The Handbook of
Brain Theory
and Neural Networks

The Handbook of
Brain Theory
and Neural Networks

EDITED BY
Michael A. Arbib

A Bradford Book
THE MIT PRESS
Cambridge, Massachusetts
London, England

This book was set in Times Roman by Asco Trade Typesetting Ltd., Hong Kong, and was printed and bound in the United States of America.

Library of Congress Cataloging-in-Publication Data

The handbook of brain theory and neural networks / Michael A. Arbib, editor.
    p.   cm.
   "A Bradford book."
   Includes bibliographical references and index.
   ISBN 0-262-01148-4
    1. Neural networks (Neurobiology)—Handbooks, manuals, etc.
  2. Neural networks (Computer science)—Handbooks, manuals, etc.
I. Arbib, Michael A.
QP363.3.H36 1995
612.8′2—dc20   94-44408
                                           CIP

Contents

Preface

This volume is inspired by two great questions: "How does the brain work?" and "How can we build intelligent machines?" It provides no simple, single answer to either question because no single answer, simple or otherwise, exists. However, in hundreds of articles it charts the immense progress made in recent years in answering many related, but far more specific, questions.

The term *neural networks* has been used for a century or more to describe the networks of biological neurons that constitute the nervous systems of animals, whether invertebrates or vertebrates. Since the 1940s, and especially since the early 1980s, the term has also been used for a technology of parallel computation in which the computing elements are "artificial neurons" loosely modeled on simple properties of biological neurons, usually with some adaptive capability to change the strengths of connections between the neurons.

Brain theory is centered on "computational neuroscience," the use of computational techniques to model biological neural networks, but also includes attempts to understand the brain and its function through a variety of theoretical constructs and computer analogies. In fact, as the following pages reveal, much of brain theory is not about neural networks per se, but focuses on structural and functional "networks" whose units are at scales both coarser and finer than that of the neuron. Computer scientists, engineers, and physicists have analyzed and applied artificial neural networks inspired by the adaptive, parallel computing style of the brain, but this *Handbook* will also sample non-neural approaches to the design and analysis of "intelligent" machines. In between the biologists and the technologists are the connectionists. They use artificial neural networks in psychology and linguistics and make related contributions to artificial intelligence, using neuron-like units which interact "in the style of the brain" at a more abstract level than that of individual, biological neurons.

Many texts have described limited aspects of one subfield or another of brain theory and neural networks, but no truly comprehensive overview is available. The aim of this *Handbook* is to fill that gap, presenting the entire range of the following topics: detailed models of single neurons; analysis of a wide variety of neurobiological systems; "connectionist" studies; mathematical analyses of abstract neural networks; and technological applications of adaptive, artificial neural networks and related methodologies. The excitement, and the frustration, of these topics is that they span such a broad range of disciplines, including mathematics, statistical physics and chemistry, neurology and neurobiology, and computer science and electrical engineering, as well as cognitive psychology, artificial intelligence, and philosophy. Much effort, therefore, has gone into making this book accessible to readers with varied backgrounds (an undergraduate education in one of the above areas, for example, or the frequent reading of related articles at the level of *Scientific American*) while still providing a clear view of recent, specialized research.

The heart of the book comes in Part III, in which the breadth of brain theory and neural networks is sampled in 266 articles, presented in alphabetical order by title. Each article meets the following requirements:

1. It is authoritative within its own subfield, yet accessible to students and experts in a wide range of other fields.
2. It is comprehensive, yet short enough that its concepts can be acquired in a single sitting.
3. It includes a list of references, limited to about 15, to give the reader a well-defined and selective list of places to go to initiate further study.

4. It is as self-contained as possible, while providing cross-references to allow readers to explore issues of related interest.

Despite the fourth requirement, some articles are more self-contained than others. Some articles can be read with almost no prior knowledge; some can be read with a rather general knowledge of a few key concepts; others require fairly detailed understanding of material covered in other articles. For example, many articles on applications will make sense only if one understands the "backpropagation" technique for training artificial neural networks; and a number of studies of neuronal function will make sense only if one has at least some idea of the Hodgkin-Huxley equation. Whenever appropriate, therefore, the articles include advice on background articles.

Parts I and II of the book provide a more general approach to helping readers orient themselves. Part I: Background presents a perspective on the "landscape" of brain theory and neural networks, including an exposition of the key concepts for viewing neural networks as dynamic, adaptive systems. Part II: Road Maps then provides an entrée into the many articles of Part III, with "road maps" for 23 different themes. The "Meta-Map," which introduces Part II, groups these themes under eight general headings which, in and of themselves, give some sense of the sweep of the *Handbook*:

Connectionism: Psychology, Linguistics, and Artificial Intelligence
Dynamics, Self-Organization, and Cooperativity
Learning in Artificial Neural Networks
Applications and Implementations
Biological Neurons and Networks
Sensory Systems
Plasticity in Development and Learning
Motor Control

A more detailed view of the structure of the book is provided in the introductory section "How to Use This Book." The aim is to ensure that readers will not only turn to the book to get good brief reviews of topics in their own specialty, but also find many invitations to browse widely—finding parallels among different subfields, or simply enjoying the discovery of interesting topics far from familiar territory.

Acknowledgments

My foremost acknowledgment is to Prue Arbib, who served as Editorial Assistant during the long and arduous process of eliciting and assembling the many, many contributions to Part III; we both thank Paulina Tagle for her help with our work. The initial plan for the book was drawn up in 1991, and it benefited from the advice of a number of friends, especially George Adelman, who shared his experience as Editor of the *Encyclopedia of Neuroscience*. Refinement of the plan and the choice of publishers occupied the first few months of 1992, and I thank Fiona Stevens of The MIT Press for her support of the project from that time onward.

As can be imagined, the plan for a book like this has developed through a time-consuming process of constraint satisfaction. The first steps were to draw up a list of about 20 topic areas (similar to, but not identical with, the 23 areas surveyed in Part II), to populate these areas with a preliminary list of over 100 articles and possible authors, and to recruit the first members of the Editorial Advisory Board to help expand the list of articles and focus the search for authors. A very satisfying number of authors invited in the first round accepted my invitation, and many of these added their voices to the Editorial Advisory Board in suggesting further topics and authors for the *Handbook*.

I was delighted, stimulated, and informed as I read the first drafts of the articles; but I have also been grateful for the fine spirit of cooperation with which the authors have responded to editorial comments and reviews. The resulting articles are not only authoritative and accessible in themselves, but have also been revised to match the overall style of the *Handbook* and to meet the needs of a broad readership. With this I express my sincere thanks to the editorial advisors, the authors, and the hundreds of reviewers who so constructively contributed to the final polishing of the articles that now appear in Part III; to Doug Gordon and the copy editors and typesetters who transformed the diverse styles of the manuscripts into the style of the *Handbook*; and to the graduate students who helped so much with the proofreading.

Finally, I want to record a debt that did not reach my conscious awareness until well into the editing of this book. It is to Hiram Haydn, who for many years was editor of *The American Scholar*, which is published for general circulation by Phi Beta Kappa. In 1971 or so, Phi Beta Kappa conducted a competition to find authors to receive grants for books to be written, if memory serves aright, for the Bicentennial of the United States. I submitted an entry. Although I was not successful, Mr. Haydn, who had been a member of the jury, wrote to express his appreciation of that entry, and to invite me to write an article for the *Scholar*. What stays in my mind from the ensuing correspondence was the sympathetic way in which he helped me articulate the connections that were at best implicit in my draft, and find the right voice in which to "speak" with the readers of a publication so different from the usual scientific journal. I now realize that it is his example I have tried to follow as I have worked with these hundreds of authors in the quest to see the subject of brain theory and neural networks whole, and to share our work with readers of diverse interests and backgrounds.

Michael A. Arbib
Los Angeles and La Jolla
January 1995

How to Use This Book

Most of this book is taken up by Part III, where 266 separately authored articles cover a vast range of topics in brain theory and neural networks, from language to motor control and from the neurochemistry of memory to the applications of artificial neural networks in steelmaking. Each article has been made as self-contained as possible, but the very breadth of topics means that few readers will be expert in a majority of them. To help the reader new to certain areas of the *Handbook*, I have prepared Part I: Background and Part II: Road Maps. This brief introductory section describes these aids to comprehension and offers more information about the structure of articles in Part III.

Part I: Background

Part I provides background material for readers new to computational neuroscience or theoretical approaches to neural networks considered as dynamic, adaptive systems. Section I.1, "Introducing the Neuron," conveys the basic properties of biological neurons, introduces several simple neural models, and provides pointers to the more detailed analysis of biological neurons. Section I.2, "Levels and Styles of Analysis," explains the interdisciplinary nexus in which the present study of brain theory and neural networks is located, with historical roots in cybernetics and with current work involving brain theory, artificial intelligence, and cognitive psychology. We also review the different levels of analysis involved, with schemas providing functional units intermediate between an overall task and neural networks. Finally, Section I.3, "Dynamics and Adaptation in Neural Networks," provides a tutorial on the concepts essential for understanding neural networks as dynamic, adaptive systems. We close by stressing that the full understanding of the brain and the improved design of intelligent machines will require not just improvements of the learning methods presented in Section II.3, but also fuller understanding of architectures based on networks of networks, with initial structures well constrained for the task at hand.

Part II: Road Maps

The reader who wants to survey a major theme, rather than seeking articles in Part III one at a time, will find in Part II a set of 23 *road maps* which place each article in Part III in a thematic perspective. Part II opens with a "Meta-Map" which briefly surveys all these themes, grouping them under eight general headings:

Connectionism: Psychology, Linguistics, and Artificial Intelligence
    Connectionist Psychology
    Connectionist Linguistics
    Artificial Intelligence and Neural Networks
Dynamics, Self-Organization, and Cooperativity
    Dynamic Systems and Optimization
    Cooperative Phenomena
    Self-Organization in Neural Networks
Learning in Artificial Neural Networks
    Learning in Artificial Neural Networks, Deterministic
    Learning in Artificial Neural Networks, Statistical
    Computability and Complexity
Applications and Implementations
    Control Theory and Robotics

This ordering of the themes has no special significance. It is simply one way to approach the richness of the *Handbook*, making it easy for the reader to identify one or two road maps of especial interest at a particular time. By the same token, the order of articles in each of the 23 road maps that follow the Meta-Map is one among many possible orderings. Each road map starts with an alphabetical listing of the articles most relevant to its theme. The road map itself will provide suggestions for *interesting* traversals of articles, but this need not imply that an article provides *necessary* background for the articles it precedes.

Part III: Articles

The articles of Part III are arranged in alphabetical order by title, both to make it easier to find a specific topic (though a full subject index is provided as well) and because a given article may be relevant to more than one of the themes of Part II, a fact which would be hidden were the article to be relegated to a specific section devoted to a single theme. Most of these articles assume some prior familiarity with neural networks, whether biological or artificial, and so the reader is encouraged to master the material in Part I, especially Section I.3 on "Dynamics and Adaptation in Neural Networks," before tackling Part III.

Most articles in Part III have the following structure: The introduction provides a nontechnical overview of the material covered in the whole article; the final section provides a discussion of key points, open questions, and linkages with other areas of brain theory and neural networks. The intervening sections may be more or less technical depending on the nature of the topic, but the first and last sections should give most readers a basic appreciation of the topic irrespective of such technicalities. The bibliography for each article contains about 15 references. People who find their favorite papers omitted from the list should blame my editorial decision, not the author's judgment. The style I chose for the *Handbook* was *not* to provide exhaustive coverage of research papers for the expert. Rather, references are there primarily to help readers who look for an *introduction* to the literature on the given topic, including background material, relevant review articles, and original research citations. Expository and survey items in the references are marked with the symbol ◆. In addition to formal references to the literature, each article contains numerous cross-references to other articles in the *Handbook*. In the text, these take the form of THE TITLE OF THE ARTICLE IN SMALL CAPITALS; they may also appear at the end of the article designated as "Related Reading." In addition to suggestions for related reading, the reader will find, just prior to the list of references in each article, a mention of the road map(s) in which the article is discussed, as well as "Background" material which will usually come from Part I.

In summary: Turn directly to Part III when you need information on a specific topic. Read sections of Part I to gain a general perspective on the basic concepts of brain theory and neural networks. For an overview of a certain theme, read the Meta-Map in Part II to choose a road map; then read the road map to choose articles in Part III. A road map can also be used as an explicit guide for systematic study of the area under review. Then continue your exploration by further use of the road maps, by following cross-references in Part III, by looking up terms of interest in the index, or simply by letting serendipity take its course as you browse through Part III at random.

Part I: Background

Michael A. Arbib

How to Use Part I

Part I provides background material. As such, it is designed to hold few, if any, surprises for readers with a fair background in either computational neuroscience or theoretical approaches to neural networks considered as dynamic, adaptive systems. *Part I is not for them.* Rather, it is designed for the many readers—be they neuroscience experimentalists, psychologists, philosophers, or technologists—who are sufficiently new to *Brain Theory and Neural Networks* that they can benefit from a compact overview of basic concepts prior to reading the "road maps" of Part II and the articles in Part III. Of course, much of what is covered in Part I is also covered at some length in articles in Part III, and cross-references will steer the reader to these articles for alternative expositions and reviews of current research.

The three sections of Part I have been written so that they may be read independently. In many articles in Part III, just one of the sections of Part I will be suggested as background. The following paragraphs briefly summarize each section:

Section I.1, "Introducing the Neuron," conveys the simplest biological properties of neurons, receptors, and effectors to motivate several simple models of single neurons, including the discrete-time McCulloch-Pitts model, and the continuous-time leaky integrator model. References to Part III alert the reader to more detailed properties of neurons which are essential for the neuroscientist and provide interesting hints about future design features for the technologist.

Similar background (without the cross-references) may be obtained from the Part III article SINGLE-CELL MODELS.

Section I.2, "Levels and Styles of Analysis," relies on the reader's general intuitions concerning neurons and neuron-like units and the fact that the connections between them (synapses) may change with experience ("synaptic plasticity"), but does not require prior knowledge of the formal models presented in Sections I.1 and I.3. It presents the interdisciplinary nexus in which the *Handbook* is situated. The attempt to understand the mind and to build intelligent machines includes, but is in no sense restricted to, the study of neural networks, and so the section begins with a historical fragment which traces our federation of disciplines back to their roots in *cybernetics*, the study of control and communication in animals and machines. We look at the way in which the research addresses brains, machines, and minds, going back and forth between brain theory, artificial intelligence, and cognitive psychology. We then review the different levels of analysis involved—whether we study brains or intelligent machines—and the use of schemas to provide functional units that bridge the gap between an overall task and the neural networks which implement it.

Similar background may be obtained from the following Part III articles:

ARTIFICIAL INTELLIGENCE AND NEURAL NETWORKS
COGNITIVE MODELING: PSYCHOLOGY AND CONNECTIONISM
MOTOR CONTROL, BIOLOGICAL AND THEORETICAL
NEUROETHOLOGY, COMPUTATIONAL
PHILOSOPHICAL ISSUES IN BRAIN THEORY AND CONNECTIONISM
SCHEMA THEORY

Section I.3, "Dynamics and Adaptation in Neural Networks," contains more material than will be required in many of the Part III articles in which it will be referred to for background. However, it conveys two ideas that are so basic to the modern theory of neural networks—*neural networks are dynamic systems*; and *neural networks are adaptive systems that change with experience*—that all but the most cursory readers of the *Handbook* should be conversant with all the topics it contains. I stress again that this material is offered as an elementary introduction, and is not intended for readers with prior experience in neural network theory. To keep the section self-contained, it repeats the definitions of the McCulloch-Pitts and leaky integrator neurons, but now in the context of dynamic systems rather than biological modeling. It introduces the basic dynamic systems concepts of stability, limit cycles, and chaos, and relates Hopfield nets to attractors and optimization. It then introduces a number of basic concepts concerning adaptation in neural nets, with discussions of pattern recognition, associative memory, Hebbian plasticity, network self-organization, perceptrons, network complexity, gradient descent, credit assignment, and backpropagation. This section, and with it Part I, closes with a cautionary note. The basic learning rules and adaptive architectures of neural networks have already illuminated a number of biological issues and led to useful technological applications. However, these networks must have their initial structure well constrained (whether by evolution or technological design) to yield approximate solutions to the system's tasks—solutions that can then be efficiently and efficaciously shaped by experience. Moreover, the full understanding of the brain and the improved design of intelligent machines will not only require improvements of these learning methods and their initialization, but also a fuller understanding of architectures based on networks of networks. Cross-references to articles in Part III will set the reader on the path to this fuller understanding.

Similar background may be obtained from the following Part III articles:

ASSOCIATIVE NETWORKS
BACKPROPAGATION: BASICS AND NEW DEVELOPMENTS
COMPUTING WITH ATTRACTORS
DYNAMICS AND BIFURCATION OF NEURAL NETWORKS
PERCEPTRONS, ADALINES, AND BACKPROPAGATION
SELF-ORGANIZATION AND THE BRAIN

A Note on Mathematics. Many articles in Part III are nonmathematical, and many of the others (as in Part I) make use of little more than the elements of differential equations and linear algebra. Many of the more mathematical articles are described in the Part II road maps on **Cooperative Phenomena** and **Learning in Artificial Neural Networks, Statistical** and require the reader to have some familiarity with probability theory, or information theory, or statistical mechanics. Here, a basic bookshelf might hold the following works:

Berger, J. O., 1985, *Statistical Decision Theory and Bayesian Analysis*, 2nd ed., New York: Springer-Verlag.
Cover, T. M., and Thomas, J. A., 1991, *Elements of Information Theory*, New York: John Wiley.
Hertz, J., Krogh, A., and Palmer, R. G., 1991, *Introduction to the Theory of Neural Computation*, Redwood City, CA: Addison-Wesley.
Reif, F., 1982, *Fundamentals of Statistical and Thermal Physics*, New York: McGraw-Hill.

The other body of mathematics used in a number of articles involves ideas to do with nonlinear systems analysis and its applications to neural and biological systems, as exemplified, for instance, in Section I.3 and the Part III articles DYNAMICS

AND BIFURCATION OF NEURAL NETWORKS and PATTERN FORMATION, BIOLOGICAL. Useful textbooks for this area include these:

Hale, J. K., and Kocak, H., 1991, *Dynamics and Bifurcations*, New York: Springer-Verlag.

Hirsch, M. W., and Smale, S., 1974, *Differential Equations, Dynamical Systems, and Linear Algebra*, New York: Academic Press.
Murray, J. D., 1989, *Mathematical Biology*, Berlin: Springer-Verlag.

I.1. Introducing the Neuron

There are radically different types of neurons in the human brain, and endless variations in neuron types of other species. In *brain theory*, the complexities of real neurons are abstracted in many ways to aid in understanding different aspects of neural network development, learning, or function. In *neural computation* (technology based on networks of "neuron-like" units), the artificial neurons are designed as variations on the abstractions (usually the simpler ones) of brain theory and are implemented in software, VLSI, or other media (see the Part II road map on **Implementation of Neural Networks**). There is no such thing as a "typical" neuron; yet this section will present examples and models which provide a starting point, an essential set of key concepts, for the appreciation of the many variations on the theme of neurons and neural networks presented in Part III.

An analogy to the problem we face here might be to define *vehicle* for a *handbook of transportation*. A vehicle could be a car, a train, a plane, a rowboat, or a forklift truck. It might or might not carry people. The people could be crew or passengers, and so on. The problem would be to give a few key examples of form (such as car versus plane) and function (to carry people or goods, by land, air, or sea, etc.). In a similar fashion, Part III offers diverse examples of neural form and function, in both biology and technology.

Basic Properties of Neurons

Here, we start with the observation that a brain is made up of a network of cells called *neurons*, coupled to receptors and effectors. The input to the network is provided by *receptors* which continually monitor changes in the external and internal environment. Cells called *motor neurons* (or *motoneurons*), governed by the activity of the neural network, control the movement of muscles and the secretion of glands. In between, the intricate network of neurons (a few hundred neurons in some simple creatures; hundreds of billions in a human brain) continually combines the signals from the receptors, including receptors which provide feedback on the state of the effectors, with signals encoding past experience to barrage the motor neurons with signals which will yield adaptive interactions with the environment. In animals with backbones (*vertebrates*, including mammals in general and humans in particular), the *brain* constitutes the most headward part of the *central nervous system* (CNS), and is linked to the receptors and effectors of the body and limbs via the spinal cord. Invertebrate nervous systems (neural networks) provide astounding variations on the vertebrate theme, thanks to eons of divergent evolution. Thus, while the human brain may be the source of rich analogies for technologists in search of "artificial intelligence," both invertebrates and vertebrates will provide endless ideas for technologists designing neural networks for sensory processing, robot control, and a host of other applications. (A few of the relevant examples may be found in the Part II road maps on **Control Theory and Robotics**, **Vision**, and **Motor Pattern Generators and Neuroethology**. See also Table 1 in this section.)

The brain provides far more than a simple stimulus-response chain from receptors to effectors (although there are such reflex paths). Rather, the vast network of neurons is interconnected in loops and tangled skeins so that signals entering the net from the receptors interact there with the billions of signals already traversing the system, not only to yield the signals which control the effectors but also to modify the very properties of the network itself so that future behavior will reflect prior experience.

To understand the processes that intervene between receptors and effectors, we study the "basic neuron" shown in Figure 1, which is abstracted from a motor neuron of mammalian spinal cord. From the *soma* (cell body) protrudes a number of ramifying branches called the *dendrites*; the soma and dendrites constitute the input surface of the neuron. There also extrudes from the cell body, at a point called the *axon hillock* a long fiber called the *axon* whose branches form the *axonal arborization*. The tips of the branches of the axon, called *nerve terminals* or *boutons*, impinge upon other neurons or upon effectors. The locus of interaction between a bouton and the cell upon which it impinges is called a *synapse*, and we say that the cell with the bouton *synapses upon* the cell with which the connection is made. In fact, axonal branches of some neurons can have many varicosities corresponding to synapses along their length, not just at the end of the branch.

We can imagine the flow of information as shown by the arrows in Figure 1. Although "conduction" can go in either direction on the axon, most synapses tend to "communicate" activity to the dendrites or soma of the cell they synapse upon, whence activity passes to the axon hillock and then down the axon to the terminal arborization. The axon can be very long indeed. For instance, the cell body of a neuron that controls the big toe lies in the spinal cord and thus has an axon that runs the complete length of the leg. We may contrast the immense length of the axon of such a neuron with the very small size of many of the neurons in our heads. For example, amacrine cells in the RETINA (q.v.) have branchings which cannot appropriately be labeled dendrites or axons, for they are short and may well communicate activity in either direction to serve as local modulators of the surrounding network; while mitral and granule cells in the olfactory bulb can interact directly via their dendrites (see the discussion of dendro-dendritic interactions in OLFACTORY BULB and PERSPECTIVE ON NEURON MODEL COMPLEXITY).

To understand more about the "typical" communication along a neuron's axon, note that the cell is enclosed by a membrane, across which there is a difference in electrical charge. If we change this potential difference between the inside and outside, the change can propagate in much the same passive way that heat is conducted down a rod of metal: a normal change in potential difference across the cell membrane can propagate in

Figure 1. A "basic neuron" abstracted from a motor neuron of mammalian spinal cord. The dendrites and soma (cell body) constitute the major part of the input surface of the neuron. The axon is the "output line." The tips of the branches of the axon form synapses upon other neurons or upon effectors (although synapses may occur along the branches of an axon as well as at the ends). (Reprinted by permission of John Wiley & Sons, Inc., from M. A. Arbib, *The Metaphorical Brain 2: Neural Networks and Beyond*, New York: Wiley-Interscience, p. 52. Copyright © 1989 by John Wiley & Sons, Inc.)

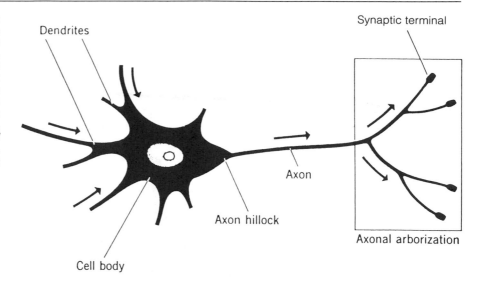

a passive way so that the change occurs later, and becomes smaller, the further away we move from the site of the original change. For "short" cells (such as the rods, cones, and bipolar cells of the retina), the passive propagation suffices to signal a potential change from one end to the other; but if the axon is long, this mechanism is completely inadequate since changes at one end will decay almost completely before reaching the other end. Fortunately, most nerve cells have the further property that if the change in potential difference is large enough (we say it exceeds a *threshold*), then in a cylindrical configuration such as the axon, a pulse can be generated which will actively propagate at full amplitude instead of fading passively.

If propagation of various potential differences on the dendrites and soma of a neuron yields a potential difference across the membrane at the axon hillock which exceeds a certain threshold, then a regenerative process is started: the electrical change at one place is enough to trigger this process at the next place, yielding a *spike* or *action potential,* an undiminishing pulse of potential difference propagating down the axon. After an impulse has propagated along the length of the axon, there is a short *refractory period* during which a new impulse cannot be propagated along the axon.

The propagation of action potentials is now very well understood. Briefly, the change of membrane potential is mediated by the flow of ions, especially sodium and potassium, across the membrane. Hodgkin and Huxley (1952) showed that the *conductance* of the membrane to sodium and potassium ions— the ease with which they flow across the membrane—depends on the transmembrane voltage. They developed elegant equations describing the voltage and time dependence of the sodium and potassium conductances. These equations (see AXONAL MODELING) have given us great insight into cellular function. Much mathematical research has gone into studying Hodgkin-Huxley–like equations, showing, for example, that neurons can support rhythmic pulse generation even without input (see OSCILLATORY AND BURSTING PROPERTIES OF NEURONS), and explicating triggered long-distance propagation. Hodgkin and Huxley used curve fitting from experimental data to determine the terms for conductance change in their model. Subsequently, much research has probed the structure of complex molecules that form *channels* which selectively allow the passage of specific ions through the membrane (see ION CHANNELS: KEYS TO NEURONAL SPECIALIZATION). This research has demonstrated how channel properties not only account for the terms in the

Hodgkin-Huxley equation, but also underlie more complex dynamics which may allow even small patches of neural membrane to act like complex computing elements. At present, most artificial neurons used in applications are very simple indeed, and much future technology will exploit these "subneural subtleties."

An impulse traveling along the axon from the axon hillock triggers new impulses in each of its branches (or *collaterals*), which, in turn, trigger impulses in their even finer branches. Vertebrate axons come in two varieties: myelinated and unmyelinated. The myelinated fibers are wrapped in a sheath of *myelin* (Schwann cells in the periphery, oligodendrocytes in the CNS). The small gaps between successive segments of the myelin sheath are called *nodes of Ranvier.* Instead of the somewhat slow active propagation down an unmyelinated fiber, the nerve impulse in a myelinated fiber jumps from node to node, thus speeding passage and reducing energy requirements (see AXONAL MODELING).

Surprisingly, at most synapses the direct cause of the change in potential of the postsynaptic membrane is not electrical but chemical. When an impulse arrives at the presynaptic terminal, it causes the release of *transmitter* molecules (which have been stored in the bouton in little packets called *vesicles*) through the presynaptic membrane. The transmitter then diffuses across the very small *synaptic cleft* to the other side, where it binds to receptors on the postsynaptic membrane to change the conductance of the postsynaptic cell (for modeling of the details, see SYNAPTIC CURRENTS, NEUROMODULATION, AND KINETIC MODELS).

The effect of the "classical" transmitters (later we shall talk of other kinds, the neuromodulators) is of two basic kinds: either *excitatory*, tending to move the potential difference across the postsynaptic membrane in the direction of the threshold (*depolarizing* the membrane), or *inhibitory*, tending to move the polarity away from the threshold (*hyperpolarizing* the membrane). There are some exceptional cell appositions which are so large, or have such tight coupling (the so-called gap junctions), that the impulse affects the postsynaptic membrane without chemical mediation.

Most neural modeling to date focuses on the excitatory and inhibitory interactions that occur on a fast time scale (a millisecond, more or less), and most biological (as distinct from technological) models assume that all synapses *from* a neuron have the same "sign." However, neurons may also secrete transmitters which modulate the function of a circuit on some quite extended time scale (see NEUROMODULATION IN INVERTE-BRATE NERVOUS SYSTEMS). Modeling which takes account of this *neuromodulation* will become increasingly important, since it allows cells to change their function, enabling a neural network to switch its overall mode of activity dramatically.

The excitatory or inhibitory effect of the transmitter released when an impulse arrives at a bouton generally causes a sub-threshold change in the postsynaptic membrane. Nonetheless, the cooperative effect of many such subthreshold changes may yield a potential change at the axon hillock which exceeds threshold—and if this occurs at a time when the axon has passed the refractory period of its previous firing, then a new impulse will be fired down the axon.

Synapses can differ in shape, size, form, and effectiveness. The geometrical relationships between the different synapses impinging upon the cell determine what patterns of synaptic activation will yield the appropriate temporal relationships to excite the cell (see DENDRITIC PROCESSING). A highly simplified example (Figure 2) shows how the properties of nervous tissue

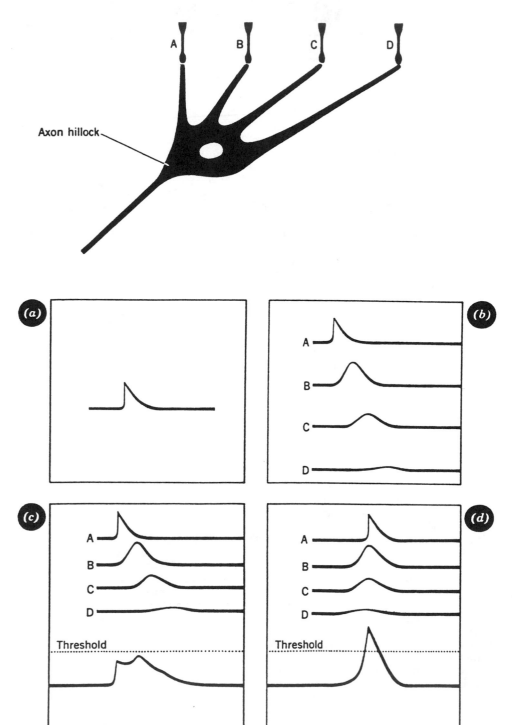

Figure 2. An example, conceived by Wilfrid Rall, of the subtleties that can be revealed by neural modeling when dendritic properties (in this case, length-dependent conduction time) are taken into account. As shown in part C, the effect of simultaneously activating all inputs may be subthreshold, yet the cell may respond when inputs traverse the cell from right to left (D). (Reprinted by permission of John Wiley & Sons, Inc., from M. A. Arbib, *The Metaphorical Brain 2: Neural Networks and Beyond*, New York: Wiley-Interscience, p. 60. Copyright © 1989 by John Wiley & Sons, Inc.)

just presented would indeed allow a simple neuron, by its very dendritic geometry, to compute some useful function. Consider a neuron with four dendrites, each receiving a single synapse from a visual receptor, so arranged that synapses A, B, C, and D (from left to right) are at increasing distances from the axon hillock. (This is not meant to be a model of a neuron in the retina of an actual organism; rather it is designed to make vivid the potential richness of single neuron computations.) We assume that each receptor reacts to the passage of a spot of light above its surface by yielding a generator potential which yields, in the postsynaptic membrane, the same time course of depolarization. This time course is propagated passively, and the further it is propagated, the later and the lower is its peak. If four inputs reach A, B, C, and D simultaneously, their effect may be less than the threshold required to trigger a spike there. However, if an input reaches D before one reaches C, and so on, in such a way that the peaks of the four resultant time courses at the axon hillock coincide, it could well exceed threshold. This, then, is a cell which, although very simple, can detect direction of motion across its input. It responds only if the spot of light is moving from right to left, and if the velocity of that motion falls within certain limits. Our cell will not respond to a stationary object, or one moving from left to right, because the asymmetry of placement of the dendrites on the cell body yields a preference of one direction of motion over others (for a more realistic account of biological mechanisms, see DIRECTIONAL SELECTIVITY IN THE CORTEX and DIRECTIONAL SELECTIVITY IN THE RETINA). This simple example illustrates that the *form* (i.e., the geometry) of the cell can have a great impact upon the *function* of the cell, and we thus speak of *form-function* relations. When we note that neurons in the human brain may have 10,000 or more synapses upon them, we can understand that the range of functions of single neurons is indeed immense. The model is adapted from Rall (1964: 90); see PERSPECTIVE ON NEURON MODEL COMPLEXITY for a fuller appreciation of the subtleties of neuronal function that can only be revealed by *compartmental models* which subdivide a cell into a number of compartments (such as the four dendrites of Figure 2) whose properties must be separated if the full function of the neuron is to be revealed.

Receptors and Effectors

Rod and cone receptors in the eyes respond to light (RETINA); hair cells in the ears respond to pressure (AUDITORY PERIPHERY AND COCHLEAR NUCLEUS); and other cells in the tongue and the mouth respond to subtle traces of chemicals. In addition to touch receptors, there are receptors in the skin that are responsive to movement or to temperature, or that signal painful stimuli (PAIN NETWORKS). These external senses may be divided into two classes: (1) the proximity senses, like touch and taste, which sense objects in contact with the body surface, and (2) the distance senses, like vision and hearing, which let us sense objects distant from the body. Olfaction (OLFACTORY BULB; OLFACTORY CORTEX) is somewhere in between, using chemical signals "right under our noses" to sense nonproximate objects. Moreover, even the proximate senses can yield information about nonproximate objects, as when we feel the wind, or the heat of a fire.

The appropriate activity of the effectors must depend on comparing where the system should be with where it is now. Thus, in addition to the external receptors, there are receptors which monitor the activity of muscles, tendons, and joints to provide a continual source of feedback (MOTOR CONTROL, BIOLOGICAL AND THEORETICAL) about the tensions and lengths of muscles and the angles of the joints, as well as their velocities.

The vestibular system in the head monitors gravity and accelerations (VESTIBULO-OCULAR REFLEX: PERFORMANCE AND PLASTICITY). Here, the receptors are hair cells monitoring fluid motion.

There are also receptors to monitor the chemical level of the bloodstream and the state of the heart and the intestines. Cells in the liver monitor glucose, while others in the kidney check water balance. Receptors in the *hypothalamus*, itself a part of the brain, also check the balance of water and sugar. The hypothalamus then integrates these diverse messages to direct behavior or other organs to restore the balance. If we stimulate the hypothalamus, an animal may drink copious quantities of water, or eat enormous quantities of food, even though it is already well supplied; the brain has received a signal that water or food is lacking, and so it instructs the animal accordingly, irrespective of whatever contradictory signals may be coming from a distended stomach.

Receptors share with neurons the property of generating potentials which are transmitted to various synapses upon neurons. However, the input surface of a receptor does not receive synapses from other neurons, but can transduce environmental energy into changes in membrane potential, which may then propagate either passively or actively. (Visual receptors do not generate spikes; touch receptors in the body and limbs use spike trains to send their message to the spinal cord.) For instance, the rods and cones of the eye contain various pigments which react chemically to light in different frequency bands, and these chemical reactions, in turn, lead to local potential changes, called generator potentials, in the membrane. If the light falling upon an array of rods and cones is appropriately patterned, then their potential changes will induce interneuron changes to, in turn, fire certain ganglion cells (retinal output neurons whose axons course toward the brain). Properties of the light pattern will thus be signaled further into the nervous system as trains of impulses (see RETINA).

At the receptors, increasing the intensity of stimulation will increase the generator potential. If we go to the first level of neurons which generate pulses, the axons "reset" each time they fire a pulse and then have to get back to a state where the threshold and the input potential meet. The higher the generator potential, the shorter the time until they meet again and, thus, the higher the frequency of the pulse. Thus at the "input" it is a useful first approximation to say that intensity or quantity of stimulation is coded in terms of pulse frequency (more stimulus → more spikes), whereas quality or type of stimulus is coded by different lines carrying signals from different types of receptors. As we leave the periphery and move toward more "computational" cells, we no longer have such simple relationships, but rather interactions of inhibitory cells and excitatory cells, with each inhibitory input moving a cell away from, and each excitatory input moving it toward, threshold.

To discuss the "output side," we must first note that a muscle is made up of many thousands of muscle fibers (MUSCLE MODELS). The motoneurons which control the muscle fibers lie in the spinal cord or the brainstem, whence their axons may have to travel vast distances (by neuronal standards) before synapsing upon the muscle fibers. The smallest functional entity on the output side is thus the *motor unit*, which consists of a motor neuron cell body, its axon, and the group of muscle fibers the axon influences.

A muscle fiber is like a neuron to the extent that it receives its input via a synapse from a motor neuron. However, the response of the muscle fiber to the spread of depolarization is to contract. Thus, the motor neurons which synapse upon the muscle fibers can determine, by the pattern of their impulses, the extent to which the whole muscle comprised of those fibers contracts, and can thus control movement. (Similar remarks apply to those cells which secrete various chemicals into the bloodstream or gut or those which secrete sweat or tears.)

Synaptic activation at the *motor endplate* (i.e., the synapse of a motor neuron upon a muscle fiber) yields a brief "twitch" of the muscle fiber. A low repetition rate of action potentials arriving at a motor endplate causes a train of twitches, in each of which the mechanical response lasts longer than the action potential stimulus. As the frequency of excitation increases, a second action potential will arrive while the mechanical effect of the prior stimulus still persists. This causes a mechanical summation or fusion of contractions. Up to a point, the degree of summation increases as the stimulus interval becomes shorter, although the summation effect decreases as the interval between the stimuli approaches the refractory period of the muscle, and maximum tension occurs. This limiting response is called a *tetanus*. To increase the tension exerted by a muscle, it is then necessary to recruit more and more fibers to contract. For more delicate motions, such as those involving the fingers of primates, each motor neuron may control only a few muscle fibers. In other locations, such as the shoulder, one motor neuron alone may control thousands of muscle fibers. As descending signals in the spinal cord command a muscle to contract more and more, they do this by causing motor neurons with larger and larger thresholds to start firing. The result is that fairly small fibers are brought in first, and then larger and larger fibers are recruited. The result, known as Henneman's Size Principle, is that at any stage, the increment of activation obtained by recruiting the next group of motor units involves about the same percentage of extra force being applied, aiding smoothness of movement (see MOTONEURON RECRUITMENT).

Since there is no command which a neuron may send directly to a muscle fiber that will cause it to lengthen itself—all the neuron can do is stop sending it commands to contract—the muscles of an animal are usually arranged in pairs. The contraction of one member of the pair will then act around a pivot to cause the expansion of the other member of the pair. For example, one set of muscles *extends* the elbow joint, while another set *flexes* the elbow joint. To extend the elbow joint, motoneurons do not signal the *flexors* to lengthen; rather they simply stop signaling them to contract, while other motoneurons signal the *extensor* muscles to contract, thus both extending the elbow and stretching the now relaxed flexor muscles. For convenience, we often label one set of muscles as the "prime mover" or *agonist*, and the opposing set as the *antagonist*. However, in such joints as the shoulder, which are not limited to one degree of freedom, many muscles, rather than an agonist-antagonist pair, participate. Most real movements involve many joints. For example, the wrist must be fixed, holding the hand in a position bent backward with respect to the forearm, for the hand to grip with its maximum power. *Synergists* are muscles which act together with the main muscles involved. A large group of muscles work together when one raises something with one's finger. If more force is required, wrist muscles may also be called in; if still more force is required, arm muscles may be used. In any case, muscles all over the body are involved in maintaining posture.

Neural Models

The brain (of, say, a typical mammal) may certainly be considered as a single vast neural network. However, anatomical inspection clearly reveals that the brain may better be viewed as a network of networks, where each constituent network has a distinctive "architecture" (e.g., pattern of layering, connectivity, and input-output connections) based on a set of neurons

with distinctive properties characteristic of this specific part of the brain. The articles described in the road map on **Mammalian Brain Regions** in Part II gives some sense of this diversity, treating (in alphabetical order) the following distinctive brain regions, many of which can themselves be further subdivided into distinctive subnetworks: auditory cortex, basal ganglia, cerebellum, hippocampus, olfactory bulb, olfactory cortex, retina, somatosensory system, superior colliculus, thalamus, and visual cortex. As noted in the article OLFACTORY BULB, it has been possible to associate with many of these regions a *basic circuit* which defines the irreducible minimum of neural components necessary to model the neural basis of the region's functions. Shepherd (1990) provides a thorough exposition of the basic circuits for a number of these regions.

By contrast to this view of the brain as a network of networks, with each subnetwork (brain region) based on the manifold repetition of a basic circuit with distinctive properties, much of neural computation is based on analysis of a single circuit with a relatively simple architecture (such as the fully connected networks and feedforward networks discussed in Section I.3). The reader will find in Part III many examples of the design of such networks, their self-organization and training. However, I want to alert the reader to the idea that more and more work in neural computation (see MODULAR AND HIERARCHICAL LEARNING SYSTEMS for one of the best-known examples) is becoming more "brain-like" in showing how complex problems are best solved by networks of networks rather than by a single network of some standard architecture. It is my expectation that future work in neural computation will also become more "brain-like" in a complementary way, by incorporating more of the subtleties of function exhibited in biological neurons. Nonetheless, we now turn to an exposition of a range of simplified models of single neurons which have been successfully applied in high-level models of brain function, in connectionist modeling, and in the artificial neural networks used in technology. The concluding subsection of Section I.1, "More Detailed Properties of Neurons," will then provide useful pointers to articles in Part III.

We first review the work of McCulloch and Pitts (1943), which combined neurophysiology and mathematical logic, using the all-or-none property of neuron firing, to model the neuron as a binary discrete-time element. They showed how excitation, inhibition, and threshold might be used to construct a wide variety of "neurons." Theirs was the first model to tie the study of neural nets squarely to the idea of computation in its modern sense (see "A Historical Fragment" in Section I.2). The basic idea is to divide time into units comparable to a refractory period so that, in each time period, at most one spike can be generated at the axon hillock of a given neuron. The McCulloch-Pitts neuron (Figure 3A) thus operates on a discrete time scale, $t = 0, 1, 2, 3, \ldots$ where the time unit is (in biology) on the order of a millisecond. We write $y(t) = 1$ if a spike does appear at time t, and $y(t) = 0$ if not. Each connection or *synapse*, from the output of one neuron to the input of another, has an attached *weight*. Let w_i be the weight on the ith connection onto a given neuron. We call the synapse *excitatory* if $w_i > 0$, and *inhibitory* if $w_i < 0$. We also associate a *threshold* θ with each neuron, and assume (contrary to the example of Figure 3) exactly one unit of delay in the effect of *all* presynaptic inputs on the cell's output, so that a neuron "fires" (i.e., has value 1 on its output line) at time $t + 1$ if the weighted value of its inputs at time t is at least θ. Formally, if at time t the value of the ith input is $x_i(t)$ and the output one time step later is $y(t + 1)$, then

$$y(t + 1) = 1 \quad \text{if and only if} \quad \sum_i w_i x_i(t) \geq \theta$$

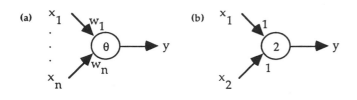

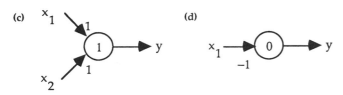

Figure 3. *A,* A McCulloch-Pitts neuron operating on a discrete time scale. Each input has an attached weight w_i, and the neuron has a *threshold* θ. The neuron "fires" at time $t + 1$ if and only if the weighted value of its inputs at time t is at least θ. *B,* Settings of weights and threshold for neurons that function as an AND gate (i.e., the output fires if x_1 and x_2 both fire). *C,* An OR gate (the output fires if x_1 or x_2, or both, fire). *D,* A NOT gate (the output fires if x_1 does not fire).

[Some authors in this *Handbook* use the variant form $y(t + 1) = 1$ if and only if $\Sigma_i w_i x_i(t) > \theta$, and some authors use states $\{-1, +1\}$ rather than $\{0, 1\}$ to increase the analogy with the statistical mechanics of systems whose units are characterized by "spins" (see STATISTICAL MECHANICS OF NEURAL NETWORKS), but these variations make no essential difference in the conclusions they reach.]

Parts B through D of Figure 3 show how weights and threshold can be set to yield neurons which realize the logical functions AND, OR, and NOT. As a result, McCulloch-Pitts neurons are sufficient to build networks which can function as the control circuitry for a computer carrying out computations of arbitrary complexity. This discovery played a crucial role in the development of automata theory (see AUTOMATA AND NEURAL NETWORKS) and in the study of learning machines. While the McCulloch-Pitts neuron no longer plays an active part in *computational neuroscience*, it is still widely used in *neural computation*, especially when it is generalized so that the input and output values can lie anywhere in the range [0, 1], and the function $f(\Sigma_i w_i x_i(t))$, which yields $y(t + 1)$, is a continuously varying function rather than a step function. As we shall see in Section I.3 (and as evidenced by the many articles discussed in the Part II road maps on **Learning in Artificial Neural Networks, Deterministic** and **Learning in Artificial Neural Networks, Statistical**), the great interest is in the use of such units when the values of the weights w_i may change through experience.

However, it is one thing to define, as McCulloch and Pitts did, neurons with sufficient logical power to subserve any discrete computation; it is quite another to understand how the neurons in actual brains perform their tasks. More generally, the problem is to select just which units to model, and to decide how such units are to be represented. Thus, when we turn from neural computation to computational neuroscience, we must turn to more realistic models of neurons. On the other hand, we may say that neural computation cannot reach its full power without applying new mechanisms based on current and future study of biological neural networks (see the following section, "More Detailed Properties of Neurons," and the road map on **Biological Neurons**).

Modern brain theory no longer uses the binary model of the neuron, but instead uses continuous-time models that either represent the variation in average firing rate of the neuron or actually capture the time course of membrane potentials. It is only through such correlates of measurable brain activity that brain models can really feed back to biological experiments. Such models also require the brain theorist to know a great deal of detailed anatomy and physiology, as well as behavioral data. Hodgkin and Huxley (1952) have shown us how much can be learned from analysis of membrane properties about the propagation of electrical activity along the axon: Rall (1964; cf. Figure 2) was a leader in showing that the study of membrane properties in a variety of connected "compartments" of membrane in dendrite, soma, and axon can help us understand small neural circuits, as in the OLFACTORY BULB (q.v.) or for DENDRITIC PROCESSING (q.v.). Nonetheless, in many cases, the complexity of compartmental analysis makes it more insightful to use a more lumped representation of the individual neuron if we are to assemble the model neurons to analyze large networks. A computer simulation of the response of a whole brain region which analyzed each component at the finest level of detail available would be too large to run on even a supernetwork of supercomputers. In addition to the importance of detailed models of single neurons in themselves, such studies can also be used to fine-tune more economical models of neurons, which can then serve as the units in models of large networks, whether to model systems in the brain or to design artificial neural networks which exploit subtle neural capabilities.

The simplest "realistic" model, consonant with the above material, is the *leaky integrator* model. Although some biological neurons communicate by the passive propagation (cable equation) of membrane potential down their (necessarily short) axons, most communicate by the active propagation of "spikes." The generation and propagation of such spikes has been described in detail by the Hodgkin-Huxley equations. However, the leaky integrator model omits such details. It is a continuous-time model in which the *internal state* of the neuron is described by a single variable, the membrane potential at the spike initiation zone.

The time evolution of the cell's membrane potential is given by a differential equation. Consider first the simple equation

$$\tau \frac{dm(t)}{dt} = -m(t) + h \tag{1}$$

We say that $m(t)$ is in an *equilibrium* if it does not change under the dynamics described by the differential equation. However, $dm(t)/dt = 0$ if and only if $m(t) = h$, so that h is the *unique* equilibrium of Equation 1. To get more information, we now integrate Equation 1 to get

$$m(t) = e^{-t/\tau} m(0) + (1 - e^{-t/\tau})h$$

which tends to the *resting level* h with *time constant* τ with increasing t so long as τ is positive. We now add synaptic inputs to obtain

$$\tau \frac{dm(t)}{dt} = -m(t) + \sum_i w_i X_i(t) + h \tag{2}$$

where $X_i(t)$ is the firing rate at the ith input. Thus an excitatory input ($w_i > 0$) will be such that increasing it will increase $dm(t)/dt$, while an inhibitory input ($w_i < 0$) will have the opposite effect. A neuron described by Equation 2 is called a *leaky integrator neuron*. This is because the equation

$$\tau \frac{dm(t)}{dt} = \sum_i w_i X_i(t) \tag{3}$$

would simply integrate the inputs with scaling constant τ,

$$m(T) = m(0) + \frac{1}{\tau} \int_0^T \sum_i w_i X_i(t) \, dt$$

but the $-m(t)$ term in Equation 2 opposes this integration by a "leakage" of the potential $m(t)$ as it tries to return to its input-free equilibrium h.

Whereas the basic discrete-time binary model has a well-defined origin—the 1943 paper of McCulloch and Pitts—continuous time modeling has a more "distributed" history, which we now sample. A simple model of a spiking cell—the *integrate and fire* model—was introduced by Lapicque (1907), coupling the above model of membrane potential to a threshold. A spike would be generated each time the neuron reached threshold. This model captures the two key aspects of neuronal excitability: a passive, integrating response for small inputs and a stereotyped impulse once the input exceeds a particular amplitude. (DIFFUSION MODELS OF NEURON ACTIVITY describes what happens to this "first passage time" model of neuron firing when noise is added to the dynamics of the membrane potential.) Hill (1936) used *two* coupled leaky integrators, one of them representing membrane potential, and the other representing the fluctuating threshold. The leaky integrator also provides the basic model of the neuronal membrane in Eccles' account of *The Physiology of Nerve Cells* (Eccles, 1957, Section I.3).

What I shall call the leaky integrator model per se does not compute spikes on an individual basis—firing when the membrane potential reaches threshold—but rather defines the *firing rate* (e.g., the number of spikes traversing the axon in the most recent 20 ms) as a continuously varying *output* measure of the cell's activity. The firing rate is approximated by a simple, sigmoid function of the membrane potential. That is, we introduce a function σ of the membrane potential m such that $\sigma(m)$ increases from 0 to some maximum value as m increases from $-\infty$ to $+\infty$ (e.g., the sigmoidal function $k/[l + \exp(-m/\theta)]$, increasing from 0 to its maximum k), then the firing rate $M(t)$ of the cell is given by the equation

$$M(t) = \sigma(m(t))$$

It should be noted that, even at this simple level of modeling, there are alternative models. In the foregoing model, we have used subtractive inhibition. But there are inhibitory synapses which seem better described by *shunting* inhibition which, applied at a given point on a dendrite, serves to divide, rather than subtract from, the potential change passively propagating from more distal synapses. Again, the "lumped frequency" model cannot model the relative timing effects crucial to our motion detector example (see Figure 2). These might be approximated by introducing appropriate delay terms

$$\tau \frac{dm(t)}{dt} = -m(t) + \sum_i w_i x_i(t - \tau_i) + h$$

All this reinforces the observation that there is no modeling approach which is automatically appropriate. Rather, we seek to find the simplest model adequate to address the complexity of a given range of problems. The articles in Part III of the *Handbook* will provide many examples of the diversity of neural models appropriate to different tasks.

More Detailed Properties of Neurons

In Part III, the emphasis will shift from single neurons to neural networks. An important focus will then be on how networks change with experience. Much current research, both biological and technological, focuses on the synapses as the loci of change. In Section I.3, the only details we will add to the neuron models just presented will be various, relatively simple, rules of *synaptic plasticity*. These are the rules which describe how the weights w_i for the connections of a cell, and thus the response of the cell to its inputs, and thus the overall behavior of the network, may change to yield self-organization or various forms of learning. This level of detail (though with many variations) will suffice for a fair range of models of biological neural networks, and for most current work on artificial neural networks.

Here, however, I want to reiterate the point that biological neurons are far more subtle than can be captured in, say, the leaky integrator or integrate-and-fire model. This has the corollary, little explored at present, that neural plasticity may take many forms other than synaptic plasticity. The road map on **Biological Neurons** in Part II surveys a set of articles which demonstrate the complexity of biological neurons. Other road maps show the structures revealed in "special-purpose" neural circuitry in different species of animals. Table 1 lists some of the relevant articles on such circuits, together with the animal species on which the studies were based. The point is that much is to be learned from features specific to many different types of nervous systems, as well as from studies in humans, monkeys, cats, and rats which focus on commonalities with the human nervous system.

As Part III makes abundantly clear, an appreciation of this complexity is necessary for the computational neuroscientist wishing to address the increasingly detailed database of experimental neuroscience, but it should also prove important for the technologist looking ahead to the incorporation of new capabilities into the next generation of artificial neural networks (ANNs). Nonetheless, much can be accomplished with simple models, as we shall see in Section I.3. For an introduction to subtleties of function of biological neurons, the reader may start with the following four articles:

Table 1. Partial List of Part III Articles Describing "Special-Purpose" Circuitry in Various Species

Article Titles	Species
CRUSTACEAN STOMATOGASTRIC SYSTEM	Crabs and lobsters
DEVELOPMENT AND REGENERATION OF EYE-BRAIN MAPS	Frogs
ECHOLOCATION: CREATING COMPUTATIONAL MAPS	Bats
ELECTROLOCATION	Electric fish
FROG WIPING REFLEXES	Frogs
HALF-CENTER OSCILLATORS UNDERLYING RHYTHMIC MOVEMENTS	Various
INVERTEBRATE MODELS OF LEARNING	*Aplysia* and *Hermissenda*
LOCOMOTION, INVERTEBRATE	Various insects
LOCUST FLIGHT: COMPONENTS AND MECHANISMS IN THE MOTOR	Locusts
NEUROMODULATION IN INVERTEBRATE NERVOUS SYSTEMS	Various
OSCILLATORY AND BURSTING PROPERTIES OF NEURONS	Various
SCRATCH REFLEX	Turtles
SOUND LOCALIZATION AND BINAURAL PROCESSING	Owls
SPINAL CORD OF LAMPREY: GENERATION OF LOCOMOTOR PATTERNS	Lampreys
VISUOMOTOR COORDINATION IN FLIES	Flies
VISUOMOTOR COORDINATION IN FROGS AND TOADS	Frogs and toads
VISUOMOTOR COORDINATION IN SALAMANDERS	Salamanders

and then follow the cross-references from there. The Part II road maps on **Biological Neurons** and **Biological Networks** may also prove helpful. Table 1 of this section shows the diversity of nervous systems in different creatures that is studied in the *Handbook*; while the road map on **Mammalian Brain Regions** indicates the diversity of neural networks within the brains of mammals.

Readers wanting comprehensive textbook treatments to back up the necessarily cursory treatments of individual *Handbook* articles may find the following works a useful start for a "neuroscience bookshelf": Levitan and Kaczmarek (1991) for a detailed treatment of the mechanisms of (primarily single) neuron function with a molecular biology emphasis; Johnston and Wu (1995) for a detailed treatment of the same topic with a computational emphasis; Nieuwenhuys, Voogd, and van Huijzen (1988) both for pictures showing cross sections of the human brain and for illuminating diagrams showing how the different neural systems fit together; and, last but not least, Kandel, Schwartz, and Jessell (1991), which is the most comprehensive, multiauthored, textbook of (noncomputational) neuroscience, and which may serve as the ideal companion volume to the present *Handbook*.

Acknowledgments. This section is based in part on material contained in Section 2.3, "The Brain as a Network of Neurons," of M. A. Arbib, *The Metaphorical Brain 2: Neural Networks and Beyond*, New York: Wiley-Interscience, Copyright © 1989 by John Wiley & Sons, Inc., and is used here by permission of John Wiley & Sons, Inc. I also want to thank those *Handbook* authors who provided references for the historical notes on the leaky integrator model.

References

Eccles, J. C., 1957, *The Physiology of Nerve Cells*, Baltimore: Johns Hopkins Press.

Hill, A. V., 1936, Excitation and accommodation in nerve, *Proc. R. Soc. Lond. B*, 119:305–355.

Hodgkin, A. L., and Huxley, A. F., 1952, A quantitative description of membrane current and its application to conduction and excitation in nerve, *J. Physiol. Lond.*, 117:500–544.

Johnston, D., and Wu, S. M.-S., 1995, *Foundations of Cellular Neurophysiology*, with simulations and illustrations by R. Gray, Cambridge, MA: Bradford/MIT Press.

Kandel, E. R., Schwartz, J. H., and Jessell, T. M. (Eds.), 1991, *Principles of Neural Science*, 3rd ed., New York: Elsevier.

Lapicque, L., 1907, Recherches quantitatifs sur l'excitation electrique des nerfs traitée comme une polarisation, *J. Physiol. Paris*, 9:620–635.

Levitan, I. B., and Kaczmarek, L. K., 1991, *The Neuron: Cell and Molecular Biology*, New York: Oxford University Press.

McCulloch, W. S., and Pitts, W. H., 1943, A logical calculus of the ideas immanent in nervous activity, *Bull. Math. Biophys.*, 5:115–133.

Nieuwenhuys, R., Voogd, J., and van Huijzen, C., 1988, *The Human Central Nervous System: A Synopsis and Atlas*, 3rd rev. ed., Berlin: Springer-Verlag.

Rall, W., 1964, Theoretical significance of dendritic trees for neuronal input-output relations, in *Neural Theory and Modeling* (R. Reiss, Ed.), Stanford, CA: Stanford University Press, pp. 73–97.

Shepherd, G. M. (Ed.), 1990, *The Synaptic Organization of the Brain*, 3rd ed., New York: Oxford University Press.

I.2. Levels and Styles of Analysis

Many articles in this book show the benefits of interplay between biology and technology. Nonetheless, it is essential to distinguish between studying the brain and building an effective technology for intelligent systems and computation, and to distinguish among the various levels of investigation that exist (from the molecular to the system level) in these related, but by no means identical, disciplines. The present section provides a fuller sense of the disciplines that come together in *Brain Theory and Neural Networks*, and of the different levels of analysis involved in the study of complex biological and technological systems. In particular, it emphasizes that the study of neural networks, while central to the *Handbook*, by no means exhausts the themes of this volume.

A Historical Fragment

The simplest history of *Neural Networks* would start with three items: McCulloch and Pitts (1943), Hebb (1949), and Rosenblatt (1958). These publications introduced the first model of neural networks as "computing machines," the basic model of network self-organization, and the perceptron model of "learning with a teacher," respectively. Many newcomers to the field would then pass over the 1960s and 1970s and see the field as "starting over" with the publication of, to pick just two papers, the work of Hopfield (1982) relating symmetric networks to optimization, and that of Rumelhart, Hinton, and Williams (1986) which introduced the powerful method of backpropagation to a wide audience. (Section 1.3 provides a semitechnical introduction to this work and a key set of currently central ideas which build upon it.) However, to fill in the story we would have to review the neural network models of vision, memory, motor control, and self-organization studied in the 1960s and 1970s by Amari, Anderson, Cooper, Cowan, Fukushima, Grossberg, Kohonen, von der Malsburg, and Widrow—to mention a few of those researchers who continue to be active, and who have contributed articles to Part III of the Handbook.

A history of *Brain Theory* would overlap that of neural networks but only to a small extent, for it would emphasize two themes little visited in the studies briefly mentioned above: the development of increasingly detailed models of single neurons (see the "Brief Historical Notes" in PERSPECTIVE ON NEURON MODEL COMPLEXITY); and the modeling of interacting networks closely modeled on the neuroanatomy and neurophysiology of specific brain regions (as discussed in the road map on **Mammalian Brain Regions** and elsewhere).

However, the present section offers neither of these histories. Rather, it emphasizes that the study of neural networks, whether biological or artificial, can only be fully understood in the broader context of such disciplines as cognitive psychology, linguistics, perception, and motor control, as well as those of computer science and control theory. To this end, the present historical fragment only takes us up to 1948, the year preceding the publication of Hebb's book, but in doing so reveals the federation of disciplines reviewed in *The Handbook of Brain Theory and Neural Networks* as the current incarnation of what emerged in the 1940s as *Cybernetics: Or Control and Communication in the Animal and the Machine* (Wiener, 1948). The articles in Part III will make abundantly clear how far we have come since 1948, and also how many problems remain. Whereas Wiener's view of cybernetics was dominated by concepts of control and communication, our subject is dominated by notions of parallel and distributed computation, with special attention to learning in neural networks. On the other hand, notions of information and statistical mechanics championed by Wiener have re-emerged as a strong strand in the study of neural networks today (see, e.g., VISUAL CODING, REDUNDANCY, AND "FEATURE DETECTION" and STATISTICAL MECHANICS OF NEURAL NETWORKS). My intent in the present "fragment" is to enrich the reader's understanding of current contributions by using a selective historical tour to place them in context.

Noting that the Greek word *cybernetics* ($\kappa\upsilon\beta\epsilon\rho\nu\epsilon\tau\epsilon\sigma$) means the helmsman of a ship (cf. the Latin word *gubernator*, which gives us the word *governor* in English), Wiener (1948) used the term for a subject in which feedback played a central role. Feedback is the process whereby, e.g., the helmsman notes the "error," the extent to which he is off course, and "feeds it back" to decide which way to move the rudder. We can see the importance of this concept in endowing automata ("self-moving" machines) with flexible behavior. In 1748, in *L'Homme Machine*, La Mettrie had suggested that such automata as the mechanical duck and flute player of Vaucanson indicated the possibility of one day building a mechanical man that could talk. While these clockwork automata were capable of surprisingly complex behavior, they lacked a crucial aspect of animal behavior, let alone human intelligence: they were unable to adapt to changing circumstances. In the following century, machines were built that could automatically counter disturbances to restore desired performance. Perhaps the best known example of this is Watt's governor for the steam engine, which would let off excess steam if the velocity of the engine became too great. This development led to Maxwell's (1868) paper, "On Governors," which laid the basis for both the theory of negative feedback and the study of system stability (discussed in Section I.3). Negative feedback was feedback in which the error (in Watts' case, the amount by which actual velocity exceeded desired velocity) was used to counteract the error; stability occurred if this feedback was apportioned to reduce the error toward zero. Bernard (1878) brought these notions back to biology with his study of what Cannon (1939) would later dub *homeostasis*, observing that physiological processes often form circular chains of cause and effect that could counteract disturbances in such variables as body temperature, blood pressure, and glucose level in the blood. In fact, following publication of Wiener's 1948 book, the Josiah Macy, Jr., Foundation conferences, in which many of the pioneers of cybernetics were involved, became referred to as *Cybernetics: Circular Causal and Feedback Mechanisms in Biological and Social Systems*.

The nineteenth century also saw major developments in the understanding of the brain. At an overall anatomical level, a crucial achievement was the understanding of localization in the cerebral cortex (see Young, 1970, for a history of brain localization). Magendie and Bell had discovered that the dorsal roots of the spinal cord were sensory, carrying information from receptors in the body, while the ventral roots (on the belly side) were motor, carrying commands to the muscles. Fritsch and Hitzig, and then Ferrier, extended this principle to the brain proper, showing that the rear of the brain contains the primary receiving areas for vision, hearing, and touch, while the motor cortex is located in front of the central fissure. All this understanding of localization in the cerebral cortex led to the 19th century neurological doctrine, perhaps best exemplified in Lichtheim's (1885) development of the insights of Broca and Wernicke, which viewed different mental "faculties" as being localized in different regions of the brain. Thus, neurological deficits were to be explained as much in terms of lesions of the connections linking two such regions as of lesions to the regions themselves. We may also note a major precursor of the connectionism of this volume, where the connections are those between neuron-like units rather than anatomical regions: the associationist psychology of Alexander Bain (1868), who represented associations of ideas by the strengths of connections between "neurons" representing those ideas.

Around 1900, two major steps were taken in revealing the finer details of the brain. In Spain, Santiago Ramón y Cajal (e.g., 1906) gave us exquisite anatomical studies of many regions of the brain, revealing the particular structure of each as a network of neurons. In England, the physiological studies of Charles Sherrington (1906) on reflex behavior provided the basic physiological understanding of synapses, the junction points between the neurons. Somewhat later, in Russia, Ivan Pavlov (1927), extending associationist psychology and building on the Russian studies of reflexes by Sechenov in the 1860s, established the basic facts on the modifiability of reflexes by conditioning (see Fearing, 1930, for a historical review).

A very different setting of the scene for cybernetics came from work in mathematical logic in the 1930s. Kurt Gödel published his famous *Incompleteness Theorem* in 1931 (see Arbib, 1987, for an accessible proof and a debunking of the claim that Gödel's theorem sets limits on machine intelligence). The "formalist" program initiated by David Hilbert, which sought to place all mathematical truth within a single formal system, had reached its fullest expression in the *Principia Mathematica* of Whitehead and Russell. But Gödel showed that, if one used the approach offered in *Principia Mathematica* to set up consistent axioms for arithmetic and prove theorems by logical deduction from them, the theory *must* be incomplete no matter which axioms ("knowledge base") one started with —there would be true statements of arithmetic that could not be deduced from the axioms.

Following Gödel's 1931 study, many mathematical logicians sought to formalize the notion of an effective procedure, of what could *and could not* be done by explicitly following an algorithm or set of rules. Kleene (1936) developed the theory of partial recursive functions; Turing (1936) developed his machines; Church (1941) developed the lambda calculus, the forerunner of McCarthy's list-processing language, LISP, a favorite of AI workers; and Emil Post (1943) introduced systems for rewriting strings of symbols, of which Chomsky's early formalizations of grammars in 1959 were a special case. Fortunately, these methods proved to be equivalent. Whatever could be computed by one of these methods could be computed by any other method if it were equipped with a suitable "program." It thus came to be believed (Church's thesis) that if a function could be computed by any machine at all, it could be computed by each one of these methods.

Turing (1936) helped chart the limits of the computable with his notion of what is now called a *Turing machine*, a device that followed a fixed, finite set of instructions to read, write, and move upon a finite but indefinitely extendable tape, each square of which bore a symbol from some finite alphabet. As one of the ingredients of Church's thesis, Turing offered a "psychology of the computable," making plausible the claim that any effectively definable computation— that is, anything that a human could do in the way of symbolic manipulation by following a finite and completely explicit set of rules—could be carried out by such a machine equipped with a suitable program. Turing also provided the most famous example of a noncomputable problem, "the unsolvability of the halting problem." Let p be the numerical code for a Turing machine program, and let x be the code for the initial contents of a Turing machine's tape. Then the halting function $h(p, x) = 1$ if Turing machine p will eventually halt if started with data x; otherwise it is 0. Turing showed that there was no "computer program" that could compute h.

And so we come to 1943, the key year for bringing together the notions of control mechanism and intelligent automata:

In "A Logical Calculus of the Ideas Immanent in Nervous Activity," McCulloch and Pitts (1943) united the studies of neurophysiology and mathematical logic. Their formal model of the neuron as a threshold logic unit (see Section I.1) built on the neuron doctrine of Ramón y Cajal and the excitatory and inhibitory synapses of Sherrington, using notation from the mathematical logic of Whitehead, Russell, and Carnap. McCulloch and Pitts provided the "physiology of the computable" by showing that the control box of any Turing machine, the essential formalization of symbolic computation, could be implemented by a network (with loops) of their formal neurons. The ideas of McCulloch and Pitts influenced John von Neumann and his colleagues when they defined the basic architecture of stored program computing. Thus as electronic computers were built toward the end of World War II, it was understood that whatever they could do could be done by a network of neurons.

Craik's (1943) book, *The Nature of Explanation*, viewed the nervous system "as a calculating machine capable of modeling or paralleling external events," suggesting that the process of forming an "internal model" that paralleled the world is the basic feature of thought and explanation. In the same year, Rosenblueth, Wiener, and Bigelow published "Behavior, Purpose and Teleology." Engineers had noted that if feedback used in controlling the rudder of a ship were too brusque, the rudder would overshoot, compensatory feedback would yield a larger overshoot in the opposite direction, and so on and so on as the system wildly oscillated. Wiener and Bigelow asked Rosenblueth whether there was any corresponding pathological condition in humans and were given the example of intention tremor associated with an injured cerebellum. This evidence for feedback within the human nervous system (see MOTOR CONTROL, BIOLOGICAL AND THEORETICAL) led the three scientists to advocate that neurophysiology move beyond the Sherringtonian view of the central nervous system as a reflex device adjusting itself in response to sensory inputs. Rather, setting reference values for feedback systems could provide the basis for analysis of the brain as a purposive system explicable only in terms of circular processes, that is, from nervous system to muscles to the external world and back again via receptors.

Such studies laid the basis for the emergence of cybernetics which in turn gave birth to a number of distinct new disciplines—such as artificial intelligence, biological control theory, cognitive psychology, and neural modeling—which went their separate ways in the 1970s. The next subsection introduces a number of these disciplines and the relations between them; this analysis will continue in many articles in Part III of the *Handbook*.

Brains, Machines, and Minds

The range of *Brain Theory and Neural Networks* covers *brains* (the biological mechanisms), *machines* (the technological mechanisms), and *minds* (a subset of the function or phenomenology of the brain's activity). Thus it overlaps with a large number of related disciplines, some involving neural networks and others not, as the present section will show.

Brains. *Brain theory* comprises many different theories as to how the structures of the brain can subserve such diverse functions as perception, memory, control of movement, and higher mental function. As such, it includes both attempts to extend notions of computing, as well as applications of modern electronic computers to explore the performance of complex models. An example of the former is the study of *cooperative computation* between different structures in the brain as a new paradigm for computing that transcends classical notions associated with serial execution of symbolic programs. For the latter, *computational neuroscience* makes systematic use of mathematical analysis and computer simulation to model the structure and function of living brains, building on earlier work in both neural modeling and biological control theory.

Machines. *Artificial intelligence (AI)* studies how computers may be programmed to yield "intelligent" behavior without necessarily attempting to provide a correlation between structures in the program and structures in the brain. *Robotics* is related to AI, but emphasizes the flexible control of machines (robots) which have receptors (e.g., television cameras and force sensors) and effectors (e.g., wheels, legs, arms, grippers) that allow them to interact with the world.

Brain theory has spawned a companion field of *neural computation*, which involves the design of machines with circuitry inspired by, but which need not faithfully emulate, the neural networks of brains. Many technologists usurp the term "neural networks" for the study of *artificial* neural networks, but we will use it as an umbrella term which may, depending on context, describe biological nervous systems, models thereof, or the artificial networks which (sometimes at great remove) they inspire. When the emphasis is on "higher mental functions," neural computation may be seen as a new branch of AI (see the road map **Artificial Intelligence and Neural Networks** in Part II), but it also contributes to robotics (especially to those robot designs inspired by analysis of animal behavior), and to a wide range of technologies, including those based on image analysis, signal processing, and control (see the road maps **Control Theory and Robotics** and **Applications of Neural Networks**).

Much work on artificial neural networks emphasizes adaptive neural networks which, without specific programming, can adjust their connections through self-organization or to meet specifications given by some teacher. However, the systematic design of neural networks, especially for applications in low-level vision (such as stereopsis, optic flow, and shape-from-shading), is also important. Moreover, complex problems cannot, in general, be solved by the tuning or the design of a single unstructured network. For example, robot control may integrate a variety of low-level vision networks with a set of competing and cooperating networks for motor control and its planning. Brain theory and neural computation thus have to address the analysis and design, respectively, of networks of networks (see, e.g., MODULAR NEURAL NET SYSTEMS, TRAINING OF and MODULAR AND HIERARCHICAL LEARNING SYSTEMS).

Minds. Here, I want to distinguish the brain from the mind (the realm of the "mental"). In great part, brain theory seeks to analyze how the brain guides the behaving organism in its interactions with the dynamic world around it, but much of the control of such interactions is not mental, and much of what is mental is subsymbolic and/or unconscious (see PHILOSOPHICAL

ISSUES IN BRAIN THEORY AND CONNECTIONISM and CONSCIOUSNESS, THEORIES OF). People can agree on examples of mental activity (perceiving a visual scene, reading, thinking, etc.) even if they take the diametrically opposite philosophical positions of dualism (mind and brain are separate) or monism (mind is a function of brain). They would then agree that some mental activity (e.g., contemplation) need not result in overt "interactions with the dynamic real world," and that much of the brain's activity (e.g., controlling normal breathing) is not mental. Face recognition seems to be a mental activity which we do not carry out through symbol manipulation.

Cognitive psychology attempts to explain the mind in terms of "information processing" (a notion which continues to change). It thus occupies a middle ground between brain theory and AI. A cognitive model must explain psychological data (such as what tasks are hard for humans; people's ability at memorization; the development of the child; patterns of human errors; etc.), but the units of the model need not correspond to actual brain structures. In the 1960s and 1970s, the majority of cognitive psychologists formulated their theories in terms of information theory and/or symbol manipulation, while theories of biological organization were ignored. However, workers in both AI and cognitive psychology now pay increasing attention to the cooperative computation paradigm. The term *connectionism* has come to be used for studies which model human thought and behavior in terms of parallel distributed networks of neuron-like units, with learning mediated by changes in strength of the connections between these elements (see COGNITIVE MODELING: PSYCHOLOGY AND CONNECTIONISM).

Brain theory seeks to enhance our understanding of human thought and the neural basis of human and animal behavior. It requires empirical data to shape and constrain modeling; and in return provides concepts and hypotheses to shape and constrain experimentation. *Neural computation* develops new strategies for building "intelligent" machines or adaptive robots. Here, the criterion for success is the design of a machine which can perform a task cheaply, reliably, and effectively, even if, in the process of making the best use of available (e.g., silicon) technology, the final design departs radically from the biological neural network that inspired it. It will be important in reading this *Handbook*, then, to be clear as to whether a particular study is an exercise in brain theory/computational neuroscience or in AI/neural computation. What will not be in doubt is not only that brain mechanisms can inspire new technology, but also that new technologies provide metaphors to drive new theories of brain function. To this it must be added that, at present, most workers in neural computation know little of brain function, and relatively few neuroscientists know much of neural computation beyond the basic ideas of Hebbian plasticity and, perhaps, backpropagation (see Section I.3). This *Handbook* is designed to increase the flow of information between these scientific communities.

Levels of Analysis

The study of brain and mind offers insight at many different levels of analysis—from large information processing blocks down to the finest details of molecular structure:

- Much of *psychology* and *linguistics* looks at human behavior "from the outside," whether studying overall competence or attending to details of performance.
- *Neuropsychology* relates behavior to the interaction of various brain regions.
- *Neurophysiology* studies the activity of neurons in networks whose structure is revealed by *neuroanatomy*, both to understand the intrinsic properties of the neurons and to help understand their role in the subsystems dissected out by the neuropsychologist, such as networks for pattern recognition or for visuomotor coordination.
- *Molecular* and *cell biology* and *biophysics* correlate the structure and connectivity of the membranes and subcellular systems which constitute cells with the way these cells transform incoming patterns or subserve memory by changing function with repeated interactions.

Much study of the brain is guided by evolutionary and comparative studies of animal behavior and brain function (cf. EVOLUTION OF THE ANCESTRAL VERTEBRATE BRAIN). The information about the function of the human brain that is gained in the neurological clinic or during neurosurgery can thus be supplemented by humane experimentation upon animals. (However, as evidenced by Table 1 of Section I.1, we can learn a great deal by studying the *differences*, as well as the similarities, between the brains of different species.) We learn by stimulating, recording from, or excising portions of an animal's brain and seeing how the animal's behavior changes. We may then compare such results with observations using such techniques as positron emission tomography imaging (PET) or functional magnetic resonance imaging (MRI) (Martin, Brust, and Hilal, 1991) of the relative activity of different parts of the human brain during different tasks. The grand aim of *cognitive neuroscience* (as neuropsychology has now become) is to use clinical data and brain imaging to form a high-level view of the involvement of various brain regions in human cognition, using single-cell activity recorded from animals engaged in analogous behaviors to suggest the neural networks underlying this involvement. The catch, of course, is that the "analogous behaviors" of animals are not very analogous at all when it comes to such symbolic activities as language and reasoning. In Part III, we will see that "higher mental functions" tend to be modeled in connectionist terms constrained (if at all) by psychological or psycholinguistic data (cf. the Part II road maps on **Connectionist Psychology** and **Connectionist Linguistics**). If a connectionist model succeeds in describing some psychological input/output behavior, it may become an important hypothesis that its internal structure is "real" (see DYNAMIC MODELS OF NEUROPHYSIOLOGICAL SYSTEMS); in general, however, connectionist "neurons" are formal units which do not actually represent biological neurons in the brain. The greatest successes to date in seeking the neural underpinnings of human behavior have come in areas such as vision, memory, and motor control where we can make neural network models of animal analogs of human capabilities (cf. the road maps on **Vision, Other Sensory Systems, Learning in Biological Systems, Motor Pattern Generators and Neuroethology**, and **Primate Motor Control**).

We also learn from the attempt to reproduce various aspects of human behavior in a robot, even though human action, memory, learning, and perception are far richer than those of any machine yet built or likely to be built in the near future. To the extent that they address "higher mental function," the studies presented in this *Handbook* suggest that there is no single "thing" called *intelligence*, but rather a plexus of properties which varies from case to case. Thus, when we suggest that the brain can be thought of in some ways as a (highly distributed) computer, we are not trying to reduce humans to the level of extant machines, but rather to understand ways in which machines give us insight into human attributes.

These differing levels make it possible to focus individual research studies, but they are ill-defined, and a scientist who works on any one level needs to make occasional forays, both downward to find mechanisms for the functions studied, and upward to understand what role the studied function can play in the overall scheme of things. *Top-down* modeling starts from some overall behavior and explains it in terms of the interaction of high-level functional units, while *bottom-up* modeling starts from the interaction of individual neurons (or even smaller units) to explain network properties.

Perhaps the most cited view of top-down modeling is afforded by Marr's (1982) "three levels" at which information processing is to be understood:

- Level 1, computational theory, specifies a function to be computed.
- Level 2, representation and algorithm, describes a set of computational steps to apply to a representation of the inputs and outputs to compute that function.
- Level 3, implementation, concerns how the level 2 structures can be carried out in some real-world device such as a brain or a computer.

These levels are certainly useful in structuring our thoughts about a complex system. However, Marr went further, suggesting that scientific analysis should proceed purely top-down, with work at level 2 directed by the specification at level 1 but totally unconstrained by work at level 3. However "the underlying idea that studies at each level can be independently pursued is itself highly dubious. For example, our top-level decomposition of a task into subtasks apt for computational modeling may be challenged once we become familiar with the distribution of information-processing resources in the brain" (from PHILOSOPHICAL ISSUES IN BRAIN THEORY AND CONNECTIONISM). As will be shown in the discussion of approach and avoidance behavior, what we originally though of as two distinct functions may turn out to share circuitry in the brain. Conversely, the modeling studies of House (1989, cf. NEUROETHOLOGY, COMPUTATIONAL) suggest that (in frogs and toads at least) an apparently unitary process—namely that of depth perception—may invoke a number of different implementations, with the depths to prey and to barriers extracted from the optic array by different neural systems distinguished on the grounds of "level 3" data on neurophysiology which thus drive, rather than being driven by, the "algorithmic" analysis of level 2. Thus, top-down modeling cannot be guaranteed to succeed unless constrained by the insights of bottom-up modeling. It requires a judicious blend of the two to connect the clear overview of crucial questions to the hard data of neuroscience or, in the case of neural engineering, to the details of implementation. Most successful modeling will be purely bottom-up or top-down only in its initial stages, if at all. Constraints on an initial top-down model will be given, for example, by the data on regional localization offered by the neurologist, or the circuit-cell-synapse studies of much current neuroscience.

However, even if we agree on the need for bottom-up, as well as top-down, considerations to constrain a level of analysis intermediate between abstract function and neural implementation, a question remains: What is the appropriate language for this intermediate representation? Our consensus in *Brain Theory and Neural Networks* is that algorithms expressed in the style of serial computation will seldom be adequate. The further point to be emphasized here is that it will in general be unrealistic to expect the analysis of a complex system to jump directly from the highest level decomposition of tasks to neural networks, be they connectionist or biological. To make sense of the brain, we often divide it into functional systems—such as the motor system, the visual system, and so on—as well as into structural systems—from the spinal cord and the hippocampus to the various subdivisions of the prefrontal cortex. Similarly,

in DISTRIBUTED ARTIFICIAL INTELLIGENCE, the solution of a task may be distributed over a complex set of interacting agents, each with their dedicated processors for handling the information available to them locally. Thus, both neuroscience and artificial intelligence require a language that complements connectionism by providing a bridging language between functional description and neural networks. One such approach is based on a theory of the concurrent activity of interacting functional units called *schemas*. Perceptual schemas are those used for perceptual analysis, while motor schemas are those which provide the control systems which can be coordinated to effect a wide variety of movement. Other schemas compete and cooperate to meld action, internal state, and perception in an ongoing action-perception cycle (see SCHEMA THEORY).

Schemas have been little studied within the field of neural network analysis per se, since their role is to analyze function without a necessary commitment to the mode of implementation such as that involved in the design or adaptation of specific neural networks. Rather, they have played an important role (as has control theory) where an attempt is made to define units of behavior as a basis for analyzing their development, interaction, and/or mechanism. Thus, the term *schema* arises in related, though not always formally congruent, uses in such fields relevant to this *Handbook* as cognitive psychology (Mandler, 1985; see Rumelhart, Smolensky, McClelland, and Hinton, 1986, for a rapprochement with connectionism), developmental psychology (Piaget, 1971), motor control and kinesiology (Schmidt, 1988), neuroethology (Ewert, 1989), and neuropsychology (Shallice, 1988).

Figure 4A provides a multilevel view of brain theory in which schemas provide an intermediate level of description for interactions mediated by biological neural networks. We may model the brain either functionally, analyzing some behavior in terms of interacting schemas, or structurally, through the interaction of anatomically defined units, such as brain regions (cf. the examples in the road map on **Mammalian Brain Regions**) or substructures of these regions, such as layers or columns (see CORTICAL COLUMNS, MODULES, AND HEBBIAN CELL ASSEMBLIES, which includes a possible bridge to the "schema-level" functional units called cell assemblies). In brain theory, we ultimately seek an explanation in terms of neural networks, since the neuron may be considered the basic unit of function as well as of structure. Other work in computational neuroscience seeks to explain the complex functionality of real neurons in terms of "subneural" units, such as membrane compartments, channels, spines, and synapses (see, e.g., DENDRITIC SPINES, ION CHANNELS: KEYS TO NEURONAL SPECIALIZATION, and PERSPECTIVE ON NEURON MODEL COMPLEXITY, as well as the road map on **Biological Neurons**). The article SPINAL CORD OF LAMPREY: GENERATION OF LOCOMOTOR PATTERNS illustrates several of the more detailed levels of modeling when study is restricted to a specific structure (spinal cord of lamprey) and function (generation of oscillations involved in swimming). Connectionist modeling is used to investigate whether a generic network structure can yield a particular class of oscillations, while biophysical modeling is essential to determine the importance of particular cellular properties to function, such as the mechanism by which locomotor bursts may be terminated. At all levels, biological data are used to build a model, and investigation of the model then leads to predictions for biological experiments—and so the theory-experiment cycle continues.

Figure 4B offers a similar picture for distributed AI, but with the essential difference that schemas provide a programming language for the *design* of a complex system, and that the implementation of different schemas may be distributed across a variety of different mechanisms, whether as conventional com-

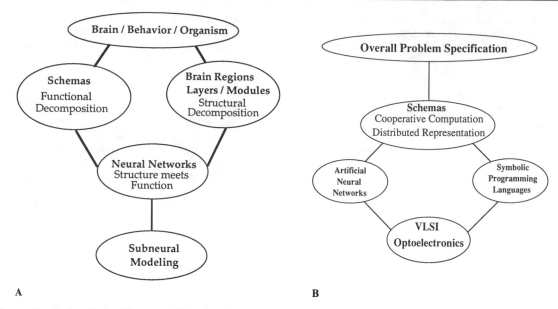

Figure 4. Views of level of analysis of brain and behavior (*A*) and a distributed technological system (*B*), highlighting the role of schemas as an intermediate level of functional analysis in each case.

puter programs, ANNs, or special-purpose devices. These different approaches then rest upon effective design of VLSI "chips" or other computing materials (cf. the road map on **Implementation of Neural Networks**).

Our analysis of Marr's views on top-down analysis has already stated that, in general, a functional analysis proceeding top-down from some overall behavior need not map directly into a bottom-up analysis proceeding upward from the neural circuitry (brain theory) or basic set of processors (distributed AI). Rather, several iterations from the "middle out" may be required to bring the structural and functional accounts into consonance. To illuminate the notion of experimental insight modifying an initial top-down analysis, we consider an example from *Rana computatrix*, a set of models of visuomotor coordination in the frog and toad (cf. VISUOMOTOR COORDINATION IN FROGS AND TOADS). Frogs and toads snap at small moving objects and jump away from large ones (to oversimplify somewhat). Thus, a simple schema-model of the frog brain might simply postulate four schemas: two perceptual schemas (processes for recognizing objects or situations) and two motor schemas (processes for controlling some structured behavior). One perceptual schema would recognize small moving objects and activate a motor schema for approaching the prey; the other would recognize large moving objects and activate a motor schema for avoiding the predator. Lesion experiments can put such a model to the test. It was thought that the tectum (a key visual region in the animal's midbrain) was the locus for recognizing small moving objects, while the pretectum (a region just in front of the tectum) was the locus for recognizing large moving objects. Based on these hypotheses on localization of schemas in the brain, the model described would predict that an animal with a lesioned pretectum would be unresponsive to large objects, but would respond normally to small objects. However, the facts are quite different. A pretectum-lesioned toad will approach moving objects, both large and small, and does not exhibit avoidance behavior. This has led to a new schema model in which a perceptual schema to recognize large moving objects is still localized in the pretectum but the tectum now contains a perceptual schema for *all* moving objects. We then add that activity of the pretectal schema

not only triggers the avoidance motor schema but also inhibits approach. This new schema model still yields the normal behavior to large and small moving objects, but also fits the lesion data, since removal of the pretectum removes inhibition, meaning that the animal will now approach any moving object (Ewert and von Seelen, 1974). The language of schemas lets us express hypotheses about the various functions that the brain performs without assuming localization of any one function in any one region, but also allows us to express the way in which many regions participate in a given function, or a given region participates in many functions.

The style of *cooperative computation* exhibited in both SCHEMA THEORY (q.v.) and connectionism is far removed from serial computation and the symbol-based ideas that have dominated conventional AI. As we shall see in example after example in Part III, the brain has many specialized areas, each with a partial representation of the world (for the case of the primate visual system, see the figure in VISUAL SCENE PERCEPTION: NEUROPHYSIOLOGY). It is only through the interaction of these regions that the unity of behavior of the animal emerges, and the human is no different in this regard. The representation of the world is *the pattern of relationships between all its partial representations*.

References

Arbib, M. A., 1987, *Brains, Machines, and Mathematics*, 2nd ed., New York: Springer-Verlag.

Bain, A., 1868, *The Senses and the Intellect*, 3rd ed.

Bernard, C., 1878, *Leçons sur les phénomènes de la Vie*.

Brooks, R. A., 1986, A robust layered control system for a mobile robot, *IEEE Robot. Automat.*, RA-2:14–23.

Cannon, W. B., 1939, *The Wisdom of the Body*, New York: Norton.

Chomsky, N., 1959, On certain formal properties of grammars, *Inform. and Control*, 2:137–167.

Church, A., 1941, *The Calculi of Lambda-Conversion*, Annals of Mathematics Studies 6, Princeton, NJ: Princeton University Press.

Craik, K. J. W., 1943, *The Nature of Explanation*, New York: Cambridge University Press.

Ewert, J.-P., 1989, The release of visual behavior in toads: Stages of parallel/hierarchical information processing, in *Visuomotor coordi-*

nation: *Amphibians, Comparisons, Models, and Robots* (J.-P. Ewert and M. A. Arbib, Eds.), New York: Plenum Press, pp. 39–120.

Ewert, J.-P., and von Seelen, W., 1974, Neurobiologie and System-Theorie eines visuellen Muster-Erkennungsmechanismus bei Kroten, *Kybernetik*, 14:167–183.

Fearing, F., 1930, *Reflex Action*, Baltimore: Williams and Wilkins.

Gödel, K., 1931, Uber formal unentscheidbare Stze der *Principia Mathematica* und verwandter Systeme, I, *Monats. Math. Phys.*, 38:173–198.

Hebb, D. O., 1949, *The Organization of Behavior*, New York: Wiley.

Hopfield, J., 1982, Neural networks and physical systems with emergent collective computational properties, *Proc. Natl. Acad. Sci. USA*, 79:2554–2558.

House, D., 1989, *Depth Perception in Frogs and Toads: A Study in Neural Computing*, Lecture Notes in Biomathematics 80, Berlin: Springer-Verlag.

Kleene, S. C., 1936, General recursive functions of natural numbers, *Math. Ann.*, 112:727–742.

La Mettrie, J., 1953, *Man a Machine* (trans. by G. Bussey from the French original of 1748), La Salle, IL: Open Court.

Lichtheim, L., 1885, On aphasia, *Brain*, 7:433–484.

Mandler, G., 1985, *Cognitive Psychology: An Essay in Cognitive Science*, Hillsdale, NJ: Lawrence Erlbaum Associates.

Marr, D., 1982, *Vision: A Computational Investigation into the Human Representation and Processing of Visual Information*, New York: W.H. Freeman and Co.

Martin, J. H., Brust, J. C. M., and Hilal, S., 1991, Imaging the living brain, in *Principles of Neural Science*, 3rd ed., (E. R. Kandel, J. H. Schwartz, and T. M. Jessel, Eds.), New York: Elsevier.

Maxwell, J. C., 1868, On governors. *Proc. R. Soc. Lond.*, 16:270–283.

McCulloch, W. S., and Pitts, W. H., 1943, A logical calculus of the ideas immanent in nervous activity. *Bull. Math. Biophys.*, 5:115–133.

Pavlov, I. P., 1927, *Conditioned Reflexes: An Investigation of the Physiological Activity of the Cerebral Cortex* (Trans. from the Russian by G. V. Anrep), New York: Oxford University Press.

Piaget, J., 1971, *Biology and Knowledge*, Edinburgh: Edinburgh University Press.

Post, E. L., 1943, Formal reductions of the general combinatorial decision problem, *Am. J. Math.*, 65:197–268.

Ramón y Cajal, S., 1906, The structure and connexion of neurons, reprinted in *Nobel Lectures: Physiology or Medicine, 1901–1921*, New York: Elsevier, 1967, pp. 220–253.

Rosenblatt, F., 1958, The perceptron: A probabilistic model for information storage and organization in the brain, *Psychol. Rev.*, 65:386–408.

Rosenblueth, A., Wiener, N., and Bigelow, J., 1943, Behavior, purpose and teleology, *Philos. Sci.*, 10:18–24.

Rumelhart, D. E., Hinton, G. E., and Williams, R. J., 1986, Learning internal representations by error propagation, in *Parallel Distributed Processing: Explorations in the Microstructure of Cognition*, vol. 1 (D. Rumelhart, J. McClelland, and PDP Research Group, Eds.), Cambridge, MA: MIT Press, pp. 318–362.

Rumelhart, D. E., Smolensky, P., McClelland, J. L., and Hinton, G. E., 1986, Schemata and sequential thought processes in PDP models, in *Parallel Distributed Processing: Explorations in the Microstructure of Cognition*, vol. 2 (J. L. McClelland and D. E. Rumelhart, Eds.), Cambridge, MA: Bradford/MIT Press, chap. 14.

Schmidt, R. A., 1988, *Motor Control and Learning: A Behavioral Emphasis*, Champaign, IL: Human Kinetics.

Shallice, T., 1988, *From Neuropsychology to Mental Structure*, Cambridge: Cambridge University Press.

Sherrington, C., 1906, *The Integrative Action of the Nervous System*, New York: Oxford University Press.

Turing, A. M., 1936, On computable numbers with an application to the entscheidungsproblem, Proc. Lond. Math. Soc. (Series 2), 42:230–265.

Turing, A. M., 1950, Computing machinery and intelligence, *Mind*, 59:433–460.

Wiener, N., 1948, *Cybernetics: Or Control and Communication in the Animal and the Machine*, New York: The Technology Press and Wiley (2nd ed., Cambridge, MA: MIT Press, 1961).

Young, R. M., 1970, *Mind, Brain and Adaptation in the Nineteenth Century: Cerebral Localization and Its Biological Context from Gall to Ferrier*, New York: Oxford University Press.

I.3. Dynamics and Adaptation in Neural Networks

Section I.1 introduced a number of key concepts from the biological study of neurons, stressing the diversity of neurons both within the human CNS and across species. It presented several simple models of neurons, noting that computational neuroscience has gone on to produce more subtle and complicated neuronal models, while neural computation tends to use simple neurons augmented by "learning rules" for changing connection strengths on the basis of "experience." The purpose of this section is to introduce two key approaches that dominate the modern study of neural networks:

• The study of neural networks as dynamic systems
• The study of neural networks as adaptive systems

To make this section essentially self-contained, we start by presenting the definitions of the McCulloch-Pitts and leaky integrator neurons from Section I.1, but we do this in the context of a general, semiformal, introduction to dynamic systems.

Dynamic Systems

We motivate the notion of dynamic systems by considering how to abstract the interaction of an organism (or a machine) with its environment. The organism will be influenced by aspects of the current environment—the *inputs* to the organism—while the activity of the environment will be responsive, in turn, to aspects of the current activity of the organism—the *outputs* of the organism. The inputs and outputs that actually enter into a *theory* of the organism (or machine) are a small sampling of the flux of its interactions with the rest of the universe. Our abstraction of any real system contains five elements:

1. The set of *inputs*: those variables of the environment which we believe will affect the system behavior of interest to us.
2. The set of *outputs*: those variables of the system which we choose to observe, or which we believe will significantly affect the environment.
3. The set of *states*: those internal variables of the system (which may or may not also be output variables) which determine the relationship between input and output. Essentially, the state of a system is the system's "internal residue of the past"; when we know the state of a system, no further information about the past behavior of the system will enable us to refine predictions of the way in which future inputs and outputs of the system will be related.
4. The *state-transition function*: that function which determines how the state will change when the system obtains various inputs.

5. The *output function*: that function which determines what output the system will yield with a given input when in a given state.

Any system in which the state-transition function and output function uniquely determine the new state and output from a specification of the initial state and subsequent inputs is called a *deterministic* system. If, no matter how carefully we specify subsequent inputs to a system, we can at best specify a probability distribution on the subsequent states and outputs, we say the system is *probabilistic* or *stochastic*. A stochastic treatment may be worthwhile either because we are analyzing systems which are "inescapably" stochastic (e.g., at the quantum level) or because we are analyzing macroscopic systems which lend themselves to a stochastic description by ignoring "fine details" of microscopic variables. For example, it is usually more reasonable to describe a coin in terms of a 0.5 probability of coming up heads than to measure the initial placement of the coin on the finger and the thrust of the thumb in sufficient detail to determine whether the coin will come up heads or tails.

Continuous-Time Systems

In Newtonian mechanics, the state of the system comprises the positions of its components, which are directly observable, and their velocities, which can be estimated from the observed trajectory over a period of time. Time is continuous (i.e., characterized by the set $\mathbb{R}$ of real numbers), and the way in which the state changes is described by a differential equation: classical mechanics provides the basic example of *continuous*-time systems in which the present state and input determine *the rate at which the state changes*. This requires that the input, output, and state spaces be continuous spaces in which such continuous changes can occur. Consider the simple example of a point mass undergoing rectilinear motion. At any time, its position $y(t)$ is the observable output of the system, and the force $u(t)$ acting upon it is the input applied to the system. Newton's third law says that the force applied to the system equals the mass times the acceleration: $\ddot{y}(t) = mu(t)$, where the acceleration $\ddot{y}(t)$ is the second derivative of $y(t)$. According to Newton's laws, the state of the system is given by the position and velocity of the particle. We call the position-velocity pair, at any time, the *instantaneous state* $q(t)$ of the system. In fact, the earlier equation gives us enough information to deduce the rate of change $dq(t)/dt$ of this state. Using standard matrix formalism, we have

$$\frac{d}{dt}\begin{bmatrix} y(t) \\ \dot{y}(t) \end{bmatrix} = \begin{bmatrix} \dot{y}(t) \\ \ddot{y}(t) \end{bmatrix} = \begin{bmatrix} \dot{y}(t) \\ mu(t) \end{bmatrix} = \begin{bmatrix} 0 & 1 \\ 0 & 0 \end{bmatrix}\begin{bmatrix} y(t) \\ \dot{y}(t) \end{bmatrix} + \begin{bmatrix} 0 \\ m \end{bmatrix}u(t)$$

while

$$y(t) = \begin{bmatrix} 1 \\ 0 \end{bmatrix}q(t)$$

This is an example of a *linear* system in which the rate of change of state depends linearly on the present state and input, and the present output depends linearly on the present state. That is, there are matrices F, G, and H such that

$$\frac{dq(t)}{dt} = Fq(t) + Gu(t)$$

$$y(t) = Hq(t)$$

More generally, a physical system can be expressed by a pair of equations:

$$\frac{dq(t)}{dt} = f(q(t), u(t))$$

$$y(t) = g(q(t))$$

The first expresses the rate of change $dq(t)/dt$ of the state as a function of both the state $q(t)$ and the input or control vector $u(t)$ applied at any time t.

We now present the definition of a *leaky integrator neuron* as a continuous-time system. The internal state of the neuron is its membrane potential, $m(t)$, and its output is the firing rate, $M(t)$. The *state transition function* of the cell is expressed as

$$\tau\frac{dm(t)}{dt} = -m(t) + \sum_i w_i X_i(t) + h \tag{1}$$

while the *output function* of the cell is given by the equation

$$M(t) = \sigma(m(t)) \tag{2}$$

Thus if there are m inputs $X_i(t)$, $i = 1, \ldots, m$, then the *input space* of the neuron is $\mathbb{R}^m$, with current value $(X_1(t), \ldots, X_m(t))$, while the *state* and *output* spaces of the neuron both equal $\mathbb{R}$, with current values $m(t)$ and $M(t)$, respectively.

Let us now briefly (and semiformally) see how a neural network comprised of leaky integrator neurons can also be seen as a continuous-time system in this sense. As typified in Figure 5, we characterize a neural network by selecting N neurons (each with specified input weights and resting potential) and by taking the axon of each neuron, which may be split into several branches carrying identical output signals, and either connecting each line to a unique input of another neuron or feeding it outside the net to provide one of the K network output lines. Then every input to a given neuron must be connected either to an output of another neuron or to one of the (possibly split) L input lines of the network. Then the input set $X = \mathbb{R}^L$, the state set $Q = \mathbb{R}^N$, and the output set $Y = \mathbb{R}^K$. If the ith output line comes from the jth neuron, then the *output function* is determined by the fact that the ith component of the output at time t is the firing rate $M_j(t) = \sigma_j(M_j(t))$ of the jth neuron at time t. The state-transition function for the neural network follows from the state-transition functions of each of the N neurons

$$\tau\frac{dm_i(t)}{dt} = -m_i(t) + \sum_j w_{ij} X_{ij}(t) + h_i \tag{3}$$

as soon as we specify whether $X_{ij}(t)$ is the output $M_k(t)$ of the kth neuron or the value $x_l(t)$ currently being applied on the lth input line of the overall network.

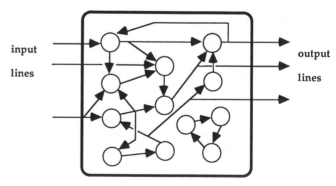

Figure 5. A neural network viewed as a system. The input at time t is the pattern of firing on the input lines; the output is the pattern of firing on the output lines; and the internal state is the vector of firing rates of all the neurons of the network.

Discrete-Time Systems

In contrast to continuous-time systems, which *must* have continuous-state spaces on which the differential equations for the state transition function can be defined, *discrete-time* systems may have either continuous- or discrete-state spaces. (A *discrete*-state space is just a set with no specific metric or topological structure.) For example, a McCulloch-Pitts neuron is considered to operate on a discrete time scale, $t = 0, 1, 2, 3, \ldots$, and have m binary input lines and one binary output line. Such a neuron has input set $\{0, 1\}^m$ and state and output set $\{0, 1\}$. The neuron is characterized by connection weights w_i and threshold θ. If at time t the value of the ith input is $x_i(t)$, then the state and output one time step later, $y(t + 1)$, equals 1 if and only if $\Sigma_i w_i x_i(t) \geq \theta$.

The important learning scheme known as backpropagation (defined later) is based on neurons which are discrete-time, but with both input, state, and output taking continuous values in some range, say [0, 1].

In computer science, an *automaton* is a discrete-time system with discrete input, output, and state spaces. Formally, we describe an automaton by the sets X, Y, and Q of inputs, outputs, and states, respectively, together with the *next-state function* $\delta: Q \times X \to Q$ and the *output function* $\beta: Q \to Y$. If the automaton is in state q and receives input x at time t, then its next state will be $\delta(q, x)$ and its next output will be $\beta(q)$. A McCulloch-Pitts neural network is a network like that shown in Figure 5, but a discrete-time network with each neuron a McCulloch-Pitts neuron. Such a network functions like a finite automaton whose state, given by the firing pattern of all the constituent neurons, changes synchronously on each tick of the time scale $t = 0, 1, 2, 3, \ldots$ Conversely, it can be shown (McCulloch and Pitts, 1943; see AUTOMATA AND NEURAL NETWORKS) that any finite automaton can be simulated by a suitable McCulloch-Pitts neural network.

Stability, Limit Cycles, and Chaos

With the previous discussion, we now have more than enough material to understand the crucial dynamic systems concept of *stability* and the related concepts of limit cycles and chaos (see COMPUTING WITH ATTRACTORS and CHAOS IN NEURAL SYSTEMS; as well as Hirsch and Smale, 1974, and Abraham and Shaw, 1981 1983). We want to know what happens to an "unperturbed" system, i.e., one for which the input is held constant (possibly with some specific "null input," usually denoted by 0, the "zero" input in X). An *equilibrium* is a state q in which the system can stay at rest, i.e., such that $\delta(q, 0) = q$ (discrete time) or $dq/dt = f(q, 0) = 0$ (continuous time). The study of stability is concerned with the issue of whether or not this rest point will be maintained in the face of slight disturbances. To see the variety of equilibria, we use the image of a sticky ball rolling on the "hillside" of Figure 6. We say that point A on the "hillside" in this diagram is an *unstable equilibrium* because a slight displacement from A will tend to increase over time. Point B is in a region of *neutral equilibrium* because slight displacements will tend not to change further, while C is a point of *stable equilibrium*, since small displacements will tend to disappear over time. Note the word "small": in a nonlinear system like that of Figure 6, a large displacement can move the ball from the *basin of attraction* of C (the set of states whose dynamics tend towards C) to another. Clearly, the ball will not tend to return to C after a massive displacement which moves the ball to the far side of A's hilltop.

Nonlinear systems also have another interesting property: they may exhibit *limit cycles*. These are closed trajectories in

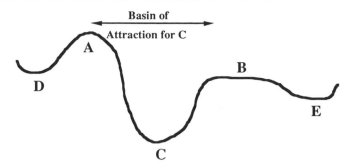

Figure 6. An energy landscape: For a ball rolling on the "hillside," point A is an *unstable equilibrium*, point B lies in a region of *neutral equilibrium*, and point C is a point of *stable equilibrium*. Point C is called an attractor: the basin of attraction of C comprises all states whose dynamics tend toward C.

the state space, and thus may be thought of as "dynamic equilibria." If the state of a system follows a limit cycle, we may also say it oscillates or exhibits periodic behavior. A limit cycle is *stable* if a small displacement from it will be reduced as the trajectory of the system comes closer and closer to the original limit cycle. By contrast, a limit cycle is *unstable* if such excursions do not die out. Research in nonlinear systems has also revealed what are called *strange attractors*. These are attractors which, unlike simple limit cycles, describe such complex paths through the state space that, although the system is deterministic, a path which approaches the strange attractor gives every appearance of being random. Very small differences in initial state that at first are not even noticeable will yield, in due course, states that are very different indeed. Such a trajectory has become the accepted mathematical model of *deterministic chaos*, and it is used to describe a number of physical phenomena, such as the onset of turbulence in a weather system, as well as a number of phenomena in biological systems (see CHAOS IN NEURAL SYSTEMS).

Hopfield Nets

Many authors have treated neural networks as dynamic systems, employing notions of equilibrium, stability, and so on, to classify their performance (see, for example, Grossberg, 1967, and Amari and Arbib, 1977, as well as COMPUTING WITH ATTRACTORS). However, it was a paper by John Hopfield (1982) that was the catalyst in attracting the attention of many physicists to this field of study. In a McCulloch-Pitts network, every neuron processes its inputs to determine a new output at each time step. By contrast, a *Hopfield net* is a net of such units with (a) *symmetric* weights ($w_{ij} = w_{ji}$) and no self-connections ($w_{ii} = 0$), and (b) *asynchronous* updating: For instance, let s_i denote the state (0 or 1) of the ith unit. At each time step, pick just one unit at random. If unit i is chosen, s_i takes the value 1 if and only if $\Sigma w_{ij} s_j \geq \theta_i$. Otherwise s_i is set to 0. Note that this is an *autonomous* (input-free) network: there are no inputs (although instead of considering θ_i as a threshold, we may consider $-\theta_i$ as a constant input, also known as a bias).

Hopfield defined a measure called the *energy* for such a net (see ENERGY FUNCTIONS FOR NEURAL NETWORKS),

$$E = -\frac{1}{2} \sum_{ij} s_i s_j w_{ij} + \sum_i s_i \theta_i \qquad (1)$$

This is not the physical energy of the neural net, but a mathematical quantity that, in some ways, does for neural dynamics what the potential energy does for Newtonian mechanics. In general, a mechanical system moves to a state of lower potential energy just as, in Figure 6, the ball tends to move downhill. Hopfield showed that his symmetrical networks with asynchronous updating had a similar property:

For example, if we pick a unit and the foregoing firing rule does not change its s_i, it will not change E. However if s_i initially equals 0, and $\Sigma w_{ij}s_j \geq \theta_i$, then s_i goes from 0 to 1 with all other s_j constant, and the "energy gap," or change in E, is given by

$$\Delta E = -\frac{1}{2}\sum_j (w_{ij}s_j + w_{ji}s_j) + \theta_i$$

$$= -\sum_j w_{ij}s_j s_j + \theta_i \quad \text{by symmetry}$$

$$\leq 0 \quad \text{since } \sum w_{ij}s_j \geq \theta_i$$

Similarly, if s_i initially equals 1, and $\Sigma w_{ij}s_j < \theta_i$, then s_i goes from 1 to 0 with all other s_j constant, and the energy gap is given by

$$\Delta E = \sum w_{ij}s_j - \theta_i < 0$$

In other words, with every asynchronous update, we have $\Delta E \leq 0$. Hence the dynamics of the net tends to move E toward a minimum. We stress that there may be different such states—they are *local* minima—just as, in Figure 6, both D and E are local minima (each of them is lower than any "nearby" state) but not global minima (since C is lower than either of them). Global minimization is not guaranteed.

The expression just presented for ΔE depends on the symmetry condition, $w_{ij} = w_{ji}$, for, without this condition, the expression would instead be $\Delta E = -\frac{1}{2}\Sigma_j(w_{ij}s_j + w_{ji}s_j) + \theta_i$. In this case, Hopfield's updating rule would not yield a passage to energy minimum, but might instead yield a limit cycle, which could be useful in, e.g., controlling rhythmic behavior (see, e.g., RESPIRATORY RHYTHM GENERATION). In a control problem, a link w_{ij} might express the likelihood that the action represented by i would precede that represented by j, in which case $w_{ij} = w_{ji}$ is normally inappropriate.

The condition of asynchronous update is crucial, too. If we consider the simple "flip-flop" with $w_{12} = w_{21} = 1$ and $\theta_1 = \theta_2 = 0.5$, then the McCulloch-Pitts network will *oscillate* between the states $(0, 1)$ and $(1, 0)$ or will sit in the states $(0, 0)$ or $(1, 1)$; in other words, there is no guarantee that it will converge to an equilibrium. However, with $E = -\frac{1}{2}\Sigma_{ij}s_i s_j w_{ij} + \Sigma_i s_i \theta_i$, we have $E(0, 0) = 0$, $E(0, 1) = E(1, 0) = 0.5$, and $E(1, 1) = 0$, and the Hopfield network will *converge* to the minimum at either $(0, 0)$ or $(1, 1)$.

Hopfield also aroused much interest because he showed how a number of optimization problems could be "solved" using neural networks. (The quotes around "solved" acknowledge the fact that the state to which a neural network converges may represent a local, rather than a global, optimum of the corresponding optimization problem.) Such networks were similar to the "constraint satisfaction" networks that had already been studied in the computer vision community. (In most vision algorithms—see, for instance, STEREO CORRESPONDENCE AND NEURAL NETWORKS and MOTION PERCEPTION—constraints can be formulated in terms of symmetric weights, so that $w_{ij} = w_{ji}$ is appropriate.) The aim, given a "constraint satisfaction" problem, is to so choose weights for a neural network so that the energy E for that network is a measure of the overall constraint violation. A famous example is the Traveling Salesman Problem (TSP): There are n cities, with a road of length l_{ij} joining city i to city j. The salesman wishes to find a way to visit the cities that is optimal in two ways: each city is visited only once, and the total route is as short as possible. We express this as a constraint satisfaction network in the following way: Let the activity of neuron N_{ij} express the decision to go straight from city i to city j. The cost of this move is simply l_{ij}, and so the total "transportation cost" is $\Sigma_{ij}l_{ij}N_{ij}$. It is somewhat more challenging to express the cost of violating the "visit a city only once" criterion, but we can re-express it by saying that, for city j, there is one and only one city i from which j is directly approached. Thus $\Sigma_j(\Sigma_i N_{ij} - 1)^2 = 0$ just in case this constraint is satisfied; a nonzero value measures the extent to which this constraint is violated. The sum

$$\sum_{ij} l_{ij}N_{ij} + \sum_j (\Sigma_i N_{ij} - l)^2$$

can then be regarded as an energy E to yield the setting of weights and thresholds for a Hopfield network with that energy. Hopfield and Tank (1986) constructed chips for this network which do indeed settle very quickly to a local minimum of E. Unfortunately, there is no guarantee that this minimum is globally optimal. The articles NEURAL OPTIMIZATION and CONSTRAINED OPTIMIZATION AND THE ELASTIC NET present this and a number of other neurally based approaches to optimization. The articles SIMULATED ANNEALING and BOLTZMANN MACHINES show how noise may be added to "shake" a system out of a local minimum and let it settle into a global minimum. (Consider, for example, shaking that is strong enough to shake the ball from D to A, and thus into the basin of attraction of C, in Figure 6, but not strong enough to shake the ball back from C toward D.)

Adaptation in Dynamic Systems

In the previous discussion of neural networks as dynamic systems, the dynamics (i.e., the state-transition function) has been fixed. However, just as humans and animals learn from experience, so do many important applications of artificial neural networks (ANNs) depend on the ability of these networks to adapt to the task at hand by, e.g., changing the values of the synaptic weights to improve performance. We now introduce the general notion of an adaptive system as background to some of the most influential "learning rules" used in adaptive neural networks. The key motivation for using learning networks is that it may be too hard to program explicitly the behavior that one sees in a black box, but one may be able to drive a network by the actual input/output behavior of that box, or by some description of its trajectories, to cause it to adapt itself into a network which approximates that given behavior. However, as we will stress at the end of this section, a learning algorithm may not solve a problem within a reasonable period of time unless the initial structure of the network is suitable.

Adaptive Control

A key problem of technology is to control a complex system so that it behaves in some desired way—whether getting a space shuttle into orbit, or a steel mill to produce high-quality steel. A common situation which complicates this *control problem* is that the controlled system may not be known accurately—it may even change its character somewhat with time. For example, as fuel is depleted, the mass and moments of inertia of the shuttle may change in unpredicted ways. The *adaptation problem* involves determining, on the basis of interaction with a given system, an appropriate "model" of the system which the controller can use in solving the control problem.

Suppose we have available an *identification procedure* which can find an adequate parametric representation of the controlled system (see IDENTIFICATION AND CONTROL). Then, rather than build a controller specifically designed to control this one system, we instead build a general-purpose controller which can accommodate to a "family" of systems characterized by a set of parameters. The controller then uses the parameters which the identification procedure provides as the best estimate of the controlled system's parameters at that time. If the identification procedure can make accurate estimates of the system's parameters more quickly than they actually change, the controller will be able to act efficiently despite fluctuations in controlled system dynamics. The controller, when coupled to an identification procedure, is an *adaptive controller*—that is, it adapts its control strategy to changes in the dynamics of the controlled system. However, the use of an explicit identification procedure is only one way of building an adaptive controller. Adaptive neural nets may be used to build adaptive procedures which may directly modify the parameters in some control rule, or identify the system *inverse* so that desired outputs can be automatically transformed into the inputs that will achieve them. (See ADAPTIVE CONTROL: NEURAL NETWORK APPLICATIONS for more information on adaptive control: our task in what follows is to introduce the types of adaptive neural networks on which they are based.)

Pattern Recognition

With x_j a "measure of confidence" that the jth of a set of features occurs in some input *pattern x*, the *preprocessor* shown in Figure 7 converts x into the *feature vector* $(x_1, x_2, \ldots, x_d)$ in a d-dimensional Euclidean space $\mathbb{R}^d$ called the *pattern space*. The pattern recognizer takes the feature vector and produces a response that has the appropriate one of K distinct values; points in $\mathbb{R}^d$ are thus grouped into at least K different categories (see CONCEPT LEARNING and PATTERN RECOGNITION). However, a category might be represented in more than one connected region of $\mathbb{R}^d$. To take an example from visual pattern recognition (although the theory of pattern recognition networks applies to any classification of $\mathbb{R}^d$), *a* and *A* are members of the category of the first letter of the English alphabet, but they would be found in different connected regions of a pattern space. In such cases, it may be necessary to establish a hierarchical system involving a separate apparatus to recognize each subset, and a further system that recognizes that the subsets all belong to the same set (a related idea was originally developed by Selfridge, 1959; for adaptive versions, see COULOMB POTENTIAL LEARNING and MODULAR AND HIERARCHICAL LEARNING SYSTEMS). Here we avoid this problem by concentrating on the case in which the decision space is divided into exactly two connected regions.

We call a function $f : \mathbb{R}^d \to \mathbb{R}^d$ a *discriminant function* if the equation $f(x) = 0$ gives the *decision surface* separating two regions of a pattern space. A basic problem of pattern recognition is the specification of such a function. It is virtually impossible for humans to "read out" the function they use (not to mention *how* they use it) to classify patterns. Thus, a common strategy in pattern recognition is to provide a classification machine with an adjustable function and to "train" it with a set of patterns of known classification that are typical of those with which the machine must ultimately work. The function may be linear, quadratic, polynomial, or even more subtle yet, depending on the complexity and shape of the pattern space and the necessary discriminations. The experimenter chooses a class of functions with parameters which, it is hoped, will, with proper adjustment, yield a function that will successfully classify any given pattern. For example, the experimenter may decide to use a linear function of the form

$$f(x) = w_1 x_1 + w_2 x_2 + \ldots + w_d x_d + w_{d+1}$$

(i.e., a McCulloch-Pitts neuron) in a two-category pattern classifier. The equation $f(x) = 0$ gives a hyperplane as the decision surface, and training involves adjusting the coefficients $(w_1, w_2, \ldots, w_d, w_{d+1})$ so that the decision surface produces an acceptable separation of the two classes. We say that two categories are *linearly separable* if an acceptable setting of such linear weights exists.

In summary, pattern recognition poses (at least) the following challenges to neural networks:

(a) Find a "good" set of preprocessors. Competitive learning based on Hebbian plasticity (see COMPETITIVE LEARNING, as well as the text that follows) provides one way of finding such features by extracting statistically significant patterns from a set of input patterns. For example, if a network were exposed to many, but only, letters of the Roman alphabet, then it would find that certain line segments and loops occurred repeatedly, even if there were no teacher to tell it how to classify the patterns.

(b) Given a set of preprocessors and a set of patterns which have already been classified, adjust the connections of a neural network so that it acts as an effective pattern recognizer. That is, its response to a preprocessed pattern should usually agree well with the classification provided by a teacher.

(c) Of course, if the neural network has multiple layers with adaptable synaptic weights, then the early layers can be thought of as preprocessors for the later layers, and we have a case of supervised, rather than Hebbian, formation of these "feature detectors"—emphasizing features which are not only statistically significant elements of the input patterns but which also serve to distinguish usefully to which class a pattern belongs.

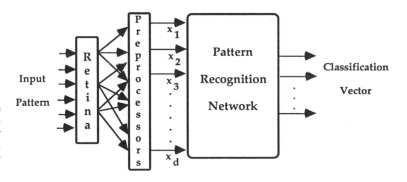

Figure 7. One strategy in pattern recognition is to precede the adaptive neural network by a layer of "preprocessors" or "feature extractors" which replace the image by a finite vector for further processing. In other approaches, the functions defined by the early layers of the network may themselves be subject to training.

Associative Memory

In pattern recognition, we associate a pattern with a "label" or "category." More generally, an associative memory takes some "key" as input and returns some "associated recollection" as output (see ASSOCIATIVE NETWORKS). For example, given the sound of a word, we may wish to recall its spelling. Given a misspelled word, we may wish to recall the correctly spelled word of which it is most plausibly a "degraded image." There are two major approaches to the use of neural networks as associative memories.

In *nonrecurrent* neural networks, there are no loops (i.e., we cannot start at any neuron and "follow the arrows" to get back to that neuron). We use such a network by fixing the pattern of inputs as the key, and holding them steady. Since the absence of loops ensures that the input pattern uniquely determines the output pattern (after the new inputs have time to propagate their effects through the network), this uniquely determined output pattern is the recollection associated with the key.

In *recurrent* networks, the presence of loops implies that the input alone may not determine the output of the net, since this will also depend on the initial state of the network. Thus, recurrent networks are often used as associative memories in the following way. The inputs are only used transiently to establish the initial state of the neural network. After that, the network operates autonomously (i.e., uninfluenced by any inputs). If and when it reaches an equilibrium state, that state is read out as the recollection associated with the key.

In either case, the problem is to set the weights of the neural network so that it associates keys as accurately as possible with the appropriate recollections. With this, we turn to the two classic learning schemes for McCulloch-Pitts-type formal neurons due to Hebb and Rosenblatt, followed by a discussion of backpropagation, a currently popular method for training multilayer loop-free networks. These and many other methods are discussed in the road maps on **Self-Organization in Neural Networks; Learning in Artificial Neural Networks, Deterministic;** and **Learning in Artificial Neural Networks, Statistical**.

Hebbian Plasticity and Network Self-Organization

In Hebb's (1949) learning scheme (see HEBBIAN SYNAPTIC PLASTICITY), the connection between two neurons is strengthened if both neurons fire at the same time. The simplest example of such a rule is to increase w_{ij} by the following amount:

$$\Delta w_{ij} = k y_i x_j$$

where synapse w_{ij} connects a presynaptic neuron with firing rate x_j to a postsynaptic neuron with firing rate y_i. The trouble with the original Hebb model is that every synapse will eventually get stronger and stronger until they all saturate, thus destroying any selectivity of association. Von der Malsburg's (1973) solution was to normalize the synapses impinging on a given neuron. To accomplish this, one must first compute the Hebbian "update" $\Delta w_{ij} = k x_i y_j$ and then divide this by the total putative synaptic weights to get the final result, which replaces w_i by

$$\frac{w_{ij} + \Delta w_{ij}}{\sum_k (w_{kj} + \Delta w_{kj})}$$

where the summation k extends over all inputs to the neuron. This new rule not only increases the strengths of those synapses with inputs strongly correlated with the cell's activity, but also decreases the synaptic strengths of other connections in which such correlations did not arise.

Von der Malsburg was motivated by the pattern recognition problem, and was concerned with how individual cells in his network might come to be tuned so as to respond to one particular input "feature" rather than another (see OCULAR DOMINANCE AND ORIENTATION COLUMNS for background as well as a review of more recent approaches). This exposed another problem with Hebb's rule: a lot of nearby cells may, just by chance, all have initial random connectivity which makes them easily persuadable by the same stimulus; alternatively, the same pattern might occur many times before a new pattern is experienced by the network. In either case, many cells would become tuned to the same feature, with not enough cells left to learn important and distinctive features. To solve this, von der Malsburg introduced *lateral inhibition* into his model. In this connectivity pattern, activity in any one cell is distributed laterally to reduce (partially inhibit) the activity of nearby cells. This ensures that if one cell—call it A—were active, its connections to nearby cells would make them less active, and so make them less likely to learn, by Hebbian synaptic adjustment, those features that most excite A.

In summary, then, when the Hebbian rule is augmented by a normalization rule, it tends to "sharpen" a neuron's predisposition "without a teacher," getting its firing to become better and better correlated with a cluster of stimulus patterns. This performance is improved when there is some competition between neurons so that if one neuron becomes adept at responding to a pattern, it inhibits other neurons from doing so (COMPETITIVE LEARNING). Thus, the final set of input weights to the neuron depends both on the initial setting of the weights and on the pattern of clustering of the set of stimuli to which it is exposed (see DATA CLUSTERING AND LEARNING). Various "post-Hebbian" rules, motivated both by technological efficiency and by recent biological findings, are discussed in several articles in Part III, including HEBBIAN SYNAPTIC PLASTICITY: COMPARATIVE AND DEVELOPMENTAL ASPECTS; BCM THEORY OF VISUAL CORTICAL PLASTICITY; and POST-HEBBIAN LEARNING RULES.

In the adaptive architecture just described, the inputs are initially randomly connected to the cells of the processing layer. As a result, none of these cells is particularly good at pattern recognition. However, by sheer statistical fluctuation of the synaptic connections, one will be slightly better at responding to a particular pattern than others are; it will thus slightly strengthen those synapses which allow it to fire for that pattern, and, through lateral inhibition, this will make it harder for cells initially less-well-tuned for that pattern to become tuned to it. Thus, without any teacher, this network automatically organizes itself so that each cell becomes tuned for an important cluster of information in the sensory inflow. This is a basic example of the kind of phenomenon treated in the road map on **Self-Organization in Neural Networks**.

Perceptrons

Perceptrons are neural nets that change with "experience," using an *error-correction rule* designed to change the weights of each response unit when it makes erroneous responses to stimuli that are presented to the network (see PERCEPTRONS, ADALINES, AND BACKPROPAGATION). We refer to the judge of what is correct as the "teacher," although this may be another neural network, or some environmental input, rather than a signal supplied by a human teacher in the usual schoolroom sense. Consider the case in which a set **R** of input lines (in the case of visual pattern recognition, we think of **R** as a "retina") feeds a single layer of preprocessors whose outputs feed into a McCulloch-Pitts neuron (called the *output unit* of the perceptron) with adjustable weights $(w_1, \ldots, w_d)$ and threshold θ to effect a twofold classification. A *simple perceptron* is one in which the preprocessors are not interconnected, *which means that the network has no short-term memory.* (If such connections are present, the perceptron is called *cross-coupled.* A cross-coupled perceptron may have multiple layers and loops back from an "earlier" to a "later" layer.) If the preprocessors feed the pattern $x = (x_1, \ldots, x_d)$ to the output unit, then the response of that unit will be to provide the pattern discrimination with discriminant function $f(x) = w_1 x_1 + \ldots + w_d x_d - \theta$. In other words, the simple perceptron can only compute a linearly separable function of the pattern as provided by the associator units. Rosenblatt (1958) showed that if the patterns of the training set (i.e., a set of feature vectors, each one coupled with a classification of 0 or 1) are linearly separable, then there is a learning scheme which will eventually yield a satisfactory setting of the weights. The best known perceptron learning rule strengthens an active synapse if the efferent neuron fails to fire when it should have fired, and weakens an active synapse if the neuron fires when it should not have done so:

$$\Delta w_{ij} = k(Y_i - y_i)x_j$$

As before, synapse w_{ij} connects a presynaptic neuron with firing rate x_j to a postsynaptic neuron with firing rate y_i, but now Y_i is the "correct" output supplied by the "teacher." (This is similar to the Widrow-Hoff [1960] least mean squares model of adaptive control; see PERCEPTRONS, ADALINES, AND BACKPROPAGATION.) Notice that the rule does change the response to x_j "in the right direction." If the output is correct, $Y = y$ and there is no change, $\Delta w_j = 0$. If the output is too small, then $Y - y > 0$, and the change in w_j will add $\Delta w_j x_j = k(Y - y)x_j x_j > 0$ to the output unit's response to $(x_1, \ldots, x_d)$. Similarly, if the output is too large, then $Y - y < 0$, Δw_j will add $k(Y - y)x_j x_j < 0$ to the output unit's response. Thus, there is a sense in which the new setting $w + \Delta w$ classifies the input pattern x "more nearly correctly" than w does. Unfortunately, in classifying x "more correctly" we run the risk of classifying another pattern "less correctly." However, the *perceptron convergence theorem* (see Nilsson, 1965, for a proof) shows that Rosenblatt's procedure does not yield an endless seesaw, but will eventually converge to a correct set of weights if one exists, albeit perhaps after many iterations through the set of trial patterns.

Network Complexity

The perceptron convergence theorem states that, if a linear separation exists, the perceptron error-correction scheme will find it. Minsky and Papert (1969) revivified the study of perceptrons (although some AI workers thought they had killed it!) by responding to such results with questions like "Given a pattern-recognition problem, how much of the retina must each associator unit 'see' if the network is to do its job?" Of course, we can always get away with using a single preprocessor computing an arbitrary Boolean function that is connected to all the units of the retina. So the question that really interests us is whether we can get away with a small number of response units connected to each of the preprocessors to make a global decision by synthesizing an array of local views.

Consider XOR, the simple Boolean operation of addition modulo 2, also known as the exclusive-or. If we imagine the square with vertices $(0, 0)$, $(0, 1)$, $(1, 1)$, and $(1, 0)$ in the Cartesian plane, with (x_1, x_2) being labeled by $x_1 \oplus x_2$, we have 0s at one diagonally opposite pair of vertices and 1s at the other diagonally opposite pair of vertices. It is clear that there is no way of interposing a straight line such that the 1s lie on one side

and the 0s lie on the other side. However, we shall prove it mathematically to gain insight into the techniques used by Minsky and Papert.

Consider the claim that we wish to prove wrong: that there actually exists a neuron with threshold θ and weights α and β such that $x_1 \oplus x_2 = 1$ if and only if $\alpha x_1 + \beta x_2 \geq \theta$. The crucial point is to note that the function of addition modulo 2 is symmetric; therefore, we must also have $x_1 \oplus x_2 = 1$ if and only if $\beta x_1 + \alpha x_2 \geq \theta$, and, so, adding together the two terms, we have $x_1 \oplus x_2 = 1$ if and only if $\frac{1}{2}(\alpha + \beta)(x_1 + x_2) \geq \theta$. Writing $\frac{1}{2}(\alpha + \beta)$ as γ, we see that we have reduced three putative parameters α, β, and θ to just two, namely γ and θ.

We now set $t = x_1 + x_2$ and look at the polynomial $\gamma t - \theta$. It is a degree 1 polynomial, but note: at $t = 0$, $\gamma t - \theta$ must be less than zero ($0 \oplus 0 = 0$); at $t = 1$, it is greater than or equal to zero ($0 \oplus 1 = 1 \oplus 0 = 1$); and at $t = 2$, it is again less than zero ($1 \oplus 1 = 0$). This is a contradiction—a polynomial of degree 1 cannot change sign from positive to negative more than once. We conclude that there is no such polynomial, and thus that there is no threshold element which will add modulo 2.

The general method starts with a pattern-classification problem and observes that certain symmetries leave it invariant. For instance, for the parity problem (is the number of active elements even or odd?), which includes the case of addition modulo 2, any permutation of the points of the retina would leave the classification unchanged. Use this to reduce the number of parameters describing the circuit. Then lump items together to get a polynomial and examine actual patterns to put a lower bound on the degree of the polynomial, fixing things so that this degree bounds the number of inputs to the response unit of a simple perceptron.

Minsky and Papert prove that the parity function requires preprocessors big enough to scan the whole retina if the preprocessors can only be followed by a single McCulloch-Pitts neuron. By contrast, to tell whether the number of active retinal inputs reaches a certain threshold only requires two inputs per neuron in the first layer of a simple perceptron. Of course, if the fan-in (the number of input lines to each neuron) is fixed, then more and more layers of processing will be required to compute a given function. For a very general perspective on such results, see STRUCTURAL COMPLEXITY AND DISCRETE NEURAL NETWORKS. Other complexity results are reviewed in articles in the road map on **Computability and Complexity**.

Gradient Descent and Credit Assignment

The implication of the results on "network complexity" is clear: if we limit the complexity of the units in a neural network, then in general we will need many layers, rather than a single layer, if the network is to have any chance of being trained to realize many "interesting" functions. This conclusion motivates the study of training rules for multilayer perceptrons of which the most widely used is *backpropagation.* Before describing this method, we first discuss two general notions of which it is an important exemplar: *gradient descent* and *credit assignment.*

In discussing Hopfield networks, we introduced the metaphor of an "energy landscape" (see Figure 6). The asynchronous updates move the state of the network (the vector of neural activity levels) "downhill," tending toward a local energy minimum. Our task now is to realize that the metaphor works again on a far more abstract level when we consider learning. In learning, the dynamic variable is not the network state, but rather the vector of synaptic weights (or whatever other set of network parameters is adjusted by the learning rules). We now conduct *gradient descent in weight space* (see Learning as Hill-Climbing in Weight Space). At each step, the weights are adjusted in such a way as to improve the performance of the network. (As in the case of the simple perceptron, the improvement is a "local" one based on the current situation. It is, in this case, a matter for computer simulation to prove that the cumulative effect of these small changes is a network which solves the overall problem.)

But how do we recognize which "direction" in weight space is "downhill"? Suppose success is achieved by a complex mechanism after operating over a considerable period of time (for example, a chess-playing program wins a game). To what particular decisions made by what particular components should the success be attributed? And, if failure results, what decisions deserve blame? This is closely related to the problem known as the "mesa" or "plateau" problem (Minsky, 1961). The performance evaluation function available to a learning system may consist of large level regions in which gradient descent degenerates to exhaustive search, so that only a few of the situations obtainable by the learning system and its environment are known to be desirable, and these situations may occur rarely.

One aspect of this problem, then, is the *temporal* credit assignment problem. The utility of making a certain action may depend on the sequence of actions of which it is a part, and an indication of improved performance may not occur until the entire sequence has been completed. This problem was attacked successfully in Samuel's (1959) learning program for playing checkers. The idea is to interpret predictions of future reward as rewarding events themselves. In other words, neutral stimulus events can themselves become reinforcing if they regularly occur before events that are intrinsically reinforcing. Such *temporal difference learning* (see Reinforcement Learning) is like a process of erosion: the original uninformative mesa, where only a few sinkholes allow gradient descent to a local minimum, is slowly replaced by broader valleys in which gradient descent may successfully proceed from many different places on the landscape.

Another aspect of credit assignment concerns structural factors. In the simple perceptron, only the weights to the output units are to be adjusted. This architecture can only support linearly separable maps (based on the patterns presented by the preprocessors), and we have seen that many interesting problems require preprocessing units of undue complexity to achieve linear separability. We thus need multiple layers of preprocessors, and—since one may not know a priori the appropriate set of preprocessors for a given problem—these units should be trainable too. This raises the question "How does a neuron deeply embedded within a network 'know' what aspect of the outcome of an overall action was 'its fault'?" This is the *structural* credit assignment problem.

Backpropagation

We now study *backpropagation*, which trains a *multilayer perceptron*—a loop-free network which has its units arranged in layers, with a unit providing input only to units in the next layer of the sequence—by propagating back some measure of "responsibility" to a hidden (nonoutput) unit. Backpropagation is an "adaptive architecture": it is not just a local rule for synaptic adjustment; it also takes into account the position of a neuron in the network to indicate how the neuron's weights are to change. (In this sense, we may see the use of lateral inhibition to improve Hebbian learning as the first example of an adaptive architecture in these pages.)

The first layer of a multilayer perceptron comprises fixed input units; there may then be several layers of trainable "hidden units" carrying an internal representation, and finally, there is the layer of output units, also trainable. (In a simple perceptron, we view the input units as fixed associator units—i.e., they deliver a preprocessed, rather than a "raw," pattern—which connect directly to the output units without any hidden units in between.) For what follows, it is crucial that each unit *not* be binary: it has both input and output taking continuous values in some range, say $[0, 1]$. The response is a sigmoidal function of the weighted sum. Thus, if a unit has inputs x_k with corresponding weights w_{ik}, the output x_i is given by $x_i = f_i(\Sigma w_{ik}x_k)$, where f_i is a sigmoidal function, say

$$f_i(x) = \frac{1}{1 + e^{-(x+\theta_i)}}$$

with θ_i being a bias or threshold for the unit.

The environment only evaluates the output units. We are given a training set of input patterns p and corresponding desired target patterns t^p for the output units. With o^p the actual output pattern elicited by input p, the aim is to adjust the weights in the network to minimize the error

$$E = \sum_{\substack{\text{patterns} \\ p}} \sum_{\substack{\text{output} \\ \text{neurons} \\ k}} (t_k^p - o_k^p)^2$$

Rumelhart, Hinton, and Williams (1986) were among those (see Backpropagation: Basics and New Developments for a broader perspective) who devised a formula for propagating back the gradient of this evaluation from a unit to its inputs. This process can continue by backpropagation through the entire net. The scheme seems to avoid many false minima. At each trial, we fix the input pattern p and consider the corresponding "restricted error"

$$E = \sum_k (t_k - o_k)^2$$

where k ranges over designated "output units." The net has many units interconnected by weights w_{ij}. The learning rule is to change w_{ij} so as to reduce E by *gradient descent*:

$$\Delta w_{ij} = -\frac{\partial E}{\partial w_{ij}} = 2 \sum_k (t_k - o_k) \frac{\partial o_k}{\partial w_{ij}}$$

Consider a net divided into $m + 1$ layers, with layer 0 comprising the input units, with nets in layer $g + 1$ receiving all their inputs from layer g, and with layer m comprising the output units. The following theorem can then be proved by induction on how many layers back we must go to reach a unit:

Theorem. Consider a layered loop-free net with error $E = \Sigma_k(t_k - o_k)^2$, where k ranges over designated "output units." Then changing the weights w_{ij} according to the gradient descent rule

$$\Delta w_{ij} = -\frac{\partial E}{\partial w_{ij}} = 2 \sum_k (t_k - o_k) \frac{\partial o_k}{\partial w_{ij}}$$

is equivalent to changing the weights, working back from the output units, by the rule

$$\Delta w_{ij} \text{ is proportional to } \delta_i o_j$$

where:

Basis Step: $\delta_i = (t_i - o_i)f_i'$ for an output unit.
Induction Step: If i is a hidden unit, and if δ_k is known for all units which receive unit i's output, then $\delta_i = (\Sigma_k \delta_k w_{ki})f_i'$, where k runs over all units which receive unit i's output.   □

Thus the "error signal" δ_i propagates back layer by layer from the output units. In $\Sigma_k \delta_k w_{ki}$, unit i receives error propagated back from a unit k to the extent to which i affects k. For output units, this is essentially the *delta rule* given by Widrow and Hoff (1960) (see PERCEPTRONS, ADALINES, AND BACKPROPAGATION).

The theorem just presented tells us how to compute Δw_{ij} for gradient descent. It does not guarantee that the above step-size is appropriate to reach the minimum, nor does it guarantee that the minimum, if reached, is global. The backpropagation rule defined by this proposition is, thus, a heuristic rule, not one guaranteed to find a global minimum. However, the road map for **Applications of Neural Networks** contains a table which pairs applications with appropriate subsets of some 26 different neural and non-neural mechanisms and adaptive architectures. Backpropagation is the most diversely used adaptive architecture. This architecture is an example of "neurally inspired" modeling, not modeling of actual brain structures: there is no evidence that backpropagation represents actual brain mechanisms of learning.

A Cautionary Note

The previous subsections have introduced a number of techniques which can be used to make networks more adaptive. In a typical training scenario, we are given a network N which, in response to the presentation of any x from some set of input patterns, will eventually settle down to produce a corresponding y from the set Y of the network's output patterns. A *training set* is then a sequence of pairs (x_k, y_k) from $X \times Y$, $1 \leq k \leq n$. The foregoing results say that, in many cases (and the bounds are not yet well defined), if we train the net with repeated presentations of the various (x_k, y_k), it will converge to a set of connections which cause N to compute a function $f : X \to Y$ with the property that, for each k from 1 to n, the $f(x_k)$ "correlate fairly well" with the y_k. Of course, there are many other functions $g : X \to Y$ such that the $g(x_k)$ "correlate fairly well" with the y_k, and they may differ wildly on those x in X that do not equal an x_k in the training set. The view that one may simply present a trainable net with a few examples of solved problems, and it will then adjust its connections to be able to solve all problems of a given class, glosses over three main issues:

a. *Complexity:* Is the network complex enough to encode a solution method?
b. *Practicality:* Can the net achieve such a solution within a feasible period of time?
c. *Efficacy:* How do we guarantee that the generalization achieved by the machine matches our conception of a useful solution?

Part III provides many "snapshots" of the research underway to develop answers to these problems (for the "state of play" for the three questions listed see, for example, STRUC-TURAL COMPLEXITY AND DISCRETE NEURAL NETWORKS; TIME COMPLEXITY OF LEARNING; and GENERALIZATION AND REGULARIZATION IN NONLINEAR LEARNING SYSTEMS). Nonetheless, it is clear that these training techniques will work best when training is based on an adaptive architecture and an initial set of weights appropriate to the given problem. Future work on the neurally inspired design of intelligent systems will involve many domain-specific techniques for system design, such as those exemplified in the road maps for **Vision** and **Primate Motor Control**, as well as general advances in adaptive architectures.

References

Abraham, R. H., and Shaw, C. D., 1981–1983, *Dynamics—The Geometry of Behavior*: Part 1: *Periodic Behavior*, and Part 2: *Chaotic Behavior*, Santa Cruz, CA: Ariel Press.

Amari, S., and Arbib, M. A., 1977, Competition and cooperation in neural nets, in *Systems Neuroscience* (J. Metzler, Ed.), New York: Academic Press, pp. 119–165.

Grossberg, S., 1967, Nonlinear difference-differential equations in prediction and learning theory, *Proc. Natl. Acad. Sci. USA*, 58:1329–1334.

Hebb, D. O., 1949, *The Organization of Behavior*, New York: Wiley.

Hirsch, M. W., and Smale, S., 1974, *Differential Equations, Dynamical Systems, and Linear Algebra*, New York: Academic Press.

Hopfield, J., 1982, Neural networks and physical systems with emergent collective computational properties, *Proc. Natl. Acad. Sci. USA*, 79:2554–2558.

Hopfield, J. J., and Tank, D. W., 1986, Neural computation of decisions in optimization problems, *Biol. Cybern.*, 52:141–152.

McCulloch, W. S., and Pitts, W. H., 1943, A logical calculus of the ideas immanent in nervous activity, *Bull. Math. Biophys.*, 5:115–133.

Minsky, M. L., 1961, Steps toward artificial intelligence, *Proc. IRE*, 49:8–30.

Minsky, M. L., and Papert, S., 1969, *Perceptrons: An Essay in Computational Geometry*, Cambridge, MA: MIT Press.

Nilsson, N., 1965, *Learning Machines*, New York: McGraw-Hill.

Rosenblatt, F., 1958, The perceptron: A probabilistic model for information storage and organization in the brain, *Psychol. Rev.*, 65:386–408.

Rumelhart, D. E., Hinton, G. E., and Williams, R. J., 1986, Learning internal representations by error propagation, in *Parallel Distributed Processing: Explorations in the Microstructure of Cognition*, vol. 1 (D. Rumelhart and J. McClelland, Eds.), Cambridge, MA: MIT Press/Bradford Books, pp. 318–362.

Samuel, A. L., 1959, Some studies in machine learning using the game of checkers, *IBM J. Res. Dev.*, 3:210–229.

Selfridge, O. G., 1959, Pandemonium: A paradigm for learning, in *Mechanisation of Thought Processes*, London: Her Majesty's Stationery Office, pp. 511–531.

von der Malsburg, C., 1973, Self-organization of orientation-sensitive cells in the striate cortex, *Kybernetik*, 14:85–100.

Widrow, B., and Hoff, M. E., Jr., 1960, Adaptive switching circuits, in *1960 IRE WESCON Convention Record*, vol. 4, pp. 96–104.

Part II: Road Maps

Michael A. Arbib

The Meta-Map

As explained in the introductory section "How to Use This Book," Part II offers 23 *road maps* which provide guided tours to different sets of articles in Part III. The reader will find that some of the roads cross, to give some small sample of the interconnectedness of our themes; these connections are further developed in the cross-references provided in the articles of Part III. In the road maps, we depart from the convention used elsewhere in this text whereby TITLES IN SMALL CAPITALS are used for *all* cross-references to other articles. Rather, in Part II we reserve TITLES IN SMALL CAPITALS for articles on a given tour, and we use "Titles in Quotation Marks" to refer to related articles which are not primary to the current road map. We will use **boldface** type to refer to road maps in Part II and to the Meta-Map. The 23 different road maps are grouped into eight sections:

II.1. Connectionism: Psychology, Linguistics, and Artificial Intelligence

Connectionist Psychology (page 31)
Connectionist Linguistics (page 32)
Artificial Intelligence and Neural Networks (page 33)

II.2. Dynamics, Self-Organization, and Cooperativity

Dynamic Systems and Optimization (page 34)
Cooperative Phenomena (page 35)
Self-Organization in Neural Networks (page 36)

II.3. Learning in Artificial Neural Networks

Learning in Artificial Neural Networks, Deterministic (page 37)
Learning in Artificial Neural Networks, Statistical (page 38)
Computability and Complexity (page 40)

II.4. Applications and Implementations

Control Theory and Robotics (page 41)
Applications of Neural Networks (page 42)
Implementation of Neural Networks (page 43)

II.5. Biological Neurons and Networks

Biological Neurons (page 45)
Biological Networks (page 46)
Mammalian Brain Regions (page 48)

II.6. Sensory Systems

Vision (page 50)
Other Sensory Systems (page 52)

II.7. Plasticity in Development and Learning

Mechanisms of Neural Plasticity (page 53)
Development and Regeneration of Neural Networks (page 54)
Learning in Biological Systems (page 54)

II.8. Motor Control

Motor Pattern Generators and Neuroethology (page 55)
Biological Motor Control (page 56)
Primate Motor Control (page 57)

The **Meta-Map** follows the same structure as these sections of Part II. However, there is no one best path to the study of *Brain Theory and Neural Networks*, and you should use the Meta-Map simply to help you choose a path that is pleasing, or useful, to you.

Connectionism: Psychology, Linguistics, and Artificial Intelligence

One could start with biological neurons and then proceed to the artificial neural networks they have inspired, but I will start this survey by noting three topics—**Connectionist Psychology, Connectionist Linguistics,** and **Artificial Intelligence and Neural Networks**—which are driven more by a desire to understand human psychology than to understand the details of biology. In the networks used in these so-called "connectionist" models, the "neurons" rarely correspond to the actual biological neurons of the human brain (the underlying structure). Rather, the driving idea is that the functioning of the human mind (the functional expression of the brain's activity) is best explored through a parallel processing methodology in which large populations of elements are simultaneously active, passing messages back and forth between each other, and "adapting" by changing the strength of their connections as they do so. This is in contrast to the serial computing methodology that was dominant in computer design from the 1940s through the 1970s and that persists, to a great extent, even now.

In dividing this introduction to connectionism into three themes, I have first distinguished those aspects of **Connectionist Psychology** that relate to perception, memory, emotion, and other aspects of cognition in general from those specifically involved in **Connectionist Linguistics**. The section on connectionist psychology also contains articles which address philosophical issues in brain theory and connectionism, including the notion of consciousness, as well as articles which approach psychology from a developmental perspective. The central idea in **Connectionist Linguistics** is that rich linguistic representations can emerge from the interaction of a relatively simple learning device and a structured linguistic environment, rather than requiring the details of grammar to be innate, captured in a genetically determined Universal Grammar.

The next road map presents articles on **Artificial Intelligence and Neural Networks** which are similar in what they explain, but are part of artificial intelligence (AI) because the attempt is to get a machine to exhibit some intelligent-like behavior, without necessarily meeting the constraints imposed by experimental psychology or psycholinguistics. The reader will, of course, find here a number of models of equal interest to psychologists and to AI researchers. "Classical" symbolic AI is contrasted with a number of methods for flexible information processing as well as with neural network approaches. The point is that, whereas brain theory seeks to know "how the brain does it," AI must weigh the value of artificial neural networks (ANNs) as a powerful technology for parallel, adaptive computation against that of other technologies on the basis of efficacy in solving practical problems on available hardware.

The articles gathered in these first three road maps do not exhaust the scope of their subject matter for at least two reasons. First, in addition to connectionist models of psychological phenomena, there are many biological models which embody genuine progress in relating the phenomena to known parts of the brain, perhaps even grounding a phenomenon in the behavior of identifiable classes of biological neurons. Models of this kind will be treated in later sections specifically focusing on memory or motor control or vision, for example. Second, while **Artificial Intelligence and Neural Networks** will focus on broad thematic issues, a number of these will reappear in applying neural networks in computer vision, or speech recognition, or elsewhere.

Dynamics, Self-Organization, and Cooperativity

The next three road maps view the dynamics of neural networks considered as general information processing structures, rather than as models of a particular biological or psychological phenomenon, or as solutions to specific technological problems. First, we examine how neural networks behave when they have fixed connections, considering **Dynamic Systems and Optimization**. To what extent will the dynamics of a neural network settle down to an equilibrium state, and to what extent can that state be seen as the solution of some problem of optimization? Under what circumstances will the network exhibit a dynamic pattern of oscillatory behavior (a limit cycle), and under what circumstances will it undergo chaotic behavior (traversing what is known as a strange attractor)? This theme continues, but also begins to get into the area of learning, in the study of **Cooperative Phenomena**. In a gas or a magnet, we do not know the behavior of any single atom with precision, but can infer the overall "cooperative" behavior—the pressure, volume, and temperature of a gas, or the overall magnetization of a magnet—through statistical methods, methods which even extend to the analyses of such dramatic phase transitions as that of a piece of iron from an unmagnetized lump to a magnet, or of a liquid to a gas. So, too, can statistical methods provide insight into the large-scale properties of neural networks, abstracting away from the detailed function of individual neurons, when our interest is in statistical patterns of behavior rather than the fine details of information processing. This, then, leads us into the next topic, **Self-Organization in Neural Networks**, in which we ask for ways in which the interaction between elements in a neural network can lead to the spontaneous expression of pattern, whether this be constituted by the pattern of activity of individual neurons or by the pattern of synaptic connections which records earlier experience.

Learning in Artificial Neural Networks

With this question of earlier experience, we have fully made the transition to the study of learning, and we turn to two road maps which address the topic, one of which emphasizes deterministic behavior, and the other of which uses the methods of probability theory and statistics: **Learning in Artificial Neural Networks, Deterministic**, and **Learning in Artificial Neural Networks, Statistical**. For many people attracted to the study of neural networks by the technological adaptability (some would even say the "self-programming") of ANNs—whether in unsupervised, supervised, or reinforcement learning—it is these road maps, together with those gathered under the heading "Applications and Implementations," that will constitute the heart of this *Handbook*.

For those interested in the formal analysis of neural network capabilities, the road map on **Computability and Complexity** provides a rapprochement between ANNs and a number of ideas developed within the mainstream of computer science, especially those arising from the study of complexity of computational structures. The extensive work on modeling *biological* mechanisms of learning in neural assemblies is surveyed in the

set of road maps introduced in Section II.7 under the general heading of "Plasticity in Development and Learning."

Applications and Implementations

In the study of **Control Theory and Robotics**, the adaptive properties of neural networks play a special role, enabling a control system, through experience, to become better and better suited to solve a given repertoire of control problems, guiding a system along a desired trajectory, through the use of feedback or model-based feedforward. These general control strategies are exemplified in a number of different approaches to robot control. We then turn to a diverse set of **Applications of Neural Networks**, which include signal processing, astronomy, speech recognition, high-energy physics, steel making, telecommunications, and visual processing. A table relates each application to the subset of some 26 different neural network architectures and alternative methodologies. Since a neural network cannot be applied unless it is implemented in software or hardware, we close this trio of road maps with the study of **Implementation of Neural Networks**, whether it be by simulation on a general-purpose computer, emulation on a specially designed neurocomputer, or implementation in a device built with electronic or photonic materials or even biomaterials.

Biological Neurons and Networks

The next three road maps—**Biological Neurons, Biological Networks**, and **Mammalian Brain Regions**—are ones that, for readers whose primary interest is brain theory rather than ANNs, may provide the appropriate entry point for the book as a whole, namely an understanding of neural networks from a biological point of view. The essay on **Biological Neurons** gives us some sense of how sophisticated real biological neurons are, with each patch of membrane being itself a subtle electrochemical structure. An appreciation of this complexity is necessary for the computational neuroscientist wishing to address the increasingly detailed database of experimental neuroscience on how signals can be propagated, and how individual neurons interact with each other. But such complexity may also provide an eye-opener for the technologist planning to incorporate new capabilities into the next generation of ANNs.

Studying the ways in which networks of biological neurons can act together, the road map for **Biological Networks** points to a variety of structures that occur in neural networks, and a variety of dynamic properties that they have been shown to exhibit. These properties include the phenomenon of neuromodulation, whereby the action of specific substances can change the nature of individual neurons, and thus the dynamics of networks of which they are a part, for seconds or even minutes at a time. Finally, the survey of **Mammalian Brain Regions** introduces the essential structure of a key set of such regions, of the functions they may serve, and of the state of modeling of this relation between structure and function. Clearly, a number of these brain regions will be met again in other road maps in which we look at specific functions, such as vision, memory, or motor control. The reader may also wish to contrast these mammalian systems with the neural systems which underlie a variety of crucial behaviors in nonmammalian species; a number of these are introduced in the road map on **Motor Pattern Generators and Neuroethology** in Section II.8 on "Motor Control" (see also Table 1 in Part I, page 10).

Sensory Systems

Vision has been the most widely studied of all sensory systems, both in brain theory and in applications and analysis of artificial neural networks, and thus has a special road map of its own which exhibits a rich interplay and overlap between the studies of technological and biological visual systems. The **Other Sensory Systems**, treated at much less length in the next road map, include audition, touch, and pain, electrolocation in electric fish, echolocation in bats, time perception, and sensor fusion. One may also note the related work on speech recognition included in the road map on **Applications of Neural Networks** which appears in Section II.4 on "Applications and Implementations."

Plasticity in Development and Learning

Having earlier looked at learning from an abstract and/or technological point of view, we now turn to the road maps which chart learning in a more biological manner. **Mechanisms of Neural Plasticity** presents the specific mechanisms at the level of synapses, or even finer-grain molecular structures, which allow the changes in the strength of connections that underlie both learning and development. With this, we have the basic mechanisms which underlie both **Development and Regeneration of Neural Networks** and **Learning in Biological Systems**. However, I again stress to the reader that one may approach the road maps, and the articles in Part III of this *Handbook*, in many different orders. For example, in discussing **Learning in Biological Systems**, we may take advantage of rather simple descriptions of the changes that occur in individual synapses during learning without concerning ourselves with the molecular mechanisms which allow synapses to change in the indi-

cated way. Thus, the road maps for "Plasticity in Development and Learning" may be followed in any subset and in any order.

Motor Control

The final set of road maps in Part II addresses the control of movement by neural networks. **Motor Pattern Generators and Neuroethology** focuses on subconscious functions such as breathing or locomotion in vertebrates, and on a wide variety of pattern generation activity in invertebrates. The road map also discusses other studies in animal behavior (neuroethology) which show how sensory input, especially visual input, and motor behavior are integrated in a cycle of action and perception. **Biological Motor Control** places increased emphasis on the interaction between neural control and the kinematics or dynamics of limbs, and also looks at various forms of motor-related learning. In showing how the goals of movement can be achieved by a neural network through the time course of activity of motors or muscles, this road map overlaps some of the issues taken up in the more applications-oriented road map on **Control Theory and Robotics**. Most of the material on biological motor control is of general relevance, but **Primate Motor Control** emphasizes movements of the eyes, head, arm, and hand which are studied in a variety of mammals but are most fully expressed in primates and humans. Of course, there will be some readers for whom this last road map will be an excellent starting place for study of the *Handbook* since—by showing how visual and motor systems are integrated in a number of human behaviors—it motivates the study of the specific neural network mechanisms required to achieve these behaviors.

II.1. Connectionism: Psychology, Linguistics, and Artificial Intelligence

Connectionist Psychology

Much classical psychology was grounded in notions of association—of ideas or of stimulus and response—which were well developed in the philosophy of Hume, but with roots going back as far as Aristotle. The article on ASSOCIATIVE NETWORKS shows how these old ideas gain new power because neural networks can provide *mechanisms* for the formation of associa-

tions which automatically yield many further properties. One of these is that neural networks have similar responses to similar inputs, a property which is exploited in the study of ANALOGY-BASED REASONING and whose less fortunate side effects are studied in DISTORTIONS IN HUMAN MEMORY.

Connectionism can apply many different types of ANN techniques to explain psychological phenomena, and the article COGNITIVE MODELING: PSYCHOLOGY AND CONNECTIONISM places a sample of these in perspective. The general theme here is that much of our psychology is better understood in terms of parallel networks of adaptive units than in terms of serial symbol processing, and that connectionism gains much of its power from using very simple units with explicit learning rules. However, these "units of thought" may be quite high level compared to the fine-grain computation of the myriad neurons in the human brain, and their properties may hence be closer to those of entire neural networks than of single neurons. Thus, SCHEMA THEORY offers a strategy that is complementary to connectionism in which the schemas are functional units which bridge between the overall specification of a task and its implementation as a network (or network of networks) of (possibly biologically constrained) neurons. PHILOSOPHICAL ISSUES IN BRAIN THEORY AND CONNECTIONISM reviews these issues from a more philosophical perspective, highlighting the distinction between connectionism and "sequential" syntax (the language of thought), and analyzing ways in which many different branches of psychology and neuroscience may influence the development of a connectionist psychology. Turning to the role of connectionism and brain science in illuminating old philosophical questions, the article on CONSCIOUSNESS, THEORIES OF addresses the "mind-brain problem" by reviewing ways consciousness has been defined, then presenting relevant neuropsychological data and the very preliminary state of neural modeling of consciousness.

Much of the early work on artificial neural networks was inspired by the problem of pattern recognition. The article on CONCEPT LEARNING provides a general introduction to recent work, placing such ideas in a psychological perspective; the mechanisms discussed briefly here are developed at greater length in many articles in the various road maps on learning. The psychology of concept learning receives a special application in the study of FACE RECOGNITION. Complementing the study of concept formation, the article on MENTAL ARITHMETIC USING NEURAL NETWORKS applies neural networks to a special form of reasoning. Other approaches to reasoning are closer to the symbolic style of artificial intelligence, though progress is being made in seeing how certain symbolic processes may be implemented in connectionist style (see, for example, the articles on "Artificial Intelligence and Neural Networks," "Connectionist and Symbolic Representations," "Semantic Networks," and "Structured Connectionist Models" which are described in the road map on **Artificial Intelligence and Neural Networks**).

Many of the concepts of connectionist psychology are strongly related to work in behaviorism, but neural networks provide a stronger "internal structure" than stimulus-response probabilities. Connectionist research has enriched a number of concepts that seemed "anticognitive" by embedding them in mechanisms, namely, neural networks, which can both support internal states and yield stimulus-response pairs as part of a general input-output map. This is shown in the articles on CLASSICAL LEARNING THEORY AND NEURAL NETWORKS and CONDITIONING, and by the exemplification of internal structure in the article on COGNITIVE MAPS, which shows how to represent knowledge of the spatial structure of the world as a basis for action.

Learning in neural networks can be either supervised or unsupervised, and supervision can be in terms of a specific error signal or some general reinforcement. However, in real animals, these signals seem to have some "heat" to them, which brings us to the issues of motivation and emotion. The specific effects of such internal "signals" as thirst and hunger are modeled in the article on MOTIVATION. Motivated behavior is usually goal-oriented. The goal may be associated with a drive, such as hunger or thirst (called primary motivation). However, motivation is also closely tied to sensory stimuli: an animal will not usually exhibit eating behavior unless food is presented, and motivation can be learned (in which case it is called secondary motivation). The relevant data are analyzed in terms of Hull's drive reduction theory and of neural network models which expand upon it. EMOTION AND COMPUTATIONAL NEUROSCIENCE and EMOTION-COGNITION INTERACTIONS provide complementary analyses of the nature of emotion, the role of brain structures such as the amygdala, the interaction of body and cognitive states, the facial expression of the emotions, and the status of neural modeling.

Finally, we turn to mental development, a theme of special concern in connectionist linguistics (see the next road map). Connectionist studies of COGNITIVE DEVELOPMENT tend to employ relatively general learning devices in specific problem domains to see to what extent the structure of a domain constrains development. In particular, although some "stages" in

development may reflect the maturation of distinct information processors, connectionist models show that learning in networks can result in periods of stable behavior interrupted by sudden discontinuities, even though the basic mechanism for learning is one of small continuous change.

Discussions about the modular structure of the brain have frequently been based on selective brain damage in adults, with the assumption that characteristic breakdown in the adult implies a built-in modular organization in the neonate brain. Two articles question this conclusion from different perspectives. Because abnormal development often results in varied mixes of proficiency and deficit, DEVELOPMENTAL DISORDERS uses the comparison of different abnormal phenotypes to explore constraints on the *emergence* of modularity (at a suitably fine grain) in the developing mind/brain. LESIONED ATTRACTOR NETWORKS AS MODELS OF NEUROPSYCHOLOGICAL DEFICITS provides further insights from the clinical observations of neuropsychology. Simulated damage to connectionist models of language processing can yield deficits which qualitatively mimic some of the deficits seen following human brain damage. In particular, such studies suggest how there may be an appearance of functional modularity (i.e., two functions may be differentially impaired by network damage) even when the functions are implemented by a single network.

Connectionist Linguistics

The traditional grounding of linguistics is in grammar, a systematic set of rules for structuring the sentences of a particular language. Much modern work in linguistics has been dominated by the ideas of Noam Chomsky, who placed the notion of grammar in a mathematical framework. His ideas have gone through successive stages in which the formulation of grammars has changed radically. However, two themes have remained stable in the "generative linguistics" that has grown from his work:

1. There is a Universal Grammar which defines what makes a language human, and each human language has a grammar which is a parametric variation of the Universal Grammar.
2. Language is too complicated for a child to learn from scratch; instead, a child has Universal Grammar as an innate mental capacity. When the child hears example sentences of a language, they set parameters in the Universal Grammar so that the child can then acquire the grammar of the particular language.

Connectionist linguistics attacks this reasoning on two fronts. First, it says that language processing is better understood in terms of connectionist processing which, as a *performance* model (i.e., a model of behavior, as distinct from a *competence* model which gives a static representation of a body of knowledge), can give an account of errors, as well as regularities, in language use. Second, it notes that connectionism has powerful learning tools which Chomsky has chosen to ignore. With those tools, connectionism can model how children could acquire language on the basis of far less specific mental structures than those posited by Chomsky in his Universal Grammar.

The article on LANGUAGE PROCESSING presents a number of models, especially networks providing some equivalent of syntactic and semantic processes. Among the issues that arise is the need to understand the representational and processing properties of networks (particularly recurrent networks) in order to be able to compare them with more traditional models. Another issue is the scaling problem, whose solution for larger and larger search spaces will require new ways of constraining the search space during learning. This may lead to a new understanding of the extent to which knowledge (or the mechanisms for self-organization of knowledge) can be innate, a central issue in cognitive science. The article on LANGUAGE ACQUISITION presents models with which developmental connectionists support the claim that rich linguistic representations can emerge from the interaction of a relatively simple learning device and a structured linguistic environment (compare the article on "Cognitive Development"). The article reviews connectionist models of lexical development, inflectional morphology, and syntax acquisition, stressing that these models use similar learning algorithms to solve diverse linguistic problems. The article on LINGUISTIC MORPHOLOGY then sets forth in greater detail the prime example of the debate between a rule-based and a connectionist account of language processing, the forming of regular and irregular past tenses of verbs in English.

The connectionist approach to language, which emphasizes the learning processes of each new speaker over the existence of an immutable Universal Grammar shared by all humans, seems well equipped to approach the issue of how a language changes from generation to generation. The article on LANGUAGE CHANGE introduces this theme, showing how the inherent generalization ability of neural networks makes certain errors in language transmission from one generation to the next more likely than others.

At present, the state of play may be summarized as follows: Generative linguistics has shown how to provide grammatical rules which explain many subtle sentence constructions of English and many other languages, revealing commonalities and differences between languages, with the differences in some cases being reduced to very elegant and compact formulations in terms of general rules with parametric variations. However, in offering the notion of Universal Grammar as the substrate for language acquisition, generative linguistics ignores issues of learning that must, in any case, be faced in explaining how children acquire the large and idiosyncratic vocabulary of their native tongue. Connectionist linguistics, on the other hand, has made great strides in bringing learning to the center, not only showing how specific language skills (e.g., use of the past tense) may be acquired, but also providing insight into psycholinguistics, the study of language *behavior*. However, connectionist linguistics still faces three major hurdles: it lacks the systematic overview of language provided by generative linguistics; the problem of finding how complex structural relationships such as constituency can be implemented in neural networks remains open; and little progress has been made in developing a neurolinguistic theory of the contributions of specific brain regions to language capabilities. It is one thing to train an artificial neural network (ANN) to yield a convincing model of performance on the past tense; it is quite another to offer an account of how this skill interfaces with all the other aspects of language, and what neural substrates are necessary for their acquisition by the human child.

Artificial Intelligence and Neural Networks

In the 1950s, the precursors of today's fields of artificial intelligence and neural networks were still subsumed under the general heading of cybernetics. Much of the work in the 1960s sought to distance AI from its cybernetic roots, emphasizing models of, e.g., logical inference, game playing, and problem solving that were based on explicit symbolic representations manipulated by serial computer programs. However, work in computer vision and in robotics (discussed in the road maps on **Vision** and **Control Theory and Robotics,** respectively) showed that this distinction was never entirely convincing, since these were areas of AI that made use of parallel computation and numerical transformations. For a while, a case could be made that use of parallelism might be appropriate for peripheral sensing and motor control but not for the "central" processes involved in "real" intelligence. However, work from at least the mid-1970s onward has made this fallback position untenable. For example, work on the HEARSAY speech understanding system introduced the methodology of DISTRIBUTED ARTIFICIAL INTELLIGENCE by grounding language understanding in the action (implicitly distributed, though implemented on a serial computer) of knowledge sources to update numerical confidence levels of multiple hypotheses distributed across a set of "levels" in a data structure known as a blackboard. This stood in contrast to "classical" AI's use of serial manipulation of symbolic structures. In the following years, work on expert systems provided an important application success in which numerical confidence values played a role, but with the emphasis still on manipulation of hypotheses through the serial application of explicit rules abstracted from the knowledge of an expert solving problems in some specific domain. As shown in EXPERT SYSTEMS AND DECISION SYSTEMS USING NEURAL NETWORKS, we now see many cases in which the application of separate rules is replaced by transformations effected in parallel by (trainable) neural networks. The article on LEARNING BY SYMBOLIC AND NEURAL METHODS continues this theme, looking at how knowledge may be acquired both by the learning of symbolic structures and by the training of neural networks.

As the general overview article on ARTIFICIAL INTELLIGENCE AND NEURAL NETWORKS makes clear, there are many problems for which the (not necessarily serial) manipulation of symbolic structures can still outperform connectionist approaches, at least with today's software running on today's hardware. Nonetheless, if we define AI by the range of problems it is to solve—or the "packets of intelligence" it is to implement—then it is no longer useful to define it in opposition to connectionism. In general, the technologist facing a specific problem should choose between, or should combine, connectionist and symbolic approaches on the basis of efficacy, not ideology. On occasion, for rhetorical purposes, authors will use the term AI for a serial symbolic methodology distinct from connectionism. However, we will generally use it in an extended sense of a technology which seeks to realize aspects of intelligence in machines by whatever methods work best. The term *symbolic*

AI will then be used for the "classical" approach. In brain theory, everything, whether symbolic or not, is, in the final analysis, implemented in a neural network. But even here (and the methodology of schema theory, which has some points in common with distributed AI, attests to this), an analysis of the brain will often best be conducted in terms of interacting subsystems which are not all fully explicated in neural network terms.

The notion of representation plays a central role in AI. As discussed in SEMANTIC NETWORKS, one classic form of representation in AI is the "semantic network," in which nodes represent concepts and links represent relations between them. The article on CONNECTIONIST AND SYMBOLIC REPRESENTATIONS analyzes two ways of relating these representational styles. In one, the translation from symbolic to neural is fairly direct: nodes become "neurons," but now processing is done by neural interactions rather than by an "inference engine" acting on a passive representation. At the other extreme, certain neural networks (connectionist, rather than biological) may transform input "questions" to output "answers" via the distributed activity of neurons whose "firing conditions" have no direct relationship to the concepts that might normally arise in a logical analysis of the problem. The discussion of these varying styles of representation is continued in STRUCTURED CONNECTIONIST MODELS. In symbolic AI, two concepts can be linked by providing a pointer between them. In a neural net, the problem of "binding" the two patterns of activity that represent the concepts is a more subtle one, and several models address the use of rapidly changing synaptic strengths to provide temporary "assemblages" of currently related data. This theme is developed not only in the article on STRUCTURED CONNECTIONIST MODELS, but also in the articles on COMPOSITIONALITY IN NEURAL SYSTEMS (how can inferences about a structure be based on the way it is composed of various elements?), DYNAMIC LINK ARCHITECTURE (the basic methodology), and OBJECT RECOGNITION (combining elements of an object into a recognizable whole).

Complementing the theme of representation in symbolic AI has been that of planning: i.e., that of going from (representations of) the current state and some desired state to a sequence of operations that will transform the former to the latter. The article on PLANNING, CONNECTIONIST provides an overview of this area, while PROBLEM SOLVING, CONNECTIONIST gives a specific approach to some subproblems. This is clearly related to a number of problems addressed in other road maps, such as "Cognitive Maps" (**Connectionist Psychology**), "Hippocampus: Spatial Models" (**Mammalian Brain Regions**), and "Reinforcement Learning in Motor Control" (**Learning in Artificial Neural Networks, Deterministic**).

The final three papers in this road map are not on neural networks per se, but instead provide related methods which add to the array of techniques extending AI beyond the serial, symbol-based approach. BAYESIAN NETWORKS provides an explicit method for following chains of probabilistic inference such as those appropriate to expert systems. FUZZY LOGIC SYSTEMS AND QUALITATIVE KNOWLEDGE provides a means of defining and applying rules which are based not on probability but on some "fuzzy membership function." (For example, people whose heights are 1 m, 1.5 m, and 2 m might belong to the set of tall people with membership values 0, 0.2, and 0.99, respectively. What can one infer about each of them on the basis of one's knowledge of tall people?) Finally, MEMORY-BASED REASONING applies massively parallel computing to answer questions about a new situation by searching for data on the most similar stored instances. In each case, the reader may ponder whether these methods are alternatives to connectionist AI, or whether they can contribute to the emergence of a technologically efficacious hybrid. Brain theory seeks to know "how the brain does it," but, as stated before, AI must weigh the value of ANNs as a powerful technology for parallel, adaptive computation against that of other technologies on the basis of efficacy in solving practical problems on available hardware.

II.2. Dynamics, Self-Organization, and Cooperativity

Dynamic Systems and Optimization

AUTOMATA AND NEURAL NETWORKS
CELLULAR AUTOMATA
CHAOS IN NEURAL SYSTEMS
COMPUTING WITH ATTRACTORS
CONSTRAINED OPTIMIZATION AND THE ELASTIC NET
DYNAMICS AND BIFURCATION OF NEURAL NETWORKS
ENERGY FUNCTIONS FOR NEURAL NETWORKS
FRACTAL STRATEGIES FOR NEURAL NETWORK SCALING
MODULAR AND HIERARCHICAL LEARNING SYSTEMS
NEURAL OPTIMIZATION
PATTERN FORMATION, BIOLOGICAL
PHASE-PLANE ANALYSIS OF NEURAL ACTIVITY
PRINCIPAL COMPONENT ANALYSIS
SELF-REPRODUCING AUTOMATA
WINNER-TAKE-ALL MECHANISMS

Neural networks without loops, especially those structured as a series of layers of neurons in which each neuron feeds its outputs forward to the next layer from input to output, are called *feedforward* networks. Much interest in artificial neural networks has been based on the use of trainable feedforward

networks as universal approximators for functions $f: X \to Y$ from the input space X to the output space Y. These functions are "memoryless" in that the output depends only on the present input. However, the provenance of neural networks was more general. The founding paper of Pitts and McCulloch established the result that, by the mid-1950s, could be rephrased as saying that any finite automaton (in which behavior is state-dependent) could be simulated by a network of McCulloch-Pitts neurons. The article on AUTOMATA AND NEURAL NETWORKS spells out this result. It also shows that, if we extend formal neurons to have real-valued states adjustable with infinite precision, then neural networks not only can simulate Turing machines but can even "compute the noncomputable." For more articles related to automata and theory of computation, see the road map on **Computability and Complexity** in Section II.3.

Von Neumann's interest in biological automata led him to derive inspiration from McCulloch and Pitts for the design of stored-program digital computers (the so-called von Neumann architecture), but it also led him to a formal account of biological reproduction, which included the theory of SELF-

REPRODUCING AUTOMATA. The theory introduced the notion of "non–von Neumann" cellular automata, a tessellation of identical automata, i.e., a tiling pattern with each tile being a finite automaton connected to its neighbors in the same way. Computations of arbitrary complexity could be achieved by "programs" encoded as "patterns in the tiles," and patterns could be designed which were not only able to compute but also to reproduce themselves. The article on CELLULAR AUTOMATA shows how this form of parallel computation has burgeoned in recent years, complementing our study of neural networks.

A finite automaton is a discrete-time dynamic system: that is, on some suitable time scale, it specifies the *next state* $q(t + 1)$ as a function $\delta(q(t), x(t))$ of the current state and input. But (as in the leaky integrator model of Sections I.1 and I.3), a neuron can be modeled as a continuous-time system (with, e.g., the membrane potential as the state variable). A network of continuous-time neurons can then be considered as a continuous-time system with the *rate of change of the state* (which could, for example, be a vector whose elements are the membrane potentials of the individual neurons) defined as a function $\dot{q}(t) = f(q(t), x(t))$ of the current state and input. When the input is held constant, the network (whether discrete- or continuous-time) may be analyzed by dynamic system theory. COMPUTING WITH ATTRACTORS shows some of the benefits of such an approach. In particular, a net may go to equilibrium (providing a state from which the answer to some problem may be read out), enter a limit cycle (undergoing repetitive oscillations which are useful in control of movement and in other situations in which a "clock cycle" is of value), or exhibit chaotic behavior (acting in an apparently random way, even though it is deterministic). The article on DYNAMICS AND BIFURCATION OF NEURAL NETWORKS treats similar mathematical issues from a somewhat different perspective, with especial emphasis on chaos and bifurcations. PHASE-PLANE ANALYSIS OF NEURAL ACTIVITY gives specific examples of how dynamic system considerations of stability may be used to reveal qualitative patterns of neural network behavior with remarkable clarity.

Two articles provide mathematical complements to the study of neural networks as dynamic systems. CHAOS IN NEURAL SYSTEMS provides a view of the appearance of this phenomenon of "deterministic randomness" in a variety of physical and biological systems and broaches the vexing question of whether chaos plays a useful role in neural networks. PATTERN FORMATION, BIOLOGICAL presents a general methodology, based on analysis of the largest eigenvalue, for tracing the asymptotic behavior of a dynamic system, applies it to the problem of biological pattern formation (the zebra's stripes and the leopard's spots), and closes with applications of the methodology to neural networks.

WINNER-TAKE-ALL MECHANISMS presents a number of designs for neural networks which solve the following problem: given a number of networks which each provide as output some "confidence measure," find in a distributed manner the network whose output is strongest. The article on MODULAR AND HIERARCHICAL LEARNING SYSTEMS addresses a somewhat more general problem: given a complex problem, find a set of networks which each provide an approximate solution in some region of the state space together with a network which can combine these approximations to yield a globally satisfactory solution (i.e., "blend" the "good" solutions rather than extract the "best" solution).

Hopfield contributed much to the resurgence of interest in neural networks in the 1980s by associating an "energy function" with a network, showing that if only one neuron changed state at a time (the so called asynchronous update), a symmetri-

cally connected net would settle to a local minimum of the energy, and that many optimization problems could be mapped to energy functions for symmetric neural networks. The article on ENERGY FUNCTIONS FOR NEURAL NETWORKS shows how the definition of energy function and the conditions for convergence to a local minimum can be broadened considerably. (Of course, a network undergoing limit cycles or chaos will *not* have an energy function which is minimized in this sense.)

The remaining articles develop the theme of using neural networks to solve optimization problems. They all use, to some extent, the methods of statistical mechanics for which the article on "Statistical Mechanics of Neural Networks" (see the next road map) serves as background. NEURAL OPTIMIZATION and CONSTRAINED OPTIMIZATION AND THE ELASTIC NET provide complementary perspectives on this theme, while FRACTAL STRATEGIES FOR NEURAL NETWORK SCALING shows cases in which these methods can be improved by using multiple scales of analysis (like a fractal pattern which repeats itself at each scale). A particular application of such optimization is described in PRINCIPAL COMPONENT ANALYSIS. In data compression applications, like image or speech coding, a distribution of input vectors may be economically encoded, with small expected values of the distortions, in terms of eigenvectors of the largest eigenvalues of the correlation matrix that describes the distribution of these patterns (these eigenvectors are the "principal components"). However, it is usually not possible to find the eigenvectors on-line. The ideal solution is then replaced by a neural network learning rule embodying a constrained optimization problem which converges to the solution given by the principal components.

Cooperative Phenomena

A major theme in physics for the last century has been the study of statistical mechanics: showing how, for example, to average out the individual variations of position and velocity of the myriad molecules in a gas to understand the relationship between pressure, volume and temperature, or to see how variations in temperature can yield dramatic *phase transitions*, such as that from ice to water, or from water to steam. The article on COOPERATIVE PHENOMENA places these ideas in a general setting, stressing the notions of *control parameter* (such as temperature in the previous example) whose variation can yield qualitative changes in system behavior, and *order parameters* which measure some degree of order during phase transitions (e.g., spontaneous magnetization in a ferromagnet, or the density of a substance during a liquid-gas transition). STATISTICAL MECHANICS OF NEURAL NETWORKS introduces the reader to some of the basic methods of statistical mechanics, and shows that they can be applied to systems made up of large numbers of (formal) neurons. Statistical mechanics has studied magnets as lattices with an atomic magnet (modeled as, e.g., a spin that can be up or down) at each lattice point, and this has led to the statistical analysis of neural networks as "spin glasses," where firing and nonfiring corresponds to "spin up" and "spin down," respectively. It has also led to the study of MARKOV RANDOM FIELD MODELS IN IMAGE PROCESSING, in which the initial information at each lattice site represents some local features of the raw image, while the final state allows one to read off a processed image. (Readers who wish to extend their appreciation of the application of statistical mechanics to neural networks should consult "Constrained Optimization and the Elastic Net" and "Neural Optimization," described in the preceding road map, as well as the articles on statistical mechanics discussed later in the present road map.)

As was noted in the preceding road map, **Dynamic Systems and Optimization**, the passage of the "energy" of a Hopfield network to a local minimum can be construed as a means for solving an optimization problem. The catch is the word "local" in local minimum—the solution may be the best in the neighborhood, yet far better solutions may be located elsewhere. The resolution of this, at the expense of great increases in time to convergence, is SIMULATED ANNEALING, adding noise which is then gradually reduced ("lowering the temperature"). The initially high temperature (i.e., noise level) stops the system from getting trapped in "high valleys" of the energy landscape; the lowering of temperature allows optimization to occur in the "deepest valley" once it has been found. The article on BOLTZMANN MACHINES then applies this method to design a class of neural networks. In simulated annealing, the controlled use of noise proves beneficial in stopping a search from getting stuck in the wrong place. But noise may also be harmful in causing neurons to "make mistakes." The article on FAULT TOLERANCE thus looks at ways to insert redundancy into neural networks so that the performance of the overall network is more reliable than the performance of its elements.

Hopfield networks fall within the earlier tradition of associative networks and synaptic matrices, but in a manner inviting precise analysis by physicists exploiting the analogy with spin glasses. The storage of specific patterns relies on a Hebbian learning rule whereby the weight of the connection between any two neurons is increased or decreased depending on whether these neurons are simultaneously active or not. Questions concerning the learning dynamics include: Do networks exist which are able to perform the desired task, and if so does the learning process converge? The STATISTICAL MECHANICS OF LEARNING provides analytic answers to these questions in the case of large networks with simple architecture processing random patterns. This study continues in the article on STATISTICAL MECHANICS OF GENERALIZATION, which asks how well, after a network has been adapted to a training set, will the trained network be able to classify an input that it has not seen before. In order to give a statistical answer, the analysis assumes that all inputs are produced at random from the network's environment. The generalization error is defined as the probability that the network will make a mistake on the new input, and this serves as the basis for studying the learning curve of the network. In the original Hopfield model, the patterns were chosen randomly, with half of the neurons active on average in each pattern (this is called dense coding). However, with SPARSELY CODED NEURAL NETWORKS, the maximum number of patterns that can be stored in a network as fixed points of its dynamics increases with the degree of sparseness (i.e., fewer active neurons in each pattern). Note in all these articles the strong statistical assumptions which provide true insight for some learning situations, but ignore the deterministic structure of the training set in other cases (cf. the contrast between the first two road maps in Section II.3, **Learning in Artificial Neural Networks, Deterministic** and **Learning in Artificial Neural Networks, Statistical**).

The last two articles of this road map return us from learning to dynamics of a network with fixed connections. COLLECTIVE BEHAVIOR OF COUPLED OSCILLATORS explains the use of phase models to help understand how temporal coherence arises over populations of densely interconnected oscillators, even when their frequencies are randomly distributed, whether these oscillators are the periodic flashing of fireflies or the pacemaker cells in the heart. By contrast, the article on COOPERATIVE BEHAVIOR IN NETWORKS OF CHAOTIC ELEMENTS analyzes conditions under which the dynamics of an ensemble of elements with complex dynamics and complex coupling may yield chaotic behavior of varying degrees of coherence, and discusses the relevance of this to models of neural information processing.

Self-Organization in Neural Networks

COMPETITIVE LEARNING
DYNAMIC LINK ARCHITECTURE
LEARNING VECTOR QUANTIZATION
OSCILLATORY ASSOCIATIVE MEMORIES
PATTERN FORMATION, BIOLOGICAL
SELF-ORGANIZATION AND THE BRAIN
SELF-ORGANIZATION IN THE TIME DOMAIN
SELF-ORGANIZING FEATURE MAPS: KOHONEN MAPS
SPATIOTEMPORAL ASSOCIATION IN NEURAL NETWORKS
TEMPORAL PATTERN PROCESSING

The article on PATTERN FORMATION, BIOLOGICAL presents general methods for the analysis of how patterns can emerge from the interactions of cells in a biological system. These may be the patterns of stripes on a zebra or, as shown in the overview article on SELF-ORGANIZATION AND THE BRAIN, they may be exhibited in patterns of synaptic connectivity, such as those involved in the regular "retinotopic" mapping of one part of the visual system to another (cf. "Development and Regeneration of Eye-Brain Maps"). The latter article stresses the statistical role of fluctuations in self-organization.

The input patterns to a neural network define a continuous vector space. Vector quantization provides a means to "quantize" this space by forming a "code book" of significant vectors linked to useful information—we can then analyze a novel vector by looking for the vector in the code book to which it is most similar. LEARNING VECTOR QUANTIZATION provides a means whereby a neural network can self-organize, both to provide the code book (one neuron per entry) and to find (by a winner-take-all technique) the code associated with a novel input vector. If this methodology is augmented by constraints which force nearby neurons to become associated with similar codes, the result is SELF-ORGANIZING FEATURE MAPS: KOHONEN MAPS, whereby a high-dimensional feature space is mapped quasi-continuously onto the neural manifold. Another approach is based on "clustering," as described in the article on COMPETITIVE LEARNING, a form of unsupervised learning in which each input pattern comes, through learning, to be associated with the activity of one, or at most a few, neurons, leading to sparse representations of data that are easy to decode. Competitive learning algorithms employ some sort of competition between neurons in the same layer via lateral connections.

Whereas the previous three articles emphasize the role of long-term modifications in changing the information processing of a network, DYNAMIC LINK ARCHITECTURE shows how synaptic plasticity may be used on a fast time scale to bind together structures coded in different parts of a neural network into a unified whole for further processing. The article on OSCILLATORY ASSOCIATIVE MEMORIES reviews a related approach. Storage and retrieval in neural networks with fixed point attractors are limited to one pattern at a time, but oscillatory neural network dynamics can associate different patterns with different phases of the oscillation.

The last three articles in this road map are concerned with the processing of temporal patterns, rather than the time-sharing of spatial patterns in relation to some oscillation. SELF-ORGANIZATION IN THE TIME DOMAIN tackles this by passing a sequence of inputs through an array of leaky integrator neurons with a sufficient spread of time constants to get an output at any time that is in almost one-to-one correspondence with the sequence that generated it. Similarly, SPATIOTEMPORAL ASSOCIATION IN NEURAL NETWORKS approaches its problem by extending the statistical mechanics of neural networks to those which incorporate signal delays into the neural dynamics and allow for a distribution of time lags. TEMPORAL PATTERN PROCESSING uses neurons with decaying activity so that a neuron's output codes the time since it was last activated.

These findings are related to a number of the articles that appear in the three road maps on learning—the boundary between self-organization and learning depends more on rhetoric than on subject matter! Perhaps it can be said that a study of self-organization emphasizes the changing patterns of connections, whereas a study of learning emphasizes changing values of strengths of connections that are already in place to improve the capability of the network to process information, but these concerns will not always be clearly separated.

II.3. Learning in Artificial Neural Networks

Learning in Artificial Neural Networks, Deterministic

ADAPTIVE RESONANCE THEORY (ART)
ASSOCIATIVE NETWORKS
BACKPROPAGATION: BASICS AND NEW DEVELOPMENTS
CONVOLUTIONAL NETWORKS FOR IMAGES, SPEECH, AND TIME SERIES
COULOMB POTENTIAL LEARNING
KOLMOGOROV'S THEOREM
LEARNING AS ADAPTIVE CONTROL OF SYNAPTIC MATRICES
LEARNING AS HILL-CLIMBING IN WEIGHT SPACE
LEARNING BY SYMBOLIC AND NEURAL METHODS
MODULAR NEURAL NET SYSTEMS, TRAINING OF
NEOCOGNITRON: A MODEL FOR VISUAL PATTERN RECOGNITION
NEUROSMITHING: IMPROVING NEURAL NETWORK LEARNING
NONMONOTONIC NEURON ASSOCIATIVE MEMORY
PATTERN RECOGNITION
PERCEPTRONS, ADALINES, AND BACKPROPAGATION
RECURRENT NETWORKS: SUPERVISED LEARNING
REINFORCEMENT LEARNING
TOPOLOGY-MODIFYING NEURAL NETWORK ALGORITHMS

Much of our concern is with supervised learning, getting a network to behave in a way which successfully approximates some specified pattern of behavior or input-output relationship. In particular, much emphasis has been placed on feedforward networks, that is, networks which have no loops, so that the output of the net depends on its input alone, since there is then no internal state defined by reverberating activity. The most direct form of this is a synaptic matrix, a one-layer neural network for which input lines directly drive the output neurons and a "supervised Hebbian" rule sets synapses so that the network will exhibit specified input-output pairs in its response repertoire. This is addressed in the article on ASSOCIATIVE NETWORKS, which notes the problems that arise if the input patterns (the "keys" for associations) are not orthogonal vectors. Association also extends to recurrent networks obtained from one-layer networks by feedback connections from the output to the input, but in such systems of "dynamic memories" (e.g., Hopfield networks) there are no external inputs as such. Rather the "input" is the initial state of the network, and the "output" is the "attractor" or equilibrium state to which the network then settles. Unfortunately, the usual "attractor network" memory model, with neurons whose output is a sigmoid function of the linear combination of their inputs, has many spurious memories, i.e., equilibria other than the memorized patterns, and there is no way to decide a memorized pattern is recalled or not. The article on NONMONOTONIC NEURON ASSOCIATIVE MEMORY shows that, if the output of each neuron is a nonmonotonic function of its input, the capacity of the network can be increased, and the network does not exhibit spurious memories: when the network fails to recall a correct memorized pattern, the state shows a chaotic behavior instead of falling into a spurious memory.

Historically, the earliest forms of supervised learning involved changing synaptic weights to oppose the error in a neuron with a binary output (the perceptron error-correction rule), or to minimize the sum of squares of errors of output neurons in a network with real-valued outputs (the Widrow-Hoff rule). This work is charted in the article on PERCEPTRONS, ADALINES, AND BACKPROPAGATION, which also charts the extension of these classic ideas to multilayered feedforward networks. Multilayered networks pose the *structural credit assignment problem*: when an error is made at the output of a network, how is credit (or blame) to be assigned to neurons deep within the network? One of the most popular techniques is called *backpropagation*, whereby the error of output units is propagated back to yield estimates of how much a given "hidden unit" contributed to the output error. These estimates are used in the adjustment of synaptic weights to these units within the network. The article on BACKPROPAGATION: BASICS AND NEW DEVELOPMENTS places this idea in a broader mathematical and historical framework in which backpropagation is seen as a general method for calculating derivatives to adjust the weights of nonlinear systems, whether or not they are neural networks. The underlying theoretical grounding is that, given any function $f: X \rightarrow Y$ for which X and Y are codable as input and output patterns of a neural network, then, as shown in the article on KOLMOGOROV'S THEOREM, f can be approximated arbitrarily well by a feedforward network with one layer of hidden units. The catch, of course, is that many, many hidden units may be required for a close fit. It is often an empirical question whether there exists a sufficiently good approximation achievable *in principle* by a network of a given size—an approximation which a given learning rule may or may not find (it may, for example, get stuck in a *local* optimum rather than a global one). The article on NEUROSMITHING: IMPROVING NEURAL NETWORK LEARNING provides a number of "rules of thumb" to be used in applying backpropagation in trying to find effective settings for network size and for various coefficients in the learning rules.

One useful perspective for supervised learning views LEARN-ING AS HILL-CLIMBING IN WEIGHT SPACE, so that each "experience" adjusts the synaptic weights of the network to climb (or descend) a metaphorical hill for which "height" at a particular point in "weight space" corresponds to some measure of the performance of the network (or the organism or robot of which it is a part). When the aim is to minimize this measure, one of the basic techniques for learning is what mathematicians call "gradient descent"; optimization theory also provides alternative methods such as, e.g., that of conjugate gradients, which are also used in the neural network literature. REINFORCEMENT LEARNING describes a form of "semi-supervised" learning where the network is not provided with an explicit form of error at each time step but rather receives only generalized reinforcement ("you're doing well"; "that was bad!") which yields little immediate indication of how any neuron should change its behavior. Moreover, the reinforcement is intermittent, thus raising the *temporal* credit assignment problem: how is an action at one time to be credited for positive reinforcement at a later time? One solution is to build an "adaptive critic" which learns to evaluate actions of the network on the basis of how often they occur on a path leading to positive or negative reinforcement.

Another perspective on supervised learning is presented in LEARNING AS ADAPTIVE CONTROL OF SYNAPTIC MATRICES, which views learning as a control problem (controlling synaptic matrices to yield a given network behavior) and then uses the adjoint equations of control theory to derive synaptic adjustment rules. Gradient descent methods have also been extended to adapt the synaptic weights of recurrent networks, as discussed in RECURRENT NETWORKS: SUPERVISED LEARNING, where the aim is to match the time course of network activity, rather than the (input, output) pairs of some training set.

The task *par excellence* for supervised learning is pattern recognition, the problem of classifying objects, often represented as vectors or as strings of symbols, into categories. Historically, the field of pattern recognition started with early efforts in neural networks (see PERCEPTRONS, ADALINES, AND BACKPROPAGATION). While neural networks played a less central role in pattern recognition for some years, recent progress has made them the method of choice for many applications. As PATTERN RECOGNITION demonstrates, multilayer networks, when properly designed, can learn complex mappings in high-dimensional spaces without requiring complicated hand-crafted feature extractors. To rely more on learning, and less on detailed engineering of feature extractors, it is crucial to tailor the network architecture to the task, incorporating prior knowledge to be able to learn complex tasks without requiring excessively large networks and training sets.

Many specific architectures have been developed to solve particular types of learning problem. ADAPTIVE RESONANCE THEORY (ART) bases learning on internal expectations. When the external world fails to match an ART network's expectations or predictions, a search process selects a new category, representing a new hypothesis about what is important in the present environment. The *neocognitron* (see NEOCOGNITRON: A MODEL FOR VISUAL PATTERN RECOGNITION) was developed as a neural network model for visual pattern recognition which addresses the specific question "how can a pattern be recognized despite variations in size and position?" by using a multilayer architecture in which local features are replicated in many different scales and locations. More generally, as shown in CONVOLUTIONAL NETWORKS FOR IMAGES, SPEECH, AND TIME SERIES, shift invariance in convolutional networks is obtained by forcing the replication of weight configurations across space. Moreover, the topology of the input is taken into ac-

count, enabling such networks to force the extraction of local features by restricting the receptive fields of hidden units to be local. COULOMB POTENTIAL LEARNING derives its name from its functional form's likeness to a coulomb charge potential, replacing the linear separability of a simple perceptron with a network that is capable of constructing arbitrary nonlinear boundaries for classification tasks.

We have already noted that networks that are too small cannot learn the desired input to output mapping. However, networks can also be too large. Just as a polynomial of too high a degree is not useful for curve-fitting, a network that is too large will fail to generalize well, and will require longer training times. Smaller networks, with fewer free parameters, enforce a smoothness constraint on the function found. For best performance, it is, therefore, desirable to find the smallest network that will "properly" fit the training data. The article TOPOLOGY-MODIFYING NEURAL NETWORK ALGORITHMS reviews algorithms which adjust network topology (i.e., adding or removing neurons during the learning process) to arrive at a network appropriate to a given task.

The last two articles in this road map take a somewhat different viewpoint from that of adjusting the synaptic weights in a single network. MODULAR NEURAL NET SYSTEMS, TRAINING OF presents the idea that, although single neural networks are theoretically capable of learning complex functions, many problems are better solved by designing systems in which several modules cooperate together to perform a global task, replacing the complexity of a large neural network by the cooperation of neural network modules whose size is kept small. The article on LEARNING BY SYMBOLIC AND NEURAL METHODS focuses on the distinction between symbolic learning based on producing discrete combinations of the features used to describe examples and neural approaches which adjust continuous, nonlinear weightings of their inputs. The article not only compares but also combines the two approaches, showing, for example, how symbolic knowledge may be used to set the initial state of an adaptive network.

Learning in Artificial Neural Networks, Statistical

Whereas the foregoing road map for **Learning in Artificial Neural Networks, Deterministic** places most emphasis on the learning of a fixed training set or the approximation of a given behavioral function, the present road map emphasizes situations where learning from examples is stochastic in the sense that examples are randomly generated and/or the neurons themselves are modeled as stochastic computing elements so that network behavior is to be analyzed from a statistical point of view. Many of the articles in the present road map rest on some fluency in the language of probability and statistics; they thus tend to be more technical (on average!) than those in the previous road map. However, it should be stressed that the division between deterministic and statistical studies was designed to help the reader approach the many, many articles on learning, rather than reflecting a strict dichotomy in subject matter.

LEARNING AND STATISTICAL INFERENCE studies learning by using such statistical notions as Fisher information, Bayesian loss, and sequential estimation. GENERALIZATION AND REGULARIZATION IN NONLINEAR LEARNING SYSTEMS sets forth the essential relationship between multivariate function estimation in a statistical context and supervised machine learning. Given a training set consisting of (input, output) pairs (x_i, y_i), the task is to construct a map which generalizes well in that, given a new value of x, the map will provide a reasonable prediction for the unobserved output associated with this x. Regularization simplifies the problem by applying constraints to the construction of the map which reduce the generalization error. Ideally, these constraints embody a priori information concerning the true relationship between input and output, though various ad hoc constraints have sometimes been shown to work well in practice. EXPLORATION IN ACTIVE LEARNING contrasts active and passive learning. In the passive learning paradigm, a learner simply observes its environment, which is assumed to generate a stream of training data. In the active learning paradigm, on the other hand, the learner can execute actions which have an impact on the generation of training data. This raises an important challenge: How can a learner efficiently explore its environment?

The article on MODULAR AND HIERARCHICAL LEARNING SYSTEMS presents ways to solve a complex learning problem by dividing it into a set of subproblems. In the context of supervised learning, modular architectures arise when the data can be described by a collection of functions, each of which works well over a relatively local region of the input space, allocating different modules to different regions of the space. The expectation maximization (EM) algorithm, an alternative to gradient methods based on the asymptotic theory of estimation, can be used to estimate the parameters of both the modular system and its extension to a hierarchical system. Another use of modularity results from the finding that averaging over the outputs of a population of neural network estimates can lead to improved network performance. AVERAGING/MODULAR TECHNIQUES FOR NEURAL NETWORKS describes the basic method and presents a firm theoretical basis for understanding why averaging improves performance for a wide class of tasks. The work in modular and hierarchical learning systems may be seen as extending the notion of averaging to interpret the component networks as generic modules which can be combined using sophisticated nonlinear techniques.

BAYESIAN METHODS FOR SUPERVISED NEURAL NETWORKS shows how to apply Bayes's rule for the use of probabilities to quantify inferences about hypotheses given data. The idea is to take the predictions $P(D|H_i)$ made by alternative models H_i about data D, and the prior probabilities of the models $P(H_i)$, and obtain the posterior probabilities $P(H_i|D)$ of the models given the data, using Bayes's rule in the form $P(H_i|D) = P(D|H_i)P(H_i)/P(D)$. To apply this to neural networks, we regard a supervised neural network as a nonlinear parameterized mapping from an input x to an output $y = y(x; w)$, which depends continuously on the "weight" parameters w. Clearly, the idea is to choose weights w_o from a weight space with some given probability distribution $P(w_i)$ so as to maximize the likelihood of the net's yielding the given set of (input, output) observations. The article on RADIAL BASIS FUNCTION NETWORKS applies Bayesian methods to the case where the approximation to the given $y = y(x; w)$ is based on a network using combinations of "radial basis" functions, each of which is "centered" around a weight vector w, so that the response to input x depends on some measure of "distance" of x from w, rather than on the dot product $w \cdot x = \Sigma_i w_i x_i$, as in many formal neurons. The distribution of the w's may be determined by some form of clustering (see DATA CLUSTERING AND LEARNING discussed below). Further learning adjusts the connection strengths to a neuron whose outputs give an estimate of, e.g., the posterior probability $p(c|x)$ that class c is present given the observation (network input) x. However, it is easier to model other related aspects of the data, such as the unconditional distribution of the data $p(x)$ and the likelihood of the data, $p(x|c)$, and then recreate the posterior from these quantities according to Bayes's rule, $p(c_i|x) = p(c_i)p(x|c_i)p(x)$.

MINIMUM DESCRIPTION LENGTH ANALYSIS shows how ideas relating to minimum description length (MDL) have been applied to neural networks, emphasizing the direct relationship between MDL and Bayesian model selection methods. The classic MDL approach defined the information in a binary string to be the length of the shortest program with which a general-purpose computer could generate the string. The Bayes bridge is obtained by replacing the Bayesian goal of inferring the "most likely" model M from a set of observations by minimizing the length of an encoded message which describes M as well as the data D expressed in term of M. MDL and Bayesian methods both formalize Occam's razor in that a complex network is preferred only if its predictions are sufficiently more accurate.

The Bayesian articles stress the "global" statistical idea of "finding the weights which maximize some expectation, according to given probability distributions," as distinct from the deterministic idea of adjusting the weights at each time step to provide a local increment in performance on the current input. However, gradient descent provides an important tool for finding the weight settings which decrease some stochastic expectation of error, too. The article on STOCHASTIC APPROXIMATION AND NEURAL NETWORK LEARNING shows that gradient descent has a long tradition in the literature of stochastic approximation. Any stochastic process which can be interpreted as minimizing a cost function based on noisy gradient measurements in a sequential, recursive manner may be considered to be a stochastic approximation. "Sequential" means that each estimate of the location of a minimum is used to make a new observation which, in turn, immediately leads to a new estimate; "recursive" means that the estimates depend on past gradient measurements only through a fixed number of scalar statistics. The article states general theorems on learning rate schedules applicable to the special case of neural networks.

The article on UNSUPERVISED LEARNING WITH GLOBAL OBJECTIVE FUNCTIONS makes the point that even unsupervised learning involves an *implicit* training signal based on the network's ability to predict its own input, or on some more general measure of the quality of its internal representation. It reviews three types of unsupervised neural network learning procedures: information-preserving algorithms, density estimation techniques, and invariance-based learning procedures. The first method is based on the preservation of mutual information $I_{x;y} = H(x) - H(x|y)$ between the input vector x and output vector y where $H(x)$ is the entropy of random variable x and $H(x|y)$ is the entropy of the conditional distribution of x given y. The second approach is to assume a priori a class of models which constrains the general form of the probability density function and then to search for the particular model parameters defining the density function (or mixture of density functions) most likely to have generated the observed data (cf. the earlier discussion of Bayesian methods). Finally, invariance-based learning extracts higher-order features and builds more abstract representations. Once again, the approach is to make constraining assumptions about the structure that is being sought, and to build these constraints into the network's architecture and/or objective function to develop more efficient, specialized learning procedures.

The problem with Hopfield networks is that they can get trapped in local, rather than global, minima of the energy function. "Simulated Annealing" (q.v.) provides a general solution to this problem of entrapment, which finds its neural network realization in BOLTZMANN MACHINES. These machines use stochastic computing elements to extend discrete Hopfield networks in two ways: they replace the deterministic, asynchronous dynamics of Hopfield networks with a randomized local search dynamics; and they replace the Hebbian learning rule with a more powerful stochastic learning algorithm.

COMPETITIVE LEARNING is a form of unsupervised learning in which each input pattern comes, through learning, to be associated with the activity of one, or at most a few, neurons, leading to sparse representations of data that are easy to decode. Competitive learning algorithms employ some sort of competition between neurons in the same layer via lateral connections. *Hard* competition allows the final activity of only one neuron, the strongest one to start with; whereas in *soft* competition, the activity of the lateral neurons does not necessarily drive all but one to zero. DATA CLUSTERING AND LEARNING emphasizes the related idea of data clustering, discovering, in an unsupervised fashion, structure which is hidden in a data set, e.g., the pronounced similarity of groups of data vectors. There is a delicate tradeoff: not to superimpose too much structure, and yet not to overlook structure. The article discusses two methods; parameter estimation of mixture models (cf. UNSUPERVISED LEARNING WITH GLOBAL OBJECTIVE FUNCTIONS) and vector quantization of a data set by combinatorial optimization (cf. "Learning Vector Quantization").

Nearing its end, this road map takes us to two articles which apply the methods introduced in the earlier article on "Statistical Mechanics of Neural Networks": STATISTICAL MECHANICS OF LEARNING and STATISTICAL MECHANICS OF GENERALIZATION. As usual in supervised learning, a network is adapted to a training set by adjusting its parameters w so that it responds fairly correctly on the examples. A common statistical model assumes that all inputs are produced independently, at random, with the same probability density from the network's environment. The probability that the network will make a mistake on the new input defines its generalization error for a given training set. Its average over many realizations of the training set, as a function of the number of examples, gives the so-called learning curve. Its calculation requires the knowledge of the network weights generated by the learning process, for which an explicit form will not be available in most cases. The methods of statistical mechanics provide an approach to this problem, in many cases yielding an exact calculation of learning curves in the "thermodynamic limit" of a very large network in which the network size increases in proportion to the number of training examples, while the statistical or information-theoretic approach is applicable to the learning curve of a medium-size network (cf. LEARNING AND STATISTICAL INFERENCE).

Finally, "GENOTYPES" FOR NEURAL NETWORKS adds another temporal dimension to the biological process of adaptation, namely that of evolution. Rather than adapt the weights of a single network to solve a problem in the network's "lifetime," the evolutionary approach applies the methodology of genetic algorithms to evolve a population of neural networks over several generations so that the population becomes better and better suited to some computational ecology.

Computability and Complexity

The 1930s saw the definition of an abstract notion of computability when it was discovered that the set of functions on the natural numbers, $f : \mathbb{N} \to \mathbb{N}$, computable by a Turing machine (an abstraction from following a finite set of rules to calculate on a finite but extendable tape, each square of which could hold a fixed number of symbols), lambda functions (which later came to be better known as functions computable by programs written in LISP), and general recursive functions (a class of functions obtained from very simple numerical functions by repeated application of composition, minimization, etc.) were identical. As general-purpose electronic computers were developed and used in the 1940s and 1950s, it was firmly established that these *computable functions* were precisely the functions that could be computed by such computers with suitable programs *provided there were no limitations on computer memory or computation time*. This set the stage for the development of *complexity theory* in the 1960s and beyond: to chart the different subsets of the computable functions that would be obtained when restrictions were placed on computing resources.

STRUCTURAL COMPLEXITY AND DISCRETE NEURAL NETWORKS measures how the set of functions computed by a network of simple "neurons" (linear threshold units; McCulloch-Pitts neurons) varies with the size (number of units) and the depth (the number of units in the longest path from an input to an output) of the network. PARALLEL COMPUTATIONAL MODELS explores similar issues, placing cellular automata and neural networks within the setting of conventional complexity theory, but then looking at the implications of analyses obtained when the discrete operations of conventional automata theory are replaced by a computing model in which operations on real numbers are treated as basic. Recent results have shown that neural networks with real weights are *more powerful* than traditional models of computation (see also "Automata and Neural Networks") in that they can compute more functions within given time bounds. This property, combined with their intrinsic parallelism, makes these networks very appealing, but the practicality of an approach based on infinite precision real operations remains to be seen. Nonetheless, the revival of real numbers in computability theory has renewed complexity theory and introduced many open problems in computational learning theory and neural network theory.

TIME COMPLEXITY OF LEARNING studies the issue of how long it takes a given learning rule to adjust the weights of a network (if, indeed, it is capable of doing so) to correctly process a set of examples. The article also presents results concerning the number of mistakes that will be made during a learning process. Other issues to be addressed in more or less detail are how "neural" the algorithms are, how exactly correct they need to be, how dependable they need to be, how they get their data, and how helpful the teacher is. PAC LEARNING AND NEURAL NETWORKS discusses the "probably approximately correct" (PAC) learning paradigm as it applies to artificial neural networks. Roughly speaking, if a large enough sample of randomly drawn training examples is presented, then it should be likely that, after learning, the neural network will classify most other randomly drawn examples correctly. The PAC model formalizes the terms "likely" and "most." The two main issues in PAC learning theory are how many training examples should be presented, and whether learning can be achieved using a fast algorithm. PAC learning makes use of the Vapnik-Chervonenkis dimension (VC-dimension) as a combinatorial parameter which measures the "expressive power" of a family of functions, and the utility of this parameter is explored more fully in the article on VAPNIK-CHERVONENKIS DIMENSION OF NEURAL NETWORKS. Bounds for the VC-dimension of a neural network **N** provide estimates for the number of random exam-

ples that are needed to train **N** so that it has good generalization properties (i.e., so that the error of **N** on new examples from the same distribution is very small, with probability very close to 1). These bounds tend to be large, since they provide such generalization-guarantees simultaneously for any probability distribution on the examples and for any training algorithm that minimizes disagreement on the training examples. Tighter bounds are available for average behaviors over probability distributions and for specific training algorithms. This theme is further developed in the article on LEARNING AND GENERALIZATION: THEORETICAL BOUNDS in relation to three learning problems: pattern recognition, regression estimation, and density estimation.

II.4. Applications and Implementations

Control Theory and Robotics

As noted in the "Historical Fragment" section of Part I, the interchange between biology and technology that characterizes the study of neural networks is an outgrowth of work in *cybernetics* in the 1940s. One of the keys to cybernetics was control (the other was communication of the kind studied in information theory). It is thus appropriate that control theory should have become a major application area for neural networks. MOTOR CONTROL, BIOLOGICAL AND THEORETICAL sets forth the basic cybernetic concepts. A motor control system acts by sending motor commands to a controlled object, often referred to as "the plant," which in turn acts on the local environment. The plant or the environment has one or more variables which the controller attempts to regulate. If the controller bases its actions on signals which are not affected by the plant output, it is said to be a feedforward controller. If the controller bases its actions on a comparison between desired behavior and the controlled variables, it is a feedback controller.

The major advantage of negative feedback control is that it is a very simple, robust strategy. The controller constantly seeks to cancel the feedback error, and so operates adequately without exact knowledge of the controlled object, and despite internal or external disturbances. The advantage of feedforward control is that it can, in the ideal case, give perfect performance with no error between the reference and the controlled variable. The main disadvantages are the practical difficulties in developing an accurate controller, and the lack of corrections for unexpected disturbances. IDENTIFICATION AND CONTROL explores the major strategy for developing an accurate model of the plant, namely to "identify" the plant as belonging to (or more precisely, being well approximated by) a system

obtained from a general family of systems by setting key parameters (e.g., the coefficients in the matrices of a linear system). By coupling a controller which can control a plant when supplied with its parameters to an identification procedure which supplies them, one obtains an adaptive controller which can handle an unknown plant even if its dynamics is (slowly) changing.

ADAPTIVE CONTROL, GENERAL METHODOLOGY and ADAPTIVE CONTROL: NEURAL NETWORK APPLICATIONS introduce adaptive control. As the titles suggest, the first gives a broad review of the subject, while the second focuses particularly on the use of neural networks to provide the adaptation. Since neural networks are able to approximate arbitrary nonlinear functions, their use extends adaptive control far beyond the range of linear systems. MODEL-REFERENCE ADAPTIVE CONTROL explains the approach to adaptive control in which, given a plant with some unknown parameters, one first develops a reference model that describes the desired behavior of the plant; the adaptation mechanism then generates compensatory mechanisms to make the real plant behave like the model plant for any given reference input signal.

REINFORCEMENT LEARNING IN MOTOR CONTROL shows the utility in motor control of the general theory introduced in the article on "Reinforcement Learning." Instead of providing detailed error information, as in supervised learning, the "reinforcement" (or evaluative feedback) provides an indication of success or failure or tells the learner whether or not, and possibly by how much, its behavior has improved. To maximize the goodness of behavior as indicated by evaluative feedback, a reinforcement learning system has to actively try alternatives, compare the resulting evaluations, and use some kind of selection mechanism to guide behavior toward the better alternatives. The article on SENSORIMOTOR LEARNING develops several related ideas. It explains how neural networks can acquire "models" of some desired sensorimotor transformation. A forward model is a representation of the transformation from motor commands to movements, in other words, a model of the controlled object. An inverse model is a representation of the transformation from desired movements to motor commands, and so can be used as the controller for the controlled object. The article explains the crucial role of coarse coding in the design of computational systems capable of learning sensorimotor mappings. It also notes that, while a special purpose machine may be controlled by a single control system, an animal or "intelligent" robot is able to deploy any of a variety of "coordinated control programs," linking perceptual schemas and motor schemas (controllers) as the occasion demands. The article on ROBOT CONTROL addresses a similar set of issues. The availability of precise mappings from physical space or sensor space to joint space or motor space is a crucial issue in robot control. Three main approaches are discussed: Correlational procedures carry out feature discovery or clustering and are often used to represent a given state space in a compact and topology-preserving manner, using procedures such as those described in the articles "Self-Organizing Feature Maps: Kohonen Maps" and "Adaptive Resonance Theory (ART)." Error-minimization procedures require explicit data on input/output pairs; their goal is to build a mapping from inputs to outputs that generalizes adequately using, e.g., the least mean squares (LMS) rule and backpropagation. In between both extremes lie procedures that use reinforcement learning to build a mapping that maximizes reward.

REACTIVE ROBOTIC SYSTEMS provides a conceptual framework for robotics that is rooted in schema theory rather than symbolic AI. Here, robots are instructed to perform through the activation of a collection of low-level primitive behaviors (schemas), and complex behavior emerges through the interaction of these schemas and the complexities of the environment in which the robot finds itself. This work was inspired in part by studies of animal behavior (see, e.g., "Neuroethology, Computational" and related articles discussed in the road map on **Motor Pattern Generators and Neuroethology**, Section II.8). However, the article not only shows the power of reactive robots in many applications; it also notes the utility of hybrid systems capable of using deliberative reasoning as well as reactive execution (which fits in with an evolutionary view of the human brain in which reactive systems handle many functions but can be overruled or orchestrated by, e.g., the deliberative activities of prefrontal cortex). VISION FOR ROBOT DRIVING gives a dramatic example of the successful use of a neural network in learning the task of vision-based autonomous driving. The objective is to steer a robot vehicle, using input from an onboard video camera to decide how to control motors on the steering wheel. An artificial neural network has been trained to process images from the video camera to generate a steering command which will keep the vehicle on the road while driving at high speed.

EXPLORATION IN ACTIVE LEARNING develops a learning paradigm very suitable for the analysis of animals and mobile robots exploring their environment: in *active learning*, the learner can execute actions which have an impact on the generation of training data. Among the strategies presented for exploration are one in which actions are favored unless they have repeatedly been found to be disadvantageous, one in which actions that have not been executed for a long time are favored for exploration, and one in which exploration is achieved through a Bayesian prior that expresses uncertainty as a function of how often and when an action has been executed. These issues are illustrated with a robot navigation task. A similar task is addressed using a different approach in the article on POTENTIAL FIELDS AND NEURAL NETWORKS. The robot's task is to navigate in a two-dimensional world from a starting position to a goal while avoiding collisions with the walls or obstacles. The underlying concept is that of a potential function which combines an "attraction" to the goal and a "repulsion" from the obstacles, such that steepest descent yields a collision-free path to the goal location from arbitrary starting positions. Both the potential function and the means to exploit it are initially unknown, and the goal of learning is thus to acquire the potential function and the state transition function. There are

links here to the more biological concerns of the articles on "Cognitive Maps" and "Hippocampus: Spatial Models."

Applications of Neural Networks

ADAPTIVE FILTERING
ADAPTIVE SIGNAL PROCESSING
APPLICATIONS OF NEURAL NETWORKS
ASTRONOMY
AUTOMATIC TARGET RECOGNITION
EXPERT SYSTEMS AND DECISION SYSTEMS USING NEURAL NETWORKS
FORECASTING
HANDWRITTEN DIGIT STRING RECOGNITION
HIGH-ENERGY PHYSICS
INVESTMENT MANAGEMENT: TACTICAL ASSET ALLOCATION
NOISE CANCELING AND CHANNEL EQUALIZATION
PROCESS CONTROL
PROSTHETICS, NEURAL
PROTEIN STRUCTURE PREDICTION
SPEAKER IDENTIFICATION
SPEECH RECOGNITION: A HYBRID APPROACH
SPEECH RECOGNITION: FEATURE EXTRACTION
SPEECH RECOGNITION: PATTERN MATCHING
STEELMAKING
TELECOMMUNICATIONS
VISUAL PROCESSING OF OBJECT FORM AND ENVIRONMENT LAYOUT

The preceding road map, **Control Theory and Robotics**, has already presented a number of applications of neural networks. Here we offer a representative (but by no means exhaustive) set of other applications, a list which can be augmented by the study of many other road maps. The article on APPLICATIONS OF NEURAL NETWORKS gives a selective overview of applications, but is not a road map of *Handbook* articles. In fact, little in the way of a road map is required for these articles, since their topics are so diverse that they can be read in almost any order, although the quadruplet on speech recognition might best be read in the order (1) SPEECH RECOGNITION: A HYBRID APPROACH, (2) SPEAKER IDENTIFICATION, (3) SPEECH RECOGNITION: FEATURE EXTRACTION, and (4) SPEECH RECOGNITION: PATTERN MATCHING.

Several articles review the various contributions of adaptive neural networks to signal processing. These are ADAPTIVE SIGNAL PROCESSING, ADAPTIVE FILTERING, and NOISE CANCELING AND CHANNEL EQUALIZATION. Two very different biomedical applications are presented in PROSTHETICS, NEURAL and PROTEIN STRUCTURE PREDICTION. The former challenges us to understand the neural circuitry of both sensory and motor processes well enough to replicate key functions in silicon and other devices. In the author's words, "It is difficult to imagine a more appropriate application of electronic neural networks than in the repair of the biological systems that have inspired them." Applications in physics include ASTRONOMY and HIGH-ENERGY PHYSICS. The applications discussed in FORECASTING are related to those in INVESTMENT MANAGEMENT: TACTICAL ASSET ALLOCATION, while EXPERT SYSTEMS AND DECISION SYSTEMS USING NEURAL NETWORKS shows how neural-based expert systems have been applied in stock trading, credit scoring, fraud detection, clinical and epidemiological applications, and factory production and maintenance.

PROCESS CONTROL exemplifies how neural networks are used in the chemical processing and related industries not only for control per se, but also for related applications, such as virtual sensors and fault detection and diagnosis. Other industrial applications include STEELMAKING and TELECOMMUNICATIONS. Finally, visual processing (see also the road map on **Vision**, Section II.6) has applications all the way from the military to the post office: AUTOMATIC TARGET RECOGNITION, VISUAL

PROCESSING OF OBJECT FORM AND ENVIRONMENT LAYOUT, and HANDWRITTEN DIGIT STRING RECOGNITION.

The crucial question I posed to the authors of these articles was "Which neural network architecture is best suited for your problem, and why?" Some answered this in detail, and others answered it at best implicitly. Moreover, some argued for a pure neural network architecture (possibly modular), whereas others argued for a hybrid architecture in which some tasks were allocated to neural networks whereas others were addressed by non-neural methods. The article APPLICATIONS OF NEURAL NETWORKS addresses the question in part by providing a table in which different methods are linked to different applications. The reader who wishes to pursue this linkage further may turn to Table 1 of the present section which lists the various architectures applied in the other articles of this road map. (PROSTHETICS, NEURAL does not appear in the table, since that article focuses on the biological systems the prostheses are to replace, rather than the use of artificial neural networks as an approach to their replacement—an interesting opening for future research.) Column numbers in the table refer to the methods listed in the key provided in the table footnote.

As the reader is well aware, many of the methods listed in the table are analyzed, discussed, or applied in other articles in the *Handbook*. But rather than offer an unwieldy listing of all these articles, I refer the reader to the index for further references to methods of particular interest. Finally, to close this road map, note that several authors of articles on applications stress the importance of using special-purpose VLSI chips to gain the full efficiency of artificial neural network approaches. Such chips are among the methods for **Implementation of Neural Networks** discussed in the next road map.

Implementation of Neural Networks

ANALOG VLSI FOR NEURAL NETWORKS
BIOMATERIALS FOR INTELLIGENT SYSTEMS
COMPUTER MODELING METHODS FOR NEURONS
DIGITAL VLSI FOR NEURAL NETWORKS
MULTIPROCESSOR SIMULATION OF NEURAL NETWORKS
NEUROSIMULATORS
NSL: NEURAL SIMULATION LANGUAGE
OPTICAL ARCHITECTURES FOR NEURAL NETWORK IMPLEMENTATIONS
OPTICAL COMPONENTS FOR NEURAL NETWORK IMPLEMENTATIONS

Table 1. Neural Networks: Applications and Methods

Article Title	1	2	3	4	5	6	7	8	9	10	11	12	13	14	15	16	17	18	19	20	21	22	23	24	25	26
																										Methods
Adaptive Filtering	✓	✓																								
Adaptive Signal Processing		✓	✓	✓		✓																				
Astronomy					✓								✓	✓												
Automatic Target Recognition		✓			✓		✓	✓	✓	✓	✓															
Expert Systems and Decision Systems...					✓																					
Forecasting						✓								✓												
Handwritten Digit String Recognition					✓										✓	✓	✓									
High-Energy Physics					✓	✓																				
Investment Management...					✓	✓						✓		✓				✓								
Noise Canceling...	✓	✓	✓																							
Protein Structure Prediction					✓															✓						
Speaker Identification														✓								✓				
Speech Recognition: A Hybrid Approach																				✓	✓					
Speech Recognition: Feature Extraction					✓							✓								✓	✓					
Speech Recognition: Pattern Matching					✓	✓															✓					
Steelmaking					✓																					
Telecommunications		✓	✓		✓																			✓	✓	
Visual Processing of Object Form...										✓										✓	✓					✓

Key to column numbers: **1**, Single-layer perceptron; adaline; **2**, least-mean-square (LMS) adaptive algorithm; Widrow-Hoff learning rule; **3**, Wiener filter; Kalman filter; recursive least-squares (RLS) algorithm; **4**, finite-duration impulse response (FIR) multilayer perceptron; **5**, multilayer (loop-free) perceptron (MLP) trained by backpropagation; **6**, backpropagation in recurrent networks; **7**, neocognitron; **8**, wavelets; **9**, projection pursuit; **10**, Adaptive Resonance Theory (ART) network; **11**, classification figure of merit (CFM) and minimum misclassification error (MME) functions; **12**, principal components analysis (PCA); **13**, Bayesian neural network classifier; **14**, auto-regressive moving average (ARMA) models; multiple linear regression (MLR) models (non-neural); **15**, feedforward convolutional networks; **16**, nonlinear programming (non-neural); **17**, Viterbi algorithm to find the shortest path traversing a given number of nodes (non-neural); **18**, additive nonlinear regression models such as additive conditional expectation (ACE) (non-neural); **19**, topographic (Kohonen) maps (Tmaps); **20**, learning vector quantization (LVQ); **21**, hidden Markov models (HMM) (non-neural); **22**, time-delay neural networks (TDNN); **23**, spectral analysis, including linear predictive coding (LPC), perceptual linear prediction (PLP) and log power spectral or cepstral coefficients (non-neural); **24**, reinforcement learning; Q-learning; **25**, optimization via a Hopfield neural network; **26**, radial basis function (RBF) networks.

PROGRAMMABLE NEUROCOMPUTING SYSTEMS
SILICON NEURONS

Briefly, a neural network (whether an artificial neural network for technological application or a simulation of a biological neural network in computational neuroscience) can be implemented in three main ways: by programming a general-purpose electronic computer; by programming an electronic computer designed for neural network implementation; or by building a special-purpose device to emulate a particular network or parametric family of networks. We discuss these three approaches in turn.

The article on NEUROSIMULATORS briefly reviews a selection of about 40 of the currently available programming languages for neural networks. These are reviewed under four main headings: neurosimulators for novices; neurosimulators available from academia; tools for single-neuron modeling; and neurosimulators designed for business and industry. The author concludes that the tools available for neural network simulations still fall short of the standards achieved in more mature fields, such as statistics or computer-aided design, with most current neurosimulators mainly useful for prototyping small- to mid-sized standard applications, although a number are designed for "general-purpose" neural network programming. For research purposes, a neurosimulator requires, at the very least, a highly developed interface, a scalable design (e.g., through parallel hardware), and extendability with new neural network paradigms. The article on COMPUTER MODELING METHODS FOR NEURONS presents some of the numerical methods necessary for efficient simulation of detailed models of single neurons (see, e.g., the articles on "Axonal Modeling" and "Dendritic Processing" in the **Biological Neurons** road map, Section II.5), focusing on the Crank-Nicholson method for solving the cable equation with voltage-dependent ion conductances. By contrast, NSL: NEURAL SIMULATION LANGUAGE provides methods for simulating very large networks of relatively simple (artificial or biological simulation) neurons. One of the most exciting challenges for neurosimulation in the next decade is to provide multilevel neurosimulation environments in which one can move effortlessly between the levels of schemas (functional decomposition of an overall behavior), large neural networks, detailed models of single neurons, and neurochemical models of synaptic plasticity. To be fully effective, such an environment will also need visualization tools, and the ability to access a database to provide experimental results for comparison with model-based predictions.

The next three articles address the digital, parallel implementation of neural networks. MULTIPROCESSOR SIMULATION OF NEURAL NETWORKS offers a theoretical analysis of the advantages of parallel implementation, with special attention to parallelizing the matrix product computations and to parallelizing by blocks of examples. DIGITAL VLSI FOR NEURAL NETWORKS first discusses the differences between digital and analog design techniques, focusing primarily on cost-performance tradeoffs in flexibility (Amdahl's law). It then discusses (as does PROGRAMMABLE NEUROCOMPUTING SYSTEMS) the RAP machine designed for ANN emulation using traditional processors, as well as two custom digital ANN processors: the Adaptive Solutions CNAPS and the Intel/Nestor Ni1000. In each case, the medium used is CMOS VLSI, the material used in the chips for essentially all today's commercially available electronic computers because of both its price performance and the availability of

design tools that are relatively easy to use. PROGRAMMABLE NEUROCOMPUTING SYSTEMS takes up a related set of topics from a somewhat different perspective. System design for neurocomputers must pay attention to scalar and other non-neural operations, communication costs, input/output between the host/disk subsystem(s) and the neurocomputer, software, and diagnostic capabilities. The article focuses on operations other than dot-product computation, since matrix arithmetic is dealt with adequately elsewhere. Networks with a million neurons are unlikely to be fully connected, and so a major concern for future neurocomputers will be efficient implementation of sparsely interconnected networks.

To put it simply, digital VLSI assigns a different circuit to each bit of information that is to be stored and processed. Each circuit is driven to the limit so that it settles into a 0-state or a 1-state, passing through a linear voltage-current region to get from one saturation state to the other. Thus, if a synaptic weight is to be stored with eight-bit precision in digital VLSI, it requires eight such circuits. By contrast, the linear region of a single circuit element can store data with about three bits of precision with far less "real estate" on the chip, and with far less power loss. The price, of course, is that precision cannot be guaranteed on the same scale as for digital circuits. However, analog precision is adequate in many neural network applications. ANALOG VLSI FOR NEURAL NETWORKS provides an overview of the implementation of circuitry in analog VLSI, and then summarizes a number of technological implementations of ANNs on analog chips. By contrast, SILICON NEURONS takes the same implementation methodology into the realm of computational neuroscience. As we will see in the next road map, **Biological Neurons**, biological neural networks are difficult to model in detail because they are composed of large numbers of nonlinear elements and have a wide range of time constants. Where simulation on a general-purpose digital computer slows dramatically as the number and coupling of elements increase, silicon neurons operate in real time and the speed of the network is independent of the number of neurons or their coupling. On the other hand, high connectivity still poses problems in 2D chip layouts, and the design of special-purpose hardware is a significant investment, since analog VLSI still lacks a general set of easy-to-use design tools.

Finally, we turn from "today's" implementation medium, VLSI, to media whose widespread application lies further in the future. The success of optic fibers as media for telecommunications has been complemented by the use of holograms and spatial light modulators as mechanisms for storing and processing information via patterns of light (photonics) rather than patterns of electrons (electronics). The current state of photonic approaches to neural network implementation is charted in OPTICAL ARCHITECTURES FOR NEURAL NETWORK IMPLEMENTATIONS and OPTICAL COMPONENTS FOR NEURAL NETWORK IMPLEMENTATIONS. The article on BIOMATERIALS FOR INTELLIGENT SYSTEMS takes an even bolder leap into the future, presenting the possibility of utilizing biomaterials for information technology, building on the study of the relation between the computational capabilities of organisms and their material organization. In some cases, biomaterials may complement inorganics in the context of conventional information-processing architectures, for example by providing novel optical interconnects, but they also open up the possibility of information processing that specifically exploits the unique characteristics of carbon macromolecules.

II.5. Biological Neurons and Networks

Biological Neurons

Nearly all the articles of the previous road maps discuss networks made of very simple neurons of the kind explained at length in Section I.3, "Dynamics and Adaptation in Neural Networks." These are describable by a single internal variable, either binary or real-valued (the "membrane potential"), and communicate with other neurons by a simple (generally nonlinear) function of that variable, sometimes referred to as the firing rate. Incoming signals are usually summed linearly via "synaptic weights," and these weights in turn may be adjusted by simple learning rules, such as the Hebbian rule, the perceptron rule, or a reinforcement learning rule. Such simplifications remain valuable both for technological application of ANNs and for approximate models of large biological networks. Nonetheless, the time has come to acknowledge that biological neurons are vastly more complex than these single-compartment models suggest. An appreciation of this complexity is necessary for the computational neuroscientist wishing to address the increasingly detailed database of experimental neuroscience. I would argue that it is also important for the technologist looking ahead to the incorporation of new capabilities into the next generation of ANNs.

PERSPECTIVE ON NEURON MODEL COMPLEXITY discusses the range from very simple to rather complex neuron models. Which model to choose depends, in each case, upon the context, such as how much information we already have about the neurons under consideration, and what questions we wish to answer. The use of more realistic neuron models when seeking functional insights into biological nervous systems does *not* mean choosing the most complex model, at least in the sense of including all known anatomical and physiological details. Rather, the key is to preserve the *most significant* distinctions between regions (soma, proximal dendritic, distal dendritic, etc.), using "compartmental modeling" whereby one compartment represents each functionally distinct region.

SINGLE-CELL MODELS starts by reviewing the "simple" models of Section I.3 (the McCulloch-Pitts, perceptron, and Hopfield models) and the slightly more complex polynomial neuron. It then turns to more realistic biophysical models, most of which are explored in detail further along this road map. These include the Hodgkin-Huxley model of squid axon, integrate-and-fire models, modified single-point models, cable and compartmental models, and models of synaptic conductances. Before turning to the detailed analysis of mechanisms of neuronal function, we first consider three articles that offer a so-called high-level view of the neuron, but this time a stochastic one.

Most nerve cells encode their output as a series of action potentials, or spikes, which originate at or close to the cell body and propagate down the axon at constant velocity and amplitude. DIFFUSION MODELS OF NEURON ACTIVITY studies the membrane potential of a single neuron as engaged in a stochastic process which will eventually bring it to the threshold for spike initiation. This leads to the first-passage-time problem, inferring the distribution of neuronal spiking based on the "first passage" of the membrane potential from its resting value to threshold. SYNAPTIC CODING OF SPIKE TRAINS continues this stochastic approach by viewing spike trains in terms of "point processes" which describe the statistics of spike timing, rather than the detailed electrical shaping of each spike. Trains have interval and instantaneous rate "averages," and exhibit various patterns ("pacemaker," "bursty," etc.). Of particular concern is "entrainment" in the sense of the influence of one cell's spike train on another through synaptic coupling. SENSORY CODING AND INFORMATION THEORY continues the study of spike trains by using information theory to analyze explicitly how the spacing of spikes may code sensory information. The article analyzes vision in flies, hearing in frogs, mechanoreception in crickets, and spatial pattern vision in monkeys.

Now for the details of neuronal function. The ionic mechanisms underlying the initiation and propagation of action potentials were elucidated in the squid giant axon by a number of workers, most notably by Hodgkin and Huxley. Variations on the Hodgkin-Huxley equation still underlie the vast majority of contemporary biophysical models. AXONAL MODELING describes this model and its assumptions, introduces the two classes of axons (myelinated and nonmyelinated) found in most animals, and concludes by briefly commenting on the possible functions of axonal branching in information processing. The article on CHAOS IN AXONS provides a specialized supplement to this article, noting that squid giant axons and the Hodgkin-Huxley equation which describes them can exhibit chaotic behavior (cf. "Chaos in Neural Systems"). It then offers a simple neuron model which can qualitatively reproduce such chaotic responses in a building block for neural networks.

The Hodgkin-Huxley equation was brilliantly inferred from detailed experiments on conduction of nerve impulses. Much research since then has revealed that the basis for these equations is provided by "channels," structures built from macromolecules and embedded in the neuron membrane which, in a voltage-dependent way, can selectively allow different ions to pass through to change the neuron's membrane potential. Similarly, channels (also known in this case as receptors) in the postsynaptic membrane can respond to neurotransmitters, chemicals released from the presynaptic membrane, to change the neuron's local membrane potential in response to presynaptic input. These changes, local to the synapse, must propagate down the dendrites and across the cell body to help determine whether or not the axon will "pass threshold" and generate an action potential. All of these mechanisms are addressed in the next five articles.

ION CHANNELS: KEYS TO NEURONAL SPECIALIZATION notes that, besides the channels described by the Hodgkin-Huxley equation, there are many other types of channels which differ in many respects. Each class of neurons is endowed with a different set of channels, and the diversity of channels between different types of neurons explains, in great part, the functional classes of neurons found in the brain. Some neurons fire spontaneously, some show adaptation, some fire in bursts, and so on. Therefore, a channel-based cellular physiology is relevant to questions about the role of different brain regions in overall function.

SYNAPTIC CURRENTS, NEUROMODULATION, AND KINETIC MODELS provides kinetic models of how synaptic currents arise from ion channels whose opening and closing is controlled (gated) directly or indirectly by the binding of neurotransmitters. The article shows how these different types of synaptic interactions can be modeled efficiently using a formalism similar to that pioneered by Hodgkin and Huxley. It compares several models of synaptic interaction, focusing on simple models based on the kinetics of postsynaptic receptors, and shows how these models capture the time courses of postsynaptic currents of several types of synaptic responses, as well as synaptic summation, saturation, and desensitization. Some neurotransmitters do not gate the channel by binding directly to it, but modulate the channel through an intracellular molecule which links the activated receptor to the opening or closing of the channel. Such a molecule is called a "second messenger" to contrast it with the case in which the transmitter itself provides a "primary message" which acts directly on the channel, which is then called "ligand-gated." Second-messenger-based synaptic interaction occurs at a slower time scale than ligand-gated interaction and is called *neuromodulation* since it may modulate the behavior of the postsynaptic neuron over a time scale of seconds or minutes rather than milliseconds (cf. "Neuromodulation in Invertebrate Nervous Systems").

OSCILLATORY AND BURSTING PROPERTIES OF NEURONS offers a dynamic systems analysis of the linkage between a fascinating variety of endogenous oscillations (neuronal rhythms) and appropriate sets of channels. However, membrane potential oscillations with apparently similar characteristics can be generated by different ionic mechanisms, and a given cell type may display several different firing patterns under different neuromodulatory conditions. Here, membrane dynamics are described by coupled differential equations, the behavior modes by attractors (cf. "Computing with Attractors"), and the transitions between modes by bifurcations. The rest state is represented by a time-independent steady state and repetitive firing by a limit cycle. The article on SILICON NEURONS shows how such differential equations can be directly mapped into an electronic circuit built using analog VLSI, to allow real-time exploration of the behavior of quite realistic neural models. The article DYNAMIC CLAMP: COMPUTER-NEURAL HYBRIDS carries the use of modeling a surprising step further: a model of a neuron that is part of a small (invertebrate, say) neural network is used to replace the real neuron by electrical activity generated by the computer model to reveal to what extent the remaining (biological) neurons retain their normal activity as the model of the substituted neuron is varied.

Roughly a dozen different types of ion channels contribute to the membrane conductance of a typical neuron. The article ACTIVITY-DEPENDENT REGULATION OF NEURONAL CONDUCTANCES takes as its starting point the fact that the electrical characteristics of a neuron depend on the number of channels of each type active within the membrane and on how these channels are distributed over the surface of the cell. A complex array of biochemical processes controls the number and distribution of ion channels by constructing and transporting channels, modulating their properties, and inserting them into and removing them from the neuron's membrane. The point to note here is that channels are molecules synthesized on the basis of genetic instructions in the cell nucleus. Thus, changing which genes are active (i.e., regulating gene expression) can change the set of channels in a cell, and thus the characteristics

of the cell. In fact, electrical activity in the cell can affect a range of processes, from activity-induced gene expression to activity-dependent modulation of assembled ion channels. Channel synthesis, insertion, and modulation are much slower than the usual voltage- and ligand-dependent processes that open and close channels. Thus, consideration of activity-dependent regulation of conductances introduces a dynamics acting at a new, slower time scale into neuronal modeling, a feedback mechanism linking a neuron's electrical characteristics to its activity.

DENDRITIC PROCESSING focuses on dendrites as electrical input-output devices that operate on a time-scale range of several to a few hundred milliseconds. The input is temporal patterns of synaptic inputs spatially distributed over the dendritic surface, whereas the output is (except, for example, in the case of dendrodendritic interactions) an ionic current delivered to the soma for transformation there, via a threshold mechanism, to a train of action potentials at the axon. The article discusses how the morphology, electrical properties, and synaptic inputs of dendrites interact to perform their input-output operation. It uses cable theory and compartmental modeling to model the spread of electric current in dendritic trees. The variety of excitable (voltage-gated) channels that are found in many types of dendrites enrich the computational capabilities of neurons. COMPUTER MODELING METHODS FOR NEURONS offers numerical methods for solving the equations describing branched cables. DENDRITIC SPINES are short appendages found on the dendrites of many different cell types. They are composed of a bulbous "head" connected to the dendrite by a thin "stem"; an excitatory synapse is usually found on the spine head, and some spines also have a second, usually inhibitory, synapse located on or near the spine stem. Models in which the spine is represented as a passive electric circuit show that the large resistance of a thin spine stem can attenuate a synaptic input delivered to the spine head.

Biological Networks

CORTICAL COLUMNS, MODULES, AND HEBBIAN CELL ASSEMBLIES
DISEASE: NEURAL NETWORK MODELS
DYNAMIC MODELS OF NEUROPHYSIOLOGICAL SYSTEMS
EEG ANALYSIS
EPILEPSY: NETWORK MODELS OF GENERATION
EVOLUTION OF THE ANCESTRAL VERTEBRATE BRAIN
LAYERED COMPUTATION IN NEURAL NETWORKS
NEUROANATOMY IN A COMPUTATIONAL PERSPECTIVE
NEUROMODULATION IN INVERTEBRATE NERVOUS SYSTEMS
THALAMOCORTICAL OSCILLATIONS IN SLEEP AND WAKEFULNESS
TRAVELING ACTIVITY WAVES
WAVE PROPAGATION IN CARDIAC MUSCLE AND IN NERVE NETWORKS

We now turn to biological neural networks, a study which will be extended in the next road map on **Mammalian Brain Regions** and in other road maps on sensory systems, memory, and motor control. The article on EVOLUTION OF THE ANCESTRAL VERTEBRATE BRAIN reminds us of the variability of vertebrate brains from one species to another and challenges us to understand both the similarities and the differences. Problems with basing evolutionary hypotheses on comparative and functional analyses of adult vertebrate brains are, to some extent, mitigated by studies of comparative neuroembryology, i.e., relating adult structures to their specific development. The article introduces a possible prototype of the ancestral vertebrate brain, followed by a scenario for mechanisms that may have diversified the ancestral vertebrate brain. The layered structure of cerebral cortex is a distinctive evolutionary feature of the mammalian brain. CORTICAL COLUMNS, MODULES, AND HEBBIAN

CELL ASSEMBLIES notes that cells in a single trajectory orthogonal to the layers of mammalian cerebral cortex have similar properties. This allows related cells to be grouped into "columns," "stripes," and "modules." These anatomically defined structures are related to hypotheses about Hebbian cell assemblies: Hebb's theory postulates that meaningful events are represented by groups of neurons which are connected to each other more strongly than to other neurons. These groups are formed by way of an associative learning process which is assumed to strengthen the connections between those neurons which are often activated together. NEUROANATOMY IN A COMPUTATIONAL PERSPECTIVE notes that a vertebrate brain contains so many neurons, and each neuron has so many connections, that the task of neuroanatomy cannot be to study all the connections in detail, but rather to reveal the structural properties typical of each region of the brain to provide clues to the understanding of its specific function. Large brains have comparatively more white matter (i.e., regions containing only axons) than small brains. Nonetheless, in a large brain, distant elements may not be able to collaborate efficiently because of the delays in the transmission from one point to the other. The way out of this problem may be the higher degree of functional specialization of cortical regions in larger brains. The article provides many data on cortical structure (my favorite is that there are 3.6 km of axons in each cubic millimeter of mouse cortex!) and argues that the data fit Hebb's theory of cell assemblies remarkably well: precisely predetermined connections are not required since, as a result of learning, the patterns of interactions between neurons will be different for each brain. What is crucial, however, is an initial connectivity sufficiently rich to allow many constellations of neuronal activity to be detected and "learned" in the connections.

LAYERED COMPUTATION IN NEURAL NETWORKS presents a general framework for modeling computations performed in layered structures (which occur in many parts of the vertebrate and invertebrate brain, including the optic tectum, the avian visual wulst, and the cephalopod optic lobe, as well as mammalian cerebral cortex). A general formalism is presented for the connectivity between layers and the dynamics of typical units of each layer. Information processing capabilities of neural layers include filter operations; lateral cooperativity and competition that can be used in, e.g., stereo vision and winner-take-all mechanisms; topographic mapping that underlies the allocation of cortical neurons to different parts of the visual field (fovea/periphery), or the processing of optic flow patterns; and feature maps and population coding, which may be applied both to sensory systems and to "motor fields" of neurons so that the flow of activity in motor areas can predict initiated movements. The article thus provides a framework for many of the specific topics treated in later road maps.

Traveling waves of neural activity with velocities of 10–90 cm/s have been recorded from hippocampal slices, and waves with velocities of 20–30 cm/s have been recorded from the cortical surface. One mechanism for the spread of neural activity involves the diffusion of K^+ ions which produces waves of spreading depression that travel at much lower velocities (about 1–3 mm/min). WAVE PROPAGATION IN CARDIAC MUSCLE AND IN NERVE NETWORKS provides a comparative dimension by presenting methods for the study of waves in an "excitable medium" very different from a neural network. Excitable cells in heart muscle connect to their immediate neighbors through electrically conductive gap junctions; the mechanism is described by "reaction-diffusion equations." The article focuses on cardiac activation wave and fibrillation. Returning to the central nervous system, TRAVELING ACTIVITY WAVES focuses on activity waves which spread by nerve conduction, empha-

sizing the role played by the relative refractory state of a neuron in shaping the spatiotemporal dynamics. It demonstrates how a relative refractory state can be incorporated into a neural network model and shows how a migrating "traveling" wave pattern can arise in such networks.

The organization of large masses of neurons into synchronized waves of activity lies at the basis of phenomena such as the electroencephalogram (EEG) and evoked potentials. The EEG consists of the electrical activity of relatively large neuronal populations that can be recorded from the scalp. To extract information from these signals that may be relevant for the diagnosis of brain diseases, and to obtain a better understanding of the brain processes underlying psychophysical and cognitive functions, a large number of quantitative analysis methods has been applied to the EEG. The article on EEG ANALYSIS analyzes the EEG as a time series, as a spatiotemporal signal, and as a signal that provides information about the state of complex neuronal networks considered as nonlinear dynamical systems. The article on EPILEPSY: NETWORK MODELS OF GENERATION presents related issues. Epilepsy is a neurological disorder characterized by the occurrence of seizures, sudden changes of neuronal activity that interfere with the normal functioning of neuronal networks, resulting in disturbances of sensory and/or motor activity, and of the flow of consciousness. During an epileptic seizure, the neuronal network exhibits typical oscillations that usually propagate throughout the brain, involving progressively more brain systems. These oscillations are revealed in the EEG. Certain brains are much more likely to generate epileptic activity, raising the question of what changes in the dynamics of neuronal populations are responsible. The article considers theoretical aspects of the behavior of neuronal populations as generators of epileptic activity and briefly discusses physiological/biophysical changes in a neuronal network that can cause the formation of an epileptogenic focus.

Oscillations are, of course, part of normal behavior (as in walking and breathing) as well as signs of disease. A variety of such normal oscillations are treated in the article on THALAMOCORTICAL OSCILLATIONS IN SLEEP AND WAKEFULNESS. Slow wave sleep is characterized by synchronized events in billions of synaptically coupled neurons in thalamocortical systems. The early stage of quiescent sleep is associated with EEG spindle waves, which occur at a frequency of 7–14 Hz; as sleep deepens, waves with slower frequencies appear on the EEG. The other sleep state, associated with rapid eye movements (REM sleep) and dreaming, is characterized by abolition of low-frequency oscillations and an increase in cellular excitability, very much like wakefulness, although motor output is markedly inhibited. Activation of a series of neuromodulatory transmitter systems during arousal blocks low-frequency oscillations, induces fast rhythms, and allows the brain to recover full responsiveness. The article analyzes cortical and thalamic networks at multiple levels from molecules to single neurons to large neuronal assemblies, with techniques ranging from intracellular recordings to computer simulations, to illuminate the generation, modulation, and function of brain oscillations.

The last three articles on this road map lead off in three separate directions: the use of high-level models of neural networks to shed insight into neural diseases with cognitive deficits; the use of backpropagation as a tool to suggest hypotheses for investigation by neurophysiologists; and the use of invertebrate neural networks to probe the dramatic effects of neuromodulation on neural function. DISEASE: NEURAL NETWORK MODELS views mental illness or malfunction of a brain area as a diminished function in part of a neural network designed to perform some set of cognitive tasks. While an integrated system model of schizophrenia has not been constructed, there are simulations focused on subclasses of negative symptoms, e.g., disturbances in processing context. Depressive and manic-depressive disorders have been studied qualitatively in terms of several principles used in neural network models: associative learning, lateral inhibition, opponent processing, neuromodulation, and interlevel resonant feedback. Neural network models of frontal lobe dysfunction focus on such frontal damage effects as deficits on standard tests from clinical neuropsychology, such as the Wisconsin Card Sorting, Stroop, and verbal fluency tests; deficits on delayed response; excessive attraction to novelty; and inability to learn flexible motor sequences.

It is a truism that similarity of input-output behavior is no guarantee of similarity of internal function in two neural networks. In particular, a recurrent neural network trained by backpropagation to mimic some biological function may have little internal resemblance to the neural networks responsible for that function in the living brain. Nonetheless, DYNAMIC MODELS OF NEUROPHYSIOLOGICAL SYSTEMS demonstrates that dynamic recurrent network models can provide useful tools to help systems neurophysiologists understand the neural mechanisms mediating behavior. Biological experiments typically involve bits of the system; neural network models provide a method of generating working models of the complete system. Confidence in such models is increased if they not only simulate dynamic sensorimotor behavior but also incorporate anatomically appropriate connectivity. The utility of such models is illustrated in the analysis of four types of biological function: oscillating networks, primate target tracking, short-term memory tasks, and the construction of neural integrators.

The above two articles use the simplified neurons described in Section I.3. We now return to the importance of subtle details of the mechanisms within individual neurons. As noted earlier in this road map, neurons are not all alike: they show a rich variety of conductances that endow them with different functional properties. These properties and, hence, the collective activity of interacting groups of neurons, are not fixed, but are instead subject to modulation. The effects of neuromodulators are long-lasting, affecting neurons or circuits for seconds, minutes, or even longer. NEUROMODULATION IN INVERTEBRATE NERVOUS SYSTEMS provides a number of examples of these effects in invertebrates. The sensory information an animal needs depends on a number of factors, including its activity patterns and motivational state. The modulation of the sensitivities of many sensory receptors is shown for a stretch receptor in crustaceans. Modulators can activate, terminate, or modify rhythmic, pattern-generating networks. One example of such "polymorphism" is that neuromodulation can reconfigure the same network to produce either escape swimming or reflexive withdrawal in the nudibranch mollusk *Tritonia*. Mechanisms and sites of neuromodulation include alteration of intrinsic properties of neurons, alteration of synaptic efficacy by neuromodulators, and modulation of neuromuscular junctions and muscles. All this makes clear the subtlety of neuronal function that must be addressed by computational neuroscience and which may inspire the design of a new generation of artificial neurons.

Mammalian Brain Regions

AUDITORY CORTEX
AUDITORY PERIPHERY AND COCHLEAR NUCLEUS
BASAL GANGLIA
CEREBELLUM AND CONDITIONING
CEREBELLUM AND MOTOR CONTROL
COLLICULAR VISUOMOTOR TRANSFORMATIONS FOR SACCADES
GRASPING MOVEMENTS: VISUOMOTOR TRANSFORMATIONS
HIPPOCAMPUS: SPATIAL MODELS

OLFACTORY BULB
OLFACTORY CORTEX
RETINA
SOMATOSENSORY SYSTEM
THALAMUS
VISUAL CORTEX CELL TYPES AND CONNECTIONS
VISUAL SCENE PERCEPTION: NEUROPHYSIOLOGY

This road map introduces the conceptual analysis and neural network modeling of a variety of regions of the mammalian brain. These regions recur in many articles not included in the foregoing list—a fact that illustrates the complementary ways of exploring the mammalian brain in a top-down fashion. We may start either from the gross anatomy (what a given region of the brain does, which is the approach of this road map) or from some function (sensory, perceptual, memory, motor control, etc., as in other road maps). These top-down approaches may themselves be contrasted with bottom-up approaches, which may start from neurons and seek to infer properties of circuits, or from biophysics and neurochemistry and seek to infer properties of neurons (as in much of the two previous road maps). It must be stressed that whole books can be, and have been, written on each of the brain regions discussed in this section. The aim of each article is to get the reader started by seeing how a selection of biological data can be confronted with models which seek to illuminate them. In some cases (especially near the sensory or motor periphery), the function of a region is clear, and there is little question as to which phenomena the brain theorist is to explain. However, in more central regions, what the experimentalist observes may vary wildly with the questions that are asked, and what the modeler has to work with is more like a Rorschach blot than a well-defined picture.

Most of the articles in this road map are associated with sensory systems: vision, body sense (the somatosensory system), hearing (the auditory system), and smell (the olfactory system). Several articles then discuss brain regions associated more with motor control, learning, and cognition.

Five articles take us through the visual system. We start with the RETINA, the outpost of the brain which contains both light-sensitive receptors and several layers of neurons which "pre-process" these responses. Different ganglion cells (the output cells of the retina) specialize in coding one or more special properties of the signal, such as contrast, motion, bright, dark, or colored light flashes, or even signals related to the ecological niche of the organism. In mammals, the retinal output branches into a variety of pathways, of which the collicular pathway and the geniculostriate pathway receive most attention in this volume. There is a trend to increasing the geniculostriate contingent in mammalian species with increasing neocortex, and a relative decrease in the collicular pathway, which nonetheless remains of great importance. The destination of the collicular pathway is the midbrain region known as the superior colliculus. COLLICULAR VISUOMOTOR TRANSFORMATIONS FOR SACCADES charts one of the important ways in which this region generates orientation responses, namely, its role in controlling saccades, rapid eye movements to bring visual targets onto the fovea. Even this basic activity involves the cooperation of many brain regions; conversely, the function of the superior colliculus is not restricted to eye movements (the article on "Visuomotor Coordination in Frogs and Toads" charts the role of the tectum, which is homologous to the superior colliculus, in approach and avoidance behavior). In mammals, the geniculostriate pathway travels from the retina via a specialized region of the thalamus, called the lateral geniculate nucleus (LGN), to the primary visual cortex, which is also called the striate cortex because of its somewhat striated appearance. The thalamus has many divisions, not only those involved with sensory pathways, but also those involved in loops linking the cortex to other brain regions like the cerebellum and basal ganglia. However, after some general preliminaries, the article on the THALAMUS focuses on the visual system, offering some thoughtful speculations inspired by the fact that there are more axons descending from the striate cortex to the LGN than there are ascending axons from the LGN to the striate cortex. Hypotheses include a gating role, an iterative algorithm carried out in the thalamocortical loop, and the notion that the LGN is an active "blackboard" on which various patterns can be written, misleading patterns can be suppressed, and a best reconstruction can thereby be generated. The last two articles on the visual system in this road map offer complementary views of the visual cortex. VISUAL CORTEX CELL TYPES AND CONNECTIONS reviews features of the microcircuitry of the primary visual cortex, area V1, and physiological properties of cells in its different laminae. It then outlines several hypotheses as to how the anatomical structure and connections might develop and how they might serve the functional organization of the region. Among these are the origin of orientation specificity, and the possible role of lateral interactions in such low-level vision processes as the inference of patterns of depth or motion. VISUAL SCENE PERCEPTION: NEUROPHYSIOLOGY moves beyond V1 to chart the bifurcation of V1 output in monkeys and humans into a pathway that ascends to the parietal cortex (the "where/how" system involved in object location and setting of parameters for action) and a pathway that descends to the inferotemporal cortex (the "what" system involved in object recognition).

The SOMATOSENSORY SYSTEM article argues that the tactile stimulus representation changes from an original form (more or less isomorphic to the stimulus itself) to a completely distributed form (underlying perception) in a series of partial transformations in successive subcortical and cortical networks. At the level of the primary somatosensory cortex (SI), the neural image of the stimulus is sensitive to the shape and temporal features of peripheral stimuli, rather than simply reflecting the overall intensity of local stimulation. Of particular interest are cortical *barrels*, discrete neuronal aggregates in rodent somatosensory cortex, each of which receives its principal input from one of the facial whiskers. The picture that emerges is that somatosensory information processing is modular on at least two different scales, macrocolumnar (barrels or segregates) and minicolumnar, with each minicolumn in a segregate receiving afferent connections from a unique subset of the thalamic neurons projecting to that segregate.

The auditory system is introduced in two articles. AUDITORY PERIPHERY AND COCHLEAR NUCLEUS spells out how the auditory periphery transforms a very high-information-rate acoustic stimulus into a series of lower-information-rate auditory nerve firing. The incoming acoustic information must thus be split across hundreds of nerve fibers to avoid loss of information. The transformation involves complex mechanical-to-electrical transformations. The cochlear nucleus continues this process of parallelization by creating multiple representations of the original acoustic stimulus, with each representation presumably emphasizing different acoustic features. The article emphasizes monaural signal processing although the cochlea and cochlear nucleus receive input from descending pathways which are binaural in nature, and the cochlear nuclei provide input to each other (cf. "Sound Localization and Binaural Processing"). AUDITORY CORTEX notes that relatively little is known of the detailed connectivity and function of mammalian auditory cortex save in the case of the highly specialized echolocating bat (cf. "Echolocation: Creating Computational Maps"). Nevertheless, in other mammals, a few auditory tasks have been broadly accepted as vital for all species, such as sound localization, timbre recognition, and pitch perception. The article discusses a few of the functional and stimulus feature maps that have been found or postulated, and relates them to the more intuitive and better understood case of the echolocating bats.

The olfactory system is distinctive in that paths from the periphery to the cortex do not travel via a thalamic nucleus. The study of this system offers prime examples of seeking a "basic circuit" that defines the irreducible minimum of neural components necessary for a model of the functions carried out by a region. OLFACTORY BULB offers examples of information processing without impulses and of output functions of dendrites (dendrodendritic synapses); the article on the OLFACTORY CORTEX explains that this is the earliest cortical region to differentiate in the evolution of the vertebrate forebrain, and it is the only region within the forebrain to receive direct sensory input. Models of olfactory cortex emphasize the importance of cortical dynamics, including the interactions of intrinsic excitatory and inhibitory circuits and the role of oscillatory potentials in the computations performed by the cortex.

We now turn to three systems related to motor control and to visuomotor coordination in mammals (cf. the road map on **Primate Motor Control** in Section II.8): cortical areas involved in grasping; the basal ganglia; and the cerebellum. GRASPING MOVEMENTS: VISUOMOTOR TRANSFORMATIONS shows the tight coupling between (specific subregions of) parietal and premotor cortex in controlling grasping. The anterior region of the inferior parietal lobe (AIP) appears to play a fundamental role in extracting intrinsic visual properties from the object for organizing grasping movements (cf. "Gaze Coding in the Posterior Parietal Cortex" for the role of a parietal region in the control of saccades). The extracted visual information is then sent to the F5 region of premotor cortex, there activating neurons that code grip types congruent to the size, shape, and orientation of the object. In addition to visually activated neurons in AIP, there are cells whose activity is linked to motor activity, possibly reflecting corollary discharges sent by F5 back to the parietal cortex. The hypothesis is that, if F5 motor commands correctly match the visual signals, the parietal cells are fully activated and send a positive feedback signal to the premotor cortex to continue the movement until the object is successfully grasped and held. Otherwise, the parietal activity is suppressed and the movement is interrupted or modified.

The basal ganglia include the striatum, the globus pallidus, the substantia nigra, and the subthalamic nucleus. The article on BASAL GANGLIA stresses that all of these structures are functionally subdivided into skeletomotor, oculomotor, associative, and limbic territories. The basal ganglia can be viewed as a family of loops, each taking its origin from a particular set of functionally related cortical fields (skeletomotor, oculomotor, etc.), passing through the functionally corresponding portions of the basal ganglia, and returning to parts of those same cortical fields by way of specific zones in the dorsal thalamus. Dopamine has an important role in synaptic plasticity within the striatum. Since dopamine neurons discharge in relation to conditions involving the probability and imminence of behavioral reinforcement, dopamine neurons might be seen as playing a role in striatal information processing analogous to that of an "adaptive critic" in connectionist networks (cf. "Reinforcement Learning").

It has been argued that the basal ganglia play a role in determining when to initiate one phase of movement or another, and that the cerebellum adjusts the metrics of movement, allowing fast feedforward-like movements, rather than slow feed-back-dominated movements. CEREBELLUM AND MOTOR CONTROL reviews a number of models for cerebellar mechanisms underlying the learning of motor skills. The cerebellum can be decomposed into cerebellar nuclei and a cerebellar cortex. The only output cells of the cerebellar cortex are the Purkinje cells, and their only effect is to provide varying levels of inhibition on the cerebellar nuclei. Each Purkinje cell receives two types of input: a single climbing fiber and many tens of thousands of parallel fibers. The most influential model of the cerebellar cortex has been the Marr-Albus model of the formation of associative memories between particular patterns on parallel fiber inputs and Purkinje cell outputs, with the climbing fiber acting as a "training signal." Later models place more emphasis on the relation between the cortex and nuclei, and on the way in which the subregions of this coupled cerebellar system can adapt and coordinate the activity of specific motor pattern generators. The plasticity of the cerebellum is approached from a different direction in CEREBELLUM AND CONDITIONING. Many

experiments indicate that the cerebellum is involved in learning and performance of classically conditioned, somatic motor reflexes; the present article focuses on rabbit eyelid conditioning. The article reviews a number of models of the role of the cerebellum in such conditioning, providing a useful complement to the models in CEREBELLUM AND MOTOR CONTROL.

Finally, we turn to the hippocampus, a region which has been implicated in a variety of memory functions (see "Short-Term Memory") both as working memory and as a basis for long-term memory. In this road map, however, we visit just one article, HIPPOCAMPUS: SPATIAL MODELS. Single-unit recordings in freely moving rats have revealed "place cells" in subfields of the hippocampus whose firing is restricted to small portions of the rat's environment (the corresponding "place fields"). These data underlie the seminal idea of the hippocampus as a spatial map (cf. "Cognitive Maps"). The article reviews the data and describes some models of hippocampal place cells and of their role in circuits controlling the rat's navigation through its environment.

II.6. Sensory Systems

Vision

ACTIVE VISION
BINDING IN THE VISUAL SYSTEM
COLLICULAR VISUOMOTOR TRANSFORMATIONS FOR SACCADES
COLOR PERCEPTION
DIRECTIONAL SELECTIVITY IN THE CORTEX
DIRECTIONAL SELECTIVITY IN THE RETINA
DISSOCIATIONS BETWEEN VISUAL PROCESSING MODES
FIGURE-GROUND SEPARATION
GABOR WAVELETS FOR STATISTICAL PATTERN
    RECOGNITION
ILLUSORY CONTOUR FORMATION
LOCALIZED VERSUS DISTRIBUTED REPRESENTATIONS
MOTION PERCEPTION
OBJECT RECOGNITION
PERCEPTION OF THREE-DIMENSIONAL STRUCTURE
PERCEPTUAL GROUPING
REGULARIZATION THEORY AND LOW-LEVEL VISION
RETINA
ROUTING NETWORKS IN VISUAL CORTEX
SELECTIVE VISUAL ATTENTION
SPARSE CODING IN THE PRIMATE CORTEX
STEREO CORRESPONDENCE AND NEURAL NETWORKS
SYNCHRONIZATION OF NEURONAL RESPONSES AS A PUTATIVE BINDING
    MECHANISM
TEXTURED IMAGES: MODELING AND SEGMENTATION
THALAMUS
VISION: HYPERACUITY
VISUAL CODING, REDUNDANCY, AND "FEATURE DETECTION"
VISUAL CORTEX CELL TYPES AND CONNECTIONS
VISUAL SCENE PERCEPTION: NEUROPHYSIOLOGY
VISUAL SCHEMAS IN OBJECT RECOGNITION AND SCENE ANALYSIS
WAVELET DYNAMICS

The topic of vision has provided one of the most fertile fields of investigation both for brain theory and for technologists constructing artificial neural networks. In view of the large number of articles included in this road map, I will keep the discussion of each article to a bare minimum. Five articles in the previous road map on **Mammalian Brain Regions**— RETINA; COLLICULAR VISUOMOTOR TRANSFORMATIONS FOR SACCADES; THALAMUS; VISUAL CORTEX CELL TYPES AND CONNEC-

TIONS; and VISUAL SCENE PERCEPTION: NEUROPHYSIOLOGY— have already introduced regions associated with vision. Rather than discuss these further, we now turn to the other articles. The article on ACTIVE VISION emphasizes the role of vision in gaining information relevant for animals and robots considered as real-time perception-action systems. This is a theme that is further developed in Section II.8 in the road maps on **Motor Pattern Generators and Neuroethology** and **Primate Motor Control**. Nonetheless, many articles in the present road map will analyze vision as the process of discovering from images what is present in the world: we may see active vision as more like the mode of vision employed by the "where/how" system described in VISUAL SCENE PERCEPTION: NEUROPHYSIOLOGY than that provided by the "what" pathway. DISSOCIATIONS BETWEEN VISUAL PROCESSING MODES extends the discussion of this dichotomy, especially in its analysis of relevant psychophysical evidence. However, the choice of the term "active vision" is unfortunate, since it suggests that other forms of vision are "passive." (The term "action-oriented vision" is more appropriate.) Determining the identity of an object need not be a passive process. Context and expectation can influence recognition, and the extent of shape and pattern analysis may be determined on a "need-to-know" basis; i.e., those aspects of the object that are critical for distinguishing it from other similar objects may be analyzed in more detail than characteristics that are less discriminating. Moreover, attentional mechanisms are constantly moving the eyes in response to the needs of both "visual systems" to foveate on items of particular relevance to the current interests of the organism. The article on COLLICULAR VISUOMOTOR TRANSFORMATIONS FOR SACCADES presents mechanisms for these overt shifts of attention (i.e., eye movements); the articles on SELECTIVE VISUAL ATTENTION and ROUTING NETWORKS IN VISUAL CORTEX offer data and hypotheses for cortical mechanisms which complement these by shifting attention covertly (i.e., without eye movement), and by routing visual input to the appropriate part of the evolving representation of the visual scene, respectively.

The issue of how information about objects in the world may be coded by neural activity in the visual systems of animals is addressed in a number of articles. LOCALIZED VERSUS DISTRIB-

UTED REPRESENTATIONS asks whether the final neural encoding of visual recognition of one's grandmother, say, involves neurons that respond selectively to "grandmothers"—so-called "grandmother cells"—or whether the sight of grandmother is never made explicit at the single neuron level, with the representation instead distributed across a large number of cells, none of which responds selectively to grandmothers alone. SPARSE CODING IN THE PRIMATE CORTEX marshals theoretical reasons and experimental evidence suggesting that the brain adopts a compromise between these extremes, which is often referred to as "sparse coding." This thesis is illustrated with data on object recognition and face recognition in the inferotemporal cortex (the "what" pathway) in monkeys. The issue of visual encoding is approached from a more general theoretical viewpoint in VISUAL CODING, REDUNDANCY, AND "FEATURE DETECTION" (see also the related article on "Information Theory and Visual Plasticity"). This article emphasizes the redundancy of the visual signal and the transformation of the signal as it passes along the visual pathway. Describing a particular cell as an "X detector" implies that the cell responds when and only when that particular feature is present (e.g., an edge detector responds only in the presence of an edge) but the article argues that describing cells in the early visual system as "detectors" of any type of feature is both misleading and inaccurate. Nonetheless, the response properties of many cells in primary visual cortex are well described by what are called "Gabor wavelets" which respond best to patterns of a given spatial frequency and orientation within a given neighborhood. GABOR WAVELETS FOR STATISTICAL PATTERN RECOGNITION relates this notion to both biology and technology, while WAVELET DYNAMICS places these ideas in a technological framework.

Given that the characteristics of objects in the visual scene are encoded by activity distributed across many cells in the visual system, the issue arises of how the activity of these cells is bound together so that these attributes will be associated with a single object in the visual scene. TEXTURED IMAGES: MODELING AND SEGMENTATION provides one very basic answer: to conduct a statistical analysis to see whether local patches of the image can be classified according to a small set of statistical descriptors. Patches of the same texture are then candidates for aggregation. PERCEPTUAL GROUPING fills the gulf between early vision (edge detection) and high-level vision (object recognition). This article surveys the range of processes that have been implicated in perceptual grouping, as well as the different functional roles attributed to them. It shows that perceptual grouping involves inferences over different types of objects, ranging from point sets indicating (projected) locations of interest on solid objects, through curve-like objects, to surface coverings. FIGURE-GROUND SEPARATION offers a theory of 3D emergent boundary completion, relating filled-in surface properties to connected boundaries, but with near boundaries obstructing the filling-in of occluded regions. The theory rests on cooperation between the boundary contour system (BCS) and the feature contour system (FCS). The same cooperation plays a crucial role in ILLUSORY CONTOUR FORMATION to explain a variety of optical illusions based on seeing brightness edges where none exists in the image.

By contrast, SYNCHRONIZATION OF NEURONAL RESPONSES AS A PUTATIVE BINDING MECHANISM argues that the "binding" of cells that correspond to a given visual object may exploit another dimension of cellular firing, namely, the phase at which a cell fires within some overall rhythm of firing in a neural population. The article presents data consistent with the proposal that the synchronization of responses on a time scale of milliseconds provides an efficient mechanism for response selection and binding of population responses (cf. "Dynamic Link Ar-

chitecture"). Synchronization also increases the saliency of responses because it allows for effective spatial summation in the population of neurons receiving convergent input from synchronized input cells. BINDING IN THE VISUAL SYSTEM places this in a broader philosophical perspective, and argues that binding mechanisms based on synchronization must be complemented by a fundamental reanalysis of the notion of cell assembly.

Beyond the basic issue of how the visual scene is segmented (how visual elements are grouped) into possibly meaningful wholes lies the question of determining for such a region its motion, color, distance, shape, etc. These issues are addressed in the next seven articles. DIRECTIONAL SELECTIVITY IN THE RETINA reviews models of retinal direction selectivity (which contributes to oculomotor responses rather than motion perception). Older models depend on the way in which amacrine and other cells of the retina are connected to the ganglion cells, the retinal output cells. A newer model is based on the directionality of synaptic interactions on the dendrites of amacrine cells, involving a spatial asymmetry in the inputs and outputs of a dendrodendritic synapse, and its shunting inhibition. It is argued that this latter mechanism's development might involve Hebbian processes driven by spontaneous activity and light. DIRECTIONAL SELECTIVITY IN THE CORTEX shows that cortical directional selectivity (which does contribute to motion perception as well as the control of eye movements) involves many cortical regions. Directionally sensitive cells in the primary visual cortex (V1) project to middle temporal cortex (MT), where directional selectivity becomes more complex, as MT cells typically having larger receptive fields. From the MT, the motion pathway projects to the middle superior temporal cortex. Cortical directional selectivity has been modeled in three manners: as a spatially asymmetric excitatory drive followed by multiplication or squaring; via a spatially asymmetric nonlinear inhibitory drive; and through a spatially asymmetric linear inhibitory drive followed by positive feedback. It is argued that development of this selectivity might involve Hebbian processes driven by spontaneous activity and binocular interactions. The issues in this article have some overlap with those presented in MOTION PERCEPTION, but the latter emphasizes two stages in motion perception and provides models of each. The first is to measure the direction and speed of movement of features in the 2D image, linking successive views to infer *optic flow*, which is the pattern of image velocities that is projected onto the retina. The second is to interpret the optic flow in terms of the three-dimensional structure of the scene and the motion of the observer and/or objects in the scene.

Stereoscopic vision exploits the fact that points in a 3D scene will, in general, project to different positions in the images formed in the left and right eyes. The differences in these positions are termed disparities. The stereo correspondence problem is: how can a pair of stereo images be mapped into a single representation, called a disparity map, that makes explicit the disparities of all points common to both images, thus revealing the distance of various visual elements from the observer. STEREO CORRESPONDENCE AND NEURAL NETWORKS notes that various constraints have been used to help determine which features on the two eyes should be matched in inferring depth. These include compatibility of matching primitives, cohesivity, uniqueness, figural continuity, and the ordering constraint. Various neural network stereo correspondence algorithms are then reviewed, and the problems of surface discontinuities and uncorrelated points, and of transparency, are addressed. The article closes with a brief review of neurophysiological studies of disparity mechanisms. PERCEPTION OF THREE-DIMENSIONAL STRUCTURE reviews various computational models for inferring an object's 3D structure from different types of optical information, such as shading, texture, motion and stereo, and examines how the performance of these models compares with the capabilities and limitations of human observers in judging different aspects of 3D structure under varied viewing conditions. By contrast, REGULARIZATION THEORY AND LOW-LEVEL VISION offers a general mathematical framework, regularization theory, to deal with the fact that the problem of inferring 3D structure from 2D images is ill-posed: there are many spatial configurations compatible with a given 2D image or set (motion sequence, stereo pair, etc.) of images. The issue then becomes to find which spatial configuration is most "plausible." We have already seen a number of constraints associated with stereo vision. The general approach of regularization theory is to define a "cost function" which combines a measure of how close a spatial configuration comes to yielding the given image (set) with a measure of the extent to which the configuration violates the constraints—and then to seek that configuration which minimizes this cost.

The next two articles address two other aspects of "low-level" vision, while the last two look at "high-level" vision. COLOR PERCEPTION stresses that color is not a local property inferred from the wavelength of light hitting a patch of retina, but is a property of regions of space that depends both on the light they reflect and on the surrounding context. Our visual system "recreates" the world in the form of boundaries that contain surfaces, and color perception involves the perception of aspects of these surfaces. Hyperacuity refers to visual judgments, such as the vernier task of aligning two parallel line segments, which can be completed with stunning accuracy. For the vernier task, the threshold may be as low as 5 arcsec, which is equivalent to one fifth of the size of the most sensitive photoreceptor. Thus, hyperacuity cannot depend on the precision of single receptors, and VISION: HYPERACUITY shows how it may depend on the ability of neural networks to infer immensely precise information from the activity of a set of such receptors.

There is a vast literature on pattern recognition in neural networks, but the article on OBJECT RECOGNITION focuses on models of viewpoint invariant object recognition which are constrained by psychological data on human object recognition. The four main approaches reviewed are direct template matching; hierarchical template matching; transform-and-match; and structural description. The strengths of each of these in modeling different aspects of human perception are analyzed. By contrast to this emphasis on the psychology of recognizing a single object, VISUAL SCHEMAS IN OBJECT RECOGNITION AND SCENE ANALYSIS provides a less psychological approach which focuses on mechanisms which integrate schemas for recognition of different objects into the perception of some overall scene.

Other Sensory Systems

AUDITORY CORTEX
AUDITORY PERIPHERY AND COCHLEAR NUCLEUS
ECHOLOCATION: CREATING COMPUTATIONAL MAPS
ELECTROLOCATION
OLFACTORY BULB
OIFACTORY CORTEX
PAIN NETWORKS
SENSOR FUSION
SOMATOSENSORY SYSTEM
SOUND LOCALIZATION AND BINAURAL PROCESSING
TIME PERCEPTION: PROBLEMS OF REPRESENTATION AND PROCESSING

In this volume, there are so many articles on vision that they warrant the separate road map which precedes this one. Here,

we analyze other sensory systems—e.g., touch, audition, pain—in addition to considering such general issues as sensor fusion and time perception. A number of the following articles were discussed in the road map on **Mammalian Brain Regions** in Section II.5, but we will now present some related and additional topics.

SOMATOSENSORY SYSTEM shows how the somatosensory system changes the tactile stimulus representation from a form more or less isomorphic to the stimulus to a completely distributed form via a series of partial transformations in successive subcortical and cortical networks. The article on PAIN NETWORKS adds a new dimension to body sensation. The pain system encodes information on the intensity, location, and dynamics of strong, tissue-threatening stimuli, but differs from other sensory systems in its "emotional-motivational" factors. In the pain system, these factors strongly modulate the relation between stimulus and felt response. At one extreme is "allodynia," a state where the slightest touch with a cotton wisp is agonizing. People display wide individual and trial-to-trial variability in the amount of pain reported following administration of calibrated noxious stimuli; that is, pain sensation is subject to ongoing modulation by a complex of extrinsic (stimulus-generated) and intrinsic (CNS-generated) state variables. The article spells out how these act in the CNS as well as the periphery.

The earlier road map introduced the auditory system in two articles: AUDITORY PERIPHERY AND COCHLEAR NUCLEUS explains how the auditory periphery parcels out acoustic stimulus across hundreds of nerve fibers, and how the cochlear nucleus continues this process by creating multiple representations of the original acoustic stimulus. The second article discusses the functional and stimulus feature maps that have been found or postulated in AUDITORY CORTEX. We now turn to SOUND LOCALIZATION AND BINAURAL PROCESSING to understand how information from the two ears is brought together. The article focuses on the use of interaural time difference (ITD) as one way to estimate the azimuthal angle of a sound source. It describes one biological model (ITD detection in the barn owl's brainstem) and two psychological models. The underlying idea is that the brain attempts to match the sounds in the two ears by shifting one sound relative to the other, with the shift that produces the best match assumed to be the one that just balances the real ITD. AUDITORY CORTEX notes the importance of such tasks as sound localization, timbre recognition, and pitch perception; discusses a few of the relevant functional and stimulus feature maps; and relates them to the echolocating system of bats. ECHOLOCATION: CREATING COMPUTATIONAL MAPS provides us with a more detailed understanding of this system. Mustached bats emit echolocation (ultrasonic) pulses for navigation and for hunting flying insects. On the basis of the echo, prey must be detected and distinguished from the background clutter of vegetation, characterized as appropriate for consumption, and localized in space for orientation and prey capture. The bats emit ultrasonic pulses that consist of a long constant-frequency component followed by a short frequency-modulated component. Each pulse-echo combination provides a discrete sample of the continuously changing acoustic scene. The auditory network contains two key design features: neurons that are sensitive to combinations of pulse and echo components, and computational maps that represent systematic changes in echo parameters to extract the relevant information.

Electrolocation is another sense that helps certain animals locate themselves in their world, but this time the animals are electric fish rather than bats, and the signals are electric rather than auditory. The article on ELECTROLOCATION relates its topic to the general issues of mechanisms which facilitate the processing of relevant signals while rejecting noise, and of attentional processes which select which stimuli to attend to. Many of these adaptive features of sensory processing networks are thought to be implemented by descending or feedback pathways which enable higher centers to shape the processing characteristics of their lower-order targets (cf. "Thalamus"). Possible mechanisms for attentive processes within the electrosensory system are described and related to the concept of "sensory searchlights," and mechanisms are proposed to explain these animals' abilities to selectively process "behaviorally relevant" information in the presence of naturally occurring patterns of masking stimuli.

The two articles on the olfactory system have already been presented in the road map on **Mammalian Brain Regions**. The article OLFACTORY BULB notes the special circuitry involved in the basic preprocessing, while OLFACTORY CORTEX presents a dynamical system analysis of further olfactory processing.

We next turn from individual senses to senses acting in concert. When one sense cannot provide all the necessary information, complementary observations may be provided by another sense. For example, haptic sensing complements vision during the act of placing a peg in a hole when the effector occludes the agent's view. Also, senses may offer competing observations, such as the competition between vision and the vestibular system in maintaining balance (and its occasional side effect of seasickness). Another type of interplay between the senses is the use of information extracted by one sense to focus the attention of another sense, coordinating the two, as in audition cueing vision. The article on SENSOR FUSION discusses this diverse set of phenomena, bridging the neurophysiological and cognitive science viewpoints, and analyzing recent efforts in sensor fusion for autonomous mobile robots.

The final article, on TIME PERCEPTION: PROBLEMS OF REPRESENTATION AND PROCESSING, addresses a unique constellation of perceptual phenomena. The most basic phenomena are the experience of simultaneity versus nonsimultaneity, the integration of events into a subjective present, and the experience of duration. Subjective time is analyzed within a hierarchical taxonomy of elementary temporal experiences which is related to neuronal mechanisms of information processing.

II.7. Plasticity in Development and Learning

Mechanisms of Neural Plasticity

BCM THEORY OF VISUAL CORTICAL PLASTICITY
HEBBIAN SYNAPTIC PLASTICITY
HEBBIAN SYNAPTIC PLASTICITY: COMPARATIVE AND DEVELOPMENTAL ASPECTS
LONG-TERM DEPRESSION IN THE CEREBELLUM
NMDA RECEPTORS: SYNAPTIC, CELLULAR, AND NETWORK MODELS
POST-HEBBIAN LEARNING RULES

We have already devoted two road maps to **Learning in Artificial Neural Networks** in Section II.3. Such learning involves a variety of learning rules, inspired in great part by the psychological hypotheses of Hebb and Rosenblatt (cf. Section I.3) about rules by which synaptic connections may change their strength as a result of experience. In recent years, much progress has been made in tracing the processes that underlie the plasticity of synapses of biological neurons. The present road map samples this research and related modeling. The following two road maps abstract away from many of these details to show how plausibly biological rules of synaptic plasticity may be used in models explaining a variety of phenomena in **Development and Regeneration of Neural Networks** and **Learning in Biological Systems**.

Hebb's idea was that a synapse (what we would now call a *Hebbian synapse*) strengthens when the presynaptic and postsynaptic elements tend to be coactive. The plausibility of this hypothesis has been enhanced by the neurophysiological discovery of a synaptic phenomenon in the hippocampus known as long-term potentiation (LTP) which is induced by a Hebbian mechanism. HEBBIAN SYNAPTIC PLASTICITY reviews recent facts and hypotheses about LTP, as well as evidence regarding variations and extensions of Hebb's original postulate for learning. Interestingly, the type of LTP seen in the CA1 region of the hippocampus is Hebbian (i.e., depending on near coincidence of presynaptic and postsynaptic activity), while that seen in the CA3 region is not. The article notes the crucial role played in the CA1 form of LTP by channels called NMDA receptors in the synapses. This role is further explained in the article on NMDA RECEPTORS: SYNAPTIC, CELLULAR, AND NETWORK MODELS which discusses the molecular structure of NMDA receptors, provides kinetic models of their behavior, and charts the role of these receptors in short- and long-term synaptic plasticity and synaptic integration, as well as in oscillators and epileptiform activity.

The article on HEBBIAN SYNAPTIC PLASTICITY also notes a classic problem with Hebb's original rule, namely that it only strengthens synapses. But this would mean that all synapses would eventually saturate, depriving the cell of its pattern separation ability. A number of biologically inspired responses to this problem are described in the articles on POST-HEBBIAN LEARNING RULES and BCM THEORY OF VISUAL CORTICAL PLASTICITY, while HEBBIAN SYNAPTIC PLASTICITY: COMPARATIVE AND DEVELOPMENTAL ASPECTS offers a broad biological perspective rooted in the conceptual framework of cell assemblies. Experimental networks ranging from the abdominal ganglion in the invertebrate *Aplysia* to the hippocampus and visual cortex offer converging validation of the viability of Hebb's postulate, but with suitable modifications to address, e.g., the saturation problem. Similar algorithms of potentiation can be implemented using different cascades of second messengers triggered by activation of synaptic and/or voltage-dependent conductances. What changes according to a correlational rule is not so much the efficacy of transmission at a given synapse, it is argued, but rather a more general coupling term mixing the influence of polysynaptic excitatory and inhibitory circuits linking the two cells, modulated by the diffuse network background activation.

The final article presents a mode of synaptic plasticity that is more like the perceptron rule (but in this case with reversed sign): synaptic plasticity controlled by a training signal. As noted in the road map on **Mammalian Brain Regions** in Section II.5, many cerebellar models have been influenced by the Marr-Albus hypothesis on the formation of associative memories between particular patterns on parallel fiber inputs and Purkinje cell outputs, with the climbing fiber acting as a "training signal." Neurophysiological research eventually showed that coincidence of climbing fiber and parallel fiber activity on a Purkinje cell led to long-term depression (LTD) of the synapse from parallel fiber to Purkinje cell. The article on LONG-TERM DEPRESSION IN THE CEREBELLUM summarizes the data on the neurochemical mechanisms underlying this form of plasticity.

Development and Regeneration of Neural Networks

BCM THEORY OF VISUAL CORTICAL PLASTICITY
DEVELOPMENT AND REGENERATION OF EYE-BRAIN MAPS
INFORMATION THEORY AND VISUAL PLASTICITY
MOTION PERCEPTION: SELF-ORGANIZATION
OCULAR DOMINANCE AND ORIENTATION COLUMNS
SOMATOTOPY: PLASTICITY OF SENSORY MAPS

There is no hard and fast line between the cellular mechanisms underlying development of the nervous system and those involved in learning. Nonetheless, the former emphasizes the questions of how one part of the brain comes to be connected to another and how overall patterns of connectivity are formed, while the latter tends to regard the connections as in place, and asks how their strengths can be modified to improve the network's performance. Studies of regeneration—the reforming of connections after damage to neurons or cell tracts—are thus associated more with developmental mechanisms than with learning per se. Another significant area of research that complements development is that of aging, but there is still too little work relating aging to neural modeling.

The study of the regeneration of *retinotopic* eye-brain maps in frogs (i.e., the mapping of neighboring points in the frog retina to neighboring points in the optic tectum) has been one of the most fruitful areas for theory-experiment interaction in neuroscience. Roger Sperry demonstrated that following optic nerve section, even if this is accompanied by eye rotation, the regenerating fibers tend to form connections with those target neurons to which they were connected before surgery. This suggested a one-to-one form of "neural specificity" in which each cell in both the retina and tectum has a unique chemical marker signaling 2D location, the idea being that retinal axons seek out tectal cells with the same positional information. DEVELOPMENT AND REGENERATION OF EYE-BRAIN MAPS reviews the subsequent interaction between theory and experiment. In experiments in which lesions were made in goldfish retina or tectum, it was found that topographic maps regenerate in conformance with whatever new boundary conditions are created by the lesions: e.g., the remaining half of a retina will eventually connect in a retinotopic way to the whole of the tectum, rather than just to the half to which it was originally connected. Theory and experiment paint a subtle view in which genetics sets a framework for development, but the final pattern of connections depends both on boundary conditions and on patterns of cellular activity. This view is now paradigmatic for our understanding of how patterns of neural connectivity are determined.

OCULAR DOMINANCE AND ORIENTATION COLUMNS studies two issues that go beyond basic map formation to provide further insight into activity-dependent development. When cells in layer IVc of visual cortex are tested to see which eye drives them more strongly, it is found that ocular dominance takes the form of a zebra-stripe-like pattern of alternating dominance. Model and experiment support the view that the stripes are not genetically specified but instead form through network self-organization. Another classic example is the formation of orientation specificity. A number of models are reviewed in light of current data, both theoretical analysis based on the idea that leading eigenvectors dominate (cf. "Pattern Forma-

tion, Biological"), as well as computer simulations. BCM THEORY OF VISUAL CORTICAL PLASTICITY provides additional detail on experiments on the development of orientation specificity, and develops a model which hinges on a variant of the Hebbian learning rule. Where the Hebbian rule associates any coactivity of presynaptic and postsynaptic activity with an increase in synaptic strength, the BCM rule only does so if the coactivity passes some threshold; for low coactivity, the synapse is weakened. INFORMATION THEORY AND VISUAL PLASTICITY treats the issue of plasticity from an informational-theoretical perspective.

MOTION PERCEPTION: SELF-ORGANIZATION models possible mechanisms for development of the kind of networks presented in the article on "Motion Perception." A set of motion detector neurons, with a spectrum of preferred directions, speeds, and locations, can self-organize because local correlations across time exist between the activation of pairs of nearby "feature detectors." Standard Hebbian-type learning rules in a competitive learning framework can use such correlations (plus a range of signal transmission time delays) to shape the receptive fields of the neurons in a higher stage. The article goes on to show how other motion-processing stages combine the outputs of these short-range motion-detector neurons to produce usable representations of object motion, combining the local 2D motion information gathered across the image to produce representations of 3D object motion.

As can be seen from the earlier discussions, neural network models of development and regeneration have been dominated by studies of the visual system. The last article on this road map, however, takes us to the somatosensory system. Research in the past decade has demonstrated plastic changes at all levels of the adult somatosensory system in a wide range of mammalian species. Changes in the relative levels of sensory stimulation as a result of experience or injury produce modifications in sensory maps. SOMATOTOPY: PLASTICITY OF SENSORY MAPS discusses which features of somatotopic maps change and under what conditions, the mechanisms that may account for these changes, and the functional consequences of sensory map changes.

Learning in Biological Systems

CEREBELLUM AND CONDITIONING
CEREBELLUM AND MOTOR CONTROL
CONDITIONING
HABITUATION
HIPPOCAMPUS: SPATIAL MODELS
INVERTEBRATE MODELS OF LEARNING: *APLYSIA* AND *HERMISSENDA*
REINFORCEMENT LEARNING IN MOTOR CONTROL
SENSORIMOTOR LEARNING
SHORT-TERM MEMORY

Just as the squid giant axon provided the invaluable insights into the active properties of neural membrane summarized in the Hodgkin-Huxley equation, so have invertebrates provided many insights into other basic mechanisms (see "Neuromodulation in Invertebrate Nervous Systems" and "Crustacean Stomatogastric System" for two examples). INVERTEBRATE MODELS OF LEARNING: *APLYSIA* AND *HERMISSENDA* describes the role of these two invertebrates in the study of basic learning mechanisms. A ganglion (localized neural network) of these invertebrates can control a variety of different behaviors, yet a given behavior, such as a withdrawal response, may be mediated by 100 neurons or less. Moreover, many neurons are relatively large and can be uniquely identified, functional properties of an individual cell can be related to a specific behavior, and changes in cellular properties during learning can be related to specific changes in behavior. Biophysical and molecular events underlying the changes in cellular properties can then be determined and mathematically modeled. The present article illustrates this with studies of associative and nonassociative modifications of defensive siphon and tail withdrawal reflexes in *Aplysia* and associative learning in *Hermissenda*.

Habituation is a decrease in the strength of a behavioral response that occurs when an initially novel stimulus is presented repeatedly. Habituation of defensive reflexes was among the first types of learning explained successfully at the cellular level. The article on HABITUATION recalls the basic mechanisms studied in *Aplysia* and then places habituation in a vertebrate context by studying habituation of prey-catching behavior in toads. Early models of habituation concentrated on describing changes of response strength under repeated stimulation and spontaneous recovery. More recent studies have modeled habituation at the systems level, molecular mechanisms underlying synaptic plasticity, and the relationship between short-term and long-term habituation.

During conditioning, animals modify their behavior as a consequence of their experience of the contingencies between environmental events. The article on CONDITIONING delineates formal theories and neural network models that have been proposed to describe *behavioral* experiments in classical and operant conditioning. CEREBELLUM AND CONDITIONING presents and models data which implicate a specific brain region, the cerebellum, in classical conditioning of the rabbit eye blink reflex. As we have already seen in the road map on **Mammalian Brain Regions** in Section II.5, plasticity depends, at least in part, on the plasticity of parallel fiber–Purkinje cell synapses under the control of climbing fiber input, and this same mechanism plays a key role in the functions discussed in CEREBELLUM AND MOTOR CONTROL.

The next two articles concern the role of the hippocampus. There is now strong evidence for "working memory," a process of short-term memory (STM) involved in performing tasks requiring temporary storage and manipulation of information to guide appropriate actions. The article on SHORT-TERM MEMORY addresses two main issues: how neural information is selected for storage and is temporarily stored in STM; and how and under what conditions information selected by STM can be transferred to long-term memory (LTM). The article discusses the roles of both the hippocampus and the prefrontal cortex in STM at the systems level, and provides a model of short-term working memory based on bistable units which can implement STM and exhibit long-term changes. HIPPOCAMPUS: SPATIAL MODELS turns from the role of hippocampus in memory to neuronal models of spatial processing in the rat hippocampus. Data on the spatial correlates of hippocampal cell firing and the idea of the hippocampus as a spatial map provide the background for several models of hippocampal place cells and navigation.

The last two articles of this road map were treated at length in the road map on **Control Theory and Robotics** in Section II.4. Briefly, REINFORCEMENT LEARNING IN MOTOR CONTROL recalls the general theory introduced in the article on "Reinforcement Learning" and demonstrates its utility in understanding how motor skills can be attained in the absence of explicit feedback about muscle contractions or joint angles. SENSORIMOTOR LEARNING explains how neural networks can acquire "models" of some desired sensorimotor transformation, and explores the role of coarse coding in the learning of sensorimotor mappings.

II.8. Motor Control

Motor Pattern Generators and Neuroethology

The article on MOTOR PATTERN GENERATION provides an overview of the basic building blocks of behavior, to be expanded upon in many of the following articles. The emphasis is on rhythmic behaviors (such as flight or locomotion), but a variety of "on-off" motor patterns (as typified by a frog snapping at its prey) are also studied. The crucial notion is that a central pattern generator (CPG), an autonomous neural circuit, can yield a good "sketch" of a movement, but that the full motor pattern generator (MPG) augments the CPG with sensory input which can adjust the motor pattern to changing circumstances (e.g., varying the pattern of locomotion when going uphill rather than on level terrain, or when the animal carries a heavy load).

The two articles on FROG WIPING REFLEXES and the SCRATCH REFLEX look at two behaviors (the former studied in frogs, the latter studied primarily in turtles) elicited by an irritant applied to the animal's skin. In each case, the position at which the limb is aimed varies with the position of the irritant; i.e., there is somatotopic control of the reflex. More importantly, the reflex has different "modes." To understand this, just think of scratching your lower back. As the scratch site moves higher, the positioning of the limb changes continuously with the position of the irritant until the irritant moves up so much that you make a discontinuous switch to the "over-the-shoulder mode" of back-scratching. Such a mode change may be compared to the GAIT TRANSITIONS discussed later. In any case, we see here two important issues: how is an appropriate pattern of action chosen, and how is the chosen pattern parameterized on the basis of sensory input?

With this, we switch to nine articles in which the emphasis is on rhythmic behavior, with rather little concern for the spatial structure of the movement. (For example, the discussion of locomotion will focus on coordinating the rhythms of the legs when the animal progresses straight ahead, rather than on how these rhythms are modified when the animal traverses uneven terrain or turns to avoid an obstacle.) The article on HALF-CENTER OSCILLATORS UNDERLYING RHYTHMIC MOVEMENTS looks at a set of minimal circuits for generating rhythmic behavior, while the CRUSTACEAN STOMATOGASTRIC SYSTEM shows how specific circuits of identified neurons controlling the chewing (by teeth inside the stomach!) of lobsters and crabs have been modeled. Of particular interest is the finding that neuropeptides (see "Neuromodulation in Invertebrate Nervous Systems") can change the properties of cells and the strengths of connections so that, e.g., a cell can become a pacemaker or a previously ineffective connection can come to exert a strong influence and with this, a network can dramatically change its overall behavior. Thus, the change of "mode" may be under the control of an explicit chemical "switch" of underlying cellular properties. Of course, in other systems, the maintenance of patterns of excitation and inhibition may enable a given circuit to act in one mode or another, while in other cases, the change of mode may involve the transfer of control from one neural circuit to another.

The next three articles transfer the emphasis to oscillators seen in vertebrate systems. The article on RESPIRATORY RHYTHM GENERATION presents several alternative models of breathing which can be evaluated against mammalian data. The article on the SPINAL CORD OF LAMPREY: GENERATION OF LOCOMOTOR PATTERNS marks an important transition: from seeing how one network can oscillate to seeing how the oscillation of a series of networks can be coordinated. Experiments show that neural circuitry in isolated pieces of the spinal cord of a lamprey (a primitive type of fish) can exhibit oscillations. When these pieces are combined in an intact spinal cord, they all oscillate with the same frequency, but with a phase relationship that has a wave of bending progressing down the fish from head to tail, yielding the coordinated pattern of swimming. The article on CHAINS OF COUPLED OSCILLATORS then abstracts from the specific circuitry to show how oscillators and their coupling can be characterized in a way which allows the proof of mathematical theorems about patterns of coordination.

The study of animal progression is continued in the next four articles. Descriptions of experiments on, and models of, LOCUST FLIGHT: COMPONENTS AND MECHANISMS IN THE MOTOR and LOCOMOTION, INVERTEBRATE are followed by two more abstract articles. WALKING reviews data on human walking and then analyzes ANNs rather than biological networks, with a special interest in how a set of networks may be *trained* to exhibit coordinated behavior, while GAIT TRANSITIONS studies its topic (e.g., the transition from walking to running) from the abstract perspective of dynamical systems.

All the previous articles may be seen as addressing the question of how a neural network (whether biologically identified, artificial, or abstractly characterized as a dynamical system) controls a single type of motor pattern, or a set of closely related modes of movement. The last five articles on this road map look at the broader study of ethology (i.e., animal behavior), in which our concern is not simply with generating a motor pattern but rather with the circumstances under which a particular motor pattern will be deployed as an appropriate part of the animal's behavior. NEUROETHOLOGY, COMPUTATIONAL provides a general overview of models of neural networks for coordinated animal behavior, while COMMAND NEURONS AND COMMAND SYSTEMS analyzes the extent to which an MPG may be activated alone or in concert with others through perceptual stimuli mediated by a single "command neuron" or by more diffuse "command systems." The final articles discuss three species in which experiments on visual control of behavior have a strong linkage to neural network modeling. VISUOMOTOR COORDINATION IN FLIES explains the mechanisms underlying the extraction of retinal motion patterns in the fly and their transformation into the appropriate motor activity. Rota-

tory large-field motion can signal unintended deviations from the fly's course; image expansion signals can indicate that the animal is approaching an obstacle; and discontinuities in the retinal motion field can indicate the presence of nearby stationary or moving objects. VISUOMOTOR COORDINATION IN FROGS AND TOADS stresses that visuomotor integration implies a complex transformation of sensory data, since the same locus of retinal activation might release behavior directed toward the stimulus (as in prey catching) or directed to another part of the visual field (as in predator avoidance). The article also shows how the efficacy of visual stimuli to release a response is determined by many factors, including the stimulus situation, the motivational state of the organism itself, and previous experience with the stimulus (habituation and conditioning). Finally, VISUOMOTOR COORDINATION IN SALAMANDERS discusses the anatomical basis for a range of behaviors in salamanders, presenting models of prey catching, prey-directed saccades, and gaze stabilization to show how low-level mechanisms add up to produce complicated behaviors, such as the devious approach of monocular salamanders to their prey.

Biological Motor Control

COROLLARY DISCHARGE IN VISUOMOTOR COORDINATION
EQUILIBRIUM POINT HYPOTHESIS
GEOMETRICAL PRINCIPLES IN MOTOR CONTROL
HUMAN MOVEMENT: A SYSTEM-LEVEL APPROACH
LIMB GEOMETRY: NEURAL CONTROL
MOTONEURON RECRUITMENT
MOTOR CONTROL, BIOLOGICAL AND THEORETICAL
MUSCLE MODELS
OPTIMIZATION PRINCIPLES IN MOTOR CONTROL
REINFORCEMENT LEARNING IN MOTOR CONTROL
SENSORIMOTOR LEARNING

MOTOR CONTROL, BIOLOGICAL AND THEORETICAL characterizes biological motor control as a problem of controlling nonlinear systems whose states are monitored with slow, sometimes low-quality, sensors. In response to changing sensory inputs, internal goals, or motor errors, the motor system must select an appropriate action, transforming control signals from sensory to motor coordinate frameworks; coordinate the selected movement with other ongoing behaviors and with postural reflexes; and monitor the movement to ensure its accuracy. If the controller bases its actions on signals which are not affected by the plant output, it is said to be a feedforward controller. If the controller bases its actions on a comparison between the reference values (desired behavior) and the controlled variables, it is a feedback controller. To complicate matters, the various stages and processes may be interrelated, so that no strict decomposition may be possible.

Muscle transduces chemical energy into force and motion, thereby providing power to move the skeleton. Because of the intricacies of muscle microstructure and architecture, no comprehensive models are yet able to predict muscle performance completely. The article on MUSCLE MODELS thus reviews three classes of models, each fulfilling a more narrowly defined objective, ranging from attempts to understand the molecular level to simulations of whole muscle behavior as part of a broader study of basic musculoskeletal biomechanics or issues of neural control. A motoneuron together with the muscle fibers that it innervates constitutes a motor unit, and each muscle is a composite structure, whose force-generating components, the motor units, are typically heterogeneous and partly neuronal. Such aggregates can produce much larger forces than a single motor unit. The article on MOTONEURON RECRUITMENT shows how the motor units can be recruited in the service of reflexes, voluntary movement, and posture. HUMAN MOVE-MENT: A SYSTEM-LEVEL APPROACH identifies many of the inherent characteristics of the physiology and anatomy of human movement that must be recognized in an integrated model of movement which couples neural network modeling with analysis of the underlying biomechanics. These characteristics include the converging nature of the neuromusculoskeletal system, neural delays and muscle mechanical lag characteristics, interlimb coupling, and total limb and body synergies. Examples from human gait illustrate these characteristics, as well as the challenge of modeling a total body movement.

As we have seen, the full understanding of movement must rest on analysis of the integration of neural networks with the biomechanics of the skeletomuscular system. Nonetheless, much has been learned about limb control from a more abstract viewpoint, as the next four articles show. Optimization theory has become an important aid to discovering organizing principles that guide the generation of goal-directed motor behavior, specifying the results of the underlying neural computations without requiring specific details of the way those computations are executed. OPTIMIZATION PRINCIPLES IN MOTOR CONTROL illustrates the successes of, and challenges to, this approach in studying arm trajectory formation, comparing the purely kinematic minimum-jerk model with the more dynamics-based minimum-torque-change model. How does one go from a kinematic description of the movement of the hand to the pattern of muscle control that yields it? There are still many competing hypotheses. One approach starts from the observation that a muscle is like a controlled-length spring: set its length, and it will naturally return to the equilibrium length that was set. Upon this premise, the EQUILIBRIUM POINT HYPOTHESIS builds a systems-level description of how the nervous system controls the muscles so that a stable posture is maintained or a movement is produced. In this framework, the "controller" is composed of muscles and the spinal-based reflexes, and the "plant" is the skeletal system. The controller defines a force field which is meant to capture the mechanical behavior of the muscles and the effect of spinal reflexes. Motion is viewed as a gradual transition from posture, and it is suggested that, for the case of multijoint arm movements, one can predict the hand's motion if the supraspinal system smoothly shifts the equilibrium point from the start point to a target location. GEOMETRICAL PRINCIPLES IN MOTOR CONTROL argues that the transition from the spatial representation of a motor goal to a set of appropriate neuromuscular commands is, in many respects, similar to a coordinate transformation. The article describes three types of coordinate system—endpoint coordinates, generalized coordinates, and actuator coordinates—each representing a particular "point of view" on motor behavior. It then examines the geometrical rules that govern the transformations between these classes of coordinates. It shows how a proper representation of dynamics may greatly simplify the transformation of motor planning into action. LIMB GEOMETRY: NEURAL CONTROL offers another perspective, starting from a discussion of the role of extrinsic and intrinsic coordinates when a human makes a movement. The article analyzes coordinate transformations in goal-directed human movements, kinematic representations in the brain, and coordinate transformations in the control of cat posture.

Fast, coordinated movements depend on the nervous system's ability to use copies of motor control signals (the corollary discharge) to compute expectations of how the body will move, rather than always awaiting sensory feedback to signal the current state of the body. Corollary discharge is thus a general mechanism for reducing mismatch between desired, often sensory-based, commands and actually performed actions.

COROLLARY DISCHARGE IN VISUOMOTOR COORDINATION describes three functions of corollary discharge: the stabilization of the visual world during eye movements, the determination of an egocentric reference for spatial localization, and the proactive control of visually goal-directed movements. Clearly, the effective use of corollary discharge rests on the brain having learned the relation between current state, motor command, and the movement that ensues. With this, the road map returns us to two articles treated in the road map on **Control Theory and Robotics**, SENSORIMOTOR LEARNING and REINFORCEMENT LEARNING IN MOTOR CONTROL, which present general learning strategies based on adaptive neural networks.

Primate Motor Control

BASAL GANGLIA
CEREBELLUM AND MOTOR CONTROL
COLLICULAR VISUOMOTOR TRANSFORMATIONS FOR SACCADES
DYNAMIC REMAPPING
EYE-HAND COORDINATION IN REACHING MOVEMENTS
GAZE CODING IN THE POSTERIOR PARIETAL CORTEX
GRASPING MOVEMENTS: VISUOMOTOR TRANSFORMATIONS
HEAD MOVEMENTS: MULTIDIMENSIONAL MODELING
PURSUIT EYE MOVEMENTS
REACHING: CODING IN MOTOR CORTEX
REACHING MOVEMENTS: IMPLICATIONS OF CONNECTIONIST MODELS
SACCADES AND LISTING'S LAW
VESTIBULO-OCULAR REFLEX: PERFORMANCE AND PLASTICITY

This final road map is primarily concerned with visually controlled behaviors for which neurophysiological data are available from the monkey brain, as well as behavioral and, in some cases, imaging data for humans. The map takes us from basic unconscious behaviors to those involving skilled action. The vestibulo-ocular reflex (VOR) serves to stabilize the retinal image by producing eye rotations that counterbalance head rotations. Vestibular nuclei neurons are much more than a simple relay; their functions include multimodality integration, temporal signal processing, and adaptive plasticity. VESTIBULO-OCULAR REFLEX: PERFORMANCE AND PLASTICITY reviews the empirical data, as well as control-theoretic and neural network models, for these functions. HEAD MOVEMENTS: MULTIDIMENSIONAL MODELING describes a "tensorial" approach (essentially, one based on applying the Moore-Penrose generalized inverse) to explore how the vestibulocollic reflex transforms head motions sensed by the six vestibular semicircular canals into patterns of neck muscle activation that would counter those motions, stabilizing the head in space. The model takes into account the precise pulling actions of 30 neck muscles and shows that predicted optimal activation directions typically differ from muscle-pulling directions.

The next five articles are concerned with different aspects of eye movements. COLLICULAR VISUOMOTOR TRANSFORMATIONS FOR SACCADES analyzes the role of the superior colliculus in the control of the rapid movements of the eyes toward a target, called saccades. The article touches on afferent and efferent mapping, target selection, visuomotor transformations in motor error maps, remapping of target representations after a saccade, and coding of dynamic motor error. SACCADES AND LISTING'S LAW spells out the interaction of the torsional component with the horizontal and vertical components of saccades. The theme of remapping is further pursued in the article on DYNAMIC REMAPPING, which distinguishes "one-shot" remapping (updating the internal representation in one operation to compensate for an entire movement) from a continuous remapping process based on the integration of a velocity signal or the relaxation of a recurrent network. The article uses data on arm movements as well as saccades. A related issue is the extent to which the internal representation of a visual target varies with head position as well as the place it projects to the retina. GAZE CODING IN THE POSTERIOR PARIETAL CORTEX reviews evidence that a portion of the posterior parietal cortex of the primate brain participates in sensory to motor coordinate transformations for the purpose of programming the direction of gaze. The article reviews a set of neural network models developed to study how such coordinate transformations might be achieved, presenting both a backpropagation and a reinforcement learning model. PURSUIT EYE MOVEMENTS takes us from saccadic "jumps" to those smooth eye movements involved in following a moving target. Current models of pursuit vary in their organization and in the features of pursuit that they are designed to reproduce. Three main types of pursuit models include "image motion" models, "target velocity" models, and models which address the role of prediction in pursuit. However, these models make no explicit reference to the neural structures that might be responsible. The article thus analyzes the neural pathways for pursuit, stressing the importance of both the visual areas of the cerebral cortex and the oculomotor regions of the cerebellum, to set goals for future modeling.

The final five articles are concerned with reaching and grasping. REACHING: CODING IN MOTOR CORTEX spells out the relation between the direction of reaching and changes in neuronal activity that have been established for several brain areas, including the motor cortex, the premotor cortex, area 5, the cerebellar cortex, and the deep cerebellar nuclei. The cells involved each have a broad tuning function, the peak of which denotes the "preferred" direction of the cell, i.e., the direction of movement for which the cell's activity would be highest. Typically, cell activity varies as a linear function of the cosine of the angle formed between the preferred direction of the cell and the direction of reaching. A movement in a particular direction will engage a whole population of cells. A unique code for the direction of movement regards this population as an ensemble of vectors; the weighted vector sum of these neuronal contributions is the "population vector" which points in the direction of the movement for discrete movements in 2D and 3D space. Further observations link this population encoding to speed of movement, as well as to preparation for movement. REACHING MOVEMENTS: IMPLICATIONS OF CONNECTIONIST MODELS reviews a number of connectionist models for the control of arm trajectories and analyzes their implications for the analysis of neurophysiological data on the control of reaching. EYE-HAND COORDINATION IN REACHING MOVEMENTS focuses on possible mechanisms responsible for visually guiding the hand toward a point within the prehension space and argues for a synthesis between positional coding and vectorial coding models. It notes the strong reliance on predicted sensory "reafference" (corollary discharge), based on the idea of an internal model of the controlled limb predicting the consequences of a command at the level of the task space. Redundancy at the level of sensory encoding may also be crucial, a point that has been little studied. In any case, while this article focuses on how the hand is brought to a target, GRASPING MOVEMENTS: VISUOMOTOR TRANSFORMATIONS emphasizes the neural mechanisms which control the hand itself in grasping an object. The latter article, which notes the crucial preshaping of the hand during reaching prior to grasping the object, emphasizes the cooperation of visual mechanisms in the parietal cortex with motor mechanisms in the premotor cortex. This pattern of cooperative computation integrates sensing and corollary discharge throughout the movement. The article on CEREBELLUM AND MOTOR CONTROL reviews a number of models of the role of the cerebellum in building "internal models" to improve motor skills. Finally, the article on BASAL GANGLIA reviews the structure of this system in terms of multiple loops, with especial emphasis on those involved in skeletomotor and oculomotor functions, reviewing the role of dopamine in motor learning, as well as mechanisms underlying Parkinson's disease.

Part III: Articles

Active Vision

Andrew Blake

Introduction

Active vision is a new paradigm in machine vision, recently emerged from the study of computational vision, that emphasizes the role of vision as a sense for robots and for other real-time perception-action systems. This is a marked departure from the orthodoxy that has held sway for much of the past 15 years. Methodology, both in machine vision and to a considerable extent in biological vision, has been dominated by the influence of David Marr, who states at the beginning of his book (Marr, 1982) that "vision is the process of discovering from images what is present in the world, and where it is"—a prescription for the seeing "couch potato." In contrast, an active vision system is far more selfish. It picks out the properties of images which it needs to perform its assigned task, and ignores the rest. In this context, there is no clear need for the sort of detailed reconstructions of the visible world that have been an accepted, traditional goal of machine vision (Horn, 1986). The general appeal of goal-oriented perception was pointed out more than 20 years ago (Arbib, 1972), but specific developments, particularly in the analysis of visual motion, are what have actually made progress in active vision possible. Elements of the emerging goal-oriented approach have been described by numerous authors over the past five years or so, under a variety of roughly synonymous titles including "Active Vision," "Dynamic Vision," and "Animate Vision"—see Blake and Yuille (1992) for a review.

Active vision mirrors the recent trend in experimental robots, pioneered by Brooks (1986), toward building animal-like behavior into the lowest levels of a robot's control system. The essence of the idea is captured by Braitenberg (1984) in an elegantly simple setting, a "bug" with two wheels and a few single-channel sensors. Appropriate excitatory and inhibitory connections generate surprisingly effective and diverse ecological behavior: food seeking, reacting to light, and cooperation and competition with other creatures. A somewhat more sophisticated bug has been realized physically, with surprising success, by Franceschini and his coworkers (Franceschini, 1992), who built an entire two-dimensional (floor-bound) fly, encircled by a planar compound eye with 24 elements. Direct couplings between the eye's elements and wheel motors lead to low-level behavior that is well adapted to obstacle avoidance and goal seeking. Neural perceptuomotor couplings were hardwired in this fly-robot; but with modern computing technology, reprogrammable systems can be built that allow experiments in the learning of those couplings.

Those systems employed somewhat primitive vision systems, but they are, in some senses, precursors of active vision. Modern Charge Coupled Device (CCD) cameras have high resolution, which brings opportunities and drawbacks. High resolution enhances the degree of perceivable detail and potentially facilitates visual guidance for tasks such as fine motion planning and manipulation (Latombe, 1991). However, the rate of flow of data becomes large, typically 50 million bits per second. Economy of data processing is clearly crucial here. It becomes advantageous, rather than processing snapshots, or sequences of snapshots treated individually, to allow the observer and sensor to interact continually. Visual sensory data then can be analyzed purposefully over relatively narrow regions of interest to answer specific queries posed by the observer. The observer can constantly adjust its vantage point to allow the sensor to uncover the piece of information that is immediately most pressing. This parsimonious, opportunistic, mobile observer operates at a distinct advantage for several reasons, which are laid out in the following four sections.

Structure from Controlled Motion

It seems natural to expect that if the analysis of single images is hard, then the analysis of sequences of images should be harder, but, as a consequence of the richness of geometric information in image flow, the opposite is actually true. A robot is able, therefore, to perceive *more* while it is moving. It is often worthwhile for the robot to generate deliberate exploratory motions precisely in order to induce motion of the image. In terms of information processing theory, the corresponding advances in understanding have come over the last decade or so. The theory of the computation of "egomotion" and of surface structure from "optic flow" has been developed extensively (e.g., Longuet-Higgins and Prazdny, 1980). To explain these terms a little, the *image-motion field* is the pattern of velocities relative to the observer of points on visible surfaces, projected onto the image. It can be thought of as a dense field of velocity vectors covering the image. However, the image-motion field is not generally observable in its entirety for several reasons; for instance, the component of a velocity vector parallel to an iso-intensity line is unobservable, the so-called aperture problem (Horn, 1986). The portion of the image-motion field that is observable is termed the *optic flow* field. Principles have been established and algorithms have been developed for analyzing the optic flow field to infer the motion of the observer relative to a rigid world (egomotion) and the shapes of surfaces in that world ("structure from motion"). (For fuller discussion, see MOTION PERCEPTION and PERCEPTION OF THREE-DIMENSIONAL STRUCTURE.)

Tracking

The complexity of the structure-from-motion problem is considerably reduced by fixation: tracking a point in the middle of a moving object and maintaining it in the center of the field of view. Of course, in human vision, this brings the considerable advantage of directing the fovea, with its high spatial resolution, at the area of interest. A further gain is that optic flow can be measured in a frame in which the fixated flow is at rest (Aloimonos, Weiss, and Bandyopadhyay, 1987). The rotational component of the optic flow field is therefore canceled out by an equal eye rotation. The dominant but relatively uninformative component of the optic flow field is removed, leaving just *parallax*—image motion relative to the fixation point—the component containing information about egomotion, surface slope (Longuet-Higgins and Prazdny, 1980), and surface curvature (Cipolla and Blake, 1992). A further benefit of tracking is that the temporal continuity of the optic flow field can be exploited to reduce the computational complexity of image matching problems. In stereoscopic vision, the accuracy of recovered surface shape increases as the baseline (distance between the two eyes in binocular stereo) increases, but there is a serious computational problem—the *correspondence problem*—of establishing matches between features in pairs of views. However, if feature motion is tracked continuously over time, for a moving observer, the identity of features can be maintained by assuming that their image motion is continuous. After some time has elapsed, the observer has moved a finite distance which becomes an effective baseline, separating the first and last views. The correspondence problem between those views is now computationally straightforward because feature identity has been maintained throughout the observer's motion. (Note that although the correspondence problem is effectively sidestepped for continuous motion, the *aperture problem* remains; see MOTION PERCEPTION.)

Focused Attention

Attention mechanisms are well known in human vision (Treisman and Gelade, 1980), and it is reasonable to expect, on grounds of computational efficiency, that they should be needed in machine vision systems too. An autonomous low-level mechanism for tracking low-level features liberates higher-level processes to analyze the features relevant to a particular task, switching freely between them as required. This is particularly effective in binocular vision (Ballard and Ozcandarli, 1988), where tracking fixates a region not merely of an image but of 3D space. Control of fixation in binocular vision heads has been studied a good deal lately (Blake and Yuille, 1992), and effective eye-control systems have been achieved that approach the agility of the human oculomotor system.

Prediction

A compelling reason that a temporal image sequence should be *easier* to process than a single image is that the images in a sequence are not mutually independent. Features found in one image are strongly correlated with those in the image immediately previous. A weak form of this correlation is the temporal continuity mentioned in the context of tracking. A stronger form can be invoked if prior models of 3D motion are available. The Kalman filter (Gelb, 1974) is the classical mechanism for doing this, capable of combining a model for the process of measurement of optic flow in an image and its reliability with a model of object motion or of observer motion. The filter combines the two components, weighing the confidence of measurements against the accumulated confidence of current estimated position and motion.

As an example, consider how a filter to track a cube translating through space would work. The simplest reasonable model for the cube's motion would be that it has approximately constant velocity in the world. This is conveyed by assuming that the cube's acceleration has a zero mean but with a certain random component. This has the effect that the cube's velocity is assumed to be constant over relatively small time scales but can vary over longer times. Armed with this model, the filter is effectively a computational simulation that aims to "mirror" the true motion. At a given instant, the position of the cube in the 3D simulation is converted, using conventional projective geometry of the kind routinely used to generate line drawings in computer graphics, into a *prediction* of how the image of the cube should appear at that instant. Then image measurements are compared with that prediction, and the difference measure or *innovation* is used, as a feedback signal, to correct the computational simulation. The cycle of prediction and measurement is repeated indefinitely to track motion over time.

Some very impressive practical demonstrations have been made of the power of active vision; a notable example is the vehicle navigation system of Dickmanns and Graefe (1988). All the components of active vision mentioned here are present in their system:

Structure from motion: The forward movement of the vehicle constitutes a controlled motion so that world structure can be elicited from image motion.

Focusing of attention, for example, on road edges or on the outlines of other vehicles, means that the necessary visual processing is within the scope of currently realistic computing resources.

Tracking: Within the focus of attention, a straightforward measurement process using templates and correlation, akin to tuned receptive fields in a biological context, localizes visible features, generating the innovation signal needed for tracking.

Prediction: A comprehensive feature predictor forms the backbone of the system. The model includes both the dynamics of the vehicle and the geometry, assumed piecewise smooth, of the road edge.

Conclusions

Tests of the viability of active vision theories have examined their effectiveness in machine seeing systems. Systems are being demonstrated, in several cases operating at real-time rates, that perform navigation, recognition, and surface analysis. Of course, these systems are nowhere near the general-purpose vision systems that researchers were dreaming of 10 or 15 years ago, and much remains to be done in enhancing versatility and generality. From the point of view of biological modeling, however, these systems constitute thorough simulations, strenuously tested. Active vision theories recommend themselves, therefore, to be adapted as experimentally testable, hypothetical models for biological vision.

Consideration of biological implementation of active vision models is less clear and is a little premature while the computational principles are still being established. In the case of oculomotor control, biology and machine vision are content to use a common language, that of control system theory; see, for example, the section on control of vision heads in Blake and Yuille (1992). It also has been demonstrated that simple visual navigation systems can be implemented in neural-like hardware (Franceschini, 1992). However, in the case of the relatively complex visual tracking systems that are being developed, assessment of the feasibility of biological implementation must wait for further advances in several areas, for instance, effective simplification of trackers by approximation, adaptivity and the learning of motions, and the integration of feature detectors.

Road Map: Vision
Related Reading: Reactive Robotic Systems; Vision for Robot Driving

References

Aloimonos, J., Weiss, I., and Bandyopadhyay, A., 1987, Active vision, in *Proceedings of the 1st International Conference on Computer Vision*, Washington, DC: IEEE Computer Society Press, pp. 35–54.

Arbib, M. A., 1972, *The Metaphorical Brain: An Introduction to Cybernetics as Artificial Intelligence and Brain Theory*, New York: Wiley-Interscience.

Ballard, D. H., and Ozcandarli, A., 1988, Eye fixation and early vision: Kinetic depth, in *Proceedings of the 2nd International Conference on Computer Vision*, Washington, DC: IEEE Computer Society Press, pp. 524–531.

Blake, A., and Yuille, A., Eds., 1992, *Active Vision*, Cambridge, MA: MIT Press. ◆

Braitenberg, V., 1984, *Vehicles*, Cambridge, MA: MIT Press.

Brooks, R., 1986, A robust layered control system for a mobile robot, *IEEE Trans. Robot. Automat.*, 2:14–23.

Cipolla, R., and Blake, A., 1992, Surface shape and the deformation of apparent contours, *Int. J. Comput. Vis.*, 9:83–112.

Dickmanns, E., and Graefe, V., 1988, Dynamic monocular machine vision, *Machine Vis. Appl.*, 1:223–240.

Franceschini, N., 1992, Real-time visual motor control: From flies to robots, *Proc. R. Soc. Lond. B*, 337:282–293. ◆

Gelb, A., Ed., 1974, *Applied Optimal Estimation*, Cambridge, MA: MIT Press.

Horn, B., 1986, *Robot Vision*, New York: McGraw-Hill.

Latombe, J.-C., 1991, *Robot Motion Planning*, Norwell, MA: Kluwer.

Longuet-Higgins, H., and Prazdny, K., 1980, The interpretation of a moving retinal image, *Proc. R. Soc. Lond. B*, 208:385–397.

Marr, D., 1982, *Vision*, San Francisco: Freeman.

Treisman, A. M., and Gelade, G., 1980, A feature-integration theory of attention, *Cognit. Psychol.*, 12:97–136.

Activity-Dependent Regulation of Neuronal Conductances

L. F. Abbott and Eve Marder

Introduction

The activity of a neuron in a neural network depends both on the synaptic input it receives and on its own intrinsic electrical characteristics. An enormous amount of both theoretical (Gluck and Rumelhart, 1990) and experimental (Byrne and Berry, 1989) work has focused on the implications of activity-dependent synaptic plasticity for network function. Less attention has been paid to the fact that the intrinsic characteristics of individual neurons can also be modified by activity; yet this intrinsic plasticity has important implications for neural network behavior (Levy, Colbert, and Desmond, 1990; LeMasson, Marder, and Abbott, 1993; Abbott and LeMasson, 1993; Siegel, Marder, and Abbott, 1994) .

Roughly a dozen different types of ion channels contribute to the membrane conductance of a typical neuron (McCormick, 1990). The electrical characteristics of a neuron depend on the number of channels of each type active within the plasma membrane and on how these channels are distributed over the surface of the cell. A complex array of biochemical processes controls the number and distribution of ion channels by constructing and transporting channels, modulating their properties, and inserting them into and removing them from the plasma membrane. Many of these processes are affected by electrical activity, ranging from activity-induced gene expression (Morgan and Curran, 1991; Armstrong and Montminy, 1993) to activity-dependent modulation of assembled ion channels (Kaczmarek and Levitan, 1987).

Channel synthesis, insertion, and modulation are much slower than the usual voltage- and ligand-dependent processes that open and close channels. Thus, consideration of activity-dependent regulation of conductances introduces a dynamics acting at a new, slower time scale into neuronal modeling. What are the implications of slow, activity-dependent plasticity of intrinsic neuronal properties for neuronal and network function? Fortunately, we can begin to address this question without modeling all of the complex processes responsible for maintaining and modifying ion channels. The essential element we need to understand is the feedback mechanism linking a neuron's electrical characteristics to its activity so that we can uncover the "rules" by which that activity modulates these characteristics. A similar approach is used to study synaptic plasticity, wherein the implications of a synaptic "learning rule" can be studied without a precise knowledge of the mechanisms of synaptic modification.

Activity-dependent regulation could, in principle, affect many channel properties, including the kinetics, voltage dependence, and open conductance of single channels; the number of active channels present in the membrane; and the distribution of channels over the surface of the neuron. In conductance-based neuron models, each membrane current is written in the form $I = \bar{g}m^{p}h^{q}(V - E_{r})$, where E_{r} is the equilibrium potential, p and q are integers, and $\bar{g}$ is the maximal conductance for the current. The dynamic variables m and h are determined by first-order rate equations (Hodgkin and Huxley, 1952; see Abbott, 1994, for a review). Regulation of channel kinetics in these models can be represented by allowing the voltage-dependent rate constants in these equations to be modified by activity. Similarly, the regulation of channel conductances, numbers, and/or distributions is modeled by allowing the maximal conductances $\bar{g}$ to vary with activity. The behavior of model neurons is particularly sensitive to the values of the maximal conductances, so considering activity-dependent $\bar{g}$ parameters is an interesting approach to the study of activity-dependent conductances.

The feedback element giving rise to activity-dependent conductances should be sensitive to the electrical activity of the neuron and must be capable of regulating numerous biochemical processes. Intracellular calcium is a prime candidate for such an element. Because calcium enters the neuron through voltage-dependent channels, the average level of intracellular calcium is highly correlated with the level of electrical activity of the neuron (Ross, 1989; LeMasson, Marder, and Abbott, 1993). Calcium is a ubiquitous regulator of biochemical pathways, and it appears to play a role in many processes affecting membrane conductances. Changes in the intracellular calcium concentration are associated with both modifications of channel properties (Kaczmarek and Levitan, 1987) and long-term changes in gene expression (Armstrong and Montminy, 1993).

Models

Models that use intracellular calcium as the feedback element linking neuronal characteristics to electrical activity have been constructed and studied (LeMasson, Marder, and Abbott, 1993; Abbott and LeMasson, 1993). In these models, the maximal conductances of each of the membrane currents are dynamic variables, whereas in conventional models they are fixed constants. Because intracellular calcium is used as a feedback element in these models, the steady-state values of the maximal conductances $\bar{g}$ depend on the intracellular calcium concentration. The maximal conductances approach their steady-state values exponentially with a long time constant, on the order of many minutes or even hours. The long time constant of exponential approach reflects the fact that the processes that are responsible for changing neuronal conductances, like channel synthesis or degradation, are slow.

In the models studied, the steady-state maximal conductances are sigmoidal functions of the intracellular calcium concentration and have the property that outward currents are increasing sigmoidal functions and inward currents are decreasing sigmoidal functions of the calcium concentration. Specifically,

$$\tau\frac{d\bar{g}}{dt} = \frac{G}{1 + \exp[\pm([\text{Ca}] - C_T)/\Delta]} - \bar{g} \tag{1}$$

where [Ca] is the intracellular calcium concentration, τ is the long time constant discussed earlier, and G, C_T, and Δ are parameters controlling the shape of the sigmoidal function. We use the plus sign in this equation for inward currents and the minus sign for outward currents. This ensures that calcium will provide a negative, stabilizing feedback loop. Excessive activity tends to increase the intracellular calcium concentration. According to Equation 1, this has the effect of increasing outward currents and decreasing inward currents. Both increased outward and decreased inward currents tend to reduce the activity of the neuron, thereby compensating for the excessive activity. A decrease in the activity of the neuron will have the opposite effect, decreasing outward and increasing inward currents. This has the stabilizing effect of increasing activity. The equilibrium or target activity of the neuron, which is the level of activity at which the maximal conductances will remain roughly constant, is determined by the parameter C_T in Equation 1.

Model neurons with calcium-dependent regulation of conductances are much more stable than their unregulated counterparts. A simple model in which this can be seen is the Morris-Lecar (1981) model, a model with voltage-dependent calcium and potassium currents. (For a review of this model, see Rinzel and Ermentrout, 1989.) In the Morris-Lecar model, the active component of the membrane current is written as

$$I = \bar{g}_{\text{Ca}}M(V)[V - E_{\text{Ca}}] + \bar{g}_K n[V - E_K] \tag{2}$$

where $M(V)$ is a sigmoidal function of V and n is a dynamic variable determined by a first-order equation with voltage-dependent parameters. In the original, unregulated form of the model, the maximal conductances $\bar{g}_{\text{Ca}}$ and $\bar{g}_K$ are constants. In the regulated version, they are allowed to vary slowly as a function of the intracellular calcium concentration, as in Equation 1, with different values of the parameter G (G_{Ca} and G_K), but the same values of τ, C_T, and Δ. The plus sign in the $\pm$ in Equation 1 is used for $\bar{g}_{\text{Ca}}$, whereas the minus sign is for $\bar{g}_K$.

Results

Figure 1A contrasts the behavior of the regulated and unregulated Morris-Lecar models for different values of the potassium and calcium equilibrium potentials E_K and E_{Ca}. The firing frequency of the unregulated model is extremely sensitive to these values. For the regulated model, the adjustment of maximal conductances maintains approximately the same frequency over a wide range of E_K and E_{Ca}. This occurs because the regulatory mechanism adjusts the maximal conductances of the calcium and potassium currents whenever the internal calcium concentration changes. Because the intracellular calcium concentration is closely correlated with the activity of the neuron, maintaining a constant [Ca] automatically stabilizes the firing frequency of the neuron.

Similar results were obtained for a regulated version of a model of the LP neuron of the crustacean stomatogastric ganglion (Buchholtz et al., 1992). This is a much more complex model neuron with seven active conductances, and it is capable of firing action potentials either tonically or in periodic bursts. In Figure 1B, the neuron initially shows bursting activity. When the extracellular potassium concentration is modified, the pattern of activity initially changes, but then the dynamic regulation mechanism restores the activity to bursting once again.

Activity-dependent regulation will change membrane conductances whenever the activity of the neuron is modified for

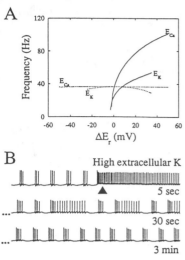

Figure 1. A, The firing frequency for regulated (dashed lines) and unregulated (solid lines) Morris-Lecar models for different values of potassium and calcium equilibrium potentials. ΔE_r is the difference between the reversal potential for either calcium (E_{Ca}) or potassium (E_K) and its control value. Changing the equilibrium potentials simulates the effect of changing the extracellular concentration of these ions. The regulated model maintains a fairly constant firing rate over a wide range of equilibrium potentials. B, Regulation of the model LP neuron discussed in the text. Initially, the neuron exhibits bursting behavior. At the time marked by the triangle, the potassium equilibrium potential is shifted. At first, this changes the activity of the neuron, but ultimately, the regulatory mechanism shifts the balance of conductances to produce behavior similar to that seen initially. In Figures 1 and 2, the regulatory process has been accelerated to speed up the simulations.

a long enough period of time to produce a sustained change in the level of intracellular calcium. This modification could be caused by changes in the extracellular environment, as in Figure 1, or by sustained synaptic inputs, as in Figure 2. The adjustments will always be in the direction of restoring or maintaining the original level of activity and holding the intracellular calcium concentration relatively constant.

Figure 2A shows how the regulated LP model neuron responds to sustained pulses of injected current. The period and duration of current pulses is much less than the time constant τ for regulation of conductances. Without any current injection, the neuron shown is a burster. However, after prolonged stimulation by injected current, the neuron changes so that it is silent. This activity-dependent shift is attributable to the changes in the intracellular calcium concentration caused by the external input. The depolarizing pulses cause the calcium concentration to rise so that the regulation mechanism reduces the inward current and increases the outward current of the model neuron. This is what produces the reduced activity seen immediately after current injecting is terminated.

Figure 2B illustrates the effects of activity-dependent modulation of intrinsic properties in a two-cell network. Two identical model LP neurons are symmetrically coupled through an electrical synapse. Figure 2B shows that the symmetric state seen initially, in which both neurons fire identically, is unstable (Abbott and LeMasson, 1993). In the stable configuration, one neuron is an intrinsic burster and the other is a follower neuron that fires tonically when isolated. Thus, the two identical neurons have spontaneously differentiated to play different roles in producing the network output.

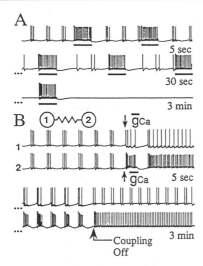

Figure 2. *A*, Injection of current pulses shifts the conductances of a model LP neuron. Initially, the model shows bursting behavior. After current pulses (shown by bars), the balance of conductances shifts so that the neuron is silent when the current pulses are terminated. *B*, Two model LP neurons are coupled electrically. The initial state in which both neurons fire identically is unstable. To show this, a small perturbation is applied at the time marked by the arrows. After a relaxation period, the network establishes stable steady-state behavior which is not symmetrical. This is seen at the beginning of the lower two traces. At the time marked "coupling off," the neurons are decoupled, revealing that the stable state of this symmetrical circuit consists of a pacemaker neuron that is an intrinsic burster, coupled to a follower neuron that fires tonically when isolated.

Discussion

The model studies have revealed several interesting consequences of activity-dependent regulation of conductances:

1. Conductance regulation stabilizes the model neuron against activity shifts caused by extracellular perturbations.
2. Intrinsic properties of model neurons are modified by sustained activity.
3. Regulated model neurons in networks can spontaneously differentiate, thus developing different intrinsic properties and playing different roles in the network.

Ion channels are not spread uniformly over the surface of a neuron but rather reveal a specific spatial distribution. The calcium regulation scheme described here, applied as a local regulator of channel density in multicompartmental models, can produce a realistic spatial distribution of channels (Siegel, Marder, and Abbott, 1994; for a different scheme, see Bell, 1992). This distribution depends on the activity of the neuron, its morphology, and the pattern of synaptic input that it receives. The intrinsic plasticity produced by the activity-dependent distribution of conductances tends to balance the synaptic contribution coming from different dendritic branches and keeps the activity of the neuron in the middle of its range of firing frequencies (Siegel, Marder, and Abbott, 1994).

One of the most significant messages provided by models of conductance regulation is that the same mechanisms that develop and maintain membrane conductances are likely to modify these conductances in response to long-lasting changes in the activity of the neuron. There is initial experimental evidence for the type of activity-dependent changes in the intrinsic properties of neurons predicted by our models (Turrigiano, Ab-

bott, and Marder, 1994). Most of the work done on activity-dependent plasticity in networks has focused on synaptic modification. As more is learned about the processes that control channel conductances, activity-dependent plasticity of intrinsic neuronal properties is likely to play an increasing role in our understanding of network function.

Acknowledgments.   Research was supported by National Institute of Mental Health grant MH–46742, by National Science Foundation grant DMS–9208206, and by the W. M. Keck Foundation.

Road Map: Biological Neurons
Background: I.1. Introducing the Neuron; Ion Channels: Keys to Neuronal Specialization
Related Reading: Synaptic Currents, Neuromodulation, and Kinetic Models

References

Abbott, L. F., 1994, Single neuron dynamics: An introduction, in *Neural Modelling and Neural Networks* (F. Ventriglia, Ed.), Oxford: Pergamon Press, pp. 57–78. ◆

Abbott, L. F., and LeMasson, G., 1993, Analysis of neuron models with dynamically regulated conductances, *Neural Comp.*, 5:823–842.

Armstrong, R. C., and Montminy, M. R., 1993, Transsynaptic control of gene expression, *Annu. Rev. Neurosci.*, 16:17–30. ◆

Bell, A., 1992, Self-organization in real neurons: Anti-Hebb in "channel space"?, in *Neural Information Processing Systems 4* (J. E. Moody and S. J. Hanson, Eds.), San Mateo, CA: Morgan Kaufmann, pp. 59–66.

Buchholtz, F., Golowasch, J., Epstein, I., and Marder, E., 1992, Mathematical model of an identified stomatogastric neuron, *J. Neurophysiol.*, 67:332–340.

Byrne, J. H., and Berry, W. O., 1989, *Neural Models of Plasticity*, San Diego: Academic Press. ◆

Gluck, M. A., and Rumelhart, D. E., 1990, *Neuroscience and Connectionist Theory*, Hillsdale, NJ: Erlbaum. ◆

Hodgkin, A. L., and Huxley, A. F., 1952, A quantitative description of membrane current and its application to conduction and excitation in nerve, *J. Phys.*, 117:500–544. ◆

Kaczmarek, L. K., and Levitan, I. B., Eds., 1987, *Neuromodulation: The Biochemical Control of Neuronal Excitability*, New York: Oxford University Press ◆

LeMasson, G., Marder, E., and Abbott, L. F., 1993, Dynamic regulation of conductances in model neurons, *Science*, 259:1915–1917.

Levy, W. B., Colbert, C. M., and Desmond, N. L., 1990, Elemental adaptive processes of neurons and synapses: A statistical/computational perspective, in *Neuroscience and Connectionist Theory* (M. A. Gluck and D. E. Rumelhart, Eds.), Hillsdale, NJ: Erlbaum, pp. 187–236. ◆

McCormick, D. A., 1990, Membrane properties and neurotransmitter actions, in *The Synaptic Organization of the Brain* (G. M. Shepherd, Ed.), New York: Oxford University Press. ◆

Morgan, J. I., and Curran T., 1991, Stimulus-transcription coupling in the nervous system: Involvement of the inducible proto-oncogenes fos and jun, *Annu. Rev. Neurosci.*, 14:421–451. ◆

Morris, C., and Lecar, H., 1981, Voltage oscillations in the barnacle giant muscle fiber, *Biophys. J.*, 35:193–213.

Rinzel, J., and Ermentrout, G. B., 1989, Analysis of neural excitability and oscillations, in *Methods in Neuronal Modeling* (C. Koch and I. Segev, Eds.), Cambridge, MA: MIT Press. ◆

Ross, W. M., 1989, Changes in intracellular calcium during neuron activity, *Annu. Rev. Physiol.*, 51:491–506. ◆

Siegel, M., Marder, E., and Abbott, L. F., 1994, Activity-dependent current distribution in model neurons, *Proc. Natl. Acad. Sci. USA* (in press).

Turrigiano, G., Abbott, L. F., and Marder, E., 1994, Activity-dependent changes in the intrinsic properties of cultured neurons, *Science*, 264:974–977.

Adaptive Control: General Methodology

Karl J. Åström

Introduction

Adaptive control is a natural extension of feedback control. It may be regarded as an attempt to achieve a higher degree of automation by adjusting parameters of the controller in an ordinary feedback loop. The adjustments are made by learning the characteristics of the process and its environment and changing the controller accordingly. Learning is done by estimating parameters of a model with given structure. Adaptation can also be viewed as automation of modeling and control design.

Research on adaptive control was very active in the 1950s. The motivation was to develop flight control systems for supersonic aircraft. Many ideas for adaptive control were conceived at that time. The development was, however, hampered by hardware problems. It was difficult to obtain the precision and reliability required with analog hardware.

Adaptive systems are fundamentally nonlinear and may exhibit very complex behavior. It has taken a considerable time to develop theory that helps us to understand the behavior of adaptive systems and to design adaptive controllers. Adaptive control has benefited significantly from the development of control theory that has taken place since the 1950s. Progress in nonlinear systems, stability theory, and system identification have been particularly important. This led to an increased interest in adaptive control in the 1970s and 1980s. There are still many interesting research problems in the field.

The development was stimulated by the advent of the microprocessor in the early 1970s. This made it possible to implement adaptive techniques in a cost-effective manner. The first commercial adaptive systems for process control appeared in the early 1980s. Most of the applications are in automatic tuning of simple controllers. The benefit from using adaptive techniques in these applications is that the engineering effort required for tuning of a controller is reduced drastically. Adaptive techniques are also used in systems for process control and in special products like autopilots for ship steering and in biomedical systems. Adaptation is also an important ingredient of the emerging field of intelligent control.

Techniques for Adaptation

A block diagram of a plant with an adaptive controller is shown in Figure 1. The system has two loops. An ordinary feedback loop is composed of the controller and the plant. The parameters of the controller are then adjusted by the adaptation loop.

Model-Reference Adaptive Systems (MRAS)

MODEL-REFERENCE ADAPTIVE CONTROL (q.v.) was one of the first approaches to adaptive control. It was originally conceived as a solution to the flight control problem, which is naturally formulated as a servo problem. The system is based on the idea of characterizing the desired closed-loop behavior by a model and adjusting the parameters using a gradient method. Figure 2 shows a block diagram of the adjustment mechanism used in a model-reference adaptive system.

The block labeled "Reference model" in Figure 2 gives the desired response to applied command signals. Let y be the process output and y_m the output of the reference model. The parameters of the controller are adjusted so that the error $e = y_m - y$ becomes small. This is accomplished by making the rate

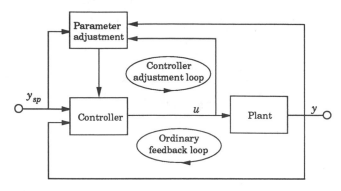

Figure 1. Block diagram of an adaptive system.

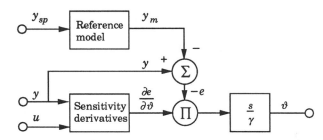

Figure 2. Parameter adjustment mechanism in a model-reference adaptive system (MRAS).

of change of the controller parameters proportional to the product of the error and the sensitivity derivative of the error, i.e.,

$$\frac{d\vartheta}{dt} = -\gamma e \frac{de}{d\vartheta} \qquad (1)$$

where ϑ is a vector of controller parameters, e is the error, and $de/d\vartheta$ is the sensitivity derivative. This scheme, which is called the MIT rule, is illustrated by the block diagram in Figure 2.

Self-Tuning Regulators (STR)

The model-reference adaptive system shown in Figure 2 is called a *direct* adaptive system because the adjustment rules tell directly how the controller parameters should be updated. A different scheme is obtained if the plant parameters are updated and the controller parameters are obtained by solving a design problem. A block diagram of the parameter adjustment mechanism for such a system is shown in Figure 3. A controller of this type is called a *self-tuning regulator* (STR). The unknown process parameters are estimated in the block labeled "Parameter estimator." The estimates are then used to design a controller in the block labeled "Design." The data filters shown in Figure 3 are used to improve the robustness of the estimates.

The system shown in this figure may be viewed as automation of plant modeling and design. The STR scheme is very flexible with respect to the choice of the underlying design and estimation methods.

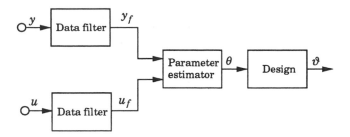

Figure 3. Parameter adjustment mechanism in a self-tuning regulator (STR).

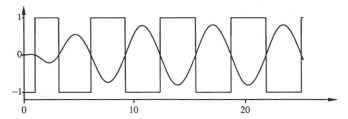

Figure 4. Process input (piece-wise constant signal) and output (continuous) signal for a process subject to relay feedback.

Gain Scheduling

Another method for adjusting the parameters of a controller, called *gain scheduling*, can be used when the required changes in controller parameters can be related to signals from the plant that can be measured directly. The controller parameters are then determined as functions of the scheduling variable.

Gain scheduling is commonly used in systems for flight control. In this case, the controller parameters are determined from measurements of Mach number and dynamic pressure. Systems with gain scheduling are implemented by designing controllers for a wide range of operating conditions, which requires a significant engineering effort. Other adaptive techniques can be used to reduce the effort substantially, as will be discussed in the next section.

Detailed presentations of different adaptive control techniques are found in books by Åström and Hägglund (1988), Narendra and Annaswamy (1989), and Sastry and Bodson (1989); and in survey papers by Åström (1991a) and Ioannou and Datta (1991).

Automatic Tuning

There are many situations where plant characteristics do not change too frequently. Both plant dynamics and disturbance characteristics may, however, be unknown. Industrial experience thus indicated a need for methods for automatic tuning of controllers. All adaptive techniques can be used for such a task. An adaptive algorithm is simply used until parameters have settled; from then on, the parameters are kept constant. It is advantageous to combine the schemes with diagnostic methods that indicate that retuning is required.

Proportional, Integral, and Derivative Control

Most industrial control problems are solved by so-called PID (proportional, integral, and derivative) controllers, where the control signal is a sum of three terms, which are proportional to the error, the time integral of the error, and the derivative of the error. In the process control field, they account for more than 90% of the control loops. It is a significant incentive to provide PID controllers with facilities for automatic tuning. A particular criterion for this application is that the tuner should be robust and easy to use for those with little knowledge of control theory. There are many methods for automatic tuning of PID controllers (see Åström and Hägglund, 1988).

Manual tuning of PID controllers is typically based on a heuristic procedure developed in Ziegler and Nichols (1942). This method is based on knowledge of the ultimate frequency ω_{180}, which is the frequency where the plant has a phase lag of $180°$, and the plant gain at that frequency. If $GP(i\omega)$ is the

transfer function of the process we have

$$k_{180} = -GP(i\omega_{180}) \tag{2}$$

Relay feedback is a convenient method for determining the parameters ω_{180} and k_{180}. Many systems will oscillate when subject to relay feedback. This is illustrated in Figure 4, which shows the input and output obtained for a process with relay feedback.

After a transient, the plant output is almost sinusoidal and has opposite phase to the plant input (the relay signal). The frequency of this oscillation is approximately equal to ω_{180}. If a is the amplitude of the plant output and d is the relay amplitude, we also find that

$$k_{180} = \frac{a\pi}{4d} \tag{3}$$

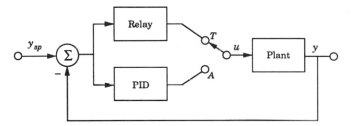

Figure 5. Block diagram of an auto-tuner based on relay feedback.

Relay Auto-Tuning

Relay feedback can be used to automate the empirical tuning procedures for PID controllers. A block diagram of an automatic tuner based on this idea is shown in Figure 5. The system works as follows: Tuning is activated by setting the switching to position T. The system then operates under relay feedback. The zero crossings and the peak amplitudes are determined when a steady-state oscillation is obtained. The frequency ω_{180} is determined from the zero crossings, and k_{180} is given by Equation 3. The controller parameters are then computed from the Ziegler-Nichols rules, and the system is automatically switched to PID control (position A in Figure 5).

Relay tuning works well for systems that are encountered in process control where plant dynamics is well damped. An attractive feature of relay feedback is that it requires little prior information about the plant. Another advantage is that the method automatically generates an input signal with much energy at the unknown ultimate frequency. A detailed discussion of PID controllers and their tuning is given in Åström and Hägglund (1988).

Industrial Products

From the early 1980s, adaptive techniques have been used in industrial products in many different ways. If plant and disturbance characteristics are known, use of a constant-gain controller is sufficient. Auto-tuning can be used to tune a controller automatically. Gain scheduling can be applied if plant characteristics of relevance for control can be related to a measurable signal. Auto-tuning can be applied to generate the gain schedule automatically. Continuous adaptation is used when plant dynamics are changing in an unpredictable manner. Notice that in the first two cases, adaptive techniques are used primarily to make engineering more cost-effective.

Industrial products based on adaptive control can be classified as off-line tuning devices, single-loop controllers, distributed-control systems (DCS), and special-purpose systems.

Off-Line Tuning Devices

The off-line tuning devices are intended to help users tune PID controllers. There are a few products of this type. One example is the Protuner, which consists of software and a special card with process IO that connects to a personal computer. There are algorithms for determining process transfer function from operational data based on frequency domain techniques and tools for controller tuning. Other systems of similar type are the Supertuner and the system SIEPID from Siemens.

Single-Loop Controllers

Single-loop controllers are the largest category of adaptive controllers in industrial use. Practically all new single-loop controllers have some form of automatic tuning or adaptation (see Åström et al., 1993). There are many systems of this type using a variety of adaptive techniques. Automatic tuning is most common, but gain scheduling and continuous adaptation are also used.

The EXACT controller from Foxboro, where adaptation is based on pattern recognition, is widely used. Features of the transient response of the closed-loop system are used to modify the controller parameters. The modification is based on empirical rules (see Bristol, 1977; Kraus and Myron, 1984). To use the system, it is necessary to start with a PID controller which gives a closed-loop system with reasonable properties.

The relay auto-tuner is also widely used. It is available from several vendors in many different forms as single-loop controllers as well as DCS systems. Tuning can be performed simply by pushing a button. Automatic tuning and gain scheduling are most common. There are also more sophisticated systems with continuous adaptation of feedback and feedforward parameters. Feedforward adaptation is very useful because it is difficult to tune feedforward loops manually. An overview of some applications is given in Hägglund and Åström (1991). This paper also describes the algorithms for continuous adaptation.

Distributed Control Systems for Process Control

The first distributed control system that employed adaptive techniques was the Novatune by ABB in Sweden, which was announced in 1982. This system was a small system that had blocks for adaptive control. The adaptive algorithm was based on least-squares estimation and minimum variance control. This system was followed by FIRSTLOOP by FirstControl AB in Sweden, which used an algorithm based on least-squares estimation and pole-placement control. In 1984 SattControl in Sweden introduced a system SDM20 based on relay auto-tuning

and gain scheduling. ABB has transferred Novatune technology to its distributed control system, ABB Master. However, the introduction of adaptive techniques into distributed control systems has been quite slow. The reason for this is probably that the systems are large, with complicated software that is difficult and costly to modify.

Special-Purpose Systems

There are many dedicated controllers that use adaptive techniques. They are used in autopilots for ship steering and for adaptive roll damping. The spacecraft Gemini has adaptive friction compensation and an adaptive notch filter. There are several adaptive systems for motor drives, and many engine control systems for cars use gain scheduling. There are systems for control of paper machines and cement kilns. Adaptive techniques are also used in biomedical applications. (A comprehensive presentation is found in Åström and Wittenmark, 1989.)

Intelligent Control

There is a strong trend to increase the level of automation and autonomy of control systems. Autonomy is needed in systems that operate in difficult environments, and it can lead to improved performance. Robotics, autonomous guided vehicles, and advanced process control systems are typical applications. The ideas and techniques come from many different fields, such as automatic control, artificial intelligence, computer engineering, and neural networks. The field is often referred to as *intelligent control* (see Passino and Antsaklis, 1992). Similar developments have occurred in quite diverse areas. For example, the learning systems considered in the AI community have strong similarities to adaptive control systems, both conceptually and algorithmically. Another is that the backpropagation algorithm in neural networks is very similar to a model-reference adaptive system (see LEARNING AS ADAPTIVE CONTROL OF SYNAPTIC MATRICES).

Expert Control

Expert control is one approach to intelligent control (see Åström and Årzén, 1992). Figure 6 shows a block diagram of such a system, which may be considered an extension of adaptive control. Instead of having one control algorithm and one estimation algorithm, there are several algorithms. The system performs control, tuning, adaptation, and scheduling, as well as diagnosis, performance assessment, control-loop assessment, and loop auditing.

It is natural to perform diagnosis in the primary control loop, since malfunctions of sensors, actuators, processors, and controllers all show up there. It is important to detect both slow degradation and catastrophic failures. Many different methods can be used for diagnosis. Parameter estimation, pattern recognition based on neural networks, and qualitative methods have been used. The results of the diagnosis can be used for alarms and reconfiguration of the control system.

Performance assessment is used to determine what can be achieved and the factors that limit performance. Current performance can also be compared with historical data. Time-series analysis is a useful tool. The aim of loop assessment is to determine the key qualitative and quantitative features of the control loop. For example, does the system have long time delays and severe nonlinearities? Some methods are outlined in Åström (1991b). Neural networks have many uses in these tasks, which involve classification of processes and disturbances.

Figure 6. Block diagram of an expert control system.

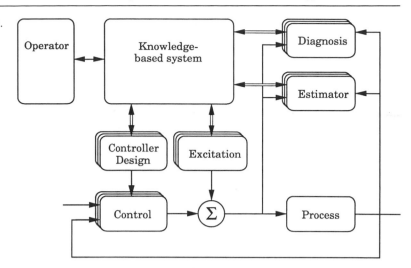

An expert control system has diverse tasks. It has to perform algorithmic as well as symbolic data processing. Also, the system can more conveniently be implemented as a real-time, knowledge-based system using a blackboard architecture (see Årzén, 1989).

Discussion

The closed-loop systems obtained with adaptive control are nonlinear because of the parameter adjustment mechanism; their behavior may be very complex. Much current research is devoted to understanding the behavior in detail. Issues such as parameterization, excitation, unmodeled dynamics, and transient properties are of significant interest. Much theoretical work on adaptive control has been devoted to adapting parameters in a controller in the form of a generic linear system. There are many interesting practical and theoretical reasons for considering controllers that have other structures. One example of this type is friction compensation; another is adaptive control of systems with backlash or asymmetric response; and a third is systems where the nonlinearity is modeled by a neural network. The algorithms used to update parameters in adaptive control are similar to other learning algorithms (for example, those used to train neural networks).

Road Map: Control Theory and Robotics
Background: Motor Control, Biological and Theoretical
Related Reading: Adaptive Control: Neural Network Applications; Process Control

References

Årzén, K.-E., 1989, An architecture for expert system based feedback control, *Automatica*, 25:813–827.
Åström, K. J., 1991a, Adaptive feedback control, *Proc. IEEE*, 75:185–217.
Åström, K. J., 1991b, Assessment of achievable performance of simple feedback loops, *Int. J. Adapt. Control & Sig. Proc.*, 5:3–19.
Åström, K. J., and Årzén, K.-E., 1992, Expert control, in *An Introduction to Intelligent and Autonomous Control* (K. M. Passino and P. J. Antsaklis, Eds.), Norwell, MA: Kluwer, pp. 163–189.
Åström, K. J., and Hägglund, T., 1988, *Automatic Tuning of PID Controllers*, Research Triangle Park, NC: Instrument Society of America.
Åström, K. J., Hägglund, T., Hang, C. C., and Ho, W. K., 1993, Automatic tuning and adaptation for PID controllers—A survey, *Control Engrg. Practice*, 4:699–714.
Åström, K. J., and Wittenmark, B., 1989, *Adaptive Control*, Reading, MA: Addison-Wesley.
Bristol, E. H., 1977, Pattern recognition: An alternative to parameter identification in adaptive control, *Automatica*, 13:197–202.
Hägglund, T., and Åström, K. J., 1991, Industrial adaptive controllers based on frequency response techniques, *Automatica*, 27:599–609.
Ioannou, P. A., and Datta, A., 1991, Robust adaptive control: A unified approach, *Proc. IEEE*, 79:1736–1768.
Kraus, T. W., and Myron, T. J., 1984, Self-tuning PID controller uses pattern recognition approach, *Control Engrg.*, June: 106–111.
Narendra, K. S., and Annaswamy, A. M., 1989, *Stable Adaptive Systems*, Englewood Cliffs, NJ: Prentice Hall.
Passino, K. M., and Antsaklis, P. J., Eds., 1992, *An Introduction to Intelligent and Autonomous Control*, Norwell, MA: Kluwer.
Sastry, S., and Bodson, M., 1989, *Adaptive Control*, Englewood Cliffs, NJ: Prentice Hall.
Ziegler, J. G., and Nichols, N. B., 1942, Optimum settings for automatic controllers, *Trans. ASME*, 64:759–768.

Adaptive Control: Neural Network Applications

Kumpati S. Narendra

Introduction

Although it is well known that multilayer feedforward networks can approximate arbitrary continuous functions on a compact set to any degree of accuracy, most of the applications in the 1980s dealt with static systems. Since dynamics constitutes an essential part of all physical systems, it was proposed by Narendra and Parthasarathy (1990) that neural networks should be used as components in dynamical systems and that a study of such networks should be undertaken within a unified framework of systems theory. A great deal of progress has been made during the past five years in both the theory and the practice of control using neural networks (or neural control). Many dynamical systems have been successfully identified and controlled in computer simulations, and a few have been practically implemented. In this article, the current status of adaptive control using neural networks is reviewed, and its potential for the future is assessed.

Why Neural Networks?

Ever since its birth five decades ago (Brown and Campbell, 1948), automatic control theory has been strongly influenced by the demands of a growing technology for faster, more accurate, and robust controllers. As control applications become more complex, the processes to be controlled are increasingly characterized by poor models, distributed sensors and actuators, multiple subsystems, multiple time scales, multiple performance criteria, high noise levels, complex information patterns, and decision spaces of high dimensions. These difficulties can be broadly classified under three headings. These are (i) computational complexity, (ii) presence of nonlinearities, and (iii) uncertainty. Artificial neural networks are capable, at least in theory, of coping with all three categories of difficulties. Their massively parallel architecture permits computation to be performed at high rates. Since they can approximate nonlinear maps to any desired degree of accuracy, they possess at least the capacity to identify (and hence control) nonlinear dynamical systems. The fact that various rules are currently available for the adjustment of the parameters of the network on the basis of observed behavior implies that the networks can deal with uncertainty. In view of this versatility, neural networks are emerging as ideal candidates for the building blocks of complex adaptive control systems.

Even in a purely deterministic context, the presence of nonlinearities in a dynamical system generally makes the control problem very complex. To use neural networks in such situations, well-known results in linear control theory have to be extended to the nonlinear domain. Quite often, the equations describing the process are unknown, making the problem one of nonlinear adaptive control. The use of the neural networks in such situations has been described in a companion article in this volume (see IDENTIFICATION AND CONTROL). This provides the mathematical basis for using neural networks.

Prior Information, Modeling Aspects, and Design Considerations

The problem of controlling a plant P using a controller C is conceptually a straightforward one. In Figure 1, P and C are shown in a feedback configuration. Given the characteristics of P, the objective is to determine C, which uses all the available data to generate a control input u to the plant. The desired input, u^*, is such that the output y of the plant is either maintained at a desired constant value (set point regulation), or follows a desired time-varying signal y^* with a small error (tracking).

In most practical applications, the plant P has multiple inputs and multiple outputs and hence u, y, and y^* in Figure 1 are vectors in general. Since the control of complex systems is effected using digital computers, the plant and controllers can be considered (with little or no loss of generality) as discrete-time dynamical systems. Further, since powerful techniques exist for the control of linear plants with known (Chen, 1984) or unknown (Narendra and Annaswamy, 1989) coefficients (adaptive control techniques in the latter case), we assume that the plant characteristics are nonlinear in the domain of interest. It is for the adaptive control of such nonlinear multivariable systems whose dynamics are unknown that neural networks are particularly suited.

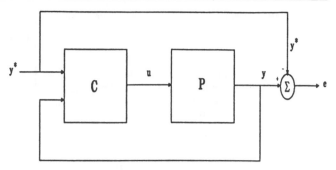

Figure 1. Feedback control of plant P using controller C. C is designed to minimize the output error $e = y - y^*$, where y^* is the desired output.

If the equations describing the plant are known, only the controller C in Figure 1 has to be determined. When the plant equations are unknown, or only partially known (i.e., the adaptive case), a model $\hat{P}$ of the plant has to be determined using all available data, and the design of the controller C in turn has to be based on $\hat{P}$. These constitute the identification and control problems of indirect adaptive control (Narendra and Annaswamy, 1989).

Adaptive control implies the ability of the system to adjust its own characteristics on-line to improve its performance. The fundamental problem of linear adaptive control, which was solved in 1980 (Narendra, Lin, and Valavani, 1980), is to stabilize an unstable linear plant with unknown parameters using only input-output data. The corresponding nonlinear problem is significantly more complex and has not been solved thus far. Hence, for the successful implementation of neural networks for adaptively controlling complex nonlinear systems, considerable prior information regarding the plant is needed; the kinds of information are briefly described below.

To identify the plant using neural networks, its inputs and outputs must be bounded functions. Hence the plant must be stabilized before using neural networks to improve its performance. Even assuming that the above conditions are satisfied, a knowledge of the bounds is needed for the choice of the neural networks used for identification and control. Since, in practical systems, the input of the controlled process cannot be altered at the will of the designer, identification has to be carried out off-line using recorded input-output data from the operating plant.

Although the above information is needed for the identification of the plant, further information is needed to undertake its control. This includes the delay of the signals from all the inputs to all the outputs, and the knowledge that the input signals of the plant cannot grow in an unbounded fashion even while the outputs remain bounded.

In the companion article (IDENTIFICATION AND CONTROL), a detailed description of a neural controller which will stabilize the feedback loop and track a reference input is given. Even though such a controller can be justified mathematically, it is rather difficult to realize using gradient methods such as back-propagation. This is because the plant together with the controller can be considered as a recurrent dynamical system, and the adjustment of the controller parameters within such a system has to take into account the effect of system dynamics. In view of this, many approximation schemes which are substantially simpler have been suggested in the literature to determine the control parameters. To a large extent, most of them avoid the dynamical questions involved. The most popular among these are briefly described in the following sections.

Supervised Control

There are many tasks which humans can perform but which are hard to imitate using standard control techniques. If neural networks are used, it may be possible to mimic the behavior of the human. This has been called supervised control. Using the same sensory inputs that the human uses, the neural network generates an output which approximates the human's control input to the system. In those cases where this is successful, the neural network can replace the human in the control tasks.

The same idea can also be used to mimic the behavior of an existing controller in a dynamical system. In this case, the neural controller can be trained off-line before being implemented in the system to replace the existing one. The objective in this case is to choose appropriate initial conditions for the parameters of the neural controller so that the on-line training time is reduced substantially. The aim of using a neural controller when a controller already exists is to achieve increased robustness and improved performance.

Inverse Control

Assuming that a plant P is given, inverse control involves the determination of its approximate inverse using a neural network. It should be noted that P in this case is not a mapping of the instantaneous value of the input into the instantaneous value of the output (i.e., a function) but the mapping of an input sequence into the value of the output at an instant (i.e., a functional). Strictly speaking, the entire past history of the input determines the current value of the output, but from a practical point of view it is assumed that a finite number, N, of past values suffices. The determination of an inverse operator P^{-1} consequently implies a mapping which, given an output sequence of length N, determines the input sequence that produced it.

It should be further mentioned that a unique input-output mapping exists only if the plant is in a state of rest when the input is applied (i.e., the plant is in zero state). If the state of the plant is non-zero (which is invariably the case), the output is also affected. Hence, the input-output mapping is changed when the initial state of the system is non-zero. The rate at which the effect of the initial values of the plant tends to zero (in the absence of an input) as the state tends to zero consequently affects the choice of the length of the interval over which mapping is defined.

All these factors make the determination of the inverse both theoretically imprecise and quite often practically not very effective. Even assuming that an inverse is found, its domain of validity is confined to the range of the outputs for which it was determined. If and when such an operator is used to control the system, it must be verified that the range of validity is preserved. In Figures 2A and 2B, two different ways of determining the inverse of P using neural networks are shown. These are referred to in this article as *Direct Inverse Control* and *Feedforward Inverse Control*.

Given a plant P and a desired output r, the objective of control is to determine the input u to P such that its output y approximates r in some sense (u, y, and r are sequences). In Figure 2A, the input to NN_1 is y, the output of the plant. The network is then trained to approximate the input to the plant. The squared errors summed over the entire interval could be used as a measure of performance. Denoting the composition of two operators by the symbol $\circ$, we attempt to determine NN_1 such that $[(NN_1 \circ Pu) - u]$ is minimized in some sense. After such a network is obtained, it is used in series with the plant to realize the output (i.e., the control input u is $NN_1 r$).

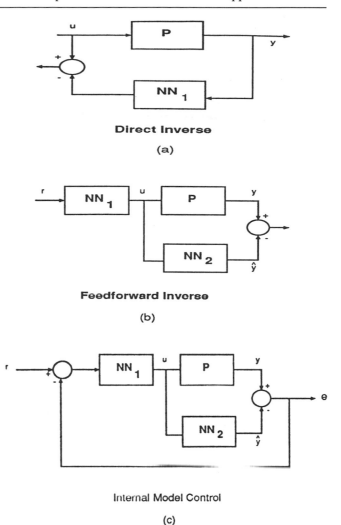

Direct Inverse

(a)

Feedforward Inverse

(b)

Internal Model Control

(c)

Figure 2. Determination of inverse of plant. In Part A, $NN_1 y$ approximates u. In Part B, NN_2 approximates P, and NN_1 approximates the inverse of P. In Part C, NN_1 and NN_2 form part of a feedback loop.

This is defined as a Direct Inverse Controller. From a mathematical point of view, in this case, NN_1 is trained as a left inverse of P but used in the control system as its right inverse.

The procedure adopted in Figure 2B overcomes the difficulty by using a second network NN_2 to identify the plant. During this stage a random input is used for both P and NN_2. Once NN_2 has been determined, NN_1 is trained using the reference input r so that $NN_2 \circ NN_1 r \cong r$, or $NN_2 \circ NN_1$ is approximately the unit operator over the desired range. The control using NN_1, derived in this fashion, is called Feedforward Inverse Control. It has proved considerably more successful than Direct Inverse Control in control applications. Both methods, however, suffer from lack of robustness since no direct feedback of error is used.

Internal Model Control

Internal model control is shown in Figure 2C. In contrast to the two types of control described in the preceding section, the identification model (NN_2) and the controller (NN_1) now form part of a feedback loop. The essential difference in this case is that the error *e* between plant and identification model is fed back and forms an input to the controller.

This model has been extensively studied and is widely used by chemical engineers (Garcia and Morari, 1982). It should be noted that if the controller is trained off-line without feedback, conventional backpropagation can be used. However, if the training is carried out within the feedback loop, the adjustment of the parameters has to be carried out in a dynamic context, and dynamic backpropagation has to be used as described in the companion article (see IDENTIFICATION AND CONTROL).

Model-Reference Adaptive Control

An extensive literature exists on MODEL-REFERENCE ADAPTIVE CONTROL. The desired performance is specified by a stable reference model with a specified reference input. The objective is to determine the controller parameters adaptively so that the output of the system follows the output of the model with minimum error. Recently, a theory of nonlinear adaptive control for multiple input, multiple output plants was presented (Narendra and Mukhopadhyay, 1994) in which the nonlinear plant, linearized around the equilibrium state, is assumed to satisfy all the conditions mentioned earlier. Simulation results have clearly demonstrated that nonlinear adaptive systems using neural networks result in a performance which is far superior to that achieved using linear adaptive control.

Predictive Control

Another commonly used method of control is predictive control, which is based on the predicted values of the system output over a finite interval of time in the future. Assuming that the desired outputs over this horizon are also known, the future control inputs are determined using a neural controller by minimizing a performance criterion.

Some Engineering Applications

Based on the principles described in the previous section, identification and control schemes have been attempted for a large number of applications. Since the field is in its infancy and since no universally accepted set of design procedures exists, the methods used in the various cases are tailored to the needs of the specific system. Further, because the approach must be validated in each case before being implemented in hardware, many of the adaptive control schemes have been evaluated using only computer simulations. The superior performance obtained over conventional methods indicates that such schemes may be ready for use in real systems. In a few cases, neural control has been successfully implemented in real time. These provide reason to believe that neural networks may become an important technology for many industrial adaptive control applications. In this section we review briefly some of the many applications that have been reported in the literature to provide the reader with some idea of the status of the field.

Furnace Control

A neural network–based adaptive control scheme has been given for a three-input three-output furnace used in plastic molding (Khalid, Omatu, and Yusof, 1993). The objective is to control three output temperatures accurately. An identification model of the system was trained off-line using data obtained from the operating system. Since no guidelines exist for choosing the number of layers and the number of nodes in each layer of the neural network used in the identifier, several networks were used in parallel, and the one that performed best on-line was selected. The controller chosen was of the inverse type described earlier in the section "Inverse Control." In all cases, backpropagation was used to adjust the parameters of the identifier and the controller, and the latter continued to adapt on-line to take into account time variations. The authors conclude that the simplicity and reliability of neural control over traditional control techniques will make such control more attractive in practical applications.

Neural Control of Steel Rolling Mill

Sbarbaro-Hofer, Neumerkel, and Hunt (1993) attempt the problem of controlling the strip thickness of a steel rolling mill using neural networks (see also STEELMAKING). The nonlinear nature of the problem, the varying time delays, and varying gains make the use of neural networks attractive in this case. Neural controllers based on internal model control and predictive control were developed and compared with a conventional PI controller (where the control signal is a sum of two terms, one Proportional to the error, and the other the time Integral of the error—PID controllers add the Derivative; see ADAPTIVE CONTROL: GENERAL METHODOLOGY). In all cases, the neural controllers achieved great control precision. The authors attribute it to the greater accuracy of the neural network model in the different operating regions of the plant. Since neural networks require more computational effort than conventional controllers, an attempt was made to combine the two types of controllers. The best results from an engineering viewpoint were obtained using an inverse model in parallel with a simple integral controller in which an integral of the error was also used as a component of the input.

The authors believe that the viability of the neural network schemes has been established by the simulations and that additional technical aspects such as memory and computational requirements should be studied before attempting hardware implementation.

Spacecraft and Aircraft Systems

The use of the theory and technology of artificial neural networks for problems of control, identification, and diagnosis in large, complex space systems is proposed in Rauch and Schaecter (1992). Improvement of performance without interrupting the high-speed control loop, evaluation of the output sensors to determine the existence of spurious structural vibrations, and implementation of health monitoring which allows the system to recognize long-term degradation as well as immediate faults are some of the problems considered in this article. The principal advantage of neural networks, according to the authors, is the tremendous computational speed achieved by massively parallel analog hardware. For illustrative purposes, they consider the control of a single segment of a six-segment mirror.

The idea of using stored information to react rapidly to changing environments was proposed by Narendra and Mukhopadhyay (1992). For example, in an aircraft system, if a finite set of possible failures that can occur is known, the controller corresponding to each failure can be determined in advance and stored. If a pattern recognition system can be used to identify the specific fault that has occurred, the precomputed controller can be used for fast and accurate response.

Autonomous Navigation

A truly interesting and novel application of neural adaptive control is described by Pomerleau (1991) (see also VISION FOR

ROBOT DRIVING). Since artificial neural networks take a prohibitively long time for training, very few attempts have been made to apply them to complex real-world perception problems, which are computationally very demanding. Pomerleau designed a neural network to drive a van automatically. Using real-time techniques, the system learns to control the vehicle in an autonomous fashion by watching a driver's reaction. This can consequently be classified as supervised control. A van trained in such a fashion has been able to drive under a variety of road conditions at speeds up to 20 mph!

Pitch Attitude Control

The pitch attitude of an underwater telerobot at the Space Systems Laboratory at MIT has been controlled using neural networks (Sanner and Akin, 1990). The problem is a relatively simple one in which the system is of second order and both states are accessible. The authors used state feedback, and the neural network was trained with static backpropagation. The success of the method using only a simple network leads the authors to believe that more sophisticated networks would improve the performance substantially. However, the one major drawback, according to them, is the large number of iterations needed to train the network, which makes on-line training infeasible. They also believe that real-time control with neural networks will be truly efficient only when special-purpose hardware takes advantage of the parallel processing capabilities of neural networks.

Refining PID Controllers

A problem that frequently confronts designers of neural controllers is the topology of the network to be used and the number of nodes to be included in the network. It is also generally accepted that, qualitatively speaking, the greater the prior information concerning the plant, the shorter the time needed for designing the controller. Scott, Shavlik, and Ray (1992) claim that the design of a neural controller for a two-output system (temperature and outflow of a water tank) could be substantially improved by basing the initial configuration of the network on a simple PID controller. Such a neural network improved performance in terms of accuracy, variability, and rate of learning over a conventionally designed neural network.

Conclusions

The prospects for adaptive control with neural networks are well expressed in the views of Werbos, McAvoy, and Su (1992) concerning the use of neural networks in chemical process industries. According to them, artificial neural networks are already in use in real-world applications by many major chemical manufacturers. Most of the applications involve pattern recognition, or soft-sensing, or feedforward control systems, where safety and stability are not a major issue. Some of the applications include real-time sensing of concentrations of products (which cannot be directly measured), nonlinear modeling, and fault detection. The authors believe that greater efficiency in chemical processing through advanced control could save millions of dollars and could have a significant impact on major global environmental problems.

A very large number of applications of neural networks for the adaptive control of nonlinear dynamical systems have been reported in the numerous journals that exist in this field. The six applications discussed in this article represent a small fraction of them. Yet they indicate the wide variety of problems that can be efficiently dealt with, and they provide a sense of the difficulties involved. The authors are unanimous in their conviction that neural networks will have a wider application in the technology of the future.

Biological systems, which we have not considered in this article, contain millions of examples of both simple and complex nonlinear multivariable systems which are controlled adaptively. In fact, the tremendous efficiency of control strategies to be found in the biological world has inspired engineers for centuries and has led in recent decades to the appropriation of numerous terms, such as *adaptation*, *learning*, and *self-organization*, from psychology and life sciences, for use in engineering. A controller in a biological system is not a single black box but rather a multiplicity of parallel and serial stages, each working on a very different part of the problem. The analysis and synthesis of such systems will require all the tools in the arsenal of the control engineer, including adaptive and learning techniques, stochastic control, optimal control, automata theory, and game theory. It is anticipated that the increasing interaction between adaptive control theorists and neurophysiologists will lead to better and novel design procedures for the design of neural adaptive controllers for nonlinear, multi-input, multi-output systems.

Road Map: Control Theory and Robotics
Background: I.3. Dynamics and Adaptation in Neural Networks; Motor Control, Biological and Theoretical
Related Reading: Adaptive Control: General Methodology; Learning as Adaptive Control of Synaptic Matrices; Process Control; Reinforcement Learning in Motor Control; Sensorimotor Learning

References

Brown, G. S., and Campbell, D. P., 1948, *Principles of Servomechanisms*, New York: Wiley.

Chen, C. T., 1984, *Linear System Theory and Design*, New York: Holt, Rinehart and Winston.

Garcia, C. E., and Morari, M., 1982, Internal model control: I. A unifying review and some new results, *Ind. Eng. Chem. Process. Des. Dev.*, 21:308–323.

Khalid, M., Omatu, S., and Yusof, R., 1993, MIMO furnace control with neural networks, *IEEE Trans. Control Sys. Technol.*, 1:238–245.

Narendra, K. S., and Annaswamy, A. M., 1989, *Stable Adaptive Systems*, Englewood Cliffs, NJ: Prentice Hall.

Narendra, K. S., Lin, Y. H., and Valavani, L. S., 1980, Stable adaptive controller design: Part II. Proof of stability, *IEEE Trans. Autom. Control*, 25:440–448.

Narendra, K. S., and Mukhopadhyay, S., 1992, Intelligent control using neural networks, *IEEE Control Sys. Mag.*, 12(2):11–18.

Narendra, K. S., and Mukhopadhyay, S., 1994, Adaptive control of nonlinear multivariable systems using neural networks, *Neural Netw.*, 7:737–752.

Narendra, K. S., and Parthasarathy, K., 1990, Identification and control of dynamical systems using neural networks, *IEEE Trans. Neural Netw.*, 1:4–27. ◆

Pomerleau, D. A., 1991, Efficient training of artificial neural networks for autonomous navigation, *Neural Computat.*, 3:88–97.

Rauch, H. E., and Schaecter, D. B., 1992, Neural networks for control, identification and diagnosis, *Proceedings of the World Space Congress 1992*, F4.4–M1.06.

Sanner, R. M., and Akin, D. L., 1990, Neuromorphic pitch attitude regulation of an underwater telerobot, *IEEE Control Sys. Mag.*, 10(3):62–68.

Sbarbaro-Hofer, D., Neumerkel, D., and Hunt, K. J., 1993, Neural control of a steel rolling mill, *IEEE Control Sys. Mag.*, 13(3):69–75.

Scott, G. M., Shavlik, J. W., and Ray, W. H., 1992, Refining PID controllers using neural networks, *Neural Computat.*, 4:746–757.

Werbos, P. J., McAvoy, T., and Su, T., 1992, Neural networks, system identification and control in the chemical process industries, in *Handbook of Intelligent Control* (David A. White and Donald A. Sofge, Eds.), New York: Van Nostrand Reinhold, pp. 283–356. ◆

Adaptive Filtering

John J. Shynk

Introduction

Adaptive filtering has been an active area of research for over thirty years and has found widespread use in numerous signal processing applications (Widrow and Stearns, 1985; Haykin, 1991). These include adaptive channel equalization and echo cancellation for telecommunications, array processing and adaptive beamforming, adaptive noise canceling and signal enhancement, adaptive control, system identification, and pattern recognition. Adaptive filters are particularly useful for problems where the underlying statistics are unknown or nonstationary, which is usually the case in many practical situations. An adaptive filter attempts to learn or track signal and system variations according to a well-defined performance criterion.

The field of adaptive filtering was derived from work on adaptive pattern recognition conducted in the 1950s and 1960s. In pattern recognition applications, such as speech recognition and image classification, a neural network is trained on a wide range of representative input patterns (called the training set). If the training set is sufficiently large, the neural net is capable of successfully classifying new patterns; e.g., distorted patterns corrupted by noise will be mapped to the most similar pattern in the training set, as defined by an error criterion. Classification is performed by a threshold or decision device that quantizes the net output to one of many different levels representing the various classes. In contrast, an adaptive filter usually does not incorporate a decision device; instead, the linear (unquantized) output is used directly in any subsequent processing. For this reason, and because the filter itself is a linear operator, it is often referred to as an *adaptive linear filter*.

An adaptive filter is a *linear combiner* comprised of adjustable weights (coefficients) and an algorithm (learning rule) that modifies these weights, as shown in Figure 1. To illustrate the connection between an adaptive filter and an artificial neuron, we also show in the figure a nonlinear device that processes the filter output $y(n)$, yielding $y_{nl}(n)$. The nonlinearity might be a hard limiter such as the signum function, or it might be a soft nonlinearity having a sigmoidal shape. This nonlinear system is essentially a single-layer perceptron (excluding the bias term), and it is often referred to as *adaline* (for *ada*ptive *li*near *neu*ron) (Widrow and Lehr, 1990). It also may be viewed as the basic element of a multilayer perceptron (Lippmann, 1987). Thus, the adaptive filter is a subcomponent of the adaline struc-

ture, although the nonlinearity may be included to make decisions in certain applications (e.g., pattern recognition).

We should emphasize that although the structure of an adaptive filter is quite basic, having much less complexity than that of a multilayer perceptron, it has been extremely successful for solving many signal processing and communications problems. There now exists a variety of adaptive filter configurations and learning algorithms, and many of their statistical properties are well understood, both analytically and from extensive computer simulations. The filter itself may be a tapped delay line, corresponding to a finite impulse response digital filter. Alternatively, the filter may have feedback, in which case it is referred to as an adaptive infinite impulse response filter. The most commonly used adaptive algorithms are based on the method of gradient descent, although several "fast" algorithms based on least-squares criteria have been developed recently (Haykin, 1991).

It is beyond the scope of this article to cover in any depth all adaptive filter configurations or the many adaptive algorithms. Instead, we will focus on the most widely used adaptive filter and describe in some detail the least-mean-square (LMS) adaptive algorithm, also known as the Widrow-Hoff learning rule (Widrow and Stearns, 1985). This algorithm is based on the linear output $y(n)$ shown in Figure 1, and the filter is a simple tapped delay line (i.e., a shift register with adjustable coefficients). We briefly show how an adaptive filter is configured for applications in system identification and channel equalization.

For comparison, we also describe a learning rule based on the nonlinear output known as Rosenblatt's perceptron procedure (Rosenblatt, 1962), which apparently was developed prior to the LMS algorithm. A specific implementation of the algorithm is presented which is very similar to the LMS algorithm (Lippmann, 1987), so that the two algorithms may be easily compared. As we shall see, both algorithms are gradient-descent methods that iteratively search the surface of a specific cost ("performance") function. Examples of their performance surfaces will be presented, along with computer simulations of the weight trajectories during learning.

Adaptive Filter Configuration

An adaptive filter is comprised of two components: (i) an *adjustable set of weights* that perform the filtering and (ii) a *learn-*

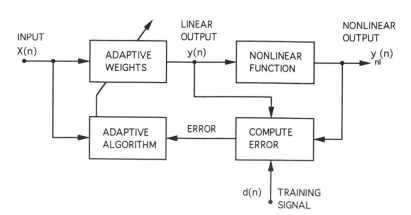

Figure 1. Adaptive filter configuration.

ing algorithm that modifies the weights to satisfy a specific performance criterion. To be more precise, define the following weight vector

$$W(n) = [w_1(n), w_2(n), \dots, w_N(n)]^T \quad (1)$$

and input signal vector

$$X(n) = [x_1(n), x_2(n), \dots, x_N(n)]^T \quad (2)$$

where the argument n denotes discrete time and the superscript T is matrix transpose. This particular form of the signal vector contains N distinct inputs and would be appropriate for multidimensional applications such as pattern recognition and array processing. However, there are many signal processing systems involving only one input signal, such as channel equalization and echo cancellation. For these cases, the input signal vector instead would be

$$X(n) = [x(n), x(n-1), \dots, x(n-N+1)]^T \quad (3)$$

where delayed versions of the input signal $x(n)$ are stored in a shift register or delay line.

The filter output is given by the inner product

$$y(n) = W^T(n)X(n) = X^T(n)W(n) \quad (4)$$

As mentioned in the previous section, we will also discuss Rosenblatt's algorithm, which is based on the nonlinear output $y_{nl}(n) = f(y(n))$. This nonlinearity could have a sigmoidal shape, such as the hyperbolic tangent function

$$\tanh(y(n)) = \frac{e^{\gamma y(n)} - e^{-\gamma y(n)}}{e^{\gamma y(n)} + e^{-\gamma y(n)}} \quad (5)$$

where $\gamma > 0$ determines the steepness of the function at the origin, or it might be the signum function (1 bit quantizer)

$$\text{sgn}(y(n)) = \begin{cases} +1 & y(n) \geq 0 \\ -1 & y(n) < 0 \end{cases} \quad (6)$$

where we have arbitrarily assigned $\text{sgn}(0) = +1$. The signum function is used for Rosenblatt's algorithm (Rosenblatt, 1962), while the hyperbolic tangent might be used in the BACKPROPAGATION algorithm for a multilayer perceptron (Rumelhart, McClelland, and PDP Research Group 1986; Werbos, 1988).

We will defer our discussion of the error signal in Figure 1 until we cover the adaptive algorithms. However, at this point we should mention that the training signal is usually called the *desired response* in the adaptive filtering literature. The reason for this is that often the goal of the adaptive filter is to generate an output that matches some desired signal, as in a system identification application. In this article we will use the term *training signal* since in other applications it may not be the desired response of the adaptive filter output.

Applications

Two representative applications of adaptive filtering are described in this section. Many signal processing applications fall into one of these general models. For example, adaptive echo cancellation is a specific case of the system identification formulation, and adaptive channel equalization belongs to the field of deconvolution (see also NOISE CANCELING AND CHANNEL EQUALIZATION).

System Identification

The configuration for system identification is shown in Figure 2 (Ljung and Söderström, 1983). Observe that the adaptive filter is placed in *parallel* with the unknown system, which could be

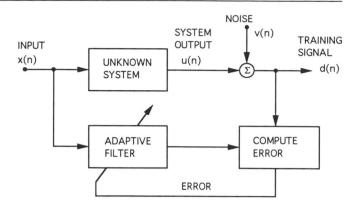

Figure 2. System identification.

an arbitrary linear or nonlinear function and may be time-varying. The adaptive filter has direct access to the system input $x(n)$ and a noisy version of the system output, i.e., $d(n) = u(n) + v(n)$, where $u(n)$ is the actual system output and $v(n)$ is additive noise. The process $v(n)$ is usually uncorrelated with $x(n)$, and it is often attributed to measurement noise. Since both $x(n)$ and $d(n)$ are available, an adaptive algorithm may be used to adjust the weights such that the filter and the unknown system statistically have similar input/output characteristics.

The system identification configuration also can be used to study the convergence properties of learning algorithms (Shynk and Bershad, 1991). In fact, many of the convergence results developed for more complex versions of the LMS algorithm require this structure in order to obtain analytically tractable results. This framework is used to derive the convergence results presented later in this article.

Adaptive Channel Equalization

The configuration for adaptive channel equalization is shown in Figure 3 (Qureshi, 1985). Observe that the adaptive filter is placed in *cascade* with the unknown channel. The goal of the adaptive filter is to mitigate the channel distortion (e.g., multipath propagation) and in effect estimate the channel *inverse*. The input of the adaptive filter is obtained from the channel output, which may be corrupted by an additive noise process $v(n)$. The training signal is the transmitted signal $x(n)$. Since the channel and the adaptive filter have a finite nonzero delay associated with each, it is necessary to delay the transmitted signal such that $d(n) = x(n - \Delta)$, where Δ is an appropriate delay.

Clearly, since the purpose of any communication system is to transmit *unknown* information across the channel, the above configuration for training the adaptive filter is not always feasible. Typically, there is a short training period at the beginning of transmission whereby a signal known at the receiving end is transmitted and used by the adaptive filter to adjust its weights. When the error rate is sufficiently reduced at the end of training, information (unknown to the receiver) then can be transmitted across the channel. During this time, the adaptive algorithm stops updating, or instead a *blind* adaptive algorithm may be employed which requires knowledge of only the received signal and certain statistical properties of the transmitted signal (Haykin, 1991).

Learning Algorithms

In this section, we consider two performance functions that depend on the filter output, either $y(n)$ or $y_{nl}(n)$, and the train-

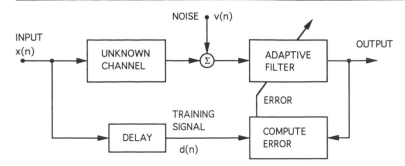

Figure 3. Adaptive channel equalization.

ing signal $d(n)$. In essence, the goal of the learning algorithm is to adjust the weight vector $W(n)$ such that the output matches the training signal $d(n)$, thereby finding a minimum (stationary point) of the performance function. The algorithms discussed here have the following general form of *gradient descent*

$$W(n + 1) = W(n) + \alpha[-\nabla(n)] \tag{7}$$

where $\nabla(n)$ is the *gradient* with respect to $W(n)$ of the corresponding performance function. The positive step size α controls the transient and steady-state convergence properties of the weight update. In the following discussion, we consider criteria that are based on two error signals: (i) one that is a function of the linear output

$$e(n) = d(n) - y(n) \tag{8}$$

and (ii) one that depends on the nonlinear output

$$e_{nl}(n) = d(n) - y_{nl}(n) \tag{9}$$

Both of these errors can be handled by the configuration in Figure 1.

Least-Mean-Square Algorithm (Widrow-Hoff)

Consider the following cost function

$$C_{LMS}(n) = e^2(n) \tag{10}$$

which is a stochastic variable fluctuating around the mean square error (MSE)

$$\xi_{LMS} = E[e^2(n)] \tag{11}$$

where $E[\cdot]$ denotes statistical expectation. For this reason, the weight update in Equation 7 is referred to as a *stochastic gradient* algorithm. Note that Equation 10 is a very simple estimate of the ensemble average in ξ_{LMS}. In this article, we will interchangeably refer to ξ and its stochastic estimate $C(n)$ as performance (or cost) functions. Using Equation 4 and Equation 8, and noting that $\partial(W^T(n)X(n))/\partial W(n) = X(n)$, the gradient of Equation 10 is

$$\nabla_{LMS}(n) = -2X(n)e(n) \tag{12}$$

and the LMS algorithm is given by (Widrow and Stearns, 1985)

$$W(n + 1) = W(n) + 2\alpha X(n)e(n) \tag{13}$$

Thus, with each new data $\{X(n), d(n)\}$, the error $e(n)$ is computed and the filter weights are updated in an attempt to learn the underlying signal statistics. The algorithm converges on average when $E[X(n)e(n)] = 0$ (the zero vector), which, after substituting $e(n)$, yields the following unique stationary point (Widrow and Stearns, 1985):

$$W_{LMS} = R^{-1}P \tag{14}$$

where $R = E[X(n)X^T(n)]$ and $P = E[X(n)d(n)]$. Clearly, this convergence point is influenced by the correlation between $X(n)$

and $d(n)$. Observe that the algorithm itself does not incorporate the nonlinear output $y_{nl}(n)$, although this output may be used for other purposes depending on the particular application (Widrow, Winter, and Baxter, 1988).

Rosenblatt's Algorithm (Perceptron)

Rosenblatt's algorithm, also called the perceptron convergence procedure, is one of the earliest algorithms developed to train a network of adaptive weights (Rosenblatt, 1962). We present a specific formulation of Rosenblatt's algorithm that is similar to the LMS algorithm and is slightly different from its original formulation (Lippmann, 1987). If we view Rosenblatt's algorithm as a gradient-descent method as in Equation 7, then the corresponding performance function is (Shynk, 1990)

$$C_{RO}(n) = 2|y(n)| - 2d(n)y(n) \tag{15}$$

where again $d(n)$ is the training signal, and we assume that the nonlinearity is given by the signum function in Equation 6. Differentiating Equation 15 with respect to $W(n)$ yields

$$2\frac{\partial|y(n)|}{\partial W(n)} - 2d(n)\frac{\partial y(n)}{\partial W(n)} = 2\,\mathrm{sgn}(y(n))X(n) - 2d(n)X(n)$$

$$= -2e_{nl}(n)X(n) \tag{16}$$

where we have substituted Equation 9. Thus, the weight update for Rosenblatt's algorithm is

$$W(n + 1) = W(n) + 2\alpha X(n)e_{nl}(n) \tag{17}$$

which is similar to the LMS algorithm except that the *nonlinear error* is used in the gradient. This algorithm converges when $E[X(n)e_{nl}(n)] = 0$, yielding (Shynk, 1990)

$$W_{RO} = c\sigma_y R^{-1}P \tag{18}$$

where R and P are as defined above, $c = \sqrt{\pi/2}$, and $\sigma_y^2 = E[y^2(n)] = W^T RW$, which is evaluated at W_{RO}. It should be mentioned that this result assumes $X(n)$ is a zero-mean Gaussian vector (which explains the appearance of π in the expression). Note that Equation 18 is actually an implicit function in W_{RO}, and it is possible that more than one stationary point exists, unlike the unique solution of the LMS algorithm (see the section on performance surfaces later in this article).

Example Convergence Results

In this section, we present some examples of the performance surfaces for the LMS algorithm and Rosenblatt's algorithm, as well as typical trajectories of the connection weights during learning. A *performance surface* is simply a three-dimensional plot of the performance function versus two of the weight vector components. We also give closed-form expressions for the performance functions and their stationary points using a model for $d(n)$ based on the system identification configuration (Shynk,

A

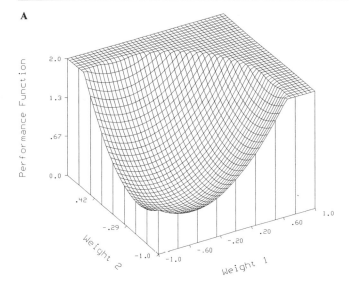

B

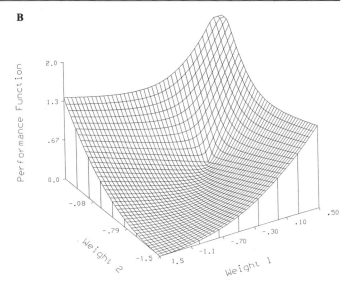

Figure 4. Performance surfaces (truncated at 2.0). A, ξ_{LMS}. B, ξ_{RO}. (From J. J. Shynk, Performance surfaces of a single-layer perceptron,

IEEE Transactions on Neural Networks, © 1990 IEEE. Reprinted with permission.)

1990). Recall that the performance functions are computed as $\xi = E[C(n)]$, where $C(n)$ is a stochastic estimate. Assume that $X(n)$ is a Gaussian random vector with zero mean and correlation matrix R, and that

$$d(n) = \text{sgn}(X^T(n)F) \qquad (19)$$

where F is an unknown fixed weight vector defined in a manner similar to $W(n)$ in Equation 1.

This structure allows one to use a theorem by Price (1958) which relates a correlation of nonlinear functionals to a correlation of the underlying linear functionals. Specifically, one form of Price's theorem states that for $G = E[f_1(z_1)f_2(z_2)]$,

$$\frac{\partial^k G}{\partial g^k} = E[f_1^{(k)}(z_1)f_2^{(k)}(z_2)] \qquad (20)$$

where $\{z_i\}$ are zero-mean, unit variance Gaussian variates, $\{f_i(z_i)\}$ are zero-memory nonlinearities, and $g = E[z_1 z_2]$ is the underlying cross-correlation. Thus, the desired expectation G is obtained by integrating Equation 20 k times (where, typically, $k = 1$ or 2), yielding an expression in terms of g. The usefulness of this theorem is that an expectation involving complicated nonlinearities may be replaced by a simpler expectation involving their derivatives. For example, the derivative of the signum function is a delta function, for which the expectation in Equation 20 is trivial to evaluate.

Performance Surfaces

In our performance surface examples, we have chosen $N = 2$, $F = [-1, -1]^T$, and $R = I$ (the identity matrix). The performance function for the LMS algorithm is given by (Shynk, 1990)

$$\xi_{LMS} = \sigma_y^2 - 2\rho_{dy}/(c\sigma_d) + 1 \qquad (21)$$

where σ_y^2 is as previously defined, $\sigma_d^2 = E[d^2(n)] = F^TRF$, and $\rho_{dy} = E[d(n)y(n)] = F^TRW$. The corresponding stationary point is

$$W_{LMS} = F/(c\sigma_d) \qquad (22)$$

Thus, the optimal weight vector is collinear with F, which is the weight vector of the underlying model generating $d(n)$. An ex-

ample of this result is shown in Figure 4A. Observe that the surface has a quadratic curvature ("bowl shape") with a *unique* minimum point. This property is characteristic of the MSE surface for a linear combiner, and it is a desirable feature of the LMS algorithm.

By comparison, the performance function of Rosenblatt's algorithm is

$$\xi_{RO} = (2/c)(\sigma_y - \rho_{dy}/\sigma_d) \qquad (23)$$

and the stationary points are given by

$$W_{RO} = (\sigma_y/\sigma_d)F \qquad (24)$$

Because $\sigma_y = \sqrt{W^TRW}$ (evaluated at W_{RO}), it is clear that ξ_{RO} has *infinitely many* stationary points. Observe in Figure 4B that the minimum points are determined by the straight line starting at the origin and extending out such that $w_1 = w_2$ for $w_1 < 0$ (as expected because of the form chosen for F). This nonuniqueness is caused by the signum function whose output $y_{nl}(n)$ is amplitude independent.

It should be mentioned that these results are specific for the adaptive filter configuration in Figure 1 and the system identification model in Figure 2 with $d(n)$ defined in Equation 19. If, for example, a bias term is included in the adaptive filter (i.e., an additional weight $w_0(n)$ with fixed input $+1$), Rosenblatt's algorithm may have a unique minimum point depending on the data. However, its performance surface will not in general have the quadratic shape in Figure 4A. On the other hand, the LMS algorithm always has a unique convergence point for any configuration of the data provided that R is nonsingular.

Weight Trajectories

Figure 5 shows some convergence examples of both algorithms for the conditions described in the previous section, except that here $F = [-1, 1]^T$. The step size was chosen to be $\alpha = 0.01$, the initial weights were $W(0) = [0, 0]^T$, and the weight trajectories were averaged over 25 independent computer runs to obtain smooth curves. Observe that the LMS algorithm converges to a unique minimum (Figure 5A) as predicted by the previous results. On the other hand, Rosenblatt's algorithm has not yet converged (Figure 5B); in fact, the weights continue to grow

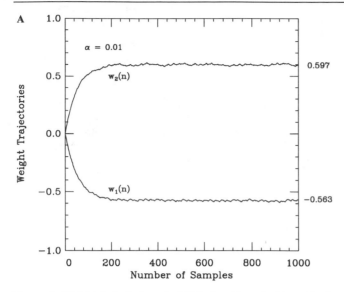

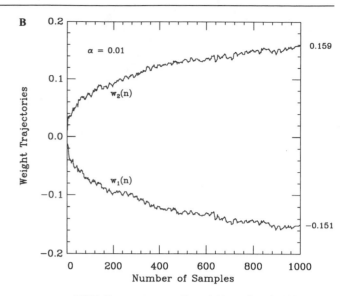

Figure 5. Weight trajectories. *A*, LMS algorithm. *B*, Rosenblatt's algorithm. (From J. J. Shynk, Performance surfaces of a single-layer perceptron, *IEEE Transactions on Neural Networks*, © 1990 IEEE. Reprinted with permission.)

without bound. The reason for this follows from the shape of the performance surface in Figure 4*B*, where we see that it slopes outward away from the origin. Since this algorithm is gradient-based, it is not surprising that the trajectories tend toward infinity. Furthermore, even if we initialized the weights at a stationary point as defined by Equation 24, the weights would still grow unbounded because the gradient in Equation 16 is a noisy estimate of the true gradient.

Discussion

Adaptive filters are now in widespread use for a variety of signal processing applications. Although several configurations are possible, the linear combiner shown in Figure 1 is the most common, and the least-mean-square algorithm is the most popular means of adjusting the adaptive weights. There are several variations of the LMS algorithm, including those that have less complexity or improved convergence properties. Although Rosenblatt's algorithm was one of the earliest learning algorithms to be developed, it is not widely used today because the nonlinearity removes information from the filter output which otherwise would be helpful in the learning process. This was evident from the example of its performance surface (Figure 4*B*), which is not quadratic and has infinitely many stationary points. Its convergence properties are not as well understood as those of the LMS algorithm.

The adaptive filter configuration described in this article is the basic component of a multilayer perceptron. The additional layers provide greater nonlinear modeling capabilities (Lippmann, 1987), which are necessary for complex applications such as speech and image processing. Gradient-descent adaptive algorithms are typically used to adjust the weights of a multilayer perceptron. They are similar to the adaptive algorithms described in this article, but they have an additional degree of complexity due to the cascade of layers. One such algorithm, known as the backpropagation algorithm (Rumelhart, McClelland, and PDP Research Group 1986; Werbos, 1988), has been successfully applied to a number of signal processing problems. For further discussion of multilayer perceptrons and

the corresponding training algorithms, the interested reader is referred to Widrow and Lehr (1990).

Road Map: Applications of Neural Networks
Background: I.3. Dynamics and Adaptation in Neural Networks
Related Reading: Adaptive Signal Processing; Perceptrons, Adalines, and Backpropagation

References

Haykin, S., 1991, *Adaptive Filter Theory*, 2nd ed., Englewood Cliffs, NJ: Prentice-Hall. ◆
Lippmann, R. P., 1987, An introduction to computing with neural nets, *IEEE ASSP Mag.*, 4:4–22. ◆
Ljung, L., and Söderström, T., 1983, *Theory and Practice of Recursive Identification*, Cambridge, MA: MIT Press.
Price, R., 1958, A useful theorem for nonlinear devices having Gaussian inputs, *IRE Trans. Inf. Theory*, IT-4:69–72.
Qureshi, S. U. H., 1985, Adaptive equalization, *Proc. IEEE*, 73:1349–1387. ◆
Rosenblatt, F., 1962, *Principles of Neurodynamics*, New York: Spartan.
Rumelhart, D. E., McClelland, J. L., and PDP Research Group, Eds., 1986, *Parallel Distributed Processing: Explorations in the Microstructure of Cognition*, Cambridge, MA: MIT Press. ◆
Shynk, J. J., 1990, Performance surfaces of a single-layer perceptron, *IEEE Trans. Neural Netw.*, 1:268–274.
Shynk, J. J., and Bershad, N. J., 1991, Steady-state analysis of a single-layer perceptron based on a system identification model with bias terms, *IEEE Trans. Circuits Sys.*, 38:1030–1042.
Werbos, P. J., 1988, Backpropagation: Past and future, in *Proceedings of the IEEE International Conference on Neural Networks*, vol. I, San Diego, CA, pp. 343–353. ◆
Widrow, B., and Lehr, M. A., 1990, 30 years of adaptive neural networks: Perceptron, madaline, and backpropagation, *Proc. IEEE*, 78: 1415–1441. ◆
Widrow, B., and Stearns, S. D., 1985, *Adaptive Signal Processing*, Englewood Cliffs, NJ: Prentice-Hall. ◆
Widrow, B., Winter, R. G., and Baxter, R. A., 1988, Layered neural nets for pattern recognition, *IEEE Trans. Acoust., Speech, Signal Process.*, 36:1109–1118.

Adaptive Resonance Theory (ART)

Gail A. Carpenter and Stephen Grossberg

Introduction: ART Models

Adaptive Resonance Theory, or ART, was introduced as a theory of human cognitive information processing (Grossberg, 1976). The theory has since led to an evolving series of real-time neural network models that perform unsupervised and supervised category learning, pattern recognition, and prediction. ART networks create stable memories in response to arbitrary input sequences with either fast or slow learning. Models that carry out unsupervised learning include ART 1 (Carpenter and Grossberg, 1987a) for binary input patterns, ART 2 (Carpenter and Grossberg, 1987b) for analog and binary input patterns, and ART 3 (Carpenter and Grossberg, 1990) for parallel search of distributed recognition codes in a multilevel network hierarchy. ARTMAP and fuzzy ARTMAP models (Carpenter et al., 1992) join two ART modules to carry out supervised learning. Fuzzy ARTMAP represents a computational synthesis of ideas from neural networks, expert production systems, and fuzzy logic. Researchers have also developed many ART model variations adapted to individual applications.

Match-Based Learning and Error-Based Learning

A *match-based* learning process is the basis of ART stability. Match-based learning allows memories to change only when attended portions of the external world match internal expectations, or when something completely new occurs. When the external world fails to match an ART network's expectations or predictions, a search process selects a new category, representing a new hypothesis about what is important in the current environment. Match-based learning, with its intrinsic stability feature, makes ART and ARTMAP well suited to problems that require on-line learning of a large and evolving database. Conversely, *error-based* learning is more naturally suited to other classes of problems, such as the learning of sensorimotor maps, that require adaptation to present statistics rather than the construction of a knowledge system. Error-based learning responds to a mismatch by sending the difference between a target output and an actual output toward zero, rather than by initiating a search for a better match. Neural networks that employ error-based learning include backpropagation (Werbos, 1974) and other descendants of the perceptron (Figure 1).

ART Search

Figure 1 illustrates the ART 1 model, and Figure 2 illustrates a typical ART search cycle. Level F_1 in Figure 1 contains a network of nodes, each of which represents a particular combination of sensory features. Level F_2 contains a second network that represents recognition codes selectively activated by patterns at F_1. The activity of a node in F_1 or F_2 is called a *short-term memory* (STM) trace. STM is the type of memory that can be rapidly reset without leaving an enduring trace. For example, a distracting unexpected event can reset the STM of a list of numbers that a person has heard once. STM is distinct from LTM, or *long-term memory*, the type of memory that we usually associate with learning. For example, we do not forget our parents' names when an unexpected event distracts us. In an ART model, adaptive weights in both bottom-up (F_1-to-F_2) paths and top-down (F_2-to-F_1) paths store LTM. Auxiliary gain con-

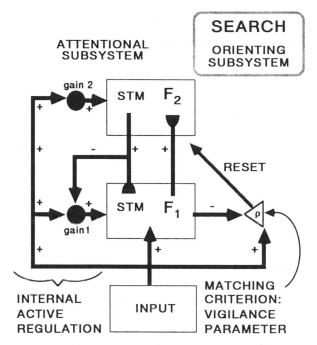

Figure 1. Typical ART 1 neural network (Carpenter and Grossberg, 1987a).

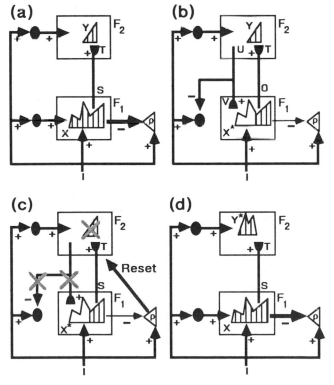

Figure 2. ART search cycle.

trol and orienting processes regulate category selection, search, and learning, as described in the following section.

An ART input vector I registers itself as a pattern X of activity across level F_1 (Figure 2A). The F_1 output vector S is then transmitted through the multiple converging and diverging adaptive filter pathways emanating from F_1. This transmission event multiplies the vector S by a matrix of adaptive weights, or LTM traces, to generate a net input vector T to level F_2. The internal competitive dynamics of F_2 contrast-enhance vector T. The resulting F_2 activity vector Y represents only one or a few active nodes, the ones that receive maximal filtered input from F_1. In ART 1, the competition is sharply tuned to select only the F_2 node that receives the maximal $F_1 \rightarrow F_2$ input, leaving just one positive component of Y. Activation of such a *winner-take-all* node defines the category, or symbol, of the input pattern I (see WINNER-TAKE-ALL MECHANISMS). Such a category represents all the F_1 inputs I that send maximal input to the corresponding F_2 node. These rules governing bottom-up activation and learning constitute SELF-ORGANIZING FEATURE MAPS: KOHONEN MAPS and the related COMPETITIVE LEARNING or LEARNING VECTOR QUANTIZATION, as introduced by Grossberg (1972, 1976), von der Malsburg (1973), and Kohonen (1989).

Activation of an F_2 node may be interpreted as "making a hypothesis" about an input I. When Y becomes active, it sends an output vector U top-down through the second adaptive filter. After multiplication by the adaptive weight matrix of the top-down filter, a net vector V becomes the input to F_1 (Figure 2B). Vector V plays the role of a learned top-down expectation. Activation of V by Y may be interpreted as "testing the hypothesis" Y, or "reading out the category prototype" V. The ART 1 network matches the "expected prototype" V of the category against the active input pattern, or exemplar, I. Nodes in F_1 that were activated by I are now suppressed if they do not correspond to large LTM traces in the prototype pattern V. Thus F_1 features that are not "expected" by V are suppressed. This matching process changes the F_1 activity pattern X by suppressing activation of all the features in I that are not "confirmed" by hypothesis Y. The resultant matched pattern X^* encodes the cluster of features in I that are relevant to hypothesis Y based on the network's learned experiences. Pattern X^* is, at that moment, the pattern of features to which the network "pays attention."

Resonance, Attention, and Learning

If expectation V is close enough to input I, then a state of *resonance* develops as the attentional focus takes hold. The pattern X^* of attended features reactivates hypothesis Y, and Y, in turn, reactivates X^*. The network locks into a resonant state through the mutual positive feedback that dynamically links X^* with Y. The resonant state persists long enough for learning to occur, hence the term *adaptive resonance* theory. ART systems learn prototypes, rather than exemplars, because memories encode the attended feature vector X^*, rather than the input I itself.

This attentive matching process combines three different types of inputs at level F_1 (Figure 1): bottom-up inputs, top-down expectations, and attentional gain control signals. The attentional gain control channel sends the same signal to each F_1 node; it is a "nonspecific," or modulatory, channel. Attentive matching obeys a $\frac{2}{3}$ *rule* (Carpenter and Grossberg, 1987a) that permits an F_1 node to reach its output threshold only if two of the three input sources that converge on it are large. The $\frac{2}{3}$ rule shows how an ART system can be "primed" by a previous event to expect a subsequent event that may or may not occur,

and why priming is "unconscious." A top-down expectation activates only to subthreshold levels the F_1 nodes in its prototype. No F_1 node can generate suprathreshold output signals in the absence of a "second third." Nodes are nonetheless "primed," or ready, to fire rapidly and resonantly if a bottom-up input does match the prototype well enough. Thus, ART systems are "intentional" or "goal-oriented" systems that can selectively process expected events.

The $\frac{2}{3}$ rule also allows an ART system to react to inputs in the absence of prior priming, since a bottom-up input both directly activates its target features and indirectly activates them through the nonspecific gain control channel to satisfy the $\frac{2}{3}$ rule (Figure 2A). After the input instates itself at F_1, an active hypothesis Y both transmits a top-down expectation V to F_1 and inhibits the F_1 gain control signal. The $\frac{2}{3}$ rule then ensures that only F_1 nodes that are confirmed by the expectation remain active in STM.

A dimensionless parameter called *vigilance* defines the criterion of an acceptable $\frac{2}{3}$ rule match. Vigilance weighs how close the input exemplar I must be to the top-down prototype V for resonance to occur. Because vigilance can vary across learning trials, a single ART system can learn recognition categories capable of encoding widely differing degrees of generalization, or morphological variability. Low vigilance leads to broad generalization and abstract prototypes. High vigilance leads to narrow generalization and to prototypes that represent few input exemplars. At very high vigilance, prototype learning reduces to exemplar learning. Thus, a single ART system may learn to recognize abstract categories of faces and dogs, as well as individual faces and dogs. In supervised ARTMAP systems, which prototypes are learned depends on the predictive success of their categories in a particular task environment.

If the top-down expectation V and the bottom-up input I are too novel, or unexpected, to satisfy the vigilance criterion, then the *orienting subsystem* (Figure 1) triggers a bout of hypothesis testing, or memory search. Search allows the network to select a better recognition code, symbol, category, or hypothesis to represent input I at level F_2. The orienting subsystem interacts with the attentional subsystem to enable the ART system to learn new representations of novel events without risking unselective forgetting of previous knowledge. ART 3 (Carpenter and Grossberg, 1990) implements search by means of a *medium-term memory* (MTM), or transmitter process, within the bottom-up paths. MTM temporarily diminishes signals in those paths that project to the most active F_2 nodes. Then, if a signal from the orienting subsystem resets F_2 activity, MTM selectively biases the next choice process against the previously active nodes, which the ART system sees as having caused some kind of error.

The search process prevents associations from forming between Y and X^* if X^* is too different from I to satisfy the vigilance criterion. As shown in Figure 2C, the search process resets Y before such an association can form. Search may select a familiar category if its prototype is similar enough to the input I to satisfy the vigilance criterion. Learning then may refine the prototype in light of new information carried by I. If I is too different from any of the previously learned prototypes, then the ART network selects an uncommitted F_2 node that establishes a new category. A network parameter controls the depth of search, before the system chooses an uncommitted node.

Over learning trials, the system establishes stable recognition categories. This process corresponds to making the inputs "familiar" to the network. All inputs that are familiar to one category access it directly in a one-pass fashion, and search is automatically disengaged. The category selected is, moreover, the one whose prototype provides the globally best match to the input pattern. Learning can continue on-line, and in a stable fashion, with familiar inputs directly accessing their categories and novel inputs triggering adaptive searches for better categories, until the network fully uses its memory capacity.

Scientists have used ART systems to explain and predict a variety of cognitive and brain data (e.g., Carpenter and Grossberg, 1991). For example, a formal lesion of the orienting subsystem creates a memory disturbance that mimics properties of medial temporal amnesia after lesions of the hippocampal formation (Carpenter and Grossberg, 1993). Dynamics of the ART F_1 and F_2 levels correspond to data concerning the prestriate visual cortex and the inferotemporal cortex during visual object recognition (Desimone, 1992), and the action of the attentional gain control pathway corresponds to functions of the pulvinar in the thalamus.

Supervised Learning and Prediction

The supervised ARTMAP architecture learns a map that associates categories of one input space (e.g., visual features) learned by one ART subsystem with categories of another input space (e.g., auditory features) learned by a second ART subsystem. Figure 3 illustrates how ARTMAP can organize dissimilar patterns into distinct recognition categories that then make the same prediction. ARTMAP systems include a *match-tracking* control process. After a predictive failure, match tracking raises vigilance just enough to trigger memory search for a better category. ARTMAP thereby creates multiple scales of generalization, from fine to coarse, as needed. Match tracking realizes a Minimax Learning Rule that conjointly minimizes predictive error and maximizes generalization, using only information that is locally available under incremental learning conditions in a nonstationary environment.

Rules and Applications

The learned expertise of an ARTMAP system translates to IF-THEN "rules." Suppose, for example, that the input vectors encode biochemicals and that the output vectors encode drug effects on behavior. Various biochemicals may achieve the same clinical effect on behavior for different chemical reasons. Correspondingly, at any time during learning, the operator of an ARTMAP system can test how many recognition categories give rise to a desired clinical effect by checking which LTM traces are large in the pathways from input recognition categories to the desired output node. Within each recognition category, the prototype characterizes a particular "rule," or bundle of biochemical features, that predicts the desired clinical effect. A list of these prototype vectors provides a transparent set of rules that predict the desired outcome. Many such rules may coexist without mutual interference because of the competitive interactions that compress each F_2 hypothesis. Associative networks such as backpropagation often mix multiple rules among the same LTM traces because they do not have the competitive dynamics to separate them.

ARTMAP is one of a rapidly growing family of attentive self-organizing prediction systems that have evolved from the biological theory of cognitive information processing of which ART forms an important part. ART modules have found their way into such diverse applications as the control of mobile robots, a Macintosh system that adapts to user behavior, diagnostic monitoring systems for nuclear plants, learning and search of airplane part inventories, face recognition, remote sensing land-cover classification, target recognition, medical diagnosis, electrocardiogram analysis, protein/DNA analysis, 3D visual object recognition, musical analysis, seismic recognition, sonar recognition, and laser radar recognition (e.g., Bachelder, Waxman, and Seibert, 1993; Escobedo, Smith, and Caudell, 1993; Gopal, Sklarew, and Lambin, 1993). All of these applications exploit the ability of an ART system to learn rapidly to classify large databases in a stable fashion, to calibrate its confidence in a classification, and to focus attention on those featural groupings that it deems to be important based on experience. The supervised ARTMAP system promises to find an even broader range of applications because of its ability to adapt the number, shape, and scale of its category boundaries to meet the on-line demands of nonstationary databases.

Acknowledgments. This research was supported in part by the Air Force Office of Scientific Research (AFOSR F49620–92–J–0225), ARPA (N00014–92–J–4015), the National Science Foundation (NSF IRI–94–01659), and the Office of Naval Research (ONR N00014–91–J–4100).

Road Map: Learning in Artificial Neural Networks, Deterministic
Background: I.3. Dynamics and Adaptation in Neural Networks
Related Reading: Short-Term Memory; Visual Processing of Object Form and Environment Layout

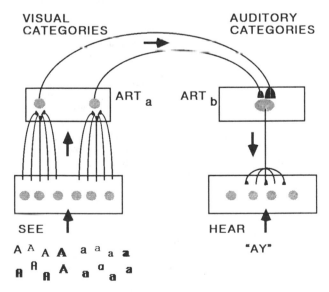

Figure 3. ARTMAP many-to-one learning combines categorization of many exemplars into one category by one ART subsystem, and labeling of many categories with the same name or prediction at a second ART subsystem.

References

Bachelder, I. A., Waxman, A. M., and Seibert, M., 1993, A neural system for mobile robot visual place learning and recognition, in *Proceedings of the World Congress on Neural Networks (WCNN-93)*, vol. 1, Hillsdale, NJ: Erlbaum, pp. 512–517.
Carpenter, G. A., and Grossberg, S., 1987a, A massively parallel architecture for a self-organizing neural pattern recognition machine, *Comput. Vis., Graph., Image Proc.*, 37:54–115.
Carpenter, G. A., and Grossberg, S., 1987b, ART 2: Stable self-organization of pattern recognition codes for analog input patterns, *Appl. Opt.*, 26:4919–4930.
Carpenter, G. A., and Grossberg, S., 1990, ART 3: Hierarchical search using chemical transmitters in self-organizing pattern recognition architectures, *Neural Netw.*, 3:129–152.
Carpenter, G. A., and Grossberg, S. (Eds.), 1991, *Pattern Recognition by Self-Organizing Neural Networks*, Cambridge, MA: MIT Press.

Carpenter, G. A., and Grossberg, S., 1993, Normal and amnesic learning, recognition, and memory by a neural model of cortico-hippocampal interactions, *Trends Neurosci.* 16:131–137.

Carpenter, G. A., Grossberg, S., Markuzon, N., Reynolds, J. H., and Rosen, D. B., 1992, Fuzzy ARTMAP: A neural network architecture for incremental supervised learning of analog multidimensional maps, *IEEE Trans. Neural Netw.*, 3:698–713.

Desimone, R., 1992, Neural circuits for visual attention in the primate brain, in *Neural Networks for Vision and Image Processing* (G. A. Carpenter and S. Grossberg, Eds.), Cambridge, MA: MIT Press, pp. 343–364.

Escobedo, R., Smith, S. D. G., and Caudell, T. P., 1993, The ART of design retrieval, in *Adaptive Neural Systems* (T. P. Caudell, Ed.), Technical Report BCS–CS–ACS–93–008, Seattle, WA: Boeing Company, pp. 149–160.

Gopal, S., Sklarew, D. M., and Lambin, E., 1993, Fuzzy-neural network classification of land cover change in the Sahel, *Proceedings of the DOSES/EUROSAT Workshop on New Tools for Spatial Analysis*, Lisbon, Portugal, 1993.

Grossberg, S., 1972, Neural expectation: Cerebellar and retinal analogs of cells fired by learnable or unlearned pattern classes, *Kybernetik*, 10:49–57.

Grossberg, S., 1976, Adaptive pattern classification and universal recoding: I. Parallel development and coding of neural feature detectors. II. Feedback, expectation, olfaction, and illusions, *Biol. Cybern.*, 23:121–134, 187–202.

Kohonen, T., 1989, *Self-Organization and Associative Memory*, 3rd ed., Berlin: Springer-Verlag.

von der Malsburg, C., 1973, Self-organization of orientation sensitive cells in the striate cortex, *Kybernetik*, 14:85–100.

Werbos, P., 1974, Beyond regression: New tools for prediction and analysis in the behavioral sciences, PhD Dissertation, Harvard University, Cambridge, MA.

Adaptive Signal Processing

Simon Haykin

Introduction

Adaptive signal processing is an important part of the ever-expanding discipline of signal processing. The term *adaptive* refers to the ability of a signal processing system to respond to statistical variations of the input signal in a continuous fashion. The traditional approach to adaptive signal processing has been based on the use of linear techniques (Haykin, 1991; Widrow and Stearns, 1985). However, with the resurgence of interest in neural networks, it is only natural that adaptive signal processing is being reexamined on many fronts in light of some unique properties offered by neural networks.

We begin this brief overview of adaptive signal processing by discussing the filtering problem that is central to the subject matter.

The Filtering Problem

The term *filter* is used to describe a device in the form of a piece of physical hardware or computer software that is applied to noisy data for the purpose of extracting information about a prescribed quantity of interest. The noise may arise from a variety of factors. For example, the sensor used to gather the data may be noisy, or the data available for processing may represent the output of a noisy channel. In any event, the filter is used to perform three basic information-processing tasks:

- *Filtering*, which refers to the extraction of useful information at some time t by employing data measured up to and including time t.
- *Smoothing*, which differs from filtering in that information about the quantity of interest need not be available at time t, and data measured later than time t can be used to obtain this information. Smoothing is expected to be more accurate than filtering, since it involves the use of more data than is available for filtering.
- *Prediction*, which is the forecasting side of information processing. The aim here is to derive information about the quantity of interest at time $t + \tau$ for some $\tau > 0$ by using data up to and including time t.

Consider, for example, a *pulse radar tracking system*. In such a system, a sequence of uniformly spaced radar pulses is transmitted, giving rise to a received signal that consists of target echoes contaminated by additive receiver noise. The requirement is to process the received signal to determine the position of a maneuvering airborne target visible to the radar. The processor may involve prediction, filtering, and smoothing, depending on the particulars of target tracking information required.

Two classes of filters may be identified, with linearity as the basis for classification. In a *linear filter*, the filter output is a linear function of the observations applied to the filter input. A *nonlinear filter*, however, is characterized by an output that is a nonlinear function of its input. Clearly, the design of a nonlinear filter is more demanding than that of a linear one.

Much has been written on the classical approach to the *linear optimum filtering problem*, assuming the availability of second-order statistical parameters (i.e., mean and covariance functions) of the useful signal and unwanted additive noise. In this statistical approach, an *error signal* is usually defined between some desired response and the actual filter output, and the filter is designed to minimize a *cost function*, defined as the mean-square value of the error signal. For stationary inputs, the resulting optimum filter is commonly known as the *Wiener filter*, in recognition of the pioneering work done by Norbert Wiener in the 1940s (Wiener, 1949). A more general solution to the linear optimum filtering problem is provided by the *Kalman filter*, so named after its originator (Kalman, 1960). Compared with the Wiener filter, the Kalman filter is more powerful, with a wider variety of engineering applications.

The Kalman Filter

A distinguishing feature of the Kalman filter is that it is based on the notion of the *state* of a system. Specifically, a linear dynamical system (assumed to be operating in discrete time) is modeled by a pair of coupled state equations:

- The *process (state-update) equation*, which defines the state vector $\mathbf{x}(n + 1)$ at time $n + 1$ in terms of the state vector $\mathbf{x}(n)$ at the previous interation n.

- The *measurement (output) equation*, which defines the observation (output) vector at time n in terms of the corresponding state vector $\mathbf{x}(n)$.

Another distinguishing feature of the Kalman filter is that the *estimate* of the state vector is computed *recursively*, as shown by the formula:

$$\begin{bmatrix} \text{Estimate of} \\ \text{the state at} \\ \text{time } n+1 \end{bmatrix} = \begin{bmatrix} \text{old estimate of} \\ \text{the state at} \\ \text{time } n \end{bmatrix} + (\text{gain matrix})$$
$$\times \begin{pmatrix} \text{innovations vector} \\ \text{received at time } n \end{pmatrix}$$

Thus, as the filter receives a new piece of information, estimate of the state is updated in "real time" in light of the "old" estimate of the state and the *innovations* vector (comprising the new information) amplified by the so-called *Kalman gain* or *gain matrix* of the filter. It is this recursive computation that makes the Kalman filter so well suited for implementation on a digital computer.

Adaptive Filters

The design of Wiener filters and Kalman filters requires *a priori* knowledge of the second-order statistics of the data to be processed. These filters are therefore optimum in their individual ways only when the assumed statistical model of the data exactly matches the actual statistical characteristics of the input data. In most practical situations, however, we do not have the statistical parameters needed to design a Wiener filter or Kalman filter, in which case we have to resort to the use of an *adaptive filter*. By such a device we mean one that is *self-designing*, relying for its operation on a *recursive algorithm* that makes it possible for the filter to operate satisfactorily in an environment where knowledge of the relevant signal characteristics is not available. The algorithm starts from some prescribed *initial conditions* representing complete ignorance about the environment; thereafter it proceeds to adjust the free parameters of the filter in a step-by-step fashion, such that after each step the filter becomes more knowledgeable about its environment (Haykin, 1991; Widrow and Stearns, 1985). Consequently, an adaptive filter is *time-varying with data-dependent parameters*; this, in turn, means that an adaptive filter is in reality a *nonlinear device*, in that it does not obey the principle of superposition.

Notwithstanding this nonlinear property, adaptive filters are commonly classified into linear and nonlinear adaptive filters. In a *linear adaptive filter*, the estimate of a quantity of interest is computed at the filter output as a linear combination of the available set of observations applied to the filter input. The ubiquitous *least-mean-square (LMS) algorithm*, originated by Widrow and Hoff (1960), is an example of linear adaptive filtering. Indeed, the LMS algorithm is the workhorse of traditional forms of adaptive signal processing. Another example is the *recursive least-squares (RLS) algorithm*, which may be viewed as a special case of the Kalman filter. The LMS algorithm is *stochastic*, providing an approximation to the Wiener filter formulated in accordance with the method of steepest descent. Conversely, the RLS algorithm is *exact*, providing a recursive solution to the linear filtering problem formulated in accordance with the method of least squares that goes back to Gauss.

The LMS algorithm offers simplicity of implementation and a robust performance in tracking statistical variations of a nonstationary environment. The RLS algorithm, however, offers a faster rate of convergence (defined in a later section) than the LMS algorithm at the expense of increased computational complexity. Various schemes, known collectively as *fast algorithms* (Haykin, 1991), have been devised to improve the computational efficiency of RLS estimation. In any event, a major limitation of all linear adaptive filtering algorithms is the inability to exploit higher-order statistics of the input data, which, in turn, restricts the scope of their practical applications. In contrast, *nonlinear adaptive filters* involve the use of some form of nonlinearity, which makes it possible to extract the full information content of the input data. However, the presence of nonlinearity makes it difficult to analyze mathematically the behavior of nonlinear adaptive filters in a way comparable to their linear counterparts.

The issues of concern in the design of adaptive filters, be they linear or nonlinear, include the following:

- *Stability*, which imposes certain constraints on the learning-rate parameter involved in computing the adjustments applied to the free parameters of the filter from one iteration to the next.
- *Rate of convergence*, which refers to the number of iterations needed by the adaptive filtering algorithm to reach steady state.
- *Tracking* of statistical variations, which applies when operating in a nonstationary environment.
- *Robustness*, which refers to the ability of the adaptive filter to continue to perform satisfactorily under adverse operating conditions.
- *Structure*, which refers to the form of information flow in the adaptive filter.

Four Classes of Applications

The ability of an adaptive filter to operate satisfactorily in an unknown environment and to track time variations of input statistics makes it a powerful device for signal processing and control applications. In particular, we may identify four distinct classes of applications (Haykin, 1991):

(a) *Identification*, where the adaptive filter is used to provide a *model* that represents the best fit to an unknown plant (e.g., control system). In this class of applications, the unknown plant and the adaptive filter share a common input, and the output of the plant provides the desired response for the adaptive filter; both the input and output represent *histories over time*.

(b) *Inverse modeling*, where the function of the adaptive filter is to provide an *inverse model* connected in cascade with the unknown plant; in this case, some practical mechanism has to be found to supply the desired response. For example, in *adaptive equalization* of an unknown communication channel, a known sequence is transmitted over the channel whose output supplies the input signal to the equalizer (see NOISE CANCELING AND CHANNEL EQUALIZATION). During training, a replica of the transmitted sequence is generated in the receiver to supply the desired response for the equalizer. The purpose of the equalizer is to accommodate the highest possible rate of digital data transmission through the channel, subject to a specified reliability measured in terms of the *error rate* (i.e., average probability of symbol error). Typically, data transmission through the channel is limited by *intersymbol interference* (ISI) caused by dispersion in the communication system. Adaptive equalization provides a powerful method to control ISI and also to combat the effects of channel noise. Indeed, every modem (modulator-demodulator), designed to facilitate the transmission of computer data over a voice-grade telephone channel, employs some form of adaptive equalization (Qureshi, 1985).

(c) *Prediction*, where the function of the adaptive filter is to provide an optimum prediction of the current value of an input signal, given a past record of the input. In this application, the current value of the input signal serves as the desired response. The procedure described herein constitutes the basis of building a *model* for an unknown physical phenomenon that manifests itself by producing an observable time series. The difference between prediction and identification should be carefully noted. Prediction deals with a *single* time series produced by a phenomenon of interest, whereas system identification involves both the input and output signals of the system (plant) under study.

(d) *Interference cancellation*, where the adaptive filter is used to suppress an unknown interference contained (alongside an information-bearing signal of interest) in a *primary signal* that serves as the desired response. The input to the adaptive filter is represented by a *reference signal* that is dominated essentially by the unknown interference. To achieve these functional requirements, an interference cancellation system uses two separate sensors, one supplying the primary signal and the other designed specifically to supply the reference signal. *Adaptive beamforming* (Compton, 1988) is an example of interference cancellation; this operation is a form of spatial filtering, the purpose of which is to distinguish between the spatial properties of a target signal of interest and an interfering signal. Another example is that of *echo cancellation* (Sondhi and Berkley, 1980) used in long-distance telephone communications over a satellite channel; in this latter application, adaptive (temporal) filtering is used to synthesize a replica of the echo experienced on the satellite channel caused by unavoidable impedance mismatch at the receiving point and then to subtract it from the received signal.

For these applications, the adaptive filter may be linear or nonlinear. In practice, however, the use of a nonlinear adaptive filter improves on the performance attainable with a linear filter, particularly when the underlying mechanism responsible for generating the input data is inherently nonlinear.

Supervised and Self-Organized Forms of Error-Correction Learning

The design of an adaptive filter must include a mechanism for the provision of an error signal, the minimization of which in the mean-square sense is basic to the underlying learning process. In this context, we may distinguish two forms of error-correction learning, namely, *supervised learning* and *self-organized learning*. In supervised learning, the input signal and desired response originate from different sources (locations), whereas in self-organized learning, they originate from the same source. Examples of supervised learning include system identification, adaptive equalization, and interference cancellation as described in the preceding section. Conversely, adaptive prediction is an example of self-organized learning in that all the training examples are obtained from the incoming time series.

The need for self-organized learning also arises in certain applications where it is simply not feasible to have a desired response. A class of adaptive signal processing tasks that fall under this later category is that of *blind deconvolution* (Haykin, 1991, 1993). The need for blind deconvolution arises in a hands-

free telephone system, blind equalization, seismic deconvolution, and image restoration, just to mention a few examples. The use of a *hands-free telephone* is severely limited by the barrel effect caused by an acoustic reverberation component produced in the surrounding environment of the near-end talker. *Blind equalization* provides a practical method for dealing with anomalous propagation conditions that arise in line-of-sight digital radio because of natural phenomena. The *seismic deconvolution* problem is complicated by the fact that the exact waveform of the actual excitation responsible for the generation of the received signal is usually unknown. In *image restoration*, we have an unknown system that represents blurring effects caused by photographic or electronic imperfections or both. In all of these situations, the system of interest is unknown, and its input is inaccessible; hence, precise knowledge of the actual signal applied to the input of the system is not available for processing. To perform blind deconvolution, we are permitted to invoke reasonable assumptions about the statistics of the input signal. Most importantly, a solution of the blind deconvolution problem requires the use of nonlinear adaptive filtering (Haykin, 1991, 1993).

The Role of Neural Networks in Nonlinear Adaptive Filtering

Neural networks are well suited for the design of nonlinear adaptive filters by virtue of three inherent characteristics:

(a) Ability of the network to *learn* from training examples representative of the working environment and then generalize.

(b) Sigmoidal *nonlinearity* in the form of a logistic function or hyperbolic tangent function built into the model of each neuron in the network, making it possible for the network to account for nonlinear behavior of the physical phenomenon responsible for generating the input data.

(c) *Universal approximation*, whereby the network can approximate an input-output continuous mapping to any desired accuracy.

These characteristics are exemplified by a multilayer perceptron trained with the backpropagation algorithm. However, a multilayer perceptron in its conventional form is a static network and is therefore unable to respond to statistical variations of the environment. To remedy this deficiency, the learning process must account for *time* (i.e., nonstationary nature of the input) in its mechanization. This requirement may be satisfied in various ways (see Haykin, 1994, chap. 13). One network configuration that deserves to be mentioned here is the *finite-duration impulse response (FIR) multilayer perceptron* (Wan, 1994). In this discrete-time network, each synapse of a neuron is made up of a low-pass FIR filter, consisting of a tapped-delay-line with adjustable weights connected to its taps. To train the network, a temporal version of the backpropagation algorithm is used, the derivation of which achieves two important objectives (Wan, 1994):

(a) Preservation of symmetry between the forward propagation of states and the backward propagation of error terms.

(b) Improved computational efficiency by eliminating the redundant use of terms in the backpropagation learning process.

The FIR multilayer perceptron accounts for time by equipping the synapses of each neuron in the network with short-term memory. Another way of accounting for time in a neural network is to use *feedback*, exemplified by the *real-time recur-*

rent learning (RTRL) algorithm (Williams and Zipser, 1989). This algorithm provides *continuous learning* or *learning-on-the-fly*. A practical limitation of the RTRL, however, is that its computational complexity increases as N^4, where N is number of neurons in the network. We may overcome this limitation by using the *pipelined recurrent neural network (PRNN)*, whose design follows the *principle of divide and conquer:*

> *To solve a complex problem, break it into a set of simpler problems.*

According to Van Essen, Anderson, and Felleman (1992), this same principle is also reflected in the design of the brain, as summarized here in the context of the primate visual system:

- Separate modules are created for different subtasks, permitting the neural architecture to be optimized for particular types of computation.
- The same module is replicated several times over.
- A coordinated and efficient flow of information is maintained between the modules.

In a loose sense, all three elements of this principle feature in the PRNN, as described here:

- The PRNN is composed of separate modules, each designed to perform nonlinear adaptive filtering on an appropriately delayed version of the input signal.
- The modules are identical, each designed as a fully connected recurrent network with a single output neuron; the output of one module feeds the next.
- Information flow into and out of the modules proceeds in a synchronized fashion.

The PRNN is nonlinear, with infinite memory, by virtue of feedback built into the design of each module. Its synaptic weights are adjusted continuously by using an extension of the RTRL algorithm (Haykin and Li, 1993) while the filtering of the input signal is performed.

A useful application of the PRNN is described in Haykin and Li (1993), where it is employed as the nonlinear predictor for *adaptive differential pulse-code modulation* (ADPCM) of speech signals. Differential pulse-code modulation achieves a reduction in the bit rate of the transmitted data, compared with standard pulse-code modulation, by using prediction to remove redundant bits from the input data stream. A nonlinear ADPCM continues to perform satisfactorily at rates as low as 16 kilobits per second, providing a significant improvement over the traditional method.

Discussion

To sum up, much can be gained from the use of neural networks in the design of nonlinear adaptive filters for important signal processing applications in such diverse fields as communications, control, radar, sonar, astronomy, seismology, and biomedical engineering. In the material presented in this article, we focused on a class of neural networks well suited for system identification, inverse modeling, prediction, and interference cancellation, where nonlinearity is of paramount importance. We briefly mentioned the blind deconvolution problem that requires nonlinear adaptive filtering of a self-organized kind. Here, too, neural networks can be of great help, though it must be said that, in the literature, the use of neural networks for solving the blind deconvolution problem has not been featured as prominently. But, then, the whole subject of nonlinear adaptive filtering using neural networks is (relatively speaking) still in its infancy stage. For more information on adaptive signal processing applications of neural networks, see Haykin (1994, chap. 6, 7, 13, and 14). For more information on dynamic recurrent neural networks, their theory, and applications, see Giles, Kuhn, and Williams (1994) (see also RECURRENT NETWORKS: SUPERVISED LEARNING).

Road Map: Applications of Neural Networks
Background: I.3. Dynamics and Adaptation in Neural Networks
Related Reading: Adaptive Control: Neural Network Applications; Adaptive Filtering

References

Compton, R. T., Jr., 1988, *Adaptive Antennas: Concepts and Performance*, Englewood Cliffs, NJ: Prentice-Hall.

Giles, C. L., Kuhn, G. M., and Williams, R. J., 1994, Special issue on dynamic recurrent neural networks, *IEEE Trans. Neural Netw.*, 5: 153–337.

Haykin, S., 1991, *Adaptive Filter Theory*, 2nd ed., Englewood Cliffs, NJ: Prentice-Hall.

Haykin, S., Ed., 1993, *Blind Deconvolution*, Englewood Cliffs, NJ: Prentice-Hall.

Haykin, S., 1994, *Neural Networks: A Comprehensive Foundation*, New York: Macmillan.

Haykin, S., and Li, L., 1993, 16 kb/s adaptive differential pulse-code modulation of speech, in *Applications of Neural Networks to Telecommunications* (J. Alspector, R. Goodman, and T. X. Brown, Eds.), Hillsdale, NJ: Erlbaum, pp. 132–137.

Kalman, R. E., 1960, A new approach to linear filtering and prediction theory, *Trans. ASME J. Basic Eng.*, 83:95–108.

Qureshi, S. U. H., 1985, Adaptive equalization, *Proc. IEEE*, 73:1349–1387.

Sondhi, M., and Berkley, D. A., 1980, Silencing echoes in the telephone network, *Proc. IEEE*, 68:948–963.

Van Essen, D. C., Anderson, C. H., and Felleman, D. J., 1992, Information processing in the primate visual system: An integrated systems perspective, *Science*, 255:419–423.

Wan, E. A., 1994, Time series prediction by using a connectionist network with internal delay lines, in *Time Series Prediction: Forecasting the Future and Understanding the Past* (A. S. Weigend and N. A. Gershenfeld, Eds.), Reading, MA: Addison-Wesley, pp. 195–217.

Widrow, B., and Hoff, M. E., Jr., 1960, Adaptive switching circuits, *WESCON Convention Record*, 4:96–104.

Widrow, B., and Stearns, S. D., 1985, *Adaptive Signal Processing*, Englewood Cliffs, NJ: Prentice-Hall.

Wiener, N., 1949, *Extrapolation, Interpolation, and Smoothing of Stationary Time Series, with Engineering Applications*, Cambridge, MA: MIT Press.

Williams, R. J., and Zipser, D., 1989, A learning algorithm for continually running fully recurrent neural networks, *Neural Computat.*, 1: 270–280.

Analog VLSI for Neural Networks

Robert W. Newcomb and Jason D. Lohn

Introduction

One of the most promising strategies for implementing neural networks is through the use of electronic analog VLSI (Very Large Scale Integration) circuits. An analog circuit is one that processes a continuum of real-valued signals in continuous time, in contrast to a digital circuit which processes integer-valued (most often binary) signals in discretized time. VLSI refers to an integrated circuit design and manufacturing technology whereby hundreds of thousands to millions of active components (most often transistors) are placed on a chip on the order of 100 mm² in area and 0.5 mm thick. Because Artificial Neural Networks (ANNs) attempt to behave similarly to the brain with its millions of neurons, VLSI is the most appropriate presently available technology for their hardware implementations. Furthermore, both VLSI circuits and biological neurons are of the same class, that is, fundamentally analog.

Although during the 1980s digital ANNs held more interest, the first neural-like circuits were analog ones constructed by Dr. Otto Schmitt in the late 1930s using vacuum tube analog computer circuits (Schmitt, 1937). These were extended to rather cumbersome transistor circuits after the Second World War to obtain artificial neurons, in which much of the emphasis was placed on the initiation and propagation of action potentials. The circuits developed to accomplish these tasks relied on nonlinearities for implementing amplitude saturation, pulse repetition saturation, threshold effects, and dynamics for effecting time-domain changes on the action potentials (Reiss, 1964). But, because of the large size of the circuits used for just one neuron, very little was done to make full ANN systems until the advent of integrated circuits (ICs). In a number of research centers around the world during the mid-1960s, considerable interest began to develop in the design of analog IC neurons and systems built from them. Of importance to the signal processing capabilities in this development has been the recent em-

phasis on the synaptic combining of signals via weight-matrix summations, as opposed to the axon propagation of action potentials. Present analog ANNs consist of synaptic weights, implemented by amplifier gains; summation of the weighted signals, implemented by the use of Kirchhoff's laws (most conveniently the current law, denoted KCL); activation functions, realized by amplifier nonlinearities; and in many cases dynamics, via capacitors, for smoothly transitioning from an initial state to a desired equilibrium.

The workhorses of analog VLSI ANNS are the Differential Voltage Controlled Current Source (DVCCS) and capacitors. The DVCCs is used for making synaptic weights and activation functions, and capacitors are used for dynamics. A DVCCS takes a voltage difference as input, and gives an output current as a function of that difference. DVCCS gains can realize the weights when operating on small signals in a linear fashion and can also realize saturation nonlinearities when operating on large signals. In both cases, the DVCCS output currents can be conveniently summed by KCL. By using capacitors in conjunction with DVCCSs, any linear circuit can be realized (Bialko and Newcomb, 1971), so that any desired filtering of ANN signals is available. Along with the DVCCS and the capacitor, it is also convenient to have resistors for conversion of currents to voltage and voltage divisions, as well as devices for creating and scaling currents (called current sources and current mirrors, respectively). Except for passive resistors and capacitors (both of which are generally avoided in VLSI because of large area or nonideal characteristics), all of these devices can basically be constructed from VLSI transistors, which are discussed later in this article.

Overview of an Analog ANN Implementation

First we present a complete ANN analog circuit to give an overview of the circuits discussed in later sections. Figure 1

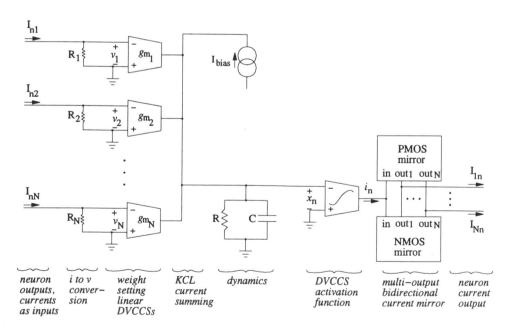

Figure 1. The *n*th neuron for a VLSI continuous-type ANN of *N* neurons.

shows an analog circuit suitable for the VLSI realization of the continuous-time neuron equations which, for the nth neuron of a set of N, are (for simplicity of notation we omit subscripts n on device parameters but not on the output currents and state)

$$C dx_n/dt + Gx_n = \sum_{i=1\ldots N} g_{mi} \cdot R_i I_{ni} + I_{\text{bias}} \quad n = 1, \ldots, N \quad (1a)$$

$$i_n = g(x_n) \quad (1b)$$

In Equation 1, $g(\cdot)$ is any of the activation functions available (see Equations 2–4), with i_n being its current output; x_n is the nth neuron's state variable; I_{ni} is the current output of the ith neuron, which is fed to the input of the nth neuron; the $g_{mi} \cdot R_i$ are the synaptic weights; and I_{bias} is the bias input. With reference to Figure 1 and Equation 1, it will be shown below that VLSI circuits can be constructed to make this Hopfield class of analog neural networks, as well as any other analog ANN, such as ART 2 (see ADAPTIVE RESONANCE THEORY), pulsed Hebbian, and biological mimics. The nonlinear function $g(\cdot)$ in Equation 1b can be realized via a DVCCS (see Figure 5) exhibiting square-law, exponential, or sigmoidal processing. In this simple model of a neuron, these nonlinearities can be thought to correspond to the activation processes in the cell body. The weighted inputs from the synapses to the cell body can be thought to correspond to varying amounts of currents linearly summing, via KCL, at the input node to the left of the activation function DVCCS in Figure 1. On the right side of Equation 1a the weights $g_{mi} \cdot R_i$ are the current gains of DVCCSs operating as linear amplifiers with resistor inputs, the resistors (of resistance R_i) being used to convert the neuron output currents to voltages (constructed using direct layouts, see Figure 3, or DVCCS connections, depending upon their Ohmic value). The bias input I_{bias}, also on the right of Equation 1a, is constructed as a constant current source made of a transistor. Using a resistor-capacitor branch connected to this same input node of the activation function amplifier, we obtain the dynamics of the analog circuit, shown on the left side of Equation 1a incorporating the derivative. Because each neuron output current i_n (which can be positive or negative) needs to be sent to each of the other neurons, it needs to be repeated N times, this being accomplished by the bidirectional current mirror (see Figure 4B) in multioutput form on the right of Figure 1. This simply reproduces N copies of the neuron output current i_n irrespective of what load is presented to it. For adjustments, as may be needed for adaptive ANNs, the transconductances g_{mi} can be made voltage variable by variation of the gain of the associated DVCCS (via the tail current introduced in Figure 5 below).

Transistors and VLSI Layouts

The key circuit component in analog VLSI is the transistor, a three- or four-terminal device that can behave as a switch in digital circuits and as an amplifier in analog ones. Transistors may be fabricated using a variety of technologies: BJT (bipolar junction transistor), MOS (metal oxide semiconductor), CMOS (complementary MOS), and others. For neural network implementations, MOS and BJT devices have been used the most; when both occur together, the process is called BiCMOS and is the most prevalent present-day analog VLSI technology.

Figure 2 shows the circuit symbols for those transistors of most interest to ANN VLSI, along with a top view of an IC layout of each. The fabrication details can be found in Geiger, Allen, and Strader (1990); however, for our purposes, it is enough to know only a few aspects of their operation.

In the MOS transistor the drain current, I_D, which flows from the outside into the drain, D, and then through the device to the source, S, is controlled by the voltage at the gate, G, with respect to the source, V_{GS}, when the latter is "above" threshold voltage, V_{th}. As the threshold voltage can be used as a fine control on the ANN weights, we note that it is dependent on the bulk-to-source voltage, V_{BS}, where the bulk material (B) is that of the substrate into which the transistor is embedded. The two types of MOS transistors, NMOS and PMOS, are distinguished by their conduction mechanisms, with the currents and voltages of the latter being ideally the negative of the former in the complementary case desired for CMOS fabrications. Since the channel can be formed by enhancing or depleting charge, we have enhancement- and depletion-mode transistors of each of the NMOS and PMOS types; the distinction is that the threshold voltages of depletion-mode devices are generally of opposite sign to those of enhancement-mode devices. Depletion-mode transistors are not as common in analog VLSI because of the extra fabrication steps needed, but they can be used to obtain more flexible designs. MOS transistors can be operated such that between the drain and source a resistor is seen whose value depends on the gate-to-source voltage, giving a voltage-variable resistor useful for adaptation. More commonly the MOS transistor is operated in its saturation mode where, instead of a resistor between drain and source, a current source is seen. This current source depends on the gate-to-source voltage in a square-law fashion, conveniently allowing for quadratic weights. By operating an MOS transistor at very low (subthreshold) gate-to-source voltages, exponential behavior is obtained; subthreshold operation is convenient for low power designs but is not too robust. In

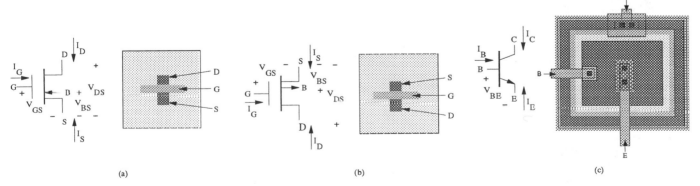

Figure 2. VLSI transistors and layouts. *A*, NMOS. *B*, PMOS. *C*, NPN BJT.

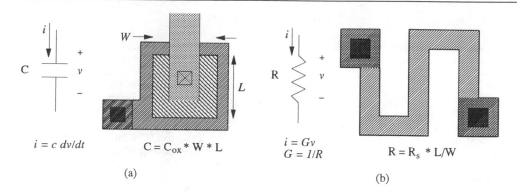

Figure 3. Passive components and layouts. *A*, MOS capacitor. *B*, Snake resistor.

$$i = c\, dv/dt$$

$$C = C_{ox} * W * L$$

$$i = Gv$$
$$G = 1/R$$

$$R = R_s * L/W$$

(a)   (b)

all of these cases, the drain current is proportional to the width-to-length ratio, W/L, of the channel; this acts as a design parameter that is very easily set in a VLSI layout. For bipolar transistors, exponential nonlinearities are obtained, and the VLSI design parameter is the emitter area to which the collector current is proportional. For more details on these devices, see Geiger et al. (1990).

Besides the active transistors, passive capacitors are used to obtain dynamics needed for realization of those ANNs that are described by differential equations, and for the derivatives needed for backpropagation. Several types of capacitors are available in VLSI. The primary one is realized by an oxide between two conductive layers (presently polysilicon, but possibly metal on top and doped silicon on the bottom), as shown in Figure 3A. These capacitors are linear time-invariant capacitors and satisfy $i = C\,dv/dt$ with capacitance $C = C_{ox}WL$ for which C_{ox} is the capacitance per unit area of the (gate) oxide used. The area, WL, of the polysilicon plate serves as a design constant. Unfortunately, to obtain capacitance of useful values requires considerable area, and, hence, capacitors often take up a good portion of analog VLSI neural networks. At times, one also needs linear resistors, for transformation of currents to voltages or for biasing, in which case the most common means of VLSI implementation is via strips of polysilicon, often in snake form to optimize layout (Figure 3B). The conductance G is given through the sheet resistance R_s (in ohms/square micron, a material constant) by $G = W/(LR_s)$, in which L is the length (distance between contact pads) and W is the width of the polysilicon. These resistors also take up considerable area and thus are avoided, but for small values of resistance they are sometimes invaluable (for values of 10 to 100 ohms). Larger-valued resistors are constructed using transistors.

Primary Circuits

Two of the key components in an ANN are the weights and the nonlinear activation functions. A weight can be realized by a DVCCS operating in its linear region, while a sigmoidal nonlinear activation function can be realized by operating a DVCCS over its full nonlinear range. We consider, as background, current sources, current mirrors, and resistors constructed as diode-connected transistors. These are all used for biasing the transistors, that is, setting the modes of operation of transistors, while the current mirrors and sources are used for various adjustments, as in adapting weights.

A current source can be constructed as the drain to source, of current I, of an MOS transistor operating in its saturation region (see Geiger et al., 1990, p. 49) with a voltage source of voltage V attached gate-to-source. We note that (1) the current I can be adjusted by varying the above-threshold voltage V; (2) current sources of one polarity are changed into ones

of opposite polarity by reversing the attachment points or by interchanging NMOS and PMOS; and (3) one needs to maintain the saturation mode of operation (by application of sufficient voltage across the current source nodes). If the transistor is operated in its ohmic region (see Geiger et al., 1990:49), with small V_{DS}, then the same circuit gives a voltage-variable resistor of conductance $G(V)$, which is useful for making small-area resistors (10–1000 ohms) as well as adaptive adjustments.

Figure 4A shows current mirrors which allow the current in one section of an ANN to determine that in another, perhaps for adjusting weights. These mirrors use a diode connection of one transistor to set the gate-source voltages for the input and output transistors to be equal. The current mirrors of Figure 4A allow current to flow in only one direction. However, by placing a P-mirror on top of an N-mirror, as in Figure 4B, we can get a bidirectional current mirror. By replacing all of the MOS devices by BJTs, similar BJT devices can be constructed. Furthermore, by placing several output transistors on one input transistor, multiple-output current mirrors are easily constructed, and these are of considerable use for distributing current in current-mode VLSI ANNs (as in Figure 1).

Figure 5 shows the basic configuration of a DVCCS. The tail current, I_T, is steered between I_1 and I_2 by the differential pair consisting of identical transistors T_1 and T_2 (of NMOS or NPN types), with the steering controlled by the voltage difference of the input voltages, $V_d = V_1 - V_2$. The difference of the transistor currents, $I_d = I_1 - I_2$, is designed to be a function of V_d and I_T, independent of any loads or the current mirror. To obtain the current output as this difference, the current mirror is used along with KCL at the output node so that $I_{out} = -I_d$. The function of I_{out} versus V_d realized depends upon the NMOS or NPN transistors and their modes of operation used to form the current difference. In all cases, the gates/bases are the leads to the left (in T_1) and right (in T_2), the drains/collectors are at the top, and the sources/emitters are at the bottom. In practice, there is some loading by whatever is attached, in which case other current mirrors are attached for isolation.

For possible nonlinearities of I_d versus V_d, there are several design alternatives. For NMOS we can obtain the sigmoidal function

$$I_d = (KW/L)[(2I_T/(KW/L)) - V_d^2]^{1/2}\,V_d \qquad V_d^2 < [I_T/(KW/L)] \tag{2}$$

which is linearized to

$$I_d = g_m V_d \qquad g_m = [2I_T(KW/L)]^{1/2} \tag{3}$$

Typical orders of magnitude are $O(I_T) = 10^{-3}$, $O(K) = 10^{-4}$, $10^{-2} < O(W/L) < 10^2$, giving $10^{-5} < O(g_m) < 10^{-2}$ over a

Figure 4. MOS current sources resistors and current mirrors. *A*, NMOS and PMOS unidirectional current mirrors, $I_{out} = ((L_2/W_2)/(L_1/W_1))I_{in}$. *B*, Bidirectional; current source.

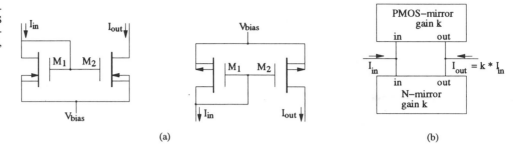

(a)   (b)

Figure 5. DVCCS. Basic configuration and circuit symbol, $T = M$ (when MOS) or Q (when BJT).

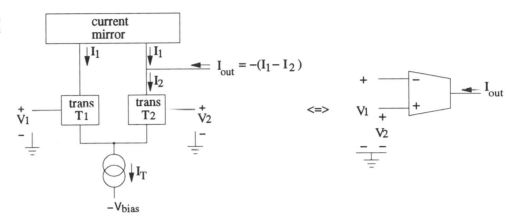

limited range of input V_d. For NPN transistors in Figure 5, or NMOS in the subthreshold range,

$$I_d = I_T \tanh[V_d/(2V_T)] \qquad (4)$$

where $V_T \approx 0.025$ is the thermal voltage at room temperature. This characteristic is quite nicely sigmoidal, leading to the BJT DVCCS having considerable importance for VLSI construction of ANN activation functions, especially for backpropagation circuits.

In some instances, it is necessary to convert the output of a DVCCS into a voltage (producing a Differential Voltage Controlled Voltage Source, DVCVS), as when voltage output for an activation function is desired. This can be accomplished by directing DVCCS output current into a resistor. However, one of the best ways to do this is to attach the gates of a CMOS pair to the output of the DVCCS. Since the CMOS pair allows no current at its input, the DVCCS can of course no longer act as a current source, but its output voltage is determined by other factors (specifically the channel-length modulation effect through the Early voltage). Other voltage amplifiers are available in the literature (Geiger et al., 1990), but high-gain operational amplifiers are not generally reasonable in VLSI for ANNs.

Since the DVCCS and the capacitor are sufficient to generate all linear circuits, we can construct many of the components of ANNs using them; see Kardontchik (1992) for filter examples. However, ANNs also require nonlinearities and, as we have seen, several nonlinearities are available, such as square-law and sigmoidal tanh ones. To build other nonlinearities, it is convenient to have multipliers, which can also be constructed from the DVCCS. An excellent multiplier is based upon the four-quadrant Gilbert multiplier (Geiger et al., 1990, p. 737), which uses two DVCCSs with the transistors of their tail current sources forming another differential pair. Assuming linear operation in the saturation region of the transistors and inputs V_x and V_y bounded by $V_x^2 \ll 2I_1(KW/L)$, $\ll 2I_2(KW/L)$, $V_y^2 \ll$

$2I_T(KW/L)$, with K, I_1 and I_2 as in Equation 2, the Gilbert multiplier gives $I_{out} \approx V_x V_y$. This Gilbert multiplier has successfully been used to multiply voltage-determined weights with neuron output voltages (Linares et al., 1993, p. 416). Dividers are possible but even less recommended than multipliers in VLSI.

Applications

Neural modeling implementations using analog VLSI have produced numerous systems. In this section, we briefly highlight some of these applications and invite the interested reader to learn more by way of the following references: Mead (1989), Zornetzer (1990), and Sánchez (1992, 1993). Other applications include constrained optimization (Tank and Hopfield, 1986), Fourier transform computations (Culhane, in El-Leithy and Newcomb, 1989), oscillators (Linares, in El-Leithy and Newcomb, 1989), Hebbian learning (Meador, Watola, and Nintunze, 1991), A/D conversion (Yuh, in Sánchez and Newcomb, 1993), data compression (Fang, in Sánchez and Newcomb, 1992), pattern recognition (Salam and Wang, 1991), and fuzzy controllers (Yamakawa, in Sánchez and Newcomb, 1993).

Analog VLSI circuits have been applied to the task of modeling neurophysiological phenomena (see SILICON NEURONS). One approach in which nerve cell characteristics were modeled in hardware is the silicon neuron of Mahowald and Douglas (1991). In the circuits comprising the silicon neuron, the neuron's ability to self-generate electrochemical impulses is emulated. Other approaches have been taken by Moon (in Sánchez and Newcomb, 1992). Circuit realizations for a set of low-level electrochemical processes occurring within synapses have also been constructed. Using dynamics derived from actual neurophysiological data, second-messenger chemical "pools" (Hartline, in El-Leithy and Newcomb, 1989) were simulated (Tsay, 1993) using VLSI analog multipliers and DVCCSs.

One of the best matches to date between analog VLSI circuitry and a biologically based application is the silicon retina of Mead (1989). This chip implements the first stages of invertebrate retinal processing and produces signals similar to those found in real retinas. Another silicon retina implementation which includes tuned pixels is discussed by Delbrück (in Sánchez and Newcomb, 1993).

ANN associative memories (such as the Hopfield net) store patterns in weights such that when a "noisy" pattern is presented, the complete pattern is produced from the memory. Boahen and his co-workers describe the implementation of a three-layer, 46-neuron heteroassociative memory (in El-Leithy and Newcomb, 1989). Using current-mode circuits operating in subthreshold conduction, the chip contains a regular array of cells, each cell containing two synapses and a 1-bit weight memory cell. Inverters are used for thresholding neurons, current sources are used in the bias circuit, and a multiplier circuit is used in the synapse. A class of adaptable associative memories can also be realized using DVCCSs, incorporating Gilbert multipliers for transconductance weights (Linares, in Sánchez and Newcomb, 1993).

Discussion

Compared to digital technology, analog VLSI offers the ANN world the distinct advantages of speed and real-time processing, though it suffers from relatively large size requirements and lack of standard cells. It also offers the ability to make continuous and speedy adjustments for adaptive neural networks and those needing efficient calculations of derivatives, as in backpropagation ANNs. Although the absolute error for analog components is typically larger than 5%, the relative precision can usually be controlled to be under 0.1% when implemented in VLSI. Roughly, this is the equivalent of 8-bit digital resolution at hundreds to thousands of mHz. When working with the primary circuits discussed here, such as the DVCCS and the current mirrors, voltage and current differences matter most, so that it is the relative tolerance that is critical. In any event, because ANNs are by conception fault tolerant, precision is not usually of concern.

VLSI neurons can have their dimensions comparable to those of biological neurons with considerably faster signal processing. However, real neurons take full advantage of their 3D nature, whereas most present-day VLSI structures are essentially planar. Although connection wires can be routed under other wires in multiple metal VLSI constructs, and there do exist some prototype 3D processes, the technology is still quite limited. Pulse-coded ANNs are amenable to a mixture of analog and digital realizations; the action potentials can be standardized and then realized by digital pulses, while the synaptic effects can be most conveniently realized by analog devices, since real-valued weights are involved.

A large number of other devices of interest to specialized areas of ANNs are presently available for VLSI. Among such devices are charge-coupled devices (CCDs), possibly for axon-like propagation or enzyme-effect mimicking, floating-gate devices for long-term storage of weights, and JFETs for less delicate fabrications. It should be noted that the MOS devices take minimal area but they are subject to damage by static charge that can puncture the very thin gate oxide. For the future, there are the very small resonant tunneling devices, which use a different substrate than the silicon of present VLSI, and molecular devices, which probably show the greatest long-range potential because of their minimal size and general signal handling.

Road Map: Implementation of Neural Networks
Background: I.1. Introducing the Neuron; I.3. Dynamics and Adaptation in Neural Networks
Related Reading: Silicon Neurons; Digital VLSI for Neural Networks

References

Bialko, M., and Newcomb, R. W., 1971, Generation of all finite linear circuits using the integrated DVCCS, *IEEE Trans. Circuit Theory*, CT-18(6):733–736.

El-Leithy, N., and Newcomb, R. W., Eds., 1989, Special issue on neural networks, *IEEE Trans. Circuits Sys.*, 36.

Geiger, R. L., Allen, P. E., and Strader, N. R., 1990, *VLSI Design Techniques for Analog and Digital Circuits*, New York: McGraw-Hill. ◆

Graf, H. P., and Jackel, L. D., 1989, Analog electronic neural network circuits, *IEEE Circuits Devices Mag.*, 5:44–55.

Kardontchik, J. E., 1992, *Introduction to the Design of Transconductor-Capacitor Filters*, Boston, MA: Kluwer.

Lee, B. W., and Sheu, B. J., 1991, *Hardware Annealing in Analog VLSI Neurocomputing*, Norwell, MA: Kluwer.

Linares, B., Sanchez, E., Rodriguez, A., and Huertas, J., 1993, A CMOS analog adaptive BAM with on-chip learning and weight refreshing, *IEEE Trans. Neural Netw.*, 4:445–455.

Mahowald, M., and Douglas, R., 1991, A silicon neuron, *Nature*, 354: 515–518.

Mead, C. A., 1989, *Analog VLSI and Neural Systems*, Reading, MA: Addison-Wesley. ◆

Meador, J., Watola, D., and Nintunze, N., 1991, VLSI implementation of a pulse Hebbian learning law, *Proc. 1991 IEEE Int. Sympos. Circuits Sys.*, Singapore, pp. 1287–1290.

Mueller, P., van der Spiegel, J., Blackman, D., et al., 1989, Design and fabrication of VLSI components for a general purpose analog neural computer, in *Analog VLSI Implementation of Neural Systems* (C. A. Mead and M. Ismail, Eds.), Boston: Kluwer.

Reiss, R. F. Ed., 1964, *Neural Theory and Modeling*, Stanford, CA: Stanford University Press.

Salam, F., and Wang, Y., 1991, A real-time experiment using a 50-neuron CMOS analog silicon chip with on-chip digital learning, *IEEE Trans. Neural Netw.*, 2:461–464.

Sánchez, E., and Newcomb, R. W., Eds., 1992, 1993, Special issues on neural network hardware, *IEEE Trans. Neural Netw.*, 3, 4.

Schmitt, O. H., 1937, An electrical theory of nerve impulse propagation and mechanical solution of the equations of nerve impulse propagation, *Am. J. Physiol.*, 119:399–400.

Tank, D. W., and Hopfield, J. J., 1986, Simple "neural" optimization networks: An A/D converter, signal decision circuit, and a linear programming circuit, *IEEE Trans. Circuits Sys.*, CAS-33:533–541.

Taylor, G. W., 1978, Subthreshold conduction in MOSFET's, *IEEE Trans. Electron Devices*, ED-3:337–350.

Tsay, S.-W., 1993, Implementation of Hartline pools and neural-type cells by VLSI circuits, in *Advances in Control Networks and Large-Scale Parallel Distributed Processing Models*, vol. 2, (M. Fraser, Ed.), Norwood, NJ: Ablex.

Zornetzer, S. F., Davis, J. L., and Lau, C., Eds., 1990, *An Introduction to Neural and Electronic Networks*, San Diego: Academic Press. ◆

Analogy-Based Reasoning

Dedre Gentner and Arthur B. Markman

Introduction

Analogy is important to connectionism by reasons of affinity and difference: *affinity* because analogical reasoning lays claim to connectionist advantages, like representation completion, similarity-based generalization, graceful degradation, and learning (Barnden, 1994); *difference* because analogy is a structure-sensitive process that involves the comparison of *systems of relations* between items in a domain, whereas connectionist models have concentrated on *correlational* representations that are sensitive to the statistical structure of the input. For both reasons, modeling analogy will be a key challenge for connectionist systems.

This chapter outlines four principles of analogy that have emerged from psychological studies. These serve to constrain models of comparison and can act as benchmarks for the development of connectionist models. After summarizing the principles, we discuss connectionist models of analogy. Our discussion takes place at Marr's *informational* and *algorithmic levels*, at which cognition is explained in terms of representations and associated processes. We will not evaluate the models in terms of brain functioning, not only because this seems premature, but because we believe a computational model should first justify itself as a cognitive account.

Four Tenets of Analogical Reasoning

Analogy is the perception of relational commonalities between domains that are dissimilar on the surface. These correspondences may suggest inferences about the target domain. Analogy has been widely studied in humans, and four general principles can be derived from this work. Analogy involves (1) structured pattern matching; (2) structured pattern completion; (3) a focus on common relational structure rather than on common object descriptions; and (4) flexibility in that (a) the same domain may yield different interpretations in different comparisons, and (b) a single comparison may yield multiple distinct interpretations. Any model of analogy must account for these phenomena.

Structured Pattern Matching

The defining characteristic of analogy is the alignment of relational structure. Alignment involves finding *structurally consistent* matches (those observing parallel connectivity and one-to-one correspondence). *Parallel connectivity* requires that matching relations have matching arguments; *one-to-one correspondence* limits any element in one representation to, at most, one matching element in the other representation (Gentner, 1989; Holyoak and Thagard, 1989). For example, when comparing the atom (a *target*) to the solar system (a *base*), the sun corresponds to the nucleus and the planets to the electrons because they play similar roles in a common relational structure: e.g., **revolve**(sun, planets) and **revolve**(nucleus, electron). The sun is not matched to both the nucleus and the electron, as that violates one-to-one correspondence. Another characteristic of analogy is *relational focus*: objects correspond by virtue of playing like roles and need not be similar (e.g., the nucleus need not be hot).

There is considerable evidence that people can align two situations, preserving connected systems of commonalities and

making the appropriate lower-order substitutions. For example, Clement and Gentner (1991) showed people analogous stories and asked them to state which of two assertions shared by base and target was most important to the match. Subjects chose the assertion connected to matching causal antecedents. Their choice was based on both the goodness of the local match and its connection to a larger matching system. This finding demonstrates that analogies seek *connected systems of relations* rather than isolated relational matches.

Effects of structure are also seen in perceptual comparisons. Lockhead and Pomerantz (1991) have reviewed studies demonstrating that higher-order perceptual relations, like symmetry, aid perceptual discriminations. Subjects could more easily discriminate between the stimuli) and (when they were presented in the context of a third identical element,), yielding the pair)) versus (). The added component introduced a higher-order symmetry relation in the right pattern, making the two patterns more discriminable. These configural effects are evidence for the role of relational structure in perceptual similarity.

Structured Pattern Completion

Analogical reasoning also involves the mapping of inferences from one domain to another. Thus, a partial representation of the target is completed based on its structural similarity to the base. For example, Clement and Gentner (1991) extended the findings described earlier by deleting some key matching facts from the target story and asking subjects to make a new prediction about the target based on the analogy with the base story. Consistent with the previous result, subjects mapped just those predicates that were causally connected to other matching predicates.

Flexibility (1): Same Term → Different Interpretations

Flexibility, a hallmark of connectionist models, is also a virtue of analogy. Barnden (1994) suggests that analogical reasoning may reconcile connectionism's flexibility with symbolic AI's rigidity. One way that analogical reasoning is flexible is that the same item can take part in many comparisons, and different aspects of its representation can participate in each comparison. Spellman and Holyoak (1992) compared politicians' analogies for the Gulf War. Some likened it to World War II, implying that the United States was acting to stop a tyrant, whereas others likened it to Vietnam, implying that the United States had entered into a potentially endless conflict between two other opponents. Different comparisons highlighted different features of the target, the Gulf War. Flexibility is also evident when the same base term is combined with different targets. The comparison statement, "The atom is like a solar system," conveys a central attractive force causing its satellites to revolve, whereas "The campfire is like a solar system" conveys a central source of heat and light.

Flexibility (2): Same Comparison → Multiple Interpretations

A second kind of flexibility is that a single base-target comparison may give rise to multiple distinct interpretations. For a comparison like "Cameras are like tape recorders," people can readily provide an object-level interpretation ("Both are small mechanical devices,") or a relational interpretation ("Both record events for later replay").

Despite this flexibility, people generally maintain structural consistency within an interpretation. In one study, Spellman and Holyoak (1992) asked subjects to map Operation Desert Storm onto World War II (WWII). They asked, "If Saddam Hussein corresponds to Hitler, who does George Bush correspond to?" Some subjects chose Franklin Delano Roosevelt, whereas others chose Winston Churchill. The key finding was that, when asked to make a further mapping for the United States in 1991, subjects chose structurally consistent correspondences. Those who mapped Bush to Roosevelt usually mapped the U.S.-1991 to the U.S.-during-WWII, and those who mapped Bush to Churchill mapped the U.S.-1991 to Britain-during-WWII.

An extreme case of conflicting interpretations is *cross-mapping*, in which the object similarities suggest different correspondences than do the relational similarities. For example, in the comparison between "Spot bit Fido" and "Fido bit Rover," Fido is cross-mapped. When presented with cross-mapped comparisons, people can compute both alignments. Research suggests that adding higher-order relational commonalities increases people's preference for the relational alignment (Gentner and Toupin, 1986; Markman and Gentner, 1993). This ability to compute relational interpretations (even for cross-mappings) is central to human analogizing.

This flexibility and the ability to process cross-mappings have significant implications for the comparison process because they mean that simulations cannot simply be trained to generate a particular kind of interpretation. Rather, the comparison process must be able to determine both object matches and structural matches and to attend selectively to one or the other.

Connectionism and Analogical Mapping

As connectionist techniques for structured representation have become available (Hinton, 1991; also see STRUCTURED CONNECTIONIST MODELS), several models of analogy have emerged. The first model, ACME (Holyoak and Thagard, 1989), is a localist system that combines the constraints of structural consistency, semantic matching (via a table of predicate similarities), and pragmatic centrality (by activating nodes related to goals and correspondences known in advance) in a constraint satisfaction network. The analogy's interpretation reflects the best solution satisfying these constraints. The interpretation need not maintain structural consistency, so spontaneous inferencing is not generally possible. Hummel et al. (1994) point out that the implementation of the pragmatic constraint often causes the important node(s) to map to everything in the other analog. Finally, because ACME settles on a single interpretation of an analogy, its solution to cross-mappings merges the object and relational interpretations.

To escape the difficulty with pragmatic bindings, Hummel et al. (1994) propose a distributed connectionist model—IMM—that makes use of temporal synchrony in unit firing to encode relations. In IMM, the connections between relations and their arguments are maintained by having individual units, which represent concepts, fire in phase with units that represent particular relational bindings (see STRUCTURED CONNECTIONIST MODELS). For example, to represent **kiss**(John, Mary), nodes for kiss, John, and *agent* fire in phase. Nodes for kiss, Mary, and *patient* also fire in phase. To map a base to target, trainable connections are set up between object nodes and relational binding nodes. As activation spreads through the system,

corresponding relations in the base and target begin to fire in phase as connections between analogically corresponding binding nodes and object nodes are strengthened. These connection strengths determine the corresponding items in base and target. IMM finds only a single interpretation of a comparison. If there are two competing interpretations, the model gives equal connection strengths to correspondences consistent with both interpretations. Thus, instead of distinguishing between two possible interpretations, the model finds a single merged interpretation that it considers to be weak.

A model of analogy has also been constructed using *tensor product representations* (Smolensky, 1991). In a tensor product, two vectors $\mathbf{X}$ and $\mathbf{Y}$ are bound by taking the outer product of these vectors, $\mathbf{YX}^\mathrm{T}$. The outer product normally forms a matrix, but a vector can be constructed from this matrix by concatenating its columns. Given $\mathbf{X}$, the vector $\mathbf{Y}$ can be obtained as $\mathbf{YX}^\mathrm{T}\mathbf{X}$ if $\mathbf{X}$ is a unit vector. Variable bindings can thus be captured by using one vector to represent a predicate and the other to represent its argument.

Tensor products have been used in a distributed connectionist model—STAR—that performs a:b::c:d analogies (Halford et al., in press). STAR represents binary relations ($\mathbf{R(a,b)}$) using tensor products of rank 3 (which are like the binary tensor products just described except that three vectors are bound together). In this model, long-term memory consists of a matrix of tensor products corresponding to various relations the system knows about. To process an analogy, the model takes the **a** and **b** terms and probes long-term memory to find a relation between them. It then takes this relation and the **c** term of the analogy and finds a fourth term that shares that relation with the **c** term. This model successfully uses a distributed connectionist representation to perform a one-relation analogical reasoning task. Presently, STAR cannot generate multiple distinct interpretations of a comparison. If the system knows many different items that could be the answer to the analogy, the output vector is a combination of them all. In addition, this model does not make use of higher-order relational structure to constrain its matches. Efforts are being made to remedy this problem.

Related Approaches

Other connectionist models have addressed some of the challenges that analogy poses to connectionism. The first challenge is the representation of higher-order relations. A promising approach to higher-order relations is holographic reduced representations (HRR) (Plate, 1994) whereby two vectors are associated using a circular convolution of an outer product matrix, yielding a vector with the same dimensionality as the original vector. A relation can be stored by adding a vector corresponding to the relation to convolutions of vectors representing the arguments of the relation and the relational roles they fill. This representation can be augmented by adding contextual information to the fillers. Hierarchical relational systems can be represented by using the vector corresponding to a whole relation as an argument to another relation. In this system, vectors corresponding to similar relations over similar objects have a high dot product. Another advantage is that vectors corresponding to similar relational structures are detected as similar, even if they contain cross-mappings.

A second challenge is structured pattern completion. Pattern matching and completion are central features of many connectionist models. In the simple, feature-based Brain State in a Box (BSB) model of Anderson et al. (1977), partial vectors are completed based on their similarity to learned vectors. It is not clear whether the pattern completion done in these models is the same as that done in analogical mapping. In many connectionist models, completion of a partial vector is based on the geometric projection of that vector onto the space of connection weights (i.e., the distance of the vector from an attractor state). This contrasts to the structured pattern completion common in analogical reasoning that we described earlier.

A third challenge is producing two or more distinct interpretations of a single analogy. Most current connectionist models of analogy yield one interpretation. Interestingly, models of the perception of the Necker cube have been able to maintain multiple stable states in the same network. In one localist model, units represent possible interpretations of corners of the Necker cube (Rumelhart et al., 1986). Excitatory connections link nodes that yield a consistent interpretation of the cube, and inhibitory connections link nodes for inconsistent interpretations. Given different random starting configurations, the system tends to settle in one of the two states representing consistent interpretations of the cube. This network displays two human-like features: (1) if one interpretation is significantly better than the other, it usually settles on the better interpretation; and (2) if multiple interpretations of the cube are equally good, its choice depends on its starting state.

Kawamoto and Anderson's (1985) model of the Necker cube is a distributed autoassociative system that learns the consistent interpretations of the cube. After learning, the network is allowed to settle on a stable state. This stable state is then *anti-learned* by subtracting the outer product matrix of the vector representing the stable state XX^T from the connection matrix (thereby flattening the basin of attraction). When the network is allowed to settle again, it settles on a different stable state. These techniques may be useful for capturing humans' ability to process multiple interpretations of a single comparison.

Discussion

Analogical reasoning relies heavily on correspondences in the structural relations of two domains. This leads to the seven hallmark phenomena, summarized in Table 1, that pose a challenge to connectionist models. No existing model exhibits the kind of structural alignment and mapping that people do. Three possibilities remain. First, connectionist models may

be ill-suited to model analogy and should be augmented by symbolic representations. Second, advances in models of structured representations and comparison may enable connectionist models to succeed in addressing the seven empirical challenges. Third, a hybrid connectionist/symbolic system like Barnden's *implementational connectionism* (Barnden, 1994) may be most appropriate. The empirical evidence described here will be helpful in constraining the development of these models.

Acknowledgment.   This work was supported by ONR grant N00014-89-J1272, awarded to Dedre Gentner and Kenneth Forbus.

Road Map: Connectionist Psychology
Background: Associative Networks
Related Reading: Semantic Networks

References

Anderson, J. A., Silverstein, J. W., Ritz, S. A., and Jones, R. S., 1977, Distinctive features, categorical perception and probability learning: Some applications of a neural model, *Psychol. Rev.*, 84:413–451.

Barnden, J. A., 1994, On using analogy to reconcile connections and symbols, in *Neural Networks for Knowledge Representation and Inference* (D. S. Levine and M. Aparicio, Eds.), Hillsdale, NJ: Erlbaum.

Clement, C. A., and Gentner, D., 1991, Systematicity as a selection constraint in analogical mapping, *Cognit. Sci.*, 15:89–132.

Gentner, D., 1989, The mechanisms of analogical learning, in *Similarity and Analogical Reasoning* (S. Vosniadou and A. Ortony, Eds.), New York: Cambridge University Press, pp. 199–241.

Gentner, D., and Toupin, C., 1986, Systematicity and surface similarity in the development of analogy, *Cognit. Sci.*, 10:277–300.

Halford, G. S., Wilson, W. H., Guo, J., Wiles, J., and Stewart, J. E. M, in press, Connectionist implications for processing capacity limitations in analogies, in *Advances in Connectionist and Neural Computation Theory*, vol. 2: *Analogical Connections*, (K. J. Holyoak and J. Barnden, Eds.), Norwood, NJ: Ablex.

Hinton, G. E., 1991, *Connectionist Symbol Processing*. Cambridge, MA: MIT Press. ◆

Holyoak, K. J., and Thagard, P., 1989, Analogical mapping by constraint satisfaction. *Cognit. Sci.*, 13:295–355.

Hummel, J. E., Burns, B., and Holyoak, K. J., 1994, Analogical mapping by dynamic binding: Preliminary investigations in *Advances in Connectionist and Neural Computation Theory*, vol. 2: *Analogical Connections* (K. J. Holyoak and J. A. Barnden, Eds.), Norwood, NJ: Ablex.

Kawamoto, A. H., and Anderson, J. A., 1985, A neural network model of multistable perception, *Acta Psychol.*, 59:35–65.

Lockhead, G. R., and Pomerantz, J. R., 1991, *The Perception of Structure*, Washington, DC: American Psychological Association. ◆

Markman, A. B., and Gentner, D., 1993, Structural alignment during similarity comparisons. *Cognit. Psychol.*, 23:431–467.

Plate, T. A., 1994, Estimating structural similarity by vector dot products of Holographic Reduced Representations, in *Advances in Neural Information Processing Systems 6* (J. D. Cowan, G. Tesauro, and J. Alspector, Eds.), San Mateo, CA: Morgan Kaufmann.

Rumelhart, D. E., Smolensky, P., McClelland, J. L., and Hinton, G. E., 1986, Schemata and sequential thought processes in PDP models, in *Parallel Distributed Processing: Explorations in the Microstructure of Cognition* (D. E. Rumelhart, J. L. McClelland, and PDP Research Group, Eds.), Cambridge, MA: MIT Press, pp. 7–57.

Smolensky, P., 1991, Tensor product variable binding and the representation of symbolic structures in connectionist systems, in *Connectionist Symbol Processing* (G. E. Hinton, Ed.), Cambridge, MA: MIT Press.

Spellman, B. A., and Holyoak, K. J., 1992, If Saddam is Hitler then who is George Bush? Analogical mapping between systems of social roles, *J. Pers. Soc. Psychol.*, 62:913–933.

Table 1. Seven Phenomena of Analogy

1. Structural consistency	Analogical mapping involves one-to-one correspondences and parallel connectivity.
2. Candidate inferences	Analogy inferences are generated via structured completion.
3. Systematicity	People prefer connected relations rather than collections of isolated relations.
4. Relational focus	Relational matches are made whether or not the objects making up the relations also match.
5. Flexibility (1)	Analogy allows multiple interpretations of a single item in different comparisons.
6. Flexibility (2)	Analogy allows multiple interpretations of a single comparison.
7. Cross-mapping	Though difficult, cross-mappings are generally interpreted relationally although the competing object similarities are perceived.

Applications of Neural Networks

Françoise Fogelman-Soulié

Introduction

Neural networks (NNs) can be used to solve a wide variety of problems found in varied fields of applications: classification, clustering, identification, diagnosis, signal or image analysis, structure prediction, optimization, and so on. There exist other techniques as well (e.g., pattern recognition and statistics, data analysis, mathematical programming, expert systems) to solve such problems. However, NNs have proved easy to implement and extremely efficient, especially in those areas where explicit knowledge of the problem is not available, data are very noisy, or the problem could change in time. These advantages have recently led to the deployment of NNs in many different industrial domains.

Most NN applications developed so far use a very restricted set of algorithms: multilayer networks (or multilayer perceptrons, MLP) and RADIAL BASIS FUNCTION NETWORKS (RBF) with the BACKPROPAGATION learning algorithm, WINNER-TAKE-ALL MECHANISMS with their learning algorithms of LEARNING VECTOR QUANTIZATION (LVQ), and topological maps (see SELF-ORGANIZING FEATURE MAPS) probably account for more than 90% of all applications, mostly in the supervised learning mode for classification, identification, diagnosis, and prediction. Although early developments of meaningful applications can be traced back to the 1950s, the introduction of NNs into industry on a large scale started in the mid-1980s with fully connected MLPs. By the end of the 1980s, scientists started to incorporate knowledge about the problem into their NNs (e.g., structured MLPs). Finally, in the last few years, it has been realized that systems with only one network are insufficient for complex applications; for real-size applications, multimodular, hybrid systems are required. NNs have proved fairly easy to integrate into complex information processing chains, in comparison to, for example, conventional artificial intelligence techniques, where knowledge representation schemes at the symbolic level do not merge easily within systems handling essentially numerical data.

The Methods

Developing an application which involves an NN requires the same methodology and ingenuity as usual in engineering: one must clearly identify and analyze the application, gather data, evaluate costs, and check the statistical validity of results. Usually, the analysis phase will end up with an architecture composed of successive modules (typically data acquisition, preprocessing, low-level processing, and high-level integration for final decision: see Figure 1). One should not expect a unique NN to be capable of handling all of these modules: a crucial matter for the success of any development lies in the designer's capacity to clearly identify the different modules as well as the optimal technique to be implemented in each one. No automatic decomposition is available today, and, in that respect,

NNs are no different from other techniques; however, recent work on "competing experts" (Jacobs et al., 1991) could lead to some automatization of the decomposition of a task (see MODULAR AND HIERARCHICAL LEARNING SYSTEMS). The analysis phase is indeed crucial; in addition to providing the usual benefits of modularity, it aims at reducing the costs: development costs (some modules can be reutilized from previous developments), computing time (decomposition into modules leads to smaller NNs, faster to train than larger ones), and required number of examples (smaller NNs require fewer examples to train to a given level of performance).

Sometimes, feedback between tasks may be necessary to recover from early errors: e.g., recognition results might hint at early segmentation errors. Such feedback is usually hard to build into a complex application. An interesting issue here is the capacity to cooperatively optimize the different modules, i.e., allow some feedback, so that the overall architecture will be globally optimal. NNs might be one simpler and more efficient way to handle such a problem, since there exists a global training algorithm (Bottou and Gallinari, 1991).

When trying out an NN implementation of one module in an application, one should always compare its performances to those obtained with a reference conventional technique, e.g., a linear classifier in a classification problem. There are many instances of applications in which adequately preprocessed data can be further analyzed by conventional techniques, with both speed and accuracy. NNs should not be assumed to be systematically better than other techniques; in fact, experience shows that, on most problems, NNs happen to be just as good as—but no better than—your usual favorite technique. However, on problems where no solution is known yet, where the only existing solutions are extremely simple (e.g., linear models), where continuous adaptation is needed or execution speed is required (e.g., through an NN chip), NNs have indeed proved most valuable. These are precisely the problems for which one should use NNs for maximal benefit.

Application Domains

Problem Areas Where NNs Can Be Used

Neural techniques can deal with problems falling into the following four main categories.

1. *Classification, diagnosis.* A pattern is presented to the network, which has to decide to which one of a predefined set of classes that pattern belongs. This situation typically occurs in pattern recognition problems.
2. *Function approximation.* The network has to associate to the input pattern a value, close to that computed by an underlying, unknown function representing the "real" phenomenon. This problem will be found in time-series prediction, identification, and process control.

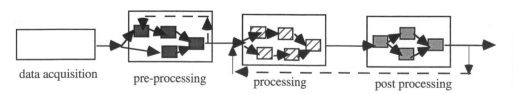

data acquisition   pre-processing   processing   post processing

Figure 1. Decomposition of a problem into tasks (here, e.g., data acquisition, preprocessing, processing, and post-processing). Possible feedback between tasks is shown by dashed arrows.

Table 1. The NN Algorithms of Choice for Each Class of Problem

Type of Problem	Type of Learning		
	Supervised	Unsupervised	None
Classification, diagnosis	Adaline, perceptron, MLP, TDNN, LVQ, RBF, Boltzmann machine	TMaps and labeling	
Function approximation	MLP, RBF, recurrent network		
Compression, quantization, feature extraction	MLP, LVQ, RCE	MLP, TMaps, ART	
Optimization			Hopfield, mean field

Abbreviations as follows: MLP (Multilayer Perceptron), TDNN (Time-Delay Neural Network), LVQ (Learning Vector Quantization), RBF (Radial Basis Function), TMaps (Topological Maps), RCE (Reduced Coulomb Energy), ART (Adaptive Resonance Theory). All these algorithms are described in this *Handbook*.

3. *Compression, feature extraction, quantization.* In most applications, raw data, such as those coming from sensors, are of high dimensionality and noisy. Neural nets can be used to reduce dimensionality and noise, either by numerical compression or through extraction of relevant features. Quantization, both supervised and unsupervised, can also be used to represent the data through a limited set of references.

4. *Optimization.* NNs can be set so that their weights embody both the objective function to be optimized and the constraints on the variables. A solution (usually approximate) is then obtained as the limit state of the network dynamics.

Depending on the problems, different networks have to be used; the most common choices are shown in Table 1. They are motivated by the bulk of experiments which have now been run on different applications.

Application Domains

NNs are now being used in many industrial domains. Descriptions can be found in proceedings of specialized meetings (of the various scientific societies, e.g., INNS, ENNS, or IEEE); however, it should be stressed that, quite often, companies tend to keep their NN methods secret, to avoid leaking their advantages to competitors. Precise details about successful applications are thus rather scarce. Some specialized newsletters are good sources for information about industrial NN applications and patents; see, for example, Rosenfeld (1993).

The main application domains where NNs have been used are the following:

- *Military sector:* signal processing (radar, sonar, infrared) for target or source identification, data fusion, image analysis for target localization and identification.
- *Image processing:* object recognition, scene analysis, satellite image analysis, teledetection.
- *Telecommunications:* routing and control, coding, filtering, channel assignment/equalization.
- *Speech:* speech recognition, speaker identification.
- *Manufacturing:* prediction of energy consumption (water, electricity), nondestructive testing, quality control.
- *Process control:* identification and control of industrial processes.
- *Robotics:* autonomous vehicles, robot-arm control.
- *Biology and medicine:* image analysis, signal processing, automatic diagnosis, genome analysis.
- *Oil:* seismic image analysis, monitoring of well productivity.

- *High-energy physics:* detection of events, triggers, analysis of particle trajectories.
- *Tertiary sector:* automatic reading of texts (printed, handwritten), signature authentification, fraud detection, marketing.
- *Finance:* prediction of stock prices, currency exchange rates, economic indices; risk analysis.

In the next few sections, I will review the most significant developments for various application domains, and give a few examples of applications, to present the reader with typical methods presently used by developers. For more detailed descriptions, the reader should refer to specific articles elsewhere in this *Handbook*.

There are also applications of NNs in biology, where scientists try to model particular subsystems so as to better understand them, make predictions about their behaviors, or even suggest experiments. We will not discuss such applications here. These are described in articles throughout the *Handbook*.

Signal Processing

Overview

NNs have been used in various areas of signal processing, at either the theoretical or the application level. In his pioneering work in the 1960s, Widrow developed the adaline algorithm to train an adaptive equalizer (used in modems, which probably makes it the most commonly used NN) (Widrow and Winter, 1988; see also NOISE CANCELING AND CHANNEL EQUALIZATION). Recent work has established a systematic comparison of NN algorithms and ADAPTIVE FILTERING (Nerrand et al., 1993): a general framework is given which includes algorithms for training NNs and for estimating filter parameters. Within that framework, NNs appear as adaptive nonlinear filters, non-recursive (if the NN is feedforward) or recursive (when the NN is recurrent). The capacity for adaptivity (i.e., continually training along time) and nonlinearity indeed provides NNs with improved performances with respect to conventional linear filters.

At the application level, there exists a wide variety of contributions. Major areas were noted in the earlier section on application domains.

Time-Series Prediction

A linear multilayer network with no hidden layer is equivalent to an Auto-Regressive (AR) model (where signal at time *t*,

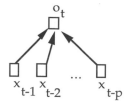
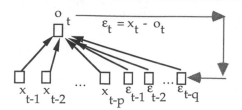

Figure 2. Linear neural network equivalent to an AR (left) or an ARMA (right) process: x_t is estimated by o_t, a linear combination of past values x_{t-i} (AR), or of past values x_{t-i} and past noise ε_{t-i} (ARMA).

Figure 3. Multimodular architecture for time-series analysis.

x_t, is a function of past values x_{t-i}, $i = 1 \ldots p$) or an Auto-Regressive Moving Average (ARMA) model (where x_t is a function of both past values and noise ε_{t-i}, $i = 1 \ldots q$) (Figure 2). One could hope that using nonlinearities and hidden layers in the NN would automatically improve upon the performances of AR or ARMA models. This is not so. Conventional techniques usually incorporate some preprocessing of the time series before performing the actual prediction: for example, differentiation (to obtain a stationary process) or trend elimination. Detailed experiments have been run on various well-known time series to demonstrate that such preprocessing could also be used to enhance NN performances, and that, in such a case, an NN was indeed better than ARMA models (see FORECASTING). We describe below an example of such results (Kouam, Badran, and Thiria, 1992), which could be used to process real-world time series (e.g., in finance).

The method described below has been tested on the classical international air-travel time series (Kouam, Badran, and Thiria, 1992), for the years 1949 to 1960. Conventional techniques led to an ARMA model with a preprocessing of the form

$$y_t = \nabla_{12} \nabla_1 \log x_t \qquad \text{with } \nabla_i z_t = z_t - z_{t-i}$$

where x_t is the number of travelers of year t and ∇_i is a shift operator which works on any series z_t as shown.

The trend—or smoothed evolution of the series—can be estimated by conventional techniques (Box and Jenkins, 1970) or by an NN trained to perform smoothing (Bishop, 1990). Multilayer networks were used (Kouam et al., 1992) and compared to ARMA after various types of preprocessing.

A global Multimodular Architecture (MMA; see Figure 3) was designed, by integrating three successive networks, to estimate the trend (denoted by s in the figure) and to predict. The first network is trained to estimate the trend $\hat{s}_j$ of the time series

X_j as a function of values $X_{j-d}, \ldots, X_j, X_{j+1}, \ldots, X_{j+d}$ (where $2d$ is the size of the so-called smoothing window). A new time series Y_j is thus obtained from X_j by subtracting the estimated trend $\hat{s}_j$. Then prediction network 1 is trained to predict Y_j one-step-ahead. Prediction network 2 is trained to predict X_j one-step-ahead and reiterated, thus generating predictions $\hat{X}_{j+1}, \ldots, \hat{X}_{j+d}$.

When the system is used for prediction, past observations $X_{j-d}, \ldots, X_j$ are passed through prediction network 2 generating predictions $(\hat{X}_{j+1}, \ldots, \hat{X}_{j+d})$; then $(X_{j-d}, \ldots, X_j)$ and $(\hat{X}_{j+1}, \ldots, \hat{X}_{j+d})$ are used by the smoothing network to produce $\hat{s}_{j+1}$, and finally prediction network 1 produces $\hat{Y}_{j+1}$. Predictions produced by prediction network 2 are not very accurate (since prediction k-steps-ahead, $k = 1 \ldots d$, is obtained from estimated past values, because of reiteration), but they are sufficient for the smoothing network to produce a good estimate of the trend. All networks are MLPs with one hidden layer, fully connected, trained by backpropagation.

Results (Kouam et al., 1992) are shown in Table 2, where MLP x-y-z represents a multilayer network with an input (respectively, hidden, output) layer of x (respectively, y, z) neurons, all layers being fully connected. Performances are measured by the arv (average relative variance) defined (Weigend, Rumelhart, and Huberman, 1991) as

$$\text{arv}(S) = \frac{\sum_{t \in S}(x_t - o_t)^2}{\sum_{t \in S}(x_t - m)^2} = \frac{\sum_{t \in S}(x_t - o_t)^2}{N\sigma^2}$$

where S is the learning set or the test set (of size N), o_t is the computed output when the input is x_t, m is the estimated mean, and σ^2 is the estimated variance of the time series.

These results indeed show that using an NN can improve upon a conventional model, but only after careful preprocessing of the data presented to the network: a simple MLP 12-5-1

Table 2. Time-Series Prediction with ARMA Model and MLPs, for Various Types of Preprocessing

International Air Travelers Traffic	Arv on Learning Set	Arv on Test Set
ARMA	0.0064	0.0066
MLP 12-5-1 on $\log X_t$	0.015 ± 0.0012	0.017 ± 0.0017
MLP 13-5-1 on $\nabla_1 \log X_t$	0.0068 ± 0.00068	0.0071 ± 0.00063
MLP 24-5-1 on $\nabla_{12} \nabla_1 \log X_t$	0.0052 ± 0.00083	0.0054 ± 0.00060
MMA	0.0048 ± 0.00041	0.0049 ± 0.00047

Note: ∇_i is the shift operator by i years.

on log X_t has an arv three times larger than the ARMA model, and it is only the MMA (see Figure 3) which produces a smaller arv.

TDNN architecture (except from the output layer, which now has one neuron per person to be identified) is then trained to identify faces; the learning set has 150 images/person, for 10 different persons. The overall system was tested on 120 new ons. Further work on larger ., 1993) led to a system which 'C-486, with 8% nondetection)y the system) and 1% identi- iose computed identity is not)wn faces.

)ecause of their capability to linear transformation (Hertz, 50 KOLMOGOROV'S THEOREM);)erform identification and/or Applications are varied: in stacle avoidance, direct and ng), in autonomous vehicles rial process control. Narendra : a general framework for NN)ook edited by Miller, Sutton, irious applications (see also TWORK APPLICATIONS). re of the NN approach is the hich integrates an NN for the another NN for the control gether. An NN controller can ontinuously in time, allowing roccss modifications. This fea- itrol, where machines can de-)ehavior to meet new require-

Arbib, *The Handbook of Brain Theory and Neural Networks*

The following figure should appear on p. 96:

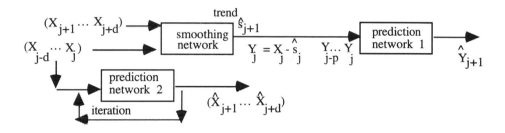

Figure 3. Multimodular architecture for time-series analysis.

Ns have already demonstrated theoretical ideas are presently l lead to significant improve-

1992) provides means to inte- cial intelligence systems, over- i sometimes undesirable char- r cannot easily be understood ter applications of NNs in the nilitary) where explanations of interesting line of work is the s for improved performance.)ed *committee* or *cooperation*, in statistics.

recognition

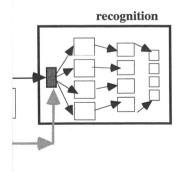

Active investigation in statistics and the theory of regularization will certainly produce more efficient NNs in the near future. In a recent competition for energy prediction, MacKay, the prize winner, used a Bayesian regularization approach (MacKay, 1994), combined with committee ideas. Conferences have been organized to bring the ideas and methods of statistics into the NN community, and vice versa (Cherkassky, Friedman, and Wechsler, 1994). (See AVERAGING/MODULAR TECHNIQUES FOR NEURAL NETWORKS; BAYESIAN METHODS FOR SUPERVISED NEURAL NETWORKS; and GENERALIZATION AND REGULARIZATION IN NONLINEAR LEARNING SYSTEMS.)

These are but a few of the promising avenues presently being explored in the NN community. Certainly, exploitation of such research will produce, in the near future, even more successful NN applications.

Road Map: Applications of Neural Networks
Background: I.3. Dynamics and Adaptation in Neural Networks
Related Reading: See the Road Map

References

Bishop, C., 1990, Curvature driven smoothing in backpropagation neural networks, in *International Neural Network Conference (INNC'90)*, vol. 2, Paris: Kluwer, pp. 749–752.

Bottou, L. Y., and Gallinari, P., 1991, A framework for the cooperation of learning algorithms, in *Neural Information Processing Systems* (R. P. Lippmann, J. E. Moody, and D. S. Touretzky, Eds.), vol. 3, San Mateo, CA: Morgan Kaufmann, pp. 781–788.

Box, G., and Jenkins, G., 1970. *Time Series Analysis: Forecasting and Control*, San Francisco: Holden-Day.

Cherkassky, V., Friedman, J. H., and Wechsler, H. Eds., 1994, *From Statistics to Neural Networks: Theory and Pattern Recognition Applications*, NATO ASI Series F, vol. 136, New York: Springer-Verlag. ◆

Fogelman-Soulié, F., Lamy, B., and Viennet, E., 1993, Multi-modular neural network architectures for pattern recognition: Applications in optical character recognition and human face recognition, in *Advances in Pattern Recognition Using Neural Network Technologies* (I. Guyon and P. Wang, Eds.), River Edge, NJ: World Scientific.

Goodman, R. M., Higgins, C. M., Miller, J. W., and Smyth, P., 1992, Rule-based neural networks for classification and probability estimation, *Neural Comput.*, 4:781–804.

Hertz, J., Krogh, A., and Palmer, R. G., 1991, *Introduction to the Theory of Neural Computation*, Lecture Notes of the Santa Fe Institute Studies in the Sciences of Complexity, vol. 1, Reading, MA: Addison-Wesley. ◆

Jacobs, R. A., Jordan, M. I., Nowlan, S. J., and Hinton, G. E., 1991, Adaptive mixtures of competing experts, *Neural Comput.*, 3:79–87.

Kouam, A., Badran, F., and Thiria, S., 1992, Approche méthodologique pour l'étude de la prévision à l'aide de réseaux de neurones, *Neuro-Nîmes '92*, EC2 ed., 117–127.

MacKay, D. J. C., 1994, Bayesian nonlinear modeling for the 1993 energy prediction competition, in *Maximum Entropy and Bayesian Methods: Santa Barbara 1993* (G. Heidbreder, Ed.), Dordrecht: Kluwer.

Matan, O., Burges, C. J. C., LeCun, Y., and Denker, J. S., 1992, Multi-digit recognition using a space displacement neural network, in *Neural Information Processing Systems* (J. E. Moody, S. J. Hanson, and R. P. Lippmann, Eds.), vol. 4, San Mateo, CA: Morgan Kaufmann, pp. 488–495.

Miller, W. T., Sutton, R. S., and Werbos, P. J., Eds., 1990, *Neural Networks for Control*, Cambridge, MA: MIT Press. ◆

Narendra, K. S., and Parthasarathy, K., 1990, Identification and control of dynamical systems using neural networks, *IEEE Trans. Neural Netw.*, 1:4–27.

Nerrand, O., Roussel-Ragot, P., Personnaz, L., Dreyfus, G., and Marcos, S., 1993, Neural networks and nonlinear adaptive filtering: Unifying concepts and new algorithms, *Neural Comput.*, 5:165–199.

Rosenfeld, E., Ed., 1993, *Intelligence* (a periodical). To subscribe, call 800-NEURALS, or send e-mail to ier@aol.com.

Weigend, A. S., and Rumelhart, D. E., and Huberman, B. A., 1991, Generalization by weight elimination with application to forecasting, in *Neural Information Processing Systems* (R. P. Lippmann, J. E. Moody, and D. S. Touretzky, Eds.), vol. 3, San Mateo, CA: Morgan Kaufmann, pp. 875–882.

Widrow, B., and Winter, R. G., 1988, Neural nets for adaptive filtering and adaptive pattern recognition, *IEEE Comput.*, 21:25–39.

Artificial Intelligence and Neural Networks

John A. Barnden

Introduction

This article surveys the distinctions between symbolic artificial intelligence (AI) systems and neural networks (NNs), their relative advantages, some differences of purpose, and ways of attempting to bridge the gap between the two. Such a review is hampered by the great variety of aims and approaches that are covered by the terms *artificial intelligence* and *neural networks* (or *connectionism*). We will therefore start with some brief comments on AI. (For an introductory textbook on AI, see for example Rich and Knight, 1991.)

It is notoriously difficult to define AI, partly because there is no clear definition of human intelligence, let alone artificial cousins of it. For this review, we can take AI to consist of the development, analysis, and simulation of computationally detailed, efficient systems for performing complex tasks, where the tasks are broadly defined, involve considerable flexibility and variety, and are typically similar to aspects of human cognition or perception. These broad tasks include, for instance: natural language understanding and generation; expert problem solving; common-sense reasoning; visual scene analysis; action planning; and learning. Notice that our description of AI does not prevent the studied systems from being psychological models, but AI researchers typically do not make strong claims of psychological realism for their systems.

There is nothing in the above description of AI that prevents the computational systems from being neural networks. Nevertheless, it is fair to say that the bulk of AI can be called "traditional" or "symbolic" AI, relying on computation over symbolic structures (e.g., logic formulae). The rest of the review will therefore discuss relationships between symbolic AI and NNs.

Further relevant discussion and information can be found in Sun (1994) and various other parts of this *Handbook*. In addition to the *Handbook* sections cited throughout this article, see CONNECTIONIST AND SYMBOLIC REPRESENTATIONS; DISTRIBUTED ARTIFICIAL INTELLIGENCE; and SEMANTIC NETWORKS.

Relative Advantages

Advantages of Neural Networks

One of the main benefits claimed for NNs is *graceful degradation*. A computational system is said to exhibit graceful degradation when it can tolerate imperfect input or significant corruption of its internal workings. The toleration consists of the system's continuing to perform usefully, though not necessarily perfectly. Imperfect input is input that is corrupted, excessive, or deficient.

It is reasonably claimed that neural networks, especially "distributed" ones, often exhibit graceful degradation. (A neural network is distributed, with respect to a given level of conceptual description of a domain, if the concepts at that level are represented by patterns of activation rather than activations at single units, and the patterns for different concepts can overlap in the units they use.) Input corruption takes the form of noise in an input activation vector. Excessive or deficient input can arise when a training set has, respectively, too many or too few members. Internal corruption usually takes the form of deletions of nodes or links or corruptions of the link weights.

On the other hand, symbolic AI systems tend not to degrade gracefully. For definiteness, consider a simple rule-based system. A small corruption of an input data structure is likely to make it fail to match the precise form expected by the rules which would have applied to the uncorrupted data structure, so that those rules totally fail to be enabled. Equally, other rules might erroneously be enabled. Similarly, even minor damage to some component of a rule can have very large effects on how the system operates.

NNs often exhibit a *pattern completion* property. An incomplete pattern of activation on an input bank of units can lead to a known completion of the pattern appearing on that same bank or on an output bank. NNs are also widely noted for *automatic similarity-based generalization*, whereby previously unseen inputs that are sufficiently similar to training inputs lead to behavior that is usefully similar to (or captures central tendencies in) the behavior elicited by the training inputs. Clearly, pattern completion and similarity-based generalization can be viewed as types of graceful degradation.

A strongly related property of many NNs is their ability to *learn* generalizations or category prototypes by exposure to instances, by virtue of fairly straightforward, uniform weight-modification procedures. These generalizations or prototypes are implicit in the weights. Although learning is intensively studied in symbolic AI, it is fair to say that certain types of learning are more naturally supported by NNs.

NNs can have (merely) *emergent rule-like behavior*. Such behavior can be described, approximately at least, as the result of following symbolic rules, even though the system exhibiting the behavior does not operate by means of the interpretation of explicit rules (Rumelhart and McClelland, 1986; Smolensky, 1988). Emergent rules are a central issue in the application of neural networks to high-level cognitive tasks, and the topic is pursued further in LINGUISTIC MORPHOLOGY. More generally, NNs tend to be more sensitive to subtle contextual effects than symbolic AI systems are, because multiple sources of information can more easily be brought to bear in a gracefully interacting and parallel way.

Many NNs allow efficient *content-based access* (or *associative access*) to long-term memory, in two different senses. First, let us assume, as usual, that a neural net's long-term memory is its set of weights. The manipulation of an input vector by the network can be thought of as the bringing to bear of particular content-relevant long-term memories on that vector. Secondly, in any NN that learns a map from particular inputs to particular outputs, an output can be thought of as a particular long-term memory recalled directly on the basis of the content of the input. Content-based access is not so easily provided in symbolic systems, as implemented on conventional computers, although it can be obtained to some useful degree by sophisticated indexing schemes (see, e.g., Bonissone et al. in Barnden and Holyoak, 1994), associative computer memories (e.g., Hwang and Briggs, 1984), or hashing (see Touretzky in Hinton, 1991, for discussion).

NNs facilitate *soft constraint satisfaction*. That is, it is possible to arrange for some hypotheses to compete and cooperate with each other, gradually influencing each other's levels of confidence until a stable set of hypotheses is found. Each hypothesis is represented by a node or group of nodes in the neural network, and the constraints are encoded by links joining those nodes or groups. The constraint satisfaction is soft because no individual constraint absolutely needs to be satisfied. By contrast, although many symbolic AI systems are designed to do constraint satisfaction, the symbolic framework provides no special support for it, particularly when the constraints are soft.

Advantages of Symbolic AI Systems

The symbolic framework is better than NNs at encoding and manipulating the *complex, dynamic structures* of information that, as far as we know, must be brought into play in cognition. These structures can, for instance, be interpretations of natural language utterances, descriptions of complex scenes, complex plans of action, or conclusions drawn from other information. The encodings of such structures, whether these encodings are symbolic or otherwise, need to have the following important properties:

1. The encodings must often be highly temporary—for instance, encodings of interpretations of natural language sentences and encodings of intermediate conclusions during reasoning need to be rapidly created, modified, and destroyed (or at least taken out of consideration).
2. The encoding technique must allow the encoded structures to combine information items (e.g., word senses) that have never been combined before, or never been combined in quite the same way. For instance, a natural language sentence interpretation can combine concepts never before combined in the experience of the understander.
3. The encodings must allow for the encoded information to be of widely varying structural complexity. Natural language sentence interpretations also provide illustrations of this point.
4. In particular, the encoded structures can be multiply nested. In the sentence "John believes that Peter's anger with Mary caused her to write him a strongly worded letter," the anger description is nested within a causation description that is nested within a belief report.
5. A given type of information can appear at different levels of nesting. A system might have to represent a room that has a wall that bears a picture that itself depicts a room (an example due to G. E. Hinton; cf. CONNECTIONIST AND SYMBOLIC REPRESENTATIONS). A belief might be about a hope that is about a belief.
6. A given type of information may also have to be multiply instantiated in other ways, as when, for instance, there are three love relationships that need to be simultaneously represented.

Turning to manipulations, cognitive systems must exhibit strong properties of *systematicity* of processing—each information structure *J* that one cares to mention has an extremely large class of variants that must be able to be subjected to the same sort of processing as *J* is; and the class of variants is far too large to imagine that each variant is processed by a separate piece of neural network or a separate symbolic module. So, we must have symbolic AI systems and NNs capable of very flexible and general processing. Systematicity is championed by Fodor and Pylyshyn as a strike against connectionism in Pinker and Mehler (1988). Although many neural network researchers feel that those authors overstated the case, and are wrong in supposing that systematicity and related desiderata force us to adopt traditional symbolic structures and manipulations, there is general recognition that some strong form of systematicity is important.

The variable binding problem for neural networks is one manifestation of the need for systematicity of processing. Suppose one wishes a neural network to make inferences that obey the following rule: *X* is jealous of *Z* whenever *X* loves *Y*; *Y* loves *Z*; and *X*, *Y*, and *Z* are distinct people. In this statement of the rule, the variables *X*, *Y*, and *Z* can be replaced by any suitable people-descriptions, such as "Joe Bloggs's father's boss." The systematicity issue in this example, irrespective of whether a neural network directly implements this rule or merely works in accordance with it, is that of avoiding having replication of machinery for all the different possible combinations of values for these variables. (Each such combination is a *J* in the terms of the previous paragraph.) In a broad sense, this issue can also be thought of as the variable binding problem for the example. Now consider the special case in which a neural network implements the rule as a subnetwork and has particular units, subnetworks, or activation patterns that play the role of the three variables. The variable binding problem, in a narrower sense now, is the problem of how the network is to be able to "bind" such a unit, subnetwork, or activation pattern to a particular value at any given moment, and how the binding is to be used in processing.

Cognitive systems must also exhibit a high degree of *structure sensitivity* in their processing (Fodor and Pylyshyn, ibid.). Pieces of information that have complex structure must be processed in ways that are heavily dependent on their structure as such, not (just) on the nature of constituents taken individually. For example, consider the operation of inferring from "not both *A* and *B*" that "not *A* or not *B*." The operation is independent of what *A* and *B* are—it is only the "not both . . . and . . ." structure that is important. Structure sensitivity is a special but central aspect of systematicity of processing.

These features of information structure encodings and manipulations combine to distinguish the types of information that neural networks for reasoning, natural language understanding, etc., must deal with from the types that typical neural networks cater to. Traditional NN techniques were originally developed largely for specific "low-level" applications, such as restricted forms of visual pattern-recognition, or for limited forms of pattern association. Because of the limitations in such applications of NNs, it has been sufficient for NNs to adhere, by and large, to the following restrictions (although almost every restriction is violated by some NN subparadigm). These restrictions, however, cause difficulty in trying to apply NNs to natural language understanding, common-sense reasoning, and the like.

1. There is typically no dynamic, rapid creation and destruction of nodes and links. Therefore, temporary information cannot be encoded in temporary network topology changes. (Some of this effect can, however, be obtained by techniques such as dynamic links, described in DYNAMIC LINK ARCHITECTURE, or by higher-order units, whose activation is sensitive to weighted sums of products of input values, rather than just to weighted sums of individual input values.)

2. Links in NNs are not differentiated by labeling, unlike the links in symbolic structures such as SEMANTIC NETWORKS. Therefore, in an NN, information that could otherwise be put into link labels has to be encoded somehow in activation values, weights, extra links, or other features of network topology, adding significantly to the cumbersomeness of the net and its processing. (See Barnden and Srinivas, 1991, for more discussion.)

3. The resolution of NN activation values is generally not fine enough to allow them individually to encode complex symbolic structures. Most typically, activation values merely encode confidence levels of some sort.

4. Pointers are usually not allowed. That is, activation values, or groups or sequences of them, are not allowed to act as names or addresses of parts of the network.

5. Stored programs are not allowed, in any conventional sense. Activation value groups or sequences cannot act as instructions (names of internal computational actions).

Further Comparative Remarks

Some of the advantages claimed above for NNs are not clear-cut. For instance, there are types of AI systems that readily exhibit forms of graceful degradation, pattern completion, and similarity-based generalization. In particular, as Barnden (in Barnden and Holyoak, 1994) argues in detail and other researchers have noted, these benefits are natural properties of suitably designed symbolic analogy-based reasoning systems (see ANALOGY-BASED REASONING). Equally, just as NNs support some types of learning more readily than symbolic AI systems do, the reverse holds equally well. Symbolic AI systems, by virtue of the ability to handle complex temporary information structures, are in a better position to perform various types of rapid learning, proceeding in large steps rather than lengthy, gradual weight modification. For instance, a symbolic AI system is in a good position to reason about why some plan of action failed, and thus quickly and greatly amend relevant parts of its knowledge base or planning strategies.

NNs are good at allowing hypotheses to be held with varying degrees of confidence, the degrees being realized as activation levels. However, it is commonplace also in symbolic AI to have numerical degrees of confidence. These appear in EXPERT SYSTEMS AND DECISION SYSTEMS and elsewhere. The relevant difference between NNs and symbolic AI lies in the fact that the normal properties of activation spread and activation combination in NNs support confidence levels *in a natural way*. In symbolic AI systems, the computations have to be specially and explicitly designed.

The above contrasts between NNs and symbolic AI are clouded by the possibility of *implementational* neural networks. These are neural networks that are implementations of symbol processing systems. That is, momentary states of the network (usually activation states) can be regarded as encoding symbolic objects, and changes in the state can be regarded as encoding symbolic manipulation steps. See, for example, the chapters by Barnden and Shastri (in Barnden and Pollack, 1991), Touretzky (in Hinton, 1991), and Lange and Wharton (in Barnden and Holyoak, 1994).

Nonimplementational NNs avoid any direct implementation of logical formulae or symbolic structures. More exactly, the states of the network cannot accurately be viewed at any convenient level of description as traditional symbolic objects. In an extreme form of nonimplementational NN, we would not be able to regard the network states even approximately as being traditional symbolic objects. However, the nonimplementational style also includes NNs which can, though only under some significant degree of approximation, be usefully viewed as manipulating symbolic objects in a traditional, rule-like way (cf. Smolensky, 1988). Such networks can be called *subsymbolic* (although this term tends to be used more loosely by some authors).

There is some slipperiness in the notion of implementational NNs, depending on how large or abstract the symbolic processing steps are. For instance, in a low-level case, a step might be a minute change to a list of symbols, whereas in a more high-level case, a step might be the splitting of a logic formula into two parts. Also, a step could, in principle, be one that would not normally be regarded as a single *unanalyzed* step in the symbolic arena. An example would be a complex inference step applied to a logic formula or the spreading of some markers in a semantic network.

The more closely that a neural network merely implements small, conventional symbol-processing steps, the more it runs the danger of inheriting the disadvantages (e.g., relative lack of graceful degradation) of symbolic AI.

Bridging the Gap

In recent years, there has been an increasing amount of work on bridging the gap between symbolic AI systems and NNs (see, e.g., Barnden and Pollack, 1991; Bookman and Sun, 1993; Dinsmore, 1992; Gallant, 1993; Hinton, 1991; Holyoak and Barnden, 1994). The gap consists of the relative advantages of (nonimplementational) NNs and symbolic AI systems, as explained above. We would like systems that combine both sets of advantages.

The chapters by Hinton, Pollack, and St. John, and Mc Clelland (in Hinton, 1991) are promising initial investigations in the use of a sophisticated technique for extending conventional types of NN processing to handle complex dynamic structures. A structure is encoded into a *reduced representation* (or *compressed encoding*). This is a single activation vector that is created from the several activation vectors that encode the constituents of the structure in such a way that the resulting vector is of roughly the same size as each of the constituents' vectors. For example, the constituents could be words, and a sequence of word encodings could represent a sentence. The compressed encoding is then a roughly word-sized vector for the whole sentence.

In Pollack's Recursive Auto-Associative Memory (RAAM), a three-layer autoassociative backpropagation network is used to create the compressed encodings. The input and output layers are divided into segments that hold constituent encodings. The net is trained to map sequences of constituent encodings to themselves. The activation pattern that appears on the hidden layer of the trained network in response to a particular sequence of constituent encodings on the input layer is the compressed encoding for the sequence. (And a compressed encoding can be decoded by placing it in the hidden layer: a close—one hopes—approximation to the sequence of constituent encodes appears on the output layer.) Also, during training, a hidden layer pattern can be copied into one of the segments in the input and output layer, leading to the ability of the network to handle recursive structures, some of whose constituents are themselves sequences of constituents. An example of such a structure is the sentence "John knows that Sally is clever," thought of as having the sentence "Sally is clever" as a constituent.

There are some indications that compressed encodings can support *holistic* structure-sensitive processing by means of

conventional NN techniques, such as feedforward association networks (see, e.g., Chalmers, 1990; Pollack in Hinton, 1991; Blank et al. in Dinsmore, 1992). The processing is holistic in that the encodings are not uncompressed into the activation vectors that encode their notional constituents. For example, Chalmers successfully trained a three-layer backpropagation network to transform compressed encodings of active English sentences into compressed encodings of their passive counterparts. The hidden layer had the same size as the input and output layers—i.e., the size of a compressed encoding—and the trained network therefore cannot have been working by first decoding the input compressed encodings into the corresponding sequence of constituent encodings. (However, further work must be done to show that, in general, holistic transformation networks do not effect an approximate decoding, by producing separate hidden-layer activation subpatterns that are compressed versions of the constituent encodings.)

In a different approach to bridging the gap, Barnden (in Barnden and Holyoak, 1994) capitalizes on the comment made above that symbolic analogy-based reasoning possesses many of the main advantages of nonimplementational NNs. The claim is that an implementational NN that implements a symbolic analogy-based reasoning system inherits those advantages, as well as the symbolic AI advantage with respect to complex dynamic information structures.

The above approaches assume that it is worthwhile to develop gap-bridging systems that are neural networks in their entirety, rather than developing systems that are some combination of NN machinery with symbolic AI machinery (where the latter machinery is given no NN realization). This hybrid strategy is currently the most common approach to bridging the gap. The simpler types of hybridization occur in systems that have largely separate neural and symbolic modules (see, e.g., Gallant, 1993, and Hendler in Barnden and Pollack, 1991). But more intimate hybridizations have been developed, for instance in networks where an individual node or link can act partially like those in neural networks and partially like those in symbolic networks (see, e.g., the Eskridge and Kokinov chapters in Holyoak and Barnden, 1994). The hybrid strategy has obvious advantages, but it also has disadvantages. Most important, a researcher's system development may (ultimately) be directed at biological realism, in which case it is eventually necessary to replace symbolic AI machinery by NN machinery.

Although the more strongly implementational a neural network is, the more it is at risk of inheriting disadvantages of symbolic AI systems, it could still be that some of the techniques on which implementational NNs are founded could be adapted for use in gap-bridging systems that escape those disadvantages. Therefore, we will now look at some of the techniques.

A crucial aspect of implementational NNs is the way in which they allow representational items to be rapidly and temporarily combined so as to form encodings of temporary complex information structures. One form of this *dynamic combination* (or *temporary association*) issue is the variable-binding problem, and a closely related form is the role-binding problem. The variable-binding problem was described above. The role-binding problem is concerned with giving specific values to the roles (slots) in predicates, frames, schemas, and the like.

An immediately obvious, and somewhat natural, approach to dynamic combination is to combine network nodes or assemblies by adding new links or giving non-zero weights to existing zero-weight links. However, this method is highly cumbersome because network structure is not directly analyzable by the network itself. Another rather similar approach is to facilitate existing (non-zero-weight) connection paths, between nodes/assemblies that are to be combined, by activating intermediate nodes on the paths. These nodes are called *binding nodes*. Since the dynamic combination structure is now encoded in the activation levels of binding nodes, the net can more easily analyze that structure. However, there is still a considerable degree of cumbersomeness (Barnden and Srinivas, 1991).

A distinctly different approach is to deem nodes/assemblies to be bound together when they have oscillating activation levels that are in phase with each other. (See STRUCTURED CONNECTIONIST MODELS and OBJECT RECOGNITION.) It remains to be seen whether enough different phases can be realistically postulated for the oscillation of a biological neuron to account for the number of different bindings needed in cognitive tasks. The oscillation phase method is an important special case of the more general notion of binding nodes together by giving them similar spatiotemporal activation patterns. Similarity could take many forms other than equality of oscillation phase.

Distinctly different again is the use of positional encodings of dynamic combinations. In the more developed forms of this idea (see Barnden in Barnden and Pollack, 1991), activation patterns (rather than nodes or assemblies) are dynamically combined by being put into suitable relative positions with respect to each other, much as bit-strings in computer memory can be put into contiguous memory locations to form records.

A somewhat pointer-like technique has been implemented (see Lange and Wharton in Barnden and Holyoak, 1994). Different parts of the network are capable of emitting activation patterns that are thought of as their "signatures." Other parts can then temporarily hold signatures and thereby point, in a sense, to the parts that possess the signatures.

Discussion

One theme of this review has been that the relative advantages of symbolic AI and NNs are less clear-cut than is usually implied. In particular, while NNs have been successful for some purposes, including some purposes shared with symbolic AI, and NNs can have advantages such as graceful degradation, most NN research has not addressed the complex information processing issues routinely tackled in symbolic AI research. The latter field has contributed much more, for instance, to the study of how natural language discourse can be understood and common-sense reasoning performed. Nevertheless, some NN researchers have pointed the way toward NNs that may ultimately be able to cope (indeed, cope better) with cognitive processes for which only symbolic approaches are currently available.

Some of the open questions in the area of this review are: Is it actually necessary to go beyond symbolic AI in order to account for complex cognition? If it is, should symbolic AI be dispensed with entirely, or is some amount of complex symbol-processing unavoidable? How can reasoning, natural language understanding, etc., be effected by neural networks without just implementing conventional symbol processing? How can different styles of system (e.g., implementational and nonimplementational neural networks, or neural networks and nonneural systems) be gracefully combined into hybrid systems?

Road Map: Artificial Intelligence and Neural Networks
Background: I.2. Levels and Styles of Analysis
Related Reading: Associative Networks; Cognitive Modeling: Psychology and Connectionism

References

Barnden, J. A., and Holyoak, K. J., Eds., 1994, *Advances in Connectionist and Neural Computation Theory*, vol. 3, *Analogy, Metaphor and Reminding*, Norwood, NJ: Ablex.
Barnden, J. A., and Pollack, J. B., Eds., 1991, *Advances in Connectionist and Neural Computation Theory*, vol. 1, *High Level Connectionist Models*, Norwood, NJ: Ablex.
Barnden, J. A., and Srinivas, K., 1991, Encoding techniques for complex information structures in connectionist systems, *Connection Sci.*, 3:263–309.
Bookman, L., and Sun, R., 1993, *Connection Sci.*, special issue on architectures for integrating neural and symbolic processes.
Chalmers, D. J., 1990, Syntactic transformations on distributed representations, *Connection Sci.*, 2:53–62.
Dinsmore, J., Ed., 1992, *The Symbolic and Connectionist Paradigms: Closing the Gap*, Hillsdale, NJ: Erlbaum.
Gallant, S. I., 1993, *Neural Network Learning and Expert Systems*, Cambridge, MA: MIT Press.
Hinton, G. E., 1991, *Connectionist Symbol Processing*, Cambridge, MA: MIT Press.
Holyoak, K. J., and Barnden, J. A., Eds., 1994, *Advances in Connectionist and Neural Computation Theory*, vol. 2, *Analogical Connections*, Norwood, NJ: Ablex.
Hwang, K., and Briggs, F. A., 1984, *Computer Architecture and Parallel Processing*, New York: McGraw-Hill.
Pinker, S., and Mehler, J., Eds., 1988, *Connections and Symbols*, Cambridge, MA: MIT Press.
Rich, E., and Knight, K., 1991, *Artificial Intelligence*, 2nd ed., New York: McGraw-Hill. ◆
Rumelhart, D. E., and McClelland, J. L., 1986, On learning the past tenses of English verbs, in *Parallel Distributed Processing: Explorations in the Microstructure of Cognition* (D. E. Rumelhart, J. L. McClelland, and the PDP Research Group, Eds.), vol. 2, *Psychological and Biological Models*, Cambridge, MA.: MIT Press.
Smolensky, P., 1988, On the proper treatment of connectionism, *Behav. Brain Sci.*, 11:1–74.
Sun, R., 1994, *Integrating Rules and Connectionism for Robust Commonsense Reasoning*, New York: Wiley.

Associative Networks

James A. Anderson

Introduction

The operation of *association* involves the linkage of information with other information. Although the basic idea is simple, association gives rise to a particular form of computation, powerful and idiosyncratic. The mechanisms and implications of association have a long history in psychology and philosophy. Association is also the most natural form of neural network computation. This article will discuss association as realized in neural networks as well as association in the more traditional senses.

Neural networks are often justified as abstractions of the architecture of the nervous system. They are composed of a number of computing units, roughly modeled on neurons,

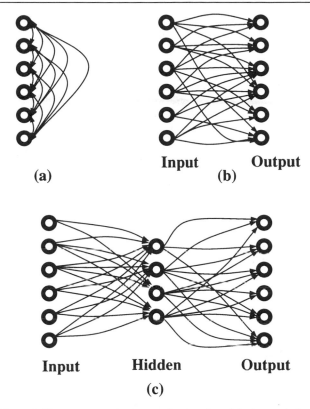

(a)

Input   Output

(b)

Input   Hidden   Output

(c)

Figure 1. Three common basic neural network architectures. *A*, A set of units connects recurrently to itself by way of modifiable connections. (The connections are drawn as reciprocal.) *B*, A feedforward network in which an input pattern is transformed to an output pattern by way of a layer of modifiable connections. *C*, A more general feedforward network. An input layer projects to an intermediate layer of units. The intermediate layer is often called a *hidden layer* because it may not be accessible from outside the network. The hidden layer then projects to the output units.

joined together by connections that are roughly modeled on the synapses connecting real neurons together. The basic computational entity in a neural network is related to the pattern of activity shown by the units in a group of many units.

Because of the use of activity patterns—mathematized as state vectors—as computational primitives, the most common neural network architectures are pattern transformers which take an input pattern and transform it into an output pattern by way of system dynamics and a set of connections with appropriate weights. In a very general sense, therefore, neural networks are frequently designed as *pattern associators*, which link an input pattern with the "correct" output pattern. Learning rules are designed to construct accurate linkages. The most common feedforward neural network architectures realize this linkage by way of connections between layers of units (Figure 1). There may be a single set of modifiable connections between input and output (Figure 1*B*), or multiple layers of connections (Figure 1*C*). Another common architecture is realized by a single layer of units where the units in the layer are recurrently interconnected (Figure 1*A*).

One common design goal of a feedforward associator (Figures 1*B* and 1*C*) is to realize what Kohonen (1977) has labeled *heteroassociation*, that is, to link input and output patterns that need have no relation to each other. Another possibility is to realize what Kohonen has called *autoassociation*, where the input and output patterns are identical. Recurrent networks (Figure 1*A*) are well suited to autoassociation.

Because the input and output patterns must correspond to information about the real world, the *data representation* is of critical importance at all levels of network operation. For example, simple pattern recognizers are often realized by neural networks as a special form of pattern associator by assuming a particular output representation, one where a single active output unit corresponds to the category of the input. Different categories correspond to different active output units. This highly localized representation is sometimes called a *grandmother cell* representation, because it implies that only when one particular unit is active is "grandmother" being represented. The alternative representation is called a *distributed representation* where representation of a concept like "grandmother" may contain many active units. Choice of representation makes a major difference in how networks are used and how well they work, and is usually more important than the exact choice of network architecture and learning rule. A common situation in engineering applications of neural networks is to have a distributed representation at the input of the network and a grandmother cell representation at the output. In the vertebrate nervous system there is little evidence for this output representation; essentially all normal motor acts involve the coordinated discharge of large groups of neurons. Distributed activity patterns are associated with distributed patterns from one end of most biological networks to the other in vertebrates (but see LOCALIZED VERSUS DISTRIBUTED REPRESENTATIONS), though there are some examples of extreme selectivity in invertebrates. The degree of distribution is a matter for experimental investigation.

Neural Network Associators

Let us give an example of how easily neural network learning rules and architectures give rise to associative behavior. Consider the two-layer network diagrammed in Figure 1*B*. Consider a situation where a pattern of activity, a state vector *f*, is present at the input set of units and another pattern, state vector *g*, is shown by the output set of units.

We want to link two patterns so that when *f* is presented to the input of the network, *g* will be generated at the output. In this two-layer network (two layers of units, one layer of connections), we will assume that the connections initially are zero and we want to change them to make the association between patterns *f* and *g*. We will also assume that all connection strengths are potentially changeable and the set of connection strengths form a *connection matrix* (or *synaptic matrix*) which we will call *W*, for "weights."

We have to propose a learning rule, but we also have to make some additional assumptions about the entire system. For example, virtually all artificial neural network learning assumes that the network is learning discrete pairs of patterns, that is, learning takes place only occasionally, when the time is ripe. One could speculate that learning in animals is a dangerous operation—after all, the nervous system is being rewired—and is kept under tight control. Primates are unusual in the degree of learned flexibility their nervous system allows. There is physiological evidence that amount of learning is controlled by diffuse biochemical processes. Dangerous and striking events, causing a biochemical upheaval, give rise to what have been called "flashbulb memories" where everything, including totally irrelevant detail, is learned. ("Where were you when John F. Kennedy was assassinated?" is practically guaranteed to involve a flashbulb memory in those old enough to remember it.) Presumably this corresponds to an undiscriminating "learn" command. In terms of modeling, these observations mean that the decision to learn is decoupled from the act of learning.

Let us assume that we have an input pattern and an output pattern and we wish to associate them for good and sufficient reasons. We assume that we can impress pattern f on the input set of units and pattern g on the output set of units. By far the most common network learning rule used is one or another variant of what is called the "Hebb synapse," described in Hebb (1949). Perhaps the most quoted sentences in the neural network literature (and certainly in this *Handbook*! [MAA]) are from Hebb: "When an axon of cell A is near enough to excite a cell B and repeatedly or persistently takes part in firing it, some growth process or metabolic change takes place in one or both cells, such that A's efficiency as one of the cells firing B, is increased." (Hebb, 1949:62). The essence of the Hebb synapse is that there has to be a *conjunction* of activity on the two sides of the connection.

There is good physiological evidence for the existence of some form of Hebb synapse in parts of the mammalian central nervous system, in particular the hippocampus (see HEBBIAN SYNAPTIC PLASTICITY). However, there are a number of "technical" problems involved in mathematically describing the resulting system. The orginal formulation by Hebb was concerned with coincident excitation. Nothing was said about coincident inhibition or about coincident excitation and inhibition. Also, the exact function determining strength of modification was not given, and, in fact, is not known. A common assumption in *artificial* network theory is to assume some version of what is called the *generalized Hebb rule* or the *outer product rule*. This states that the change in strength of a connection during learning is given by the *product* of activities on the two sides of the connection, that is, if W_{ij} is the strength of the connection, then the change in strength ΔW_{ij} is proportional to the product, $f_j g_i$, where f_j is the activity of the jth input unit and g_i is the activity of the ith output unit. This convenient expression may have only a weak relationship to physiological reality.

Given the generalized Hebb rule, if we have only a single pair of vectors to associate, the results can be written compactly as

$$W = \eta g f^T$$

where η is a learning constant and W is the connection (or weight) matrix.

By making an additional assumption about the properties of the individual neural elements, this rule leads almost immediately to a simple pattern associator called the *linear associator*. Suppose the elementary computing units are linear, so that the output is given by the inner product between input activity and connection strengths. Then the output pattern is given by the matrix product of an input pattern f and the connection matrix W; that is, the output of the network is Wf. Because we know what W is—it was constructed by the generalized Hebb rule—we can compute the output pattern,

$$\text{(output pattern)} = Wf = \eta g f^T f = \text{(constant)}g$$

since $f^T f$ is a constant, the squared length of f. The output pattern is a constant multiple of g and, except for length, we have reconstructed the learned associate of f, that is, g.

Suppose we have a whole set of associations $\{f^i \to g^i\}$ that we want to teach the network. (Superscripts stand for individual pattern vectors.) If we assume that the overall strength of a connection is the algebraic *sum* of its past history (an unsupported assumption), then we have the weight matrix W given by

$$W = \sum_i \eta g^i f^{iT}$$

Notice that in the special case where the input patterns $\{f^i\}$ are *orthogonal*, that is, $f^i f^j = 0$ if $i \neq j$,

$$Wf^i = \text{(constant)}g^i$$

because the contributions to the output pattern from the other terms forming W are identically zero since they involve the inner product $[f_i, f_j] = f_i^T f_j$. This model, and in fact most simple network models, make the prediction that outer product associators will work best and most reliably with representations where different input associations are as orthogonal as possible. For this reason, some cortical models in the neuroscience literature have explicitly discussed aspects of cortical processing in terms of orthogonalization. The most complete reference for the linear associator and related models is Kohonen (1977, 1984).

It is possible to change almost any assumption and still have an associator. *Hebb learning rules of virtually any kind give rise to associative systems*. As only one example, the nonlinear Hebbian associator proposed by Willshaw, Buneman, and Longuet-Higgins (1969) used binary connections—with strengths either one or zero—and the resulting system still worked nicely as a pattern associator.

Supervised Networks

The outer product associator is less accurate with nonorthogonal patterns. However, observed distortions and human performance are sometimes remarkably similar. (See Anderson, 1995, chap. 11, for a model of "concept formation" that emerges when correlated inputs are stored in the linear associator.)

Most designers of artificial networks prefer networks to produce accurate reproductions of learned associations, rather than interesting distortions (but see DISTORTIONS IN HUMAN MEMORY). (This seemingly natural assumption is not necessarily a good one.) *Supervised* network algorithms can perform more accurate association. Examples of such algorithms would include the Widrow-Hoff (LMS) algorithm, the perceptron, backpropagation, and many others. The basic mechanism employed is *error correction*. Suppose we have an initial training set of patterns to be learned. This means we know what the output patterns are for a number of input patterns. We take an input from the training set and let the network generate an output pattern. We then compare the desired output pattern and the actual output pattern in some way. This process generates an *error signal*. The network is then modified using a learning rule so as to *reduce* the error signal.

The most commonly used error signal is based on the distance between the actual and desired output; however, other error signals can be more desirable. For example, one could incorporate a term penalizing large numbers of connections or large values of connection strength. The network learning problem reduces to a minimization problem where the space formed by the connection strengths (*weight space*) is searched to find the point where error is reduced to as low a value as possible. This process requires the use of control structures that can be complex; for example, there is assumed to be an omniscient *supervisor* who compares desired and actual network output and computes the error term as well as implements the mechanisms to change connection strengths appropriately. The structure of these algorithm is designed to produce good pattern association whether or not this is the aim of the network architects. (See LEARNING AS HILL-CLIMBING IN WEIGHT SPACE; PERCEPTRONS, ADALINES, AND BACKPROPAGATION.)

Autoassociative Models

We have described association as pattern linkage. However, there are alternative descriptions in the neural network literature. For example, in the initial line of the second chapter, the textbook by Hertz, Krogh, and Palmer (1991) says, "Associative memory is the 'fruit fly' or 'Bohr atom' problem of the field" (p. 11). Their definition of association is: "Store a set of patterns ξ... in such a way that when presented with a new pattern ζ_i, the network responds by producing whichever one of the stored patterns most closely resembles ζ_i" (p. 11). This is not, however, a description of association but of a *content addressable memory* where input of partial or noisy information is used to retrieve the correct stored information. The source of this limited view of association lies in the ability of autoassociative systems to reconstruct missing or noisy parts of learned patterns.

Consider the autoassociative version of the linear associator. Suppose we learn one pattern, f, of length 1, with learning constant $\eta = 1$. Then

$$W = ff^{\mathrm{T}} \quad \text{and} \quad Wf = f$$

Suppose we take vector f, with n elements, and set to zero some of the elements, forming a new vector, f'. Let us make a second vector, f'', from only the elements that were set to zero in f'. Then $f' + f'' = f$ and $f'f''^{\mathrm{T}} = 0$. If f' is input to the autoassociator,

$$Wf' = (f' + f'')(f' + f'')f'^{\mathrm{T}} = (\text{constant})f$$

where the constant is related to the length of f'. In operation, by putting a part of f, f', into the network, we retrieve all of f, bar a constant. This behavior is often referred to as the *reconstructive* or *holographic* property of neural networks. Of course, more subtle problems arise when W stores multiple vectors. Anyway, this type of memory is associative because if, for example, the state vector was meaningfully partitioned, then f' is associatively linked to f'' and vice versa in the sense that input of one pattern will produce the other. This kind of associator produces intrinsically bidirectional links (i.e., $f' \rightarrow f''$ and $f'' \rightarrow f'$), unlike feedforward heteroassociators ($f \rightarrow g$).

Some nonlinear "attractor" neural networks with dynamics that minimize energy functions develop their associative abilities largely from their autoassociative architecture. The best-known examples of this kind of associator are Hopfield networks and parallel feedback networks such as the BSB (Brain State in a Box) model (Anderson, 1995, chap. 15). For a general review of attractor networks, see Amit (1989) and COMPUTING WITH ATTRACTORS. Multilayer autoassociators are also possible. The multilayer *encoder networks*, which require the output pattern to be as accurate a reconstruction as possible of the input pattern, also have this form. Many autoassociative networks have close ties to known statistical techniques such as PRINCIPAL COMPONENT ANALYSIS.

A related associative attractor model called a *bidirectional associative memory* or BAM (Kosko, 1988), is a nonlinear dynamical system with a reciprocal feedback structure. It assumes two layers of units, as well as pairs of associations to be learned, as in a heteroassociator. There are connections from both input to output and output to input. Given f and g patterns to be learned, assumed to be binary vectors, we can form both a forward and a backward connection matrix. If f is input, then g will be given as the output; g at the output will give rise to f at the input because of the backward connections. Suppose the input is not exactly what was learned. After a few passes back and forth through the system, it can be shown that the network will stabilize, in the noise-free case, to the learned f and g.

Psychological Association

We have shown how neural networks easily form associators of many different kinds. We will now discuss a little of the history of association in psychology to show how associators form a style of computation with considerable power as well as severe limitations.

The major outlines of one way to use an associative computer can be found clearly expressed in Aristotle in the fourth century B.C. Aristotle made two important claims about memory structure: First, the elementary unit of memory is a *sense image*, that is, a sensory-based set of information. Second, links between these elementary memories serve as the basis for higher-level cognition. A recent English translation by Richard Sorabji (1968) used the term *memory* for the elementary memory unit and *recollection* for reasoning by associations between elementary units. Aristotle discussed at length how one "computes" with memorized sense images. The word *recollection* was used in the translation to denote this process: "Acts of recollection happen because one change is of a nature to occur after another." That is, Aristotle proposed a linkage mechanism between memories. He suggested several ways that linkage could occur: by temporal succession or by "something similar, or opposite, or neighboring." This list of the mechanisms for the formation of associations is approximately what would be given today by psychologists.

Recollection in Aristotle's sense was computation. It was a dynamic and flexible process: "recollecting is, as it were, a sort of reasoning." Aristotle argued that properly directed recollection is capable of discovering new truths, using memorized sense images as the raw material and learning to traverse new paths through memory (Figure 2).

A practical problem with such an associative net is branching, that is, what to do if there is more than one link leaving an elementary memory. Aristotle was aware of this problem: "it is possible to move to more than one point from the same starting point." A general solution to the branching problem requires a nonlinear mechanism to select one or the other branch.

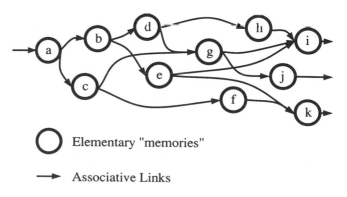

Possible Paths: abdhi, acgi, abei, ...

Figure 2. A simple model of associative computation. Elementary memories ("sense images" according to Aristotle) are associatively linked (arrows) to other sense images. Branches are possible, and they present some difficulties. There are many possible paths through the network. Forming and traversing links between elementary memories is the basis of mental computation.

The most influential psychologists in this century were the behaviorists, in particular B. F. Skinner, the Harvard psychologist whose ideas about reinforcement learning unfortunately dominated much of the theoretical discussion in psychology for several decades. This school held that learning formed an associative link between a stimulus and a specific response. The link could be strengthened by positive reinforcement (to a first approximation, something useful or pleasant, or the cessation of something unpleasant) or weakened by negative reinforcement (either absence of something pleasant or something actively unpleasant) when the response followed the stimulus. A number of careful experiments showed that there were accurate quantitative "laws of learning" that were followed by animals in some simple situations.

It was debatable whether this view of association is useful in more complex situations. From the beginning, human behavior has seemed to humans to be far richer than stimulus-response ($S \rightarrow R$) association. In the 1950s Skinner wrote a book attempting to explain language behavior using associative rules. In a famous book review, Chomsky (1957) pointed out that simple $S \rightarrow R$ association cannot do some kinds of linguistic computation. The argument used was that Skinner was proposing a well-defined computing machine with his associative model and that this computing machine was not powerful enough to do the computations we know language users perform. The simple $S \rightarrow R$ models of Skinner had about as much computing power as the simplest heteroassociative neural networks which no one claimed were a general-purpose computer. However, supervised network learning algorithms applied without insight may produce systems with only this degree of overall computational power.

"Connectionist" Models

Much modern work using association assumes that the entities linked, and the links themselves, can have complex internal structure. Flexible systems capable of complex reasoning can be produced by using labeled links: for example, a robin IS-A bird, an IS-A link, or "Fred is the father of Herb," meaning that there is an associative link between Fred and Herb and that the link carries the relationship "Father-of." Complex and sophisticated computational models, *semantic neworks*, can be built from these pieces.

In the 1980s, many of those interested in semantic network models started working with neural networks. The term *connectionism* was often used to indicate the application of neural networks to high-level cognition. Recently there have been many attempts to apply networks to reasoning, to complex concept structures, and, in particular, to language understanding. A heated but illuminating debate arose from an early connectionist paper by Rumelhart and McClelland (1986) that used a neural network to simulate the way young children learn past tenses of verbs. Past tense learning had always been considered to be a good example of the application and misapplication of a specific rule, suggesting symbolic processing. Rumelhart and McClelland's neural network acted as if it were using rules, but the rule-like behavior was the result of generalizing from examples and learning specific cases. (See LINGUISTIC MORPHOLOGY.) Perhaps because this model was such a direct attack on the existence of rules in language, a vigorous counterattack developed. As one example, a long paper by Pinker and Prince (1988) finished its abstract with the sentence, "We conclude that connectionists' claims about the dispensability of rules in explanations in the psychology of language must be rejected, and that, on the contrary, the linguistic and developmental facts provide good evidence for such rules"

(p. 74). The vigor of the attack is perhaps due in part to the authors' feeling that the connectionists had violated the "central dogma of modern cognitive science, namely that intelligence is the result of processing symbolic expressions" (Pinker and Prince, p. 74). Many other cognitive scientists feel that the "central dogma" is actually more like a central, and open, question.

Less well known outside psychology are several associative neural network models that were constructed to explain the fine structure of experimental data in more traditional areas of psychology such as verbal learning. An interesting example of such a model is the TODAM model of Murdock (see CLASSICAL LEARNING THEORY AND NEURAL NETWORKS). TODAM and variants blur the distinction between the network and the representation. In the associative networks we have discussed, there are two formally distinct entities, state vectors and connection matrices. In the TODAM class of models, the association is stored with the items themselves and is therefore the same type of entity. TODAM makes a number of testable qualitative predictions about a wide range of data from the classical verbal learning literature. Recently, models assuming networks composed of large numbers of local networks (a "network of networks") suggest networks like TODAM might be realizable with neural networks.

Discussion and Open Questions

An often proclaimed virtue of neural networks is their ability to generalize effectively and to do computation based on similarity. Having learned example associations from a training set, the network can then generate correct answers to new examples. Many have pointed out the formal similarity of neural networks to approximation and interpolation as studied in numerical analysis. A properly designed neural network can act as a useful adaptive interpolator with good, even optimal, generalization around the region of the learned examples. However, it is not easy for neural networks to make good generalizations other than by approximation and interpolation. On this basis, Fodor and Pylyshyn (1988) made some telling arguments against the promiscuous application of connectionism to cognition. The essential criticism they made is one that an engineer would be happy to make: associative neural networks are such an inefficient way to compute that it would be foolish to build a cognitive system like that. Neural networks do not generalize well outside of a restricted definition based on mathematical interpolation, they cannot reason effectively, and they cannot extrapolate in any meaningful sense. These criticisms are part of a battle involving the limitations of association that has been going on for centuries. Fodor and Pylyshyn commented, "It's an instructive paradox that the current attempt to be thoroughly modern and 'take the brain seriously' should lead to a psychology not readily distinguishable from the worst of Hume and Berkeley" (p. 64).

Fodor and Pylyshyn contrasted neural network associators with what they call the classical view of mental operation. In essence, this view postulates "a language of thought"; that is, "mental representations have *a combinatorial syntax and semantics*" (p. 12). The classical view is dominant in virtually all branches of traditional artificial intelligence and linguistics. The power of the digital computer arises in part from the fact that it is designed to be an extreme example of this organization: a programming language operating on data is the prototype of the classical view.

Suppose we have a sentence of the form *A and B* that we hold is true. An example Fodor and Pylyshyn used is *John went to the store and Mary went to the store*. The truth of this sentence logically entails the truth of *Mary went to the store*. This conclusion arises from the rules of logic and of grammar. It is not easy for an associative neural network to handle this problem. Such a network could easily learn that *John went to the store and Mary went to the store* is associated with *Mary went to the store*. But the power of the classical approach arises from the fact that *every* sentence of this form gives rise to the same result. Given the huge number of possible sentences, *it makes practical sense* to assume that some kind of logical syntax exists. It would be hard to figure out how language could function without some global rule-like operations, however implemented.

The ability to understand and answer sentences or phrases that are new to the listener is hard to explain purely with association. To give one example (see MENTAL ARITHMETIC USING NEURAL NETWORKS), consider number comparisons such as "Is seven bigger than five?" There are nearly 100 such single-digit comparisons, nearly 10,000 two-digit comparisons, and so on. Children cannot possibly learn them as individual cases.

If there is a qualitative difference between human and animal cognition, it lies right here. There have been attempts to build neural networks that realize parts of the classical account, with indifferent success (see Hinton, 1991). Is it possible to build a neural network based largely on natural associators that can reproduce the kind of rule-governed behavior—even in limited domains—that does in fact seem to be part of human cognition? A neural network with this ability would allow for much more powerful and useful generalization than current networks provide. It may not be easy to find this solution. There are many animals with complex nervous systems capable of associative learning, but only our own species, one out of millions of species, is really effective at using these powerful extensions to association.

Road Maps: Connectionist Psychology; Learning in Artificial Neural Networks, Deterministic
Background: I.3. Dynamics and Adaptation in Neural Networks

Related Reading: Artificial Intelligence and Neural Networks; Cognitive Modeling: Psychology and Connectionism

References

Amit, D. J., 1989, *Modelling Brain Function: The World of Attractor Neural Networks*, Cambridge, Eng.: Cambridge University Press.
Anderson, J. A., 1995, *Introduction to Neural Networks*, Cambridge, MA: MIT Press. ◆
Anderson, J. R., 1983, *The Architecture of Cognition*, Cambridge, MA: Harvard University Press.
Chomsky, N., 1957, A review of Skinner's *Verbal Behavior, Language*, 35:26–58.
Fodor, J. A., and Pylyshyn, Z. W., 1988, Connectionism and cognitive architecture: A critical analysis, in *Connections and Symbols* (S. Pinker and J. Mehler, Eds.), Cambridge, MA: MIT Press.
Hebb, D. O., 1949, *The Organization of Behavior*, New York: Wiley.
Hertz, J., Krogh, A., and Palmer, R. G., 1991, *Introduction to the Theory of Neural Computation*, Redwood City, CA: Addison-Wesley. ◆
Hinton, G. E., 1991, *Connectionist Symbol Processing*, Cambridge, MA: MIT Press. ◆
Kohonen, T., 1977, *Associative Memory: A System Theoretic Approach*, Berlin: Springer-Verlag. ◆
Kohonen, T., 1984, *Self-Organization and Associative Memory*, Berlin: Springer-Verlag. ◆
Kosko, B., 1988, Bidirectional associative memories, *IEEE Trans. Sys., Man Cybern.*, 18:49–60.
Pinker, S., and Prince, A., 1988, On language and connectionism: Analysis of a parallel distributed processing model of language acquisition, in *Connections and Symbols* (S. Pinker and J. Mehler, Eds.), Cambridge, MA: MIT Press.
Rumelhart, D. E., and McClelland, J. L., 1986, On learning the past tenses of English verbs, in *Parallel Distributed Processing: Explorations in the Microstructure of Cognition* (D. E. Rumelhart, J. L. McClelland, and PDP Research Group, Eds.), vol. 2, *Psychological and Biological Models*, Cambridge, MA: MIT Press.
Sorabji, R., 1969, *Aristotle on Memory*, Providence, RI: Brown University Press.
Willshaw, D. J., Buneman, O. P., and Longuet-Higgins, H. C., 1969, Non-holographic associative memory, *Nature*, 222:960 962.

Astronomy

Michael C. Storrie-Lombardi and Ofer Lahav

Introduction

A recent review of neural network (NN) experiments in astronomy (Miller, 1993) identifies some two dozen applications including adaptive optics, telescope guidance, star pattern recognition for spacecraft guidance, feature detection in satellite images, scheduling Hubble Space Telescope (HST) time, detection of cosmic ray hits on HST, automated star/galaxy discrimination, faint object classification, morphological classification of galaxies, and stellar spectra classification. NNs' ability to handle nonlinear data interactions, their speed, and their robustness have allowed their utilization in diverse problem areas ranging from real-time control to classification.

NN algorithms used in astronomical work have included both supervised and unsupervised strategies. Supervised NNs learn to replicate the decisions of human experts (if the task is classification) or reproduce machine parameters for control applications. Training usually proceeds by progressively minimizing (with respect to free parameters called *weights*) the

square of the difference between the NN output and the objective "true" answer. Rediscovery of a mechanism for training multiple layers of NN nodes sparked intense investigation of supervised backpropagation NNs (BPNNs). BPNNs consist of a series of nodes arranged in layers, with each node connected to nodes in the subsequent layer by a weight. For easily describing the general configuration of a BPNN, we can use, e.g., {13;12,11,2} to represent 13 input nodes, 12 nodes in the first hidden layer, 11 in the second, and 2 nodes for output.

Unsupervised NNs do not use prior classification. An analogous technique in classical multivariate statistics is principal component analysis (PCA). Interestingly, PCA itself can be implemented as an NN (see PRINCIPAL COMPONENT ANALYSIS). Unsupervised NNs require only a set of input vectors and an algorithm for adjusting the weights connecting the nodes of the network. Many of these NNs rely on the insight of D. O. Hebb: that the likelihood of a neuron firing depends on the level of input from neighboring neurons. Unsupervised self-organizing networks as developed by Teuvo Kohonen can produce an objective mapping of target objects as described by the input parameters, allowing us to search objectively for previously unrecognized classes and class boundaries.

Control and Tracking with Neural Networks

Controlling mirror alignment to minimize atmospheric turbulence and tracking astronomical objects pose specific nonlinear problems amenable to NN strategies.

Adaptive Optics

Atmospheric turbulence compromising instrument performance remains a major concern in optical and infrared astronomy. The extended nature of the backscatter source renders laser-beacon wavefront correction techniques relatively insensitive to large-scale errors. Work initiated at the Steward Observatory (Angel et al., 1990; Lloyd-Hart et al., 1992) indicated the possibility of utilizing a BPNN in wavefront correction for the multiple mirror telescope (MMT) by controlling tilt and path difference for each telescope beam. The MMT combines light from six co-mounted 1.8 m mirrors into a single coherent focus. Simultaneous recording of an in-focus and out-of-focus pair of focal plane images permits derivation of the wavefront distortion secondary to atmospheric turbulence. The images can be superimposed on two detector arrays. The normalized image intensity from each pixel serves as input to a neural network. Performance in real time demands that atmospheric distortion not change significantly between detection and correction (about 10–30 ms cycle time at 2.2 μm wavelength). Computer simulations indicating a BPNN-controlled system could reduce the mean square wavefront phase error were verified by an in-line prototype on the MMT. Subsequent work has investigated the utilization of transputer technology to enhance system performance for large BPNN configurations (e.g., {338;144,18} for the six-mirror configuration).

Remote Tracking

For remotely operated telescopes, an automated system for monitoring field star shapes and radial positions would provide significant backup if primary guidance systems failed. Investigators at the Dominion Astrophysical Observatory (Ozard and Morbey, 1993) have explored the ability of NNs to control such systems for the Lyman Far-Ultraviolet Spectrographic Explorer (FUSE). The fine-error sensor camera in FUSE has residual field curvature aberration, significant enough to alter star image across the field of view. A computer model simulated star images as they might appear at different radii and a simulated, noise-free, 16-pixel CCD provided the inputs to a {16:9,16} BPNN. The NN successfully learned 90% of the training patterns and correctly classified 89% of the test set. While this effort will require refinement with CCD optimization, increased image resolution, and larger training sets from real-world data, it appears a most promising line of investigation.

Meteor Trails in Real Time

A study at the University of Natal (Fraser, Khan, and Levy, 1992) used a BPNN for classifying meteor type. The ionized trail left by a meteor can reflect transmitted microwave radiation. High-density trails provide better communication conditions than low-density ones. The NN used 12 trail parameters and 20 normalized sample points from the first 100 ms of the signal as input. A {32;32,2} BPNN correctly classified 97% of the test set as trails of high or low density. The trained NN could identify target objects in millisecond time intervals, a significant improvement over the several seconds required using a rule-based expert system.

Identification and Classification of Astronomical Objects

Astronomy has experienced an exponential growth in data generation. Translation of human expertise into automated classification systems for images and spectra would provide rapid screening, standardization, and the extension of expertise beyond a single site or one expert's life span. Efforts to develop rule-based expert systems for the automated analysis of images and spectra have met with mixed success (Adorf, 1989). Supervised NNs can learn directly from the previous decisions of human experts. NNs trained according to decisions of Hubble or de Vaucouleurs classify new data in a manner similar to the original expert. In addition, the NN can provide an estimate of the accuracy of the classification. Such NN techniques supply one of the few practical strategies for classifying the enormous amount of data produced by machine scans of Schmidt plates at the Cambridge Automated Plate Measuring facility and CCD data from the Sloan Digital Sky Survey projects. NN efforts to date have included star/galaxy separation, galaxy image classification, and stellar spectral classification. A classical approach to classification problems is to utilize Bayes' theorem. The a posteriori probability for a class C_j given the data vector $\mathbf{x}$ is

$$P(C_j|\mathbf{x}) \propto P(\mathbf{x}|C_j)P(C_j)$$

where $P(C_j)$ is the prior frequency distribution for the classes, and $P(\mathbf{x}|C_j)$ is the conditional probability for the data given a class (as deduced from the training set). The difficulty in applying Bayes' theorem directly to realistic problems is that it usually requires the parameterization of the probability functions involved (e.g., to Gaussians, although usually this is only a crude approximation). NNs in multiple output configuration can be shown to produce the a posteriori probability $P(C_j|\mathbf{x})$ without this complication. A practical diagnostic for NN performance as a probabilistic classifier is that the output values (per object) sum to unity. If training and testing sets both derive from the same parent distribution, then the frequency distribution $P(C_j)$ for the objects as classified by NN will be similar to the one seen in the training set.

Star/Galaxy Separation

Investigators at the University of Minnesota have pioneered the utilization of NNs for automated star/galaxy separation (Odewahn et al., 1992). An automated plate scanner digitizes images from the Palomar Sky Survey. A simple linear perceptron and a variety of BPNNs were tested using 14 image features as input vector and two output nodes for the star/galaxy classification. A {14;14,13,2} BPNN produced a 99% correct classification rate for galaxy images with blue magnitude $B \leq 18.5$ and 95% for images $18.5 \leq B \leq 19.5$. After training on all 14 input features, 14 separate test data sets were prepared. Each test set had a percentage of random noise added to one component of the input vector prior to testing. Assessing the effect of varying noise levels on classification accuracy provided an indication of both the significance of the feature and the robustness of the NN in handling signal distortion. Except for the most dominant feature (average transmission), the features important in star/galaxy separation for large diameter objects were not the same as those for small diameter images. While light gradients were useful in large objects, central transmission and image area were more dominant when classifying small images. The system also seemed relatively impervious to noise. These findings and other work in progress indicate that NNs will find a significant role in star/galaxy image separation during automated plate scanning and CCD data acquisition.

Galaxy Image Classification

Investigations carried out at the Institute of Astronomy in Cambridge, England (Storrie-Lombardi et al., 1992) have demonstrated that a BPNN can learn to classify galaxies. The origin of the "Hubble sequence," i.e., the morphological classification of galaxies according to visual image, remains a fundamental problem in understanding galaxy formation, evolution, and large-scale structure of the universe. In spite of several attempts to automate galaxy classification, current efforts remain human-intensive. Using 13 galaxy parameters measured by machine from the European Southern Observatories (ESO) galaxy survey catalog of Lauberts and Valentijn as input, a {13;13,5} BPNN was trained according to human visual classification. During training, the BPNN transformed the 13-dimensional input vector into a five-dimensional output vector classifying galaxies as ellipticals, lenticulars, early-spirals, late-spirals, and irregulars. When tested on > 3000 galaxies not used in training, the NN highest probability choice agreed with catalog classification for over 64% of galaxies as opposed to 56% for the linear automated procedure. Over 90% of the time, the NN first or second highest probability choice agreed with human experts. The BPNN produced a classification distribution similar to that of experts. The classification probabilities did sum to unity. The value of the maximal output component was significantly higher for galaxies correctly classified than for those wrongly identified. In a sense, the BPNN issued a warning of its uncertainty about wrongly classified objects. This ability to generate classification probabilities makes NNs extremely useful for assigning a reliability index to observational target lists. A subsequent prospective experiment has now demonstrated that NNs can replicate the classification of a human expert as accurately as can another human expert (Lahav et al., in preparation).

Stellar Spectral Classification

A BPNN given analog inputs can generate an analog output similar to a nonlinear fit. The NN cannot attach a classification probability to each object, but it can produce a continuous classification instead of adhering to artificial discrete classification boundaries. A project already underway to automatically classify stellar spectra from IIaO high dispersion objective prism plates uses such a single output BPNN trained on the visual classification work of N. Houk (von Hippel et al., 1994). The study is investigating NN ability to classify spectra by line information alone, continuum alone, and using full spectra. The line-only case replicates the general method employed by human classifiers with continuum-only equivalent to classification based on photometric colors. Pilot data from six plates (575 stars) were split into six training/testing sets with six different nets trained on five plates of data, then tested on the sixth unknown plate. Using this rotational scheme allowed testing of all stars as unknowns and allowed assessment of interplate training-set variability. Using 382 spectral bins as input to a {382;3,1} BPNN produced a mean rms classification uncertainty of ~1.7 spectral subtypes (line-only ~2.1 and continuum-only ~2.8). Classification accuracy is clearly sensitive to size and adequate representation of stellar subtypes in the training sample. Increasing training set size to several thousand stars across all temperature classes should produce a system approaching human expert accuracy of <1 spectral subtype. Larger data sets will also make it possible to employ a more complex NN. Since an NN can be trained to predict multiple functions simultaneously, NN output expansion to three linear nodes can provide temperature, luminosity, and metallicity classification.

Discussion

NNs have several practical advantages compared to traditional techniques. They require no prior assumptions about the statistical distribution of objects except that the training set distribution be a representative sample of expected test sets. They invoke no heuristics to define class membership. They reflect the assumptions we make about our input features. In our efforts to date, we find the choice of input features and adequate sampling far more important than choice of a particular learning parameter or network topology. When we encounter unexpected NN classification decisions, we generally do not find the problem one of entrapment in local minima or a pathological error surface. Instead, the NN has usually helped elucidate unexpected interaction effects, scaling anomalies, errors in data acquisition, and/or difficulty with feature orthogonality. Network weight analysis using the often-forgotten original BPNN configuration (with input nodes directly connected to output as well as to a hidden layer) makes it possible for us to determine that neither higher-order interaction nor nonlinear capability will save us from an ill-chosen set of input features inadequately characterizing our test objects. We consider this capacity for driving us back to the fundamentals of our data a major advantage of NN strategies.

We find it encouraging that NN algorithms can be shown to generalize and/or complement classical data analysis techniques, e.g., Fourier analysis, Wiener filtering, PCA, and vector clustering algorithms (Hertz, Krogh, and Palmer, 1991). As familiarity with NN techniques increases in astronomy, we anticipate more studies exploring this complementarity. One such study already completed (Hernandez-Parajes et al., 1992) compares principle component analysis (PCA), hierarchical clustering, Kohonen self-organizing map, BPNN, and genetic algorithms in stellar classification. Acceptance of novel search and classification strategies in a scientific community often depends on our ability to describe the new techniques in terms of familiar algorithms. With more awareness that we can view BPNNs as simply a gradient-descent-search strategy and most unsupervised NNs as vector analysis variants, then we can expect increased investigation of more powerful NN algorithms sharing much with efforts in statistical physics, e.g., simulated annealing, Boltzmann machines, and the Ising spin-glass models.

Astronomy will need to face one of the major NN problems enumerated by Minsky and Papert (1969) concerning scalability in NN paradigms. For NNs to move away from using small numbers of parameters as input and try to directly utilize pixel input from large CCDs or handle real-time control problems, we shall need to find more efficient algorithms and configurations. Several possibilities exist. Combinations of unsupervised and supervised NNs may offer the rapid learning characteristic of unsupervised systems and the precision of supervised BPNNs. Aleksander and Morton (1990) have attacked speed and complexity by suggesting use of probabilistic logic node pyramids in hardware. Pao (1989) has demonstrated that eliminating hidden layers in BPNNs while using higher-order input vectors considerably speeds convergence but at considerable cost to system memory. Here, the higher-order interaction inputs function as hidden layer nodes with input to hidden weights held constant. Success depends on the intelligent choice of interactions to provide a sufficiently rich set of input features. Genetic algorithm strategies may provide a method for evolving more computationally efficient learning rules (Jones, 1993). In our own efforts on star and galaxy classification, preprocessing input data by PCA and then training on the PCA factors can significantly decrease the size of the input vector and speed both training and testing. Weight analysis can then provide information about the orthogonality, linearity, and information content of input features.

Over the next decade, we anticipate NN technology will play a significant role in adaptive optics, star/galaxy separation, automated stellar and galaxy classification, and the development of survey target lists complete with classification probabilities.

Acknowledgments. MCSL thanks the Harrison Watson Fund and Clare College for financial support and the Aspen Center for Physics for assistance in the preparation of this work.

Road Map: Applications of Neural Networks
Background: I.3. Dynamics and Adaptation in Neural Networks

References

Adorf, H.-M., 1989, Connectionism and neural networks, in *Knowledge-Based Systems in Astronomy* (A. Heck and F. Murtagh, Eds.), Heidelberg: Springer-Verlag, pp. 215–245. ◆

Aleksander, I., and Morton, H., 1990, *An Introduction to Neural Computing*, London: Chapman and Hall. ◆

Angel, J. R. P., Wizinowich, P., Lloyd-Hart, M., and Sandler, D., 1990, Adaptive optics for array telescopes using neural network techniques, *Nature*, 348:221–224.

Fraser, D. D., Khan, Z., and Levy, D. C., 1992, A neural network for meteor trail classification, in *Artificial Neural Networks* (I.

Aleksander and J. Taylor, Eds.), Amsterdam: Elsevier, pp. 1155–1158.

Hernandez-Parajes, M., Commellas, F., Monte, E., and Floris, J., 1992, Classifying stars: A comparison between classical and neural network algorithms, in *Proc. Astron. from Large Databases II*, (A. Heck and F. Murtagh, Eds.), Heidelberg: Springer-Verlag, pp. 325–330. ◆

Hertz, J., Krogh, A., and Palmer, R. G., 1991, *Introduction to the Theory of Neural Computation*, Redwood City, CA: Addison-Wesley. ◆

Jones, A. J., 1993, Genetic algorithms and their applications to the design of neural networks, *Neural Comput. Applic.*, 1:32–45.

Lahav, O., Naim, A., Buta, R. J., Corwin, H. G., de Vaucouleurs, G., Dressler, A., Huchra, J. P., van den Bergh, S., Raychaudhury, S., Sodré Jr., L., and Storrie-Lombardi, M. C. (in preparation), Galaxies, human eyes and artificial neural networks.

Lloyd-Hart, M., Wizinowich, P., McLeod, B., Wittman, D., Colucci, D., Dekany, R., McCarthy, D., Angel, R., and Sandler, D., 1992, First results of an on-line adaptive optics system with atmospheric wavefront sensing by an artificial neural network, *Astron. J. Let.*, 390:L42–44.

Miller, A. S., 1993, A review of neural network applications in astronomy, *Vistas Astronom.*, 36(2):141–161. ◆

Minsky, M. L., and Papert, S., 1969, *Perceptrons: An Essay in Computational Geometry*, Cambridge, MA: MIT Press.

Odewahn, E. B., Stockwell, R. L., Pennington, R. M., Humphreys, R. M., and Zumach, W., 1992, Automated star/galaxy discrimination with neural networks, *Astron. J.*, 103:318–331.

Ozard, S., and Morbey, C., 1993, The application of artificial neural networks for telescope guidance: A feasibility study for Lyman FUSE, *Astron. Soc. Pac.*, 105:625–629.

Pao, Y.-H., 1989, *Adaptive Pattern Recognition and Neural Networks*, New York: Addison-Wesley.

Storrie-Lombardi, M. C., Lahav, O., Sodré, L., and Storrie-Lombardi, L. J., 1992, Morphological classification of galaxies by artificial neural networks, *Mon. Not. Royal Astron. Soc.*, 259:8p–12p.

von Hippel, T., Storrie-Lombardi, L. J., Storrie-Lombardi, M. C., and Irwin, M., 1994, Automated classification of stellar spectra, I: Initial results with artificial neural networks, *Mon. Not. Royal Astron. Soc.*, 269:97–104.

Auditory Cortex

Shihab A. Shamma

Introduction

The auditory cortex plays a critical role in the perception and localization of complex sounds. It is the last station in a long chain of processing centers beginning with the cochlea of the inner ear and passing through the cochlear nuclei (CN) (see AUDITORY PERIPHERY AND COCHLEAR NUCLEUS), the superior olivary complex (SOC), the lateral lemniscus, the inferior colliculus (IC), and the medial geniculate body (MGB) (Figure 1). Recent studies have rapidly expanded our knowledge of the neuroanatomical structure, the subdivisions, and the connectivities of all central auditory stages (Winer, 1992). However, apart from the midbrain cochlear and binaural SOC nuclei, relatively little is known about the functional organization of the central auditory system, especially compared to the visual and motor systems. Consequently, modeling cortical auditory networks is complicated by the uncertainty of what the cortical machinery is exactly trying to accomplish.

One exception to this state of affairs is the highly specialized echolocating bat, in which these uncertainties are much re-lieved by the existence of a stereotypical behavioral repertoire that is closely linked to its acoustic environment (see ECHOLOCATION). This has made it possible to construct a functional map of the auditory cortex, revealing the specific acoustic features extracted and represented in the cortex. In turn, these cortical maps have acted as a guide to discovering the organization and nature of the transformations occurring in lower auditory centers, such as the MGB, IC, and the SOC. Thus, it has become meaningful in these species to investigate and model cortical and other central auditory neural networks.

In other mammals, it is more difficult to isolate an auditory behavior and its associated stimulus features with comparable specificity. Nevertheless, a few tasks have been broadly accepted as being vital for all species, such as sound localization, timbre recognition, and pitch perception. For each, evidence of various functional and stimulus feature maps has been found or postulated, a significant number of them being found in the last few years. In this review, a few examples of such maps are elaborated, along with their relationships to the more intuitive and better understood case of echolocating bats. In

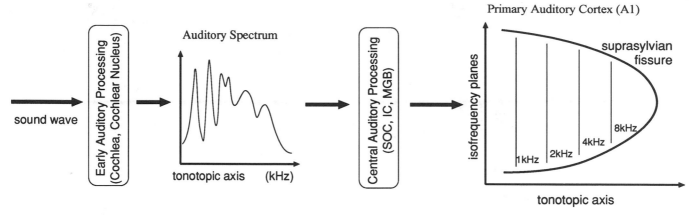

Figure 1. Schematic representation of the multiple stages of processing in the mammalian auditory pathway. Sound is analyzed in the cochlea, and an estimate of the acoustic spectrum (an auditory spectrum) is known to be extracted at the cochlear nucleus (Blackburn and Sachs, 1990). The tonotopic organization of the cochlea is preserved all the way up to the cortex, where it has a two-dimensional layout. The isofrequency plane encodes perhaps other features of the stimulus.

each example, our goal is to determine how and whether models of the underlying neural networks can advance our understanding of the auditory cortex.

Parcellization and Neuroanatomy of the Auditory Cortex

The layout and neural structure of the auditory cortex is, in many respects, similar to that of other sensory cortices (Winer, 1992). For instance, based on cytoarchitectonic criteria and patterns of connectivity, the auditory cortex is subdivided into a primary auditory field (AI) and several other surrounding fields, e.g., the anterior auditory field (A) and the secondary auditory cortex (AII). The number and specific arrangement of the surrounding fields vary among different species, presumably reflecting the complexity of the animal's acoustic environment. The AI (and possibly other fields) is further subdivided into smaller regions, serving perhaps different functional roles, such as echo delay and amplitude measurements in the bat (see ECHOLOCATION).

The anatomical parcellization of the auditory cortex into different fields is mirrored by physiologically based divisions. Most important is the systematic frequency organization in different fields, or the so-called tonotopic maps. For example, AI cells are spatially ordered based on the tone frequency to which they best respond, i.e., their best frequency (BF). They also respond vigorously to the onset of a tone and exhibit little evidence of adaptation to its repeated presentations. In other fields, cells may be less frequency-selective, respond more adaptively, or be totally unresponsive to single tones, preferring more spectrally or temporally complex stimuli. A sudden change in these response patterns or in the gradual spatial order of the tonotopic map is usually taken to signify a border between different fields. In the cat, which has the most extensively mapped auditory cortex, four well-ordered tonotopic fields have been described, together with many other less precisely delineated secondary areas (Clarey, Barone, and Imig, 1992).

Timbre: Models for the Encoding of Spectral Profiles

Recognizing and classifying environmental sounds is critical for the survival and propagation of many animals. Although a multitude of cues are responsible, the single most important one is the shape of the so-called spectral envelope (or the spec-

tral profile) of the sound (Green, 1988). It is largely this cue which allows us to distinguish between speech vowels or between different instruments playing the same note. The spectral profile emerges early in the auditory system (Blackburn and Sachs, 1990) as the sound is analyzed into different frequency bands, in effect distributing its energy across the tonotopic axis (the auditory sensory epithelium) (see Figure 1). As far as the central auditory system is concerned, the spectral profile is a one-dimensional pattern of activation analogous to the two-dimensional distribution of light intensity upon the retina.

An important organizational feature of the central auditory system is the expansion of the one-dimensional tonotopic axis of the cochlea into a two-dimensional sheet, with each frequency represented by an entire sheet of cells (see Figure 1). An immediate question thus arises as to the functional purpose of this expansion and the nature of the acoustic features that might be mapped along these isofrequency planes. For example, one might conjecture that the amplitude or the local shape of the spectrum is explicitly represented along this new dimension.

In general, there are two ways in which the spectral profile can be encoded in the central auditory system. The first is *absolute*: that is, the encoding of the spectral profile in terms of the absolute intensity of sound at each frequency, in effect combining both the shape information and the overall level. The second is *relative*, whereby the spectral profile shape is encoded separately from the overall loudness of the stimulus. Examples of each of these two hypotheses are discussed in the following sections.

The Best-Intensity Model

The first hypothesis is motivated primarily by the strongly nonmonotonic responses as a function of stimulus intensity observed in many cortical and other central auditory cells (Clarey, Barone, and Imig, 1992). In a sense, one can view such a cell's response as being selective to (or encoding) a particular intensity. Consequently, a population of such cells, tuned to different frequencies and intensities, can provide an explicit representation of the spectral profile by their spatial pattern of activity (Figure 2). This scheme is not a true transformation of the spectral features represented, but rather is strictly a change in the means of the representation. The most compelling example of such a representation is that in the DSCF (Doppler-

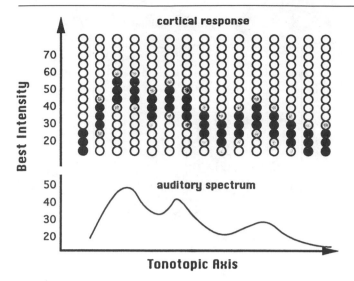

Figure 2. Schematic diagram of the way in which the spectral profile (lower plot) can be encoded by arrays of nonmonotonic cells (circles) tuned to different BFs (along the tonotopic axis) and best intensities (BIs). The black circles signify strongly activated cells, whereas the white circles indicate weakly activated cells. Thus, a peak in the input pattern located at a given BF and at an intensity of 40 dB would best activate cells with the same BF and BI.

Shifted Constant Frequency processing) area of AI in the mustache bat (see ECHOLOCATION). However, an extension of this hypothesis to multicomponent stimuli (i.e., as depicted in Figure 2) has not been demonstrated in any species.

The Ripple Analysis Model

The second hypothesis, in which the relative shape of the spectrum is encoded, is supported by recent physiological experiments in cat and ferret AI, coupled with psychoacoustical studies in human subjects. The data reveal a substantial

transformation of the way the spectral profile is represented centrally. Specifically, besides the tonotopic axis, two additional features of the response areas of AI neurons (the analog of the receptive fields in the visual system) are found to be topographically mapped across the isofrequency planes. They are the bandwidth and symmetry of the response areas, depicted schematically in Figure 3 as the scale and symmetry axes, respectively, and discussed in greater detail as follows.

Changes in cell response areas. Cell response areas, i.e., the excitatory and inhibitory responses they exhibit to a tone of various frequencies and intensities, change their bandwidth in an orderly fashion along the isofrequency planes (Mendelson and Schreiner, 1990; Shamma, Vranic, and Versnel, 1995). Near the center of AI, cells are narrowly tuned. Toward the edges, they become more broadly tuned. This orderly progression occurs at least twice, and it correlates with several other response parameters, such as increasing response thresholds toward the edges.

An intuitively appealing implication of this finding is that response areas of different bandwidths are selective to spectral profiles of different widths. Thus, broad spectral profiles (e.g., broad peaks or gross trends, such as spectral tilts secondary to preemphasis) would best drive cells with wide response areas. Similarly, narrower spectral profiles (e.g., sharp peaks or edges, or fine details of the spectral profile) would best be represented in the responses of cells with more compact response areas. In effect, having a range of response areas at different widths allows us to encode the spectral profile at different scales or levels of detail (resolution). From a mathematical perspective, this is basically equivalent to analyzing the spectral profile into different scales or "bands," much like performing a Fourier transform of the profile, hence representing it as a weighted sum of elementary sinusoidal spectra, usually known as *ripples* (Shamma, Vranic, and Versnel, 1995). Coarser scales then correspond to the low-frequency ripples, whereas finer scales correspond to the high-frequency ripples.

Physiological responses in AI support this model in that AI cells act as selective filters for ripple frequency, with each tuned around a specific (characteristic) ripple frequency, Ω_o

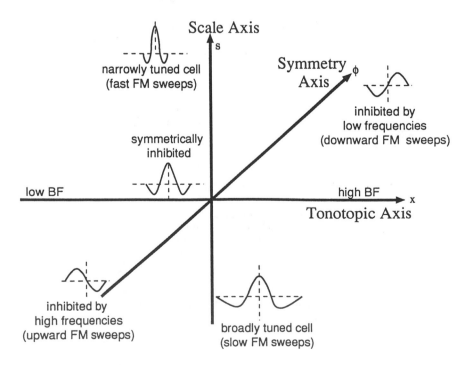

Figure 3. Schematic diagram of the three representational axes thought to exist in AI: the tonotopic (BF) axis, the scale (or bandwidth) axis, and the symmetry axis.

(Shamma, Vranic, and Versnel, 1995). If this ripple transfer function is inverse Fourier transformed, we obtain the so-called response area function (RF), which essentially resembles the outlines of the usual response area measured with tones. In general, Ω_o reflects the width or scale of the RF: a higher Ω_o usually implies a finer or narrower RF. Furthermore, AI units exhibit a uniform columnar organization with respect to their characteristic ripple, much like that seen for the BF (Shamma, Vranic, and Versnel, 1995). They also have an ordered and clustered topographic distribution along the isofrequency planes of AI, with Ω_o usually highest near the center of AI and lowest near the edges, thereby mirroring the response area bandwidth maps described earlier.

Changes in inhibitory response areas. The response areas exhibit systematic changes in the symmetry of their inhibitory response areas. For instance, cells in the center of AI have sharply tuned excitatory responses around a BF, flanked by symmetric inhibitory response areas. Toward the edges, the inhibitory response areas become significantly more asymmetrical, with inhibition dominated by either higher- or lower-than-BF frequencies. This trend is repeated at least twice across the length of the isofrequency plane (Shamma et al., 1993).

It is intuitively clear that response areas with different symmetries would respond best to input profiles that match their symmetry. For instance, an odd-symmetrical response area would respond best if the input profile had the same local odd-symmetry, and worst if it had the opposite odd-symmetry. As such, one can state that a range of response areas of different symmetries (represented by the symmetry axis in Figure 3) is capable of encoding the shape of a local region in the profile. From an opposite perspective, it can be shown mathematically that the local symmetry of a pattern can be changed by manipulating only the phase of its Fourier transform (Shamma, Vranic, and Versnel, 1995). Therefore, the axis of response area symmetries, in effect, is able to encode the phase of the profile transform, thus providing a complementary description to that of the magnitude along the scale axis discussed earlier.

Once again, physiological responses in AI support this model in that cells are usually tuned to specific or characteristic phase (Φ_o) of a rippled spectrum, and hence exhibit various asymmetrical response areas (Shamma, Vranic, and Versnel, 1995). Furthermore, Φ_os are both columnarly and topographically organized in a manner which closely mirrors that of the response area asymmetries described earlier.

In summary, according to the ripple analysis model, properties of response areas in AI are conceptually organized along three organizational axes (see Figure 3): (1) the tonotopic axis (as reflected by the systematic change in BFs); (2) the symmetry axis, as reflected by the change in response area asymmetries along the isofrequency planes; and (3) the scale axis, as reflected by the systematic change in bandwidths along the isofrequency planes.

An arbitrary spectral profile, e.g., a speech vowel /aa/, would then be analyzed and detected along all three dimensions, as illustrated in the AI model response shown in Figure 4 (Wang and Shamma, in press). Thus, at each point along the surface of AI, the response can be computed by convolving the input pattern with an RF that has the BF and the scale (Ω_o) indicated by the tonotopic and scale axes. The symmetry index is represented by the clockwise direction of the arrows, with zero phase pointing upward. The strength of the response is denoted by the gray scale. Note that peaks in the envelope of the spectrum (called *formants*) are relatively broad in bandwidth and thus are represented in the low scale regions, generally <2 cycles/octave. In contrast, the fine structure of the spectral harmonics

Cortical Response to the Vowel /aa/

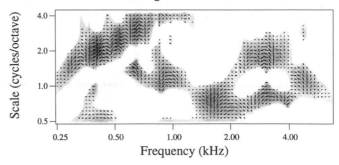

Auditory Spectrum of Vowel /aa/

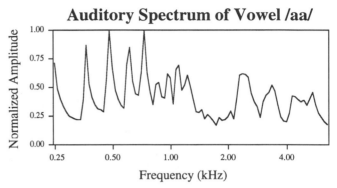

Figure 4. The auditory spectrum of naturally spoken vowel /aa/ (top) and the corresponding sharpened cortical model output (bottom). At each point, the output is computed by convolving the input pattern with an RF that has the BF and bandwidth indicated along the tonotopic and scale axes. The scale axis is labeled by the characteristic ripple of the cortical cells. The strength of the response is denoted by the gray scale. The symmetry index is represented by the clockwise direction of the arrows, with zero phase pointing upward.

is only visible at high scales (usually >1.5–2 cycles/octave; top left corner of the plot).

Finally, it should be recognized that the ripple analysis model in AI implies a functional organizational that is closely analogous to the well studied organization of receptive fields in the primary visual cortex (De Valois and De Valois, 1988).

Models of AI Responses to Frequency-Modulated Tones

Another feature of cortical responses that has been the subject of neural network modeling is that which is due to frequency-modulated (FM) tones. These stimuli are important because they mimic the dynamic aspects of many natural vocalizations, as in speech consonant-vowel combinations or in the trills of many birds and other animal sounds. The effects of manipulating two specific parameters of the FM sweep—its direction and rate—have been well studied. In several species, and at almost all central auditory stages, cells can be found that are selectively sensitive to the FM direction and rate. Most studies have confirmed a qualitative theory in which directional selectivity arises from an asymmetrical pattern of inhibition in the response area of the cell (Shamma et al., 1993; Wang and Shamma, in press), whereas rate sensitivity is correlated to the bandwidth of the response area (Heil, Langner, and Scheich, 1992).

Specifically, FM responses may be modeled as a temporal sequence activating the excitatory and inhibitory portions of the response area. For example, if an FM sweep first traverses the excitatory response area, discharges will be evoked which cannot be influenced by the inhibition activated later by the ongoing sweep. Conversely, if an FM sweep first traverses the inhibitory area, the inhibition may still be effective at the time the tone sweeps through the excitatory area. If the response is assumed to be a temporal summation of the instantaneous triggers, then it follows that it will be smaller in this latter direction of modulation. This theory also explains why the response area bandwidth is correlated with the FM rate preference (cells with very broad response area respond best to very fast sweeps), and why FM directional selectivity decreases with the FM rate.

Models of Pitch Representation in the Central Auditory System

A sound complex consisting of several harmonics is heard with a strong pitch at the fundamental frequency of the harmonic series, even if there is no energy at all at that frequency. This percept has been variously called the missing fundamental, virtual pitch, or residue pitch (Moore, 1989). A large number of psychoacoustical experiments have been carried out to elucidate the nature of this percept and its relationship to the physical parameters of the stimulus. Basically, all models fall into one of two camps. The first believes that the pitch is extracted explicitly from the harmonic spectral pattern. This can be accomplished in a variety of ways, for instance, by finding the best match between the input pattern and various harmonic templates assumed to be stored in the brain (Goldstein, 1973). The second group claims that the pitch is extracted from the periodicities in the time-waveform of responses in the auditory pathway which can be estimated, for example, by computing their autocorrelation functions (Moore, 1989). In these latter models, some form of organized delay lines are assumed to exist in order to do the computations, much like those that seem to exist in the FM-FM area of the mustached bat (see ECHOLOCATION for details).

In all pitch models, however, the extracted pitch is assumed to be represented finally as a *spatial* map in higher auditory centers. This is because many studies have confirmed that neural synchrony to the repetitive features of a stimulus, be it the waveform of a tone or its AM modulations, becomes progressively worse toward the cortex (Langner, 1992). It is a remarkable aspect of pitch that, despite its fundamental and ubiquitous role in auditory perception, only a few reports exist of physiological evidence of spatial pitch maps, and none has been independently confirmed. One source is human subjects who have undergone nuclear magnetic resonance (NMR) scans of the primary auditory cortex. The other is from multiunit mappings in various central auditory structures (Schreiner and Langner, 1988).

Of course, the difficulty in finding spatial pitch maps in the auditory cortex may be attributable to the fact that they do not exist! This possibility is counterintuitive, given the results of ablation studies that show that bilateral cortical lesions in the auditory cortex severely impair the perception of pitch of complex sounds but do not affect the fine discrimination of frequency and intensity of simple tones. Another more likely possibility is that the maps sought are not at all as straightforward as we imagine.

As an example, consider the representation of the pitch of the vowel complex /aa/ in the ripple analysis model representation shown in Figure 4. Specifically, note the distinctive regular appearance of the harmonic portion of the spectrum (< 1 kHz) in the cortical representation. On a logarithmic frequency axis, a harmonic series appears as a train of peaks with progressively closer spacing or, in effect, a logarithmically increasing spatial frequency. Since the scale is a logarithmic mapping of spatial frequency, this trend is resolved as a straight line in a tonotopic-scale plane, which is outlined by the dashed lines. A change in

the fundamental frequency, corresponding to a change in the pitch of the vowel, simply results in a translation of the harmonic peak pattern along the tonotopic axis. Therefore, an harmonic series with a higher fundamental frequency appears again as a straight line with the *same* slope as before, but horizontally shifted to the right. If a harmonicity detection mechanism is based on such a straight line with this particular slope, it should be able to interpolate or extrapolate the missing components to determine the fundamental frequency, much like the auditory perception of pitch (Moore, 1989).

Models of Sound Localization

It has been recognized for many years that the auditory cortex (and especially the AI) is involved in sound localization (see also SOUND LOCALIZATION AND BINAURAL PROCESSING). Detailed physiological studies have further confirmed that AI cells are rather sensitive to all kinds of manipulations of the binaural stimulus (Clarey, Barone, and Imig). For instance, changing either of the two most important binaural cues—interaural level difference (ILD) or interaural time difference (ITD)—causes substantial changes in their firing rate patterns. This sensitivity to interaural cues has its origins early in the auditory pathway, at the superior olivary complex, where the first convergence of binaural inputs occurs. However, despite this diversity, two elements typical of a functional organization of AI have been lacking. The first of these is the lack of a significant transformation of the single-unit responses. For example, if ILD-sensitive cells are to encode the location of a sound source based on this cue, then they ought to become uniformly more stable with overall sound intensity. This, however, does not seem to be the case (Semple and Kitzes, 1993). The second element that is lacking is a topographical distribution of the responses with respect to these cues or to a more complex combination of features, e.g., a map of acoustic space derived from ILD and ITD cues, as in the barn owl (Sullivan and Konishi, 1986).

A map of auditory space has indeed been found in the superior colliculus of several mammals. No such map, however, has yet been detected in AI or other cortical fields despite intensive efforts (Clarey, Barone, and Imig, 1992). What has been found, however, is a topographic order of certain binaural responses along the isofrequency planes of AI. Specifically, cells excited equally well by sounds from both ears (called EE cells), and others inhibited by ipsilateral sounds (called EI cells) are found clustered in alternating bands that parallel the tonotopic axis. One possible functional model that utilizes such maps assumes that EI cells are tuned to particular ILDs and, hence, encode the location of a sound source based on this cue. In contrast, EE cells would encode the absolute level of the sound. However, there is little evidence to support this hypothesis in the sense that neither EE nor EI cells are particularly stable encoders of specific ILDs or absolute sound levels (Semple and Kitzes, 1993). An alternative hypothesis that has recently been proposed is that these cells encode the absolute levels of the stimulus at each ear, rather than the difference and average binaural levels, as previously postulated (Semple and Kitzes).

Summary

The study of central auditory function has recently reached a sufficiently advanced stage to allow for meaningful quantitative and neural network models to be formulated. In most mammals, these models are still systemic in nature, with a primary focus on understanding the overall functional organization of the cortex and other central auditory structures. In the bat and other specialized animals, the models are somewhat more detailed, addressing specific neural mechanisms, such as the coincidences and the delay lines of the FM-FM areas. The auditory system, with its multitude of diverse functions, and its combination of temporal and spatial processes, should thus prove to be a valuable window into the brain, and an effective vehicle for understanding its underlying mechanisms.

Road Maps: Mammalian Brain Regions; Other Sensory Systems
Related Reading: Sound Localization and Binaural Processing; Auditory Periphery and Cochlear Nucleus

References

Blackburn, C. C., and Sachs, M. B., 1990, The representation of steady-state vowel sound /e/ in the discharge rate patterns of cat anteroventral cochlear nucleus, *J. Neurophysiol.*, 63:1191–1212.

Clarey, J., Barone, P., and Imig, T., 1992, Physiology of thalamus and cortex, in *The Mammalian Auditory Pathway: Neurophysiology* (D. Webster, A. Popper, and R. Fay, Eds.), New York: Springer-Verlag, pp. 153–334.

De Valois, R., and De Valois, K., *Spatial Vision*, Oxford: Oxford University Press, 1990.

Goldstein, J., 1973, An optimum processor theory for the central formation of the pitch of complex tones, *J. Acoust. Soc. Am.*, 54: 1496–1516.

Green, D., 1988, *Profile Analysis*, New York: Oxford Press.

Heil, P., Langner, G., and Scheich, H., 1992, Processing of FM stimuli in the chick auditory cortex analogue: Evidence of topographic representations and possible mechanisms of rate and directional sensitivity, *J. Comp. Physiol.* [A], 171:583–600.

Langner, G., 1992, Periodicity coding in the auditory system, *Hear. Res.*, 6:115–142.

Mendelson, J., and Schreiner, C., 1990, Functional topography of cat primary auditory cortex: Distribution of integrated excitation, *J. Neurophysiol.*, 64:1442–1459.

Moore, B., 1989, *An Introduction to the Psychology of Hearing*, London: Academic Press, chap. 5.

Schreiner, C., and Langner, G., 1988, Periodicity coding in the inferior colliculus of the cat. II. Topographical organization, *J. Neurophysiol.*, 60:1823–1840.

Semple, M., and Kitzes, L., 1993, Binaural processing of sound pressure level in cat primary auditory cortex: Evidence for a representation based on absolute levels rather than interaural level differences, *J. Neurophysiol.*, 69:449–461.

Shamma, S., Fleshman, J., Wiser, P., and Versnel, H., 1993, Organization of response areas in ferret primary auditory cortex, *J. Neurophysiol.*, 69:367–383.

Shamma, S., Vranic, S., and Versnel, H., 1995, Representation of spectral profiles in the auditory system: Theory, physiology, and psychoacoustics, in *Advances in Hearing Research: Proceedings of the 10th International Symposium on Hearing* (G. Manley, G. Klump, G. Koppl, H. Fastl, and H. Occkinghaus, Eds.), Singapore: World Scientific.

Sullivan, W., and Konishi, M., 1986, Neural map of interaural phase difference in the owl's brainstem, *Proc. Natl. Acad. Sci. USA*, 83: 8400–8404.

Wang, K., and Shamma, S., in press, Representation of spectral profiles in the primary auditory cortex, *IEEE Trans. Audio Speech*.

Winer, J., 1992, The functional architecture of the medial geniculate body and the primary auditory cortex, in *The Mammalian Auditory Pathway: Neuroanatomy* (D. Webster, A. Popper, and R. Fay, Eds.), New York: Springer-Verlag, pp. 222–410.

Auditory Periphery and Cochlear Nucleus

David C. Mountain

Introduction

The auditory periphery transforms a very high information rate signal into a group of lower information rate signals. This process of parallelization is essential because the potential information rate in the acoustic stimulus is on the order of 0.5 MB per second, and yet typical auditory nerve (AN) fibers have maximum sustained firing rates of 200 per second. If we assume the information content of a single action potential is 1 bit, then the incoming acoustic information will need to be split across hundreds of nerve fibers to avoid loss of information. The cochlear nucleus (CN) continues this process of parallelization by creating multiple representations of the original acoustic stimulus, with each representation presumably emphasizing different acoustic features.

This article reviews the anatomy from a systems perspective and emphasizes what is known about the signal processing functions performed by the auditory periphery and the cochlear nucleus. The emphasis of this review is on monaural signal processing, although the cochlea and cochlear nucleus receive input from descending pathways which are binaural in nature (Warr, in Webster et al., 1992). In addition, there is evidence that the cochlear nuclei provide input to each other.

System Overview

The first stages of auditory processing are summarized in block diagram form in Figure 1. The cochlea takes a serial time signal and converts it into parallel signals in the form of the mechanical stimuli to the hair bundles of the hair cells located in the organ of Corti. This process acts as a bank of mechanical filters, with each filter only responding to a limited range of frequencies and with the center frequency of the filter changing as a function of position along the length of the cochlea. The outer hair cells (OHCs) appear to act collectively as a nonlinear electromechanical feedback system which enhances cochlear sensitivity for low-level sounds, but which provides little enhancement for high-level sounds.

The major output of the second stage consists of the receptor potentials of the inner hair cells (IHCs). The IHC receptor potential is a rectified and low-pass filtered version of the mechanical stimulus, with an amplitude roughly proportional to the log of the stimulus power. This logarithmic relationship is attributable to a combination of nonlinearity in the mechanics and in the IHC transduction process, and gives the hearing apparatus its very large dynamic range.

A typical mammalian cochlea will have around 2000 IHCs. The center frequencies of the mechanical filters driving the IHCs are closely spaced (approximately 0.005 octave per IHC), and yet the filter functions are relatively broad (approximately 0.1 to 0.3 octave). These facts mean that the representation of the acoustic signal (as seen in the IHC receptor potentials) is similar to a highly smoothed spectrum plotted on log-log coordinates.

The output of the IHCs becomes the input to the next level of processing—the type I spiral ganglion (SG) cells—which results in the firing patterns of fibers in the auditory nerve. At this level, the processing takes the form of automatic gain con-

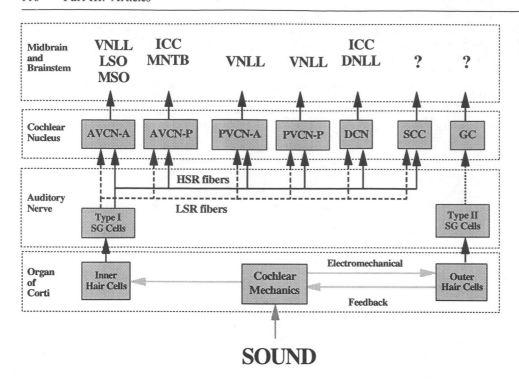

Figure 1. Block diagram of early processing in the auditory system. The gray lines indicate mechanical signals, whereas the black lines indicate neural signals. (See text for an explanation of the abbreviations.)

trol and further parallelization into high (HSR) and low (LSR) spontaneous-rate pathways. A third pathway comes from the type II spiral ganglion cells which receive their input from the OHCs.

The auditory nerve provides the input to the cochlear nucleus where the process of feature extraction begins. These features include onsets, temporal fine structure, and amplitude fluctuations. The cochlear nucleus is the first level at which there is significant involvement of inhibitory pathways and interactions across frequency channels. The output of the cochlear nucleus consists of multiple pathways originating from different cell types within the nucleus.

Cochlear Mechanics

The cochlea (Ryugo, in Webster et al., 1992) consists of a spiral-shaped, fluid-filled tube embedded in the temporal bone. It is separated into three longitudinal compartments by two membranes: the basilar membrane and Reissner's membrane. From a hydromechanical and sensory-physiological point of view, the basilar membrane is the more important structure of the two. It supports the organ of Corti, which contains the sensory hair cells. The combination of the basilar membrane and the organ of Corti is often referred to as the cochlear partition. Acoustic stimuli are coupled to the cochlear fluids by the middle ear ossicles, resulting in a pressure difference across the cochlear partition. The partition is a resonant structure with resonant frequencies (characteristic frequencies) that are graded from high to low along its length.

Direct measurements of basilar membrane motion demonstrate that, at low sound levels, the response can be highly tuned, with each cochlear location only responding to a narrow range of frequencies (Hubbard and Mountain, in Hawkins et al., in press) which widens as sound level is increased. The tuning of the cochlear partition appears to involve the presence of a group of sensory cells, the OHCs. The OHCs respond to mechanical stimuli with voltage changes which, in turn, cause the OHCs to change their length (Hubbard and Mountain, in

Hawkins et al., in press). These voltage-dependent length changes appear to be attributable to voltage-sensitive transmembrane proteins. This novel form of electromotility is piezoelectric in nature, allowing the length changes to achieve very high velocities (Mountain and Hubbard, 1994).

Many hydromechanical models have been proposed to explain these findings (Hubbard and Mountain, in Hawkins et al., in press). The passive properties of the cochlea can be modeled functionally as a transmission line with a propagation velocity and a local resonant frequency which decreases exponentially along its length. Current models differ from one another mostly in the level of anatomical detail and in how the active role of the OHCs is represented. In most models, the OHCs are assumed to act locally by modifying the partition impedance through regenerative feedback. An alternative approach is to assume that there are two modes of energy propagation which are coupled through the OHCs to create a traveling-wave amplifier.

All of these hydromechanical models are computationally intense. As a result, it is common practice for modeling studies that attempt to represent multiple levels of the auditory pathway to represent cochlear mechanics with a bank of digital band-pass filters which capture the salient features of mechanical frequency response (Hubbard and Mountain, in Hawkins et al., in press). One of the most commonly used approaches is the gamma-tone filter bank, so called because its impulse response is a sinusoid which is modulated by a gamma function. The impulse response, $h(t)$, of a gamma-tone filter may be determined by the following equation:

$$h_j(t) = t^{n-1} e^{-2\pi b_j t} \cos(2\pi f_j t + \phi_j) \tag{1}$$

where f_j is the characteristic (center) frequency of the jth filter in Hertz, b_j is the bandwidth of the jth filter, ϕ_j is the phase offset of the jth filter, and n is a constant (typically, $n = 4$).

Filters of this type reproduce the magnitude of the cochlear frequency response reasonably well, but have difficulties capturing some of the features of the phase response. Also, since this is a linear filter, it cannot reproduce the changes in cochlear tuning which occur with changes in stimulus level. Recently, Carney (1993) has used a feedback approach in order to replicate some of the nonlinear features of cochlear mechanics.

Inner Hair Cells

Although much progress has been made in measuring basilar membrane motion, little direct data exist to explain how this motion gets coupled to the IHC hair bundle. The hair bundles protrude out of the organ of Corti and are in contact with, or at least in close proximity to, the tectorial membrane. The tectorial membrane is a flap of extracellular matrix which lies over the top of the organ of Corti. When the organ of Corti is displaced, a shearing motion is believed to result between the organ and the tectorial membrane that deflects the hair bundles. Comparisons of IHC receptor potentials to inferred basilar membrane motion have led to the hypothesis that the hair bundle motion is a high-pass filtered version of basilar membrane motion. Alternatively, Mountain and Cody (1989) have proposed that the OHCs, through their electromotility, displace the IHC hair bundles via movements of the tectorial membrane. Since the details of the coupling process are at present unclear, a simple high-pass filter is usually used to model the process, with cutoff frequencies around 500 Hz being typical.

The mechanical-to-electrical transduction process in hair cells is believed to be the result of tension-gated channels located in the hair bundle (Mountain and Hubbard, in Hawkins et al., in press). The relationship between stereocilia displacement x and the mechanically induced conductance change $G(x)$ is most commonly modeled as

$$G(x) = \frac{G_{\max}}{1 + e^{-S_x(x-x_o)}} \qquad (2)$$

where $G_{\max}$ is the maximum conductance, S_x is the sensitivity, and x_o is an offset constant.

Although inner hair cells also contain voltage-dependent conductances, most models only include the mechanically sensitive conductance coupled to a linear leakage conductance and a linear membrane capacitance. In fact, most models simplify things further by treating the mechanically sensitive conductance as a soft rectifier and then low-pass filtering the rectified signal using a cutoff frequency of around 1 kHz (Mountain and Hubbard, in Hawkins et al., in press).

If a linear filter bank is used to represent cochlear mechanics, then it is often desirable to use a rectification function which includes considerable compression to accommodate the large dynamic range of many acoustic signals. Since the DC receptor potentials of IHCs measured using best-frequency tones appear to grow as a logarithmic function of sound pressure, a combination of a half-wave rectifier followed by a logarithmic compressor provides a reasonable model (Mountain and Hubbard, in Hawkins et al., in press).

Auditory Nerve

Each IHC synapses with 10–30 type I AN fibers, the cell bodies of which are located in the SG (Ryugo, in Webster et al., 1992). AN fibers exhibit spontaneous activity in the absence of sound, and they are often segregated into low (LSR) and high (HSR) spontaneous rate categories. Spontaneous rate (SR) tends to correlate with threshold, with HSR fibers being more sensitive to sound stimuli. A single IHC will synapse with ganglion cells of both SR categories. A comparison of IHC and SG cell responses (Mountain and Hubbard, in Hawkins et al., in press)

has revealed that the steady-state firing rate of low-threshold SG cells has a dynamic range which corresponds to only about 1–2 mV of IHC receptor potential, whereas that of high-threshold cells has a dynamic range of approximately 8 mV of IHC receptor potential.

The firing rate of AN fibers in response to sustained tones exhibits an initial rapid increase followed by adaptation to a lower steady-state rate (Ruggero, in Popper and Fay, 1992). The steady-state response has only a limited dynamic range, typically saturating at sound levels of approximately 20 dB above the fiber's threshold. There are three components to the adaptation. The fastest component, usually called rapid adaptation, has a time constant of a few milliseconds and creates an onset response with a large dynamic range. The second component, usually called short-term adaptation, has a time constant of a few tens of milliseconds. It creates a slower component immediately after the onset response which has a smaller dynamic range similar to that of the steady-state response. The third component of adaptation operates on a time scale of seconds and is rarely included in auditory models.

The instantaneous firing rate (IFR) of AN fibers can be modulated on a cycle-by-cycle basis by the acoustic stimulus (phase-locking) up to about 4 kHz (Ruggero, in Popper and Fay, 1992). The fast dynamics of the AN IFR, coupled with only modest frequency resolution, suggests that we should think of the AN representation as that of a spectrogram which has been optimized more for temporal resolution than for spectral resolution. This excellent temporal resolution plays an important role in sound-source localization which relies heavily on cues from interaural time delays.

Little biophysical data are available for the IHC synapse, but since adaptation is not observed in the IHC receptor potentials, adaptation must be taking place in the IHC-AN synapse. The adaptation processes are most commonly assumed to be the result of synaptic vesicle depletion (Mountain and Hubbard, in Hawkins et al., in press). Synaptic vesicles are typically divided into two or more pools. One of these pools represents vesicles which are docked at the active zones and is often referred to as the releasable pool or the immediate pool. Additional vesicles, which are located near the release sites but appear to be tethered to the cytoskeleton, are not available for immediate release. In many current models, these reserve vesicles are further divided into two pools, one of which contains a constant number of vesicles. This multicompartment approach leads to at least two state variables which are essential if the model is to exhibit both rapid and short-term adaptation. Current models differ largely in the number of pools, how vesicles are transferred between pools, and whether neurotransmitter recycling is included in the model.

Cochlear Nucleus

Fibers of the AN travel through the core of the cochlear spiral and enter the ventral cochlear nucleus where they branch. The ascending branch innervates the anteroventral cochlear nucleus (AVCN) and the descending branch travels through the posteroventral cochlear nucleus (PVCN) and enters the dorsal cochlear nucleus (DCN). The cochlear nucleus appears to consist of many subdivisions. For the purposes of this article, we will follow the approach of Cant (in Webster et al., 1992) and divide both AVCN and PVCN into anterior and posterior divisions (AVCN-A, AVCN-P, PVCN-A, PVCN-P). These ventral regions are surrounded by the small-cell cap and marginal layer (SCC) and by granule-cell domains (GC). Within a subdivision, the low-frequency fibers project to more ventral regions and the high-frequency fibers project to more

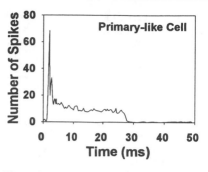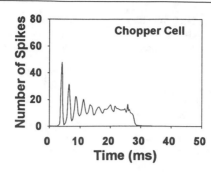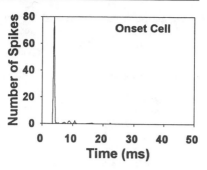

Figure 2. Examples of peristimulus-time histograms (PSTHs) for pure-tone responses from cells in the cochlear nucleus.

dorsal regions. This orderly arrangement is referred to as a "cochleotopic" or tonotopic projection.

The most commonly used physiological classification scheme in the cochlear nucleus is based on the peristimulus-time histogram (PSTH). These histograms are derived by averaging the responses to short tone bursts (25–50 ms in duration) presented at the cell's characteristic frequency (Rhode and Greenberg, in Popper and Fay, 1992). Figure 2 illustrates three of the most common PSTH types found in the CN. The primary-like cell PSTHs are similar to the PSTHs recorded from AN fibers. The primary-like cells appear to follow the fine structure of the acoustic stimulus extremely well. The onset-cell PSTHs all have large responses to the stimulus onset, followed by reduced or nonexistent activity during the remainder of the stimulus. These cells also respond well to the peaks of amplitude-modulated stimuli. The chopper-cell PSTHs exhibit periodically modulated activity at the beginning of the histogram which is the result of the regular firing pattern of these cells becoming synchronized to the stimulus onset. The chopper cells do not synchronize as well to the temporal fine structure, but may be important in encoding amplitude fluctuations.

The different PSTH types are believed to be the result, in part, of differences in intrinsic membrane properties and different degrees of AN fiber convergence. This has led many modelers to focus on membrane and convergence parameters in an effort to gain a better understanding of the diversity of PSTH shapes. Most of the effort to date has concentrated on the primary-like and chopper cells. For example, Hewitt, Meddis, and Shackleton (1992) used a single-compartment model in which the effects of dendritic filtering were approximated by a low-pass filter. Their model compared favorably to data from chopper cells over a variety of experimental conditions.

An alternative classification scheme is to measure the cell's response to tones by stepping the acoustic frequency and intensity over a wide range and then mapping out the borders of excitatory and inhibitory areas to create a response map (RM) (Rhode and Greenberg, in Popper and Fay, 1992). RMs are classified according to the configuration of the excitatory and inhibitory areas. A type I RM has a v-shaped excitatory area similar to that of AN fibers and no signs of inhibition. Type II RMs have a v-shaped excitatory area and exhibit little spontaneous activity, which makes assessment of inhibition with pure tones difficult. However, cells with type II RMs respond poorly to wide-band noise, indicating that they receive broadly tuned inhibitory input. Type III RMs have a v-shaped excitatory area flanked by inhibitory sidebands. Type IV RMs exhibit broad inhibitory regions and only small islands of excitation, including one at low stimulus levels. The firing rate measured with tones at the characteristic frequency is a nonmonotonic function of intensity for these cells. Cells with type IV RMs, how-

ever, respond well to broad-band noise. The type V RM is similar to that of type IV except that no low-level regions of excitation exist. Type II, IV, and V RMs are found almost exclusively in the DCN whereas type I RMs are found principally in the VCN and type III RMs are found throughout the cochlear nucleus.

The complex nature of many of the RM types naturally leads to more complex models. The DCN has attracted particular attention owing to the complex nature of type II, IV, and V RMs. Type IV RMs require a model with at least two layers of interneurons and a layer of output neurons. The cells with type IV RMs project to the inferior colliculus. Simultaneous recording studies have shown that the cells with type IV RMs are inhibited by cells with type II RMs. Since cells with type II RMs are excited by tones but inhibited by wide-band noise, they must receive input from another set of inhibitory interneurons (Rhode and Greenberg, in Popper and Fay, 1992).

Pont and Damper (1991) have developed this three-layer hypothesis into a full network model of the DCN designed to produce the RMs just described. In their model, the AN fibers project to all three layers. The first layer receives the greatest degree of convergence, the third layer the least. The first layer inhibits the second layer and the second layer inhibits the third. The degree of inhibitory convergence increases as one progresses through the layers. Recently, a more detailed DCN model has been proposed by Davis and Voigt (1994) which emphasizes the temporal properties of type II and type IV cells.

Discussion

Despite many modeling and experimental studies of the processes involved in early audition, we know little about the functional significance of even the major CN output pathways. The problem, in part, lies in the difficulty of studying the complex interactions between inhibition and excitation with such simple stimuli as tones and such simple analysis tools as the PSTH. It may be that, in the future, the use of more natural stimuli or the use of pseudorandom stimuli coupled with higher-order analysis techniques will yield additional insights (Eggermont, 1993). More detailed anatomical knowledge is needed, as well, in order to work out the intrinsic CN circuitry.

Of the major pathways shown in Figure 1, only the ventral acoustic stria (VAS) can be assigned some definite function. A large fraction of this pathway comes from primary-like cells in the AVCN and projects to the superior olivary complex which appears to be specialized for the computation of binaural features. The medial superior olive (MSO) computes a representation which is effectively the cross-correlation function for the signals arriving at the two ears (Irvine, in Popper and Fay, 1992). This computation is believed to be very important for the estimation of interaural time delays (see SOUND LOCALIZA-

TION AND BINAURAL PROCESSING). The lateral superior olive (LSO) receives an excitatory input from the ipsilateral AVCN and an inhibitory input from the medial nucleus of the trapezoid body (MNTB). The MNTB is excited by the contralateral AVCN. This circuit creates responses in the LSO which are sensitive to interaural intensity differences, another important feature for the computation of sound source location. An AVCN pathway of unknown function arises from chopper cells and projects to the central nucleus of the inferior colliculus (ICC). The PVCN presents an even greater mystery. Many of its cells produce onset-type PSTHs, suggesting that it may be involved in processing temporal features. Many of these cells project to the ventral nucleus of the lateral lemniscus (VNLL), the cells of which synchronize well to amplitude-modulated stimuli.

The DCN has attracted considerable attention owing to the complex response properties of its neurons as well as the fact that, in many species, it is a layered structure similar to the cerebellum. The output pathway from the DCN is via the dorsal acoustic stria (DAS) and projects to the ICC and the dorsal nucleus of the lateral lemniscus (DNLL). Neither physiological nor computational studies have, however, shed much light on the function of the DCN. One hint comes from the lesion studies of Sutherland (1991) who found that cats could not orient effectively in the vertical dimension to unexpected noise stimuli when the DAS was lesioned. The same group showed, however, that lesions of the DAS had no effect on the animal's ability to discriminate between sources with different elevations (Masterton and Sutherland, 1994). In fact, for most fine discrimination tasks, lesion studies imply that the VAS is the important pathway (Masterton, Granger, and Glendenning, 1994).

One simple function that is becoming better understood is the acoustic startle reflex. This simple reflex appears to be triggered by very large cells located where the AN enters the ventral cochlear nucleus. These cells receive input from the full range of AN characteristic frequencies and project to giant cells in the reticular formation. These giant cells, in turn, project directly to motor neurons which control head movements. Although giant cells appear to be triggered by cells in the ventral cochlear nucleus (VCN), they also receive extensive input from the DCN and LSO (Lingehhöl and Friauf, 1994). It may be that the DCN and, perhaps, even the LSO are largely involved in orienting reflexes and not in tasks requiring fine discrimination. It is likely that there has been significant evolutionary pressure to create very fast reflexes, with the result that some acoustic features are computed twice, once in a fast but crude way for reflexive tasks and again a second time in a slower, more precise fashion to facilitate subtle discriminations.

Acknowledgment. The preparation of this chapter was supported in part by ONR.

Road Maps: Mammalian Brain Regions; Other Sensory Systems
Related Reading: Auditory Cortex

References

Carney, L. H., 1993, A model for low-frequency auditory-nerve fibers in cat, *J. Acoust. Soc. Am.*, 93:401–417.

Davis, K. A., and Voigt, H. F., 1994, Neural modeling of the dorsal cochlear nucleus: Cross-correlation analysis of short-duration toneburst responses, *Biol. Cybern.*, 71:511–521.

Eggermont, J. J., 1993, Wiener and Volterra analyses applied to the auditory system, *Hear. Res.*, 66:177–201. ◆

Hawkins, H. L., McMullen, T. A, Popper, R. R., and Fay, R. R., in press, *The Springer Handbook of Auditory Research: Auditory Computation*, Springer-Verlag: New York.

Hewitt, M. J., Meddis, R., and Shackleton, T. M., 1992, A computer model of a cochlear-nucleus stellate cell: Responses to amplitude-modulated and pure-tone stimuli, *J. Acoust. Soc. Am.*, 91:2096–2109.

Lingehhöl, K., and Friauf, E., 1994, Giant neurons in the rat reticular formation: A sensorimotor interface in the elementary acoustic startle circuit? *J. Neurosci.*, 14:1176–1194.

Masterton, R. B., and Sutherland, D. P., 1994, Discrimination of sound source elevation in cats: I. Role of dorsal/intermediate and ventral acoustic striae, *ARO Abstr.*, 17:84.

Masterton, R. B., Granger, E. M., and Glendenning, K. K., 1994, Role of acoustic striae in hearing: Mechanism of enhancement of sound detection in cats, *Hear. Res.*, 73:209–222.

Mountain, D. C., Cody, A. R., 1989, Mechanical coupling between inner and outer hair cells in the mammalian cochlea, in *Cochlear Mechanisms: Structure, Function and Models* (J. P. Wilson and D. T. Kemp, Eds.), New York: Plenum Press, pp. 153–160.

Mountain, D. C., and Hubbard, A. E., 1994, A piezoelectric model of outer hair cell function, *J. Acoust. Soc. Am.*, 95:350–354.

Pont, M. J., and Damper, R. I., 1991, A computational model of afferent neural activity from the cochlea to the dorsal acoustic stria, *J. Acoust. Soc. Am.*, 89:1213–1228.

Popper, A. N., and Fay, R. R., 1992, *The Springer Handbook of Auditory Research*, vol. 2, *The Mammalian Auditory Pathway: Neurophysiology*, New York: Springer-Verlag.

Sutherland, D. P., 1991, A role of the dorsal cochlear nucleus in the localization of elevated sound sources, *ARO Abstr.*, 14:33.

Webster, D. B., Popper, A. N., and Fay, R. R., 1992, *The Springer Handbook of Auditory Research*, vol. 1, *The Mammalian Auditory Pathway: Neuroanatomy*, New York: Springer-Verlag.

Automata and Neural Networks

Eduardo D. Sontag

Introduction

An automaton, or sequential machine, is a device which evolves in time, reacting to external stimuli and in turn affecting its environment through its own actions. In computer science and logic, *Automata Theory* deals with various formalizations of this concept.

Artificial *Neural Networks*, as understood in this article, are systems obtained from a finite number of memory-free scalar elements or "neurons" by means of weighted interconnections. The complete system is updated synchronously, in discrete time steps, and the transmission of information among neurons requires a unit time delay. The term *recurrent* or *feedback* network is sometimes used in order to emphasize the fact that possible loops in signal paths may allow neurons to mutually affect each other.

In this formal sense, neural networks constitute a (very) particular type of automaton. It is therefore natural to analyze the information processing and computational power of neural networks through their comparison with the more abstract, general models of automata classically studied in computer science. This permits a characterization of neural capabilities in unambiguous mathematical terms.

Several variants of the notion of automata are possible, depending on the type and availability of "external" secondary memory storage. Similarly, many different types of neural networks can be conceived of, depending on the time scales of operation and the types of signals transmitted among neurons. This article will present some basic background about automata and will briefly explain how such objects can be compared to, and simulated by, neural networks.

The area of relations between automata and neural nets is an old one, dating back at least to the work of the neurophysiologists McCulloch and Pitts (1943), but it is also one of active current research, so only a small fraction of the topic can be covered in this short article; pointers to relevant literature are given in the references. For the same reason, most ideas will be only discussed in an intuitive and sketchy fashion, with mathematical details omitted.

Automata

The components of actual automata may take many physical forms, such as gears in mechanical devices, relays in electromechanical ones, integrated circuits in modern digital computers, or neurons. The behavior of such an object will depend on the applicable physical principles. From the point of view of automata theory, however, all that is relevant is the identification of a set of *internal states* which characterize the status of the device at a given moment in time, together with the specification of rules of operation which predict the next state on the basis of the current state and the inputs from the environment. Rules for producing output signals may be incorporated into the model as well.

Although the beginnings of the mathematical formalization of automata took place prior to the advent of digital computers, it is useful to think of computers as a paradigm for automata in order to explain the basic principles. In this paradigm, the state of an automaton, at a given time t, corresponds to the specification of the complete contents of all random access memory locations as well as of all other variables that can affect the operation of the computer, such as registers and instruction decoders. The symbol $x(t)$ will be used to indicate the state at the time t. At each instant (clock cycle) the state is updated, leading to $x(t + 1)$. This update depends on the previous state, as instructed by the program being executed, as well as on external inputs like keyboard strokes and pointing-device clicks. The notation $u(t)$ will be used to summarize the contents of these inputs. (It is mathematically convenient to consider "no input" as a particular type of input.) Thus one postulates an update equation of the type

$$x(t + 1) = f(x(t), u(t)) \qquad (1)$$

for some mapping f. Also at each instant, certain outputs are produced: update of video display, characters sent to printer, and so forth; $y(t)$ symbolizes the total output at time t. (Again, it is convenient to think of "no output" as a particular type of output.) A mapping

$$y(t) = h(x(t)) \qquad (2)$$

provides the output at time t associated to the internal state at that instant.

Abstractly, an automaton is defined by the above data. As a mathematical object, an *automaton* is simply a quintuple

$$\Sigma = (X, U, Y, f, h)$$

consisting of sets X, U, and Y (called, respectively, the state, input, and output spaces), as well as two functions

$$f: X \times U \to X \qquad h: X \to Y$$

(called the next-state and the output maps, respectively). A *finite automaton* is one for which each of the sets X, U, and Y is finite.

The Finiteness Assumption

It would appear on first thought that it is sufficient in practice to restrict studies to finite automata. After all, only a finite amount of memory is available in any computer. However, even for digital computation, finiteness imposes theoretical constraints which are undesirable when one is interested in the understanding of ultimate computational capabilities. As a trivial illustration, assume that one wishes to design a program which reads an input string of 0s and 1s and, *after* this string ends, displays the same string in its output. A finite automaton cannot accomplish this task, obviously, since the task requires an unbounded amount of memory (unless one knows in advance that the strings to be memorized and repeated will be of no more than a certain predetermined length, in which case enough memory, represented by a certain number of states, can be preallocated for storage). On the other hand, one could certainly write a computer program, in any modern programming language, to perform this task. The program will instruct the computer to write the string into a file as it is being received, to be retrieved later when $u(t) = \#$ is encountered. This program will execute correctly as long as enough external storage (e.g., in the form of disk drives) is potentially available.

A mathematical model more general than finite automata allows for "external" storage in addition to the information represented by the current "internal" state of the system. This is the *Turing Machine* model, introduced by the English mathematician Alan Turing in 1936, and it forms the basis of most of modern computer science. In a Turing machine, a finite automaton is used as a "control" or main computing unit, but this unit has access to a potentially infinite read/write storage device. The entire system, consisting of the control unit and the storage device, together with the rules that specify access to the storage, can be seen as a particular type of infinite automaton, albeit one with a very special structure. It is widely accepted today in the computer science community that no possible digital computing device can be more powerful, except for relative speedups due to more complex instruction sets or parallel computation, than a Turing machine.

It has been known at least since McCulloch and Pitts (1943) that finite automata can be simulated by (recurrent) neural networks. This fact is stated formally later in this article. One may wish to ask as well about neural simulation of infinite automata, such as Turing machines. Such a simulation can be approached in at least two different ways. The first is to make a distinction between "RAM" and secondary storage for neural networks, in which case the simulation of a finite automata (the control part) is all that is needed, and details of the secondary storage implementation are not studied. An alternative possibility, perhaps more reasonable from a biological standpoint, is to blur this distinction and to allow the extra memory to be represented by quantities also associated with neurons, such as activation levels or concentrations of neurotransmitters at synapses. In this latter case, a more sophisticated mathematical analysis is required.

Types of Machines Not Covered in This Article

In this article it is assumed that all behavior is deterministic. The field of *probabilistic* automata (Paz, 1971) deals with random effects and will not be studied here.

Note that the definition of automata implicitly assumes that all attention is restricted to operation at discrete instants of time, or *epochs*: $t = 0, 1, 2, \ldots$ On the other hand, many physical devices (and "analog computers") are most naturally described by means of continuous quantities such as voltages or forces, evolving perhaps in continuous time, and are subject to continuous-valued external signals. Such more general devices can be studied as well, leading to connections between automata theory and control systems theory; see Padulo and Arbib (1974) and Sontag (1990). Various areas of continuous mathematics, including differential equation theory, are relevant in that study.

Recurrent Neural Nets

As mentioned earlier, this article deals with recurrent neural networks. These are devices built by linearly combining a finite number n of memory-free scalar processing units, each of which performs the same nonlinear transformation $\sigma : \mathbb{R} \to \mathbb{R}$ on a linear combination of its inputs. The units are interconnected through unit delays. One may describe such a system by introducing n real-valued variables x_i, $i = 1, \ldots, n$ which represent the internal state of the ith processor respectively, subject to update equations such as

$$x_i(t + 1) = \sigma\left(\sum_{j=1}^{n} a_{ij} x_j(t) + \sum_{j=1}^{m} b_{ij} u_j(t) \right) \qquad (3)$$

where each u_i, $i = 1, \ldots, m$ is an external input signal. The coefficients a_{ij}, b_{ij} denote the weights, intensities, or "synaptic strengths" of the various connections. The function $\sigma : \mathbb{R} \to \mathbb{R}$, which appears in all the equations, is called the *activation function*. It is often taken to be a sigmoidal-type map, as discussed below; it characterizes how each neuron responds to its aggregate input. One also assumes a certain number p of probes, or measurement devices, whose outputs signal to the environment the collective response of the net. Each such device averages the activation values of many neurons. Mathematically, this is modeled by adding a set of functions

$$y_i(t) = \sum_{j=1}^{n} c_{ij} x_j(t) \qquad i = 1, \ldots, p \qquad (4)$$

The coefficient c_{ij} represents the effect of the jth neuron on the ith measurement.

As defined, a recurrent neural net is a particular type of automaton, with state space $X = \mathbb{R}^n$, input space $U = \mathbb{R}^m$, and output space $Y = \mathbb{R}^p$.

In many studies, starting with the classical work by McCulloch and Pitts (1943), the function σ is taken to be the hard-limiter, threshold, or *Heaviside* function $\mathcal{H}(x)$, which equals 1 if $x > 0$ and 0 for $x \le 0$. Often one wants a differentiable saturation. For this, especially in contemporary neural network studies, it is customary to consider the *standard sigmoid* $\sigma(x) = (1 + e^{-x})^{-1}$ or equivalently, up to translations and change of coordinates, the hyperbolic tangent, $\tanh(x)$. Also common is a piecewise linear function, $\pi(x) := x$ if $|x| < 1$ and $\pi(x) = \text{sign}(x)$ otherwise; this is sometimes called a "semilinear" or "saturated linearity" function. In this article, it will only be assumed that σ is a *sigmoid* (sometimes called a "squashing" function) in the following sense: σ is monotone (not necessarily continuous), bounded, and not constant. Note

that both $\lim_{x \to -\infty} \sigma(x)$ and $\lim_{x \to +\infty} \sigma(x)$ necessarily exist and are distinct. The above activation functions are all sigmoids.

Simulations

In order to precisely state that neural nets can do everything that a finite automaton can, one needs to introduce the notion of simulation.

In general, given an automaton $\Sigma = (X, U, Y, f, h)$, the map f can be extended by induction to arbitrary input sequences. That is, for any sequence $u_1, \ldots, u_k$ of values in U,

$$f_*(x, u_1, \ldots, u_k)$$

is defined as the iterated composition $f(f(\ldots f(f(x, u_1), u_2), \ldots, u_{k-1}), u_k)$. Suppose now that two automata are given, $\Sigma = (X, U, Y, f, h)$ and $\overline{\Sigma} = (\overline{X}, U, Y, \overline{f}, \overline{h})$, which have the same input and output sets. The automaton $\overline{\Sigma}$ *simulates* Σ if there exist two maps

$$\text{ENC}: X \to \overline{X} \qquad \text{and} \qquad \text{DEC}: \overline{X} \to X$$

called the *encoding* and *decoding* maps, respectively, such that for each $x \in X$ and each sequence $\omega = u_1, \ldots, u_k$ of elements of U,

$$f_*(x, \omega) = \text{DEC}[\overline{f}_*(\text{ENC}[x], \omega)] \qquad h(x) = \overline{h}(\text{ENC}[x])$$

Assume that for some integer m the input value set U consists of the vectors $e_1, \ldots, e_m$ in $\mathbb{R}^m$, where e_i is the ith canonical basis vector, that is, the vector having a 1 in the ith position and zero in all other entries. Similarly, suppose that Y consists of the vectors $e_1, \ldots, e_p$ in $\mathbb{R}^p$. (The assumption that U and Y are of this special "unary" form is not very restrictive, as one may always encode inputs and outputs in this fashion.) There holds then the following *simulation theorem*:

Theorem. Every finite automaton can be simulated by a neural network with activation function $\sigma = \mathcal{H}$.

See Alon, Dewdney, and Ott (1991) for a study of the minimal number of neurons (the dimension n) required for the simulation; the theory of threshold functions, as covered by Muroga (1971), is useful in this context. The implementation of finite automata by networks has been suggested in some application areas; see, for instance, Cleeremans, Servan-Schreiber, and McClelland (1989). For more on automata, their behaviors, and relations to neural networks, the reader may wish to consult Kleene (1956), Rosenblatt (1962), Hopcroft and Ullman (1979), Minsky (1967), and Arbib (1987).

A sketch of a proof of the theorem is as follows. Assume that the states of the finite automaton Σ to be simulated are $\{\xi_1, \ldots, \xi_N\}$. A neural network that simulates Σ has $n = Nm + 1$ neurons and is built as follows. Denote the coordinates of the state vector $x \in \mathbb{R}^n$ by x_{ij}, $i = 1, \ldots, N, j = 1, \ldots, m$, and x_0. (This last coordinate will be identically equal to 1 in the simulation.) In terms of these coordinates, the update equations are $x_0^+ = \mathcal{H}(x_0)$ and, for $r = 1, \ldots, N, s = 1, \ldots, m$,

$$x_{rs}^+ = \mathcal{H}\left(\sum_{j=1}^{m} \sum_{i \in S_{rs}} x_{ij} - x_0 + u_s \right)$$

where

$$S_{rs} := \{l \mid f(q_l, e_s) = q_r\}$$

For each $l = 1, \ldots, p$, the lth coordinate of the output is defined as

$$y_l = \sum_{j=1}^{m} \sum_{i \in T_l} x_{ij}$$

where $T_l := \{i | h_l(q_i) = 1\}$ and h_l is the lth coordinate of h. The encoding map is

$$\text{ENC}[q_r] := e_{r1} + e_0$$

where e_{ij} and e_0 denote the canonical basis vectors in the coordinates x_{ij} and x_0 respectively. The decoding map is

$$\text{DEC}[e_{ij} + e_0] = q_i$$

for all i, j, and is arbitrary on all other elements of $\mathbb{R}^n$. The proof that this is indeed a simulation is immediate, based on the observation that, provided that the starting state has the form $e_{ij} + e_0$, at each instant the state of the network has the same special form, where q_i is the corresponding state of the original automaton. Note that the expression $\sum_{j=1}^{m} \sum_{i \in S_r} x_{ij} - x_0 + u_s = \sum_{j=1}^{m} \sum_{i \in S_r} x_{ij} + u_s - 1$ can only take the values $-1, 0$, or 1, because only one of the x_{ij} can be equal to 1 at any given time. Moreover, the value 1 can only be achieved for this sum if both $u_s = 1$, and there is some i so that $x_{ij} = 1$, that is, if the current state of the original machine is q_i and $f(q_i, e_s) = q_r$. Thus the next state is $e_{rs} + e_0$ precisely if the next state of Σ is q_r and the input applied is e_s.

Unbounded Machines

As remarked earlier, it follows as a simple consequence of the theorem just stated that every possible Turing machine (i.e., any digital computer) can also be simulated by a neural network, provided that an external memory is assumed. It suffices to simulate the control part of the Turing machine by a net. However, this begs the question of what are the capabilities of neural networks when no such *additional* memory is allowed. This issue is discussed next.

It is not very difficult to establish that the above simulation theorem is true whatever sigmoid σ one wishes to use, not merely the Heaviside map $\mathcal{H}$. If this sigmoid happens to be $\mathcal{H}$, which is used in most of the classical literature on neural networks, then no more than finite automata can be realized without additional memory. This is because after the first time step the states of the system are forced to lie in the finite set $\{0, 1\}^n$. Observe, however, that neural networks might be more powerful than finite automata provided that the next-state mapping be allowed to have an infinite range, that is, if the sigmoid σ itself can take infinitely many possible values.

This latter observation was exploited by Siegelmann and Sontag (1992). (See also Cosnard, Garzon, and Koiran, 1993, which expanded on this work.) It was shown that, if one uses the piecewise linear sigmoid $\sigma(x) = \pi(x)$ introduced earlier, a novel simulation result is possible. It turns out that all Turing machines can be simulated by neural networks, with no recourse to an artificial secondary storage device. The basic idea is to take advantage of the unbounded precision of the state variables x_i to store information. The construction results in a simulation of a universal Turing machine by a recurrent network which employs roughly 1000 neurons. The main mathematical difficulty in the proof lies in the fact that a continuous sigmoid does not allow, in principle, for discontinuous logical decisions of the "if $x \leq 0$ then do $(\ldots)$ else do $(\ldots)$" type, and such decisions are required in general programs. This drawback was overcome by the use of a Cantor set representation for the storage of activation values. Cantor sets permit the making of binary decisions by means of finite precision devices, taking advantage of the fact that no values may ever appear in the "middle" range.

The simulation of Turing machines by sigmoidal networks in effect substitutes one assumption, namely that a potentially infinite external storage device exists, by another, namely the hypothesis that unlimited precision is possible in neural computation and in the storage of activations. Neither assumption (infinitely many finite precision storage units nor finitely many infinite precision cells) is valid in the real world, but the two models represent different idealizations of the potential unlimited availability of resources. By studying these mathematical idealizations, one gains a better understanding of the power and limitations of real devices.

While classical computation theory essentially stops with Turing machines, in the context of neural networks, seen as analog computing devices, there is no need to do so. If weights and activations are allowed to assume arbitrary real values, neural networks with continuous sigmoidal σ are able to "compute" more than digital computers. A precise study of such analog computations was carried out in a paper by Siegelmann and Sontag (1993), which characterized the information-processing power of neural networks under resource constraints, in particular in terms of computation time. One of the main results in that study is that the class of languages recognized in polynomial time using recurrent nets, called there "circuit P" or "analog P," is exactly the same as a class also studied in computer science, namely the class of languages recognized in polynomial time by Turing machines which consult oracles, where the oracles are sparse sets. This gives a precise characterization of the power of recurrent nets in terms of a known (noncomputable in the classical sense) complexity class. In summary, even though networks, as analog devices, can "compute" far more than digital computers, they also give rise to a rich theory of computation. The ultimate implications of this characterization for analog implementations of neural computers are as yet unclear, but the result implies that many problems, namely those *not* in the class analog-P, cannot be expected to be solved efficiently even in analog computers such as neural networks.

Sontag (1993) provides an approximation type of result. He shows (with a proof similar to the one sketched above) that arbitrary machines $\Sigma = (X, U, Y, f, h)$—even infinite, but supposing now that X as well as U and Y have a topological structure and the maps f and h are continuous—can be *approximated* arbitrarily well, on compact subsets, by neural networks with sigmoidal activations. Various theoretical aspects of neural network theory are explored in that paper as well.

Road Map: Dynamic Systems and Optimization
Related Reading: Parallel Computational Models

References

Alon, N., Dewdney, A. K., and Ott, T. J., 1991, Efficient simulation of finite automata by neural nets, *J. Assoc. Comput. Mach.*, 38:495–514.

Arbib, M. A., 1987, *Brains, Machines, and Mathematics*, 2nd ed., New York: Springer.

Cleeremans, A., Servan-Schreiber, D., and McClelland, J. L., 1989, Finite state automata and simple recurrent networks, *Neural Computat.*, 1:372–381.

Cosnard, M., Garzon, M., and Koiran, P., 1993, Computability properties of low-dimensional dynamical systems, in *Proceedings of the 10th Symposium on Theoretical Aspects of Computer Science*, Berlin: Springer-Verlag.

Hopcroft, J. E., and Ullman, J. D., 1979, *Introduction to Automata Theory, Languages, and Computation*, Reading, MA: Addison-Wesley.

Kleene, S. C., 1956, Representation of events in nerve nets and finite automata, in *Automata Studies* (C. E. Shannon and J. McCarthy, Eds.), Princeton, NJ: Princeton University Press, pp. 3–41.

McCulloch, W. S., and Pitts, W., 1943, A logical calculus of the ideas immanent in nervous activity, *Bull. Math. Biophys.*, 5:115–133.

Minsky, M. L., 1967, *Computation: Finite and Infinite Machines*, Englewood Cliffs, NJ: Prentice Hall.

Muroga, S., 1971, *Threshold Logic and Its Applications*, New York: Wiley.

Padulo, L., and Arbib, M. A., 1974, *System Theory: A Unified State-Space Approach to Continuous and Discrete Systems*, Philadelphia: W. B. Saunders (Publishing continued by Hemisphere Publ., Washington, DC.)

Paz, A., 1971, *Introduction to Probabilistic Automata*, New York: Academic Press.

Rosenblatt, F., 1962, *Principles of Neurodynamics*, New York: Spartan.

Siegelmann, H. T., and Sontag, E. D., 1992, On the computational power of neural nets, in *Proceedings of the 5th ACM Workshop on Computational Learning Theory*, Pittsburgh, pp. 440–449. To appear in *J. Computer Sys. Sci.*

Siegelmann, H. T., and Sontag, E. D., 1993, Analog computation via neural networks, in *Proceedings of the 2nd Israel Symposium on Theory of Computing and Systems (ISTCS93)*, Los Alamitos, CA: IEEE Computer Society Press. Also, Analog computation, neural networks, and circuits, *Theoret. Comput. Sci.*, to appear.

Sontag, E. D., 1993, Neural networks for control, in *Essays on Control: Perspectives in the Theory and Its Applications* (H. L. Trentelman and J. C. Willems, eds.), Boston: Birkhauser, pp. 339–380.

Sontag, E. D., 1990, *Mathematical Control Theory: Deterministic Finite Dimensional Systems*, New York: Springer.

Automatic Target Recognition

Harold Szu and Brian Telfer

Introduction

Automatic Target Recognition (ATR) involves detecting and classifying targets that may be moving, partially occluded, or camouflaged in a real-world unstructured environment. ATR thus differs in several respects from commercial pattern recognition applications, such as character or speech recognition. ATR encompasses a wide variety of sensors, such as ladar, radar, sonar, and electro-optical, in both passive and active modes, and producing a wide variety of signal types, e.g., one-dimensional signals, two-dimensional images, and range images. The ATR environment is uncontrolled, with the hostile targets often actively attempting to avoid detection/recognition, in contrast to commercial applications which allow control over the environment to enhance recognizability. The uncontrolled environment leads to a tremendous variety of background clutter, and to partially occluded targets that can appear in a wide range of scales and aspects. Targets can also vary greatly owing to environmental effects, e.g., the diurnal and climatic variations in infrared (IR) imagery. Immense amounts of data must be processed in real time. Because of the difficulties and expense involved in collecting real data, simulated data must often be used during system development. All of these factors cause ATR to be a very difficult recognition application. (For a further introduction to the difficulties of ATR, see Roth, 1990; Sadjadi, 1991; *MIT Lincoln Laboratory Journal*, 1993).

Despite the differences from commercial applications, ATR systems generally follow the classic sequence of operations used for any classification task: (1) preprocessing to remove noise and clutter, (2) segmentation and detection of possible targets, (3) feature extraction to reduce the dimensionality and include invariances, and (4) classification as target/nontarget or according to the class of target. Variations on this sequence occur, with one of the most common being to eliminate the segmentation and detection phase, instead correlating feature extraction/classification filters with the entire signal and image. In such cases, outputs exceeding a threshold indicate a target. For moving targets, tracking is an additional challenge, but we consider this outside the scope of ATR. Depending on the scenario, the output of the ATR system is a decision aid that assists a human operator or an input to a weapons system.

Many biological neural networks excel at ATR, and their inspiration has played an important role in much of the ATR artificial neural network development. Examples of the most closely biologically motivated work include that of Grossberg and also that of Waxman (see VISUAL PROCESSING OF OBJECT FORM AND ENVIRONMENT LAYOUT).

System Profiles

Overviews of a few neural network ATR systems and system components are given as representative samples that are readily accessible in open literature. Many other organizations have been investigating neural networks for ATR. Though ATR deals with a wide variety of sensors and targets, these accessible examples mainly consider electro-optical or radar and ground targets. To illustrate the type of challenge faced in these applications, Figure 1 shows a view of a partially obscured target model on a Naval Surface Warfare Center terrain board (a model of terrain used for testing ATR algorithms), together with various types of clutter.

Sahtirn^{TM}

The Self Adaptive Hierarchical Target Identification and Recognition Neural Network (SAHTIRN^{TM}) was developed by the Hughes Aircraft Company (Daniell et al., 1992) and demonstrated on real and terrain board IR targets (in addition to character recognition). The early processing applies the Boundary Contour System (BCS) (see FIGURE-GROUND SEPARATION) to enhance contrast and remove varying illuminance, followed by a Canny edge detector. BCS employs a center-surround shunting network. The hybrid BCS/Canny segmentor was shown to produce better results than either system alone. To remove clutter, several combinations are formed of the largest connected-edge components within a detection gate, and each combination is input to the next stage as a tentative target. The next stage extracts features with a modified NEOCOGNITRON (q.v.), a hierarchical feedforward network that provides a degree of shift, scale, and distortion tolerance. This produces a set of features that are used as input to a multilayer perceptron trained by BACKPROPAGATION (q.v.). Multiple targets in the same image are handled by detecting each at the segmentation stage and classifying each detection separately. The system has been implemented on specialized high-throughput hardware. A three-class real-time test with terrain board imagery resulted in

Figure 1. Example image of a target and clutter taken from a Naval Surface Warfare Center terrain board (courtesy of N. Caviris). The type of data generated poses numerous challenges, including detection of the target in the midst of target-like clutter, discrimination between types of targets, and recognition of the target regardless of aspect, position, scale, in-plane rotation, varying illumination, and partial occlusion.

91% correct, 6% rejected, and 3% error. This was a low-clutter environment, but the system is being extended to handle more clutter.

Modals

The Multiple Object Detection and Location System (MOD-ALS) was developed by Booz-Allen and Hamilton for low angle-of-attack air-to-surface ATR and was tested on visible-wavelength terrain board imagery (Thoet et al., 1992). The system consists of three processing layers. The first computes a similarity measure between local spatial features (each within an 11 × 11 window for the case of 300–700 pixels on target) and the input image, for each pixel in the input image. This measure is invariant to constant illumination changes. The local features are computed by an unspecified neural network learning technique during the training phase. The second layer treats each pixel in the image as a centroid of a tentative target and finds the best match of each feature within a window where that feature could appear in a target with respect to the centroid. The third layer searches for good matches (according to another similarity measure) of a group of features indicating a target. Any third-layer matches exceeding a threshold indicate

a target of a particular class. Receiver operating characteristics (ROCs) have been measured by varying the output threshold. ROCs measure the tradeoff between detection probability and false alarm rate. ROCs are important for ATR because to detect targets from clutter, the probability of detection (the probability of correctly classifying a target) is commonly traded off against the probability of false alarm (probability of misclassifying clutter as target), forming an ROC curve. Tolerance to scale and distortion is built into the multilayer system by allowing the feature locations to vary somewhat. Good ROC test results were obtained for five target classes for varying scales, resolutions, and clutter levels.

Other Tests

At a more preliminary level, other systems or system components have been tested for a variety of sensors. We mention a few here.

In one of the earliest neural network ATR demonstrations, sonar returns from different aspects of a metal cylinder and a similarly shaped rock were discriminated, with features computed by integrating the short-time Fourier transform over time for different frequency bands, and by using a multilayer perceptron classifier (Gorman and Sejnowski, 1988). The best test set results were 90.4%, as compared to 82.7% for the conventional nearest-neighbor classifier.

Recognizing targets regardless of position, scale, rotation, and aspect is one of the challenges central to ATR. Although the examples in the immediately preceding sections provide a degree of in-plane invariance through tolerance accumulated over several processing layers, other studies have used feature spaces, such as moment invariants (combinations of moments) and polar-log and Fourier-Mellin transforms. Out-of-plane invariance is invariably achieved through the training set, i.e., by incorporating numerous out-of-plane views in the training set.

Tests have been made of one-class recognition, in which the goal is to train on only one class and, in testing, to recognize that class while rejecting all others (Fogler et al., 1992). This is an attractive approach, since it is unrealistic to expect to train on all possible types of clutter that a system may confront. A neocognitron was used to extract features and ART2-A (see ADAPTIVE RESONANCE THEORY) was used as a classifier, using synthetic aperture radar data (real and synthetic), with a particular tank as the class of interest.

Wavelet features (see WAVELET DYNAMICS) have been shown to be attractive for target detection. Wavelet features chosen to match an object or parts of an object, inputted to a multilayer perceptron, have been shown to improve detection accuracy in multispectral electro-optical images and in sidescan sonar images (Telfer, Szu, et al., 1994). A more powerful approach is to adaptively compute wavelet features in conjunction with the neural network to maximize detection accuracy. This idea has been successfully demonstrated for acoustic backscatter (Telfer, Szu, et al., 1994) and IR imagery (Casasent and Smokelin, 1994) detection applications.

Multilayer perceptrons trained with standard backpropagation and variants incorporating BCM (see BCM THEORY OF VISUAL CORTICAL PLASTICITY), projection pursuit, and lateral inhibition have also been tested for recognizing simulated inverse synthetic aperture radar imagery of ships (Bachmann et al., 1994). This work has demonstrated that the projection pursuit feature extraction improves results for noisy and shifted inputs.

Sensor and Temporal Fusion

SENSOR FUSION (q.v.) and temporal fusion have received much attention because they offer means for improving classification results by combining information from different sources and times. Information can be fused at three levels: data, features, and classifier outputs. As one example of feature-level fusion (Ruck et al., 1988), Doppler and relative range images from a laser radar were segmented using histogram-based algorithms. Moment invariant features (invariant to position, in-plane rotation, and scale) were measured from targets extracted from the segmented images. Features from both image types were concatenated and classified by a multilayer perceptron with good results for five target classes (86.4% correct vs. 76.5% correct for a nearest-neighbor classifier). Neural network outputs can be fused in a variety of ways, e.g., voting schemes, and those networks whose outputs estimate class probabilities seem especially well suited to fusing classifier outputs.

In one recent novel approach to temporal fusion, target aspect views are learned by an ART network, and an aspect network is computed to encode the allowed transitions between aspect views (see Waxman et al. in *MIT Lincoln Laboratory Journal*, 1993, pp. 77–116). During recognition, both the aspect views of an input object and the aspect transitions are used to build confidence in a classification.

Increasing Classification Rates and ROC Performance

From the numerous examples summarized thus far, the general roles of neural networks in current ATR systems should be clear. Improvements are being made in a variety of areas. In this section, we highlight recent developments in neural network objective functions to improve ATR performance. This is a basic issue because finding network weights that minimize an objective function is a technique that is used to train a variety of network architectures.

The role and importance of the proper choice of objective functions in ATR (and other pattern recognition) performance are becoming increasingly understood. The classic mean-squared-error (MSE) function results in network outputs that approximate a posteriori probabilities (Ruck et al., 1990). This is a very desirable property for classifier outputs that are subject to further processing, such as when outputs are combined from multiple classifiers, or when multiple outputs are combined over time.

Although MSE is attractive from the probability standpoint, a disadvantage is that the network must have sufficient functional complexity (i.e., enough weights) to approximate the probabilities accurately. For many ATR applications, only an absolute classification, and not a probability, is desired. In this case, only the class decision boundaries, not entire probability functions, need be learned by the network. Less network complexity is needed to represent class boundaries than to represent class probabilities. This is very important because networks with fewer weights tend to generalize better to test data. This is especially important in ATR, in which large training sets are difficult and expensive to obtain, meaning that frequently there are insufficient data to estimate class probabilities accurately, even if that is the desired goal.

Objective functions have been developed to minimize misclassification error and to optimize class boundaries rather than to estimate probabilities. Such functions have been investigated under the names *classification figure of merit* (CFM) (Hampshire and Waibel, 1990; Hampshire, 1993) and the *minimum misclassification error* (MME) function (Telfer and Szu, 1994). Hampshire has demonstrated that such functions require fewer training samples than MSE to obtain the same error, and also that they require fewer weights than MSE to obtain the same error—in one example, about one-fifth as many (Telfer and Szu, 1994). Minimizing the misclassification error for a given network complexity and given training set size is vital for many ATR applications, for which a misclassification can have far-reaching consequences.

Especially for ATR, classification goals other than simply minimizing misclassification error—such as producing various operating points along an ROC—are very important. Although an MSE-trained network can easily produce an ROC by varying the output threshold, an improved ROC results from computing networks with specific operating points on the ROC, using a modified MME function. The function can be modified to assign different costs to errors from different classes, or to maximize the probability of detection for a fixed false alarm rate, which is a Neyman-Pearson criterion (see Telfer and Szu, 1994, for references). The Neyman-Pearson criterion is desirable because a certain low false alarm rate (probability of misclassifying clutter as target) is acceptable, and the classification problem is then to maximize the detection probability (probability of correctly classifying a target) given the acceptable false alarm rate. The computational tradeoff is that it is very easy to vary a threshold to generate an ROC, but the results are suboptimal. Generating each operating point separately requires more computations, but these are normally performed off-line, and a better result is obtained.

Discussion

After surveying recent neural network efforts in ATR, the natural question is: what have neural networks contributed? Compared to the quadratic and nearest-neighbor statistical classifiers and fixed template methods that were in vogue earlier, the various neural network architectures have displayed clear advantages in classification accuracy and, often, in computational speed. In addition, neural networks have been developing into special-purpose architectures of weights tailored to ATR applications using techniques that improve results, such as weight sharing and weight pruning. New objective functions suited to ATR and ROCs further improve results. Additional capability will be realized as the information from time-varying imagery becomes more widely exploited.

Another competitor to neural networks in ATR has been expert systems (see EXPERT SYSTEMS AND DECISION SYSTEMS USING NEURAL NETWORKS). Because expert systems are coded by hand rather than trained on data, they tend to extract only simple relationships among data, and they tend not to generalize well to new environments. A common complaint about neural networks from proponents of rule-based AI is that neural networks are black boxes whose interior workings are not understood and cannot be extracted; thus, the assumption is that they offer no insight into further improvements in performance. From the advances surveyed in this and other articles, that is clearly not the case. Objective functions are well understood in terms of probabilities, misclassification rates, and ROCs. The importance of individual features can be measured, and the importance of individual weights can be computed.

Further advances in neural network ATR can be expected in improved classification performance and robustness, and also in neural network hardware for real-time implementation.

Acknowledgment. Support from the NSWCDD Independent Research Program is gratefully acknowledged.

Road Map: Applications of Neural Networks
Background: I.3. Dynamics and Adaptation in Neural Networks

References

Bachmann, C. M., Musman, S. A., Luong, D., and Schultz, A., 1994, Unsupervised BCM projection pursuit algorithms for classification of simulated radar presentations, *Neural Netw.*, 7:709–728.

Casasent, D. P., and Smokelin, J.-P., 1994, Neural net design of macro Gabor wavelet filters for distortion-invariant object detection in clutter, *Optical Engineering*, 33:2264–2271.

Daniell, C. E., Kemsley, D. H., Lincoln, W. P., Tackett, W. A., and Baraghimian, G. A., 1992, Artificial neural networks for automatic target recognition, *Opt. Engrg.*, 31:2521–2531.

Fogler, R. J., Koch, M. W., Moya, M. M., Hostetler, L. D., and Hush, D. R., 1992, Feature discovery via neural networks for object recognition in SAR imagery, *Proceedings of the International Joint Conference on Neural Networks*, 4:408–413.

Gorman, R. P., and Sejnowski, T. J., 1988, Analysis of hidden units in a layered network trained to classify sonar targets, *Neural Netw.*, 1:75–89.

Hampshire, J., 1993, A differential theory of learning for efficient statistical pattern recognition, PhD Dissertation, Carnegie Mellon University, Electrical Engineering Deptartment.

Hampshire, J., and Waibel, A., 1990, A novel objective function for improved phoneme recognition using time delay neural networks, *IEEE Trans. Neural Netw.*, 1:216–228.

MIT Lincoln Laboratory Journal, Spring 1993, Special Issue on Automatic Target Recognition, vol. 6. ◆

Roth, M. W., 1990, Survey of neural network technology for automatic target recognition, *IEEE Trans. Neural Netw.*, 1:28–43. ◆

Ruck, D. W., Rogers, S. K., Kabrisky, M., and Miller, J. P., 1988, Multisensor target detection and classification, *Proc. SPIE*, 931:14–21.

Ruck, D. W., Rogers, S. K., Kabrisky, M., Oxley, M. E., and Suter, B. W., 1990, The multilayer perceptron as an approximation to a bayes optimal discriminant function, *IEEE Trans. Neural Netw.*, 1:296–298.

Sadjadi, F., 1991, Automatic object recognition: Critical issues and current approaches, *Proceedings of the SPIE*, 1471:303–313. ◆

Telfer, B., and Szu, H., 1994, Energy functions for minimizing misclassification error with minimum-complexity networks, *Neural Netw.*, 7:809–818.

Telfer, B., Szu, H., Dobeck, G., Garcia, J., Ko, H., and Dubey, A., 1994, Adaptive wavelet classification of acoustic backscatter and imagery, *Opt. Engrg.*, 33:2192–2203.

Thoet, W. A., Rainey, T. G., Brettle, D. W., Slutz, L. A., and Weingard, F. S., 1992, ANVIL neural network program for three-dimensional automatic target recognition, *Opt. Engrg.*, 31:2532–2539.

Averaging/Modular Techniques for Neural Networks

Michael P. Perrone

Introduction

Recent neural network simulations have shown that averaging over the outputs of a population of neural network estimates can lead to improved network performance (Baxt, 1992; Hansen, Liisberg, and Salamon, 1992; Perrone, 1993; Perrone and Cooper, 1993; Neal, 1993; Xu, Krzyzak, and Suen, 1992). In this article, the basic averaging method is described and a firm theoretical basis for its performance is presented. This analysis enables us to understand why averaging improves performance and to identify a wide class of common optimization tasks for which averaging can be used to improve performance.

Furthermore, one can extend the notion of averaging to interpret the component networks as generic modules that can be combined using more sophisticated nonlinear techniques (Jacobs et al., 1991; Reilly et al., 1988; Wolpert, 1992). These nonlinear methods can potentially improve performance over simple averaging. However, they do not lend themselves to mathematical analysis, and therefore motivations for these methods are at present only heuristic.

Simple Averaging

Given a population of n neural network estimates of some function $f(x)$, the ith of which is denoted by $f_i(x)$, it can be shown (Perrone and Cooper, 1993), assuming the following error orthogonality condition for $i \neq j$,

$$E[(f_i(x) - f(x))(f_j(x) - f(x))] = 0 \qquad (1)$$

that the mean square error (MSE) of some function g relative to f, given by

$$MSE[g] = E[(g(x) - f(x))^2] \qquad (2)$$

satisfies

$$MSE[\bar{f}] = \frac{1}{N}\overline{MSE[f]} \qquad (3)$$

where a bar indicates an average over the population of estimates. For example, $\bar{f}$ is defined as

$$\bar{f} \equiv \frac{1}{n}\sum_i f_i(x) \qquad (4)$$

This result implies that the MSE can be made arbitrarily small by increasing the size of the population sufficiently. This result is at the heart of improvement in performance seen experimentally. Thus, one should average over neural networks whose errors are as orthogonal as possible. One could, for example, attempt to maximize the orthogonality while minimizing the MSE during network training.

In practice, the orthogonality condition is a good approximation when the number of networks averaged is small and the networks differ significantly. Unfortunately, the orthogonality condition does not hold in general and must be relaxed to find a more accurate formulation of the effects of averaging.

In the case of MSE minimization, we can relax the orthogonality assumption by using the *Cauchy inequality* (Beckenbach and Bellman, 1965):

$$\left(\sum_{i=1}^{n} x_i y_i\right)^2 \leq \left(\sum_{i=1}^{n} x_i^2\right)\left(\sum_{i=1}^{n} y_i^2\right) \qquad (5)$$

We now set $y_i = 1$ for all i, replace the x_i with $f_i(x) - f(x)$, and average over the data to find

$$MSE[\bar{f}] \leq \overline{MSE[f]} \qquad (6)$$

Equation 6 indicates that the average function is never worse than the population average. Note, however, that by relaxing the orthogonality assumption, the $1/N$ behavior from Equation 3 is lost.

The equality in Equation 6 holds only when all of the estimators in the population are identical on all of the data; otherwise, the average is better. This implies that averaging would have no effect on a method which always finds the same solution (optimal or not). It also implies that finding a variety of solutions can be beneficial. In particular, it is well known that the optimal solution for a finite data set will be biased to that data and will in general perform worse on new data. One is therefore faced with the problem of choosing from a multitude of other possible solutions. In this case, averaging is a reasonable choice.

Furthermore, Equation 6 is independent of the data in the sense that it holds for any data set regardless of how it is chosen, including previously unseen test data. Thus, Equation 6 is not biased in any way toward the training data.

It is important to note that Equation 6 does not imply that the average estimator, $\bar{f}$, will perform better than each estimator in the population. It only performs better than the average of the performances. Thus, it is possible that a certain estimator in the population could perform better on some data than the average estimator. If one has a reliable method for selecting the best estimator from a population, then averaging may not be necessary; however, when such a method is not available (or is not sufficiently reliable) or when several estimators all have similar performance on some test set, then averaging will do better than the expected performance of a random choice of one estimator from the population.

Averaging can be interpreted as a generalization of standard Monte Carlo integration techniques. The difference lies in the fact that averaging places no restrictions on the process used to generate the estimates. Although relaxing this restriction abandons the guarantee of estimating the Bayes optimal solution, it has several benefits: averaging avoids the computational overhead of stochastic sampling, including achieving equilibrium and assuring that the samples drawn from the distribution are independent; averaging can be used to combine networks generated from more than one learning algorithm and with more than one network architecture; and averaging can be used to combine networks generated from different data.

Variance Reduction and Smoothing

The averaging process is inherently a smoothing operation, as can be seen by the following derivation. Fixing x, we can write that

$$\begin{aligned}
E[(\hat{f}(x) - f(x))^2] &= E[(\hat{f} - E[\hat{f}] + E[\hat{f}] - f(x))^2] \\
&= E[(\hat{f} - E[\hat{f}])^2] + E[(E[\hat{f}] - f(x))^2] \\
&\quad + 2E[(\hat{f} - E[\hat{f}])(E[\hat{f}] - f(x))] \\
&= VAR(\hat{f}) + BIAS^2(\hat{f})
\end{aligned}$$

where $VAR(\hat{f}) \equiv E[(\hat{f} - E[\hat{f}])^2]$ and $BIAS(\hat{f}) \equiv E[\hat{f}] - f(x)$. Now, since $E[\bar{f}] = E[\hat{f}]$, we find that the bias term is the same for both $MSE[\hat{f}]$ and $MSE[\bar{f}]$; therefore, when we reduce the MSE by averaging, it is because we have reduced the variance term. This variance reduction corresponds to smoothing our estimate of $\hat{f}$.

Weighted Averaging

We now generalize to weighted averaging by defining our average estimator to be given by

$$g(x) \equiv \sum_{i=1}^{N} \alpha_i f_i(x) \qquad (7)$$

for α_i real. Thus simple averaging sets $\alpha_i = 1/N$ for all i.

If we impose no other constraints, then the α_i which minimize the MSE, given in Equation 2, are given by

$$\alpha = (XX^t)^{-1}XY \qquad (8)$$

where α is the vector of the α_i, $X_{ij} = f_j(x_i)$, and $Y_i = f(x_i)$. This solution may have problems with the inversion of the XX^t matrix if two or more of the networks being averaged have highly correlated (i.e., nonorthogonal) errors. One common solution to this problem is to add the term $\lambda\alpha^t\alpha$ to the MSE and minimize both simultaneously. This is equivalent to assuming that the α_i are distributed according to a Gaussian distribution of mean zero and variance $(2\lambda)^{-1}$; it is commonly called *weight decay regularization* or *ridge regression*.

With this modification, the optimal solution becomes

$$\alpha = (XX^t + \lambda I)^{-1}XY \qquad (9)$$

where the optimal λ can be chosen by cross-validation or by generating several estimates of α with different values of λ and then averaging over the different α's.

If we impose the constraint that $\Sigma_i \alpha_i = 1$, then the optimal α_i's are given by

$$\alpha = \frac{C^{-1}\mathbf{1}}{\mathbf{1}^T C^{-1}\mathbf{1}} \qquad (10)$$

where $\mathbf{1}$ is the vector with all elements equal to 1, and

$$C_{ij} \equiv E[(f(x) - f_i(x))(f(x) - f_j(x))] \qquad (11)$$

Again adding weight decay results in an optimal α of

$$\alpha_i = \frac{(C + \lambda)^{-1}\mathbf{1}}{\mathbf{1}^T(C + \lambda)^{-1}\mathbf{1}} \qquad (12)$$

Other constraints may also be imposed, such as positivity (i.e., $|\alpha_i| \geq 0$ for all i), but a closed form is not readily computed.

Note that the process described here may introduce over-fitting because the C we calculate is the sample correlation matrix not the true correlation matrix, so C is a random variable, as are MSE[g] and the optimal α_i's. Thus, noise in the estimate of C can lead to bad estimates of the optimal α_i's.

The results depend on two assumptions: the rows and columns of C are linearly independent, and we have a reliable estimate of C. In certain cases, where we have nearly duplicate networks in the population $\mathscr{F}$, we will have nearly linearly dependent rows and columns in C, which can make the inversion process unstable, so our estimate of C^{-1} will be unreliable. In these cases, we can use heuristic techniques to sub-sample the population $\mathscr{F}$ to assure that C has full rank. In practice, the increased stability produced by removing near-degeneracies outweighs any information lost by discarding nets.

Convexity and Averaging

The fundamental principle behind the improvement seen from averaging is the notion of convexity. A function $\Phi(x)$ is convex if for all x_1 and x_2

$$\Phi\left(\frac{x_1 + x_2}{2}\right) \leq \frac{\Phi(x_1) + \Phi(x_2)}{2} \qquad (13)$$

Such a function is bounded below and has only one minimum. For convex $\Phi(u)$, the following inequality holds (Perrone, 1994):

$$\Phi(\bar{f}) \leq \overline{\Phi(f)} \qquad (14)$$

This result indicates that averaging will improve the performance of any algorithm which attempts to minimize a convex cost.

We must be careful to distinguish between the cost function's dependence on the network (i.e., through x_i above) and its dependence on the parameters of the network. Except in the most trivial cases, the cost functions for neural networks generally have numerous local minima as a function of the parameter space. However, the cost functions are generally convex relative to the network. For example, the most common cost function, MSE, is the sum of squares of the errors, which can be easily shown to be convex.

This is a powerful result, since a wide variety of commonly used minimization techniques are convex (Perrone, 1994). These include the L_p-norms, E_2 (i.e., the MSE) and E_p; the Regression Spline cost, S; the Kullback-Leibler Information, K; and the Penalized Maximum Likelihood, P:

$$E_2(g) = \sum_i (g(x_i) - f(x_i))^2 \qquad (15)$$

$$E_p(g) = \sum_i |g(x_i) - f(x_i)|^p \qquad \text{for all } p \geq 1 \qquad (16)$$

$$S(g) = \frac{1}{n}\sum_i (g(x_i) - f(x_i))^2 + \lambda \int (g'')^2\,dx \qquad (17)$$

$$K(g,f) = \int g \ln\left(\frac{g}{f}\right) \qquad (18)$$

$$P(p) = -\ln \prod_i p(x_i) + \lambda \int (p'')^2\,dx \qquad (19)$$

where g is the estimate function, f is the target function, and $p(x)$ is a probability distribution. In addition, the sum of convex functions is convex; so the averaging method can be used with a combination of any of these costs. One can also include any convex regularization term.

Several very important optimization costs are negative convex, including Maximum Entropy, H; Maximum Mutual Information, I; and Maximum Likelihood Estimation, L:

$$H(p) = -\sum_i p(x_i) \ln p(x_i) \qquad (20)$$

$$I(p,q) = H(p) + H(q) - H(pq) \qquad (21)$$

$$L(p) = \prod_i q(x_i) \qquad (22)$$

where p and q are probability distributions. For these costs, the inequality in Equation 13 is reversed, which in turn reverses the inequality in Equation 14. Thus, for negative convex functions, averaging increases the total cost. However, as the names of the negative convex costs imply, we seek to maximize them, and therefore averaging is working in the correct direction.

Variations

In addition to the methods described above, averaging can be combined with the jackknifing, bootstrapping, and cross-validation resampling techniques (Efron, 1982; Miller, 1974; Stone, 1974, for review) without further modification of the averaging algorithms. The basic idea behind these resampling techniques is to improve one's estimate of a given statistic by combining multiple estimates generated by subsampling or resampling of a finite data set. Thus, using a single data set, one can generate numerous estimators by finding different local minima or by using various subsets of the original data set to generate distinct estimates.

Also, averaging can be used to combine networks with different architectures and/or with different training algorithms.

Discussion

Related methods exist which construct more sophisticated, nonlinear combinations of networks. For example, the competing expert algorithm (Jacobs et al., 1991; see MODULAR AND HIERARCHICAL LEARNING SYSTEMS) uses weighted averaging controlled by a nonlinear gating network that is trained to learn which network(s) perform(s) best in various regions of the input space. In this way, the competing expert network can specialize on a subset of the total problem, and it is used only in that region for which it is well trained. Stacked generalization (Wolpert, 1992) trains a population of networks and then uses the errors generated by these networks to train new networks to correct for these errors. This process can be repeated as needed and combined with a variety of different methods for treating the errors. The GENSEP algorithm (Reilly et al., 1988) also uses the errors of previously trained networks to train new networks; but it trains new networks only on local tasks, and it includes the incorporation of new/subset feature selection to focus successive networks on particular tasks. The Boosting algorithm (Drucker, Schapire, and Simard, 1993) uses successive networks as data filters. Only a fraction of the data set (including all the incorrectly classified data points) is passed to the next network for training. The estimates of these networks can then be combined via averaging.

Although these nonlinear methods have been shown to improve performance, these algorithms are motivated heuristically and currently do not lend themselves easily to theoretical analysis analogous to that presented for averaging. One very interesting line of research is to extend the analysis presented here to these more sophisticated methods.

Road Map: Learning in Artificial Neural Networks, Statistical
Related Reading: Bayesian Methods for Supervised Neural Networks

References

Baxt, W. G., 1992, Improving the accuracy of an artificial neural network using multiple differently trained networks, *Neural Computat.*, 4:772–780.

Beckenbach, E. F., and Bellman, R., 1965, *Inequalities*, New York: Springer-Verlag. ◆

Drucker, H., Schapire, R., and Simard, P., 1993, Improving performance in neural networks using a boosting algorithm, in *Advances in Neural Information Processing Systems 5* (S. J. Hanson, J. D. Cowan, and C. L. Giles, Eds.), San Mateo, CA: Morgan Kaufmann, pp. 42–49.

Efron, B., 1982, *The Jackknife, the Bootstrap and Other Resampling Plans*, Philadelphia, PA: SIAM. ◆

Hansen, L. K., Liisberg, C., and Salamon, P., 1992, Ensemble methods for handwritten digit recognition, in *Neural Networks for Signal Processing II: Proceedings of the 1992 IEEE Workshop*, (S. Y. Kung, F. Fallside, and C. A. Kamm, Eds.), Piscataway, NJ: IEEE, pp. 333–342.

Jacobs, R. A., Jordan, M. I., Nowlan, S. J., and Hinton, G. E., 1991, Adaptive mixtures of local experts, *Neural Computat.*, 3(2).

Miller, R. G., 1974, The jackknife—A review, *Biometrika*, 61:1–16. ◆

Neal, R. M., 1993, Bayesian learning via stochastic dynamics, in *Advances in Neural Information Processing Systems 5*, (S. J. Hanson, J. D. Cowan, and C. L. Giles, Eds.), San Mateo, CA: Morgan Kaufmann, pp. 475–482.

Perrone, M. P., 1993, *Improving Regression Estimation: Averaging Methods for Variance Reduction with Extensions to General Convex Measure Optimization*, PhD thesis, Brown University, Institute for Brain and Neural Systems. ◆

Perrone, M. P., 1994, General averaging results for convex optimization, in *Proceedings of the 1993 Connectionist Models Summer School*, Hillsdale, NJ: Erlbaum, pp. 364–371.

Perrone, M. P., and Cooper, L. N., 1993, When networks disagree: Ensemble method for neural networks, in *Artificial Neural Networks for Speech and Vision* (R. J. Mammone, Ed.), New York: Chapman and Hall, chap. 10.

Reilly, D. L., Scofield, C. L., Cooper, L. N., and Elbaum, C., 1988, Gensep: A multiple neural network learning system with modifiable network topology, in *Abstracts of the First Annual International Neural Network Society Meeting*, INNS.

Stone, M., 1974, Cross-validatory choice and assessment of statistical predictions (with discussion), *J. R. Statist. Soc. Ser. B*, 36:111–147.

Wolpert, D. H., 1992, Stacked generalization, *Neural Netw.*, 5:241–260.

Xu, L., Krzyzak, A., and Suen, C. Y., 1992, Methods of combining multiple classifiers and their applications to handwriting recognition, *IEEE Trans. Syst. Man Cybern.*, 22:418–435.

Axonal Modeling

Christof Koch and Öjvind Bernander

Introduction

Axons are highly specialized "wires" that conduct the neuron's output signal to target cells—in the case of cortical pyramidal cells, up to 10,000 other cortical neurons. As such, they are highly specialized, with a relatively stereotypical behavior. Most authors agree that their role in signaling is limited to making sure that whatever pulse train is put into one end of the axon is rapidly and faithfully propagated to the other end. This is in contrast to the complexity of electrical events occurring at the cell body and in the dendritic tree, where the information from thousands of synapses is integrated.

Despite the uniformity of electrical behavior, there is great morphological variability that largely reflects a tradeoff between propagation speed and packing density. Axonal size varies over four orders of magnitude: diameters range from 0.2 μm fibers in the mammalian central nervous system to 1 mm in squid; lengths range from a few hundred microns to over a meter for motoneurons (Kandel, Schwartz, and Jessell, 1991). Some axons are bound only by the thin cellular membrane while others are wrapped in multiple sheaths of myelin. An example of an axonal arbor is shown in Figure 1.

The majority of nerve cells encode their output as a series of brief voltage pulses. These pulses, also referred to as *action potentials* or *spikes*, originate at or close to the cell body of nerve cells and propagate down the axon at constant velocity and amplitude. Their shape is relatively constant across species and types of neurons. Common to all is the rapid upstroke (depolarization) of the membrane above 0 mV and the subsequent, somewhat slower, downstroke (repolarization) toward the resting potential and slightly beyond to more hyperpolarized potentials. At normal temperatures, the entire sequence occurs within one millisecond. A minority of cell types are axonless and appear to use graded voltage as output, such as cells in the early part of the retina, or interneurons in invertebrates (Roberts and Bush, 1981). Action potentials are such a dominant feature of the nervous system that for a considerable amount of time it was widely held—and still is in parts of the neural network community—that all neuronal computations only involve these all-or-none events. This belief provided much of the impetus behind the neural network models originating in the late 1930s and early 1940s (see SINGLE-CELL MODELS).

The ionic mechanisms underlying the initiation and propagation of action potentials were elucidated in the squid giant axon by a number of workers, most notably by Hodgkin and Huxley (1952). For this work they shared, together with Eccles, the 1963 Nobel Prize in physiology and medicine (for a historical overview see Hodgkin, 1976). Their model has played a paradigmatic role in biophysics; indeed, the vast majority of contemporary biophysical models use essentially the same mathematical formalism Hodgkin and Huxley introduced forty years ago. This is all the more surprising since the kinetic description

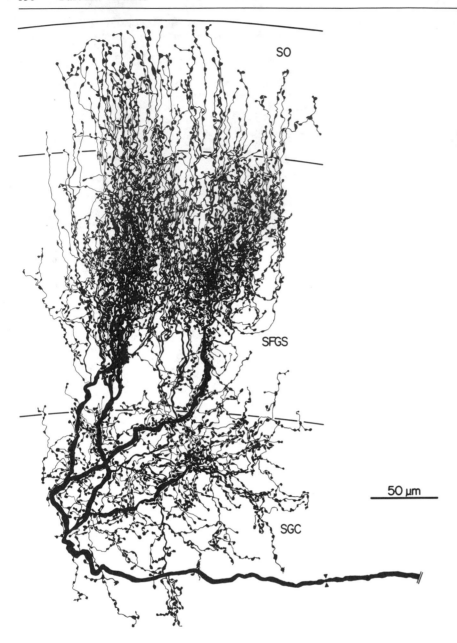

SO

SFGS

50 μm

SGC

Figure 1. Axonal terminations. An axon from nucleus isthmi terminating in turtle tectum was labeled with horseradish peroxidase and reconstructed from a series of parallel sections. The thick parent trunk (3 μm) is wrapped in myelin and shows a node of Ranvier (triangles). The thin (<1 μm) branches are nonmyelinated and are home to approximately 3600 synaptic boutons (bulbous thickenings), where contact is made onto other cells. The boutons vary greatly in size but average about 1.5 μm in diameter. (From Sereno, M. I., and Ulinski, P. S., 1987, Caudal topographic nucleus isthmi and the rostral nontopographic nucleus isthmi in the turtle, *Pseudemys scripta, Journal of Comparative Neurology*, 261:319–346. Copyright © 1987 by Wiley-Liss. Reprinted by permission of John Wiley & Sons, Inc.)

of the continuous, deterministic, and macroscopic membrane permeability changes within the framework of the Hodgkin-Huxley model was achieved without any knowledge of the underlying all-or-none, stochastic, and microscopic ionic channels.

Given its importance, we will describe the Hodgkin-Huxley model and its assumptions in some detail in the following section. We will then introduce the two classes of axons found in most animals, myelinated and nonmyelinated, and describe their differences. Axons possess heavily branched axonal trees. We will conclude the overview by briefly alluding to additional complications that arise when attempting to understand the role and function of these trees in information processing.

The Hodgkin-Huxley Model of Action Potential Generation

Electric current in nerve cells is carried by the flow of ions through membrane proteins called *channels*. Figure 2A shows a schematic of a membrane patch containing a sodium-selec-

tive channel. The concentration of sodium ions is high in the extracellular fluid and low in the intracellular axoplasm. This *concentration gradient* gives rise to an osmotic pressure on sodium ions to flow into the cell. At some membrane potential, termed the *reversal potential*, this pressure will be canceled by the *electrical gradient* and so the net flow of sodium ions will be zero at that point. The channel can be blocked if, by a conformational change, one of the four diagramed *gating particles* covers the pore opening. A similar situation holds for potassium ions flowing through separate potassium-selective channels, except that the concentration gradient is reversed.

In the squid giant axon, the membrane potential is determined by three conductances: a voltage-independent (passive) leak conductance, g_l, a voltage-dependent (active) sodium conductance, g_{Na}, and an active potassium conductance, g_K. The equivalent circuit used to model the membrane is shown in Figure 2. The conductances are in series with batteries, the values of which correspond to the respective reversal potentials of the ionic currents, E_l, E_{Na}, and E_K. The outside is connected

A)

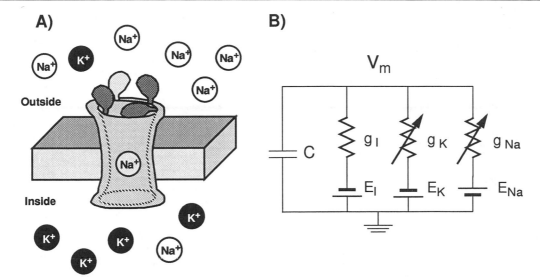

B)

V_m

Figure 2

Figure 2. *A,* Schematic of ionic flow through channel proteins. Membrane channels are highly selective for different ionic species. The difference in ion concentration between the inside and outside determines the reversal potential and thus whether the current will flow inward or outward at a given membrane potential. This sodium channel has four specialized areas called *particles,* each one of which can block the channel pore through a conformational change. The probability of blockage increases with membrane potential for the *inactivation* particle (light gray) and decreases for the three *activation* particles (dark grey). *B,* Equivalent circuit of axonal membrane. The Hodgkin-Huxley model of squid axon incorporates a capacitance and three conductances. Two of the conductances are voltage-dependent (active), g_{Na} and g_K, while the third is a passive "leak" conductance, g_l. The maximal conductances are 120, 36, and 0.3 ms/cm^2, respectively. Each conductance is in series with a battery that defines the *reversal potential* for each conductance type. The values are $E_{Na} = 50$, $E_K = -77$, and $E_l = -54.3$ mV. See text for the voltage dependences of g_{Na} and g_K. The top rail corresponds to the axoplasm (inside) of the axon, while the bottom rail, grounded, is the external medium. When a *membrane action potential* or *space clamp* is modeled, only one compartment is used as shown and the spatial structure of the membrane is ignored. When *propagating action potentials* are modeled, the specific resistivity of the axoplasm, $R_a = 34.5$ Ωcm, cannot be ignored. R_a is then modeled as a series of resistors connecting identical compartments that correspond to different spatial locations along the axon. The membrane capacitance is 1 μF/cm^2.

to ground under the assumption that the resistivity of the external medium is negligible.

The time course of an action potential is illustrated in Figure 3. In this simulation of a membrane patch, a brief current pulse initiates an action potential. Before stimulation, the membrane voltage is at rest, $V_m = -65$ mV. At this potential, g_{Na} and g_K are almost fully inactivated; g_K is still much larger than g_{Na}, and the membrane is dominated by the leak current and the residual potassium current. The applied current slowly depolarizes the membrane by charging up the capacitance. As V_m approaches threshold ($V_t \approx -50$ mV), sodium channels begin to open up, allowing for the influx of Na$^+$ ions which further depolarize the membrane. About 1 ms later, two events occur to bring the voltage back toward and somewhat beyond the resting value: the sodium conductance inactivates, that is, the sodium channels slowly close again, and potassium channels open up, causing an outward current to flow. This outward current forces the membrane potential below the resting value of −65 mV (hyperpolarization), but it too eventually deactivates, allowing g_l to pull V_m back to rest.

Mathematical Formulation

The equation describing the circuit in Figure 2 is

$$C\frac{dV_m}{dt} = g_l(E_l - V_m) + g_{Na}(E_{Na} - V_m) + g_K(E_K - V_m) \quad (1)$$

While g_l is constant, g_{Na} and g_K are time- and voltage-dependent:

$$g_{Na} = \overline{G}_{Na} \cdot m(t)^3 h(t)$$
$$g_K = \overline{G}_K \cdot n(t)^4$$

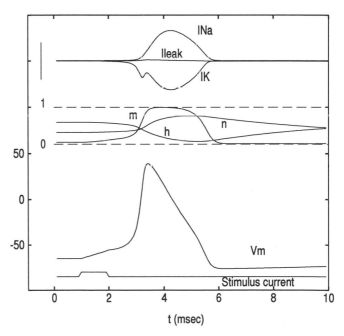

Figure 3. Membrane (nonpropagating) action potential. Simulation of a single compartment, corresponding to either a small patch of membrane or a "space-clamped" axon (see circuit in Figure 2). A brief (1 ms) current stimulus was applied at $t = 1$ ms. The top set of curves shows the membrane current density at the stimulated end (scale bar = 1 mA/cm^2). The initial hump in I_{Na} is due to a reduction in the driving force $E_{Na} - V_m$ that occurs during the rapid upstroke of the spike. The fractional activations of the m, h, and n channels are shown second from top (the two dashed lines correspond to activations of 0 and 1). Temperature = 6.3°C.

where the constants $\bar{G}_{Na}$ and $\bar{G}_K$ are the maximal conductances and the time dependence and voltage dependence reside in the so-called *gating particles*, described by the state variables m, h, and n. These variables follow first-order kinetics, relaxing exponentially toward a steady-state value x_∞ with a time constant τ_x:

$$\frac{dm}{dt} = \frac{m_\infty(V_m) - m}{\tau_m(V_m)}$$

$$\frac{dh}{dt} = \frac{h_\infty(V_m) - h}{\tau_h(V_m)}$$

$$\frac{dn}{dt} = \frac{n_\infty(V_m) - n}{\tau_n(V_m)}$$

The steady state activations (m_∞, h_∞, and n_∞) have a sigmoidal dependence on voltage. The *activation* variables m and n have the asymptotes $\lim_{V_m \to -\infty} m_\infty, n_\infty = 0$, $\lim_{V_m \to \infty} m_\infty, n_\infty = 1$, while the reverse holds for the *inactivation* variable h. That is, for very negative voltages, the m and n particles shut off current flow through both channels types, while at very positive potentials, the h particle shuts off the sodium current. The time "constants" (τ_m, τ_h, and τ_n) are not constant with respect to voltage, but rather have a roughly bell-shaped dependence with peaks in the -80 to -40 mV range. The x_∞ and τ_x values were the ones actually measured by Hodgkin and Huxley using a series of voltage clamp steps. Instead of fitting these curves directly with mathematical functions, which would be sufficient for simulation purposes, they chose to express x_∞ and τ_x in terms of the variable α_x and β_x:

$$x_\infty = \frac{\alpha_x}{\alpha_x + \beta_x}$$

$$\tau_x = \frac{1}{\alpha_x + \beta_x}$$

where α_x and β_x depend on V_m as follows:

$$\alpha_m = \frac{-0.1(V_m + 40)}{e^{-(V_m + 40)/10} - 1} \qquad \beta_m = 4e^{-(V_m + 65)/18}$$

$$\alpha_h = 0.07 e^{-(V_m + 65)/20} \qquad \beta_h = \frac{1}{e^{-(V_m + 35)/10} + 1}$$

$$\alpha_n = \frac{-0.01(V_m + 55)}{e^{-(V_m + 55)/10} - 1} \qquad \beta_n = 0.125 e^{-(V_m + 65)/80}$$

These rate constants assume a temperature of 6.3°C. At higher temperatures, they should be multiplied by a factor of 3 per 10°C. The functional forms were chosen for two reasons. First, they were among the simplest that fit the data, and second, they resemble the equations that govern the movement of a charged particle in a constant field.

While this set of equations may seem daunting at first, they correspond to hypothetical biophysical correlates. The *gating particles* can be thought of as charged portions of the channel protein that we now know can undergo a conformational change and flip between physically covering and uncovering the channel opening. Since the channel will be blocked if any of the four particles is in the closed position, multiplicative dependences of the form $m^3 h$ and n^4 are obtained. $x_\infty(V_m)$ then becomes the probability that an x particle will be in the open state at potential V_m. Each particle thus follows a two-state Markov model, where α_x is the rate constant from the closed to the open state, and β_x is the rate constant from the open to the closed state. The time courses of the three variables are graphed in Figure 3.

This mathematical formalism was laid down in 1952. Since then, most models of voltage-dependent conductances—not only in axons, but also in cell bodies and dendrites—have used the same formalism with only minor modifications (Koch and Segev, 1989).

Action Potential Propagation

Equation 1 describes a patch of membrane with no spatial extent. This corresponds to the original experiments where the axon was "space clamped": a long electrode was inserted into the axon along its axis, removing any spatial dependence. In response to stimulation, the whole membrane would fire simultaneously as a single isopotential unit. More commonly, one end of the axon is stimulated and an action potential propagates to the other end. The equation that governs extended structures is the cable equation

$$C \frac{\partial V_m}{\partial t} = \frac{d}{R_a} \frac{\partial^2 V_m}{\partial x^2} + g_l(E_l - V_m) + g_{Na}(E_{Na} - V_m)$$
$$+ g_K(E_K - V_m) \qquad (2)$$

where d is the axon diameter. The equation rests on the assumption of radial symmetry, i.e., radial current flow can be neglected, leaving only one spatial dimension, the distance x along the cable, in addition to t. If the last two (active) terms are dropped from the right-hand side, we are left with the classical cable equation for passive cables (see DENDRITIC PROCESSING). Associated with that equation is the *space constant* $\lambda = 1/\sqrt{g_l R_a}$, which is the distance across which the membrane potential decays a factor e in an infinite cable.

Figure 4 shows the result of a simulation of a 100-cm-long axon of diameter $d = 1$ mm. One end was stimulated with a brief current pulse, and the voltage was graphed for five positions along the axon. The form of the action potential is very similar to that in Figure 3; furthermore, the action potential is self-similar as it propagates, showing no signs of dispersion.

The total delay from one end to the other is about 5 ms, giving an average velocity of about 20 m/s. By assuming a constant conduction velocity, that is, by postulating the existence of a wave $V_m(x, t) = V_m(x - vt)$, Equation 2 shows that the velocity is proportional to the square root of axon diameter: $v \propto \sqrt{d}$ (Rushton, 1951). Indeed, in a truly remarkable test of their model, Hodgkin and Huxley estimated the velocity to be 18.8 m/s, a value within 10% of the experimental value of 21.2 m/s. This is surprisingly accurate, considering that the model was derived from a space-clamped axon. This represents one of the rare instances in which a neurobiological model makes a successful quantitative prediction. The square root relationship had previously been found experimentally in the squid (Pumphrey and Young, 1938).

Myelinated and Nonmyelinated Fibers

The principle of action potential generation and propagation appears to be very similar across neuronal types and species. One important evolutionary invention is that of myelination in the vertebrate phylum. Myelin sheaths are white fatty extensions of Schwann cells or neuroglial cells that are wrapped in many layers around axons. Myelin is a major component of the white matter of the brain, as opposed to the gray matter of the neocortex, which has a high concentration of cell bodies and dendrites. The myelin sheaths extend for up to 1–2 mm along the axon (the *internodes*) and are separated by *nodes of Ranvier* that are only a few μm long. The internodal distance appears to be approximately linear in fiber diameter (Hursh, 1939).

Figure 4. Action potential propagation. Simulation of a 100 cm axon of diameter $d = 1$ mm. At $t = 1.5$ ms, a current pulse stimulus (3.5 mA, 1 ms) is applied to one end (bottom trace). The voltage is shown at 5 locations: 0, 25, 50, 75, and 100 cm. Temperature = 6.3°C.

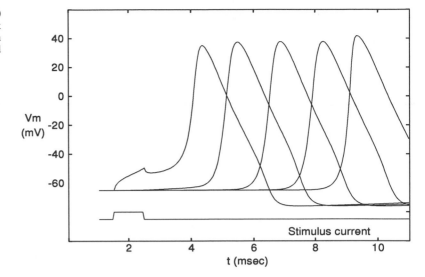

Figure 5. Spike propagation velocity and diameter for myelinated and nonmyelinated fibers. The propagation velocity has a square root dependence on diameter for nonmyelinated axons. For myelinated axons, the dependence is linear or slightly sublinear. Myelination increases velocity for axons as thin as 0.2 μm, which are among the smallest found in the brain. Adapted from Waxman and Bennett (1972).

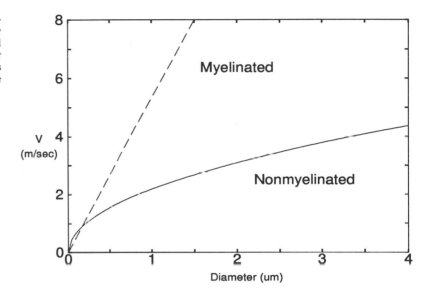

Myelin insulates the axon from the surrounding medium, increasing the membrane resistance and decreasing the capacitance. This reduces the electrotonic length of the axon for both DC and AC signals, making the cable electrically shorter, thereby significantly increasing the propagation speed. While a 1 mm nonmyelinated axon in the squid has an associated propagation speed of only about 20 m/s (Hodgkin and Huxley, 1952), a myelinated 20 μm vertebrate axon can reach over 100 m/s (Hursh, 1939). For a nonmyelinated axon to reach that velocity, it would have to be an inch thick! This reduction in axon diameter allows for a much higher packing density while conserving speed.

It has been shown experimentally (Hursh, 1939) and theoretically (Rushton, 1951) that the velocity of propagation is linear or slightly sublinear in the fiber diameter for myelinated axons. Figure 5 compares the spike propagation velocity for myelinated and nonmyelinated axons for small diameters. The myelinated axons already overtake nonmyelinated ones in the sub-micrometer range.

As opposed to their uniform distribution in nonmyelinated nerve, the voltage-gated channels in myelinated nerve are highly segregated between node and internode (Waxman and Ritchie, 1985; Hille, 1992). The nodal membrane has a high concentration of sodium channels and voltage-independent leak channels. The internodal membrane has a low concentration of potassium and leak channels, and is virtually devoid of sodium channels. Here, the repolarization of the membrane following the initial phase of the spike is via the leak channels and sodium inactivation. This low density of channels in the internodal membrane, which makes up more than 99% of the axonal membrane, reduces the average current density across the membrane, resulting in great savings in metabolic energy. Most of the activity occurs at the nodes of Ranvier, while the propagation along the internodes is chiefly passive.

In summary, myelin provides three advantages: propagation speed and packing density are both dramatically increased, and power consumption is decreased.

The Axonal Tree

Some axons branch profusely in the vicinity of the cell body. Others send off one or a few branches that course through the body for up to a meter before branching. Others extend for a few millimeters, giving rise to axonal arbors at regular intervals. The axon often arises at the "axon hillock," a somatic bulge opposite from the trunk of the dendritic tree. However, other arrangements are possible, such as the axon's emanating from the dendrite rather than the soma.

Figure 1 shows an example of a terminal arbor in turtle tectum from a cell originating in nucleus isthmi. This particular axon has a 3 μm myelinated parent trunk and initial daughter branches. These give rise to hundreds of thin, highly varicosed daughter branches that lack myelin. The varicosities are usually the location of synaptic *boutons*, a local thickening where action potentials trigger the release of neurotransmitter, which in turn induces a conductance change in the postsynaptic target neuron. Boutons of some neurons may receive synaptic input that can inhibit this signal transmission, a process known as presynaptic inhibition. The 3600 boutons on this arbor average 1.5 μm in diameter, though the size is highly variable, with a few boutons being as large as 7 μm.

The propagation speed along an unbranched axon depends on the diameter, as discussed earlier. In addition, a delay might be introduced at branch points, at varicosities at presynaptic terminals, and at locations where the diameter changes abruptly (Manor, Koch, and Segev, 1991). The delay may be negative (a speed-up) depending on the geometrical aspects, in particular the diameter of the parent branch in relation to that of the daughter branches. In a simulation of a 3.5-mm-long branched terminal axonal tree, Manor, Koch, and Segev (1991) found that the total axonal delay from the cell body to the synaptic terminals ranged from about 3 to 6 ms. Most of this delay (67–78%) arose from the properties of unbranched, uniform cables; 16–26% resulted from branch point delays, and 6–7% from the presence of varicosities. In theory, the delay at a single branch point may be as large as 2 ms or more, if the temperature is low and the impedance mismatch is large. If the mismatch is too large, however, *branch point failure* may occur, a condition in which the action potential fails to propagate beyond the branch point. The concept of branch point filtering has been put forth by Chung, Raymond, and Lettvin (1970): the branch point may constitute a point of control where selective transmission occurs, allowing the axonal tree to distribute action potentials only to a subset of nerve terminals.

While the axonal propagation delay may seem an unavoidable fact of life that slows down neural communication, it may also have important computational advantages. For instance, sound localization (see SOUND LOCALIZATION AND BINAURAL PROCESSING) in the barn owl depends on interaural time differences as small as a tenth of a millisecond and is apparently obtained by using the axon as a delay line (Carr and Konishi, 1988). Also, several models of brain function depend critically on the exact timing of inputs from different sources. While delays may be imposed by the dendritic trees at the input end of the neuron, the axons are also important candidates for this function.

Road Map: Biological Neurons
Background: I.1. Introducing the Neuron
Related Reading: Ion Channels: Keys to Neuronal Specialization

References

Carr, C. E., and Konishi, M., 1988, Axonal delay lines for time measurement in the owl's brainstem, *Proc. Natl. Acad. Sci. USA*, 85: 8311–8315.

Chung, S.-H., Raymond, S. A., and Lettvin, J. Y., 1970, Multiple meaning in single visual units, *Brain Behav. Evol.*, 3:72–101.

Hille, B., 1992, *Ionic Channels of Excitable Membranes*, 2nd ed., Sunderland, MA: Sinauer Associates. ◆

Hodgkin, A. L., 1976, Chance and design in electrophysiology: An informal account of certain experiments on nerve carried out between 1934 and 1952, *J. Physiol.*, 263:1–21. ◆

Hodgkin, A. L., and Huxley, A. F., 1952, A quantitative description of membrane current and its application to conduction and excitation in nerve, *J. Physiol.*, 117:500–544.

Hursh, J. B., 1939, Conduction velocity and diameter of nerve fibers, *Amer. J. Physiol.*, 127:131–139.

Jack, J. J. B., Noble, D., and Tsien, R. W., 1975, *Electric Current Flow in Excitable Cells*, Oxford: Oxford University Press. ◆

Kandel, E. R., Schwartz, J. H., and Jessell, T. M., 1991, *Principles of Neural Science*, 3rd ed., New York: Elsevier. ◆

Koch, C., and Segev, I., Eds., 1989, *Methods in Neuronal Modeling*, Cambridge, MA: MIT Press.

Manor, Y., Koch, C., and Segev, I., 1991, Effect of geometrical irregularities on propagation delay in axonal trees, *Biophys. J.*, 60:1424–1437.

Pumphrey, R. J., and Young, J. Z., 1938, The rates of conduction of nerve fibres of various diameters in cephalopods, *J. Exp. Biol.*, 15: 453–467.

Roberts, A., and Bush, B. M. M., 1981, *Neurones Without Impulses*, Cambridge, Eng.: Cambridge University Press. ◆

Rushton, W. A. H., 1951, A theory of the effects of fibre size in medullated nerve, *J. Physiol.*, 115:101–122.

Sereno, M. I., and Ulinski, P. S., 1987, Caudal topographic nucleus isthmi and the rostral nontopographic nucleus isthmi in the turtle, *Pseudemys scripta, J. Comp. Neurol.*, 261:319–346.

Waxman, S. G., and Bennett, M. V. L., 1972, Relative conduction velocities of small myelinated and non-myelinated fibres in the central nervous system, *Nature*, 238:217–219.

Waxman, S. G., and Ritchie, J. M., 1985, Organization of ion channels in the myelinated nerve fiber, *Science*, 228:1502–1507. ◆

Backpropagation: Basics and New Developments

Paul J. Werbos

Introduction and Discussion

Backpropagation is the most popular algorithm for Artificial Neural Networks (ANNs). It is used in about 70% of real-world ANN applications, though estimates range from 40% to 90%. Sophisticated forms of backpropagation may be crucial to understanding Biological Neural Networks (BNNs) as well, contrary to conventional wisdom.

The most popular form of backpropagation today is the simplest version, popularized by Rumelhart, Hinton, and Williams (1986). That version applies backpropagation to problems of supervised learning or pattern recognition. It adapts a particular form of multilayer ANN by propagating "error signals" or derivatives backwards from layer to layer of the network. In this article I describe a more general version, closer to the original version I developed in the 1970s (Werbos,

1994a), with their version presented as a special case, which I call "basic backpropagation." (However, other articles in the *Handbook* will almost always use the term *backpropagation* to refer to the "basic" version. [MAA])

Most useful applications of backpropagation require further refinements, discussed in later sections of this article. The greatest benefits of backpropagation involve *prediction* and *control* (including optimal planning and reinforcement learning). The most significant control applications are *not* examples of supervised learning; for example, backpropagation in its general form has helped to speed up reinforcement learning systems used to optimize manufacturing and flight control systems (White and Sofge, 1992; Werbos, 1993b). Recent empirical results in biology have begun to converge with engineering-based reinforcement learning designs down to the circuit level; Werbos (1994b) provides citations and details. Werbos (1994a) suggests links to humanistic psychology.

Definitions and History

Because backpropagation has had so many diverse applications, the word itself has been used in different ways, some precise and some confusingly sloppy. There are two standard definitions, both equally acceptable.

1. Backpropagation is a procedure for *efficiently* calculating the derivatives of some output quantity of a nonlinear system, with respect to all inputs and parameters of that system, through calculations proceeding *backwards* from outputs to inputs. It permits "local" implementation on parallel hardware (or wetware).
2. Backpropagation is any technique for adapting the weights or parameters of a nonlinear system by somehow using such derivatives or the equivalent.

In basic backpropagation—a special case—the weights are adapted by gradient descent (see the next section), but useful applications of backpropagation often use the derivatives in more sophisticated ways (see later sections). Backpropagation is critical when working with *large* networks or systems—even when large computers are available—because it reduces the cost of computing derivatives by a factor of N, compared with traditional methods, where N is the number of derivatives to be calculated.

Backpropagation can be applied to *any* differentiable, sparse, nonlinear system; it is not restricted to ANNs or to particular forms of ANNs. Therefore, even if BNNs are more complex than most ANN designs, this does not rule out backpropagation (as distinct from basic backpropagation) in biological systems.

There is no such thing as a "backpropagation network." In basic backpropagation, we use an ANN design called a *multilayer perceptron* (MLP). Intuitively, MLPs are feedforward networks constructed by interconnecting model neurons defined by the McCulloch-Pitts neuron model, with variations. The idea of adapting MLPs by minimizing square error is older than backpropagation; see the classic papers by Widrow, Rosenblatt, and Amari in Anderson and Rosenfeld (1988). The word *backpropagation* was coined by Rosenblatt to describe a method of his which has gone out of use.

The original impetus for my approach to backpropagation came from Sigmund Freud. Freud is most famous for the clinical ideas which brought him his funding. However, Freud (1954) also worked hard to develop a theory of "psychodynamics," incorporating explicit ideas about flows of information between neurons which I translated into methods for adapting MLPs. The first working applications of backpropagation, in my 1974 Harvard PhD thesis, were to *general* nonlinear systems for political forecasting, not ANNs. Related work prior to Rumelhart, Hinton, and Williams (1986) includes papers by Parker and LeCun and efforts by Charles Smith; see Werbos (1994a) for more details. Haykin (1994) also provides in-depth information on backpropagation. DeBoeck (1994) contains additional history and financial applications.

Basic Backpropagation

Supervised Learning

Basic backpropagation is the most popular method for performing the task called *supervised learning*. In supervised learning, we try to adapt an ANN so that its actual outputs ($\hat{\mathbf{Y}}$) approximate some target outputs ($\mathbf{Y}$) for a training set which contains T patterns. The goal is to adapt the parameters of the network so that it performs well for patterns outside the training set.

Pattern recognition is the most common application of supervised learning today. For example, suppose that we want to build an ANN which learns to recognize handwritten ZIP codes. Assume that we already have a camera and preprocessor which can digitize the image, locate the five digits, and provide a 19×20 grid of bits (1s and 0s) representing the image of each digit. We want the ANN to input the 19×20 image and output a classification; for example, we might ask the network to output four binary digits which, together, identify which decimal digit is being observed.

Before adapting the parameters of the ANN, we must obtain a training database of actual handwritten digits and correct classifications. Suppose, for example, that this database contains 2000 examples of preclassified handwritten digits. In that case, $T = 2000$. We may give each example a label t between 1 and 2000. For each sample t, we have a record of the input pattern and the correct classification. Each input pattern consists of 380 numbers, which may be viewed as a vector with 380 components; we may call this vector $\mathbf{X}(t)$. The desired classification consists of four bits, which form a vector $\mathbf{Y}(t)$. The actual output of the network will be $\hat{\mathbf{Y}}(t)$, which may differ from the desired output $\mathbf{Y}(t)$, especially in the period *before* the network has been adapted. (I use the term *adapted* since the term *trained* is not appropriate when backpropagation is used for unsupervised learning.) To solve the supervised learning problem, there are two steps.

1. We must specify the "topology" (connections and equations) for a network which inputs $\mathbf{X}(t)$ and outputs a four-component vector $\hat{\mathbf{Y}}(t)$, an approximation to $\mathbf{Y}(t)$. The relation between inputs and outputs must depend on a set of weights (parameters) $\mathbf{W}$ which can be adjusted.
2. We must specify a "learning rule"—a procedure for adjusting the weights $\mathbf{W}$ so as to make the actual outputs $\hat{\mathbf{Y}}(t)$ approximate the desired outputs $\mathbf{Y}(t)$.

In basic backpropagation, we use a very simple network design, described in the next section.

Even when we use a simple network design, the vectors $\mathbf{X}(t)$ and $\mathbf{Y}(t)$ need not be made up of ones and zeros. They can contain any values which the network is capable of inputting and outputting. Let us denote the components of $\mathbf{X}(t)$ as $X_1(t), \ldots, X_m(t)$, so that there are m inputs to the network. Let us denote the components of $\mathbf{Y}(t)$ as $Y_1(t), \ldots, Y_n(t)$, so that we have n outputs. Throughout this article, the components of a vector will be represented by the same letter as the vector itself, in the same case; this convention is convenient because it lets us work with two different vectors, $\mathbf{x}(t)$ and $\mathbf{X}(t)$, which are closely related to each other.

The Generalized Feedforward MLP

Before we specify a learning rule, we must define how the outputs of the ANN depend on its inputs and weights. In basic backpropagation, we assume the following:

$$x_i = X_i \qquad 1 \leq i \leq m \qquad (1)$$

$$v_i = \sum_{j=1}^{i-1} W_{ij} x_j \qquad m < i \leq N + n \qquad (2)$$

$$x_i = s(v_i) \qquad m < i \leq N + n \qquad (3)$$

$$\hat{Y}_i = x_{i+N} \qquad 1 \leq i \leq n \qquad (4)$$

where Equation 1 presents the inputs and Equation 4 reads the outputs. The function s in Equation 3 is usually the sigmoidal function

$$s(v) = 1/(1 + e^{-v}) \qquad (5)$$

and where N is a constant which can be any integer you choose larger than m. The value of $(N + n)$ decides how many neurons are in the network (counting inputs as neurons). Intuitively, v_i represents the total level of voltage exciting a neuron, and x_i represents the intensity of the resulting output from the neuron (x_i is called the *activation level* of the neuron). Many authors assume that there is a threshold or constant weight W_{io} added to the right side of Equation 2; however, we achieve the same effect by assuming that one of the inputs (like X_1) is always 1.

Figure 1 illustrates the significance of these equations. There are $N + n$ circles, representing all the neurons in the network, *including* the input neurons. The first m circles are copies of the inputs $X_1, \ldots, X_m$; they are included in the vector $\mathbf{x}$ only as a way of simplifying the notation. Every other neuron in the network—such as neuron number i, which calculates v_i and x_i—takes input from every cell which *precedes* it in the network. Even the last output cell, which generates $\hat{Y}_n$, takes input from other output cells, such as the one which outputs $\hat{Y}_{n-1}$.

In ANN terminology, this network is "fully connected" in the extreme. In practical applications, it is usually desirable to limit the connections between the neurons. This can be done by fixing some of the weights W_{ij} to zero so that they drop out of all calculations. Many researchers prefer to use "layered" networks, in which all connection weights W_{ij} are zeroed out except for those going from one "layer" (subset of neurons) to the next layer. In general, we may zero out as many or as few of the weights as we choose, based on our understanding of

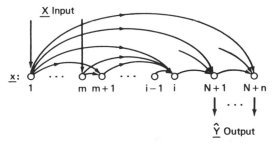

Figure 1. Generalized MLP used in basic backpropagation. For a 3-layer MLP, zero out connections within the hidden layer ($m + 1$ to N), within the output layer ($N + 1$ to $N + n$), or direct from the input layer to the output layer.

individual applications. Most beginners define only three layers—an input layer, a single "hidden" layer, and an output layer, as in Rumelhart et al. (1986). This section assumes the full range of *allowed* connections, for the sake of generality.

Adapting the Network: Approach

In basic backpropagation, we choose the weights W_{ij} so as to minimize square error over the training set:

$$E = \sum_{t=1}^{T} E(t) = \sum_{t=1}^{T} \sum_{i=1}^{n} (1/2)(\hat{Y}_i(t) - Y_i(t))^2 \qquad (6)$$

This is a special case of the method of least squares, used routinely in statistics, econometrics, and engineering; the uniqueness of backpropagation lies in *how* this expression is minimized.

In basic backpropagation, there are two popular alternative approaches: pattern learning (or observation-by-observation learning) and batch learning. This section discusses pattern learning, the most popular approach.

We start out with arbitrary values for the weights $\mathbf{W}$. (Most people choose random numbers from -0.1 to 0.1, but it may be better to guess the weights based on prior information, when prior information exists. The best results often come from reusing weights adapted to solve a similar problem with a similar connectivity.) Next, we calculate the outputs $\hat{\mathbf{Y}}(t)$ and the errors $E(t)$ for that set of weights. Then we calculate the derivatives of E with respect to all the weights, using a backwards sweep of calculations. If increasing a given weight would lead to more error, we adjust that weight downward. If increasing a weight leads to less error, we adjust it upward. After adjusting all the weights, we start all over, and keep on going through this process until the weights and the error settle down. (Some people iterate until error approaches zero; however, if the number of training patterns exceeds the number of weights in the network—as recommended by studies of generalization—this should usually be impossible.) The uniqueness of backpropagation lies in the method used to calculate the derivatives exactly for all the weights in only one pass through the system.

Calculating Derivatives: Theoretical Background

Many authors suggest that we need only use the conventional chain rule for partial derivatives to calculate the derivatives of E with respect to all the weights; we do this by first calculating the derivatives of error with respect to the outputs of the network, and then working back layer-by-layer to the input layer and the weights. Under certain conditions, this can be a rigorous approach, but its generality is limited, and it requires extraordinary care with the side conditions (which are rarely spelled out); calculations like this often become confused and erroneous when networks and applications grow complex. Even when using Equation 7, it is best to test one's gradient calculations using explicit perturbations, to be sure that there is no bug in one's code. If test results appear suspicious, one may develop a spreadsheet version of a very simple test problem to help identify the error.

In order to prove the validity of these calculations and expand their generality, I proved a new chain rule for ordered derivatives in 1974 (Werbos, 1994a):

$$F_{-}z_i = \frac{\partial \text{TARGET}}{\partial z_i} + \sum_{j>i} F_{-}z_j * \frac{\partial z_j}{\partial z_i} \qquad (7)$$

where we write the ordered derivative of TARGET with respect to z_i as $F_{-}z_i$, which represents "F_eedback to z_i." This chain rule is valid only for *ordered* systems where the values to be calculated could be calculated one by one (if necessary) in the order $z_1, z_2, \ldots, z_n$, TARGET. The simple partial derivatives represent the *direct* impact of z_i on z_j *through the system equation* which determines z_j. The ordered derivative represents the *total* impact of z_i on TARGET, accounting for both the direct and indirect effects. For example, suppose that we had a simple system governed by the following two equations, in order:

$$z_2 = 4 * z_1$$

$$z_3 = 3 * z_1 + 5 * z_2 \qquad (8)$$

The "simple" partial derivative of z_3 with respect to z_1 (the *direct* effect) is 3; to calculate the simple effect, we look *only* at the equation which determines z_3. However, the ordered derivative of z_3 with respect to z_1 is 23 because of the indirect impact by way of z_2. The simple partial derivative measures what happens when we increase z_1 (e.g., by 1 in this example) and assume that everything else (like z_2) in the equation which determines z_3 remains constant. The ordered derivative not only measures what happens when we increase z_1, but also recalculates all other quantities—like z_2—which are later than z_1 in the causal ordering we impose on the system.

This chain rule provides a straightforward, plodding, "linear" procedure to calculate the derivatives of a given TARGET variable with respect to *all* of the inputs (and parameters) of an ordered differentiable system *in only one pass through the system*. For purposes of debugging, one can calculate the true value of any ordered derivative simply by perturbing z_i at the point in the program where z_i is calculated; this is particularly useful when applying backpropagation to a complex network of functions other than ANNs.

Adapting the Network: Equations

For a given set of weights **w**, it is easy to use Equations 1 through 6 to calculate $\hat{Y}(t)$ and $E(t)$ for each pattern t. The trick is in how we then calculate the derivatives.

Let us use the prefix $F\_$ to indicate the ordered derivative of E with respect to whatever variable the $F\_$ precedes. (This is a special case of the preceding section, where we choose E to be our TARGET.) Thus, for example,

$$F\_\hat{Y}(t) = \frac{\partial E}{\partial \hat{Y}_i(t)} = \hat{Y}_i(t) - Y_i(t) \qquad (9)$$

which follows simply by differentiating Equation 6. By the chain rule for ordered derivatives as expressed in Equation 8,

$$F\_x_i(t) = F\_\hat{Y}_{i-N}(t) + \sum_{j=i+1}^{N+n} W_{ji} * F\_v_j(t)$$
$$i = N+n, \ldots, m+1 \qquad (10)$$

$$F\_v_i(t) = s'(v_i) * F\_x_i(t) \qquad i = N+n, \ldots, m+1 \qquad (11)$$

$$F\_W_{ij} = \sum_{t=1}^{T} F\_v_i(t) * x_j(t) \qquad (12)$$

where s' is the derivative of $s(v)$ as defined in Equation 5 and $F\_\hat{Y}_k$ is assumed to be zero for $k \leq 0$. Note how Equation 10 requires us to run *backwards* through the network to calculate the derivatives; this backwards propagation of information gives backpropagation its name. A little calculus and algebra, starting from Equation 5, shows that

$$s'(v) = s(v) * (1 - s(v)) \qquad (13)$$

which we can use when we implement Equation 11. Finally, to adapt the weights, people usually set

$$\text{New } W_{ij} = W_{ij} - \text{learning\_rate} * F\_W_{ij} \qquad (14)$$

where the learning_rate is some small constant chosen on an ad hoc basis. (People usually make it as large as possible, up to 1, until the error begins to diverge; the next section describes alternatives.)

Equations 9–14 do not directly tell us how to calculate derivatives through *recurrent* networks, but they are easily generalized, as explained later.

Improving Basic Backpropagation (MLPS)

Basic backpropagation is powerful enough for many applications. However, researchers have tried to understand and improve three characteristics of the method: (1) *generalization*, the ability of a network to predict **Y** accurately outside of the original training set; (2) *learning speed* or *convergence rate*, which is vital for systems learning from real-time experience (rather than fixed training sets); and (3) *fault tolerance*, the ability to perform well despite noise or breakage. I will discuss the first two topics.

Generalization

The key to generalization is to keep one's network "simple." One must seek networks with fewer connections and smaller weights. Three procedures are commonly used: pruning, weight sharing, and growing.

In pruning, we start out with highly connected networks, like generalized MLPs (as in Figure 1) or layered MLPs. (The former eliminates the need to choose the number of layers a priori; see Krishnakumar, 1993.) Instead of minimizing $E(t)$ from Equation 6, one minimizes

$$E'(t) = E(t) + f(W, k) \qquad (15)$$

where f is some function of the weights, representing network complexity, and k is some arbitrary parameter. For example, Werbos (1987) proposed that we generalize "ridge regression," a statistical method, by choosing

$$f(W, k) = k \sum_{ij} W_{ij}^2 \qquad (16)$$

White and Sofge (1992, chap. 10) and Weigend, Rumelhart, and Huberman (1991) discuss current variations.

When weights approach zero, they can be removed or "pruned." One can also expect improved generalization owing to smaller weights. As in ridge regression, the choice of k is itself a subject for research. One way to choose k is to split one's data into two parts, a training set and a test set; we try different values of k on the training set and choose whatever works best on the test set, by some measure (White and Sofge, 1992). Theoreticians have studied other complexity measures such as the VC-dimension (see VAPNIK-CHERVONENKIS DIMENSION OF NEURAL NETWORKS); however, practical results demand more attention to computational cost, robustness, and the early stages of learning (White and Sofge, 1992).

Weight sharing exploits the old principle that you should exploit prior information. Weight sharing is common in image processing, character recognition, and other applications where the input is a rectangle of pixels. If your input data, **X**(t), forms a matrix of values $X_{ij}(t)$, representing the gray-scale level at spatial coordinates (i, j), you can define an array of hidden units (also forming a rectangle) by

$$v_{lk} = \sum_{i,j} W_{l-i, k-j} X_{i,j} \qquad (17)$$

Equation 2 would have required an *array* of weights $W_{l,k,i,j}$. Equation 17 "reuses" the same weights for different neurons (l, k). Weight sharing was popularized by Guyon et al. (1989) of AT&T, whose ZIP code recognizer matches the best in existence. Simpler versions of backpropagation, without refinements, perform much worse on this application.

From a theoretical viewpoint, weight sharing is simply a way of exploiting our a priori knowledge that visual images are symmetric with respect to spatial translation.

In *growing networks*, people start with minimal networks and *add* neurons "as needed" (see TOPOLOGY-MODIFYING NEURAL NETWORK ALGORITHMS). The most popular method is Fahlman's cascade-correlation method, which does well on small-scale classification problems on fixed databases. Ultimately, to manage huge brain-like networks, we must combine pruning and growing—and weight sharing too, if we can find a workable way to learn usable symmetries without overcommitting ourselves to them. We must create and test new connections "at random," and whole new neurons and even circuits; such exploratory subsystems would serve a function similar to that of university research in a market economy, to help keep the system from being caught in a rut, a local minimum (Werbos, 1987).

Learning Speed

Popular wisdom states that basic backpropagation generalizes better than the popular alternatives—RBF networks (see RADIAL BASIS FUNCTION NETWORKS) and the Cerebellar Model Articulation Controller, or CMAC (Albus, 1975)—but learns more slowly in real time. Equation 14 represents the well-known method of *gradient descent*, which is extremely slow when used naively.

According to the usual linearized analysis of steepest descent, there are two common sources of slow learning: (1) *scaling problems*, where weights vary dramatically in size and should be adapted at different rates (different learning rates for different weights); (2) *correlation problems*, where inputs are highly dependent on each other, so that adapting one disturbs the other. (Shanno, 1990, cites several texts containing standard linearized analysis.) There are many ad hoc methods which help with these problems. Good practitioners often employ tricks like using tanh instead of the sigmoidal function, developing learning-rate multipliers for groups of weights based on trial and error, rescaling input variables, and using the generalized adaptive learning rate (Werbos, 1993a; White and Sofge, 1992). (BHI has patents pending on the latter, and a version working on large-sample pattern learning.) Older conjugate gradient methods—Polak-Ribiere and Fletcher-Reeves—are popular but inferior to recent methods. Werbos (1993b) discusses more complex approaches for research and neuroscience connections.

Backpropagation with Alternative Functional Forms

Using Equation 7 or 8, we can apply backpropagation to feedforward networks other than ANNs, including econometric-style models and fuzzy logic (Werbos, 1993a). Slow speed and good generalization are due to MLPs, *not* to backpropagation. Backpropagation is much faster with other functional forms (DeClaris and Su, 1991; Narayan, 1993). The usual methods to adapt RBFs and CMACs are mathematically equivalent to backpropagation, for those networks. Even with MLPs, one can adapt *two separate learning rates*—a fast one for the output layer (as in RBFs) and a slow one (faster than the zero used in most RBFs!) for hidden layers. See Werbos (1993b) for details, and for ideas on combining the capabilities of local and global networks.

Backpropagation and Prediction in Recurrent Networks

The use of backpropagation for recurrent networks is straightforward, but confused in present literature, because of communications problems between disciplines. By 1981, backpropagation was generalized to yield an *exact* calculation of derivatives for recurrent networks; as in basic backpropagation, the cost of calculating a complete set of derivatives is similar to the cost of a single forward pass through the system—or, in some cases, far less—if the correct algorithm is used for the task at hand.

To avoid the common pitfalls in using recurrent networks, it is crucial to distinguish two different types of recurrent networks, essential to performing two different types of task:

1. *Simultaneous Recurrent Networks* (SRNs) have an output conceptually defined by

$$\hat{Y}(t) = y_\infty(t) \tag{18}$$

$$y_{n+1}(t) = f(X(t), W, y_n(t)) \tag{19}$$

where f represents the operation (inputs and outputs, in effect) of a feedforward "inner" network, and where $y_n(t)$ represents the nth iterate of y as the network settles down at time t. SRNs have superior capabilities in function approximation, critical to higher-order control and planning (Werbos, 1993b; White and Sofge, 1992, chap. 13). They can solve language processing tasks which are too hard for MLPs (Werbos, 1994a, chap. 10). They may even be crucial to efficient scene segmentation and binocular vision.

2. *Time-Lagged Recurrent Networks* (TLRNs) can be represented as

$$\hat{Y}(t) = f(X(t), R(t-1), W) \tag{20}$$

$$R(t) = g(X(t), R(t-1), W) \tag{21}$$

White and Sofge (1992, chap. 10) compare TLRNs against other networks for forecasting or spatiotemporal pattern recognition. TLRNs provide capabilities like short-term memory, adaptation to parameter drift or pattern reversals, filtering, etc.

For TLRNs, we can use Equation 8 directly for *backpropagation through time*. Rumelhart et al. (1986) proposed something similar for SRNs, but for SRNs it is far better to use *simultaneous backpropagation*, combining methods bin Werbos (1993b) and White and Sofge (1992, chap. 3).

Backpropagation through time is *not biologically plausible*, because it does not allow true forward-time, real-time learning. Other methods exist which do calculate derivatives in forward time but are still implausible, because they require computational structures which grow as the square of the network size (Werbos, 1994a). Note that the terms *backpropagation through time* and *recurrent backpropagation* have been used very loosely, without standardization, in much of the literature; thus it is common to read assertions about one algorithm, based on an implementation of a different algorithm or a misunderstanding of someone else's work.

White and Sofge (1990, chap. 10) explain how backpropagation through time has been used to cut errors in half in predicting real-world chemical plants, relative to basic backpropagation, even when the same feedforward ANN was being used. The same authors provide a generic review of the hundreds of working applications of neural networks in the chemical process industries today.

Many authors use the term BEP—Backwards Error Propagation—instead of backpropagation. The specific examples given earlier *do* involve a backwards propagation of error, a calculation of the derivatives of *prediction errors*. However, backpropagation can also calculate the derivatives of *utility*, for a measure which represents success or goal achievement in the external world. This could be called BUP.

There have been several applications of BUP recently, far surpassing conventional control and conventional neurocontrol for the same problems. For example, F-15s can learn to stabilize themselves in 2 seconds after having parts shot off, using BUP as part of more complex reinforcement-based control systems (far more sophisticated than conventional scalar reinforcement systems). White and Sofge (1992) explain several examples in detail. Werbos (1993a, 1994a) gives a few more recent examples and an introductory overview. These sources survey other control methods as well, and applications ranging from outer space to pollution abatement.

Road Map: Learning in Artificial Neural Networks, Deterministic
Background: I.3. Dynamics and Adaptation in Neural Networks
Related Reading: Neurosmithing: Improving Neural Network Learning; Perceptrons, Adalines, and Backpropagation; Recurrent Networks: Supervised Learning

References

Albus, J. S., 1975, A new approach to manipulator control: The cerebellar model articulation controller (CMAC), *Trans. ASME J. Dyn. Syst. Meas. Control*, 97:220–227.

Anderson, J., and Rosenfeld, E., Eds., 1988, *Neurocomputing: Foundations of Research*, Cambridge, MA: MIT Press.

Baum, E., and Haussler, D., 1989, What size net gives valid generalization? *Neural Comput.*, 1(1).

DeBoeck, G., Ed., 1994, *Trading on the Edge: Neural, Genetic and Fuzzy Systems for Chaotic Financial Markets*, New York: Wiley.

DeClaris, N., and Su, M., 1991, A novel class of neural networks with quadratic junctions, in *Proceedings of the IEEE Conference on Systems, Man, and Cybernetics, 1991*, New York: IEEE, pp. 1557–1562.

Freud, S., 1954, *The Origins of Psychodynamics: Letters to Wilhelm Fliess, 1887–1902*, New York: Basic Books.

Guyon, I., Poujaud, I., Personnaz, L., Dreyfus, G., Denker, J., and

LeCun, Y., 1989, Comparing different neural network architectures for classifying handwritten digits, in *Proceedings of the International Joint Conference on Neural Networks, 1989*, vol. 2, New York: IEEE, pp. 127–132.

Haykin, S., 1994, *Neural Networks: A Comprehensive Foundation*, New York, Macmillan. ◆

Krishnakumar, K., 1993, Optimization of the neural network connectivity pattern using a backpropagation algorithm, in *Proceedings of the 1993 World Congress on Neural Networks*, vol. 3, Hillsdale, NJ: Erlbaum, pp. 498–502.

Narayan, S., 1993, ExpoNet, in *Proceedings of the 1993 World Congress on Neural Networks*, vol. 3, Hillsdale, NJ: Erlbaum, pp. 494–497.

Rumelhart, D. E., Hinton, G. E., and Williams, R. J., 1986, Learning internal representations by error propagation, in *Parallel Distributed Processing: Explorations in the Microstructure of Cognition* (D. E. Rumelhart, J. L. McClelland, and PDP Research Group, Eds.), vol. 1, *Foundations*, Cambridge, MA: MIT Press, pp. 318–362. ◆

Shanno, D., 1990, Recent advances in numerical techniques for large-scale optimization, in *Neural Networks for Control* (W. Miller, R. Sutton, and P. Werbos, Eds.), Cambridge, MA: MIT Press, pp. 171–178.

Weigend, A., Rumelhart, D., and Huberman, B., 1991, Generalization by weight elimination with application to forecasting, in *Proceedings of the International Joint Conference on Neural Networks, 1991*, vol. 1, New York: IEEE, pp. 837–841.

Werbos, P., 1987, Learning how the world works, in *Proceedings of the IEEE Conference on Systems, Man, and Cybernetics, 1987*, New York: IEEE, pp. 302–310.

Werbos, P., 1989, Maximizing long-term gas industry profits in two minutes in Lotus using neural network methods, *IEEE Trans. Syst. Man Cybern.*, March/April:315–333.

Werbos, P., 1993a, Elastic fuzzy logic, *J. Intelligent Fuzzy Syst.*, 1:365–377.

Werbos, P., 1993b, Supervised learning, in *Proceedings of the 1993 World Congress on Neural Networks*, vol. 3, Hillsdale, NJ: Erlbaum, pp. 358–363.

Werbos, P., 1994a, *The Roots of Backpropagation: From Ordered Derivatives to Neural Networks and Political Forecasting*, New York: Wiley. ◆

Werbos, P., 1994b, The brain as a neurocontroller: New hypotheses and experimental possibilities, in *Origins: Brain and Self-Organization* (K. Pribram, Ed.), Hillsdale, NJ: Erlbaum, pp. 680–706.

White, D., and Sofge, D., Eds., 1992, *Handbook of Intelligent Control*, New York: Van Nostrand.

Basal Ganglia

Garrett E. Alexander

Introduction

Enormous progress has been made in characterizing the structure and functional organization of the basal ganglia, and yet a comprehensive understanding of how these nuclei contribute to behavioral control remains elusive. Although the functional topography of these structures is now understood in some detail, we still do not understand precisely what the basal ganglia do or how they do it. What is missing is a comprehensive theory of basal ganglia operations, and this is likely to require functional models of the basal ganglia that are able both to assimilate the constraints imposed by neurobiological data and to simulate various candidate behavioral functions in which these structures are believed to participate.

Functional Organization of the Basal Ganglia

The basal ganglia include the striatum (comprising the putamen, the caudate nucleus, and the ventral striatum), the globus pallidus, the substantia nigra, and the subthalamic nucleus. All of these structures are functionally subdivided into skeletomotor, oculomotor, associative, and limbic territories based on their physiological properties and on their interconnections with cortical and thalamic territories having the same functions (Alexander, Crutcher, and DeLong, 1990). As indicated in Figure 1, the large-scale organization of the basal ganglia can be viewed as a family of reentrant loops that are organized in parallel, each taking its origin from a particular set of functionally related cortical fields (skeletomotor, oculomotor, etc.),

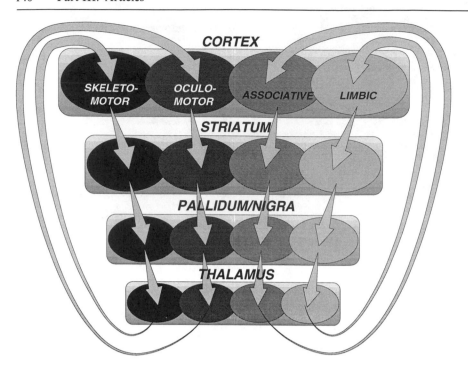

Figure 1. Parallel, functionally segregated, basal ganglia–thalamocortical loops.

passing through the functionally corresponding portions of the basal ganglia, and returning to parts of those same cortical fields by way of specific basal ganglia recipient zones in the dorsal thalamus.

Owing to the segregation that is maintained along each of these cortico–basal ganglia–thalamocortical circuits, there is little direct communication among the separate functional domains. Virtually the entire cortical mantle is topographically mapped onto the striatum, which represents the "input" portion of the basal ganglia. The cortically directed signals that emerge from the basal ganglia's output nuclei (internal pallidum and substantia nigra pars reticulata) are returned exclusively to foci within the frontal lobe, after first passing through select portions of the thalamus. The mapping of corticostriatal projections is such that functionally interconnected cortical fields (e.g., frontal motor and premotor areas and parietal somatosensory areas) tend to project to overlapping or contiguous domains within the striatum. Because of their parallel organization, corresponding stations along each of the functional loops are thought to carry out similar operations.

Motor Circuitry of the Basal Ganglia

Key features of the motor circuitry of the basal ganglia are depicted in Figure 2. Like the cerebellum, the skeletomotor portions of the basal ganglia receive inputs from most of the sensorimotor territories of the cerebral cortex, including primary and secondary somatosensory areas, primary motor cortex, and a variety of premotor areas, including the supplementary motor area, the dorsal and ventral premotor areas, and the cingulate motor areas. Directed to the putamen, which is the skeletomotor portion of the striatum, these coordinated corticostriatal inputs impose a well-defined somatotopic organization upon their target nucleus. Most of the putamen-projecting sensorimotor areas also send excitatory projections directly to the spinal cord (Dum and Strick, 1991), as well as to the subthalamic nucleus (STN).

Most striatal neurons are medium spiny, GABAergic, projection neurons (Wilson, 1992), known by the acronym MSNs

(for Medium Spiny Neurons). In primates they comprise two distinct populations: one projecting to the external segment of the globus pallidus (GPe), and the other projecting either to the internal segment of the globus pallidus (GPi) or to the substantia nigra pars reticulata (SNr) (Parent, 1990). At rest, putamen projection neurons are nearly silent, but most have well-defined sensorimotor fields and discharge selectively in relation to specific parameters of movement or specific aspects of the preparation for movement (Alexander and Crutcher, 1990). MSNs have two stable subthreshold states—one hyperpolarized, silent, and relatively resistant to unsynchronized excitatory inputs, and the other depolarized and more responsive to excitatory afferents (Wilson, 1992).

GPi and SNr constitute the output nuclei of the basal ganglia, sending GABAergic projections to their respective targets in both the ventrolateral and intralaminar thalamus. Anatomically, GPi and SNr share many common features, and from a functional standpoint, they can be viewed as two subdivisions of a single-output nucleus. The large projection neurons in the sensorimotor territories of GPi/SNr have movement-related receptive fields that are similar to those of the putamen's MSNs, but GPi/SNr neurons differ in having relatively high spontaneous discharge rates (60–80 Hz). The discharge properties and sensorimotor fields of GPe neurons are comparable to those in GPi/SNr.

The output nuclei receive not only *direct* inputs from the striatum via the GPi- and SNr-projecting MSNs, but also an important converging source of information via what has come to be termed the *indirect* pathway (Alexander and Crutcher, 1990). As indicated in Figure 2, the indirect pathway takes its origin from GPe-projecting MSNs. In turn, the GABAergic neurons of GPe project mainly to the STN, whose excitatory, glutamatergic neurons send feedforward connections to GPi/SNr, completing one arm of the indirect pathway. Subthalamic neurons also send glutamatergic connections back to GPe and the putamen.

The GPe also sends projections to the output nuclei themselves, thereby completing a second arm of the indirect pathway. Activation of MSNs associated with either arm of the

Figure 2. Basal ganglia skeletomotor circuitry. Excitatory pathways are indicated by light shading, whereas inhibitory pathways are indicated by dark shading. Connections between the thalamus and cortex are reciprocal. Connections to and from the SNr (which parallel those of GPi) are not shown. A further simplification shows the thalamostriate projection arising from ventrolateral thalamus, rather than from the centromedian nucleus.

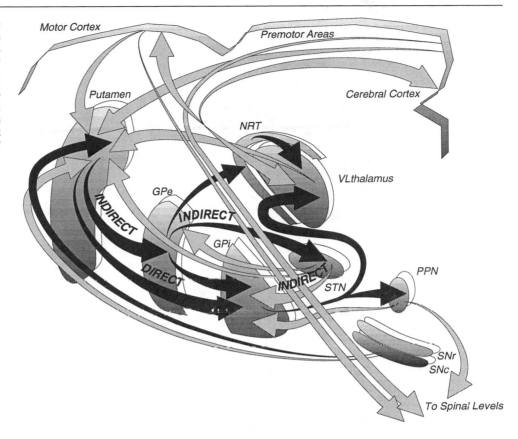

indirect pathway will tend to increase basal ganglia output by increasing neuronal activity at the level of the output nuclei—in one case by disinhibiting the STN with its excitatory projections to GPi/SNr, and in the other by directly disinhibiting GPi/SNr. In contrast, activation of MSNs associated with the direct pathway tends to decrease basal ganglia output by directly suppressing activity at the level of GPi/SNr. Given the reentrant nature of basal ganglia–thalamocortical connections, cortically initiated activation of the direct pathway will tend to be reinforced by disinhibition of excitatory thalamocortical projection neurons in basal ganglia recipient zones of the thalamus. Conversely, cortically initiated activation of the indirect pathway will tend to suppress reentrant thalamocortical excitation by increasing the inhibitory outflow from basal ganglia to thalamus.

The GPe sends an inhibitory, feedforward projection to the nucleus reticularis of the thalamus (NRT) which, in turn, imposes a robust, GABAergic modulation upon the basal ganglia recipient nuclei of the ventrolateral thalamus. There appears to be a functional consistency among the various GPe projections. By enhancing synaptic inhibition within the ventrolateral thalamus, all of the projections from GPe (including both arms of the indirect pathway as well as the GPe projections to NRT) tend to reduce corticothalamocortical interactions within the same functional loop.

On the other hand, the role of the NRT in basal ganglia operations may be much more complicated. The NRT is known to play an important role in gating thalamocortical transmission from the dorsal thalamus, and it does so in a roughly topographic manner. The NRT receives collateral, excitatory inputs from corticothalamic as well as thalamocortical fibers that must pass through this nucleus en route to their respective termination zones in the thalamus and cortex. In addition, the NRT may also receive collateral inputs from both the inhibitory pallidothalamic pathway and the excitatory thalamostriate projections.

While the pedunculopontine nucleus (PPN) is not generally considered to be part of the basal ganglia, this structure has strong, reciprocal connections with the basal ganglia output nuclei and with the STN, sending excitatory, probably glutamatergic projections to all of these structures, as well as to the striatum and to the dopaminergic neurons of the substantia nigra pars compacta (SNc) (see the next section). Until recently, descending projections of the basal ganglia output nuclei had received relatively little attention (save in the oculomotor system, where projections to superior colliculus are emphasized); it had been generally assumed that directly descending basal ganglia influences on the skeletomotor system extended no further caudally than the PPN. Recent studies in rats, however, have also revealed substantial PPN outflow to the reticulospinal system. These findings have not yet been extended to primates.

One of the remarkable features of basal ganglia circuitry is that functionally discrete channels of information processing are maintained throughout the various cortico–basal ganglia–thalamocortical pathways in the face of layer-to-layer connectivity that is highly convergent. The convergence ratio along the corticostriatal pathways, which provide approximately half of the excitatory input to the neostriatum, may be on the order of 5000:1. Cell counts in the human striatum and globus pallidus suggest minimum convergence ratios on the order of 300:1 and 100:1, respectively, for the striatal projections to GPe and GPi. Nevertheless, basal ganglia neurons involved in skeletomotor and oculomotor control show highly refined sensorimotor fields that are comparable in their somatotopic and behavioral specificity to the sensorimotor fields of neurons at cortical levels (Alexander and Crutcher, 1990).

Role of Dopamine

Dopaminergic input to the putamen consists of nigrostriatal projections that originate in the SNc. The cortical motor and premotor areas receive separate dopaminergic projections from the ventral tegmental area. At the network level, dopamine appears to have differential effects on the direct and indirect pathways, tending to activate striatal MSNs that project directly to GPi while suppressing those that project to GPe. Given the reciprocal reentrant effects associated with differential activation of the direct versus indirect pathways, the differential effects of dopamine on these two pathways could be viewed as resulting in the net enhancement of positive feedback, and suppression of negative feedback, returned to the various cortical areas that receive basal ganglia influences.

Dopamine has also been shown to have a role in synaptic plasticity within the striatum, being implicated in both long-term potentiation (LTP) and long-term depression (LTD) (Calabresi et al., 1992). Much of the evidence gathered thus far suggests that dopamine plays a permissive role in striatal synaptic plasticity. The presence of dopamine at striatal synapses appears to be necessary but not sufficient for LTP and/or LTD to occur, the other required factor being the activation of corticostriatal inputs with (or, in some cases, without) associated depolarization of the postsynaptic MSN (see article by Wickens in Houk, Davis, and Beiser, 1995). The nigrostriatal pathway provides an extraordinarily dense dopaminergic input to each MSN, comparable in magnitude to the thousands of corticostriatal synapses that each cell also receives. Dopamine neurons do not show activity changes in relation to movement per se, but discharge instead in relation to conditions involving the probability and imminence of behavioral reinforcement (Schultz and Romo, 1990).

Taken together, these findings suggest the possibility that dopamine neurons may play an important role in determining when striatal synapses should be strengthened or weakened. In this respect, dopamine neurons might be seen as playing a role in striatal information processing, analogous to that of an "adaptive critic" in connectionist networks (see REINFORCEMENT LEARNING). Adaptive critic circuitry learns to predict the future reward potential of current sensory conditions. The output of an adaptive critic network, which represents the imminence-weighted sum of expected rewards (primary reinforcements), can be employed in conjunction with an executive network as a global, secondary reinforcement signal that determines when it has become appropriate (because of successful behaviors) to update the adaptive synapses in both networks. The secondary reinforcement signals generated by an adaptive critic network show some of the same reward-related contingencies manifested by dopamine neurons (see articles by Schultz and by Barto in Houk, Davis, and Beiser, 1995).

With their combined voltage dependency and ligand specificity, N-methyl-D-aspartate (NMDA) receptors (see NMDA RECEPTORS: SYNAPTIC, CELLULAR, AND NETWORK MODELS) are widely assumed to play a role in at least one form of activity-dependent synaptic plasticity, that is, LTP. Evidence for the existence of NMDA receptors has been found in a number of basal ganglia nuclei that are believed to receive significant glutamatergic input, including the striatum, the STN, and the SNc. NMDA receptors have also been demonstrated at cortical levels, where the phenomenon of LTP has been documented in many regions, including some of the sensorimotor fields (Iriki

et al., 1991). Adaptive synapses thus appear to be distributed across multiple stations along the basal ganglia–thalamocortical pathways.

Limits of Our Current Understanding of Basal Ganglia Function

Disinhibition plays an important role in many current models of basal ganglia operation (Chevalier and Deniau, 1990). In the case of the basal ganglia's oculomotor pathways, for example, there is compelling evidence that striatally induced, phasic inhibition of SNr neurons leads to the generation of saccadic eye movements through the release of command neurons, in the superior colliculus, from the tonic, GABAergic inhibition they receive via the nigrotectal pathway (Hikosaka and Wurtz, 1983). The elegant simplicity of this model has helped to motivate similar models of how the basal ganglia's skeletomotor circuitry may contribute to movement control, the chief difference being the emphasis in most skeletomotor models on disinhibition at the level of the thalamus rather than brainstem (Alexander and Crutcher, 1990).

These models of the basal ganglia's skeletomotor circuitry assume that voluntary movements are facilitated in the context of focused disinhibition of the basal ganglia recipient portions of the ventrolateral thalamus. This focused thalamic disinhibition may be generated either by phasic enhancement of transmission through the direct pathway, by phasic suppression of transmission through the indirect pathway, or by a combination of both processes. Conversely, according to this same scheme, decreased transmission through the direct pathway, or increased transmission through the indirect pathway, would have the effect of suppressing voluntary movements.

This relatively simplistic model raises questions of its own, however. It is not known, for example, whether a given corticostriatal neuron (or functional group of such neurons) engages both the direct and indirect pathways in a balanced manner. Nor is it evident whether the known convergence of inputs from the direct and indirect pathways onto individual GPi/SNr neurons results in a functional interaction between the two pathways that is antagonistic or complementary. A possibility at one extreme would be a functionally antagonistic, push/pull system that could be used to scale or brake the intended movement. At the other extreme would be a functionally complementary, center-surround system that might serve to facilitate the intended movement pattern while suppressing potentially conflicting ones.

Studies of primate models of basal-ganglia–induced movement disorders have tended to confirm many of the predictions of the thalamic disinhibition models (DeLong, 1990). These models predict that excessive basal ganglia outflow should be associated with hypokinetic states, as seen in the various akinetic/rigid disorders of movement control. Because of dopamine's reciprocal effects on the direct and indirect pathways, striatal dopamine deficiency should and does result in excessive discharge of neurons in the STN as well as in the GPi/SNr. In primates with parkinsonian akinesia and bradykinesia that has been induced experimentally by treatment with the toxic contaminant 1-methyl-4-phenyl-1,2,3,6-tetrahydropyridine (MPTP), the discharge rates of GPi neurons, as well as those neurons in the STN, have been found to be abnormally high, consistent with predictions of the thalamic disinhibition models. Voluntary movements can be restored in these akinetic animals by selective lesions of the STN, which results in the return of GPi neurons to more normal rates of discharge (Bergman, Wichmann, and DeLong, 1990). Also consistent with thalamic disinhibition models is the fact that akinesia in humans with idiopathic Parkinson's disease can be relieved by selective lesions of the GPi.

Such models are also able to account for hyperkinetic disorders (dyskinesias) on the basis of reduced basal ganglia outflow (DeLong, 1990). Such reduced outflow is thought to result in excessive disinhibition of the basal ganglia-receiving territories within the thalamus, with consequent release of involuntary movements. This explanation is consistent with various clinical and experimental data, including the fact that lesions of the STN result in dyskinesias, and that such lesions are associated with reduced levels of spontaneous discharge in the GPi.

There are, however, a number of observations that are difficult to reconcile with simple models of basal ganglia function based primarily on the concept of thalamic disinhibition. One of the most serious is that lesions of the basal ganglia output nuclei do not result in dyskinesias as the models predict they should (although dyskinesias have been reported in association with muscimol-induced suppression of GPi output).

A possible explanation for this discrepancy may be that the voltage-dependent, bistable properties of thalamic neurons (Steriade and Llinás, 1988) may cause them to respond to downward modulations of inhibitory basal ganglia input in a nonmonotonic manner. When a thalamic relay neuron has been hyperpolarized sufficiently long to deinactivate its low-threshold Ca^{2+} conductance, a sufficient depolarizing stimulus will elicit a low-threshold Ca^{2+} spike, superimposed on which will be a burst of conventional Na^+ action potentials (see THALAMOCORTICAL OSCILLATIONS IN SLEEP AND WAKEFULNESS for the basic concepts used here). Even a phasic reduction in the level of tonic inhibition may be sufficient to elicit a low-threshold spike. In the absence of sustained depolarization, hyperpolarizing conductances initiated by the low-threshold spike return the cell to a hyperpolarized state, where further transient depolarizations may initiate additional low-threshold spikes and their associated bursts of action potentials. With sustained depolarization near threshold, however, the low-threshold Ca^{2+} conductance is inactivated and the neuron enters a nonbursting relay mode in which its discharge pattern will tend to follow that of its predominant excitatory inputs.

Moderate reductions of GABAergic input from the basal ganglia, as would be expected with subthalamic lesions, may leave thalamic neurons with sufficient hyperpolarizing inputs that the low-threshold Ca^{2+} conductance is not inactivated, and at the same time, may reduce the level of tonic inhibition to the point where low-threshold spikes are readily generated even by small, phasic depolarizations. This could lead to frequent bursts of inappropriate thalamocortical activity and resultant dyskinesias. On the other hand, severe reductions in GABAergic input, as would be expected with GPi lesions, may actually restore some stability by increasing the level of tonic depolarization in thalamic neurons to the point where they shift into the nonbursting, relay mode.

Another difficulty for thalamic disinhibition theories is that lesions of the ventrolateral thalamus do not result in akinesia, and yet according to such theories, parkinsonian akinesia is generally attributed to increased basal ganglia outflow that results in excessive inhibition at the level of the ventrolateral thalamus (DeLong, 1990). One possible explanation for this apparent discrepancy is that we may have underestimated the functional significance of descending basal ganglia outflow to the PPN (and from there to the reticulospinal system), in which case lesions of the thalamus that block only the reentrant (cortically projected) influences of the basal ganglia may leave intact those descending influences that are conveyed more directly to the segmental motor apparatus.

The Need for Functional Models

In the absence of a comprehensive, testable theory of basal ganglia operations, neurobiological techniques alone are unlikely to provide a complete answer to the question of what functions are subserved by the basal ganglia. The behavioral effects of selective lesions of neural structures only hint at the underlying functions. Lesions reveal only the function of the remaining system, and in systems as complex as the brain, with its rich possibilities for redundancy and functional compensation, residual function may give us relatively little direct insight into the functions subserved by the lesioned structure.

The neurobiologist's other principal tools for examining functional issues comprise neurophysiological recordings (especially single neuron recordings in behaving animals) and, more recently, various functional imaging techniques (such as positron emission tomography and functional magnetic resonance imaging). Unfortunately, these approaches also have their interpretational limits. Both types of studies are correlative in nature and, as such, they generally cannot, by themselves, establish causal linkages.

On the other hand, if it were possible to develop functional models of basal ganglia circuitry—models that captured essential aspects of the known anatomy and physiology and were able to simulate some of the behavioral functions that have been imputed to these structures—the validity of those models could be tested interactively by comparing their dynamical properties with those of the basal ganglia themselves. Specifically, such comparisons between real and simulated networks could be made by using many of the same tools that the neurobiologist now uses to examine function.

Adaptive models of basal ganglia circuitry are only now beginning to be developed. A recently described model of the basal ganglia oculomotor circuit incorporates several important biological details and uses reinforcement learning to control sequential, target-directed eye movements (see article by Arbib and Dominey in Houk, Davis, and Beiser, 1995). The model comprises multiple, functionally differentiated cortical fields (including frontal eye fields and posterior parietal cortex) that send convergent projections to a layer of neurons representing the caudate nucleus. The caudate layer sends inhibitory projections to an SNr layer, which in turn sends its own inhibitory projections to separate layers representing the superior colliculus and the oculomotor portion of the thalamus. Corticostriatal projections include modifiable synapses, and learning depends on dopamine-mediated reinforcement signals that reflect reward contingencies.

Other adaptive models of basal ganglia circuitry have also been reported recently (see Houk, Davis, and Beiser, 1995), but for the most part, these have not yet attempted to incorporate significant neurobiological details into the modeling schemes. There are, to be sure, certain *cellular* models of the basal ganglia neurons that are well constrained by neurobiological data and are providing important new insights into the neurodynamics of basal ganglia circuitry. For example, Wilson has modeled the bistable dynamics of striatal MSNs and shown that the processing of corticostriatal inputs may differ sharply depending on whether each neuron is in the hyperpolarized ("down") or depolarized ("up") state (see article by Wilson in Houk, Davis, and Beiser, 1995). What is still generally lacking, however, are adaptive *network* models that capture essential aspects of basal ganglia neurodynamics and are also able to mimic some of the behavioral functions attributed to these structures. The advent of such models is likely to prove crucial for a full understanding of how the basal ganglia contribute to behavioral processing.

Acknowledgments. This work was supported by grants from the Office of Naval Research (N00014–92–J–1132) and from the National Institutes of Health (NS–17678).

Road Map: Mammalian Brain Regions
Related Reading: Thalamus

References

Alexander, G. E., and Crutcher, M. D., 1990, Functional architecture of basal ganglia circuits: Neural substrates of parallel processing, *Trends Neurosci.*, 13:266–271.

Alexander, G. E., Crutcher, M. D., and DeLong, M. R., 1990, Basal ganglia-thalamocortical circuits: Parallel substrates for motor, oculomotor, "prefrontal" and "limbic" functions, *Prog. Brain Res.*, 85:119–146.

Bergman, H., Wichmann, T., and DeLong, M. R., 1990, Reversal of experimental parkinsonism by lesions of the subthalamic nucleus, *Science*, 249:1436–1438.

Calabresi, P., Pisani, A., Mercuri, N. B., and Bernardi, G., 1992, Long-term potentiation in the striatum is unmasked by removing the voltage-dependent magnesium block of NMDA receptor channels, *Eur. J. Neurosci.*, 4:929–935.

Chevalier, G., and Deniau, J. M., 1990, Disinhibition as a basic process in the expression of striatal functions, *Trends Neurosci.*, 13:277–280.

DeLong, M. R., 1990, Primate models of movement disorders of basal ganglia origin, *Trends Neurosci.*, 13:281–285.

Dum, R. P., and Strick, P. L., 1991, The origin of corticospinal projections from premotor areas in the frontal lobe, *J. Neurosci.*, 11:667–689.

Hikosaka, O., and Wurtz, R. H., 1983, Visual and oculomotor functions of monkey substantia nigra pars reticulata, IV. Relation of substantia nigra to superior colliculus, *J. Neurophysiol.*, 49:1285–1301.

Houk, J., Davis, J., and Beiser, D. G., Eds., 1995, *Models of Information Processing in the Basal Ganglia*, Cambridge, MA: MIT Press.

Iriki, A., Pavlides, C., Keller, A., and Asanuma, H., 1991, Long-term potentiation of thalamic input to the motor cortex induced by co-activation of thalamocortical and corticocortical afferents, *J. Neurophysiol.*, 65:1435–1441.

Parent, A., 1990, Extrinsic connections of the basal ganglia, *Trends Neurosci.*, 13:254–258.

Schultz, W., and Romo, R., 1990, Dopamine neurons of the monkey midbrain: Contingencies of responses to stimuli eliciting immediate behavioral reactions, *J. Neurophysiol.*, 63:607–624.

Steriade, M., and Llinás, R. R., 1988, The functional states of the thalamus and the associated neuronal interplay, *Physiol. Rev.*, 68: 649–742.

Wilson, C. J., 1992, Dendritic morphology, inward rectification, and the functional properties of neostriatal neurons, in *Single Neuron Computation*, (T. McKenna, J. Davis, and S. F. Zornetzer, Eds.), Boston: Academic Press, pp. 141–171.

Bayesian Methods for Supervised Neural Networks

David J.C. MacKay

Introduction

Bayesian probability theory provides a unifying framework for data modeling. This article first explains the Bayesian interpretation of neural network learning, then describes three particular benefits of applying Bayesian methods to neural networks. First, the *overfitting problem* can be solved by using Bayesian methods to control model complexity. Bayesian model comparison can be used, for example, to optimize weight decay rates and the number of hidden units in a multilayer perceptron. Second, probabilistic modeling handles uncertainty in a natural manner. There is a unique prescription, *marginalization*, for incorporating uncertainty about parameters into predictions; this procedure yields better predictions. Third, we can define more sophisticated probabilistic models which are able to extract more information from data.

Thomas Bayes was an eighteenth-century English clergyman who suggested the use of probabilities to quantify inferences about hypotheses given data. The Bayesian use of probability to represent degrees of belief contrasts with the more restrictive use within orthodox statistics, where probabilities may refer only to the frequencies of random variables, not to hypotheses. Bayes's theorem asserts the relationship between conditional probabilities:

$$P(A|B, C) = \frac{P(B|A, C)P(A|C)}{P(B|C)} \tag{1}$$

Here $P(A|B, C)$ denotes a *conditional probability*: the probability of A, given B and C. Bayes's theorem can be used to take the predictions $P(D|\mathscr{H}_i)$ made by alternative models $\mathscr{H}_i$ about data D, and the *prior probabilities* of the models $P(\mathscr{H}_i)$, and obtain the *posterior probabilities* $P(\mathscr{H}_i|D)$ of the models given the data, $P(\mathscr{H}_i|D) = P(D|\mathscr{H}_i)P(\mathscr{H}_i)/P(D)$. For a good text on Bayesian methods, see Box and Tiao (1973).

Neural Network Learning

For the purposes of this article, a supervised neural network is a nonlinear parameterized mapping from an input $\mathbf{x}$ to an output $\mathbf{y} = \mathbf{y}(\mathbf{x}; \mathbf{w}, \mathscr{A})$. The output is a continuous function of the parameters $\mathbf{w}$. The architecture of the net, i.e., the functional form of the mapping, is denoted by $\mathscr{A}$. The details of the architecture do not matter; Bayesian methods can be applied for example to multilayer perceptrons and RADIAL BASIS FUNCTION NETWORKS (q.v.). Such networks can be "trained" to perform *regression* and *classification* tasks.

A *regression network* has *targets* that are real numbers. It is trained using a data set $D = \{\mathbf{x}^{(m)}, \mathbf{t}^{(m)}\}$ by adjusting $\mathbf{w}$ so as to minimize an *error function*: for example, ignoring $\mathscr{A}$,

$$E_D(\mathbf{w}) = \frac{1}{2} \sum_m \sum_i (t_i^{(m)} - y_i(\mathbf{x}^{(m)}; \mathbf{w}))^2 \tag{2}$$

This minimization is based on repeated evaluation of the gradient of E_D using BACKPROPAGATION (q.v.). The details of the learning algorithm do not concern us here; what matters is the objective function that is optimized. Often, *regularization* (also known as "weight decay") is included, modifying the objective function to

$$M(\mathbf{w}) = \beta E_D + \alpha E_W \tag{3}$$

where, for example, the regularizer could be $E_W = \frac{1}{2} \sum_i w_i^2$. This additional term favors small values of $\mathbf{w}$ and can, if α is appropriately set, decrease the tendency of a model to "overfit" noise in the training data.

For a *classification network* the targets are classifications, and a different objective function is appropriate. I assume that the model is a "softmax" network (see WINNER-TAKE-ALL MECHANISMS) having coupled outputs which sum to 1 and are interpreted as class probabilities $y_i = P(t_i = 1|\mathbf{x}, \mathbf{w}, \mathscr{A})$. We can represent the targets by a vector, $\mathbf{t}$, in which a single element is set to 1, indicating the correct class, and all other elements are set to 0. The error function βE_D is replaced by the cross entropy:

$$G = \sum_m \sum_i t_i^{(m)} \log y_i(\mathbf{x}^{(m)}; \mathbf{w}) \tag{4}$$

The function $-G$ is similar to the sum-squared error in that it is smallest if $\mathbf{y}$ is close to $\mathbf{t}$ for all examples. The total objective function is then $M = -G + \alpha E_W$.

The Overfitting Problem in Neural Networks

A problem with neural networks is that an over-flexible network can be duped by stray correlations in the data into "discovering" nonexistent structure. This problem is illustrated in Figure 1a–d. Consider a control parameter which influences the complexity of a model, for example, a regularization constant like the weight decay parameter α mentioned earlier. As the control parameter is varied to increase the complexity of the model (descending from Figure 1a to Figure 1c and going from left to right across Figure 1d), the best fit to the training data that the model can achieve becomes increasingly good. However, the empirical performance of the model, the *test error*, has a minimum as a function of the control parameters. *An over-complex model overfits the data and generalizes poorly.* This problem may also complicate the choice of the number of hidden units in a multilayer perceptron, as well as the radius of the basis functions in a radial basis function network.

The first message of this article is illustrated in Figure 1e. If we give a probabilistic interpretation to the model, then we can evaluate the posterior probability distribution of the control parameters given the training data; we find that the data's preferences automatically embody Occam's razor, the principle that states a preference for simple models. The mechanism of this razor is explained in MacKay (1992a, 1994). Over-complex models are, in fact, less probable, and the "evidence" P (Training Data|Control Parameters) can be used as an objective function for optimization of model control parameters. No test set is needed. This can yield savings in computational resources, and it allows better use to be made of valuable data.

Probabilistic Interpretation of Neural Network Learning

The neural network learning process can be given the following probabilistic interpretation. The error function is interpreted as defining a noise model:

$$P(D|\mathbf{w}, \beta, \mathscr{H}) = \frac{1}{Z_D(\beta)} \exp(-\beta E_D) \tag{5}$$

where $Z_D(\beta) = \int d^N D \exp(-\beta E_D)$ is the normalizing constant of this probability over the N-dimensional data space. Thus, the use of the sum-squared error E_D in Equation 2 corresponds to an assumption of Gaussian noise on the target variables, and the parameter β defines a noise level $\sigma_v^2 = 1/\beta$.

Similarly, the regularizer is interpreted in terms of a log prior probability distribution over the parameters:

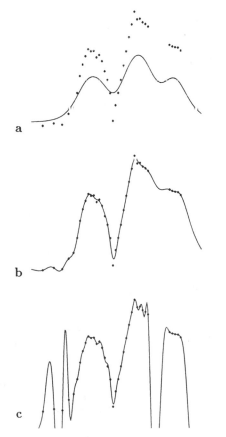

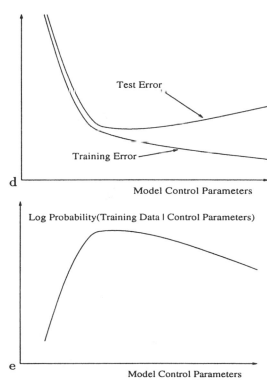

Figure 1. Optimization of model complexity. Parts *a* through *c* show a radial basis function model interpolating a data set with one input variable and one output variable. As the regularization constant is varied to increase the complexity of the model (from *a* to *c*), the interpolant is able to fit the training data increasingly well, but beyond a certain point the generalization ability (test error) of the model deteriorates, as shown in part *d*. Probability theory (part *e*) allows us to optimize the control parameters without a test set.

$$P(\mathbf{w}|\alpha, \mathcal{H}) = \frac{1}{Z_W(\alpha)} \exp(-\alpha E_W) \qquad (6)$$

where $Z_W(\alpha) = \int d^k \mathbf{w} \exp(-\alpha E_W)$. Thus the use of "weight decay" is interpreted as a prior belief that mappings involving small $|\mathbf{w}|$ are more probable than those that involve large parameter values. The probabilistic model $\mathcal{H}$ specifies the functional form $\mathcal{A}$ of the network, the noise model in Equation 5, and the prior of Equation 6.

The minimization of $M(\mathbf{w})$ then corresponds to the *inference* of the parameters $\mathbf{w}$, given the data. Using Bayes's theorem,

$$P(\mathbf{w}|D, \alpha, \beta, \mathcal{H}) = \frac{P(D|\mathbf{w}, \beta, \mathcal{H}) P(\mathbf{w}|\alpha, \mathcal{H})}{P(D|\alpha, \beta, \mathcal{H})} \qquad (7)$$

$$\propto \exp(-M(\mathbf{w})) \qquad (8)$$

Since minimizing $M(\mathbf{w})$ is equivalent to maximizing $\exp(-M(\mathbf{w}))$, the $\mathbf{w}$ found by (locally) minimizing $M(\mathbf{w})$ is now interpreted as the (locally) most probable parameter vector, $\mathbf{w}^{\mathrm{MP}}$. The quantity $P(D|\alpha, \beta, \mathcal{H})$ in the denominator here is simply the normalizing constant for this inference, but it assumes an important role later if we wish to infer the hyperparameters α and β.

Why is it natural to interpret the error functions as *log* probabilities? Error functions are usually additive. For example, E_D is a sum of squared errors. Probabilities, on the other hand, are usually multiplicative: for example, the joint probability of independent events A and B is $P(A, B) = P(A)P(B)$. The logarithmic mapping maintains this correspondence.

Note that this interpretation refers only to the objective function minimized by the learning algorithm, and not to the details of the learning algorithm itself. Bayesian methods can be applied to supervised neural networks whatever learning algorithm is used to train them.

The interpretation of $M(\mathbf{w})$ as a log probability adds little new at this stage. But new tools will emerge when we proceed to other inferences. Bayesian inference for neural networks may be implemented by Monte Carlo sampling (Neal, 1993) or by deterministic methods employing Gaussian approximations (MacKay, 1992b). The latter are reviewed here.

Bayesian Solution to the Overfitting Problem

Setting Regularization Constants α and β

The control parameters α and β determine the complexity of a model. Setting α is a nontrivial problem because, as illustrated in Figure 1, the best fit to the training data (the minimum training error) is achieved by the most complex and flexible model of all. However, probability theory can tell us how to infer α and β given the data. We apply Bayes's theorem.

$$P(\alpha, \beta|D, \mathcal{H}) = \frac{P(D|\alpha, \beta, \mathcal{H}) P(\alpha, \beta|\mathcal{H})}{P(D|\mathcal{H})} \qquad (9)$$

The data-dependent factor $P(D|\alpha, \beta, \mathcal{H})$ is the normalizing constant from our previous inference in Equation 7; we call this factor the "evidence" for α and β.

Assuming we have only weak prior knowledge about the noise level and the smoothness of the interpolant, the evidence framework optimizes the constants α and β by finding the maximum of the evidence for α and β. If we can approximate the posterior probability distribution in Equation 8 by a single Gaussian (Taylor-expanding $\log P(\mathbf{w}|D, \alpha, \beta, \mathcal{H})$),

$$P(\mathbf{w}|D, \alpha, \beta, \mathcal{H})$$

$$\simeq \frac{1}{Z_M'} \exp\left(-M(\mathbf{w}^{\mathrm{MP}}) - \frac{1}{2}(\mathbf{w} - \mathbf{w}^{\mathrm{MP}})^{\mathrm{T}} \mathbf{A}(\mathbf{w} - \mathbf{w}^{\mathrm{MP}})\right) \qquad (10)$$

where $\mathbf{A} = -\nabla\nabla \log P(\mathbf{w}|D, \mathcal{H})$, then the evidence for α and β can be written down:

$$\log P(D|\alpha, \beta, \mathcal{H}) \simeq \log \frac{Z_M'}{Z_W(\alpha) Z_D(\beta)} \qquad (11)$$

$$\simeq -M(\mathbf{w}^{\mathrm{MP}}) + \frac{1}{2} \log \det[2\pi \mathbf{A}^{-1}]$$

$$- \log Z_W(\alpha) - \log Z_D(\beta) \qquad (12)$$

The maximum of the evidence has elegant properties, found by differentiating the log evidence with respect to α:

$$1/\alpha_{\mathrm{MP}} = \sum_i w_i^{\mathrm{MP}2}/\gamma \qquad (13)$$

where $\mathbf{w}^{\mathrm{MP}}$ is the parameter vector which minimizes the objective function $M = \beta E_D + \alpha E_W$, and γ is the "number of well-determined parameters," given by

$$\gamma = k - \alpha \mathrm{Trace}(\mathbf{A}^{-1}) \qquad (14)$$

Here k is the total number of parameters, and the matrix $\mathbf{A}^{-1}$ is the variance-covariance matrix that defines error bars on the parameters $\mathbf{w}$. Thus $\gamma \to k$ when the error bars on the parameters are small relative to their prior range, defined by $\sigma_w^2 = 1/\alpha$.

Similarly, in a regression problem with a Gaussian noise model, the maximum evidence value of β satisfies

$$1/\beta_{\mathrm{MP}} = 2E_D/(N - \gamma) \qquad (15)$$

Equations 13 and 15 can be used as re-estimation formulas for α and β; that is, one can alternately optimize $\mathbf{w}$ given α and β, and then update the hyperparameters using these equations. The computational overhead for these Bayesian calculations is not severe: it is only necessary to evaluate properties of the error bar matrix, $\mathbf{A}^{-1}$.

Relationship to ideal hierarchical Bayesian modeling.   Bayesian probability theory has been used above to *optimize* the hyperparameters α and β by maximizing the evidence. Ideally we would *integrate over* these nuisance parameters in order to obtain the posterior distribution of the parameters $P(\mathbf{w}|D, \mathcal{H})$ and the predictive distributions $P(\mathbf{t}^{(N+1)}|D, \mathcal{H})$. The larger the number of well-determined parameters, the more accurate the given approximation becomes.

Multiple regularization constants.   For simplicity, it has been assumed that there is only a single class of weights, which are modeled as coming from a single Gaussian prior with $\sigma_w^2 = 1/\alpha$. However, the parameters usually fall into three or more distinct dimensional classes. It is appropriate to give these classes separate weight decay constants α_c (MacKay, 1992b). The methods described here generalize to such models in a straightforward manner.

Model Comparison

Just as the evidence for α and β can be used to optimize these control parameters, we can also evaluate the evidence $P(D|\mathcal{H})$ to compare models $\mathcal{H}$ that have different network architectures or different regularizers, or to compare alternative local minima of a single model. As demonstrated in MacKay (1992b), the evidence can be strongly correlated with the generalization error of a model, so that the overfitting problem in model selection is solved.

In a practical application, Thodberg (1993) has applied these methods to the prediction of pork fat content from spectroscopic data, and has obtained better performance than standard techniques involving cross-validation. This improvement is attributed to the fact that Bayesian methods need no

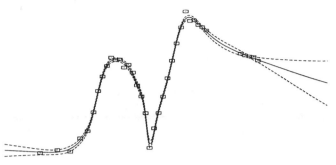

Figure 2. Error bars on the predictions of a trained regression network.

$$\mathbf{g} = \frac{\partial y}{\partial \mathbf{w}} \qquad (17)$$

This vector measures the sensitivity of the output to the parameters and can be found by backpropagation. The error bars $\mathbf{A}^{-1}$ associated with $\mathbf{w}$ produce a variance in y equal to

$$\sigma_y^2 = \mathbf{g}^T \mathbf{A}^{-1} \mathbf{g} \qquad (18)$$

Figure 2 illustrates these error bars on the predictions of a multilayer perceptron with ten hidden units trained on 37 data points. Notice that the error bars become larger where the data are sparse.

validation set: all the available data can be used for parameter fitting, for optimization of model complexity, and for model comparison.

Error Bars and Predictions

We will now see how the Bayesian approach allows us to quantify the uncertainty in neural network predictions.

It is common practice simply to use the most probable values of $\mathcal{H}$, $\mathbf{w}$, etc., when making predictions, but this is not optimal. Bayesian prediction of a new datum $\mathbf{t}^{(N+1)}$ at a given input vector $\mathbf{x}^{(N+1)}$ involves *marginalizing* or *integrating over* our uncertainty about the variables $\mathbf{w}$, α, β, and $\mathcal{H}$:

$$P(\mathbf{t}^{(N+1)}|D) = \sum_{\mathcal{H}} \int d\alpha \, d\beta \int d^k \mathbf{w} P(\mathbf{t}^{(N+1)}|\mathbf{x}^{(N+1)}, \mathbf{w}, \alpha, \beta, \mathcal{H})$$
$$\times P(\mathbf{w}, \alpha, \beta, \mathcal{H}|D) \qquad (16)$$

Typically, marginalization over $\mathbf{w}$ and $\mathcal{H}$ affects the predictive distribution significantly, and integration over α and β has a lesser effect.

Error Bars in Regression

It is simple to turn Gaussian error bars on $\mathbf{w}$ into error bars on the output, if we assume that the output is a locally linear function of the parameters, $y(\mathbf{x}^{N+1}, \mathbf{w}) \simeq y(\mathbf{x}^{N+1}; \mathbf{w}^{MP}) + \mathbf{g} \cdot (\mathbf{w} - \mathbf{w}^{MP})$, where

Integrating over Models: Committees

If we have multiple regression models $\mathcal{H}$, then the predictive distribution is obtained by summing together the predictive distribution of each model, weighted by its posterior probability.

It should be noted that the Bayesian methods described here prescribe only how one should perform *inference* and make predictions; they do not specify how to make *decisions*. Choosing between possible actions requires in addition that a *cost function* be specified. It is then in principle a trivial exercise in decision theory to choose the action that minimizes the expected cost. If a single estimator is required and the loss function is quadratic, the optimal answer is a weighted mean of the models' predictions, $y(\mathbf{x}^{N+1}; \mathbf{w}^{MP}, \mathcal{H})$. The weighting coefficients are the posterior probabilities, which are obtained from the evidences $P(D|\mathcal{H})$. If we cannot evaluate these accurately, then alternative pragmatic prescriptions for the weighting coefficients can be used (Thodberg, 1993; MacKay, in press). The mean of the predictions of several models can give performance superior to that of any of the individual models (see AVERAGING/MODULAR TECHNIQUES FOR NEURAL NETWORKS).

Error Bars in Classification

In a classification problem, the best-fit parameters give overconfident predictions. This is illustrated for a simple two-class problem in Figure 3. Figure 3a shows a binary data set, which in Figure 3b is modeled with a "network" consisting of a single neuron. The best-fit parameter values give predictions shown by three contours. Are these reasonable predictions? Consider

Figure 3. Integrating over error bars in a classifier. Part *a* shows a binary data set. The two classes are denoted by the point styles $\times = 1$, $o = 0$. In part *b*, the best-fit model is shown by its 0.1, 0.5, and 0.9 predictive contours. This model assigns probability 0.9 of being in class 1 to both inputs A and B. Part *c* graphs the posterior probability distribution of the model parameters, $P(\mathbf{w}|D, \mathcal{H})$. Two typical samples from the posterior are indicated by the points labeled 1 and 2. Parts *d* and *e* show the corresponding classification contours. Point B is classified differently by these different plausible classifiers, whereas the classification of A is relatively stable. In part *f*, Bayesian predictions are obtained by integrating over the posterior distribution of $\mathbf{w}$. The probability that point B is in class 1 is closer to 0.5, in accordance with our intuition.

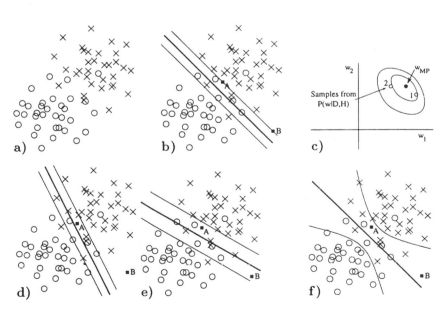

new data arriving at points *A* and *B*. The best-fit model assigns both of these examples the probability 0.9 of being in class 1. But intuitively we might be inclined to assign a less confident probability (closer to 0.5) at *B* than at *A*, since point *B* is far from the training data.

Precisely this result is obtained by marginalizing over the parameters, whose posterior probability distribution is depicted in Figure 3*c*. Two random samples from the posterior define two different classification surfaces, which are illustrated in Figure 3*d* and 3*e*. The point *B* is classified differently by these equally plausible classifiers, whereas the classification of *A* is relatively stable. We obtain the Bayesian predictions (Figure 3*f*) by averaging together the predictions of the plausible classifiers. The width of the resulting decision region increases as we move away from the data, in accordance with intuition. The Bayesian approach therefore better identifies the points whose classification is uncertain.

Automatic Relevance Determination

Because Bayesian data modeling forces one to make explicit all one's modeling assumptions, it also makes it easy to design useful modifications to one's models. The automatic relevance determination (ARD) model is designed to capture the concept that, in a regression problem, some of the input variables may in fact be irrelevant to the prediction of the output variable. Now, because a finite data set shows random correlations between the irrelevant inputs and the output, any conventional neural network (even with weight decay) fails to set the weights for these junk inputs to zero. Thus, the irrelevant variables hurt the model's performance, particularly when the variables are many and the data are few.

The uncertain relevance of the inputs can be represented by introducing multiple weight decay constants, one α associated with each input. When these control parameters are optimized using the methods described earlier, the decay rates for junk inputs are inferred to be large, so these inputs are automatically switched off.

The ARD model has been used to create the winning entry in a recent prediction competition, involving modeling of the energy consumption of a building (MacKay, in press). For further discussion and references, see the review by MacKay (1994).

Discussion: Relationship to Other Theories of Generalization

Conventional wisdom in learning theory and statistics states that the complexity of the model should be matched to the amount of data. Thus, it is believed that, when there are few data, it is better to use a model with few parameters, even if that model is not believed to be capable of representing the true function. For a Bayesian, in contrast, the choice of which models to consider is a matter of prior belief, and should not depend on the amount of data collected. It is now common practice for Bayesians to fit models that have more parameters than the number of data points.

I now discuss the relationship between the evidence and quantities arising in learning theory.

Relationship to Generalized Prediction Error

Moody's generalized prediction error or GPE (Moody, 1992) is an estimator of generalization error which can be derived for regression models under minimal assumptions about the distribution of the residuals and the true interpolant. The predicted error per data point is

$$\mathrm{GPE} = (E_D + \sigma_v^2 \gamma)/N \qquad (19)$$

where γ was defined in Equation 14. Like the log evidence, the GPE has the form of the data error plus a term that penalizes complexity. However, the penalty term in the evidence does not have the same scaling behaviour as γ, and, empirically, the GPE is not always a good predictor of generalization. One reason is that, in the derivation of GPE, it is effectively assumed that test samples will be drawn only at the **x** locations at which we have already received data. The consequences of this false assumption are most serious for over-parameterized and over-flexible models.

Relationship to Structural Risk Theory

Recent work on "structural risk minimization" uses empirical expressions of the form

$$E_{\mathrm{gen}} \simeq E_D/N + c_1 \frac{\log(N/\gamma) + c_2}{N/\gamma} \qquad (20)$$

where γ, the "effective VC-dimension" of the model (see Vapnik-Chervonenkis Dimension of Neural Networks), was defined in Equation 14. The constants c_1 and c_2 are determined by experiment. Interestingly, the scaling behavior of this expression (Equation 20) is similar to the scaling behavior of the log evidence (Equation 12).

The structural risk theory provides no equivalent to Bayesian marginalization over uncertain parameters.

Relationship to Minimum Description Length (MDL)

A complementary view of Bayesian model comparison is obtained by replacing probabilities of events by the lengths in bits of messages which communicate the events. Message lengths $L(\mathbf{x})$ correspond to a probability distribution $P(\mathbf{x})$ via the relations

$$P(\mathbf{x}) = 2^{-L(\mathbf{x})} \qquad L(\mathbf{x}) = -\log_2 P(\mathbf{x}) \qquad (21)$$

The MDL principle (see Minimum Description Length Analysis) states that one should prefer models which can communicate the data in the smallest number of bits. This conveys an intuitive picture of model comparison (Figure 4). One can imagine that each model $\mathcal{H}_i$ communicates the data D by sending the name of the model, sending the best-fit parameters of the model $\mathbf{w}_{(i)}^{\mathrm{MP}}$, then sending the data relative to those parame-

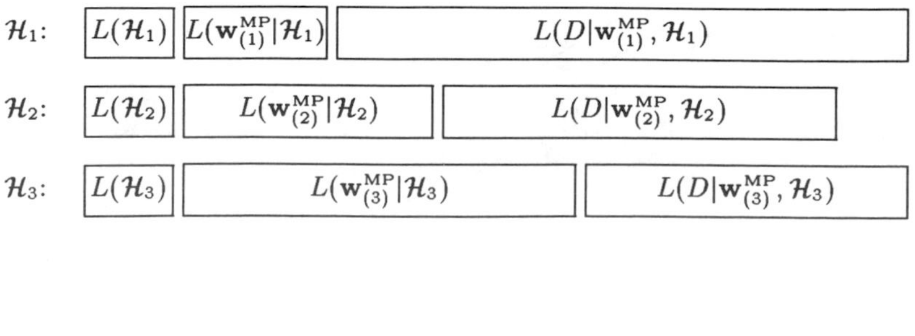

$\mathcal{H}_1$: $\boxed{L(\mathcal{H}_1)}$ $\boxed{L(\mathbf{w}_{(1)}^{\mathrm{MP}}|\mathcal{H}_1)}$ $\boxed{\qquad\qquad L(D|\mathbf{w}_{(1)}^{\mathrm{MP}}, \mathcal{H}_1) \qquad\qquad}$

$\mathcal{H}_2$: $\boxed{L(\mathcal{H}_2)}$ $\boxed{\qquad L(\mathbf{w}_{(2)}^{\mathrm{MP}}|\mathcal{H}_2) \qquad}$ $\boxed{\qquad L(D|\mathbf{w}_{(2)}^{\mathrm{MP}}, \mathcal{H}_2) \qquad}$

$\mathcal{H}_3$: $\boxed{L(\mathcal{H}_3)}$ $\boxed{\qquad L(\mathbf{w}_{(3)}^{\mathrm{MP}}|\mathcal{H}_3) \qquad}$ $\boxed{\qquad L(D|\mathbf{w}_{(3)}^{\mathrm{MP}}, \mathcal{H}_3) \qquad}$

Figure 4. A popular view of model comparison by minimum description length.

ters. As we proceed to more complex models, the length of the parameter message increases. On the other hand, the length of the data message decreases, because a complex model is able to fit the data better, making the residuals smaller. In this example, the intermediate model $\mathcal{H}_2$ achieves the optimum trade-off between these two trends.

Road Map: Learning in Artificial Neural Networks, Statistical
Related Reading: Generalization and Regularization in Nonlinear Learning Systems

References

Box, G. E. P., and Tiao, G. C., 1973, *Bayesian Inference in Statistical Analysis*, Reading, MA: Addison–Wesley.

MacKay, D. J. C., 1992a, Bayesian interpolation, *Neural Computat.*, 4(3):415–447.

MacKay, D. J. C., 1992b, A practical Bayesian framework for backpropagation networks, *Neural Computat.*, 4(3):448–472.

MacKay, D. J. C., 1994, Bayesian methods for backpropagation networks, in *Models of Neural Networks III* (E. Domany, J. L. van Hemmen, and K. Schulten, Eds.), New York: Springer-Verlag, chap. 6.

MacKay, D. J. C., in press, Bayesian non-linear modelling for the 1993 energy prediction competition, in *Maximum Entropy and Bayesian Methods: Santa Barbara 1993* (G. Heidbreder, Ed.), Dordrecht: Kluwer.

Moody, J. E., 1992, The *effective* number of parameters: An analysis of generalization and regularization in nonlinear learning systems, in *Advances in Neural Information Processing Systems 4* (J. E. Moody, S. J. Hanson, and R. P. Lippmann, Eds.), San Mateo, CA: Morgan Kaufmann, pp. 847–854.

Neal, R. M., 1993, Bayesian learning via stochastic dynamics, in *Advances in Neural Information Processing Systems 5* (C. L. Giles, S. J. Hanson, and J. D. Cowan, Eds.), San Mateo, CA: Morgan Kaufmann, pp. 475–482.

Thodberg, H. H., 1993, *Ace of Bayes: Application of Neural Networks with Pruning*, Technical Report 1132E, Roskilde, Den.: Danish Meat Research Institute (submitted to *IEEE Trans. Neural Netw.*).

Bayesian Networks

Judea Pearl

Introduction

This article surveys the historical development of Bayesian networks, summarizes their semantical basis, and assesses their properties and applications vis-à-vis those of neural networks.

Bayesian networks are directed acyclic graphs (DAGs) in which the nodes represent variables of interest (e.g., the temperature of a device, the gender of a patient, a feature of an object, the occurrence of an event) and the links represent causal influences among the variables. The strength of an influence is represented by conditional probabilities that are attached to each cluster of parents-child nodes in the network.

Figure 1 illustrates a simple yet typical Bayesian network. It describes the causal relationships among the season of the year (X_1), whether rain falls (X_2) during the season, whether the sprinkler is on (X_3) during that season, whether the pavement would get wet (X_4), and whether the pavement would be slippery (X_5). All variables in this figure are binary, taking a value of either true or false, except the root variable X_1, which can take one of four values: Spring, Summer, Fall, or Winter. Here,

the absence of a direct link between X_1 and X_5, for example, captures our understanding that the influence of seasonal variations on the slipperiness of the pavement is mediated by other conditions (e.g., the wetness of the pavement).

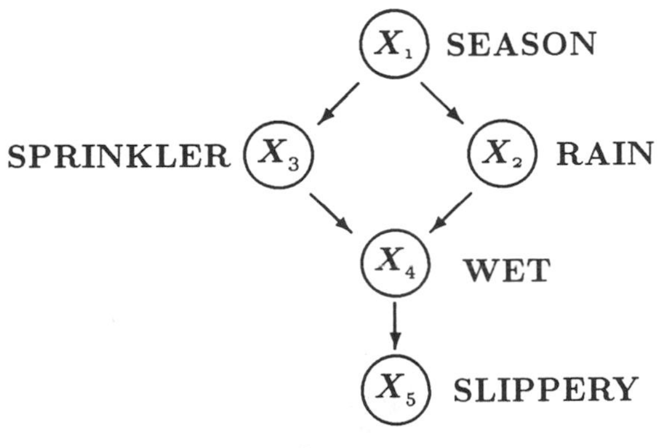

As this example illustrates, a Bayesian network constitutes a model of the environment rather than, as in many other knowledge representation schemes (e.g., rule-based systems and neural networks), a model of the reasoning process. It simulates, in fact, the causal mechanisms that operate in the environment, and thus allows the investigator to answer a variety of queries, including the following types: associational queries, such as "Having observed A, what can we expect of B?"; abductive queries, such as "What is the most plausible explanation for a given set of observations?"; and control queries, such as "What will happen if we intervene and act on the environment?" Answers to the first type of query depend only on probabilistic knowledge of the domain, while answers to the second and third types rely on the causal knowledge embedded in the network. Both types of knowledge, associative and causal, can effectively be represented and processed in Bayesian networks.

The associative facility of Bayesian networks may be used to model cognitive tasks such as object recognition, reading comprehension, and temporal projections. For such tasks, the probabilistic basis of Bayesian networks offers a coherent semantics for coordinating top-down and bottom-up inferences, thus bridging information from high-level concepts and low-level percepts. This capability is important for achieving selective attention, that is, selecting the most informative next observation before actually making the observation. In certain structures, the coordination of these two modes of inference can be accomplished by parallel and distributed processes that communicate through the links in the network.

However, the most distinctive feature of Bayesian networks, stemming largely from their causal organization, is their ability to represent and respond to changing configurations. Any local reconfiguration of the mechanisms in the environment can be translated, with only minor modification, into an isomorphic reconfiguration of the network topology. For example, to represent a disabled sprinkler, we simply delete from the network

Figure 1. A Bayesian network representing causal influences among five variables.

all links incident to the node "Sprinkler." To represent a pavement covered by a tent, we simply delete the link between "Rain" and "Wet." This flexibility is often cited as the ingredient that marks the division between deliberative and reactive agents, and that enables the former to manage novel situations instantaneously, without requiring retaining or adaptation. Thus, Bayesian networks can model a wide spectrum of cognitive activities, ranging from low-level perception (reaction) to planning and explaining (deliberation).

Historical Background

Networks employing directed acyclic graphs (DAGs) have a long and rich tradition, which began with the geneticist Sewall Wright (1921). He developed a method called *path analysis*, which later became an established representation of causal models in economics, sociology, and psychology. *Recursive models* is the name given to such networks by statisticians seeking meaningful and effective decompositions of contingency tables. *Influence diagrams* represent another application of DAGs developed for decision analysis. The primary role of a DAG in these applications is to provide an efficient description of the probability functions; once the network is configured, all subsequent computations are pursued by symbolic manipulation of probability expressions.

The potential for the network to work as a computational architecture, and hence as a model of cognitive activities, was noted in Pearl (1982), where a distributed scheme was demonstrated for probabilistic updating on tree-structured networks. The motivation behind this particular development was the modeling of distributed processing in reading comprehension, where both top-down and bottom-up inferences are combined to form a coherent interpretation. This dual mode of reasoning is at the heart of Bayesian updating, and in fact motivated Reverend Bayes's original 1763 calculations of posterior probabilities (representing explanations), given prior probabilities (representing causes), and likelihood functions (representing evidence).

Bayesian networks have not attracted much attention in cognitive modeling circles, but they did in expert systems. The ability to coordinate bidirectional inferences filled a void in expert systems technology of the late 1970s, and it is in this area that Bayesian networks truly flourished. Over the past ten years, Bayesian networks have become a tool of great versatility and power, and they are now the most common representation scheme for probabilistic knowledge (Shafer, 1990; Shachter, 1990). They have been used to aid in the diagnosis of medical patients and malfunctioning systems, to understand stories, to filter documents, to interpret pictures, to perform filtering, smoothing, and prediction, to facilitate planning in uncertain environments, and to study causation, nonmonotonicity, action, change, and attention. Some of these applications are described in a tutorial article by Charniak (1991); others can be found in Pearl (1988) and Shafer (1990).

Formal Semantics

Bayesian Networks as Carriers of Probabilistic Information

Given a DAG Γ and a joint distribution P over a set $X = \{X_1, \ldots, X_n\}$ of discrete variables, we say that Γ *represents* P if there is a one-to-one correspondence between the variables in X and the nodes of Γ, such that P admits the recursive product decomposition

$$P = (x_1, \ldots, x_n) = \prod_i P(x_i | \mathbf{pa}_i) \qquad (1)$$

where $\mathbf{pa}_i$ are the direct predecessors (called *parents*) of X_i in Γ. For example, the DAG in Figure 1 induces the decomposition

$$P(x_1, x_2, x_3, x_4, x_5)$$
$$= P(x_1)P(x_2|x_1)P(x_3|x_1)P(x_4|x_2,x_3)P(x_5|x_4) \qquad (2)$$

The recursive decomposition in Equation 1 implies that, given its parent set $\mathbf{pa}_i$, each variable X_i is conditionally independent of all its other predecessors $\{X_1, X_2, \ldots, X_{i-1}\}\backslash\mathbf{pa}_i$. Using Dawid's notation (Dawid, 1979), we can state this set of independencies as

$$X_i \perp\!\!\!\perp \{X_1, X_2, \ldots, X_{i-1}\}\backslash\mathbf{pa}_i | \mathbf{pa}_i \qquad i = 2, \ldots, n \qquad (3)$$

Such a set of independencies is called *Markovian*, since it reflects the Markovian condition for state transitions: each state is rendered independent of the past, given its immediately preceding state. For example, the DAG of Figure 1 implies the following Markovian independencies:

$$X_2 \perp\!\!\!\perp \{0\}|X_1 \qquad\qquad X_3 \perp\!\!\!\perp X_2|X_1$$
$$X_4 \perp\!\!\!\perp X_1|\{X_2, X_3\} \qquad X_5 \perp\!\!\!\perp \{X_1, X_2, X_3\}|X_4 \qquad (4)$$

In addition to these, the decomposition of Equation 1 implies many more independencies, the sum total of which can be identified from the DAG using the graphical criterion of *d-separation* (Pearl, 1988):

Definition. If X, Y, and Z are three disjoint subsets of nodes in a DAG Γ, then Z is said to *d-separate* X from Y, denoted $d(X, Z, Y)_\Gamma$, if and only if there is no path from a node in X to a node in Y along which the following two conditions hold:

(1) every node with converging arrows either is or has a descendant in Z, and
(2) every other node is outside Z.

A path satisfying these two conditions is said to be *active*; otherwise, it is said to be *blocked* (by Z). By *path* we mean a sequence of consecutive edges (of any directionality) in the DAG.

In Figure 1, for example, $X = \{X_2\}$ and $Y = \{X_3\}$ are d-separated by $Z = \{X_1\}$; the path $X_2 \leftarrow X_1 \rightarrow X_3$ is blocked by $X_1 \in Z$, while the path $X_2 \rightarrow X_4 \leftarrow X_3$ is blocked because X_4 and all its descendants are outside Z. Thus $d(X_2, X_1, X_3)$ holds in Γ. However, X and Y are not d-separated by $Z' = \{X_1, X_5\}$, because the path $X_2 \rightarrow X_4 \leftarrow X_3$ is rendered active by virtue of X_5, a descendant of X_4, being in Z'. Consequently, $d(X_2, \{X_1, X_5\}, X_3)$ does not hold in Γ; in words, learning the value of the consequence X_5 renders its causes X_2 and X_3 dependent, as if a pathway were opened along the arrows converging at X_4.

The d-separation criterion has been shown to be both necessary and sufficient relative to the set of distributions that are represented by a DAG Γ (see Geiger et al., in Shachter, 1990). In other words, there is a one-to-one correspondence between the set of independencies implied by the recursive decomposition of Equation 1 and the set of triples (X, Z, Y) that satisfy the d-separation criterion in Γ. Furthermore, the d-separation criterion can be tested in time linear in the number of edges in Γ. Thus, a DAG can be viewed as an efficient scheme for representing Markovian independence assumptions and for deducing and displaying all the logical consequences of such assumptions. Additional properties of DAGs and their applications to evidential reasoning in expert systems are discussed in Pearl (1988), Shachter (1990), and Spiegelhalter et al. (1993).

Bayesian Networks as Carriers of Causal Information

The interpretation of DAGs as carriers of independence assumptions does not necessarily imply causation and will in fact be valid for any set of Markovian independencies along any ordering (not necessarily causal or chronological) of the variables. However, the patterns of independencies portrayed in a DAG are typical of causal organizations, and some of these patterns can be given meaningful interpretation only in terms of causation. Consider, for example, two independent events, E_1 and E_2, that have a common effect E_3. This triple represents an intransitive pattern of dependencies: E_1 and E_3 are dependent, E_3 and E_2 are dependent, yet E_1 and E_2 are independent. Such a pattern cannot be represented in undirected graphs because connectivity in undirected graphs is transitive. Likewise, it is not easily represented in neural networks, because E_1 and E_2 should turn dependent once E_3 is known. The DAG representation provides a perfect language for intransitive dependencies via the converging pattern $E_1 \rightarrow E_3 \leftarrow E_2$, which implies the independence of E_1 and E_2 as well as the dependence of E_1 and E_3 and of E_2 and E_3. The distinction between transitive and intransitive dependencies is the basis for the causal discovery systems of Pearl and Verma (1991) and Spirtes, Glymour, and Schienes (1993) (see the later section on recent developments).

However, the Markovian account still leaves open the question of how such intricate patterns of independencies relate to the more basic notions associated with causation, such as influence, manipulation, and control, which reside outside the province of probability theory. The connection is made in the mechanism-based account of causation.

The basic idea behind this account goes back to H. Simon, and it was adapted by Pearl and Verma (1991) for defining probabilistic causal theories, as follows. Each child-parents family in a DAG Γ represents a deterministic function

$$X_i = f_i(\mathbf{pa}_i, \varepsilon_i) \qquad (5)$$

where $\mathbf{pa}_i$ are the parents of variable X_i in Γ, and ε_i, $0 < i < n$, are mutually independent, arbitrarily distributed random disturbances. Characterizing each child-parent relationship as a deterministic function, instead of the usual conditional probability $P(x_i|\mathbf{pa}_i)$, imposes equivalent independence constraints on the resulting distributions and leads to the same recursive decomposition that characterizes DAG models (see Equation 1). However, the functional characterization $X_i = f_i(\mathbf{pa}_i, \varepsilon_i)$ also specifies how the resulting distributions would change in response to external interventions, since each function is presumed to represent a stable mechanism in the domain and therefore remains constant unless specifically altered. Thus, once we know the identity of the mechanisms altered by the intervention and the nature of the alteration, the overall effect of an intervention can be predicted by modifying the appropriate equations in the model of Equation 5 and using the modified model to compute a new probability function of the observables.

The simplest type of external intervention is one in which a single variable, say X_i, is forced to take on some fixed value x_i'. Such *atomic* intervention amounts to replacing the old functional mechanism $X_i = f_i(\mathbf{pa}_i, \varepsilon_i)$ with a new mechanism $X_i = x_i'$ governed by some external force that sets the value x_i'. If we imagine that each variable X_i could potentially be subject to the influence of such an external force, then we can view each Bayesian network as an efficient code for predicting the effects of atomic interventions and of various combinations of such interventions, without representing these interventions explicitly.

This function-replacement operation yields a simple and direct transformation between the pre-intervention and the post-intervention distributions:

$$P(x_1, \ldots, x_n | do(X_i = x_i')) = \begin{cases} \dfrac{P(x_1, \ldots, x_n)}{P(x_i | \mathbf{pa}_i)} & \text{if } x_i = x_i' \\ 0 & \text{if } x_i \neq x_i' \end{cases} \qquad (6)$$

which reflects the removal of the term $P(x_i|\mathbf{pa}_i)$ from the product decomposition of Equation 1, since $\mathbf{pa}_i$ no longer influence X_i. Graphically, the removal of this term is equivalent to removing the links between $\mathbf{pa}_i$ and X_i while keeping the rest of the network intact. Transformations involving conjunctive actions can be obtained by straightforward generalization of Equation 6.

The transformation in Equation 6 exhibits all the properties we normally associate with actions, and it was therefore proposed as a solution to the frame problem and its two satellites, the ramification problem and the concurrency problem (Darwiche and Pearl, 1994; Pearl, 1994a). For example, to represent the intervention "turning the sprinkler ON" in the network of Figure 1, we delete the link $X_1 \rightarrow X_3$ and fix the value of X_3 to ON. The resulting joint distribution on the remaining variables will be

$$P(x_1, x_2, x_4, x_5 | do(X_3 = \text{ON}))$$
$$= P(x_1) P(x_2 | x_1) P(x_4 | x_2, X_3 = \text{ON}) P(x_5 | x_4) \qquad (7)$$

Note the difference between the action $do(X_3 = \text{ON})$ and the observation $X_3 = \text{ON}$. The latter is encoded by ordinary Bayesian conditioning, the former by conditioning a mutilated graph with the link $X_1 \rightarrow X_3$ removed. This indeed mirrors the difference between seeing and doing: after observing that the sprinkler is ON, we wish to infer that the season is dry, that it probably did not rain, and so on; no such inferences should be drawn in evaluating the effects of the deliberate action "turning the sprinkler ON."

Properties and Algorithms

By providing graphical means for representing and manipulating probabilistic knowledge, Bayesian networks overcome many of the conceptual and computational difficulties of rule-based systems (Pearl, 1988). Their basic properties and capabilities can be summarized as follows:

1. Graphical methods make it easy to maintain consistency and completeness in probabilistic knowledge bases. They also prescribe modular procedures of knowledge acquisition which significantly reduce the number of assessments required.
2. Independencies can be dealt with explicitly. They can be articulated by an expert, encoded graphically, read off the network, and reasoned about; yet they forever remain robust to numerical imprecision.
3. Graphical representations uncover opportunities for efficient computation. Distributed updating is feasible in knowledge structures that are rich enough to exhibit intercausal interactions (e.g., "explaining away"). And, when extended by clustering or conditioning, tree-propagation algorithms are capable of updating networks of arbitrary topology (Pearl, 1988; Shafer, 1990).
4. The combination of predictive and abductive inferences resolves many problems encountered by first-generation ex-

pert systems and renders belief networks a viable model for cognitive functions requiring both top-down and bottom-up inferences.

5. The causal information encoded in Bayesian networks facilitates the analysis of action sequences, their consequences, their interaction with observations, and their expected utilities, and hence the synthesis of plans and strategies under uncertainty (Dean and Wellman, 1991; Pearl, 1994a).

6. The isomorphism between the topology of Bayesian networks and the stable mechanisms that operate in the environment facilitates modular reconfiguration of the network in response to changing conditions, and permits deliberative reasoning about novel situations.

The first algorithms proposed for probability updating in Bayesian networks used message-passing architecture and were limited to trees (Pearl, 1982) and singly connected networks (Kim, 1983). The idea was to assign each variable a simple processor, forced to communicate only with its neighbors, and to permit asynchronous back-and-forth message-passing until equilibrium was achieved. Coherent equilibrium can indeed be achieved in this way, but only in singly connected networks, where an equilibrium state occurs in time proportional to the diameter of the network.

Many techniques have been developed and refined to extend the tree-propagation method to general, multiply connected networks. Among the most popular are Shachter's method of node elimination, Lauritzen and Spiegelhalter's method of clique-tree propagation, and the method of loop-cut conditioning (see Pearl, 1988; Shafer, 1990).

Clique-tree propagation, the most popular of the three methods, works as follows. Starting with a directed network representation, the network is transformed into an undirected graph that retains all of its original dependencies. This graph, sometimes called a Markov network, is then triangulated to form local clusters of nodes (cliques) that are tree-structured. Evidence propagates from clique to clique by ensuring that the probability of their intersection set is the same, regardless of which of the two cliques is considered in the computation. Finally, when the propagation process subsides, the posterior probability of an individual variable is computed by projecting (marginalizing) the distribution of the hosting clique onto this variable.

While the task of updating probabilities in general networks is NP-hard (see PAC LEARNING AND NEURAL NETWORKS for the definition of NP), the complexity for each of the three methods cited above is exponential in the size of the largest clique found in some triangulation of the network. It is fortunate that these complexities can be estimated prior to actual processing; when the estimates exceed reasonable bounds, an approximation method such as stochastic simulation (Pearl, 1988) can be used instead. Learning techniques have also been developed for systematic updating of the conditional probabilities $P(x_i|\mathbf{pa}_i)$ so as to match empirical data (see Spiegelhalter and Lauritzen, in Shachter, 1990).

Recent Developments

Causal discovery.   One of the most exciting prospects in recent years has been the possibility of using Bayesian networks to discover causal structures in raw statistical data. Several systems have been developed for this purpose (Pearl and Verma, 1991; Spirtes, Glymour, and Schienes, 1993). Technically, such discovery is feasible only if one is willing to accept forms of guarantees that are weaker than those obtained through controlled randomized experiments: minimality and

stability (Pearl and Verma, 1991). *Minimality* guarantees that any other structure compatible with the data is necessarily more redundant, and hence less trustworthy, than the one(s) inferred. *Stability* ensures that any alternative structure compatible with the data must be less stable than the one(s) inferred; namely, slight fluctuations in experimental conditions will render that structure no longer compatible with the data. With these forms of guarantees, the theory provides criteria for identifying genuine and spurious causes, with or without temporal information, and yields algorithms for recovering causal structures with hidden variables from empirical data.

Plain beliefs.   In mundane decision making, beliefs are revised not by adjusting numerical probabilities but by tentatively accepting some sentences as "true for all practical purposes." Such sentences, often named *plain beliefs*, exhibit both logical and probabilistic character. As in classical logic, they are propositional and deductively closed; as in probability, they are subject to retraction and to varying degrees of entrenchment (Goldszmidt and Pearl, 1992).

Bayesian networks can be adopted to model the dynamics of plain beliefs by replacing ordinary probabilities with nonstandard probabilities, that is, probabilities that are infinitesimally close to either zero or one. This amounts to taking an "order of magnitude" approximation of empirical frequencies and adopting new combination rules tailored to reflect this approximation. The result is an integer-addition calculus, very similar to probability calculus, with summation replacing multiplication and minimization replacing addition. A plain belief is then identified as a proposition whose negation obtains an infinitesimal probability (i.e., an integer greater than zero).

This combination of infinitesimal probabilities with the causal information encoded by the structure of Bayesian networks facilitates linguistic communication of belief commitments, explanations, actions, goals, and preferences, and serves as the basis for current research on qualitative planning under uncertainty (Darwiche and Pearl, 1994; Goldszmidt and Pearl, 1992; Pearl, 1994b).

Discussion

The most distinctive characteristics of Bayesian networks are their abilities to faithfully represent causal relationships, to combine top-down and bottom-up inferences, and to adapt to changing conditions by updating the probability measures attached to the links. Although Bayesian networks can model a wide spectrum of cognitive activity, their greatest strength is in causal reasoning, which, in turn, facilitates reasoning about actions, explanations, counterfactuals, and preferences. Such capabilities are not easily implemented in neural networks. Except for the common ability to perform distributed inferencing, the relation between Bayesian networks and neural networks is rather tenuous. For example, there are very few neural features in Bayesian networks: weights, sums, and sigmoids play no significant role; all computational units represent familiar linguistic notions; and deployment of bidirectional messages in acyclic structures has yet to find a biological basis.

Some questions arise: Does an architecture resembling that of Bayesian networks exist anywhere in the human brain? If not, how does the brain perform those cognitive functions in which Bayesian networks excel? The answer is, I speculate, that nothing resembling Bayesian networks actually resides permanently in the brain. Instead, fragmented structures of causal organizations are constantly being assembled on the fly, as needed, from a stock of functional building blocks. Each such building block is specialized to handle a narrow context of experience and is probably embodied in a neural network architecture. For example, the network of Figure 1 may be assembled from several neural networks, one specializing in the experience surrounding seasons and rains, another in the properties of wet pavements, and so forth. Such specialized networks are probably stored permanently in some mental library, from which they are drawn and assembled into the structure shown in Figure 1 only when a specific problem presents itself, for example, to determine whether an operating sprinkler could explain why a certain person slipped and broke a leg in the middle of a dry season.

I believe the properties of Bayesian networks will be useful to scientists studying higher cognitive functions, where the problem of organizing and supervising large assemblies of specialized neural networks becomes important.

Acknowledgments. The research was partially supported by Air Force grant #F49620–94–1–0173, NSF grant #IRI–9200918, and Northrop-Rockwell Micro grant #93–124.

Road Map: Artificial Intelligence and Neural Networks
Related Reading: Expert Systems and Decision Systems Using Neural Networks

References

Charniak, E., 1991, Bayesian networks without tears, *AI Mag.*, 12(4): 50–63. ◆

Darwiche, A., and Pearl, J., 1994, Symbolic causal networks for planning under uncertainty, in *Symposium Notes of the 1994 AAAI Spring Symposium on Decision-Theoretic Planning*, Stanford, CA, pp. 41–47.

Dawid, A. P., 1979, Conditional independence in statistical theory, *J. R. Statist. Soc. Ser. A.*, 41:1–31.

Dean, T. L., and Wellman, M. P., 1991, *Planning and Control*, San Mateo, CA: Morgan Kaufmann. ◆

Goldszmidt, M., and Pearl, J., 1992, Default ranking: A practical framework for evidential reasoning, belief revision and update, in *Proceedings of the Third International Conference on Knowledge Representation and Reasoning*, San Mateo, CA: Morgan Kaufmann, pp. 661–672.

Kim, J. H., and Pearl, J., 1983, A computational model for combined causal and diagnostic reasoning in inference systems, in *Proceedings of the 1983 International Joint Conference on Artificial Intelligence*, Karlsruhe, Germany, pp. 190–193.

Pearl, J., 1982, Reverend Bayes on inference engines: A distributed hierarchical approach, in *Proceedings of the AAAI National Conference on AI*, Pittsburgh, pp. 133–136.

Pearl, J., 1988, *Probabilistic Reasoning in Intelligence Systems*, San Mateo, CA: Morgan Kaufmann, 1988. (Revised 2nd printing, 1991.) ◆

Pearl, J., 1994a, A probabilistic calculus of actions, in *Proceedings of the Tenth Conference on Uncertainty in Artificial Intelligence (UAI-94)*, San Mateo, CA: Morgan Kaufmann.

Pearl, J., 1994b, From Adams' conditionals to default expressions, causal conditionals, and counterfactuals, in *Probability and Conditionals* (E. Eells and B. Skyrms, Eds.), New York: Cambridge University Press. ◆

Pearl, J., and Verma, T., 1991, A theory of inferred causation, in *Principles of Knowledge Representation and Reasoning: Proceedings of the Second International Conference* (J. A. Allen, R. Fikes, and E. Sandewall, Eds.), San Mateo, CA: Morgan Kaufmann, pp. 441–452.

Shachter, R. D., Ed., 1990, Special issue on influence diagrams, *Networks: Internat. J.*, 20(5).

Shafer, G., and Pearl, J., Eds., 1990, *Readings in Uncertain Reasoning*, San Mateo, CA: Morgan Kaufmann. ◆

Spiegelhalter, D. J., Lauritzen, S. L., Dawid, P. A., and Cowell, R. G., 1993, Bayesian analysis in expert systems, *Statist. Sci.*, 8(3):219–247.

Spirtes, P., Glymour, C., and Schienes, R., 1993, *Causation, Prediction, and Search*, New York: Springer-Verlag.

Wright, S., 1921, Correlated and causation, *J. Agric. Res.*, 20:557–585.

BCM Theory of Visual Cortical Plasticity

Nathan Intrator and Leon N. Cooper

Visual Cortical Plasticity

The BCM theory of cortical plasticity (called BCM for Bienenstock, Cooper, and Munro, 1982) was created to account for the striking dependence of the selectivity of cat visual cortex neuron responses on the animal's visual environment during the critical period. It has been successful in explaining the great variety of experience-dependent responses of cortical neurons. Among these are monocular and binocular deprivation and reverse suture. Recently, some of the underlying assumptions of this theory have been verified in direct physiological experiments in visual cortex. This has led to a great deal of current work on the details of long-term potentiation and depression as well as the underlying basis for these phenomena. It is expected that such work will result in a much improved understanding of the molecular basis of learning and memory storage. We present a brief account of the BCM theory as well as some of the recent experimental work.

Neurons in the primary visual cortex, area 17, of normal adult cats are sharply tuned to the orientation of an elongated slit of light, and most are activated by stimulation of either eye (Hubel and Wiesel, 1959). Both of these properties—orienta-
tion selectivity and binocularity—depend on the type of visual environment experienced during a critical period of early postnatal development. For example, monocular deprivation (MD) has profound and reproducible effects on the functional connectivity of striate cortex during the critical period, extending from approximately 3 weeks to 3 months of age in the cat (Frègnac and Imbert, 1984; Sherman and Spear, 1982). Brief periods of MD will result in a dramatic shift in the ocular dominance (OD) of cortical neurons so that most will be responsive exclusively to the open eye. The OD shift after MD is the best-known and most intensively studied type of visual cortex plasticity.

When MD is initiated late in the critical period, or after a period of rearing in the dark, it will induce clear changes in cortical OD without a corresponding anatomic change in the geniculocortical projections. Long-term recordings from awake animals also indicate that OD changes can be detected within a few hours of monocular experience; this seems too rapid to be explained by the formation or elimination of axon terminals. Moreover, deprived-eye responses in visual cortex may be restored within minutes to hours under some conditions, which suggests that synapses deemed functionally disconnected are nonetheless physically present. Therefore, it is reasonable to assume that changes in the functional binocularity may be explained by changes in the efficacy of individual cortical synapses.

The consequences of binocular deprivation (BD) on visual cortex stand in striking contrast to those observed after MD. Although seven days of MD during the second postnatal month leave few neurons in the striate cortex responsive to stimulation of the deprived eye, most cells remain responsive to stimulation through either eye after a comparable period of BD (Wiesel and Hubel, 1965). Thus, it is not merely the absence of patterned activity in the deprived geniculate projection that causes the decrease in synaptic efficacy after MD.

The result of a reversed suture (RS) experiment is even more striking. In this experiment, the kitten is first exposed to a normal visual environment; then one eye is sutured closed for a few days until the sutured eye becomes functionally disconnected. At that time the sutured eye is opened and exposed to a normal visual environment again, and the previously opened eye is closed. The result from this experiment is that in general the newly opened eye does not recover before the previously opened eye becomes disconnected.

These results are explained by the BCM theory for visual cortical plasticity, which will briefly be described below.

Single-Cell Theory

A typical neuron in striate cortex receives thousands of afferents from other cells. Most of these afferents derive from the lateral geniculate nucleus (LGN) and from other cortical neurons. We have approached the analysis of this complex network in several stages. In the first stage, we consider a single neuron with inputs from both eyes (i.e., via LGN) but without intracortical interactions.

The output of this neuron (in the linear region) can be written

$$c = \mathbf{m}^l \cdot \mathbf{d}^l + \mathbf{m}^r \cdot \mathbf{d}^r \tag{1}$$

where $\mathbf{d}^l(\mathbf{d}^r)$ are the LGN inputs coming from the left (right) eye to the vector of synaptic junctions $\mathbf{m}^l$ ($\mathbf{m}^r$). Thus the neuron integrates signals from the left and right eyes. (For simplicity, whenever possible we shall omit the left and right superscripts.) According to the theory presented by Bienenstock, Cooper, and Munro (1982), the synaptic weight changes over time as a function of local and global variables: its change in time is

$$\dot{m}_j = F(d_j, m_j; d_k \ldots, m_k \ldots, c; \bar{\bar{c}}; X, Y, Z) \tag{2}$$

Here variables such as d_j, m_j are designated *local*. These represent information (such as the incoming signal, d_j, and the strength of the synaptic junction, m_j) available locally at the synaptic junction, m_j. Variables such as $d_k, \ldots, c$ are designated *quasi-local*. These represent information (such as c, the *activity* or the depolarization of the postsynaptic cell, or d_k, the incoming signal to another synaptic junction) that is not locally available to the junction m_j but is physically connected to the junction by the cell body itself—thus necessitating some form of internal communication between various parts of the cell and its synaptic junctions. Variables such as $\bar{\bar{c}}$ (the time-averaged output of the cell) are averaged local or quasi-local variables. *Global* variables are designated $X, Y, Z, \ldots$ These latter represent information (e.g., presence or absence of neurotransmitters such as norepinephrine or the average activity of large numbers of cortical cells) that is present in a similar fashion for all or a large number of cortical neurons (distinguished from local or quasi-local variables presumably carrying detailed information that varies from synapse to synapse or cell to cell).

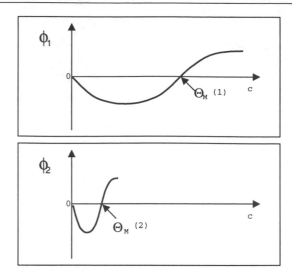

Figure 1. The ϕ function for two different Θ_M's.

In a form relevant to this discussion, BCM modification can be written

$$\dot{m}_j = \phi(c, \bar{\bar{c}}; X, Y, Z, \ldots)d_j \tag{3}$$

so that the jth synaptic junction, m_j, changes its value in time as the product of the input activity (the local variable d_j) and a function ϕ of quasi-local and time-averaged quasi-local variables, c and $\bar{\bar{c}}$, as well as global variables X, Y, Z. Neglecting global variables, one arrives at the following form of synaptic modification equation:

$$\dot{m}_j = \phi(c, \Theta_M)d_j \tag{4}$$

where Θ_M is a nonlinear function of some time-averaged measure of cell activity.

The shape of the function ϕ is given in Figure 1 for two different values of the threshold Θ_M. Of particular significance is the change of sign of ϕ at the modification threshold Θ_M and the nonlinear variation of Θ_M with the average output of the cell $\bar{c}$. In the form originally proposed by Bienenstock, Cooper, and Munro (1982), this was written as

$$\Theta_M \sim (\bar{c})^2 \tag{5}$$

A recently proposed form

$$\Theta_M \sim (\bar{c}^2) \tag{6}$$

displays very similar qualitative properties but improved stability. The precise form of Θ_M is, of course, expected to be more complex and remains an outstanding experimental question. For simplicity, in stationary environments, the time average is often replaced by a spatial average over the environmental inputs ($\bar{c} \to \mathbf{m} \cdot \bar{\mathbf{d}}$).

The occurrence of negative and positive regions for ϕ results in the cell's becoming selectively responsive to subsets of stimuli in the visual environment. This happens because the response of the cell is diminished to those patterns for which the output, c, is below threshold (ϕ negative) while the response is enhanced to those patterns for which the output, c, is above threshold (ϕ positive). The nonlinear variation of the threshold Θ_M with the average output of the cell contributes to the development of selectivity and the stability of the system (Bienenstock et al., 1982; Intrator and Cooper, 1992). Similarly, in a Hebbian rule (see HEBBIAN SYNAPTIC PLASTICITY), postsynaptic activity above a predefined threshold, along with strong input activity, will lead to increase in synaptic strength, in a manner similar to BCM; in contrast, however, this threshold does not move in a Hebbian rule.

Information-extraction properties of this modification formula have been studied and compared with those of other cortical plasticity theories as well as other feature-extraction methods related to projection pursuit (see INFORMATION THEORY AND VISUAL PLASTICITY).

Cortical Network: Mean-Field Theory

The actual cortical network is very complex. It includes different cell types, intracortical interactions, and recurrent collaterals. In what follows we present a method of analyzing this complex system. The first step is to divide the inputs to any cell into those from the LGN and those from all other sources. The activity of neuron i is affected by its input vector d from the LGN, and by the adjacent cortical neurons:

$$c_i = \mathbf{m}_i \cdot \mathbf{d} + \sum_j L_{ij} c_j \qquad (7)$$

where L_{ij} are the cortico-cortical synapses, which are assumed to be inhibitory. Scofield and Cooper (1985) have analyzed a network extension of the single-cell theory. In a mean-field approximation to the full network, the lateral inhibition can be replaced by a constant shift in the synaptic weight vector of the form $\bar{c} = (1/N) \Sigma_i c_i$, where N is the number of neurons in the network, so that c_i becomes

$$c_i = \mathbf{m}_i \cdot \mathbf{d} + \bar{c} \sum_j L_{ij} \qquad (8)$$

From a consistency condition it follows that $\bar{c} = \bar{\mathbf{m}} \cdot \mathbf{d} + \bar{c} L_0 = (1 - L_0)^{-1} \bar{\mathbf{m}} \cdot \mathbf{d}$, where $\bar{\mathbf{m}} = (1/N) \Sigma_i \mathbf{m}_i$, and $L_0 = (1/N) \Sigma_{ij} L_{ij}$, so that $c_i = (\mathbf{m}_i + (1 - L_0)^{-1} \bar{\mathbf{m}} \Sigma_j L_{ij}) \mathbf{d}$.

If we assume that the lateral connection strengths are a function only of the relative distance $i - j$, then L_{ij} becomes a circular matrix so that $\Sigma_i L_{ij} = \Sigma_j L_{ij} = L_0$, and

$$c_i = (\mathbf{m}_i + L_0 (1 - L_0)^{-1} \bar{\mathbf{m}}) d \qquad (9)$$

Thus, in the mean-field approximation, the effect of lateral inhibition is a shift in the vector of synaptic weights. Since $L_0 < 0$, the shift reduces m_i.

When analyzing the position and stability of the fixed points using this approximation, it follows, under some mild assumptions about the evolution of the average synaptic weights, that there is a mapping

$$m_i' \leftrightarrow \mathbf{m}_i(\alpha) - \alpha$$

such that, for every neuron in such a network with synaptic weight vector m_i, there is a corresponding neuron with weight vector m_i' that undergoes the same evolution (around the fixed points) subject to a translation α.

Although the averaged inhibition assumption used in the mean-field theory is an approximation, the mean-field network just described provides a powerful tool for analyzing a certain type of network architecture in great detail, and for gaining an intuitive understanding of a complex network in terms of the behavior of a single neuron. In addition, it provides justification for the study of single-cell theories which allow both positive and negative synapses. More detailed lateral inhibition networks which do not rely on the mean-field assumption are described in Intrator and Cooper (1992).

Comparison of Theory and Experiment

While the theory aims to provide a physiologically plausible account of synaptic plasticity, it does not address the mecha-

nism by which plasticity diminishes at the end of the critical period. A number of possible mechanisms have been proposed to account for the short duration of the plastic period, but at present it is not clear that the length of the critical period is determined by the same mechanism as that underlying synaptic change.

The validity of the BCM theory, as with any theory, can be tested in two ways. The first is to derive predictions or consequences of the theory in various situations that can be compared with experimental results. There is a considerable experimental literature on visual cortical plasticity, reaching back 30 years, which facilitates such comparisons with the BCM theory. The second approach is to attempt to verify the underlying assumptions of the theory, particularly those assumptions that distinguish it from others. In the case of BCM, the most important and unique assumptions concern the form of the synaptic modification function ϕ and the movement of the modification threshold. Over the last five years, we have made significant progress using both of these approaches, and this work is summarized briefly below.

In a recent simulation study, the consequences of the BCM theory were compared in detail with the results of experiments on what were called "classical" rearing conditions (Clothiaux, Cooper, and Bear, 1991). These conditions include normal binocular vision, monocular deprivation, reverse suture, strabismus, binocular deprivation, as well as the restoration of normal binocular vision after various forms of deprivation. The theory is in excellent agreement with observations. In addition, various new predictions, such as the correlation between selectivity and ocular dominance, seem to be confirmed. Comparisons with the pharmacological manipulations that affect visual cortical plasticity (Reiter and Stryker, 1988; Bear et al., 1990, for review) were not considered and remain an area that is ripe for further work. The modifications considered by Clothiaux et al. were those that occur in kitten visual cortex during the second postnatal month after brief ($\simeq 2$ weeks) changes in visual experience. Particular attention was given to the manner in which the theory predicts that changes in visual experience should affect the binocularity of cortical neurons and the selectivity of these neurons for the stimulus pattern (e.g., its orientation). It is these properties of binocularity and selectivity which distinguish cortical neurons from those in the retina and thalamus. A review of the experimental literature as it relates to the modification of these properties may be found in Clothiaux et al. (1991). For the relevance of information theory and comparison with other cortical theories, see INFORMATION THEORY AND VISUAL PLASTICITY.

Neurobiological Foundations for the Assumptions of the BCM Theory

Recent advances in our understanding of excitatory amino acid (EAA) receptors have suggested a possible physiological basis for the BCM form of synaptic modification. In 1987, Bear et al. proposed that the modification threshold Θ_M of BCM related to the membrane potential at which the N-methyl-D-aspartate (NMDA) receptor dependent Ca^{2+} flux reached the threshold for inducing synaptic long-term potentiation (LTP). (See NMDA RECEPTORS: SYNAPTIC, CELLULAR, AND NETWORK MODELS for background concepts.) In support of the hypothesis that NMDA receptor mechanisms play a role in synaptic plasticity, Bear and co-workers have found that the pharmacological blockade of NMDA receptors with the competitive antagonist AP5 disrupts the physiological and anatomical consequences of monocular deprivation in striate cortex. Although the interpretation of these experiments is compromised by the

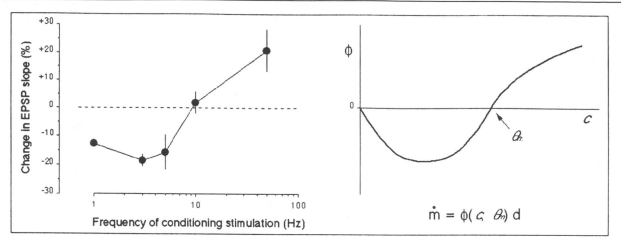

Figure 2. Comparison of experimental observations with BCM ϕ function for synaptic modification. Data replotted from Dudek and Bear (1992).

finding that AP5 reduces visually evoked responses, the data indicate that activity evoked in visual cortex in the absence of NMDA receptor activation is not sufficient to produce loss of closed-eye responsiveness in MD.

In the past several years, to better investigate the assumptions of the BCM theory and to address possible underlying mechanisms, our work has been focused on the synaptic plasticity that can be evoked in brain slices (Bear, Press, and Connors, 1992; Dudek and Bear, 1992; Kirkwood et al., 1993, and the references therein). Hippocampus, particularly CA1 and dentate gyrus, is an advantageous preparation because robust and long-lasting experience-dependent synaptic modifications can be evoked in this structure. Dudek and Bear (1992) recently tested a theoretical prediction that patterns of excitatory input activity that consistently fail to activate target neurons sufficiently to induce synaptic potentiation will instead cause a specific synaptic depression. To realize this situation experimentally, the Schaffer collateral projection to CA1 in rat hippocampal slices was stimulated electrically at frequencies ranging from 0.5 to 50 Hz. 900 pulses at 1–3 Hz consistently yielded a depression of the CA1 population EPSP that persisted without signs of recovery for more than 1 hour after cessation of the conditioning stimulation. This long-term depression was specific to the conditioned input and could be prevented by application of NMDA receptor antagonists. This result was surprising in that NMDA receptors are known to participate in the induction of long-term potentiation, an increase in synaptic effectiveness. Indeed, at higher stimulation frequencies the depression was replaced by a potentiation. If the effects of varying stimulation frequency in the experiments of Dudek and Bear are explained by different values of postsynaptic response (perhaps the integrated postsynaptic depolarization or Ca^{2+} level) during the conditioning stimulation, then it can be seen from Figure 2 that their data are in striking agreement with assumptions of the BCM theory.

Of course, as striking as this similarity is, this work was performed in hippocampus and the BCM theory was developed for visual cortex. And, although these two forms of synaptic plasticity (depression and potentiation) have been reported previously in the sensory neocortex (cf. Artola, Bröcher, and Singer, 1990), evidence to date has indicated that they occur with far lower probability, usually require pharmacological treatments for their induction, and are elicited by stimulation patterns that differ dramatically from those that are effective in hippocampus (see discussion in Bear, Press, and Connors,

1992). Together, these data have been taken as support for the view that hippocampus and sensory neocortex may be quite distinct with respect to their capability for synaptic change. However, a direct comparison of plasticity of synaptic responses evoked in adult rat hippocampal field CA1 with those evoked in adult rat and immature cat visual cortical layer III has now been carried out by Kirkwood et al. (1993). They found that very similar forms of plasticity, LTP and LTD, are evoked with precisely the same types of stimulation in the three types of cortex without the use of pharmacological treatments. Furthermore, in all three preparations, both LTP and LTD depend on activation of NMDA receptors. These data suggest, first, that hippocampus should not be considered as a privileged site for plasticity in the adult brain and, second, that a common principle may govern experience-dependent synaptic plasticity, both in CA1 and throughout the superficial layers of the neocortex. We believe that this work represents an important advance toward a general theory of experience-dependent synaptic plasticity in the mammalian brain.

It is our opinion that, in its entirety, this work gives strong justification for a form of modification similar to that assumed by BCM. However, still open is the question of the sliding modification threshold. Although more work remains to be done on this question, we note that, according to two recent studies, the sign and magnitude of a synaptic modification in both hippocampus (Huang et al., 1992) and the Mauthner cell of goldfish (Yang and Faber, 1991) have been shown to depend on the recent history of synaptic activation.

Acknowledgments. Research was supported by the Office of Naval Research, the Army Research Office, and the National Science Foundation.

Road Maps: Development and Regeneration of Neural Networks; Mechanisms of Neural Plasticity
Background: I.3. Dynamics and Adaptation in Neural Networks
Related Reading: Hebbian Synaptic Plasticity: Comparative and Developmental Aspects; Ocular Dominance and Orientation Columns

References

Artola, A., Bröcher, S., and Singer, W., 1990, Different voltage dependent thresholds for the induction of long-term depression and long-term potentiation in slices of rat visual cortex, *Nature*, 347:69–72.
Bear, M. F., Cooper, L. N., and Ebner, F. F., 1987, The physiological basis of a theory for synapse modification, *Science*, 237:42–48. ◆

Bear, M. F., Gu, Q., Kleinschmidt, A., and Singer, W., 1990, Disruption of experience-dependent synaptic modification in the striate cortex by infusion of an NMDA receptor antagonist, *J. Neurosci.*, 10:909–925.

Bear, M. F., Press, W. A., and Connors, B. W., 1992, Long-term potentiation of slices of kitten visual cortex and the effects of NMDA receptor blockade, *J. Neurophysiol.*, 67:841–851.

Bienenstock, E. L., Cooper, L. N., and Munro, P. W., 1982, Theory for the development of neuron selectivity: Orientation specificity and binocular interaction in visual cortex, *J. Neurosci.*, 2:32–48. ◆

Clothiaux, E. E., Copper, L. N., and Bear, M. F., 1991, Synaptic plasticity in visual cortex: Comparison of theory with experiment, *J. Physiol*, 66:1785–1804.

Cooper, L. N., and Scofield, C. L., 1988, Mean-field theory of a neural network, *Proc. Nat. Acad. Sci. USA*, 85:1973–1977.

Dudek, S. M., and Bear, M. F., 1992, Homosynaptic long-term depression in area CA1 of hippocampus and the effects on NMDA receptor blockade, *Proc. Natl. Acad. Sci. USA*, 89:4363–4367. ◆

Frègnac, Y., and Imbert, M., 1984, Development of neuronal selectivity in primary visual cortex of cat, *Physiol. Rev.*, 64:325–434.

Huang, Y. Y., Colino, A., Selig, D. K., and Malenka, R. C., 1992, The influence of prior synaptic activity on the induction of long-term potentiation, *Science*, 255:730–733.

Hubel, D. H., and Wiesel, T. N., 1959, Integrative action in the cat's lateral geniculate body, *J. Physiol.*, 148:574–591. ◆

Intrator, N., and Cooper, L. N., 1992, Objective function formulation of the BCM theory of visual cortical plasticity: Statistical connections, stability conditions, *Neural Netw.*, 5:3–17. ◆

Kirkwood, A., Gold, S.M.D.J.T., Aizenman, C., and Bear, M. F., 1993, Common forms of synaptic plasticity in hippocampus and neocortex *in vitro, Science*, 260:1518–1521. ◆

Reiter, H. O., and Stryker, M. P., 1988, Neural plasticity without action potentials: Less active inputs become dominant when kitten visual cortical cells are pharmacologically inhibited, *Proc. Natl. Acad. Sci. USA*, 85:3623–3627.

Scofield, C. L., and Cooper, L. N., 1985, Development and properties of neural networks, *Contemp. Phys.*, 26:125–145.

Sherman, S. M., and Spear, P. D., 1982, Organization of visual pathways in normal and visually deprived cats, *Physiol. Rev.*, 62:738–855.

Wiesel, T. N., and Hubel, D. H., 1965, Comparison of the effects of unilateral and bilateral eye closure on cortical unit responses in kittens, *J. Neurophysiol.*, 28:1029–1040. ◆

Yang, X., and Faber, D. S., 1991, Initial synaptic efficacy influences induction and expression of long-term changes in transmission, *Proc. Natl. Acad. Sci. USA*, 88:4299–4303.

Binding in the Visual System

J. I. Nelson

Introduction: Many Areas, No Master

The trouble with visual cortex areas is that there are so many of them: 18 in the cat visual system alone (Payne, 1993), and over 30 in the monkey (Felleman and Van Essen, 1991). An everyday object, with its edges, color, surface texture, and movement, will excite cells in most of these areas. If the responses are so widespread, how can we identify the representation of a single object? When there are two similar objects in the field, to which representation does a feature detector's response belong? In vision, *binding* is the synthesis of a multi-area response into a unified, single-object representation. Bressler, Coppola, and Nakamura (1993) put it more generally: the binding problem is "the way in which the brain integrates fragmentary neural events at multiple locations to produce unified perceptual experience and behavior."

It is believed that binding is *not* achieved by convergence onto a master area or cell because anatomical pathways do not form an organizational chart which supports a "chief-of-staff" (Boussaoud et al., 1990; Distler et al., 1993; Felleman and Van Essen, 1991; also see Figure 1 in VISUAL SCENE PERCEPTION: NEUROPHYSIOLOGY) and because destruction of so-called integrative areas beyond the sensory systems does not destroy the unity of sensory experience, a point argued by Damasio (1989).

Interest in binding today is a belated acknowledgment that the analysis of a stimulus into its components and the scattering of the responses to those components has left neurophysiology bereft of principles for synthesis and coherence. There is a feeling that success on some fronts is carrying us away from an understanding of perception as the imposition of order on sensation. This carries with it the nagging doubt that binding is a *deus ex machina*, called upon to fix something which should not have become a problem in the first place.

The binding problem arises from an unspoken assumption: namely that, with activity spread across the cortical mantle, processing for visual perception cannot occur until some mech-anism intervenes to label the activity which arose from one object, and to make the relevant neurons act like a functional unit.

The visual response must be transmitted out of the visual system for many purposes: to allow the object it represents to be named (otherwise we have an anomia), to be understood for its utility (otherwise we have associative agnosia), to be understood for its motivational significance and reward value (otherwise we have sociopathy; see the case of patient Boswell of Damasio, Tranel, and Damasio, 1990), and to be linked to the motor programs for using it (otherwise we have ideational apraxia; see De Renzi and Lucchelli, 1988, also see VISUAL SCENE PERCEPTION: NEUROPHYSIOLOGY). The visual system is a major exporter; binding must make information more compact and easier to transmit.

In what I call the *transmission-first view*, processing activity from just the right neurons is invoked to reconcentrate scattered activity and to undo the damage which transmission along the many known pathways has done. By and large, the anatomical basis of intracortical connectivity—hierarchies and streams—is taken as the model for the functional nature of intracortical information exchange. Information is transmitted from one neuron to another, from one area to a higher one, and successively refined. After a higher area has processed the information, feedback may occur, but it is optional. Feedback today is not an essential accompaniment of forward transmission. If feedback occurs at all, it is as an error signal by which a higher area corrects a false response already received from a lower area.

There are several problems with a transmission-first approach. The transmission-first, process-later view permits incomplete, uncompensated (e.g., no constancies of size, color, etc. in perception), and otherwise erroneous information to be passed on to higher centers. If feedback is used to correct the earlier area's error, we start over and lose time. A better system would be to get it right the first time: if it does not fit, do not transmit.

In order to get it right the first time using better mechanisms of spike transmission, we must realize that the goal of sensory processing in cortex is *transmission*. Sensation is changed to perception by organizing it; perception is the imposition of order on sensation. What fits is transmitted onward. If we assume transmission works and then go looking for the processing, we have already missed the chance to find it.

The transmission-first view creates the binding problem. It assumes transmission always works and spikes can get everywhere without regard to the information they carry. Binding is missing until something special happens.

The transmission-first view is a product of a neuroscience unleavened by Gestalt psychology. By contrast, in what I term the *processing-first view*, transmission is impossible unless something happens. Long-range transmission in cortex is impossible on a cell-to-cell basis because the divergence from one pyramidal cell to many targets (500 to 2000 other cells) leaves too few synapses per target (1 to 8, whereas 20 to 30 would be more adequate) to fire it. This is the *sparse connectivity problem*.

The sparse connectivity problem can be solved by assemblies and timing: the cell which cannot transmit on its own must be aided by additional cells, and these must fire in synchrony with it. If these cells lie in multiple areas, then transmission becomes contingent upon activity in cells with multiple stimulus specializations. The involvement of distributed cell assemblies in seemingly local transmission guarantees that multiple cells concerned with multiple cue systems must concur about something before transmission can proceed. The power of the cortex can be brought to bear on the issue of where one cell should send one spike.

The First Mechanism: 40-Hz Oscillation

It was obvious some time ago to people who thought deeply about the problem of binding that it must involve changes in timing, because timing is the key to the construction of neuronal assemblies and to the support of cooperativity in cortex (von der Malsburg and Schneider, 1986). The first and currently only prominent neurophysiological binding mechanism is 40-Hz oscillation, observed most easily in the local field potential (LFP) (reviewed by Singer, 1993; Engel et al., 1992; see also SYNCHRONIZATION OF NEURONAL RESPONSES AS A PUTATIVE BINDING MECHANISM). The oscillation arises during the presentation of moving stimuli whose elements move together, not with separate trajectories or opposite motion directions, and so its occurrence signals perceptual grouping (Gestalt theory's "common fate") or the presence of a rigid object. The phase of the oscillation is the same in widely separated places, suggesting that some kind of communication is occurring among the areas involved. Binding would arise if the neurons encoding one object were forced into synchrony and so could communicate more effectively.

Indeed, 40-Hz oscillation has several attributes of a binding mechanism—notably, multi-area involvement and temporal synchrony to facilitate neural transmission—but it is unsatisfying in other respects.

First, what is the circuit? The network underlying the observed synchrony must embrace multiple areas because the synchrony embraces multiple areas. In a direct test of this supposition, it has been shown that cortical connectivity is necessary for interhemispheric synchrony, because cutting the callosum abolishes the oscillation. Endogenous pacemaker activity is not the synchronizer (Jagadeesh, Gray, and Ferster, 1992).

Specifying the neuron-to-neuron circuits and their relationship to the pathways between areas—namely, the reciprocal projection systems of cortex—remains unfinished business. Are projections between areas reciprocal because both directions are essential somehow? If the circuit operates synchronously, do we have to discard integrate-and-fire neuronal principles and assume cells operate as spike coincidence detectors? (see SINGLE-CELL MODELS). How is synchrony to be reconciled with the latency inherent in all neural transmission? This latency is attributable to conduction delays down the axon (up to 60 ms for callosal fibers in the rabbit; Swadlow, 1983) and to transmission delays at the synaptic cleft between cells (1–2 ms) and is unavoidable. These are provocative findings, and at least we are moved to ask good questions.

Second, what do we do with the field potential? It is clear that rhythmicity, or at least phase coherence, arises in cortex in association with stimulus coherence and intentional states (Bressler, Coppola, and Nakamura, 1993; Murthy and Fetz, 1992). However, mass action potentials take us a step away from spikes and the circuits we must unravel for transmitting them. We do not know how to integrate LFP oscillations with spike transmission. Spike activity causes the LFP (e.g., depolarization and synaptic current flow at a target neurons' dendritic tree; Engel et al. 1992), but we cannot rule out the possibility that the LFP in return is a "shadow government," a feedback channel onto spike transmission which travels 200 μm or more entirely outside the constraints of anatomical pathways, using volume conduction of potentials summated from everywhere. Not that this is always bad (epilepsy, migraine), but we would have to understand it.

Third, oscillations are too distributed in space to be a powerful binding mechanism. Phase coherence extends from striate to motor cortex. True, oscillations arise when two bars move in unison as if belonging to a single rigid object. When oscillations are present, the spikes' increased synchrony enhances transmission to the next stage of processing. These mechanisms for enhancing and redirecting information exchange among clusters of cortical areas must be better understood at the circuitry level. The circuits must be specific to selected neurons, because many separate bindings would need to be maintained while a complex scene was being recognized. Oscillations which spread over a great deal of cortical surface seem too crude a mechanism for perception.

Finally, oscillations are too diffuse temporally. The shift into correlation can be abrupt (30–40 ms), but on other trials, this shift may not occur for 500 ms or more after stimulation has started. For example, a monkey has already done stimulus processing and made an express saccade by 70 ms. The time of onset for correlated oscillation is too variable with respect to the time of stimulus onset, and the periods of oscillation are too spotty and variable in duration (Gray et al., 1992) for oscillations to constitute a fundamental mechanism for transmitting drive of visual activity in cortex. Long after perception is over (hundreds of milliseconds), oscillations might trigger memory consolidation or plasticity mechanisms because *other* mechanisms had transmitted the drive and accomplished the binding.

The Next Mechanism of Binding

The next mechanism of binding will come from a more fundamental grasp of spike transmission in cortex. This mechanism must be assembly-based (see CORTICAL COLUMNS, MODULES, AND HEBBIAN CELL ASSEMBLIES). The circuits could be very specifically targeted on selected cells, very stereotypically structured, and deterministic in their operation. Multiple areas must be embraced. Spikes could be synchronous in all the areas; we need not necessarily rely on probabilistic firing and recruitment. If we had a grasp of these circuits, then we would see that there is no bottom-up transmission without a top-down return of EPSPs (excitatory post-synaptic potentials), so that dispersed activity is not only labeled by synchrony, it is also physically returned, as so many models of pattern recognition require (see VISUAL SCHEMAS IN OBJECT RECOGNITION AND SCENE ANALYSIS). We would see that, steeped in principles developed in squid and at the motor endplate, we have been wrongly trained to assume that transmission works, and then processing starts. After all, information cannot be processed in a given cortical area until it arrives. Information continues to arrive in one area after another, waiting for processing, until we wake from the nightmare, calling for a binding mechanism. From out of the cacophony, the right responses must somehow be found and labeled, with a supplementary oscillation or some other timing code. But in cortex, as in perception, personality, or political prejudice, everything must fit or little is transmitted. The unbound drive will never arrive. The first spike discriminates the face. I myself took transmission for granted, searching tuning curves and receptive fields for the processing. But the processing is in the transmission. This processing, as we would expect, is distributed and massively parallel (see VISUAL SCENE PERCEPTION: NEUROPHYSIOLOGY). Binding is a nonproblem, but we have others to take its place.

Road Map: Vision

References

Boussaoud, D., Ungerleider, L. G., and Desimone, R., 1990, Pathways for motion analysis: Cortical connections of the medial superior temporal and fundus of the superior temporal visual areas in the macaque, *J. Comp. Neurol.*, 296:462–495.

Bressler, S. L., Coppola, R., and Nakamura, R. N., 1993, Episodic multiregional cortical coherence at multiple frequencies during visual task performance, *Nature*, 366:153–155.

Damasio, A. R., 1989, Time-locked multiregional retroactivation: A systems-level proposal for the neural substrates of recall and recognition, *Cognition*, 33:25–62.

Damasio, A. R., Tranel, D., and Damasio, H., 1990, Individuals with sociopathic behavior caused by frontal damage fail to respond autonomically to social stimuli, *Behav. Brain. Res.*, 41:81–94.

De Renzi, E., and Lucchelli, F., 1988, Ideational apraxia, *Brain*, 111:1173–1185. ◆

Distler, C., Boussaoud, D., Desimone, R., and Ungerleider, L. G., 1993, Cortical connections of inferior temporal area TEO in Macaque monkeys, *J. Comp. Neurol.*, 334:125–150.

Engel, A. K., König, P., Kreiter, A. K., Schillen, T. B., and Singer, W., 1992, Temporal coding in the visual cortex—New vistas on integration in the nervous system, *Trends Neurosci.*, 15:218–226.

Felleman, D. J., and Van Essen, D. C., 1991, Distributed hierarchical processing in the primate cerebral cortex, *Cerebral Cortex*, 1:1–47.

Gray, C. M., Engel, A. K., Konig, P., and Singer, W., 1992, Synchronization of oscillatory neuronal responses in cat striate cortex—Temporal properties, *Visual Neurosci.*, 8:337–347.

Jagadeesh, B., Gray, C. M., and Ferster, D., 1992, Visually evoked oscillations of membrane-potential in cells of cat visual-cortex, *Science*, 257:552–554.

Murthy, V. N., and Fetz, E., 1992, Coherent 25- to 35-Hz oscillations in the sensorimotor cortex of awake behaving monkeys. *Proc. Natl. Acad. Sci. USA.*, 89:5670–5674.

Payne, B. R., 1993, Evidence for visual cortical area homologs in cat and macaque monkey, *Cerebral Cortex*, 3:1–25.

Singer, W., 1993, Synchronization of cortical activity and its putative role in information-processing and learning, *Annu. Rev. Physiol.*, 55:349–374. ◆

Swadlow, H. A., 1983, Efferent systems of primary visual cortex: A review of structure and function, *Brain Res. Rev.*, 6:1–24.

von der Malsburg, C., and Schneider, W., 1986, A neural cocktail-party processor, *Biol. Cybern.*, 54:29–40.

Biomaterials for Intelligent Systems

Michael Conrad and Klaus-Peter Zauner

Introduction

Advances in the biomolecular and chemical sciences have opened up the possibility of utilizing biomaterials for information technology. In some cases, these materials may complement inorganics in the context of conventional information processing architectures, e.g., by providing novel optical interconnects. More importantly, they open up the possibility of studying the relationship between the computational capabilities of organisms and their material organization, and of implementing modes of information processing that specifically utilize the unique characteristics of carbon macromolecules.

This article briefly reviews the essential facts about molecular-level information processing as it occurs in biological systems or as it may be modified in an artificial context. It then considers the pertinence of these principles to computing in both artificial and natural systems.

Molecular Essentials

The function of biological systems depends on an enormous number of macromolecular structures with a wide variety of highly specific functions. These structures are assembled from a small set of common building blocks. Proteins provide the most prominent example. In nature, these are built up from 20 types of amino acids, which are themselves built up from between 10 and 27 atoms. The amino acids differ in shape and size as well as in their electrical and chemical properties. Proteins are linear chains of, typically, a few hundred amino acids. These fold up to form a characteristic three-dimensional shape (in the size range of 10 nm) that largely determines the ability to specifically recognize other molecular structures. The image of a key fitting into a lock is often used to characterize this shape-based mode of recognition. Recognition can lead to the highly selective making or breaking of covalent bonds (catalytic switching) or to the formation of stable polymacromolecular structures (principle of self-assembly). Other functions performed by proteins include chemical, optical, and electrical sensing, as well as mechanical actuation.

Nucleic acids, like proteins, are built up from a small set of building blocks. The building block principle also applies to membranes, but at a higher level, since these are formed from a variety of lipids and proteins. In the laboratory, it is possible to enlarge the set of building blocks, both at the amino acid and at the macromolecular level, and to organize them into structures that are not found in natural systems.

Complex molecular structures of the types just described cannot be produced by a classical chemical approach to the sequence of reactions, since the yield in each step is limited and separation of the product from the very similar but nonfunctional byproducts becomes unfeasible. The common phenomenon of self-assembly noted earlier—that is, the spontaneous formation of well-defined supramolecular structures on the basis of free energy minimization—provides one way of overcoming this (Lehn, 1988). Genetic engineering, another important technique, makes it possible to produce complicated structures that are covalently bonded. Cells produce proteins with the aid of a sophisticated synthesizing mechanism according to the genetic information encoded in the sequences of nucleotide bases in DNA. It is possible to manipulate the pattern of bases so that the cell will produce an altered version of a specific protein. Biotechnological methods make it possible to produce the modified protein on a large scale.

Biomolecules in Optical Computing

Materials with controllable optical properties, such as photochromics that change their color on illumination, are necessary for many real-time optical information processing architectures (see OPTICAL COMPONENTS FOR NEURAL NETWORK IMPLEMENTATIONS). Pioneering work in the former USSR during the 1980s demonstrated that the protein bacteriorhodopsin (BR), found in the light harvesting bacterium *Halobacterium salinarium* (until recently called *Halobacterium halobium*), exhibits interesting features in this respect. BR, functioning as an individual molecule, utilizes solar energy to maintain a proton gradient across the cell membrane. The pumping of a proton from the cytoplasm to the extracellular side of the membrane is accompanied by a series of photochemical and thermal transitions of the protein which altogether involve eight states with different absorption characteristics. Yellow light excites the transition from the B state to the M state, and exposure to blue light resets the protein to the B state. BR is capable of repeating this cycle several hundred thousand times, a feature that is unrivaled by conventional materials and allows applications in real-time processing. The modification of certain critical amino acids in BR has led to specifically tailored variants with improved state lifetime and recording sensitivity (Hampp et al., 1992).

The application of BR in holographic films is based on the shift in the absorption spectrum accompanying the conformational transitions of the molecule. The transitions also entail a rapid charge displacement (Hong, 1986), leading to a photocurrent that has been employed by Haronian and Lewis (1991) in a prototype for an opto-electronic hybrid neural network. In their architecture (Figure 1), analog amplifiers are interconnected through a matrix of oriented BR patches which stores the synaptic weights. This allows for extremely fast reprogramming with optical signals. BR, which is a highly stable molecule, also has applications as an optical sensor.

Molecular Matrices

Maintaining the spatial relationship between components without compromising desired interactions is essential for the proper function of a molecular device. In nature, biological membranes play an essential role. These are lipid bilayers that are self-assembled in an aqueous medium from molecules with a polar (water-attracting) and a nonpolar (water-repelling) region. Langmuir-Blodgett (LB) films are promising matrix materials that resemble biomembranes in a number of respects, but are not restricted to two layers. Multilayer films are de-

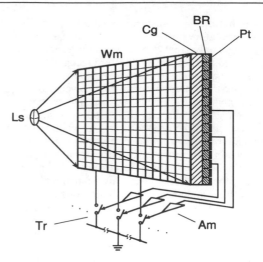

Figure 1. The neural network design suggested by Haronian and Lewis (1991) is based on the protein bacteriorhodopsin (BR). In this hybrid architecture, the summation, amplification (Am), and thresholding (Tr) functions of the neuron are realized with conventional silicon technology. (For clarity, only three such units are shown.) Optical accessibility of the synaptic weight matrix (Wm) permits fast reprogramming, even for a large number of connections. The weight of a connection is represented in the state distribution of the molecules in the corresponding BR patch and is set with a pattern of blue light from the light source (Ls). On illumination with a yellow flash, each BR patch yields an electric potential between the conducting glass surface (Cg) and the platinum electrode (Pt); if the patch is selected by the threshold circuit (Tr), the resulting photocurrent is fed to the amplifiers (Am). A subsequent red flash restores the states of the BR molecules that gave rise to the photocurrent.

posited on a solid surface by stacking monomolecular layers at an air-water interface. The resulting films possess interesting electrical and nonlinear optical properties, but more importantly, may also serve to embed ordered arrangements of macromolecules. The thermal, mechanical, and chemical stability of LB films is currently subject to some debate, and improvements that may make them more suitable for certain applications, such as cross-linking with the surface or photopolymerization, are being investigated. The anticipated scope of LB films includes piezoelectric and pyroelectric detectors, photoelectric devices, and optical sensors (reviewed in Ulman, 1991). Burrows and Wilson (1990) suggest a dynamic memory design based on LB multilayers deposited on the gate electrode of a field effect transistor (FET). Molecular layers with low and high electron affinities form a multiple quantum well structure. By shifting electrons with an electric field across the layers, it is possible to store several bits in serial fashion on the area occupied by one FET, whereas conventional memory technology allows only the storage of 1 bit.

Interface Schemes

Commercial use of biomaterials requires that they compete with silicon-based devices in terms of production costs as well as functional performance. Niches for which present-day technologies are not well suited are clearly the most accessible. Interfaces that support the integration of biocomponents into existing architectures are the key to filling these niches. Biosensors provide a link. These are composed of a molecular recognition unit and a transducer that converts a chemical signal into electrical or optical output. Biosensors can react selectively to the presence of specific molecules in a complex medium, whereas ordinary sensors require follow-on data processing to discriminate different molecules (Blum and Coulet, 1991). Recognizing different combinations of molecules should also be possible in active media if the signal-carrying molecules interact with each other. Light-sensitive active media have been used in optical pattern processing, though not yet in conjunction with biosensors (Rambidi, 1992).

The ability of biomolecules to discriminate signal-carrying molecules in a complex medium is also important because it provides a mechanism for interconnection-free transmission at a local level. This mechanism has a biological analog in intracellular communication in the form of messenger molecules and receptors. Other possibilities include voltage- or pH-sensitive dyes for optical readout, charge transfer from proteins to the gate electrode of an FET, and attempts to directly connect to proteins by means of a conducting polymer or a metallic contact.

The conceptual significance of interface technologies is more general than indicated earlier. In order to utilize molecular-level mechanisms for information processing, it is necessary to connect them to macroscopic input and output. This type of cross-scale interface is not required for silicon technologies, since the latter possess no processors whose size is in the range of individual molecules and whose function is controlled by the unaveraged motions of electrons. The nonclassical quantum behavior of the electrons is pertinent to the speed and reliability of silicon switches, but is otherwise masked by averaging so far as the input-output behavior of the switch is concerned. In the case of proteins and other molecular processors, quantum features are also pertinent to the speed and specificity of the recognition process owing to the importance of electronic-conformational interactions (Conrad, 1992). Cross-scale interfaces are required to exploit this recognition capability fully.

Transduction-amplification cascades that serve as interfaces across levels of scale are ubiqitous in biology. These enable biological cells to convert macroscopic signals impinging on the external membrane to chains of molecular-level events and to amplify the end events for the control of macroscopic activities. Second-messenger mechanisms involving cyclic nucleotides are prominent examples. Individual hormone or transmitter molecules ("first messengers") bind to a receptor protein on the cell surface. The receptor activates enzymes that amplify the single molecule signal by catalytic formation of second-messenger molecules (e.g., cyclic nucleotides) inside the cell. Effects of second messengers include the opening and closing of ion channels that alter the membrane potential of neurons and, hence, influence macroscopic firing behavior (Schwartz and Kandel, 1991).

Information Processing Within the Neuron

The view of the neuron as exclusively an integrating threshold device is no longer tenable. Here, we concentrate on the role of the cytoskeleton, a network of filamentary structures assembled from protein units. The components of the cytoskeleton are actin filaments (diameter of 6 nm), microtubules (diameter of 24 nm), and intermediate filaments (diameter of 8–10 nm). The latter, called neurofilaments in neurons, seem to play a static roll as microskeleton in the literal sense. The cytoskeleton plays a major role in the movement and location of proteins and intracellular organelles and in the motions of the cell as a whole.

In early anatomical studies of neuronal tissue, the cytoskeleton —revealed, in part, by staining with silver salts—was considered to be an excitation-transmitting structure within the neuron. This hypothesis, which appeared to be disproven by developments in membrane electrophysiology, has been revived by recent experimental evidence. For example, Matsumoto et al. (1984) have concluded from their experiments with squid giant axons that the conditions for microtubule assembly and membrane excitability are closely related. Liberman et al. (1985) have shown that neuron potentials can be evoked by mechanical distention in the same way that they are evoked by microinjection of second messengers. Results such as these have suggested the possibility that the cytoskeleton can participate in the control of nerve impulse activity, either as a subneuronal signal integration network or as a memory-supporting medium (Hameroff, 1989; Conrad, 1990).

The learning capability of many artificial neural network architectures depends on information feedback. Some of the successful architectures require feedback along forward connections, as in backpropagation. Since the information flow along the axonal membrane is generally unidirectional and the feedback is less time-critical than the forward information flow, the suggestion has been made that the cytoskeleton provides a backward transmission line (Werbos, 1992).

Discussion

It is often noted that the firing behavior of neurons is slow when compared to electronic switches, but that the connectivity can be enormous. Such connectivity is not possible on a two-dimensional surface. Building up a highly interconnected, three-dimensional structure with high packing density almost certainly requires a self-organizing growth process that is ultimately based on the recognition and self-assembly properties of macromolecules. At the present time, we are not aware of any technological efforts in this direction. However, ongoing attempts to control the growth of neurons in a prescribed manner on two-dimensional templates may contribute to an understanding of the biologically pertinent mechanisms.

The fabrication of molecular information-processing devices is helping to clarify the manner in which the properties of materials—in particular, the shape-based specificity of proteins and other biological macromolecules—relate to functional capabilities at the system level. By so doing, it may help to elucidate the relationship between the material organization of the brain and its computational capabilities.

Acknowledgment. The preparation of this paper was supported in part by grant ECS–9109860 from the U.S. National Science Foundation.

Road Map: Implementation of Neural Networks
Related Reading: Silicon Neurons

References

Blum, L. J., and Coulet, P. R., Eds., 1991, *Biosensor Principles and Applications*, vol. 15 of *Bioprocess Technology*, New York: Marcel Dekker. ◆

Burrows, P. E., and Wilson, E. G., 1990, The inchworm memory—A new molecular electronic device, *J. Mol. Electronics*, 6:209–220.

Conrad, M., 1990, Molecular computing, in *Advances in Computers* (M. C. Yovits, Ed.), Boston: Academic Press, pp. 235–324. ◆

Conrad, M., 1992, Molecular computing: The lock-key paradigm, *Computer (IEEE)*, 25:11–20. ◆

Hameroff, S. R., 1989, *Ultimate Computing*, Amsterdam: North Holland. ◆

Hampp, N., Thoma, R., Oesterhelt, D., and Bräuchle, C., 1992, Biological photochrome bacteriorhodopsin and its genetic variant Asp96 → Asn as media for optical pattern recognition, *Appl. Opt.*, 31:1834–1841.

Haronian, D., and Lewis, A., 1991, Elements of a unique bacterior-hodopsin neural network architecture, *Appl. Opt.*, 30:597–608.

Hong, F. T., 1986, The bacteriorhodopsin model membrane system as a prototype molecular computing element, *Biosystems*, 19:223–236.

Lehn, J.-M., 1988, Supramolecular chemistry—Scope and perspectives. Molecules, supermolecules, and molecular devices (Nobel lecture), *Angew. Chem. Int. Ed. Engl.*, 27:89–112. ◆

Liberman, E. A., Minina, S. V., Mjakotina, O. L., Shklovsky-Kordy, N. E., and Conrad, M., 1985, Neuron generator potentials evoked by intracellular injection of cyclic nucleotides and mechanical distension, *Brain Res.*, 338:33–44.

Matsumoto, G., Ichikawa, M., Tasaki, A., Murofushi, H., and Sakai, H., 1984, Axonal microtubules necessary for generation of sodium current in squid giant axons: I. Pharmacological study on sodium current and restoration of sodium current by microtubule proteins and 260K protein, *J. Membr. Biol.*, 77:77–91.

Rambidi, N. G., 1992, Nondiscrete biomolecular computing, *Computer (IEEE)*, 25:51–54.

Schwartz, J. H., and Kandel, E. R., 1991, Synaptic transmission mediated by second messengers, chap. 12, in *Principles of Neural Science* (E. R. Kandel, J. H. Schwartz, and T. M. Jessell, Eds.), New York: Elsevier, pp. 182–185, 187–189. ◆

Ulman, A., 1991, *An Introduction to Ultrathin Organic Films—From Langmuir-Blodgett to Self-Assembly*, Boston: Academic Press. ◆

Werbos, P. J., 1992, The cytoskeleton: Why it may be crucial to human learning and to neurocontrol, *Nanobiology*, 1:75–95.

Boltzmann Machines

Emile H. L. Aarts and Jan H. M. Korst

Introduction

Neural network models are based on the hypothesis that massive parallelism and the adaptive adjustment of neural states and interneural connection strengths are what produce self-organization and learning in the neural network of the human brain. A Boltzmann machine is a neural network model that tries to realize these brain functions through the use of stochastic computing elements. The model is interesting since

- It provides a generalized approach to optimization, classification, and learning tasks.
- There exist mathematical models that support the analysis of the network's dynamics and learning properties.
- It is relatively easy to implement in hardware.

Boltzmann machines were introduced by Hinton and Sejnowski (1983) and can be viewed as an extension of discrete Hopfield networks (Hopfield, 1982) in the following two ways. First, they replace the greedy local search dynamics of Hopfield networks with a randomized local search dynamics. Second, they replace the relatively simple Hebbian learning rule with a more powerful stochastic learning algorithm. These extensions are both achieved through the use of stochastic computing elements.

Here, we present a mathematical model and briefly discuss the theory of self-organization and learning in Boltzmann machines. For a more extensive discussion and an overview of the applications, we refer to our book (Aarts and Korst, 1989).

Network Structure

As in a Hopfield network, a Boltzmann machine consists of a number of two-state units that are connected in some way. The network can be represented by a pseudograph $\mathscr{B} = (\mathscr{U}, \mathscr{C})$, where $\mathscr{U}$ denotes a finite set of units and $\mathscr{C} \subseteq \mathscr{U} \times \mathscr{U}$ a set of symmetric connections. A *connection* $\{u, v\} \in \mathscr{C}$ joins the units u and v. The set of connections may contain *loops*, i.e., $\{u, u\} \in \mathscr{C}$, for some $u \in \mathscr{U}$. If two units are connected, they are called *adjacent*. A *unit* u can be in one of two *states*, state "0" or state "1."

Definition 1. A *configuration* k is a global state which is uniquely defined by a sequence of length $|\mathscr{U}|$, whose uth com-ponent $k(u)$ denotes the *state* of unit u in configuration k. $\mathscr{R}$ denotes the set of all configurations.

Definition 2. A connection $\{u, v\} \in \mathscr{C}$ is *activated* in a given configuration k if both u and v have state "1," i.e., if $k(u) \cdot k(v) = 1$.

Definition 3. With each connection $\{u, v\} \in \mathscr{C}$ a *connection strength* $s_{\{u,v\}} \in \mathbb{R}$ is associated as a quantitative measure of the *desirability* that $\{u, v\}$ be activated. By definition, $s_{\{u,v\}} = s_{\{v,u\}}$. If $s_{\{u,v\}} > 0$, it is desirable that $\{u, v\}$ is activated; if $s_{\{u,v\}} < 0$, it is undesirable. Connections with a positive (negative) strength are called *excitatory* (*inhibitory*).

Definition 4. The *consensus function* $C: \mathscr{R} \to \mathbb{R}$ assigns to each configuration k a real number, called the *consensus*, which equals the sum of the strengths of the activated connections, i.e.,

$$C(k) = \sum_{\{u,v\} \in \mathscr{C}} s_{\{u,v\}} k(u) k(v) \qquad (1)$$

The consensus is a global measure indicating to what extent the units in the network have reached a consensus about their individual states. In general, the measure will be large if many excitatory connections are activated, and small if many inhibitory connections are activated.

Network Dynamics

Self-organization in a Boltzmann machine is achieved by allowing units to change their states, from "0" to "1" or the reverse. This is similar to the self-organization in Hopfield networks. However, in a Boltzmann machine the acceptance of a proposed state change is stochastic, whereas in a Hopfield network it is deterministic. Let the network be in configuration k; then a *state change* of unit u results in a configuration l, with $l(u) = 1 - k(u)$ and $l(v) = k(v)$ for each $v \neq u$. Furthermore, let $\mathscr{C}_u$ denote the set of connections incident with unit u, excluding $\{u, u\}$. Then the *difference in consensus*, induced by a state change of unit u in configuration k, is given by

$$C_k(u) = (1 - 2k(u)) \left(s_{\{u,u\}} + \sum_{\{u,v\} \in \mathscr{C}_u} s_{\{u,v\}} k(v) \right) \qquad (2)$$

The effect on the consensus, resulting from a state change of unit u, is completely determined by the states of its adjacent units and the corresponding connection strengths. Consequently, each unit can locally evaluate its state change since no global calculations are required.

Definition 5. In a Boltzmann machine, the response of an individual state change of unit u to its adjacent units in a configuration k is a stochastic function given by

$$\mathbb{P}_c(u|k) \equiv \mathbb{P}_c\{\text{accept a state change of unit } u|k\}$$

$$= \frac{1}{1 + \exp(-C_k(u)/c)} \tag{3}$$

where $C_k(u)$ is given by Equation 2, and $c \in \mathbb{R}^+$ denotes a control parameter.

Implementation of state changes in a Boltzmann machine is done by SIMULATED ANNEALING (q.v.), which is a randomized local search algorithm that derives from the simulation of physical annealing processes. Generally speaking, the algorithm allows the acceptance of state changes that deteriorate the consensus to escape from poor locally maximal configurations. The probability of accepting a state change is controlled by the parameter c, and is given by Equation 3. In most cases, the algorithm is implemented such that initially the value of c is large, in which case the probability of accepting deteriorations is large. Subsequently, the value of c is decreased to eventually become 0, in which case no deteriorations are accepted.

Simulated annealing can be mathematically modeled by Markov chains (Aarts and Korst, 1989). This model can also be used to describe the state changes of the units in a Boltzmann machine. To this end, we distinguish between two models, *sequential Boltzmann machines* and *parallel Boltzmann machines*.

Sequential Boltzmann Machines

In a sequential Boltzmann machine, units may change their states only one at a time. The resulting iterative procedure can be described as a sequence of Markov chains, where each chain consists of a sequence of trials and the outcome of a given trial depends probabilistically only on the outcome of the previous trial. A *trial* consists of the following two steps. Given a configuration k, a neighboring configuration k_u is generated, determined by a unit $u \in \mathscr{U}$ that proposes a state change. Next, it is evaluated whether k_u is accepted or not. More specifically, the outcome of the trial is k_u with probability $\mathbb{P}_c(u|k)$, and otherwise k. For the acceptance probability of Equation 3, we obtain the following result.

Theorem 1 (Aarts and Korst, 1989). For a sequential Boltzmann machine with a response function given by Equation 3, the following statements hold.

(i) The probability $q(c)$ of obtaining a configuration k after a sufficiently large number of trials carried out at a fixed value of c is given by

$$q_k(c) = \frac{\exp(C_k/c)}{\sum_{l \in \mathscr{R}} \exp(C_l/c)} \tag{4}$$

(ii) For $c \downarrow 0$, Equation 4 reduces to a uniform distribution over the set of configurations with maximum consensus.

The expression in Equation 4 is often referred to as the *stationary distribution* of the corresponding Markov chain, and it is known in statistical physics as the Boltzmann distribution, which explains the name "Boltzmann machine." The process of reaching the stationary distribution is called *equilibration*. The first part of the theorem states that configurations with a higher consensus have a larger probability of occurring than configurations with a lower consensus. (This result plays an important role in the learning algorithm presented later in this article.) The second part states that if c approaches 0 slowly enough to allow equilibration, the Boltzmann machine finds with probability 1 a configuration with maximum consensus. This result plays an important role in optimization and classification, as discussed later.

In practical implementations, the asymptoticity conditions cannot be attained; thus, convergence to a configuration with maximum consensus is not guaranteed. In those cases, the Boltzmann machine finds a locally optimal configuration, i.e., a configuration with consensus no worse than that of any of its neighboring configurations. The convergence of the Boltzmann machine is determined by a set of parameters, known as the cooling schedule, which determine the convergence of the simulated annealing algorithm. These parameters are: a start value of c, a decrement rule to lower the value of c, the length of the individual Markov chains, and a stop criterion. In our book (Aarts and Korst, 1989), we discuss some cooling schedules for Boltzmann machines.

Parallel Boltzmann Machines

We distinguish between synchronous and asynchronous state changes. A *synchronous state change* consists of a set of state changes of individual units that are evaluated all at the same time. The synchronization is done by a global clocking scheme. Zwietering and Aarts (1991) use Markov chains to analyze synchronously parallel Boltzmann machines. Their main result is a conjecture that under certain mild conditions a stationary distribution different from Equation 4 is attained, converging as c approaches 0 to a distribution over the set of configurations for which the so-called *extended consensus* is maximal.

Asynchronous state changes are state changes of individual units that are evaluated concurrently and independently. Units generate state changes and accept or reject them on the basis of information that is not necessarily up to date, since the states of adjacent units may have changed in the mean time. Asynchronous parallelism does not require a global clocking scheme—an advantage in hardware implementations. This type of parallelism is hard to analyze, and only few results are known (Aarts and Korst, 1989).

Combinatorial Optimization and Classification

The ability of a Boltzmann machine to obtain configurations with a large consensus can be used to handle combinatorial optimization and classification problems.

Combinatorial Optimization

A combinatorial optimization (CO) problem can be formalized as a pair $(\mathscr{S}, \mathscr{S}', f)$ where $\mathscr{S}$ denotes a set of *solutions*, $\mathscr{S}' \subseteq \mathscr{S}$ a set of *feasible solutions*, and $f: \mathscr{S} \to \mathbb{R}$ a *cost function*. The problem now is finding a feasible solution with optimal cost. A Boltzmann machine can be used to solve CO problems by defining a correspondence between the configurations of the Boltzmann machine and the solutions of the CO problem in such a way that the cost function of the CO problem is transformed into the consensus function associated with the Boltzmann machine. This can be done by formulating the CO problem as a 0-1 integer programming problem in which the decision variables assume values equal to 0 or 1. The values of the 0-1 variables correspond to the states of the units. The cost function and the constraints of the CO problem are implemented by choosing the appropriate connections and their strengths. In this way, maximizing the consensus in the Boltzmann machine is equivalent to solving the corresponding CO problem. More specifically, it is often possible to construct a Boltzmann machine for which

1. Each locally maximal configuration of the Boltzmann machine corresponds to a feasible solution.
2. The higher the consensus of the corresponding configuration, the better the cost of the corresponding feasible solution.

These properties imply that feasible solutions can be obtained and that configurations with near-maximal values of the consensus function correspond one-to-one to near-optimal solutions of the combinatorial optimization problem. This feature enables Boltzmann machines to be used for approximation purposes, which we have demonstrated for several well-known problems, including the traveling-salesman problem, graph coloring, independent set, and others (Aarts and Korst, 1989). It was found that for graph theoretical problems, such as graph coloring and independent set, the performance was good in the sense that high-quality solutions were obtained within small running times. The performance for the traveling-salesman problem was poor.

Classification

A classification problem can be formalized as a pair $(\mathcal{O}, \mathcal{S})$ where $\mathcal{O}$ denotes a set of *objects* and $\mathcal{S}$ a collection of disjoint subsets $\mathcal{S}_1, \dots, \mathcal{S}_l$ that partitions $\mathcal{O}$. The problem is to determine the subset $\mathcal{S}_j \subset \mathcal{S}$ to which a given object $o \in \mathcal{O}$ belongs. In practice, the set of objects is usually very large, and providing an explicit description of each subset is impracticable. The subsets are therefore often implicitly described by specifying a number of typical examples for each subset. To use a Boltzmann machine for solving classification problems, the set of units is often subdivided into three disjoint subsets, $\mathcal{U}_i, \mathcal{U}_h$ and $\mathcal{U}_o$, denoting the sets of *input*, *hidden*, and *output units*, respectively. Some *input pattern* representing an object is clamped into the states of the input units, i.e., the states of the input unit are fixed to values given by the input pattern. The remaining units then adjust their states to maximize the consensus, subject to the fixed states of the input units. After maximization of the consensus, the states of the output units represent the subset to which the object is thought to belong. In this way, a Boltzmann machine can implement a given input-output function. The hidden units are usually required to implement correctly a given input-output function. The minimum number of hidden units that is needed strongly depends on the intrinsic complexity of the input-output function that is to be implemented (Aarts and Korst, 1989). Choosing appropriate connection strengths for classification problems is often difficult, since different items may give rise to conflicting connection strengths. However, a Boltzmann machine can often acquire a given input-output function by learning from examples, as shown in the next section.

Learning

Learning in a Boltzmann machine takes place by examples that are clamped into the states of the *environmental units*, i.e., the

input and output units. The hidden units are used to construct an internal representation that captures the regularities of the examples clamped into the environmental units. The learning algorithm we discuss starts off by setting all connection strengths equal to zero. Next, a sequence of learning cycles is completed, each consisting of two phases. In the first, the *clamped phase*, examples of a given input-output function are clamped successively into the states of the environmental units, and for each example the Boltzmann machine is equilibrated using the current set of connection strengths; see the earlier section on network dynamics. In the second phase, the *free-running phase*, all units are free to adjust their state, and again the Boltzmann machine is equilibrated. At the end of each learning cycle, the connection strengths are modified using statistical information obtained from the two phases. This process is continued until the average change, over a number of learning cycles, of the connection strengths approaches zero. If the learning is successfully completed, then the Boltzmann machine is able to complete a partial example, i.e., to find the states of all units, given that the states of a subset of the environmental units are clamped to values given by the partial example, by maximizing the consensus.

Extension of the Network Structure

The set of environmental units is denoted by $\mathcal{U}_{io} = \mathcal{U}_i \cup \mathcal{U}_o$. We assume that $|\mathcal{U}_{io}| = m$. An *environmental configuration l* is determined by the states of the units $u \in \mathcal{U}_{io}$. The state of an environmental unit u in an environmental configuration l is denoted by $q_l(u)$. $\mathcal{Q}$ denotes the set of all environmental configurations. With each environmental configuration l, a set $\mathcal{Q}_l \subset \mathcal{R}$ can be associated, consisting of all configurations for which the states of the environmental units are given by l.

The *learning set $\mathcal{T}$* consists of the environmental configurations that can be clamped into the environmental units during learning. Clearly we have $\mathcal{T} \subseteq \mathcal{Q}$. An *input* determines the states of the input units $u \in \mathcal{U}_i$. The set of all possible inputs is denoted by $\mathcal{X}$. Similarly, an *output* is determined by the states of the output units $u \in \mathcal{U}_o$. The set of all possible outputs is denoted by $\mathcal{Y}$. By definition, we have $\mathcal{Q} = \mathcal{X} \times \mathcal{Y}$.

Toward a Learning Algorithm

In this section we discuss the basic elements of the Boltzmann machine learning algorithm proposed by Ackley, Hinton, and Sejnowski (1985). The objective is to modify the connection strengths such that a Boltzmann machine in a free-running phase will have a high probability of being in those environmental configurations that belong to the learning set. To this end, we introduce two probability distributions d and d' defined over the set of environmental configurations. d_l is the probability of obtaining an environmental configuration l in a clamped phase. d'_l is the probability of obtaining l in a free-running phase. The probability distribution d is determined by the environmental configurations in the learning set and by the frequency at which they are clamped into the environmental units. d_l is large if l belongs to the learning set and l is frequently clamped into the environmental units. The probability distribution d' depends on the connection strengths and, as is pointed out below, d' can be defined by using the stationary distribution of Equation 4. The objective can be reformulated as follows: *modify the connection strengths of the Boltzmann machine such that d' is close to d*. If $d \approx d'$, then the Boltzmann machine can determine for a given input the corresponding output. More generally, if a subset of the environmental units are clamped, then the Boltzmann machine can determine the most probable corresponding environmental configuration.

An information-theoretic measure of the distance between the two probability distributions d and d' is the divergence G (Kullback, 1959), which is given by

$$G = \sum_{l \in \mathcal{Q}} d_l \ln \frac{d_l}{d'_l} \tag{5}$$

It can be shown that $G = 0$ if and only if $d = d'$, and $G > 0$ otherwise.

The objective of the learning algorithm can be rephrased as: *minimize G by modifying the connection strengths.* But before we describe how G is minimized, we must readdress the network dynamics discussed earlier. Using Equation 4, we know that after equilibration in a free-running phase the Boltzmann machine tends to be in configurations corresponding to large values of the consensus. If equilibrium is achieved, d'_l is given by

$$d'_l(c) = \sum_{k \in \mathcal{Q}_l} q_k(c) \tag{6}$$

where the $q_k(c)$ are the components of the stationary distribution given by Equation 4. The partial derivative of G with respect to $s_{\{u,v\}}$ can be written as

$$\frac{\partial G}{\partial s_{\{u,v\}}} = \frac{p'_{\{u,v\}} - p_{\{u,v\}}}{c} \tag{7}$$

where $p_{\{u,v\}}$ and $p'_{\{u,v\}}$ denote the probabilities of connection $\{u, v\}$ being activated at equilibrium in the clamped and the free-running phases, respectively; see, e.g., Aarts and Korst (1989). To minimize G, it suffices to collect statistics on $p_{\{u,v\}}$ and $p'_{\{u,v\}}$ and to iteratively change the connection strength proportionally to the difference between the probabilities, i.e.,

$$s_{\{u,v\}} := s_{\{u,v\}} + \eta(p'_{\{u,v\}} - p_{\{u,v\}}) \tag{8}$$

where $\eta \in \mathbb{R}$ is called the *learning parameter.*

Informally speaking, the learning algorithm consists of the following steps:

1. Set all connection strengths to zero.
2. Complete a number of learning cycles until $d \approx d'$, where each learning cycle consists of the following steps:
   (a) Clamped phase: clamp a number of examples into the environmental units, equilibrate, and collect statistics on $p_{\{u,v\}}$.
   (b) Free-running phase: equilibrate and collect statistics on $p'_{\{u,v\}}$.
   (c) Modify connection strengths.

By definition, environmental configurations l not included in the learning set $\mathcal{T}$ have a corresponding probability $d_l = 0$. Consequently, G is not properly defined for these environmental configurations unless $d'_l = 0$, a condition that can only be realized by infinitely large connection strengths, which is unrealistic. To avoid this problem, a *noise ratio* is introduced, allowing the environmental units in the clamped phase to change their states with a certain probability. Furthermore, the use of noise suppresses the unlimited growth of connection strengths. This was found to be essential for obtaining good results.

The average-case performance and time complexity of the learning algorithm depend on the following issues:

1. The parameters of the learning algorithm: the cooling schedule, the learning parameter, the noise ratio, and the number of examples per cycle.
2. The ratio between the environmental and the hidden units.
3. The way the units are connected.
4. The intrinsic hardness of a given classification problem.

Many studies presented in the literature have considered the influence of these issues on the performance of the learning algorithm. Most of the studies are based on experimental evaluations. The overall conclusion is that learning in a Boltzmann machine is rather slow. This is essentially due to the time needed to equilibrate and collect statistics.

Discussion

Compared to Hopfield networks, Boltzmann machines have two major advantages. First, they can escape from poor locally optimal configurations; and second, a powerful learning algorithm exists that can handle hidden layers. A disadvantage is the slow convergence of the self-organization. Speed-up can be achieved by hardware implementations or model extensions. Examples of VLSI implementations are given by Hirai (1993). An optoelectronic implementation is given by Lalanne et al. (1993). Recent model extensions concentrate on deterministic learning algorithms (Meir, 1991) and asynchronous and asymmetric Boltzmann machines (see Ferscha and Haring, 1991, and Apolloni and de Falco, 1991, respectively). Furthermore, extensions of the binary states to multivalued states (Lin and Lee, 1991) and even to continuous states (Beiu et al., 1992) have been investigated.

Road Map: Learning in Artificial Neural Networks, Statistical
Background: I.3. Dynamics and Adaptation in Neural Networks
Related Reading: Computing with Attractors

References

Aarts, E. H. L., and Korst, J. H. M., 1989, *Simulated Annealing and Boltzmann Machines*, Chichester, Eng.: Wiley. ◆
Ackley, D. H., Hinton, G. E., and Sejnowski, T. J., 1985, A learning algorithm for Boltzmann machines, *Cognit. Sci.*, 9:147–169.
Apolloni, B., and De Falco, D., 1991, Learning by asymmetric parallel Boltzmann machines, *Neural Computation*, 3:402–408.
Beiu, V., Ioan, D. C., Dumbrava, M. C., and Robciuc, O., 1992, *Physical Fields Determination Using Continuous Boltzmann Machines*, Zurich: Acta.
Ferscha, A., and Haring, G., 1991, Asynchronous parallel Boltzmann machines for combinatorial optimization: Parallel simulation and convergence, *Meth. Oper. Res.*, 64:545–555.
Hinton, G. E., and Sejnowski, T. J., 1983, Optimal perceptual inference, in *Proceedings of the IEEE Conference on Computer Vision and Pattern Recognition*, New York: IEEE, pp. 448–453.
Hirai, Y., 1993, Hardware implementation of neural networks in Japan, *Neurocomputing*, 5:3–16.
Kullback, S., 1959, *Information Theory and Statistics*, New York: Wiley.
Lalanne, P., Rodier, J.-C., Chavel, P., Belhaire, E., and Garda, P., 1993, Optoelectronic devices for Boltzmann machines and simulated annealing, *Opt. Engrg.*, 32:1904–1914.
Lin, C. T., and Lee, C. S. G., 1991, A multivalued Boltzmann machine, in *Proceedings of the International Joint Conference on Neural Networks*, vol. 3, New York: IEEE, pp. 2546–2552.
Meir, R., 1991, On deriving deterministic learning rules from stochastic systems, *Int. J. Neural Syst.*, 2:283–293.
Zwietering, P. J., and Aarts, E. H. L., 1991, Parallel Boltzmann machines: A mathematical model, *J. Parallel Distrib. Comput.*, 13:65–75.

Cellular Automata

Tommaso Toffoli

Introduction

Cellular automata are abstract dynamical systems that, in discrete mathematics, play a role comparable to that played in the mathematics of the continuum by partial differential equations. In terms of structure as well as applications, they are the computer scientist's counterpart to the physicist's concept of a "field" governed by "field equations." It is not surprising that they have been reinvented innumerable times under different names and within different disciplines; the canonical attribution is to Stanislaw Ulam and John von Neumann, circa 1950; much early material is collected in Burks (1970).

Alan Turing showed that, no matter how complex a computation, it can always be reduced to a sequence of elementary operations chosen from a fixed catalog. For a more complex computation, it is not necessary to expand the catalog with new "specialty" items; a longer sequence of the same old items will suffice. In this sense, Turing had reduced thought to simple, well-understood operations.

Von Neumann was interested in doing the same for life. Conventional models of computation such as the Turing machine make a distinction between the structural part of a computer, which is fixed, and the data on which the computer operates, which are variable. The computer cannot operate on its own matter; it cannot extend or modify itself, or build other computers. In a cellular automaton, objects that may be interpreted as passive data and objects that may be interpreted as computing devices are both assembled out of the same kind of structural elements and subject to the same fine-grained laws; computation and construction are just two possible modes of activity. Using cellular automata, von Neumann was able to show that movement, growth according to a plan, self-reproduction, evolution—life, in brief—can be achieved within a toy world governed by simple local rules (somewhat like a game of checkers); in that world at least, life is in principle reducible to well-understood mechanisms given once and for all. (See Arbib, 1987, for the connections between Turing's and von Neumann's programs.)

The most important role for cellular automata—in biology, computer science, complex systems theory, etc., as well as in physics itself—still remains that of a versatile conceptual tool for reductionism based on physically realistic constraints. Besides this theoretical role, a complementary, practical role is rapidly developing: cellular automata architectures are among the most efficient ways to exploit the technological potential for fine-grained, massively parallel computation.

Structure and Operation

In a cellular automaton, space is represented by a uniform array. To each site of the array, or cell (whence the name *cellular*), there is associated a state variable ranging over a finite set—typically just a few bits' worth of data. Time advances in discrete steps, and the dynamics is given by an explicit rule—say, a lookup table—through which at every step each cell determines its new state from the current state of its neighbors.

Thus, the system's laws are local (no action-at-a-distance) and uniform (the same rule applies to all sites); in this respect, they reflect fundamental aspects of physics. Moreover, they are finitary: even though one may be dealing with an indefinitely extended array (in which case the state space has the cardinality of the continuum, just as with differential equations), the evolution over a finite time of a finite portion of the system can be computed exactly by finite means (unlike with differential equations). A more formal characterization of cellular automata may be given in terms of topological dynamics on the Cantor set (Toffoli, 1984).

As a simple example, consider a two-dimensional array where each cell can be in one of two states—black or white—and can see as a neighborhood the entire 3×3 window centered on and including the cell itself. At each step, let the new state of the cell follow the majority of its neighbors—i.e., become white if at least five of the nine neighbors are currently white, and black otherwise. If we start the array in a random state, with this rule the cellular automaton will in a few steps converge to the state of Figure 1A; after that, no further changes will take place: the system has attained a limit point and is, as it were, in a frozen state.

Freezing may be overcome by injecting some "thermal noise" into the dynamics: when the vote is $4:5$ or $5:4$, i.e., close to a tie, toss a coin instead of following the majority. With this variant of the rule, the separation of the black and white phases will progress indefinitely, yielding ever larger and more rounded domains, as shown in Figure 1. Indeed, there emerge, on a macroscopic scale, well-defined continuum features such as curvature and surface tension. (Substantially the same effect can be achieved with a deterministic variant of the rule: when you are close to a tie as above, let the party with 4 votes—rather than the one with 5—win the contest.)

Our next example is more directly relevant to brain sciences. Let each cell have three states: ready, firing, and recovering. At time $t + 1$, a ready cell will fire with a probability p close to 1 if any of the four adjacent cells (to the north, south,

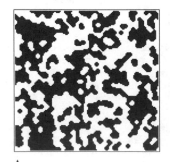

A   B   C

Figure 1. From a random mix of black and white cells, in a few steps the majority rule leads to an attractor consisting of two clearly separated phases (*A*). By adding thermal noise to the majority rule, phase separation proceeds indefinitely (*B* and *C*).

Figure 2. Excitation waves in an excitable medium ("prairie fire"), for different values of the *p* parameter (decreasing from *A* to *C*).

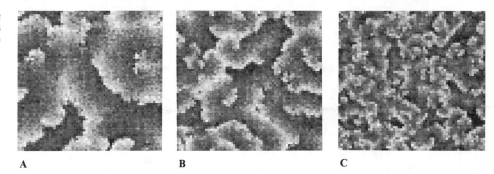

A   B   C

east, and west) was firing at time *t*. After firing, the cell will go into the recovering state, from which at each step it has a probability *q* of returning to the ready state (thus, for small *q*, the average recovery time is of the order of $1/q$ steps). This is a bit like a brain with all neurons and no axons—all action is by immediate physical contact. The rule yields excitation patterns that spread, die out, and revive much like prairie fires, as shown in Figure 2; in this metaphor, *p* represents the "flammability" and *q* the "rate of regrowth" of grass (Schroeder, 1991).

Another excitable medium with a rich phenomenology is Conway's game of "life" (Gardner, 1970), a cellular automaton that in the 1970s attained the status of a cult.

Behavior and Programming

Cellular automata supply useful models for many investigations in natural sciences, combinatorial mathematics, computer science, and artificial intelligence (Farmer, Toffoli, and Wolfram, 1984). One of the most natural and successful applications for cellular automata is the modeling of macroscopic physical phenomena (e.g., fluid dynamics, flow through porous media, phase separation) by reduction to extremely simple fine-grained mechanisms (Vichniac, 1984). Indeed, with current progress in computer technology, reductionistic models of this kind—which are in any case conceptually very attractive—are also becoming computationally accessible. Another natural application is in the early stages of image processing, where the dominant features one has to deal with are short-range spatial correlations, with attendant demands on fine-grained, massively parallel computational methods.

The example of Figure 2 gives evidence that simplicity of structure is not inconsistent with complex behavior. But how does one come up with a cellular automaton that will exhibit a desired behavior? Are there any correspondence rules that allow one to predict the global consequences of given local laws? This is a fundamental issue, on which substantial progress has started accumulating only rather recently.

The first approach, which may be termed *microprogramming*, is a natural extension of the Turing machine philosophy to a parallel computing context. Since each cell of a cellular automaton contains some data and some data-processing machinery, the entire cellular automaton can be viewed as a "tape" covered by a swarm of "heads." In particular, a few bits in the initial state of each cell can be used to instruct the cell to use only a specific part of its machinery, thus effectively yielding different types of cells. In a trivial limit, this approach is equivalent to saying, "Let this cell act as a bit of wire, this one as a NAND gate, this one as empty space; only three or four different kinds of behavior will be necessary. Give me the sche-

matics of any computer, and I will build a similar one within the cellular automaton. Once I've in this way 'transliterated' your schematics, whatever your computer can do, mine will too" (cf. Codd, 1968). Most early work on cellular automata follows this approach.

This approach may be complemented by modulating the activity of the cells in time as well as in space—turning the automaton into an SIMD (single-instruction, multiple-data) computer (Margolus, 1995).

The second approach, emergent computation, is more challenging and, conceptually at least, more rewarding. Here, the basic question is, what aspects of a cellular automaton's microscopic dynamics will be obliterated by coarse-grain averaging, and what aspects instead will make themselves felt all the way to the macroscopic level? For example, with two-state cells, consider a rule that under certain conditions may change a 0 into a 1 or a 1 into a 0, but conserves the overall number of 0s and 1s. A combinatorial argument shows that, for almost all (in a sense that can be made technically precise) such rules, the macroscopic variable "density of 1s" obeys the diffusion equation; in this sense, diffusive behavior is one of the universality classes of emergent computation.

Even though their original motivation was understanding natural systems rather than programming artificial ones, analytical and statistical mechanics have accumulated an enormous expertise with questions of this kind. The fundamental role of conservation laws in physics is well known: they allow one to reach certain universal conclusions regardless of much lack of knowledge in matters of detail. A more recent development is renormalization-group theory (Pfeuty and Toulouse, 1977), which deals with what "you can take with you" and what "you must leave behind" as you gradually move upscale from a microscopic description to a macroscopic one. The reader familiar with the analysis and programming of neural networks is certainly aware of the inroads that statistical mechanical methods have made in that field; with cellular automata, this "technology transfer" is even easier, since the structure of cellular automata has much more to share with the character of physical law.

The scalar quantity "number of 1s" mentioned above is, in the given context, analogous to energy; more specialized forms of detailed balance (analogous to conservation of molecular species, charge, vectorial quantities such as momentum, etc.) lead to a panoply of universality classes of behavior, characterized by power laws, critical exponents, fractal dimensions, and phase transitions. The behaviors that one can "program" in this fashion probably span only a fraction of the repertoire one might wish to achieve; on the other hand, they are accompanied by a guarantee of robustness (against various kinds of noise) and a wealth of theoretical understanding—which is not always the case with more conventional, "microprogrammed"

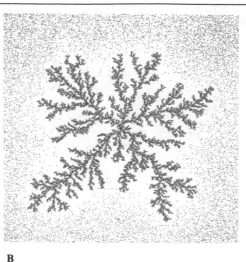

A   B

Figure 3. *A*, Phase separation in three dimensions; the 3D rendering was achieved by simulating the trajectories of light "particles" within the cellular automaton itself. *B*, Dendritic growth fed by diffusing particles, starting from a nucleation center.

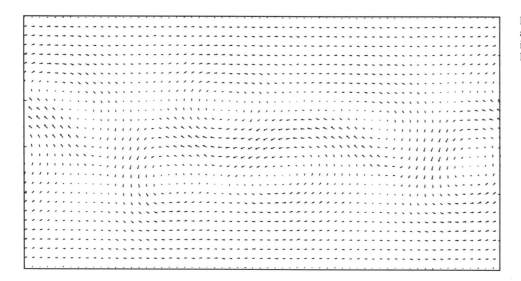

Figure 4. Turbulent counterflow in a channel (Kelvin-Helmholtz shear instability) modeled using the FHP lattice gas.

behavior. Much recent work on cellular automata stresses this approach (Toffoli, 1984; Wolfram, 1986; Toffoli and Margolus, 1987; Gutowitz, 1991).

To illustrate this approach, Figure 3 shows a three-dimensional annealing process analogous to that of Figure 1, and the dendritic growth achieved by a process of diffusion-limited aggregation. In the last decade, fluid dynamics modeling by means of lattice-gas cellular automata has become a burgeoning scientific business (Doolen et al., 1990). Figure 4 shows the macroscopic behavior, rendered by coarse-grain averaging, of a cellular-automaton model of turbulent flow (Margolus, 1995).

Implementation and Efficiency

The cellular-automaton scheme of computation is unique in the way it lends itself to extremely efficient concrete implementations.

From a logic designer's viewpoint, a cellular automaton is a uniform sequential network with polynomial interconnectivity. That is, the number of new nodes one can reach in n steps grows as n^a; by contrast, in an exponential network (e.g., hypercube, binary tree), this number grows as a^n. These con-

straints make possible very direct and efficient physical implementations of the sequential network. Specifically, polynomial growth (with $a = 3$) permits an isometric embedding of the network in ordinary three-dimensional space: nodes that are directly connected are also physically close. With that, since all connections are local, one can clock the network at a rate that is independent of network size; thus, the implementation is indefinitely scalable.

The next few decades will see the number of cells that can be updated at a useful rate by a dedicated cellular automata machine slowly climb from 10^8, now commonplace (Margolus, 1995), to 10^{16}—the Avogadro number "in two dimensions." Thus, we can look forward to simulations that directly span the gap between the microscopic and the macroscopic world. Of course, with such a large number of cells, the system itself will have to be used as a preprocessor for initialization and readout tasks (cf. Figure 3).

Discussion

In brief, cellular automata are a playground where computational theory may discover what aspects of physics are relevant to it, may come to terms with them, and eventually may incorporate them in its axioms. Among these aspects, we have already stressed the connectivity of space-time (which already provides qualitative guidelines for the theory of computational complexity). Another basic aspect of physics that has stimulated much theoretical work in cellular automata is the reversibility of physical law (Toffoli and Margolus, 1990); a computation can be carried out efficiently by a microscopic mechanism only if it matches physics in this respect—i.e., if it is invertible. Finally, interest is currently mounting with respect to another question, namely, to what extent the quantum nature of physics is relevant to computational theory, and to what extent cellular automata may be able to act as a bridge between the two.

More abstractly, cellular automata are a fertile playground for the theory of complex systems, collective phenomena, and emergent computation. This field deals with the aggregate behavior of an astronomical number of simple, identical subsystems; it is really combinatorics with a time dimension added to it—a generalization of statistical mechanics.

Finally, cellular automata are a paradigm for parallel computation much as Turing machines arc for sequential computation. However, while the Turing machine was never expected to be more than a conceptual device, cellular automata may well turn out to provide the most practical blueprint for a parallel computer architecture. Such fine-grained computers may be programmed with "microalgorithms" that explicitly prescribe the state of every cell at every step, as in conventional computers; alternatively, they can be used as collective computation engines, where state variables and interactions are distributed over a large number of cells.

Much as computers can be thought of as unspecialized, programmable machines, perhaps the best characterization of cellular automata is as unspecialized, programmable matter.

Acknowledgment. This work was funded in part by ARPA, under contract N00014−93−1−0660.

Road Map: Dynamic Systems and Optimization
Background: I.3. Dynamics and Adaptation in Neural Networks
Related Reading: Markov Random Field Models in Image Processing; Parallel Computational Models; Pattern Formation, Biological; Self-Reproducing Automata

References

Arbib, M., 1987, *Brains, Machines, and Mathematics*, New York: Springer-Verlag. ◆
Burks, A., 1970, *Essays on Cellular Automata*, Chicago: University of Illinois Press.
Codd, E. F., 1968, *Cellular Automata*, New York: Academic Press.
Doolen, G., et al., Eds., 1990, *Lattice Gas Methods for Partial Differential Equations*, Redwood City, CA: Addison-Wesley.
Farmer, D., Toffoli, T., and Wolfram, S., 1984, *Cellular Automata*, Amsterdam: North-Holland.
Gardner, M., 1970, The fantastic combinations of John Conway's new solitaire game "life," *Sci. Am.*, 223(April):120–123. ◆
Gutowitz, H., 1991, *Cellular Automata: Theory and Experiment*, Cambridge, MA: MIT Press.
Manneville, P., et al., 1989, *Cellular Automata and Modeling of Complex Physical Systems*, Berlin: Springer-Verlag.
Margolus, N., 1995, CAM-8: A computer architecture based on cellular automata, in *Pattern Formation and Lattice-Gas Automata* (A. Lawniczak and R. Kapral, Eds.), Providence, RI: American Mathematical Society.
Pfeuty, P., and Toulouse, G., 1977, *Introduction to the Renormalization Group and Critical Phenomena*, New York: Wiley.
Schroeder, M., 1991, *Fractals, Chaos, Power Laws*, New York: Freeman. ◆
Toffoli, T., 1984, Cellular automata as an alternative to (rather than an approximation of) differential equations in modeling physics, *Physica D*, 10:117–127. ◆
Toffoli, T., and Margolus, N., 1987, *Cellular Automata Machines—A New Environment for Modeling*, Cambridge, MA: MIT Press. ◆
Toffoli, T., and Margolus, N., 1990, Invertible cellular automata: A review, *Physica D*, 45.229–253.
Vichniac, G., 1984, Simulating physics with cellular automata, *Physica D*, 10:96–116.
Wolfram, S., Ed., 1986, *Theory and Applications of Cellular Automata*, Singapore: World Scientific.

Cerebellum and Conditioning

Gábor T. Bartha and Richard F. Thompson

Introduction

A large number of experiments now indicate that the cerebellum is involved in learning and performance of classically conditioned, somatic motor reflexes (Thompson, 1990). These results appear to be general for mammals, including humans. Conditioning studies have been done for many movements, including limb flexion, neck turns, and facial movements. Most studies have used rabbit eyelid conditioning, so we focus on this. For the rabbit preparation, the conditioning stimulus (CS) is usually a tone or light, and the unconditioned stimulus (US) is usually a corneal airpuff or shock. The US evokes a defensive blink reflex or unconditioned response (UR) that includes both eyelid closure and nictitating membrane (third eyelid) extension. The CS does not evoke a blink initially, but after repeated pairings of the CS followed by the US, rabbits eventually learn to blink in response to the CS, and this is called the conditioned response (CR).

By varying stimulus parameters, a large number of behavioral-conditioning phenomena can be observed (Gormezano, Kehoe, and Marshall, 1983). Blink conditioning can occur with CS-US interstimulus intervals (ISIs, defined by stimulus onsets) ranging from 100 milliseconds to well over 1 second. The CS may overlap the US (delay conditioning) or there may be an interval between the end of the CS and the start of the US (trace conditioning). The rate of learning and asymptotic response level are optimal at an ISI of about 250 ms for delay conditioning. After 100–200 training trials at the optimal ISI, rabbits give CRs on more than 90% of the trials. The CR onset initially develops near the US onset and gradually begins earlier in the trial over the course of training. The CR generally peaks near the US onset (Figure 1). If a trained rabbit is repeatedly given CS-alone trials, CRs will decrease and eventually extinguish altogether. Reacquisition after extinction is fast, usually occurring in less than 20 trials. More complex conditioning phenomena such as blocking and conditioned in-

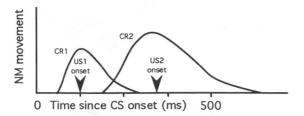

Figure 1. Simulated examples of how nictitating membrane (NM) CRs change as a function of training with different ISIs.

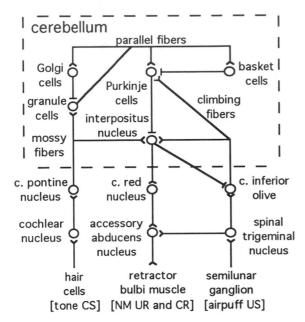

Figure 2. Neural network for nictitating membrane (NM) URs and CRs. The US-UR pathway consists of the semilunar ganglion, spinal trigeminal nucleus, and accessory abducens nucleus. The CS-CR pathway consists of auditory hair cells, cochlear nucleus, pontine nucleus, cerebellum, red nucleus, and again the accessory abducens nucleus. The accessory abducens nucleus contains retractor bulbi motoneurons. The reinforcing pathway begins with US information in the spinal trigeminal nucleus and consists of the inferior olive and its projections to the cerebellum. The interpositus nucleus provides inhibitory feedback to the inferior olive. Angled synapses are excitatory; straight synapses are inhibitory. The abbreviation *c.* means contralateral.

hibition can be observed when more than one CS is used (see CONDITIONING).

The behavioral data inspire the following questions: What brain structures are responsible for each of the behavioral phenomena? What neural computations are being performed by each brain structure? Where is plasticity occurring? and What are the mechanisms of plasticity? To answer these questions, extensive anatomical, lesion, and physiological studies have been undertaken (Thompson, 1990). This research has identified much of the neural network subserving URs and CRs in rabbit, and this is summarized in Figure 2. Note that the US and CS information must have pathways that bring them together for associative learning of the CR to occur. Evidence points to the cerebellum as the site of convergence and to the olivary climbing fiber system as the essential reinforcing pathway. Also note that the US elicits the UR and provides reinforcement using distinct pathways. The interpositus provides

inhibitory feedback to the inferior olive that increases during the course of learning, which is consistent with the view that the inferior olive represents an error signal. It is worth emphasizing that, for blink and related forms of conditioning, the neural substrates shown in Figure 2 constrain what can be learned.

Although a large body of data has been collected, none of the preceding questions have been completely answered. While experiments appear to have answered most questions concerning the neural network subserving delay conditioning, much less is known about structures responsible for other conditioning phenomena. For example, the hippocampus appears to be important for long ISI trace conditioning (> 500 ms) but not for delay conditioning or short ISI trace conditioning. The data also strongly suggest that the cerebellum is the primary site of plasticity; however, the respective contributions of the cortex and deep nuclei are not known. Least is known about the neural computations performed and mechanisms of plasticity. The complexity of behavioral and neurobiological data makes modeling a valuable tool.

Modeling the Role of the Cerebellum in Classical Conditioning

Next, we describe several models of cerebellum that attempt to account for classical conditioning. In particular, we discuss what data the models account for, what mechanisms they use, and what limitations they have. Most models focus on accounting for CR timing because it is adaptive and provides clues to the underlying neural computations.

Marr (1969) and Albus (1971) proposed similar theories of cerebellar cortex that attempt to account for its intriguing anatomy and physiology. Their most controversial idea, that parallel fiber synapses on Purkinje cells are modifiable, has considerable experimental support. (For the basic cell types and fibers of cerebellum, see Figure 2, and CEREBELLUM AND MOTOR CONTROL.) Albus also anticipated the discovery of long-term depression (LTD) by predicting a decrease in parallel fiber synaptic strength (see LONG-TERM DEPRESSION IN THE CEREBELLUM). Both theories assume some form of expansion encoding of mossy fiber activity in the Golgi–granule cell network (Figure 2), and that Purkinje cells learn to respond to these patterns, thereby generating movements. Although the theories focus mainly on learning voluntary movements, they also include learning of conditioned reflexes. Marr considered reflexes for maintenance of posture. Albus predicted explicitly that climbing fiber spikes are the US and mossy fiber activity patterns are the CS for Purkinje cell conditioning. This prediction is supported by eyelid conditioning studies. However, neither theory accounts for conditioned reflex phenomena such as rate of acquisition, optimal ISI, and CR timing.

Moore, Desmond, and Berthier (1989) use the approach of taking a neural network model derived from functional considerations and mapping its computations onto the cerebellum. The major goal of their model is to account for CR timing, but it also addresses other conditioning phenomena. In their model, each CS stimulus trace is internally represented as a spectrum of time delays using a tapped delay-line. Stimulus traces are activated by both CS onsets and offsets and make contact by modifiable connections with so-called V- and E-units. The CS also generates a *CR image*, which is a short-latency increase to a plateau of activity that lasts as long as the US. The V-unit is the output element, and the E-unit acts to gate learning in the V-unit near the expected time of the US. In the cerebellar mapping of Moore et al., Purkinje cells are the V-units and Golgi cells are the E-units; both units can learn with conjoint parallel-climbing fiber activity. The delay-line structures are not specified, but delays occur before the parallel fibers outputs. Moore et al. suggest that the CR image is learned in the brainstem in parallel with learning in the cerebellar cortex. The Golgi cell gates the association of the CR image with the Purkinje cell. Moore et al. assume that release of interpositus nucleus inhibition by Purkinje cells is responsible for generation of CRs.

Purkinje and Golgi cell plasticity combined with the model's tapped delay-lines yields anticipatory CRs for delay and trace conditioning at all ISIs. Conditioned responses initially develop near the US onset, begin earlier in trial over the course of training, and peak near the US onset as observed experimentally. The model also reproduces discrete shifts in CR timing and double-peaked CRs with mixed ISI training. The learning rules follow the form of Rescorla and Wagner's model (1972) and thus account for some of the same stimulus context effects, such as blocking and conditioned inhibition. The Rescorla-Wagner formula is $\Delta w_i = \alpha_i(\lambda - \Sigma w_i x_i)$, where w_i are CS associative weights, α_i are CS-dependent learning rate parameters, λ indicates the presence of the US, and x_i are CS activities. Moore et al. have added time dependency and other complexities to the α_i in their learning rules. Symmetric treatment of CS onsets and offsets accounts for observations of CS onset and offset conditioning. In trace conditioning, the model predicts that, if CS duration is lengthened on probe trials, then offset associations will generate a second CR peak delayed by the amount of time that the CS was lengthened. This prediction has been confirmed experimentally (Desmond and Moore, 1991).

The model does not include learning of the hypothesized CR image, nor is there experimental support for its existence. Contrary to observations, in the model, the duration of the CR after training is independent of ISI because the duration of CS onset and offset processes are fixed at 100 ms. This also limits the amount of time that the CR may anticipate the US. Experimental evidence suggests that the hippocampus is required for long ISI trace conditioning, but this model accounts equally well for delay and trace conditioning without including a model of the hippocampus. Finally, the proposed delay-lines must be more than 1 second long, and no physiologically plausible mechanism has been suggested for this.

Gluck et al. (in press) present a lumped model of the cerebellum and the inferior olive. The main approach of this model, as for the model by Moore et al. (1989), is to take an algorithm of conditioning and map its computations onto the cerebellum. The algorithm is also based on the Rescorla-Wagner model and adds higher-order input. The mapping includes negative feedback from cerebellum to inferior olive and a recurrent cerebellar connection. Cerebellar output is identified with the CR. It is assumed that Purkinje cell dendrites transform CS input and CR information from the recurrent connection into a higher-order combination of CS and CR input. Negative feedback to the inferior olive implements the subtractive term, $(\lambda - \Sigma w_i x_i)$, in the Rescorla-Wagner formula. This allows plasticity for the lumped cerebellum node to follow a Hebbian rule conjoining cerebellar inputs and inferior olive activity.

The model qualitatively accounts for cerebellar output after delay conditioning for a short range of ISIs. Peak CRs occur near US onset, but this does not account for output pathway delays. Consistent with behavioral data, CRs are broader for longer ISIs, but the delay in CR onset does not increase at longer ISIs as it should. The model accounts for extinction and reacquisition that is faster than initial acquisition, but reacquisition is slower than in rabbits. The model also accounts for blocking and conditioned inhibition but is unable to account for trace conditioning or even long ISI delay condition-

ing. Gluck et al. (in press) assume, in accord with lesion data, that the hippocampus is required for long ISI trace conditioning, but this is not the case for shorter ISIs. Since the CR is driven by the CS, removal of the CS quickly terminates the CR, a result inconsistent with experiment. Early in training, the model appears to produce CRs that peak too early in the trial and only later peak near the US.

Jaffe (1990) proposed that single interpositus neurons generate the full range of delays for CRs by adjusting their input weights and by exploiting the phenomenon of delayed inhibitory rebound. Delayed inhibitory rebound is similar to post-inhibitory rebound except that there may be a delay as long as several hundred milliseconds before rebound firing. This phenomenon depends on a low-threshold calcium-dependent conductance (Llinás and Mühlethaler, 1988). In Jaffe's model, the cerebellar cortex is represented as lumped populations of on- and off-beam Purkinje cells that project to a single interpositus nuclear cell. (A "beam" is a set of parallel fibers whose width equals the dendritic spread of a Purkinje cell.) The delayed response is approximately an exponential function of the connection strength between the off-beam Purkinje cell population and the interpositus neuron.

Perhaps the most significant problem is that Jaffe's interpositus cell model must be quiescent at CS onset and cannot fire throughout the delay period for the delayed rebound mechanism to work. Most neurons in the interpositus are spontaneously active, including neurons that show behaviorally related activity. It seems unlikely that a few neurons silent during the delay can account for most of the CR by relaying their rebound activity to other neurons in the interpositus. Finally, in contrast to the discrete shift observed experimentally (Gormezano, Kehoe, and Marshall, 1983), the continuous range of each cell would result in a continuous shift in CR timing when the ISI is shifted.

Buonomano and Mauk (1994) present a semirealistic neural network model of cerebellar cortex consisting of 10,000 granule cells, 900 Golgi cells, and one Purkinje cell. A conductance-based neuron model is used for all cells, and convergence and divergence ratios are approximated. The main idea of the network model is that different granule cell populations code for different times. A CS constantly activates a subset of mossy fibers and these excite a subset of granule cells. Both mossy fibers and granule cells excite Golgi cells. Inhibitory feedback from Golgi cells to granule cells is instrumental in varying the subset of granule cells active at a given time. Simulations demonstrate that by weakening granule cell synapses active during a US, Purkinje cells learn to decrease their activity around the time of the US for a wide range of ISIs. Lesion studies also provide experimental support for a cerebellar cortical role in CR timing (Perrett, Ruiz, and Mauk, 1993).

This model's greatest problem is extreme sensitivity to noise or input variability. For example, any change in the subset of active mossy fibers or their pattern of activity would disrupt Purkinje cell timing by changing the active subsets of granule cells. The model also depends on mossy fibers to drive the granule cells throughout the period of the CR and so cannot account for trace conditioning. There is some evidence that CSs are primarily represented as onset or offset responses on the mossy fibers even for delay conditioning. To account for delays in the motor output pathway, Purkinje cell activity should decrease further in advance of the US than demonstrated by the model. Although the neural model used produces spike trains, comparisons with single-unit recordings of cerebellar neurons were not made.

Bartha (1992) stresses the importance of neural input/output representations by (a) use of neural recordings to constrain mossy fiber input and (b) studies of a detailed CR pathway and oculomotor plant model to constrain cerebellar output. The oculomotor plant model shows that the relation between neural firing and movement can be highly nonlinear (Bartha and Thompson, 1992). The cerebellar cortex model consists of a Purkinje cell, Golgi cell, 200 granule cells, 50 mossy fiber inputs, and 10 basket cell inputs. The divergence and convergence of the connections are scaled in proportion to the level of scaling of the neural populations. Neural activity is modeled as spike trains using conductance-based, single-compartment neuron models.

The primary hypothesis of Bartha (1992), supported by early experimental work (Eccles, Ito, and Szentágothai, 1967), is that Golgi cells inhibit granule cells for periods of tens to hundreds of milliseconds. Purkinje cell plasticity such as LTD can selectively decrease the synaptic weights from granule cells that are active around the time of the US. In this way, Purkinje cells can learn to reduce their firing near the time of the US and thus release inhibition on interpositus cells, which in turn may drive the CR. With realistic single-neuron parameters, the model is able to reproduce many aspects of CR timing and form, including CR onset, time to peak, peak amplitude, and time-amplitude area, as well as single neural unit activity for ISIs less than 500 ms. A major limitation of the model is that it offers little control over CR onset. This makes it difficult for the Purkinje cell to select appropriately timed granule cells because it may select all cells that depress for a longer duration than the ISI. In addition, the model cannot account for the decrease in CR onset time during conditioning.

Conclusions and Future Directions

No model accounts for all the data on cerebellar neurobiology and conditioning, even when restricted to delay conditioning. Nevertheless, computer models are beginning to influence experimental work (Desmond and Moore, 1991; Perrett, Ruiz, and Mauk, 1993; Tracy and Thompson, unpublished data). As models become more accurate, the cycle of interaction with experiment will tighten. To test detailed models, more technically challenging experiments are needed. For example, both Buonomano and Mauk (1994) and Bartha (1992) have suggested single-unit recordings from granule cells. These recordings are difficult because granule cells are small and closely spaced, and many samples are needed. Buonomano and Mauk (1994) predict that a subset of granule cells will exhibit non-periodic activity, and Bartha (1992) predicts that the subset of granule cells will primarily show depressions in activity at all conditionable intervals after the presentation of a CS. We look forward to experiments that will test these and future model predictions and to models that incorporate more neurobiological data.

Acknowledgments.   This work was supported by NRC Associateship Award to G. T. Bartha and ONR Grant (N00014–88–K–0112) to R. F. Thompson.

Road Maps: Learning in Biological Systems; Mammalian Brain Regions
Related Reading: Cerebellum and Motor Control

References

Albus, J. S., 1971, A theory of cerebellar function, *Math. Biosci.*, 10: 25–61.
Bartha, G. T., 1992, A computer model of oculomotor and neural contributions to conditioned blink timing, Doctoral Dissertation, University of Southern California.
Bartha, G. T., and Thompson, R. F., 1992, Control of rabbit nictitating membrane movements: II. Analysis of the relation of motoneuron activity to behavior, *Biol. Cybern.*, 68:145–154.
Buonomano, D. V., and Mauk, M. D., 1994, Neural network model of the cerebellum: Temporal discrimination and the timing of motor responses, *Neural Computat.*, 6:38–55.
Desmond, J. E., and Moore, J. W., 1991, Altering the synchrony of stimulus trace processes: Tests of a neural-network model, *Biol. Cybern.*, 65:161–169.
Eccles, J. C., Ito, M., and Szentágothai, J., 1967, *The Cerebellum as a Neuronal Machine*, New York: Springer-Verlag.
Gluck, M. A., Goren, O. A., Myers, C. E., and Thompson, R. F., in press, A higher-order recurrent network model of the cerebellar substrates of response timing in motor-reflex conditioning, *J. Cognit. Neurosci.*
Gormezano, I., Kehoe, E. K., and Marshall, B. S., 1983, Twenty years of classical conditioning with the rabbit, *Prog. Psychobiol. Physiol. Psychol.*, 10:197–275. ◆
Jaffe, S., 1990, A neuronal model for variable latency response, in *Analysis and Modeling of Neural Systems* (F. H. Eeckman, Ed.), Boston: Kluwer, pp. 405–410.
Llinás, R., and Mühlethaler, M., 1988, Electrophysiology of guinea-pig cerebellar nuclear cells in the *in vitro* brain stem–cerebellar preparation, *J. Physiol.*, 404:241–258.
Marr, D., 1969, A theory of cerebellar cortex, *J. Physiol.*, 202:437–470.
Moore, J. W., Desmond, J. E., and Berthier, N. E., 1989, Adaptively timed conditioned responses and the cerebellum: A neural network approach, *Biol. Cybern.*, 62:17–28.
Perrett, S. P., Ruiz, B. P., and Mauk, M. D., 1993, Cerebellar cortex lesions disrupt learning-dependent timing of conditioned eyelid responses, *J. Neurosci.*, 13:1708–1718.
Rescorla, R. A., and Wagner, A. R., 1972, A theory of Pavlovian conditioning: Variations in the effectiveness of reinforcement and nonreinforcement, in *Classical Conditioning II* (A. Black and W. Prokasy, Eds.), New York: Appleton-Century-Crofts, pp. 64–99.
Thompson, R. F., 1990, Neural mechanisms of classical conditioning in mammals, *Philos. Trans. R. Soc. Lond. [Biol.]*, 329:161–170. ◆

Cerebellum and Motor Control

Mitsuo Kawato

Introduction

Fast, smooth, and coordinated movements in humans and animals cannot be achieved by feedback control alone because delays associated with feedback loops are long (about 200 ms for visual feedback and 100 ms for somatosensory feedback) and feedback gains are low in biological motor control systems (see MOTOR CONTROL, BIOLOGICAL AND THEORETICAL). Additionally, feedback controllers, such as the commonly used PID (proportional, integral, and derivative) controllers, do not incorporate predictive dynamic and/or kinematic knowledge of controlled objects or environments. Feedforward control ex-

Figure 1. Schematic diagram of a neural circuit for voluntary-movement learning control by the cerebro-cerebellar communication loop. The lateral part of the cerebellum hemisphere is shown. Although inputs and outputs of other cerebellar regions are vastly different, the neural circuit of the cerebellar cortex is rather uniform. It must be emphasized that only one climbing fiber makes contact with a single Purkinje cell, whereas parallel fibers account for 200,000 synapses on a single Purkinje cell. Abbreviations: CF, climbing fiber; BC, basket cell; GO, Golgi cell; GR, granule cell; MF, mossy fiber; PC, Purkinje cell; PF, parallel fiber; ST, stellate cell; DE, dentate nucleus; IO, inferior olivary nucleus; PN, pontine nuclei; RNp, parvocellular red nucleus; VL, ventrolateral nucleus of the thalamus.

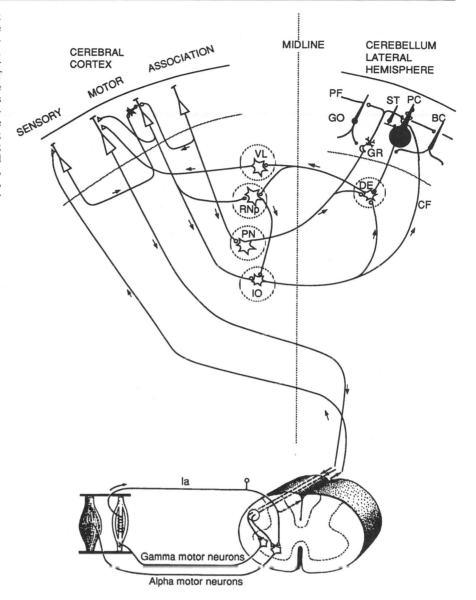

plicitly incorporates predictive internal models and appears to be essential at least for relatively fast movements. The internal models in the brain must be acquired through motor learning in order to accommodate the changes that occur with the growth of controlled objects such as hands, legs, and torso, as well as the unpredictable variability of the external world.

Where in the brain are internal models of the motor apparatus likely to be stored? First, the locus should exhibit a remarkable adaptive capability, essential for acquisition and continuous update of internal models of the motor apparatus. A number of physiological studies have suggested important functional roles of the cerebellum in motor learning and remarkable synaptic plasticity in the cerebellar cortex (Ito, 1984, 1989; see also CEREBELLUM AND CONDITIONING). Second, biological objects of motor control by the brain, such as arms, speech articulators, and the torso, possess many degrees of freedom and complicated nonlinear dynamics. Correspondingly, neural internal models should receive a broad range of sensory inputs and possess a capacity high enough to approximate complex dynamics.

Extensive sensory signals carried by mossy fiber inputs and an enormous number of granule cells in the cerebellar cortex seem to fulfill these prerequisites for internal models. (See Figure 1 for cerebellar circuitry and its connection to the cerebellar nucleus in the case of the lateral cerebellum.) The cerebellar cortex is divided into longitudinal microzones 200 μm wide and more than 50 mm long. Finally, the cerebellar symptoms usually classified as the "triad" of hypotonia, hypermetria, and intention tremor could be understood as degraded performance when control is forced to rely solely on negative feedback after internal models are destroyed and/or cannot be updated. This is because precise, fast, and coordinated movements can be executed if accurate internal models of the motor apparatus can be utilized during trajectory planning, coordinate transformation, and motor control, while pure feedback controllers with long feedback delays and small gains can attain only poor performance in these computations and usually lead to oscillatory instability for forced fast movements.

Miall et al. (1993) identified several classes of theories regarding the role of the cerebellum in the control of visually guided movements (such as coordination by Flourens, comparators by Holmes, gain controllers, and associative learning). The most complete class of theories, these authors suggested, comprises theories in which the cerebellum forms an internal model of the motor system; such theories can encompass all the alternative theories, while fitting many of the known facts of cerebellar organization. These theories require that the cerebellum be an adaptive system capable of learning and of updating a model as the behavior of the motor system changes. They also require that the cerebellum store relevant parameters of the motor system, as these parameters form part of the description of the motor system's behavior. Another requirement is timing capabilities: the motor system is dynamic, so a useful model will also need dynamic, i.e., time-dependent, behavior. Such models also account for the connectivity of the cerebellum at a reasonable level of detail, and they can account as well for many of the symptoms of cerebellar disorder (Miall et al., 1993). How might internal models be acquired in the cerebellum through motor learning?

Marr-Albus Model and Synaptic Plasticity

Marr (1969) and Albus (1971) proposed a detailed model of the cerebellum, according to which the cerebellum can form associative memories between particular patterns on parallel fiber inputs and Purkinje cell outputs. The basic idea is that the parallel fiber–Purkinje cell synapses can be modified by input from the climbing fibers. In the perceptron models, the efficacy of a parallel fiber–Purkinje cell synapse is assumed to change when there exists a parallel fiber and climbing fiber input conjunction. The presence of the putative heterosynaptic plasticity of Purkinje cells was demonstrated as a long-term depression (LTD) (Ito, 1989). An associative LTD found in Purkinje cells can be modeled as the following heterosynaptic plasticity rule: the rate of change of the synaptic efficacy of a single parallel fiber synapse is proportional to the negative product of the firing rate of that synapse's input and the increment of the climbing fiber firing rate from its spontaneous level:

$$\tau \, dw_i/dt = -x_i(F - F_{\text{spont}}) \tag{1}$$

where τ is the time constant, w_i is the synaptic weight of the ith parallel fiber–Purkinje cell synapse, x_i is the firing frequency of the ith parallel fiber–Purkinje cell synapse, F_{spont} is the firing frequency of the climbing fiber input, and F is its spontaneous level. This single rule reproduces both the LTD and the long-term potentiation (LTP) experimentally found in Purkinje cells. When the climbing fiber and the parallel fiber are simultaneously stimulated, the parallel fiber synaptic efficacy decreases. In contrast, the parallel fiber synaptic efficacy increases when only the parallel fiber is stimulated (the climbing fiber firing frequency is lower than its spontaneous level).

Model for Adaptive Modification of VOR

Cerebellar motor learning has been most intensively studied in the vestibulocerebellum using the vestibulo-ocular reflex (VOR). When the head is turned, the VOR normally acts to stabilize retinal images by generating smooth eye movements that are equal and opposite to the rotary head movements. Under experimental manipulations of retinal slips using magnifying spectacles, inversion prisms, or a rotating visual screen, the gain of the VOR (the ratio of eye to head movement) changes (see VESTIBULO-OCULAR REFLEX: PERFORMANCE AND PLASTICITY). Such VOR adaptation is abolished when the

flocculus is destroyed in cats, rabbits, and monkeys, or when only the visual climbing fiber pathway is destroyed in rabbits (see Ito, 1984, 1989).

Fujita (1982) expanded the basic Marr-Albus model to incorporate dynamic responses, and he proposed that the cerebellum is an adaptive filter that can learn to compensate for the dynamical characteristics of the oculomotor plant. He simulated adaptive modification of the VOR based both on LTD and on the hypothesis that the synapses between parallel fibers and Purkinje cells of the cerebellar flocculus are the sites of modification. Climbing fiber activities are assumed to reflect motor performance error information conveyed by retinal slip velocity.

Until recently, there was no direct evidence as to the existence of inverse dynamics models of controlled objects (Atkeson, 1989; see also SENSORIMOTOR LEARNING) in the cerebellum. One of the most compelling pieces of circumstantial evidence is the known neural circuit for the VOR.

The horizontal VOR is regulated by a three-neuron reflex arc as well as by the microzone of the flocculus called the H-zone (Ito, 1984), which also participates in the control of the optokinetic response (OKR). The OKR moves the eye in the direction of visual field motion to stabilize the retinal image, and it shares major neural mechanisms with the VOR. When the head is rotated to the left, the semicircular canal sends the head rotational velocity to the vestibular nucleus and the flocculus, and the eye is rotated to the right. If the VOR is not perfect, images move on the retina, and the retinal slip information is sent back to both the vestibular nucleus and the flocculus by the climbing and mossy fibers. This retinal slip information is the sensory input used by the OKR.

Figure 2A is a schematic neural circuit, and Figure 2B is its block diagram. This is a model of only the microzone of the flocculus (the H-zone) which is related to the horizontal VOR and OKR. Here, θ_h, θ_e, and θ_{ext} denote the head rotation angle, the eye-rotation angle, and the rotation angle of the external world, while u is the motor command sent to the muscles. Superscripts w, r, c, and m indicate variables represented in world, retinal, canal, and muscle coordinates. The term s indicates the Laplace operator, and $1/s$ corresponds to integration. T_{wc} and T_{wr} represent coordinate transformations from the world frame to the canal frame and to the retinal frame, respectively. The vestibular organ sends the head-rotation velocity in canal coordinates, θ_h^c, to the vestibular nucleus and the flocculus. The forward dynamics of the eyeballs are represented by the operator P.

The semicircular canal outputs the head velocity signal, which is the negative of the desired eye velocity for a perfect VOR. On the other hand, in motor neurons of extraocular muscles, firing frequencies represent necessary muscle forces which are roughly linear combinations of the eye position, eye velocity, and eye acceleration signals. Thus, between the vestibular output and the motor neuron, the inverse dynamics computation P^{-1} must be solved. The three-neuron reflex arc constitutes the main part of this inverse dynamics computation. Because the cerebellar flocculus constitutes a side path within the neural circuit that executes inverse dynamics transformation, it must provide a part of the inverse dynamics model of the eye plant. Furthermore, because characteristics of the side path are variable because of Purkinje cell plasticity, the side path can provide additional and necessary transfer characteristics (e.g., decrease or increase of VOR gain in the simplest case) so that the summation of the side path and the main reflex arc provides the complete inverse dynamics transformation (e.g., gain 1 in the simplest case).

Figure 2. A neural circuit diagram and a block diagram for horizontal VOR and OKR. *A*, A schematic diagram of a neural circuit for horizontal VOR and OKR. *B*, A block diagram of adaptive modification of VOR and OKR by the flocculus. Information flow via the external world is shown by dashed lines. Neural pathways are shown by solid lines. The motor command positive feedback pathway is shown by a broken line (long and short dashes) to indicate that origins in the brainstem are not clear.

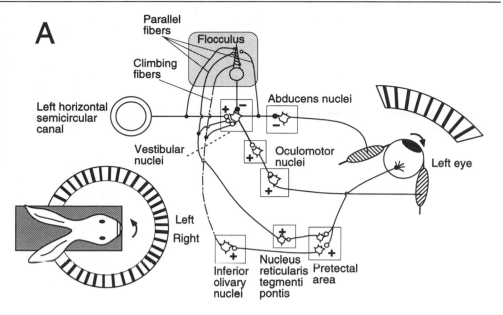

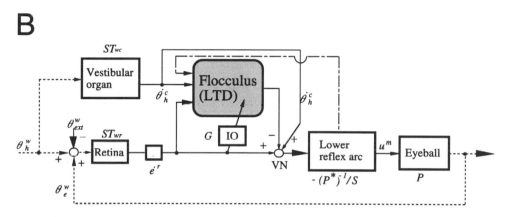

Figure 3. Components of cerebellar-related circuitry and their interconnections, as conceived in Boylls's theory of cerebellar function in mesencephalic locomotion (cited from Szentágothai and Arbib, 1975). Topographic interconnections between a microzone of the cerebellar cortex and the patch of cerebellar nucleus and precerebellar nuclei (microcomplex of Ito, 1984) are schematically depicted by limited divergence and convergence of neural connections (arrows).

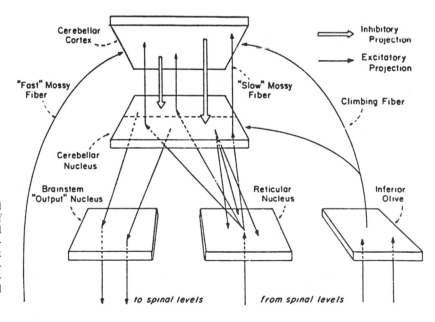

Models of Limb Motor Control in the Cerebellum

Boylls (1975) proposed that the spatiotemporal neural firing patterns formed by the excitatory loop of the cerebellar reverberating circuit and the inhibitory loop via the Purkinje cells are computationally beneficial for generation of rhythmic interlimb coordination patterns in locomotion. Tsukahara et al. (1983) physiologically and anatomically demonstrated positive feedback loops formed between cerebellar deep nuclei and brainstem nuclei, naming them the *cerebellar reverberating circuit.* In Boylls' theory the purpose of cerebellar computation is to create synergically meaningful excitation profiles on a cerebellar nucleus; these are subsequently transmitted via an "output nucleus" to spinal levels (Figure 3). The principal instrument of this pattern sculpting is spatiotemporally significant inhibition from the cerebellar cortex (Szentágothai and Arbib, 1975).

Boylls' model was recently expanded to accommodate motor learning in the cerebellum by Houk and Barto (1992) as an adjustable pattern generator (APG) model of the cerebellum. Temporal patterns of movements are acquired through motor learning, based on the LTD of Purkinje cells in combination with the reverberating circuit. Artificial neural network models with recurrent connections that can learn and generate arm trajectories (see SENSORIMOTOR LEARNING) were the computational base of Houk and Barto's model. The learning scheme proposed is mathematically based on associative reward-penalty learning. One of the model's attractive features is that a motor pattern is selected and generated—an impossibility with a simple internal forward or inverse model. Correspondingly, however, learning is more difficult.

Miall et al. (1993) proposed that the cerebellum forms two types of internal models. One is a forward model of the motor apparatus. The second is a model of the transport time delays in the control loop (attributable to receptor and effector delays, axonal conductance, and cognitive processing delays). The second model delays a copy of the prediction made by the first model, so that it can be compared in temporal registration with actual sensory feedback from the movement. The second model resolves the difficulty of temporal mismatching between the sensory signal delayed by the feedback loop and the output calculated by internal models. In manual tracking of visual targets by humans and primates, the control performance became noisier and more unstable when an extra time delay was inserted before the visual target presentation. Several experimental phenomena like this were reproduced by the model.

My colleagues and I proposed a model of cerebellar motor learning (Figure 4) based on the feedback-error-learning scheme (Kawato and Gomi, 1992). The premotor networks correspond to a feedback controller, the cerebellar cortex to an inverse model, climbing fiber inputs to the error signal, and, finally, the combination of motor units, the environment, and the sensory receptors to the controlled object. The parallel fiber inputs are assumed to carry the desired motor pattern information as well as the current state of the motor apparatus. We assume that climbing fiber responses represent motor commands generated by some of the premotor networks, i.e., networks that are upstream of the motor neurons and include feedback controllers at the spinal, brainstem, and cerebral levels (including the motor cortex). This assumption by no means contradicts the known fact that many climbing fibers carry sensory information. It is assumed that sensory information is represented in the motor-command-error coordinates. We do not necessarily assume that every premotor network literally compares a "desired trajectory" with an "actual trajectory." It is required, however, that the premotor network calculate the motor error in motor-command coordinates which vanishes

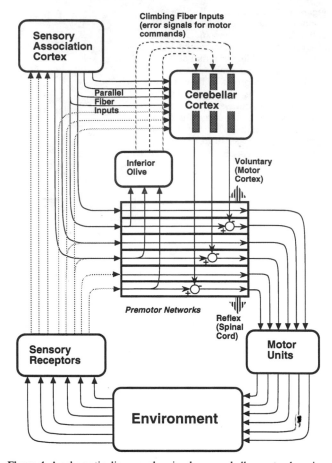

Figure 4. A schematic diagram showing how cerebellar motor learning might be incorporated in sensorimotor control. Premotor networks are motor control networks that are upstream of the motor neurons. They range in complexity from simple spinal reflexes (bottom of large rectangle) to motor cortical circuits controlling voluntary movements (top of large rectangle). Thus, the top rectangle is assumed to contain the motor cortex. Some of the premotor networks are under the inhibitory influence of the cerebellar cortex. Parallel fibers to Purkinje cells carry vast amounts of information both from the sensory receptors and from the cerebrum; this information is necessary for coordinative and predictive control. Some of the parallel fibers represent desired motor pattern information and some represent current state information of the motor apparatus, regardless of whether they originate from the sensory association cortex (solid curves) or from the sensory receptors (dashed curves). The climbing fiber input is assumed to carry motor error signals represented in motor-command coordinates. This is achieved by a closed loop and a one-to-one anatomical correspondence between each premotor network, the small region in the inferior olive, and the microzone of the cerebellar cortex.

when the resultant movement is desirable. Based on the LTD in Purkinje cells, each longitudinal microzone in the cerebellar cortex, in conjunction with a small portion of the deep cerebellar nucleus that is connected to the microzone, learns to execute predictive and coordinative control of different types of movements. This is achieved by a closed loop and one-to-one anatomical correspondence between each premotor network, a small region in the inferior olive, and a microzone of the cerebellar cortex. If one premotor network is regulated by one microzone of the cerebellar cortex, then the latter must receive climbing fiber inputs from the specific part of the inferior olive that receives inputs from the earlier part of the same premotor network. With this anatomical organization, the cerebellar microzone is trained by feedback error, which represents the copy of the motor-error command generated by the corresponding premotor network. Ultimately, each microzone acquires an inverse model of a specific controlled object, and it complements the relatively crude feedback control provided by the premotor networks. Thus, the activity of the corresponding earlier part of the premotor network decreases as cerebellar learning proceeds, if it works in feedback-control mode. However, the latter half of the premotor circuit is quite active. If the premotor network constitutes a part of the feedforward controller, as in the VOR case, the premotor network activity does not decrease to zero. Moreover, it must be noted that other premotor networks which are not connected to the cerebellum (see Figure 4 for such independent premotor networks) may be active even after learning.

Inverse-Dynamics Encoding of Ocular-Following Eye Movements by the Ventral Paraflocculus

Recently, Shidara et al. (1993) examined an inverse-dynamics representation of the firing pattern of ventral paraflocculus Purkinje cells during ocular-following responses (OFR) in monkeys. They found that movements of the visual scene evoke tracking movements of the eyes, termed *ocular following*. When the monkey turns its head, the eyes counterrotate under the influence of the VOR, but because this reflex is not perfect, some disturbance of gaze, and hence also of the retinal image, often occurs. By working to stabilize the eye with respect to the environment, ocular following helps to compensate for the animal's own movements.

Simple spike activities of a Purkinje cell in the left ventral paraflocculus were recorded together with the ocular-following response to 50–200 presentations of a downward or ipsilateral test ramp (20°–160° per second) of a large-field random dot pattern. The responses were aligned with stimulus onset. The ensemble average spike response over the trials, eye acceleration, eye velocity, eye position, and stimulus velocity were calculated. The temporal pattern of the ensemble mean firing frequency of cells was reconstructed as a linear weighted superposition of the position, velocity, and acceleration of the actual eye movement according to the following equation:

$$f(t - \Delta) = M \cdot \ddot{\theta}(t) + B \cdot \dot{\theta}(t) + K \cdot \theta(t) + f_{spont} \qquad (2)$$

Here, $f(t), \theta(t), \dot{\theta}(t), \ddot{\theta}(t), \Delta, f_{spont}$ are the firing frequency at time t, the eye position, velocity, acceleration at time t, the time delay, and the spontaneous firing level, respectively. Four coefficients and the time delay were estimated to minimize the squared error between the average and reconstructed firing frequencies. Shidara et al. (1993) found that the cell firing frequency is fairly well reproduced by this expression: the determination coefficient was larger than 0.7 for more than 80% of the neurons studied (19/23). They used the parameters of the 19 well-fitted Purkinje cells for the following statistical analysis. The mean ratio of acceleration coefficient to velocity coefficient (M/B) of the Purkinje cells was 72.1, which was close to that of motoneurons, 67.39. On the other hand, while the mean ratio of acceleration coefficient to position coefficient ($M/K = -294.5$) of the Purkinje cells was of a similar size, it was different from that of motoneurons (344.8) and had the reverse sign. These results indicate that the activity of the Purkinje cells encodes not the static but the dynamic part of the motor command for the required eye movement.

For a satisfactory OFR, the desired trajectory information given by the ramp velocity profile must be converted into a motor command which produces the desired eye movement. Although this inverse-dynamics problem could be solved at multiple sites in the brain, these results support the hypothesis that the cerebellum may be the major site of the inverse-dynamics model for controlled movements. There are three primary reasons. First, the Purkinje cell firing profiles were well reconstructed by the inverse-dynamics representation. Second, the ratio of acceleration and velocity coefficients of the Purkinje cells was very close to that of the motoneurons, but the ratio for mossy fiber inputs was only about half that of the Purkinje cells. That is, while the outputs of Purkinje cells could constitute the dynamic part of the final motor command, their inputs could not. Third, the preferred directions of the Purkinje cells could be classified into only two groups, downward and ipsilateral. This indicates that the Purkinje cells encode coordinates suitable for generating the motor output, whereas the mossy fibers do not.

Analog-Firing-Frequency Coding Versus Local Coding for Parallel Fiber Inputs

For previous cerebellar models, it has been repeatedly proposed that parallel fiber inputs represent information about the dynamical state of the controlled object in the manner of a table lookup (e.g., the codon theory of Marr, 1969, and the Cerebellar Model Articulation Controller of Albus, 1975). To clarify the argument in a very simple case, let us assume that a single Purkinje cell constitutes an inverse-dynamics model of a single-degree-of-freedom controlled object while receiving necessary information about the desired trajectory θ_d (position, velocity, and acceleration). Furthermore, we simply assume that the output of a Purkinje cell y is a linear weighted summation of its parallel fiber inputs x_i weighted by the synaptic efficacy w_i of the ith parallel fiber–Purkinje cell synapse. Then we obtain the following:

$$\tau(t) = y = \sum_n w_i x_i = \sum_n w_i h_i(\ddot{\theta}_d(t), \dot{\theta}_d(t), \theta_d(t)) \qquad (3)$$

Here, τ is the necessary motor command, θ is the state variable, and h_i is a (nonlinear) function of the position, velocity, and acceleration of the desired trajectory which gives the firing frequency of the ith parallel fiber.

There are two possible classes of representation by the parallel fiber inputs. The first class could be called local representation, table-lookup representation, codon theory, CMAC, or the radial basis function network. In the first class of representations, the function h_i has local support in the three-dimensional input space, that is, h_i is non zero only within a small finite domain in the input space. The other class of representation could be called global representation, analog-firing-frequency coding, or the MADALINE network. The second class is a slightly narrower class of representations. In the second class, the function h_i is a rather simple function of the desired trajectory input such as the component itself (e.g., $h_i = \theta_d$) or a linear combination of the three inputs $h_i = a\ddot{\theta}_d(t) + b\dot{\theta}_d(t) + c\theta_d(t)$. It is clear that the two classes are mutually exclusive.

The cerebellar model for the first class of representations is inspired by the enormous number of granule cells in the cerebellar cortex. Because the number is so large, probably on the order of 10^{11}, some researchers expect that the state space of the controlled object may be covered by small boxes in which only a small number of granule cells fire. However, this coding scheme requires too many neurons even for a single arm which has 13 degrees of freedom. In order to calculate the necessary torque for each joint, we need to consider a 39-degree-of-freedom state space because each joint contains three variables (position, velocity, and acceleration) and there exist interactional forces among all degrees of freedom. Even if we simply divide each degree of freedom into only 5 segments, we still need 5^{39} boxes to cover the full space—an excessive number, even compared with the number, 10^{11}, of granule cells. There are, of course, several schemes to decrease the number of necessary nodes (e.g., Albus's coarse coding scheme, CMAC), but the order of difference between 5^{39} and 10^{11} is almost 10^{15} for the above two figures, five segments per degree of freedom is clearly an underestimate, and the cerebellum should control not only a single arm but also many body parts. Thus, we believe that the local coding for parallel fibers is not adopted by the cerebellum for motor control.

The flocculus for the VOR gives the best-known experimental data supporting our argument. The mossy fibers for this system code the head velocity θ_h^c as the firing frequency. This is analog coding in contrast to local coding.

More compelling evidence was recently obtained for the ventral paraflocculus during OFR. Shidara et al. (1993) recorded parallel and mossy fiber firing frequencies and reconstructed them as a linearly weighted summation of eye position, velocity, and acceleration, using Equation 2 just as for Purkinje cells. Although the reconstruction coefficients were different from the Purkinje cell outputs described earlier and the fitting was less dramatic, the firing frequency was fairly well explained by linear combinations of the eye position, velocity, and acceleration. This finding is again in accordance with the analog coding scheme, and it contradicts the local coding scheme.

Acknowledgments. This work was supported by a Human Frontier Science Project Grant.

Road Maps: Learning in Biological Systems; Mammalian Brain Regions; Primate Motor Control
Background: Motor Control, Biological and Theoretical
Related Reading: Long-Term Depression in the Cerebellum; Pursuit Eye Movements

References

Albus, J. S., 1971, A theory of cerebellar functions, *Math. Biosci.* 10: 25–61.
Albus, J. S., 1975, A new approach to manipulator control: The cerebellar model articulation controller (CMAC), *Trans. ASME J. Dyn. Syst. Meas. Control*, 97:220–227.
Atkeson, C. G., 1989, Learning arm kinematics and dynamics, *Annu. Rev. Neurosci.*, 12:157–183. ◆
Boylls, C. C., 1975, A theory of cerebellar function with applications to locomotion, I: The physiological role of climbing fiber inputs in anterior lobe operation, in *COINS Tech. Rep.: Computer and Information Science*, Amherst: University of Massachusetts.
Fujita, M., 1982, Adaptive filter model of the cerebellum, *Biol. Cybern.*, 45:195–206.
Houk, J. C., and Barto, A. G., 1992, Distributed sensorimotor learning, in *Tutorial in Motor Behavior II* (G. E. Stelmach and J. Requin, Eds.), Amsterdam: Elsevier, pp. 71–100.
Ito, M., 1984, *The Cerebellum and Neural Control*, New York: Raven. ◆
Ito, M., 1989, Long-term depression, *Annu. Rev. Neurosci.*, 12:85–102. ◆
Kawato, M., and Gomi, H., 1992, The cerebellum and VOR/OKR learning models, *Trends Neurosci.*, 15:445–453. ◆
Marr, D., 1969, A theory of cerebellar cortex, *J. Physiol.*, 202:437–470.
Miall, R. C., Weir, D. J., Wolpert, D. M., and Stein, J. F., 1993, Is the cerebellum a Smith predictor? *J. Motor Behav.*, Special issue on computational studies of arm control, 25:203–216.
Shidara, M., Kawano, K., Gomi, H., and Kawato, M., 1993, Inverse-dynamics encoding of eye movements by Purkinje cells in the cerebellum, *Nature*, 365:50–52.
Simpson, J. I., and Alley, K. E., 1974, Visual climbing fiber input to rabbit vestibulo-cerebellum: A source of direction-specific information, *Brain Res.*, 82:302–308.
Szentágothai, J., and Arbib, M. A., 1975, *Conceptual Models of Neural Organization*, Cambridge, MA: MIT Press. ◆
Tsukahara, N., Bando, T., Murakami, F., and Oda, Y., 1983, Properties of cerebello-precerebellar reverberating circuits, *Brain Res.*, 274:249–259.

Chains of Coupled Oscillators

Nancy Kopell

Introduction

The mathematical analysis discussed in this article is directed to understanding the patterns of phases that emerge from the interaction of a collection of oscillators. Much of the work was motivated by the desire to understand the origin and regulation of spatiotemporal patterns in neural networks that govern undulatory locomotion. The main experimental preparation for this work is the lamprey central pattern generator (CPG) for locomotion (see SPINAL CORD OF LAMPREY: GENERATION OF LOCOMOTOR PATTERNS). However, the mathematics is considerably more general and can be used to gain insights into other pattern generators (Friesen and Pearce, 1993), as well as waves of activity in other neural tissue (Kleinfeld et al., 1994) and in smooth muscle (see Ermentrout and Kopell, 1984, for references). It is also related to synchronization and waves of activation that occur during sensory processing in the cortex, e.g., in the olfactory system, as well as to other kinds of patterns in the nervous system. (See SYNCHRONIZATION OF NEURONAL RESPONSES AS A PUTATIVE BINDING MECHANISM; OLFACTORY BULB; OLFACTORY CORTEX; OSCILLATORY ASSOCIATIVE MEMORIES.)

For the lamprey swimming CPG, the pattern in question is a constant speed wave (i.e., with phase-lag between adjacent oscillators independent of position along the chain); for some waves in neural tissue, or in peristalsis, the phase lags vary in space. In some situations, notably the formation of cell assemblies, the desired pattern is close to synchrony. For any such pattern, we are interested in what properties of the oscillators and the interactions between them are important in producing that kind of pattern.

There is also a significant literature analyzing the behavior of chains of oscillators in physical problems. However, much of that work depends on central theoretical notions, such as physical energy. In biological oscillators, however, there are other structures that are important to the behavior of collections of such oscillators. For example, unlike models of mechanical oscillators, models of biological oscillators have stable limit cycles (see PHASE-PLANE ANALYSIS OF NEURAL ACTIVITY). In the context of undulatory locomotion, each "oscillator" is likely to be a local subnetwork of neurons that produces rhythmic patterns of membrane potentials. Since the details of these oscillators are not yet known and are difficult to obtain, the object of the mathematics is to investigate the consequences of that part of the structure that is known and to generate sharper questions to motivate further experimentation.

The most striking aspect of the structure of the lamprey CPG for undulatory locomotion is the linear geometry of the network. That is, this network can be crudely described as a chain of oscillators (Cohen, Holmes, and Rand, 1982). As we discuss later, the linear geometry provides important constraints on the behavior of the network, making it possible to draw general conclusions under qualitative assumptions about the behavior of the local oscillators and their interactions. A pair of previous review articles discussing the mathematics of chains of oscillators are Kopell (1987) and Rand, Cohen, and Holmes (1987). See also Murray (1989) for related phenomena and some analysis of systems of oscillators. References for further related papers can be found in Ermentrout and Kopell (1994b) and SPINAL CORD OF LAMPREY: GENERATION OF LOCOMOTOR PATTERNS.

Physical Models and Phase Models

A very primitive question about the CPG for undulatory locomotion is the origin of the traveling waves of neural activity that pass down the spinal cord. A mathematical version of this question is: what properties of oscillators and their interactions suffice to produce constant-speed traveling waves? As is well known, interacting oscillators can produce almost unlimited complexity (Guckenheimer and Holmes, 1983), making analysis in complete generality an impossibility. Thus, further hypotheses are added, at least temporarily, to get insight into the possible mathematical mechanisms for producing the waves.

In the absence of knowledge of details, it is desirable to keep the oscillator description as general as possible. Thus, we assume only that the local network can be described by some (possibly high-dimensional) system of ordinary differential equations having a stable limit cycle, with no concern for the mechanistic origin of that oscillation. The further hypotheses are on the nature of the interactions. The first is that the interactions are local, with direct connections only among neighboring units (though not just nearest neighbors). Later in the article, we discuss current understanding of network behavior when direct connections between distant units are allowed.

The second hypothesis is more technical but has important consequences for the behavior of the chain. The hypothesis is that the behavior of the chain is qualitatively the same as one in which the interactions are averaged over a cycle of the rhythm. [Technically, this means that the use of the Averaging Theorem (Guckenheimer and Holmes, 1983) is valid.] This is known to be true for arbitrary kinds of coupling, provided that the interactions are sufficiently weak. It also holds, even with strong coupling, under circumstances more likely to be relevant to CPGs, namely, when each oscillator is itself a composite of cells, emitting coupling signals several times during the cycle (Ermentrout and Kopell, 1991). It need not hold for general strong coupling, as qualitatively different phenomena often occur. See the section "Phase Oscillators and Relaxation Oscillators" below and Guckenheimer and Holmes (1983).

Under these hypotheses, the system can be reduced to one in which all of the original physical variables are determined by the behavior of the phases of the set of oscillators. Furthermore, with no more assumptions, the behavior is qualitatively the same as if the interactions are through phase differences, even though the physical interactions may be quite different. With nearest neighbor coupling, the reduced equations are

$$\dot{\theta}_j = \omega_j + H^+(\theta_{j+1} - \theta_j) + H^-(\theta_{j-1} - \theta_j) \qquad (1)$$

Here, θ_j is the phase of the jth oscillator and ω_j is the frequency of that oscillator in the absence of interactions. H^+ and H^- are periodic functions representing the composite effects of oscillators $j + 1$ and $j - 1$ on the jth oscillator. (See Ermentrout and Kopell, 1984, for a derivation.) These functions are numerically computable, given the equations for the uncoupled oscillators and the mathematical description of the original physical interactions (Ermentrout and Kopell, 1991). They depend not only on the interactions but on properties of the oscillators as well. The existence of the reduction procedure allows us to come back to questions about how detailed structure of the oscillators or coupling can affect the chain behavior through their effects on the functions $H^\pm$.

Mechanisms for the Production of Traveling Waves

When this reduction is valid, the question about the generation of traveling waves reduces to the question of the roles of the uncoupled frequencies $\{\omega_j\}$ and the coupling functions $H^\pm$ in determining the behavior of the chain. There are at least two different mechanisms that can produce waves in Equation 1. One of these relies on differences in natural frequency along the chain and the other on properties of the coupling.

The first mechanism is easy to illustrate, using the simple choice of coupling functions $H^+(\phi) = \sin \phi = H^-(\phi)$, where ϕ denotes the relevant phase difference. Such functions arise from the reduction procedure when the oscillators have symmetrical limit cycles, and the interaction is through the standard mathematical description of diffusion between compartments (Ermentrout and Kopell, 1984). In this case, if all the ω_j are equal, the synchronous solution ($\theta_j \equiv \theta_k$ for all j, k) is the stable output of the chain. If the frequencies are not all the same, however, other behavior is produced. For example, if there is a gradient in frequencies, there is a traveling wave of activity from the oscillator with the highest frequency to that with the lowest. Waves induced by frequency differences *are* important in some physical situations, such as peristalsis (see Ermentrout and Kopell, 1984, and their references). There is also evidence that waves in some neural tissue, such as the limax olfactory organ (Kleinfeld et al., 1994), are produced by gradients of frequency. Such gradients are also known to exist in the leech CPG for swimming (Friesen and Pearce, 1993).

In the lamprey, there is no evidence for a gradient in frequency. Furthermore, unlike the observed wave in the lamprey CPG, waves produced by frequency gradients do not have constant phase lag along the chain (Cohen et al., 1982). It *is* possible to produce constant phase lags with this choice of $H^\pm$ and *some* collection of uncoupled frequencies: this occurs if all of the oscillators but the first and last have the same frequency, the first is incremented by some amount Δ, and the last is decremented by the same amount (Cohen et al., 1982). However, these hypotheses are very implausible for the lamprey network, because the waves can be obtained for any sufficiently long stretch of spinal cord anywhere along the cord. This motivated the search for mechanisms that do not depend on differences in frequencies between the ends of the chain and the other oscillators.

There is such a mechanism, one that produces essentially constant speed waves (or, equivalently, phase differences between adjacent oscillators independent of position) in a chain of identical oscillators. (There is a small *boundary layer* with different phase lags at one end of the chain.) The mechanism relies on the use of coupling functions that have one property very different from that of the sine coupling. An oscillator forced by another oscillator of the same frequency with forcing function a multiple of $\sin(\theta_2 - \theta_1)$ phaselocks with a zero-phase difference. The essential property for H^+ or H^- (which we will call the *edge property*) is that, if used as a forcing function (i.e., with only one-way coupling), the oscillators would lock with a non-zero difference. A simple example of such an H^+ or H^- is $A \sin \phi + B \cos \phi$, $B \neq 0$. There are several points to make about this kind of coupling and the patterns of phase lags that it induces:

1. The edge property holds for almost all kinds of coupling between two oscillators, and in particular for models of chemical synapses. An important special exception is the mathematical description of simple diffusion across a membrane, which is a standard model for electrical synapses (*gap junctions*). Indeed, as noted previously, such models can give rise to sine coupling, which does not have the property.

2. The edge property ensures that if a chain of identical oscillators is coupled in *only one direction*, a stable solution is a constant-speed traveling wave with phase lag between any two oscillators determined by that pair alone. For example, if

$$\dot{\theta}_1 = \omega + H(\theta_2 - \theta_1)$$
$$\dot{\theta}_2 = \omega \tag{2}$$

then the phase lag between the oscillators is the value $\phi \equiv \theta_2 - \theta_1$ at which $H(\phi) = 0$, $H'(\phi) > 0$. In the case of one-way coupling, the analysis for the chain is no more difficult than for the pair and produces a wave with phase lag constant along the chain. We can think of this wave as "edge driven," because of the difference in input at one edge of the chain.

If the coupling is in *both* directions, as in the lamprey CPG, the analysis does not reduce to that of a single forced oscillator. However, at least for sufficiently long chains of oscillators and $H^+ \neq H^-$, in general the system behaves like that of a system coupled in only one direction. In particular, either H^+ or H^- "dominates" the other and (except in a small boundary layer) determines the phase lag as if the other one were absent. This can be shown (Kopell and Ermentrout, 1986) using a continuum limit analysis. A simpler version using linearized equations to display the dominance effect is discussed in Spinal Cord of Lamprey: Generation of Locomotor Patterns; this simpler version is adequate for understanding the production of traveling waves and the idea of dominance, but does not reproduce other effects, described in the following sections, that are elicited by experimental perturbations.

3. With identical oscillators in Equation 1, the phase lags produced are independent of the frequency ω of the uncoupled oscillators. This helps to provide an explanation for how it is possible for undulatory swimming to occur at many different swimming speeds (which are proportional to the frequency of the electrical activity at each of the local oscillators) while keeping phase lag constant. However, this is a solution to that question only at an abstract level; at a more biophysical level, there is still an issue. The difficulty is that physical parameters that change oscillator features, such as frequency, also may change properties of the coupling; thus, it is not clear at the biophysical level how to change the frequencies ω without changing the coupling functions $H^\pm$. We note, though, that constancy of the phase lags under change of oscillator frequency does not, in this set of models, require that the coupling functions $H^\pm$ be unchanged; it only requires that the zero of the dominant function, which determines the lag between successive oscillators, be independent of parameters that are used to vary oscillator frequency. This raises a question at the level of oscillator structure, which will be addressed briefly at the end of this article, and more fully in Spinal Cord of Lamprey: Generation of Locomotor Patterns.

4. *Constant*-speed waves are produced when the oscillators are identical. However, the analysis that produces these results also holds more generally, e.g., if there is a gradient in frequencies in addition to coupling with the edge property. In that case, the wave pattern produced results from a balance of the inhomogeneities in frequencies and the properties of the coupling (Kopell and Ermentrout, 1986). Thus, frequency gradients may be an important mechanism in producing wave behavior, as in peristalsis, whenever constancy of phase lag along the chain is not required for function.

Mathematics and Experimental Perturbations

If there are differences in frequencies of the uncoupled oscillators or input to the oscillators from outside the chain, other spatiotemporal patterns of phases can be generated. This suggests experimental perturbations that can be used to distinguish the mechanisms discussed in earlier sections for producing waves from a mechanism that relies only on differences in the uncoupled frequencies of the oscillators.

Perturbations of Frequency

The first such experimental protocol is the *split-bath* paradigm, in which different parts of the spinal cord are exposed to different concentrations of an excitatory amino acid whose presence affects frequency in a dose-dependent manner. The mathematical analog of this protocol is a chain of oscillators in which the uncoupled frequencies change abruptly across one or more points in the chain. The mathematics used to analyze this situation is very similar to that used to study the case of identical oscillators (Kopell and Ermentrout, 1986). Under the same hypotheses on the coupling as in the previous section, the result of the analysis is that the phase lags are constant within each compartment, with a change in lag across a boundary between compartments. Furthermore, there are relationships between the lags within the compartments in the split-bath protocol and the lags that occur when the uncoupled frequencies are identical. These relationships, to be explained below, do not occur in models in which the lags are caused only by frequency differences, and hence they allow for experimental differentiation between the two kinds of mechanisms.

We consider the case of two compartments. The major result concerns the role of dominance in the behavior of a chain having *piecewise constant* frequencies. If the coupling from last to first (i.e., in the caudal to rostral direction) is dominant, we shall say that the back (or *caudal*) compartment is dominant; if the opposite coupling is the dominant one, we say that the front (or *rostral*) compartment is dominant. It follows from the analysis that, for small enough differences in frequency across the boundary, the phase difference between two points within the dominant compartment is the same as that in the chain of identical oscillators. The phase difference between two points within the other compartment changes in the same direction as the frequency: if the frequency in the second compartment is lower, the phase lag is also smaller. Note that if the lag in the dominant compartment is negative, the lag in the other compartment will be *more* negative. For larger differences in frequency, other behavior can emerge.

The behavior just described for not-too-large differences in frequencies between the compartments is what is observed in the lamprey data (see Spinal Cord of Lamprey: Generation of Locomotor Patterns). By contrast, if the split-bath protocol for frequencies is used with sine coupling in both directions, the results are quite different: the phase lags are not constant within a compartment for fixed frequency differences, and in neither compartment are the lags independent of frequency.

Perturbations of Input: Forcing Experiments

The second experimental perturbation directly affects only the oscillators near one of the ends of the cord. A small motor attached to one end creates periodic motions of the segments near that end, and this creates periodic electrical perturbations of the end oscillators (see SPINAL CORD OF LAMPREY: GENERATION OF LOCOMOTOR PATTERNS for references to the experimental and mathematical literature). The mathematical analog of this situation is a chain of oscillators that receives periodic input at one end. The mathematics that governs such a situation is not the same as that of the unforced chain. The analysis is somewhat easier, partly because, if the chain is to phase lock at all, it must lock at the forcing frequency (i.e., $\dot{\theta}_j \equiv \Omega$, the forcing frequency). This contrasts with the case of the unforced chain, in which the nondominant coupling acts to alter the ensemble frequency of the chain from that of the uncoupled lags.

The equations for the forced chain are the same as Equation 1, except that the first or last equation has one more term, corresponding to the forcing. Assume for definiteness that the last oscillator is forced. The calculations of the phase lags can then be done one equation at a time, starting from the first, which determines ϕ_1; each new equation, if solvable, inductively determines one new ϕ_j. (The equation can be solved for $H^+(\phi_j)$, but there need not be any ϕ_j that solves the equation.) The last equation, if solvable, determines the phase lag between the last oscillator of the chain and the forcing oscillator. If the frequency of the forcing is quite different from the ensemble frequency of the chain, it may not be possible to solve for all the $\{\phi_j\}$. This can occur either because the final equation cannot be solved or because some previous equation not dependent on the forcing cannot be solved. If the forcing is relatively large, it is not the size (or nature) of the forcing but properties of the coupling in the unforced chain that cause locking to be lost at frequencies sufficiently distant from the ensemble frequency. We shall refer to the interval of forcing frequencies compatible in this sense with the internal structure of the chain as the *internal entrainment range*. Within this range, for oscillators and coupling satisfying the hypotheses we have been using, it turns out that locking is possible independent of the length of the chain.

The mathematical analysis relates the internal entrainment range to the ensemble frequency of the chain and to which of the two coupling functions is dominant. A chain that is forced at the dominant end (i.e., in the direction of the dominant coupling) has an internal entrainment range that extends on both sides of the ensemble frequency. By contrast, forcing in the nondominant direction leads to an internal entrainment range that has the ensemble frequency as one boundary.

Unlike the existence of constant-speed waves, the reaction of the model to forcing depends crucially on the nonlinearities of the coupling functions. A substantial amount of the qualitative behavior of the coupled chain depends on the function $f(\phi) \equiv [H^+(\phi) + H^-(-\phi)]/2$, and in particular on the sign of f'' in an interval of phases containing the zeros of the functions H^+ and H^-. If $f'' > 0$, then the internal entrainment range for forcing in the nondominant direction extends above the ensemble frequency; if $f'' < 0$, it extends below the latter. These conclusions, like those from the mathematical analog of the split-bath protocol, allow inferences from the lamprey data about the existence of a dominant coupling and which direction the dominant coupling takes; the forcing data also specify the sign of f''. See Kopell and Ermentrout (1986) for further discussion of the significance of f'' in the determination of the ensemble frequency.

These conclusions may be compared with those drawn by analyzing the isotropic sine-coupled model with forcing at one end of a chain of identical oscillators. The same method of analysis just described shows that, for *any* forcing frequency not equal to that of the uncoupled oscillators, locking is lost for a sufficiently long chain, and the larger the difference between the forcing frequency and the natural frequency, the shorter the length of the chain that can sustain entrainment. If the chain with isotropic sine coupling has a *leading oscillator*, e.g., the first oscillator has higher intrinsic frequency than the rest (which are identical), the locking is still lost for long chains for forcing at any frequency other than that of the nonleading oscillators.

Long-Range Coupling

All of this theory has dealt with the spatiotemporal behavior of phase lags that are produced by nearest neighbor coupling. Similar results hold for multiple-neighbor coupling. However, there are known to be long fibers in the lamprey CPG that could presumably affect intersegmental coordination. This has motivated attempts to understand the effects of long-range coordination on the production of phase lags.

The subject is potentially a very large one, since the effects of a combination of long and short connections appear to depend in a complicated way on many things, including the phase lags that would be induced by the local coupling or long coupling alone and the lengths over which the long-range coupling operates. Thus, there is not yet any general theory. However, several special cases have been investigated in some detail, and from these cases, some answers and some new questions arise. The two main "architectures" that have been considered are translation-invariant coupling (Kiemel, 1990) and sparse coupling between the ends of the chain and the oscillators near the middle of the chain (Ermentrout and Kopell, 1994a). [Another architecture, with long coupling only between the two ends of the chain (Cohen, Holmes, and Rand, 1982), turns out to be analyzable as a special case of Ermentrout and Kopell (1994a).]

Translation-invariant coupling means that the connections between oscillator 1 and oscillator k are repeated between oscillator $1 + j$ and oscillator $k + j$, for all j that are relevant. For such an architecture, with long coupling in only one direction, but short coupling in both, Kiemel investigated long-range connections that are roughly "tuned" to the short ones; by tuning, we mean that the phase lag between a pair of separated oscillators that would be produced directly by the long-range fibers in the absence of local coupling is close to the phase lag that would be produced between those oscillators by the local coupling. For example, if the local coupling produces equal phase lags of ϕ, and a pair of oscillators are k points apart on the chain, then a direct long-range connection between those oscillators should produce a stable phase lag close to $k\phi$ if it is to be tuned to the local coupling. Intuitively, tuned long-range coupling does not drastically fight the effects of the local coupling. Kiemel found that the long coupling has very different effects on the phase lags, depending on whether the long coupling is in the direction of the dominant short coupling or not. He also showed that the effects of the long coupling are roughly proportional to the distance along the chain of the oscillators directly connected; thus, long coupling is strong coupling.

The work on *sparse long-range coupling* between the ends of the chain and oscillators near the middle was motivated by the question of why the wavelength of the traveling waves in the lamprey and other similar animals is approximately one body length (Ermentrout and Kopell, 1994a). Local coupling alone, as described in the previous sections, provides no intrinsic wave-length; it specifies the lag per pair of oscillators rather than the total end-to-end phase lag. Hence, it is not apparent from such a theory why the end-to-end phase lag should be one body length. By contrast, long-range coupling connecting end oscillators to points near the middle does provide an intrinsic length scale.

The investigation was also motivated by descriptions of locomotion in other species, as well as motor behavior of general vertebrates during very early development (Beckoff, 1985). Spatiotemporal patterns that occur during early development include rhythmic contractions that are synchronous along the body (C-coils), traveling waves with wavelength equal to body length, and S-waves, in which the upper and lower halves of the body each contract in unison, and oscillate in antiphase with each other (with a node between the halves). Some adult animals, including salamanders, use undulatory locomotion when they swim and S-wave locomotion when they walk.

All of these behaviors can be obtained with short-range coupling plus long-range coupling as just discussed (Ermentrout and Kopell, 1994a). In this study, the local coupling is sinusoidal, which produces synchrony in the absence of long-range coupling; this was done to obtain solutions that mimic the C-coils, which are the most primitive of the rhythmic behaviors. The particular long-range coupling chosen is an abstraction of inhibitory coupling: like the phase models obtained using the reduction procedure from biophysical models with inhibitory synapses (Ermentrout and Kopell, 1991), the long-range coupling alone would produce antiphase behavior between the oscillators directly coupled.

With the sparse connectivity and assumptions on the local coupling described here, subsets of the chain shorter than half the length have only local coupling, and hence synchronize. However, such subsets of the lamprey cord self-organize into traveling waves with phase lag per segment relatively independent of the size or portion of the chain. The latter observation is entirely consistent with the *local* mechanism for producing waves from identical oscillators described in a previous section. To account for this observation, plus the constraint on the wavelength and observations about early development, we have suggested that the waves could be produced by sparse connectivity in early development; those waves could then act as teaching signals to tune (in the sense of Kiemel, 1990) the local connections to the long-range connections, making the long connections redundant in the adult. A model showing that such learning of phase lags is possible is found in Ermentrout and Kopell (1994b).

Variations on a Mathematical Theme

H-Functions and C-Functions

The phases of oscillators we have been discussing are abstract quantities that can be precisely defined from equations describing the oscillator dynamics and the interactions among the oscillators; however, the definition we have been using of $H^{\pm}$ does not lend itself easily to measurement in the biological preparation, where the equations are not known. For this reason, it is desirable to recast at least some of the theory in terms of quantities that can be experimentally measured. We now discuss a variation of the $H^{\pm}$ functions that is so measurable.

We shall describe a coupling function C^+ that is an analog of H^+; a similar description holds for C^-, an analog of H^- (see SPINAL CORD OF LAMPREY: GENERATION OF LOCOMOTOR PAT-

TERNS). Recall that the equations for oscillator 1 forced by oscillator 2 are

$$\dot{\theta}_1 = \omega_1 + H^+(\theta_1 - \theta_1)$$
$$\dot{\theta}_2 = \omega_2$$

The phase lag $\phi \equiv \theta_2 - \theta_1$ that occurs if there is locking is determined as the solution (if it exists) of

$$\omega_2 - \omega_1 = H^+(\phi). \qquad (3)$$

This equation has solutions for a range of ω_2 near ω_1; as ω_2 is varied, the solution ceases to exist when the solution ϕ reaches a point at which $(H^+)' = 0$. Within that range, Equation 3 gives a relationship between the frequency difference $\Delta \equiv \omega_2 - \omega_1$ of the two oscillators and the resultant phase difference. The significance of this *interpretation* of the function H^+ is that frequency differences can in principle be experimentally manipulated, and phase differences can be measured at some important point in the cycle, such as a burst onset or offset; thus, this gives a potential means of measuring H^+, or at least the part of H^+ that has positive slope, which is the only relevant part for stable locked solutions. To distinguish a function so constructed, we shall call it $\Delta = C^+(\phi)$. This definition is also useful for working with mathematical oscillators themselves described by complicated networks in which it is hard to compute H^+, but much easier to compute C^+ by simulation. We note that if the function is experimentally measured, the result may depend on the marker (such as phase onset) chosen to measure phase difference; for one constructed from an H-function, it is independent of choice of marker. We conjecture that the theory developed for the $H^{\pm}$ functions holds for the $C^{\pm}$ as well if the latter are relatively independent of the marker. Because the $C^{\pm}$ functions are not rigorously known to behave exactly like the $H^{\pm}$ functions, it is useful to have numerical simulations to back up intuition. Simulations using $C^{\pm}$ functions, including some which address questions about phase regulation under change of frequency, are discussed in SPINAL CORD OF LAMPREY: GENERATION OF LOCOMOTOR PATTERNS, in the context of particular cellular mechanisms for production of the local oscillations.

Phase Oscillators and Relaxation Oscillators

Equation 1 is derived from more physical descriptions of oscillators and their interactions by a reduction procedure described in an earlier section. When the reduction procedure is not valid, the emergent behavior of the collection of oscillators can be very different from that of the lamprey system, which is the inspiration of this work (Guckenheimer and Holmes, 1983).

One such situation in which the reduction procedure is not valid involves relaxation oscillators coupled using models of excitatory chemical synapses (see PHASE-PLANE ANALYSIS OF NEURAL ACTIVITY). Such oscillators do not behave like oscillators interacting through phase differences. For systems of relaxation oscillators, thresholds play an important role in locking, which occurs by a different mechanism, referred to as *fast threshold modulation* (FTM) (Somers and Kopell, 1993).

The different mechanisms lead to some important properties. One is that differences in frequency or in inputs among such collections of oscillators need not lead to phase differences among the oscillators, as occurs for Equation 1. This provides a potential mechanism to be used in constructing a network whose phase lags are well regulated even if the frequencies are not (as is true of the lamprey spinal cord). A second difference concerns speed of locking. Long chains of oscillators of the form in Equation 1 can take many, many cycles to approach the locked solution, with the time increasing with the length of the chain. By contrast, a chain of relaxation oscillators can lock within a few cycles (Somers and Kopell, 1993).

It is difficult to obtain traveling waves in a chain of such oscillators; the usual outcome is close to synchrony unless the coupling is very weak (in which case the averaging procedure is relevant). Indeed, changes in gating of synapses or voltage range of the presynaptic cell can change the mechanism of the interaction from that of phase oscillator–like behavior to that of FTM, changing waves into almost synchronous cell assemblies. Thus, a generalized signal in the network that changes, e.g., the threshold or sharpness of the synapses, can provide a stimulus-induced change in mechanism, and hence in emergent behavior. This is not the only way in which stimuli can change the pattern behavior; in limax, stimuli turn frequency-induced waves into approximate synchrony (Kleinfeld et al., 1994), possibly by eliminating the frequency gradient.

It is possible that some more complicated local circuit, with several different elements within a local oscillator, may have some features of each type. For example, individual elements may be of relaxation type, whereas the composite local circuit may have some features of phase oscillators (Ermentrout and Kopell, 1991). Such composite oscillators are not yet well understood, and current analytic techniques need to be expanded.

Acknowledgment. This work was partially supported by NIMH Grant MH47150.

Road Map: Motor Pattern Generators and Neuroethology
Background: I.3. Dynamics and Adaptation in Neural Networks
Related Reading: Collective Behavior of Coupled Oscillators; Pattern Formation, Biological

References

Bekoff, A., 1985, Development of locomotion in vertebrates: A comparative perspective, in *The Comparative Development of Adaptive Skills: Evolutionary Implications* (E. S. Gallin, Ed.), Hillsdale, NJ: Erlbaum, pp. 57–94. ◆

Cohen, A. H., Holmes, P. J., and Rand, R. H., 1982, The nature of the coupling between segmental oscillators of the lamprey spinal generator for locomotion: A mathematical model, *J. Math. Biol.*, 13: 345–369.
Ermentrout, G. B., and Kopell, N., 1984, Frequency plateaus in a chain of weakly coupled oscillators: I, *SIAM J. Math. Anal.*, 15: 215–237.
Ermentrout, G. B., and Kopell, N., 1991, Multiple pulse interactions and averaging in coupled neural oscillators, *J. Math. Biol.*, 29: 195–217.
Ermentrout, G. B., and Kopell, N., 1994a, Inhibition-produced patterning in chains of coupled nonlinear oscillators, *SIAM J. Appl. Math.*, 54:478–507.
Ermentrout, G. B., and Kopell, N., 1994b, Learning of phase lags in chains of neural oscillators, *Neural Comp.*, 6:225–241.
Friesen, W. O., and Pearce, R. A., 1993, Mechanisms of intersegmental coordination in leech locomotion, *Semin. Neurosci.*, 5:41–47. ◆
Guckenheimer, J., and Holmes, P., 1983, *Nonlinear Oscillations, Dynamical Systems and Bifurcations of Vector Fields*, New York: Springer-Verlag.
Kiemel, T., 1990, Three problems from the mathematics of neural oscillators, PhD Thesis, Cornell University.
Kleinfeld, D., Delaney, K. R., Fee, M. S., Flores, J. A., Tank, D. W., and Gelperin, A., 1994, Dynamics of propagating waves in the olfactory network of a terrestrial mollusc: An electrical and optical study, *J. Neurophysiol.*, 72:1402–1419.
Kopell, N., 1987, Toward a theory of modelling central pattern generators, in *Neural Control of Rhythmic Movements in Vertebrates* (A. H. Cohen, S. Grillner, and S. Rossignol, Eds.), New York: Wiley, pp. 369–413. ◆
Kopell, N., and Ermentrout, G. B., 1986, Symmetry and phaselocking in chains of weakly coupled oscillators, *Commun. Pure Appl. Math.*, 39:623–660.
Murray, J. D., 1989, *Mathematical Biology*, New York: Springer-Verlag. ◆
Rand, R. H., Cohen, A. H., and Holmes, P.J., 1987, Systems of coupled oscillators as models of central pattern generators, in *Neural Control of Rhythmic Movements in Vertebrates* (A. H. Cohen, S. Grillner, and S. Rossignol, Eds.), New York: Wiley, pp. 369–413. ◆
Somers, D., and Kopell, N., 1993, Rapid synchronization through fast threshold modulation, *Biol. Cybern.*, 68:393–407.

Chaos in Axons

Kazuyuki Aihara

Introduction

Information in biological neural networks is carried by nerve impulses called *action potentials*, though it is an important future problem to clarify how much and what kind of information each action potential carries. An axon, or nerve fiber, is a constituent part of a neuron which generates and propagates the action potentials (Hodgkin and Huxley, 1952; Cole, 1968).

The process of generating and propagating action potentials in axons is not as simple as it might seem when modeled by linear threshold elements. In fact, the process is supported by nonlinear dynamics peculiar to the axonal nerve membranes, which produce abundant nonlinear phenomena, such as self-sustained oscillation, synchronized oscillation, quasiperiodic oscillation, and chaotic oscillation (Aihara and Matsumoto, 1986).

In this article, the deterministic chaos in the axons is reviewed both experimentally with squid giant axons and numerically with nerve equations (Aihara, Matsumoto, and Ikegaya, 1984; Aihara and Matsumoto, 1986).

Chaos in Squid Giant Axons

Squid giant axons are excellent materials for studying nonlinear dynamics of the axonal membranes because they are literally giant, with diameters between 400 μm and 900 μm, and structually simple, without myelin. Therefore, squid giant axons have been widely used in electrophysiological experiments.

Under normal physiological conditions, the axons are in a state of rest, stably maintaining a constant membrane potential of about −60 mV, called the *resting potential*. It is also possible, by changing external ionic concentrations around the axons, to induce self-sustained oscillation with spontaneous and repetitive firing of action potentials, which emulates the

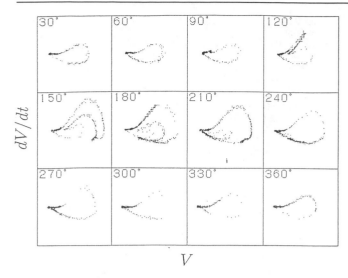

dV/dt

V

Figure 1. Poincaré sections of a chaotic oscillation experimentally observed when a squid giant axon in the state of self-sustained oscillation with the natural frequency of 183 Hz is forced by a sinusoidal current with a frequency of 270 Hz and an amplitude of 2.35 µA. The number in each section designates the phase of the sinusoidal force. (From Aihara, K., Numajiri, T., Matsumoto, G., and Kotani, M., Structures of attractors in periodically forced neural oscillators, 1986, *Phys. Lett. A*, 116:313. Reprinted with permission from Elsevier Science.)

behavior of pacemaker neurons. From the viewpoint of dynamic system theory, nerve membranes are nonlinear dissipative systems. Dynamic evolution of a nonlinear dissipative system can be asymptotically characterized by a concept of attractors, such as equilibrium points, limit cycles, tori, and strange attractors. The resting and oscillatory states of squid giant axons correspond to a stable equilibrium point and a stable limit cycle, respectively.

Dynamic behavior described by a strange attractor is classified as deterministic chaos (Ruelle, 1989). *Deterministic chaos* represents deterministic dynamic behavior with sensitive dependence on initial conditions, although it is still a challenging mathematical problem to define deterministic chaos rigorously. It has been experimentally shown with squid giant axons (Aihara and Matsumoto, 1986) and other cells (Glass and Mackey, 1988) that deterministic chaos can be easily and reproducibly observed in both resting and oscillatory cells when forced by periodic currents.

Figure 1 shows an example of chaotic oscillation experimentally recorded with a squid axon forced by a sinusoidal current. Each picture in Figure 1 is the Poincaré section obtained first by reconstructing a possible attractor in a space of $\mathbf{R}^2$ (the membrane potential V and its temporal differential dV/dt) × S^1 (sinusoidal force) from time-series data of the membrane potential, and then by observing its cross section at a fixed phase of the force. Figure 1 demonstrates that a part of the surface of the reconstructed attractor is stretched, folded, and compressed during one forcing period. Such transformation with stretching, folding, and compressing is typical of chaotic systems.

Different routes to chaos have been observed experimentally with biological cells (e.g., Aihara and Matsumoto, 1986; Glass and Mackey, 1988), including (1) *successive period-doubling bifurcations*, in which the period increases in the form of 2^n times, (2) *intermittency*, whereby chaotic bursts occur intermittently among apparently periodic phases, and (3) *collapse of quasi-*

periodicity, whereby chaotic behavior is produced through the collapse of a two-dimensional torus representing quasiperiodicity.

Chaos in Nerve Equations

It has been confirmed that various phenomena in squid giant axons can be well described quantitatively with the Hodgkin-Huxley equations (Hodgkin and Huxley, 1952) and qualitatively with the FitzHugh-Nagumo equations (Nagumo, Arimoto, and Yoshizawa, 1962; FitzHugh, 1969). Chaotic oscillation and the routes to chaos in the squid giant axons are no exceptions. These chaotic phenomena can be essentially understood with deterministic dynamics described by these nerve equations, even if the behavior seems erratic and complex (Aihara, Matsumoto, and Ikegaya, 1984; Aihara and Matsumoto, 1986).

A fundamental property of deterministic chaos is orbital instability owing to sensitive dependence on initial conditions. The orbital instability of the chaotic neural oscillation is produced by threshold phenomena of nerve membranes; namely, trajectories near a threshold veer sharply toward either a response with an action potential or a subthreshold response without any action potentials. The threshold separatrix delimiting the two kinds of nerve responses (i.e., the boundary *separating* them) is composed of a fuzzy trajectory, and the property of sharp veering is not strictly all-or-none, as in a saddle-point threshold model (FitzHugh, 1969). It is this continuous type of threshold separatrix inherent in these excitable dynamics that brings about the orbital instability of the neural chaos.

From Chaos in Axons to Chaotic Neural Networks

As explained in the previous section, chaos in axons has been well described by the Hodgkin-Huxley and FitzHugh-Nagumo equations. However, these equations are too complicated to be used as constituent elements for large-scale artificial neural networks. For this reason, the following simple neuron model (Aihara, Takabe, and Toyoda, 1990) which can qualitatively reproduce such chaotic responses, has been proposed on the basis of Caianiello's neuronic equation (Caianiello, 1961) and the biological properties of graded responses and relative refractoriness:

$$x(t + 1) = f\left[s(t) - \alpha \sum_{r=0}^{t} k^r g\{x(t - r)\} - \theta \right] \quad (1)$$

where $x(t + 1)$ is the graded analog output at the discrete time $t + 1$, f is the continuous output function, assumed to be the logistic function $f(y) = 1/\{1 + \exp(-y/\varepsilon)\}$ with the steepness parameter ε, $s(t)$ is the strength of the stimulation at t, α is a positive parameter, k is the damping factor of the refractoriness, g is the refractory function, assumed to be the identity function here, and θ is the threshold.

As Equation 1 is a model of an axon hillock or a trigger zone, the output function f is electrophysiologically equivalent to a stimulus-response curve obtained by plotting response strength of the membrane potential to a single pulse stimulation versus strength of the stimulation. The continuous type of threshold phenomena, explained in the previous section, is represented by a continuous stimulus-response curve (FitzHugh, 1969).

Equation 1 can be transformed to the following simpler one-dimensional mapping (Aihara, Takabe, and Toyoda, 1990):

$$y(t + 1) = ky(t) - \alpha g[f\{y(t)\}] + a \quad (2)$$

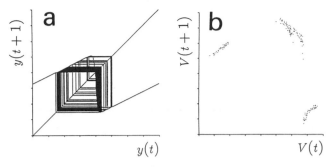

Figure 2. *A*, A chaotic solution to Equation 2, with $k = 0.5$, $\varepsilon = 0.015$, $a = 0.315$, and $y(0) = 0.3$. *B*, Corresponding return plots of the membrane potential experimentally obtained from a chaotic response of a resting squid axon stimulated by periodic pulses. The distribution of the dots in part *B* shows that the data have an approximately one-dimensional structure which is qualitatively similar to part *A*, in which the chaotic trajectory is demonstrated by shifting squares. The dots in *B* correspond to the vertices of the shifting squares in *A*, except on the diagonal line.

where $y(t + 1)$ is the internal state defined by $y(t + 1) = s(t) - \alpha \sum_{r=0}^{t} k^r g\{x(t - r)\} - \theta$, $a = (A - \theta)(1 - k)$, and A is the strength of the stimulation supposed to be temporally constant. An example of a chaotic solution to Equation 2 is shown in Figure 2*A*, with corresponding return plots of the experimentally observed membrane potential in Figure 2*B*.

The chaotic neuron model can be generalized as a fundamental element of Chaotic Neural Networks to investigate spatiotemporally chaotic neurodynamics and their possible functional roles in neural computing (Aihara, Takabe, and Toyoda, 1990). The dynamics of the *i*th chaotic neuron in a simple version of chaotic neural networks, with M constituent neurons and N externally applied inputs, are described by the following discrete-time equation under the assumption of exponential temporal-decay of both spatiotemporal inputs and refractoriness (Aihara, Takabe, and Toyoda, 1990):

$$y_i(t + 1) = ky_i(t) + \sum_{j=1}^{M} W_{ij} h_j\{f_j(t)\} + \sum_{j=1}^{N} V_{ij} A_j(t)$$
$$- \alpha g_i\{f_i(y_i(t))\} - \theta_i(1 - k) \qquad (3)$$

where for the *i*th neuron, y_i is the internal state, k is the decay factor, W_{ij} is the connection weight from the *j*th neuron, h_j is the transfer function of the axon from the *j*th neuron, f_j is the continuous output function of the *j*th neuron given the output $x_i(t + 1) = f_i\{y_i(t + 1)\}$, V_{ij} is the connection weight from the *j*th external input, $A_j(t)$ is the strength of the *j*th external input, g_i is the refractory function, and θ_i is the threshold. The model is a natural extension of conventional discrete-time neural network models, such as the McCulloch-Pitts model (McCulloch and Pitts, 1943) and multilayer perceptrons, because the model includes these conventional networks as special cases by adjusting values of the model parameters. In other words, we can introduce chaotic dynamics into conventional neural networks to study possible parallel distributed processing with spatiotemporal chaos. Moreover, various dynamic neural network models have been proposed to realize computational chaotic neurodynamics (Yamaguti, 1994).

Discussion

Simple linear threshold elements are often adopted as artificial neuron models. Although such simple neuron models have been useful for deriving some fundamental principles in artificial neurocomputing, real neurons are far more complicated (Segundo, 1986). In fact, real axons have chaotic dynamics which most usual neuron models lack.

On the other hand, experimental results implying the existence of deterministic chaos at the level of neural networks have also been reported. In particular, Freeman and colleagues proposed, on the basis of their experiments, an intriguing hypothesis that chaotic neural activity contributes to functional roles, such as rapid and unbiased access to learned patterns and classification and learning of novel patterns in perceptual processes (Skarda and Freeman, 1987). The nonlinear spatiotemporal dynamics of neural networks composed of chaotic elements (Moore, 1990) is an attractive subject not only for models of biological neural networks, but also for models for a new generation of artificial analog computing.

Road Map: Biological Neurons
Background: I.3. Dynamics and Adaptation in Neural Networks
Related Reading: Chaos in Neural Systems; Epilepsy: Network Models of Generation; Synaptic Coding of Spike Trains

References

Aihara, K., and Matsumoto, G., 1986, Chaotic oscillations and bifurcations in squid giant axons, in *Chaos* (A. V. Holden, Ed.), Manchester: Manchester University Press, and Princeton, NJ: Princeton University Press, pp. 257–269.
Aihara, K., Matsumoto, G., and Ikegaya, Y., 1984, Periodic and nonperiodic responses of a periodically forced Hodgkin-Huxley oscillator, *J. Theoret. Biol.*, 109:249–269.
Aihara, K., Takabe, T., and Toyoda, M., 1990, Chaotic neural networks, *Phys. Lett. A.*, 144:333–340.
Caianiello, E. R., 1961, Outline of a theory of thought processes and thinking machines, *J. Theor. Biol.*, 1:204–235.
Cole, K. S., 1968, *Membranes, Ions and Impulses*, Berkeley: University of California Press. ◆
FitzHugh, R., 1969, Mathematical models of excitation and propagation in nerve, in *Biological Engineering*, (H. P. Schwan, Ed.), New York: McGraw-Hill, pp. 1–85. ◆
Glass, L., and Mackey, M. C., 1988, *From Clocks to Chaos*, Princeton, NJ: Princeton University Press. ◆
Hodgkin, A. L., and Huxley, A. F., 1952, A quantitative description of membrane current and its application to conduction and excitation in nerve, *J. Physiol. (Lond.)*, 117:500–544.
McCulloch, W. S., and Pitts, W. H., 1943, A logical calculus of the ideas immanent in neural nets, *Bull. Math. Biophys.*, 5:115–133.
Moore, C., 1990, Unpredictability and undecidability in dynamical systems, *Phys. Rev. Lett.*, 64:2354–2357.
Nagumo, J., Arimoto, S., and Yoshizawa, S., 1962, An active pulse transmission line simulating nerve axon, *Proc. IRE*, 50:2061–2070.
Ruelle, D., 1989, *Chaotic Evolution and Strange Attractors*, Cambridge, Eng.: Cambridge University Press. ◆
Segundo, J. P., 1986, What can neurons do to serve as integrating devices?, *J. Theoret. Neurobiol.*, 5:1–59.
Skarda, C. A., and Freeman, W. J., 1987, How brains make chaos in order to make sense of the world, *Behav. Brain Sci.*, 10:161–195.
Yamaguti, M., Ed., 1994, *Towards the Harnessing of Chaos*, Amsterdam: Elsevier Science.

Chaos in Neural Systems

Leon Glass

Introduction

Chaotic dynamics, i.e., aperiodic dynamics in deterministic systems displaying acute sensitivity to initial conditions, has emerged from a relatively obscure topic of mathematics to an area that is of great current interest amongst scientists as well as the general public.

The important characteristics of chaos are the apparent irregularity of time traces and the divergence of the trajectories over time (starting from two nearby initial conditions) in a system that is deterministic. Although the rhythm is irregular, the underlying deterministic equations can lead to structure in the dynamics.

Deterministic chaotic dynamics is different in principle from what is commonly known as random dynamics. In random dynamics, prediction is intrinsically impossible, except in a statistical sense. A natural system believed to be random is the radioactive decay of isotopes leading to random time intervals between decay events.

Chaotic dynamics is now well documented in a variety of different mathematical models, physical systems, and biological systems (Cvitanovic, 1989). Although the technical aspects of the field can be presented using very elementary methods (Devaney, 1992), mathematical rigor is not always easy to translate to the biological domain.

Experimental recordings of neural activity are notable for complex dynamic behavior. Examples include: fluctuations in the electrical current through ionic channels observed during patch clamp recordings, the spike trains recorded from individual neuron cells using intracellular electrodes, and the activities of collections of nerve cells reflected in electroencephalograms. In this article, I discuss the origin of these complex behaviors from a nonlinear dynamics perspective.

In the next section, I describe basic methods that have been used to identify chaos. Then I describe attempts to apply these methods to neural systems at the subcellular and cellular level, and finally I turn to neural networks. I emphasize that solid demonstrations of chaotic dynamics in spontaneous activity are very difficult to obtain.

Identifying Chaos

Nonlinear Equations

Chaotic dynamics occurs in deterministic systems. In a deterministic system, there is a definite rule governing the dynamics, which is usually expressed as an equation. For example, there are now many well-known model systems formulated as difference equations, differential equations, difference-differential equations, and partial differential equations that display chaotic dynamics.

As long as one is dealing with deterministic equations, the signature of deterministic chaos is clear. It is easy to numerically integrate the equations starting from two nearby initial conditions and to watch the evolution. To argue that the system is chaotic, one should observe aperiodic dynamics with the same statistical properties (such as density histograms) for both trajectories. In addition, the trajectories should diverge but remain in a bounded region. Moreover, many of these equations show characteristic bifurcations (changes in qualitative dynamics) as parameters change that are representative of well-studied routes to chaos. For example, many different theoreti-

cal and experimental systems display successive doublings of the period of an oscillation before the establishment of chaotic dynamics (Cvitanovic, 1989).

Operational Tests for Chaos

When dealing with dynamics in experimental systems or in natural systems, it is much more problematic to figure out the dynamic origins of the complex rhythms. Many different methods have been developed to characterize time series. I briefly mention these methods and describe their relevance to the identification of chaotic dynamics. The bottom line is that there is not a good operational definition for deterministic chaos. Consequently, many of the claims for chaotic dynamics in biology must be viewed with skepticism. An excellent critical review is given by Grassberger, Schreiber, and Schaffrath (1991).

Power spectrum. The power spectrum of a time series can be determined using fast Fourier transform algorithms. Many natural processes show a power spectrum that goes as $1/f^\alpha$ where $\alpha \approx 1$ for low frequencies. Although some workers have interpreted $1/f$ power spectra as a signature for chaos, many random models also show this behavior. Since one can always construct random models that have the same power spectrum as a deterministic model, the power spectrum is not a good measure for defining chaos.

Dimension. Grassberger and Procaccia (1983) defined the correlation dimension of a set of points. Call r the distance from a given point. The point correlation function, $C_i(r)$ is proportional to the number of points lying within a radius r of a point i. The spatial correlation function, $C(r)$, is an average of the $C_i(r)$ over all points i. Then the correlation dimension is v if $C(r) = kr^v$ for small values of r. For random points distributed in d dimensions, $C(r) = kr^d$, so that the dimension is an integer equal to the dimension of the space in which points are distributed. Thus a line has dimension 1, a plane has dimension 2, and so forth. However, some sets of points, called *fractals*, have a fractional dimension. Some workers take a fractional correlation dimension as a definition for deterministic chaos, since many chaotic dynamical systems have attractors with a fractional dimension. However, there are many pitfalls associated with determining the dimension, and not all chaotic systems have a fractional dimension. Consequently, a fractional correlation dimension cannot be accepted as a definition for chaos.

Lyapunov exponent. The Lyapunov exponent measures the rate at which trajectories diverge. A negative Lyapunov exponent indicates convergence of trajectories, and a positive Lyapunov exponent indicates divergence of trajectories. Therefore, for a time series, a positive Lyapunov exponent is a necessary condition for chaotic dynamics. However, the numerical algorithms for determination of the Lyapunov exponent may yield a positive exponent for periodic oscillations with large first derivatives, or in noisy systems. Therefore, a positive Lyapunov exponent found using numerical methods from naturally occurring data cannot be taken as a definition for chaos.

Prediction. Since chaotic dynamics are generated by deterministic equations, it should be possible to predict dynamics that will occur in the proximate future. Several methods have been developed to try to make predictions based on the past dynamics without necessarily knowing the equations of motion, by estimating local properties of the dynamics in some appropriate phase space. A confounding factor of prediction tests is that random inputs to linear systems lead to linear correlations for short times that enable prediction even though the system is not deterministic and hence not chaotic.

Surrogate data.   None of these measures can be simply applied to determine if a given data set displays deterministic chaos. Consequently, several people have advocated generating "surrogate" data sets that have similar statistical properties to a given data set but are generated from a random process (Theiler et al., 1992). For example, it is a simple matter to generate a random time series that has an identical power spectrum to a given time series. Statistical analyses are then carried out using the time series under investigation and surrogates. The results of the analysis of the surrogate data are compared with the analysis of the original data. The origins of significant differences are then analyzed. They may be attributable to deterministic chaos, though other hypotheses, such as failure to capture some important aspect in the surrogate data set, need to be entertained. Application of this technique has led to the recognition that many earlier claims for deterministic chaos were in error.

Chaos at Subcellular and Cellular Levels

Carefully controlled experiments, in which it is possible to generate large amounts of high-quality data while varying critical control parameters, have demonstrated chaotic dynamics in chemical, electronic, and hydrodynamics systems (Cvitanovic, 1989). Observations of chaotic dynamics in biological and in particular neural systems has been much more problematic. Sensible people disagree in their interpretation of the same sets of experimental data. I will review critically some of the areas in which there have been claims for chaotic dynamics.

Subcellular Dynamics: Ion Channels

The basis for neural activity lies in the changes of conductance of specialized protein molecules called ion channels that allow ions to pass between the intracellular and extracellular medium. For example, action potentials of nerve cells are usually associated with the opening of sodium ion channels. Classical intracellular electrophysiological techniques average the activity of the thousands of ion channels present on a single nerve cell. Macroscopic ionic models, the earliest and most famous of which is the Hodgkin-Huxley equation, model this average activity and its dependence on the membrane voltage. An important technological advance, called the patch clamp, enabled researchers to measure directly the conductance of a single ion channel during its open state. Experimental recording of a single ion channel typically shows a complex switching behavior in which the channel makes transitions between open and closed states in a seemingly random manner. Most researchers believe that this switching is random and have developed theoretical descriptions based on Markov-type models (see Synaptic Currents, Neuromodulation, and Kinetic Models). However, deterministic chaotic models may show the same statistical features as the Markovian models (for review, see Liebovitch, 1994).

Molecular biological methods are giving us increasing insight into the molecular structure of the ion channels and enabling a classification of families of different channels. The

structural information has yet to be translated into an understanding of the control of the conformational changes that underly the channel activity. In view of the molecular scale of the ion channels, it seems unlikely that it will be necessary to invoke deterministic chaos to interpret the opening and closing of ion channels; rather, it seems that Markovian models will prevail.

The observation that subcellular ion channel events are following a random (or perhaps chaotic) dynamics raises a conceptual problem: "Why is it possible to study cellular or supercellular processes using deterministic models such as the Hodgkin-Huxley equations?" Presumably, the thousands of channels in a single cell, and millions and billions in collections of cells, enable an averaging so that macroscopic equations are valid, with fluctuations being small. We can therefore deal with ion kinetics in nerve cells the same way that physicists carry out deterministic computations as a consequence of averaging over suitably large ensembles.

Cellular Activity in Single Cells

The electrical activity of single cells can be easily measured using intracellular or extracellular microelectrodes, either in vitro or in vivo. The activity can be measured during spontaneous activity or in response to manipulation environmental parameters. In my opinion, the most convincing demonstrations of chaotic dynamics in biological systems have been carried out in this scale of preparation.

Periodic stimulation of oscillatory or excitable biological systems, such as nerve or cardiac tissue, has generally identified chaotic dynamics over restricted ranges of stimulation parameters (Glass and Mackey, 1988; see also Chaos in Axons). Over other stimulation ranges, there are regular rhythms observed in which there are repeating sequences of stimuli and action potentials. This work is placed on a firm footing by extensive theoretical analyses and numerical simulation mathematical models of the experimental systems. The mathematical models are formulated as simplified maps, simplified differential equations, or realistic ionic models. Depending on the stimulation parameters, the theoretical models display periodic behavior or deterministic chaos in agreement with experimental observations. Because one has deterministic equations in which chaos can be demonstrated, the identification of chaos in the biological system is reasonably secure.

Spontaneous activity from single cells can also show complex fluctuations of activity. An example is in the work by Mpitsos et al. (1988), recording spontaneous activity from molluscan pacemaker neurons. Detailed analysis and data manipulation claimed that activity from a single cell was well described by a one-dimensional map known to generate chaotic dynamics. However, better controls and reproducibility are needed to lend credibility to these claims.

There is no clear functional role for chaos at a cellular level. However, a recent intriguing speculation is that stochastic dynamics may be useful in some detection tasks. *Stochastic resonance* describes the concept that, in some transducers, such as stretch or mechanoreceptors, the signal-to-noise ratio is optimal at an intermediate level of noise (Moss, Bulsara, and Shlesinger, 1993). For example, imagine a subthreshold oscillation (induced by a weak input) of a nerve membrane. In the presence of noise, the oscillation might be raised to a firing threshold by an additive noise. Because of the underlying oscillation, firing times would occur at multiples of the period of the underlying oscillation. The concept of stochastic resonance and the mechanisms that have been suggested for stochastic resonance would be equally applicable if deterministic chaos rather than random noise were generating the additive signals. Consequently, chaos might play a role in optimizing the performance of such transducers, but clear demonstrations of operation of this phenomenon in intact organisms have not yet been made. In this mechanism, either chaotic or random inputs would be effective.

Chaos in Neural Networks

Networks of neurons regulate major bodily functions in animals, ranging from breathing and control of heartbeat variability to the cognitive and intellectual functions of humans. A clear theoretical understanding of the dynamics that underlie neural functioning is still not at hand. In particular, although deterministic chaos has occasionally been invoked as an essential feature in the functioning of neural networks, the experimental evidence supporting the existence of chaos in neural networks is still meager. I discuss separately deterministic chaos in neural control circuits and chaos in cortical networks and cognitive processes.

Chaos in Neural Control Circuits

Respiration.   Respiration is a good example of a system that displays complex rhythms. Although early studies demonstrated a broad power spectrum associated with respiration, more recent work has measured the dimension and Lyapunov exponent of time series from respiration (Sammon and Bruce, 1991). There have been claims that respiration in animal models and in humans can display chaotic dynamics. Although respiratory rhythms can be irregular, I believe that it is nevertheless difficult to assert chaos based on the analyses carried out to date.

If there were chaotic dynamics in respiration, the origin of this behavior would be important to elucidate. Most attempts at deriving theoretical models for respiration have assumed that the respiratory rhythm can be associated with a stable limit cycle oscillation yielding periodic dynamics (see RESPIRATORY RHYTHM GENERATION). However, it is possible that the feedback systems that regulate respiration give rise to deterministic chaos. Since nonlinear feedback can give rise to deterministic chaos as a consequence of nonmonotonic feedback control or multiple feedback loops (Glass and Mackey, 1988; Glass and Malta, 1990), the operation of the respiratory control system might lead to deterministic chaotic dynamics.

Heart rate variability.   Similar considerations apply to the analysis of heart rate variability. The cardiac pacemaker (a specialized bit of cardiac muscle) is modulated through the sympathetic and parasympathetic nerve activity, as well as the level of circulating hormones. Heart rate varies as a consequence of coupling to the respiratory system (respiratory sinus arrhythmia) and also as a consequence of a whole host of feedback control circuits governing blood pressure control. Numerous claims for chaotic dynamics of heart rate variability have been based on computation of measures such as dimension, Lyapunov exponent, broad band power spectra, and $1/f$ power spectra (Goldberger, Rigney, and West, 1990). These are not conclusive tests for deterministic chaos. Thus, although there is significant variability in normal heart rate, there is not a clear demonstration that this variability is deterministic chaos.

Chaos and health.   Goldberger et al. (1990) correlate diminished heart rate variability with a higher risk of sudden cardiac death. However, since patients do not die of reduced heart rate variability, it is important to clarify the underlying mechanisms. One possibility relates to sympathetic activity. Impaired cardiac function leads to increased sympathetic activity associated with elevated levels of catecholamines. This in turn increases the heart rate, leading to less variability and possibly an increase in the risk of arhythmia. Patients with impaired heart function are also more likely to take drugs that blunt fluctuations of heart rate or even to have implanted artificial pacemakers that might lead to a more regular rhythm. Clinical studies that associate higher risk for sudden cardiac death with decreased heart rate variability do not always address these potentially confounding factors.

Oversimplifications (e.g., "it is healthy to be chaotic") must be viewed warily. Many different types of serious respiratory and cardiac arhythmias are associated with highly irregular dynamics, so there is no a priori reason why irregular dynamics should be associated with improved function. Of course, failure to respond appropriately to a fluctuating environment is a sign of poor health.

Chaos in Cortical Function

Chaotic dynamics has been demonstrated in simplified theoretical models (Hopfield-type models) of neural networks (Lewis and Glass, 1992). However, the experimental evidence for chaotic dynamics in neural networks underlying higher cortical functions is scanty.

An enormous amount of effort has been directed toward carrying out analysis of electroencephalogram (EEG) signals. Over the past decade, numerous studies computed the dimension and Lyapunov exponent of EEG traces, concluding that the EEG signal was chaotic (Elbert et al., 1994). This approach was criticized at an early stage (Glass and Mackey, 1988). There is now a growing recognition of the difficulties associated with application of these statistical measures to assert deterministic chaos in the EEG (Albano and Rapp, 1993; Glass, Kaplan, and Lewis, 1993). Nevertheless, different states of alertness are reflected in differences in the dimension of the EEG. Although the functional significance of these variations is still obscure, it is possible that the dimension will be useful to characterize EEG dynamics.

Another attempt to analyze higher neural function used recordings from the olfactory bulb of rabbits (Skarda and Freeman, 1987). In the resting stage, the irregular fluctuation of neural activity was associated with deterministic chaos. In response to an odor, there was a transient reproducible waveform that depended on the odor delivered to the animal. Freeman believes that chaos plays an important role in information processing. The basic idea is that many different types of cycles associated with different odors emerge from the chaotic background. The criticisms raised earlier with respect to identification of chaotic dynamics apply equally to Freeman's work. Despite attempts to formulate theoretical models of the observed system dynamics, it seems premature to claim that the observed behavior is deterministic chaos. Consequently, Freeman's claims that chaotic dynamics plays an essential role in cognitive function and information processing must be viewed as hypothesis rather than established fact.

Discussion

Theoretical studies have demonstrated chaotic dynamics in models of ion channels, single cells in spontaneous activity, single cells subjected to periodic stimulation, feedback control networks, and spontaneously active neural networks. These theoretical demonstrations of the possibility for chaotic dynamics have not yet been matched by equally convincing experimental studies. The only definite observation of chaotic dynamics in neural systems is found in the response of nerve cells to periodic stimulation (see CHAOS IN AXONS).

The high dimensions of neural systems, the fluctuating environment leading to nonstationarity, and the subtleties of the numerical methods are factors that make convincing demonstration of chaos *in vivo* a difficult matter. Therefore, though it might not be surprising that clear demonstrations of deterministic chaos in neurobiology are rare, it is surprising that exaggerated claims have been founded on such weak experimental evidence, and that so many have accepted claims for chaos without sufficiently critical analysis.

Two fundamental issues concerning complex neural rhythms remain. One issue is to understand the physical basis of the rhythm, using methods from mathematics and physics to help interpret the observed activity. There are well-developed methods for studying deterministic chaotic dynamics and for studying random systems. Is a deterministic or a random model more suitable to describe some complex time series? Since real biological systems (as opposed to computer models of these systems) are exposed to random thermal changes and other fluctuations, and are composed of very large number of interacting elements (i.e., they are high-dimensional), it is still not easy to decide the appropriate class of theoretical model for a given system.

Another difficult issue is to understand the biological significance, if any, of the observed fluctuations. Consider complex fluctuations in the timing between spikes of a neuron: does the neuron function well despite these irregularities, or are the irregularities essential to its task? And if the irregularities are essential for the task, is there any reason to expect that deterministically chaotic irregularities would be better than random ones?

Probably what drives many in brain theory is to understand the neural correlates of the human mind: higher cognitive function, originality, and free will. We neither behave in a random manner nor in a totally predictable manner. Although it seems inevitable that the mathematical concept of chaos will help us interpret the human brain, the definite results in that direction are still meager.

Acknowledgments.   This work was partially supported by the Natural Sciences and Engineering Research Council (Canada).

Road Map: Dynamic Systems and Optimization
Background: I.3. Dynamics and Adaptation in Neural Networks

Related Reading: Dynamics and Bifurcation of Neural Networks; Epilepsy: Network Models of Generation

References

Albano, A. M., and Rapp, P. E., 1993, On the reliability of dynamical measures of EEG signals, in *Proceedings of the Second Annual Conference on Nonlinear Dynamical Analysis of the EEG* (B. H. Jansen and M. E. Brandt, Eds.), Singapore: World Scientific, pp. 117–139.

Cvitanovic, P., Ed., 1989, *Universality in Chaos*, 2nd ed., Bristol: Adam Hilger. ◆

Devaney, R. L., 1992, *A First Course in Chaotic Dynamical Systems: Theory and Experiment*, Reading, MA: Addison-Wesley. ◆

Elbert, T., Ray, W. J., Kowalik, A. J., Skinner, J. E., Graf, K. E., and Birbaumer, N., 1994, Chaos and physiology: Deterministic chaos in excitable cell assemblies, *Physiol. Rev.*, 74:1–47. ◆

Glass, L., Kaplan, D. T., and Lewis, J. E., 1993, Tests for deterministic dynamics in real and model neural networks, in *Proceedings of the Second Annual Conference on Nonlinear Dynamical Analysis of the EEG* (B. H. Jansen and M. E. Brandt, Eds.), Singapore: World Scientific, pp. 233–249.

Glass, L., and Mackey, M. C., 1988, *From Clocks to Chaos: The Rhythms of Life*, Princeton, NJ: Princeton University Press. ◆

Glass, L., and Malta, C. P., 1990, Chaos in multi-looped negative feedback systems, *J. Theor. Biol.*, 145:217–223.

Goldberger, A. L., Rigney, D. R., and West, B. J., 1990, Fractals and chaos in human physiology, *Sci. Am.*, February:43–49. ◆

Grassberger, I., and Procaccia, I., 1983, Measuring the strangeness of strange attractors, *Physica D*, 9:189–208.

Grassberger, T., Schreiber, T., and Schaffrath, C., 1991, Nonlinear time series analysis, *Int. J. Bif. Chaos*, 1:521–547. ◆

Lewis, J. E., and Glass, L., 1992, Nonlinear dynamics and symbolic dynamics of neural networks, *Neural Computat.* 4:621–642.

Liebovitch, L. S., 1994, Single channels: From Markovian to fractal models, in *Cardiac Electrophysiology: From Cell to Bedside*, 2nd ed. (D. P. Zipes and J. Jalife, Eds.) (in press). ◆

Moss, F. M., Bulsara, A., and Shlesinger, M. (Guest Eds.), 1993, *J. Stat. Phys.*, nos. 1/2, Proceedings of the NATO Advanced Research Workshop on Stochastic Resonance in Physics and Biology.

Mpitsos, G. J., Burton, R. M., Creech, H. C., and Soinila, S. O., 1988, Evidence for chaos in spike trains of neurons that generate rhythmic motor patterns, *Brain Res. Bull.*, 21:529–538.

Sammon, M. P., and Bruce, E. N., 1991, Vagal afferent activity increases dynamical dimension of respiration in rats, *J. Appl. Physiol.*, 70:1748–1762.

Skarda, C. A., and Freeman, W. J., 1987, How brains make chaos in order to make sense of the world, *Behav. Brain Sci.*, 10:161–195.

Theiler, J., Galdrikian, B., Longtin, A., Eubank, S., and Farmer, J. D., 1992, Using surrogate data to detect nonlinearity in time series, in *Nonlinear Modeling and Forecasting* (E. Casdagli and S. Eubank, Eds.), Boston, MA: Addison-Wesley, pp. 163–188.

Classical Learning Theory and Neural Networks

Bennet B. Murdock

Introduction

Classical learning theory refers to a body of theory and data that has tried to elucidate the fundamental principles of learning and memory. It represents the best efforts of experimental psychologists to understand learning and memory. As psychologists, they rely on behavioral data to test their theories and models. Both learning theorists and neural modelers view elementary associative processes as the building blocks of learning.

A common assumption shared by many neuroscientists is that one must understand the structure and function of the nervous system before gaining an understanding of how the brain works. Psychologists have taken a different approach. They maintain that behavior is the end result of brain functioning, and the consequences of hypothetical brain processes can be observed directly in this behavior. Their approach has been to postulate conceptual brain processes and then derive or simulate the behavioral consequences. The success or failure of this enterprise depends on the correspondence between the predicted outcome and the behavioral data obtained under controlled laboratory conditions.

Classical Learning Theory

Historically (and neglecting the contribution of a few pioneers before 1940), there have been three major learning theories: the reinforcement theory of Hull, Miller, and Spence; the stimulus-sampling theory of Estes, Suppes, and Atkinson; and the interference theory of Underwood and Postman. Reinforcement theory dealt with principles of animal learning and first appeared in Hull's *Principles of Behavior*. The stimulus-sampling theory of Estes was not species-specific, but tried to provide a general framework for behavioral change. Interference theory dealt with principles of human learning and memory.

The reinforcement theory of Hull followed in the tradition of the animal learning work of Thorndike and the classical conditioning work of Pavlov. This theory had as its two major variables drive strength and habit strength. Drive strength was a motivational variable, whereas habit strength was a result of learning, and the two combined multiplicatively to determine reaction potential. There were various measures of reaction potential (response amplitude, probability, latency, and resistance to extinction). Behavioral oscillation (variability in the underlying processes) introduced variability in responding. Inhibitory factors worked against excitatory factors (reaction potential), and discrimination was the algebraic sum of the two. Stimulus generalization was another important concept; similarity between the test and training stimulus was another determinant of the observable behavior.

Stimulus-sampling theory postulated a set of abstract elements that were conditioned (associated to some response) or not conditioned to a response, and that were available or not available (two subsets) at the time of test. Only the number (proportion) of conditioned elements in the active (available) subset determined the probability of a response. Over time, there was random interchange between the two subsets, and this interchange provided a natural interpretation for changes in context that occurred over time. It also provided a natural explanation for forgetting—the stimulus-sampling theory of encoding variability of Bower assumed a binomial distribution of conditioned elements (stimulus attributes), and this theory was able to deduce a large number of empirical effects in verbal learning and memory.

The basic premise of interference theory was that forgetting occurs through interference, not through decay. Two types of interference were retroactive interference (RI) and proactive interference (PI). RI is forgetting of some specified material owing to the acquisition (learning) of some subsequent material. PI is forgetting of some specified material as a result of the acquisition (learning) of some prior material. If we designate items as A, B, and C, then an associated pair would be A-B or A-C. In RI, we learn A-B, then learn A-C, then try to recall A-B. In PI, we learn A-B, then learn A-C, then try to recall A-C. One of the major achievements of interference theory was to document beyond all reasonable doubt the existence of these two phenomena (RI and PI), but it was never able to provide a satisfactory theoretical synthesis.

Empirical Effects

A large number of empirical effects have been documented in experimental studies of learning and memory (see Benjafield, 1992; Broadbent, 1958; Greene, 1992). These can be organized into the storage and retrieval of item information, associative information, and serial-order information. *Item information*

underlies our ability to recognize single objects or events, such as names, faces, words, tastes, odors, etc. *Associative information* binds two ideas together—a name and a face, a word and its meaning, or abstract concepts. *Serial-order information* preserves the order in a sequence of items—a telephone number, an access code, the spelling of words. There is much empirical evidence to justify the claim that item, associative, and serial-order information are different (Murdock, 1974).

Memory for item information drops off exponentially with time, even for periods as long as a year. Recognition memory for pictures is better than recognition memory for words, but recognition memory for rare words is better than recognition memory for common words. The mirror effect shows that, for a number of different variables, if A is better remembered than B, then the hit rate (proportion of "yes" responses to old items) will be higher for A than B and the correct rejection rate (proportion of "no" responses to new items) will also be higher for A than B. This mirror effect rules out a simple strength-theory model of item recognition based on the principles of signal-detection theory. (Signal-detection theory is a precursor of Anderson's matched-filter model, to be described later.)

Memory for associative information does not show much forgetting, at least over short intervals of time. However, working-memory effects are very pronounced; one or two intervening pairs (either studied or tested) markedly reduce the recall of other pairs. In RI and PI, similarity effects are pronounced, but the locus of the similarity is critical. If we change our symbolism slightly, we can write the RI paradigm as A-B, A'-D, where the similarity of A and A' can vary from maximum (identity, so A' = A) to minimum (none, so A' is a different item; call it C). Now we learn A-B, learn A'-D, then try to recall A-B. The more similar A and A' (up to identity), the worse the retention of A-B. (PI is the reverse of this; it is the second item that must be recalled.) However, if the two lists are A-B, C-B', there is not much effect of the B-B' similarity (i.e.,

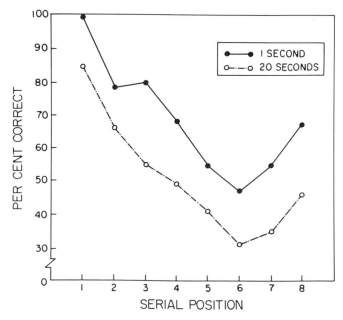

Figure 1. Probability of correct recall of letters in an eight-item list after a 1- or 20-second delay as a function of serial position. (From Parkinson, S. R., 1972, Short-term memory while shadowing, *J. Exp. Psychol*, 92:262. Copyright 1972 by the American Psychological Association. Reprinted by permission.)

retention of A-B after C-B is not much worse than retention of A-B after C-D).

Serial-order effects are quite different from item and associative effects. If we ask a person to recall a short list of items in order, the data show pronounced primacy effects (Figure 1). This almost never happens with item or associative information. On the other hand, item recognition or probe recall (what item followed item *i*?) show marked recency effects.

Memory for items and memory for order are dissociable; for instance, similarity facilitates memory for item information, whereas it impairs memory for order information. Transpositions (juxtaposition of two items) are rare events but very regular; their frequency drops off as the distance between the two items increases. Chunks are formed with repetition; a chunk is a "glob" of information (Miller, 1956) which can function as a higher-order unit or can be unpacked into its constituents if necessary. (We use words in sentences, but we can spell them if asked to do so.)

This brief discussion does not begin to exhaust the list of important empirical effects (and there is not space to discuss recent developments in priming and implicit memory), but it does provide a set of core data that current memory models must consider.

Current Memory Models

Current memory models began with the buffer model of Atkinson and Shiffrin (1968), which highlighted the distinction between short-term and long-term memory. The short-term store contained a rehearsal buffer with a small number (2–4) of storage registers, generally one per item. While items were resident in the buffer, they built up strength in long-term store through a transfer process, but once an item was "bumped" from the buffer (like musical chairs), decay in long-term memory started. A schematic representation of this model is shown in Figure 2. The model was supported by subsequent evidence of the experimental separation of short-term and long-term effects. For instance, in free recall of lists of unrelated words, the modality of presentation (auditory or visual) only affected the short-term component, but the presentation rate or list length only affected the long-term component (Murdock, 1974).

Holographic memory models of the 1970s (see Hinton and Anderson, 1981), along with work by Kohonen and his collaborators (see Kohonen, Oja, and Lehtiö, in Hinton and Anderson, 1981), pioneered the notion of distributed representation. They did not have much impact on cognitive psychology, perhaps because we could not see any testable implications. However, the seminal work of Anderson (Anderson et al., 1977) changed all that. While his matched-filter and linear-associator models were prototypical connectionist models, they also had clear-cut empirical consequences as well. This was brought out in his applications of the matched-filter model to data from reaction-time experiments and to categorical perception and probability learning in the linear-associator model. Anderson also presented some very convincing arguments as to why von Neumann–type computer models (e.g., Atkinson and Shiffrin's buffer model) were less plausible biologically than models which used distributed representation and parallel processing.

The matched-filter and linear-associator models were early examples of what are now generally referred to as global-matching models. By *global-matching models*, we mean models in which the comparison process (in recognition) or the retrieval process (in recall) operates globally, not locally. In recognition, the probe does not just make contact with the target

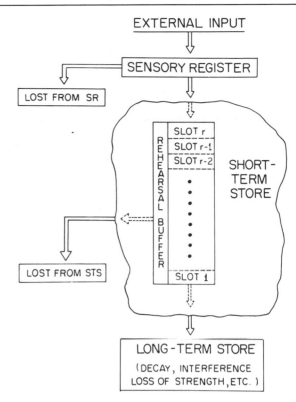

Figure 2. A schematic representation illustrating the operation of the buffer model of Atkinson and Shiffrin (1968). Items in the short-term store can be maintained by rehearsal, and the longer they reside in the short-term store, the greater the build-up of strength in the long-term store. Items are lost from the sensory register (SR) by decay, but they are displaced from the short-term store (STS) by succeeding items. [From Atkinson, R. C., and Shiffrin, R. M., 1968, Human memory: A proposed system and its control processes, in *The Psychology of Learning and Motivation: Advances in Research and Theory*, vol. 2 (K. W. Spence and J. T. Spence, Eds.), New York: Academic Press, p. 113. Reprinted with permission.]

item; instead, one way or another (and the models differ as to how), a number of items enter into the comparison process, but the system resonates most strongly to the probe. In recall, the retrieval operation also operates on more than a single trace, but again the target item dominates over its possible competitors.

The matched-filter model is a simple but elegant model for item recognition which represents items as *N*-dimensional vectors and stores them in a common memory vector. On a probe test, the subject is assumed to compare an encoded version of the probe item to the memory vector by taking the dot product and responding "yes" or "no" depending upon the output of the decision system. Because it is a linear model, it is mathematically tractable, and its predictions are in general accord with much experimental data (but not the mirror effect).

The Search of Associative Memory (SAM) model, proposed by Shiffrin, Raaijmakers, and their colleagues (e.g., Gillund and Shiffrin, 1984), is an updated version of the buffer model. Like the matched-filter model and the linear-associator model, it is a global-matching model. There is some elaboration of the long-term store and a wealth of parameters to explain the myriad experimental effects in item and associative recognition and recall. The MINERVA2 model of Hintzman (1988) is a multiplex model whereby items are stored separately but come together at the time of retrieval. It is particular strong in explaining some of the data from studies of prototypes and judgments of frequency. The matrix model of Humphreys, Bain, and Pike (1989) is an expanded version of Anderson's matched-filter model and is designed to encompass item recognition and associative recognition and recall in a single theoretical framework. It is particularly good at handling the diversity of experimental paradigms which have been employed in recent studies of cognitive processing. Finally, convolution-correlation models—e.g., the CHARM model of Metcalfe-Eich (1982) and the TODAM2 model of Murdock (1993)—use a vector representation rather than a matrix representation for associative information. The TODAM2 model of Murdock attempts to subsume item, associative, and serial-order information in a common framework.

The basic idea of TODAM2 is that we store labeled chunks; a chunk is a sum of n-grams, where n-grams are n-way multiple autoconvolutions of sums of n item vectors. Labels and lebals (a *lebal* is the involution or mirror image of a label) give us two ways to access chunks and the information they contain. They also allow converging operations so the same retrieval task can be done in several ways, and this forms the basis for confidence judgments. Item and associative information are just special cases of chunks where $n = 1$ and $n = 2$, respectively. The TODAM2 model goes beyond TODAM1 or the CHARM model to explain a wider range of phenomena.

Discussion

Where are we now? The current cognitive revolution has considerably dimmed the luster of reinforcement theory and interference theory, but stimulus-sampling theory is still alive and well. The current global-matching memory models are a very active area of investigation, and they not only can explain many aspects of our current data base, but they have also uncovered some interesting new results (e.g., the list-strength effect). (The *list-strength effect* refers to the fact that recognition accuracy is not affected by the strength of other items in the list.) Rather than recount their successes, though, let me consider some of the major problems now facing these models.

1. *The mirror effect.* As noted, the mirror effect refers to the fact that the hit rate is higher and the correct-rejection rate is lower for the more memorable material. None of the current models can yet explain this. Glanzer et al. (1993) have proposed a model which does account for the mirror effect, but how it will account for other data remains to be seen.

2. *Catastrophic unlearning.* Some of the connectionist models show much faster forgetting (unlearning) of an A-B list after learning an A-D list than human subjects do, and the same effect holds for item recognition (Ratcliff, 1990). Various patches have been proposed, but the implications of these modifications have not yet been determined.

3. *Receiver-operating characteristic (ROC) slopes.* An ROC curve is a plot of hit rate as a function of false-alarm rate (the complement of the correct-rejection rate). The slope tells us the ratio of the standard deviations of the underlying old- and new-item strength distributions. Ratcliff and his colleagues have reported that it is consistently about 0.8 (hence, the old-item variance is about 50% greater than the new-item variance), regardless of the base strength itself. None of the current memory models predict this.

4. *Item-associative interaction.* As discovered by Hockley (1992), recognition for item information declines over short intervals of time (e.g., 30 seconds) more rapidly than does recognition of associative information, as determined by comparing intact pairs with rearranged pairs. This interaction is a real puzzle for global-matching models, especially those which posit a common storage for item and associative information. If these two kinds of information are stored in a common memory system, how can there be different rates of forgetting?

These models thus have their strengths and weaknesses, but the enterprise is still young and promising. They illustrate the active interplay between theory and data, long the hallmark of the experimental psychologist's approach to cognitive processes. They are process models, and they suggest what sorts of abstract processes must be involved in brain functioning. There are no real links to brain function yet, but their implementation in neural networks could be a very exciting next step.

Acknowledgment. Preparation of this paper was supported by NSERC operating grant 146 from the Natural Sciences and Engineering Research Council of Canada.

Road Map: Connectionist Psychology
Related Reading: Associative Networks; Distortions in Human Memory

References

Anderson, J. A., Silverstein, J. W., Ritz, S. A., and Jones, R. S., 1977, Distinctive features, categorical perception, and probability learning: Some applications of a neural model, *Psychol. Rev.*, 84:413–451.

Atkinson, R. C., and Shiffrin, R. M., 1968, Human memory: A proposed system and its control processes, in *The Psychology of Learning and Motivation: Advances in Research and Theory*, vol. 2 (K. W. Spence and J. T. Spence, Eds.), New York: Academic Press, pp. 89–195.

Benjafield, J. G., 1992, *Cognition*, Englewood Cliffs, NJ: Prentice-Hall.

Broadbent, D. E., 1958, *Perception and Communication*, New York: Pergamon Press. ◆

Gillund, G., and Shiffrin, R. M., 1984, A retrieval model for both recognition and recall, *Psychol. Rev.*, 91:1–67.

Glanzer, M., Adams, J. K., Iverson, G. J., and Kim, K., 1993, *Psychol. Rev.*, 100:546–567.

Greene, R. L., 1992, *Human Memory: Paradigms and Paradoxes*, Hillsdale, NJ: Erlbaum. ◆

Hinton, G. E., and Anderson, J. A., Eds., 1981, *Parallel Models of Associative Memory*, Hillsdale, NJ: Erlbaum. ◆

Hintzman, D. L., 1988, Judgments of frequency and recognition memory in a multiple-trace memory model, *Psychol. Rev.*, 95:528–551.

Hockley, W. E., 1992, Item versus associative information: Further comparisons of forgetting rates, *J. Exp. Psychol. [Learn. Mem. Cogn.]*, 18:1321–1330.

Humphreys, M. S., Bain, J. D., and Pike, R., 1989, Different ways to cue a coherent memory system: A theory for episodic, semantic, and procedural tasks, *Psychol. Rev.*, 96:208–233.

Metcalfe-Eich, J. M., 1982, A composite holographic associative recall model, *Psychol. Rev.*, 89:627–661.

Miller, G. A., 1956, The magical number seven, plus or minus two: Some limits on our capacity for processing information, *Psychol. Rev.*, 63:81–96. ◆

Murdock, B. B., 1974, *Human Memory: Theory and Data*, Potomac, MD: Erlbaum. ◆

Murdock, B. B., 1993, TODAM2: A model for the storage and retrieval of item, associative, and serial-order information, *Psychol. Rev.*, 100:183–203.

Ratcliff, R., 1990, Connectionist models of recognition memory: Constraints imposed by learning and retention functions, *Psychol. Rev.*, 97:285–308.

Cognitive Development

James L. McClelland and Kim Plunkett

Introduction

Research in cognitive development seeks to understand how it is that the impressive cognitive capacities of the human mind can develop, and how this process is controlled. A central question has always been, Is the main source of control the innate endowment of the organism? Is it experience that arises from the interaction of the organism with its environment? Or is it an intricate interplay between genetic endowment and experience? These issues have been under consideration by philosophers and cognitive scientists for centuries. While no one today would deny the importance of either innate factors or experience, the exact nature of the mechanisms that exploit these factors, and the extent to which these mechanisms are pre-programmed with specific domain content, remains a source of intense debate within developmental psychology and cognitive neuroscience. In this article we consider how connectionist models may contribute to the evolution of thinking about these crucial matters.

Many connectionist models of cognitive development have been driven by two primary considerations:

1. *An interest in the role of environmental structure.* Developmental connectionists have tended to adopt a minimalist strategy by employing relatively general learning devices in specific problem domains in order to determine the extent to which the inherent structure of a domain constrains the construction of internal representations and thereby the developmental process.
2. *An attempt to specify the nature of the mechanisms involved in the developmental process.* In particular, connectionist models have tended to focus on the issue of whether discontinuities in behavioral development reflect the maturation of distinct information processors or nonlinear changes resulting from learning in a more homogeneous system.

Most connectionist models of cognitive development employ a general learning algorithm (such as backpropagation) that computes small changes to the connection strengths in a network so as to reduce the output error for any given input pattern. It is, therefore, appropriate to conceive of learning in these networks as a process of gradient descent on a multi-dimensional error landscape. Although the device that drives learning (the learning algorithm) usually only promotes gradual change in the weight matrix, the uneven surface of the error landscape can result in relatively sudden, dramatic qualitative shifts in network performance. Conversely, the error surface may be relatively flat at a particular configuration of the weight matrix, and the behavioral consequences of weight changes may be comparatively minor. Learning in networks can easily result in periods of stable behavior interrupted by sudden discontinuities, even though the basic mechanism for learning is one of small, continuous change.

The interpretation of change in a connectionist network as a process of movement through an even landscape is closely connected to Waddington's conception of development as epigenesis. Waddington (e.g., 1975) suggested that development can be viewed as a trajectory through a landscape in which stable states are achieved by an organism when occupying relatively homogeneous regions of the landscape, whereas change is observed during periods in which the trajectory crosses a downward sloping surface. The trajectory along the epigenetic landscape is determined by an interaction of the constitution of the organism itself and the environment in which it is required to survive. The same can be said of connectionist network models. What connectionist models add to Waddington's picture is a framework for constructing explicit models of the process of developmental change. Some of the properties of connectionist models, such as their use of graded parameters (connection weights) and gradient-based learning rules for adjusting these weights (such as backpropagation and many other connectionist learning rules) make them particularly well suited to the study of these issues. For fuller reviews of connectionist models and their importance for cognitive development, see Plunkett and Sinha (1992) and McClelland (1994).

It should be noted that connectionist models discussed in what follows apply to the development of implicit knowledge—knowledge that can govern behavior without itself being accessible to overt report. Knowledge in connections is implicit knowledge in just this sense. One key issue in cognitive development is the extent to which such implicit knowledge actually underlies performance in particular tasks. Clearly explicit rules and explicit reasoning strategies are sometimes used—a case in point will be suggested below. Developmental connectionists, however, have tended to stress that the behavior that others may have accounted for in terms of explicit rules might in fact be captured implicitly in connection weights.

Rethinking the Need for Innate Knowledge

We now consider the first general issue noted above, namely the extent to which cognitive processes must rely on innate domain knowledge. There can be no doubt that there is some initial structuring of the nervous system before birth that strongly influences what is experienced and the form this experience takes, and thus how this experience initiates changes in the structure of the system. However, there are several arguments that have been given by nativists that lead them to postulate innate knowledge of specific concepts or principles (see McClelland, in press, for fuller discussion). Here, we consider what we take to be the most crucial of these arguments, namely, the argument that a general-purpose learning mechanism that does not exploit domain-specific constraints is insufficient to account for the knowledge children acquire. This argument is based on the false assumption that the general-purpose learning mechanisms offered by connectionist learning rules adhere to the same principles as classical associationist models of learning: that learning occurs by contiguity, and application of what is learned to new cases depends on a similarity-based generalization process. Given this assumption, evidence that generalization depends on anything other than surface similarity is taken as evidence that there must be some domain-specific knowledge that is brought to bear. While there is considerable debate about the exact role that input or surface similarity might play and the sources of input that might count as providing data relevant to assessing input similarity, it nevertheless seems reasonably clear that the immediate perceptual characteristics of stimuli do not always determine the basis for correct generalization (Keil, 1987).

Where this argument goes wrong is in its initial assumption. Connectionist learning algorithms such as backpropagation are capable of gradually discovering task-appropriate representations through the learning process. Hinton's (1989) family tree model is one of the best examples of this: Before learning, this network treats all members of each of two "families" as approximately equally similar (with random initial variations). But after learning the kinship relations among these individuals through training with examples, the acquired internal representations use one dimension of the representational space to distinguish the two families, and use other dimensions to position the individuals within each family so that those individuals who play structurally analogous roles have similar representations. In other words, the similarity structure changes over time, gradually becoming appropriate to the domain through the course of the learning process. Another example that makes somewhat closer contact with the developmental literature is Rumelhart's model of learning the structure of the semantic domain of living things (Rumelhart and Todd, 1993). In this example, the representations of concepts gradually differentiate to capture a conceptual hierarchy that first distinguishes plants from animals and then later distinguishes among subclasses of these major categories (e.g., birds and fish). This process of conceptual differentiation is reminiscent of the developmental process of categorical differentiation discussed by Keil (1979).

Stages in Cognitive Development

We now consider the developmental process itself, in particular the issue of stage-like transitions in cognitive development. Stage theories of cognitive development characterize distinct stages of development in terms of qualitatively different underlying principles and operations. Transitions between stages are explained in terms of endogenous factors and/or a complex interaction between developing internal structures and the environmental niches which regulate the unfolding of those structures. In this section, we review a connectionist model of cognitive development whose behavior develops in a stage-like fashion. The properties of representation in such a model are interesting because input does not change over the course of training, and a single learning algorithm (backpropagation) with a fixed learning rate is used as the mechanism driving change throughout the simulation. An important finding from this simulation work is that a single mechanism can exhibit stage-like behavioral properties (i.e., there are accelerations and decelerations in the behavior and indeed in the connection weights in the system) despite the fact that the representational changes in the underlying mechanism are governed by a simple homogeneous process.

The domain of the model is the balance scale problem, introduced by Inhelder and Piaget (1958) and later studied extensively by Siegler (1976, 1981) and others (particularly Ferretti and Butterfield, 1986). Children are shown a balance scale with varying weights on either side and at varying distances from the fulcrum (Figure 1). They are asked to judge which side will go down when the scale is released. Siegler has shown that children pass through a series of stages in which their responses appear to be determined by a succession of procedures that make differential reference to the dimensions of weight and distance. Children below the age of 4 respond relatively haphazardly in the task. By 5, nearly all children are in the first stage, where they focus exclusively on the number of weights on each side in making a decision. In the second stage, they incorporate the distance dimension, but only under those conditions where the weights are equal. In the third stage, their responses are confused under those conditions where both weights and distance differ. Finally, some individuals eventually behave in accordance with a procedure that amounts to

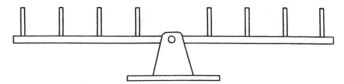

Figure 1. A balance scale of the type used first by Inhelder and Piaget and later by Siegler (1976, 1981). (Reprinted by permission from Siegler, 1976, fig. 1.)

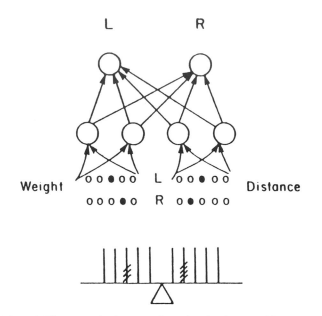

Figure 2. The connectionist network used to simulate acquisition of the balance scale task by McClelland (1989). Different input units code the number of weights placed on the Left and Right beams, and the positions at which they are placed. (Reprinted by permission from McClelland, 1989, fig. 2.7.)

computing the torque exerted by the weights on each side of the scale by multiplication of weight times distance. Above the age of 5, a large proportion (approximately 93%) of children's responses to the balance scale task fitted into one of these four categories (see Siegler, 1976, 1981, for details).

The connectionist model that captured the developmental progression seen in the balance scale task was presented by McClelland (1989; Shultz, Mareschal, and Schmidt, 1994, have a different connectionist model of the same task). The input to the model (Figure 2) is divided into two channels—in this case, a channel that represents the weights of the two objects on either side of the fulcrum and a channel that represents the distance of two objects from the fulcrum. There are five possible weight values for each object and five possible distance positions for each object. Weight and distance values are represented for each object by a five-place input field, in which there is a single unit assigned to each integer value of weight or distance on each side of the scale. However, the weight and distance units are themselves unstructured—i.e., weight and distance values are arbitrarily assigned to single input units. The network is not told explicitly which units correspond to large weights or distances, nor which units represent the weight or position of the left or the right object. The network must discover these correspondences through experiencing outcomes of balance scale problems—i.e., by discovering connection weights that allow it to predict which side will go down for various combinations of object weights and distances.

The network is trained on random samples from the possible combinations of object weights and distances on the balance scale. Training proceeds gradually through connection weight changes made in response to a random sequence of such problems. On each trial, the network's output is compared to the correct output for that problem, and the discrepancy between the actual output and the desired output is used to generate an error signal which is used by a backpropagation learning algorithm to adjust the connection strengths in both channels of the network. The network is tested at regular intervals on both trained and novel weight/distance combinations.

Two factors are crucial to the performance of the network. First, the network must be structured so as to treat the dimensions of weight and distance as separate. Second, some differential treatment of weight and distance is necessary to reflect the fact that children rely on the weight cue earlier in development than the distance cue. In McClelland (1989), this differential treatment amounts to incorporating more examples involving weight variations than distance variations in the training set, on the assumption that children may have more experience with weight than with distance as a factor in determining balance. In McClelland (in press), it is shown that differential initial use of weight versus distance can arise if the

weight on each side is a unary predicate (depending on the weight alone) while distance is a binary relation (a property arising from the relation between the weight and the fulcrum of the scale). Attention here focuses on the simpler case of more frequent exposure to variations in weight relative to distance. In both variants, the training sequence is stable over time. Discontinuity in development arises from the learning process, not from changes in the input.

At the beginning of training, the weights in the network are randomly set. Once training commences, the weight matrix gradually moves away from its random state, and after a time systematic behavioral patterns begin to appear. After an initial period of near-null output, there is a relatively rapid transition into conformity with the first of four rules developed by Siegler to characterize the stages of development of performance on the balance scale task. From this point on, the network's performance is impressive in that it can can be classified 85% of the time according to one of the first three of the four rules outlined by Siegler, and passes through the rules in the same sequence that characterizes children's development. The model does not reliably achieve the final stage of learning on the balance scale problem, which we take to be based on the use of an explicit multiplication strategy. (Siegler provides considerable evidence that performance in stage 4 is qualitatively different for crucial problems that rely on multiplication of weight times distance.)

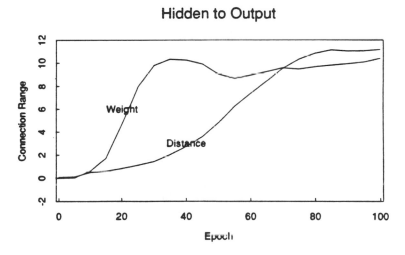

Figure 3. Sensitivity of the McClelland (1989) model to variations in weight and distance as cues to which side should go down on balance scale problems, as reflected by the input to hidden weights and by the hidden to output weights in the network (see Figure 2). The figure illustrates the periods of acceleration and deceleration seen in both pathways over the course of training. The index of sensitivity is based on absolute values of differences between connection weights from input to hidden units and from hidden to output units. For example, for sensitivity to object weight in the input to hidden connections, the graph shows the absolute value of the following difference: (the connection weight to the left-most hidden unit from the input unit coding for one weight placed on the left of the scale) minus (the connection weight to the same hidden unit from the input unit coding for five weights placed on the left of the scale). (Reprinted by permission from McClelland, 1989, fig. 2.12.)

Three aspects of the model commend the general approach it exemplifies for accounting for transitions in development in this and other domains:

1. The model shows periods of stasis followed by relatively abrupt transitions (Figure 3). Learning on the weight and then on the distance dimension is initially quite slow, then gradually accelerates until further changes fail to lead to further improvement in performance. The reason for the acceleration lies in the fact that the network must learn both how to encode the relevant dimension in the input to hidden weights, and how to use the results of this encoding to respond correctly (hidden to output weights). Changes at either level are effectively incoherent until the other level starts to become organized. This illustrates a crucial fact about development stressed extensively by Piaget: that the progress occasioned by a particular experience is sensitive to the existing state of knowledge, and that there must be a firm foundation before progress can be made (see Flavell, 1963, for discussion).

2. The model shows differential readiness to progress from one stage to the next at different points within the stage. Stage 1 typically lasts for quite a while in children, and the same thing is observed in the McClelland (1989) model. Both in the model and in children, exposure to a well-chosen set of problems leads to a transition to stage 2 or 3 if it occurs toward the end of stage 1, but either to no change or to regression to random performance if it occurs near the beginning. This differential readiness is a reflection of two factors in the model: first, the simple fact that weight changes accumulate, bringing the model closer to the point at which its behavior exhibits some sensitivity to distance; and second, the fact that the transition out of stage 1 corresponds to a point in development where weight changes in the distance pathway within the network accumulate rapidly. Very small changes build up during the early part of the time the network spends in a stage, leading to relatively abrupt transitions between stages.

3. Children's responses are not in fact exactly in line with the specific rules enumerated by Siegler, and the network captures many aspects of these discrepancies. Most interestingly, Siegler's rules are discrete, in the sense that they are not sensitive to different degrees of variation of either the weight or distance cue: The child either uses a cue (weight, distance) or not, according to Siegler's procedures. However, Ferretti and Butterfield (1986) and others have shown that children's behavior is strongly affected by the actual magnitude of the difference between the two sides on both the weight and the distance cues. McClelland (in press) shows that the model accounts quite well for these effects, whose relevance to the predictions of connectionist models was first pointed out by Shultz, Mareschal, and Schmidt (1994).

Finally, it may be noted that the model provides fairly clear links to the earlier work of Piaget (e.g., Piaget, 1952) in that it provides a simple but precise illustration of how a continuous developmental process of gradual adaptation can lead to periods of behavior that can be characterized as stage-like. The structural assumptions of the model require the assimilation of the input data to separate representations of weight and distance. The input assumptions of the model require the accommodation of the network's weights to sets of combinations of weights and distances which are repeatedly presented to the network. The interaction between structural and input assumptions can be seen as an embodiment of what Piaget would have called an *equilibration process*, that coordinates the representations of weight and distance in relation to the network's performance on the balance scale problem.

Discussion

The previous sections have shown that connectionist models provide powerful learning mechanisms that are capable of discovering appropriate internal representations. These mechanisms are more powerful than the learning mechanisms proposed by the classical associationists: in particular, they are not doomed to rely on raw input similarity as the basis for generalization. Because of their nonlinear, multilayer structure, connectionist models that learn are capable of exhibiting stage-like developmental progressions and other phenomena reminiscent of children's performance and of their developmental progress, as illustrated by the McClelland (1989) simulation of development in the balance scale task.

Connectionist models have also been applied to a number of other aspects of cognitive development. These include the development of "object permanence" (Munakata et al., 1994), the ability to order elements sequentially (Mareschal and Shultz, 1993), and the development of visually guided tracking and reaching (Mareschal and Plunkett, 1994). For a review of related connectionist-inspired models of language acquisition, see LANGUAGE ACQUISITION. Taken together, these other models illustrate the potential breadth of application of the approach, and lead to the expectation that there will be more such models in the future.

The main issues that need to be addressed in further work are as follows:

1. How much prestructuring is actually needed to account for cognitive development? The McClelland (1989) model was prestructured to reflect the dimensional distinctions of weight and distance, and it will be important to consider whether such prestructuring is necessary and under what circumstances. It seems certain that we should view the brain as a developing network of networks, rather than a single, unstructured homogeneous system, and it seems likely that this network of networks receives some initial structuring prior to exposure to inputs from the external environment.

2. What is the relationship between the acquisition of implicit knowledge, captured by connectionist models, and explicit cognitive functions, including the ability to describe the basis for responses in tasks such as the balance scale task, and the ability to use this explicit knowledge as the basis for further processes? Many examples of one-trial learning would seem to rely on this form of explicit knowledge. (See DEVELOPMENTAL DISORDERS for a discussion of this issue.)

It seems likely that these questions will motivate a great deal of future work, and that future developments in our understanding of learning processes in neural networks will shed considerable light on these issues.

Road Map: Connectionist Psychology
Related Reading: Developmental Disorders; Development and Regeneration of Eye-Brain Maps; Ocular Dominance and Orientation Columns

References

Ferretti, R. P., and Butterfield, E. C., 1986, Are children's rule-assessment classifications invariant across instances of problem types? *Child Dev.*, 57:1419–1428.
Flavell, J. H., 1963, *The Developmental Psychology of Jean Piaget*, Princeton, NJ: Van Nostrand. ◆

Hinton, G. E., 1989, Learning distributed representations of concepts, in *Parallel Distributed Processing: Implications for Psychology and Neurobiology* (R. G. M. Morris, Ed.), Oxford: Clarendon, pp. 46–61.

Inhelder, B., and Piaget, J., 1958, *The Growth of Logical Thinking from Childhood to Adolescence*, New York: Basic Books.

Keil, F. C., 1979, *Semantic and Conceptual Development: An Ontological Perspective*, Cambridge, MA: Harvard University Press.

Keil, F. C., 1987, Conceptual development and category structure, in *Concepts and Conceptual Development: Ecological and Intellectual Factors in Categorization* (U. Neisser, Ed.), Cambridge, Eng.: Cambridge University Press, pp. 175–200.

McClelland, J. L., 1989, Parallel distributed processing: Implications for cognition and development, in *Parallel Distributed Processing: Implications for Psychology and Neurobiology* (R. G. M. Morris, Ed.), Oxford: Clarendon, pp. 9–45.

McClelland, J. L., 1994, The interaction of nature and nurture in development: A parallel distributed processing perspective, in *International Perspectives on Psychological Science*, vol. 1, *Leading Themes* (P. Bertelson, P. Eelen, and G. d'Ydewalle, Eds.), Hillsdale, NJ: Erlbaum, pp. 57–88. ◆

McClelland, J. L. (in press), A connectionist perspective on knowledge and development, in *Developing Cognitive Competence: New Approaches to Process Modeling* (G. Halford and T. Simon, Eds.), Hillsdale, NJ: Erlbaum.

Mareschal, D., and Plunkett, K., 1994, Developing object permanence: Connectionist insights, Paper presented at the Annual Conference of the Developmental Section of the British Psychological Society, September, Portsmouth, UK.

Mareschal, D., and Shultz, T. R., 1993, A connectionist model of the development of seriation, in *Proceedings of the 15th Annual Conference of the Cognitive Science Society*, Hillsdale, NJ: Erlbaum, pp. 676–681.

Munakata, Y., McClelland, J. L., Johnson, M. H., and Siegler, R. S., 1994, *Now You See It, Now You Don't: A Gradualistic Framework for Understanding Infants' Successes and Failures in Object Permanence Tasks*, Technical Report PDP.CNS.94, Pittsburgh: Carnegie Mellon University.

Piaget, J., 1952, *Origins of Intelligence in Children*, New York: International Universities Press. ◆

Plunkett, K., and Sinha, C. G., 1992, Connectionism and developmental theory, *Br. J. Dev. Psychol.*, 10:209–254. ◆

Rumelhart, D. E., and Todd, P. M., 1993, Learning and connectionist representations, in *Attention and Performance XIV: Synergies in Experimental Psychology, Artificial Intelligence, and Cognitive Neuroscience* (D. E. Meyer and S. Kornblum, Eds.), Cambridge, MA: MIT Press, pp. 3–30. ◆

Shultz, T. R., Mareschal, D., and Schmidt, W. C., 1994, Modeling cognitive development on balance scale phenomena, *Machine Learn.*, 16:59–88.

Siegler, R. S., 1976, Three aspects of cognitive development, *Cog. Psychol.*, 8:481–520.

Siegler, R. S., 1981, Developmental sequences between and within concepts, *Monogr. Soc. Res. Child Dev.*, 46 (189).

Waddington, C. H., 1975, *The Evolution of an Evolutionist*, Edinburgh: Edinburgh University Press. ◆

Cognitive Maps

Nestor A. Schmajuk

Introduction

Cognitive maps store information about the relationships between contiguous temporal events or proximal spatial locations and combine this information to determine the relationships between remote temporal events or distant spatial locations. For example, during classical conditioning (see CONDITIONING), animals learn to predict what temporally contiguous events might follow other events. By linking several contiguous predictions, cognitive maps allow organisms to predict what temporally remote events might be expected. Similarly, during maze learning, animals learn to predict what spatially proximal locations are connected to other spatial locations. By linking several proximal predictions, cognitive maps allow organisms to predict what spatially remote locations are connected to other locations. This article outlines different formal theories and neural network models that have been proposed to describe both temporal and spatial cognitive mapping.

According to Tolman (1932), animals acquire an *expectancy* that the performance of response R1 in a situation S1 will be followed by a change to situation S2 (S1-R1-S2 expectancy). Tolman hypothesized that a large number of local expectancies can be combined, through inferences, into a *cognitive map*. Tolman proposed that place learning, latent learning, and detour learning illustrate the animals' capacity for reasoning by generating inferences. In place learning, animals learn to approach a given spatial location from multiple initial positions, independently of any specific set of responses. In latent learning, animals are exposed to a maze without being rewarded at the goal box. When a reward is later presented, animals demonstrate knowledge of the spatial arrangement of the maze, which remains "latent" until reward is introduced. Detour problems are maze problems that can be solved by integrating separately learned pieces of local knowledge into a global depiction of the environment.

When seeking reward in a maze, organisms compare the expectancies evoked by alternative paths. For Tolman, *vicarious trial-and-error behavior*, i.e., the active scanning of alternative pathways at choice points, reflects the animal's generation and comparison of different expectancies. At choice points, animals sample different stimuli before making a decision. For example, a rat often looks back and forth between alternative stimuli before approaching one or the other. According to Tolman's *stimulus-approach* view, organisms learn that a particular stimulus situation is appetitive, and therefore it is approached. Supporting this assumption, Mackintosh (1974:554) suggested that, in the presence of numerous intra-maze and extra-maze cues, animals typically learn to approach a set of stimuli associated with reward and to avoid a set of stimuli associated with punishment. However, in a totally uniform environment, animals learn to make the correct responses that lead to the goal.

Interestingly, Tolman (1932:177) suggested that the relations between initial and goal positions can be represented by a directed graph, and he called this graph a means-end field. Many years later, artificial intelligence theories described problem solving as the process of finding a path from an initial to a desired state through a directed graph (see PLANNING, CONNECTIONIST).

Place Learning

Place learning has been studied in the "water maze," requiring rats to escape from a pool filled with opaque water (Morris,

1981) by navigating to a submerged, invisible platform. Since the pool is a homogeneous environment with no intra-maze cues, the animal must learn a combination of distances to distal landmarks outside the tank to discriminate the platform's location (place learning). Morris (1981) found that if rats were trained to navigate to the submerged platform from a given starting point, they could swim directly to the goal from any other starting location in the pool perimeter during testing.

Barto and Sutton (1981) characterized a model that described place learning. They suggested that the association between one of four distal landmarks and a response (a movement in one of four spatial directions) increases whenever the response moves the animal toward a more appetitive location. In contrast to the Barto and Sutton model, in which visual patterns trigger sterotyped responses, Arbib and House (1987:153) suggested that spatial trajectories that characterize detour behavior could be described in terms of the interaction between vector fields that represent the preferred direction of motion at different spatial locations. One appetitive field represents motion toward the location of the goal, a second aversive field represents motion away from the position of a fence, and a third field represents motion from the location of the animal. When the goal appetitive field is combined with the animal's position field, the resulting field shows various paths starting at the position of the animal and ending at the location of the goal. When the three fields are combined, the net field shows a set of paths that are diverted around the fence ends (see POTENTIAL FIELDS AND NEURAL NETWORKS).

Zipser (1986) proposed a neural model capable of recognizing a given spatial location. The model comprises two layers of neural elements. Each unit in the first layer is specific for a given distal landmark and fires maximally when the current visual angle of the landmark equals a previously stored visual angle. Units in the second layer generate an output proportional to the sum of the outputs of all units in the first layer and consequently respond maximally at a specific location.

Extending some of Zipser's notions, Schmajuk and Blair (1993) presented a real-time neural network capable of describing place learning and the dynamics of spatial navigation. The network incorporates detectors that can be tuned to the values of visual angles of different distal landmarks as perceived from the spatial location where a reinforcing event is encountered. After a detector has been tuned, its output generates an effective stimulus which peaks at the distance from the landmark where positive or negative reinforcement was encountered before. At all other distances from the landmarks, the output becomes smaller, so that the detector describes spatial generalization with a peak at the location of the goal.

The outputs of the tuned detectors become associated with the reinforcing event, and therefore, each tuned detector activates the network's output unit in proportion to its association with the output unit. The output of the network at all points in a spatial environment yields a generalization surface, which can guide navigation from any location that is within view of familiar landmark cues, even if that location has never been visited before, and independently of the responses used to approach the goal. Spatial navigation is accomplished by adopting a *stimulus-approach* principle, that is, by approaching appetitive places and avoiding aversive places. When generalization surfaces are assumed to represent the forces exerted by the animal to move in the environment, the dynamics of spatial movements can be described. Schmajuk and Blair (1993) showed that the network correctly describes the navigational trajectories and dynamics of many spatial learning tasks.

Figure 1*A* shows an appetitive place learning simulation. The simulated environment consists of a rectangular matrix

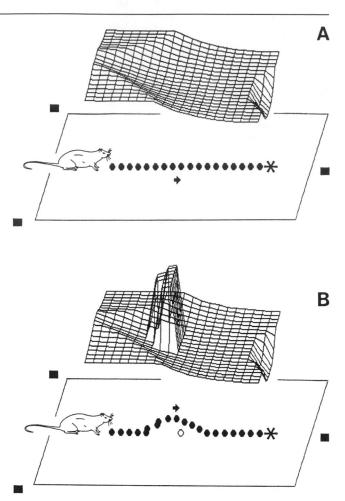

Figure 1. Appetitive generalization surface for combined appetitive and aversive place learning. *A*, Appetitive generalization surface, $G^{\text{appetitive}}(x, y)$, for an appetitive unconditioned stimulus (US) encountered at the point indicated with the asterisk. *B*, Composite generalization surface for an appetitive US encountered at the point indicated with the asterisk and an aversive US encountered at the point indicated with the open circle. Squares indicate the location of the spatial landmarks. Solid circles indicate different routes to the goal from different starting places. Arrows indicate the direction of movement.

(20 × 20), representing the pool. Simulated acquisition consists of 10 trials, during which the animal is rewarded at one location. During training, the animal perceives three landmarks in the environment. The animal builds a generalization surface for the appetitive location, $G^{\text{appetitive}}(x, y)$. Using this appetitive generalization surface, the subject finds the goal from any start point.

Figure 1*B* shows the simulated testing of a goal-directed behavior in which an invisible aversive location has been incorporated. Simulated acquisition consisted of 10 trials, during which the animal is rewarded at the same location as in Figure 1*A* and punished at another location. During training, the animal perceives three landmarks in the environment. The animal builds two generalization surfaces, one for the appetitive and another for the aversive location. Figure 1*B* shows the appetitive surface where the net generalization surface is appetitive, i.e., $G^{\text{appetitive}}(x, y) > G^{\text{aversive}}(x, y)$; and the aversive generalization surface where the net generalization surface is aversive, i.e., $G^{\text{appetitive}}(x, y) < G^{\text{aversive}}(x, y)$. Using these generalization surfaces, the subject finds the goal from any start point, avoiding the aversive place.

Maze Learning

Deutsch (1960) presented a formal description of cognitive mapping that incorporates many of Tolman's cognitive concepts. Deutsch assumed that when an animal explores a given environment, it learns that stimuli follow one another in a given sequence. Internal representations of the stimuli are linked together in the order they are encountered by the animal. Deutsch suggested that a given drive activates its goal representation, which in turn activates the linked representations of stimuli connected to it. When the animal is placed in the maze, it searches for stimuli stimulated by the goal representation. When a stimulus activated by the goal is perceived, the activation of lower stimuli in the chain is cut off, and behavior is controlled by the stimulus that is closer to the goal. Deutsch's theory can account for latent learning in the following terms. When animals are exposed to the maze without being rewarded at the goal box, they learn about the connections between different places in the maze. When a reward is subsequently presented, activation of the goal representation activates the representations of the stimuli connected to it.

Milner (1960) proposed a system capable of building a Tolmanian spatial map and of using it to control the animal's movements in a spatial environment. The model has nodes that are active only when a particular response (R_i) has been made in a particular location (S_j). The output of these nodes can be associated with nodes representing the location (S_k) that results from making response R_i at location S_j. When the organism is next placed in location S_j, random responses are generated. When response R_i appears, the node with inputs S_j and R_i is active and, in turn, activates the node representing S_k. If S_k is associated to an appetitive stimulus, it activates a mechanism that holds R_i in the response generator, and location S_k can be reached. Some views of hippocampal function (e.g., McNaughton, 1989) suggest that S_j-R_i-S_k associations would be stored in the hippocampus (see HIPPOCAMPUS: SPATIAL MODELS).

Lieblich and Arbib (1982) addressed the question of how animals build a cognitive model of the world. They posited that spatial representations take the form of a directed graph, the World Graph, in which nodes represent a *recognizable situation* in the world. In their scheme, a node represents not only a place but also the motivational state of the animal. Consequently, a place in the world might be represented by more than one node if the animal has been there under different motivational states. Lieblich and Arbib postulated that each node in the world graph is labeled with *learned* vectors, **R**, that reflect the drive-reduction properties for multiple motivations of the place represented by that node. Based on the value of **R**, the animal moves to the node most likely to reduce its current drive.

Hampson (1990) analyzed maze navigation in terms of a model that combines stimulus-response and stimulus-stimulus mechanisms into a system capable of assembling action sequences to reach a final goal (see PROBLEM SOLVING, CONNECTIONIST). Schmajuk and Thieme (1992) presented a real-time, biologically plausible neural network approach to purposive behavior and cognitive mapping. The system is composed of (a) an action system that implements a stimulus-approach principle; and (b) a cognitive system, involving a neural cognitive map. The goal-seeking mechanism displays exploratory behavior until either (a) the goal is found or (b) an adequate prediction of the goal is generated. The cognitive map built by

the network is a *topological map*, i.e., it represents only the adjacency, but not distances or directions, between places and reward. The network has recurrent and nonrecurrent properties that allow the reading of the cognitive map without modifying it. Schmajuk and Thieme (1992) described two types of predictions: fast-time and real-time predictions. Fast-time predictions are produced in advance of what occurs in real time, when the information stored in the cognitive map is used to predict the remote future. Real-time predictions are generated simultaneously with the occurrence of environmental events, when the information stored in the cognitive map is being updated. Computer simulations show that the network successfully describes latent learning and detour behavior in rats. To ascertain the power of the network in problem solving, Schmajuk and Thieme (1992) applied the network to the Tower of Hanoi task. Simulations show that the network takes a few trials to solve the problem in the minimal number of movements.

Schmajuk, Thieme, and Blair (1993) incorporated a route system to the network described by Schmajuk and Thieme (1992). Whereas the cognitive map stores associations between places and reward, the route system establishes associations between cues and reward. Both systems compete with each other to establish associations with the reward, with the cognitive system generally overshadowing the route system. In agreement with O'Keefe and Nadel (1978), after hippocampal lesions (see HIPPOCAMPUS: SPATIAL MODELS), animals navigate through mazes by making use of the route system.

Discussion

In sum, several formal and neural network approaches seem to be successful at describing some aspects of cognitive mapping in both animals and machines. Future research may combine two issues that have been separately addressed so far: (a) how to define recognizable temporal or spatial situations in the environment (such as in place learning), and (b) how to connect these situations to conform a cognitive map (such as in maze learning).

Road Map: Connectionist Psychology

References

Arbib, M. A., and House, D. H., 1987, Depth and detours: An essay on visually guided behavior, in *Vision, Brain and Cooperative Computation* (M. A. Arbib and A. R. Hanson, Eds.), Cambridge, MA: MIT Press, pp. 129–163.

Barto, A. G., and Sutton, R. S., 1981, Landmark learning: An illustration of associative search, *Biol. Cybern.* 42:1–8.

Deutsch, J. A., 1960, *The Structural Basis of Behavior*, Cambridge, Eng.: Cambridge University Press.

Hampson, S. E., 1990, *Connectionistic Problem Solving*, Boston: Birkhauser.

Lieblich, I., and Arbib, M. A., 1982, Multiple representations of space underlying behavior, *Behav. Brain Sci.*, 5:627–659.

Mackintosh, N. J., 1974, *The Psychology of Animal Learning*, London: Academic.

McNaughton, B. L., 1989, Neuronal mechanisms for spatial computation and information storage, in *Neural Connections and Mental Computations* (L. Nadel, L. Cooper, P. Culicover, and R. Harnish, Eds.), New York: Academic.

Milner, P. M., 1960, *Physiological Psychology*, New York: Holt, Rinehart and Winston.

Morris, R. G. M., 1981, Spatial location does not require the presence of local cues, *Learn. Motiv.*, 12:239–260.

O'Keefe, J., and Nadel, L., 1978, *The Hippocampus as a Cognitive Map*, Oxford: Clarendon.

Schmajuk, N. A., and Blair, H. T., 1993, Place learning and the dynamics of spatial navigation: An adaptive neural network, *Adapt. Behav.*, 1:355–387.

Schmajuk, N. A., and Thieme, A. D., 1992, Purposive behavior and cognitive mapping: An adaptive neural network, *Biol. Cybern.*, 67: 165–174.

Schmajuk, N. A., Thieme, A. D., and Blair, H. T., 1993, Maps, routes, and the hippocampus: A neural network approach, *Hippocampus*, 3:387–400.

Tolman, E. C., 1932, Cognitive maps in rats and men, *Psychol. Rev.*, 55:189–208.

Zipser, D., 1986, Biologically plausible models of place recognition and goal location, in *Parallel Distributed Processing: Explorations in the Microstructure of Cognition*, vol. 2, *Psychological and Biological Models* (J. L. McClelland, D. E. Rumelhart, and PDP Research Group, Eds.), Cambridge, MA: MIT Press, pp. 432–470.

Cognitive Modeling: Psychology and Connectionism

Amanda J. C. Sharkey and Noel E. Sharkey

Models of Cognition

In 1986, a two-volume work edited by McClelland and Rumelhart presented a number of connectionist models of different aspects of cognition that had been trained by exposure to samples of the required tasks. This work was indebted to earlier pioneering neural network research related to cognitive processing and memory (e.g., Anderson, 1972; Grossberg, 1973), but it was these two volumes that set the agenda for connectionist cognitive modelers and offered a methodology that has become the standard. The number of connectionist cognitive models are now legion. The domains that have been simulated include memory retrieval; category formation; in language: phoneme recognition, word recognition, speech perception, acquired dyslexia, and language acquisition; and, in vision: edge detection and object and shape recognition (for reviews, see McClelland and Rumelhart, 1986; Quinlan, 1991; and articles cited in the Road Map on **Connectionist Psychology** in Part II of this *Handbook*).

We have chosen to discuss only supervised learning techniques, as these have been the most commonly used recent cognitive modeling. In the simplest case of supervised learning, a net consists of a set of input units, a layer of hidden units, and a set of output units, each layer being connected to the next via modifiable weights. This is a feedforward net. When the net is trained on a set of input-output pairs, the weights are adjusted via a learning algorithm (e.g., backpropagation) until the required output is produced in response to each input in the training set. When tested on a set of previously unseen inputs, the net will, to a greater or lesser extent, display an ability to generalize—that is, to go beyond the data it was trained on, and to produce an appropriate response to some of the test inputs. The ability of the net to generalize depends on the similarity between the function extracted as a result of the original training, and the function that underlies the test set. Given a training set that is sufficiently representative of the required function, generalization results are likely to be good. Where the inputs and outputs of such a net are given an interpretation relevant to the performance of a cognitive task, the net may be seen as a model of that task.

As well as their basic architecture and mode of operation, four typical features characterize connectionist models of cognition, and these features account for much of the popularity of the approach: (1) Connectionist models can be used both to model mental processes and to simulate the actual behavior involved; (2) they can provide a "good fit" to the data from psychology experiments; (3) the model, and its fit to the data, is achieved without explicit programming, or "hand-wiring"; and

(4) the models often provide new accounts of the data. We discuss each of these features in turn.

The first two—namely, the way in which connectionist nets can both provide a model of a cognitive process and simulate a related task, and their ability to provide a good fit to the empirical data—can be seen as combining some of the characteristics of two earlier routes to modeling. One of these, taken by the cognitive psychology community, involved building models that could account for the results from psychology experiments with human subjects, but which did not incorporate simulations of experimental tasks. The second route, followed by the artificial intelligence community, was to build computer models that actually perform the task in ways that resemble human performance without regard to detailed psychological evidence. The connectionist approach, as described here, provides the benefits of both simulating the performance of human tasks and, at the same time, fitting the data from psychological investigations.

As an example of the latter, consider Seidenberg and McClelland's (1989) model of word pronunciation. This simulates a number of experimental behaviors including: pronunciation of novel items, differences in performance on lexical decision and naming tasks, and ease of pronunciation with relation to variables such as frequency of occurrence, orthographic redundancy, and orthographic-phonological regularity. At the same time, it is generally accepted to provide a good fit to the experimental data. This is achieved by deriving an analog to the data from the performance of the model. In the case of Seidenberg and McClelland's account of lexical decisions, the comparison to lexical decision data is accomplished by computing an "orthographic error score," based on the sum of squares of the differences between the feedback pattern computed by the network and the actual input to the orthographic units. Here, a low error score is equated with a fast reaction time (RT). Others have assumed different data analogs. For example, Cohen, Dunbar, and McClelland (1990) modeled reaction time by incorporating a time-dependent constant into the activation equation. Yet another approach has been to equate RTs with movement through lexical space (Sharkey and Sharkey, 1992). One may also use convergence time in attractor networks to model RTs more directly.

The third feature identified above is that of achieving the model and the fit to the data without explicit hand-wiring, and is one that can be favorably contrasted to the symbolic programming methodology employed in artificial intelligence, where the model must be programmed step by step, leaving room for ad hoc modifications and kludges. The fourth characteristic, perhaps the most scientifically exciting, is the possibility of providing a novel explanation of the data. Seidenberg and McClelland showed that their network provided an integrated (single mechanism) account of data on both regular and exception words where previously the old cognitive modeling conventions had forced an explanation in terms of a dual route.

On first consideration, the four features discussed above seem to provide support in favor of a connectionist approach to cognitive modeling, as opposed to the approaches taken in the two preceding routes to cognition, by AI and psychological modelers. When the result has been the development of new, and simpler, accounts of the data, there are obvious benefits. The connectionist approach would also seem to be preferable inasmuch as it makes it possible to both simulate the tasks and provide a good fit to the data; and to do this without having to program these abilities in by hand. However, we must tread warily here in claiming a clear distinction between the connectionist approach and those that preceded it.

Let us first reexamine the idea that connectionist network models are not explicitly programmed by hand. While this is strictly true, there are still a number of decisions that have to be made by the researcher and that affect the subsequent performance of the model. The following factors, for example, both govern the creation of a model and *are* in the hands of the researcher: the architecture of the net, or nets, and its initial structure; the learning technique; the learning parameters (e.g., the learning rate); the input and output representations; and the training sample. The term *extensional programming* can be used to refer to the manipulation of these factors. By means of extensional programming, the clever experimenter can determine the ultimate form of a model and hence its performance in terms of any data set.

Control of the content, and presentation of the training sample, is an important aspect of extensional programming. Its potential influence on the performance of a connectionist model can be illustrated by a consideration of some of the criticisms of McClelland and Rumelhart's past-tense model (McClelland and Rumelhart, 1986). Their model is said to mirror several aspects of human learning of past-tense verb endings. However, when the model was criticized by Pinker and Prince (1988), an important element of their criticism was that the experimenters had unrealistically tailored the environment to produce the required results, and that the results were an artifact of the training data. More specifically, the results indicated a U-shaped curve in the rate of acquisition, as occurs with children, but, Pinker and Prince argued, this curve occurred only because the net was exposed to the verbs in an unrealistically structured order. These criticisms have largely been answered as a result of further research (see LINGUISTIC MORPHOLOGY), but the point remains. Selection of the input and control of the way that it is presented to the net affect what the net learns.

If the performance of a connectionist net is determined by its extensional programming, this means that there are a number of parameters which can be altered until a good data fit is achieved. Not only that, but an acceptance that a good fit to the data has been achieved relies on the acceptance of a number of other assumptions—e.g., about the way that the relevant tasks have been simulated—and the consequent way that analogs to the data have been derived from the performance of the model. Thus, when a model is said to explain the empirical data from lexical decision tasks, the model does not actually perform a lexical decision task like humans do, outputting a "yes" when the input is a word, and a "no" when it is not. An equivalence must be drawn between the task as it is performed by humans and the actual input-output relationships encoded by the net. In fact, the actual task in question is rarely performed by the model. What is performed is often something that approximates the task and is assumed to capture its essential characteristics. Thus, in a model of word pronunciation, the output of the model does not take the form of spoken words but rather a set of phonological features. Here, the approximation is fairly straightforward; in other cases, the relationship is less obvious. Ratcliff (1990), for instance, equates the phenomenon of recognition memory with that of auto-association, assuming that the ability of a net to reproduce its inputs as outputs corresponds to the human ability to recognize inputs that have been seen before.

From the foregoing discussion, we can see that some of the advantages of connectionist models of cognition are not entirely straightforward. Nonetheless, their potential for developing new accounts that go beyond the data is an important justification for their employment. Much of the excitement in the psychology community has been about the ability of connectionist models to handle apparently rule-governed phenomena without any explicit rules. For example, learning to produce the appropriate past tense of verbs was always held as an example of a rule-governed phenomena until McClelland and Rumelhart's (1986) model showed that the behaviors could be trained in a model that did not contain any explicit rules. This is part of a new approach to cognition that exploits the novel style of computation afforded by connectionism. We now turn to discuss some of the recent work that is examining the potential of connectionism as a new theory of mind.

An Emerging Theory of Cognition

There has been a certain amount of discussion about the extent to which connectionism constitutes the emergence of a new paradigm for psychology (Schneider, 1987). Whether or not the changes actually constitute a paradigm shift, it is certain that the adoption of connectionist modeling techniques has led to changes. One of these is the emergence of a new breed of cognitive modeler who is concerned with exploring the implications of connectionist computation for a new theory of cognition, with the ultimate aim of providing an account of the way the mind emerges from the brain. The methodology employed by these "cognitive connectionists" represents a move away from the detailed modeling of cognitive psychology experiments toward more abstract computational experimentation. It is to be hoped this will broaden the scope of their investigations, for while there are benefits to models that provide a new explanation of a wide spectrum of data, the maintenance of a close relationship between model and data can limit investigations to previously identified questions from an older paradigm.

The promise of a new connectionist theory of cognition is that it will further our understanding of the relationship between brain and mind. However, this does not necessarily mean that such a theory will incorporate details of neural processing. It is important to make a clear distinction between the use of brain-style computation and neuropsychological modeling. While connectionist models are sometimes described as being "neurally inspired," their relationship to neural processing is delicate and tenuous. In a model of cognitive processes, the computation performed is unlikely to correspond to what goes on at the neural level. Indeed, it may take units on the order of several thousand neurons to encode stimulus categories of significance to animals. Clearly, where the inputs to a net are things like noun phrases, or disease symptoms, or even the phonological representations of words, the inputs cannot be equated with neural inputs, but must represent substantially preprocessed stimuli. In fact, there are few cases where actual facts about the nervous system are used to constrain the architecture and design of a model. It is, of course, important to build computational models of real brain circuits in all of their glorious detail (see, e.g., articles cited in the Part II Road Map on **Mammalian Brain Regions**). But if one is concerned with cognition rather than the details of neural processes, an appropriate research strategy is to use broader brushstrokes, relying on computational abstractions.

In support of a movement away from the details of neurophysiology and psychology experiments, we can cite an example of progress from the history of connectionism itself. Arguably, it was two major simplifying assumptions made by McCulloch and Pitts (1943) that enabled them to develop their theoretical analyses without getting bogged down in the physical and chemical complexity of the nervous system. Their first simplification was based on the observation that neural communication is thresholded, and thus the spike activation potential is all or none: it either fires or does not fire. Thus, the neuron could be conceived of as a binary computing device. The second simplification was to view synapses as numerical weightings between simple binary computing elements. This meant that computation proceeded by summing the weighted inputs to an element and using the binary threshold as an output function (see Sharkey and Sharkey, 1994, for further historical details). The position of the cognitive connectionist is that we can best proceed by being constrained by simplifying, rather than fully biological, assumptions about neural computation and cognition so long as these assumption are formed in a principled way.

Nonetheless, if the use of connectionist computation (as opposed to symbolic computation) is to lead to the development of a new theory of cognition, it must first be shown to be capable, in principle, of supporting higher mental processes, *and* to do so in a novel manner: mere implementation of symbolic architectures will not do. The question of whether or not connectionism is capable of supporting a cognitive architecture has mainly been addressed in the context of discussions about the novelty and value of connectionist representation. In a review of representational formalisms in the connectionist literature, Sharkey (1991) pointed out that the most novel representational class is that comprising distributed representations developed over the internal hidden units of a multilayer network. These are vectors of continuously valued activations, normally in the range 0 to 1.

Proponents of the classical symbolic tradition have claimed that such representations are, in principle, incapable of supporting a cognitive architecture because in order to account for the systematic nature of human thought, representations must be able to support structure-sensitive processes, and this requires compositional representations. The assumption is that there is only one kind of compositionality: namely, the concatenative compositionality of symbolic strings. This permits structure-sensitive operations, because in their mode of combination the constituents of complex expressions are tokened whenever the complex expression is tokened. For example, in order to develop an expression from a sentence such as "John kissed Mary," arbitrary symbols representing the constituents *John*, *kissed*, and *Mary* are combined in a contextually independent concatenation to produce the propositional representation *kiss(John, Mary)*. Whenever this latter complex expression is tokened, its constituents—*kiss*, *Mary*, and *John* —are also tokened. This makes the manipulation of the representations by a mechanism sensitive to the syntactic structure resulting from concatenative compositionality relatively easy.

Distributed representations, on the other hand, do not exhibit this kind of compositionality. Instead, cognitive connectionists have identified an alternative form of compositionality, one that has been described as merely functional, nonconcatenative compositionality (van Gelder, 1990). Distributed representations combine tokens without those tokens appearing in the complex expression, since the tokens of the input constituents are destroyed in their combination. The point is that such representations can still be shown to be *functionally* compositional because there exist general and reliable procedures for combining constituents into complex expressions and for decomposing those expressions back into the constituents. It is possible, for example, to encode simple syntactic trees in terms of connectionist distributed representations, and to decode them back into the same syntactic trees (Pollack, 1990). Thus, the constituents of the tree have been combined into a form of representation that is nonconcatenative, but which preserves the necessary information.

A considerable and growing body of research has shown that not only are distributed representations compositional, but they can also enable *systematic* structure-sensitive operations. Chalmers (1990), for example, found that it was possible to use connectionist nets to transform distributed representations for active sentences into distributed representations for passive sentences. Thus, distributed representations allow at least a limited form of systematicity without emergence onto the symbol surface. Moreover, this is not just an example of "old wine in new bottles"—i.e., a mere re-implementation of the classical account. These uniquely connectionist representations operate in a different manner.

Unlike classic representations, the structural similarity relations in networks cannot be characterized in terms of *syntactic* systematicity. It has been argued that, instead, the representations in connectionist nets exhibit *spatial* systematicity. In other words, if the distributed representations are conceived as points in a multidimensional hidden unit space, the Euclidean distances between the representations provide a metric of similarity. It is the precise distances between these representational points that captures structural and semantic similarity. However, to complicate matters, the assumption of spatial systematicity has been challenged recently on computational grounds (e.g., Sharkey, Sharkey, and Jackson, 1994), because exclusive concentration on the relative position of the hidden unit representations ignores the important functional role of the output weights of a net in using these representations. In their analysis of weight representations, Sharkey et al. (1994) showed that it is possible for spatially similar representations to have different computational roles, and for distant representations to have similar roles.

The distinction between hidden unit and weight representations is quite a subtle one for our current purposes, and we include it only to bring the paper to the present state of the art. The essential point of the foregoing discussion is that connectionist representations are capable of forms of both compositionality and systematicity, and that they accomplish them by very different means from those of classical symbolic representations. This means that there are now strong indications that connectionist research is capable of supporting a new style of cognitive architecture.

Discussion

Connectionist representations have been shown to be capable of much of the systematicity and compositionality required of them by their critics. They do not merely implement symbolic computation, but operate in a novel manner. Whether it will be possible to provide a connectionist account of all aspects of cognition remains to be seen. There are certain aspects of thought, particularly those associated with logical deduction and metaknowledge (knowledge about one's own knowledge) which at present seem to pose a challenge. In the meantime, the connectionist approach is sufficiently novel and interesting to warrant further investigation. The tantalizing promise is that its use of "neurally inspired" computation will increase our understanding of the elusive relationship between brain and mind.

Road Map: Connectionist Psychology
Related Reading: Connectionist and Symbolic Representations; Developmental Disorders; Language Processing; Lesioned Attractor Networks as Models of Neuropsychological Deficits

References

Anderson, J. A., 1972, A simple neural network generating an interactive memory, *Math. Biosci.*, 8:137–160.

Chalmers, D. J., 1990, Syntactic transformations on distributed representations, *Connection Sci.*, 2:53–62.

Cohen, J. D., Dunbar, K., and McClelland, J. M., 1990, On the control of automatic processes: A parallel distributed processing account of the Stroop effect, *Psychol. Rev.*, 97:332–361.

Fodor, J. A., and Pylyshyn, Z., 1988, Connectionism and cognitive architecture: A critical analysis, *Cognition*, 28:3–71.

Grossberg, S., 1973, Contour enhancement, short-term memory, and constancies in reverberating neural networks, *Stud. Appl. Math.*, 52: 213–257.

McClelland, J. L., and Rumelhart, D. E., and PDP Research Group, Eds., 1986, *Parallel Distributed Processing: Explorations in the Microstructure of Cognition*, vol. 2, *Psychological and Biological Models*, Cambridge, MA: MIT Press. ◆

McCulloch, W. S., and Pitts, W. H., 1943, A logical calculus of the ideas immanent in nervous activity, *Bull. Math. Biophys.*, 5:115–133.

Pinker, S., and Prince, A., 1988, On language and connectionism: Analysis of a parallel distributed processing model of language acquisition, in *Connections and Symbols* (S. Pinker and J. Mehler, Eds.), Cambridge, MA: MIT Press, pp. 73–194.

Pollack, J., 1990, Recursive distributed representations, *Artif. Intell.*, 46:77–105.

Quinlan, P. T., 1991, *Connectionism and Psychology: A Psychological Perspective on New Connectionist Research*, London: Harvester Wheatsheaf. ◆

Ratcliff, R., 1990, Connectionist models of recognition memory: Constraints imposed by learning and forgetting functions, *Psychol. Rev.*, 96:523–568.

Schneider, W., 1987, Connectionism: Is it a paradigm shift for psychology? *Behav. Res. Methods, Instrum. & Comput.*, 19:73–83.

Seidenberg, M. S., and McClelland, J. L., 1989, A distributed developmental model of word recognition and naming, *Psychol. Rev.*, 96: 523–568.

Sharkey, A. J. C., and Sharkey, N. E., 1992, Weak contextual constraints in text and word priming, *J. Memory and Language*, 31:543–572.

Sharkey, N. E., 1991, Connectionist representation techniques, *Artif. Intell. Rev.*, 5:143–167. ◆

Sharkey, N. E., and Sharkey, A. J. C., 1994, Emergent cognition, in *Handbook of Neuropsychology*, vol. 9, *Computational Modeling of Cognition* (J. Hendler, Ed.), Amsterdam: Elsevier, pp. 347–360.

Sharkey, N. E., Sharkey, A. J. C., and Jackson, S. A., 1994, Opening the black box of connectionist nets: Some lessons from cognitive science, *Comput. Stand. Interfaces*, 16:279–293.

van Gelder, T., 1990, Compositionality: A connectionist variation on a classical theme, *Cognit. Sci.*, 14:355–364.

Collective Behavior of Coupled Oscillators

Yoshiki Kuramoto

Introduction

Limit-cycle oscillations are particularly relevant to the functions of living organisms. They are nonlinear and dissipative, in contrast to linear and conservative type oscillations that are commonly met in statistical physics. A limit-cycle oscillator comes to have a unique amplitude and wave form as $t \to \infty$ independently of initial conditions. Consequently, a representative point chosen initially in a suitable domain in the phase space is eventually attracted to a unique closed orbit C. Mathematically, a limit-cycle oscillator is modeled by a set of nonlinear differential equations, a simple example of which is the van der Pol equation proposed originally for electronic oscillators. Repetitive spike discharges of an excitable membrane may also be regarded as a form of limit-cycle oscillations for which the Hodgkin-Huxley equation is widely known.

Limit-cycle oscillators are generally difficult to treat analytically, and this remains true even for a simple van der Pol oscillator. Therefore, theoretical study of *collective* behavior of multi-oscillator systems would be hopelessly difficult without resorting to some drastic simplifications of the models. In this article, we are mainly concerned with the phase model as a particularly useful tool for such purposes. We also discuss, but only briefly, the complex Ginzburg-Landau model, which removes some drawbacks inherent in the phase model.

Phase Model

The great utility of the phase model was first realized by Winfree (1967, 1980). The theoretical basis of the phase model was presented by Kuramoto (1984), and it is generally applicable to weakly perturbed or weakly coupled oscillators. In the phase model, each oscillator is represented by just one degree of freedom θ, i.e., the phase (mod 2π) which is defined in a suitable way along C (and also outside C, if necessary). The free motion of the oscillator on C may then be described by

$$\frac{d\theta}{dt} = f(\theta) \qquad (1)$$

where $f(\theta)$ is a 2π-periodic function of θ with definite sign. When the oscillator is perturbed only weakly, it will stay close to C, but the right-hand side of Equation 1 has to be modified with an additional term depending on θ itself. If the perturbation is caused by the coupling from another oscillator supposedly almost identical in nature to the first one, then this correction term will depend also on the phase θ' of the second oscillator. Thus the corresponding equation has the form

$$\frac{d\theta}{dt} = f(\theta) + G(\theta, \theta') \qquad (2)$$

where G is 2π-periodic in both θ and θ'. Equation 2 can be further simplified. First, we employ a particular definition of the phase on C in such a way that $f(\theta)$ may take a constant value ω. We assume weakness of the coupling, i.e., $|G/\omega| \ll 1$, so that on putting $\theta = \omega t + \phi$ and $\theta' = \omega t + \phi'$, we have slowly varying variables ϕ and ϕ'. Then the change of ϕ and ϕ' over the period of oscillation $T \equiv 2\pi/\omega$ will be so small that the equation for ϕ, i.e., $d\phi/dt = G(\omega t + \phi, \omega t + \phi')$ (and also the equation for ϕ') may safely be time-averaged over T under fixed ϕ and ϕ'. This leads to $d\phi/dt = \Gamma(\phi' - \phi)$ or

$$\frac{d\theta}{dt} = \omega + \Gamma(\theta' - \theta) \qquad (3)$$

where

$$\Gamma(x' - x) = (2\pi)^{-1} \int_0^{2\pi} G(\lambda + x, \lambda + x')\,d\lambda$$

One may apply this averaging method even to pulsatile coupling such as described in the next paragraph. Generalizing the above idea to a large population of N oscillators with weak distribution of frequencies, we obtain

$$\frac{d\theta_i}{dt} = \omega_i + \sum_{j=1}^N \Gamma_{ij}(\theta_j - \theta_i) \qquad i = 1, 2, \ldots, N \qquad (4)$$

There are also other forms of the phase model depending on different choices for $f(\theta)$ and $G(\theta, \theta')$. In an important variant relevant to neurodynamics, $f(\theta)$ has a sawtooth form like $f(\theta) = a - \theta$ $(0 < \theta \leq 2\pi)$. This is equivalent to the so-called integrate-and-fire oscillator model (Peskin, 1975). If a is slightly less than 2π, the system actually does not represent an oscillator, but it changes into an excitable element. This is because after "firing" at $\theta = 0$, the system comes back to a resting state at $\theta = a$, which is close to the phase of firing. The coupling between such elements is usually assumed to be pulsatile. Specifically, it is reasonable to replace the coupling G in Equation 2 by $\varepsilon \sum_n \delta(t - t_n)$, where t_n $(n = 1, 2, \ldots)$ represent the moments at which the second oscillator fires or $\theta'(t_n) = 0$. Thus, the coupling from the second to the first oscillators is excitatory or inhibitory if $\varepsilon > 0$ or $\varepsilon < 0$, respectively. There exists a mathematical study showing how perfect mutual synchronization is established in a population of globally coupled elements of identical integrate-and-fire oscillators.

Collective Oscillations

Some simple phase models aid our theoretical understanding of how temporal coherence arises over entire populations of densely interconnected oscillators even when their frequencies are randomly distributed (Winfree, 1967, 1980; Kuramoto, 1984; Strogatz, Mirollo, and Matthews, 1992). Among the frequently cited examples of collective oscillations, the most spectacular is periodic flashing of swarms of fireflies in Southeast Asia (Buck, 1988). As a biochemical example, suspensions of yeast cells undergoing glycolitic oscillations also exhibit synchronous collective oscillations (Ghosh, Chance, and Pye, 1971). The assembly of pacemaker cells in the sinoatrial node of the heart also generates rhythms at a collective level (DeHaan and Sachs, 1972).

As a very simple population model, suppose the pair coupling $\Gamma_{ij}(x)$ in Equation 4 to be identical over all pairs and simply proportional to $\sin x$. Then we have

$$\frac{d\theta_i}{dt} = \omega_i + \frac{K}{N} \sum_{j=1}^N \sin(\theta_j - \theta_i) \qquad (5)$$

where K is assumed positive and the factor N^{-1} ensures physically meaningful results in the limit $N \to \infty$. How collective rhythms emerge from Equation 5 will be outlined below (for more details, see Kuramoto, 1984). In order to describe the collective behavior, it is helpful to introduce a complex variable $w = \sigma \exp(i\Theta)$ as a measure for the amplitude and phase of collective oscillations. It is defined by

$$w = \frac{1}{N} \sum_{j=1}^N e^{i\theta_j} = \int_0^{2\pi} d\theta\, \rho(\theta, t) e^{i\theta} \qquad (6)$$

where ρ denotes the normalized density distribution $N^{-1} \sum_j \delta(\theta - \theta_j(t))$. Imagine that the oscillators are distributed like point masses along the unit circle, centered at the origin in the complex plane in such a way that their arguments coincide with their respective phases θ_j. Then w represents the center of mass of these running "particles." Uniform ρ gives $w = 0$, implying the absence of collective oscillations, whereas polarized ρ which will necessarily propagate along the circle gives rotating w, implying collective oscillations. Note that Equation 5 can be expressed as

$$\frac{d\theta_i}{dt} = \omega_i + K\sigma(t) \sin(\Theta(t) - \theta_i(t)) \qquad (7)$$

and in this form the oscillators look mutually independent, except that they are driven by a common force from the internal field $w(t) = \sigma \exp(i\Theta)$.

By taking advantage of this fact, it can be shown analytically that for a simple class of distributions $g(w)$ of the natural frequencies, the above model exhibits a phase transition-like behavior from unpolarized to polarized distributions. We can explain briefly. Suppose that ρ propagates steadily or $\rho(\theta, t) = \rho(\theta - \Omega t)$, which means $\sigma = $ constant and $\Theta = \Omega t$. Then Equation 7 can be solved for each $\theta_j(t)$ with yet unknown parameters σ and Ω, and the entirety of these solutions determines the distribution ρ (again σ and Ω included). From the definition of w in terms of ρ (Equation 6), this means that w is now given by a certain function of w itself. In the special case in which $g(w)$ is unimodal and symmetric about $\bar{\omega}$, the problem is simplified by virtue of the obvious property $\Omega = \bar{\omega}$. Then a self-consistent equation for σ is obtained in the form

$$\sigma = K\sigma \int_{-\pi/2}^{\pi/2} dx\, g(\Omega + K\sigma \sin x) \cos x\, e^{ix} \equiv S(\sigma) \qquad (8)$$

$S(\sigma)$ is an S-shaped odd function saturating to 1 as $\sigma \to \infty$. Therefore, besides the trivial solution $\sigma = 0$ which always exists, a nontrivial solution appears when K exceeds the critical value $K_c = (2/\pi)g(\Omega)$ at which $S'(0) = 1$. The distribution of the true frequencies $\tilde{\omega}$, which is denoted by $\tilde{g}(\tilde{\omega})$, has a peculiar feature above K_c. Each $\tilde{\omega}_j$ is obtained by solving Equation 7, and substitution into the identity $\tilde{g}(\tilde{\omega}) = g(\omega)\, d\omega/d\tilde{\omega}$ gives $\tilde{g}(\tilde{\omega})$. It is clear that $\tilde{g}$ is identical to g below K_c where $\sigma = 0$. Above K_c, the oscillators are divided into two groups, one satisfying $|K\sigma/(\Omega - \omega_j)| > 1$ and the other $|K\sigma/(\Omega - \omega_j)| < 1$. The oscillators in the first group have an identical frequency Ω, which means that they are perfectly entrained to the collective oscillation. The onset of collective oscillation thus implies nucleation in the frequency space. The ratio r of the size of the frequency nucleus to the system size is given by

$$r = \int_{|\omega - \Omega| < K\sigma} g(\omega)\, d\omega \qquad (9)$$

The oscillators in the second group fail to be entrained to the collective oscillations, their frequencies being still distributed and given by

$$\tilde{\omega}_j = \Omega + (\omega_j - \Omega)\{1 - (K\sigma/(\Omega - \omega_j))^2\}^{1/2}$$

Putting these facts together, we obtain

$$\tilde{g}(\tilde{\omega}) = r\delta(\tilde{\omega} - \Omega)$$
$$+ g(\Omega + \sqrt{(\tilde{\omega} - \Omega)^2 + (K\sigma)^2}) \frac{|\omega - \Omega|}{\sqrt{(\tilde{\omega} - \Omega)^2 + (K\sigma)^2}}$$

$$(10)$$

This spectrum is composed of a sharp central peak and a broad background. The intensity of the latter component exhibits a conspicuous drop near the central peak, as is seen from its asymptotic behavior: $(K\sigma)^{-1}g(\Omega + K\sigma)|\tilde{\omega} - \Omega|$. Similar spectral characteristics aroused the interest of Wiener (1958) in his observation of electroencephalographic recordings of human brain waves. Although we will not show it here, the exact analytic form for the phase distribution $\rho(\theta - \Omega t)$ can also be obtained explicitly.

Wave Phenomena

Equation 4 is also useful for the study of pattern dynamics in self-oscillatory media. For example, imagine a long chain of oscillators with regular spacing d, coupled only with their nearest neighbors. Then Equation 4 becomes

$$\frac{d\theta_i}{dt} = \omega_i + \Gamma(\theta_{i+1} - \theta_i) + \Gamma(\theta_{i-1} - \theta_i) \quad (11)$$

A remarkable feature of such a system is that the chain is segmented into a number of clusters, in each of which the oscillators are perfectly in mutual entrainment. Thus the pattern of the true frequencies is composed of plateaus with a discontinuity between neighboring clusters. If we assume a linear gradient in the natural frequencies ω_i, we obtain a staircase-like pattern of $\tilde{\omega}_i$, and such results have been used to explain some physiologically observed facts in mammalian small intestine (Ermentrout and Koppel, 1984).

The continuum limit of Equation 10 is also worth studying (Kuramoto, 1984). On the assumption of a small local gradient in phase and natural frequencies, Equation 11 is reduced to a partial differential equation,

$$\frac{\partial \theta}{\partial t} = \omega(x) + v\frac{\partial^2 \theta}{\partial x^2} + \mu\left(\frac{\partial \theta}{\partial x}\right)^2 \quad (12)$$

where $v = -d^2\Gamma'(0)$ and $\mu = -d^2\Gamma''(0)$. These arguments may easily be extended to higher dimensions, and the corresponding partial differential equation takes the form

$$\frac{\partial \theta}{\partial t} = \omega(x) + v\nabla^2\theta + \mu(\nabla\theta)^2 \quad (13)$$

Detailed analysis of Equation 12 or 13 is possible because they are reduced to a linear equation

$$\frac{\partial Q}{\partial t} = \left(v^{-1}\mu\omega(x) + v\frac{\partial^2}{\partial x^2}\right)Q \quad (14)$$

and its high-dimensional analogs via the transformation $\theta = v\mu^{-1}\ln Q$. In what follows, we assume $v > 0$, excluding the possibility of *phase instabilities* (a notion to be explained shortly). If ω is uniform, Equation 12 admits a 1-parameter family of solutions $\theta = (\omega + \mu k^2)t + kx$. Since any physical quantity should be a 2π-periodic function of θ, this class of solutions represents undamped traveling waves with various frequencies and velocities. Such waves are called *phase waves*. When two plane waves collide, they do not interfere with each other, but

a shock front is formed. There is a corresponding class of solutions of Equation 12 with the form

$$\theta = \omega t + v\mu^{-1}\ln\left(\cosh\left(\frac{q\mu}{v}(x - x_s(t))\right)\right) + px + \mu(p^2 + q^2)t$$

$$(15)$$

where $x_s = -2p\mu t$, and p and q are arbitrary constants. The asymptotic form of the above solutions far from the shock front is identical to the plane wave solutions; it is given by

$$\theta \cong \omega_{\pm}t + k_{\pm}x \qquad x \gg x_s(t) \quad \text{or} \quad x \ll x_s(t) \quad (16)$$

where $k_{\pm} = p \pm |q|$ and $\omega_{\pm} = \omega + \mu k_{\pm}^2$. Suppose $\mu > 0$, which represents a normal situation. Then the shock front moves in such a way that the higher frequency domain expands at the expense of the lower frequency domain. When the natural frequency ω is not perfectly uniform, but there is a small high-frequency region (or pacemaker region, as is often called) in an otherwise homogeneous medium, then a synchronizing shock front is formed spontaneously from a uniform initial phase distribution. Analysis based on Equation 13 in two-dimensional systems shows that the high-frequency domain with a pacemaker at its center appears as an expanding disc. In this domain, the isophase contour looks like concentric rings (or a target pattern) expanding faster than the domain expansion and vanishing at the domain boundary. When there are multiple pacemakers with different frequencies, the target pattern corresponding to the highest frequency encroaches step by step on the other domains and eventually dominates the entire medium, leaving a single target pattern. Similar phenomena are known in the Belousov-Zhabotinsky reaction (the best-studied chemical reaction system for wave patterns), although the waves in the oscillating Belousov-Zhabotinsky reaction are more akin to trigger waves characteristic of excitable media than to phase waves, i.e., waves generated by smooth oscillations.

Complex Ginzburg-Landau Model

Neglecting amplitude degree of freedom in the phase model sometimes causes difficulty. This happens when one tries to understand a certain class of phenomena involving phase singularities. Another drawback of the phase model is that it cannot explain the Benjamin-Feir phase instability, i.e., the instability of uniform oscillations against long-wave phase disturbances. The most extensively studied model for self-oscillatory media involving amplitude degree of freedom is the complex Ginzburg-Landau equation:

$$\frac{\partial A}{\partial t} = (1 + ic_0)A + (1 + ic_1)\nabla^2 A - (1 + ic_2)|A|^2 A \quad (17)$$

Here the individual oscillator is represented by a complex variable A, and its free motion is given by a perfectly smooth rotation along the unit circle in the complex A-plane. The oscillators are mutually coupled through diffusion with a complex diffusion constant. The complex Ginzburg-Landau model is generally derived from the reaction-diffusion equation near the onset of limit-cycle oscillation (Kuramoto, 1984).

Besides the family of plane wave solutions already discussed in relation to the phase model, the two-dimensional complex Ginzburg-Landau model admits a solution of rigidly rotating spiral waves (Hagan, 1982). In polar coordinates (r, θ), the spiral-wave solution has the general form

$$A(r, \theta) = R(t)e^{i(\pm\theta + S(r) + \omega t)} \quad (18)$$

Far from the origin, S behaves like kr and R tends to a constant $R_\infty(<1)$, so that the isophase contour forms an Archimedean spiral. Depending on the sign before θ, the pattern rotates clockwise or counterclockwise. Near the origin, R is proportional to r, so that there is a phase singularity right at the origin. Since the phase singularity cannot be eliminated without global change of the wave pattern, this kind of pattern, once created, persists very stably. Similar rotating waves also appear in excitable media such as heart muscle, and their relevance to a high-frequency irregularity of heartbeat has been suggested (Winfree, 1980).

Under certain conditions, spatially uniform oscillations $A(t) = A_0(t) \equiv \exp(i(c_0 - c_2)t)$ loses stability, and then the system immediately gets into a turbulent state (Kuramoto, 1984). The standard linear stability analysis shows that the instability of $A_0(t)$ (i.e., the Benjamin-Feir instability) occurs if $1 + c_1 c_2 < 0$. The turbulence near its onset is weak where $|A|$ has almost its full value of 1 everywhere, while the phase of A exhibits long-scale spatiotemporal irregularity. Thus it is called *phase turbulence*. As the turbulence becomes stronger, the amplitude fluctuation becomes significant, and spontaneous creation of phase singularities becomes more and more frequent. This is called *defect-mediated turbulence*.

Discussion

In conclusion, some simple many-oscillator models presented here (or possibly their variants) may find many applications to real and artificial neural networks as well as other biological systems. There are some experimental indications that the mechanism of synchronization in oscillatory neuronal activities could be responsible for feature linking in the primary visual cortex of cat (Gray et al., 1990). As a corresponding theory, we mention the work by Sompolinsky, Golomb, and Kleinfeld (1991), which attempts an explanation of such experiments in terms of a neural oscillator model similar to our phase model.

Road Map: Cooperative Phenomena
Background: I.3. Dynamics and Adaptation in Neural Networks
Related Reading: Chains of Coupled Oscillators; Traveling Activity Waves; Wave Propagation in Cardiac Muscle and in Nerve Networks

References

Buck, J., 1988, Synchronous rhythmic flashing of fireflies, II, *Quart. Rev. Biol.*, 63:265–289.
DeHaan, R. L., and Sachs, H. G., 1972, Cell coupling in developing systems—The heart-cell paradigm, *Curr. Top. Dev. Biol.*, 7:193–228.
Ermentrout, G. B., and Kopell, N., 1984, Frequency plateaus in a chain of weakly coupled oscillators, I, *SIAM J. Math. Anal.*, 15:215–237.
Ghosh, A. K., Chance, B., and Pye, E. K., 1971, Mitabolic coupling and synchronization of NADH oscillations in yeast cell populations, *Arch. Biochem. Biophys.*, 145:319–331.
Gray, C. M., Engel, A. K., König, P., and Singer, W., 1990, Stimulus-dependent neuronal oscillations in cat visual cortex—Receptive field properties and feature dependence, *Eur. J. Neurosci.*, 2:607–619.
Hagan, P. S., 1982, Spiral waves in reaction diffusion equation, *SIAM J. Appl. Math.*, 42:762–786.
Kuramoto, Y., 1984, *Chemical Oscillations, Waves, and Turbulence*, Berlin: Springer-Verlag. ◆
Kuramoto, Y., 1990, Collective synchronization of pulse-coupled oscillators and excitable units, *Physica D*, 50:15–30.
Peskin, C. S., 1975, *Mathematical Aspects of Heart Physiology*, New York: Courant Institute of Mathematical Science, pp. 268–278.
Sompolinsky, H., Golomb, D., and Kleinfeld, D, 1991, Cooperative dynamics in visual processing. *Phys. Rev. A*, 43:6990–7011.
Strogatz, S. H., Mirollo, R. E., and Matthews, P. C., 1992, Coupled nonlinear oscillators below the synchronization threshold—Relaxation by generalized Landau damping, *Phys. Rev. Lett.*, 68:2730–2733.
Wiener, N., 1958, *Nonlinear Problems in Random Theory*, Cambridge, MA: MIT Press.
Winfree, A. T., 1967, Biological rhythms and the behavior of populations of coupled oscillators, *J. Theor. Biol.*, 16:15–42.
Winfree, A. T., 1980, *The Geometry of Biological Time*, New York: Springer-Verlag. ◆

Collicular Visuomotor Transformations for Saccades

J. A. M. Van Gisbergen and A. J. Van Opstal

Introduction

Based on studies of brainstem structures involved in movement execution, Robinson (1975) proposed that saccades are generated by a nonlinear pulse generator in the brainstem, embodied by burst cells, which is driven by the difference between a signal-specifying target location in head coordinates and a local feedback signal coding current eye position. In this scheme, pause cells suppress the saccadic controller during fixations through strong inhibition of burst cells.

The emphasis in more recent research has shifted to include more central areas. For example, Scudder (1988) modified Robinson's model by incorporating the superior colliculus (SC) which, in his scheme, provides a static desired displacement signal with some capability to affect the dynamic properties of saccades. We review these developments by discussing both recent theoretical concepts and the experimental data on which they are based. For a broader orientation, the reader may consult earlier reviews (Robinson, 1975; Sparks and Mays, 1990; Van Gisbergen and Van Opstal, 1989).

Fixation (FIX) cells in the rostral pole of the SC are active during attentive fixation and pause during large saccades. In the caudal zone, the pattern of activity is just the opposite. A group of saccade-related (SAC) cells becomes active prior and during the saccade and has low levels of activity during active fixation. A similar contrasting pattern of activity characterizes omnipause neurons and saccadic burst cells in the brainstem. These cells pause and burst during saccades; it is thought that they are driven by FIX and SAC cells, respectively. As Figure 1 indicates, pause cells and burst cells have mutual inhibition; supposedly, this is also the case for FIX and SAC cells in the SC. Target position on the retina is topographically coded. At the level of burst cells and motoneurons, saccade size is coded temporally. This transition from spatial to temporal coding is a classical problem in the study of the saccadic system. Early studies of SAC cells suggested that the SC contained a topographical map specifying desired saccade displacement. This research showed that each SAC neuron is recruited only for a limited range of saccade amplitudes and directions, denoted as its movement field. As one moves from the fixation zone to the

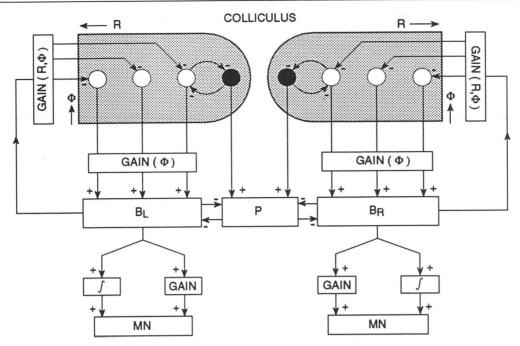

Figure 1. Scheme of putative connections of collicular neurons with saccade-related neurons in the brainstem, including recurrent signals responsible for local feedback. The SC consists of two neural maps, each subserving one half-field, which contain FIX cells (solid circles) in the rostral zone and SAC cells (open circles) in the caudal zone. FIX cells and SAC cells have mutual inhibitory connections (only partly shown). The same applies to omnipause cells (P) and right-(B_R) and left-going (B_L) medium-lead burst cells. When the pause cells have been silenced, the burst cells are driven by the population of SAC cells recruited at the location in the SC map that corresponds to target location (R, Φ). According to the model proposed by Van Opstal and Kappen (1993), the strength of the connections to horizontal burst cells depends only on the Φ coordinate of the SAC neuron in the SC. To account for the findings of Waitzman et al. (1991), the integrated burst cell signal inhibits SAC cells through topographically weighted connections such that caudal SAC neurons can maintain their burst longer, thereby ensuring larger saccades. The velocity command signal carried by burst cells has access to motoneurons (MN) both by a direct path (gain) and an indirect path through the neural integrator ($\int$) to create the pulse-step command signal driving the muscles.

caudal border, the movement field of local SAC neurons becomes more eccentric and larger. Since saccade direction is coded along the perpendicular dimension, the colliculus can be considered to be a map of saccade amplitude and direction which is organized in polar coordinates. By virtue of its topographical organization, the SC has become a key area for experimental and modeling approaches to the question of how sensory signals can be transformed into goal-directed movements. In fact, the SC is a layered structure that contains a stacked array of topographically organized maps. When the electrode passes from the superficial through the intermediate to the deep layers, the signals change from purely or mainly visual to mainly movement-related, with many cells carrying both types of signals (see Sparks and Mays, 1990). Because of this neat organization, it has been possible to deduce underlying neural computations from observed spatio temporal patterns of neural activity.

Afferent and Efferent Mapping

An early electrical stimulation study by Robinson showed, for the first time, that the SC motor map is highly nonhomogeneous: relatively more space is devoted to the representation of small saccades. Based upon these results, Ottes has modeled the collicular motor representation in the monkey by a complex logarithmic mapping function. This description characterizes the relationship between retinal target location and the site of maximum SAC cell activity (afferent mapping), as well as the inverse relationship between the locus of movement cell activity and the resulting saccade vector (efferent mapping). The population activity in the model resembles a Gaussian function centered around the point defined by the afferent mapping. This activity profile is translation-invariant, so that the number of active cells is constant, independent of saccade size. To explain how such a topographical representation may generate saccades, Van Gisbergen, Van Opstal, and Tax (1987) assumed that each cell has fixed connections with the horizontal and vertical burst cells downstream. Thereby, each cell generates a small movement contribution (minivector), proportional to its firing rate, in the direction of the retinal location to which it is connected by the afferent mapping. It was shown that the population activity generates appropriate movements for any target location if all minivectors generated by individual cells are summed vectorially. A problem with this scheme is that it does not assign any role to feedback connections implied by experimental data (see later sections). A recent feedback version is discussed later in this article.

Target Selection

Primary visual cortex neurons cannot distinguish between relevant and irrelevant stimuli. By contrast, a recent study indicates that visual-motor neurons in the frontal eye field (FEF) are involved in target selection (Schall and Hanes, 1993). Schall and Hanes propose a center-surround type of organization that causes a facilitated response in neurons processing the target and suppression at adjacent locations. Schlag-Rey and coworkers found evidence that the projection from the FEFs to

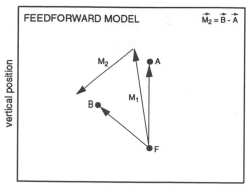
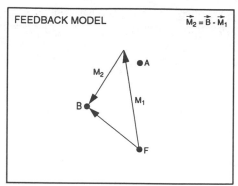

Figure 2. Feedforward and feedback schemes explaining control of the second response in a double-step task. Note that only the feedback model can correct the second movement (M_2) to the final target (**B**) for an error in the first movement (M_1) elicited by stimulus **A**. The point of fixation prior to the first movement is denoted by **F**.

the SC excites neurons involved in similar saccade vectors and inhibits all others. These mechanisms may help to weed out all stimulus representations except one.

Interestingly, under certain conditions involving a double stimulus, the saccadic system does not select one or the other target but makes a compromise response. This phenomenon has been studied extensively at a behavioral level. Two studies, first by Van Opstal and then by Glimcher, have investigated the collicular involvement in averaging. The findings are compatible with the idea that there is only a single mountain of activity during such saccades. In fact, the neural activity associated with averaging saccades could be predicted quite well from the movement field data collected in single-target trials. Thus, the SC seems to show only a single peak of activity in its motor map during the saccade, irrespective of the visual stimulus pattern.

Visuomotor Transformations in Motor Error Maps

Much about the visuomotor transformations in the oculomotor system has been learned by the use of paradigms that dissociate the metrics of the resulting saccade from the retinal signals of the eliciting stimulus by requiring an intermittent movement. For example, execution of the second movement in a double-step paradigm, relying on two briefly flashed consecutive targets A and B which are both extinguished before the first movement begins, requires a computation that accounts for the effects of the first movement. Clearly, the second movement cannot be based solely on the retinal coordinates of target B at the time when it was presented. The required computation to specify saccade B, necessitated by the intervening movement, could be done by relying on either a feedforward or a feedback strategy (Figure 2).

Behavioral studies have shown that humans and monkeys can perform the double-step task quite well, which means that the system can predict, or take into account, the effects of the first movement. One way to do this is to compute in advance the required second movement (M_2) from the relative retinal locations of the two flashed targets A and B by taking $M_2 = B - A$ (feedforward model). This is a visuospatial transformation, requiring the computation of a retinal difference vector, which may well be within reach of the capacities of the visual system. However, it only leads to satisfactory performance if the first movement is executed accurately ($M_1 = A$): Any error in the first movement, which is not taken into account, will propagate into the execution of the second.

An alternative strategy, based on the use of nonretinal feedback about the first movement (M_1), is to take the actual movement into account by vector subtraction: $M_2 = B - M_1$ (feedback model). Behavioral studies have shown only that the

feedback hypothesis performs better than the so-called retinal error hypothesis ($M_2 = B$). This does not prove that the first movement was taken into account using feedback, so the feedforward model should also be tested against the feedback model.

An interesting variation on the double-step task is the perturbation paradigm. The monkey has to make a saccade to a remembered target, but in some trials, the normal execution of this task is perturbed by a wrong saccade provoked by electrical SC stimulation before the remembered-target saccade begins. Strikingly, the animal can subsequently make a corrective movement, in full darkness, toward the location of the remembered target to compensate for the artificial saccade. Since the latter was created unexpectedly and outside the visual system, it cannot be explained by a feedforward visual mechanism of the type just discussed; rather, a feedback mechanism seems more likely in this case.

It is interesting to consider neuron behavior at various levels in the system during a double-step task. The first study to investigate the neural substrate for the execution of double steps was performed by Mays and Sparks. Quasi-visual cells, thought to carry a precursor signal for SAC cells, show a new locus of activity in the SC, appropriate for the new saccade, after the first movement has been made. This means that the system somehow updates the stimulus information about the final target which is made obsolete by the intervening movement. This work could not distinguish whether feedforward or feedback mechanisms were used, as this would have required an analysis of the effect of variability in the first movement on the precise locus of the new collicular activity. A later study, using the perturbation paradigm to elicit the second saccade, also found updating of collicular neural activity preceding the corrective saccade. As explained earlier, the need for feedback in the latter case is much more obvious.

Schiller and Sandell have found that, after ablation of either SC or FEF, monkeys are still able to make saccades and to perform the compensation necessitated by an electrical stimulus in the remaining area. Their results indicate that neither the SC nor the FEF can be held uniquely responsible for the compensation. In the FEF, the double-step experiment has been used to investigate the updating of neural activity after a previous eye movement. The results reported by Goldberg and Bruce are compatible with the possibility of remapping based on feedback, but do not rule out the feedforward model (see also a later section). Further work, using the perturbation paradigm, may clarify whether compensation by local feedback signals is possible already at this level.

The use of a novel paradigm has recently challenged the idea that the FEF can create motor error signals. In their so-called colliding-saccade paradigm, Schlag-Rey and co-workers applied local microstimulation during an ongoing natural saccade. In the deep SC layers, the electrically evoked saccade summed vectorially with the ongoing natural saccade, a result judged to be consistent with the notion that these layers code motor error. By contrast, at more superficial SC sites, the electrically elicited movement was compensated for by a later corrective movement that brought the fovea to the retinal location represented at the location of the electrode, even if it was heading in a very different direction before the stimulation began. It has been proposed that, at these sites, the electrode creates an artificial retinotopic goal, defined with respect to the position of the eye at the time of stimulation, which is compensated for by subsequent changes in eye position in downstream structures computing motor error. Remarkably, the compensatory result characteristic of the superficial SC layers was also consistently obtained in the FEF, even at sites where the microelectrode encountered movement-related cells. The latter finding is unexpected if one assumes that these FEF cells already code motor error, as one would then predict the same result as obtained in the deep SC sites (i.e, vector summation, rather than compensation). To explain this discrepancy, Dassonville and colleagues proposed that the FEF provides only the equivalent of a retinotopic signal and that the required compensation for eye movement occurs downstream to yield the saccade vector. But why, then, did Goldberg and Bruce find that movement-related cells in FEF could do the vector subtraction in double-step target experiments? According to Dassonville and co-workers, this paradox could be resolved if the FEF unit activity "occurred in relation to a saccade based on the exocentric relationship of the two stimuli in the double-step task," an interpretation which is in fact equivalent to the feedforward hypothesis formulated earlier.

Single-unit activity in the lateral intraparietal area (LIP) of the parietal cortex has been studied with the double-step paradigm by Goldberg, Colby, and Duhamel (1990) and by Duhamel, Colby, and Goldberg (1992). Their data suggest that LIP neurons represent the motor error predicted on the basis of retinal information, assuming correct execution of the first movement. Interestingly, in this case, there are data from trials in which the monkey made an incorrect first movement. Yet, in these cases, the cell showed the same activity as in a correct trial, suggesting that, at least in these cells and under these experimental conditions, the remapping was by a feedforward mechanism. Andersen and his group, who also studied LIP neurons, have concluded from their results that these neurons code motor error (see GAZE CODING IN THE POSTERIOR PARIETAL CORTEX). Again, one of the critical experiments that remains to be done to resolve this issue is to study the effect of perturbation saccades on the activity patterns of LIP neurons. Finding perturbation-induced remapping would strongly indicate a role for feedback signals (without ruling out a feedforward scheme in the double-step paradigm).

Discussion

Remapping Models Based on Feedback

According to Robinson's (1975) model, the recomputation of motor error necessary to generate the correction saccade after a perturbation is done by comparing target location in head coordinates and current eye position. Recently, two models have shown how remapping can be understood without resorting to craniotopic coding (Droulez and Berthoz, 1991; Dominey and Arbib, 1992; and DYNAMIC REMAPPING). In this alternative approach, motor error is represented in an oculocentric map which is continually updated for changes in eye position by an elegant shifting mechanism driven either by an efference copy of eye velocity (Droulez and Berthoz, 1991) or by eye position signals (Dominey and Arbib, 1992). When the eye moves, the map is modified by moving both mountains of activity by this intrinsic process which creates transitory activity in intermediate local neurons. In both schemes, the neural network representing motor error also serves as the memory for storing retinal target location. This nicely accounts for the events in a perturbation experiment, but becomes more problematical when the map has to store multiple remembered targets, as in a double-step experiment. First of all, to our knowledge, there is no clear neurophysiological evidence that sequential targets are represented simultaneously. Furthermore, having two simultaneous mountains of activity in the motor error map representing two sequential targets would require some way of tagging them to indicate which mountain should be translated into movement first.

An alternative possibility, perhaps more in line with the neurophysiological literature, is that during the execution of a movement sequence, the motor error map represents only a single movement at a time until the target is foveated and the activity is erased, making the map available for handling the next movement. This requires that the location of the final double-step target be stored outside the motor error map (see, e.g., Krommenhoek et al., 1993).

Coding of Dynamic Motor Error

So far, we have discussed remapping after a previous saccade at various stages in the system by concentrating on spatial aspects and ignoring dynamic changes. Until recently, it was thought that the precise time course of the SC population activity had no direct relevance for the speed and the size of the resulting saccade. According to this notion, the code for saccade metrics in the SC is purely spatial, providing a static desired displacement signal that is transformed into a temporally coded velocity command signal downstream. In this vein, Van Gisbergen and colleagues looked at the relationship between instantaneous firing rate of short-lead burst cells in the brainstem and instantaneous motor error. They found a strong nonlinear relation: initially, the burst cell starts off at a high rate, then decays gradually as the saccade proceeds, falling sharply to zero as the eye nears its final position. This behavior was interpreted in terms of local internal feedback (Robinson, 1975) which reduces the net driving signal at the level of the burst cells as the eyes move. In Robinson's model, the nonlinear relationship between saccade peak velocity and saccade amplitude (called the main sequence of saccades) is a consequence of a saturation-type nonlinearity in the burst cells. Recent recordings by Berthoz shed some doubt on the entrenched notion that the SC signal specifies only metrical saccade properties. It has also became clear that the effects of electrical stimulation, once regarded as all-or-nothing, are in fact graded with stimulus intensity and frequency. Furthermore, local reversible SC inactivation may cause visually guided saccades to become abnormally slow, but the eye still arrives on target. Since work in Sparks' group provides evidence that internal feedback signals enclose the SC, the question arises as to whether feedback may underlie these effects. Recent intriguing experimental data indicate that the collicular saccade representation may be modified at a saccadic time scale during the ongoing movement. This may take the form of a decay in firing rate without a concurrent shift in population activity (Waitzman et al., 1991) or a rapid shift of movement cell activity in the map (Munoz, Pélisson, and Guitton 1991). On this basis, it has been suggested that collicular neurons may code saccadic motor error dynamically, and that fast efference copy feedback may underlie these phenomena (Droulez and Berthoz, 1991). Waitzman et al. (1991) found that a subpopulation of saccade-related cells (clipped cells) showed a significant temporal relationship with saccade offset, indicating that their discharge may terminate as a result of internal negative feedback. Subsequently, Van Opstal and Kappen (1993) proposed a quantitative model incorporating a realistic SC motor map whereby negative feedback from saccadic burst cells was deemed to be responsible for this decline see (Figure 1). The strength of inhibitory connections to SC neurons is graded with their rostrocaudal position: caudal neurons, representing large saccades, receive weaker burst cell signals, allowing them to continue their activity for a longer period of time. Interestingly, the main sequence of saccades then becomes an emerging property of this connectivity without requiring a nonlinearity in the burst cells. This model can also explain saccade averaging and duration stretching in the components of oblique saccades. The latter phenomenon entails that a given movement component lasts longer when it is part of an oblique saccade than when it is executed alone. It occurs in the model because the SC in this scheme acts as a vectorial pulse generator followed by a decomposition stage at the level of the burst cells, an arrangement proposed earlier in more abstract terms on the basis of behavioral data.

Munoz, Pélissor, and Guitton (1991) studied the firing patterns of collicular output neurons for a wide amplitude range of gaze saccades in head-free cats. Before a small saccade some cells fire early, but before large saccades, they fire later, more near the end, suggesting that the mountain of collicular activity may expand and shift toward the fixation zone during the saccade. Munoz and co-workers suggest that the instantaneous position of the mountain of activity codes current gaze motor error and that efference copy signals may be responsible for shifting the activity during the movement. When the activity invades the foveal zone, recruitment of FIX cells provides a potential mechanism for terminating the saccade. It remains to be seen, however, whether FIX cells really subserve this role. It has been noticed, for example, that they hardly become active in intervals between spontaneous saccades. This suggests that their main function may be to ensure active fixation by suppressing SAC cells (see Figure 1), thereby allowing foveal analysis of interesting stimuli. Since earlier studies on SAC neurons in monkey SC have never shown the intrasaccadic shift in population activity, the hypothesis that the SC codes dynamic motor error by a mountain of activity shifting with saccadic speed is still quite controversial. However, a preliminary report from work in the monkey suggests that movement-related cells with a gradually increasing burst (build-up of cells) may show wave-shifting.

Acknowledgment.   The authors acknowledge support from EC project BRA 6615 (MUCOM).

Road Maps: Mammalian Brain Regions; Primate Motor Control; Vision
Related Reading: Saccades and Listing's Law

References

Dominey, P. F., and Arbib, M. A., 1992, A cortico-subcortical model for generation of spatially accurate sequential saccades, *Cereb. Cortex*, 2:153–175. ◆
Droulez, J., and Berthoz, A., 1991, A neural network model of sensoritopic maps with predictive short-term memory properties, *Proc. Natl. Acad. Sci. USA*, 88:9653–9657.
Duhamel, J.-R., Colby, C. L., and Goldberg, M. E., 1992, The updating of the representation of visual space in parietal cortex by intended eye movements, *Science*, 255:90–92.
Goldberg, M. E., Colby, C. L., and Duhamel, J.-R., 1990, Representation of visuomotor space in the parietal lobe of the monkey, *Cold Spring Harbor Symp. Quant. Biol.*, 55:729–739. ◆
Krommenhoek, K. P., Van Opstal, A. J., Gielen, C. C. A. M., and Van Gisbergen, J. A. M., 1993, Remapping of neural activity in the motor colliculus: A neural network study, *Vis. Res.*, 33:1287–1298.
Munoz, D. P., Pélisson, D., and Guitton, D., 1991, Movement of neural activity on the superior colliculus map during gaze shifts, *Science*, 251:1358–1360.
Munoz, D. P., and Wurtz, R. H., 1993, Fixation cells in monkey superior colliculus. I. Characteristics of cell discharge, *J. Neurophysiol.*, 70:559–575.
Robinson, D. A., 1975, Oculomotor control signals, in *Basic Mechanisms of Ocular Motility and Their Clinical Implications* (G. Lennerstrand and P. Bach-y-Rita, Eds.), Oxford: Pergamon, pp. 337–374. ◆
Schall, J. D., and Hanes, D. P., 1993, Neural basis of saccade target selection in frontal eye field during visual search, *Nature*, 366:467–469.
Scudder, C. A., 1988, A new local feedback model of the saccadic burst generator, *J. Neurophysiol.*, 59:1455–1475.
Sparks, D. L., and Mays, L. E., 1990, Signal transformations required for the generation of saccadic eye movements, *Annu. Rev. Neurosci.*, 13:309–336. ◆
Van Gisbergen, J. A. M., and Van Opstal, A. J., 1989, Models, in *The Neurobiology of Saccadic Eye Movements* (R. H. Wurtz and M. E. Goldberg, Eds.), Amsterdam: Elsevier, pp. 69–101. ◆
Van Gisbergen, J. A. M., Van Opstal, A. J., and Tax, A. A. M., 1987, Collicular ensemble coding of saccades based on vector summation, *Neuroscience*, 21:541–555.
Van Opstal, A. J., and Kappen, H., 1993, A two-dimensional ensemble coding model for spatial-temporal transformation of saccades in monkey superior colliculus, *Network*, 4:19–38.
Waitzman, D. M., Ma, T. P., Optican, L. M., and Wurtz, R. H., 1991, Superior colliculus neurons mediate the dynamic characteristics of saccades, *J. Neurophysiol.*, 66:1716–1737.

Color Perception

Jules Davidoff

Introduction

The objects of the world either reflect or emit light of different wavelengths into our eyes. It is not these light rays that are colored. Rather, *color* is the name we give to our perception of aspects of surfaces (or, more generally, regions of space). To understand this definition, we need to explore the operations of our visual system and to be clear about two points. The first of these is the difference between boundaries and surfaces; the second is whether computational procedures are based on wavelength differences or on color differences. Our experience of color has a reliable but complex relationship to wavelength (Walsh and Kulikowski, 1993), but wavelength differences alone do not produce the sensation of color.

Our visual system "recreates" the world in the form of boundaries that contain surfaces (Grossberg and Mingolla, 1985). The perception of the recovered boundaries is what we call shape; the perception of aspects of surfaces we call color. It is somewhat a matter of convention as to which aspects of the perceived surfaces are included under the term *color*. We normally only consider surfaces of red, green, blue, etc. to be colored, but it is not the same in all cultures. In Japan, for example, black-and-white photography is known as two-color photography, and it is at least conceivable that some languages include a surface's luster as part of their color vocabulary. This role of color as surface appearance may appear to make it "secondary" and relatively unimportant, but it corresponds to its grammatical form in English. Color words are adjectives. We speak of seeing an apple or a green apple, but not of seeing a green. However, not all surface properties are colors. Unlike colored properties, other properties of surfaces (e.g., textural) are not affected by the wavelength composition of incident illumination; we do not call these surface properties color.

We need to make one further distinction before considering the computational problems in recovering the surface appearance that we call color. That distinction is between color and object-color (Davidoff, 1991). It will prevent us from making the mistake of Hering, who believed that color has a necessary connection to our past experience with the colors of objects (object-colors). Color, as defined here, has no connection to our memory of the surface appearance of particular objects. Object-colors are the representations in memory for the surfaces of shapes we know to be objects. Of course, color can be compared with object-color. We can even compare the surface appearance of a patch with a surface that forms part of an object, even when only one is present. Thus, we might say that a particular red surface reminds us of a ripe apple or that an apple does not look ripe because it is not sufficiently red.

The distinction between color and effects attributable to wavelength is illustrated by some recent reports claiming that reading-disabled children are helped by viewing text through low-pass (blue) filters. These filters are effective for some reading-disabled children, but not by virtue of perceived color. With the filters, surfaces of the page will look blue; but it is not the blueness that improves the reading. The cause of the improvement lies in the differential effect of chromatic (wavelength) information on the boundary-forming components of the visual system. To better understand these differences between boundaries and surfaces, we need to consider the emergence of chromatic differences and boundary differences within the visual system.

Chromatic Differences Within the Visual System

The main stages of our visual system up to the recovery of boundaried surfaces are as follows: the retina, the lateral geniculate nucleus (LGN) of the thalamus, and the visual cortex. Each of these stages may be subdivided.

The human retina has three different receptors for photopic vision, but some animals have four or more (see Thompson, Palacios, and Varela, 1992 for the potential consequences for color vision). The three receptors are the L cones (with a peak sensitivity of 560 nm (Figure 1), the M cones that have peak sensitivities of around 530 nm, and the short-wave (S) cones. The S cones, which have a peak sensitivity of around 430 nm, constitute only 3%–10% of the total. A light source of 560 nm would be seen as red, one of 530 nm as green, and one of 430 nm as blue. However, it would be a mistake to identify the percepts of red, green, and blue with the outputs of individual

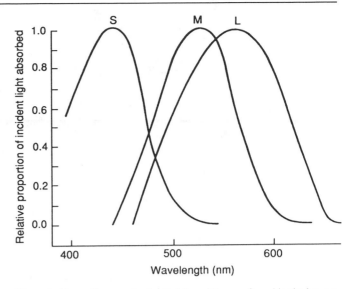

Figure 1. Absorption spectra for L, M, and S cones found in the human retina. The heights have been adjusted so that peak sensitivity equals 1.0 for each cone type.

cone types; this is why we now refer to them as L, M, and S cones.

Loss of one or more of these cone types as a result of a genetic disorder produces forms of what is commonly referred to as color blindness. Cone-deficient individuals (usually male) confuse colored surfaces that normal trichromats distinguish without error. Individual differences in normal observers with trichromatic vision may result from a genetic variation in the L and M cones that produces different peak sensitivities.

To perceive color differences, a system requires a comparison between at least two types of receptor with different absorption spectra. The reason is that the receptors transmit information, not only as a result of their sensitivity to a particular wavelength, but also as a result of the intensity of the illumination at that wavelength. So, our long-wave (L) receptor would not distinguish between equally intense monochromatic light sources of 550 nm and 570 nm. But more importantly, any monochromatic light within the range of the cone's absorption spectrum could be made sufficiently intense to equal the cone's output to a 560-nm source (peak sensitivity) at a lower intensity.

The two absorption spectra of L and M cones (see Figure 1), which are apparently quite similar, are sufficient to produce the basis of a chromatically sensitive system. A spatial antagonism based on wavelength arises at the retina in the bipolar cells and their output to retinal ganglion cells. The receptive field of the neuron is divided into a center and surround with different spectral sensitivities; it is possible for one part of the receptive field to cancel the output of the other. The cells, therefore, have chromatic opponency. These cells are divided into those that respond to the onset of stimulation and those that respond to the offset (on and off channels). In order to model the interactions of the cone types, it is necessary to take into account the adaptation state of the receptors. The model of Guth (1991), for example, combines noise and a gain control between the cone receptors and cone opponency stages.

At the LGN, parvocellular cells have the potential to distinguish between luminance and wavelength information because of their different receptive field properties; there is a much greater spatial resolution for luminance-defined stimuli.

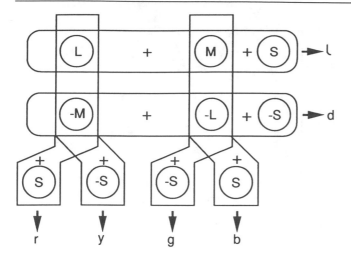

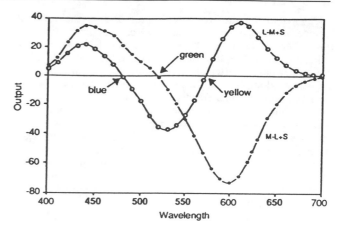

Figure 2. A model (adapted from DeValois and DeValois, 1993) for the production of achromatic and chromatic opponency. In the horizontal rows, cone-opponent units with the same center sign are added together $(L + M + S)$ and $(-L - M - S)$ to yield achromatic units (light [l] and dark [d]). In the vertical columns, cone-opponent units of different center signs are added; this cancels luminance signals. The outputs from these systems are labeled red (r), yellow (y), green (g), and blue (b) (but see also Figure 3).

Figure 3. The output of red $(L - M + S)$ and Blue $(M - L + S)$ units. Green and yellow units would produce mirror images. Unique blue and yellow occur at zero output from the red (and green) system. (Adapted from DeValois and DeValois, 1993.)

DeValois and DeValois (1993) have demonstrated that an LGN cell may, for example, fire with a shift toward long wavelengths if excited over all its receptive field, but it responds to increments in luminance with an excitatory center and an inhibitory surround. So, at this stage, the same neurons are transmitting both chromatic and luminance information. The separation of luminance and wavelength information is achieved more fully at the cortex and, for this reason, monocular information and cells of the same polarity (on and off channels) are kept separate at the LGN (see Cronley-Dillon, 1991 vol. 6).

The primary reception area for visual information in the primate cortex is known as V1. From V1, information passes in series and parallel to many visual areas containing neurons that respond relatively selectively to different aspects of the visual input (e.g., motion, wavelength, or location). The motion pathway is, for example, through V3 and V5; it is fed by information from the magnocellular layers of the LGN. The recovery of chromatically defined boundaried surfaces involves V4 and subsequent locations in the temporal lobe.

The most important difference between the receptive fields of cells at V1 and those earlier in the visual system is, according to Hubel and Wiesel, that some of them are orientation-specific. Cells without orientation specificity but with wavelength specificity were discovered more recently. Livingstone and Hubel (1987) called the areas containing these cells blobs, based on their appearance in tangential section after staining with cytochrome oxidase. The role of the blob neurons in color vision is not fully understood. (There are, for example, species with poor color vision that possess many blob neurons.)

At V1, there is an increasing selectivity for wavelength. DeValois and DeValois (1993) argue that it takes a combination from cone-opponent cells (including those from the S cones) to produce opponent color mechanisms. Their proposed neural network also extracts luminance information independent of wavelength. The opponent color processes are a red-green (r-g) system and a yellow-blue (y-b) system. The production of the opponent-color systems from the three cone

types is shown in detail in Figure 2 and schematically in Figure 4. Different combinations of cone output produce the r-g and y-b systems. Note that the S cells serve a modulatory role, as they are added to or subtracted from each combination of L and M outputs. DeValois and DeValois (1993) add a half-wave rectification to the combination of cone outputs.

The cells proposed by DeValois and DeValois (1993) would produce the unique red, green, blue, and yellow primary colors that do not perceptually resemble any of the other primaries. As can be seen from Figure 3, the output of the red system has two null points as wavelength varies; these appear as unique blue and yellow to our perception. The output of the blue system has only one null point; this is unique green (see Figure 3). Unique red is obtained from a combination of the extraspectral long and short wavelengths that would cancel the y-b function. Another proposal (Krauskopf, Williams, and Heeley, 1982) contrasts yellow (y) with violet (v). In their study, two directions of perceived color were taken to be mutually independent if color discriminations along each direction were not affected by prolonged exposure to a field temporally modulated along the other direction. For the r-g and y-v axes, thresholds were raised in the direction of habituation but were virtually unchanged in the orthogonal direction. DeValois and DeValois (1993), however, claim that prolonged exposure affects a precortical cone opponent system rather than that producing perceived color.

Boundaries

Boundaries can only be achieved from the recovery of orientation information; the surface-forming networks are not sensitive to orientation. On approaching a blob with their microelectrodes, Livingstone and Hubel (1987) found that cells showed a decline in orientation selectivity; this decline may have been exaggerated, however (Ts'o and Gilbert, 1988). In the interblob region, cells have relatively small receptive fields and are precisely tuned for orientation but not obviously selective to direction of motion or wavelength. Nevertheless, it is clear that some of these cells also respond to chromatic contrasts. If cells can be both orientation- and wavelength-specific, it means that there may be boundary-forming mechanisms based on chromatic as well as luminance information. Thus, as in machine vision models, regions may be defined according to either color or luminance (and any other property, such as depth or texture, that can define orientation).

Cells with responses that would be effective in the production of boundary-forming mechanisms become more pronounced as information is passed from V1 to V2. Nevertheless, it may be that cells responsive to the direction of chromatic contrast do not emerge prior to V4. The particular cells emphasized by Grossberg and Mingolla (1985) form boundaries without recognizing the direction of the contrast (polarity) on either side of the border. There are, in fact, five so-called domains for the production of shapes: luminance, motion, stereopsis, texture, and color. According to Livingstone and Hubel (1987), boundary contours are produced primarily in the luminance domain, although they are also strongly achieved through motion. Moreover, as the Gestalt psychologist Koffka pointed out in 1935, the figure in perception is characterized by distinct form and fine detail. Figural perceptions are, therefore, likely to be based on the luminance boundary-forming procedures of the parvocellular pathway (Weisstein, Maguire, and Brannan, 1992) and not on chromatic boundaries. Under conditions that do not require fine detail, it is usually important to form boundary segmentation at a rapid rate, so, again, nonchromatic (i.e., magnocellular pathways) would be used. However, in natural conditions, lighting variations throughout the day produce vastly different luminance contrasts between shadow and nonshadow areas, so chromatic differences are more reliable. Thus, it would also be beneficial to be able to form boundaries from chromatic differences. In fact, chromatic contrasts are actually superior for low spatial frequencies. If one wanted to form a rough segmentation of the visual scene at a distance, albeit one insufficient for detailed object recognition, it would only be effective to use chromatic, rather than luminance, contours.

Domain Independence and Integration

Many studies, using a variety of behavioral techniques, have shown that luminance and chromatic domains are not independent. Chromatic differences may also contribute to stereopsis and texture-forming domains. A contribution of wavelength information to the motion domain would be more surprising because the latter is supposed to rely on the magnocellular pathway, indeed, original reports indicated that chromatically defined stimuli appear to be stationary. However, there is now evidence that chromatic information does contribute to the motion domain. To make effective use of the color domain, we would need cooperative algorithms (Poggio, Gamble, and Little, 1988) to integrate it with boundary-forming domains based on luminance, motion, stereopsis, and textural differences. Nevertheless, for our understanding of color, it may not be important to know how the boundary-forming domains integrate (for a description of the problems, see Savoy, 1987). By considering individual domains in neurophysiology or psychophysics, we may be getting a false picture of neuronal activity. A division of load between neurons does not mean that they are separate in the whole train of neuronal firing. The engine and carriages are separate parts of another sort of train, but they both proceed with the same speed.

Color

Studies of chromatic adaptation indicate the need for narrowly selective neurons to explain perceived color. We do not have a clear idea of the neurophysiological organization that provides these neurons (but see Cronley-Dillon, 1991, vol. 6; DeValois and DeValois, 1993). Zeki originally proposed V4 as the site

for these cells with narrow wavelength selectivities, but later identified V4 as the substrate for color constancy because of the large receptive field of the neurons. Color constancy allows us to know the true surface color, independent of the spectral qualities of the illuminant. To achieve color constancy, we need to see a background of a different spectral reflectance. Land also realized the importance of nearby surfaces in his Retinex theory, and used the lightness of surrounding and distant surfaces in the computation of surface color. Surrounding surfaces also dramatically affect the color of a surface by inducing contrast determined by color and lightness opponency. In some cases, the effect can be dramatic enough to change the name we wish to give to a surface. There is, for example, no surface that we would always wish to call brown; any such surface may be called orange or yellow, depending on the background. So, a major problem for providing the neurophysiological substrate is that perceived surface color is affected by spatial factors. It is also affected by the temporal and luminance characteristics of the display (see Davidoff, 1991). All of these require a combination of information from both magnocellular and parvocellular pathways.

Another important aspect of perceived surface colors is that they have a systematic relationship to each other; that is to say, the individual colors fall into categories (reds, oranges, blues, etc.). One could say that we will only *understand* color (while ignoring what it means to *experience* color) when we have a system for measuring color appearance that produces a common output for surfaces that we wish to label with the same color name. At the moment, there are many systems for unambiguously coding color samples, but none are psychologically perfect. For example, the Munsell system codes color by a hue range (e.g., 2.5R [Red]), saturation (chroma), and lightness (value). Unfortunately, subjects have great difficulty perceiving stimuli in the same hue range (e.g., 2.5R) as similar. Another color appearance system is the Natural Color System (NCS), widely in use in Scandinavia. The NCS codes colors according to three components: hue, blackness, and chromaticness. Hue is based on the four unique hues (red, green, yellow, blue) arranged in a circle; colors are designated as proportions of two unique hues. As in the Munsell system, opponent colors are separated because we cannot perceive a reddish-green or a yellowish-blue. The NCS differs from the Munsell system by scaling whiteness-blackness rather than value. The outcome is to divide color space into 10 major areas that have the same "nuance." Details of these and other measuring systems are included in Cronley-Dillon (1991, vol. 6). However, there are several limitations with all of them. First, none of these systems include all colors (e.g., chrome or fluorescent yellow). Second, the enormous perceptual difference between a colored light source (film color) and a surface color cannot be shown.

Achromatopsia

Cerebral achromatopsia is an impairment that causes the affected individual to describe the world as being in shades of grey. Two recent historical and theoretical reviews document this fascinating disorder, which is caused by damage to the posterior regions of the lingual and fusiform gyri in the ventromedial extrastriate cortex (Zeki, 1990; Cronley-Dillon, 1991, vol. 12). But, as a matter of fact, there is no account of cerebral achromatopsia in which grey discrimination has been found to be normal; this is consonant with an impairment to the perception of surface appearance.

It is most interesting that, while achromatopsic patients have markedly degraded surface appearances, in other ways, they make normal use of chromatic information. For example, pa-

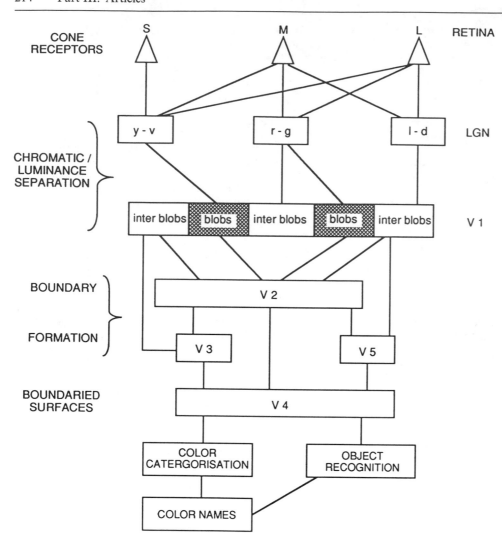

CONE RECEPTORS

CHROMATIC / LUMINANCE SEPARATION

BOUNDARY

FORMATION

BOUNDARIED SURFACES

S   M   L   RETINA

y - v   r - g   l - d   LGN

inter blobs | blobs | inter blobs | blobs | inter blobs   V 1

V 2

V 3   V 5

V 4

COLOR CATERGORISATION   OBJECT RECOGNITION

COLOR NAMES

Figure 4. A scheme for the production of surface colors at V4 from the chromatic pathways. Subsequent to V4 are areas concerned with color memory and color names. Sparse connections are omitted, including those from S cones to the r-g and l-d systems (but see also Figure 2) and those from V1 to V4. The model is a predominantly feedforward one.

tients may derive shape from isoluminant borders. The consequence is that, at some distances, patients can read the figures in the Ishihara Test for color blindness. Zeki (1990) suggested that these patients are using the boundary-forming systems of the magnocellular pathways. Shape discerned from movement would be normal. However, Heywood, Cowey, and Newcombe (1991) found that such patients could distinguish chromatically defined edges. They showed that an achromatopsic patient used edge detection rather than a color comparison; when the samples were separated by a short distance, the patient could no longer make the discrimination. Heywood et al. (1991) argue that the surviving boundary-forming system in achromatopsic patients codes chromatic borders without polarity, and that their normal cone functions (especially, the function of the S cones) militates against the use of the magnocellular pathway. However, the issues are far from settled. Also unsettled is whether surface colors are finally recovered at the visual area V4. Although rare, some achromatopsic patients have normal shape discrimination that would be impossible with V4 lesions. Heywood et al. (1991) suggest that achromatopsic patients sustain damage to the white matter under the fusiform gyrus (supposed homologue for V4 in humans) but have the grey matter preserved, a proposal originally put forward by Lenz in 1921.

Unanswered Questions

A scheme for the neuronal pathways and processing stages involved in color is shown in Figure 4. Its aim is to describe the production of narrow color categories; clearly, it is incomplete and imprecise. For example, there is no consideration of the ways attention might affect segmentation from chromatic differences or memory for color (Davidoff, 1991). Also, there ought to be directional arrows in Figure 4. To change the procedures currently in use for boundary segmentation, we require top-down mechanisms that operate on V1, at least from V4 and probably from many other loci. We may have a moderately clear understanding of the neural correlates of opponent processes, but we still await a description of how these processes may combine to produce the nuances of color sensation.

Road Map: Vision
Related Reading: Figure-Ground Separation; Visual Cortex Cell Types and Connections; Visual Scene Perception: Neurophysiology

References

Cronley-Dillon, J. R., Ed., 1991, *Vision and Visual Dysfunction*, London: Macmillan Press. ◆

Davidoff, J., 1991, *Cognition Through Color*, Cambridge, MA: MIT Press. ◆

DeValois, R. L., and DeValois, K. K., 1993, A multi-stage color model, *Vis. Res.*, 33:1053–1065. ◆

Grossberg, S., and Mingolla, E., 1985, Neural dynamics of form perception: Boundary completion, illusory figures, and neon color spreading, *Psychol. Rev.*, 92:173–211.

Guth, S. L., 1991, Model for color vision and light adaptation, *J. Opt. Soc. Am.*, 8:976–993.

Heywood, C. A., Cowey, A., and Newcombe, F., 1991, Chromatic discrimination in a cortically blind observer, *Eur. J. Neurosci.*, 3: 802–812.

Krauskopf, J., Williams, D. R., and Heeley, D. W., 1982, Cardinal directions of color space, *Vis. Res.*, 22:1123–1131.

Livingstone, M. S., and Hubel, D. H., 1987, Psychophysical evidence for separate channels for the perception of form, color, movement and depth, *J. Neurosci.*, 7:3416–3468.

Poggio, T., Gamble, E. B., and Little, J. J., 1988, Parallel integration of visual modules, *Science*, 242:436–440. ◆

Savoy, R. L., 1987, Contingent after effects and isoluminance: Psychophysical evidence for separation of color, orientation, and motion, *Comput. Vis. Graph. Image Proc.*, 37:3–19.

Thompson, P., Palacios, A., and Varela, F. J., 1992, Ways of coloring, *Behav. Brain Sci.*, 15:1–74.

Ts'o, D. Y., and Gilbert, C. C., 1988, The organization of chromatic and spatial interactions in the primate striate cortex, *J. Neurosci.*, 8:1712–1727.

Walsh, V., and Kulikowski, J. J., 1993, Seeing color: The non-Newtonian brain, in *The Artful Brain* (R. L. Gregory, J. Harris, D. Rose, and P. Heard, Eds.), Oxford: Oxford University Press. ◆

Weisstein, N., Maguire, W., and Brannan, J., 1992, M and P pathways and the perception of figure and ground, in *Applications of Parallel Processing in Vision* (J. Brannan, Ed.), Amsterdam: North Holland.

Zeki, S. M., 1990, A century of cerebral achromatopsia, *Brain*, 113: 1721–1777.

Command Neurons and Command Systems

J.-P. Ewert

Introduction

If a certain interneuron is stimulated electrically in the brain of a marine slug, the animal then displays a species-specific escape swimming behavior, although a predator is absent. If in a toad a certain brain area of the optic tectum is stimulated in this manner, snapping behavior is triggered, although no prey is present. In both cases, a commanding trigger recalls a motor program. We address the notion "command" to certain sensorimotor processes involving sensory pattern recognition and localization on the one side and motor pattern generation on the other; command functions provide the interface (Ewert et al., 1994). The operation translates a specific pattern of sensory input into an appropriate spatiotemporal pattern of activity in motoneurons to coordinate the muscle contractions for the adequate action pattern which—in various animal groups and depending on the task—are either rather stereotyped (fixed action pattern) or leave considerable room for variability (modal action pattern). This correspondence can be innate, modified innate, or acquired and strategy-related. Once an action pattern is commanded, it tends to carry on to completion, although there may be cases in which a command is "countermanded."

How does this interface operate? Before we tackle the question, a few comments on pattern generation are in order. A *motor pattern generator* (MPG) is an internuncial network that, in response to a commanding trigger input, coordinates appropriate muscle contractions. The network is activated if and only if a specific (combination of) input occurs. An intrinsic pattern of neuronal connectivity in the network assures the generation of a consistent spatiotemporal distribution of excitation and inhibition. The output of the network, mediated by premotoneurons, has privileged access to the requisite motoneuronal pools. Proprioceptive and internal feedback—or positive feedback from components of the motor network to the command—can play a role in the coordination and maintenance of a motor pattern. Recurrent negative feedback from the motor system to the command and from the command to the sensory afferents terminates the command and prevents sensory input during the animal's movement.

The property of an MPG-commanding trigger is correlated with the conditions under which an action pattern is released: (1) the locus of the stimulus; (2) the presence of adequate stimulus features; (3) motivation (variable on a relatively long time scale), attention, and arousal (variable in the short term); (4) gating input; and (5) evaluation of the stimulus in connection with experience. The first two listed concern aspects of the releasing stimulus, whereas the latter three refer to modulatory functions.

The Command Neuron Concept

The question whether information about the conditions just listed can be mediated by a command neuron is controversial (Kupfermann and Weiss, 1978; Eaton and DiDomenico, 1985; Ewert, 1980, 1989). According to Kupfermann and Weiss (1978), a command neuron is an interneuron whose excitation is both necessary and sufficient to activate the corresponding MPG. Test criteria include: (a) the recording of the activity from the putative command neuron during the presentation of a stimulus signal in registration with the corresponding action pattern (link between stimulus and motor activity); (b) electrical stimulation of the putative command neuron with demonstration that this action pattern is executed (sufficiency criterion); and (c) removal of the putative command neuron and demonstration that this action pattern is no longer elicited by the stimulus signal (necessity criterion).

The best candidates for this approach are among the identified neurons of various invertebrates, such as the neuron initiating escape swimmeret movements in the crayfish, for which the term "command" was first applied by C. A. G. Wiersma and K. Ikeda (for a review, see Gillette, 1987). An example in which the processes that contribute to command functions are particularly well understood concerns feeding in the sea slug *Pleurobrachea*. The ravening behavior commanding neuron, in response to chemosensory stimulation by food in the buccal cavity, displays a burst episode whose maintenance and inhibition involve specific intrinsic electrochemical processes; these are independent of the performance of the motor act, but leave open the possibility that (also) an element of the MPG feeds back to the command neuron. Among vertebrates, the Mauthner cell of teleost fish—via a spike in one cell of the bilateral pair—commands the fast startle escape response to vibratory stimulation (Eaton and DiDomenico, 1985).

However, even in these cases, it was technically not feasible to test the necessary-and-sufficient condition, and in the case where it was possible (the Mauthner cell), the necessity criterion was not fulfilled. After lesioning of both Mauthner cells in the larval stage, zebra fish displayed the startle response, although at a longer lantecy.

Is the command concept, therefore, obsolete? Regarding dynamic, plastic, backup, and feedback-guided properties that probably exist, the necessary-and-sufficient condition is too rigorous. If more than a few neurons cooperate in command functions, it is technically impossible to test these criteria. Depending on the task of a sensorimotor process, we envision a spectrum of possibilities by which command functions can be executed (Eaton and DiDomenico, 1985; Ewert, 1989). Hence, the idea behind the redefined command concept is fruitful in the neurophysiological analysis of behavior (neuroethology) and in perceptual robotics (neural engineering) (Ewert, 1991; Ritzmann, 1993).

Command Systems

A command system (Kupfermann and Weiss, 1978) consists of collectively operating neurons of the same type called a command element (CE), such as the pair responsible for the ravening behavior of feeding in *Pleurobrachea* (Gillette, 1987). For a CE, the necessary-and-sufficient condition cannot be fulfilled: if CEs are connected to the MPG like an AND-gate, a CE fulfills the necessity but not the sufficiency criterion; in an OR-gate, a CE fulfills the sufficiency but not the necessity criterion.

The concept of a command-releasing system (CRS) (Ewert, 1980, 1989) is suggested for functions requiring a more complex perception of the sensory input. Different types of CEs (a set of each) cooperatively trigger an MPG. Each type of CE evaluates a certain aspect of the releasing stimulus signal, such as features, feature combinations, and their location in space, respectively; we call this a *sensorimotor code*. The nature of a CRS leads to the analogy of "neural working groups," whose members (CE), acting in different parts of the brain, exhibit goal-specific cooperation (Figure 1). On the one hand, the same goal can be reached by differently combined groups; on the other hand, certain members can be shared by different groups

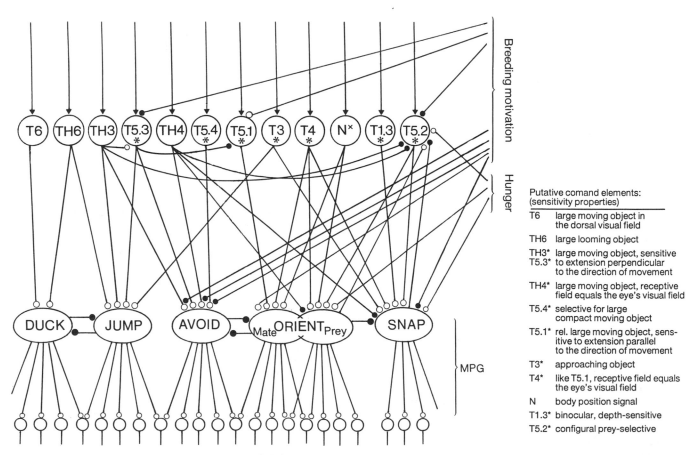

Figure 1. Sensorimotor codes implemented by CRSs in common toads. Different subsets of functionally identified types of pretectal thalamic (TH) and tectal (T) neurons may contribute to commanding and parameterizing motor patterns of feeding (prey-orienting, snapping), escape (ducking, jumping, avoidance turning), and orienting by the male toad toward the female in the mating reason. Breeding motivation and state of hunger, for example, deliver different modulatory influences. Each circle represents a set of neurons; each oval stands for a neuronal circuit of the MPG with access to motoneuronal pools (motoneuronal and proprioceptive feedback from muscles are not considered here). The open dots refer to excitatory influence, and the solid dots to inhibitory influence; arrows refer to sensory inputs, e.g., from retinal ganglion cells and pretectal/tectal interneurons. The asterisks indicate that the projective character of the neuron type is evidenced by means of the antidromic stimulation/recording method. (Modified from Ewert, 1980, 1989, including the jump/duck model of Liaw and Arbib, 1993.)

for different goals. The concept of coded commands stresses that firing of various types of efferent neurons in certain combinations characterizes an object in space and selects the appropriate goal-directed action pattern. This concept is supported by neurophysiological and neuroanatomical results in amphibians (Ewert, 1989; Ewert et al., 1994) and mammals (Dean and Redgrave, 1992).

The behavioral choice of toads in terms of so-called hierarchies of behavior may depend on dual excitatory/inhibitory actions of the respective CEs and/or MPGs, yielding predator-avoiding > prey-catching > mate-approaching = 0. This is modified by hormonal influences in spring: mate-approaching > predator-avoiding > prey-catching = 0 (see Figure 1).

A CRS can be regarded as the neurobiological correlate of the ethological concept of an *(innate) releasing mechanism*, originally called the *(innate) releasing schema*. The prefix "command" points to the initiating, primarily ballistic function, "releasing" denotes its coded property, and "code" relates to the combinative cooperative nature. The CRS concept also considers dynamic and plastic properties which allow broad applicability.

Schema theory (Arbib, 1989) offers an interdisciplinary science that allows one to treat principles of neuroethology or neural engineering in the same language. In the language of schema theory, the sensorimotor code of a CRS embodies a *perceptual schema* that exists for only one purpose: namely, to determine the conditions for the activation of a specific MPG embodying a *motor schema*. The CRS must also ensure that the resultant movement is directed in relation to the target. A schema and its instantiation usually are coextensive, i.e., instantiation of a schema appears to be identifiable with appropriate activity in certain populations of neurons of the brain, whereby each schema may involve several cell types or brain regions, while a given cell type or brain region may be involved in several schemas. The motor schemas of directed appetitive behaviors and consummatory behavior need not occur in a fixed order; rather, each may proceed to completion, followed by perceptual schemas that will determine which motor schema is to be executed next. Schemas may be linked by so-called coordinated control programs. Motor schemas, for example, can take the form of compound motor coordinations (such as a frog's programmed jump-snap-gulp sequence), which comprise a set that will proceed to completion without intervening perceptual tests, e.g., in such a manner that schema *A* proceeds to completion, and completion of schema *A* triggers the initiation of schema *B*, or that schema *A* passes a parameter *x* to schema *B*. It is also possible that two or more motor schemas are coactivated simultaneously and interact through competition and cooperation to yield a more complicated motor pattern (Cobas and Arbib, 1992).

Command Properties

With respect to requirements 1 through 5 listed earlier, the properties of the commanding trigger can be many-fold. Modulatory influences (numbers 3 to 5) must not arrive at the MPG directly, but rather by a (feature-selective) CE of the CRS.

Localization

The system must monitor the stimulus in space to select and direct the behavior in relation to the target, for example, in the field of vision. How is visual space, the *x-y-z* position of an object, translated into appropriate motor space, the amplitude of an orienting movement? In cats, the superior colliculus is involved in visual orientation of the eye, head, and trunk. Pre-

vious authors have argued that information about the position of these movable segments must be integrated in the topographic correspondence between input and output. For example, in cats, the sensory map in the collicular superficial layer is transformed into a motor map in the deep layers, in which a vector from an initial eye position to a goal eye position is represented (Sparks, 1988). The segment vector situation is somewhat less complex in amphibians, which display no target-oriented eye movements.

Information on visual depth can be obtained in various ways, e.g., by binocular vision or motion parallax. Such information is not only necessary to select appropriate behavior (e.g., planning a route) but also to estimate the real size of the target.

Feature Analysis

Recognition of the sensory input is necessary for the release of adequate target-oriented ballistic behavior. This calls for an analysis of the releasing key stimulus. For example, studies in common toads *Bufo bufo* show that the visual prey key stimulus for prey-catching is not represented by a specific feature; the specificity of the key lies in an object-features–relating algorithm (Ewert, 1980, 1989). It determines the prey schema through an analysis of the geometry of an object in relation to its direction of movement by linking spatial and spatiotemporal features. More specifically, within behaviorally relevant limits of size, area extension parallel to the direction of movement (spatiotemporal effect) increases the prey-value, while extension perpendicular to the direction of movement (spatial effect) decreases it. A bug or a millipede is thus not recognized explicitly; rather, they are implicit in the prey schema. This prey schema is *species-universal*, meaning that it is common to the members of the same species (*bufo*). But species-universal does not mean *species-specific*. In fact, such different predatory animals as the toad (vertebrate) and praying mantis (invertebrate) take advantage of a comparable object-features–relating algorithm for their prey schema (Prete, 1992), showing that such an algorithm can be implemented by entirely different neural networks.

The relational principle of object classification—approximateness versus exactness—is of general importance. For gaping blackbird nestlings, the feeding parent is characterized by a 1:3 ratio of a head-rump schema. The mating call of the male treefrog is characterized by certain low- and high-frequency tone components which, in order to be optimal for the female, must occur in a particular ratio. In songbirds, it is suggested that an innate "song template" crudely defines the song of the species, leaving considerable freedom for imitation, improvisation, and invention. The processes underlying face recognition in primates probably involve so-called broad tuning, which has the advantage of interference between different memories stored in the neural network (for a review, see Ewert, 1989).

Given the discussion on the function of neurons which display complex response properties, there is no doubt that feature-selective cells exist, such as face-selective neurons in monkeys, biosonar-coding neurons in bats, song-specific neurons in songbirds, prey-selective neurons in toads, and δ-F-sign-stimulus–selective neurons in weakly electric fish. The complexity of these neurons does not simply arise from serial information processing. For example, it can be shown that prey-selective neurons in toads express the subtractive interactions of tectal and pretectal network modules sensitive to spatiotemporal and spatial object features, respectively (Arbib, 1989; Ewert, 1989, 1991). The features-relating algorithm of

Input

visual field

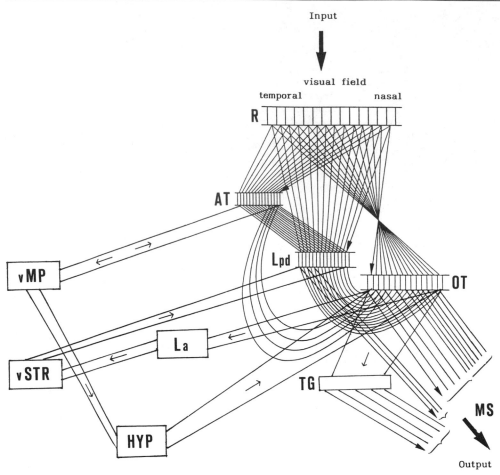

Figure 2. Macro-network of the toad's brain. It is suggested that the species-specific configural prey schema takes advantage of retino (R)–tectal (OT) and retino (R)–pretectal (Lpd) filtering processes; the former evaluates the dimension of an object in the direction of its movement (spatiotemporal features), whereas the latter evaluates the dimension perpendicular to the direction of movement (spatial features). Pretecto-tectal inhibitory influences determine the prey-selective properties of tectomedullary projecting neurons. Loop operations modulate this network with respect to *learning* (involving the vMP), *gating* (involving the vSTR), and *motivational states* (involving the hypothalamus, HYP). Other abbreviations are: TG, tegmentum; MS, medullary/spinal system; AT, anterior thalamus; La, lateroanterior thalamus; Lpd, pretectal lateroposterodorsal thalamus. (For details, see Ewert, 1989, 1991.)

the toad's prey schema is implemented by this circuitry, integrated in a macro-network that allows modulation, e.g., with respect to raising or lowering the classification threshold in satiated versus hungry toads, respectively, or to a generalization or specification of the prey schema involving learning (Figure 2). This processing structure fulfills Bullock's (1961) concept of the multiple-input metastable feedback loop. From a systems-theoretical point of view, such a neuronal filter system compares (e.g., cross-correlates) information related to the stimulus signal with stored information provided by the neural circuitry, including the pattern of neurochemical transmission.

This brings up an interesting point. Prey-selective neurons do not respond in an all-or-none fashion; rather, their response activity is correlated with the probability that the signal fits the perceptual schema they are expressing. However, the behavioral response, e.g., snapping, *is* all-or-none, but the probability of the occurrence of this behavior—the response frequency—is correlated with the degree of resemblance of the stimulus signal with the schema (e.g., see Ewert, 1989).

Although we know that feature-selective neurons exist, controversy surrounds their function. Of course, in a prey-motivated toad faced with a prey object, many types of neurons will display various degrees of activity, particularly prey-selective cells, but also other types of cells relating to the function of the visuomotor task. Since prey-selective cells recorded from the toad's optic tectum in an area representing the frontal visual field evidently project their axons toward the bulbar/spinal motor systems close to the tongue that control hypoglossal motoneurons, one is led to the conclusion that these are

feature-selective CEs of a CRS for the tongue flip, given that tectal electrostimulation activates the tongue motoneurons and snapping (Ewert, 1989; Ewert et al., 1994).

In toads, rats, and cats, it has been shown that tectal (superior collicular) feature-selective neurons belong to a tectoreticulospinal system which, in addition to other influences, represents a premotor substrate for goal-specific head movements (Dean and Redgrave, 1992; Grantyn et al., 1992). For example, a toad's prey-selective neurons display a strong premotor warming-up activity before the animal turns toward the prey, but they are silent during a spontaneous head movement (Ewert et al., 1994). Turning proceeds in the declining phase of the neuron's activity. The firing pattern of a prey-selective neuron toward prey within snapping distance is different: warming-up until snapping, then silence during the snap.

Motivation and Attention

A commanding trigger cannot operate if state-dependent inputs are not appropriate. Neither in food-deprived, occasionally unresponsive toads, nor in toads that have been fed until prey neglect, do the prey-selective neurons display the warming-up typical for a subsequent prey-catching response (Ewert et al., 1994). Comparably, in a study of the association cortex of monkeys, a neuron type discharged strongly shortly before the animal reached forward with his arm to grasp a rewarding object, such as a banana, however, the neuron did not respond when the arm was used for other tasks; if the monkey was satiated, the banana hardly activated the neuron and did not elicit grasping (Mountcastle et al., 1975).

Gating

An MPG is normally suppressed, so that triggering presumes gating. For example, sensorimotor transfer in the mammalian colliculus is controlled by inhibitory neurons of the substantia nigra, which in turn receive inhibitory inputs from neurons of the telencephalic basal ganglia (modeled by Dominey and Arbib, 1992). The execution of a collicular-mediated motor response requires temporal coincidence of striatal (basal-ganglionic) disinhibitory influences with collicular sensorimotor commands. Generally, such disinhibitory effects are prerequisites for the translation of perception into action.

An analogous disinhibitory striato-pretecto-tectal pathway is found in amphibians, (Ewert, 1991). Toads or frogs seldom react toward prey immediately. Initially, they hesitate safely. Clearly, their tectal prey-catching releasing system is under pretectal suppression. Owing to tectal and internal influences, structures of the caudal ventral striatum vSTR (see Figure 2) inhibit structures of the pretectum to shape the filter property of prey-selective neurons and to gate their appropriate activity. The premotor warming-up of prey-selective tectal neurons toward prey (before turning or snapping) could be interpreted as a gating disinhibition. The prey neglect observed in experiments involving telencephalic ablation or a lesion to the vSTR, then, would express a permanent hesitation, whereas the enhanced response to prey after pretectal damage would be a permanent gating connected with a loss of prey-selective properties.

Present and Past

The concept of CRS is applicable both to innate, modified innate, and acquired properties. Various parts of the brain involve the capability of information storage related to associative and nonassociative learning. Among vertebrates, the hippocampus of mammals and the ventromedial pallium vMP (primordium hippocampi; see Figure 2) of amphibians play a prominent role in the evaluation of an ongoing stimulus signal to past experience. The hippocampus is thus suitable for mediating between present and past to shape future behavior. The species-universal, prey-selective properties in toads can be modified by individual experience (Ewert, 1991) involving changes in sensorimotor codes. After a lesion of the vMP, the original species properties again emerge—and persist.

Discussion

Operations between sensory analysis and motor response represented by sensorimotor codes are of general importance. The tuning of *sensorimotor instruments* could be regarded as a primordial response form of an organism's adaptation to environmental constraints, preparing the way for the emergence of *cognitive instruments* dealing with the control of perception and action. Cognition, as the process by which knowledge is gained about the world, involves internal activities in response to antecedent signals that are classified, transformed, and coordinated prior, at least potentially, to initiating actions.

The way in which fundamental algorithms of cognition (related to perception, memory, retention, revival, and retrieval) are implemented in the human brain may become understandable in terms of the evolution of related structures and functions of vertebrate brains. Asking for phyletic roots of cognitive processes at *algorithmic* levels (constructed by processes that enable adaptive behavior) and *implementational* levels (defined as the mechanisms that implement those processes), we speculate that the rules according to which significant coincidences of events determine an animal's behavioral *strategy* are inherent in sensorimotor codes, and that the implementational level refers to CRSs, including decoding premotor circuitry. Functional units of interconnected neurons that allow the animal to select behaviorally relevant objects from irrelevant ones may implement their function according to a "wired-in" algorithm. In this context, it is important to note that the same goal-specific algorithm can be implemented by entirely different network structures (cf. the configural prey schemas of the common toad and praying mantis), which "biologically" justifies the artificial neural network approach. Modulatory circuits make CRSs adaptive in relation to past and present events, thus opening further algorithmic levels. This implies comparable basic neural mechanisms in the animal kingdom and suggests that the capabilities of neuronal networks—and also of neural engineering—may turn out to be a matter of degree of complexity.

Road Map: Motor Pattern Generators and Neuroethology
Related Reading: Motor Pattern Generation; Neuroethology, Computational; Visuomotor Coordination in Frogs and Toads

References

Arbib, M. A., 1989, Visuomotor coordination: Neural models and perceptual robotics, in *Visuomotor Coordination: Amphibians, Comparisons, Models, and Robots* (J.-P. Ewert and M. A. Arbib, Eds.), New York: Plenum, pp. 121–171. ◆

Bullock, T. H., 1961, The problem of recognition in an analyzer made of neurons, in *Sensory Communication* (W. A. Rosenblith, Ed.), Cambridge, MA.: MIT Press, pp. 717–724. ◆

Cobas, A., and Arbib, M. A., 1992, Prey-catching and predator avoidance in frog and toad: Defining the schemas, *J. Theor. Biol.*, 157:271–304. ◆

Dean, P., and Redgrave, P., 1992, Approach and avoidance systems in the rat, in *Visual Structures and Integrated Functions* (M. A. Arbib and J.-P. Ewert, Eds.), Berlin/Heidelberg/New York: Springer-Verlag, pp. 191–204. ◆

Dominey, P. F., and Arbib, M. A., 1992, A cortico-subcortical model for the generation of spatially accurate sequential saccades, *Cereb. Cortex*, 2:153–175. ◆

Eaton, R. C., and DiDomenico, R., 1985, Command and the neural causation of behavior: A theoretical analysis of the necessity and sufficiency paradigm, *Brain Behav. Evol.*, 27:132–164. ◆

Ewert, J.-P., 1980, *Neuroethology*, Berlin/Heidelberg/New York: Springer-Verlag. ◆

Ewert, J.-P., 1989, The release of visual behavior in toads: Stages of parallel/hierarchical information processing, in *Visuomotor Coordination: Amphibians, Comparisons, Models, and Robots* (J.-P. Ewert and M. A. Arbib, Eds.), New York: Plenum, pp. 39–120. ◆

Ewert, J.-P., 1991, A prospectus for the fruitful interaction between neuroethology and neural engineering, in *Visual Structures and Integrated Functions* (M. A. Arbib and J.-P. Ewert, Eds.), Berlin/Heidelberg/New York: Springer-Verlag, pp. 31–56. ◆

Ewert, J.-P., Beneke, T. W., Schürg-Pfeiffer, E., Schwippert, W. W., and Weerasuriya, A., 1994, Sensorimotor processes that underlie feeding behavior in tetrapods, in *Advances in Comparative and Environmental Physiology*, vol. 18, *Biomechanics of Feeding in Vertebrates* (V. L. Bels, M. Chardon, and P. Vandevalle, Eds.), Berlin/Heidelberg/New York: Springer-Verlag, pp. 119–161. ◆

Gillette, R., 1987, The role of neural command in fixed action patterns of behaviour, in *Aims and Methods in Neuroethology* (D. M. Guthrie, Ed.), Manchester: Manchester University Press, pp. 46–79. ◆

Grantyn, A., Berthoz, A., Hardy, O., and Gourdon, A., 1992, Contribution of reticulospinal neurons to dynamic control of head movements: Presumed neck bursters, in *The Head-Neck Sensory Motor System* (A. Berthoz, W. Graf, and P. P. Vidal, Eds.), New York: Oxford University Press, pp. 318–329. ◆

Kupfermann, I., and Weiss, K. R., 1978, The command neuron concept, *Behav. Brain Sci.*, 1:3–39. ◆

Liaw, J. S., and Arbib, M. A., 1993, Neural mechanisms underlying direction-selective avoidance behavior, *Adapt. Behav.*, 1:227–261.

Mountcastle, V. B., Lynch, J. C., Georgopoulos, A., Sakata, H., and Acuna, C., 1975, Posterior parietal association cortex of the monkey: Command functions for operations within extrapersonal space, *J. Neurophysiol.*, 38:871–908. ◆

Prete, F. R., 1992, Discrimination of visual stimuli representing prey versus nonprey by the praying mantis *Sphodromantis lineola* (Burr.), *Brain Behav. Evol.*, 39:285–288. ◆

Ritzmann, R. E., 1993, The neuronal organization of cockroach escape and its role in context-dependent orientation, in *Biological Neural Networks in Invertebrate Neuroethology and Robotics* (R. D. Beer, R. E. Ritzmann, and T. McKenna, Eds.), New York: Academic Press, pp. 113–137. ◆

Sparks, D. L., 1988, Neural cartography: Sensory and motor maps in the superior colliculus, in *Neural Cartography: How Does the CNS Use Sensory Maps?* (T. E. Finger, Ed.), Basel: Karger, pp. 49–55. ◆

Competitive Learning

Nathan Intrator

Introduction

Competitive learning is a family of algorithms that uses some sort of competition between lateral neurons during learning. In a typical competitive network architecture, neurons in each layer are connected to the layer above, but in addition, there are lateral connections between neurons in the same layer which cause the competition. Competitive learning includes a wide variety of algorithms performing different tasks such as encoding, clustering, and classification (Hertz, Krogh, and Palmer, 1991, for review). After briefly reviewing some early competitive learning algorithms, we shall concentrate on several issues concerning competitive learning and mention some less discussed, and some recent algorithms in this context.

What Is Competitive Learning?

In recent years, competitive learning has received considerable attention due to its demonstrated applicability and its biological plausibility. The idea of competition between neurons leads to sparse representations of data which are easy to decode and which conserve energy (see SPARSELY CODED NEURAL NETWORKS).

Competitive learning algorithms employ some sort of competition between neurons in the same layer via lateral connections. In early models, the competition, which was called hard competition, led to final activity of a single neuron, the strongest one to start with (von der Malsburg, 1973). In more recent models, the competition affects the activity of the lateral neurons but does not necessarily drive all but one to zero (Rumelhart and Zipser, 1986). Such competition, which is called *soft competition*, has nice mathematical properties, as we shall see below. In a general competitive learning architecture, there are several layers of neurons. Neurons in each layer are connected via lateral inhibition to adjacent neurons and via excitatory connections to higher layers. Current competitive learning algorithms can be distinguished by their learning rule (which is driven by their desired objective function) or by the form and role of the competition during learning. One form of competitive learning algorithm can be described as an application of a successful single neuron learning algorithm in a network with lateral connections between adjacent neurons. The lateral connections are needed so that each neuron will extract a different feature from the data. For example, if one has an algorithm for finding the principal component in a data set— i.e., the first eigenvector of the correlation matrix of the data— then a careful application of this learning rule in a lateral inhi-

bition (competitive) network gives an algorithm that finds the first few principal components of the data (Oja, 1989; Sanger, 1989; see also PRINCIPAL COMPONENT ANALYSIS). This set of algorithms can be characterized by the fact that if the lateral connections are turned off, then neurons will learn the most easily extracted feature in the data (based on its objective function). For example, in the principal component network, all neurons will learn the first principal component of the data. In such algorithms, unless the lateral inhibition has a complex form (such as that of a Mexican hat), it is sufficient to study the learning rule of a single neuron and deduce from it the performance of the network as a whole.

A second family of algorithms is characterized by the fact that turning off the lateral connections will result in great loss of ability to extract useful features. Thus, the competition between neurons has an important role in improving, or sharpening, the features extracted from the data by each single neuron. This is an important distinction which suggests that competitive learning may be useful in cases where other forms of learning will have difficulties. A simple example of such a network is one that searches for clusters. Clustering is the process of grouping together data points based on some measure of distance. When no inhibition exists between neurons, it is likely that neurons will converge to the mean of the data distribution. This follows from the fact that based on a simple Hebbian rule, neuronal weights (of radial basis neurons) tend to converge to the center of the distribution in order to maximize correlation of input activity with output activity. A smart algorithm that looks for tight clusters will probably find those that are occurring with highest probability, or those cluster centers which are closest to the initial weight values. When inhibition is turned on, the cluster centers will move to new locations and will become sharper and more distinct.

There is evidence that inhibition plays a similar role in various brain functions, such as the creation of orientation columns in the visual cortex (Ramoa, Paradiso, and Freeman, 1988) or sharpening receptive fields in the sensory cortex (von der Malsburg, 1973; Merzenich et al., 1988).

The Usefulness of Competitive Learning

Frequently, in a classification problem, there are regions in pattern space in which classification is easier, while there are other regions in which classification is not that simple and thus requires a larger network architecture. A simple example would be regions in which the different classes are linearly separable and others in which the boundary between classes is

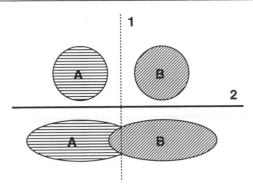

Figure 1. The ability of an unsupervised splitting rule to reduce confusion.

highly nonlinear. Thus, one of the key ideas in machine learning and statistical parameter estimation is recursive partitioning of the observation space so as to separate between such regions. Motivation comes from the desire to study structure of high dimensional space by search for homogeneous subregions which present simpler structures than the original whole input space. Statistical theory (Barron, 1991) suggests that the variance of the estimator (which directly influences generalization performance) is affected by the complexity of the estimator (polynomial degree, number of hidden units, etc.). Thus, performing the estimation with a simpler architecture is likely to improve generalization performance. However, if the network architecture is too simple, in the sense that the architecture is not rich enough to allow a function flexible enough to fit the data, large training errors are unavoidable. One solution to this trade-off is to recursively partition the data into several homogeneous regions so that a smaller network (simpler architecture) will suffice for each region.

Motivation for Recursive Partitioning

In this section, we will discuss simple examples to illustrate the need for recursive partitioning.

Consider the example presented in Figure 1. It shows a subregion in space in which two classes are strongly mixed. A supervised classification algorithm will split according to hyperplane 1. Alternatively, a split along hyperplane 2, with a subsequent split of the top part along hyperplane 1, will increase the purity of the classifier. The bottom region can then be transferred to another classifier (possibly more complex), or it may be transferred to a feature extractor which, for improved performance, will look for more features in that region.

It turns out that in many real-world problems there are distinct regions in which the classification is much simpler than in others. One reason is the fact that different features are needed for the classification in different regions, and another is that different regions possess various levels of complexity. This is probably one of the reasons for the apparent success of recursive partitioning methods in classification and regression. In speech, for example, some phonemes have a different structure in male and in female voices, and thus may require different features for the recognition. Hampshire and Waibel (1990) achieved improved performance for stop-consonant recognition by separately training networks on voiced and unvoiced phonemes, and combining the hidden structure found by either net into a single classifier. In face recognition, it has been shown that even with the absence of trivial cues like hair style, facial hair, etc., female faces are distinct from male faces

(Golomb, Lawrence, and Sejnowski, 1991). Thus, it is conceivable that classification of each face by a separate network for each gender may require a smaller classifier than classification of both genders together by a single classifier.

Hard and Soft Competition and Gaussian Mixtures

In this section, the connection between hard and soft competition and the probabilistic model of Gaussian mixtures is briefly discussed.

A cluster center (the mean of a Gaussian) is associated with a neuronal weight vector, and neurons compete among themselves to add each of the input patterns to their cluster. For a given pattern, the winner (in hard competition) is the neuron with highest probability for that pattern to belong to the Gaussian distribution represented by its center. In soft competition, the cluster centers are organized in such a way that the distance of each data point from each of the cluster centers corresponds to the probability of this point under each of the Gaussian distributions (Nowlan, 1991)

If an input $x \in R^n$ is drawn with probability π_i from one of J independent Gaussians (for simplicity, assume that the Gaussians are radially symmetric), its a posteriori probability to belong to the jth Gaussian is $\pi_j p_j(x)$, for

$$p_j(x) = \frac{1}{M\sigma_j} \exp\left(-\frac{\|x - \mu_j\|^2}{2\sigma_j^2}\right)$$

where $\sigma_j^2 I$ is the covariance matrix of Gaussian j, and M is a normalizing constant. The likelihood which measures the probability of input x_k under this model is $P(x_k) = \Sigma_j \pi_j p_j(x_k)$. The likelihood of a pattern set $\{x_k\}_{k=1}^K$ is

$$L = \prod_k P(x_k)$$

assuming K independent patterns. Parameter estimation of such a model involves adjusting the centers $\{\mu_j\}$, the covariances $\{\sigma_j^2\}$, and the prior probabilities $\{\pi_j\}$, so as to maximize the likelihood of the model for a given training set. It is mathematically equivalent but more convenient to maximize the log likelihood, since then the maximization is on summation and not multiplication. In a Radial Basis Function (RBF) network realization of this model, one has J radial basis hidden units and a linear output unit which gives the likelihood function under the model (see RADIAL BASIS FUNCTION NETWORKS).

Assuming for simplicity that the σ_j are equal, the hard competition approach for solving this problem would assume that each input can belong to only one cluster center (thus the competition between neurons will lead to only one active neuron at a time); therefore the above definition of $P(x)$ is replaced by

$$P(x) = \max_i \pi_i p_i(x)$$

The analytic solution for cluster centers (under the assumptions of radial symmetry of the σ_i and π_i being independent of i) is

$$\mu_i = \frac{\sum_{k \in C_i} x_k}{N_i}$$

where C_i is the set of training patterns that were assigned to the cluster center i, and N_i is the number of these patterns. Note that this is not an explicit solution since the C_i depend on all of the μ_k's. A simple learning rule that converges to this solution is given by:

$$\Delta\mu_{ij} = \eta(x_{ki} - \mu_{ij})p_i(x)$$

for the winning cluster i and no change for the other clusters.

This learning rule moves the closest cluster center in the direction of each data point which belongs to that cluster, thus converging to the mean of the cluster. During this process, input patterns may "cross" from one cluster to another, due to the movement of the cluster center; however, the process will converge to a stable solution simply because it is easy to see that the mean squared distance of the patterns from their corresponding cluster centers goes down.

Under the soft competition paradigm, which assumes that each pattern can belong to any of the clusters, the solution which maximizes the likelihood of the model is given by

$$\mu_i = \frac{\sum_k p(i|x_k)x_k}{\sum_k p(i|x_k)}$$

where

$$p(i|x_k) = \frac{p_i(x_k)}{\sum_j p_j(x_k)}$$

A learning rule for soft competition is similar to that for hard competition, the difference being that each of the cluster centers is modified in the direction of the line connecting the cluster center and pattern x_k in proportion to the probability for it to belong to that cluster. This follows from the formulation of the data as a Gaussian mixture model. The soft competition approach is closely related to k means clustering (Devijver and Kittler, 1982).

Competing Experts

The competing experts idea (Jacobs et al., 1991; Nowlan, 1991; see also MODULAR AND HIERARCHICAL LEARNING SYSTEMS) is a direct extension of the above competitive learning idea. In this architecture, each of the hidden units of the previous RBF model is replaced by an expert net, while the linear summation output unit is replaced by a controller network that receives inputs not only from the hidden layer but also directly from the input layer. The competition between neurons is replaced by a competition between experts, all competing to perform well on the data in various parts of the input space. The underlying assumption is that the data were generated by several generators, each with its own set of parameters. Therefore, the best performance is achieved if an expert is trained, and performs best, on a single generator rather than on the whole collection of data. The decision on how to split the space is based on the performance of these experts. The hard competition in this case assumes that each pattern could have been generated by a single generator, and the soft competition assumes that the data could have been generated by any one of the generators with some appropriate probability. The architecture contains a controller network which gives the relative weight to the decision (or regression) suggested by each expert. The experts need not be of the same architecture or complexity, and the controller net is usually of a different architecture from the experts; for example, Nowlan (1991) has used an RBF controller and sigmoidal feedforward net as experts.

During training of the hybrid architecture, the controller network is trained to direct input patterns to those experts that perform best on them. Consequently, each expert is trained on subpopulations of the training data on which it performs best. The soft competition version directs inputs to all experts, but measures the output of each expert based on its relative previous performance. Similarly, the error used for training the experts is affected by the amount of "belief" in each expert as given by the controller. This approach is different from the Bayesian method (see BAYESIAN METHODS FOR SUPERVISED NEURAL NETWORKS) for choosing a model, since different experts are only trained on parts of the input space on which they perform best as suggested by the controller.

Hierarchical Mixture of Experts

There are various ways one can construct controller networks, and they resemble various statistical approaches. One can either construct a wide, shallow network in which input patterns are split only once into one of several experts performing the task, or one can construct a deep tree with many splits so that the regression or classification task is performed only by a bottom (terminating) node of the tree. This approach has gained considerable attention with the appearance of the classification and regression trees (CART) methodology (Breiman et al., 1984). This methodology constructs a tree leading to different decisions in different regions of pattern space. Experimental evidence from the olfactory cortex suggests the performance of hierarchical clustering by a biological network (Lynch, 1986) and has led to the proposal of a model for hierarchical clustering (Ambros-Ingerson, Granger, and Lynch, 1990). It uncovered the potential of recursive partitioning in complex pattern encoding and classification.

A direct extension of competitive learning ideas motivated by the success of the above recursive partitioning methods is the hierarchies of experts approach (Jordan and Jacobs, 1993; MODULAR AND HIERARCHICAL LEARNING SYSTEMS). In this tree-structured network, each expert can again be a competitive mixture of experts. Learning in this model is more involved and therefore the different mixture components are simplified to be generalized linear models. The use of the expectation-maximization (EM) algorithm accelerates the learning.

Road Map: Self-Organization in Neural Networks
Background: I.3. Dynamics and Adaptation in Neural Networks
Related Reading: Coulomb Potential Learning; Learning Vector Quantization; Ocular Dominance and Orientation Columns

References

Ambros-Ingerson, J., Granger, R., and Lynch, G., 1990, Simulation of paleocortex performs hierarchical clustering, *Science*, 247:1344–1348.
Barron, A. R., 1991, Complexity regularization with application to artificial neural networks, in *Nonparametric Functional Estimation and Related Topics* (G. Roussas, Ed.), Dordrecht: Kluwer, pp. 561–576. ◆
Breiman, L., Friedman, J. H., Olshen, R. A., and Stone, C. J., 1984, *Classification and Regression Trees*, Belmont, CA: Wadsworth. ◆
Devijver, P. A., and Kittler, J., 1982, *Pattern Recognition: A Statistical Approach*, London: Prentice Hall. ◆
Golomb, B. A., Lawrence, D. T., and Sejnowski, T. J., 1991, SEXNET: A neural network identifies sex from human faces, in *Advances in Neural Information Processing Systems 3* (R. P. Lippmann, J. E. Moody, and D. S. Touretzky, Eds.), San Mateo, CA: Morgan Kaufmann, pp. 572–577.
Hampshire, J. B., and Waibel, A., 1990, Connectionist architectures for multi-speaker phoneme recognition, in *Advances in Neural Information Processing Systems 2* (D. S. Touretzky, and R. P. Lippmann, Eds.), San Mateo, CA: Morgan Kaufmann, pp. 203–210.
Hertz, J., Krogh, A., and Palmer, R. G., 1991, *Introduction to the Theory of Neural Computation*, New York: Addison-Wesley. ◆
Jacobs, R. A., Jordan, M. I., Nowlan, S. J., and Hinton, G. E., 1991, Adaptive mixtures of local experts, *Neural Computat.*, 3(1):79–87.
Jordan, M. I., and Jacobs, R. A., 1993, Hierarchical mixture of experts and the EM algorithm, submitted to *Neural Computat.*
Lynch, G., 1986, *Synapses, Circuits and the Beginnings of Memory*, Cambridge, MA: MIT Press. ◆

Merzenich, M. M., Recanzone, G., Jenkins, W. M., Allard, T. T., and Nudo, R. J., 1988, Cortical representation plasticity, in *Neurobiology of Neocortex* (P. Rakic and W. Singer, Eds.), New York: Wiley, pp. 41–68.

Nowlan, S. J., 1991, Soft competitive adaptation: Neural network learning algorithms based on fitting statistical mixtures, PhD Dissertation, Carnegie Mellon University. ◆

Oja, E., 1989, Neural networks, principal components, and subspaces, *Int. J. Neural Syst.*, 1:61–68.

Ramoa, A. S., Paradiso, M. A., and Freeman, R. D., 1988, Blockade of intracortical inhibition in kitten striate cortex: Effects on receptive field properties and associated loss of ocular dominance plasticity, *Exp. Brain Res.*, 73:285–296.

Rumelhart, D., and Zipser, D., 1986, Feature discovery by competitive learning, in *Parallel Distributed Processing* (D. E. Rumelhart, J. L. McClelland, and PDP Research Group, Eds.), vols. 1 and 2, Cambridge, MA:. MIT Press.

Sanger, T., 1989, Optimal unsupervised learning in a single-layer linear feedforward neural network, *Neural Netw.*, 2:459–473.

von der Malsburg, C., 1973, Self-organization of orientation sensitivity cells in the striate cortex, *Kybernetik*, 14:85–100. ◆

Compositionality in Neural Systems

Elie Bienenstock and Stuart Geman

Compositionality in Cognition

Compositionality refers to our ability to construct mental representations, hierarchically, in terms of parts and their relations. The so-called rules of composition are that (1) we have at our disposal an infinite repertoire of hierarchically constructed entities built from relatively small numbers of lower-level constituents, a phenomenon sometimes referred to as (infinite) *productivity* (also known as *creativity* or generativity), and (2) allowable constructions nevertheless respect specific constraints, whereby most combinations are made meaningless.

Example 1. In *language*, no more than 26 characters and a half-dozen additional symbols—alternatively, no more than a few dozen phonemes—are required to compose a story on any subject one can possibly imagine. Moreover, the collection of all texts that have ever been spoken or written in the English language is an infinitesimal fraction of what could possibly be composed. At the same time, most arrangements of symbols that are possible a priori from a mere combinatorial point of view are illegitimate as linguistic constructions. The number of character strings of length 1000 that make up a proper English text is vanishingly small when compared to the number of all possible strings of such length. Thus, while infinitely productive, language is at the same time severely constrained.

On the surface, the composition mechanism in language appears simple. Individual characters are assembled into syllables, which are then assembled into words, then further composed into phrases, sentences, etc. One text differs from another text in the same language only by the *relative positioning* (relations) among the constituents (symbols), and not, for instance, by the frequencies of occurrence of each symbol; these frequencies are about the same for any sufficiently long text. Yet, encoded within this apparently simple surface structure is a considerable amount of highly complex organization: syntax, semantics, pragmatics, etc. Linguistic constraints are partly arbitrary and conventional (a fact sometimes viewed as a violation to the principle of compositionality), and partly obey some language-specific regularities. Moreover, Chomsky (e.g., 1986) maintains that some so-called universal, *language-independent* regularities may result from specific properties of our brains that are largely genetically determined (but see LANGUAGE ACQUISITION).

Example 2. In the perception and production of *visual imagery*, primitive elements, which are analogous to letters or phonemes, are combined in a highly specific relational manner to form composite entities suitable themselves for relational bindings into even more specific higher-level structures. Edge elements combine, in accordance with rules governing gradient magnitudes and directions, to form curve elements, which in turn can combine, end-to-end, to build the cartoon-like boundary description of a scene. Surface elements piece smoothly together, in a manner consistent with boundary-determined discontinuities, to form three-dimensional shapes, which are themselves combined into the objects of everyday imagery. There is, furthermore, an infinite, but nevertheless topologically and logically restricted, repertoire of object placements to produce a meaningful scene. Grenander (1993) has built a mathematical theory of the composition of patterns, characterized by these basic principles of productivity and restrictedness. Biederman (1987) has exploited a compositional description of objects using a small repertoire of volumetric shapes, called geons, in his psychological theory of OBJECT RECOGNITION (q.v.).

Example 3. *Procedures* can be recursively decomposed from broadly defined goals, which are achieved over minutes or hours, into simple motor actions with durations of fractions of a second. These basic units, analogous to phonemes or simple shapes, can be combined effectively to generate an infinite variety of goal-achieving activities.

Organization by composition is, in fact, so ubiquitous as to suggest that it is fundamental to cognition. This being the case, there are several implications worth highlighting:

Disambiguation. The problems of interpreting an image or understanding a spoken language are sometimes said to be ill-posed. Yet they really are very well-posed, as attested to by the spectacular recognition performances achieved by humans under ordinary circumstances. It is true, however, that auditory and visual data are often ambiguous at all but the most global levels of interpretation. Isolated phonemes, or even words and phrases spliced from a continuous speech signal, can be impossible to interpret. In fact, the mere segmentation of a speech signal into phonemes or words is difficult, if not impossible, in the absence of a simultaneous global interpretation. This apparent ambiguity may persist at any given level, or at several levels at once (acoustic, phonetic, lexical, syntactic). Analogous considerations apply to scene analysis.

Evidently, despite the richness of possible scenes or utterances, there are severe constraints restricting plausible interpretations. Recalling that compositional constructs are themselves highly restricted, we may interpret our remarkable auditory and visual perceptual skills as exploiting, in a fundamental way, the restrictedness of compositionality.

Invariance.   Relational descriptions are invariant. In *computer vision*, this is the basis for many object-recognition algorithms (see Dickinson, Pentland, and Rosenfeld, 1992): to define the objects of interest relationally, and often hierarchically, in terms of the relative positionings of identifiable subparts. Identification becomes, essentially, a matter of relational graph matching, and is, a fortiori, invariant. One may, for example, identify a chair as a planar rectangular surface with four attached, more-or-less-identical, cylindrical parts that are situated near corners and roughly perpendicular to the plane, and a further plane, attached perpendicular to the first plane, on the side opposite the cylinders. The parts, furthermore, may themselves be defined as relational compositions of still more primitive elements.

Analogous, in language, is the invariance of meaning over a multitude of almost equivalent expressions: many different word strings can adequately evoke the mental objects and the (partly metaphorical) relations between them that constitute a given intended message. Speech demonstrates other invariances as well: the simple linear relation of constituents confers an invariance with respect to the *rate* of articulation as well as other phonetic parameters, which can be altered in many ways without affecting meaning. By and large, it is the *order* (in English and most other natural languages) and not the duration of constituents that is the primary vehicle of meaning.

In general then, to the extent that cognitive entities are relationally organized, they can be identified and/or described in multiple equivalent ways. Invariance is thereby related to the relational and productive properties of compositionality.

Computation.   Artificial interpretation of speech and image data are daunting engineering tasks; difficulties generally manifest themselves as overwhelming computational requirements. A few successes, however, have been realized—mostly in speech recognition—and these have relied on divide-and-conquer strategies that exploit the hierarchical organization of data. Compositional models may be based upon primitive grammars that restrict word sequences, phonetic models for allowable word pronunciations, and acoustic models for the articulation of the phonetic units. This hierarchy is the basis for computationally feasible algorithms that infer a word sequence from a raw acoustic signal (Rabiner, 1989; SPEECH RECOGNITION: PATTERN MATCHING).

Such successes, although sparse, strongly suggest that our brains, too, avoid explosive combinatorial search by exploiting in a recursive manner the compositional organization of mental representations.

Compositionality in Neural Systems

Since compositionality is so central to cognition, it appears reasonable to construe it as an observable manifestation of a property of compositionality in *neural activity*. Drawing from the previous section, we can identify several features that neural mechanisms for compositionality would likely possess.

Compositional representation through dynamic binding.   The neural representation of a composite entity should include the suitably defined composition of those patterns of neural activity that make up the representations of the constituents of this composite entity. A popular simple example is the problem of representing a scene containing a red triangle and a blue square. The mere *coactivation* of four cells (or groups of cells) representing the four elementary features "red," "blue," "triangle," and "square" would lead to a "superposition catastrophe" (von der Malsburg, 1987); that is, in this case, the inability to distinguish a scene containing a red triangle and a blue square from a scene containing a red square and a blue triangle. Thus, composition is more than coactivation; a *binding* mechanism is required to attach with each other the neural representations of the entities "red" and " triangle." Binding needs to be *dynamical*, i.e., reversible, to allow the representation of other constructs at different times.

Relational binding.   Binding further needs to be *relational*, that is, qualified in terms of a collection of domain-specific relations among constituents. For example, to account for our *linguistic* ability to assemble six lexical items, such as "feed, carve, Elsa, Sophie, pumpkin, cat" into a string, such as "Sophie feeds the cat and Elsa carves the pumpkin," a compositional model should use bonds that are qualified in terms of *predicate roles*. Thus, it will be specified that the bond between the neural representation of the item "Sophie" and the neural representation of the item "feed" is of the *subject* type, i.e., that it is Sophie who does the feeding. Only then will the representation of one particular string be distinguishable from the representation of alternative strings, constructed from the same constituents. Note that these alternative constructs can be (1) syntactically and semantically legitimate, such as "Elsa feeds the cat and Sophie carves the pumpkin;" (2) syntactically correct but semantically/contextually unacceptable, such as "Sophie carves the cat and Elsa feeds the pumpkin;" or (3) syntactically wrong, such as "Feeds Elsa Sophie carves and the cat pumpkin."

Hierarchical computation.   A basic tenet of compositionality is that cognitive representations are hierarchically organized. Likewise, it is natural to expect that the computational mechanisms that elicit the sequence of neural events corresponding to a perceptual or motor act—e.g., in visual pattern recognition or in the interpretation or production of spoken language—are hierarchically organized.

It is hardly disputable that, in the sense of the features just outlined, no satisfactory encompassing treatment of neural compositionality is available to date. In particular, models of the cell-assembly type, inasmuch as they address the issue of compositionality, represent each new composite entity by allocating for it *separate* neural machinery, rather than by composing the representations of its constituents. This has led some authors (e.g., Fodor and Pylyshyn, 1988) to the strong and highly controversial conclusion that modern "connectionism" is wholly inadequate to model cognition at the representational level, the level discussed here.

Nevertheless, models do exist that provide some elements of a theory of neural compositionality (e.g., von der Malsburg, 1987; Shastri and Ajjanagadde, 1993; Smolensky, 1990; Gindi, Mjolsness, and Anandan, 1991; Hummel and Biederman, 1992; see also DYNAMIC LINK ARCHITECTURE, SYNCHRONIZATION OF NEURONAL RESPONSES AS A PUTATIVE BINDING MECHANISM, and STRUCTURED CONNECTIONIST MODELS). An important common feature of most of these models is the use of mechanisms through which a number of neural activity patterns are combined into a composite pattern that preserves, as subpatterns, the original constituent activities. Binding is dynamic: the constituent patterns can either appear by themselves, representing isolated entities, or they can be bound explicitly to represent a composite entity.

Thus, a compositional model will, in general, *not* allocate a specific cell—or group of cells—for a composite entity, such as "red triangle," as a typical feedforward-net model would. Rather, it will employ the machinery already available, that is, the elementary-feature cells or cell groups, and posit the existence of an *additional degree of freedom* in neural activity. This new degree of freedom will be used to dynamically express the bond between "red" and "triangle," thereby avoiding the superposition catastrophe. Similarly, in the linguistic example just cited, the primary activity patterns associated with the six lexical items will be preserved, and an additional degree of freedom will be used to express syntactical dynamical bonds. In short, composite patterns will be constructed by suitably arranging constituent activities, thereby providing an explicit representation of parts and their relations. Productivity will then arise, fundamentally, from combinatorics in a space of neural activity patterns; refer to Damasio (1989) for a similar picture based on neuroanatomical data and lesion studies.

Most compositional models to date, following a suggestion of von der Malsburg (1987), use *fine temporal structure* of neural activity as the medium for expressing dynamic binding. One currently popular implementation of this idea—not a part of von der Malsburg's original theory—posits that the neurons whose activities are to be bound fire, for some time, periodically or nearly periodically. Each neuron is then viewed as carrying two *independent* variables, a level of firing and a phase. The latter is the additional degree of freedom used to express dynamic binding. At this point, there exists no conclusive neurophysiological evidence for this mechanism. Some support, however, comes from recent findings in the visual cortex (see SYNCHRONIZATION OF NEURONAL RESPONSES AS A PUTATIVE BINDING MECHANISM).

Shastri and Ajjanagadde (1993; see also STRUCTURED CONNECTIONIST MODELS) propose a linguistic model along these lines, in which the representation of a predicate, such as "carve," would include the oscillatory activity of two distinct neurons (or neuronal populations) for the two roles: subject ("person carving") and object ("thing carved"). When representing "Elsa carves the pumpkin," the person-carving neuron is phase-locked with the oscillating neuron whose activity represents Elsa. The other entity occurring in the representation—namely, the instantiation of the "thing carved" as "pumpkin"—uses a different phase. Bindings are propagated between predicates along hard wired *phase-preserving* lines, which embody long-term rules. For instance, the rule that maintains "a person (or animal) being fed eats" uses a phase preserving line from the object neuron of the predicate "feed" to the subject neuron of the predicate "eat;" it allows the system to *infer*, from the short-term fact that Sophie feeds the cat, another short-term fact—namely, that the cat eats. Long-term facts may also be encoded in the system, e.g., "Sophie loves animals," or "children carve pumpkins for Halloween." Shastri and Ajjanagadde show that such a system can perform simple, "reflexive," reasoning. Albeit limited in several ways, this reasoning may access a virtually unlimited store of long-term rules and facts.

In the same spirit of locking the phases of oscillators to express binding, although with the additional assumption that specialized fast links are used for signal synchronization, Hummel and Biederman (1992) propose a model of object recognition from line drawings based upon the compositional approach to object representation of Biederman (1987) mentioned in the first section of this article (see OBJECT RECOGNITION for further details).

Although these attempts exhibit most of the features outlined in the beginning of the section, they do not add up to a fully coherent theory of neural compositionality. Most notably, they use rigid architectures, some of which are hierarchically structured, as in Hummel and Biederman (1992). They suggest no convincing hypothesis for the mechanisms underlying the extreme versatility manifested by our brains in linguistic behavior, e.g., in the handling of recursive constructions, or of metaphors, or, more generally, of analogical discourse or reasoning. Furthermore, they fail to address the important issue of how compositional representations are learned and modified, e.g., during language acquisition. This stands in sharp contrast with the wealth of ideas about learning advanced for feedforward connectionist networks.

Some efforts have been made to tackle these issues as well. In particular, von der Malsburg (1987) has suggested adopting a developmental/epigenetic approach, stressing the role of processes of self-organization and natural selection in neural compositionality. In this approach, one investigates mechanisms of brain development that could result in the formation of specific spatiotemporal activity patterns providing a suitable medium for highly versatile compositional operations. Bienenstock (1994) has proposed that "synfire chains" (Abeles, 1991, and references therein) may be relevant here, more so than oscillating circuits. *Synfire chains* are, roughly, large networks that are wired in such a fashion as to support wavelike patterns of activity specified with a millisecond accuracy. Electrophysiological data collected in frontal cortical areas of behaving monkeys are suggestive of the existence of (reverberating) synfire chains (Abeles et al., 1993).

The hypothesis is that these structures could be dynamically bound via weak synaptic couplings; the wavelike activities of two synfire chains could be synchronized in much the same way as coupled oscillators lock their phases. More complex spatiotemporal patterns could arise from reverberation of activity, involving "folding" of the chains upon themselves. Such complex patterns could exhibit highly specific binding properties (think of the highly specific interactions between folded proteins), providing a suitable medium for both productive and restricted composition. Recursiveness of compositionality could, in principle, arise from the further binding of these composite structures. Here again, however, we are a long way from a completely coherent—much less a comprehensive—theory.

In sum, neural compositionality remains among the most challenging issues in brain theory. Particularly vexing are the computational aspects—for instance, those related to the problem of graph matching, e.g., for object recognition—in compositional neural models. Although one may expect significant progress in theoretical investigations, such progress is bound to remain largely speculative until it becomes possible to map cortical activity with high spatial *and* temporal resolution, and to process in a useful way the overwhelming amounts of data that will result.

Acknowledgments. This work was supported by Army Research Office contract DAAL03–92–G–0115 to the Center for Intelligent Control Systems, National Science Foundation grant DMS–8813699, and Office of Naval Research contract N00014–91–J–1021.

Road Map: Artificial Intelligence and Neural Networks

References

Abeles, M., 1991, *Corticonics: Neuronal Circuits of the Cerebral Cortex*, Cambridge, Eng.: Cambridge University Press.
Abeles, M., Bergman, H., Margalit, E., and Vaadia, E., 1993, Spatiotemporal firing patterns in the frontal cortex of behaving monkeys, *J. Neurophysiol.*, 70:1629–1638.
Biederman, I., 1987, Recognition-by-components: A theory of human image understanding, *Psychol. Rev.*, 94:115–147.

Bienenstock, E., 1994, A model of neocortex, *Network: Computation in Neural Systems* (in press).

Chomsky, N., 1986, *Knowledge of Language: Its Nature, Origin, and Use*, New York: Praeger.

Damasio, A. R., 1989, Time-locked multiregional retroactivation: A systems-level proposal for the neural substrates of recall and recognition, *Cognition*, 33:25–62.

Dickinson, S. J., Pentland, A. P., and Rosenfeld, A., 1992, From volumes to views: An approach to 3-D object recognition, *Comput. Vis. Graph. Image Proc.: Image Understanding*, 55:130–154.

Fodor, J. A., and Pylyshyn, Z. W., 1988, Connectionism and cognitive architecture: A critical analysis, *Cognition*, 28:3–71.

Gindi, G., Mjolsness, E., and Anandan, P., 1991, Neural networks for model based recognition, in *Neural Networks: Concepts, Applications and Implementations* (P. Antognetto and V. Milutinovic, Eds.), Englewood Cliffs, NJ: Prentice-Hall, pp. 144–173.

Grenander, U., 1993, *General Pattern Theory: A Study of Regular Structures*, New York: Oxford University Press.

Hummel, J. E., and Biederman, I., 1992, Dynamic binding in a neural network for shape recognition, *Psychol. Rev.*, 99:480–517.

Rabiner, L. R., 1989, A tutorial on hidden Markov models and selected applications in speech recognition, *Proc. IEEE*, 77:257–286.

Shastri, L., and Ajjanagadde, V., 1993, From simple associations to systematic reasoning: A connectionist representation of rules, variables and dynamic bindings, *Behav. Brain Sci.*, 16:417–494.

Smolensky, P., 1990, Tensor product variable binding and the representation of symbolic structures in connectionist networks, *Artif. Intell.*, 46:159–216.

von der Malsburg, C., 1987, Synaptic plasticity as a basis of brain organization, in *The Neural and Molecular Bases of Learning* (J. P. Changeux and M. Konishi, Eds.), New York: Wiley, pp. 411–432. ◆

Computer Modeling Methods for Neurons

Michael Hines and Nicholas T. Carnevale

Introduction

This article attempts to provide an intuitive rationale for the most-used numerical method for solving the cable equation with voltage-dependent ion conductances: the Crank-Nicholson method (Crank and Nicholson, 1947). Problems involving chemical diffusion in one dimension are also susceptible to the same computational strategy (Oran and Boris, 1987).

Of the many previous articles that discuss numerical methods for solving the cable equation, Mascagni (1989) is notable for a reasonably complete explanation and bibliography. Douglas (1961) gives a fairly rigorous account of numerical methods for solving parabolic partial differential equations. Hines (1984) discusses special techniques for fast simulations of neurons. A textbook with good general coverage of numerical methods for differential equations has been written by Dahlquist and Björck (1974).

The Basic Approach

The easiest way to deal with quantities that vary continuously with position and time is to approximate the differential equations that describe them with difference equations. This involves replacing continuous derivatives by finite differences in a manner analogous to the definition of the derivative

$$\left.\frac{df}{dx}\right|_{x=x_0} \approx \frac{f(x_0 + h) - f(x_0)}{h} \tag{1}$$

However, in a difference equation we do not take the limit as h goes to zero; rather, we treat it as a small constant. This produces a set of algebraic equations (one for each discrete value of x and t) that can be solved by computer. The practical issue is to choose a difference replacement that optimizes the accuracy of the simulation and the computation time.

The Cable Equation

The physical principle of conservation of charge is combined with Ohm's law to derive the cable equation. We focus on these separately to provide insight into the process of spatial discretization and the meaning of boundary conditions.

Conservation of charge requires that the sum of currents flowing into any region from all sources (e.g., adjacent interior regions, transmembrane ionic fluxes, microelectrodes) must equal zero.

$$\sum i_a - \int i_m \, dA = 0 \tag{2}$$

where the sum is over all the axial currents i_a in units of charge/time (e.g., milliamperes) flowing into the region through cross-section boundaries; i_m is the transmembrane current density (mA/cm^2); and the integral is taken over the membrane area A of the region. This is illustrated in Figure 1. The usual convention is that outward transmembrane current flow is positive and axial current flow into a region is positive.

The standard approach in computer simulation is to divide the neuron into regions or compartments small enough that the spatially varying i_m in any compartment j is well approximated by its value i_{m_j} at the center of compartment j. Therefore Equation 2 becomes

$$i_{m_j} A_j = \sum i_a \tag{3}$$

where A_j is the surface area of compartment j.

From Ohm's law, the axial current between adjacent compartments j and k is approximated by the voltage drop between the centers of the compartments divided by the resistance of the path between them. This transforms Equation 3 into

$$i_{m_j} A_j = \sum_k \frac{V_k - V_j}{r_{jk}} \tag{4}$$

The total membrane current $i_{m_j} A_j$ is the sum of capacitive and ionic components

$$c_j \frac{dV_j}{dt} + I(V_j, t)$$

where c_j is the membrane capacitance of the compartment and $I(V_j, t)$ includes the effects of ionic conductances. In summary, the spatial discretization of branched cables yields a set of ordinary differential equations of the form

$$c_j \frac{dV_j}{dt} + I(V_j, t) = \sum_k \frac{V_k - V_j}{r_{jk}} \tag{5}$$

Injected currents would be added to the right-hand side of this equation.

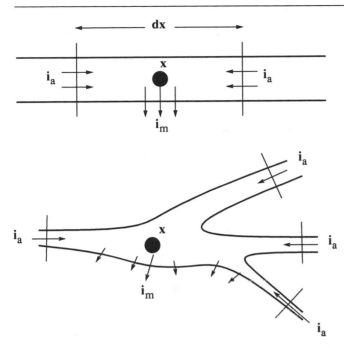

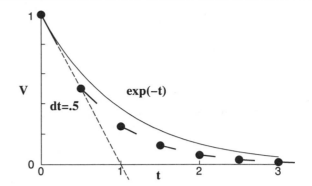

Figure 2. Simulation of the behavior of Equation 6 by successively marching forward by a fixed interval, assuming the current is constant within each interval. The current value used for a given interval is found from the value of the voltage at the beginning of the interval. The voltage values are shown by filled circles. The slope of the line segment emanating from the voltage value depends on the current *at that time step*. The dashed line shows the value of the voltage after the first time step as a function of Δt.

Figure 1. The sum of axial currents flowing into a region equals the current flowing out of the membrane area. Note that the axial current i_a varies with position.

In deriving this equation we made two approximations: representing spatially varying i_m by its value at the center of each compartment, and specifying axial current i_a in terms of the voltage drop between the centers of adjacent compartments. If the compartments are of equal size, it is easy to use Taylor's series to show that both of these approximations have errors proportional to the square of compartment length. Thus, doubling the number of compartments reduces the error by a factor of four.

Forward Euler: Simple, Inaccurate, Unstable

Spatial discretization reduced the cable equation to a set of equations with first-order derivatives in time. Euler's method is the simplest possible method for solving these equations. While it can suffer from low accuracy and can be numerically unstable, it has the advantage of being the easiest to understand, provides concrete examples of the concepts of accuracy and stability, and is a platform from which we can branch out to more complicated methods that have fewer pitfalls.

We illustrate Euler's method with models of a passive neuron (constant membrane resistance) that have only one or two compartments. In the one-compartment model the right-hand side of Equation 5 disappears, so

$$\frac{dV}{dt} + kV = 0 \qquad (6)$$

The membrane capacitance and conductance have been subsumed in the constant k—the inverse of the membrane time constant. In this easy linear case, we can readily compare the results of our computer methods to the analytic solution of this equation, which is

$$V(t) = V(0)e^{-kt} \qquad (7)$$

The numerical methods that we use to understand and control the error are immediately generalizable to the nonlinear case.

Euler's method says that, since we know the initial value $V(0)$ of our dependent variable and its initial slope $-kV(0)$ (Equation 6), we will assume the slope is constant for a short period of time and extrapolate to a new value a brief interval into the future. In Figure 2, we start with the initial condition $V(0) = 1$ and use a rate parameter $k = 1$. The time interval over which we extrapolate is $\Delta t = 0.5$.

Notice that the absolute error increases at first, but then decreases as the analytic solution and the simulation solution approach the same steady state ($V = 0$). This error is a combination of errors from two sources. First is the local error, which is due to the extrapolation within a time step. This is easily analyzed with Taylor's theorem truncated at the term proportional to Δt^2:

$$V(t + \Delta t) = V(t) + \Delta t V'(t) + \frac{\Delta t^2}{2} V''(t^*) \qquad t \leq t^* \leq t + \Delta t$$

$$(8)$$

Euler's method ignores the second-order term, so the local error at each step is proportional to Δt^2. Integrating over a fixed time interval T requires $T/\Delta t$ steps, so the cumulative local error is on the order of $\Delta t^2 \cdot T/\Delta t$, i.e., proportional to Δt. We can always decrease this error as much as we like by reducing Δt.

The second source of error has to do with the effect of past errors on the future trajectory of the simulation. Thus, if our computer solution has a small error at time t_1, then even a subsequent exact solution may have a very large error at time t_2 simply because it is on a different trajectory.

The magnitude of simulation error is not as important as whether the simulation error puts us on trajectories that are different from the set of trajectories defined by the error in our parameters. There may be some benefit in treating the model equations as sacred runes which must be solved to an arbitrarily high precision—removal of any source of error has value. But judgment is required in order to determine the meaning of a simulation. For example, consider the Hodgkin-Huxley membrane action potential elicited by a short but strong current stimulus and one elicited by a much weaker stimulus. The top panel of Figure 3 compares these action potentials with those calculated by Euler's method using a time step of 25 µs. While the voltage hovers near threshold, a little bit of error

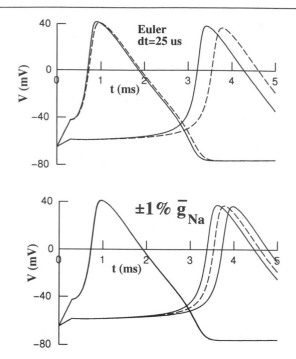

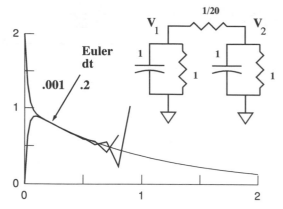

Figure 4. Euler's method is numerically unstable whenever Δt is greater than twice the smallest time constant. The analytic solution is the sum of two exponentials with time constants 1 and 1/41. The solution step size Δt is 0.001 ms for the first 0.2 ms, after which it is increased to 0.2 ms.

Figure 3. Hodgkin-Huxley membrane action potentials elicited by a current stimulus of duration 0.3 ms, and amplitudes 0.08 mA/cm^2 and 0.022 mA/cm^2. In the top panel, the solid lines are for Euler's method, with $\Delta t = 0.025$ ms; the dashed line is computed using a very much smaller Δt. All action potentials are calculated with $\bar{g}_{Na} = 0.12$ mA/cm^2. The bottom panel shows very accurate simulations with $\bar{g}_{Na} \pm 1\%$. In this panel, the three simulations that involve the large stimulus superimpose.

resulting from our time step is amplified into a considerable error in the actual time of occurrence of the spike. However, the behavior around the threshold is highly sensitive to almost any parameter, as is seen by changing the sodium channel density by only 1%. Clearly, it is crucial to know the sensitivity of our results to every parameter of the model, and the time step is just one more parameter which is added as a condition of being able to simulate a model on the computer.

It might seem that using extremely small Δt would be the best way to reduce error. However, computers represent real numbers with only a fixed number of digits. Operations that involve the difference of similar numbers, as when differences are substituted for derivatives, are especially prone to roundoff error. Consequently, there is a limit to the accuracy improvement that can be achieved by decreasing Δt.

Generally speaking, it would be nice to be able to use "physiological" values of Δt, i.e., values that allow representation of the state trajectories, with numerical accuracy commensurate with the accuracy of our physiological measurements.

Numerical Stability

Suppose the time step is too large for the example of Equation 6, e.g., $\Delta t = 3$ when $k = 1$. Now the first iteration extrapolates down to $V = -2$, the second ends up at $V = -2 + 6 = 4$, and each successive step oscillates with geometrically increasing magnitude—the simulation is numerically unstable. An important aspect of instability is most easily illustrated with a model in which two compartments are connected by a small axial resistance so that the membrane potentials are normally in quasi-equilibrium and at the same time are decaying fairly

slowly. Figure 4 shows the time course of the simulation when the initial condition is $V = 0$ in one compartment and $V = 2$ in the other. One might expect that a small Δt would be needed initially because the voltages are changing rapidly, but that a larger Δt would suffice when they are changing slowly. Figure 4 illustrates the results of this strategy as well. After 0.2 units of time with $\Delta t = 0.001$, the two voltages have nearly come into equilibrium. Then we changed Δt to 0.2—still small enough to follow the slow decay closely. Unfortunately, what happens is that, no matter how small the difference between the voltages (even if it consists only of roundoff error), the difference grows geometrically at each time step. For Euler's method, the time step must never be more than twice the smallest time constant in the system.

The notion of "time constant" and its relationship to stability are clarified by appealing to linear algebra. For a linear system with N compartments, there are exactly N spatial patterns of voltage over all compartments, such that only the amplitude of the pattern changes with time, and the shape of the pattern is preserved. These patterns (eigenvectors) have the property that the time course of change of the ith pattern is given by $e^{\lambda_i t}$, where λ_i is called the eigenvalue of the ith eigenvector. If the real part of λ_i is negative, then the ith pattern decays exponentially to 0; if the real part is positive, the amplitude grows catastrophically. If λ_i has an imaginary component, then the pattern oscillates with frequency $\omega_i = \mathrm{Im}(\lambda_i)$.

Our two-compartment model has two such patterns. In one pattern, the voltage is the same in both compartments and decays with the time course e^{-t}. The other pattern has equal but opposite voltages that decay with time course e^{-41t}. The key idea is that a problem involving N coupled differential equations can always be transformed into a set of N independent equations, each of which is solved separately as in the single compartment of Equation 6. It is essential to use a Δt small enough that the solution of each equation is stable. This is why stability criteria depend on the smallest time constant.

Systems that have a very large ratio between their slowest and fastest time constants are said to be stiff. Stiffness is a serious problem because we may need to use a small Δt to follow changes resulting from the fast time constant, while running the simulation for a very long time in order to observe changes governed by the slow time constant.

Whether an imposed driving force changes the stability properties depends on whether it alters the time constants that

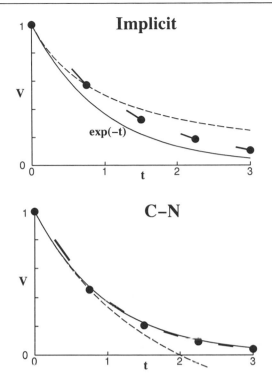

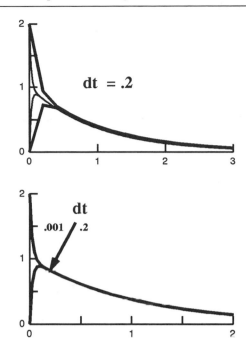

Figure 6. Two compartments as in Figure 4, simulated with the fully implicit method. Top: Δt is much larger than the fast time constant. Bottom: for the first 0.2 time units, Δt is small enough to accurately follow the fast time constant. Thereafter, Δt is increased to 10 times the fast time constant; yet the simulation remains numerically stable.

Figure 5. First-order, fully implicit (backward Euler) and second-order, Crank-Nicholson methods. At the end of each implicit step, the slope at the new value points back to the beginning of the step. In the Crank-Nicholson method, the slope at the midpoint of each step is used to determine the new value. The dashed lines show the voltage after the first time step as a function of Δt.

describe the system. A current source (perfect current clamp) will not change the time constants and therefore will not affect stability. Any other signal source introduces a load into the compartment to which it is attached, changing the time constants and their corresponding eigenvectors. The more closely it approximates a voltage source (perfect voltage clamp), the greater this effect will be.

Implicit (Backward Euler) Method: Inaccurate but Stable

We can avoid the numerical stability problems of Euler's method by evaluating the equations at time $t + \Delta t$, i.e.,

$$V(t + \Delta t) = V(t) + \Delta t f(V(t + \Delta t), t + \Delta t) \tag{9}$$

This is called the implicit method, or sometimes "backwards Euler," since it is derived from Taylor's series truncated at the Δt term but with $t + \Delta t$ in place of t. For our simple example, we have

$$V(t + \Delta t) = \frac{V(t)}{1 + k\Delta t} \tag{10}$$

Several iterations are shown in the top half of Figure 5. At each step, we move to a new value such that the slope there points back to the beginning of the step. If Δt is very large, then instead of geometrically increasing error oscillations, we get an exponential approach to the steady state.

The attractive stability properties of the implicit method are illustrated in Figure 6, where we simulate the two-compartment model. Although a large Δt does not allow us to follow fast changes, it does give a qualitative understanding of the be-

havior. Also, the artifice of changing the step size depending on how quickly the states are changing now works without blowing up. As in Euler's method, the global error is proportional to Δt for small Δt. Unfortunately, we now have to solve a set of nonlinear simultaneous equations at each step. This takes a lot of extra work, and we will want to use a step size as large as possible while retaining good quantitative accuracy. It is safest to use the first-order implicit method for initial exploratory simulations because its robust stability properties give fast simulations that are almost always qualitatively correct for reasonable values of Δt, and one does not have to worry about large error oscillations between tightly coupled compartments.

Central Difference (Crank-Nicholson) Method: Stable and More Accurate

Alternating the implicit and explicit methods with a time step of $\Delta t/2$ (Crank and Nicholson, 1947) yields a simulation with a much smaller error, proportional to the square of the step size. The bottom half of Figure 5 illustrates the idea. The value at the end of a full step Δt is along the line determined by the estimated slope at the midpoint of the step.

It is proper to wonder, though, what effect the Euler half-step has on numerical stability. Figure 7 shows the two-compartment model using this central difference method with a Δt much larger than the fast time constant. We say the method is stable because the error oscillations eventually decay away. Clearly, this method will also work with a variable time step approach. This method approximates an exponential decay by

$$V(t + \Delta t) = V(t) \frac{1 - \dfrac{k\Delta t}{2}}{1 + \dfrac{k\Delta t}{2}} \tag{11}$$

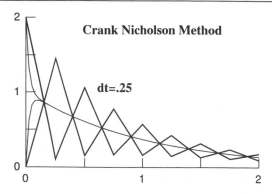

Figure 7. The Crank-Nicholson method can have significant error oscillations when there is a large amplitude component in the simulation that has a time constant much smaller than Δt. However, the oscillation amplitude decreases at each step, so the simulation is numerically stable.

As Δt, gets very large, the step multiplier approaches -1 from above, so the solution oscillates with decreasing amplitude.

One interesting feature of the central difference method is that the amount of computational work for the extra accuracy, compared to the backward difference method, is trivial since, after computing $V(t + \Delta t/2)$, we just have

$$V(t + \Delta t) = 2V\left(t + \frac{\Delta t}{2}\right) - V(t) \tag{12}$$

In other words, the extra accuracy does not require extra computations of the model functions.

Efficiency

The ideas presented so far are very general, and the problem of solving the large sets of nonlinear algebraic equations that re-sult from the implicit methods has not been addressed. Here, efficiency is all-important, and a great deal of work is ongoing in iterative and direct methods. The oldest direct method, and still the primary workhorse for stiff equations, is Newton's method, which produces a sequence of linear systems that are typically solved by Gauss elimination (Dahlquist and Björck, 1974). Although advancing N equations by one time step generally requires $O(N^3)$ arithmetic operations, branched architectures with no loops are computationally equivalent to unbranched cables and need only $O(N)$ operations. Furthermore, voltage-dependent membrane properties, which are typically formulated in analogy to Hodgkin-Huxley channels, allow the cable equation to be cast in a linear form, still second-order correct, that can be solved without iterations (Hines, 1984).

Road Map: Implementation of Neural Networks
Background: Axonal Modeling
Related Reading: Dendritic Processing

References

Crank, J., and Nicholson, P., 1947, A practical method for numerical evaluation of solutions of partial differential equations of the heat-conduction type, *Proc. Cam. Phil. Soc.*, 43:50–67.
Dahlquist, G., and Björck, Å., 1974, *Numerical Methods* (N. Anderson, Trans.), Englewood Cliffs, NJ: Prentice-Hall. ◆
Douglas, J., 1961, A survey of numerical methods for parabolic differential equations, in *Advances in Computers*, vol. 2 (F. Alt, Ed.), New York: Academic Press, chap. 1. ◆
Hines, M., 1984, Efficient computation of branched nerve equations, *Int. J. Biomed. Comput.*, 15:69–76.
Mascagni, M., 1989, Numerical methods for neuronal modeling, in *Methods in Neuronal Modeling* (C. Koch and I. Segev, Eds.), Cambridge, MA: MIT Press, pp. 439–484. ◆
Oran, E. S., and Boris, J. P., 1987, *Numerical Simulation of Reactive Flow*, New York: Elsevier. ◆

Computing with Attractors

John Hertz

Introduction

This article describes how to compute with networks with feedback that exhibit complex dynamical behavior. In order to compute with any machine, we need to know how data are to be fed into it and how the result of a computation is to be read out. These questions are trivial for layered feedforward networks, but not for networks with feedback. A natural proposal is to wait until the network has "settled down" and then read the answer off a suitably chosen set of units. The state a dynamical system settles into is called an *attractor*, so this paradigm is called computing with attractors.

The term "settling down" is not meant to restrict this picture to cases in which the dynamical state of the network stops changing. This is one kind of attractor, but it is also possible to settle down into periodic or even chaotic patterns of activity.

This way of computing is appealing because it does not require the person reading the result to observe the entire evolution of the network. Neither need she know when the compu-tation was started or how long it took. All that is required is a way of recognizing that the network has reached the attractor. Therefore, this survey will begin with a brief description of the kinds of attractors one can meet: fixed points, limit cycles, and strange attractors. We will illustrate the paradigm with a simple example: the Hopfield model for associative memory. We then indicate briefly both how the connections necessary to embed desired patterns can be learned and how the paradigm can be extended to time-dependent attractors. Finally, we examine the possible relevance of attractor computation to the functioning of the brain.

Networks, Attractors, and Stability

We will focus our attention on networks described by systems of differential equations like

$$\tau_i \frac{du_i}{dt} + u_i(t) = \sum_{j \neq i} w_{ij} g[u_j(t)] + h_i(t) \tag{1}$$

Here $u_i(t)$ is the net input to unit i at time t and $g(\ )$ is a sigmoidal activation function ($g' > 0$), so that $V_i = g(u_i)$ is the activation of unit i. The connection weight to unit i from unit j is denoted w_{ij}; $h_i(t)$ is an external input, and τ_i is a relaxation time.

We can also consider discrete-time systems governed by

$$V_i(t + 1) = g\left[\sum_j w_{ij} V_j(t) + h_i(t) \right] \qquad (2)$$

Here it is understood that all units are updated simultaneously.

These models are deterministic. Noisy networks can be analyzed by the methods of statistical mechanics and are treated extensively elsewhere in this *Handbook* (see STATISTICAL MECHANICS OF NEURAL NETWORKS).

Viewing such a network as a computer, we see that data can be read into it in two ways. One is as the $h_i(t)$ on a subset of the units, which we can call input units. The $h_i(t)$ values might be held fixed or varied in time, depending on the problem. This way of loading data is, of course, carried over directly from the conventional input-output paradigm as we normally apply it to feedforward networks. Alternatively, we can load the data by setting the values of the initial activations $V_i(0)$. For layered feedforward networks this procedure is equivalent to the previous one, but for recurrent nets it is fundamentally different. We can also use both these input mechanisms simultaneously. The first is appropriate when we want to have the output vary with the input (e.g., continuous mapping); we use the second when we want an output to be insensitive to small changes in the input (e.g., error correction). As the second one is intrinsic to recurrent networks, we will focus most of our attention on it.

The program of such a computer is its connection weights w_{ij}. Some of them may be zero, but the questions we address here are rather trivial unless there is some feedback, i.e., unless our networks are *recurrent*. We will not restrict our attention to symmetric connections; w_{ij} need not equal w_{ji}.

Finding the correct weights to implement a particular computation is a highly complex problem. However, for recurrent networks, we must first understand something about the attractors which represent the results of computations, so we now turn our attention to this problem.

To describe the dynamics of our networks, we make use of a picture in which the activation of each unit in the network is associated with a direction in a multidimensional space, called the *configuration space*. Every point in this space represents a possible state of the network, called the *state vector*, and the motion of this vector represents its evolution in time. For all recurrent networks of interest, there are just three possibilities for the asymptotic state.

1. The state vector comes to rest, i.e., the unit activations stop changing. This is the simplest case to analyze, and it is called a *fixed point*. Different results of a computation (owing to different input data) are characterized by settling into different fixed points. The region of initial states which settles into a single fixed point is called its *basin of attraction*. Most of the examples of recurrent networks in the literature, such as the Hopfield model and many related ones, compute with fixed points.
2. The state vector settles into a periodic motion, called a *limit cycle*.
3. The state vector moves chaotically, in the sense that two copies of the system which initially have nearly identical states will grow more and more dissimilar as they evolve: the two state vectors diverge from each other. However, the way they diverge is restricted. At any time, the two state vectors

are actually growing closer together in many directions in the configuration space; the divergence occurs only in some (typically a few) directions. A Poincaré map showing, for instance, the states of some of the units every time the state vector passes through some hyperplane in configuration space, will be a fractal object with a dimensionality greater than zero (Schuster, 1989). This kind of attractor is called *strange* (see CHAOS IN NEURAL SYSTEMS).

Which kind of attractor we obtain will depend on the connections in the network and, possibly, on the input data. Suitable learning algorithms make it possible to design the desired type of attractor. In simple applications, fixed points are naturally easiest to deal with. However, it may sometimes be advantageous to exploit the richer dynamical possibilities available in nonstationary attractors. For example, limit cycles allow the timing of the network's response to be controlled.

There are conditions under which the attractors will always be fixed points. For nets described by the continuous dynamics of Equation 1, a sufficient (but not necessary) condition is that the connection weights be symmetric: $w_{ij} = w_{ji}$.

General results concerning the stability of recurrent nets were obtained by Cohen and Grossberg (1983). They showed, for static external input h_i, that if the connection matrix w_{ij} is symmetric, the attractors of Equation 1 are always fixed points, even if the activation function is allowed to differ from unit to unit, and even if the $u_i(t)$ on the left-hand side is replaced by a general monotonic function $b_i(u_i)$ and τ_i is a (positive) function of u_i.

The proof illustrates the basic mathematical strategy for proving stability. Suppose we can find some quantity, a nontrivial function of the state variables u_i, which always decreases under the dynamics in Equation 1 except for special values of the u_i at which it does not change. These values are fixed points. For values of the u_i near such a point, the system will evolve either toward it (an attractor) or away from it (a repellor). For almost all starting states, the dynamics will end at one of the attractor fixed points. Furthermore, these are the only attractors. If there were a limit cycle, for example, our function would decrease everywhere on a closed curve, which is impossible. Thus, whether such a quantity exists for a given network is very important. A function with this property is called a *Lyapunov function*.

There is indeed such a function for the Cohen-Grossberg extension of the dynamics of Equation 1:

$$L(\mathbf{u}) = \sum_i \int^{u_i} [b_i(u) - h_i] g_i'(u)\, du - \frac{1}{2} \sum_{ij} w_{ij} g_i(u_i) g_j(u_j) \qquad (3)$$

We can show directly that this is a Lyapunov function by computing its time derivative. Making use of the equations of motion and the symmetry of w_{ij}, we find

$$\dot{L} = -\sum_i g_i'(u_i) \tau_i(u_i) \dot{u}_i^2 \le 0 \qquad (4)$$

Thus L is always decreasing, except at the fixed points $\dot{u}_i = 0$.

When a Lyapunov function exists, we may think about the dynamics in terms of sliding downhill on a surface in configuration space, the height of which is given by $L(\mathbf{u})$. The motion is not simple gradient descent, since $\dot{u}_i$ is not exactly proportional to $-\partial L/\partial u_i$ (there is an extra factor $g_i'(u_i) \tau_i(u_i)$). Nevertheless, the motion is always downhill, and the bottoms of the valleys correspond to the fixed points.

If we know the form of the Lyapunov function for a particular kind of network, this picture gives us a clue about how to program desired fixed-point attractors: we try to choose the connection weights w_{ij} and biases h_i so that L has minima at these points in configuration space.

To gain a little more insight, we restrict ourselves to the case $b_i(u) = u$ and an activation function $g(u) = \tanh(\beta u)$. Using the activation variables $V_i = g(u_i)$ instead of u_i, we find that we can write L in the form

$$L(\mathbf{V}) = \frac{1}{\beta} \sum_i \int^{V_i} dy \tanh^{-1} y - \sum_i h_i V_i - \frac{1}{2} \sum_{ij} w_{ij} V_i V_j \quad (5)$$

For large gain β, the first term is small. It is natural to think of the other two simply as a "potential energy" which the system tries to minimize. The w_{ij} and h_i should thus be chosen so that their sum has minima at or near the desired fixed points. The main effect of the first term is just to prevent the activations from reaching 1 or -1, since its derivatives diverge there.

Sometimes it is simple to construct a potential energy with the desired minima, at least approximately. In other problems, this strategy may be inadequate, and we have to resort to iterative learning algorithms to determine the network parameters.

Limit cycles and strange attractors are harder to handle mathematically. Often, however, it is possible to proceed in some kind of analogy with the fixed-point case.

Associative Memory

The most celebrated application of computing with fixed points is associative memory (see ASSOCIATIVE NETWORKS). Here we follow the treatment due to Hopfield (1984) (see also Hertz, Krogh, and Palmer, 1991, chaps. 2 and 3). There is a set of patterns to be stored somehow by the computer. Given as input a pattern which is a corrupted version of one of these, the result of the computation should be the corresponding uncorrupted one.

The strategy for solving this problem is to try to guess a form for the potential energy part of Equation 5 which will have minima at the configurations corresponding to the patterns to be stored. We take patterns $\xi_i^\mu = \pm 1$. The subscript i labels the N elements of the pattern (e.g., pixel values), and the superscript μ labels the p different patterns in the set. The patterns are assumed random and independent for both different i and different μ. If we wanted to store just one such pattern ξ_i, a natural choice would be to take the potential energy function proportional to $-(\sum_i \xi_i V_i)^2$. The quantity $\sum_i \xi_i V_i$ measures the similarity between the pattern and the state of the network. It achieves its maximal value at $+N$, and therefore $-(\sum_i \xi_i V_i)^2$ is minimal if and only if every V_i coincides with ξ_i. For more than one pattern we try one such term for each pattern, yielding a total potential energy

$$H = -\frac{1}{2N} \sum_{\mu=1}^{p} \left(\sum_i \xi_i^\mu V_i \right)^2 \quad (6)$$

Multiplying out the square of the sum, we can identify the connection weights in Equation 5 as

$$w_{ij} = \frac{1}{N} \sum_\mu \xi_i^\mu \xi_j^\mu \quad (7)$$

The form of this equation suggests a Hebbian interpretation (see HEBBIAN SYNAPTIC PLASTICITY). For each pattern, there is a contribution to the connection weight proportional to the product of the activities of the sending units (ξ_j^μ) and the receiving units (ξ_i^μ) when the network is in the state $V_i = \xi_i^\mu$. This is just the form of synaptic strength proposed by Hebb (1949) as the basis of animal memory, so this is sometimes called a Hebbian storage prescription. This matrix is symmetric and has positive definite eigenvalues, so our earlier results guaran-

tee that the attractors of our network dynamics are fixed points for both continuous and discrete-time dynamics.

The hope is that this will produce a fixed point of the network dynamics at or near each ξ_i^μ. (Because our H is purely quadratic in the V_i, we also expect fixed points near $-\xi_i^\mu$.) We can see how well this works by examining the stationary points of the Lyapunov function of Equation 5, which are

$$V_i = \tanh\left(\beta \sum_j w_{ij} V_j \right) \quad (8)$$

We would first like to know whether there are solutions of Equation 8 which vary across the units like the individual patterns ξ_i^μ.

The quality of retrieval of a particular stored pattern ξ_i^μ is measured by the quantity $m_\mu = N^{-1} \sum_i \xi_i^\mu V_i$. Using Equation 8, with the weight formula of Equation 7, we obtain

$$m_\mu = \frac{1}{N} \sum_i \xi_i^\mu \tanh\left(\beta \sum_\nu \xi_i^\nu m_\nu \right) \quad (9)$$

The kind of solution we are looking for should describe a state of the network which is correlated with only one of the stored patterns, i.e., just one of the $m_\mu \neq 0$. With this restriction, Equation 9 reduces to

$$m = \tanh(\beta m) \quad (10)$$

where m is the value of the one nonzero m_μ. This equation has nontrivial solutions whenever $\beta > 1$. Next, we have to inquire whether these solutions are truly attractors, i.e., whether they are stable. At this point the story gets mathematically involved, so we simply survey the results (Kühn, Bös, and van Hemmen, 1991). Everything here is derived in the limit of a large network, i.e., $N \to \infty$.

The story is quite simple when p (the number of stored patterns) is a negligible fraction of N, the number of units in the network. Then the nontrivial solutions are globally stable, while the solution $m = 0$ is unstable, whenever $\beta > 1$.

If the gain is high enough, there are other attractors in addition to the ones we have tried to program into the network with the choice represented by Equation 7. In the simplest of these, the state of the network is equally correlated with three of the ξ_i^μ, say, $\xi_i^{\mu_1}$, $\xi_i^{\mu_2}$, and $\xi_i^{\mu_3}$. These other attractors are thus mixtures of three of our desired attractors. Such solutions exist whenever $\beta > 1$, but they are locally stable only when $\beta > 2.17$. Turning the gain up higher still, combinations of greater numbers of the desired memories also become stable. Thus, by keeping the gain between 1 and 2.17, we can limit the attractor set to the desired states.

When p is of the same order as N, the analysis is more involved. The root of the problem is that the different terms in the weight formula (Equation 7) interfere with each other, even for independent random patterns. (The overlap between two patterns is of order $N^{-1/2}$, but as the number of such overlaps is of order N^2, the net effect is of order 1.) This cross-talk has three important effects. First, it induces small mismatches, which grow with increasing $\alpha = p/N$, between the original patterns ξ_i^μ and the attractors. Second, and more dramatically, it destroys the pattern-correlated attractors completely above a critical value of α, α_c. This critical value depends on the gain β, and in the limit $\beta \to \infty$, α_c approaches 0.14. Finally, one finds that whenever the gain exceeds $\beta_s = (1 + 2\sqrt{\alpha})^{-1} \leq 1$ there are infinitely many fixed points, all completely uncorrelated with the patterns ξ_i^μ.

Nevertheless, as long as we are not trying to store too many patterns ($\alpha < \alpha_c$), there will be attractors which are strongly correlated with the patterns. The unwanted other attractors can no longer be completely eliminated by suitable tuning of the gain, but as they are uncorrelated with the patterns, they do not have much effect on the retrieval of a pattern, starting from an initial configuration not too far from the attractor.

Thus, attractor computation works in this system over a wide range of model parameters. It can be shown to be robust with respect to many other variations, as well. These include dilution (random removal of connections), asymmetry (making some of the $w_{ij} \neq w_{ji}$), and quantization or clipping of the weight values.

A number of other problems, in particular in optimization theory, have been treated using the same strategy of choosing the connections so that the potential energy has minima in the appropriate places (see NEURAL OPTIMIZATION). The features we have noted in the associative memory problem appear to be universal. It is possible to obtain the desired attractors (at least approximately), but other, undesired attractors are generally also created. These can be controlled to some degree by suitable adjustment of the gain or other parameters.

Learning

The weight formula of Equation 7 was only an educated guess. It is possible to obtain better weights, which reduce the cross-talk and increase α_c, by employing systematic learning algorithms.

One of the simplest of these is *Boltzmann learning* (see BOLTZMANN MACHINES and Hertz et al., 1991, chap. 7). Originally, Boltzmann learning was formulated for stochastic binary units. Here, we use a formulation for continuous-valued deterministic units. Suppose, as earlier, that we want to make an attractor of the configuration in which the unit activations are proportional to pattern ξ_i^μ. Now, if we start the network in the configuration $V_i = \xi_i^\mu$, it will settle into some fixed point V_i^μ. The algorithm is to change the weights according to

$$\Delta w_{ij} = \eta(\xi_i^\mu \xi_j^\mu - V_i^\mu V_j^\mu) \tag{11}$$

The first term is a Hebb-like learning term like Equation 7, and the second term ensures that learning stops when the fixed point V_i^μ coincides with the ξ_i^μ. This is then performed for every pattern and repeated until the attractors converge to the desired locations. It is evident from Equation 11 that the resulting connections will be symmetric if the initial ones are. Boltzmann learning can also be used when there are hidden units. In that case, when i or j is a hidden unit, the patterns ξ_i^μ or ξ_j^μ are replaced by the stationary values those units take when the nonhidden units are clamped at the pattern values.

The other simple learning rule that can be used to learn particular attractors is the delta rule. When there are no hidden units, its weight updating rule is

$$\Delta w_{ij} = \eta(\xi_i^\mu - V_i^\mu)\xi_j^\mu \tag{12}$$

As in Equation 11, the role of the second or "unlearning" term in the parentheses is to turn learning off when the desired fixed points are achieved. With hidden units, the delta rule becomes what is known as BACKPROPAGATION (q.v.). There is not space here to go into the mathematical description of backpropagation in recurrent networks. We only remark that it is describable in terms of a network like the original one but with all the directions of the connections reversed (see Hertz et al., 1991, chap. 7).

Nonstationary Attractors

So far, we have worked with networks with first-order dynamics (Equations 1 and 2) and a symmetric weight matrix. If we relax either of these conditions, nonstationary attractors are possible.

By including suitable delays in the discrete-time dynamics (Equation 2), it is possible to extend the Hopfield model to store pattern sequences. This problem is treated in detail in SPATIOTEMPORAL ASSOCIATION IN NEURAL NETWORKS. The problem can be mapped onto the one with static patterns, and much of the analysis for that model can be carried over to the dynamic case.

Iterative learning algorithms can also be brought to bear to stabilize specific desired limit cycles. In particular, the recurrent backpropagation algorithm mentioned earlier for learning fixed points can be extended rather straightforwardly to learning arbitrary time-dependent patterns (see Hertz et al., 1991) chap. 7.

If networks can learn arbitrary periodic attractors, the obvious next question is whether they can learn strange attractors. The answer to this is also affirmative. The initial work on this problem was done by Lapedes and Farber (1987), who succeeded in teaching a network a strange attractor generated by a nonlinear differential-delay equation known as the Mackey-Glass equation, which was originally introduced in a model of blood production (see CHAOS IN NEURAL SYSTEMS).

Discussion: Attractors in the Brain?

Both local neural activity (see MOTOR PATTERN GENERATION) and macroscopic neural activity (see EPILEPSY: NETWORK MODELS OF GENERATION) can be described in terms of attractors. However, the most interesting questions about attractors in the brain are about their functional roles in processes such as perception, recognition, and memory. That is, are particular attractors associated with, for example, the recognition of particular objects? Can the settling of the brain's activity into such an attractor be identified with the recognition process?

W. Freeman (1991; see also Skarda and Freeman, 1987) has argued that cortical computation may follow this paradigm. This hypothesis is based largely on the analysis of multielectrode recordings in the olfactory bulb and piriform cortex (the next stage in the olfactory processing pathway) in rabbits and cats. The claim is that the resting state of the system is a strange attractor, and that the system builds a set of distinct attractors corresponding to different odors.

Does a similar picture apply in the rest of the brain? Some very interesting evidence in favor of attractor computation comes from a combination of experiments by Miyashita (1988) and theoretical modeling by Griniasty, Tsodyks, and Amit (1993). In these experiments, monkeys learn to identify a set of visual patterns. During training, the patterns are presented in a particular order. After learning, the patterns are presented in random order and multicellular recordings are made in a small region of anterior ventral temporal cortex during the period between stimuli. The spatial pattern of the mean firing rates across the electrode array is found to be stimulus-specific. The interesting finding is that, although the stimuli are not correlated with each other, the resulting firing patterns are. Strong correlations are found only between pairs of activity patterns evoked by stimuli which were close together in the training sequence.

In the theoretical analysis, the firing rate patterns are identified with attractors of the cortical dynamics. Griniasty et al. suppose there are pre- and postsynaptic delays in the learning of the patterns, with the result that new terms proportional to $N^{-1} \sum_\mu \xi_i^\mu \xi_j^{\mu \pm 1}$ are added to the weight formula of Equation 7. Consequently, the attractor corresponding to a particular pattern is correlated with those of nearby patterns in the sequence.

The form of this correlation can be calculated and is quite similar to that observed in Miyashita's experiments. This result lends credence to both the idea of attractor computation and the Hebbian learning picture.

Abeles (1991) has proposed another picture, in which the brain computes using precisely timed sequences of patterns, called *synfire chains*, rather than stationary attractors. There is accumulating evidence that particular sequences of neuronal firings, accurate to one or two milliseconds, occur statistically far above chance level in multicell recordings. The function of these sequences is not yet known, but this intriguing conjecture offers another, quite different, kind of attractor for further experimental and theoretical investigation.

It is evident that the attractor hypothesis at least provides a nontrivial framework within which to address the problem of cortical computation, both theoretically and experimentally. It is too soon to know in what degree it is correct or how it will have to be elaborated in the future. However, at present, there is no other paradigm on the market that offers a comparably fertile ground for the interplay of theory and experiment that will be necessary to achieve an understanding of the nervous system.

Road Map: Dynamic Systems and Optimization
Background: I.3. Dynamics and Adaptation in Neural Networks
Related Reading: Constrained Optimization and the Elastic Net; Dynamics and Bifurcation of Neural Networks; Energy Functions for Neural Networks

References

Abeles, M., 1991, *Corticonics*, Cambridge, Eng.: Cambridge University Press. ◆

Cohen, M., and Grossberg, S., 1983, Absolute stability of global pattern formation and parallel memory storage by competitive neural networks, *IEEE Trans. Sys. Man Cybern.*, 13:815–826.

Freeman, W. J., 1991, The physiology of perception, *Sci. Am.*, 264(2): 78–85. ◆

Griniasty, M., Tsodyks, M. V., and Amit, D. J., 1993, Conversion of temporal correlations between stimuli to spatial correlations between attractors, *Neural Computat.*, 5:1–17.

Hebb, D. O., 1949, *The Organization of Behavior*, New York: Wiley. ◆

Hertz, J. A., Krogh, A. S., and Palmer, R. G., 1991, *Introduction to the Theory of Neural Computation*, Redwood City, CA: Addison-Wesley. ◆

Hopfield, J. J., 1984, Neurons with graded responses have collective computational properties like those of two-state neurons, *Proc. Natl. Acad. Sci. USA*, 79:3088–3092.

Kühn, R., Bös, S., and van Hemmen, L., 1991, Statistical mechanics of graded-response neurons, *Phys. Rev. A*, 43:2084–2087.

Lapedes, A., and Farber, R., 1987, *Nonlinear Signal Processing Using Neural Networks: Prediction and Signal Modelling*, Technical Report LA–UR–87–2662, Los Alamos, NM: Los Alamos National Laboratory.

Miyashita, Y., 1988, Neuronal correlate of visual associative long-term memory in the primate temporal cortex, *Nature*, 335:817–820.

Schuster, H. G., 1989, *Deterministic Chaos*, 2nd ed., Weinheim, Ger.: VCH Verlagsgesellschaft mbH. ◆

Skarda, C. A., and Freeman, W. J., 1987, How brains make chaos in order to make sense of the world, *Behav. Sci.*, 10:161–195. ◆

Concept Learning

Philippe G. Schyns and Luc Rodet

Introduction

Imagine that on a particular morning, while you are brushing your teeth, your ability to perceive similarities between objects slowly dissipates. You first notice that the red object in your hand behaves strangely. Each time your hand moves, the object looks completely different. Then you look in the mirror, at the object you were sure was your face a few minutes ago. You tilt your head to the left, then to the right, but each time your own reflection in the mirror looks like a different face. *Toothbrush* and *face* as permanent entities have just vanished from your mental life. Your mind does not "see" through your eyes anymore. As you contemplate this continuous flow of unrelated experiences, you suddenly realize you are less and less able to forget a difference, to generalize, to abstract . . . to think.

The study of *concept learning*, within the field of cognitive science, considers how people segment the world into categories of similar events or objects. Concepts are mental representations of object categories. For example, the concept *my-toothbrush* embraces different views of my toothbrush, and the concept *bathroom* represents the rooms in which one finds the objects toothbrush, glass, mirror, razor, brush, bathtub, and so forth. Categorization is the operation by which an object is identified as a member of a particular category. To illustrate, your face from a three-quarters view should be categorized as a view of your face, and a toothbrush might be classified as a bathroom object. The mechanisms of concept acquisition are probably among the most fundamental processes of cognition: they authorize the categorization of a potentially unlimited number of objects and events with limited resources.

As a computational metaphor, connectionist modeling has deep implications for its domains of application. On the one hand, the connectionist framework provides mechanistic tools to interpret and to formulate the workings of cognitive capacities. On the other hand, and perhaps more importantly, network formulations may stretch theories of cognitive capacities beyond their current stages of development. Mechanistic formulations may reveal implicit assumptions of old formulations that become explicit issues of new formulations. This article briefly discusses some aspects of network modeling which illustrate important issues in human categorization theory. The selected issues reflect our own biases as to what we think are promising cross-fertilizations between human and connectionist categorization theories.

Features, Feature Spaces, and the Psychological Relevance of Features

Categorization theories conceive of an object as a composition of attributes, such as its color, texture, height, weight, whether it has wings, legs, or fins, whether it lives in the air, on the ground, or in the sea, and so forth. In the connectionist metaphor, such description translates into a *feature vector*: each slot of an n-dimensional vector encodes the presence/absence or the value of the n properties. Geometrically, a particular point in space encodes an object, and categories of similar objects form tight clouds of points.

Consider the problem of distinguishing Martians from other beings. Martians are green and have pointed ears. Other beings are pink with rounded ears, green with rounded ears, or even pink with pointed ears. We can represent Martians in a two-dimensional feature space by a binary encoding of their *color* and *ear-shape* properties.

A simple one-layer network, with two input units and one output unit solves this categorization task. For each exemplar $\mathbf{x}^p$, we want the network output o^p to match the expected output t^p (1 if $\mathbf{x}^p$ is a Martian exemplar, -1 otherwise). In short, for every p,

$$o^p = \mathrm{sgn}(\mathbf{w} \cdot \mathbf{x}^p - \theta) = t^p$$

A weight vector $\mathbf{w}$ must be found so that its inner product is positive (for a Martian input) or negative (for all other inputs). As $\mathbf{w} \cdot \mathbf{x} = |\mathbf{w}||\mathbf{x}|\cos(\gamma)$, where γ is the angle between the two vectors, the inner product is positive, negative, or equal to 0 when γ is, respectively, less than, greater than, or equal to 90 degrees. Therefore, $\mathbf{w} \cdot \mathbf{x} = 0$ is a line through the origin, perpendicular to $\mathbf{w}$, which segregates Martians from non-Martians. The addition of a threshold θ moves the separating line a distance θ away from the origin.

The *perceptron learning rule* (Rosenblatt, 1962) finds the weight vector $\mathbf{w}$ to distinguish between linearly separable categories (categories that are separable by a line, a plane, or a hyperplane):

$$\mathbf{w}^{\mathrm{new}} = \mathbf{w}^{\mathrm{old}} + \eta \Theta(\theta - \mathbf{w} \cdot \mathbf{x}^p)\mathbf{x}^p$$

With learning, the weight vector $\mathbf{w}$ rotates in the direction of the input pattern $\mathbf{x}^p$ by a proportion η when there is a large mismatch between the input and $\mathbf{w}$. When the Θ threshold function returns 0 for all input patterns, the line perpendicular to $\mathbf{w}$ partitions the input space. The *perceptron convergence theorem* guarantees that a simple perceptron partitions the input space in a finite number of steps, when the linear separability requirement is met (Minsky and Papert, 1969).

The Martian example illustrates concept learning in a *fixed space*: the space spanned by the feature set. A fixed feature set approach has the major advantage of specifying the aspects of an object that are considered for categorization, but fixed features have the drawback of limiting possible concepts to combinations of the features of the set. If a feature external to the set distinguishes between two objects, this distinction is not representable in the learning space. So, the fixed set should be as large as possible to model the learning of complex concepts, but it should not be so large that all aspects of the objects are always considered for categorization. The issue of *feature relevance* is particularly salient in fixed set models because each feature constitutes the design decision of including (versus excluding) a feature in the learning space. In principle, infinitely many features can characterize an object, but only a subset of the features are psychologically relevant. Task constraints, sophisticated high-level knowledge, attentional mechanisms, and perceptual organization could constrain the selection of a set of psychologically relevant features. But it is not clear that such a fixed set exists, as new features are often needed to represent new categorizations. Thus, the fixed set approach may face serious difficulties in providing a general account of human categorization. These difficulties do not preclude interesting modeling. Many interesting categorization problems do occur once the features of the space are known (e.g., Gluck and Bower, 1988; Kruschke, 1992).

Unsupervised Learning and the Importance of Similarity Measures

In the Martian example, no learning occurs if exemplars are presented independently of the exemplar's category. This assumption is quite limiting when one attempts to model the varieties of observed human concept learning. For example, when looking for the first time through a microscope, researchers were exposed to a world in which no a priori categorizations existed. Researchers had to discover order by themselves. This is the task of *unsupervised* algorithms: they discover similarities or structures in the input without feedback about the relevance of their discoveries.

The simplest form of unsupervised learning involves a one-layer network with *competitive* output units—units that compete to respond to the input exemplars (see COMPETITIVE LEARNING). Each output o_i computes the projection of the input on the weight vector $\mathbf{w}_i$:

$$o_i = \mathbf{w}_i \cdot \mathbf{x}^p$$

In a winner-take-all algorithm (see WINNER-TAKE-ALL MECHANISMS), only the winner unit o_w fires after the competition; all other outputs are inhibited. After learning, each output unit o_i responds to a particular category, the category represented by the weight vector $\mathbf{w}_i$. The weights $\mathbf{w}_i$ are changed to insure that the winning unit fires more intensively the next time it sees an exemplar from the same category. Geometrically, this translates into a reduction of the Euclidean distance between the weight vector and the exemplar; o_w is maximum when $\mathbf{x}^p$ is parallel to $\mathbf{w}$:

$$\Delta\mathbf{w}_i = \eta o_i(|\mathbf{w}_i - \mathbf{x}^p|)$$

Feature maps are similar to competitive networks, but maps have the property that nearby output units respond to nearby input patterns. To illustrate, consider a simple problem: We want a network to classify different views of a teapot which rotates in depth. Ideally, we would like nearby views of the teapot to be mapped onto nearby output units.

Figure 1 shows the self-organization of the weights of a seven-output-unit feature map. Although teapot views were presented in a random order, the weight vectors are perfectly ordered. Straightforward modifications of the simple competitive learning rule implement a feature map algorithm (see SELF-ORGANIZING FEATURE MAPS: KOHONEN MAPS). The most important modification is the neighborhood function $N(o_w)$, which denotes the units around the winner:

$$\Delta\mathbf{w}_i = \eta N(o_w)(|\mathbf{w}_i - \mathbf{x}^p|)$$

This learning rule updates the weight vectors in $N(o_w)$ so that they all rotate in the same direction. When $N(o_w)$ is large, this operation imposes a global order on the weights. In Figure 1, a global organization of the weights is shown in the third panel from the top. If $N(o_w)$ and η decrease with time, the order proceeds from global to local (see the last panel of Figure 1). Feature maps have the advantage that nearby output units represent the similarity of the concepts encoded in their weights (Schyns, 1991).

Unsupervised algorithms rely on a similarity metric to partition the input space. Similar featural encodings form tight clouds which are captured by the weights. Similarity can be very versatile. To illustrate, consider a kite, a parrot, and a painting of the parrot. Perceptually, the parrot and its portrait look more alike than the parrot and the kite. However, the parrot and the kite fly, the painting does not. From this perspective, parrot and kite are more similar. The similarity metric itself may change the similarity measures of the two objects

Figure 1. Evolution of the weights of a self-organizing network which orders automatically randomly presented two-dimensional views of three-dimensional objects. The endpoint of the process is a set of two-dimensional views which categorizes the input view according to angle of rotation. The inputs are high-dimensional vectors encoding the gray-level values of the two-dimensional projections of the teapot.

being compared. For example, if $M_1 = \mathbf{w} \cdot \mathbf{x}^p$ and $M_2 = |x_1 - x_2| + |y_1 - y_2|$ is another metric (with x_i and y_i as distinct feature values), nearby points under M_1 become distant points under M_2, and vice versa. In principle, infinitely many features can enter similarity comparisons. The justification of feature use in similarity estimates is therefore intertwined with the justification of feature use in object descriptions. A complete theory of concept learning must justify the features that are the basis of the learning space.

Learning the Basis of the Feature Space as a Side-Effect of Categorization

The issue of feature learning has recently come under closer scrutiny in psychology (Schyns and Murphy, 1994). It has become more and more apparent that many tasks require the creation of new dimensions of categorization—dimensions that were not used prior to the experience with the classification system.

To illustrate how this might be done, consider a "communication game" involving an observer, a narrow communication channel, and a very large set of brand new "Martian objects" (objects whose shapes are completely new). The observer's task is to describe all the objects in as few words as possible. A good strategy is probably to recode the object set so that each word of the code represents a major difference of the set. Assume another observer who experiences a completely different object set, except one object which is identical to one of the first set. The contrasts and similarities within the two object sets will probably be quite different, resulting in different object description vocabularies. In particular, the object which intersects both sets will be coded differently.

This communication game assigns a *functional role* to each feature, and the objects of the set form a context which constrains feature extraction (Schyns and Murphy, 1994). Networks and their ability to *recode* a data set may provide most useful analogies for understanding the learning of the relevant features of categorization.

Supervised Recoding

Simple perceptrons do not recode their input space and are therefore limited in the range of problems they can solve. A *multilayer* feedforward network with nonlinear units is a much more powerful learning device. With only one layer of *hidden units*, such networks have enough power to recode an input space and solve nonlinearly separable problems—more generally, any Boolean function (Hertz, Krogh, and Palmer, 1991).

The *backpropagation* algorithm (see BACKPROPAGATION: BASICS AND NEW DEVELOPMENTS) prescribes the weight change of a multilayer feedforward network in learning complex input/output mappings (Rumelhart, Hinton, and Williams, 1986); $\Delta \mathbf{w}$ directly depends on a measure of the mismatch between the desired output and the network's response to an exemplar:

$$E[\mathbf{w}] = \frac{1}{2} \sum_{p,i} (t_i^p - o_i^p)^2$$

In a multilayer perceptron, the activation rule of units is a nonlinear differentiable function. Since the cost function is differentiable for every weight, a gradient descent algorithm can learn the appropriate weights for the mapping.

Categorization in a multilayer perceptron is a two-step process: the input-to-hidden layer recodes the input, and the hidden-to-output assigns a category label to the recoded input. Two-layer perceptrons present interesting instances of featural recoding of the input space. For example, Gorman and Sejnowski's (1988) architecture learned a mapping between the frequency components of sonar returns (*rock* or *explosive mines* of the same size) and their category labels. They found that explosive echoes (wide-band signals) would turn off all hidden units, and rock echoes (narrow-band signals) would turn them on. Further analysis revealed that hidden units responded to three correlated dimensions of the echoes (their bandwidth, onset, and decay). These dimensions were not explicitly represented on the input vectors (as *color* and *has-pointed-ears* were in the Martian example). Rather, they were implicit features of the pattern set that were discovered during learning and recoded on the hidden layer.

Unsupervised Recoding

A robust classification of vectors in a high-dimensional space requires many data points to obtain good estimators of the categories, but high-dimensional spaces are almost empty. Although stimuli may in principle vary along each of the n dimensions, it is expected that most of the structure of a categorization problem lies in a much smaller dimensional space—because, for example, the categories can be distinguished from only a few covarying variables. *Dimensionality reduction* techniques transform high-dimensional problems to low-dimensional spaces and preserve most of their intrinsic information. This is interesting for concept learning, because the dimensions of low-dimensional spaces may be useful categorization features, if the search for relevant dimensions is properly constrained.

Principal components analysis. The major sources of variation of a data set are called the *m principal components* of the data set. If there are redundancies in the input data, there should be fewer principal components than dimensions (i.e., $m \ll n$). Because principal components account for most of the variability of the data set, recoding the input stimuli with principal components should preserve the intrinsic information of the data.

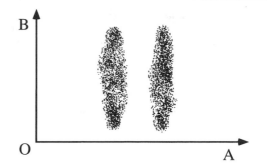

Figure 2. The top panel presents two simple categories projected on the OA and OB axes (from Linsker, 1988). The OA axis points in the direction of highest variance, and the OB axis points in the next orthogonal direction of highest variance. A projection onto OA preserves the structure of the categorization task. The bottom panel also presents two simple categories projected onto the OA and OB axes (from N. Intrator, private communication). Again, the projection onto OA preserves the structure of the categorization problem, but here OA is not the direction of highest variance (OB is).

This is the goal of PRINCIPAL COMPONENT ANALYSIS (PCA) (q.v.).

To illustrate PCA, consider the simple example of two categories presented in the top panel of Figure 2. The projection of the clouds onto OA preserves more information than the projection onto OB. The first principal component OA points in the direction of highest variance. The second principal component OB points in the direction of highest variance in the subspace perpendicular to OA. This shows a successful dimensionality reduction with the discovery of a new dimension onto which exemplars are projected and classified.

Sanger's (1989) learning rule directly implements PCA in a one-layer feedforward network. This architecture extracts exactly the m first components of the input set and represents the mapping on the m output units. Sanger's learning rule is

$$\Delta w_{ij} = \eta o_i \left(x_j - \sum_{k=1}^{i} o_k w_{kj} \right)$$

Projection pursuit and minimal code description length. The featural interpretation of principal components is often difficult because orthogonal directions of highest variance have no connections in principle to the best projections for categorization (though see Linsker, 1988). For example, the bottom panel of Figure 2 shows that the direction of maximal variance is not necessarily an interesting projection for categorization. The two clouds can be separated on the OA axis, but not on the OB axis. Maximal variance, however, is in the OB direction.

In general, it is difficult to know what a good projection (or feature) is. Recent approaches have incorporated objective functions of "feature goodness" to constrain the search. For example, Intrator and Gold (1993) (see also Information Theory and Visual Plasticity) discuss a technique in which the algorithm searches directions of multimodal projections of the categories (to illustrate, in the bottom panel of Figure 2, the projection of the clouds is unimodal on OB and multimodal on OA). Another approach is taken by Zemel and his collaborators. Their learning algorithm searches a coding scheme that maximizes the uniqueness and the distinctiveness of the individual features by minimizing their mutual information, or the way features predict each other (see Minimum Description Length Analysis).

Discussion

The psychology of categorization is complex: Some exemplars are more typical of a category than others are (e.g., a robin versus a penguin as an exemplar of *bird*), and categories are hierarchically nested (e.g., a robin is a bird, which is an animal). One level of categorization (called the middle, or basic, level of a hierarchy, *bird* in the previous example) is often accessed faster than more specific or more general levels of categorization (Murphy and Smith, 1982). These effects illustrate some of the important aspects of the agenda of concept modeling.

Another important aspect of the agenda concerns the nature of the features of categorization. Schyns and Murphy (1994) argued that the features defining the concept learning space are not fixed, but continuously adjustable to task demands. A flexible, context-sensitive (as opposed to a fixed, context-independent) feature space is a desirable property of a concept learning system. Experience with the environment could result in people learning finely tuned features—features that could enable the encoding of sophisticated hierarchical representations, often needed to represent various degrees of domain-specific conceptual expertise.

Connectionist learning procedures have already demonstrated their capabilities of extracting varities of simple low-level features from raw data. There is still a large gap between learning those features and the extraction of high-level features useful for object classification (features such as *hood, trunk, has-legs, wheel*, and so forth). The integration of different sources of constraints (bottom-up and top-down) is probably the way to achieve complex feature extraction.

Of course, there is also the learning of complex concepts, such as "living beings," "healthy versus nonhealthy behaviors," "politically correct versus incorrect attitudes," and so forth. These probably require a substantial expansion of the positional framework of a vector space to a framework in which relationships between conceptual entities are better represented. Whether the connectionist framework can be extended to provide powerful representations of relationships between concepts is an open empirical question (but see Analogy-Based Reasoning).

Road Map: Connectionist Psychology
Background: Perceptrons, Adalines, and Backpropagation
Related Reading: Pattern Recognition

References

Gluck, M. A., and Bower, G. H., 1988, From conditioning to category learning: An adaptive network model, *J. Exp. Psychol. Gen.*, 117: 227–247.

Gorman, R. P., and Sejnowski, T. J., 1988, Analysis of hidden units in a layered network trained to classify sonar targets, *Neural Netw.*, 1:75–89.

Hertz, J., Krogh, A., and Palmer, R. G., 1991, *Introduction to the Theory of Neural Computation*, Redwood City, CA: Addison-Wesley.

Intrator, N., and Gold, J., 1993, Three dimensional object recognition using an unsupervised BCM network: The usefulness of distinguishing features, *Neural Computat.*, 5:61–74.

Kruschke, J. K., 1992, ALCOVE: An exemplar-based connectionist model of category learning, *Psychol. Rev.*, 99:22–44.

Linsker, R., 1988, Self-organization in a perceptual network, *Computer*, March:105–117.

Minsky, M. L., and Papert, S. A., 1969, *Perceptrons*, Cambridge, MA: MIT Press.

Murphy, G. L., and Smith, E. E., 1982, Basic-level superiority in picture classification, *J. Verb. Learn. Verb. Behav.*, 77:353–363.

Nosofsky, R. M., 1984, Choice, similarity, and the context theory of classification, *J. Exp. Psychol. Learn. Mem. Cogn.*, 10:104–114.

Rosenblatt, F., 1962, *Principles of Neurodynamics*, New York: Spartan.

Rumelhart, D. E., Hinton, G., and Williams, R. J., 1986, Learning internal representations by back-propagating errors, *Nature*, 323: 533–536.

Sanger, T. D., 1989, Optimal unsupervised learning in a single-layer linear feedforward network, *Neural Netw.*, 2:459–473.

Schyns, P. G., 1991, A neural network model of conceptual development, *Cognit. Sci.*, 15:461–508.

Schyns, P. G., and Murphy, G. L., 1994, The ontogeny of part representation in object concepts, *Psychol. Learn. Motiv.*, 31:301–349.

Conditioning

Nestor A. Schmajuk

Introduction

During conditioning, animals modify their behavior as a consequence of their experience with the contingencies between environmental events. This article delineates different formal theories and neural network models that have been proposed to describe classical and operant conditioning.

Classical Conditioning

During classical conditioning, animals change their behavior as a result of the contingencies between conditioned stimuli (CSs) and unconditioned stimuli (USs). Contingencies may vary from very simple to extremely complex. For example, in simple acquisition, animals are exposed to CS_1, followed by the US. Although, at the beginning of training, animals generate only an unconditioned response (UR) when the US is presented, with an increasing number of CS_1-US pairings, CS_1 presentations elicit a conditioned response (CR). In general, the CR is considerably analogous to the UR (but see the later discussion of opponent processes), CR onset precedes the US onset, and the peak CR amplitude tends to be located around the time of occurrence of the US. When acquisition is followed by presentations of CS_1 alone, the CR extinguishes.

Associations, Predictions, and Connections

Modern learning theories assume that the association, $V_{i,k}$, between events CS_i and CS_k represents the *prediction* that CS_i will be followed by CS_k. Neural network, or connectionist, theories frequently assume that the association between CS_i and CS_k is represented by the efficacy of the synapses, $V_{i,k}$, that connect a presynaptic neural population excited by CS_i with a postsynaptic neural population excited by CS_k. (Event k might be another CS or the US.) When CS_k is the US, this second population controls the generation of the CR. At the beginning of training, synaptic strength $V_{i,\text{US}}$ is small, and therefore CS_i is incapable of exciting the second neural population and generating a CR. As training progresses, synaptic strengths gradually increase and CS_i comes to generate a CR.

Although some models of conditioning describe changes in $V_{i,k}$ on a trial-to-trial basis, real-time networks describe the unbroken, continual temporal dynamics of $V_{i,k}$. In general, real-time neural networks assume that CS_i gives rise to a trace, $\tau_i(t)$, in the central nervous system that increases over time to a maximum and then gradually decays to zero. The increment in $V_{i,k}$ is a function of the intensity of the CS_i trace at the time the US is presented. Figure 1 illustrates the time courses of the CS_i, τ_i, and the US in a delay conditioning experiment. Obviously, only real-time networks can describe the effects of CS duration, US duration, interstimulus interval (ISI), or intertrial interval (ITI).

Following Hebb (1949; cf. HEBBIAN SYNAPTIC PLASTICITY), changes in synaptic strength $V_{i,k}$ might be described by $\Delta V_{i,k} = f(CS_i)f(CS_k)$, where $f(CS_i)$ represents the presynaptic activity and $f(CS_k)$ the postsynaptic activity. Different $f(CS_i)$ and $f(CS_k)$ functions have been proposed. Learning rules for $V_{i,k}$ assume variations in the effectiveness of either CS_i, $f(CS_i)$, or the US, $f(CS_k)$, or both. The following sections describe how different types of models deal with the many experimental results. In general, new models improve on older models by describing additional and more complicated paradigms. As shown later, models that assume variations in both the effectiveness of the CS_i and the US are able to describe a wide variety of complex conditioning paradigms.

Variations in the Effectiveness of the CS: Attentional Models

Attentional theories assume that the formation of CS_i-US associations depend on the magnitude of an "internal representation" of CS_i, $f(CS_i)$. In neural network terms, attention may be interpreted as the modulation of the CS representation that activates the presynaptic neuronal population involved in associative learning. When focused on a particular CS, selective attention enhances the internal representation of that specific CS.

According to Grossberg's (1975) neural attentional theory, the pairing of CS_i with a US causes both an association of $f(CS_i)$ with the US and an association of the US with $f(CS_i)$. Sensory representations $f(CS_i)$ compete among themselves for a limited-capacity short-term memory activation that is reflected in CS_i-US associations. In addition to latent inhibition, blocking, and overshadowing, the Grossberg model correctly predicts that latent inhibition might be obtained after training with a weak US.

Variations in the Effectiveness of the US: Simple and Generalized Delta Rules

A popular rule, proposed independently in psychological (Rescorla and Wagner, 1972) and neural network domains, has

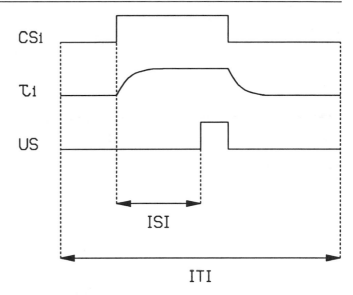

Figure 1. Time courses of the CS_i, τ_i, and the US in a delay conditioning experiment. ISI: interstimulus interval. ITI: intertrial interval.

been termed the *delta rule* (see PERCEPTRONS, ADALINES, AND BACKPROPAGATION). The delta rule describes changes in the synaptic connections between the two neural populations by way of minimizing the squared value of the difference between the output of the population controlling the CR generation and the US. According to the "simple" delta rule, CS_i-US associations are changed until $f(\text{US}) = \text{US} - \Sigma_j V_{j,\text{US}}CS_j$ is zero. In neural network terms, $f(\text{US})$ can be construed as the modulation of the US signal that activates the postsynaptic neural population involved in associative learning. Rescorla and Wagner (1972) showed that the model describes acquisition, extinction, conditioned inhibition, blocking, and overshadowing.

Sutton and Barto (1981) presented a temporally refined version of the Rescorla-Wagner model. In the model, the effectiveness of CS_i is given by the temporal trace $\tau_i = f(CS_i(t)) = Af(CS_i(t)) + BCS_i(t)$, which does not change over trials. The effectiveness of the US changes over trials according to $f(\text{US}(t)) = (y(t) - y'(t))$, where the output of the model is

$$y(t) = f\left[\sum_j V_{j,\text{US}}f(CS_j) + f(\text{US}) \right]$$

where $f(\text{US})$ is the temporal trace of the US and $y'(t) = Cy'(t) - (1 - C)y(t)$. Computer simulations show that the model correctly describes acquisition, extinction, conditioned inhibition, blocking, overshadowing, primacy effects, and second-order conditioning. More recently, Barto and Sutton (1990) proposed a new rendering of the Sutton and Barto (1981) model, designated the temporal difference model, in which $f(\text{US}) = (\text{US} + \Gamma y'(t + 1) - y'(t))$. The temporal difference model correctly describes ISI effects, serial-compound conditioning, no extinction of conditioned inhibition, second-order conditioning, and primacy effects.

Klopf (1988) introduced an interesting extension of the Sutton-Barto model, termed a drive-reinforcement model of single neuron function. According to this neuronal model, changes in pre- and postsynaptic activity levels are correlated to determine changes in synaptic efficacy. Changes in presynaptic signals, $p(CS_i) = f(\Delta CS_i)$, are correlated with changes in postsynaptic signals, $p(\text{US}) = \Delta y$. Changes in the efficacy of a synapse are also proportional to the current efficacy of the synapse. The

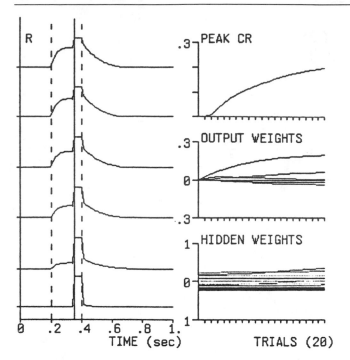

Figure 2. Acquisition of classical conditioning. *Left panels:* Real-time simulated conditioned and unconditioned response on trials 1, 4, 8, 12, 16, and 20. Vertical dashed lines indicate CS onset and offset. Vertical solid line indicates US onset. Trial 1 is represented at the bottom of the panel. *Right panels:* Peak CR is shown as a function of trials. Output weights are average VSs and VNs as a function of trials. Hidden weights are average VHs as a function of trials.

model can describe delay and trace conditioning, conditioned and unconditioned stimulus duration and amplitude effects, partial reinforcement effects, interstimulus interval effects, second-order conditioning, conditioned inhibition, extinction, reacquisition effects, backward conditioning, blocking, overshadowing, compound conditioning, and discriminative stimulus effects.

Kehoe (1988) presented a network that incorporates a hidden-unit layer, trained according to a delta rule. In addition to the paradigms described by the Rescorla-Wagner model, the network describes stimulus configuration, learning to learn, savings effects, and positive and negative patterning.

Schmajuk and DiCarlo (1992) introduced a model that, by employing a "generalized– delta rule to train a layer of hidden units that "configure" simple CSs, is able to solve negative and positive patterning. Interestingly, the biologically plausible, real-time rendition of backpropagation offered by Schmajuk and DiCarlo (1992) differs from the original version in that the error signal used to train hidden units, instead of including the derivative of the activation function of the hidden units, simply contains their activation function. The network provides correct descriptions of acquisition of delay and trace conditioning, extinction, acquisition-extinction series, blocking, overshadowing, discrimination acquisition and reversal, compound conditioning, feature-positive discrimination, conditioned inhibition, negative patterning, positive patterning, generalization, and occasion setting. Figure 2 shows real-time simulations on trials 1, 4, 8, 12, 16, and 20 in a delay conditioning paradigm with a 200-ms CS, 50-ms US, and 150-ms ISI. As CR amplitude increases over trials, output weights VS_i and VN_j, and hidden weights VH_{ij} may increase or decrease.

Variations in the Effectiveness of Both the CS and the US

In order to account for a wider range of classical conditioning paradigms, some theories have combined variations in the effectiveness of both the CS and the US. For example, Wagner (1978) suggested that CS_i-US associations are determined by (a) $f(US) = (US - \Sigma_j V_{j,US}CS_j)$ as in the Rescorla-Wagner model, and (b) $f(CS_i) = (CS_i - V_{i,CX}CX)$, where CX represents the context and $V_{i,CX}$ the strength of the CX-CS_i association.

Schmajuk and Gray (1993) introduced a theory that assumes that $f(CS_i)$ is modulated by the association of the internal representation of CS_i with the total environmental novelty, z_i. Total environmental novelty is given by $\Sigma_j|\bar{\lambda}_j - \bar{B}_j|$, that is, the sum of the absolute values of the differences beween the average predicted and the average observed event j. Schmajuk and Gray (1993) showed that the model correctly describes latent inhibition of excitatory and inhibitory conditioning, stimulus specificity, latent inhibition increases with the number of CS preexposures, increasing ITI durations, increasing CS intensities, and increasing CS durations. The model also accounts for other features of conditioning: the unexpected presence or absence of a stimulus previous to conditioning reduces latent inhibition; latent inhibition shows overshadowing and blocking; context presentations following CS preexposure do not extinguish latent inhibition; context changes following CS preexposure decrease latent inhibition; preexposure to the apparatus prior to CS preexposure facilitates latent inhibition; latent inhibition is obtained after training with a weak US; context changes that produce little dishabituation of the orienting response can produce severe attenuation of latent inhibition; and CS preexposure is able to yield perceptual learning.

When combined with the Schmajuk and DiCarlo (1992) model (Figure 3), the Schmajuk and Gray (1993) approach can describe a wide variety of classical conditioning data. Figure 3 shows that simple stimuli CS_i become *configured* with other CSs in a hidden unit that represents configural stimulus CN_j. As before, it is assumed that $f(CS_i)$ and $f(CN_i)$ are modulated by the association of the internal representations of CS_i and CN_j with the total environmental novelty. Both $f(CS_i)$ and $f(CN_i)$ become associated with the US.

Opponent Processes

In agreement with Pavlov's stimulus subtitution theory, most theories presented in this article suggest that repeated CS-US pairing leads the CS to substitute for the US in generating the UR. However, although in many cases the CR is considerably analogous to the UR, there are some cases in which the CR might be the opposite of the UR.

In order to predict whether a CR will be analogous or opposite to the UR, Wagner (1981) proposed a theory called a *sometimes opponent process* (SOP). The theory suggests that presentation of the US drives the US node into primary activity A1, which rapidly grows and decays back to zero, followed by secondary activity A2, which grows and decays at a slower rate. A CS paired with the US can trigger only activity A2 in the US node. For instance, rats receiving a brief foot-shock US develop a UR that consists of hyperactivity (A1) followed by freezing (A2). A CS paired with the foot-shock US elicits a CR that consists of freezing. In contrast, rabbits receiving a corneal-air-puff US develop a UR that consists of an eyeblink (A1 and A2). A CS paired with the air-puff US elicits an eyeblink CR. According to the SOP model, excitatory CS-US associations increase whenever the CS and the US node are in the A1 state. Conversely, inhibitory CS-US associations increase whenever the CS node is in state A1 and the US node is in state A2.

CONFIGURAL STIMULI

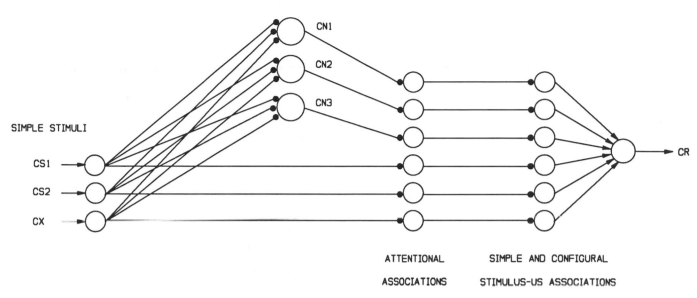

Figure 3. Diagram of a network that incorporates both a layer of hidden units capable of describing stimulus configuration and a layer of attentional units capable of describing latent inhibition. CS: conditioned stimulus; CN: configural stimulus; CR: conditioned response. Arrows represent fixed synapses. Solid circles represent variable synapses.

In a conditioned inhibition paradigm, CS_1-US presentations will cause the US node to assume state A2 during CS_1-CS_2 presentations. Because during CS_1-CS_2 presentations node CS_2 is in state A1, and the US node in state A2, inhibitory CS_2-US associations are formed. In sum, SOP is able to predict when the CR will resemble and when it will differ from the UR, and when excitatory or inhibitory conditioning is obtained.

Grossberg and Schmajuk (1989) presented a real-time neural network model, the READ (REcurrent Associative Dipole) circuit, which combines a mechanism of associative learning with an opponent-processing circuit. CSs form excitatory associations with direct activations of a dipole, and inhibitory associations with the antagonistic rebounds of a previously habituated dipole. Conditioning can be actively extinguished, and associative saturation prevented, by a process called opponent extinction, even if no passive memory decay occurs. Opponent extinction exploits a functional dissociation between read-in and read-out of associative memory, which may be achieved by locating the associative mechanism at specific locations in the neuronal dendrites. The READ architecture is able to explain conditioning and extinction of a conditioned excitor, second-order conditioning, conditioning and nonextinction of a conditioned inhibitor, and properties of conditioned inhibition as a "slave" process.

Multiple Representations of the CS: Timing

In classical (and operant) conditioning, the peak CR amplitude tends to be located around the time of the occurrence of the US, a fact that suggests that animals learn about the temporal relationship between the CS and the US.

Commonly, timing theories assume that presentation of CS_i generate multiple time-dependent $f(CS_i)$'s. Desmond and Moore (1988) proposed a neural network model that simulates the temporal characteristics of a CR in classical conditioning (cf.

CEREBELLUM AND CONDITIONING). The model assumes that the CS onset activates an onset process in one set of neural elements, and the CS offset activates an offset process in another set of elements. Each process consists of sequential and overlapping activation of neural elements, X. Once activated, each element X remains activated for a certain time. Each element X makes contact, through a modifiable synapse, with two neural elements, V and E. Element V receives input from the US and from element E. Element E receives input only from the US. The output of the V element describes the CR. The model is able to describe ISI curves with single and multiple USs. However, the model does not describe a Weber's law for temporal generalization.

Grossberg and Schmajuk (1989) proposed a neural network, called the *spectral timing model*, capable of learning the temporal relationships between two events. The model consists of three layers of neural elements. A step function, activated by the CS presentation, excites the first layer that contains many elements, each one having a different reaction time. The output of each element in the first layer is a sigmoid function that activates a second layer of habituating transmitter gates. In turn, the output of each transmitter gate activates an $f(CS_i)$ element. Those $f(CS_i)$ elements active at the time of the US presentation become associated with the US in proportion to their activity. All $f(CS_i)$ elements activate their corresponding $V_{i,US}$ weights and are added in order to generate the CR. During testing, the CR shows a peak at the time when the $f(CS_i)$ elements that have been active simultaneously with the US are active again. The model is able to describe ISI curves with single and multiple USs, as well as a Weber's law for temporal generalization. Grossberg and Schmajuk (1989) showed that the model can explain the effects of increasing CS and US intensity, an inverted U in learning as a function of ISI, multiple timing peaks, the effect of increasing US duration, and the effect of drugs on timed motor behavior.

Church and Broadbent (1991) presented a connectionist version of Church's *scalar timing theory*. The model consists of the following components: (1) a pacemaker that emits pulses; (2) a switch that is opened at the onset of the event to be timed, and closed at its offset; (3) a counter that accumulates pulses; (4) a "reference memory" that accumulates pulses of reinforced times, and a "working memory" that stores the total number of pulses accumulated in a particular trial; and (5) a comparator that compares the values stored in both memories. Values stored in working memory are compared to values stored in reference memory. If they are similar, a response is produced. Notice that the number of stored pulses increases with the measured time. The model is able to describe timing with single but not with multiple USs, and it describes a Weber's law for temporal generalization.

Sensory Preconditioning

By including CS-CS associations, some models of classical conditioning are able to describe sensory preconditioning. For instance, Gelperin, Hopfield, and Tank (1985) proposed a model that simulates many of the learning abilities of the snail *Limax*. The network incorporates a highly interconnected network that can display different stable states of activity. These stable states, which can be regarded as memories stored in the network, are determined by the strength of the synaptic connections between the neurons. The network is able to simulate first- and second-order conditioning, extinction, and blocking. Similarly, Schmajuk and Gray's (1993) model is able to describe sensory preconditioning by incorporating an autoassociative, recurrent network.

Operant Conditioning

During operant (or instrumental) conditioning, animals change their behavior as a result of a triple contingency between their responses (R), discriminative stimuli (S_D), and the reinforcer (S_R). Animals are exposed to the reinforcer (S_R) in a relatively close temporal relationship with the discriminative stimuli (S_D) and a response (R_i). At the beginning of training animals generate only a few R_i when S_D is presented; with an increasing number of experiences with the S_D-R_i-S_R contingency, however, they start emitting R_i when S_D is presented. Four classes of S_D-R_i-S_R contingencies are possible: (1) positive reinforcement, in which R_i is followed by the presence of an appetitive stimulus; (2) punishment, in which R_i is followed by the presence of an aversive stimulus; (3) omission, in which R_i is followed by the absence of an appetitive stimulus; and (4) negative reinforcement (escape and avoidance), in which R_i is followed by the absence of an aversive stimulus. As in REINFORCEMENT LEARNING (q.v.), during operant conditioning animals learn by trial and error from feedback that evaluates their behavior but does not indicate the correct behavior.

Operant conditioning procedures can be divided into discrete trial procedures, in which the operant response occurs only once on a given trial (runways, mazes, shuttle box), and free operant procedures, in which the operant response can occur repeatedly (bar pressing, key pecking). Operant spatial tasks such as place learning and maze learning are described in the article on COGNITIVE MAPS.

Positive Reinforcement

Staddon and Zhang (1991) proposed a nonassociative model (no S_D-R_i associations are formed) that describes the context-free properties of operant conditioning, selection, delay of reward, contingency, and unsignaled avoidance. The model assumes that different responses are represented by a filtered noise, V, that is augmented or decremented by pleasant or unpleasant events in an amount proportional to its value. The response of the highest V value is the one that actually occurs.

Maki and Abunawas (1991) presented a neural network that, employing a generalized delta rule, describes matching-to-sample tasks (MTS). In the MTS paradigm, animals are first presented with the sample stimulus and later presented with the comparison stimulus. Animals are rewarded for choosing the comparison stimulus that matches the preceding sample. The network correctly describes code sampling stimuli in terms of anticipated events (prospective coding) and improvement with delay training (rehearsal). Its accuracy decreases when compound samples are presented (shared attention).

Negative Reinforcement

Discrete-trial escape and avoidance paradigms are usually run in a two-way shuttle box. The shuttle box is a chamber with two compartments separated by a barrier with a door. Each compartment has a metal-grid floor that can deliver a shock US. Lights above the chambers provide warning signals (WS) for the US. The experiment starts with both compartments being illuminated. At time zero, the light above the compartment where the animal is located turns off (WS) and the door separating both compartments opens. If the animal has not crossed to the opposite side after a given time (that may vary between 2 and 40 s), the shock US is applied. If the animal has crossed to the opposite side before that time, it avoids the US, and the separating door closes behind it. After a constant or an average intertrial interval that varies from 15 s to 4 min, the whole sequence restarts.

Grossberg (1972) proposed a neural theory of punishment and avoidance that combines classical and operant conditioning (response-selection) mechanisms and describes avoidance in terms of an architecture that provides a rebound mechanism from fear to relief, similar to Grossberg and Schmajuk's antagonistic-rebound mechanism mentioned earlier. According to Grossberg, if an S is paired with the US, it becomes associated with fear and its termination produces a rebound that provides a relief signal which, in turn, controls the association of S with the avoidance response. The model describes phenomena such as the lesser effect of reducing J units of shock to $J/2$ units than of reducing $J/2$ to 0 units, persistent nonspecific fear that biases the interpretation of specific cues, different effects of gradual and abrupt shock on response suppression, reduction of pain in the presence of a loud noise, and the influence of drugs on conditional emotional and avoidance responses.

Schmajuk and Urry (1992) presented a real-time, two-process theory of avoidance that combines elements of classical and operant conditioning. The network incorporates two processes, classical and operant conditioning. Whereas the classical conditioning process controls US-US, S-US, and R-US associations, the operant conditioning process controls US-R_{escape} and S-$R_{avoidance}$ associations. Whereas classical conditioning is regulated by a delta rule, $f(US) = US - \Sigma_j V_{j,US} X_j$, where X represents S, R, or the US; operant conditioning is regulated by a novel algorithm that mirrors the classical conditioning algorithm, $f(US) = -(US - \Sigma_j V_{j,US} X_j)$. Schmajuk and Urry (1992) applied the network to the description of escape and avoidance behavior with a shuttle box, a running wheel, a leg flexion, or a lever-pressing paradigm. Schmajuk and Urry (1992) demonstrated through computer simulations that the model describes many of the features that characterize avoidance behavior: fear of the US decreases as the animal masters the response; techniques that decrease fear decrease ongoing avoidance responses; techniques that increase fear increase ongoing avoidance responses; the amount of time needed to generate the avoidance response decreases with increasing number of trials; in some situations avoidance has negligible extinction; trace avoidance is less resistant to extinction than delay avoidance; acquisition of trace avoidance is slower and extinction faster with increasing WS-US intervals; extinction of avoidance is obtainable by blocking the animal's capability to elicit the avoidance response without delivering the US (by shocking the animal when it emits the avoidance response, or by shocking the animal whether or not it emits the avoidance response); when the avoidance response terminates the WS but does not prevent the US presentation, animals show a slower acquisition and lower asymptotic levels of avoidance; when the avoidance response prevents the US presentation but does not terminate the WS, animals learn less than in the normal avoidance situation; when the avoidance response has no effect either on the WS duration or the delivery of the US (classical conditioning), animals do not acquire consistent avoidance, and the avoidance response may be different from that required to escape the US.

Discussion

Neural networks might be regarded as abstract, albeit biologically motivated, computational architectures that describe different types of conditioning. At this level, new models improve on older models by describing additional and more complicated paradigms, in general through the introduction of more elaborate mechanisms and internal representations of environmental events.

When regarded as models of specific neural circuits, network models of conditioning provide theories that extend to anatomical and physiological levels. Neural networks are usually evaluated at different levels. At the behavioral level, simulated behavioral results are compared with experimental data. At the computational level, simulated activity of the neural elements of the model are compared with the activity of single neurons or of neural populations. At the anatomical level, interconnections among neural elements in the model are compared with neuroanatomical data. At the brain manipulation level, model performance is correlated with animal performance after lesioning, modifying neural connectivity, or blocking changes in neural connectivity in the network. Simultaneous evaluation at behavioral, computational, anatomical, and manipulation levels requires the selection of a specific animal preparation. For example, Schmajuk and DiCarlo (1992) modeled conditioning of the nictitating membrane in rabbit, assuming that CS-US associations were stored in the cerebellum (see CEREBELLUM AND CONDITIONING) and that this storage was modulated by the hippocampal formation.

Road Maps: Connectionist Psychology; Learning in Biological Systems

References

Barto, A. G., and Sutton, R. S., 1990, Time-derivative models of Pavlovian conditioning, in *Learning and Computational Neuroscience: Foundations of Adaptive Networks* (M. Gabriel and J. Moore, Eds.), Cambridge, MA: MIT Press, pp. 497–537.

Church, R. M., and Broadbent, H. A., 1991, A connectionist model of timing, in *Neural Network Models of Conditioning and Action* (M. Commons, S. Grossberg, and J. E. R. Staddon, Eds.), Hillsdale, NJ: Erlbaum, pp. 225–240.

Desmond, J. E., and Moore, J. W., 1988, Adaptive timing in neural networks: The conditioned response, *Biol. Cybern.*, 58:405–415.

Gelperin, A., Hopfield, J. J., and Tank, D. W., 1985, The logic of Limax learning, in *Model Neural Networks and Behavior* (A. I. Selverston, Ed.), New York: Plenum, pp. 237–261.

Grossberg, S., 1972, A neural theory of punishment and avoidance, II: Quantitative theory, *Math. Biosci.*, 15:253–285.

Grossberg, S., 1975, A neural model of attention, reinforcement, and discrimination learning, *Int. Rev. Neurobiol.*, 18:263–327.

Hebb, D. O., 1949, *The Organization of Behavior*, New York: Wiley.

Kehoe, E. J., 1988, A layered network model of associative learning: Learning to learn and configuration, *Psychol. Rev.*, 95:411–433.

Klopf, A. H., 1988, A neuronal model of classical conditioning, *Psychobiology*, 16:85–125.

Maki, W. S., and Abunawas, A. M., 1991, A connectionist approach to conditional discriminations: Learning, short-term memory, and attention, in *Neural Network Models of Conditioning and Action* (M. Commons, S. Grossberg, and J. E. R. Staddon, Eds.), Hillsdale, NJ: Erlbaum, pp. 241–278.

Rescorla, R. A., and Wagner, A. R., 1972, A theory of Pavlovian conditioning: Variation in the effectiveness of reinforcement and nonreinforcement, in *Classical Conditioning II: Theory and Research* (A. H. Black and W. F. Prokasy Eds.), New York: Appleton-Century-Crofts.

Schmajuk, N. A., and DiCarlo, J. J., 1992, Stimulus configuration, classical conditioning, and the hippocampus, *Psychol. Rev.*, 99:268–305.

Schmajuk, N. A., and Gray, J. A., 1993, Latent inhibition: A neural network approach, paper presented at Thirty-fourth Annual Meeting of the Psychonomic Society, Washington, DC.

Schmajuk, N. A., and Urry, D. W., in press, The frightening complexity of avoidance: An adaptive neural network, in *Models of Action* (J. E. R. Staddon and C. Wynne, Eds.), Hillsdale, NJ: Erlbaum.

Staddon, J. E. R., and Zhang, Y., 1991, On the assignment-of-credit problem in operant learning, in *Neural Network Models of Conditioning and Action* (M. Commons, S. Grossberg, and J. E. R. Staddon, Eds.), Hillsdale, NJ: Erlbaum, pp. 279–293.

Sutton, R. S., and Barto, A. G., 1981, Toward a modern theory of adaptive networks: Expectation and prediction, *Psychol. Rev.*, 88:135–170.

Wagner, A. R., 1978, Expectancies and the priming of STM, in *Cognitive Processes in Animal Behavior* (S. H. Hulse, H. Fowler, and W. K. Honig, Eds.), Hillsdale, NJ: Erlbaum, pp. 177–209.

Wagner, A. R., 1981, SOP: A model of automatic memory processing in animal behavior, in *Information Processing in Animals: Memory Mechanisms* (N. E. Spear and R. R. Miller, Eds.), Hillsdale, NJ: Erlbaum, pp. 5–47.

Connectionist and Symbolic Representations

David S. Touretzky

Introduction

Knowledge representation is the subfield of artificial intelligence in which epistemology meets algorithm design. Many early knowledge representation systems were built on first-order predicate calculus and performed only limited deductive reasoning. Others used more ad hoc inference methods, and sometimes radically different notations, but still operated on discrete symbol structures. In some sense, all such systems are logics.

Research on neural network or *connectionist-style* knowledge representation attempts to address three problems yet unsolved by the logical approach. The first is the problem of flexible reasoning. Human cognition is often more like constraint satisfaction than theorem proving. What mechanisms underlie this style of inference? The second problem is neural plausibility: how does symbolic behavior emerge from the parallel activity of a large number of primitive, slow, computing elements? The third problem is self-organization: how do reasoners construct appropriate representations based on their experience? Some progress has been made on each of these, but none is close to being solved. There are also some problems in which the symbolic approach currently has the upper hand. One that is considered here is the representation of structured relationships among concepts.

Flexible Reasoning

In a classical symbolic representation, symbols are arbitrary tokens composed into structures that encode relationships among objects, such as *loves (John, Mary)*. Formal rules for manipulating these symbol structures determine the kinds of inference that take place. Newell's paper on physical symbol systems (Newell, 1980) is a definitive statement of the symbolic artificial intelligence (AI) framework.

A problem with systems that follow this approach is *brittleness*: if the input does not match perfectly what the inference rules expect, the system cannot function. One may try to add rules to provide for partial matches, but this complicates the rule system, and it is difficult to anticipate and formalize all the ways the input might differ from what is expected.

Some types of connectionist networks handle partial matching automatically, as a side effect of their normal functioning. For example, ASSOCIATIVE NETWORKS (q.v.) can match a stored pattern given a small portion as input. Errors or noise in the input are cleaned up automatically during the retrieval process. Retrieval normally maps an input to the stored pattern closest to it in Hamming space. (The Hamming distance between two binary patterns is the number of positions where they differ.) Such models mimic certain aspects of human memory in that any portion of the pattern may serve as a retrieval cue, and retrieval is error tolerant. But partial matching based on Hamming distance is overly simplistic. A more realistic model would have to weight different parts of the pattern differently, and would be invariant over transformations appropriate to the domain. For example, when recalling images, we would expect retrieval to be invariant over translations and small rotations of the cue.

Apart from matching, another component of reasoning in which brittleness is a problem is inference. For more robust reasoning, strict deductive rules can be replaced with some form of weighted evidence combination. Then, if there is incomplete or inconsistent input, the reasoner can still jump to a plausible conclusion rather than having to give up. The symbolic approach attempts to devise formal accounts of evidence combination and jumping to conclusions (e.g., Dempster-Shafer theory, and nonmonotonic and default logics; see Pearl, 1988), but important unsolved problems remain. For example, inference in standard nonmonotonic logic is uncomputable; restricted sublanguages must be found that are sufficiently expressive yet have tractable inference algorithms.

In the connectionist approach, some types of flexible inference emerge from the model's normal operation. *Interactive activation* models are a good example. Unlike the associative memory models mentioned earlier, they use a *local representation* in which concepts are identified with individual nodes, and excitatory or inhibitory links encode relations between them. Computation is by spreading activation and lateral inhibition.

"Strength of evidence" is reflected in a node's activation level, and evidence combination occurs by nodes summing their inputs. Competing hypotheses connected by inhibitory links form WINNER-TAKE-ALL MECHANISMS (q.v.) in which only the node with the greatest evidential support remains active when the network settles.

In the Rumelhart and McClelland letter perception model (see STRUCTURED CONNECTIONIST MODELS), there are three layers of structure corresponding to letter features (strokes), letter instances, and letter sequences (words). The first two layers are replicated fourfold to provide a parallel representation of four-letter words. Bidirectional excitatory connections between layers allow bottom-up perceptual processing to be guided by top-down expectations, which conversely are primed by bottom-up cues. It is not necessary to anticipate every possible noisy or incomplete cue; the population of stroke, letter, and word units collectively guides the model to a state representing a valid interpretation of the input.

As in the case of partial matching, the flexible inference provided by connectionist models suggests a more humanlike mode of information processing.

Flexibility and Formality

One advantage of SEMANTIC NETWORKS (q.v.) over formal logic is that they allow knowledge to be represented informally, by assigning arbitrary meanings to nodes and links. Localist connectionist models, in which each concept is represented by a distinct node, have much in common with early AI semantic networks that used spreading activation for inference. For example, an excitatory link between an *astronomer* node and a *stellar body* node in such a system would indicate that astronomers are associated with stellar bodies in some way, without saying precisely how. Let us suppose there are also excitatory links from *star* to *stellar body* and *famous performer*. Inhibitory links between the latter two nodes indicate that they are competing senses of the word "star" that should not be active at the same time. Spreading activation simultaneously from both *astronomer* and *star* will select the *stellar body* sense of the word, since it receives more total activation than *famous performer* receives from *star* alone. But in a sentence such as "The astronomer married the star," evidence from "married" would override this heuristic. Waltz and Pollack (1985) showed how lexical ambiguity and garden path effects can be resolved in localist connectionist networks using this informal style of representation.

A formal specification of the relationship between astronomers and stars would be difficult to construct and would involve a great deal of general knowledge, such as the concepts of scientific disciplines and their objects of study. There are efforts under way in the AI community to specify this kind of world knowledge, but there is as yet no consensus that such an approach will ultimately succeed.

Neural Plausibility

One of the supposed advantages of connectionist-style representations is their neural plausibility. One such representation, coarse coding of points in a continuous space (discussed in the next section), is known to be used in the brain. Examples include place cells in hippocampus that represent an animal's location in the environment, and populations of cells in motor cortex that encode direction (see REACHING: CODING IN MOTOR CORTEX). But the representation of purely symbolic information in the brain remains a mystery; we cannot be certain that coarse coding plays a significant role. Other approaches to representation also have some claims to neural plausibility; they are considered at the end of this section.

Distributed Representations and Coarse Coding

Coarse coding is a representation strategy that relies on populations of broadly tuned units, many of which are simultaneously active, to collectively encode a value with high precision. Each unit covers a range of values in some continuum, and units are active when the input to be encoded falls within their *receptive field*. Given enough units with overlapping receptive fields, an input value can be recovered from the population's activity pattern. For binary-valued units, the value is obtained by averaging the receptive field centers of all active units. For continuous-valued units whose activity is a function of distance of the input from the receptive field center, a weighted average is used.

This representation is robust against noise or damage, an important property of real neural systems. Coarse coding also makes it possible to accurately represent points in high-dimensional spaces, using far fewer units than would be required in a localist scheme. Multiple points can even be represented simultaneously, provided there are not too many and they do not cluster too tightly.

Touretzky and Hinton (1988) showed how coarse coding could be applied to symbolic representations. Their DCPS (Distributed Connectionist Production System) model manipulated triples of symbols that constituted the working memory of a production rule reasoner. The contents of working memory were represented as a distributed pattern of activity over a collection of 2000 units, and rule matching and variable binding were done by simulated annealing. Touretzky (1990) subsequently showed in the BoltzCONS model that the same techniques could be used to represent and manipulate tree structures as coarse-coded linked lists.

Another sort of distributed representation is the feature vector, which could be processed by either symbolic or connectionist methods. Hinton, McClelland, and Rumelhart (1986) proposed the term *microfeature* to denote learned low-level features in a connectionist net with no high-level conceptual correlates. For example, a microfeature might encode some complex statistical regularity of the domain, whereas conventional "features" would correspond to semantic categories such as *soft* or *animate*. Smolensky (1988) uses the term *sub-symbolic* to refer to microfeature representations of this type, but today's connectionist models intended to operate on microfeatures often use hand-constructed feature vectors, and these contain ordinary symbolic features.

One problem with using feature vectors to represent symbol structures is that it is difficult to simultaneously represent multiple similar concepts within the same population of units. Suppose we represent *elephant* as an activity pattern. A particular elephant would then be encoded as some variant of that pattern. But how could we represent two distinct elephants simultaneously? In a localist scheme, one would simply create a separate node for each instance and link them both to the *elephant* node. In distributed schemes, one needs a way to represent composite structures, so that the set of two elephants has a representation that is different from the individuals themselves. Composite representations are discussed later.

Localist Neworks

Localist networks, although not coarse coded, also express certain aspects of what mental representations are thought to be like. Their nodes have graded activity levels and can send acti-

vation to other nodes, which can be seen as analogs of attention (or confidence) and priming. But individual nodes are not the right level at which to make a neural analogy. The way localist representations could be realized in neural tissue is an open problem. Current thinking is that such nodes might be implemented as *cell assemblies* composed of a few dozen to a few thousand neurons, bound together in some way such that the overlap between distinct assemblies is not too great. Dynamic recruitment schemes that can form cell assemblies on demand have been proposed, but there remains the problem of how to achieve powerful processing in such a dynamically constructed network.

Self-Organization

Perhaps the most intriguing property of connectionist networks is their ability to construct new representations. Often this is done using a gradient descent learning algorithm. Hinton (1990) described a five-layer backpropagation network trained on a near exhaustive set of kinship relations: triples such as (*Colin father-of James*) generated from two isomorphic three-generation family trees. The network's hidden layer developed a binary feature representation that included features for the family, gender, and generation level of each individual. Similarly, the FGREP architecture of Miikkulainen and Dyer (1991) constructs feature vector representations for lexical entries (nouns and verbs) using backpropagation learning, based on the context in which each word occurs in a set of training sentences.

Still more intriguing is the ability of recurrent networks to learn grammars, or equivalently, finite state automata. Unlike symbolic learning algorithms that construct discrete nodes for possible automaton states, and add or subtract nodes and links as they learn, the connectionist networks go through intermediate states that have no concise symbolic description. Pollack's (1990) Recursive Auto-Associative Memory (RAAM) for representing trees, and Elman's (1991) experiments with simple recurrent networks learning context-free sentences, indicate that such networks can construct representations having a distributed recursive quality, albeit with limited depth. Figure 1 shows the structure of a RAAM network. It is a 2N-N-2N encoder, trained by backpropagation. Two N-element input tokens A and B are compressed into an N-element hidden representation, and then expanded back into their original representation. The hidden unit pattern $H_{A,B}$ can be thought of as an encoding of the nonterminal node whose children are the terminal symbols A and B. This hidden pattern can then be used as an input token and combined with others to build more complex structures, such as $H_{H_{(A,B)}, H_{(C,D)}}$. To go in the opposite direction, from a nonterminal node to its two children, the

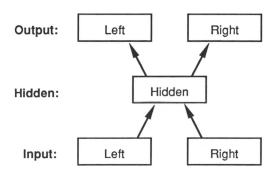

Figure 1. Pollack's RAAM (Recursive Auto-Associative Memory) architecture.

pattern for the nonterminal is latched into the hidden layer; forward propagation from there to the output layer produces the patterns for the left and right children.

Examination of the hidden unit states of the RAAM shows interesting feature structures, but, as in most connectionist work, it is difficult except in the simplest cases to give a definitive symbolic interpretation to individual hidden unit activity levels.

Representing Structured Information

A major problem in knowledge representation is the handling of structured information. Logic-based approaches are particularly strong in this regard. In network models where a concept corresponds to a node, a link between two nodes can only indicate that *some* relationship exists between them. To represent precise multiplace relations, such as "John gave Mary a book," additional machinery is required.

Various binding mechanisms and node replication techniques have been proposed to allow labels to be dynamically associated with John, Mary, and book, and then propagated to a node representing an instance of the relation "*x* gave *y* a *z*." See STRUCTURED CONNECTIONIST MODELS for more information on this approach. Unfortunately, current proposals for connectionist inference from structured symbolic information suffer from the same brittleness problems as do logic-based approaches, which diminishes their appeal relative to the less ambitious, less formal, spreading activation theories.

Distributed Representations and Structured Information

Models based on distributed representations also have problems representing structured information. Smolensky's (1990) tensor product representation constructed framelike knowledge structures using vector representations both for slot names and their fillers, but this system did not perform serious inference, only retrieval. In contrast, Derthick's (1990) μKLONE system, a connectionist version of the KLONE knowledge representation language, produced both ordinary associative retrieval/inference and limited counterfactual reasoning through the same simulated annealing search mechanism. μKLONE encoded classes and individuals as complex feature vectors. Its major drawback was that it required highly structured, dedicated connections representing constraints between the many units that encoded each frame. Thus, unlike Smolensky's proposal, it could not construct new frames on the fly.

Further progress on framelike and graphlike encodings comes from the recent work of Plate (1995), whose "holographic reduced representations" (HRRs), constructed by circular convolution, handle nested structures more elegantly than earlier proposals. Using an HRR encoding, semantic and structural similarity among frames can be estimated by a simple dot product.

Compositionality and Reduced Descriptions

Van Gelder (1990) points out that, while compositionality in symbolic systems is based on some form of concatenation, this need not be the case in connectionist networks (see COMPOSITIONALITY IN NEURAL SYSTEMS). The Pollack and Elman models, for example, represent both atomic and composite structures as points in a high-dimensional real-valued state space. Composition and decomposition are simply mappings between points in the space. There is no analog of concatenation here because the representations of the components are not "included" in the representation of the composite structure.

Merely replicating symbolic functions in an unusual way is not the point of this work. Connectionist representations could potentially exhibit unique properties that support more flexible forms of symbol processing than the classical model. One example of a uniquely connectionist view of compositionality is Hinton's (1990) notion of *reduced descriptions*. Consider a frame describing a *room*, with slots for the floor, walls, doors, and windows. Each of these parts could be considered as a whole, in which case it would be represented as a frame in its own right, e.g., the *door* frame might have slots for the doorknob, lintel, and hinges. Classical representation formalisms give us two choices for how to represent the *room* frame.

First, we could fold in all the subparts, sub-subparts, etc., of rooms. This would make the *room* frame very large and complex. And sometimes we will want to refer to a door as an independent concept, which this approach would not support. We would also like to be able to refer to the size of both rooms and doors, but we cannot use the same name, "size," to access both attributes if both are included directly in the *room* frame. If we want to represent a painting on one wall of the room depicting another room with a door, then there are even more opportunities for confusion. The alternative taken by most knowledge representation formalisms is to give *room* a "door" slot whose filler is some arbitrary symbol, such as *room-door*, that points to a frame for doors. This hierarchical approach allows an unbounded number of levels of description without individual frames becoming too large; it also avoids conflicts between identically named slots by keeping them at different levels. The drawback is that high-level frames like *room* become impoverished; the real information is down below, accessible only by extensive pointer following.

In a connectionist architecture, pointer following interferes with the process of parallel inference. So basic knowledge about rooms must be available in the *room* frame itself. But it should also be possible to "zoom in" on components such as doors or walls to access more detailed information. The suggested solution is for slot fillers to not be arbitrary pointers to other frames, but rather, meaningful reduced descriptions of them. If necessary, the full frame could be accessed from the reduced description by associative retrieval. Automatic construction of useful reduced descriptions remains an open problem.

Discussion

Research on connectionist models is motivated by a desire for insight into the workings of our brains, and a need for more humanlike machine reasoning systems. But these models are not yet sophisticated enough to handle large, real-world applications, where symbolic AI techniques still reign. The development of reduced descriptions, and the flexibility they promise in processing structured information, may be the first step in taking connectionist representations significantly beyond the powers of the classical logic framework.

Road Map: Artificial Intelligence and Neural Networks
Background: I.2. Levels and Styles of Analysis
Related Reading: Artificial Intelligence and Neural Networks; Cognitive Modeling: Psychology and Connectionism; Localized Versus Distributed Representations

References

Derthick, M. A., 1990, Mundane reasoning by settling on a plausible model, *Artif. Intell.*, 46:107–157.
Elman, J. L., 1991, Distributed representations, simple recurrent networks, and grammatical structure, *Machine Learn.*, 7:195–225.

Hinton, G. E., 1990, Mapping part-whole hierarchies into connectionist networks, *Artif. Intell.*, 46:47–75.

Hinton, G. E., McClelland, J. L., and Rumelhart, D. E., 1986, Distributed representations, in *Parallel Distributed Processing: Explorations in the Microstructure of Cognition* (D. E. Rumelhart, J. L. McClelland, and PDP Research Group, Eds.), Cambridge, MA: MIT Press, chap. 3. ◆

Miikkulainen, R., and Dyer, M. G., 1991, Natural language processing with modular neural networks and distributed lexicon, *Cognit. Sci.*, 15:343–399.

Newell, A., 1980, Physical symbol systems, *Cognit. Sci.*, 4:135–183.

Pearl, J., 1988, *Probabilistic Reasoning in Intelligent Systems: Networks of Plausible Inference*, San Mateo, CA: Morgan Kaufmann.

Plate, T. A., 1995, Holographic reduced representations, *IEEE Trans. Neural Netw.*, in press.

Pollack, J. B., 1990, Recursive distributed representations, *Artif. Intell.*, 46:77–105.

Smolensky, P., 1988, On the proper treatment of connectionism, *Behav. Brain Sci.*, 11:1–74. ◆

Smolensky, P., 1990, Tensor product variable binding and the representation of symbolic structures in connectionist systems, *Artif. Intell.*, 46:159–216.

Touretzky, D. S., 1990, BoltzCONS: Dynamic symbol structures in a connectionist network, *Artif. Intell.*, 46:5–46.

Touretzky, D. S., and Hinton, G. E., 1988, A distributed connectionist production system, *Cognit. Sci.*, 12:423–466.

van Gelder, T. J., 1990, Compositionality: Variations on a classical theme, *Cognit. Sci.*, 14:355–384.

Waltz, D. L., and Pollack, J. B., 1985, Massively parallel parsing: A strongly interactive model of natural language interpretation, *Cognit. Sci.*, 9:51–74.

Consciousness, Theories of

Max Velmans

Introduction

Given the centrality of consciousness to human life, it is surprising that it is often regarded as nature's deepest mystery. We have made considerable progress in understanding the events and entities in the world that we are conscious *of*, but find there to be something about consciousness itself that is unfathomable. In recent years there has nevertheless been a scientific renaissance in this area.

One reason for this is the recognition that consciousness and the "mind/body problem" present not one problem, but many. By way of introduction, these can be arranged into four interconnected groups, each focused on one central question:

1. What is consciousness?
2. Is consciousness causally efficacious?
3. What is the function of consciousness?
4. What is the relationship of consciousness to the brain?

Definitions

What consciousness is taken to be (Question 1 above) is partly a matter of arbitrary definition—and some confusion in the literature has resulted from differences in definition. In some writings, *consciousness* is synonymous with *mind*. However, given the extensive evidence for nonconscious mental processing (Velmans, 1991), this definition of consciousness is too broad. In other writings, *consciousness* is synonymous with *self-consciousness*. As one can be conscious of many things other than oneself (other people, the external world, and so on), this definition is too narrow. Here, we follow the common usage in which the term *consciousness* is synonymous with *awareness* or *conscious awareness* (sometimes *phenomenal awareness*). The contents of consciousness encompass all that we are conscious of, aware of, or experience. These include not only experiences that we commonly associate with ourselves (such as thoughts, feelings, images, dreams, body experiences, and so on), but also the experienced three-dimensional world (the phenomenal world) beyond the body surface.

However, what consciousness is taken to be is not entirely arbitrary. Once a given reference for the term *consciousness* is fixed, the investigation of its nature can begin, and this may, in time, transmute the meaning (or sense) of the term. As noted by the philosopher John Dewey, to grasp the meaning of a thing, an event, or situation is to see it in its relations to other things; to note how it operates or functions, what consequences follow from it, what causes it, what uses it can be put to. Thus, to understand what consciousness is, we need to understand what causes it, what its function(s) may be, how it relates to nonconscious processing in the brain, and so on. As our scientific understanding of these matters deepens, our understanding of what consciousness *is* will also deepen. A similar transmutation of meaning (with growth of knowledge) occurs with basic terms in physics such as *energy* and *time*.

Consciousness and Cognitive Psychology

One of the main aims of cognitive psychology is to provide a functional description of the human mind in information processing terms. But how does conscious awareness relate to human information processing? Is consciousness causally active in certain forms of information processing (Question 2 above)? If so, what is its function (Question 3)? According to William James (1890), the current contents of consciousness define the "psychological present" and are contained in "primary memory" (a form of short-term working store). The contents of "secondary memory" (a long-term memory store) define the "psychological past," and while they remain in secondary memory, they are unconscious. James also suggested that stimuli which enter consciousness are at the "focus of attention," having been selected from competing stimuli to enable effective interaction with the world. Stimuli at the focus of attention are also given significance and value by their contextual surround —a conscious "fringe" or flowing consciousness "stream." These ideas, developed around a hundred years ago, are still the focus of much psychological research.

James's linking of consciousness to primary memory, for example, is developed in more recent attempts to specify with more precision *how* consciousness relates to such a short-term store. It is generally thought that only information at the focus of attention enters primary memory (and consciousness), in which case it may later be consciously recognized or recalled. However, there is also much evidence for implicit memory—

the ability to acquire and use long-term information in advance of, or in the absence of, any ability to consciously remember that information (cf. Schacter, 1989, and the chapter by Gardiner in Velmans, in press).

The processes which govern how stimuli are selected for focal attention and consciousness have been extensively researched in the context of the *cocktail party problem*. In a cocktail party situation, the conversation one attends to enters consciousness, while the competing conversations seem to form a relatively undifferentiated background noise. However, all speech waveforms arrive concurrently at the eardrum. So how does the brain select the required message from such complex auditory stimuli? It is thought that some form of preattentive, preconscious analysis must be involved, and much research has been devoted to the nature and extent of such analysis.

Initial studies in this area used a shadowing task in which one set of stimuli is presented (via earphones) to a subject's left ear and other stimuli to his right ear, while the subject is asked to attend to one ear and repeat out loud what he hears. Early findings indicated that subjects in this situation could not report the identity or meaning of stimuli on the nonattended ear, although they could report some of their physical features. This suggested that preattentive analysis is restricted to physical cues, while analysis for meaning requires focal attention. Only stimuli at the focus of attention enter short-term memory, become conscious, and can be subsequently remembered. However, later experiments demonstrated that the meaning of nonattended stimuli affected the processing of attended stimuli, indicating that analysis of meaning of nonattended stimuli can take place without conscious awareness or subsequent recall—for example, words in the nonattended ear were found to affect the interpretation of ambiguous messages in the attended ear. Subsequently, it was thought that the meaning and significance of input stimuli in different channels are analyzed in a parallel, automatic, preconscious fashion, without mutual interference, up to the point where each stimulus is matched to its previous traces in long-term memory. More complex processing (for example, the identification of novel stimuli or novel stimulus combinations) was assumed to require focal attention and consciousness (see review chapter by Underwood in Velmans, in press).

Kahneman and Treisman (1984) noted, however, that different *forms* of attention may have to be devoted to different *stages* of input analysis. Attention may, for example, be devoted to physical features if one is searching for a target input stimulus, but other resources may be required to integrate the set of features at the location found by the search. In addition, the results of input analysis need to be disseminated to other processing modules. Kahneman and Treisman cautioned against the general tendency to identify focal-attentive processing with consciousness. Velmans (1991) also argued against taking such identities for granted, and questioned whether consciousness is necessary for the processing of novel, complex input. Sentences, for example, may be both novel and complex, and to read a sentence one must focus one's attention on it. But in what sense is the information processing involved in reading conscious? If one silently reads the sentence, "The forest ranger did not permit us to enter the park without a permit," one experiences phonemic imagery ("inner speech") in which the stress pattern on the word *permit* changes appropriately—from per*mit* (a verb) to *per*mit (a noun). But for this to happen, the semantic and syntactic analysis required to determine the role of this word in the sentence must take place *before* its phonemic image enters consciousness. Nor is one consciously aware of any of the detailed information processing involved in input analysis. In this instance, consciousness (of the input) appears to *follow* input analysis rather than enter into it.

Such considerations force a reexamination of the way consciousness relates to focal-attentive processing, and similar questions can be raised about the many other forms of information processing which have been thought to require attention and consciousness for their operation—for example, stimulus selection and choice, learning and memory, and the production of complex, flexible, or novel responses (see Baars, 1989; Farthing, 1992; Velmans, 1991, in press, for reviews). It may be, for example, that consciousness relates closely not to the *detailed operation* of attentional processing, but to the *results* of such processing which need to be disseminated throughout the processing system to enable its integrated functioning, based on continuously updated, pertinent information. This possibility has been suggested by Baars (1989) and also by Navon, Van Gulick, and Velmans (see commentaries accompanying Velmans, 1991).

It is also clear that one needs to define the meaning of a "conscious process" more precisely. A process may be said to be conscious in the sense that one is conscious of its operations (certain aspects of problem solving are conscious in this sense, but not, for example, the processing involved in visual perception); a process may also be said to be conscious in the sense that its *output* enters consciousness (as in silent reading and the other forms of attentional processing mentioned above); a process may also be said to be conscious insofar as consciousness causally affects or enters into that process. Evolutionary theory suggests that consciousness ought to affect mental processing. Whether it actually does so, however, is a much debated issue (see Velmans, 1991, 1993, and accompanying commentaries).

James (1890) stressed that the significance and value of conscious material at the focus of attention is indicated by the relatively vague feelings that surround it. Mangan (1993) argues that such feelings provide *contextual* information about conscious material at the focus of attention, in a highly condensed form. For example, the goodness-of-fit of currently focused-on material with prior material stored in long-term memory may be manifest in consciousness as a simple feeling of its "rightness" or "wrongness." According to Mangan, the unconscious process which produces such feelings may resemble the computation discovered by Hopfield (1982), in which the goodness-of-fit of an immense number of interacting, neuronlike nodes is condensed into a single metric or index. In their present form, connectionist networks do not require such computations in order to settle into stable states. Consequently, Mangan suggests that there may be a second network in the human brain that generates a goodness-of-fit metric (or its analog) thereby overseeing the operations of a more basic network that settles into stable states in the normal way. Another possibility is a more sophisticated network which generates a global, state-of-the-system metric as an emergent property. Whether or not the existence of such neural nets are confirmed by experiment, it is clear that contextual processing and its accompanying conscious "fringe" need to be incorporated into any complete model of brain function and accompanying experience.

The relation of consciousness to focal-attention and primary memory is fundamental to its understanding in information processing terms. However, there are many other areas of psychology in which the contents of consciousness have been investigated. For example, studies of perception and psychophysics have extensively charted the way specific contents of consciousness (percepts) relate to environmental stimuli and brain processing. Other prominent areas of investigation include the study of consciousness in imagery, emotion, and dreams. There is also a continuing debate about methodological issues—for example, over how to determine when a process is conscious, about the value and limitations of introspection, and about the necessary conditions for a science of consciousness.

The Neuropsychology of Consciousness

The cognitive issues outlined above provide a context for the more detailed investigation of how consciousness relates to the workings of the brain (Question 4). The contents of consciousness are, of course, immensely varied and are likely to draw on information that is widely distributed throughout the brain. But what happens to this information to make it conscious? Cognitive psychological theory suggests that the neural conditions for consciousness are closely linked to processes (and structures) responsible for attention and short-term memory (see, for example, the chapter by Posner and Rothbart in Milner and Rugg, 1992). Cognitive studies of input analysis also suggest that processing is initially preconscious, and this is consistent with neuropsychological evidence that direct microelectrode stimulation of the cortex has to proceed for at least 200 ms before neural conditions develop that are adequate to support conscious experience (see review chapters by Libet in Ciba Foundation Symposium 174, 1993, and in Velmans, in press). Libet and his co-workers also found evidence that the brain compensates for this preconscious processing time by "marking" the time of arrival of stimuli at the cortical surface with an early evoked potential, and then referring experienced time of occurrence "backwards in time" to this early time marker. In short, experienced time of arrival corresponds to actual time of arrival of input at the cortical surface (rather than to the time when a conscious representation has had time to develop).

Studies with brain-damaged patients have revealed many different forms of *dissociation* between psychological functioning and conscious awareness. A classic example is the case of "blindsight." Weiskrantz (1986), for example, reviews studies with a patient with lesions in his right hemisphere striate cortex who had no awareness of stimuli projected to his blind (left) hemifield. When persuaded to guess, he could nevertheless point accurately to spots flashed in the blind area, discriminate horizontal from vertical lines, discriminate color, movement, and, to a limited extent, simple forms. Dissociation of awareness from ability to discriminate has also been found for touch ("blindtouch") produced by somatosensory cortex damage. Similarly, in amnesia produced by lesions in the medial temporal lobe (including the hippocampus), patients may learn new motor skills and have implicit acquisition of semantic information although they have no (overt) conscious recognition or recall of what they have learned (see review by Schacter, 1989, and the chapter by Schacter in Milner and Rugg, 1992).

Given the close theoretical linkage of consciousness to focal attention (discussed above), it is significant that in some of these conditions awareness is, in part, dissociated from aspects of focal attention. Weiskrantz (1986) found, for example, that a blindsighted subject can direct his attention to a screen and discriminate between stimuli projected to his blind hemifield without having any visual awareness of the stimuli; there is also evidence that amnesiacs can attend to and respond appropriately to a sequence learning task without having any explicit awareness of the sequence being learned (see Posner and Rothbart in Milner and Rugg, 1992). Such findings are troublesome for functionalist theories of consciousness, in that they demonstrate conscious awareness to be dissociable from many of the functions with which it is commonly identified (cf. Velmans, 1991). At the same time, neural differences between functions accompanied or not accompanied by consciousness may provide clues about the neural conditions which support conscious awareness (see Baars, 1989, and the chapters by Baars and by Libet in Velmans, in press); the striate cortex, for example, appears to be an important component of neural systems which support visual awareness (cf. review chapter by Cowie and Stoerig in Milner and Rugg, 1992).

For Plato, Descartes, and modern dualists such as Eccles (1980), the unity of consciousness is indivisible. However, dissociations also appear to occur *within* the contents of consciousness. For example, patients with *prosopagnosia* (usually caused by bilateral lesions to occipital-temporal cortical regions) are able to perceive familiar faces, but do not consciously recognize them. They may, however, show evidence of unconscious recognition. One technique used to demonstrate such dissociations is a variant of the "guilty knowledge test," sometimes used in criminal investigations, in which a guilty person may have an involuntary physiological response to stimuli related to the crime. For example, one prosopagnosic subject studied by Bauer was shown photographs of familiar faces, each of which was accompanied by five names (one correct name and four distractors). When the subject was asked to choose the correct name, his performance was no better than chance (22%). But his skin conductance to the correct name rose to maximal levels on 61% of the trials, well above chance (see review chapter by Young and De Haan in Milner and Rugg, 1992, and the chapter by Young in Velmans, in press). In short, such patients consciously experience faces and *unconsciously* recognize them. However, tacit recognition is not accompanied by the usual conscious feelings of familiarity—an apparent dissociation of focal consciousness from "fringe" consciousness (see Mangan, 1993). In *anosognosia*, patients appear to be unaware of their impairments and may actively deny them. Patients with Anton's syndrome, for example, may deny their own blindness, and other patients deny their deafness; some patients with prosopagnosia deny their recognition problems—and lack of insight into memory impairments is also common. Anosognosia may be thought of as impaired self-monitoring, or (from a first-person perspective) as a deficit in self-awareness. Schacter (1989) accounts for dissociations of consciousness in terms of disconnections of specific processing modules from a general conscious awareness system. However, the anosognosias are generally specific to a given impairment rather than reflecting any global lack of self-awareness. This opens up the possibility that self-monitoring (and accompanying self-awareness) may not rely on just one centralized monitoring mechanism—although it might simply be that self-monitoring is not always accurate.

A further, dramatic dissociation of consciousness occurs within "split brain" patients, whose corpus callosum has been sectioned in an operation to relieve focal epilepsy (in some patients the anterior commissure is also sectioned—see the extensive review in Springer and Deutsch, 1993). Sectioning the corpus callosum prevents direct communication between the left and right cerebral cortices although subcortical connections remain between the two halves of the brain. Surprisingly, such patients seem outwardly normal. Typically, however, they report being aware of stimuli (such as pictures or names of objects) projected to their left hemisphere (which controls speech), but deny awareness of stimuli projected to the right hemisphere. On the other hand, if objects corresponding to the stimuli are hidden behind a screen, the right hemisphere can identify them by touch, picking them out from other objects with the left hand, which it controls. The right hemisphere also appears able to understand concrete concepts, being able to identify (with the left hand) an object that "goes with" the stimulus it receives—for example, picking out a lighter when shown a cigarette (Gazzaniga, 1985); it is also generally superior to the left hemisphere in visuo-spatial tasks, although inferior in verbal and conceptual tasks.

What such findings imply for the unity of consciousness and the relation of consciousness to the brain has been much debated. According to Eccles (1980), the ability to introspect and verbally communicate one's experience is required for full self-consciousness. While the right hemisphere has a form of consciousness associated with its ability to perceive, learn, and so on, this may be likened to the consciousness of nonhuman animals. For Eccles, only the left hemisphere is *fully* conscious. Gazzaniga (1985) develops a similar view, arguing that consciousness is associated with a specialized "interpreter system" which nearly always exists in the left hemisphere. This makes sense of input from left and right hemisphere modules to produce a coherent narrative, which is fed to a language module (for report), thereby producing the *illusion* of a unified consciousness.

On the other hand, Sperry (1985) maintains that in the normal brain, conscious unity is not an illusion. Rather, it is an emergent property of brain functioning that *supervenes* (regulates, coordinates) the brain activity from which it emerges. In the bisected brain (according to Sperry), each hemisphere has an associated consciousness of its own. Not only can the right hemisphere perceive, learn, engage in intelligent action, and so on, but may also have self-consciousness and a sense of social awareness. Striking evidence of right hemisphere ability was, for example, found by Gazzaniga and his co-workers in one exceptional patient whose right hemisphere was able to communicate its experiences by spelling out answers to questions with Scrabble letters, using the left hand (see Springer and Deutsch, 1993, for further evidence, and discussion in Farthing, 1992).

Given the distributed nature of information processing in the brain and the many dissociations of consciousness discussed above, it is remarkable that the contents of consciousness generally appear bound into a coherent, relatively well-integrated stream. But how does the brain solve this "binding" problem? Some theorists argue for the existence of specific regions of the brain where all currently relevant information comes together and becomes conscious (see, for example, the chapter by Posner and Rothbart in Milner and Rugg, 1992). However, it may be that a fixed consciousness location is ruled out by the distributed nature of human information processing, both in location and time. An alternative binding process may be the synchronous or correlated firing of diverse neuron groups representing currently attended-to objects or events. Although this possibility remains tentative, evidence for the existence of such binding processes (involving rhythmic frequencies in the 30 to 80 Hz region; see SYNCHRONIZATION OF NEURONAL RESPONSES AS A PUTATIVE BINDING MECHANISM) has recently been reviewed by Llinás and Paré (1991) and Shastri and Ajjanagadde (1993). The latter (see also STRUCTURED CONNECTIONIST MODELS) give a detailed, innovative account of how such variable bindings might propagate over time, as attended-to representations change, within neural networks.

Theories of consciousness are diverse—ranging from global theories about the nature of consciousness to micro-theories about the way aspects of consciousness relate to specific processes within the brain. The issues addressed often cross the boundaries between philosophy, psychology, neuropsychology, AI, and neural networks. Comprehensive reviews are given in Farthing (1992), Ciba Foundation Symposium 174 (1993), and Velmans (in press).

Road Map: Connectionist Psychology
Related Reading: Binding in the Visual System; Dissociations Between Visual Processing Modes; Visual Scene Perception: Neurophysiology

References

Baars, B. J., 1989, *A Cognitive Theory of Consciousness*, Cambridge, Eng.: Cambridge University Press. ◆
Ciba Foundation Symposium 174, 1993, *Theoretical and Experimental Studies of Consciousness*, Chichester, Eng.: Wiley. ◆
Eccles, J. C., 1980, *The Human Psyche*, New York: Springer International. ◆
Farthing, W., 1992, *The Psychology of Consciousness*, Englewood Cliffs, NJ: Prentice Hall. ◆
Gazzaniga, M. S., 1985, *The Social Brain: Discovering the Networks of the Mind*, New York: Basic Books. ◆
Hopfield, J., 1982, Neural networks and physical systems with emergent collective computational abilities, *Proc. Natl. Acad. Sci. USA*, 79:2554–2558. ◆
James, W., 1890 (reprinted 1950), *The Principles of Psychology*, New York: Dover. ◆
Kahneman, D., and Treisman, A., 1984, Changing views of attention and automaticity, in *Varieties of Attention* (R. Parasuraman and D. R. Davies, Eds.), New York: Academic Press. ◆
Llinás, R. R., and Paré, D., 1991, Of dreaming and wakefulness, *Neuroscience*, 44:521–535. ◆
Mangan, B., 1993, Taking phenomenology seriously: The "fringe" and its implications for cognitive research, *Conscious. Cognit.*, 2(2):89–108. ◆
Milner, A. D., and Rugg, M. D., Eds., 1992, *The Neuropsychology of Consciousness*, San Diego, CA: Academic Press. ◆
Schacter, D. L., 1989, On the relation between memory and consciousness: Dissociable interactions and conscious experience, in *Varieties of Memory and Consciousness* (H. L. Roediger and F. I. M. Craik, Eds.), Hillsdale, NJ: Erlbaum, pp. 355–389. ◆
Shastri, L., and Ajjanagadde, V., 1993, From simple associations to systematic reasoning: A connectionist representation of rules, variables and dynamic bindings using temporal synchrony, *Behav. Brain Sci.*, 16(3):417–494. ◆
Sperry, R. W., 1985, *Science and Moral Priority: Merging Mind, Brain and Human Values*, New York: Praeger. ◆
Springer, S. P., and Deutsch, G., 1993, *Left Brain, Right Brain*, New York: Freeman. ◆
Velmans, M., 1991, Is human information processing conscious? *Behav. Brain Sci.*, 14(4):651–726. ◆
Velmans, M., 1993, Consciousness, causality, and complementarity, *Behav. Brain Sci.*, 16(2):404–416. ◆
Velmans, M., Ed., in press, *The Science of Consciousness: Psychological, Neuropsychological and Clinical Reviews*, London: Routledge. ◆
Weiskrantz, L., 1986, *Blindsight: A Case Study and Implications*, London: Open University Press. ◆

Constrained Optimization and the Elastic Net

Alan L. Yuille

Introduction

This chapter describes neural network approaches to solving optimization problems. Many problems in neural networks and related disciplines can be formulated in terms of minimizing cost functions, or equivalently, of finding the most probable estimates of probability distributions. Thus there is a lot of interest in finding algorithms for solving such problems. For some of these problems, computer scientists have developed highly effective discrete algorithms. By contrast, workers in neural networks have concentrated more on analog continuous-time algorithms, or dynamical systems, which are believed to be more biologically plausible and to have other attractions, such as inherent parallelism and the feasibility of being implemented in VLSI. Recent results show that such algorithms are competitive with more traditional approaches to a variety of problems.

There have been a number of different neural network methods for optimization. In recent years a theoretical framework adapted from statistical mechanics has been able to unify three of the most important approaches. The first approach, commonly associated with Hopfield and Tank (1985), attempts to solve optimization problems by using neural network models of a type also studied by Grossberg and his collaborators (see COMPUTING WITH ATTRACTORS). The second approach is the elastic net developed by Durbin and Willshaw (1987), which is closely related to self-organizing networks (see SELF-ORGANIZING FEATURE MAPS: KOHONEN MAPS). The third approach, statistical mechanics, first appeared in the guise of stochastic algorithms such as SIMULATED ANNEALING (q.v.) and BOLTZMANN MACHINES (q.v.); but then deterministic methods, in particular deterministic annealing or mean-field annealing, were developed (see references in Hertz, Krogh, and Palmer, 1991, and Yuille, 1990). It was then realized that most examples of the first two approaches could be considered as special cases of the third (Simic, 1990; Yuille, 1990).

For pedagogical reasons, we will introduce everything from the statistical mechanics perspective (see also NEURAL OPTIMIZATION). We will be particularly concerned with problems involving global constraints, for example, combinatorial problems. The issue of how these constraints are imposed is very important.

Formalizing Optimization in Terms of Energy Functions

We can express optimization problems in terms of minimizing energy, or cost, functions. Thus, for example, we can write the Traveling Salesman Problem (TSP) in terms of minimizing an energy function:

$$E_{TSP}[V] = \sum_{i=1}^{N} \sum_{a,b=1}^{N} d_{ab} V_{ia} V_{(i+1)b} \qquad (1)$$

where $a, b = 1, \ldots, N$ labels the cities; d_{ab} is the distance between the ath and bth cities; $i = 1, \ldots, N$ is the stage at which a city is visited on a tour (the tour is a closed loop so the $(N+1)$th stage is the first stage); and the $\{V_{ia}\}$ are binary decision units so that $V_{ia} = 1$ if city a is the ith city on the tour, and zero otherwise. The problem is solved by finding the $\{V_{ia}\}$ that correspond to the shortest tour passing through each of the cities. In the following, we will represent the decision variables by a vector V with N^2 elements or, equivalently, by an $N \times N$ matrix with elements $\{V_{ia}\}$.

Not all $\{V_{ia}\}$ correspond to legal tours. We must impose global constraints to restrict the allowable configurations. For this problem, we must impose the condition that in each row and each column of the matrix V_{ia}, there is exactly one element which takes the value 1. This is equivalent to the algebraic conditions $\Sigma_i V_{ia} = 1$, for all a, and $\Sigma_a V_{ia} = 1$, for all i.

There are two main strategies for imposing these constraints. The first strategy, known as soft constraints, corresponds to adding energy-bias terms which penalize configurations that do not obey the constraints. For example, one could add to $E_{TSP}[V]$, defined in Equation 1, an energy-bias term (Hopfield and Tank, 1985). One possibility is:

$$E_b[V] = A \sum_a \left\{ \sum_i V_{ia} - 1 \right\}^2 + A \sum_i \left\{ \sum_a V_{ia} - 1 \right\}^2 \qquad (2)$$

where A is a large positive constant. The second strategy, called

hard constraints, reduces the configuration space by eliminating configurations which do not satisfy the constraints. One way of doing this is to re-parameterize the space so that the constraints are implicitly imposed. For example, when minimizing a cost function $f(x, y)$ on the circle $x^2 + y^2 = 1$, one can switch to polar coordinates and minimize the function $g(\theta) = f(\cos\theta, \sin\theta)$. Another standard approach is the method of Lagrange multipliers. This is performed by introducing additional parameters, the Lagrange multipliers, for each constraint to be satisfied (see Gill et al., 1986). The constraints are multiplied by these parameters and added to the original cost function. Extremizing the modified cost function with respect to the new parameters, as well as the original variables, will yield a solution with the constraints automatically satisfied. For our example, we write a modified cost function $g(x, y, \lambda) = f(x, y) + \lambda\{x^2 + y^2 - 1\}$, where λ is the additional parameter. We now extremize with respect to x, y, and λ. Observe that extremizing with respect to λ will ensure that the constraint $x^2 + y^2 = 1$ is automatically satisfied.

Statistical mechanics will suggest an alternative way of imposing these constraints, but in all known cases it reduces to Lagrange multipliers or re-parameterization.

Now that the energy function has been defined and the constraints have been imposed, a procedure is still needed to find the minimum. In contrast to the discrete algorithms studied by computer scientists, neural network theorists are predominantly interested in continuous-time dynamical systems. More specifically, they seek to define equations of the form

$$\frac{dX}{dt} = F(X) \qquad (3)$$

which converge to minima of the energy function.

For equations of this type to make sense, however, the variables must be continuous. Thus they cannot be directly applied to combinatorial optimization problems of the type described by Equation 1. One solution is to replace the binary decision variables $\{V_{ia}\}$ by continuous variables $\{S_{ia}\}$ which take values in the range $[0, 1]$ and to set up dynamics so that the $\{S_{ia}\}$ will converge to values close to 0 or 1, which can then be rounded off to the nearest integer. Such a method was proposed by Hopfield and Tank (1985). They allowed the $\{S_{ia}\}$ to take values in $[0, 1]$ and introduced additional variables $\{U_{ia}\}$ related to them by $S_{ia} = 1/\{1 + e^{-\beta U_{ia}}\}$, for all i, a. They then defined dynamics

$$\frac{dU_{ia}}{dt} = -U_{ia} - \frac{\partial E[S]}{\partial S_{ia}} \quad \text{for all } i, a \qquad (4)$$

and showed that this converges to a locally optimal solution to the problem. Since E is quadratic, the second term on the right-hand side is linear in the $\bar{V}$'s. This can be thought of as a re-derivation of the leaky integrator neuron (see section I.1, "Introducing the Neuron").

In the next subsection, we will describe a more general approach, which includes Hopfield and Tank as a special case, and involves interpreting the $\{S_{ia}\}$ as representing approximations to the means of the variables $\{V_{ia}\}$ with respect to a probability distribution. In this approach, it is shown that Hopfield and Tank's algorithm converges to a minimum of an effective energy, $E_{\text{eff}}[S]$, which can be derived from $E[V]$. The effective energy is one of the fundamental concepts of neural net optimization, and statistical mechanics will provide ways of deriving it.

Suppose we have transformed the problem into minimizing an effective energy $E_{\text{eff}}[S]$ with continuous variables S. An infinite number of dynamical systems can be defined which converge to a minimum of $E_{eff}[S]$. The simplest example is gradient descent:

$$\frac{dS_{ia}}{dt} = -\frac{\partial E_{eff}[S]}{\partial S_{ia}} \quad \text{for all } i, a \tag{5}$$

By applying the chain rule, we obtain $dE_{eff}/dt = -\Sigma_{ia}\{\partial E_{eff}/\partial S_{ia}\}^2$, so the energy always decreases until we get to a minimum. But many other algorithms are possible. The key idea is to define a *Lyapunov function* $L[S]$ and a dynamical system $dS_{ia}/dt = F_{ia}[S]$ for all a, i so that $dL/dt \leq 0$ along a trajectory and $dL/dt = 0$ if and only if we are at a minimum of $E_{eff}[S]$. In this example $E_{eff}[S]$ acts as the Lyapunov function.

The effective energy $E_{eff}[S]$ may still have many local minima, and the dynamical system may converge to any one of them. One final ingredient, therefore, is to generalize the effective energy function $E_{eff}[S]$ to a one-parameter family of effective energy functions $E_{eff}[S; T]$, with the property that increasing T corresponds to smoothing the effective energy, so that at large T there is a single global minimum. The strategy is to minimize the effective energy at large T and to track this solution down as T decreases. This strategy will not always guarantee a correct solution, but it has empirically been a good heuristic. In the next section we see how statistical mechanics provides a natural way of obtaining such a one-parameter family with T being identified as the temperature of the system.

Statistical Mechanics and the Gibbs Distribution

Statistical mechanics was first invented to model the behavior of gases and related systems. Such systems continually undergo random fluctuations whose magnitude depends on the temperature T. Thus, it is practically impossible to specify the precise configuration of the system. But instead, using statistical mechanics, one can specify the probability that the system is in a specific state. As the temperature increases, the fluctuations become very strong and all configurations become equally likely. As the temperature tends to zero, the fluctuations diminish and the system tends to the configuration with lowest energy.

The probability of the system being in a specific state is given by the Gibbs distribution (Reif, 1982). This is derived by assuming that the *entropy*, or disorder, of the system is as large as possible while requiring that the average energy is fixed. Strictly speaking, the Gibbs distribution specifies a one-parameter family of probability distributions parameterized by the temperature T.

Suppose we have formulated a problem in terms of minimizing an energy $E[V]$. The associated Gibbs distribution is

$$P[V; \beta] = \frac{e^{-\beta E[V]}}{Z} \tag{6}$$

where Z is a normalization factor also known as the *partition function*. The constant $\beta = 1/T$, where T is the temperature of the system.

It can be seen directly from Equation 6 that maximizing the probability P with respect to V is equivalent to minimizing the energy E. This gives a nice philosophical interpretation based on Bayesian ideas (see, for example, Durbin, Szeliski, and Yuille, 1989).

It can also be seen that as $\beta \to 0$ ($T \to \infty$) all states become equally likely, but as $\beta \to \infty$ ($T \to 0$) only the lowest energy state has nonzero probability of occurring (although the probabilistic ordering of states does not change as we vary β).

This temperature-dependent behavior suggests *simulated annealing* (Kirkpatrick, Gelatt, and Vecchi, 1983): find the thermal equilibrium of the system (i.e., make the states satisfy the Gibbs distribution) at large T (small β) and keep the system in

thermal equilibrium as T decreases to 0 ($\beta \to \infty$). The intuition is that it is easy to get to the Gibbs distribution for high temperatures (since all states are roughly equally probable) and that thermal equilibrium at low temperatures will correspond to the lowest energy state. Indeed, simulated annealing can be proven to obtain the optimal solution provided the system is cooled at a logarithmic rate. Unfortunately, this rate of cooling is prohibitively slow, and so the bounds are only of theoretical interest. Nevertheless, simulated annealing has been shown empirically to be an effective heuristic optimization algorithm when the system is cooled far faster than the theoretical bounds.

Observe, however, that to solve the optimization problem we need only know the means of the V variables with respect to the Gibbs distributions, i.e., $\bar{V}(\beta) = \Sigma_V V P[V; \beta]$, and not the full distribution. At zero temperature, only the lowest energy state can occur, and so these means will correspond to the state that minimizes the energy. In a few cases, the means can be calculated explicitly and hence the solution found directly (see WINNER-TAKE-ALL MECHANISMS for an example). If not, we can use deterministic annealing, or mean-field annealing. This corresponds to approximating the means of the distribution at high temperature and tracking these solutions as the temperature decreases. Unlike simulated annealing, there is usually no guarantee of convergence to the optimal solution. Still, the method is usually very quick and empirically gives good results. As we will show, it is also the basis of the neural network and elastic net approaches to optimization.

The Mean-Field Approximation

It can be shown (see Hertz et al., 1991) that if the partition function in Equation 6 can be computed analytically, then it is possible to solve directly for the means. (This procedure involves introducing additional auxiliary terms in the energy function.) Unfortunately, this is often impractical. Physicists, however, have developed (see the appendix to this article) a set of techniques for approximating the partition function and obtaining approximate consistency conditions, the *mean-field equations*, for the means of variables of interest. Because the variables we are considering, the V, are spatially indexed, they are called fields, and we can refer to their means as the *mean fields*. (One must be careful with this terminology, however, because physicists—and some neural network workers—often reserve the term *mean fields* for a specific type of field. Moreover, *mean* may correspond to spatial averaging rather than to the Gibbs distribution.)

We will briefly describe the main results and refer elsewhere for proofs (Reif, 1982). The free energy of the system is $F = -T \log Z$. By standard thermodynamic identities, it can be written as $F = E - TS$, where E is the energy and S is the entropy, a measure of the disorder of the system. We can obtain an effective energy $E_{eff}[\bar{V}; T]$ by taking the expectation of F with respect to the Gibbs distribution and making some approximations (see appendix). This gives an effective energy

$$E_{eff}[\bar{V}; T] = E[\bar{V}] + T \sum_{ia} \{\bar{V}_{ia} \log \bar{V}_{ia} + (1 - \bar{V}_{ia}) \log(1 - \bar{V}_{ia})\} \tag{7}$$

where the second term is an approximation to the expected entropy. It is straightforward to show, by elementary calculus, that the entropy term has a single minimum at $\bar{V}_{ia} = 0.5$ for all a, i.

The global minimum of this energy corresponds to the mean-field equations. For large T, the entropy term will force the effective energy to become convex and hence have a single unique minimum. As $T \to 0$, the global minimum of E_{eff} will tend to the global minimum of E, provided technical conditions are imposed to ensure that E_{eff} has no minima in the interior of the hypercube.

Consider the specific example of the Traveling Salesman Problem, in which the global constraints are imposed by energy-bias terms. The energy becomes

$$E_{\text{eff}}[\bar{V}; T] = \sum_{iab} \bar{V}_{ia} \bar{V}_{(i+1)b} d_{ab} + A \sum_a \left\{ \sum_i \bar{V}_{ia} - 1 \right\}^2$$
$$+ A \sum_i \left\{ \sum_a \bar{V}_{ia} - 1 \right\}^2 + T \sum_{ia} \{ \bar{V}_{ia} \log \bar{V}_{ia}$$
$$+ (1 - \bar{V}_{ia}) \log(1 - \bar{V}_{ia}) \} \qquad (8)$$

It is easy to check that the gradient of the entropy term becomes infinite as $\bar{V} \to 0$ or $\bar{V} \to 1$. This makes it energetically unfavorable for the $\bar{V}$'s to leave the region $[0, 1]$.

There are a large number of algorithms that can be used to find minima of E_{eff}. Perhaps the simplest is to use gradient descent to find a minimum of E_{eff} at large T. By tracking this solution down to small T, we may hope to find a good approximation to the global minimum at small T.

Surprisingly, the Hopfield and Tank neural network model will also converge to a minimum of E_{eff}. To see this, we perform the coordinate change $S_{ia} = 1/\{1 + e^{-\beta U_{ia}}\}$ for all a, i to eliminate the U's from Equation 4. This yields the equation

$$\frac{dS_{ia}}{dt} = -S_{ia}(1 - S_{ia}) \frac{\partial E_{\text{eff}}}{\partial S_{ia}} \quad \text{for all } i, a \qquad (9)$$

which converges to a minimum of E_{eff} because $\{dE_{\text{ef}}/dt\} = -\Sigma_{ia} \{\partial E_{\text{eff}}/\partial S_{ia}\}^2 S_{ia}(1 - S_{ia}) \leq 0$, so E_{eff} acts as a Lyapunov function for the system.

This shows that neural network dynamics will converge to a minimum of the effective energy, hence giving a solution to the mean-field equations, provided that the energy $E[V]$ is quadratic and the constraints are soft constraints imposed by energy biases.

There are many possible algorithms for solving the mean-field equations (see Yuille and Kosowsky, 1994), and it is by no means certain that neural network dynamics is the best choice. Moreover, the use of soft constraints has been criticized because the Hopfield and Tank model has poor behavior on large-scale problems (although it can be shown that clever choices of the constants can significantly improve performance; see Aiyer, Niranjan, and Fallside, 1990). If hard constraints are used, then these problems can still be solved by dynamical systems, though these are no longer of the Hopfield/linear threshold type.

It is unclear, at least to this author, why neural network researchers should restrict themselves to the linear threshold type of model. If the goal is to model the brain, then it is clear that such models are oversimplifications and that real neurons have far richer dynamics. If the goal is to design a working engineering system, then one should pick the network that best solves the problem.

The Elastic Net

We now look at a model which imposes some of its constraints in a hard manner: the *elastic net* model (Durbin and Willshaw, 1987). (For another model using hard constraints, see WINNER-TAKE-ALL MECHANISMS.) To show how the elastic net model uses hard constraints, we will first introduce it by a slightly different formulation (Yuille, 1990).

Let $\{x_a\}$, $a = 1, \ldots, N$, be the positions of the cities that the traveling salesman wants to visit. We propose to formulate the TSP as a problem of matching the cities to an elastic net of cities $\{y_j\}$, $j = 1, \ldots, M$, with $M \geq N$, connected together by elastic springs joining y_j to y_{j+1} (mod M) for $j = 1, \ldots, M$. These springs can be interpreted as a prior distribution biasing the length of the net to the minimum-length tour. We can write this in the form

$$E[\{V_{ia}\}, \{y_i\}] = \sum_{i=1}^M \sum_{a=1}^N V_{ia} |x_a - y_i|^2 + v \sum_{j=1}^M |y_j - y_{j+1}|^2 \qquad (10)$$

where the $\{V_{ia}\}$ are binary matching elements as before. V_{ia} is 1 if x_a is matched to y_i and is 0 otherwise. We restrict this sum to impose the condition that each city is matched to exactly one city on the net, hence $\Sigma_i V_{ia} = 1$, for all a. The first term is the matching term and gives the penalty paid if there is no city on the net close to each data city. The second term on the right-hand side corresponds to stretching the springs and minimizes the square of the distances. The net is periodically identified so that $y_{M+1} = y_1$. The variables $\{V_{ia}\}$ and $\{y_j\}$ denote a state of the system.

We define the Gibbs distribution for the system:

$$P[V, y; \beta] = \frac{e^{-\beta E[V, y]}}{Z} \qquad (11)$$

The "solutions" to the TSP will correspond to the means of the fields $\{V_{ia}\}$ and $\{y_i\}$ as $\beta \to \infty$. We simply round off the $\{V_{ia}\}$ to the nearest integers and read off the tour. Alternatively, the final positions of the $\{y_i\}$ will determine the tour. We must, however, impose constraints to ensure that we obtain legal tours. Observe that if the $\{V_{ia}\}$'s are fixed, then the probability distributions for the $\{y_i\}$'s are products of Gaussians.

We can write the partition function as

$$Z = \sum_V \int [dy] e^{-\beta E[V, y]} \qquad (12)$$

where the sum is taken over all the possible states of the $\{V_{ia}\}$. We restrict this sum to impose the condition that each city is matched to exactly one city on the net. This restricts us to configurations $\{V_{ia}\}$ such that $\Sigma_i V_{ia} = 1$, for all a. Observe that this restriction will allow undesirable configurations where two cities are matched to the same city on the net. Such situations are, however, biased against by the first term in the energy function (Equation 10). Thus the elastic net imposes constraints in a hybrid soft/hard manner.

Obtaining the Durbin and Willshaw Algorithm

We now show how to derive the original, and more standard, formulation of the elastic net.

Writing the partition function as

$$Z = \sum_V \int [dy] \left\{ \prod_a e^{-\beta \sum_j V_{ja} |x_a - y_j|^2} \right\} e^{-\beta v \sum_k |y_k - y_{k+1}|^2} \qquad (13)$$

we perform the sum over the $\{V_{ia}\}$'s using the constraint that for each a there exists exactly one i such that $V_{ia} = 1$. This gives (Yuille, 1990):

$$Z = \int [dy] \prod_a \left\{ \sum_j e^{-\beta |x_a - y_j|^2} \right\} e^{-\beta v \sum_k |y_k - y_{k+1}|^2} \qquad (14)$$

This can be written as

$$Z = \int [dy] e^{-\beta E_{\text{eff}}[y; \beta]} \qquad (15)$$

where

$$E_{\text{eff}}[y; \beta] = \frac{-1}{\beta} \sum_a \log \left\{ \sum_j e^{-\beta |x_a - y_j|^2} \right\} + v \sum_k |y_k - y_{k+1}|^2 \qquad (16)$$

is the energy function for the original elastic net algorithm given by Durbin and Willshaw (1987).

Finding the global minimum of $E_{\mathrm{eff}}[y; \beta]$ with respect to the y in the limit as $\beta \to \infty$ will give the shortest possible tour (assuming the square norm of distances between cities).

This can be performed by a gradient descent algorithm

$$\frac{dy_j}{dt} = -\frac{\partial E_{\mathrm{eff}}}{\partial y_j} \quad \text{for all } j \tag{17}$$

while increasing β (reducing T).

By substituting Equation 16 into Equation 17, we obtain

$$\frac{dy_j}{dt} = 2 \sum_a w_{ja}(\beta)(x_a - y_j) - 2v\{(y_j - y_{j-1})$$
$$+ (y_j - y_{j+1})\} \quad \text{for all } j \tag{18}$$

where

$$w_{ja}(\beta) = \frac{e^{-\beta|x_a - y_j|^2}}{\sum_k e^{-\beta|x_a - y_k|^2}} \quad \text{for all } j, a \tag{19}$$

We can interpret this by saying that each city y_j on the net is pulled toward each data city x_a by a force weighted by the factor $w_{ja}(\beta)$, and is pulled as well toward its neighboring cities, y_{j-1} and y_{j+1}, on the net. At high temperature, the weighting factor is the same for all data cities, but as the temperature goes to zero it becomes strongly biased toward the data city closest to y_j. See Figure 1 for an illustration of the solution.

What values of T (or β) should the system start out with? It was observed empirically that at sufficiently high temperatures the elastic net collapses to a point at the center of mass of the cities. Mathematical analysis of the effective energy showed that there was a critical temperature T_c, which could be computed analytically, at which this configuration became unstable (Durbin et al., 1989). Thus T_c gives a natural upper bound for starting the algorithm. The existence of such critical temperatures is related to phase transitions in the underlying physical system (Reif, 1982). At the critical temperature, properties of the system change discontinuously and the system bifurcates.

The elastic net has been shown (see references given in Yuille, 1990) to perform significantly better than the Hopfield and Tank algorithm. It is generally believed that this is attributable to its use of hard constraints.

Additional Properties of the Elastic Net

We briefly mention three important properties of the elastic net which we do not have sufficient space to describe in detail.

First, there are close, and somewhat surprising, relations between the Hopfield and Tank formulation and the elastic net (Simic, 1990; Yuille, 1990). Although some approximations are involved, it appears that the key difference lies in the way the global constraints are imposed, while the fact that the elastic net uses hard constraints enables it to avoid the bad scaling behavior sometimes associated with the Hopfield and Tank model.

Second, the elastic net can be given a simple interpretation in terms of Bayesian probability theory (Durbin et al., 1989). This makes it a straightforward matter to generalize the elastic net to solve other problems.

Third, the EM algorithm, developed by statisticians, gives a very successful way of minimizing $E_{\mathrm{eff}}[y; \beta]$ (R. Durbin, personal communication). See Yuille, Stolorz, and Utans (1994) for a description of this approach.

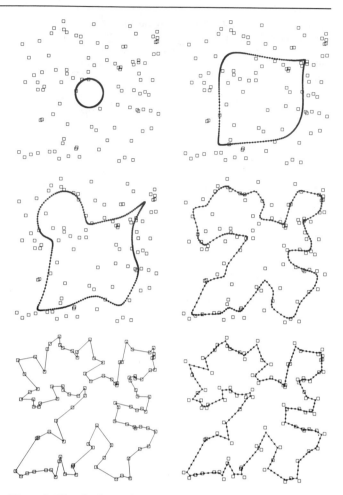

Figure 1. The elastic net in action. The large boxes represent cities through which the traveling salesman must pass. The net is the set of small boxes linked by a closed contour. The initial configuration is shown at the top left. The next three frames, from left to right and top to bottom, show minima of the effective energy at decreasing temperatures. The final, and correct, solution, shown at the bottom left, corresponds to the net configuration at the bottom right. Observe that the number of data cities is far smaller than the number of cities on the net. (This figure is courtesy of Richard Durbin.)

Hard Constraints and Mean-Field Theory

It is also possible to impose both global constraints in a hard manner (see review by Yuille and Kosowsky, 1994), and several equivalent effective energies can be found. A particularly intuitive one is

$$E_{\mathrm{eff}}[S, P, Q; \beta] = E[S] + T \sum_{ia} S_{ia} \log S_{ia}$$
$$+ \sum_a P_a \left\{ \sum_i S_{ia} - 1 \right\} + \sum_i Q_i \left\{ \sum_a S_{ia} - 1 \right\} \tag{20}$$

where the P's and Q's can be interpreted as Lagrange multipliers which impose the global constraints.

Conclusion

The Hopfield and Tank paper (1985) stimulated a lot of excitement in neural network optimization. Reports of the model's poor scaling behavior somewhat dampened this enthusiasm, and this behavior was traced to the use of soft constraints. The next generation of networks, using hard constraints, seemed to scale well (Peterson, 1990). Moreover, it also seems possible for algorithms with soft constraints to scale well, provided attention is paid to the values of the coefficients in the bias terms (Aiyer et al., 1990).

Neural network optimization has developed into a good general-purpose heuristic algorithm that is competitive with other optimization procedures. On well-studied problems like the TSP, however, it still remains inferior to the best computer science algorithms.

There is some difficulty, however, in comparing neural network algorithms fairly to existing computer algorithms. The neural network algorithms are designed to work on special-purpose analog hardware, and they suffer when these equations are discretized for serial computation. A new generation of neural network algorithms—discrete in time, though parallel and using analog variables (Waugh and Westervelt, 1993)—may be more directly comparable.

Finally, recent theoretical analysis (Yuille and Kosowsky, 1994) shows some precise relations between neural network optimization and more traditional optimization methods using barrier functions and interior point methods (Gill et al., 1986).

Appendix: Desirable Background Knowledge

The mean-field equations can be obtained in at least three different ways (see Reif, 1982, and Hertz et al., 1991):

1. *The saddle-point approximation.* This is a standard technique in statistical mechanics. The partition function $Z = \Sigma_V e^{-\beta E[V]}$ is transformed into an integral form $Z = \int [dS][dU] e^{-\beta E_{\text{eff}}[S, U; \beta]}$, where the S variables correspond to the mean fields of the V and the U are auxiliary variables which impose constraints on the S. $E_{\text{eff}}[S, U : \beta]$ is extremized with respect to S and U; then the mean fields are approximated by the resulting values S^*.
2. *Approximating the free energy.* The free energy $F = -T \log Z$. From thermodynamic relations, it can be expressed as $U - TS$, where U is the energy and S is the entropy. The expectation of the free energy, with respect to the Gibbs distribution, is approximated to give the effective energy.
3. *Mean-field techniques.* When evaluating the partition function, one evaluates the contribution from each variable by fixing the remaining variables at their mean values. This leads to consistency conditions for the mean fields, the mean-field equations.

Note that physicists have established conditions under which these approximations are valid. Unfortunately, these situations typically involve making the system infinitely large and imposing undesirable conditions on the form of the energy function. For our purposes, however, we are interested only in the weaker result that these approximations are valid at zero temperature.

Acknowledgments.   I would like to thank DARPA and the Air Force for support under contract F49620–92–J–0466.

Road Map: Dynamic Systems and Optimization
Background: I.3. Dynamics and Adaptation in Neural Networks
Related Reading: Energy Functions for Neural Networks; Statistical Mechanics of Neural Networks

References

Aiyer, S. V. B., Niranjan, M., and Fallside, F., 1990, A theoretical investigation into the performance of the hopfield model, *IEEE Trans. Neural Netw.*, 1:204–215.

Durbin, R., and Willshaw, D., 1987, An analog approach to the travelling salesman problem using an elastic net method, *Nature*, 326:689–691.

Durbin, R., Szeliski, R., and Yuille, A. L., 1989, An analysis of the elastic net approach to the travelling salesman problem, *Neural Computat.*, 1:348–358.

Gill, P., Murray, W., Saunders, M., Tomlin, J., and Wright, M., 1986, On projective newton barrier methods for linear programming and an equivalence to Karmarkar's projective method, *Math. Programming*, 36:183–209.

Hertz, J., Krogh, A., and Palmer, R. G., 1991, *Introduction to the Theory of Neural Computation*, Redwood City, CA: Addison-Wesley.

Hopfield, J. J., and Tank, D. W., 1985, Neural computation of decisions in optimization problems, *Biol. Cybern.*, 52:141–152.

Kirkpatrick, S., Gelatt, C., Jr., and Vecchi, M., 1983, Optimization by simulated annealing, *Science*, 220:671–680.

Peterson, C., 1990, Parallel distributed approaches to combinatorial optimization problems—benchmark studies on T.S.P., *Neural Computat.*, 2:261–270.

Reif, F., 1982, *Fundamentals of Statistical and Thermal Physics*, New York: McGraw-Hill.

Simic, P., 1990, Statistical mechanics as the underlying theory of "elastic" and "neural" optimization, *Network: Computat. Neural Syst.*, 1:1–15.

Waugh, F., and Westervelt, R., 1993, Stability of analog networks with local competition, *Phys. Rev. E* (in press).

Yuille, A. L., 1990, Generalized deformable models, statistical physics and matching problems, *Neural Computat.*, 2:1–24.

Yuille, A. L., and Kosowsky, J. J., 1994, A mathematical analysis of deterministic annealing, in *Artificial Neural Networks with Applications in Speech and Vision*, New York: Chapman and Hall.

Yuille, A. L., Stolorz, P., and Utans, J., 1994, Statistical physics, mixtures of distributions, and the EM algorithm, *Neural Computat.*, 6:334–340.

Convolutional Networks for Images, Speech, and Time Series

Yann LeCun and Yoshua Bengio

Introduction

The ability of multilayer backpropagation networks to learn complex, high-dimensional, nonlinear mappings from large collections of examples makes them obvious candidates for image recognition or speech recognition tasks (see PATTERN RECOGNITION). A typical approach is to feed the network with "raw" inputs (e.g., normalized images) and to rely on backpropagation to turn the first few layers into an appropriate feature extractor. While this can be done with an ordinary fully connected feedforward network with some success for tasks such as character recognition, there are problems.

First, typical images, or spectral representations of spoken words, are large, often with several hundred variables. A fully connected first layer with, say, a few hundred hidden units, would already contain tens of thousands of weights. Overfitting problems may occur if training data are scarce. In addition, the memory requirement for that many weights may rule out certain hardware implementations. But the main deficiency of unstructured nets for image or speech aplications is that they have no built-in invariance with respect to translations, or local distortions of the inputs. Before being sent to the fixed-size input layer of a neural net, character images, spoken word spectra, or other 2D or 1D signals must be approximately size normalized and centered in the input field. Unfortunately, no such preprocessing can be perfect: handwriting is often normalized at the word level, which can cause size, slant, and position variations for individual characters; words can be spoken at varying speed, pitch, and intonation. This causes variations in the position of distinctive features in input objects. In principle, a fully connected network of sufficient size could learn to produce outputs that are invariant with respect to such variations. However, learning such a task would probably result in multiple units with identical weight patterns positioned at various locations in the input. Learning these weight configurations requires a very large number of training instances to cover the space of possible variations. Conversely, in convolutional networks (which will be defined in the next section), shift invariance is automatically obtained by forcing the replication of weight configurations across space.

Second, a deficiency of fully connected architectures is that the topology of the input is entirely ignored. The input variables can be presented in any (fixed) order without affecting the outcome of the training. On the contrary, images, or spectral representations of speech, have a strong 2D local structure, and time series have a strong 1D structure: variables (or pixels) that are spatially or temporally nearby are highly correlated. Local correlations are the reasons for the well-known advantages of extracting and combining *local* features before recognizing spatial or temporal objects. Convolutional networks force the extraction of local features by restricting the receptive fields of hidden units to be local.

Convolutional Networks

Convolutional networks combine three architectural ideas to ensure some degree of shift and distortion invariance: local receptive fields, shared weights (or weight replication), and, sometimes, spatial or temporal subsampling. A typical convolutional network for recognizing characters is shown in Figure 1 (from LeCun et al., 1990). The input plane receives images of characters that are approximately size normalized and centered. Each unit of a layer receives inputs from a set of units located in a small neighborhood in the previous layer. The idea of connecting units to local receptive fields on the input goes back

to the perceptron in the early 1960s, and was almost simultaneous with Hubel and Wiesel's discovery of locally sensitive, orientation-selective neurons in the cat's visual system. Local connections have been reused many times in neural models of visual learning (see Mozer, 1991; LeCun, 1986; and NEOCOGNITRON). With local receptive fields, neurons can extract elementary visual features such as oriented edges, endpoints, or corners (or similar features in speech spectrograms). These features are then combined by the higher layers. As stated earlier, distortions or shifts of the input can cause the position of salient features to vary. In addition, elementary feature detectors that are useful on one part of the image are likely to be useful across the entire image. This knowledge can be applied by forcing a set of units, whose receptive fields are located at different places on the image, to have identical weight vectors (Rumelhart, Hinton, and Williams, 1986). The outputs of such a set of neurons constitute a *feature map*. At each position, different types of units in different feature maps compute different types of features. A sequential implementation of this, for each feature map, would be to scan the input image with a single neuron that has a local receptive field and to store the states of this neuron at corresponding locations in the feature map. This operation is equivalent to a convolution with a small-size kernel, followed by a squashing function. The process can be performed in parallel by implementing the feature map as a plane of neurons that *share* a single weight vector. Units in a feature map are constrained to perform the same operation on different parts of the image. A convolutional layer is usually composed of several feature maps (with different weight vectors), so that multiple features can be extracted at each location. The first hidden layer in Figure 1 has four feature maps with 5 × 5 receptive fields. Shifting the input of a convolutional layer will shift the output but will leave it unchanged otherwise. Once a feature has been detected, its exact location becomes less important, as long as its approximate position relative to other features is preserved. Therefore, each convolutional layer is followed by an additional layer which performs a local averaging and a subsampling, reducing the resolution of the feature map, and reducing the sensitivity of the output to shifts and distortions. The second hidden layer in Figure 1 performs 2 × 2 averaging and subsampling, followed by a trainable coefficient, a trainable bias, and a sigmoid. The trainable coefficient and bias control the effect of the squashing nonlinearity (for example, if the coefficient is small, then the neuron operates in a quasi-linear mode). Successive layers of convolutions and subsampling are typically alternated, resulting in a *bi-pyramid*: at each layer, the number of feature maps is increased as the spatial resolution is decreased. Each unit in the third hidden layer in Figure 1 may have input connections from several feature maps in the previous layer. The convolution/subsampling combination, inspired by Hubel and Wiesel's notions of "simple"

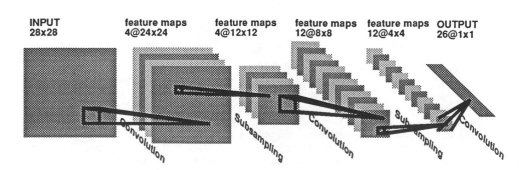

INPUT
28x28

feature maps
4@24x24

feature maps
4@12x12

feature maps
12@8x8

feature maps
12@4x4

OUTPUT
26@1x1

Convolution   Subsampling   Convolution   Subsampling   Convolution

Figure 1. Convolutional neural network for image processing, e.g., handwriting recognition.

and "complex" cells, was implemented in the neocognitron model (see NEOCOGNITRON), although no globally supervised learning procedure such as backpropagation was available then.

Since all the weights are learned with backpropagation, convolutional networks can be seen as synthesizing their own feature extractor. The weight-sharing technique has the interesting side effect of reducing the number of free parameters, thereby reducing the "capacity" of the machine and improving its generalization ability (see LeCun, 1989, on weight sharing, and LEARNING AND GENERALIZATION: THEORETICAL BOUNDS for an explanation of generalization). The network in Figure 1 contains approximately 100,000 connections, but only approximately 2600 free parameters because of the weight sharing. Such networks compare favorably with other methods on handwritten character recognition tasks (Bottou et al., 1994; see also HANDWRITTEN DIGIT STRING RECOGNITION), and they have been deployed in commercial applications.

Fixed-size convolutional networks that share weights along a single temporal dimension are known as Time-Delay Neural Networks (TDNNs). TDNNs have been used in phoneme recognition (without subsampling) (Lang and Hinton, 1988; Waibel et al., 1989), spoken word recognition (with subsampling) (Bottou et al., 1990), and on-line handwriting recognition (Guyon et al., 1991).

Variable-Size Convolutional Networks: SDNNs

While characters or short spoken words can be size normalized and fed to a fixed-size network, more complex objects such as written or spoken words and sentences have inherently variable size. One way of handling such a composite object is to segment it heuristically into simpler objects that can be recognized individually (e.g., characters, phonemes). However, reliable segmentation heuristics do not exist for speech or cursive hand writing. A brute-force solution is to scan (or replicate) a recognizer at all possible locations across the input. While this can be prohibitively expensive in general, convolutional networks can be scanned or replicated very efficiently over large, variable-size input fields. Consider one instance of a convolutional net and its *alter ego* at a nearby location. Because of the convo-

lutional nature of the networks, units in the two nets that look at identical locations on the input have identical outputs; therefore, their output does not need to be computed twice. In effect, replicating a convolutional network can be done simply by increasing the size of the field over which the convolutions are performed, and replicating the output layer, effectively making it a convolutional layer. An output whose receptive field is centered on an elementary object will produce the class of this object, while an in-between output may be empty or contain garbage. The outputs can be interpreted as evidence for the categories of object centered at different positions of the input field. A postprocessor therefore is required to pull out consistent interpretations of the output. Hidden Markov models (HMMs) or other graph-based methods are often used for that purpose (see SPEECH RECOGNITION: PATTERN MATCHING, and PATTERN RECOGNITION in this volume). The replicated network and the HMM can be trained simultaneously by backpropagating gradients through the HMM. Globally trained, variable-size TDNN/HMM hybrids have been used for speech recognition (see PATTERN RECOGNITION for a list of references) and on-line handwriting recognition (Schenkel et al., 1993). Two-dimensional replicated convolutional networks, called *Space Displacement Neural Networks* (SDNNs) (Figure 2), have been used in combination with HMMs or other elastic matching methods for handwritten word recognition (Keeler, Rumelhart, and Leow, 1991; Matan et al., 1992; Bengio, LeCun, and Henderson, 1994). Another interesting application of SDNNs is object spotting (Wolf and Platt, 1994).

An important advantage of convolutional neural networks is the ease with which they can be implemented in hardware. Specialized analog/digital chips have been designed and used in character recognition and in image-preprocessing applications (Boser et al., 1991). Speeds of more than 1000 characters per second were obtained with a network with approximately 100,000 connections.

The idea of subsampling can be turned around to construct networks that are similar to TDNNs but can generate sequences from labels. These networks are called reverse-TDNNs because they can be viewed as upside-down TDNNs: temporal resolution increases from the input to the output, through alternated oversampling and convolution layers (Simard and LeCun, 1992).

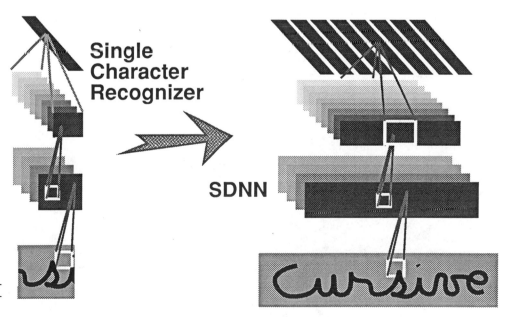

Figure 2. A variable-size replicated convolutional network, called a Space Displacement Neural Network (SDNN).

Discussion

Convolutional neural networks are a good example of an idea inspired by biology that resulted in competitive engineering solutions that compare favorably with other methods (Bottou et al., 1994). Although applying convolutional nets to image recognition removes the need for a separate hand-crafted feature extractor, normalizing the images for size and orientation (if only approximately) is still required. Shared weights and subsampling bring invariance with respect to small geometric transformations or distortions, but fully invariant recognition is still beyond reach. Radically new architectural ideas, possibly suggested by biology, will be required for a fully neural image or speech recognition system.

Road Map: Learning in Artificial Neural Networks, Deterministic
Background: I.3. Dynamics and Adaptation in Neural Networks

References

Bengio, Y., LeCun, Y., and Henderson, D., 1994, Globally trained handwritten word recognizer using spatial representation, space displacement neural networks and hidden Markov models, in *Advances in Neural Information Processing Systems 6* (J. Cowan, G. Tesauro, and J. Alspector, Eds.), San Mateo, CA: Morgan Kaufmann, pp. 937–944.

Boser, B., Sackinger, E., Bromley, J., LeCun, Y., and Jackel, L., 1991, An analog neural network processor with programmable topology, *IEEE J. Solid-State Circuits*, 26:2017–2025.

Bottou, L., Cortes, C., Denker, J., Drucker, H., Guyon, I., Jackel, L., LeCun, Y., Muller, U., Sackinger, E., Simard, P., and Vapnik, V., 1994, Comparison of classifier methods: A case study in handwritten digit recognition, in *Proceedings of the International Conference on Pattern Recognition*, Los Alamitos, CA: IEEE Computer Society Press.

Bottou, L., Fogelman-Soulié, F., Blanchet, P., and Lienard, J. S., 1990, Speaker independent isolated digit recognition: Multilayer perceptrons vs dynamic time warping, *Neural Netw.*, 3:453–465.

Guyon, I., Albrecht, P., LeCun, Y., Denker, J. S., and Hubbard, W., 1991, Design of a neural network character recognizer for a touch terminal, *Pattern Recognition*, 24(2):105–119.

Keeler, J., Rumelhart, D., and Leow, W., 1991, Integrated segmentation and recognition of hand-printed numerals, in *Advances in Neu-*

ral Information Processing Systems 3 (R. P. Lippman, J. M. Moody, and D. S. Touretzky, Eds.), San Mateo, CA: Morgan Kaufmann, pp. 557–563.

Lang, K., and Hinton, G., 1988, *The Development of the Time-Delay Neural Network Architecture for Speech Recognition*, Technical Report CMU–CS–88–152, Carnegie-Mellon University, Pittsburgh.

LeCun, Y., 1986, Learning processes in an asymmetric threshold network, in *Disordered Systems and Biological Organization* (E. Bienenstock, F. Fogelman-Soulié, and G. Weisbuch, Eds.), Les Houches, France: Springer-Verlag, pp. 233–240.

LeCun, Y., 1989, *Generalization and Network Design Strategies*, Technical Report CRG–TR–89–4, Department of Computer Science, University of Toronto.

LeCun, Y., Boser, B., Denker, J., Henderson, D., Howard, R., Hubbard, W., and Jackel, L., 1990, Handwritten digit recognition with a back-propagation network, in *Advances in Neural Information Processing Systems 2* (D. Touretzky, Ed.), San Mateo, CA: Morgan Kaufmann, pp. 396–404.

Matan, O., Burges, C., LeCun, Y., and Denker, J., 1992, Multi-digit recognition using a space displacement neural network, in *Advances in Neural Information Processing Systems 4* (J. Moody, S. Hanson, and R. Lipmann, Eds.), San Mateo, CA: Morgan Kaufmann, pp. 488–495.

Mozer, M., 1991, *The Perception of Multiple Objects, A Connectionist Approach*, Cambridge, MA: MIT Press.

Rumelhart, D., Hinton, G., and Williams, R., 1986, Learning representations by back-propagating errors, *Nature*, 323:533–536.

Schenkel, M., Weissman, H., Guyon, I., Nohl, C., and Henderson, D., 1993, Recognition-based segmentation of on-line hand-printed words, in *Advances in Neural Information Processing Systems 5* (C. Hanson and L. Giles, Eds.), San Mateo, CA: Morgan Kaufmann, pp. 723–730.

Simard, P., and LeCun, Y., 1992, Reverse TDNN: An architecture for trajectory generation, in *Advances in Neural Information Processing Systems 4* (J. Moody, S. Hanson, and R. Lipmann, Eds.), San Mateo, CA: Morgan Kaufmann, pp. 579–588.

Waibel, A., Hanazawa, T., Hinton, G., Shikano, K., and Lang, K., 1989, Phoneme recognition using time-delay neural networks, *IEEE Trans. Acoustics, Speech, Signal Processing*, 37:328–339.

Wolf, R., and Platt, J., 1994, Postal address block location using a convolutional locator network, in *Advances in Neural Information Processing Systems 6* (J. Cowan, G. Tesauro, and J. Alspector, Eds.), San Mateo, CA: Morgan Kaufmann, pp. 745–752.

Cooperative Behavior in Networks of Chaotic Elements

Kunihiko Kaneko

Chaos in the Brain

The importance of chaotic dynamics in the brain has been appreciated both theoretically and experimentally. Chaos is an irregular motion appearing in a deterministic system governed, for example, by some differential equations or maps. The necessity of chaotic dynamics in the brain can be inferred in the following contexts. So far it is the only scientific mechanism to connect deterministic and probabilistic views at a macroscopic level. By chaotic dynamics, a tiny perturbation can be amplified to a macroscopic level. Second, chaos is the only mechanism so far to create complexity from simple rules. For other examples of complex behavior, such as the spin glass model or cellular automaton, we need a huge number of degrees of freedom or rules (often made from "random couplings," externally assigned, which also require much information). The

necessity of chaotic dynamics for biological information processing has been discussed by Tsuda (1992).

Experimentally, the existence of chaotic dynamics in the brain has been confirmed. Even a single neuron or a small ensemble of neurons can show a chaotic time series or frequency locking (Hayashi, Nakao, and Hirakawa, 1982). Some data of EEG time series show irregular motion, which is found to be chaotic with the aid of reconstruction of attractors or dimension estimates (Basar, 1990; Freeman, 1986). Partial synchronization of nonlinear oscillations has been discovered in the visual cortex (Gray et al., 1989; Eckhorn, 1991; see also SYNCHRONIZATION OF NEURONAL RESPONSES AS A PUTATIVE BINDING MECHANISM). Although the nature of their nonlinear oscillation has not yet been clarified, the dynamics strongly suggests the existence of chaos (but see CHAOS IN NEURAL SYSTEMS). Indeed, the significance of chaos can be deduced there; both the synchronization and desynchronization mechanisms are necessary for the processing of external inputs, by some balance between chaos and entrainment (see the discussion of the globally coupled map later in this article).

Consequently, the neural dynamics consists of an ensemble of elements with complex dynamics and complex coupling. Most current neural network studies, however, use oversimplified local dynamics (0-1 or a sigmoid function). Therefore, it is natural and important to ask, "What happens if we use moderately simplified elements with a chaotic response with oversimplified couplings, as a different kind of simplification from traditional neural dynamics?" This article offers an answer to this question by providing a novel phenomenology in a network of chaotic elements. Note that each element does not necessarily correspond to a single neuron; rather, it represents the activity of an ensemble of neurons which shows (chaotic) oscillations.

The dynamics of coupled chaotic elements has recently been investigated in various topologies. In particular, coupled systems on a simple topology have been intensively studied. Here we will review two of these: the coupled map lattice with nearest-neighbor interaction; and globally coupled maps. We will also briefly review the relevance of the results to neural information processing.

The Coupled Map Lattice

The coupled map lattice (CML) is a dynamical system with discrete time ("map"), discrete space ("lattice"), and a continuous state (Kaneko, 1993). It usually consists of dynamical elements on a lattice, each interacting ("coupled") with suitably chosen sets of other elements. It was originally introduced to study spatiotemporal chaos and turbulent phenomena in general. Our modeling with CML is based on the following steps:

1. Choose a (set of) field variable(s) on a lattice. This set of variable(s) is not at a microscopic, but at a macroscopic, level.
2. Decompose the processes underlying the phenomena into independent components (e.g., reaction, diffusion, and so on).
3. Replace each component by a simple parallel dynamics on a lattice. The dynamics consists of nonlinear transformation of the field variable on each lattice point, and/or a coupling term among suitably chosen neighbors.
4. Carry out each unit dynamics procedure successively.

A process with a local chaotic process and diffusion has extensively been investigated. As the simplest choice, we take a local (one-dimensional) map for chaos, and a discrete Laplacian operator for the diffusion. The former process is given by $x'_n(i) = f(x_n(i))$, where $x_n(i)$ is a variable at time n and lattice site i, and $x'_n(i)$ is introduced as the intermediate value. The discrete Laplacian operator for diffusion is given by $x_{n+1}(i) = (1 - \varepsilon)x'_n(i) + (\varepsilon/2)(x'_n(i+1) + x'_n(i-1))$. Combining these two processes, our dynamics is given by

$$x_{n+1}(i) = (1 - \varepsilon)f(x_n(i)) + \frac{\varepsilon}{2}(f(x_n(i+1)) + f(x_n(i-1))) \quad (1)$$

For the local map $f(x)$, the logistic map $f(x) = 1 - ax^2$ is often adopted, since it is the best-examined model in the studies of chaos.

In contrast with CELLULAR AUTOMATA (CA) (q.v.), or other systems of on-off units, elements at the lattice points take continuous, rather than discrete, values. A cell represents not a microscopic but a semimacroscopic state. This is the reason why a moderate number of cells is sufficient for simulations of CML, while a huge number of cells is often required for CA.

Because of the sensitive dependence on initial conditions that is characteristic of chaos, a homogeneous state becomes unstable in our CML with chaotic components. In the model (Equation 1) with the logistic map with weak nonlinearity (e.g., $1.42 < a < 1.55$), domains of various sizes are spontaneously formed in which the oscillations are highly correlated. Oscillations of elements are out of phase between the neighboring domains, while they are in phase for elements in the same domain, although they cannot be completely synchronized because of chaos. The nature of dynamics of a site depends on the domain size it belongs to (a feature called *spatial bifurcation*). In a large domain, the motion is chaotic, while in smaller domains it is periodic, with the periods depending on the domain sizes.

In the *frozen random state*, domain positions are frozen in space. With increasing nonlinearity in each element, domains "melt," and a transition to fully developed spatiotemporal chaos begins. In this spatiotemporal intermittency (STI) transition, ordered motion and turbulent bursts coexist in space-time. The ordered region forms a large cluster whose size distribution obeys the power-law distribution, leading to a long-range correlation in space-time (Kaneko, 1989). Other discoveries in the CML include the chaotic traveling wave and the Brownian motion of chaotic defects (Kaneko, 1993). The relevance of these discoveries to the information processing will be discussed later in this article.

The Globally Coupled Map

In neural dynamics, the interaction among elements is not necessarily local; rather, it is global with a rather complicated structure. The simplest case is given by the *globally coupled map* (GCM) of chaotic elements (Kaneko, 1990). An example is

$$x_{n+1}(i) = (1 - \varepsilon)f(x_n(i)) + \frac{\varepsilon}{N}\sum_{j=1}^{N} f(x_n(j)) \quad (2)$$

where n is a discrete time step, i is the index of an element ($i = 1, 2, \ldots, N$ = system size), and $f(x) = 1 - ax^2$. The model is a mean-field theory type of extension of the CML. The dynamics consists of parallel nonlinear transformation and a feedback from the "mean field." It is equivalent to

$$y_{n+1}(i) = f\left[(1 - \varepsilon)y_n(i) + \frac{\varepsilon}{N}\sum_{j=1}^{N} y_n(j)\right]$$

with the aid of transformation $y_n(i) = f(x_n(i))$. In this form, one can see a clear correspondence with neural nets: if one chooses a sigmoid function (e.g., $\tanh(\beta x)$) as $f(x)$ and a random or coded coupling $\varepsilon_{i,j}$ instead of the above homogeneous coupling, a typical neural net is obtained.

Through the interaction, some elements oscillate synchronously, while chaotic instability leads to the destruction of coherence. Attractors in GCM are classified by the number of synchronized clusters k and the number of elements for each cluster N_k. Here a cluster is defined as the set of elements in which $x(i) = x(j)$. For each clustering condition $(N_1, N_2, \ldots, N_k)$, there are $N!/(N_1!N_2!\cdots N_k!)$ partitions. We have exponentially many attractors per clustering condition.

In a globally coupled chaotic system in general, the following phases appear successively with the increase of nonlinearity in the system:

1. *Coherent phase.* A coherent attractor ($k = 1$) is obtained from almost all initial conditions.
2. *Ordered phase.* All initial conditions lead to attractors with few clusters ($k = o(N)$).

3. *Partially ordered phase.* Coexistence of attractors with many clusters ($k = O(N)$) and of attractors with few clusters.
4. *Turbulent phase.* All attractors have many ($k = N$) clusters.

In the turbulent phase, $x(i)$ takes almost random values almost independently. One might then expect that the mean field $h \equiv (1/N) \Sigma_j f(x(j))$ obeys the law of large numbers and the central limit theorem. This is not the case. The mean square deviation of the mean field, $\langle h^2 \rangle - \langle h \rangle^2$, decreases with $1/N$ up to some crossover size, but the decrease stops with the further increase of N. The distribution of mean field h approaches a slightly non-Gaussian form. This observation means that *globally coupled chaos violates the law of large numbers*, which implies the emergence of hidden correlation (coherence) among elements. This is a general statement in a globally coupled chaotic system (Kaneko, 1992). Since EEG is an average of some neuronal activities, one may expect that this statement holds for EEG time series.

In the partially ordered phase, orbits visit several ordered states successively via highly chaotic states. In the ordered states, elements split into a few "effective" clusters, whose elements are almost synchronized, i.e., they take the same value up to a given level of precision, say 10^{-5}. Our system exhibits *intermittent change between self-organization toward a coherent structure and its collapse to a high-dimensional disordered motion*. The total dynamics consists of residences at ordered states interspersed with a high-dimensional chaotic state. This dynamics, called *chaotic itinerancy*, has been found in a model of neural dynamics, (Tsuda, 1991), in optical turbulence (Ikeda, Matsumoto, and Ohtsuka, 1989), and in GCM (Kaneko, 1990).

Relevance to Neural Information Processing

The relevance of chaos to neural information processing has recently been emphasized by Tsuda (1992), while high-dimensional chaos has a much larger potential. Here we briefly list some of the capabilities of a network of chaotic elements and discuss the relevance to neural dynamics.

1. *Hierarchical memory storage at many attractors and switching.* In the frozen random states and the partially ordered states discussed earlier, a huge number of attractors (c^N with $c > 2$) coexist, and they are hierarchically organized as a tree or a domain structure. Chaos is essential to this hierarchical complexity, since it leads to successive splittings of clusters. These attractors can serve as basic hierarchical memory units, and they can be related to dynamic categorization in the brain. By applying an input to a single element, one can make a switch from one attractor to another. Depending on inputs, different attractors are retrieved as different memories. During the course of switching, intermittent chaos is often observed, which may be useful: the system partially retains its previous history, but a search for a novel state is also possible through the destruction of structure by chaos.
2. *Spontaneous transition among local structures with spatiotemporal intermittency or chaotic itinerancy.* Spontaneous switching among ordered states is possible through chaotic states. The switch is deterministic, but because of the chaotic motion of the system it is not rigidly fixed. The importance of chaotic itinerancy in neural networks has been stressed by Tsuda (1992) with respect to spontaneous recall of memories.
3. *Information creation and transmission by chaotic traveling wave.* The ability to create information through chaos was first noted by Shaw (1981). Through chaotic dynamics, microscopic information is amplified to a macroscopic level. By measuring an orbit successively, information on initial states is gained. In CML, the created information can be transmitted to other elements through the interaction. In the chaotic traveling wave discovered in CML, for example, attractors with different velocities coexist. Switching among attractors by single (local) inputs is again possible, to change the speed of transmission. By means of the traveling wave, some information is transmitted to the whole space. Thus a transformation from local to global information is possible through this switching.

For example, assume that some information is encoded as a real number. By applying an input of this real number to one site, the CML dynamics changes its motion and transmits the information. By measuring the time series of the dynamics at a distant lattice site, this information can be decoded again.

4. *Search with the use of spatiotemporal chaos.* Freeman has discovered chaotic transients in the activity of neurons of the olfactory bulb when a rabbit encounters novel odors. He has proposed that chaos may be useful in the search for memorized states (Freeman, 1986). The search involving chaos is based on its two faces: randomness and temporal order. The former is necessary for the search, while the latter eliminates unnecessary wandering in the phase space. In the course of the search, some spatial and temporal structures are thus preserved. Indeed, Nozawa (1992) has proposed a GCM model combined with a neural network, which is more powerful in optimization of the Traveling Salesman Problem than conventional neural networks of the Hopfield type. Besides the efficiency, we note that the searching proceeds in a similar manner to that of a human being. In the model, the strength of chaos at an element decreases as a path is fixed, which may remind us of a kind of simulated annealing. However, the decrease is not given in advance (as in simulated annealing) but occurs spontaneously, and it is not always monotonic.
5. *Separated procedures as different functions.* If each procedure in our CML is related to some task in processing (instead of chaos or diffusion), it may be possible to compromise multiple different requests by a suitable choice of CML. In image processing, for example, the preservation of relevant structure (e.g., patterns with clear edges) and smoothing to eliminate a noise may be carried out by a CML with a suitably chosen local mapping, and a diffusion process (for smoothing).
6. *Partial coherence as a mechanism of grouping and feature detection.* In GCM, partitions into clusters are spontaneously created, while domains of partial synchronization are formed in CML. In information processing, grouping of many inputs is required, based on some condition such as continuity (Gray et al., 1989; Eckhorn, 1991). Our network of chaotic elements can provide such a capability, since the coupling term leads to the grouping of elements with (partial) coherent oscillations. In neural processing, a (partially) synchronized cluster changes its members according to inputs to the system (Vaadia and Aertsen, 1992). In GCM, this request is attained by the orbital instability of chaos.

Change of coherence is often important in neural dynamics. In epilepsy, an ensemble of neurons exhibits a large spike because of the oversynchronized oscillation of neurons, and information-processing ability is often lost. Vaadia and Aertsen (1992) have found that effective coupling among neurons varies temporally in a rather short time scale, and that the degree of synchronization of multineuron activity changes both temporally and by the choice of pairs. Such change of synchronization is typically observed in the partially ordered phase of GCM.

Acknowledgments. This work was partially supported by a Grant-in-Aid for Scientific Research from the Ministry of Education, Science, and Culture of Japan.

Road Map: Cooperative Phenomena
Background: I.3. Dynamics and Adaptation in Neural Networks
Related Reading: Chains of Coupled Oscillators; Collective Behavior of Coupled Oscillators

References

Basar, E., 1990, *Chaos in Brain Function*, New York: Springer-Verlag.

Eckhorn, R., 1991, Stimulus-specific synchronizations in the visual cortex, in *Neuronal Cooperativity* (J. Kruger, Ed.), Berlin: Springer-Verlag.

Freeman, W., 1986, Petit mal seizure spikes in olfactory bulb and cortex by caused runaway inhibition after exhaustion of excitation, *Brain Res. Rev.*, 11:259–284.

Gray, C. M., Koenig, P., Engel, P., and Singer, W., 1989, Oscillatory responses in cat visual cortex exhibit inter-columnar synchronization which reflects global stimulus properties, *Nature*, 338:334–337.

Hayashi, H., Nakao, M., and Hirakawa, K., 1982, Chaos in the self-sustained oscillation of an excitable membrane under sinusoidal stimulus, *Phys. Lett. A*, 88:265–269.

Ikeda, K., Matsumoto, K., and Ohtsuka, K., 1989, Maxwell-Bloch turbulence, *Prog. Theoret. Phys. Suppl.*, 99:295–324.

Kaneko, K., 1989, Pattern dynamics in spatiotemporal chaos, *Physica D*, 34:1–41.

Kaneko, K., 1990, Clustering, coding, switching, hierarchical ordering, and control in network of chaotic elements, *Physica D*, 41:137–172. ◆

Kaneko, K., 1992, Mean field fluctuation in network of chaotic elements, *Physica D*, 55:368–384.

Kaneko, K., Ed., 1993, *Theory and Applications of Coupled Map Lattices*, New York: Wiley. ◆

Nozawa, H., 1992, A neural network model as a globally coupled map and applications based on chaos, *Chaos*, 2:377–386.

Shaw, R., 1981, Strange attractors, chaotic behavior, and information flow, *Z. Naturforsch.*, 36a:80–112. ◆

Tsuda, I., 1991, Chaotic itinerancy as a dynamical basis of hermeneutics in brain and mind, *World Future*, 32:167–184.

Tsuda, I., 1992, Dynamic link of memory: Chaotic memory map in non-equilibrium neural networks, *Neural Netw.*, 5:313–326.

Vaadia, E., and Aertsen, A., 1992, Coding and computation in the cortex: Single-neuron activity and cooperative phenomena, in *Information Processing in the Cortex: Experiments and Theory* (A. Aertsen and V. Braitenberg, Eds.), Berlin: Springer-Verlag.

Cooperative Phenomena

Hermann Haken

Introduction

Most objects of scientific study in physics, chemistry, biology, and many other fields, are composed of many individual parts that interact with each other. By their interaction, the individual parts may produce cooperative phenomena that are connected with the emergence of new qualities that are not present at the level of the individual subsystems. For instance, the human brain consists of a network of some 100 billion neurons that, through their cooperation, bring about pattern recognition, associative memory, steering of locomotion, speech, etc. Physical systems may serve as model systems or as test grounds for the development of new concepts and mathematical tools. Examples are provided by phase transitions of systems in thermal equilibrium, such as the water-ice transition or the evaporation of water. Ferromagnets are composed of many individual elementary magnets that change their orientation randomly, but interact to align each other below a critical temperature, the so-called Curie temperature, and may thus produce a macroscopic magnetization. In metals at a somewhat elevated temperature, the electrons move entirely independently of each other; however, in superconducting metals below a critical temperature, the electrons form pairs and are then able to carry electric current without any resistance. Quite evidently, in each of these cases the macroscopic properties of the systems change dramatically when the temperature, changed from the outside, passes through a critical value. Theories to deal with these phase transitions were originally developed by Landau (Landau and Lifshitz, 1959), and later in a more adequate fashion by Wilson (1971) and others. In contrast to biological systems, whose functioning is maintained by a continuous flux of energy and/or matter through them, the physical systems just mentioned are truly dead. There are, however, physical systems whose spatial, temporal, or spatio-temporal structures are maintained by a continuous influx of energy and/or matter, and which may show phase-transition-like phenomena. Thus, they seem suited to act as model systems for biological systems, and they may also guide the development of new types of neural nets. Typical examples are provided by lasers and fluids.

Consider a solid-state laser (Haken, 1983). In a solid matrix, laser-active atoms are embedded. The material has the form of a transparent rod with mirrors mounted at each end. The mirrors reflect the light that is emitted from the laser-active atoms and that runs in the axial direction; thus the mirrors allow the light to interact with the laser-active atoms for an extended period. The atoms can be excited by light from other lamps. When the excitation level is low, the individual atoms emit their light independently of each other so that microscopically chaotic light waves emerge. If the pump power exceeds a critical value, however, the properties of the light change dramatically. It comes to be composed of a single, practically infinitely long, wave track that shows only minor fluctuations in phase and amplitude. In the laser, the emission acts of the individual electrons have become correlated to produce the collective phenomenon of laser light. The basic mechanism for the emergence of a single coherent light wave is as follows: When a light wave has been emitted from an excited atom, it may hit another excited atom and force that atom to enhance the impinging light wave by the process of stimulated emission (as studied by Einstein). When a number of atoms are excited, an avalanche of that light wave is generated. Again and again such avalanches are generated, and they start to compete with each other. The one that has the largest growth rate wins the competition and establishes the laser light wave. Because the energy supply to the system is limited, the light wave eventually saturates. Thus, the established light wave forces the individual electrons of the atoms into its rhythm, which in turn forces them to support it. In the terminology of synergetics (Haken, 1983), the light wave acts as the order parameter. This is a variable that describes the macroscopic order of the system and gives orders to the individual parts of it. In the laser, the order parameter enslaves the electrons of the atoms. When the pump power to the laser is further increased, new effects may appear; for instance, several coherent waves may be produced simultaneously, ultrashort regular light pulses may occur, or light may show deterministic chaos. Thus, an input variable, namely the pump power, controls the self-organization of the system to produce entirely new temporal structures. This input variable is called a control parameter. Note the difference between the concepts of order parameter and control parameter: a *control parameter* is a (physical) quantity that is imposed on the system from the outside, whereas the *order parameter* is established by the system itself via self-organization.

When a fluid is heated from below, beyond a critical temperature difference ΔT between the lower and upper surfaces, macroscopic patterns may occur—in the form of rolls or hexagons, for instance. When the temperature difference ΔT is increased, the rolls may start several kinds of oscillatory motions. Macroscopic spatiotemporal patterns may also be formed in special chemical reactions, such as in the Belousov-Zhabotinsky reaction.

Outline of the Mathematical Approach

In order to make visible the profound analogies in the formation of patterns by quite different systems, and to prepare the ground for establishing an important analogy between pattern formation and pattern recognition, we have to adopt a rather abstract formulation (Haken, 1983). To describe a system at the microscopic level, we introduce the state vector

$$\mathbf{q} = (q_1, q_2, \ldots, q_n) \tag{1}$$

In the example of a laser, the individual components q_j may stand for the time-dependent field amplitudes used in a decomposition of the electric field strengths of the laser light into so-called modes. The modes are typically standing or running sinusoidal waves that fit between the mirrors of the laser. Further components q_j may stand for the dipole moments of the individual atoms, and for the inversion (i.e., degree of excitation) of the atoms. In fluids, q_j may denote the density, the components of the velocity field, and the temperature field. In this case, the components are both space- and time-dependent. In models of chemical reactions, q_j may stand for the concentration of a chemical of kind j. In all these cases and many others, the state vector develops in the course of time; this time evolution is described by so-called evolution equations, which are of the form

$$\dot{\mathbf{q}} = \mathbf{N}(\mathbf{q}, \nabla, \alpha) + \mathbf{F}(t) \tag{2}$$

The left-hand side is the temporal derivative of the state vector $\mathbf{q}$, which is determined by the right-hand side of this equation. $\mathbf{N}$ is a nonlinear function of the state vector. The state vector may be subjected to spatial differential operations $\nabla = (\partial/\partial x, \partial/\partial y, \partial/\partial z)$, and the system may be controlled from the outside by control parameters α. In the case of the laser, a typical control parameter is the power input into the laser. Finally, $\mathbf{F}$ represents stochastic forces that stem from internal or external fluctuations. In the case of the laser, this may include the spontaneous emission process. In chemical reactions, the typical reaction-diffusion equations are of the form

$$\dot{\mathbf{q}} = \mathbf{N}(q, \alpha) + \mathbf{F}(t) + D\Delta \mathbf{q} \tag{3}$$

where D is a diffusion matrix, and $\Delta = \nabla^2$ is the Laplace operator.

Equation 2, in this general form, covers an enormous range of phenomena, and at first sight it appears impossible to devise a general method of solution. From the experimental point of view, however, we are quite often confronted with the following situation. Below a certain pump power threshold, a laser acts as a lamp, not emitting coherent light. When we slowly increase the pump power, suddenly the laser forms coherent laser light. In other words, the former state has become unstable and has been replaced by a new state. This suggests the following strategy: We assume that, for a given control parameter value α_0, a solution $\mathbf{q}_0$ of Equation 2 (with $\mathbf{F} \equiv 0$) is known. The general procedure allows us to treat all kinds of $\mathbf{q}_0$ as such initial states; $\mathbf{q}_0$ may be time-independent (representing a fixed point), time-periodic (representing a limit cycle), time-quasi-periodic (forming a torus), or even time-chaotic (forming a chaotic attractor).

Common features and differences with respect to bifurcation theory, an important branch of dynamic systems theory, are worth mentioning (Guckenheimer and Holmes, 1990). That theory considers bifurcation from a fixed point and from time-periodic solutions; it neglects, however, the important impact of fluctuations. It also does not treat relaxation processes. Synergetics, in contrast, includes these phase-transition effects, as well as the bifurcation from quasi-periodic and chaotic reference states $\mathbf{q}_0$.

Here we explicitly treat the case of a fixed point $\mathbf{q}_0$. To check the stability of the state $\mathbf{q}_0$ when we change the control parameter, we make the hypothesis that, for α close to α_0, the state q_α can be written as

$$\mathbf{q}_\alpha = \mathbf{q}_0 + \mathbf{w}(\mathbf{x}, t) \tag{4}$$

where $\mathbf{w}$ is assumed to be a small quantity. We may thus insert Equation 4 into Equation 2. In the resulting equation for $\mathbf{w}$ (still with $\mathbf{F} \equiv 0$), we keep only the linear terms and obtain

$$\dot{\mathbf{w}} = L(\mathbf{q}_0)\mathbf{w} \tag{5}$$

where L is a linear operator. It can be shown quite generally in all the previously mentioned cases of $\mathbf{q}_0$ that the solutions $\mathbf{w}$ can be written in the form

$$\mathbf{w}(\mathbf{x}, t) = \begin{cases} e^{\lambda t}\mathbf{v}_u(\mathbf{x}, t) & Re\lambda \geq 0 \\ e^{\lambda t}\mathbf{v}_s(\mathbf{x}, t) & Re\lambda < 0 \end{cases} \tag{6}$$

where we distinguish between two sets of modes: *unstable modes* with $Re\lambda \geq 0$ and *stable modes* with $Re\lambda < 0$ (Re = real part of). Here $\mathbf{v}$ depends on t in a way that is weaker than an exponential growth or decay. The λ's are the eigenvalues of Equation 5. Any linear combination of Equation 6 is, of course, again a solution of Equation 5. For what follows, however, it will be crucial to treat the solutions of Equation 6 individually; we must distinguish between those eigenvalues λ whose real part is positive or zero, and those whose real part is negative. It is our goal to solve the nonlinear stochastic Equation 2 exactly, i.e., not only in a linear approximation. To this end, we expand the unknown function $\mathbf{q}$ into a superposition of the individual eigenfunctions $\mathbf{v}$ of the linear operator L in Equation 5:

$$\mathbf{q}(\mathbf{x}, t) = \mathbf{q}_0 + \sum_u \xi_u(t)\mathbf{v}_u(\mathbf{x}) + \sum_s \xi_s(t)\mathbf{v}_s(\mathbf{x}) \tag{7}$$

In the mathematical sense, this is a complete superposition representing $\mathbf{q}$ as a function of $\mathbf{x}$. The amplitudes ξ_u and ξ_s are still unknown functions of time. To obtain equations for ξ_u, ξ_s, we insert the expansion (Equation 7) into

$$\dot{\mathbf{q}} = \mathbf{N}(\mathbf{q}, \alpha) + \mathbf{F}(t) \tag{8}$$

On the right-hand side of Equation 8, we expand the nonlinear function $\mathbf{N}$ that has become a function of ξ_u and ξ_s into a power series of ξ_u and ξ_s. The *linear* terms in ξ_u and ξ_s may be written $\xi_u L(\mathbf{q}_0)\mathbf{v}_u$ and $\xi_s L(\mathbf{q}_0)\mathbf{v}_s$, respectively. Because of Equation 6, in the case of a fixed point $\mathbf{q}_0$, we may replace, for instance, $L\mathbf{v}_u$ by $\lambda_u\mathbf{v}_u$.

Keeping these and all higher-order terms and projecting both sides of Equation 8 on the modes $\mathbf{v}$, we obtain the following two equations:

$$\dot{\xi}_u = \lambda_u\xi_u + \hat{N}_u(\xi_u, \xi_s) + \hat{F}_u(t) \qquad (9)$$

$$\dot{\xi}_s = \lambda_s\xi_s + \hat{N}_s(\xi_u, \xi_s) + \hat{F}_s(t) \qquad (10)$$

The first term on the right-hand side of Equations 9 and 10 stems from the linear part of N; use was made of the fact that $\mathbf{v}_u$ and $\mathbf{v}_s$ are eigenfunctions of the linear operator L in Equation 5 with the eigenvalues λ that occur on the right-hand side of Equation 6. These equations are entirely equivalent to Equation 8. However, provided the inequality

$$|Re\,\lambda_s| \gg |Re\,\lambda_u| \qquad (11)$$

holds, the *slaving principle* of synergetics (Haken, 1983) may be applied. According to this principle, the mode amplitudes ξ_s are uniquely determined (enslaved) by ξ_u. The possibility of expressing ξ_s by ξ_u can be made plausible in the following fashion: Let us assume that according to Equation 11 the mode amplitudes ξ_s relax much faster than the mode amplitudes ξ_u. Consider Equation 10 with slowly varying ξ_u that act as driving forces on ξ_s. When we neglect transients, ξ_s, being driven by ξ_u, changes much more slowly than it normally would because of the first term on the right-hand side in Equation 10. In other words, this means that $\dot{\xi}_s$ can be neglected against $\lambda_s\xi_s$, or that $\dot{\xi}_s = 0$. This turns Equation 10 into an algebraic equation that can be solved for ξ_s, expressing ξ_s by ξ_u. This approximation is called *adiabatic approximation*. The slaving principle assures us that this procedure is a first step toward a systematic procedure by which ξ_s can be expressed uniquely and exactly by ξ_u and $\hat{F}_s$:

$$\xi_s(t) = f_s(\xi_u(t), t) \qquad (12)$$

The explicit dependence of f_s on t stems from the time dependence of the fluctuating forces, but not that of the amplitudes ξ_u. In most cases of practical interest, f_s can be approximated by a low-order polynomial in ξ_u.

In practically all cases that have been treated in the literature, the systems are of a very high dimension, i.e., they contain very many variables, but the number of unstable mode amplitudes ξ_u is very small. The amplitudes ξ_u are called *order parameters*, whereas the variables ξ_s can be called *enslaved variables*. The order-parameter concept allows for an enormous reduction of the degrees of freedom. Think of a single-mode laser in which we have one mode and, say, 10^{16} degrees of freedom stemming from the dipole moments and inversions of the individual laser atoms. The order parameter in the single-mode laser is identical with the lasing mode, i.e., a single degree of freedom. Once we have expressed the enslaved modes by means of the order parameters ξ_u via Equation 12, we may insert Equation 12 into Equation 9, and thus find equations for the order parameters alone:

$$\dot{\xi}_u = \lambda_u\xi_u + \hat{N}_u(\xi_u, f_s(\xi_u, t)) + \hat{F}_u(t) \qquad (13)$$

In a number of cases, these equations can be put into *universality classes* describing the similar behavior of otherwise quite different systems. The term *universality classes* derives from the theory of phase transitions of systems in or close to thermal equilibrium, such as superconductors or ferromagnets, though the classes treated here are of a more general character. In the present context, the term means that Equation 13 can be put into specific general forms. For instance, a single-mode laser, the formation of a roll pattern in a fluid, and the formation of a stripe pattern in chemical reactions obey the same basic

order-parameter equation. Such universality classes can be established because of the following facts:

1. When we are dealing with a *soft transition* of a system, its order parameters are small close to the instability point. In analogy to conventional phase transition theory, we call a transition a soft transition if the order parameter(s) change(s) smoothly with the control parameter. This allows us to expand $\hat{N}_u$ into a power series with respect to the order parameters, where we may keep only a few leading terms.
2. Furthermore, we may exploit symmetries: for instance, when there is a term $\beta\xi_u^2$, and there is an inversion symmetry of the system, it follows that $\beta = 0$. Symmetries also lead in a number of cases to relationships between coefficients.
3. Finally, we may invoke so-called normal form theory to simplify the nonlinear functions on the right-hand side of Equation 13.

Some Examples for the Formation of Spatial Patterns

In the case of a single-order parameter with inversion symmetry, where the properties of a system remain unchanged when we replace ξ_u by $-\xi_u$, the order-parameter equation reads

$$\dot{\xi}_u = \lambda\xi_u - \beta\xi_u^3 + F(t) \qquad (14)$$

Equation 14 is an example of cases in which order-parameter equations can be treated in terms of Thom's catastrophe theory (Thom, 1975). There are, indeed, some special cases in which the right-hand side of an order-parameter equation can be derived from a potential function. The properties of such a potential function, and in particular the changes of its shape as a function of control parameters, can then be classified according to catastrophe theory. Note, however, that in quite a number of cases such potential functions cannot be found for systems far from thermal equilibrium.

Let us consider the general case. The state vector may be written as

$$\mathbf{q}(\mathbf{x}, t) = \mathbf{q}_0 + \xi_u(t)\mathbf{v}_u(\mathbf{x}) + \sum_s \xi_s(t)\mathbf{v}_s(\mathbf{x}) \qquad (15)$$

Because the sum over the enslaved modes is generally small, the evolving pattern is determined by the second term on the right-hand side, which is called the mode skeleton since it represents the basic features of the evolving pattern. While $\mathbf{v}_u$ determines the spatial variation of $\mathbf{q}$, ξ_u determines the temporal rise and saturation. In the case of two-order parameters, the basic equations read

$$\dot{\xi}_1 = \lambda\xi_1 + \beta_1\xi_1^2 + \beta_{12}\xi_1\xi_2 + \beta_2\xi_2^2 + \cdots + F_1(t) \qquad (16)$$

$$\dot{\xi}_2 = \lambda\xi_2 + \beta_1^1\xi_1^2 + \beta_{12}^1\xi_1\xi_2 + \beta_2^1\xi_2^2 + \cdots + F_2(t) \qquad (17)$$

and the total state vector is determined by

$$\mathbf{q}(\mathbf{x}, t) = \mathbf{q}_0 + \xi_1(t)\mathbf{v}_1(\mathbf{x}) + \xi_2(t)\mathbf{v}_2(\mathbf{x}) \qquad (18)$$

Depending on the coefficients in Equations 16 and 17, modes may either coexist or compete. In the case of competition, either ξ_1 or ξ_2 vanish. In the case of coexistence, both are nonvanishing, so that the total pattern becomes a superposition of the patterns $\mathbf{v}_1$ and $\mathbf{v}_2$.

Let us now consider the case of three-order parameters where the modes are represented by plane waves

$$v_1(\mathbf{x}) = A\exp\{i\mathbf{k}_1\mathbf{x}\}$$
$$v_2(\mathbf{x}) = A\exp\{i\mathbf{k}_2\mathbf{x}\}$$
$$v_3(\mathbf{x}) = A\exp\{i\mathbf{k}_3\mathbf{x}\} \qquad (19)$$

When hexagons are formed, $\mathbf{k}_1, \mathbf{k}_2, \mathbf{k}_3$ form a triangle, and the coexistence of the corresponding three-order parameters is

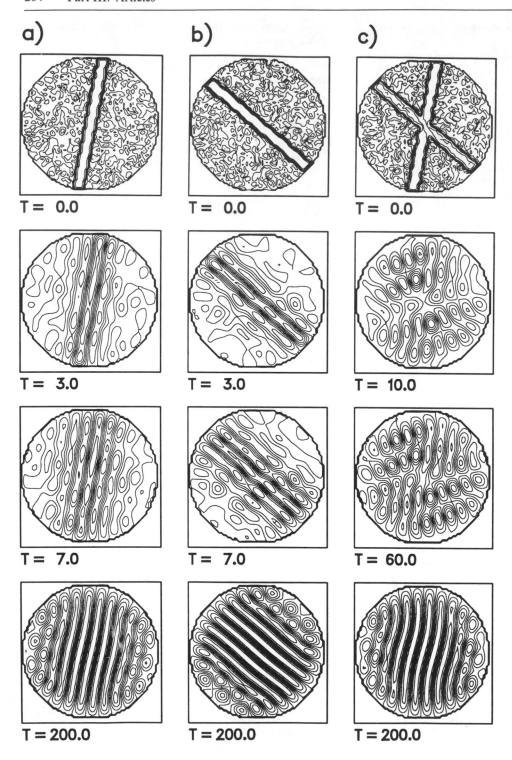

Figure 1. Computer simulation of a fluid heated from below. Left column: Completion of a single upwelling stripe to a full stripe pattern in the course of time. Middle column: The same as before, but with a different orientation of the initial stripe. Right column: Two initially given upwelling stripes, with one stripe somewhat stronger than the other one. In the course of time, the originally stronger stripe wins the competition and determines the whole stripe pattern.

guaranteed by quadratic terms in Equation 13. In some fluids the coexistence of three-order parameters may give way to competition if the energy input into the system is enhanced. Also, the possibility of regimes of coexistence and competition is possible.

A further example for pattern formation in physics is provided by the computer simulation shown in Figure 1, where a liquid is heated from below in a circular vessel. Depending on a prescribed initial state, different stripe patterns of upwelling

fluid may evolve. The fluid acts as an associative memory; that is, an incomplete set of data (one upwelling stripe) is supplemented automatically by the system to a full stripe pattern. The mechanism is as follows: Once a stripe is provided, it establishes its corresponding order parameter, which competes with all other order parameters governing other stripe directions. The originally given order parameter wins this competition and, eventually, forces the whole system into the selected ordered state.

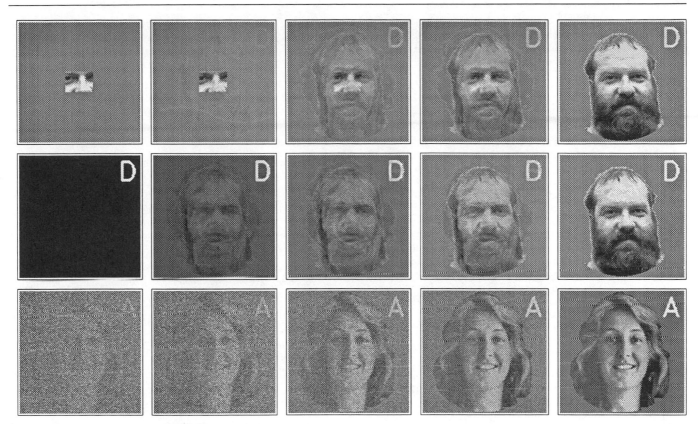

Figure 2. Upper row: Restoration of a full face from part of it. Middle row: Restoration of a full face from the single-letter coding for the name. Lower row: Restoration of the face from an originally noisy face.

A Synergetic System for Associative Memory

We have just seen that synergetic systems may act as an associative memory (Kohonen, 1987). This allows us to devise an algorithm for pattern recognition based on an analogy between pattern formation and pattern recognition (Haken, 1991). In both cases, an incomplete set of data (fluid: a single stripe; pattern recognition: part of a pattern) is completed to a full set of data (fluid: a complete stripe pattern; pattern recognition: a complete pattern) by means of order parameters and the slaving principle. The algorithm can be formulated as follows:

We consider a set of prototype patterns stored in our system. These patterns are represented by vectors $\mathbf{v}^{(k)}$ of a high-dimensional space, where one component $v_i^{(k)}$ corresponds to the gray value or color value of a specific pixel of a pattern. In the same way, we encode a starting or test vector $\mathbf{q}$ (pattern to be recognized).

By means of these vectors $\mathbf{v}^{(k)}$, we construct the dynamics for the test vector $\mathbf{q}$ in the following sense: The test vector $\mathbf{q}$ becomes a time-dependent quantity, undergoing a gradient dynamics in a potential field, which may be visualized as a mountainous landscape. This potential field possesses those and only those minima that correspond to the stored prototype pattern vectors $\mathbf{v}^{(k)}$. Note that this approach avoids the well-known difficulty of a number of neural networks, in particular the Hopfield type (Hopfield, 1982), in which the system can be trapped in spurious minima that do not correspond to any stored patterns (see STATISTICAL MECHANICS OF NEURAL NETWORKS). Here, the dynamical system leads to an identification of prototype patterns without the need to introduce SIMULATED ANNEALING (q.v.), i.e., a statistical pushing of the test vector $\mathbf{q}$.

In Figure 2 the associative property of the dynamical system is shown using three different initial conditions: A part of a face, the name of a pattern, and a pattern that is disturbed by noise. In every case, there is a complete restoration of the original prototype.

The dynamics of a *synergetic computer* can be interpreted or realized by means of a parallel network in which each component q_j of $\mathbf{q}$ represents the activity of a model neuron j. By means of the hypothesis

$$\mathbf{q}(t) = \sum_{k'=1}^{M} \xi_{k'}(t)\mathbf{v}^{(k')} + \mathbf{w}(t) \tag{20}$$

where $\mathbf{w}$ is a residual vector, the dynamics of $\mathbf{q}$ can be transformed into the order-parameter equations

$$\dot{\xi}_k = \lambda_k \xi_k - B \sum_{k' \neq k}^{M} \xi_{k'}^2 \xi_k - C \sum_{k'=1}^{M} \xi_{k'}^2 \xi_k \tag{21}$$

The attention parameters λ_k serve for an amplification of each order parameter; the second term serves for a discrimination, and the last term for a saturation of the order parameters. In this way, to each perceived pattern, an individual order parameter ξ_k is attached. The formalism allows one to treat time-dependent attention parameters and the properties of the perception of ambiguous patterns (Ditzinger and Haken, 1990).

Discussion

Synergetics, as outlined in this article, may be considered a strategy for coping with complex systems in physics, chemistry, economics, and, it is hoped, some important aspects of the brain. In this approach, a connection is established between the microscopic level, described by the individual parts of a system, and the macroscopic level, described by order parameters. This separation is made possible by a separation of time scales for the reaction of the parts and of the order parameters. While the typical time constants of neurons in the brain are on the order of milliseconds, those of the brain's macroscopic performance, such as recognition, speech, etc., are on the order of hundreds of milliseconds. At the microscopic level we may mention, in particular, the neural network model by McCulloch and Pitts (McCulloch, Pitts, and Warren, 1943), who modeled the individual neurons as two-state elements. A further step was the recognition of the analogy between the McCulloch and Pitts network and spin glasses; this insight allowed Hopfield (1982) to introduce an energy function for that network, and it gave rise to an avalanche of studies by theoretical physicists in particular. An important and very comprehensive modeling of brain action in terms of neurons was established by Wilson and Cowan (1973), who solved their equations numerically. To the best of this author's knowledge, the order-parameter concept and slaving principle have not been applied yet to these models. Thus we may ask whether there is any evidence for the occurrence of adequate order parameters in brain activities. Here we list a few types of evidence:

1. The pattern recognition model described in the preceding section.
2. The modeling of oscillations in the perception of ambiguous patterns cited earlier (Ditzinger and Haken, 1990).
3. The identification of low-dimensional chaos (describable by order parameters) by Babloyantz (1985).
4. The low-dimensional attractors in the olfactory bulb noted by Freeman (1975).
5. The analysis of petit mal epilepsy, describable by Shilnikov chaos (Friedrich and Uhl, 1992).
6. The MEG analysis of finger-tapping by Kelso Fuchs, and Haken (1992).
7. The analysis of movement coordination by Haken, Kelso, and Bunz (1985).

Because in some EEG and MEG measurements, multi-electrode or squid derivations were possible, the spatiotemporal patterns could be determined and, in particular, the basic modes in the sense of a mode decomposition could be identified. In a number of experiments, a surprisingly low number of dominating modes and thus order parameters could be found.

In conclusion, we may state that the strategy of searching for order parameters describing brain functions has found some justification, but considerable work remains to be done—for instance, to connect the microscopic neuronal level with the macroscopic level of brain functions.

Road Map: Learning in Artifical Neural Networks, Deterministic
Related Reading: Associative Networks; Development and Regeneration of Eye-Brain Maps; Dynamics and Bifurcation of Neural Networks; Epilepsy: Network Models of Generation; Pattern Formation, Biological; Self-Organization and the Brain

References

Babloyantz, A., 1985, Evidence of chaotic dynamics of brain activity during the sleep cycle, in *Dimensions and Entropies in Chaotic Systems* (G. Mayer-Kress, Ed.), New York and Berlin: Springer.
Ditzinger, T., and Haken, H., 1990, The impact of fluctuations on the recognition of ambiguous patterns, *Biol. Cybern.*, 63:453–456. ◆
Freeman, W., 1975, *Mass Action in the Nervous System: Examination of the Neurophysiological Basis of Adaptive Behavior Through the EEG*, New York: Academic Press.
Friedrich, R., and Uhl, C., 1992, Synergetic analysis of human electroencephalograms: Petit-mal epilepsy, in *Evolution of Dynamical Structures in Complex Systems* (R. Friedrich and A. Wunderlin, Eds.), Berlin and New York: Springer.
Guckenheimer, J., and Holmes, P., 1990, *Nonlinear Oscillations, Dynamical Systems, and Bifurcations of Vector Fields*, 3rd. ed., Berlin and New York: Springer.
Haken, H., 1983, *Synergetics: An Introduction*, 3rd ed., Berlin and New York: Springer. ◆
Haken, H., 1991, *Synergetic Computers and Cognition*, Berlin and New York: Springer.
Haken, H., Kelso, J. A. S., and Bunz, H., 1985, A theoretical model of phase transitions in human hand movements, *Biol. Cybern.*, 51:347–356.
Hopfield, J. J., 1982, Neural networks and physical systems with emergent computational abilities, *Proc. Natl. Acad. Sci. USA*, 79:2554–2558.
Kelso, J. A. S., Fuchs, A., and Haken, H., 1992, Phase transitions in the human brain: Spatial mode dynamics, *Internat. J. Bifur. Chaos Appl. Sci. Engrg.*, 2:917–939.
Kohonen, T., 1987, *Associative Memory and Self-Organization*, 2nd ed., Berlin and New York: Springer.
Landau, D., and Lifshitz, I. M., 1959, *A Course of Theoretical Physics*, vol. 5, *Statistical Physics*, London and Paris: Pergamon.
McCulloch, W. S. and Pitts, W. H., 1943, A logical calculus of the ideas immanent in nervous activity, *Bull. Math. Biophys.*, 5:115–133.
Thom, R., 1975, *Structural Stability and Morphogenesis: An Outline of a General Theorie of Models*, Reading MA: Benjamin.
Wilson, H. R., and Cowan, J. D., 1973, A mathematical theory of the functional dynamics of cortical and thalamic nervous tissue, *Kybernetik*, 13:55–80. ◆
Wilson, K. G., 1971, Renormalization group and critical phenomena: I. Renormalization group and the Kadanoff scaling picture; II. Phase-space cell analysis of critical behavior, *Phys. Rev. B*, 4:3174–3205.

Corollary Discharge in Visuomotor Coordination

M. Jeannerod

Introduction

The concept of Corollary Discharge (CD) has been used to explain visuomotor functions. We will examine three of these functions, the stabilization of the visual world during eye movements, the determination of an egocentric reference for spatial localization, and the proactive control of visually goal-directed movements.

Visual Stabilization

During visuomotor behavior, many factors are likely to introduce visual instability, particularly during eye movements, yet, the visual scene appears stationary. By contrast, passive displacement of the eyeball (by exerting a slight pressure on the outer canthus) produces the impression of displacement of the surrounding objects. This difference can be explained by

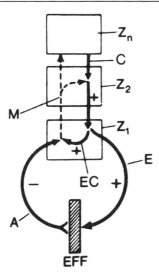

Figure 1. Von Holst and Mittelstaedt's conception of the so-called reafference principle. A high-level center (Z_n) issues a command (C) which reaches executive center Z_1. From Z_1 an efference (E) and an efference copy (EC) simultaneously arise. E modifies the effector (EFF) and this modification is signaled by the reafference (A) to Z_1. The residual difference between A and EC is transmitted as a message (M) to higher centers.

the fact that, in the case of a self-produced movement, the intentional effort carries a corrective mechanism that cancels its visual effects. Indeed, when the effort is not associated with a movement (e.g., if one eye muscle is paralyzed), visual objects appear to move in the direction of the intended movement. Sperry (1950) observed that fish with inverted vision caused by surgical 180° eye rotation tended to turn continuously in circles when placed in a visual environment. He interpreted this circling behavior as the result of a disharmony between the retinal input generated by movement of the animal and a compensatory mechanism for maintaining the stability of the visual field. This mechanism was a centrally arising discharge that reached the visual centers as a corollary of any excitation generated by the motor centers and normally resulting in movement (hence, the term *corollary discharge*, as used by Sperry). In this way, visual changes produced by a movement of the animal were normally "cancelled" by a CD of a corresponding size and direction, and had no effect on behavior. If, however, the CD did not correspond to the visual changes (e.g., after inversion of vision), these changes were not cancelled and were read by the motor system as having their origin in the external world. The animal thus moved in the direction of this apparent visual displacement.

Contemporary to Sperry's observations, von Holst and Mittelstaedt (1950) independently came to the same conclusion, using the term *efference copy* to designate the cancellation mechanism (Figure 1). Thus, corollary discharge and efference copy represent two nearly identical formulations of the same general principle that the nervous system can inform itself about its own (motor) activity. Experiments in invertebrate and vertebrate animals have indeed demonstrated the existence of motor-to-sensory projections in several sensory systems (see review in Jeannerod, Kennedy, and Magnin, 1979). In the visual domain, retinal and extraretinal signals would thus combine to produce a perceived position of the visual world during eye movements. Besides this explanation based on the processing of neural outflow signals, the existence of an inflow signal arising from eye muscle proprioceptors has been advo-

cated by several authors (see Bridgeman and van der Heijden, 1994). A CD type of regulation should, in principle, be more advantageous because of its timing: centrally propagating motor-to-sensory discharges are expected to be available earlier than discharges arising from the periphery, as discussed later in this article.

Visual Localization

The function of visual stabilization should imply that the extraretinal signal (whether from an outflow or an inflow origin) carries information as to spatial location of the target. The retinal position of the target is not, by itself, sufficient for localization because, as the eyes move in the head and the head moves on the trunk, several different retinal positions correspond to the same spatial locus. Remapping the spatial location of the target, therefore, requires that the eye/head position signals be combined with the retinal position signals. Provided this information is stored appropriately, it can be used for guiding the arm to that location.

Demonstration of the role of CD in this remapping process requires that motor commands for eye position be dissociated from actual eye position. If the arm is found to be directed according to the motor command, this would favor the CD hypothesis; if the arm is directed according to the actual eye position, however, this would favor a proprioceptive hypothesis. Perenin, Jeannerod, and Prablanc (1977) tested the ability of patients with partial paralysis of extrinsic eye muscles to localize targets by hand. The targets were presented in the dark to the paralyzed eye, whereas the normal eye was covered. When the target appeared in an area of the visual field that corresponded to a paralyzed muscle, the hand erred beyond target location (past-pointing). Recording the movements of the normal (covered) eye showed that they were an exaggerated amplitude, so that they would have clearly overreached target position. Thus, because of the weakness of the paretic muscle, the neural command for moving the eyes was stronger than normally required. This made the subject overestimate the rotation angle made by his eyes. Past-pointing directly reflects this perceived increase in effort.

Although these data clearly favor a mechanism based on CD for monitoring eye position, proprioceptive signals may also contribute. Gauthier, Nommay, and Vercher (1990) experimentally deviated one eye by pulling it with a suction cap. The deviated eye was covered. Pointing with the arm at targets presented to the open (nondeviated) eye was systematically biased (by 13% to 16% of the angular deviation) in the direction of the pull. As the innervation sent to the two eyes was likely to be the same, mislocalization of objects was partly attributable to signals arising from the muscle stretched by the pulling (proprioception). In the normal condition, both the efferent signals and those of proprioceptive origin concur for spatial localization.

The relationships between the retinal error signal and the extraretinal positional signal were first formalized in a model (Robinson, 1975) in which the efference copy from eye position was derived from the output of the neural integrator which maintains eye position during fixation. It is this signal of actual eye position which is combined with retinal error (the difference between where the fovea is and where the target is) to provide other motor systems (e.g., the arm) with target location information. If the eye were driven only by retinal error, the eye position signal would have to be subtracted from the target position signal (in head coordinates) to reobtain the retinal error. In fact, Hallett and Lightstone (1976) demonstrated that saccades are not generated on the basis of retinal error. Instead, the driving signal for the eye saccadic generator to reach the desired eye position relative to the head is the eye motor error signal. This signal is constructed by "subtracting" the actual change in eye position in the orbit from the desired position. The movement stops when the motor error equals zero. The CD signaling the actual change in eye position is derived from executive structures, such as brainstem saccadic burst generators. As these neurons code eye velocity, the corollary signal has to be integrated to provide eye position.

Recent work by Guitton and colleagues has directly demonstrated the dynamic nature of the remapping process by showing that output neurons from the superior colliculus—the tectoreticular (TR) neurons—code the change in eye motor error during movement (see Guitton, 1992). Before the movement takes place, a TR neuron coding the vector to the desired eye position is activated and drives the saccadic generator. As the movement progresses and motor error is reduced, other TR neurons coding for smaller vectors are activated until the error is zero. At this point, a TR neuron coding for a zero vector is activated and fixation is maintained. An internal representation of change in gaze position is generated and compared with the desired gaze position to yield instantaneous gaze motor error. If this gaze motor error is the parameter represented topographically on the collicular map, one can conceive how this signal will activate the proper sequence of TR neurons—hence, Guitton's hypothesis of a moving "hill of activity" shifting across the collicular map. There are some difficulties with this model, however, most notably with the timing of discharges in the superior colliculus which, in order to be suitable for coding motor error, should precede those of the saccadic generator (Sparks, 1993).

The CD concept used in this model implies that the driving signal changes during the course of the movement, a concept which departs from that of a static combination of vectors, as postulated by earlier models.

Representation of Goals of Movements

A CD type of mechanism has also been used to explain the control of skeletal movements. Animals with deafferented limbs are able, in the absence of vision of their limbs, to learn to repeat movements until they become conditioned. Since no information about limb position is available from the periphery, it must be provided by purely central mechanisms. Miles and Evarts (1979) proposed that signals generated by the motor centers provide information about future movements before they have reached the periphery, a mechanism now known as internal feedback. The existence of collaterals arising from the pyramidal tract and distributed to subcortical structures involved in motor control lends evidence to this mechanism.

At variance with eye movements, visually goal-directed skeletal movements require that incoming signals provide information on both the current position of the limb with respect to the body (the role of proprioception) and its position with respect to the visual target. It is this combined signal (the position of the limb relative to the target) which will be used by the comparator to compute the error of the movement with respect to the desired trajectory (Jeannerod, 1988). This mode of internal feedback regulation based on CD has a major advantage in minimizing correction delays when a perturbation occurs during execution of a movement. When the visual target of an arm movement is suddenly displaced during the movement, a correction can be generated within 100 ms or less, so that the target is correctly acquired with only a minimal increase in movement time. This correction is achieved through a kinematic rearrangement during the early phase of the movement (Paulignan et al., 1991). Abbs and Gracco (1984) also de-

scribed rapid compensation (within 50 ms or less) of perturbations applied to articulators during speech, by unexpectedly pulling down the lower lip. A similar result was obtained by Cole and Abbs (1987) when mechanical perturbations were applied to fingers during rapid grasps. This type of correction also implies a dynamic comparison between the CD, which represents the desired movement, and incoming signals, which monitor the current state of execution. Since reafferent signals are delayed with respect to the command signal, the comparator must look ahead in time and produce an estimate of the movement velocity corresponding to the command. The image of this estimated velocity is used at the input level for computing the actual position of the limb with respect to the target. Simulation experiments (Hoff and Arbib, 1992) were able to generate accurate corrections when target position was perturbed, without a notable increase in movement duration, which is compatible with the observed data.

In addition to matching the movement trajectory to the represented movement, this mechanism also has other potential functions. It might be used to produce a correspondence between the motor command and the amount of muscular contraction in conditions in which the muscular plant is not linear. Other nonlinearities may also arise from interaction of the moving limb with external forces, especially if it is loaded (for a review, see Bullock and Grossberg, 1988). This problem, which is critical for producing accurate limb movements, is less important for eye movements, where interactions with the external force field are minimal and where the load of the moving segment is constant. In this case, the pattern of command issued by the saccadic pulse generator should unequivocally reflect the final desired position of the eye, that is, the position where the retinal error is zero.

Road Map: Biological Motor Control
Related Reading: Dynamic Remapping; Gaze Coding in the Posterior Parietal Cortex

References

Abbs, J. H., and Gracco, V. L., 1984, Control of complex motor gestures: Orofacial muscle responses to load perturbations of lip during speech, *J. Neurophysiol.*, 51:705–723.

Bridgeman, B., and van der Heijden, A. H. C., 1994, A theory of visual stability across saccadic eye movements, *Behav. Brain Sci.*, 17:247–292.

Bullock, D., and Grossberg, S., 1988, Neural dynamics of planned arm movements: Emergent invariants and speed-accuracy properties during trajectory formation, *Psychol. Rev.*, 95:49–90.

Cole, K. J., and Abbs, J. H., 1987, Kinematic and electromyographic responses to perturbation of a rapid grasp, *J. Neurophysiol.*, 57:1498–1510.

Gauthier, G. M., Nommay, D., and Vercher, J. L., 1990, The role of ocular muscle proprioception in visual localization of targets, *Science*, 249:58–61.

Grüsser, O. J., 1986, Interaction of efferent and afferent signals in visual perception: A history of ideas and experimental paradigms, *Acta Psychol.*, 63:3–21.

Guitton, D., 1992, Control of eye-head coordination during orienting gaze shifts, *Trends Neurosci.*, 15:174–179.

Hallett, P. E., and Lightstone, A. D., 1976, Saccadic eye movements towards stimuli triggered by prior saccades, *Vis. Res.*, 16:99–106.

Hoff, B., and Arbib, M. A., 1992, A model of the effects of speed, accuracy and perturbation on visually guided reaching, in *Control of Arm Movement in Space* (R. Caminiti, P. B. Johnson, and Y. Burnod, Eds.), *Experimental Brain Research*, series 22, Berlin and New York: Springer-Verlag, pp. 285–306.

Jeannerod, M., 1988, *The Neural and Behavioural Organization of Goal-Directed Movements*, Oxford: Clarendon.

Jeannerod, M., Kennedy, H., and Magnin, M., 1979, Corollary discharge: Its possible implications in visual and oculomotor interactions, *Neuropsychologia*, 17:241–258.

Miles, F., and Evarts, E. V., 1979, Concepts of motor organization, *Annu. Rev. Psychol.*, 30:327–362.

Paulignan, Y., MacKenzie, C. L., Marteniuk, R., and Jeannerod, M., 1991, Selective perturbation of visual input during prehension movements, I: The effects of changing object position, *Exper. Brain Res.*, 83:502–512.

Perenin, M. T., Jeannerod, M., and Prablanc, C., 1977, Spatial localization with paralysed eye muscles, *Ophthalmologica*, 175:206–214.

Robinson, D. A., 1975, Oculomotor control signals, in *Basic Mechanisms of Ocular Motility and Their Clinical Implications* (G. Lennerstrand and P. Bach-y-Rita, Eds.), Oxford: Pergamon, pp. 337–374.

Sparks, D., 1993, Are gaze shifts controlled by a moving hill of activity in the superior colliculus? *Trends Neurosci.*, 16:214–216.

Sperry, R. W., 1950, Neural basis of the spontaneous optokinetic response produced by visual inversion, *J. Comp. Physiol. Psychol.*, 43:482–489.

von Holst, E., and Mittelstaedt, H., 1950, Das Reafferenzprinzip: Wechselwiskungen zwischen Zentralnervensystem und Peripherie, *Naturwissenschaften*, 37:464–476.

Cortical Columns, Modules, and Hebbian Cell Assemblies

William H. Calvin

Introduction

Our cerebral cortex sits atop the white matter like a thin layer of icing, about 2 mm thick. About 148,000 neurons lie under each square millimeter; however, it is this "icing" that is horizontally layered, not the "cake" beneath. Neurons with similar interests tend to be vertically arrayed in the cortex, forming cylinders known as *cortical columns*, although they sometimes appear as elongated, curtain-like bands. *Minicolumns* are about 30 μm in diameter, whereas *macrocolumns* are 0.4–1.0 mm.

The superficial cortical layers of the primary visual cortex in primates (V1) have regions that contain neurons that are particularly sensitive to color; these *blobs* selectively project to color-sensitive zones in area V2 called *stripes*. The blobs are separated by macrocolumnar distances (Bartfeld and Grinvald, 1992), with surrounding regions containing neurons more sensitive to visual form than to color; however, only 30%–35% of blob neurons are color-sensitive, and many animals with poor color vision nonetheless have blobs. Besides the color stripes of V2, there are stripes in parallel which specialize in form, as well as others involved in binocular aspects (Livingstone and Hubel, 1988).

Columns, barrels, blobs, and stripes have all been called *cortical modules*, and the term is frequently applied to any segmentation or repeated patchiness (Purves, Riddle, and LaMantia, 1992). By so loose a definition, both a dried-up river bed and the fur of a "marmalade" cat would also be considered modular. Frequently, there is no evidence of either function or detailed internal circuitry, just premature talk of "repeating patterns of circuitry" and "iterated modular units." *Module* is also loosely used by theoreticians to mean any functional grouping in the brain, which also violates the notion of repeating standardized units. Some columns may indeed be modular based on the usual conventions of modular furniture and modular electronics, but a favorite candidate for a higher-order module, the so-called hypercolumn, has now been shown not to be modular at all (Blasdel, 1992).

Why should neurons with similar functional specializations cluster together? Why should some clusters extend vertically through most of the cortical layers? The reasons could be functional (Bullock, 1980; Shaw, Harth, and Scheibel, 1982) or merely an epiphenomenon of development (Purves et al., 1992). Here, the problem will be approached via clustering tendencies provided by the cortex's horizontal connections (Katz and Callaway, 1992).

Layering and Its Functional Correlates

Traditionally, six neocortical layers have been distinguished by neuroanatomists, but this has been subject to some lumping and splitting. Layers II and III can usually be lumped together, as when one talks of the superficial pyramidal neurons. But layer IV has had to be subdivided repeatedly in the visual cortex (IVa, IVb, IVcα, IVcβ).

Layer IV neurons send most of their outputs up to layers II and III. Some superficial neurons send messages down to layers V and VI, although most connect, either laterally or via U-fibers in white matter, in their own layers. Layer VI sends messages back down to the thalamus via the white matter, while V sends signals to other deep and distant neural structures, sometimes even the spinal cord. A simple path would come into layer IV, then up to III, down to V or VI, and then back out of the cortex to some subcortical structure (White, 1989). So for any column of cortex, the bottom layers are like a subcortical "out" box, the middle layer like an "in" box, and the superficial layers somewhat like an "interoffice" box connecting the columns and different cortical areas. These are not, of course, exclusive roles, e.g., layer VI pyramids also have axon branches extending horizontally to terminate in layer IV.

Traditionally, the association cortex has simply been considered *terra incognita*, the 90% of the human cerebral cortex that is not motor strip or a primary sensory area. However, Diamond (1979) argues that the motor cortex is not restricted to the motor strip, but is the fifth layer of the entire cerebral cortex. That's because V, whatever the area, contains neurons that, at some stage of development, send their outputs down to the spinal cord, with copies to the brainstem, basal ganglia, and hypothalamus. Diamond likewise argues that the fourth layer everywhere is the sensory cortex, and that the second and third layers everywhere are the true association cortex. Calvin (1994) argues that the superficial layers have the right properties to implement Darwinian processes on the time scale of thought and action.

Columnar Clustering

A given neuron, however, may have dendrites spanning a few layers, especially if it is a pyramidal neuron; these taproot-shaped cells are the excitatory neurons of neocortex. The other neuron types are thought to be inhibitory (except that layer IV's spiny stellate neuron is excitatory). In Nissl stains that

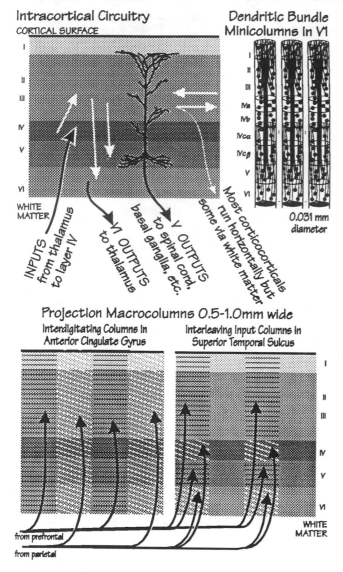

Intracortical Circuitry

CORTICAL SURFACE

INPUTS from thalamus to layer IV

VI OUTPUTS to thalamus

V OUTPUTS to spinal cord, basal ganglia, etc.

WHITE MATTER

Dendritic Bundle Minicolumns in V1

Most corticocorticals run horizontally but some via white matter

0.031 mm diameter

Projection Macrocolumns 0.5-1.0mm wide

Interdigitating Columns in Anterior Cingulate Gyrus

Interleaving Input Columns in Superior Temporal Sulcus

WHITE MATTER

from prefrontal

from parietal

Figure 1. A diagrammatic representation of cortical layering, dendritic bundle minicolumns, and the structure of macrocolumns. (Adapted from Calvin and Ojemann, 1994; Peters and Yilmaz, 1993; and Goldman-Rakic, 1990.)

show only cell bodies of neurons, Ramón y Cajal saw narrow columns (hereafter called *minicolumns*) running from white matter to the cortical surface; the cell-sparse gaps were about 30 μm apart in human cortex. It now appears that the gaps are vertical bundles of axons and apical dendrites. In monkeys, these minicolumns are 31 μm in diameter, but in cats, they are about twice that diameter (Peters and Yilmaz, 1993). In monkeys, there are about 100 neurons in such a minicolumn (143 in V1), 39 of which are superficial pyramidal neurons.

The human cerebral cortex totals 2200 cm²; unfolded, it would fit on four sheets of letter-sized paper. By comparison, the cerebral cortex of apes would fit on a single sheet, that of cats and monkeys would fit on a postcard, and that of a rat would fit on a small postage stamp. On the basis of layering differences, there are 52 areas in each human hemisphere; a Brodmann area averages 21 cm² and 250 million neurons grouped into several million minicolumns. Physiologically, there are aggregations (hereafter called *macrocolumns*) which would seem to contain, at most, a few hundred minicolumns;

this may be secondary to an organization of the input wiring into projection macrocolumns (Figure 1).

In 1957, Mountcastle and co-workers discovered a tendency for somatosensory strip neurons responsive to skin stimulation (hair, light touch) to alternate, at distances of about every 0.5 mm, with those specializing in joint and muscle receptors (see Mountcastle, 1957). It now appears that there is a mosaic organization of similar dimensions, with the neurons within each macrocolumn (or "segregate") having a receptive field optimized for the same patch of body surface (see SOMATO-SENSORY SYSTEM). Hubel and Wiesel (1977), recording in monkey visual cortex, saw curtain-like clusters ("ocular dominance columns") which specialized in the left eye, with an adjacent cluster at a distance of about 0.4 mm specializing in the right eye.

Orientation columns are of minicolumn dimensions; within these, the neurons prefer lines and edges that are tilted about the same angle from the vertical. There are many such minicolumns specializing in various angles within an ocular dominance macrocolumn (Hubel and Wiesel, 1977). The visual cortex provides us with our best insights into cortical circuitry (see VISUAL CORTEX CELL TYPES AND CONNECTIONS). The relationships between minicolumns and macrocolumns are best seen there (see OCULAR DOMINANCE AND ORIENTATION COLUMNS), although it is hazardous to generalize because ocular dominance columns themselves are less than universal, e.g., they are not a typical feature of New World monkeys.

Horizontal Organization in Neocortex

Lateral connections in an array are similar to the relaxation algorithms used for neural-like problem solving. In cortex as elsewhere, *recurrent inhibition* (or *lateral inhibition*) is thought to provide an antagonistic organization that sharpens responsiveness to an area far smaller than would be predicted from the anatomical funneling of inputs. *Recurrent excitation* may be especially prominent in the superficial layers of primate neocortex. The superficial pyramids send myelinated axons out of the cortical layers into the white matter, their eventual targets being typically the superficial layers of other cortical areas when of the "feedback" type; when "feedforward," they terminate in layer IV and deep in layer III. Long corticocortical terminations are often organized into interdigitating macrocolumns.

However, superficial pyramidal neurons also send unmyelinated collaterals to adjacent superficial pyramids. Indeed, roughly 70% of the excitatory synapses on any superficial pyramid (but less than 1% of those on layer V pyramids) are derived from pyramidal neurons less than 0.3 mm away. In general, the average cortical neuron contacts fewer than 10% of all the neurons in that radius, so superficial pyramids may be said to have an unusually strong propensity to excite one another.

There is also an unusual pattern to such superficial connections that suggests a columnar organizing principle. The collateral axon travels a characteristic lateral distance without giving off any terminal branches; then it produces a tight terminal cluster (see figure 5 in Gilbert, 1993). The distance to the center of the terminal patch is about 0.43 mm in primary visual cortex, 0.65 mm in the secondary visual areas, 0.73 mm in the sensory strip, and 0.85 mm in the motor cortex of monkeys (Lund, Yoshioka, and Levitt, 1993). The axon may then continue for an identical distance and then produce another cluster, and this occasionally continues for several millimeters.

Because of this local standard for axon length, recurrent excitation becomes probable among some cell pairs. Horizontal connections are also found among the pyramidal neurons of layers V and VI, but the regular spacing has been noted only for the pyramids of the superficial layers. In the absence of simultaneous recordings at appropriate spacings, we can only guess at the consequences of the mutual reexcitation. Even if the synaptic strengths were high, much longer axons would be required to give conduction times that escape refractoriness and create (even briefly) the proverbial reverberating circuit (corticocortical conduction velocities are about 0.3–0.5 mm/ms).

Synchrony-Shaped Connectivity

Recurrent excitation, however, can induce synchrony, as even weak coupling between relaxation oscillators is known to quickly produce entrainment. Were the cells otherwise active, they would soon tend to produce some spikes at about the same time. Cells in cortical minicolumns often fire in synchrony; more widespread synchronized firing has been a recent theme in cortical neurophysiology. Such synchrony occurs more frequently during difficult tasks and is the best-known correlate of perceptual binding (see SYNCHRONIZATION OF NEURONAL RESPONSES AS A PUTATIVE BINDING MECHANISM).

Synchronized firing in the context of the superficial layers of neocortex has some important implications for synaptic enhancement. The local field potentials associated with synchronized firing have their source in the superficial layers. The long-term potentiation (LTP) demonstrated in motor cortex is dependent on sufficient postsynaptic depolarization by other inputs and is confined to the superficial layers; that is where most N-methyl-D-aspartate (NMDA) receptors for glutamate are located.

In such a system, near-simultaneous arrivals may enhance the synaptic strength of the coincident inputs. Even more important, though, repeated coincidences should be particularly effective in NMDA-like synapses. One can imagine synchronous "test patterns" during development serving to shape up adult cortical connectivity via use dependence and selective survival. The superficial pyramid's lateral connections suggest that they could organize synchronous clusters about 0.4 mm apart in primary visual cortex, which happens to be the approximate size of the macrocolumns of area 17, the ocular dominance columns. It is thought that ocular dominance columns are organized by a gradual segregation of geniculate afferents into layer IV, but such selective survival might be secondary to neural activity during development and thus dependent on connectivity in the superficial layers.

Discussion

Permanent and Temporary Macrocolumns

Given that both are about half a millimeter in diameter, what is the relationship between the permanent macrocolumns and the ephemeral entrained pairs, whose synchrony can be destroyed in an instant by a wave of inhibition?

Entrained cells probably form triangular mosaics on occasion, given that a superficial pyramid sends axon collaterals in many directions. Two entrained cells may send axon terminals to an equidistant point, with impulses arriving simultaneously, and so entrain it as well. This equilateral triangle forms the basis of a triangular mosaic; the spacing will be the local metric, whether 0.43 mm or 0.85 mm (hereafter, simply approximated as 0.5 mm). Because the basis of recruitment and entrainment is conduction time, not distance per se, various distortions of the triangular mosaic might be seen if conduction velocity (typically, 0.3–0.5 mm/ms within the cortex) and synaptic delay are not constant.

Just from the geometry, the overall impression should be like that of wallpaper, where corresponding points in a repeating pattern can be identified, but where the origin of the pattern cannot necessarily be identified. One can, however, artificially inscribe a boundary such that homologous points are approached but not included. If one does this with many neighboring points, each of which is part of a different triangular mosaic, the largest possible "unit area" without repetitions will be hexagonal in shape. The distance between parallel sides of the hexagon will be 0.5 mm as well. Like the wallpaper's unit pattern, this hexagon need not have a unique origin, and it would be meaningless to speak of a hexagonal boundary unless underlying resonances with synaptic strengths were so organized. One could speak (Calvin and Ojemann, 1994), however, of this hexagonal unit pattern being laterally copied or cloned, thanks to the triangular mosaic tendencies of recurrent excitation among superficial pyramids.

Because a pair of synchronized neurons could form a triangular mosaic that is not parallel to the others, hexagonal mosaic cloning depends, in effect, on two identical and adjacent hexagons of spatiotemporal patterning. This parent pair could be the result of evoked activity in an input pathway, or the initial spatiotemporal pattern could be generated by cortical connectivity just as the gaits of locomotion are generated by spinal cord circuitry. The spatial-only connectivity patterns in cortex that generate the spatiotemporal firing patterns could themselves be contained in adjacent hexagons; were this visible to anatomical or physiological techniques, we might consider it a permanent macrocolumn. Otherwise, we would merely see transient physiological ensembles of macrocolumnar dimensions with hexagonal repeats.

Modular Aspects

It is widely expected that cerebral cortex will turn out to have circuits which, in different cortical patches, are merely repeats of a standard pattern. However, *module* has instead been loosely applied to a wide range of functional or anatomical collectives. A true module would be a cortical patch which, whatever the origin of its inputs or the destination of its outputs, nonetheless has an internal architecture which is the same from one instance to another, with only minor differences in local tuning. The best candidate for a true module was the "hypercolumn" (Hubel and Wiesel, 1977): two adjacent ocular dominance columns, each containing a full set of orientation columns. Adjacent hypercolumns would merely represent different patches of the visual field; inputs and outputs might differ, but internal wiring would be similar from one hypercolumn to the next.

However, newer mapping techniques have shown that ocular dominance repeats are somewhat independent of orientation column repeats (Blasdel, 1992), making adjacent hypercolumns internally nonidentical, i.e., not iterated. For this reason, "hypercolumn" now appears with scare quotes around it in the visual system literature. No current use of "cortical module" bears any relationship to the technological use of the term; *cortical cluster* would be a more appropriate term.

The Hebbian Cell Assembly

Representations via a pattern are familiar from the trichromatic theory of color perception and the similar aspects of taste. Evoking the memory of your grandmother's face is probably not a matter of activating a single specialized neuron; rather, it is thought to involve the activity pattern in an ensemble of cortical neurons, each of which helps to implement other memories as well. Different sensory or motor schemas might be characterized by different firing patterns in time and space, with each characteristic spatiotemporal pattern serving as a code for an item from an individual's memory.

Memory recall may consist of the creation of a spatiotemporal sequence of neuron firings, probably one similar to that present at the time of the input to memory, just shorn of some of the nonessential frills that promoted it. It would be like a message board in a stadium, with lots of little lights flashing on and off, but creating a pattern. A somewhat more general version of a Hebbian cell assembly (Calvin and Ojemann, 1994) avoids anchoring the spatiotemporal pattern in particular cells, making it more like a message scrolling on a message board. The pattern continues to mean the same thing, even though it is implemented by different lights. Although we tend to focus on the lights which turn on, the lights which remain off also contribute to the pattern.

The notion of convergence zones for associative memories raises the issue of maintaining the identity of a spatiotemporal code during long-distance corticocortical transmission, such as through the corpus callosum. Distortions of the spatiotemporal pattern by a lack of precise topographic mappings may be unimportant when the information flows in only the one direction. However, because the connections between distant cortical regions are typically reciprocal, any distortions of the original spatiotemporal firing pattern during forward transmission would need to be compensated for in the reverse path in order to maintain the characteristic spatiotemporal pattern as the local code for a sensory or motor schema. In addition to inverse transforms or error-correction mechanisms, degenerate codes seem possible (as when six different RNA triplets all code for leucine).

Error correction is particularly interesting because reliable copying capabilities often provide insight into possible codes, e.g., the genetic code puzzle was solved after identifying which physical patterns could be readily duplicated. Are there physiological clones of cerebral schemas that might help us identify a cerebral code such as the relevant Hebbian cell assembly?

Spatiotemporal pattern copying was inferred from the need to reduce timing jitter during precision throwing using law-of-large-numbers averaging. The triangular mosaic of electrical activity predicted from standard-length axons has an inherent error-correction property that can maintain spatiotemporal patterns during copying, so long as they are small enough to fit inside a macrocolumn-sized hexagon of cerebral cortex. This suggests that hexagonal Hebbian cell assemblies could implement a cerebral code (Calvin, 1994).

Road Map: Biological Networks
Related Reading: Fault Tolerance; Hebbian Synaptic Plasticity: Comparative and Developmental Aspects; NMDA Receptors: Synaptic, Cellular, and Network Models

References

Bartfeld, E., and Grinvald, A., 1992, Relationships between orientation-preference pinwheels, cytochrome oxidase blobs, and ocular-dominance columns in primate striate cortex, *Proc. Natl. Acad. Sci. USA*, 89:11905–11909.

Blasdel, G. G., 1992, Orientation selectivity, preference, and continuity in monkey striate cortex, *J. Neurosci.*, 12:3139–3161. ◆

Bullock, T. H., 1980, Reassessment of neural connectivity and its specification, in *Information Processing in the Nervous System* (H. M. Pinsker and W. D. Willis, Jr., Eds.), New York: Raven, pp. 199–220.

Calvin, W. H., 1994, The emergence of intelligence, *Sci. Am.*, 271:89–96.

Calvin, W. H., and Ojemann, G. A., 1994, *Conversations with Neil's Brain: The Neural Nature of Thought and Language*, Reading, MA: Addison-Wesley.

Diamond, I., 1979, The subdivisions of neocortex: A proposal to revise the traditional view of sensory, motor, and association areas, in *Progress in Psychobiology and Physiological Psychology 8* (J. M. Sprague and A. N. Epstein, Eds.), New York: Academic Press, pp. 1–43.

Gilbert, C. D., 1993, Circuitry, architecture, and functional dynamics of visual cortex, *Cereb. Cortex*, 3:373–386.

Goldman-Rakic, P., 1990, Parallel systems in the cerebral cortex: The topography of cognition, in *Natural and Artificial Parallel Computation* (M. A. Arbib and J. A. Robinson, Eds.), Cambridge, MA: MIT Press, pp. 155–176.

Hubel, D. H., and Wiesel, T. N., 1977, Functional architecture of macaque visual cortex, *Proc. R. Soc. Lond. B*, 198:1–59.

Katz, L. C., and Callaway, E. M., 1992, Development of local circuits in mammalian visual cortex, *Annu. Rev. Neurosci.*, 15:31–56.

Livingstone, M. S., and Hubel, D. H., 1988, Segregation of form, color, movement, and depth: Anatomy, physiology, and perception, *Science*, 240:740–749. ◆

Lund, J. S., Yoshioka, T., and Levitt, J. B., 1993, Comparison of intrinsic connectivity in different areas of macaque monkey cerebral cortex, *Cereb. Cortex*, 3:148–162.

Mountcastle, V. B., 1957, Modality and topographic properties of single neurons of cat's somatic sensory cortex, *J. Neurophysiol.*, 20:408–434.

Mountcastle, V. B., 1979, An organizing principle for cerebral function: The unit module and the distributed system, in *The Neurosciences: Fourth Study Program* (F. O. Schmitt and F. G. Worden, Eds.), Cambridge, MA: MIT Press, pp. 21–42. ◆

Peters, A., and Yilmaz, E., 1993, Neuronal organization in area 17 of cat visual cortex, *Cereb. Cortex*, 3:49–68.

Purves, D., Riddle, D. R., and LaMantia, A.-S., 1992, Iterated patterns of brain circuitry (or how the cortex gets its spots), *Trends Neurosci.*, 15:362–368 (also see letters in 16:178–181). ◆

Shaw, G. L., Harth, E., and Scheibel, A. B., 1982, Cooperativity in brain function: Assemblies of approximately 30 neurons, *Exp. Neurol.*, 77:324–358.

White, E. L., 1989, *Cortical Circuits*, Boston: Birkhauser.

Coulomb Potential Learning

Michael P. Perrone and Leon N. Cooper

Introduction

The Coulomb Potential Learning (CPL) algorithm (Bachmann et al., 1987), which derives its name from its functional form's likeness to a coulomb charge potential, was originally motivated by the shortcomings of the simple perceptron (Rosen-blatt, 1962) and the original Hopfield net (Hopfield, 1982). In the case of the perceptron, it was clear almost from the outset that the linear separability provided by the simple perceptron would not be sufficient to perform complex tasks. In the case of the original Hopfield model, the recall capacity of the network is low because of nonorthogonal memories and the existence of spurious memories. The CPL algorithm addresses both of these problems by providing a simple network that is capable of storing an arbitrarily large number of memories with perfect recall and no spurious memories, as well as constructing arbitrary nonlinear boundaries for classification tasks. In addition, the CPL algorithm is easy to implement in hardware and is readily adaptable to parallel computation.

Perfect Memory Recall

The CPL algorithm constructs a network in the following way: Suppose that we are given a set of memories, $\mathcal{M} = \{\mathbf{m}_i | \mathbf{m}_i \in \mathbb{R}^N,$ for all $i\}$. For each memory, construct a neuron, $n_i(\mathbf{x})$, with the activation function $n_i(\mathbf{x}) = -\|\mathbf{x} - \mathbf{m}_i\|_2^{-L}$, where $\|z\|_2 = \Sigma_i z_i^2$, and combine all of these neurons with a single perceptron whose output, $E(\mathbf{x}) = \Sigma_i Q_i n_i(\mathbf{x})$, is given by

$$E(\mathbf{x}) = -\sum_i Q_i \|x - \mathbf{m}_i\|_2^{-L}$$

where $N > 3$, $L > N - 2$, and $Q_i > 0$. Clearly, each memory corresponds to a minimum of the network activation. Thus, we have stored all of the memories. Now, we need to define a method for retrieving memories from this network. We retrieve memories by relaxing from an arbitrary initial state to a minimum of the $E(x)$ function; to do so, we perform gradient descent on $E(x)$ to find a fixed point of the following differential equation:

$$\dot{\mathbf{x}} = -\sum_i Q_i \|\mathbf{x} - \mathbf{m}_i\|_2^{-(L+2)}(\mathbf{x} - \mathbf{m}_i)$$

Note that relaxation to a memory may be slow if the initial point is far from all of the memories, since in that case the gradient will be very small. Another practical consideration is that when the gradient-descent process is implemented, the gradients near memories will be very large, and so care must be taken to terminate computer implementations of the relaxation process before overflow errors occur.

No Spurious Memories

In order to gain more insight into the process of memory recall, we consider the case in which $L = 1$ and $N = 3$. In this case, we can make a direct analogy with electrostatics. The memories can be interpreted as negative electric point charges and the relaxation equation corresponds to the motion of a positive particle in a Coulomb field—thus the name of the algorithm. In this special case, it is easy to see why the network has no spurious memories. Suppose a spurious memory exists. Then, by definition, it is distinct from the true memories. In a small neighborhood of the spurious memory, the gradient must point inward, toward a spurious minimum; but Gauss's law (Jackson, 1975) tells us that a net inward gradient implies that there must be charge enclosed in the neighborhood. Thus, we arrive at a contradiction, so the spurious memory cannot exist. It is possible to prove in general that the CPL network has no spurious memories (Dembo and Zeitouni, 1988).

Note that we can control the size of the basins of attraction for each memory by adjusting the strength of the "charge," Q_i, and the exponential power of the potential, L. This flexibility gives us a natural method for controlling the relative importance of various memories and how likely they are to be recalled. Thus, the network will not necessarily relax to the closest memory in the Euclidean sense; rather, it will relax to the most prominent memory, which implies that the network is constructing nonlinear boundaries between memories. Under certain conditions, however (Dembo and Zeitouni, 1988), these nonlinear effects can be minimized.

Deterministic Classification

It is also possible to modify the CPL network to function as a classification network. If we have two or more memory sets, $\mathcal{M}_1, \mathcal{M}_2, \ldots,$ which correspond to distinct classes, and we construct a CPL network, $E_i(\mathbf{x})$, for each memory set $\mathcal{M}_i$, then we can compare the total activations of each network (without using gradient descent) and choose the network with the smallest activation, weighted by the number of corresponding memories, as the winner. In this way, the CPL network $E_i(\mathbf{x})$ can be thought of as approximating the negative of the probability density of class $\mathcal{M}_i$, and we are simply choosing the class with the highest probability. Note, however, that the $E_i(\mathbf{x})$ are not density functions because they are not in general normalizable. It is possible to show that under certain conditions the $E_i(\mathbf{x})$'s will converge in the infinite memory limit to the correct distributions (Duda and Hart, 1973).

Alternatively, we can construct a CPL network from all classes and use gradient descent to relax to a memory. The classification in this case is given by the class of the memory to which the network relaxes.

In practice, however, it is not feasible to store all of the memories associated with a given classification task, since testing new patterns may be unacceptably slow for large memory sets. In this case, it is helpful to sub-sample the memories. A simple yet efficient method for sub-sampling has been outlined by the Reduced Coulomb Energy (RCE) algorithm (Reilly, Cooper, and Elbaum, 1982). The RCE algorithm creates networks of neurons analogous to the CPL algorithm, except the neurons have a bounded activity function given by $n_i(\mathbf{x}) = 1 - \Theta(\|\mathbf{x} - \mathbf{m}_i\|_2 - t_i)$, where $\Theta(\cdot)$ is a step function. Thus, the activity of RCE neuron i is 1 if the input is within a distance t_i of $\mathbf{m}_i$, and 0 otherwise. The RCE neuron can be viewed as a "clipped" version of the CPL neuron. Classification of a given input is determined by choosing the class of memories which has the largest total output. No relaxation process is used, and therefore spurious memories are not an issue. In its simplest version, the RCE algorithm builds a network in the following manner. For each memory in the data set:

1. If the classification is correct, make no changes.
2. If the network activity is zero (no classification), add a new neuron to the network using the new memory as the center, and set the neuron's threshold equal to the distance to the nearest memory of a different class.
3. If the classification is incorrect or confused,
   (a) Shrink the thresholds of the neurons which were responsible for the error.
   (b) Pass the memory through the network again.

This process is repeated until all the data are correctly classified. The network generated attempts to cover each class with hyperspheres as depicted in Figure 1. Given enough resources, this algorithm can cover arbitrarily complex boundaries between classes for a deterministic classification problem.

The algorithm lends itself to many natural extensions, which include additional knowledge. For example, one can include one's confidence in m_i or knowledge about the relevance of m_i by choosing the Q_i accordingly; or one may choose the initial threshold from step 2 more or less conservatively.

Once such a network has been constructed, it can be used as it is, or the RCE neurons can be replaced with CPL neurons, where the Q_i are set equal to the number of members from the same class which activate the corresponding RCE neuron. Combining CPL with RCE results in a Radial Basis Function (RBF) network (Powell, 1987) which efficiently uses the data.

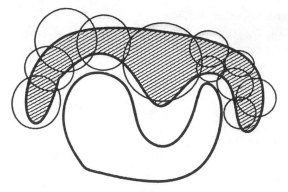

Figure 1. Two deterministic class regions are shown (shaded and clear) for an arbitrary two-dimensional classification problem. The receptive fields of the RCE neurons are shown for the shaded class.

Note that such a CPL network can also be trained by gradient descent on the Q_i, the t_i, and the $\mathbf{m}_i$.

Probabilistic Classification

For probabilistic classification in which class membership is a random variable, we can improve performance by using a variation of the algorithm. In principle, memories from different classes can be arbitrarily close for probabilistic classification. In practice, this fact results in many neurons having very small thresholds, t_i, and therefore much of the feature space may be left uncovered, particularly near class boundaries or anywhere where the class probabilities are nearly equal. This uneven covering can lead to poor classification performance. One way around this problem is the Probabilistic RCE algorithm (Scofield et al., 1987). The probabilistic version is the same as the deterministic version, except that a minimum value for the threshold, t, is set. After training, any cell whose threshold is below the minimum value is automatically reset to the minimum value. During testing, the probabilistic regions will give multiple responses from one or more classes. The number of cells active from a given class divided by the total number of cells active can be used as an estimate of the class probability. Also, these cells can keep pattern and classification counts that allow them to estimate the probability that a pattern falling within their influence field is an example of the class they represent (Scofield et al., 1987).

Discussion

In high-dimensional classification problems, networks such as CPLs, kernel estimators, and RBFs all suffer from the *curse of dimensionality* (Bellman, 1961), which implies that the amount of data required to construct a reliable estimate of the true solution increases exponentially with dimensionality. Thus, a CPL network which functions well for low-dimensional classification problems may perform no better than the level of chance for high-dimensional problems, unless there is a ridiculously large amount of data. In practice, we rarely, if ever, have this much data. We run out of physical space to store all of the data long before we can offset the exponential factor!

One way to lessen the effects of the curse of dimensionality is to construct hybrid neural networks which can divide a problem into subtasks that are handled in lower-dimensional spaces and are therefore less troubled by the curse of dimensionality. In one variation of the RCE algorithm (Reilly et al., 1987;

Reilly et al., 1988), multiple RCE networks are generated to solve subtasks of a large multiclass task in the following way:

1. Train a network on the full task.
2. Identify which classes are being confused.
3. For each pair of confused classes:
   (a) Use the previously trained network to select patterns which are confused.
   (b) Train a new network on the confused patterns.

Note that in both training phases, the networks do not have to be trained with the full dimensionality of the task. In the first phase (step 1), this fact is justified because we can correct mistakes in the second phase; and in the second phase (step 3), we are dealing with less complex tasks, and therefore the full dimensionality may not be needed. In operation, this hybrid network classifies a new pattern using the main network unless there is confusion between classes, in which case the pattern is classified by the appropriate subtask network. This method has been used to improve classification performance significantly on real-world optical character recognition tasks (Scofield, Kenton, and Chang, 1991).

In practice, the subtask selection procedure can be iterated as long as there are data available on which to train. However, as the data become sparse, our estimates become more and more noisy, and we run into the problem of overfitting. We can lessen this problem by averaging over several hybrid network solutions from several different training runs (see AVERAGING/MODULAR TECHNIQUES FOR NEURAL NETWORKS). In this way, we can reduce the variance of our estimate as much as we like (Perrone and Cooper, 1993). This method can significantly improve performance.

Road Map: Learning in Artificial Neural Networks, Deterministic
Background: I.3. Dynamics and Adaptation in Neural Networks
Related Reading: Digital VLSI for Neural Networks; Modular and Hierarchical Learning Systems; Radial Basis Function Networks; Topology-Modifying Neural Network Algorithms

References

Bachmann, C. M., Cooper, L. N., Dembo, A., and Zeitouni, O., 1987, A relaxation model for memory with high storage density, *Proc. Natl. Acad. Sci. USA*, 84:7529–7531.

Bellman, R. E., 1961, *Adaptive Control Processes*, Princeton, NJ: Princeton University Press. ◆

Dembo, A., and Zeitouni, O., 1988, General potential surfaces and neural networks, *Phys. Rev. A*, 37:2134–2143. ◆

Duda, R. O., and Hart, P. E., 1973, *Pattern Classification and Scene Analysis*, New York: Wiley. ◆

Hopfield, J. J., 1982, Neural networks and physical systems with emergent collective computational abilities, *Proc. Natl. Acad. Sci. USA*, 79:2554–2558.

Jackson, J. D., 1975, *Classical Electrodynamics*, New York: Wiley. ◆

Perrone, M. P., and Cooper, L. N., 1993, Learning from what's been learned: Supervised learning in multi-neural network systems, in *Proceedings of the World Conference on Neural Networks*, vol. 3, Hillsdale, NJ: Erlbaum, pp. 354–357.

Powell, M. J. D., 1987, Radial basis functions for multivariable interpolation: A review, in *Algorithms for Approximation* (J. C. Mason and M. G. Cox, Eds.), Oxford: Clarendon, pp. 143–167.

Reilly, D. L., Cooper, L. N., and Elbaum, C., 1982, A neural model for category learning, *Biol. Cybern.*, 45:35–41.

Reilly, D. L., Scofield, C. L., Cooper, L. N., and Elbaum, C., 1988, Gensep: A multiple neural network learning system with modifiable network topology, in *Abstracts of the First Annual International Neural Network Society Meeting*, INNS.

Reilly, R. L., Scofield, C. L., Elbaum, C., and Cooper, L. N., 1987, Learning system architectures composed of multiple learning modules, in *Proceedings of the IEEE First International Conference on Neural Networks*, vol. 2, New York: IEEE.

Rosenblatt, F., 1962, *Principles of Neurodynamics*, New York: Spartan. ◆

Scofield, C., Kenton, L., and Chang, J.-C., 1991, Multiple neural net architectures for character recognition, in *Proceedings of Compcon, San Francisco, CA, February 1991*, Los Alamitos, CA: IEEE Computer Society Press, pp. 487–491.

Scofield, C. L., Reilly, D. L., Elbaum, C., and Cooper, L. N., 1987, Pattern class degeneracy in an unrestricted storage density memory, in *Neural Information Processing Systems* (D. Z. Anderson, Ed.), New York: American Institute of Physics.

Crustacean Stomatogastric System

Scott L. Hooper

Introduction

The stomatogastric nervous system (STNS) of decapod crustacea generates the rhythmic motor patterns of the esophagus and the three areas of the crustacean stomach: the cardiac sac, the gastric mill, and the pylorus. The STNS has been intensely studied, and contains some of the best understood central pattern-generating networks in biology. This work has identified three general characteristics of STNS networks that are relevant to biological and artificial neural networks. First, rhythmicity in these highly distributed networks arises emergently from the combination of the network's synaptic connectivity and the inherent cellular properties of its neurons. Second, modulatory influences can induce individual STNS networks to assume different functional configurations that produce qualitatively different rhythmic outputs. Third, modulatory influences can "switch" individual neurons between different neural networks or fuse individual networks into a single large network. The experimental advantages of the STNS have allowed many of the cellular mechanisms underlying these characteristics to be described. A repeated result of these investigations has been the observation of neurons with long-lasting (tens to hundreds of milliseconds) synaptic and active membrane properties. Model neurons with such complex electrical properties are not used in most artificial neural networks. The widespread presence of such neurons in the STNS and other biological neural networks indicates that (1) the relevance of neural network simulations with simple neurons to real network function is unclear, and (2) the introduction of these types of neurons into distributed neural networks may afford significant functional advantages.

Background

The STNS is an anatomically separate part of the crustacean nervous system that, under most conditions, contains four distinct neural networks that respectively generate the rhythmic motions of the esophagus and the three stomach areas. Four general characteristics of STNS neurons have particularly important functional consequences.

Morphology

All known STNS neurons are monopolar and have inexcitable cell bodies. Synaptic contacts and the ion channels that underlie the slow active properties of these neurons are located in neuropil areas that are physically distant from the cell bodies. Spike initiation zones are electrically distinct from the neuropil regions, and spike generation occurs in response to the summed membrane potential of the neuropil region (Selverston et al., 1976).

Synapses

Each electrophysiologically observed interneuronal interaction (both chemical and electrical) results from the summed activities of hundreds of anatomical synaptic contacts. Regional segregation of input and output synapses has not been observed, i.e., the neurons are not divided into specific presynaptic and postsynaptic regions. Thus, the relationship between neuronal input and output results from local integrative processes in the neuropil instead of the whole cell integration typical of neurons divided into distinct presynaptic and postsynaptic regions (Selverston et al., 1976). Finally, in the two best studied STNS networks—the pyloric and gastric networks —synaptic release occurs as a graded function of membrane potential; thus, these networks are analog devices (Graubard, Raper, and Hartline, 1983).

Several different transmitter substances are used in STNS networks and, depending on transmitter and receptor type, different synapses have different ionic bases and time courses. In addition to rapid onset rapid decay synaptic contacts, many STNS synapses induce slowly rising and decaying postsynaptic responses with characteristic times as long as 100 ms (Marder and Eisen, 1984). The presence of synapses with different time constants is very important in determining neuronal firing order in STNS output patterns.

Cellular Properties

STNS neurons have multiple time- and voltage-dependent membrane conductances that endow individual neurons with specific, active, long-duration, intrinsic membrane properties. Some are endogenous oscillators (Selverston and Moulins, 1985), i.e., neurons that, as a result of inherent membrane conductances, spontaneously rhythmically depolarize and fire bursts of action potentials. Others have so-called plateau properties. A plateauing neuron has two quasi-stable membrane potentials (a hyperpolarized "rest" potential and a depolarized "plateau" potential). The neurons make transitions between the two states in response to synaptic input, postinhibitory rebound, or current injection. These transitions are regenerative, i.e., depolarization above a certain threshold voltage from rest activates voltage-dependent conductances that drive the neurons to the plateau, and relatively small hyperpolarizations from the plateau induce an active repolarization to the rest state (Russell and Hartline, 1978). These properties actively transform inputs in both amplitude and time, in that small amplitude inputs can induce a full plateau, and brief inputs (tens of milliseconds) induce long-lasting responses (plateaus can last for seconds). Finally, many show postinhibitory rebound, i.e., they depolarize and fire after inhibition (Selverston et al., 1976).

These three properties—endogenous oscillation, plateau properties, and postinhibitory rebound—result in neurons that have very nonlinear input–output characteristics in sign, amplitude, and time. Multiconductance stomatogastric neuronal models suggest that these properties are particularly important for reproducing the dynamic properties of the STNS networks (Marder and Selverston, 1992). Neurons expressing these properties are also present in vertebrate nervous systems (Hounsgaard and Kjaerulff, 1992); computer simulations using simple integrate-and-fire neurons are likely to prove inadequate for biological networks containing such neurons.

Neuromodulation

Both synaptic strength and expression of inherent cellular properties in STNS networks are under the control of modulatory influences. These influences can, for instance, induce individual neurons to express plateau properties or transform non-oscillatory neurons into spontaneous bursters (Harris-Warrick, Nagy, and Nusbaum, 1992); as such, they dramatically increase the functional repertoire of individual neurons and of the networks of which they are a part.

Mechanisms of Central Pattern Generation

The pyloric network is the best understood STNS neural network, and will be used to illustrate the general principles that underlie rhythmic pattern generation in it and in the other well-understood STNS network, the gastric mill. These networks generate the motor patterns of the gastric mill (which chews food) and the pylorus (a filter that separates chewed food into three streams, one for further chewing, one for absorption, and one for excretion). Each of these networks are central pattern generators, i.e., they can endogenously generate their respective motor patterns without timing cues from the rest of the nervous system. Figure 1 (right) shows simultaneous intracellular recordings from all six pyloric network neuron types. Each pyloric cycle consists of a sequence of bursts of action potentials from the pyloric neurons; this sequence of bursts then rhythmically repeats (two cycles of the rhythm are shown in Figure 1). The pyloric dilator (PD) motorneurons define the

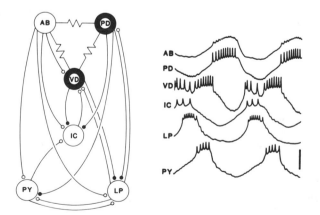

Figure 1. Pyloric network synaptic connectivity (left), and output (right). Small circles represent chemical synapses, whereas resistor symbols indicate electrical coupling. Calibration bar: 20 mV for PD and LP neurons, 10 mV for other neurons. [Modified, with permission, from Miller, J. P., 1987, Pyloric mechanisms, in *The Crustacean Stomatogastric System* (A. I. Selverston and M. Moulins, Eds.), Berlin: Springer-Verlag, p. 112.]

beginning of the sequence; they and the anterior burster (AB) interneuron fire together. Lateral pyloric (LP), inferior cardiac (IC), and weak ventricular dilator (VD) motorneuron firing occurs next, followed by pyloric (PY) neuron and strong VD motorneuron firing. The pyloric network synaptic connectivity (Miller, 1987) is shown in the left-hand portion of Figure 1. The clear neurons are glutamatergic and induce rapid, short-lasting inhibition in their follower neurons; the black neurons are cholinergic and induce slow, long-lasting inhibitions (Marder and Eisen, 1984).

The combined effects of active cellular properties, the network's interconnectivity pattern, and the different time courses of the glutamatergic and cholinergic inhibitions qualitatively explain the observed pyloric pattern (Miller, 1987). All the pyloric neurons, depending on the state of the network, can be endogenous bursters (Bal, Nagy, and Moulins, 1988), but the AB neuron is generally the fastest oscillator and drives the PD neurons through their electrical coupling. These neurons thus rhythmically inhibit all the other pyloric neurons. After the AB/PD neuron burst, postinhibitory rebound triggers plateau potentials in the VD, LP, and IC neurons, and they fire bursts of action potentials. The VD neuron fires first because only the glutamatergic AB neuron (with its short-lasting IPSP) inhibits it; the LP and IC neurons also receive a late, slow inhibitory postsynaptic potential (IPSP) from the PD neurons. VD neuron inhibition of the LP and IC neurons is insufficient to block their plateaus, and they reduce VD neuron firing during their bursts. Finally, the PY neurons, which have cellular properties that delay their postinhibitory rebound (Miller, 1987), fire (again through the induction of plateau potentials), inhibiting the IC and LP neurons and releasing the VD neuron from inhibition. The PY and VD neurons are turned off by the next AB/PD neuron burst, and the cycle repeats.

No single rhythmic-pattern-generating mechanism "explains" the pyloric network. It has both endogenous oscillators *and* multiple half-center oscillators and neuronal rings that, in the presence of postinhibitory rebound and plateau potentials, could drive rhythmic pattern production (see Selverston and Moulins, 1985, for a discussion of mechanisms underlying rhythm generation). Indeed, in network states (see the section that follows) in which none of the neurons are endogenous oscillators, a correctly ordered pyloric pattern (of dramatically reduced frequency) can be achieved by tonic depolarization (by current injection) of various pyloric neurons (Miller, 1987). This multiplicity of mechanisms makes it impossible to ascribe specific aspects of network output to specific network neurons. Network rhythmicity and pattern instead arise emergently from a combination of neuronal cellular properties and the network's distributed synaptic connectivity.

Multifunctional Networks

This complexity seems excessive for a network that produces the simple output pattern seen in Figure 1. In fact, however, the pyloric network is capable of producing many qualitatively different output patterns in response to various modulatory inputs (more than 20 modulatory substances are known to be present in inputs to the stomatogastric ganglion, STG) that modify neuronal cellular properties or synaptic strengths (Harris-Warrick et al., 1992). (The pattern shown in Figure 1, and the explanation provided, applies to one such state, the one that is most commonly seen in the isolated STNS.) Pyloric network complexity may thus exist to provide a substrate from which multiple, functionally different network configurations can be constructed. An example of the pyloric outputs in the presence of the neuromodulators serotonin and proctolin is shown in Figure 2. A detailed examination of these different

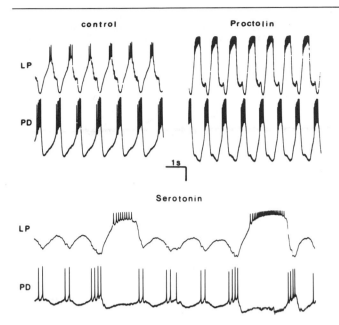

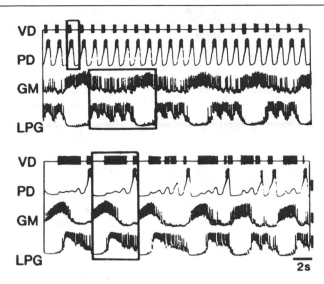

Figure 2. Proctolin (top right panel, 10^{-6} M) and serotonin (bottom panel, 10^{-4} M) applications induce the pyloric network to produce qualitatively different motor patterns. Calibration bar: 6 mV. [Modified, with permission, from Marder, E., and Hooper, S. L., 1985, Neurotransmitter modulation of the stomatogastric ganglion of decapod crustaceans, in *Model Neural Networks and Behavior* (A. I. Selverston, Ed.), New York: Plenum, pp. 319–337.]

Figure 3. A defined modulatory input can induce networks that are generally distinct (top panel, pyloric [VD and PD neuron traces] and gastric mill [GM and LPG neuron traces]) to form a fused network that produces a novel output (bottom panel). Calibration bars: 10 mV. (Modified, with permission, from Meyrand, P., Simmers, J., and Moulins, M., 1991, Construction of a pattern-generating circuit with neurons of different networks, *Nature*, 351:60–63, fig. 3. Copyright 1991 Macmillan Magazines Limited.)

patterns is beyond the scope of this article; it is only important to note that modulator application results in the network producing very different outputs. Proctolin dramatically increases LP neuron (top trace) activity. Serotonin induces an output pattern in which the LP and PD neuron bursts no longer strictly alternate; instead, the LP neuron bursts once every two or three PD neuron bursts. Network multifunctionality is not limited to the pyloric network; the gastric mill network also produces multiple output patterns (Turrigiano and Heinzel, 1992). Research on in vivo and semi-intact preparations shows that these networks produce a variety of motor patterns in the animal, and thus the multiple output patterns observed in vitro (and the network switches and fusions noted later in this article) may be behaviorally relevant (Turrigiano and Heinzel, 1992).

The cellular targets and changes in intrinsic membrane properties involved are known for several of these modulators (Harris-Warrick et al., 1992). This work has two results that are particularly relevant to artificial neural networks. First, many modulators alter the expression of long-lasting cellular properties (endogenous bursting, plateau properties) of the directly affected neurons, which again demonstrates the importance of long-lasting cellular properties in these biological networks. Second, owing to the highly interconnected nature of the networks and the active responses of follower neurons to the inputs they receive, changes in network activity cannot be explained by considering solely the neurons directly affected by a modulator. This is because the neurons directly affected by a modulator alter the activity of the rest of the network's neurons, and these in turn feed back onto, and modify the responses of, the directly affected neurons. The network rearrangements are, therefore, global responses of the network as a whole, and so not only network rhythmicity, but also network response to modulatory input, are *distributed* functions in these networks.

Neuron Switching and Network Fusion

Up to now, the four STNS neural networks have been treated as though they were separate entities. However, the stomach is an anatomically compact organ, and it is likely that the motor patterns these networks produce are coordinated. Indeed, considerable internetwork connectivity exists, and, depending on the species and the animal's behavioral state, the boundaries between networks are more or less variable. The extent of these interactions ranges from simple coordination of two networks, to cases in which one or more neurons are switched between different networks, to instances where two or more networks fuse to form a new network that produces a novel neural output (Dickinson, Mecsas, and Marder, 1990; Hooper and Moulins, 1989; Meyrand, Simmers, and Moulins, 1991). An example of fusion is shown in Figure 3. The upper panel shows simultaneous recordings from the VD (top trace, extracellular recording) and PD (second trace, intracellular recording) neurons from the pyloric network and the GM and LPG (bottom two intracellular traces) neurons of the gastric mill network under control conditions. The bottom panel shows the activity of these four neurons after discharge of an identified modulatory neuron that alters the activity of these networks. It is apparent that the two formerly separate networks now produce a single conjoint rhythm that is different from either the pyloric or gastric rhythms.

The changes in directly affected neurons responsible for these internetwork switches and fusions have been described in two cases. In one case (a switch), the change was attributable to a suppression of plateau property expression in the switching neuron (Hooper and Moulins, 1989); in the other case (a fusion), it was attributable to increased strength of an internetwork synaptic connection (Dickinson et al., 1990). Nevertheless, the effects on total network activity again could be understood only by considering the entire network. Thus, long-lasting cellular properties and distributed network architecture are not only crucial to central pattern generation and multifunctionality in the STNS, but also underlie many aspects of internetwork interactions and restructuring.

Discussion

Work in a well-defined invertebrate system of distributed biological neural networks has revealed several cellular and synaptic properties that are essential for biological function but generally have not been included in artificial neural network studies. Chief among these properties are the following:

1. Synapses with many different characteristic time courses are present in the same network.
2. Many neurons have, in addition to typical short-duration active membrane properties, long-lasting (tens to hundreds of milliseconds) regenerative membrane conductances that give rise to long-duration active intrinsic membrane characteristics (endogenous oscillations, plateau potentials, postinhibitory rebound).
3. Neuromodulatory inputs can profoundly alter both synaptic strength and inherent cellular properties.

The presence of these properties in biological networks indicates that neural network simulations with simple model neurons often may be inadequate to explain neurobiological function. These properties are largely responsible for the remarkable functional characteristics (rhythmic pattern production, multiple pattern production from single networks, neuron switching between different networks, and network fusion) of STNS networks. It is thus possible that inclusion of model neurons with such properties can result in considerable enhancement of artificial neural network capabilities, and a deeper understanding of both artificial and biological neural network function.

Road Map: Motor Pattern Generators and Neuroethology
Background: I.1. Introducing the Neuron
Related Reading: Half-Center Oscillators Underlying Rhythmic Movements; Neuromodulation in Invertebrate Nervous Systems

References

Bal, T., Nagy, F., and Moulins, M., 1988, The pyloric central pattern generator in crustacea: A set of conditional neuronal oscillators, *J. Comp. Physiol.*, 163:715–727.

Dickinson, P. S., Mecsas, C., and Marder, E., 1990, Neuropeptide fusion of two motor pattern generator circuits, *Nature*, 344:155–158.

Graubard, K., Raper, J. A., and Hartline, D. K., 1983, Graded synaptic transmission between identified spiking neurons, *J. Neurophysiol.*, 50:508–521.

Harris-Warrick, R. M., Nagy, F., and Nusbaum, M. P., 1992, Neuromodulation of stomatogastric networks by identified neurons and transmitters, in *Dynamic Biological Networks: The Stomatogastric Nervous System* (R. M. Harris-Warrick, E. Marder, A. I. Selverston, and M. Moulins, Eds.), Cambridge, MA: MIT Press, pp. 87–138. ◆

Hooper, S. L., and Moulins, M., 1989, A neuron switches from one network to another by sensory induced changes in its membrane properties, *Science*, 244:1587–1589.

Hounsgaard, J., and O. Kjaerulff, 1992, Ca^{2+}-mediated plateau potentials in a subpopulation of interneurons in the ventral horn of the turtle spinal cord, *Eur. J. Neurosci.*, 4:183–188.

Marder, E., and Eisen, J. S., 1984, Transmitter identification of pyloric neurons: Electrically coupled neurons use different transmitters, *J. Neurophysiol.*, 51:1362–1373.

Marder, E., and Hooper, S. L., 1985, Neurotransmitter modulation of the stomatogastric ganglion of decapod crustaceans, in *Model Neural Networks and Behavior* (A. I. Selverston, Ed.), New York: Plenum, pp. 319–337. ◆

Marder, E., and Selverston, A. I., 1992, Modeling the stomatogastric nervous system, in *Dynamic Biological Networks: The Stomatogastric Nervous System*, (R. M. Harris-Warrick, E. Marder, A. I. Selverston, and M. Moulins, Eds.), Cambridge, MA: MIT Press, pp. 161–196.

Meyrand, P., Simmers, J., and Moulins, M., 1991, Construction of a pattern-generating circuit with neurons of different networks, *Nature*, 351:60–63.

Miller, J. P., 1987, Pyloric mechanisms, in *The Crustacean Stomatogastric System* (A. I. Selverston and M. Moulins, Eds.), Berlin: Springer-Verlag, pp. 109–136. ◆

Russell, D. F., and Hartline, D. K., 1978, Bursting neural networks: A reexamination, *Science*, 200:453–456.

Selverston, A. I., and Moulins, M., 1985, Oscillatory neural networks, *Annu. Rev. Physiol.*, 47:29–48. ◆

Selverston, A. I., Russell, D. F., Miller, J. P., and King, D. G., 1976, The stomatogastric nervous system: Structure and function of a small neural network, *Prog. Neurobiol.*, 7:215–290. ◆

Turrigiano, G. G., and Heinzel, H.-G., 1992, Behavioral correlates of stomatogastric network function, in *Dynamic Biological Networks: The Stomatogastric Nervous System* (R. M. Harris-Warrick, E. Marder, A. I. Selverston, and M. Moulins, Eds.), Cambridge, MA: MIT Press, pp. 197–220. ◆

Data Clustering and Learning

Joachim M. Buhmann

Introduction

Data clustering (Jain and Dubes, 1988) aims at discovering and emphasizing structure which is hidden in a data set. The structural relationships among individual data vectors, e.g., pronounced similarity of groups of data vectors, have to be detected in an unsupervised fashion. This search for prototypes poses a delicate tradeoff: not to superimpose too much structure, which does not exist in the data set, and not to overlook structure by a too-simplistic modeling philosophy. Conceptually, there exist two different approaches to data clustering which are discussed and compared in this paper:

- *Parameter estimation of mixture models* by parametric statistics.
- *Vector quantization* of a data set by combinatorial optimization.

Parametric statistics assumes that noisy data have been generated by an unknown number of qualitatively similar stochastic processes. Each individual process is characterized by a univariate probability density, which is sometimes called a "natural" cluster of the data set. The density of the full data set is described by a parametrized *mixture model*; Gaussian mix-

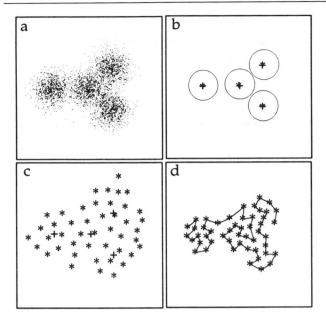

Figure 1. Clustering of multivariate data (*A*) generated by four Gaussian sources. The estimated Gaussian mixture model is depicted in part *B*; plus signs (+) represent the centers of the Gaussians and asterisks (∗) represent cluster centers. The circles denote the covariance estimates. Part *C* shows a data partitioning using a logarithmic complexity measure. Data clustering by a self-organizing chain is shown in part *D*.

tures are frequently used (McLachlan and Basford, 1988). This model-based approach to data clustering requires us to estimate the mixture parameters, e.g., the mean and variance of four Gaussians for the data set depicted in Figure 1*A*. Bayesian statistics provides a conceptual framework to compare and validate different mixture models.

The second approach to data clustering, which has been popularized as vector quantization in communication theory, partitions the data into clusters according to a suitable cost function. The resulting assignment problem is known to be a deterministic, NP-hard, combinatorial optimization problem (see PAC LEARNING AND NEURAL NETWORKS). This second strategy can be related to the statistical model-based approach in two different ways: (i) a stochastic optimization strategy such as SIMULATED ANNEALING (q.v.) is employed to search for solutions with low costs; (ii) the data set constitutes a randomly drawn instance of the combinatorial optimization problem. The optimization approach to clustering aims at partitioning the data set into disjunct cells according to a cost function. This approach, applied to the data set in Figure 1*A*, does not search for decision boundaries of exactly those four partitions which correspond to the univariate components of the underlying probability distribution.

The most difficult problem in mixture density estimation and in vector quantization deals with the complexity of the clustering solution, e.g., how many clusters should be chosen and how we should compare and validate solutions with different degrees of complexity. Learning, which is unsupervised in both cases, addresses the questions of how to estimate the model parameters and how to estimate the data assignments to clusters, respectively. The parameter values can be adjusted in an iterative or on-line fashion, if a steady stream of data is available. Another preferred choice for small data sets is batch

learning, in which parameters are estimated on the basis of the full data set.

Gaussian Mixture Models

Clusters in data sets are modeled by a mixture of stochastic data sources. Each component of this mixture, a data cluster, is described by a univariate probability density which is the stochastic model for an individual cluster. The sum of all component densities forms the probability density of the mixture model

$$P(x|\Theta) = \sum_{v=1}^{K} p(x|\theta_v)\pi_v \qquad (1)$$

Let us assume that the functional form of the probability density $P(x|\Theta)$ is completely known up to a finite and presumably small number of parameters $\Theta = (\theta_1, \dots, \theta_K)$. For the most common case of Gaussian mixtures, the parameters $\theta_v = (y_v, \Sigma_v)$ are the coordinates of the mean vector and the covariance matrix. The a priori probability π_v of the component v is called the mixing parameter.

Adopting this framework of parametric statistics, the detection of data clusters reduces mathematically to the problem of how to estimate the parameters Θ of the probability density for a given mixture model. A powerful statistical tool for finding mixture parameters is the maximum likelihood (ML) method (Duda and Hart, 1973), in which one maximizes the probability of the independently, identically distributed data set $\{x_i | i = 1, \dots, N\}$ given a particular mixture model. For analytical purposes, it is more convenient to maximize the log-likelihood

$$L(\Theta) = \sum_{i=1}^{N} \log P(x_i|\Theta) = \sum_{i=1}^{N} \log\left(\sum_{v=1}^{K} p(x_i|\theta_v)\pi_v\right) \qquad (2)$$

which yields the same maximum likelihood parameters because of the monotonicity of the logarithm. The straightforward maximization of Equation 2 results in a system of transcendental equations with multiple roots. The ambiguity in the solutions originates from the lack of knowledge as to which mixture component v has generated a specific data vector x_i and, therefore, which parameter θ_v is influenced by x_i.

For the computational problem of estimating parameters of mixture models with the maximum likelihood method, an efficient bootstrap solution is provided by the expectation maximization (EM) algorithm (Dempster, Laird, and Rubin, 1977). The EM algorithm estimates the unobservable assignment variables $\{M_{iv}\}$ in a first step. $M_{iv} = 1$ denotes that x_i has been generated by component v; otherwise, $M_{iv} = 0$. On the basis of these maximum likelihood estimates $\{\hat{M}_{iv}\}$, the parameters Θ are calculated in a second step. An iteration of these two steps renders the following algorithm for Gaussian mixtures:

- *E-step.* The expectation value of the complete data log-likelihood is calculated, conditioned on the observed data $\{x_i\}$ and parameter estimates $\hat{\Theta}$. This yields the expected assignments of data to mixture components, i.e.,

$$\hat{M}_{i\alpha}^{(t+1)} = \frac{p(x_i|\hat{y}_\alpha^{(t)}, \hat{\Sigma}_\alpha^{(t)})\pi_\alpha}{\sum_{v=1}^{K} p(x_i|\hat{y}_v^{(t)}, \hat{\Sigma}_v^{(t)})\pi_v}$$

$$= \frac{|\hat{\Sigma}_\alpha^{(t)}|^{-1/2} \exp(-\frac{1}{2}(x_i - \hat{y}_\alpha^{(t)})^T (\hat{\Sigma}_\alpha^{(t)})^{-1}(x_i - \hat{y}_\alpha^{(t)}))}{\sum_{v=1}^{K} |\hat{\Sigma}_v^{(t)}|^{-1/2} \exp(-\frac{1}{2}(x_i - \hat{y}_v^{(t)})^T (\hat{\Sigma}_v^{(t)})^{-1}(x_i - \hat{y}_v^{(t)}))}$$

$$(3)$$

- *M-step.* The likelihood maximization step estimates the mixture parameters, e.g., the centers and the variances of Gaussians (Duda and Hart, 1973:47):

$$\hat{y}_\alpha^{(t+1)} = \frac{\sum_{i=1}^{N} \hat{M}_{i\alpha}^{(t+1)} x_i}{\sum_{i=1}^{N} \hat{M}_{i\alpha}^{(t+1)}} \tag{4}$$

$$\hat{\Sigma}_\alpha^{(t)} = \frac{1}{\sum_{i=1}^{N} \hat{M}_{i\alpha}^{(t+1)}} \sum_{i=1}^{N} \hat{M}_{i\alpha}^{(t+1)} (x_i - \hat{y}_\alpha^{(t+1)})(x_i - \hat{y}_\alpha^{(t+1)})^T \tag{5}$$

Note that the Equations 4 and 5 have a unique solution after the expected assignments $\{\hat{M}_{i\alpha}^{(t+1)}\}$ have been estimated. The monotonic increase of the likelihood up to a local maximum guarantees the convergence of the EM algorithm. In our example of four Gaussians (Figure 1A), the EM algorithm estimates centers and variances as depicted by asterisks and circles in Figure 1B.

An important question has not yet been raised: How do we estimate the correct number of components in the data set? To differentiate between structure and noise of a data set, the complexity of the mixture model has to be constrained in the spirit of Occam's razor. Such a preference for simple models with few components is mathematically implemented by such techniques as MINIMUM DESCRIPTION LENGTH ANALYSIS (q.v.; see also Rissanen, 1989). The reader should note that ML estimation without constraints yields a singular solution of one component with zero variance for each data vector (Duda and Hart, 1973:198).

Gaussian mixture models have attracted a lot of attention in the neural network community for three primary reasons: (1) Networks of neural units with Gaussian-shaped receptive fields, so-called radial basis functions, compute function approximations in a robust and efficient way (Poggio and Girosi, 1990; see also RADIAL BASIS FUNCTION NETWORKS). (2) The a priori assumption that synaptic weights in neural networks are generated by a Gaussian mixture model has considerably reduced the generalization error of layered neural networks in time-series prediction tasks (Nowlan and Hinton, 1992), e.g., predicting sun spots and stock market indicators. (3) A neural network architecture based on Gaussian mixtures, the Hierarchical Mixture of Experts (Jacobs et al., 1991; see also MODULAR AND HIERARCHICAL LEARNING SYSTEMS), is able to solve efficiently a real-world classification or regression task in a divide-and-conquer fashion.

Data Clustering as a Vector Quantization Process

The second approach to data clustering uses an optimization principle to partition a set of data vectors which are characterized either by coordinates $\{x_i | x_i \in \mathbb{R}^d; i = 1, \ldots, N\}$ or by pairwise distances $\{D_{ik} | D_{ik} \in \mathbb{R}; i, k = 1, \ldots, N\}$. Data sets which are explicitly represented by vectors in a d-dimensional Euclidean space occur in almost all areas of science, whereas dissimilarity data are analyzed in fields such as the social sciences, economics, molecular biology, and linguistics. The goal of data clustering is to determine a partitioning of a data set which minimizes either the average distance of data vectors to their cluster centers or the average distance between data vectors of the same cluster. The two cases are referred to as *central* and *pairwise clustering*, respectively. Note that the dissimilarities D_{ik} do not necessarily qualify as a distance measure: dissimilarity or confusion matrices of protein, genetic, psychometric, or linguistic data frequently violate the triangular inequality, and the self-dissimilarity D_{ii} is not necessarily zero. For the following discussion, we require only the symmetry $D_{ik} = D_{ki}$.

Central Clustering

Solutions to central clustering are represented by a set of K data *prototypes* $\{y_v | y_v \in \mathbb{R}^d; v = 1, \ldots, K\}$. The Boolean assignments $\{M_{iv} | M_{iv} \in \{0, 1\}; v = 1, \ldots, K; i = 1, \ldots, N\}$ indicate that data vector i is uniquely assigned to cluster v ($\sum_{v=1}^{K} M_{iv} = 1$). Such an assignment gives rise to the distortion/quantization costs D_{iv} because of information loss. The functional form of D_{iv} depends on the weighting of data distortions, in which the quadratic costs $D_{iv} = (x_i - y_v)^2$ and K means $y_\alpha = \sum_{i=1}^{N} M_{i\alpha} x_i / \sum_{i=1}^{N} M_{i\alpha}$ are the most common choices for a clustering model. The cost function for this clustering is

$$\mathcal{E}_K^{cc}(\{M_{iv}\}) = \sum_{i=1}^{N} \sum_{v=1}^{K} M_{iv}(x_i - y_v)^2 \tag{6}$$

The size K of the cluster set, i.e., the complexity of the clustering solution, has to be determined by a problem-dependent complexity measure (Buhmann and Kühnel, 1993) which monotonically grows with the number of clusters. Simultaneous minimization of the distortion costs and the complexity costs yields an optimal number K^* of clusters. Constant complexity costs per cluster or logarithmic complexity costs $\log(N/\sum_{i=1}^{N} M_{i,v})$ (Shannon information) are utilized in various applications, such as signal processing, image compression, and speech recognition. A clustering result with logarithmic complexity costs is shown in Figure 1C. It is important to emphasize that the vector quantization approach to clustering optimizes a data partitioning and, therefore, is not suited to detect univariate components of a data distribution. The splitting of the four Gaussians in Figure 1A into 45 clusters (Figure 1C) results from low complexity costs and is not a defect of the method. Note also that the distortion costs as well as the complexity costs determine the position of the prototypes $\{y_v\}$. For example, the density of clusters in Figure 1C is independent of the density of data, which is a consequence of the logarithmic complexity costs. The cost function in Equation 6 has to be minimized by varying the assignments M_{iv}, and this amounts to a search in a discrete space with exponentially many states. The optimization procedure implicitly yields the cluster means $\{y_v\}$ by estimating optimized assignments $\{M_{iv}\}$.

A well-known algorithm for on-line estimation of K prototypes is the K-means algorithm (MacQueen, 1967). The K means $\{y_v | v = 1, \ldots, K\}$ are initialized by the first K data vectors $\{x_i | i = 1, \ldots, K\}$. A new data vector $x_{t+1}, t \geq K$ is assigned to the closest mean y_α:

$$M_{t+1,\alpha} = \begin{cases} 1 & \text{if } \|x_{t+1} - y_\alpha\| < \|x_{t+1} - y_v\| \text{ for all } v \neq \alpha \\ 0 & \text{otherwise} \end{cases} \tag{7}$$

The new mean vector is adjusted in response to the data vector x_{t+1} according to the learning rule

$$y_\alpha^{(t+1)} = y_\alpha^{(t)} + \frac{1}{\sum_{i=1}^{t+1} M_{i\alpha}} M_{t+1,\alpha}(x_{t+1} - y_\alpha^{(t)}) \tag{8}$$

The learning rule in Equation 8 adjusts the closest mean y_α proportional to the deviation $(x_{t+1} - y_\alpha^{(t)})$. The total adjustment is normalized to the number of data vectors which have already been assigned to this cluster center.

The deterministic optimization strategy for data clustering can be related to the stochastic mixture model estimation by using a Monte Carlo search method with a controlled noise level to minimize the clustering costs in Equation 6. The maximum entropy framework (Tikochinsky, Tishby, and Levine, 1984) allows us to estimate the most stable probability distribution for cluster assignments if the average clustering costs are fixed at a particular noise level. A careful decrease of the noise level, called SIMULATED ANNEALING (q.v.), yields optimized assignments with a high probability. One of the frequently encountered drawbacks of Monte Carlo processes is their slow relaxation to the equilibrium. An analytical calculation of the expectation values for cluster assignments and cluster centers would be clearly preferable for efficiency reasons. *Deterministic annealing* (Rose, Gurewitz, and Fox, 1990; Buhmann and Kühnel, 1993), an approximation method which is exact for central clustering in the limit $N \to \infty$, yields robust estimates of the expected (rather than the randomly fluctuating) assignments

$$\langle M_{i\alpha} \rangle = \frac{\exp(-\beta(\boldsymbol{x}_i - \boldsymbol{y}_\alpha)^2)}{\sum_{v=1}^{K} \exp(-\beta(\boldsymbol{x}_i - \boldsymbol{y}_v)^2)} \qquad (9)$$

The expectation values by the maximum entropy method are denoted by $\langle \cdot \rangle$. The estimated cluster centers are

$$\langle \boldsymbol{y}_\alpha \rangle = \frac{\sum_{i=1}^{N} \langle M_{i\alpha} \rangle \boldsymbol{x}_i}{\sum_{i=1}^{N} \langle M_{i\alpha} \rangle} \qquad (10)$$

The noise parameter β controls the fuzziness of the clustering problem, i.e., in the limit $\beta \to \infty$ the solution of Equation 9 corresponds to hard clustering with Boolean assignments $\langle M_{i\alpha} \rangle \in \{0, 1\}$ of a data vector $\boldsymbol{x}_i$ to the closest cluster center. Small β values represent the fuzzy extreme with partial assignments of data vectors to several clusters ($0 \leq \langle M_{i\alpha} \rangle \leq 1$). Equations 9 and 10, display a striking similarity to the Gaussian mixture estimation by the EM approach (Equations 3 and 4) since the maximum entropy estimate of the assignment probability distribution can be interpreted as an (admittedly implausible) Gaussian mixture model with fixed variances $1/\beta$.

In the neural network community, much attention has been paid to the biologically inspired data clustering systems known as SELF-ORGANIZING FEATURE MAPS (q.v.; see also Kohonen, 1984). In addition to the distance between data vector $\boldsymbol{x}_i$ and the nearest cluster $\boldsymbol{y}_\alpha$, the cost function measures the distances to cluster centers $\boldsymbol{y}_{\alpha \pm 1}$ which are neighbors in index space ($|v - \mu| = 1$). Such a self-organization principle generates a topologically ordered arrangement of nodes as shown in Figure 1D.

Pairwise Clustering

Clustering nonmetric data which are characterized by a dissimilarity matrix and not by explicit Euclidean coordinates have to be formulated as an optimization problem with quadratic assignment costs. Whenever two data items i and k are assigned to the same cluster v, the costs are increased by the amount D_{ik}/p_v, with $p_v = (1/N)\Sigma_{k=1}^{N} M_{kv}$ being the weight of cluster v. Then the pairwise clustering costs are

$$\mathscr{E}^{\mathrm{pc}}(\{M_{iv}\}) = \frac{1}{2} \sum_{i,k=1}^{N} D_{ik} \left(\sum_{v=1}^{K} \frac{M_{iv} M_{kv}}{p_v} - 1 \right) \qquad (11)$$

with $M_{iv} \in \{0, 1\}$ and $\Sigma_{v=1}^{K} M_{iv} = 1$. The constant cost term $\Sigma_{i,k=1}^{N} D_{ik}/2$ has been subtracted to gauge the cost scale, i.e., to have zero average costs for random cluster assignments. The cost function (Equation 11) is similar to energies of disordered systems in statistical physics, e.g., spin glasses and Potts glasses. Note that the correct scaling of the dissimilarities has to be $D_{ik} \sim 1/\sqrt{N}$ to get total expected clustering costs $\mathscr{E}^{\mathrm{pc}} \sim N$.

Minimization of the quadratic cost function (Equation 11) turns out to be algorithmically complicated because of pairwise, potentially conflicting correlations between assignments. The deterministic annealing technique, which produces robust reestimation equations for central clustering in the maximum entropy framework, is not directly applicable to pairwise clus-

tering since there is no analytical technique known to capture correlations between assignments M_{iv} and $M_{\kappa v}$ in an exact way. The expected assignments $\langle M_{iv} \rangle$, however, can be approximated by calculating the average influence $\mathscr{E}_{iv}$ exerted by all M_{kv}, $k \neq i$ on the assignment M_{iv} (Hertz, Krogh, and Palmer, 1991:29), thereby neglecting pair correlations ($\langle M_{iv}M_{\kappa v} \rangle$). A maximum entropy estimate of $\mathscr{E}_{iv}$ yields these transcendental equations:

$$\langle M_{iv} \rangle = \frac{\exp(-\beta \mathscr{E}_{iv})}{\sum_{\mu=1}^{K} \exp(-\beta \mathscr{E}_{i\mu})} \qquad (12)$$

$$\mathscr{E}_{iv} = \frac{1}{p_v N} \sum_{\kappa=1}^{N} \langle M_{\kappa v} \rangle \left(D_{ik} - \frac{1}{2 p_v N} \sum_{l=1}^{N} \langle M_{lv} \rangle D_{\kappa l} \right) \qquad (13)$$

The variables $\mathscr{E}_{iv}$ depend on the given distance matrix D_{ik}, the averaged assignment variables $\{\langle M_{iv} \rangle\}$, and the cluster weights $\{p_v\}$. Equation 13 suggests an algorithm for learning the optimized cluster assignments which resembles the EM algorithm: In the E-step, the assignments $\{M_{iv}\}$ are estimates for given $\mathscr{E}_{iv}$. In the M-step, the $\mathscr{E}_{iv}$ are reestimated on the basis of new assignment estimates. This iterative algorithm converges to a consistent solution of assignments for the pairwise data clustering problem.

Discussion

Data clustering can be formulated either as a parameter estimation problem or as a combinatorial optimization problem for data partitioning. In the first approach, rooted in parametric statistics, an explicit goal is to find the correct multivariate probability distribution of the data set, i.e., the correct number of mixture components. Vector quantization, in contrast, partitions a data set according to a quality criterion, which often compromises between the precision of the data representation and other constraints like data coding or hardware considerations. Both strategies for clustering, although apparently quite different, are related on an algorithmic level by the stochastic nature of the data source. The parameter estimation for probability densities of mixture models and the maximum entropy estimation of stochastic data assignments in the optimization problem yield similar equations for the cluster centers and the expected assignments. The differences between the two clustering approaches are most significant when complexity constraints are introduced and solutions with different numbers of clusters have to be compared.

Road Map: Learning in Artificial Neural Networks, Statistical
Related Reading: Competitive Learning; Neural Optimization

References

Buhmann, J., and Kühnel, H., 1993, Vector quantization with complexity costs, *IEEE Trans. Inform. Theory*, 39:1133–1145.
Dempster, A. P., Laird, N. M., and Rubin, D. B., 1977, Maximum likelihood from incomplete data via the EM algorithm, *J. R. Statist. Soc. Ser. B*, 39:1–38.
Duda, R. O., and Hart, P. E., 1973, *Pattern Classification and Scene Analysis*, New York: Wiley, sect. 3.2. ◆
Hertz, J., Krogh, A., and Palmer, R. G., 1991, *Introduction to the Theory of Neural Computation*, New York: Addison-Wesley.
Jacobs, R. A., Jordan, M. I., Nowlan, S. J., and Hinton, G. E., 1991, Adaptive mixtures of local experts, *Neural Computat.*, 3:79–87.
Jain, A. K., and Dubes, R. C., 1988, *Algorithms for Clustering Data*, Englewood Cliffs, NJ: Prentice Hall. ◆
Kohonen, T., 1984, *Self-Organization and Associative Memory*, Berlin: Springer.
MacQueen, J., 1967, Some methods for classification and analysis of multi-variate observations, in *Proceedings of the 5th Berkeley Sym-*

posium on Mathematical Statistics and Probability, Berkeley: University of California Press, pp. 281–297.

McLachlan, G. J., and Basford, K.-E., 1988, Mixture Models, New York and Basel: Dekker. ◆

Nowlan, S. J., and Hinton, G. E., 1992, Simplifying neural networks by soft weight-sharing, Neural Computat., 4:473–493.

Poggio, T., and Girosi, F., 1990, Networks for approximation and learning, Proc. IEEE, 79:1481–1497.

Rissanen, J., 1989, Stochastic Complexity in Statistical Inquiry, Singapore: World Scientific.

Rose, K., Gurewitz, E., and Fox, G., 1990, A deterministic annealing approach to clustering, Pattern Recognition Lett., 11:589–594.

Tikochinsky, Y., Tishby, N. Z., and Levine, R. D., 1984, Alternative approach to maximum-entropy inference, Phys. Rev. A, 30:2638–2644.

Dendritic Processing

Idan Segev

Introduction

If Gertrude Stein, who is famous for using repetition in her writings, had written about the nervous system, she might observe that "dendrites are dendrites are dendrites." In fact, *dendrites* are strikingly exquisite and unique structures that are used to classify neurons into types: pyramidal, Purkinje, amacrine, stellate, etc. (Figure 1); they are also the largest component in both surface area and volume of the brain. But most meaningful is that most of the synaptic inputs occur on the dendritic tree, and it is there where this information is processed. Indeed, dendrites are the elementary computing devices of the brain.

This article presents a summary of our present ideas about and understanding of the computational function of dendrites. The focus here is on dendrites as electrical input-output devices that operate on a time scale ranging from several to a few hundred milliseconds. The input comprises temporal patterns of synaptic inputs impinging over the dendritic surface, whereas the output is either an ionic current that is delivered to the axon (where, via a threshold mechanism, a train of action potentials is produced) or, for dendrodendritic interactions, a pattern of synaptic potentials at neighboring dendrites. How do the machinery of dendrites, their morphology, electrical properties, and synaptic inputs interact to perform their input-output operation?

We first introduce dendrites, then we explain Rall's cable theory and his compartmental modeling approach, which provided the basic insights regarding dendritic processing. Recent ideas on the computational role of the variety of excitable (voltage-gated) channels in dendrites are discussed, and the epilogue suggests several directions for future research, including new technologies for viewing how dendrites process synaptic information.

Dendrites and Their Synapses

Table 1 concisely captures the functionally important facts about dendrites. Because dendrites come in many shapes and sizes, such a summary unavoidably gives only a rough range of values. Nonetheless, several important functional conclusions can be drawn from this table. (For reviews, see Braitenberg and Schüz, 1991; Shepherd, 1990; McKenna, Davis, and Zornetzer, 1992; and Rall et al., 1992.)

1. Dendrites tend to ramify, creating large and complicated trees. Dendrites are thin processes, starting with a diameter of a few microns near the soma; subsequently, branch diameter typically decreases to less than 1 μm as they successively branch. Many (but not all) types of dendrites are studded with abundant tiny branches, or appendages, called the dendritic spines. When present, DENDRITIC SPINES (q.v.) are the major postsynaptic target for *excitatory* synaptic inputs.

2. Synapses are not randomly distributed over the dendritic surface. In general, inhibitory synapses are more proximal than excitatory synapses (see legend in Figure 1, and Shepherd, 1990).

3. Synaptic inputs, both excitatory and inhibitory, typically operate by locally changing (increasing) the conductance of the postsynaptic membrane (opening specific ion channels). Synaptic input perturbs the electrical properties of dendrites, and so the dendrites are inherently nonlinear devices. The time course of the synaptic conductance change associated with the various types of inputs in a given neuron may vary by 1–2 orders of magnitude. The fast excitatory (non-NMDA) and inhibitory (GABA$_A$) inputs operate on a time scale of 1 ms and have a peak conductance on the order of 1 nS; this conductance is approximately 10 times larger than the slow excitatory (NMDA) and inhibitory (GABA$_B$) inputs that both act on a time scale of 10–100 ms.

4. The dendritic membrane is a relatively good electrical insulator (R_m is on the order of 10,000–50,000 Ωcm^2 and more). With capacitance, C_m, of approximately 1 μF/cm^2, the membrane (integration) time window at the soma is on the order of $\tau = R_m C_m = 10$–50 ms. The large R_m and the small dimensions of dendrites imply large (DC) input resistance ($\sim 10^3$ MΩ) as well as a large (AC) input impedance at distal arbors and on spines.

5. Many types of dendrites, most notably the cerebellar Purkinje cell, are covered with nonlinear excitable (voltage-dependent) channels. These nonlinear channels have important consequences for the computational capabilities of dendrites.

The Theoretical Foundation for Dendritic Modeling

In two ingenious studies, Rall (1959, 1964) established the theoretical foundation that has allowed us to functionally link the three columns of Table 1. His passive cable theory for dendrites, complemented by his compartmental modeling approach, laid the groundwork for a quantitative exploration of the integrative function of dendrites.

Passive Cable Theory for Dendrites

The underlying equation that describes the flow of electrical current (and the spread of the resultant voltage) in morphologically and physiologically complicated dendritic trees is the one-dimensional cable equation:

Figure 1. Dendrites have unique shapes that are used to characterize neurons into types. In many neuron types, synaptic inputs from a given source are preferentially mapped into a particular region of the dendritic tree *A*, Cerebellar Purkinje cell of the guinea pig (reconstructed by M. Rapp). *B*, CA1 pyramidal neuron from the rat (reconstructed by B. Claiborne). *C*, Neostriatal spiny neuron from the rat (from Wilson, 1992). *D*, Axonless interneurons of the locust (reconstructed by G. Laurent). Typically, synaptic inputs are distributed nonrandomly over the dendritic surface.

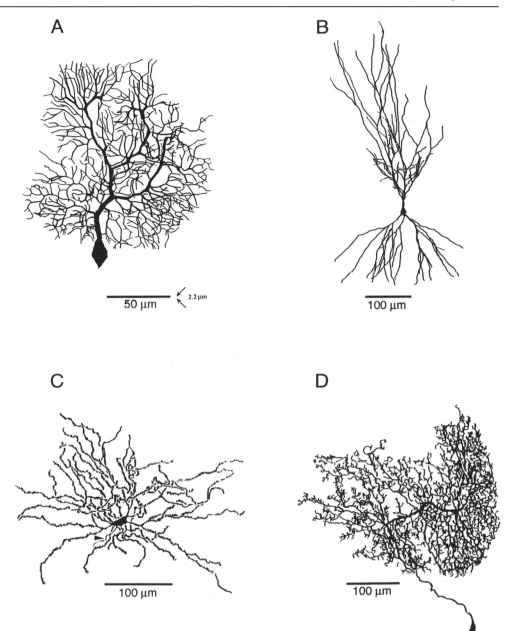

$$\frac{\partial^2 V(X,T)}{\partial X^2} = \frac{\partial V}{\partial T} + V(X,T) \tag{1}$$

where V is the voltage across the membrane (interior minus exterior, relative to the resting potential); $X = x/\lambda$, where x is the distance along the core conductor (cm) and the *space constant*, λ, is defined as $\sqrt{r_m/r_i}$; and $T = t/\tau_m$, where the *time constant*, τ_m, is $r_m c_m$. Further, r_m is the membrane resistance per unit length (in Ωcm), c_m is the membrane capacitance per unit length (in F/cm), and r_i is the cytoplasm resistance per unit length (in Ω/cm). Complete derivation of the cable equation can be found in the chapter by Rall (in Koch and Segev, 1989; see also Jack, Noble, and Tsien, 1975).

The solution for Equation 1 depends, in addition to the electrical properties of the membrane and cytoplasm, on the boundary condition at the ends of the segment toward which the current flows. Rall (1959) showed that Equation 1 could be

solved analytically also for passive dendritic trees with arbitrary branching. He modeled the dendritic tree as a collection of short cylindrical segments (Figure 2*B*), where the tree attached at the end of each segment acts as a sink for the longitudinal current (i.e., a "leaky end"). An example of such a solution for the steady-state case is shown in Figure 3*A*, and one for the transient case is shown in Figure 3*B*. Several important implications of this figure are discussed later in this article, and a recent extension of cable theory that allows an explicit definition of the notion of input synchronicity in dendrites can be found in Agmon-Snir and Segev (1993).

Compartmental Modeling Approach

The compartmental modeling approach complements cable theory by overcoming the assumption that the membrane is passive and the input is current. Mathematically, the com-

Table 1. Recent Ranges of Values for Dendritic Machinery

Morphology	Physiology	Synaptology
Diameter near soma: 1–6 μm Tip diameter: 0.3–1 μm Average dend. length: 0.15–1.5 mm Total dend. length: 1–10 mm Dendrite area: 2000–750,000 μm^2 Dendritic trees/neuron: 1–16 Dendritic tips/neuron: 10–400 Dend. spines/neuron: 300–200,000 Spines dens./1-μm dendrite: 0.5–14 Spine neck length: 0.1–2 μm Spine neck diameter: 0.04–0.5 μm Spine head diameter: 0.3–1 μm Spine volume: 0.005–0.3 μm^3	*Passive properties*[a] Mem. resistivity (R_m): 5–50,000 Ωcm^2 Axial resistivity (R_i): 70–250 Ωcm Mem. capacitance (C_m): 1–2 μF/cm^2 Time constant (τ_m): 7–50 ms Dendrite space const. (λ): 0.2–1 mm Elect. length ($L = x/\lambda$): 0.2–2 Soma input resist. (R_N): 1–10^3 MΩ Tip input resist. (R_T): 10^2–10^3 MΩ SS V attenuation (soma → tip): 1.1–2 SS attenuation (tip → soma): 2–15 *Excitable channels*[b] High-threshold Ca^{+2} (L, N, P type) Fast Na$^+$ current K$^+$ current (?)	No. synapses/neuron: 500–200,000 Type I[c] (excitatory): 60%–90%; distributed, with majority on spines Type II[c] (inhibitory): 10%–40%; near soma; some on spines *Excitatory synaptic input*[d] non-NMDA: g_{peak}: 0.1–0.3 ns; t_{peak}: 0.3–1 ms NMDA: g_{peak}: 0.05–0.5ns; t_{peak}: 5–50 ms *Inhibitory synaptic input*[e] GABA$_A$: g_{peak}: 0.4–1 ns; t_{peak}: 0.2–1.2 ms GABA$_B$: g_{peak}: 0.1–0.3 ns; t_{peak}: 40–150 ms

[a] Passive properties of dendrites can be strongly modulated by synaptic activity as well as by voltage-gated channels. Steady state (SS) voltage attenuation and input resistance at dendritic tips were estimated from detailed models of reconstructed neurons (e.g., Rapp et al., 1994, *J. Physiol.*, 474:101–118).

[b] Characterization (i.e., type, distribution, density, and kinetics) of the voltage-gated channels in dendrites is only starting to emerge, using antibodies against ion channels as well as patch clamp techniques. For references on dendritic Ca^{+2} channels, see Llinás, *Science*, 242:1654–1664; Westenbroek et al., 1992, *Neuron*, 9:109–115; and Markram et al., *J. Physiol.* (in press). Information on voltage-gated Na channels in dendrites can be found in Stuart and Sakmann, 1994, *Nature*, 367:69–72. Information on K channels in dendrites is still missing.

[c] Based on data from cortical and hippocampal pyramidal neurons.

[d] See, e.g., Stern et al., 1992, *J. Physiol.*, 449:247–278.

[e] See, e.g., Thomas et al., 1993, *J. Physiol.*, 463:391–407.

partmental approach is a finite-difference (discrete) approximation to the (nonlinear) cable equation. It replaces the continuous cable equation by a set, or a matrix, of ordinary differential equations and, typically, numerical methods are employed to solve this system (which can include thousands of compartments and thus thousands of equations) for each time step. In the compartmental model, dendritic segments that are electrically short are assumed to be isopotential and are lumped into a single R-C (either passive or active) membrane compartment (Figure 2C). Compartments are connected to each other via a longitudinal resistivity according to the topology of the tree. Hence, differences in physical properties (e.g., diameter, membrane properties, etc.) and differences in potential occur between compartments rather than within them. It can be shown that when the dendritic tree is divided into sufficiently small segments (compartments), the solution of the compartmental model converges to that of the continuous cable model. A compartment can represent a patch of membrane with a variety of voltage-gated (excitable) and synaptic (time-varying) channels. A review of this very popular modeling approach can be found in a chapter by Segev et al. (in Koch and Segev, 1989).

Main Insights for Passive Dendrites with Synapses

The theoretical background just outlined and the many results obtained from modeling and experimental studies of dendrites conducted during the last 40 years provide important insights into the input-output properties of passive dendrites. These properties can be assembled as follows:

1. Dendritic trees are electrically distributed (rather than isopotential) elements. Consequently, voltage gradients exist over the tree when synaptic inputs are applied locally. Because of inherent nonsymmetric boundary conditions in dendritic trees, voltage attenuates much more severely in the

dendrites-to-soma direction than in the opposite direction (Figure 3A).

2. The large voltage attenuation from dendrites to soma (which may be a few hundredfold for brief synaptic inputs at distal sites) implies that many (several tens) of excitatory inputs should be activated within the integration time window, τ_m, in order to build-up depolarization of 10–20 mV and reach threshold for firing of spikes at the soma and axon.

3. Although severely attenuated in *peak* values, the attenuation of the time integral of transient potentials (as well as the attenuation of charge) is relatively small. Thus, the "cost" (in terms of area under the voltage transient, or charge) of placing the synapse at the dendrites rather than at the soma is minimal (Figure 3A).

4. Synaptic potentials are delayed, and they become significantly broader, as they spread away from the input site (Figure 3B). The large sink provided by the tree at distal arbors implies that, locally, synaptic potentials are very brief. This change in width of the synaptic potential implies multiple time windows for synaptic integration in the tree (Figure 4). At the soma level the time window for synaptic integration is primarily governed by τ_m, whereas at distal dendritic arbors it may be as short as 0.1 τ_m and less (see Agmon-Snir and Segev, 1993).

5. Inhibitory synapses (whose conductance change is associated with a battery near the resting potential) are more effective when located on the path between the excitatory input and the "target" point (soma) than when placed distal to the excitatory input. Thus, when strategically placed, inhibitory inputs can specifically veto parts of the dendritic tree and not others (Rall, 1964; Koch, Poggio, and Torre, 1982; Jack et al., 1975).

6. Because of dendritic delay, the somatic depolarization that results from activation of excitatory inputs at the dendrites is very sensitive to the temporal sequence of the synaptic

Figure 2. Dendrites (*A*) are modeled either as a set of cylindrical membrane cables (*B*) or as a set of discrete, isopotential, R-C compartments (*C*). In the cable representation (*B*), the voltage at any point in the tree is computed from Equation 1, and the appropriate boundary conditions are imposed by the tree. In compartmental representation (*C*), the tree is discretized into a set of interconnected R-C compartments; each is a lumped representation of a sufficiently small dendritic segment. Membrane compartments are connected via axial, cytoplasmic resistances. Here, the voltage can be computed in each compartment for any nonlinear input and voltage- and time-dependent membrane properties.

A. Physiologically & Morphologically Characterized Neuron

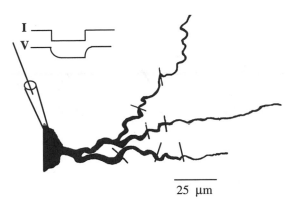

25 μm

B. Cable Model

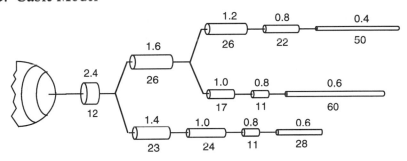

C. Compartmental Model

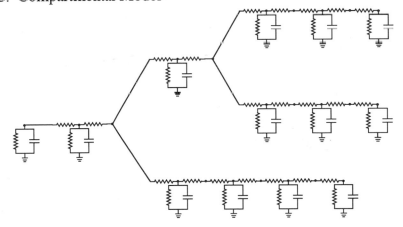

A

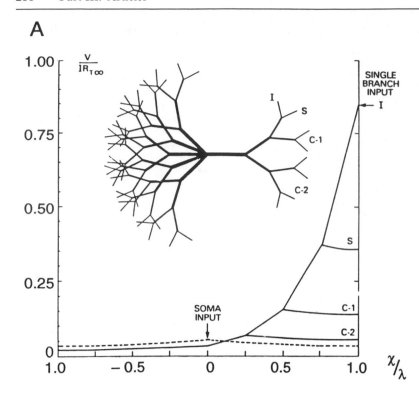

Figure 3. The voltage spread in passive dendritic trees is asymmetrical (*A*); its time-course changes are broadened, and the peak is delayed as it propagates away from the input site (*B*). The solid curve in *A* shows the steady-state voltage computed for current input to the terminal branch (I). The dashed line corresponds to the same current when applied to the soma. Note the small difference, at the soma, between the solid curve and the dashed curve, indicating the negligible "cost" of placing this input at the distal branch rather than at the soma. (From chapter by W. Rall in Koch and Segev, 1989). In *B*, a brief transient current is applied to the terminal branch (I), and the resultant voltage is shown on a logarithmic scale. Note the several hundredfold attenuation from the input site to the soma and the broadening of the transient as it spreads away from the input site (from Rinzel, J., and Rall, W., 1974, *Biophys. J.*, 14:759–790).

B

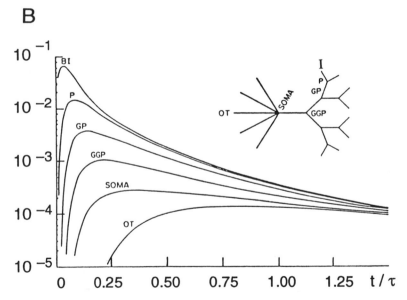

activation. The somatic depolarization is largest when the synaptic activation starts at distal dendritic sites and progresses proximally. Activation of the same synapses in the reverse order in time will produce decreased somatic depolarization. Thus, the output of neurons with dendrites is inherently *directionally selective* (Figure 5).

Excitable Dendrites

In recent years, it has been clearly demonstrated that the membrane of many types of dendrites is endowed with voltage-gated (nonlinear) ion channels, including the NMDA channels, as well as voltage-activated inward (Ca^{+2} and Na^{+}) and out-

ward (K^{+}) conductances (Ross and Werman, 1987; Stuart and Sakmann, 1994; Laurent, 1993; McKenna et al., 1992; Wilson, in McKenna et al., 1992). These channels are responsible for a variety of subthreshold electrical nonlinearities and, under favorable conditions, they can generate full-blown action potentials. In contrast to axonal trees, the regenerative phenomenon from input into excitable dendrites tends to spread only locally.

What is the electrical behavior to be expected from dendrites with voltage-gated membrane ion channels? First, these channels can modulate the input-output properties of the neuron. For example, they can *amplify* the excitatory synaptic current locally, and the regenerative activity can spread ("chain reaction"), indirectly activating nearby dendritic regions and thereby further enhancing the excitatory synaptic inputs (Rall and

Figure 4. Dendrites provide multiple time windows for input integration. *A*, In the reference case of an isopotential R-C circuit, this time window (measured by the local delay, LD) is governed by the membrane time constant, $\tau_m = RC$ (τ_m was set to 20 ms). *B*, In the electrically distributed structure of soma coupled to a sealed-end cylinder (at $x/\lambda = 1$), the local time window at intermediate locations along the cylinder is somewhat smaller than τ_m (it is exactly $\tau_m/2$ in an infinite cylinder). *C*, In complicated trees, however, the time windows at distal arbors are 10 times briefer than τ_m, which remains the time window for input integration at the soma. The soma behaves more as an integrator, whereas distal dendrites act more like a coincidence detector for local inputs. The time window is defined here as the difference between the centroid of the input current input and the centroid of the resultant voltage transient at the input site (Agmon-Snir and Segev, 1993). Layer V cortical pyramidal cell was reconstructed by R. Douglas.

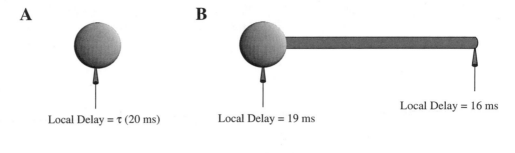

A   Local Delay = τ (20 ms)

B   Local Delay = 19 ms   Local Delay = 16 ms

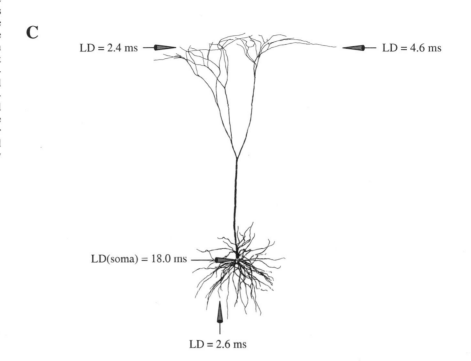

C   LD = 2.4 ms   LD = 4.6 ms

LD(soma) = 18.0 ms

LD = 2.6 ms

Segev, 1987). Because of the asymmetry of voltage attenuation in dendritic trees (see Figure 3*A*), regenerative activity in dendrites tends to spread more securely in the centrifugal (soma-to-dendrites) direction than in the centripetal direction.

Second, unlike the passive case, in the excitable situation a certain degree of input clustering implies an increased charge transfer to the soma (owing to the extra-active inward current). In this case, the output at the axon depends sensitively on the size (and site) of the "clusteron," and this may serve as a mechanism for implementing multidimensional discrimination tasks of input patterns via multiplication-like operations (Mel, 1993).

Third, active dendritic currents (both inward and outward) may serve as a mechanism for synaptic gain control. As a result of active currents, the integrative capabilities of the neuron (e.g., its input resistance and electrotonic length) are dynamically controlled by the membrane potential; thus, the neuron output depends on its state (membrane potential). Active currents (e.g., outward K^+ current) can act to counterbalance excitatory synaptic inputs (negative feedback), thus stabilizing the input-output characteristics of the neuron. At other voltage regimens, active currents might effectively increase the input resistance and reduce the electrotonic distance between synapses (positive feedback), with the consequence of nonlinearly boosting a specific group of coactive excitatory synapses (Wilson, in McKenna et al., 1992; Laurent, 1993).

Computational Function of Dendrites

It seems appropriate to conclude this article by asking what kind of computations could be performed by a neuron with dendrites. Several answers have already been discussed earlier; they are succinctly highlighted as follows:

1. Neurons with dendrites can compute the direction of motion (Rall, 1964; Koch, Poggio, and Torre, 1982).
2. Neurons with dendrites can simultaneously function on multiple time windows. Distal arbors act more as coincidence detectors, whereas the soma acts more as an integrator when brief synaptic inputs (i.e., non-NMDA GABA_A) are involved (Agmon-Snir and Segev, 1993).
3. Neurons with dendrites can implement a multidimensional classification task (Mel, 1993).
4. Neurons with dendrites can function as many, almost independent, functional subunits. Each unit can implement a rich repertoire of *logical* operations (Koch et al., 1982; Rall and Segev, 1987), as well as other local computations (e.g., local synaptic plasticity), and they can function as semi-autonomous input-output elements (e.g., via dendrodendritic synapses).
5. Neurons with slow ion currents in the dendrites that are partially electrically decoupled from fast spike-generating

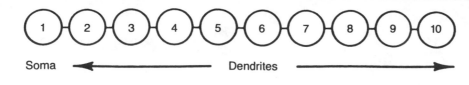

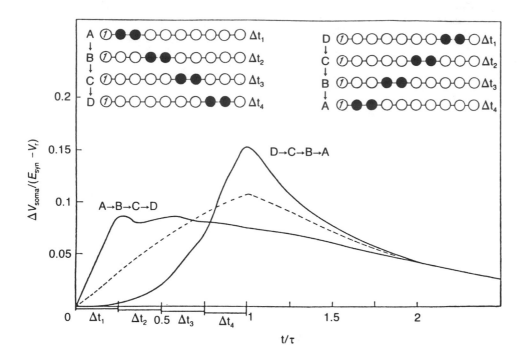

Figure 5. Neurons with dendrites and synapses can compute the direction of motion. The depolarization (ΔV_{soma}) at the soma (compartment 1) is sensitive to the spatiotemporal sequence of synaptic activation. When only excitatory inputs are involved (solid circles), the depolarization at the soma is larger and more delayed when the synaptic activation starts distally and progresses proximally. The dotted curve shows the effect when the same total synaptic conductance is spread uniformly over compartments 2–9 for the full duration $t = 0 - \tau$. ($E_{syn} - V_r$) is the difference between the excitatory battery and the resting potential. The y-axis could be interpreted as firing probability (from Rall, 1964).

currents at the soma/axon hillock can produce a large repertoire of frequency pattern, including regular, high-frequency spiking, as well as bursting, as is thought to occur in experimental and theoretical models of epileptic seizures (Pinsky and Rinzel, 1994).

Road Map: Biological Neurons
Background: Axonal Modeling; Ion Channels: Keys to Neuronal Specialization
Related Reading: Directional Selectivity in the Retina

Epilogue

In her critical writings, Gertrude Stein said that "Science, well they never are right about anything, not right enough so that science cannot go enjoying itself as if it is interesting, which it is...." Interesting—and very promising—are several novel technologies that may prove useful for probing, physically emulating, and precision-altering parts of the nervous system. Among these are voltage-dependent dyes that enable one to view the electrical activity of neurons when the system carries out specific computations. VLSI technology potentially makes it possible to emulate the electrical (and chemical) activity of synapses, dendrites, and axons and to construct realistic neural networks in chips that operate in real time (see SILICON NEURONS). These, and molecular biological methods, including antibodies against specific ion channels, combined with high-resolution optical probes, may serve as the essential link between the single neuron level and the system levels, a step needed for understanding how the neuronal machinery implements in vivo, rather than in theory, its computational functions.

Acknowledgments. This work was supported by a grant from the BSF and ONR.

References

Agmon-Snir, H., and Segev, I., 1993, Signal delay and input synchronization in passive dendritic structures, *J. Neurophysiol.*, 70:2066–2085.

Braitenberg, V., and Schüz, A., 1991, *Anatomy of the Cortex*, Berlin: Springer-Verlag. ◆

Jack, J. J. B., Noble, D., and Tsien, R. W., 1975, *Electric Current Flow in Excitable Cells*, Oxford: Clarendon. ◆

Koch, C., Poggio, T., and Torre, V., 1982, Retinal ganglion cells: A functional interpretation of dendritic morphology, *Philos. Trans. R. Soc. Lond. [Biol.]*, 298:227–264.

Koch, C., and Segev, I., Eds., 1989, *Methods in Neuronal Modeling: From Synapses to Networks*, Boston: MIT Press.

Laurent, G., 1993, A dendritic gain control mechanism in axonless neurons of the locust, *Schistocerca americana, J. Physiol. (Lond.)*, 370:45–54.

McKenna, T., Davis J., and Zornetzer, S. F., Eds., 1992, *Single Neuron Computation*, Boston: Academic Press.

Mel, B. W., 1993, Synaptic integration in excitable dendritic trees, *J. Neurophysiol.*, 70:1086–1101.

Pinsky, P. F., and Rinzel, J., 1994, Intrinsic and network rhythmogenesis in a reduced model of CA3 neurons, *J. Computat. Neurosci.*, 1:39–60.

Rall, W., 1959, Branching dendritic trees and motoneuron membrane resistivity, *Exp. Neurol.*, 2:503–532.

Rall, W., 1964, Theoretical significance of dendritic trees for neuronal input-output relations, in *Neural Theory and Modeling* (R. F. Reiss, Ed.), Stanford, CA: Stanford University Press, pp. 73–97.

Rall, W., Burke, R. E., Holmes, W. R., Jack, J. J. B., Redman, S. J., and Segev, I., 1992, Matching dendritic neuron models to experimental data, *Physiol. Rev.* (Suppl.), 72:S159–S186.

Rall, W., and Segev, I, 1987, Functional possibilities for synapses on dendrites and on dendritic spines, in *Synaptic Function* (G. M. Edelman., E. E. Gall., and W. M. Cowan., Eds.), New York: Wiley, pp. 605–636.

Ross, W. N., and Werman, R., 1987, Mapping calcium transients in the dendrites of Purkinje cells from the guinea-pig cerebellum in vitro, *J. Physiol.*, 389:319–336.

Shepherd, G. M., 1990, *The Synaptic Organization of the Brain*, New York: Oxford University Press. ◆

Stuart, G. J., and Sakmann B., 1994, Active propagation of somatic action potentials into neocortical pyramidal cell dendrites, *Nature*, 367:69–72.

Dendritic Spines

William R. Holmes and Wilfrid Rall

Introduction

The function of dendritic spines has been debated ever since they were discovered by Ramon y Cajal. Although it was widely believed that spines were important for intercellular communication, this was not demonstrated until 1959 when Gray, using the electron microscope, showed that synapses are present on spines. Why should synapses exist on spines, rather than on dendrites? What role does spine morphology play in their function? Early theoretical studies of these questions focused on the large electrical resistance provided by the thin spine stem and suggested that changes in stem diameter might be important for synaptic plasticity. Later investigations found that if voltage-dependent conductances were present on spines, then spines could increase the computational possibilities of a cell. Recent models suggest that spines are isolated compartments where highly localized chemical reactions can take place.

Spine Morphology

Dendritic spines are short, appendage-like structures found on many different cell types. Spines are composed of a bulbous "head" connected to the dendrite by a thin "neck" or "stem." An excitatory synapse is usually found on the spine head, and some spines also have a second, usually inhibitory, synapse located near or on the spine neck. Spines typically are small in size, but because they occur in densities of 1–2 spines/μm or more, spine membrane area can comprise 40–60% of the total neuron membrane area.

Attempts have been made to classify spines based on their size, shape, and dendritic location. Jones and Powell (1969) categorized spines as sessile (stemless) or pedunculated (having a peduncle, or stem), with sessile spines more common in proximal regions and pedunculated spines in distal regions. Peters and Kaiserman-Abramof (1970) classified spines as (1) stubby, (2) mushroom-shaped, and (3) thin, or long-thin (Figure 1). Stubby spines were most numerous in proximal regions, long-thin spines dominated distal regions, and mushroom-shaped spines were distributed almost uniformly. These categories are arbitrary, since spine shape varies continuously and all types of spines are found in all areas. In some brain regions, it has not been possible to categorize spines in any systematic manner, but categories and a range of dimensions are useful for models.

Passive Models of Spine Function

Models in which the spine is represented as a passive electric circuit show that the large resistance of a thin spine stem can

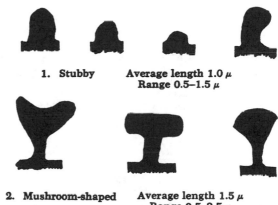

1. Stubby   **Average length 1.0 μ**
Range 0.5–1.5 μ

2. Mushroom-shaped   **Average length 1.5 μ**
Range 0.5–2.5 μ
Average stalk length 0.8 μ
Average bulb dimensions 1.4 × 0.6 μ

3. Thin   **Average length 1.7 μ**
Range 0.5–4.0 μ
Average stalk length 1.1 μ
Average bulb dimensions 0.6 μ

Figure 1. Variety of spine shapes in the parietal cortex. (Adapted from Peters, A., and Kaiserman-Abramof, I. R., 1970, The small pyramidal neuron of the rat cerebral cortex: The perikaryon, dendrites and spines, *American Journal of Anatomy*, 127:325, table 1. Copyright © 1970 by the Wistar Institute of Anatomy and Biology. Used by permission of John Wiley & Sons, Inc.)

attenuate a synaptic input delivered to the spine head. This can be seen by considering the circuit pictured in Figure 2.

Assuming a constant synaptic conductance, currents flowing in the spine head can be described by Kirchhoff's law as:

$$V_{SH}/R_{SH} + g_{syn}(V_{SH} - V_{EQ}) + V_{SH}/(R_{SS} + R_{BI}) = 0 \qquad (1)$$

where the first term is leakage current across the spine head membrane, the second is the synaptic current, and the third is the flow of current through the spine stem to ground. Because R_{SH} is large, the first term in Equation 1 is negligible compared to the other two terms and can be ignored. With this simplifica-

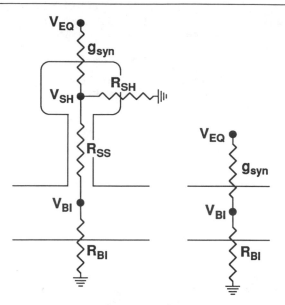

Figure 2. Electrical circuit of a dendritic spine. V_{EQ} is the synaptic reversal potential, V_{SH} is the voltage in the spine head, and V_{BI} is the voltage in the dendrite at the base of the spine. R_{SH} is the spine head resistance, R_{SS} is the spine stem resistance, and R_{BI} is the branch input resistance at the base of the spine. g_{syn} is the synaptic conductance. The corresponding circuit for an input on a dendrite is shown on the right.

tion, the steady-state spine head potential is

$$V_{SH} = V_{EQ}/[1 + 1/\{g_{syn}(R_{SS} + R_{BI})\}]. \qquad (2)$$

Similarly, the currents in the dendrite are described as

$$V_{BI}/R_{BI} + (V_{BI} - V_{SH})/R_{SS} = 0 \qquad (3)$$

which can be rearranged as

$$V_{BI} = V_{SH}R_{BI}/(R_{BI} + R_{SS}) \qquad (4)$$

Combining Equations 2 and 4, the voltage at the dendrite is

$$V_{BI} = V_{EQ}/[1 + 1/(g_{syn}R_{BI}) + (R_{SS}/R_{BI})] \qquad (5)$$

If, however, the synapse were on the dendrite, Kirchhoff's current law says that

$$g_{syn}(V_{BI} - V_{EQ}) + V_{BI}/R_{BI} = 0 \qquad (6)$$

Rearranging, we have

$$V_{BI} = V_{EQ}/[1 + 1/(g_{syn}R_{BI})] \qquad (7)$$

The only difference between Equations 5 and 7 is the presence of the ratio R_{SS}/R_{BI} in the denominator of Equation 5, and this term accounts for voltage attenuation when the synapse is on the spine instead of the dendrite.

Spines also attenuate synaptic current. The synaptic currents for spine and dendritic inputs can be computed by substituting expressions for V_{SH} and V_{BI} in Equations 2 and 7 in place of V in $g_{syn}(V - V_{EQ})$. The resulting currents are given by the right side of Equations 5 and 7 divided by R_{BI}. The size of the synaptic current entering the spine is attenuated because, for identical g_{syn}, the voltage in the spine head owing to an input there is closer to V_{EQ} than the voltage change with a dendritic input.

The R_{SS}/R_{BI} ratio in these equations suggests another possible function for dendritic spines. If a neuron can adjust spine stem morphology (and hence R_{SS}), then spines provide a mechanism to allow synaptic weights to be modified (Rall, 1978).

For this mechanism to be important, the value of R_{SS}/R_{BI} should lie between 0.1 and 10 times $1 + 1/(g_{syn}R_{BI})$, because within this range a small change in R_{SS}/R_{BI} can have a significant effect on V_{BI} (see Equation 5). Early experimental estimates of R_{SS}/R_{BI} suggested that it might lie in this effective operating range. This encouraged investigators to look for spine dimension changes in various experimental situations. However, recent estimates of g_{syn}, R_{SS}, and R_{BI} suggest that R_{SS}/R_{BI} may not fall in this range. For example, if $g_{syn} = 0.5$ nS, and $R_{BI} = 200$ MΩ, then $1/(g_{syn}R_{BI}) = 10$. This means that R_{SS} should be 220–2200 MΩ; morphological measurements indicate, however, that R_{SS} of long-thin spines is less than 100 MΩ in many neuron types.

Although it does not appear that the function of spines can be explained by changes in passive electrical properties caused by changes in spine dimensions, prudence suggests that one should keep an open mind to this possibility. Even today we remain unsure of actual spine stem dimensions, especially in a dynamic, living spine. We also remain unsure of the cytoplasmic resistivity. Estimates of intracellular resistivity, R_i, have risen in recent years, and R_i may be larger in spines than elsewhere in the cell. While it is true that estimates of spine stem resistance for some neurons have turned out to be too small to fit the hypothetical operating range, there is too much uncertainty to justify an assertion that this holds true for all spines in all varieties of neurons.

Models of Excitable Spines

If voltage-dependent conductances exist on spines, then spines might exist to amplify synaptic inputs. Theoretical studies by Miller, Rall, and Rinzel (1985) and Perkel and Perkel (1985) showed that postsynaptic potential and charge transfer were five-to-ten-fold larger for synaptic input on spines with voltage-dependent conductances than for input to passive spines. Results were sensitive to voltage-dependent channel densities and spine stem resistance. Maximal amplification occurred for R_{SS} in a particular range. Below this range, voltage-dependent conductances could not be activated, and the spine behaved as a passive spine. Above this range, the large R_{SS} reduced current flow at the spine head, and this reduced the amplification.

Subsequent studies explored interactions among neighboring spines. Interactions among excitable spines create a number of interesting possibilities for information transfer. It is theoretically possible to get a chain reaction of spine head action potentials, spreading a certain distance proximally and distally. Rall and Segev (1988) explored signal amplification of this sort in models of "excitable spine clusters." Considerable amplification of the initial input occurs even if only a small percentage of spines possess voltage-dependent channels. Placing voltage-dependent conductances in the spine head rather than on dendrites means that fewer channels are needed to create dendritic "hot spots." Alternatively, activation of voltage-dependent channels in one spine could prevent an action potential from occurring later on a neighboring spine because of sodium inactivation caused by subthreshold depolarization.

Shepherd and Brayton (1987) have shown that basic logic operations such as AND gates, OR gates and AND-NOT gates can arise from interactions of excitable spines. (AND refers to action potential in a spine A only if spine A *and* spine B are activated; OR means action potential in spine A if spine A *or* spine B is activated; AND-NOT indicates action potential in spine A if spine A is activated *and* inhibition is *not* activated.) The OR gates might be more common in thin dendrites, and the AND gates might be more common proximally. The AND-NOT function might be local or more global depending on the location of the inhibition. Inhibition on the spine might veto input from that spine only, whereas inhibition on the dendrite might veto input from activated spines distal to the inhibition. Complex logic operations could be developed depending on the pattern of synaptic input.

If spines possess voltage-dependent conductances, then spines might exist to amplify synaptic input and to allow special kinds of information processing to occur. However, it is not known if nature takes advantage of these possibilities. The models assume high densities of voltage-dependent sodium and potassium channels in spines; but, as yet, there is no experimental evidence for this. The models also require an R_{SS} that is much larger than present estimates to get the interesting interaction effects. However, smaller R_{SS} values would be required if the spine head action potentials were due to calcium instead of sodium (because of slower kinetics).

Models of Calcium Diffusion in Spines

Previous theoretical studies searched for a function for spines that depended on the *electrical* resistance of the spine neck. Besides providing an electrical resistance to current flow, the thin spine neck provides a *diffusional* resistance to the flow of ions and molecules. Koch and Zador (1993) refer to the spine neck as having a small *diffusive space constant*. The spine neck, by restricting the flow of materials out of the spine head, might effectively isolate the spine head and provide a localized environment in which reactions specific to a particular synapse can occur.

Calcium is a prime candidate for a substance that might be selectively concentrated in the spine head. Calcium is important for a large number of metabolic processes and has been shown to be necessary for the induction of long-term potentiation, but high concentrations can lead to cell death. Spines might provide isolated locations where high concentrations of calcium can be attained safely without disrupting other aspects of cell function.

Compartmental models of dendritic spines have been developed to include the effects of diffusion, calcium buffering, and a calcium pump (Gamble and Koch, 1987; Holmes and Levy, 1990; Zador, Koch, and Brown, 1990). The major difficulty with these models is that parameter values for calcium influx, buffer kinetics, buffer concentration, pump rate, and pump capacity are largely unknown for dendritic spines. Nevertheless, these models show that calcium influx at the spine head can lead to large transient increases in spine head calcium concentration. If, instead, the calcium influx occurs on the dendrite, the local calcium concentration change is much smaller. Large increases occur in the spine head because of its small volume and because incoming calcium cannot be buffered or pumped out instantaneously.

Large transient increases in calcium concentration are restricted to the spine head because of the diffusional resistance of the spine neck. The thin spine neck acts as a constriction that slows calcium diffusion. Any calcium that does enter the spine neck is further hindered from diffusing to the dendrite by the presence of calcium buffer and pumps in the spine neck. Calcium that does diffuse through the spine neck to the dendrite has little effect on dendritic calcium concentration because of the large volume of the dendrite.

There are a number of factors that can affect the magnitude of the spine head calcium transient. The most important of these seem to be buffer concentration, the magnitude and duration of the calcium current, and spine shape (Holmes, 1990). Buffer concentration and calcium currents in spine heads have been difficult to quantify experimentally, but the range of spine shapes is known for several neuron types. Simulations with different spine shapes suggest that spines with fat necks are not able to concentrate calcium as quickly or to the same high levels as spines with long, thin necks. Thus, spine morphology may affect the rates and types of calcium-dependent reactions that can occur near a synapse on a particular spine.

Discussion

Early theoretical work with passive spine models showed the importance of a large electrical spine stem resistance for synaptic transmission and demonstrated how synaptic weights might be modified by changes in this resistance. Later modeling showed that input could be amplified or transformed in a variety of interesting ways if spines have excitable membrane. Although spine stem measurements suggest that spine stem *electrical* resistance is too small to play a role in plasticity in most cases, and although the existence of excitable spines has yet to be demonstrated, we cannot discard hypotheses of spine function based on the electrical properties of spines without better estimates of spine stem resistance and without knowing whether voltage-dependent ion channels exist in sufficient densities on spine membrane.

If spine stem resistance is indeed small, and if the required voltage-dependent channels do not exist on spines, then one must look elsewhere to explain spine function. The current hypothesis is that spines, by restricting the diffusion of substances away from the synapse, provide a local, isolated environment in which specific biochemical reactions can occur. In particular, the spine stem may provide a *diffusional* resistance that allows calcium to become concentrated in the spine head and calcium-dependent reactions to be localized to the synapse. This could be very important for plasticity changes, such as those that occur with long-term potentiation. Spine morphology may determine the magnitude of the diffusional resistance and play a role in determining, or restricting, the types of biochemical reactions that can take place at a synapse.

Road Map: Biological Neurons
Background: I.1. Introducing the Neuron; Ion Channels: Keys to Neuronal Specialization
Related Reading: Hebbian Synaptic Plasticity; Olfactory Bulb; Olfactory Cortex

References

Brown, T. H., Chang, V. C., Ganong, A. H., Keenan, C. L., and Kelso, S. R., 1988, Biophysical properties of dendrites and spines that may control the induction and expression of long-term synaptic potentiation, in *Long-Term Potentiation: From Biophysics to Behavior* (P. W. Landfield and S. A. Deadwyler, Eds.), New York: Alan R. Liss, pp. 201–264.

Fifková, E., and Morales, M., 1992, Actin matrix of dendritic spines, synaptic plasticity, and long-term potentiation, *Int. Rev. Cytol.*, 139:267–307. ◆

Gamble, E., and Koch, C., 1987, The dynamics of free calcium in dendritic spines in response to repetitive synaptic input, *Science*, 236:1311–1315.

Harris, K. M., and Kater, S. B., 1994, Dendritic spines: Cellular specializations imparting both stability and flexibility to synaptic function, *Annu. Rev. Neurosci.*, 17:341–371. ◆

Holmes, W. R., 1990, Is the function of dendritic spines to concentrate calcium? *Brain Res.*, 519:338–342.

Holmes, W. R., and Levy, W. B., 1990, Insights into associative long-term potentiation from computational models of NMDA receptor-mediated calcium influx and intracellular calcium concentration changes, *J. Neurophysiol.*, 63:1148–1168.

Jones, E. G., and Powell, T. P. S., 1969, Morphological variations in the dendritic spines of the neocortex, *J. Cell. Sci.*, 5:509–529.

Koch, C., and Zador, A., 1993, The function of dendritic spines: Devices subserving biochemical rather than electrical compartmentalization, *J. Neurosci.*, 13:413–422. ◆

Miller, J. P., Rall, W., and Rinzel, J., 1985, Synaptic amplification by active membrane in dendritic spines, *Brain Res.*, 325:325–330.

Perkel, D. H., and Perkel, D. J., 1985, Dendritic spines: Role of active membrane in modulating synaptic efficacy, *Brain Res.*, 325:331–335.

Peters, A., and Kaiserman-Abramof, I. R., 1970, The small pyramidal neuron of the rat cerebral cortex: The perikaryon, dendrites and spines, *Am. J. Anat.*, 127:321–356.

Rall, W., 1978, Dendritic spines and synaptic potency, in *Studies in Neurophysiology* (R. Porter, Ed.), New York: Cambridge University Press, pp. 203–209. ◆

Rall, W., and Segev, I., 1988, Synaptic integration and excitable dendritic spine clusters: Structure/function, in *Intrinsic Determinants of Neuronal Form and Function* (R. J. Lasek and M. M. Black, Eds.), New York: Alan R. Liss, pp. 262–282.

Shepherd, G. M., and Brayton, R. K., 1987, Logic operations are properties of computer-simulated interactions between excitable dendritic spines, *Neuroscience*, 21:151–165.

Zador, A., Koch, C., and Brown, T. H., 1990, Biophysical model of a Hebbian synapse, *Proc. Natl. Acad. Sci. USA*, 87:6718–6722.

Developmental Disorders

Annette Karmiloff-Smith

Introduction

Compared to classical artificial intelligence (AI) and its focus on on-line processing, neural networks are of particular interest to the cognitive developmentalist because of their emphasis on emergent representations and learning. Networks embody several principles inherent to the major developmental theory of the twentieth century (Piaget, 1971). There have been several attempts to provide a computational perspective on Piagetian theory, with varying amounts of connectionist involvement (Elman et al., 1995; see also COGNITIVE DEVELOPMENT and LANGUAGE ACQUISITION). Networks also capture the notion of a *gradual* acquisition of fine-grained functional units (Arbib, 1987) more akin to Piaget's schemas than to the innately fixed, broad modules proposed by Fodor (1983).

Discussions about the modular structure of the brain have frequently been based on selective brain damage in adults, with the assumption that characteristic breakdown in the adult implies a built-in modular organization in the neonatal brain. The actual process of acquisition and progressive modularization is rarely taken into account. Yet the path of both normal and abnormal development can be particularly informative about the adult end state.

The purpose of this article is to use the comparison of different abnormal phenotypes to explore constraints on modeling the *developing* mind/brain. Because abnormal development often results in peaks and valleys of proficiences and deficits, certain issues can be more clearly pinpointed than in the normal case: the nature of the predispositions in the neonatal brain, the extent to which learning is domain-specific/domain-general, and the passage from implicit to explicit knowledge.

Developmental disorders arise during embryogenesis and postnatal development. Although they often give rise to specific impairments which are thought to imply damage to a module, the connectionist framework suggests that such a conclusion is not always warranted. A model of learning regular/irregular past tense in English investigated the constraints on dissociation and plasticity in a network undergoing *random* lesions, both prior to and during training (Marchman, 1993). *General* lesions to the network resulted in *selective* impairments, because of an overall reduction in the ability to find a single-mechanism solution when resolving the competition between two classes of mappings. This holds at the fine-grained

level of *within-domain* dissociations in learning. In the rest of this article, we focus on *across-domain* dissociations.

The Starting State of Networks

A problem with early connectionist simulations was that little attention was paid to what is known about the neonatal brain and cognitive development in infancy. The neonatal brain has some functional specialization in the form of a large number of fine-grained schemas (Arbib, 1987), and very young infants demonstrate surprising cognitive capacities (Spelke, 1994). By contrast, most network models of development give theoretical prominence to a seamless, domain-general view of learning. They have invoked starting states with random weights and connections, based on the assumption that the human child starts life simply with domain-general learning mechanisms. However, a close examination of networks reveals that this so-called randomness is only random with respect to an already highly constrained architecture and learning algorithm. Different architectures (feedforward, interactive activation, recurrent, autoassociative, etc.) are chosen for different domains and, within these, precise specifications are made with respect to number of layers, number of units within layers, fast versus slow weight changes, and so forth. These architectural choices are not arbitrary: recurrent networks lend themselves particularly well to processing sequential input like language, whereas autoassociative networks are especially suited for certain forms of categorization. Despite the theoretical insistence on general learning and random starting states, in practice, networks are often architecturally and computationally domain-specific in approach.

Different architectures could emerge from interactions among specific task constraints and a domain-general architecture. However, if networks are to model human development, then consideration must be given to what is known about domain-specific predispositions. Predispositions can be architecture-specific, computation-specific, and/or representation-specific. They can also involve a maturational timetable, in that different parts of the brain become sensitive to different types of stimuli at different times. Rather than denying the importance of the starting state, networks would lend themselves particularly well to an exploration of the minimum type(s) of domain-specific predispositions that get different types of learning rapidly off the ground, as well as the effects on abnormal development of the absence of one or more types of predisposition.

Developmental Disorders: Clues to Domain-Specific Development

In contrast to adult neuropsychological deficits (see LESIONED ATTRACTOR NETWORKS AS MODELS OF NEUROPSYCHOLOGICAL DEFICITS and DISTORTIONS IN HUMAN MEMORY), developmental disorders have rarely been modeled. Yet comparisons between disorders provide a particularly clear window on the issue of domain-general and domain-specific learning. Studies of single syndromes can be misleading. For example, a conclusion drawn from research in subjects with Down syndrome, who score poorly across most domains, is that the functioning of a domain-general learning mechanism is impaired. However, comparisons between different phenotypes, as well as a fact about the brain of those with Down syndrome, challenge the domain-general view of development. When processing data, the brains of subjects with Down syndrome are more active across the entire brain than those of normal age-matched controls. This indicates that the brain of subjects with Down syndrome is less specialized than the normally developing brain, i.e., it has not become progressively modularized by postnatal loss of connectivity of synapses (Dehaene and Changeux, 1993; Johnson and Karmiloff-Smith, 1992). Domain-general processing seems to be the exception, not the rule. Many developmental disorders show uneven cognitive profiles that suggest complex interactions between domain-specific and domain-general capacities.

The example of Williams syndrome, a developmental disorder caused by a microdeletion on one of the alleles of chromosome 7 and contingent genes, provides a telling case in point. This syndrome presents a behavioral profile of relative proficiency in language, face processing, and theory-of-mind (attributing mental states to others), which coexists with severe deficits in visuospatial processing, number, and problem solving (for a full neuropsychological/cognitive profile and references, see Karmiloff-Smith et al., 1994). In another phenotype—hydrocephalus with associated myelomeningocele (a protrusion of the membranes of the brain or spinal cord through a defect in the skull or spinal column)—language can be the only area of proficiency. Subjects suffering from specific language impairment show the opposite pattern, performing within the normal range in all domains except language. In yet another phenotype—autism—even individuals with normal IQs are selectively impaired in tasks that require judging another's mental states (Baron-Cohen, Tager-Flusberg, and Cohen, 1993). Like adult brain damage, abnormal phenotypes rarely show across-the-board deficits. A purely domain-general theory of learning cannot explain these different, uneven cognitive profiles.

A comparison between different phenotypes shows that domain specificity does not imply simple brain-to-behavior relations, however. For example, subjects with autism are socially aloof and particularly deficient in theory-of-mind tasks (Baron-Cohen et al., 1993). By contrast, subjects with Williams syndrome are socially friendly and show proficiency in the theory-of-mind domain. It is known that the processing of socially relevant information depends on intact limbic structures. Brain imaging studies indicate that the limbic structures in the temporal lobe of subjects with Williams syndrome are normal, and that the reverse is true for those with autism (for details and

references, see Karmiloff-Smith et al., 1994). A plausible conclusion, then, on both psychological and neurological grounds, is that socially relevant computations form a module which is preserved in those with Williams sydrome and impaired in autistic subjects. However, comparisons with other phenotypes challenge this conclusion. For example, subjects with Down syndrome show impairment of the limbic structures in the temporal lobe and yet perform significantly better on theory-of-mind tasks than do subjects with autism (Baron-Cohen et al., 1993). Phenotypical variations point to a more fine-grained level of processing than that hypothesized by an impaired or intact module (Arbib, 1987). The comparison of psychological data and different brain abnormalities could serve as a particularly useful tool for network explorations of the possible architectural, representational, and computational variations that result in different phenotypical outcomes in particular domains.

Representation-Specific Predispositions: The Case of Face Processing

Subjects with Williams syndrome show proficiency in face-processing tasks, despite their impaired performance on all other visuospatial tasks. They demonstrate a domain-specific proficiency. By contrast, those with Down syndrome, matched according to mental age and chronological age, score very poorly on both face-processing and visuospatial tasks. Yet the face input to infants with Down syndrome is similar to that received by infants with Williams syndrome. Exposure to relevant stimuli, then, is necessary but not sufficient. Indeed, the differences between phenotypical patterns indicate that development does not simply start with general predispositions conducive to any form of visuospatial input, but with some skeletal representation-specific predispositions with respect to faces. The data available from normal neonates and from other species point to similar conclusions.

Recent research on normal infants was inspired by work with another species, the chick, in which it was shown that two independent neural systems underlie filial preference (see review in Johnson and Morton, 1992). On the basis of neural development of the visual system and of behavioral changes at 2 months, Johnson and Morton argued for two independent systems in the human infant also. Neonates preferentially track a stimulus with three high-contrast blobs over other stimuli. This initial behavior is mediated subcortically (through the colliculus and pulvinar). Around 2 months of age, a cortical system takes over which inhibits the subcortical pathway. The cortical system does not require built-in representations, but it is primed for its learning by the preferential inputs set up by the first system. The subcortical system, by contrast, does rely on a skeletal representation-specific predisposition. Once the cortical system takes over, infants no longer preferentially track the high-contrast blobs, but prefer stimuli that resemble real faces, suggesting that they are learning increasingly more about the facial characteristics of conspecifics rather than relying on the skeletal predisposition that makes them attend to face-like stimuli in the first place.

A strong nativist position would argue for a detailed template of the human face. There are several clues to suggest that evolution's aim is not to specify modules in detail. First, infants show no preference for detailed faces over schematic ones prior to 2 months. They need to learn about the details. Second, up to the age of 9 months, infants are able to discriminate between different nonhuman primate faces, whereas adults fail such tasks (Sargent and Nelson, 1992). In other words, specialized pathways in the adult brain have become modularized and dedicated to *human* face recognition, whereas the infant brain is still in the process of specialization.

These studies indicate that recognition progressively *becomes* species-specific with experience, an argument against any strong nativist view and dear to the connectionist heart. However, the very same findings, together with the differences among abnormal phenotypes, argue against a unitary, seamless view of the mind/brain. Evolution has provided a skeletal representation-specific predisposition for face processing and allocated specific subsystems to the task.

Why is a domain-general pattern recognition device insufficient? Actually, it would suffice if the *only* thing the child (or a network) learns about is faces, but real environments are exceedingly rich. Infants are not presented with faces neatly cordoned off from other visual or auditory input, with syntax independent of meaning, or with physics input independent of number input, and so forth. Second, we should recall that subjects with Williams syndrome, Down syndrome, and hydrocephalus with associated myelomeningocele all perform poorly on visuospatial processing, yet only those with Williams syndrome score well on face-processing tasks. A domain-general pattern recognition device and adequate experience will not explain these differences.

Simulations of development should, therefore, place less focus on showing that structured learning *can* result from random starting states. Rather, the connectionist framework should be viewed as a particularly privileged lens through which to explore and rethink development in terms of minimal architectural, representational and/or computational predispositions and their effect on subsequent learning, particularly with respect to the absence of such predispositions in different developmental disorders.

The Passage from Implicit to Explicit Representations

We have focused on areas of proficiency and deficiency in developmental behavior, but little has been said about the representations underlying such behavior. Yet the status of such representations is important, because in normal development, children gradually build up *explicit* representations of their implicit knowledge. Modifications in input patterns or internal recruitment of new hidden units can cause networks to develop more generalizable representations, but such representations remain implicit in the network's functioning. While behavioral proficiency is the *endpoint* of learning in a connectionist network, in the normal human case, it is the *starting point* for generating explicit representations of implicitly defined knowledge (Karmiloff-Smith, 1992). In the normal case, then, knowledge *in* the mind progressively becomes knowledge *to* the mind. This rarely happens in abnormal development, even in areas of proficiency.

Connectionist models provide interesting accounts of how children reach proficiency in a domain. However, they do not model subsequent representational change (Karmiloff-Smith, 1992; Clark and Karmiloff-Smith, 1993; see also CONNECTIONIST AND SYMBOLIC REPRESENTATIONS). Although the modeler can demonstrate the existence of *potentially* explicit knowledge via cluster or principal component analyses, and although clusters and principal components can be extracted by neural networks (see DATA CLUSTERING AND LEARNING and PRINCIPAL COMPONENT ANALYSIS), it is the modeler who labels the resulting clusters (e.g., Elman, 1990). Network representations are highly task-specific and not transportable to other networks. Yet the importance, in normal development, of progressively forming explicit representations that can be used in contexts outside the special-purpose procedures for which the knowledge was originally generated cannot be overestimated. The use of networks to explore the ways in which transportable (explicit) representations can emerge may help to pinpoint the *representational* limits on abnormal development, even in areas of *behavioral* proficiency.

Discussion

Comparisons of uneven cognitive profiles in different phenotypes should be considered an important source of constraint in neural network modeling of the development of the human mind/brain. Although networks have demonstrated that structured learning *can* occur from random starting states and general learning algorithms, consideration must be given to the psychological data on domain-specific predispositions and resulting fine-grained modularization. In general, it is essential to focus on normal and abnormal *developing* mind/brains, rather than solely on the final product of adult normal or lesioned mind/brains.

Road Map: Connectionist Psychology
Background: I.2. Levels and Styles of Analysis
Related Reading: Linguistic Morphology

References

Arbib, M. A., 1987, Modularity and interaction of brain regions underlying visuomotor coordination, in *Modularity in Knowledge Representation and Natural Language Understanding* (J. L. Garfield, Ed.), Cambridge, MA: MIT Press, pp. 333–363.

Baron-Cohen, S., Tager-Flusberg, H., and Cohen, D. J., 1993, *Understanding Other Minds: Perspectives from Autism*, Oxford: Oxford University Press.

Clark, A., and Karmiloff-Smith, A., 1993, The cognizer's innards: A psychological and philosophical perspective on the development of thought (with peer commentary and response), *Mind Lang.*, 8:487–581.

Dehaene, S., and Changeux, J.-P., 1993, Neuronal models of cognitive functions, in *Brain Development and Cognition: A Reader* (M. H. Johnson, Ed.), Oxford: Blackwell Scientific, pp. 363–402. ◆

Elman, J. L., 1990, Finding structure in time, *Cognit. Sci.*, 14:179–211.

Elman, J. L., Bates, E., Johnson, M. H., Karmiloff-Smith, A., Parisi, D., and Plunkett, K., in press, *Connectionism in a Developmental Framework*, Cambridge, MA: MIT Press. ◆

Fodor, J. A., 1983, *The Modularity of Mind*, Cambridge, MA: MIT Press.

Johnson, M. H., and Karmiloff-Smith, A., 1992, Can neural selectionism be applied to cognitive development and its disorders? *New Ideas Psychol.*, 10:35–46.

Johnson, M. H., and Morton, J., 1991, *Biology and Cognitive Development: The Case of Face Recognition*, Oxford: Blackwell Scientific. ◆

Karmiloff-Smith, A., 1992, *Beyond Modularity: A Developmental Perspective on Cognitive Science*, Cambridge, MA: MIT Press/Bradford Books. ◆

Karmiloff-Smith, A., Klima, E., Bellugi, U., Grant, J., and Baron-Cohen, S., in press, Is there a social module? Language, face processing and theory-of-mind in subjects with Williams syndrome, *J. Cognit. Neurosci.*

Marchman, V. A., 1993, Constraints on plasticity in a connectionist model of the English past tense, *J. Cognit. Neurosci.*, 5:215–234.

Piaget, J., 1971, *Biology and Knowledge*, Edinburgh: Edinburgh University Press.

Sargent, P. L., and Nelson, C. A., 1992, Cross-species recognition in infant and adult humans: ERP and behavioral measures. Poster presentation, International Conference on Infant Studies, Miami Beach, FL, May 1992.

Spelke, E., 1994, Initial knowledge: Six suggestions, *Cognition*, 50:431–445.

Development and Regeneration of Eye-Brain Maps

Jack D. Cowan and A. Edward Friedman

Introduction

The brain is a biological computer of immense complexity comprised of highly specialized neurons and neural circuits. Such neurons are interconnected with high precision in many regions of the brain, if not in all. How is this precision or *specificity* achieved? There are many observations which indicate that there is also considerable *plasticity*. How does this work? Sperry (1943) first demonstrated specificity in the regeneration of *topographic* eye-brain maps in frogs (i.e., neighboring points in the frog retina map into neighboring points in the optic tectum). Essentially he found that after optic nerve section and eye rotation, optic nerve fibers tended to regenerate connections with those target neurons to which they were connected before surgery. Sperry suggested that some kind of chemical marker was involved in the process. Many years later, Gaze and Sharma (1970) and others found evidence for plasticity in experiments in which lesions are made in goldfish retina or tectum. Following the lesions, topographic maps regenerated, conforming to whatever new boundary conditions were created by the lesions. In effect, these researchers found that optic nerve fibers can change their connections; in fact, a whole bundle of fibers can completely reorganize its pattern of connections. There are now many experiments indicating that the formation of connections involves both specificity and plasticity.

Eye-Brain Maps and Models

Figure 1 shows the structure of a typical topographic map from the goldfish retina to its optic tectum. How does such a map develop? Initially, there is considerable disorder in the pathway: retinal ganglion cells make contacts with many widely dispersed target neurons. However, the mature pathway shows a high degree of topographic order. How is such a map organized? One answer was suggested by Prestige and Willshaw (1975): retinal axons and their target neurons (in the optic tectum of lower vertebrates, or via the lateral geniculate nucleus in the visual cortex in mammals) are *polarized* by contact adhesion molecules distributed so that axons from one end of the retina are stickier than those from the other end, and neurons at one end of their target are correspondingly stickier

than those at the other end. Of course, this means that isolated retinal fibers will all tend to stick to one end of the target. However, if such fibers *compete* with each other for terminal sites, and if such sites compete for retinal axon terminals, less sticky axons will be displaced, and eventually a topographic map will form. The Prestige-Willshaw theory explains many observations indicating neural specificity. It does not provide for plasticity—the ability of axonal fiber systems to adapt to changed target conditions, and vice versa. It was Willshaw and von der Malsburg (1976) who first provided a theory for the plasticity of map reorganization, by postulating that synaptic growth in development is *Hebbian*. Such a mechanism provides self-organizing properties in retino-tectal and retino-geniculo-cortical map formation and reorganization.

Map Formation

In this article we describe the basic structure of many of the models which have been proposed. We follow the formulation introduced by Häussler and von der Malsburg (1983). Let $r_j(t)$ be the activity of the jth retinal cell and $t_i(t)$ the corresponding activity of the ith tectal cell. Let $w_{ij}(t)$ be the strength or weight of the synaptic contact made by the jth retinal fiber and the ith tectal cell, where both i and j run from 1 through N, each in a one-dimensional array with periodic boundary conditions. Thus we are considering map formation from one one-dimensional ring or chain to another. Later, we will look at two-dimensional maps. We assume that $0 \le w_{ij}$, so that the weights define a real-valued function w on the domain $\{(i,j)|1 \le i,j \le N\}$ which can easily be displayed in an $N \times N$ array, called the *weight matrix*.

Tectal activity is expressed in terms of retinal activity by the simple equation

$$t_i(t) = \sum_{j=1}^{N} w_{ij}(t)r_j(t) \tag{1}$$

and the following update equation expresses the changes in w_{ij}.

$$\dot{w}_{ij} = f_{ij}(\mathbf{w}) - \frac{1}{2}w_{ij}\left(\frac{1}{N}\sum_k f_{kj}(\mathbf{w}) + \frac{1}{N}\sum_l f_{il}(\mathbf{w})\right) \tag{2}$$

with $f_{ij}(\mathbf{w}) = \alpha + \beta \cdot w_{ij} \cdot C_{ij}(\mathbf{w})$, and $C_{ij}(\mathbf{w}) = \sum_{kl} c(ik,jl)w_{kl}$. This is a system of cubic equations in N^2 variables. The growth rate of the weight w_{ij}, $f_{ij}(\mathbf{w})$, consists of a positive constant α representing the rate of formation of retino-tectal synapses, and a rate for cooperation between neighboring tectal synapses given by the function $C_{ij}(\mathbf{w})$ multiplied by another positive constant β. The function $C_{ij}(\mathbf{w})$ is an approximation of the effects produced by Hebbian growth of tectal synapses. What is really taking place is that the synaptic weight w_{ij} increases at a rate $\beta \cdot w_{ij} \cdot r_j t_i = \beta \cdot w_{ij} \cdot r_j \sum_{k=1}^{N} w_{ik}(t)r_k(t)$ by Equation 1. If retinal stimulation occurs as random correlated patches, the net effect of this rule is to produce synaptic weights w_{ij} which show some degree of local cooperativity. It is such cooperativity that is represented by the term $C_{ij}(\mathbf{w})$. Häussler and von der Malsburg make the further assumption that $c(ik, jl) = c_T(|i - k|) \cdot c_R(|j - l|)$, so that retinal fiber cooperativity acts independently of tectal synapse cooperativity. These assumptions define $c_T(m)$ and $c_R(m)$ for the integer m, $1 \le m \le N$ as nonnegative, even, and monotone-decreasing functions of m

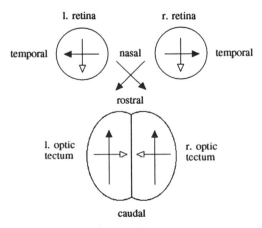

Figure 1. The retino-tectal map in goldfish.

for $m \leq N/2$ such that

$$\sum_m c_T(m) = \sum_n c_R(n) = 1 \tag{3}$$

Sum Rules

Examination of Equation 2 indicates that the total weight of all contacts made by a single retinal fiber, and the total weight of all contacts on one tectal cell, are held close to the saturation value N by the negative or competition term

$$-\frac{1}{2} w_{ij} \left(\frac{1}{N} \sum_k f_{kj}(\mathbf{w}) + \frac{1}{N} \sum_l f_{il}(\mathbf{w}) \right)$$

To see this, suppose only retinal fiber competition is present, so that Equation 2 reduces to

$$\dot{w}_{ij} = f_{ij}(\mathbf{w}) - \frac{1}{2} w_{ij} \left(\frac{1}{N} \sum_k f_{ij}(\mathbf{w}) \right) \tag{4}$$

Let $W_j^R = \Sigma_k w_{kj}$ (the sum over all contacts made by a single retinal fiber). Then Equation 4 gives

$$\dot{W}_j^R = -\frac{1}{N} \left(\sum_k f_{kj}(\mathbf{w}) \right) (W_j^R - N)$$

It follows that since all weights are bounded, $0 \leq w_{ij}$, they remain bounded and the sum W_j^R converges to the value N. Thus Equation 3 embodies the sum rule

$$\sum_k w_{kj} = N \tag{5}$$

In similar fashion the last term in Equation 1 can be thought of as embodying the tectal sum rule

$$\sum_l w_{ij} = N \tag{6}$$

Of course, both terms are present in Equation 2 and the rules interfere, but the overall effect is to limit synaptic weights in both directions. It follows that if any synaptic weight increases, all others from the same retinal fiber, and on the same tectal cell, will decrease. This is an example of WINNER-TAKE-ALL MECHANISMS (q.v.). Its overall effect is precisely what is needed to produce a one-to-one map.

Competition and Pattern Selection

Further insights into the map-forming process can be obtained by introducing in W, the linear space of all weight patterns, the pointwise multiplication operator $(x \cdot y)_{ij} = x_{ij} y_{ij}$. Then Equation 2 can be rewritten in the more concise form

$$\dot{\mathbf{w}} = -\alpha(\mathbf{w} - 1) + \beta \mathbf{w}(C(\mathbf{w}) - B(\mathbf{w}C(\mathbf{w}))) \tag{7}$$

where the linear operator B is given as

$$B_{ij}(\mathbf{x}) = \frac{1}{2} \left(\frac{1}{N} \sum_k x_{kj}(\mathbf{w}) + \frac{1}{N} \sum_l x_{il}(\mathbf{w}) \right)$$

The first term in Equation 7 forces $\mathbf{w}$ to converge to the homogeneous state $\mathbf{w} = 1$ (all weights equal to 1) at a rate α. Because of its winner-take-all effect, the second term forces $\mathbf{w}$ to converge to a state in which one weight is equal to N in each row and column of the weight matrix $\mathbf{w}$, and all the others equal 0. Some configurations correspond to one-to-one maps. However, most of these configurations are unstable. Evidently, it is the combination of the two terms, together with the initial conditions, which gives rise to stable topographic maps. Further analysis of the process can be carried out using bifurcation theory (see PATTERN FORMATION, BIOLOGICAL). The essence of the method is contained in the analysis of small deviations from the homogeneous state $\mathbf{w} = 1$. Linearizing about this state, one obtains a linear operator $L(v)$, the eigenfunctions of which determine, in large part, the nature of the solutions to Equation 7. They take the form

$$e_{ij}^{kl} = \exp\left(i \frac{2\pi}{N} (ik + jl) \right)$$

where $i, j, k,$ and l are all integers in the range from 0 to N. Evidently, when $k = l = 0$, $e^{kl} = 1$. The corresponding eigenvalues satisfy the equation

$$C(e^{kl}) = \gamma^{kl} e^{kl} = \sum_{ij} c(-i, -j) \exp\left(i \frac{2\pi}{N} (ik + jl) \right) e^{kl} \tag{8}$$

from which one can show that the eigenvalues λ^{kl} of $L(v)$ satisfy

$$\lambda^{kl} = \begin{cases} -\alpha - 1 & k = l = 0 \\ -\alpha + \frac{1}{2}(\gamma^{kl} - 1) & k = 0, l \neq 0, \text{ or } k \neq 0, l = 0 \\ -\alpha + \gamma^{kl} & \text{otherwise} \end{cases} \tag{9}$$

These results for eigenfunctions and eigenvalues are similar to those of reaction-diffusion equations on a ring (see PATTERN FORMATION, BIOLOGICAL). The essential point is that, by a suitable choice of the decay parameter α, the largest of these eigenvalues can be selected to be the only nonnegative one. Only diagonal patterns in the weight matrix (corresponding to topographic maps) belong to such an eigenvalue: all other patterns belong to negative eigenvalues, and therefore decay. The process is shown in Figure 2.

Feature Maps and Elastic Nets

The Häussler–von der Malsburg analysis deals with the processes of map formation in competitive Hebbian learning models. It emphasizes the point that such learning can generate topographic maps, provided there is a mechanism to enforce continuity. A more abstract but still closely related approach to map formation was introduced by Kohonen (1984). A variant of this motivated further work on the mapping problem by Durbin and Mitchison (1990) and others.

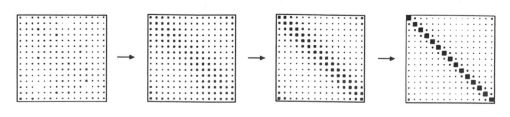

Figure 2. Diagonal pattern formation. Retinal fibers are represented by the columns of the matrix $\mathbf{w}$, and tectal cells by its rows. (Redrawn from figure 4 of Häussler, A., and von der Malsburg, C., 1983, Development of retinotopic projections: An analytical treatment, *Journal of Theoretical Neurobiology*, 2:47–73; used with permission.)

Feature Maps

Kohonen essentially reformulated the Willshaw–von der Mals-burg model in a way that emphasized neighboring correlations. We recall Equation 1, which we rewrite in the vector form

$$t = \mathbf{w}r = \begin{bmatrix} w_1 \\ w_2 \\ \cdot \\ \cdot \\ w_N \end{bmatrix} = r \qquad (10)$$

where r and t are, respectively, vectors of retinal and tectal activity, and where w_i is the ith weight vector.

As Kohonen noted, the essential feature of any self-organizing map is that such weight vectors adapt so as to provide a *match* to the input vectors t. If one *normalizes* them (instead of using sum rules as we did earlier), then if c is the location of the best match, it can be determined by the condition

$$\|r - w_c\| = \min_i \|r - w_i\| \qquad (11)$$

where $\|x - y\|$ is the Euclidean distance between x and y. An appropriate weight adaptation or update equation can be formulated which achieves such matches. It is essentially a competitive Hebbian learning equation with weight decay, similar to that described earlier. The details can be found in SELF-ORGANIZING FEATURE MAPS: KOHONEN MAPS. The net effect is that the distribution of weight vectors $M(w_i)$ converges essentially to match the probability distribution of the input data $P(r)$. In addition, the w_i become ordered according to their mutual *similarity*.

An important consequence is that similar input features map into neighboring units. This is called *feature mapping*. Since local interactions tend to preserve the continuity of sequences of w_i, it follows that differing effects can be obtained that depend on the *dimensionality* and *variance* of $P(r)$. This is relevant to the formation of topographic maps, especially in cases where the input dimensionality is greater than that of the network topology itself. Figure 3 shows such an example, in which a three-dimensional distribution maps onto a two-dimensional rectangular topology. The result is striking: if the variance in

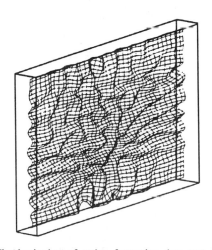

Figure 3. The beginning of stripe formation (see text for details). (Redrawn from figure 5.31 of Kohonen, T., 1984, *Self-Organization and Associative Memory*, New York: Springer-Verlag, © by Springer-Verlag Berlin Heidelberg 1984; used with permission.)

the third input dimension is sufficiently large, $M(w_i)$ becomes corrugated, the retinotopic map becomes distorted, and *stripes* develop. This suggests an elegant explanation of the formation of ocular dominance stripes in superimposed binocular maps of the visual field in higher vertebrates, and of similar stripes in three-eyed frogs (Law and Constantine-Paton, 1982).

Elastic Nets

Another algorithm, introduced by Durbin and Willshaw (1987), is called the *Elastic Net*. In this approach, a topographic map is considered to be an elastic surface or net (with its ends connected). The algorithm works by attaching units to inputs, and neighboring units to each other. It updates units in terms of their proximity to an input, normalized across all units, and by forcing continuity of unit representations. Thus it is closely related to the feature map. Durbin and Mitchison (1990) suggested that the principles underlying the elastic net operate in the formation of cortical maps, and that the length of cortical wiring is minimized in map development. However, they noted that it is computationally infeasible to extend such a minimization procedure to maps of high-dimensional data onto two-dimensional cortex; instead, they used what is essentially a feature map to investigate the mapping problem, but with a different method of ensuring continuity. (See CONSTRAINED OPTIMIZATION AND THE ELASTIC NET for details.)

Plasticity Versus Specificity

It is evident that the mechanisms described so far can generate topographic maps, and that such maps will form as long as input fibers and their targets are electrochemically active. Such mechanisms can clearly account for the plasticity seen in Gaze and Sharma's (1970) experiment. However, experiments such as eye rotation, in which retinal fibers find their way back to more or less where they were connected before optic nerve sectioning, cannot be explained only by Hebbian learning mechanisms. Additional cues seem to be involved, such as those provided by chemical markers that polarize nerve fibers and their targets. Prestige and Willshaw's analysis (1975) provided a conceptual framework and motivation for investigations—by Willshaw and von der Malsburg (1979), Fraser (1980), Whitelaw and Cowan (1981), and many others—of models which combine competitive Hebbian learning mechanisms with chemical markers. In the following sections we briefly describe the model introduced by Cowan and Friedman (1990), and some of the regeneration experiments which it addresses.

Group I and Group II Affinities

Prestige and Willshaw's analysis provides a framework for understanding markers and how they might act. They introduced two types of marker affinities, labelled "Group I" and "Group II." A single fiber with Group I markers has maximal affinity for a single tectal location. Conversely, any single fiber with Group II markers will always stick preferentially to one end of the tectum, no matter what its retinal origin: i.e., it is *polarized*. The Whitelaw-Cowan model employs single retinal and tectal homophilic Group II markers. Conversely, Fraser used what is essentially a homophilic (i.e., self-attracting) Group I marker.

Friedman and Cowan postulated that retinal fibers display both types of markers—the *tips* display Group II markers, and the *sides* display Group I markers. However, these Group I markers are specific to retinal fiber interactions; they convey information about retinal position, so that retinal fibers tend to stick to their retinal neighbors. Conversely, the Group II tip markers are such that retinal fibers are *repelled* by inappropriate tectal surfaces, and tend to stick only at one end or another of the tectum. In addition, it is postulated that there is not one but two varieties of tip marker, and two corresponding tectal markers. The Friedman-Cowan model combines these more elaborate marker properties with the cubic nonlinearities of the Häussler–von der Malsburg formulation. The resulting equation is very similar to Equations 1 and 2, but with an explicit Hebbian growth term $f_{ij}(\mathbf{w}) = \alpha_{ij} + \beta_{ij} \cdot w_{ij} \cdot [\mu_{ij} + (r_j - \lambda)t_i]$. However, the $\mathbf{b} = \{\beta_{ij}\}$ matrix is complicated and comprises several terms which express the presence of the various markers. The model which results from all these modifications and extensions is much more complex in its mathematical structure than any of the previous models. But computer simulations show it to be capable of correctly reproducing the observed details of almost all the experiments cited earlier.

The Schmidt et al. Experiment

Figure 4 shows a simulation of the retinal induction experiment of Schmidt, Cicerone, and Easter (1978) in which a topographic map from a half-retina is allowed to form on an intact goldfish tectum. The map is said to be *expanded* in that the retinal fibers expand their coverage of the tectum until it is completely innervated. After this, the expanded map is sectioned, and fibers from the other intact retina are diverted onto the same tectum. The result is that two maps form: an expanded map which mirrors the previous map from the other tectum, and a nearly normal map. The final result is a patchy map composed of competing portions of the two maps. The simulation reproduces this result in striking fashion.

At generation time 0, half a retina is ablated (the corresponding right half of the matrix $\mathbf{w}$ is shown as dark), and an expanded map forms, complete at generation time 499, with the remaining half-retina mapping topographically onto an entire tectum. At generation time 500, the expanded map is sectioned and fibers from another retina are diverted onto the same tectum. Almost immediately, at generation time 510, expanded and nearly normal maps can be seen, which later on compete and intercalate to form the final patchy map seen at generation time 2000. In terms of the Friedman-Cowan model, these effects occur because some incoming retinal fibers stick to debris left over from the previous expanded map, and other fibers stick to nonoccluded tectal markers. The fiber-fiber

markers control the regeneration of the expanded map, whereas the retino-tectal markers control the formation of the nearly normal map.

The Meyer "Polarity Reversal" Experiment

The Meyer polarity reversal experiment (Meyer, 1979) also shows the effects of fiber-fiber interactions. In this experiment, the left half of a goldfish eye and its attached retinal fibers are removed, leaving an intact normal half-eye map to the right optic tectum. At the same time, the right half of the other eye and its attached fibers are also removed, and fibers from the remaining half-eye are allowed to innervate the same tectum. The result is that fibers from the right half-retina, which normally would make contact with the corresponding cells in the left tectum, instead make connections with inappropriate neurons in the right tectum, and in a *reversed* direction, i.e., their polarity is reversed. Meyer interpreted this result to mean that optic nerve fibers show a tendency to aggregate with their retinal neighbors. A simulation of this experiment makes this evident. If f_{ik}, the mutual stickiness of neighboring fibers, is not strong enough, retino-tectal markers dominate, and the mismatched map forms with normal polarity. However, if f_{ik} is large enough, Meyer's result is found: the mismatched map forms with a reversed polarity.

Binocular and Compound Eye Maps

Several extensions of the model are of considerable complexity and interest. Direct binocular maps of the visual field are found in the visual cortex of many mammalian species, but rarely in lower vertebrates. However, in a remarkable experiment, Law and Constantine-Paton (1982) created a three-eyed frog via embryonic manipulations, and were therefore able to investigate the situation in which one tectum receives direct binocular input. The results are spectacular: fibers from the third, supernumerary eye intercalate into the normal map in a stripe pattern. Do there exist markers which distinguish between eyes? To test this, Ide, Fraser, and Meyer (1982) created a compound frog eye made up of two fragments from the same eye, the so-called isogenic compound eye. Such an eye consists essentially of two half-eyes with mirror symmetry. Interestingly, the two halves each expanded over the same tectum, again in a stripe representation. This indicates that eye-specific markers are not necessary for stripe formation. It is possible to simulate

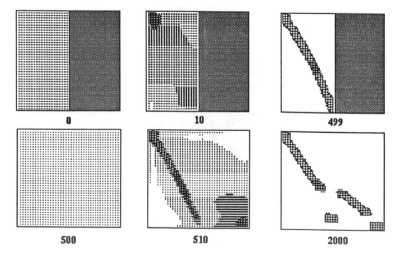

Figure 4. Simulation of the Schmidt et al. retinal induction experiment in which a nearly normal map is intercalated into an expanded map (see text for details).

this experiment within the framework of the one-dimensional Friedman-Cowan model. Two intercalated, expanded patchy maps—the one-dimensional equivalent of stripes—are formed from the two halves of a compound eye projecting to an entire tectum. Uncorrelated, not anticorrelated, noise in the two retinal halves is sufficient to produce the result. Once again, it is the fiber-fiber markers which act to produce the patchiness.

Discussion

It is evident that models which combine Hebbian plasticity with intrinsic, genetically determined eye-brain and fiber-fiber markers can generate correctly oriented retinotopic maps. The Friedman-Cowan model is one such example. It permits the simulation of a large number of experiments and provides a consistent explanation of almost all of them. In particular, it shows how the apparent induction of central markers by peripheral effects, as seen in the Schmidt et al. (1978) experiment, can be produced by the effects of debris. The model also shows how the polarity reversal seen in Meyer's experiment can be produced by fiber-fiber interactions, In addition, it demonstrates how these interactions can lead to patchy map formation, as seen in the third-eye graft performed by Law and Constantine-Paton and in the Ide et al. compound eye experiment.

In summary, much of the complexity of the many regeneration experiments carried out in the last fifty years can be understood in terms of the effects produced by contact adhesion molecules with differing affinities, acting to control an activity-dependent self-organizing mechanism. It remains to be seen whether such models can be extended to account for the full complexity of the binocular maps seen in higher vertebrates, or even for the complexities still to be investigated in the goldfish and the frog. It also remains to be seen if alternative formulations of competitive Hebbian learning, for example, feature maps or elastic nets, when provided with additional markers, can reproduce the details of such experiments and many others.

Road Map: Development and Regeneration of Neural Networks
Background: I.3. Dynamics and Adaptation in Neural Networks
Related Reading: Cooperative Phenomena; Ocular Dominance and Orientation Columns

References

Cowan, J. D., and Friedman, A. E., 1990, Development and regeneration of eye-brain maps: A computational model, *Advances in Neural Information Processing Systems 2* (D. Touretzky, Ed.), San Mateo, CA: Morgan Kaufmann, pp. 92–99.

Durbin, R. M., and Mitchison, G., 1990, A dimension reduction framework for understanding cortical maps, *Nature*, 343:644–647.

Durbin, R. M., and Willshaw, D., 1987, An analogue approach to the Travelling Salesman Problem using an Elastic Net method, *Nature*, 326:689–691.

Fraser, S. E., 1980, A differential adhesion approach to the patterning of nerve connections, *Dev. Biol.*, 79:453–464.

Gaze, R. M., and Sharma, S. C., 1970, Axial differences in the re-innervation of the goldfish optic tectum, *Exp. Brain Res.*, 10:171–181.

Häussler, A., and von der Malsburg, C., 1983, Development of retinotopic projections: An analytical treatment, *J. Theor. Neurobiol.*, 2:47–73.

Ide, C. F., Fraser, S. E., and Meyer, R. L., 1982, Eye dominance columns from an isogenic double-nasal frog eye, *Science*, 221:293–295.

Kohonen, T., 1984, *Self-Organization and Associative Memory*, New York: Springer-Verlag.

Law, M. I., and Constantine-Paton, M., 1982, A banded distribution of retinal afferents within layer 9A of the normal frog optic tectum, *Brain Res.*, 247:201–208.

Meyer, R. L., 1979, Retinotectal projection in goldfish to an inappropriate region with a reversal in polarity, *Science*, 205:819–821.

Prestige, M. C. and Willshaw, D. J., 1975, On a role for competition in the formation of patterned neural connections, *Proc. R. Soc. Lond. B*, 190:77–98.

Schmidt, J. T., Cicerone, C. M. and Easter, S. S., 1978, Expansion of the half retinal projection to the tectum in goldfish: An electrophysiological and anatomical study, *J. Comp. Neurol.*, 177(2):257–278.

Sperry, R. W., 1943, Effect of 180 degree rotation of the retinal field on visuomotor coordination, *Quart. J. Exp. Physiol.*, 92.263–279.

Whitelaw, V. A., and Cowan, J. D., 1981, Specificity and plasticity of retino-tectal connections: A computational model, *J. Neurosci.*, 1: 1369–1387.

Willshaw, D. J., and von der Malsburg, C., 1976, How patterned neural connections can be set up by self-organization, *Proc. R. Soc. Lond. B*, 194:431–445.

Willshaw, D. J., and von der Malsburg, C., 1979, A maker induction mechanism for the establishment of ordered neural mappings: Its application to the retino-tectal problem, *Philos. Trans. R. Soc. Lond. B*, 287:203–243.

Diffusion Models of Neuron Activity

Luigi M. Ricciardi

Introduction

We offer a survey of one-dimensional stochastic diffusion models for the membrane potential of a single neuron, with emphasis on the related first-passage-time (FPT) problems, namely, the determination of the neuronal output. While we make use of the formalism of stochastic differential equations (SDEs), we shall also sketch, in a couple of paradigmatic instances, how diffusion models can be constructed from first principles, that is, without resorting to SDEs. We shall thus indicate how the much-celebrated Wiener and Ornstein-Uhlenbeck (OU) neuronal models can be obtained as the limit of a Markov process with discrete state spaces. Similarly, suitable limit procedures

can be seen to lead to more general neuronal models, such as those that embody the notion of reversal potentials. Some more references, and numerous expanded and complementary considerations, can be found in a review article (Ricciardi, 1994).

Diffusion Processes and Neuronal Modeling

Let us look at a neuron as a black box characterized by an input and an output. We may then pose two distinct problems: (1) to determine the output for a given input, and (2) to guess the input by analyzing the output.

In the present framework, problem (1) is an *FPT problem*, while problem (2) is an *inverse problem*, that is, the inverse of an FPT problem. In both cases, the class of input functions must be specified first to make these problems mathematically sound. We shall assume that the neuron's membrane potential is modeled by a one-dimensional diffusion process.

From a biophysical point of view, neuronal models of single cells reflect the electrical properties of the membrane via electric circuit models that contain energy storage elements. Such circuit models can be described by means of differential equations for the membrane potential. When the input to the cell includes a random component, then the differential equation includes a *noisy* term, and hence becomes itself a stochastic differential equation whose solution can sometimes be approximated by a diffusion process. In other cases—for instance, when the neuron has very few synaptic inputs near the trigger zone—a Poisson-driven differential equation may be a biologically more appropriate model.

An action potential (spike) is produced when the membrane voltage exceeds a threshold voltage. This situation corresponds to an FPT problem for the associated diffusion process.

Let us now turn to the mechanisms responsible for the production of stationary spike trains. In this case, we are interested in modeling the "spontaneous" or "resting" activity of a neuron, or its steady-state response to a constant stimulus. Such an analysis is usually required before attempting to model the neuron's response to time-varying stimuli. We shall not explore the derivation of models. The interested reader can refer to the several available reviews of stochastic neuronal models (see, for instance, Ricciardi, 1977, 1994, and references therein), in which a detailed construction of some of the diffusion approximations considered later in this article can also be found.

We shall assume that the changes in the membrane potential between two consecutive neuronal firings (spikes) are represented by a scalar diffusion process $X(t)$. Such a process can be described by the stochastic differential equation (SDE)

$$dX(t) = \mu[X(t), t] \, dt + \sigma[X(t), t] \, dW(t) \qquad (1)$$

where μ and σ are real-valued functions of their arguments satisfying certain regularity conditions, and $W(t)$ is a standard Wiener process. (See Ricciardi and Sato, 1989, for an expository discussion of stochastic differential equations.) The Wiener process has historically been exploited as a first mathematical model of Brownian motion. It can be defined as the process $W(t)$ such that

(a) $W(0) = 0$.
(b) $W(t)$ has stationary independent increments.
(c) For every $t > 0$, $W(t)$ is normally distributed.
(d) For all $t > 0$, $E[W(t)] = 0$.

In neuronal modeling, the reference level for the membrane potential is usually taken to be equal to the resting potential. The initial voltage, namely the reset value following a spike, is often assumed to be equal to the resting potential: $X(t_0) = x_0$, where $t_0 \in \mathbb{R}$ denotes the initial time. The threshold potential, denoted by $S = S(t)$ with $S(t_0) > x_0$, is customarily assumed to be a deterministic function of time. Hereafter, we shall mainly be interested in the FPT random variable

$$T = \inf_{t \geq t_0} \{t: X(t) > S(t) | X(t_0) = x_0\} \qquad (2)$$

which is the theoretical counterpart of the interspike interval. The importance of interspike intervals is due to the generally accepted hypothesis that the information transferred within the nervous system is usually encoded by the timing of neuronal

spikes. Therefore, the reciprocal relationship between the firing frequency and the interspike interval naturally leads to the problem of determining the probability density function (pdf) of T, namely the function

$$g[S(t), t | x_0, t_0] = \frac{\partial}{\partial t} P\{T \leq t\} \qquad (3)$$

When this density cannot be obtained analytically (which is practically the rule), or when it is too difficult to give sufficiently precise estimations of it, the analysis is restricted to its moments, primarily mean and variance. The coefficient of variation (standard deviation/mean) is also used, since it measures the relative spread of the distribution and its deviation from exponentiality. In some cases, it is useful to work out asymptotic estimates of the FPT pdf for large times or large thresholds.

An alternative description of the process $X(t)$, modeling the membrane potential changes, is sometimes obtained via the so-called *diffusion equations approach*. First of all, one defines the pdf of $X(t)$ conditional on $X(t_0) = x_0$:

$$f(x, t | x_0, t_0) = \frac{\partial}{\partial x} P\{X(t) \leq x | X(t_0) = x_0\} \qquad (4)$$

It can then be seen (cf. Ricciardi, 1977) that f satisfies the Fokker-Planck (FP) equation

$$\frac{\partial f}{\partial t} = -\frac{\partial}{\partial x}[A_1(x, t) f] + \frac{1}{2}\left(\frac{\partial^2}{\partial x^2}\right)[A_2(x, t) f] \qquad (5)$$

and the Kolmogorov (K) equation

$$\frac{\partial f}{\partial t_0} + A_1(x_0, t_0)\frac{\partial f}{\partial x_0} + \frac{1}{2} A_2(x_0, t_0)\frac{\partial^2 f}{\partial x_0^2} = 0 \qquad (6)$$

where the coefficients $A_1(x, t)$ and $A_2(x, t)$ are the *infinitesimal moments*, or the *drift* and *infinitesimal variance*, defined as

$$A_i(x, t) = \lim_{\Delta t \downarrow 0} \frac{1}{\Delta t} \int (y - x)^i f(y, t + \Delta t | x, t) \, dy$$

$$\equiv \lim_{\Delta t \downarrow 0} \frac{1}{\Delta t} E\{[X(t + \Delta t) - X(t)]^i | X(t) = x\} \quad (i = 1, 2)$$
$$\qquad (7)$$

It is essential to mention that the quantities $A_1(x, t)$ and $A_2(x, t)$ just defined are related to the functions μ and σ that we considered earlier when talking about the differential equations approach. Within the so-called Ito approach (Ricciardi and Sato, 1989), the relation is the following:

$$\begin{cases} A_1(x, t) = \mu(x, t) \\ A_2(x, t) = \sigma^2(x, t) \end{cases} \qquad (8)$$

Equations 5 and 6 characterize the class of diffusion processes, which are special Markov processes.

Neuronal Models

The year 1964 marks the beginning of the history of neuronal models based on diffusion processes. Indeed, in a much celebrated article, Gerstein and Mandelbrot (1964) postulated that for a number of experimentally monitored neurons subject to spontaneous activity, the firing pdf could be modeled by the FPT pdf of a Wiener process. Actually, these authors were able to show that, by suitably choosing the parameters of the model, the experimentally recorded interspike interval histograms of many units could be fitted to an excellent degree of approximation by means of the FPT pdf of a diffusion process, characterized by the constant infinitesimal moments

$$A_1(x) = \mu \qquad \mu \in \mathbb{R} \qquad (9a)$$

$$A_2(x) = \sigma^2 \qquad \sigma \in \mathbb{R}^+ \qquad (9b)$$

As is well known, in this case the transition pdf (Equation 4) is normal, with mean $\mu(t - t_0) + x_0$ and variance $\sigma^2(t - t_0)$. Since this is a temporally homogeneous diffusion process, without loss of generality we can take $t_0 = 0$. By means of the methods outlined by Ricciardi (1994) and others, one can then prove that the FPT pdf of such a process is given by

$$g(S, t | x_0) = \frac{S - x_0}{\sigma \sqrt{2\pi t^3}} \exp\left[-\frac{(S - x_0 - \mu t)^2}{2\sigma^2 t} \right] \qquad x_0 < S \quad (10)$$

For $\mu \geq 0$, neuronal firing is a sure event as Equation 10 is normalized to unity. If one takes $\mu < 0$, the FPT pdf (Equation 10) can be interpreted as the firing pdf conditional upon the event "firing occurs." The case $\mu = 0$ is also of interest since Equation 10 is a stable law (Feller, 1966). This case provides an interpretation of numerous experimental results indicating that the shapes of histograms are sometimes preserved when the adjacent interspike intervals are summed.

To provide an interpretation of the neuronal model based on the assumptions in Equation 9, let us imagine that the neuron's membrane potential undergoes a simple random walk under the effect of excitatory and inhibitory synaptic actions. For simplicity, let us thus assume that the neuronal dynamics develop on a discrete time scale $0, \tau, 2, \tau, \ldots$, with $\tau > 0$ as an arbitrary time unit. Passing to the limit as $\tau \to 0$, it can be shown (Ricciardi, 1977) that the random walk can be made to converge to the diffusion process having drift μ and infinitesimal variance σ^2—in other words, to a Wiener process.

This procedure provides the simplest example in which the Wiener model for neuronal activity can be constructed. Although the underlying assumptions are undoubtedly oversimplified and some well-known electrophysiological properties of neuronal membrane are not taken into account, it must be stressed that the fitting of some experimental data by the FPT pdf is truly remarkable (Gerstein and Mandelbrot, 1964).

If a time-varying threshold is introduced to account for relative refractoriness—which is relevant in high firing rate conditions—the determination of the firing pdf for the Wiener neuronal model cannot generally be accomplished analytically. Hence, ad hoc numerical methods had to be devised.

In conclusion, despite the excellent fitting of some data, the neuronal model based on the Wiener process has been the object of various criticisms. We limit ourselves to pointing out that it does not include the well-known spontaneous exponential decay of the neuron's membrane potential that occurs between successive PSPs. A diffusion model for neuronal activity that includes this specific feature is the Ornstein-Uhlenbeck (OU) neuronal model. This is defined as the diffusion process characterized by the following drift and infinitesimal variance:

$$A_1(x) = -\frac{x}{\vartheta} + \mu \qquad (11a)$$

$$A_2(x) = \sigma^2 \qquad (11b)$$

where $\mu \in \mathbb{R}$, and ϑ and σ are positive constants. While μ and σ depend on the neuron's input, ϑ represents the membrane time constant (this model is referred to as the "leaky integrator neuronal model"). Comparing Equation 11 with Equation 9, we see that now the drift is state-dependent. However, in the limit as $\vartheta \to \infty$, the moments of Equation 11 identify with those in 9, meaning that the OU model yields the Wiener (or "perfect integrator") model in the limit of an infinitely large time constant. Recalling Equations 1 and 8, we can interpret the OU

model as generated by the following SDE:

$$dX(t) = \left(-\frac{X}{\vartheta} + \mu \right) dt + \sigma \, dW(t) \qquad (12)$$

where $W(t)$ is a standard Wiener process.

Although Equation 12 can be taken as defining the OU model, such a model can also be obtained from first principles by using the formalism of diffusion equations. To this purpose, let us initially assume that the neuron is subject to a sequence of excitatory and inhibitory PSPs of constant magnitudes $e > 0$ and $i < 0$, occurring in time with Poisson laws of parameters α_e and α_i, respectively. The membrane potential is thus viewed as a stochastic process $X(t)$ in continuous time with a discrete space consisting of the lattice $x_0 + ki + he$ ($k, h = 0, \pm 1, \ldots$), with the points of discontinuity randomly occurring in time. Denoting by x_0 ($x_0 < S$) the resting potential, we assume that all sample paths start at x_0 at the fixed initial time $t_0 = 0$. The firing pdf is the pdf of the instants at which the sample paths for the first time reach, or cross in the upward direction, the threshold S.

Ricciardi (1977) shows that if the input rates are taken larger and larger while simultaneously taking smaller and smaller PSPs with a suitable constraint, the membrane potential "converges" to the diffusion process with infinitesimal moments

$$\begin{cases} A_1(x) = -\dfrac{x}{\vartheta} \\[2mm] A_2 = k_i^2 \sigma_i^2 \left[\left(\dfrac{\sigma_i}{\sigma_e} \right)^2 + 1 \right] \equiv \sigma^2 \\[2mm] A_n = 0 \quad (n = 3, 4, \ldots) \end{cases} \qquad (13)$$

These identify with moments in Equation 11 if $\mu = 0$. The case $\mu \neq 0$ can be obtained by a slightly more complicated model in which multiple Poisson-distributed excitatory and inhibitory inputs impinge on the neuronal membrane.

Let us now return to the OU model with the infinitesimal moments of Equation 11. Its transition pdf is obtained by solving either the Fokker-Planck equation (5) or the Kolmogorov equation (6) with the initial conditions

$$\lim_{t \downarrow t_0} f(x, t | x_0, t_0) = \delta(x - x_0) \qquad (14a)$$

$$\lim_{t_0 \uparrow t} f(x, t | x_0, t_0) = \delta(x_0 - x) \qquad (14b)$$

expressing the circumstance that initially the whole probability mass is concentrated at the initial value x_0. One thus finds (cf. Ricciardi, 1977):

$$f(x, t | x_0) = [2\pi V(t)]^{-1/2} \exp\left\{ -\frac{[x - M(t|x_0)]^2}{2V(t)} \right\} \qquad (15)$$

where

$$M(t | x_0) = \mu \vartheta (1 - e^{-t/\vartheta}) + x_0 e^{-t/\vartheta} \qquad (16a)$$

$$V(t) = \frac{\sigma^2 \vartheta}{2} (1 - e^{-2t/\vartheta}) \qquad (16b)$$

Hence, at each time t, the transition pdf is normal with mean $M(t|x_0)$ and variance $V(t)$. It must be pointed out that the OU model differs from the Wiener model in some relevant respects. For instance, an equilibrium regime exists, since in the limit as $t \to \infty$ the pdf (Equation 15) becomes normal with mean $\mu \vartheta$ and variance $\sigma^2 \vartheta / 2$. Furthermore, the crossing of the firing threshold is a sure event. However, in contrast to the Wiener model, the FPT problem is in general very complicated, even in the case of constant thresholds.

We conclude this section by stressing an essential difference between the Wiener and the OU neuronal models. The former is unrealistical in lacking the presence of the finite time constant of the neuronal membrane, but it is susceptible to a closed form solution. Indeed, the firing pdf has the simple expression of Equation 10 (from which one can also calculate mean, variance, etc., of the firing time). The latter, although more realistically depicting the neurophysiological reality, does not allow us to obtain any closed form expression for the firing pdf. Moreover, although the moments of the firing time for the case of a constant threshold can be obtained analytically, their expressions are extremely complicated, so that rather tricky and cumbersome computations are required to come to any evaluation of the statistics of the firing time (Nobile, Ricciardi, and Sacerdote, 1985).

Numerical Evaluation of the Firing pdf

In the preceding section we mentioned the lack of a closed form solution to the firing pdf for the OU neuronal model. However, some efficient procedures have been made available to obtain accurate numerical evaluations for the general case of time-varying thresholds and for arbitrary diffusion models (not necessarily of the OU type). This is an important target, because to calculate the FPT pdf one would have to solve equations of the type shown in Equations 5 and 6, not only under the initial conditions of Equation 14, but also in the presence of complicated boundary conditions—a very difficult task that only rarely leads one to analytical solutions. Efficient numerical algorithms are thus especially desirable if one has to deal with time-varying neuronal thresholds. Without exploring any mathematical details, for which the reader is referred to the review by Ricciardi (1994) and to the references in it, it must be mentioned that the first fundamental contribution to the solution of this problem is due to Durbin (1971), who studied the case of a standard Wiener process and a time-dependent boundary. A purely numerical method for the same process was developed by Anderssen, De Hoog, and Weiss (1973). Their procedure is based on the remark that the FPT pdf of Equation 3 can be proved to be a solution of the following integral equation:

$$f[S(t), t|x_0, t_0] = \int_0^t f[S(t), t|S(\tau), \tau]g[S(\tau), \tau|x_0, t_0]\,d\tau \quad (17)$$

with $x_0 < S(t_0)$. This is a first-kind Volterra integral equation whose solution is made complicated by the circumstance that the kernel $f[S(t), t|S(\tau), \tau]$ exhibits a singularity of the type $1/\sqrt{t - \tau}$ as $\tau \uparrow t$. Hence, the problem of determining $g[S(t), t|x_0, t_0]$ from Equation 17 via numerical methods is by no means trivial. Such a method was then modified by Favella et al. (1982) and applied to the numerical solution of the FPT problem for the OU process. However, all these algorithms necessitate the use of large computation facilities and sophisticated library programs. As a consequence, they are expensive to run and not suitable to suggest to the modeler in real time how to identify the various parameters to fit the recorded data.

An entirely different approach, which is quite general and especially suited to handle neuronal firing problems, was later developed (cf. Ricciardi, 1994). The guiding idea was to prove that, in place of the singular Equation 17, the following second-kind Volterra integral equation for g can be written down, characterized by a nonsingular kernel:

$$g[S(t), t|x_0, t_0] = -2\psi[S(t), t|x_0, t_0]$$
$$+ 2\int_{t_0}^t g[S(\tau), \tau|x_0, t_0]\psi[S(t), t|S(\tau), \tau]\,d\tau \quad (18)$$

where we have set

$$\begin{cases} \psi[S(t), t|y, \tau] = \varphi[S(t), t|y, \tau] + k(t)f[S(t), t|y, \tau] \\ \varphi(x, t|y, \tau) = \dfrac{d}{dt}\displaystyle\int_{-\infty}^x f(z, t|y, \tau)\,dz \end{cases} \quad (19)$$

with $k(t)$ an arbitrary continuous function. Apparently, Equation 18 is more complicated than Equation 17. However, it possesses an extra degree of freedom, the arbitrary function $k(t)$ whose specification can be established in such a way as to remove the singularity of the kernel or, when possible, to make the integral on the right-hand side of Equation 18 vanish, thus obtaining the closed form solution $g[S(t), t|x_0, t_0] = -2\psi[S(t), t|x_0, t_0]$. Specific applications of this method to the process describing Wiener and OU neuronal models can be found in Ricciardi (1994).

The regularization procedure of Equation 18 is not restricted to these two neuronal models. Indeed, under mild assumptions one can prove that, for any time-homogeneous diffusion processes characterized by drift $A_1(x)$ and infinitesimal variance $A_2(x)$, the kernel of Equation 18 can be made continuous at $\tau = t$ by selecting

$$k(t) = \frac{1}{2}\left\{A_1[S(t)] - S'(t) - \frac{A_2'[S(t)]}{4}\right\} \quad (20)$$

The Moments of the Firing Time

Sometimes it is necessary to obtain information on the shape of the firing pdf or on the average firing time and its standard deviation. Closed form expressions for all moments of the firing time for the OU neuronal model have been provided by Nobile, Ricciardi, and Sacerdote (1985). These turned out to be very useful for discovering some asymptotic trends of the firing pdf. For instance, after systematic computations of mean and variance of the firing time and of the skewness (a measure of the deviation from symmetry of the firing pdf) for a variety of thresholds and initial states, a striking feature emerged: for boundaries of the order of a couple of units or more above the starting point (the neuron's resting potential), the variance of the firing time equals the square of its mean value to an excellent degree of approximation. Moreover, the skewness (the ratio of the third-order central moment to the cube of the standard deviation) is equal to 2 to a very high precision. Finally, the goodness of these approximations was seen to increase with increasing values of the threshold. Putting all this information together, the conjecture emerged that the firing pdf is susceptible to an excellent exponential approximation for a wide range of thresholds and initial states. In addition, these "experimental" results have led to quantitative results concerning the asymptotic exponential behavior of the FPT pdf, not only for the OU process, but also for a wider class of diffusion processes, both for constant and for time-varying boundaries (Ricciardi, 1994).

As expected, the convergence to the exponential distribution for increasing thresholds is accompanied by a large increase of the mean firing time. This is in agreement with the finding that for some neurons the histograms of the recorded interspike intervals become increasingly better fitted by exponential functions as the firing rates decrease (see, for instance, Škvařil et al., 1971).

Neuronal Diffusion with State-Dependent Infinitesimal Variance

The OU neuronal models considered so far can be modified to include a time-dependent drift. For instance, it is conceivable that the model of Equation 11 might be changed into

$$\begin{cases} A_1(x) = -\dfrac{x}{\vartheta} + \mu + P(t) \\ A_2(x) = \sigma^2 \end{cases} \quad (21)$$

if a time-dependent extra effect were induced by some kind of DC external stimulation acting on the neuron. It is interesting, in particular, to investigate the behavior of Equation 21 when $P(t)$ is a periodic function of period T that reflects some oscillatory action of the environment. This situation quite naturally leads us to considering an FPT problem for the OU process (Equation 11) and a periodic boundary. To give a hint of why this is so, let us refer to the OU neuronal model with infinitesimal moments (Equation 21), where $P(t)$ is periodic with period T. It is then intuitive that, by a suitable transformation, an FPT problem through a constant threshold S_0 can be changed into an FPT problem for a time-independent OU model through a periodic boundary. Putting it the other way around, an FPT problem through a periodic boundary $S(t)$ for the OU process defined via

$$\begin{cases} dX(t) = -\dfrac{1}{\vartheta} X(t)\, dt + \sigma\, dW(t) \\ P\{X(0) = x_0\} = 1 \end{cases} \quad (22)$$

is equivalent to the FPT problem through the constant boundary $S(0) = S_0$ for the process

$$Y(t) = X(t) - S(t) + S_0 \quad (23)$$

that satisfies

$$\begin{cases} dY(t) = \left[-\dfrac{1}{\vartheta} Y(t)\, dt - \dfrac{S(t)}{\vartheta} + \dfrac{S_0}{\vartheta} - S'(t) \right] dt + \sigma\, dW(t) \\ P\{Y(0) = x_0\} = 1 \end{cases}$$
$$(24)$$

In other words, both $X(t)$ and $Y(T)$ start at x_0 at time 0. However, $Y(t)$ is an OU process containing a periodic drift, and the threshold is constant; $X(t)$, in contrast, is a time-independent linear-drift process, and the corresponding threshold is a periodic function. The question then naturally arises: how does the firing pdf look for an OU model and a periodic boundary? This is another paradigmatic case in which a mathematical answer is made possible by a systematic analysis of computational results. By using the numerical methods for the firing pdf evaluation, one can, for instance, prove that, for a normalized OU neuronal model, the periodicity of the threshold is reflected in an oscillatory behavior of the firing pdf. The firing pdf, in turn, approaches a time-nonhomogeneous exponential limit as the threshold moves farther and farther away from the reset value of the process.

In order to embody some other physiological features of real neurons, several alternative models have been proposed. We mention two of the diffusion models with state-dependent infinitesimal variance.

In models with *reversal potentials*, it is assumed that the input-related changes of the membrane potentials depend on the actual value of the membrane potential. This is suggested by some experimental findings (cf., for instance, Schmidt, 1985). Models embodying the concept of reversal potentials were introduced to account for the biological restriction that the state space of the process $X(t)$ must be finite, and for the assumption that a nonlinear summation of synaptic input may take place.

Hanson and Tuckwell (1983) considered two reversal potential diffusion models; the first is given by

$$\begin{cases} A_1(x) = -\dfrac{x}{\tau} + \mu(V_E - x) \\ A_2(x) = \sigma(V_E - x) \end{cases} \quad (25)$$

where $\tau > 0$, μ and $\sigma > 0$ are constant, and V_E is a constant, called the *excitatory reversal potential*, for which $V_E > S$ holds. The second model was defined by the infinitesimal moments

$$\begin{cases} A_1(x) = -\dfrac{x}{\tau} + \mu_1(V_E - x) + \mu_2(x - V_I) \\ A_2(x) = [\sigma_1^2(V_E - x)^2 + \sigma_2^2(x - V_I)^2]^{1/2} \end{cases} \quad (26)$$

where $\tau, \mu_1, \mu_2, \sigma_1$, and σ_2 are constants, and $V_I < x_0$ is an additional constant called the *inhibitory reversal potential*. The role of V_E is the same as in Equation 25. As before, the lower boundary was taken to be either $-\infty$ or a constant. Here, V_I was a reflecting boundary.

We emphasize that in diffusion models one must expect a great variety of forms for the infinitesimal variance and the linear form of the drift. In particular, the linear drift models the passive electrical circuit properties of the membrane at the trigger zone, and also the mean effect of the noisy input. The infinitesimal variance, on the other hand, must take into account not only the diversity of spatial configurations for different neurons, but also the location and type of synaptic input on that neuron as well. Hence, a variety of forms for this term in the diffusion equation appear to be appropriate. An outline of several other state-dependent neuronal models can be found in Ricciardi (1994) and in the references there.

It should be mentioned that the important topic of model verification has not yet been extensively studied, although some results are available for diffusion models (cf. references in Ricciardi, 1994).

Discussion

It must be stressed that the neuronal behavior described by diffusion models ultimately assume, that the output is a renewal process—in other words, that intervals between firings are independent and identically distributed. However, one can conceive models aimed, for instance, at simulating the clustering effect in spike generation. Serial dependence among interspike intervals can be modeled in various ways: for instance, by adjusting the reset value after each spike in the OU process.

Stochastic neuronal models are also used for the description of more complex neuronal structures. A critical review of deterministic and stochastic approaches to the analysis of neural networks can be found in Cowan and Sharp (1988). A review of several specific neuronal systems and networks, together with simulation programs and an extensive bibliography, can be found in MacGregor (1987).

Using the theory of diffusion processes, we have outlined a few stochastic models for the activity of single neurons. The prediction of the firing pdf remains a difficult problem whose solution can usually only be approached by numerical or simulation procedures. To conclude this bird's eye view of the topic, we would like to mention an alternative approach to the construction of diffusion models that fit experimental data: reversing the problem. Namely, instead of formulating a neuronal model $X(t)$ based on some reasonable assumptions, and then trying to compute the firing pdf as an FPT pdf through a preassigned threshold, one can proceed as follows. First, construct the histogram of the experimentally recorded spike train and try to fit it by a function $g(S, t|x_0)$, with S and x_0 parameters to be determined by the standard methods. Once this task has been accomplished, ask the following questions: Can $g(S, t|x_0)$ be viewed as the FPT pdf, through the threshold S, of a diffusion process conditioned on $X(0) = x_0$? If the answer is yes, can such a process be uniquely determined? This is, so to speak, the inverse of the FPT problem. Quite surprisingly, a precise answer to this question can be provided, at least in principle (Capocelli and Ricciardi, 1972).

Road Map: Biological Neurons
Background: I.1. Introducing the Neuron
Related Reading: Synaptic Coding of Spike Trains

References

Anderssen, R. S., De Hoog, F. R., and Weiss, R., 1973, On the numerical solution of Brownian motion processes, *J. Appl. Probab.*, 10:409–418.

Capocelli, R. M., and Ricciardi, L. M., 1972, On the inverse of the first passage time probability problem, *J. Appl. Probab.*, 9:270–287.

Cowan, J. D., and Sharp, D. H., 1988, Neural nets, *Q. Rev. Biophys.*, 21:365–427.

Durbin, J., 1971, Boundary-crossing probabilities for the Brownian motion and Poisson processes and techniques for computing the power of the Kolmogorov-Smirnov test, *J. Appl. Probab.*, 8:431–453.

Favella, L., Reineri, M. T., Ricciardi, L. M., and Sacerdote, L., 1982, First passage time problems and some related computational problems, *Cybern. Syst.*, 13:95–128.

Feller, W., 1966, *An Introduction to Probability Theory and Its Applications*, vol. 2, New York: Wiley.

Gerstein, G. L., and Mandelbrot, B., 1964, Random walk models for the spike activity of a single neuron, *Biophys. J.*, 4:41–68.

Hanson, F. B., and Tuckwell, H. C., 1983, Diffusion approximations for neuronal activity including synaptic reversal potentials, *J. Theor. Neurobiol.*, 2:127–153.

MacGregor, R. J., 1987, *Neural and Brain Modeling*, San Diego: Academic Press.

Nobile, A. G., Ricciardi, L. M., and Sacerdote, L., 1985, Exponential trends of Ornstein-Uhlenbeck first-passage-time densities, *J. Appl. Probab.*, 22:360–369.

Ricciardi, L. M., 1977, *Diffusion Processes and Related Topics in Biology*, Lecture Notes in Biomathematics, vol. 14, Berlin: Springer-Verlag. ◆

Ricciardi, L. M., 1994, Diffusion models of single neurons activity, in *Neural Modeling and Neural Networks* (F. Ventriglia, Ed.), Oxford: Pergamon, pp. 129–162. ◆

Ricciardi, L. M., and Sato, S., 1989, Diffusion processes and first-passage-time problems, in *Lectures in Applied Mathematics and Informatics* (L. M. Ricciardi, Ed.), Manchester: Manchester University Press, pp. 206–285. ◆

Schmidt, R. F., Ed., 1985, *Fundamentals of Neurophysiology*, 3rd ed., New York: Springer.

Škvařil, J., Radil-Weiss, T., Bohdanecký, Z., and Syka, J., 1971, Spontaneous discharge patterns of mesencephalic neurons, interval histogram and mean interval relationship, *Kybernetik*, 9:11–15.

Tuckwell, H. C., and Richter, W., 1978, Neuronal interspike time distributions and the estimation of neurophysiological and neuroanatomical parameters, *J. Theor. Biol.*, 71:167–183.

Digital VLSI for Neural Networks

Dan Hammerstrom

Introduction

The simulation of Artificial Neural Network (ANN) models, whether simplistic or realistically complex, is a key problem in the field because of their computational complexity. These simulations are required both for research into functional models and for commercial products using ANNs. Most researchers perform these simulations on standard computer technology, primarily on workstations or personal computers.

Despite advances that have increased the speed and reduced the cost of general-purpose computing, these approaches are often insufficient. An ANN model of moderate size may have thousands of nodes and millions of connections, where a *connection* is a data transfer path between two neurons. Depending on the complexity of the model, even a high-speed workstation cannot emulate such a network in real time.

For these reasons, there has been much interest in developing custom hardware for ANNs. The inherent parallelism in ANN and connectionist models suggests an opportunity to speed up the simulations. Their simple computations also suggest an opportunity to employ simpler and cheaper low-precision digital hardware.

This chapter discusses digital electronic implementations of ANNs. First, the differences between digital and analog design techniques are discussed, focusing primarily on cost-performance trade-offs in flexibility (Amdahl's law). The second topic is a special system, the RAP machine, created by the International Computer Science Institute and designed for ANN emulation using traditional processors. And finally, to convey a sense of the state of digital design, two custom digital ANN processors are discussed: the Adaptive Solutions CNAPS and the Intel/Nestor Ni1000.

Many implementation media have been proposed for ANN simulation, including technologies such as optics or gallium arsenide (GaAs) semiconductors. Currently, the best medium is CMOS VLSI—Very Large Scale Integration (VLSI) placing complex circuits on chips constructed using complementary metal oxide semiconductor (CMOS) technology (Weste and Eshraghian, 1993). Because of intense competitive pressures, and the volume produced, CMOS silicon provides unprecedented price performance. And CMOS VLSI is relatively easy to design. With few exceptions, it is therefore the technology of choice for commercial computer implementation and the only technology considered in this chapter.

Why Digital?

Cost-Performance

One commonly held belief in the ANN research community is that analog computation (where signals are transmitted and manipulated as strengths, generally voltage or current) is inherently superior to digital computation (where signals are transmitted and manipulated as serial or parallel streams of 1s and 0s). In fact, both technologies have advantages and drawbacks. The best choice depends on the application.

Why is analog appealing? Partly because it provides 10–100 times greater computational density than digital computation. *Computational density*—the amount of computation per unit area of silicon—is important because the cost of a chip is generally proportional to its total area. In analog circuitry, complex, nonlinear operations such as multiply, divide, and hyperbolic tangent can be performed by a handful of transistors (typically 5–10). Digital circuitry requires hundreds or even thousands of wires and transistors for the same operations. Analog computation also performs these operations using far less power per computation than digital computation.

So, if analog is so good, why are people still building digital chips, and why are most commercial products digital? One important reason is familiarity. People know how to build digital circuits, and they can do it reliably, no matter the size and complexity of the system. This is partly the legacy of having thousands of digital designers all over the world constantly tweaking and improving design techniques and software. It is also easier to create a digital version of a computation, where a computer program represents the algorithm, than an analog version, where the circuit itself represents the algorithm. This is particularly true if you are trying to build a system that is robust and reliable over the wide temperature and voltage ranges needed in commercial products. Analog design is an uncommon capability. And it is becoming less common as people find they can do more with digital circuitry. For example, Digital Signal Processors (DSPs) now perform most of what was once the domain of analog circuitry.

Flexibility

Another factor working in favor of digital is that analog designs are generally algorithms wired into silicon. Such designs are inflexible. Digital designs can be either hardwired or programmable. Their flexibility is a major benefit, since it allows software control as well as an arbitrary level of precision (low to high, and fixed or floating point). The price of this flexibility is reduced cost-performance, but the result is a chip that can solve a larger part of a problem. It also leads to a device that has broader applicability and that can track incremental algorithm improvements by changing the software, not by redesigning the circuitry.

The role flexibility plays in system cost-performance can be understood more clearly by examining *Amdahl's law* (Hennessy and Patterson, 1991), which describes the benefits from parallelizing a computation. Briefly, a computing task has portions or *subtasks* that often can be executed in parallel. Other, sequential, tasks cannot begin until a previous task has completed, which forces a sequential ordering of these tasks.

Amdahl's law states that no matter how many processors are available to execute subtasks, the speed of a particular task is roughly proportional to those subtasks that cannot be executed in parallel. In other words, sequential computation dominates as parallelism increases. Amdahl quantifies the relationship as follows:

$$S = 1/(op_s + (op_p/p)) \qquad (1)$$

where S is the total speed-up; op_s, the number of operations in the serial portion of the computation; op_p, the number of operations in the parallel portion; and p, the number of processors. Hence, as p gets large, S approaches $1/op_s$.

For example, suppose we have two chips to choose from. The first can perform 80% of the computation with a 20× speed-up on that 80%. The second can perform only 20% of the computation, but executes that 20% with a 1000× speed-up. Plugging into the equation, the first chip gives us a speed-up of over 4×, whereas the second—and "faster"—chip has only a 1.25× speed-up. A programmable device that accelerates several phases of an application generally offers a much larger benefit than a dedicated device.

Signal Intercommunication

One difference between silicon and biological networks is that, when compared with computation, internode communication for silicon is relatively more expensive in size than for biological systems. Although several levels of wire interconnect (three to four today) can be used, each level is restricted to two-dimensional interconnection because wires on the same level cannot pass over or touch one another.

Two-dimensional layout and large expensive wires require us to modify our biologically derived computational models to more closely match the strengths and weaknesses of the implementation technology. To show the need for such modifications, Bailey and Hammerstrom (1988) modeled a hypothetical neural circuit. This circuit, modest by biological standards, had one million "neurons" with one thousand inputs each, or one billion connections total.

Their first calculations assumed a direct implementation, that is, one connection per wire. This billion-connection ANN required more than 86 square meters of silicon for dedicated communication pathways. Since silicon averages tens of dollars per square centimeter, such a system is too costly to be practical. These costs result from a theorem showing that the metal area required by direct communication is proportional to the *cube* of the convergence of each node.

Their second calculations assumed a multiplexed or "shared" interconnect structure, that is, more than one connection uses a single wire. Wire multiplexing adds complexity at each end for arbitration, since only one requester can use a wire at a time. Likewise, an address must be sent with each data packet to identify the sender, and some decoding must be performed on that address. Bailey and Hammerstrom showed that, with the proper communication architecture, a 100× reduction in silicon area over the direct approach was possible with little impact on performance. Since these large networks will no doubt be sparsely activated, multiplexing interconnect makes even more sense; silicon area is too expensive to be wasted on rarely used, low-bandwidth communication. Analog voltages and currents are difficult to multiplex. Consequently, analog interconnection requires pulse frequency, phase, or both to represent the value of a connection, which many researchers are using in their circuits.

Digital Neural Networks: Off-the-Shelf Processors

One successful approach to high-speed ANN simulation has been to use arrays of commercial DSP or microprocessor chips. Although companies such as Intel offer general-purpose multiprocessor systems as commercial products, these systems are usually less suitable (Jackson and Hammerstrom, 1992) for ANN simulation than systems specifically designed for ANNs. Artificial neural network models have vastly different computational characteristics from the traditional computations that general-purpose parallel systems were designed to perform. For example, ANNs have large numbers of simple processes that communicate with most other processes. By optimizing the interconnect architecture to take advantage of these characteristics, special-purpose parallel systems can be built using off-the-shelf microprocessors or DSPs. These systems offer higher performance (sometimes 10× or more) than comparable general-purpose parallel systems.

One representative example, the Ring Array Processor (RAP) machine, was designed and built at the International Computer Science Institute (Morgan et al., 1992). The RAP uses as its

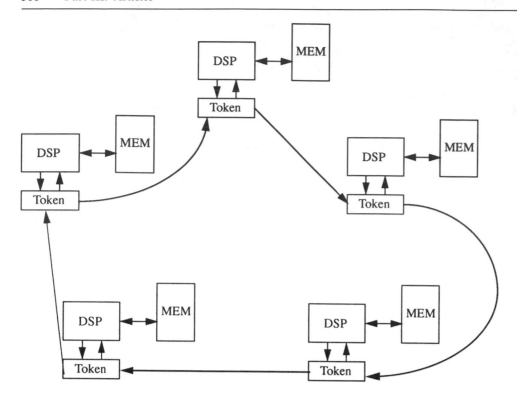

Figure 1. RAP data-passing ring. Each processor node consists of a Digital Signal Processor (DSP) with its own local memory and ring interface circuitry. For simple feedforward networks, an entire network is divided equally across all the processors.

core processor a Texas Instruments DSP, the TMS320C30. Each processor node consists of a DSP, local memory, and ring communication circuitry for interprocessor communication (Figure 1). Each processor receives data from its nearest neighbor and then copies and sends the data on the ring bus. Messages flow only in one direction, and the ring network can move data quickly at 64 MB per second from one processor to the next. The typical RAP machine has four DSPs per board, with up to 10 boards in a system. Since each TMS320C30 can compute 16 million connections per second, the largest machine can compute 640 million connections per second.

Artificial neural networks can be mapped to the RAP architecture in several ways. Generally, they are partitioned by placing one or more network nodes on each processor. For multilayered, artificial networks, each processor can simulate several nodes from each layer. All processors emulate the first layer simultaneously; when done, they transmit the node outputs to all other processors over the ring; then the computation of the next layer begins.

The RAP system has been successful in a number of ANN-based speech projects, providing high performance for ANN training and allowing research results that would have been impossible otherwise.

Digital Neural Networks: Full Custom Processors

Designing architectures customized for ANN simulation permits significant improvements in cost-performance, where the processors and their interconnect can be optimized to further improve performance and lower cost. This section discusses two custom digital processor architectures, both designed for emulating ANN models: the Adaptive Solutions CNAPS (Connected Node Architecture for Parallel Systems) and the Intel/Nestor Ni1000. Both are commercial products. They also represent the two ends of the design spectrum of digital ANN chips: relatively generalized versus specialized functionality.

CNAPS

The CNAPS architecture (Hammerstrom, 1991) has multiple processor nodes (PNs) connected in a one-dimensional structure, forming a Single-Instruction Multiple-Data (SIMD) array (Figure 2). SIMD structures have one instruction storage and decode unit and many execution units, simplifying system design and reducing costs. Unlike the RAP machine, each CNAPS PN does not have program storage and sequencing hardware, and each executes the same instruction each clock. Like the RAP, many network nodes can be allocated to each PN. Node outputs are broadcast from each PN to all the others over a single broadcast bus.

Another major simplification of the CNAPS architecture, found in other digital ANN chips, is the use of limited-precision, fixed-point arithmetic. Many researchers have shown how floating point and high precision are unnecessary in ANN simulation (Fahlman and Hoehfeld, 1992). CNAPS supports 1-, 8-, and 16-bit precision in hardware. The PNs are smaller and cheaper because they use less precision and avoid the extra hardware to support floating-point arithmetic. This reduced precision has been found to be more than adequate for the applications implemented on CNAPS.

The CNAPS architecture has 64 PNs per chip. At 25 MHz, each chip can execute at a rate of 1.6 billion connections per second. A single chip can execute backpropagation learning at a rate of 300 million connection updates per second (each update consists of reading the weight associated with the connection, modifying it, and then writing it back). Each PN (Figure 3) has 4096 bytes of on-chip local memory, used to store synaptic weight data and other local values. Hence, a 64-PN chip can store up to 256 KB of information. Multiple chips can be combined to create larger, more powerful systems. The general programmability of the device allows it to execute a large range of functions, including many non-ANN algorithms such as the discrete Fourier transform, nearest neighbor classification, and dynamic time warping.

Figure 2. CNAPS Architecture. This is a Single-Instruction Multiple-Data (SIMD) architecture, in which all processor nodes (PNs) execute the same instruction on each clock. There is a single broadcast data bus that allows efficient many-to-many and many-to-one communication.

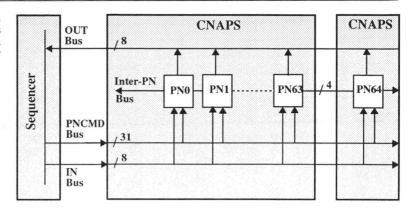

Figure 3. CNAPS PN Architecture. A single PN has a multiplier, accumulator, logic/shifter unit, register file, and separate memory address adder. Each PN also has its own memory for storing weights, lookup tables, and other data.

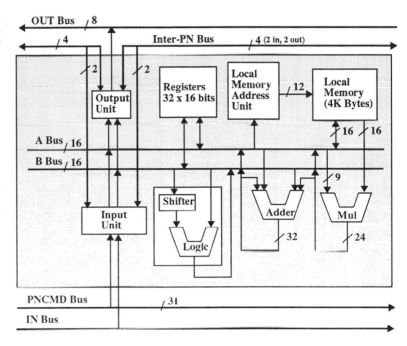

Figure 4 shows a simple two-layer network mapped to a CNAPS array. The network nodes are labeled CN0–CN7; the processor nodes, PN0–PN3. Multiple network nodes map to a single processor node; in this example, one node from each layer is mapped to a single PN. For feedforward calculation, assume that the outputs of nodes CN0–CN3 have been computed. To compute the inner product of nodes CN4–CN7, the output value of node CN0 is broadcast on the bus to all PNs in the first clock. Each PN then multiplies the CN0 output with the corresponding weight element, different for each PN. On the next clock, CN1's output is broadcast, etc. After four clocks, all 16 products have been computed—$O(n^2)$ connections in $O(n)$ time.

Ni1000

The Ni1000, developed jointly by the Intel and the Nestor Corporations (Holler et al., 1992), is a specialized neural network chip optimized to accelerate pattern recognition tasks. The architecture implements a radial basis function network. The learning algorithms supported include Nestor's RCE (Reduced Coulomb Energy; see COULOMB POTENTIAL LEARNING), and Specht's PNN (Probabilistic Neural Network). These networks

usually have a single layer of hidden nodes to represent regions in the input vector space. Region size is controlled by a single threshold value. These networks are trained by creating new "prototypes." Each hidden node represents a single prototype. Training then consists of writing the value of a particular training vector into the weight vector for a hidden node.

The Ni1000 emulates a three-layer network of 256 input nodes, 1024 hidden nodes, and 64 output nodes. It has 512 processors, and each processor emulates two hidden nodes. Input data are broadcast to all processors from the inputs. Input data and hidden node weight data use a 5-bit fixed-point representation. The hidden node output is not a sum of products, but a sum of differences, using "Manhattan" distance rather than vector inner product computation:

$$d_j = \sum_i |v_i - w_{ij}| \tag{2}$$

The output node computation is done by another set of processors in a special 16-bit floating-point format. The function computed by the output nodes is:

$$\sum_i w_j f(-t_j d_j) \tag{3}$$

where the function $f(\ )$ is nonlinear and can be either an expo-

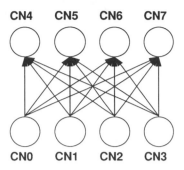

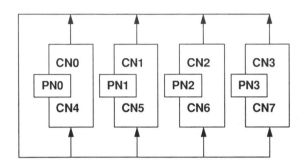

Figure 4. Mapping of a simple two-layer feedforward network to the CNAPS array. When emulating a feedforward network, each layer is spread across the PN array. The neuron outputs of one layer are broadcast sequentially to all PNs while they compute the multiply-accumulations for the next layer of neurons.

nential or signum (step) and t_j is a threshold that determines the "range" of "height" of the hidden node response.

The simple hidden-unit computation allows all 512 processors to be placed on one chip, since each processor is a small, low-precision adder and subtractor. Consequently, the Ni1000 has excellent computational density and can compute more than 10 billion connections per second, an impressive rate for digital-based computation. Weight data are stored in small, compact Electrically Erasable Programmable Read-Only Memory (EEPROM) cells that retain their state when the chip is powered down.

During feedforward operation, a three-stage pipeline is enabled which processes input vectors at a rate of 40,000 vectors/sec. The pipeline outputs class labels or probability estimates suitable for Bayesian classification, also at 40,000 per second.

Other Chips

Other digital ANN chips include the MAC-16 by Siemens (Ramacher et al., 1991) and the HNC SNAP (Means and Lisenbee, 1991). Both of these are similar to the CNAPS. The MAC-16 uses a processor architecture that restricts the range of possible computations. This restriction allows off-chip memory and thereby the emulation of larger networks. The SNAP uses floating-point processors, which are larger and more expensive, but allow it to execute applications needing 32-bit floating-point dynamic range and precision, and also uses off-chip memory.

Future Trends

It is difficult to predict what new technology will look like, but speculation is always possible. Today most ANNs are used for pattern recognition. The final stage of most pattern recognition

algorithms involves checking a series of classification results to see if they fit in the larger context of the domain in question. Including this contextual knowledge can be as simple as checking spelling; or it can be as complex as accessing high-order rules or schemas that reflect complex syntactical and semantic relationships. Since classification is imperfect, contextual processing of some kind is essential to guarantee the accuracy of the final result.

Although the results are still speculative, research (Ambros-Ingerson, Granger, and Lynch, 1990) has shown that scaling to large contexts requires networks with relatively sparse interconnect and sparse activation (only a few nodes are actively firing at a time; see SPARSELY CODED NEURAL NETWORKS). Based on the work described earlier on VLSI connectivity, digital-based systems can handle such networks much more efficiently than can analog systems. Therefore, at some point in the processing, the data will probably be converted from analog to digital representation. Today the conversion is done at or just after the transducer. Based on the analog technologies of the type described in SILICON NEURONS, systems of the future will probably take advantage of the computational density of analog VLSI to perform the feature extraction and some preliminary classification at the front end, with conversion to digital form for contextual processing and final classification.

It is a matter of open research whether these digital context engines will be heavily oriented to biological and artificial neural network architectures, or whether they will look more like more traditional semantic and connectionist network structures. Because the implementation costs and strengths of silicon differ from those of biology, the computational algorithms will differ as well. Nevertheless, some of the basic computational characteristics of massively parallel, subsymbolic computation will be retained (see CONNECTIONIST AND SYMBOLIC REPRESENTATIONS).

Road Map: Implementation of Neural Networks
Background: I.1. Introducing the Neuron
Related Reading: Analog VLSI for Neural Networks; Programmable Neurocomputing Systems

References

Ambros-Ingerson, J., Granger, R., and Lynch, G., 1990, Simulation of paleocortex performs hierarchical clustering, *Science*, 247:1344–1348.
Bailey J., and Hammerstrom, D., 1988, Why VLSI implementations of associative VLCNs require connection multiplexing, in *Proceedings of the 1988 International Conference on Neural Networks*, San Diego, IEEE, pp. 173–188.
Fahlman, S. E., and Hoehfeld, M., 1992, Learning with limited numerical precision using the cascade-correlation algorithm, *IEEE Trans. Neural Netw.*, 3:602–611.
Hammerstrom, D., 1991, A highly parallel digital architecture for neural network emulation, in *VLSI for Artificial Intelligence and Neural Networks* (J. Delgado-Frias and W. Moore, Eds.), New York: Plenum.
Hennessy, J. L., and Patterson, D. A., 1991, *Computer Architecture: A Quantitative Approach*, Palo Alto, CA: Morgan Kaufmann. ◆
Holler, M., Park, C., Diamond, J., Santoni, U., The, S. C., Glier, M., and Scofield, C. L., 1992, A high performance adaptive classifier using radial basis functions, in *Proceedings of the Government Microcircuit Applications Conference*, Las Vegas, pp. 261–264.
Jackson, D., and Hammerstrom, D., 1992, Distributing back propagation networks over the Intel iPSC/860 Hypercube, in *Proceedings of the International Joint Conference on Neural Networks, 1992*, vol. 1, New York: IEEE, pp. 575–580.
Means, R., and Lisenbee, L., 1991, Extensible linear floating point SIMD neurocomputer array processor, in Proceedings of the International Joint Conference on Neural Networks, 1991, New York: IEEE.

Morgan, N., 1990, *Artificial Neural Networks: Electronic Implementations*, Washington, DC: Computer Society Press Technology Series and Computer Society Press of the IEEE. ◆

Morgan, N., Beck, J., Kohn, P., Bilmes, J., Allman, E., and Beer, J., 1992, The ring array processor: A multiprocessing peripheral for connectionist applications, *J. Parallel Distrib. Comput.*, 14:248–259.

Ramacher, U., et al., 1991, Architecture of a general-purpose neural signal processor, in *Proceedings of the International Joint Conference on Neural Networks, 1991*, New York: IEEE.

Weste, N., and Eshraghian, K., 1993, *Principles of CMOS VLSI Design: A Systems Perspective*, 2nd ed., MA: Addison-Wesley. ◆

Directional Selectivity in the Cortex

Norberto M. Grzywacz and Anthony M. Norcia

Introduction

A visual neuron is directionally selective (DS) if back-and-forth motions, symmetric about the middle of its receptive field, elicit different responses. For the axis along which the ratio between the responses to back-and-forth motions is maximal, we call the direction eliciting the largest response the *preferred direction*, and the opposite direction the *nonpreferred direction*.

Motion perception is mediated by cortical directional selectivity. That cortical directional selectivity (CDS) contributes to motion perception is supported by correlates between perceptual decisions in motion tasks, on the one hand, and performances and numbers of DS neurons on the other (Pasternak, Merigan, and Movshon, 1981; Maunsell and Newsome, 1987). Moreover, lesions or small current injections in the middle temporal cortex (MT) affect specific motion-integration tasks (Salzman et al., 1992).

Other perceptual functions to which CDS contributes to are related to *optic flow*. Optic flow is a spatiotemporal distribution of image motion vectors that approximates the distribution obtained by the projection of the three-dimensional vectors of a moving object onto the retina (for a review, see Hildreth and Koch, 1987). One definition of optic flow assumes $dE/dt = 0$, where E is the image's brightness. This equation, which says that brightness varies slowly over time, defines optic flow $\mathbf{v}$ from brightness as

$$\nabla E \cdot \mathbf{v} + \frac{\partial E}{\partial t} = 0 \qquad (1)$$

The components of ∇E are the spatial changes of intensity in the horizontal and vertical directions, and thus can be measured from the image. Similarly, one can measure $\partial E/\partial t$. Therefore, Equation 1 has two variables (horizontal and vertical components of $\mathbf{v}$) and thus does not have a unique solution. To remove the ambiguity from this equation, one typically assumes that $\mathbf{v}$ varies smoothly over small image regions.

Important information can be extracted from optic flows. For instance, if one translates forward (or backward), the optic flow is that of an expansion (or contraction), whose focus signals the direction of heading (see Hildreth, 1992). To find this focus, DS information can be combined across the image by "higher" cortical areas. One such higher area is the middle superior temporal cortex (MST), which appears to contain neurons tuned to expansion and contraction (Maunsell and Newsome, 1987).

Cortical directional selectivity contributes to the control of eye movements. Cortical directional selectivity is also useful for eye-movement control (for reviews, see Miles and Wallman, 1993). For example, neuroanatomical data demonstrate that

MT and MST furnish a large projection to the dorsolateral pons, known to contribute to smooth-pursuit eye movements.

Mechanism

Directional Selectivity in Cortical Pathways

As a first-order approximation, motion computation is hierarchical in the cortex (Maunsell and Newsome, 1987). The hierarchy begins with simple and complex DS cells in layer IVb of primate area V1. (Primate retinal DS cells appear to project to the superior colliculus and, thus, CDS is probably computed from scratch.) The DS cells from V1 project to MT (or V5) and V2, which also projects to MT. Cortical directional selectivity becomes more complex in MT, where cells typically have larger receptive fields. From MT, the motion pathway projects to MST.

Psychophysical Models of Cortical Directional Selectivity

One class of models of the hierarchy's first stage is based on human psychophysics and is similar to the Reichardt model (van Santen and Sperling, 1985). Figure 1*A* provides an illustration. In the figure, the model has a slow input from the left and a fast input from the center. The interaction between them is multiplicative. For rightward, but not leftward motion, the sluggishness of the left pathway compensates for its being reached first by the stimulus, and the signals arrive at the multiplication site roughly simultaneously. Other models, called motion-energy models (Adelson and Bergen, 1985), use distributed spatial asymmetry and squaring nonlinearity. This distributed spatial asymmetry arises from the fact that different locations in the receptive field have different impulse responses

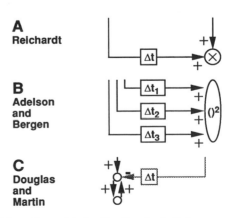

Figure 1. Schematic models for directional selectivity.

(Figure 1*B*). These differences are such that only when an object moves in the preferred direction does an area stimulated later have a faster reaction. Thus, it is only in this case that the responses of all areas occur roughly simultaneously to yield a response. One model for such space and time co-dependence uses a receptive field profile based on Gabor functions (Grzywacz and Yuille, 1990; see also GABOR WAVELETS FOR STATISTICAL PATTERN RECOGNITION). It is written as

$$G(\mathbf{r}, t) = \frac{1}{(2\pi)^{3/2}(\sigma_r)^2(\sigma_t)} \exp\left(-\frac{|\mathbf{r}|^2}{2\sigma_r^2}\right) \exp(-i\Omega_r \mathbf{u} \cdot \mathbf{r})$$
$$\exp\left(-\frac{t^2}{2\sigma_t^2}\right) \exp(-i\Omega_t t) \qquad (2)$$

where $\mathbf{r}$ and t are a spatial location in the image and time, respectively; $\sigma_r > 0$, $\sigma_t > 0$, $\Omega_r > 0$, and $\Omega_t > 0$ are scalar parameters corresponding to receptive field's size, temporal window, optimal spatial frequency, and optimal temporal frequency, respectively; and $\mathbf{u}$ is a unit-vector parameter, corresponding to the preferred direction.

Physiological Models of Cortical Directional Selectivity

Emerson et al.'s complex DS-cell data from cat cortex suggest that CDS can arise from a motion-energy mechanism (Emerson, Bergen, and Adelson, 1992). However, Reid, Soodak, and Shapley (1991) showed that for nonpreferred directions, an inhibitory mechanism suppresses responses in a manner inconsistent with motion energy models. Reid et al. suggested that this mechanism might be nonlinear inhibition similar to that of retinal directional selectivity. Reid et al.'s interpretation for cortical inhibition shares its asymmetry with the mechanism suggested by Douglas and Martin (1992). However, the nonlinearity underlying Douglas and Martin's "functional microcircuit" for cortex is positive feedback (mediated by intracortical excitation), which is prevented from being "ignited" by relatively weak and essentially linear inhibition (Figure 1*C*). In contrast, Heeger (1993) reinterpreted the Reid et al. data by saying that the nonpreferred-direction suppression might be due to a cortical inhibitory network devoted to responses' normalization (see also Albrecht and Geisler, 1991).

Another psychophysically and physiologically inspired computational model for CDS is the gradient model, which computes velocity as the ratio between temporal and spatial gradients, as suggested by Equation 1 (Hildreth and Koch, 1987).

Models of Motion Integration: Velocity

Computational models have also been proposed for how CDS signals are integrated to obtain higher-level motion information. For instance, Grzywacz and Yuille's (1990) model for the computation of velocity comprised a three-dimensional space of cells whose axes are the spatial ($\Omega_r \mathbf{u}$) and temporal (Ω_t) frequency parameters of Equation 2. These authors showed that if an arbitrary stimulus is translating with velocity $\mathbf{v}$, then the maximal responses in this space fall in the plane

$$\Omega_r \mathbf{u} \cdot \mathbf{v} + \Omega_t = 0 \qquad (3)$$

Therefore, to measure velocity, one could measure the slant and orientation of this plane. Grzywacz and Yuille proposed a plausible neural architecture to perform this measurement.

Development

Single-Cell Physiology

At least 5% of the cells in areas 17 and 18 are DS in the kitten's cortex at eye opening (see Blakemore, 1978). Consequently,

some of CDS appears to be genetically coded. However, one can modulate CDS during development by using visual environments illuminated only by brief, low-frequency, stroboscopic flashes of light. Directionally selective cells are much less common in cortices of strobe-reared animals than in normal animals. This loss results in an impairment of perceptual performance in motion tasks, which can be reversed by training in adult animals along with the generation of new DS neurons (Pasternak et al., 1981). Also, selective biases in the distribution of preferred directions of DS cells have been produced by exposing kittens to stripes moving in a particular direction.

Binocular Interactions

Studies of human infants using visual-evoked cortical potentials have suggested that a critical parameter "training" proto-DS cells during development is binocularity (Norcia et al., 1991). These studies recorded monocularly the response to vertical gratings undergoing an oscillatory apparent motion at a temporal frequency of 10 Hz. In normal infants six months old or younger, and in patients with a history of constant strabismus beginning before six months of age, the evoked potentials showed strong nasalward/temporalward direction biases. This pattern was not seen in normal adult evoked potentials. The fact that strabismus early in the critical period prevents normal development indicates that binocularity and CDS are tightly coupled.

These findings, and those from single-cell physiology, suggest that CDS in humans develops as follows: It is either present at birth or develops shortly after it. This initial CDS has a temporal/nasal neural population bias that is due, perhaps, to genetic factors controlling retinal ganglion-cell asymmetries. After six months of life, binocular interactions reduce the bias, eliminating the temporal/nasal asymmetries. In patients with early disruptions of binocularity, these neural populations remain biased, causing monocular smooth pursuit to be better for nasalward motion, and nasalward motion to be perceived as faster (Tychsen, 1993).

Models for the Development of Cortical Directional Selectivity

Models for the development of orientation selectivity (that is, selectivity to spatial orientation of anisotropic stimuli, such as lines, as depicted in Figure 2*A*, as distinct from direction of motion) might be relevant to the development of CDS. Motion energy models (see Figure 1*B*) interpret CDS as orientation in space-time. This can be seen by plotting the impulse responses

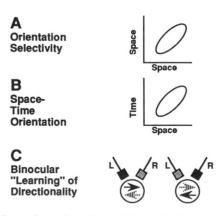

Figure 2. Space-time orientation and binocular learning of directionality. The symbols *L* and *R* stand for left and right eyes, respectively.

of different positions of the receptive field on space-time axes, as schematically illustrated in Figure 2*B*. Hence, a model of orientation selectivity might apply to CDS if one of the spatial dimensions in the orientation selectivity model is switched to a temporal axis.

Linsker (1990) proposed a model for the development of orientation selectivity that started with cells with isotropic receptive field properties, except for small statistical biases (see also INFORMATION THEORY AND VISUAL PLASTICITY). He showed that in the presence of random signals, Hebbian processes can amplify these biases to form orientation selectivity. In mathematical terms, Linsker used a multilayer network such that cell M of a particular layer at instant τ responded to the inputs from the preceding layer $(L_1, L_2, \ldots, L_N)$ in a linear manner as

$$M^\tau = a_1 + \sum_k L_k^\tau c_j \qquad (4)$$

where a_1 is a parameter and c_k is the strength of the kth input to the Mth cell. The Hebb rule used was

$$(\Delta c_k)^\tau = a_2 L_k^\tau M^\tau + a_3 L_k^\tau + a_4 M^\tau + a_5 \qquad (5)$$

where the a_k are parameters ($a_2 > 0$). To prevent c values from becoming infinite during the development process, a saturation constraint was imposed.

If time were an independent variable of M and L_i, then such a model would perhaps lead to oriented space-time receptive fields and thus to CDS.

Problems for this Hebbian hypothesis of CDS are the initial distribution bias toward temporal-to-nasal preferred directions and its elimination during development by binocular interactions. These interactions may take the form of a Hebbian mechanism, in which a binocular cell, initially selective to a nasalward motion in one eye, selects by Hebbian correlation the same direction of motion from the other eye, making the cell equally responsive to that direction in either eye (Figure 2*C*).

Discussion

Cortical directional selectivity has been modeled in three manners: as (1) spatially asymmetric excitatory drive followed by multiplication or squaring, (2) spatially asymmetric nonlinear inhibitory drive, and (3) spatially asymmetric linear inhibitory drive followed by positive feedback. The development of CDS might involve Hebbian processes driven by spontaneous activity and binocular interactions. Given that CDS is central to motion perception and control of eye movement, it might be an excellent paradigm to study the interface between cortical mechanisms and behavior.

Acknowledgments. Support for this work came from grants from the National Eye Institute (EY08921), the Office of Naval Research (N00014–91–J–1280), and the Air Force Office of Sponsored Research (F49620–92–J0156), and an award from the Paul L. and Phyllis C. Wattis Foundation to N.M.G.; from a grant from the National Eye Institute (EY06579) to A.M.N.; and from a core grant from the National Eye Institute to Smith-Kettlewell (EY06883).

Road Map: Vision
Background: I.3. Dynamics and Adaptation in Neural Networks
Related Reading: Directional Selectivity in the Retina; Motion Perception; Pursuit Eye Movements

References

Adelson, E. H., and Bergen, J. R., 1985, Spatio-temporal energy models for the perception of motion, *J. Opt. Soc. Am. A*, 2:284–299.

Albrecht, D. G., and Geisler, W. S., 1991, Motion selectivity and the contrast-response function of simple cells in the visual cortex, *Vis. Neurosci.*, 7:531–546.

Blakemore, C., 1978, Maturation and modification in the developing visual system, in *Handbook of Sensory Physiology*, vol. 8 (R. Held, H. W. Leibowitz, and H.-L. Teuber, Eds.), Berlin: Springer-Verlag, pp. 377–436. ◆

Douglas, R. J., and Martin, K. A. C., 1992, Exploring cortical microcircuits: A combined anatomical, physiological, and computational approach, in *Single Neuron Computation* (T. McKenna, J. Davis, and S. F. Zornetzer, Eds.), Orlando: Academic Press, pp. 381–412. ◆

Emerson, R. C., Bergen, J. R., and Adelson, E. H., 1992, Directionally selective complex cells and the computation of motion energy in cat visual cortex, *Vis. Res.*, 32:203–218.

Grzywacz, N. M., and Yuille, A. L., 1990, A model for the estimate of local image velocity by cells in the visual cortex, *Proc. R. Soc. Lond. B*, 239:129–161.

Heeger, D. J., 1993, Modeling simple cell direction selectivity with normalized, half-squared, linear operators, *J. Neurophysiol.*, 70:1885–1898.

Hildreth, E. C., 1992, Recovering heading for visually guided navigation, *Vis. Res.*, 32:1177–1192.

Hildreth, E. C., and Koch, C., 1987, The analysis of visual motion: From computational theory to neuronal mechanisms, in *Annual Review of Neuroscience*, vol. 10 (W. M. Cowan, E. M. Shooter, C. F. Stevens, and R. F. Thompson, Eds.), Palo Alto, CA: Annual Reviews, pp. 477–533. ◆

Linsker, R., 1990, Perceptual neural organization: Some approaches based on network models and information theory, in *Annual Review of Neuroscience*, vol. 13 (W. M. Cowan, E. M. Shooter, C. F. Stevens, and R. F. Thompson, Eds.), Palo Alto, CA: Annual Reviews, pp. 257–281. ◆

Maunsell, J. H. R., and Newsome, W. T., 1987, Visual processing in monkey extrastriate cortex, in *Annual Review of Neuroscience*, vol. 10 (W. M. Cowan, E. M. Shooter, C. F. Stevens, and R. F. Thompson, Eds.), Palo Alto, CA: Annual Reviews, pp. 363–401. ◆

Miles, F. A., and Wallman, J., Eds., 1993, *Visual Motion and Its Role in the Stabilization of Gaze*, Amsterdam: Elsevier. ◆

Norcia, A. M., Garcia, H., Humphry, R., Holmes, A., Hamer, R. D., and Orel-Bixler, D., 1991, Anomalous motion VEPs in infants and in infantile esotropia, *Invest. Ophthalmol. Vis. Sci.*, 32:436–439.

Pasternak, T., Merigan, W. H., and Movshon, J. A., 1981, Motion mechanisms in strobed reared cats: Psychophysical and electrophysical measures, *Acta Psychol.*, 48:321–332.

Reid, R. C., Soodak, R. E., and Shapley, R. M., 1991, Directional selectivity and spatiotemporal structure of receptive fields of simple cells in cat striate cortex, *J. Neurophysiol.*, 66:505–529.

Salzman, C. D., Murasugi, C. M., Britten, K. H., and Newsome, W. T., 1992, Microstimulation in visual area MT: Effects on direction discrimination performance, *J. Neurosci.*, 12:2331–2355.

Tychsen, L., 1993, Motion sensitivity and the origins of infantile strabismus, in *Early Visual Development: Normal and Abnormal* (K. Simons, Ed.), New York: Oxford University Press, pp. 364–390. ◆

van Santen, J. P. H., and Sperling, G., 1985, Elaborated Reichardt detectors, *J. Opt. Soc. Am. A*, 2:300–320.

Directional Selectivity in the Retina

Norberto M. Grzywacz, Evelyne Sernagor, and Franklin R. Amthor

Introduction

A visual neuron is directionally selective (DS) if back-and-forth motions, symmetric about the middle of its receptive field, elicit different responses. For the axis along which the ratio between the responses to back-and-forth motions is maximal, we call the direction eliciting the largest response the *preferred direction*, and the opposite direction the *null direction*.

There is no evidence that retinal directional selectivity (RDS) contributes to motion perception. However, there is evidence that RDS contributes to oculomotor responses. In vertebrates, RDS is involved in optokinetic nystagmus (for reviews, see Miles and Wallman, 1993). In this article, we discuss the mechanisms and development of RDS.

Mechanism

Theoretical Preliminaries

The theoretical work of Reichardt, Poggio, and collaborators (Poggio and Reichardt, 1976) emphasized two requirements that models of directional selectivity must fulfill. The first is a spatially-asymmetric mechanism. The second is a nonlinear-functional mechanism to yield two different single-number estimates of preferred- and null-direction responses.

Figure 1*A* illustrates the simplest model proposed by Reichardt and colleagues for insects' RDS, in which inputs from two locations converge to an interaction site where these inputs are multiplied (the nonlinearity) and integrated. The inputs are spatially asymmetric, since a slow signal comes from the left and a fast signal from the center. For rightward, but not leftward, motion, the sluggishness of the left pathway is compensated by the fact that the stimulus arrives at it first. Hence, for rightward motion, if the speed is appropriate, both signals reach the multiplication site at roughly the same time, yielding a positive multiplication. In contrast, for leftward motion, the multiplication is near zero.

The multiplication in Figure 1*A* is one of many quadratic non-linearity models supported by insect data. Poggio and Reichardt investigated the predictions common to all quadratic models of RDS. For that purpose, they used the Volterra formulation. The output under such a formulation for a smooth, time-in-variant, nonlinear interaction between the responses to stimuli in spatial locations a (z_a) and b (z_b) is

$$y(t) = h_{0,0} + \sum_{m=1}^{\infty} \sum_{j=0}^{m} h_{j,m-j} *^m z_a^{(j)} z_b^{(m-j)} \tag{1}$$

where the mth order convolution $*^m$ is

$$h_{j,m-j} *^m z_a^{(j)} z_b^{(m-j)} = \int_{-\infty}^{\infty} \cdots \int_{-\infty}^{\infty} dt_1 \ldots dt_m h_{j,m-j}(t_1, \ldots, t_m)$$
$$\prod_{i=1}^{j} z_a(t - t_i) \prod_{k=1}^{m-j} z_b(t - t_{j+k}) \tag{2}$$

and where $h_{j,m-j}$ are the mth order kernels of the interaction. A quadratic, or second-order, nonlinearity is one for which, for $m \geq 3$, $h_{j,m-j} = 0$. Two predictions of models with only quadratic nonlinearities are frequency doubling and superposition of nonlinearities. The former is the appearance in the Fourier spectrum of the response to moving sinusoidal gratings of energy at a frequency twice the fundamental, but not at frequencies higher than that. In superposition of nonlinearities, the nonlinear average response to a grating composed of two sinusoidal gratings of different frequencies, but whose ratio is a rational number, is equal to the sum of the responses to the individual gratings.

Vertebrate Spatial Asymmetry

The vertebrate retina contains five classes of neurons relevant for RDS (Dowling, 1987). Turtle-retina evidence (Marchiafava, 1979) indicates that the lateral asymmetry mediating RDS does not originate in photoreceptors, horizontal cells, or bipolar cells. Rabbit data (Amthor, Takahashi, and Oyster, 1989) show no correlation between asymmetries in dendritic trees of DS ganglion cells and their preferred-null axes. Consequently, these data suggest that the dendritic trees of amacrine cells mediate the RDS's spatial asymmetry.

Vertebrate Nonlinearities

Barlow and Levick (1965) performed apparent-motion experiments in rabbit retinas and suggested that the nonlinear interaction mediating RDS is inhibitory (Figure 1*B*). In Figure 1*B*, the spatial asymmetry is due to the fact that inputs come only from the right (inhibitory and slow) and center (excitatory and fast). The inhibition is nonlinear, and Barlow and Levick called it *veto inhibition*. The model generates RDS, because when the motion comes from the right, but not the left, the light reaches the inhibitory pathway before the excitatory pathway, compensating for the sluggishness of the former.

Torre and Poggio (1978) proposed a biophysical implementation of this nonlinear veto inhibition. Their rationale was based on RDS's being elicited by motions spanning short distances almost anywhere inside the receptive field (Barlow and Levick, 1965). Torre and Poggio suggested that this subunit-like behavior could be accounted for if each subunit corresponded to a branch of the DS ganglion cell's dendritic tree. To keep the computation constrained to each branch, they suggested that the inhibition mediating RDS works through a synapse that causes local changes of membrane conductance (shunting inhibition) and little hyperpolarization. To understand how

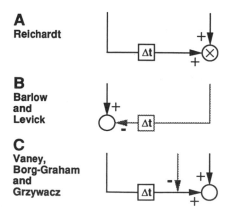

Figure 1. Schemes of models for directional selectivity.

such a synapse works, consider a patch of membrane receiving excitatory (g_e) and shunting inhibitory (g_i) synaptic conductances. Setting without loss of generality the resting and inhibitory reversal potentials to zero, the voltage V obeys

$$C\frac{dV(t)}{dt} + (g_e(t) + g_i(t) + g_{leak})V(t) = g_e(t)E_e + g_{leak}E_{leak}$$

(3)

where C is membrane capacitance, g_{leak} is the membrane's leak conductance, and E_e and E_{leak} are reversal potentials of g_e and g_{leak}, respectively. When $g_i \gg g_e, g_{leak}$, then V falls toward the following equilibrium ($dV/dt \to 0$) value:

$$V(t) \to \frac{g_e E_e + g_{leak}E_{leak}}{g_i}$$

(4)

which is small, because g_i is large. Therefore, this inhibition is division-like, rather than subtraction-like.

Torre and Poggio also argued that a shunting-inhibition mechanism for vertebrate RDS might be consistent with the insects' quadratic nonlinearity. Their argument was that for sufficiently low contrasts, one can neglect the higher-order nonlinearities (Equation 1), as in a Taylor series approximation.

Problems with Old Models for Retinal Directional Selectivity

Experiments have shown that the inhibitory mechanism mediating RDS in rabbit is division-like (Amthor and Grzywacz, 1993). However, recent experiments did not support the Torre and Poggio model. A quadratic approximation is not valid even for the smallest contrasts (Grzywacz, Amthor, and Mistler, 1990); rabbit DS ganglion cells failed the frequency-doubling and superposition-of-nonlinearities tests. Moreover, a whole-cell patch-clamp experiment by Borg-Graham and Grzywacz (1992) casts doubt that the relevant inhibition acts on the DS ganglion cell.

Recent data also challenged the Barlow and Levick model. Pharmacological evidence indicates that the inhibition mediating RDS is GABAergic via GABA$_A$ receptors (Caldwell, Daw, and Wyatt, 1978). However, when responses of turtles' DS ganglion cells to motion were recorded under picrotoxin (a GABA$_A$ antagonist), preferred and null directions would sometimes reverse as the stimulus's contrast or speed varied (Borg-Graham and Grzywacz, 1992).

Finally, the postulate of an exclusively inhibitory mechanism of RDS is incorrect. Grzywacz and Amthor (1993) expanded earlier experiments by Barlow and Levick (1965) to find that if the spatiotemporal parameters of the stimulus are appropriate, then there is a preferred-direction facilitation that can be as strong as null-direction inhibition.

A New Model

Vaney (1990) and Borg-Graham and Grzywacz (1992) independently proposed new models for RDS (Figure 1C). They are similar to the Reichardt model (Figure 1A) in that the spatial asymmetry is excitatory (the inhibitory mechanism is spatially symmetric). However, in contrast to that model, the nonlinearity is not quadratic, but inhibitory, as in the Barlow and Levick model, and of the shunting type, as in the Torre and Poggio model. Vaney, and Borg-Graham and Grzywacz, suggested that the DS signal flows from the distal end of each amacrine dendrite to the ganglion cell. Consequently, these dendrites would be autonomous units, bypassing integration at the amacrine soma. In turn, the ganglion cell would sample signals

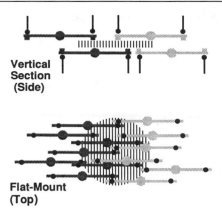

Figure 2. Model for the relationship between amacrine and ganglion dendrites in retinal directional selectivity. The vertically hatched, dark-textured, light-textured, and black elements correspond to one layer of ganglion-cell dendritic field, excitatory amacrine cells making strong synapses, excitatory amacrine cells making weak synapses, and inhibitory synapses, respectively. Preferred direction is to the right.

from several dendrites with the same preferred direction (Figure 2).

Computer simulations with this model show that it can generate RDS in individual dendrites of an amacrine cell without leading to RDS in its soma (Borg-Graham and Grzywacz, 1992). Furthermore, simulations show that the model accounts for reversals of preferred and null directions under picrotoxin as a result of saturation of the synapse between the amacrine dendrite and DS ganglion cell. Finally, the model accounts for preferred-direction facilitation as, for example, synaptic facilitation.

Development

Physiology and Anatomy

From studies in turtle (Sernagor and Grzywacz, current research), one can conclude that mature RDS emerges relatively late in retinal development. The adult incidence of DS cells ($> 20\%$) emerges after receptive fields reach mature sizes and after the disappearance of spontaneous waves of activity typical of immature retinas (Meister et al., 1991). In addition, RDS matures after orientation selectivity (that is, selectivity to spatial orientation of anisotropic stimuli, such as lines), concentric receptive fields, and surround inhibition. There are some immature DS cells at early stages of turtle embryonic development. However, this embryonic directional selectivity completely disappears at hatching. Directional selectivity is absent in turtle for at least the first 40 days of life, reappearing later on. Embryonic ganglion cells might express directional selectivity because of accidental asymmetric wiring in their polarized, undeveloped dendritic trees. Because mature RDS may emerge mainly at the expense of orientationally selective cells (Sernagor and Grzywacz, current research), it has been suggested that DS cells are modified orientationally selective cells. (RDS also emerges relatively late in rabbit, although its development is faster than that of turtle; Masland, 1977.)

The late emergence of RDS suggests two hypotheses for its development: (1) it requires light exposure, and/or (2) it requires the late emergence of an inhibitory drive onto the network mediating orientation selectivity.

If visual experience is necessary for the development of RDS, then this development probably involves synaptic plasticity. Daw and Wyatt (1974) attempted to test this hypothesis by raising rabbits from 15 to 60 days after birth such that their only visual experience was motion in one particular direction. These authors did not observe changes in the distribution of preferred directions, and they argued against plasticity in developing retinas. However, a criticism of this interpretation is that, in rabbit, the critical period for establishment of RDS might occur earlier than 15 days postnatal (Masland, 1977).

Models for the Development of Retinal Directional Selectivity

Borg-Graham and Grzywacz (1992) emphasized that the developmental problem is not to create spatial asymmetry, since amacrine dendrites are already asymmetric. Rather, the problem is to coordinate asymmetries of several dendrites (see Figure 2) to explain the subunit-like behavior in ganglion cells. Borg-Graham and Grzywacz suggested that a Hebbian correlational process (see HEBBIAN SYNAPTIC PLASTICITY) could reinforce an initial statistical bias in connections from amacrine cells to proto-DS ganglion cells (Figure 3A).

There are two problems with this model: (1) It does not account for the apparent emergence of RDS at the expense of orientationally selective cells. (2) This model predicts homogeneous distribution of preferred directions. The distribution of preferred directions is not necessarily homogeneous; for rabbit on-off cells, the distribution clusters around four directions parallel and perpendicular to the visual streak (Oyster, 1968).

A solution for the first problem would be a two-stage development for RDS (Figure 3B). In the first stage, the inhibition necessary for the model in Figures 1C and 2 would be absent and a Hebbian mechanism would reinforce statistical biases in orientation rather than direction. (The stimulus for this mechanism would be mainly spontaneous activity.) Dendrites of immature amacrine and ganglion cells could be orientationally selective despite lack of inhibition, because, for instance, a moving edge perpendicular to the dendrite would stimulate it better than a parallel edge, as a result of voltage saturation. In the second stage of the model, emergence of inhibition onto certain amacrine dendrites would turn them into DS cells, and a Hebbian process would coordinate the preferred directions, as in Figure 2.

The mechanism underlying the nonhomogeneous distribution of preferred directions might tap temporal-nasal, superior-inferior asymmetries of ganglion-cell densities, which, like directional selectivity, mature relatively late.

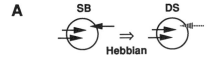

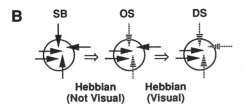

Figure 3. Two models for the development of retinal directional selectivity.

Since this mechanism's development might involve Hebbian processes driven by spontaneous activity and light, RDS may be an excellent paradigm to study dendritic computations and brain self-organization.

Acknowledgments.   Support for this work came from a grant from the Office of Naval Research (N00014–91–J–1280), a grant from the National Eye Institute (EY08921), and an award from the Paul L. and Phyllis C. Wattis Foundation to N.M.G.; from a grant from the National Eye Institute (EY05070) to F.R.A.; and from a core grant from the National Eye Institute to Smith-Kettlewell (EY06883).

Road Map: Vision
Background: Axonal Modeling
Related Reading: Directional Selectivity in the Cortex; Retina; Visuomotor Coordination in Flies

References

Amthor, F. R., and Grzywacz, N. M., 1993, Inhibition in On-Off directionally selective ganglion cells in the rabbit retina, *J. Neurophysiol.*, 69:2174–2187.

Amthor, F. R., Takahashi, E. S., and Oyster, C. W., 1989, Morphologies of rabbit retinal ganglion cells with complex receptive fields, *J. Comp. Neurol.*, 280:97–121.

Barlow, H. B., 1953, Summation and inhibition in the frog's retina, *J. Physiol., Lond.*, 119:69–88.

Barlow, H. B., and Levick, W. R., 1965, The mechanism of directionally selective units in rabbit's retina, *J. Physiol.*, 178:477–504.

Borg-Graham, L. J., and Grzywacz, N. M., 1992, A model of the direction selectivity circuit in retina: Transformations by neurons singly and in concert, in *Single Neuron Computation* (T. McKenna, J. Davis, and S. F. Zornetzer, Eds.), Orlando: Academic Press, pp. 347–375. ◆

Caldwell, J. H., Daw, N. W., and Wyatt, H. J., 1978, Effects of picrotoxin and strychnine on rabbit retinal ganglion cells: lateral interactions for cells with more complex receptive fields, *J. Physiol.*, 276:277–298.

Daw, N. W., and Wyatt, H. J., 1974, Raising rabbits in a moving visual environment: An attempt to modify direction sensitivity in the retina, *J. Physiol. (Lond.)*, 240:309–330.

Dowling, J. E., 1987, *The Retina: An Approachable Part of the Brain*, Cambridge, MA: Harvard University Press. ◆

Grzywacz, N. M., and Amthor, F. R., 1993, Facilitation in On-Off directionally selective ganglion cells in the rabbit retina, *J. Neurophysiol.*, 69:2188–2199.

Grzywacz, N. M., Amthor, F. R., and Mistler, L. A., 1990, Applicability of quadratic and threshold models to motion discrimination in the rabbit retina, *Biol. Cybern.*, 64:41–49.

Marchiafava, P. L., 1979, The responses of retinal ganglion cells to stationary and moving visual stimuli, *Vis. Res.*, 19:1203–1211.

Masland, R. H., 1977, Maturation of function in the developing rabbit retina, *J. Comp. Neurol.*, 175:275–286.

Meister, M., Wong, R. O. L., Baylor, D. A., and Shatz, C. J., 1991, Synchronous bursts of action potentials in ganglion cells of the developing mammalian retina, *Science*, 252:939–943.

Miles, F. A., and Wallman, J., Eds., 1993, *Visual Motion and Its Role in the Stabilization of Gaze*, Amsterdam: Elsevier. ◆

Oyster, C. W., 1968, The analysis of image motion by the rabbit retina, *J. Physiol.*, 199:613–635.

Poggio, T., and Reichardt, W. E., 1976, Visual control of orientation behaviour in the fly: Part II: Towards the underlying neural interactions, *Q. Rev. Biophys.*, 9:377–438. ◆

Torre, V., and Poggio, T., 1978, A synaptic mechanism possibly underlying directional selectivity to motion, *Proc. R. Soc. Lond. B*, 202:409–416.

Vaney, D. I., 1990, The mosaic of amacrine cells in the mammalian retina, in *Progress in Retinal Research*, vol. 9 (N. Osborne and J. Chader, Eds.), Oxford: Pergamon, pp. 49–100. ◆

Disease: Neural Network Models

Daniel S. Levine

Introduction

In neural networks, mental illness or malfunction of a brain area is typically modeled as diminished function in part of a network designed to perform some set of cognitive tasks. Functional diminution can be effected either by breaking a connection or by altering one or more network parameters, such as connection weights or signal strengths. Most ensuing models are offshoots of models developed for normal cognitive processes.

Since the early 1980s, data have accumulated on functional components and neurochemistry of specific mental illnesses (e.g., schizophrenia, depression, epilepsy). For none of these diseases, however, is there yet a universally accepted qualitative theory. Individual variations in symptoms limit the universality of any theory of a disease even as knowledge deepens through positron-emission tomography (PET) scanning and other clinical techniques. Also, current qualitative theories of brain areas involved in high-level cognitive function (e.g., prefrontal cortex and hippocampus) are suggestive but anatomically imprecise.

Hence, current computational models of these illnesses focus on poor performance in a few behavioral tasks or impairment of a general class of functions. (See also EPILEPSY: NETWORK MODELS OF GENERATION.) Enough success has been achieved with such models to suggest this will be a major growth area for neural networks and computational neuroscience.

Schizophrenia

Schizophrenic symptoms fall into two major categories—positive and negative (see, e.g., Frith and Done, 1988). Positive symptoms include hallucinations and paranoid delusions, whereas negative symptoms include flat affect and poor speech content. Some efforts have been made at unified qualitative theories encompassing both symptom types; Frith and Done (1988) say in their abstract:

> the negative signs of schizophrenia reflect a defect in the initiation of spontaneous action, while the positive symptoms reflect a defect in the internal monitoring of action. . . . [I]nitiation of action depends upon brain systems linking the prefrontal cortex and the basal ganglia. Internal monitoring . . . depends upon links between the prefrontal cortex and the hippocampus via the parahippocampal cortex and the cingulate cortex.

This integrated system has not been modeled. Thus far, simulations have focused on subclasses of negative symptoms, particularly disturbances in processing context. Cohen and Servan-Schreiber (1992) studied one such symptom using a backpropagation network. In the experiment that Cohen and Servan-Schreiber simulated, subjects interpret a phrase which uses a word in a sense that is not its dominant meaning but the one that fits the context. An example is "without a *pen* you cannot keep chickens." More schizophrenics than normals interpret "pen" in its dominant meaning, a writing implement, rather than its context-appropriate meaning, a fenced enclosure.

Figure 1 shows Cohen and Servan-Schreiber's network. Different modules represent words and context. Schizophrenia is mimicked by reduced gain of the sigmoid activation function of nodes within the context module. While connections between modules in this network do not reflect specific brain connec-

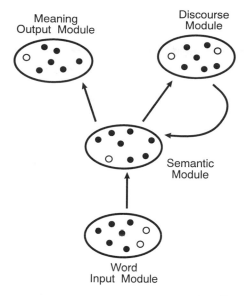

Figure 1. Network architecture that simulates lexical disambiguation by context. Schizophrenia is interpreted as loss of connections between discourse and semantic modules. (From Cohen, J. D., and Servan-Schreiber, D., 1992, Context, cortex, and dopamine: A connectionist approach to behavior and biology in schizophrenia, *Psychol. Rev.*, 99: 60. Copyright 1992 by the American Psychological Association. Reprinted by permission.)

tions, the modules representing words and concepts, however, may be analogous to functional units in association cortex. For example, Fuster (1989, chap. V) reviews evidence from single-cell studies of the prefrontal cortex in monkeys involved in a delayed matching-to-sample task. Different neurons in the frontal lobes fire in response to different aspects of the task—the original (sample) stimulus, the reward, the matching stimulus, etc.—and neurons responding to the same aspect tend to be grouped in columns. While not all neuroscientists accept this columnar organization as a general principle, it has been observed enough to provide a justification for encoding high-level entities as single functional units, as is done in all network models discussed herein.

Cohen and Servan-Schreiber (1992) interpreted gain reduction in their sigmoid function as analogous to reduced dopamine input to the prefrontal cortex. (Frontal dopamine loss goes along with *excess* of dopamine at a different brain location, probably on pathways leading to the limbic system. That dopamine excess is related to positive symptoms and often treated by antidopaminergic *neuroleptic* drugs.)

Depression and Manic-Depressive Illness

Depressive and manic-depressive disorders have not yet been simulated in neural networks but have been qualitatively studied in three chapters from the book of Levine and Leven (1992): the chapters by Banquet, Smith, and Guenther; Hestenes; and Leven. The theories discussed in those chapters are based on several overarching principles used in many other neural network models: associative learning, lateral inhibition, opponent processing, neuromodulation, and interlevel resonant feedback.

Associative learning (see ASSOCIATIVE NETWORKS) and lateral inhibition are widely used in neural networks. Opponent processing is a common idea in psychology involving representations of opposing concepts, such as pain and pleasure, light and dark, or flexion and extension. Grossberg (e.g., 1987) applies it to a network architecture, the *gated dipole*, using habituating chemical transmitters. If a representation of one concept (channel) in a gated dipole has recently been active, transmitter in that channel is more habituated than in the opposite channel, leading to transient activity of the opposite representation (a release from inhibition). Thus, reactions to novel events or unexpected reinforcement are more enhanced than reactions to familiar events or expected reinforcement.

Neuromodulation is defined by Kaczmarek and Levitan (1987) as "the ability of neurons to alter their electrical properties in response to intracellular biochemical changes resulting from synaptic or hormonal stimulation" (p. 3). In networks, modulation is suggested by cognitive data whereby contextual shifts influence the organism's selective attention. Such attention shifts have suggested to many modelers that one neural signal may modulate the strength of another signal along a different pathway (e.g., Levine and Leven, 1992, chaps. 5, 7, and 9). Modulation is also suggested by neurochemical data about monoamine transmitters (norepinephrine, dopamine, and serotonin), each produced and released by a different midbrain nucleus, then sent out broadly to the cerebral cortex and limbic system.

Interlevel resonant feedback was introduced by Carpenter and Grossberg (1987) into a computational architecture, the adaptive resonance theory (ART) network (see ADAPTIVE RESONANCE THEORY). Many other models combine associative learning and lateral inhibition into a two-layer feedforward network consisting of feature and category representation layers. If an input perturbs the feature layer, each category node receives a weighted "bottom-up" signal, and the node receiving the largest signal determines the category in which the input is classified. Carpenter and Grossberg argued, however, that in such a feedforward network, presentation of one pattern can lead to recoding of other patterns. To prevent this, they added "top-down" feedback connections from category to feature layers that indicate whether the current input matches a stored category prototype.

Hestenes, modeling manic-depressive illness, and Leven, modeling learned helplessness (a common animal model for depression), related their theories to neurotransmitter biochemistry. Hestenes posited network roles for the monoamine transmitters and related them to the network principles mentioned above. These roles are: for norepinephrine, selective enhancement of novel or significant inputs; for serotonin, matching between two patterns; for dopamine, potentiation of reward signals. He noted that the manic stage of manic-depressive (bipolar) disorders, like serotonin deficiency, leads to pattern matching deficits. This led him to suggest treating mania with dietary supplements of the amino acid tryptophan, a biochemical precursor of serotonin. He roughly correlated the depressive part with reduced dopamine and found evidence that both transmitters act on the nucleus accumbens, which links limbic system emotional regions to basal ganglia motor control regions.

Leven developed a theory whereby learned helplessness is again caused by changes in the balance between two neurotransmitters, in this case, 5-HT and NE. These two transmitters were treated as operating on two sites of a gated dipole or opponent processing network (see above).

Banquet et al. (in Levine and Leven, 1992) presented results on differing evoked potential responses to common and rare auditory stimuli. They showed that depressed patients did not exhibit as pronounced differences in responses to common and rare stimuli as did normal subjects. Their current work in progress involves simulations that add to the adaptive resonance theory network (see above) a module for processing probabilities.

Frontal Lobe Dysfunction

The prefrontal area of the neocortex, in feedback with other brain regions, is involved in a wide range of cognitive functions (Fuster, 1989). Its connections with other association cortices and with secondary sensory cortices give it a role in processing sensory data. Its connections, both directly and via the thalamus, with the limbic system and hypothalamus give it a role in processing emotions and internal states. Its connections with midbrain monoaminergic regions give it a role in modulating responses to sensory patterns. Its connections with the motor cortex and basal ganglia give it a role in planning movement.

Many neural network models of frontal lobe function focus on reproducing frontal damage effects in humans or monkeys. These effects include deficits on standard tests from clinical neuropsychology, such as the Wisconsin Card Sorting, Stroop, and verbal fluency tests; deficits on delayed response; excessive attraction to novelty; and inability to learn flexible motor sequences.

Dehaene and Changeux (1991) and Levine and Prueitt (1989) both modeled effects of frontal damage on the Wisconsin Card Sorting Test (WCST) (Milner, 1964). In the WCST, a subject is given a sequence of 128 cards, each displaying a number, color, and shape, and must match the card shown to one of four template cards. The experimenter then says the match is right or wrong, giving no reason. After ten straight correct matches to color, the experimenter (without warning) switches the criterion to shape. Then, if classification by shape is learned, the criterion shifts to number, then back to color, and so on. Most patients with damage to the dorsolateral frontal cortex learn the color criterion as quickly as normals, but cannot switch to shape. Normals and patients with other brain lesions, by contrast, usually achieve four or five criteria (color, shape, number, color, shape) over 128 trials.

Milner's results were simulated by Levine and Prueitt (1989), using the neural network of Figure 2, which builds on ART (Carpenter and Grossberg, 1987). Frontal damage is modeled as reduction in reinforcement signals to bias nodes. Nodes in F_1 code features (numbers, colors, shapes). Nodes in F_2 code template cards, each of a category prototype. To each feature class corresponds a *habit node* and *bias node*. Connection weights z_{ij} and z_{ji} are large when node x_i represents a feature present in card y_j. Attentional gating from bias nodes enhances currently relevant bottom-up signals. When an input card is presented, the template card with largest node activity is chosen as the one matched. If template and input share a feature, a signal is sent to that feature's habit node. The same signal excites or inhibits the feature's bias node, depending on the sign of reinforcement. If negative reinforcement is weak, positive feedback between habit and bias nodes, attentional gating, and category choice make the network perseverate in color classification, as seen in frontally lesioned patients.

The Levine-Prueitt network was based on principles such as associative learning, lateral inhibition, opponent processing, neuromodulation, and resonant feedback (see above). While Dehaene and Changeux (1991) derived their network from primary neurobiological considerations, their resulting network also instantiates some of these same principles. For example, there is feedback between input and memory nodes, and rule clusters modulate synapses from input to memory nodes. Dehaene and Changeux also included a process, similar to opponent

Figure 2. Network that simulates card sorting data. Frontal damage is modeled by reduced gain from reinforcement nodes to bias nodes. (Adapted from Leven, S. J., and Levine, D. S., Effects of reinforcement on knowledge retrieval and evaluation, in *Proceedings of the First International Conference on Neural Networks*, vol. 2, pp. 269–279. © 1987 IEEE; reprinted by permission.)

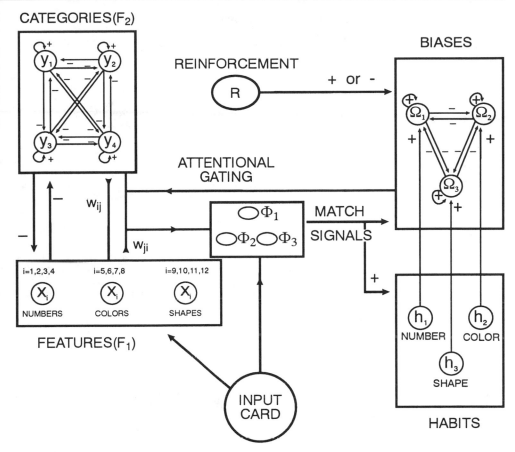

processing, for rejecting previously used but currently unrewarding rules.

In a variant of their WCST network, Levine and Prueitt (1989) also reproduced Pribram's (1961) data on attraction to novelty. Novelty preference was simulated using Grossberg's gated dipole (see above). Pribram compared normal and frontally lesioned rhesus monkeys in a scene with several junk objects. As successive objects were added, a reward (peanut) was placed each time under the new object. The peanut remained under the same cue until the animal found the peanut a certain number of times. Normal monkeys tended to go first to the previously rewarded object, whereas frontally damaged monkeys tended to go quickly to the novel object.

The same types of models have been extended to other human and animal data. Levine and Parks (1992), using a neural network similar to that of Figure 2, simulated effects of frontal damage on the *verbal fluency* test. In this test, the subject is asked to generate as many words as he or she can think of in 60 seconds beginning with a certain letter, excluding proper nouns and variants of previously uttered words. Dehaene and Changeux (1989) simulated effects of frontal damage on the *delayed response* test, whereby a monkey observes the location of a reward and then must find it after the scene is removed from view.

Cohen, Dunbar, and McClelland (1990), using a network similar to that of Figure 1, simulated effects of frontal damage on the *Stroop test*. In this test, subjects are shown a word representing one color printed in ink that might be the same color or another color, then asked to read aloud the color of the ink. This is easy when the colors of the ink and text match, but hard when the colors mismatch. If the stimulus word is, for example, "red" but that word is printed in green ink, subjects have difficulty saying the word "green." The test is sensitive to any injury or dysfunction that disturbs processing of context (see the section "Schizophrenia," above).

More biologically realistic and testable models of the prefrontal cortex, currently under development, will build on these networks but also include pathways connecting the prefrontal cortex to subcortical areas, particularly the limbic system and basal ganglia.

Discussion

Neural modeling is now sophisticated enough to construct networks whereby weakening some connections or removing some nodes has effects analogous to known mental dysfunctions. In addition to the work discussed above, analogies have been drawn of some gated dipole abnormalities with schizophrenia, Parkinsonism, and juvenile hyperactivity (Grossberg, 1987, vol. 1, chap. 2) and some ART abnormalities with classic amnesia (Carpenter and Grossberg, 1987). Primitive analogies have been drawn between parts of these networks and specific brain areas or chemical transmitter systems. Also, a modified backpropagation network to map orthographic representations into semantic ones has been constructed. Lesioning some connections in this network (see LESIONED ATTRACTOR NETWORKS AS MODELS OF NEUROPSYCHOLOGICAL DEFICITS) leads to errors similar to those found in *deep dyslexia*, whereby a patient cannot access the phonological representation of a written word without first accessing a semantic representation.

These networks point in interesting directions, but their physiological and neurochemical constraints are still crude. As the models develop, they will gradually incorporate much more detail on neuronal transmitters, messengers, and receptors in specific brain areas. Conversely, instantiations of network principles in models of particular mental processes can suggest structures for physiologists and biochemists to search for in the brain.

Road Map: Biological Networks
Related Reading: Basal Ganglia; Developmental Disorders; Short-Term Memory

References

Carpenter, G. A., and Grossberg, S., 1987, A massively parallel architecture for a self-organizing neural pattern recognition machine, *Comput. Vis. Graph. Image Proc.*, 37:54–115.
Cohen, J. D., Dunbar, K., and McClelland, J. L., 1990, A parallel distributed processing model of the Stroop effect, *Psychol. Rev.*, 97:332–361.
Cohen, J. D., and Servan-Schreiber, D., 1992, Context, cortex, and dopamine: A connectionist approach to behavior and biology in schizophrenia, *Psychol. Rev.*, 99:45–77.
Dehaene, S., and Changeux, J.-P., 1989, A simple model of prefrontal cortex function in delayed-response tasks, *J. Cognit. Neurosci.*, 1:244–261.
Dehaene, S., and Changeux, J.-P., 1991, The Wisconsin Card Sorting Test: Theoretical analysis and modeling in a neuronal network, *Cereb. Cortex*, 1:62–79.
Frith, C. D., and Done, D. J., 1988, Toward a neuropsychology of schizophrenia, *Br. J. Psychiatry*, 153:437–443.
Fuster, J. M., 1989, *The Prefrontal Cortex*, New York: Raven. ◆
Grossberg, S., 1987, *The Adaptive Brain*, vols. 1 and 2, Amsterdam: Elsevier.
Kaczmarek, L. K., and Levitan, I. B., 1987, *Neuromodulation: The Biochemical Control of Neuronal Excitability*, New York: Oxford University Press. ◆
Levine, D. S., and Leven, S. J., Eds., 1992, *Motivation, Emotion, and Goal Direction in Neural Networks*, Hillsdale, NJ: Erlbaum.
Levine, D. S., and Parks, R. W., 1992, Frontal lesion effects on verbal fluency in a network model, in *International Joint Conference on Neural Networks, Baltimore, Maryland*, vol. 2, Piscataway, NJ: IEEE, pp. 39–44.
Levine, D. S., and Prueitt, P. S., 1989, Modeling some effects of frontal lobe damage: Novelty and perseveration, *Neural Netw.*, 2:103–116.
Milner, B., 1964, Some effects of frontal lobectomy in man, in *The Frontal Granular Cortex and Behavior* (J. Warren and K. Akert, Eds.), New York: McGraw-Hill, pp. 313–334.
Pribram, K. H., 1961, A further experimental analysis of the behavioral deficit that follows injury to the primate frontal cortex, *J. Exp. Neurol.*, 3:432–466.

Dissociations Between Visual Processing Modes

Bruce Bridgeman

Introduction

The visual system has two kinds of jobs to do. One function is to support visual cognition or perception, our visually informed knowledge about the identities and relative locations of objects and surfaces in the world. Another, sensorimotor, function is to drive visually guided behavior. There is no guarantee that the two functions always use the same visual information, and indeed they need somewhat different kinds of information. The cognitive function has less need of quantitative, egocentrically based, spatial localization (in a coordinate system based on the head or body) because cognition is concerned primarily with pattern recognition and with the positions of objects relative to one another. Qualitative information about location may be adequate for this system, and there is abundant evidence that humans are not very good at quantitatively estimating distances, directions, and the like if the measures of these abilities are perceptual judgments rather than motor behaviors. The sensorimotor function, in contrast, needs quantitative, egocentrically calibrated, spatial information to guide motor acts.

Physiological Evidence

The visual pathways begin with a unified optical system. The representation of the left visual field goes through the thalamus to the right side of the primary visual cortex, at the back of the head, while the right visual field projects to the left side. From here, visual signals are relayed to a multitude of topographic maps in other visual areas (Van Essen, Newsome, and Bixby, 1982). There may be as many as 20 regions topographically coding all or part of the visual field. This characteristic of the visual system raises two questions: Do all of these maps work together in a single visual representation, or are they functionally distinct? If they are distinct, how many functional maps are there and how do they communicate with one another?

The evidence reveals that the multiple maps support at least two functionally distinct representations of the visual world; under some conditions, the two representations can simultaneously hold different spatial values. The representations do not always function independently, however, but sometimes communicate with one another. Each representation uses several of the physiological retinotopic maps; the cognitive and sensorimotor representations may correspond respectively to pathways in the temporal and parietal cortex, respectively (Mishkin, Ungerleider, and Macko, 1983). The temporal pathway consists of a number of regions in which neurons generally respond to stimuli in larger and larger regions of the visual field, but require more and more specific features or properties to excite them. This processing pathway culminates in the inferotemporal cortex, which seems to specialize in pattern recognition problems involving choices and discriminations (Pribram, 1971). Neurons in this region typically respond to very large areas of the visual field, usually including the fixation point, and their responses are highly modified by visual experience and by the nature of the visual task currently being executed.

The parietal pathway, in contrast, specializes in physical features of the visual world, such as motion and location. One area contains a map of the cortex that seems to specialize in motion of objects, while others contain neurons that respond both to characteristics of the visual world and to intended movements.

The spatial function can be further subdivided into two processing pathways that probably reflect two different modes of spatial coding. Receptive fields of neurons in the lateral intraparietal area are spatially corrected before each rapid eye movement, so that they respond to stimuli that will be in their retinotopic receptive fields (i.e., in retinally based coordinates) following a planned eye movement (Duhamel, Colby, and Goldberg, 1992). The changes in these receptive fields can also be conceived as activity that signals candidates for planned eye movements (see DYNAMIC REMAPPING).

A second coding scheme is seen in another parietal cortex area, 7a, involved in ocular fixation. Neurons in this area provide information that is sufficient to reconstruct spatial position in head-centered coordinates. The responses of these neurons depend on both eye position in the orbit and retinal location of a target, in a multiplicative interaction. No single neuron responds to targets at a constant location in space, but within the retinotopic map of the area, information is available about the head-centered location of a target. Later, simulations showed that information sufficient to derive spatiotopic output exists in such a network of cells (Zipser and Andersen, 1988). A parallel distributed processing network could be trained to respond to targets at particular locations in a visual field. After training, the response properties of the nodes in the model turned out to bear a striking resemblance to the receptive fields of the neurons in area 7a (see GAZE CODING IN THE POSTERIOR PARIETAL CORTEX).

Spatial processing is too basic a function to be limited to the parietal cortex, a relatively high-order structure that is well differentiated only in primates. The midbrain superior colliculus is involved in sensorimotor vision (see COLLICULAR VISUOMOTOR TRANSFORMATIONS FOR SACCADES). The superior colliculus also performs broad intermodal integration (see SENSOR FUSION).

Clinical Evidence

The earliest evidence for a separation of visual functions has come from neurological patients. Damage to a part of the primary visual cortex results in an effective blindness in the affected field—a scotoma. Patients have no visual experience in the scotoma—that is, they do not recognize objects or surfaces in the blind area. But when forced to point to targets located in their scotoma, which they insist they cannot see, they point accurately to the targets by using visuomotor information unavailable to their perception. Visual information coexisting with a lack of visual experience is called *blindsight* (Weiskrantz et al., 1974). It may be made possible by an alternative pathway from the retina to other parts of the visual cortex through the superior colliculus or other subcortical structures. This is an example of visually guided behavior without the experience of perception. At first it was thought that blindsight allowed only gross localization of objects, but more recent work has shown surprisingly sophisticated processing without awareness, including orientation and color discriminations.

Another example of dissociation of visuomotor and perceptual function has been found in a patient with damage to the lateral occipital and occipitoparietal regions of the cortex (Goodale and Milner, 1992). When asked to match line orientations, the patient showed gross errors, such as judging horizontal to be vertical, although she had no scotoma in which visual experience was altogether absent. When asked to put a card through a slot at varying orientations, however, she oriented the card correctly even as she raised her hand from the start position. There was a dissociation between her ability to perceive object orientation and her ability to direct accurate reaching movements toward objects of varying orientations. Her cognitive representation of space was unavailable, but her visuomotor representation remained intact. The complementary dissociation, with retention of perception accompanied by loss of motor coordination, is seen clinically under the label of optic ataxia.

Such examples demonstrate a separate cognitive and sensorimotor representation, but all of the demonstrations are in patients with brain damage. When a human suffers brain damage with partial loss of visual function, there is always the possibility that the visual system will reorganize itself, isolating fragments of the machinery that normally function as a unit. The clinical examples, then, leave open the possibility that the system may function differently in intact humans. Any rigorous proof that normal visual function also shows this cognitive/sensorimotor distinction must include psychophysical measures in intact humans.

Psychophysical Evidence

Some of the earliest evidence for a cognitive/sensorimotor distinction in normal subjects has come from studies of rapid (saccadic) eye movements. Subjects are normally unaware of sizable displacements of the visual world if the displacements occur during saccadic eye movements. This implies that information about spatial location is degraded during saccades. There is a seeming paradox to this degradation, however, for people do not become disoriented after saccades, implying that spatial information is maintained. Experimental evidence supports this conclusion. For instance, the brain can produce accurate saccadic movements in response to a target that is flashed (and mislocalized) during an earlier saccade, and hand-eye coordination remains fairly accurate following saccades. How can perceptual information be lost while visually guided behavior is preserved?

Resolution of this paradox begins by noting that the two kinds of conflicting observations use different response measures. The experiments on perception of displacement during saccades require a symbolic response, such as a nonspatial verbal report or a button press. The behavior has an arbitrary spatial relationship to the target. Orienting of the eye or hand, in contrast, requires quantitative spatial information, defined as involving a 1:1 correspondence between a target position and a motor behavior, such as directing the hand or the eyes to the target. The conflict might be resolved if these two types of measures, which can be labeled as cognitive and sensorimotor, could be combined in a single experiment. If two pathways in the visual system process different kinds of information, spatially oriented motor activities might have access to accurate position information, even when that information is unavailable at a cognitive level.

The two conflicting observations (perceptual suppression on one hand and accurate motor behavior on the other) were combined by asking subjects to point to the position of a target that had been displaced and then extinguished (reviewed in Bridgeman, 1992). In some trials, the target jump was detected, while on others, the jump went undetected owing to a simultaneous eye movement (monitored photoelectrically). As one would expect, subjects could point accurately to the position of the now-extinguished target following a detected displacement. Pointing was equally good, however, following an undetected displacement. It appeared that some information was available to the motor system but not to perception.

The results of this trial imply that quantitative control of motor activity is unaffected by the perceptual detectability of target position. One can also interpret the result in terms of signal detection theory as a high-response criterion for the report of displacement. The first control for this possibility was a two-alternative, forced-choice measure of saccadic suppression of displacement. Even this criterion-free measure showed an inability to perceive displacements under conditions where pointing was accurate, even when the target had been displaced (Bridgeman and Stark, 1979). The information was available to a motor system controlling pointing, but not to a cognitive system informing visual perception.

Dissociation of cognitive and sensorimotor function has also been demonstrated by creating conditions in which cognitive and sensorimotor systems receive opposite signals at the same time, which is a more rigorous way to separate cognitive and sensorimotor systems. A signal is inserted selectively into the cognitive system with stroboscopic-induced motion. In this illusion, a surrounding frame is displaced, creating the illusion that a target jumps, although it remains fixed relative to the subject. We know that induced motion affects the cognitive system because we experience the effect, and because subjects can make verbal judgments of it. But earlier experiments implied that the information used for motor behavior may come from sources unavailable to perception. This experiment involved stroboscopic-induced motion in an otherwise uniform field; a target spot jumped in the same direction as a frame, but not far enough to cancel the induced motion. The spot still appeared to jump in the direction opposite the frame, while it actually jumped in the same direction. At this point, the retinal position of the target was stabilized by feedback from an eye tracker, so that retinal error could not drive eye movements. Saccadic eye movements followed the actual direction, even though subjects perceived stroboscopic motion in the opposite direction (Wong and Mack, 1981). If a delay in responding was required, however, eye movements followed the perceptual illusion. These results imply that the sensorimotor system has no memory, but must rely on information from the cognitive system when responding to what had been present rather than what is currently present.

Paralleling these distinctions between perceptual and motor-oriented systems, there is also behavioral evidence that extraretinal signals are used differently for movement discrimination and for egocentric localization. During pursuit tracking, Honda (1990) found systematic errors in perceptually judged distance of target movement, even when egocentric localization was not affected by the mode of eye movements.

All of these techniques involve motion or displacement, leaving open the possibility that the dissociations are associated in some way with motion systems, rather than with representation of visual location per se. Motion and location may be confounded in some kinds of visual coding schemes. A newer design allows the examination of visual context in a situation where there is no motion or displacement at any time (Bridgeman, 1992). The design is based on the Roelofs effect (Roelofs, 1935), a perceptual illusion seen when a static frame is presented asymmetrically to the left or the right of a subject's center-line. Objects that lie within the frame tend to be mislocated in the direction opposite to the offset of the frame. For example, in an otherwise featureless field, a rectangle is presented to the subject's left. Both the rectangle and stimuli within it will tend to be localized too far to the right.

All subjects reveal a Roelofs effect for a perceptual measure, judging targets to be further to the right when a frame is on the left and vice versa. However, only half of them show the effect for pointing—the rest remain uninfluenced by frame position in pointing. If a delay in responding causes some subjects to switch from using motor information directly to using information imported from the cognitive representation, delaying the response should force all subjects to switch to using cognitive information. By delaying the response cue long enough, all subjects showed a Roelofs effect in pointing as well as judging. Thus, like Wong and Mack's induced-motion experiments reviewed earlier, this design showed a switch from motor to cognitive information in directing the motor response, revealed by the appearance of a cognitive illusion after a delay. The sensorimotor branch of the system seems to hold spatial information just long enough to direct current motor activity, and no longer.

Discussion

The general conclusion is that information about egocentric spatial location is available to the visual system despite cognitive illusions of location. Egocentric localization information is available to the sensorimotor system even while the cognitive system, relying on relative motion and relative position information, holds unreliable information about location.

Acknowledgment.   This work was supported by AFOSR grant number 90–0095.

Road Map: Vision
Related Reading: Grasping Movements: Visuomotor Transformations; Visual Scene Perception: Neurophysiology

References

Bridgeman, B., 1992, Conscious vs unconscious processes: The case of vision, *Theory Psychol.*, 2:73–88. ◆

Bridgeman, B., and Stark, L., 1979, Omnidirectional increase in threshold for image shifts during saccadic eye movements, *Percept. & Psychophys.*, 25:241–243.

Duhamel, J.-R., Colby, C. L., and Goldberg, M. E., 1992, The updating of the representation of visual space in parietal cortex by intended eye movements, *Science*, 255:90–92.

Goodale, M. A., and Milner, A. D., 1992, Separate visual pathways for perception and action, *Trends Neurosci.*, 15:20–25. ◆

Honda, H., 1990, The extraretinal signal from the pursuit-eye-movement system: Its role in the perceptual and the egocentric localization systems, *Percept. & Psychophys.*, 5:509–515.

Mishkin, M., Ungerleider, L., and Macko, K., 1983, Object vision and spatial vision: Two cortical pathways, *Trends Neurosci.*, 6: 414–417. ◆

Pribram, K. H., 1971, *Languages of the Brain*, Englewood Cliffs, NJ: Prentice-Hall, chaps. 11, 17, 18, and 19. ◆

Roelofs, C., 1935, Optische Localisation, *Arch. Augenheilkunde*, 109: 395–415.

Van Essen, D. C., Newsome, W. T., and Bixby, J. L., 1982, The pattern of interhemispheric connections and its relationship to extrastriate visual areas in the macaque monkey, *J. Neurosci.*, 2:265–283.

Weiskrantz, L., et al., 1974, Visual capacity in the hemianopic field following a restricted occipital ablation, *Brain*, 97:709–728.

Wong, E., and Mack, A., 1981, Saccadic programming and perceived location, *Acta Psychol.*, 48:123–131.

Zipser, J., and Andersen, R. A., 1988, A back-propagation programmed network that simulates response properties of a subset of posterior parietal neurons, *Nature*, 33:679–684.

Distortions in Human Memory

Janet Metcalfe

Introduction

Human memory is messy. What do the biases, blends, and mistakes signify about the nature of the system underlying human episodic memory? Two models, CHARM and MINERVA, which produce the distortions found in the human data, are reviewed. They differ in fundamental assumptions: the former is a distributed holographic model with superposition of events in a composite trace; the latter assumes separate storage of individual memory traces, like so many pictures in a photo album. Being easier to understand than memory itself, they can be used as tools to isolate the principles of human episodic memory responsible for distortions.

In the classic memory-distortion paradigm, people witness an event and then later are given a suggestion that the event was other than it was. For example, the person may see a sports car go through a yield sign at a corner and later be told it was a stop sign (Loftus, 1975, 1979). As compared to control subjects receiving no suggestion, the misled people choose the incorrect stop sign frequently in tests, when it is pitted against the yield sign. Such memory distortions may occur in situations, such as "suggested" abuse cases, far more serious than this example. If a no-parking sign is the test alternative, however, the misled and control subjects do not differ (McCloskey and Zaragoza, 1985).

CHARM

In CHARM (Metcalfe, 1990, 1991, 1993) two items represented as vectors, $\mathbf{A} = (a_{-(n-1)/2}, \ldots, a_{-1}, a_0, a_1, \ldots, a_{(n-1)/2})$ and $\mathbf{B} = (b_{-(n-1)/2}, \ldots, b_{-1}, b_0, b_1, \ldots, b_{(n-1)/2})$, may be associated by the operation of convolution, denoted $*$, to yield the vector $\mathbf{A} * \mathbf{B}$ whose mth element is defined as

$$(A * B)_m = T_m = \sum a_i b_j \qquad (i,j) \in S_+(m) \qquad (1)$$

where $S_+(m) = \{(i,j) \mid -(n-1)/2 \le i, j \le (n-1)/2, \text{ and } i + j = m\}$. The results of successive convolutions are added into a single composite memory trace $\mathbf{T}$, where

$$\mathbf{T} = \mathbf{A} * \mathbf{A} + \mathbf{A} * \mathbf{B} + \mathbf{B} * \mathbf{B} + \mathbf{C} * \mathbf{C} + \mathbf{C} * \mathbf{D} + \mathbf{D} * \mathbf{D} + \cdots$$

$$+ \text{ pre-existing noise} \qquad (2)$$

Retrieval generates a new vector $\mathbf{R}_m$ from the elements of the cue and trace vectors by cross-correlating them:

$$\mathbf{R}_m = \sum q_i t_j \qquad (i,j) \in S_-(m) \qquad (3)$$

where $\mathbf{Q}$ is the cue vector with elements q_i, $\mathbf{T}$ is the trace with elements t_i, and $S_-(m)$ is the domain of paired elements over which the correlation is attempted, i.e., $S_-(m) = \{(i,j) \mid -(n-1)/2 \le i, j < (n-1)/2, \text{ and } i - j = m\}$. $\mathbf{R}$ is matched either to a lexicon—the best match above a lower threshold is chosen—or to the test alternatives.

If the trace $\mathbf{T} = \{\text{Corner} * \text{Yield sign}\} + \{\text{Corner} * \text{Stop sign}\}$ + other associations, then when Corner is given as a cue it generates Yield sign + Stop sign, superimposed (i.e., added together feature by feature). If only (Corner $*$ Yield sign) was added into the composite trace, Corner retrieves only Yield sign. When, in the misled condition, the retrieved (superimposed) item is matched to the alternatives Yield sign and Stop sign, it matches both, and hence the choice will sometimes favor one and sometimes the other. In the control condition, since only Yield sign is retrieved, the Stop sign alternative does

not match. In neither the misled nor the control condition does the retrieved item match the No-parking sign. Hence, the model produces the data (see Metcalfe, 1990).

MINERVA

In MINERVA (Hintzman, 1986, 1987) each event vector is stored as a separate trace in secondary memory (SM). The retrieval cue resonates with all SM traces to produce a composite "echo" vector. All traces (including the missing parts that are not given in the cue) are activated, A, as a function of their similarity, S, to the cue, where S is

$$S_i = \sum_{j=1}^{N} C_j T_{ij} / N_r \qquad (4)$$

C_j is the value of feature j in the cue, T_{ij} is the value of feature j in trace i, and N_r is the number of features that are non-zero in both the trace and probe. Activation A is

$$A_i = S_i^3 \qquad (5)$$

Each trace vector is weighted by its activation value and added into the composite echo vector. The relevant part of the composite echo is matched to the alternatives presented at test and the best match is chosen.

The parts of an event are concatenated into a single vector such that in the misled condition the traces include {Corner, yield sign}, {Corner, stop sign}, and so forth. The cue {Corner, ???} produces a high S and A for both traces containing Corner. The content echo vector consists of the superposition of the stop and the yield sign traces. Therefore, when these alternatives are pitted against one another in the test, MINERVA will waver. However, in the control condition it will unambivalently choose the yield sign. Regardless of many differences between the models, MINERVA produces the distortions in these data for the same reason as does CHARM.

The Principle of Superposition (POS)

Both models produce the correct results because both retrieve a composite representation—a superposition of the relevant traces. It has been thought since Galton (1878) invented composite photography that superposition may be important in human memory. Prototype extraction, generalization, and interference all depend on the principle of superposition. Superposition gains further plausibility as a fundamental principle because, at the level of the brain, different memories reuse the same neurons.

Caveats

Despite its importance, the POS cannot be the full story. People can sometimes report the source of memories (Lindsay and Johnson, 1989). In addition, there is less generalization and interference than the POS, taken alone, would suggest. Both CHARM and MINERVA include *binding* constructs (convolution in CHARM, the cubing operation in MINERVA) that individualize events, keeping them from indiscriminately smearing into one another. As Mesulam noted (Saltus, 1994), "There are plenty of reasons to expect distortion. As you are talking to me, you are using the same brain cells that contain all the knowledge you've acquired in your life, and you're writing these new thoughts on top of that. What's mysterious and astonishing is how the brain retrieves a particular memory from the jumble" (p. 25).

Road Map: Connectionist Psychology
Background: Associative Networks
Related Reading: Classical Learning Theory and Neural Networks; Lesioned Attractor Networks as Models of Neuropsychological Deficits

References

Galton, F., 1878, Composite portraits, *Nature*, 18:97–100.
Hintzman, D. L., 1986, "Schema abstraction" in a multiple-trace memory model, *Psychol. Rev.*, 93:411–428. ◆
Hintzman, D. L., 1987, Recognition and recall in MINERVA2: Analysis of the "recognition-failure" paradigm, in *Modeling Cognition* (P. Morris, Ed.), New York: Wiley, pp. 215–229.
Lindsay, D. S., and Johnson, M. K., 1989, The eyewitness suggestibility effect and memory for source, *Memory and Cognition*, 17:349–358.
Loftus, E. F., 1975, Leading questions and the eyewitness report, *Cognit. Psychol.*, 7:560–572.
Loftus, E. F., 1979, *Eyewitness Testimony*, Cambridge, MA: Harvard University Press. ◆
McCloskey, M., and Zaragoza, M, 1985, Misleading postevent information and memory for events: Arguments and evidence against memory impairment hypotheses, *J. Exp. Psychol. Gen.*, 114:1–16.
Metcalfe, J., 1990, A composite holographic associative recall model (CHARM) and blended memories in eyewitness testimony, *J. Exp. Psychol. Gen.*, 119:145–160. ◆
Metcalfe, J., 1991, Recognition failure and the composite memory trace in CHARM, *Psychol. Rev.*, 98:529–553.
Metcalfe, J., 1993, Novelty monitoring, metacognition, and control in a composite holographic associative recall model: Implications for Korsakoff amnesia, *Psychol. Rev.*, 100:3–22. ◆
Saltus, R., 1994, Memories are made of this: Facts, fantasies, fiction, *Boston Globe*, May 20, pp. 25, 28.

Distributed Artificial Intelligence

Edmund H. Durfee

Introduction

Natural systems are based on parallel, distributed processing at many different levels. At the neural level, interconnected and concurrently acting neurons propagate signals among themselves such that coherent behavior emerges from their joint activity. Within the brain, different neural subsystems combine and exchange signals to work together to yield an overall intelligent nervous system. And beyond the boundaries of a single intelligent entity are other such entities, which together form societies that achieve more than the individual entities can. Thus, what constitutes an "individual" can be highly subjective: An individual to one researcher can, to another, be a complex distributed system composed of finer-grained agents.

In this context, an agent's "granularity" corresponds to the amount of processing it does between interactions with others. Research in brain theory and neural networks has concentrated, not surprisingly, on the finer-grained levels of intelligence evidenced in the brain, drawing on neurophysiology and psychology as sources of inspiration when trying to build computational models of such natural systems. Distributed artificial intelligence (DAI) has instead focused on coarser-grained levels of individuality and interaction, where the goal is to draw upon sociological, political, and economic insights to develop what might be more closely described as computational models of societies (Gasser, 1991).

But the roots of DAI are actually not far from brain-theoretic and neural network concerns. Some of the early research in the DAI field emerged from studies of individual intelligent systems and is more closely aligned, metaphorically speaking, with tightly coupled subsystems in the mind. In this article, we will begin with that type of metaphor, as embodied in blackboard architectures. From there, we will move on to a brief overview of current DAI research, which has gone beyond the tightly coupled and cooperative metaphors and is now exploring issues in the motivations of individuals for forming societies and the processes involved in societal formation. We conclude this article with a brief summary of the similarities and differences between DAI and the fields of brain theory and neural networks, highlighting some potentially important areas for cross-fertilization between them.

Blackboard Architectures

The blackboard architecture (Nii, 1989) brought together many of the new, popular ideas that were taking root in the 1970s, including software engineering principles of modularity, AI ideas of pattern-directed invocation (where a procedure is not directly invoked by another, but instead invokes itself in response to a particular input pattern), and hardware ideas of shared-memory multiprocessors. At the heart of the architecture is the metaphor of a blackboard, with which experts communicate with one another by posting hypotheses representing tentative partial solutions to a collective problem (such as constructing a plan, interpreting sensor data, or designing a complex artifact). By building on each other's hypotheses, the experts extend and enlarge upon the scope and completeness of the solutions being constructed. For the original designers of the blackboard architecture, this metaphor served to externalize intuitions about the interpretation processes internal to the mind, where multiple specialized processes would cooperate and compete to generate partial interpretations that would reinforce and refute each other until a winning interpretation would emerge. Not surprisingly, researchers in brain theory had been following similar intuitions in developing concepts such as *schema theory* (Arbib, 1989).

In computational terms, the blackboard is shared memory, and the experts are modular pieces of software, called *knowledge sources* (KSs) in blackboard architecture terminology (Figure 1). Each KS is an expert on processing hypotheses that satisfy a particular pattern, and when the KS sees a hypothesis matching its pattern, the KS invokes itself, does some amount of processing, and then typically posts one or more new hypotheses on the blackboard, which will then cause another KS to be invoked, and the chain goes on until either no KSs can be invoked (at which point the system is said to be quiescent) or a suitable solution has been found. Detecting and responding to a suitable solution can be the task of a dedicated KS that compares hypothetical solutions against thresholds for completeness and confidence, or it can itself be a distributed process leading to KSs hypothesizing alternative competing responses to the situation. To streamline the pattern-matching process, the blackboard is usually partitioned into several levels, such

Figure 1. Example of blackboard architecture. Knowledge sources (KSs) post hypotheses (tentative partial solutions) to the data blackboard, which trigger goal processing to generate possible goals to extend or refine the hypotheses. Each goal is rated according to the preference information. For each combination of a goal to achieve and specific hypotheses to use in achieving it, the goal processor forms a KS instantiation (KSI). KSIs are stored on a prioritized queue, ranked according to goal preferences and quality of hypotheses and knowledge, from which one is selected and the appropriate KS is executed.

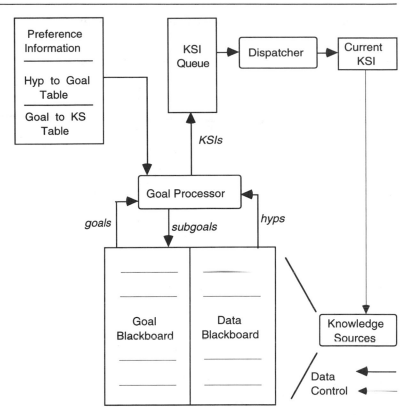

that KSs read from and write to specific levels, saving them from having to "watch" (match against) the entire blackboard.

The blackboard architecture thus has the advantage of providing an asynchronous communication link (the blackboard) among independent software modules (KSs). Expertise can be added to the overall system incrementally by introducing more and more KSs. Because KSs are invoked based on patterns of entries on the blackboard, KSs never have to be aware of other KSs.

For reasons initially stemming from the limited computer systems of the mid-1970s, however, the blackboard architecture in practice was more complicated than the metaphor would lead one to believe. Most important, because blackboard systems have typically been implemented on serial hardware, the blackboard architecture concept has had to be enlarged to include the ability to control which KS would be given permission to use the processor at any given time. A range of blackboard control paradigms have been investigated; most share the idea of having a prioritized agenda of potential KS executions, each rated by how well the KS expects to perform on the current data. More sophisticated control approaches, moreover, have the agenda biased toward pursuing particular problem-solving goals (Figure 1).

Blackboard systems have been applied to a variety of problems, ranging from construction site layout to protein interpretation, from intensive-care monitoring to errand planning. Probably the most memorable application, however, is the problem for which the blackboard architecture was initially designed: speech understanding. The Hearsay-II system (Erman et al., 1980) employed a blackboard architecture, incorporating a variety of KSs. Some were expert at forming hypotheses about syllables from sounds; others could hypothesize words from syllables; and others, phrases from words. A complete phrase that adequately accounted for the input sounds would

constitute a solution to the speech understanding task. Moreover, expectations for subsequent words in phrases could help guide the KSs at the phoneme and syllable levels to process the most relevant data sooner. Thus, Hearsay-II processed speech signals in a manner that was opportunistic (taking advantage of the best information it had at any given time) and island-driven (working outward from islands of reliability such as subphrases). While Hearsay-II's speech understanding capabilities were not always state-of-the-practice, its groundbreaking architecture did provide a degree of flexibility unmatched in its day.

Since the days of Hearsay-II, the blackboard architecture has continued to evolve while still maintaining, at its heart, the idea of cooperating KSs interacting through a blackboard. With strides in computer hardware and software technology, implementations of blackboard systems where KSs run in parallel are a current topic of investigation, as are issues in real-time computation and integrating highly diverse programming paradigms into a single blackboard system (Jagannathan, Dodhiawala, and Baum, 1989). However, even before these strides in technology, researchers had worked to achieve parallel activity in blackboard systems by employing multiple, cooperating blackboard architectures, which interact through a shared, global blackboard or through communication hardware. The architecture shown in Figure 1, from the Distributed Vehicle Monitoring Testbed (Bond and Gasser, 1988:268–284), illustrates a blackboard architecture with sophisticated goal processing, scheduling, and a communication interface.

Cooperative Problem-Solving Systems

The initial forays into viewing problem solving as a collective activity among experts triggered increasing amounts of work in this field. The principal line of research involved maintaining the view of a system as being composed of a number of co-operating experts, but dropping the assumption that these experts are able to read from and write to a shared-memory blackboard. Frameworks, such as Actors (Hewitt, 1991), were developed to account for the semantics of message passing among loosely coupled agents, and the challenge of using slow and unreliable communication channels to support cooperative activities came to the fore. This view—of agents being loosely coupled through communication channels—has been the dominant perspective of DAI since.

Cooperative distributed problem solving (Bond and Gasser, 1988; Durfee, Lesser, and Corkill, 1989) took two major forms. One of these forms was born directly out of the Hearsay-II work, in the form of a distributed Hearsay-II system. Using a number of Hearsay-II blackboard systems, this research explored how these systems could work together to identify a spoken phrase when each system received some portion of that phrase. This led to research on the Functionally-Accurate/Cooperative (FAC) paradigm (Corkill and Lesser, 1983), where problem solvers would individually solve their local subproblems and share their partial results so that, over time, increasingly complete results would be formed and the system would solve the entire problem (one or more agents would construct and share a hypothesis that satisfied the solution criteria). Of course, the wholesale exchange of all partial results can bog a system down quickly, and so substantial effort went into developing techniques by which cooperating agents could model each other to be smart about which results to share. These models included organizational structures (Corkill and Lesser, 1983) and partial global plans (Durfee and Lesser, 1991).

A complementary perspective on cooperative problem solving viewed the process as sharing tasks rather than results. As embodied in the Contract Net protocol (Bond and Gasser, 1988:357–366), the task-sharing perspective saw the coordination problem as associating tasks to be done with the right agents to do them. In Contract Net, for example, an agent with a large task to do would decompose it into smaller tasks, and then attempt to contract these out to the most suitable agents by announcing each subtask, collecting bids, and awarding the subtask. Because agents choose to bid on the tasks they receive, the assignment of tasks to agents involves mutual selection. Moreover, task sharing can be coupled with result sharing, such that organizational roles could be contracted out, FAC would follow, and then tasks to implement the solution could be contracted out.

Subsequent research in cooperative distributed problem solving has served to extend and refine the basic paradigms of task and result sharing, involving, among other things, the introduction of planning to the process, and the formulation of negotiation strategies for reaching a compromise between what different agents want that maximizes the overall system performance. Note, however, that in this discussion the emphasis has been on the overall system rather than on an individual. The assumption is that agents in these systems are built with an implicit goal of doing whatever they need to improve the performance of the entire system. While this is often reasonable when building real systems, since we can design such agents, it does neglect the issue of how to get individual agents, each with its own selfish goals, to solve problems cooperatively. This issue has more recently been the focus of a variety of research activities.

Socially Adept Individual Agents

Cooperative problem solving has been shown to be generally beneficial from a system-wide perspective, but many DAI researchers are concerned with how such cooperation might emerge when an agent can only see benefits to itself. In other words, what is the knowledge and reasoning that agents employ in making smart decisions about when to work together? Once they have adopted the goal of working together, then they can employ cooperative problem-solving techniques to actually accomplish their joint activities.

One of the most fruitful fields for insights when investigating how selfish agents can still work together for their mutual benefit is economics. Capitalist markets are examples of how selfish, profit-maximizing entities can still engage in transactions to their mutual benefit, and where the whole system somehow performs well through the local decisions of each of its many constituent parts (in fact, central control over markets, such as was formerly practiced in some Eastern European countries, has generally resulted in poor performance).

As an example of a situation where cooperation is to the mutual benefit of selfish agents, consider the well-known Prisoners' Dilemma problem, captured in the payoff matrix shown in Figure 2. In the matrix, the joint actions of the agents lead to an entry in the matrix with a payoff for each agent (agent I and agent II). For example, in Figure 2, if I takes action C and II takes action D, then I gets a payoff of 0 and II gets 5. Think of the payoffs as being years off a sentence for the player, and each has been given the choice of either cooperating with the other prisoner by not squealing (C), or of squealing on the other prisoner (D).

Each player wants to maximize his payoff and, in this symmetric game, will recognize that, whatever the other player does, he is better off with D than C. The paradox of this game is that, from a purely selfish viewpoint, the players will both choose D and both will receive a payoff of 1. Had they both cooperated, they could have each received 3. How can cooperation emerge given selfish agents?

There are several answers to this question. Axelrod (1984), for example, mapped the Prisoners' Dilemma game to a variety of real-world situations where cooperation emerged, such as across trenches in World War I. He used computer simulations to show that, given a population of strategies for playing this game considering previous encounters, a strategy where agents are cooperative initially (take action C in the first encounter with a particular opponent) but would respond "tit for tat" (take the action the opponent took the last time they met) was the fittest strategy in the sense that if strategies reproduce from generation to generation based on their relative payoffs against the entire population, then over time "tit for tat" takes over the population. Thus, eventually the population of agents will consistently cooperate.

But given the desire to avoid the learning curve required by evolution, there have been other formulations in DAI that lead to understanding how cooperation can emerge even in such situations. For example, if the agents see each other as rational, then they can make a deal allowing them to cooperate (Bond and Gasser, 1988:227–234). Alternatively, they can apply metagame techniques to evaluate the performance of alternative

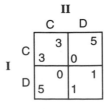

Figure 2. Prisoners' Dilemma game matrix.

strategies to select the cooperative strategy from a purely selfish view (Gmytrasiewicz, Durfee, and Wehe, 1991), leading to self-organization for their mutual benefit.

Organization theory (Bond and Gasser, 1988:151–158) has also looked at the problem of how organizations form and why individuals are willing to become parts of an organization. In joining an organization, an individual forgoes some of his free-dom because he has now committed to performing some set of actions for others in the organization. He has also accepted dependencies on others. The positive side of the resulting web of commitments (Gasser, 1991) is that, when the agents abide by their responsibilities and while the circumstances under which the organization was formed hold, the payoff to mem-bers of the successful organization is higher than those members could have gained individually. So organization, under the right circumstances, is the rational thing to do. The remain-ing challenge, then, is in developing computational mechanisms by which individuals can detect that the circumstances are right for organizing and for performing the organizational self-design.

Organizational self-design approaches come in both top-down and bottom-up varieties. In a top-down approach, one (or a few) agent(s) monitor the global performance of the orga-nization and detect when a particular organizational structure is needed. Most typically, roles in the new organization are assigned (often through a contracting style of protocol) and the new organization is adopted (Bond and Gasser, 1988:102–105; Corkill and Lesser, 1983). Alternatively, in a bottom-up ap-proach, an agent monitors its own performance, decides when a change (such as to the tasks it itself performs) is in order, and unilaterally makes that change. This in turn could cause other agents to change what they are doing, and reorganization pro-pagates throughout the network (Ishida, Gasser, and Yokoo, 1992). However, often in order to know whether it is behaving inappropriately, an agent needs more global knowledge of how it is performing in relation to others.

Discussion

Given the information above, and the larger context in which this article is written, the remaining question to address is how DAI relates to brain theory and neural networks (BTNN). We have noted that the roots of DAI, in the more tightly coupled mechanisms of the blackboard architecture, are lodged in bio-logical and social metaphors, and thus share many of the same intuitions as schema theory. However, the more recent trends in DAI toward increasingly loosely coupled and self-interested agents seems to be drifting farther from BTNN concerns.

DAI and BTNN differ in many respects. One respect is clearly the granularity of computation relative to communica-tion. DAI generally assumes that communication is time-consuming, error-prone, and costly. Thus, communication de-cisions are made judiciously, and messages among entities in DAI systems are at the symbol (rather than the signal) level, encoding much richer semantic content. Because communica-tion is at such a premium, and because envisioning the impact of a message requires a model of the hearer of that message, agents in DAI systems usually have explicit models of other agents, including their interests, abilities, and expectations. De-cision making in DAI agents is thus a complex process of map-ping potential actions (including communication acts) into explicit models of the anticipated activities of others, leading to a wide range of behaviors. Often, to anticipate the actions of another, a DAI agent will execute its inferencing processes on its model of that other agent, drawing conclusions on what the other agent could be thinking or doing by "putting itself in the other's shoes."

In BTNN, this implies that an agent should be able to ignore its current state and to instead "think" as if it were another agent. Because such projection requires an agent to be able to disconnect from its own reality and superimpose that assumed of another, this implies that models of the mind must provide a higher-level override capability to control the processing of the parallel brain regions and schemas. In this way, the same ma-chinery that an agent uses to control its own actions can be used to predict those of others.

At the level of neural networks, individual units are generally simpler, signal-processing elements, and the complexities arise from the sheer number of interconnections and the ways that those interconnections evolve over time. In DAI, the primary concern has been to endow agents with knowledge about pro-tocols, conventions, common goals, and so on, right from the start so that they can immediately interact intelligently; interac-tion is too expensive and slow to depend on "learning" how to interact. In many approaches to neural networks, the focus is precisely on learning because interaction is cheap and fast in the tightly coupled, signal-propagating architecture assumed. Thus, in neural networks, important performance criteria in-clude trainability, convergence, and robustness. DAI shares the robustness criterion, but rather than the capability to learn, it focuses on performance in terms of correctly and quickly co-ordinating activity to work as an effective team from the outset, making maximum use of resources at all times.

For example, having mobile robots learn to coordinate their movements by allowing them to learn from colliding is gener-ally infeasible, since collisions can lead to disablement of the robots. In a DAI approach (Durfee and Montgomery, 1991), the robots each plan their behaviors and then engage in a dialog to efficiently isolate and resolve conflicting actions. Following a predominant paradigm in AI, the robots essen-tially engage in a distributed search through the space of joint behaviors, using a hierarchical representation to focus their search. Through the distributed hierarchical search, the agents can balance the costs of coordination with its benefits, to attain an appropriate level of coordination. As problems scale up, moreover, the agents can employ abstractions to represent teams of agents as single entities. Mobile robots performing deliveries, for example, are clustered into geographic teams such that a robot only models the other members of its team and coordinates with other teams through the decisions of team leaders (Montgomery and Durfee, 1993). Properties of the task and of the agents, known ahead of time by the agents, dictate appropriate decomposition and task-abstraction strate-gies that, in the best case, can reduce the time to solution from exponential in the single agent case to logarithmic in the multiagent case (Montgomery and Durfee, 1993).

Yet, despite their differences, DAI and BTNN share some critical similarities. One of these is an emphasis on emerging intelligence—of the whole being more than the sum of its parts. From relatively simple neural units, complex patterns of activity can arise in neural networks; from cooperation and competition among schemas, intelligent behavior results; from rational choices among individuals, societies and civilizations (to anthropomorphize) emerge. As was alluded to at the outset of this article, all systems are distributed if you look closely enough, and it is the fact that the collection is more than the sum of its parts that allows us to call it a system rather than a collection of component parts. Whether simple or complex, communicating signals or symbols, distributed systems are a ubiquitous framework encompassing all of these studies. As such, these studies have common ground for sharing ideas and insights.

And as a specific example of the opportunities in a cross-disciplinary study, a key concern in DAI, as in BTNN, is in how to propagate global feedback such that individual entities can modify their behavior correctly. DAI suffers from the same credit/blame assignment problems as in neural networks and schema systems. If the system as a whole performs well or poorly, which entities, and which interactions among entities, were responsible? Propagation algorithms that have been developed for neural networks, analysis tools for blackboard and schema systems, and feedback loops that support reorganization in DAI systems raise many common concerns. Similarly, DAI, brain theory, and neural networks are all concerned with timely and appropriate communication among computational units; the thresholding computations used in neural networks and the cooperative/competitive links among brain regions have analogs in DAI systems that decide when a result is good enough to share and how much to believe results received from others. Such analogies are, to date, not firmly understood, and represent opportunities for cross-fertilization.

Acknowledgments. This research was sponsored in part by the NSF under grants IRI–9015423, IRI–9010645, and IRI–9158473; and by ARPA under contract DAAE–07–92–C–R012.

Road Map: Artificial Intelligence and Neural Networks
Related Reading: Planning, Connectionist; Schema Theory

References

Arbib, M. A., 1989, *The Metaphorical Brain 2*, New York: Wiley-Interscience.

Axelrod, R., 1984, *The Evolution of Cooperation*, New York: Basic Books.

Bond, A. H., and Gasser, L., Eds., 1988, *Readings in Distributed Artificial Intelligence*, San Mateo, CA: Morgan Kaufmann.

Corkill, D. D., and Lesser, V. R., 1983, The use of meta-level control for coordination in a distributed problem-solving network, in *Proceedings of the Eighth International Joint Conference on Artificial Intelligence*, Karlsruhe, Germany, pp. 748–756.

Durfee, E. H., and Lesser, V. R., 1991, Partial global planning: A coordination framework for distributed hypothesis formation, *IEEE Trans. Syst. Man Cybern.*, 21:1167–1183.

Durfee, E. H., Lesser, V. R., and Corkill, D. D., 1989, Cooperative distributed problem solving, in *The Handbook of Artificial Intelligence*, vol. IV (A. Barr, P. Cohen, and E. Feigenbaum, Eds.), Reading, MA: Addison-Wesley, pp. 83–137.

Durfee, E. H., and Montgomery, T. A., 1991, Coordination as distributed search in a hierarchical behavior space, *IEEE Trans. Syst. Man Cybern.*, 21:1363–1378.

Erman, L. D., Hayes-Roth, F., Lesser, V. R., and Reddy, R., 1980, The Hearsay-II speech-understanding system: Integrating knowledge to resolve uncertainty, *Comput. Surv.*, 12:213–253.

Gasser, L., 1991, Social conceptions of knowledge and action: DAI foundations and open systems semantics, *Artif. Intell.*, 47:107–138.

Gmytrasiewicz, P. J., Durfee, E. H., and Wehe, D. K., 1991, A decision-theoretic approach to coordinating multiagent interactions, in *Proceedings of the Twelfth International Joint Conference on Artificial Intelligence*, Sydney, Australia, pp. 62–68.

Hewitt, C., 1991, Open information systems semantics for distributed artificial intelligence, *Artif. Intell.*, 47:79–106.

Ishida, T., Gasser, L., and Yokoo, M., 1992, Organization self-design of distributed production systems, *IEEE Trans. Knowl. Data Eng.*, 4:123–134.

Jagannathan, V., Dodhiawala, R., and Baum, L., Eds., 1989, *Blackboard Architectures and Applications*, San Diego, CA: Academic Press.

Montgomery, T. A., and Durfee, E. H., 1993, Search reduction in hierarchical distributed problem solving, *Group Decision and Negotiation*, 2:301–317.

Nii, H. P., 1989, Blackboard systems, in *The Handbook of Artificial Intelligence*, vol. IV (A. Barr, P. Cohen, and E. Feigenbaum, Eds.), Reading, MA: Addison-Wesley, pp. 1–82.

Dynamic Clamp: Computer-Neural Hybrids

M. B. O'Neil, L. F. Abbott, A. A. Sharp, and E. Marder

Introduction

What role does each ionic current play in controlling the intrinsic properties of a neuron? What role does each neuron and synapse play in governing the output of a circuit? In the case of a single neuron with many voltage and time-dependent conductances, it is often difficult to predict how behavior will be influenced by any one conductance. In networks in which the individual neurons differ in their response to synaptic inputs and in which synaptic connections vary in strength and time course, it is even more difficult to assess the functional role of a given neuron or synapse in shaping network function (Marder and Selverston, 1992).

These are questions that often lead experimental neurobiologists to simulate the relevant neuron or network (Selverston, 1993; Koch and Segev, 1989). Conventional simulation techniques allow the investigator to construct models of neurons or networks that, hopefully, capture the essential features of the components. The modeler can then vary parameters at will to study their influence on the output of the system. Because the system is in a computer, the modeler can display all the variables at once, a compelling advantage over the limited data obtainable in an experimental situation.

Experimentalists utilizing computer simulations to gain insight into the functioning of biological systems are faced with a sometimes insurmountable limitation of the simulation approach. An accurate computer simulation relies on measurements of all the parameters underlying the relevant biological processes in the system. Often, measurement or even identification of these parameters is difficult or impossible. These unknowns may not be critical to the problem being addressed, or may entirely invalidate any conclusions that might be drawn from the simulation.

The voltage clamp is frequently used as an aid in characterizing the voltage and time dependence of currents in neurons. With the voltage clamp, the potential of the cell is maintained at a user-specified value. The voltage clamp injects the current necessary to clamp the cell's potential to this preset voltage. Since the voltage clamp controls the potential of the cell under study, the function of the cell is disrupted. Because of this, while voltage clamping is a powerful tool in conductance characterization, it fails to provide insight into the role of a conductance in shaping the behavior of neurons or networks.

Method

The *dynamic clamp* is a new technique that creates hybrid computer-biological circuits in which the computer simulates modeled currents and applies them, in real time, directly to a biological system (Marder et al., 1993; Renaud-LeMasson et al., 1993; Sharp et al., 1993a, b). The biological system continues to perform its own current computations, elements of which frequently confound the modeler. The dynamic clamp thus serves as an interface between theoretical and experimental work. The investigator may ask what a given current will do to the output of a cell by adding it to that cell. Circuits of arbitrary connectivity can be formed allowing the investigator to determine the role of each synapse in circuit function.

The dynamic clamp implements conductance simulations by determining and controlling current injection into a neuron through an intracellular electrode. The appropriate current is determined by the dynamic clamp software based on the membrane potential of the cell and user-specified properties of the conductance to be added (Sharp et al., 1993a, b). The injected current (I) is given by $I = gm^p h^q(E_r - V)$, where g is the maximal conductance, p and q are integers, E_r is the reversal potential of the current, V is the membrane potential, and m and h are activation and inactivation variables described by the usual Hodgkin and Huxley (1952) equations. In this paradigm, the experimenter has complete control over the activation parameters, time constants, maximal conductance, and reversal potential of the added conductance.

A schematic of a commonly used hardware configuration for an experiment using the dynamic clamp to add a conductance to a neuron is shown in Figure 1A. The dynamic clamp controls the current injected into the neuron through the use of an amplifier operating in discontinuous current clamp mode (DCC). In DCC, the amplifier employs a high-speed switching circuit so that a single electrode can alternately record the membrane potential of a neuron and pass current into it. Alternatively, two electrodes can be used, one to record membrane potential, the other to pass current. The computer running the dynamic clamp software communicates with and controls the amplifier

through computer interface hardware consisting of multichannel Analog to Digital (A/D) and Digital to Analog (D/A) converters as well as accurate timers. Once the experimenter has entered the description of the conductance using the dynamic clamp software, the real-time conductance simulation is performed via iterations of the following loop: The current clamp amplifier samples the membrane potential, V_m, of the neuron through a microelectrode inserted into the neuron. The dynamic clamp reads V_m from the current clamp amplifier via an A/D converter and determines the time from a timer. Using this information, the conductance description specified by the user and the elapsed time since the last voltage was read, the appropriate membrane current I_e is computed. The system then uses a D/A converter to convert the calculated value for the membrane current I_e to an analog voltage V_e and sends it to the amplifier, which injects the current I_e proportional to V_e into the neuron. This loop repeats until the experimenter terminates the conductance simulation.

The dynamic clamp software allows the experimenter to add multiple conductances to a neuron. That is, the computer is able to simulate and add to the neuron the composite output of multiple conductance simulations, each with independent and user definable parameters. Additionally, the system employs two independent channels which allow the introduction of the same or different membrane conductances into two neurons.

The dynamic clamp can construct artificial chemical and/or electrical synapses between two neurons. Figure 1B shows a schematic of a synapse hardware configuration using two amplifiers both operating in DCC. Artificial chemical synapses are simulated by using the membrane potential of the presynaptic neuron to control the conductance of the postsynaptic neuron. The synaptic current is given by $I_{syn} = gs(E_{syn} - V_{post})$ where E_{syn} is the reversal potential for the artificial synapse in the postsynaptic neuron and V_{post} is the membrane potential of the

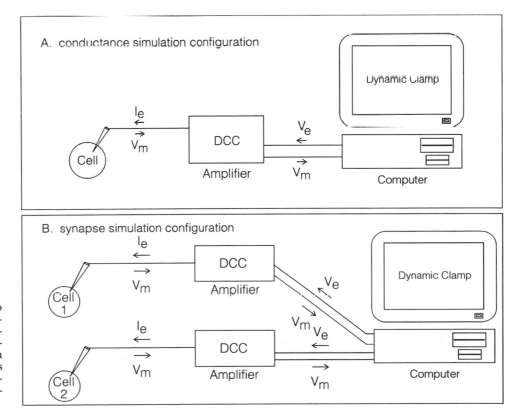

Figure 1. Schematic of dynamic clamp hardware configurations. *A*, Conductance simulation configuration. *B*, Synapse simulation configuration. The dynamic clamp will work equally well with any current clamp method, as long as the method provides accurate measurement of membrane potential and accurate current injection.

postsynaptic neuron. The synaptic activation variable *s* varies between 0 and 1 and is determined by the potential of the presynaptic neuron through a differential equation used to model the specific synapse being created. In this configuration, two independent A/D and D/A channels are used, one for the presynaptic neuron and one for the postsynaptic neuron. The user may designate which of the neurons is presynaptic and which is postsynaptic. It is also possible to construct reciprocal chemical synapses with this dynamic clamp configuration. The voltage and time dependence of these two synapses may be independently defined.

The dynamic clamp software can also simulate electrical synapses between neurons. An electrical synapse behaves like a resistor connecting the membrane potentials of the two neurons producing a synaptic current given by $I_{syn} = g(V_2 - V_1)$ in cell 1 where g is the conductance, V_1 is the potential of cell 1, and V_2 is the potential of cell 2. An equal and opposite current is injected into cell 2. While simulating synapses, the dynamic clamp uses calculation loops and current clamp amplifier control similar to those for the conductance simulation configuration described previously.

The dynamic clamp technique can be used to give insights into the role of a particular conductance in shaping electrical activity, even when the full biophysical details of the cell's behavior are not understood. This is illustrated by the example in Figure 2. The Anterior Burster (AB) neuron of the crustacean stomatogastric ganglion (STG) is an oscillator which is strongly activated by a number of modulators, including the peptide proctolin. When applied to an isolated AB neuron, proctolin increases the amplitude and frequency of the AB neuron's bursts (Hooper and Marder, 1987). The proctolin-activated conductance increases with depolarization (Golowasch and Marder, 1992). Figure 2 illustrates the importance of the voltage dependence of the proctolin conductance for the physiological activation of the AB neuron's membrane potential oscillations. When the activation curve of the simulated proctolin-activated current has a midpoint at −45 mV, the voltage dependence of the proctolin current amplifies the amplitude and frequency of the burst, without changing the baseline from which the oscillation starts. However, when the activation curve is shifted only 5 mV in the hyperpolarized direction, the proctolin current is "on" even at the most hyperpolarized levels of membrane potential. The net effect is similar to that produced by a non–voltage-dependent increase in conductance that depolarizes the oscillator, thus shifting the baseline membrane potential and decreasing the amplitude of the bursts (Figure 2). The sensitivity to the precise voltage dependence of the proctolin current seen in Figure 2 is surprising. Because we do not have detailed voltage-clamp measurements of the actual currents in the AB neuron, it would not have been possible to obtain these results by conventional simulation techniques.

The dynamic clamp can be used to form circuits from neurons in culture or in intact ganglia (Sharp et al., 1993a, 1993b). The dynamic clamp is being used to understand the role of synaptic

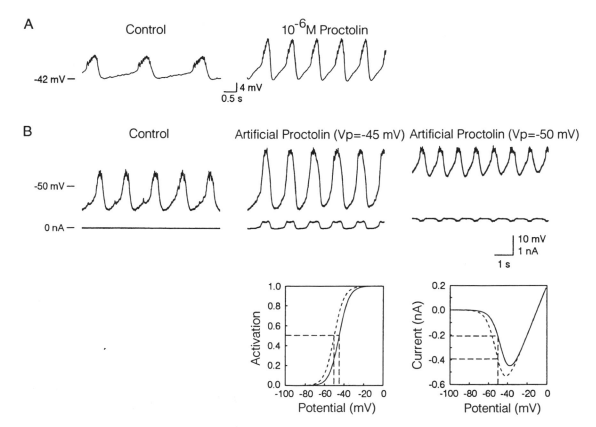

Figure 2. Using the dynamic clamp to understand the role of a voltage-dependent modulator current in oscillator function. *A*, Intracellular recordings from an isolated AB neuron in control saline and in the presence of proctolin. Proctolin application increases both the amplitude and frequency of the neuron's oscillation. Modified from Hooper and Marder (1987). *B*, Intracellular recordings from an AB neuron in the absence of a proctolin current, and with the dynamic clamp addition of the proctolin current. Bottom left: the activation curve shown by the solid line was used to give the recording labeled $Vp = -45$ mV, and the activation curve with the dashed line was used to give the recordings labeled $Vp = -50$ mV. Note the importance of the correct activation curve in properly simulating the effects of proctolin. Bottom right: current-voltage relationship of dynamic clamp proctolin current in voltage clamp.

release threshold, synaptic strength, and intrinsic membrane properties in the operation of rhythmic alternation in the reciprocally inhibitory HN interneurons in the leech heartbeat system (Ronald Calabrese, personal communication, 1994). In these experiments, the existing synaptic connections are first pharmacologically blocked and are replaced with defined synaptic connections using the dynamic clamp. This approach should allow the construction of interacting networks with both biological and computational components in numerous preparations. Hybrid circuits between a complete model neuron and a biological neuron have also been constructed (Renaud-LeMasson et al., 1993; Marder et al., 1993).

Discussion

The dynamic clamp method has several features which, depending on the circumstances, can either be significant limitations or can be exploited in novel ways. First, unlike real membrane channels, which are distributed over the surface of the cell, the conductances produced by the dynamic clamp are introduced at a single point source at the tip of the electrode. For neurons which can be modeled as a single compartment, this will reproduce the electrical effect of the distributed channels. For neurons with extensive dendritic structures, this either can be a limitation, or can be used as a way of studying the effect of channel distribution in a complex cable. Second, membrane Ca^{++} currents may often activate Ca^{++}-dependent conductances and other cellular processes that will not be replicated by the dynamic clamp addition of the current alone. Again, this will be a limitation for some investigations, or it can be exploited to determine the consequences of the current flowing through Ca^{++} channels in the absence of the other effects of intracellular Ca^{++} changes that normally accompany Ca^{++} currents.

In summary, because biological neurons often have surprising and unknown attributes not captured in conventional models, the dynamic clamp approach is likely to reveal interesting features of neural networks. The dynamic clamp serves as a new tool to aid in the understanding of the function of biological networks.

Requests for information concerning the dynamic clamp software should be addressed to Mr. Michael O'Neil at the address given in the Contributors list at the back of this volume.

Acknowledgments. The dynamic clamp was developed under the auspices of NSF 9009251, NSF IBN–9312975 to Brandeis University, and 1R43MH51988–01 to Dyna-Quest Technologies, Inc.

Road Map: Biological Neurons
Background: Ion Channels: Keys to Neuronal Specialization
Related Reading: Crustacean Stomatogastric System; Half-Center Oscillators Underlying Rhythmic Movements

References

Golowasch, J., and Marder, E., 1992, Proctolin activates an inward current whose voltage dependence is modified by extracellular Ca^{2+}, *J. Neurosci.*, 12:810–817.

Hodgkin, A. L., and Huxley, A. F., 1952, A quantitative description of membrane current and its application to conduction and excitation in nerve, *J. Physiol.*, 117:500–544.

Hooper, S. L., and Marder, E., 1987, Modulation of a central pattern generator by the peptide, proctolin, *J. Neurosci.*, 7:2097–2112.

Koch, C., and Segev, I., 1989, *Methods in Neuronal Modeling: From Synapses to Networks*, Cambridge, MA: MIT Press. ◆

Marder, E., Abbott, L. F., LeMasson, G., O'Neil, M. B., Renaud-LeMasson, S., and Sharp, A. A., in press, Biological simulators: Computer modification of neuronal conductances and formation of novel networks, in *Enabling Technologies for Cultured Neural Networks* (D. A. Stenger and T. M. McKenna, Eds.), Orlando, FL: Academic Press/Harcourt Brace Jovanovich.

Marder, E., and Selverston, A. I., 1992, Modeling the stomatogastric nervous system, in *Dynamic Biological Networks: The Stomatogastric Nervous System* (R. M. Harris-Warrick, E. Marder, A. I. Selverston, and M. Moulins, Eds.), Cambridge, MA: MIT Press, pp. 161–196. ◆

Renaud-LeMasson, S., LeMasson, G., Marder, E., and Abbott, L. F., 1993, Hybrid circuits of interacting computer model and biological neurons, in *Advances in Neural Information Processing Systems 5* (C. L. Giles, S. J. Hanson, and J. D. Cowan, Eds.), San Mateo, CA: Morgan Kaufmann, pp. 813–819.

Selverston, A. I., 1993, Modeling of neural circuits: What have we learned? *Annu. Rev. Neurosci.*, 16:531–546. ◆

Sharp, A. A., O'Neil, M. B., Abbott, L. F., and Marder, E., 1993a, The dynamic clamp: Computer-generated conductances in real neurons, *J. Neurophysiol.*, 69:992–995.

Sharp, A. A., O'Neil, M. B., Abbott, L. F., and Marder, E., 1993b, The dynamic clamp: Artificial conductances in biological neurons, *Trends Neurosci.*, 16:389–394.

Dynamic Link Architecture

Christoph von der Malsburg

Introduction

The field of neural computation rests on the belief that there is a set of basic structures and mechanisms that enable the brain to extract and process the regularities behind sensory and motor patterns. This may be called its cognitive architecture. According to classical neural networks, it rests on four central tenets: (a) Short-term memory is represented by the firing rate of neurons. A neuron stands for an elementary symbol (e.g., "triangle" or "red") that is active or dormant, depending on the neuron's state of activity. (b) Short-term memory is organized by the exchange of excitation and inhibition, arranged such as to favor stationary states. (c) Long-term memory resides in the strengths of connections between neurons. (d)

Long-term memory is organized by mechanisms of synaptic plasticity.

Classical neural networks have many strengths, but they also have some important weaknesses. Among them is their apparent inability to represent and process high-level cognitive symbols (Fodor and Pylyshyn, 1988). Another weakness is that, in spite of all their claim and promise to learn functional structures autonomously from examples, this program runs into a dilemma: either the designer of the network constructs a neural representation already well adapted to the specific problem at the outset (assigning appropriate symbolic meaning to neurons), or the system will run into a learning time problem, needing astronomical numbers of examples (Geman, Bienenstock, and Doursat, 1992).

The dilemma can be traced back to the inability of neural networks to dynamically form new symbols by binding together existing neural symbols (von der Malsburg, 1981). To give an example, assume a model visual system has four neurons, representing a triangle, a square, and the colors red and green, all neurons generalizing over position. When the system is presented with a red triangle and a green square, all four neurons go on, but in doing so they represent the situation ambiguously, failing to distinguish it from one with a green triangle and red square. Psychophysical experiments (Treisman, 1985) show such "conjunction errors" when the subject is given too little viewing time. Within neural networks, this situation can only be mended with the help of combination coding cells, which either have to be present "at birth" (needing the foresight of a designer) or have to be created by learning (leading to the learning-time problem).

Temporal Binding

The Dynamic Link Architecture (von der Malsburg, 1981) proposes to solve the binding problem by letting neural signals fluctuate in time (as indeed they do in higher levels of the brain) and by synchronizing those sets of neurons that are to be bound together into a higher-level symbol (Figure 1). This modification to the neural networks architecture raises a number of questions: How are signal correlations dynamically created? What are their dynamical consequences in the circuit? How does temporal binding help solving fundamental problems with neural networks? and, Is there experimental evidence for the existence and significance of signal correlations in the brain?

To start with the last question, signal correlations are very important in the brain, and there are experimental indications

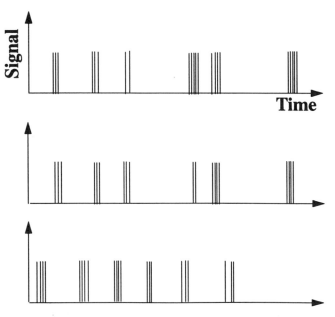

Figure 1. Temporal binding. Neurons can dynamically express grouping into composite structures by synchronizing their activity. In the situation shown, the middle neuron is bound to the upper neuron and not to the lower. If the middle neuron stood for the color red and the other two for objects, e.g., a triangle and a square, the redness would in the current situation apply to the upper object, not to the lower. At another time, the lower neuron's object might be red, which would be expressed by spike synchrony between the middle and bottom neurons.

(Sillito et al., 1994, and work reviewed in Engel et al., 1992) that some of the observed correlations encode the binding of features into more comprehensive wholes; in the examples studied, visual line elements are bound into longer edges. Some of the correlations were based on signals in the broad 40-Hz range (for review, see Engel et al., 1992; SYNCHRONIZATION OF NEURONAL RESPONSES AS A PUTATIVE BINDING MECHANISM).

How are appropriate correlations produced? Some temporal correlations are already contained in the sensory signal. In the auditory modality, for instance, modulations in amplitude and frequency tend to be common to the harmonic components of one sound source. This effect has been exploited in work by von der Malsburg and Schneider (1986) to create a model of the cocktail party effect. More important, however, are cases in which correlations are created by connections within the circuit. In the binding example, signal correlations between the red-detector and the triangle-detector can be created with the help of the indirect connections that are present between those neurons when (and only when) they are triggered from the same locus in primary visual cortex or in thalamus. In Wang, Buhmann, and von der Malsburg (1990; see also OSCILLATORY ASSOCIATIVE MEMORIES), the connections inherent in the associative memory model were used to correlate the signals of neurons within memory traces, permitting several of them to be activated by the same stimulus and still be kept separate (as when we have a mixture of several known odors in our nose). Another model (von der Malsburg and Buhmann, 1992) uses regular connections in a sensory field to correlate all neurons hit by one object, thus achieving figure-ground. In a model for object vision (Hummel and Biederman, 1992; see OBJECT RECOGNITION), structured connections (*fast enabling links*) are used to synchronize feature detector neurons, thus binding features within one compact object part (geon) and avoiding confusion between features in different geons. In Shastri and Ajjanagadde (1993; see STRUCTURED CONNECTIONIST MODELS), a logical reasoning system uses connections to temporally bind logical predicates.

What are the dynamical effects of signal correlations? We know that neurons are coincidence detectors. Signals arriving at a dendritic membrane out of sync by more than the membrane time constant cannot sum up to fire the neuron. Therefore, signal correlations have deep consequences for neural dynamics, and the common practice of ignoring them is simply a mistake. A model of invariant object recognition (von der Malsburg, 1988) demonstrates how appropriate signal correlations can be used to reliably distinguish between objects that are composed largely of the same features, although in a different spatial arrangement.

Probably the most important effect of temporal binding is on the efficiency of learning. The main problem for neural network learning arises from the drowning of significant feature constellations in input patterns by insignificant—that is, accidental—ones. All of these lead to synaptic strengthening that has to be undone in an extremely costly way by determining and exploiting the difference in frequency of occurrence between significant and accidental constellations. Since it is often possible to detect the significance of feature constellations on the basis of stored information and express it by temporal binding, this can be picked up by a refined long-term plasticity mechanism to speed up learning considerably. Such speedup has been demonstrated in a model (Konen and von der Malsburg, 1993) that could learn to recognize types of mirror symmetry in otherwise random pixel patterns after seeing single examples. (A conventional neural network version, reviewed in Konen and von der Malsburg, 1993, took dozens of thousands of examples to learn the same performance.)

Figure 2. Rapid reversible synaptic plasticity. The synaptic weight J_{ij} between neurons j and i can vary on a fast time scale. (a) If the neurons i and j fire synchronously, the dynamic weight J_{ij} is rapidly increased from a resting value to the maximal value set by the permanent synaptic strength, T_{ij}. (b) If both neurons are active but fire asynchronously, the synaptic weight is rapidly decreased to zero. Dynamic switching can take place within small fractions of a second. (c) When there is no more activity in one or both of the neurons, the dynamic weight slowly falls back to the resting value, with the time constant of short-term memory.

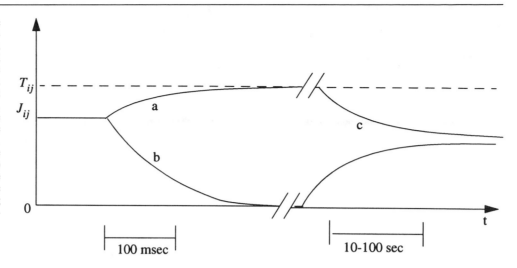

Rapid Reversible Synaptic Plasticity

The second important feature besides temporal signal binding distinguishing the Dynamic Link Architecture from conventional neural networks is rapid reversible synaptic plasticity (Figure 2). Although rapid synaptic changes have been observed experimentally (Zucker, 1989), the exact mode of control predicted here has yet to be demonstrated. The main effect of rapid synaptic plasticity is to switch off connections that do not fit the context and to stabilize connections corresponding to correct bindings. Rapid plasticity is important when the occurrence of signal correlations needs to be memorized over short periods of silence, as demonstrated in von der Malsburg and Schneider (1986), and in systems with dense permanent connections of which only a very small subsystem is required in a given situation. An example is position-invariant object recognition, where the invariant memory trace needs connections from all potential positions in primary visual cortex of the image of that object, and where many of these connections would create confusion for any given position. It has been demonstrated (Konen, Maurer, and von der Malsburg, 1994) that the irrelevant connections can be rapidly switched off by link dynamics.

Conclusion

Two features distinguish the Dynamic Link Architecture (von der Malsburg, 1981) from conventional neural networks: binding by signal correlations and rapid reversible synaptic plasticity. Some of its aspects have been demonstrated experimentally for the brain; others are predictions to be validated by future experiments. A large and rapidly growing number of models have demonstrated the power of this architecture, solving, for instance, problems like sensory segmentation, invariant pattern recognition, and reduction of learning time. It is obvious that dynamic links are required to build up large-scale neural systems which are composed of many modules. Such systems need many potential connections, but in any given context, they must be liberated from the confusing influence of temporarily irrelevant links.

Road Maps: Artificial Intelligence and Neural Networks; Self-Organization in Neural Networks

Related Reading: Binding in the Visual System; Compositionality in Neural Systems; Face Recognition; Gabor Wavelets for Statistical Pattern Recognition

References

Engel, A. K., König, P., Kreiter, A. K., Schillen, T. B., and Singer, W., 1992, Temporal coding in the visual cortex: New vistas on integration in the nervous system, *Trends Neurosci.*, 15:218–226. ◆

Fodor, J, and Pylyshyn, Z. W., 1988, Connectionism and cognitive architecture: A critical analysis, *Cognition*, 28:3–71.

Geman, S., Bienenstock, E., and Doursat, R., 1992, Neural networks and the bias/variance dilemma, *Neural Computat.*, 4:1–58.

Hummel, J. E., and Biederman, I., 1992, Dynamic binding in a neural network for shape recognition, *Psychol. Rev.*, 99:480–517.

Konen, W., Maurer, T., and von der Malsburg, C., 1994, A fast dynamic link matching algorithm for invariant pattern recognition, *Neural Netw.*, 7:1019–1030.

Konen, W., and von der Malsburg, C., 1993, Learning to generalize from single examples in the dynamic link architecture, *Neural Computat.* 5:719–735.

Shastri, L., and Ajjanagadde, V., 1993, From simple associations to systematic reasoning: A connectionist representation of rules, variables and dynamic bindings, *Behav. Brain Sci.*, 16:417–494.

Sillito, A. M., Jones, H. E., Gerstein, G. L., and West, D. C., 1994, Feature-linked synchronization of thalamic relay cell firing induced by feedback from the visual cortex, *Nature*, 369:479–482.

Treisman, A., 1985, Preattentive processing in vision, *Comput. Vis. Graph. Image Proc.*, 31:156–177.

von der Malsburg, C., 1981, The correlation theory of brain function, Internal Report 81–2, Göttingen, Germany: Max-Planck-Institut für Biophysikalische Chemie. Reprinted in *Models of Neural Networks* (K. Schulten and H.-J. van Hemmen, Eds.), 2nd ed., Springer, 1994. ◆

von der Malsburg, C., 1988, Pattern recognition by labeled graph matching, *Neural Netw.*, 1:141–148.

von der Malsburg, C., and Buhmann, J., 1992, Sensory segmentation with coupled neural oscillators, *Biol. Cybern.*, 67:233–242.

von der Malsburg, C., and Schneider, W., 1986, A neural cocktail-party processor, *Biol. Cybern.*, 54:29–40.

Wang, D. L., Buhmann, J., and von der Malsburg, C., 1990, Pattern segmentation in associative memory, *Neural Computat.*, 2:94–106.

Zucker, R. S., 1989, Short-term synaptic plasticity, *Annu. Rev. Neurosci.*, 12:13–31.

Dynamic Models of Neurophysiological Systems

Eberhard E. Fetz and Larry E. Shupe

Introduction

Dynamic recurrent network models can provide invaluable tools to help systems neurophysiologists understand the neural mechanisms mediating behavior. Biological experiments typically involve bits of the system: anatomical structures and their connections, effects of lesions on behavior, activity of single neurons in behaving animals. The missing element required to synthesize these pieces can be provided by neural network models: a method of generating working models of the complete system. New algorithms make it possible to derive networks that simulate dynamic sensorimotor behavior and incorporate anatomically appropriate connectivity. The resulting networks determine the remaining free parameters based on examples of the behavior itself.

Training procedures initially developed for feedforward networks have been extended to dynamic recurrent networks, which have three key properties (see RECURRENT NETWORKS: SUPERVISED LEARNING): First, the units are *dynamic*, meaning they can exhibit time-varying activity, which can represent the mean firing rates of single or multiple neurons, membrane potentials, or some relevant time-varying stimulus or motor parameter. Second, the networks can have *recurrent* connectivity, including feedback and cross-connections. Third, the network connections required to simulate a particular dynamic behavior can be derived from examples of the behavior by *gradient descent* methods such as backpropagated error correction. The resulting models provide complete neural network solutions of the behavior, insofar as they determine all the connections and activations of the units that simulate the behavior.

Neural networks that emulate particular dynamic behaviors basically transform spatiotemporal inputs into appropriate spatiotemporal outputs. These networks are usually composed of interconnected "sigmoidal" units, whose outputs are sigmoidal functions of their inputs; this mimics a biological neuron's property of saturating at maximal rates for large inputs, and decreasing to zero for low inputs. To train the network, the synaptic weights between units are initially assigned randomly, and the output response of the network is determined. The difference between network output patterns and the desired target output activations is the error. The backpropagation algorithm optimally modifies the weights to reduce this error. This weight change implements a gradient descent of the error as a function of the weight. The process of presenting input patterns and changing the weights to reduce the remaining error is iterated until the network converges on a solution with minimal error.

Applications

The applications for these dynamic recurrent networks fall into three general categories:

1. *Pattern recognition* applications involve identification of spatiotemporal input patterns into discrete categories. A set of input units receiving time-varying signals can represent a spatiotemporal pattern, and the output codes the categories.
2. *Pattern generation* networks produce temporal patterns in one or more output units, either autonomously or under the control of a gating input. These include oscillating networks (Williams and Zipser, 1989) and simulations of central pattern generators (Tsung, Cottrell, and Selverston, 1990; Rowat and Selverston, 1991).

3. *Pattern transformation* networks convert spatiotemporal input patterns into spatiotemporal outputs. Examples include simulations of the leech withdrawal reflex (Lockery and Sejnowski, 1992); step target tracking in the primate (Fetz and Shupe, 1990); the vestibulo-ocular reflex (Arnold and Robinson, 1991; Lisberger and Sejnowski, 1992; see also VESTIBULO-OCULAR REFLEX: PERFORMANCE AND PLASTICITY); and short-term memory tasks (Zipser, 1991). Recurrent networks can also simulate analytical transforms such as integration and differentiation of input signals (Munro, Shupe, and Fetz, 1994).

Oscillating Networks

Biological systems provide numerous examples of autonomously generated periodic motor activity: locomotion, mastication, respiration, etc. The neural circuitry underlying cyclic movements has been called a central pattern generator (CPG). Williams and Zipser (1989) trained dynamic networks to generate oscillatory activity with various frequencies. The smallest circuit that sustained quasi-sinusoidal oscillations consisted of two interconnected sigmoidal units.

Tsung, Cottrell, and Selverston (1990) trained a network with the connectivity and sign constraints of neurons in the lobster gastric mill circuit to simulate their oscillatory activity. This network replicated the correct phase relations of the biological interneurons. If its activity was perturbed, the network reverted to the original pattern, indicating that the weights found by the learning algorithm represented a strong limit cycle attractor. Rowat and Selverston (1991) developed algorithms to train oscillations in networks of units with biological properties like gap junctions and membrane currents, and trained networks with realistic constraints on connection weights to simulate known oscillatory activations. Simulations suggest that recurrent networks of sigmoidal units are quite robust in generating oscillatory activity, even to the point of meeting various constraints in phase and period. (But see CRUSTACEAN STOMATOGASTRIC SYSTEM for more subtle oscillator properties involving neuromodulation.)

Primate Target Tracking

We used dynamic networks to simulate the neural circuitry controlling forelimb muscles of the primate. In monkeys performing a step-tracking task, physiological experiments documented the discharge patterns and output connections of task-related neurons. Premotoneuronal (PreM) cells were identified by postspike facilitation of target muscle activity in spike-triggered averages of EMG. During alternating wrist movements, the response patterns of different PreM cells—corticomotoneuronal (CM), rubromotoneuronal (RM), and dorsal root ganglion afferents—as well as single motor units (MU) of agonist muscles, fall into specific classes (Fetz et al., 1989). All groups include cells that exhibit phasic-tonic, tonic, or phasic discharge, as well as cells with unique firing properties. Many MUs show decrementing discharge through the static hold period. Some RM cells fire during both flexion and extension, and some are unmodulated with the task.

To investigate the function of these diverse cells and to determine what other types of discharge patterns might be required to transform a step signal to the observed output of motoneurons, we derived dynamic networks that generated as outputs the average firing rates of motor units recorded in monkeys performing a step-tracking task. Changes in target position were represented by step inputs to the network and/or by brief transient bursts at the onset of target changes. The input signals were transformed to either MU output patterns (phasic-tonic, tonic, decrementing, and phasic flexors and extensors) by intervening hidden units consisting of interconnected excitatory and inhibitory units.

The activation patterns and connection matrix of units in such networks are illustrated elsewhere (Fetz and Shupe, 1990; Fetz, 1992 and 1993). In these simulations, the network solutions have several features which resemble biological situations but which were not explicitly incorporated:

1. Divergent connections of hidden units to different coactivated motor units are representative of divergent outputs of physiological PreM neurons (Fetz et al., 1989).
2. Many hidden units have discharge patterns resembling the outputs, but some have counterintuitive patterns that are also seen in biological neurons, e.g., bidirectional and sustained activity.
3. Different network simulations with the same architecture, but initialized with different weights often converged on different solutions, comparable to the diversity of neural relations seen in biological networks.

A useful heuristic feature of these networks is the ability to quickly probe their operation with manipulations (Fetz and Shupe, 1990; Fetz, 1993). The contributions of hidden units can be tested by making selective lesions and analyzing the behavior of the remaining network. The output effects of a given unit can also be tested by delivering a simulated stimulus and analyzing the propagated network response. Because of changing activation levels, the effect of a stimulus depends on the time it is delivered, as also observed in physiological experiments. These networks can also be trained to scale their responses—i.e., to generate output activation patterns proportional to the size of the input. Moreover, their ability to generalize across stimulus dimensions can be quickly tested by presenting different inputs.

To generate more realistic models of the primate motor system, the same approach has been used with networks incorporating additional biological features (Maier, Shupe, and Fetz, 1993): (1) the connectivity of specific neurons in motor cortex, red nucleus and spinal cord (interneurons, and motoneurons) and afferent fibers was included with appropriate relative conduction delays; (2) the activity of representative subsets of these additional elements, where known, was required to be part of the solution; and (3) in addition to the active target tracking task, the network was required to simulate reflex responses to peripheral perturbations of the limb. The resulting networks have the ability to generate both types of behaviors, and have more realistic properties. Although these networks are still highly abstracted, they reflect many of the essential features of the biological system in the monkey. Some complex activity patterns seen in PreM neurons of monkeys, such as bidirectional responses of RM cells, also appear in the networks. Even some apparently paradoxical relations seen in monkeys, such as cortical units that covary with muscles which they inhibit, appear in networks and make contributions that are understandable in terms of other units: their activity subtracts out inappropriate components of bidirectional activity patterns. Thus, network simulations have proven useful in elucidating the function of many puzzling features of biological networks, including some puzzling properties.

Short-Term Memory Tasks

Neural mechanisms of short-term memory have been investigated in many experiments by recording cortical cell activity in animals performing instructed delay tasks. A common type of instructed delay task requires remembering the value of a particular stimulus. Zipser (1991) trained recurrent networks to simulate short-term memory of an analog value during the delay; the resulting network implements a sample-and-hold function. The network has two inputs: an analog signal representing the stimulus value to be remembered, and a gate signal specifying the times to take samples. The network output is the value of the analog input at the time of the previous gate. During the delay between gate signals, the activity of many hidden units resembles the response patterns of cortical neurons recorded in monkeys performing comparable instructed delay tasks. The activity patterns of hidden units, like those of cortical neurons, fall into three main classes: sustained activation proportional to the remembered analog value, often with a decay or rise; transient modulation during the gate signal; and combinations of the two. The network simulations allow the function of the patterns observed in the animal to be interpreted in terms of its possible role in the memory task.

We investigated such short-term memory networks to further analyze their operation. To elucidate the underlying computational algorithm, we constrained units to have either excitatory or inhibitory output weights, and reduced the network to the minimal essential network. A larger network was initially trained, then reduced by (1) combining units with similar responses and connections into one equivalent unit and (2) eliminating units with negligible activation or weak connections, then (3) retraining the smaller networks to perform the same operation. A reduced network performing the sample-and-hold function is illustrated in Figure 1. It consists of three excitatory and one inhibitory unit. The two inputs are the sample gate signal (S) and the random analog variable (A); the output (O) is the value of A at the last sample gate. This reduced version reveals a computational algorithm that exploits the nonlinear sigmoidal input-output function of the units. The first excitatory unit (SA) carries a transient signal proportional to the value of A at the time of the gate. This signal is derived by clipping the sum of the analog and gating inputs with a nega-

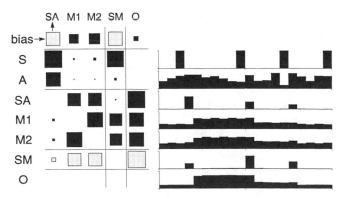

Figure 1. Reduced network performing a sample-and-hold function, simulating short-term memory. The units are indicated by abbreviations and their representative activation patterns, shown at right. The weights are indicated by squares (black = excitatory; grey = inhibitory) proportional to the connection from row unit to column unit (e.g., arrows). The two inputs are the sample signal (S) and a random analog value (A); the output (O) is the sustained value of the last sampled analog value.

tive bias, as shown by the input weights to SA in the first column. This input sample is then fed to two excitatory units ($M1$ and $M2$) that maintain their activity through reciprocal connections and also feed their summed activity to the output ($M1$ and $M2$ could also be replaced by a single self-connected M unit). The inhibitory unit (SM) carries a transient signal proportional to the previous value of A. Its value is derived from a clipped sum of the gate S and the previous values held in $M1$ and $M2$. As shown by its output weights, the function of SM is to subtract the previously held value from the integrating hidden units and from the output. This illustrates how the weights and activations of a complete network solution can reveal the underlying algorithm—in this case, an elegant use of nonlinearity and integration to yield the appropriate remembered value. It seems plausible that networks with more units implement a comparable algorithm in a distributed manner.

Neural Integration

In biological motor systems, neural integrators have been postulated to transform transient commands into sustained activity and to mediate the vestibulo-ocular reflex (VOR). Arnold and Robinson (1991) modeled the VOR integrator with a recurrent network whose connections resembled those of the vestibulo-ocular system. Two input signals represented the reciprocal responses of opposed vestibular afferents to head movement; these connected to four interneurons, which were interconnected to each other and to motoneurons. Since vestibular afferents carry tonic activity in the absence of head movement, the integrator had to be configured so as to integrate only deviations from baseline, but not the baseline activity itself. The authors used units with intrinsically sustained activity with decay and a nondifferentiable rectifying input-output characteristic. To train the networks, they tweaked individual weights and used the effect on the error to update the weights. Integration was performed through positive recurrent connections between the interneurons. Removing hidden units in a trained network reduced the time constant of integration, but the network could be readily retrained. The networks could also mimic more complex physiological responses, such as post-saccadic drift. Similarly Anastasio (see VESTIBULO-OCULAR REFLEX: PERFORMANCE AND PLASTICITY) trained dynamic networks of sigmoidal units with backpropagation to simulate velocity storage in the vestibulo-ocular system.

Lisberger and Sejnowski (1992) used dynamic networks to investigate mechanisms of learning in the vestibulo-ocular system. The network was constructed to include many anatomical and physiological constraints, including pathways through the cerebellar flocculus, with appropriate delays. The two inputs to the network—head velocity and target velocity—were converted to a single output: eye velocity. The network was initially trained to simulate three behaviors: smooth pursuit of a moving visual target; the VOR to head movement; and suppression of the VOR (when head and target move together). Then the network was required to change the gain of the VOR (as occurs after wearing magnifying or minifying goggles) and also to maintain accurate smooth-pursuit visual tracking. These requirements led to changes in the weights of connections at two specific sites: the vestibular input to the flocculus and to the brainstem neurons controlling oculomotoneurons. This study exemplifies the insights gained from a biologically constrained dynamic model that can incorporate the time course of neural activity observed under different behavioral conditions, and shows the power of such simulations to reveal novel network mechanisms.

Discussion

The unique insights provided by neural network simulations assures their continued use in elucidating the operations of neural systems. The basic limitation of conventional physiological and anatomical data is that they provide a selective sample of a complex system, leaving a wide gap between particular glimpses of neural activity or anatomical structure, and the behavior of the overall system. This gap is usually bridged by intuitive inferences, often based on selective interpretations of the data (Fetz, 1992). A more objective approach would be to derive neural network models that simulate the behavior. These models can incorporate the observed responses of units and can help explain the functional meaning of neural patterns. Thus, integrative neurophysiologists can profitably use a combination of unit recording techniques and neural modeling to elucidate network mechanisms. To the extent that models can incorporate anatomical and physiological constraints, they can provide plausible explanations of the biological neural mechanisms mediating behavior.

Acknowledgments.   This work was supported by ONR (grant N18–89–J–1240) and by NIH grants NS12542 and RR00166.

Road Map: Biological Networks
Background: I.3. Dynamics and Adaptation in Neural Networks
Related Reading: Reaching Movements: Implications of Connectionist Models; Short-Term Memory

References

Arnold, D. B., and Robinson, D. A., 1991, A learning network model of the neural integrator of the oculomotor system, *Biol. Cybern.*, 64:447–454.

Fetz, E. E., 1992, Are movement parameters recognizably coded in the activity of single neurons? *Behav. Brain Sci.*, 15:679–690. ◆

Fetz, E. E., 1993, Dynamic neural network models of sensorimotor behavior, in *The Neurobiology of Neural Networks* (D. Gardner, Ed.), Cambridge, MA: MIT Press, pp. 165–190. ◆

Fetz, E. E., Cheney, P. D., Mewes, K., and Palmer, S., 1989, Control of forelimb muscle activity by populations of corticomotoneuronal and rubromotoneuronal cells, *Prog. Brain Res.*, 80:437–449. ◆

Fetz, E. E., and Shupe, L. E., 1990, Neural network models of the primate motor system, in *Advanced Neural Computers* (R. Eckmiller, Ed.), Amsterdam: Elsevier North, pp. 43–50.

Lisberger, S. G., and Sejnowski, T. J., 1992, *Computational Analysis Suggests a New Hypothesis for Motor Learning in the Vestibulo-Ocular Reflex*, Technical Report INC-9201, Institute for Neural Computation, University of California, San Diego.

Lockery, S. R., and Sejnowski, T. J., 1992, Distributed processing of sensory information in the leech: A dynamical neural network model of the local bending reflex, *J. Neurosci.*, 12:3877–3895.

Maier, M., Shupe L. E., and Fetz, E. E., 1993, A spiking neural network model for neurons controlling wrist movement, *Soc. Neurosci. Abst.*, 19:993.

Munro, E., Shupe, L., and Fetz, E., 1994, Integration and differentiation in dynamic recurrent neural networks, *Neural Computat.*, 6: 405–419.

Rowat, P. F., and Selverston, A. I., 1991, Learning algorithms for oscillatory networks with gap junctions and membrane currents, *Network*, 2:17–41.

Tsung, F.-S., Cottrell, G. W., and Selverston, A. I., 1990, Experiments on learning stable network oscillations, in *Proceedings of the International Joint Conference on Neural Networks, 1990*, vol. 1, New York: IEEE, pp. 169–174.

Watrous, R. L., and Shastri, L., 1986, *Learning Phonetic Features Using Connectionist Networks: An Experiment in Speech Recognition*, Technical Report MS–CIS–86–78, Linc Lab 44, University of Pennsylvania, Philadelphia.

Williams, R. J., and Zipser, D., 1989, A learning algorithm for continually running fully recurrent neural networks, *Neural Computat.*, 1: 270–280.

Williams, R. J., and Zipser, D., 1990, *Gradient-Based Learning Algorithms for Recurrent Connectionist Networks*, Technical Report NU–

CCS–90–9, College of Computer Science, Northeastern University, Boston. ◆

Zipser, D., 1991, Recurrent network model of the neural mechanism of short-term active memory, *Neural Computat.*, 3:179–193.

Dynamic Remapping

Alexandre Pouget and Terrence J. Sejnowski

Introduction

The term *dynamic remapping* has been used in many different ways, but one of the clearest formulations of this concept comes from the mental rotation studies by Georgopoulos et al. (1989; see also REACHING: CODING IN MOTOR CORTEX). In these experiments, monkeys were trained to move a joystick in the direction of a visual stimulus, or 90° counterclockwise from it. The brightness of the stimulus indicated which movement was required on a particular trial: a dim light corresponding to a 90° movement and a bright light to a direct movement. An analysis of reaction time suggested that, by default, the initial motor command always pointed straight at the target and then continuously rotated if the cue indicated a 90° rotation, an interpretation that was subsequently confirmed by single-unit recordings.

The term *remapping* is also commonly used whenever a sensory input in one modality is transformed to the sensory representation in another modality. The best-known example in primates is the remapping of auditory space, which is head-centered in the early stages of auditory processing, into the retinotopic coordinates used in the superior colliculus (Jay and Sparks, 1987; Stein and Meredith, 1993). This type of remapping, equivalent to a change of coordinates, is closely related to sensorimotor transformations. It does not have to be performed over time but could be accomplished by the neuronal circuitry connecting different representations.

This review is divided into three parts: In the first part, we briefly describe the types of cortical representations typically encountered in dynamic remapping. We then summarize the results from several physiological studies in which it has been possible to characterize the responses of neurons involved in temporal and spatial remappings. Finally, in the third part, we review modeling efforts to account for these processes.

Neural Representation of Vectors

A saccadic eye movement toward an object in space can be represented as a vector, S, whose components s_x and s_y correspond to the horizontal and vertical displacement of the eyes. Any sensory or motor variable can be represented by similar vectors. There are two major ways of representing a vector in a neural population: by a map and by a vectorial representation.

The encoding of saccadic eye movements in the superior colliculus is an example of a map representation. A saccade is specified by the activity of a two-dimensional layer of neurons (see COLLICULAR VISUOMOTOR TRANSFORMATIONS FOR SACCADES). Before a saccade, a bump of activity appears on the map at a location corresponding to the horizontal and vertical displacement of the saccade.

A vectorial code appears to be used for the coding of the direction of hand movement in the primary motor cortex. Georgopoulos et al. (1989) showed that neurons in the primate motor cortex of a monkey moving its hand fire maximally for a particular direction of the hand movement and respond with a cosine tuning as a function of angle from this best direction. This suggests that each cell encodes the projection of the vector along its preferred direction.

In both cases, the original vector can be recovered from the population activity pattern through a simple transformation. In the map, the center of mass of the activity pattern codes for the direction of movement. For the vectorial representation, an estimate of the original vector, called a population vector, can be recovered by having each unit vote for its best direction by an amount proportional to its activity.

Neurophysiological Correlates of Remapping

Continuous Remappings

Georgopoulos et al. (1989) studied how the population vector varies over time in the mental rotation experiment described in the introduction. They found that, for 90° movements, the vector initially pointed in the target direction and then continuously rotated 90° counterclockwise, at which point the monkey initiated a hand movement (Figure 1A). This is consistent with the interpretation of the reaction time experiments: the monkey had initially planned to move toward the stimulus, and then it updated this command according to stimulus brightness.

Evidence for continuous remappings also has been reported in a double-saccade paradigm. In these experiments, two targets are briefly flashed in succession on the screen, and the monkey makes successive saccades to their remembered locations (Figure 1B). Monkeys can perform this task with great accuracy, demonstrating that they do not simply keep a trace of the retinotopic location of the second target, since, after the first eye movement, this signal no longer corresponds to where the target was in space. Single-unit recordings in the superior colliculus, frontal eye field, and parietal cortex have shown that, before the first saccade, the brain encodes the retinotopic location of the second target. Then, while the first eye movement is executed, this information is updated to represent where the second target would appear on the retina after the first saccade (Mays and Sparks, 1980; Gnadt and Andersen, 1988; Goldberg and Bruce, 1990). In certain cases, this update is predictive, i.e., it starts before the eye movement (Duhamel, Colby, and Goldberg, 1992).

If these remappings were continuous, a bump of activity should sweep through the superior colliculus, going from the retinal location of the second target to its new retinal location after the first saccade. Notice that if a similar mechanism operates on the hill of activity related to the first target, this hill would move toward the center of the map, since the first saccade results in the foveation of the first target. Munoz, Pelisson, and Guitton (1991) have reported evidence for this moving-hill mechanism in single-saccade paradigm, but it is an

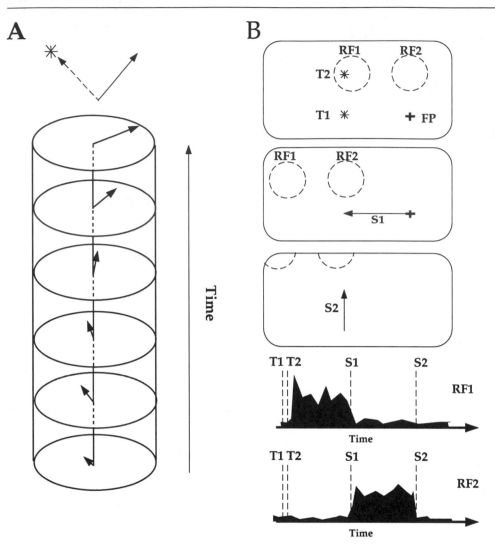

Figure 1. *A*, Rotation of population vector in the primary motor cortex when the brightness of the target (star) indicates a 90° clockwise movement. (Adapted from Georgopoulos et al., 1989.) *B*, Saccade remapping. The monkey makes a double saccade to the remembered positions of T1 and T2. The post-stimulus-time histograms at the bottom show the responses of two cells with receptive fields RF1 and RF2. The second cell (RF2) responds only after the first eye movement, encoding the new retinal location of T2, even though it is no longer present on the screen.

open question whether a moving hill is also involved in the remapping of the second target.

Although the data from Munoz et al. (1991) and from Georgopoulos et al. (1989), are consistent with continuous remappings, they do not constitute proof of this concept. Hence, the population vector rotation could be the consequence of the simultaneous decay and growth of, respectively, the initial planned hand direction and the final one, without ever truly activating intermediate directions. There are also several unresolved problems with the moving-hill hypothesis of Munoz et al. (1991), as emphasized by Sparks (1993).

One-Shot Sensory Remapping

In the inferior colliculus and primary auditory cortex, neurons have bell-shaped auditory receptive fields in space whose position is fixed with respect to the head. In contrast, in the multi-sensory layer of the superior colliculus, the position of the auditory receptive fields is fixed in retinotopic coordinates, which implies that the auditory map must be combined with eye position (Jay and Sparks, 1987). Therefore, the auditory space is remapped in visual coordinates, presumably for the purpose of allowing auditory targets to be foveated by saccadic eye movements, a function mediated by the superior colliculus.

A similar transformation has been found in the striatum and the premotor cortex, where some of the cells have visual receptive fields in somatosensory coordinates (skin-centered) (Graziano and Gross, 1993). In all cases, these remappings are thought to reflect an intermediate stage of processing in sensorimotor transformations.

These remappings can be considered as a change of coordinates, which correspond to a translation operation (see also SENSOR FUSION). For example, the auditory remapping in the superior colliculus requires the retinal location of the auditory stimulus, **R**, which can be computed by subtracting its head-centered location, **A**, from the current eye position, **E**:

$$\mathbf{R} = \mathbf{A} - \mathbf{E} \qquad (1)$$

Remapping Models

The remappings we have described so far fall into two categories: vector rotation with a vectorial code (e.g., mental rotation) and vector translation within a topographic map (e.g., auditory remapping in the superior colliculus). These transformations are very similar, for rotating a vector within a vectorial representation consists of translating a pattern of activity along a circle. Therefore, in both cases, the remapping involves

Figure 2. Remappings as a moving hill of activity in a map. In continuous remapping (left), a recurrent network dynamically moves the hill of activity according to a velocity signal, $\dot{E}$. In feedforward remapping (right), the hill is moved in one shot by the full amount of the current displacement, E, through an intermediate stage of processing in the hidden layer.

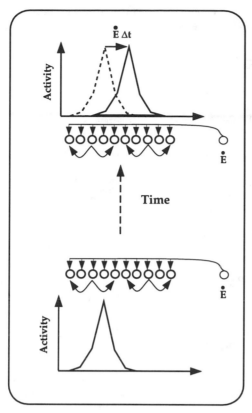

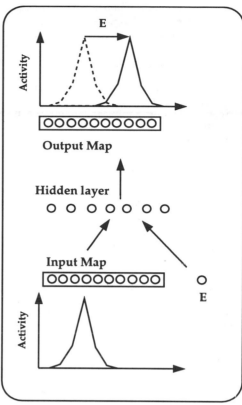

translating a bell-shaped pattern of activity across a map. Most models perform this operation either dynamically through time or in one shot through the hidden layer of a feedforward network (Figure 2).

Dynamical Models

Two kinds of mechanisms have been used in models of continuous remapping: the integration of a velocity signal or the relaxation of a recurrent network.

Integrative model for remapping. In the double-saccade paradigm just described, the retinal coordinates of the second target were updated during the first saccade, a process which might involve moving a hill of activity across the deep layer of the superior colliculus (Munoz et al., 1991). A model by Droulez and Berthoz (1991) shows how this bump of activity could be moved continuously across the map by integrating the eye velocities during the first saccade. Their model is essentially a *forward* model of motion; i.e., given a velocity signal, it generates the corresponding moving image. Interestingly, the equations are very similar to those used for *inverse* models of motion processing. In both cases, the analysis relies on the assumption that the overall gray-level profile in the image is unchanged; only the position of the image features is changed. It is possible to design a recurrent network to implement this constraint (Droulez and Berthoz, 1991). The resulting network moves arbitrary patterns of activity in response to an instantaneous velocity signal.

This model would not only update the coordinates of second target, but would also move the hill of activity corresponding to the first target toward the center of the map, consistent with Munoz et al. (1991). Finally, the model does not require the true eye velocity to work. An approximation of eye velocity,

obtained from the eye position modulated neurons found in the parietal cortex (see GAZE CODING IN THE POSTERIOR PARIETAL CORTEX), would be sufficient (Dominey and Arbib, 1992).

Recurrent networks. Mental rotation of a population vector can be reproduced by training a neural network to follow a circular trajectory over time. In this case, the population vector rotates as a consequence of the network dynamics in the absence of any input signals. This approach has been used by Lukashin and Georgopoulos (1994) to model the generation of hand trajectory, but when the trajectory is a circle, mental rotation and a circular hand trajectory are equivalent. Although the model generates a rotating vector, additional mechanisms must be specified to stop the rotation.

Single-Shot Models

Feedforward models have been used for vectorial as well as map representations. They are used whenever the amplitude of the shift is available to the brain beforehand, as for auditory remapping in the superior colliculus in which the shift is directly proportional to the current eye position (Equation 1). In contrast, in the mental rotation case, the amplitude of the shift is specified by an external stimulus.

Shifter models. As demonstrated by Touretzky, Redish, and Wan (1993), rotation within a vectorial representation can be performed by using a shifter circuit (for more details on shifter circuits, see also ROUTING NETWORKS IN VISUAL CORTEX). Their architecture uses N independent circuits, each implementing a rotation through a particular angle. This mechanism is limited in resolution since it rotates only in multiples of $360/N$ degrees. Whether such shifter circuits actually exist in the brain remains to be demonstrated.

Feedforward network models. There are multiple examples of three-layer networks, or variations thereof, which are trained or hand-crafted to perform sensory remapping. Since these remappings perform vector addition, it might appear unnecessary to deploy a fully nonlinear network for such a task. However, with a map representation, vector addition requires moving a hill of activity in a map as illustrated in Figure 2, an operation that is highly nonlinear.

Special-purpose nonlinear circuits can be designed to perform this operation (Groh and Sparks, 1992), but more biologically realistic solutions have been found with networks of sigmoidal units trained with backpropagation. Hence, the model of Zipser and Andersen (see Gaze Coding in the Posterior Parietal Cortex), which was trained to compute a head-centered map from a retinotopic input, uses hidden units with retinotopic receptive fields modulated by eye position, as in parietal neurons (see also Krommenhoek et al., 1993).

However, backpropagation networks are, in general, quite difficult to analyze, providing realistic models but not much insight into the algorithm used by the network. Pouget and Sejnowski (1995) have recently explored a way to analyze such networks using the theory of basis functions.

Basis functions. The process of moving a hill of activity in a single shot can be better understood when considered within the larger framework of nonlinear function approximation. Consider, for example, the feedforward network shown in Figure 2 applied to a remapping from retinotopic, r_x, to head-centered coordinates, a_x. It can be shown that the responses of the output units are nonlinear in the input variables, namely, the retinal position, r_x, and eye position, e_x.

Therefore, the actual goal of the network is to find an appropriate intermediate representation to approximate this output function. One possibility is to use basis functions of r_x and e_x in the hidden layer (Pouget and Sejnowski, 1995).

Perhaps the best-known set of basis functions is the set of cosine and sine functions used in the Fourier transform. Another example is the set of Gaussian or radially symmetric functions with local support (see Radial Basis Function Networks). A good model of the response of parietal neurons, which are believed to be involved in remapping, is a set of Gaussian functions of retinal position multiplied by sigmoids of eye position. The resulting response function is very similar to that of gain-modulated neurons in the posterior parietal cortex (see Gaze Coding in the Posterior Parietal Cortex).

Conclusions

Remapping can be continuous and dynamic or a single shot through several layers of neurons. In both cases, the problem amounts to moving a hill of activity in neuronal maps. Whether some models are better than others is often difficult to establish because the neurophysiological data available are still relatively sparse. Modelers might be able to constrain their models further by considering deficits that have been documented after lesions in humans. These data not only provide valuable insights into the nature of remappings, but they also might help bridge the gap between behavior and single-cell responses.

Road Maps: Primate Motor Control; Vision
Related Reading: Corollary Discharge in Visuomotor Coordination

References

Dominey, P. F., and Arbib, M. A., 1992, A cortico-subcortical model for the generation of spatially accurate sequential saccades, *Cereb. Cortex*, 2:153–175.

Droulez, J., and Berthoz, A., 1991, A neural model of sensoritopic maps with predictive short-term memory properties, *Proc. Natl. Acad. Sci.*, 88:9653–9657. ◆

Duhamel, J. R., Colby, C. L., and Goldberg, M. E., 1992, The updating of the representation of visual space in parietal cortex by intended eye movements, *Science*, 255:90–92.

Georgopoulos, A. P., Lurito, J. T., Petrides, M., Schwartz, A. B., and Massey, J. T., 1989, Mental rotation of the neuronal population vector, *Science*, 243:234–236. ◆

Gnadt, J. W., and Andersen, R. A., 1988, Memory related motor planning activity in posterior parietal cortex of macaque, *Exp. Brain Res.*, 70(1):216–220.

Goldberg, M. E., and Bruce, C. J., 1990, Primate frontal eye fields: III. Maintenance of a spatially accurate saccade signal, *J. Neurophysiol.*, 64(2):489–508.

Graziano, M. S., and Gross, C. G., 1993, A bimodal map of space: Somatosensory receptive fields in the macaque putamen with corresponding visual receptive fields, *Exp. Brain Res.*, 97(1):96–109.

Groh, J. M., and Sparks, D. L., 1992, Two models for transforming auditory signals from head-centered to eye-centered coordinates, *Biol. Cybern.*, 67(4):291–302. ◆

Jay, M. F., and Sparks, D. L., 1987, Sensorimotor integration in the primate superior colliculus: 1. Motor convergence, *J. Neurophysiol.*, 57:22–34.

Krommenhoek, K. P, Van Opstal, A. J., Gielen, C. C. A. M., and Van Gisbergen, J. A. M, 1993, Remapping of neural activity in the motor colliculus: A neural network study, *Vision Res.*, 33(9):1287–1298.

Lukashin, A. V., and Georgopoulos, A. P., 1994, A neural network for coding trajectories by time series of neuronal population vectors, *Neural Computat.*, 6(1):19–28.

Mays, L. E., and Sparks, D. L., 1980, Dissociation of visual and saccade-related responses in superior colliculus neurons, *J. Neurophysiol.*, 43(1):207–232. ◆

Munoz, D. P., Pelisson, D., and Guitton, D., 1991, Movement of neural activity on the superior colliculus motor map during gaze shifts, *Science*, 251:358–360.

Pouget, A., and Sejnowski, T. J., 1995, Spatial representation and basis functions (in preparation). ◆

Sparks, D. L., 1993, Are gaze shifts controlled by a "moving hill" of activity in the superior colliculus? *Trends Neurosci.*, 16(6):214–218.

Stein, B., and Meredith, M. A., 1993, *The Merging of the Senses*, Cambridge, MA: MIT Press.

Touretzky, D. S., Redish, A. D., and Wan, H. S., 1993, Neural representation of space using sinusoidal arrays, *Neural Computat.*, 5:869–884.

Dynamics and Bifurcation of Neural Networks

Xin Wang and Edward K. Blum

Introduction

While a feedforward neural network implements a function from its input layer to its output layer, a recurrent neural network (RNN) defines a dynamical system whose dynamics is utilized to carry out a computation task such as implementing associative memories, pattern recognition, and optimization. The practical objective of studying RNNs is to design network dynamics for performing a given task by adjusting network parameters. This involves two major issues in dynamical system theory, dynamics and bifurcation. *Dynamics* is concerned with the asymptotic behavior of the networks, which includes limit sets (e.g., fixed points, periodic orbits, and chaotic invariant sets) and their asymptotic stability (e.g., stable, unstable, and saddle), while *bifurcation* is concerned with how the dynamics changes as parameters are varied. In many applications of RNNs in information processing, such as associative memories, pattern recognition and classification, combinatorial optimization, and segmentation and binding of objects, limit sets are used to represent the computational objects, namely, memories, temporal patterns, optimal solutions, visual objects, and so forth. Asymptotic stability of these limit sets is considered as associativity, fault tolerance, generalization, and robustness; bifurcation of these asymptotic structures as parameters change is related to "learning" happening in the networks. A fundamental problem in RNNs is *to design attractors and their basins of attraction such that the dynamics of the networks (possibly driven by external inputs) leads to transitions among the attractors, so that some computation is performed.*

Neural Networks as Dynamical Systems

A general *dynamical system* is a triple $D = (X, T, \phi)$, consisting of a nonempty state (or phase) space X, a temporal domain T, and a state-transition function $\phi: X \times T \to X$ such that (i) either $T = R^+$ (the set of non-negative real numbers) or $T = N$ (the set of natural numbers) and (ii) for all $x \in X$ and $s, t \in T$,

$$\phi(x, 0) = x \qquad \phi(\phi(x, t), s) = \phi(x, s + t)$$

Essentially, the function ϕ describes how the state changes in time. The system D is a *continuous-time* (CT) system when $T = R^+$ and a *discrete-time* (DT) system when $T = N$.

Differential equations and maps are usually used to define dynamical systems. A differential equation

$$dx(t)/dt = F(x(t)) \qquad x(0) = x_0 \qquad x(t) \in X \subseteq R^n \quad (1)$$

defines, under certain conditions (Ruelle, 1989b), a CT system $D = (X, R^+, \phi)$, where ϕ is the solution of Equation 1,

$$\phi(x, t) = \int_0^t F(x(s))\, ds \qquad \phi(x, 0) = x$$

for $x \in X$, $t \in R^+$. A map $g: X \to X$ defines a DT system $D = (X, N, \phi)$, where ϕ is given by iterates of g,

$$\phi(x, k) = g^k(x) \quad (2)$$

for $x \in X$, $k \in N$, Here $g^0(x) = x$ and $g^{k+1}(x) = g(g^k(x))$.

As all biological neurons have bounded activities, only systems whose state spaces X are bounded subsets of the n-dimensional, real vector space R^n are of interest. To fix ideas, consider a network of $n \geq 1$ neurons of the following type. Let $x_i(t)$ be the state of neuron i at time t, taking values in an interval I (e.g., $I = [-1, 1]$ or $[0, 1]$); $f_i: R \to I$ is the neuron activation function, J_i is the external input of neuron i, and w_{ij} are the connection weights from neuron j to neuron i. A CT model of the network is

$$dx_i(t)/dt = -x_i(t) + f_i\left(\sum_{j=1}^n w_{ij}x_j(t) + J_i\right) \quad (3)$$

and a DT model is

$$x_i(t + 1) = f_i\left(\sum_{j=1}^n w_{ij}x_j(t) + J_i\right) \quad (4)$$

Usually the functions f_i take a *sigmoidal* form, namely, they are bounded and monotonically increasing functions, and there exists a unique value $z^* \in R$ at which the derivative $f'(z)$ attains a global maximum $f'(z^*)$. Two examples are $\tanh(z)$ with $I = [-1, 1]$ and $1/(1 + e^{-x})$ with $I = [0, 1]$, both having $z^* = 0$.

Denote $X = I^n$, $x(t) = (x_1(t), \ldots, x_n(t))$, $W = [w_{ij}]$, $J = (J_1, \ldots, J_n)$, and $f(z) = (f_1(z_1), \ldots, f_n(z_n))$. Then $X, x(t), W, J$, and f are regarded as network's state space, state at time t, connection weight matrix, external input, and activation function, respectively. The models expressed in Equations 3 and 4 have vector forms

$$dx(t)/dt = -x(t) + f(Wx(t) + J) \quad (5)$$

and

$$x(t + 1) = f(Wx(t) + J) \quad (6)$$

The main objective of studying RNNs is to understand how the dynamical systems defined by these models behave.

Note that the variables $x_i(t)$ in Equations 3 and 4 are intended to represent neuron firing rates. Similar models arise from considering membrane potentials as neuron states. Let $u_i = \Sigma_{j=1}^n w_{ij}x_j(t) + J_i$ and $x_i = f_i(u_i)$, or in vector form $u = Wx + J$ with $x = f(u)$. Then Equations 5 and 6 become

$$du(t)/dt = -u(t) + Wf(u(t)) + J$$

and

$$u(t + 1) = Wf(u(t)) + J$$

It turns out that if the matrix W is invertible, these two sets of models have the same dynamical behavior under the transformation $u(t) = Wx(t) + J$. The former has the advantage of being a dynamical system on the hypercube I^n, depending only on the range I of the function f, whereas the range of the latter varies with the matrix W as well.

There are many other models of RNNs which have been developed in the neural network literature. Some are described in the forms of Boolean networks, partial differential equations and delayed differential equations, for which the reader is referred to Hirsch (1989) and Hertz, Krogh, and Palmer (1991), and references therein.

Preliminaries Concerning Dynamical Systems

Dynamics

Let $D = (X, T, \phi)$ be a dynamical system. Define, for each $x \in X$, a map $\phi_x: T \to X$ by $\phi_x(t) = \phi(x, t)$ and, for each $t \in T$, a map $\phi^t: X \to X$ by $\phi^t(x) = \phi(x, t)$. The map ϕ_x is the *trajectory* of initial state x and the map ϕ^t is the *flow* of the system over time t. The set of states $\Gamma(x) = \{\phi_x(t) | t \in T\}$ is the *orbit* of x.

A *limit point p* of a trajectory $x(t) = \phi_x(t)$ is a point satisfying $p = \lim_{t_k} x(t_k)$ for some sequence $t_k \to \infty$. The set of all limit points of the trajectory ϕ_x, denoted by $\omega(x)$, is the *limit set* of x. As the space X is assumed to be bounded, $\omega(x)$ is nonempty, closed, and bounded. Moreover, $\omega(x)$ is *invariant*: any trajectory starting at a point y in $\omega(x)$ remains in $\omega(x)$.

Let A be a closed invariant subset of X. A is *asymptotically stable*, or A is an *attractor*, if there is a neighborhood U of A such that the limit set of any state x in U is a subset of A; that is, the trajectory ϕ_x approaches A as $t \to \infty$. If, on the other hand, for any neighborhood U of A, the limit set of any state x in U, but not in A, has empty intersection with A, A is *asymptotically unstable*. In cases where A is neither asymptotically stable nor unstable, A is *saddle*, which implies that in any neighborhood of A some trajectories tend to A, and others stay away from A. In general, the set of states whose limit sets are subsets of A is called the *basin of attraction* of A, and is denoted by $B(A)$. If $B(A)$ is equal to the entire space X, A is *globally asymptotically stable*.

There are just a few elementary types of attractor with simple geometric structures in the forms of fixed points or equilibria, or periodic and quasi-periodic trajectories. A trajectory $\phi_x(t)$ is *stationary* if $\phi_x(t) = x$ for all $t \in T$. The state x then is called a *fixed point* or an *equilibrium*. The condition for x to be a fixed point is $F(x) = 0$ in CT systems defined by Equation 1 and $g(x) = x$ in the DT systems of Equation 2. Often cited for showing existence of fixed points is the Brouwer fixed point theorem, which says that any continuous function $g: X \to X$ on a compact (bounded and closed) subset X of R^n has at least one fixed point in X. This theorem can be applied directly to DT systems, whereas for CT systems it is applicable to function $g(x) \equiv F(x) + x$, because the fixed points of such defined g satisfy $F(x) = 0$. However, the Brouwer theorem does not tell how many fixed points there are, or where they are.

Two other simple types of attractor take the forms of periodic and quasi-periodic trajectories. In the periodic case, $\phi_x(p) = x$ for some $p > 0$, and x is a *periodic state* with period p if p has the smallest value satisfying $\phi_x(p) = x$. In the quasi-periodic case, $\phi_x(t)$ is not periodic but its limit set $\omega(x)$ is a smooth curve or surface in R^n. Precisely, a trajectory $x(t) = \phi_x(t)$ is *quasi-periodic* if there exists a finite set of periodic functions $\phi_i(t)$ ($i = 1, \ldots, n$) with *linearly independent* periods p_i (i.e., there are no rational numbers $q_1, \ldots, q_k$ such that $p_i = \Sigma_{j \neq i} q_j p_j$), such that $x(t) = \Sigma_i a_i \phi_i(t)$ for some $a_i \neq 0$. This happens especially when a rotation map on the unit circle in R^2 has a rotating angle $\alpha \pi$ with α irrational: no point on the circle is periodic, but the limit set of any trajectory is the whole circle. Often associated with periodic and quasi-periodic trajectories are *limit cycles*. A limit cycle is a periodic or quasi-periodic trajectory by itself, and also a limit set of some other trajectories.

There are also complicated attractors with geometric structures characterized by a fractal Hausdorff dimension (Ruelle, 1989a). These attractors often exist in nonlinear dynamical systems that possess *sensitive dependence on initial conditions*, meaning that the distance between trajectories $\phi_x(t)$ and $\phi_y(t)$ of two close initial states x and y increases exponentially in time t. Attractors of this kind are loosely called *chaotic* or *strange attractors*, and related systems are *chaotic systems*. The most distinguishable dynamical feature is their practically unpredictable dynamics, as any small error in measuring or computing initial conditions will be magnified exponentially over a long period of time. There are also other features associated with a chaotic attractor A (Devaney, 1986). Important ones are topological transitivity on A (i.e., A cannot be decomposed into two or more independent parts) and density of a set of the system's

periodic trajectories in A (i.e., complex dynamics on A interplay with regular dynamics on the periodic trajectories). At present, there exist some rigorous approaches and results concerning chaotic systems and attractors, most of which are for low-dimensional systems: for example, symbolic dynamics analysis of the dynamics of one-dimensional unimodal maps, and findings concerning the presence of homoclinic and heteroclinic structures in two- or higher-dimensional systems. Among others, the one-dimensional logistic map, two-dimensional Hénon map, and three-dimensional Lorenz equation are often-cited examples of dynamical systems that have strange attractors (Devaney, 1986; Ruelle, 1989b).

Bifurcation

When a dynamical system depends on a set of parameters, study of bifurcation, i.e., quantitative changes of dynamic structure as the parameters are varied, becomes possible.

The bifurcation idea is formalized by means of the concept of topological conjugacy. Two systems, $D = (X, T, \phi)$ and $D' = (Y, T, \psi)$, are *topologically conjugate* if there is a continuous map $h: X \to Y$ that establishes a one-to-one correspondence between trajectories of the two systems, namely,

$$h(\phi(t, x)) = \psi(t, h(x)) \qquad x \in X, t \in T$$

Suppose that a system D_μ on R^n depends on a set of parameters $\mu \in R^r$, where R^r ($r \geq 1$) is the parameter space of the system. If in any small neighborhood U of a parameter value μ_0 there are two values $\mu_1, \mu_2 \in U$ such that D_{μ_1} and D_{μ_2} are not topologically conjugate, then a *bifurcation* is said to occur at μ_0, and μ_0 is a *bifurcation point*. For instance, a bifurcation occurs when an attracting fixed point becomes asymptotically unstable, and the parameter value at which the fixed point changes its stability is the bifurcation point.

Current research in bifurcation theory is still restricted to systems with only one or two parameters. Most results are obtained for local bifurcation around fixed points that are caused by change of the hyperbolicity of fixed points, or for global bifurcation caused by changes of the system's homoclinic and heteroclinic structures (Ruelle, 1989b; Hale and Kocak, 1991). Some examples of local bifurcation, such as Hopf, pitchfolk, and period-doubling bifurcations, will be given later in this article in the context of neural networks.

Dynamics of Neural Networks

Asymptotic Stability of Fixed Points

While detection of fixed points for a given RNN amounts to solving a certain equation such as $x = f(Wx + J)$ for the models in Equations 5 and 6, determination of asymptotic stability of fixed points relies on calculating the eigenvalues of the Jacobian of the network at the fixed points, according to the Grobman-Hartman theorem (Hale and Kocak, 1991; Ruelle, 1989b). For Equation 5, this is to calculate the eigenvalues λ of the matrix $-I + DW$ at a fixed point $x^* = (x_1^*, \ldots, x_n^*)$, where $D = \text{diag}(f_i'(\Sigma_{j=1}^n w_{ij}x_j^*(t) + J_i))$. If all λ's have negative (positive, or some negative and some positive) real parts, then x^* is asymptotically stable (unstable, or saddle). For Equation 6, the stability depends on the eigenvalues λ of the matrix DW. If all λ's have moduli smaller (larger, or some smaller and some larger) than 1, then x^* is asymptotically stable (unstable, or saddle).

Convergent Dynamics

Convergent dynamics is further classified (Hirsch, 1989) into two types: *quasi-convergence*, meaning that every trajectory approaches a set of fixed points, and *global convergence*, meaning that there is a unique fixed point to which every trajectory converges.

There are basically three approaches for showing that an RNN has convergent dynamics. They are Lyapunov, contraction, and monotonicity approaches.

The *Lyapunov approach* is the most common one to guarantee convergence. It involves constructing an energy function E, which is continuous on the state space X and nonincreasing along trajectories. Such a function is constant on the set of limit points of a trajectory. If E is strictly decreasing ($dE(x(t))/dt < 0$ in the CT case and $E(x(t+1)) - E(x(t)) < 0$ in the DT case) for all nonfixed points $x(t)$, then all limit points of any trajectory are fixed points.

Many conditions for convergence have been derived using the Lyapunov approach (Hirsch, 1989). For example, conditions on the activation functions f_i and weights w_{ij}, such as sigmoidal f_i's and zero-diagonal symmetry ($w_{ij} = w_{ji}, w_{ii} = 0$) (Hopfield, 1984), guarantee that the network of Equation 3 has an energy function

$$E = -\frac{1}{2}\left(\sum_{i,j} w_{ij}x_i x_j + \sum_i x_i^2\right)$$

For a more general class of RNNs of the form

$$dx_i(t)/dt = a_i(x_i)\left[b_i(x_i) - \sum_{j=1}^{n} w_{ij}c_j(x_j)\right] \qquad (7)$$

where a_i, b_i and c_i ($i = 1, \ldots, n$) are functions, when $a_i > 0$, $c_i' > 0$, and $w_{ij} = w_{ji}$, the function

$$E = -\sum_i \int_0^{x_i} b_i(z)c_i'(z)\,dz + \left(\sum_{i,j} w_{ij}c_i(x_i)c_j(x_j)\right)\Big/2$$

is a strict energy function and therefore the network of Equation 7 is quasi-convergent. For DT networks defined by Equation 4 with symmetric weight matrices W, the only possible limit points are shown by Marcus and Westervelt (1989) to be fixed points and periodic points of period 2. Notice that this is different from the result obtained by Hopfield (1982); the latter uses asynchronous dynamics, which will be addressed later in this article.

The *contraction approach* for convergence is based on the concept of contractive functions. A function $h: R^n \to R^n$ is *contractive* if it has some norm $\|h\| < 1$. It has been shown by Kelly (see Hirsch, 1989) that the network of Equation 5 is globally convergent if $f(Wx + J)$ is contractive, or if $\mu\|W\| < 1$ in this case where μ is the maximum of the neuron gains $\mu_i = \max_z f_i'(z)$.

Depending on which norm $\|\cdot\|$ is chosen, several sufficient conditions for Equation 5 to converge globally have been obtained. One is $\mu(w_{ii} + (\Sigma_{ij}|w_{ij}| + |w_{ji}|))/2 < 1$ (Hirsch, 1989). More detailed considerations on the norm will lead to more sophisticated conditions on convergence.

The *monotonicity approach* was formulated by Hirsch (1984) for convergence of monotone dynamical systems. Let $\leq$ be a partial order on a set $X \subseteq R^n$. For any $x, y \in X$, define $x \ll y$ when $x \leq y$ and $x_i \neq y_i$ ($i = 1, \ldots, n$). A function $f: X \to X$ is *monotone* (respectively, *strongly monotone*) if $x \leq y$ implies $f(x) \leq f(y)$ ($f(x) \ll f(y)$). A system (X, T, ϕ) is (strongly) monotone if for each $t \in T$, the flow ϕ^t is (strongly) monotone. An important class of monotone systems is composed of *cooperative systems*, namely, systems defined by Equation 1 whose off-diagonal terms of the Jacobians $DF = [\partial F_i/\partial x_j]$ are ≥ 0. If, in addition, the Jacobians are *irreducible*, then the systems are *strongly monotone*. It turns out that in a monotone system, if a trajectory $x(t) = \phi_x(t)$ satisfies $x \ll x(t_0)$ for some $t_0 > 0$, then $x(t)$ converges to a fixed point x^* with $x \ll x^*$; and in a strongly monotone system, almost all trajectories converge to a set of fixed points—almost quasi-convergence.

The convergence results for monotone systems can be applied to excitatory networks (Hirsch, 1989). The network of Equation 3 with sigmoidal functions f_i is *excitatory* if $w_{ij} \geq 0$ for all $i \neq j$. Clearly, an excitatory network is cooperative. Hence, if in addition the matrix W is irreducible, then the network is strongly monotone and therefore almost quasi-convergent.

Oscillatory Dynamics

Oscillation is a main function of some networks, e.g., spiking neurons, pacemakers or central pattern generators, and oscillatory cortices (Glass and Mackey, 1988).

Though there have been many studies of biological oscillations, the oscillatory dynamics of networks of reasonable size is in general still difficult to describe, mainly because mathematical tools for analyzing periodic orbits in high-dimensional spaces are not sufficiently developed. A practical approach to studying oscillatory dynamics is *modular*, meaning that it focuses first on oscillatory dynamics of networks of small size, and then deals with the dynamics of collections of small networks through certain functional coupling structures. Typical examples are coupled oscillators (Glass and Mackey, 1988; see also CHAINS OF COUPLED OSCILLATORS and COLLECTIVE BEHAVIOR OF COUPLED OSCILLATORS).

Chaotic Dynamics

Chaotic dynamics may play a significant computational role in biological information processing (Glass and Mackey, 1988; Yao and Freeman, 1990). Many numerical simulations have been performed to demonstrate the existence of chaos in RNNs (see Wang, 1991).

However, to show rigorously that an RNN displays chaotic dynamics or has strange attractors is not an easy matter. One

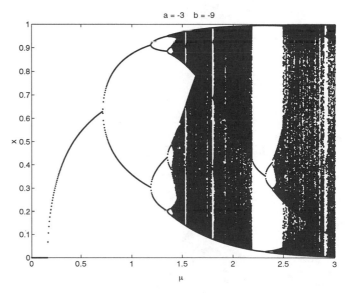

Figure 1. Period doublings to chaos.

approach is to reduce the dynamics of a network through topological equivalence to the dynamics of some familiar chaotic models, such as the logistic map, Hénon map, or Lorenz equation. In Wang (1991), the network of Equation 4 is treated as a one-parameter family of maps with the neuron gain as the parameter. For a class of weight matrices W of the form

$$\begin{bmatrix} a & -a \\ b & -b \end{bmatrix}$$

the network of two neurons is proven to be equivalent to a full family of S-unimodal maps on the interval [0, 1]. Essentially, an *S-unimodal map* is a mathematical generalization of the one-dimensional logistic map, and a full family of S-unimodal maps will go through, as the parameter varies, a sequence of period-doubling bifurcations, at each of which previously stable periodic orbits lose their stability, giving rise to new stable periodic orbits with doubled periods. The maps eventually be-

come chaotic (Devaney, 1986). Figure 1 shows the bifurcation diagram for the activity of one of the two neurons.

In Yao and Freeman (1990), a network based on Lorenz-like strange attractors is constructed to understand the role of chaos in biological pattern recognition. The network maintains a global chaotic attractor that provides a basal activity of the resting system and allows transitions driven by external inputs between chaotic and other periodic attractors.

In practice, several statistic measurements, such as Lyapunov exponents, entropy, and information dimension, are used as indications for a system to be chaotic (Ruelle, 1989a). For instance, the Lyapunov exponents of a given trajectory can be viewed as long-time averages of the real parts of the eigenvalues of the linearized system along the trajectory. They capture asymptotic information about local expansion and contraction of the state space in such a way that *positive* Lyapunov exponents indicate expansion, and hence sensitive dependence on initial conditions, while *negative* ones indicate contraction.

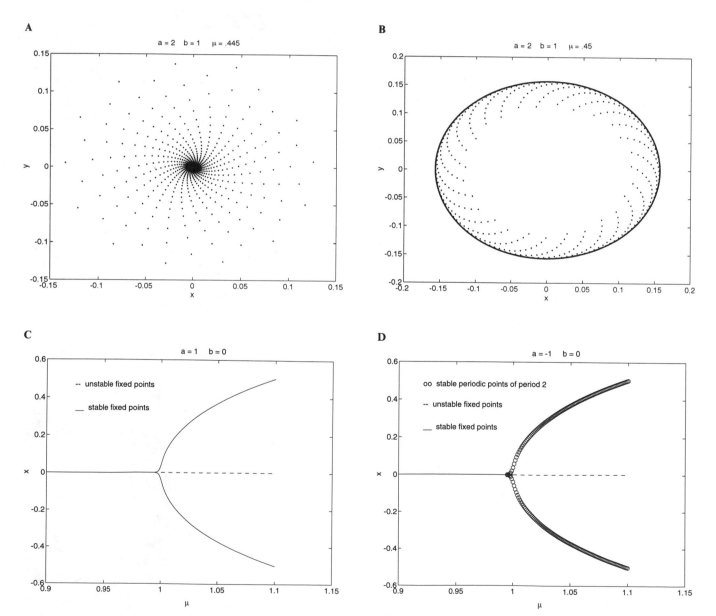

Figure 2. Various bifurcations.

Bifurcation

Bifurcation analysis provides a means to describe the qualitative behavior of a network in different parameter regions and the changes in the network dynamics as parameters are varied. For example, the Wilson and Cowan model for the dynamics of average activities of excitatory and inhibitory populations of neurons, which shows hysteresis phenomena and limit-cycle activity, is analyzed in Borisyuk and Kirillov (1992) for its bifurcation behavior. As a result, a parameter plane is partitioned into regions of dynamics which are qualitatively the same, bounded by bifurcation curves.

Possible local bifurcations in neural networks can be illustrated using the network of Equation 6, with the neuron gain as a bifurcation parameter. Consider a two-neuron case with an input ($J = 0$). Assume that all f_i's have the same maximal slope (neuron gain) μ at the origin and that W is of the form

$$\begin{bmatrix} a & -b \\ b & a \end{bmatrix}$$

for some real a, b. Then the Jacobian at the origin $Df(0, 0) = \mu W$ has two eigenvalues $\lambda_{1,2} = \mu(a \pm ib)$, and $\mu_0 = 1/(a^2 + b^2)^{1/2}$ is a bifurcation point because $|\lambda_{1,2}| = 1$. With different a and b, the following bifurcations occur as μ is varied:

1. *Hopf bifurcation* (Figure 2A–B). If a and b satisfy $[(a + ib)/(a^2 + b^2)^{1/2}]^k \neq 1$ for $k = 1, 2, 3, 4$, then as μ increases past μ_0, the origin loses its stability, and an attracting limit cycle emerges.
2. *Pitchfork bifurcation* (Figure 2C). If $a > 0$ and $b = 0$, then as μ increases past μ_0, the origin loses its stability, and two new attracting fixed points emerge.
3. *Period-doubling bifurcation* (Figure 2D). If $a < 0$ and $b = 0$, then as μ increases past μ_0, the origin loses its stability, and a new pair of attracting periodic points of period 2 emerges.

Asynchronous Dynamics

Neural networks as dynamical systems have one distinct feature: they are massively distributed computing systems. One of the major issues in parallel and distributed computation is *asynchronization*. The dynamics as described by Equations 5 and 6 require that the neurons communicate their states to all others instantaneously and synchronize their dynamics precisely all the time. Any mechanism used to enforce such synchronization may have an important effect on performance of the network when it is implemented or simulated over a real distributed system. Moreover, the biological manifestation of asynchrony is inherent, as it can be caused by delays in nerve signal propagation, variability of neuron parameters such as refractory periods, and adaptive neuron gains.

Asynchronous dynamics has been thoroughly studied in the contexts of DT dynamical systems (see Wang, Li, and Blum, 1994, for references). Among others, contractive maps on Banach spaces and continuous maps on partial ordered sets are *asynchronizable*, i.e., any asynchronous iterations of these maps will converge to the fixed points of synchronous iterations.

The asynchronization issue has also been addressed in the context of neural networks. For example, in the celebrated DT Hopfield model (Hopfield, 1982), only one randomly chosen neuron is allowed to update its state at each time, and the networks with symmetric weight matrices are always convergent to fixed points. The issue is also discussed for CT RNNs. Previous approaches have involved converting CT models into DT ones through the Euler discretization, and then applying the existing result for contractive maps.

Recently, a rigorous formulation of asynchronous dynamics was presented by Wang et al. (1994) for CT dynamical systems, based on concepts of local time scales of individual components and communication time delays between the components. The asynchronous dynamics of systems defined by the general equation

$$C \, dx/dt = -x + F(x)$$

where $C = \text{diag}(c_1, c_2, \ldots, c_n)$ with $c_i > 0$ and $F: R^n \to R^n$, is discussed for F being contractive and monotone. When the concept and results are applied to the network of Equation 5, convergent dynamics of contractive and monotone networks are, under certain conditions, asynchronizable.

Conclusion

The fundamental problem mentioned in the introduction remains open. Mathematical analysis of the dynamical and bifurcation behaviors of RNNs increases our understanding of their intrinsic mechanism, capacity, and limitation when they are used to process and generate temporal information. Analysis of this type also promotes understanding of the fundamental principles of parallel and distributed information processing as complementary to the computational principles suggested by biological experiments.

Road Map: Dynamic Systems and Optimization
Background: I.3. Dynamics and Adaptation in Neural Networks
Related Reading: Chaos in Neural Systems; Computing with Attractors; Cooperative Phenomena; Energy Functions for Neural Networks

References

Borisyuk, R. M., and Kirillov, A. B., 1992, Bifurcation analysis of a neural network model, *Biol. Cybern.*, 66:319–325.

Devaney, R. L., 1986, *An Introduction to Chaotic Dynamical Systems*, Redwood City, CA: Benjamin-Cummings. ◆

Glass, L., and Mackey, M. C., 1988, *From Clock to Chaos: The Rhythms of Life*, Princeton, NJ: Princeton University Press. ◆

Hale, J. K., and Kocak, H., 1991, *Dynamics and Bifurcations*, New York: Springer-Verlag. ◆

Hertz, J., Krogh, A., and Palmer, R. G., 1991, *Introduction to the Theory of Neural Computation*, Redwood City, CA: Addison Wesley. ◆

Hirsch, M. W., 1984, The dynamical systems approach to differential equations, *Bull. Am. Math. Soc.*, 11:451–514.

Hirsch, M. W., 1989, Convergent activation dynamics in continuous time networks, *Neural Netw.*, 2:331–349. ◆

Hopfield, J., 1982, Neural networks and physical systems with emergent computational abilities, *Proc. Natl. Acad. Sci. USA*, 79:2554–2558.

Hopfield, J., 1984, Neurons with graded response have collective computational properties like those of two-state neurons, *Proc. Natl. Acad. Sci. USA*, 81:3088–3092.

Marcus, C. M., and Westervelt, R. M., 1989, Dynamics of iterated-map neural networks, *Phys. Rev. A*, 40:501–504.

Ruelle, D., 1989a, *Chaotic and Strange Attractors: The Statistical Analysis of Time Series for Deterministic Nonlinear Systems*, Cambridge, Eng.: Cambridge University Press. ◆

Ruelle, D., 1989b, *Elements of Differentiable Dynamics and Bifurcation Theory*, New York: Academic Press. ◆

Wang, X., 1991, Period-doublings to chaos in a simple neural network: An analytical proof, *Complex Systems*, 5:425–441.

Wang, X., Li, Q., and Blum, E. K., 1994, Asynchronous dynamics of continuous-time neural networks, in *Advances in Neural Information Processing Systems 6* (J. Cowan, G. Tesauro, and J. Alspector, Eds.), San Mateo, CA: Morgan Kaufmann, pp. 493–500.

Yao, Y., and Freeman, W. J., 1990, Model of biological pattern recognition with spatially chaotic dynamics, *Neural Netw.*, 3:153–170.

Echolocation: Creating Computational Maps

Nobuo Suga and Jagmeet S. Kanwal

The Echolocation Signal

Mustached bats emit echolocation pulses for navigation and for hunting flying insects. Bats hunting insects on the wing must determine several prey characteristics accurately and dynamically for a successful interception. First and foremost, the prey must be detected and distinguished from the background clutter of vegetation in which these bats hunt. The targeted prey must be characterized in terms of its surface features and wing beat pattern to ensure that it is appropriate for consumption. At the same time, other features such as relative velocity and distance of the fluttering target must be dynamically computed, and the target must be localized in space for orientation and prey capture. This poses a huge computational problem that must be solved repeatedly in real time for sets of input parameters that continuously vary as the bat approaches its target.

To extract much of this information, mustached bats emit ultrasonic pulses that consist of a long constant-frequency (CF) component followed by a short frequency-modulated (FM) component. Each pulse contains four harmonics (H_{1-4}), so that there are eight components that can be defined (CF_{1-4} and FM_{1-4} in Figure 1). In the pulse, the second harmonic (H_2) is always predominant. By comparison, the fundamental (H_1) in the pulse is very weak. The CF_2 frequency is slightly different among individual bats and between sexes and is approximately 61 kHz. In FM_2, the frequency sweeps down from 61 kHz to approximately 49 kHz (Suga, Simmons, and Shimozawa,1974).

Each pulse-echo combination provides a discrete sample of the continuously changing acoustic scene. Different parameters of echoes reflected back from the target, in reference to the emitted pulse, carry different types of information about a target (Suga, 1973). Briefly, the auditory network contains two key design features: neurons that are sensitive to combinations of pulse and echo components and computational maps that represent systematic changes in echo parameters to extract the relevant information.

Frequency Analysis in the Periphery

The eight components (CF_{1-4} and FM_{1-4}) of the echo are analyzed in the frequency domain in different regions of the basilar membrane in the cochlea (Figure 1B), so that biosonar signals may be considered to be processed along eight different channels: CF_1 channel, CF_2 channel, and so on (Suga, 1988). Peripheral neurons function as an array of filter banks that are more sharply tuned in frequency to the CF components than to the FM components. Among the CF channels, the CF_2 channel is unique, because it is associated with an extraordinarily sharply tuned local resonator in the cochlea (Pollak and Casseday, 1989) and occupies approximately one-third of the ascending auditory system. Frequency-tuning curves of peripheral neurons are similar to a base-up, vertex-down triangle, so that the ambiguity in coding frequency by a single neuron is large at high stimulus amplitudes. All peripheral neurons respond not only to tone bursts, but also to FM sounds and noise bursts.

Frequency and Amplitude Processing Along Central Pathways

At levels higher than the auditory nerve, frequency selectivity of many neurons within the CF_1, CF_2, and CF_3 channels is

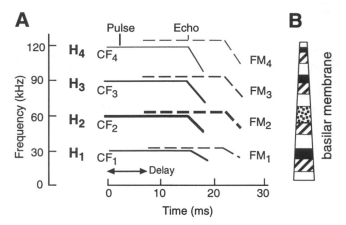

Figure 1. *A,* Schematized sonograms of a biosonar pulse and its Doppler-shifted echo. *B,* Linear representation of different CF and FM channels (denoted by different shading as indicated in Figure 3) on the basilar membrane in the cochlea. Frequency differences between the same harmonic in the pulse and echo are attributable to Doppler shift.

increased by lateral inhibition, so that the ambiguity in frequency coding remains small even at high stimulus amplitudes. Neurons with such "level-tolerant" frequency-tuning curves sandwiched between inhibitory tuning curves respond best to tone bursts (CF tones) at particular frequencies, but neither to FM sounds nor to noise bursts (Suga and Tsuzuki, 1985). The extent of neural sharpening in frequency tuning is most dramatic in the CF_2 channel, so that many neurons in this channel selectively respond to CF tones in a very narrow frequency range regardless of amplitude.

To optimize its foraging strategy, the bat must scan for small periodic changes in frequency and amplitude in the echo from the target relative to those returning from the background vegetation. For a flying bat, the returning echoes undergo a Doppler shift in their frequency. The wing beats of insects create additional small though systematic variations in the Doppler-shifted frequencies in the returning echo because of a component of wing movement that moves either away or toward the bat's ears. This systematic variation in the Doppler-shifted frequencies is important for detecting insects in a cluttered background consisting of moving leaves, small birds, and inedible insects. Similarly, echo amplitude depends on the relative size (subtended angle) and distance of a target. To facilitate the detection of appropriate targets, the CF_2 channel in the bat's auditory system projects to the Doppler-shifted CF processing (DSCF) area in the primary auditory cortex (Figures 2 and 3). The DSCF area is very large, and neurons in a single cortical column in this area have an identical best frequency and best amplitude. In other words, each column is characterized by a particular combination of best frequency and best amplitude.

At the periphery, auditory neurons show a monotonic impulse-count function: the stronger the stimulus, the larger the number of impulses discharged. Many neurons in the auditory cortex and thalamus, however, are tuned to particular amplitudes: "best" amplitudes. The creation of amplitude selectivity is prominent in the CF_2 channel compared with other CF channels. Thus, most neurons in the CF_2 channel are tuned both in frequency and amplitude (Suga, 1988).

Figure 2. Graphic summary of the computational maps in the auditory cortex of the mustached bat. The DSCF area lies within the main tonotopic axis of the primary auditory cortex and is divided into two subdivisions suited either for target detection or localization. It also has an amplitopic map for representation of subtended target angle. The FM-FM, DF, and VF areas consist of neurons that are facilitated by FM_1-FM_n combinations and represent different, overlapping ranges of a delay axis. The CF/CF area consists of two major types of CF_1/CF_n facilitation neurons that are sensitive to combinations of harmonically or quasiharmonically related CF components and has frequency-versus-frequency coordinates that represent a velocity map. The thick lines are arteries, the larger one of which is on the sulcus. All maps are generated from this simple spatially linear representation of inputs.

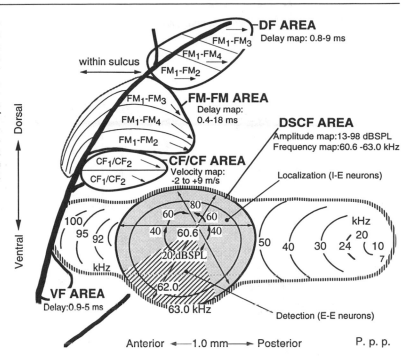

Along the cortical surface of the DSCF area, both best frequency and best amplitude are systematically mapped, forming a frequency-versus-amplitude coordinate system (Figure 2). The amplitude axis is circular, representing 13–98 dB sound pressure level (SPL), whereas the frequency axis is radial, representing 60.6–62.3 kHz. The best frequencies of single neurons often change at a rate of approximately 20 Hz/column within certain portions of the DSCF area. This allows the detection of small frequency and amplitude modulations in an echo from a flying insect that are important for characterizing the size and location of a target. Inactivation experiments with muscimol, a drug which silences large numbers of neurons, indicate that the DSCF area is necessary for fine frequency analysis at approximately 61 kHz (Riquimaroux, Gaioni, and Suga, 1991).

In addition to the frequency-versus-amplitude coordinates, in the DSCF area, approximately 50% of the neurons are sensitive to periodic frequency modulation that would be found in echoes from flying insects. The DSCF area also has "binaural" bands or subdivisions which are mainly occupied by either E-E (excitatory input from both ears) or I-E neurons (inhibitory input from ipsilateral side and excitatory input from the contralateral side). This allows integration of interaural time and intensity differences for the cortically extracted target characteristics. This computation conveys the azimuth of the acoustically defined target. Furthermore, recent studies indicate that DSCF neurons show facilitative response to the combination of a pulse FM_1 and echo CF_2 and respond best to echo delays ranging from approximately 12 to 26 ms (Fitzpatrick et al., 1993), corresponding to the transition from the search to the approach phase of insect pursuit.

Extraction of Velocity Information

As just explained, to measure relative target velocity, the frequencies of echo CF components must be compared with those of pulse CF components. The ascending auditory system creates sharply tuned, level-tolerant neurons by lateral inhibition,

mainly at subthalamic nuclei. In the dorsal division of the medial geniculate body (MGBd), part of the CF_1 channel and part of either the CF_2 or CF_3 channel are integrated, so that neurons in this region respond poorly to the CF_1, CF_2, and CF_3 tones when delivered alone, but respond strongly when the CF_1 tone is delivered together with either the CF_2 or CF_3 tone. These CF/CF neurons project separately to the CF/CF area in the auditory cortex and form the CF_1/CF_2 and CF_1/CF_3 subdivisions.

During insect pursuit, the amount of Doppler shift is dependent on the relative velocity between the source (target) and the observer (bat), and therefore a measurement of a Doppler-shifted frequency in the CF components of the echo can predict the relative velocity of the target. To enable this measurement, each CF/CF neuron has two frequency-tuning curves for facilitation: one is tuned to pulse CF_1 (~ 29 kHz), and the other is tuned to CF_2 (~ 61 kHz) or CF_3 (~ 92 kHz). A critical parameter for the facilitation of almost all CF/CF neurons is a deviation of the CF_2 or CF_3 frequency from the exact harmonic relationship with the CF_1 frequency, i.e., an amount of Doppler shift in terms of CF_2 or CF_3 (see Figure 1). Positively Doppler-shifted (raised) frequencies corresponding to decreasing source-observer distances are well represented, whereas negatively Doppler-shifted (lowered) frequencies corresponding to increasing source-observer distances are scarcely represented, as they are irrelevant to target interception. The CF/CF neurons are specialized to respond to particular combinations of two frequencies, regardless of stimulus amplitudes. The width of the facilitative frequency-tuning curves of some neurons in the CF_2 and CF_3 channels are only $+1.0\%$ of their best frequencies regardless of stimulus amplitudes. Each CF channel provides a separate measure of velocity, perhaps at a different resolution.

In the CF/CF area, each location along the cortical surface represents a particular combination of two frequencies. The combination of two frequencies systematically varies along the cortical surface, but does not vary with depth, so that there are frequency-versus-frequency coordinates (Suga, 1988).

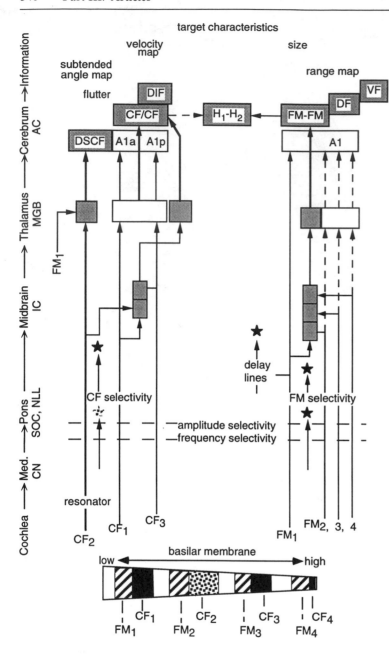

Figure 3. Creation of computational maps by parallel-hierarchical processing of different types of biosonar information. The CF_{1-4} and FM_{1-4} of the pulse and echo are analyzed at different locations on the basilar membrane in the cochlea (bottom). The signal elements coded by auditory nerve fibers are sent up to the auditory cortex (AC) through several auditory nuclei (left margin). During the ascent of the signals, frequency selectivity is increased in some neurons, and amplitude, CF, and FM selectivities are created in others (arrows with a star). Each star indicates that the addition of selectivity also takes place in the auditory nuclei and cortex as well as in the nucleus where the arrow starts. In certain portions of the IC and the MGB, two channels processing different signal elements (e.g., CF_1 and CF_2 or FM_1 and FM_2) are integrated to produce CF/CF and FM-FM combination-sensitive neurons. These combination-sensitive neurons respectively project to the CF/CF and FM-FM areas of the auditory cortex, where target velocity or range information is systematically represented. The delay lines used by FM-FM neurons are presumably created in the IC. Because of corticocortical connections, DF, VF, and VA areas also consist of combination-sensitive neurons (center top). The DSCF area has frequency-versus-amplitude coordinates to represent velocity and subtended angle information of a target (based on Suga, 1988). Triangular frequency-tuning curves at the periphery are changed into level-tolerant frequency-tuning curves in subthalamic auditory nuclei by lateral inhibition. Then, the CF_1 and CF_2 or CF_3 channels are integrated in the medial geniculate body, creating two types of CF/CF combination-sensitive neurons. These CF/CF neurons project to the CF/CF area in the auditory cortex and form the CF_1/CF_2 and CF_1/CF_3 subdivisions. In each subdivision, a cortical column is characterized by a particular combination of two frequencies. Each subdivision has frequency-versus-frequency coordinates, in which target relative velocities are systematically mapped. Stippled boxes indicate regions where combination-sensitive neurons are present. Abbreviations: CN, cochlear nucleus; SOC, superior olivary complex; NLL, nucleus of lateral lemniscus; IC, inferior colliculus; and MGB, medial geniculate body.

Within these coordinates, there is a velocity axis representing different amounts of Doppler shifts in terms of CF_2 or CF_3, i.e., relative target velocities. These range from 8.6 to -1.2 m/sec. Relative velocities between 6 and 0 m/sec that are encountered by the bat during the approach and terminal phases of insect pursuit are overrepresented.

Extraction of Distance Information

Echo delay is a primary cue for measuring target range. An FM signal has two important advantages for measuring time delays that are used for ranging. According to the physics of sonar, a short, broadband signal, e.g., a click, is the most appropriate signal for ranging. In a broadband signal emitted by the vocal tract of an animal, however, the signal amplitude per frequency channel decreases proportionately with increase in bandwidth

because of physiological constraints. Therefore, bats emit a short FM signal so that the same intensity level is maintained for each frequency channel that is restricted to a very short period within the FM sweep. A second advantage of the FM sweep is that it allows multiple comparisons between several different frequencies encoded by an array of auditory receptors for extracting distance information about the same target. These multiple measurements further increase the accuracy of the computation.

At the periphery, neurons respond to both the pulse and the echo. The time interval between the two grouped discharges is directly related to the echo delay. A 1.0-ms echo delay corresponds to a target distance of 17.3 cm at 25°C. The central auditory system creates neurons that are selective for FM sweeps so that they cannot be stimulated by the long CF tone preceeding the FM sweep in an echolocation pulse. FM selec-

Figure 4. Neural mechanisms for creating delay (target-range)-tuned neurons. Neurons tuned to frequencies swept by the FM_1 (FM_1 channel) create delay lines from 0.4 to 18 ms, perhaps in the inferior colliculus, whereas neurons tuned to frequencies swept by FM_n (FM_n channel: $n = 2$, 3, or 4) do not create delay lines (*A* and *B*). In an FM-FM neuron (denoted by an asterisk), where an echo delay is equal to the delay line associated with it, the signals from the FM_1 and FM_4 channels arrive at the same time. FM-FM neurons in the MGBd receive signals from both FM_1 and FM_n channels (*C*). The amount of coincidence is amplified by NMDA receptors at the synapse of the FM-FM neuron. Abbreviations: Ampl., amplification; DT, delay tuning; MGB, medial geniculate body.

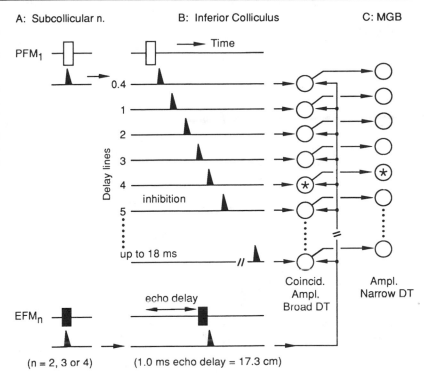

tivity is also added to some neurons by disinhibition or facilitation. This makes computations of target distance more reliable because they are unaffected by pre-FM stimulus characteristics.

In the MGBd and perhaps at the level of the inferior colliculus (IC) of the mustached bat, part of the FM_1 channel and most of either the FM_2, FM_3, or FM_4 channels are integrated to create three major types of FM-FM combination-sensitive neurons: FM_1-FM_2, FM_1-FM_3, and FM_1-FM_4 (see Figure 3). These neurons respond poorly to single FM sounds, but respond strongly to the FM_1 sound combined with either the FM_2, FM_3, or FM_4 sound (Olsen and Suga, 1991a, 1991b). The delay of the echo FM_2, FM_3, or FM_4 sound from the pulse FM_1 sound is the critical parameter for their facilitative responses. These FM-FM neurons act as *delay-dependent multipliers* for processing target range information.

For creating the response properties of these neurons, at least three mechanisms are involved: delay lines, coincidence detection, and multiplication (Figure 4). These experimentally verified mechanisms are currently being modeled. Delay lines are created to spread the neural responses over 18 ms in the FM_1 channel. Thalamic FM-FM neurons receive an input from the FM_1 channel through delay lines and the other input from the FM_n channel ($n = 2, 3$, or 4) without delay lines (Olsen and Suga, 1991a, 1991b). When an echo returns, say, with a 3.0-ms delay from a pulse, an FM-FM neuron associated with a 3.0-ms delay line receives the inputs from both the FM_1 and the FM_n channels at the same time. The extent of the coincidence is multiplied by NMDA receptors at the synapse of the FM-FM neuron (Suga, Olsen, and Butman, 1990; see also NMDA RECEPTORS: SYNAPTIC, CELLULAR, AND NETWORK MODELS). The NMDA receptor complex actually contains several binding sites and is coupled to ion channels that show voltage-dependent blockade by magnesium. Activation of one or more of these binding sites leads to allosteric interactions that strongly potentiate and amplify the neural response, e.g., by magnifying the depolarization or producing a long-lasting excitatory postsynaptic potential (EPSP).

The three types of FM-FM neurons in the MGBd separately project to the FM-FM area of the auditory cortex, forming FM_1-FM_2, FM_1-FM_3, and FM_1-FM_4 subdivisions. In each subdivision, the best echo delay that can excite a neuron systematically varies along the cortical surface but not with depth. Therefore, each subdivision in the FM-FM area has an echo-delay axis for the systematic representation of target distances, which span from 7 to 310 cm.

Parcellation of Cortical Auditory Information

The FM-FM area projects to the DF (dorsal fringe) and VF (ventral fringe) areas of the auditory cortex in addition to other regions of the brain (see Figure 2). The DF area, like the FM-FM area, consists of three subdivisions containing FM-FM neurons. In each subdivision in the DF area, target ranges of up to 160 cm are systematically represented (Suga, 1990). The DF area projects to the VF area as well as to other areas of the brain. The VF area also consists of three clusters of FM-FM neurons and represents target ranges of up to 80 cm. It is most likely that each of these three different areas is related to distance measurements in a different behavioral situation. It is suggested that the VF area is used for velocity measurements when the bat is roosting, whereas the DF and FM-FM areas are probably used for ranging during insect pursuit.

Discussion

In summary, the auditory system exhibits several important design features for processing acoustic signals for echolocation. These are listed below:

1. The input layer (periphery) of the auditory system has an array of nonlinear frequency filters. The sharpness of filters at the periphery and neural sharpening of filters in the central nervous system are remarkably different between the channels processing different types of auditory information.

2. Different parameters of acoustic signals are processed in parallel. However, hierarchical processing is incorporated within each parallel channel, so that signal processing in the auditory system is parallel and hierarchical.

3. In five layers of the central auditory system, several important operations, such as lateral inhibition, delay lines, and gates/coincidence detection and amplification, are combined to create various building blocks: e.g., latency constancy, duration sensitivity, CF and FM selectivity, amplitude selectivity, combination-sensitivity, etc. Some of these building blocks protect signal processing by reducing masking that would normally occur in a cluttered environment.

4. As a result of parallel-hierarchical processing, the auditory cortex (output layer) consists of several subdivisions (maps) specialized for systematic representation of particular types of echolocation information, such as target distance, target size, and relative target velocity.

Acknowledgments.   This work was supported by the McKnight Foundation, NIDCD research grant DC00175, and ONR research grant N00014–90–J–1068.

Road Map: Other Sensory Systems
Related Reading: Auditory Cortex; Electrolocation; Neuroethology, Computational; Sound Localization and Binaural Processing

References

Fitzpatrick, D. C., Kanwal, J. S., Butman, J., and Suga, N., 1993, Combination-sensitive neurons in the primary auditory cortex of the mustached bat, *J. Neurosci.*, 320:509–520.

Gooler, D. M., and O'Neill, W. E., 1987, Topographic representation of vocal frequency demonstrated by microstimulation of anterior cingulate cortex in the echolocating bat, *Peronotus parnelli parnelli, J. Comp. Physiol.*, 161:283–294.

Horikawa, J., and Suga, N., 1986, Biosonar signals and cerebellar auditory neurons of the mustached bat, *J. Neurophysiol.*, 55:1247–1267.

Metzner, W., 1989, A possible neural basis for Doppler-shift compensation in echo-locating horseshoe bats, *Nature*, 341:529–532.

Olsen, J. F., and Suga, N., 1991a, Combination-sensitive neurons in the medial geniculate body of the mustached bat: Encoding of relative velocity information, *J. Neurophysiol.*, 65:1254–1274.

Olsen, J. F., and Suga, N., 1991b, Combination-sensitive neurons in the medial geniculate body of the mustached bat: Encoding of target range information, *J. Neurophysiol.*, 65:1275–1296.

Pollak, G. D., and Casseday J. H., 1989, *The Neural Basis of Echolocation in Bats*, New York: Springer-Verlag.

Riquimaroux, H., Gaioni, S. J., and Suga, N., 1991, Cortical computational maps control auditory perception, *Science*, 251:565–568.

Suga, N., 1973, Feature extraction in the auditory system of bats, in *Basic Mechanisms in Hearing* (A. R. Möller, Ed.), New York: Academic Press, pp. 675–744.

Suga, N., 1988, Auditory neuroethology and speech processing: Complex sound processing by combination-sensitive neurons, in *Functions of the Auditory System* (G. M. Edelman, W. E. Gall, and W. M. Cowan, Eds.), New York: Wiley, pp. 679–720. ◆

Suga, N., 1990, Cortical computation maps for auditory imaging, *Neural Netw.*, 3:3–21. ◆

Suga N., Olsen J. F., and Butman, J. A., 1990, Specialized subsystems for processing biologically important complex sounds: Cross-correlation analysis for ranging in the bat's brain, in *The Brain, Cold Spring Harb. Symp. Quant. Biol.*, 55:585–597. ◆

Suga, N., Simmons, J. A., and Shimozawa T., 1974, Neurophysiological studies of echolocation in awake bats producing CF-FM orientation sounds, *J. Exp. Biol.*, 61:379–399.

Suga, N., and Tsuzuki K., 1985, Inhibition and level tolerant frequency tuning in the auditory cortex of the mustached bat, *J. Neurophysiol.*, 53:1109–1144.

EEG Analysis

Fernando H. Lopes da Silva and Jan Pieter Pijn

Introduction

The electroencephalogram, or EEG, consists of the electrical activity of relatively large neuronal populations that can be recorded from the scalp. Along with the EEG, the magnetic fields generated by these populations (i.e., the magnetoencephalogram or MEG) can also be recorded using very sensitive transducers. Here, we will discuss these two types of activity jointly, since the same analysis methods apply to both. A constant preoccupation of electroencephalographic research has been to develop techniques to extract information from signals, recorded at the scalp, that may be relevant for the diagnosis of brain diseases, and to obtain a better understanding of the brain processes underlying psychophysical and cognitive functions. To this aim, a large number of quantitative analysis methods applied to the EEG have been developed. Recently, the mathematical theory of dynamical nonlinear systems has started to influence the field of brain sciences, particularly by providing a framework that may contribute to obtain a better understanding of the dynamics of EEG signals in relation to brain functions.

Here, we discuss briefly the main aspects of these different endeavors, taking into consideration, first, the EEG as a time series; second, the EEG as a spatiotemporal signal; and third, the EEG as a signal that provides information about the state of complex neuronal networks considered as nonlinear dynamical systems.

Analysis of EEG Signals as Time Series

The EEG is a complex time series, the statistical properties of which depend on the state of the subject and on external factors. Even when the subject's behavioral state is almost constant, the duration of epochs that have the same statistical properties—i.e., are stationary—is usually short. Therefore, EEG signals present essential nonstationary properties. According to the interest of the researcher, the emphasis may be placed on the analysis of the EEG during steady states, or on the detection and characterization of transients. The former is important to define the properties of the so-called ongoing EEG in which the main emphasis lies on the quantification of the basic rhythmic activities; the latter is relevant to put in evidence the dynamics of EEG activities—i.e., the changes in EEG features, either transients of short duration (less than about 200 ms), such as the paroxysmal patterns that commonly occur in epileptic patients in between seizures (interictal), or alterations of the basic rhythmic activity that occur during changes of the state of alertness and in relation to cognitive or motor tasks. A special type of EEG transients is formed by the event-related, or evoked, potentials. These usually have a small amplitude relative to that of the ongoing EEG and, thus, must be detected using averaging techniques, which lie outside the scope of this article.

In the analysis of the ongoing EEG, it is customary to subdivide EEG signals into quasi-stationary epochs and to characterize them by a number of statistical parameters, such as probability distributions, correlation functions, and frequency spectra. In general, these methods of statistical analysis are based on the assumption that EEG signals are realizations of random processes without assuming a specific generation model, and are therefore nonparametric, in contradistinction to the parametric methods that assume a specific generation model.

EEG time series often present a certain degree of interdependence. In order to analyze this property, correlation functions have been commonly used. Here, we must distinguish two main questions: the determination of whether, within one EEG signal, dependency between successive time samples exists (such as in the case of brain rhythmic activity); and the determination of the degree of relationship between multiple EEG signals. The former can be approached by computing the time average of the product of the signal and a replica of itself shifted by a given time interval, i.e., the autocorrelation function (for a more extensive and formal description, see Lopes da Silva, 1993). Its Fourier transform is the power density spectrum, or simply the power spectrum. The latter gives the distribution of the (squared) amplitude of different frequency components. Currently, these spectral distributions are calculated using the Fast Fourier transform or FFT. In case the EEG signals deviate from a Gaussian distribution, higher-order spectral moments may be calculated to characterize them, such as the bispectrum or bispectral density. The latter can be useful to determine whether the system responsible for the EEG generation has nonlinear properties. The same aim can also be achieved by alternative methods based on the mathematical analysis of deterministic dynamical systems (see below) and discussed by Takens (1993). A significant bispectrum exists when there are harmonically related frequency components in an EEG signal. An example of a power spectrum and the corresponding bispectrum are shown in Figure 1.

The degree of relationship between two EEG signals can be estimated using the cross-correlation function. Similarly to the autocorrelation, the latter is the time average of the product of two signals as a function of the time delay between both. The Fourier transform of this function yields the cross-power spectrum. The latter is a complex quantity that has magnitude and phase. To quantify the degree of relationship between pairs of EEG signals as a function of frequency, the magnitude of the cross-power spectrum is usually normalized by dividing it by the value of the autospectra at that frequency of the corresponding signals. This yields a normalized quantity called the *coherence function*. Coherence functions can be used to estimate the degree of the relationship, in the frequency domain, between pairs of EEG signals. The counterpart of the coherence function is the phase function, which provides information about the time relationship between two signals as a function of frequency. The computation of phase functions has been used to estimate time delays between EEG signals in order to obtain evidence for the propagation of EEG signals in the brain. However, the existence of a time delay between two EEG signals can only be concluded with certainty when there is a linear phase relationship between phase and frequency over a certain frequency range. In this context, it should be stressed that cross-correlation and/or coherence analysis are limited to the case that the relationship between the signals is linear, which is not necessarily always the case. In particular, during

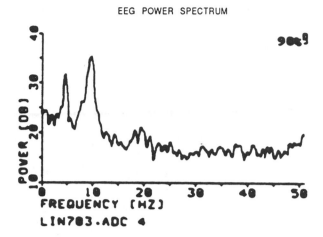

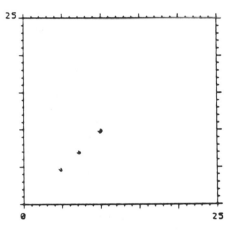

Figure 1. *Above*: Power spectrum of a segment of the scalp EEG of a subject showing an alpha rhythm variant, with a main peak at 10 Hz and a subharmonic at 5 Hz; the 90% confidence band is indicated in the top right corner. *Below*: Contour map of the corresponding normalized bispectrum; only values above 0.25 are plotted. The bispectrum shows three maxima: one at the intersection of 5 Hz and 5 Hz for the two axes, indicating phase coupling between 5 and 10 Hz; another at the intersection between 7 Hz and 7 Hz, indicating also phase coupling between 7 Hz and 14 Hz, components that cannot be clearly seen in the power spectrum above. Still another peak is at the intersection between 10 Hz and 10 Hz, indicating phase coupling between 10 Hz and the second harmonic 20 Hz. This analysis illustrates the nonlinear properties of some EEG signals. (Adapted from Lopes da Silva, 1993.)

epileptiform activity, deviations from linearity are commonly encountered. Therefore, alternative methods have been developed, such as the computation of a coefficient of nonlinear regression between two signals (Pijn et al., 1991).

EEG signals can also be analyzed using parametric methods, assuming that these signals result from the operation of linear filtering on a noise source with a flat spectrum. In this way, the EEG signal is described as a function of its own past and a noise term. This constitutes an autoregressive model (Lopes da Silva, 1993). From the coefficients of the model, one can estimate the spectral density of the signal and describe it by a relatively small number of parameters. Parametric methods can also be extended to deal with the fact that most EEG signals have time-varying properties. In this way, automatic adaptive segmentation methods have been developed (Barlow, 1984), by means of which quasi-stationary EEG segments, which can be described by the same set of parameters, are defined. These quasi-stationary EEG segments can be characterized by the corresponding parameters. This also allows use of the parameters to classify EEG segments with the same characteristics, using clustering analysis or other approaches based on artificial neural networks. More recently, another parametric method has been introduced in EEG analysis, the continuous wavelet transform, that has the advantage over spectral analysis of being able to extract the information present in the signal at a wide range of time scales (see WAVELET DYNAMICS). Using wavelet transforms, short transients can be effectively detected in EEG signals as bursts of rhythmic waves at very different frequencies. Accordingly, the wavelet transform acts as a mathematical zoom that detects signal components at wide ranges of time and frequency scales (Schiff et al., 1994).

Analysis of EEG Signals as Potential Distributions in Space

EEG signals are distributed not only in time but also in space. Indeed, a most fundamental aspect of the analysis of brain signals is to estimate from multiple scalp EEG, or MEG, recordings the distribution within the brain of the corresponding sources. This implies solving the so-called inverse problem of volume conduction theory—i.e., to determine the location within the brain tissue of the sources of electrical activity, taking into consideration the properties of both the brain and the conductive media surrounding the brain. This problem has no unique solution: it is not possible to determine a unique source current distribution in a volume conductor from measurements taken at the conductor surface. However, it is possible to solve this problem by putting constraints on the current source distributions—i.e., by defining a specific model of the source. A commonly assumed source model is the equivalent current dipole that represents an active patch of cortex. Therefore, two models are required: one of the source and another of the volume conductor. Of course, it is necessary to have a sufficient number of measurement points at the scalp to obtain a satisfactory solution. In general, to account for potential distributions with the steepest spatial gradients, it is necessary to have inter-electrode distances of at most 2.5 cm (Gevins et al., 1994), which corresponds to recording from about 128 electrodes placed on the scalp.

We consider, briefly, the main problems posed by a choice of the source and of the volume conductor models. The problem of estimating the source is a nonlinear problem that has to be solved iteratively. However, when multiple sources are active, errors in their estimation are likely to occur, since the contribution of two or more sources equals the sum of the contributions of each one. Nevertheless, this problem may be approached if additional information is available about the possible location of the sources—for example, when two symmetric sources are to be expected on both hemispheres (Scherg, 1990). A general approach that takes into account both the time functions of the activity of the sources and the corresponding spatial properties (positions and orientations) has been proposed by De Munck (1990). As regarding the volume conductor models, the most commonly used has been a set of concentric spheres that represent, from inside to outside, the brain, the cerebrospinal fluid in some cases, the skull, and the scalp. However, the head devi-

ates appreciably from a sphere. With the growing availability of magnetic resonance imaging (MRI) techniques, it became possible to reconstruct the different head compartments in a more realistic way. This is important, since several simulation studies have shown that deviations from a realistic shape of the head can influence significantly the magnetic fields and the potential distributions (Hämäläinen et al., 1993).

In most applications of EEG spatial analysis, either in medicine or in psychophysiology, one does not attempt an estimation of the brain sources using the inverse approach in view of the inherent difficulties of this methodology and the uncertainties of the estimated localizations. Most researchers are satisfied, rather, with representing the sets of multiple EEG signals projected as a map at the surface of the scalp. This is a form of *brain mapping*. In essence, an EEG feature is extracted by an appropriate analysis method from the set of EEG records at a given time sample, or within a certain time window, and a contour map of the corresponding values over the scalp is constructed. In general, such brain maps represent steady-state variables, but there are also interesting applications to dynamic changes. In the latter case, the maps may represent the change of the ongoing activity within a given frequency band induced by some event—for example, a sensory stimulus, a cognitive task, or a movement. Recently, this form of analysis has been applied with interesting results in the assessment of the cortical areas involved in the planning of specific movements (Pfurtscheller, Flotzinger, and Neuper, 1994), as illustrated in Figure 2.

Valuable additions to brain mapping have been recently developed, making use of modern computer technology. One of these consists in approximating the scalp potential distribution with a spherical spline function. Thereafter, the second spatial derivative (Laplacian) of this function is calculated. Assuming a homogeneous conductivity of the medium, this spatial second derivative can be considered equivalent to the current density in the scalp. In this way, a sharpened image of the scalp potential can be obtained (Law and Nunez, 1993). Another consists in deblurring the scalp EEG—i.e., estimating the potentials at the cortical surface by means of a realistic finite element model

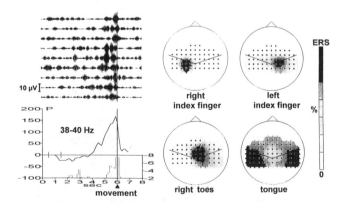

Figure 2. *Left side:* Examples of bandpass filtered (38–40 Hz) movement-synchronous EEG trials and time course of 38–40 Hz instantaneous power changes during finger movement in relation to the indicated reference interval 4.5–5.5 s prior to movement onset. Note the significant power increase in the last second prior to movement onset. *Right side:* Topographical display of narrow band power increase in the 40 Hz range during different movements with approximate location of the central sulcus. Each map represents a time window of 125 ms. Black marks the spatial location of 40 Hz power increase. (Adapted from Pfurtscheller, Flotzinger, and Neuper, 1994.)

of the subject's head made by using three-dimensional anatomical information obtained from MRI scans. A basic difference between this analysis method and the equivalent dipole localization approach is that the former makes no assumption about the number or type of generator sources but only about the properties of the passive media surrounding the sources. Therefore, this method is applicable in those cases where a simple model of the sources cannot be assumed a priori.

EEG Signals as Expression of Dynamical Nonlinear Systems

The basic elements of the theory of complex dynamical nonlinear systems, and the definition of a number of concepts relevant to the present issue, are discussed in the article EPILEPSY: NETWORK MODELS OF GENERATION. These concepts are important in distinguishing whether EEG signals are generated by random filtered processes or by deterministic nonlinear systems that may possess multiple types of attractors, including the possibility of exhibiting chaotic dynamics. Generally, the term *chaos* refers to low-dimensional irregular signals that may have quasi-periodic components, but result from deterministic systems with complex dynamics that are very sensitive to initial conditions (see CHAOS IN NEURAL SYSTEMS). One way to characterize the system's behavior is to compute its dimensionality. With the aim of distinguishing whether an EEG signal may be generated by a noise source or by a nonlinear system with deterministic dynamics, a characteristic measure is currently used—the so-called *correlation dimension* (D_2) (Grassberger and Procaccia, 1983). Several investigators have applied this form of analysis to EEG signals recorded during different behavioral states and found in some cases that these signals had a low (noninteger) correlation dimension, D_2, which was interpreted as corresponding to a chaotic attractor—for example, during sleep (Babloyantz, Salazar, and Nicolis, 1985). However, a low value of D_2 does not provide sufficient evidence to allow the conclusion that the EEG signals are generated by a process with deterministic chaotic dynamics, since noise characterized by a $f^{-\beta}$ power law (f: frequency, $\beta > 1$) can also have a low D_2 when analyzed with the current algorithms. To solve this ambiguity, one has to use surrogate signals (controls) obtained, for example, by randomizing the phase of the same EEG signals. The latter should yield a large D_2 value since the transformed signal is not distinguishable from Gaussian noise (Pijn et al., 1991). In most cases of ongoing EEGs, including sleep (Achermann et al., 1994), the difference between the real and the surrogate signals is very small, and the hypothesis that the EEG signals are generated by a deterministic chaotic process in these cases cannot be drawn. However, in the case of EEG signals recorded during epileptic seizures, the value of D_2 was shown to be much smaller than that of the corresponding surrogate signals, so in these cases, the hypothesis that the brain networks behave as nonlinear systems with chaotic dynamics during epileptic seizures (see EPILEPSY: NETWORK MODELS OF GENERATION, Figure 1) cannot be rejected (Pijn et al., 1991). Nevertheless, a more profound analysis of the dynamics of these systems is needed before definite conclusions about the underlying mathematical models may be drawn (Takens, 1993).

Discussion

EEG recordings are complex signals that may provide valuable information about the underlying brain systems, since they have unsurpassed resolution in time, although their spatial resolution is rather limited. Therefore, mapping cortical activity using EEG/MEG signals combined with realistic models of the brain, extracted from MRI scans, may yield new possibilities for functional imaging of dynamical brain states. The choice of analysis method used to extract information from EEG/MEG signals depends, in the first instance, on the question formulated by the researcher. There is no universal method that can be applied successfully under all circumstances. When the main concern is to characterize a given brain state, or a change in state, spectral analysis using either the Fast Fourier transform or the autoregressive model may be appropriate. However, if the dynamics of the underlying neuronal networks are the subject of interest, more sophisticated mathematical methods are necessary. Questions relating to the functional localization of the brain sources of certain EEG features imply the use of advanced computer models of the sources and of the volume conductor, or of methods of topographical analysis that may provide significant brain maps.

Road Map: Biological Networks

References

Achermann, P., Hartmann, R., Gunziger, A., Guggenbühl, W., and Borbély, A., 1994, All-night sleep EEG and artificial stochastic control signals have similar correlation dimensions, *Electroencephalogr. Clin. Neurophysiol.*, 90:384–387.

Babloyantz, A., Salazar, J. M., and Nicolis, C., 1985, Evidence of chaotic dynamics of brain activity during the sleep cycle, *Phys. Lett.*, 111A(3):152–156.

Barlow, J. S., 1984, Analysis of EEG changes with carotid clamping by selective analog filtering, matched inverse digital filtering and automatic adaptive segmentation: A comparative study, *Electroencephalogr. Clin. Neurophysiol.*, 58:193–204.

De Munck, J., 1990, The estimation of time-varying dipoles on the basis of evoked potentials, *Electroencephalogr. Clin. Neurophysiol.*, 77:156–160.

Gevins, A. S., Le, J., Martin, N. K., Brickett, P., Desmond, J., and Reutter, B., 1994, High-resolution EEG: 124-channel recording, spatial deblurring and MRI integration methods, *Electroencephalogr. Clin. Neurophysiol.*, 90:337–358.

Grassberger, P., and Procaccia, I., 1983, Measuring the strangeness of strange attractors, *Physica*, 9:183–208.

Hämäläinen, M., Hari, R., Ilmoniemi, R., Knuutila, J., and Lounasmaa, O. V., 1993, Magnetoencephalography—theory, instrumentation, and applications to noninvasive studies of the working human brain, *Rev. Modern Phys.*, 65:413–497. ◆

Law, S. K., and Nunez, P. L., 1993, High-resolution EEG using spline generated surface Laplacians on spherical and ellipsoidal surfaces, *IEEE Trans. Biomed. Eng.*, 40:145–153.

Lopes da Silva, F. H., 1993, EEG analysis: Theory and practice, in *Electroencephalography: Basic Principles, Clinical Applications and Related Fields* (E. Niedermeyer and F. H. Lopes da Silva, Eds.), Baltimore: Williams & Wilkins, pp. 1097–1123. ◆

Pfurtscheller, G., Flotzinger, D., and Neuper, C., 1994., Differentiation between finger, toe and tongue movements in man based on 40 Hz EEG, *Electroencephalogr. Clin. Neurophysiol.*, 90:456–460.

Pijn, J. P. M., van Nerveen, J., Noest, A., and Lopes da Silva, F. H., 1991, Chaos or noise in EEG signals: Dependence on state and brain site, *Electroencephalogr. Clin. Neurophysiol.*, 79:371–381.

Scherg, M., 1990, Fundamentals of dipole source potential analysis, in *Auditory Evoked Electric and Magnetic Fields: Advanced Audiology*, vol. 6 (F. Grandori, M. Hoke, and G.-L. Romani, Eds.), Basel: Karger, pp. 40–69.

Schiff, S. J., Aldroubi, A., Unser, M., and Sato, S., 1994, Fast wavelet transformation of EEG, *Electroencephalogr. Clin. Neurophysiol.* (in press).

Takens, F., 1993, Detecting non-linearities in stationary time series, *Internat. J. Bifur. Chaos*, 3:241–256.

Theiler, J., 1990, Estimating fractal dimension, *J. Opt. Soc. Am. [A]*, 7:1055–1073.

Electrolocation

Joseph Bastian

Introduction

Sensory systems must be adaptable; ever-changing circumstances require that the gain, dynamic range, and spatial and temporal selectivity or "tuning" of a sensory system be adjustable. Likewise, mechanisms are needed to facilitate the processing of relevant signals while rejecting noise and predictable background signals; moreover, at some level within a system, processes must select which stimuli are to be attended to. Many of these adaptive features of sensory processing networks are thought to be implemented by descending or feedback pathways which enable higher centers to shape the processing characteristics of their lower-order targets. Although the importance of descending pathways is widely recognized, clear demonstrations of their function are few, and fewer still are examples in which the cellular mechanisms underlying these functions are known.

This review summarizes recent studies of the electrosensory system of South American weakly electric fish and focuses on the roles of descending or feedback projections in shaping the information processing characteristics of electrosensory neurons. The neural circuitry underlying gain control within this sensory system is described, along with modeling studies which propose gain control mechanisms based on the regulation of total membrane conductance. Possible mechanisms for attentive processes within the electrosensory system are described and are related to the concept of so-called sensory searchlights, and mechanisms are proposed to explain these animals' abilities to selectively process behaviorally relevant information in the presence of naturally occurring patterns of masking stimuli. Additional recent reviews can be consulted for summaries of social communication in weakly electric fish (Hopkins, 1988), physiological and behavioral studies of electrolocation (Heiligenberg, 1977; Bastian, 1990), and studies of the jamming avoidance response (JAR) (Heiligenberg, 1991). Bullock and Heiligenberg (1986) provide a comprehensive review of anatomical, physiological, and behavioral studies of electrosensory systems.

Electrolocation and the Electrosensory Lateral Line Lobe

Electrosensory systems allow animals to locate and identify objects in the absence of visual cues or inputs from other sensory systems. Fish that are so equipped generate an electric signal by means of an electric organ located in their elongate tail, and current flow resulting from this electric organ discharge (EOD) causes a voltage to develop across the animal's skin. The magnitude and timing of this EOD is measured by electroreceptors (modified hair cells) scattered over the surface of the body. Objects having an impedance different from that of the water distort the EOD field, and the resulting changes in voltage across the skin are encoded by the electroreceptors, then further processed by higher-order components of the electrosensory system.

An illustration of an electrolocation signal generated by a highly conductive target moving past such an animal is shown below the fish in Figure 1. The otherwise constant amplitude of the EOD seen by a given electroreceptor is positively modulated as the target passes through a receptor's receptive field. The amplitude of this modulation depends upon the size and shape of the target, its conductivity, its distance away from the fish, and the amplitude of the discharge itself.

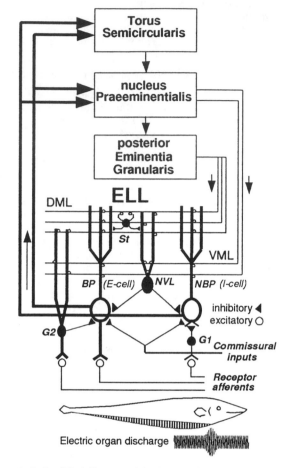

Figure 1. A simplified diagram of the ELL and its afferent and efferent connections illustrating the descending electrosensory projections to the dorsal and ventral molecular layers (DML and VML, respectively). See text for further explanation.

Factors, such as changes in the conductivity of the water, proximity to large conductive or nonconductive surfaces, as well as postural changes that occur during locomotion, can significantly alter the amplitude of an animal's discharge. The fish cannot modify the output of their electric organ; hence, compensation for changes in discharge must be accomplished within the sensory processing circuitry to avoid loss of sensitivity if EOD amplitude falls, and to avoid saturating the system if amplitude rises.

Since electroreceptors receive no efferent innervation, control of electrosensory processing must occur within the central nervous system (CNS). Much of this control is likely to occur within the first-order nucleus, the electrosensory lateral line lobe (ELL). In addition to the receptor afferent projection, the ELL receives multiple sources of descending electrosensory input, as well as a significant proprioceptive projection; moreover, ELLs are linked bilaterally by GABAergic commissural connections (Carr and Maler, in Bullock and Heiligenberg, 1986; Sas and Maler, 1987).

Figure 1 is a schematic diagram of the ELL circuitry and its afferent and efferent connections. Within this diagram, the amplitude-encoding *receptor afferents*, which project somato-topically to the ELL, fire at high constant rates owing to the continuously present EOD. These afferents make excitatory synapses (open circles) with one category of ELL output neuron—the basilar pyramidal cells (*BP*)—and with inhibitory interneurons (*G1, G2*). These interneurons provide inhibitory input to a second category of ELL output neuron, the non-basilar pyramidal cell (*NBP*). Basilar pyramidal cells (E-cells) are excited, while nonbasilar pyramidal cells (I-cells) are inhibited, by increased EOD amplitude and receptor afferent firing. The interneurons also contribute to the antagonistic surrounds of the E-cell and I-cell receptive fields. The ELL's output consists of two parallel and oppositely responding cell types having receptive fields organized in a center-surround fashion. Additionally, there is a parallel pathway consisting of receptors and higher-order neurons which encode the timing of EOD cycles (not shown in Figure 1). The amplitude-encoding efferent neurons project to two higher centers, the nucleus praeeminentialis (NP) and the torus semicircularis (TS), and outputs of both of these higher-order nuclei ultimately project back to the ELL closing feedback loops, as diagrammed in Figure 1.

The ELL pyramidal neurons and certain inhibitory interneurons (*G2* and *NVL*) have large dendrites which extend into the dorsal and ventral molecular layers of the ELL (DML and VML, respectively). The DML is composed of typical "cerebellar" parallel fibers, and the granule cells which give rise to these axons reside within the posterior eminentia granularis (EGp). The EGp granule cells receive descending electrosensory information from the NP, as well as proprioceptive and other inputs. This pathway is termed "indirect" since information destined for the ELL's DML is relayed through the EGp. The ELL's VML is made up of axons that descend "directly" from the NP. This pathway is also involved in modulating pyramidal cell activity but, as is described below, its role is quite different from that of the indirect projection. The properties of the ELL-NP-ELL feedback loop are influenced by still higher centers; the NP receives a large projection from the TS, and the TS also receives a direct projection from the ELL pyramidal cells.

In addition to the general neuroanatomical features of the electrosensory system summarized in Figure 1, the physiological characteristics of the principal cell types involved in these pathways have been described, and in many cases, these neurons have been intracellularly labeled and reconstructed (Bastian and Bratton, 1990; Bratton and Bastian, 1990; Bastian and Courtright, 1991). The distributions of neurotransmitters and of receptor types within the ELL have been studied. The synaptic inputs from DML and VML fibers to the apical dendrites of pyramidal cells and interneurons are glutamatergic, and both NMDA and non-NMDA receptors are found within these molecular layers (Maler and Monaghan, 1991). The direct descending input from the NP to the VML also contains a population of GABAergic fibers (Maler and Mugnaini, personal communication).

Global Changes in EOD Amplitude and Gain Control

Large-scale changes in EOD amplitude occur as a result of seasonal changes in the conductivity of the water, damage to the electric organ, or changes in posture. Experiments in which EOD amplitude was artificially manipulated to mimic these effects have shown that the ELL pyramidal cells maintain constant responses to electrolocation targets over a wide range of EOD amplitudes. Experiments using lesions or local anesthetic injection have demonstrated that the indirect descending pro-

jection from the NP through the EGp to the ELL's DML must be intact for this gain-control mechanism to operate (Bastian, 1986).

Adjustment of pyramidal cell gain to compensate for changing EOD amplitude requires that EOD amplitude itself be measured by the system. The typical basilar and nonbasilar pyramidal cells are unable to do this; they are rapidly adapting and do not signal long-term amplitude changes. However, a subset of pyramidal cells—deep pyramidal cells (*DBP*)—which are shown in Figure 2, are completely nonadapting. These ELL efferents project to a region within the NP where another category of nonadapting neurons, the NP multipolar cells (*Mp*), are found. The firing of both ELL deep basilar pyramidal cells and NP multipolar cells accurately tracks long-term changes in EOD amplitude. Multipolar cell axons descend bilaterally to the EGp, where they provide excitatory input to EGp granule cells. The axons of the granule cells compose the ELL's DML, providing excitatory input to the apical dendrites of pyramidal cells and inhibitory interneurons, but not to the shorter apical dendrites of the deep basilar pyramidal cells. This pathway, consisting of tonically responding cell types (stippled somata in Figure 2), provides only excitatory input to its targets within

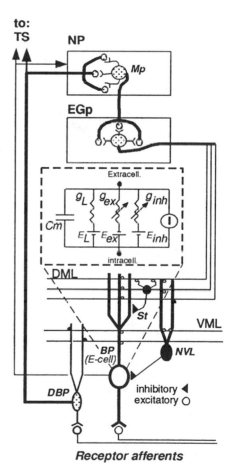

Figure 2. A diagram of the ascending and descending components of the electrosensory system that is thought to mediate gain control (stippled cells). Balanced changes in the conductances associated with the excitatory and inhibitory synapses—g_{ex} and g_{inh}—of the pyramidal cell equivalent circuit will alter the cells' input resistance, resulting in gain changes that are independent of changes in spontaneous firing rate, as described in the text. (Pyramidal cell equivalent circuit modified from Nelson, 1994.)

the ELL, yet its interruption increases pyramidal cell excitability rather than decreasing it. This suggests that descending control of ELL inhibitory neurons is involved. Additionally, minimal changes in the spontaneous activity of pyramidal cells has been shown to result from removal of descending input (Bastian, 1986).

Nelson (1994) has proposed an interesting mechanism that allows control of the pyramidal cell's gain, independent of membrane potential and spontaneous activity. This model shows that balanced changes in excitatory and inhibitory inputs can alter a neuron's input impedance while leaving resting potential unchanged. Since the cell's input impedance will determine the amplitude of membrane potential changes resulting from other synaptic currents, gain can be controlled effectively.

The implementation of this model in terms of descending inputs to the ELL is diagramed in Figure 2. A rise in EOD amplitude increases activity in the nonadapting deep basilar pyramidal (*DBP*) cells and NP multipolar cells, leading to enhanced EGp granule cell activity. The granule cell axons, DML parallel fibers, provide excitatory input to inhibitory interneurons (e.g., *NVL* cells or DML stellate cells [*St*]) as well as to ELL pyramidal cells. The pyramidal cells receive increased hyperpolarizing and depolarizing inputs, increasing the conductances associated with these synapses (g_{ex} and g_{inh} of the equivalent circuit for the pyramidal cell membrane of Figure 2). Given appropriately balanced synaptic weights, the effects of these increased conductances will cancel, leaving resting potential and spontaneous activity unchanged. However, the pyramidal cell's input resistance will be reduced and synaptic currents resulting from increased receptor afferent input (*I*) will produce smaller changes in membrane potential, thereby reducing the cell's gain. A reduction in EOD amplitude will result in oppositely signed changes in the circuit's activity and increased gain. As suggested by Nelson, this mechanism not only accounts for the observed behavior of ELL pyramidal cells, but may also be a strategy commonly used by modulatory inputs to independently control gain and spontaneous levels of activity.

Direct Descending Input and Sensory Searchlights

The anatomical and physiological properties of NP stellate and bipolar cells, which contribute to the direct descending input, differ significantly from the properties of the deep basilar pyramidal and NP multipolar cells, suggesting that the direct pathway plays a significantly different role in electroreception. The stellate cells have been studied physiologically (Bratton and Bastian, 1990) and are most likely glutamatergic. The GABAergic bipolar cells have yet to be physiologically characterized.

The projection of the NP multipolar cells to the EGp is diffuse; single multipolar cell axons ramify over long distances within the EGp, providing synaptic input to numerous, spatially separate domains of granule cells. The indirect projection is thought to affect ELL circuitry globally. In contrast, the projection of the NP stellate cells is reciprocally topographic with the ELL; stellate cells receiving input from a given region of the ELL project back to the VML of that same region (Maler and Mugnaini, 1993).

Unlike the multipolar cells which are spontaneously active and tonically responsive to changes in EOD amplitude, the stellate cells have virtually no spontaneous activity and give rapidly and completely adapting responses to transient changes in EOD amplitude. The stellate cells are unable to provide any information about the long-term status of the EOD and cannot contribute to the gain-control mechanism described earlier.

The stellate cells are, however, exceptionally responsive to the EOD distortions caused by moving electrolocation targets, and their anatomical and physiological properties suggest a function analogous to a "sensory searchlight," such as proposed by Crick (1984; see also Koch, 1987) for neurons of the corticothalamic projection. Activity in small populations of NP stellate cells may enhance the activity of subsets of ELL neurons processing "interesting" information (Bratton and Bastian, 1990).

Maler and Mugnaini (1993) have expanded on this suggestion, drawing more specific comparisons between the corticothalamic circuitry discussed by Crick and the direct NP-ELL projection summarized in Figure 3. These researchers propose that the NP bipolar cells (*Bi* in Figure 3), which probably integrate inputs from numerous stellate cells, provide a diffuse GABAergic projection to ELL pyramidal cells via the VML (dashed line in Figure 3). This would suppress pyramidal cell responses to spatially extensive, or uninteresting, stimuli

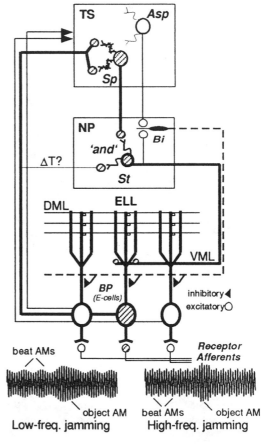

Figure 3. A diagram summarizing the proposed role of n. praeeminentialis (NP) stellate cells (*St*) acting as a "sensory searchlight," and the control of this function by inputs descending from the torus semicircularis (TS). Populations of ELL pyramidal cells that process "interesting" stimuli, indicated by the crosshatched basilar pyramidal cell (*BP*), project to a topographically related population of NP stellate cells. These, in turn, project back to the same population of ELL pyramidal cells, providing positive feedback excitation at least partly via NMDA synapses in the ELL's ventral molecular layer (VML). Torus cells with high spine density (*Sp*) may supply an input required for stellate cell activation. The low-pass characteristic of the spiny cells may facilitate their recognition of relevant stimuli, such as low-frequency EOD AMs, even when partially masked by higher-frequency beats, such as illustrated by the waveform-labeled, high-frequency jamming.

in a manner analogous to the diffuse inhibition proposed for thalamic reticular complex neurons (Crick, 1984). The projection of NP stellate cells (*St*) provides an excitatory pathway paralleling the inhibitory bipolar projection. Stellate cells project back to the VML, providing excitatory input to the same ELL pyramidal cells from which they receive input (cross-hatched cells in Figure 3). Excitation of small populations of stellate cells then acts as a searchlight, providing positive feedback excitation of pyramidal cells that are concurrently receiving certain patterns of receptor afferent inputs. NMDA receptors (see NMDA RECEPTORS: SYNAPTIC, CELLULAR, AND NETWORK MODELS) have been identified within the VML, and physiological experiments show an NMDA component of pyramidal cell excitatory postsynaptic potentials (EPSPs) resulting from VML activity (Maler and Mugnaini, 1993). Therefore, responses of a given pyramidal cell depolarized via increased receptor afferent input are expected to show highly nonlinear increases in activity owing to simultaneous activation of their NMDA receptors via inputs descending from a topographically related population of stellate cells. The facilitated responses of the selected group of ELL pyramidal cells would, in a sense, stand out more clearly against the background activity of the inhibited neighboring pyramidal cells.

Controlling the Searchlight: Dendritic Spines, Temporal Filtering, and the Rejection of Disruptive Signals

Mechanisms for locally and transiently enhancing the responsiveness of subsets of neurons in a sensory processing network must be controlled or directed so that cells processing relevant stimuli are specifically selected. Higher control of the NP stellate cells would also be needed to reduce the obvious stability problems inherent in the positive feedback loop described earlier. This control is expected to emanate from higher centers which can somehow recognize relevant stimuli while rejecting those that are irrelevant or confounding.

Clear-cut distinctions can be made between patterns of stimuli that are disruptive and should be rejected and those that carry useful information for these animals. Electrolocation targets cause transient amplitude modulations of the EOD, as shown below the fish in Figure 1, and behavioral studies show that the fish are capable of detecting and orienting to these targets based on electrosensory cues alone (Heiligenberg, 1977; Bastian, 1990). Similar, but continuous, amplitude modulations can result from the interaction of the EODs of nearby fish. When weakly electric fish having similar EOD frequencies are in close proximity, each animal perceives the sum of the individual discharges, and the summed discharges produce a beat waveform. These continuous amplitude modulations can mask EOD modulations attributable to electrolocation targets (as illustrated by the waveform showing low-frequency jamming in Figure 3), reducing an animal's electrolocation abilities. The degree to which electrolocation is jammed depends on the EOD frequencies of the animals involved. The beat frequency, or amplitude modulation (AM) repetition rate, is determined by the difference in the animals' discharge frequencies. When this difference is low (less than about 10 Hz), the spectral characteristics of the beats are similar to those of AMs caused by detectable electrolocation targets, and electrolocation is severely jammed. Higher beat frequencies are less deleterious, and several species have developed jamming avoidance responses (JARs) whereby the animals alter their EOD frequencies, enlarging the frequency difference between their EODs.

The JAR results in a shift of the beat frequency to higher values and an improvement in the animal's ability to electro-locate, presumably because the nervous system can somehow reject these higher-frequency beats while attending to the lower-frequency AMs attributable to electrolocation targets. The effect of high-frequency beats on the AM due to an electrolocation target is illustrated by the high-frequency jamming waveform in Figure 3. The rejection of high-frequency beats is not attributable simply to filtering the input signal at the periphery, as electroreceptors respond well to AM frequencies in excess of 64 Hz. Higher-order processes involving the descending electrosensory pathways may be involved in selecting relevant, i.e., low-frequency, AMs, rejecting the higher-frequency AMs that remain after the JAR is completed.

Rose and Call (1992) recently identified neurons within the TS which act as low-pass filters, preferentially responding to AMs of less than about 5 to 10 Hz. Other torus neurons have broad-band properties, and respond well to AMs in excess of 15 Hz. The low-pass characteristic of torus neurons is highly correlated with the density of DENDRITIC SPINES (q.v.) on a given cell. Patterns of postsynaptic potentials recorded from cells with high spine densities accurately reflect the envelope of low-frequency AMs, but postsynaptic potential (PSP) amplitude is significantly attenuated as AM frequency increases beyond about 10 Hz. Torus neurons devoid of spines show minimal reduction of PSP amplitude as a function of AM frequency. The low-pass filtering is thought to be a consequence of the passive effects resulting from synaptic inputs to spines having stems with high axial resistance. These low-pass neurons could facilitate the animal's analysis of low-frequency, object-caused modulations, even when partially masked by the higher-frequency modulations present after two animals' discharge frequencies have diverged owing to the JAR (Rose and Call, 1992).

Both heavily spined as well as torus neurons devoid of spines are known to receive input from ELL pyramidal cells and to project to the NP, as diagramed in Figure 3 (Carr and Maler, in Bullock and Heiligenberg, 1986). Heavily spined torus neurons (*Sp*), which are preferentially responsive to low-frequency amplitude modulations, are candidates for directing the searchlight function proposed for the NP stellate cells. Increased activity of these torus cells should signal the presence of electrolocation targets (low-frequency EOD AM), and if they provide excitatory input to the stellate cells (as proposed in Figure 3), their activity could gate the stellate cell positive feedback to the ELL pyramidal cells.

The stellate cells can be thought of as acting as an "and" gate, providing descending excitation to pyramidal cells only upon receipt of simultaneous inputs from both the torus spiny cells and the topographically related population of ELL pyramidal cells. Having stellate cell firing contingent upon, or gated by, torus spiny cells restricts stellate cell responses to only those electrosensory stimuli having appropriate low-frequency characteristics. Continuous high-frequency AMs, such as are present following an animal's JAR, are ineffective stimuli for the spiny cells and would not permit stellate cell firing; hence, pyramidal cell responses to these higher-frequency AMs would not be facilitated.

Although speculative, the suggestion that the spiny or low-pass torus neurons project to NP stellate cells would explain some of the interesting physiological properties of stellate cells. Stellate cell responses appear after a long latency, which is approximately 5 ms longer than the responses of the NP multipolar cells. This is surprising since both the stellate and multipolar cells receive direct input from the ELL. It has been suggested that the longer stellate cell latency may be a result of a 'required' input from the torus (Bratton and Bastian, 1990). Furthermore, single pyramidal cell axons branch to project to both the NP and the torus. The NP branch is notably thinner, implying a conduction delay. Such a delay (ΔT in Figure 3) could ensure the stellate cells' receipt of temporally coincident inputs ascending from the ELL and descending from the torus.

The AM tuning of the stellate cells has also been studied, and these cells show a strong low-pass characteristic. Their responses peak at AM frequencies similar to those preferred by the spiny torus cells (about 8 Hz), then cut off rapidly as AM frequencies increase further. The stellate cells are totally unresponsive to AMs above about 32 Hz. The high-frequency insensitivity of the stellate cells could result from reduced descending input from torus spiny cells, but active inhibitory mechanisms are also present (Bratton and Bastian, 1990). The low-frequency AM selectivity of the circuit diagramed in Figure 3 could be further improved if the output of the wide-band spine-free torus cells contributed to the diffuse inhibition of the ELL thought to be mediated by the NP bipolar cells.

Discussion

This review indicates specific roles for descending or feedback pathways in a sensory processing system. Anatomically and physiologically distinct pathways are likely to control two important properties of the electrosensory system. The gain of the output cells of the first-order processing station is controlled by a diffuse descending projection that may involve simultaneous modulation of excitatory and inhibitory inputs to the distal apical dendrites of these cells. Modeling studies show how balanced changes in these inputs can control gain, independent of the cell's membrane potential and resting levels of activity. A parallel descending pathway can potentially influence localized regions of the first-order processing station, effectively increasing the responses of smaller numbers of output neurons processing behaviorally relevant information. The anatomy and physiology of this pathway shares many features with a sensory searchlight proposed for the mammalian corticothalamic projection (Crick, 1984; Maler and Mugnaini, 1993).

Animals, such as electric fish, echolocating bats, and others, rely very heavily on a single sensory system. Use of these animals in studies of neural networks simplifies the problems associated with identifying behaviorally relevant stimuli and facilitates the design of experiments that determine how the nervous system accomplishes well-defined tasks. The general problems and constraints faced by animals relying on electrosensory information are commonly faced by many others. Understanding the ways that these animals solve some of these problems may indicate general solutions used by a wide variety of sensory systems.

Acknowledgment. This work was supported by NIH grant NS12337.

Road Map: Other Sensory Systems
Related Reading: Echolocation: Creating Computational Maps; Neuroethology, Computational; Thalamus

References

Bastian, J., 1986, Gain control in the electrosensory system: A role for the descending projections to the electrosensory lateral line lobe, *J. Comp. Physiol.*, 158:505–515.
Bastian, J., 1990, Electroreception, in *Comparative Perception*, vol II, *Complex Signals* (W. C. Stebbins and M. A. Berkley, Eds.), New York: Wiley. ◆
Bastian, J., and Bratton, B., 1990, Descending control of electroreception, I: Properties of nucleus praeeminentialis neurons projecting indirectly to the electrosensory lateral line lobe, *J. Neurosci.*, 10: 1226–1240.
Bastian, J., and Courtright, J., 1991, Morphological correlates of pyramidal cell adaptation rate in the electrosensory lateral line lobe of weakly electric fish, *J. Comp. Physiol.*, 168:393–407.
Bratton, B., and Bastian, J., 1990, Descending control of electroreception, II: Properties of nucleus praeeminentialis neurons projecting directly to the electrosensory lateral line lobe, *J. Neurosci.*, 10:1241–1253.
Bullock, T. H., and Heiligenberg, W., 1986, *Electroreception*, New York: Wiley. ◆
Crick, F., 1984, Function of the thalamic reticular complex: The searchlight hypothesis, *Proc. Natl. Acad. Sci. USA*, 81:4586–4590.
Heiligenberg, W., 1977, *Principles of Electrolocation and Jamming Avoidance*, vol. 1, *Studies of Brain Function*, New York: Springer-Verlag. ◆
Heiligenberg, W. F., 1991, *Neural Nets in Electric Fish*, Cambridge, MA: MIT Press. ◆
Hopkins, C. D., 1988, Neuroethology of electric communication, *Annu. Rev. Neurosci.*, 11:497–535. ◆
Koch, C., 1987, The action of the corticofugal pathway on sensory thalamic nuclei: A hypothesis, *Neuroscience*, 23:399–406.
Maler, L., and Monaghan, D., 1991, The distribution of excitatory amino acid binding sites in the brain of an electric fish, *Apteronotus leptorhynchus*, *J. Chem. Neuroanat.*, 4:36–91.
Maler, L., and Mugnaini, E., 1993, Organization and function of feedback to the electrosensory lateral line lobe of Gymnotiform fish, with emphasis on a searchlight mechanism, *J. Comp. Physiol.*, 173:667–670.
Nelson, M. E., 1994, A mechanism for neuronal gain control by descending pathways, *Neural Computat.*, 6:242–254.
Rose, G. J., and Call, S. J., 1992, Evidence for the role of dendritic spines in the temporal filtering properties of neurons: The decoding problem and beyond. *Proc. Natl. Acad. Sci. USA*, 89:9662–9665.
Sas, E., and Maler, L., 1987, The organization of afferent input to the caudal lobe of the cerebellum of the gymnotid fish *Apteronotus leptorhynchus*, *Anat. Embryol.*, 177:55–79.

Emotion and Computational Neuroscience

Joseph E. LeDoux and Jean-Marc Fellous

Introduction

Emotion is clearly an important aspect of the mind; yet it has been largely ignored by the "brain and mind (cognitive) sciences" in modern times. However, there are clear signs that this is beginning to change. In this article, we survey some issues about the nature of emotion, describe what is known about the neural basis of emotion, and consider some efforts that have been made to develop computer-based models of different aspects of emotion.

What Is Emotion?

The nature of emotion has been debated within psychology for the past one hundred years. The formal debate goes back to William James's famous question: Do we run from the bear because we are afraid, or are we afraid because we run? James suggested that we are afraid because we run. Subsequently, the psychological debate over emotion has centered on the question of what gives rise to the subjective states of awareness that we call emotions or emotional experiences. Theories of emotional experience typically seek to account for how different emotional states come about and can be grouped into several broad categories: feedback, central, arousal, and cognitive theories (for review, see LeDoux, 1989). Though very different in some ways, each of these theories proposes that emotional experiences are the result of prior emotional processes. Feedback and arousal theories require that the brain detect emotionally significant events and produce responses appropriate to the stimulus; these responses then serve as a signal that determines the content of emotional experience. Central and cognitive appraisal theories, which are in some ways different levels of description of similar processes, assume that emotional experience is based on prior evaluations of situations; these evaluations then determine the content of experience. Interestingly, the evaluative processes that constitute central and appraisal theories are also implicitly necessary for the elicitation of the peripheral responses and arousal states of feedback and arousal theories.

The disparate theories of emotional experience thus all point to a common mechanism—a central evaluative system that determines whether a given situation is potentially harmful or beneficial to the individual. Since these evaluations are the precursors to conscious emotional experiences they must, by definition, be unconscious processes. Such processes are the essence of the ignored half of James's question. That is, we run from a bear because our brain determines that bears are dangerous. Many emotional reactions are likely to be of this type: unconscious information processing of stimulus significance, with the experience of "emotion" (the subjective state of fear) coming after the fact.

Unfortunately, we have no adequate theory of how conscious experiences of the emotional or nonemotional kind emerge from prior processing. In spite of this, researchers in other areas have made progress. For example, vision researchers have achieved considerable understanding of the neural mechanisms underlying the processing of color while still knowing little about how color experience emerges from color processing (see COLOR PERCEPTION). Similarly, we can study how the brain processes the emotional significance of situations without having to first solve the problem of how those situations are experienced as conscious content.

The Neural Basis of Emotional Processing

Traditionally, emotion has been ascribed to the brain's limbic system, which is presumed to be an evolutionary old part of the brain involved in the survival of the individual and species (MacLean, 1952). Some of the areas usually included in the limbic system are the hippocampal formation, septum, cingulate cortex, anterior thalamus, mammillary bodies, orbital frontal cortex, amygdala, hypothalamus, and certain parts of the basal ganglia. However, the limbic system concept and the limbic system theory of emotion are both problematic (LeDoux, 1991). The survival of the limbic system theory of emotion is due in large part to the fact that the amygdala, a small region in the temporal lobe, was included in the concept (LeDoux, 1991).

The amygdala has been consistently implicated in emotional functions (see various chapters in Aggleton, 1992). Lesions of this region interfere with both positive and negative emotional reactions. Moreover, unit recording studies show that cells in the amygdala are sensitive to the rewarding and punishing features of stimuli and to the social implications of stimuli. Other limbic areas have been less consistently implicated in emotion, and when they have been implicated, it has been difficult to separate out the contribution of the region to emo-

tion per se as opposed to some of the cognitive prerequisites of emotion (for discussion, see LeDoux, 1991).

The contribution of the amygdala to emotion results in large part from its anatomical connectivity (for review, see the Amaral chapter in Aggleton, 1992). The amygdala receives inputs from each of the major sensory systems and from higher-order association areas of the cortex. The sensory inputs arise from both the thalamic and cortical levels. These various inputs allow a variety of levels of information representation (from raw sensory features processed in the thalamus to whole objects processed in sensory cortex to complex scenes or contexts processed in the hippocampus) to impact on the amygdala and thereby activate emotional reactions. All of these inputs converge on the lateral nucleus of the amygdala, which can be viewed as the sensory and cognitive gateway into the amygdala's emotional functions. At the same time, the amygdala sends output projections to a variety of brainstem systems involved in controlling emotional responses, such as species-typical behavioral (including facial) responses, autonomic nervous system responses, and endocrine responses. All of these outputs originate from the central nucleus of the amygdala. Within the amygdala, the stimulus input region (lateral nucleus) and the motor output region (central nucleus) are interconnected, allowing an anatomical account of how events in the world come to elicit emotional responses.

Much of the anatomical circuitry of emotion described above has been elucidated through studies of fear conditioning, a procedure whereby an emotionally neutral stimulus, such as a tone or light, is associated with an aversive event, such as a mild footshock (see chapters by LeDoux, Kapp, et al., and Davis in Aggleton, 1992). After such pairings, the tone or light comes to elicit emotional reactions that are characteristically expressed when members of the species in question are threatened. While there are other procedures for studying emotion, none has been as successfully applied to the problem of identifying stimulus-response connections in emotion. The fear conditioning model is at this point particularly attractive since it has laid out pathways from the sensory input stage to the motor output stage of processing, showing how simple stimulus features, stimulus discriminations, and contexts control the expression of behavioral, autonomic, and endocrine responses in threatening situations.

Although many emotional response patterns are hard-wired in the brain's circuitry, the particular stimulus conditions that activate these are mostly learned by association through classical conditioning. The amygdala appears to contribute significantly to this aspect of learning and memory and may be a crucial site of synaptic plasticity in emotional learning (LeDoux, 1992). This form of memory is quite different from what has come to be called *declarative memory*, the ability to consciously recall some experience from the past (Squire, Knowlton, et al., 1993). Declarative memory, in contrast to *emotional memory*, crucially requires the hippocampus and related areas of the cortex. When we encounter some stimulus that in the past had aversive consequences, we recall the details of who we were with and where we were and even that it was a bad experience. However, in order to give the declarative memory an emotional flavor, it may be necessary for the stimulus, simultaneously and in parallel, to activate the emotional memory system of the amygdala. It is likely to be this dual activation of memory systems that gives our ongoing declarative memories their emotional coloration.

At this point, we have mentioned "emotional experience" a number of times, and it may be worth speculating on just what an emotional experience is and how it might emerge. The emotion of *fear* will be used as an example. All animals, regardless of their stage of evolutionary development, must have the ability to detect and escape from or avoid danger. The widespread distribution of these behaviors in the animal kingdom makes it unlikely that the subjective experience of fear is at the heart of this ability. It may well be the case that subjective, consciously experienced fear is a mental state that occurs when the defense system of the brain (the system that detects threats and organizes appropriate responses) is activated, but only if that brain also has the capacity for consciousness. That is, by this reasoning, fear and other emotions reflect the representation of the activity of neural systems shaped by evolution and the responses they produce as conscious content. If this is true, then it is important that we focus our research efforts on these stimulus detection-and-response organizing systems, as these are the systems that generate the conscious content we call emotions. While emotional behaviors may be triggered by sensory inputs that bypass or pass through the neocortex, the experience of emotion is likely to involve the cortical representation of the emotional episode. Unfortunately, our understanding of the cortical representation of emotion episodes (or other conscious experiences) is poor at present.

Computational Models of Emotion

Using computers to understand emotions has always been a challenge. Popular beliefs define computing devices as inherently incapable of exhibiting and experiencing any emotions and, at present, no definite claims have been made that computers are (or may be) suitable for such a task. Nevertheless, consistent with the notion put forth in the introduction, computers can be and have begun to be used as tools for modeling certain aspects of emotional processing.

Models of Emotional Learning and Memory

As proposed by most central theories, many emotional responses are hard-wired in brain circuitry. Nevertheless, in humans and many other organisms, the environmental events that trigger these responses are often learned through experiences in which emotionally neutral stimuli come to be associated with emotionally charged stimuli. One important aspect of emotional processing, therefore, involves the manner in which the brain forms, stores, and uses associations between meaningless and meaningful stimuli.

Grossberg and Schmajuk (1987) developed a model of conditioned affective states based on the notion that conditioned reinforcement involves pairs of antagonistic neural processes, such as fear and relief. Computer simulations show and possibly explain various effects related to the acquisition and extinction of conditioned excitation and inhibition which have been observed experimentally. On the basis of their model, the authors suggest a mechanism by which neutral events are charged with a reinforcing value (either positive or negative) dependent on the previous activity of the model. Although none of the neurons simulated have been designed specifically to match the individual properties of actual neurons, activity in the model neurons has been related by the authors to the activity of brain structures involved in the processing of certain emotions, such as the hippocampo-amygdaloid system (described as a zone of convergence of conditioned (CS) and unconditioned (US) stimulus pathways), the septum (described as a zone in which the opposition of the processes is represented), the hypothalamus, the nucleus of the solitary tract, and the reticular formation (described as zones of visceral and somatosensory inputs).

In an extension of the basic model, Ricart added a new neural center to the US pathway, the role of which is to prolong the neural representation of the US after its actual termination (Ricart, 1992). The amount and nature of the activity of this center is related to the "unexpectedness" of the stimuli and has been suggested to be analogous to the activity of the locus coeruleus (LC). Also, the extended model attempts to account for phenomena related to the interstimulus interval (ISI) in classical conditioning (excitatory backward conditioning), which the original model did not address.

These are interesting and innovative models with strong predictive power. However, the models might be even more valuable if they were focused on findings related exclusively to one form of conditioning rather than blending different forms, such as fear conditioning and eyeblink conditioning. For example, the ISI phenomenon is important in eyeblink conditioning, but fear conditioning is fairly insensitive to this variable. Focusing on one form of conditioning would also allow one to make more direct correlations with anatomical findings, since the different forms of conditioning have different underlying neural systems, which have been precisely identified in recent years (Thompson, 1986; LeDoux, 1992).

Armony and co-workers have implemented another connectionist model of emotional learning and memory that, like the previous two conditioning models, also focuses on zones of convergence of US and CS pathways (Armony, Servan-Schreiber, et al., 1993). However, in contrast to Grossberg and Schmajuk's model and Ricart's model, this model is explicitly based on the facts known about the functional and neural mechanisms of fear conditioning. It examines processing in two parallel sensory (CS) transmission pathways to the amygdala from the auditory thalamus and the auditory cortex in a learning situation involving an auditory CS paired with a footshock US. The model is initially trained using a modified Hebb-type learning rule and, under testing conditions, reproduces data related to frequency-specific changes of the receptive fields known to exist in the auditory thalamus and amygdala.

Computational Models of Cognitive-Emotion Interactions and Appraisal

Researchers in experimental psychology, artificial intelligence (AI), and cognitive science have long recognized the mutual influences of emotions and cognition. However, these interactions are still not clearly formulated. We still do not have adequate theories defining each of these components of human mentation (emotion and cognition), much less a full understanding of how cognition and emotion might relate (see EMOTION-COGNITION INTERACTIONS).

As described above, most theories of emotion recognize the importance of evaluative or appraisal processes. Although there is considerable disagreement as to how these processes should best be viewed, most workers nevertheless see evaluative or appraisal processes as functioning by comparing sensed characteristics of the world to internal goals, standards, and attitude structures, deducing the emotional significance of the stimulus, guiding the expression of emotional behavior and other physiological responses, and influencing other modules pertaining to behavioral decisions.

In principle, it is possible to model appraisal processes using classical AI techniques (see Chwelos and Oatley, 1994, for a review and a critique). For example, using a vector space approach, one could attempt to find a plausible mapping between appraisal features (e.g., novelty, urgency, intrinsic pleasantness) and emotion categories (e.g., fear, joy, pride). Relying on a posteriori verbal reports and a predefined set of emotions, one could then derive a limited set of appraisal criteria, sufficient for emotion prediction and differentiation. Other AI approaches, such as decision trees, pattern matching, and production rules (expert systems), are also possible, although each of these methods encounters theoretical difficulties. These types of systems, however, do not generally account for any neurophysiological data.

One criticism often made of cognitive models of emotion is related to the complexity of processing involved and to the time they consequently require. From an AI point of view, the criticism has been addressed by introducing reactivity to "classical" cognitive models (e.g., Bates, Loyall, and Reilly, 1992). Classical AI approaches assume that systems possess a well-defined representation of their environment, state, actions, and goals. In contrast, reactive systems do not make such assumptions; they are mostly based on real-time, incomplete evaluations, their performance being based more on the properties of the evaluative mechanisms than on the quality and quantity of their internal representations (see Lyons and Hendricks, 1992, for a review and examples).

It is interesting to note that, as we mentioned earlier, appraisal of sensory information might be one of the most prominent functions of the amygdala, placing this structure in a key position to actually perform the mapping of the emotional value of the stimuli. In this view, the relation between amygdala activity and emotion is a computational one (in the broad sense of the term) rather than a subjective one. The existence of multiple pathways to the amygdala from input processing systems of various levels of complexity (see above) provides a biological resolution to some of the concerns that have been raised about the importance of cognition in driving emotion. The involvement of cognition can be minimal or maximal, depending on the situation.

Models of Facial Expressions of Emotion

Of interest to feedback and arousal theories, the expression of emotion in the face is an important biological aspect of emotion that has significant implications for how emotion is communicated in social situations (Darwin, 1872). Face recognition and analysis of facial expression has only recently been an active field of research in the computer vision community (for a recent review, see Samal and Iyengar, 1992). Face analysis can be computationally decomposed into three subproblems: detecting the face in a scene; identifying the face; and analyzing its expression. At present, each of these tasks uses different features of the face, and often different computational approaches. These approaches are based on psychophysical observations and are not yet explicitly based on neurophysiological data. However, a number of neurophysiological studies have been conducted (for review, see Rolls's chapter in Aggleton, 1992). These studies have shown cells selectively responsive to particular faces in areas of temporal neocortex and in the amygdala. More recently, other studies showed that there might be an influence of facial expressions on the actual neural correlates of the emotional states experienced, through modifications of blood flow characteristics (for review, see Ekman, 1992). Other approaches are more physico-mathematical, relying on image processing techniques. These implementations address exclusively the problem of emotional expression (and, possibly, communication of emotions) without relying on any theory of emotional experience.

Conclusion

It is important to distinguish between emotional experiences and the underlying processes that lead to emotional experi-

ences. One of the stumbling blocks to an adequate scientific approach to emotion has been the focus of the field on constructing theories of the subjective aspects of emotion. Studies of the neural basis of emotion and emotional learning have instead focused on how the brain detects and evaluates emotional stimuli and how, on the basis of such evaluations, emotional responses are produced. The amygdala was found to play a major role in the evaluation process. It is likely that the processing that underlies the expression of emotional responses also underlies emotional experiences, and that progress can be made by treating emotion as a function that allows the organism to respond in an adaptive manner to challenges in the environment rather than to a subjective state. While computational approaches to subjective experiences of the emotional or nonemotional kind are not likely to be easily achieved, computational approaches to emotional processing are both possible and practical. Although relatively few models currently exist, this situation is likely to change as researchers begin to realize the opportunities that are present in this too-long neglected area.

Road Map: Connectionist Psychology
Related Reading: Conditioning; Emotion-Cognition Interactions; Sparse Coding in the Primate Cortex

References

Aggleton, J. P., Ed., 1992, *The Amygdala: Neurobiological Aspects of Emotion, Memory, and Mental Dysfunction*, New York: Wiley-Liss.

Armony, J. L., Servan-Schreiber, D., et al., 1993, A connectionist model of the thalamo-amygdala network mediating conditioned fear reactions, *Soc. Neurosci. Abst.*, 19:1227.

Bates, J., Loyall, A. B., and Reilly, W. S., 1992, Integrating reactivity, goals, and emotion in a broad agent, in *Proceedings of the Fourteenth Annual Conference of the Cognitive Science Society*, Hillsdale, NJ: Erlbaum, pp. 696–701.

Chwelos, G., and Oatley, K., 1994, Appraisal, computational models, and Scherer's expert system, *Cognition and Emotion*, 8(3):245–257.

Darwin, C., 1872, *The Expression of Emotions in Man and Animals*, reprinted 1979, London: Julian Friedman.

Ekman, P., 1992, Facial expressions of emotion: New findings, new questions, *Psychol. Sci.*, 3(1):34–38.

Grossberg, S., and Schmajuk, N. A., 1987, Neural dynamics of attentionally modulated pavlovian conditioning: Conditioned reinforcement, inhibition, and opponent processing, *Psychobiology*, 15(3):195–240.

LeDoux, J. E., 1989, Cognitive-emotional interactions in the brain, *Cognition and Emotion*, 3(4):267–289.

LeDoux, J. E., 1991, Emotion and the limbic system concept, *Concepts Neurosci.*, 2:169–199.

LeDoux, J. E., 1992, Brain mechanisms of emotion and emotional learning, *Curr. Opin. Neurobiol.*, 2:191–197.

Lyons, D. M., and Hendricks, A. J., 1992, Planning, reactive, in *Encyclopedia of Artificial Intelligence*, 2nd ed., New York: Wiley, pp. 1171–1181.

MacLean, P. D., 1952, Some psychiatric implications of physiological studies on the frontotemporal portion of the limbic system (visceral brain), *Electroencephalogr. Clin. Neurophysiol.*, 4:407–418.

Ricart, R., 1992, Neuromodulatory mechanisms in neural networks and their influence on interstimulus interval effects in Pavlovian conditioning, in *Motivation, Emotion, and Goal Direction in Neural Networks* (D. S. Levine and S. J. Leven, Eds.), Hillsdale, NJ: Erlbaum, pp. 115–165.

Samal, A., and Iyengar, P. A., 1992, Automatic recognition and analysis of human faces and facial expressions: A survey, *Pattern Recognition*, 25(1):65–77.

Squire, L. R., Knowlton, B., et al., 1993, The structure and organization of memory, *Annu. Rev. Psychol.*, 44:453–495.

Thompson, R. F., 1986, The neurobiology of learning and memory, *Science*, 233:941–947.

Emotion-Cognition Interactions

Mark Beeman, Andrew Ortony, and Laura A. Monti

Introduction

The central problem of emotion modeling is to differentiate situations that elicit emotions from those that do not, and to associate those that do with appropriate emotional states. Crucial to this process is the determination of the emotional significance of emotion-inducing situations. Neuroanatomical research on emotions has necessarily focused on a simplified version of this process using relatively simple and measurable physiological responses (see EMOTION AND COMPUTATIONAL NEUROSCIENCE).

In considering interactions between cognition and emotion, a key question is whether emotion processing and cognitive processing are independent. In support of the independence position, it is known that mere exposure to a stimulus causes subjects to show increased positive affect toward the stimulus, even if they are not aware of having previously encountered it (e.g., Zajonc, 1980). This *mere exposure effect* precludes the involvement of conscious cognition, but not of unconscious cognition. Whereas some instances of some human emotions involve little or no conscious cognition, others (e.g., pride, relief) certainly require considerable cognition, much of it conscious. Cognition often plays an important role in the elicitation and intensification of emotions.

Brain Theories

Although emotions are thought to be mediated by separable subsystems (e.g., Gray, 1987), neuroanatomical studies in vertebrates often implicate the amygdala in affective processing of sensory stimuli. Some neurons in the amygdala respond preferentially to particular categories of reward-associated stimuli, while others respond selectively to novel visual stimuli, perhaps reflecting the functional equivalence of novel and rewarded stimuli: i.e., both deserve increased attention and processing (Rolls, 1995).

The amygdala receives inputs both directly from the thalamus, including from thalamic sensory nuclei, and indirectly from the cortex. The former projections may allow fast but simple processing of the affective aspects of sensory stimuli, whereas the latter allow slower but fuller (cognitive) processing of stimulus affect. Thus, anatomical evidence suggests that some emotional processing could occur independently, but that cognitive processing may influence or elicit other emotions. Neurobehavioral evidence is consistent with this view: when areas of the sensory cortex appropriate to a conditioned stimulus are lesioned, rats continue to respond to the conditioned stimulus (LeDoux, Sakuguchi, and Reis, 1984). A reasonable conclusion is that higher cognitive functions are not always necessary for organisms to respond to the affective aspects of stimuli.

Neural Networks

Neural network models can certainly assess familiarity and feed that information to emotion subsystems. For example, in the Categorizing and Learning Module (CALM) (Murre, Phaf, and Wolters, 1992), each network module includes an arousal unit that receives input from all other units and "acts as a measure of the novelty of the input pattern." Competition weeds out other activations, propagating low activation to the arousal unit, when the input pattern activates a single representation unit (which, in CALM, represents a single input pattern). Otherwise, competition is decreased and high activation is propagated to the arousal unit. In such a system, arousal—reflecting novelty—could trigger emotional responses and influence liking, as in the mere exposure effect.

However, the most interesting aspect of the mere exposure effect explained earlier is that it occurs without conscious cognition, presumably through direct thalamo-amygdaloid afferents. By contrast, in CALM, the (representation) units which detect novelty seem more closely related to cognitive processing and the cortex than to emotional processing and the amygdala. Also, it is unclear whether exposure at an unconscious level—say, when the input pattern is so weak or brief that the system cannot settle into any stable solution—would sufficiently decrease activation of the arousal units at subsequent presentations. Nevertheless, such networks can detect novelty, and it seems likely that a more biologically realistic network could be designed to modulate affective responses as a result of increasing familiarity.

Associating initially neutral stimuli with emotional responses may be a core component of emotion, and has been well described neurally (see EMOTION AND COMPUTATIONAL NEUROSCIENCE). Many neural networks have simulated various aspects of classical conditioning, particularly within the ADAPTIVE RESONANCE THEORY (ART) model.

Neural networks have also simulated reward-mediated learning of behavior, which allows emotional systems to relate so-called hardwired emotional responses to a variety of stimuli. One network (Nakamura and Ichikawa, 1990) models long-term potentiation (LTP; see HEBBIAN SYNAPTIC PLASTICITY) that requires two inputs: calcium influx caused by activation of the pyramidal cells from the sensory afferent, and catecholamine release from the reward input. Although a far cry from the complex emotional experiences of humans, the models already implemented demonstrate how neural networks can use mathematical parameters to simulate components of emotional processing.

Artificial Intelligence Models

Most artificial intelligence (AI) models of emotion processing have used symbolic representation approaches to model various aspects of the influence of cognition on emotion, and thus are based on cognitive models. Some AI models have been concerned with reasoning about emotions and emotion-related states, either in the context of language-understanding systems—where inferences are made about people's presumed goals from described emotions—or in the context of reasoning systems—where inferences are made as a means of generating plausible explanations of described emotional states. Systems have also been developed or proposed that take as their starting point the idea that the function of emotions is to control attention and behavior (Simon, 1967) in light of individuals' goals and concerns. Some of these systems model the generation of emotions in response to perceived situations in a microworld and then allow the generated emotions to play a role in the control of subsequent events in the world.

The Influence of Cognition on Emotion

Despite the possibility that emotion processing can occur without conscious cognition, cognition not only often affects emotion processing, it often determines it. Many emotion theorists have proposed that cognition contributes to emotion by shaping the way in which the subjective world is interpreted. Such theorists attempt to characterize different emotions in terms of different kinds of antecedent cognitive conditions. The goal is not simply to specify the cognitive antecedents of different emotions, but to identify more generally some systematic structural principles governing the relation between such conditions and emotions.

Brain Theories

There is considerable neurobiological evidence that emotion processing is indeed influenced by cognitive processing. For instance, in addition to subcortical inputs, brain structures involved in emotion processing, such as the amygdala, receive input from many cortical areas. These inputs come from most, if not all, input modalities, but more from secondary and tertiary sensory areas than from primary areas. In fact, the cortical areas projecting to the amygdala do not respond selectively to stimuli associated with reinforcement, which suggests that such cortical areas do not process emotions per se, but feed into and affect emotion processing in the amygdala. A more specific breakdown of how cortical and cognitive processing influence the myriad complex emotions described by psychologists remains to be described at the neural level.

Neural Networks

Given the behavioral and neuroanatomical evidence that cognition normally plays an important role in emotion processing, viable neural network models of emotion processing must allow for inputs from cognitive processing units. Several existing neural network models do provide input to arousal units or emotional learning systems from units that are presumably involved in cortical and cognitive processing (such as CALM's representation units).

There have been several attempts within neural network frameworks (albeit without simulations; see Levine and Leven, 1992) to posit dysfunctional attentional or regulatory mechanisms to explain emotional psychological disorders. One proposal has suggested that manic-depressive illness results from defective monoamine regulation, and that delusions and hallucinations result from a defect in a pattern-matching process; a similar theory addressed learned helplessness and depression. Some behavioral features of schizophrenia have been simulated in a neural network which models the enhanced signal-to-noise ratio caused by catecholamine release (Cohen and Servan-Schreiber, 1992).

These theories and models provide an important first step toward modeling cognition-emotion interactions. Such interactions have been widely examined within cognitive frameworks, but future neural networks will have to devote more attention to the way in which cognition shapes and elicits the various complex emotions that people experience.

Cognitive Models

From a cognitive perspective, the challenge is to characterize emotion differentiation and intensification. It is necessary to differentiate distinct emotions (e.g., shame, embarrassment, regret) in terms of the cognitions that give rise to them. This requires a classification of emotions according to the way in which people experience situations, while recognizing that a classification appropriate for one culture might be inappropriate for another. It is also necessary to specify the factors that influence the intensity of emotions. One attempt (Ortony, Clore, and Collins, 1988) to address both of these questions characterizes classes of human emotions in terms of reactions to appraisals of different kinds of situations (see Table 1). In

Table 1. Twenty-two Distinct Emotion Types, with Representative Emotion Words

Well-being Emotions (arising from the appraisal of events evaluated in terms of goals and interests)	
Pleased about a desirable event (HAPPY)	Displeased about an undesirable event (UNHAPPY)
Fortunes-of-Others Emotions (Well-being Subtypes)	
Pleased about an event desirable for other (HAPPY-FOR)	Displeased about an event desirable for other (RESENTMENT)
Pleased about an event undesirable for other (GLOATING)	Displeased about an event undesirable for other (PITY)
Prospect-Based Emotions (Well-being Subtypes)	
Pleased about a prospective desirable event (HOPE)	Displeased about a prospective undesirable event (FEAR)
Pleased about a confirmed desirable event (SATISFACTION)	Displeased about a confirmed undesirable event (FEARS-CONFIRMED)
Pleased about a disconfirmed undesirable event (RELIEF)	Displeased about a disconfirmed desirable event (DISAPPOINTMENT)
Attribution Emotions (arising from the appraisal of actions evaluated in terms of standards and values)	
Approving of one's own praiseworthy action (PRIDE)	Disapproving of one's own blameworthy action (SHAME)
Approving of other's praiseworthy action (ADMIRATION)	Disapproving of other's blameworthy action (REPROACH)
Well-Being/Attribution Emotions (Compounds)	
Approving of one's own praiseworthy action and pleased about the related desirable event (GRATIFICATION)	Disapproving of one's own blameworthy action and displeased about the related undesirable event (REMORSE)
Approving of someone else's praiseworthy action and pleased about the related desirable event (GRATITUDE)	Disapproving of someone else's blameworthy action and displeased about the related undesirable event (ANGER)
Attraction Emotions (arising from the appraisal of objects evaluated in terms of tastes and preferences)	
Attracted by an appealing object (LIKING)	Repelled by an unappealing object (DISLIKE)

Events are evaluated in terms of goals, the actions of agents in terms of standards and values, and the appealingness of objects in terms of tastes and preferences (based on Ortony, Clore, and Collins, 1988).

this model, each eliciting condition (e.g., the prospect of an undesirable event) is a precondition for a distinct emotion type (e.g., a fear emotion), and each emotion type has an associated set of factors that can influence its intensity. Neural network models will eventually have to include mechanisms for the kinds of classifications represented by such cognitive models.

The Influence of Emotion on Cognition

Research on mood-congruent memory suggests that people in positive moods are more likely to be able to recall positive than negative material, and vice versa, whereas (affective) state-dependent memory research indicates that memory is better if the affective state at the time of recall matches that at the time of learning. Both of these effects can be interpreted in terms of neoassociationist (or localist connectionist) models: affective units activated at the time of learning that are also activated at the time of recall provide additional connections to the to-be-recalled material, thereby increasing its availability.

Research on the influence of emotional states on evaluative judgments has shown that relatively low levels of happiness induced by trivial good fortune (e.g., finding a dime) significantly increase people's satisfaction ratings about their life in general, as well as specific aspects of it. Similarly, people made to feel afraid judge feared events to be more probable than do people made to feel angry, who themselves judge the perpetrators of potentially anger-inducing acts as more blameworthy than do people made to feel afraid.

Could the influence of affect on memory and on judgment both be explained in terms of the enhanced activation of mood-compatible material? Apparently not, because, for example, people tend to judge their level of happiness and life-satisfaction more positively on sunny days than on rainy days, but not if they articulate how they feel about the weather prior to making their judgment. The best explanation of this is that feelings which have not been consciously attributed to some external cause (e.g., the weather) are implicitly used as relevant information in making judgments about other things (Schwartz and Clore, 1983). A neural network model could accommodate such findings by utilizing a competitive binding scheme: once emotional arousal has been linked to one possible cause, it is less available to influence cognition or to be linked to other possible causes.

Brain Theories

Affect influences cognitive processing at the level of neurotransmitter systems. Depression engages a norepinephrine-driven arousal system, most prominently in the right hemisphere, biasing the brain to fill working memory with varied information, compromising attention. In contrast, anxiety may engage a dopamine-driven activation system, predominantly in the left hemisphere, biasing working memory toward redundant information and producing obsessive attentional focus (Tucker, 1990).

The outputs from brain structures involved in emotion processing suggest how neurotransmitter systems might influence cognitive processing. Activity in the amygdala may modulate the signal-to-noise ratio through direct projections to the cortex. Amygdalocortical projections could also cue a memory by providing a depolarizing input to enhance long-term potentiation to encode a significant stimulus (Rolls, 1995), similar to the neural network model of reward-mediated learning described earlier. Indeed, norepinephrine affects memory storage.

Neurophysiological evidence also suggests ways in which emotions can influence cognitive processing. The receptive fields of neurons in sensory areas can be altered through sensitization or classical conditioning (Weinberger, 1995). Sensory neurons can be sensitized rather generally, but even more interesting, the receptive field properties of sensory neurons can expand or shift toward a stimulus associated with reward or punishment, relative to changes in receptive fields of neurons coding nonrewarded stimuli. Apparently, reward and punishment alter the sensory processing of stimuli, although it is unclear whether these effects require input from emotion systems, such as the amygdala.

Finally, emotions may have enduring effects on cognition via the release of hormones. The level of adrenal steroid hormones, which may help prevent depression, is modulated by stress level and, in turn, modulates the density of synaptic spines in the hippocampus (McEwen, 1995). Therefore, emotion-modulated hormone levels could clearly affect memory functions. Likely, a wide range of mechanisms allow various emotions to exert distinct effects on cognitive processing.

Neural Networks

Various neural network architectures could allow emotions to influence cognition. In the CALM model, activation of the arousal unit increases the learning rate by globally changing the weights on intermodule connections. This might be analogous to increasing the availability of a neurochemical which facilitates learning. Alternatively, learning of a sensory input could be facilitated by coincident activation from the arousal unit onto the units encoding the stimulus.

Other emotion-driven free parameters of neural networks could affect cognitive activity in the networks. In one ART model, a "neuromodulator" parameter is "released" (from a center supposedly analogous to the locus coeruleus) when unexpected emotional patterns are inputted to the network (Ricart, 1992). This prolongs the transient activation caused by the stimulus, accounting for some effects of interstimulus interval on learning, although there are some problems with this model (see EMOTION AND COMPUTATIONAL NEUROSCIENCE).

Neural network models have employed various mechanisms to modulate the signal-to-noise ratio, one way in which emotion could influence cognitive processing. In a CALM module, an activated arousal unit outputs random noise to all representation units within the module. ART models contain feedback mechanisms that effectively boost activation in emotion-associated representations, giving them an advantage when competing for activation. Other neural network models of cognitive processing include attentional mechanisms. Although these mechanisms are interesting, a complete model would necessarily include various mechanisms which could affect cognition independently or in combination.

Conclusion

To date, neural network models of emotion processing and of emotion-cognition interactions have concentrated on the contribution of emotion processing to learning and memory. Future neural networks will have to address other aspects of emotion and emotion-cognition interactions. For instance, input from an active arousal system could selectively increase the learning rate for connections to units that participate in the representation of an input that is rewarded or punished. This might allow the representational properties of these units to shift or expand, as do receptive fields of neurons coding stimuli paired with reward or punishment. Other challenges lie in the short- and long-term effects of hormonal effects of emotion on memory and cognition. Future models will also have to expand their repertoire of emotion mechanisms, and help determine whether biological mechanisms newly discovered in animals can account for the behavioral effects of emotions observed in humans. However, in the long run, progress will be heavily dependent on models making better use of the cognitive information necessary to differentiate and intensify more complex emotions.

Road Map: Connectionist Psychology
Related Reading: Conditioning; Disease: Neural Network Models

References

Cohen, J. D., and Servan-Schreiber, D., 1992, Context, cortex, and dopamine: A connectionist approach to behavior and biology in schizophrenia, *Psychol. Rev.*, 99:45–77.

Gray, J. A., 1987, *The Psychology of Fear and Stress*, Cambridge, Eng.: Cambridge University Press.

LeDoux, J. E., Sakuguchi, A., and Reis, D. J., 1984, Subcortical efferent projections of the medial geniculate nucleus mediate emotional responses conditioned to acoustic stimuli, *J. Neurosci.*, 4:683–698.

Levine, D. S., and Leven, S. J., 1992, *Motivation, Emotion, and Goal Direction in Neural Networks*, Hillsdale, NJ: Erlbaum.

McEwen, B. S., 1995, Stressful experience, brain and emotions: Developmental, genetic and hormonal influences, in *The Cognitive Neurosciences* (M. S. Gazzaniga, Ed.), Cambridge, MA: MIT Press, pp. 1117–1135.

Murre, J. M. J., Phaf, R. H., and Wolters, G., 1992, CALM: Categorizing and learning module, *Neural Netw.*, 5:55–82.

Nakamura, K., and Ichikawa, A., 1990, Cerebral mechanism for reward-mediated learning: A mathematical model of neuropopulational network plasticity, *Biol. Cybern.*, 63:1–13.

Ortony, A., Clore, G. L., and Collins, A., 1988, *The Cognitive Structure of Emotions*, New York: Cambridge University Press. ◆

Ricart, R., 1992, Neuromodulatory mechanisms in neural networks and their influence on interstimulus interval effects in Pavlovian conditioning, in *Motivation, Emotion, and Goal Direction in Neural Networks* (D. S. Levine and S. J. Leven, Eds.), Hillsdale, NJ: Erlbaum, pp. 115–165.

Rolls, E. T., 1995, A theory of emotion and consciousness, and its application to understanding the neural basis of emotion, in *The Cognitive Neurosciences* (M. S. Gazzaniga, Ed.), Cambridge, MA: MIT Press, pp. 1091–1106.

Schwarz, N., and Clore, G. L., 1983, Mood, misattribution, and judgments of well-being: Informative and directive functions of affective states, *J. Pers. Soc. Psychol.*, 45:513–523.

Simon, H. A., 1967, Motivational and emotional controls of cognition, *Psychol. Rev.*, 74:29–39.

Tucker, D. M., 1990, Asymmetries of cortical architecture and the structure of emotional experience, in *Event Related Potentials: Basic Issues and Applications* (J. W. Rohrbaugh, R. Parasuraman, and R. Johnson, Jr., Eds.), New York: Oxford University Press.

Weinberger, N. M., 1995, Retuning the brain by fear conditioning, in *The Cognitive Neurosciences* (M. S. Gazzaniga, Ed.), Cambridge, MA: MIT Press, pp. 1071–1089.

Zajonc, R. B., 1980, Feeling and thinking: Preferences need no inferences, *Am. Psychol.*, 35:151–175. ◆

Energy Functions for Neural Networks

Eric Goles

Introduction

In this article we survey some of the most common neural network models for which an *energy* can be defined—where an energy is a quantity $E(x(t))$, depending on the current configuration of the network, $x(t)$, which does not increase when the network is updated, i.e., $E(x(t + \tau)) \leq E(x(t))$, and which is constant in steady state. Classically, these quantities appeared in the dynamical study of ordinary differential equations as what is called a Lyapunov function. In fact, the Lyapunov function approach consists in determining a positive quantity which decreases when the differential system approaches the equilibrium points. In such a case, it is possible to study the stability of the solutions (see Grossberg, 1988, and Hirsch, 1989, in the neural networks context). From the physical point of view, the possibility of associating an energy with neural networks arose from the deep analogy between them and the spin-glass magnetic model (Little and Shaw, 1975; Hopfield, 1984; see also NEURAL OPTIMIZATION). The interest in determining when such quantities exists arises from the fact that their existence allows study of the convergence rate, for specific update modes of the network, to stable or short-period configurations. Moreover, the attractors are local minima of the energy E, so this kind of network can be used to model associative memories (Hopfield, 1982) and hill-climbing optimization strategies.

Energies have been developed for discrete and continuous networks. There are three principal models: discrete transition–discrete time, continuous transition–discrete time, and continuous transition–continuous time. The first model was made famous by Hopfield (1982) for associative memories with Hebb interactions updated asynchronously. It was later extended to

sequential and parallel update (Fogelman-Soulié, Goles, and Weisbuch, 1983; Goles, Fogelman-Soulié, and Pellegrin, 1985). The discrete time–continuous function appears in the context of the BSB model (Golden, 1986). A survey of the energy approach can be found in Goles and Martínez (1990). Finally, the continuous time–continuous transition model has been studied first by Cohen and Grossberg (1983; see COMPUTING WITH ATTRACTORS) and in a particular case by Hopfield (1984). It is important to point out that Hopfield's energy approach was based on a spin-glass analogy of symmetric neural networks. This analogy as well as some preliminary results were presented in Little (1974) and Little and Shaw (1975).

In this article, we present first the linear argument model for discrete time and continuous transition. We determine, under symmetric assumptions about interconnections, the associated energy. Moreover, we extend the approach to the related class of quasi-symmetric interconnections. In a similar way, we present the discrete time–continuous transition model by taking as a local function a real, bounded, increasing function. It is important to point out that in this case symmetry is also the key hypothesis in determining the energy. Further, we extend the analysis to any increasing function.

For continuous time–continuous transition, we present the general model studied by Grossberg (1988), and we illustrate the energy determination for the particular case developed in Hopfield (1984).

The Binary State–Discrete Time Model

Suppose the neurons take values in a binary set, usually $\{0, 1\}$ or $\{-1, 1\}$. We present here the threshold case, i.e., $y_i =$

$\mathbb{1}(\Sigma_{i=1}^n w_{ij}x_j - b_i)$, for $x \in \{0,1\}^n$ and $\mathbb{1}(u) = 1$ iff $u \geq 0$ (0 otherwise). Throughout this paragraph we will assume, without loss of generality, that for any $x \in \{0,1\}^n$ and $i \in \{1, \ldots, n\}$, $\Sigma_{j=1}^n w_{ij}x_j - b_i \neq 0$. (Otherwise, it suffices to make a small change in threshold b_i without changing the dynamics; Goles and Martínez, 1990.) The threshold model is the classical one. Other binary models—for instance, states in the set $\{-1, 1\}$ with the sign transition function—can be reduced to the threshold model, with similar expressions for the energies.

Asynchronous Update

Let us consider the foregoing model with asynchronous update, i.e., the neurons are updated one by one in random order. In this context, we have the following result.

Theorem 1. (Hopfield, 1982; Fogelman-Soulié, Goles, and Weisbuch, 1983.) Let $W = (w_{ij})$ be a symmetric $n \times n$ matrix with non-negative diagonal entries (i.e., $\text{diag}(W) \geq 0$). Then the quantity $E(x) = -\frac{1}{2}\Sigma_{i,j=1}^n w_{ij}x_i x_j + \Sigma_{i=1}^n b_i x_i$ is an energy associated with the asynchronous iteration of the network.

Proof. Let $x = (x_1, \ldots, x_n) \in \{0,1\}^n$ be the current configuration. Suppose we update the kth neuron, obtaining the new configuration $\tilde{x} = (x_1, \ldots, x_{k-1}, \tilde{x}_k, \ldots, x_n)$, where $\tilde{x}_k = \mathbb{1}(\Sigma_{j=1}^n w_{kj}x_j - b_k)$. Let $\Delta_k E = E(\tilde{x}) - E(x)$. Since W is symmetric, we get:

$$\Delta_k E = -(\tilde{x}_k - x_k)\left(\sum_{j=1}^n w_{kj}x_j - b_k\right) - \frac{1}{2}w_{kk}(\tilde{x}_k - x_k)^2 \quad (1)$$

By definition of the threshold function, the first term is negative when $\tilde{x}_k \neq x_k$. Since $w_{kk} \geq 0$, we conclude that $\Delta_k E \leq 0$, with $\Delta_k E < 0$ iff $\tilde{x}_k \neq x_k$.   □

The first determination of this energy for a particular symmetric threshold model was done by Hopfield (1982) for associative memories and interactions w_{ij} defined by the Hebb rule, i.e., $w_{ij} = \Sigma_{k=1}^p x_i^{(k)} x_j^{(k)}$, ($w_{ii} = 0$) where $x = \{x^{(1)}, \ldots, x^{(p)}\}$ is the set of prototypes to be memorized. This rule was used before Hopfield in a neural model proposed by Anderson (see ASSOCIATIVE NETWORKS).

As a corollary to the previous theorem, we can state that the only stable states of the network are fixed points, i.e., configurations remaining invariant by the application of the threshold rule. In fact, it suffices to remark that between two successive different configurations the energy decreases. Furthermore, since the energy is bounded on the set $\{0,1\}^n$, the network converges in a finite number of steps to a fixed point. We will come back to this aspect in the next section. On the other hand, when $\text{diag}(W) = 0$ (as in Hopfield, 1982), it is easy to verify that the fixed points of the network are local minima of E. In fact, consider a fixed point $x \in \{0,1\}^n$ and its kth neighborhood in the hypercube, $\tilde{x} = (x_1, \ldots, 1 - x_k, \ldots, x_n)$. Since x is a fixed point and $\text{diag}(W) = 0$, one gets $x_k = \mathbb{1}(\Sigma_{j \neq k} w_{kj}x_j - b_k)$ and $\Delta_k E = E(\tilde{x}) - E(x) = -(1 - 2x_k)(\Sigma_{j \neq k} w_{kj}x_j - b_k) > 0$, so x is a local minimum of E. The previous aspect is important for modeling, by hill-climbing neural strategies, some hard combinatorial optimization problems (see NEURAL OPTIMIZATION and CONSTRAINED OPTIMIZATION AND THE ELASTIC NET).

Periodic Update

Assume now that the neurons are updated one by one in a periodic order: $\{1 \rightarrow 2 \rightarrow \cdots \rightarrow n \rightarrow 1\}$. Clearly, this update

strategy is a particular case of the asynchronous one, but the periodicity permits us to obtain bounds on the transient time. It is obvious that the network has the same energy, E, as the asynchronous iteration. Given a matrix W and a threshold vector b, we define $\tau(W,b)$ as the maximum number of steps taken by the network, for any initial condition, to reach a stable configuration.

Theorem 2. (Fogelman-Soulié, Goles, and Weisbuch, 1983.) Let W be an $n \times n$ symmetric matrix with non-negative diagonal. Then the transient time $\tau(W,b)$ for periodic update is bounded by

$$\tau(W,b) \leq \frac{\|W\|_1 + 2\|b\|_1}{2\left(e + \min_i w_{ii}\right)} \quad (2)$$

where $e = \min\{|\Sigma_{j=1}^n w_{ij}x_j - b_i| : i \in \{1, \ldots, n\}, x \in \{0,1\}^n\}$, $\|W\|_1 = \Sigma_{i,j=1}^n |w_{ij}|$, and $\|b\|_1 = \Sigma_{i=1}^n |b_i|$.

Proof. From the proof of Theorem 1, one gets, for any $k \in \{1, \ldots, n\}$, $|\Delta_k E| \geq e + \frac{1}{2}\min_i w_{ii}$. On the other hand, $|E(x)| \leq \frac{1}{2}\Sigma_{i,j=1}^n |w_{ij}| + \Sigma_{i=1}^n |b_i|$. From previous inequalities we obtain the bound directly.   □

Better bounds can be obtained with a finer analysis of $|\Delta_k E|$ and $|E(x)|$. See, for instance, Kamp and Hasler (1990) and Goles and Martínez (1990).

Parallel Update

Suppose we update the network synchronously:

$$x_i(t + 1) = \mathbb{1}\left(\sum_{j=1}^n w_{ij}x_j(t) - b_i\right) \quad 1 \leq i \leq n, x(0) \in \{0,1\}^n$$

$$(3)$$

In this context we have the following result.

Theorem 3. (Goles, Fogelman-Soulié, and Pellegrin, 1985.) Let W be an $n \times n$ symmetric matrix. Then the expression

$$E(t) = -\sum_{i=j=1}^n w_{ij}x_i(t)x_j(t-1) + \sum_{i=1}^n b_i(x_i(t) + x_i(t-1))$$

is an energy associated with the parallel update.

Proof. Let $\Delta E = E(t) - E(t-1)$. Since W is symmetric, one gets

$$\Delta E = -\sum_{i=1}^n (x_i(t) - x_i(t-2))\left(\sum_{j=1}^n w_{ij}x_j(t-1) - b_i\right) \quad (4)$$

By definition of the threshold function, $\Delta E \leq 0$ and it is strictly negative when $x(t) \neq x(t-2)$.   □

This energy can also be obtained in the framework of the statistical mechanics model proposed by Little (1974). In fact, the energy $E(t)$ with threshold $b = 0$ is the zero temperature limit of the Hamiltonian: $H(x) = -\beta^{-1} \Sigma_{i=1}^n \log_2 \cosh(\beta n^{-1} \Sigma_{j \neq i} w_{ij}x_j)$, where $\beta = 1/kT$, k being the Boltzmann constant (Peretto, 1984). More information about the physical approach can be found in CONSTRAINED OPTIMIZATION AND THE ELASTIC NET and STATISTICAL MECHANICS OF NEURAL NETWORKS.

Corollary 1. For a symmetric matrix W, the parallel iteration converges to fixed points or two-periodic configurations.

Proof. Suppose that $\{x(0), \ldots, x(T-1), x(T) = x(0)\}$ is a cycle of period T. From Theorem 3 it follows that $E(t)$ is necessarily constant on the cycle. If $T > 2$, we have $x(0) \neq x(2)$, so $E(2) < E(0)$, which is a contradiction.   □

Corollary 2. For a positive-definite matrix W, the parallel update converges to fixed points.

Proof. From Corollary 1 it is enough to prove that there are no two-cycles. Suppose $\{x(0), x(1)\}$ is a two-cycle, i.e., $x(2) = x(0)$. Since W is positive definite, $\alpha = (x(1) - x(0))^T(Wx(1) - x(0)) \geq 0$, with equality only if $x(1) = x(0)$. Further, by Equation 4,

$$\alpha = (x(0) - x(1))^T(Wx(1) - b) - (x(1) - x(0))^T(Wx(0) - b)$$
$$= \Delta_2 E + \Delta_1 E$$

From the proof of Theorem 3, we know that $\Delta_2 E \leq 0$ and $\Delta_1 E \leq 0$, so $\alpha = 0$. Hence, $x(1) = x(0)$, i.e., two-cycles do not exist.   □

The application of the foregoing result to associative memory models is straightforward. Consider the Hopfield model with generalized Hebb interconnections on the prototype set $\{x^{(1)}, \ldots, x^{(p)}\} \subseteq \{0, 1\}^n$, with interconnection matrix $W = (1/p)X^T X$, where $X = (x^{(1)}, \ldots, x^{(p)})$. Consider also the projection or pseudo-inverse interconnection model, i.e., $W = (X^T X)^{-1} X^T$. Since in both cases W is positive definite, the parallel updates of previous models converge only to fixed points (Kamp and Hasler, 1990).

From the energy given in Theorem 3, one may determine that the transient time for the parallel update is bounded by $\tau(W, b) \leq (1/e)(\|W\|_1 + 3\|2b - W\bar{1}\|_1 - 2\Sigma_{i=1}^n e_i)$ if $e > 0$ (0 if $e = 0$), where $e = \min\{-E^*(2) - E^*(1) : x(0) \neq x(2)\}$, $e_i = \min\{|\Sigma_{j=1}^n w_{ij} u_j - b_i| : u \in \{0, 1\}^n\}$, and $\bar{1} = (1, \ldots, 1)$. Clearly, if all the vectors belong to a two-cycle, then $e = 0$ and $\tau(W, b) = 0$.

It is important to remark that this bound is not necessarily polynomial. In fact, it is possible to build symmetric neural networks of size n with exponential transient time (recall that n neurons corresponds to 2^n states). When the matrix W takes values on the integers, the quantities e_i, e can be controlled and an explicit bound on $\tau(W, b)$ can be given in terms of W, b. Further, there exist symmetric neural networks where the bound is attained. More information about these topics can be found in Goles and Martínez (1990).

Another model that can be studied by using the energy approach is the Bidirectional Associative Memory (BAM) model proposed by Kosko (1988). Roughly it consists of a two-layer bidirectional network which achieves heteroassociations with a smaller correlation matrix. Given its correlation matrix W and a pair of vectors $(x, y) \in \{0, 1\}^{2n}$, the energy is $E = -x^T W y$ (Wang, Cruz, and Mulligan, 1991).

Energies for Nonsymmetric Models

When the matrix W is no longer symmetric, it is difficult, for general classes of matrices with other nontrivial regularities, to determine energies which insure convergence to fixed points (in the asynchronous and sequential update) and to fixed points and two-cycles (for the parallel update). That fact is not surprising, since if we permit arbitrary interconnections and enough neurons we can model arbitrary automata (see AUTOMATA AND NEURAL NETWORKS).

Furthermore, the dynamics (sequential or parallel) is very dependent on symmetry. In fact, arbitrarily small variations in a symmetric matrix can generate long cycles. A similar situation occurs for the sequential iteration and the non-negativity diagonal hypothesis: a negative diagonal also generates long cycles (Goles and Martínez, 1990). So one may say that the existence of energies relies very much on the hypothesis of symmetric interconnections. Of course, one may find non-symmetric matrices which accept an energy; but one could not find a nontrivial class, really different from the symmetric one, with the energy property. However, minor variation on symmetry are possible.

One can generalize the foregoing results to the quasi-symmetric class of matrices. We say that a matrix $W = (w_{ij})$ is quasi-symmetric if there exists a positive vector $(u_1, \ldots, u_n)$ such that for any $i, j \in \{1, \ldots, n\}, u_i w_{ij} = u_j w_{ji}$. We have the following result.

Theorem 4. (Goles and Martínez, 1990.) Let W be a quasi-symmetric matrix. Then the quantity $E(x) = -\frac{1}{2}\Sigma_{i,j} u_i w_{ij} x_i x_j + \Sigma_{i=1}^n u_i b_i x_i$, is an energy for the asynchronous and sequential update.

The quantity $E(t) = -\Sigma_{i,j=1}^n u_i w_{ij} x_i(t) x_j(t-1) + \Sigma_{i=1}^n u_i b_i (x_i(t) + x_i(t-1))$ is an energy for the parallel update.

The Continuous Transition–Discrete Time Model

Let us now consider a real transition function, $f : \mathbb{R} \to \mathbb{R}$, continuous, strictly increasing on an interval $S = (-a, a), f(0) = 0$, and constant outside S, i.e., $f(x) = f(-a)$ for $x \leq -a$ and $f(x) = f(a)$ for $x \geq a$. As an example, f could be a truncated "sigmoidal" similar to those used by Hopfield (1984) and classically empolyed in multilayered networks. Another example is the BSB model (Golden, 1986) where f is linear in S. In the previous context, the update function is as follows:

$$y_i = f(\arg_i(x)) \qquad 1 \leq i \leq n, x \in S^n \qquad (5)$$

where $\arg_i(x) = \Sigma_{j=1}^n w_{ij} x_j - b_i$. For this model and the iterations defined in previous paragraphs we have the following result.

Theorem 5. (Goles and Martínez, 1990.) Suppose W is an $n \times n$ symmetric matrix. Then when $\text{diag}(W) \geq 0$ the asynchronous or periodic update admits the energy

$$E(x) = -\frac{1}{2}\sum_{i,j=1}^n w_{ij} x_i x_j + \sum_{i=1}^n \left(\int_0^{x_i} f^{-1}(s)\,ds + b_i x_i \right) \qquad (6)$$

Furthermore, the quantity

$$E(t) = -\sum_{i,j=1}^n w_{ij} x_i(t) x_j(t-1)$$
$$+ \sum_{i=1}^n \left[\int_0^{x_i(t)} f^{-1}(s)\,ds + \int_0^{x_i(t-1)} f^{-1}(s)\,ds \right]$$
$$+ \sum_{i=1}^n b_i(x_i(t) + x_i(t-1)) \qquad (7)$$

is an energy associated with the parallel update.

Proof. We first give the proof for the asynchronous (analogous to the periodic) update. Suppose we update the kth neuron, so that $\tilde{x} = (x_1, \ldots, \tilde{x}_k, \ldots, x_n)$, where $\tilde{x}_k = f(\arg_k(x))$. Suppose also that each neuron has been updated at least one time, so $x_k = f(\arg_k(z))$, where $z \in S^n$. Since W is symmetric, one gets

$$\Delta_k E = -(\tilde{x}_k - x_k) \arg_k(x) - \tfrac{1}{2} w_{kk}(\tilde{x}_k - x_k)^2 + \int_0^{\tilde{x}_k} f^{-1}(s)\, ds$$
$$- \int_0^{x_k} f^{-1}(s)\, ds$$

Since $w_{kk} \geq 0$, the quadratic term is clearly negative. For the other two terms, let $u = \arg_k(x)$ and $v = \arg_k(z)$. Then $\Delta_k E = -(f(u) - f(v))u + \int_0^{f(u)} f^{-1}(s)\, ds - \int_0^{f(v)} f^{-1}(s)\, ds$.

From the definition of f, we have easily

$$\int_0^{f(\alpha)} f^{-1}(s)\, ds = \alpha f(\alpha) - \int_0^{\alpha} f(s)\, ds \qquad (8)$$

so $\Delta_k E = (u - v)f(v) + \int_0^v f(s)\, ds - \int_0^u f(s)\, ds$. Since f is an increasing function, one concludes that $\Delta_k E \leq 0$. For the parallel update the proof is similar. □

One may also prove that the only finite orbits are fixed points for the asynchronous and periodic update, while fixed points and/or two-cycles are possible for the parallel update.

The generalization to a nonsymmetric interval, and functions such that $f(0) \neq 0$, can be studied in a similar way. A more general approach would suppose that local update functions act in a high-dimensional space. In this framework, by considering $f : \mathbb{R}^p \to \mathbb{R}^p$, it has been proved that when f is positive (i.e., $\langle f(x) - f(y), x \rangle \geq 0$ for all $x, y \in \mathbb{R}^p$, where $\langle \ , \ \rangle$ is a scalar product) the sequential and parallel update for symmetric interconnections admits an energy (Goles, 1985).

Further, if f is a subgradient of a convex function g (i.e., $g(y) \geq g(x) + \langle f(x), y - x \rangle$ for all $x, y \in \mathbb{R}^p$), it can be proved that, under the symmetry hypothesis, the parallel iteration also admits an energy (Goles and Martínez, 1990).

Continuous Transition–Continuous Time Models

Several authors have introduced differential models of neural networks such that the transition function and the time steps are continuous. In this context Cohen and Grossberg (1983) proposed the nonlinear update

$$\frac{dx_i}{dt} = a_i(x_i)\left[b_i(x_i) - \sum_{j=1}^n w_{ij} d_j(x_j) \right] \qquad 1 \leq i \leq n \qquad (9)$$

To study the dynamics of Equation 9, the authors assume that $a_i(x) \geq 0$, $d_j'(x) \geq 0$, and the matrix W is symmetric. Then they determine the following energy function:

$$E(x) = -\sum_{i=1}^n \int^{x_i} b_i d_i'(x)\, dx + \tfrac{1}{2} \sum_{i,j=1}^n w_{ij} d_i(x_i) d_j(x_j) \qquad (10)$$

A particular case of this model was studied by Hopfield (1984):

$$c_i \left(\frac{dy_i}{dt} \right) = \sum_{j=1}^n w_{ij} x_j - \frac{y_i}{R_i} + I_i \qquad (11)$$

where $y_i = g_i^{-1}(x_i)$.

Theorem 6. (Hopfield, 1984.) Let W be a symmetric matrix and g_i a sigmoid. Then

$$E(x) = -\tfrac{1}{2} \sum_{i,j=1}^n w_{ij} x_i x_j + \sum_{i=1}^n \left(\frac{1}{R_i} \right) \int_0^{x_i} g_i^{-1}(s)\, ds + \sum_{i=1}^n I_i x_i \qquad (12)$$

is an energy function associated with Equation 11.

Proof. Since W is symmetric, one gets

$$\frac{dE}{dt} = -\sum_{i=1}^n \frac{dx_i}{dt} \left(\sum_{j=1}^n w_{ij} x_j - \frac{y_i}{R_i} + I_i \right)$$

From Equation 11 we have

$$\frac{dE}{dt} = -\sum_{i=1}^n c_i \frac{dx_i}{dt} \frac{dy_i}{dt} = -\sum_{i=1}^n c_i (g_i^{-1})'(x_i) \left(\frac{dx_i}{dt} \right)^2$$

Since g_i^{-1} is increasing, one concludes that $dE/dt \leq 0$. □

Good surveys of differential models of neural networks are provided by Grossberg (1988) and Hirsch (1989).

Discussion

In this article we have reviewed the principal neural models which accept an energy function. In all cases, the existence of such operators depends strongly on regularities of interconnections: symmetry and quasi-symmetry. One may also determine an energy for a class of antisymmetric matrices (Goles and Martínez, 1990), but it seems difficult to find other nontrivial classes of matrices which accept energy functions. In fact, arbitrary small perturbation of the matrices discussed in this article can induce long cycles in the network dynamics. Necessary and sufficient conditions for the existence of an energy function have been formally studied by Kobuchi (1991) under a plausible decomposition hypothesis of ΔE. But, in practice, the authors recover only the quasi-symmetry class.

Further, some of the results can be extended to high-order networks, i.e., neural networks with polynomial interactions. In this context, under a generalization of the symmetry hypothesis, one may find an energy function for the periodic update (Xu and Tsai, 1990; Kamp and Hasler, 1990).

Acknowledgments. This work was partially supported by DTI-U. de Chile, FONDECYT 1940520–94, and CEE Project CI1–CT92–0046.

Road Map: Dynamic Systems and Optimization
Background: I.3. Dynamics and Adaptation in Neural Networks
Related Reading: Dynamics and Bifurcation of Neural Networks

References

Cohen, M., and Grossberg, S., 1983, Absolute stability of global pattern formation and parallel memory storage by competitive neural networks, *IEEE Trans. Sys. Man Cyber.*, SMC-13:815–826.

Fogelman-Soulié, F., Goles, E., and Weisbuch, G., 1983, Transient length in sequential iteration of threshold functions, *Discrete Appl. Math.*, 6:95–98.

Golden, R. M., 1986, The brain-state-in-a-box neural model is a gradient descent algorithm, *J. Math. Psych.*, 30:73–80.

Goles, E., 1985, Dynamic of positive automata networks, *Theoret. Comput. Sci.*, 41:19–32.

Goles, E., Fogelman-Soulié, F., and Pellegrin, D., 1985, Decreasing energy functions as a tool for studying threshold networks, *Discrete Appl. Math.*, 12:261–277.

Goles, E., and Martínez, S., 1990, *Neural and Automata Networks*, Norwell, MA: Kluwer. ◆

Grossberg, S., 1988, Nonlinear neural networks: Principles, mechanisms, and architectures, *Neural Netw.*, 1:17–61. ◆

Hirsch, M., 1989, Convergent activation dynamics in continuous time networks, *Neural Netw.*, 2:331–349. ◆

Hopfield, J. J., 1982, Neural networks and physical systems with emergent collective computational abilities, *Proc. Natl. Acad. Sci. USA*, 79:2554–2558.

Hopfield, J. J., 1984, Neurons with graded response have collective computational properties like those of two-state neurons, *Proc. Natl. Acad. Sci. USA*, 81:3088–3092.

Kamp, Y., and Hasler, M., 1990, *Réseaux de neurones récursifs pour mémoires associatives*, Lausanne: Presses Polytechniques et Universitaires Romandes. ◆

Kobuchi, Y., 1991, State evaluation functions and Lyapunov functions for neural network, *Neural Netw.*, 4:505–510.

Kosko, B., 1988, Adaptive bidirectional associative memories, *Appl. Opt.*, 26:4947–4960.

Little, W. A., 1974, The existence of persistent states in the brain, *Math. Biosci.*, 19:101–120.

Little, W. A., and Shaw, G. L., 1975, A statistical theory of short and long term memory, *Behav. Biol.*, 14:115.

Peretto, P., 1984, Collective properties of neural networks: A statistical physics approach, *Biol. Cybern.*, 50:51–62.

Wang, Y. F., Cruz, J. B., and Mulligan, J. H., 1991, Guaranteed recall of all training pairs for bidirectional associative memory, *IEEE Trans. Neural Netw.*, 2:559–567.

Xu, X., and Tsai, W. T., 1990, Constructing associative memories using neural networks, *Neural Netw.*, 3:301–309.

Epilepsy: Network Models of Generation

Fernando H. Lopes da Silva and Jan Pieter Pijn

Introduction

Epilepsy is a neurological disorder characterized by the occurrence of *seizures* (i.e., *ictal activity*). These seizures consist of sudden changes of neuronal activity that interfere with the normal functioning of neuronal networks, resulting in disturbances of sensory and/or motor activity, as well as of the flow of consciousness. During an epileptic seizure, the neuronal network exhibits typical *oscillations* that usually propagate throughout the brain, involving progressively more brain systems. These oscillations are revealed in the local field potentials, i.e., in the local electroencephalogram (EEG), as illustrated in Figure 1, where the third trace represents an epileptic seizure. In principle, it is possible to induce an epileptic seizure in any brain, given a sufficiently strong stimulus (e.g., an electroshock can produce an epileptic seizure in a normal brain). However, certain brains have a greater tendency to generate epileptic activity, even in response to trivial stimuli. The latter are termed epileptic brains. The question is what are the changes in the dynamics of the neuronal populations that are responsible for this dramatic change in behavior.

In this article, we consider first the theoretical aspects of the behavior of neuronal populations as generators of epileptic activity. Thereafter, we briefly discuss current ideas about the physiological/biophysical changes that occur in a neuronal network and that can cause the formation of an epileptogenic focus.

Epilepsy Models

From a theoretical point of view, to understand how epileptic activity may occur in a neuronal population implies an analysis of the behavior of a system consisting of neuronal populations with complex nonlinear properties. In this respect, two main approaches may be followed. First, an EEG time series that represents the output signal of a neuronal population is analyzed in an attempt to deduce some properties of the dynamics of the underlying neuronal networks. Second, we can infer from the biophysical properties of individual neurons the behavior of the network using computer simulations of groups of neurons. Generally speaking, these two approaches are complementary, as one is analytical and the other synthetic.

Dynamical Nonlinear Models

In essence, epilepsy is manifested as the sudden occurrence of a typical oscillation of relatively long duration triggered by some initial and/or input conditions. Under normal circumstances,

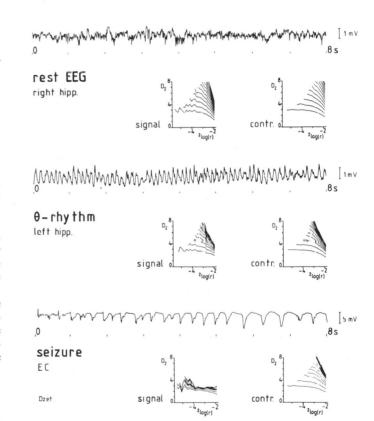

Figure 1. Three EEG signals (epochs of 8 s) recorded from the hippocampus (hipp.) and the entorhinal cortex (EC) of the freely walking rat, with the corresponding plots of the correlation dimension (D_2) for a number of different values of embedding: top trace, resting condition; middle trace, exploratory behavior; bottom trace, epileptiform seizure triggered by a series of electrical pulses (note the difference in amplification factor). The plots below each EEG trace represent two cases: on the left-hand side, the plot of D_2 is shown for the *real* signals; on the right-hand side the corresponding plots are shown for *surrogate* signals (controls, abbreviated contr.) obtained by randomizing the phase of the same EEG signals. The latter should indicate a large D_2 value since the transformed signal is not distinguishable from noise. Note that, for the resting and exploratory conditions, the plots for the EEG and the surrogate signals are similar, with both indicating a large D_2 value; thus, they cannot be distinguished from noise. In contrast, for the epileptiform seizure, D_2 converges to a value between 2 and 3, indicating the presence of low-dimensional chaos, whereas that of the surrogate signal has a large value. (Adapted from Pijn et al., 1991.)

such an input would cause no more than a transient and harmless change in brain activity; however, in the epileptic brain, it can cause a disastrous, massive series of discharges. Why this happens in the epileptic brain may be understood by assuming that some changes in the operating regimen of the networks occur, such that their dynamics undergo a qualitative change. The recently developed theory of nonlinear dynamics offers the possibility of understanding, in formal terms, the changes in state that occur in a neuronal network during epilepsy.

The basic elements of this theory, relevant to the present issue, can be summarized as follows (for a good introduction, see Ott, 1993). The dynamics of a system can be analyzed in phase space (see PHASE-PLANE ANALYSIS OF NEURAL ACTIVITY). In this space, a vector, having coordinates equal to the values of the relevant variables (e.g., a representative time series), characterizes the state of the system at any moment of time. The evolution of the vector forms a trajectory in this space. The subspace to which the trajectory may settle after transients have died out is called the *attractor*. In general, four types of evolutions can be distinguished: standing still, periodic, quasi-periodic, and chaotic. The limit behavior of these evolutions are, respectively, a point attractor, a limit cycle, a torus, and a "strange attractor." An evolution is chaotic if the distance between any two nearby points on the attractor diverges exponentially as a function of time. Under these circumstances, the behavior of the system is so strongly dependent on initial conditions that, although intrinsically deterministic, it is practically unpredictable (see CHAOS IN NEURAL SYSTEMS).

In general, a dynamical system has a relatively small number of parameters which can modify the overall dynamic structure. For some values of a parameter, the dynamic structure may change—for example, instead of a point attractor, a limit cycle or a chaotic attractor may occur—and we then say that a bifurcation has taken place (Elbert et al., 1994; see DYNAMICS AND BIFURCATION OF NEURAL NETWORKS). Thus, a bifurcation represents a qualitative change and depends on a set of control parameters that define the operating regimen of the system.

One way to characterize the system's dynamic behavior is to estimate its *correlation dimension* (D_2) (Grassberger and Procaccia, 1983). The concept of correlation dimension is presented in formal terms in Ott (1993) and in a more intuitive way in Elbert et al. (1994). It is a measure of the dimensionality of the phase space within which the attractor is embedded, and thus corresponds to the minimal number of variables that are necessary to describe the system's dynamics. A periodic sinusoidal oscillation (e.g., a limit cycle) corresponds to a D_2 of 1, whereas an oscillation consisting of two incommensurate frequencies corresponds to a D_2 of 2; if D_2 is larger than 1 and it is not an integer, the system is likely to exhibit chaotic oscillations. If used properly, and employing the necessary controls, the correlation dimension D_2 is useful in discriminating between noise and deterministic dynamics, since in the first case, its value should always increase with the number of samples and does not converge to a given value (as in the two top EEG traces in Figure 1); i.e., it does not saturate, contrary to the latter case (as in the third trace). This is an important distinction regarding the generation of EEG signals, since it has implications regarding the basic processes responsible for the generation of such signals; i.e., whether they are generated by random filtered processes or by deterministic nonlinear systems that may exhibit chaotic dynamics. From the three examples in Figure 1, it is clear that, EEG signals with different values of D_2 can be recorded from the same area of the brain, depending on the behavioral condition. This implies that a given neuronal network may have attractors which change with varying conditions. Several investigators have applied this form of analysis

to EEG signals recorded during epileptic seizures, and have found that these signals were characterized by a low-dimensional process that most likely corresponds to a chaotic attractor (Babloyantz and Destexhe, 1986; Pijn et al., 1991), although the EEG signal recorded from the same area may have a high value of D_2 when recorded outside the epileptic seizure. Thus, a neuronal network may switch from one to another state; i.e., it may present a bifurcation, characterized by different values of D_2, depending on specific conditions. We hypothesize that the main difference between a normal and an epileptic brain is essentially that the operating regimen of a given neuronal network in an epileptic brain is much closer to a bifurcation point than in a normal brain, leading to low-dimensional chaos. Compared to the normal case, in which the operating regimen is situated far from such a bifurcation point, in an epileptic neuronal network, the distance between operating and bifurcation points is so small that the system may easily switch from a stable equilibrium to a chaotic attractor, even in response to a very weak stimulus. Considered from the perspective of the theory of nonlinear dynamic systems, the basic question in the case of epilepsy is which factors are responsible for the change in operating point of the neuronal networks.

It is interesting to add that the multineuronal model of Traub and Miles (1991), described in the following section, contains a large number of degrees of freedom (9900 cells, each with a number of compartments and 5 nonlinear conductances) and generates seizure-like discharges with a D_2 of 4.2, most likely corresponding to a low-dimensional chaotic attractor. Thus, the general tendency, both in the natural case and in the simulations, is for seizure activity to correspond to a relatively low-dimensional chaotic process. Whether one can speak of the existence of a real chaotic attractor in these cases is still a matter for further investigation.

Neuronal Network Models of the Generation of Seizure Epileptiform Activity

Models of neuronal networks have been developed with the aim of understanding how the intrinsic membrane properties of neurons, in combination with the properties of the interneuronal interactions, can produce epileptic activity. Indeed, a synthetic approach may help to understand the relationship between neuronal activity at the single neuron level and at the population level. An essential characteristic of epileptic activity is the occurrence of synchronous bursting activity in a large population of neurons. In this sense, the models of Traub and collaborators (Traub and Wong, 1982; Traub, Jefferys, and Whittington, 1994) are paradigmatic. In a recent study, they modeled a hippocampal network consisting of 1000 pyramidal and 100 inhibitory neurons. Each pyramidal neuron contained five ionic conductances (nonlinear type of kinetics; see ION CHANNELS: KEYS TO NEURONAL SPECIALIZATION). The neurons were randomly interconnected by way of excitatory (glutamatergic) and inhibitory (GABAergic) synapses. The modeled seizure discharges consisted of tonic depolarizations of dendrites with repetitive Ca^{2+} bursts of action potentials, enhanced by phasically acting glutamatergic receptors (Figure 2). The conditions that lead to epileptiform activity, represented in Figure 2, correspond to changes in the balance between excitatory and inhibitory processes, either by increasing the former or decreasing the latter. This illustrates the fact that changes in the control parameters of the network can lead to epileptic behavior. Model studies (Traub, Miles, and Jefferys, 1993) have demonstrated that enhanced synchrony in the neuronal network underlies epileptiform seizure activity, and may

RESPONSES OF THE MODEL POPULATION
DIFFERENT CONDITIONS OF EXCITATION/INHIBITION

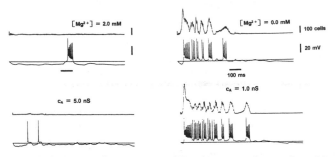

Figure 2. Representative responses of Traub's *neuronal network model* to a brief stimulus under different conditions. Each of the four sets shows two traces: the upper one represents the number of pyramidal cells depolarized by more than 20 mV from rest, whereas the lower trace illustrates the somatic membrane potential of one pyramidal neuron. The upper row shows how the responses depend on a change in an excitatory parameter. Here, the glutamatergic receptors of the NMDA subtype are activated by lowering $[Mg^{2+}]$ extracellularly; when the excitatory level is low ($[Mg^{2+}] = 2.0$ mM; left-hand side), no synchronized activity is present; when the excitation is strengthened by the reduction of $[Mg^{2+}]$ to 0 mM, synchronized "epileptiform" bursts occur. The lower row shows the dependency on the inhibitory parameter represented by the GABAergic synaptic conductance, c_A. At $c_A = 5$ nS, synchrony is minimal. Blocking of the $GABA_A$ receptors ($c_A = 1$ nS) leads to a synchronized series of "epileptiform" bursts. (Adapted from Traub, Jefferys, and Whittington, 1994.)

occur when the excitatory connectivity has a high density and is sufficiently strong. Under these circumstances, the stimulation of a small group of neurons can cause a wave of spreading excitation throughout the neuronal network. These studies emphasize the importance of recurrent excitatory interconnections in order for excessive synchronous activity to occur. This model represents a number of the properties that are characteristic of neuronal populations in epileptogenic foci, although it cannot account for all forms of epilepsy.

Other model studies have addressed the mechanisms underlying another type of epilepsy, namely *absence epilepsy*, which is characterized behaviorally by the abrupt cessation of all activities, by a state of unawareness and unresponsiveness that lasts for a few seconds, and by a typical EEG with bursts of spikes and waves at about 3 Hz. Recently, Wang (1994) constructed a model of thalamic neurons that were capable of producing two oscillatory modes: the usual 7- to 14-Hz spindles seen under normal conditions, and the 3-Hz oscillations encountered in absence epilepsy. In this model, the strength and the inactivation time constant of a calcium transient current of the thalamic neurons appear to be the most important factors in determining the burst frequency. Thus, different balances of the same set of ion channels can cause distinct oscillatory modes (corresponding to multiple attractors), and a change in the balance between different ionic processes may cause epileptiform bursts of activity.

Discussion

An essential discussion point is which physiological and biophysical phenomena are responsible for the changes in the dynamics of a neuronal network leading to the generation of epileptic activity. Currently, we have only fragmentary knowledge of these processes. Experimental evidence, obtained in vivo during the development of an epileptogenic focus (Lopes

da Silva, Kamphuis, and Wadman, 1992; Lothman, Bertram, and Stringer, 1991), and the model studies just discussed show that the balance between excitatory and inhibitory processes changes in the epileptic condition. This strongly suggests that the essential control parameter of the dynamics of a neuronal network should be seen as the ratio between two sets of parameters: i.e., those responsible for excitation and those responsible for inhibition, including synaptic and intrinsic ion currents. It is likely that multiple combinations of changes in these various parameters may lead to the same type of qualitative change in network dynamics that leads to epileptogenesis. For example, the detailed physiological changes occurring in an epileptogenic focus in the hippocampus that can cause the type of epileptic seizure activity of focal origin illustrated in Figure 1 are probably quite different from those occurring in the thalamus during absence epilepsy.

This means that models of realistic neuronal networks have to take into account the essential control parameters of a network: namely, the balance between all processes (synaptic and intrinsic) that cause excitation and inhibition, respectively, and their influence on the network dynamics.

Nonlinear dynamic analysis provides, in this context, a powerful way to define the existence and domain of possible bifurcations in the operating regime of the network which lead to epileptic behavior. This may be more important than trying to define the underlying attractors in a precise way.

Road Map: Biological Networks
Background: I.3. Dynamics and Adaptation in Neural Networks
Related Reading: Cooperative Behavior in Networks of Chaotic Elements; Oscillatory and Bursting Properties of Neurons; Thalamocortical Oscillations in Sleep and Wakefulness

References

Babloyantz, A., and Destexhe, A., 1986, Low-dimensional chaos in an instance of epilepsy, *Proc. Natl. Acad. Sci. USA*, 83:3513–3517.

Elbert, T., Ray, W. J., Kowalik, Z., Skinner, J. E., Graf, K. E., and Birbaumer, N., 1994, Chaos and physiology: Deterministic chaos in excitable cell assemblies, *Physiol. Rev.*, 74:1–47. ◆

Grassberger, P., and Procaccia, I., 1983, Measuring the strangeness of strange attractors, *Physica*, 9:183–208.

Lopes da Silva, F. H., Kamphuis, W., and Wadman, W. J., 1992, Epileptogenesis as a plastic phenomenon of the brain: A short review, *Acta Neurol. Scand.*, 86:34–40. ◆

Lothman, E. W., Bertram, E. H. III, and Stringer, J. L., 1991, Functional anatomy of hippocampal seizures, *Prog. Neurobiol.*, 37:1–82. ◆

Ott, E., 1993, *Chaos in Dynamical Systems*, New York: Cambridge University Press.

Pijn, J. P., Van Neerven, J., Noest, A., and Lopes da Silva, F. H., 1991, Chaos or noise in EEG signals: Dependence on state and brain site, *Electroencephalogr. Clin. Neurophysiol.*, 79:371–381.

Traub, R. D., Jefferys, J. G. R., and Whittington, M. A., 1994, Enhanced NMDA conductances can account for epileptiform activities induced by low Mg^{2+} in the rat hippocampal slice, *J. Physiol.*, 478:379–393.

Traub, R. D., and Miles, R., 1991, *Neuronal Networks of the Hippocampus*, New York: Cambridge University Press. ◆

Traub, R. D., Miles, R., and Jefferys, J. G. R., 1993, Synaptic and intrinsic conductances shape picrotoxin-induced synchronized afterdischarges in the guinea-pig hippocampal slice, *J. Physiol.*, 461:525–547.

Traub, R. D., and Wong, R. K. S., 1982, Cellular mechanism of neuronal synchronization in epilepsy, *Science*, 216:745–747.

Wang, X.-J., 1994, Multiple dynamical modes of thalamic relay neurons: Rhythmic bursting and intermittent phase-locking, *Neuroscience*, 59:21–31.

Equilibrium Point Hypothesis

Reza Shadmehr

Introduction

The Equilibrium Point Hypothesis (EPH) provides a systems-level description of how the nervous system controls the muscles so that a stable posture is maintained or a movement is produced. In the EPH framework, the *controller* is composed of muscles and the spinal-based reflexes, and the *plant* is the skeletal system. The controller is represented as a mapping which assigns a force to each state of the plant, i.e., a *force field*. This mapping, which is meant to capture the mechanical behavior of the muscles and the effect of the spinal reflexes, is also dependent on a set of *control inputs* from supra-spinal centers. EPH predicts certain properties for the controller. For example, the field will have an attractor, and movement is a result of supra-spinal input which causes a gradual shift of this attractor.

Mathematical Basis of the Hypothesis

Equilibrium refers to a state of a system in which the forces acting on it are zero. For example, if the dynamics of the system are

$$\dot{q} = h(q, u) \tag{1}$$

where q is the state of the system and $u(t)$ is a control input, then the equilibrium points q^* satisfy the following condition:

$$0 = h(q^*, u) \quad \text{for all } t \geq t_0 \tag{2}$$

In short, if the system reaches an equilibrium point, it will remain there.

For a mechanical system, the state is an ordered pair $q = \{\theta, \dot{\theta}\}$, where θ and $\dot{\theta}$ are the (generalized) position and velocity of the system. A change in the state occurs when there are forces acting on it. This relationship can be written in the framework of Equation 1 as

$$\ddot{\theta} = I(\theta)^{-1}(f_c(\dot{\theta}, \theta, u(t)) - f_m(\dot{\theta}, \theta) - f_p(\theta)) \tag{3}$$

where I is the system's inertia, f_c is the external force field imposed on the system due to the controller with control input $u(t)$ (e.g., by the torque motors or muscles), f_m is the force field produced due to the motion of the inertial coordinate frames (e.g., coriolis and centripetal), and f_p is the force field due to gravity or other potential fields. It follows that the system is at equilibrium at any state $\{\theta = \theta^*, \dot{\theta} = 0\}$ where the force in the net field $f_c - f_m - f_p$ is zero. Any such position θ^* is an equilibrium point.

We call each state where a field has zero force a *null point* of that field. An equilibrium point for the system exists only at those null points of the force field $f_c - f_m - f_p$ where the state has zero velocity.

Let us consider how we would go about controlling the system of Equation 3. Our objective may be to select the input u to the control field f_c in such a way that for time $t \geq t_0$, the system of Equation 3 is at a desired position θ_d. For this to occur, $f_c - f_m - f_p$ must be a converging force field about the state $\{\theta_d, 0\}$. Another objective of the control system may be to select the input u in such a way that the system follows a particular trajectory $\{\theta_d(t), \dot{\theta}_d(t)\}$. For this to occur, the EPH approach suggests that we select u in such a way that $f_c - f_m - f_p - I(\theta)\ddot{\theta}_d(t)$ is a force field converging to the state $\{\theta_d(t), \dot{\theta}_d(t)\}$. For example, choose $u(t) = \{\theta_d(t), \dot{\theta}_d(t)\}$, and set f_c to be

$$f_c = \hat{f}_m + \hat{f}_p + \hat{I}\ddot{\theta}_d(t) - B(\dot{\theta} - \dot{\theta}_d(t)) - K(\theta - \theta_d(t)) \tag{4}$$

where $\hat{x}$ is the controller's estimate (or model) of x, and B and K are positive-definite matrices representing the system's viscosity and stiffness (Slotine and Li, 1991).

In Equation 4 we get the impression that, during movement, the field f_c might also have a region where forces converge on an equilibrium point. Indeed, the transition from posture to movement could be made by a controller that gradually moved its equilibrium point from the starting state of the system to its goal state. In other words, EPH asserts that by manipulating the single equilibrium point of the control field f_c, the system's state is controlled during both posture and movement.

Biomechanical Interpretation

Let us use the model of Equation 3 to describe the mechanics of a limb's musculoskeletal system. f_c is the force field produced by the muscles, and f_m and f_p are the forces generated by motion of the skeletal system and the gravitational forces on the skeleton. The forces in the field f_c are dependent on the length and velocity of contraction of each muscle, as well as on an input vector u, which serves as a "composite control signal" (Feldman, 1966), representing the supra-spinally generated neural command that acts on the muscles and the spinal reflex circuitry. If we consider the static properties of the field f_c in Equation 4, we see that the input vector takes the form of a *threshold joint angle* θ_d, around which a converging force field develops with stiffness properties K. This threshold length can be thought of as a resting length of a nonlinear spring.

Feldman and colleagues (Feldman, 1966) were the first to suggest that the supra-spinal input to the spinal control system can be represented as a threshold length for a spring-like system. They experimented on the muscles of the elbow joint and quantified how the muscles responded to a displacement. It was found that as the muscles were stretched from some predetermined length, they produced monotonically increasing amounts of force. Feldman's thesis was that the signal sent from supra-spinal centers to the spinal reflexes and muscles could be interpreted as setting the threshold length beyond which force developed for each muscle. Neurophysiological evidence for this idea was later provided by experiments of Feldman and Orlovsky (1972).

The muscle-reflex model proposed by Feldman was compared to other muscle models by Shadmehr and Arbib (1992). It was shown that without their reflexive circuitry, muscles did not behave as spring-like elements with adjustable threshold lengths. However, from the rate of change of stiffness with respect to force in an intact muscle-reflex system (Hoffer and Andreassen, 1981), Shadmehr and Arbib derived a force-length relationship that was quite similar to the results of Feldman (1966) and Feldman and Orlovsky (1972). This suggested that one role of spinal reflexes was to functionally transform muscles into spring-like elements with an adjustable threshold length.

The observation that each muscle-reflex subsystem can be viewed as a nonlinear spring implies that the control field f_c is *conservative* (Hogan, 1985). This suggests that when one considers behavior of the arm in its multi-joint case, the curl of the static component of the field f_c should be zero. Mussa-Ivaldi, Hogan, and Bizzi (1985) measured the static

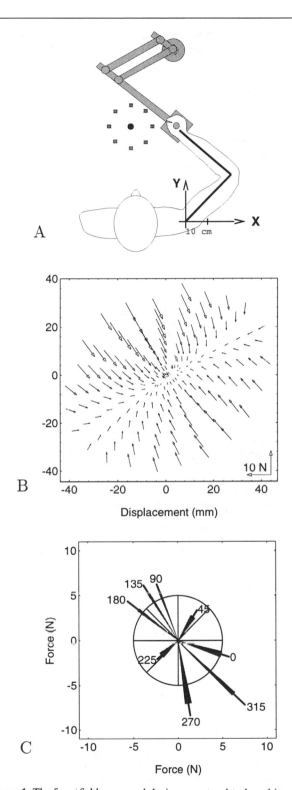

Figure 1. The force field measured during a postural task and its relation to forces generated during initiation of a reaching movement. *A,* The subject is instructed to position the hand at the "start" position (black circle) and not to intervene as the robot slowly displaces the hand from equilibrium. *B,* The measured force field during posture. *C,* Forces generated as the subject initiates a movement from the start position toward one of eight targets (gray squares in *A*). (From Shadmehr, R., Mussa-Ivaldi, F. A., and Bizzi, E., 1993, Postural force fields of the human arm and their role in generating multi-joint movements, Journal of Neuroscience, 13:45–62; reprinted with permission.)

component of this field (Figure 1*A*, 1*B*) and found the field to be conservative.

These results suggest that posture is maintained at an equilibrium position through production of a converging force field by the muscles. The muscles in turn are functionally similar to nonlinear springs with an adjustable threshold length.

Movement

The major contribution of EPH has been to describe motion as a gradual transition from posture. For slow movements, where the forces generated by the motion of the system (i.e., the field f_m in Equation 3) are small, one would expect the trajectory of this equilibrium point to closely follow the actual trajectory of the system. Experiments of Bizzi et al. (1984) on intact and deafferented monkeys making elbow movements have shown that this is indeed the case.

Flash (1987) suggested that for the case of multi-joint arm movements, one can predict the hand's motion if the supraspinal system smoothly shifts the equilibrium point from the start point to a target location. She showed that in her model the hand exhibited characteristic deviations from a straight-line trajectory, and these matched hand trajectories observed in psychophysical experiments. In this model, the control input $u(t)$ was defined as a time-varying equilibrium position $\theta_c^*(t)$ for the field f_c:

$$f_c(\theta, \dot{\theta}, t) = K(\theta - \theta_c^*(t)) + B\dot{\theta}$$

The field had the property that its static behavior about equilibrium was defined by a stiffness matrix K, as measured by Mussa-Ivaldi, Hogan, and Bizzi (1985).

The success of the simulations of Flash (1987) was debated by Katayama and Kawato (1993). Flash had assumed that, during movement, the magnitude of stiffness K was approximately three times that measured during posture. Data of Bennett et al. (1992), however, have shown that stiffness during a highly practiced movement is 25–80% of that measured during posture. Although measurements of Milner (1993) in single-joint movements seem to support the assumptions of Flash, this controversy regarding relative magnitudes of stiffness during a movement has remained unresolved. If stiffness of the moving arm is small compared to posture, then the notion that, during movement, the field f_c has an equilibrium position, and that this position is somehow close to the actual movement, is considerably challenged.

Part of the difficulty of assessing the validity of EPH through a model-based approach is that while Bennett et al. assumed that, in the field f_c, static forces are linearly related to a displacement from equilibrium, results of Shadmehr, Mussa-Ivaldi, Bizzi (1993) suggest that the relationship is highly nonlinear: the further the hand is from equilibrium, the smaller the local stiffness of the field. There are now other approaches which have attempted to test the existence of an equilibrium trajectory more directly.

Shadmehr, Mussa-Ivaldi, and Bizzi (1993) suggested that if a movement is generated through a shift of the equilibrium position of the postural field to the target, then from postural measurements of this field (Figure 1*B*), one can predict the direction and magnitude of the forces that should be produced by the muscles during the initiation of the reaching movement. Since the postural field is not isotropic, forces measured during initiation of a movement should not point toward the target. These movement initiation forces were measured (Figure 1*C*), and it was found that the pattern of forces from postural measurements agreed with the measured forces during initiation of movement.

Won (1993) suggested that, during a movement, the static component of the field f_c should be similar to that measured during posture, i.e., they should converge to an equilibrium position. In his experiment, the hand was displaced from its intended trajectory via a rigid mechanical constraint. He showed that as the arm is being displaced, it produces forces directed toward the intended trajectory (Figure 2). This was in agreement with the idea that the controller's output during movement is a force field with an equilibrium point moving roughly along the path connecting the start to the target position.

The most direct test of the hypothesis to date has been the experiments of Hodgson (1994). In this work, the aim was to measure the trajectory for which there was zero force on the hand. Such a position would be an equilibrium point for the control field f_c. The power of this approach was that it did not rely on an a priori model of f_c, but simply measured its gradient and, through an iterative procedure, found its attractor. Hodgson showed that the equilibrium position of the human

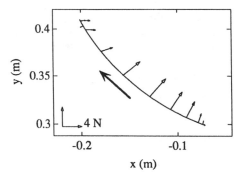

Figure 2. The trajectory that the hand follows is typically straight, but when the hand is displaced from its natural trajectory, forces at the hand point toward the path of the unperturbed movement. (From Won, 1993.)

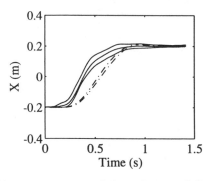

Figure 3. Direct measurement of the trajectory of the equilibrium position during a one-dimensional reaching movement. The dashed lines are typical hand trajectories, whereas the solid lines are the mean and the confidence intervals for the equilibrium trajectory. (From Hodgson, 1994.)

arm gradually shifted from the start to the target position (Figure 3). The equilibrium trajectory led the hand throughout the movement.

Recent work on the frog spinal cord has suggested a neural architecture by which the neuromuscular system produces these force fields and manipulates their equilibrium (see GEOMETRICAL PRINCIPLES IN MOTOR CONTROL and FROG WIPING REFLEXES).

Acknowledgments. Preparation of this article was supported in part by grants from the NIH (NS09343 and AR26710) and the ONR (N00014/88/K/0372). The author is supported by the McDonnell-Pew Center and by the Center for Biological and Computational Learning at MIT.

Road Map: Biological Motor Control
Related Reading: Corollary Discharge in Visuomotor Coordination; Optimization Principles in Motor Control

References

Bennett, D. J., Hollerbach, J. M., Xu, Y., and Hunter, I. W., 1992, Time varying stiffness of the human elbow joint during cyclic voluntary movement, *Exp. Brain Res.*, 88:433–442.

Bizzi, E., Accornero, N., Chapple, W., and Hogan, N., 1984, Posture control and trajectory formation during arm movement, *J. Neurosci.*, 4:2738–2744.

Feldman, A. G., 1966, Functional tuning of the nervous system with control of movement or maintenance of a steady posture, II: Controllable parameters of the muscles, *Biophysics*, 11:565–578.

Feldman, A. G., and Orlovsky, G. N., 1972, The influence of different descending systems on the tonic stretch reflex in the cat, *Exp. Neurol.*, 37:481–494.

Flash, T., 1987, The control of hand equilibrium trajectories in multi-joint arm movements, *Biol. Cybern.*, 57:257–274.

Hodgson, A., 1994, Inferring central motor plans from attractor trajectory measurements, PhD Dissertation, Health Science and Technology Program, Massachusetts Institute of Technology.

Hoffer, J. A., and Andreassen, S., 1981, Regulation of soleus muscle stiffness in premammillary cats: Intrinsic and reflex components, *J. Neurophysiol.*, 45:267–285.

Hogan, N., 1985, The mechanics of multi-joint posture and movement control, *Biol. Cybern.*, 52:315–331.

Katayama, M., and Kawato, M., 1993, Virtual trajectory and stiffness ellipse during multijoint arm movements predicted by neural inverse models, *Biol. Cybern.*, 69:353–362.

Milner, T. E., 1993, Dependence of elbow viscoelastic behavior on speed and loading in voluntary movements, *Exp. Brain. Res.*, 93:177–180.

Mussa-Ivaldi, F. A., Hogan, N., and Bizzi, E., 1985, Neural, mechanical, and geometric factors subserving arm posture, *J. Neurosci.*, 5:2732–2743.

Shadmehr, R., and Arbib, M. A., 1992, A mathematical analysis of the force-stiffness characteristics of muscles in control of a single joint system, *Biol. Cybern.*, 66:463–477.

Shadmehr, R., Mussa-Ivaldi, F. A., and Bizzi, E., 1993, Postural force fields of the human arm and their role in generating multi-joint movements, *J. Neurosci.*, 13:45–62.

Slotine, J. J. E., and Li, W., 1991, *Applied Nonlinear Control*, Englewood Cliffs, NJ: Prentice Hall, pp. 310–388.

Won, J., 1993, The control of constrained and partially constrained arm movements, Master's Thesis, Dept. of Mechanical Engineering, Massachusetts Institute of Technology.

Evolution of the Ancestral Vertebrate Brain

Bernd Fritzsch

Introduction

Unraveling the evolution of the ancestral vertebrate brain has been centered on comparative and functional analyses of adult vertebrate brains. The remarkable variations in size and shape have often been viewed as evidence for progressive increases in size and complexity. Despite dramatic progress attributable to modern tracing techniques, immunocytochemistry, and sophisticated single-cell analysis, much remains to be discovered with respect to variations in "wiring," the functionally relevant aspect of brain variation. While both stability and variation in connections have been found, it is unclear how these relate to functional constraints imposed on the organism under study. This is true, in part, because there are always several ways to implement a certain function using existing neural connections. Moreover, a given function may be served by pathways which are alike not because of ancestry but because of functional necessities. Convergent evolution of similar functions, or *homoplasy*, stands in contrast to a structural similarity resulting from common ancestral circuitry, or *homology*. Teasing apart these two aspects of evolution is a major, unresolved issue of brain evolution. Given the theoretical and practical problems of identifying evolutionary changes through comparisons of adult structures, it appears worthwhile to reinvigorate the old and recently renewed studies of comparative neuroembryology, i.e., to augment the comparison of adult structures by comparing, in addition, their specific development.

In this overview, I will first introduce a possible prototype of the ancestral vertebrate brain, then present a possible ontogenetic scenario through which the problem of homology/homoplasy may be minimized. I will then suggest possible mechanisms for diversifying the ancestral vertebrate brain, along with data about structural changes whose functional implications are, as yet, unclear.

When and How Was the First Vertebrate Brain Formed?

The origin of the vertebrate brain is tightly coupled to the origin of the vertebrate head, and the origins of both remain unresolved (Gilland and Baker, 1993; Krumlauf et al., 1993). In the past, the early evolution of the brain was often attributed to some iterative homology of hindbrain and spinal cord, i.e., the evolution of the hindbrain from a spinal-cord–like ancestral system. The events controlling this change into hindbrain remained unresolved. Equally unclear was how and when the forebrain, and in particular, the cortex (Karten, 1991), evolved and whether or not it was a modified hindbrain-like organization or was derived from an olfactory target organ. Irrespective of these aspects, three major developmental issues underlying the evolution of the brain have been clarified:

1. The vertebrate central nervous system invaginates from the dorsal ectoderm to form a neural tube. How and when the interaction between the deep layers of the embryo (i.e., the mesoderm) and the superficial neuroectoderm occurred in ancestral chordates is unknown. The molecular machinery of these interactions are currently being investigated, and may eventually help us to understand how this interaction transforms a sheath of cells into a patterned neural tube (Ruiz i Altaba, 1994).
2. The evolutionary changes that generate the vertebrate head (and brain) correlate with the appearance of novel embryonic material—neural crest and placodes—which contribute to all sensory systems of the head; the evolutionary origin of this material is unknown (Fritzsch and Northcutt, 1993).
3. During development, the nervous tissue becomes polarized and then regionalized through specific gene expression pattern. By comparing the expression of genes with comparable (homologous) nucleotide sequence in topologically comparable (homologous) parts of the developing vertebrate brain, we may eventually be able to identify homologous areas in the nervous tissue. Thus far, only rudiments of this map are known.

Based on these insights offered by comparative and molecular neuroembryology, the evolution of the brain can be viewed as a sequence of developmental variations leading to modified adult structures. These modified structures will subserve somewhat different functions which, in turn, will be chosen in the process of natural selection. Thus, the way selection of novel structures and functions is related to mutations is rather indirect. The motor system, the common output of the brain, may be the best understood part of the brain in this respect, as we know much about its evolution, molecular control of induction, regionalization, diversification, pathway selection, and trophic support (Fritzsch and Northcutt, 1993; Korzh, 1994). As the inclusion of other systems of the brain in this overview would require considerable speculation while nevertheless dealing with similar problems, this article will focus predominantly on the evolution of spinal and cranial motoneurons.

From Genes to Brain: Evolution of the Brainstem Motor System

Recent research shows that developmental selector genes play an important role in the differentiating vertebrate brain: they form a space map which may activate topologically appropriate structural genes within the rostrocaudal and ventrolateral subdivisions of the brain (Krumlauf et al., 1993; Korzh, 1994). These genes were first identified in fruit flies, in which they were found to be related to homeotic transformations, i.e., control of the developmental fate of segments. Some of them were consequently named homeotic genes. Subsequently, similar selector genes were identified in vertebrates, other animals, and plants.

In vertebrates, selector genes govern (1) the formation of longitudinal columns of motoneurons and (2) the regionalization of these columns into domains associated with a specific cranial nerve. Genetic manipulations, treatments with teratogens, and mutations show that there is a correlation between the expression of these genes and the development of colums and their regionalization, although the details have not yet been explicated (Joyner and Guillemot, 1994). It has also become clear that, in some cases, one of these genes can be eliminated without any apparent defect in the region in which they are expressed, probably because other genes provide sufficient redundancy to conserve differentiation even in the absence of the gene.

The conservation of selector genes coincides with an apparently stable pattern of motoneuron regionalization in the hindbrains of some extant vertebrates. However,

1. We do not yet know how these selector genes are themselves selected in their expression, nor how they achieve their effect through differential activation of the genes they can regulate.
2. Not all vertebrates appear to share the same complement of homeotic genes (Pendleton et al., 1993).
3. There is a fairly stable pattern of hindbrain motoneuron populations but there are also intriguing variations (Gilland and Baker, 1993; Fritzsch and Northcutt, 1993).

More vertebrates need to be screened for expression of selector genes and motoneuron regionalization before the conclusion derived from studies in some can be generalized to all vertebrates. Clarifying these relationships is paramount for providing a rational basis for any reductionistic approach to homology. In the following discussion, some examples of variations in the hindbrain motor systems are given.

The Abducens Motoneurons

A population of ocular motoneurons—the abducens—changes position with respect to other landmarks (Gilland and Baker, 1993), as do its fiber pathways in the brainstem (Fritzsch and Northcutt, 1993). A subdivision of the abducens, the accessory abducens, seems to disappear in some and reappear in other vertebrate lineages (Fritzsch, 1991), and may even innervate a muscle that is not involved in eye movement. This variation of the abducens motoneurons should be governed by selector genes comparable to those for nearby motoneurons which are more conserved. How these differences in position and pathway selection are controlled during development and how the outcome of the various developmental programs that lead to these differences are stabilized through selective advantages for a given pattern is unclear.

The Oculomotor System

The system of ocular muscles and motoneurons offers additional variations among vertebrates: four out of six eye muscles in jawed vertebrates, but only three out of six eye muscles in lampreys are innervated by the oculomotor motoneurons (Fritzsch, 1991). In addition, jawless vertebrates, such as lampreys, lack horizontal canals. This is one of the major inputs into the oculomotor system of jawed vertebrates. It is not yet known how these differences in the connections of the ocular motoneurons and the ear evolved. It is also unclear how horizontal angular acceleration is measured and transformed into motor comands for the different coordinates of ocular muscles. Thus, the vestibulo-ocular system of lampreys, which likely displays the primitive vertebrate pattern, consists of three neurons between the ear and the eye muscles. A three-neuron arc exists also in jawed vertebrates, but the detailed pattern differs in this functionally equivalent system (Figure 1).

If we are to understand the selective advantage (if any) of these differences, we must first unravel the structural/functional relationship in this vestibulo-ocular arc in lampreys and compare this to the two functionally equivalent, but structurally different, vestibulo-ocular systems of jawed vertebrates (Figure 1). We then need to find out whether each of these circuits is (1) optimally designed for all functions of the system, (2) optimized only with respect to a specific function, or (3) just sufficient to fulfill its function. Currently, we cannot exclude the fact that the major driving force behind this transformation of the vestibulo-ocular reflex system may not be the need of the system for a different function, but rather "accidental" developmental changes which may have led to a reorganization of, e.g., the subdivision of, eye muscles and their nerves (Fritzsch, 1991). Once in place, the entire system implemented the "new" structures into the preexisting pattern, changing it only enough to achieve this implementation.

How Does the Brain Change Its Pattern?

Unfortunately, there is no universal answer for this question but, rather, limited speculation, primarily because past research has been directed toward unraveling the common connections rather than the dissimilarities. Theories about how changes in the nervous system come about usually explicitly or implicitly identify functionally uncommitted neurons as the source of evolutionary novelty. Logically, there are only three possible ways these neurons can be generated in the first place:

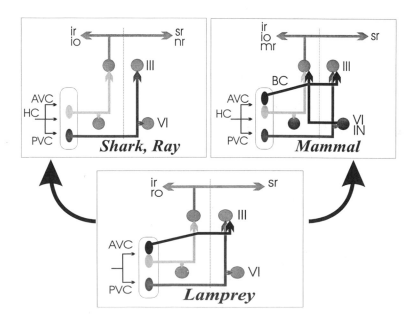

Figure 1. Reorganization of the ocular motor system among vertebrates. Lampreys have three eye muscles: inferior rectus (ir), rostral oblique (ro), and superior rectus (sr); jawed vertebrates have four eye muscles innervated by cranial nerve III. However, in sharks and rays, the nasal rectus (nr) receives a crossed innervation, whereas the functionally equivalent medial rectus (mr) of mammals receives an uncrossed innervation. All vertebrates have a rostral population in the vestibular nuclei which projects uncrossed, and a caudal population which projects crossed, to the ocular motoneurons. Lampreys and mammals have an additional rostral population that projects through the brachium conjunctivum (BC) to the cranial nerve III nucleus. Note that lampreys have no horizontal canal (HC), and that only mammals have internuclear neurons (IN) from cranial nerve VI to III which provide the basis for conjugated horizontal eye movements. This task must be performed differently in lampreys and sharks. AVC is the anterior vertical canal; PVC, the posterior vertical canal; io, the inferior oblique muscle. The dashed line indicates the midline. (Modified from Fritzsch, 1991, and Fritzsch and Northcutt, 1993.)

1. Through increased proliferation of an existing population which forms redundant and, therefore, potentially uncommitted neurons.
2. Through the loss of an input or target or both, which frees neurons of their previous functional constraints and allows them to adopt a new commitment.
3. Through the de novo formation of a novel set of neurons.

The third possibility requires that neurons somehow escape the pattern of gene activations normally mediated through the spatially restricted expression of developmental selector genes. This could happen in three ways: (a) alteration of selector gene expression, e.g., through changes in the morphogen gradients which activate these genes; (b) mutation of selector genes which alters their expression pattern; or (c) mutation of structural genes. I will not deal with these aspects because of the lack of good examples that are clearly distinct from the other two possibilities (options 1 and 2 just mentioned).

Increased proliferation and formation of more neurons (option 1) clearly occurred in the evolution of vertebrates, as exemplified by the relatively and absolutely larger brains of humans compared to those in same-sized bony fish and other tetrapods. Moreover, in some vertebrates, specific areas have become the largest part of the entire brain while other species may lack these areas altogether. For example, the valvulae cerebelli are unique to bony fish and can overgrow other parts of the brain in the same way that the forebrain has overgrown the brainstem in humans. This increased proliferation may come about through a simple mutation of genes regulating the proliferation of stem cells (Joyner and Guillemot, 1994). As such, this formation of additional neurons would resemble well-known examples in molecular evolution (e.g., the formation of different hemoglobin chains). Subsequently, some neurons of this initially identical population could assume a different function by reaching a different target or by segregating their perikarya and dendrites (and thus input) through differential migration, or both. How this new identity could be achieved is still unclear, but one possibility might be that the enlarged population of postmitotic neurons could be affected by a slightly different gradient of selector gene products, thereby achieving different phenotypes (Kaufman, 1993).

A second possible scenario for the formation of uncommitted neurons may involve the loss of either the target or the input (option 2). This would eliminate the constraints normally acting on these neurons, allowing them to develop a new function. One well-known example of this is the evolution of the middle ear ossicles of land vertebrates. There is evidence to suggest that these ossicles are derived from former jaw-supporting ossicles which have undergone a radical change in function since their original functional constraint, supporting the jaw, was lost (Webster, Popper, and Fay, 1992). Previous suggestions of a similar reorganization involved in the evolution of the auditory nuclei have been refuted, but it is still unclear what role such a reorganization of function may play in the evolution of the brain. Clearly, entire sensory systems, such as the mechanosensory lateral line and the electroreceptive system of ampullary organs of fish and many amphibians, have been lost in most land vertebrates (Fritzsch and Northcutt, 1993). Whether or not the loss of this peripheral system and the concomitant freeing of central neurons from their previous functional constraints allowed incorporation of these neurons into other functional systems is unknown. In summary, while there is evidence of the loss of both inputs and targets, it has not yet been confirmed that the neurons that are freed from their previous functional constraints by such a process are, in fact, modified to perform a novel, different function.

From New Neurons to New Functions

Regardless of how new neurons evolve, they must achieve new input/output relations to mediate any new function. Clearly, migration of neurons into a new position is a widespread phenomenon in the developing brain, and it is widely agreed that this is correlated with the formation of novel input to this differently positioned subset of neurons. Unfortunately, we do not yet understand what initiates, guides, and arrests this migration, which can bring neurons from a dorsal part of the brain to the ventral part or from one side to the other (Fritzsch and Northcutt, 1993). It is known that even those neurons that are likely to be homologous and that have comparable functions can differ, as in the cases of the laminar nucleus of birds and the medial superior olive in mammals. Both serve as relay neurons for the auditory pathway, but in birds (and reptiles), they are dorsal, next to the primary auditory nuclei, whereas in mammals, they are ventral, near the base of the hindbrain (Webster et al., 1992). These differences in position can be reconciled with the assumed homology of these populations by showing that, indeed, both nuclei arise dorsally but migrate further ventrally in mammals.

In contrast to the well-accepted role of migration for possibly achieving a novel input, it is much more controversial to explain how a novel target is reached by the axon. Some theories are based on the undeniable fact of a widespread exuberant projection of axons to many different targets during development in birds and mammals, out of which a few will be selected. Although appealing, this idea fails to integrate the available data on the peripheral nervous system, which show a rather precise selection of pathways by navigating axons. In their ultimate version, such theories assume that no new connections ever form in the brain, but that differences occur exclusively through differential loss of connections present in ancestral forms (Ebbesson, 1984). Given the broad appeal of these ideas, it appears necessary to highlight the evidence in favor of the formation of new connections.

One of the most striking examples is the growth of axons from the mammalian cortex to the spinal cord, the corticospinal tract. The pyramidal neurons are the only cortical neurons in vertebrates which project to the lumbar spinal cord; thus, they are considered by most researchers to form a novel tract. Any other interpretation would mean that this tract is ancestral to vertebrates and was lost in all lineages except for mammals. Even more compelling evidence for the de novo formation of pathways comes from fiber outgrowth into the periphery. While this theory was rejected for almost a century, it is now clear that no motoneuron axons pass through the "ventral roots" of Amphioxus, a likely ancestor of vertebrates. Instead, muscle fibers project with noncontractile processes to these ventral roots. One of the major steps in the evolution of the vertebrate brain and spinal cord was, then, for motoneurons to themselves project through the ventral root to reach their target muscles (Fritzsch and Northcutt, 1993). While it is still unclear when and how this novel invasion of axons into the periphery happened, those motoneurons found a pathway where they had never been before in prior species. Moreover, addition of novel tissue may have caused redirection of growing fibers.

For example, in rhombomeres 4 and 5 of the chicken hindbrain, motoneurons initially form two paramedian strips (Fritzsch, Christensen, and Nichols, 1993). All of these motoneurons project their axons to the facial root to exit the brain. Within the facial nerve, some axons reroute to reach the developing ear, ending there as statoacoustic efferents on a remarkably different target: instead of innervating striated muscle fibers as the facial motoneurons do, they innervate hair cells of the ear. Moreover, this pathway of efferent fibers to the ear differs among vertebrates: rerouting within the nerve occurs in chicken and frogs, but in mammals it occurs within the brain (Fritzsch et al., 1993). In addition, the initially overlapping populations of efferent neurons segregate into different positions within the hindbrain through differential migration. While this example shows formation of connections to a novel target and segregation of inputs, it is unclear to what degree the reorganization of the facial motoneurons to the ear to form the statoacoustic efferents depends on the formation of the ear and a concomitant suppression of differentiation of parts of branchiomeric muscle fibers. Such losses of a target could conceivably have forced a subset of facial motoneurons either to innervate a new target and become the efferents to the ear or else to disappear. Thus, even in fairly accessible and well-studied systems, the details underlying the revealed variations are not sufficiently understood.

In conclusion then, there is evidence both for ongoing invasion of certain fibers into novel territories and for rather precise pathway selection, precluding the idea that evolution must always pick from a completely randomized network. Nonetheless, the ongoing plasticity in certain areas like the forebrain may, in fact, benefit from a less constrained development that enables them to form a wider array of initial connections from which only a certain few will be selected in a later step.

Variation in Functionally Relevant Details

Nervous tissue not only undergoes modifications in its long-range connections and formation of new cells, but also shows numerous cellular reorganizations, as evidenced by the degree of branching of dendrites, variations in neurotransmitter(s) and postsynaptic receptors, and stratification into distinct laminae or the absence of these in topologically comparable areas of the brain (Arbas, Meinertzhagen, and Shaw, 1991). For example, the gustatory nuclei in closely related species of bony fish may or may not be laminated. Or the neurons of the cerebellar nuclei, which are all assembled in the white matter in mammals, may become stratified with Purkinje cells into a single cortical layer in bony fish. It is still unclear what, if any, functional implications these differences in cellular assembly may offer. This is particularly evident in the stratification of the tectum of many vertebrates, which apparently can be secondarily reduced in salamanders without appreciable functional deficits (Roth et al., 1993).

Another example can be found in the forebrains of birds and mammals, in which there are differences in organization despite similarities in function and long-range connections. Clearly, comparable functions can be accomplished with either a laminar, cortical organization or with a set of interconnected groups of neurons. Indeed, Karten (1991) has suggested that "laminar organization of populations is an alternative means of organizing populations of neurons." It appears, then, that some of the well-appreciated laminar networks are but one way in which the brain can solve implementation of a specific functional circuit.

One of the most striking examples in the context of laminated structures is the difference in the organization of ganglion cells in the vertebrate retina. In the retina of jawed vertebrates, these cells are almost exclusively located in a distinct ganglion cell layer. However, the likely homologous ganglion cells of the retina in lampreys are predominantly located in the inner nuclear layer, together with other neurons (Fritzsch, 1991). While the functional significance of such a reorganization of cell distribution is still unknown, the organization in lampreys is noteworthy in that it resembles the layering of cells

typical for the brain. As such, this similarity in stratification minimizes the problem of deriving the retina from brain tissue and shows that even such complicated organs as the eye can be shown to be derived by stepwise transformation in evolution. We now see that evagination of the eye and transformation of the stratification of neurons are independent steps in the evolution of the lateral vertebrate eyes (Fritzsch, 1991). Thus, any possible combination of partly evaginated eyes and partly transformed brain/retina can, theoretically, fill in gaps in the known evolution of eyes.

Discussion

This brief overview has stressed some of the emerging developmental principles that govern the remarkably stable overall pattern of the vertebrate brain but that provide enough room for plasticity in both local as well as long-range circuitry. It is proposed that complex interactive genetic cascades offer room for "developmental errors" which are the driving forces for any reorganization. Thus, the "error" of one generation serves as the source for novelty in the next generation, provided it can be integrated into the existing network and its ever-changing functional needs. Looking at the evolution of the brain as a system that has managed to implement developmental "errors" to evolve into a multitude of forms may change our view of the adaptiveness of the brain to a more dynamic perspective: that is, adaptation allows a developing system with changes directed by internal mechanisms to stay on top of its changing adaptive (epigenetic) landscape (Kaufman, 1993). Modeling such systems must take into account the fact that optimal adaptation is compromised by these internal and external dynamics, and thus may never be achieved in any living systems to the extent that it can be achieved in technical systems.

Acknowledgments.   This work was supported, in part, by a grant from the NIH (P50 DC00215–09).

Road Map: Biological Networks
Related Reading: Cognitive Development; Olfactory Cortex; Vestibulo-Ocular Reflex: Performance and Plasticity

References

Arbas, E. A., Meinertzhagen, I. A., and Shaw, S. R., 1991, Evolution in nervous systems, *Annu. Rev. Neurosci.*, 14:9–38.

Ebbesson, S. O. E., 1984, Evolution and ontogeny of neural circuits, *Brain Behav. Sci.*, 7:321–366.

Fritzsch, B., 1991, Ontogenetic clues to the phylogeny of the visual system, in *The Changing Visual System* (P. Bagnoli and W. Hodos, Eds.), London: Plenum, pp. 33–49.

Fritzsch, B., Christensen, M. A., and Nichols, D. H., 1993, Fiber pathways and positional changes in efferent perikarya of 2.5 to 7 day chick embryos as revealed with DiI and dextran amines, *J. Neurobiol.*, 24:1481–1499.

Fritzsch, B., and Northcutt, R. G., 1993, Cranial and spinal nerve organization in amphioxus and lampreys, *Acta Anat.*, 148:96–110. ◆

Gilland, E., and Baker, R., 1993, Conservation of neuroepithelial and mesodermal segments in the embryonic vertebrate head, *Acta Anat.*, 148:111–132. ◆

Joyner, A. L., and Guillemot, F., 1994, Gene targeting and development of the nervous system, *Curr. Opin. Neurobiol.*, 4:37–42.

Karten, H. J., 1991, Homology and evolutionary origins of the "neocortex," *Brain Behav. Evol.*, 38:264–272.

Kaufman, S. A., 1993, *The Origin of Order: Self-Organization and Selection in Evolution*, New York: Oxford University Press, p. 709. ◆

Korzh, V. P., 1994, Genetic control of early neuronal development in vertebrates, *Curr. Opin. Neurobiol.*, 4:21–28.

Krumlauf, R., Marshall, H., Studer, M., Nonchev, S., Sham, M. H., and Lumsden, A., 1993, *Hox homeobox* genes and regionalization of the nervous system, *J. Neurobiol.*, 24:1328–1340. ◆

Pendleton, J. W., Nagai, B. K., Murtha, M. T., and Ruddle, F. H., 1993, Expansion of the *Hox* gene family and the evolution of chordates, *Proc. Natl. Acad. Sci. USA*, 90:6300–6304.

Roth, G., Nishikawa, K. C., Naujoks-Manteuffel, C., Schmidt, A.,

and Wake, D. B., 1993, Paedomorphosis and simplification in the nervous system of salamanders, *Brain Behav. Evol.*, 42:137–170.

Ruiz i Altaba, A., 1994, Pattern formation in the vertebrate neural plate, *Trends Neurosci.*, 17:233–243.

Webster, D., Popper, A. N., and Fay, R. R., 1992, *The Evolutionary Biology of Hearing*, New York: Springer-Verlag, p. 858.

Expert Systems and Decision Systems Using Neural Networks

Stephen I. Gallant

Introduction

One direct application of neural networks is to predict some value of interest (for example, blood glucose level), given the input features that affect that value (meal size, insulin dose, etc.). Often, however, we are most interested in making a decision based on the model's inputs. For example, we may really want to know whether, given certain medical data, we should "admit patient to hospital." This article gives a short introduction to decision making using neural networks, including neural network expert systems and specific applications.

Types of Decision Systems

Some decisions involving neural networks require a "human in the loop," and some do not. For example, if we want to decide which stock trades to make based on a system that predicts next month's stock prices, we can make this decision based on information supplied by the net. Contrast this with a system that directly recommends specific stocks to buy and sell. In the latter case, the neural network produces a decision (which is subject to review by a human). For the most extreme, time-critical cases, a computer must make a decision and act upon it *without any human intervention*. For example, deciding whether to approve credit card charges requires automated processing because there is simply no time for a human in the loop. This holds similarly for time-critical arbitrage activity.

Expert Systems

An important subclass of neural network decision systems requires additional capabilities that are usually associated with the non-neural *expert systems* of traditional artificial intelligence (AI). Duda and Shortliffe (1983) give a good overview of this subject. These expert system–like characteristics include the ability to:

- Make inferences based on partial input data.
- Direct user input in an efficient manner.
- Justify inferences by means of IF-THEN rules.

Note that these three capabilities can be important. For example, in a general medical diagnostic system, it would be ludicrous to require that all possible symptoms and all possible test results be entered prior to reaching a diagnosis. Thus, a medical expert system should stop user input as soon as enough information has been entered so that a conclusion cannot be changed on the basis of additional information. Furthermore, an expert system should help order the user's input so that only relevant information need be entered.

With respect to IF-THEN rules, a bank or insurance company may be required to give such a rule as justification for any loan or insurance application that is denied. Although the underlying decision model may be complex, such credit scoring justifications are invariably simple (e.g., "salary too low," "poor credit rating," "monthly expenses too high"). Rules also help instill confidence in the system on the part of the user.

In practice, it can be fairly arbitrary whether a system is classified as an expert system or not, and this holds for both conventional and neural expert systems. We prefer to reserve the term *expert system* for systems with at least some of the above three capabilities.

The next sections look at ways to use a neural network model as a decision system and introduce the MACIE model for neural network expert systems, arguably the simplest and most direct way of incorporating expert system characteristics into a neural network model. We will also see some other neural network expert system models, and survey some actual applications. Some of the material in this article comes from Gallant (1993), where many of these issues are explored in greater detail.

Neural Network Decision Systems

The most straightforward way of using a neural network as a decision system is to threshold its output into two (or more) classes. For example, a real-valued output, u, in the range (0, 1) might be interpreted as:

$$\text{Action A if } u \geq 0.5$$
$$\text{Action B if } u < 0.5$$

or

$$\text{Action A if } u \geq 0.9$$
$$\text{Action B if } u \leq 0.1$$
$$\text{No action otherwise}$$

Of course a model with discrete outputs, e.g., $\{+1, -1\}$, is trivially interpretable as a decision model.

All decision models (including expert systems) involve some sort of loss function that quantifies the penalty for making a wrong decision. A simple way to accommodate such a loss model is to use a neural network model with discrete outputs and a learning algorithm that maximizes the probability of producing a correct output, and then to adjust the set of training examples to reflect the loss function. For example, if incorrect "yes" ($+1$) decisions are twice as expensive as incorrect "no" (-1) decisions, we would take the set of training examples and double the number of "no" examples. For an extended illustration of this type of processing, see Gallant (1993, chap. 16).

It is worth noting that mean squared error learning algorithms, including the popular BACKPROPAGATION: BASICS AND NEW DEVELOPMENTS (q.v.) algorithm, minimize mean squared error (or some other differentiable error function) and do not necessarily produce maximum likelihood estimators that minimize the probability of misclassifying a training example. For example, mean squared error objective functions overemphasize the influence of outliers in the data. Nevertheless, the similarity between these two types of objective functions is usually close enough so that it is not an issue with respect to performance. (See Gallant, 1993, chap. 6, for more on this point.)

Neural Network Expert Systems

The fact that a neural network could do things normally attributed to expert systems was shown in Gallant (1988a, 1988b, 1993). Saito and Nakano (1988) implemented an early neural network expert system for diagnosing causes of headaches, and explored rule extraction using a kind of sensitivity analysis. Fu (1989) combined rules from MYCIN, a classic expert system, with neural training.

Here we look at MACIE (MAtrix Controlled Inference Engine), a model for neural network expert systems, to give a flavor of how expert system tasks are accomplished. Consider a multilayer perceptron network model where cell activations, u_i, take on values in $\{+1, -1, 0\}$ with interpretations:

$$+1 \sim \text{True}$$
$$0 \sim \text{Unknown}$$
$$-1 \sim \text{False}$$

Forward Chaining

Now consider Figure 1 with weights $w_{i,j}$ as shown. Here, we are focusing on a single cell in a multilayer perceptron network,

and we are partway through an expert system consultation. This cell may be either an intermediate cell or an output cell. Currently, we know that inputs u_2 and u_6 are True, input u_4 is False, and input u_5 is not available for this run. (Note that the bias weight, $w_0 = -3$, is always added to the weighted sum when computing a cell's output so that the threshold may be taken to be zero. Therefore it is convenient to define a bias input, u_0, which is always $+1$ and is considered a "known" input.)

Inputs u_1 and u_3 are currently Unknown, but their values are potentially available based upon additional information from the user.

Our first task is to see whether we already have enough information to make an inference. We compute the current weighted sum and the amount this sum might change if more information becomes known:

$$\text{Current}_i = \sum \{w_{i,j} u_j | u_j \text{ not unknown}\}$$
$$\text{Change}_i = \sum \{\text{ABS}(w_{i,j}) | u_j \text{ unknown}\}$$

Notice that we include unavailable variables in the Current summation, even though all of their corresponding terms are 0. The bias is also included as the first term in the Current computation. For our example we have:

$$\text{Current} = (-3)(+1) + (6)(+1) + (-3)(-1)$$
$$+ (-4)(0) + (-2)(+1) = 4$$
$$\text{Change} = |-1| + |-2| = 3$$

Current gives the current weighted sum (including bias), and Change gives how much this weighted sum might possibly change in either direction, based upon any possible combination of unknown information. For example, if we should find

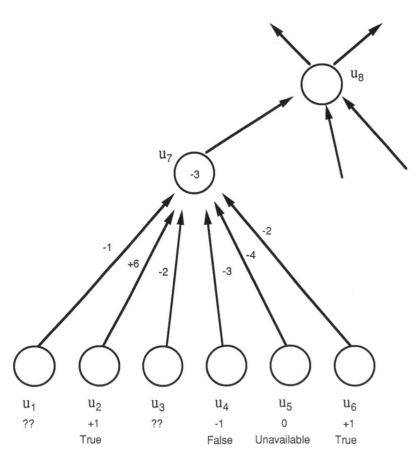

$$
\begin{array}{cccccc}
u_1 & u_2 & u_3 & u_4 & u_5 & u_6 \\
?? & +1 & ?? & -1 & 0 & +1 \\
 & \text{True} & & \text{False} & \text{Unavailable} & \text{True}
\end{array}
$$

Figure 1. A single cell, partway through a run of MACIE. The bias weight, -3, is pictured inside the cell; the bias "input" is always taken to be $+1$. (A *threshold* is equivalent to a bias with its weight negated.)

u_3 to be True, then this would change the weighted sum by -2. In our example, the weighted sum can change by no more than 3.

Because the final weighted sum will be greater than 0 regardless of eventual values for any unknown variables, we can now make the inference that u_7 must be True. More generally, whenever

$$\text{ABS}(\text{Current}_i) > \text{Change}_i$$

then we can make an inference for u_i, namely that u_i is True or False, according to whether Current_i is positive or negative, respectively. MACIE's inferencing method is *conservative*, because the inference can never be changed by finding out more information (i.e., values for unknown variables).

Note that once we infer a value for a cell, this might trigger other inferences higher up in the network. This is a type of forward chaining in expert systems.

It is also possible to justify an inference by an IF-THEN rule, giving an explanation capability for neural network expert systems. The basic idea is to take just enough of the inputs that led to a particular inference to guarantee that conclusion, and then to express this minimal set of inputs as an IF-THEN rule for the system. In the example discussed above, such a rule (see Gallant, 1993, chap. 17, for details) would be:

If u_2 is True
And u_5 is Unavailable
And u_4 is False
Then conclude that u_7 is True.

Although such rules are good enough for justifications of credit scoring decisions, they typically are not cognitively compelling, and this raises subtle issues concerning the very nature of "justifications," "explanations," and "understanding." When we speak of *explaining something*, we usually mean that the explanation is in terms of some rules in an underlying model. For example, a financial system might have a model that predicts falling stock prices if unemployment drops. A more satisfying explanation of the situation would be that decreased unemployment makes both inflation and higher interest rates more likely, which in turn reduces stock prices. Although more cognitively compelling, the latter explanation is in fact merely a collection of rules in a more complex model. The essential nature of the explanation is not different from the original simple explanation. Moreover, any attempt to completely understand the situation would lead to a large (infinite?) and fruitless regress:

Q: Why does a drop in unemployment cause more
        inflation?
A: Fewer available workers means higher wages, which
        fuels inflation.
Q: Why does fewer available workers mean higher wages?
A: Supply and demand with respect to wages.
Q: Why does supply and demand hold?
            · · ·
Q: Why do companies want profit?
            · · ·
Q: Why do individuals want more money?
            · · ·
Q: Why do individual cognitive processes lead to profit-
        seeking behavior?
            · · ·
Q: Why do neurons produce this behavior?
            · · ·
Q: Why do subatomic particles have mass?
            · · ·

Because of such difficulties, we prefer to speak of less ambitious *justifications* produced by the expert system rather than *explanations*.

Although the procedure outlined above for extracting rules is fairly simple for single-cell models, if we are given a complex-enough network, there is an open research issue on how to efficiently derive the *simplest and most compelling collection* of IF-THEN rules that justify a network output.

Another approach for extracting rules was developed by Towell and Shavlik (1993). Here, the weights are clustered by magnitude and replaced by single values. For example, weights of $\langle 5.1, 5.2, 5.3, 1.1, 0.9 \rangle$ would be replaced by $\langle 5.1, 5.1, 5.1, 1.0, 1.0 \rangle$. The groups are then interpreted by predicates of the form "at least N of the following M variables are True." This method is easily extended to include negative weights—for example, if the first weight were replaced by -5.1. Towell and Shavlik (1993) and Craven and Shavlik (1993) report good results in extracting simple rules that are easily understandable and that do not produce severe degradation in performance when used in place of the original networks.

Saito and Nakano (1990) have also developed their RN (rules from networks) method for approximating a network by rules. This approach takes advantage of the original set of training examples to construct hypercube intersections that approximate the behavior of the original network in the regions surrounding the training examples.

Fuzzy Models

Hayashi has done much work in combining neural network expert systems with fuzzy logic (Hayashi, 1991). He represents values by a group of discrete cells, with fuzzy values represented by patterns in the group. For example, "somewhat tall" might be represented as $\langle +1 +1 +1 +1 -1 \rangle$ using a thermometer code (B. Widrow's terminology) or level set representation (see also Frydenberg and Gallant, 1986). (A *thermometer code* is a vector of $+1$ values followed by -1 values, where the number of $+1$ values corresponds to the degree to which something is true or to the magnitude of a value.) Hayashi also developed ways of extracting fuzzy IF-THEN rules from such networks.

Hall and Romaniuk (1990) have also developed somewhat complex neural network expert system models with built-in fuzzy logic functions that they named SC-net and FUZZNET. Here the sign of the bias of a cell is used to signal different cell activation functions.

Applications

There have been many successful applications of neural network models as decision systems in areas such as portfolio management, credit scoring, credit card fraud detection, marketing, direct mail, debt collection, aerospace instrument and process calibration, fault detection, inventory level management, and mortgage appraisal and underwriting. This section outlines a few successful applications.

Fraud Detection

HNC Inc.'s Falcon™ system, a neural network system for credit card fraud detection, recently went on-line with First Data Resources, the largest credit card processor in the United States. Tests showed that network models were better at spotting fraud (and with fewer false alarms than with humans), and HNC claims to have saved over \$50 million in fraud losses (*AI Expert*, June 1994). The whole area of fraud detection looks promising for neural network decision models.

Stock Trading

Although financial institutions tend to be very secretive, especially concerning new technology, it is well known that many companies are using neural network models, including some marketed by NeuralWare (Burke, 1992; Mandelman, 1992; McGough, 1992). Fidelity Investments uses neural network models in at least four mutual funds; for example, with Fidelity's Small Cap Fund, a neural network is used to screen large numbers of stocks of smaller companies. Small company stocks are followed less closely than large company stocks because there are so many small stocks and because only a limited amount can be invested in any small company stock. Fidelity uses a neural network model to screen thousands of these small stocks, then a manager makes final trading decisions. Note that these are human-in-the-loop decision systems, where the net is suggesting stock trades but a human is making final decisions.

Credit Scoring

HNC has also had success with a neural network expert system for approving loan applications for a major national lender. Here, it is vital to justify loan "declines" using an IF-THEN rule involving the applicant's financial status, so this task requires expert system functionality. Accuracy of the system built by HNC is very high. One test involving 100,000 cases produced only a few errors due to corrupted data, and got the remaining cases correct (same decisions as humans had made), all with reasonable justifications of the net's actions.

Clinical and Epidemiological Applications

There is an increasing use of neural networks in clinical and epidemiological domains. For example, Baxt (1990) built a neural network for diagnosis of acute coronary occlusion, and reported excellent results in predicting whether patients required hospitalization. More recent work has analyzed the rules discovered by these networks.

Factory Production and Maintenance

Hughes Missile Systems has reported successful neural network decision systems for several tasks, including failure diagnosis and optimizing resource use. In one spectacular case, they built a neural network expert system in one day that replaced a conventional expert system that had required one person-year to construct.

In another case, a civilian aircraft company was able to demonstrate large savings in maintenance by having a neural network model determine the need for major aircraft engine overhauls.

These cases are but a tiny sample of the current uses for neural networks.

Concluding Remarks

When is a neural network expert system preferable and when is a conventional expert system preferable? Conventional expert systems (and simple programming) are usually better able to handle the procedural aspects that are often required by practical expert systems. For example, MYCIN is able to dynamically generate nodes that correspond to individual sites or treatments (but see TOPOLOGY-MODIFYING NEURAL NETWORK ALGORITHMS). If aspects of a problem are truly captured by a known collection of IF-THEN rules, then a conventional expert system is clearly best. For decisions based upon complex, noisy, and redundant data, then neural networks should be useful, either alone or to produce the key judgmental inferences as part of a combined system.

In the future, we can expect continued growth in applications of neural network decision systems and neural-network expert systems, as well as continued integration with conventional expert systems. It is encouraging to note that six European organizations have joined together to explore these issues in the MIX project (Modular Integration of Connectionist and Symbolic Processing in Knowledge-Based Systems) under the ESPRIT initiative.

In the future, we can also expect an increased spread of interest from large, high-tech industries into smaller and lower-tech enterprises, and eventually we may even see personal neural network expert systems in common use.

Road Maps: Applications of Neural Networks; Artificial Intelligence and Neural Networks
Related Reading: Bayesian Networks; Fuzzy Logic Systems and Qualitative Knowledge; Learning by Symbolic and Neural Methods

References

Baxt, W. G., 1990, Use of an artificial neural network for data analysis in clinical decision-making: The diagnosis of acute coronary occlusion, *Neural Computat.*, 2:480–489.

Bookman, L. A., and Sun, R., Eds., 1993, *Connection Sci.*, 5(3, 4), special issue on architectures for integrating neural and symbolic processes.

Burke, G., 1992, Neural networks: Brainy way to trade? *Futures*, August, pp. 34–36.

Craven, M. W., and Shavlik, J. W., 1993, Learning symbolic rules using artificial neural networks, in *Machine Learning: Proceedings of the Tenth International Conference* (P. E. Utgoff, Ed.), San Mateo, CA: Morgan Kaufmann.

Duda, R. O., and Shortliffe, E. H., 1983, Expert systems research, *Science*, 220:261–268. ◆

Frydenberg, M., and Gallant, S. I., 1986, Fuzziness and expert system generation, in *International Conference on Information Processing and Management of Uncertainty in Knowledge-Based Systems*, Paris, France, June 30–July 4. Extended paper reprinted 1987, in *Uncertainty in Knowledge-Based Systems* (B. Bouchon and R. Yager, Eds.), Berlin: Springer-Verlag, pp. 137–143.

Fu, L. M., 1989. Integration of neural heuristics into knowledge-based inference, *Connection Sci.*, 1:325–340. ◆

Gallant, S. I., 1988a, Connectionist expert systems, *Commun. ACM*, 31(2):152–169. Japanese translation 1988, in *Neurocomputer* (Nikkei Artificial Intelligence), pp. 114–136. ◆

Gallant, S. I., 1988b, Matrix Controlled Expert System Producible from Examples, United States Patent 4,730,259.

Gallant, S. I., 1993, *Neural Network Learning and Expert Systems*, Cambridge, MA: MIT Press. ◆

Hall, L. O., and Romaniuk, S. G., 1990, A hybrid connectionist symbolic learning system, paper presented at American Association for Artificial Intelligence (AAAI) annual conference, Boston, MA.

Hayashi, Y., 1991, A neural expert system with automated extraction of fuzzy if-then rules and its application to medical diagnosis, in *Advances in Neural Information Processing Systems 3* (D. S. Touretzky and R. Lippman, Eds.), San Mateo, CA: Morgan Kaufmann, pp. 578–584.

Mandelman, A., 1992, The computer's bullish! *Barron's*, 14 Dec.

McGough, R., 1992, Fidelity's Bradford Lewis takes aim at indexes with his "neural network" computer program, *The Wall Street Journal*, 27 Oct.

Saito, K., and Nakano, R., 1988, Medical diagnostic expert system based on pdp model, in *Proceedings of the IEEE International Conference on Neural Networks*, vol. 2, New York: IEEE.

Saito, K., and Nakano, R., 1990, Automatic extraction of classification rules, paper presented at International Neural Network Conference, Paris.

Towell, G. G., and Shavlik, J. W., 1993, Extracting refined rules from knowledge-based neural networks, *Machine Learn.*, 13:71–101. ◆

Exploration in Active Learning

Sebastian Thrun

Introduction

Research on machine learning has, over the last decades, produced a variety of techniques to automatically improve the performance of computer programs through experience. Approaches to machine learning can roughly be divided into two categories, passive and active, each making characteristic assumptions about the learner and its environment.

Passive Learning

In the passive learning paradigm, a learner learns purely through observing its environment. The environment is assumed to generate a stream of training data according to some unknown probability distribution. Passive learning techniques differ in the type of results they seek to produce, as well as in the way they generalize from observations. Common learning tasks are the clustering, classification, or prediction of future data.

Passive learning techniques can be subdivided into order-free and order-sensitive approaches. Order-free approaches rest on the assumption that the temporal order in which the training data arrives does not matter for the task to be learned. It is assumed that the training examples are generated independently according to a stationary probability distribution. The majority of machine learning approaches falls into this category. For example, unsupervised learning usually aims to characterize the underlying probability distribution or to cluster the data. Supervised learning, on the other hand, is concerned with approximating an unknown target function (conditional probability) from a set of observed input-output examples.

Passive learning has also been studied in order-sensitive learning scenarios, which are settings in which the temporal order of the training data carries information relevant to the learning task. This is the case, for example, if consecutive training examples are conditionally dependent on each other, and learning about these dependencies is crucial for the success of the learner. Time-series prediction or speech recognition are examples of order-sensitive learning domains.

Active Learning

The active learning paradigm allows the learner to interact with its environment. More specifically, the learner can execute actions which have an impact on the generation of training data. The freedom to execute actions imposes an important challenge that is specific to active learning: Which actions shall a learner generate during learning? How can a learner efficiently explore its environment?

In active learning, one can also distinguish between order-free and order-sensitive cases. Order-free active learning rests on the assumption that what is observed in the environment depends only upon the most recently executed action. Perhaps the best-studied approach of this kind is learning by queries (Angluin, 1988; Atlas et al., 1990; Baum and Lang, 1991). In query learning, the available actions are queries for values of an unknown target function. The environment provides immediate responses (answers) to these queries.

In order-sensitive approaches, on the other hand, observations may depend on many actions. For example, approaches to learning optimal control (like airplane control, or game playing) fall into this category. To describe the long-term dependencies between actions and observations, it has frequently proven helpful to assume that the environment possesses internal state information. Actions influence the state of the environment, and the state determines what the learner observes.

Exploration refers to the process of selecting actions in active learning. Although most of the exploration techniques reviewed in this article are applicable to active learning in general, we will primarily focus on action selection issues in order-sensitive scenarios. Indeed, most of the approaches listed here have originally been applied in order-sensitive frameworks. Notice that throughout the article, we will make the simplifying and restrictive assumption that the state of the environment is fully observable.

Action Selection Strategies

Principles

How can a learner pick the right action for learning? At first glance, it might seem appropriate to use random action selection mechanisms to generate actions (Whitehead, 1991). Random action selection is frequently used, primarily for two reasons: (a) it is simple; and (b) it usually ensures that any possible finite sequence of actions will be executed eventually. However, it has been shown, both through theoretical analyses as well as empirical findings, that more sophisticated query and exploration strategies can often drastically reduce the number of training examples required for successful learning. This is because responses to different actions typically carry different amounts of information. Random sampling often does not take full advantage of the opportunity to select the most informative query/action.

Intuitively speaking, in order to learn efficiently, one would like to execute actions that are most informative for the learning task at hand. The more one expects to learn from the outcome of an action, the better it is. Indeed, this "greedy" principle, the optimization of knowledge gain, has been employed in most approaches to action selection in active learning.

Query Selection

Recent research on query learning has led to a variety of approaches for the active selection of queries. For example, two recent approaches to learning by queries both use a neural network model of the learner's uncertainty. During learning, queries are favored that have the least predictable outcome. Uncertainty is estimated either by the difference of two models constructed from the same observations (Atlas et al., 1990), or based on an analysis of the parameters of the estimator (Cohn, 1994). Both approaches have proven superior to random sampling in empirical comparisons. Paass and Kindermann (in press) propose a method that integrates an external cost function into the active learning framework. More specifically, their approach favors queries that minimize the decision costs, which allows learning to be focused in performance-relevant areas. Their approach is computationally expensive, since it relies on explicit Monte-Carlo integration.

Exploration

In order-sensitive scenarios, exploring unknown parts of the environment requires that sequences of actions be executed. For example, if a lunar robot aims to explore the back side of the moon, it has to get there first—and getting there might require even more exploration. Techniques that employ models of the expected knowledge gain and then direct explorative actions to unknown parts of the environment are called *directed exploration techniques* (see Thrun, 1992b, for an overview). While most existing approaches share the philosophy of selecting actions through maximizing the knowledge gain, they differ in the particular way knowledge gain is estimated. Some estimate this quantity implicitly from a specific data structure; others utilize explicit models represented using separate data structures. In addition, a variety of heuristic estimators have been used for estimating the expected knowledge gain. Estimators typically use related quantities such as frequency, density, recency, or empirical prediction errors.

For example, Kaelbling (1993) suggests an approach to exploration in which actions are favored unless they have repeatedly been found to be disadvantageous. As a consequence, actions are selected that exhibit good performance or that are unexplored or both. A similar approach following the same line of thought is proposed by Koenig and Simmons (1993). Their approach bears close resemblance with heuristic search techniques for graphs (Korf, 1988), although it differs in that it does not assume the availability of a model of the environment. Koenig and Simmons also derive worst-case bounds for the complexity of exploration for deterministic shortest-path problems. Sutton (1990) assigns a so-called exploration bonus to actions. This bonus measures, for each environment state, the elapsed time since each available action was executed. As a consequence, actions that were not executed for a long time are favored for exploration. Sutton also employs a dynamic programming technique to propagate exploration utility through the state space of the environment using a model of the environment, which is easy to obtain for the environments he studied. Another approach achieves exploration through a Bayesian prior that expresses uncertainty as a function of how often and when an action has been executed. Thrun (1992a) compares several of these approaches empirically, along with a combined approach taking frequency and recency into account. Approaches specific to memory-based learning have been proposed by Moore (1990) and Schaal and Atkeson (1994). Memory-based learning memorizes all training data explicitly. In these approaches, the density of previous data points is used to assess the utility of actions for exploration. Schaal and Atkeson also take into account knowledge about the goal of learning to focus exploration.

Focusing Exploration

Most active learning techniques estimate the expected knowledge gain of the learner for each applicable action, and they select actions through maximizing knowledge gain. In order for this methodology to work efficiently, two assumptions have to be made: the heuristic for estimating the gain of knowledge must yield approximately correct action preferences; and gaining knowledge per se must be helpful for the learning task.

Both assumptions are not necessarily fulfilled in practice. Most heuristics for exploration are somewhat ad hoc, hence their effectiveness varies across environments and learning tasks. Moreover, depending on what goal the learner aims to achieve, sometimes only parts of the environment have to be known in order to perform optimally. This is typically the case, for example, in the context of REINFORCEMENT LEARNING (q.v.; see also Barto et al., in press; Sutton, 1990; Watkins and Dayan, 1992). In reinforcement learning, the learning task is to generate control, i.e., to learn action policies that maximize a given reward function. Exploring regions in state space that are irrelevant for the task of learning control is a waste of both time and memory resources.

A common strategy to focus exploration is to explore and to exploit simultaneously, by taking both knowledge gain and the task-specific utility of actions into account. Boltzmann distributions and semi-uniform distributions provide ways to combine random exploration with exploitation. These distributions explore by flipping coins, but the likelihood of individual actions is determined by the task-specific exploitation utility: In Boltzmann distributions, the likelihood of picking an action is exponentially weighted by its utility; and in semi-uniform distributions, the action with the largest utility has a distinctly high probability of being executed. Notice that most of the aforementioned approaches have indeed originally been proposed in combination with task-specific exploitation. Thrun (1992b) empirically demonstrated that the combination of exploration and exploitation can yield faster learning than either component in isolation. The fundamental dilemma of choosing the right ratio between exploration and task-specific exploitation is called the *exploration-exploitation dilemma*. Often, exploration and exploitation are traded off dynamically so that exploration fades in time.

Complexity Results

In addition to empirical studies, theoretical results emphasize the importance of exploration in active learning. Based on a result by Whitehead (1991), which shows that random walk exploration can require exponential learning time in various cases, it has been shown that directed exploration techniques can reduce the complexity of active learning from exponential training time (random exploration) to polynomial training time (Thrun, 1992a; Koenig and Simmons, 1993). Similar results exist in the query learning framework (Angluin, 1988; Baum and Lang, 1991). Although most results apply to certain deterministic environments only, in practice they often carry over to stochastic environments.

Example

This section briefly describes an artificial neural network approach to exploration in real-valued domains. This approach shares many of the ideas in the current literature: the expected knowledge gain is estimated during learning, and actions are selected greedily such as to maximize knowledge gain. Unlike most other approaches, it operates in real-valued domains and uses artificial neural networks to estimate the gain of knowledge.

Modeling Competence

Assume the learning task is to approximate an unknown target function $f: I \rightarrow O$. Here, I denotes the input space, and O denotes the output space of both f and its approximation, denoted by $\hat{f}$. Assume that at any instance in time, the learner can execute one of its actions, denoted by A. Actions influence the state of the environment and hence have some impact on the training examples given to the function approximator. Notice that in the experiments reported below, I is the product of A and the state space of the environment.

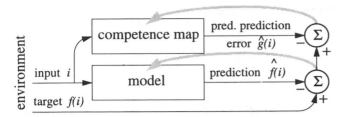

Figure 1. Training the model and the competence map.

A competence map is a function $\hat{g}: I \to \mathbb{R}$ that assesses the accuracy of $\hat{f}$. It is trained as follows. For each observed input-output example $\langle i, f(i) \rangle \in I \times O$ of f, there will be some model error $\varepsilon(i) = \|\hat{f}(i) - f(i)\|$. The competence map models this error $\varepsilon(i)$ as a function of i. Hence, each training example for $\hat{f}$ also produces a training example for $\hat{g}$, as illustrated in Figure 1. The competence map is used to direct exploration, by selecting actions that maximize $\hat{g}$. More specifically, the learner explores by picking actions for which its own competence is minimal, i.e., for which its internal models are most inaccurate. Such actions are assumed to maximize the gain of knowledge.

It should be noted that competence estimates, as they are described here, may only be approximately correct, since the dynamics of the estimators $\hat{f}$ and $\hat{g}$ are usually hard to model. In addition, if due to model limitations $\hat{f}$ fails to model the environment in sufficient detail, unmodeled effects can be a constant source of model error and perpetually provoke more exploration. This is the case, for example, if (many-to-one) function approximators like artificial neural networks are employed in highly stochastic environments.

Empirical Results

To illustrate exploration via a competence map, consider the environment depicted in Figure 2. The input to the learner is its x-y position in a two-dimensional world. Its task is to navigate from the starting position (box) to the goal position (cross) while avoiding collisions with the walls or the obstacle. Actions, which will be indicated by arrows in Figure 3, are small displacements which, when executed, are added to the current position. When the agent reaches its goal or, alternatively, when it collides with a wall or the obstacle, it is reset to its starting position.

In addition to its coordinates, the learner is able to perceive a potential function $d(x, y)$, which is depicted in Figure 2. The potential measures the "distance" to the goal and to the obstacles, such that gradient descent yields a collision-free path to the goal location from arbitrary starting positions. Both the state transition function and the potential function are initially unknown. The goal of learning is to learn a control strategy for selection actions which carry the agent to the goal. This is done by learning the state transition function and the potential function. Once a reasonable model of these functions has been identified, pure hill climbing will result in admissible paths, such that the goal can be reached without collision.

In our experiments, a multilayer network was trained with the BACKPROPAGATION: BASICS AND NEW DEVELOPMENTS (q.v.) algorithm to model the motion dynamics and the potential function values. The input to this network was the current position (x, y) and action. It was trained to predict the next position (x', y') and the corresponding potential function value $d(x', y')$. The actual network consisted of two separate components, one for predicting the next position (x', y') with no hidden units,

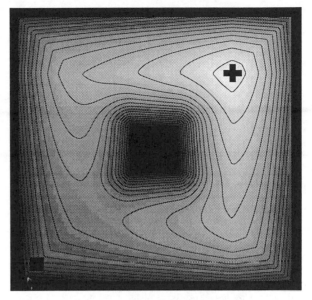

Figure 2. Potential function. The darkness indicates the combined "distance" to the goal and the obstacle/walls.

and one for predicting the potential function value. The latter component consisted of 10 units with radial basis activation functions in the first hidden layer, and eight units with a logistic activation function in the second layer.

Competence was also modeled by an artificial neural network which received the same four input values as the model, but was trained to predict the squared model prediction error, given by $\alpha[(x_{\text{pred}} - x_{\text{obs}})^2 + (y_{\text{pred}} - y_{\text{obs}})^2 + (d_{\text{pred}} - d_{\text{obs}})^2]$. Here α is an appropriate normalization constant which ensures that competence values lie in $[0, 1]$. In the actual implementation, the competence network had two hidden layers with six logistic units each. During learning, exploration and exploitation were combined using a selective attention mechanism described elsewhere (Thrun, 1992b), which traded off exploration and exploitation dynamically based on expected costs and benefits.

In an experimental study, three approaches to exploration were compared: (a) random exploration; (b) pure exploitation, i.e., always following the best known path; and (c) directed exploration based on competence. Since pure exploitation might get stuck and fail to explore exhaustively, in rare cases actions had to be generated randomly. In 15,000 learning steps, each technique learned the linear motion dynamics well. However, they produced different models of the potential function. Random exploration (cf. Figure 3A) performed most poorly. The resulting model was not accurate enough to allow the agent to navigate to the goal. When actions were generated by pure exploitation, a reasonable path was found from the start to the goal. The approach yielded good performance in terms of navigation. The model, however, was rather inaccurate and the world was poorly explored, as can easily be seen from Figure 3B. The best results in terms of both control and model accuracy were found with directed exploration using the competence map. Competence map exploration also resulted in a much smaller number of collisions during learning, yet yielded the most accurate model (Figure 3C). These findings demonstrate the advantage of directed exploration techniques over random exploration. (Further details may be found in Thrun, 1992b).

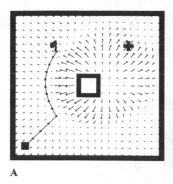 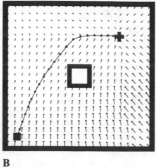 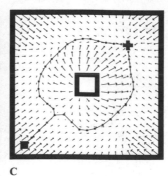

A   B   C

Figure 3. Results. *A*, random exploration; *B*, exploitation (whenever possible); *C*, exploration with a competence map.

Conclusion

In active learning, the learner is given the ability to execute actions during learning. Hence, the learner can, to a certain extent, control the stream of training data. A key challenge of active learning is to select actions so as to optimize the rate of learning.

This article reviews and discusses several heuristic approaches to the selection of actions during active learning, which can primarily be found in the literature on neural networks and reinforcement learning. In order to illustrate how these ideas work in practice, a concrete exploration mechanism based on artificial neural networks is outlined. In this approach, exploration is achieved through estimating the learner's competence. The basic philosophy behind this and most other approaches to the selection of actions is to estimate the expected gain of knowledge as a function of the actions to be executed. The learner then picks actions which maximize the expected knowledge gain.

While the area of passive, order-free learning has been studied extensively in the field of machine learning, statistics, and artificial neural networks, considerably less effort has been spent on the exploration of active learning issues. This might be attributed to the fact that passive learning is simpler, and many learning tasks do not provide the opportunity to select actions. Natural learners such as animals and humans, however, learn actively, as they make use of their capability to act and influence their environment. They are truly embedded in their environments, they act *and* observe. If artificial agents need to learn autonomously, they will probably have to follow the same learning principles.

Road Maps: Control Theory and Robotics; Learning in Artificial Neural Networks, Statistical
Background: I.3. Dynamics and Adaptation in Neural Networks
Related Reading: Potential Fields and Neural Networks; Planning, Connectionist; Reinforcement Learning in Motor Control

References

Angluin, D., 1988, Queries and concept learning, *Machine Learn.*, 2(4):319–342. ◆

Atlas, L., Cohn, D., Ladner, R., El-Sharkawi, M. A., Marks, R. J., Aggoune, M. E., and Park, D. C., 1990, Training connectionist networks with queries and selective sampling, in *Advances in Neural Information Processing Systems 2* (D. Touretzky, Ed.), San Mateo, CA: Morgan Kaufmann, pp. 567–573.

Barto, A. G., Bradtke, S. J., and Singh, S. P., in press, Learning to act using real-time dynamic programming, *Artif. Intell.*

Barto, A. G., and Singh, S. P., 1990, On the computational economics of reinforcement learning, in *Connectionist Models, Proceedings of the 1990 Summer School* (D. S. Touretzky, J. L. Elman, T. J. Sejnowski, and G. E. Hinton, Eds.), San Mateo, CA: Morgan Kaufmann, pp. 35–44.

Baum, E. B., and Lang, K. J., 1991, Constructing hidden units using examples and queries, in *Advances in Neural Information Processing Systems 3* (R. P. Lippmann, J. E. Moody, and D. S. Touretzky, Eds.), San Mateo, CA: Morgan Kaufmann, pp. 904–910.

Cohn, D., 1994, Queries and exploration using optimal experiment design, in *Advances in Neural Information Processing Systems 6* (J. Cowan, G. Tesauro, and J. Alspector, Eds.), San Mateo, CA: Morgan Kaufmann.

Kaelbling, L. P., 1993, *Learning in Embedded Systems*, Cambridge, MA: MIT Press. ◆

Koenig, S., and Simmons, R. G., 1993, Complexity analysis of real-time reinforcement learning, in *Proceedings of the Eleventh National Conference on Artificial Intelligence (AAAI-93)*, Menlo Park, CA: AAAI Press/MIT Press, pp. 99–105.

Korf, R. E., 1988, Real-time heuristic search: New results, in *Proceedings of the Sixth National Conference on Artificial Intelligence (AAAI-88)*, Computer Science Dept., University of California, Los Angeles: AAAI Press/MIT Press, pp. 139–143.

Moore, A. W., 1990, Efficient memory-based learning for robot control, PhD thesis, Trinity Hall, University of Cambridge, England.

Paass, G., and Kindermann, J., in press, Bayesian query construction for neural network models, in *Advances in Neural Information Processing Systems 7*, San Mateo, CA: Morgan Kaufmann.

Schaal, S., and Atkeson, C. G., 1994, Assessing the quality of learned local models, in *Advances in Neural Information Processing Systems 6* (J. Cowan, G. Tesauro, and J. Alspector, Eds.), San Mateo, CA: Morgan Kaufmann.

Sutton, R. S., 1990, Integrated architectures for learning, planning, and reacting based on approximating dynamic programming, in *Proceedings of the Seventh International Conference on Machine Learning*, San Mateo, CA: Morgan Kaufmann, pp. 216–224.

Thrun, S. B., 1992a, *Efficient Exploration in Reinforcement Learning*, Technical Report CMU–CS–92–102, Pittsburgh: Carnegie Mellon University.

Thrun, S. B., 1992b, The role of exploration in learning control, in *Handbook of Intelligent Control: Neural, Fuzzy and Adaptive Approaches* (D. A. White and D. A. Sofge, Eds.), Florence, KY: Van Nostrand Reinhold. ◆

Watkins, C. J. C. H., and Dayan, P., 1992, Q-learning, *Machine Learn.*, 8:279–292.

Whitehead, S. D., 1991, Complexity and cooperation in Q-learning, in *Proceedings of the Eighth International Workshop on Machine Learning* (L. A. Birnbaum and G. C. Collins, Eds.), San Mateo, CA: Morgan Kaufmann, pp. 363–367. ◆

Eye-Hand Coordination in Reaching Movements

Peter F. Dominey, Philippe Vindras, Claude Prablanc, and Denis Pelisson

Introduction

This article focuses on possible mechanisms responsible for visually guiding the hand toward a point within the prehension space, seen from the view point of modeling, in order to sketch some plausible rules of their functioning. A more detailed overview of the psychophysiological, neurophysiological, and clinical aspects can be found in Jeannerod (1988) and in the other articles on reaching movements in this volume (REACHING: CODING IN MOTOR CORTEX and REACHING MOVEMENTS: IMPLICATIONS OF CONNECTIONIST MODELS).

A simplified way to analyze the ongoing processes occurring when a subject is required to point at a visual target appearing within the prehension space is to consider how this task would be achieved by an anthropomorphic robot equipped with two cameras on its head and a nonredundant arm. An analytical approach might operate according to the following sequence. First, retinal, ocular, and cephalic angles would be measured and combined to compute a body-centered target position (extrinsic coordinates). Then, through a coordinate transformation (known as the inverse kinematics problem), target position would be represented as a set of intended joint angles (intrinsic coordinates). If there were spatial constraints on the path to follow, the robot would have to determine the trajectory. Then, taking into account the initial posture of the joints, it would compute a set of torques on each joint (known as the inverse dynamics problem) to control the duration of the movement. Finally, it would control the ongoing movement and, if necessary, correct the torques on-line to ensure that the physical movement followed the intended one.

Such a control strategy, based on a series of discrete symbolic representations performing analytical computations (such as matrix calculus or optimality computation), although biologically implausible, has driven much of the past research on movement control. However, recent neurophysiological studies have failed to find evidence for the discrete representations and for the analytic stages of computations just described. It will be seen how the different tasks described for the robot, which for the sake of clarity have been identified separately, could well be achieved by a widely parallel processing of redundant neuronal signals. An open issue is whether intermediate analytical representations needed in a Cartesian approach are necessary for motor control in living organisms. In the following sections, we describe the three stages of visuomotor transformation just discussed, as well as the more biologically valid, but functionally equivalent, distributed processes, exemplified by recent neural networks.

Representation of Target Position

The position of a visual target in the prehension space can be specified by the set consisting of retinal location, eye, and head positions. There is a many-to-one mapping from this set to a given spatial location, producing a form of "spatial equivalence" in which different combinations of retinal, eye, and head signals correspond to a single target position and lead to the same motor behavior. Within the conceptual framework of a symbolic representation of position signals, this excess of degrees of freedom is not a problem as long as we admit that a "summation" of the signals into a single variable does occur, an idea which has long been accepted by neurophysiologists

and engineering modellers. The correlate of this view is that some single cell (or a localized cluster of cells), probably in an integrative cortical area, encodes the target location, ensuring the stability of the representation of the target when the eyes and head move freely.

In fact, such a representation based on individual neurons lacks experimental evidence, and there is now a trend to consider representations distributed over large cell populations. One possible implementation of such a distributed mapping has been found by Andersen et al. (1990) in the parietal cortex of the monkey. In a head-fixed paradigm (reducing the previous set of variables to a pair), these researchers recorded visual responses in area 7a that were both retinotopically organized and modulated by eye position. This kind of target representation is perhaps typical of parallel computation in the brain: Mazzoni, Andersen and Jordan (1991) reproduced it in the middle (hidden unit) layer of a three-layered neural network that used an associative reward-penalty learning rule to transform distributed representations of retinal location and eye position into head-centered coordinates (see GAZE CODING IN THE POSTERIOR PARIETAL CORTEX).

On the other hand, the head-centered representation used by Mazzoni, Andersen, and Jordan (1991) in their output layer, and for which Andersen et al. (1990) have not reported conclusive experimental evidence, is perhaps unnecessary. Kuperstein (1988) partly addresses this issue: in a head-centered model, he showed that a hand position can be computed, without an explicit representation of space, using a distributed oculomotor representation of the target. Indeed, his distributed, self-organizing model maps extraocular muscle activation during (head-fixed) target fixation to activation of antagonist muscle pairs that correspond to an arm posture for grasping the target, without using an explicit representation of the target in body-centered space. Some recent psychophysical studies, testing the capacity of the eye and hand motor systems to compensate for a sudden change of target position during movement execution (Prablanc and Martin, 1992), could fit with that view. Despite a target perturbation triggered during the saccade (small enough to remain undetected at the conscious level), both a corrective saccade and an automatic amendment of the ongoing hand pointing were observed, as if the hand automatically reached a posture associated with the new eye refixation.

Another point specific to living systems, but not well integrated in neural network models, is the multimodal redundancy of sensory information. Any physical reality (sensory input from the target; eye, head, or arm position) can be signaled to the central nervous system by different modalities (e.g., by vestibular, proprioceptive, and efferent modalities for the head). Each sensory modality alone can provide a rough estimate, but accurate spatial localization, a characteristic of living systems evolving in an unconstrained natural situation, relies upon the integration of multimodal signals. Neurophysiological aspects of this integration have been studied during localization of a visual/auditory target in the superior colliculus of the cat (Meredith and Stein, 1986; see also SENSOR FUSION). This study showed a neuronal response to multimodal stimuli that was higher than expected from the linear addition of each unimodal response. Another example of multimodal integration is the improvement of pointing with the unseen hand at an object in space when a view of the hand is allowed prior to the movement, suggesting a perceptual enhancement effect by the integration of proprioceptive and visual cues (Prablanc et al., 1979). While there have been some attempts to build an accurate spatial representation from nonhomogeneous submodality attributes in visual perception (Poggio, Gamble, and Little, 1988), neural networks in motor control generally transform single modality target position signals into motor commands.

Mapping a Target Position Signal into a Pattern of Motor Commands

There is still a large debate about the nature of goal-directed movement motor commands: do they encode a set of final length and tension state of the muscle, or are they related to the direction and amplitude of hand displacement? In other words, is a final arm position planned (leaving our viscoelastic skeletomuscular plant to achieve the displacement), is a vectorial displacement planned, or do we use either modality according to the circumstances? While cortical recordings show directional tuning of neuronal activity and have been claimed to support the "vectorial" hypothesis, some very strong observations favor static "positional" hypotheses. For instance, consider a naive subject wearing, for the first time, prismatic goggles that shift the visual field toward the right, who tries to point quickly from left to right toward a target along a frontoparallel line. Although the perceived movement vector amplitude (defined by vision of the hand-to-target vector) is preserved by the prism shift, the subject largely overshoots the target and almost reaches its virtual image (see Jeannerod, 1988, for a review). Therefore, the movement seems to depend much more on the shifted (virtual) target position than on the visually preserved hand-to-target vector.

Most of the "positional" (or "static") neural network models consist of a transformation of a visual target's extrinsic coordinates into a set of joint postures and thus into a final limb muscular activity, using the spring-like properties of muscles. As a result, the motor commands need not know initial hand position, strongly simplifying the problem. This allows neural networks to learn a plain transformation of a three-dimensional Cartesian point into three (or more, if redundant) joint angles or final muscular activity. The idea that movement path needs to be defined in an extrinsic frame of reference has been addressed specifically. Known as the trajectory formation problem, it has been addressed by the use of a "minimum jerk" criterion (see OPTIMIZATION PRINCIPLES IN MOTOR CONTROL). It explains the production of smooth, stereotyped, human hand trajectories along straight lines rather than any arbitrary trajectory between two points. While the class of such models is generally based on the existence of a teacher with supervised learning, another class uses unsupervised learning (Bullock, Grossberg, and Guenther, 1993) with self-organizing maps. The main feature of the latter models is a coding of the instantaneous vector error between the endpoint effector and the target. They include both spatial representations of the target and endpoint effector in a body frame of reference and implement strategies to achieve motor equivalence with redundant effectors, under unexpected circumstances of limb segment perturbation. They also integrate the property of quasi-continuous modifiability of the motor response in double-step experiments, when the target may be changed prior to or during the hand movement (see Jeannerod, 1988, for a review).

An overview of some motor control models not quoted in this review can be found in Miller, Sutton, and Werbos (1990). The issues listed earlier originate in the robotics field, and most of the neural networks discussed are based on the BACKPROPAGATION: BASICS AND NEW DEVELOPMENTS (q.v.) learning technique, the biological plausibility of which has been criticized.

Among the exceptions are models which use Kohonen-like algorithms based on local recurrent inhibition (see SELF-ORGANIZING FEATURE MAPS: KOHONEN MAPS) and models based on reinforcement learning (Mazzoni, Andersen, and Jordan, 1991). A last, but not least robotic characteristic shared by many of the models discussed is their symbolic output (mainly, joint angles).

On the contrary, some neural network models based on the "vectorial" hypothesis have a distributed output, since they aim to reproduce some features of the neuronal population behavior recorded in the motor cortex (see REACHING: CODING IN MOTOR CORTEX), where neuronal activity is statistically linearly related to the cosine of the angle between preferred direction and input vector. The sum of all these preferred directions weighted by the neurons' activities results in a neuronal population vector parallel to the visual input vector. This is the case, for instance, for the models proposed by Bullock, Grossberg, and Guenther (1993) and Burnod et al. (1992), which are designed to map a three-dimensional hand-target vector into a three-dimensional vector related to the desired direction of the future hand movement, and which depend on arm posture.

Dynamics

Unlike models generating final position or kinematic trajectories, models of dynamics are aimed at transforming an intrinsic vector (i.e., the difference between initial and final joint angles, or between initial and final muscle state) into a temporal pattern of motor commands. A recent approach to learning the dynamics is Kawato's relaxation technique (1992) which allows a learning of torques based on the minimum torque change constraint. Another model of dynamics, proposed by Jordan and Rumelhart (1992), involves a sequential network that uses recurrent connections to some auxiliary input state units in order to feed the hidden units with a time-varying input. One of its main features is the existence of an action-to-sensation internal model and of a predicted performance error signal, taking into account both the kinematics and dynamics of the moving arm (see SENSORIMOTOR LEARNING).

One important feature of dynamic eye-hand coordination is the speed-accuracy tradeoff quantified by Fitts' law, which states that movement time is related to a logarithmic function of the ratio between distance to move and size of the target to hit. At least two kinds of modeling have tried to incorporate this type of timing performance criteria, one based on feedforward (Kawato, 1992), and the other on delayed feedback with lookahead (Hoff and Arbib, 1992).

The first model is based on a feedforward process linking the developed force to the noise and, therefore, to the variability of the response. It is largely independent of the vision of the moving hand. It has been modeled by Kawato (1992) using a network of N identical dynamics models linked in a cascade. Each receives joint position and velocity from its predecessor, and uses the current torque value to compute the subsequent joint angles and velocities, which are passed on to the next model. A relaxation technique adjusts the successive torques, with a least square criterion minimizing their derivative, and with a number of iterations that increases as hand position error decreases. Since the transmission delays of peripheral information are too long to allow for a feedback action, variability reduction is obtained through a time-consuming feedforward computation process.

The second model addresses the problem of feedback control with delays. It involves three concurrent mechanisms: one based on a dynamic hand-to-target retinal error, a second dealing with a visual (target)-to-kinesthetic (hand) motor error signal, and a third one comparing the efference copy and the goal, for very fast corrections in unpredictable situations, when the target may be suddenly changed. According to Hoff and Arbib (1992), a feedback control system based on the minimum jerk criterion for generation of arm trajectories can compensate for target perturbations, as in the classical target perturbation paradigms. Input to the model is the Cartesian position of the target, and output is the state (position, velocity, and acceleration) of the hand. The problems arising from the physiological sensorimotor delays in target vision and hand proprioception are addressed by the introduction of a "lookahead" unit that estimates the current state, based on delayed hand position and velocity signals. This model uses the sensory motor loop to provide an account of Fitts' law.

Discussion

Pioneering works on neuronal population coding (see REACHING: CODING IN MOTOR CORTEX) suggest a vectorial representation of movement, while many psychophysical studies support a final position encoding (see Jeannerod, 1988, for a review). A synthesis between positional coding and vectorial coding models is thus likely to be expected. For example, in the oculomotor system, cortical and collicular codes for displacement combine with brainstem codes for position, yielding a final position (Dominey and Arbib, 1992). This analogy might extend to reaching, where posture-dependent cortical codes for displacement could combine with postural codes in the spinal cord to yield equilibrium point specifications.

Another specificity of hand dynamic behavior is its strong reliance on predicted sensory "reafferences" (afferences resulting from a given motor command), like kinesthetic or visual reafferences. The idea of an internal model of the controlled limb predicting the consequences of a command at the level of the task space is a key point in human motor control and has been integrated in some NNs (Jordan and Rumelhart, 1992).

Redundancy at the level of sensory encoding may also be a crucial point on which little has been theorized. The fact that we are more capable of reproducing orientations of a hand in space than the joint angles it forms with the forearm suggests that there exists some integrated representation of proprioceptive signals derived from the endpoint effector (the hand) in a task space. Similarly, a visual representation within the same task space is likely elaborated from head, eyes, and retinal position signals. Two such higher-level maps representing the visually and proprioceptively derived hand postures in a task space could have a constructive interaction. They could enhance each other in a normal situation of sensory coherence, whereas in a situation of intersensory conflict they could produce either a weighting or a winner-take-all effect (see WINNER-TAKE-ALL MECHANISMS) when the conflict is too large. The paradigms of intersensory conflicts, such as prism-displaced vision or vibration-induced altered proprioception, dissociate normally redundant information in a random or systematic way. The data obtained from such dissociations may provide a powerful basis for comparing experimental behavior with the response of different neural network models processing multisensory information.

Road Map: Primate Motor Control
Background: I.3. Dynamics and Adaptation in Neural Networks
Related Reading: Equilibrium Point Hypothesis

References

Andersen, R. A., Bracewell, R. M., Barash, S., Gnadt, J. W., and Fogassi, L., 1990, Eye position effects on visual, memory and saccade-related activities in areas LIP and 7a of macaque, *J. Neurosci.*, 10:1176–1196.

Bullock, D., Grossberg, S., and Guenther, F. H., 1993, A self-organizing neural model of motor equivalent reaching and tool use by a multijoint arm, *J. Cognit. Neurosci.*, 54:408–435.

Burnod, Y., Grandguillaume, P., Otto, I., Ferraina, S., Johnson, P. B., and Caminiti, R., 1992, Visuomotor transformations underlying arm movements toward visual targets: A neural network model of cerebral cortical operations, *J. Neurosci.*, 12:1435–1453.

Dominey, P. F., and Arbib, M. A., 1992, A cortical-subcortical model for generation of spatially accurate sequential saccades, *Cereb. Cortex*, 2:153–175.

Hoff, B., and Arbib, M. A., 1992, A model of the speed-accuracy and perturbation on visually guided reaching, in *Control of Arm Movement in Space: Neurophysiological and Computational Approaches* (R. Caminiti, P. B. Johnson, and Y. Burnod, Eds.), Experimental British Research, Series 22, Berlin: Springer-Verlag, pp. 285–306.

Jeannerod, M., 1988, *The Neural and Behavioural Organization of Goal-Directed Movements*, Oxford: Clarendon. ◆

Jordan, M. I., and Rumelhart, D. E., 1992, Forward models: Supervised learning with a distal teacher, *Cognit. Sci.*, 16:307–354. ◆

Kawato, M., 1992, Optimization and learning in neural networks for formation and control of coordinated movement, in *Attention and Performance*, vol. XIV (D. Meyer and S. Kornblum, Eds.), Cambridge, MA: MIT Press, pp. 821–849. ◆

Kuperstein, M., 1988, Neural model of adaptive hand-eye coordination for single postures, *Science*, 239:1303–1311.

Mazzoni, P., Andersen, R. A., and Jordan, M., 1991, A more biologically plausible learning rule for neural networks, *Proc. Natl. Acad. Sci. USA*, 88:4433–4437.

Meredith, M. A., and Stein, B. E., 1986, Visual, auditory and somatosensory convergence on cells in superior colliculus results in multisensory integration, *J. Neurophysiol.*, 56:640–662.

Miller, W. T., Sutton, R. S., and Werbos, P. J., 1990, *Neural Networks for Control*, Cambridge, MA: MIT Press. ◆

Poggio, T., Gamble, E. B., and Little, J. J., 1988, Parallel integration of vision modules, *Science*, 242:436–440.

Prablanc, C., Echallier, J. F., Komilis, E., and Jeannerod, M., 1979, Optimal response of eye and hand motor systems in pointing at visual target, II: Static and dynamic cues in the control of hand movements, *Biol. Cybern.*, 35:183–187.

Prablanc, C., and Martin, O., 1992, Automatic control during hand reaching at undetected two-dimensional target displacements, *J. Neurophysiol.*, 67:455–469.

Face Recognition

Alice J. O'Toole, Hervé Abdi, and Dominique Valentin

Introduction

We provide a survey of neural network models of face recognition, identification, and classification. *Recognition* is the categorization of a face as familiar or unfamiliar, *identification* is the retrieval of a name or other semantic key (e.g., "my dentist"), and *classification* is the assignment of a face to visually based categories such as sex, race, and age. We refer to these tasks collectively as *face processing* tasks. The primary challenge for computational models of face processing is to deal effectively with the complexity of faces as visual patterns. All faces have two eyes, a nose, and a mouth, arranged in a universally recognizable configuration. Individual faces comprise unique variations on this configural theme. Thus, to recognize faces, we must be able to encode subtle variations in the form and configuration of these features. Selecting a representation that captures this information is a difficult problem. Most neural network models avoid this "feature selection" problem entirely by operating directly on viewer-centered images of faces, thus making use of the ability of neural networks to act as powerful tools for selecting the most useful information in faces for different tasks.

This image-based representation, used by most neural network models, captures both local and global information in faces, when the faces have been normalized previously in size and position in the image. This normalization requirement is generally not prohibitive since there are good algorithms for locating faces in images (Turk and Pentland, 1991). The simplest involves projecting subimages of a larger image into the face space derived from the learned faces. The distance between the subimage and its projection acts as a measure of "faceness." Motion-detecting and head-tracking algorithms can be used when information over time is available.

We begin with models viewing face processing as a pattern recognition task. These can be divided into linear, nonlinear, and dynamic link architecture models. We then present psychological models that emphasize the cognitive components of the human face processing system, including semantic memory for persons, context of encounter, and names.

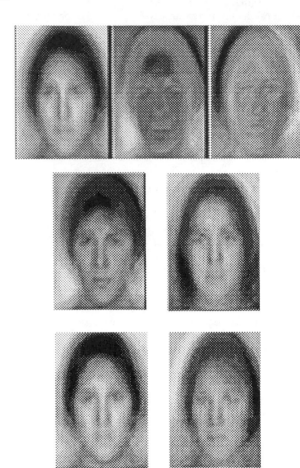

Figure 1. *Top row*: The first three eigenvectors. *Middle row*: Left, the sum of the first two eigenvectors; right, the first eigenvector minus the second eigenvector. *Bottom row*: Same as middle row, using the third eigenvector with inverse weights on the "male" and "female" face.

The Pattern Recognition Approach

Linear Models

Kohonen (1977) showed that face images stored in an autoassociative memory could be retrieved and restored even when noisy or incomplete face images were used as memory keys. Since the autoassociative memory is a cross-product matrix, it can be expressed as the weighted sum of the outer-products of its eigenvectors, where the weights are the eigenvalues of the matrix (see PRINCIPAL COMPONENT ANALYSIS). Further, since retrieving a face from the memory is implemented by applying the memory matrix as a filter to the memory key, retrieved faces can be expressed as a weighted combination of the eigenvectors. The autoassociator, therefore, implements principal component analysis (PCA). As we shall see, most neural network models of face processing make implicit or explicit use of PCA.

One interesting aspect of PCA models is that the eigenvectors are interpretable as images and are often facelike (Sirovich and Kirby, 1987). Further, these images depend on the statistical properties of the learned faces and can be shown in some

cases to relate reliably to categorical information like face sex (O'Toole et al., 1993). For example, the first three eigenvectors of an autoassociative matrix comprising 60 male and 60 female faces appear in the top row of Figure 1. The sex of a face can be predicted by its projection onto each of these eigenvectors. For the second eigenvector, male faces typically project onto positive parts of the axes, whereas female faces project onto negative parts of the axes. In face representation terms, this means that male faces generally require positive weights on this eigenvector to be reconstructed, whereas female faces require negative weights (i.e., like the positive and negative of a photograph). The relevance of these eigenvectors to the gender appearance of a face is illustrated in Figure 1. The left image of the middle row is made by adding the first and second eigenvectors, whereas the right image is made by subtracting the second eigenvector from the first. An analogous demonstration for the third eigenvector appears in the bottom row. Both of the eigenvectors seem to capture aspects related to the gender of the face.

PCA-based models have been applied to the tasks of face recognition and classification. A recognition decision can be made by projecting a face into the face space and assessing its "distance" to other "known" faces, setting a criterion for accepting the face as "known" (Turk and Pentland, 1991). For an alternative method, see O'Toole, Millward, and Anderson (1988). Both methods show excellent recognition performance. True to the image-based representation, however, novel instances of learned faces can be recognized only insofar as they are physically similar to a learned instance of the face. As such, little recognition generalization is possible with large pose changes and with other similarly disruptive changes in the image-based code. When the general properties of the global configuration (e.g., light source, spatial resolution, partial occlusion manipulations) are preserved, however, the performance of PCA-based models is more impressive, and in some cases mirrors human performance on similar tasks (e.g., O'Toole et al., 1994; Turk and Pentland, 1991).

Applied to face classification, PCA-based models have lent insight into the nature of facial category information. In general, the eigenvectors with relatively large eigenvalues contain the most reliable information for classification, whereas eigenvectors with smaller eigenvalues provide better information for recognition (O'Toole et al., 1993). These latter eigenvectors explain small amounts of variance and, hence, code what is unique to individual or small subsets of faces.

Nonlinear Models

Backpropagation (see BACKPROPAGATION: BASICS AND NEW DEVELOPMENTS) has been applied to the problems of face classification and identification. These models learn a mapping between face images and "information" about the faces, including the identity, sex, and expression of a face. Since backpropagation is computationally intensive, and since the face images are large, the task is divided into subtasks handled by two networks (e.g., Cottrell and Fleming, 1990; Cottrell and Metcalfe, 1991; Golomb, Lawrence, and Sejnowski, 1991). The first network compresses the images. This is accomplished by training a backpropagation network to reproduce face images through a narrow channel of hidden units. The nature of the representation emerging in the hidden units is related to PCA (cf. Valentin et al., in press).

The activation pattern of hidden units that emerges from the image-compression network for each face is then used as input to a second network that learns a mapping from this condensed code to the desired information about the faces. For example, Cottrell and Fleming (1990) trained a backpropagation network simultaneously to classify images as faces or nonfaces, to identify faces, and to classify faces by sex. Their network generalized reasonably well to classifying novel images as face/nonface, and less well to classifying novel faces by sex. Golomb et al. (1991) trained SexNet to classify faces only for sex. This "dedicated" network generalized better in classifying novel faces by sex.

Cottrell and Metcalfe (1991) extended this approach to classifying faces by expression. They used images of people feigning emotions, including happy, astonished, and angry, and showed that the model learned to discriminate most of the emotions reasonably well, but showed somewhat limited ability to generalize to novel faces. In general, the model distinguished positive emotions more accurately than negative emotions. It is likely that increasing the number of faces would result in better performance on the generalization tasks.

Dynamic Link Architecture

Buhmann, Lange, and von der Malsburg (1989) applied the DYNAMIC LINK ARCHITECTURE (q.v. to face recognition with

the goal of overcoming the relative intolerance of image-based models to changes in the size or pose of a face. The face representation used is a compromise between image- and feature-based codes. They begin by choosing a small number of spatial positions on each face by superimposing a regular lattice over the face image and using the grid intersections for "match points." The subimage surrounding each match point is coded with a "Gabor jet," consisting of a series of coefficients of Gabor filters of varying orientations and sizes. A novel image is classified as an exemplar face by allowing the superimposed grid positions of Gabor jets to deform using an optimization task constrained by the topology of the grid and the similarity of the match points to match points on the comparison image. The novel image is then compared to each image in the database and a cost criterion for classifying the face as "known" is set. Buhmann et al. tested their model with faces rotated in depth by 15 degrees and with grimaced faces. The model performed well in both cases. This approach differs from the classic connectionist framework in that: (1) distributed representations of faces are not used, and (2) classifying the face as "known" requires a serial search through the database.

Neural Network Models of Human Face Processing

Connectionist approaches from a psychological perspective focus on coordinating the various information processing tasks relevant for human face memory. These models generally employ arbitrary representations for faces and for the semantic information related to face processing. We present three examples that characterize this perspective.

Semantic priming and identity priming produce well-known effects on the speed with which humans recognize faces. *Semantic priming* is elicited by preceding a target face with a semantically related "prime" face. *Identity priming* is elicited when the target and prime are different pictures of the same face. Burton, Bruce, and Johnston (1990) have simulated semantic and identity priming effects using an interactive-activation and competition model (McClelland and Rumelhart, 1981). Nodes in the model are interconnected with excitatory and inhibitory links and are of three types: person identity units, semantic information units, and face recognition units. Speed to identify a face is taken as the number of processing cycles to reach the threshold for the person identity code. This model successfully simulates the time course of semantic and identity priming.

Context effects refer to the difficulty people have in "placing" faces seen out of their usual context. Schreiber, Rousset, and Tiberghien (1991) trained a backpropagation network to reconstruct its input face and context codes and to identify faces. They varied the number of contexts with which a particular face was presented and simulated several well-known findings, including the fact that a change of context is less disruptive for the identification of faces known from a variety of contexts than for those known from a single context.

Face typicality, as judged by human observers, is a reliable predictor of human performance, with atypical faces being recognized more accurately and classified (as faces) less quickly than typical faces. Using several neural networks and abstract face patterns created synthetically by varying stimulus vectors about a prototype, Valentine and Ferrara (1991) simulated the effects of typicality on face identification and categorization. Of particular interest were the time-course effects for learning typical and atypical faces. These were simulated with a backpropagation network measuring the identification and category label learning across training epochs. Using a linear auto-associator with face images, O'Toole et al. (1994) found that a model-derived measure of face distinctiveness was more strongly related to human recognition performance than were human ratings of typicality.

Discussion

Connectionist models of face processing have considered several tasks, including face recognition, identification, and classification. Typically, pattern-recognition-oriented models have operated directly on face images and have been concerned with selecting the information most useful for the various tasks. Models concerned with cognitive components of the human face processing system have generally focused on coordinating modules of the cognitive system. Although these models mimic human performance on analogous tasks, the leap to understanding the underlying neural processes is a large one. Even with prodigious research efforts into the neuropsychology of face recognition deficiencies (e.g., prosopagnosia) and the neurophysiological substrates of face recognition (see SPARSE CODING IN THE PRIMATE CORTEX), the complexity of the pattern recognition problem posed by faces still limits the utility of these data for guiding model design.

Road Map: Connectionist Psychology
Background: I.3. Dynamics and Adaptation in Neural Networks
Related Reading: Emotion and Computational Neuroscience; Gabor Wavelets for Statistical Pattern Recognition

References

Buhmann, J., Lange, J., and von der Malsburg, C., 1989, Distortion invariant object recognition by matching hierarchically labeled graphs, in *Proceedings of the International Conference on Neural Networks*, vol. I, Washington, DC, pp. 155–159.

Burton, A. M., Bruce, V., and Johnston, R. A., 1990, Understanding face recognition with an interactive activation model, *Br. J. Psychol.*, 81:361–380.

Cottrell, G. W., and Fleming, M., 1990, Face recognition using unsupervised feature extraction, in *Proceedings of the International Conference on Neural Networks*, Dordrecht, The Netherlands: Kluwer, pp. 322–325.

Cottrell, G. W., and Metcalfe, J., 1991, Empath: Face, emotion, and gender recognition using holons, in *Advances in Neural Information Processing Systems 3* (R. Lippmann, J. Moody, and D. S. Touretzky, Eds.), San Mateo, CA: Morgan Kaufmann, pp. 557–564.

Golomb, B. A., Lawrence, D. T., and Sejnowski, T. J., 1991, SexNet: A neural network identifies sex from human faces, in *Advances in Neural Information Processing Systems 3* (R. Lippmann, J. Moody, and D. S. Touretzky, Eds.), San Mateo, CA: Morgan Kaufmann, pp. 572–577.

Kohonen, T., 1977, *Associative Memory: A System Theoretic Approach*, Berlin: Springer-Verlag.

McClelland, J. L., and Rumelhart, D. E., 1981, An interactive activation model of the effect of context in perception: Part 1, An account of basic findings, *Psychol. Rev.*, 88:375–406.

O'Toole, A. J., Abdi, H., Deffenbacher, K. A., and Valentin, D., 1993, Low-dimensional representation of faces in higher dimensions of the face space, *J. Opt. Soc. Am. [A]*, 10:405–410.

O'Toole, A. J., Deffenbacher, K. A., Valentin, D., and Abdi, H., 1994, Structural aspects of face recognition and the other-race effect, *Memory and Cognition*, 2:208–224.

O'Toole, A. J., Millward, R. B., and Anderson, J. A., 1988, A physical system approach to recognition memory for spatially transformed faces, *Neural Netw.*, 1:179–199.

Schreiber, A., Rousset, S., and Tiberghien, G., 1991, Facenet: A connectionist model of face identification in context, *Eur. J. Cognit. Psychol.*, 3:177–198.

Sirovich, L., and Kirby, M., 1987, Low-dimensional procedure for the characterization of human faces, *J. Opt. Soc. Am. [A]*, 3:519–524.

Turk, M., and Pentland, A., 1991, Eigenfaces for recognition, *J. Cognit. Neurosci.*, 3:71–86.

Valentin, D., Abdi, H., O'Toole, A. J., and Cottrell, G., in press, Connectionist models of face processing: A survey, *Pattern Recognition*, 27. ◆

Valentine, T., and Ferrara, A., 1991, Typicality in categorization, recognition and identification: Evidence from face recognition, *Br. J. Psychol.*, 82:87–102.

Fault Tolerance

Jack D. Cowan

Introduction

In many cases, even the loss of substantial amounts of brain cells or tissue does not totally abolish brain function, a property known as *graceful degradation*. It is therefore not surprising that the problems of constructing fault-tolerant networks have been studied almost since the earliest days of neural networks. This review covers a variety of approaches to the problem, almost all of which can be related in some way to information theory, i.e., to Shannon's work on encoding messages for reliable transmission through noisy channels. Within this framework, we describe some of the arguments for sparse coding, superimposed random subsets, overlapping coarse codes, and their applications to fault tolerance. We also briefly describe, in terms of regularization theory, the use of injected noise during the training of perceptrons and the application of this procedure to fault tolerance.

Early Work

Probabilistic Logics

Problems related to fault tolerance motivated the work of Warren McCulloch, and through him Jon von Neumann in the early 1950s. Von Neumann's (1956) study of "Probabilistic Logics and the Synthesis of Reliable Organisms from Unreliable Components" used *redundancy* techniques to provide error detection and correction in neural networks composed of components which malfunctioned from time to time with probability ε. By using N components everywhere instead of 1, and the bundle encoding scheme in which activation of more than $(1 - \Delta)N$ lines signals "1" and less than ΔN signals "0," together with "restoring organs" which mapped bundle activity levels toward either "0" or "1," he was able to achieve arbitrarily reliable computing. More precisely, for $\varepsilon = 0.005$ and $\Delta = 0.07$, the probability of network malfunction was

$$\rho(N) \approx \frac{6.4}{N} 10^{-0.00086N}$$

This is disappointing since $\rho(N) \le \varepsilon$ if and only if $N \ge 2000$.

Von Neumann's result is not very rigorous: the action of the restoring organ depends on the use of random permutations of bundle lines, and it is not clear how to achieve them. Recent work has produced rigorous results. Consider the sum (modulo 2) of n arguments computed by a network of $O(n)$ reliable gates. Such a function can be computed by not less than $o(n \log n)$ unreliable gates, as can the n-argument functions

AND and OR. (We use the 'big-O" and "little-o" notation of asymptotic analysis: $f = O(g)$ means f/g is bounded, whereas $f = o(g)$ means $f/g \to 0$ in some limit.) Similarly, consider the parallel computation of a set of k functions, each of which is the sum (modulo 2) of some of their n arguments. It has been shown that for $n = k$ the minimum possible number of unreliable components required to compute reliably such a set is $O(n^2/\log n)$. In all these cases, then, the ratio R_c of the numbers of reliable to unreliable gates required to achieve reliability varies from $O(1/N)$ to, at best, $O(\log N/N)$.

Information Theory

All these results are at variance with what information theory tells us about the reliable transmission of messages through noisy channels. Claude Shannon showed that the noise in a channel determines a maximum rate (in bits of information per coded message symbol) at which information in the form of coded messages may be transmitted and decoded without error at the receiver. This maximum rate is known as the channel capacity C. Shannon's noisy channel coding theorem then says that coded messages may be transmitted through such a channel and decoded without error at a rate R (in bits per symbol) if and only if $R \leq C$. Shannon also showed that the probability of error in decoding $\varsigma(N) \approx O(2^{-N(C-R)}) = O(\exp(-\alpha N))$, where $\alpha = (C - R)/\log_2 e$ and where N is the length of the code words sent through the channel.

There is a big difference between a "computation ratio" R_c, which is at best $O(\log N/N)$, and a transmission rate R, which is independent of N. This motivated Winograd and Cowan (1963) to study reliable computation in the presence of noise. They focused on the nature of the decoding process. In a communication system, only the channel is assumed to be noisy, and the various computations performed by the decoder are assumed to be error-free; for computation with unreliable elements, in contrast, decoding operations cannot be assumed to be error-free. Winograd (1963) proved that if t errors are to be corrected anywhere in a network, then the average number of inputs per device in the network, n, is at least $(2t + 1)R_c$. Conversely, $R_c \leq n/(2t + 1)$, so that if R_c is required to remain constant as more and more errors are corrected, then the complexity of the elements, as measured by n, must increase. Here is the crux of the problem, for if one is provided with elements which compute only functions of $m \leq n$ arguments, then it can be shown that $p = O(2^{n-m})$ elements are required to compute each function of n arguments. Evidently ε, the probability of error per element, increases exponentially with p so that the Shannon bound cannot be reached in such a network. This possibility occurs only if ε is independent of p. This would be the case if $m = n$, and ε does not increase exponentially with n.

Winograd and Cowan introduced a scheme for the reliable parallel computation of a set of k functions, each of which is a Boolean function of some of their n arguments. The scheme incorporated an (N, k) error-correcting code into the computations, i.e., a code in which messages that are k bits long are encoded into signals that are N symbols long for transmission through a noisy channel. In a Hamming block code, $N - k$ of the received symbols in each code word are used for error detection and correction. The information transmission rate is k/N bits/symbol. In the computing case, this implies that k computations are implemented by a network comprising N/k times as many elements as would be required in the error-free case. Figure 1 shows what is meant. Here, the function to be computed has M elements, of which a fraction $\alpha = u/M$ are output elements. Then the number of elements in the k irredundant precursors is just Mk, and the number in the redundant network is $(M - u)N + uk$. Thus the computation ratio is

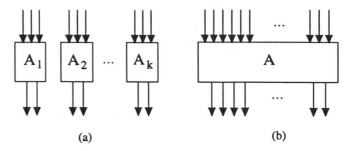

Figure 1. A, k copies of a Boolean function computed in parallel. B, One redundant Boolean function that replaces them. (Redrawn from Arbib, 1964.)

$$R = \frac{k}{N - \alpha(N - k)} = \frac{k}{N}\left[1 + \alpha\left(1 - \frac{k}{N}\right)\right] \approx \frac{k}{N} \quad \text{for } \alpha \text{ small}$$

The effect of such an encoding scheme is to *distribute* the logical functions to be implemented over the entire network. Such a scheme works most efficiently with large k and N—in effect, in a parallel distributed architecture. Thus the Winograd-Cowan scheme is an early example of *Parallel Distributed Processing* (PDP). Of course, it can be argued that the scheme is not realistic, in that all the extra coding machinery that is incorporated into the neurons is assumed to be error-free. This is not true for simple logical elements, but it may be more plausible for real neurons.

From McCulloch-Pitts Elements to Adalines and Sigmoids

Recently, work on fault and failure tolerance has been strongly influenced by the shift from synchronous binary logic networks to asynchronous analog networks, in which the weight patterns required to implement encoding and decoding can be learned, or in which decoding is performed by the dynamics of a recurrent network.

Stevenson, Winter, and Widrow (1990) calculated the effects on ε of both input errors and weight perturbations on adalines. (Adalines are single-layer perceptrons in which gradient descent is used to generate the learning rule.) The basic idea is common to any threshold device. Consider an adaline with n inputs where n is large. Let X and W be $(n + 1)$-dimensional binary input and weight vectors (the extra dimension is needed for the threshold setting). Let ΔX and ΔW be variations in their amplitudes with Hamming distance t, i.e., t of the $n + 1$ vector components are inverted. Finally, let $\delta X = \Delta X/|X|$ and $\delta W = \Delta W/|W|$. Using the geometric properties of n-dimensional vectors, the probability of an adaline decision error P is approximately $(1/\pi)[\delta X^2 + \delta W^2]$ for small errors. From this, one can calculate the probability of a madaline decision error, P_L, i.e., the output probability of error of a network comprising L adaline layers. In general this will be binomially distributed. Stevenson et al. found the remarkably accurate approximation for P_L in the case of weight errors:

$$P_L = (\delta W/\pi)(1 + \beta(1 + \beta(\ldots(1 + \beta(1 + \beta)^{1/2}\ldots)^{1/2})^{1/2})^{1/2})^{1/2}$$

where $\beta = 4/(\pi\delta W)$, the number of square roots is $L - 1$, and the number n of adalines per layer is large, i.e., at least 100.

Two conclusions follow from this formula, both of which can be seen in simulations. The first is that as long as n is large, P_L is essentially independent of n; the second is that P_1, the probability of error for a one-layer network of adalines, is also essentially independent of the number of inputs per adaline. Similar formulas have recently been derived for analog threshold devices based on the sigmoid. These results support the assumptions of the Winograd-Cowan theorem.

Associative Memories

Our discussion to this point applies to the general problem of computation, which includes pattern classification and decoding. We now turn to the storage and retrieval of information in the form of *associative memories*. Work on this problem over the past decade has focused mainly on Hopfield networks, but analysis of an older generation of associative memories provides insights into fault tolerance.

Sparse Coding

Consider, for example, the network introduced by Willshaw, Buneman, and Longuet-Higgins (1969) comprising V A-lines, R B-lines, and VR binary switches. Such switches correspond to binary synaptic weights. Using binary weights makes the analysis much simpler. Let an A-pattern correspond to the activation of n randomly chosen A-lines, and a B-pattern to m randomly chosen B-lines. Thus there will be a total of nm doubly activated switches. Let these be turned on, if they are not already on. After K pairs of patterns have been associated in this way, a fraction p of the switches will have been turned on. If the patterns are random, p will be approximated by

$$p = 1 - \exp\left(-K\frac{nm}{VR}\right)$$

To recall a pattern is simple. An A-pattern is put in, so that each B-line of the associated B-pattern is activated through n switches (which, by hypothesis, are all on). However, B-lines not belonging to the associated B-pattern will also be activated. The probability that such a B-line is activated through every one of its n intersections with the activated A-lines is just p^n. So if the firing threshold of each B-line is n, not only will all those lines fire which comprise the associated B-pattern, but a further mp^n will probably also fire. If this number is less than one, no errors will occur, so the critical value is $mp^n = 1$. Now a single B-pattern requires $\log_2\binom{m}{R}$ bits to specify it, so if K B-patterns can be accurately retrieved, the amount of available stored information is $K\log_2\binom{m}{R} \approx Km\log_2 R$ bits. The combination of these three equations leads to an expression for the number of bits stored in the network, namely, $VR\log_2 p\log_2(1-p)$ bits. This reaches its maximum value of $0.693\ VR$ when $p = 0.5$. In such a case, $n \approx \log_2 R$, i.e., the number of lines per A-pattern should be small if the network is to be an efficient information store. This is an early argument for *sparse coding*. Willshaw et al. later showed that associative networks can function reliably even if the A-patterns are noisy, and even if the network is damaged, by raising n and therefore storing fewer associations. The resulting sharp drop in the information storage density is predicted by the Winograd-Cowan theorem.

Superimposed Random Subsets

The foregoing analysis indicates that to retrieve patterns without error places a sharp bound on the number of lines per A-pattern which can be used to encode information. If there are V input lines, there are at most $\binom{V}{n}$ differing A-patterns or *descriptors* which can be fed into the store. Since n has to be small, the only way to increase the possible number of descriptors, i.e., the *vocabulary*, is to increase V. Remarkably, a paper

dealing with this issue appeared in 1965, written by P. H. Greene. Greene's paper is particularly interesting in that it is virtually identical in conceptual content with Marr's (1969) well-known paper on the cerebellum, and with later works.

Greene based his analysis on earlier work by Mooers (1949) on the storage and retrieval of information using decks of punched cards. Holes are punched in a number of locations on each card in a deck. These holes (or their absence) embody Boolean descriptions $x_1 \& \bar{x}_2 \& x_3 \& x_4 \ldots x_N$, where the x_i represent various categories: female, married, Caucasian, etc. Because of the holes, decks of such cards can be automatically sorted, either mechanically or electrically. Consider the combinatorics: if there are N locations, there are 2^N binary patterns to be used as descriptors. Suppose that D categories are allocated to each descriptor, with no overlap. Then there are $M = N/D$ subfields, and a vocabulary comprising $V = M2^D$ independent descriptors. So if $D = N$, $V = 2^N$; if $D = N/2$, $V = 2N$; and if $D = 1$, $V = 2N$.

Mooers' contribution, patented under the name Zatocoding, was to use randomly overlapping or *superimposed* subfields as descriptors. This generates a vocabulary of $V = \binom{N}{D}$ independent descriptors. Suppose that, on the average, there are K descriptors per card. Then the maximum total number of descriptions is $\binom{V}{K}$, i.e., K out of V D-tuples. The probability P of any of the N locations being in one of the K descriptors is bounded by $1 - (1 - (D/N))^K$ so that the number of locations used to form the K descriptors is

$$G \approx NP = N\left(1 - \left(1 - \frac{D}{N}\right)^K\right) \le N\left(1 - \exp\left(-\frac{X}{N}\right)\right)$$

where $X = KD$. This expression is maximized when $X \approx N\log 2$, at the value $G \approx N/2$, when $P \approx 1/2$. This is an important result, and Mooers related it to information theory: the maximum capacity is obtained when each location on a card can signal one bit of information, and this occurs when about half the locations are used. This is similar to the conclusion reached later by Willshaw et al. by a slightly different argument.

What about the decoding process—retrieving information from the card deck? Suppose there are R cards, and suppose that the joint presence of n descriptors is required to select a card. Then approximately 2^{nD} different cards can be uniquely decoded, i.e., $R \approx 2^{nD}$ or $D \approx (\log_2 R)/n$. Furthermore, the fraction of wrong cards selected does not exceed

$$\frac{\binom{G}{n}}{\binom{N}{n}} < \left(\frac{G}{N}\right)^n$$

This can be made as small as desired by adjusting G, N, and n. In the optimal case when $G \approx N/2$, it is less than 2^{-n}.

It follows that one should first choose D given R, then set a value of N large enough to generate enough descriptors, on the average K per card. For example, suppose $n = 3$ and $R = 4000$. Then $D \approx (\log_2 4000)/n = 4$. Choose $N = 40$, then $X \approx 40\log 2 \approx 28$, whence $K = X/D = 7$. This demonstrates that random subset coding is advantageous when there is a large number of descriptors, each occurring with a relatively low frequency—another argument for sparse coding. It would not be advantageous if one or two descriptors were present on every card, in addition to other descriptors. It would be more economical to use a 1-location fixed field in each card to indicate these two, rather than D locations in each card as required by the superimposed random coding method. This is essentially

Figure 2. A network implementation of the Mooers-Greene scheme (see text for details).

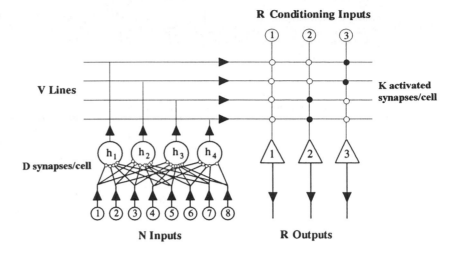

the argument for *grandmother cells* versus sparse distributed coding.

Greene suggested the network implementation shown in Figure 2. Let each card correspond to one of R output neurons, and let there be N input lines. Selecting random D-tuples from these lines will generate V secondary lines, each of which contacts all the output neurons. Let these neurons have a common threshold of n units of excitation. Suppose further that conducting excitatory synapses with these neurons are formed by a classical conditioning process: on the average about K per neuron. This can be set up in such a way as to correspond to card selection by activation of n out of K descriptors; and it corresponds to the standard architecture of the associative memory network.

Greene did not implement this scheme with circuit diagrams, nor did he develop in any detail the various inhibitory mechanisms he saw were needed to control the operation of the system. Remarkably, all these mechanisms were implemented by Marr (1969) in a virtually identical system (proposed apparently without prior knowledge of Greene's paper).

In Marr's scheme D-tuples, called *codons*, are selected by cerebellar granule cells (corresponding to the hidden units h_i of Figure 2) and decoded by cerebellar Purkinje cells (the output cells of Figure 2) following associative learning at parallel-fiber (V lines in Figure 2) Purkinje cell synapses. Marr's contribution was to introduce inhibitory interneurons: Golgi cells to control D and K, and basket and stellate cells to control n. Such controls stabilize the values of D, K, and n and allow sparse coding to take effect to provide reliable operation.

Hopfield Networks

Analyzing the fault-tolerance properties of associative memories is not easy. The associative memory studied by Hopfield (1982), based on systems of coupled analog neurons with sigmoid characteristics, is more amenable to analysis. It comprises a fully connected recurrent network of N units with a symmetric weight matrix W. Since the eigenvalues of such a W are real, the network will, with asynchronous update, settle into some stationary, nonoscillatory state, say $\sigma = \{\sigma_1, \sigma_2, \ldots, \sigma_N\}$, where each $\sigma_i = \pm 1$. To program the network, Hopfield used the standard outer-product algorithm. Let σ^P be a pattern to be stored in the network. Then choose the weights such that $w_{ij} = \sigma_i^P \cdot \sigma_j^P$. Starting from a random initial state σ^0, the network will settle into the state σ^P. Now let there be m patterns to be stored in the network, $\sigma^1, \sigma^2, \ldots, \sigma^m$. Let the weights be chosen via the algorithm:

$$w_{ij} = \sum_{P=1}^{m} \sigma_i^P \cdot \sigma_j^P$$

Thus starting from σ^0, the network will settle into the nearest σ^P. Evidently if m is too large, ambiguity will result. Thus there is an upper limit on m, a capacity, which limits the number of memories that may be reliably stored and retrieved. There is an analogy between the processes of encoding and decoding messages in a communications system and the storage and retrieval of memories in a Hopfield network. Both encoded messages and memories are N-bit strings, corresponding to points in an N-dimensional hyperspace. In order for reliable decoding or retrieval to occur, such points need to be sufficiently separated from each other, i.e., they must be the centers of sufficiently large N-dimensional hyperspheres. Both processes then involve the same calculation—*sphere packing* in N dimensions. This analogy was exploited by McEliece et al. (1987), who proved that for error-free retrieval $m \leq O(N/4 \log N)$. This result should be compared with that of Amit, Gutfreund, and Sompolinsky (1985), who exploited the analogy between Hopfield networks and spin glasses to show that if a 5% error rate can be tolerated, $m \approx 0.14N$. (Spin glasses are systems of binary spins which can interact either ferromagnetically, in which case ↑ spins tend to make their neighbors spin ↑, or anti-ferromagnetically, in which case ↑ spins tend to make their neighbors spin ↓.)

To correct errors in associative networks is straightforward. A common scheme is that suggested by Marr (1969), which we discussed earlier. A recent implementation of this for Hopfield networks uses a winner-take-all circuit (Moopen et al., 1986). In addition, slight asymmetries in the weight matrix W are introduced. Together these result in better elimination of the spurious memories that are found in all associative networks, and better tolerance of errors.

Error-Correcting Codes

Earlier we described the Winograd-Cowan construction in which a Hamming code was embedded into network structure. In recent years there has been a growing appreciation of the connection between codes and neural networks. Several papers have described the use of neural networks to implement the decoding process for a variety of codes. In this review, we concentrate on the control of errors within the decoding process itself. An interesting use of error-correcting codes for this purpose is that of Petsche and Dickinson (1990), who start out by making the point that the representations used in neural networks are equivalent to codes of various kinds, from block codes in which an entire input sequence is mapped onto an output sequence to those which do not mix any symbols. Petsche and Dickinson note that there is an intermediate family of codes which provide a *coarse-coded* or "receptive field" representation of input sequences, in terms of the action of a few semi-local encodings, namely, *convolution codes*. In a convolution code, relatively short *overlapping* subsets of an input vector are used to determine the output vector. This is what is involved in the Mooers et al. schemes, as well as in the coarse-coded representations via overlapping receptive fields introduced in recent years. An example is provided by the linear (3, 1) triplet code, which is made out of three generators, $g_0 = (111)$, $g_1 = (110)$, and $g_2 = (011)$. Let $u = (u_1, u_2, \ldots, u_b)$ be an input vector and $v = (v_1, v_2, \ldots, v_b)$ the output vector, where each $v_i = (v_{i,1}, v_{i,2}, v_{i,3})$ is a 3-bit subset. Then v can be written as the convolution of u with the generators g_i, i.e.,

$$v_i = \sum_{k=\max(1, i-2)}^{\min(i, b)} u_k g_{i-k}$$

Examination of this equation shows that each 3-bit subset v_j of **v** depends only on the *i*th bit of the input vector and the two previous bits. The number of bits of the input vector that uniquely determine each output subset is called the *constraint length K*. In the noise-free case, v_i contains no information about v_j for $|i - j| \geq K$. For a coarse-coded representation, this corresponds to nonoverlapping receptive fields.

Petsche and Dickinson constructed a neural network which incorporates a convolution code. Such a network can detect and correct errors in its input. In addition, if learning is allowed, the network can also repair itself, since failed components can be replaced by spares.

Perceptrons

The encoding schemes we have described, using either Hamming codes, superimposed random subsets, or convolution with overlapping receptive fields (coarse coding), generate a redundant representation of the input embodied in a layer of internal or "hidden" neurons. The decoding of this representation generates a further layer of output neurons. Thus there is a close correspondence between the architecture of multilayered perceptrons and the standard information theory paradigm. This correspondence led several workers to investigate the fault tolerance of perceptrons, particularly in the case when weight noise is present during training. Simulations by Judd and Munro (1993), for example, show that training in the presence of hidden-unit faults (misfirings) produces more fault tolerance: during subsequent testing, enhanced fault tolerance is found in networks trained with a higher rate of hidden-unit misfiring.

Regularization

These results indicate that *nonlinear optimization* is needed to obtain the appropriate weight settings that produce fault tolerance. Injected weight-noise adds *constraints* to the usual error or penalty functions which must be minimized in the process of optimization. As Neti, Schneider, and Young (1992) have noted, this is similar to the *regularization* processes employed to learn continuous maps from a finite number of examples—a classic "ill-posed" problem. Neti et al. formulated the fault-tolerance problem in such terms, obtaining what they called "maximally fault-tolerant" networks for several auditory encoding problems. Their results indicate that more distributed and *uniform* representations are more fault tolerant and gener-

alize better. These representations are also similar to those obtained by Petsche and Dickinson: increasing the number of hidden units, up to some limit, leads to improved fault tolerance and better generalization.

Discussion

In this review we have discussed many papers on fault tolerance which use *redundancy* techniques in a variety of ways. One approach we have not discussed is that followed by Satyanarayana, Tsividis, and Graf (1990), who constructed a reconfigurable analog VLSI neural network chip comprised of *distributed neurons*, each with N weights. One advantage gained by such a construction is that large current buildups in each neuron are avoided. Moreover, since the chip is reconfigurable, defects leading to failures can be isolated and ignored, thus producing greater fault tolerance in the neurons themselves. This provides another reason to assume that neural firing-error probabilities are largely independent of N, as discussed earlier.

It is clear that fault tolerance in neural networks can be achieved in many ways, most of which involve parallel distributed architectures. Methods include sparse coding, random subset encoding, coarse coding via overlapping receptive fields, and block coding. All such methods can be seen as attempts to encode computations to allow for error detection and correction using methods borrowed from information theory. A somewhat differing viewpoint and methodology are provided by the theory of regularization; this, too, leads to parallel distributed architectures and fault tolerance. It remains to be seen exactly what connections exist among these various methods, and whether there is, in any sense, an optimum design for fault tolerance.

Road Map: Cooperative Phenomena
Background: I.3. Dynamics and Adaptation in Neural Networks
Related Reading: Generalization and Regularization in Nonlinear Learning Systems; Sparsely Coded Neural Networks; Statistical Mechanics of Neural Networks; Visual Coding, Redundancy, and "Feature Detection"

References

Amit, D. J., Gutfreund, H., and Sompolinsky, H., 1985, Storing infinite numbers of patterns in a spin-glass model of neural networks, *Phys. Rev. Lett.*, 55:1530–1533.

Arbib, M. A., 1964, *Brains, Machines and Mathematics*, New York: McGraw-Hill.

Greene, P. H., 1965, Superimposed random coding of stimulus-response connections, *Bull. Math. Biophys.*, 27:191–202.

Hopfield, J., 1982, Neural networks and physical systems with emergent collective computational properties, Proc. Natl. Acad. Sci. USA, 79: 2554–2558.

Judd, S., and Munro, P. W., 1993, Nets with unreliable hidden nodes learn error-correcting codes, in *Advances in Neural Information Processing Systems 5*, (S. J. Hanson, J. D. Cowan, and C. L. Giles, Eds.), San Mateo, CA: Morgan Kaufmann, pp. 89–96.

Marr, D., 1969, A theory of cerebellar cortex, *J. Physiol. (Lond.)*, 202:437–470.

McEliece, R., Posner, E., Rodemich, E., and Venkatesh, S., 1987, The capacity of the Hopfield associative memory, *IEEE Trans. Inform. Theory*, 33:461–482.

Mooers, C. N., 1949, Application of random codes to the gathering of statistical information, Zator Company Technical Bulletin, no. 31.

Moopen, A., Khanna, S. K., Lambe, J., and Thakoor, A. P., 1986, Error correction and asymmetry in a binary matrix model, in *Neural Networks for Computing* (J. S. Denker, Ed.), New York: American Institute of Physics, pp. 315–320.

Neti, C., Schneider, M. H., and Young, E. D., 1992, Maximally fault-tolerant neural networks and nonlinear programming, *IEEE Trans. Neural Netw.*, 3:14–23.

Petsche, T., and Dickinson, B. W., 1990, Trellis codes, receptive fields, and fault-tolerant, self-repairing neural networks, *IEEE Trans. Neural Netw*, 1:154–166.

Satyanarayana, S., Tsividis, Y., and Graf, H. P., 1990, A reconfigurable analog VLSI neural network chip, in *Advances in Neural Information Processing Systems 2* (D. Touretzky, Ed.), San Mateo, CA: Morgan Kaufmann, pp. 758–768.

Stevenson, M., Winter, R., and Widrow, B., 1990, Sensitivity of feedforward neural networks to weight errors, *IEEE Trans. Neural Netw.*, 1:71–80.

von Neumann, J., 1956, Probabilistic logics and the synthesis of reliable organisms from unreliable components, in *Automata Studies* (C. E. Shannon and J. McCarthy, Eds.), Princeton, NJ: Princeton University Press.

Willshaw, D. J., Buneman, O. P., and Longuet-Higgins, H. C., 1969, Non-holographic associative memory, *Nature*, 222:960–962.

Winograd, S., 1963, Redundancy and complexity of logical elements, *Inform. and Control*, 5:177–194.

Winograd, S., and Cowan, J. D., 1963, *Reliable Computation in the Presence of Noise*, Cambridge, MA: MIT Press.

Figure-Ground Separation

Stephen Grossberg

Introduction

When we observe the world through one or both of our eyes, we readily perceive objects that are distinct from one another and from their scenic background. This competence is often called figure-ground separation. Figure-ground percepts are so ubiquitous and immediate that their remarkable and paradoxical nature is often not evident to a naive observer. When we reflect, however, that the three-dimensional (3D) world is projected onto the two-dimensional (2D) surface of each eye's retina, then figure-ground separation seems harder to understand. When we consider, further, that object or figures often seem to pop out from their backgrounds even when we view a 2D picture with a single eye, then the process may seem mysterious indeed.

Many factors can contribute to figure-ground separation, including differences in luminance, color, size, binocular disparity, and motion between a figure and its background. An analysis of all of these factors would require a comprehensive study of visual perception. The current article has a more limited goal. It summarizes perceptual data that clarify key issues which must be dealt with to understand figure-ground perception, including data about how luminance contrast, binocular disparity, and spatial scale contribute to figure-ground separation in response to both 3D scenes and 2D pictures. An outline is then provided of how these data may be explained by a recent model of how the visual cortex works. Although unfamiliar objects can be separated from unfamiliar backgrounds, prior knowledge about the world can also facilitate figure-ground separation. A framework for analyzing how unfamiliar figures may be separated, yet how knowledge may modulate or facilitate the separation process, also is summarized.

Size Differences Attributable to Perspective

It has been known since the Renaissance that the perspective with which a 2D drawing or painting is rendered can make a figure appear to pop out from its background. Typically, a large foreground figure (e.g., of a person) in front of small background figures (e.g., of trees, houses, and hills) makes the foreground figure appear nearby and the background figures appear farther away. A 2D picture can thereby generate a 3D percept. This type of observation is consistent with the maxim that "large size scales signal near objects."

As with many other properties of visual perception, this maxim is not always true, as is noted later. Whatever the cause, 3D percepts derived from 2D perspectives show that the points and lines of Euclidean geometry and the surface elements and normals of Gaussian geometry are insufficient to explain figure-ground separation. New geometrical ideas are needed to explain how a 2D picture can generate a 3D percept. In this new geometry, points and lines are generalized to emergent boundary segmentations, and surface elements and normals are replaced by the filling-in of surface properties. What these segmentation and surface processes are and how they work is indicated below.

Binocular Disparity and the Size-Disparity Correlation

Image size alone is not a completely reliable cue to a figure's depth with respect to its background. In particular, a nearby small object and a far away large object may both subtend the same "size" on the retina. Another cue to depth is the different relative positions, or binocular disparity, with which an object is registered on an observer's two retinas. Under a variety of viewing conditions, nearer objects generate a larger binocular disparity than farther objects. For example, objects viewed from a great distance generate an approximately zero disparity on the retinas.

By combining information about size and disparity, much more can be inferred about a figure's location relative to its background. For example, two objects may generate identical retinal image sizes, but the one that generates a larger disparity under appropriate viewing conditions will be closer, and therefore smaller. This linkage between size and disparity is often called the size-disparity correlation (Julesz and Schumer, 1981). It has often been proposed that larger receptive fields, or spatial scales, preferentially represent the size-disparity correlations of nearer, and thus larger and binocularly more disparate, objects. Although such an implementation of size-disparity correlation may abet figure-ground separation, it is not sufficient, as the following examples show.

DaVinci Stereopsis

Many figure-ground percepts can be better understood by analyzing the following type of ubiquitous experience. When we view a farther surface that is partly occluded by a nearer surface, one eye typically registers more of the farther surface than the other eye does. Our conscious percept of the farther surface is often derived from the view of the eye that registers more of this surface. For example, under the viewing conditions depicted in Figure 1, observers see surface BC at the same depth as surface CD, even though surface BC is registered by only the right eye. Thus, BC is part of the same "figure" as CD,

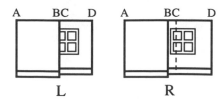

Figure 1. A DaVinci stereopsis display. (Figure reprinted with permission from Grossberg, 1994.)

even though only CD benefits from binocular disparity cues. This perceptual situation is often called *DaVinci stereopsis* (Nakayama and Shimojo, 1990). The challenging perceptual properties that subserve this percept are now illustrated under simpler stimulus conditions.

Deformable Fusion by Allelotropia

Because each eye views the world from a different position in the head, the same material point on an object is registered at a different location on the two retinas, except for that object region which is foveally fixated by both eyes. To binocularly fuse such a disparate pair of monocular images, the two images must be deformed into one percept, as in the phenomenon of *displacement* or *allelotropia*. Here, when a pattern EF G is viewed through one eye and a pattern E FG is viewed through the other eye, the letter F can be seen in depth at a position halfway between E and G. Thus, the process of binocular fusion deforms the two monocular appearances of F into one binocular percept of F whose spatial position differs from either monocular position of F with respect to E and G. This deformation of F's relative position is necessitated by the disparity of the two monocular F positions when E and G are binocularly fused.

During inspection of a 3D scene, the amount of deformation needed to achieve binocular fusion depends on how far away each object is with respect to an observer's retinas, since images of closer objects are more disparate than images of further objects. Thus, different parts of the left eye and right eye images are deformed by different amounts to generate a single binocular percept of the world. During DaVinci stereopsis, the vertical boundaries of regions AB and CD in the left eye and right eye images of Figure 1 need to be deformed by different amounts to be binocularly fused. If deformation of monocular boundaries occurs to form fused binocular boundaries, why are no "holes" created in binocular perceptual space?

Distance of Zero-Disparity Points

In particular, the retinal images of objects at optical infinity have zero disparity on the two retinas, and the disparities on the two retinas of corresponding object points tend to increase as an object approaches the observer. This is the familiar reason for assuming that larger size scales and disparities signal near objects. Conversely, when both eyes focus on a single point on a planar surface viewed in depth, the fixation point is a point of zero disparity. Points of the surface that are registered by the retinas further from the fixation point generate larger binocular disparities. Why do planar percepts not recede toward optical infinity at the fixation point and curve toward the observer at the periphery of the visual field? Why does the plane not become distorted in a new way every time our eyes fixate on a different point on its surface? For current purposes, a key fact is that a "zero disparity" condition also occurs under monocular viewing conditions, as in detecting region BC of

Figure 1. How does the monocularly viewed region BC inherit the depth of the binocularly viewed region CD?

Both the absence of "holes" in space attributable to boundary fusion and the inheritance by BC of the depth CD may be explained by a filling-in process that selectively completes a BC surface representation at a depth corresponding to that of region CD. In other words, the process that fills in the surface depth of CD in response to its binocular boundaries keeps flowing until it also fills in BC. Demonstrations that a filling-in process completes depthful surface properties include those of Nakayama, Shimojo, and Ramachandran (1990) and Watanabe and Cavanagh (1992).

Binocular and Monocular Boundary Segmentations

The surface filling-in process is activated and contained by boundary segmentations. Some boundaries are derived from binocularly viewed parts of a scene, others from monocularly viewed parts. In Figure 1, binocular fusion of the AB boundaries and the CD boundaries registers different disparities and amounts of allelotropia. The monocularly viewed boundaries in region BC do not register any binocular disparity. Nor do the horizontal image boundaries. Thus, at least three ways exist for an image to be registered with zero, or near-zero, disparity: as an occluded region during DaVinci stereopsis, as a monocularly viewed image, or as a horizontal boundary during either monocular or binocular viewing. Monocular and near-zero disparity cells are known to be separately processed by visual cortex (Poggio and Talbot, 1981). Grossberg (1994) suggested that monocular and near-zero disparity boundaries are processed in a separate pool of cortical cells for the following reasons.

Monocular and Near-Zero Disparity Cell Pools

We need to explain how the monocularly viewed vertical and horizontal boundaries in region BC are joined with the binocularly fused, large-disparity vertical boundaries and horizontal near-zero disparity boundaries in region CD to form the window frame in Figure 1. Disparity-sensitive cortical cells are tuned to a limited range of disparities. Let us assume that active near-zero disparity cells, whether they are monocularly or binocularly activated, give rise to spatially organized boundary signals that are combined with the spatially organized activations of cells that code non-zero disparities to create a more complete boundary representation (Figure 2a). The non-zero disparity cells are themselves assumed to be segregated into separate cell pools that are organized to correspond to different relative depths of an observed image feature. Thus, near-zero disparity cells add their boundary activations to multiple boundary representations, each corresponding to a differently tuned pool of non-zero disparity cells.

In response to the scene in Figure 1, consider BC boundaries added to CD boundaries at those scales and disparities that are capable of computing binocularly fused CD boundaries. These composite BCD boundaries must enclose *connected* regions, such as the connected window frame in the right eye image of Figure 1, if the following problem can be solved.

Three-Dimensional Emergent Boundary Completion

Because of allelotropia, the binocularly fused boundaries within region CD may be positionally displaced relative to the monocularly viewed boundaries within region BC. As a result, gaps may occur between the locations of cells in the visual cortex that represent binocular and monocular boundaries. When regions contain oblique contours, the binocular and

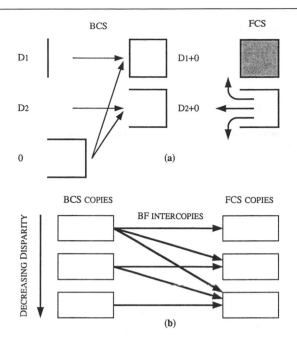

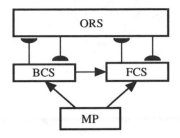

Figure 2. *A*, Near-zero disparity and monocular boundaries are added to boundaries of all the selective pools of non-zero disparity cells (with disparities D_1 and D_2). Only regions enclosed by connected boundaries can fill in. Other regions dissipate activity through uncontrolled diffusion. *B*, Multiple FCS copies exist corresponding to the BCS copies that code different ranges of relative depth from an observer. Each FCS copy contains a complete set of filling-in domains, or FIDOs, that correspond to the opponent colors (red, green), (blue, yellow), and (black, white). Near boundaries add to far boundaries in the FCS copies to prevent filling-in from occurring behind opaque surfaces. (Figure reprinted with permission from Grossberg, 1994.)

Figure 3. Completed boundaries within the boundary contour system (BCS) can be recognized within the object recognition system (ORS) through direct BCS → ORS interactions, whether or not they are seen in the feature contour system (FCS) by separating two regions with different filled-in brightnesses or colors. (Figure reprinted with permission from Grossberg, 1994.)

monocular responses of cortical cells may be both orientationally and positionally displaced. These gaps and misalignments need to be corrected by a boundary completion process (Grossberg and Mingolla, 1985; see also ILLUSORY CONTOUR FORMATION). Boundary completion is capable of generating an emergent boundary segmentation which realigns and connects the boundaries that join regions BC and CD. These completed boundaries completely enclose the window frame in Figure 1.

Capture of Filled-in Surface Properties by Connected Boundaries

The connected boundaries within region BCD form a sparse and discontinuous representation of the scene. How are the scene's continuous surface properties generated, to form a scenic figure, including its brightnesses, colors, and surface depths? Suppose that boundaries which enclose *connected* regions in BCD, and *only* these boundaries, can trigger filling-in of surface properties of the regions that form part of the final visible 3D percept (see Figure 2a). For this to work, multiple filling-in domains, or FIDOs, exist such that filling-in within each FIDO is controlled by boundaries that are sensitive to a restricted range of binocular disparities (see Figure 2b). A feature contour system (FCS) input is broadcast to all the FIDOs that code its color. Filling-in is triggered in only those FIDOs where color signals (FCS inputs) and boundary signals (boundary contour system, or BCS, boundaries) spatially coincide (Grossberg, 1987, 1994; see later discussion). These boundaries "capture" the surface color for their FIDO. Filling-in is modeled as a diffusion of featural activity across a FIDO until it

hits a boundary barrier (Grossberg and Todorović, 1988). The activity dissipates unless a connected boundary can contain it. As explained below, region BCD in Figure 1 contains a connected boundary within the FIDO, or subset of FIDOs, corresponding to the binocularly fused boundaries of region CD. Such surface representations combine position, depth, orientation, brightness, and color properties.

Near Boundaries Obstruct Filling-in of Occluded Regions

How does the filling-in of surface BC at the depth of CD stop at boundary B? Boundary B is binocularly fused at a disparity corresponding to a nearer surface than are the boundaries of region CD. Without further processing, boundary B could not form a connected boundary around region BD. Nor could it prevent filling-in of region AB within the FIDO whose depth corresponds to region CD. Filling-in would also occur within the FIDO whose depth corresponds to boundaries A and B of region AB. If both filling-in events could occur, region AB would appear transparent; it would be represented by two different filled-in surface representations at two different depths from the observer. In fact, if filling-in is the basis for many depthful figure-ground percepts, then why do not *all* figures look transparent?

This will not happen if the boundaries of closer objects are added to the boundaries of further objects in the FIDOs (see Figure 2b), so that near and far data are processed asymmetrically. Then filling-in initiated in region BD does not flow behind region AB. This restriction on surface filling-in does not prevent *boundaries* from being completed behind an occluding region. Then pathways from boundary representations to the object recognition system (Figure 3) enable partially occluded figures to be recognized by their completed boundaries, even if visible surface properties are not filled in behind the occluding object.

These properties of DaVinci stereopsis illustrate how the multiple spatial scales that are used for disparity-selective early visual filtering may interact with later boundary segmentation and surface filling-in processes to bind visual features into surface representations of figure and ground.

Large Size Scales Signal Far Objects (The Weisstein Effect)

The Weisstein effect clarifies how figure-ground percepts in depth can occur in response to pictures that are constructed from multiple spatial scales or spatial frequencies. As already noted, large size scales, or low spatial frequencies, often seem to selectively process near objects, whereas high spatial frequencies selectively process far objects. In contrast to this property, Klymenko and Weisstein (1986) demonstrated that if regions filled with relatively higher spatial frequency sinusoidal gratings are adjacent to regions containing relatively lower spatial frequency gratings, then the regions with the higher frequency appear closer in depth than those containing the lower frequency. They studied a variant of the classical Rubens faces/vase reversible figure for which, in the absence of the sinusoidal gratings, a temporally bistable percept is perceived. At one instant, two faces pop out as figures. At the next instant, a vase pops out between the faces as they recede into the background. With a higher spatial frequency sinusoid placed within the faces than the vase, the faces are perceived as figures most of the time. The Weisstein effect shows that whether a spatial frequency difference signals "near" or "far" depends on how the image is segmented into boundaries and surfaces, not merely on a size difference per se.

Three-Dimensional Percepts of Occluded and Occluding Figures in 2D Pictures

The spatial organization and relative luminance of occluding and occluded objects also has a powerful influence on figure-ground perception during inspection of 2D pictures as well as 3D scenes (Bregman, 1981; Kanizsa, 1979). Comparing Figures 4b and 4c shows that the occluding black sinewy shape in front of the occluded Bs helps us recognize them as Bs.

How does a 2D image create a 3D percept of occluding figures in front of occluded figures, as in Figure 4b? How are the gray fragments easily recognized in Figure 4b as occluded B shapes but not in Figure 4c, even though they are equally well seen in both? A comparison of Figures 4b and 4c illustrates that properties of contrast, form, and depth interact to generate a percept, and that this interaction may, as in Figure 4b, generate a 3D representation of a 2D image. This 3D repre-

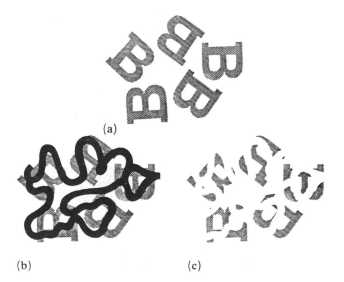

(a)

(b)   (c)

Figure 4. Role of occluding region in recognition of occluded letters: (a) Uppercase B letters. (b) The same letters partially hidden by a black snakelike occluder. (c) The same, except the occluder is white, and therefore merges with the remainder of the white background. Although the exposed portions of the letters are identical in (b) and (c), they are much better recognized in (b). (Reprinted from Nakayama, K., Shimojo, S., and Ramachandran, V. S., 1989, Stereoscopic depth: Its relation to image segmentation, grouping, and the recognition of occluded objects, *Perception*, 18:55–68, figure 1, with permission from Pion Limited, London.)

sentation enables the occluded boundaries of the B shapes to be completed for purposes of recognition, even though the occluded surfaces are not seen in either figure. How does this happen?

Suppose that the boundaries which are shared by the gray B shapes and the black occluder are assigned to the occluder and detached from the remaining B boundaries. Suppose also that these shared boundaries, along with the other occluder boundaries, are used to generate a boundary segmentation and filled-in surface representation of the black occluder "in front of" the surface on which the B fragments lie. These boundaries are also reattached to the B boundaries at a later processing stage, as in DaVinci stereopsis, to keep the gray from flowing behind the black.

Occluded Boundary Completion and Recognition Without Filling-in

Given that the shared boundaries between occluder and B shapes in Figure 4b are somehow removed from the B shapes, how does an observer so quickly recognize the incomplete B figures? As in the case of DaVinci stereopsis, a boundary completion process generates illusory contours between the (approximately) colinear line ends of the incomplete B figures. This property of illusory contour completion raises a central question in figure-ground perception: if illusory contours complete the B shapes and thereby enhance their *recognition*, why do we not *see* these illusory boundaries in the sense of detecting a perceived brightness or color contrast at their locations?

Figure 3 schematizes part of the answer that was proposed by Grossberg and Mingolla; see ILLUSORY CONTOUR FORMATION. A boundary that is completed within the segmentation system (which is called the boundary contour system, or BCS) does not generate visible contrasts within the BCS. In this sense, *all boundaries are invisible*. Visibility is a property of the surface filling-in system (the feature contour system, or FCS). The completed BCS boundary can directly activate the visual object recognition system (ORS), whether or not it is visible within the FCS. Neurophysiological data suggest that the ORS includes the inferotemporal cortex (Mishkin and Appenzeller, 1987), whereas the FCS visible surface representation includes area V4 of the extrastriate cortex (Desimone et al., 1985; Zeki, 1983). A boundary may thus be completed within the BCS, and thereby may improve pattern recognition by the ORS, without necessarily generating a visible brightness or color difference within the FCS. In the classical literature, such boundaries were said to be *amodally* completed, but the relationship between amodal completion, modal completion, and filling-in was not specified.

Three-Dimensional Kanizsa Squares

Three-dimensional Kanizsa squares provide a vivid example of how boundary and surface processes interact to define figure and ground occurs. When one inspects a 2D Kanizsa square (see ILLUSORY CONTOUR FORMATION), one perceives a square boundary that encloses a bright square region. A square illusory contour is generated by four black pac-man figures and triggers selective filling-in of the bright square region.

Three-dimensional Kanizsa square percepts illustrate how binocularly fused boundaries can generate illusory boundaries that selectively capture the brightness signals induced by the pac-man figures to fill in two surfaces at different perceived depths from the observer. In Nakayama, Shimojo, and Ramachandran (1990), the disparity of the vertical boundaries in the pac-man figures of two Kanizsa square images was varied. The image pairs were viewed through a stereoscope or free fused. In the crossed-disparity case, which corresponds to closer objects, the illusory contours that frame the square are greatly enhanced, and the Kanizsa square appears nearer. Observers recognize that the pac-man boundaries are completed into disks behind the square surface, but only the pac-men are seen as visible surfaces. When the disparity is reversed, an occluding surface is perceived, through whose four (almost) circular windows are seen the four corners of an occluded square. The illusory contours that complete the four circular windows are visible because the occluding surface fills in at the nearer depth. The Kanizsa square is recognized behind the occluding surface, but only its four pac-man regions are visible through the four circular windows.

These remarkable percepts show that binocular matching of just a few edges in a scene can trigger completion of a 3D boundary segmentation that captures figural surface percepts at the relative depths that the boundaries encode. Also, once again, the ORS may *recognize* the BCS boundaries that are completed behind the nearer occluding surface, even if they are not *seen* within the FCS. They are not seen within the FCS whenever the BCS boundaries of nearer segmentations create barriers to filling-in of farther surfaces, as in Figure 2b. These and many other figure-ground percepts can be explained by such model rules (Grossberg, 1994). The article ILLUSORY CONTOUR FORMATION introduces some of the monocular properties of the BCS, including the simple, complex, hypercomplex, and bipole cells of the visual cortex that the BCS models.

Concluding Remarks

These experimental data and theoretical concepts suggest that figure-ground separation in particular, and biological vision in general, use principles and mechanisms that are very different from those described in classical geometries and computer vision algorithms. These new ideas are naturally expressed using suitably defined neural networks in which the complementary properties of emergent boundary segmentations and filled-in surface representations are interactively combined.

Acknowledgments. Professor Stephen Grossberg was supported in part by the Air Force Office of Scientific Research (AFOSR F49620-92-J-0499), ARPA (ONR N00014-92-J-4015), and the Office of Naval Research (ONR N00014-91-J-4100).

Road Map: Vision
Related Reading: Color Perception; Perceptual Grouping; Visual Scene Perception: Neurophysiology

References

Bregman, A. L., 1981, Asking the "what for" question in auditory perception, in *Perceptual Organization* (M. Kubovy and J. R. Pomerantz, Eds.), Hillsdale, NJ: Erlbaum, pp. 99–118.

Desimone, R., Schein, S. J., Moran, J., and Ungerleider, L. G., 1985, Contour, color, and shape analysis beyond the striate cortex, *Vision Res.*, 25:441–452.

DeYoe, E. A., and Van Essen, D. C., 1988, Concurrent processing streams in monkey visual cortex, *Trends Neurosci.*, 11:219–226. ◆

Grossberg, S., 1987, Cortical dynamics of three-dimensional form, color, and brightness perception, *Percept. & Psychophys.*, 41:87–158.

Grossberg, S., 1994, 3-D vision and figure-ground separation by visual cortex, *Percept. & Psychophys.*, 55:48–120.

Grossberg, S., and Mingolla, E., 1985, Neural dynamics of form perception: Boundary completion, illusory figures, and neon color spreading, *Psych. Rev.*, 92:173–211.

Grossberg, S., and Todorović, D., 1988, Neural dynamics of 1-D and 2-D brightness perception: A unified model of classical and recent phenomena, *Percept. & Psychophys.*, 43:241–277.

Julesz, B., and Schumer, R. A., 1981, Early visual perception, *Annu. Rev. Psychol.*, 32:572–627. ◆

Kanizsa, G., 1979, *Organization in Vision: Essays in Gestalt Perception*, New York: Praeger. ◆

Klymenko, V., and Weisstein, N., 1986, Spatial frequency differences can determine figure-ground organization, *J. Exp. Psychol. [Hum. Percept.]*, 12:324–330.

Mishkin, M., and Appenzeller, T., 1987, The anatomy of memory, *Sci. Am.*, 256:80–89. ◆

Nakayama, K., and Shimojo, S., 1990, DaVinci stereopsis: Depth and subjective occluding contours from unpaired image points, *Vision Res.*, 30:1811–1825.

Nakayama, K., Shimojo, S., and Ramachandran, V. S., 1990, Transparency: Relation to depth, subjective contours, luminance, and neon color spreading, *Perception*, 19:497–513.

Poggio, G. F., and Talbot, W. H., 1981, Mechanism of static and dynamic stereopsis in foveal cortex of the rhesus monkey, *J. Physiol.* 315:469–492.

Watanabe, T., and Cavanagh, P., 1992, Depth capture and transparency of regions bounded by illusory and chromatic contours, *Vision Res.*, 32:527–532.

Zeki, S., 1983, Colour coding in the cerebral cortex: The reaction of cells in monkey visual cortex to wavelengths and colours, *Neuroscience*, 9:741–765.

Forecasting

Lyle H. Ungar

Introduction

The task of forecasting the future values of sequences of observations is, in many ways, ideally suited for neural networks. Large amounts of data may be available, and the underlying relationships are often nonlinear and unknown. Neural nets, mostly standard backpropagation networks (i.e., multilayer perceptrons trained by backpropagation; see BACKPROPAGATION: BASICS AND NEW DEVELOPMENTS), have been used with great success in many forecasting applications, but they have had less success in other applications. This article looks at the use of neural nets for forecasting with particular attention to understanding when they perform better or worse than other technologies.

The success of neural networks in forecasting depends significantly on the characteristics of the process being forecast. One may want to predict minute-by-minute progress of a chemical reaction, hour-by-hour power usage (load) for an electric power utility, daily weather, monthly prices of products and inventory levels, and quarterly or yearly sales and profits. These problems differ in the quantity and type of information available for forecasting, and hence call for different forecasting techniques.

Forecasting problems can be characterized on a number of dimensions: (i) Is a single series of measurements used, as is often done in conventional forecasting, or are multiple related measurements available? (ii) Are the data seasonal or not? Monthly or quarterly data, such as those for sales volume or energy use, often show strong seasonal variation, while annual data or data measured each second or minute do not. (iii) The number of observations and (iv) the degree of randomness (signal/noise ratio) of the process also strongly limit the complexity of the model which can be fitted. If data are available only annually for the past ten or twenty years, and no measurement is available for most of the disturbances, one should not expect to be able to fit a complex model such as a neural network. This is unfortunately the case for many forecasting problems such as those represented in the Makridakis collection (described later in this article). (v) Finally, for some forecasting problems, one only requires prediction a single time step in the future; for others, multiple time step forecasts are required. This difference has implications for the method used to train the neural network.

Before looking at neural networks, we will briefly review conventional forecasting methods. Forecasting has mostly used one of two different classes of methods, depending on whether the data are seasonal or not. For monthly data, such as those for sales or unemployment levels, the seasonal variation is often removed by dividing the series by an index representing the historical seasonal variation. For example, dividing the unemployment rate for each month (perhaps averaged over several years) by the average annual unemployment rate gives an index which indicates monthly variations. This index will have an average value of one. Dividing the actual unemployment rate in a given month by the index for that month gives the seasonally adjusted unemployment, which shows overall trends after typical monthly variations are accounted for. A linear or exponential regression (i.e., fitting the data as a linear or exponential function of time), or some form of smoothing such as a moving average, can then be used to make predictions of the deseasonalized unemployment. Actual levels are then forecast by multiplying these base predictions by the index for the month being forecast (Makridakis, Wheelwright, and McGee, 1983).

In contrast, for many complex processes, such as those involved in chemical plants, robots, or stock prices, the best prediction of the near future is obtained by using an appropriately weighted combination of recent measurements of the variable being predicted and other correlated variables. The most widely used approach employs Auto-Regressive Moving Average (ARMA) models. For example, to predict the value of a variable y (such as a temperature, a pressure, or a stock price) at time $t + 1$ using past values of y and of a second variable z, one would use a linear regression to fit a model of the form:

$$y_{t+1} = c_0 + c_1 y_t + c_2 y_{t-1} + c_1 y_{t-2} + \cdots$$
$$+ c_n z_t + c_{n+1} z_{t+1} + \cdots \qquad (1)$$

Note that ARMA models differ from the linear regression models mentioned earlier in that they are functions of previous variables rather than of time.

Neural networks can be used to learn a nonlinear generalization of ARMA models of the form

$$y_{t+1} = f(y_t, y_{t-1}, y_{t-2}, \ldots, z_t, z_{t+1}, \ldots) \qquad (2)$$

When the process is nonlinear and sufficient data are available, the neural networks will provide a more accurate model than the linear ARMA model. See Box, Jenkins, and Reinsel (1994) for extensive descriptions of conventional ARMA models and of the Box-Jenkins modeling approach, which involves picking a model of the form of Equation 1 with some subset of the coefficients set to zero. Later in this article we summarize the results of a number of studies which compare ARMA and neural network models.

Two other modeling methods are also often used by engineers: Kalman filtering and Wiener-Volterra series. Kalman filters (see ADAPTIVE SIGNAL PROCESSING) assume a known model structure in which the parameters and their covariance, which is modeled explicitly, may be changing over time. Kalman filters are good for modeling relatively simple but noisy processes but, unlike neural networks, do not form nonparametric models which can accurately forecast the behavior of nonlinear systems. Wiener-Volterra series are polynomial expansions of input-output relations fitted to past data. As such, they, like neural nets, can approximate arbitrary functions. However, for models with multiple inputs they require more data than neural networks to obtain an equal level of accuracy.

Using Neural Nets for Forecasting

Neural networks are most often used to fit ARMA-style models of raw time-series data from one or more measurements, but they can also be used as a piece of larger forecasting systems: for example, in combination with deseasonalizing (i.e., forecasting a time series from which the seasonal component has been removed, as described earlier). Even for the simpler ARMA-style models, attention to the method is required if one is making forecasts multiple time steps in the future rather than a single time step.

Direct Versus Recurrent Prediction

A simple form of multistep forecasting is direct prediction (Figure 1A), in which a network takes past values as inputs and has separate outputs for predictions that are one, two, and more time steps in the future. Alternatively, one can train a network to predict one time step in the future and then use the network recursively to make multistep predictions (Figure 1B). Such networks are sometimes called "externally recurrent networks," in contrast to networks which have internal memory. Direct forecasting networks are easier to build than externally recurrent nets because they do not require unfolding in time (described later), but their predictions are generally less accurate, since they have more parameters which must be fitted from the same limited data.

The obvious way to train a network such as the one in Figure 1B is to minimize the error on the one-time-step predictions.

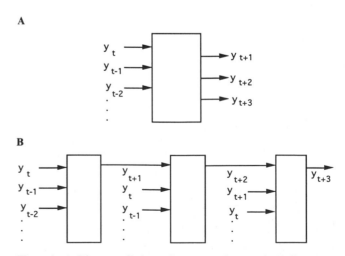

Figure 1. *A*, Direct prediction using a neural network. *B*, Recurrent one-step-ahead prediction using a neural network repeatedly.

This unfortunately does not give optimal networks for multi-step predictions. To better understand this somewhat confusing point, consider the case of a simple linear ARMA model:

$$y_{t+1} = c_0 + c_1 y_t + c_2 y_{t-1} \tag{3}$$

A two-step-ahead prediction would then take the form

$$y_{t+2} = c_0 + c_1(c_0 + c_1 y_t + c_2 y_{t-1}) + c_2 y_t \tag{4}$$

Selecting coefficients c_0, c_1, and c_2 to minimize the prediction error for the one-step-ahead error yields a different equation than selecting the same coefficients to minimize the error in the two-step prediction. (Note that the former is a linear regression problem, while the latter requires nonlinear regression because the coefficients multiply each other.) More accurate long-range predictions are obtained by training to minimize the multistep prediction error. The solution using backpropagation employs the same unfolding in time or other solution methods as for internally recurrent networks see RECURRENT NETWORKS: SUPERVISED LEARNING. This and related issues are covered in detail in books on conventional system-identification methods; see, e.g., Ljung and Torsten (1983).

Combining Neural Networks with Other Methods

There are a number of ways in which neural networks can be combined with data preprocessing techniques, with partial models, based on mechanistic first principles, of the process being forecast, and with other forecasting techniques. Most commonly, if there is a strong seasonal component to the data, the data may be deseasonalized and the neural net used to forecast the basic trend. It may appear pointless to use a seasonal index when it is well known that neural networks can approximate arbitrary functions, which should include any seasonal variation. Experience indicates that if sufficient data are available, this is true, but that for shorter time series, deseasonalizing gives more accurate forecasts.

Similarly, when modeling complex physical systems, much better forecasts can be obtained with much less data when prior knowledge (e.g., in the form of mass, energy, or kinematic constraints on the variables or in terms of monotonic relations between measured and forecast variables) is built into the network (Psichogios and Ungar, 1992). In a typical example, the equations governing a fermentation reactor are known except for the growth kinetics of the cells (e.g., yeast) in the reactor. If a neural network is used just to approximate the growth kinetics, rather than to model the whole system, models are learned which are more accurate and which extrapolate better to operating regimes where no data were available.

Neural networks can also be used in conjunction with conventional forecasting methods. For example, one can often produce more accurate forecasts by providing several conventional forecasts as input to the neural network. In this case, the network serves partly as a combining method, producing a weighted average of the different forecasts (Foster, Collopy, and Ungar, 1992). Such combining of forecasts is widely practiced in the forecasting community, mostly with relatively arbitrary combining weights (Clemons, 1989).

Assessing Neural Nets for Forecasting

There are several difficulties in assessing forecasting methods. The most serious is that the results of a single forecast tell little about whether the method will be superior for other forecasts. In testing any method, it is important to have a large set of representative time series upon which the methods will be tested. An example of such a collection of time series which has been widely used to compare forecasting methods is the M-competition (Makridakis et al., 1982). This competition included 1001 series and evaluated 24 forecasting methods. The series, taken from a variety of organizations in a number of countries, include macroeconomic, microeconomic, industrial, and demographic data for phenomena such as production levels, net sales, unemployment, spending, GNP, vital statistics, and infectious disease incidence. The series include yearly, quarterly, and monthly series, but no series arising from securities or commodities trading. These time series all involve only a single variable; they do not provide correlated variables which might enhance the predictions.

One must also decide which error criteria to use. The most obvious criterion, and the one which is optimized by standard neural networks, is minimization of the mean square prediction error. This criterion has the property that a small number of unusual series may have a large effect on the error. In looking at combined errors for different time series, one must, of course, also normalize for the different magnitudes of the series. Thus, forecasters often measure performance by using measures which are more robust for outliers or atypical time series.

Three error measures which have proved particularly robust are: the percentage of time a method had a lower absolute error than the no-change forecast (*percent-better*); the relative absolute error (RAE); and the median absolute percent error (mdAPE). The RAE is calculated as the geometric mean across all series i of

$$\text{rae}_i = \frac{\sum_{t=1}^{T} |\tilde{x}(t) - x(t)|_i}{\sum_{t=1}^{T} |x(0) - x(t)|_i} \tag{5}$$

where $\tilde{x}(t)$ is the forecast and $x(t)$ represents the true value of the series at time t. The RAE represents a comparison over the forecast horizon T for series i of the absolute error of the forecast method with the no-change or random walk forecast. One then calculates a geometric mean over all the series:

$$\text{RAE} = \left[\prod_{i=1}^{n} \text{rae}_i \right]^{1/n} \tag{6}$$

The median average percent error is defined as the median across all series i of

$$\text{ape}_i = \frac{1}{T} \sum_{t=1}^{T} 100 \frac{|\tilde{x}(t) - x(t)|_i}{|x(t)|_i} \tag{7}$$

Good forecast performance is reflected in higher percent-betters and lower RAEs and mdAPEs.

In assessing neural networks for forecasting, one must compare the accuracy of the neural networks with that of other statistical tools such as exponential smoothing (for a single time series) or linear ARMA models (for several correlated time series). Surprisingly, many studies fail to compare neural network forecasts with well-made conventional forecasts.

Table 1 shows a sample of the applications in which neural networks have been used for forecasting. Almost all of the studies use standard backpropagation networks with less than a dozen inputs and less than a dozen hidden nodes, with the exact architecture being selected by trial and error. Also, most of the studies use data from a single source and evaluate their results on the basis of the mean square error on out-of-sample forecasts (i.e., error when forecasting data other than that used for building the model). Table 1 does not include any studies

Table 1. Forecasting Using Neural Nets (NNs): Results

Application	Authors	Results	Compared with
Car sales, airline passengers	Tang et al.	NN better for longer-term forecast; Box-Jenkins better for shorter	Box-Jenkins
Currency exchange rates	Weigend et al.	NNs better	Random guessing
Electric load forecasting	Park et al.	NN better	Currently used technology (not specified precisely)
Electrochemical reaction	Hudson et al.	Prediction looks good	—
Flour prices	Chakraborty et al.	NNs better than ARMA	ARMA
Polypropylene sales	Chitra	NNs slightly better than ARMA	ARMA
Stock prices	White	NNs provide no benefit	Random walk
Widely varied (Makridakis collection)	Foster et al.	NNs better on quarterly data; worse on annual	Many exponential smoothing and deseasonalizing methods

See Vemuri and Rogers (1994) for citations of the studies listed here.

using chaotic time series such as from the Mackey-Glass equation, which give little insight into neural network forecasts of realistic data. See Vemuri and Rogers (1994) for a wide variety of neural network forecasting studies, including all the studies cited in Table 1 which are not listed in the references. For literature on neural network forecasting for process control, see Process Control in this volume. Process control and robotics applications have seen some of the most successful use of neural networks for forecasting: the processes involved are often sufficiently multivariable and nonlinear to warrant the use of neural networks, but sufficiently well characterized and free of noise to allow accurate models to be built.

Dangers in Using Forecasts

Forecasts rely on a number of assumptions. They assume that the system that is modeled remains constant: i.e., that the model which held when the model was built still applies when the forecast is made. If the system structure is evolving over time, techniques from adaptive control may be more appropriate. (See the two articles on adaptive control earlier in this volume.) It is also implicitly assumed, when forecasting using neural networks with multiple inputs, that the covariance structure of the inputs will remain constant. This presents a major difficulty when modeling systems that have feedback in them, if the feedback structure is variable. For example, consider a house controlled by a thermostat. One will typically find that the heater will be on more often when the house is cold (this is, after all, what the heating system is designed to do). Forecasts of future house temperature can be accurately made using historic temperature measurements. If, however, these forecasts are used as part of the control scheme (the thermostat), then instability often results, since the forecasts fail to account for the new thermostat behavior. (See Process Control for more details.) Similar situations often occur in economics and marketing, where forecasts can result in new laws being passed or in new prices being charged (and resulting actions by competitors), thus invalidating the original forecast. Unfortunately, there is generally little that one can do other than monitoring forecasts and distrusting them or collecting more data. (This is true in linear regression as well, where it is impossible to tell which of two highly correlated inputs is responsible for changes in an output; but in linear regression one can easily detect the problem by examining the uncertainty on the regression coefficients, whereas the problem is usually concealed in neural nets.)

Discussion

Neural networks have a fair number of demonstrated successes as forecasting tools and a smaller number of documented failures. All the usual warnings about model building apply. In particular, one needs good data to build a good model. When the data are noisy and occur in short series, neural networks often fail to do better than simple forecasting techniques. For example, the 181 yearly series of the M-competition, which have a mean length of 19 data points on which to base a prediction, do not provide a good basis for complex nonlinear models. Neural networks generally give significant improvements over conventional forecasting methods when applied to monthly data in the M-competition set, but not to the yearly data (Marquez et al., 1992). This is probably due to the high ratio of noise to data in the yearly data.

It may also be the case that the data are truly random or that the key independent variables are not being measured. Many people claim that this is true of the stock market (White, 1989). If this is true, then neural networks will not produce useful market forecasts, although they may help sell forecasting products. Several fund managers claim that they are getting superior predictions using neural networks, but for obvious competitive reasons they generally do not provide enough information to test the claims.

Neural networks have proven successful in forecasting such phenomena as prices (Chakraborty et al., 1992), product demand (Chitra, 1993), electric utility loads (Yu, Moghaddamjo, and Chen, 1992), and inventory levels. (See Table 1.) Such problems are characterized by ample measurements with a relatively high signal-to-noise ratio. In most cases, substantially better performance is obtained by using several related inputs to the network. For example, in forecasting wheat prices in three cities, superior performance was found by using recent wheat prices and measures of the local earning power. Similarly, in forecasting demand for polypropylene production, several macroeconomic variables were fed into the network. For longer, more deterministic time series, as in measuring the progress of a chemical reaction, neural networks have been shown to be a relatively accurate means of forecasting even chaotic series (Lapedes and Farber, 1987; Hudson et al., 1990).

All of these applications use standard backpropagation networks, occasionally with some degree of structure built into the network. For example, in the currency exchanges, "excess" weights were eliminated; likewise, for forecasting wheat prices, past values of prices in three different cities were used to predict the logarithm of flour prices. All such methods have demonstrated better performance than conventional forecasting methods, except when only short time series were available (10–30 data points) or when it was unclear whether there was an underlying model other than a biased random walk (e.g., stock prices). However, the gains in accuracy over conventional forecasting methods are often relatively small. Also, many of the studies fail to examine enough different series to prove that the neural network–based methods will work on different problems. They do, however, provide sufficient encouragement that many companies are starting to use neural networks for problems such as demand forecasting. The future will probably bring not only more use of neural networks in practical forecasting problems, but also more use of other network types, including unsupervised learning combinations of the measured variables (features) which allow accurate forecasts with simpler models.

Road Map: Applications of Neural Networks
Related Reading: Applications of Neural Networks; Averaging/Modular Techniques for Neural Networks; Investment Management: Tactical Asset Allocation

References

Box, G. E. P., Jenkins, G. M., and Reinsel, G. C., 1970, *Time Series Analysis: Forecasting and Control*, 3rd ed., Englewood Cliffs, NJ: Prentice Hall. ◆

Chakraborty, K., Mehrotra, K., Mohan, C. K., and Ranka, S., 1992, Forecasting the behavior of multivariate time series using neural networks, *Neural Netw.*, 5:961–970. ◆

Chitra, S. P., 1993, Use neural networks for problem solving, *Chem. Eng. Prog.*, April:44–52. ◆

Clemons, R., 1989, Combining forecasts: A review and annotated bibliography, *Int. J. Forecasting*, 5:559–583.

Foster, B., Collopy, F., and Ungar, L. H., 1992, Neural network forecasting of short noisy time series, *Comput. Chem. Engrg.*, 16(4):293–298. ◆

Hudson, J. L., et al., 1990, Nonlinear signal processing and system identification: Applications to time series from electrochemical reactions, *Chem. Eng. Sci.*, 45:2075–2981.

Lapedes, A. S., and Farber, R. M., 1987, Nonlinear signal processing using neural networks: Prediction and system modelling, in *Los Alamos National Laboratory Technical Report LA–UR–87–266*; reprinted in Vemuri and Rogers (1994).

Ljung, L., and Torsten, S., 1983, *Theory and Practice of Recursive Identification*, Cambridge, MA: MIT Press.

Makridakis, S., et al., 1982, The accuracy of extrapolation (time series) methods: Results of a forecasting competition, *J. Forecasting*, 1:111–153.

Makridakis, S., Wheelwright, S., and McGee, V., 1983, *Forecasting: Methods and Applications*, New York: Wiley. ◆

Marquez, L., Hill, T., O'Connor, M., and Remus, W., 1992, Neural network models for forecasting: A review, in *Proceedings of the Twenty-fifth Hawaii International Conference on System Sciences*, Los Alamitos, CA: IEEE Computer Society Press, pp. 494–497.

Psichogios, D. C., and Ungar, L. H., 1992, A hybrid neural network: First principles approach to process modeling, *Am. Inst. Chem. Eng. J.*, 38:1499–1512.

Vemuri, V. R., and Rogers, R. D., 1994, *Artificial Neural Networks: Forecasting Time Series*, Los Alamitos, CA: IEEE Computer Society Press. ◆

White, H., 1989, Economic prediction using neural networks: The case of IBM daily stock returns, in *Proceedings of the IEEE International Conference on Neural Networks*, San Diego, CA: IEEE TAB Neural Network Committee, vol. II, p. 451.

Yu, D. C., Moghaddamjo, A. R., and Chen, S.-T., 1992, Weather sensitive short-term load forecasting using nonfully connected artificial neural network, *IEEE Trans. Power Syst.*, 7:1098–1105.

Fractal Strategies for Neural Network Scaling

Raymond Lister

Introduction

Many neural network optimization algorithms contain a mechanism analogous to SIMULATED ANNEALING (q.v.). This algorithm exhibits *asymptotic convergence*. That is, as run time approaches infinity, the probability of finding an optimal solution approaches 1. Of more practical interest, however, is the rate at which this probability approaches 1. In many neural network applications of annealing, the rate of convergence appears to be extremely slow. Thus, although these algorithms work well on small problems, they do not scale well to interesting problem sizes.

Sorkin (1990) has offered a proof that the rate of convergence can be fast for the special case where the energy landscape is a fractal. Figure 1 illustrates the following intuitive explanation of his proof. The landscape shown is the second generation of a fractal. The first generation simply contains a line between points *C* and *D*. The third-generation fractal would add, between points *c* and *d*, a structure that is self-similar to the first generation. At the first temperature iteration, the annealing algorithm moves only on the first-generation structure of the fractal energy landscape; the finer second-generation fractal structure is not part of this restricted solution space. This restricted space is small, so the algorithm will quickly find

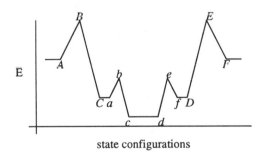

state configurations

Figure 1. A fractal energy landscape.

the lower energy region between *C* and *D*. The temperature for the second temperature iteration is chosen so that transitions from *C* to *B* or from *D* to *E* are very unlikely, restricting the algorithm to the second-generation structure between *C* and *D*. This is once again a small solution space, and so the algorithm will quickly find the lower energy region between *c* and *d*. With each succeeding temperature iteration, the algorithm explores an increasingly restricted region of progressively higher generations of the fractal. Sorkin proved that, with this approach, a solution of expected energy no more than ε above the global minimum can be found in time polynomial in $1/\varepsilon$.

A major determinant of the shape of an energy landscape is the *changing mechanism* by which candidate solutions are generated from the current solution. A *recursive changing mechanism* is one that can be applied to a solution at different spatial or temporal scales, with similar effect at each scale. Recursive changing mechanisms often give landscapes that are approximately fractal. We formally define such *quasi-fractal* landscapes in the next section. We then present empirical evidence that simulated annealing can also be relatively fast on quasi-fractal energy landscapes. We illustrate that with the Traveling Salesman Problem (TSP). The empirical evidence also offers an explanation for the mediocre performance of Hopfield and Tank's (1985) neural network, and its descendants (Van den Bout and Miller, 1989; Peterson and Söderberg, 1989; see NEURAL OPTIMIZATION): those networks do not have quasi-fractal energy landscapes.

Quasi-fractal Energy Landscapes

A metric space is a set X with a distance function $d(x, y)$, such that for any x, y, z in X, three conditions apply: that $d(x, y) = 0$ if and only if $x = y$; that $d(x, y) = d(y, x)$; and that $d(x, z) \leq d(x, y) + d(y, z)$. The Euclidean plane is a metric space, and for this reason, the last of these three conditions is known as the *triangle inequality*. The minima of an energy landscape also define a metric space, when $d_c(x, y)$ is the minimum number of applications of the changing mechanism required to turn locally minimal solution x into locally minimal solution y.

An *ultrametric space* satisfies an additional condition stronger than the triangle inequality: that $d(x, z) \leq \text{maximum}\{d(x, y), d(y, z)\}$. This condition implies that all triangles in the space are isosceles, with the third side shorter than or equal to the other two sides. The archetypal example is a tree, where the distance between nodes is the height one must ascend the tree to reach a common predecessor. Any ultrametric space can be mapped onto a tree structure (Baldi and Baum, 1986), so all ultrametric spaces exhibit a recursive structure.

Energy landscapes are also ultrametric spaces, when the metric $d_E(x, y)$ is the minimax energy between locally minimal solutions x and y. The minimax energy is defined thus. Consider the set of all possible paths from x to y. Each of those paths has a highest energy maximum. The minimax energy is the lowest of these maxima. It follows trivially that d_E defines an ultrametric space, as one can always move from x to any other locally minimal solution z, via y.

In some metric spaces, the two longer sides of triangles are approximately equal, with high probability. This can be quantified by a statistical correlation function, ranging from -1 to 1. When the correlation is exactly 1, the space is ultrametric. When the correlation is near 1, the space is said to exhibit a high degree of *ultrametricity*.

Solla, Sorkin, and White (1986) conjectured that annealing works well in landscapes where $d_c(x, y)$ and $d_E(x, y)$ are highly correlated, that is, in landscapes where solutions close to each other are separated by low minimaxima, and solutions far apart are separated by high minimaxima. Such landscapes are *quasi-fractal*.

To empirically determine whether $d_c(x, y)$ and $d_E(x, y)$ are highly correlated, it is sufficient to determine whether $d_c(x, y)$ exhibits a high degree of ultrametricity. That is, for triples of locally minimal solutions x, y, z, where $d_c(x, y) \leq \text{minimum}\{d_c(x, z), d_c(y, z)\}$, determine the correlation between $d_c(x, z)$ and $d_c(y, z)$.

The Traveling-Salesman Problem

Segment Reversal Versus Moving Single Cities

Perhaps the most obvious changing mechanism for the TSP is the removal of a single city from its current position in the path and its reinsertion at some other position. That mechanism, however, does not scale well to large problems. This is illustrated in Figure 2. The 32 cities are grouped into four clusters of eight cities. The path in Figure 2A contains a suboptimal cross-over (i.e., intercluster edges A → C and B → D cross). Transforming the path in Figure 2A into the path in Figure 2B requires the movement of all eight cities forming one cluster, with no improvement in path length until all eight cities have been moved.

Segment reversal is a well-known way to change a path (Lin and Kernighan, 1973), and it scales well to large problems. The transition between Figure 2A and 2B requires a single segment reversal. The direction of travel along the segment through clusters C and B is reversed, with intercluster edges A → C and B → D in Figure 2A replaced by edges A → B and C → D in Figure 2B. Segment reversal scales well because it can be applied to segments of arbitrary length, and thus it may be applied at different spatial scales. In Figure 2, it is as if the four clusters were "super cities" (hence the four circles around the clusters, to emphasize the impression). Segment reversal operates equally on changing the order of individual cities and the order of clusters of cities.

Kirkpatrick and Toulouse (1985) numerically studied the distribution of minima in energy landscapes defined by segment reversal. They found that the distances between the minima exhibited a high degree of ultrametricity.

Performance Comparison

Figure 3 contrasts the final solutions generated by simulated annealing when using the above two changing mechanisms. The simulations differed in no respect other than the changing mechanism, and they consumed the same amount of CPU time. The TSP used is Peterson's (1990) 200-city problem, in which the cities are enclosed by a square of unit side length. The path on the left, generated by segment reversal, is 1.4 units shorter than the path on the right—a large difference, since 1.4 is approximately the length of the diagonal of the enclosing square.

The final solution obtained by moving single cities contains many locations where the path crosses itself. This is the same ill-structure that Wilson and Pawley (1988) noted in their final solutions from Hopfield and Tank networks. The two algorithms exhibit this same ill-structure for the same fundamental reason: their respective energy surfaces are related. Moving sin-

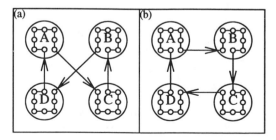

Figure 2. Two Traveling Salesman paths. The path on the left can be changed into the better path on the right by a single segment reversal.

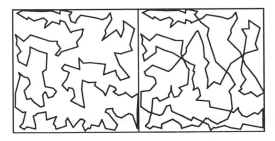

Figure 3. The final solutions from two annealing runs on Peterson's (1990) 200-city TSP, using segment reversal (left, length equals 10.7) and moving a single city at a time (right, length equals 12.1). Each solution is enclosed by a square of unit length sides.

gle cities is the nearest discrete approximation to the changing mechanisms implemented by existing analog networks. These analog networks can be pictured intuitively as moving portions of cities with each path update, and so they are also slow to move clusters of cities.

Table 1 summarizes results by Peterson (1990) and Lister (1993) on the same 200-city problem. The least effective algorithm was Peterson and Söderberg's (1989) neural network. The length of its best run is approximately equal to the average performance when simulated annealing moves single cities at a time; note that the right path in Figure 3 has approximately the same length. The elastic net (Durbin and Willshaw, 1987; see CONSTRAINED OPTIMIZATION AND THE ELASTIC NET) is a deterministic algorithm, so only one run is required. The elastic net is significantly better than annealing by moving single cities, but annealing with segment reversal performs better still.

Whether moving single cities, or reversing segments, a parallel implementation of annealing only requires $O(\log N)$ time per path update (Lister, 1994). That probably compares well with the time required by Peterson and Söderberg's network and the elastic net. However, such time comparisons are tangential to the thrust of this article: the annealing runs with segment reversal and moving single cities require exactly the same amount of time; thus they support the hypothesis that quasi-fractal energy landscapes give better solutions.

Table 1. Performance of Several Algorithms on Peterson's (1990) 200-City TSP

		Path Length	
Algorithm	No. Runs	Best	Average
1. Mean-field network	5	12.1	12.7
2. Moving single cities	100	11.2	12.1
3. Elastic net	1	11.1	11.1
4. Segment reversal	100	10.6	10.8

The statistics for algorithms 1 and 3 are from Peterson (1990).

Discussion

Hopfield and Tank (1985) conjectured that analog neural networks have a fundamental computational advantage over discrete-state networks, because analog networks can move through intermediate states not available to discrete networks. However, in this article, we have surveyed more recent annealing work concerning fractal and quasi-fractal landscapes, and these results suggest that analog computation alone is not sufficient; analog networks need to implement recursive changing mechanisms. Perhaps such recursive mechanisms will be found to emerge naturally in networks of coupled oscillators (see COLLECTIVE BEHAVIOR OF COUPLED OSCILLATORS).

Road Map: Dynamic Systems and Optimization
Background: I.3. Dynamics and Adaptation in Neural Networks

References

Baldi, P., and Baum, E., 1986, Caging and exhibiting ultrametric structures, in *Neural Networks for Computing* (J. Denker, Ed.), American Institute of Physics Conference Proceedings 151, New York: American Institute of Physics, pp. 35–40. ◆

Durbin, R., and Willshaw, D., 1987, An analogue approach to the travelling salesman problem using an elastic net method, *Nature*, 326:689–691.

Hopfield, J., and Tank, D., 1985, Neural computation of decisions in optimization problems, *Biol. Cybern.*, 52:141–152.

Kirkpatrick, S., and Toulouse, G., 1985, Configuration space analysis of travelling salesman problems, *J. Phys.*, 46:1277–1292. ◆

Lin, S., and Kernighan, B., 1973, An effective heuristic algorithm for the Traveling-Salesman Problem, *Oper. Res.*, 21:498–516.

Lister, R., 1993, Annealing networks and fractal landscapes, in *Proceedings of the IEEE International Conference on Neural Networks*, San Francisco, vol. 1, pp. 257–262. ◆

Lister, R., 1994, Multiprocessor rejectionless annealing, in *PCAT'93: Parallel Computing and Transputers* (D. Arnold et al., Eds.), Washington, DC: IOS Press, pp. 199–206. ◆

Peterson, C., 1990, Parallel distributed approaches to combinatorial optimization: Benchmark studies on traveling salesman problem, *Neural Computat.*, 2:261–269.

Peterson, C., and Söderberg, B., 1989, A new method for mapping optimization problems onto neural networks, *Int. J. Neural Sys.*, 1:3–22.

Solla, S., Sorkin, G., and White, S., 1986, Configuration space analysis for optimization problems, in *Disordered Systems and Biological Organization* (E. Bienenstock et al., Eds.), NATO ASI Series F20:283–293, Berlin: Springer-Verlag. ◆

Sorkin, G., 1990, Simulated annealing on fractals: Theoretical analysis and relevance for combinatorial optimization, in *Advanced Research in VLSI* (W. Dally, Ed.), Cambridge, MA: MIT Press, pp. 331–351.

Van den Bout, D. E., and Miller, T. K. III, 1989, Improving the performance of the Hopfield-Tank neural network through normalization and annealing, *Biol. Cybern.*, 62:129–139.

Wilson, G., and Pawley, G., 1988, On the stability of the travelling salesman problem algorithm of Hopfield and Tank, *Biol. Cybern.*, 58:63–70.

Frog Wiping Reflexes

Simon F. Giszter

Introduction

Frog wiping reflexes have been studied since the nineteenth century (Stein, Mortin, and Robertson, 1986). Their advantages are the documented adjustments, the simplicity of aspects of frog motor organization, and the robustness of the behaviors. The frog's wiping responses are largely organized in the spinal cord: they persist unaltered following spinal transection.

Experiments have been broadly organized around three conceptual themes: (1) understanding the biological implementation of sensorimotor transformations in motor planning, (2) understanding the modularity of vertebrate movement and its organization, and (3) understanding the control strategies stabilizing limb motions. Each of these broad themes is elaborated in the following sections.

The task of wiping consists of locating a stimulus on the body surface and executing an action which removes it. The animal subdivides the task as follows: (1) it closes the kinematic chain; this moves the effector limb to the body segment with the target stimulus, and (2) it executes a movement that removes the irritant. These subtasks require the frog to transform skin location, limb configuration, and body scheme to select an appropriate effector, to select a set of postures and limb trajectories for the effector, and then to generate a set of appropriate muscle activations to control and move the effector through these postures and trajectories. Each stage involves ill-posed problems, i.e., there are many possible solutions to the problem.

A Description of Wiping Kinematics

The frog body surface can be divided into skin regions within which different wiping strategies are used. The wipe strategies that have been best examined are those to the back, forelimb, hindlimb and cloaca. Some of the strategy skin regions overlap. In the region of overlap (termed a transition zone by Stein; cf. SCRATCH REFLEX), both strategies are competent to remove the irritant. This *motor equivalence* was first described by Stein and colleagues in the turtle (1986) and replicated in the frog (Berkinblit, Gelfand, and Feldman, 1986b). When multiple stimuli occur in different skin zones, switching between or blending of strategies can occur. In transition zones, the strategies chosen on a trial depend both on history and context of stimulation. All of these phenomena can be duplicated in intact or spinal frogs.

Most of the postures and transitions observed in wipes to the back are summarized in Figure 1, with switching or blending examples. Some phases are optional and need not be executed each cycle. Thus, in wiping to the back, extension may be omitted, flexion may be omitted, and whisk and flexion may be blended in both intact or spinal frogs.

It is clear that body scheme information is used to control wiping (Fookson, Berkinblit, and Feldman, 1980). This frequently involves active posturing of target limbs (Giszter, McIntyre, and Bizzi, 1989; Sergio and Ostry, 1993).

Solution of Ill-Posed Problems

Berkinblit et al. (1986a, 1986b) first pointed out that, in wiping to the back, the frog adjusts limb position on the basis of stimulus position, and the selection of limb configuration is ill-posed. Individual frogs exhibit fixed solutions to the problem

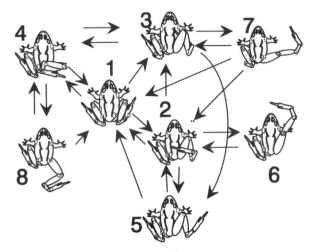

Figure 1. Examples of wiping postures (1–8) and the possible transitions among them. *1*, Flexion. *2*, Back placing (first strategy). *3*, Shoulder placing. *4*, Cloacal placing (second strategy). *5*, Whisk/wipe. *6*, Forward extension. *7*, Lateral extension. *8*, Caudal extension. A transition zone exists between the cloacal wipe (strategy 2) and the back wipe (strategy 1) which overlap in the areas of skin covered by the movements. A typical wipe might follow a sequence of 1, 2, 5, 1. Other rapid abbreviated sequences, such as 1, 2, 1, also occur. For two stimuli on the back and shoulder, a blend sequence (1, 2, 3, 7, 1) could occur. For stimuli on the back and cloacal area, a switching sequence (1, 2, 5, 1, 4, 8, 1) would be typical. Adjustments shown in Figure 2 are for the back-placing posture (2).

of choosing joint angles to position the limb in the placing phase (Giszter et al., 1989); see Figure 2. While each frog's strategy is fixed and linear, the parameterization of the strategy used can vary from frog to frog (see Sergio and Ostry, 1993).

Work on the hindlimb-to-hindlimb wipe by Giszter et al. (1989) has shown that the spinal frog is incapable of solving the problem of hindlimb-to-hindlimb wiping for the whole work space, but only wipes successfully in a restricted zone. This restricted zone allows a simple fixed strategy that is roughly equivalent to joint angle matching between the target and effector limbs. Hindlimb-to-hindlimb wiping throughout the work space would necessarily involve a nonlinear transformation of configuration information between the target and effector limbs. Although intact frogs can do this, this is apparently beyond the abilities of the spinal frog. However, frogs in which half of the brainstem underlying the fourth ventricle is spared exhibit wipes that are very flexible, extending into areas where spinal frogs fail or abort their wipes. Regarding actuation of the limb and the ill-posed problems posed by the musculature, little information is available. Schotland and Rymer (1993) found both stereotypy and adjustment of muscle activation in wiping.

Control of the Limb

A study of kinematic control by Ostry, Feldman, and Flanagan (1991) supports joint-based planning of limb motion. Force-field approaches have also been used. These associate a force vector generated under isometric conditions with a position in the frog limb's work space. Forces were examined both prior

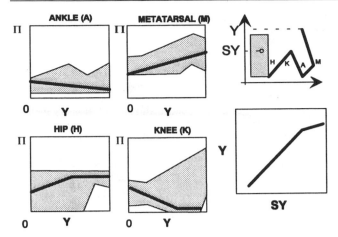

Figure 2. Example of an ill-posed kinematic problem solved by the spinal cord during wiping. To wipe to the back, the frog takes the base or body-centered coordinate system rostrocaudal to coordinate location SY, then chooses a solution for angles H (hip), K (knee), A (ankle), and M (metatarsal angle) to set the limb tip position (Y) close to the stimulus (SY). This problem is ill-posed. Legitimate combinations of angles providing a solution may be chosen from a volume which depends on joint limits and link parameters. Monte Carlo models show that permissible angles to set tip to Y lie in the shaded regions. The solution chosen by the frog spinal cord is a simple straight line parameterized by SY (bold lines). This solution only deviates from an ideal match in extreme locations owing to joint limits. (Redrawn from Giszter, McIntyre, and Bizzi, 1989.)

to and during an attempted spinal behavior in frogs, or as a result of some stimulation in the central nervous system (CNS) (Giszter, Mussa-Ivaldi, and Bizzi, 1993). By obtaining a set of such records across the work space, a force-field description may be constructed (see Bizzi, Mussa-Ivaldi, and Giszter, 1991). In frogs, the force fields examined in this way have several interesting properties which suggest they are primitive control entities. First, the field at any time can be viewed as a scaled version of the field at other times. Second, the fields elicited by stimulation of the skin and stimulation in the grey matter of the spinal cord correspond very closely. Third, the total force field of the limb in the isometric case usually converges on a posture. This posture would be the final posture of the limb were it free to move in wiping. That this is indeed the case has been confirmed experimentally. The force field represents an attractor in configuration space which is apparently simply turned on and scaled up in strength.

Evoked force fields fall into two distinct categories. The first category consists of convergent force fields, which probably represent stable postures. The second category consists of force fields that are parallel, either in joint space or hip-centered polar coordinates. These latter fields probably represent force control. One manner in which force fields might be combined in a limb is by linear vector summation. In principle, this would be expected to be difficult to achieve, given the limb and muscle nonlinearities. However, a body of evidence supports this mechanism for field combinations in the frog spinal cord (Bizzi et al., 1991).

Modularity of Wiping

Both kinematic work (Ostry et al., 1991; Berkinblit et al., 1986a) and force-field descriptions (Giszter et al., 1993) support the notion of a modular organization of the wiping behaviors. Berkinblit et al. (1986a) proposed that flexion could form

an initial and intermediate phase of wiping movements and might be utilized in several different behaviors. The decomposition of movement yielded serially activated behavioral elements: postures and movement fragments were combined sequentially through time. In order to have separate dynamics in different phases, the limb must come to rest between phases. Such pauses were observed, either in normal motion or after cooling the animal. A fixed timing of phases was also noted.

The force-field descriptions discussed earlier also support a movement primitive, i.e., a self-contained motor element, or module. Vector summation mechanisms and winner-take-all mechanisms observed in experiments provide the adjustment, blending, and selection mechanisms needed for modular control of wiping.

Path Planning and Sensorimotor Transformations

Planning was addressed by Berkinblit and co-workers (1986b), and Shadmehr (1991). The goal of their models has been to examine wiping and computational issues involved in general positioning. In this regard, they succeed well, although it is likely that the wiping mechanisms are not general-purpose positioning mechanisms. The work is summarized by Shadmehr (1991), who has also provided a model. Both the goal and the general value of these models have been to specify algorithms operating locally at the joint or muscle level which can solve ill-posed problems of kinematic planning.

The existing models assume that a body-centered coordinate representation is available to the frog; they then proceed to the issue of obtaining a pattern of joint motions of a serially redundant linkage in order to lead the endpoint trajectory to a target. They are thus suitable for wiping to the back (whereas a body-centered coordinate representation is presumably directly available from skin location). Current models do not address how the initial body-centered stimulus representation is derived for stimuli on limbs in various configurations. Both the Berkinblit and Shadmehr models use some minimization or endpoint compliance criterion to provide a particular choice of solution for the path planning problem.

The Berkinblit model (1986b) is a gradient descent based on a simple independent rule for the motion of each joint (see Figure 2). Shadmehr points out that this can be compactly expressed as

$$\Delta\theta = B \cdot J_s^T \cdot e$$

where B is a constant diagonal matrix, J_s is the forward Jacobian of the linkage, and e is the endpoint error vector. This represents a special (nonintegrable) case of the Mussa-Ivaldi backdriving algorithm (Mussa-Ivaldi et al., 1991). In the Berkinblit algorithm, the joint stiffness matrix is constant and diagonal. Thus, the Berkinblit algorithm is a particular pseudoinverse. One goal of the algorithm proposed by Berkinblit and colleagues was to replicate the (within-strategy) motor equivalence which they had suggested. The lack of integrability was thus viewed as a feature of the model, not a drawback, since it provided motor equivalence: different trajectories to the same endpoint could arrive there using very different configurations. The model also has the attractive feature of finding its way to the target location in the event of a joint constraint.

In order to utilize the backdriving algorithm, Shadmehr (1991) chose a constant joint stiffness matrix that included double joint muscles, and thus had off-diagonal terms. He chose straight-line endpoint paths to reach the target stimulus, although this did not affect the outcomes he observed. A variant of the algorithm was used to drive the endpoint motion to sets of target locations arranged rostrocaudally along the model "back." He observed two types of folding of the limb as the target was moved along the back, matching the response types seen in the frog. He was also able to obtain variations of these two folding strategies in the transition zone (i.e., different folding of the limb or configuration, as discussed earlier) by beginning model trajectories from a single endpoint position with several different null space motions of the limb's configuration. The model leads to the limb utilizing different solution manifolds for each different initial configuration variation in the null space. In effect, the boundary in the set of target positions at which the model switches between the two folding strategies is moved back and forth in the transition zone by the null space motion of the initial position configuration. However, while the limb is operating on each solution manifold, the chosen behavior remains fully integrable.

The Role and Organization of Force-Field Modules

The notion of force-field modules simplifies motor control provided they can be combined in a simple way in different arrangements. This implies that there cannot be very much nonlinear interaction among modules. Vector summation of force-field primitives is thus central. Vector summation seems to be sufficiently common in frog stimulation work (80% of cases) that its inclusion in various models is well warranted. Modularity notions can be divided into (1) decomposition into basis fields, and (2) decompositions into behavioral elements. Data obtained from frog studies support both types of decomposition.

Basis fields form general-purpose elements from which all postures and movements can be constructed. In contrast, behavioral force fields are conceived of as special-purpose elements designed for participation in a few types of motor tasks. This distinction is useful for examining how the spinal cord might be used by other structures. Nonetheless, it remains possible that the fields obtained so far fulfill both roles in different contexts, i.e., perhaps the most common behaviors may represent a set of basis fields for arbitrary or general-purpose positioning.

Basis-Field Models

Mussa-Ivaldi (1992) showed that arbitrary fields were readily approximated using a linear combination of conservative and circulating (or rotational) radial vector field primitives. This concept has now been applied in the frog, as well as in human biomechanics (see EQUILIBRIUM POINT HYPOTHESIS and GEOMETRICAL PRINCIPLES IN MOTOR CONTROL). The coefficients of these basis-field elements can be found by least squares or minimum norm methods.

Mussa-Ivaldi and Giszter (1992) applied the Mussa-Ivaldi vector field primitive work to the motor system in the nonredundant limb. Since this framework allows approximation of arbitrary smooth vector fields, it follows that the planning and choice of the control fields to be approximated must be constructed in some other manner.

Planning and calculating how to combine primitives in order to approximate a desired field structure is a task which the CNS must perform for each behavior and perhaps each individual instance. In principle, arbitrary field structures and field time courses can be generated by the techniques of basis-field approximation (Mussa-Ivaldi, 1992). The clear advantage of this framework is that several of the processes necessary for actual movement execution can be supported by basis-field approximation. However, this flexibility implies that planning

the details of these processes must be deferred to other mechanisms.

Behavioral Force-Field Models

The modeling of a wiping system using behavioral primitives has been explored by Giszter (1994). Planning, decision making, and control of execution are all resident in and generated by a single network. The models did not deal explicitly with redundancy issues. A serially redundant limb was either viewed as a single behavioral entity or as two connected and individually nonredundant behavioral entities. A set of so-called behavioral force-field primitives was used to drive the virtual trajectory of a limb in order to simulate wiping to various regions of body surface.

The model structure chosen consisted of three parts. The first represented areas of skin and associated sensory systems. These areas had associated overlapping sensory fields which transformed skin stimulation into the activations of a network layer representing the conditions of stimulation. The units in this network layer acted as conditions or propositions for input to an augmented Maes network which formed the second part of the simulation. In its original form, the Maes network is similar to a spreading activation network. The basic behavior network comprises a collection of nodes. These represent behaviors, or actions. Biologically, these nodes may be thought of as (1) pattern generators for movement fragments or (2) primitive actions. The behavior nodes pass activation back and forth among themselves. This activation is passed through links. Flow through links is controlled by a layer of units representing conditions derived from the environment and by the states of the behavior nodes themselves. For a behavior node to execute, all of the (pre)condition units necessary for its execution must be on. The condition units thus provide the direct activation of behavior nodes, control the spread of activity through the links among behavior nodes, and gate execution of nodes. Behavior nodes have three types of outgoing links (see Giszter, 1994, for details and references). *Successor links* activate behavior nodes with a precondition which will be turned on by the action of the originating behavior node. *Predecessor links* activate other behavior nodes which will turn on a precondition of the originating behavior node if they execute. *Conflictor links* inhibit behavior nodes which will turn off a precondition of the originating behavior node if they execute. Thus, there are two different types of processes between nodes: a cooperation and positive feedback between nodes likely to activate in succession and a competition among nodes likely to conflict with one another in their actions and requirements. The full model operation is diagrammed in Figure 3. Activation at any given time flows in the network according to a first-order differential equation, with parameters governed by the condition units. As conditions become true or false, the parameters of this differential equation change according to the rules of operation of the Maes network. Behaviors execute when above threshold and when all necessary (pre)conditions are on. The network selects sequences of actions based on the global situation, which acts on the network by modulating local interactions. Thus, adaptive behavior is an emergent property of the interaction of network and environment. The overall flow of activation propagated by the links leads to increasing activation (i.e., preparation) of those behavior nodes that are currently executable, as well as increasing activation of those nodes that will most likely shortly become executable as a result of environmental variations or current actions.

The final portion of the simulation consisted of the limb model and a set of behavioral primitives. This provided vector

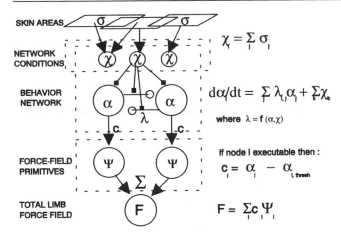

$$\chi_i = \sum_j \sigma_j$$

$$d\alpha/dt = \sum_i \lambda_{i,j}\alpha_j + \sum_k \chi_k$$

where $\lambda = f(\alpha, \chi)$

If node I executable then :

$$c_i = \alpha_i - \alpha_{i,\,thresh}$$

$$F = \sum_i c_i \Psi_i$$

Figure 3. A combined Maes and force-field model applied to frog wiping behavior. Skin stimuli set the condition units state (network conditions). These units both activate and configure flow in the Maes network layer (behavior network). Execution of nodes in the Maes network layer occurs for nodes with values above threshold and with all preconditions true. The execution recruits and controls the strength of force-field primitives. These primitives are combined by approximate vector summation to generate the total limb field (F) that determines limb dynamics and wiping motions.

summation of executing behavioral force fields and limb dynamics. Force-field strength was determined by the level of activation of the associated executing behavior node.

The choice of sequences and combinations of force fields and the strength of activation of the force-field sequences (and hence, resulting limb motions) were an emergent property determined by the network state and the sensory context. The model was successful in replicating kinematic and force-field blending, as well as switching and transition zone phenomena, and successfully dealt with multiple stimuli by choosing switches and blends. "Random" exploration of allowed force-field combinations was exhibited by the network in response to nearly unreachable stimuli. This usually led to eventual success.

Discussion

Frog spinal cord and reflex behaviors form a relatively simple system in which theoretical, biomechanical, and neurophysiological issues can be explored and interact. They also form an ideal framework for examining the interactions of spinal decision making and trajectory control, with control and planning generated by descending systems. The experimental studies in this area have resulted in several unexpected findings: e.g., the frog spinal cord contains a limited set of body scheme information. The sensory motor transformations examined closely have involved use of strategies allowing simple linearization or matching, rather than general solutions using all portions of the solution space. Static force-field descriptions seem adequate to capture the qualitative behavior of the frog limb, although perhaps not all quantitative aspects of it. Finally, vector summation appears to be a valid mechanism for combining at least a subset of spinal responses. The picture of frog wiping reflexes that emerges is one of a restricted and special-purpose collection of modular mechanisms. These are competent to generate

apparently purposive and relatively complex behaviors. It remains to be discovered whether these mechanisms may form a "bootstrap," or even perhaps the substrate, for generating more general position control of the limb.

Acknowledgments. This work was supported by NIH NS09343, AR26710, ONRN00014/90/J/1946, and the Medical College of Pennsylvannia Department of Anatomy.

Road Map: Motor Pattern Generators and Neuroethology
Related Reading: Limb Geometry: Neural Control; Motor Pattern Generation

References

Berkinblit, M. B., Feldman, A. G., and Fookson, O. I., 1986a, Adaptability of innate motor patterns and motor control mechanisms, *Behav. Brain Sci.*, 9:585–638. ◆

Berkinblit, M. B., Gelfand, I. M., and Feldman, A. G., 1986b, A model of the aiming phase of the wiping reflex, in *Neurobiology of Vertebrate Locomotion* (S. Grillner, P. S. G. Stein, D. G. Stuart, H. Forssberg, and R. M. Herman, Eds.), London: Macmillan, pp. 217–227.

Bizzi, E., Mussa-Ivaldi, F. A., and Giszter, S. F., 1991, Computations underlying the execution of movement: A biological perspective, *Science*, 253:287–291. ◆

Fookson, O. I., Berkinblit, M. B., and Feldman, A. G., 1980, The spinal frog takes into account the scheme of its body during the wiping reflex, *Science*, 209:1261–1263. ◆

Giszter, S. F., 1994, Reinforcement based tuning of action synthesis and selection in a virtual frog, in *From Animals to Animats 3: Proceedings of the Third International Conference on the Simulation of Adaptive Behavior* (D. Cliff, P. Husbands, J.-A. Meyer, and S. W. Wilson, Eds.), Cambridge, MA: MIT Press, pp. 291–301.

Giszter, S. F., McIntyre, J., and Bizzi, E., 1989, Kinematic strategies and sensorimotor transformations in the wiping movements of frogs, *J. Neurophysiol.*, 62:750–767.

Giszter, S. F., Mussa-Ivaldi, F. A., and Bizzi, E., 1993, Convergent force fields organized in the frog spinal cord, *J. Neurosci.*, 13:467–491.

Mussa-Ivaldi, F. A., 1992, From basis functions to basis fields: Using vector primitives to capture vector patterns, *Biol. Cybern.*, 67:479–489.

Mussa-Ivaldi, F. A., Bizzi, E., Morasso, P., and Hogan, N., 1991, Network models of motor systems with many degrees of freedom, in *Advances in Control Networks and Large-Scale Parallel Distributed Processing Models* (M. D. Fraser, Ed.), Norwood, NJ: Ablex.

Mussa-Ivaldi, F. A., and Giszter, S. F., 1992, Vector field approximation. A computational paradigm for motor control and learning, *Biol. Cybern.*, 67:491–500.

Ostry, D. J., Feldman, A. G., and Flanagan R. F., 1991, Kinematics and control of frog hindlimb movements, *J. Neurophysiol.*, 65:547–562.

Schotland, J. L., and Rymer, W. Z., 1993, Wiping and flexion reflexes in the frog, II: Response to perturbations, *J. Neurophysiol.*, 69:1736–1748.

Sergio, L., and Ostry, D. J., 1993, Three-dimensional kinematic analysis of frog hindlimb movement in reflex wiping, *Exp. Brain Res.*, 94:53–70.

Shadmehr, R., 1991, *A Computational Theory for Control of Posture and Movement in a Multi-joint Limb*, PhD Dissertation, Technical Report 91–07, Center for Neural Engineering, University of Southern California, Los Angeles.

Stein, P. S. G., Mortin, L. I., and Robertson, G. A., 1986, The forms of a task and their blends, in *Neurobiology of Vertebrate Locomotion* (S. Grillner, P. S. G. Stein, D. G. Stuart, H. Forssberg, and R. M. Herman, Eds.), London: Macmillan, pp. 201–216. ◆

Fuzzy Logic Systems and Qualitative Knowledge

Jerry M. Mendel

Introduction

For many problems two distinct forms of problem knowledge exist: (1) *quantitative knowledge*, which is used all the time in engineering problem formulations (e.g., mathematical models), and (2) *qualitative knowledge*, which represents linguistic information that is usually impossible to quantify using traditional mathematics (e.g., rules, expert information, design requirements). The latter knowledge is usually ignored at the front end of engineering designs, but it is frequently used to evaluate such designs. Both types of knowledge should and can be utilized to solve real problems. They can be coordinated using *fuzzy logic* (FL).

In this chapter we focus on a *model-free approach* in which rules are extracted from numerical data and are then combined with linguistic information (collected from experts). Both steps use FL. The model-free approach can be applied to the same class of problems as feedforward neural networks (FFNN); hence, we can compare the FL and FFNN approaches. For an expanded version of this article, see Mendel (1994).

A Short Primer on Fuzzy Logic

A *fuzzy set* F (Zadeh, 1965) in a universe of discourse U is characterized by a grade of *membership function* $\mu_F(x)$ which takes on values in the interval [0, 1]. We assume that $U \subseteq R$, the set of real numbers. A fuzzy set is a generalization of a *crisp set* (i.e., a set in the usual sense) whose membership function takes on only two values, zero or unity. Consequently, a fuzzy set F in U may be represented as a set of ordered pairs of a generic element x and its grade of membership function, i.e., $F = \{(x, \mu_F(x)) | x \in U\}$. A membership function provides a *measure of the degree of similarity* (degree of membership) of an element in U to the fuzzy set. For example, suppose U is the set of all automobiles in Los Angeles, and we want to classify specific cars as "foreign" or "domestic." Using crisp-set theory, we could do this simply by examining the brand name (e.g., a Toyota is "foreign," whereas a Ford is "domestic"). A different approach is to determine the percentage of parts for a specific car made in the United States and then to assign a degree of similarity to what we perceive to be a domestic or foreign car. For example, if 25% of our car's parts are made in the USA, then our car is similar to a domestic car to degree 0.25, and it is similar to a foreign car to degree 0.75. In this case, we would say our car is "foreign." The point of this example is to demonstrate that in fuzzy logic an element can reside, with different degrees of similarity, in sets that crisp-set theory would insist were nonoverlapping.

Linguistic variables (Zadeh, 1973), whose values are not numbers but words in sentences or artificial language, are heavily used in FL. The name of the variable is denoted u; numerical values of a linguistic variable are denoted x. In engineering applications of fuzzy logic, membership functions, are associated with terms that appear in the antecedents or consequents of rules (e.g., Wang, 1994). The most commonly used shapes for membership functions are triangular, trapezoidal, piecewise linear, and Gaussian. Until very recently, membership functions were chosen by the user arbitrarily, on the basis of the user's experience. More recently, membership functions have been designed using optimization procedures, including neural network learning methods. Greater resolution is achieved by using more membership functions, at the price of greater computational complexity. Membership functions are usually scaled between zero and unity, and they overlap (as in the case of foreign cars and domestic cars). This overlapping is one of the great strengths of FL because it allows an input to be distributed across a number of rules.

The fundamental operations in crisp-set theory are union, intersection, and complement. Let A denote a set and $\overline{A}$ denote its complement. The two fundamental (Aristotelian) laws of crisp-set theory are: (1) *Law of Excluded Middle*: $A \cup \overline{A} = U$ (i.e., a set and its complement must comprise the universe of discourse), and (2) *Law of Contradiction*: $A \cap \overline{A} = \phi$ (i.e., an element can be either in a set or in its complement; it cannot simultaneously be in both). *These two laws are broken in FL.* We have already seen this fact in the automobile example. In FL, union, intersection, and complement are defined in terms of their membership functions, either by extending or generalizing comparable membership functions from crisp-set theory (Zadeh, 1965; Zimmermann, 1991). Let fuzzy sets A and B be described by their membership functions $\mu_A(x)$ and $\mu_B(x)$. One definition of fuzzy union leads to the membership function $\mu_{A \cup B}(x) = \max[\mu_A(x), \mu_B(x)]$, and one definition of fuzzy intersection leads to the membership function $\mu_{A \cap B}(x) = \min[\mu_A(x), \mu_B(x)]$. Additionally, fuzzy complement is given by $\mu_{\overline{B}}(x) = 1 - \mu_B(x)$. Other definitions are possible.

Other important notions from fuzzy-set theory are: fuzzy relations (e.g., "u is considerably larger than v"), composition of fuzzy relations on the same product space (e.g., "u is considerably larger than v" or "v is very close to u"), composition of fuzzy relations on different product spaces (e.g., "u is smaller than v, and v is close to w"), and the special case of the latter when one relation is just a fuzzy set (e.g., "v is medium large and w is smaller than v").

Compositions are handled mathematically by the *sup-star composition* (e.g., Zimmermann, 1991; Lee, 1990; Wang, 1994). Let R and S be fuzzy relations in $U \times V$ and $V \times W$, respectively. $R \circ S$ denotes the fuzzy composition of R and S; it is a fuzzy set in $U \times W$, and its membership function is

$$\mu_{R \circ S}(x, z) = \sup_{y \in V}[\mu_R(x, y) \star \mu_S(y, z)] \tag{1}$$

where $x \in U$, $y \in V$, $z \in W$, and $\star$ is usually chosen, in engineering applications, to be either the minimum or the product. When R is just a fuzzy set in U, so that $\mu_R(x, y)$ just becomes $\mu_R(x)$, then $V = U$ and Equation 1 becomes

$$\mu_{R \circ S}(z) = \sup_{x \in U}[\mu_R(x) \star \mu_S(x, z)] \tag{2}$$

In crisp logic there are two very important inference rules, *modus ponens* and *modus tollens*. To date, only a generalized modus ponens rule has been used in controls and signal processing applications of FL. The premises and consequences of modus ponens rules can be stated as follows:

- *Modus ponens rule*

  Premise 1: *u* is *A*.
  Premise 2: IF *u* is *A*, THEN *v* is *B*.
  Consequence: *v* is *B*.

  Modus ponens is associated with the implication "*A* implies *B*" $[A \rightarrow B]$.
  The extension of modus ponens to FL treats *A* and *B* as fuzzy sets and *u* and *v* as linguistic variables.

- *Generalized modus ponens rule*
  Let A^*, A, B^*, and B be fuzzy sets, and u and v be linguistic variables; then the following premises and consequence obtain:

  Premise 1: u is A^*.
  Premise 2: IF u is A, THEN v is B.
  Consequence: v is B^*.

The generalization from crisp-set theory is two-fold: we use "similar" sets A and A^* (B and B^*) rather than repeated use of the single set A (B); and the consequence comes with a grade of membership given by

$$\mu_{B^*}(y) = \sup_{x \in A^*} [\mu_{A^*}(x) \star \mu_{A \to B}(x, y)] \tag{3}$$

Example: IF a man is *short* (A), he will *not make* a *very good professional basketball player* (B). This guy is *under five feet tall* (A^*); he will (to some specific degree of membership) make a *poor professional basketball player* (B^*).

In controls and signal processing applications of FL, a *fuzzy implication* $A \to B$ is a fuzzy IF-THEN rule. A fuzzy implication, known as a fuzzy conjunction, $A \to B$, is a special type of fuzzy relation in $U \times V$ with the following membership function:

$$\mu_{A \to B}(x, y) = \mu_A(x) \star \mu_B(y) \tag{4}$$

where $\star$ is the "star" composition. The two most widely used engineering choices for the star composition are: (1) the *min-operation rule of fuzzy implication* (Lee, 1990): $\mu_{A \to B}(x, y) = \min\{\mu_A(x), \mu_B(y)\}$; and (2) the *product-operation rule of fuzzy implication* (Lee, 1990): $\mu_{A \to B}(x, y) = \mu_A(x)\mu_B(y)$. These implications do *not* coincide with the commonly used implication from propositional logic, but do preserve *cause and effect*, whereas the latter does not; this is very important in engineering applications of FL.

A Fuzzy Logic System

Figure 1 depicts a fuzzy logic system (FLS) that is widely used in control and signal processing applications (e.g., Wang, 1994). An FLS maps crisp inputs into crisp outputs. It contains four components: rules, fuzzifier, inference engine, and defuzzifier. Once the rules have been established, an FLS can be viewed as a mapping from inputs to outputs (the solid path in Figure 1, from "crisp inputs" to "crisp outputs"), and this mapping can be expressed quantitatively.

Rules may be provided by experts or can be extracted from numerical data. In either case, engineering rules are expressed as a collection of IF-THEN statements, e.g., "IF u_1 is very warm and u_2 is quite low, THEN turn v somewhat to the right." A *fuzzy rule base* consists of a collection of M IF-THEN rules:

$$R^{(l)}: \text{IF } u_1 \text{ is } F_1^l \text{ and } u_2 \text{ is } F_2^l \text{ and } \cdots \text{ and } u_p \text{ is } F_p^l, \text{ THEN } v \text{ is } B^l \tag{5}$$

where $l = 1, 2, \ldots, M$; F_i^l and B^l are fuzzy sets in $U_i \subset R$ and $V \subset R$, respectively; and $\mathbf{u} = (u_1, \ldots, u_p)^T \in U_1 \times \cdots \times U_p$ and $v \in V$ are linguistic variables. Note that a multi-input, multi-output rule can always be considered as a group of multi-input, single-output rules (Lee, 1990:426), which is why everyone concentrates on the latter rules. Note, also, that it is possible to cast "nonobvious" rules into the form of Equation 5 (Wang, 1994). Rules are connected by the term "also." Their membership functions can be combined in different ways, e.g., (1) maximum or (2) weighted addition (Kosko, 1992). Sometimes they are combined by the defuzzifier (see below).

The *inference engine* of the FLS maps fuzzy sets into fuzzy sets via $\mu_{A \to B}(\mathbf{x}, y)$. It handles the way in which rules are combined. Just as we humans use many different types of inferential procedures to help us understand things or to make decisions, there are many different fuzzy logic inferential procedures. Only a very small number of them are actually used in signal processing or control applications. FL principles are used to combine fuzzy IF-THEN rules from the fuzzy rule base into a mapping from fuzzy input sets in $U = U_1 \times \cdots \times U_p$ to fuzzy output sets in V. $R^{(l)}$ is interpreted as a fuzzy implication $F_1^l \times F_2^l \times \cdots \times F_p^l \to B^l$ in $U \times V$. In this case,

$$\mu_{B^*}(y) = \sup_{\mathbf{x} \in A^*} [\mu_{A^*}(\mathbf{x}) \star \mu_{A \to B}(\mathbf{x}, y)] \tag{6}$$

Let A^* be an arbitrary fuzzy set in U; then each rule $R^{(l)}$ determines a fuzzy set $A_x \circ R^{(l)}$ in R, such that

$$\mu_{B^l}(y) = \mu_{A_x \circ R^{(l)}} = \sup_{\mathbf{x} \in U} [\mu_{A^*}(\mathbf{x}) \star \mu_{F_1^l \times F_2^l \times \cdots \times F_p^l \to B^l}(\mathbf{x}, y)] \tag{7}$$

This equation is the most important one so far in our development of an FLS.

The *fuzzifier* maps a crisp point $\mathbf{x} = (x_1, \ldots, x_p)^T \in U$ into a fuzzy set A^* in U. This is needed in order to activate rules that are defined in terms of linguistic variables which have fuzzy sets associated with them. The singleton fuzzifier is most widely used. In a *singleton fuzzifier*, A^* is a fuzzy singleton with support $\mathbf{x}$ if $\mu_A(\mathbf{x}^*) = 1$ for $\mathbf{x} = \mathbf{x}^*$ and $\mu_{A^*}(\mathbf{x}^*) = 0$ for all other $\mathbf{x} \in U$ with $\mathbf{x} \neq \mathbf{x}^*$. In a *nonsingleton fuzzifier*, $\mu_{A^*}(\mathbf{x}^*) = 1$ and $\mu_{A^*}(\mathbf{x})$ decreases from unity as $\mathbf{x}$ moves away from $\mathbf{x}^*$. The singleton fuzzifier does not let the user account for uncertainty about $\mathbf{x} = \mathbf{x}^*$, whereas the nonsingleton fuzzifier does. The main reason for using the singleton fuzzifier is that the supremum operation in Equation 7 disappears, because fuzzy input set A^* contains only a single element $\mathbf{x}^*$. In that case,

$$\mu_{B^l}(y) = \mu_{A_x \circ R^{(l)}} = \mu_{A^*}(\mathbf{x}^*) \star \mu_{F_1^l \times F_2^l \times \cdots \times F_p^l \to B^l}(\mathbf{x}^*, y) \tag{8}$$

Fuzzy Logic System

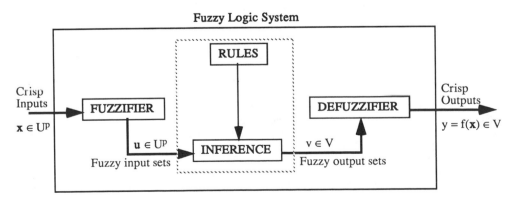

Figure 1. A fuzzy logic system (FLS).

The *defuzzifier* maps output sets into crisp numbers. In many applications, crisp numbers must be obtained at the output of an FLS. In a controls application, such a number corresponds to a control action. In a signal processing application, such a number could correspond to the prediction of next year's sunspot activity. Let $\bar{y}^l$ be the center of gravity of the fuzzy set B^l. The *height defuzzifier*, which is computationally simple, first evaluates $\mu_{B^l}(y)$ at $y = \bar{y}^l$ and then computes the output of the FLS as

$$y = \sum_{l=1}^{M} \bar{y}^l \mu_{B^l}(\bar{y}^l) \bigg/ \sum_{l=1}^{M} \mu_{B^l}(\bar{y}^l) \tag{9}$$

It is very easy to use Equation 9 because the centers of gravity of commonly used membership functions are known ahead of time. For discussions of other defuzzifiers, see Lee (1990) and Wang (1994).

We are now able to interpret the FLS in Figure 1 as a *nonlinear system*, one that maps a vector of inputs $\mathbf{x}$ into an output $y = f(\mathbf{x})$. Suppose, for example, we use singleton fuzzification, product inference, and height defuzzification (as described by Equation 9). Then (see Wang, 1994; Wang and Mendel, 1992b) we have

$$f(\mathbf{x}) = \sum_{l=1}^{M} \bar{y}^l \left[\prod_{i=1}^{p} \mu_{F_i^l}(x_i) \right] \bigg/ \sum_{l=1}^{M} \left[\prod_{i=1}^{p} \mu_{F_i^l}(x_i) \right] \tag{10}$$

In this formula $\bar{y}^l$ is the point at which μ_{B^l} achieves its maximum value, and it is assumed that $\mu_{B^l}(\bar{y}^l) = 1$. If, in addition, the membership functions are all Gaussian, i.e.,

$$\mu_{F_i^l}(x_i) = \exp[-(x_i - \bar{x}_i^l)/(\sigma_i^l)^2] \tag{11}$$

where $\bar{x}_i^l$ and σ_i^l are design parameters, then

$$f(\mathbf{x}) = \sum_{l=1}^{M} \bar{y}^l \left[\prod_{i=1}^{p} \exp[-(x_i - \bar{x}_i^l)/(\sigma_i^l)^2] \right]$$
$$\bigg/ \sum_{l=1}^{M} \left[\prod_{i=1}^{p} \exp[-(x_i - \bar{x}_i^l)/(\sigma_i^l)^2] \right] \tag{12}$$

Equations 10 and 12 can also be expressed as

$$f(\mathbf{x}) = \sum_{l=1}^{M} \bar{y}^l \phi_l(\mathbf{x}) \tag{13}$$

where $\phi_l(\mathbf{x})$ are called *fuzzy basis functions* or FBFs (Wang and Mendel, 1992b), which are given by

$$\phi_l(\mathbf{x}) = \prod_{i=1}^{p} \mu_{F_i^l}(x_i) \bigg/ \sum_{l=1}^{M} \left[\prod_{i=1}^{p} \mu_{F_i^l}(x_i) \right] \tag{14}$$

or, when membership functions are Gaussian, by

$$\phi_l(\mathbf{x}) = \prod_{i=1}^{p} \exp[-(x_i - \bar{x}_i^l)/(\sigma_i^l)^2]$$
$$\bigg/ \sum_{l=1}^{M} \left[\prod_{i=1}^{p} \exp[-(x_i - \bar{x}_i^l)/(\sigma_i^l)^2] \right] \tag{15}$$

Each rule provided by an expert, or each rule extracted from data, can lead to an FBF in Equation 13, and each FBF is affected by all of the M rules because of its denominator; hence, there is a sharing of rule information across all of the FBFs.

Equation 15 is a direct result of the fuzzifier–inference–defuzzifier path in Figure 1. Although crisp inference can in general involve *iterative chains of reasoning*, fuzzy inferences, especially as used in the Figure 1 FLS, have not yet been extended to iterative chains of reasoning.

Finally, FLSs are capable of approximating a wide range of nonlinear functions, just as FFNNs are. They are able to incor-

porate both linguistic and numerical data, whereas an FFNN is able to incorporate only numerical data. Wang and Mendel (1992b) have proven that FLSs of the form given in Equation 12 are capable of uniformly approximating any real continuous function to any degree of accuracy. Kosko (1994) has also done this, but for an additive FLS. These results justify applying FLSs to many nonlinear problems. They also help explain why FLSs are so successful in engineering applications. This "existence" result is important; but it does not tell us how to specify an FLS. Design methods tell us how to do this (see Wang, 1994).

Fuzziness and Probability

A lot has been written about fuzzy logic and its relation to probability (see Laviolette and Seaman, 1994, and the references therein for a very comprehensive discussion and debate on probability versus fuzziness). Many fuzzy-logic theorists maintain that FL is quite different from probability, for a wide variety of reasons, including the facts that: (a) the laws of excluded middle and contradiction are broken in FL, but are not broken in probability; and (b) conditional probability must be defined in probability theory, but can be derived from first principles using FL (Kosko, 1992). Others maintain that FL subsumes probability. Subjective probabilists, on the other hand, maintain that anything one can do with FL can also be done with probability, and that the latter is to be preferred because it has an axiomatic basis, whereas FL does not. They bemoan the fact that engineers, who are the largest users of FLSs, are not adequately trained in subjective probability.

There is some truth to both sides of the argument. While it is of great intellectual interest to establish the proper connections between FL and probability, this author does not believe that doing so will change the ways in which we solve problems, because both probability and FL should be in the arsenal of tools used by engineers. FL will not solve all problems, nor will probability.

Probability builds on crisp-set theory, but so does fuzzy logic. Fuzzy logic measures the degree of similarity of an object to imprecisely defined properties; probability measures the number of occurences in an experiment, i.e., "probabilities convey information about relative frequencies" (Bezdek and Pal, 1992: 5). A new measurement has no effect on the value of a membership function, whereas it can change an a priori probability value to an a posteriori probability value. See Bezdek and Pal (1992) for a very interesting example that further clarifies this point.

Fuzzy logic and FLSs should not be viewed as replacements for existing successful approaches; they should be viewed as enhancements, or as Bezdek and Pal (1992: 5) state, "the idea of fuzziness is one of enrichment, not of replacement."

Discussion

We have demonstrated that a fuzzy logic system (FLS) is a nonlinear system that maps a crisp input vector into a crisp scalar output. We have provided mathematical formulas that describe this system; we have shown that it can be expressed as a linear combination of fuzzy basis functions; and we have explained that an FLS is a universal function approximator, which makes it a competitor to all other function approximators that share this property (e.g., FFNNs). The fuzzy basis function expansion is unique among all other basis function expansions, in that it can derive its basis functions in a unified manner from either numerical data (as can the other expansions) or linguistic knowledge (as can none of the other expansions). Finally, *FBFs are not radial basis functions* (Kim and Mendel, 1993); they are nonlinear functions of radial basis functions, and they can be created from a combination of linguistic and numerical information.

The architecture of an FLS is determined by a careful understanding of fuzzy sets and fuzzy logic, and it is rich with possibilities: there is no one FLS, there are many. As users of an FLS, we must decide on the type of fuzzification, on functional forms for membership functions, on parameters of membership functions, and on composition, inference, and defuzzifier.

We conclude with a short comparison of FLSs and FFNNs, because they can both be used to solve the following problem: given a set of N desired input-output pairs (the training set), $(\mathbf{x}^{(1)}: y^{(1)}), (\mathbf{x}^{(2)}: y^{(2)}), \ldots, (\mathbf{x}^{(N)}: y^{(N)})$, where $\mathbf{x}$ is the $p \times 1$ input and y is the output, determine an FLS or FFNN $f: \mathbf{x} \to y$.

Both the FLS and FFNN approaches are model free. They are both given training information contained in some examples from which they are required to learn so as to give correct or successful outputs when new inputs are presented.

The parameters of an FFNN are its weights. The parameters of an FLS are (Equation 12) the $M \bar{y}^l$, $M p \bar{x}_i^l$, and $M p \sigma_i^l$. FFNNs and FLSs can be trained using backpropagation learning algorithms, which are gradient descent algorithms constrained by the architecture of either the FFNN or the FLS. For the former, the backpropagation algorithm adjusts the weights, whereas for the latter the backpropagation algorithm (Wang and Mendel, 1992a) adjusts the $M + 2Mp \bar{y}^l$, $\bar{x}_i^l$, σ_i^l parameters.

Backpropagation (BP) FFNNs and FLSs have been applied to identification of nonlinear dynamical systems (Narendra and Parthasarathy, 1990; Wang and Mendel, 1992b), and in every case the BP FLS identifier outperformed the BP FFNN. Either the BP FLS was able to converge to results that the BP FFNN identifier was unable to, or it converged at a faster rate. The BP FLS and BP FFNN have also been applied to first-break picking in reflection seismology (Chu and Mendel, 1994). Although very similar processed results were obtained by the two first-break pickers, there was a huge difference in the training times: 14 min–54 min for the BP FFNN versus 6 s–21 s for the BP FLS! The reason for the fast convergence of the BP FLS was established, by simulations, to be the excellent initialization procedure for its design parameters. The weights of the BP FFNN are initialized randomly, whereas the parameters of the BP FLS are initialized in a smart way. This is possible because the parameters of a BP FLS have physical meanings, whereas the weights of the BP FFNN do not.

The construction of a numerical fuzzy decision maker can range from a simple one-pass design procedure (Wang, 1994) to training procedures (Wang and Mendel, 1992a), whereas the construction of a feedforward neural decision maker always requires time-consuming iterative training.

The choice of the number of hidden-layer neurons of an FFNN is very difficult and quite problem dependent; an inaccurate choice of the number of hidden-layer neurons may cause either too large a network or a network which cannot match all of the desired patterns. The question of how many fuzzy sets we decide to partition a variable into is somewhat analogous, as is the choice of the number of rules; but FLS results are

excellent even for rather coarse partitions and a modest number of rules.

An FLS is more general than an FFNN. In other words, our numerical-fuzzy approach can use not only desired input-output patterns but also linguistic rules from human experts, whereas the FFNN can presently utilize only examples.

An FLS is more flexible than an FFNN. That is, we can choose the membership functions in a wide variety of forms and divide the domain interval into different regions, and it is very easy to update the FLS. Hence, an FLS can be used easily for adaptive time-series prediction, whereas any new data shown to an FFNN may require time-consuming retraining.

All of these comparisons suggest that incorporation of prior knowledge, as in a BP FLS, should be of prime concern for supervised learning problems, especially in cases where training time is limited.

Road Map: Artificial Intelligence and Neural Networks
Background: Perceptrons, Adalines, and Backpropagation
Related Reading: Expert Systems and Decision Systems Using Neural Networks; Learning by Symbolic and Neural Methods

References

Bezdek, J. C., and Pal, S. K., 1992, *Fuzzy Models for Pattern Recognition*, New York: IEEE Press. ◆

Chu, P., and Mendel, J. M., 1994, First break refraction event picking using fuzzy logic systems, *IEEE Trans. Fuzzy Sys.*, 2:255–266.

Kim, Y. M., and Mendel, J. M., 1993, *Fuzzy Basis Functions: Comparisons with Other Basis Functions*, SIPI Report #229, Los Angeles: University of Southern California.

Kosko, B., 1992, *Neural Networks and Fuzzy Systems: A Dynamical Systems Approach to Machine Intelligence*, Englewood Cliffs, NJ: Prentice-Hall.

Kosko, B., 1994, Fuzzy systems as universal approximators, *IEEE Trans. Comput.*, 43:1329–1333.

Laviolette, M., and Seaman, J. W., Jr., 1994, The efficacy of fuzzy representations and uncertainty, *IEEE Trans. Fuzzy Sys.*, 2:4–15.

Lee, C. C., 1990, Fuzzy logic in control systems: Fuzzy logic controller, Part II, *IEEE Trans. Syst. Man Cybern.*, 20:419–435.

Mendel, J. M., 1994, Fuzzy logic systems for engineering: A tutorial, submitted for publication. ◆

Narendra, K. S., and Parthasarathy, K., 1990, Identification and control of dynamical systems using neural networks, *IEEE Trans. Neural Netw.*, 1:4–27.

Wang, L. X., 1994, *Adaptive Fuzzy Systems and Control: Design and Stability Analysis*, Englewood Cliffs, NJ: PTR Prentice-Hall.

Wang, L. X., and Mendel, J. M., 1992a, Back-propagation fuzzy systems as non-linear dynamic system identifiers, in *Proceedings of the IEEE International Conference on Fuzzy Systems*, San Diego, pp. 1409–1418.

Wang, L. X., and Mendel, J. M., 1992b, Fuzzy basis functions, universal approximation, and orthogonal least squares learning, *IEEE Trans. Neural Netw.*, 3:807–814.

Zadeh, L. A., 1965, Fuzzy sets and systems, in *Proceedings of the Symposium on Systems Theory*, Brooklyn, NY: Polytechnic Institute of New York, pp. 29–37. ◆

Zadeh, L. A., 1973, Outline of a new approach to the analysis of complex systems and decision processes, *IEEE Trans. Syst. Man Cybern.*, 3:28–44. ◆

Zimmermann, H.-J., 1991, *Fuzzy Set Theory—and Its Applications*, 2nd ed., Boston: Kluwer Academic.

Gabor Wavelets for Statistical Pattern Recognition

John Daugman and Cathryn Downing

Introduction: Signal-to-Symbol Converters

A central problem both for machine intelligence and for the modeling of biological vision and cognition, is provided by the concept of a *signal-to-symbol converter*. The external world presents itself only as physical signals at the sensory surfaces, which explicitly express very little of the information required for intelligent interaction with the environment. These signals must be converted into symbolic representations (but see Ac-TIVE VISION for a complementary view), whose manipulation allows the organism or machine to bring an appropriate model of its external environment into contact with its internal goals and purposes.

Bridging this gap from signal reception to symbolic representation remains a major barrier to progress in machine intelligence, neural network theories, and neuroscience. This article reviews aspects of the neurobiology of spatial vision that reveal the degrees of freedom used in visual codes and that indicate certain strategies for signal-to-symbol conversion. We describe work on *quadrature demodulating networks*, whose input is a raw sensory signal, and whose output is a few bits' worth of information which, for a given scale of analysis, makes an abstraction from signal structure in a way that lends itself to symbolic (Boolean) operations appropriate and useful for pattern recognition. Finally, we illustrate this approach in connection with two practical systems for automatic visual recognition of personal identity.

Degrees of Freedom in Spatial Visual Codes

Over the past 30 years, electrophysiological recordings from the brain's primary visual cortex have generated a list of coding dimensions that underlie spatial vision. The key functional concept is that of a neuron's *receptive field*, which specifies that region of two-dimensional visual space in which image structure can influence the neuron's response. In particular, the neuron's *receptive field profile* indicates the relative degree to which the cell is excited or inhibited by the distribution of light as a function of spatial position in the image. These concepts are illustrated in the top portion of Figure 1, both for the case of linearly responding *simple cells* (indicated by the ovals) and for the case of nonlinear *complex cells* (represented by the outputs emerging at the top of the diagram).

It is useful to characterize the nature of the spatial image code at this early cortical level. Ignoring color, motion, and stereoscopic selectivities, there are arguably five major degrees of freedom (i.e., independent dimensions or forms of variation) spanned by the primary cortical visual neurons, the simple cells. These can be regarded as defining the dimensions of the spatial visual code. The first two degrees of freedom are the *location* of a neuron's receptive field, defined by retinotopic coordinates (x, y). The third is the *size* of its receptive field (which can be described using a single scalar diameter, provided we view variation in the field width/length aspect ratio as secondary population structure). The fourth is the *orientation* of the boundaries separating excitatory and inhibitory regions, as indicated in Figure 1, normally also corresponding to the direction of receptive field elongation. The fifth is the *symmetry*, which may be even or odd, or some linear combination of these two canonical states. (Since any function can be decomposed into the sum of an even function plus an odd function,

their relative strengths define a continuum that allows this fifth dimension to be regarded as *phase*.)

These five degrees of freedom in the visual code also correspond to some dimensions of the "cortical architecture" (rules of topographic and modular organization), although such structure is more clear for some variables than for others. The (x, y) position coordinates of receptive fields in visual space form systematic (although nonconformal) topographic maps of these two dimensions across the cortical surface. For the third degree of freedom, subpopulations of neurons that share the same orientation preference are grouped together into columns, and successive columns rotate systematically across angles (*sequence regularity*). These rather crystalline organizational principles were originally documented in papers by Hubel and Wiesel (1962, 1974). Of the remaining two degrees of freedom, there is some evidence for pairwise grouping by quadrature (90-degree) phase relationship (Pollen and Ronner, 1981), and also for anatomical grouping by field size, either in different cortical layers (Maffei and Fiorentini, 1973) or in adjacent columns similar to those for orientation (Tootell, Silverman, and DeValois, 1981).

The benefit of identifying the primary degrees of freedom in a spatial image code is that it allows us to characterize the coding strategy in information-theoretic terms (Shannon and Weaver, 1949). The information-carrying capacity associated with each degree of freedom is a function of the number of individually resolvable states for that dimension. These define a kind of "information budget" that can be allocated in alternative ways among the different available degrees of freedom. Certain inescapable conflicts arise, however, which limit the degree to which some combinations of information can be simultaneously resolved. These conflicts take the form of an *uncertainty principle*, whose mathematical form (Daugman, 1985) is just a two-dimensional (2D) generalization of the one familiar from quantum physics in the famous work of Weyl and Heisenberg. One such conflict, or tradeoff, is intuitively clear from considering the oval receptive fields in Figure 1. Orientation resolution (the "sharpness" of orientation tuning) would be enhanced by making the ovals more elongated, but this would reduce their resolution for spatial location in that direction. Similarly, increasing the field width by adding more cycles of undulation would sharpen the tuning for spatial frequency, but at the cost of lost resolution for spatial location in this direction. The optimal solution to these tradeoffs, achieving maximal *conjoint* resolution of image information in both 2D spatial and 2D spectral terms, is the family of complex-valued 2D Gabor filters (Daugman, 1980, 1985), to be described in the next section.

The 2D Gabor Wavelet Representation

The optimal solution to the conjoint uncertainty relation is a family of complex-valued filters having the following functional form in the (x, y) space domain:

$$G(x, y) = e^{-\pi[(x-x_0)^2/\alpha^2 + (y-y_0)^2/\beta^2]} e^{-2\pi i[u_0(x-x_0) + v_0(y-y_0)]} \quad (1)$$

where (x_0, y_0) specify position in the image, (α, β) specify the filter's effective width and length, and (u_0, v_0) specify the filter's modulation wave-vector, which can be interpreted in polar coordinates as spatial frequency $\omega_0 = \sqrt{u_0^2 + v_0^2}$ and orientation (or direction) $\theta_0 = \arctan(v_0/u_0)$. The real and imaginary parts of this complex filter function (both of which are, of course,

Figure 1. The 2D Gabor phasor–based quadrature demodulation network, from Daugman (1993b). Even- and odd-symmetric receptive fields, of the kind associated with cortical "simple cells," subserve a phasor resolution of continuous image structure in the complex plane as illustrated in the lower portion.

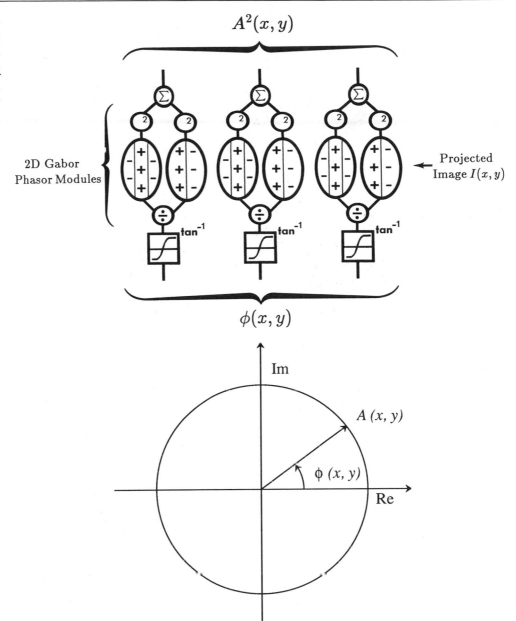

real quantities; see Daugman, 1985, 1993b) describe associated pairs of simple cells in *quadrature phase* (90-degree phase relation), as discovered by Pollen and Ronner (1981), and as portrayed by the paired ovals in Figure 1. The 2D Fourier transform $F(u, v)$ of a 2D Gabor filter, which reveals its spectral response selectivity in the Fourier plane, has exactly the same functional form as the space-domain function (i.e., it is "self-Fourier") but with the parameters interchanged or inverted:

$$F(u, v) = e^{-\pi[(u-u_0)^2\alpha^2+(v-v_0)^2\beta^2]}e^{-2\pi i[x_0(u-u_0)+y_0(v-v_0)]} \quad (2)$$

Thus the 2D Fourier power spectrum of this filter, $F(u, v)F^*(u, v)$, is simply a bivariate Gaussian centered on (u_0, v_0). Hence, its peak response occurs for an orientation θ_0 and spatial frequency ω_0 as defined above, matching the excitatory/inhibitory structure of the receptive fields, as one would expect.

This family of 2D filters was introduced by Daugman (1980) as a model for the structure of simple cell receptive field pro-

files, and they are termed *2D Gabor functions* because they are the 2D generalization of a class of one-dimensional (1D) functions proposed by Dennis Gabor for representing time-varying signals (Gabor, 1946). Experimental work by Jones and Palmer (1987) confirmed that this family of functions provided good fits to the receptive field profiles of about 97% of the simple cells whose 2D profiles the researchers recorded in cat visual cortex. (Other investigators, such as Parker and Hawken [1988], have preferred other functions, such as differences of several offset Gaussians, which, having additional fitting parameters, offered better fits to their data.) The top row of Figure 2 illustrates three of the 131 measured 2D receptive field profiles from the work of Jones and Palmer. For each case, the second row shows the best-fitting 2D Gabor filter (minus the associated quadrature component, since this experiment did not record from quadrature pairs). The third row shows the residuals obtained by subtracting the fitted model from the measured receptive field profiles. For 97% of the cells studied,

2D Receptive Field Profiles

Fitted 2D Gabor Phasors

Residuals

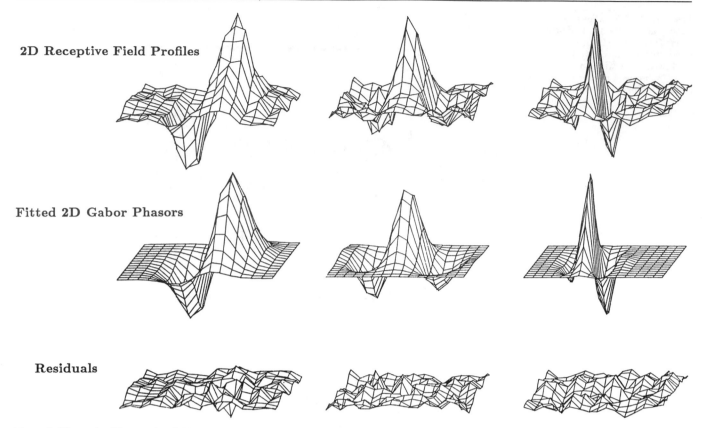

Figure 2. Illustrative 2D receptive field profiles of simple cells found in cat visual cortex, from the work of Jones and Palmer (1987). The raw receptive field measurements (top row) are well described by the 2D Gabor phasor model (middle row) in 97% of the cells studied, yielding residuals (bottom row) indistinguishable from random error in chi-square tests.

these residuals were indistinguishable from random error in chi-square tests. Although other analytic forms could be chosen to fit the available 2D receptive field data, there can be no doubt that the 2D Gabor filter model specifies an efficient set of coding primitives that capture 2D spatial location, orientation, size, and phase (or symmetry) in a natural way.

Besides their optimality in terms of the uncertainty relation, 2D Gabor functions have other useful properties. They can be used as a complete self-similar 2D *wavelet* expansion basis (Meyer, 1986), with the requirement of orthogonality relaxed (Daugman, 1988), by appropriate parameterization for dilation, rotation, and translation. If we take $\Psi(x, y)$ to be a chosen generic 2D Gabor wavelet as specified in Equation 1, which will be called a *mother wavelet*, then we can generate from this one function a complete self-similar family of *daughter wavelets*, which will constitute a complete expansion basis, through the generating function

$$\Psi_{mpq\theta}(x, y) = 2^{-2m}\Psi(x', y') \tag{3}$$

where the substituted variables (x', y') incorporate dilations of the wavelet in size by 2^{-m}, translations in position (p, q), and rotations through angle θ:

$$x' = 2^{-m}[x\cos(\theta) + y\sin(\theta)] - p \tag{4}$$

$$y' = 2^{-m}[-x\sin(\theta) + y\cos(\theta)] - q \tag{5}$$

It is noteworthy that, as consequences of the similarity, shift, and modulation theorems of 2D Fourier analysis, together with the rotation isomorphism of the 2D Fourier transform, all of these effects of the generating function of Equation 3, ap-

plied to a 2D Gabor mother wavelet $\Psi(x, y) = G(x, y)$ in generating a 2D Gabor daughter wavelet $\Psi_{mpq\theta}(x, y)$, will have corresponding or reciprocal effects on its 2D Fourier transform $F(u, v)$ without any change in functional form (Daugman, 1985). This family of 2D wavelet filters and their 2D Fourier transforms is closed under the transformation groups of dilations, translations, rotations, and convolutions.

Any image can be represented completely in terms of such a basis of elementary expansion functions. Interesting issues arise concerning how the "information budget" should be allocated. For example, in exchange for indexing all Gabor wavelets by their 2D location, the parameters that specify each wavelet's orientation and spatial frequency can be sampled less frequently. Whereas, for example, the 2D Fourier basis must sample the frequency plane along a Cartesian grid, a 2D Gabor expansion can sample the frequency plane on just a log-polar grid. This great reduction in sampling density at the higher frequencies is paid for by the use of positioning parameters. Illustrations of self-similar 2D Gabor representations of images obtained with varying numbers of wavelet orientations (6, 4, 3, and 2 orientations in the sampling set) may be found in Daugman (1988). Further information-theoretic and computational issues are raised by the fact that these expansion functions are *nonorthogonal* (i.e., the integral of the product of any two receptive field profiles is, in general, non-zero); therefore the 2D Gabor coefficients cannot be obtained simply by taking the inner product of the image with each localized elementary function. A relaxation network solution to this problem was presented by Daugman (1988). The consequences of nonorthogonality include paradoxes for the classical interpretation of what it is that a neuron's receptive field actually enables it to encode about the image; in particular, the classical view that a linear neuron's response signifies the "relative presence" of its receptive field profile in local image structure cannot be true when the ensemble of neurons is nonorthogonal.

The Quadrature Demodulator as Signal-to-Symbol Converter

A purpose of a signal-to-symbol converter is to create a succinct and useful representation of signal structure having much lower dimension than the raw signal itself. In a sense, standard machine-vision strategies for edge extraction are examples of such devices, since edge maps can signify object structure, and

they clearly have lower dimension than the pixel array. However, many naturally occurring objects, such as bodies and faces, lack the planar or geometrical forms of manufactured objects that generate simple data trees of edges and vertices; rather, they are defined by continuous-tone structure and undulations which are not well extracted by edge maps.

The quadrature demodulator network seen in Figure 1 offers a means to extract and encode such subtler undulations as define faces, textures, and natural forms. The $\phi(x, y)$ function emerging from below the bank of 2D Gabor modules represents continuous local image structure in terms of the rotating *phasor* angle indicated in the bottom diagram. Further details about the use of complex variables in such a coding network

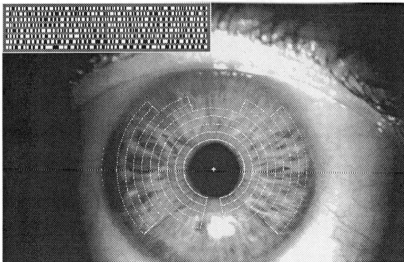

Figure 3. Illustration of face and eye codes using the scheme summarized in Figure 1. The bit arrays indicate in which quadrant the Gabor phasor resides, for different wavelet positions, orientations, and sizes. The graphics superimposed upon the face indicate where facial features (eyes, nasal ridge, and mouth) were located by algorithms that are also based on 2D Gabor phasors.

may be found in Daugman (1993b). The complex phasor angle that is resolved by the arctangent-like *squashing function* in the network output nodes can be coarsely quantized to construct extremely compact image codes. Even with as few as two bits preserved, signifying merely in which of the four quadrants the phasor resides for a local patch of the image at a given scale of analysis and orientation of variation, the network can function as a very effective signal-to-symbol converter. In the next section we illustrate applications of such coarse phasor codes in two different approaches to face recognition.

Applications: Recognizing Facial "Signatures"

Pattern demodulation codes constructed by quantizing the local phasor angle into just two bits are illustrated in Figure 3, both for an entire face and for an isolated iris pattern. The detailed texture visible in an iris is a spatial stochastic process having high dimensionality but little genetic penetrance except for color. Details about the automatic algorithms for locating and isolating the facial features are given in Daugman (1993a); doubly dimensionless coordinate systems are established for size-, translation-, and rotation-invariant representations of the face or facial feature. Image structure is then resolved into phasors in the complex plane with just the quadrant encoded, yielding two bits. The different bit arrays shown are derived at

different scales of analysis, and the position of a bit within an array represents the wavelet's location and orientation. The face code contains 146 bytes, and the iris code contains 256 bytes. An application of similar ideas for coding and recognizing handwritten characters is described in Shustorovich (1994).

Recognizing personal identity by comparing face codes or iris codes in a decision network is illustrated in Figure 4. The extracted 2D Gabor phasor quadrants are simply compared by parallel (integer) Exclusive-OR, which detects disagreement between corresponding bit pairs. The total strength of disagreement is summed and compared with a threshold to generate a decision about "same" or "different." Thus, in a sense, this network transforms problems of pattern recognition into a simple test of statistical independence on the Gabor phasor angles derived from the patterns in question.

Finally, performance in the case of iris recognition is illustrated in Figure 5, which shows "same-different" separation performance using the 2D Gabor demodulation phasor network of Figure 1 and the decision network of Figure 4. Since XOR comparisons of phasor bits are Bernoulli trials, whose values of p and q depend on whether a pair of iris codes come from the same or from different eyes, the confidence levels associated with recognition decisions can be estimated from the binomial distribution for the likelihood of observing a fraction

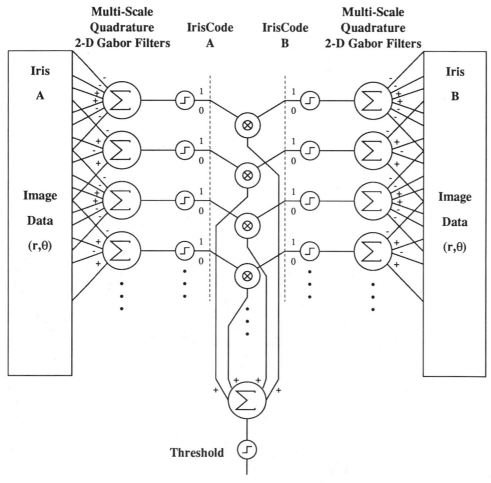

Figure 4. The comparison and recognition network used to make decisions about the identity of eyes or faces. In effect, this network transforms the problem of pattern recognition into a statistical test of independence.

Figure 5. Recognition performance for the real-time iris recognition system. The white histogram shows the measured Hamming distances among different images of same eyes ("authentics"), while the black histogram shows this distribution among different images of different eyes ("impostors").

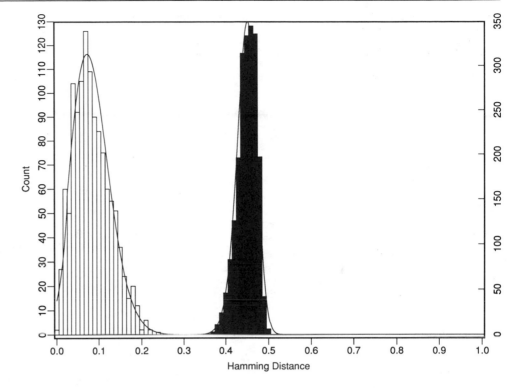

$x = m/N$ True XOR outcomes in N comparisons:

$$f(x) = \frac{N!}{m!(N-m)!}p^m q^{(N-m)} \tag{6}$$

The solid curves that have been fitted to the two observed distributions in Figure 5 are derived from this binomial expression, and they provide good fits to the data. The best-fitting values for p, q, and N indicate that when different irises are compared using this method of phase-quadrant texture demodulation at this resolution, there are approximately 173 orthogonal degrees of freedom spanned by such stochastic signatures. The formal decidability of this recognition task, analogous to detectability in a signal detection task, is measured at $d' = 8.41$ for the two distributions shown. The crossover point between the two fitted distributions (where the theoretical false accept rate equals the theoretical false reject rate, around an operating criterion of 0.321 Hamming distance) corresponds to an error probability of 1 in 131,000. Thus, it appears that this complex yet stable textural signature could provide a very reliable basis for automatically recognizing personal identity (in lieu of using PINs, cards, or keys) for purposes such as access control, passports, ATMs, computer login, and authentication or security generally.

Discussion

Many open questions in neuroscience are related to building a bridge from signals to symbols. In this article we reviewed some well-established facts about early visual coding that can be implemented in a simple 2D Gabor wavelet network, which functions as an elementary sort of signal-to-symbol converter. Although there is no neural evidence for actual pattern recognition based on such coarsely quantized phasor codes, we demonstrated practical applications of this network in automatic visual recognition of personal identity.

Acknowledgments. This work was supported by U.S. National Science Foundation PYI Award IRI–8858819, by U.S. Air Force Office of Scientific Research subcontract 90–0175, and by research grants from the Eastman Kodak Corporation.

Road Map: Vision

Related Reading: Ocular Dominance and Orientation Columns; Visual Coding, Redundancy, and "Feature Detection"; Visual Cortex Cell Types and Connections; Wavelet Dynamics

References

Daugman, J. G., 1980, Two-dimensional spectral analysis of cortical receptive field profiles, *Vis. Res.*, 20:847–856.

Daugman, J. G., 1985, Uncertainty relation for resolution in space, spatial frequency, and orientation optimized by two-dimensional visual cortical filters, *J. Opt. Soc. Am. A*, 2:1160–1169. ◆

Daugman, J. G., 1988, Complete discrete 2D Gabor transforms by neural networks for image analysis and compression, *IEEE Trans. Acoust. Speech Signal Process.*, 36:1169–1179.

Daugman, J. G., 1993a, High confidence visual recognition of persons by a test of statistical independence, *IEEE Trans. Pattern Analysis Machine Intell.*, 15:1148–1161.

Daugman, J. G., 1993b, Quadrature-phase simple-cell pairs are appropriately described in complex analytic form, *J. Opt. Soc. Am. A*, 10:375–377.

Gabor, D., 1946, Theory of communication, *J. Inst. Electr. Eng.*, 93:429–457.

Hubel, D. G., and Wiesel, T. N., 1962, Receptive fields, binocular interaction, and functional architecture in the cat's visual cortex, *J. Physiol. (Lond.)*, 160:106–154. ◆

Hubel, D. G., and Wiesel, T. N., 1974, Sequence regularity and geometry of orientation columns in the monkey striate cortex, *J. Comp. Neurol.*, 158:267–293. ◆

Jones, J. P., and Palmer, L. A., 1987, An evaluation of the 2D Gabor filter model of simple receptive fields in cat striate cortex, *J. Neurophysiol.*, 58:1233–1258. ◆

Maffei, L., and Fiorentini, A., 1973, The visual cortex as a spatial frequency analyzer, *Vis. Res.*, 13:1255–1267.

Meyer, Y., 1986, Principe d'incertitude, bases Hilbertiennes et algèbres d'opérateurs, *Sém. Bourbaki*, 662:209–223.

Parker, A. J., and Hawken, M. J., 1988, Two-dimensional spatial structure of receptive fields in monkey striate cortex, *J. Opt. Soc. Am. A*, 5:598–605.

Pollen, D. A., and Ronner, S. F., 1981, Phase relationships between adjacent simple cells in the visual cortex, *Science*, 212:1409–1411.

Shannon, C., and Weaver, W., 1949, *Mathematical Theory of Communication*, Urbana: University of Illinois Press. ◆

Shustorovich, A., 1994, A subspace projection approach to feature extraction: The 2D Gabor transform for character recognition, *Neural Netw.*, 7:1295–1301.

Tootell, R., Silverman, M., and DeValois, R. L., 1981, Spatial frequency columns in primary visual cortex, *Science*, 214:813–815.

Gait Transitions

J. J. Collins

Introduction

Legged animals typically employ multiple gaits for terrestrial locomotion. Bipeds, for example, walk, run, and hop, whereas quadrupeds commonly walk, trot, and bound. Animals make transitions between different gaits depending on their speed and the terrain. Experimental studies have demonstrated that animal locomotion is controlled, in part, by a central pattern generator (CPG), which is a network of neurons in the central nervous system (CNS) capable of producing rhythmic output (Shik and Orlovsky, 1976; Grillner, 1981; Pearson, 1993). (The control of locomotion, however, is not purely central; e.g., the output of a locomotor CPG is modulated by feedback from the periphery.) Shik and colleagues, for instance, showed that mesencephalic cats could exhibit a walking gait on a treadmill when the midbrain was electrically stimulated. Moreover, they found that such preparations could switch between different gaits if either the stimulation strength or the treadmill speed was varied.

Although the aforementioned studies established the existence of rhythm-generating networks in the CNS, a vertebrate CPG for legged locomotion remains to be identified or isolated. As a result, little is known about the specific characteristics of the neurons and interconnections making up such systems. Consequently, researchers have resorted to using modeling techniques to gain insight into the possible functional organization of these networks. The most popular approach has involved the analysis of systems of coupled oscillators. Coupled-oscillator models have been used to control the gaits of bipeds (Bay and Hemami, 1987; Taga, Yamaguchi, and Shimizu, 1991), quadrupeds (Stafford and Barnwell, 1985; Schöner, Jiang, and Kelso, 1990; Collins and Stewart, 1993a; Collins and Richmond, 1994), and hexapods (Beer, 1990; Collins and Stewart, 1993b).

The neural mechanisms underlying gait changes are not well understood. A key question in this regard is whether gait transitions involve (1) switching between different CPGs, or (2) bifurcations of activity in a single CPG. In this article, we discuss a number of modeling approaches that have been developed to explore the feasibility of using either one or the other of these mechanisms to generate gait transitions in coupled-oscillator networks.

A Neuromodulatory Approach

As a model for legged-locomotion control, Grillner (1981) proposed that each limb of an animal is governed by a separate CPG, and that interlimb coordination is achieved through the actions of interneurons which couple together these CPGs. Within this scheme, gait transitions are produced by switching between different sets of coordinating interneurons, i.e., a locomotor CPG is reconfigured to produce different gaits.

Grillner's proposed strategy has been adopted, in spirit, by several CPG modeling studies. Stafford and Barnwell (1985), for example, used a similar approach in a study of quadrupedal locomotion. They considered a CPG model that was composed of four coupled networks of oscillators. Each network controlled the muscle activities of a limb of a model quadruped. Stafford and Barnwell showed that this model could produce the walk, trot, and bound. In addition, they demonstrated that the walk-to-trot and walk-to-bound transitions could be generated by changing the relative strength of certain interoscillator connections or by eliminating others altogether. (Transitions in the reverse direction, e.g., bound-to-walk, were not reported.) Along similar lines, Bay and Hemami (1987) used a CPG network of four coupled van der Pol oscillators to control the movements of a segmented biped. Each limb of the biped was composed of two links, and each oscillator controlled the movement of a single link. Bipedal walking and hopping were simulated by using the oscillators' output to determine the angular positions of the respective links. Transitions between out-of-phase and in-phase gaits were generated by changing the nature of the interoscillator coupling; e.g., the polarities of the network interconnections were reversed to produce the walk-to-hop transition.

This approach is, in principle, physiologically reasonable. For instance, the notion that supraspinal centers may call on functionally distinct sets of coordinating interneurons to generate different gaits is plausible but not yet experimentally established. In addition, from a different but relevant perspective, it has been shown that rhythm-generating neuronal networks can be modulated, e.g., reconfigured, through the actions of neuroamines and peptides, and thereby they are enabled to produce several different motor patterns (see Pearson, 1993, and CRUSTACEAN STOMATOGASTRIC SYSTEM), at least in invertebrate preparations.

A Synergetic Approach

Synergetics deals with COOPERATIVE PHENOMENA (q.v.) in nonequilibrium systems (Haken, Kelso, and Bunz, 1985). In synergetics, the macroscopic behavior of a complex system is characterized by a small number of collective variables, which in turn govern the qualitative behavior of the system's components.

Schöner et al. (1990) used a synergetic approach in a study of quadrupedal locomotion. They analyzed a network model that was made up of four coupled oscillators. Each oscillator represented a limb of a model quadruped. Three relative phases— the phase differences between the right-front and the left-front, left-hind, and right-hind oscillators, respectively—were used as collective variables to characterize the system's interlimb-coordination patterns. Gait transitions were modeled as non-equilibrium phase transitions, which, in this case, could also be interpreted as bifurcations in a dynamical system (see the next section). Schöner et al. demonstrated that various four-component networks could produce and switch (abruptly or gradually) between different gaits, such as the gallop, trot, and pace, if the coupling terms that operated on the relative phases were varied. Importantly, this work predicted that gait transitions should be accompanied by loss of stability; i.e., signs of instability, such as spontaneous gait transitions, should arise near a switching point. Phenomena of this sort have been observed experimentally; e.g., the decerebrate cats in the Shik study could, near the trot-gallop transition point, switch back and forth spontaneously between the trot and gallop.

This approach is significant in that it relates system parameter changes and stability issues to gait transitions. Its primary weakness, however, is that the physiological relevance of the aforementioned relative-phase coupling terms is unclear. This remains an open issue.

A Group-Theoretic Approach

The traditional approach for modeling a locomotor CPG has been to set up and analyze, either analytically or numerically, the parameter-dependent dynamics of a hypothesized neural circuit. Collins and Stewart (1993a, 1993b), however, approached this problem from the perspective of group theory. Specifically, they considered various networks of symmetrically coupled nonlinear oscillators and examined how the symmetry of the respective systems leads to a general class of phase-locked oscillation patterns. Within this approach, the onset of a given pattern is modeled as a symmetric Hopf bifurcation, and transitions between different patterns are modeled as symmetry-breaking bifurcations of various kinds. In standard Hopf bifurcation, the dynamics of a nonlinear system changes as some parameter is varied, and a stable steady state becomes unstable, "throwing off" a limit cycle (or periodic solution). At a symmetric analog of a Hopf bifurcation, which is appropriate for symmetric dynamical systems, one or more periodic solutions, usually several, bifurcate. There may also be secondary branches of solutions and other more complicated bifurcations. Successive bifurcations tend to break more and more symmetry; i.e., they lead to states with less and less symmetry. Importantly, the pattern of bifurcations that can occur and the nature of the periodic states that arise through such bifurcations are controlled primarily by the symmetries of the system.

The theory of symmetric Hopf bifurcation thus predicts that symmetric oscillator networks with invariant structure can sustain multiple patterns of rhythmic activity. From the standpoint of CPGs, this prediction challenges the notion that a network's coupling architecture *needs* to be altered to produce different oscillation patterns. Importantly, the symmetry-breaking analysis is independent of the details of the oscillators' intrinsic dynamics and the interoscillator coupling. (The production of periodic states through symmetric Hopf bifurcation, however, does depend on the variation of *some* suitable system parameter.) This approach thus provides a framework for distinguishing model-independent features (attributable to symmetry alone) from model-dependent features.

Collins and Stewart used this approach to study the dynamics of symmetric networks of two, four, and six coupled oscillators. These networks were considered as models for bipedal, quadrupedal, and hexapodal locomotor CPGs, respectively. They demonstrated that many of the generic phase-locked os-

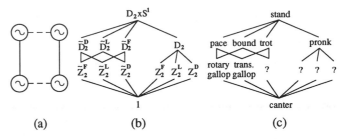

Figure 1. *A*, A rectangularly symmetric network of four coupled oscillators. The solid and dashed lines represent two forms of coupling. *B*, Patterns of symmetry-breaking for the network in *A*. The respective group-theoretic symbols are described in Collins and Stewart (1993a). *C*, Quadrupedal gaits corresponding to the patterns in *B*.

cillation patterns for these models correspond to animal gaits. They also showed that transitions between these gaits could be modeled as symmetry-breaking bifurcations occurring in such systems. These studies led to natural hierarchies of gaits, ordered by symmetry, and to natural sequences of gait bifurcations (Figure 1). This work thus related observed gaits and gait transitions to the organizational structures of the underlying CPGs.

This approach is significant in that it provides a novel mechanism for generating gait transitions in locomotor CPGs. Its primary disadvantage, however, is that its model-independent features cannot provide information about the internal dynamics of individual oscillators. In particular, the stability of the predicted gait patterns and the conditions under which one is selected over another depend on the specific parameters of the model under investigation.

A Hardwired Network Approach

Motivated by the predictions of the above group-theoretic approach, Collins and Richmond (1994) conducted a series of computer experiments with a symmetric, hardwired locomotor CPG model that consisted of four coupled oscillators. They demonstrated that it was possible for such a network to produce multiple phase-locked oscillation patterns that correspond to three quadrupedal gaits: the walk, trot, and bound. Transitions between the different gaits were generated by varying the driving signal or by altering internal oscillator parameters. As observed in real animals (Alexander, 1989), transitions between the walk and trot, which were generated by varying the intrinsic frequency of the CPG oscillators and the amplitude of the driving signal, could be either gradual or abrupt, depending on the nature of the parameter variation. Similar parameter changes could also shift the CPG model from either the walk or trot into the bound. However, once the CPG model was in the bound, it maintained that gait even if the system parameters were returned to their original values for either the walk or trot, i.e., there was "total" hysteresis in the network's dynamics. To produce transitions from the bound, it was necessary to subject two of the CPG oscillators to an increased driving stimulus before the system parameters were changed to those of the desired gait. (Experimental data which indirectly support such a strategy for generating transitions from the bound were provided by Afelt, Blaszczyk, and Dobrzecka, 1983. Specifically, they found that the initiation of the gallop-to-trot transition in dogs was characterized by kinematic changes in a *single pair* of diagonal limbs.) Importantly, the above *in numero* results were obtained without changing the relative strengths or polarities of the system's interconnections;

i.e., the network maintained an invariant coupling architecture. Collins and Richmond (1994) also showed that the ability of the hardwired CPG network to produce and switch between multiple gaits was, in essence, a model-independent phenomenon: three different oscillator models—the Stein neuronal model, the van der Pol oscillator, and the FitzHugh-Nagumo model—and two different coupling schemes were incorporated into the network without impeding its ability to produce the three gaits and the aforementioned gait transitions. This general finding was likely attributable to the symmetry of the network, which was maintained in all the numerical experiments.

Earlier, Beer (1990) designed a hardwired CPG network for controlling hexapodal locomotion (see LOCOMOTION, INVERTEBRATE). In Beer's model, each leg of a model cockroach was controlled by a circuit made up of one pacemaker neuron, two sensory neurons, and three motor neurons. The pacemaker neurons of adjacent leg-controller circuits inhibited one another. If the pacemaker neurons of the network were identical, then the model could generate the tripod gait. To produce metachronal-wave gaits (in which waves of leg movements sweep from the back of the animal to the front), Beer varied the intrinsic frequencies of the pacemaker neurons such that the natural frequency of the back-leg pacemakers was lower than that of the middle-leg pacemakers, which was lower than that of the front-leg pacemakers. With this arrangement, the progression speed of the model cockroach could be changed, and transitions between different gaits could be produced by varying the tonic level of activity of a single command neuron, which was connected to every leg-controller circuit. This model's ability to generate and switch between different gaits was a direct consequence of the interactions between its coupled pacemaker neurons and their respective central and afferent inputs.

A similar model, made up of six coupled unit oscillators, was developed by Taga et al. (1991) to control bipedal locomotion. In this case, each unit oscillator controlled a single joint, i.e., an ankle, knee, or hip, of a multi-link biped. As with Beer's model, the CPG network was driven by a tonic activation signal, and each unit oscillator received feedback about the state of the system's limbs. With this arrangement, the biped's speed could be changed, and abrupt transitions between walking and running could be generated by varying the amplitude of the network's activation signal. Interestingly, these gait transitions exhibited hysteresis; i.e., the walk-to-run transition occurred at a faster progression speed than did the reverse transition. Similar hysteretic behavior has been observed in humans (Alexander, 1989). Taga et al., unfortunately, did not report on the model's ability to switch between out-of-phase gaits (i.e., walking and running) and in-phase gaits (i.e., hopping).

In these studies, gait transitions were produced by varying the CPG's driving signal. From a physiological standpoint, this pattern-switching mechanism is reasonable; e.g., experimental studies have shown that the output of a locomotor CPG can be modified by changes to its descending inputs. Nonetheless, it is important to note that the exact form of the driving signal or signals acting on a locomotor CPG is unknown. Similarly, it is unclear how externally applied stimulation signals are transmitted to locomotor CPGs. For instance, although the stimulation signal in the Shik study was amplitude modulated (to produce gait transitions), this does not necessarily mean that the resulting descending signals were also amplitude modulated. In addition, although the results of the Shik study were largely independent of the stimulation frequency, there is evidence that frequency-modulated stimulation signals can affect the output of locomotor CPGs. Lennard and Stein (1977), for example, electrically stimulated the dorsolateral funiculus in spinal and intact turtles and found that an increase in the stimulus frequency resulted in an increased repetition rate of hindlimb swimming movements. Finally, it should be reiterated that it is most likely erroneous to assume (as it has been in several CPG modeling studies) that the net driving signal of a locomotor CPG consists only of descending influences from supraspinal centers. The results from several experimental studies indicate that a CPG "driving" signal may also consist of afferent inputs from peripheral sensory organs (Pearson, 1993).

Discussion

The discussed modeling studies fall into two camps: (1) gait transitions are produced by changing the relative strength or polarity of the interoscillator coupling in a CPG; i.e., "different" CPGs are used to produce different gaits; or (2) gait transitions are generated by changing the CPG's driving signal; i.e., bifurcations in a single CPG are used to generate different gaits. Both of these pattern-switching mechanisms are physiologically plausible, and they each lead to realistic locomotor patterns; e.g., in most of these studies, the stepping frequency or progression speed of the model animal increased when the CPG network switched to "faster" gaits. However, for a consistent theory of gait transitions to emerge, additional experimental data about the functional organization and operation of locomotor CPGs will have to be obtained. In particular, work is needed: (1) to determine whether a locomotor CPG uses functionally distinct sets of coordinating interneurons to produce different motor patterns, (2) to establish the extent to which neuromodulatory mechanisms are employed in vertebrate motor systems, and (3) to clarify the nature of the peripheral and descending inputs that influence the output of a locomotor CPG. Further experimentation is also needed to examine the possible role of bifurcation in gait transitions. In this regard, future investigations should explore the extent of hysteresis in gait transitions and the occurrence of increased instabilities near switching points, as well as consider more extensively the effects of system-parameter variation on gait-transition dynamics.

Road Map: Motor Pattern Generators and Neuroethology
Background: I.3. Dynamics and Adaptation in Neural Networks
Related Reading: Dynamics and Bifurcation of Neural Networks; Spinal Cord of Lamprey: Generation of Locomotor Patterns

References

Afelt, Z., Blaszczyk, J., and Dobrzecka, C., 1983, Speed control in animal locomotion: Transitions between symmetrical and nonsymmetrical gaits in the dog, *Acta Neurobiol. Exp.*, 43:235–250.

Alexander, R. McN., 1989, Optimization and gaits in the locomotion of vertebrates, *Phys. Rev.*, 69:1199–1227. ◆

Bay, J. S., and Hemami, H., 1987, Modeling of a neural pattern generator with coupled nonlinear oscillators, *IEEE Trans. Biomed. Eng.*, 34:297–306.

Beer, R. D., 1990, *Intelligence as Adaptive Behavior: An Experiment in Computational Neuroethology*, San Diego: Academic Press.

Collins, J. J., and Richmond, S. A., 1994, Hard-wired central pattern generators for quadrupedal locomotion, *Biol. Cybern.*, 71:375–385.

Collins, J. J., and Stewart, I. N., 1993a, Coupled nonlinear oscillators and the symmetries of animal gaits, *J. Nonlin. Sci.*, 3:349–392. ◆

Collins, J. J., and Stewart, I., 1993b, Hexapodal gaits and coupled nonlinear oscillator models, *Biol. Cybern.*, 68:287–298.

Grillner, S., 1981, Control of locomotion in bipeds, tetrapods and fish, in *The Handbook of Physiology, Section 1: The Nervous System, Vol. II, Motor Control* (V. B. Brooks, Ed.), Bethesda, MD: American Physiological Society, pp. 1179–1236.

Haken, H., Kelso, J. A. S., and Bunz, H., 1985, A theoretical model of phase transitions in human hand movements, *Biol. Cybern.*, 51:347–356.

Lennard, P. R., and Stein, P. S. G., 1977, Swimming movements elicited by electrical stimulation of turtle spinal cord: I. Low-spinal and intact preparations, *J. Neurophysiol.*, 40:768–778.

Pearson, K. G., 1993, Common principles of motor control in vertebrates and invertebrates, *Annu. Rev. Neurosci.*, 16:265–297. ◆

Schöner, G., Jiang, W. Y., and Kelso, J. A. S., 1990, A synergetic theory of quadrupedal gaits and gait transitions, *J. Theoret. Biol.*, 142:359–391.

Shik, M. L., and Orlovsky, G. N., 1976, Neurophysiology of locomotor automatism, *Phys. Rev.*, 56:465–501.

Stafford, F. S., and Barnwell, G. M., 1985, Mathematical models of central pattern generators in locomotion: III. Interlimb model for the cat, *J. Motor Behav.*, 17:60–76.

Taga, G., Yamaguchi, Y., and Shimizu, H., 1991, Self-organized control of bipedal locomotion by neural oscillators in unpredictable environment, *Biol. Cybern.*, 65:147–159.

Gaze Coding in the Posterior Parietal Cortex

Pietro Mazzoni and Richard A. Andersen

Introduction

An issue of active debate in the area of biological motor control is what coordinate frames the nervous system uses to represent sensory stimuli and planned movements. Much evidence suggests that a portion of the posterior parietal cortex (PPC) of the primate brain participates in sensory to motor coordinate transformations. Specific areas within the PPC appear to compute such transformations to program the direction of gaze. We review a set of neural network models developed in our laboratory to study how such coordinate transformations might be achieved by neurons in the PPC.

Neuronal Properties and Presumed Function of the Primate's Posterior Parietal Cortex

Because an animal's sensory and motor organs can move relative to one another, a requirement of sensorimotor integration is the transformation of spatial locations across coordinate frames. Early studies of the monkey's PPC suggested that this area plays a role in the integration of visual perception and motor behavior because neurons were found that responded to visual stimuli and to changes in eye position (reviewed in Andersen, 1987). The portions of the visual field in which luminous stimuli elicited responses—i.e., their receptive fields—corresponded to particular retinal locations. As the monkey looked in different directions, the receptive fields maintained their retinal location but were modulated by eye position. These were called *spatial gain fields* because eye position acted as a gain on the visual response. For most neurons, the modulation had a planar component—i.e., proportional to the horizontal and/or vertical eye position.

The Zipser-Andersen Model

The properties of PPC neurons suggested that individual neurons were unlikely to subserve spatial computations. Being sensitive to both retinal location over a large area and to eye position, a single neuron's activity is an ambiguous signal of stimulus location. This location could in principle be retrieved, however, from the pooled activity of a group of such neurons. Zipser and Andersen developed a neural network to study how an ensemble of neuron-like model units might solve the coordinate transformation problem (Zipser and Andersen, 1988). The aim was to examine the properties of individual units that were trained to solve the problem as a group. If the brain was indeed encoding spatial locations in the distributed pattern of activity of many parietal neurons, then some features of the brain's algorithm might emerge in the model network too.

The Zipser-Andersen network model (Figure 1) is a three-layer feedforward network whose input units carry signals

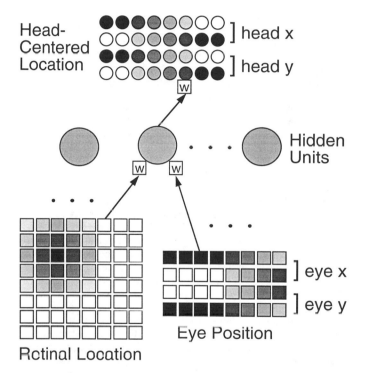

Figure 1. The Zipser-Andersen network model. The input layer consists of 64 units encoding retinal position topographically and 32 units encoding eye position linearly. Each hidden unit has a logistic activation function and projects to all output units. These are 32 logistic units trained to encode the vector sum of the retinal and eye position locations (see Zipser and Andersen, 1988, for details).

known to be available to the PPC. The output layer was trained to encode the head-centered location of the visual stimulus. The task was to perform vector addition of the stimulus's retinal position and the eye position. After the network was trained using backpropagation, its hidden units were found to respond to visual stimuli and to eye position very much like PPC neurons. They had retinotopic visual receptive fields whose activity profiles were modulated by eye position—i.e., spatial gain fields—and the modulation was largely planar. The receptive fields also had shapes similar to those of PPC neurons.

It was thus shown that a layered network can learn to transform retinal coordinates into head-centered ones using the input signals available to the PPC. This result is consistent with the adaptability of spatial behavior. Moreover, the hidden units' representation of spatial information was very similar to that found in PPC neurons. These neurons can thus play a similar role in the organism—i.e., build up an intermediate representation between input and output stages that is part of the coordinate transformation computation. The network demonstrated explicitly that units with PPC neurons' properties contain a distributed representation of space that is sufficient for accurate localization. Finally, the similarity of response properties suggested that the network and the brain may employ a common strategy in solving the coordinate transformations problem.

Body-Centered Coordinates

Having the problem's solution programmed in a network model made it possible to further investigate what algorithms the PPC may indeed be using through analysis and manipulations of the model. An immediate question was how locations could be encoded in other coordinate frames. The transformation from retinal to head-centered coordinates has a natural application in the programming of eye movements, the eyes having to move to particular positions relative to the head. Large gaze shifts, however, are achieved by coupled movements of the eyes and head; in this case, the target's position must be calculated in body-centered coordinates. Evidence from lesion studies suggests that the PPC is necessary for the proper execution of not only eye movements but other forms of spatial behavior as well. Could the Zipser-Andersen network be modified to compute body-centered coordinates; and if so, what predictions would it make about the PPC?

Goodman and Andersen (1990) added a group of units encoding head position to the input layer of the Zipser-Andersen network and trained this new network to produce body-centered locations at the output layer. The hidden units were found to be sensitive to all three input types. They had retinotopic visual receptive fields modulated by both eye and head position, each in a planar fashion. In other words, they developed planar "gaze fields," that is, linear modulation of visual responses along a particular direction of gaze, which is the sum of eye and head positions. Moreover, the "eye" gain field of a given hidden unit was always aligned (with the same direction and slope) with the same unit's "head" gain field. This was a natural solution for the network given the constraints of its architecture (the eye and head position inputs produced signals in very similar formats) and of the problem (eye and head position are indeed coupled for a given spatial position). The result suggested, however, that if the PPC subserves coordinate transformations beyond the head-centered reference frame and does so with an algorithm analogous to the neural network's strategy, then it should contain units with gaze fields similar to those of the network just described. Such units have recently been identified in the PPC (Brotchie and Andersen, 1991). Brotchie and Andersen trained monkeys to look in various directions by moving their eyes alone or by moving both their eyes and their head. A population of PPC neurons had visual responses modulated equivalently by eye or head position. These gaze fields were largely planar and the direction of eye and head position modulation was the same.

How the Neural Network Transforms Coordinates

Goodman and Andersen (1990) outlined a simple explanation of how the network performs coordinate transformations. Over the course of learning, each hidden unit develops a "preferred direction," that is, a direction in its input space along which to maximally modulate its activity. By maximal modulation, we mean that an input vector parallel to the preferred

direction produces the unit's largest activation, and a vector in the opposite direction produces the smallest activation (or largest inhibition from the resting activity level). The hidden units of the Zipser-Andersen network align their sensitivity in retinal space and in eye position space, and develop an eye position response field that approximates a plane oriented along what becomes the unit's preferred direction (direction $\mathbf{a}_i$ for the ith hidden unit in the network). A hidden unit effectively collapses the multidimensional signal of the retinal and eye position units into two two-dimensional vectors, one for retinal and one for eye position ($\mathbf{r}$ and $\mathbf{e}$, respectively). The goal is to add these two vectors to obtain the head-centered position vector, $\mathbf{h}$:

$$\mathbf{r} + \mathbf{e} = \mathbf{h}$$

A hidden unit's activation is proportional to the dot product of its input vectors and its preferred direction (i.e., its input weight vector):

$$\alpha_i \approx \mathbf{r} \cdot \mathbf{a}_i + \mathbf{e} \cdot \mathbf{a}_i$$

Each hidden unit thus extracts the components of the retinal and eye position vectors along its preferred direction and adds them. Because these vectors' components are added at the hidden unit's input, the output of each hidden unit effectively consists of the component of the head-centered vector along the unit's preferred direction. Formally, because

$$\mathbf{r} \cdot \mathbf{a}_i + \mathbf{e} \cdot \mathbf{a}_i = \mathbf{h} \cdot \mathbf{a}_i$$

then

$$\alpha_i \approx \mathbf{h} \cdot \mathbf{a}_i$$

The preferred directions of the hidden units span the two-dimensional input space so that the retinal and eye position vectors are decomposed without losing information. These components are combined again at the output layer to give the vector that is the sum of the retinal and eye position vectors. A single hidden unit's operation can thus be described as a sum of dot products, and is an elegant way of adding two vectors that are encoded in the activity of many input units.

A notable feature of the distributed representation of the Zipser-Andersen network is the absence of topography in the hidden layer. Maintenance of topographic relationships across processing stages can be an effective mechanism for processing spatial information. Several models of saccade generation, for example, use representations with well-defined spatial relationships in order to generate an appropriate saccadic command to look at a sensory stimulus (e.g., Droulez and Berthoz, 1991; Dominey and Arbib, 1992). In these models, the saccade vector is determined by which units are active within a given stage. Units in the hidden layer of the Zipser-Andersen network, on the other hand, are connected to every input and every output unit, and encode the head-centered position vector without regard to any input or output topography. The output vector is determined not by which units are active but by the activity level of every unit in the hidden layer. It is not clear whether PPC areas are topographically organized. The Zipser-Andersen model demonstrates that PPC neurons can transmit to other cortical areas the head-centered position of a stimulus, encoded in their collective firing rate, without any topographic organization.

Perturbing the Model

Stimulation of the Lateral Intra-Parietal (LIP) area in the PPC —a region that directly projects to eye movement centers and that is active during the programming of saccadic eye movements—elicits saccadic eye movements (Shibutani, Sakata, and Hyvärinen, 1984). Goodman and Andersen (1989) simulated the effect of electrically stimulating the PPC by setting the output of a hidden unit to its maximum possible value and interpreting the new position encoded by the output layer as the endpoint of the simulated gaze shift. This process was repeated for many initial eye positions. Because PPC neurons encode spatial locations in neither a retinotopic nor a head-centered reference frame, the effect of varying eye position on the elicited eye movement is not intuitively predictable. Knowing how the network computes coordinate transformations, however, allows us to predict what the effect should be.

Because a hidden unit's activity encodes the component of the head-centered vector along the unit's preferred direction, maximal activation of that unit will shift the network's output along the unit's preferred direction. This direction is encoded in the unit's weights and so should not be affected by the values of the inputs; i.e., the direction of the eye movement will be independent of the starting eye position. The movement's amplitude, on the other hand, depends on how far the unit's initial activation level is from the unit's maximum possible activation. The initial activation level is determined by the unit's input, and thus by the initial eye position. Thus, we expect the amplitude of the simulated saccades to depend on the initial eye position. In particular, as the initial eye position changes along directions orthogonal to the unit's preferred direction, the elicited saccade should be affected very little, because the unit's initial activation will remain the same. As the initial eye position changes along the same direction as the unit's preferred one, on the other hand, the unit's initial activation will vary, and therefore so will the change in the position encoded at the network's output layer when stimulation brings the hidden unit's activation to maximum.

The pattern of eye movements just described was indeed obtained by stimulation of most hidden units in a trained network (Goodman and Andersen, 1989). The elicited saccades had very similar directions from all starting eye positions, but their amplitude decreased as the eye position was shifted along one direction. The direction of this amplitude decrease was very similar to the direction of the elicited eye movement, indicating that the saccades were getting smaller as the eye moved along the unit's preferred direction, as predicted.

Thier and Andersen (1991) found that stimulation of area LIP elicited a pattern of saccades similar to that obtained from the neural network. The saccades evoked from various initial eye positions were all in the same direction. Their amplitude decreased as the starting eye position was moved in one direction, and remained the same as the initial position varied along the orthogonal direction.

Encoding the head-centered location of a stimulus is not the only way in which a saccade to that stimulus can be programmed. Another commonly proposed scheme maps the sensory vector falling on the retina (from the fovea to the stimulus's image) directly into a motor command encoding the required saccade vector, without ever computing the head-centered location of the stimulus. This method still requires some mechanism for keeping track of eye position, so that an appropriate saccade can be made to targets that appeared before one or more intervening eye movements. One such mechanism updates the planned saccade vector base on the last eye movement made. This method has been postulated as a cortical mechanism for saccade planning (Goldberg, Colby, and Duhamel, 1990) and has been used in saccade-generation models. In these models, the future saccade vector is remapped (see DYNAMIC REMAPPING) based either on the integrated eye velocity signal from the intervening saccade (Droulez and

Berthoz, 1991) or on a damped copy of the intervening saccade's eye position signal (Dominey and Arbib, 1992). The Zipser-Andersen model does not address the issue of multiple saccade plans. Extending the model to handle sequences of saccades, however, would not require a remapping scheme that kept track of intervening saccades. All saccade targets would be directly encoded in head-centered coordinates as they appear, and a saccade to each could be planned based only on the current eye position, independently of past eye movements.

Biological Plausibility of the Learning Algorithm

The biological plausibility of the Zipser-Andersen model was an issue of concern because the backpropagation algorithm is an unlikely candidate as a biological learning mechanism. To address this issue, Mazzoni, Andersen, and Jordan (1991) trained a neural network to perform the retinal-to-head-centered coordinate transformation using a reinforcement learning rule developed by Barto and Jordan (1987). This algorithm adjusts the network's connections based on a single error signal computed from the network's overall performance and on the local presynaptic and postsynaptic activation for each connection. Because it combines a reinforcement signal with Hebbian updating of connection strength, it is biologically more plausible than backpropagation (see REINFORCEMENT LEARNING). The hidden units of this network developed gain fields and receptive fields virtually identical to those of the backpropagation-trained networks. The networks' algorithm for computing coordinate transformations, therefore, did not depend on the specific learning mechanism used. The fact that this model learned the computed coordinate transformations based on a simple reinforcement signal also supported the idea that PPC neurons can learn to solve this task from simple feedback signals directly available to the nervous system. Moreover, because the properties of the hidden units are not specific to backpropagation training, the use of backpropagation in the original model does not invalidate its role as a model of PPC function.

Conclusion

The Zipser-Andersen network has been a valuable tool in the study of the PPC. It helped put into an explicit theoretical context many experimental results, and predicted a few additional ones. The original data was not easily summarized by an intuitive coding scheme, partly because the experiments addressed how neurons encode more than one variable. The neural network paradigm provided a framework for developing an intuition about the distributed representation of several variables. As more experiments address the encoding and interactions of several parameters in the nervous system, we expect neural networks to continue to fruitfully assist our investigations of nervous system functions.

Road Map: Primate Motor Control
Background: Perceptrons, Adalines, and Backpropagation
Related Reading: Grasping Movements: Visuomotor Transformations

References

Andersen, R. A., 1987, The role of the inferior parietal lobule in spatial perception and visual-motor integration, in *The Handbook of Physiology, Section 1: The Nervous System, Vol. IV, Higher Functions of the Brain, Part 2* (F. Plum, V. B. Mountcastle, and S. T. Geiger, Eds.), Bethesda, MD: American Physiological Society, pp. 483–518. ◆
Barto, A. G., and Jordan, M. I., 1987, Gradient following without backpropagation in layered networks, in *Proceedings of the IEEE*

International Conference on Neural Networks, vol. 2, New York: IEEE, pp. 629–636.

Brotchie, P. R., and Andersen, R. A., 1991, A body-centered coordinate system in posterior parietal cortex, *Soc. Neurosci. Abst.*, 17:1281.

Dominey, P. F., and Arbib, M. A., 1992, A cortical-subcortical model for generation of spatially accurate sequential saccades, *Cereb. Cortex*, 2:153–175.

Droulez, J., and Berthoz, A., 1991, A neural network model of sensoritopic maps with predictive short-term memory properties, *Proc. Natl. Acad. Sci. USA*, 88:9653–9657.

Goldberg, M. E., Colby, C. L., and Duhamel, J. R., 1990, Representation of visuomotor space in the parietal lobe of the monkey, *Cold Spring Harbor Symp. Quant. Biol.*, 55:729–739. ◆

Goodman, S. J., and Andersen, R. A., 1989, Microstimulation of a neural-network model for visually guided saccades, *J. Cog. Neurosci.*, 1:317–326.

Goodman, S. J., and Andersen, R. A., 1990, Algorithm programmed by a neural network model for coordinate transformation, in *Proceedings of the International Joint Conference on Neural Networks, 1990*, vol. 2, New York: IEEE, pp. 381–386.

Mazzoni, P., Andersen, R. A., and Jordan, M. I., 1991, A more biologically plausible learning rule for neural networks, *Proc. Natl. Acad. Sci. USA*, 88:4433–4437.

Shibutani, H., Sakata, H., and Hyvärinen, J., 1984, Saccade and blinking evoked by microstimulation of the posterior parietal association cortex of the monkey, *Exp. Brain Res.*, 55:1–8.

Thier, P., and Andersen, R. A., 1991, Electrical microstimulation delineates 3 distinct eye-movement related areas in the posterior parietal cortex of the rhesus monkey, *Soc. Neurosci. Abst.*, 17:1281.

Zipser, D., and Andersen, R. A., 1988, A backpropagation programmed network that simulates response properties of a subset of posterior parietal neurons, *Nature*, 331:679–684. ◆

Generalization and Regularization in Nonlinear Learning Systems

Grace Wahba

Introduction

In this article we will describe generalization and regularization from the point of view of multivariate function estimation in a statistical context. Multivariate function estimation is not, in principle, distinguishable from supervised machine learning. However, until fairly recently supervised machine learning and multivariate function estimation had fairly distinct groups of practitioners and little overlap in language, literature, and the kinds of practical problems under study.

In any case, we are given a *training set*, consisting of pairs of input (feature) vectors and associated outputs $\{\mathbf{t}(i), y_i\}$, for n training or example subjects, $i = 1, \ldots n$. From these data, it is desired to construct a map which *generalizes well*, that is, given a new value of $\mathbf{t}$, the map will provide a reasonable prediction for the unobserved output associated with this $\mathbf{t}$.

Most applications fall into one of two broad categories, which might be called nonparametric regression and classification. In *nonparametric regression*, y may be (any) real number or a vector of r real numbers. In *classification*, y is usually represented as a q-dimensional vector of zeros and ones, with a single 1 in the kth position if the example (subject) came from category k. In some classification applications, the desired algorithm will, given $\mathbf{t}$, return a vector of zeros and ones indicating a category assignment ("hard" classification). In other applications, it may be desired to return a q-vector of probabilities (that is, non-negative numbers summing to 1) which represent a forecast of the *probabilities* of an object with predictor vector $\mathbf{t}$ being in each of the q categories ("soft" classification).

In some problems the feature vector $\mathbf{t}$ of dimension d contains zeros and ones (as in a bitmap of handwriting); in other problems, it may contain real numbers representing physical quantities. In this article we will be generally concerned with the latter case, since the ideas of generalization and regularization are easiest to discuss when there is a convenient topology (for example, that determined by distance in Euclidean d-space) so that "closeness" and "smoothness" can be easily defined. *Regularization*, loosely speaking, means that some constraints are applied to the construction of the map with the goal of reducing the generalization error (see also REGULARIZATION THEORY AND LOW-LEVEL VISION). Ideally, these constraints em-

body a priori information concerning the true relationship between input and output; alternatively, various ad hoc constraints have sometimes been shown to work well in practice.

Generalization and Regularization in Nonparametric Regression

Single-Input Spline Smoothing

We will use Figure 1 to illustrate the ideas of generalization and regularization in the simplest possible nonparametric regression setup, that is, $d = 1$, $r = 1$, with $\mathbf{t} = t$ any real number in some interval of the real line. The boxed points (which are identical in each of the three panels of Figure 1) represent $n = 100$ (synthetically generated) input-output pairs $\{t(i), y_i\}$, generated according to the model

$$y_i = f_{TRUE}(t(i)) + \epsilon_i \qquad i = 1, \ldots, n \qquad (1)$$

where $f_{TRUE}(t) = 4.26(e^{-t} - 4e^{-2t} + 3e^{-3t})$, and the ϵ_i came from a pseudo-random number generator for normally distributed random variables with mean 0 and standard deviation 0.2. These figures are from Wahba and Wold (1975). Given these training data $\{t(i), y_i, i = 1, \ldots, n\}$, the learning problem is to create a map which, if given a new value of t, will predict the response $y(t)$. In this case, the data are noisy, so that even if the new t coincides with some predictor variable $t(i)$ in the training set, merely predicting y as the response y_i is not likely to be satisfactory. Also, this does not yet provide any ability to make predictions when t does not exactly match any predictor values in the training set. It is desired to generate some sort of curve, which will allow a reasonable prediction of the response for any t within a reasonable vicinity of the set of training predictors $\{t(i)\}$. The dashed line in each panel of Figure 1 is $f_{TRUE}(t)$; the three solid black lines in the three panels of Figure 1 are three solutions to the variational problem: Find f in the (Hilbert) space W_2 of functions with continuous first derivatives and square integrable second derivatives which minimizes

$$\frac{1}{n} \sum_{i=1}^{n} (y_i - f(t(i)))^2 + \lambda \int (f^{(2)}(u))^2 \, du \qquad (2)$$

for three different values of λ. The parameter λ is known as the

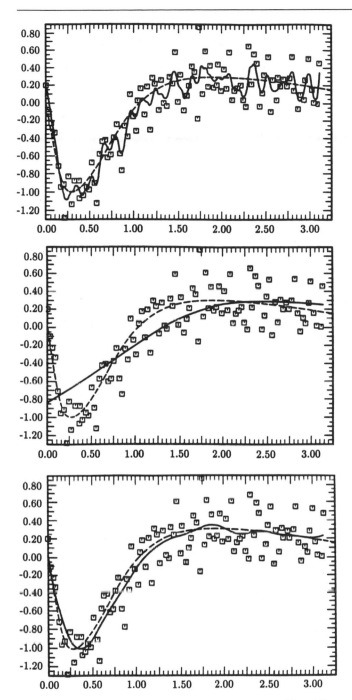

Figure 1. Training data (boxed points) have been generated by adding noise to $f_{TRUE}(t)$, shown by the dashed curve in each panel. All three panels have the same data. *Top*, Solid curve is fitted spline with λ too small. *Middle*, Solid curve is fitted spline with λ too large. *Bottom*, Solid curve is fitted spline with λ obtained by leaving-out-one cross validation.

regularization or smoothing parameter. As $\lambda \to \infty$, f_λ tends to the least squares straight line best fitting the data, and as $\lambda \to 0$ the solution tends to that curve in W_2 which minimizes the penalty functional $J(f) = \int (f^{(2)}(u))^2 \, du$ subject to interpolating the data (provided the $\{t(i)\}$ are distinct). This latter interpolating curve is known as a *cubic interpolating spline*, and minimizers of Equation 2 are known as *smoothing splines*. See Wahba (1990) and references cited there for further informa-

tion concerning these and other properties of splines, and for further references.

These splines have been studied at least since they were discussed by I. Schoenberg in the 1940s. Schoenberg gave the interpolating spline its name after the mechanical spline (a thin, flexible rod with weights attached at selected points), which was used by draftsmen for drawing smooth curves to represent cross sections of ships' hulls. In the top panel of Figure 1, λ has been chosen too small, and the wiggly solid line is attempting to fit the data too closely. It can be seen that using the wiggly curve in the top panel is not likely to give a good prediction of y, assuming that future predictor-response data is generated by the same mechanism as the training data. In the middle panel, λ has been chosen too large; the curve has been forced to flatten out, and again it can be seen that the solid line will not give a good prediction of y. In the bottom panel, λ has been chosen by a leaving-out-one cross validation (see Wahba and Wold, 1975), and it can be seen that the λ obtained in this way does a good job of choosing the right amount of smoothing to best recover f_{TRUE} of Equation 1. The f_{TRUE} of Equation 1 would provide the best predictor of the response in an expected mean-square-error sense if future data were generated according to Equation 1. The curve in the bottom panel has a reasonable ability to *generalize*, that is, to predict the response given a new value t of the predictor variable, at least if t is not too far from the training predictor set $\{t(i)\}$.

For each positive λ, there exists a unique $\kappa = \kappa(\lambda)$ so that the minimizer f_λ of Equation 2 is also the solution to the problem: Find f in W_2 to minimize

$$RSS(f) = \frac{1}{n} \sum_{i=1}^{n} (y_i - f(t(i)))^2 \qquad (3)$$

subject to the condition

$$J(f) = \int (f^{(2)}(u))^2 \, du \leq \kappa \qquad (4)$$

As λ becomes large, the associated $\kappa(\lambda)$ becomes small, and conversely. In general, the term *regularization* refers to solving some problem involving best-fitting, subject to some constraint(s) on the solution. These constraints may be of various forms. When they include a quadratic penalty involving derivatives, like $J(f)$, the method is commonly referred to as *Tikhonov regularization*. The "tighter" the constraints (i.e., the smaller κ, and the larger λ), the further away the solution f_λ will generally be from the training data; that is, RSS will be larger. As the constraints get weaker and weaker, ultimately (if there are enough degrees of freedom in the method) the solution will interpolate the data. However, as is clear from Figure 1, a curve which runs through all the data points is *not* a good solution.

A fundamental problem in machine learning with noisy or incomplete data is to balance the "tightness" of the constraints with the "goodness of fit" to the data, in such a way as to minimize the generalization error, that is, the ability to predict the unobserved response for new values of t (or $\mathbf{t}$). This trade-off, called the *bias-variance tradeoff* in the statistical literature, has been discussed in the context of machine learning by Geman, Bienenstock, and Doursat (1992). Numerous methods for curve fitting, other than smoothing splines, in the case $d = 1$, $r = 1$ have been proposed in the statistics literature. Popular methods include Parzen kernel estimates, nearest-neighbor estimates, orthogonal series estimates, and least squares spline estimates. See Eubank (1988) and the references cited there. Each method has one or more regularization parameters, either explicit or implicit, that control the bias-variance tradeoff.

Single-Input, Single-Hidden-Layer, Feedforward Neural Net

A single-input, single-hidden-layer, feedforward neural net (NN) predictor for the learning problem of the preceding section is typically of the form

$$f_{NN}(t) = \sigma_o\left(b_0 + \sum_{j=1}^{N} w_j\sigma_h(a_j t + b_j)\right) \tag{5}$$

where σ_h is the so-called "activation function" of the hidden layer and σ_o is the activation function for the output. σ_h is generally a sigmoidal function, for example, $\sigma_h(\tau) = e^\tau/(1 + e^\tau)$, while σ_o may be linear, sigmoidal, or a threshold unit. In the learning problem we have described, best results would likely be obtained with σ_o linear. Here N is the number of hidden units, and the w_j, a_j, and b_j are "learned" from the training data by some appropriate iterative gradient descent algorithm that tries to steer these values toward minimizing some distance measure, typically $RSS = \Sigma_{i=1}^n (f_{NN}(t(i)) - y_i)^2$. It is clear that if N is sufficiently large, and the descent algorithm is run long enough, it should be possible to drive the RSS as close as one likes to 0. (In practice it is possible to get stuck in local minima.) However, it is also clear intuitively from Figure 1 that driving RSS all the way to zero is not a desirable thing to do. Regularization in this problem may be done by controlling the size of N, by imposing penalties on the w_j, by stopping the descent algorithm early, that is, not driving down RSS as far as it can go, or by various combinations of these strategies. Each strategic choice will influence how closely f_{NN} will fit the data, how "wiggly" it will be, and how well it will be able to predict unobserved data that are generated by a mechanism similar to that of the observed data.

Multiple-Input, Single-Hidden-Layer, Feedforward Neural Net

For d greater than 1, the single-hidden-layer, feedforward neural net with a d-dimensional input and one-dimensional output is of the form

$$f_{NN}(\mathbf{t}) = \sigma_o\left(b_0 + \sum_{j=1}^{N} w_j\sigma_h(\mathbf{a}_j'\mathbf{t}(i) + b_j)\right) \tag{6}$$

where σ_o and σ_h are as before, but now the $\mathbf{a}_j$ and $\mathbf{t}$ are d-vectors. All the previous remarks concerning the regularizing of the network by controlling N, penalizing the w_j, and stopping short of driving RSS to a minimum continue to apply here.

Multiple-Input Radial Basis Function and Related Estimates

Radial basis functions are a popular method for nonparametric regression. (See the article in this volume on RADIAL BASIS FUNCTION NETWORKS.) We first describe a general form of nonparametric regression which will specialize to radial basis functions and other methods of interest. Let $R(\mathbf{s}, \mathbf{t})$ be *any* symmetric, strictly positive definite function on $E^d \times E^d$. Here, *strictly positive definite* means that, for any $K = 1, 2, \ldots$, the $K \times K$ matrix with j, kth entry $R(\mathbf{s}(j), \mathbf{s}(k))$ is strictly positive definite whenever the $\mathbf{s}(1), \ldots, \mathbf{s}(K)$ are distinct. (A symmetric $K \times K$ matrix M is said to be positive definite if for any K-dimensional column vector x, $x^\mathrm{T}Mx$ is greater than or equal to 0, and is said to be strictly positive definite if $x^\mathrm{T}Mx$ is always strictly greater than 0.) Positive definiteness will play a key role in the following discussion because (among other reasons) any positive definite matrix can be the covariance matrix of a random vector and any positive definite function $R(\mathbf{s}, \mathbf{t})$ can be the

covariance function of some stochastic process, $X(\mathbf{t})$. That is, there exists $X(\cdot)$ such that $\mathrm{Cov}\, X(\mathbf{s})X(\mathbf{t}) = R(\mathbf{s}, \mathbf{t})$. Given training data $\{\mathbf{t}(i), y_i\}$, it is always possible in principle to obtain a (regularized) input-output map from these data by letting the model $f_{R,\lambda}$ be of the form

$$f_{R,\lambda}(\mathbf{t}) = \sum_{j=1}^{N} c_j R(\mathbf{t}, \mathbf{s}(j)) \tag{7}$$

where the $\mathbf{s}(j)$ are $N \le n$ "centers" which are placed at distinct values of the $\{\mathbf{t}(i)\}$ and $c = (c_1, \ldots, c_N)^\mathrm{T}$ is chosen to minimize $RSS(f) + \lambda J(f)$. Here

$$RSS(f_{R,\lambda}) = \sum_{i=1}^{n} (f_{R,\lambda}(\mathbf{t}(i)) - y_i)^2 \tag{8}$$

and the regularizing penalty $J(\cdot)$ is of the form

$$J(f_{R,\lambda}) = \sum_{j,k=1}^{n} c_j c_k J_{jk} \tag{9}$$

where J_{jk} are the entries of a non-negative definite quadratic form. The (strict) positive definiteness of R guarantees that

$$RSS(f_{R,\lambda}) + \lambda J(f_{R,\lambda}) \tag{10}$$

always has a unique minimizer in c, for any non-negative λ. This follows by substituting Equation 7 into Equation 10, and using the fact that the columns of the $n \times N$ matrix with i, j entry $R(\mathbf{t}(i), \mathbf{s}(j))$ are linearly independent since they are just N columns of the $n \times n$ positive definite matrix with i, j entry $R(\mathbf{t}(i), \mathbf{t}(j))$.

Radial basis function estimates are obtained for the special case where $R(\mathbf{s}, \mathbf{t})$ is of the special form

$$R(\mathbf{s}, \mathbf{t}) = r(\|W(\mathbf{s} - \mathbf{t})\|) \tag{11}$$

where W is some linear transformation on E^d and the norm is Euclidean distance. That is, $R(\mathbf{s}, \mathbf{t})$ depends only on some generalized distance in E^d between $\mathbf{s}$ and $\mathbf{t}$. The regularization, that is, the effecting of the tradeoff between goodness of fit to the data and "smoothness" of the solution, is performed by reducing N and/or increasing λ. The choice of W will also affect the "wiggliness" of $f_{R,\lambda}$ in the radial basis function case. Alternatively, a model can be obtained by choosing N small and minimizing $RSS(f)$. In that case N and W are the smoothing parameters.

In the special case $N = n$, $\mathbf{s}(i) = \mathbf{t}(i)$, the $f_{R,\lambda}$ can (for *any* positive definite R) be shown to be Bayes estimates (see Kimeldorf and Wahba, 1970; Wahba, 1990, 1992; Girosi, Jones, and Poggio, 1994). Arguments can be given to show that if n is large and $N < n$ is not too small, then they are good approximations to Bayes estimates (see Wahba, 1990, chap. 7). In the special case $J_{i,j} = R(\mathbf{t}(i), \mathbf{t}(j))$, the Bayes model is easy to describe:

$$y_i = X(\mathbf{t}(i)) + \epsilon_i \tag{12}$$

with $X(\mathbf{t})$ a zero-mean Gaussian stochastic process with covariance $EX(\mathbf{s})X(\mathbf{t}) = bR(\mathbf{s}, \mathbf{t})$ and the ϵ_i independent zero-mean Gaussian random variables with common variance σ^2, and independent of $X(\mathbf{t})$. In this case, the minimizer $f_{R,\lambda}$ of $RSS(f) + \lambda J(f)$, evaluated at $\mathbf{t}$, is the conditional expectation of $X(\mathbf{t})$, given $y_1, \ldots, y_n$, provided that λ is chosen as σ^2/nb. In general, pretending that one has a prior and computing the posterior mean or mode will have a regularizing effect. However, the degree of regularization (choice of λ here) may not be the same as what one would get by attempting to minimize the bias-variance tradeoff if the prior is not correct. (See the later section on choosing how much to regularize.)

Thin plate splines in d variables (of order m) consist of radial basis functions plus polynomials of total degree less than m in

d variables. ($2m - d > 0$ is required for technical reasons.) Letting $\mathbf{t} = (t_1, \ldots, t_d)$, the thin plate splines are minimizers (in an appropriate function space) of

$$\frac{1}{n} \sum_{i=1}^{n} (y_i - f(\mathbf{t}(i)))^2 + \lambda \sum_{\alpha_1 + \cdots + \alpha_d = m} \frac{m!}{\alpha_1! \cdots \alpha_d!}$$
$$\int_{-\infty}^{\infty} \cdots \int_{-\infty}^{\infty} \left(\frac{\partial^m f}{\partial t_1^{\alpha_1} \cdots \partial t_d^{\alpha_d}} \right)^2 dt_1 \cdots dt_d \qquad (13)$$

Setting $d = 1$, $m = 2$ gives the cubic spline case discussed earlier. Note that there is no penalty on polynomials of total degree less than m; the thin plate splines with a particular choice of λ are Bayes estimates with an improper prior (that is, infinite variance) on the polynomials of total degree less than m (see Wahba, 1990, and the references cited there).

There are a number of variations on regularized estimates. Additive smoothing splines are of the form

$$f(\mathbf{t}) = \mu + \sum_{\alpha=1}^{d} f_\alpha(t_\alpha) \qquad (14)$$

where μ and the f_α are the solution to a variational problem of the form: Find μ and $f_1, \ldots, f_d$ in a certain function space to minimize

$$\sum_{i=1}^{n} (f(\mathbf{t}(i) - y_i))^2 + \sum_{\alpha=1}^{d} \lambda_\alpha J_\alpha(f_\alpha) \qquad (15)$$

The J_α may be of the form of J in Equation 4. Here, there is a *regularization parameter* for each component. See Hastie and Tibshirani (1990) and Wahba (1990). These additive models generalize to smoothing spline analysis of variance models (SS-ANOVA), whereby terms of the form $f_{\alpha\beta}, f_{\alpha\beta\gamma}$, etc. are added to the representation in Equation 14, and corresponding penalty terms with regularization parameters are added in Equation 15. The $f_\alpha, f_{\alpha\beta}$, etc. may be generalized to themselves being radial basis functions. See Gu and Wahba (1993).

Regression spline ANOVA models can be obtained by setting the $f_\alpha, f_{\alpha\beta}$, etc. as linear combinations of a (relatively small) number of basis functions (usually splines). In this case the number of the basis functions is probably the most influential regularization parameter. (See Hastie and Tibshirani, 1990; Friedman, 1991; and the references cited in those sources.) These and similar methods all have either explicit or implicit regularization parameters which govern the balance between the complexity of the model and the fit to the data—the bias-variance tradeoff. Most of them use some form of cross validation to make this tradeoff. Other references may be found in Ripley (1994) and Wahba (1992).

Generalization and Regularization in Classification

The top panel of Figure 2 is a scatterplot from Wahba et al. (1994) of body mass index (bmi) and age for 669 subjects from the Wisconsin Epidemiologic Study of Diabetic Retinopathy. Subjects who had diabetic retinopathy that progressed are indicated with a plus sign; others, with a dot. We can then consider each plus sign as a 1 and each dot as a 0; thus y_i is 1 or 0. If the plusses and dots were reasonably separable by some simple partition of the square in the top panel of Figure 2, the classification problem would be one of finding a reasonably simple description of this partition. Then a "hard" classification would be made by assigning 1 or 0 to points according to which part of the partition they are in. In this kind of data, it is frequently of interest to make a "soft" classification, that is, to estimate the *probability* $p(\mathbf{t})$ that a subject with predictor vector $\mathbf{t} =$ (bmi, age) will become a 1. (This argument carries over to sev-

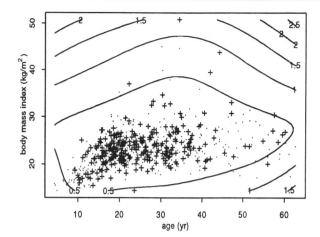

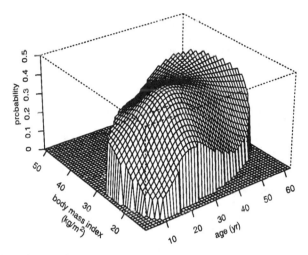

Figure 2. *Top,* Scatterplot of body mass index versus age. A + sign indicates subjects whose diabetic retinopathy progressed; the dots indicate other subjects. *Bottom,* Estimate of probability of progression as a function of body mass index and age.

eral classes, but for ease of discussion, we will only consider two classes.) In this case, some regularized (that is, "smooth") estimate for $p(\mathbf{t})$ is desirable, since it would be highly unreasonable for the estimate to pass through the data, that is, to take on the value 1 at the plusses and 0 at the dots. Regularized estimates can be obtained as follows. First, define

$$f(\mathbf{t}) = \log[p(\mathbf{t})/(1 - p(\mathbf{t}))] \qquad (16)$$

where f is known in the statistics literature as the *logit*. Then $p(\mathbf{t})$ is a sigmoidal function of $f(\mathbf{t})$, that is, $p(\mathbf{t}) = e^{f(\mathbf{t})}/(1 + e^{f(\mathbf{t})})$. (Other sigmoidal functions may be used.) We will get a regularized estimate for f. $RSS(f)$ of Equation 3 will be replaced by an expression more suitable for 0-1 data, by using the likelihood for these data. To describe the likelihood, note that if y is a random variable with $\text{Prob}[y = 1] = p$ and $\text{Prob}[y = 0] = (1 - p)$, then the probability density (or likelihood) $P(y, p)$ for y, when p is true, is just $P(y, p) = p^y(1 - p)^{(1-y)}$. This merely says that $P(1, p) = p$ and $P(0, p) = (1 - p)$. Thus, the likelihood for $y_1, \ldots, y_n$ (assuming that the y_i are independent) is

$$P(y_1, \ldots, y_n; p(\mathbf{t}(1)), \ldots, p(\mathbf{t}(n)))$$
$$= \prod_{i=1}^{n} p(\mathbf{t}(i))^{y_i}(1 - p(\mathbf{t}(i)))^{(1-y_i)} \qquad (17)$$

Substituting f for p in Equation 17, taking the negative logarithm, and suppressing the $\{y_i\}$ in the notation gives the negative log likelihood $L(f)$ in terms of f:

$$-\log P(y_1, \ldots, y_n; p(\mathbf{t}(1)), \ldots, p(\mathbf{t}(n))) \equiv L(f)$$

$$= \sum_{i=1}^{n} [\log(1 + e^{f(t(i))}) - y_i f(\mathbf{t}(i))] \tag{18}$$

It is natural for $L(f)$ to replace $RSS(f)$ of Equation 3 when y_i is restricted to 0 or 1, since $RSS(f)$ is just (a linear function of) the negative log likelihood for y_i generated according to Equation 1. A neural net implementation of soft classification would consist of finding $f_{NN}(\mathbf{t}) = \text{logit } p_{NN}(\mathbf{t})$ of the form of Equation 6 to minimize $L(f)$ of Equation 18. If N is large enough, then, in principle, f_{NN} may be driven so that $p_{NN}(\mathbf{t}(i))$ is close to 1 if y_i is 1, and is close to 0 if y_i is 0. Again, it is intuitively clear that this is not desirable. As before, a regularized or smooth f_{NN} can be obtained by controlling N, penalizing the w_i, stopping the iterative fitting early, or some combination of these actions. The bottom panel in Figure 2 gives the estimated probability of progression of diabetic retinopathy (estimated probability of a 1), as a function of bmi and age. This figure (from Wahba et al., 1994) is actually a cross section of a model of the form

$$f(\text{age, gly, bmi}) = \mu + f_1(\text{age}) + b \cdot \text{gly} + f_3(\text{bmi})$$

$$+ f_{13}(\text{age, bmi}) \tag{19}$$

where gly (glycosylated hemoglobin) was held fixed at the median value of the training set data for plotting purposes. μ, b, f_1, f_3, and f_{13} were obtained by finding f of the form of Equation 19 to minimize

$$L(f) + \lambda_1 J_1(f_1) + \lambda_3 J_3(f_3) + \lambda_{13} J_{13}(f_{13}) \tag{20}$$

in an appropriate function space. The component f's are smoothing splines, and the regularization or smoothing parameters λ_1, λ_3, and λ_{13} have been chosen according to a multivariate version of the iterative unbiased risk criteria in Gu (1992). Larger values of the λ's would have caused the plot to flatten out, and smaller values would have caused it to be "wiggly."

Choosing How Much to Regularize

At the time of this writing, it is a matter of lively debate and much research how to choose the various regularization parameters. Leaving out a large fraction of the training sample for this purpose and tuning the regularization parameter(s) to best predict the set-aside data (according to whatever criteria of best prediction is adopted) is conceptually simple, defensible, and widely used (this is called out-of-sample tuning). Successively leaving-out-one, successively leaving-out-10%, and generalized cross validation are all popular. If the variance of the observational error is known, then unbiased risk estimates become available. See Li (1986), Wahba (1990) and references cited there, and Gu (1992) for these "in-sample" tuning methods. When there is a Bayesian model behind the regularization procedure, maximum likelihood estimates may be derived (see, for example, Wahba, 1985), although in order for these and other Bayes estimates to do a good job of minimizing the generalization error in practice, it is usually necessary that the priors on which they are based be realistic.

Which Method Is Best?

Feedforward neural nets, radial basis functions, and various forms of splines all provide regularized or regularizable methods for estimating "smooth" functions of several variables, given a training set $\{\mathbf{t}(i), y_i\}$. Which approach is best? Unfortunately, there is not, nor is there likely to be, a single answer to that question. The answer most surely depends on the particular nature of the underlying but unknown "truth," the nature of any prior information that might be available about this "truth," the nature of any noise in the data, the ability of the experimenter to choose the various smoothing or regularization parameters well, the size of the data set, the use to which the answer will be put, and the computational facilities available. From a mathematical point of view, the classes of functions well approximated by neural nets, radial basis functions, and sums and products of splines (ANOVA splines) are not the same, although all of these methods have the capability of approximating large classes of functions. Of course, if a large enough data set is available, models utilizing all of these approaches may be built, and tuned, and then compared on data that have been set aside for this purpose.

Acknowledgments. This work was supported by NSF Grant DMS 9121003 and NIH Grant R01 EY09946.

Road Map: Learning in Artificial Neural Networks, Statistical
Related Reading: Bayesian Methods for Supervised Neural Networks

References

Eubank, R., 1988, *Spline Smoothing and Nonparametric Regression*, New York: Marcel Dekker. ◆

Friedman, J., 1991, Multivariate adaptive regression splines, *Ann. Statist.*, 19:1–141.

Geman, S., Bienenstock, E., and Doursat, R., 1992, Neural networks and the bias/variance dilemma, *Neural Computat.*, 4:1–58. ◆

Girosi, F., Jones, M., and Poggio, T., 1994, Regularization theory and neural networks architectures, *Neural Computat.* (to appear). ◆

Gu, C., 1992, Cross-validating non-Gaussian data, *J. Comput. Graph. Stats.*, 1:169–179.

Gu, C., and Wahba, G., 1993, Semiparametric analysis of variance with tensor product thin plate splines, *J. R. Statist. Soc. Ser. B*, 55:353–368.

Hastie, T., and Tibshirani, R., 1990, *Generalized Additive Models*, New York: Chapman and Hall. ◆

Kimeldorf, G., and Wahba, G., 1970, A correspondence between Bayesian estimation of stochastic processes and smoothing by splines, *Ann. Math. Statist.*, 41:495–502.

Li, K. C., 1986, Asymptotic optimality of C_L and generalized cross validation in ridge regression with application to spline smoothing, *Ann. Statist.*, 14:1101–1112.

Ripley, B., 1994, Neural networks and related methods for classification, *J. R. Statist. Soc. Ser. B*, 56:409–437. ◆

Wahba, G., 1985, A comparison of GCV and GML for choosing the smoothing parameter in the generalized spline smoothing problem, *Ann. Statist.*, 13:1378–1402.

Wahba, G., 1990, *Spline Models for Observational Data*, CBMS-NSF Regional Conference Series in Applied Mathematics, vol. 59, Philadelphia: Society for Industrial and Applied Mathematics. ◆

Wahba, G., 1992, Multivariate function and operator estimation, based on smoothing splines and reproducing kernels, in *Nonlinear Modeling and Forecasting* (M. Casdagli and S. Eubank, Eds.), SFI Studies in the Sciences of Complexity, Proc. vol. XII, Reading, MA: Addison-Wesley, pp. 95–112.

Wahba, G., Wang, Y., Gu, C., Klein, R., and Klein, B., 1994, Structured machine learning for "soft" classification with smoothing spline ANOVA and stacked tuning, testing and evaluation, in *Advances in Neural Information Processing Systems 6* (J. Cowan, G. Tesauro, and J. Alspector, Eds.), San Mateo, CA: Morgan Kaufmann, pp. 415–422.

Wahba, G., and Wold, S., 1975, A completely automatic French curve, *Commun. Statist.*, 4:1–17. ◆

"Genotypes" for Neural Networks

Stefano Nolfi and Domenico Parisi

Introduction

Neural networks are computational models of nervous systems. However, organisms do not possess only nervous systems and other phenotypic traits but also genetic information stored in the nucleus of their cells (genotype). The nervous system is part of the phenotype which is derived from this genotype through a process called *development*. The information specified in the genotype determines aspects of the nervous system which are expressed as innate behavioral tendencies and predispositions to learn. When neural networks are viewed in the broader biological context of "artificial life" (Langton, 1992), they tend to be accompanied by genotypes and to become members of evolving populations of networks in which genotypes are inherited from parents to offspring.

To study evolution computationally, genetic algorithms can be used (Holland, 1975). An initial population of artificial organisms is tested to determine the fitness of each individual, and individuals reproduce as a function of their fitness. To maintain population variability, an offspring's genotype can be a new combination of parts of its two parents' genotypes (sexual reproduction) and/or random modifications can be introduced in the reproduction process (mutations). If organisms are modeled by neural networks, what an offspring inherits from its parent(s) is a blueprint for the offspring's neural network.

The inherited genotype can completely specify the phenotypic network; i.e., both the network's architecture and connection weights are genetically determined. In this case, the behavior of the network is entirely innate and there is no learning (Parisi, Cecconi, and Nolfi, 1990). Or the genotype specifies the network's architecture, but the weights are learned (Miller, Todd, and Hedge, 1989). In still other cases, what is selected during evolution are good initial weights for learning or good learning rates and momentums (Belew, McInerney, and Schraudolph, 1991).

Development as Mapping from Genotypes to Phenotypes

To apply genetic algorithms to neural networks, it is necessary to codify the neural network (phenotype) into a string (genotype). However, the way in which this coding should be realized is not straightforward. In most current models, the representations of the genotypic and phenotypic forms coincide. That is, the inherited genotype directly and literally describes the phenotypical neural network. This approach encounters problems of scalability (Kitano, 1990). The number of bits of information necessary to encode a network increases exponentially with the number of neurons of the network and, as a consequence, the space to be searched by the evolutionary process also increases exponentially.

Aside from scalability problems, a direct genotype-to-phenotype mapping is biologically implausible. In real life, we cannot predict which phenotype will emerge from a given genotype because of the large nonlinearities present in the mapping process. If the genotype is viewed as a set of instructions, it is not the case that each of these instructions will result in a single network property. Rather, the properties of the network emerge as the result of many interactions among the various genetic instructions and their products (see SELF-ORGANIZATION AND THE BRAIN).

Another problem concerns the nature of the network which is encoded in the genotype. In conformity to standard practice, almost all proposed schemes for the genotypic encoding of network architecture (Miller et al., 1989; Kitano, 1990) view this architecture as a topological structure or connectivity matrix. On the contrary, real nervous systems are physical objects in three-dimensional space and neurons that are responsible for a single function tend to be located in the same part of the brain. Developmentally, the probability that two neurons will end up as being connected depends on the physical location of the two neurons. (For an account of development that emphasizes the role of spatial information, cf. DEVELOPMENT AND REGENERATION OF EYE-BRAIN MAPS.)

In this section, we will describe a method for codifying a neural network architecture into a genetic string which is inspired by the characteristics of neural development in real animals. Inherited genetic material specifies developmental instructions that control the axonal growing and branching process of a set of neurons. During the growth process (Figure 1), connections between neurons with their synaptic weights are established. When a growing axonal branch of a particular neuron reaches another neuron, a connection between the two neurons is established. Hence, the spatial location of neurons is critical in determining the resulting connectivity.

We are interested in the evolution of organisms (Os) that can efficiently capture food elements randomly distributed in a two-dimensional environment. Os have a facing direction and a rudimentary sensory system that allow them to receive as input a coding of the angle (measured clockwise relative to where Os are currently facing) and distance of the nearest food element. Os are also equipped with a simple motor system that provides them with the possibility, in a single action, to turn through any angle from 90° left to 90° right and then move from zero to five steps forward. Finally, when Os happen to step on a food cell, they eat the food element, which then disappears.

To evolve Os that are able to reach food elements, a genetic algorithm is applied to the Os' "genotypes." We begin with 100 randomly generated genotypes. This is Generation 0 (G0). G0 networks are allowed to "live" for eight epochs, an epoch consisting of 250 actions in five different environments (50 actions in each). The environment is a grid of 40 × 40 cells with 10 pieces of food randomly distributed in it. The 20 individuals that have accumulated the most food in the course of their life are allowed to reproduce agametically by generating five copies of their genotype. These 20 × 5 = 100 new organisms constitute the next generation (G1). Random mutations are introduced in the copying process (by replacing 20 bits with a randomly selected value), resulting in possible changes of the architecture and of the values of the weights. The process is repeated for 1,000 generations.

Os' genotypes are bit strings divided up into 40 blocks, each block corresponding to the growth instructions for a single neuron. The first five blocks correspond to sensory neurons, the last five blocks to motor neurons, and the 30 intermediate blocks to internal neurons. The standard logistic function is used to compute the activation value of the neurons. Each block of the genotype contains instructions that determine the properties of the corresponding neuron. A neuron-expression gene determines if the corresponding neuron will be present or

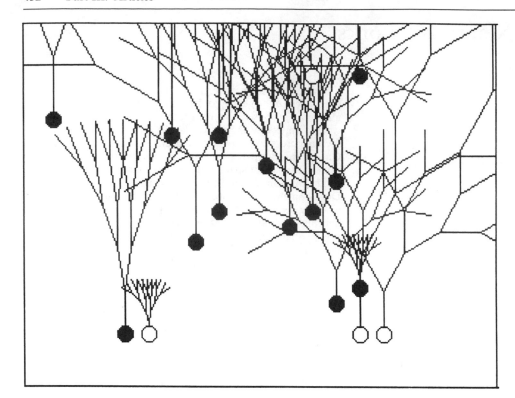

Figure 1. Development of a neural network from a randomly generated genetic string.

not in Os' nervous system. Two physical-position genes specify the Cartesian coordinates of the neuron in two-dimensional space. A branching-angle gene and a segment-length gene determine the angle of branching of the neuron's axon and the length of the branching segments. A synaptic-weight gene determines the synaptic coefficients of all the connections that will be established by the neuron (all connections departing from the same neuron have the same weight). A bias gene represents the activation bias or threshold of the neuron. A neuron-type gene specifies, in the case of a sensory neuron, if the sensory neuron codifies angle or distance of food, and in the case of a motor neuron, if the motor neuron codifies the angle of turn or the amplitude of the motor step.

By looking at Os' fitness (i.e., number of food elements eaten) throughout the 1000 generations, we observe the evolutionary emergence of Os that are increasingly able to approach food elements. After 1000 generations, an average of 72% of the available food is eaten in 10 replications of the simulation. In the best of these 10 simulations, a value of 89% is reached.

If we analyze how network architecture changes in these individuals during the course of evolution, we notice that evolved architectures are extremely simple. Similar simulations conducted with networks with an externally decided fixed architecture require many more connections and result in poorer performance.

By analyzing how network architecture changes in these individuals during the course of evolution, we observe regularities in the evolutionary development of the Os' sensory and motor systems. In the very first generations (cf. G0 of Figure 2), Os have neural architectures which take into account only food direction (angle) on the sensory side, and control only direction of movement on the motor side. Some generations later (from G5), we see the emergence of architectures that control the amplitude of the motor step in addition to direction of movement, but still rely on sensory information which is restricted to food direction. Only from G225 do Os become

sensitive to food distance as well. This type of progression in the selection of sensory and motor neurons across generations is observed in most of the simulations performed.

It is also interesting that, despite the simplicity of the task, neural architectures progressively structure themselves into functional subnetworks or neural modules. (For the importance of structure in the architecture of "what" and "where" neural networks, cf. Rueckl, Cave, and Kosslyn, 1989). At G999, the division of Os' neural system into two functional subnetworks is evident. There is a larger subnetwork which computes Os' direction of movement as a function of the state of the neuron sensitive to food angle, and a smaller subnetwork that controls Os' movement speed as a function of food distance.

Finally, we observe that the complexity of the subnetworks corresponds to the complexity of the mapping performed by these subnetworks. The mapping between the angle of the nearest food element and the angle of turn should be relatively complex in order to ensure high performance. On the contrary, the mapping between the distance of the nearest food element and the amplitude of the motor step can be much simpler. What is observed in the simulations is that the evolved subnetworks that are responsible for computing the angle of turn are always more complex than the subnetworks that are responsible for computing the amplitude of the motor step.

Learning and Evolution

The networks described so far do not learn during their lifetime. Their behavior (searching for food) is entirely genetically transmitted. To investigate the interaction between learning and evolution, a new set of simulations was run using genotypes that encode the weights of a fixed network architecture. The architecture includes two parts: a standard network and a teaching network (cf. Figure 3). The two networks share the same input units but have separate sets of internal and output

Figure 2. Functional architectures of individuals of the most successful simulation. Within the sensory layer, filled circles represent neurons sensitive to the angle of the nearest food element; empty circles represent neurons sensitive to its distance. In the motor layer, filled circles represent motor neurons that codify the angle of turn; empty circles represent neurons that codify the amplitude of the motor step.

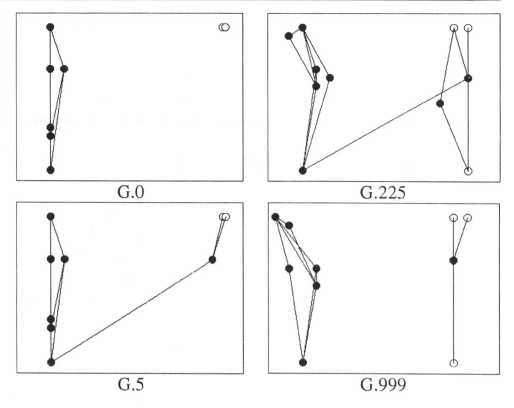

G.0   G.225

G.5   G.999

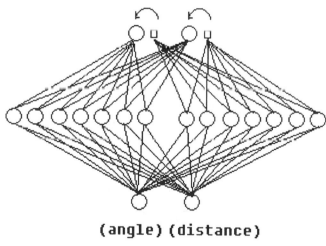

motor action

(angle) (distance)

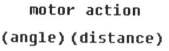

(angle) (distance)

sensory input

Figure 3. Os' architecture.

units. The output units of the standard network are interpreted as generating motor actions, while the output units of the teaching network (i.e., the teaching units) are interpreted as generating teaching input for the motor output units of the standard network.

When an O is placed in the environment that has been described above, a sequence of events will occur. Sensory input is received on the input units. Activation flows from the input units to both the hidden units of the standard network and to those of the teaching network, and then from these hidden units separately to the output units of the standard network (encoding motor actions) and to the output units of the teaching network (encoding teaching input for the standard network). The activation values of the motor output units are used to move O in the manner specified by these values, thereby changing the sensory input for the next cycle. At the same time, the backpropagation algorithm uses the discrepancy between the two sets of output units to change the connection weights of the standard network (a learning rate of 0.15 was used).

As in the preceding simulations, Os should be able to capture food elements randomly distributed in the environment. Os are allowed to "live" for 20 epochs. Mutations are introduced in the copying process by selecting at random four weights of the standard subnetwork and four weights of the teaching subnetwork, and adding a random value between $+1.0$ and -1.0 to their current value. All other parameters are those of the preceding simulations.

By looking at Os' fitness across generations, it is observed that Os that are increasingly able to approach food elements tend to evolve. Interestingly, evolution does not select Os which perform well at birth but selects Os learning to perform well—i.e., Os that have a predisposition to learn. Os' performance at birth does not increase across generations. What increases is the ability to learn the desired performance. This genetically transmitted ability to learn is incorporated in the weights of both the teaching network and the standard network. The teaching weights generate increasingly good teaching input for the standard network. The standard weights are selected for their ability to let a good behavior emerge as a result of learning. By randomly changing either sets of weights, no useful behavior emerges.

Conclusions

For neural networks to be credible as models of the nervous system of real organisms (aside from the inevitable simplifications of simulations), they must allow us to study not only the functioning of the adult nervous system, but also neural evolution and neural development. To this purpose, what is needed are distinct representations for inherited genotypes and for phenotypic networks and some model of the complex mapping from genotypes to phenotypes (development).

This article has presented some models of genotypes for neural networks and of the genotype-to-phenotype mapping. Genotypes can directly incorporate innate behaviors that require no learning, or they can incorporate innate predispositions to learn some behaviors. Innate behaviors can be present at birth (congenital) or they can develop during life under the control of genetically specified information. With regard to mapping the information contained in the genotype into the phenotypic network, it can be direct and one-to-one or, with more biological plausibility, it may be more indirect and many-to-many.

In multicellular organisms, other important processes involved in the mapping between genotype and phenotype are cell division (one cell becomes two daughter cells) and the migration of neurons to reach their terminal position in the brain. These processes are ignored in the models described in this article but have been examined in other models (Wilson, 1989). In Cangelosi, Parisi, and Nolfi (in press), both cell division and migration and the growth of neurites are included in the same model. Another crucial property of the genotype/phenotype mapping in biological organisms is its temporal character. Biological development (or maturation) does not simply yield an "individual"; rather, the phenotypic individual is a succession of different phenotypic forms which are sequentially generated by the genotype in interaction with the environment. A model in which the mapping between genotype and phenotype (i.e.,

ontogeny) takes place during the individual's lifetime and is influenced both by the genotype and by the external environment is described in Nolfi, Miglino, and Parisi (in press).

Road Map: Learning in Artificial Neural Networks, Statistical

References

Belew, R. K., McInerney, J., and Schraudolph, N. N., 1992, Evolving networks: Using the Genetic Algorithm with connectionist learning, in *Artificial Life II* (C. G. Langton, Ed.), Reading, MA: Addison-Wesley, pp. 511–547.

Cangelosi, A., Parisi, D., and Nolfi, S., in press, Cell division and migration in "genotypes" for neural networks, *Network*.

Hinton, G. E., and Nowlan, S. J., 1987, How learning guides evolution, *Complex Systems*, 1:495–502. ◆

Holland, J. J., 1975, *Adaptation in Natural and Artificial Systems*, Ann Arbor, MI: University of Michigan Press. ◆

Kitano, H., 1990, Designing neural networks using genetic algorithms with graph generation system, *Complex Systems*, 4:461–476.

Langton, C. G., 1992, Artificial life, in *Lectures in Complex Systems* (L. Nadel and D. Stein, Eds.), Reading, MA: Addison-Wesley, pp. 189–242. ◆

Miller, G. F., Todd, P. M., and Hedge, S. U., 1989, Designing neural networks using genetic algorithms, in *Proceedings of the Third International Conference on Genetic Algorithms* (L. Nadel and D. Stein, Eds.), San Mateo, CA: Morgan Kaufmann, pp. 379–384.

Nolfi, S., Miglino O., and Parisi, D., in press, Phenotypic plasticity in evolving neural networks, in *Proceedings of the International Conference from Perception to Action* (D. P. Gaussier and J.-D. Nicoud, Eds.), Los Alamitos, CA: IEEE Press.

Parisi, D., Cecconi, F., and Nolfi, S., 1990, Econets: Neural networks that learn in an environment, *Network*, 1:149–168. ◆

Rueckl, J. G., Cave, K. R., and Kosslyn, S. M., 1989, Why are "what" and "where" processed by separate cortical visual systems? A computational investigation, *J. Cognit. Neurosci.*, 1:171–186.

Wilson, S. W., 1989, The genetic algorithm and simulated evolution, in *Artificial Life* (C. G. Langton, Ed.), Reading, MA: Addison-Wesley, pp. 157–166.

Geometrical Principles in Motor Control

Ferdinando A. Mussa-Ivaldi

Introduction

The central role played by geometry in the control of motor behaviors was recognized several decades ago by the Russian investigator Nikolai Bernstein. Using only the tool of logical reasoning applied to common observations, Bernstein reached the conclusion that "there exist in the higher levels of the C.N.S. projections of space, and not projections of joint and muscles" (Bernstein, 1967). This intuition led other investigators to consider motor planning and execution as separate processes. The transition from the spatial representation of motor goal to a set of appropriate neuromuscular commands is in many respects similar to a *coordinate transformation*. This analogy is indeed the focus of this article. We will begin by describing three main types of coordinate systems, each one representing a particular point of view on motor behavior. Then, we will examine the geometrical rules that govern the transformations between these classes of coordinates. Finally, we will show how a proper representation of dynamics may

greatly simplify the transformation of motor planning into action.

Coordinate Systems for Motor Control

Endpoint Coordinates

Consider a monkey in the act of reaching for an apple with a wooden stick. The free extremity of the stick is the site at which the monkey interacts with its environment. Borrowing some terminology from robotics, we call such a site an *endpoint*. The position of the stick is fully determined by six coordinates. These coordinates are measured with respect to three orthogonal axes originating, for example, from the monkey's shoulder. Why six coordinates? Because this is the smallest set of numbers needed to specify unambiguously the location and orientation of a rigid object in three-dimensional space. Then, in our example, a position in endpoint coordinates is a point $p = (x, y, z, \theta_x, \theta_y, \theta_z)$. The coordinates x, y, and z determine a translation with respect to the orthogonal axes. The angular coordinates, θ_x, θ_y, and θ_z, determine an orientation with respect to the same axes. Consistent with this notation, a *force* in endpoint coordinates is a vector with a system of three linear and three angular components:

$$F = (F_x, F_y, F_z, \tau_x, \tau_y, \tau_z)$$

Generalized Coordinates

A different way of describing the position of the monkey's arm is to provide the set of joint angles that define the orientation of each skeletal segment, either with respect to fixed axes in space or with respect to the neighboring segments. Joint angles are a particular instance of *generalized coordinates*. According to the standard definitions of analytical mechanics, generalized coordinates are *independent variables* that are suitable for describing the dynamics of a system (Goldstein, 1980).

Once we have defined a set of generalized coordinates, we may also define a *generalized force* vector. For example, if we use joint angles as generalized coordinates, the corresponding generalized forces are the torques measured at each joint. The dynamics of any mechanical system with N generalized coordinates are described by N second-order differential equations, relating the generalized coordinate to their first and second time derivatives and to the generalized forces.

In vectorial notation, the dynamics equations for the skeletal system of the monkey's arm can be written as:

$$\tau(q, \dot{q}, u(t)) = I(q)\ddot{q} + C(q, \dot{q}) \tag{1}$$

where $q = (q_1, q_2, \ldots, q_n)$ is the arm configuration in joint-angle coordinates, $\dot{q}$ and $\ddot{q}$ are respectively the first (velocity) and second (acceleration) time derivatives of q, I is an $N \times N$ matrix of inertia (that is, configuration-dependent), and $C(q, \dot{q})$ is a vector of centripetal and Coriolis torques (Raibert, 1978). The whole right-hand side of Equation 1 represents the torque resulting from the inertial properties of the arm. The term $\tau(\cdot)$ on the left-hand side stands for the net torque generated nonlinearly by the muscles, by the environment (e.g., the gravitational torque), and by other dissipative elements, such as friction. The variable u is a control vector—for example, a set of neural signals directed to the motoneurons—whose dimension may far exceed the number of generalized coordinates (see EQUILIBRIUM POINT HYPOTHESIS).

Actuator Coordinates

Actuator coordinates afford the most direct representation for the motor output of the central nervous system. A *position* in this coordinate system may be, for example, a collection of muscle lengths, $l = (l_1, l_2, \ldots, l_M)$. Accordingly, a *force* would be a collection of muscle tensions, $f = (f_1, f_2, \ldots, f_M)$. The number, M, of actuator coordinates depends upon how detailed is the model of control under consideration. For instance, one may be dealing with a macroscopic representation of muscles (biceps, triceps, etc.) as the elements being controlled ($M \sim 30$). In an accurate representation of the neuromuscular system, one should consider each motor unit as an independently controlled element ($M \sim 10,000$). *Unlike generalized coordinates, actuator coordinates do not constitute a system of independent variables*; we cannot set arbitrary values to l_i without eventually violating some kinematic constraint. Nevertheless, actuators are independent units of control. Indicating with u the state of activation of a muscle, one may say that the central nervous system determines the tension generated by a muscle at any given state $(q, \dot{q})$ of the skeletal system.

The Workspace and Its Transformations

The transformations from generalized coordinates to endpoint coordinates, and from generalized coordinates to actuator coordinates, are, in general, nonlinear mappings. In the case of the monkey's arm, the transformation from joint to hand coordinates is a nonlinear function

$$r = \mathscr{L}(q) \tag{2}$$

where r indicates the position of the hand in endpoint coordinates and q is the joint configuration. The transformation from joint to muscle coordinates is another nonlinear mapping

$$l = \mathscr{M}(q) \tag{3}$$

This section deals with transformations between different representations of the workspace in the framework of differential geometry. A more extensive tutorial on this discipline and its applications can be found in Bishop and Goldberg (1980).

The Transformation of Vectors and Vector Fields

A function that associates a vector to each point of a multidimensional domain, M, is called a *vector field over M*. For example, we may rearrange the terms of Equation 1, the dynamics equation, so as to represent the arm's acceleration, $\ddot{q}$, as a time-varying vector field over the state space described by q and $\dot{q}$:

$$\ddot{q} = I^{-1}(q)(\tau(q, \dot{q}, u(t)) - C(q, \dot{q})) \tag{4}$$

Another vector field describes the viscoelastic behavior of the arm muscles. This behavior can be measured by stimulating each muscle and recording the resulting tension, f, at different muscle lengths, l, rates of shortening, $\dot{l}$, and times, t. Then, the collective mechanical output of the skeletomotor system is summarized by a force field in muscle coordinates

$$f = \alpha(l, \dot{l}, t) \tag{5}$$

Vector fields such as those expressed in Equations 4 and 5 determine the way in which a system reacts to its environment on one hand, and to its control signals on the other. To investigate the transformation from planning to control of actions, we must understand how such mechanical fields are affected by a change of coordinates.

Let us begin by considering the laws that govern the transformation of a point from a set of coordinates, x, into a new set of coordinates, $\bar{x}$. The coordinate transformation is a nonlinear function

$$\bar{x} = T(x) \tag{6}$$

We assume that this function is continuous and sufficiently differentiable (the existence and continuity of second partial derivatives is enough for most practical purposes). However, we do not require the existence of an inverse mapping, $x = T^{-1}(\bar{x})$. In many biologically relevant cases—such as the transformation from joint to muscle coordinates—the two coordinate systems have different dimensions, and the inverse mapping does not exist.

Next, consider how a vector field, $v(x)$, is related to the corresponding vector field, $\bar{v}(\bar{x})$, in the new coordinate system. First, assume that $v(x)$ is a field of velocity vectors, $\dot{x} = v(x)$, and apply the chain rule:

$$\dot{\bar{x}} = \frac{d\bar{x}}{dt} = \frac{\partial \bar{x}}{\partial x}\frac{dx}{dt} = \frac{\partial \bar{x}}{\partial x}\dot{x}$$

As we are dealing with multivariate functions, the expression $\partial\bar{x}/\partial x$ represents the functional derivative, or *Jacobian*, of the transformation T. This is a position-dependent matrix, $J(x)$, whose elements are

$$J_{i,j}(x) = \frac{\partial \overline{x}_i}{\partial x_j}$$

and the transformation for the velocity field can be rewritten as

$$\dot{\overline{x}} = J(x)\dot{x} \qquad (7)$$

A vector that changes according to this law is called *contravariant*.

Does Equation 7 provide a rule for transforming the whole velocity field $\dot{x}(x)$ into $\dot{\overline{x}}(\overline{x})$? The answer is affirmative only in the hypothesis—*to which we are not committing ourselves*—that the mapping T is invertible. In fact, in this case we may write $\dot{\overline{x}}$ as a function of $\overline{x}$:

$$\dot{\overline{x}} = J(T^{-1}(\overline{x}))\dot{x}(T^{-1}(\overline{x})) = \overline{v}(\overline{x})$$

In contrast, if T is not invertible, we are stuck with the fact that the terms on the right side of Equation 7, J and $\dot{x}$, both depend on x and not on $\overline{x}$. Summing up, in general we know how to transform a contravariant vector *at a given point* but we do not know how to transform a contravariant *field*.

The situation is quite different when dealing with vectors that, in a change of coordinates, transform like the gradient operator:

$$\frac{\partial}{\partial x} = \frac{\partial \overline{x}}{\partial x}\frac{\partial}{\partial \overline{x}}$$

That is,

$$\frac{\partial}{\partial x} = J^T(x)\frac{\partial}{\partial \overline{x}} \qquad (8)$$

A vector following this type of transformation is said to be *covariant*. Note the dual or reciprocal nature of Equations 7 and 8. An infinitesimal displacement (or a velocity) is transformed by the Jacobian into the new coordinate system. The same Jacobian maps a covariant vector the other way around —from the new to the old coordinate system.

An example of a covariant vector is force, F. The covariance of force derives from the tensor invariance of work and power. *Work and power are indeed true scalar variables whose value cannot be modified by a change of coordinates.* In the original coordinate system, power is calculated as

$$F^T\dot{x}$$

and in the new coordinate system as

$$\overline{F}^T\dot{\overline{x}}$$

By equating these two expressions and using Equation 7, we obtain

$$\overline{F}^T J(x)\dot{x} = F^T\dot{x}$$

Hence, the transformation of a force field, $\overline{F}(\overline{x})$, is

$$F(x) = J(x)^T\overline{F}(T(x)) \qquad (9)$$

Then we reach the important conclusion that, unlike contravariant vectors, *covariant vectors transform globally, as fields.* No inverse transformation is required.

Transforming Plans into Action

We now consider how the discussion of the preceding section may shed some light on the problem of implementing planned behaviors by a pattern of motor commands.

A number of investigations (Morasso, 1981; Hogan, 1984; Flash and Hogan, 1985) have suggested that actions are planned in endpoint coordinates. To some extent, this is a rather obvious concept that does not require much experimental testing. It is clear, for example, that we can mentally formulate (and execute) a command such as "trace the shape of a circle with the left hand" without being concerned about the set of muscles that are involved in this behavior. However, once we have decided to trace a circle, we must somehow choose which muscles to activate and in what temporal sequence. In carrying out this task, our brain must face the challenges associated with *kinematic redundancy*—the imbalance between the number of muscles, joints and endpoint coordinates that is typical of any biological system.

The purpose of this section is to show how a proper representation of dynamics and of coordinate transformations leads to a straightforward solution to the problems associated with redundancy. The key for this solution lies in the representation of both the "high-level" plans and the "low-level" actuator actions as covariant fields of force.

The Transformation of Dynamics

The dynamics of a wide variety of systems may be represented by a mapping

$$\ddot{q} = D(q, \dot{q}, t) \qquad (10)$$

from the current "state," $(q, \dot{q})$, and time, t, to the acceleration, $\ddot{q}$ (as in Equation 4). Effectively, this is a mapping from the current to the next state because the acceleration determines the velocity at the next instant of time, $(\dot{q} + \ddot{q}dt)$, and the velocity determines the next configuration, $(q + \dot{q}dt)$.

The dependence of $\ddot{q}$ upon time in Equation 10 represents the action of a control signal whose presence may be made more explicit by an input function $u(t)$:

$$\ddot{q} = D(q, \dot{q}, u(t))$$

With this representation, one may describe the execution of different tasks through the selection of different input functions.

The foregoing expressions state that a mechanical system's dynamics is a time-varying field of generalized acceleration over the system's state space. Like position and velocity, acceleration can be transformed from generalized to endpoint coordinates, and from generalized to actuator coordinates, but not, in general, the other way around. Hence, the earlier discussion on contravariant vector fields applies to the particular case of a dynamics field. In a kinematically redundant system, we may map acceleration from generalized coordinates either to endpoint or to actuator coordinates. However, in a kinematically redundant system there is not a well-posed rule for mapping an acceleration *field* to either endpoint or actuator coordinates. Essentially, we have just stated in another way that dynamics can be properly represented only in a system of generalized coordinates. *One cannot achieve a consistent description of dynamics in terms of muscle states and muscle forces, or in terms of endpoint states and endpoint forces.* Then how can we represent dynamic plans and translate them into actions?

This problem may be solved by specifying motor plans and actuator actions as force fields in the respective coordinate systems. To this end, we begin by reformulating the dynamics as a mapping

$$\ddot{q} = D(q, \dot{q}, \Phi(q, \dot{q}, t)) \qquad (11)$$

whose arguments are the current state and *a time-dependent field of generalized forces*, Φ, defined over the system's state space. This is a quite general formulation that includes, as a particular case, the torque function τ of Equation 1.

The next step consists of expressing the motor plans as fields of force in endpoint coordinates:

$$F = \pi(x, \dot{x}, t) \qquad (12)$$

For example, a reaching movement of the hand may be planned, as proposed by Flash, by specifying a time-varying field whose equilibrium point moves along a smooth trajectory (see Equilibrium Point Hypothesis). In contrast, a constant and uniform force field, $F = F_0$, encodes the task of exerting a force of given direction and amplitude, regardless of the uncertainty with which the hand location may be determined. Following a similar approach, Khatib (1986) proposed to plan the movements of a robotic arm in an obstacle-ridden environment by associating a field of repulsive forces to each obstacle and a field of attractive forces to the target location (see Potential Fields and Neural Networks).

From our earlier considerations of vector fields, we may conclude that the planned field, $\pi(x, \dot{x}, t)$, has a unique image, $\Phi_\pi(q, \dot{q}, t)$, in configuration space. This is true regardless of kinematic redundancy. The only condition for Φ_π to be defined is that the kinematic transformation from generalized to endpoint coordinates is defined with its first partial derivatives. Operationally, we may construct Φ_π as a combination of three mappings:

$$\Phi_\pi = l_2 \circ \pi \circ l_1 : (q, \dot{q}, t) \xrightarrow{l_1} (x, \dot{x}, t) \xrightarrow{\pi} F \xrightarrow{l_2} Q$$

The first mapping, l_1, has three components: the direct kinematics transformation ($q \to x$), its Jacobian ($q, \dot{q} \to \dot{x}$), and the identity function ($t \to t$). The second mapping is the planned endpoint field, $\pi(x, \dot{x}, t)$. Finally, the third mapping, l_2, is the transformation from endpoint to generalized force, which again is provided by the Jacobian of the direct kinematics. Once we have expressed the planning field in generalized coordinates, we obtain the dynamics

$$\ddot{q} = D(q, \dot{q}, \Phi_\pi(q, \dot{q}, t))$$

On the other end of the planning/execution problem, one must deal with a number of actuators that, in any biological system, exceeds the number of generalized coordinates. As in Equation 5, we represent the mechanical action of the actuators under a particular pattern of control as a force field

$$f = \alpha(l, \dot{l}, t) \qquad (13)$$

This field has a well-defined image in generalized coordinates, $\Phi_\alpha(q, \dot{q}, t)$, which can be derived from the combination of three direct transformations:

$$\Phi_\alpha = m_2 \circ \alpha \circ m_1 : (q, \dot{q}, t) \xrightarrow{m_1} (l, \dot{l}, t) \xrightarrow{\alpha} f \xrightarrow{m_2} Q$$

The first transformation, m_1, has as components the actuator kinematics ($\mathcal{M}: q \to l$), together with its Jacobian ($q, \dot{q} \to \dot{l}$) and the identity mapping ($t \to t$). The second transformation is the actuator force field, $\alpha(l, \dot{l}, t)$, and the third transformation, m_2, is again given by the Jacobian of $\mathcal{M}$.

The generalized force field, Φ_α, induces the dynamics

$$\ddot{q} = D(q, \dot{q}, \Phi_\alpha(q, \dot{q}, t))$$

We may consider this expression as the description of a "pattern generator" corresponding to a particular setting of the control signals. On the other hand, the field Φ_π and its associated dynamics correspond to some planned behavior. Clearly, if $\Phi_\pi = \Phi_\alpha$, this planned behavior is exactly implemented by the given pattern of actuators. However, such a fortunate situation may not be attainable, and one must be content with an *approximation* of the planned goal. The argument that we have

illustrated so far indicates that such an approximation can be properly represented in a system of generalized coordinates, regardless of kinematics redundancy. The view of motor control as a form of functional approximation is consistent with other aspects of neural information processing. For example, Poggio and Edelman (1990) have reduced the task of recognizing a three-dimensional object to the approximation of a function that maps two-dimensional views into a fixed "standard view" of the object. These investigators have shown that the approximation may be obtained by a linear combination of Gaussian radial functions.

A set of experiments in the spinalized frog have suggested that the neural circuits enclosed within the lumbar region of a frog's spinal cord may be organized into a few distinct modules of control (Bizzi, Mussa-Ivaldi, and Giszter, 1991; Frog Wiping Reflexes). Each one of these modules generates a field of forces acting on the ipsilateral hindlimb. How can the central nervous system generate a significant repertoire of behaviors starting from such a simple organization? An answer to this question comes from the empirical observation that the simultaneous stimulation of two spinal sites results in the vectorial sum of the fields obtained from the isolated stimulation of each site. This finding suggests that the combination of k independent control modules generates a net field

$$\Phi(q, t, \alpha_1, \ldots, \alpha_k) = \sum_{i=1}^{k} \alpha_i \phi_i(x, t) \qquad (14)$$

where α_i are activation coefficients.

This approach was explored by Mussa-Ivaldi and Giszter (1992), who concluded that a wide repertoire of field features can indeed be accurately reproduced by adding only a few given fields, as in Equation 14. Following this approach, a control system may approximate a *target field*, $\Phi_\pi(q, t)$, by determining a set of parameters, α_i^*, such that the norm

$$\|\Phi(q, t, \alpha_1^*, \ldots, \alpha_k^*) - \Phi_\pi(q, t)\| \qquad (15)$$

is at a minimum. This least-squares method requires the definition of an inner-product structure for the linear space of vector fields. Given two fields, $f(x, t)$ and $g(x, t)$, one may define an inner product of f and g as

$$\langle f, g \rangle = \iint f^T(x, t) \cdot g(x, t) p(x, t) \, dx \, dt$$

where $p(x, t)$ is a weighting function. Then the norm of a field, $\|f\|$, is

$$\|f\| = \langle f, f \rangle^{1/2}$$

The norm of the error field, Equation 15, is minimized by solving for α_i the K equations

$$\langle \phi_j, \Phi_\pi \rangle = \sum_i \langle \phi_j, \phi_i \rangle \alpha_i \qquad (j = 1, \ldots, K)$$

In essence, each coefficient is obtained by taking the projections, $\langle \phi_j, \Phi_\pi \rangle$, of the desired vector field along the system of fields generated by the control modules.

In this geometrical framework, the execution of an arbitrary task can be regarded as a "negotiation" between planning and control processes. On one hand, the planning processes may require a field on the basis of the features of the task and without too much of a concern for the properties of the hardware. On the other hand, the control processes may simply try to execute the best possible approximation to the planned field. As the quality of this approximation depends on the structure of the fields, ϕ_i, the adaptation of the control modules to the statistics of required behaviors is likely to play a crucial role in motor learning.

Acknowledgments. This work was supported by NIH grants NS09343 and AR26710 and by ONR grant N00014/88/K/0372.

Road Map: Biological Motor Control
Related Reading: Limb Geometry: Neural Control; Optimization Principles in Motor Control

References

Bernstein, N., 1967, *The Co-ordination and Regulation of Movements*, Oxford: Pergamon Press.

Bishop, R. L., and Goldberg, S. I., 1980, *Tensor Analysis on Manifolds*, New York: Dover.

Bizzi, E., Mussa-Ivaldi, F. A., and Giszter, S. F., 1991, Computations underlying the execution of movement: A biological perspective, *Science*, 253:287–291.

Flash, T., and Hogan, N., 1985, The coordination of arm movements: An experimentally confirmed mathematical model, *J. Neurosci.*, 5:1688–1703.

Goldstein, H., 1980, *Classical Mechanics*, 2nd ed., Reading, MA: Addison-Wesley.

Hogan, N., 1984, An organizing principle for a class of voluntary movements, *J. Neurosci.*, 4:2745–2754.

Khatib, O., 1986, Real-time obstacle avoidance for manipulators and mobile robots, *Intl. J. Robotics Res.*, 5:90–99.

Morasso, P., 1981, Spatial control of arm movements, *Exp. Brain Res.*, 42:223–227.

Mussa-Ivaldi, F. A., and Giszter, S. F., 1992, Vector field approximation: A computational paradigm for motor control and learning, *Biol. Cybern.*, 67:491–500.

Poggio, T., and Edelman, S., 1990, A network that learns to recognize 3D objects, *Nature*, 343:263–266.

Raibert, M. H., 1978, A model for sensorimotor control and learning, *Biol. Cybern.*, 29:29–36.

Grasping Movements: Visuomotor Transformations

G. Rizzolatti, G. Luppino, and M. Matelli

Introduction

When one attempts to pick up an object, one executes two distinct motor operations. One operation consists in bringing the hand toward object location in space, the other consists in shaping the hand and fingers in anticipation of the object's size, shape, and orientation. These two operations were defined by Jeannerod (1988) as the transport and grasp components of prehension, respectively. An analysis of the two components shows that they rely on markedly different visuomotor transformations. Transport requires establishing the location of objects in space with respect to the body. It implies the formation of a stable frame of reference, independent of eye or head position, and the encoding of visual information in body-centered coordinates. Visual information is then transformed to a pattern of proximal (shoulder, elbow) muscle activity which allows one to bring the hand in contact with the object. By contrast, grasping deals with the intrinsic qualities of the object, that is, with its shape and size. The coordinate system in which grasping movements are generated relates to these properties. Once the object size and shape are appropriately coded, they are transformed into a pattern of distal (finger, wrist) movements which permit the object to be grasped (Arbib, 1981; Jeannerod, 1988).

This article focuses on the second component of prehension, grasping. Its aim is to examine where in the cerebral cortex the visuomotor transformations underlying it occur, and which are the neural mechanisms responsible for them.

Somatomotor Areas in the Inferior Parietal Lobe and the Agranular Frontal Cortex

The parietal lobe of primates consists of three main sectors: the postcentral gyrus, the superior parietal lobule, and the inferior parietal lobule (IPL). This last sector is related to vision. Visual inputs to it arrive from occipitotemporal areas DP, PO, MT, MST, V3A, the areas located in the anterosuperior temporal sulcus, and from the visual field periphery of V3 and V2 (Andersen et al., 1990; Baizer, Ungerleider, and Desimone, 1991).

The IPL is functionally subdivided into several different areas (Figure 1). Three of them are buried in the intraparietal sulcus (IPs). They include the *lateral intraparietal area* (LIP), the *ventral intraparietal area* (VIP), and the *anterior intraparietal area* (AIP). Two areas are located on the crown of the inferior parietal lobule: area 7a and area 7b. Finally, the parietal operculum is occupied, at least in part, by the *second somatosensory area* (SII).

Physiological studies have shown that each of these areas is involved in specific sensorimotor transformations. Thus, the IPL can be conceived as a set of largely independent modules, each of which is responsible for the organization of a specific motor action. These actions include: saccadic eye movements (LIP), ocular fixation (7a), reaching (mostly 7b), and grasping (AIP).

A modular organization similar to that of the parietal lobe is present also in the motor sector of the frontal lobe (see Figure 1; see also Matelli and Luppino, 1992). This sector is characterized by a particular cytoarchitectonic structure that is devoid of granular cells (agranular cortex). Functionally, it is related to body movements, the only exception being a small rostral area controlling eye movements (supplementary eye fields, F7).

Which of the many areas forming the agranular frontal cortex is involved in grasping? A priori, in order to mediate grasping, a motor area must satisfy two logical requisites: (1) to have an adequate representation of distal movements, and (2) to receive an appropriate visual input. The area that possesses these requisites is area F5 (Rizzolatti and Gentilucci, 1988).

Area F5

Area F5 forms the rostral part of inferior area 6. Its main anatomical connections are illustrated in Figure 2. Particularly important are the connections with AIP and F1. The connections with F1, the precentral motor area, are within the hand field of this area (Muakkassa and Strick, 1979; Matsumura and Kubota, 1979; Matelli et al., 1986). Intracortical microstimulation and single neuron studies have confirmed that F5 is specif-

Figure 1. Lateral and mesial views of the monkey cerebral cortex showing the location of the frontal and parietal areas referred to in the text. The intraparietal (IPs) and lateral (Ls) sulci are opened. AIs, inferior arcuate sulcus; ASs, superior arcuate sulcus; Cs, central sulcus; Cgs, cingulate sulcus; MIP, medial intraparietal area; Ps, principal sulcus; STs, superior temporal sulcus.

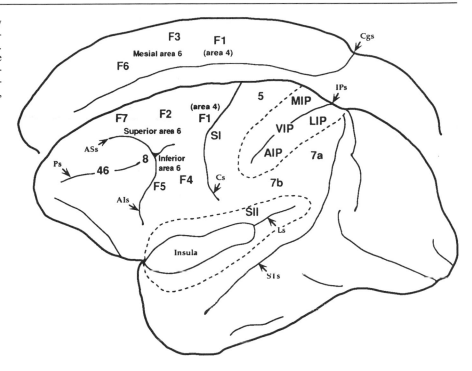

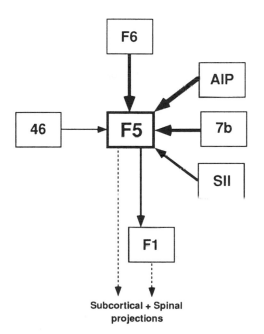

Figure 2. Schematic diagram of the main cortical connections of F5. All connections are reciprocal. Line thickness is roughly proportional to the connection richness. The arrows indicate the presence of subcortical and spinal cord projections. The richness of these projections is not indicated.

ically related to distal movements (Kurata and Tanji, 1986; Rizzolatti et al., 1988).

Rizzolatti et al. (1988) have described various classes of F5 neurons, each of which discharges during specific hand movements (e.g., grasping, holding, tearing, manipulating). The largest class is related to grasping. The temporal relations between neuron discharge and grasping movements vary among neurons. Some of them fire only during the last part of grasp-ing, i.e., during finger flexion. Others start to fire with finger extension and continue to fire during finger flexion. Finally, others are activated in advance of the movement initiation and often cease to discharge only when the object is grasped.

Many grasping neurons show selectivity according to the type of hand prehension. Three basic grip types have been extensively tested: precision grip, finger prehension, and whole-hand prehension. Eighty-five percent of grasping neurons were found to be selective for one of these types of grasping, the most represented being the precision grip (Rizzolatti et al., 1988).

Rizzolatti and co-workers have concluded that F5 contains a sort of "vocabulary" of motor acts related to prehension (Rizzolatti and Gentilucci, 1988). The "words" of the vocabulary indicate populations of neurons related to different motor acts. There are various categories of "words." Some indicate very general commands, e.g., "grasp," "hold," and "tear." Others indicate how the objects have to be grasped, held, or torn (e.g., by precision grip, finger prehension, whole-hand prehension, or their subtypes). Finally, a third group of "words" is concerned with the temporal segments of the actions.

The presence in F5 of a store of "words" or motor schemas (see Arbib, 1981) has two important consequences. First, since information is concentrated in relatively few abstract elements, the number of variables to be controlled is much less than it would be if the movements were described in terms of moto-neurons or muscles. This solution to the high degree of freedom of hand movement is remarkably similar to that proposed theoretically (see Arbib, Iberall, and Lyons, 1985). Second, the retrieval of the appropriate movement is simplified. Both for internally generated actions and for those emitted in response to an external stimulus, a schema (Arbib, 1981) or a small ensemble of schemas only, have to be selected. In particular, the retrieval of a movement in response to a visual object is reduced to the task of matching its size and orientation with the appropriate schema.

How can the motor vocabulary of F5 be addressed? The simplest way to examine this issue is to present different types of stimuli—for example, different visual stimuli—and to establish whether a neuron studied fires in the absence of movements, and, if this is the case, in response to which stimuli.

Using this approach, "visual" responses are observed in about 20%–30% of F5 neurons. According to the type of effective stimuli, two separate classes of neurons can be distinguished. Neurons of the *first class* respond to the presentation of graspable objects (Rizzolatti et al., 1988). Often, there is a relation between the type of prehension coded by the cell and the size of the stimulus effective in triggering the neurons. Neurons of the *second class* (so-called mirror neurons) respond when, for instance, monkeys see movements executed by the experimenters or by another monkey which are similar to those coded by the neurons. For example, many F5 neurons fire when a monkey grasps a piece of food and when the experimenter does it (di Pellegrino et al., 1992).

Anterior Intraparietal Area

F5 is strongly connected with AIP (see Figure 2). Physiological studies of IPL have shown that parietal neurons discharge in relation to arm projection and hand movements (Mountcastle et al., 1975). The functional properties of neurons related to hand movements located in the AIP were analyzed by Sakata and his co-workers (Taira et al., 1990; Sakata et al., 1992).

In monkeys trained to grasp objects requiring different types of grip, about half of the task-related neurons fired almost exclusively during one type of grip. Precision grip was the grip type most represented.

About 40% of the AIP neurons discharged equally well if the appropriate grasping movement was made in the light, with the monkey looking at the object, or in the dark. These neurons appear to be essentially motor neurons. They have been described as "motor-dominant neurons." The remaining neurons discharged more strongly ("visual and motor neurons"), or even exclusively ("visual-dominant neurons"), in the light. Half of the visual-dominant neurons and some of the visual and motor neurons became active when the animal fixated the object in the absence of any movement. For these last neurons, the visually effective object and the type of grip coded (assessed in the dark) coincide. They appear, therefore, to match the visual representation of the objects with the way in which the objects can be grasped.

While some neurons respond to the visual aspects of the object to be grasped, others respond to visual stimuli other than graspable objects. For example, about half of the visual-dominant neurons did not fire during object fixation in the absence of movement, and yet these neurons became active when the monkey grasped objects in the light. There is recent evidence that, at least for some neurons, the visual feature that activates the neurons is the sight of the monkey's own hand approaching and grasping the food.

If one compares these data with those relating to F5, several analogies, but also some important differences, emerge. Let us start with motor organization. Both areas appear to have a "vocabulary" of motor acts for manipulating objects. Precision grip, whole-hand prehension, and finger prehension are represented in both areas. However, whereas in F5 the discharge of a given neuron is usually limited to one motor act (e.g., grasping or holding) or even to a segment of the act (e.g., hand closure), parietal neurons most often start to discharge during hand shaping and continue discharging during object holding. The available evidence suggests, therefore, that AIP neurons code actions in a more global way, whereas F5 neurons parcel these actions into smaller segments.

Visual responses to three-dimensional objects are more frequent in AIP than in F5. Both areas respond to complex biological stimuli, like the vision of the hand. Mirror neurons, however, have not yet been reported in AIP. Thus, it is likely that the hand neurons described by Sakata et al. (1992) could have a role in hand control during manipulation, rather than in recognition of biologically or socially relevant movements.

In conclusion, the IPL appears to play a fundamental role in extracting the intrinsic visual properties from an object for organizing grasping movements. The extracted visual information is then sent to F5, where it is used to activate neurons that code grip types congruent to the size, shape, and orientation of the object. Two different roles can be attributed to the motor-dominant neurons. In keeping with the proposal by Mountcastle et al. (1975), they can act (in a broad sense) as command neurons. An alternative point of view is that the motor responses in AIP are the reflection of corollary discharges sent by F5 back to the parietal cortex. If an F5 motor command correctly matches the visual signals, the parietal cells are fully activated and send a positive feedback signal to the premotor cortex to carry on the correct movement until the object is successfully grasped and held. Otherwise, the parietal activity is suppressed and the movement is interrupted or modified (Sakata et al., 1992).

Discussion

If one compares the data on grasping with those on the functional organization of primary sensory or motor cortices, it is clear that the emerging picture is still rather sketchy and by no means satisfactory. Many fundamental points are unclear and some have not yet been addressed. We have no idea, for example, as to how complex stimulus parameters, like object weight or fragility, are coded by grasping neurons. We do not understand how grasping neuron activity is integrated with that of reaching neurons. We do not understand at all how and where the processes described in this article are transformed into movement parameters. Yet, in spite of all this, some interesting principles of cellular encoding of complex actions have begun to emerge.

The first is that motor information, as well as visual information for action, is centrally coded in abstract terms. A motor act is coded in terms of its goal ("grasp," "hold"), regardless of whether this goal is to be achieved by the muscles of the right hand, left hand, or even the mouth. Congruently, any relation with the basic visual receptive field organization is lost. Grasping neurons respond to the appropriate visual stimuli, independently of where in space the stimuli are located. A second principle is that information for complex acts is compressed in populations of specific units. It is unclear at present whether there is a set of neurons (genetically preformed or determined by learning) for any type of grip, or whether different types of grips result from a joint action of a limited series of fundamental grip neuron types (coarse coding; see, e.g., FAULT TOLERANCE). What is clear, however, is that grip movements are not organized on the basis of a direct link between visual neurons and motor neurons controlling individual muscles or simple finger movements. A third principle is that general commands (e.g., grasp with the mouth or hands) are often not distinct from a precise specification of how action has to be performed (e.g., precision grip). It appears that what to do and how to do it are processed conjointly.

The data on grasping that we have presented are certainly very preliminary. They just begin to piece together the cellular encoding at more abstract levels. They appear, however, to be very promising for achieving a better understanding of sensori-motor transformations and for future insights into the neurophysiology of abstract reasoning in the brain.

Acknowledgment. This work was supported by the Human Frontier Science Program.

Road Maps: Mammalian Brain Regions; Primate Motor Control
Related Reading: Corollary Discharge in Visuomotor Coordination; Eye-Hand Coordination in Reaching Movements; Gaze Coding in the Posterior Parietal Cortex

References

Andersen, R. A., Asanuma, C., Essick, G., and Siegel, R. M., 1990, Corticocortical connections of anatomically and physiologically defined subdivisions within the inferior parietal lobule, *J. Comp. Neurol.*, 296:65–113.

Arbib, M. A., 1981, Perceptual structures and distributed motor control, in *Handbook of Physiology*, sect. 1, vol. 2, part 2 (V. B. Brooks, Ed.), Bethesda, MD: American Physiological Society, pp. 1449–1480. ◆

Arbib, M. A., Iberall, T., and Lyons, D., 1985, Coordinated control programs for movements of the hand, in *Hand Function and the Neocortex* (A. W. Goodman and I. Darian-Smith, Eds.), *Experimental Brain Research*, supplement 10, Berlin: Springer-Verlag, pp. 111–129. ◆

Baizer, J. S., Ungerleider, L. G., and Desimone, R., 1991, Organization of visual inputs to the inferior temporal and posterior parietal cortex in macaques, *J. Neurosci.*, 11:168–190.

di Pellegrino, G., Fadiga, L., Fogassi, L., Gallese, V., and Rizzolatti, G., 1992, Understanding motor events: A neurophysiological study, *Exp. Brain Res.*, 91:176–180.

Jeannerod, M., 1988, *The Neural and Behavioral Organization of Goal-Directed Movements*, Oxford: Clarendon. ◆

Kurata, K., and Tanji, J., 1986, Premotor cortex neurons in macaques: Activity before distal and proximal forelimb movements, *J. Neurosci.*, 6:403–411.

Matelli, M., Camarda, R., Glickstein, M., and Rizzolatti, G., 1986, Afferent and efferent projections of the inferior area 6 in the macaque monkey, *J. Comp. Neurol.*, 251:281–298.

Matelli, M., and Luppino, G., 1992, Anatomo-functional parcellation of the agranular frontal cortex, in *Control of Arm Movement in Space* (R. Caminiti, P. B. Johnson, and Y. Burnod, Eds.), *Experimental Brain Research*, supplement 22, Berlin: Springer-Verlag, pp. 85–102.

Matsumura, M., and Kubota, K., 1979, Cortical projection of hand-arm motor area from postarcuate area in macaque monkey: A histological study of retrograde transport of horseradish peroxidase, *Neurosci. Lett.*, 11:241–246.

Mountcastle, V. B., Lynch, J. C. G. A., Sakata, H., and Acuna, C., 1975, Posterior parietal association cortex of the monkey: Command functions for operations within extrapersonal space, *J. Neurophysiol.*, 38:871–908.

Muakkassa, K. F., and Strick, P. L., 1979, Frontal lobe inputs to primate motor cortex: Evidence for four somatotopically organized "premotor" areas, *Brain Res.*, 177:176–182.

Rizzolatti, G., Camarda, R., Fogassi, L., Gentilucci, M., Luppino, G., and Matelli, M., 1988, Functional organization of inferior area 6 in the macaque monkey, II: Area F5 and the control of distal movements, *Exp. Brain Res.*, 71:491–507.

Rizzolatti, G., and Gentilucci, M., 1988, Motor and visual-motor functions of the premotor cortex, in *Neurobiology of Neocortex* (P. Rakic and W. Singer, Eds.), Chichester, Eng.: Wiley, pp. 269–284. ◆

Sakata, H., Taira, M., Mine, S., and Murata, A., 1992, Hand-movement related neurons of the posterior parietal cortex of the monkey: Their role in visual guidance of hand movements, in *Control of Arm Movement in Space* (R. Caminiti, P. B. Johnson, and Y. Burnod, Eds.), *Experimental Brain Research*, supplement 22, Berlin: Springer-Verlag, pp. 185–198. ◆

Taira, M., Mine, S., Georgopoulos, A. P., Murata, A., and Sakata, H., 1990, Parietal cortex neurons of the monkey related to the visual guidance of hand movement, *Exp. Brain Res.*, 83:29–36.

Habituation

DeLiang Wang

Introduction

Habituation is a decrease in the strength of a behavioral response that occurs when an initially novel stimulus is presented repeatedly. It is probably the most elementary and ubiquitous form of plasticity, and its underlying mechanisms may provide the basis for understanding other forms of plasticity and more complex learning behaviors. As a simple and fundamental form of learning, habituation has been extensively studied in different animals. Habituation of defensive reflexes that have simple neural circuitry was among the first types of learning explained successfully at the cellular level. To distinguish habituation from other types of behavioral decrement (like fatigue), Thompson and Spencer (1966) summarized nine criteria of habituation, among which are (1) exponential decay of response strength with the number of stimulus presentations, (2) spontaneous recovery with rest, (3) more rapid and pronounced habituation with repeated series of habituation training, (4) generalization of habituation to similar stimuli, and (5) restoration of the habituated response with presentation of a different stimulus. The last property is often referred to as *dishabituation*.

In 1967, when studying habituation of prey-catching behavior in anuran amphibians, Ewert observed that the forgetting seen in habituation exhibits two phases: a short-term one that lasts for a few minutes, and a long-term process that lasts for at least 6 hours (see Ewert, 1984, and VISUOMOTOR COORDINATION IN FROGS AND TOADS, for a review). The two forms of habituation were studied in *Aplysia* at the cellular level, and it has been found that long-term memory (LTM) is stored at the same site as short-term memory (STM), but requires new mechanisms of synaptic plasticity (Bailey and Chen, 1988).

Since it is well described behaviorally and its cellular mechanisms have been revealed in simple systems, habituation has been studied computationally for quite some time. Early models concentrated on describing changes of response strength with repeated stimulation and spontaneous recovery. Since most data at the cellular level were obtained from small neural networks, simple computational models often proved to be sufficient. More recent studies have modeled habituation at the systems level, frequently linking it to pattern recognition, while others have modeled molecular mechanisms underlying synaptic plasticity. Also, the computational relationship between short-term and long-term habituation has been investigated recently.

Animal Systems

Higher Vertebrate Systems

The orienting reflex in higher vertebrates is subject to habituation on repeated presentation. On the basis of a wide variety of experiments, Sokolov (1960) showed that habituation of the orienting reflex is stimulus-specific: that is, changes in the parameters of the stimuli readily recover the response. Furthermore, he proposed a comparator theory to account for the observations. His theory asserts that the cortex creates a model of the stimulus that is continuously compared with the object currently presented. When the model matches the current stimulus, habituation develops; otherwise the orienting reflex reappears. Vinogradova (1975) later found that cells in the rabbit hippocampus are highly sensitive to various parameters of the stimuli typically used in the orienting reflex, and changes in the responses of these cells correlate well to behavioral alterations in habituation of the reflex.

In order to understand the neurophysiological mechanisms of habituation, W. Spencer, R. Thompson, and D. Nielson in 1967 published a series of studies using the flexion reflex in spinal cats (see Hawkins, Clark, and Kandel, 1987, for a review). After repeated stimulation, the response was found to habituate. However, habituation recovered with rest or after strong stimulation at a different site (dishabituation). These researchers found that dishabituation does not simply recover the habituated response, but is an independent process. Based on this evidence and other data from intact animals, Groves and Thompson (1970) put forward the dual process theory, which claims that the net behavioral outcome can be best understood as a summation of the two independent processes of habituation and sensitization. Owing to the relative simplicity of the spinal cord (compared to the brain), Spencer and coworkers also used the preparation to explore the cellular mechanisms of habituation. They concluded that habituation must occur in the interneuron chain connecting sensory inputs and motoneurons.

Lower Vertebrate Systems

Although isolated spinal preparations proved invaluable in elucidating neurophysiological substrates of habituation, they provided little information about the involvement of the brain in habituation. Visually induced prey-catching orienting behavior in anuran amphibians turned out to be suitable for studying habituation at the systems level. This behavior has long been known to be habituatable. Besides common characteristics, habituation of the orienting behavior in toads also exhibits (1) locus specificity, whereby after habituation at a given location, the response can be released by the same stimulus applied at a different locus; and (2) hierarchical stimulus specificity, whereby another stimulus given at the same locus may restore the response habituated by a previous stimulus. Only certain stimuli can dishabituate a previously habituated response, and this dishabituation forms a hierarchy where a stimulus can dishabituate the habituated response of another stimulus if the latter is lower in the hierarchy (see Ewert, 1984). Wang and Ewert found that, like higher vertebrates, dishabituation in toads also involves an independent process (1992).

Neuronal adaptation in retinal, tectal, or thalamic neurons lasted no longer than 90 to 120 seconds (Ewert, 1984). Since behavioral studies show that habituation effects may last 24 hours or longer, habituation substrates must be located elsewhere. Anatomically, the medial pallium (MP) has been thought to be a homolog of the mammalian hippocampus. After bilateral lesions of MP, toads showed no progressive decrease of the orienting activity to repetitive stimulation. In addition, both effects of conditioning and associative learning abilities in naive toads are abolished as a result of MP lesions. Finkenstädt (1989) identified two types of visually sensitive neurons in MP whose activities strongly increase or decrease in response to repetitive stimulation. Taken together, these data point to the MP as the locus of neural substrates of habituation.

Invertebrate Systems

None of the vertebrate preparations provided evidence about the cellular basis of habituation at the same level of definiteness as the marine snail *Aplysia*. Kandel and his colleagues have extensively analyzed habituation of a specific defensive reflex whereby the animal withdraws its gill in response to tactile stimulation of the siphon (see Hawkins et al., 1987, for a review). They demonstrated that both short-term and long-term habituation share a common locus, namely the synapses of the sensory neurons on the motor neurons, and habituation results from synaptic depression. Furthermore, synaptic depression accompanying short-term habituation was found to be caused by the reduction of neurotransmitter release in the presynaptic terminals. In addition, Bailey and Chen showed that long-term habituation is caused by structural changes of presynaptic terminals, such as the number of presynaptic terminals (varicosities) and the number and size of active zones (Bailey and Chen, 1988). See the article INVERTEBRATE MODELS OF LEARNING: *APLYSIA* AND *HERMISSENDA* for more discussion.

Computational Models of Habituation

Modeling Habituation as an Isolated Process

With a specific behavior, habituation can often be measured quantitatively by changes in the intensity of the behavior. It is thus interesting to explain the observed quantitative data by a computational/mathematical model. This type of model generally views habituation as an isolated process of synaptic plasticity, and directly links modification of model synapses to behavioral changes. In simple systems, this type of model may be readily applied to explain cellular processes of habituation.

Stanley (1976) proposed a model that described the decrease of synaptic efficacy y by a first-order differential equation, and used the model to simulate the habituation data obtained from the cat spinal cord. The model basically takes on the following form:

$$\tau \frac{dy(t)}{dt} = \alpha(y_0 - y(t)) - S(t) \tag{1}$$

where y_0 is the normal, initial value of y; $S(t)$ represents external stimulation; τ, the time constant, governs the rate of habituation; and α regulates the rate of recovery. This linear differential equation exhibits the exponential curve of habituation and spontaneous recovery, as shown in Figure 1. The figure displays the value of y with different values for τ and α during two habituation sessions separated by a recovery interval. Similar models are used to simulate habituation of the orienting reflex and prey-catching behavior (Lara and Arbib, 1985), and the behavioral decrement related to associative learning (Gluck and Thompson, 1987).

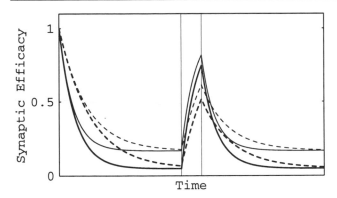

Figure 1. Decrease of synaptic efficacy with different parameters, as defined by Equation 1. Two continuous habituation sessions with a recovery interval were simulated. For solid lines, $\tau = 33$, whereas for dashed lines, $\tau = 66$. For thick lines, $\alpha = 1.05$; for thin lines, $\alpha = 1.2$. $S(t)$ is set to 1 during training, and it is set to 0 during recovery, which corresponds to an increase in synaptic efficacy in the middle of the figure. $y_0 = 1.0$.

While Equation 1 successfully models a number of properties of habituation, it can only explain short-term habituation. Wang and Hsu in 1990 proposed a model to incorporate both short- and long-term habituation (see Wang and Arbib, 1992, for the detailed reference). The main idea was to use an inverse S-shaped curve to describe two forms of memory. Wang and Arbib (1992) adjusted the same idea to model stimulus-specific habituation in toads in the following form:

$$\tau \frac{dy(t)}{dt} = \alpha z(t)(y_0 - y(t)) - \beta y(t)S(t) \qquad (2)$$

$$\frac{dz(t)}{dt} = \gamma z(t)(z(t) - l)S(t) \qquad (3)$$

where y, y_0, τ, and α are as described in the previous equation. The second term in Equation 2 regulates habituation with parameter β. The presynaptic stimulus $S(t)$ multiplied by $y(t)$ forms activity-gated input, in contrast to direct input, in Equation 1. Variable $z(t)$, as defined in Equation 3, controls the rate of recovery, evolving much slower than $y(t)$. With constant stimulation $S(t)$, $z(t)$ exhibits a typical inverse S-shaped curve. When $z(t)$ remains in the higher half of the inverse S-shaped curve, recovery is relatively fast; when $z(t)$ is in the lower half, recovery is relatively slow. These two phases of recovery are used to model two forms of memory: STM and LTM. Wang (1993) tested the model on the long-term habituation data relating to the toad's prey-catching behavior obtained by Ewert (see Ewert, 1984). Figure 2 displays the experimental data and the model outputs. The simulation results are quantitatively similar to the experimental data. The discrepancy in the 24-hour curves suggests that the recovery (forgetting) should occur more quickly in the model when it enters profound habituation, a topic of future study. Linking to biological synapses, Equation 2 would be a description of changes in neurotransmitter release, and Equation 3 would describe morphological changes that follow long-term habituation.

Modeling Habituation as Embedded in Complex Behaviors

As a form of negative learning, habituation plays a significant role in adaptive behaviors. It helps to direct an animal's atten-

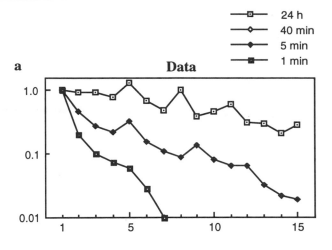

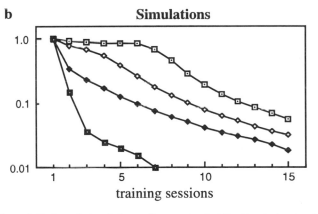

Figure 2. Data and simulations of successive habituation in toads. *A*, Habituation effects in successive training sessions separated by constant recovery times of 1 minute, 5 minutes, and 24 hours (adapted from Ewert, 1984). Each data point represents the normalized cumulative number of orienting turns during one session. *B*, The corresponding simulation results (from Wang, 1993). It is assumed that instantaneous response of the model is proportional to the synaptic weight, $y(t)$. For comparison, the results obtained with a 40-minute rest are also included.

tion to novel and potentially more useful stimuli while ignoring repetitive stimuli with little significance. Modeling habituation at the cellular level is not sufficient for understanding complex behaviors which often involve large networks of neurons. As is well known to the neural network community, a network of neurons can exhibit emergent behaviors unique to the systems level. Thus, the challenge is to bridge habituation behaviors and synaptic depression.

Kohonen has studied a type of network, called the novelty filter, which can demonstrate habituation behaviors (Kohonen, 1989). It is a one-layer network with recurrent projections whose strengths are subject to modification. It can be shown that, if an input pattern is kept constant, the induced output of the filter tends to zero exponentially fast, reminiscent of behavioral habituation. This network can be used to detect novelty in input patterns in the following way. A set of standard patterns is stored in the network using the synaptic modification rule. During novelty filtering, an input pattern is presented to the network, and the output of the network will be the part of the input that cannot be constructed from a linear combination of the stored patterns. Apparently, this type of network can be used for tracking missing parts of standard images, or abnormality diagnosis.

As introduced earlier, anuran amphibians exhibit a dishabituation hierarchy in the response to the repeated presentation of a set of visual patterns, such as different types of triangles, rectangles, etc. Wang and Arbib (1992) proposed a neural network model for the MP which consists of a network of MP columns. The model includes the visual pathway from the retina to the anterior thalamus (AT), and from there to MP, where habituation occurs. After a transformation, different AT responses activate different cell groups in a layer of an MP column. Synapses in MP columns exhibit depression upon repeated presynaptic stimulation, as described in Equations 2 and 3. There are unilateral connections in the layer so that a cell group corresponding to one stimulus can dishabituate other cell groups in only one direction, thus forming the dishabituation hierarchy. Furthermore, the model provides a number of specific predictions concerning mechanisms of habituation and cellular organization of the medial pallium.

Conclusion

In the last 30 years, neuroscientists have made remarkable advances in elucidating cellular mechanisms of habituation. It is generally agreed that habituation behaviors are caused by depression of synaptic transmission, which occurs at the presynaptic terminals owing to reduced neurotransmitter release. Despite these advances, gaps clearly exist between different levels of studies, particularly between synaptic depression and habituation behaviors. Although habituation behaviors at the basic level are known to be caused by some elementary rules of synaptic depression, these rules are far from sufficient to understand complex adaptive behaviors. For example, associative memory and backpropagation both use the Hebbian learning rule, but their behaviors at the systems level are vastly different (see ASSOCIATIVE NETWORKS and BACKPROPAGATION: BASICS AND NEW DEVELOPMENTS). For this reason, it is important to relate behaviors to basic synaptic plasticity. Because emergent properties of neural networks require advanced computational analysis, interactions between experimental and computational studies will be more important than ever.

Acknowledgment.   This work was supported in part by NSF grant IRI–9211419.

Road Map: Learning in Biological Systems
Background: I.1. Introducing the Neuron

References

Bailey, C. H., and Chen, M. C, 1988, Long-term memory in *Aplysia* modulates the total number of varicosities of single identified sensory neurons, *Proc. Natl. Acad. Sci. USA*, 85:2373–2377.

Ewert, J.-P., 1984, Tectal mechanisms that underlie prey-catching and avoidance behaviors in toads, in *Comparative Neurology of the Optic Tectum* (H. Vanegas, Ed.), New York: Plenum, pp. 246–416. ◆

Finkenstädt, T., 1989, Stimulus-specific habituation in toads: 2DG studies and lesion experiments, in *Visuomotor Coordination: Amphibians, Comparisons, Models, and Robots* (J.-P. Ewert and M. A. Arbib, Eds.), New York: Plenum, pp. 767–797.

Gluck, M. A., and Thompson, R. F., 1987, Modeling the neural substrates of associative learning and memory: A computational approach, *Psychol. Rev.*, 94:1–16.

Groves, P. M., and Thompson, R. F., 1970, Habituation: A dual-process theory, *Psychol. Rev.*, 77:419–450.

Hawkins, R. D., Clark, G. A., and Kandel, E. R., 1987, Cell biological studies of learning in simple vertebrate and invertebrate systems, in *Handbook of Physiology*, section 1, *The Nervous System* (F. Plum, Ed.), Bethesda, MD: American Physiological Society, pp. 25–83. ◆

Kohonen, T., 1989, *Self-Organization and Associative Memory*, 3rd ed., New York: Springer-Verlag. ◆

Lara, R., and Arbib, M. A., 1985, A model of the neural mechanisms responsible for pattern recognition and stimulus specific habituation in toads, *Biol. Cybern.*, 51:223–237.

Sokolov, E. N., 1960, Neuronal models and the orienting reflex, in *The Central Nervous System and Behavior*, vol. III (M. A. B. Brazier, Ed.), New York: Macy Foundation, pp. 187–276.

Stanley, J. C., 1976, Computer simulation of a model of habituation, *Nature*, 261:146–148.

Thompson, R. F., and Spencer, W. A., 1966, Habituation: A model phenomenon for the study of neuronal substrates of behavior, *Psychol. Rev.*, 73:16–43.

Vinogradova, O., 1975, The hippocampus and the orienting reflex, in *Neuronal Mechanisms of the Orienting Reflex* (E. N. Sokolov and O. Vinogradova, Eds.), New York: Erlbaum, pp. 128–154.

Wang, D. L., and Arbib, M. A., 1992, Modeling the dishabituation hierarchy in toads: The role of the primordial hippocampus, *Biol. Cybern.*, 67:535–544.

Wang, D. L., and Ewert, J.-P., 1992, Configurational pattern discrimination responsible for dishabituation in common toads *Bufo bufo* (L.): Behavioral tests of the predictions of a neural model, *J. Comp. Physiol. [A]*, 170:317–325.

Wang, D. L., 1993, A neural model of synaptic plasticity underlying short-term and long-term habituation, *Adapt. Behav.*, 2:111–129.

Half-Center Oscillators Underlying Rhythmic Movements

Ronald L. Calabrese

Introduction

The half-center oscillator model was first proposed by Brown (1914) to account for the observation that spinal cats could produce stepping movements even when all dorsal roots were severed, thereby eliminating sensory feedback from the animal's motion. He envisioned pools of interneurons that could control flexor and extensor motor neurons, respectively (the half-centers), that had reciprocal inhibitory connections, and that were capable of sustaining alternating activity if properly activated. In the model, he assumed that the duration of reciprocal inhibition was limited by some intrinsic factor, e.g.,

synaptic fatigue, and that the neuron pools showed rebound excitation. We now know that most rhythmic movements, including locomotory movements, are programmed in part by central pattern-generating networks that comprise neural oscillators (Getting, 1989). In many of these pattern-generating networks, in both vertebrates and invertebrates, reciprocal inhibitory synaptic interactions between neurons or groups of neurons are found, and are thought to form the basis of oscillation in the network.

We are only beginning to understand how membrane properties interact with reciprocal inhibition to initiate and sustain oscillation in these networks, i.e., the extrinsic and intrinsic neuronal factors that promote oscillation (Calabrese and De Schutter, 1992; Arshavsky et al., 1993). Modeling studies will be important in sorting out these interactions and identifying critical parameters. Detailed models which consider all identified voltage-gated and synaptic currents should be particularly important in understanding individual cases, but simplified models will play a role in elucidating general principles of organization (Wang and Rinzel, 1992).

In physiological analyses of half-center oscillators, particular attention has been paid to plateau potentials (Pearson and Ramirez, 1992) and *N*-methyl-*D*-aspartate (NMDA) receptor channels (Roberts and Sillar, 1990; Grillner et al., 1991; Arshavsky et al., 1993). Plateau potentials are regenerative depolarizing potentials caused by non-inactivating or slowly inactivating cation channels. They can support bursts of action potentials and promote graded transmitter release. Plateau formation upon release from hyperpolarization may account for the *postinhibitory rebound* which is often observed in these networks (e.g., see Satterlie, 1989). Plateau properties can be activated by neuromodulators (see NEUROMODULATION IN INVERTEBRATE NERVOUS SYSTEMS), and their activation is associated with the onset of oscillation in some networks. NMDA channels—cation channels which open in response to glutamate but are voltage-sensitive—can provide, in the presence of glutamate, the *tonic excitatory drive* that initiates and sustains oscillation, and, in some cases, may confer on neurons *intrinsic oscillatory properties* (see NMDA RECEPTORS: SYNAPTIC, CELLULAR, AND NETWORK MODELS). *Sag potentials* (DiFrancesco and Noble, 1989) are slow depolarizations produced by a hyperpolarization-activated inward current (I_h), activated by inhibition (or hyperpolarization). They may play a role in pacing oscillation in these networks (Calabrese, Angstadt, and Arbas, 1989). Unlike plateau potentials and NMDA channel-mediated depolarizations, which are regenerative and depend on inhibition, inactivation, or activation of outward currents for their termination, sag potentials move the membrane potentials into the range where they are deactivated.

Some Specific Examples

A few examples of identified half-center oscillators illustrate how these properties contribute to oscillation.

Xenopus *and Lamprey Swimming*

The spinal pattern-generating networks that produce swimming in early-stage tadpoles of the frog *Xenopus* and in adult lampreys are among the best understood of vertebrate motor networks (Roberts, 1990; Grillner et al., 1991; Arshavsky et al., 1993). In *Xenopus* tadpoles, swimming movements (10–20 Hz) are initiated by mechanical stimulation that activates Rohon-Beard sensory neurons (Roberts, 1990). In paralyzed embryos, similar stimulation elicits a fictive swim motor program. The sensory neurons produce long-lasting, dual, transient and sustained, component excitatory potentials in spinal interneurons resulting from the activation of non-NMDA and NMDA excitatory amino acid receptors, respectively (Roberts and Sillar, 1990). Similar excitatory connections among the excitatory interneurons produce a prolonged excited state within the excitatory interneuronal pool. The excitatory interneurons likewise activate reciprocally inhibitory pools of inhibitory interneurons, which can then produce oscillation. The intrinsic properties of the inhibitory interneurons that produce oscillation are not well characterized, but are thought to include *postinhibitory rebound* and *spike accommodation*, the waning of spike frequency with sustained depolarization. Computer sim-

ulations based on the known membrane properties and synaptic connections reproduce the major properties of the system, including initiation by brief excitatory input and accurate voltage waveforms in the oscillator neurons (Arshavsky et al., 1993). The swim oscillator of the pelagic mollusk *Clione* is organized in a remarkably similar fashion (Satterlie, 1989; Arshavsky et al., 1993).

In lampreys, swimming movements (0.25–10 Hz) can be initiated by mechanical stimulation of the tail fin (Grillner et al., 1991). In the isolated brainstem, stimulation of reticulospinal neurons can produce the fictive swim motor program. In the isolated spinal cord, stimulation of tail fin sensory neurons or spinal tracts, or the application of excitatory amino acids, will also produce the fictive motor pattern. Since even very small pieces of spinal cord (1–2 segments long) can produce the fictive motor pattern in response to excitatory amino acids, it is assumed that each spinal segment has its own swim pattern generator, which are then tied together by coordinating fibers.

The descending reticulospinal system and applied excitatory amino acids activate fictive swimming through both non-NMDA and NMDA excitatory amino acid receptors on all the types of spinal interneurons that form the segmental pattern generators (Grillner et al., 1991). Within the pattern generators, the excitatory interneurons act through the same receptors. Tail fin sensory neurons activate the swim pattern generator through the excitatory interneurons. A reciprocally inhibitory network of both crossed and uncrossed inhibitory interneurons receives excitatory amino-acid–mediated excitation from the excitatory interneurons. Both excitatory and inhibitory interneurons have outputs onto motor neurons.

Activation of non-NMDA receptors provides increased general network excitability leading to rhythm generation (Grillner et al., 1991). Activation of NMDA receptors can induce rhythmic plateau formation in spinal neurons that are also thought to contribute to rhythmicity. Modeling experiments with the known network connections and cellular properties indicate that reciprocal inhibitory synaptic interactions are important in burst termination within the pattern-generating network, but they also point out that Ca^{2+}-activated K^+ currents may contribute (Grillner et al., 1991).

Leech Heartbeat

The rhythmic constrictions (at about 0.1 Hz) of the bilateral heart tubes of the leech *Hirudo medicinalis* are paced and coordinated by rhythmically active segmental heart motor neurons (Calabrese et al., 1989). A network of seven bilateral pairs of heart interneurons, one pair of which is located in each of the first seven segmental ganglia of the nerve cord, paces activity in the heart motor neurons. This network is continuously active in the isolated nervous system and produces a fictive motor program that can account for the constriction pattern of the hearts observed in situ. The synaptic connections among the interneurons, and from the interneurons to the motor neurons, are inhibitory.

Two foci of oscillation occur within the pattern generator in the third and fourth segmental ganglia of the nerve cord, where the heart interneurons form reciprocally inhibitory connections across the midline (Calabrese et al., 1989). The normal activity cycle of each heart interneuron of these reciprocally inhibitory pairs consists of an active burst phase, during which it inhibits its contralateral partner, and an inhibited phase, during which firing is suppressed by synaptic inhibition from the contralateral partner.

The synaptic interaction between reciprocally inhibitory heart interneurons consists of a graded component in addition

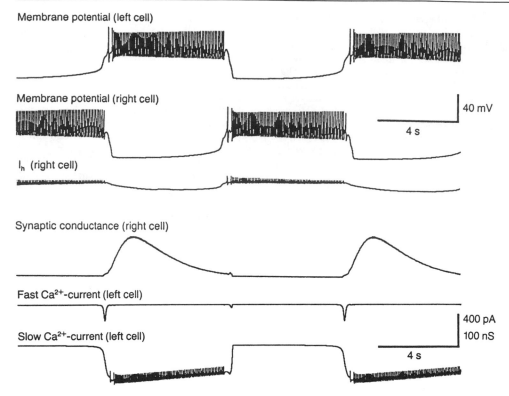

Membrane potential (left cell)

Membrane potential (right cell)

I_h (right cell)

Synaptic conductance (right cell)

Fast Ca²⁺-current (left cell)

Slow Ca²⁺-current (left cell)

40 mV

4 s

400 pA
100 nS

4 s

Figure 1. Simulated oscillations in a pair of leech heart interneurons. Membrane potential is shown in both cells. Note that the synaptic conductance in the right cell is controlled by Ca^{2+} influx (Ca^{2+} concentration) in the left cell (and vice versa). Plateau and associated burst formation, and Ca^{2+} influx, result mainly from the slow Ca^{2+} current (I_{CaS}). The contribution of the fast Ca^{2+} current (I_{CaF}) is limited to the period of transition to the burst phase. The small blip of I_{CaF} that occurs during the transition to the inactive state results from the rapid deinactivation and subsequent activation of I_{CaF} as the cell is inhibited by its contralateral partner. Spike-mediated transmission is not explicitly modeled, so spike-mediated inhibitory postsynaptic potentials (IPSPs) are not simulated. The role of I_h is explained in the text. See Calabrese and De Schutter (1992) for model description. (From Calabrese, R. L., and De Schutter, E., 1992, Motor-pattern-generating networks in invertebrates: Modeling our way toward understanding, *Trends in Neurosciences*, 15:439–445. Reprinted with permission.)

to spike-mediated synaptic transmission (Calabrese et al., 1989). Both of these components are blocked by bicuculline, and bath application of bicuculline reversibly halts oscillation of a reciprocally inhibitory heart interneuron pair, demonstrating the importance of reciprocal inhibition for rhythm generation (Schmidt and Calabrese, 1992).

The hyperpolarization-activated inward current (I_h), appears to be important in pacing oscillation because it promotes escape from inhibition and transition to the burst phase (Calabrese et al., 1989). Blockade of this current with external Cs^+ reversibly disrupts normal oscillation. Low-threshold Ca^{2+} currents (I_{CaS} [slowly inactivating] and I_{CaF} [rapidly inactivating]) underlie graded transmission and promote burst formation by sustaining prolonged depolarized plateaus (Angstadt and Calabrese, 1991).

Perturbation experiments have identified the transition from the inhibitory to the burst phase as the critical timing transition in the activity cycle of the heart interneurons (Calabrese et al., 1989). The inhibited cell escapes from the inhibitory influence of the active cell during this transition. Two factors appear to contribute to this escape: 1) the buildup of I_h in the inhibited cell that overcomes the inhibition, and (2) the waning of the inhibitory influence of the active cell. Both graded transmission and spike-mediated inhibition decline as the low-threshold Ca^{2+} currents that support the burst formation (spike frequency) and mediate the graded transmission are inactivated.

Modeling studies in this system (Figure 1) indicate that the former factor predominates (Calabrese and De Schutter, 1992). In the heart interneuron model, identified voltage-gated currents were represented by Hodgkin-Huxley–like equations generated from voltage clamp data, and graded synaptic transfer was an explicit function of presynaptic Ca^{2+} entry (De Schutter, Angstadt, and Calabrese, 1993). The voltage-gated currents included in the model were two low-threshold Ca^{2+} currents, one rapidly (I_{CaF}) and one slowly inactivating (I_{CaS}); a fast Na^+ current (I_{Na}); a delayed rectifier current (I_K); a tran-

sient outward current (I_{KA}); and a hyperpolarization-activated inward current (I_h). Varying the conductance g_h in the model indicates that it regulates the period of the oscillation and that the transition to the active state occurs when I_h has developed sufficiently to overcome the waning synaptic conductance (see Figure 1) and other outward currents. However, no specific formulation for spike-mediated transmission is present in the model, and this transmission is not well simulated. Thus, the model, while informative, is incomplete.

Discussion

A general theoretical framework for understanding how reciprocally inhibitory neurons oscillate was recently developed by Wang and Rinzel (1992). Their model neurons are minimal. Each contains a synaptic conductance which is a sigmoidal function of presynaptic membrane potential with a set threshold and instantaneous kinetics, a constant leak conductance, and a voltage-gated postinhibitory rebound conductance (g_{pir}). The g_{pir} is derived from a quantitative model of a T-type calcium current in thalamic neurons; it activates rapidly and inactivates slowly, and is strongly inactivated at rest so that "hyperpolarization of sufficient duration and amplitude is required to deinactivate" g_{pir} to produce "rebound excitation after removal of hyperpolarization." (This current corresponds to the low-threshold Ca^{2+} currents found in the leech heart interneurons.) The authors note that an h-conductance "would have an expression similar to" g_{pir}, and thus their model should be relevant to oscillators that employ h-conductance.

Two fundamentally different modes of oscillation appear in the model: "release" and "escape" (Wang and Rinzel, 1992). For the release mode to occur, the synaptic threshold must be above the steady-state voltage of the neurons when uninhibited. In the release mode, the inactivation of g_{pir} erodes the depolarized or active phase of a neuron so that it falls below threshold for synaptic transmission. Consequently, its partner is released from inhibition and rebounds into the active depolarized state. For the escape mode to occur, the synaptic threshold must be below the steady-state voltage of the neurons when uninhibited. This condition can be accomplished simply by increasing g_{pir}. In the escape mode, once g_{pir} becomes deinactivated by the hyperpolarization associated with inhibition, it activates and overcomes the maintained synaptic current so that the neuron escapes into the active phase. In the release mode, the transition from the inactive state to the active state is controlled by the active presynaptic neuron, and the period of the oscillation is critically dependent on the threshold for synaptic transmission. The synaptic threshold determines the fraction of time during rebound in which the neuron can inhibit its partner. In the escape mode, the transition from the inactive state to the active state is controlled by the inactive postsynaptic neuron, and the period of the oscillation is relatively insensitive to the synaptic threshold. The period is roughly twice the time it takes for g_{pir} to deinactivate and subsequently activate sufficiently to overcome the inhibition.

It seems clear that the heart interneuron oscillator operates in the escape mode; whenever I_h is sufficiently activated to overcome the synaptic inhibition, a transition from the inactive state to the active state occurs. In heart interneuron oscillator and its model, g_{pir} is represented by at least three conductances —g_h, g_{CaF} and g_{CaS}—with differing activation ranges and inactivation properties (Calabrese and De Schutter, 1992). By contrast, the *Xenopus* swim oscillator appears to operate in the release mode. Perhaps this mode is more suited to the operational frequency range of this oscillator, which is some 10 times faster than the leech heartbeat oscillator.

Road Map: Motor Pattern Generators and Neuroethology
Background: Ion Channels: Keys to Neuronal Specialization
Related Reading: Crustacean Stomatogastric System; Oscillatory and Bursting Properties of Neurons; Spinal Cord of Lamprey: Generation of Locomotor Patterns

References

Angstadt, J. D., and Calabrese, R. L., 1991, Calcium currents and graded synaptic transmission between heart interneurons of the leech, *J. Neurosci.*, 11:746–759.

Arshavsky, Y. I., Orlovsky, G. N., Panchin, Y. V., Roberts, A., and Soffe, S. R., 1993, Neuronal control of swimming locomotion: Analysis of the pteropod mollusc *Clione* and embryos of the amphibian *Xenopus*, *Trends Neurosci.*, 16:227–233. ◆

Brown, T. G., 1914, On the nature of the fundamental activity of the nervous centres; together with an analysis of the conditioning of rhythmic activity in progression, and a theory of the evolution of function in the nervous system, *J. Physiol. (Lond.)*, 48:18–46.

Calabrese, R. L., Angstadt, J. D., and Arbas, E. A., 1989, A neural oscillator based on reciprocal inhibition, in *Perspectives in Neural Systems and Behavior* (T. J. Carew and D. Kelley, Eds.), New York: Alan R. Liss, pp. 33–50.

Calabrese, R. L., and De Schutter, E., 1992, Motor-pattern-generating networks in invertebrates: Modeling our way toward understanding, *Trends Neurosci.*, 15:439–445. ◆

De Schutter, E., Angstadt, J. D., and Calabrese, R. L., 1993, A model of graded synaptic transmission for use in dynamic network simulations, *J. Neurophysiol.*, 69:1225–1235.

DiFrancesco, D., and Noble, D.,1989, Current I_f and its contribution to cardiac pacemaking, in *Cellular and Neuronal Oscillators* (J. W. Jacklet, Ed.), New York: Marcel Dekker, pp. 31–57.

Getting, P. A., 1989, Emerging principles governing the operation of neural networks, *Annu. Rev. Neurosci.*, 12:184–204. ◆

Grillner, S., Wallén, P., Brodin, L., and Lansner, A., 1991, Neuronal network generating locomotor behavior in lamprey: Circuitry, transmitters, membrane properties, and simulation, *Annu. Rev. Neurosci.*, 14:169–199.

Pearson, K. G., and Ramirez, J.-M., 1992, Parallels with other invertebrate and vertebrate motor systems, in *Dynamic Biological Networks: The Stomatogastric Nervous System* (R. M. Harris-Warrick, E. Marder, A. I. Selverston, and M. Moulins, Eds.), Cambridge, MA: MIT Press, pp. 263–281.

Roberts, A., 1990, How does a nervous system produce behavior? A case study in neurobiology, *Sci. Prog.*, 74:31–51.

Roberts, A., and Sillar, K. T., 1990, Characterization and function of spinal excitatory interneurons with commissural projections in *Xenopus laevis* embryos, *Eur. J. Neurosci.*, 2:1051–1062.

Satterlie, R. A., 1989, Reciprocal inhibition and rhythmicity: Swimming in a Pteropod mollusk, in *Cellular and Neuronal Oscillators* (J. W. Jacklet, Ed.), New York: Marcel Dekker, pp. 151–172.

Schmidt, J., and Calabrese, R. L., 1992, Evidence that acetylcholine is an inhibitory transmitter of heart interneurons in the leech, *J. Exp. Biol.*, 171:329–347.

Wang, X.-J., and Rinzel, J., 1992, Alternating and synchronous rhythms in reciprocally inhibitory model neurons, *J. Neural Comp.*, 4:84–97.

Handwritten Digit String Recognition

C. J. C. Burges

Introduction

In this article, we consider the problem of machine reading of handwritten digits. The handwriting is assumed unconstrained, so that the digits may touch or even overlap. The raw image data are usually lifted from paper by a camera or scanner and then coded as a bit map. Systems must first locate the area of interest in the image and remove noise (variegated backgrounds, underlines, etc.). We will ignore such problems here and explore the core problem: given an image of a string of digits, determine the corresponding ascription.

This problem may be considered a first step toward reading any unconstrained handwriting, whether cursive, printed, or mixed. In fact, the approaches considered here are all generalizable to such problems, and workers are actively pursuing such generalizations.

We will consider systems that use neural networks for the recognition function. Considerable work has also been done using more classical recognition techniques (USPS, 1990 and 1992).

To solve the problem, one might consider the following three-step approach: (1) cut the image into pieces, where each piece contains only one character; (2) read each character separately; and (3) combine the results to form the answer. But how would one know how to cut the image into the correct pieces before doing any recognition? This puzzle leads immediately to the kinds of approaches described below. They fall into two broad categories: *oversegmentation* and *integrated segmentation and recognition* (ISR) techniques.

Oversegmentation is summarized as follows: First, construct a set of cuts (a *cut* is an imaginary curve dropping from the top to the bottom of the image). Cuts are generated using heuristics in such a way that the *correct* cuts (those that delineate characters) are contained in the set. Second, use these cuts to construct segments which are likely to contain single characters. Third, score the segments, using a neural network, and determine the set of segments that both comprises a plausible segmentation of the image and has the highest overall score. In this way, incorrect segments may occur, but the system is trained to reject them.

The segmentation step is difficult, and ISR techniques attempt to forgo it altogether. The idea is to detect directly both the presence of characters in the image and their approximate location.

Oversegmentation has the advantage that the segmentation algorithms can be designed to incorporate knowledge directly about the problem domain. Its disadvantage is that the cuts are generated by heuristics which may not generalize well to other, similar problems. ISR techniques have the advantage that they bypass difficult segmentation problems, and the disadvantage that more of the work is done by the neural network "black box," and that it is therefore more difficult to build in domain knowledge.

For problems where the total number of possible words that the user can write is small (for example, on checks), a third approach may be preferable: that of recognizing the whole word at once, by building one or more models for each word that can occur, and then comparing the data against each model.

In the following sections, we will briefly describe four systems and provide sufficient references for further reading. For those readers unacquainted with the basic notions of feedforward neural nets, the article PATTERN RECOGNITION provides an excellent introduction.

Integrated Segmentation/Recognition Networks

One of the earliest ISR systems (Keeler, Rumelhart, and Leow, 1991; Keeler and Rumelhart, 1992) is based on a feedforward convolutional network (see CONVOLUTIONAL NETWORKS FOR IMAGES, SPEECH, AND TIME SERIES). The approach is characterized by an arrangement of the hidden units into three-dimensional "blocks," where each horizontal "sheet" in a block comprises sigmoidal units sharing the same feature-extracting kernel, and where the units in a column of a given block all have the same receptive field. In this way, positional information is maintained throughout the network, and in test mode, each output unit has a fixed piece of the image that contributes to its activation, enabling the system to determine not only the presence but also the location of digits in the image.

The network takes as input a gray-scale image. Above the feature-extracting blocks, a third block of hidden units, also connected via local receptive fields, uses an exponential rather than a sigmoidal transfer function. During training, a final summing layer of 10 units is used, where each output unit sums all outputs from a given horizontal sheet (corresponding to a particular digit class) from the exponential layer. This summing layer therefore throws away all positional information present in the previous layer. This choice of the last two layers has a self-organizing effect in training, during which the network is told which digits are present, but not where they are in the image. The exponential units introduce a "winner takes all," competitive learning effect. As a result, the exponential units themselves turn into "smart histograms" which give sharp peaks above the corresponding characters in the input image.

Results

The authors trained two networks with slightly different architectures in a two-step process, starting with single characters and then training on two- and three-digit fields. They then combined these systems, accepting the answer only when the two networks agreed. They trained and tested on the NIST Special Database 1. They quote results (averaged over fields of length 2 through 6, with average length 4) of an error rate of 0.7% at 83% acceptance. At the same acceptance rate, the individual nets had performances of roughly 2% error and 5% error. As other workers have found, combining results from different networks, even when similar architecture networks trained on similar data are used, can give a marked boost in performance (see AVERAGING/MODULAR TECHNIQUES FOR NEURAL NETWORKS).

The Saccade System

The guiding philosophy of this ISR system is to mimic the processes used by humans to read text. The system is called *Saccade* (Martin and Rashid, 1992; Martin et al., 1993) after the ophthalmological term given to small, rapid eye movements when changing focus from one point to another. The idea is to teach a neural network not only to classify individual digits, but also to estimate the position of the current character, and of the next character to the right, from the current position. The net has four sets of output units, which perform distinct roles: one is used to make corrective saccades so as to better center the digit being recognized; another is used to make saccades to the next character to be recognized; the third is used to determine whether or not the input field of view is correctly centered over a character; and the fourth set of units indicates the class of the character being recognized.

Training

Training images (consisting of several characters) must first be classified by humans both for the ascription and for the position of the center of each digit. The network is then trained using backpropagation. The maximum jump size is constrained never to exceed half the input field of view width, since the net has no information available to make a judgment about positions of characters which lie outside of its field of view.

Results

Martin et al. (1993) report results on the NIST Special Database 1. 80,000 characters (300 ppi) were used for training and 20,000 for testing. The training and test sets were drawn from handwriting samples from different groups of people. Images where the number of characters determined by the Saccade system did not match the number in the ascription were rejected. For five-digit images, the authors report field rates of 23.2% reject at 1.1% error. The system requires an average of 1.3 forward passes per character recognized.

Future Directions

The authors point out that this biology-inspired approach suggests several methods for improvement, in particular, foveal transforms, in which the network would be given a higher resolution in the center of its field of view by analogy with the human eye, and two-dimensional saccades, in which the system would be able to move vertically as well as horizontally across the image, in order to better center the characters.

Spatiotemporal Networks

The spatiotemporal connectionist approach (Fontaine and Shastri, 1992, 1993) is more rooted in computer science than in biology. The image is scanned through the network, one column at a time. Horizontal spatial relationship information is kept by using propagation delays in the connections between units, and by giving some units delayed connections to themselves, so that each unit's activation depends on a spatial window.

Spatiotemporal Network Architecture

The architecture has a hierarchical structure. The fundamental unit is the Local Receptive Field (LRF) subnetwork, which consists of four inputs, four hidden units, and one output. These are used to detect lines at one of four angles, and are pretrained. LRFs are then replicated and tesselated to form a Feature Detection Module (FDM), which can then act on larger fields. A typical FDM might comprise three LRFs. A few FDMs are then combined, along with some unstructured hidden layers, to form a Single Scan Network (SSN), which is used to scan the image in one particular direction. Four SSNs (scanning left-to-right, right-to-left, top-to-bottom, and bottom-to-top) are then combined into a single digit recognition network. The complete recognition network comprises 10 single digit networks, one for each digit. The above system is extended to recognize fields by combining one left-to-right SSN, the full network described above, and a procedural controller, into which domain knowledge can be incorporated.

Network Size and Training

Although the architecture may seem complex, it results in 17,108 weights, which is quite small for a digit recognizer; in addition, the LRF's pretrained connections (which amount to 3,888 weights) can be frozen during training of the aggregate network. Training is done using the BFGS quasi-Newton optimization method (McCormick, 1983). Since the network is temporal, the targets are functions in time; classification decisions are made based on the time integral of the outputs.

Results

Test sets consisted of 500 images of pairs of characters, in which touching and overlapping digits were synthesized and the fraction of touching and overlapping digits was therefore controllable. For 5% overlapping and 45.6% touching digits, for example, an accuracy of 73.3% at 7.2% rejection rate was obtained.

Shortest Path Segmentation

In this oversegmentation system, the recognition engine is a convolutional feedforward neural net single character recognizer. The idea is to represent the recognition/segmentation problem as a graph in which there is a one-to-one correspondence between paths and segmentation choices for the image (Burges et al., 1992; Burges et al., 1993). The approach has been extended to read images of bit-mapped cursive script (Burges et al., 1993; Matan et al., 1992). The system has five basic sequential operations: image preprocessing, cut construction, graph generation, recognition, and graph evaluation.

Graph Generation

Once a set of candidate cuts has been computed, the segmentation graph is constructed. The nodes in the graph correspond to image pieces, while the arcs connect nodes if and only if the corresponding image pieces are "legal neighbors." The definition of *legal neighbor* depends on the application: skip between or overlap of image pieces can be allowed.

Recognition and Graph Evaluation

Nodes in the graph corresponding to image pieces that are deemed unlikely to be digits are removed. This is important because calls to the recognizer are time-expensive, and each image piece corresponding to a surviving node is sent to the neural net for scoring. The number of calls per character after pruning is approximately 1.4.

Graph Evaluation

The graph is populated in such a way that the shortest path through the graph corresponds to the best combined segmentation/recognition of the full image. The Viterbi algorithm (Forney, 1973) is used to find the shortest path that traverses a given number N of nodes (corresponding to N digits). This graphical approach offers many advantages: a given list of answers can easily be scored, the Kth shortest path algorithm (Gondran and Minoux, 1990) can be used to generate the K best answers, and the graph can be combined with lexicon or grammar graphs to incorporate contextual information.

Training

The graphical approach allows "in-loop" training. The idea is to use correctly (incorrectly) segmented image pieces as positive (negative) examples, and to use the system itself to generate the counterexample data. The system is thus able to learn from its mistakes as it develops during training.

Results

On a test set of 2,642 handwritten ZIP codes, the system gets 85.4% recognition rate at 0% rejection, and 3.1% error at 27.6% rejection rate.

Road Map: Applications of Neural Networks
Background: Pattern Recognition
Related Reading: Competitive Learning; Convolutional Networks for Images, Speech, and Time Series

References

Burges, C. J. C., Ben, J. I., LeCun, Y., Denker, J. S., and Nohl, C. R, 1993, Off-line recognition of handwritten postal words using neural networks, *Int. J. Pattern Recog. Artif. Intell.*, 7:689–703.

Burges, C. J. C, Matan, O., LeCun, Y., Denker, J. S., Jackel, L. D., Stenard, C. E., Nohl, C. R., and Ben, J. I., 1992, Shortest path segmentation: A method for training a neural network to recognize character strings, in *Proceedings of the International Joint Conference on Neural Networks*, New York: IEEE, vol. 3, pp. 165–172.

Fontaine, T., and Shastri, L., 1992, Handprinted digit recognition using spatiotemporal connectionist models, in *Proceedings of the IEEE Conference on Computer Vision and Pattern Recognition*, Los Alamitos, CA: IEEE Computer Society Press, pp. 169–175.

Fontaine, T., and Shastri, L., 1993, Recognizing handprinted digits strings: A hybrid connectionist/procedural approach, in *15th Annual Meeting of the Cognitive Science Society*.

Forney, G. D., Jr., 1973, The Viterbi algorithm, *Proc. IEEE*, 61(3):268. ◆

Gondran, M., and Minoux, M., 1990, *Graphs and Algorithms*, trans. S. Vadja, New York: Wiley. ◆

Keeler, J., and Rumelhart, D. E., 1992, A self-organizing integrated segmentation and recognition neural net, in *Advances in Neural Information Processing Systems 4* (J. E. Moody, S. J. Hanson, and R. P. Lippmann, Eds.), San Mateo, CA: Morgan Kaufmann, pp. 496–503.

Keeler, J., Rumelhart, D. E., and Leow, W., 1991, Integrated segmentation and recognition of hand-printed numerals, in *Advances in Neural Information Processing Systems 3* (R. P. Lippmann, J. E. Moody, and D. S. Touretzky, Eds.), San Mateo, CA: Morgan Kaufmann, pp. 557–563.

McCormick, G. P., 1983, *Nonlinear Programming: Theory, Algorithms, and Applications*, New York: Wiley. ◆

Martin, G. L., and Rashid, M., 1992, Recognizing overlapping hand-printed characters by centered-object integrated segmentation and recognition, in *Advances in Neural Information Processing Systems 4*

(J. E. Moody, S. J. Hanson, and R. P. Lippmann, Eds.), San Mateo, CA: Morgan Kaufmann, pp. 504–511.

Martin, G. L., Rashid, M., Chapman, D., and Pittman, J., 1993, Learning to see where and what: Training a net to make saccades and recognize handwritten characters, in *Advances in Neural Information Processing Systems 5* (S. J. Hanson, J. D. Cowan, and C. L. Giles, Eds.), San Mateo, CA: Morgan Kaufmann, pp. 441–447.

Matan, O., Burges, C. J. C., LeCun, Y., and Denker, J. S., 1992, Multi-digit recognition using a space displacement neural network, in *Advances in Neural Information Processing Systems 4* (J. E. Moody, S. J. Hanson, and R. P. Lippmann, Eds.), San Mateo, CA: Morgan Kaufmann, pp. 488–495.

United States Postal Service (USPS), November 1990 and 1992, *Advanced Technology Conference Proceedings*, Washington, DC.

Head Movements: Multidimensional Modeling

Barry W. Peterson

The head movement system has been modeled using biomechanical, linear systems analysis and neural network techniques (see Peterson and Richmond, 1988, chaps. 2, 12, 14, 16, and 19). This article concentrates on the latter, specifically on models derived from Pellionisz and Llinás's (1979, 1980) tensorial approach to explore how the vestibulocollic reflex (VCR) transforms head motions sensed by the six vestibular semicircular canals (SCCs) into patterns of neck muscle activation that would counter those motions, stabilizing the head in space.

Tensor network theory is designed to represent the stages of sensorimotor transformation in multidimensional, non-orthogonal systems. It recognizes that an external action (i.e., a sensory stimulus or the movement it produces) can be represented in two different ways in a non-orthogonal system such as the VCR. Figure 1*A* illustrates the *contravariant* (parallelogram) representation that corresponds to how forces generated by muscles combine to generate a net force *X* or *Y*. Each neck muscle exerts a torque on the head with a unique direction, and the angle between two such directions could be 120° as in Figure 1*A*. Then, the net force produced by activating these muscles is the *vector sum* of their torques as shown. In Figure 1*B*, points *X* and *Y* are represented by their perpendicular projections on each of the coordinate axes. This *covariant* type of

representation corresponds to sensory systems, such as the vestibular SCCs, where each canal responds independently to a head rotation in proportion to the projection of that rotation on the canal's direction of maximal sensitivity. These two representations are not identical, and thus neural transformations underlying a reflex like the VCR must involve conversion from covariant to contravariant representations.

Pellionisz and Peterson's (1988) tensorial model of the feline VCR treats this reflex as a three-stage matrix transformation of

A CONTRAVARIANT (PARALLELOGRAM) REPRESENTATION

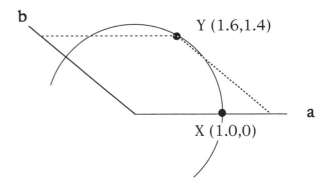

B COVARIANT (PROJECTION) REPRESENTATION

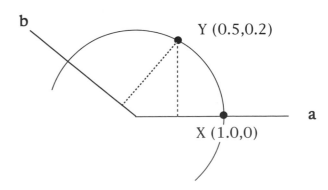

Figure 1. Illustration of contravariant and covariant representations of a sensory stimulus or movement in a non-orthogonal coordinate frame. Part *A* shows contravariant (parallelogram) representations of points *X* and *Y* which correspond to physical combination of muscle actions. Here two muscles, **a** and **b**, are required to produce movements in the *X* and *Y* directions. Parallelograms representing summation of muscle actions are constructed to determine how much activity is required in **a** and **b** to generate each movement. In this non-orthogonal frame, **a** must be maximally activated for a movement not along **a**'s direction, but for movement along the 1-o'clock direction (30° from **a**'s direction). Similarly, **b** is maximally activated not for a movement along **b**'s direction, but along the 3-o'clock direction. Part *B* shows the covariant (projection) representation of the same two movements, corresponding to responses of sensors with optimal response-directions aligned with **a** and **b**. Here responses are determined by perpendicular projection of each point on the two sensor axes. Thus, each sensor is maximally activated when the stimulus is along its own axis. Numbers in parentheses give values of activation components of **a** and **b**, respectively, in response to movement to *X* and *Y*.

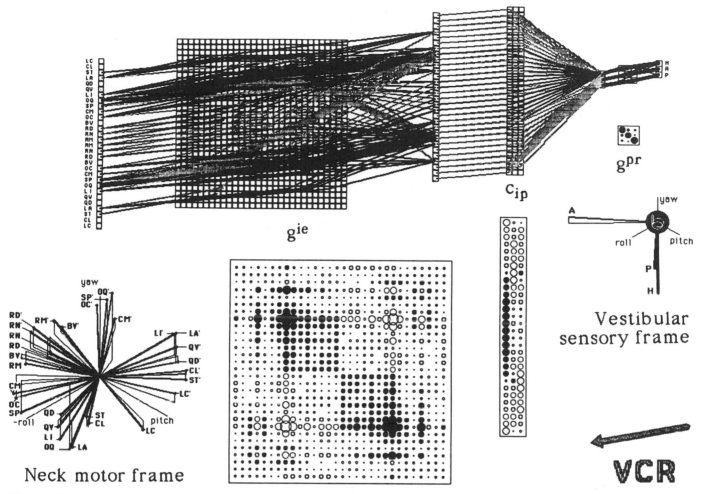

Figure 2. Tensor network model of the feline VCR (Pellionisz and Peterson, 1988). Sensorimotor transformation occurs in three stages: g^{pr}, sensory metric tensor; c_{ip}, sensorimotor embedding tensor; and g^{ie}, neck motor metric tensor. These general tensorial operations are expressed by matrices, represented by patch diagrams where positive and negative components are shown by filled and open circles, respectively, with surface area proportional to size of the component. The CNS can implement these matrices by neuronal networks as indicated at the top, where strengths of connectivities from H, A, and P canals to the 30 neck muscles are represented by line shadings (black for excitation, gray for inhibition), with line thickness proportional to the value of the interconnecting matrix element.

covariant vestibular sensory input to contravariant neck motor output (Figure 2). For simplicity, the neck was treated as a single, 3-degree-of-freedom "joint" about which head rotations occur. However, the model takes into account the precise pulling actions of the 30 neck muscles (Wickland, Baker, and Peterson, 1991) and axes of maximum sensitivity of the six SCCs (Blanks, Curthoys, and Markham, 1972). The latter define the coordinate frame of input signals to the VCR; the former, the coordinate frame of its motor output. Each transformation is implemented by a matrix that expresses a general tensorial operation in the particular coordinate frames of the VCR. Stage 1, the vestibular metric tensor, converts covariant signals from the three pairs of SCCs into contravariant form. It is implemented by a 3×3 matrix, which is the inverse of the matrix composed of cosines of the angles between SCCs. In stage 2, sensorimotor embedding, the contravariant sensory signals are projected on neck coordinate axes defined as pulling directions of the 30 neck muscles. The 3×30 matrix involved is a table of cosines of angles between SCC sensitivity axes and each of the 30 muscle pulling axes. Its output is a covariant representation of the VCR motor command.

Stage 3, the 30×30 neck metric tensor, converts the covariant motor command from stage 2 into the contravariant form required to generate a pattern of muscle activity that will physically sum in parallelogram fashion to produce the desired movement. Unlike the other matrices, this matrix cannot be uniquely computed because the neck motor system is functionally overcomplete because 30 muscles control just 6 degrees of freedom of head motion. As a result, any given head motion could be produced by an infinite number of muscle activation patterns, and correspondingly, the 30×30 matrix of cosines has no unique inverse: an infinite number of solutions exist. In considering this problem, which arises for most motor systems, Pellionisz (1984) hypothesized that the central nervous system (CNS) chooses a matrix corresponding to the Moore-Penrose generalized inverse (Albert, 1972) of the 30×30 array of cosines, a solution equivalent to a least squares fit in which unwanted muscle co-contractions are minimized, thus generating efficient motor patterns. Once this inverse is chosen, the model predicts a unique direction of head rotation for which each muscle will be maximally activated. Activation during rotation about other axes declines as the cosine of the angle

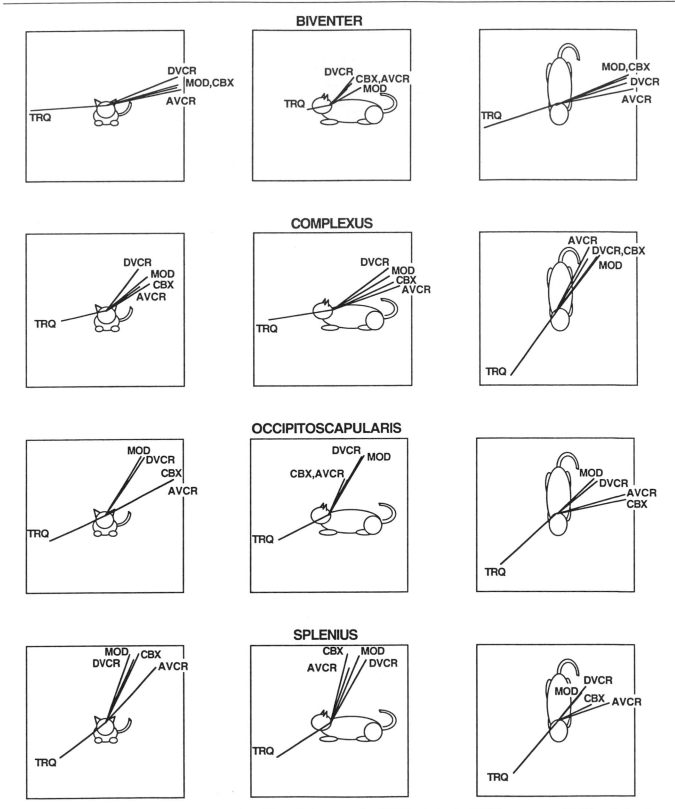

Figure 3. Comparison of torques of cat neck muscles with predictions and electromyographic measurements of their excitation. Vector representations are as in Figure 2, except that vector lengths are normalized to 1. AVCR = average of measurements of neck muscle excitation during VCR in alert cats; DVCR = average of VCR excitation in same cats after decerebration; CBX = VCR measurements after cerebellectomy in decerebrate cats; MOD = model predictions of neck muscle excitation; TRQ = torques of neck muscles.

between those axes and the optimal axis. As in Figure 1A, predicted optimal activation directions typically differ from muscle pulling directions.

An advantage of the model is that it can be tested experimentally. My colleagues and I have observed neck muscle electromyographic (EMG) activity in cats during three-dimensional head or body movements. Interestingly, different animals employ quite different muscle activations when generating a particular movement voluntarily (Keshner et al., 1992), indicating that the head-neck motor system is indeed overcomplete. In contrast, the VCR employs a single unique pattern of muscle activation in both decerebrate (Baker, Goldberg, and Peterson, 1985) and alert (Banovetz et al., 1987) cats, which is well predicted by the tensorial model. Figure 3 compares activation directions of four muscles with those predicted by the model and with vectors representing the torque produced by each muscle. The model predictions "MOD" agree quite closely with the decerebrate VCR measurements, "DVCR," which were made with paradigms that primarily activated semicircular canal receptors. Torque vectors, "TRQ," do not fall exactly opposite the VCR activation directions, as would occur if muscles were maximally activated for rotations that required a compensatory head movement aligned with their pulling directions. Alert VCR data, "AVCR," were obtained with a paradigm that induced significant activation of otolith organs. Since the model does not include otolith input, it is not surprising that the AVCR and MOD vectors are less well aligned.

The model thus passes its first test, replicating VCR input/output kinematics. However, a motivation in constructing the model was that it would also predict internal stages of signal processing within VCR neural networks. One prediction, based on the original concepts of Pellionisz and Llinás (1979, 1980), was that the metric tensors constituting model stages 1 and 3 are implemented by the cerebellum (see Arbib and Amari, 1985, for a critique). To test this idea, Banovetz et al. (1987) made partial cerebellar lesions, destroying vermis lobules 5 through 10 and much of the deep nuclei in four cats. While VCR amplitude increased postlesion, no significant shift in maximal activation directions occurred, as indicated by CBX vectors in Figure 3. However, it is still possible that critical neck metric tensor transformations are mediated by a portion of the cerebellum that was spared by Banovetz and co-workers, possibly lobule 1, which has extensive interconnections with neck and vestibular system (Hirai, 1987).

A more direct test is to compare spatial properties of neurons at each stage of the VCR with those predicted by the model. Our recordings of spatial properties of vestibulospinal neurons failed to confirm model predictions and suggested why the internal representations of the model may not reflect the processing used by the CNS. Perlmutter et al. (1989) recorded from second-order vestibular neurons projecting either selectively to the neck (VC neurons) or to neck and oculomotor nuclei (VOC neurons). As shown in Figure 4, while activation directions of VOC neurons were closely aligned with those of the ipsilateral SCCs, VC neurons received a great deal of convergent input from horizontal and vertical canals and often responded best to rotations that would primarily activate contralateral SCCs. Despite this diversity of responses, activation directions aligned with pitch rotation were conspicuously absent. Graf, Baker, and Peterson (1993) observed a similar lack of pitch response in vestibuloocular (VO) neurons and concluded that vestibular circuits generate pitchlike activations required by vertical rectus eye muscles by combining signals from relay neurons in the left and right vestibular nuclei on motoneurons. Apparently, a similar mechanism accounts for activa-

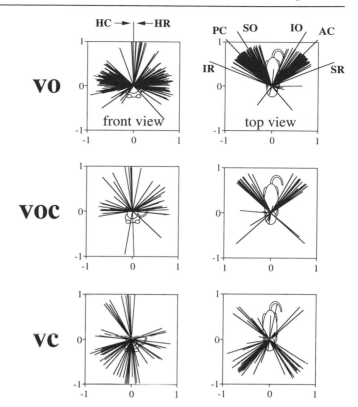

Figure 4. Maximal activation directions of vestibulo-ocular neurons (VO), vestibulo-collic neurons (VC), and neurons with both oculomotor and spinal projections (VOC), from Perlmutter et al. (1989). SCC and extraocular muscle maximal activation direction vectors (from Baker and Peterson, 1991) are shown for comparison. HC = horizontal canals; AC, PC = vertical canal pairs; HR = average of medial and lateral rectus extraocular muscles; IO = inferior oblique extraocular muscle; IR = inferior rectus extraocular muscle; SO = superior oblique extraocular muscle; SR = superior rectus extraocular muscle.

tion of neck muscles like biventer cervicis, whose activation directions are near pitch.

These observations indicate that the model in its current form does not incorporate a key restriction governing projections within the VCR neural network. It is therefore not surprising that it fails to predict spatial patterns of activation of vestibulocollic neurons or changes in EMG activity that underlie cerebellar dysmetria of head movements. These findings should not be taken as an indictment of the tensorial modeling approach but rather as confirmation of a principle embraced by its authors: that models will only be successful if they incorporate all aspects of the geometry of sensors, effectors, and neural projections. Given the linearity of the vestibulocollic system, I am confident that spatial transformations underlying the VCR can be represented as a series of matrix transformations, as exemplified by the Pellionisz and Peterson (1988) model. It remains to be seen if other postulates of the tensorial approach, such as using the Moore-Penrose generalized inverse for covariant-to-contravariant transformations in overcomplete systems, prove useful in predicting actual VCR neural network transformations once a new iteration of the model incorporating the findings of our neural recordings is constructed.

Finally, it is appropriate to note some limitations of this modeling approach. First, the internal structure of the model is arbitrarily determined by the modeler. Pellionisz and Peterson

(1988) chose three serial transformations since these most closely conform to the tensorial concepts on which the model is based. Ideally, however, the transformations should more closely match known neuroanatomy of the modeled system (parallel paths with two and three transformations for the VCR, for instance; Peterson and Richmond, 1988). Also, in its current form, the model does not represent VCR networks as distributed systems with many elements at each stage, as required to predict the diversity of neural responses in Figure 4. Other neural network models of vestibular reflexes (Anastasio and Robinson, 1989) capture this aspect of the system, but unless they are very closely constrained by known neuroanatomy and neurophysiology, they tend to "find" solutions that provide little insight into how the brain actually functions. The advantage of the tensorial model is that its core principles and assumptions are clear, and its predictions are testable—valuable features for models that aspire to explain brain functioning.

Road Map: Primate Motor Control
Related Reading: Geometrical Principles in Motor Control; Vestibulo-Ocular Reflex: Performance and Plasticity

References

Albert, A., 1972, *Regression and the Moore-Penrose pseudoinverse*, New York: Academic Press.

Anastasio, T. J., and Robinson, D. A., 1989, The distributed representation of vestibulo-oculomotor signals by brain-stem neurons, *Biol. Cybern.*, 61:79–88. ◆

Arbib, M. A., and Amari, S.-I., 1985, Sensori-motor transformations in the brain (with a critique of the tensor theory of cerebellum), *J. Theor. Biol.*, 112:123–155.

Baker, J., Goldberg, J., and Peterson, B., 1985, Spatial and temporal response properties of the vestibulocollic reflex in decerebrate cats, *J. Neurophysiol.*, 54:735–756.

Baker, J. F., and Peterson, B. W., 1991, Excitation of the extraocular muscles in decerebrate cats during the vestibulo-ocular reflex in three-dimensional space, *Exp. Brain Res.*, 84:266–278.

Banovetz, J. M., Rude, S. A., Perlmutter, S. I., Peterson, B. W., and Baker, J. F., 1987, A comparison of neck reflexes in alert and decerebrate cats, *Soc. Neurosci. Abstr.*, 13:1312.

Blanks, R., Curthoys, I., and Markham, C., 1972, Planar relationships of semicircular canals in the cat, *Am. J. Physiol.*, 223:55–62.

Graf, W., Baker, J., and Peterson, B. W., 1993, Sensorimotor transformation in the cat's vestibuloocular reflex system, I: Neuronal signals coding spatial coordination of compensatory eye movements, *J. Neurophysiol.*, 70:2425–2441.

Hirai, N., 1987, Input-output relations of lobules I and II of the cerebellar anterior lobe vermis in connection with nack and vestibulospinal reflexes in the cat, *Neurosci. Res.*, 4:167–184.

Keshner, E. A., Baker, J. F., Banovetz, J., and Peterson, B. W., 1992, Patterns of neck muscle activation in cats during reflex and voluntary head movements, *Exp. Brain Res.*, 88:361–374.

Pellionisz, A., 1984, Coordination: A vector-matrix description of transformations of overcomplete CNS coordinates and a tensorial solution using the Moore-Penrose generalized inverse, *J. Theor. Biol.*, 110:353–375. ◆

Pellionisz, A., and Llinás, R., 1979, Brain modeling by tensor network theory and computer simulation. The cerebellum: Distributed processor for predictive coordination, *Neuroscience*, 4:323–348.

Pellionisz, A., and Llinás, R., 1980, Tensorial approach to the geometry of brain function: Cerebellar coordination via a metric tensor, *Neuroscience*, 5:1761–1770.

Pellionisz, A. J., and Peterson, B. W., 1988, A tensorial model of neck motor activation, in *Control of Head Movement* (B. W. Peterson and F. J. R. Richmond, Eds.), New York: Oxford University Press, pp. 178–186. ◆

Perlmutter, S. I., Iwamoto, Y., Baker, J. F., and Peterson, B. W., 1989, Contribution of medial vestibulospinal neurons (VSNs) to spatial transformation in the vestibulocollic reflex (VCR), *Soc. Neurosci. Abstr.*, 15:516.

Peterson B. W., and Richmond, F. J. R., Eds., 1988, *Control of Head Movement*, New York: Oxford University Press.

Wickland, C. R., Baker, J. F., and Peterson, B. W., 1991, Torque vectors of neck muscles in the cat, *Exp. Brain Res.*, 84:649–659.

Hebbian Synaptic Plasticity

Thomas H. Brown and Sumantra Chattarji

Introduction

Interest in the idea that certain forms or aspects of learning and memory result from use-dependent synaptic modifications has a long history (Brown, Kairiss, and Keenan, 1990). Here we review some of the conceptual and experimental aspects of the type of synaptic learning mechanism originally suggested by the Canadian psychologist Donald Hebb (1949). We first summarize the contemporary concept of a Hebbian synaptic learning mechanism. Theoretical studies suggest that useful and potentially powerful forms of learning and self-organization can emerge in networks of elements that are interconnected by various formal representations of a Hebbian modification (Kohonen, 1984). Interest in the computational aspects of Hebbian modification algorithms has been enhanced by the neurophysiological discovery of a synaptic phenomenon in the hippocampus known as long-term potentiation (LTP), which is induced by a Hebbian mechanism. We review recent facts and hypotheses about LTP that are pertinent to contemporary interpretations of a Hebbian synaptic modification. Finally we review more recent evidence regarding variations and extensions of Hebb's original postulate for learning.

Concept of a Hebbian Synapse

In his 1949 book *The Organization of Behavior*, Donald Hebb proposed that an important condition for triggering enhanced synaptic efficacy would be the repeated conjunction of presynaptic activity and the firing of the cell on to which this activity was afferent (Hebb, 1949:62):

> When an axon of cell A is near enough to excite a cell B and repeatedly or persistently takes part in firing it, some growth process or metabolic change takes place in one or both cells such that A's efficiency, as one of the cells firing B, is increased.

This idea has come to be known as *Hebb's postulate* for learning. Hebb proposed this change as the basis of a memory formation and storage process that would cause enduring modifications in the elicited activity patterns of spatially distributed "nerve cell assemblies." Thus, it specifies the *location* of the modification and provides a qualitative statement of the *conditions* for change. In brief, a *Hebbian synapse* strengthens when the presynaptic and postsynaptic elements tend to be coactive.

Hebb's postulate can be seen as the synthesis of two earlier ideas about memory. In 1890 the American psychologist William James proposed that the most basic law of association was the *law of neural habit*, which maintained that if two "elementary brain processes" are active at the same time, then each will tend to "propagate its excitement into the other." James did not suggest a cellular correlate or substrate for these elementary brain processes. The second idea concerns the physical nature of the changes underlying memory. In 1893 the Italian neuroanatomist Eugenio Tanzi proposed that the synapse might be the locus of the modification, an idea that was advanced by Santiago Ramón y Cajal in 1911. Thus, Hebb's postulate can be seen as the synthesis of the law of neural habit and the synaptic hypothesis for memory.

Contemporary Concept of a Hebbian Synaptic Modification

Since 1949 the concept of a Hebbian synapse has undergone considerable evolution to include several defining characteristics. These key features (Brown et al., 1990), which form the basis of the contemporary understanding of Hebbian synaptic modification, are as follows:

1. *Local mechanism.* The process is *local*; i.e., all the information that is necessary for the modification is available to the synapse as a consequence of ordinary synaptic transmission. Although a Hebbian mechanism uses only local information, the modification process may be subject to global control signals. Such signals may enable the induction or consolidation of changes at synapses that have met the criteria for a Hebbian modification. A global "reinforcement signal" can thus control Hebbian plasticity in a large population of activated synapses. Substances that have been demonstrated to act as neuromodulators (such as catecholamines and acetylcholine) are plausible candidates for this role. Such global modulation of a local process is different from an external "teacher" signal that explicitly "instructs" selective modification on a synapse-by-synapse basis independent of local activity.

2. *Interactive mechanism.* Whether a change occurs at a Hebbian synapse depends on activity on *both* sides of the synaptic cleft. It involves a true interaction, in the statistical sense, between presynaptic and postsynaptic activity. Thus, mechanisms that depend on either presynaptic activity alone (e.g., posttetanic potentiation or PTP) or postsynaptic activity alone, or on a linear superposition of the consequences of presynaptic and postsynaptic activity, are termed noninteractive and would not qualify as Hebbian. This interactive requirement makes the mechanism fundamentally associative. The nature of the interaction can involve a *conjunctive-type* rule, where the simple co-occurrence of (some level) of pre- and postsynaptic activity (within some short time interval) is sufficient to cause synaptic enhancement, or a *correlational-type* rule, where the synaptic enhancement depends on the covariance, or some other statistical measure of the association, between pre- and postsynaptic activity.

3. *Time-dependent mechanism.* Modifications in a Hebbian synapse depend on the exact time of occurrence of pre- and postsynaptic activity. The timing of pre- and postsynaptic events plays an essential role in determining the change. This requirement for temporal specificity is best illustrated in the phenomenon called associative LTP between two separate afferent pathways. In area CA1 of the hippocampus, Brown and co-workers (Brown et al., 1990) demonstrated that tetanic stimulation of either a "weak" pathway (i.e., one that is unable to induce potentiation on its own) or a "strong" pathway (i.e.,

one capable of inducing potentiation on its own) *alone* failed to induce LTP in the weak input. Only temporally overlapping pairing of the two inputs produced associative LTP. However, several studies (reviewed in Brown et al., 1990) have reported the occurrence of LTP even if there is a small temporal gap—a brief "trace period"—of up to 40 ms.

Thus, a Hebbian synapse may be defined as one that uses a time-dependent, highly local, and strongly interactive mechanism to increase synaptic efficacy as a function of the conjunction or correlation between pre- and postsynaptic activity. Neither Hebb nor his colleagues of the day had the technology to conduct experiments to study Hebbian synaptic modifications. Thus, situation apart from providing a conceptual framework, Hebb's ideas had little impact on neurobiology. This situation changed with two important developments during the 1960s and 1970s. First, even though experimental tools and preparations amenable to testing Hebb's ideas were lacking, this lack did not prevent theoretical explorations of their computational implications. Second, emergence of more sophisticated physiological techniques eventually led to the discovery of Hebbian synaptic plasticity mechanisms in the hippocampus.

Experimental Evidence for Hebbian Synaptic Mechanisms

Hebbian synapses have now been shown to exist in the sense that one can demonstrate experimentally a use-dependent form of synaptic enhancement that is governed by a time-dependent, highly local, and strongly interactive mechanism. This mechanism appears to be responsible for at least one form of hippocampal synaptic plasticity called long-term potentiation (LTP). LTP is a use-dependent and persistent increase in synaptic strength that can be rapidly induced by brief periods of synaptic stimulation. It has been reported to last for hours in vitro, and for days and weeks in vivo (Baudry and Davis, 1991). LTP was first discovered and is very prominent in a brain structure—the hippocampus—which has long been implicated in learning and memory.

However, it should be noted that even though LTP has several features that make it an attractive candidate for memory mechanisms, LTP does not last indefinitely. Experiments in which the time course of LTP has been studied have typically found that, once induced, LTP seems to involve a heterogeneous family of synaptic changes with dissociable time courses lasting a few hours or several days (for discussion, see Morris et al. and Laroche et al. in Baudry and Davis, 1991). Thus, one can question whether the duration of hippocampal LTP is sufficient to sustain long-term memory. On the other hand, there appear to be multiple LTP mechanisms, one of which is dependent on protein synthesis and may be important for long-term memory. The application of various inhibitors of protein synthesis such as anisomycin and cycloheximide disrupts long-term memory formation, while it leaves short-term memory almost intact. A similar approach to examining involvement of protein synthesis in LTP suggests that, in agreement with behavioral data, protein synthesis inhibition disrupts the maintenance of LTP, leaving the induction of LTP relatively or totally intact. Furthermore, the neocortex rather than the hippocampus is likely to be the final depository of long-term memory storage.

The form of LTP that has been most extensively studied is the variety that occurs in the Schaffer collateral/commissural (Sch/com) synaptic input to the pyramidal neurons of hippocampal region CA1. Several groups have shown that the induction of LTP in the CA1 Sch/com synapses satisfies the definition of a Hebbian modification (Bliss and Collingridge, 1993;

Brown et al., 1990; Brown et al., 1988). The LTP observed at the mossy-fiber synapse in area CA3 of the hippocampus has properties that are fundamentally different from that found in CA1 and dentate gyrus (Johnston et al., 1992). In the following subsection we summarize some of the key aspects of the physiology of LTP induction in the Sch/com synapses of the hippocampus.

Induction of Hippocampal LTP

Glutamate, the major excitatory neurotransmitter in the hippocampus, exerts its action through at least two major classes of receptors—NMDA (N-methyl-D-aspartate) and non-NMDA —named for the specific exogenous agonists which selectively activate them (see NMDA RECEPTORS: SYNAPTIC, CELLULAR, AND NETWORK MODELS). The non-NMDA category can be further subdivided into at least two classes of receptors—the AMPA receptor (selectively activated by α-amino-3-hydroxy-5-methyl-4-isoxazole) and the metabotropic receptor (a second messenger-coupled receptor). The fast component of the excitatory postsynaptic current (EPSC) at the Sch/com synapses in the CA1 region is mediated primarily via the AMPA receptors, while the slow component is mediated by NMDA receptors. The fast AMPA-receptor component lasts a few milliseconds and has a fast rise time, and its duration is limited by the time constant of the cell membrane. The postsynaptic potentials mediated by the AMPA receptors are caused by an increase in a mixed cation conductance (mainly Na^+ and K^+). In contrast, the slow NMDA component has a slower rise time (~ 20 ms) and a longer decaying phase (about 200–300 ms). Furthermore, stimulation of NMDA receptors results in the activation of a voltage-dependent current that is carried not only by Na^+ and K^+ but also importantly by Ca^{2+}.

There is general agreement that the induction of the most commonly studied form of LTP at Sch/com synapses is controlled by the NMDA subclass of glutamate receptor. The common working hypothesis is that the Ca^{2+} influx through the NMDA receptor-gated channel and the resultant increase in postsynaptic Ca^{2+} are partly responsible for triggering the induction of LTP (Bliss and Collingridge, 1993). How does Ca^{2+} influx through the NMDA receptor-gated channel relate to the Hebbian nature of LTP induction? It turns out that this channel has just the kind of gating properties needed for a Hebb-type conjunctive mechanism.

The NMDA conductance is a function of both glutamate binding to the receptor site and voltage across the channel (Bliss and Collingridge, 1993). The voltage dependence results from the fact that the NMDA channels are normally partially blocked by Mg^{2+} and the fact that this channel block is gradually relieved as the membrane is depolarized. Thus, the conjunction of presynaptic activity (to cause glutamate release) and a sufficient amount of postsynaptic depolarization (to relieve the channel block) allows Ca^{2+} influx into the postsynaptic cell.

Just as there are many ways of implementing a Hebbian algorithm theoretically, nature may have more than one way of designing a Hebbian synapse. As mentioned earlier, the induction of LTP at the mossy-fiber synapses in field CA3 is different from that observed in the CA1 region and dentate gyrus. In particular, induction of mossy-fiber LTP does not require the activation of NMDA receptors (Johnston et al., 1992). Nevertheless, there is evidence that mossy-fiber LTP depends on activation of postsynaptic voltage-dependent Ca^{2+} channels *and* high-frequency presynaptic activity (Johnston et al., 1992). Thus, at the molecular level there may be more than one way to build a Hebbian synapse.

Biophysical Models of LTP Induction

A quantitative model of the NMDA-dependent postsynaptic mechanism for LTP induction has been developed for the Sch/com synapses (for a detailed review see Zador, Koch, and Brown, 1990; Brown et al., 1990). In this model, the trigger for synaptic enhancement is Ca^{2+} influx through NMDA receptor-gated channels that are located on the dendritic spine head (see DENDRITIC SPINES). The model maintains that (a) the peak transient increase in intracellular Ca^{2+} is localized within the dendritic spine, (b) the spine amplifies the local change in intracellular Ca^{2+}, and (c) the relationship between the peak transient increase in Ca^{2+} and the amount of LTP is nonlinear. Two other fundamental components of this model are calcium buffering and extrusion mechanisms throughout the spine head and neck and longitudinal calcium diffusion between the spine and the dendrite. Voltage-gated Ca^{2+} channels were assumed in this model to be located only on the dendritic shaft.

New technology is enabling predictions of the model to be tested experimentally. Several lines of recent evidence call into question some of the features of the model. There is now evidence that in many synaptic systems LTP can be induced by mechanisms that do not require the activation of NMDA receptors (Johnston et al., 1992). Furthermore, the relative amount of Ca^{2+} influx through voltage-gated Ca^{2+} channels is large (Jaffe, Fisher, and Brown, 1994). Finally, recent evidence suggests that NMDA receptor-mediated currents are not sufficient for inducing LTP. Activation of the metabotropic subtype of glutamate receptor may also be necessary (Bortolotto et al., 1994).

More recently, experiments using fluorescence imaging techniques with confocal scanning laser microscopy to visualize dendritic spines in hippocampal slices have raised further questions about some of the assumptions underlying the model described above (Jaffe et al., 1994). These studies suggest that most spines are not isolated from changes in dendritic Ca^{2+}; equivalent changes in Ca^{2+} were observed in both spines and dendrites in response to membrane depolarization. Thus, future models of dendritic spines not only need to incorporate influx of Ca^{2+} through NMDA receptors, but also need to consider the contribution of voltage-gated Ca^{2+} channels and the role of metabotropic receptors in releasing Ca^{2+} from intracellular stores.

These and other recent findings have led to the development of a second-generation biophysical model (Jaffe et al., 1994) that incorporates a scheme in which multiple sources of calcium and calcium-dependent processes interact within different spatiotemporal domains of the soma, dendritic shaft, and spine. Different locations, magnitudes, and durations of intracellular $[Ca^{2+}]$ changes may play a critical role in determining the amplitude, time course, or even polarity of the synaptic change, giving rise to more complex computations than are possible with a strictly Hebbian learning rule that is constrained by the properties of the NMDA receptor-gated channel. One of the key features that will be of importance in the second-generation spine model is the issue of going beyond the strictly Hebbian learning rule toward a more generalized Hebbian rule that allows bidirectional regulation of synaptic strength.

Bidirectional Regulation of Synaptic Strength

According to the preceding discussion, a strictly Hebbian synaptic modification makes use of a local, time-dependent, and interactive mechanism to *increase* synaptic strength as a function of correlation or conjunction between presynaptic and postsynaptic activities. Hebb, however, did not discuss the consequence of uncorrelated or negatively correlated pre- and postsynaptic activity. Hebb's original postulate did not explore the conditions under which synaptic efficacy could *decrease*. It is clear that, in the absence of a mechanism for synaptic weakening, the synaptic strength could tend to increase without bound, or else saturate at some asymptotic value.

One obvious problem with synapses that can only increase in efficacy is that the range of associative relationships which the memory system can encode is limited, an example being whether memory can be updated when the relationship between two events changes. A second problem is that memory systems of this type can reliably store only a certain amount of information, and one question is whether the facility to decrease as well as increase synaptic strengths can improve performance. A third point is that LTP does not last indefinitely; i.e., experiments in which the time course of LTP has been studied have typically found that, once induced, LTP seems to involve a heterogeneous family of synaptic changes with dissociable time courses lasting a few hours or several days (for discussion see Morris et al. and Laroche et al. in Baudry and Davis, 1991). Thus, processes that cause synaptic weakening could be relevant to all these points (Sejnowski, Chattarji, and Stanton, 1990). The existence of such a capacity in the hippocampus has long been suspected, and indeed is required by certain neural-network models of associative memory and synaptic connectivity. Thus, this final section provides a brief review of the theoretical justifications as well as experimental evidence for synaptic depression.

Theoretical Representations of Generalized Hebbian Modifications

The emergence of theoretical studies involving "neural network" models brings us—decades after Cajal's work and Hebb's book—to a third and equally important conceptual issue. After *where* and *when* comes the issue of *how* synaptic plasticity enables networks of neurons or neuron-like processing elements to perform various cognitive operations. Most models of information storage in neural networks rely on changing the synaptic strengths, or weights, between model neurons (Kohonen, 1984). The weights in these simplifying models are altered by learning rules or algorithms so that the network can later retrieve the stored information or perform the desired task.

Probably the most thoroughly explored use of the Hebb rule in neural network models is the formation of associations between one stimulus or pattern of activity in one neural population and another. The Hebb rule and variations on it have been desirable for use in such neural network models, because they provide a way of forming global associations between large-scale patterns of activity in assemblies of neurons using only the local information available at individual synapses (Sejnowski et al., 1990).

The earliest models of associative memory were based on network models in which the output of a model neuron was assumed to be proportional to a linear sum of its inputs, each weighted by a synaptic strength. Thus,

$$V_B = \sum W_{BA} V_A \tag{1}$$

where V_B are the firing rates of a group of M output cells, V_A are the firing rates of a group of N input cells, and W_{BA} is the synaptic strength between input cell A and output cell B.

The transformation between patterns of activity on the input vectors to patterns of activity on the output vectors is determined by the synaptic weight matrix, W. This matrix can be constructed during repeated presentation of pairs of associated input and output vectors using the simplest version of the Hebb rule:

$$\Delta W_{BA} = \varepsilon V_B V_A \tag{2}$$

where the strength of the learning ε can be adjusted to scale the outputs to the desired values. This equation states that the variables relevant to synaptic change are the co-occurring activity levels, and that increases in synaptic strength are proportional to the product of the presynaptic and postsynaptic values. It should also be emphasized that this simple rule admits of many variations that still qualify as Hebbian.

This model of associative storage is simple and has several attractive features: first, learning occurs at each trial; second, the information is distributed over many synapses, so that recall is relatively immune to noise or damage; and third, input patterns similar to stored inputs will give output similar to stored outputs, a form of generalization.

The matrix model of associative storage has some obvious limitations: First, items with input vectors that are similar, that is, with subpopulations of activated neurons having a significant overlap, will produce outputs that are mixtures of the stored outputs; but discriminations must often be made among similar inputs. Second, the matrix model of associative memory cannot respond contingently to pairs of inputs, that is, have an output that is other than the sum of the individual outputs. Last, any learning system that uses a mechanism that can *only* increase the strengths of synapses will eventually degrade as all the synapses begin to saturate at their maximum values.

One way to prevent saturation of the synaptic strengths is to reduce the weights by nonspecific decay, but this method results in information loss at the same decay rate. Another approach is to renormalize the total synaptic weight of the entire terminal field from a single neuron to a constant value. Sejnowski (1977) proposed a learning algorithm that accomplishes this purpose by using the selective *decrease* of synaptic strength to accomplish optimal error-correction learning based on storing the covariances between the pre- and postsynaptic neurons.

According to this *covariance rule*, the change in strength of a plastic synapse should be proportional to the covariance between presynaptic and postsynaptic firing:

$$\Delta W_{BA} = \varepsilon(V_B - \langle V_B \rangle)(V_A - \langle V_A \rangle) \tag{3}$$

where $\langle V_B \rangle$ are the average firing rates of the output neurons and $\langle V_A \rangle$ are the average firing rates of the input neurons. Thus, the strength of the synapse should *increase* if the firing of the presynaptic and postsynaptic elements are *positively correlated*, *decrease* if they are *negatively correlated*, and remain *unchanged* if they are *uncorrelated*. The covariance rule is an extension of the Hebb rule, and it is easy to show that traditional Hebbian synapses can be used to implement it. Taking a time average of the change in synaptic weight in Equation 3:

$$\langle \Delta W_{BA} \rangle = \varepsilon(\langle V_B V_A \rangle - \langle V_B \rangle \langle V_A \rangle) \tag{4}$$

The first term on the right-hand side has the same form as the simple Hebbian synapse in Equation 2. The second term is a learning *threshold* that varies with the product of the time-averaged pre- and postsynaptic activity levels. This learning threshold ensures that no change in synaptic strength should occur if the average correlation between the pre- and postsynaptic activities is at chance level, i.e., when there is no net covariance.

Both *heterosynaptic depression* and *homosynaptic depression* of synaptic strength are required by the covariance rule, but

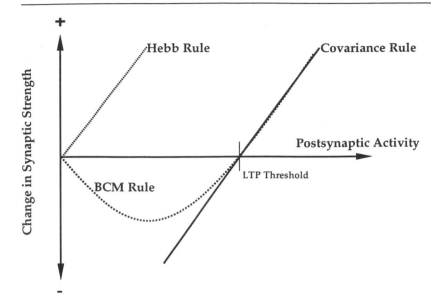

Figure 1. Schematic drawing of the change in synaptic strength as a function of the postsynaptic activity for the Hebb rule, the covariance rule, and the BCM rule. Both covariance and BCM rules are variants of the Hebb rule and postulate a threshold above which there is LTP (positive change in synaptic strength) and below which there is LTD (negative change in synaptic strength).

performance almost as good can be obtained by using only one or the other. However, homosynaptic depression, by preserving synaptic specificity, should be more sensitive to higher-order correlations in synaptic activity than heterosynaptic depression.

Hebbian synaptic plasticity has been used to model the development of OCULAR DOMINANCE AND ORIENTATION COLUMNS (q.v.) in visual cortex. A network model of cortical development incorporating Hebbian potentiation as well as depression was proposed by Bienenstock, Cooper, and Munro (1982). Their model of synaptic plasticity, now known as the *BCM rule* (Figure 1), strengthens the synapse when the average postsynaptic activity exceeds a threshold and weakens the synapse when the activity falls below the threshold level for potentiation, as in the covariance rule (see BCM THEORY OF VISUAL CORTICAL PLASTICITY). But the BCM rule has an additional threshold that must be reached to produce depression (see Figure 1). This threshold for depression gives the network model desirable stability properties.

Experimental Evidence for Long-Term Synaptic Depression

In spite of the obvious usefulness of a generalized Hebbian modification that includes both Hebb's original enhancement process and one or more processes causing synaptic weakening, except for studies in the cerebellum there has been a relative paucity of studies on long-term synaptic depression (LTD) compared to those on LTP. More recently, several laboratories have provided evidence for several different forms of LTD in the hippocampus and other cortical areas (Linden, 1994; Brown and Chattarji, in press; Sejnowski et al., 1990).

There is now evidence for both homosynaptic and heterosynaptic LTD in area CA1 of the hippocampus. Furthermore, an associative form of LTD has been reported in hippocampal field CA1 and dentate gyrus. Moreover, there is increasing evidence for similarities in properties of LTD observed in the hippocampus and other cortical regions. Recent findings suggest that Ca^{2+} may also play an important role in the induction of LTD. Thus, future models of generalized Hebbian plasticity will have to include a more complete picture of the various different sources of Ca^{2+} within dendritic spines as well as the complex spatiotemporal dynamics of changes in the levels of Ca^{2+} within the cell (Jaffe et al., 1994; Brown and Chattarji, 1994). Recent developments in fluorescence imaging techniques with confocal scanning laser microscopy will play a crucial role in shaping the next generation of theoretical and experimental investigations on Hebbian plasticity.

Acknowledgment. This work was supported by the Office of Naval Research.

Road Map: Mechanisms of Neural Plasticity
Background: I.3. Dynamics and Adaptation in Neural Networks; Ion Channels: Keys to Neuronal Specialization
Related Reading: Associative Networks; Hebbian Synaptic Plasticity: Comparative and Developmental Aspects; Long-Term Depression in the Cerebellum

References

Baudry, M., and Davis, J. L., 1991, *Long-Term Potentiation: A Debate of Current Issues*, Cambridge, Eng., and London: Bradford. ◆
Bienenstock, E. L., Cooper, L. N., and Munro, P. W., 1982, Theory for the development of neuron selectivity: Orientation specificity and binocular interaction in visual cortex, *J. Neurosci.*, 2:32–48.
Bliss, T. V. P., and Collingridge, G. L., 1993, A synaptic model of memory: Long-term potentiation in the hippocampus, *Nature*, 361: 31–39.
Bortolotto, Z. A., Bashir, Z. I., Davies, C. H., and Collingridge, G. L., 1994, A molecular switch activated by metabotropic glutamate receptors regulates induction of long-term potentiation, *Nature*, 368: 740–743.
Brown, T. H., Chapman, P. F., Kairiss, E. W., and Keenan, C. L., 1988, Long-term synaptic potentiation, *Science*, 242:724–728.
Brown, T. H., and Chattarji, S., in press, Hebbian synaptic plasticity: Evolution of the contemporary concept, in *Models of Neural Networks*, vol. 2 (L. van Hemmen, Ed.), New York: Springer-Verlag. ◆
Brown, T. H., Kairiss, E. W., and Keenan, C. L., 1990, Hebbian synapses: Biophysical mechanisms and algorithms, *Annu. Rev. Neurosci.*, 13:475–511. ◆
Hebb, D. O., 1949, *The Organization of Behavior*, New York: Wiley.
Jaffe, D. B., Fisher, S. A., and Brown, T. H., 1994, Confocal laser scanning microscopy reveals voltage-gated calcium signals within hippocampal dendritic spines, *J. Neurobiol.*, 25:220–223.

Johnston, D., Williams, S., Jaffe, D. B., and Gray, R., 1992, NMDA-receptor independent long-term potentiation, *Annu. Rev. Physiol.*, 54:489–505. ◆

Kohonen, T., 1984, *Self-Organization and Associative Memory*, Berlin and Heidelberg: Springer-Verlag.

Linden, D. J., 1994, Long-term synaptic depression in the mammalian brain, *Neuron*, 12:457–472. ◆

Sejnowski, T. J., 1977, Storing covariance with nonlinearly interacting neurons, *J. Math. Biol.*, 4:303–321.

Sejnowski, T. J., Chattarji, S., and Stanton, P. K., 1990, Homosynaptic long-term depression in hippocampus and neocortex, *Semin. Neurosci.*, 2:355–363. ◆

Zador, A., Koch, C., and Brown, T. H., 1990, Biophysical model of a Hebbian synapse, *Proc. Natl. Acad. Sci. USA*, 87:6718–6722.

Hebbian Synaptic Plasticity: Comparative and Developmental Aspects

Yves Frégnac

Introduction

Appropriate levels of description must be chosen in order to analyze dynamic changes in brain function during perception, learning, or development. One approach is to go from simple phenomenological rules to complex mechanistic scenarios of synaptic plasticity, and evaluate progressively how each level of complexity affects the processing and adaptive capacities of the overall network. This article will briefly summarize the historical foundations and subsequent elaboration by theoreticians and experimenters of a simple activity-dependent algorithm of synaptic plasticity proposed by Hebb in 1949. We will then review biological validations of predictions of Hebb's postulate at different levels of neural integration (i.e., synaptic efficacy, functional coupling, adaptive change in behavior).

Five major issues will be dealt with:

1. Should the definition of Hebbian plasticity refer to a simple positive correlational rule of learning, or are there biological justifications for including additional "pseudo-Hebbian" terms (such as synaptic depression owing to disuse or competition) in a generalized phenomenological algorithm?
2. Do the predictions of Hebbian-based algorithms account for most forms of activity-dependent dynamics in synaptic transmission throughout phylogenesis? How do the predictions depend on the complexity of the neural network (e.g., direct sensory-motor connections in *Aplysia* versus associative networks in neocortex)?
3. On which time scales (perception, learning, epigenesis) and at which stage of development of the organism (embryonic life, "critical" postnatal developmental periods, adult age) are up- and down-regulations of functional links induced by neuronal activity?
4. What are the spatiotemporal constraints (e.g., input specificity, temporal associativity) that characterize induced changes in synaptic efficacy?
5. Are there anti-Hebbian modifications which would contradict Hebb's postulate?

The Conceptual Framework of Cell Assemblies

The foundations of Hebbian theories can be traced as early as 1890: according to William James, the adaptive capacities of our brain depend on mechanistic laws of association that operate under the guidance of cerebral cortex: "When two elementary brain-processes have been active together or in immediate succession, one of them, on re-occurring tends to propagate its excitement into the other."

Building Assemblies

The major contribution of Hebb was to incorporate a physiological association principle in a multilevel model of cerebral functioning during perception and learning (Hebb, 1949). His theory was based on an entirely new concept, i.e., that of "cell assembly," an activity process which reverberates in "a set of closed pathways." The neurophysiological postulate of Hebbian synapses—if cell A synapses upon cell B, then coactivity of A and B increases the strength of the synapse from A to B—was introduced as a way of reinforcing coupling between coactive cells and thus of growing assemblies. Similar hypotheses were developed at a higher hierarchical level of organization, which allowed the linking between cognitive events and their recall under the form of a temporally organized series of activations of assemblies. Hebb referred to this binding process as a "phase sequence."

The initial formulation of Hebb's postulate requires spatial convergence of one neuron onto another, and provides a specific prediction: a period of maintained temporal correlation between pre- and postsynaptic activity will lead to an increase in the efficacy of synaptic transmission. His postulate referred exclusively to excitatory synapses, but a symmetric version of Hebb's postulate is obtained by reducing the strength of inhibitory synapses activated at the same time as the postsynaptic cell (Stent, 1973).

Linking Assemblies

Peter Milner was probably the first theoretician to propose explicit rules for the compositionality of assemblies (Milner, 1957): the repeated activation of a given cell assembly would reinforce synaptic links within this assembly, and in addition "prime" a restricted number of cells. Their latent synapses would remain transiently eligible for further potentiation by the contiguous firing of other assemblies. As early as 1974, Milner predicted that temporal correlation could be used to bind elementary representations into a cognitive "whole" on a much faster time scale than initially proposed by Hebb. A related formalism, revived seven years later, postulated fast binding processes during visual shape recognition depending on the temporal correlation of firing between costimulated cells (von der Malsburg, 1981; see DYNAMIC LINK ARCHITECTURE): the hypothesis of "fast Hebbian synapses" offered a new field of validation for Hebbian associative theories, on the millisecond time scale rather than the classical developmental scale.

Theoretical Predictions and Neurobiological Tests of Hebb's Postulate

Ten years after the publication of his book *The Organization of Behavior*, Hebb was doubtful whether his theory was definite enough to be testable, not so much because of technical constraints but rather because of the number of postulates that were hypothesized, each addressing a different level of integration.

Hebbian Analogs of Cellular Learning

Hebb's postulate was initially applied by cyberneticians and electrophysiologists in the context of supervised learning. In a similar way to the external teacher of the gamma perceptron, which imposes an increase of the gains of active synapses that participated in the "correct" answer, electrophysiogical tests of Hebbian synapses impose depolarization of the postsynaptic element concomitantly with afferent activity. The control of postsynaptic activity is achieved using various technical means (electrical stimulation of an unmodifiable pathway, iontophoresis of excitatory neurotransmitters, intracellular current injection), in order to elicit a "positive reinforcement" of the modifiable test response. This strategy has been attempted at a variety of sites in the central nervous system ranging from molluscan neuronal ganglia to the mammalian forebrain (reviews in Brown et al., 1990; Glanzman, 1994; and Frégnac and Shulz, 1994). Most successful results in demonstrating the role of postsynaptic factors have been observed in the vertebrate cortex, where synaptic potentiation of various duration could be induced under the cooperative influence of other inputs (as is the case during high-frequency tetanus of afferent pathways) or by forcing the postsynaptic cell to an artificially "high" level of activity (using intracellular current or juxtacellular iontophoresis). Long-lasting enhancement of monosynaptic transmission was first obtained in vivo in motor cortex, before being replicated in vitro in CA1 pyramidal cells of hippocampus. As we will discuss later, results opposite to the Hebbian prediction have been found in the striatum, in the cerebellum, and in a related structure in electric fish, the electrosensory lobe.

In spite of some reports that invertebrate synapses possess the capacity for long-term potentiation, until recently it has been assumed that nonassociative and associative forms of behavioral plasticity in *Aplysia* are the result of an activity-dependent presynaptic modulation of the efficacy of sensorimotor synapses. However, a recent report by Glanzman (1994) provides strong evidence for a specific influence of the target cell in inducing spatial competition and segregation of the locus of termination of the afferent axons corresponding to different presynaptic axons. It also demonstrated that Hebbian pairing protocols induced potentiation of identified sensorimotor synapses.

By which cellular machinery does coactivity exert a control of synaptic gain? In vertebrate hippocampus as well as in *Aplysia* sensorimotor cocultures, evidence implicates the NMDA receptor and its invertebrate homologous form, respectively (see NMDA RECEPTORS: SYNAPTIC, CELLULAR, AND NETWORK MODELS). These receptors are ideally suited to operate conjunctive mechanisms, since they require depolarization of the postsynaptic neuron above a critical level to free their ionophore channel from the magnesium block. Under certain conditions of afferent stimulation (or with the concomitant help of an external teacher), the presynaptic release of glutamate will activate non-NMDA receptors (and possibly of NMDA receptors) colocalized at the same synapse, and induce a postsynaptic response above that threshold. This response will allow entry of free calcium through the NMDA channel in the postsynaptic cell, which triggers a chain of intracellular events acting both post- and presynaptically, eventually resulting in the expression of a long-lasting increase in synaptic efficacy. However, this mechanism is certainly not unique, since Hebbian forms of plasticity can be observed even during the pharmacological blockade of NMDA receptors by its antagonist APV (2-amino-5-phosphonovaleric acid).

A Theoretical Need for Synaptic Depression: "Pseudo-Hebbian" Rules

Most algorithms of synaptic plasticity use rules of normalization which require depression of certain synapses in addition to Hebbian reinforcement of active connections: some assume forgetting mechanisms slowly activated by disuse, or complementary plasticity rules operating at the level of synapses fed by the active pathway ("homosynaptic depression") and by neighboring inactive afferents ("heterosynaptic depression"). In the latter case, the decay may depend only on local variables, or it may be the result of a global constraint which maintains a constant sum (or sum of the squares) of all the synaptic weights converging onto the same neuron. Gunther Stent proposed a biophysical mechanism inducing a selective decrease in the efficacy of synaptic transmission of afferent fibers which were inactive at the time when the postsynaptic neuron was discharging under the influence of other inputs (Stent, 1973). The prediction of his own postulate finds strong support from cross-depression studies in visual pathways, and from the observation of heterosynaptic depression in the CA1 field and the dentate area of the hippocampus.

An interesting nonassociative depression rule, which also does not require the simultaneous knowledge of modifications occurring elsewhere in the neuron, assumes that at each update of synaptic weight, the efficiency of all synapses would be decremented by a term proportional to the square of the postsynaptic response. The predictions of this rule may be compared with experimental nonassociative depression revealed in cortical neurons subjected to repeated antidromic activation.

The assumption of global constraints in maintaining constant the total synaptic weight onto a recipient neuron has also been made on the basis of more biological grounds, such as the theory of selective stabilization. Correlates have been obtained recently in cocultures of specified numbers of pre- and postsynaptic partners, suggesting that the total capacity of the target neuron for synaptic interaction is fixed and divided among the different input lines (Glanzman, 1994). Related arguments can also be found during synaptogenesis of neuromuscular junctions and of vertebrate visual pathways.

In spite of their diversity, these different rules have a common implication: they predict spatial and temporal competition between active fibers which impinge on a common target cell. They are referred to as being *pseudo-Hebbian*.

Experimental Support for a Generalized Hebbian Algorithm

Most algorithms that are currently used to model synaptic plasticity in self-organizing networks or behavioral learning are surprisingly uniform, and are based on coactivity of pre- and postsynaptic neurons. They may be summarized by a general equation in which the change of synaptic efficacy with respect to time is equal to the product of a presynaptic term and a postsynaptic term (review in Frégnac and Shulz, 1994). The so-called "covariance hypothesis" (Sejnowski, 1977; Bienenstock, Cooper, and Munro, 1982) replaces the pre- and postsynaptic terms by the departure of instantaneous pre- and postsynaptic activities from their respective average values over a certain time window. These average values constitute pre- and postsynaptic thresholds that determine the sign of the modification (potentiation versus depression) and can be replaced by nonlinear functions of past activity (Bienenstock et al., 1982). The "floating threshold" hypothesis predictions agree with the observation of an increased rate of cortical specification in previously deprived animals that are reexposed to a visually structured environment, when compared with the normal process observed in nondeprived animals.

In addition to the straightforward Hebbian condition, the covariance hypothesis predicts two forms of depression. The first one is an associative heterosynaptic depression at the level of synapses whose activity was uncorrelated with that of the tetanized pathway (Levy and Steward, 1983; Debanne, Gahwiler, and Thompson, 1994). The second form is a homosynaptic depression, when presynaptic activity is associated with repetitive failure in synaptic transmission (Frégnac et al., 1988; Frégnac et al., 1994).

Constraints in Spatial and Temporal Specificity

Input Specificity and Cooperativity

A first limitation in the locality of learning depends on the minimal neuroanatomical convergence of input necessary to induce a functional change. In the case when activity of one afferent alone is sufficient, convergence should be considered as related to the density and the spatial distribution of boutons made by a single presynaptic axon onto the target cell. Using simultaneous dual intracellular recordings, Redman and collaborators (Friedlander, Sayer, and Redman, 1990) failed to observe significant potentiation of individual synaptic connections, whereas their compound activation revealed an increased postsynaptic response. These authors hypothesized the existence of a postsynaptic threshold mechanism controlling the *expression* of synaptic potentiation: a critical level of depolarization had to be achieved so that the enhancement at the "primed" synapse would be revealed in response to the test input. This nonlinear behavior of the postsynaptic neuron would greatly increase the spatial input specificity of long-term potentiation (LTP) by making it conditional on the strength and the convergence of multiple inputs. It could thus prevent temporally unstructured or spatially disperse afferent information from benefiting from the potentiation.

Volume Plasticity

The input specificity of Hebbian schemes of plasticity, i.e., their restriction to active synapses, might suffer strong limitations when the release of retrograde factors and the spatial diffusion of second messengers are taken into consideration. Since quantal analysis studies have implicated presynaptic factors in the maintenance of LTP, it is admitted that some feedback signal indicates to the presynaptic terminal that the correlation operation has been accomplished and that potentiation is authorized. Various retrograde messengers have been proposed, including arachidonic acid, nitric oxide, carbon monoxide, and platelet-activating factor.

Since these messengers diffuse in the extracellular medium, the correlation between high levels of the released molecule and active axon terminals could then become the key factor controlling which synapses should be potentiated. This scheme accounts for the observed generalization of potentiation to neighboring synapses belonging to the axon which has initiated the retrograde messenger process, independently of the target neuron. The consequence of this "volume plasticity" is that correlation will be reinforced between elements which are coactive within a given time window and are within some critical volume without being necessarily physically connected. Reasonable estimates of the space constant on which retrograde messenger-induced changes occur are in the order of 50–150 μm, based on dual intracellular recordings of a conditioned cell and a neighbor, both of which receive parent branches from the same input fiber. Functional effects could operate on a larger scale if a permanent imbalance in activity is introduced between competing axons, for example, altering the spatial grouping of bands of ocular dominance driven by the open eye in monocular deprived visual cortex.

Temporal Associativity

Hebbian rules are correlational, and in their simplest form are symmetric in time; i.e., no temporal ordering is required between pre- and postsynaptic activation. This symmetry does not account for long delays such as optimal interstimulus intervals of association (in the range of one second) in classical conditioning (see CEREBELLUM AND CONDITIONING). Similar values (± 100 ms) have been found for homosynaptic potentiation and depression induced by pairing test input with intracellular depolarizing and hyperpolarizing pulses, respectively (Frégnac et al., 1994).

An overlooked consequence of additional pseudo-Hebbian rules is that their interplay with a purely Hebbian scheme predicts a loss of symmetry in the temporal domain, and a possible narrowing of the critical interval of association. In vitro studies of heterosynaptic plasticity in cocultures of embryonic spinal neurons and myotomal muscles show that synchronous activation of two presynaptic pathways protects them from depression, whereas a delay as short as 100 ms is sufficient to depress one or both pathways. Other forms of long-term depression have been observed. Among them, associative LTD has been observed during contiguous dual-pathway stimulation paradigms or when the test input follows a postsynaptic depolarization induced by a brief current pulse (Debanne et al., 1994). The exact temporal window during which a recurrent input remains eligible for potentiation depends, however on the strength of the last unconditioned activation of the cell. These results agree partially with the temporal order requirement in associative heterosynaptic depression reported in the early 1980s in the study of the crossed (weak) and uncrossed (strong) enthorinal cortex projection to the dentate gyrus, which described much shorter intervals (20 ms) enabling associative LTP (Levy and Steward, 1983).

For cases in which associativity is met when postsynaptic activity (equivalent to the unconditioned stimulus of Pavlovian terminology) follows presynaptic activity (the conditioned stimulus) with a several-hundred-millisecond delay, predictive rules have been modeled with ad hoc correlation functions. However, no experimental evidence so far has been obtained to account for the buildup of optimal interstimulus intervals through Hebbian mechanisms. This lack of evidence does not negate the implication of ionic mechanisms in lagged excitability changes or the slow buildup of a second-messenger mediated intracellular response.

From Hebbian Synapses to Behavioral Learning

The present section will deal with the functional and behavioral consequences that Hebbian rules of plasticity induce in biological self-organizing systems.

Unsupervised Learning

Three possible applications of Hebbian processes acting on a long-term scale can be found in the early development of retinofugal pathways in lower vertebrates, mammals, and primates. These exemplify unsupervised learning, or self-organization:

1. The intrinsic synchronous bursting activity which arises prenatally from the retina before rods and cones are even formed ("dark discharge") exerts a structuring influence on the developing retinofugal pathway. The correlated firing among neighboring retinal ganglion cells within one eye, and the lack of synchronous firing between ganglion cells of different eyes, are conditions that allow competition between geniculate afferents according to their ocularity. This correlated input is present prenatally, taking the form of spatially organized waves of activity which spread intermittently at random directions across the whole retina. These synchronizing waves could provide the local correlations necessary for the sorting out and topographic refinement of retinal projections onto the lateral geniculate nucleus (LGN) before visual experience. Afterward, this intrinsic pattern-generating mechanism will give way to correlated inputs under the guidance of vision.

2. Similar activity-dependent rules might apply to the development of intracortical connectivity: the validity of correlational Hebbian rules seems to hold throughout ontogenesis, if one does not restrict the choice of the postsynaptic control variable to spike activity. Indeed, subthreshold calcium activity and electrical coupling could act prenatally as a substitute for synaptic transmission to ensure assembly formation in the absence of conventional fast Na^+ action potentials. A more classical form of Hebbian plasticity responsible for the progressive maturation of the horizontal intracortical network occurs during a few weeks following birth in the cat; the process results in a selective activity-dependent pruning and stabilization of horizontal connectivity.

3. Evidence has been obtained for the implication of Hebbian mechanisms during the functional reorganization of cortical processing following anomalous visuomotor behavior. A neuroanatomical and electrophysiological study in divergent strabismic kittens—which compensate misalignment of their eyes by alternate fixation—showed that only territories with the same ocular dominance are linked by tangential intracortical connections, and that synchronized activity is achieved only between cell groups dominated by the same eye. Furthermore, in the case of convergent strabismus—which results behaviorally in a loss of acuity through the eye which is not used for fixation—neurons dominated by the "amblyopic" eye exhibited much weaker synchronized activity than cells driven by the "good" eye. The observed correspondence between alterations in intracortical horizontal connectivity topology, the selective impairment of response synchronization, and the perceptual deficit is probably the best evidence so far for a role of temporal correlation in the functional organization of cortical domains.

In summary, the grouping and sorting out of fibers, the morphological tuning in the spatial distribution of the terminal boutons of intrinsic and extrinsic axons, and the functional expression and possible silencing of synapses could all be, at some stage of postnatal development, under the influence of temporal correlation between presynaptic fibers converging onto the same target, or between pre- and postsynaptic partners. This essential role of coactivity in self-organization was foreseen by theoreticians such as Linsker and Miller who proposed a unifying role for activity, whether triggered endogenously by the nervous system or evoked by interaction with the environment (see OCULAR DOMINANCE AND ORIENTATION COLUMNS).

Supervised Learning

Support for a functional implication of Hebb's postulate has also been found when studying the physiological effects of forced patterns of activity, which *simulate* the functional effects of anomalous visual experience during critical periods of development (epigenesis in Frégnac et al., 1988). A differential supervised association protocol was used to test specific predictions of the covariance rule by imposing opposite changes in the temporal correlation between two test parameters characterizing afferent visual activity and the output signal of the cell. Here, an external supervisor (i.e., the experimenter) helped the cell to respond to one input and blocked the cell's response to another, different input. The common outcome was that the relative preference between the two test stimulus characteristics was generally displaced toward that which had been paired with imposed increased visual responsiveness. These pairing-induced modifications of specificity of the visual response have been considered as cellular analogs of epigenesis since they reproduce functional changes occurring during development or following early manipulation of the visual environment (monocular deprivation, rearing in an orientation-biased environment, or optically induced interocular orientation disparity). Surprisingly, the probability of inducing functional changes was comparable in the kitten during the critical period and in older kittens and adults, suggesting that the cellular potential for plasticity might extend well beyond the classical extent of the critical period.

Gating Signals and Attentional Processes

Both Hebb and Milner were aware of the fact that the expression of synaptic changes could largely depend on the level of preactivation of nonspecific projection systems and arousal. These factors could influence the likelihood of summation at the synapse, and thus affect the amount of correlated input needed to induce synaptic changes. Because of methodological and technical difficulties, the role of the "behavioral context" (i.e., attention, reinforcement) has often been ignored in the study of synaptic mechanisms underlying learning in higher vertebrates. Recently, Ahissar and collaborators applied cross-correlation techniques for studying plasticity of "functional connectivity" between pairs of neurons in the auditory cortex in awake monkeys performing a sensory discrimination task (Ahissar et al., 1992). The correlation of activity between two neurons was artificially controlled by activating the target cell of the pair (the postsynaptic cell) by the presentation of its preferred auditory stimulus every time (and immediately after) the other cell fired spontaneously. Under these Hebbian conditions, reversible changes in functional coupling could be induced only when the animal was attentive to the tone used to control the activity of the postsynaptic cell: these changes lasted for a few minutes and followed the covariance hypothesis predictions. The results indicate that Hebb's requirement is necessary, but not sufficient, for cortical plasticity in the adult monkey to occur: internal signals indicating the behavioral relevance are also required.

Anti-Hebbian Forms of Learning

Depending on the neural structure under study (i.e., cerebellum versus cortex), or the time course of the functional effect looked for (i.e., sensory adaptation versus learning), forms of plasticity have been observed which are contrary to the predictions of Hebb's postulate. Such changes are called *anti-Hebbian* (or reverse Hebbian), and should be unambiguously distinguished from pseudo-Hebbian modifications (see POST-HEBBIAN LEARNING RULES).

Theoretical Implications of Anti-Hebb Rules

The theoretical interest of anti-Hebbian learning is best shown when associated with specific assumptions about connectivity organization, as was first done in formal nets with linear neural response functions. When used with normalization procedures, Hebbian learning is known to be equivalent to the extraction of the first principal component from the distribution of presented input patterns (see PRINCIPAL COMPONENT ANALYSIS). Anti-Hebbian rules limited to connections across the output layer allow in addition the extraction of higher-order principal components.

Experimental Evidence for Anti-Hebbian Plasticity

The best-known example of anti-Hebbian plasticity is LONG-TERM DEPRESSION IN THE CEREBELLUM (q.v.), which, by its time-course and induction requirements, appears similar to Hebbian associative potentiation. The trigger mechanism appears to be the same in both cases: free calcium entry in the postsynaptic cell. In order to unite the two processes, it could be assumed that the sign in the change of the synaptic modification depends on the type of neurotransmission (excitatory/inhibitory) that the postsynaptic neuron will exert on other cells. Indeed, the Purkinje cells for which Hebbian protocols induce depression are the inhibitory output neurons of the cerebellar cortex. Although no biological basis has been found to support the hypothesis that excitatory and inhibitory cells undergo Hebbian potentiation and depression, respectively, its implications sound very attractive in terms of system theory: forced coactivity would produce the same type of global positive gain control whatever the neuronal structure under study, either by increasing the transmission of the selected input through a purely excitatory loop or by reducing the excitation fed into the inhibitory efferent pathway.

The application of a sign-inverted Hebbian rule to excitatory networks has by itself a straightforward prediction: the output of the association neurons will tend to be reduced in response to input patterns to which the neural system is exposed frequently. Evidence in the visual system has been found for gain control processes acting in the range of hundreds of milliseconds, which is too long to be accounted for by direct inhibitory action and too short to be compared with the effects of behavioral learning.

A final example of anti-Hebbian plasticity links both cerebellar LTD and fast adaptation processes in perception. In the mormyrid electric fish, the neural structure homologous to cerebellum is the electrosensory lobe. Its Purkinje-like cells are contacted at their soma by primary sensory projections which inform the fish of its electrical environment and provide as well a reafferent response resulting from its own electric discharge (see ELECTROLOCATION). The dendrites of these cells also receive a copy of the motor command responsible for electric discharge. Remarkably, pairing an electrosensory stimulus 0–100 ms after the electric organ discharge motor command results in altered corollary discharge responses intervening at the same delay. The change is opposite in sign to the effect on firing pattern evoked by the paired stimulus: the modifiable corollary discharge elicits in the electrosensory lobe the transient storage of a negative image of the temporal and spatial pattern of sensory input that has followed the motor command. The response to the natural association of command and its reafferent stimulus will consequently decrease because of the pairing-induced development of an inverted-sign response to the command. The synaptic nature of the change is demonstrated by replacing the sensory input (affecting the whole structure) by an intracellular current pulse (affecting only the cell under study) (Bell et al., 1993). Similar mechanisms based on anti-Hebbian plasticity could potentially act in the vertebrate brain to filter out modification of sensory input caused by a motor exploration of the environment, and thus optimize the detection of new events.

Discussion

The study of memory formation benefits from the use of simple putative elementary principles of plasticity, operating at a local level (the synapse) and uniformly across the cell assembly. The large number of experimental attempts to demonstrate the validity of Hebb's postulate prediction during the last 40 years should have narrowed its fields of application. Surprisingly, however, Hebbian schemes have survived to become the symbol of an ever-renewed concept of synaptic plasticity, open for further generalization. A variety of experimental networks ranging from the abdominal ganglion in the invertebrate *Aplysia* to the hippocampus and visual cortex offer converging validation of the prediction of Hebb's postulate. In these networks, similar algorithms of potentiation can be implemented using different cascades of second messengers triggered by activation of synaptic and/or voltage-dependent conductances. Classes of processes occurring on different time scales—development or epigenesis (days and weeks), learning (minutes or hours), and even perception (milliseconds)—could have similar phenomenological outcomes.

When followed literally, Hebb's postulate refers to the set of direct excitatory synaptic contacts, originating from one presynaptic neuron, onto a postsynaptic neuron that may participate in triggering its activity, and to the correlational rules that predict an increase in efficacy in synaptic transmission. Modelers have often simplified this view to the extreme, using ideal connections between pairs of neurons and ignoring much of the complexity of different biological implementations of the so-called Hebbian synapse in invertebrates and vertebrates. Most cellular data supporting Hebb's predictions have been derived from electrophysiological measurements of composite postsynaptic potentials or synaptic currents, or of short-latency peaks in cross-correlograms, which cannot always be interpreted simply at the synaptic level. The basic conclusion of these experiments is that covariance between pre- and postsynaptic activity up- and down-regulates the "effective" connectivity between pairs of functionally coupled cells.

It may be concluded that what changes according to a correlational rule is not so much the efficacy of transmission at a given synapse, but rather a more general coupling term mixing the influence of polysynaptic excitatory and inhibitory circuits linking the two cells, modulated by the diffuse network background activation. Replacing this composite interaction by a single coupling term defines an ideal Hebbian synapse, and it has the additional interest for the modeler of providing a weighting function of the input, which can even change sign when inhibition overcomes excitation.

Acknowledgment.   Experimental data from my lab have been collected with the support of the Human Frontier Science Program.

Road Map: Mechanisms of Neural Plasticity
Background: I.3. Dynamics and Adaptation in Neural Networks
Related Reading: Hebbian Synaptic Plasticity; Invertebrate Models of Learning: *Aplysia* and *Hermissenda*; Post-Hebbian Learning Rules

References

Ahissar, E., Vaadia, E., Ahissar, M., Bergman, H., Arieli, A., and Abeles, M., 1992, Dependence of cortical plasticity on correlated activity of single neurons and on behavioral context, *Science*, 257:1412–1415.

Bell, C., Caputi, A., Grant, K., and Serrier, J., 1993, Storage of a sensory pattern by anti-Hebbian synaptic plasticity in an electric fish, *Proc. Natl. Acad. Sci. USA*, 90:4650–4654.

Bienenstock, E., Cooper, L. N., and Munro, P., 1982, Theory for the development of neuron selectivity: Orientation specificity and binocular interaction in visual cortex, *J. Neurosci.*, 2:32–48.

Brown, T. H., Ganong, A. H., Kairiss, E. W., and Keenan, C. L., 1990, Hebbian synapses: Biophysical mechanisms and algorithms, *Annu. Rev. Neurosci.*, 13:475–511. ◆

Debanne, D., Gahwiler, B. H., and Thompson, S. M., 1994, Asynchronous presynaptic and postsynaptic activity induces associative long-term depression in area CA1 of the rat hippocampus in vitro, *Proc. Natl. Acad. Sci. USA*, 91:1148–1152.

Frégnac, Y., Burke, J., Smith, D., and Friedlander, M. J., 1994, Temporal covariance of pre- and postsynaptic activity regulates functional connectivity in the visual cortex, *J. Neurophysiol.*, 71:1403–1421.

Frégnac, Y., and Shulz, D, 1994, Models of synaptic plasticity and cellular analogs of learning in the developing and adult vertebrate visual cortex, in *Advances in Neural and Behavioral Development* (V. Casagrande and P. Shinkman, Eds.), Norwood, NJ: Neural Ablex, vol. 4, pp. 149–235. ◆

Frégnac, Y., Shulz, D., Thorpe, S., and Bienenstock, E., 1988, A cellular analogue of visual cortical plasticity, *Nature*, 333:367–370.

Friedlander, M. J., Sayer, R. J., and Redman, S. J., 1990, Evaluation of long-term potentiation of small compound and unitary EPSPs at the hippocampal CA3-CA1 synapse, *J. Neurosci.*, 10:814–825.

Glanzman, D. L., 1994, Postsynaptic regulation of the development and long-term plasticity of *Aplysia* sensorimotor synapses in cell culture, *J. Neurobiol.*, 25:666–693.

Hebb, D. O., 1949, *The Organization of Behavior*, New York: Wiley. ◆

James, W., 1890, *Psychology: Briefer Course*, Cambridge, MA: Harvard University Press. ◆

Levy, W. B., and Steward, O., 1983, Temporal contiguity requirements for long-term associative potentiation/depression in the hippocampus, *Neuroscience*, 8:791–797.

Milner, P. M., 1957, The cell assembly: Mark II, *Psych. Rev.*, 64:242–252.

Milner, P. M., 1974, A model for visual shape recognition, *Psychol. Rev.*, 81:521–535.

Sejnowski, T. J., 1977, Statistical constraints on synaptic plasticity, *J. Theor. Biol.*, 69:387–389.

Stent, G., 1973, A physiological mechanism for Hebb's postulate of learning, *Proc. Natl. Acad. Sci. USA.*, 70:997–1001.

von der Malsburg, C., 1981, The correlation theory of brain function, Internal report, Max-Planck Institute for Biophysical Chemistry, Göttingen, FRG. ◆

High-Energy Physics

Bruce Denby

Introduction

High-energy physics (HEP) is the study of the basic constituents of matter and the fundamental forces through which they interact. In the past few years a wide variety of applications of neural networks to pattern recognition problems in experimental high-energy physics have appeared. Most of the applications to date make use of multilayer perceptrons (MLP) trained by backpropagation for identification of different types of particles produced in high-energy particle collisions. There have also been applications of recurrent networks to the reconstruction of the trajectories of charged particles. In HEP, much of the pattern recognition must be performed on-line, in a few microseconds or less. There have already been some very interesting applications using hardware neural networks to do pattern classification directly in the readout hardware of HEP experiments. An in-depth review of HEP applications can be found in Denby (1993).

HEP Research

Current research in HEP is involved with the verification of the "standard model" of particle physics, which describes the basic constituents of matter. During a collision, these constituents, called *quarks* and *leptons*, interact by "exchanging" one of the force-carrying particles W, Z, γ (the *photon*), or g (the *gluon*). These "elementary" particles are not directly observable. The physical process which occurred in the collision must be determined from an examination of the stable daughter particles in the debris of the collision, and the ways in which they interact in the detectors.

Some of the major areas of research in HEP today are the study of the production and decay properties of the "heavy" (i.e., massive) quarks, called c and b; the search for the heaviest quark, called top, or simply t, which is postulated but as yet undiscovered; and the study of the characteristics of the so-called *jets* of particles into which quarks and gluons disintegrate.

HEP Accelerators and Detectors

HEP data are produced in experiments at accelerator centers. The major centers today are Fermilab in Illinois, CERN in Switzerland, and HERA in Germany. Each site features an underground *ring*, with diameter up to several miles, in which opposing beams of particles are made to collide at one or more *interaction regions* (there are also the so-called fixed-target experiments, but for simplicity we shall not discuss them here). In the collisions, daughter particles of many kinds are produced,

Figure 1. Particle bunches in the accelerator are brought together in interaction regions which house detector systems. A portion of the cylindrically symmetric system is shown, along with tracks from an electron, pion, muon, and jet (see text for an explanation of these terms). The small histograms accompanying the particle tracks indicate the distribution of the energy of the particle in the calorimeter cells (one cell per histogram bin).

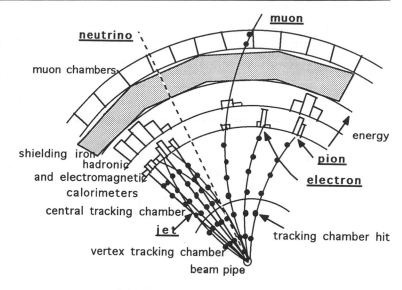

and these are detected in cylindrical arrays of particle detectors surrounding the interaction region. Figure 1 shows a cross section of part of such a detector system. These detectors have dimensions of tens of meters and weigh many tons.

Most particle detectors can be classified as one of two main types, *tracking chambers* and *calorimeters*. Tracking chambers are used to detect the trajectories of electrically charged particles emerging from a collision. Calorimeters are designed to cause most particles incident upon them to interact and deposit all of their energy, and are highly segmented into cells in order to give information on the spatial extent of the energy deposit from the particle.

Figure 1 shows a generic HEP detector system with central and vertex (inner) tracking chambers, electromagnetic and hadronic calorimeters, and muon-shielding iron, followed by another set of tracking chambers called muon chambers. The figure illustrates the behavior of the detectors for the four most commonly encountered types of particles and for a jet. A muon passes through the calorimeters, depositing only a small amount of energy in each, and through the shielding iron, to be detected in the muon chambers. An electron deposits its energy in a localized region of the electromagnetic calorimeter, and a pion deposits its energy over a region of both electromagnetic and hadronic calorimeters. A neutrino does not interact and passes undetected through the apparatus. A jet is composed of many particles and deposits energy in both calorimeters over a broad region. The different behaviors of these particles in the detectors are used to determine particle type.

A typical experiment may record hundreds of thousands of individual detector channels, corresponding to about 1 million bits of information, for each collision, or *event*, as they are usually called, and it is not uncommon to record many millions of events during a data-taking run. At Fermilab currently, particle collisions occur every 4 microseconds; at the Large Hadron Collider (LHC), the new accelerator which will be built at CERN toward the end of the decade, they will occur about every 16 *nano*seconds.

Triggering

It is neither desirable nor feasible to log all of the events to permanent storage media. On-line pattern recognition, called *triggering*, is required to reject background events and retain

the interesting events. This is usually performed by three separate modules called *levels*. The data from the detectors pass into the trigger as a stream of events. Each level of trigger scans the incoming events for certain features, rejects most of the events it receives, and passes the remainder on to the next level. In level 1, simple tests on global event information are performed; for example, events with too little energy transverse to the beam direction correspond to glancing collisions and are rejected. In the level 2 trigger, somewhat more sophisticated tests can be done—for example, looking for the presence of clusters of energy in the calorimeter which correspond to *jets*. While levels 1 and 2 are implemented in special hardware, level 3 is executed on a "farm" of conventional processors operating in parallel on separate events. The processing done by level 3 can be quite sophisticated—for example, full reconstruction of trajectories of charged particles. Events passing all three levels of trigger are logged onto magnetic tape. Typical processing times for levels 1, 2, and 3 are 1 microsecond, 20 microseconds, and 1 millisecond, respectively.

The use of neural network technology in HEP triggering is still experimental. Although more standard technologies are currently favored, neural networks may be able to make triggers which are more efficient and less costly by moving to level 1 or level 2 the complex pattern recognition normally done in level 3, thus reducing the requirements placed on the level 3 processor farm as well as the amount of data recorded on tape for later analysis. Present-day silicon neural networks, with settling times of the order of 1–10 microseconds, are too slow for most level 1 triggers (much faster chips may eventually be possible) but are sufficiently fast to be a competitive technique for level 2 triggers. Since most level 2 triggers are special purpose hardwired devices, a change of algorithm implies rebuilding or rewiring. With a neural network, the algorithm can be changed by downloading a different set of weights, making neural network triggers more flexible.

Off-Line Reconstruction

Off-line reconstruction, in which the final physics analyses are carried out, is applied after all the selected events have been logged to tape. Historically, high-energy physicists have eschewed "complicated" data analysis techniques in favor of simple one-dimensional cuts on variables selected "by hand."

However, such ad hoc selection procedures are often inefficient. The key to the value of neural networks in off-line HEP analyses is creating efficient cuts to retain events from physics processes to be studied while rejecting as many as possible of the background events.

Applications to Triggering

First Application: Muon Trigger

The first trigger application of a neural network in HEP was accomplished recently at Fermilab (Lindsey et al., 1992). Identification of a muon track is a useful trigger for detecting decays of Ws, Zs, and top and b quarks, since each will often decay to a muon. Typical conventional trigger systems can only crudely measure the slopes and intercepts of muon tracks, and as a result often give false identifications. In an experiment at Fermilab, the slopes and intercepts of muon tracks traversing a small tracking chamber were calculated accurately, in real time, using a commercial VLSI neural network chip incorporated into the standard data acquisition system. Twelve analog wire signals from the chamber were coupled with three latch signals to form the 15 inputs to the neural network chip, configured as an MLP. The network was trained, using gradient backpropagation, on 10,000 tracks generated with a simple simulation, and the weights obtained were downloaded into an Intel ETANN chip (Intel, 1991). Simulations of simple entities such as tracks or energy deposits are well understood and will not bias classification results. The intercept position resolution using the conventional trigger technique was 5 centimeters. The neural network trigger had a position resolution of 1.2 millimeters, which is only a factor of two worse than the best obtainable off-line using the complete reconstruction algorithm, and is available in a few microseconds. There is a plan to incorporate such an ETANN readout into the muon trigger of one of the two major detectors (called D0) at the Fermilab proton antiproton collider in its next run (Fortner, 1992).

First Application in a Major Collider Experiment: CDF

Neural network trigger hardware was already installed for the 1992–93 run of the other major detector at Fermilab, CDF (Collider Detector at Fermilab) (Denby et al., in press). The calorimeter trigger for CDF operates upon an array of voltages, of dimension 24 (azimuthal angle) by 42 (polar angle) by 2 (electromagnetic/hadronic section), which represent the energies in the calorimeter cells. In the existing trigger, the total energy of every cluster of energy, the number of cells it contains, and its width are computed. This information is adequate for a great many triggers, but for some purposes a more sophisticated analysis is required. For every cluster, the new, neural, level 2 trigger selects a 5-by-5-cell region of interest (in hadronic and in electromagnetic compartments) centered on the cluster and passes these 50 analog signals to neural network chips configured as MLPs (Intel, 1991). The chips are programmed to execute three different algorithms which determine if the cluster could be (1) an isolated photon in the central calorimeter (i.e., a photon surrounded by a region containing very little energy); (2) an isolated electromagnetic shower in the end-plug calorimeter; or (3) the semileptonic decay of a b quark. Data from triggers 1 and 2 taken in the 1992–93 run of the CDF experiment have now been analyzed. Both triggers have efficiency close to 100% and have been verified to select clean isolated clusters. The chips have proven to be highly reliable. Results from trigger 3 are still under study.

Off-Line Applications

Numerous groups have used neural networks for identifying reactions containing b quarks (see in particular Brandl et al., 1993), usually referred to as b tagging, which is of considerable interest since many of the particles containing b quarks have yet to be studied in detail. A group at CERN has used an MLP to classify decays of the Z into c quarks and b quarks (Abreu et al., 1992a), and this procedure has permitted an improved measurement of the probabilities of the Z to decay into these particles. We will use this work as an illustrative example of the types of applications of neural networks typically found in off-line analyses.

The standard technique for distinguishing heavy quarks from light quarks is through their so-called semileptonic decays, in which a particle containing a heavy quark decays to a lepton plus other particles. This method is inefficient because semileptonic decays account for only 20 percent of heavy quark decays. A technique which allows the use of all types of heavy quark decays is thus desirable. In this work, 19 jet and event-shape variables were created as inputs to an MLP. The network architecture used had 25 hidden units and 3 output units to encode the three classes. The outputs are normalized to sum to one so that they will represent probabilities of the three classes. The training data for the network were generated with a standard simulation program and another which simulates the response of the apparatus. In off-line applications, some parameters in the simulations are unknown because of an incomplete knowledge of the underlying physics; dependence of the final answer on this uncertainty must be taken into account and incorporated into the quoted uncertainties. Six thousand events were used for training and an independent set of 200,000 events for testing.

The trained network was used to determine the relative fractions F_b, F_c, of b and c quark decays in a sample of 123,475 real events. The results are

$$F_c = 0.151 \pm 0.008_{stat.} \pm 0.041_{syst.}$$

$$F_b = 0.232 \pm 0.005_{stat.} \pm 0.017_{syst.}$$

That is, the Z decays some 15.1% of the time into c quarks and 23.2% into b quarks. For comparison, the best results to date (Abreu et al., 1990; Abreu et al., 1992b) using semileptonic decays are

$$F_c = 0.162 \pm 0.030_{stat.} \pm 0.050_{syst.}$$

$$F_b = 0.215 \pm 0.017_{(stat.+syst.)}$$

The two results are consistent with each other. In the case of F_c, however, both statistical and systematic errors are better for the neural net result than for the semileptonic decay result. In the case of F_b, although the neural network result has a slightly larger overall error, it is obtained with a much smaller number of events. This difference occurs because the neural net analysis allows all the data to be used in the analysis, while in the standard analysis, only the rarer semileptonic decays can be used. The neural network approach also has the advantage of providing a probability on an event-by-event basis, whereas the semileptonic decay technique relies upon global distributions for the entire data set.

Neural Nets and Charged Track Reconstruction

Recurrent networks (see RECURRENT NETWORKS: SUPERVISED LEARNING) have been used in HEP for track reconstruction (Denby, 1988; Peterson, 1989; Stimpfl-Abele and Garrido,

1990). The neurons in the network in this case represent directed links (i.e., arrows) connecting two hits in a tracking detector. In Figure 2, two such links are denoted by the label of their respective neurons, i and j. The weight, $w_{ij} = w_{ji}$, between neurons i and j is determined by the angle θ_{ij} between their links:

$$w_{ij} = \frac{A \cos^n \theta_{ij}}{l_i l_j}$$

where n is some integer (a typical optimal value is $n = 5$), l_i and l_j are the lengths of the links. This form is used if the links do not point into or out of the same hit; if they are head to head or tail to tail, $w_{ij} = -B$ is used. Thus there will be cooperation between pairs of neurons whose links share a hit, are head to tail, and point in nearly the same direction; and competition between neurons whose links share a hit, are head to head (or tail to tail) and point in different directions. An energy function is defined, $E = -\frac{1}{2} \Sigma w_{ij} o_i o_j$, where o_i is the output of neuron i. This function will be most negative when the angles between connected links are small (cosine approximately equal to one),

their corresponding neurons have activation o equal to one, and the activations of other neurons are zero. This situation favors neurons lying along smooth trajectories such as those of particles moving in a magnetic field. First, the recurrent network is set up, using the existing tracking chamber hits in a particular event, to define the links and hence the neurons. Then, the evolution of the system is obtained by iteratively solving the update equations:

$$du_i/d\tau = \sum_j w_{ij} o_j - u_i \qquad o_i = \text{sigmoid}(u_i)$$

On each iteration, $d\tau$ is kept much smaller than 1.

The method has been used on real data at CERN (Stimpfl-Abele and Garrido, 1990). Figure 3 shows a view transverse to the beam line of a reconstructed event after settling of the network. The algorithm is as efficient as the conventional track reconstruction program and somewhat faster to execute, and it has been shown that the speed advantage increases with the number of tracks. The algorithm is not in widespread use but may prove to be important in the future.

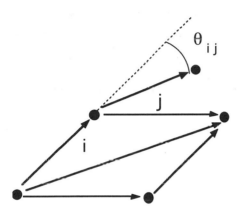

Figure 2. Definition of neurons in the Denby-Peterson recurrent neural network used for reconstruction of charged tracks. Neurons i and j are represented by the links with those labels in the figure. The weight between neurons i and j is determined by the cosine of the angle between the corresponding links, using the formula given in the text.

Conclusion

Neural network techniques in HEP are showing their worth. The decay probabilities of the Z boson into b and c quarks have been measured with higher precision than before using a technique based upon an MLP. Recurrent networks provide a faster way of performing charged track reconstruction. One of the most exciting promises of neural network technology is in the realm of triggering. A VLSI neural network used in the data acquisition system of a drift chamber provided, in only a few microseconds, track intercept resolution 50 times more accurate than that previously obtainable on-line. Neural network triggers at the CDF experiment are in place, and their performance to date is excellent; further applications at HERA (Fent et al., 1992) and elsewhere (Baldanza et al., 1994; Fortner, 1992) are planned.

Road Map: Applications of Neural Networks
Background: I.3. Dynamics and Adaptation in Neural Networks

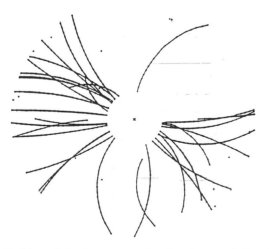

Figure 3. Charged track reconstruction based on real data from CERN, using a recurrent neural network.

References

Abreu, P., et al., 1990, Measurement of the partial width of the decay of the Z^0 into charm quark pairs, *Phys. Lett. B*, 252:140.

Abreu, P., et al., 1992a, Measurement of the partial width of the Z^0 into b$\bar{b}$ final states using their semileptonic decays, *Z. Phys. Chem.*, 56:47–62.

Abreu, P., et al., 1992b, Classification of the hadronic decays of the Z^0 into b and c quark pairs using a neural network, the DELPHI collaboration, *Phys. Lett. B*, 295:383–395.

Baldanza, C., Bisi, F., Cotta-Ramusino, A., D'Antone, I., Malferrari, L., Mazzanti, P., Odorici, F., Odorico, R., and Zuffa, M., 1994, The WA92 collaboration: Results from an on-line neural trigger within a fixed target experiment for the production of beauty particles, in *New Computing Techniques in Physics Research III* (K. H. Becks and D. Perret-Gallix, Eds.), La Londe les Maures, France: World Scientific, pp. 391–409.

Brandl, B., Falvard, A., Guicheney, C., Henrard, P., Jousset, J., and Proriol, J., 1993, Multivariate analysis methods to tag b quark events at LEP/SLC, *Nucl. Instrum. Methods A*, 324:307. ◆

Denby, B., 1988, Neural networks and cellular automata in experimental high-energy physics, *Comput. Phys. Commun.*, 49:429. ◆

Denby, B., 1993, The use of neural networks in high-energy physics, *Neural Computat.*, 5:505–549. ◆

Denby, B., Lindsey, C. S., Dickson, M., Konigsberg, J., Pauletta, G.,

Badgett, W., and Burkett, K., in press, Performance of the CDF neural network electron isolation trigger, *Nucl. Instrum. Methods.*

Fent, J., Gruber, A., Kiesling, C., Oberlack, H., and Ribarics, P., 1992, A second level neural network trigger in the H1 esperiment at HERA, *Int. J. Neural Syst.*, 3(supp.):277–284.

Fortner, M., 1992, Analog neural networks in an upgraded muon trigger for the D0 detector, in *Proceedings of the Second International Workshop on Software Engineering, Artificial Intelligence, and Expert Systems for Nuclear and High-Energy Physics*, La Londe les Maures, France: World Scientific, pp. 381–385.

Intel, 1991, Intel 80170NX Electrically Trainable Analog Neural Network, Intel Corporation, Santa Clara, California.

Lindsey, C. S., Denby, B., Haggerty, H., and Johns, K., 1992, Real time track finding in a drift chamber with a VLSI neural network, *Nucl. Instrum. Methods A*, 317:346–356. ◆

Peterson, C., 1989, Track finding with neural networks, *Nucl. Instrum. Methods A*, 279:537.

Stimpfl-Abele, G., and Garrido, L., 1990, Fast track finding with neural nets, *Comput. Phys. Commun.*, 64:45–56.

Hippocampus: Spatial Models

Neil Burgess, Michael Recce, and John O'Keefe

Introduction

The hippocampus is one of the most studied areas of the brain. It has attracted interest for several reasons: its position, which is many synapses removed from sensory transducers or motor effectors; the putative role that hippocampal damage plays in human amnesia; the discovery of long-term potentiation (LTP) (see HEBBIAN SYNAPTIC PLASTICITY); and the discovery that cell firing is spatially coded. It has also been studied for its role in neurological disorders, including epilepsy, schizophrenia, and Alzheimer's disease. Bilateral damage to the hippocampus and nearby structures in patient H. M., intended as treatment for epilepsy, produced a profound retrograde and anterograde amnesia, prompting extensive cross-species research to uncover the specific memory deficit that results from hippocampal damage (the most prominent of which, in the rat, appears to be a deficit in spatial navigation).

In short, the hippocampus has become the primary region in the mammalian brain for the study of the synaptic basis of memory and learning. Structurally, it is the simplest form of cortex (Figure 1). It contains one projection cell type, which is confined to a single layer (compared with the large number of

A)

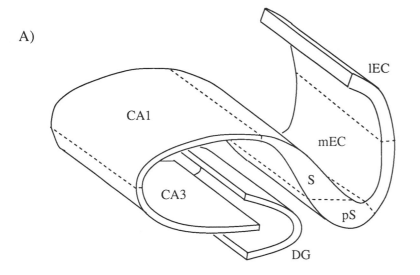

B)

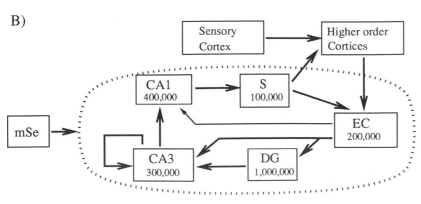

Figure 1. The hippocampus is formed from sheets of cells. *A*, A schematic section cut perpendicular to the longitudinal axis of the hippocampus. EC, entorhinal cortex; mEC, medial EC; lEC, lateral EC; S, subiculum; pS, pre- and para-subiculum; DG, dentate gyrus. *B*, The major projections between subfields (mSe, medial septum), and the approximate numbers for the major cell type in each subfield (i.e., pyramidal cells, except for the DG, in which the major cell type is granule cells) in the rat. The human hippocampus contains one order of magnitude more cells. In the DG–CA3 projection, a single mossy fiber projects from each granule cell, making very large synapses onto only 14 or so pyramidal cells. All the other projections have large divergence and convergence (many thousands to one), and involve the type of synapse in which Hebbian LTP (Long-Term Potentiation) has been observed. A variety of interneurons provide feedforward and feedback inhibition. Cells in the mSe project into DG, CA3, and (less strongly) CA1, playing a role in producing the θ rhythm of the hippocampal EEG. Cells in CA3 and CA1 also project out to the lateral septum (S) via the fornix. (Adapted from McNaughton, 1989.)

cell types and six principal layers of the neocortex), and it receives inputs from all sensory systems and association areas (see Figure 1 for more details).

Attempts to model the hippocampus differ both in the level of anatomical detail and in the function they seek to reproduce. Marr (1971) proposed a theory for how the hippocampus could function as an associative memory, from which have grown many extensions, usually focusing on the role of the CA3 recurrent collaterals (see Figure 1; ASSOCIATIVE NETWORKS; McNaughton and Nadel, 1990; and Treves and Rolls, 1992). However, the discovery that much of the deficit in short-term memory tasks for monkeys is attributable to lesions of the nearby perirhinal cortex (Suzuki et al., 1993) is prompting a revision of the idea of the hippocampus as a general store for declarative memories. Here, we specifically consider neuronal models of spatial processing in the rat hippocampus: the domain in which the least controversial experimental data are available. Models at this level, which are based on repeatable data, are the most likely to provide insight into hippocampal function with respect to higher levels of description and wider functionality. We introduce data on the spatial correlates of hippocampal cell firing and the seminal idea of the hippocampus as a spatial map, and we describe some models of hippocampal place cells and navigation.

Single-Unit Recordings

Single-unit recordings in freely moving rats have revealed place cells (PCs), in fields CA3 and CA1 of the hippocampus, whose firing is restricted to small portions of the rat's environment (the corresponding "place fields"). Recently, PCs have been identified in monkeys. There appears to be little topographic organization of PCs relating their positions in CA3 or CA1 to the positions of their firing fields, and different subsets of place cells fire in different environments. The firing properties of PCs can be manipulated by changing the rat's environment: e.g. rotating the major cues in an environment can cause the place fields to rotate; increasing the environment's size can cause some of the place fields to expand correspondingly. In environments in which direction of movement is restricted (e.g., mazes with narrow arms), PC firing rates appear to depend on the rat's direction of travel as well as its location.

Cells in the entorhinal cortex (the main cortical input to the hippocampus; see Figure 1) also have spatially correlated firing, but tend to have larger, less well-defined place fields than those in CA3 or CA1. Cells whose primary behavioral correlate is "head direction" have also been found in the (dorsal) presubiculum (see Figure 1). They fire when the rat points its head in a specific direction relative to the cues in the environment, and independently of its location (see Muller et al., 1991, and Barnes et al., 1990, for reviews of these data).

The electroencephalogram (EEG) recorded in the hippocampus is the largest electrical signal in the brain. One form of the EEG, called the theta (θ) rhythm, is a sinusoidal oscillation of 7–12 Hz. O'Keefe and Nadel (1978) have suggested that, in the rat, the θ rhythm coincides with displacement movements, e.g., walking. Recently, PC firing has been found to have a systematic phase relationship to θ (O'Keefe and Recce, 1993): when a rat on a linear track runs through a place field, the PC tends to fire groups of spikes, with each successive group occurring at an earlier phase of the θ cycle. Consistent with these data, initial analysis of the phase of firing of PCs as a rat runs in an open field environment indicates that cells firing at a late phase tend to have place fields centered ahead of the rat, whereas those firing at an early phase tend to have place fields centered behind the rat (Burgess, Recce, and O'Keefe, 1994).

There are two features of PC firing that raise immediate problems for their use in navigation: (1) information about a place in an environment (i.e., the firing of the corresponding PCs) can only be accessed locally (by actually visiting that place); and (2) place fields appear to be no more affected by the location of the goal (which is obviously essential for navigation) than by the location of any other cue. Unfortunately, there are no reports to date of the existence of the postulated destination cells.

Cognitive Maps

Cognitive maps were first introduced by Tolman to explain place learning in rats, including, for example, their ability to take short-cuts (see COGNITIVE MAPS). An alternative view, suggested by Hull, is that navigation is achieved by following a list of stimulus-response-stimulus steps. O'Keefe and Nadel (1978) proposed that independent neural systems exist in the brain to support a "taxon" system for route navigation and a "locale" system for map-based navigation (for a current synopsis, see O'Keefe, 1991). The "map" was taken to be a Euclidean description of the environment in a coordinate system based on the world and not on some part of the animal's body. They proposed that the locale navigation system resides in the hippocampus, based on: (1) the firing properties of hippocampal PCs; (2) the presence of θ rhythm during displacement movement; (3) spatial tasks, including the Morris water maze and the Olton eight-arm maze, that show clear deficits in performance after hippocampal lesions; and (4) the interpretation of the amnesic syndrome as the loss of episodic memory (memory for specific events set in a spatiotemporal context).

Place Cell Firing

In this section, we describe a model of how the spatial firing of place cells might develop as the rat explores an environment. Following an earlier, mathematical model of PC firing proposed by Zipser in 1985, Sharp (1991) used a simple network with an input layer and two layers of cells governed by "competitive learning" dynamics (a simple model of the effect of Hebbian synaptic increment combined with inhibitory interneurons; see COMPETITIVE LEARNING). In this model, there are two types of input (or "cortical") cell that respond to cues placed around the environment: a type 1 cell that fires whenever a particular cue is at a given distance from the rat, and a type 2 cell that likewise responds to a particular cue being at a given distance, but only if the cue is within a certain range of angle from the rat's head.

As the simulated rat moves around its environment, competitive learning in the synaptic weights (initially random) leads to unsupervised clustering of the input vectors: a PC learns to fire in a portion of the environment in which the inputs (i.e., the distances and angles of cues) are similar. Cells in the same competitive group will fire in different places, whereas cells from different groups can have overlapping firing fields. The model generates realistic-looking place fields after 64 minutes of simulated exploration. Interestingly, if the simulated rat's exploration is restricted to movements consistent with being on an eight-arm maze, then PC firing tends to be much more strongly correlated with the orientation of the rat (as well as its location) than in the case of unrestricted exploration. This fits well with the available experimental data.

Navigation

This section presents models that address how the hippocampus can be used to enable navigation. The simplest map-based strategies are based on defining a surface, over the whole environment, on which gradient following leads to the goal (see REINFORCEMENT LEARNING; see also Barto and Sutton, 1981, for a simple related mechanism which learns to navigate on the basis that it can "smell" how far away the goal is). These strategies have the disadvantage that, to build up the surface, the goal must be reached very many times, from different points in the environment. A new surface must be computed if the goal is moved, and multiple goals, as in the eight-arm maze, cannot be handled. Learning in these models is much slower and more goal-dependent than in rats, and they are unable to perform "latent learning" (e.g., in rats, exploration in the absence of any goals improves subsequent navigation). Interestingly, the performance of these models improves somewhat when a spatially diffuse representation (like place fields), rather than a punctate representation, is built up during exploration (Dayan, 1990). The following sections describe some recent models that have related navigation to the action of individual cells in the hippocampus.

Graph Model

A role in navigation was proposed for the CA3 recurrent collaterals (the axonal projections by which each CA3 PC contacts approximately 5% of the other CA3 PCs) by Muller, Kubie, and Saypoff (see Traub et al., 1992). Given a model of LTP in which presynaptic and postsynaptic firing within a short time interval leads to a small increase in synaptic "strength," it is possible to construct a model of CA3 in which PC firing (as the rat moves around) leads to the synaptic strength of a connection between two PCs, depending on the proximity of their place fields. If two PCs have closely overlapping fields, they will both fire near-coincidentally relatively often, whereas if they have well-separated place fields, they will very rarely fire simultaneously.

After brief exploration, the synaptic strengths represent distances along the paths taken by the rat; however, to build up a true distance metric would take a long time, as with reinforcement learning approaches. Another problem involves accessing the distance information (i.e., getting projection cells to fire so that information can be passed out of region CA3). The model also falls short of being a model of navigation in that it does not provide a direction in which to move.

Local View Model

McNaughton and Nadel (1990) proposed that the hippocampus functions as an associative memory (see ASSOCIATIVE NETWORKS). According to this model, as the rat explores, it learns to associate each local view and movement made with the local view from the place visited as a result of the movement. Thus, routes through an environment can be stored as a chain of local view/movement associations. The model is supported by the fact that, in some situations, PC firing depends on the rat's direction (and, therefore, its local view). Some major problems with this theory are that (1) simple route-following strategies appear to be the kind of navigation of which hippocampectomized rats are capable (O'Keefe and Nadel, 1978); (2) it is difficult to know which particular route will lead to the desired goal, and solving this problem leads one back to the reinforcement learning approach; (3) the model is not capable of more sophisticated navigation, such as taking short-cuts; and (4) in open fields, PC firing does not seem to depend on direction.

Inertial/Directional Systems

The existence of a directional system in rodents is supported by the head-direction cells and by experiments in which (1) a goal location seems to be stored in terms of a distance and absolute direction from one symmetrical cue, and (2) inertial homing (i.e., an internal sense of direction) is used for navigation rather than external cues (e.g., in complete darkness). It is also interesting to note that, once the rat has oriented itself from the array of cues, PCs can continue to fire in the correct places after all the salient cues in an environment have been removed, or after the lights have been switched off.

Various aspects of this system have been modeled (McNaughton, Knierim, and Wilson, 1995). A model of "dead-reckoning" describes how the rodent's current heading direction can be calculated as it moves about. Briefly, it is proposed that the combinations of all heading directions ($H(t)$) and all possible angular displacements ($\Delta H(t)$) over a short time step (τ) are associated with the resulting heading directions ($H(t + \tau)$) by an associative neural network. Evidence for this includes cells (in the posterior parietal cortex) with responses that depend both on the rat's orientation ($H(t)$) and on the sense of its angular velocity ($\Delta H(t)/\tau$).

The hippocampus is invoked to correct cumulative errors in direction by associating the so-called local views at each location with the appropriate heading direction (H). However, it is suggested that, if the local view–heading direction associations cannot be learned, then the heading direction system is the fundamental determinant of the orientation of the PCs' spatial map. Evidence comes from experiments in a symmetrical environment with one polarizing cue; in normal circumstances, when the rat is familiar with the environment, the position of place fields can be rotated by moving the cue between sessions. However, this is not the case if (from the very first session) the rat is disoriented between each session.

It is also proposed that each PC codes for a single cue that is at a particular angle (provided by the H system) and distance. The rat could then navigate by comparing the current vectors $\mathbf{p}_i(t)$ to individual cues with the memory $\mathbf{m}_i$ of their respective values from the goal location. The vector back to the goal is: $\mathbf{p}_i(t) - \mathbf{m}_i$ for any cue i. This has the advantage that navigation would be possible using only one cue, but is hard to reconcile with the robustness of place cell firing to removal of individual cues.

Centroid Model

O'Keefe (1991) has proposed a navigational mechanism whereby environments are characterized by two parameters: the centroid and the slope of the positions of environmental cues, which can be used as the origin and $0°$ direction of a polar coordinate framework. The firing of a PC could represent the average position of a small number of cues (their minicentroid), while head direction cells could represent the translation vector between pairs of cues. The environmental centroid and slope are then found by averaging the mini-centroids and slopes. It was proposed that single cells could represent two-dimensional vectors as phasor: that is, taking the θ rhythm of the EEG as a clock cycle, the amplitude of firing would code for proximity, and the phase of firing within a clock cycle would code for angle. In this formalism, summing the output of several neurons results in vector addition, and subtraction is equivalent to a phase inversion followed by addition. Thus, the summed PC activity could provide a vector $\mathbf{v}(t)$ continually pointing to the centroid of the environment, so that, if the "goal" was encountered at time t_g, storing $\mathbf{v}(t_g)$ (outside the hippocampus) would enable the translation vector $\mathbf{v}(t) - \mathbf{v}(t_g)$ from the rat to the goal to be calculated whenever the rat wanted to go to the goal.

There are some points in favor of this model: (1) it provides a powerful alternative set of mechanisms to those usually considered by so-called connectionists; (2) it gives a plausible interpretation of PC firing; (3) if a PC's phase of firing with respect to θ was given by $360° - 2\alpha$, where α is the angle between the rat's direction of travel and the direction of the mini-centroid, then the model would be consistent with the phase shift data presented by O'Keefe and Recce (1993); (4) the rat would be able to take short-cuts. The model's disadvantages include the sensitivity of a unique reference direction to movements or occlusion of cues.

Place Cell Firing and Navigation

The population vector model (Burgess et al., 1994) is implemented at the neuronal level but also generates actual movement trajectories for the simulated rat, aiming to surmount some of the difficulties discussed earlier. The input to the model is a set of sensory cells with tuning curve responses to the distance of cues from the rat. The output is able to guide the rat back to a goal that has been visited one or more times in a single environment (Figure 2). The output representation chosen is analogous to that found in primate motor cortex (see REACHING: CODING IN MOTOR CORTEX): that is, a *population vector* (i.e., the vector sum of cells' preferred directions weighted by their firing rates). In the model, it is instantiated by a set of "goal" cells, such that each cell has a conical firing rate map whose peak is displaced from the goal position in a particular absolute direction (the "preferred direction" for that cell). The intermediate layers, entorhinal cells (ECs), PCs, and subicular cells (SCs) map the sensory input (from 4 to 16 intramaze and extramaze cues) to the population vector output. A type of competitive learning governs the dynamics of the PCs and SCs, similar to that in Sharp (1991).

Each EC receives connections from a pair of sensory cells, although not all possible pairs are represented. The choice of cues and distances at which the cues in a pair are represented results in ECs with large spatial firing fields that tend to be centered near to the centroid of the pair of cues. The model relies on the phase firing of SCs (relative to θ) being such that those firing late in a cycle have place fields centered ahead of the rat. This is achieved, somewhat arbitrarily, by making the phase of firing of ECs depend on the angle between the rat's heading direction and the direction of the centroid of the corresponding pair of cues. (That is, if the centroid is ahead, the cell fires at a late phase, whereas if it is behind, it fires early.) This property propagates throughout the PC and SC layers. Less severe competition among SCs compared to that among PCs leads to larger firing fields, which avoids the locality of information access problem (so that goal cell firing fields cover the whole environment).

The rat's direction of motion is updated at the end of each θ cycle; during "exploration," it is simply a random variable within 30° of the previous direction; during "searching," it is determined by the goal cells. A set of eight goal cells codes for each new goal location that is encountered (representing north, northeast, east, etc.). When the rat is at the goal, the goal cell with preferred direction closest to the rat's heading direction (given by head-direction cells) receives a strong input, allowing connections to it to be switched on. Information from the goal is only received at the goal location (cf. Barto and Sutton, 1981). This signal arrives at a late phase of the θ rhythm, and connections are switched on from those SCs that are active at that time (which tend to have firing rate maps that are peaked ahead of the rat). When a goal is encountered, the rat looks around in all directions so that connections are switched on to all eight goal cells.

The direction and proximity (represented by the net firing of the group of goal cells) of interesting objects is the output of the hippocampal map, and allows the simulated rat to navigate. A small number of obstacles can be avoided during navigation by subtracting the population vector of cells coding for an obstacle from the population vector for the goal. The model assumes that the role of the dentate gyrus (DG) and CA3 recurrent collateral system is to cause different subsets of PCs to be active in different environments (e.g., via mossy fiber "detonator" synapses in CA3 and an associative network; see McNaughton and Nadel, 1990; Treves and Rolls, 1992). The advantage of this model over those discussed earlier is that reasonable trajectories, including short-cuts, are performed after one visit to the goal following brief exploration (during which the goal need not be present)—that is, the model learns very quickly, and shows latent learning. Future work includes modeling the discrimination of many different environments.

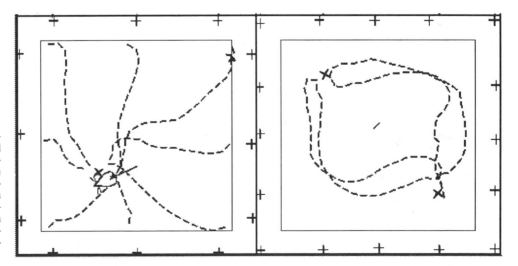

Figure 2. On the left are shown simulated trajectories, from eight novel starting positions, to a goal encountered after 30 seconds of exploration (at 60 cm/s in a 135 × 135 cm² environment; the rat is shown to scale. On the right is shown simulated navigation between two goals (×) with an obstacle (/) in between; cues are represented by + signs. (Adapted from Burgess et al., 1994.)

Discussion

There has been much development of spatial models of hippocampal processing in recent years. We are now close to being able to construct a simple neuronal model of the hippocampus that takes in sensory information from environmental cues, turns it into a place cell representation of space, and uses that spatial representation to create a representation of the direction in which interesting things are located, relative to it, as it moves about an environment. Such a model would allow systematic examination of the role of individual cells in the hippocampus in terms of their effect on (spatial) behavior, and would provide direction for experiments.

The preceding section describes a possible starting point for this type of model, following the ideas of Sharp (1991) and O'Keefe (1991). Inertial/directional navigation systems should also be integrated with the hippocampal place system, as in McNaughton et al. (1995). Finally, a model should be able to distinguish between many different environments, and to "remember" where things are in each. There is evidence that distinct PC representations are formed to code for different environments. Incorporating this into a model leads us back to consideration of the role of the DG and the CA3 recurrent collaterals in providing an associative memory (Marr, 1971; McNaughton and Nadel, 1990; Treves and Rolls, 1992). Exciting future possibilities include application of these models to related tasks, allowing consideration of experiments involving monkeys and humans, and to the navigation of autonomous mobile robots.

Acknowledgments.   Neil Burgess was supported by a Joint Councils Initiative on Human Computer Interaction, and as a Royal Society University Research Fellow.

Road Map: Mammalian Brain Regions
Background: I.3. Dynamics and Adaptation in Neural Networks
Related Reading: Potential Fields and Neural Networks; Short-Term Memory

References

Barnes, C. A., McNaughton, B. L., Mizumori, S. J. Y., Leonard B. W., and Lin, L.-H., 1990, Comparison of spatial and temporal characteristics of neuronal activity in sequential stages of hippocampal processing, *Prog. Brain Res.*, 83:287–300.

Barto, A. G., and Sutton, R. S., 1981, Landmark learning: An illustration of associative search, *Biol. Cybern.*, 42:1–8.

Burgess, N., Recce, M., and O'Keefe, J., 1994, A model of hippocampal function, *Neural Netw.*, 7:1065–1081.

Dayan, P., 1990, Navigating through temporal difference, in *Advances in Neural Information Processing Systems 3* (R. P. Lippmann, J. E. Moody, and D. S. Touretzky, Eds.), San Mateo, CA: Morgan Kaufmann, pp. 464–470. ◆

Marr, D., 1971, Simple memory: A theory for archicortex, *Philos. Trans. R. Soc. Lond. [Biol.]*, 262:23–81.

McNaughton, B. L., 1989, Neuronal mechanisms for spatial computation and information storage, in *Neural Connections, Mental Computation* (L. Nadel, L. A. Cooper, P. Culicover, and R. M. Harnish, Eds.), Cambridge, MA: MIT Press, pp. 285–350.

McNaughton, B. L., Knierim, J. J., and Wilson, M. A., 1995, Vector encoding and the vestibular foundations of spatial cognition: A neurophysiological and computational hypothesis, in *The Cognitive Neurosciences* (M. Gazzaniga, Ed.), Cambridge, MA: MIT Press.

McNaughton, B. L., and Nadel, L., 1990, Hebb-Marr networks and the neurobiological representation of action in space, in *Neuroscience and Connectionist Theory* (M. A. Gluck and D. E. Rumelhart, Eds.), Hillsdale, NJ: Erlbaum, pp. 1–63. ◆

Muller, R. U., Kubie, J. L., Bostock, E. M., Taube, J. S., and Quirk, G. J., 1991, Spatial firing correlates of neurons in the hippocampal formation of freely moving rats, in *Brain and Space* (J. Paillard, Ed.), Oxford: Oxford University Press, pp. 296–333. ◆

O'Keefe, J., 1991, The hippocampal cognitive map and navigational strategies, in *Brain and Space* (J. Paillard, Ed.), Oxford: Oxford University Press, pp. 273–295. ◆

O'Keefe, J., and Nadel, L., 1978, *The Hippocampus as a Cognitive Map*, Oxford: Clarendon Press. ◆

O'Keefe, J., and Recce, M., 1993, Phase relationship between hippocampal place units and the EEG theta rhythm, *Hippocampus*, 3:317–330.

Sharp, P. E., 1991, Computer simulation of hippocampal place cells, *Psychobiology*, 19:103–115.

Suzuki, W. A., Zola-Morgan, S., Squire, L. R., and Amaral, D. G., 1993, Lesions of the perirhinal and parahippocampal cortices in the monkey produce long-lasting memory impairment in the visual and tactual modalities, *J. Neurosci.*, 13:2430–2451.

Traub, R. D., Miles, R., Muller, R. U., and Gulyas, A. I., 1992, Functional organization of the hippocampal CA3 region: Implications for epilepsy, brain waves and spatial behavior, *Network*, 3:465–488. ◆

Treves, A., and Rolls, E. T., 1992, Computational constraints suggest the need for two distinct input systems to the hippocampal CA3 network, *Hippocampus*, 2:189–200.

Human Movement: A System-Level Approach

David A. Winter

Introduction

For researchers in neural networks who wish to make a meaningful contribution to the study of human movement, there are a large number of neurological, anatomical, and biomechanical considerations that must be accounted for in their models. Human gait analysis is particularly attractive to model because of its importance in everyday life and its complexity as a total body movement. Also, human gait has been analyzed more than any other movement; thus the validity of many of the neural models can be tested.

The purpose of this article is to identify the various inherent characteristics of the physiology and anatomy of human movement that must be recognized in any meaningful model. At this point in time, these include: the converging nature of the neuromusculoskeletal system, the neural delays and muscle mechanical lag characteristics, the interlimb coupling, and the total limb and body synergies. Examples from human gait will be presented to illustrate those characteristics, as well as to illustrate the challenge of modeling a total body movement.

Converging Nature of the Neuromusculoskeletal System

The structure of the neural system has many excitatory and inhibitory synaptic junctions, all converging on a final synaptic junction in the spinal cord to control individual motor units (see MUSCLE MODELS). Figure 1 depicts the various stages of

Figure 1. A schematic diagram showing four levels of convergence in the neuromusculoskeletal system of human movement. The converging nature of the system leads to considerable variability (redundancy) at the neural and muscular levels.

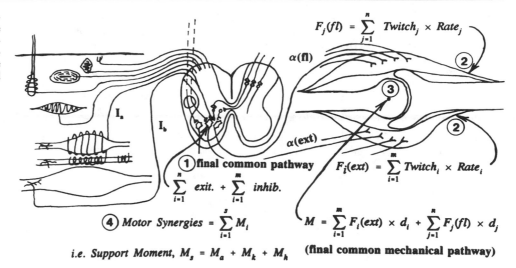

$$F_j(fl) = \sum_{j=1}^{n} Twitch_j \times Rate_j$$

$\alpha(fl)$

$\alpha(ext)$

① **final common pathway**

$$F_i(ext) = \sum_{i=1}^{m} Twitch_i \times Rate_i$$

$$\sum_{i=1}^{n} exit. + \sum_{i=1}^{m} inhib.$$

④ $Motor\ Synergies = \sum_{i=1}^{s} M_i$

$$M = \sum_{i=1}^{m} F_i(ext) \times d_i + \sum_{j=1}^{n} F_j(fl) \times d_j$$

i.e. Support Moment, $M_s = M_a + M_k + M_h$ **(final common mechanical pathway)**

this convergence, beginning with the α motoneuron (labeled 1), which is commonly described as the *final common pathway*. Each α neuron, via its synapse on the muscle, controls one motor unit. At the individual muscle level, we see a second summation of all motor unit tensions at the level of the tendon (labeled 2). This summation results from the neural recruitment of motor units, and has been the subject of scores of research papers (cf. Henneman and Olson, 1965; see also MOTONEURON RECRUITMENT). The resulting tension is a temporal superposition of the twitches of all active motor units (modulated by the length and velocity of each motor unit). A third level of musculoskeletal integration takes place at each joint, where the moment-of-force (labeled 3) is the algebraic sum of all muscle force/moment-arm-length products. The calculated moments-of-force are net effects of all agonist and antagonist muscles crossing that joint, and in spite of their mechanical units (Newton meters) they must be considered to be neurological signals. Finally, an intersegment integration is evident when the moments at two or more joints collaborate toward a common goal. One such synergy (labeled 4), reported over a decade ago, quantifies the integrated defense of all three joints of the

weight-bearing lower limb against a gravity-induced vertical collapse (Winter, 1980).

One of the by-products of these many levels of integration is that there is considerably more variability at the motor level than at the kinematic level. Figure 2 illustrates this point. A total of 15 major muscles at three joints can act to control the knee angle during weight bearing. Many contributions of muscle forces could yield the same muscle moment. In turn, many combinations of ankle, knee, and hip moments could yield the same knee angle. This redundancy can frustrate neural researchers, but there is a positive interpretation—the neuromuscular system is very plastic, and therefore very adaptable. This adaptability is very purposeful in pathological gait to compensate for pathological motor or skeletal deficits. For example, Winter (1989) reports on a major adaptation that took place in a patient who had a knee joint replaced because of its degeneration due to osteoarthritis. For years prior to the surgery, the patient had refrained from using her quadriceps muscles to support her during walking: the resultant increased bone-on-bone forces induced pain in her arthritic knee. She compensated by using her hip extensors instead of her knee extensors

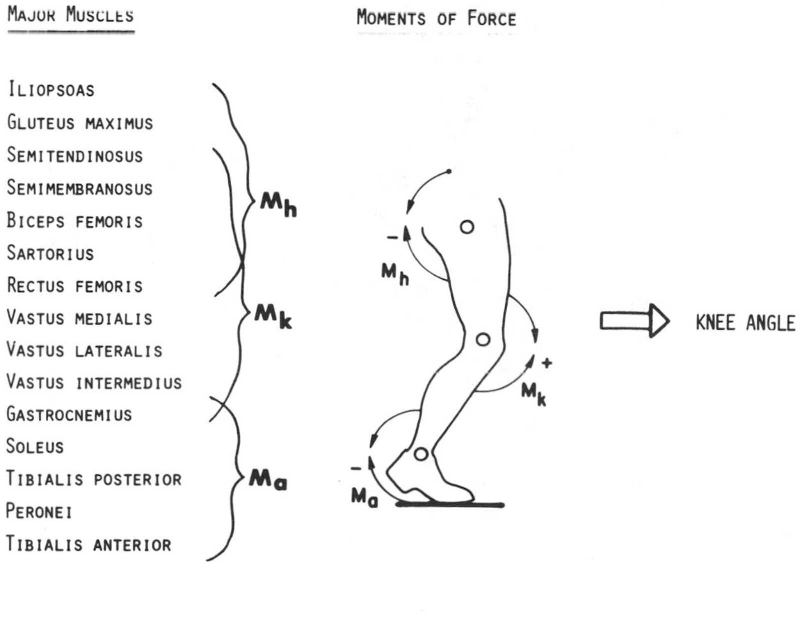

MAJOR MUSCLES   MOMENTS OF FORCE

ILIOPSOAS
GLUTEUS MAXIMUS
SEMITENDINOSUS
SEMIMEMBRANOSUS
BICEPS FEMORIS M_h
SARTORIUS
RECTUS FEMORIS
VASTUS MEDIALIS M_k
VASTUS LATERALIS
VASTUS INTERMEDIUS
GASTROCNEMIUS
SOLEUS
TIBIALIS POSTERIOR M_a
PERONEI
TIBIALIS ANTERIOR

M_h

M_k

M_a

KNEE ANGLE

Figure 2. Major lower limb muscles and joint muscles that control the knee angle during stance. The complex contributions of muscle forces would lead us to predict considerable indeterminacy at the kinetic level for the same outcome for the knee angle. Such indeterminacy allows considerable plasticity (adaptability) at the neural level to achieve the same outcome with different combinations of motor patterns.

and still maintained a near-normal walking pattern. This author would speculate that the spinal network is hardwired so that the total extensor pattern at both hip and knee was kept constant. Thus, a decrease in the knee extensor pattern was instantaneously replaced with a hip extensor pattern. More examples and discussion of this synergy will be presented later.

Essential Neural Delays and Muscle Lag Characteristics

Compared to robotic locomotion, which has no delays in its electrical signals and negligible delays in its torque motors, the human system is severely disadvantaged. Although the axon velocity is quite high (50 m/s), the minimal reflex response time from a foot sensor (afferent + efferent pathway ≈ 2 m) is about 40 ms. Similarly, a centrally initiated command will also take about 40 ms to reach a distal muscle. This means that triggering of the first motor unit takes 40 ms with a further electromechanical delay in the start of the muscle twitch, followed by the lag due to the low-pass characteristics of the muscle twitch itself (Milner-Brown, Stein, and Yemm, 1973). Thus, the peak tension from the very first motor unit might not be reached for a further 50 to 110 ms (Buchthal and Schmalbruch, 1970), followed by delays in the recruitment of further motor units (De Luca et al., 1982). The reverse is true for turn-off of activation except that, because of the characteristic shape of the twitch waveform, the fall in tension is slower than the rise. Thus, the neural system must program its neural patterns somewhat in advance of the time-course of the desired muscle tension and even further in advance of the displacements controlled by those tensions. The relation between these neural and mechanical events has been studied in a simulation model of motor unit recruitment which has been used to predict EMG and muscle tension (Fuglevand, Winter, and Patla, 1993).

Example of neural delays and muscle lag characteristics. Figure 3 is an electromyographic (EMG) signal from a surface

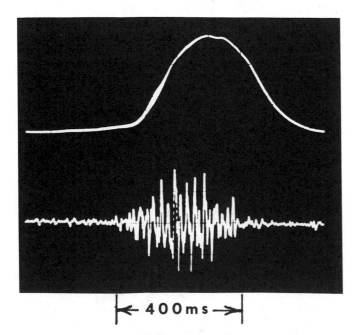

← 400ms →

Figure 3. Phasic relationship between the EMG and tension profiles from a muscle during a moderate length contraction. The tension waveform lags behind the EMG in both rise and fall of tension and is due to the low-pass characteristics of the muscle.

electrode of the biceps brachii muscle (lower trace) along with the moment-of-force (upper trace) that this elbow flexor generated during an isometric contraction. The EMG is an activation profile and is slightly delayed behind the neural pulse trains in the active α motoneurons. The electrode location was over the middle of the muscle while the moment-of-force reflects the tension seen at tendons. There is initially an electromechanical delay between the time that the EMG signal is recorded and the time it takes for the electrical wave to propagate along the muscle fibers to each tendon at a velocity of about 4 m/s. The mechanical wave of force from each crossbridge coupling within each sarcomere must also propagate from each active crossbridge to the tendon. These electromechanical delays can add another 20 ms to the neurological delays. The propagating force wave arrives at the tendon in the form of a twitch, which has been analyzed as a critically damped, second-order, low-pass system with cut-off frequency of around 3 Hz (Milner-Brown et al., 1973). Such a transfer function of the muscle will cause a further lag in the buildup of tension. The 400-ms burst of activity for this biceps muscle is seen to be somewhat in advance of its associated force waveform. The rise in force lags the rise in EMG by about 100 ms, the peak in force is delayed about 100 ms behind the EMG peak, and the tension wave lasts at least 150 ms after the end of its neural activation.

Anatomical Knowledge of the CNS-Interlimb Coupling and Total Body Synergies

The central nervous system (CNS) is not stupid about the anatomy it is controlling. Each muscle has its unique functional characteristics that influence the tension it generates: physiological cross-sectional area, force-length and force-velocity characteristics, and twitch characteristics. The CNS must know the moment arm of each muscle at each joint at both the proximal and distal ends, and, finally, it must know the inertial characteristics of all segments it may control and their interactions with adjacent segments. Our knowledge of the various control strategies allows us to see how the CNS is thinking, and invariably we see that the most appropriate strategy is used. Such interactions require in-depth understanding of the anatomy, the detailed goals of the movement, and the mechanics of multisegment systems. Finally, the CNS is acutely aware of the gravitational load acting on every body segment, and it knows when to stabilize against gravity or to use gravity to advantage. We will now examine the complex total body movement known as *gait* to illustrate the CNS's knowledge about the anatomy and mechanics of the system it is controlling.

Gait as an Example Total Movement

The challenge of gait analysis is its complexity, because of the large number of segments involved and the fact that each joint has three degrees of angular rotation, especially the hip joint. There is a continuous interaction between segments (interlimb coupling), which means that a rather simple event such as toe clearance during midswing can be seen to be under the control of at least six different muscle groups acting at separate joints of the stance and swing limbs (Winter, 1992). The second reason for the complexity is the large number of apparently redundant muscles that can be controlled. As was discussed earlier, this redundancy not only results in considerable variability at the motor level but also signals the tremendous plasticity of the CNS. The third characteristic of gait as a total movement is the continuously changing and competing tasks that must all be satisfied in order to achieve this apparently simple movement. Not only does gait involve single and double support phases (with critical transitions four times per stride), but the same muscles can at the same time be involved in two or three independent tasks.

Support Synergy: An Example of Redundancy and Interlimb Coupling

The control of a collapse of the body in the vertical direction during walking or running involves all three joints of the support limb. In order to collapse, not only must the knee flex, but also the ankle and hip. Thus, the extensor muscles at each of those joints are the main line of defense against a vertical collapse. This response during gait has been quantified as a total limb synergy called *support moment* (Winter, 1980). With all extensor moments reported as positive, the support moment $M_s = M_a + M_k + M_h$, where M_a, M_k, and M_h are the sagittal plane moments at the ankle, knee, and hip, respectively.

M_s has been seen to be positive during weight bearing and is very consistent during repeat intrasubject analyses. Figure 4 demonstrates those profiles for nine repeat walking trials across days on the same subject, who was asked to walk with her natural cadence while the kinetic analyses were done. The details of the biomechanical measurements and analyses are the subject of textbooks; the important neurological information contained in these motor patterns is how all three joints collaborate toward the common goal of total limb support (Winter, 1991).

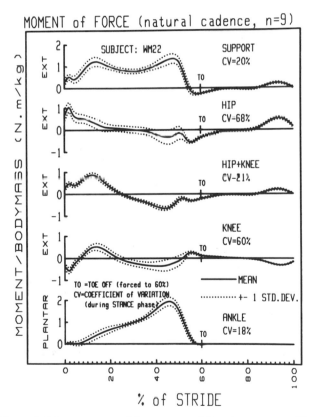

Figure 4. Joint moments of force averaged for nine repeat walking trials across separate days. The ankle, knee, and hip angles (not shown) had consistent joint angle patterns which did not vary more than $\pm 1/ 1/2°$, respectively, over the nine days. However, considerable variability is evident in the hip moments (CV = 68%) and knee moments (CV = 60%).

As can be seen in Figure 4, the ensemble average of the nine repeat walking trials is normalized to the same time-base, with one stride = 100% with heel contact at 0% and next heel contact at 100%. Extensor moments at all three joints are positive. The M_s profile represents a net extensor motor pattern pushing upward away from the floor, so it is not surprising that the peaks and valley in M_s are mimicked in the vertical ground reaction profile. What is surprising is the large variability in the hip and knee patterns, as indicated in the coefficient of variation (CV) scores of 60 to 70% (Winter, 1989). Here, this high variability is not seen in the hip, knee, and ankle angle profiles for these same trials (hip CV = 21%, knee CV = 8%, ankle CV = 16%). This reduced variability is a result of the considerable redundancy in the motor control system that would be predicted by the converging nature of the system and its ability to accomplish the same kinematic output. We see the CNS taking advantage of that adaptability to accomplish more than one goal with the same muscle groups, as is now discussed.

Balance control of the upper body in the anterior posterior (A/P) direction is the parallel task that must be satisfied along with the support task. Because of the large inertial load of the head, arms, and trunk (HAT) during each stride, the hip extensors and flexors are required to balance HAT against large perturbing forces acting at the hip (Winter, Ruder, and MacKinnon, 1990). During the first half of stance, there is a large deceleration of the stance hip joint followed by a large forward acceleration during the latter half of stance. These hip accelerations create a large couple (≈ 40 Nm) at the hip joint which unbalances the large inertial HAT segment. Fortunately, the CNS anticipates these acceleration perturbations and provides a counteracting hip extensor moment for the first half of stance and flexor moment during the second half of stance. These neurally controlled moments virtually cancel out the unbalancing moments so that the HAT remain erect within a few degrees. The question remains, how does the CNS know these accelerations before they occur? The matter is puzzling because any reflex responses from the vestibular system are not only too late but of the wrong polarity (Winter et al., 1990). There is no doubt that the CNS has in its memory an approximate template of the desired hip motor pattern (extensor for the first half of stance, flexor for the latter half). However, how does it know exactly when to start the pattern and allow for changes related to our speed of walking? Such proactive control could be triggered from earlier events in the gait cycle that are functionally and temporally related. This author would speculate that the vigor of the previous push-off could satisfy those criteria. Push-off starts about 10% of the gait cycle (≈ 100 ms) prior to the start of the pattern at the next heel contact, and the magnitude of the balance perturbation increases with the power of push-off.

However, in our assessment of the strides that were selected days apart, this hip motor pattern varies considerably because of small variations in posture of the large mass of HAT and because of a certain looseness in the balance control system (e.g., overcorrection may take place on one stride followed by an undercorrection on successive strides). One example of the range of that variability is a 66% CV of the hip moment. While this variability is desired in the balance control of HAT, it causes considerable problems in the support task. A reduced hip extensor moment during weight acceptance on one stride would cause the knee to collapse more than normal during that time. However, we know from the consistency of the knee angle that this does not happen. This is because the knee extensors compensate by increasing their activity. Thus, the net moment acting on the thigh (at proximal and distal ends) remains quite consistent over the stance phase. In Figure 4, this consistency is seen in the summation of hip plus knee moments, which shows a CV of 21%. There is almost a one-for-one tradeoff between the hip and the knee moments. The best single measure of the collaboration between the knee and hip is a covariance measure, which, for these nine repeat trials, has been calculated to be 89% (Winter, 1989). A 100% covariance would mean that the sum of the hip plus knee moments would have no variability in spite of the high individual variance at each joint. In summary, the support moment must be consistent to result in invariable knee, ankle, and hip angles; all this is accomplished in the presence of the necessarily high variable hip joint moments. One could speculate as to where the neural control of this synergy is located. This author suggests it is a hardwired spinal network, with an interneuron network which accomplished the one-for-one tradeoff. The total neural drive from this network represents the support moment, which is partitioned over the stance period into separate drives to the ankle, knee, and hip. An independent input to this network would be the anticipatory hip balance control which varies on a stride-to-stride basis, depending on the state of balance of HAT. The reader is referred to a recent text that has addressed the issue of adaptability by the CNS in gait (Patla, 1991).

Are There Any General Approaches to Discover the Neural Control?

Optimization methods search for the "best" controls based on an assumed criterion (see OPTIMIZATION PRINCIPLES IN MOTOR CONTROL). Biomechanics researchers have used intuitively attractive criteria such as minimum energy expenditure and minimum muscle exertion. Linear programming has been the most popular optimization technique. Seireg and Arvikar (1975) modeled 31 lower extremity muscles in a seven-segment, 3D model with inertial forces assumed to be negligible. Their objective function minimized a linear combination of muscle forces and joint moments; however, the force profiles for each muscle bore little resemblance to the EMG profiles reported by others. Crowninshield et al. (1978) used a more realistic criterion: to minimize the total tensile stress in their 30-muscle model. Their results were somewhat better, but for most muscles the predicted profiles had significant disagreement with established EMG profiles. Hardt (1978), using a 31-muscle model, optimized his solution based on a minimum sum of all forces; the resultant force profiles not only differed drastically from previously reported EMG profiles, but the model utilized square wave (on-off) activation profiles. All these studies suffered from the assumption that human walking satisfied a single criterion. A multiplicity of criteria is evident if we look at the biomechanics of walking (Winter, 1989; Yamaguchi, 1990). Several suboptimal goals are being sought simultaneously but with hierarchical priorities: balance control of HAT plus support against vertical collapse are the first priority. Forward propulsion with a safe toe clearance during swing and a gentle landing are the second priority. In order to achieve these goals, the muscles involved have periods of coactivation; there are also simultaneous periods of energy generation and absorption (Winter, 1991) within the same limb. Such evidence refutes any suggestion that a minimum energy criterion applies. Thus, any general approach toward discovering the framework of the neural control must be based on hierarchical priorities of subgoals, all of which must be satisfied.

Current research in neural networks has presented researchers with techniques to "learn" the correct neural (motor) output for any given set of desired kinematic patterns (see PERCEPTRONS, ADALINES, AND BACKPROPAGATION). Here, the experimental angles of the joints of the lower limbs and trunk over time would be inputs and the joint moments of both limbs would be outputs. The configuration of nodes in the input, hidden, and output layers combined with the connection weights defines an optimal solution. However, the network solution that satisfies those sets of inputs and outputs may fail to predict the motor patterns for a modified set of kinematics (Hammerstrom, 1993). This author believes that the architecture of the neural networks is important and must reflect the anatomy and synergies as described in this article. Only then will such techniques result in meaningful identification of valid neural networks.

Discussion

The major general conclusion is to emphasize the role of biomechanics as a window onto the net goals of the CNS in human movement. Biomechanics allows us to examine the fundamental characteristics of the final neural control and to see evidence of the integration at the muscle, joint, and total body level. We also quantify the neuromuscular delays and muscle lag characteristics that the CNS is programmed to account for. Biomechanics also allows researchers to identify total limb synergies and the integration of total body movement. All this must be programmed into the neural networks, along with the capability to adapt depending on the changing demands of the movement. It is hoped that this article, through its generalizations and its specific examples, gives insight into how neural networks might be modeled to take these factors into account.

Road Map: Biological Motor Control
Related Reading: Limb Geometry: Neural Control; Walking

References

Buchthal, F., and Schmalbruch, H., 1970, Contraction times and fibre types in intact human muscle, *Acta. Physiol. Scand.*, 79:435–452.
Crowninshield, R. D., Johnston, R. C., Andrews, J. G., and Brand, R. A., 1978, A biomechanical investigation of the human hip, *J. Biomech.*, 11:75–85.
DeLuca, C. J., LeFever, R. A., McCue, M. P., and Xenakis, A. P., 1982, Control scheme governing concurrently active motor units during voluntary contractions, *J. Physiol.*, 329:129–142.
Fuglevand, A. J., Winter, D. A., and Patla, A. E., 1993, Models of recruitment and rate coding organization in motor-unit pools, *J. Neurophysiol.*, 70:2470–2488. ◆
Hammerstrom, D., 1993, Neural networks at work, *IEEE Spectrum*, June:26–32.
Hardt, D. E., 1978, Determining muscle forces in the leg during normal human walking, an application and evaluation of optimization methods, *Trans. ASME J. Biomech. Eng.*, 100:72–78.
Henneman, E., and Olson, C. B., 1965, Relations between structure and function in the design of skeletal muscle, *J. Neurophysiol.*, 28:581–598.
Milner-Brown, H. S., Stein, R. B., and Yemm, R., 1973, The contractile properties of human motor units during voluntary isometric contractions, *J. Physiol.*, 230:359–370.
Patla, A. E., 1991, *Adaptability of Human Gait: Implications for the Control of Locomotion*, Advances in Psychology, vol. 78, Amsterdam: Elsevier. ◆
Seireg, A., and Arvikar, R. J., 1975, The predication of muscular load sharing and joint forces in the lower extremities during walking, *J. Biomech.*, 8:89–102.
Winter, D. A., 1980, Overall principle of lower limb support during stance phase of gait, *J. Biomech.*, 13:923–927.
Winter, D. A., 1989, Biomechanics of normal and pathological gait: Implications for understanding human locomotor control, *J. Motor Behav.*, 21:337–355.

Winter, D. A., 1991, *The Biomechanics and Motor Control of Human Gait: Normal, Elderly and Pathological*, 2nd ed., Waterloo, Ont.: University of Waterloo Press. ◆

Winter, D. A., 1992, Foot trajectory in human gait: A precise and multifactorial motor control task, *Phys. Ther.*, 72:45–56.

Winter, D. A., Ruder, G. K., and MacKinnon, C. D., 1990, Control of balance of upper body during walking, in *Multiple Muscle Systems:* *Biomechanics and Movement Organization* (J. Winters and S. Woo, Eds.), New York: Springer-Verlag, pp. 534–541.

Yamaguchi, G. T., 1990, Performing whole body simulation of gait with 3D, dynamic musculoskeletal models, in *Multiple Muscle Systems: Biomechanics and Movement Organization* (J. Winters and S. Woo, Eds.), New York: Springer-Verlag, pp. 663–679.

Identification and Control

Kumpati S. Narendra

Introduction

The past five decades have witnessed major advances in systems theory made through a combination of mathematics, modeling, computation, and experimentation. During the past ten years, there has also been an explosive growth in pure and applied research related to neural networks. In this article we attempt to indicate how the concepts and methods developed in these areas are being combined to generate general principles for the identification and control of complex nonlinear dynamical systems (Narendra and Parthasarathy, 1990).

Systems can be classified as either continuous-time (in which the variables of the system are defined for all values of time $t \in \mathbb{R}$) or discrete-time (in which they are defined at integer values, i.e., $t = 0, 1, 2, \ldots$). In many cases of interest, the latter class is obtained by sampling the variables of continuous-time systems at discrete time intervals. The sampling time intervals are invariably chosen to obtain the desired accuracy.

A very general method of representing multi-input multioutput continuous and discrete-time dynamical systems is using vector differential and difference equations as follows:

Continuous-Time Systems	Discrete-Time Systems	
$\dot{x}(t) = f[x(t), u(t)]$	$x(k + 1) = f[x(k), u(k)]$	(1)
$y(t) = h[x(t)]$	$y(k) = h[x(k)]$	

where $u(t)$ $(u(k)) \in \mathbb{R}^r$ is an input vector, $x(t)(x(k)) \in \mathbb{R}^n$ is the state vector, and $y(t)(y(k)) \in \mathbb{R}^m$ is an output vector. From Equation 1, it follows that given the state at time t_0 (k_0) and the input (input-sequence) for $t \geq t_0$ (k_0), the corresponding state and output can be determined. Equation 1 emphasizes the central role played by the state of the system, since if the state at time t_0 is known, the past history of the system is not relevant and the output is determined uniquely by the input from time t_0.

A second fact worth noting is that the dynamical systems defined in Equation 1 do not map the input at any time instant to the output at the same time instant but rather the inputs applied over time intervals into outputs over time intervals (see ADAPTIVE CONTROL: NEURAL NETWORK APPLICATIONS). More precisely, the state at time t_0 and the entire input $u(\tau)$, $t_0 \leq \tau \leq t_1$, defines the output at time t_1. In this article we can confine our attention to discrete-time dynamical systems of the form given in Equation 1.

A special case of the system of Equation 1 is one described by an equation of the following form where the functions f and h are linear:

$$x(k + 1) = Ax(k) + Bu(k)$$
$$y(k) = Cx(k) \qquad (2)$$

This system in which A, B, and C are constant $n \times n, n \times r$, and $m \times n$ matrices, respectively, is said to be linear and time invariant, and parametrized by the triple $\{A, B, C\}$. The former property implies that the law of superposition applies; i.e., if two input sequences $\{u_1(k)\}$ and $\{u_2(k)\}$ applied to the system, with zero initial conditions, result in output sequences $\{y_1(k)\}$ and $\{y_2(k)\}$, respectively, an input sequence $\{\alpha u_1(k) + \beta u_2(k)\}$ for any constant $\alpha, \beta \in \mathbb{R}$ will result in an output sequence $\{\alpha y_1(k) + \beta y_2(k)\}$. The second property of time invariance implies that the properties of the system do not vary with time for the same initial condition x_0.

Control theory deals with the process of influencing the behavior of a dynamical system so as to achieve a desired objective. Generally, this objective is to maintain the outputs of a system (e.g., altitude of an aircraft, glucose level in the blood) around prescribed constant levels (regulation), or follow predetermined time functions, e.g., the trajectory of a rocket in space (tracking). The best developed part of control theory deals with linear time-invariant systems for which design methods are currently well established. In recent years it has become evident that in many applications, linear models are not adequate for the representation of the process to be controlled. This discovery has led to an increased interest in the representation, identification, and control of nonlinear systems. Although much effort has been expended on the mathematical properties of nonlinear systems, very few constructive procedures currently exist for the design of controllers for such systems. We describe in the following sections efforts made in recent years to identify and control nonlinear systems using artificial neural networks.

Neural Networks in Dynamical Systems

The ability of neural networks to approximate large classes of nonlinear maps sufficiently accurately is well known (see KOLMOGOROV'S THEOREM). Since any finite-dimensional discrete-time dynamical system can be represented using summation, delay, and nonlinear maps (see Equation 1), it follows that neural networks are prime candidates for the identification of nonlinear plants. Assuming that a specific representation is chosen for the plant, identification consists of approximating the unknown nonlinear maps in such a representation. This purpose is accomplished by using suitably interconnected artificial neural networks. The parameters of the networks are then adjusted using gradient methods.

Artificial Neural Networks

Motivated by different networks in biological systems, various artificial neural network structures have been proposed for the representation of nonlinear systems. The most commonly used network structures for approximating nonlinear maps are the multilayer feedforward network (also known as a multilayer perceptron, MLP) and the radial basis function network (RBFN). These two classes of networks (which are described in greater detail in other articles in this volume) form the principal building blocks of the dynamical systems considered in this article. In an MLP with one hidden layer, the weight matrices associated with the two layers, W_1 and W_2, are the adjustable parameters (or weights) of the network. An alternative to the MLP is the RBFN, which can be considered as a two-layer network in which the hidden layer performs a nonlinear transformation on the inputs. (See RADIAL BASIS FUNCTION NETWORKS.) The output layer then combines the outputs of the first layer linearly so that the output is described by the equation

$$y = f(u) = \sum_{i=1}^{N} w_i R_i(u) + w_0$$

The functions R_i are termed activation functions or radial basis functions and, typically, these are Gaussian functions.

It has been shown by various authors that both MLPs and RBFNs can approximate continuous functions on a compact set arbitrarily closely. It has also been shown by Barron (1993) that the effectiveness of MLPs (in which the output depends nonlinearly on the weights) for approximating a general class of nonlinear functions increases with the dimension of the input space, as compared to networks in which the output depends linearly on the weights.

In the control of complex dynamical systems, the dimension of the inputs used for identification and control is generally quite high. In such cases, in view of the advantages assured by Barron (1993), MLPs are found to be very attractive for identification and control. The parameters of the network are adjusted using backpropagation or related methods, and convergence is generally quite slow. On the other hand, RBFNs generate linear maps in the parameters which can be adjusted very rapidly.

Characterization and Identification of Systems

System characterization and system identification are fundamental problems in systems theory. The problem of characterization is concerned with the mathematical representation of a system as an operator S which maps input signals into output signals. The problem of identification is to approximate S in some sense using an identification model.

If the state vector representation given in Equations 1 and 2 is used, the problem of identification arises when the functions f and h in Equation 1 or the matrices A, B, and C in Equation 2 are unknown. In the past three decades, numerous methods have been suggested for the efficient identification of linear systems, which involve the estimation of the parameters of A, B, and C. While several convenient canonical representations of the linear plants are known which can be used in the choice of models, this is not the case with nonlinear systems. Hence, proper parametrizations for the approximation of the functions f and h have to be chosen, and suitable methods must be developed for the adjustment of the parameters.

State Variables of Plant Accessible

When the state variables of the plant are accessible, a model of the plant can be set up as

$$\hat{x}(k + 1) = N_f[\hat{x}(k), u(k)]$$
$$\hat{y}(k) = N_h[\hat{x}(k)]$$
(3)

where $\hat{x}(k)$ and $\hat{y}(k)$ are the estimates of $x(k)$ and $y(k)$, and N_f and N_h are neural networks used to approximate the maps f and h, respectively. The neural networks N_f and N_h are then trained using the errors $x(k) - \hat{x}(k)$ for N_f and $y(k) - \hat{y}(k)$ for N_h. An alternative model used for system identification has the form

$$\hat{x}(k + 1) = N_f[x(k), u(k)]$$
$$\hat{y}(k) = N_h[x(k)]$$
(4)

It is seen that this model uses the accessible state vector $x(k)$ in the place of $\hat{x}(k)$ in the right-hand side of Equation 4. The models of Equations 3 and 4 are generally referred to as parallel and series-parallel models, respectively. From the point of view of practical identification, the series-parallel model is found to be substantially better. The reasons for its superiority are related to the methods that have to be used to adjust the parameters of the network and are briefly described in the next paragraph.

In Equation 4, since $x(k)$ is accessible, the output $N_f[x(k), u(k)]$ can be compared to $x(k + 1)$ using static backpropagation. Similarly, N_h can also be trained in a straightforward manner. In Equation 3, however, determination of the partial derivative of the error (in gradient methods) with respect to parameters is no longer straightforward, since the equation describing $\hat{x}(k)$ is not a function but a difference equation (or equivalently, Equation 3 describes a recurrent network). Hence, static backpropagation can no longer be used. The method of determining partial derivatives in recurrent networks is called dynamic backpropagation and is computationally intensive.

State Variables Not Accessible

The identification problem using a state representation becomes substantially more complex when the state variables of the plant are unknown and identification has to be carried out using only input-output data. It is well known even in the linear case that a unique parameterization of the plant no longer exists (Kalman, Falb, and Arbib, 1969). The success of nonlinear identification techniques therefore strongly depends upon specific parameterizations used.

Motivated by autoregressive moving average (ARMA) models which have been extensively used in linear system identification, a new model has recently been suggested for nonlinear system identification. The various properties of this model are not yet completely understood and at present are the subject of considerable theoretical investigation. It is my opinion that this model is bound to play an important role in the future in the identification and control of complex nonlinear dynamical systems using neural networks.

From the state Equations 1, the outputs at instants k, $k + 1, \ldots$, can be determined as $y(k) = h[x(k)]$, $y(k + 1) = h[f[x(k), u(k)]], \ldots$. If the linearized system around the equilibrium state of the system of Equation 1 is observable, it has been shown (Levin and Narendra, 1993) using the implicit function theorem that $x(k)$ and hence $y(k)$ can be determined using n past values of the input and output as follows:

$$x(k) = G[y(k), y(k - 1), \ldots, y(k + n - 1),$$
$$u(k), u(k - 1), \ldots, u(k + n - 1)]$$
$$y(k + 1) = F[y(k), y(k - 1), \ldots, y(k + n - 1),$$
$$u(k), u(k - 1), \ldots, u(k + n - 1)]$$
(5)

where G and F are functions mapping $\mathbb{R}^n \times \mathbb{R}^n$ into $\mathbb{R}^n$ and $\mathbb{R}$ respectively. Equation 5 is an exact mathematical representation of the given system (from Equation 1) in a domain around the equilibrium state. It provides a rigorous mathematical basis for the synthesis of identification models using input-output data.

The problem of identification in the present case reduces to the approximation of the function F in Equation 5. A neural network model of this system is given by the equation

$$\hat{y}(k + 1) = N_F[y(k), y(k - 1), \ldots, y(k + n - 1),$$
$$u(k), u(k - 1), \ldots, u(k + n - 1)] \qquad (6)$$

where $N_F[\cdot]$ is a neural network and the parameters of the latter are adjusted using the error $e_i(k)$ between $\hat{y}(k)$ and $y(k)$. The n past values of the output and input $y(k), y(k - 1), \ldots, y(k + n - 1)$, $u(k), u(k - 1), \ldots, u(k + n - 1)$ are used as inputs to the network N_F. This model has been used extensively for the identification of practical systems.

Comments on Identification Models

System identification using neural networks is at present still very much of an art. The number of layers to be used in the network as well as the number of nodes contained in each layer are essentially chosen by trial and error. There are also numerous methods (generally gradient based) for the adjustment of the parameters, and the performance criterion also varies from design to design. A criterion function which has proved effective in applications is one in which the output errors as well as the incremental inputs are weighted over a finite interval of time (i.e., $\Sigma_k[e_i^2(k) + (u(k) - u(k - 1))^2]$).

In many applications, some authors have used a neural network directly to approximate a given dynamical system (as a function rather than as a functional) but such an approach lacks rigorous justification. In the following section, while discussing different architectures which have been proposed for the control of dynamical systems using neural networks, "identification" invariably implies identification using the model described by Equation 6.

Recently, efforts have been made to obtain approximate models of Equation 6 in which the control terms appear linearly. They have the following form:

$$\hat{y}(k + 1) = F_0[y(k), y(k - 1), \ldots, y(k + n - 1)]$$
$$+ \sum_{j=0}^{n-1} F_j[y(k), y(k - 1), \ldots, y(k + n - 1)]u(k - j)$$
$$(7)$$

The importance of such models for the practical synthesis of controllers is described in the following section.

Control

Regulation in a control system involves the generation of a control input which will stabilize the system around a desired constant equilibrium value. In the tracking problem, a reference output $y^*(k)$ is specified and the output of the plant is to approximate it in some sense, i.e., $\lim \sup_{k \to \infty} \| y(k) - y^*(k) \| \le \varepsilon$. For theoretical analysis, ε is assumed to be zero so that asymptotic tracking is achieved.

In the following, we assume that the linearization of the system to be controlled around the equilibrium state is both controllable and observable. Controllability implies that any state of the linearized system can be transferred to any other state by the application of a suitable control input. Observability implies that by observing the input and output of the system over a finite interval of time, the state of the system can be determined. This condition on the linearization assures the controllability and observability of the nonlinear system in some domain around the equilibrium. This is a reasonable assumption, since almost all control systems in operation, designed using linear control principles, satisfy this criterion. This conclusion in turn implies that the results which follow can be applied to the control of these systems even when their domains of operation are increased and the resulting dynamics is nonlinear.

Regulation

Consider the dynamical system

$$x(k + 1) = f[x(k), u(k)] \qquad f[0, 0] = 0$$

which has to be regulated around the origin. If the linearized system around the origin is controllable, it has been shown (Levin and Narendra, 1993) that $u = g[x]$ (i.e., nonlinear state feedback) exists which stabilizes the system. A neural network, in principle, can be used to approximate $g[x]$. Assuming that such a network is used, we have

$$x(k + 1) = f[x(k), N_c[x(k)]] \triangleq F[x(k), \alpha] \qquad (8)$$

where α is the adjustable weight vector of the network. Equation 8 represents a dynamical system, and hence α has to be adjusted using dynamic backpropagation. In Levin and Narendra (1993), different methods are described for determining α to stabilize the system, based on the measured state $x[k]$. Simulation results indicate that this method is far superior to that obtained using linear theory.

Tracking

The problem of tracking a reference signal, when the dynamics of the nonlinear system is unknown, poses a real challenge to the control engineer. In this case, the plant dynamics is first identified using an identification model as described in the previous section. The principal question that has to be addressed is whether an input $u(\cdot)$ exists so that the output of the plant can asymptotically follow the reference signal. Assuming that such an input exists, the next problem is to determine a controller whose output is this desired control input.

The delay of the plant (also known as its relative degree d) is the number of time steps which elapse before the effect of an input is observed at the output. This must be known for a precise formulation of the control problem. The reference signal $y^*(k)$ is then defined as the output of a reference model described by the equation

$$y^*(k + d) = r(k)$$

where $r(k)$ is any bounded reference input. The delay d of the reference model is seen to be the same as that of the plant, and its input $r(k)$ is assumed to be specified. This implies that at time k the designer knows the value of the reference signal which the plant has to follow at time $k + d$.

Since the plant has a delay d, it follows that $u(k)$ and the state $x(k)$ determine the output at time $k + d$. Given $y^*(k + d)$ (or alternately $r(k)$), and the fact that the linearized system is controllable, it can be shown that the desired control input $u(k)$ can be expressed as

$$u(k) = g[x(k), r(k)]$$

or expressing $x(k)$ in terms of inputs and outputs,

$$u(k) = G[y(k), y(k - 1), \ldots, y(k + n - 1),$$
$$r(k), u(k - 1), \ldots, u(k + n - 1)]$$

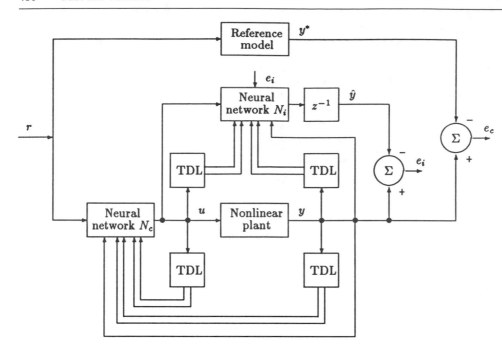

Figure 1. A general architecture for identification and control using neural networks. N_i and N_c represent the neural networks used for identification and control respectively; TDL denotes a tapped delay line which yields past values of signals; z^{-1} refers to a delay of one time unit; e_i and e_c are, respectively, the identification and control errors.

It is this function $G(\cdot)$ which, for the given control problem, is approximated by a neural network N_c.

Design of the Controller

The structure of the identifier and controller are shown in Figure 1. The past values of the input $u(k)$ and output $y(k)$, obtained using tapped delay lines (TDL), form the input to the network N_i used for identification. The same inputs (with $u(k)$ replaced by $r(k)$) are also fed back to the network N_c used as the controller. In the figure, e_i and e_c are the identification and control errors, respectively. It is readily apparent that the parameters of N_i can be adjusted using static or conventional backpropagation, while the parameters of N_c (which is in the feedback loop) have to be adjusted using dynamic backpropagation. Since, as mentioned earlier, the latter is computationally intensive, efforts have been made by different authors to approximate it in some sense.

From both theoretical and practical viewpoints, the best approach in the author's experience is to use the approximate model of Equation 7, in which the control input $u(k)$ appears linearly. In this case, the input $u(k)$ can be computed at every instant using its past values (i.e., $u(k-1), \ldots, u(k+n-1)$) as well as $y(k)$ and its past values. In the control of many complex dynamical systems, this model has been found to perform very satisfactorily.

Conclusions

Various control methods have been suggested by different authors for the practical control of dynamical systems. These include supervised control, inverse control, internal model control, and model reference control (see ADAPTIVE CONTROL: NEURAL NETWORK APPLICATIONS). In these the authors have attempted to realize identifiers and controllers as static maps using neural networks, to improve convergence and simplify analysis. To our knowledge, the identification and control models described in this article are the only ones which are based on theoretical results from nonlinear and adaptive control.

The general approach used in the previous sections and the specific structure of the identifier and controller shown in Figure 1 have been extended in numerous directions. These include control of systems with multiple inputs and multiple outputs, decoupling of such systems, and disturbance rejection. Decoupling refers to the design of a controller such that the effects of the different reference inputs are decoupled and each input affects only one output. The problem of disturbance rejection is to determine a modified control input to compensate for the effect of a disturbance at the output. More recently, the same approach has been used in the context of control using multiple identification models. At any instant, the "best" identification model is chosen according to an error criterion based on the identification error, and this, in turn, is used to choose the controller to be used. The control consequently involves switching and tuning. Considerable research is currently in progress on the stability of such systems, as well as in the choice of schemes which result in improved performance.

Road Map: Control Theory and Robotics
Background: I.3. Dynamics and Adaptation in Neural Networks
Related Reading: Reinforcement Learning in Motor Control; Sensorimotor Learning

References

Barron, A. R., 1993, Universal approximation bounds for superpositions of a sigmoidal function, *IEEE Trans. Inform. Theory*, 39:930–945.

Irwin, G. W., Warwick, K., and Hunt, K. J., Eds., 1992, *Neural Networks for Control Systems*, IEEE Control Engineering Series 46, London: Peter Peregrinus. ◆

Kalman, R. E., Falb, P. L., and Arbib, M. A., 1969, *Topics in Mathematical System Theory*, New York: McGraw-Hill.

Levin, A. U., and Narendra, K. S. 1993, Control of nonlinear dynamical systems using neural networks: Controllability and stabilization, *IEEE Trans. Neural Netw.*, 4:192–206.

Narendra, K. S., and Parthasarathy, K., 1990, Identification and control of dynamical systems using neural networks, *IEEE Trans. Neural Netw.*, 1:4–27. ◆

Illusory Contour Formation

Gregory W. Lesher and Ennio Mingolla

Introduction

Although first described by Frederich Schumann in 1904, illusory contours (ICs)—defined as the percept of clear boundaries in regions with no corresponding luminance gradient—did not spark research until Walter Ehrenstein, in 1941, and Gaetano Kanizsa, in 1955, presented their eponymous figures, which exhibit three important characteristics: a sharp IC delineating the border of a figure, an illusory brightening of the figure with respect to its background, and the appearance that the figure is floating above the inducing stimuli (Figure 1). Subsequent research has included: phenomenology (the classification of the perceptual qualities associated with ICs), psychophysics (the quantification of how variations in inducing stimuli affect IC clarity or illusory figure brightness), and modeling (the specification by algorithms or equations of formal analogs of the neural substrate of IC formation). This article considers only those ICs available in static, monocular stimuli, as described in Petry and Meyer's (1987) excellent introduction to IC research.

Why Neural Models?

Ehrenstein was primarily interested in illusory brightness, and some subsequent researchers claimed that figural illusory brightness *drove* IC formation. Kanizsa stressed the importance of understanding illusory forms as a whole, downplaying such perceptual attributes as brightness and clarity, which he believed followed from the instantiation of the illusory form. In subsequent cognitive theories, illusory figures were the brain's solution to "problematic" stimulus configurations. Brightness theories emphasized low-level explanatory mechanisms, while cognitive theories relied on high-level explanations. Gestalt theory occupied the middle, stressing that autonomous processes for visual grouping, perhaps influenced by but different from cognitive factors, discerned the "wholes" in visual stimulation.

By the early 1980s, shortcomings in both the brightness and cognitive camps had become evident. Brightness theories could not account for the demonstration of ICs without illusory

brightening. The data of von der Heydt, Peterhans, and Baumgartner (1984) most strongly undermined the claim that cognitive factors *must* be invoked to explain IC formation. They found many V2 cells in alert macaque monkeys that responded to stimuli yielding ICs in humans. Soon thereafter, Redies, Crook, and Creutzfeldt (1986) obtained responses to ICs from recordings in V1 of cats.

While early instantiation of ICs could result from either bottom-up neural interactions or top-down priming of the visual cortex, subsequent research has shown evidence of anatomical support in the cortex for the long-range interactions necessary for bottom-up contour formation. The absence of V1 response to ICs in the von der Heydt et al. (1984) study and in simple cells in the Redies et al. (1986) study, on the other hand, indicates that retinally generated brightness contrast effects are insufficient to explain ICs. Note, however, that whatever the mechanisms of IC generation, there are undeniable modulatory effects on IC perception by such cognitive factors as perceptual set and memory, as well as retinal effects, such as simultaneous brightness contrast.

Neural Models of Illusory Contour Formation

One critical factor in IC formation is the interaction of the thickness and orientation of inducers relative to the orientation of resulting illusory contours. While the Ehrenstein stimulus of Figure 1A contains four thin inducing lines with ICs *perpendicular* to these lines, the Kanizsa stimulus of Figure 1B consists of solid inducers with ICs *collinear* with the edges of the inducers. These two types are called *line-end* and *edge inducers*, respectively. Do the two types of inducers imply the existence of two separate mechanisms, or are they endpoints of a continuum handled by a single mechanism?

Ullman (1976) presented the first computational model of IC formation. His model is neural in that it consists of several layers of cells, locally interconnected so as to prefer straight-line completion, but with an allowance for curved completion. Cells excite neighbors that are roughly collinear and of similar orientation. Two initial layers compute a number of fuzzy

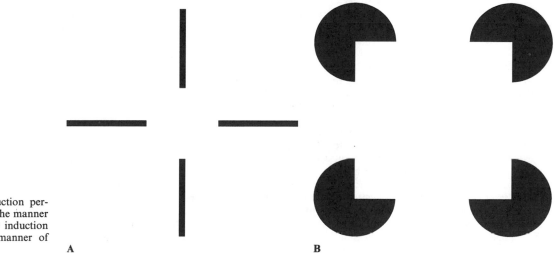

Figure 1. *A*, Contour induction perpendicular to line-ends, in the manner of Ehrenstein. *B*, Contour induction parallel to edges, in the manner of Kanizsa.

A   B

curves; competition in a third layer chooses a specific curve. Ullman's model received little attention from the IC research community upon publication, probably because (a) it only explained edge-type ICs, (b) it was too computationally complex to be widely simulated, and (c) it had little basis in real neural structures.

Peterhans, von der Heydt, and Baumgartner (1986) proposed a model of IC formation in which separate real and IC representations are unified in early visual cortex (V2). ICs are generated by end-stopped cells lying along a line roughly perpendicular to their preferred orientation. Outputs from pairs of these cells are multiplicatively gated, ensuring that IC formation cannot occur with single inducers. The output of each gated pair is summed with all other gated outputs, as well as with the activity of complex cells whose positions and preferred orientations match the line that the end-stopped cells fall along. Complex cells signal the presence of real contours. A unified representation is achieved by summing real and IC units. This dichotomy of contour representations is shared by most models of illusory contour formation, but (as discussed later) Grossberg and Mingolla (1985) showed it is not required to model psychophysical or physiological data.

Peterhans et al. posit that the corners of edge inducers trigger end-stopped cell activity. Since the same end-stopped cells that respond to line-ends can be triggered by the corners of edge inducers, contour completion can occur with either type of inducer in a direction perpendicular to the preferred orientation of the end-stopped cells.

The model of Finkel and Edelman (1989) is based upon that of Peterhans et al. (1986). As in the earlier model, end-stopped cell activity drives IC formation. Finkel and Edelman pool end-stopped cell activity over similar orientations at the same location, then specify that termination discontinuities—ecological cues to occlusion—arise only if at least three such cell pools are active along a line perpendicular to the mean orientation of these "wide-angle" pools. Perpendicular contour completion between termination discontinuities occurs at the occlusion module, leading to straight-line illusory contours. Finkel and Edelman claim their "re-entrant" (feedback) system mediates conflicts between real and illusory contours, sharpening or eliminating contours as appropriate.

In Kellman and Shipley's (1991) line-end theory, first-order spatial discontinuities are the generating features of ICs, provided they are "relatable"; that is, if they satisfy certain constraints concerning the intersection of the extensions of lines perpendicular to the inducing line or parallel to the edge associated with inducing corners. The degree of relatability determines IC strength, a hypothesis supported by psychophysical experiments. Shipley and Kellman (1992) present data indicating that for edge-type stimuli, contour clarity is a linear function of the support ratio—the fraction of the total length of an illusory figure's edge that has corresponding luminance discontinuities. Kellman and Shipley (1991; Shipley and Kellman, 1992) note that support ratio must be incorporated into the mechanics of the IC interpolation process, but provide no further computational details.

Grossberg and Mingolla (1985) propose the existence of two parallel systems, a boundary contour system (BCS) responsible for establishing boundaries of objects, and a feature contour system (FCS) responsible for establishing the color and brightness of object surfaces (see FIGURE-GROUND SEPARATION). The BCS represents the orientation of boundaries at every position and thus performs the completion required for IC formation. Note that the visibility (induced brightness, as opposed to clarity) of illusory figures involves processing by the FCS as well.

Units sensitive to oriented image contrasts may be activated by stimuli falling anywhere within extended receptive fields, so positional and orientational uncertainty are inevitable in their responses to luminance discontinuities. The BCS includes two competitive mechanisms for overcoming these uncertainties. These processes establish "end cuts" at line terminations— activity at cells with orientational preferences that are roughly perpendicular to the line itself. In the first competitive stage, cells with similar orientational preference compete across locations, functionally generating end-stopping, in that "survivors" of the competition respond most strongly to the ends of lines or corners. In the second stage, cells at the same location but of different orientations compete. Thus, strong vertical activations along the edges of a vertical line inhibit the weaker verticals near the line end. Then, weakening of vertical activations disinhibits spatially overlapping horizontal activations, generating a perpendicular end cut.

Grossberg and Mingolla believe completion occurs in directions approximately parallel (collinear) to the orientation preferred by the inducing units of the second competitive stage— whether activated by stimulus edges or end cuts—rather than in directions perpendicular to the orientation preferred by inducing end-stopped cells. To accomplish IC completion, bipole filters with bowtie-shaped receptive fields take inputs from spatially disparate cells with similar orientational preferences and whose receptive field centers fall roughly along a common line or arc. The contribution of each oriented cell is a function of its relation to bipole cell position and orientation. A bipole fires only if both sides of its receptive field are excited, and it feeds back to earlier stages, providing orientational information to locations where no signals supported by image contrast exist. For edge inducers, bipole cells use contrast-driven activity along the edge directly; for line-end inducers, end cuts are signals for completion. The long range AND completion employed by Peterhans et al. (1986) and Finkel and Edelman (1989) can be accomplished by a bipole cell, as can a measure approximating the relatability of Kellman and Shipley (1991) and the strength of "association field" interactions of Field, Hayes, and Hess (1992).

Heitger and von der Heydt (1993) have recently elaborated the model of Peterhans et al. (1986), using the edge and discontinuity detectors of Heitger et al. (1992). Unlike the earlier model, however, illusory contour completion can be triggered by both edge and line-end detectors. *Para* grouping proceeds in directions parallel to edges, while *ortho* grouping is limited to directions perpendicular to line-ends. The relative contributions of para and ortho systems are determined by a parameter that encodes the "cornerness" of the stimuli at each location. The explicit distinction between para and ortho differs from the approach taken in the BCS.

Signals from para and ortho units are combined by a "grouping field" consisting of two lobes, similar to the bipole filter of the BCS. After grouping, fuzzy bands of activity indicate the presence of illusory (and real) contours. The maximal contours of these bands correspond to human IC perceptions in several examples. Heitger and von der Heydt's model also contains mechanisms for assignment of figure and ground (cf. Grossberg, 1994, and FIGURE-GROUND SEPARATION).

Finkel and Sajda (1992) have addressed illusory contour formation within the context of a feedback network which segments objects, including illusory figures, into distinct depth planes. As a model of intermediate vision, this system addresses many of the higher-level IC determinants. For example, completion strength is functionally mediated by the figure's consistency with being an occluding object. However, the model does not appear capable of reproducing psychophysical results for parametric variations in low-level inducer determinants.

Discussion

Models based strictly on line-ends and corners for determination of IC strength predict that the extent of an edge-type inducer should have little effect on contour strength—only the corners participate in induction. Shipley and Kellman (1992) demonstrate that increasing the extent of edge inducers, while maintaining a constant gap between inducers, causes a monotonic increase in contour strength. Lesher and Mingolla (1993) found that increasing the width of line-end-type inducers resulted in monotonic increase of contour strength to a plateau. These data are not reconcilable with models which include no mechanism for collinear IC induction, or at least modulation of contour strength by edge length. Line-end models also predict that ICs cannot exist in the absence of line-ends or corners. However, stimuli containing no such discontinuities can produce salient ICs (Lesher and Mingolla, 1993).

A final problem with most of the line-end models is that they fail to make concrete predictions about the effect of the number or proximity of inducers on IC strength. However, none contain any mechanism that would predict anything other than a constant or increasing contour strength with increasing numbers of inducers. Although this has proven to be the case when increasing the number of lines in an Ehrenstein figure, Lesher and Mingolla (1993) have found that contour strength is an inverted U as a function of the number of inducing lines when Kanizsa pacmen are replaced by concentric arcs.

The BCS includes interactions in its first competitive stage that can explain these inverted-U results. As lines approach one another, mutual inhibition lowers overall responses, thereby decreasing each line's contribution to contour completion. The end-stopped cells employed in the model of Heitger and von der Heydt (1993), as specified by Heitger et al. (1992), also contain inhibitory mechanisms that may allow corresponding modeling of these results. Their model's "hard" choice of a maximum within a broad spatial distribution of potential IC signals precludes modeling of fuzzy ICs, however, and their use of a purely feedforward architecture is at odds with temporal data on IC formation and persistence. The BCS/FCS is a more mature system, capable of dynamically modulating between choice and contrast enhancement ("leader take most") modes, as required by sharp and fuzzy ICs, respectively. This model has recently been employed to model complex illusory phenomena that no other system can reproduce. Francis, Grossberg, and Mingolla (1994) have used BCS mechanisms to explain why visual persistence is longer for ICs than for real contours, and Gove, Grossberg, and Mingolla (1993) have extended BCS and FCS mechanisms to explain brightness enhancement at line-ends, as in Ehrenstein figures, making this theory the most comprehensive treatment of ICs and related phenomena.

Acknowledgments. Gregory W. Lesher was supported in part by the Air Force Office of Scientific Research (AFOSR F49620–92–J–0334), a National Science Foundation Graduate Fellowship, and Enkidu Research. Ennio Mingolla was supported in part by the Air Force Office of Scientific Research (AFOSR 90–0175 and AFOSR F49620–92–J–0334), the Northeast Consortium for Engineering Education (NCEE A303–21–93), and the Office of Naval Research (ONR N00014–91–J–4100 and ONR N00014–94–1–0597).

Road Map: Vision
Related Reading: Perceptual Grouping

References

Field, D. J., Hayes, A., and Hess, R. F., 1992, Contour integration by the human visual system: Evidence for a local "association field," *Vision Res.*, 33:173–193.

Finkel, L. H., and Edelman, G. M., 1989, Integration of distributed cortical systems by reentry: A computer simulation of interactive functionally segregated visual areas, *J. Neurosci.*, 9:3188–3208.

Finkel, L. H., and Sajda, P., 1992, Object discrimination based on depth-from-occlusion, *Neural Computat.*, 4:901–921.

Francis, G., Grossberg, S., and Mingolla, E., 1994, Cortical dynamics of feature binding and reset: Control of visual persistence, *Vision Res.*, 33:2253–2270.

Gove, A., Grossberg, S., and Mingolla, E., 1993, Brightness perception, illusory contours and corticogeniculate feedback, in *Proceedings of the World Congress on Neural Networks, vol. 1*, Hillsdale, NJ: Erlbaum, pp. 25–28.

Grossberg, S., 1994, 3-D vision and figure-ground separation by visual cortex, *Percept. & Psychophys.*, 55:48–120.

Grossberg, S., and Mingolla, E., 1985, Neural dynamics of form perception: Boundary completion, illusory figures, and neon color spreading, *Psychol. Rev.*, 92:173–211.

Heitger, F., Rosenthaler, L., von der Heydt, R., Peterhans, E., and Kubler, O., 1992, Simulation of neural contour mechanisms: From simple to end-stopped cells, *Vision Res.*, 32:963–978.

Heitger, F., and von der Heydt, R., 1993, A computational model of neural contour processing: Figure-ground segregation and illusory contours, in *IEEE 4th International Conference on Computer Vision*, Los Alamitos, CA: Computer Society Press, pp. 32–40.

Kellman, P. J., and Shipley, T. F., 1991, A theory of visual interpolation in object perception, *Cognit. Psychol.*, 23:141–221.

Lesher, G. W., and Mingolla, E., 1993, The role of edges and line-ends in illusory contour formation, *Vision Res.*, 33:2253–2270.

Peterhans, E., and von der Heydt, R., 1989, Mechanisms of contour perception in monkey visual cortex, II: Contours bridging gaps, *J. Neurosci.*, 9:1749–1763.

Peterhans, E., von der Heydt, R., and Baumgartner, G., 1986, Neuronal responses to illusory contour stimuli reveal stages of visual cortical processing, in *Visual Neuroscience* (J. D. Pettigrew, K. J. Sanderson, and W. R. Levick, Eds.), Cambridge, Eng.: Cambridge University Press, pp. 343–351.

Petry, S., and Meyer, G. E., Eds., 1987, *The Perception of Illusory Contours*, New York: Springer-Verlag. ◆

Redies, C., Crook, J. M., and Creutzfeldt, O. D., 1986, Neuronal responses to borders with and without luminance gradients in cat visual cortex and dorsal lateral geniculate nucleus, *Exp. Brain Res.*, 61:469–481.

Shipley, T. F., and Kellman, P. J., 1992, Strength of visual interpolation depends on the ratio of physically specified to total edge length, *Percept. & Psychophys.*, 52:97–106.

Ullman, S., 1976, Filling-in the gaps: The shape of subjective contours and a model for their generation, *Biol. Cybern.*, 25:1–6.

von der Heydt, R., Peterhans, E., and Baumgartner, G., 1984, Illusory contours and cortical neuron responses, *Science*, 224:1260–1262.

Information Theory and Visual Plasticity

Nathan Intrator and Leon N. Cooper

Introduction

When we try to understand the nature of synaptic learning rules, we should be concerned with possible goals that may underlie synaptic changes. It is conceivable that knowing what could be a useful goal under different input environments could serve to distinguish between synaptic plasticity theories. Only after this distinction between goals has been indicated can one continue further and distinguish between learning rules aimed at achieving the same objective, on the basis of their detailed mathematical properties, computational complexity, or other factors.

In this article we discuss information theory as related to information relays in cortex. Based on information theory, we motivate the need for, and distinguish between, two main learning rules: rules which are based on second-order correlations between inputs, and rules which are based on higher-order correlations. For the second family of rules, we present a mathematical technique which we believe can lead to a better distinction between the large family of rules based on higher-order correlations and give a more accurate definition for such synaptic plasticity rules.

In comparing learning rules, one should isolate the contributions of three different components to the overall goal of a synaptic model: the input environment, the learning rule, and the network architecture. In our discussion, we will be concerned with the relation between different input environments and learning rules.

We start with a brief review of information theory.

Information Theory and Synaptic Modification Rules

A general class of learning rules stems from the information theoretic idea of maximizing the mutual information between a network's output and input, or as it is sometimes called, minimizing the information loss across the network (Barlow, 1961: 217–234). This idea has been suggested as a general principle for cortical and artificial network models (Linsker, 1986; Bichsel and Seitz, 1989; Atick and Redlich, 1992, see also VISUAL CODING, REDUNDANCY, AND "FEATURE DETECTION").

Information theory was developed for the study of communication channels (Shannon, 1948). Shannon considered information as a loss in uncertainty. He considered a set of words as his data and measured uncertainty using the entropy of the distribution of words given by

$$H(P) = -\sum_i P_i \log P_i$$

where P_i is the probability of occurrence of the ith word. An intuitive way to look at this function is by considering the average number of bits (0s and 1s) that is needed to produce efficient code. It is desirable to use a small number of bits for sending those words which appear with high probability, and to use more bits to send words which appear with much lower probability.

Shannon extended this idea to the problem of information flow through a noisy communication channel, seeking to optimize the code so as to send the smallest number of bits on average while still achieving reliable communication (to the extent possible). This effort led to questions such as, How does the receiver, on the basis of the transmitted information, maximize his knowledge about the data known to the sender? The mutual information maximization principle formulates these ideas: If M^{in} is the message that enters the channel, and M^{out} is the message received from the channel, then the average information conveyed by the output message about the input is given by the difference in uncertainties before we know M^{out} and after we know it:

$$I[M^{out}, M^{in}] = H[P(M^{in})] - H[P(M^{in} | M^{out})]$$

where $P(M^{in})$ is the probability that message M^{in} enters the channel, and $P(M^{in} | M^{out})$ is the probability of having message M^{in} enter the channel, knowing that message M^{out} was received from the channel. This difference is also known as the relative entropy.

In simple words, the idea behind information theory is to find new representations of data which are simpler in some sense—e.g., can be described with a smaller number of bits, or reside in a lower dimensional space—but, on the other hand, contain as much information as possible about the original data.

Synaptic modification rules can be derived from solving the mutual information maximization problem under different assumptions about the probability distribution of input words. The solution to such learning rules will be based on gradient ascent or a more sophisticated optimization algorithm, e.g., conjugate gradient. A layered network based on this principle conveys information from layer to layer in such a way as to reduce the uncertainty in the system. The network's inputs are messages M^{in}, and its desired output is M^{out} which maximizes the mutual information. The mathematical properties of this information criterion are worked out for simple distributions, although the general solution is not given in a closed form.

Mathematical analysis of mutual information in linear layered networks leads to the conclusion that linear layered networks receiving Gaussian-distributed inputs should extract the principal components from the data in order to maximize the mutual input/output information. In the next section the conclusion will be contrasted with the goals of mutual information maximization under non-Gaussian distributions. In the following sections we discuss information theory as related to early and higher visual cortex regions.

The Statistics of Natural Images and Optimal Retina-Cortex Maps

The Gaussian distribution maximizes entropy among all distributions on the real line (with the above entropy definition extended to an integral). The amount of information needed to reduce the uncertainty under Gaussian-distributed data is largest; in other words, data that are not Gaussian distributed are redundant. It is thus natural to ask whether the pixel images of visual scenes are Gaussian distributed.

Ruderman (1994) demonstrated that the single log intensity of pixel distribution of a small collection of natural images is not Gaussian. This finding indicates that the multidimensional distribution of images is not Gaussian, since otherwise every (linear) projection (including single pixel projections) should have been Gaussian. Based on this finding, Ruderman suggests a new transformation to the pixel intensity, based on its local variance, which Gaussianizes the distribution. While the optimality of such a transformation is demonstrated from information theory considerations, the biological basis of such transformation has yet to be found.

Information Theory Principles in Data Transfer from Retina to Visual Cortex

Atick and Redlich's (1992) hypothesis is that the main goal of retinal transformations is to eliminate redundancy in input signals, particularly that due to pairwise correlations among pixels (second-order correlation). They discuss the optimal response of ganglion cells from the viewpoint of information maximization.

Field (1987) suggests an interesting match between the spectrum of natural images and the log polar retinotopic mapping from retina to cortex (see VISUAL CODING, REDUNDANCY, AND "FEATURE DETECTION"). Based on a small number of images, he observes that the power spectrum decreases as $1/f^2$ where f is the frequency of the changes of gray level in the image. Assuming that the coding is done similarly in each of the frequency bands, this observation implies that each frequency band does not carry the same amount of information, thus leading to suboptimal coding of the information. He suggests that log polar retinotopic mapping in which the bandwidth of each frequency band is a fraction of the central frequency causes each frequency band to carry the same amount of information, and is thus optimal from the information theory viewpoint.

Information Theory Principles in Data Transfer Within Visual Cortex

Principal Components and Maximum Information Preservation

Networks that extract principal components from data are numerous (see PRINCIPAL COMPONENT ANALYSIS). Linsker (1986) presented a set of equations stemming from a simple Hebbian rule with the addition of some weight constraints for stability, as a general framework for synaptic modification in early layers of visual cortex. Using computer simulations, Linsker showed that with a Gaussian arbor function (a function that controls the initial synaptic strengths as a function of spatial location) and a particular set of parameters related to the width of the arbor function and to the constraints on the weights, receptive fields that resemble simple and complex cells emerge. It is important to note here that the main properties attributed to this model are due to the network architecture, especially the form of the arbor function. For example, using a uniform arbor function would not have led to any meaningful result. A simplified form of these equations which ignores saturation constraints on the weights leads to extraction of principal components in each layer. Principal components are optimal solutions of the maximum information preservation (mutual information maximization) *under a Gaussian distribution assumption on the input data.* Thus, Linsker presented the rule guiding synaptic plasticity in his layered network as conforming to maximum information preservation guidelines. We emphasize that the goal of mutual information maximization—which is a general and very feasible learning rule—coincides with principal components extraction only in the special case of Gaussian data distribution.

Principal components are optimal when the goal is to accurately reconstruct the inputs. They are not optimal when the goal is classification, as can be seen by the simple example in Figure 1 (see also Duda and Hart, 1973:212). Two clusters each belonging to a different class are presented. As before, the goal is to simplify the representation with minimal loss in information, and in this case it amounts to finding a single dimensional projection that will capture the structure that is exhibited in the data. In Figure 1B clusters have different variance in either direction, whereas in Figure 1A the variance in both directions is equal. Clearly, the structure in the data is conveyed in the x projection, but in Figure 1 the variance is maximized in the y projection. This projection onto y also minimizes the mean square error (MSE) of the reconstructed image, since this is also the direction of the principal component of the data. We therefore have a simple example in which the goal of maximum information preservation stands in conflict with the goal of finding the principal component of the data. Furthermore, the example also demonstrates the superiority of the goal of maximum information preservation over the goal of extracting principal components.

One can then ask why it is that the principal component misses the important structure in the data, while another projection does not. The answer lies in the fact that principal components are concerned with first- and second-order moments of the data. When there is important information in higher-order moments, it cannot be revealed with principal components.

Another way to view what principal components do to the data is by observing that they define a new system of coordinates in which the covariance matrix is diagonal; namely, they eliminate the second-order correlation in the data, i.e., correlation between pairs of pixels. This procedure, however, does not eliminate higher-order correlations in the data.

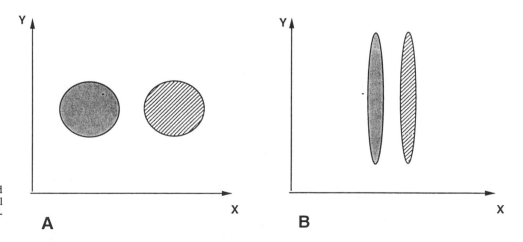

Figure 1. Principal components find useful structure in data (*A*) and fail when the variance of each cluster is different in each direction (*B*).

Projection Pursuit and Cortical Plasticity

Now that we have discovered that information preservation may be very different from extraction of principal components, and that principal components may not be useful in the case of non-Gaussian distribution, it becomes relevant to ask, What constitutes an interesting structure (important information) in a high-dimensional non-Gaussian distribution?

Projection pursuit methods seek features which emphasize the non-Gaussian nature of distributions (see Huber, 1985, for review). They seek structure that is exhibited by (semi)linear projections of the data. The relevance to neural network theory is clear, since the activity of a neuron is largely believed to be a semilinear function of the projection of the inputs on the vector of synaptic weights. Diaconis and Freedman (1984) show that for most high-dimensional clouds (of points), most low-dimensional projections are approximately Gaussian. This finding suggests that important information in the data is conveyed in those directions whose single dimensional projected distribution is far from Gaussian. Polynomial moments are good candidates for measuring deviation from Gaussian distribution; for example, skewness and kurtosis, which are functions of the first four moments of the distribution, are frequently used in this respect.

Intrator (1990) has shown that a BCM neuron (see BCM THEORY OF VISUAL CORTICAL PLASTICITY) can find structure in the input distribution that exhibits deviation from Gaussian distribution in the form of multimodality in the projected distributions. Since clusters cannot be found directly in the data because of the data's sparseness, this type of deviation, which is measured by the first three moments of the distribution, is particularly useful for finding clusters in high-dimensional data and is thus useful for classification or recognition tasks.

Principal Components Versus Projection Pursuit Learning

In this section we compare two single-neuron visual plasticity models based on the information they extract from the environment: Miller, Keller, and Stryker (1989) and BCM (Bienenstock, Cooper, and Munro, 1982; Intrator and Cooper, 1992). The former is a representative of principal components algorithms, and the latter is representative of projection pursuit learning. We use correlation of activity matrices to simplify the comparison. Let $C(\alpha, \beta)$ represent the correlation of activity between input pixel α and β; Miller's rule for a single neuron is given by

$$\frac{dS(\alpha, t)}{dt} = \lambda A(-\alpha) \left\{ \sum_{\beta} S(\beta, t) C(\alpha, \beta) - c_1 E[\alpha_\sigma(\alpha, t)] \right\}$$
$$- \gamma S(\alpha, t) - \varepsilon' A(-\alpha) \qquad (1)$$

which is a simple case of Equation 1 in Miller et al. (1989).

An analogous reformulation of the BCM modification equation (Intrator et al., 1993) leads to the following synaptic modification rule:

$$\frac{dm(\alpha, t)}{dt} = \lambda \left\{ \sum_{\beta, \gamma} \tilde{C}(\alpha, \beta, \gamma) m(\beta) m(\gamma) \right.$$
$$\left. - \sum_{\beta \gamma \delta} C(\alpha, \beta) C(\gamma, \delta) m(\beta) m(\gamma) m(\delta) \right\} \qquad (2)$$

where $\tilde{C}(\alpha, \beta, \gamma)$ is the third-order correlation of the input activity given by the expectation $E[a(\alpha)a(\beta)a(\gamma)]$. (For simplicity, we assume zero mean input activity.)

A number of important observations follow. Miller et al. (1989) use only first- and second-order statistical information

(the mean and correlation matrix of input activity) of the data, whereas BCM utilizes in addition the third-order statistics of the input activity. Since higher-order statistics carry additional information, this difference suggests that the BCM model may be more sensitive to the input environment.

If we assume that the retina performs decorrelation of the inputs (Atick and Redlich, 1992), then the covariance matrix of pixel intensities $C(\alpha, \beta)$ is diagonal (assuming that the inputs have zero mean), and so for eigenvalues $e(\alpha)$, Equation 2 becomes

$$\frac{dm(\alpha, t)}{dt} = \lambda \left\{ \sum_{\beta, \gamma} \tilde{C}(\alpha, \beta, \gamma) m(\beta) m(\gamma) - e(\alpha) m(\alpha) \sum_{\gamma} e(\gamma) m^2(\gamma) \right\}$$
$$(3)$$

In the special case of no second-order correlations, the BCM synaptic modification equation performs third-order decorrelation of the inputs, subject to some penalty related to the size of the weights. When the second-order statistics of the input data are not decorrelated, then the modification equation can be thought of as finding some balance between the second- and third-order correlation in the data.

Summary

We have demonstrated the important role of information theory in conveying information throughout the visual cortex. We have presented cases in which information theory considerations led people to seek methods for Gaussianizing the input distribution, and in other cases led people to seek learning goals for non-Gaussian distributions. We have argued that while information theory is an important and general framework for optimal information flow, the special case of linear feature extraction from Gaussian distribution, which might be sufficient for information flow through such channels as the optic nerve, may not be general enough for development of higher-order maps of (visual) cortex. We presented the framework of projection pursuit as a possible alternative.

We have demonstrated the difference of one representative of each of these models, the Miller-type correlation model and BCM learning.

Road Map: Development and Regeneration of Neural Networks
Background: I.3. Dynamics and Adaptation in Neural Networks
Related Reading: Fault Tolerance; Ocular Dominance and Orientation Columns

References

Atick, J. J., and Redlich, N., 1992, What does the retina know about natural scenes, *Neural Computat.*, 4:196–210.

Barlow, H. B., 1961, *Possible Principles Underlying the Transformations of Sensory Messages*, Cambridge, MA: MIT Press. ◆

Bichsel, M., and Seitz, P., 1989, Minimum class entropy: A maximum information approach to layered networks, *Neural Netw.*, 2:133–141.

Bienenstock, E. L., Cooper, L. N., and Munro, P. W., 1982, Theory for the development of neuron selectivity: Orientation specificity and binocular interaction in visual cortex, *J. Neurosci.*, 2:32–48. ◆

Diaconis, P., and Freedman, D., 1984, Asymptotics of graphical projection pursuit, *Ann. Statist.*, 12:793–815. ◆

Duda, R. O., and Hart, P. E., 1973, *Pattern Classification and Scene Analysis*, New York: Wiley. ◆

Field, D. J., 1987, Relations between the statistics of natural images and the response properties of cortical cells, *J. Opt. Soc. Am.*, 4:2379–2394.

Huber, P. J., 1985, Projection pursuit (with discussion), *Ann. Statist.*, 13:435–475. ◆

Intrator, N., 1990, A neural network for feature extraction, in *Advances in Neural Information Processing Systems 2* (D. S. Touretzky and R. P. Lippmann, Eds.), San Mateo, CA: Morgan Kaufmann, pp. 719–726.

Intrator, N., Bear, M. F., Cooper, L. N., and Paradiso, M. A., 1993, Theory of synaptic plasticity in visual cortex, in *Synaptic Plasticity: Molecular, Cellular and Functional Aspects* (M. Baudry, R. Thompson, and J. Davis, Eds.), Cambridge, MA: MIT Press, pp. 147–167. ◆

Intrator, N., and Cooper, L. N., 1992, Objective function formulation of the BCM theory of visual cortical plasticity: Statistical connections, stability conditions, *Neural Netw.*, 5:3–17. ◆

Linsker, R., 1986, From basic network principles to neural architecture (series), *Proc. Natl. Acad. Sci. USA*, 83:7508–7512, 8390–8394, 8779–8783. ◆

Miller, K. D., Keller, J. B., and Stryker, M. P., 1989, Ocular dominance column development: Analysis and simulation, *Science*, 245: 605–615.

Ruderman, D. L., 1994, Natural ensembles and sensory signal processing, PhD thesis, University of California at Berkeley.

Shannon, C. E., 1948, A mathematical theory of communication, *Bell Syst. Tech. J.*, 27:379–423, 623–656. ◆

Invertebrate Models of Learning: *Aplysia* and *Hermissenda*

John H. Byrne and Terry Crow

Introduction

Invertebrates are attractive animals for the study of learning and memory because of the relative simplicity of their central nervous systems. In many cases, a fairly complete "wiring diagram" can be specified and modeled. Many neurons are relatively large and can be uniquely identified in each member of a species, permitting one to examine the functional properties of an individual cell and to relate those properties to a specific behavior mediated by the cell. Biophysical and molecular events underlying the changes in cellular properties can be determined and mathematically modeled. This article summarizes progress that has been made toward a mechanistic analysis of learning in the gastropod molluscs *Aplysia* and *Hermissenda*.

Associative and Nonassociative Modifications of Defensive Siphon and Tail Withdrawal Reflexes in *Aplysia*

Behaviors and Neural Circuits

The siphon-gill and tail-siphon withdrawal reflexes of *Aplysia* have been used to analyze the neuronal mechanisms contributing to nonassociative and associative learning (Byrne et al.,

1993; Hawkins, Kandel, and Siegelbaum, 1993). The siphon-gill withdrawal reflex is elicited when a stimulus is delivered to the siphon and results in withdrawal of the siphon and gill (Figure 1*A*). A second behavior is the tail-siphon withdrawal reflex. Stimulation of the tail elicits a coordinated set of defensive responses, two components of which are a reflex withdrawal of the tail and the siphon (Figure 1*B*).

Defensive reflexes in *Aplysia* exhibit three forms of nonassociative learning: habituation, dishabituation, and sensitization. *Habituation* refers to a response decrement as a result of repeatedly delivering a stimulus (see HABITUATION). *Dishabituation* refers to restoration of a habituated (decremented) response by delivery of another stimulus. Finally, *sensitization* refers to an enhancement of a nondecremented response by delivery of another stimulus to the animal. With repeated stimulation, the reflexes undergo both short-term (minutes) and long-term (days) habituation. Restoration of a habituated response (dishabituation) or sensitization of a nonhabituated response can be produced by applying a noxious stimulus to the head or tail. Short-term sensitization lasts minutes, whereas long-term sensitization lasts days to weeks, depending on the type of sensitization training.

A. Siphon-Gill Reflex

B. Tail-Siphon Reflex

Figure 1. Siphon-gill and tail-siphon withdrawal reflexes of *Aplysia*. *A*, Siphon-gill withdrawal. Dorsal view of *Aplysia* showing the relaxed position (1) and the withdrawn position (2) following application of a stimulus (e.g., a water jet, brief touch, or weak electric shock) to the siphon. In the latter, the siphon and the gill have withdrawn into the mantle cavity. *B*, Tail-siphon withdrawal reflex. Shown are the relaxed position (1) and the withdrawn position (2), which occurs when a stimulus (e.g., a touch or weak electric shock) is applied to the tail, eliciting a reflex withdrawal of the tail and siphon.

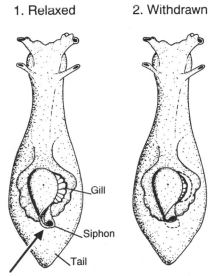
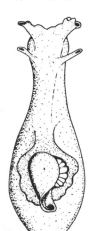
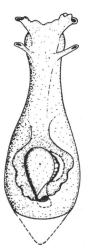

1. Relaxed   2. Withdrawn   1. Relaxed   2. Withdrawn

Gill

Siphon

Tail

Stimulus

Stimulus

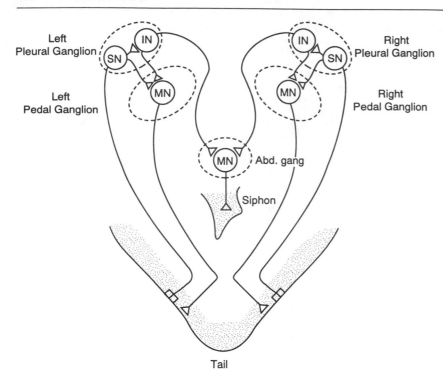

Figure 2. Simplified circuit diagram of the tail-siphon withdrawal reflex (see text for details). SN, sensory neuron; MN, motor neuron; IN, interneuron; Abd. gang, abdominal ganglion.

The afferent limb of the siphon-gill withdrawal reflex consists of sensory neurons with somata in the abdominal ganglion. The sensory neurons monosynaptically excite gill and siphon motor neurons, which are also located in the abdominal ganglion. Excitatory, inhibitory, and modulatory interneurons in the withdrawal circuit have also been identified.

The afferent limb of the tail-siphon withdrawal reflex consists of a cluster of 200 sensory neurons that are located in the pleural ganglion. These sensory neurons make monosynaptic excitatory connections with motor neurons in the adjacent pedal ganglion (Figure 2). The motor neurons produce withdrawal of the tail. In addition to their connections with tail motor neurons, sensory neurons form synapses with various identified interneurons. Some of these interneurons provide a parallel pathway to activate the tail motor neurons. These same interneurons activate motor neurons in the abdominal ganglion, which control reflex withdrawal of the siphon (see Figure 2). Aspects of the neural circuit for tail withdrawal and its plasticity have been mathematically modeled and simulated (White et al., 1993).

The sensory neurons for both the siphon-gill and tail-siphon withdrawal reflexes are similar, and appear to be key plastic elements in the neural circuits. Changes in their membrane properties and synaptic efficacy are associated with sensitization and the procedures that mimic short- and long-term sensitization training (see the next section and Byrne et al., 1993; Hawkins et al., 1993).

Cellular Mechanisms in Sensory Neurons Associated with Short- and Long-Term Sensitization in Aplysia

Short-term sensitization is induced when a single brief train of shocks to the body wall results in the release of modulatory transmitters, such as serotonin (5-HT), from facilitatory neurons that innervate the sensory neurons (Figure 3). The binding of 5-HT to receptors activates adenylyl cyclase, raising the level of the second messenger cAMP in sensory neurons. The in-

crease in cyclic adenosine monophosphate (cAMP) activates cAMP-dependent protein kinase (protein kinase A, or PKA) which adds phosphate groups to specific substrate proteins and, hence, alters their functional properties. One consequence of this protein phosphorylation is an alteration of the properties of membrane channels. Specifically, the increased levels of cAMP lead to a decrease in the S-K^+ current ($I_{K,s}$) and the calcium-activated K^+ current ($I_{K,Ca}$). These changes in membrane currents lead to depolarization of the membrane potential, enhanced excitability, and a small increase in the duration of the action potential. Moreover, cAMP appears to activate a membrane-potential and spike-duration–independent process of facilitation, which is represented in Figure 3 (large open arrow at far right) as the translocation of transmitter vesicles from a storage pool to a releasable pool, making more vesicles available for release with subsequent action potentials in the sensory neuron. These combined effects contribute to the short-term cAMP-dependent enhancement of transmitter release. Serotonin also appears to act through another receptor to increase the level of second messenger diacylglycerol (DG). DG activates protein kinase C (PKC). PKC, like PKA, activates the spike-duration–independent process of facilitation. In addition, a nifedipine-sensitive Ca^{2+} channel ($I_{Ca,Nif}$) and the delayed K^+ channel ($I_{K,V}$) are dually regulated by PKA and PKC. This modulation of $I_{K,V}$ and $I_{Ca,Nif}$ would also contribute to the increase in duration of the action potential. The consequences of the activation of these multiple messenger systems and multiple modulations of cellular processes occur when test stimuli elicit action potentials in the sensory neuron at various times after the presentation of the sensitizing stimuli. The enhanced release of transmitter from the sensory neuron leads to an enhanced activation of follower interneurons and motor neurons, and an enhanced behavioral response (i.e., sensitization).

Aspects of the modulation of membrane channels and the dynamics of second messenger systems, calcium regulation, and transmitter storage and release have been mathematically modeled and simulated (Gingrich and Byrne, 1987). The details

Figure 3. Model of heterosynaptic facilitation that contributes to short- and long-term sensitization in *Aplysia*. Sensitizing stimuli led to the release of modulatory transmitters (e.g., 5-HT), which leads to the transient activation of the second messengers DG and cAMP which affect cellular processes, the combined effects of which lead to enhanced transmitter release when a subsequent spike is fired in the sensory neuron. Long-term alterations are achieved through regulation of protein synthesis and growth. Positive (+) and negative (−) signs indicate enhancement and suppression of cellular processes, respectively (see text for additional details).

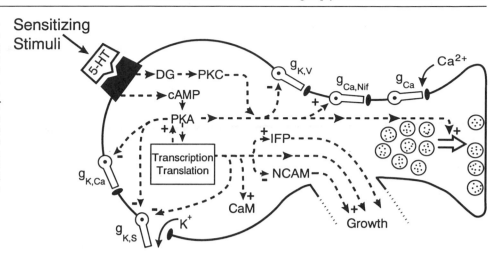

of these biophysical and biochemical processes were necessary to simulate the features of the empirical data, which could not be captured by less detailed models (see Baxter and Byrne, 1993).

Repeating the sensitizing stimuli leads to the induction of long-term facilitation. Repeated pulses of 5-HT lead to further, or more prolonged, phosphorylation and activation of nuclear regulatory proteins by PKA. Such proteins affect the regulatory regions of DNA and lead to increased transcription of RNA and, hence, increased synthesis of specific proteins. Some of the resulting proteins may be transcription factors which can activate other genes, some of which may be able to maintain their own activation. One of the newly synthesized proteins initiates the internalization and degradation of neuronal cell adhesion molecules (NCAMs), altering the interaction of the neuron with other cells and allowing the restructuring of the axon arbor. The sensory neuron can then form additional connections with the same postsynaptic target or make new connections with another. Other newly synthesized proteins, such as intermediate filament proteins (IFPs), are also likely to contribute to the new growth. Increased synthesis of calmodulin (CaM) also occurs, but the functional significance of this effect has not been determined.

Prolonged stimulation and increased cAMP levels also activate a process that decreases the level of PKA regulatory subunits, further prolonging PKA activation (Greenberg et al., 1987). With fewer regulatory subunits to bind to catalytic subunits, the catalytic units may be persistently active, contributing to long-term facilitation of transmitter release through the same cAMP-dependent processes seen in the short term. Some of these cAMP-PKA–induced changes include a decrease in $I_{K,S}$ and enhanced excitability, as well as, perhaps, a change in the synthesis of an $I_{K,S}$ channel protein or protein associated with the channel. As with short-term sensitization, the enhanced release of transmitter from existing contacts of sensory neurons onto motor neurons and interneurons underlies the long-term enhanced responses of the animal to test stimuli (i.e., sensitization). However, uniquely to long-term sensitization, increases in axonal arborization and synaptic contacts may contribute to the enhanced activation of follower interneurons and motor neurons (e.g., see Figure 2).

Associative Learning in Aplysia

Classical conditioning of withdrawal reflexes of *Aplysia* has been demonstrated (see Byrne et al., 1993; Hawkins et al.,

1993). Studies using a neuronal analog of differential classical conditioning indicate that a cellular mechanism called activity-dependent neuromodulation (see NEUROMODULATION IN INVERTEBRATE NERVOUS SYSTEMS) contributes to short-term classical conditioning observed at the behavioral level. Delivering a reinforcing or unconditioned stimulus (US), such as an electric shock, to the tail or a peripheral nerve activates a modulatory system, such as 5-HT, that nonspecifically enhances transmitter release from the sensory neurons. This nonspecific enhancement contributes to short-term sensitization (see earlier section). The temporal specificity characteristic of associative learning results from the pairing of a conditioned stimulus (CS), such as spike activity in one sensory neuron, with the US, which causes a selective amplification of the modulatory effects in that specific sensory neuron. Unpaired activity does not amplify the effects of the US. The amplification of the modulatory effects in the paired sensory neuron leads to a pairing-specific enhancement of the excitatory postsynaptic potential (EPSP) in the motor neuron.

The proposed mechanism suggests that increased Ca^{2+} levels resulting from spike activity alter adenylyl cyclase via CaM and increase the cAMP level produced by 5-HT. Thus, both Ca^{2+} and CaM appear to play a role in the activity-dependent neuromodulation that may underlie associative conditioning of the tail and gill withdrawal reflexes.

An important conclusion is that this mechanism for short-term associative learning is an elaboration of the cAMP-dependent mechanisms that contribute to a simpler form of learning, sensitization. This finding raises the interesting possibility that even more complex forms of learning may be achieved by using these simpler forms as building blocks. Indeed, a mathematical description of the learning rule (activity-dependent neuromodulation) for simple classical conditioning, when incorporated into appropriate neural circuits, has the capability to simulate higher-order features of classical conditioning, such as second-order conditioning of blocking, as well as features of operant conditioning (Byrne et al., 1990).

Associative Learning in *Hermissenda*

Classical Conditioning

Pavlovian conditioning of *Hermissenda* involves phototactic suppression and foot contraction (the conditioned responses, CRs) produced by paired stimulation of the visual and vestibular systems with their adequate stimuli (Crow, 1992). The conditioning procedure consists of light (the conditioned stimulus, CS) paired with high-speed rotation (the unconditioned stimulus, US). Retention of conditioned behavior elicited by the CS persists from days to weeks, depending upon the number of initial conditioning trials (Crow and Alkon, 1978; Alkon, 1989). The change in phototactic behavior exhibits CS specificity and is dependent upon the association of the two sensory stimuli involving both contiguity and contingency. Nonassociative contributions to behavior are expressed in the initial trials of the conditioning session and decrement rapidly following the termination of multi-trial training. In addition to multiple-trial conditioning of reduced locomotion and foot contraction, a one-trial conditioning procedure also modifies phototaxis (Crow and Forrester, 1986). Light (CS), when paired with direct application of 5-HT (nominal US) to the exposed nervous system of otherwise intact *Hermissenda*, produces significant suppression of phototactic behavior when the animals are tested 24 hours after the end of the one-trial training session. One-trial conditioning also produces cellular correlates in identified type B photoreceptors that are sites of cellular plasticity produced by multi-trial Pavlovian conditioning (see the section that follows).

Cellular Correlates of Classical Conditioning

One storage site for the memory of the associative experience is the primary sensory neurons of the pathway mediating the CS (Crow, 1992; Alkon, 1989). Cellular correlates of conditioning in identified cells of the CS pathway (type B photoreceptors) are expressed by a significant increase in CS-elicited spike frequency, enhanced excitability, an increase in input resistance, both increased and decreased amplitudes of light-elicited generator potentials, a decrease in spike-frequency accommodation, and reductions in the peak amplitudes of several diverse K^+ currents (see Crow, 1992, for a review). The enhanced excitability, expressed by significant increases in both the amplitude of CS-elicited generator potentials and spike frequency elicited by the CS, may be a major contributor to changes in the duration and amplitude of complex postsynaptic potentials and enhancement of spike frequency in postsynaptic targets. Changes in the strength of synaptic connections between the identified sensory neurons have been reported following conditioning. The amplitude of unitary inhibitory postsynaptic potentials (IPSPs), recorded from medial type A photoreceptors elicited by spikes in the medial type B photoreceptor, is enhanced by conditioning. However, the amplitude of IPSPs recorded from lateral type A photoreceptors that are elicited by spikes in the lateral type B photoreceptor is not enhanced in conditioned animals. A second site of cellular plasticity in conditioned animals is the type A photoreceptors. Lateral type A photoreceptors of conditioned animals exhibit an increase in CS-elicited spike frequency, a decrease in generator potential amplitude, and enhanced excitability and decreased spike-frequency accommodation to extrinsic current (Frysztak and Crow, 1993). The evidence for localization of cellular changes in the CS pathway indicates that multiple sites of plasticity involving changes in excitability and synaptic strength exist in the visual system of conditioned animals. Modeling studies will provide important insights into the determination of the relative contribution of changes in excitability and synaptic strength to conditioned modification of phototaxis.

Mechanisms of Classical Conditioning

Two K^+ currents—I_A and $I_{K,Ca}$—of type B photoreceptors are reduced following conditioning (Alkon, 1989). Reductions in several diverse K^+ currents could account for both the enhanced excitability and enhancement of IPSPs observed in conditioned animals. Recently, a Hodgkin-Huxley description of the time- and voltage-dependent currents was used to model the generator potential of the type B photoreceptor (Sakakibara et al., 1993).

Studies of the mechanisms responsible for the reduction in the K^+ currents of type B photoreceptors have implicated several second messenger systems (see Crow, 1992). The most recent evidence suggests that the phosphoinositide system may contribute to the reductions in K^+ currents observed in conditioned *Hermissenda*. Activation of PKC by phorbol esters and DG analogs and intracellular injection of PKC into type B photoreceptors reduce both the A-type K^+ current (I_A) and $I_{K,Ca}$ (Alkon, 1989). Activation of PKC appears to be initiated by the actions of 5-HT and GABA released by stimulation of the US pathway.

Studies of one-trial conditioning have provided insights into the mechanisms related to the induction or acquisition of conditioned behavior. The one-trial conditioning procedure (discussed earlier) produces both short- and long-term enhancement of generator potentials recorded from identified B photoreceptors (Crow and Forrester, 1991). Long-term enhancement depends on protein synthesis, is expressed only in lateral B photoreceptors, and is dependent upon the contiguity of the CS and US (Figure 4). The essential feature of the proposed mechanism for one-trial conditioning is that activation of PKC by stimulation of the US pathway occurs together with elevated intracellular Ca^{2+} levels produced by the presentation of the CS.

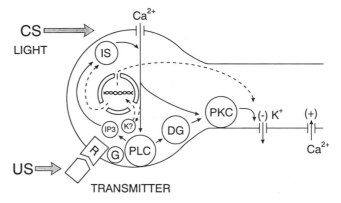

Figure 4. A cellular model for associative memory in B photoreceptors of *Hermissenda*. Short-term enhancement produced by one-trial conditioning involves activation and amplification of protein kinase C (PKC). Transmitter released by stimulation of the US pathway binds to a specific receptor (R). The receptor-activated signal is transmitted through a G protein to the enzyme phospholipase C (PLC). PLC splits PIP$_2$ (not shown) into inositol trisphosphate (IP3) and diacylglycerol (DG). The DG, Ca^{2+} influx produced by the conditioned stimulus (CS) and the Ca^{2+} released by IP3 from internal stores (IS) activate PLC and PKC, which leads to a reduction of K^+ currents and enhanced excitability of the photoreceptors. The presentation of the CS results in increased levels of intracellular Ca^{2+}. This is produced by the depolarizing generator potential serving to activate voltage-dependent Ca^{2+} channels, as well as by light-induced release of Ca^{2+} from IS. Pairing specificity results from the amplification provided by Ca^{2+} acting on PLC and PKC. Long-term memory in this model is dependent upon protein synthesis and gene products dependent upon Ca^{2+} activating an unidentified kinase (K?) or long-term changes in Ca^{2+} buffering. The evidence suggests that both short-term memory and long-term memory in this system involve independent parallel pathways (see text).

The induction of short-term enhancement by one-trial conditioning is attributable to activation of PKC, since protein kinase inhibitor and a reduction in the activity of PKC secondary to prolonged exposure to phorbol esters (down-regulation) block short-term enhancement (Crow, 1992). However, the same protein kinase inhibitors do not block the expression of established enhancement in identified type B photoreceptors (Crow, 1992). These results indicate that kinase activation may be important only for acquisition of one-trial conditioning, but not for the long-term expression of conditioned behavior. Moreover, recent evidence indicates that short- and long-term enhancement are parallel processes. The conditions that are sufficient to block short-term enhancement—down-regulation of PKC and kinase inhibition—do not block long-term enhancement (Crow and Forrester, 1993). These results suggest that short- and long-term enhancement are not sequential processes, but rather are parallel processes that involve independent mechanisms.

Discussion

The possibility of relating cellular changes to complex behavior in invertebrates is encouraged by the progress that has already been made in examining the neural mechanisms of simple forms of nonassociative and associative learning. The results of these analyses have shown that (1) learning involves changes in existing neural circuitry (as, at least for the short-term, the growth of new synapses and the formation of new circuits for learning and memory are not necessary); (2) learning involves the activation of second messenger systems; (3) the second messenger affects multiple subcellular processes to alter the responsiveness of the neuron (e.g., at least one locus for the storage of memory is the alteration of specific membrane currents); (4) long-term memory requires new protein synthesis, whereas short-term memory does not; and (5) long-term memory may be associated with structural changes in the nervous system.

Acknowledgments.   This work was supported by NIMH Award MH00649 (JHB), NIMH grant MH408060 (TC), NIH grant NS19895 (JHB), and AFOSR grant 93 NL090 (JHB).

Road Map: Learning in Biological Systems
Background: Ion Channels: Keys to Neuronal Specialization

References

Alkon, D. L., 1989, Memory storage and neural systems, *Sci. Am.*, 261(1):42–50. ◆

Baxter, D. A., and Byrne, J. H., 1993, Learning rules for neurobiology, in *The Neurobiology of Neural Networks* (D. Gardner, Ed.), Cambridge, MA: MIT Press, pp. 71–105.

Byrne, J. H., Baxter, D. A., Buonomono, D. U., and Raymond, J. L., 1990, Neuronal and network determinants of simple and higher-order features of associative learning: Experimental and modeling approaches, *Cold Spring Harbor Symp. Quant. Biol.*, 40:175–186.

Byrne, J. H., Zwartjes, R., Homayouni, R., Critz, S., and Eskin, A., 1993, Roles of second messenger pathways in neuronal plasticity and in learning and memory: Insights gained from *Aplysia*, in *Advances in Second Messenger and Phosphoprotein Research*, vol. 27 (A. C. Nairn and S. Shenolikar, Eds.), New York: Raven, pp. 47–108. ◆

Crow, T., 1992, Analysis of short- and long-term enhancement produced by one-trial conditioning in *Hermissenda*: Implications for mechanisms of short- and long-term memory, in *Neuropsychology of Memory* (L. Squire and N. Butters, Eds.), New York: Guilford, pp. 575–587. ◆

Crow, T., and Alkon, D. L., 1978, Retention of an associative behavioral change in *Hermissenda*, *Science*, 201:1239–1241.

Crow, T., and Forrester, J., 1986, Light paired with serotonin mimics the effects of conditioning on phototactic behavior in *Hermissenda*, *Proc. Natl. Acad. Sci. USA*, 83:7975–7978.

Crow, T., and Forrester, J., 1991, Light paired with serotonin in vivo produces both short- and long-term enhancement of generator potentials in identified B photoreceptors in *Hermissenda*, *J. Neurosci.*, 11:608–617.

Crow, T., and Forrester, J., 1993, Down-regulation of protein kinase C and kinase inhibitors dissociate short- and long-term enhancement produced by one-trial conditioning of *Hermissenda*, *J. Neurophysiol.*, 69:636–641.

Frysztak, R. J., and Crow, T., 1993, Differential expression of correlates of classical conditioning in identified medial and lateral type A photoreceptors of *Hermissenda*, *J. Neurosci*, 13:2889–2997.

Gingrich, K. J., and Byrne, J. H., 1987, Single-cell neuronal model for associative learning, *J. Neurophysiol.*, 57:1705–1715.

Greenberg, S. M., Castellucci, V. F., Bayley, H., and Schwartz, J. H., 1987, A molecular mechanism for long-term sensitization in *Aplysia*, *Nature*, 329:62–65.

Hawkins, R. D., Kandel, E. R., and Siegelbaum, S., 1993, Learning to modulate transmitter release: Themes and variations in synaptic plasticity, *Annu. Rev. Neurosci.*, 16:625–665. ◆

Sakakibara, M., Ikeno, H., Usui, S., Collin, C., and Alkon, D. L., 1993, Reconstruction of ionic currents in a molluscan photoreceptor, *Biophys. J.*, 65:519 527.

White, J. A., Ziv, I., Cleary, L. J., Baxter, D. A., and Byrne, J. H., 1993, The role of interneurons in controlling the tail-withdrawal reflex in *Aplysia*: A network model, *J. Neurophysiol.*, 70:1777–1786.

Investment Management: Tactical Asset Allocation

A. D. Zapranis and A. N. Refenes

Introduction

The prevailing wisdom among financial economists is that price fluctuations not due to external influences are dominated by noise and can be modeled by stochastic processes. Consequently, they try to understand the nature of noise and develop tools for predicting its effects on asset prices. It is, however, possible that these remaining price fluctuations are, to a large extent, due to nonlinear processes at work in the marketplace. Therefore, given appropriate tools, we can understand much of the market's price structure on the basis of completely or partially deterministic but nonlinear dynamics.

Nonlinear modeling techniques are the subject of increasing interest from practitioners in quantitative asset management, with neural networks assuming a prominent role. They are being applied to a number of "live" systems in financial engineering, but there is rarely a comprehensive investigation of the nature of the relationship that has been captured between asset prices and their determinants. In the section "Neural Networks and Nonparametric Regression," we formulate neural learning in a framework similar to additive nonlinear regression. This framework provides an explicit representation of the estimated models and enables modelers to use a rich collection of analytic and statistical tools to test the significance of the various parameters in the estimated neural models. The methodology can be applied to modeling asset returns in various markets and uses modern financial economics theory on market dynamics to investigate the plausibility of the estimated models and to analyze them in order to separate the nonlinear components of the models which are invariant through time from those that reflect temporary (and probably unrepeatable) market imperfections.

In the section "Active Investment Management and Neural Networks," we review the process of quantitative investment management and explain how and where neural networks can be applied to enhance the process. The key idea here is that a particular portfolio will depend on the universe of assets under consideration and the properties of those assets at that time. The main proposition of modern theories such as the arbitrage pricing theory (APT) is that the return of an asset is a *linear* combination of the asset's exposure to a set of factors. Such theories have been very useful in expanding our understanding of the capital markets, but many financial anomalies have remained unexplainable. Because of their inductive nature, neural networks have the ability to infer complex *nonlinear* relationships between an asset price and its determinants. Although this approach may lead to better nonparametric estimators, neural networks are not always easily accepted in the financial economics community because they bypass the step of theory formulation. This article gives an introduction to investment management, provides a formulation of neural learning which is synergetic rather than competitive to theory formulation, reviews some ongoing research and applications in investment management, and compares neural networks with linear models in the context of tactical asset allocation. Our empirical results challenge the linearity assumption.

Finally, we introduce the topic of tactical asset allocation, which will serve as our benchmark for evaluating linear against nonlinear modeling methodologies, and we describe the data and setup for our experimental analysis. We evaluate linear regression against neural modeling in the context of estimating differential returns between asset classes, on the basis of a universe of economic and financial factors.

Neural Networks and Nonparametric Regression

We provide a theoretical framework for analyzing the computational properties of neural networks by making use of statistical estimation theory giving a formulation of neural computation which is directly analogous to additive nonparametric nonlinear regression models and for which there is a rich collection of analytic and statistical tools.

Nonparametric Statistical Inference

Much of the work on neural networks has been compared to nonparametric statistical inference (e.g., Amari, 1990; see LEARNING AND STATISTICAL INFERENCE). One important area of statistical estimation theory which, despite its striking resemblance to neural learning, has been overlooked is recent developments in the field of nonparametric nonlinear regression. Much of the work in so-called additive nonlinear regression models such as ACE (Hardle, 1989) can be shown to be similar to feedforward neural networks. The main strength of these models derives from the fact that they provide an explicit repre-

sentation of the structure of the estimated models which is very important in analyzing the properties of the estimators. Consider for example the simple ACE (additive conditional expectation) model for nonparametric nonlinear regression. Given a set of observations (i.e., training set), ACE attempts to estimate ("learn") a function of the form

$$\theta(y_i) = \phi_1(x_{ip}) + \phi_2(x_{ip}) + \ldots + \phi_p(x_{ip}) + r_i \tag{1}$$

where $\theta, \phi_1, \ldots, \phi_p$ are smooth nonlinear transformations estimated from the data $(y_i, x_{i1}, \ldots, x_{ip}); i = 1, \ldots, p$, and r_i are the residuals. The estimation procedure may choose the nonlinear transformations $\theta, \phi_1, \ldots, \phi_p$ in a way that maximizes the correlation between $\theta(y_i)$ and $\phi_i(x_{ip}) + \phi_2(x_{ip}) + \ldots + \phi_p(x_{ip}) + r_i$ or minimizes the residual mean square error. In either case, the end effect bears a striking resemblance to learning by error backpropagation with ordinary least squares and linear output units. This similarity is explored in the next section.

Neural Networks as Nonparametric Multiplicative Regression

Consider the family of neural networks with asymmetric sigmoids as the nonlinear transfer function. For simplicity, we consider networks with one layer of hidden units. It can be shown that, under certain conditions, the function that the error backpropagation algorithm estimates is given by

$$y = \gamma_0 \frac{A^{\alpha_0} B^{\beta_0} \ldots M^{\mu_0}}{A^{\alpha_0} B^{\beta_0} \ldots M^{\mu_0} + 1} + \gamma_1 \frac{A^{\alpha_1} B^{\beta_1} \ldots M^{\mu_1}}{A^{\alpha_1} B^{\beta_1} \ldots M^{\mu_1} + 1} + \ldots$$
$$+ \gamma_n \frac{A^{\alpha_n} B^{\beta_n} \ldots M^{\mu_m}}{A^{\alpha_n} B^{\beta_n} \ldots M^{\mu_m} + 1} \tag{2}$$

where y denotes the output from the network (dependent variable); $A, B, \ldots, M$ denote m input (independent) variables; $\alpha_j, \beta_j, \ldots, \mu_j$ denote connection weights from input $A, B, \ldots, M$ (respectively) to the hidden layer; and γ_j denote connections from the hidden units to the output unit.

Thus, neural learning is analogous to searching the function space defined by the terms of Equation 2 and the range of the permissible values for the parameters. This formulation is strikingly similar to the formulation of additive nonlinear nonparametric regression (see Hardle, 1989) and it allows us to apply the analytic and statistical tools that have been developed in the field of additive nonparametric regression for the class of neural networks of a similar structure.

Active Investment Management and Neural Networks

The ultimate goal of any investment strategy is to maximize returns with minimum risk. In the framework of modern portfolio management theory, this is achieved by constructing a portfolio of investments which is weighted in a way that achieves some tradeoff between maximum return and minimum risk. The construction of such an optimal portfolio requires a priori estimates of asset returns and risk.

Traditionally, in line with the efficient market hypothesis, it has been assumed that returns are random and that the best prediction for tomorrow's return is today's return. Over a longer period, expected returns are estimated by computing the average of historical returns. The prediction error is considered as unpredictable noise and so asset risks were estimated by the standard deviation of historical returns. More recently, with multiple-factor APT, the idea has emerged that asset returns might be influenced by a number of factors. This new approach involves three stages:

1. *Factor analysis.* Identify factors which have an influence on asset prices (and/or returns).
2. *Estimating returns of the different assets.* Estimate asset returns on the basis of the above factors.
3. *Portfolio construction and optimization.* Given estimates of returns, find portfolio weights which optimize the global return/risk.

Neural networks have been applied extensively to all three stages of investment management. Comprehensive reviews can be found in DeBoeck (1994), Refenes and Azema-Barac (1994), Refenes, Zapranis, and Francis (1994), and Trippi and Turban (1993). Among the numerous contributions, notable are applications in bonds by Dutta and Shashi (1988), in stocks by Shoenenburg (1990), in foreign exchange by Weigend, Rumelhart, and Huberman (1991), and in corporate economics and macroeconomics by Sen, Oliver, and Sen (1992) and White (1994).

Ongoing research and application development at the Neuroforecasting Unit, London Business School, includes projects in tactical asset allocation, futures price sensitivity to volume and open interest, tactical intraday currency trading, factor models for equity investment, modeling and trading concurrent futures indices, and forecasting volatility for option pricing. An example of investment management based on neural networks is described, in the next section. In this description, we deal with the first two phases in the investment management process: factor analysis and the estimation of expected returns.

Tactical Asset Allocation: Neural Networks Versus Regression

Tactical asset allocation refers to the task of allocating funds between different asset classes. The main asset classes considered here are equities, bonds, and cash. This is done on the basis of forecasted (or expected) differences in returns between equities and cash versus (estimated) differences in returns between bonds and cash. Expected returns for $t + 1$ are computed on the basis of the values of several economic (and other) variables at time $t - \tau$ (a lag). We use monthly data on economic variables to estimate differential returns between equities and cash one month ahead. These economic variables are to be selected from a universe of 17 variables and we use stepwise regression (as opposed to discriminant analysis, principal components analysis, autoassociative networks, or other methods) to reduce them to a manageable level. The variables selected through a stepwise regression process are then used to estimate differential returns. For the estimation, we compare the performance of linear regression against a feedforward network. Clearly this comparison is rather unfair to the network since the variables have already been selected in a way that best suits the linear regression model. It is therefore probable that these factors only explain the linear part of the relationship. Nevertheless, since our purpose is to show that even in these restrictive conditions there are considerable nonlinearities, we shall ignore this underlying bias.

Factor Analysis: Stepwise Variable Selection with Backward Regression

In the factor (data) analysis phase, we use multiple linear regression to estimate the linear relationship between the independent variables (x_j^i) and the dependent variable (y). The least-squares technique is used to estimate the coefficients (b_i) in an equation of the form

$$y_{t+1} = b_0 + b_1 x_{t-\tau_1}^1 + b_2 x_{t-\tau_2}^2 + \ldots + b_n x_{t-\tau_n}^n + \epsilon_t \qquad (3)$$

where ϵ denotes a random disturbance term and τ a time lag. The regression coefficient b_i represents the expected change in y associated with a one-unit change in the ith independent variable.

Backward stepwise variable selection was used to determine which of the 17 independent variables to include in the final regression equation which estimates future returns. This procedure starts by including all the variables and then deleting variables one at a time. The criteria for choosing variables are whether they reduce the sum of squared errors the most and have the highest statistical significance.

The regression statistics for the whole universe of 17 input variables are given in Zapranis and Refenes (1994). The t values and significance levels are used to eliminate the low sensitivity variables. After repeated applications, this stepwise procedure retains only three independent variables (X_1, X_7, and X_{14}), which account for 3.13% of the variability in the dependent variable. Clearly, the percentage of explained variability is disappointing although the estimates of the coefficients of the regression equation are statistically significant (the significance levels are ranging from 0 for the constant to 0.1751 for X_{14}). This, however, is not unusual in financial engineering applications, where the benchmark is the random walk.

Before proceeding further (or indeed giving up), it is desirable to test the assumptions of regression to see if it is possible to make any improvements. This is a procedure which should be followed for all estimation methods including neural networks. The main areas of investigations include: (1) testing for serial correlation in the residuals (the presence of which may indicate that there is a systematic error component which could be modeled separately through an error correction term); (2) testing for possible nonlinear transformations of the input variables (the presence of which may lead to an additive regression approach as shown in Equation 1); (3) investigating the effect of influential observations in the data set (the presence of which can be dealt with in various ways, e.g., robust regression methods or simple removal); and (4) investigating the effects of possible nonstationarities in the data.

A detailed description of these tests can be found in Zapranis and Refenes (1994). Briefly, the tests indicate: (1) that there is strong evidence of nonstationarity in the time series; (2) that there are several influential observations both in the form of outliers and in the form of leverage points; and (3) that there is no evidence of nonlinearities in the data or models. However, because the strength of these tests is rather weak, we proceed with the modeling after removing the influential observations and modeling different periods. This leads to improved statistics for the regression. Having identified a reduced set of variables, we next proceed into the second phase of the process: estimating differential returns for $t + 1$ on the basis of factor exposures at time $t - \tau$. Since our primary purpose is to test the hypothesis of linearity, we compare multiple linear regression on the selected variables against a similar estimation with a simple backpropagation network.

Forecasting Differential Returns with Regression and Neural Networks

The evaluation of the two approaches (multiple linear regression and neural networks) is done on the data for the period December 1990 to November 1992, which was withheld for out-of-sample testing. For several multiple linear regression (MLR) and neural network (NN) models (REG0 to REG4 and NET0 to NET4, respectively) we obtained forecasts for the dependent variable in that period. Performance statistics for that period are reported in Table 1.

Table 1. Performance Statistics for Multiple Linear Regression and Neural Networks Models in the Validation Period

	Neural Networks					Multiple Linear Regression			
Model	Training Period	Independ. Variables	Theil's Coeff.	Correl. Coeff.	Model	Model Fit Period	Independ. Variables	Influent. Observ. Removed	Correl. Coeff.
NET0	98 months	all	0.71	0.42	REG0	all except X_6	98 months	no	0.094
NET1	24 months	X_8	0.70	0.43	REG1	X_1, X_7, X_{14}	98 months	no	0.275
NET2	24 months	X_1, X_8, X_{10}, X_{15}	0.64	0.53	REG2	X_3, X_7, X_9, X_{14}	98 months	yes	−0.07
NET3	48 months	X_1, X_3, X_5, X_8	0.61	0.59	REG3	X_1, X_3, X_5, X_8	48 months	no	0.42
NET4	48 months	X_1, X_8, X_{10}, X_{15}	0.61	0.63	REG4	X_8	24 months	no	0.38

In terms of explained variability, the full regression model (REG0) is disappointing (less than 1%). The REG1 model does better out-of-sample than in-sample, but this is a mere fluke because the cross-correlation for X_1 (historical returns) against y is almost 0.6 in the cross-validation period and much smaller elsewhere (in-sample performance statistics are given in Zapranis and Refenes, 1994).

The removal of the influential observations (model REG2) has improved the fit in-sample but resulted in a negative correlation out-of-sample. Models REG3 and REG4 (shorter periods) did much better, explaining what in financial engineering terms is a satisfactory 14.5% to 17.5% of the variability in y.

Since our purpose is to provide a rather conservative comparison between MLR and NN, we deliberately choose a sim-

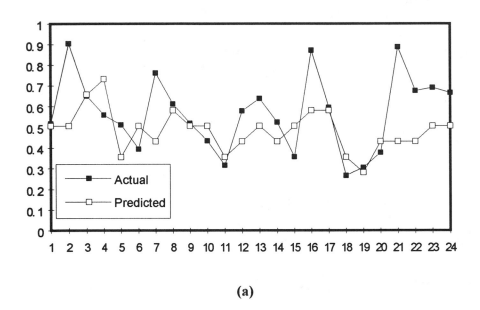

(a)

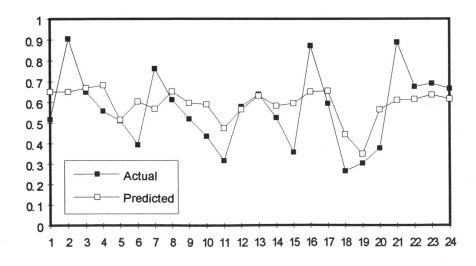

(b)

Figure 1. Actual versus predicted in the validation period. Predictions made by REG4 (*A*) and by NET4 (*B*).

ple network with no error correction terms, ordinary least squares (OLS) cost function, and no robustness against influential observations. Some of these extensions to the OLS procedure are described in Refenes et al. (1994). As with any other gradient descent/ascent algorithm, backpropagation can potentially be trapped in local minima/maxima. Only extensive experimentation can reasonably safeguard against that event. The major parameters of the algorithm are: initial conditions; training time; transfer function; gradient descent control terms; cost function; and topology.

We experimented extensively with training topology and training time, while the rest of the above parameters were fixed. The transfer function was the asymmetric sigmoid with range $(0, 1)$. The cost function was the ordinary quadratic cost function. Learning rate was 0.1 and momentum term also 0.1. The weights were updated after each presentation of a training pattern (continuous update as opposed to batch update).

We trained networks for many different topologies for the three periods. Network performance was evaluated using the following measures: (1) root mean square error (RMS); (2) percentage of change in direction correctly predicted (POCID); (3) Pearson's correlation coefficient between actual and forecasted values; and (4) Theil's coefficient of inequality (T-Coefficient). Detailed figures are given in Zapranis and Refenes (1994).

The problem we are dealing with is characterized by structural instabilities and short system memory. Nonrobust techniques are problematic when applied on long periods and, clearly, ordinary backpropagation is no exception. This becomes apparent by looking at Table 1. In that table, we see measures for out-of-sample performance for networks trained on Periods II and III (Period II: 48-month period ending November 1990; Period III: 24-month period ending November 1990) and tested on Period IV (December 1990 to November 1992). As in the case of MLR, performance is clearly related to the period used for training; but unlike regression it produces much higher correlation coefficients (see Table 1); in fact $\overline{R}^2_{net\,3} = 0.34$ and $\overline{R}^2_{net\,4} = 0.39$. Figure 1 depicts actual versus predicted figures out-of-sample for REG4 and NET4.

Because of the small number of training patterns (either 24 or 48), the number of inputs is kept to a minimum. Although the ratio of patterns to connections is considered to be very low (0.4 to 0.7), this does not seem to be a problem. The inputs are either the ones selected by the stepwise variable elimination procedure, or they were selected on the basis of their cross-correlations with the dependent variable.

In terms of explained variability (correlation squared), the networks trained on short periods do much better than the network trained on the 98-month period (NET0). Especially, the networks with four independent variables give correlations out-of-sample ranging from 0.53 to 0.63. This is equivalent with explained variability in the range of 28% to 39.5%.

In summary, neural network models far outperform multiple linear regression in- and out-of-sample. In this case study, in terms of R^2, we have $\overline{R}^2_{reg} \cong 0.17$ and $\overline{R}^2_{net} \cong 0.35$. Due to nonstationarities in the input-output relationships, the appropriate period for model fitting for both techniques should be kept relatively short (not exceeding 48 months). This restricts the number of variables that can be used as inputs to the model to

a possible maximum of five or six. However, neural networks do very well with as little as four inputs.

Discussion

There are strong reasons to believe that the relationships between asset prices and their determinants are determined by complex nonlinear processes. Neural networks provide a suitable methodology for modeling this type of relationship. However, the development of successful applications is not a straightforward task. It requires the synergetic combination of expertise in financial engineering and network engineering. Many problems of neural modeling remain unresolved. These include robustness to statistical outliers, robustness to discontinuous data and "weak nonstationarity," dealing with serial correlation and multicolinearities in the independent variables, etc. These areas are currently the subject of intensive research in the statistical, mathematical, and econometric sciences. Nevertheless, even in its current state of development, neurotechnology has demonstrated that with careful use, it is capable of providing more accurate models.

Road Map: Applications of Neural Networks
Background: Backpropagation: Basics and New Developments
Related Reading: Applications of Neural Networks; Forecasting

References

Amari, S., 1990, The mathematical foundations of neural computing, *IEEE Trans. Neural Netw.*, 78:1443–1463.

DeBoeck, G., Ed., 1994, *Trading at the Edge*, New York: Wiley.

Dutta, S., and Shashi, S., 1988, Bond rating: A non-conservative application of neural networks, in *Proceedings of the International Conference on Neural Networks, 1988*, 2:443–450.

Hardle, W., 1989, *Applied Non-parametric Regression*, Econometric Society Monographs, Cambridge, Eng.: Cambridge University Press.

Refenes, A. N., and Azema-Barac, M., 1994, Neural network applications in financial asset management, *J. Neural Comput. Applic.*, 2(1): 13–39.

Refenes, A. N., Zapranis, A. D., and Francis, G. , 1994, Stock performance modelling using neural networks, *Neural Netw.*, 7:375–388.

Schoenenburg, E., 1990, Stock price prediction using neural networks: A project report, *Neurocomputing*, 2:17–27.

Sen, T., Oliver, R., and Sen, N., 1994, Predicting corporate mergers using backpropagation neural networks: A comparative study with logistic models, in *Neural Networks in the Capital Markets* (A. N. Refenes, Ed.), New York: Wiley.

Trippi, R. R., and Turban, E., Eds. 1993, *Neural Networks in Finance and Investing*, Chicago: Probus.

Weigend, A. S., Rumelhart, D., and Huberman, B., 1991, Generalisation by weight elimination applied to currency exchange rate prediction, in *Proceedings of the International Joint Conference on Neural Networks*, New York: IEEE Press.

White, H., 1994, Economic prediction: Can neural networks outperform the pros? in *Proceedings of the 14th International Symposium on Forecasting*, Stockholm, Sweden, June 12–15, 1994, p. 65.

Zapranis, A. D., and Refenes, A. N., 1994, Neural networks in Tactical Asset Allocation: Towards a methodology for hypothesis testing and confidence intervals, Technical Report, London Business School, Department of Decision Science.

Ion Channels: Keys to Neuronal Specialization

José Bargas and Elvira Galarraga

Introduction

Voltage spikes, or impulses called "action potentials" (APs), are the major means by which neurons code information and communicate via their axons. Hodgkin and Huxley (1952) established a mathematical model (the HH model; see AXONAL MODELING) to explain how ion conductances in neuron membranes generate APs which propagate along the axon. However, aside from the conductances of the HH model, there are many other types of conductances, and these differ in many respects. Each neuronal class is endowed with a set of conductances.

Complexity has two levels: (1) Ion conductance diversity in a single type of neuron explains intrinsic firing properties. (2) Ion conductance diversity between different types of neurons explains the functional classes of neurons found in the brain (Llinás, 1988). Some neurons fire spontaneously, some show adaptation, some fire in bursts, and so on. With respect to firing (Llinás, 1988) and morphology (Cajal, 1899), there is a *neuronal specialization*. There are many classes of brain neurons (Figure 1).

Operation of Ion Conductances

Ion conductances produce APs. A second role of ion conductances is to set *a particular firing pattern, latency, rhythm, or oscillation* for the firing of these APs. Different classes of neu-

rons have different balances of ion conductances. The result is that *neurons in different regions of the brain have different intrinsic firing properties* (see Figure 1). Therefore, cellular physiology is relevant to questions about the roles of different cerebral nuclei in cerebral function.

If neurons show different firing properties (i.e., threshold, adaptation, pacemaking, bursting, etc.), how does the firing property relate to the class of information processing a given nucleus performs? Do different firing patterns mean different release arrangements at the synaptic terminals and different ways of generating a distribution of synaptic weights? For example, burst firing is found during slow-wave sleep, and single spiking is found during arousal in thalamic relay neurons (see THALAMOCORTICAL OSCILLATIONS IN SLEEP AND WAKEFULNESS). Different brain functions relate to different response patterns of the same neurons in the same nets. Thus, distributed processing of brain functions may be enhanced by channel variety.

As ion channels are responsible for firing patterns, a *functional classification* of ion channels might be based on firing (see Figures 1 and 3). Using modifications of the HH model, different types of ion conductances and firing patterns generated by them may be reproduced. A functional classification does not replace a molecular classification (Hille, 1992). However, it postulates a framework to understand how channel variety gives rise to different firing patterns *within* a single class of neuron, and different firing properties *between* different classes of neurons. Modeling examples populate the literature

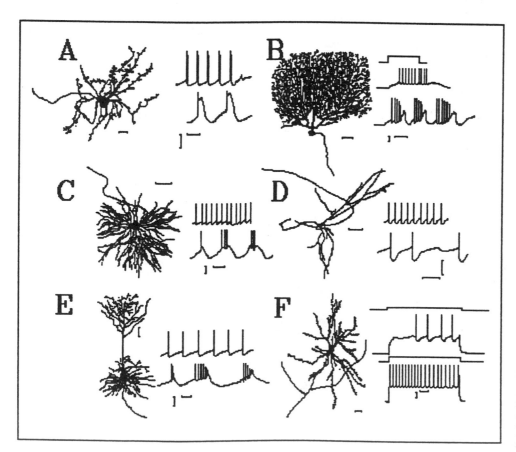

Figure 1. Neurons differ in morphology and function. *A*, Inferior olive neurons fire rhythmic broad spikes. *B*, Cerebellar Purkinje cells fire calcium and sodium spikes. *C*, Thalamic relay neurons fire tonically or in bursts. *D*, Nigra compacta neurons fire spontaneously. *E*, Cortex pyramidal neurons may fire in bursts. *F*, Principal neurons of the neostriatum only fire upon stimulation. (Modified from Bargas, Galarraga, and Surmeier, 1993.)

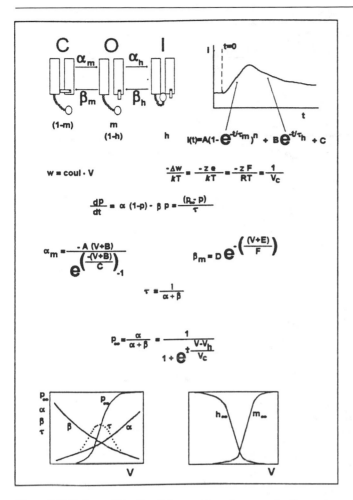

$$K(t) = A(1 - e^{-t/\tau_m})^n + B e^{-t/\tau_h} + C$$

$$w = coul \cdot V$$

$$\frac{-\Delta w}{kT} = \frac{-z e}{kT} = \frac{-z F}{RT} = \frac{1}{V_c}$$

$$\frac{dp}{dt} = \alpha (1-p) - \beta p = \frac{(P_\infty - p)}{\tau}$$

$$\alpha_m = \frac{-A(V+B)}{e^{\left(\frac{-(V+B)}{C}\right)} - 1} \qquad \beta_m = D \, e^{-\left(\frac{(V+E)}{F}\right)}$$

$$\tau = \frac{1}{\alpha + \beta}$$

$$P_\infty = \frac{\alpha}{\alpha + \beta} = \frac{1}{1 + e^{+\frac{V - V_h}{V_c}}}$$

Figure 2. HH-channel kinetics. Scheme at top: A nonconducting molecule (C) with a closed gate (m) may open (O) ("activation") following first-order kinetics with rate constant α_m. After a variable time it may become nonconductive either by "deactivating," i.e., closing m with rate constant β_m, or by "inactivating," i.e., closing h with rate constant α_h. Gate h is depicted as a "ball and chain." "De-inactivation" (I to O) has rate constant β_h. Rate constants can be fitted to exponential solutions multiplied by constants (see equations and plots). Record at top right shows that upward and downward swings of current can be fitted by a sum of exponentials. One of these exponentials describes activation and has τ_m. The other describes inactivation and contains τ_h. Each τ is directly related to the rate constants, and both τ and rate constants can be used to integrate dp/dt. If a steady state is reached ($dp/dt = 0$), p becomes p_∞, which is directly related to rate constants and can be fitted by a cumulative Boltzmann distribution (p = probability). The probabilities for m and h are similar functions except that one is ascending and the other descending.

(e.g., Mironov, 1983; Yamada, Koch, and Adams, 1989; McCormick, Huguenard, and Strowbridge, 1992). Here the emphasis will be on concepts. According to HH, an ion current can be described by

$$I_i = (G_i^* \cdot mh) \cdot (E - E_i) \qquad (1)$$

which is Ohm's law, where I_i is current, $(G_i^* \cdot mh)$ is conductance, and $(E - E_i)$ is voltage (ΔV). The top of Figure 2 shows that channels may be closed (C), open (O), or inactive (I). Only the O state is conductive and contributes to I_i in Equation 1. Any state (C, O, I) depends on membrane potential and time. At certain voltages, the most probable state is closed and $I_i \approx$

0. A voltage change modifies the probability of being in that state. Then, some channels may open, and I_i is recorded (Figure 2, top right).

Any change of state follows rate constants α and β, which define the velocity of the reaction represented by dp/dt in Figure 2, where p stands for m, h, n, etc. in the HH model (Equation 1). Reaction velocity is assessed by recording ion currents and extracting time constants (τ_s). The value τ_m is the time constant for activation, and τ_h is the time constant for inactivation. They can be extracted, for each applied voltage, by fitting exponential functions to current records (Figure 2). Then $\tau_s \approx 1/(\alpha + \beta)$, because state changes are reversible (Figure 2). Small or large τ_s indicates fast or slow activation (or inactivation) of the current, respectively. Instead of using rate constants multiplied by concentrations as in a chemical reaction, the state of a channel population is assessed by conductance measurements. Conductances are measured from the records ($G = \Delta I_i / \Delta V$), normalized, and equated to probabilities with values between 0 (all channels are closed) and 1 (all channels are open; see below). If the probability of the channels being open is indicated by m, then the probability of the channels being closed is $1 - m$. The same applies for the h parameter, i.e., the probability of the channel being inactivated. Voltage-dependent rate constants (α, β) define first-order kinetics and represent simple exponential solutions multiplied by constants (Figure 2). They can be fitted to Boltzmann-type distributions (Jack, Noble, and Tsien, 1975). Also, steady-state functions (p_∞; when $t \to \infty$, $dp/dt = 0$) can be fitted to cumulative Boltzmann distributions. They represent the probability that a number of molecules is in a given state as a function of voltage (see the plots and equation for p_∞, m_∞, and h_∞ in Figure 2). Functions of p_∞ are extracted as a function of conductance from the experimental records, e.g., as permeabilities (P) in terms of the steady-state I–V plot by using the Goldman-Hodgkin-Katz equation for current:

$$P(I, V) = -A \cdot (e^{\Delta V / V_c} - 1) \cdot G \qquad (2)$$

where $A = (RT/z^2 F^2)/(C_o - C_i)$, with C_i and C_o being internal and external ion concentrations, respectively. $G = \Delta I_i / \Delta V$ is obtained in one of several ways, e.g., by differentiating the I–V plot (dI_i/dV) (Jack et al., 1975). The Boltzmann exponent, $-w/kT$, translated to unitary charge or moles (Figure 2), becomes the slope factor for p_∞ ($1/V_c$). Except for a voltage called V_h, in which both states are equally favored, all the other values favor one of the states. The variables V_h, V_c, and τ are used to characterize and classify ion conductances. They vary between conductance types. From p_∞ and τ, rate constants as a function of voltage can be approximated: $p_\infty/\tau = \alpha$. Neurons employ ion currents with different time constants depending upon their functional needs. In both C and I states, channels are nonconductive. To go back and forth from C to O, i.e., for the channel to "activate" and "deactivate," voltage has to change. In contrast, when entering the I state, channels shut off even if voltage is maintained. Thus, conductances are *transient*, or conduct transiently, when they "inactivate." Transient conductances, which "inactivate" and "de-inactivate," have the h parameter or gate. Conductances that do not have the h parameter have only the C and O states; they *persist* until the voltage changes (see Figure 2). The activation exponential function is raised to a power n (which would make activation probability $= m^n$). Thus, the HH model assumes that complexity arises by the parallel action of n simple first-order processes; i.e., n equals the number of particles that have to move, following a first-order process, for the complete channel molecule to open. In fact, each channel molecule is composed of four similar peptides with several additional subunits. Thus, the sigmoidal

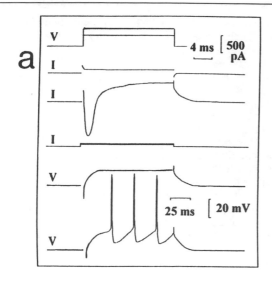

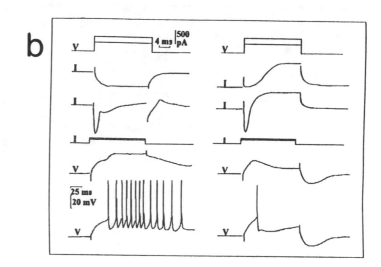

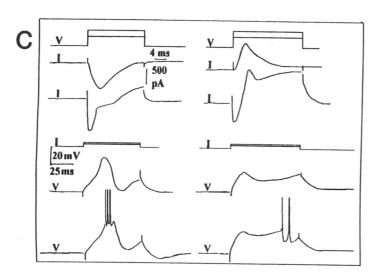

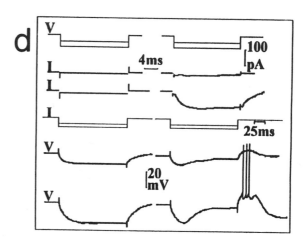

Figure 3. Basic firing machinery ($I_{IT} + I_O + I_{leak}$) in part A, to compare with part B (addition of I_P's) and Part C (addition of I_T's). The variable I_h is depicted in part D. From top to bottom there are six traces in parts A–C: (1) Sub- and suprathreshold V commands. (2) Corresponding subthreshold I responses: I_{leak} in part A, $+I_{IP}$ in part B left , $+I_{OP}$ in part B right, $+I_T$ in part C left, $+I_A$ in part C right. (3) Suprathreshold I responses. (4) Sub- and suprathreshold intracellular I steps. (5) Corre-

sponding subthreshold V responses: RC in part A, + effects of I_{IP} in part B left, + effects of I_{OP} in part B right, + effects of I_T in part C left, + effects of I_A in part C right. (6) Suprathreshold V responses. Note that addition of currents to the basic firing machinery generates a variety of firing patterns. Both passive control and response with I_h are in part D left and right, respectively. See text for abbreviations.

delay produced by n may be viewed as "closed states (C_1, $C_2, \ldots, C_n$) that have to be crossed before entering the open state."

For a given current to be inward or outward depends on the value of E_i or equilibrium potential of the ion i (Equation 1). A value more negative than the threshold for AP firing makes a hyperpolarizing outward current which will tend to arrest firing. A value more positive than threshold makes a depolarizing inward current that will increase or provoke firing. For outward currents, charge carriers are K$^+$ and Cl$^-$ with negative E_i. For inward currents, charge carriers are Na$^+$ and Ca^{2+}, with positive E_i. A neuron has several ion conductances (transient, persistent, inward, outward). The HH model is one of a set. (In Figure 3, compare part A to parts B and C.)

Firing impulses involve inward currents to depolarize, outward currents to repolarize, and leak currents for passive RC

properties. Thereafter, ion conductances are responsible for modifying the threshold, rhythm, rate of adaptation, pace, and frequency of the impulses. The kinetics of a given ion conductance may bestow it with a role in both the firing of an impulse and the afterpotential or pacemaking mechanisms which set the rhythm and frequency of firing. The change in transmembrane potential during firing is due to a sum of conductances, for example:

$$C\frac{dV}{dt} = -\sum_i^n (I_i) = -(\sum I_{OT} + \sum I_{IT} + \sum I_{OP} + \sum I_{IP} + \sum I_{IR})$$

$$(3)$$

where each I_i follows Equation 1, C is cell capacitance, and I_{OT}, I_{IT}, I_{IP}, I_{OP}, and I_{IR} represent basic conductances that one may encounter (see below); e.g., I_{OT} means "outward transient currents." Many neurons have representatives of all these functional "families" in their membranes. However, some neurons lack some of them, and others have several representatives of the same family. In any case, some current or currents dominate the firing behavior at a given moment, and firing might resemble one of the basic types illustrated in Figure 3.

Firing can be seen as a modified HH model with additional conductances. Conductances can be added with HH formalisms, differing in kinetic parameters ($\alpha, \beta, p_\infty, V_c, V_h$, or Ca dependency) (e.g., Mironov, 1983; Yamada et al., 1989; McCormick et al., 1992). Examples are pictured in Figure 3. (Figure 3A is a basic firing mechanism to which parts B and C should be compared.)

Ion Conductances Found in Neurons

A neuron may possess 5 to 10 significant conductances in terms of firing pattern (Llinás, 1988). For illustrative purposes conductances are divided here into five "families" according to the way they are activated, the direction of charge flow, and the firing contribution most mentioned in the experimental literature (we abstract from ion species, biochemical structure, etc.). For simplicity, currents are illustrated as a part of the subthreshold response (see Figure 3) (see Bargas, Galarraga, and Surmeier, 1993, for a list of examples of conductances).

Persistent or Slowly Inactivating Inward Currents

Slowly inactivating or *persistent* means that these currents either do not inactivate (do not have the h gate), or, if they do, the inactivation time constant τ_h is very large (hundreds of milliseconds or seconds). A persistent inward current (I_{IP}, being I_{Na}, I_{Ca}, etc.) is elicited by a depolarizing stimulus (cf. Figure 3, parts A and B left). The value of I_{IP} increases with the stimulus and produces a depolarization that adds to stimulus depolarization and enhances it. This action produces a higher frequency response. The additional depolarization persists after the stimulus and can contribute to *bistability* and negative slope resistance regions (Jack et al., 1975).

The neuron may stay for long periods in a *plateau potential* more depolarized than the *resting potential*. This potential can maintain firing with no stimulus. There is a gain in input-output function. There are several examples of I_{IP}'s.

In Purkinje cells of the cerebellum, there is a TTX-sensitive sodium current ($I_{Na\,P}$). In motoneurons there is a low-threshold, dihydropyridine-sensitive calcium current ($I_{Ca(L)}$). In motoneurons bistability is a "modulated" function; i.e., it can be turned on or off according to brain chemistry. A positive feedback during I_{IP} may be obtained: Each depolarization activates more I_{IP} which causes more depolarization and so on, until all-or-nothing spikes different from somatodendritic action potentials can be obtained. Accordingly, dendritic or high-threshold spikes are attributed to I_{IP}'s.

There are many classes of I_{IP}'s, and they have a complex pharmacology. Among the most intriguing are several classes of calcium conductances. Entry of calcium has significance to cell dynamics in transmitter release, muscle contraction, channel activation or inactivation, cytoskeleton function, enzyme activation, and gene activation. So persistent Ca^{2+} currents play numerous roles in neuronal function.

Persistent or Slowly Inactivating Outward Currents

Persistent outward currents (I_{OP}) oppose depolarizing stimuli and inward currents. They decrease excitability by augmenting firing threshold (cf. Figure 3, parts A and B right). They comprise a negative feedback system: Depolarization activates the current; the current hyperpolarizes the membrane; and then the current shuts down. Persistent outward currents are associated with prolonged afterhyperpolarizations and with lowering and adaptation of the firing frequency.

Tonic firing becomes adapting firing if a slow I_{OP} is superimposed, as it is the case in sympathetic neurons. Conversely, if I_{OP} is blocked, a frequency gain for the same stimulus and a decrease in the firing threshold are expected. One example is the block, by acetylcholine, of a persistent outward current (I_M). Therefore, I_{OP}'s are mainly responsible for *frequency and threshold control*. Gating of I_{OP}'s may depend on membrane voltage, intracellular calcium, both voltage and $[Ca^{2+}]_i$, intracellular sodium, and other factors (Schwindt, 1992). A purely voltage-dependent gating opens and closes a fraction of channels with each action potential. A purely calcium-dependent gating opens a fraction of channels depending on the number of spikes fired (digital to analog conversion). With each spike, calcium enters the cell and accumulates in the cytoplasm. Hence, dependency on intracellular calcium confers cumulative properties, and firing is influenced by previous history. Transmitters use intracellular signaling pathways that involve changes in intracellular calcium and thus modulate firing pattern.

Afterhyperpolarizations (AHPs) after a spike, a train, or a burst are sensitive to modulators. Afterhyperpolarizations end episodes of increased excitability, set the pace for rhythmic bursting, and are responsible for frequency gain and adaptation (Connor, 1985). Several I_{OP}'s may participate together in AHPs. This fact indicates the importance of modifying spike threshold, input-output functions, and adaptation. As a result, old classifications as "slow adapters" and "fast adapters" are being revised: The same neuron can do both at different moments depending on modulation and previous membrane potential (Nicoll, 1988; Schwindt, 1992). Therefore, network states depend not only on synaptic arrangements but also on dynamic properties of neurons (see "Discussion").

Many I_{OP}'s, as potassium currents (I_K's), are being discovered. Many I_K's link cell metabolism with membrane potential in several ways. Channel classification on a genomic and molecular basis would be as voltage-gated, ligand-gated, and metabolically gated, because gating of many channels depends on transmembrane gradients of $Ca^{2+}, Na^+, K^+, H^+, H_2O_2$, NO, free radicals, and ATP. A change in some of these gradients changes gene transcription.

Transient or Fast-Inactivating Inward Currents

Transient currents inactivate quickly, meaning that τ_h is small. Transient inward currents (I_{IT}'s) are among the first active conductances to be described: they produce spikes. A transient I_{Na} makes up the upward swing of the action potential (HH model). A transient I_{Ca} (I_T) has been described (see Figure 3C left). This conductance produces low-threshold spikes (cf. Figure 3, parts A and C left) that may trigger sodium action potentials and burst firing. Varieties of I_T have been reported. Thalamic, nigral, pontine, cortical, and other neurons use I_T to fire in bursts. Moreover, I_T is in part responsible for *rebound excitation* or postinhibitory afterdischarges. When the membrane potential is rather positive, T channels inactivate and do not participate. Neurons with I_T fire single spikes at depolarized potentials and with bursts at hyperpolarized potentials. *Thus, neurons respond differently depending on previous membrane potential, and may not fire at potentials in between. These differences allow a net to behave differently at different moments, while using the same neuronal elements. This process may be a substrate for the ability to produce different tasks with the same* net (multitask networks; see also NEUROMODULATION IN INVERTEBRATE NERVOUS SYSTEMS).

Note that if full-amplitude action potentials can be evoked from two widely separated membrane potentials, and not from potentials in between, then the neuron has two different spike firing thresholds. Thus, *a neuron may have more than one threshold*, and different firing patterns may be evoked from each threshold. Transient currents may not produce all-or-nothing Ca^{2+} spikes but, instead, regularly spaced depolarizations which contribute to pacemaking. A single cell may exhibit a variety of firing patterns due to I_T.

Transient or Fast-Inactivating Outward Currents

Transient outward currents (I_{OT}'s or I_A's) represent the opposite of I_{IT}'s or I_T's. Accordingly, rhythmic firing and spike repolarization are also main functions envisioned for I_A's. They inactivate between spike intervals, and this inactivation suits them for setting the firing pace. They oppose the depolarization at the beginning of the stimulus (cf. Figure 3, parts *A* and *C* right) and "delay" the membrane trajectory toward threshold. If stimulus strength is weak, the cell may not respond. However, if the same stimulus lasts a certain amount of time, the cell begins to respond because I_A inactivates when the stimulus lasts long enough. Therefore, with this inactivating current *the cell acquires a sort of "time threshold."*

Facilitation is the increase in the amplitude of a postsynaptic response caused by activity in the presynaptic neuron. It is caused by an increase in calcium inside the presynaptic terminals. However, a "conditioning," repetitive, weak stimulation would progressively inactivate I_A, and this response by itself would increase the amplitude of the postsynaptic response. Thus, a previously subthreshold stimulus can reach threshold after I_A inactivation as a result of repeated stimulation. The response would be "facilitated" by an intrinsic membrane property of the postsynaptic element.

Because neurons may have a variety of I_O's with different τ_h's for inactivation, *neurons may have more than one time constant for latency, pacemaking, frequency adaptation*, and so on. The same ion conductance may possess fast ("N") and slowly inactivating ("C") kinetics.

Inward Rectifiers

Anomalous or *inward rectifiers* (I_{IR}'s) are activated by *hyperpolarization*. Inward rectifiers generate inward current to oppose a hyperpolarizing stimulus. They comprise a negative feedback system "correcting" extreme hyperpolarization. They deactivate upon depolarization. There are different classes. The *fast* or *instantaneous* inward rectifier (I_{FIR}) uses K^+ as charge carrier. Its current becomes inward once the hyperpolarization surpasses E_K, but significant conductance change occurs at potentials positive to E_K. These currents contribute to resting potential. Their activation depends on membrane potential and $[K^+]_o$, and these variables depend on the network's level of excitability. Thanks to this conductance, excitability may produce dynamic changes in neuronal input resistance and, hence, in the electrotonic distance (L) of the dendritic arbor (Wilson, 1992; see DENDRITIC PROCESSING). Hence, synaptic integration is a function of the excitability level of the network.

A second type is *time dependent*. It is clearly detected as an inward current upon hyperpolarization (I_h) (cf. Figure 3*D* left and right). It is carried by Na^+ and K^+. It is manifested as a "sag" that opposes the hyperpolarization and as a rebound "hump" when the hyperpolarization is over. This rebound may attain firing threshold. In many cells, I_h and I_T collaborate (McCormick et al., 1992), producing strong bursts. Another type is carried by Cl^-.

The I_h current is ideal for pacemaking. Afterhyperpolarizations after a spike or burst activate I_h. Then, I_h opposes these same AHPs and returns the membrane potential to inward current threshold. This response keeps the cell firing. Therefore, the whole phase-plane orbit, including the spike, can be initiated by a hyperpolarization. Then, the cell has a threshold in the hyperpolarizing direction (another unstable singular point that initiates the entire cycle; see Rinzel and Ermentrout, 1989).

Discussion

Cellular physiology brings complexity to our picture of neurons and nets. Multiple ion conductances endow distinct neuronal types with peculiar firing properties: multiple changing firing thresholds, various firing patterns associated with each threshold, the possibility of switching between different firing patterns, "time thresholds," multiple time constants for spike frequency adaptation and latency, a dynamically changing electrotonic length, pacemaking, intrinsic facilitation, and so on. A given ion conductance contributes to one or more properties. However, a change in afferent inputs, membrane potential, or chemical environment may be sufficient to transform the firing characteristics of a neuron. A new subset of conductances shaping a completely different firing pattern may become dominant because of prevailing activity or brain chemistry; for example, a slowly adapting neuron may become a rhythmic burster. Therefore, *neurons in neuronal nets should be seen as dynamically changing between different states giving rise to different net configurations*. Such changes increase the number of possible net states and open the possibility that some forms of learning do *not* occur as changes in synaptic weights.

Once the net is set to work and synaptic weights begin to change, the input-output functions of the neurons also change. Neuronal nets would "evolve" from one configuration to another depending on input regularity, previous history, or neuromodulation.

New formal neurons with *n* thresholds due to *n* ion conductances activated at each threshold would trigger a variety of firing patterns. Each firing pattern would encompass a given input-output relationship. To favor a given input-output relationship one would favor one of the *n* thresholds by driving the working potential near to that threshold. This purpose may be achieved using external potential sources (*neuromodulation*). A given input-output relation may favor learning, while another may favor consolidation. Some configurations may link some nets because of similar or resonant frequencies; other configurations may disconnect those nets. Some would be preferred by the sensory neurons, others by intermediate or motor neurons, and so on. Second messengers and slow changes in intrinsic firing properties could be simulated by "learning-evolving rules." As second-messenger cascades are disclosed, together with their time frames and effects on ion conductances, these rules will be made explicit.

Acknowledgment. This work was supported by DGAPA-UNAM (México).

Road Map: Biological Neurons
Background: Axonal Modeling
Related Reading: Dynamic Clamp: Computer-Neural Hybrids; Oscillatory and Bursting Properties of Neurons; Synaptic Currents, Neuromodulation, and Kinetic Models

References

Bargas, J., Galarraga, E., and Surmeier, D. J., 1993, Neuronal networks of the mammalian brain have functionally different classes of neurons, in *Neuroscience: From Neural Networks to Artificial Intelligence* (P. Rudomín, M. A. Arbib, F. Cervantes, and R. Romo, Eds.), Heidelberg: Springer-Verlag, pp. 3–32. ◆

Cajal, S. R., 1899, *Textura del sistema nervioso del hombre y de los vertebrados*, Madrid: N. Moya. ◆

Connor, J. A., 1985, Neural pacemakers and rhythmicity, *Annu. Rev. Physiol.*, 47:17–28.

Hille, B., 1992, *Ionic Channels of Excitable Membranes*, Sunderland, MA: Sinauer Associates. ◆

Hodgkin, A. L., and Huxley, A. F., 1952, A quantitative description of membrane current and its application to conduction and excitation in nerve, *J. Physiol. (Lond.)*, 117:500–544.

Jack, J. J. B., Noble, D., and Tsien, R. W., 1975, *Electric Current Flow in Excitable Cells*, London: Oxford University Press. ◆

Llinás, R., 1988, The intrinsic electrophysiological properties of mammalian neurons: Insights into central nervous system function, *Science*, 242:1654–1664. ◆

McCormick, D. A., Huguenard, J., and Strowbridge, B. W., 1992, Determination of state-dependent processing in thalamus by single neuron properties and neuromodulators, in *Single Neuron Computation* (T. McKenna, J. Davis, and S. F. Zornetzer, Eds.), Boston: Academic Press, pp. 259–290.

Mironov, S. I., 1983, Model investigation of firing mechanisms of mammalian neurons, *Neirofiziologiia*, 16:445–451.

Nicoll, R. A., 1988, The coupling of neurotransmitter receptors to ion channels in the brain, *Science*, 241:545–551. ◆

Rinzel, J., and Ermentrout, G. B., 1989, Analysis of neural excitability and oscillations, in *Methods in Neuronal Modeling* (C. Koch and I. Segev, Eds.), Cambridge, MA: MIT Press, pp. 135–169.

Schwindt, P., 1992, Ionic currents governing input-output relations in Betz cells, in *Single Neuron Computation* (T. McKenna, J. Davis, and S. F. Zornetzer, Eds.), Boston: Academic Press, pp. 235–258.

Wilson, C. J., 1992, Dendritic morphology, inward rectification, and the functional properties of neostriatal neurons, in *Single Neuron Computation* (T. McKenna, J. Davis, and S. F. Zornetzer, Eds.), Boston: Academic Press, pp. 141–171.

Yamada, W. M., Koch, C., and Adams, P. R., 1989, Multiple channels and calcium dynamics, in *Methods in Neuronal Modeling* (C. Koch and I. Segev, Eds.), Cambridge, MA: MIT Press, pp. 97–133.

Kolmogorov's Theorem

Věra Kůrková

Introduction

In a famous 1900 lecture, David Hilbert conjectured that a multivariable continuous function need not, in general, be decomposable into a finite superposition of continuous functions of a lesser number of variables. This conjecture was disproved a half century later by A. N. Kolmogorov (1957) in his superposition theorem. De Figueiredo (1980) suggested an engineering implication for this deep mathematical result in overcoming the "curse of dimensionality." In studying the capabilities of multilayer feedforward networks to approximate continuous functions, this article will survey the following applications of this powerful mathematical tool: a proof of the "universal approximation capability," an estimate of the number of hidden units, and a noniterative learning algorithm.

Kolmogorov's Theorem and Neural Networks

It has been more than 20 years since Minsky and Papert (see Section I.3) conclusively demonstrated that the simple perceptron (with no hidden layer) is incapable of usefully representing or approximating functions outside a very narrow and special class. Although Minsky and Papert left open the possibility that multilayer networks (having at least one hidden layer) might be capable of better performance, it has only been in the last several years that researchers have begun to seriously explore the ability of multilayer perceptron-type networks to approximate general mappings from one finite dimensional space to another. The apparent ability of sufficiently elaborate hierarchies of perceptrons to approximate quite well nearly any function encountered in applications leads one to wonder about the ultimate capabilities of such networks.

As a way of proving the capability of multilayer perceptron networks to approximate arbitrarily well any continuous function, Hecht-Nielsen (1987) suggested using the superposition theorem in Kolmogorov (1957). This theorem answers a conjecture of Hilbert formulated as the 13th of his famous list of 23 open problems presented at the Second International Congress of Mathematics held in Paris in 1900.

The 13th problem, although formulated as a somewhat minor concrete hypothesis, can be interpreted analogously with a basic problem of algebra—representing the solution of polynomial equations in terms of radicals. Instead, Hilbert wondered whether roots of a general algebraic equation of higher degree can be expressed by sums and compositions of one-variable functions of some suitable type. Perhaps influenced by Niels Henrik Abel's results, Hilbert believed that this was not possible in general and conjectured specifically that the roots of the equation $x^7 + ax^3 + bx^2 + cx + 1 = 0$ as functions of the three coefficients a, b, and c are not representable by sums and superpositions even of functions of two variables.

This specific case was refuted by Arnold (1957), and the problem is completely settled by the general representation theorem of Arnold's teacher, Kolmogorov (1957), stating that any continuous function f defined on an n-dimensional cube is representable by sums and superpositions of continuous functions of only one variable. Kolmogorov's formula

$$f(x_1, \ldots, x_n) = \sum_{q=1}^{2n+1} \varphi_q \left[\sum_{p=1}^{n} \psi_{pq}(x_p) \right]$$

readily brings to mind perceptron-type networks with the reservation that the one-variable functions φ_q $(q = 1, \ldots, 2n + 1)$ and ψ_{pq} $(p = 1, \ldots, n; q = 1, \ldots, 2n + 1)$ are far from being any of the types of functions currently used in neurocomputing. In fact, having even fractal graphs, they are highly nonsmooth. (This fact may also explain the failure of Hilbert's intuition—functions with fractal graphs were supposed to be pathological then.)

Highly nonsmooth functions encountered in mathematics are mostly constructed as limits or sums of infinite series of smooth functions. For example, the classical Weierstrass function has no derivative at any point, and there are many other functions with fractal graphs, including Kolmogorov's functions.

Staircase-like functions constructed using any increasing continuous sigmoidal function have the pleasant property that they can approximate any continuous function on any closed interval with an arbitrary accuracy. Using this argument, Funahashi (1989) derived from Kolmogorov's theorem the capability of a class of networks with two hidden layers containing perceptrons (in the sense of "neurons" with any continuous increasing sigmoidal response function) to approximate continuous functions.

Kůrková (1992a) showed that with the goal of approximation, Kolmogorov's construction can be simplified and based on functions constructed from generalized sigmoidals which need not even be continuous or increasing. To illustrate her argument, consider for n staircase-like functions ψ_p $(p = 1, \ldots, n)$, constructed using a sigmoidal function σ, the function Ψ defined on the n-dimensional unit cube by $\Psi(x_1, \ldots, x_n) = \Sigma_{p=1}^n \psi_p(x_p)$. The function Ψ defines on the cube a Rubik's cube–like structure with small boxes having edges corresponding to the steps of ψ_p and gaps corresponding to the slopes of ψ_p. Choosing a suitable set of heights of steps, we can guarantee that Ψ maps small boxes into closed mutually disjoint subintervals of the real line. This allows us to ascribe to these intervals values of the function $f: I^n \to \mathbb{R}$ being approximated at some chosen points in the small boxes. So we obtain a finite family of steps that can be approximated arbitrarily well by a staircase-like function φ constructed using σ.

The function $\varphi \circ \Psi$ approximates f on the union of all small boxes. To extend the approximation to include the gaps, it is sufficient to take more Rubik's cube–like structures that are mutually shifted. The finer the small boxes, the better the approximation. Note that, in fact, the size of the small boxes needed for a given accuracy of approximation depends on the modulus of continuity of f which is the function $\omega_f : (0, \infty) \to \mathbb{R}$ such that $\omega_f(\delta) = \sup\{|f(\mathbf{x}) - f(\mathbf{y})|: \mathbf{x}, \mathbf{y} \in I^n$ and $|x_p - y_p| < \delta$ for $p = 1, \ldots, n\}$.

Since each step of one of the staircase-like functions involved in such a construction corresponds to one hidden unit, we can estimate their number. Moreover, the construction provides a universal set of weights and biases for a network capable of approximating all functions from a certain set (characterized by a bounded norm and modulus of continuity) with a given accuracy so that these weights could be hardwired and only weights corresponding to the output units need to be learned. Nees (1993) developed and tested a noniterative learning algorithm based on these ideas. She estimated both modulus of continuity and norm from a finite set of input/output data.

Discussion

Approximation capabilities for classes of networks with only one hidden layer containing perceptrons with more or less general activation function were proved elegantly using advanced theorems from functional analysis by several authors (e.g., Cybenko, 1989; Hornik, Stinchcombe, and White, 1989). Recently, Leshno et al. (1993) characterized activation functions having the so-called universal approximation property as nonpolynomial. Also proofs of similar approximation capabilities for networks with types of units other than perceptrons, such as radial basis function units and semilocal units, were derived (Girosi and Poggio, 1990; Park and Sandberg, 1991; Kůrková, 1992b).

Most of these proofs are not direct and so cannot be used to estimate the number of hidden units needed for a given level of accuracy in approximation. However, the direct approach in Kůrková (1992a), being based on Kolmogorov's theorem which is constructive, leads to estimates on the rate of approximation which depend on the modulus of continuity. Her estimates grow exponentially with the dimension of the input space as in traditional series expansions.

Recently Jones (1992) proposed the application of his *relaxed algorithm* to neural networks, guaranteeing an integrated square error independent of the input dimension. The relaxed algorithm is a way to quickly approximate functions in the closure of the convex hull of a bounded subset of a Hilbert space. Barron (1993) described spaces of functions that can be approximated by Jones's algorithm using functions computable by single-hidden-layer networks with perceptrons. For such functions, integrated square error of $O(1/k)$ can be achieved by networks with k hidden units independently of the input dimension.

However, the apparent avoidance of the "curse of dimensionality" is due to the fact that these function spaces are more and more constrained as the input dimension increases. Also, a constant may dominate the $1/k$ factor. The relationship between one- and two-hidden-layer networks is not yet well understood, but there are simple examples like XOR, where exact implementation by a two-hidden-layer network, based on Kolmogorov's construction, uses very few hidden units.

Acknowledgment.   The author was supported by GACR grant 201/93/0427.

Road Map: Learning in Artificial Neural Networks, Deterministic
Background: I.3. Dynamics and Adaptation in Neural Networks
Related Reading: Identification and Control

References

Arnold, V. I., 1957, On functions of three variables, *Dokl. Akad. Nauk SSSR*, 114:679–681.

Barron, A. R., 1993, Universal approximation bounds for superpositions of a sigmoidal function, *IEEE Trans. Inform. Theory*, 39:930–945. ◆

Cybenko, G., 1989, Approximation by superpositions of a single function, *Math. Control Signals Systems*, 2:303–314.

de Figueiredo, R. J. P., 1980, Implications and applications of Kolmogorov's superposition theorem, *IEEE Trans. Automat. Control*, 25:1227–1230. ◆

Funahashi, K., 1989, On the approximate realization of continuous mappings by neural networks, *Neural Netw.*, 2:183–192.

Girosi, F., and Poggio, T., 1990, Networks and the best approximation property, *Biol. Cybern.*, 63:169–176.

Hecht-Nielsen, R., 1987, Kolmogorov's mapping neural network existence theorem, in *Proceedings of the International Conference on Neural Networks*, New York: IEEE Press, vol. 3, pp. 11–14.

Hornik, K., Stinchcombe, M., and White, H., 1989, Multilayer feedforward networks are universal approximators, *Neural Netw.*, 2:359–366. ◆

Jones, L. K., 1992, A simple lemma on greedy approximation in Hilbert space and convergence rates for projection pursuit regression and neural networks training, *Ann. Statist.*, 20:608–613.

Kolmogorov, A. N., 1957, On the representations of continuous functions of many variables by superpositions of continuous functions of one variable and addition, *Dokl. Akad. Nauk SSSR*, 114:953–956.

Kůrková, V., 1992a, Kolmogorov's theorem and multilayer neural networks, *Neural Netw.*, 5:501–506.

Kůrková, V., 1992b, Universal approximation using feedforward neural networks with Gaussian bar units, in *Proceedings of ECAI'92*, Chichester, Eng.: Wiley, pp. 193–197.

Leshno, M., Lin, V., Pinkus, A., and Schocken, S., 1993, Multilayer feedforward networks with a non-polynomial activation function can approximate any function, *Neural Netw.*, 6:861–867.

Nees, M., 1993, Chebyshev approximation of multivariate functions by discrete superposition, Research report, Erlangen, Ger.: Mathematisches Institut Erlangen.

Park, J., and Sandberg, I. W., 1991, Universal approximation using radial-basis-function networks, *Neural Computat.*, 3:246–257.

Language Acquisition

Kim Plunkett

Introduction

Much of the language acquisition literature has been more descriptive than explanatory. Hypotheses concerning the mechanisms that drive linguistic development have not generally been sufficiently well formulated to test against performance data.

Most explanatory work has concentrated on specifying the nature of the grammars that children acquire, to the virtual neglect of the lexicon and sociocognitive dimensions of development. Connectionist modeling of language acquisition offers an opportunity to rectify this situation (see Plunkett, 1995). How much innate structure is needed to get the language learning process off the ground? Advocates of the "poverty of the stimulus" argument propose that the stimulus to which children are exposed when learning a language does not contain sufficient explicit information to account for the rich internal representations which underlie mature adult usage of language. Consequently, the learning process must be constrained by factors, endogenous to the child, which have a specifically linguistic character—e.g., a Universal Grammar (Chomsky, 1965). In contrast, developmental connectionists suggest that rich linguistic representations can emerge from the interaction of a relatively simple learning device and a structured linguistic environment. In this article, I review some connectionist models of inflectional morphology, lexical development, and syntax acquisition.

Inflectional Morphology

Language acquisition researchers have long been aware that children make overregularization errors on forms that do not take the regular inflection in the language. For example, in learning the English plural, children may spontaneously overregularize the /s/ suffix to irregular nouns to produce errors like /sheeps/. This erroneous behavior has been interpreted as resulting from children's discovery that the plural or past tense form obeys a rule, such as "Add an /s/" to the singular forms. On making this discovery, children apply the rule to irregular forms that they have previously produced correctly, hence, the onset of overregularization errors. This process of development leads to a U-shaped profile in children's performance. They start producing irregular forms correctly (presumably because they have memorized them), they discover the regular rule and begin producing overregularization errors, and finally the children sort out which forms are not governed by the rule and thereby eliminate the overregularization errors.

Rumelhart and McClelland (1986, referred to as R&M) implemented a verb learning model that mimicked this behavior (see LINGUISTIC MORPHOLOGY). The model consisted of a feedforward network that transformed a phonological representation of the stem of a verb into a phonological representation of the past tense of the same verb. The network learns the relation between stems and past tense forms through a training process. The model produced the well-known U-shaped curve that characterizes children's performance on irregular verbs in which they move from early successful performance to erroneous performance and finally to the mature adult state. After initial enthusiasm from acquisitionists, the R&M model came under detailed scrutiny in an important review by Pinker and Prince (1988), who pointed out a number of shortcomings of the model. From a developmental perspective, the most important shortcoming seemed to be that R&M had introduced a

discontinuity in the training set to which the network was exposed. Initially, they trained the network on just 10 verbs, eight of which were irregular and just two regular. This manipulation was deliberately designed to capture frequency characteristics of verbs in the language: Irregular verbs are, on the whole, more frequent than regular verbs. After the network had received some training on this initial set of 10 verbs, the training vocabulary was expanded by 410 verbs. Most of the new verbs were regular verbs. It was at this point in training that many of the overregularization errors were observed. However, Pinker and Prince argued that there was no evidence to indicate that children experienced a similar discontinuity in their verb vocabulary development. Hence the assumptions of the model were undermined and the findings brought into question.

It is relatively easy to show (as did Plunkett and Marchman, 1991, referred to as P&M) that a network exposed to a constant diet of regular and irregular verbs will still produce overregularization errors. Plunkett and Marchman found that their model produced a micro U-shaped profile of overregularization errors. This result may seem to contradict the textbook story of the developmental profile of overregularization errors. More recently, however, Marcus et al. (1992) have shown that overregularization errors are far more infrequent in children than originally supposed and that they are spread out over a much wider period of development, essentially confirming the micro U-shaped predictions of the P&M model. In a later model of past tense acquisition which attempted to replicate the conditions of development more closely, Plunkett and Marchman (1993) exposed a network to an incremental training regime in which the number of verbs increased gradually over time. This manipulation was intended to capture the fact that children do not seem to try to learn the language all at once and that the number of verbs in their productive vocabularies increases gradually over time (Marchman and Bates, 1994). The profile of overregularization errors produced in this later simulation was similar to that observed by Marcus et al. (1992) for many of the 83 children whose data were analysed. P&M discovered that the network exhibited a critical mass effect—i.e., although the number of verbs fed into the network was increased only gradually, the tendency of the network to generalize to novel stems with an /ed/ suffix increased in a nonlinear fashion.

P&M showed that the timing of the critical mass effect was directly related to a composite measure of the absolute number of regular verbs and the balance of regular and irregular verbs in the current training regime. Hence, the model predicted that the onset of overregularization errors in children should be closely related to the composition of their verb vocabularies, e.g., the relative number of regular and irregular verbs. Although Marcus et al. (1992) found no evidence for such a relationship in their study, Marchman and Bates (1994) analyzed data from a much larger number of children over a wider developmental span and confirmed the critical mass predictions of the model.

Pinker and Prince (1988) also pointed out a number of other difficulties with the original R&M model, including the implausibility of the Wickelphone representation used and the difficulty the model has dealing with verb homophones such as the verb /ring/, which can take several past tense forms (/rang/, /rung/). Since the model is exposed only to phonological information, it has no way of knowing which meaning of /ring/ is intended and hence is unable to determine which past tense form is required. MacWhinney and Leinbach (1991) constructed a connectionist model of past tense acquisition which addressed many of the problems raised by Pinker and Prince. In a similar vein, Cottrell and Plunkett (in press) implemented a connectionist model of past tense acquisition which took a semantic representation of the verb and transformed it into a phonological representation of either the stem or the past tense form. They showed that connectionist models could take verbs with similar semantic representations (e.g., /hit/, /strike/, /slap/) and yet apply quite different processes of past tense formation to these verbs. Other issues raised by Pinker and Prince (1988) and others (Kim et al., 1991), such as the fate of denominal verbs or compound forms with regular and irregular constituents, have not yet been adequately addressed by connectionist models.

One final issue that arises from the application of connectionist networks to learning inflectional morphology, concerns the problem of minority default mappings. We have already seen that connectionist models of the English past tense exploit the fact that the number of regular verbs in English greatly outnumbers the irregular verbs. The process of past tense formation associated with regular verbs becomes the default associated with novel forms and best characterizes the type of errors observed. However, there exist languages with inflectional systems where no single inflectional process dominates the system. For example, both Arabic and German have plural systems in which the default that characterizes speakers' responses to novel forms and the errors they make constitute only a minority of forms in the system—the so-called "sound plural" in Arabic and the /-s/ plural in German. At first blush, this fact would appear to embarrass any connectionist approach to explaining its acquisition: novel forms are less likely to resemble these minority inflectional processes than others in the language and so should not be treated in a minority default fashion. It turns out that connectionist networks are perfectly capable of implementing the acquisition of minority default mappings (Forrester and Plunkett, 1994). Multilayered networks can, through training, construct internal representations of the input which radically transform the similarity structure of the training set. For example, inputs which look very similar need not be similar at the level of internal representation.

Consequently, novel forms can be transformed to resemble the minority class in the language at the level of internal representation. This nonlinear property of multilayered networks enables them to learn minority default mappings. Furthermore, network models set to learn such inflectional systems generate empirical predictions concerning their acquisition in children which can then be tested.

Lexical Development

One of the most widely cited facts of early language acquisition concerns the dramatic change in rate of vocabulary growth observed in many children around 21 months of age—the so-called vocabulary spurt. Since most of the new words acquired by children at this point are names for objects, the vocabulary spurt has often been interpreted as a naming insight, though other explanations relating to children's developing phonological skills or semantic representations have also been proposed. Child language researchers have also focused on over- and underextension errors produced by children. Mastery of, say, the word *dog* does not necessarily indicate that children have correctly represented its meaning. They may use the word not only to talk about dogs, but about other creatures such as cats, cows, and horses (overextension errors), or they may restrict themselves to using the word *dog* only in relation to dogs which

are barking, or the family dog (underextension errors). These patterns of behavior have received a variety of theoretical interpretations.

Another fact of language acquisition concerns the comprehension/production asymmetry. Children seem to be able to understand words that they do not themselves actively produce. A similar asymmetry holds for adults' receptive and expressive vocabularies. The existence of this asymmetry has prompted some researchers to postulate the existence of separate cognitive mechanisms for the processes of comprehension and production. Other researchers maintain that comprehension and production exploit the same underlying knowledge base and that any asymmetry can be explained in terms of peculiarities of the task being performed. For example, production involves an encoding process whereas comprehension only seems to involve decoding.

Expanding on a model originally implemented by Yves Chauvin, Plunkett et al. (1992) trained a network to associate images with labels. The images consisted of random dot patterns generated from a set of prototypes. A total of 32 prototypes was used and six distortions (low, medium, and high level) were generated from each prototype. The distortions were of such a size that the high-level distortions generated from one prototype could be more similar to another prototype than its own. Hence, although the image space consisted of a "natural" clustering of distortions, these clusters were often ambiguous as far as the high-level distortions were concerned. All the distortions generated from a given prototype were labeled identically. The ambiguity inherent in the image space could thereby be resolved by learning the appropriate label for the distortion.

The network used to learn these image/label associations was an autoassociator. It took image/label pairs as input and was trained to reproduce these pairs at the output. Separate processing channels were used for the image and label inputs. These channels then converged on a single layer of hidden units, which were thereby forced to construct a composite representation of the image and label pairs. Two sets of output units (image and label) were driven directly by these composite representations. Use of a trained autoassociator permits evaluation of network performance using pattern completion. For example, it is possible to present the network with just the image and observe what label is generated; this is "production" in the model. Alternatively, one can measure comprehension by presenting the network with just the label and observing which image is generated. The network was trained only on the image distortions and their associated labels. The network was never trained on the prototype images themselves. These were used to test generalization in the network.

Network performance was evaluated continuously throughout training for comprehension and production. The main findings included:

1. A prototype effect-accuracy in labeling a prototype image was greater than in labeling a distortion throughout training. This result indicates that the network is not merely memorizing the input patterns but is generalizing to novel forms.
2. Performance in comprehension exceeds performance in production throughout training. Input labels, presented alone, are better at producing the correct output images than input images are at producing the correct output labels.
3. Performance in comprehension and production undergoes a vocabulary spurt. There is a distinct discontinuity in the rate of learning of labels and images which occurs earlier for images (comprehension) than for labels (production).
4. Underextension and overextension errors are apparent in both production and comprehension. However, in produc-

tion, underextension errors are most likely to occur before the vocabulary spurt and overextension errors after the spurt. In comprehension, overextension errors are also prevalent before the vocabulary spurt.

These results do more than mimic some well-known facts about early lexical development. They predict new results. For example, the models suggest that there should be a comprehension spurt in lexical development as well as a production spurt. Only recently has the comprehension spurt been confirmed as typical of children's early lexical development. Similarly, the timing of over- and underextension errors is well documented in the literature for production but not for comprehension. The model's prediction that overextension errors in comprehension should be prevalent prior to the vocabulary spurt has yet to be evaluated against children's comprehension skills. The model also suggests novel theoretical interpretations of the familiar facts. The comprehension/production asymmetry can be shown to be a direct outcome of the representational formats used for the images and the labels. Rather than invoking separate processes or mechanisms to account for the asymmetry, its explanation may lie in the differences between modes of representation of the visual system and the auditory/linguistic system. The model also provides an account of how the timing of over- and underextension errors are related to the vocabulary spurt in different ways. In the model, the extension of a given label is closely linked to the internal representation of the image space and the accuracy with which other labels are processed. Underextensions naturally go together with a period in the training of the network when labels are only imperfectly projected onto limited regions of the image space—i.e., when only a few labels are used appropriately and then only with an underextended repertoire of tokens.

This type of architecture, which combines both visual and auditory processing, would seem to have considerable potential for the modeling of more complex relations that hold between language and perception, such as learning about verbs in relation to dynamic perceptual events or complex noun phrases that incorporate adjectives to distinguish objects of different types.

Syntax Acquisition

The demise of behaviorist approaches to language under Chomskyan scrutiny in the late 1950s was fueled by twin concerns: the creativity and stimulus independence of speaker utterances on the one hand, and, on the other, the complexity of the structural relations in language, such as long-distance dependencies, which speakers honor in their productions. Any attempt to account for the acquisition of the grammatical structures governing productive language usage needs to provide an explanation of these abilities. Despite the fact that a considerable body of knowledge has accumulated about the milestones which children pass in their acquisition of syntactic abilities, there is still considerable controversy concerning the nature of the mechanisms involved in the acquisition process. Roughly speaking, two camps can be distinguished: One view supposes that children's grammatical knowledge does not change dramatically over developmental (the continuity hypothesis). From this view, grammatical knowledge is constrained and supported from the outset by specific linguistic universals specified via some universal grammar. In contrast, grammatical knowledge may be viewed as discontinuous over developmental time and its discovery highly sensitive to the structure latent in the linguistic stimulus and other sources of nonlinguistic information available to the child. The disconti-

nuity hypothesis is thus directly at odds with the "poverty of the stimulus" argument.

Connectionist approaches to the acquisition and representation of syntactic knowledge have tended to focus on evaluating the feasibility of the discontinuity hypothesis and have met with some success (Elman, 1990, 1993).

Elman (1991) trained a simple recurrent network on a prediction task. Recurrent networks maintain a history of their previous internal states and are thereby able to use their memory to make predictions about the future (see LANGUAGE PROCESSING). The prediction task was simply to predict the next word in a sequence which is the start of a grammatical string generated by a simple phrase-structure grammar. The network was given explicit information about word boundaries but no direct information about sentence boundaries. The accuracy with which the network learned to predict the next word in the sequence was quite low, as is true for humans. However, the network was very accurate at predicting what type of word (which grammatical category) would come next. So if the network encountered the definite article *the*, it could be confident that the next word would be a noun or an adjective. Effectively, the network performed a distributional analysis of the input strings in the language. Elman conducted a cluster analysis of the activation of individual words at the hidden unit level in the network and found that they had been assigned a representation which fitted their grammatical role in the sentence. So nouns were clustered together in the similarity tree on a branch different from verbs. Similarly, within a branch, verbs were subdivided into those which demanded a direct object and those which were intransitive. Of course, the network itself was not using these linguistic terms to govern the prediction task. The linguistic labels assigned to the tree structure produced by the cluster analysis are purely the constructs of the analyst. Nevertheless, these analyses show that information very much like that used by linguists to describe grammatical categories is exploited by the network in the task of predicting the next word in the sequence. Furthermore, Elman's result suggests that analysis processes akin to that performed by a simple recurrent network permit the bottom-up learning of abstract grammatical categories like noun and verb, even though these categories may not be represented symbolically in the processing system.

A second result important for language acquisition is Elman's (1993) proposals for the conditions under which these simple recurrent networks can learn long-distance dependencies. Initially, Elman had difficulty training the network to remember the correct form of the main verb (singular or plural) when it was separated from its subject by a relative clause. However, he found that either of two manipulations allowed the network to achieve final success. First, Elman showed that restricting the syntactic complexity of the sentences in the initial training set and then gradually introducing increasingly complex forms into the training set resulted in correct number agreement of verbs with their distant subjects. This strategy is reminiscent of the view in acquisition studies that caregivers provide their children with a set of language lessons which are tuned to the child's current level of development. Second, Elman manipulated the amount of memory available to the recurrent network. He started the network off with a small amount of memory effectively restricting its capacity to use the sequence information in the word strings to predict subsequent words. As training progressed, Elman expanded the memory capacity of the network. This manipulation had the same effect as restricting the complexity of the initial sentence strings: The network was able to learn long-distance dependencies. This result suggests that nonlinguistic factors internal to the child (in this case, memory span) can have important consequences for the learning of syntax. Specifically, the fact that we start off with limited memories may be an asset, not a hindrance, for our capacity to acquire complex syntactic processing skills.

Conclusions

Few child language researchers would deny that the young child comes particularly well endowed for the task of learning language. Disagreement arises when considering the exact nature of this endowment and the role of the environment in constraining the developmental process. None of the connectionist models described above start off life uninitiated in the structure of language. They are all given predefined tasks (e.g., predicting the next word), and they are all provided with information concerning the representations they are expected to manipulate (e.g., the phonological structure of verb stems). These connectionist models are therefore not tabula rasa approaches to language acquisition in the strict empiricist sense. Nevertheless, they all share the property of attempting to demonstrate how the interaction of a relatively simple computational mechanism with a structured environment can lead to the emergence of rich linguistic representations. They attempt to identify a set of minimal conditions that could get the process of language acquisition off the ground. Although future research will no doubt reveal that children exploit far richer neural structures than those described here, this minimalist approach to language acquisition helps us to establish a baseline of sufficiency conditions for development. This is not an easy task. It is often far easier for the researcher to attribute prior knowledge of a skill to explain its development than to show how that skill can emerge out of more primitive abilities. Sometimes knowledge attribution of this kind is warranted. The connectionist models described in this article suggest that the prior linguistic knowledge that children bring to the language learning situation need not consist of prewired specifications of parameters to be set, or constraints on which grammars may be learned. Language learning is multidimensional in character and is best described as emerging from a complex constraint satisfaction process.

Road Map: Connectionist Linguistics
Background: I.3. Dynamics and Adaptation in Neural Networks
Related Reading: Developmental Disorders

References

Chomsky, N., 1965, *Aspects of the Theory of Syntax*, Cambridge, MA: MIT Press.
Cottrell, G. W., and Plunkett, K., in press, Acquiring the mapping from meanings to sounds, *Connection Sci.*
Elman, J. L., 1990, Finding structure in time, *Cognit. Sci.*, 14:179–211.
Elman, J. L., 1993, Learning and development in neural networks: The importance of starting small, *Cognition*, 48:71–99. ◆
Forrester, N., and Plunkett, K., 1994, Learning the Arabic plural: The case for minority default mappings in connectionist networks, in *Proceedings of the Sixteenth Annual Cognitive Science Society Conference*, Hillsdale, NJ: Erlbaum, pp. 319–323.
Kim, J. J., Pinker, S., Prince, A., and Prasada, S., 1991, Why no mere mortal has ever flown out to center field, *Cognit. Sci.*, 15:173–218.
MacWhinney, B., and Leinbach, J., 1991, Implementations are not conceptualizations: Revising the verb learning model, *Cognition*, 40: 121–157.
Marchman, V., and Bates, E., 1994, Continuity in lexical and morphological development: A test of the critical mass hypothesis, *J. Child Lang.*, 21:339–366.
Marcus, G. F., Ullman, M., Pinker, S., Hollander, M., Rosen, T. J., and Xu, F., 1992, Overregularization in language acquisition, *Monogr. Soc. Res. Child Dev.*, 57(4).
Pinker, S., and Prince, A., 1988, On language and connectionism: Analysis of a Parallel Distributed Processing Model of language acquisition, *Cognition*, 29:73–193.
Plunkett, K., 1995, Connectionist approaches to language acquisition, in *Handbook of Child Language* (P. Fletcher and B. MacWhinney, Eds.), Oxford: Blackwell, pp. 36–72. ◆
Plunkett, K., and Marchman, V., 1991, U-shaped learning and frequency effects in a multi-layered perceptron: Implications for child language acquisition, *Cognition*, 38:1–60.
Plunkett, K., and Marchman, V. A., 1993, From rote learning to system building: Acquiring verb morphology in children and connectionist nets, *Cognition*, 48:21–69.
Plunkett, K., Sinha, C., Moller, M. F., and Strandsby, O., 1992, Symbol grounding or the emergence of symbols? Vocabulary growth in children and a connectionist net, *Connection Sci.*, 4(3, 4):293–312.
Rumelhart, D. E., and McClelland, J. L., 1986, On learning the past tense of English verbs, in *Parallel Distributed Processing: Explorations in the Microstructure of Cognition* (D. E. Rumelhart, J. L. McClelland, and PDP Research Group, Eds.), Cambridge, MA: MIT Press. ◆

Language Change

Mary Hare

Introduction

The goal of this paper is to show how certain universal patterns of language change can be explained using basic principles of connectionist learning. The paper will focus on change in morphology, the system of sound-meaning correspondences that form the subparts of words, and particularly on inflectional morphology, or the way in which a language marks nouns and verbs to indicate number, tense, aspect, subject agreement, and so on. A strong source of morphological change is the tendency to eliminate apparent irregularities and complications in this system. A clear example can be found in English verb inflection. In Old English (ca. 870) there were at least 10 forms of past tense inflection: Four subclasses (the weak verbs) took variants of the suffix /-d/ or /-t/, while seven others (the strong classes) marked the past through a stem vowel change, like the modern *give–gave*. Over the past 1000 years this system has simplified dramatically. The three Old English (OE) affixed past tenses coalesced into one, which spread through the strong classes. The result is the modern system, with a single regular suffix /-d/ now applying to all but perhaps 150 verbs.

Morphological regularization processes are extremely common cross-linguistically and raise a number of questions. First, the complex system of OE existed for hundreds of years—what permitted it to overcome the drive toward simplification? What eventually disrupted the system and caused it to change? Can the direction of change be predicted? Furthermore, many irregular forms never change: What factors contribute to this immunity? This article attempts to answer these questions, and to show the extent to which the linguistic facts result from the way the language is learned.

The Organization of Inflectional Classes

The more complex a morphological system, the more difficult it is to acquire, and forms that are difficult to acquire and process are those most likely to be changed (Kiparsky, 1968; Slobin, 1977). On the other hand, the less arbitrary the system is, the easier it is to learn, and even a highly complex morphology can be acquired with relative ease if it is structured appropriately. To illustrate this point, consider again the OE past tense. Learning the language involved assigning each of several thousand verbs to one of ten inflection classes. Furthermore, one weak class (the ancestor of the modern regular verbs) was much larger than the others, and very productive. This more regular class could have dominated the system, making the smaller classes very difficult to acquire. However, two factors conspired to make the system learnable.

The first was relative token frequency. Although the weak past tenses included approximately 75% of all OE verbs, individual weak verbs tended to be infrequent while individual strong verbs generally had a high token frequency (Quirk and Wrenn, 1975), making them easier to learn.

Second, verbs were not arbitrarily assigned to past tense classes: most classes developed phonological cues to class membership. Each strong verb class, for example, displayed a class characteristic, or phonological feature, that served as the basis for categorization. This characteristic was generally the vowel of the present tense stem; occasionally there were constraints on the consonants of the stem coda. Table 1 gives the phonological characteristics for each of the strong verb classes previous to 870 A.D.

The existence of such characteristics meant that rather than learning each past form individually, the learner could use a much smaller number of generalizations of the type "long *i* in the present → Class I vowels in the past." These could be pivotal in learning classes with low type and token frequency, and the data suggest that speakers made use of them when categorizing verbs. As examples, when the verbs *strive* (Old French) and *thrive* (Norse) were borrowed into the language, they entered strong Class I since they fit the *i.* + *fricative* pattern, while *take* (Norse) entered Class VI where the most common pattern was *a* + *k*.

Morphological Change

The tendency to link membership in the smaller morphological classes to extramorphological features like phonological form made the classes easier to learn, but it also motivated morphological change. In the example of the OE strong verbs, phono-

logical changes eventually disrupted the class characteristics. When this happened, the class structure was lost and the morphological system changed as well, to reestablish a set of valid generalizations on which the classes could be based. The most striking example involves OE strong Class III. Originally this was a class characterized by *e + sonorant + consonant* in the present tense, but a series of phonological changes affected the original shape of its members. One such process altered *e* to *i* before a nasal, as in the word *sing*. A second process changed the *e* to *ie* after a palatal consonant (resulting in an earlier form of the verb *yield*). The two changes had different consequences, due to differences in the number of verb types involved. The first affected a large number of verbs, which became a new class. The second affected only a few items, and there were no grounds for differentiating these from other verbs with *ie* in the present. These verbs soon regularized.

Token frequency also played a role in how an individual verb responded to the loss of a phonological generalization. As an illustration, three members of Class IV developed variations on the characteristic stem vowel *e*: *shieran* (shear), *niman*, (take), and *cuman* (come). Over time, *shieran* has regularized and *niman* has disappeared from the language. The highly frequent *cuman*, however, has survived as a strong verb.

In summary, the strong verb classes survived if they displayed clear phonological templates for their members to match. If this within-class similarity was lost, however, the verb classes changed. Some verbs were classified together under a different generalization (as in the *sing* subclass), others moved to join an existing class (like *yield* or *shear*). Still others, whose token frequencies were sufficiently high, resisted change altogether.

Modeling Account

In developing an account of these facts, I will argue that they exemplify the effects of frequency and regularity in learning. Seidenberg and McClelland (1989) discuss in detail these effects in networks implementing backpropagation learning (Rumelhart, Hinton, and Williams, 1986). In networks using this algorithm, the output response to an input stimulus is compared to a "teacher," the expected output, and then the discrepancy between the two (referred to as the error) is calculated. The algorithm then adjusts the connection weights of the network in a way that reduces the overall error. For frequent patterns, a reduction in error on one presentation of the pattern entails error reduction on all other presentations. Similarly, a weight change that leads to better performance on a highly regular mapping will improve performance on all other patterns that share in that regularity. Thus, both regular patterns and frequent patterns make large contributions to error reduction. Since the backpropagation algorithm changes weights in proportion to the effect that the change will have in reducing total error, greater weight changes will be associated both with frequent and with regular patterns. As a result, these items are mastered more quickly, while items that are irregular and infrequent are learned with more difficulty.

Hare and Elman (1993) showed that this aspect of network behavior can account for the patterns of morphological change in Old English verbs. The paper describes a feedforward backpropagation network that was taught the 10-class verb system of OE, with a number of large classes showing phonological cues to class membership, and a number of scattered exceptions to these cues. Frequency differences cut across the consistent and exceptional items.

The network was trained for 15 sweeps through the data set, at which point the total summed squared error was 2.5, since some verbs had been learned perfectly while others were still

Table 1. Identifying Characteristics of Early OE Verb Classes (Prior to 870 A.D.)

Class	Present/ Infinitive Vowel	Coda Consonant	Example
I	i:	one C (generally a fricative)	dri:fan (*drive*)
II	e:o	any C	fre:osan (*freeze*)
III	e	sonorant + C	helpan (*help*)
IV	e	one sonorant	bearan (*bear*)
V	e	C (usually stop or spirant)	etan (*eat*)
VI	a	one C (frequently /k/)	scacan (*shake*)

incorrect. Specifically, 70% of the verbs from phonologically coherent classes took the correct past tense inflection, and high-frequency verbs (those presented up to 30 times per sweep through the data set) were also well learned, whether or not they fit the phonological characteristic of their class. By contrast, low-frequency items (presented < 5 times per sweep) that did not fit the template of their class were learned poorly.

At this point, the output of the network was taken and used as the teacher in training a second network. This second network was trained for 15 sweeps through the data set, and its output was then used to train a third network. This procedure was repeated for a total of five networks, metaphorically viewed as "generations." The goal was to reproduce the effects of imperfect transmission in language change. Because learning was stopped before perfect performance was reached, and only the information available at that point was passed on, the errors in learning in the first network became part of the data set for subsequent networks. The assumption was that patterns most affected by generational learning would be those that were the most difficult to learn, that is, patterns with poor type and token frequency, or lacking the phonological similarity characteristics of their class.

The results supported this assumption. After five generations of learning, verbs with high token frequency continued to show lower error than other verbs. The weak verbs had high type frequency, and with few exceptions were learned as weak even though their token frequency was low. The fate of the low frequency strong verbs depended on the phonological consistency of their class: verbs from classes with clear phonological characteristics were learned much better than those that lacked phonological similarity characteristics. These results were highly consistent with the actual historical development of these verbs.

The direction of change was also consistent with historical trends. The weak classes (with their high type frequency) attracted new members from the strong classes. The larger, more consistent strong classes attracted new members as well. The characteristics of the victims were equally important in determining the direction of change: verbs that changed class most readily were low frequency items from sparse phonological neighborhoods; they also showed a high degree of phonological similarity to the members of the aggressor class.

Discussion

In summary, this article has argued that both frequency and formal similarity are important to the organization of the English verb inflection system, and that change occurs when the similarity basis of an irregular class breaks down. These are effects that are easily understood in terms of connectionist networks, where frequency and regularity play similar roles in learning. The argument is strengthened by the fact that frequency-regularity effects have also been found in the synchronic system (Seidenberg, 1992; Seidenberg and Bruck, 1990).

As a final point, these results bear on an issue that is currently of great interest in the psycholinguistic literature: Can the English verbal system be the product of a single mechanism, or are distinct mechanisms required to account for the regular-irregular distinction? The fact that the simple properties of connectionist learning offer a rationale for the shape of attested historical changes argues in favor of the single-mechanism account.

Road Map: Connectionist Linguistics
Related Reading: Language Acquisition; Linguistic Morphology

References

Bybee, J., and Modor, C., 1993, Morphological classes as natural categories, *Language*, 59:251–270.
Hare, M., and Elman, J. L., 1993, From weared to wore: A connectionist model of language change, in *Proceedings of the 15th Annual Meeting of the Cognitive Science Society*, Princeton, NJ: Erlbaum, pp. 528–533.
Kiparsky, P., 1968, Linguistic universals and linguistic change, in *Universals of Linguistic Theory* (E. Bach and R. Harns, Eds.), New York: Holt, Rinehart and Winston, pp. 170–202.
Plunkett, K., and Marchman, V., 1991, U-shaped learning and frequency effects in a multilayered perceptron: Implications for child language acquisition, *Cognition*, 38:73–193.
Quirk, R., and Wrenn, C. L., 1975, *An Old English Grammar*, London: Methuen. ◆
Rumelhart, D. E., Hinton, G., and Williams, R., 1986, Learning internal representations by error propagation, in *Parallel Distributed Processing: Explorations in the Microstructure of Cognition*, vol. 1, *Foundations* (D. E. Rumelhart, J. L. McClelland, and PDP Research Group, Eds.), Cambridge, MA: MIT Press, pp. 318–362.
Seidenberg, M., 1992, Connectionism without tears, in *Connectionism: Theory and Practice* (S. Davis, Ed.), New York: Oxford University Press. ◆
Seidenberg, M., and Bruck, M., 1990, Consistency effects in the generation of past tense morphology, paper presented at the 31st meeting of the Psychonomics Society, New Orleans, November.
Seidenberg, M., and McClelland, J., 1989, A distributed, developmental model of word recognition and naming, *Psychol. Rev.*, 97: 447–452.
Slobin, D., 1977, Language change in childhood and in history, in *Language Learning and Thought* (J. Macnamara, Ed.), New York: Academic Press.

Language Processing

Jeffrey L. Elman

Introduction

One of the most fruitful—and contentious—areas of neural network research has been in the modeling of natural language phenomena. Language has always been a central area of cognitive science. It is a behavior which is often taken as the hallmark of the human species, and has been extensively studied by scientists in a wide range of fields. Because of this, there is a rich empirical base to be accounted for. But, as importantly, there are well-developed formal models of linguistic behavior which have played a major role in shaping general theories of cognition. Thus, insofar as connectionist models may suggest alternative ways of thinking about language, they also can be seen as more fundamental challenges to the traditional theories of cognition.

Since the early 1980s, three major issues have tended to dominate the field of connectionist models of language: (1) how to deal with the parallel processing of information, including the effects of higher-level information on lower-level processing; (2) how to represent the knowledge which underlies patterned behavior, whether or not explicit rules are involved, and how this knowledge (whatever its form) is acquired; and (3) whether connectionist networks possess sufficient representational power to deal with higher-level grammatical processes.

These issues have primarily been explored within two specific areas of language behavior: morphology and syntax. (*Morphology* is the process of word formation, including things such as inflections and derivations; *syntax* deals with sentence formation, e.g., word order, constraints on possible sentence structures, etc.) Although there has been important work in other areas as well, including especially speech recognition and phonology, this article will focus on the issues and domains mentioned above.

Context Effects and Parallel Processing

It is logical to think of the flow of information in language processing as proceeding in a bottom-up manner, beginning with input (acoustic or visual) and culminating in a semantic representation. The alternative possibility, that meaning might be extracted before the input has been processed, seems prima facie unlikely. For this reason, up until the late 1970s most accounts of language understanding assumed that language was processed in a serial flow, and that there were minimal interactions between levels of processing and no feedback from higher to lower levels. This assumption also seemed warranted by the rapid rates at which humans process language, since extensive interactions or feedback loops would appear to introduce undesirable delays in processing.

By the 1970s, however, a large body of empirical data had begun to accumulate, which suggested that human listeners do in fact process multiple knowledge sources simultaneously and that the integration of this information involves complex interactions and top-down effects. One of the best-studied examples of this is the so-called word superiority effect in letter perception. When observers are asked to name letters that are presented for extremely brief exposure durations, perception is improved when those letters occur in words, relative to performance when either isolated letters or letters in nonword contexts are presented. The effect seems to be genuinely perceptual, rather than simply reflecting a post-processing bias to name letters which are consistent with the word.

In 1981, McClelland and Rumelhart developed an account of the word superiority effect based on what they called an "interactive activation model." The model involved a localist network which contained distinct nodes for visual features, letters, and words. Input to the network caused visual features nodes to be active to varying degrees, depending on the duration and quality of the input. Feature nodes activated those letter nodes which contained them; and letter nodes activated those word nodes which contained them. In addition to excitatory bottom-up connections, there were excitatory top-down connections and also inhibitory connections between inconsistent letters and words, and between all words (reflecting the fact that only one word ought to be present at one time). The top-down connections allowed the model to use lexical knowledge to assist in perception. The model also demonstrated that statistical patterns of letter occurrences present in the lexicon could exert an effect. Thus, pseudo-words (e.g., "mave") would benefit from the presence of many visually similar words even though there exists no such word in the model's lexicon. "Gang effects" were also obtained, such that letters contained in words in dense neighborhoods (that is, with many nearby similar words) benefited, compared to letters in words which were in more sparsely populated regions of the lexicon.

The word perception model was highly influential in demonstrating the power of what at first seemed to be a relatively simple architecture. In addition to the basic word-superiority effect, McClelland and Rumelhart were able to account for a range of additional experimental phenomena. Especially impressive was the model's ability to predict human data which had not yet been observed, but which Rumelhart and McClelland verified through subsequent experiments with humans. The model thus served not only as an account of existing data, but also as a source of new hypotheses.

A related model, called TRACE, was proposed by McClelland and Elman (1986) to account for speech perception. Speech exhibits many of the same top-down phenomena found in written word perception but presents a number of other problems. These include segmentation (the fact that what are perceived as distinct units, e.g., phonemes or words, are in fact not obviously marked in the time-varying speech signal), significant variability due to coarticulation (the speech articulators often move in advance of the current sound, causing a blending of features from one sound into adjacent sounds) and to interspeaker differences, and variability in the rate of speech. The fact that listeners process speech with extreme rapidity places a further constraint on models.

The TRACE model was designed to deal with a subset of these problems by having, in addition to the sorts of connections found in the word perception model, multiple networks which processed successive time slices and connections between nodes in different time slices. These additional connections allowed the network to compensate for coarticulation. Segmentation occurred as a result of lateral inhibition; thus, it was an outcome of the perceptual process rather than a prerequisite for it. The most important point made by TRACE, however, was that lawful variability in a signal can be a source of information rather than a source of noise (by providing redundancy). This perspective was supported by the experimental literature which shows that listeners are able to exploit this variability.

Learning and the Representation of Knowledge

These early models provided a satisfying answer to the question of how to model top-down effects and complex perceptual interactions, but they raised in turn several new problems. First was the question of whether there might be some automatic procedure by which a network's knowledge might be acquired, other than through programming on the part of the designer. Second, to what extent were the highly localized representations of these models realistic approximations of the human data? The empirical evidence is contradictory. Although no deficits have been reported in which specific words are lost as a consequence of damage (as might be predicted from the localist word representations used in many early models), cases of loss have been reported involving double dissociations of information from general categories (e.g., semantic categories, or regular morphology versus irregular morphology). These latter data, along with a variety of results from normal speakers, have been used to argue in favor of a more modular structure in which, for example, regular patterns are stored separately and processed by a different mechanism than irregular patterns.

Beginning in the mid-1980s, a number of investigators began to focus on models in which various learning algorithms (most commonly, backpropagation of error) were used to train networks. The results of these simulations led to interesting proposals regarding the learning and representation of linguistic knowledge in humans.

The Past Tense of English

One of the most influential models in the area of connectionist language processing has been Rumelhart and McClelland's (1986) simulation of a two-layer network which was trained to produce the past tense of English verbs, given as input the present tense. The problem is interesting for several reasons. First, the past tense morphology in English is only partially regular. Most verbs form the past with the /-ed/ morpheme, but there are a small number of highly common verbs whose past tense morphology is irregular. The present-past mapping is thus not entirely straightforward. Secondly, in the course of learning the past tense morphology, many children go through a period in which the regular inflection is applied more generally than is warranted (i.e., to irregulars as well). This has been taken as strong evidence for the psychological reality of rules, since the behavior might be interpreted as reflecting the incomplete acquisition of a rule (children know the rule, but not when it does not apply). Rumelhart and McClelland's network exhibited similar behavior. Because the network does not learn rules in any obvious sense, this result prompted Rumelhart and McClelland to suggest that children (and presumably adults) as well might not have explicitly represented rules. Their results also suggested that a single mechanism was sufficient to account for both regular and irregular morphology.

These conclusions led to a detailed critique by Pinker and Prince (1988) of Rumelhart and McClelland's model. One of the most trenchant objections was the artificiality of the training regime. The network was initially trained on a data set consisting only of high-frequency verbs (mostly irregular); this was subsequently amplified by the addition of a larger number of mid-frequency verbs (mostly regular). The appearance in the network of overgeneralization coincided with this discontinuity in training, which has no plausible correspondence with the child's experience. Pinker and Prince also raised a number of other concerns, including the representation of the verbs, and the inability to represent as distinct homophonous verbs with different past tenses.

These criticisms prompted Plunkett and Marchman (1991) to train a three-layer network on the same task, but using more naturalistic training and a less controversial representational scheme. Plunkett and Marchman's result in fact bolstered the original conclusions by Rumelhart and McClelland, and led to some important results regarding the ways in which class size and token frequency of items in the training set interact in the learning process. Plunkett and Marchman observed that in a network trained on multiple mappings, the number of items which participate in a mapping (i.e., the class size) affects the likelihood of that class's mapping being taken on by other smaller classes. On the other hand, individual token frequency of specific patterns affects the ability of an item to resist such encroachment. Plunkett and Marchman pointed out that the precise class size and token frequencies of the English regular and irregular classes seem to reflect a stable equilibrium between these forces. These results are important because they reflect basic learning characteristics of networks using gradient descent training.

Dissociated Behaviors from Single Mechanisms

In further work, Marchman (1993) addressed another criticism of the network single-mechanism account. It has been reported that in some cases following brain damage, there is selective loss of regular verb morphology. Marchman demonstrated that in fact a single network, trained on both regular and irregular mappings, also showed preferential maintenance of irregular forms following lesioning of the network. This result follows from Plunkett and Marchman's earlier work, and reflects the fact that the same conditions which are necessary for learning irregular mappings in the first place (i.e., high token frequency) also protect these forms from loss following damage.

Hinton and Shallice (1991) also considered the question of whether a dissociation of behavior necessarily implies a dissociation of mechanism. Their work focused on deep dyslexia, which is a language deficit which may occur following brain damage. It is a syndrome characterized both by the presence of semantically based errors in reading (*peach* read as *apricot*), but also many visually based errors (*mat* read as *rat*). In addition, there are more joint visual and semantic errors (*cat* read as *rat*) than would be predicted from the base rates of each type alone.

This is a perplexing phenomenon. What sort of damage might simultaneously affect two systems (visual and semantic), which are putatively handled by distinct brain regions? Hinton and Shallice attempted to simulate this syndrome by constructing a network which took word input in the form of letters, and produced the appropriate semantic output for that word. The presence of recurrent connections within the network allowed it to develop a dynamics which it employed in order to deal with cases in which visually similar inputs ultimately mapped onto semantically very different outputs. Hinton and Shallice showed that one of the properties of networks with these kinds of attractor dynamics is that lesion virtually anywhere in the system resulted in both types of loss observed in deep dyslexics. That is, the network produced not only visually and semantically based reading errors, but also a higher than expected percentage of errors which combined both types. As with Marchman's model, this work underscores the fact that behaviors which are functionally modular (regular versus irregular morphology; visual versus semantic processing) can be carried out in mechanisms whose structure is not modular in the same way.

German Noun Gender

MacWhinney et al. (1989) investigated the ability of network models to learn the morphology of the German definite article (*the* in English). There are a large number of morphologic variants (allomorphs) of the word in German, and the choice of the correct allomorph depends on the gender, number, and case of the noun. Gender in German is grammatical more than semantic, which means that whether a given noun is masculine, feminine, or neuter cannot easily be determined and often has no correspondence with the semantic gender. Many linguists have argued that gender in German is therefore an arbitrary marking which must be memorized individually for every noun.

MacWhinney et al. showed, however, that there are subtle cues which reflect joint interactions between the phonological shape of a word, morphological characteristics, and semantic gender; and that networks—like native speakers—learn to be sensitive to these cues. MacWhinney et al. interpreted their results as evidence that networks provide a superior account of speakers' tacit knowledge of language, and that language processing is better viewed as an instance of cue integration rather than symbolic rule application.

Word Recognition and Naming

The McClelland and Rumelhart (1981) model provided an account for the way in which orthographic input might be used to access words, where the words were considered as abstract entities with no phonological content. Seidenberg and McClelland (1989) developed a more complete model of word recognition in which orthographic input was used to access both a word's phonological as well as semantic representation (the actual simulation, however, was limited to the mapping from orthography to phonological representation).

The Seidenberg and McClelland model was able to simulate a number of phenomena characteristic of human readers. These phenomena include the ability to name novel forms; the transition from beginning to skilled reading; individual differences among readers; differences between words, in terms of processing difficulty; and differences in carrying out lexical decision versus naming tasks. Seidenberg and McClelland point out that these data can be handled without recourse to the dual-route mechanism which has been proposed for the reading task. Instead, the model deals with both regular and irregular words (and with known as well as novel words) using a single mechanism. The model also suggests a view of the lexicon which is a radical departure from traditional views. The lexicon in their model is not a list of entries for individual words. Instead, the lexicon is embedded, in a distributed manner, across the patterns of connection weights that have been learned in order to carry out the reading tasks. Reading thus is not so much a matter of access of information as it is of the activation of appropriate outputs, given particular inputs.

High-Level Linguistic Processing: Semantics and Syntax

The third general question which has been raised in the connectionist language processing literature is whether connectionist networks provide the necessary representational power in order to capture certain higher level syntactic and semantic behaviors. In particular, there is strong evidence that the mental representation of sentences is not simply as a linear string of words, but instead requires complex and abstract hierarchical structures. Furthermore, some grammatical processes appear to involve a degree of recursion.

Assigning Roles to Sentence Constituents

Although word processing and morphological knowledge are obviously crucial aspects of language processing, they only capture part of what listeners and speakers know. Ultimately, language is a matter of communicating meanings. This involves the use of sentence structure to link words together appropriately, and the ability of listeners to do such things as match words with roles (such as who was the agent of an action, what was the instrument used, etc.).

St. John and McClelland (1990) addressed this latter problem in their "sentence gestalt" model. In this model, words are presented to a network one at a time. As the network processes each word, it is simultaneously given a probe input which takes the form of a question about the sentence that is being presented. Questions might be about the roles (e.g., agent, object, instrument, etc.) played by one of the words in the sentence (e.g., "What role does the bus driver fill in this sentence?"). Or questions might ask which word filled a certain role (e.g., "Who is the agent in this sentence?"). The network's task was to produce the answer which was correct for that sentence. The network had internal recurrent connections, which it used to build up an ongoing representation of the sentence (the "sentence gestalt").

By requiring the network to answer questions from the very beginning, even before it had seen the entire sentence, the network was encouraged to make early inferences about the roles filled by the words it was processing. These inferences could be overridden by later information in cases where the inferences were mistaken; but nonetheless, the network learned to anticipate things which were generally true of the training corpus. Given the partial input "the bus driver ate...," for example, and the probe "What was the object of the sentence?" the network might answer "steak" since this was the typical fare of bus drivers (at least in this simulation).

The network showed a number of other interesting characteristics, including the ability to infer the fillers of roles which had not been explicitly named (but which were implied by the meaning of the sentence) and the ability to disambiguate homonyms on the basis of context.

Learning Word Meaning

In the St. John and McClelland model, semantic roles were predetermined; the model's task was to learn which words filled which roles in any given sentence. But one can also ask where such roles come from. More generally, one might ask if a network can infer something about lexical semantics from the way in which words are used.

Using a simple recurrent network architecture (an architecture in which hidden unit activation values are saved in another layer and then fed back to the hidden units on the next time slice), Elman (1990) trained a network to process simple sentences a word at a time, and to predict at every point in time which word would occur next. Words were represented by localist encodings in which a single bit was turned on. Inputs were therefore orthogonal to each other, so that the form of the input contained no information which the network could use to distinguish words along category dimensions. Given the grammar that was used to produce the sentences in this simulation, the prediction task was nondeterministic. In such cases, optimal performance is usually achieved by activating outputs in a way which approximates the conditional probability of occurrence of each possible successor. The network in fact learned to do this; it not only activated nouns and verbs where appropriate, but also distinguished between subclasses of nouns, such as inanimate, human, small animals, etc. Elman examined the hidden unit activation vectors which were produced by the different words and found that the network had partitioned the hidden unit state space into broad regions corresponding to grammatical category (nouns in one subspace and verbs in another), with finer-grained subdivisions being used to capture subtler semantic distinctions between categories of words. Although the network cannot be said to have learned semantics per se, it did learn a representational scheme which reflects the semantics that was implicit in the lexicon for this task.

Representing Hierarchical Structure

Both the St. John and McClelland and the Elman models assumed monoclausal sentences, and the problem of *parsing* (figuring out the constituents in the sentence) is avoided by having a relatively simple sentence structure. In more complex sentences, the problem arises of how to represent structure and assign words to that structure.

Fodor and Pylyshyn (1988) argued that connectionist models are incapable of representing the sort of structure which would be needed to process complex sentences. Natural languages are characterized by hierarchical structure, in which constituents may recursively be embedded in other constituents. In sentences such as "the letters that the girl writes are going to Paris," listeners must be able to recognize that "that the girl writes" is an embedded sentence, and that the embedding modifies "the letters." These structural relationships are crucial, because they account for the fact that the embedded clause verb is in the singular—to agree with "girl," not "letters" —and that listeners know that it is the letters that are going to Paris, not the girl, even though pragmatically nothing rules this out. It doesn't take much imagination to conceive of similar sentences in which the structural relations both are more complex and contribute essentially to determining meaning.

Fodor and Pylyshyn maintained that things like constituent structure (the notion that one element is part of another) are beyond the ability of connectionist nets to represent. (Actually, they argued that any connectionist network which succeeded in doing this would merely be an implementation of the classical physical symbolic system.)

There is another problem which arises when complex sentences are considered, and that is the issue of learnability. Sentences with center-embeddings (such as "The letters that the girl writes are going to Paris") schematically take the form $A^n B^n$, in which As and Bs stand for nouns and verbs; in such sentences, n As are followed by n Bs. These languages fall into a class of languages (context-free) which are known to be not learnable solely on the basis of positive-only evidence, that is, evidence which consists only of grammatical examples. And yet to all intents and purposes, this is the sort of input which language learners are given. (Learners may hear ungrammatical sentences, but they are not given ungrammatical sentences on purpose, with the information that the sentences are in fact ungrammatical.) This observation is one of the things which has prompted some linguists to make relatively strong claims about the necessity for at least some linguistic knowledge to be innately specified on the grounds that if the necessary information cannot be learned, but is exhibited in adults, then it must be prespecified from the start. Thus, complex sentences pose problems not only for representation, but also learning.

Elman (1993) addressed these two questions by training a simple recurrent network on the same word prediction described above, but using sentences which contained relative clauses. These sentences thus had precisely the properties which might require the network to develop some sort of hierarchical representation, and which might be difficult or impossible to learn. There were two results of interest from this work. First, under certain conditions, the network did learn to process these complex sentences. By looking at trajectories through the hidden unit state space, Elman determined that basic sentence types were represented by the same characteristic trajectory, and that the embedding relationship was represented by a displacement in the state space. These embeddings utilized the same canonical trajectory patterns, but were shifted along another dimension in order to distinguish the levels of embedding. This allowed the network to process long-distance dependencies between heads of relative clauses and their distant verbs, and between subject and verbs which were interrupted by multiple subclauses.

Success was not achieved under normal training. Instead, Elman found that it was necessary to use one of two incremental training strategies (what he termed "the importance of starting small"). Either the training data had to be presented in ordered format, with simple sentences first; or alternatively, the network could be given the entire corpus from the outset (including complex sentences), but had to begin training with limited working memory. This latter was achieved by disrupting the feedback of hidden unit activation patterns every two or three words, and had the effect of making it impossible for the network to attend to all but the simplest sentences. Both strategies were necessary because the network was unable to learn

an initial representational scheme to encode inputs (distinguishing nouns from verbs, singulars from plurals, etc.) when it was given overly complex sentences. By focusing on simple sentences first, however, the network could develop a set of internal representations which improved the temporal memory so that the network could then deal with inputs that were both longer and more complex. This result can be interpreted as consistent with the need for some form of innate knowledge, but the knowledge takes the form of a developmental bias which turns out to have a language-specific effect, rather than being language-specific to begin with.

Discussion

Despite impressive and exciting results, it is clear that connectionist models have really only scratched the tip of the linguistic iceberg. It is surprising that networks have been as successful as they have been in modeling this complex human behavior; but the reality of the behavior is still considerably beyond the scope of any existing model.

Two issues seem particularly pressing. The first is to understand the representational and processing properties of networks (particularly recurrent networks) in order to be able to compare them with more traditional models. What seemed like important differences at first (rules versus connections; implicit versus explicit knowledge) on closer inspection will probably turn out to be the wrong dimensions for comparison.

The second challenge is the scaling problem. Current network technology limits most models to toy versions of real problems, and the scaling up from toy to reality is not straightforward. The solution will undoubtedly require new ways of constraining the search space during learning. This may lead to a new understanding of the ways in which knowledge can be innate—which itself is a central issue in cognitive science.

Road Map: Connectionist Linguistics
Background: I.3. Dynamics and Adaptation in Neural Networks
Related Reading: Connectionist and Symbolic Representations; Language Acquisition; Lesioned Attractor Networks as Models of Neuropsychological Deficits; Speech Recognition: Pattern Matching; Structured Connectionist Models

References

Elman, J. L., 1990, Finding structure in time, *Cognit. Sci.*, 14:179–211.
Elman, J. L., 1993, Learning and development in neural networks: The importance of starting small, *Cognition*, 48:71–99.
Fodor, J., and Pylyshyn, Z., 1988, Connectionism and cognitive architecture: A critical analysis, in *Connections and Symbols* (S. Pinker and J. Mehler, Eds.), Cambridge, MA: MIT Press, pp. 3–71. ◆
Hinton, G. E., and Shallice, T., 1991, Lesioning an attractor network: Investigations of acquired dyslexia, *Psychol. Rev.*, 98:74–95.
MacWhinney, B., Leinbach, J., Taraban, R., and McDonald, J., 1989, Language learning: Cues or rules? *J. Memory and Language*, 28:255–277.
Marchman, V. A., 1993, Constraints on plasticity in a connectionist model of the English past tense, *J. Cognit. Neurosci.*, 5:215–234.
McClelland, J. L., and Elman, J. L., 1986, Interactive processes in speech perception: The TRACE model, in *Parallel Distributed Processing: Explorations in the Microstructure of Cognition* (D. E. Rumelhart, J. L. McClelland, and PDP Research Group, Eds.), vol. 2, *Psychological and Biological Models*, Cambridge, MA: MIT Press, pp. 58–121.
McClelland, J. L., and Rumelhart, D. E., 1981, An interactive activation model of context effects in letter perception, Part 1: An account of basic findings, *Psychol. Rev.*, 88:375–407.
Pinker, S., and Prince, A., 1988, On language and connectionism: Analysis of a parallel distributed processing model of language

acquisition, in *Connections and Symbols* (S. Pinker and J. Mehler, Eds.), Cambridge, MA: MIT Press, pp. 73–193.

Plunkett, K., and Marchman, V., 1991, U-shaped learning and frequency effects in a multilayered perceptron: Implications for child language acquisition, *Cognition*, 38:43–102.

Rumelhart, D. E., and McClelland, J. L., 1986, On learning the past tenses of English verbs, in *Parallel Distributed Processing: Explorations in the Microstructure of Cognition* (D. E. Rumelhart, J. L.

McClelland, and PDP Research Group, Eds.), vol. 2, *Psychological and Biological Models*, Cambridge, MA: MIT Press, pp. 216–271. ◆

Seidenberg, M. S., and McClelland, J. L., 1989, A distributed, developmental model of word recognition and naming, *Psychol. Rev.*, 96: 523–568.

St. John, M., and McClelland, J. L., 1990, Learning and applying contextual constraints in sentence comprehension, *Artif. Intell.*, 46:217–257.

Layered Computation in Neural Networks

Hanspeter A. Mallot

Introduction

Layering is a common architectural feature of many neural subsystems, both in vertebrate and invertebrate brains. It is best studied in the mammalian neocortex, but can be found in regions as diverse as the optic tectum, the avian visual wulst, or the cephalopod optic lobe. In a broader sense, layered neural areas with strong vertical connectivity and topographic organization of input and output may be called *cortical* in all these different structures.

Neural layers are characterized by various anatomical and physiological parameters, such as relative abundance of cell classes, soma size, pharmacology, and both intrinsic and interarea connectivity (Braitenberg and Schüz, 1991; see also NEURO-ANATOMY IN A COMPUTATIONAL PERSPECTIVE). These parameters remain constant within a two-dimensional sheet, but vary between sheets. In contrast, the "layers" of artificial neural networks are defined by topology only (block-structure of the connectivity matrix), leaving no room for geometrical concepts such as two-dimensional extent. Some important properties of cortical layering are as follows:

1. Both intralayer and interlayer connectivity are largely determined by spatial constraints, such as nearness. In particular, the two-dimensional topology of layers is important.
2. Connections between two neurons can be mediated by multiple synapses located in different layers. If detailed timing is considered, this multiplicity of connections can be functionally significant owing to differences in propagation time.
3. Layers can be void of nerve cell somata, mediating fiber contacts of neurons whose somata are located elsewhere (e.g., the molecular layer of the neocortex).
4. Feedback connections can occur even within each layer.

This article deals with three aspects of layering: quantitative descriptions of layered or cortical organization, the activation dynamics of network models incorporating this organization, and applications to problems of information processing and computation. Although the concepts are general, examples will repeatedly be drawn from (primate) visual cortex.

Quantitative Anatomy

Uniformity and Continuous Models

Given the vast numbers of cells found in layered cortical areas, it seems appropriate to build spatially continuous models (neural "fields") where each point in space corresponds to a neuron (Korn and von Seelen, 1972; Amari, 1977; see also the references given in Mallot and Giannakopoulos, 1992). Besides being a good approximation of large neuron numbers, the continuous description allows for a natural modeling of position and distance of neurons. In a continuous model, a "neuron" consists of (1) a point on the sheet at the position of its soma, (2) a cloud or density function of postsynaptic (dendritic) sites specifying the input sensitivity for each point on the sheet; (3) a density function of postsynaptic (axonal) sites, and (4) an appropriate activation function (see the discussion that follows). The usefulness of these model features rests on two assumptions:

- The strength (or likelihood) of a connection between two neurons is proportional to the overlap of their dendritic and axonal clouds and the efficiencies of the presynaptic and postsynaptic sites involved. This assumption is discussed at length by Braitenberg and Schüz (1991), who have termed it "Peters's rule" (see also Peters, 1985:64ff).
- Intrinsic connectivity is largely uniform; i.e., the fiber clouds of the neurons are shifted versions of each other.

While space variance (nonuniformity) presents a problem to the sketched continuous approach, some common cases can be dealt with rather easily: e.g., by topographic maps between brain areas and modulations of neuron density within single layers. Topographic mapping can be modeled in terms of piecewise, continuous, point-to-point coordinate transforms. Explicit mathematical functions have been derived from known or assumed distributions of areal magnification by integrating the "mapping-magnification equation" in one or two dimensions (Schwartz, 1980, $\log z$; Mallot and Giannakopoulos, 1992). Point-to-point models have also been proposed for columnar input patterns, such as ocular dominance stripes (Mallot, von Seelen, and Giannakopoulos, 1990). An important case of intrinsic space variance is the columnar pattern of cell densities such as the one revealed by the cytochrome oxidase stain. In the continuous model, this can be accounted for by a space-variant density factor.

Populations, Layers, and Areas

In the continuous approach, the unit of modeling is the *neural population*, i.e., a set of neurons from the same anatomical class with a space-invariant connectivity pattern. Examples of such populations in the visual system (see VISUAL CORTEX CELL TYPES AND CONNECTIONS) are the spiny stellate cells in layer 4a, the GABAergic cells in layer 3, the pyramidal cells of layer 6 connecting to the lateral geniculate nucleus, etc. The neural population is characterized by a number of variables that fall into three groups. *Anatomical variables* include the dendritic and axonal fiber clouds, $\delta(\mathbf{x})$, $\alpha(\mathbf{x})$; cell density, $\varrho(\mathbf{x})$; and topographic output maps, $\mathscr{R}(\mathbf{x})$. *Physiological variables* include a nonlinear compression function, $f(u)$; time delays for the propagation of activity, T; synaptic integration time, τ; and gain factors. Activity is described by three *state variables*: input, $s(\mathbf{x}, t)$; potential, $u(\mathbf{x}, t)$; and output, $e(\mathbf{x}, t)$, where $\mathbf{x} \in \mathbb{R}^2$.

When connecting neural populations into networks, it is convenient (although slightly redundant) to keep the two separate state variables for input and output, the somatic activity $e(\mathbf{x}, t)$, and the "synaptic" activities $s(\mathbf{x}, t)$. The idea is that the output of one population is not the immediate input to some other population. Rather, several outputs from different populations are accumulated into one distribution of presynaptic activity, which then feeds into all neural populations with an appropriate dendritic port. We call the support of the presynaptic activity *connection planes*; they are indexed by $l \in \{1, \ldots, L\}$ in Equations 1 and 3 in the next section. Connection planes are reminiscent of the *blackboard* structure in multi-agent computer systems (see DISTRIBUTED ARTIFICIAL INTELLIGENCE) in that they collect activity from several populations without keeping track of the original source of the activity. When in turn a neural population "reads" from the connection plane, it cannot know whose activity it is reacting to; this lack of labeling of the activities is a direct consequence of Peters's rule.

Using the idea of connection planes, we can now give a definition of the terms layer and area. A *layer* is a connection plane together with all neuron populations whose somata are located in that plane. (The number of populations in a layer may be zero, as is the case in the molecular layer of the cortex. There may also be just one population allowing for specific circuitry; in this case, the pooling effect of the connection planes is bypassed and the distinction between layers and populations becomes obsolete.) An *area* is a set of layers connected without topographic maps (other than the identity). That is to say, in the case of modeling a visual system, just one retinotopic map, $\mathscr{R}_i$, is assigned to each visual area A_i, which is valid for all its layers. For projections between different areas—e.g., from A_i to A_j—a mapping of the form $\mathscr{R}_j \circ \mathscr{R}_i^{-1}$ is required to connect points representing the same retinal location.

Activation Dynamics

Network Equations

Consider a network of P neural populations (index p, state variables $e_p(\mathbf{x}, t)$ and $u_p(\mathbf{x}, t)$) connected via L connection planes (index l, state variable $s_l(\mathbf{x}, t)$). The activation dynamics

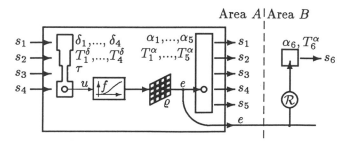

Figure 1. Activation transfer function of a neural population modeling dendritic summation, somatic point operations, and axonal spreading of activity. In Equations 1–3 (explained in the text), an additional index p is used to distinguish between different populations.

of the resulting network can be formulated in three steps (cf. Figure 1 and Mallot and Giannakopoulos, 1992):

1. Dendritic summation ($s_l \to u_p$): For each population p, inputs from different connection planes $s_1, \ldots, s_L$ are accumulated according to dendritic arborizations, δ_{pl}; delays, T_l^δ, and synaptic integration time, τ:

$$\frac{\partial}{\partial t} u_p(\mathbf{x}, t) = -\frac{u_p(\mathbf{x}, t)}{\tau} + \sum_{l=1}^{L} \int s_l(\mathbf{x}', t - T_l^\delta) \delta_{pl}(\mathbf{x} - \mathbf{x}') \, d\mathbf{x}' \tag{1}$$

2. Somatic point operations ($u_p \to e_p$): The resulting intracellular potential is passed through a nonlinearity f_p and locally weighted with the density of the cell population $\varrho_p(\mathbf{x})$. For example, if we consider a cell population in the magnocellular stream, $\varrho_p(\mathbf{x})$ reflects the pattern of cytochrome-oxidase blobs (see VISUAL CORTEX CELL TYPES AND CONNECTIONS).

$$e_p(\mathbf{x}, t) = \varrho_p(\mathbf{x}) f_p(u_p(\mathbf{x}, t)) \tag{2}$$

3. Axonal spread ($e_p \to s_l$): The resulting excitation is spread over the axonal densities α_{lp} and added to the activity of the connection layer to which the axon projects, again with appropriate delays (propagation times) T_l^α. For axons from population p projecting to a connection layer l in another cortical area, a point-to-point mapping $\mathscr{R}_{lp}$ has to be considered.

$$s_l(\mathbf{x}, t) = \sum_{p=1}^{P} \int e_p(\mathbf{x}', t - T_l^\alpha) \alpha_{lp}(\mathbf{x} - \mathscr{R}_{lp}(\mathbf{x}')) \, d\mathbf{x}' \tag{3}$$

Integro-differential equations of the leaky integrator type, such as the equations just given, have been studied extensively. One important special case is the interaction of two populations —one excitatory and one inhibitory—with all of the space-variant terms in Equations 1–3 omitted (e.g., Amari, 1977; Chipalkatti and Arbib, 1988; Ermentrout and Cowan, 1979; Murray, 1989). The formulation given here allows for space variances, both in the interarea connections (topographic mapping functions $\mathscr{R}_{lp}$ in Equation 3) and in the cell densities (ϱ_p in Equation 2). In addition, cell populations can be multiply connected via different cortical layers so that each path has its own spatiotemporal characteristic (Krone et al. 1986; Mallot et al., 1990; Mallot and Giannakopoulos, 1992).

Receptive Fields and Point Images

When stimulated with an external signal, $s_{ext}(\mathbf{x}, t)$, the network reacts with a distribution of activity $e(\mathbf{y}, t)$ that corresponds to the neural representation of the stimulus. As an example, let $\mathbf{x}$ denote retinal and $\mathbf{y}$ cortical coordinates. In neurophysiological experiments, the relation between stimulus and excitation is often described by two so-called characteristic functions which can easily be modeled in continuous neural networks.

- The point image, point spread function, or impulse response, $p_{ps}(\mathbf{y}, t)$, is the distribution of activity resulting from stimulation with a spatiotemporal Dirac function, $\delta(\mathbf{x}, t)$, e.g., a briefly flashed spot of light in the visual system. If the system were linear, space-invariant, and stationary, responses to arbitrary stimuli could be predicted by superposition of appropriately shifted, delayed, and weighted impulse responses (convolution).
- The receptive field profile $p_{rf}(\mathbf{x}, t)$ of a cortical unit at position $\mathbf{y}$ describes the influence that each input site $\mathbf{x}$ at each instant in time has on the unit in question. In linear, space-

invariant and stationary systems, point spread function and receptive field are identical up to a mirroring in spatial and temporal coordinates (e.g., while p_{rf} "looks backward in time," p_{ps} "looks forward"). In general linear systems, they are the kernels of adjoint operators (Mallot et al., 1990).

Point images and receptive fields are most useful in linear systems, where they completely describe the stimulus-response behavior by way of superposition. In order to interpret neurophysiological measurements of these functions, nonlinear approaches are required. Possible choices are (1) cascades of linear systems with stationary nonlinearities, (2) Wiener-Volterra expansions (usually terminated after order 2), and (3) nonlinear network equations, such as presented earlier in Equations 1–3.

Receptive Field Properties

Figure 2 shows four steps for increasing realism in the modeling of receptive fields. Figure 2A illustrates the space- and time-invariant feedforward system, where the spatiotemporal version of linear systems theory can be applied (Korn and von Seelen, 1972). Since the early work on lateral inhibition in the compound eye of the chelicerate *Limulus*, a number of filter functions have been discussed as models of both receptive fields and spatial vision (difference of Gaussians, various derivatives of Gaussians, Gabor functions). Many of these are now used as

filters in image processing. In the feedforward case, the spatial and temporal parts of the neural "filter function" are usually considered separable, equivalent to a cascade of two steps, one of which is temporal only and the other of which is spatial only. Simple nonseparability can be introduced by adding several of these cascades (e.g., one for the on-center and another for the off-surround of a retinal ganglion cell; see Dinse, Krüger, and Best, 1990). One important computational application of the resulting spatiotemporal filters is the processing of visual motion (Korn and von Seelen, 1972). Interestingly, many more can be found, if receptive fields specifically responding to other stimulus parameters (orientation, velocity, spectra, etc.) are considered (Adelson and Bergen, 1991, Mallot et al., 1990).

Parts *B* and *C* of Figure 2 show simple extensions of the space-invariant feedforward situation. In Figure 2B, many layers with feedback connections are considered (Krone et al., 1986). The main effects of this architecture include (1) increased width of receptive fields since point stimuli can be signaled through the entire network by feedback connections and (2) full nonseparability of spatial and temporal aspects of the receptive field. The first of these effects has been used by Horn (1974) in the deconvolution step of the "retinex" scheme for recovering lightness from image intensities. This article also introduces the idea of resistive networks for image processing which links the continuous Equations 1–3 to discrete implementations, as well as to diffusion-type equations in which spatial interaction is modeled by partial derivatives rather than by integral kernels.

In Figure 2C, the feedforward situation is extended by allowing for space variance by retinotopic mapping (Mallot et al., 1990). The combination of mapping and feedback illustrated in Figure 2D has not yet been studied in detail. Its activation dynamics is described by Equations 1 3 cited earlier.

Information Processing Capabilities of Neural Layers

The filter operations discussed earlier exploit the neighborhood relations in a neural layer. Other features that can be used for computational purposes are nonlinear activation dynamics and neural maps. Some examples include the following:

1. *Lateral cooperativity.* Lateral interactions between neural activities in a layer can have cooperative and/or inhibitory effects leading to filling-in or related kinds of shaping of the activity pattern (Murray, 1989). One important application of this principle is the solution of the correspondence problem in stereovision by means of cooperative dynamics in a disparity map (for a review, see Blake and Wilson, 1991; Chipalkatti and Arbib, 1988; STEREO CORRESPONDENCE AND NEURAL NETWORKS). Other examples of nonlinear lateral interactions include various winner-take-all or nonmaximum-suppression schemes that are widely used in artificial neural networks.

2. *Topographic mapping.* While the continuity of neural representations is a prerequisite for neighborhood operations, such as filtering, the smooth distortions introduced by topographic mapping can simplify subsequent information processing tasks. Examples include the allocation of cortical neurons to different parts of the visual field (fovea/periphery), and the simplified processing of images with systematic space variances, such as optic flow patterns. The optic flow resulting from translation in a plane can be compensated for by so-called inverse perspective mapping of the input images. Obstacles in the way of the observer lead to uncompensated changes in the flow field which are easily detected. A review

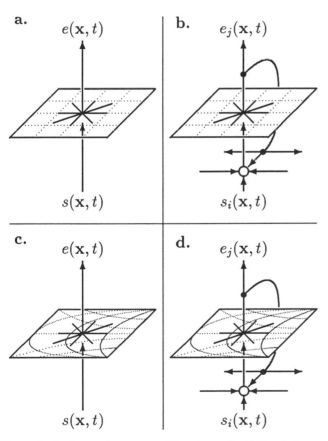

Figure 2. Continuous layers as models of receptive fields. *A*, Space-invariant, feedforward (spatiotemporal convolution). *B*, Space-invariant network of continuous layers. *C*, Space-variant, feedforward. *D*, Space-variant network of continuous layers.

of applications of topographic mapping to image processing problems is given elsewhere (Mallot et al., 1990).

3. *Feature maps and population coding.* While the examples presented so far apply to sensory input, analogous results have been obtained for motor pathways. Here, the distribution of activity on a neural layer has to be interpreted in terms of the "motor fields" of its active neurons. If this is done, the flow of activity in the appropriate motor areas predicts the initiated movements.

Discussion

The type of cortex model sketched in this article has the advantage of modeling a prominent structural unit of the vertebrate nervous system, i.e., the neural layer, on a rather high level. It can easily deal with geometric features, such as maps, columns, dendritic and axonal arborization patterns, varying cell densities, and the like. This level of modeling is required to understand large-scale activation dynamics of cortical networks as have been made accessible by recently developed imaging techniques. The continuity limit seems appropriate when entire cortical areas are to be represented in a neural network model. On the other hand, it is not very well suited to model properties which differ from one cell to the next. For example, synaptic plasticity is not easily included. It is, therefore, most useful for the modeling of rather short time scales, where plasticity may be excluded, and for systems in steady states.

Acknowledgment.   This work was supported by the *Deutsche Forschungsgemeinschaft* Grant Ma1038/5–1.

Road Map: Biological Networks
Background: I.1. Introducing the Neuron
Related Reading: Directional Selectivity in the Cortex; Gabor Wavelets for Statistical Pattern Recognition; Thalamus; Traveling Activity Waves

References

Adelson, E. H., and Bergen, J. R., 1991, The plenoptic function and the elements of early vision, in *Computational Models of Visual Processing* (M.S. Landy and J.A. Movshon, Eds.), Cambridge, MA: MIT Press, pp. 3–20. ◆

Amari, S.-I., 1977, Dynamics of pattern formation in lateral-inhibition type neural fields, *Biol. Cybern.*, 27:77–87.

Blake, R., and Wilson, H. R., 1991, Neural models of stereoscopic vision, *Trends Neurosci.*, 14:445–452. ◆

Braitenberg, V., and Schüz, A., 1991, *Anatomy of the Cortex: Statistics and Geometry*, Berlin: Springer-Verlag. ◆

Chipalkatti, R., and Arbib, M. A., 1988, The cue integration model of depth perception: A stability analysis, *J. Math. Biol.*, 26:235–262.

Dinse, H. R., Krüger, K., and Best, J., 1990, A temporal structure of cortical information processing, *Concepts Neurosci.*, 1:199–238.

Ermentrout, G. B., and Cowan, J. D., 1979, A mathematical theory of visual hallucination patterns, *Biol. Cybern.*, 34:137–150.

Horn, B. K. P., 1974, Determining lightness from an image, *Comput. Vis. Graph. Image Proc.*, 3:277–299.

Korn, A., and von Seelen, W., 1972, Dynamische Eigenschaften von Nervennetzen im visuellen System, *Kybernetik*, 10:64–77.

Krone, G., Mallot, H. A., Palm, G., Schüz, A., 1986, The spatio-temporal receptive field: A dynamical model derived from cortical architectonics, *Proc. R. Soc. Lond. B Biol. Sci.*, 226:421–444.

Mallot, H. A., and Giannakopoulos, F., 1992, Activation dynamics of space-variant continuous networks, in *Neural Network Dynamics* (J. G. Taylor, E. R. Caianiello, R. M. J. Cotterill, and J. W. Clark, Eds.), Berlin: Springer-Verlag, pp. 341–355.

Mallot, H. A., von Seelen, W., and Giannakopoulos, F., 1990, Neural mapping and space-variant image processing, *Neural Netw.*, 3:245–263.

Murray, J. D., 1989, *Mathematical Biology*, Berlin: Springer-Verlag, chap. 16. ◆

Peters, A., 1985, Visual cortex of the rat, in *Cerebral Cortex*, vol. 3, *Visual Cortex* (A. Peters and E. G. Jones, Eds.), New York: Plenum, pp. 19–80. ◆

Schwartz, E. L., 1980, Computational anatomy and functional architecture of striate cortex: A spatial mapping approach to perceptual coding, *Vis. Res.*, 20:645–669.

Learning and Generalization: Theoretical Bounds

V. Vapnik

The Learning Problem

Function Estimation Model

The general model of learning from examples can be described through three components:

1. A generator of random vectors x, drawn independently from a fixed but unknown distribution $P(x)$.
2. A supervisor that returns an output vector y to every input vector x, according to a conditional distribution function $P(y|x)$, also fixed but unknown. (This includes the case in which the supervisor uses a function $y = f(x)$.)
3. A learning machine capable of implementing a set of functions $f(x, \alpha)$, $\alpha \in \Lambda$.

The problem of learning is that of choosing from the given set of functions $f(x, \alpha)$, $\alpha \in \Lambda$, the one which best approximates the supervisor's response. The selection is based on a training set of l independent identically distributed (i.i.d.) observations

$$(x_1, y_1), \ldots, (x_l, y_l) \tag{1}$$

drawn according to $P(x, y) = P(x)P(y|x)$.

Problem of Risk Minimization

In order to choose the best available approximation to the supervisor's response, one measures the *loss* or discrepancy $L(y, f(x, \alpha))$ between the response y of the supervisor to a given input x and the response $f(x, \alpha)$ provided by the learning machine. Consider the expected value of the loss, given by the *risk functional*

$$R(\alpha) = \int L(y, f(x, \alpha)) \, dP(x, y) \tag{2}$$

The goal is to find the function $f(x, \alpha_0)$ which minimizes the risk functional $R(\alpha)$ (over the class of functions $f(x, \alpha)$, $\alpha \in \Lambda$) in the situation where the joint probability distribution $P(x, y)$ is unknown and the only available information is contained in the training set (Equation 1).

Three Main Learning Problems

This formulation of the learning problem is rather broad. It encompasses many specific problems. Consider the main ones: the problems of pattern recognition, regression estimation, and density estimation (Vapnik, 1979, 1995).

Pattern recognition.   Let the supervisor's output y take only two values $y = \{0, 1\}$ and let $f(x, \alpha)$, $\alpha \in \Lambda$, be a set of *indicator* functions (functions which take only two values, 0 and 1). Consider the following loss function:

$$L(y, f(x, \alpha)) = \begin{cases} 0 & \text{if } y = f(x, \alpha) \\ 1 & \text{if } y \neq f(x, \alpha) \end{cases} \quad (3)$$

For this loss function, the functional (Equation 2) determines the probability of the different answers: the answers y given by the supervisor and the answers given by the indicator function $f(x, \alpha)$. We call the case of different answers a *classification error*. The problem, therefore, is to find a function which minimizes the probability of classification errors when the probability measure $P(x, y)$ is unknown, but the data (Equation 1) are given.

Regression estimation.   Let the supervisor's answer y be a real value, and let $f(x, \alpha)$, $\alpha \subset \Lambda$, be a set of real functions which contains the *regression function*

$$f(x, \alpha_0) = \int y \, dP(y | x)$$

It is known that the regression function is the one which minimizes the functional (Equation 2) with the following *least squares* loss function:

$$L(y, f(x, \alpha)) = (y - f(x, \alpha))^2 \quad (4)$$

Thus the problem of regression estimation is the problem of minimizing the risk functional (Equation 2) with the loss function (Equation 4) in the situation where the probability measure $P(x, y)$ is unknown but the data (Equation 1) are given.

Density estimation.   Finally, consider the problem of density estimation from the set of densities $p(x, \alpha)$, $\alpha \in \Lambda$. For this problem we consider the following loss function.

$$L(p(x, \alpha)) = -\log p(x, \alpha) \quad (5)$$

It is known that the desired density minimizes the risk functional (Equation 2) with the loss function (Equation 5). Thus, again, to estimate the density from the data one has to minimize the risk functional under the condition where the corresponding probability measure $P(x)$ is unknown but i.i.d. data $x_1, \ldots, x_n$ are given.

The general setting of the learning problem.   The general setting of the learning problem can be described as follows. Let the probability measure $P(z)$ be defined on the space Z. Consider the set of functions $Q(z, \alpha)$, $\alpha \in \Lambda$. The goal is to minimize the risk functional

$$R(\alpha) = \int Q(z, \alpha) \, dP(z) \quad \alpha \in \Lambda \quad (6)$$

if probability measure $P(z)$ is unknown but an i.i.d. sample

$$z_1, \ldots, z_l \quad (7)$$

is given.

The learning problems considered above are particular cases of this general problem of *minimizing the risk functional* (Equa-

tion 6) *on the basis of empirical data* (Equation 7), where z describes a pair (x, y) and $Q(z, \alpha)$ is the specific loss function (for example, one of Equations 3, 4, or 5). Below we will describe the result obtained for the general statement of the problem. To apply it for specific problems one has to substitute the corresponding loss functions in the formulas obtained.

Empirical Risk Minimization Induction Principle

In order to minimize the risk functional (Equation 6), with an unknown probability measure $P(z)$, the following induction principle is usually applied.

The risk functional $R(\alpha)$ is replaced by the *empirical risk* functional

$$R_{\text{emp}}(\alpha) = \frac{1}{l} \sum_{i=1}^{l} Q(z_i, \alpha) \quad (8)$$

constructed on the basis of the training set (Equation 7).

The principle is to approximate the function $Q(z, \alpha_0)$ which minimizes risk (Equation 6) by the function $Q(z, \alpha_l)$ minimizing empirical risk (Equation 8). This principle is called the *Empirical Risk Minimization* induction principle (ERM principle).

The ERM principle is quite general. The classical methods for the solution of a specific learning problem, such as the least-squares method in the problem of regression estimation or the maximum likelihood method in the problem of density estimation, are realizations of the ERM principle for the specific loss functions considered above.

Four Parts of Learning Theory

Learning theory has to address the following four questions (Vapnik, 1995):

1. *What are (necessary and sufficient) conditions for consistency of the learning process based on the ERM principle?*
   This requires the specification of necessary and sufficient conditions for convergence in probability of the values of risks $R(\alpha_l)$ (for the functions $Q(z, \alpha_l)$ minimizing the empirical risk) to the minimal possible value of the risk $R(\alpha_0)$ as the number of observations increases:

$$R(\alpha_l) \to_{l \to \infty}^{P} R(\alpha_0) \quad (9)$$

2. *How fast is the rate of convergence of the learning process?*
3. *How can one control the generalization ability of the learning process?*
4. *How can one construct algorithms that can control the generalization ability?*

The answers to these questions form the four parts of learning theory:

1. Theory of consistency of the learning processes.
2. Nonasymptotic theory of the rate of convergence of the learning processes.
3. Theory of controlling the generalization ability of the learning processes.
4. Theory of constructing the learning algorithms.

Each of these four parts will be discussed in the following sections.

Remark on PAC model.   A particular case of the classical definition of consistency (Equation 9) where $R(\alpha_0) = 0$ (in this case $R_{\text{emp}}(\alpha_l) = 0$) is used as the definition of the *Probably Approximately Correct* (PAC) model of learning (Valiant, 1984; see PAC LEARNING AND NEURAL NETWORKS). The restriction

$R(\alpha_0) = 0$, however, is too strong. This restriction makes it impossible to analyze the most important cases of the learning problem, namely the case in which the set of functions of the learning machine does not contain the supervisor's rule and the case in which the data contain noise. The results obtained in the framework of the PAC model are a corollary of the results of statistical learning theory, where $R(\alpha_0) = R_{\mathrm{emp}}(\alpha_l) = 0$ (Saitta and Bergadano, 1993).

Theory of Consistency of the Learning Processes

The Key Assertion of the Learning Theory

The key assertion of the learning theory is the following (Vapnik and Chervonenkis, 1989):
Let $Q(z, \alpha)$, $\alpha \in \Lambda$, be a set of totally bounded functions

$$A \le Q(z, \alpha) \le B$$

Then for the ERM principle to be consistent it is necessary and sufficient that the empirical risk $R_{\mathrm{emp}}(\alpha)$ converge *uniformly* to the actual risk $R(\alpha)$ over the set $Q(z, \alpha)$, $\alpha \in \Lambda$.
Uniform (one-sided) convergence is defined as

$$\lim_{l \to \infty} \mathrm{Prob} \left\{ \sup_{\alpha \in \Lambda} (R(\alpha) - R_{\mathrm{emp}}(\alpha)) > \varepsilon \right\} = 0 \quad \text{for all } \varepsilon \quad (10)$$

In other words, according to the key assertion, the conditions for consistency of the ERM principle are equivalent to the conditions for uniform one-sided convergence (Equation 10).

The Necessary and Sufficient Conditions

To describe the necessary and sufficient conditions for uniform convergence (Equation 10), it is necessary to introduce a new concept which is called *the entropy of the set of functions* $Q(z, \alpha)$, $\alpha \in \Lambda$, *on the sample of size l*.
Let $A \le Q(z, \alpha) \le B$, $\alpha \in \Lambda$, be the set of bounded loss functions. Using this set of functions and the training set (Equation 7), one can construct the following set of l-dimensional vectors:

$$q(\alpha) = (Q(z_1, \alpha), \ldots, Q(z_l, \alpha)) \quad \alpha \in \Lambda \quad (11)$$

This set of vectors belongs to the l-dimensional cube and has a finite ε-net in the metric C. Let $N = N^{\Lambda}(\varepsilon; z_1, \ldots, z_l)$ be a number of elements of the minimal ε-net of this set of vectors $q(\alpha)$, $\alpha \in \Lambda$.
The logarithm of the (random) value $N^{\Lambda}(\varepsilon; z_1, \ldots, z_l)$

$$H^{\Lambda}(\varepsilon; z_1, \ldots, z_l) = \ln N^{\Lambda}(\varepsilon; z_1, \ldots, z_l)$$

is called the *random VC-entropy* of the set of functions $A \le Q(z, \alpha) \le B$ on the sample $z_1, \ldots, z_l$. The expectation of the random VC-entropy

$$H^{\Lambda}(\varepsilon; l) = EH^{\Lambda}(\varepsilon; z_1, \ldots, z_l)$$

is called the *VC-entropy* of the set of functions $A \le Q(z, \alpha) \le B$, $\alpha \in \Lambda$, on a sample of size l. Here expectation is taken with respect to the product measure $P(z_1, \ldots, z_l) = P(z_1) \cdot \ldots \cdot P(z_l)$.
The main results of the theory of uniform convergence are connected with the equality

$$\lim_{l \to \infty} \frac{H^{\Lambda}(\varepsilon, l)}{l} = 0 \quad \text{for all } \varepsilon \quad (12)$$

This equality describes the necessary and sufficient conditions for uniform one-sided convergence (Equation 10) over the set of bounded functions.
According to the key assertion, Equation 12 implies the necessary and sufficient conditions for consistency of the ERM principle. These necessary and sufficient conditions of consis-

tency of the ERM principle were found in Vapnik and Chervonenkis (1989). However, it was previously found that Equation 12 forms the necessary and sufficient condition for uniform two-sided convergence

$$\lim_{l \to \infty} \mathrm{Prob} \left\{ \sup_{\alpha \in \Lambda} |R(\alpha) - R_{\mathrm{emp}}(\alpha)| > \varepsilon \right\} = 0 \quad \text{for all } \varepsilon \quad (13)$$

over the set of indicator functions $Q(z, \alpha)$, $\alpha \in \Lambda$ (Vapnik and Chervonenkis, 1968, 1971) and forms the necessary and sufficient conditions for two-sided uniform convergence (Equation 13) over the set of real bounded functions as well (Vapnik and Chervonenkis, 1981).

Bounds for the Rate of Convergence

The nonasymptotic theory of the bounds for the rate of convergence of the learning processes is based on the concept of *VC-dimension* (abbreviation for Vapnik-Chervonenkis dimension; see VAPNIK-CHERVONENKIS DIMENSION OF NEURAL NETWORKS) of the set of functions $Q(z, \alpha)$, $\alpha \in \Lambda$. It was developed in the 1970s and 1980s (Vapnik and Chervonenkis, 1968, 1971; Vapnik, 1979, 1995).

VC-Dimension of the Set of Functions

Below we give an definition of VC-dimension of sets of indicator functions and then we generalize this definition for the sets of real functions.

The VC-dimension of a set of indicator functions. The *VC-dimension of a set of indicator functions* $Q(z, \alpha)$, $\alpha \in \Lambda$, is the maximum number h of vectors $z_1, \ldots, z_h$ which can be separated in all 2^h possible ways using functions of this set (*shattered* by this set of functions). If for any n there exists a set of n vectors which can be shattered by the set $Q(z, \alpha)$, $\alpha \in \Lambda$, then the VC-dimension is equal to infinity.

The VC-dimension of the set of real functions (Vapnik, 1979). Let $a \le Q(z, \alpha) \le A$, $\alpha \in \Lambda$, be a set of real functions bounded by constants a and A (a can be $-\infty$ and A can be ∞). Let us consider along with the set of real functions $Q(z, \alpha)$, $\alpha \in \Lambda$, the set of indicator functions

$$I(z, \alpha, \beta) = \theta\{Q(z, \alpha) - \beta\} \quad \alpha \in \Lambda \quad \beta \in (a, A) \quad (14)$$

where $\theta(z)$ is the step-function

$$\theta(z) = \begin{cases} 0 & \text{if } z < 0 \\ 1 & \text{if } z \ge 0 \end{cases}$$

The *VC-dimension of the set of real functions* $Q(z, \alpha)$, $\alpha \in \Lambda$, is defined to be the VC-dimension of the set of indicator functions (Equation 14).

Example. The VC-dimension of the set of *linear indicator functions*

$$Q(z, \alpha) = \theta\left\{ \sum_{p=1}^{n} \alpha_p z_p + \alpha_0 \right\}$$

in n-dimensional coordinate space $Z = (z_1, \ldots, z_n)$ is equal to $h = n + 1$, since using functions of this set one can shatter at most $n + 1$ vectors.

Example. The VC-dimension of the set of *linear functions*

$$Q(z, \alpha) = \sum_{p=1}^{n} \alpha_p z_p + \alpha_0 \quad \alpha_0, \ldots, \alpha_n \in (-\infty, \infty)$$

in n-dimensional coordinate space $Z = (z_1, \ldots, z_n)$ is equal to $h = n + 1$, because the VC-dimension of the corresponding linear indicator functions is equal to $n + 1$ (using $\alpha_0 - \beta$ instead of α_0 does not change the set of indicator functions).

Distribution Independent Bounds

Consider sets of functions which possess a finite VC-dimension h. We distinguish between two cases.

1. The case where the set of loss functions $Q(z, \alpha)$, $\alpha \in \Lambda$, is a set of *totally bounded functions*.
2. The case where the set of loss functions $Q(z, \alpha)$, $\alpha \in \Lambda$, is *not necessarily a set of totally bounded functions*.

Case 1: The set of totally bounded functions. Without loss of generality, we assume that

$$0 \leq Q(z, \alpha) \leq B \qquad \alpha \in \Lambda \qquad (15)$$

The main result in the theory of learning machines with the set of totally bounded functions is the following assertion (Vapnik, 1979, 1995):

- With probability at least $1 - \eta$, the inequality

$$R(\alpha) \leq R_{\text{emp}}(\alpha) + \frac{B\varepsilon}{2}\left(1 + \sqrt{1 + \frac{4R_{\text{emp}}(\alpha)}{B\varepsilon}}\right) \qquad (16)$$

holds true simultaneously for all functions of the set in Equation 15, where

$$\varepsilon = 4\frac{h\left(\ln\frac{2l}{h} + 1\right) - \ln\frac{\eta}{4}}{l} \qquad (17)$$

For the set of indicator functions, $B = 1$.

This inequality bounds the risks for all functions of the set given in Equation 15, including the function $Q(z, \alpha_l)$ which minimizes empirical risk (Equation 8).

Case 2: The set of unbounded functions. Consider the set of (nonnegative) unbounded functions $0 \leq Q(z, \alpha)$, $\alpha \in \Lambda$. It is easy to show (by constructing examples) that without additional information about the set of unbounded functions and/ or probability measures it is impossible to obtain an inequality of the type in Equation 16. Below we use the following information: we are given a pair (p, τ) such that inequality

$$\sup_{\alpha \in \Lambda} \frac{\left(\int Q^p(z, \alpha)\, dP(z)\right)^{1/p}}{\int Q(z, \alpha)\, dP(z)} \leq \tau < \infty \qquad (18)$$

holds true, where $p > 1$. (This inequality characterizes the *tails of distributions* of the random variables $\xi_\alpha = Q(z, \alpha)$, generated by the $P(z)$.)

The main result of the theory of learning machines with unbounded set of functions is the following assertion, which for simplicity we will describe for the case $p > 2$ (the results for case $p > 1$ one can be found in Vapnik, 1979, 1995):

- With probability at least $1 - \eta$ the inequality

$$R(\alpha) \leq \frac{R_{\text{emp}}(\alpha)}{(1 - a(p)\tau\sqrt{\varepsilon})_+} \qquad a(p) = \sqrt[p]{\frac{1}{2}\left(\frac{p-1}{p-2}\right)^{p-1}} \qquad (19)$$

holds true simultaneously for all functions, satisfying Equation 18, where ε is determined by Equation 17 and $(a)_+ = \max(a, 0)$.

This inequality bounds the risks for all functions (including the function $Q(z, \alpha_l)$ which minimizes the empirical risk (Equation 8).

These bounds cannot be significantly improved.

Theory for Controlling the Generalization Ability

The theory for controlling the generalization ability of a learning machine is devoted to constructing an induction principle for minimizing the risk functional using a *"small" sample size* of the training set. (The sample size l for a set with Vapnik-Chervonenkis dimension h is considered to be small if the ratio l/h is small, say $l/h < 20$.)

Structural Risk Minimization Induction Principle

The ERM principle can be justified by considering the inequalities in Equations 16 or 19. When l/h is large, the second summand on the right-hand side of Equation 16 becomes small. The actual risk is then close to the value of the empirical risk. In this case, the small value of the empirical risk guarantees a small value of the (expected) risk.

However, if l/h is small, even small $R_{\text{emp}}(\alpha_l)$ does not guarantee a small value of risk. In this case the minimization for $R(\alpha)$ requires a new principle, based on the simultaneous minimization of both terms in Equation 16 (Equation 19), one of which depends on the value of the empirical risk while the second depends on the VC-dimension of the set of functions.

The following principle, called the *structural risk minimization* (SRM) principle, is intended to minimize the risk functional with respect to both the empirical risk and the VC-dimension of the set of functions (Vapnik and Chervonenkis, 1974; Vapnik, 1979, 1995).

Let the set S of functions $Q(z, \alpha)$, $\alpha \in \Lambda$, be provided with a *structure*: nested subsets of functions $S_k = \{Q(z, \alpha), \alpha \in \Lambda_k\}$, such that

$$S_1 \subset S_2 \subset \ldots \subset S_n \ldots \qquad (20)$$

and $S^* = \bigcup_k S_k$.

An *admissible structure* is one satisfying the following three properties:

1. The set S^* is everywhere dense in S.
2. The VC-dimension h_k of each set S_k of functions is finite.
3. Any element S_k of the structure contains either:
   a set of totally bounded functions $0 \leq Q(z, \alpha) \leq B_k$, $\alpha \in \Lambda_k$, or
   a set of functions satisfying the inequality

$$\sup_{\alpha \in \Lambda_k} \frac{\left(\int Q^p(z, \alpha)\, dP(z)\right)^{1/p}}{\int Q(z, \alpha)\, dP(z)} \leq \tau_k \qquad p > 2 \qquad (21)$$

for some pair (p, τ_k).

For a given set of observations $z_1, \ldots, z_l$ the SRM principle suggests choosing the element of structure S_n with function $Q(z, \alpha_l^n)$ for which the guaranteed risk of Equation 16 (or risk of Equation 19) is minimal.

The SRM principle actually suggests a *tradeoff between the quality of the approximation and the complexity of the approximating function*. As n increases, the minima of empirical risk are decreased; however, the term responsible for the confidence interval (summand in Equation 16 or multiplier in Equation 19) is increased. The SRM principle takes both factors into account.

Asymptotic Rate of Convergence

For asymptotic analysis of SRM principle one should consider a law determining for a given l the number

$$n = n(l) \qquad (22)$$

of element S_n of the structure in Equation 20. In this case the method of structural risk minimization provides approximations $Q(z, \alpha_l^{n(l)})$ for which the sequence of risks $R(\alpha_l^{n(l)})$ converges to the best one $R(\alpha_0)$ with *asymptotic rate of convergence*

$$V(l) = r_{n(l)} + T_{n(l)} \sqrt{\frac{h_{n(l)} \ln l}{l}} \qquad (23)$$

if the law $n = n(l)$ is such that

$$\lim_{l \to \infty} \frac{T_{n(l)}^2 h_{n(l)} \ln l}{l} = 0 \qquad (24)$$

In Equation 23 $T_n = B_n$ if one considers a structure with totally bounded functions in subsets S_n, and $T_n = \tau_n$ if one considers a structure with elements satisfying Equation 25. To provide the best rate of convergence one has to know the *rate of approximation r_n for the chosen structure*:

$$r_n = \inf_{\alpha \in \Lambda_n} \int Q(z, \alpha) \, dP(z) - \inf_{\alpha \in \Lambda} \int Q(z, \alpha) \, dP(z) \qquad (25)$$

In this case by minimizing the right-hand side of Equation 23 one can a priori find the law $n = n(l)$ which gives the best asymptotic rate.

Example. Let $Q(z, \alpha)$, $\alpha \in \Lambda$, be a set of functions satisfying Equation 19 for $p > 2$. Consider a structure for which $n = h_n$. Let the asymptotic rate of approximation be described by the law

$$r_n = \left(\frac{1}{n}\right)^c$$

(This law describes the main classical results in approximation theory.) Then the asymptotic rate of convergence reaches its maximum value if

$$n(l) = \left[\frac{l}{\ln l}\right]^{1/2c+1}$$

where $[a]$ is integer part of a. The asymptotic rate of convergence is

$$V(l) = \left(\frac{\ln l}{l}\right)^{c/2c+1} \qquad (26)$$

Examples of the Structure for Neural Nets

The general SRM principle can be implemented in many different ways. Here we consider three different examples of structures built for the set of functions implemented by a neural network.

1. *Structure given by the architecture of the neural network.* Consider an ensemble of fully connected neural networks in which the number of units in one of the hidden layers is monotonically increased. The set of implementable functions defines the structure as the number of hidden units is increased.

2. *Structure given by the learning procedure.* Consider the set of functions $S = \{f(x, w), w \in W\}$ implementable by a neural net of fixed architecture. The parameters $\{w\}$ are the weights of the neural network. A structure is introduced through $S_p = \{f(x, w), \|w\| \le C_p\}$ and $C_1 < C_2 < \dots < C_n$. Under very gen-

eral conditions on the set of loss functions, the minimization of the empirical risk within the element S_p of the structure is achieved through the minimization of

$$E(w, \gamma_p) = \frac{1}{l} \sum_{i=1}^{l} L(y_i, f(x_i, w)) + \gamma_p \|w\|^2$$

with appropriately chosen Lagrange multipliers $\gamma_1 > \gamma_2 > \dots > \gamma_n$. The well-known "weight decay" procedure refers to the minimization of this functional.

3. *Structure given by preprocessing.* Consider a neural net with fixed architecture. The input representation is modified by a transformation $z = K(x, \beta)$, where the parameter β controls the degree of degeneracy introduced by this transformation (for instance β could be the width of a smoothing kernel).

A structure is introduced in the set of functions $S = \{f(K(x, \beta), w), w \in W\}$ through $\beta \ge C_p$ and $C_1 > C_2 > \dots > C_n$.

The method for controlling the generalization ability of the learning machine using preprocessing is important for problems of image recognition. It reflects the idea of developing the structure by increasing the detail used in describing the input image.

Constructing the Learning Algorithms

Generalization of the two ideas of minimizing the empirical risk on the set of linear indicator functions developed in the 1960s constitutes two main branches in the theory of constructing learning algorithms for the pattern recognition problem.

Idea of Sigmoid Approximation of Indicator Functions

Consider first the problem of minimizing empirical risk

$$R_{\mathrm{emp}}(w) = \frac{1}{l} \sum_{j=1}^{l} (y_i - f(x_j, w))^2 \qquad (27)$$

on the set of *linear indicator functions* $f(x, \alpha)$, $\alpha \in \Lambda$. Since y and $f(x, \alpha)$ take only two values 0 and 1, it is impossible to use regular *gradient-based* methods for minimizing Equation 27. (The gradient of the function $R_{\mathrm{emp}}(w)$ is either equal to zero or undefined.) The idea is to approximate the set of indicator functions by so-called *sigmoid functions* $\overline{f}(x, w)$—smooth monotonic functions such that $\overline{f}(-\infty, w) = 0$, $\overline{f}(+\infty, w) = 1$.

For the set of sigmoid functions $\overline{f}(x, \alpha)$, $\alpha \in \Lambda$, the empirical risk functional has a gradient $\operatorname{grad} R_{\mathrm{emp}}(w)$ and therefore can be minimized using standard gradient-based methods. Thus, the idea is to use the sigmoid approximation at the stage of estimating the coefficients and use the indicator functions with these coefficients at the stage of recognition.

The generalization of linear machines leads to Learning Neural Nets. In this generalization, instead of directly using linear indicator functions (single neuron), one considers a set of functions which are the superposition of several linear indicator functions (neural nets). All indicator functions in this superposition are replaced by sigmoid functions.

The method for calculating the gradient of the empirical risk for the sigmoid approximation of neural nets is called the *backpropagation method* (Rumelhart, Hinton, and Williams, 1986; Le Cun, 1986). Using this gradient, one can iteratively modify the coefficients (weights) of a neural net on the basis of standard gradient-based procedures.

Idea of Optimal Separating Hyperplane

The second idea for constructing learning algorithms is developing the idea of constructing the *optimal separating hyper-*

plane (Vapnik and Chervonenkis, 1974; Vapnik, 1979). The optimal hyperplane is one which among all separating hyperplanes has a maximal distance between hyperplane and the closest vector.

The idea is as follows: a learning machine maps the input vectors into a very high-dimensional feature space Z through some nonlinear mapping chosen a priori. In this space a separating hyperplane is constructed.

Example. To construct a decision surface corresponding to a polynomial of degree 2, one can create a feature space Z which has $N = [n(n + 3)]/2$ coordinates of the form:

$$z_1 = x_1, \ldots, z_n = x_n \qquad n \text{ coordinates}$$

$$z_{n+1} = x_1^2, \ldots, z_{2n} = x_n^2 \qquad n \text{ coordinates}$$

$$z_{2n+1} = x_1 x_2, \ldots, z_N = x_n x_{n-1} \qquad \frac{n(n-1)}{2} \text{ coordinates}$$

where $x = (x_1, \ldots, x_n)$. The separating hyperplane constructed in this space is a second-degree polynomial in the input space.

Two problems arise in this approach, one conceptual and one technical. The conceptual problem is how to find a separating hyperplane that will generalize well. The technical problem is how to treat computationally such high-dimensional spaces.

The conceptual part of this problem was solved in 1965 (Vapnik and Chervonenkis, 1974). It was shown that to construct the optimal hyperplanes one has only to take into account a small amount of the training data, the so-called *support vectors*, which determine the margin. It was shown also that if the training vectors are separated without errors by an optimal hyperplane, the expectation value of the probability of committing an error on a test example is bounded by the ratio

$$E[\Pr(\text{error})] \leq \frac{E[\text{number of support vectors}]}{\text{number of training vectors}} \qquad (28)$$

This bound does not depend on dimensionality of the space. Therefore, if the optimal hyperplane can be constructed from a small number of support vectors, the generalization ability will be high—even in an infinite-dimensional space. This property uses Support Vectors Networks (described later) to construct decision rules in high-dimensional space.

However, even if the optimal hyperplane generalizes well and can theoretically be found, the technical problem of how to treat the high-dimensional feature space remains. In 1992 an effective way was found for constructing a hyperplane in some high-dimensional spaces, including the space of polynomials in high-dimensional input space, the space of radial basis functions, and two-layer neural nets (Boser, Guyon, and Vapnik, 1992; Cortes and Vapnik, 1995).

Why Can Neural Networks and Support Vectors Networks Generalize?

To control the generalization ability of the learning process, one has to construct a structure (Equation 20) on the set of decision functions $S = Q(z, \alpha), \alpha \in \Lambda\}$ and then chose both an appropriate element S_k of the structure and a function $Q(z, \alpha_l^k) \in S_k$ in this element that minimizes the bound in Equation 16. The bound in Equation 16 can be rewritten in the simple form

$$R(\alpha_l^k) \leq R_{\text{emp}}(\alpha_l^k) + \Omega\left(\frac{l}{h_k}\right) \qquad (29)$$

Table 1. Handwritten Digit Recognition with Support Vectors Networks, Showing Effect of Increasing Order of Polynomials

Degree of Polynomial	Raw Error, %	Support Vectors	Dimensionality of Feature Space
1	12.0	200	256
2	4.7	127	~ 33000
3	4.4	148	$\sim 1 \times 10^6$
4	4.3	165	$\sim 1 \times 10^9$
5	4.3	175	$\sim 1 \times 10^{12}$
6	4.2	185	$\sim 1 \times 10^{14}$
7	4.3	190	$\sim 1 \times 10^{16}$

Designing a neural network, one determines a set of admissible functions with some VC-dimension h^* which for a given amount l of training data determines the confidence interval. During the learning process this network minimizes the first term in the bound of Equation 29 (number of errors on the training set). This approach to minimizing the right-hand side of Equation 29 can be described as follows: *Keep the confidence interval fixed (by choosing an appropriate network) and minimize the empirical risk.*

Support Vectors Networks (SVN) is another approach, which can be described as follows: *Keep the value of empirical risk fixed (say, equal to zero) and minimize the confidence interval.* The SVN takes advantage of the fact that the generalization ability of learning machines depends on the VC-dimension of the set of functions rather than on the number of parameters determining the set of functions. The VC-dimension of the element of structure used in the SVN depends on the number of support vectors and is independent of the dimensionality of space.

Example: Handwritten Digit Recognition with Learning Machines

Experiments were conducted with the U.S. Postal Service digits database, which contains 7300 training patterns and 2000 test patterns with 16×16 pixel resolution. To construct the decision rules for digit recognition, a five-layer neural network with architecture providing feature extraction for handwritten digits was constructed. The performance of this neural net (LeNet 1) is 5.1% raw error (LeCun et al., 1990).

Table 1 describes the results of the experiments with the same database using ten decision polynomials (one per class) constructed by an SVN (Cortes and Vapnik, 1995). It shows that the number of support vectors (Table 1 shows mean value per classifier) increases very slowly with the order of the polynomials, whereas the dimensionality of the feature space increases extremely fast. Note that performance does not decline with increasing dimensionality of the space—indicating no overfitting problems.

Road Map: Computability and Complexity
Background: I.3. Dynamics and Adaptation in Neural Networks
Related Reading: Generalization and Regularization in Nonlinear Learning Systems; Statistical Mechanics of Generalization

References

Boser, B., Guyon, I., and Vapnik, V. N., 1992, A training algorithm for optimal margin classifiers, in *Fifth Annual Workshop on Computational Learning Theory*, San Mateo, CA: Morgan Kaufmann, pp. 144–152.
Cortes, C., and Vapnik, V., 1995, Support vector networks, submitted to *Machine Learning.*

LeCun, Y., 1986, Learning processes in asymmetric threshold network, in *Disordered Systems and Biological Organizations*, Les Houches, France: Springer-Verlag, pp. 233–240.

LeCun, Y., Boser, B., Denker, J. S., Henderson, D., Howard, R. E., Hubbard, W., and Jackel, L. J., 1990, Handwritten digit recognition with back-propagation network, in *Advances in Neural Information Processing Systems* (D. S. Touretzky, Ed.), San Mateo, CA: Morgan Kaufman, pp. 396–404.

Rumelhart, D. E., Hinton, G. E., and Williams, R. J., 1986, Learning internal representations by error propagation, in *Parallel Distributed Processing: Explorations in the Microstructure of Cognition* (D. E. Rumelhart, J. L. McClelland, and PDP Research Group, Eds.), vol. 1, *Foundations*, Cambridge, MA: MIT Press, pp. 318–362.

Saitta, L., and Bergadano, F., 1993, Pattern recognition and Valiant's learning framework, *IEEE Trans. Pattern Analysis Machine Intell.*, 15:145–155.

Valiant, L. G., 1984, A theory of learnability, *Commun. ACM*, 27:1134–1142.

Vapnik, V. N., 1979, *Estimation of Dependencies Based on Empirical Data* [in Russian], Moscow: Nauka. (English translation, 1982, New York: Springer-Verlag.)

Vapnik, V. N., 1995, *Statistical Learning Theory*, New York: Wiley.

Vapnik, V. N., and Chervonenkis, A. Ja., 1968, On the uniform convergence of relative frequencies of events to their probabilities, *Rep. Acad. Sci. USSR*, 181, no. 4.

Vapnik, V. N., and Chervonenkis, A. Ja., 1971, On the uniform convergence of relative frequencies of events to their probabilities, *Theory Probab. Appl.*, 16:264–280.

Vapnik, V. N., and Chervonenkis, A. Ja., 1974, *Theory of Pattern Recognition* [in Russian], Moscow: Nauka. (German translation: Wapnik, W. N., and Cherwonenkis, A. Ja., 1979, *Theorie der Zeichenerkennung*, Berlin: Akademia-Verlag.

Vapnik, V. N., and Chervonenkis, A. Ja., 1981, Necessary and sufficient conditions for the uniform convergence of the means to their expectations, *Theory Probab. Appl.*, 26:532–553.

Vapnik, V. N., and Chervonenkis, A. Ja., 1989, The necessary and sufficient conditions for consistency of the method of empirical risk minimization [in Russian], in *Yearbook of the Academy of Sciences of the USSR on Recognition, Classification, and Forecasting*, vol. 2, Moscow: Nauka, pp. 217–249. (English translation, 1991, *Pattern Recognition Image Anal.*, 1:284–305.)

Learning and Statistical Inference

Shun-ichi Amari

Introduction

Behaviors of neural networks are often subject to stochastic laws. Even for a deterministic neuron, a stochastic interpretation is sometimes very useful. The discipline of *statistical inference* studies stochastic phenomena, and it has a long tradition of research. Learning from examples in neural networks is of a stochastic nature in the sense that examples are randomly generated and a network's behavior is intrinsically fluctuating or is stochastically interpreted. Promising approaches to the study of learning problems from the statistical point of view include such notions as Fisher information, Bayesian loss, and sequential estimation. Recent developments in neural learning have been influenced by many statistical ideas (White, 1989).

Nonlinear neurodynamics, learning, and self-organization, among others, are key concepts leading to new developments of statistical science. Consequently, many statisticians have recently become interested in neural network technology (see, e.g., Ripley, 1994). The present article reviews various aspects of neural learning from the statistical-inference point of view.

Stochastic Neurons

We first describe stochastic behaviors of single neurons and then those of networks. A mathematical neuron receives a number of input signals $\{x_1, \ldots, x_n\}$, summarized in an input vector $\mathbf{x} = (x_0, x_1, \ldots, x_n)$, and emits an output z. Here, $x_0 = 1$ is added to set the bias or the threshold term. The neuron calculates the weighted sum of inputs

$$u = \mathbf{w} \cdot \mathbf{x} = \sum w_i x_i \tag{1}$$

where w_i are the synaptic efficacies or connection weights. The output z is determined stochastically depending on u. In the simple additive noise case, the output is written as

$$z = f(u) + n \tag{2}$$

where f is a sigmoidal nonlinear function and n is a stochastic noise term. Let $r(n)$ be the probability density function of the noise. The expectation of n is assumed to be zero.

$$E[n] = 0$$

Then, the conditional probability of z when $\mathbf{x}$ is inputed is given by

$$p(z \mid \mathbf{x}) = r\{z - f(\mathbf{w} \cdot \mathbf{x})\}$$

and the expectation of z is

$$E[z \mid \mathbf{x}] = \int z p(z \mid \mathbf{x}) \, dz = f(\mathbf{w} \cdot \mathbf{x}) \tag{3}$$

where $E[z \mid \mathbf{x}]$ denotes the conditional expectation of z under the condition that the input is $\mathbf{x}$.

When z takes on the binary values 0 and 1, a widely used stochastic neuron model has the following probability:

$$p(z \mid \mathbf{x}) = \frac{\exp\{\beta z \mathbf{w} \cdot \mathbf{x}\}}{1 + \exp\{\beta \mathbf{w} \cdot \mathbf{x}\}} \qquad z = 0, 1 \tag{4}$$

so that

$$E[z \mid \mathbf{x}] = f(\mathbf{w} \cdot \mathbf{x})$$

with

$$f(u) = \frac{\exp\{\beta u\}}{1 + \exp\{\beta u\}}$$

When $\beta \to \infty$, $f(u)$ converges to the step function, implying that z becomes deterministic, taking the value 1 when $\mathbf{w} \cdot \mathbf{x}$ is positive and 0 otherwise. On the other hand, when $\beta \to 0$, $f(u)$ becomes almost uniform, implying that stochastic fluctuation dominates. The parameter β is called sometimes the inverse of the "temperature" in the statistical mechanics analogy.

A multilayer perceptron is a neural network with feedforward connections. It consists of input neurons, hidden neurons, and output neurons. It receives an input $\mathbf{x}$ from the input neurons and emits an output vector signal $\mathbf{z} = (z_1, \ldots, z_m)$ from the m output neurons. In many cases, the hidden neurons behave deterministically, but the behaviors can be interpreted stochastically. In some cases, noise is added to the output neurons. The behavior of a network is again represented by the conditional probability distribution $p(\mathbf{z}|\mathbf{x})$. Since the network includes a large number of modifiable parameters (synaptic efficacies and thresholds of component neurons), we summarize all of them in a vector $\mathbf{w} = (w_1, \ldots, w_k)$. The conditional probability is expressed in the form $p(\mathbf{z}|\mathbf{x}; \mathbf{w})$, showing that it depends on the parameters $\mathbf{w}$.

Let $(\mathbf{x}_t, \mathbf{z}_t)$, $t = 1, \ldots, T$, be examples of input-output pairs. Learning is carried out based on the training set

$$D_T = \{(\mathbf{x}_1, \mathbf{z}_1), \ldots, (\mathbf{x}_T, \mathbf{z}_T)\} \tag{5}$$

In many cases, $\mathbf{x}_t$ are generated independently subject to an unknown probability distribution $q(\mathbf{x})$, and $\mathbf{z}_t$ is the desired output from the true or ideal neural network. In the case of self-organization or unsupervised learning, the desired outputs $\mathbf{z}_t$ are missing, so that D_T consists of only $\mathbf{x}_t$'s. Learning is a procedure to modify (usually sequentially) the network parameters $\mathbf{w}$ based on the training set D_T such that the trained network behaves sufficiently well to simulate the true or ideal network from which the training data are obtained. From the statistical point of view, this looks like statistical estimation of the unknown parameters $\mathbf{w}$ from observed data D_T. However, many interesting problems arise from the point of view of learning.

For example, learning is a sequential estimation procedure where examples $(\mathbf{x}_t, \mathbf{z}_t)$ are obtained one by one. Moreover, the neural network model we use is merely an approximation of the true or desired input-output behavior; that is, the statistical model is unfaithful. This case leads us to the problem of model selection, which is closely related to generalization and training errors. Dynamic aspects of learning or learning curves are also interesting topics.

Information Measures

It is useful to summarize here information measures for later use. Let X be a random variable with probability density function $p(x)$. The *entropy*

$$H(X) = -\int p(x) \log p(x)\, dx$$

is the measure of uncertainty of X. When two random variables X and Y are correlated, we can obtain some information concerning X if we can observe Y. The amount of such information is given by the *Shannon information*:

$$I(X:Y) = H(X) - H(X|Y)$$

where $H(X|Y)$ is the *conditional information* (Cover and Thomas, 1991). In learning, we can evaluate the amount of information included in an example $(\mathbf{x}_t, \mathbf{z}_t)$ in Shannon's sense.

Let $p(x, \mathbf{w})$ be the probability density function parameterized by $\mathbf{w}$. The family $M = \{p(x, \mathbf{w})\}$ is called a *statistical model*. Let $x_1, \ldots, \mathbf{x}_T$ be T independent observations, and let $\hat{\mathbf{w}}$ be an estimator of the true $\mathbf{w}$. How well can we estimate $\mathbf{w}$? The *Fisher information* matrix $G = (g_{ij})$ is defined in this connection by

$$g_{ij} = E\left[\frac{\partial}{\partial w_i} \log p(x, \mathbf{w}) \frac{\partial}{\partial w_j} \log p(x, \mathbf{w})\right] \tag{6}$$

where E denotes the expectation with respect to $p(x, \mathbf{w})$. When we measure the accuracy of an estimator $\hat{\mathbf{w}}$ by its error covariance matrix $V = (v_{ij})$,

$$v_{ij} = E[(\hat{w}_i - w_i)(\hat{w}_j - w_j)] \tag{7}$$

the Cramér-Rao theorem shows that

$$V \geq \frac{1}{T} G^{-1} \tag{8}$$

implying that the error is at best as small as the inverse of the Fisher information divided by the number T of observations. Moreover, the maximum likelihood estimator $\hat{\mathbf{w}}_{mle}$ that maximizes the likelihood $p(x_1, \mathbf{w}) \ldots p(x_T, \mathbf{w})$ attains the bound asymptotically. (See standard textbooks such as Cox and Hinkley, 1974.)

Let $p(x)$ and $q(x)$ be two probability distributions. How different are they? A frequently used measure is the *Kullback-Leibler divergence* or the *relative entropy* defined by

$$KL(p \| q) = \int p(x) \log \frac{p(x)}{q(x)}\, dx \tag{9}$$

It is related to both the Shannon and the Fisher information as follows. Let $p_{XY}(x, y)$ be the joint probability of X and Y, and let $p_X(x)$ and $p_Y(y)$ be their marginal distributions. Then

$$I(X:Y) = KL(p_{XY} \| p_X p_Y) \tag{10}$$

When $\mathbf{w}$ and $\mathbf{w} + d\mathbf{w}$ are infinitesimally close,

$$KL(p(x, \mathbf{w}) \| p(x, \mathbf{w} + d\mathbf{w})) = \tfrac{1}{2} \sum g_{ij}\, dw_i\, dw_j \tag{11}$$

so that the KL divergence is measured by the Fisher information.

Sequential Estimation and the Bayesian Standpoint

The behavior of a parameterized network is given by the conditional probability $p(\mathbf{z}|\mathbf{x}; \mathbf{w})$. When $\mathbf{x}$ is generated subject to $q(\mathbf{x})$, the input-output joint distribution is $q(\mathbf{x})p(\mathbf{z}|\mathbf{x}; \mathbf{w})$. Therefore, the probability that the training set D_T is generated from a network specified by $\mathbf{w}$ is

$$P(D_T; \mathbf{w}) = \prod_{t=1}^{T} q(\mathbf{x}_t) p(\mathbf{z}_t | \mathbf{x}_t; \mathbf{w}) \tag{12}$$

Given the data D_T, this is called the *likelihood function*, showing the likelihood that the data are generated from a network specified by $\mathbf{w}$. The maximum likelihood estimator $\hat{\mathbf{w}}_{mle}$ is the one that maximizes the likelihood $P(D_T; \mathbf{w})$ or its logarithm

$$\log P(D_T; \mathbf{w}) = \sum \log q(\mathbf{x}_t) + \sum \log p(\mathbf{z}_t | \mathbf{x}_t; \mathbf{w})$$

The $\hat{\mathbf{w}}_{mle}$ is a consistent estimator, i.e.,

$$\lim_{T \to \infty} \hat{\mathbf{w}}_{mle} = \mathbf{w}_0$$

where $\mathbf{w}_0$ is the true parameter from which the training data are derived. More precisely, $\hat{\mathbf{w}}_{mle}$ is asymptotically normally distributed with mean $\mathbf{w}_0$ and covariance matrix G^{-1}/T (Rao, 1973; Cox and Hinkley, 1974), where G is the Fisher information matrix defined in this case by

$$G = E[\nabla \log p(\mathbf{z}|\mathbf{x}, \mathbf{w}) \{\nabla \log p(\mathbf{z}|\mathbf{x}, \mathbf{w})\}^T] \tag{13}$$

The expression $\nabla f(\mathbf{w})$ represents the gradient column vector whose components are $(\partial/\partial w_i) f(\mathbf{w})$, and $\mathbf{a}^T$ is the transposition of vector $\mathbf{a}$.

Since the Fisher information plays a fundamental role in the accuracy of estimation and learning, some examples are shown here. The first case is that the output $\mathbf{z}$ is a noise-contaminated version of $\mathbf{f}(\mathbf{x}; \mathbf{w})$, which is the ideal output of a multilayer perceptron with analog activation function:

$$\mathbf{z} = \mathbf{f}(\mathbf{x}; \mathbf{w}) + \mathbf{n}$$

where $\mathbf{n}$ is Gaussian noise with mean 0 and covariance matrix $\sigma^2 I$, I being the identity matrix. From the statistical viewpoint, this is the nonlinear regression problem (Rao, 1973) of observed data D_T to the nonlinear model $\mathbf{f}(\mathbf{x}, \mathbf{w})$. The log probability is

$$\log p(\mathbf{z}; \mathbf{w}) = -\frac{1}{2\sigma^2}\{\mathbf{z} - \mathbf{f}(\mathbf{x}, \mathbf{w})\}^T\{\mathbf{z} - \mathbf{f}(\mathbf{x}, \mathbf{w})\} + \text{const.}$$

so that

$$\nabla \log p = \frac{1}{\sigma^2}(\mathbf{z} - \mathbf{f})^T\nabla\mathbf{f}$$

Since $E\{(\mathbf{z} - \mathbf{f})(\mathbf{z} - \mathbf{f})^T\} = \sigma^2 I$, the Fisher information is given by

$$G = \frac{1}{\sigma^2}E[\nabla\mathbf{f}(\mathbf{x}; \mathbf{w})\{\nabla\mathbf{f}(\mathbf{x}; \mathbf{w})\}^T] \qquad (14)$$

where the expectation is taken over the distribution $q(\mathbf{x})$. This shows that the Fisher information tends to infinity and the estimation error tends to 0 as the noise term σ^2 becomes 0.

Another example is the discrete case, given by Equation 4. The Fisher information is calculated as

$$G = \frac{2\beta e^{\beta f}}{1 + e^{\beta f}}\nabla\nabla f + \frac{\beta^2 e^{\beta f}}{(1 + e^{\beta f})^2}\nabla f(\nabla f)^T \qquad (15)$$

It should also be noted that G tends to ∞ as the temperature β^{-1} tends to 0.

Neural learning is usually sequential in the sense that all the data D_T are not used as a batch. Instead, the previous estimate $\hat{\mathbf{w}}_t$ is modified to $\hat{\mathbf{w}}_{t+1}$ when a new datum $(\mathbf{x}_{t+1}, \mathbf{z}_{t+1})$ comes in. The best learning algorithm is derived from statistical analysis. The new estimator $\hat{\mathbf{w}}_{t+1} = \hat{\mathbf{w}}_t + \Delta\mathbf{w}_t$ should minimize the *log likelihood* $l(D_{t+1}; \mathbf{w}) = \log P(D_{t+1}; \mathbf{w})$, where we have

$$l(D_{t+1}; \mathbf{w}) = l(D_t; \mathbf{w}) + l(\mathbf{z}_{t+1}|\mathbf{x}_{t+1}; \mathbf{w})$$

$$l(\mathbf{z}_{t+1}|\mathbf{x}_{t+1}; \mathbf{w}) = \log p(\mathbf{z}_{t+1}|\mathbf{x}_{t+1}; \mathbf{w})$$

Since $l(D_{t+1}; \mathbf{w})$ should be maximized at $\mathbf{w} = \hat{\mathbf{w}}_t + \Delta\mathbf{w}_t$, by expanding

$$0 = \nabla l(D_{t+1}; \hat{\mathbf{w}}_t + \Delta\mathbf{w}_t)$$

and assuming that $l(D_t; \mathbf{w})$ is maximized at $\mathbf{w} = \hat{\mathbf{w}}_t$, we have

$$\Delta\mathbf{w}_t = -\{\nabla\nabla l(D_{t+1}; \hat{\mathbf{w}}_t)\}^{-1}\nabla l(\mathbf{z}_{t+1}|\mathbf{x}_{t+1}, \hat{\mathbf{w}}_t)$$

Here, $\nabla\nabla l$ is called the *observed Fisher information*, and its expectation over D_{t+1} is $(t + 1)$ times the negative of the Fisher information G. Since this converges to its expectation, we may use

$$\Delta\mathbf{w}_t = -\frac{1}{t + 1}G^{-1}\nabla l(\mathbf{z}_{t+1}|\mathbf{x}_{t+1}; \hat{\mathbf{w}}_t) \qquad (16)$$

obtaining the stochastic-approximation-type algorithm.

Theoretically speaking, this process gives the efficient $\hat{\mathbf{w}}_{mle}$. However, it might be difficult to calculate $(\nabla\nabla l)^{-1}$ or G^{-1}, and moreover, the statistical model might be inaccurate. So a more flexible, robust, and practical approach is often used. The Bayesian approach (Berger, 1985) assumes a prior distribution $p_{pr}(\mathbf{w})$ of the true parameter $\mathbf{w}$. Then, when data D_T are observed, the Bayes rule gives the posterior distribution of $\mathbf{w}$ based on D_T

$$p_{post}(\mathbf{w}|D_T) = \frac{p_{pr}(\mathbf{w})P(D_T|\mathbf{w})}{P(D_T)} \qquad (17)$$

where

$$P(D_T) = \int p_{pr}(\mathbf{w})P(D_T|\mathbf{w})\,d\mathbf{w} \qquad (18)$$

is the marginal distribution of D_T under the prior $p_{pr}(\mathbf{w})$.

The estimator which maximizes the posterior distribution is called the maximum posterior estimator $\hat{\mathbf{w}}_{mp}$. However, when T is large and the prior $p_{pr}(\mathbf{w})$ satisfies $p_{pr}(\mathbf{w}) > 0$ in the parameter space, it is known that the $\hat{\mathbf{w}}_{mp}$ is asymptotically equivalent to $\hat{\mathbf{w}}_{mle}$.

The predictive distribution of $\mathbf{z}_{t+1}$ given $\mathbf{x}_{t+1}$ is given by

$$p(\mathbf{z}_{t+1}|\mathbf{x}_{t+1}; D_T) = \frac{P(D_{T+1})}{P(D_T)} \qquad (19)$$

$D_{T+1} = \{D_T, (\mathbf{z}_{T+1}, \mathbf{x}_{T+1})\}$, which is the distribution of the output $\mathbf{z}_{T+1}$ under the condition that D_T is observed previously and that a new input $\mathbf{x}_{T+1}$ has arrived. This relation (Equation 19) can be used for evaluating the learning curve which shows how fast the generalization error decreases by learning from examples. (See Levin, Tishby, and Solla, 1990; Amari and Murata, 1993.)

Loss Functions and Stochastic Descent Learning

The aim of learning is not to estimate $\mathbf{w}$ but to achieve good information-processing performance. The Bayesian standpoint suggests using the notion of a risk or loss function which is to be minimized through learning.

Let $r(\mathbf{x}, \mathbf{z}_{\text{true}}, \mathbf{w})$ be a risk when an input signal $\mathbf{x}$ is processed by a network of parameter $\mathbf{w}$, where $\mathbf{z}_{\text{true}}$ is the true or desired output. A simplest example is the squared error,

$$r(\mathbf{x}, \mathbf{z}_{\text{true}}; \mathbf{w}) = \frac{1}{2}|\mathbf{z}_{\text{true}} - \mathbf{z}|^2 \qquad (20)$$

where $\mathbf{z}$ is the output from the network specified by $\mathbf{w}$. Sometimes, a function $R(\mathbf{w})$ of $\mathbf{w}$ is added to the risk in order to penalize a complex network. The function $R(\mathbf{w})$, called the regularization term, takes a large value when the network with parameter $\mathbf{w}$ is complex (Poggio, Torre, and Koch, 1985; see GENERALIZATION AND REGULARIZATION IN NONLINEAR LEARNING SYSTEMS). A typical example is

$$R(\mathbf{w}) = \sum \frac{w_i^2}{1 + w_i^2}$$

This function is nearly equal to w^2 when w is small and nearly equal to 1 when w is large. So $R(\mathbf{w})$ penalizes w's the more they depart from 0 when w is small, but penalizes each large w the same amount (by 1), implying that penalty for those large w's is just the number of them. The loss function $L(\mathbf{w})$ is the expected risk

$$L(\mathbf{w}) = E[r(\mathbf{x}, \mathbf{z}_{\text{true}}; \mathbf{w})] \qquad (21)$$

where expectation is taken with respect to the distribution $q(\mathbf{x})p(\mathbf{z}_{\text{true}}|\mathbf{x}; \mathbf{w}_0)$. This is called the generalization loss or error, since the behavior of the network specified by $\mathbf{w}$ is evaluated by the expectation with respect to a new example. The best network is supposed to be the one that minimizes $L(\mathbf{w})$. However, the point is that we do not know the loss function $L(\mathbf{w})$, since the true $\mathbf{w}_0$ which defines $\mathbf{z}_{\text{true}}$ is unknown. Instead, we have a training set D_T generated from the true distribution. This gives us the empirical loss function

$$L_{\text{train}}(\mathbf{w}) = \frac{1}{T}\sum_{t=1}^{T}r(\mathbf{x}_t, \mathbf{y}_t; \mathbf{w}) \qquad (22)$$

which is an evaluation of $L(\mathbf{w})$ using the training data themselves. It is called the training error, and it converges to $L(\mathbf{w})$ as T tends to infinity. The practical learning procedure is given by minimizing $L_{\text{train}}(\mathbf{w})$ sequentially.

The optimal learning procedure is similarly obtained by

$$\Delta\mathbf{w}_t = -A^{-1}\nabla r(\mathbf{x}_{t+1}, \mathbf{z}_{t+1}; \hat{\mathbf{w}}_t) \tag{23}$$

$$A = \sum_{i=1}^{t} \nabla\nabla r(\mathbf{x}_i, \mathbf{z}_i; \hat{\mathbf{w}}_t) \tag{24}$$

If the risk r is put equal to the negative of the log likelihood $l(\mathbf{z}|\mathbf{x}, \mathbf{w})$, we have

$$L_{\text{train}}(D_t; \mathbf{w}) = -T\log P(D_T; \mathbf{w}) \tag{25}$$

Hence, the maximum likelihood estimator is a special case of this scheme. However, a risk can be more general, including the regularization term $R(\mathbf{w})$ which penalizes complicated networks in order to avoid overfitting.

It is again not easy to evaluate A^{-1}. There may be accumulations of errors. Therefore, a more robust stochastic gradient algorithm

$$\Delta\mathbf{w}_t = -C_t\nabla r(\mathbf{x}_{t+1}, \mathbf{z}_{t+1}; \hat{\mathbf{w}}_t) \tag{26}$$

is widely used (see STOCHASTIC APPROXIMATION AND NEURAL NETWORK LEARNING), where the learning constant C_t may be put equal to a small constant ε or may be a positive matrix. This simple method is called the *stochastic descent learning rule* (Amari, 1967). It is also called the *generalized delta rule*. Rumelhart, Hinton, and Williams (1986) showed that, when it was applied to the multilayer perceptron, the calculation of the ∇r is performed through error propagation in the backward direction. So the algorithm is called the *error backpropagation* method. Many acceleration methods have been proposed so far. White (1989) formulated this perspective in clear statistical terms.

This type of learning is applicable to any parameterized family $p(\mathbf{z}|\mathbf{x}, \mathbf{w})$ of stochastic behaviors as well as deterministic behaviors. The idea is also used in the self-organization scheme, where there are no desired outputs $\mathbf{z}_{\text{true}}$ but the network modifies its structure depending only on the input data $D_T = \{\mathbf{x}_1, \ldots, \mathbf{x}_T\}$. For example, if the risk function is put equal to

$$r(\mathbf{x}, \mathbf{w}) = a|\mathbf{w}|^2 - \tfrac{1}{2}|\mathbf{w}\cdot\mathbf{x}|^2 \tag{27}$$

for a single neuron, the connection weight vector $\mathbf{w}$ of the neuron converges to the principal eigenvector of the correlation matrix

$$V = \int q(\mathbf{x})\mathbf{x}\mathbf{x}^T \, d\mathbf{x} \tag{28}$$

of the input signals, if $|\mathbf{w}|$ is normalized. This fact was pointed out in more general perspectives of neural learning (Amari, 1977) and studied by Oja (1982) in detail (see PRINCIPAL COMPONENT ANALYSIS).

For another example, consider a set of k neurons whose connection weights are $\mathbf{w}_1, \ldots, \mathbf{w}_k$. The neurons receive a common input $\mathbf{x}$ and calculate the distance $|\mathbf{w}_i - \mathbf{x}|$. The neuron whose weight $\mathbf{w}_i$ is closest to $\mathbf{x}$ wins, and output $z_k = 1$, while all the other $z_j = 0$. This is called the *winner-take-all rule*. Let us put

$$r(\mathbf{x}, \mathbf{w}) = \tfrac{1}{2}\min_t |\mathbf{w}_i - \mathbf{x}|^2$$

where $\mathbf{w} = (\mathbf{w}_1, \ldots, \mathbf{w}_k)$. Then the loss is given by

$$L(\mathbf{w}) = \tfrac{1}{2}\int \min_i |\mathbf{w}_i - \mathbf{x}|^2 q(\mathbf{x}) \, d\mathbf{x} \tag{29}$$

The learning rule (Equation 26) in this case leads to Kohonen's

LEARNING VECTOR QUANTIZATION (q.v.; see also Kohonen, 1988).

Learning Curves and Generalization Errors

The learning curve shows how quickly a learning network improves behavior that is evaluated by the generalization error. This process is related to the dynamical behavior of neural learning and the complexity of neural networks.

The stochastic approximation also guarantees that $\hat{\mathbf{w}}_t$ converges to the optimal parameter $\mathbf{w}_0$ with probability 1 when the learning constant C_t tends to 0 at an adequate rate. However, when C_t is too small, learning becomes ineffective. The purpose of learning in many cases is to adjust the network parameters in a changing environment. In this case, C_t should be kept to a small constant ε. With C_t set equal to a small constant ε, the dynamical behavior of $\hat{\mathbf{w}}_t$ was studied in an early paper (Amari, 1967) in which the stochastic descent learning rule was proposed for the multilayer perceptron from the Bayesian standpoint.

Let us consider the case that the initial value $\mathbf{w}_1$ is in a neighborhood of the (local) optimal value $\mathbf{w}_0$. Let us define two matrices,

$$A = \nabla\nabla L(\mathbf{w}_0) \tag{30}$$

$$B = E[\nabla r(\nabla r)^T] \tag{31}$$

Then, the expected value of $\hat{\mathbf{w}}_t$ converges to $\mathbf{w}_0$ exponentially, and the covariance of the error $\hat{\mathbf{w}}_t - \mathbf{w}_0$ also converges exponentially to εV, where V is a matrix obtained from A and B (Amari, 1967). The dynamic behavior of $\mathbf{w}_t$ has also been studied when the environment, that is, the optimal $\mathbf{w}_0$, is changing slowly (Amari, 1967; Heskes and Kappen, 1991).

Since the behavior of the net is evaluated by $L(\hat{\mathbf{w}}_t)$, but not directly by $\hat{\mathbf{w}}_t$, it is important to know how fast $L(\hat{\mathbf{w}}_t)$ approaches its optimum value. Here, $L(\hat{\mathbf{w}}_t)$ is the expectation of the risk $r(\mathbf{x}, \mathbf{z}; \hat{\mathbf{w}}_t)$ with respect to the new example pair $(\mathbf{x}, \mathbf{z})$. This is the generalization loss or error, evaluating the behavior of the net by a new example $(\mathbf{x}, \mathbf{z})$ which is not included in the training set D_T. This is different from the training error (Equation 22), which is an evaluation of $\hat{\mathbf{w}}_t$ based on the training data D_t. We can calculate the latter, but it is difficult to know the generalization error $L(\hat{\mathbf{w}}_t)$ because we do not know the function $L(\mathbf{w})$ itself. If we know the relation between L_{train} and L, we can then evaluate L through L_{train}.

A standard technique of statistical inference (Cox and Hinkley, 1974) can be applied to this problem. Let us define the training data as D_t, and let $\hat{\mathbf{w}}_t$ now be the best estimator obtained therefrom. It maximizes L_{train}, so that it satisfies

$$0 = \nabla L_{\text{train}}(\hat{\mathbf{w}}_t)$$

or

$$0 = \sum_{i=1}^{t} \nabla r(\mathbf{z}_i, \mathbf{x}_i, \mathbf{w}_0) + \sum_{i=1}^{t} \nabla\nabla r(\mathbf{z}_i, \mathbf{x}_i, \mathbf{w}_0)(\hat{\mathbf{w}}_t - \mathbf{w}_0)$$

By the law of large numbers, when t is large

$$\frac{1}{t}\sum_{i=1}^{t} \nabla\nabla r(\mathbf{z}_i, \mathbf{x}_i, \mathbf{w}_0) \to E[\nabla\nabla l] = A$$

On the other hand,

$$E[\nabla r(\mathbf{y}, \mathbf{x}; \mathbf{w}_0)] = 0$$

because $\mathbf{w}_0$ is optimal. Therefore, by the central limit theorem,

$$\frac{1}{\sqrt{t}}\sum_{i=1}^{t} \nabla r(\mathbf{z}_i, \mathbf{x}_i; \mathbf{w}) \to \varepsilon$$

where ε is a normally distributed vector random variable with mean 0 and covariance matrix

$$B = E[\nabla r(\nabla r)^T]$$

Hence, we have

$$\hat{\mathbf{w}}_t - \mathbf{w}_0 = \frac{1}{\sqrt{t}} A^{-1} \varepsilon \qquad (32)$$

Therefore, the covariance of the error $\Delta \mathbf{w} = \hat{\mathbf{w}}_t - \mathbf{w}_0$ is given by

$$E[(\hat{\mathbf{w}}_t - \mathbf{w}_0)(\hat{\mathbf{w}}_t - \mathbf{w}_0)^T] = \frac{1}{t} A^{-1} E[\varepsilon \varepsilon^T] A^{-1}$$

$$= \frac{1}{t} A^{-1} B A^{-1}$$

Now the generalization error is calculated as follows:

$$L(\hat{\mathbf{w}}_t) = L(\mathbf{w}_0) + + \nabla L(\mathbf{w}_0) \nabla \mathbf{w}^T + \tfrac{1}{2} (\nabla \mathbf{w})^T A \nabla \mathbf{w}$$

From $\nabla L(\mathbf{w}_0) = 0$, we have

$$E[L(\hat{\mathbf{w}}_t)] = L(\mathbf{w}_0) + \frac{1}{2t} \mathrm{tr}(B^{-1}A) \qquad (33)$$

On the other hand,

$$tL_{\text{train}}(\hat{\mathbf{w}}_t) = \sum r(\mathbf{x}_i, \mathbf{z}_i; \mathbf{w}_0) + \sum \nabla r(\mathbf{x}_i, \mathbf{z}_i; \mathbf{w}_0) \nabla \mathbf{w}_0$$
$$+ \tfrac{1}{2} \sum (\nabla \mathbf{w})^T \nabla \nabla l \nabla \mathbf{w}$$

The first term converges to $tL(\mathbf{w}_0)$. The second term is further expanded as

$$\sum \nabla l(\mathbf{x}_i, \mathbf{z}_i, \mathbf{w}_0) = \sum \nabla l(\mathbf{x}_i, \mathbf{z}_i; \hat{\mathbf{w}}_t) - \sum \nabla \mathbf{w}^T \nabla \nabla l(\mathbf{x}_i, \mathbf{z}_i; \hat{\mathbf{w}}_t)$$

where the first term of the right-hand side vanishes. Hence,

$$E[L_{\text{train}}(\hat{\mathbf{w}}_t)] = L(\mathbf{w}_0) - \frac{1}{2t} \mathrm{tr}(B^{-1}A) \qquad (34)$$

The relation between the training error and the generalization error is given from Equations 32 and 33 by

$$E[L(\hat{\mathbf{w}}_t)] = E[L_{\text{train}}(\hat{\mathbf{w}}_t)] + \frac{1}{t} \mathrm{tr}(B^{-1}A)$$

The term $\mathrm{tr}(B^{-1}A)$ shows the difference between the training error and the generalization error. When this term is large, L_{train} is much smaller than L because of overfitting. If a complex network is used, observed data D_t are easily overfitted, so this term represents the complexity of the network. This is a generalization of the AIC (Akaike information criterion; Sakamoto, Ishiguro, and Kitagawa, 1986) widely used in statistical inference. When the loss is the negative log likelihood and the model is faithful, both A and B are equal to the Fisher information matrix, so that we have an evaluation of complexity by the number of modifiable parameters,

$$\mathrm{tr}(B^{-1}A) = \mathrm{tr}(I) = \text{Number of modifiable parameters}$$

This relationship is universal in the sense that the complexity (overfitting factor) depends only on the number of parameters independently of the architecture of the network. Universal properties of this type are more or less known in various situa-

tions concerning learning machines (Amari and Murata, 1993). One more remarkable fact is the t^{-1} convergence of the learning error in learning curves. This was first remarked by T. Cover in his Ph.D. thesis in 1964. A universal result is proved in Amari and Murata (1993). The problem of learning curves attracts researchers from the algorithmic, information-theoretic, and physics points of view.

Discussion

We have studied learning in neural networks from the viewpoint of statistical inference. The accuracy of learning is shown in terms of the Fisher information. Learning is sequential estimation from the statistical point of view. However, we are interested in the behaviors of the trained network under a general loss criterion. The behavior of learning curves is shown. Moreover, the complexity of a neural network is defined to elucidate the discrepancy between the training and generalization errors.

Road Map: Learning in Artifical Neural Networks, Statistical
Background: I.3. Dynamics and Adaptation in Neural Networks
Related Reading: Bayesian Methods for Supervised Neural Networks; Statistical Mechanics of Generalization

References

Amari, S., 1967, Theory of adaptive pattern classifiers, *IEEE Trans. Electron. Comput.*, 16:299–307.
Amari, S., 1977, Neural theory of association and concept-formation, *Biol. Cybern.*, 26:175–185.
Amari, S., and Murata, N., 1993, Statistical theory of learing curves under entropic loss, *Neural Comp.*, 5:140–153.
Berger, J. O., 1985, *Statistical Decision Theory and Bayesian Analysis*, 2nd ed., New York: Springer-Verlag. ◆
Cover, T. M., and Thomas, J. A., 1991, *Elements of Information Theory*, New York: Wiley. ◆
Cox, D. R., and Hinkley, D. V., 1974, *Theoretical Statistics*, London and New York: Chapman and Hall. ◆
Heskes, T. M., and Kappen, B., 1991, Learning processes in neural networks, *Phys. Rev. A*, 440:2718–2726.
Kohonen, T., 1988, *Self-Organization and Associative Memory*, 2nd ed., Berlin: Springer-Verlag.
Levin, E., Tishby, N., and Solla, S. A., 1990, A statistical approach to learning and generalization in layered neural networks, *Proc. IEEE*, 78:1568–1574.
Oja, E., 1982, A simplified neuron model as a principal component analyzer, *J. Math. Biol.*, 15:267–273.
Poggio, T., Torre, V., and Koch, C., 1985, Computational vision and regularization theory, *Nature*, 317:314–319.
Rao, C. R., 1973, *Linear Statistical Inference and Its Applications*, 2nd ed., New York: Wiley.
Ripley, B. D., 1994, Neural networks and related method for classification, *J. R. Statist. Soc. Ser. B*, 56:409–456. ◆
Rumelhart, D. E., Hinton, G. E., and Williams, R. J., 1986, Learning internal representation by error backpropagation, in *Parallel Distributed Processing: Explorations in the Microstructure of Cognition* (D. E. Rumelhart, J. L. McClelland, and PDP Research Group, Eds.), vol. 1, *Foundations*, Cambridge, MA: MIT Press, pp. 318–362.
Sakamoto, Y., Ishiguro, M., and Kitagawa, G., 1986, *Akaike Information Criterion Statistics*, Dordrecht: Reidel. ◆
White, H., 1989, Learning in artificial neural networks: A statistical perspective, *Neural Comp.*, 1:425–464. ◆

Learning as Adaptive Control of Synaptic Matrices

Jean-Pierre Aubin

Introduction

Control theory allows us to present many examples of neural networks in a unified way by regarding them as discrete or continuous control systems, the states of which are the *signals* and the controls of which are the *synaptic matrices* (pattern classification problems, including time series in forecasting). See Aubin (1995) and that source's bibliography for an approach using this viewpoint.

The usual classification problem for which most neural networks have been designed is to find synaptic matrices with which the network maps inputs of a prescribed sequence of patterns to outputs, through one, several, or a continuous set of layers. One can use several results on control of nonlinear systems to obtain learning algorithms converging to solutions of this problem, such as the backpropagation formula (which is nothing other than the *adjoint equation* of control theory). Since the mathematical structure of the space of synaptic matrices involves tensor products, we will use *pure tensor products* formulas describing the learning algorithms. The corrections of the synaptic weights of the synaptic matrix are the products of the presynaptic and postsynaptic activities. This is the reinforcement characteristic of *Hebbian rules* in learning mechanisms.

Adaptive Systems

The general form of an adaptive network is given by a map $\Phi : X \times U \to Y$, where X is the input space, Y the output space, and U a control (or parameter) space. In this article, the spaces X, Y, and U are finite dimensional vector spaces.

A *pattern* is an input-output pair $(a_p, b_p) \in X \times Y$. The inputs a_p are often called *keys* (or search arguments) and the outputs b_p the *memorized data*.

In the case of neural networks, or more generally, of connectionist networks, the control space is a matrix space (synaptic matrices, fitness matrices, activation matrices, etc., depending on the contexts and the fields of motivation). A *pattern* $(x, y) \in X \times Y$ made of an input-output pair is *recognized* (or *discovered, generalized*) by the adaptive system programmed by such a control u if $y = \Phi(x, u)$ is the signal produced by the network excited by the input x.

The Learning Problem

A *training set* is a finite set (a_p, b_p), $p \in \mathscr{P}$, of patterns.

The choice of a control is made by learning a given training set, i.e., by finding a control $u_{\mathscr{P}}$ satisfying

$$\Phi(a_p, u_{\mathscr{P}}) = b_p \quad \text{for all } p \in \mathscr{P} \tag{1}$$

We say that a control $u_{\mathscr{P}}$ has *learned the training set*. With a control u, the system generalizes from this training set by associating with any input x the output $y = \Phi(x, u_{\mathscr{P}})$. This is called the generalization phase of the operation of the adaptive system.

This phase includes the *forecasting problem*, when $X = Y$ and when the patterns associated with a time series $a_1, \ldots, a_{T+1}$ are defined by $a_p = a_t$ and $b_p = a_{t+1}$ (the input is the present state and the output the future state).

$$\Phi(a_t, u_{\mathscr{P}}) = a_{t+1} \quad \text{for all } t = 1, \ldots, T$$

The record of past states constitutes the training set.

A feature common to adaptive systems is that they are "programmed" not by a sequence of instructions, but by making a control u (a synaptic matrix W in the case of neural networks) learn a set of patterns which they can reproduce, and thus, hopefully, discover new patterns associated with such a synaptic matrix. Such a system, after all, is nothing but an extrapolation procedure set in a new framework.

Any algorithm converging to a solution $u_{\mathscr{P}}$ to the learning problem (Equation 1) is regarded as a *learning algorithm*.

Since Equation 1 is generally a nonlinear problem, the first algorithm we can think of is the *Newton algorithm*. However, we can view the learning problem as an optimization problem and use familiar gradient algorithms.

Gradient Descent

We observe that the learning problem (Equation 1) is a nonlinear problem, which can be mapped to a minimization problem of the form

$$0 = \inf_u \left(\sum_{p \in \mathscr{P}} E_p(\Phi(a_p, u) - b_p)^\alpha \right)^{1/\alpha}$$

where $\alpha \in [1, \infty]$ and where $E_p : Y \to \mathbb{R}$ are *evaluation functions* vanishing (only) at the origin.

This approach has two advantages. The first is the possibility of defining a solution to such a minimization problem even when there is no solution to the above system of equations: When the infimum is not equal to 0, the minimal solution can be regarded as a quasi-solution. The second advantage is the possibility of using the whole family of minimization algorithms (including variants of the gradient method) converging to a solution of the minimization problem.

The simplest example of distance is naturally provided by the case when $\alpha = 2$. But we also include the important (although nonsmooth) case when $\alpha = \infty$, where we have to solve the minimization problem

$$0 = \inf_u \sup_{p \in \mathscr{P}} E_p(\Phi(u_p, u) - b_p)$$

even though the functional is no longer differentiable in the usual sense.

Once we have transformed the learning problem into an optimization problem of the form $\inf_{u \in U} H(u)$, the natural algorithms which may lead to the minima are in most cases variants of *gradient descent*. When H is differentiable, it can be written in the form

$$u^{n+1} - u^n = -\varepsilon_n H'(u^n)$$

Neural Networks

In this article, a neural network denotes an adaptive system in which the set of controls consists of (synaptic) matrices, sequences of (synaptic) matrices, or even time-dependent synaptic matrices.

Before being more specific about examples of neural networks, we pose the question, What then is special to neural networks?

We shall have to apply optimization algorithms (Newton's algorithm, gradient descent, heavy algorithms) to adaptive systems controlled by synaptic matrices, and thus to use specific properties of the spaces of matrices (regarded as tensor products).

Since Hebb's 1949 classic book *Organization of Behavior*, most of the studies of neural networks have dealt with numerous variations of learning rules which prescribe a priori the evolution of the synaptic matrix $W^n = (w_{i,j}^n)$ through an algorithm of the form

$$w_{i,j}^{n+1} - w_{i,j}^n = \alpha_i^n \beta_j^n$$

where the correction of the synaptic weight $w_{i,j}^n$ is proportional to the product of presynaptic and postsynaptic activity. Such matrices are tensor products of the vectors α and β. We shall see that the algorithms mentioned above, when applied to neural networks, i.e., adaptive systems parametrized by synaptic matrices, yield learning rules which involve tensor products of vectors.

Classes of Neural Networks

For simplicity, we consider the learning problem of one pattern only.

In the case of a one-layer neural network described by an output function $g : Y \to Y$, the map Φ is defined by

$$\Phi(\alpha, W) = g(Wa)$$

where the input $a \in X$ and the synaptic matrix W in the space $\mathscr{L}(X, Y)$ of linear functions from X to Y are given. Given an evaluation function E on the output space Y and a pattern (a, b), we look for a synaptic matrix W which minimizes

$$H(W) = E(g(Wa) - b)$$

In the case of a multilayer neural network described by the vector spaces (layers) $X_0 = X$, $X_l (l = 1, \dots, L - 1)$ and $X_L = Y$ and output functions $g_l : X_l \to X_l$ from one layer to the next, the map Φ_L associates with the sequence $(W_1, \dots, W_L)$ of synaptic matrices $W_l \in \mathscr{L}(X_{l-1}, X_l)$ the final state x_L of the sequence of states starting from $x_0 = a$ according to

$$x_l = g_l(W_l x_{l-1}) \quad \text{for all } l = 1, \dots, L \tag{2}$$

Here, the transition of the signal through the layers of the neural network is also the transition through discrete time $l = 1, \dots, L$. Therefore, we can regard a multilayer neural network as a time-varying *discrete control problem* where layers play the role of time. At time (layer) 1, one chooses the control W_1 for mapping the initial state x_0 to $g_1(W_1 x_0)$, and so on: at time (layer) l, we associate with the preceding state x_{l-1} and the control W_l the new state $g_l(W_l x_{l-1})$. Hence the final state x_L is obtained from the initial state x_0 through the time-dependent discrete dynamical system of Equation 2 controlled by the sequence of synaptic matrices W_l.

We have thus defined adaptive systems controlled by synaptic matrices. Devising gradient descent algorithms (or other algorithms) for minimizing such functions requires the computation of the gradients of such functions defined on spaces of matrices. The tensor structure of the spaces of linear operators makes its appearance here, since such gradients involve tensor products of vectors, which yield "reinforcement" learning rules and justify the specific role of neural networks compared to general adaptive systems. Hence the task at hand is to compute the derivatives of these maps Φ and their transpose, in order to write down the algorithms.

Tensor Products

Let $X = \mathbb{R}^n$, $Y = \mathbb{R}^m$ be finite dimensional vector spaces, and let $X^* = \mathscr{L}(X, \mathbb{R})$ denote the *dual space* of X, which is the vector space of linear functionals $p : X \to \mathbb{R}$ on X. We recall that the *transpose* $W^* \in \mathscr{L}(Y^*, X^*)$ is defined by

$$\langle W^* q, x \rangle = \langle q, Wx \rangle \quad \text{for all } x \in X, \text{ for all } q \in Y^*$$

where $\langle y, x \rangle$ is simply $y(x)$.

Let p belong to X^*. We associate with any $p \in X^*$ and $y \in Y$ their *tensor product* $p \otimes y \in \mathscr{L}(X, Y)$, which is the linear operator defined by

$$p \otimes y : x \to (p \otimes y)(x) = \langle p, x \rangle y$$

the matrix of which is

$$(p^i y_j)_{i=1,\dots,m; j=1,\dots,n}$$

Its transpose $(p \otimes y)^* \in \mathscr{L}(Y^*, X^*)$ maps $q \in Y^*$ to

$$(p \otimes y)^*(q) = \langle q, y \rangle p$$

because $\langle q, \langle p, x \rangle y \rangle = \langle \langle q, y \rangle p, x \rangle$. Hence, we shall identify from now on the transpose $(p \otimes y)^*$ with $y \otimes p$.

Let us consider two pairs (X, X_1) and (Y, Y_1) of finite dimensional vector spaces. Let $A \in \mathscr{L}(X_1, X)$ and $B \in \mathscr{L}(Y, Y_1)$ be given. Since $A^* \in \mathscr{L}(X^*, X_1^*)$, we denote by $A^* \otimes B$ the linear operator from $X^* \otimes Y = \mathscr{L}(X, Y)$ to $X_1^* \otimes Y_1 = \mathscr{L}(X_1, Y_1)$ defined by

$$(A^* \otimes B)(W) = BWA \quad \text{for all } W \in \mathscr{L}(X, Y)$$

We observe that when $W = p \otimes y$, we have

$$(A^* \otimes B)(p \otimes y) = A^* p \otimes By$$

and that

$$(A^* \otimes B)^* = A \otimes B^* \tag{3}$$

Let $A_1 \in \mathscr{L}(X_2, X_1)$ and $B_1 \in \mathscr{L}(Y_1, Y_2)$. Then it is easy to check that

$$(A_1^* \otimes B_1)(A^* \otimes B) = (A_1^* A^*) \otimes (B_1 B)$$

One main reason for introducing the concepts of tensor products is given by the following proposition.

Proposition 1. Let us consider three spaces X, Y, and Z, an element $x \in X$, a differential map $g : Y \to Z$, with which we associate the map $\Psi : W \to \Psi(W) = g(Wx)$.

The derivative of Ψ at $W \in \mathscr{L}(X, Y)$ is given by

$$\Psi'(W) = x \otimes g'(Wx)$$

Furthermore, if E be a differentiable functional from Z to $\mathbf{R}$, setting $H(W) = E(g(Wx))$, the gradient of H at $W \in \mathscr{L}(X, Y)$ is given by

$$H'(W) = x \otimes g'(Wx)^* E'(g(Wx)) \in \mathscr{L}(X^*, Y^*)$$

Backpropagation Algorithms

Gradient Methods

Because multilayer neural networks are special discrete control problems, we can derive from control theory the formula for the derivatives of the map Φ_L and its transposes involving the *adjoint equation*. The formulas on tensor products are then used when the controls are synaptic matrices.

In the case of a one-layer neural network described by an output function $g : Y \to Y$, the map Φ is defined by

$$\Phi(a, W) = g(Wa)$$

and given an evaluation function E on the output space Y and a pattern (a, b), we look for a synaptic matrix W which minimizes

$$H(W) = E(g(Wa) - b)$$

Then the gradient of H is given by

$$H'(W) = a \otimes g'(Wa)^*E'(g(Wa) - b)$$

and the gradient method can then be written

$$W^{n+1} - W^n = -\varepsilon_n a \otimes g'(W^n a)^* E'(g(W^n a) - b)$$

When the output function g_i depends only upon the signal arriving at neuron i, we obtain

$$w_{ij}^{n+1} = w_{ij}^n - \varepsilon_n a_j \left(\frac{\partial g_i}{\partial y_i}(W^n a) \frac{\partial E}{\partial y_i} g_i((W^n a)_i - b_i) \right)$$

This equation belongs to the class of *pseudo-Hebbian learning rules*: the synaptic weight from a neuron j to a neuron i should be *strengthened* whenever the connection is highly active, in proportion to the activities a_j of the presynaptic neuron and $(\partial g_i / \partial y_i)(W^n a)(\partial E / \partial y_i) g_i((W^n a)_i - b_i)$ of the postsynaptic neuron of the synapse.

In the case of a multilayer network described by the vector spaces $X_0 = X$, X_l ($l = 1, \ldots, L - 1$), and $X_L = Y$ and differentiable propagation rules $g_l : X_l \to X_l$, the map Φ_L associates with the sequence $(W_1, \ldots, W_L)$ of synaptic matrices the final state x_L of the sequence of states starting from $x_0 = a$ according to

$$x_l = g_l(W_l x_{l-1}) \quad \text{for all } l = 1, \ldots, L \qquad (4)$$

Given an evaluation function E defined on the output space X_L, we are looking for a sequence $(W_l)_{l=1, \ldots, L}$ of synaptic matrices minimizing

$$H(W_1, \ldots, W_L) = E(\Phi_L(a, W_1, \ldots, W_L) - b)$$

The backpropagation learning rule is nothing other than the gradient method applied to this function H defined in the previous equation. When the maps g_l and the evaluation function E are differentiable, the gradient of H is given by the formula

$$H'(W) = (x_{l-1} \otimes g_l'(x_l)^* p_l)(1 \le l \le L)$$

where p_l is given by the adjoint equation

$$p_l = W_{l+1}^{n*} g_{l+1}'(W_l x_l)^* p_{l+1}$$

backpropagating the final costate $p_L = E'(x_L - b)$ which is the gradient of the error on the final state. The adjoint equation is actually the adjoint (or transpose) of the linearized control system (Equation 4) defined by

$$y_l = g_l'(W_l x_{l-1}) W_l y_{l-1} \quad \text{for all } l = 1, \ldots, L$$

We check that

$$p_l = W_{l+1}^* g_{l+1}'(x_{l+1})^* \cdots W_L^* g_L'(x_L)^* E'(x_L - b)$$

In other words,

$$\frac{\partial H(W)}{\partial W_l} = x_{l-1} \otimes g_l'(x_l)^* p_l$$

(see Chapter 4 of Aubin, 1995). The gradient method provides the celebrated *backpropagation algorithm*: Starting with a synaptic matrix W^0, we define W^{n+1} from the synaptic matrix W^n according to the rule

$$W_l^{n+1} - W_l^n = -\varepsilon_n x_{l-1}^n \otimes g_l'(x_l^n)^* p_{l+1}^n$$

where $x_l^n = g_l(W_l^n x_{l-1}^n)$ (starting at $x_0^n = a$) and where p_l^n is obtained from the final condition $E'(x_L^n - b)$ through the adjoint equation

$$p_l^n = W_{l+1}^{n*} g_{l+1}'(x_{l+1}^n)^* p_{l+1}^n$$

which evaluates to

$$p_l^n = W_{l+1}^{n*} g_{l+1}'(x_{l+1}^n)^* p_{l+1}^n \cdots W_L^{n*} g_L'(x_L^n)^* E'(x_L^n - b)$$

We begin by modifying the synaptic weights of the synaptic matrix of the last layer L by setting $p_L^n = E'(x_L^n - b)$ and then computing $p^{L-1} = W_L^{n*} g_L'(x_L^n)^* p_L^n$. The new matrix is obtained through the *pseudo-Hebbian learning rule*

$$W_L^{n+1} - W_L^n = -\varepsilon_n x_{L-1}^n \otimes g_L'(W_L^n x_L^n)^* p_L^n$$

Then the gradient is backpropagated to modify successively the synaptic matrices W_l^n of each layer l from the last one to the first one through the adjoint equation and, at each layer, according to a reinforcement learning rule.

The Need for Nonsmooth Optimization

We may not always have the possibility of choosing the simplest evaluation functions E as we do in the case of *least square methods*. When the criterion is nonsmooth, the gradient of the criterion is replaced by a *generalized gradient*. This happens when the function to be minimized is written $H(u) = \sup_{p \in P} H_p(u)$. Even when the functions H_p are differentiable, taking their supremum destroys the usual differentiability. In another instance—the learning problem of control systems—the problem is to regulate the system by finding the controls which govern the evolution of the state in order to obey prescribed *state* (or *viability*) *constraints*. Such evolutions are called *viable*. Many control problems—in particular, the tracking of a trajectory, stabilization, and the like—fall into this category. One can obtain viable evolutions by taking for evaluation the function measuring the distance from a point to the constrained set, which is not necessarily differentiable. However, nonsmooth analysis, in particular, convex analysis, provides ways to define *generalized gradients* of any function. In general, one defines in nonsmooth analysis the concept of the generalized gradient of H at u, which is generally a subset $\partial H(u)$. Generalized gradients (which boil down to *subdifferential* in the case of convex functions) are no longer elements, but subsets. This is not the right place to recall precisely how generalized gradients are rigorously defined. See Aubin (1993) or Chapter 6 of Aubin and Frankowska (1990) for an introduction to nonsmooth analysis.

When E is locally Lipschitz and g is differentiable, we deduce that the generalized gradient of the function H is given by

$$\partial H(W) = a \otimes g'(Wa)^* \partial E(g(Wa) - b)$$

since the generalized gradient of the function $\Psi : y \to \Psi(y) = E(g(y))$ is equal to $g'(y)^* \partial E(g(y))$. The gradient algorithm takes the form

$$W^{n+1} - W^n \in -\varepsilon_n a \otimes g'(W^n a)^* \partial E(g(W^n a) - b)$$

The convergence of these algorithms holds true under convexity assumptions both in the smooth and nonsmooth cases. See Seube (1993) and Aubin (1995) for further details on this topic.

Learning Without Forgetting

An algorithm which learns without forgetting is defined as follows for the class of neural networks of the form

$$\Phi(x, W) = c(x) + G(x)W\psi(x)$$

where ψ maps the input space to the input layer, the synaptic matrix W weights the processed signal $\psi(x)$, and $G(x)$ maps the weighted processed signal to the output space. They are affine with respect to the synaptic matrices. Consider a finite training set of patterns $(a_p, b_p) \in X \times Y$. Hence we have to find a synaptic matrix W_P learning the training set in the sense that

$$c(a_p) + G(a_p)W_P\psi(a_p) = b_p \quad \text{for all } p = 1, \ldots, P \qquad (5)$$

Assuming that W_{P-1} has been obtained for learning the $P - 1$ input-output pairs (a_p, b_p) $(1 \leq p \leq P - 1)$, we want to find a new synaptic matrix W_P which learns the whole new training set (a_p, b_p) $(1 \leq p \leq P)$ according to the *heavy rule*

$$W_P \text{ is the } W \text{ which minimizes } \| W - W_{P-1} \|$$

under the constraints of Equation 5.

For that purpose, we need the concept of *pseudo-inverse* $C^{\dagger} \in \mathscr{L}(Y, X)$ of a matrix $C \in \mathscr{L}(X, Y)$. It maps any $y \in Y$ to the closest solution with minimal norm $\bar{x} = C^{\dagger} y$, i.e., the solution with minimal norm to the equation $Cx = \bar{y}$ where $\bar{y}$ is the projection of y onto the image of C. The pseudo-inverse coincides with the usual inverse when C is invertible, is involved in quadratic minimization problems under linear equality constraints, and enjoys many properties (which are often used in statistical analysis). See Kohonen (1984) or Aubin (1995) for presentations of pseudo-inverses.

Since we define the synaptic matrix W_P as a solution to a quadratic minimization problem under linear equality constraints, we have to use the pseudo-inverse of a tensor product of matrices. However, one can prove that *the pseudo-inverse of a tensor product is the tensor product of pseudo-inverses*. This is a very useful property of tensor products, since pseudo-inverses of linear operators occur in many applications—statistics and data analysis, for instance. We shall solve our problem in the following way:

We posit these assumptions:

1. The functions $c : X \to Y$ and $\psi : X \to Z_1$ are continuous.
2. The function $G : X \to \mathscr{L}(Z_2, Y)$ is continuous.
3. The function $G(x) \in \mathscr{L}(Z_2, Y)$ is surjective for all $x \in X$.
4. The elements $\psi(a_p)$ are mutually orthonormal.

Then the heavy algorithm associates with the synaptic matrix W_{P-1} the new synaptic matrix W_P defined by the formula

$$W_P = W_{P-1} - \psi(a_P) \otimes G(a_P)^{\dagger}(\Phi(a_P, W_{P-1}) - b_P)$$

This is also a pseudo-Hebbian learning rule, in the sense that the correction of the synaptic matrix W_{P-1} is obtained by adding to it a matrix $p \otimes u$ whose entries $p_i e^j$ are proportional to the product of activities in the presynaptic and postsynaptic neurons. It uses only the former synaptic matrix W_{P-1} and the last pattern (a_P, b_P) for making the correction.

Discussion

In classification problems, synaptic matrices which learn a given set of patterns are equilibria of a nonlinear problem, which are computed through gradient- and Newton-type algorithms (see Aubin, 1995). By the same token, one can also extend these results to *"continuous-layer" neural networks* in which the evolution of the signals is governed by a differential equation

$$x'(t) = g(W(t)x(t)) \tag{6}$$

(where the time here is regarded as an "infinitesimal layer").

The point was to illustrate, using the basic backpropagation algorithm, that powerful mathematical techniques could be used not only to provide other proofs, but also to cast more light on problems of neuroscience. Control theory touches many issues addressed by neuroscience.

However, by looking only at static or asymptotic problems (mapping inputs onto outputs, or finding equilibria and, more generally, attractors), the evolution mechanism involved in neural networks is neglected and the "real" time used to model the neural networks is often replaced by artificial times. For instance, the *algorithmic time*, describing the iterations of a given algorithm, is often interpreted as a learning rule, although it may not be related to the time involved in the modeling of the network. Or, in the case of multilayer networks, the layers of the network may represent a second time scale.

By choosing the route used by most neural networks, one may bypass the basic question, Why and how do synaptic matrices evolve? and the basic answer, To adapt to constraints through learning laws which are feedbacks of the neural network regarded as a control system, associating at each instant with any signal a synaptic matrix allowing the adaptation to viability constraints. Among the attempts to answer such questions, we refer to a dynamical point of view presented in Aubin (1995) and Seube (1993).

Modern mathematical techniques are needed to handle some of these issues. If we accept that physics studies much simpler phenomena than the ones investigated by neuroscience, and that physicists have been motivated to use a more and more complex mathematical apparatus, we have to accept also that neuroscience could require a new and dedicated mathematical arsenal which goes beyond what is currently available. Paradoxically, the very fact that the mathematical tools I think useful for neuroscience have to be quite sophisticated impairs their acceptance by many specialists, and the gap threatens to widen.

Among the recent mathematical developments, *set-valued analysis*, which deals with the extension of the differential and integral calculus to set-valued maps which occur naturally in artificial intelligence (to the extent that "intelligence" is sometimes grasped by the multi-valued nature of the input-output maps or of their inverses), may become a "must" (see, for instance, Aubin and Frankowska, 1990). I mention also the use of differential inclusions and *viability theory* to model Darwinian evolution by encapsulating the concept of (contingent) chance by differential inclusions instead of stochastic differential equations, subject to appropriate state (or "viability") constraints (see, for instance, Aubin and Frankowska, 1990; Aubin, 1991; Frankowska, 1995).

Road Map: Learning in Artificial Neural Networks, Deterministic
Background: I.3. Dynamics and Adaptation in Neural Networks
Related Reading: Adaptive Control: Neural Network Applications

References

Aubin, J.-P., 1991, *Viability Theory*, Boston: Birkhäuser.
Aubin, J.-P., 1993, *Optima and Equilibria*, Heidelberg: Springer-Verlag. ◆
Aubin, J.-P., 1995, *Neural Networks and Qualitative Physics: A Viability Approach*, Cambridge, Eng.: Cambridge University Press. ◆
Aubin, J.-P., and Frankowska, H., 1990, *Set-Valued Analysis*, Boston: Birkhäuser.
Frankowska, H., 1995, *Set-Valued Analysis and Control Theory*, Boston: Birkhäuser.
Kohonen, T., 1984, *Self-Organization and Associative Memory*, Series in Information Sciences, vol. 8, Heidelberg: Springer-Verlag. ◆
Seube, N., 1993, A state feedback control law learning theorem, in *Neural Networks in Robotics* (G. A. Bekey and K. Goldberg, Eds.), Dordrecht: Kluwer Academic, pp. 175–191.

Le... ing in Weight Space

An...

Le... h experience. At-
ter... artificial learning
sys... improvement strat-
eg... be pictured as the
pr... tep to be one lead-
in... wing this strategy,
th... ore steps that lead
u... o a possible config-
u... ng system, and the
a... "measure of good-
... ble in each location
... an be made to the

... se it requires so little
... ing on each step re-
... ndscape and does not
... past steps and their
... haustive examination
... a major shortcoming
... Just as a mountain
... randed on the top of a
... e hill-climbing strategy
... perior to its immediate
... configuration possible.
... at are *locally maximal*
... r of equal goodness to,
... hood. *Globally maximal*
... ior to any possible con-
... is an analogous process
... m. While the term *local*
... for this case, the distinc-
... wo problems exactly mir-
... *imization* procedures for
... y maximal or minimal, as

... an take many forms, de-
... are defined. For example,
... ," and what are "simple
... ge about the landscape is
... n of hill-climbing requires

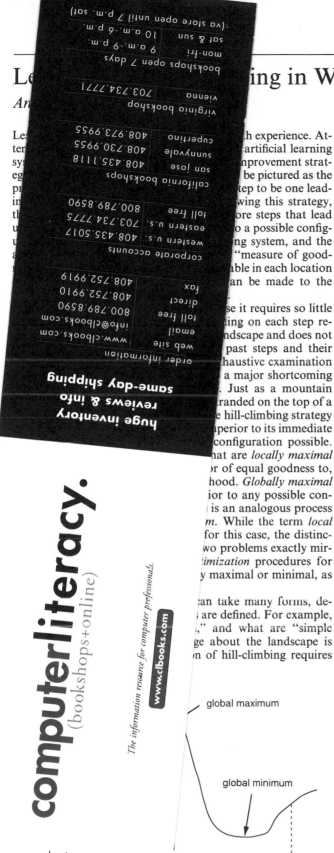

Figure 1. Local and global maxima and minima. The bold region of the abscissa contains all possible configurations, or points, x, with $f(x)$ representing the "goodness" of x.

the following components (see, e.g., Papadimitriou and Steiglitz, 1982): (1) a set of *feasible points*, (2) an *objective function*, and (3) a *neighborhood function*. The set of feasible points, also called the *search space*, contains a point representing each configuration that could possibly be a solution of the optimization problem. Let us denote the set of feasible points F. The objective function, denoted here by f, assigns a real number to each feasible point:

$$f : F \to \mathbb{R}$$

For any feasible point $x \in F$, $f(x)$ is the measure of the goodness of x. A global maximum is any $x \in F$ for which

$$f(x) \geq f(y) \quad \text{for all } y \in F$$

The neighborhood function formalizes what we mean by the simple changes that can be made to each configuration. It is a function, denoted here by N, that maps any feasible point x to the set of all feasible points, $N(x)$, that are close to x in the sense that each can be produced from x by some simple change. A local maximum is any $x \in F$ for which

$$f(x) \geq f(y) \quad \text{for all } y \in N(x)$$

Given these components, the hill-climbing procedure can be expressed using a function that for any feasible point x returns either an improved feasible point or an indication that local improvement is not possible:

improve (x)

$$= \begin{cases} \text{any } y \in N(x) & \text{with } f(y) > f(x) \text{ if such a } y \text{ exists} \\ \text{'no'} & \text{otherwise} \end{cases}$$

Then the hill-climbing procedure is as follows:

> **procedure** hill-climb
>
> **begin**
>
> $\quad x =$ some initial feasible point;
>
> $\quad$ **while** improve(x) $\neq$ 'no' **do**
>
> $\quad\quad x =$ improve(x);
>
> $\quad$ **return** x
>
> **end**

Different versions of this hill-climbing procedure exist, depending on how the function **improve** selects an improved point y from $N(x)$. If it sequentially searches $N(x)$ for an improved point, returning the first one found, it is called *first improvement* hill-climbing. In contrast, if it returns the best point in $N(x)$, it is called *steepest ascent* hill-climbing (Papadimitriou and Steiglitz, 1982).

Although the procedure just described clearly implements the basic idea of hill-climbing, it applies literally only if the search space consists of a finite number of discrete points, because only in this case can one exactly implement the function **improve**. However, the idea of hill-climbing also applies to problems with continuous search spaces, such as the n-dimensional Euclidean space $\mathbb{R}^n$. When the objective function f is a differentiable function of a region of $\mathbb{R}^n$, hill-climbing can take the form of a *gradient ascent* method. A point x in $\mathbb{R}^n$ is an n-dimensional vector $(x_1, x_2, \ldots, x_n)$. The *gradient* of f at x is the vector $(\partial f/\partial x_1(x), \ldots, (\partial f/\partial x_n(x)))$, where $(\partial f/\partial x_i(x))$ is the partial derivative of f with respect to the ith dimension evaluated at the point x. This vector, often denoted $\nabla f(x)$, points in the direction of the steepest increase of the objective function at the point x.

In gradient ascent, the function improve works by moving x some distance in the direction of the gradient vector:

$$\text{improve}(x) = x + \eta \nabla f(x), \qquad (1)$$

where η is often a nonnegative scalar that, together with the magnitude of $\nabla f(x)$, determines the distance along the gradient direction at which the new point is chosen. When $\nabla f(x) = 0$, x is a local maximum, and hill-climbing stops. The analogous procedure of *gradient descent* applies when the objective function is to be minimized. When a gradient ascent or descent method is applied to a learning system, such as an artificial neural network, it is common to call η the "learning rate parameter" because larger steps in the search space often—but not always—produce faster learning.

In fact, the choice of η is critical to the operation of the method: if it is too large, the new point may actually be worse than the old point because improve overshot the hill's peak in the gradient direction; if it is too small, however, hill-climbing progress may be unacceptably slow. More sophisticated versions of gradient ascent select a different η at each step, or even replace multiplication by η by a more complex operation, to avoid these difficulties. Many variations of basic gradient ascent have been studied (e.g., Luenberger, 1984). Obviously, to apply a gradient ascent method, it must be possible to compute the gradient of the objective function at each point generated in the search. This is possible in many optimization problems, but in others, an approximation of the gradient must be used instead.

The role hill-climbing plays in a learning system depends on what the search space represents and how one defines the objective function. Since learning involves improving aspects of behavior, each point in the search space must somehow represent a possibility for how the system might behave, and the objective function must indicate what it means to achieve varying degrees of behavioral success. A common approach to specifying a search space for a learning system is to devise a mathematical description of the system's behavior that depends on a set of numbers called *parameters*. As the values of the parameters change, the system's behavior changes.

For example, the connection weights of an artificial neural network are parameters, and learning is said to occur as the weights change according to one of many different learning algorithms, most of which implement a form of hill-climbing. In this case, the search space is the *weight space*, consisting of all feasible combinations of weight values. Usually *all* combinations of weight values are feasible, but there are some network learning methods in which some combinations are excluded from the search space (e.g., networks with weight sharing, e.g., LeCun et al., 1990; see CONVOLUTIONAL NETWORKS FOR IMAGES, SPEECH, AND TIME SERIES). The most commonly employed objective functions for artificial neural networks provide measures of the error in a network's behavior as compared with its desired behavior. Networks designed for *supervised learning* (see PERCEPTRONS, ADALINES, AND BACKPROPAGATION), for example, usually perform gradient descent on an objective function, giving the average over a set of training examples of the squared error between actual and desired network outputs. REINFORCEMENT LEARNING (q.v.) networks, however, try to maximize an objective function, giving (as one example) the probability of reward.

Hill-climbing in continuous search spaces can take forms other than the gradient ascent strategy just described. In fact,

one form of hill-climbing that works in continuous spaces is practiced by certain bacteria as they "swim" in their continuous fluid surroundings. A bacterium such as *Bacillus subtilis*, for example, propels itself along a straight path by rotating hairlike appendages called flagella. Whenever it reverses the rotational direction of its flagella, the bacterium stops and tumbles in place before heading off in a new, randomly determined, direction. This bacterium tends to move toward higher concentrations of certain chemical substances, called attractants, by regulating the frequency with which it stops and tumbles according to changes in the amount of attractant it senses. When it is moving toward higher attractant concentrations, it reduces the frequency with which it stops and tumbles. Thus, the straight segments of its path that lead uphill tend to be longer than the downhill segments, where the hill is formed by the spatial distribution of the attractant. As a consequence, the bacterium tends to find and remain near a local maximum of the attractant concentration. This behavior, which is a special kind of bacterial chemotaxis known as *klinokinesis*, is discussed by Lackie (1986).

Klinokinesis is a hill-climbing strategy that works somewhat differently from the methods already described because it neither systematically searches the neighborhoods of feasible points nor computes a gradient at each point. The bacterium effectively estimates the slope of the attractant distribution only in the direction it is moving, whereas a gradient estimate would effectively estimate the slope in all directions from each point. In fact, klinokinesis shows that hill-climbing is possible even when it is impossible to consider any alternative situation without actually committing to it. Instead of our mountain climber being able somehow to examine the neighboring terrain to determine—before it takes a step—the uphill direction, this mountain climber has to actually move to a point before its altitude can be determined. Although each step under these conditions cannot always produce an improvement, the overall trend of movement still leads uphill. An example of a strategy like klinokinesis conducted in the weight space of an artificial neural network is provided by Erich Harth's *Alopex* algorithm (Unnikrishnan and Venugopal, 1994) for adjusting a network's weights.

We have pointed out that hill-climbing is a local optimization procedure because it finds local instead of global optima. It is sometimes said of a learning system using hill-climbing that it "got stuck" in a local minimum or maximum. Another—and probably more serious—shortcoming of hill-climbing is that it needs a hill to climb! The most natural objective functions for many problems are essentially flat for large regions of the search space. On such "plateaus," hill-climbing is useless. For example, imagine a search space consisting of all possible lists of some fixed length of cooking instructions and an objective function that rates each such "recipe" according to how good you think its product tastes. One can imagine using hill-climbing to find a recipe for some delicious treat, but in reality, hill-climbing may be practically useless because only very rarely will such a random recipe produce anything edible at all. Hill-climbing works well for improving configurations having neighborhoods full of interesting alternatives, but it is not useful for finding such configurations in the first place. Minsky and Selfridge (1961) discussed this issue with regard to experimentation with randomly connected neural networks and the automatic synthesis of computer programs.

This shortcoming of hill-climbing can be addressed in several ways. First, one can try to define the search space and its neighborhood structure so that plateaus are small or few. This requires using prior knowledge about the problem to find a search-space representation that leads to a search space rich in attractive alternatives. In fact, designing appropriate representations is an essential aspect of any attempt to construct a learning system. A second way to address hill-climbing's plateau problem is to use other search methods to find regions of the search space in which hill-climbing can be useful. The *heuristic search* methods of artificial intelligence (AI) (e.g., Pearl, 1984) were developed to work when hill-climbing does not. According to AI terminology, hill-climbing is a *greedy* and *irrevocable* search strategy. It is greedy because it selects the best alternative in the immediate neighborhood of the current configuration without considering the possibility that such a selection may prevent future access to even better alternatives. Hill-climbing is irrevocable because it does not permit attention to return to a configuration considered earlier, even if that configuration showed more promise than the current one. Search strategies that are neither greedy nor irrevocable can be useful when simple hill-climbing is not.

One way to modify hill-climbing so that it is neither greedy nor irrevocable is to sometimes allow moves to configurations that are inferior to the current configuration, i.e., to sometimes allow moves downhill on the objective function landscape. For example, the SIMULATED ANNEALING (q.v.) algorithm does this using a probabilistic rule for selecting configurations (although simulated annealing is usually presented as a descent procedure). A computational temperature, T, determines how likely an inferior configuration will be accepted as the next configuration. When $T = 0$, the algorithm reverts to a greedy, irrevocable form of hill-climbing that always rejects moving to an inferior configuration.

Finally, methods exist for changing the objective function itself as a result of the learning system's experiences to make it more informative in evaluating alternatives. These methods have reached a high state of development in the field of REINFORCEMENT LEARNING (q.v.), where they are sometimes called *adaptive critic* methods.

Road Map: Learning in Artificial Neural Networks, Deterministic
Related Reading: Backpropagation: Basics and New Developments; Reinforcement Learning in Motor Control; Unsupervised Learning with Global Objective Functions

References

Lackie, J. M., 1986, *Cell Movement and Cell Behavior*, London: Allen & Unwin.

LeCun, Y., Boser, B., Denker, J. S., Henderson, D., Howard, R. E., Hubbard, W., and Jackel, L. D., 1990, Handwritten digit recognition with a back-propagation network, in *Advances in Neural Information Processing Systems 2* (D. S. Touretzky, Ed.), San Mateo, CA: Morgan Kaufmann, pp. 396–404.

Luenberger, D. G., 1984, *Linear and Nonlinear Programming*, 2nd ed., Reading, MA: Addison-Wesley. ◆

Minsky, M. L., and Selfridge, O., 1961, Learning in random networks, *Information Theory: Fourth London Symposium*, London: Butterworths, pp. 335–347.

Papadimitriou, C. H., and Steiglitz, K., 1982, *Combinatorial Optimization: Algorithms and Complexity*, Englewood Cliffs, NJ: Prentice-Hall. ◆

Pearl, J., 1984, *Heuristics: Intelligent Search Strategies for Computer Problem Solving*, Reading MA: Addison-Wesley. ◆

Unnikrishnan, K. P., and Venugopal, K. P., 1994, Alopex: A correlation-based learning algorithm for feed-forward and recurrent neural networks, *Neural Computat.*, 6:469–490.

Learning by Symbolic and Neural Methods

Jude W. Shavlik

Introduction

The last 10 or so years have produced an explosion in the amount of research on machine learning. This rapid growth has occurred, largely independently, in both the symbolic and connectionist (neural network) machine learning communities. Fortunately, over the last few years, these two communities have become less separate, and there has been an increasing amount of research that can be considered a hybrid of the two approaches. This article reviews some of the research that compares and combines the symbolic and neural-network approaches to creating artificial intelligence.

It is hard to define precisely the essential differences between the symbolic and connectionist approaches, but for the purposes of this article, one can make the coarse approximation that symbolic approaches focus on producing discrete combinations of the features used to describe examples, while neural approaches adjust continuous, nonlinear weightings of their inputs.

The history of both the symbolic and connectionist approaches to machine learning dates back to the beginnings of the field of artificial intelligence. In the 1950s, Rosenblatt began working on the perceptron (see PERCEPTRONS, ADALINES, AND BACKPROPAGATION). Also during that decade, Samuel (1959; see also REINFORCEMENT LEARNING) started his development of a program that learned to play checkers. One of the learning methods in Samuel's system improved state-based search, while another adjusted the weights assigned to various board-description features; thus, one can view his system as the first program that combined symbolic and connectionist learning methods.

To directly compare symbolic and connectionist learning algorithms, it is useful to consider the general task of learning from examples (frequently called *inductive learning*). In this common form of machine learning, an algorithm is given a collection of examples and counterexamples of some concept, and its task is to infer a general procedure for classifying future examples. For example, consider providing a computer program with a database of medical records: the age, weight, temperature, etc. of a clinic's patients, along with a physician's diagnosis for each person. An inductive learning algorithm would produce an automated procedure for predicting the diagnosis of new patients, given their physical characteristics and symptoms.

More precisely, an inductive learning program is given examples of the form (x_i, y_i), and it is supposed to learn a function f such that $f(x_i) = y_i$ for all (or most) i. Furthermore, the function f should capture the "general patterns" in the training data, so that f can be applied to predict y values for new, previously unseen, x values. Typically, each x_i is a composite de-

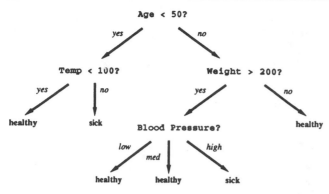

Figure 1. A sample (and hypothetical) decision tree for classifying patients on the basis of their physical properties. Interior nodes contain questions about the features of the patient being classified, while leaf nodes record the diagnosis.

scription of some object, situation, or event (e.g., a patient entering a medical clinic), and each y_i is a simpler description (e.g., of the disease of the patient).

One of the most successful symbolic learning algorithms is Quinlan's (1993) ID3, which represents the above function f as a *decision tree*. These are trees in which the nodes contain tests (e.g., of patient attributes) and the leaves indicate which class to assign the patient (e.g., "healthy" or "sick"). Figure 1 shows a sample decision tree for medical diagnosis.

Quinlan's ID3 algorithm (and its successor, the C4.5 algorithm) is designed to produce small decision trees. Ideally, it would find the smallest decision tree consistent with the training examples. Unfortunately, this is known to be computationally intractable. Hence, ID3 employs a heuristic algorithm based on information theory that constructs a reasonably small decision tree.

Neural networks are an alternative way to represent the function f, and training methods such as backpropagation (see BACKPROPAGATION: BASICS AND NEW DEVELOPMENTS) can use the provided examples to select good weight values for a network. Since one can choose to represent f as either a decision tree or neural network (many other representations are also possible), a question arises: Which approach works best? The next section describes some empirical studies that investigate this question.

This article also addresses another question: namely, is there any advantage to *combining* symbolic and connectionist approaches? (The section "Combining Symbolic AI and Neural Learning" discusses this question, as well as two more specific versions of it. It discusses methods that use neural networks to refine symbolic knowledge and also covers the task of extracting symbolic knowledge from trained neural networks.

Readers interested in obtaining a broader understanding of symbolic approaches to machine learning should consider an introductory textbook on artificial intelligence (e.g., Lugar & Stubblefield, 1993; Winston, 1992). Weiss and Kulikowski's (1991) textbook describes a range of approaches to the task of learning from examples and includes an excellent discussion of proper experimental methodology. Quinlan (1993) has written a textbook on decision tree induction, which is accompanied by computer programs written in the C programming language. Those desiring direct exposure to the research literature should check the collection of papers edited by Shavlik and Dietterich (1990), as well as the journal *Machine Learning*. Finally, the field of *computational learning theory* investigates the theoretical foundations of machine learning, addressing both the

symbolic and connectionist approaches. Natarajan's (1991) textbook further discusses computational learning theory (also see the Part II road map **Computability and Complexity**).

Comparing Decision Tree Induction with Neural Network Training

There have been several studies that empirically compared decision tree induction with neural network training, using data sets of "real-world" examples. Of the many published studies, this section reviews the results of three of the earliest ones, which used data sets from a variety of domains.

Atlas et al. (1990) used training examples from three tasks; two involved the prediction of aspects of power systems, and the third addressed speaker-independent vowel recognition. They used backpropagation to train the neural networks and used the CART algorithm, a technique closely related to ID3, to induce decision trees. Shavlik, Mooney, and Towell (1991) compared the ID3 and backpropagation algorithms on five commonly used datasets; three involved diseases, one addressed an aspect of chess, and the fifth was the NETTALK data set for pronunciation of written words. They also described experiments that measured accuracy as a function of noise in the data, size of the data sets, and amount of unspecified ("missing") feature values. Weiss and Kulikowski (1991) compared CART and backpropagation, as well as another half-dozen learning methods, on three medical test beds plus the famous Iris data set of the statistician Fisher that dates back to 1936. (Many of the data sets used in these studies are available through anonymous ftp—ftp.ics.uci.edu—from the UC-Irvine Repository of Machine Learning Databases.)

Atlas et al. reported that on all three data sets backpropagation produced an accuracy—on examples not seen during training—that was better than CART's. However, on only one data set was the difference statistically significant. Shavlik, Mooney, and Towell (1991) stated that backpropagation performed better than ID3 on four of their five data sets, but for only two of them was the difference significant. On the one data set where ID3 generalized better than backpropagation, the difference was not statistically significant. Finally, Weiss and Kulikowski (1991) reported that CART did better than backpropagation on two of their four datasets, while backpropagation generalized better on the other two. They did not discuss the statistical significance of the differences.

A conclusion from these studies is that one can expect slightly better generalization (i.e., accuracy on examples not seen during training) from backpropagation. However, these studies did not produce a definitive answer to the question of whether one should prefer decision trees or neural networks. Only a small sample of data sets were studied, algorithms and experimental methodology are continually improving for both decision tree induction and neural network training, and the magnitudes of the accuracy differences between the two techniques were small.

While generalization ability is probably the most important property of an inductive learning algorithm, there are several additional important aspects of an algorithm that learns from examples. One of these is the computer time required for training, and on this measure decision tree induction is significantly better—by about a factor of 100 in one experiment (Shavlik et al., 1991)—than backpropagation training. A gap still exists even when one considers parallel implementations.

Another important property of a learning algorithm is the interpretability of its representation of the induced concept. While some induced decision trees can be large and hard to understand, in most cases they provide less opaque representations than do neural networks. (The next section describes how one can re-represent trained neural networks as symbolic rule sets.)

In conclusion, these studies demonstrate that symbolic and connectionist approaches to machine learning are not diametric opposites. Rather, they can often be seen as addressing the same learning task, and to a first approximation they learn equally well.

Combining Symbolic AI and Neural Learning

There are many ways to combine symbolic and connectionist AI techniques, and the literature is growing rapidly. For example, Utgoff (1988) developed an algorithm that closely integrates decision trees and perceptrons. In his approach, the leaf nodes of the induced decision tree are perceptrons instead of category names. Others have created loosely coupled hybrid systems in which "high"-level decisions are made by an expert system, while "low"-level decisions are made by neural networks. For instance, in the system for autonomous driving developed by Pomerleau, Gowdy, and Thorpe (1991; see also VISION FOR ROBOT DRIVING), a rule-based technique integrates the predictions of several neural networks, which were trained on several subtasks of driving. This section describes a framework that organizes a large body of the research that combines aspects of both symbolic AI and neural learning.

In the framework illustrated by Figure 2, the learner first *inserts* symbolic information of some sort into a neural network. In this step, the symbolic knowledge is re-represented as a neural network; for example, as will be shown later, a hidden unit can be configured to make the same deductions as a conjunctive rule. It is becoming increasingly clear that a learner must make effective use of prior knowledge to perform well, and "knowledge-insertion" techniques provide a mechanism for incorporating existing information into neural networks.

Once in a neural representation, the learner uses training examples to *refine* the initial knowledge. When the backpropagation algorithm adjusts the weights to reduce the error, it is indirectly refining the initial symbolic knowledge. For example, it can delete an unnecessary antecedent ("precondition") of a rule by setting to zero the corresponding weight in the network.

Finally, the learner *extracts* symbolic information from the trained network. In this step, it converts the refined knowledge contained in the network back into a symbolic representation, thereby facilitating human inspection of what was learned.

The research of several groups fits nicely into this framework, and promising results have been achieved. The remainder of this article discusses some of this research and points out open issues in each of the three phases. (Shavlik, 1994, further describes this research and provides a lengthy bibliography.) Before continuing, it should be noted that these three steps are somewhat independent, and researchers have studied various combinations of them.

Insertion

The KBANN algorithm (Shavlik, 1994) is one technique for inserting prior knowledge into a neural network. This algorithm creates *knowledge-based artificial neural networks* by producing neural networks whose topological structure matches the dependency structure of the rules in an approximately correct "domain theory" (a collection of inference rules about the current task).

Figure 3 contains a simple example of the KBANN approach to mapping a domain theory into a neural network. KBANN converts a rule set into a network by re-representing the final conclusions of the rules as output units and creating hidden units that represent the intermediate conclusions. It connects units with highly weighted links and sets unit biases in such a manner that the (non-input) units emulate AND and OR gates, as appropriate.

Various research groups have found that knowledge-based neural networks train faster than do "standard" neural networks, presumably because the initial information is used to choose a good starting point for gradient-descent training of the network. More importantly, experiments have shown that knowledge-based networks generalize better to future examples than do standard networks, as well as several other learning methods. One can attribute this improved generalization to two aspects of the insertion process. The domain theory produces a useful inductive bias by (1) focusing attention on relevant input features and (2) indicating useful intermediate conclusions (which suggest a good network topology).

In addition to the simple, propositional rules shown in Figure 3 and used in much of the early KBANN work, researchers

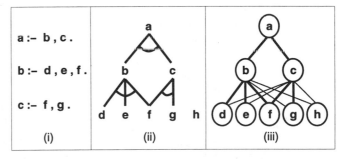

Figure 3. A sample application of the KBANN rule-insertion algorithm. Frame (i) contains a simple domain theory, while frame (ii) shows the dependency structure of these rules. (The rules are expressed using Prolog notation. The top rule can be read as "if *b* and *c* are true, then *a* can be concluded.") The third frame shows the network KBANN creates. The thick lines in frame (iii) correspond to the dependencies in the rules; KBANN sets the weights on these links in such a manner that nodes are highly active only when the domain theory supports the corresponding deduction. Thin lines in frame (iii) represent zero-weighted links that KBANN adds to the network to allow refinement of the domain theory during neural training. (Reprinted by permission from Shavlik, J., 1994, Combining symbolic and neural learning, *Machine Learning*, 14:321–331.)

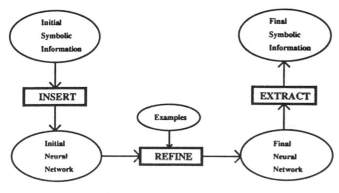

Figure 2. A framework for combining symbolic and neural learning. Prior knowledge is first re-represented as a neural network, which is then trained using a set of examples. In the last step, a symbolic description of the trained network is extracted. (Reprinted by permission from Shavlik, J., 1994, Combining symbolic and neural learning, *Machine Learning*, 14:321–331.)

have produced techniques for mapping several other forms of prior knowledge into networks. These include decision trees, rules containing certainty factors, fuzzy-logic rules, production rules, mathematical equations, and finite-state grammars. The approaches of the other researchers differ from KBANN to various degrees, but the essential idea is the same: use prior knowledge to decide how to initialize a neural network.

Refinement

Once prior knowledge is inserted into a network, the knowledge has to be refined and enhanced. A simple way of doing this is to run backpropagation, or some other standard connectionist training procedure, on the training examples. However, there are two ways to use symbolic information to improve training: (1) one could use symbolic learning methods and ideas to focus the adjustment of the network, both its weights and topology, and (2) one might alter backpropagation to better match the symbolic nature of a given problem.

Symbolic problems have motivated several changes to standard connectionist learning. Rather than minimizing mean square error, the cross-entropy error function is a better choice for knowledge-based networks (Shavlik, 1994). Refining rules with certainty factors requires the use of a different activation function for hidden and output units (Fu, 1989). Finally, networks often decay their weights toward zero during training. Weights in knowledge-based networks should decay toward their initial values, thereby encouraging the network to preserve the knowledge in the initial domain theory.

Extraction

The third phase of Figure 2's framework involves extracting symbolic information (e.g., rules) from a trained network, which need not originally be knowledge based. Why is this important? Rule extraction can help one understand what the "black box" network has learned. For example, if the network produced a scientifically interesting discovery, it would be nice if this were made explicit. Also, one may wish that a trained system could produce explanations of its future decisions. Finally, one may want to manipulate the results of learning in another AI system, such as a planner.

Several people have developed methods for extracting symbolic rules from trained neural networks. Among others, Gallant (1993) proposed an algorithm that considers various ways that a unit's weighted input can exceed its threshold and converts each of these situations into a rule. Another approach is to first cluster the weights going into a unit, next, reset each weight to the average value of the cluster it is in, and lastly, adjust the unit's bias (Shavlik, 1994); this reduces the degrees of freedom (since there is only one weight value per cluster) in the combinatorial search for rules. Others simply project trained units to the closest valid rule or express the extracted rules using a representation that is closer to that used by neural networks (e.g., fuzzy-logic rules).

Discussion

In addition to reviewing empirical comparisons of symbolic and connectionist learning methods, this article describes work that addresses three central questions about their synthesis:

1. How can one re-represent symbolic prior knowledge so that it can be refined by powerful numeric-optimization methods (e.g., backpropagation)?

2. How can symbolic knowledge about the task at hand guide neural network training?
3. How can one extract a small and comprehensible symbolic version of a trained network without losing (much) accuracy?

Several research groups have demonstrated the value of converting approximately correct symbolic information into a neural network representation, followed by connectionist learning and then extraction of the refined symbolic knowledge.

Regarding future work on the knowledge-insertion process, one would like to increase the types of prior knowledge that can be inserted into networks. For example, techniques are lacking for inserting domain theories expressed using predicate calculus. To do so, one must devise methods for dealing with unbounded symbolic structures in neural networks (whose size is usually fixed after training). Recurrent networks (see RE-CURRENT NETWORKS: SUPERVISED LEARNING) provide one such method, and Pollack's (1990) recursive autoassociative memories provide another. Also relevant is research on teaching networks to recognize context-free grammars by having them learn how to use a stack.

There are several open questions regarding the use of symbolic information to aid the network refinement step. How can one detect when extra hidden units are needed to generalize well, and where are the best places to add them? Folk wisdom says that backpropagation does not work well in networks with many layers of hidden units, because the error signal becomes too diffuse. Can one use symbolic information to focus the backpropagated error signal, especially in deep networks? Deep networks often occur when basing the network topology on the dependency structure of a rule base, so this problem is exacerbated in knowledge-based networks.

A major question with rule extraction is: how does one measure comprehensibility? An extraction algorithm must produce reasonably comprehensible rules, but without a good measure it is hard to compare alternative approaches. Most researchers resort to simply counting the number of symbols in the extracted rule sets, but a less syntactically oriented measure would be better. A second open issue relates to the refinement phase. The hidden units in knowledge-based networks generally have symbolic names attached to them, and if one is going to use these labels for the extracted rules, one needs to ensure that the symbol-unit correspondence is not severely altered during training. However, the cost of this might be an unacceptable reduction in accuracy. Thus, there is likely to be an accuracy-comprehensibility tradeoff in the design and use of rule-extraction algorithms.

Acknowledgments.   This work was partially supported by ONR Grant N00014-93-1-0998 and NSF Grant IRI-9002413.

Road Maps: Artificial Intelligence and Neural Networks; Learning in Artificial Neural Networks, Deterministic
Related Reading: Connectionist and Symbolic Representations; Expert Systems and Decision Systems Using Neural Networks; Fuzzy Logic Systems and Qualitative Knowledge

References

Atlas, L., Cole, R., Muthusamy, Y., Lippman, A., Connor, J., Park, D., El-Sharkawi, M., and Marks, R., 1990, A performance comparison of trained multilayer perceptrons and trained classification trees, *Proc. IEEE*, 78:1614-1619.
Fu, L., 1989, Integration of neural heuristics into knowledge-based inference, *Connection Sci.*, 1:325-340.

Gallant, S., 1993, *Neural Network Learning and Expert Systems*, Cambridge, MA: MIT Press, chaps. 14–17. ◆

Lugar, G., and Stubblefield, W., 1993, *Artificial Intelligence: Structures and Strategies for Complex Problem Solving*, 2nd ed., Redwood City, CA: Benjamin/Cummings, chap. 12. ◆

Natarajan, B., 1991, *Machine Learning: A Theoretical Approach*, San Mateo, CA: Morgan Kaufmann.

Pollack, J., 1990, Recursive distributed representations, *Artif. Intell.*, 46:77–105.

Pomerleau, D., Gowdy, J., and Thorpe, C., 1991, Combining artificial neural networks and symbolic processing for autonomous robot guidance, *Engrg. Appl. Artif. Intell.*, 4:279–285.

Quinlan, J., 1993, *C4.5: Programs for Machine Learning*, San Mateo, CA: Morgan Kaufmann. ◆

Samuel, A., 1959, Some studies in machine learning using the game of checkers, *IBM J. Res. Develop.*, 3:211–229. (Reprinted in Shavlik and Dietterich, 1990.)

Shavlik, J., 1994, Combining symbolic and neural learning, *Machine Learn.*, 14:321–331. ◆

Shavlik, J., and Dietterich, T., Eds., 1990, *Readings in Machine Learning*, San Mateo, CA: Morgan Kaufmann. ◆

Shavlik, J., Mooney, R., and Towell, G., 1991, Symbolic and neural network learning algorithms: An experimental comparison, *Machine Learn.*, 6:111–143.

Utgoff, P., 1988, Perceptron trees: A case study in hybrid concept representations, in *Proceedings of the Seventh National Conference on Artificial Intelligence*, St. Paul, MN: Morgan Kaufmann.

Weiss, S., and Kulikowski, C., 1991, *Computer Systems That Learn: Classification and Prediction Methods from Statistics, Neural Nets, Machine Learning, and Expert Systems*, San Mateo, CA: Morgan Kaufmann. ◆

Winston, P., 1992, *Artificial Intelligence*, 3rd ed., Reading, MA: Addison-Wesley, chaps. 16–25. ◆

Learning Vector Quantization

Teuvo Kohonen

Introduction

Neural network models are often applied to statistical pattern recognition problems, in which the class distributions of pattern vectors usually overlap and one must pay attention to the optimal location of the decision borders. The *Learning Vector Quantization* (*LVQ*) algorithms discussed in this article will be shown to define very good approximations for the optimal decision borders. These algorithms are computationally very light.

In the basic competitive learning neural networks to which the LVQ belongs, all cells may be thought to form an input layer, while there also exist mutual feedbacks or other types of lateral interaction between the cells. All cells receive the same external input, and by means of comparisons made in the lateral direction of the layer, an active response is switched on at the cell with the highest activation (the "winner"), while the responses of all the other cells are suppressed; this is called the *winner-take-all* (*WTA*) *function* (cf. Didday, 1970; Didday, 1976; Amari and Arbib, 1977; Grossberg, 1976).

Although many competitive learning neural networks have been based on nonlinear dynamic neural models, the decision or classification functions that ensue from their collective behavior are simply described by a formalism that was originally developed for signal analysis, namely, *vector quantization* (*VQ*). (For a general review of VQ, see Makhoul, Roucos, and Gish, 1985.) Thus VQ and the neural network models of competitive learning are not alternative methods: the former is an idealized description of the latter on the signal-space level.

Vector Quantization

As in simple neural-network models, we assume a signal vector $x = (\xi_1, \xi_2, \ldots, \xi_n) \in \mathbb{R}^n$ and a set of units or cells, each provided with a parametric vector (called *codebook vector*) $m_i = (\mu_{i1}, \mu_{i2}, \ldots, \mu_{in})^T \in \mathbb{R}^n$. The winner in the category of VQ problems is usually defined as the unit c whose codebook vector has the smallest Euclidean distance from x:

$$\|x - m_c\| = \min_i \left\{ \|x - m_i\| \right\} \qquad (1)$$

If x is a natural, stochastic, continuous-valued vectorial vari-

able, we need not consider multiple minima: the probability for $\|x - m_i\| = \|x - m_j\|$ for $i \neq j$ is then zero.

The VQ methods were originally developed to compress information. The m_i had to be placed into the input signal space such that the average expected quantization error E is minimized:

$$E = \int \|x - m_{c(x, m_1, \ldots, m_k)}\|^2 p(x)\, dx = \min! \qquad (2)$$

where $p(x)$ is the probability density function of x, and dx is a hypervolume differential in the signal space. Notice that c, the index of the winner, depends on x and all the m_i.

It has been shown by Zador (1982), for example, that the point density of the m_i values that minimize E is proportional to $[p(x)]^{n/(n+2)}$. Since in practical problems usually $n \gg 2$, it can then be said that the distribution of the m_i approximates $p(x)$.

Optimal Decision

Assume that all samples of x are derived from a finite set of classes $\{S_k\}$, the distributions of which are allowed to overlap. The problem of optimal decision or statistical pattern recognition is usually discussed within the framework of the Bayes theory of probability (for a textbook account, see, e.g., Kohonen, 1989, chap. 7.2). Let $P(S_k)$ be the a priori probability of class S_k, and $p(x|x \in S_k)$ be the conditional probability density function of x on S_k, respectively. In this method the so-called *discriminant functions* are defined as

$$\delta_k(x) = p(x|x \in S_k)P(S_k) \qquad (3)$$

It can be shown that the average rate of misclassifications is minimized if the sample x is determined to belong to class S_c according to

$$\delta_c(x) = \max_k \left\{ \delta_k(x) \right\} \qquad (4)$$

Learning Vector Quantization Algorithms

The LVQ1

Consider now Figure 1. In the LVQ approaches we assign *a subset of codebook vectors to each class S_k* and then search for

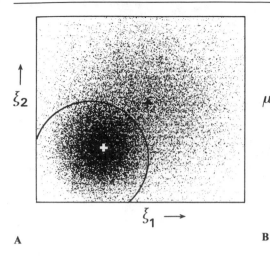

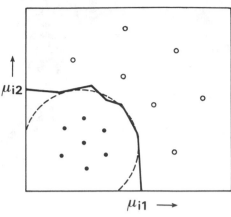

A   B

Figure 1. *A*, The probability density function of $x = [\xi_1, \xi_2]^T$ is represented here by its samples, the small dots. The superposition of two symmetric Gaussian density functions corresponding to two different classes S_1 and S_2, with their centroids shown by the white and dark cross, respectively, is shown. Solid curve: the theoretical optimal Bayes decision surface. *B*, Large black dots: codebook vectors of class S_1. Open circles: codebook vectors of class S_2. Solid curve: decision surface obtained by LVQ1. Broken curve: Bayes decision surface.

the codebook vector m_i that has the smallest Euclidean distance from x. This assignment can be made in such a way that codebook vectors belonging to different classes are not intermingled, although the class distributions overlap. The sample x is then thought to belong to the same class as the closest m_i. As only codebook vectors closest to the class borders define the decision borders, a good approximation of $p(x \mid x \in S_k)$ is not necessary everywhere. We must place the m_i into the signal space in such a way that the nearest-neighbor rule (Equation 1) minimizes the average expected misclassification probability.

Notice that in considering the average expected classification accuracy, the quantization errors can be large in regions where $p(x)$ has small values.

Let

$$c = \arg\min_i \{\|x - m_i\|\} \tag{5}$$

define the index of the nearest m_i to x, denoted by m_c; x is then determined to belong to the same class to which the nearest m_i belongs.

Let $x = x(t)$ now be a time-series sample of input, and let the $m_i(t)$ represent sequential values of the m_i in the discrete-time domain, $t = 0, 1, 2, \ldots$, obtained in the following process. Starting with properly defined initial values $m_i(0)$, Equations 6 shall define the basic learning vector quantization process (Kohonen, 1988; Kohonen, Barna, and Chrisley, 1988); this particular algorithm is called LVQ1.

$$m_c(t + 1) = m_c(t) + \alpha(t)[x(t) - m_c(t)]$$

if x and m_c belong to the same class

$$m_c(t + 1) = m_c(t) - \alpha(t)[x(t) - m_c(t)] \tag{6}$$

if x and m_c belong to different classes

$$m_i(t + 1) = m_i(t) \quad \text{for } i \neq c$$

It will be shown that the asymptotic values of m_i obtained in this process define a vector quantization for which the rate of misclassifications is approximately minimized. Here $0 < \alpha(t) < 1$, and $\alpha(t)$ (learning-rate factor) is usually made to decrease monotonically with time. It is recommended that α should initially be rather small, say, smaller than 0.1. If only a restricted time is available for learning, the exact law $\alpha = \alpha(t)$ is not crucial, especially also if only a restricted set of training samples is available; they may be applied cyclically, and $\alpha(t)$ may even be made to decrease linearly to zero.

It is in general difficult to show what the exact convergence limits of Equation 6 are. The following discussion is based on the idea that VQ tends to approximate density functions such as $p(x)$. Instead of $p(x)$, we may also consider *any* nonnegative function $f(x)$ in Equation 2.

The Bayes decision borders defined by Equations 3 and 4 divide the signal space into class regions B_k such that the rate of misclassifications is minimized. All such borders together are defined by the condition $f(x) = 0$, where, for $x \in B_k$ and $h \neq k$,

$$f(x) = p(x \mid x \in S_k)P(S_k) - \max_h \{p(x \mid x \in S_h)P(S_h)\} \tag{7}$$

Notice that $f(x)$ is piecewise continuous and nonnegative. For each $x \in B_k$, $f(x)$ has a positive hump, and these humps are separated by the Bayes borders at which $f(x) = 0$.

If we approximate $f(x)$ by the point density of codebook vectors defined by classical VQ, this point density must then also tend to zero at all borders.

In order to find the minimum of E in Equation 2 and the optimal values for the m_i in VQ by gradient descent, we need an expression for the gradient of E. From Kohonen (1991, Equations A1 through A14) we obtain the result

$$\nabla_{m_i} E = -2 \int \delta_{ci} \cdot (x - m_i)p(x)\,dx \tag{8}$$

where δ_{ci} is the Kronecker delta, and c is the index of the m_i that is closest to x (i.e., the winner). The gradient step of vector m_i is

$$m_i(t + 1) = m_i(t) - \lambda \cdot \nabla_{m_{i(t)}} E \tag{9}$$

where λ defines the step size, and

$$\nabla_{m_{i(t)}} E = -2\delta_{ci}[x(t) - m_i(t)] \tag{10}$$

If $p(x)$ in E is now replaced by $f(x)$ we get by substitution:

$$\nabla_{m_i} E = -2 \int \delta_{ci}(x - m_i)f(x)\,dx$$

$$= -2 \int \delta_{ci}(x - m_i)[p(x \mid x \in S_k)P(S_k)$$

$$- \max_h \{p(x \mid x \in S_h)P(S_h)\}]\,dx \tag{11}$$

The gradient steps must be computed separately in the event that the sample $x(t)$ belongs to S_k, and in the event that $x(t) \in S_h$.

In the event that $x(t) \in S_k$ we obtain

$$\nabla_{m_i(t)} E = -2\delta_{ci}[x(t) - m_i(t)] \tag{12}$$

with the a priori probability $P(S_k)$.

The class with $\max_h\{p(x|x \in S_h)P(S_h)\}$ is the runner-up class signified by index r. In the event that $x(t) \in S_r$ the following expression for $\nabla_{m_i} E$ is obtained with the a priori probability $P(S_r)$:

$$\nabla_{m_i(t)} E = +2\delta_{ci}[x(t) - m_i(t)] \tag{13}$$

The different cases are collected into the following set of equations, rewritten with $\alpha(t) = 2\lambda$:

$$m_c(t + 1) = m_c(t) + \alpha(t)[x(t) - m_c(t)]$$
$$\text{for } x(t) \in B_k \text{ and } x(t) \in S_k$$
$$m_c(t + 1) = m_c(t) - \alpha(t)[x(t) - m_c(t)]$$
$$\text{for } x(t) \in B_k \text{ and } x(t) \in S_r \tag{14}$$
$$m_c(t + 1) = m_r(t) \quad \text{for } x(t) \in B_k \text{ and } x(t) \in S_h, h \neq r$$
$$m_i(t + 1) = m_i(t) \quad \text{for } i \neq c$$

If the m_i of class S_k are already within B_k, the VQ will further attract them to the hump corresponding to B_k, at least if the learning steps are small. With a sufficiently large number of codebook vectors in each class region B_k, the closest codebook vectors in adjacent regions B_k will be arbitrarily close to the Bayes border. Thus VQ and Equation 7 have been shown to define the Bayes borders with arbitrarily good accuracy.

Near equilibrium, close to the borders at least, Equations 6 and 14 can be seen to define almost similar corrections; notice that in Equation 6, *the classification of* x *was approximated by the nearest-neighbor rule*, and this approximation will be improved during learning. However, notice too that in Equation 6 the minus-sign corrections were made every time when x was classified incorrectly, whereas Equation 14 only makes the corresponding correction if x is exactly in the runner-up class. The error thereby made is often insignificant. As a matter of fact, the algorithms called LVQ2 and LVQ3 (Kohonen, 1990) are even closer to Equation 14 in this respect.

The Optimized Learning Rate LVQ1 (OLVQ1)

If an individual learning rate $\alpha_i(t)$ is assigned to each m_i, we obtain the following modified learning process (Kohonen, 1992). Let c be defined by Equation 5. Then

$$m_c(t + 1) = m_c(t) + \alpha_c(t)[x(t) - m_c(t)]$$
$$\text{if } x \text{ is classified correctly}$$
$$m_c(t + 1) = m_c(t) - \alpha_c(t)[x(t) - m_c(t)] \tag{15}$$
$$\text{if } x \text{ is classified incorrectly}$$
$$m_i(t + 1) = m_i(t) \quad \text{for } i \neq c$$

We may try to determine the $\alpha_i(t)$ for fastest convergence of Equations 15. Let us express Equations 15 in the shorter form

$$m_c(t + 1) = [1 - s(t)\alpha_c(t)]m_c(t) + s(t)\alpha_c(t)x(t) \tag{16}$$

where $s(t) = +1$ if the classification is correct, and $s(t) = -1$ if the classification is wrong. It may be obvious that the *statistical accuracy* of the learned codebook vector values is approximately optimal if all samples have been used with equal weight, i.e, if the effects of the corrections made at different times, when referring to the end of the learning period, are of approximately equal magnitude. Notice that $m_c(t + 1)$ contains a trace of $x(t)$ through the last term in Equation 16, and traces of the

earlier $x(t')$, $t' = 1, 2, \ldots, t - 1$, through $m_c(t)$. In a learning step, the magnitude of the last trace of $x(t)$ is scaled down by the factor $\alpha_c(t)$, and, for instance, during the same step the trace of $x(t - 1)$ becomes scaled down by $[1 - s(t)\alpha_c(t)] \cdot \alpha_c(t - 1)$. Now we first stipulate that these two scalings must be identical:

$$\alpha_c(t) = [1 - s(t)\alpha_c(t)]\alpha_c(t - 1) \tag{17}$$

If this condition is made to hold for all t, by induction it can be shown that the traces collected up to time t of all the earlier $x(t')$ will be scaled down by an equal amount at the end, and thus the "optimal" values of $\alpha_i(t)$ are determined by the recursion

$$\alpha_c(t) = \frac{\alpha_c(t - 1)}{1 + s(t)\alpha_c(t - 1)} \tag{18}$$

For fast learning, the OLVQ1 algorithm can be started with the $\alpha_i(0)$ in the range 0.3 to 0.5.

General Considerations

Initialization of the Codebook Vectors

A rather good strategy is to start with the same number of codebook vectors in each class. An upper limit to the total number of codebook vectors is set by the restricted recognition time and computing power. An identical number of codebook vectors per class is justifiable, since for optimal approximation of the borders the average distances between the adjacent codebook vectors (which depend on their numbers per class) ought to be the same in each class. Because the final placement of the codebook vectors and thus their distances are not known until the end of the learning process, equalization of the distances in the various classes ought to be made iteratively.

Once the numbers of codebook vectors have been fixed, one may use first samples of the real training data for their initial values. Referring to the derivation of Equation 9, however, the codebook vectors should always remain inside the respective class regions. For the initial values one can then accept only samples that are not misclassified. In other words, a sample is first tentatively classified against all the other samples in the training set, for instance by the traditional k-nearest-neighbor (kNN) method, and accepted for a possible initial value only if this tentative classification is the same as the class identifier of the sample. (In the learning algorithm itself, however, no samples must be excluded; they are applied independently of whether they fall on the correct side of the class border or not.)

Overall Learning Strategy

One may start the learning with the OLVQ1 algorithm, which has fast convergence; its final recognition accuracy will approximately be achieved after a number of learning steps that is about 30 to 50 times the total number of codebook vectors. In an attempt to ultimately improve recognition accuracy, one may try to continue with the basic LVQ1, or with the other LVQ versions, using a low initial value of learning rate, which is then the same for all classes.

The neural-network algorithms often "overlearn"; i.e., when the learning and test phases are alternated, the recognition accuracy is first improved until an optimum is reached. After that point, the accuracy often starts to decrease slowly. A possible explanation in the present case is that when the codebook vectors become very specifically tuned to the training data, the ability of the algorithm to generalize with respect to new data suffers. It is therefore necessary to stop the additional learning process after some optimal number of steps, say, 50 to 200 times the total number of the codebook vectors. Such a stopping rule can only be found by experience.

Comparison with Other Methods

LVQ and SOM

Another related algorithm, the *Self-Organizing Map* (*SOM*) (Kohonen, 1989, 1990, 1993; see also SELF-ORGANIZING FEATURE MAPS: KOHONEN MAPS) should be mentioned in this context. It is an unsupervised learning method, whereas supervised training is used in the LVQ. In LVQ, only one or two winner cells are updated during each adaptation step, whereas in the SOM, a block of neighboring cells around the winner *relating to the physical network* is updated simultaneously. The main application areas of these algorithms are also different: LVQ is used for the classification of stochastic data, whereas the SOM is more useful for the visualization of high-dimensional data on a two-dimensional display.

Relative Performance of LVQ as a Classifier

The number of different applications of LVQ (and SOM) is for the present at least on the order of many hundreds, and it is very difficult to make any comparative survey. Just to give a feeling of the relative performance, Table 1 (Kohonen, 1990) compares the classification accuracies achievable by a couple of classical methods as well as by LVQ1. The data, 15-channel acoustic spectra representing 19 different phonemic classes, were collected from Finnish speech. A total of 1550 samples were used for training, and another set of 1550 independent samples for testing, respectively. There were in total 117 codebook vectors.

The term *parametric Bayes* in Table 1 means a Bayesian classification method in which the discriminant functions are approximated by multivariate normal distributions, and kNN is the classical *k*-nearest-neighbor method, i.e., comparison of each test sample against all the training samples and voting over *k* closest ones (here $k = 5$). It is generally known that the kNN algorithm gives a very good approximation of the theoretical Bayes limit; but LVQ1 is still better, and here more than ten times faster computationally.

Table 1. Error Percentages in Phonemic Classification

Parametric Bayes	kNN	LVQ1
12.1	12.0	10.2

Road Map: Self-Organization in Neural Networks
Background: I.3. Dynamics and Adaptation in Neural Networks
Related Reading: Competitive Learning; Coulomb Potential Learning; Data Clustering and Learning

References

A list of over 1000 literature references to analyses and applications of LVQ and SOM algorithms is available on the Internet and can be accessed using the anonymous-user file transfer protocol (FTP). The address of the computer on the Internet is cochlea.hut.fi (or 130.233.168.48). This same address also contains extensive software packages of LVQ and SOM algorithms, diagnostic programs, and exemplary data.

The following works have been referred to in this article:

Amari, S., and Arbib, M. A., 1977, Competition and cooperation in neural nets, in *Systems Neuroscience* (J. Metzler, Ed.), New York: Academic Press, pp. 119–165. ◆

Didday, R. L., 1970, The simulation and modelling of distributed information processing in the frog visual system, PhD Dissertation, Stanford University.

Didday, R. L., 1976, A model of visuomotor mechanisms in the frog optic tectum, *Math. Biosci.*, 30:169–180.

Grossberg, S., 1976, Adaptive pattern classification and universal recoding, I:Parallel development and coding of neural feature detectors, *Biol. Cybern.*, 23:121–134; II: Feedback, expectation, olfaction, illusions, *Biol. Cybern.*, 23:187–202.

Kohonen, T., 1988., An introduction to neural networks, *Neural Netw.*, 1:3–16. ◆

Kohonen, T., 1989, *Self-Organization and Associative Memory*, 3rd ed., Berlin: Springer-Verlag.

Kohonen, T., 1990, The Self-Organizing Map, *Proc. IEEE*, 78:1464–1480. ◆

Kohonen, T., 1991, Self-Organizing Maps: Optimization approaches, in *Articial Neural Networks* (T. Kohonen, K. Mäkisara, O. Simula, and J. Kangas, Eds.), Amsterdam: Elsevier, vol. 2, pp. 1677–1680.

Kohonen, T., 1992, New developments of Learning Vector Quantization and the Self-Organizing Map, in *Symposium on Neural Networks: Alliances and Perspectives in Senri 1992 (SYNAPSE'92)*, Osaka, Japan.

Kohonen, T., 1993, Physiological interpretation of the Self-Organizing Map algorithm, *Neural Netw.*, 6:895–905.

Kohonen, T., Barna, G., and Chrisley, R., 1988, Statistical pattern recognition with neural networks: Benchmarking studies, in *Proceedings of the IEEE International Conference on Neural Networks*, vol. 1, New York: IEEE, pp. 61–68.

Makhoul, J., Roucos, S., and Gish, H., 1985. Vector quantization in speech coding, *Proc. IEEE*, 73:1551–1588. ◆

Zador, P. L., 1982. Asymptotic quantization error of continuous signals and the quantization dimensions, *IEEE Trans. Inform. Theory*, IT-28:139–149.

Lesioned Attractor Networks as Models of Neuropsychological Deficits

David C. Plaut

Introduction

A highly controversial issue concerning the neural implementation of cognition is the degree to which cognitive functions are localized to particular brain regions as opposed to distributed throughout large areas of cortex. Connectionist models using distributed representations provide a natural formalism for expressing how cognitive processes can be distributed over large numbers of neuron-like processing units. In fact, many characteristics of these models echo Karl Lashley's original ideas about mass action and equipotentiality. However, the possibility that cognitive functions are distributed widely in the cortex is seriously challenged by the remarkably selective cognitive deficits than can occur in brain-injured patients (see Shallice, 1988, for a review). Can the nature of distributed computation in networks be reconciled with these selective deficits? In fact, as the work described below illustrates, connectionist networks are leading to new insights about how disorders of brain function can give rise to disorders of cognition, challenging traditional assumptions about the modular organization of cognitive functions (see Farah, 1994, for discussion).

Attractors and Damage

Researchers have begun to explore the degree to which the effects of damage in connectionist models of cognitive processes reproduce certain types of neuropsychological deficits. Many of these investigations use recurrent, interactive networks that develop attractors for familiar patterns of activity (see COMPUTING WITH ATTRACTORS). In an attractor network, units interact in such a way that the initial activity pattern generated by an input gradually settles to the nearest attractor pattern. If the state of each unit is represented along a separate dimension in a multidimensional *state space*, then each attractor corresponds to a particular point within this space, and the set of patterns that settle to it corresponds to a region around it called its *basin* of attraction. Although recurrent networks can also exhibit more complex dynamic behavior, such as "limit cycle" and "chaotic" attractors, virtually all existing attempts to model cognitive and neuropsychological phenomena have relied on "point" attractors.

Lesioning a network, by removing some proportion of its units or connections, alters the settling behavior of the remaining units. In state space, the effects of damage amount to distortions of the shapes and positions of attractor basins. As a result, the initial pattern of activity for an input now may fall within a neighboring basin, giving rise to an incorrect response. The detailed pattern of correct and incorrect performance depends on specifics of the layout of attractor basins (as a function of the task, network architecture, and training procedure) and how the basins are distorted by the lesion (as a function of the location, severity, and type of damage).

Models of Neuropsychological Deficits

Brain damage can produce selective impairments in a wide range of cognitive domains, including high-level vision, attention, written and spoken language, learning and memory, planning, and motor control (Shallice, 1988). In many of these domains, lesioning a connectionist model of the normal process leads to analogous deficits, suggesting that the models capture important properties of the normal cognitive system.

Deep Dyslexia

The class of neuropsychological impairments which have received perhaps the greatest theoretical attention are those that involve word reading, the so-called acquired dyslexias. Of these, *deep dyslexia* is among the most perplexing (Coltheart, Patterson, and Marshall, 1980). The hallmark characteristic of the reading behavior of deep dyslexic patients is the occurrence of *semantic* errors, such as reading the word *river* as "ocean" or *dark* as "night." These patients also make *visual* errors (e.g., *scandal* ⇒ "sandals"), suggesting a second impairment. However, if two separate lesions are involved, why do visual errors virtually always co-occur with semantic errors?

Hinton and Shallice (1991) demonstrated that the co-occurrence of visual and semantic errors is a natural consequence of a single lesion to an attractor network trained to derive the meanings of written words. They trained a recurrent backpropagation network to map from the visual form (orthography) of 40 three- or four-letter words to a simplified representation of their semantics, described in terms of 68 predetermined seman-

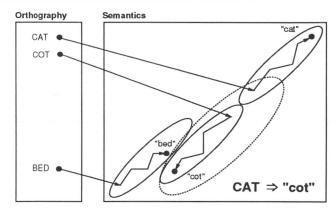

Figure 1. How damage to semantic attractors can cause visual errors. The solid ovals depict the normal basins of attraction; the dotted oval depicts a basin after semantic damage. (Reprinted from Plaut and Shallice, 1993a, p. 393.)

tic features (e.g., *brown, made-of-wood, for-cooking*). After training, lesions throughout the network resulted in both semantic errors (e.g., *cat* ⇒ "dog") and visual errors (e.g., *cat* ⇒ "cot"), similar to those observed in deep dyslexia. Essentially, for the network to solve the task, the layout of attractor basins must be sensitive to both visual and semantic similarity; as a result, these metrics are reflected in the types of errors that occur as a result of damage (Figure 1).

More recently, Plaut and Shallice (1993a) have extended these initial findings in two ways. First, they established the generality of the co-occurrence of visual and semantic errors by showing that it does not depend on particular characteristics of the network architecture, the training procedure, or the way responses are generated from semantic activity. Second, they extended the approach to account for many of the remaining characteristics of deep dyslexia, including the effect of concreteness on reading accuracy (e.g., *table* is more likely to be read correctly than *truth*) and its interaction with visual errors; the occurrence of other types of errors (e.g., *visual-then-semantic*; *sympathy* ⇒ "orchestra," presumably via *symphony*); greater confidence in visual than in semantic errors; relatively preserved lexical decision; and the existence of different subvarieties of deep dyslexia. The only major additional assumption in these extensions is the hypothesis that abstract words have fewer semantic features than concrete words, which causes the network to form weaker attractors for them. The same general approach has also been used to account for the perseverative and semantic influences on the errors that optic aphasic patients make in naming visually presented objects (Plaut and Shallice, 1993b), and the degree of recovery and generalization in cognitive rehabilitation studies with acquired dyslexic patients (Plaut, in press).

Neglect Dyslexia

Mozer and Behrmann (1990) have accounted for another reading disorder, known as *neglect dyslexia*, based on principles very similar to those used in the deep dyslexia simulations. Patients with neglect dyslexia often ignore the contralesional (typically left) portion of written material, even when it falls entirely within the intact portions of their visual fields (Riddoch, 1991). Incorrect responses to letter strings typically consist of letter omissions (e.g., *chair* ⇒ "hair"), substitutions (e.g., *house* ⇒ "mouse"), or additions (e.g., *love* ⇒ "glove"). The severity of the deficit is influenced by both peripheral and central manipulations. Thus, the accuracy of reading a letter string is better when the stimulus is presented further to the right, or when it forms a word.

In the model that Mozer and Behrmann used (Mozer, 1991), bottom-up letter information interacts with top-down lexical/semantic knowledge to form attractors for words. The bottom-up input is gated by an Attentional Mechanism (AM) that forms a spatially contiguous "spotlight" on the basis of where letter features occur. Letter features that fall outside the spotlight are much less likely to be transmitted to the word recognition system. Mozer and Behrmann model the attentional impairment in neglect dyslexia by introducing a monotonic gradient of damage to the connections from the letter features to the AM, with damage most severe on the left. This damage biases the AM toward forming a spotlight that includes only the rightmost letters of an input string, resulting in corrupted letter input to the word recognition system. The lexical attractors can often reconstruct the correct pattern of activity, particularly when the entire input forms a word. However, when this process fails, the result is often another word that differs from the presented word only on the left. Reading accuracy is better if the letter string is presented further to the right because the damage from these positions to the AM is less severe.

Hemispatial Neglect

In traditional cognitive neuropsychological accounts, investigators have often stipulated the existence of a specialized module in the cognitive system whenever a brain-damaged patient exhibits a selective deficit in some specific aspect of cognitive function. For example, patients with *hemispatial neglect*—a more general attentional deficit than neglect dyslexia—are abnormally slow to shift their attention from a pre-cued ipsilesional location to a contralesional stimulus. This has been interpreted in terms of damage to a specific "disengage" module (Posner et al., 1984). However, Cohen et al. (1994) have reproduced this deficit in shifting attention by unilaterally damaging a connectionist model with no special disengage module. Rather, in the model, attention is allocated based on competitive interactions among units representing different spatial locations. The unilateral damage causes an imbalance in this competition, making it difficult for attention to be captured by contralesional locations.

Prosopagnosia

Another example of the influence of connectionist models on the interpretation of neuropsychological deficits concerns patients with "prosopagnosia." These patients fail to name familiar faces and have no conscious recollection of them, and yet they often show evidence of recognition on tasks, such as priming or name relearning, that measure covert knowledge of faces. This behavioral dissociation has led some researchers to propose separate mechanisms of overt and covert recognition. But Farah, O'Reilly, and Vecera (1993) have demonstrated that this dissociation between overt and covert face recognition can arise in a single system. Partial damage can virtually eliminate overt naming performance while still leaving sufficient residual information available to support above-chance performance on more indirect tests of semantic knowledge.

Category-Specific Semantic Deficits

As a final example, key connectionist models are contributing to the question of whether knowledge is organized by modality or by category. Apparent evidence for category specificity comes from the finding that some brain-injured patients are selectively impaired in recognizing and recalling information about animate objects (e.g., animals) versus inanimate objects (e.g., tools). Warrington and Shallice (1984) suggested that these deficits could be accounted for if semantics were instead organized by modality, under the dual assumption that visual semantics is impaired and the animate objects rely more heavily on visual than on functional semantics. This explains why the patients were also impaired on inanimate categories, such as gemstones and fabrics, they rely primarily on visual attributes. However, in the absence of a computational model, this account was rejected in favor of categorical organization when it was noted that patients had difficulty with both functional and visual aspects of animate objects. But Farah and McClelland (1991) revived the original hypothesis by accounting for the entire pattern of deficits in an interactive, distributed model in which visual and functional semantics interact. The simulation accounts for the patients' paradoxical impairment in recalling functional information about animate objects because functional semantics normally relies on interactions with intact visual semantics to settle to the correct pattern.

Discussion

Connectionist networks would appear a priori to be an appropriate formalism within which to develop computational models of neuropsychological disorders. Although the specific relationship between these networks and neurobiology is far from clear, the belief that representation and computation in these networks resembles neural computation at some level remains one of their strongest attractions. As the research reviewed here illustrates, the computational formalism of attractors, and their behavior under damage, can lead to a deeper understanding of a wide range of neuropsychological phenomena. However, most simulations have addressed impairments only within very specific processing domains. A central challenge for future work is to extend these preliminary findings in developing more comprehensive simulations of broader aspects of normal and impaired cognitive processing. Nonetheless, even at this early stage of research, the finding that the behavior of attractor networks after damage resembles that of neurological patients supports the claim that the apparent similarity of artificial and biological neural networks is, in fact, substantive.

Road Map: Connectionist Psychology
Background: I.3. Dynamics and Adaptation in Neural Networks
Related Reading: Developmental Disorders; Distortions in Human Memory; Schema Theory

References

Cohen, J. D., Romero, R. D., Servan-Schreiber, D., and Farah, M. J., 1994, Mechanisms of spatial attention: The relation of macrostructure to microstructure in parietal neglect, *J. Cognit. Neurosci.*, 6:377–387.

Coltheart, M., Patterson, K. E., and Marshall, J. C., Eds., 1980, *Deep Dyslexia*, London: Routledge and Kegan Paul.

Farah, M. J., 1994, Neuropsychological inference with an interactive brain: A critique of the locality assumption, *Behav. Brain Sci.*, 17:43–104. ◆

Farah, M. J., and McClelland, J. L., 1991, A computational model of semantic memory impairment: Modality-specificity and emergent category-specificity, *J. Exp. Psychol. Gen.*, 120:339–357.

Farah, M. J., O'Reilly, R. C., and Vecera, S. P., 1993, Dissociated overt and covert recognition as an emergent property of a lesioned neural network, *Psychol. Rev.*, 100:571–588.

Hinton, G. E., and Shallice, T., 1991, Lesioning an attractor network: Investigations of acquired dyslexia, *Psychol. Rev.*, 98:74–95.

Mozer, M. C., 1991, *The Perception of Multiple Objects: A Connectionist Approach*, Cambridge, MA: MIT Press.

Mozer, M. C., and Behrmann, M., 1990, On the interaction of selective attention and lexical knowledge: A connectionist account of neglect dyslexia, *J. Cognit. Neurosci.*, 2:96–123.

Plaut, D. C., in press, Relearning after damage in connectionist networks: Toward a theory of rehabilitation, *Brain Lang*.

Plaut, D. C., and Shallice, T., 1993a, Deep dyslexia: A case study of connectionist neuropsychology, *Cognit. Neuropsychol.*, 10:377–500. ◆

Plaut, D. C., and Shallice, T., 1993b, Perseverative and semantic influ-

ences on visual object naming errors in optic aphasia: A connectionist account, *J. Cognit. Neurosci.*, 5:89–117.

Posner, M. I., Walker, J. A., Friedrich, F. J., and Rafal, R. D., 1984, Effects of parietal injury on covert orienting of visual attention. *J. Neurosci.*, 4:1863–1874.

Riddoch, M. J., Ed., 1991, *Cognitive Neuropsychology: Neglect and the Peripheral Dyslexias*, Hillsdale, NJ: Erlbaum.

Shallice, T., 1988, *From Neuropsychology to Mental Structure*, Cambridge, Eng.: Cambridge University Press. ◆

Warrington, E. K., and Shallice, T., 1984, Category specific semantic impairments, *Brain*, 107:829–853.

Limb Geometry: Neural Control

Francesco Lacquaniti and Claudio Maioli

Introduction

How is spatial information for limb movement encoded in the brain? To move, we must know *where* to go and *how* to get there. Because movements unfold simultaneously in the space extrinsic to the subject (*where* to go) and in the intrinsic space corresponding to the joints and muscles (*how* to go), the brain represents spatial information for both extrinsic and intrinsic spaces. Let us consider the example of drawing movements. The goal of drawing and handwriting is to produce geometrical shapes conforming to their cognitive representation. Neural processing underlying their generation can be epitomized by the transformation of an intended drawing into an actual movement. The cognitive representation of drawing and handwriting includes complex symbolic and analogic elements (Viviani and Terzuolo, 1983). Ultimately, however, it translates into an intention to produce a drawn figure in a predefined area of the world space, in congruity with visual inputs or imagery. Thus, geometrical properties (form and location) of intended drawing are defined in world (extrinsic) coordinates.

The transformation of intended drawing into actual movement must solve the problem of the spatiotemporal coordination of a large number of joints and muscles. How does the brain represent the complexity of actual limb movement? Several representations in different reference frames might be used. For instance, the movement could be coded in an earth-fixed frame, using the coordinates of the pen's tip. Limb movement could be also described in intrinsic frames of reference, such as those associated with the angles of the limb joints, with the torques required to produce the movement, or with the activities of the muscles. Each of these sets of variables defines a hypothetical space of representation, endowed with a variety of coordinate systems; for instance, Cartesian or polar coordinates might be used in world space, and absolute or relative angles might be used in joint space.

Three different approaches have been followed to understand the representations of limb geometry in the brain: (1) behavioral approach (cf. Jeannerod, 1988; Lacquaniti, 1989; Soechting and Flanders, 1992), (2) electrophysiological recording (cf. Caminiti and Johnson, 1992; Georgopoulos, Taira, and Lukashin, 1993; Schwartz 1994); and (3) computational studies (cf. Arbib and Amari, 1985; Jordan, 1990).

Coordinate Transformations in Goal-Directed Human Movements

Geometrical shape is one global property coded in the internal representation of drawing or handwriting. The form of a given letter or word can be preserved through wide changes in size and speed of execution, and even in the specific effector (Viviani and Terzuolo, 1983). Although kinematic and figural aspects of drawing are a priori completely independent, they are strictly coupled by neural control. Instantaneous tangential velocity is inversely related to curvature, according to a power law (Lacquaniti, 1989). The gain factor of this law is piecewise constant, depending on the linear extent of each curve segment and on the average speed of movement. Apparently continuous movements are, therefore, segmented. Superposition of harmonic oscillations in two orthogonal directions of the plane of motion predicts the power law. The shape of the figure is determined by modulating amplitude and phase of the oscillations in a segmented fashion. The power law is accounted for by the minimum-jerk theory, according to which the kinematics of natural movements are bound to be maximally smooth (Hogan, 1988; see OPTIMIZATION PRINCIPLES IN MOTOR CONTROL).

How is the motion of different limb segments coordinated? According to the *equilibrium point theory*, the brain codes only a sequence of equilibrium positions of the limb endpoint; there is no need, instead, to specify the time course of joint angles and torques explicitly because these variables are determined automatically by tracking the equilibrium trajectory of the endpoint by means of the muscle viscoelastic properties (cf. EQUILIBRIUM POINT HYPOTHESIS). This theory, however, does not account for the existence of specific laws of covariation of joint angles, suggesting an intermediate processing stage that transforms endpoint coordinates into the angular coordinates of the joints (inverse kinematics, cf. Lacquaniti, 1989; Soechting and Flanders, 1992).

In an intermediate representation, the configuration of the arm is defined by the angular coordinates of elevation and yaw describing the absolute orientation of the upper arm and forearm with respect to the vertical and the sagittal plane. This frame is used to code kinesthetic information and to control joint torques (Soechting and Flanders, 1992). The neural algorithm used to generate drawing movements involves oscillatory changes in the orientation angles, the phase relations of which specify azimuth, elevation, and slant of the figure drawn. Because the algorithm involves linear relations between joint angles and endpoint coordinates, whereas the exact relations are nonlinear, the algorithm predicts characteristic distortions for drawing in specific regions of the workspace. A similar kind of inverse mapping accounts for pointing errors in response to visually presented, remembered targets (Soechting and Flanders, 1992). Target location, defined in shoulder-centered spherical coordinates (azimuth, elevation, and distance), is linearly mapped into the final joint angles of the arm. Actual limb movement is generated by subtracting the values of these final angles from the initial angles as derived by kinesthesia.

Kinematic Representations in the Brain

Neural encoding of geometric parameters of limb movement has been discovered at several sites in the central nervous system (CNS) (Bosco and Poppele, 1993; Caminiti and Johnson, 1992; Georgopoulos et al., 1993; Soechting and Flanders 1992). A vector code of movement direction has been described at the level of motor and premotor cortex (see REACHING: COPING IN MOTOR CORTEX). Neurons are broadly tuned to the direction of hand movement. Activity is maximal along a preferred direction specific to each neuron, and decreases proportionally to the cosine of the angle between this direction and the direction of actual movement. Preferred directions are distributed uniformly throughout space. The direction of actual hand movement is prescribed by the overall activity of the ensemble population. Each neuron contributes a vector in its preferred direction with an amplitude proportional to the level of activity during actual movement direction. Population vectors predict movement trajectory during reaching (Georgopoulos et al., 1993) and drawing (Schwartz, 1994). In drawing, the time-varying changes in length and direction of the population vectors in motor cortex parallel the corresponding changes in the vector of tangential velocity. Population vectors obey the same power law that has been described behaviorally in humans (Schwartz, 1994).

As reviewed earlier, psychophysical studies indicate that trajectory planning involves the specification of hand position relative to the body to afford a comparison between the kinesthetic reafferences and the commands that generate limb movement. A neural substrate for a body-centered positional code has recently been discovered in the superior parietal lobule of the monkey (Lacquaniti et al., in press). Neural activity is monotonically tuned in a frame of reference whose coordinates define the azimuth, elevation, and distance of the hand relative to the shoulder. Each spatial coordinate is encoded in a different subpopulation of neurons. This parcellation is, perhaps, a neural correlate of the psychophysical observation that these spatial parameters are processed in parallel and largely independent of each other.

Coordinate Transformations in the Control of Cat Posture

The existence of kinematic representations is not surprising in the case of intentional limb movements. It may appear more surprising, however, in the case of the control of body posture. According to a classical viewpoint, postural control is equated to the stabilization of the body against gravity, that is, to a problem of statics. Because animals are statically balanced when the vertical projection of their center of mass (*pcm*) falls within the support area, it is often assumed that *pcm* is the global variable regulated in stance. Because this variable reflects the weight distribution among the limbs, it can be regulated by monitoring the contact forces at the feet. It has recently been shown, however, that the maintenance of postural geometry takes precedence in the control hierarchy, *pcm* being determined only in a subordinate manner (Lacquaniti and Maioli, 1994a). The CNS is primarily concerned with maintaining the spatial relations of body segments relative to each other and to the external environment according to an internal image of the preferred posture, the so-called *body scheme*. Two global geometric variables are high-order parameters in the control of cat posture: the length and the angle of orientation relative to the vertical of each limb axis. These variables are accurately controlled when the supporting platform is tilted in the sagittal plane, and when cat posture is perturbed by the application of external loads. There is electrophysiological evidence that spinal sensory neurons projecting to the cerebellum are broadly tuned to limb length and orientation in cats (Bosco and Poppele, 1993).

Limb length and orientation specify the position of the limb endpoints in body-centered polar coordinates. As in the case of human goal-directed movements, there is strong evidence in favor of an intermediate processing stage that transforms endpoint coordinates into the angular coordinates of the joints (Lacquaniti and Maioli, 1994b). Postural changes in cats involve three joints in either the forelimbs (scapula, shoulder, and elbow) or the hindlimbs (hip, knee, and ankle). Thus, the overall geometrical configuration of a limb can be depicted as one point in the three-dimensional space defined by the angular coordinates of these joints. Data points, measured in different trials and under different experimental conditions (variable platform tilt, weight distribution, head and trunk orientation, and interfeet distance), are not scattered throughout the anatomically permissible joint space, but are confined within a small volume close to one plane, under both static and dynamic conditions.

This constraint of covariation of the joint angles does not depend on biomechanical factors, but is attributable to neural control of limb geometry. The orientation of the plane is approximately constant in all subjects, and is the same at the forelimbs and hindlimbs, despite the large biomechanical differences. The kinematic invariance contrasts with the variable, idiosyncratic control of the contact forces at the feet and of the joint torques. Limb geometry and kinetics are controlled by means of independent neural modules (Lacquaniti and Maioli, 1994a). Interestingly, the covariation of joint angles holds in an absolute reference frame fixed to the vertical; in other words, it is contingent upon expressing one joint angle as an absolute angle (orientation angle with respect to the vertical). The application of anomalous somesthetic stimuli to the trunk results in a substantial tilt of the orientation of the plane in joint space, because it interferes with the measurement of the orientation of proximal limb segments relative to the vertical.

The processing stage that transforms endpoint coordinates into the angular coordinates of the joints is reminiscent of that described earlier for goal-directed movements. The algorithm for the coordinate transformation that is involved in the control of postural geometry, however, is different. Most algorithms considered for the inverse kinematics (e.g., Moore-Penrose generalized inverse, linear mappings, etc.) solve the mapping problem by eliminating the redundancy in the system. In other words, they establish a one-to-one correspondence between a given desired position of the limb endpoint and a restricted set of angular coordinates at all joints. Elimination of redundancy, however, may not always be desirable. The concept of *motor equivalence* makes reference to the adaptability of the specific motor patterns implemented by the CNS to achieve an invariant global goal (Berkinblit, Feldman, and Fukson, 1986; Jeannerod, 1988; Viviani and Terzuolo, 1983).

Inverse kinematics for the control of cat posture involves one-to-many mappings. Indeed, the planar constraint imposes well-defined boundaries to the admissible covariations of limb joint angles, but at the same time it allows a high degree of flexibility of the specific geometrical configurations (i.e., joint angles). The execution of a commanded change in limb orientation is minimally sensitive to the joint configuration that is specifically chosen. In other words, a given change in orientation can be produced by means of a wide range of changes in joint angles. The criterion that is effectively optimized is opposite that of Moore-Penrose, which tends to minimize the changes in joint angles. Accordingly, the plane predicted by Moore-Penrose is orthogonal to the experimental plane (Lacquaniti and Maioli, 1994b). The coordinate transformation for postural control involves two distinct mapping rules for the two global variables describing endpoint position: limb orientation maps in broad regions of joint space, but limb length maps in more discrete, linear strips.

These observations have been generalized to the dynamic responses evoked by random ramp rotations of the support platform in the nose-down or nose-up direction (Lacquaniti and Maioli, 1994b). It has been found that the trajectories described in the three-dimensional space of joint angles may diverge in different directions, but remain confined within the 95% tolerance limits of the plane of static angular covariation. Moreover, we have been able to prove that dynamic perturbations may evoke postural responses that involve a given trajectory in joint angle space in one trial but a completely different trajectory in another trial, even though endpoint position remains essentially the same in both cases.

This behavior is characteristic of dynamical systems governed by *chaotic attractors*. Dynamics of nonlinear systems with multiple degrees of freedom can be identified by the motion along specified trajectories in state-space. We have shown that the postural control system effectively reduces the degrees of freedom in the state-space of the angular coordinates of limb joints. Accordingly, limb dynamics are confined to a subspace of lower dimensionality that identifies an attractor space. It is customary to associate an *energy function* (Lyapunov function, cf. COMPUTING WITH ATTRACTORS) to the state-space of a dynamical system. In the case of postural control, the decrease of this function during dynamical evolution may tend toward several different loci owing to the existence of multi-stable basins of attraction. Indeed, we have seen that similar initial states (defined by the set of angular coordinates at the limb joints) may evolve toward different final states. Arbitrarily small deviations (attributable to ubiquitous biological noise) from the initial state may lead to arbitrarily large deviations of the final state after sufficient time has elapsed. Practically speaking, then, the behavior of the system is quantitatively nondeterministic; in other words, one cannot predict its detailed subsequent evolution. Nevertheless, as the state of the system tends to be confined in the proximity of the attractor plane, its qualitative behavior becomes predictable.

The constraint expressed by the plane of angular covariation does not have an autonomous, permanent existence, but is continuously recreated by successive postural states. The essence of motor-equivalent behavior is the continuous exploration of a wide region of joint angle space that is compatible with stability, and the reinforcement of the neural constraint. The planar constraint may emerge as an isomorphism between the internal model of the body scheme and the actual limb movement and its perception. Therefore, the specific orientation of the plane in joint space appears significant, not only from a motor-control viewpoint, but also from a sensory perspective. The multiplicity of different geometrical configurations of the limb that are enforced by the inverse mapping tend to generate a correspondingly wide range of different configurations in sensory space. In sum, motor-equivalent postural dynamics affords the continuous recalibration of sensorimotor space by means of variable sensorimotor associations.

Discussion

We have made three main points: (1) there exist neural representations of limb movement in different frames of reference, both extrinsic and intrinsic to the body; (2) the brain is endowed with different mapping rules to solve inverse kinematics; and (3) motor-equivalent solutions are predicted by specific attractor dynamics. With regard to neural network implementations of attractor dynamics, it is well known that many connectionist networks behave as nonlinear dynamical systems with multiple attractors (see COMPUTING WITH ATTRACTORS). In particular, the state-space of networks with symmetric connections is partitioned in regions that are basins of attraction for limit points. The network proposed by Jordan (1990; see also SENSORIMOTOR LEARNING) is especially relevant to the present discussion. In this network, a forward model of the controlled system is learned by monitoring both the input and the output of the controlled system. In the case of postural control, the forward model maps the expected relationship between the set of joint angles and the resulting endpoint position. Remarkably, this forward model would not need to be learned in its most general and exact manner by the postural system. In fact, although the forward mapping from joint angles to limb length and orientation is generally nonlinear, we found that the latter two parameters can be estimated simply and accurately using linear compounds of the joint angles, within the experimental range of postures. After the forward model has been learned in Jordan's network, the desired movement trajectory (sequence of endpoint positions) is fed to the inverse model to derive the feedforward motor command (sequence of joint angles). The resulting error in endpoint position is propagated through the forward model to derive the corresponding error in the motor command space (joint angle space). The latter error represents the signal to train the inverse model. Parameterized constraints, such as the planar constraint on the joint angles for the postural control, are easily incorporated into the learning procedure, and can thereby bias the choice of a particular inverse function.

Road Map: Biological Motor Control
Related Reading: Geometrical Principles in Motor Control; Head Movements: Multidimensional Modeling; Human Movement: A System-Level Approach

References

Arbib, M. A., and Amari, S., 1985, Sensori-motor transformations in the brain (with a critique of the tensor theory of cerebellum), *J. Theor. Biol.*, 112:123–155. ◆

Berkinblit, M. B., Feldman, A. G., and Fukson, O. I., 1986, Adaptability of innate motor patterns and motor control mechanisms, *Behav. Brain Sci.*, 9:585–638.

Bosco, G., and Poppele, R. E., 1993, Broad directional tuning in spinal projections to the cerebellum, *J. Neurophysiol.*, 70:863–866.

Caminiti, R., and Johnson, P. B., 1992, Internal representations of movement in the cerebral cortex as revealed by the analysis of reaching, *Cerebr. Cortex*, 2:269–276.

Georgopoulos, A. P., Taira, M., and Lukashin, A., 1993, Cognitive neurophysiology of the motor cortex, *Science*, 260:47–52. ◆

Hogan, N., 1988, Planning and execution of multijoint movements, *Can. J. Physiol Pharmacol.*, 66:508–517.

Jeannerod, M., 1988, *The Neural and Behavioural Organization of Goal-Directed Movements*, Oxford: Clarendon Press. ◆

Jordan, M. I., 1990, Motor learning and the degrees of freedom problem, in *Attention and Performance*, vol. XIII (M. Jeannerod, Ed.), Hillsdale, NJ: Erlbaum, pp. 796–836.

Lacquaniti, F., 1989, Central representations of human limb movements as revealed by studies on drawing and handwriting, *Trends Neurosci.*, 12:287–291.

Lacquaniti, F., Guigon, E., Bianchi, L., Ferraina, S., and Caminiti, R., in press, Representing spatial information for limb movement: Role of area 5 in the monkey, *Cereb. Cortex*.

Lacquaniti, F., and Maioli, C., 1994a, Independent control of limb position and contact forces in cat posture, *J. Neurophysiol.*, 72:1476–1495.

Lacquaniti, F., and Maioli, C., 1994b, Coordinate transformations in the control of cat posture, *J. Neurophysiol.*, 72:1496–1515. ◆

Schwartz, A. B., 1994, Direct cortical representation of drawing, *Science*, 265:540–542. ◆

Soechting, J. F., and Flanders, M., 1992, Moving in three-dimensional space: Frames of reference, vectors, and coordinate systems, *Annu. Rev. Neurosci.*, 15:167–191. ◆

Viviani, P., and Terzuolo, C., 1983, The organization and control of movement in handwriting and typing, in *Language Production*, vol. 2 (B. Butterworth, Ed.), London: Academic Press, pp. 103–146.

Linguistic Morphology

Mark S. Seidenberg

Introduction

The subject matter of linguistic morphology is the structure of words and the principles that govern word formation. Morphology is an important aspect of language and also a complex one insofar as syntax, phonology, and semantics are all implicated in the phenomena. It also happens to have provided the domain in which theoretical linguistics clashed with the emerging neural network or "connectionist" paradigm in the 1980s.

The goal of traditional morphological theory was to identify the smallest meaning-bearing units in language, called morphemes, and to characterize word structure in terms of operations on these units. Thus, both *houseboat* and *boathouse* consists of two morphemes, *house* and *boat*, and the principles governing the morphological process of compounding in English dictate that a houseboat is a kind of boat, whereas a boathouse is a kind of house. Natural languages subvert the simple identity between units of form and units of meaning in several ways, however. There are semantic units that are not overtly realized in a word; for example, the fact that *books* is plural is indicated by the inflectional morpheme /s/, but the fact that *book* is singular does not receive overt morphological expression. Conversely, there are units that participate in morphological processes that have meaning only as etymological traces which lie outside the scope of our theories of lexical representation: for example, the *-mit* in *permit*, *submit*, and *remit*, and the *cran-* in *cranberry*. Moreover, the meaning of a word is not always transparent from the combination of its parts: a *redhead* is a person with red hair but a *blackhead* is a kind of pimple; *flatbreads* are breads that are flat but *sweetbreads* are neither bread nor sweet. Thus, like other aspects of language, the morphological system appears to be rule-governed at some level of description, but it also admits many exceptions. These characteristics of morphology raise interesting issues concerning how this knowledge is represented, acquired, and used.

Morphological theories typically differentiate between inflectional and derivational types. Inflectional morphology is, roughly, aspects of word structure that mark grammatical information such as number, tense, gender, and case. Languages differ with respect to the extent to which information is inflectionally encoded. Whereas the English inflectional system marks tense (*like-liked-liking*) and number (*boy-boys*), Mandarin Chinese marks neither. The English system is relatively impoverished, however, in contrast to agglutinative languages such as Turkish, which has a much richer inflectional component. Whereas inflectional morphology can be seen as creating sets of related forms (or paradigms) from a root form (or stem), derivational morphology concerns the creation of words from other words through processes such as affixing, compounding, and reduplication. For example, in English the suffix *-ness* attaches to adjectives such as *happy* and *friendly* to form nouns such as *happiness* and *friendliness*. Inflectional and derivational morphology differ in several respects, although the exact boundary between them has been difficult to isolate. The use of inflectional morphology is grammatically governed: if a language encodes a particular type of information inflectionally (e.g., tense), the appropriate inflection must be present for a sentence to be grammatical. Thus, every verb in English must have the tense appropriate to its context. The system is highly productive insofar as new verbs that come into the language inherit the regular or "rule-governed" inflections for each tense. For example, once the verb *to fax* came into the language, it could immediately be used in the past (*faxed*) and other tenses. Some irregular inflectional forms are tolerated (e.g., *take-took*, *go-went*), but there are no gaps (i.e., verbs that cannot occur in specific tenses). In contrast, use of particular derivational forms is not demanded by the syntax, and derivational morphology is less fully productive, with many seemingly arbitrary gaps (e.g., *nice-nicety*, *grammatical-grammaticality*, but not *old-oldity* or *large-largeity*). Whereas all of the words in an inflectional paradigm are of the same grammatical category (e.g., for number, nouns; for tense, verbs), derivational morphology can create words that differ in grammatical category from their roots (e.g., de-adjectival nouns such as *nicety*, deverbal nouns such as *teacher*, etc.). Derivational processes can change stress patterns, whereas inflectional morphology does not; compare *photograph/photographer* (derivational) to *photograph/photo-graphs* (inflectional). In general, inflectional morphology is more closely involved with syntax, and derivational morphology with phonology.

Modern morphological theory attempts to systematize these phenomena in terms of abstract explanatory principles. Much of this work is undertaken within the generative framework of Chomsky and his followers (see Spencer, 1991, for a linguistically oriented overview, and Sproat, 1991, for a computationally oriented one). The goal is the development of a "competence" theory, i.e., a theory that provides a characterization of morphological knowledge in terms of a set of general principles governing the well-formedness of words in languages, abstracting away from details about such things as how this knowledge varies across individuals, how it is put to use in linguistic performance, and how it is represented in the brain. These general principles are thought to be revealing about the nature of the linguistic capacity which, in the Chomskyan approach, is thought to be a domain- and species-specific form of knowledge. In practice, there is not as yet a dominant theory in morphology analogous to Chomsky's "principles and parameters" theory in syntax. Within the generative approach, there has been considerable debate about the scope of the morphological domain and its relationship to other aspects of language (see, for example, Anderson, 1993).

The Past Tense Debate

An alternative view is that language reflects more general principles concerning perception, learning, memory, and action of the sort explored in connectionist or "neural network" modeling. It was in the morphological domain that these opposing views happened to collide. Morphological phenomena provided the vehicle for Pinker and Prince's (1988) critique of the role of neural network models in explaining language. Linguists have taken morphological phenomena as providing powerful evidence for the fundamental claim that knowledge of a language consists of a set of generative rules. In Berko's (1958) classic experiment, for example, children were shown an animal called a *wug*. They were then shown a second wug and told, "Here is another wug. Now there are two...," to which the children reliably answered "wugs" even though they would not have heard this word before. Most theoretical linguists take the ability to generalize to new instances as evidence for an internalized rule (here, of plural formation). Instances in which the rule does not apply (e.g., *man-men*; *sheep-sheep*) are treated as special cases that must be learned by rote. Pinker (1991) depicted the formation of the past tense of verbs in English as a quintessential example of a rule-governed linguistic process. According to this view, one of the central goals of linguistic theory is to identify the rules that underlie linguistic competence.

Rumelhart and McClelland (1986) described a neural network (connectionist) model of the acquisition of the past tense that was said to provide an alternative to this standard linguistic account. Their model took the phonological form of a present tense verb as input and generated a phonological representation of the past tense as output. The model learned to perform this mapping on the basis of exposure to a large set of present-past tense pairings, using a simple error-correcting learning algorithm. The model was therefore said to illustrate how the ability to generalize can arise in a network that does not represent rules explicitly.

The Pinker and Prince (1988) paper was a response to this challenge to linguistic orthodoxy. The paper had two main parts. First, it showed that the Rumelhart and McClelland model was flawed insofar as it failed to capture many known behavioral phenomena concerning past tense acquisition. Worse, the nominal successes of the model appeared to derive from aspects of the simulation that were psychologically unrealistic. For example, the model was said to exhibit "U-shaped learning," a developmental phenomenon wherein the child acquires a certain type of knowledge (e.g., of some past tense forms) on a case-by-case basis and then experiences a decrement in performance in the course of discovering the rule. Thus children will correctly produce irregular forms such as *went* and *took*, having learned them as individual instances, and later make regularization errors such as *goed* and *taked* as they discover the rule that governs most forms. The child eventually learns both the rule and the cases where the rule does not apply. To the extent that Rumelhart and McClelland's model exhibited this behavior, Pinker and Prince argued, it was because the outcome was preordained by using an unnatural training regime. The model was trained on a small set of irregular verbs and then exposed to a large set of regular forms, which resulted in unlearning of the irregulars and regularization errors. Children's exposure to verbs is not ordered in this way, however. In short, Pinker and Prince suggested that because of these kinds

of flaws, the Rumelhart and McClelland model could not be taken as having provided an alternative to the standard linguistic account.

The second part of the Pinker and Prince critique was the presentation of a more general view of the role of connectionist models in explaining linguistic phenomena. Pinker and Prince acknowledged that many of the inadequacies of the Rumelhart and McClelland model reflected idiosyncratic aspects of their implementation rather than general properties of neural networks. For example, their phonological representation could not represent words that occur in some languages; because the model did not include any semantics, it could not distinguish *ring* (to sound a bell), which has the past tense *rang*, from *ring* (to form a circle), which has the past tense *ringed*. Of course, there is nothing about the connectionist framework that precludes improving the phonological representation or implementing a semantic component. The analysis of this model nonetheless suggested to Pinker and Prince some broader implications concerning the utility of the entire connectionist approach. Their view was that a connectionist model that (unlike the Rumelhart and McClelland model) was capable of simulating detailed aspects of linguistic performance could only do so by "implementing" the linguistic theory of the phenomena. Insofar as linguistic theory suggests that knowledge of the past tense is represented in terms of a rule, a connectionist network could only succeed by implementing the rule using the inventory of units, connections, weights, and other mechanisms provided by connectionist theory. This might be a useful thing to do, Pinker and Prince observed, because it could provide some insight as to how linguistic knowledge is realized in the brain. However, the important generalizations about the phenomena qua language would be captured by the rule-based theory. A purely connectionist treatment would therefore fail to capture important generalizations about this aspect of human knowledge.

Fallout

Pinker and Prince's critique had a number of repercussions. First, at just the point when neural network modeling was entering a spectacular renaissance, it convinced a broad audience of linguists that the approach is antithetical to their interests. Following Pinker and Prince's lead, many linguists interpreted the inadequacies of Rumelhart and McClelland's model as reflecting problems inherent in the broader approach that it exemplified. Moreover, the Rumelhart and McClelland work seemed to recapitulate a familiar pattern in which nonlinguists develop a theory claimed to offer an alternative to the standard linguistic approach that only succeeds in demonstrating a lack of understanding or even awareness of basic aspects of the phenomena to be explained. All the worse was that this approach came packaged with an emphasis on learning principles, raising for many linguists the specter of the behaviorist approach, thought to have been buried by Chomsky many years ago.

A second consequence of the Pinker and Prince critique was that it stimulated the development of several other network models addressing the same morphological phenomena. MacWhinney and Leinbach (1991) developed a model that learned both aspects of inflectional morphology in English, the past tense and the plural. The Plunkett and Marchman (1991) models addressed the acquisition issues, particularly questions concerning U-shaped learning. Daugherty and Seidenberg (1992) described a simple feedforward net model that provided a good fit to data from behavioral experiments that examined subjects' use of their knowledge of the past tense. Hoeffner and McClelland's (1993) model mapped from a semantic representation of the present tense to a phonological representation of the past tense. This work addresses most of the limitations of the Rumelhart and McClelland model and suggests that the problems that Pinker and Prince identified were in fact specific to it rather than inherent in the connectionist approach.

Pinker and Prince have not commented on these later models, but their reaction would not be hard to anticipate. Each of the models addresses only some of the concerns raised in their critique; each of them is limited in some respects; no single model simulates in detail all aspects of past tense morphology. This case illustrates a striking difference between the linguistic and network modeling approaches. The linguistic theory can claim to have broad scope but only because specific details about how knowledge is represented, acquired, and used can be omitted. This vagueness makes the theory difficult to refute. In contrast, the models are specific about such questions, but necessarily limited in scope. They are in a sense too easy to refute because there invariably are flaws in the models that are theoretically uninteresting. The question of phonological representation provides a simple illustration of the contrast. Whereas it was easy to identify flaws in the phonological representation employed in the Rumelhart and McClelland model, Pinker and Prince offer no specific proposals of their own in this domain. Moreover, detailed properties of the phonological representation were not central to Rumelhart and McClelland's broader claims about morphological knowledge.

Further Developments

While connectionists have been playing catch-up with Pinker and Prince (1988), Pinker has been accumulating additional evidence thought to demonstrate the psychological reality of the past tense rule (Pinker, 1991; Marcus et al., 1992). Some of the data derive from studies of language-impaired populations. There is a developmental form of dysphasia called "specific language impairment" (SLI) which Pinker interprets as causing an inability to acquire some types of linguistic rules. For example, some SLI children exhibit difficulties with inflectional morphology. Pinker interprets these deficits as reflecting a congenital inability to formulate morphological rules. The fact that the impairment is heritable has been taken as evidence that there may be genetic encoding of specific aspects of grammar such as inflectional morphology (Pinker, 1994). Conversely, children with Williams syndrome (see DEVELOPMENTAL DISORDERS), which causes impairments in perception and cognition, are said to be able to formulate the past tense rule but have difficulty mastering the irregular cases. This "double dissociation" is taken as evidence that the rule and the exceptions are represented separately in the brain. While suggestive, the neurolinguistic data admit other interpretations: the SLI children have a broad range of cognitive, perceptual, and motoric problems, including serious articulatory deficits that may create difficulties in perceiving and producing unstressed speech elements, which happen to include inflectional morphemes in English. Similarly, the Williams syndrome children, far from being a pure case, are able to learn some irregular forms. It is also unclear whether these patterns of impairment necessarily implicate separate "rule" and "exception" knowledge representations. There is a flourishing field of connectionist neuropsychology that attempts to explain patterns of cognitive impairment following brain injury in terms of "lesions" to neural network models (see LESIONED ATTRACTOR NETWORKS AS MODELS OF NEUROPSYCHOLOGICAL DEFICITS). Experiments with models in other domains have demonstrated that different types of behavioral anomalies can arise from different types of damage to a single network. This research suggests that impair-

ments such as an inability to generalize or master irregular forms might be realized in connectionist networks that do not employ explicit representations of rules or lists of exceptions.

The Rumelhart and McClelland model stimulated a large amount of research that has resulted in the identification of a wealth of facts about inflectional morphology, its acquisition, and the ways in which the system is impaired in cases of developmental anomaly or brain injury. Disagreements persist about the implications of these data concerning the nature of morphological knowledge and knowledge of language more generally. Pinker interprets the data as providing compelling evidence concerning a fundamental claim about linguistic knowledge, that it consists of grammatical rules. The past tense rule is taken as a clear example of this type of knowledge representation. Neural network models of the past tense are therefore parasitic on linguistic theory insofar as they can be adequate only to the extent that they implement such rules. An alternative view holds that the Pinker and Prince account is backwards (Seidenberg, 1992). The connectionist network is not in fact an implementation of a rule-based theory because its behavior deviates from what would be expected on the basis of such theories. These deviations represent novel predictions about behavior not identical to those of the rule-based theory. The function of the rule-based theory is to provide a folk psychological account of the network's behavior. This level of description is useful but misses certain generalizations that can only be stated at the level of network behavior.

The Daugherty and Seidenberg (1992) model, for example, does not embody the dichotomy between rule-governed cases and exceptions that is central to the linguistic account. The model learned the correspondences between present and past tense forms, using a single mechanism to encode all items. The linguistic and connectionist theories, therefore, make different predictions about the behavior of items such as *bake-baked*. Whereas the linguistic approach treats this as a simple case of rule application, the network model exhibits interference from related irregular forms such as *take-took* and *make-made*. This occurs because the values of the weights on connections mediating this computation reflect the aggregate effects of exposure to all verbs, not merely the rule-governed ones. Behavioral experiments indicate that human subjects do in fact show the same kind of interference when asked to generate the past tense under speeded naming conditions (Seidenberg, 1992). Thus, it takes longer to generate the past tense of *bake* compared to *like* even though both are "rule-governed." The model, then, cannot be said to be merely "implementing" a linguistic theory that did not predict these effects.

Although considerable progress has been made, questions concerning the proper treatment of inflectional morphology are far from settled; many technical issues remain unresolved. However, even if it turns out that inflectional morphology represents a set of phenomena for which neural networks happen to be well suited, it could be argued that nothing substantive follows about the nature of language. The past tense in English is a rather simple system with characteristics that are not shared by more interesting aspects of language. In particular, it involves a single morphological rule, and while there are exceptions to it, they are finite in number and can be learned by rote, at least in principle. Thus, Pinker's (1991) revised account of the past tense involves two components: the rule and a separate "associative net" for representing the exceptions. That past tense formation is rule-governed is claimed as a major discovery, but this could not fail to be true given that the rule does not have to apply in all cases because a second mechanism is invoked to explain the exceptions to it (see Seidenberg, 1992, for discussion). There are two types of data to be explained (rule-governed cases and exceptions) and two explanatory devices in the theory (a rule and an associative net). These observations suggest that the past tense might not provide the most interesting domain in which to explore the role of neural network modeling concepts in explaining complex aspects of linguistic knowledge. Certainly derivational morphology provides a much more challenging and complex domain in which to explore these issues, insofar as the phenomena seem to derive from interactions among several types of "soft constraints," including syntactic, lexical, and phonological.

Discussion

The debate over the past tense has revealed some of the tensions that exist between the neural network modeling approach and linguistic theory. In expressing the view that connectionist modeling is antithetical to the interests of theoretical linguists, the Pinker and Prince (1988) article reflected the view of many mainstream theoretical linguists. More recently, connectionist concepts have begun to filter into linguistic theory, entering through phonology and morphology. In an important work certain to have wide impact within linguistics, Prince and Smolensky (in press) incorporate concepts drawn from connectionism into what is otherwise a standard theory of linguistic competence. Whereas Pinker and Prince (1988) saw it as a basic claim that linguistic knowledge consists of rules, Prince and Smolensky (in press) suggest that it involves the parallel satisfaction of multiple constraints of varying strengths. It is likely that this trend will continue, as linguists begin to absorb many of the insights concerning knowledge representations, processing mechanisms, and learning that have emerged from the neural network paradigm.

Acknowledgment. This article was prepared with support from NIMH grant 47566.

Road Map: Connectionist Linguistics
Related Reading: Language Acquisition: Language Processing

References

Anderson, J., 1993, *A-morphous Morphology*, Cambridge, Eng.: Cambridge University Press.
Berko, J., 1958, The child's learning of English morphology, *Word*, 14:150–177.
Daugherty, K., and Seidenberg, M. S., 1992, Rules or connections? The past tense revisited, in *Proceedings of the 14th Annual Meeting of the Cognitive Science Society*, Hillsdale, NJ: Erlbaum, pp. 259–264.
Hoeffner, J. H., and McClelland, J. L., 1993, Can a perceptual processing deficit explain the impairment of inflectional morphology in developmental dysphasia? A computational investigation, in *Proceedings of the 25th Annual Child Language Research Forum* (E. V. Clark, Ed.), Stanford, CA: Center for the Study of Language and Information, pp. 38–49.
MacWhinney, B., and Leinbach, J., 1991, Implementations are not conceptualizations: Revising the verb learning model, *Cognition*, 40: 121–158.
Marcus, G., Pinker, S., Ullman, M., Hollander, M., Rosen, T. J., and Xu, F., 1992, Overregularization in language acquisition, *Monogr. Soc. Res. Child Dev.*, serial no. 228, 57(4).
Pinker, S., 1991, Rules of language, *Science*, 253:530–534.
Pinker, S., 1994, *The Language Instinct*, New York: Morrow. ◆
Pinker, S., and Prince, A., 1988, On language and connectionism: Analysis of a parallel distributed processing model of language acquisition, *Cognition*, 28:73–194.
Plunkett, K., and Marchman, V., 1991, U-shaped learning and frequency effects in a multilayered perceptron, *Cognition*, 39:43–102.
Prince, A., and Smolensky, P., in press, *Optimality Theory*, Cambridge, MA: MIT Press.
Rumelhart, D. E., and McClelland, J. L., 1986, On learning the past tenses of English verbs, in *Parallel Distributed Processing: Explorations in the Microstructure of Cognition*, vol. 2, *Psychological and Biological Models* (J. L. McClelland, D. E. Rumelhart, and PDP Research Group, Eds.), Cambridge, MA: MIT Press.
Seidenberg, M. S., 1992, Connectionism without tears, in *Connectionism: Advances in Theory and Practice* (S. Davis, Ed.), Oxford: Oxford University Press. ◆
Spencer, A., 1991, *Morphological Theory*, Oxford: Blackwell. ◆
Sproat, R., 1991, *Morphology and Computation*, Cambridge, MA: MIT Press. ◆

Localized Versus Distributed Representations

Simon Thorpe

Introduction

What happens in the brain when one recognizes a familiar stimulus such as the face of one's grandmother? Most researchers accept that the act of perception involves the activation of some form of internal representation, but there is little agreement about how such representations are implemented in neural hardware. While it is clear that many millions of neurons in the visual system are activated during the processing of each retinal image, many of them are involved in generic visual processing. Nevertheless, it seems likely that at some level of the visual system there exists a population of neurons whose activity in some way represents the concept *grandmother*. The key issue to be addressed in this article concerns the nature of this neural representation. Specifically, could it be that the final representation involves neurons that respond selectively to grandmothers—so-called *grandmother cells*? An alternative suggestion is that the presence of *grandmother* is never made explicit at the single-neuron level, but that instead the final representation is distributed across a large number of cells, none of which responds selectively to grandmothers alone.

The way in which concepts are represented is a critical question in brain theory and has been the subject of much debate, although there is little sign of consensus. *Local representations*, in which individual nodes in a network are used to represent particular concepts, were already a part of the semantic network models proposed in the late 1960s and are used by many of the connectionist models proposed during the last decade (see SEMANTIC NETWORKS). Given the "neurally inspired" nature of connectionist models, it is natural to ask what the relationship is between nodes in a connectionist model and neurons in the brain. Could neurons behave like nodes in a semantic network, with a direct mapping between neurons and concepts? Connectionists generally avoid this question by insisting that the nodes in a connectionist network are not to be thought of as real neurons. However, the question must be tackled if we are to develop a realistic theory of the brain.

A Connectionist Example

To clarify what we mean by the terms *local* and *distributed*, consider the following very simple example. Suppose that an

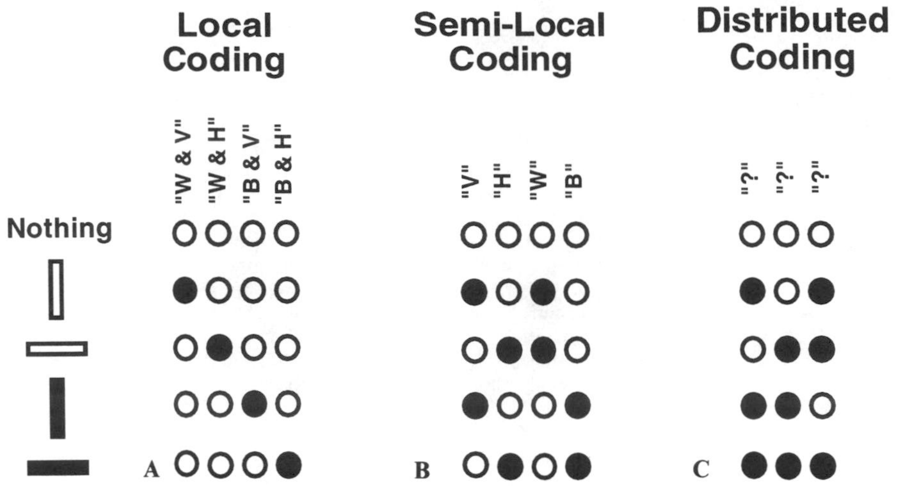

Local Coding

Semi-Local Coding

Distributed Coding

Figure 1. Three ways of representing four stimuli. In the left column (local coding), each stimulus has a corresponding unit. In the middle column (semi-local coding), the individual features —White (W), Black (B), Vertical (V), and Horizontal (H)—are coded with separate units. In the right column, the stimuli are represented using a distributed code.

animal faced with a set of stimuli has to learn which of two responses to make (X or Y) to get a reward. The stimuli are white (W) or black (B) bars that can be either vertical (V) or horizontal (H) in orientation, resulting in a total of four stimuli (WV, WH, BV, and BH).

We could model this situation with a simple three-layer neural network with an input layer for processing the stimuli, an output layer for generating motor responses, and an intermediate hidden layer of binary units. Suppose that the network was trained to perform the same task as the animal. Figure 1 illustrates just three of the many ways in which the four stimuli might be encoded by the hidden layer. One possibility, illustrated in Figure 1A, uses a separate unit to code each stimulus ("local coding"). A second possibility (Figure 1B) is a semi-local scheme in which color and orientation are encoded separately, with the result that two units need to be activated to encode each stimulus. The third possibility (Figure 1C) illustrates what might happen if only three hidden units were available. The network is forced to represent the information with fewer units, with the result that a form of distributed coding is used.

There are obviously many other ways to represent the same stimuli, especially if the units can have graded responses instead of being binary. However, the example illustrates the essential difference between local and distributed coding, namely, that with a local representation, activity in individual units can be interpreted directly: the units effectively have labels such as "White and Horizontal" or "Vertical." In contrast, with distributed coding, individual units cannot be interpreted without knowing the state of the other units in the network.

Note that all three types of representation can mediate the task. Indeed, one conclusion of much connectionist modeling is that most if not all input-output transformations can be performed without using local representations. Nevertheless, local coding schemes have certain properties difficult to reproduce in distributed representations. Only the completely local scheme responds unambiguously when multiple stimuli are simultaneously presented. For example, presenting WV and BH together would simply activate both the "WV" and the "BH" units in the local representation. In contrast, all four units in the semi-local scheme would be activated, a pattern indistinguishable from that produced by WH and BV. This illustrates the well-known binding problem that affects all representations that are not local (see BINDING IN THE VISUAL SYSTEM).

One point deserves underlining. For some authors, local coding requires that only one unit is active; Figure 1B would

thus be considered as a case of distributed coding. Here we would argue that even representations requiring many simultaneously active units can be termed local if the activation of a unit can be interpreted without reference to other units in the network. Thus, a representation of *sad grandmother*, which involves the simultaneous activation of units coding *sad* and *grandmother*, can still be considered as local if the individual units can be interpreted in isolation.

Local and Distributed Coding in the Neurosciences

How have local and distributed representations fared in the literature? Most neurophysiologists who have analyzed the properties of single cells effectively assume a degree of local coding, since the search for functional correlates at the single-unit level makes no sense in a fully distributed system. Despite this fact, remarkably few neuroscientists have gone down in print in support of local coding schemes. Jerzy Konorski and Horace Barlow are two notable exceptions. Konorski proposed that "perceptions experienced in humans' and animals' lives, are represented not by the assemblies of units but by single units in the highest levels of particular analyzers" (Konorski, 1967). He used the term *gnostic units* for such cells. Barlow's classic paper "Single Units and Sensation: A Neuron Doctrine for Perceptual Psychology?" (Barlow, 1972) proposed that "Perception corresponds to the activity of a small selection from the very numerous high-level neurons, each of which corresponds to a pattern of external events of the order of complexity of a word" and that "high impulse frequency in such neurons corresponds to a high certainty that a trigger feature is present."

In contrast, many scientists have argued in favor of distributed coding schemes. The list includes authors such as Donald Hebb, Karl Lashley, Karl Pribram, E. Roy John, Dalbir Bindra, Walter Freeman, and Gerald Edelman. In addition, distributed representations have generated considerable interest in the neural network literature; the so-called Hopfield net is a clear example of distributed encoding, since all units are involved in representing all the various input patterns (see COMPUTING WITH ATTRACTORS).

Evidence for Distributed Coding

Some of the clearest evidence for distributed coding comes from work on sensory and motor systems. Both cases require the coding of continuous variables such as movement direction or stimulus orientation. It would appear that rather than using neurons tuned to particular values, the brain uses populations of relatively coarsely tuned units with overlapping properties. For instance, neurons in the motor cortex have responses that are roughly tuned for movement direction, but an animal can accurately control the direction of movements in three-dimensional space. Georgopoulos and his co-workers found that movement direction can be predicted by calculating a "population vector" for a set of neurons; each neuron "votes"

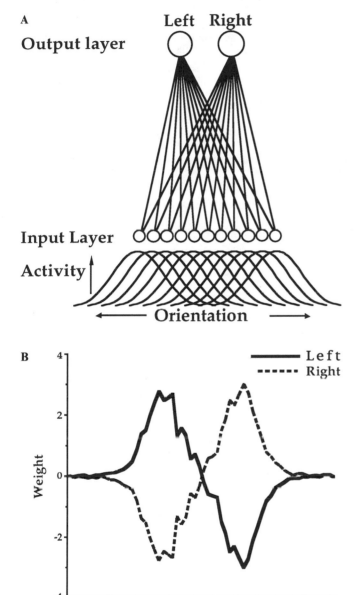

Figure 2. *A*, Structure of the neural net used to simulate performance in the two-orientation task. The input layer is composed of orientation-tuned units, each with a Gaussian filter shape. The output layer contains units trained to respond to the different test stimuli. *B*, Connection strengths between 60 input layer units and two output units A and B after training with two test orientations separated by 2°. Note that the most important units are not those "tuned" to the test stimuli, but rather those responding best to orientations 10° to 20° to either side.

for its own preferred movement direction, but the strength of the vote is weighted by the strength of the neuron's response (see REACHING: CODING IN MOTOR CORTEX).

Population vector coding also has been suggested for encoding sensory parameters such as line orientation. However, the picture may not be so simple. Pouget and Thorpe (1991) investigated orientation coding using a connectionist model with two layers, an input layer, composed of orientation-tuned filters with properties similar to cells of the visual cortex, and an output layer, containing just two units (Figure 2*A*).

After training the two output units to respond to orientations 1° to the left or right of vertical, it was found that the most important units in the input layer were not the most active ones (i.e., those tuned to ±1°). Instead, the strongest weights were from units tuned to orientations offset by about 20° to either side (Figure 2*B*). The explanation lies in the bell-shaped tuning functions of the input units; at "optimal orientations," the units are actually very insensitive to changes in orientation, but are much more useful at the point where the tuning curve is steepest. Such findings suggest that orientation is encoded using a truly distributed code, since the "meaning" of a cell tuned to −20° depends on the state of other units tuned to +20°. Similar arguments for distributed coding can be found in the study of stereoscopic depth by Lehky and Sejnowski (1990).

Thus, continuous parameters such as orientation, stereo disparity, and movement direction appear to be represented in a distributed form. However, it is not clear just how far this coarse coding strategy can be pushed. What about high-level concepts like *apple*, *chair*, and *grandmother*, which cannot be represented as continuous variables? Could such entities be explicitly encoded at the single-unit level? Many scientists regard this possibility as ridiculous, although often their reasons for rejection are not made explicit. In the following section, we analyze five criticisms of local coding commonly thought to rule out local coding.

Criticisms of Local Coding

"There Are Not Enough Neurons in the Brain"

It has often been argued that a local coding scheme with one neuron for every possible stimulus would require too many neurons. This is certainly true. Take, for example, the retinal image, which is encoded by the activity of roughly 1 million ganglion cells. Even if we consider only binary images, the total number of possible images is $2^{1,000,000}$ (roughly $10^{300,000}$), which, given a density of 100,000 cells/mm³, would require a brain larger than the known universe—not a very practical solution!

But local coding does not need a neuron for every possible stimulus. Imagine a machine designed to read the alphabet. How many different "grandmother cells" would it need using a local coding scheme? The sensible answer is clearly 26: one for each letter of the alphabet. The critical number is not the number of stimuli (which may be astronomic) but rather the number of perceptual categories. So, how many perceptual categories does a typical human have? Thorpe and Imbert (1989) suggested a value of around 100,000 for the number of visual categories, and although only a very rough guess, it is nevertheless several orders of magnitude below the number of neurons available in the temporal lobe.

"Local Coding Is Too Risky"

Using just one neuron to represent each identifiable object would result in a very fragile system; cell death would mean that representations would be continually being lost. However, as Feldman (1988) pointed out, the problem effectively disappears if, instead of having just one neuron, there was a small amount of redundancy. With 100 billion cells and a death rate of 100,000 per day, the probability of any one cell dying is 10^{-6}, but the chances of two prespecified cells failing is extremely low.

"No One Has Ever Found a Grandmother Cell"

While it is certainly true that no one has ever demonstrated the existence of a "grandmother cell," many studies have reported neurons with remarkably selective responses. Reports of visual responses selective to complex visual forms such as monkey paws in the early 1970s were dismissed by many as inconclusive, but it is much harder to ignore the increasingly detailed descriptions of neurons selective for faces in the temporal lobe (Desimone, 1991; Perrett et al., 1992; Rolls, 1992; see also Tanaka, 1993). Other studies have described neurons in the orbitofrontal cortex with independent visual and gustatory responses selective for the same food items (Thorpe, Rolls, and Maddison, 1983), and recent work by Miyashita (1993), using arbitrary fractal patterns, indicates that temporal lobe neurons become "tuned" to stimuli that the animal is trained with. While none of these researchers has claimed that such neurons are "grandmother cells," such data clearly contradict the view that individual neurons cannot show highly selective tuning properties.

"Individual Neurons Are Too Unreliable"

The responses of individual neurons are affected by noise, and many have claimed that population coding is needed to overcome this limitation. In fact, recent data indicate that, when tested with the appropriate stimuli, the most sensitive cells can rival or even exceed the performance of a psychophysical observer in the same conditions. Newsome, Britten, and Movshon (1989) recorded from movement-sensitive neurons in Middle Temporal cortex (MT) while the monkey was performing a perceptual discrimination and found that the responses of some single cells contained as much information about the stimulus as the animal's behavioral response.

"Local Representations Do Not Generalize"

It is often claimed that distributed representations are particularly interesting because of their ability to generalize to novel situations (Hinton, McClelland, and Rumelhart, 1986; Mozer, 1991) and that this feature is difficult to achieve in systems using local representations. While generalization may well be problematic for networks relying entirely on local representations, systems based on a combination of local and distributed representations can avoid such limitations. Furthermore, even in a local coding scheme, generalization is relatively easy to obtain as long as unit responses are graded rather than binary.

Discussion

We have considered five arguments commonly used to refute local coding. None seems totally convincing. Of course, the absence of a clear-cut case against local coding is not an argument in its favor. But it is clearly too soon to rule out the idea of grandmother cells. It seems that many scientists have re-

jected local coding largely on the basis of intuition or a deep-felt feeling that local coding is somehow too reductionist. Others have opted for distributed coding because it offers a real alternative to traditional symbolic approaches (see CONNECTIONIST AND SYMBOLIC REPRESENTATIONS). However, it may be that one of the advantages of the connectionist framework is the fact that there is room for both distributed and local ("symbolic") forms of representation.

Connectionist models have shown that distributed representations do work, and it may be that many organisms survive perfectly well without ever making concepts explicit at the single-neuron level. Nevertheless, we should not lose sight of the fact that local representations do have some clear advantages, in particular, their ability to handle several concepts at the same time. In addition, there may be significant computational advantages in keeping the neural code sparse (see SPARSELY CODED NEURAL NETWORKS and SPARSE CODING IN THE PRIMATE CORTEX). Given the absence of conclusive evidence against local coding, we should surely keep our options open.

Road Maps: Artificial Intelligence and Neural Networks; Vision
Related Reading: Dynamic Link Architecture; Structured Connectionist Models; Visual Coding, Redundancy, and "Feature Detection"

References

Barlow, H. B., 1972, Single units and sensation: A neuron doctrine for perceptual psychology? *Perception*, 1:371–394. ◆
Desimone, R., 1991, Face-selective cells in the temporal cortex of monkeys, *J. Cognit. Neurosci.*, 3:1–8.
Feldman, J. A., 1988, Connectionist representation of concepts, in *Connectionist Models and Their Implications* (D. Waltz and J. A. Feldman, Eds.), New York: Ablex.
Hinton, G. E., McClelland, J. L., and Rumelhart, D. E., 1986, Distributed representations, in *Parallel Distributed Processing: Explorations in the Microstructure of Cognition* (D. E. Rumelhart, J. L. McClelland, and PDP Research Group, Eds.), vol. 1, *Foundations*, Cambridge, MA: MIT Press, pp. 77–109.
Konorski, J., 1967, *Integrative Activity of the Brain: An Interdisciplinary Approach*, Chicago: University of Chicago Press.
Lehky, S. R., and Sejnowski, T. J., 1990, Neural models of stereoacuity and depth interpolation based on a distributed representation of stereo disparity, *J. Neurosci.*, 10:2281–2299.
Miyashita, Y., 1993, Inferior temporal cortex: Where visual perception meets memory, *Annu. Rev. Neurosci.*, 16:245–263. ◆
Mozer, M. C., 1991, *The Recognition of Multiple Objects*, Boston: MIT Press.
Newsome, W. T., Britten, K. H., and Movshon, J. A., 1989, Neuronal correlates of a perceptual decision, *Nature*, 341:52–54.
Perrett, D. I., Hietanen, J. K., Oram, M. W., and Benson, P. J., 1992, Organization and functions of cells responsive to faces in the temporal cortex, *Philos. Trans. R. Soc. Lond. Biol.*, 335:23–30. ◆
Pouget, A., and Thorpe, S. J., 1991, Connectionist models of orientation identification, *Connection Sci.*, 3:127–142.
Rolls, E. T., 1992, Neurophysiological mechanisms underlying face processing within and beyond the temporal cortical visual areas, *Philos. Trans. R. Soc. Lond. Biol.*, 335:11–20.
Tanaka, K., 1993, Neuronal mechanisms of object recognition, *Science*, 262:685–688. ◆
Thorpe, S. J., and Imbert, M., 1989, Biological constraints on connectionist models, in *Connectionism in Perspective* (R. Pfeifer, Z. Schreter, F. Fogelman-Soulié, and L. Steels, Eds.), Amsterdam: Elsevier, pp. 63–92. ◆
Thorpe, S. J., Rolls, E. T., and Maddison, S. P., 1983, The orbitofrontal cortex: Neuronal activity in the behaving monkey, *Exp. Brain Res.*, 49:93–115.

Locomotion, Invertebrate

Randall D. Beer and Hillel J. Chiel

Introduction

Locomotion can be defined as an animal's ability to move its body along a desired path, making it fundamental to many other animal behaviors. Given the diversity of ecological niches that animals inhabit, and the variety of body plans that they possess, it is not surprising that their modes of locomotion are equally diverse. Types of locomotion include walking, swimming, flying, crawling, and burrowing.

Despite this diversity, certain common principles can be discerned. All locomotion systems must solve the twin problems of *support* and *progression*. The problem of support arises because in many modes of locomotion (e.g., flight), the gravitational attraction of the earth must be overcome. The problem of progression arises because an animal must generate propulsive forces that overcome not only its body's inertia, but also any drag from the density and viscosity of the medium or the friction of the substrate.

Both support and progression involve the generation of forces. This is accomplished by the contraction of muscles attached to either flexible hydrostatic skeletons or rigid skeletons. In addition, many animals have specialized body structures and appendages that facilitate locomotion, such as fins, wings, and legs. Thus, the detailed design of an animal's body is a crucial component of its locomotion system. Because of the nature of these specializations, the problems of support and progression are rarely independent. Wings, for example, are used to generate both lift and propulsion in flying animals.

To provide support and progression, the movements of these specialized body structures must be coordinated by an animal's nervous system. The diverse modes of locomotion and the variety of body plans lead to equally diverse neural circuitry mediating locomotion. However, once again, certain basic principles can be discerned. Underlying many forms of locomotion are basic oscillatory patterns of movement generated by neural circuits that are referred to as *pattern generators* (see MOTOR PATTERN GENERATION). Sometimes these circuits consist of dedicated neurons that autonomously produce rhythmic outputs, and are thus referred to as *central* pattern generators. However, this central pattern is often strongly shaped by sensory feedback, fundamentally involving the body and environment, in the generation of a locomotory pattern. In fact, sensory feedback can play such a fundamental role that it sometimes makes no sense to speak of a distinct central pattern generator.

Researchers have begun to use computer modeling to understand the neural basis of locomotion. However, unlike most work in computational neuroscience, models of an animal's body are playing an increasingly important role in understanding locomotion systems. Modeling of both an animal's body and the neural circuitry underlying its behavior has been termed *computational neuroethology* (see NEUROETHOLOGY, COMPUTATIONAL). In this article, we focus on invertebrate locomotion systems for which quantitative modeling has been done, reviewing several computer models of swimming and walking.

Swimming

In swimming, support is less of a problem than for other modes of locomotion. However, unless an animal is neutrally buoyant, effort is still required to keep it from either sinking or rising. Progression, however, requires much more effort because of the drag from water's density and viscosity. Thus, the bodies of swimming animals are streamlined. Swimming invertebrates use one of two mechanisms, either hydraulic propulsion or rhythmic undulations of the body or of specific projections.

Several computer models of swimming invertebrates have been constructed. Friesen and Stent (1977) constructed a model of the pattern generator underlying swimming in the medicinal leech *Hirudo medicinalis*. Using electronic neural analogs, they demonstrated that a pattern generator composed of intersegmental inhibitory rings of identified neurons could generate the undulatory motion observed in leech swimming. However, this model predicted that individual segments could not generate the swimming rhythm, a prediction later shown experimentally to be false. Subsequently, Pearce and Friesen (1988) simulated a more abstract model of the leech swim generator, using a chain of coupled harmonic oscillators with differing natural frequencies. It reproduced the experimentally observed phase lags between adjacent segments.

Another important model of invertebrate swimming is Getting's simulation of the pattern generator for escape swimming in the mollusc *Tritonia diomedea* (Getting, 1989). Using an integrate and fire model neuron (see SINGLE-CELL MODELS) augmented with additional membrane currents, Getting simulated the network of identified cells underlying the escape swim. The model predicted the existence of a previously unidentified population of interneurons which were later identified.

Niebur and Erdös (1991) modeled creeping and swimming in the nematode *Caenorhabditis elegans*. They modeled the forces acting on its body (a hydrostatic skeleton) during locomotion, the forces generated by its longitudinal body wall muscles, and the electrotonic network of interneurons and motor neurons activating these muscles. These muscles deform the body wall and activate cuticle stretch receptors, which in turn excite motor neurons for muscles a quarter of a body length behind the point of stretch, leading to a backward-propagating wave of muscular contraction that propels the animal forward. Stretch receptors thus play a primary role in generating locomotion in *C. elegans*, rather than simply modulating an ongoing pattern as they do in the lamprey (see SPINAL CORD OF LAMPREY: GENERATION OF LOCOMOTOR PATTERNS).

Walking

In legged animals, the body is raised above the ground and propelled by a sequence of leg movements. During walking, each leg cycles between a *stance phase*, in which the leg is providing support and propulsion, and a *swing phase*, in which the leg is off the ground and swinging forward. Swing phase duration is often nearly constant, while stance phase duration varies considerably with the speed of progression. Because the legs provide both support and propulsion and must be lifted after each stance, their movements must be coordinated so that the center of mass of the body remains within a polygon of support formed by the stancing legs (static stability). Otherwise, the animal must dynamically stabilize its body. Another coordination problem arises because adjacent legs must not interfere with one another. In many-legged animals, avoiding interference between adjacent legs is the crucial coordination problem, while the maintenance of stability is more important for animals with fewer legs.

Insect locomotion is remarkably flexible and robust (Graham, 1985). Insects can walk over a variety of terrains, as well as vertically and upside down. In addition, they can also adapt their gait to the loss of up to two legs without severe degradation of performance (Delcomyn, chap. 2 in Beer, Ritzmann, and McKenna, 1993) and sometimes even use dynamically stable gaits (Full, chap. 1 in Beer et al., 1993). Most modeling has focused on statically stable walking across flat, horizontal surfaces. Even under these conditions, insects exhibit different gaits, depending on their speed of locomotion.

Slowly walking insects show distinct *metachronal waves* on each side of the body: each leg begins its swing immediately after the termination of the swing of the leg behind it, with a 180° phase relationship between the pair of legs in each segment. Fast-walking insects use a *tripod gait*, in which the front and back legs on each side of the body step in unison with the middle leg on the opposite side. In one of the earliest theoretical models of insect walking, Wilson (1966) suggested that the

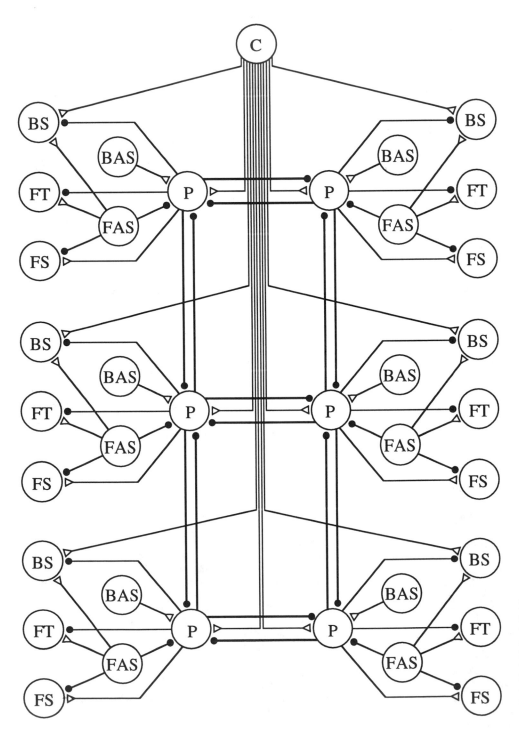

Figure 1. Neural network model of cockroach walking. Excitatory connections are denoted by open triangles, and inhibitory connections are denoted by filled circles. Abbreviations: C, command neuron; P, pacemaker neurons; FT, foot motor neurons, FS and BS, forward swing and backward swing motor neurons; FAS and BAS, forward and backward angle sensors.

entire range of observed insect gaits could be explained by assuming that fixed, antiphasic metachronal waves on each side of the body increasingly overlap as walking speed increases.

The neural organization of the American cockroach's walking system (*Periplaneta americana*) has been studied by Pearson and his colleagues (see Pearson, Fourtner, and Wong, 1973). We developed a neural network model based on Pearson's work (Beer, 1990; see Figure 1). Each of the model neurons is a leaky integrator whose integrated inputs are passed through a piecewise linear function with a threshold and saturation. Each leg controller has a pacemaker neuron whose output rhythmically oscillates because of a voltage-dependent intrinsic current. The interval between bursts depends linearly on the tonic input to the pacemaker, with excitation decreasing and inhibition increasing the interburst interval. In addition, a strong excitatory pulse between bursts or a strong inhibitory pulse during a burst can reset a pacemaker's burst rhythm. These pacemakers implement the swing burst-generators that Pearson hypothesized.

A pacemaker burst initiates a swing by inhibiting the foot and backward swing motor neurons and exciting the forward swing motor neurons, causing the foot to lift and the leg to swing forward. Between bursts, the foot is down and tonic excitation from a command neuron moves the leg backward. Feedback from two sensors that signal when a leg is nearing its extreme forward or backward position fine-tunes pacemaker output. Forward angle sensor inhibition encourages burst termination, whereas backward angle sensor excitation encourages burst initiation. The forward angle sensor also makes direct connections to the motor neurons, modeling leg reflex pathways described by Pearson.

To generate statically stable gaits, the swings of the individual legs must be coordinated in some way. Following Pearson, we inserted mutually inhibitory connections between the pacemaker neurons of adjacent legs. We also added an entrainment mechanism for generating metachronal waves: slightly increasing the angle ranges of the rear legs lowers the burst frequency of the rear pacemakers, causing the pattern generators on each side of the body to phase-lock into a stable metachronal relationship.

In simulations of this circuit in a kinematic hexapod body model, a continuous range of statically stable gaits similar to those described by Wilson (1966) were observed. This range of gaits was produced simply by varying the tonic level of excitation of the command neuron. Smooth transitions between gaits could be generated by continuously varying this excitation. We found that the ability of this circuit to generate statically stable gaits was quite robust to lesions. For example, removing any single sensor or interpacemaker connection did not generally disrupt locomotion. These studies also demonstrated that sensory feedback was crucial for the maintenance of the slower metachronal gaits but was relatively unimportant in the tripod gait.

The stick insect *Carausius morosus* has also been a major focus of research. Cruse (1990) reviewed leg coordination influences in both the stick insect and the crayfish *Astacus leptodactylus*. In the stick insect, there are three major influences: (1) a swinging leg inhibits the swing of a more anterior leg; (2) when a leg begins its stance phase, it excites the swing of a more anterior leg; and (3) as a stancing leg nears the end of its stance, it increasingly excites the swing of a more posterior leg. Some of these influences also operate between pairs of legs in the same segment.

Dean (1991) simulated these and other coordination mechanisms. The pattern generator for each leg was modeled as a relaxation oscillator, with two states corresponding to stance and swing. The positions of each of the six legs were the state variables for a kinematic model of walking. The coordination mechanisms modify the position at which an affected leg begins its swing, with inhibitory influences producing a posterior shift and excitatory influences producing an anterior shift. Dean's simulations demonstrated that these coordination mechanisms were sufficient to generate a continuous range of gaits, including the wave gait at low stepping frequencies and the tripod gait at high stepping frequencies. The model also exhibited distinct asymmetries in stepping pattern observed in the stick insect, in which the phase relationship between legs in the same segment is consistently lower or higher than 180°. A good review of earlier models of stick insect walking can also be found in Dean (1991).

Dean (1992) also explored the robustness of these coordinating mechanisms to various perturbations, including variations in starting configurations, perturbations of individual leg velocities, and obstructions to the swing of individual legs. He found that the gaits generated by these mechanisms were quite robust to such perturbations, and that, in most cases, the model's responses were similar to those of the insect. Discrepancies between the model and the insect could be traced to the need for dynamic variables in addition to kinematic variables. Dean varied the strength and form of the coordination mechanisms. He found that influence 3 was the most important in maintaining proper coordination because of its graded nature, though the model was quite robust to substantial variations in the strengths of individual mechanisms.

Discussion

We have touched on several successful examples of quantitative modeling of locomotion. It is notable that the different simulations use very different neural models. More fundamentally, it is striking that very different neural architecture can be used to generate locomotion. Undoubtedly, this variety is attributable to the diverse body plans of animals and the many different ecological niches that they occupy. One consistent theme that does emerge, however, is the complex interplay of sensory input and central circuitry in the generation of locomotion. This complex interplay is responsible for the adaptive flexibility of animal locomotion.

The remarkable flexibility and robustness of insect walking has intrigued roboticists. Biologically inspired locomotion controllers offer a number of advantages over more classical approaches, including their distributed nature, their robustness and their computational efficiency. Thus, a number of researchers have explored the application of biological ideas to legged robots. Raibert (chap. 14 in Beer et al., 1993) has argued for the importance of leg and actuator design in locomotion, designing a series of dynamically stable hopping and running robots based on the biomechanical design of animal limbs. Donner (1987) implemented a distributed locomotion controller directly inspired by Wilson's (1966) work on insect walking for a hydraulically actuated hexapod. More recently, Brooks (chap. 15 in Beer et al., 1993) developed a partially distributed hexapod controller using a network of finite state machines. The swing/stance cycle of each leg was driven by a chain of peripheral reflexes, and the movements of the individual legs were coordinated by a single, centralized machine which told each leg when to lift. Finally, we implemented the locomotion circuit just described in a hexapod robot and found that it could generate a range of gaits similar to those observed in simulation (Quinn and Espenschied, chap. 16 in Beer et al.,

1993) and that it was equally robust to perturbations (Chiel et al., 1992). The stick insect coordination rules have also been implemented in this robot (Espenschied et al., 1993). Thus, models of animal locomotion not only yield insights into the neural control of motor behavior, but they also may have significant technological applications (see also WALKING).

Acknowledgments. R.D.B. acknowledges ONR grant N00014–90–J–1545, and H.J.C. acknowledges NSF grant BNS–8810757.

Road Map: Motor Pattern Generators and Neuroethology
Background: Single-Cell Models
Related Reading: Gait Transitions; Locust Flight: Components and Mechanisms in the Motor; Neuromodulation in Invertebrate Nervous Systems

References

Beer, R. D., 1990, *Intelligence as Adaptive Behavior: An Experiment in Computational Neuroethology*, San Diego: Academic Press.
Beer, R. D., Ritzmann, R. E., and McKenna, T., Eds., 1993, *Biological Neural Networks in Invertebrate Neuroethology and Robotics*, San Diego: Academic Press. ◆
Chiel, H. J., Beer, R. D., Quinn, R. D., and Espenschied, K., 1992, Robustness of a distributed neural network controller for a hexapod robot, *IEEE Trans. Robot. Automat.*, 8:293–303.
Cruse, H., 1990, What mechanisms coordinate leg movement in walking arthropods? *Trends Neurosci.*, 13:15–21. ◆

Dean, J., 1991, A model of leg coordination in the stick insect, *Carausius morosus*, II: Description of the kinematic model and simulation of normal step patterns, *Biol. Cybern.*, 64:393–402.
Dean, J., 1992, A model of leg coordination in the stick insect, *Carausius morosus*, Parts III and IV, *Biol. Cybern.*, 66:335–355.
Donner, M., 1987, *Real-Time Control of Walking*, Cambridge, MA: Birkhäuser Boston.
Espenschied, K. S., Quinn, R. D., Chiel, H. J., and Beer, R. D., 1993, Leg coordination mechanisms in the stick insect applied to hexapod robot locomotion, *Adapt. Behav.*, 1:455–468.
Friesen, W. O., and Stent, G. S., 1977, Generation of a locomotory rhythm by a neural network with recurrent cyclic inhibition, *Biol. Cybern.*, 28:27–40.
Getting, P. A., 1989, Reconstruction of small neural networks, in *Methods in Neuronal Modeling* (C. Koch and I. Segev, Eds.), Cambridge, MA: MIT Press, pp. 171–194. ◆
Graham, D., 1985, Pattern and control of walking in insects, *Adv. Insect Physiol.*, 18:31–140. ◆
Niebur, E., and Erdös, P., 1991, Theory of the locomotion of nematodes: Dynamics of undulatory progression on a surface, *Biophys. J.*, 60:1132–1146.
Pearce, R. A., and Friesen, W. O., 1988, A model for intersegmental coordination in the leech nerve cord, *Biol. Cybern.*, 58:301–311.
Pearson, K. G., Fourtner, C. R., and Wong, R. K., 1973, Nervous control of walking in the cockroach, in *Control of Posture and Locomotion* (R. B. Stein, K. G. Pearson, R. S. Smith, and J. B. Redford, Eds.), New York: Plenum, pp. 491–514. ◆
Wilson, D. M., 1966, Insect walking, *Annu. Rev. Entomol.*, 11:103–122. ◆

Locust Flight: Components and Mechanisms in the Motor

R. Meldrum Robertson

Introduction

Locusts have certain characteristics which make them one of the best species for studying the neural bases for flight behaviors (see Kammer, 1985). The neural elements involved in generating the patterns of motor activity are idiosyncratic enough to be individually identifiable, and they are large enough to be penetrated by intracellular microelectrodes. Thus, in the locust, it is possible to describe the operation of networks of identified neurons, connected by identified synapses, and to determine how these networks contribute to the computational task of this portion of the nervous system. That task is to produce rhythmical motor patterns capable of keeping the locust aloft in an unpredictable environment.

The Motor Output

The locust flight system (Figure 1) creates a spatiotemporal pattern of electrical activity in about 80 flight motoneurons. These motoneurons activate muscles controlling the paired forewings and hindwings, causing beating at around 22 cycles/s. A slower version of the motor pattern (around 12 cycles/s) can be generated by a central nervous system deafferented from phasic timing information emanating from wing proprioceptors. The extent to which sensory feedback supercedes the role of the central pattern generator in normal intact flight remains unclear. Nevertheless, it is quite clear that proprioceptive feedback is necessary for appropriate timing of the wingbeat phase transitions. The *tegulae* are external sense organs (see Figure 1)

stimulated by depression of each wing, and they can initiate the subsequent elevator phase by excitation of elevator motoneurons and interneurons. The stretch receptors are internal at the wing base and are activated by wing elevation. They promote the occurrence of the subsequent wing depression by opposing the hyperpolarization between the bursts of action potentials in depressor motoneurons. A simple model describing how the stretch receptors regulate frequency of wingbeat has been described (Figure 2; Pearson and Ramirez, 1990) but would benefit from a quantitative implementation.

The Neuronal Components and Their Organization

Wing muscle motoneurons do not participate in timing the motor pattern, and they serve simply as output elements. Thus, the centrally generated rhythm arises as a result of the cellular properties and interactions of interneurons in the three thoracic ganglia: prothoracic, mesothoracic, and metathoracic. Numerous interneurons have been described, and they are connected into circuits via standard, short-latency, synaptic interactions that are probably mediated by gamma-aminobutyric acid (GABA) (inhibitory) and glutamate (excitatory) (Robertson, 1989). A simple model of the circuitry described to date has, at its heart, a circuit of delayed excitation and feedback inhibition which would result in an elevator-depressor burst sequence (Figure 3; Robertson and Pearson, 1985). However, this model does not take into consideration more recent findings about the operation of the flight network.

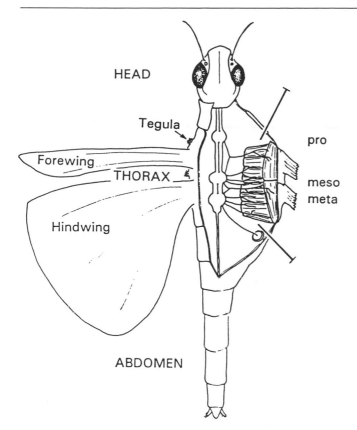

Figure 1. The locust flight motor. This dorsal view of a locust shows, on the left side, the form of the forewing and hindwing and the position of the fore and hind tegulae (only the forewing tegula is labeled). On the right side, the thorax has been pinned open to reveal the bank of flight muscles that powers the wings, as well as the three thoracic ganglia (prothoracic, mesothoracic, and metathoracic) that contain the moto-neurons and interneurons involved in generating flight motor patterns.

One of the most interesting recent findings is that longitudinal hemisection of the mesothoracic ganglion can have surprisingly little effect on the ability of an animal to fly (Ronacher, Wolf, and Reichert, 1988). A similar hemisection of the metathoracic ganglion is more disruptive—precluding free flight by impairing coordination—but it still does not abolish wing-beating. The system must be constructed with much redundancy of elements, and peripheral feedback loops probably compensate for much of the damage to central networks, particularly those of the mesothoracic ganglion. More importantly, these results question our understanding of the fundamental organization of the system.

It had been proposed that the central circuit operated essentially as a unit that was distributed throughout six serially homologous, segmental neuromeres (Robertson, Pearson, and Reichert, 1982). The hemisectioning experiments would have damaged most of the interneurons described as part of this distributed flight network, but rhythms were still generated. One suggestion is that the central oscillator is composed of hemiganglionic subunits. However, the evidence is weak that a completely isolated hemiganglion can produce an oscillatory output. Furthermore, the generation of a few cycles of rhythmical activity is not sufficient support for the idea that the system is organized in such a fashion. There is a large conceptual difference between rhythmical neural activity produced in

a hemiganglion and hemiganglionic premotor centers for the control of coordinated wingbeating.

Other investigations have attempted to demonstrate the existence of segmental oscillators for the separate control of the two pairs of wings by cutting the connectives between the ganglia. There is some tantalizing evidence in support of this proposal. In the presence of octopamine, an isolated metathoracic ganglion can produce recognizable flight rhythms. This is not true for an isolated mesothoracic ganglion, although ragged mesothoracic rhythms can be produced by an otherwise intact animal with a transection between the mesothoracic and metathoracic ganglia. The first result can be explained by most of the proposed unitary oscillator being located in the metathoracic ganglion. The second may arise from a contribution of the peripheral feedback loops.

None of the characteristic behavior predictable from coupled oscillator theory has been described. Neither hemisection nor transection experiments have demonstrated two separate, simultaneous rhythms from the same preparation. It is thus impossible to demonstrate different coordination modes, or relative coordination. The delay from hindwing to forewing depressor bursts is fixed at 5–10 ms (i.e., it is not phase-constant), and remains the same even when the output rhythm frequency is halved by deafferentation.

These results demonstrate multiple patterning elements which may aid in stabilizing the output pattern, and an extremely important role for sensory elements in timing and coordination of the four wings during intact flight. Furthermore there is a pre-eminent role for the metathoracic ganglion in central pattern generation, compared with the role of the mesothoracic ganglion. There are many oscillator mechanisms contributing to the generation of the rhythm, with the result that rhythm generation survives much experimental manipulation. However, there is not yet strong evidence to support the notion that each wing, or pair of wings, is controlled by a separate central oscillator with their relative phasing determined by the nature of the coupling between them. This makes the locust flight system unique among locomotor pattern generators, which are normally organized as coupled central oscillators.

Circuitry Underlying Steering

Much has been written about the mechanisms underlying steering in the locust flight system (see Rowell, 1988, for a review). Although behaviors subserving oriented flight are currently receiving more attention than previously, most information relates to course correction behaviors, mechanisms of the "autopilot." It is only for the course correction circuitry that there is a model at the cellular level (Figure 4) that explains how multimodal sensory input which signals deviation from course can be integrated into the operation of the central circuitry to cause the asymmetries in the motor output necessary to compensate for an unintended change in the direction of flight (e.g., Reichert, Rowell, and Griss, 1985). The basis of the model is that continuous signals from exteroceptors, such as the ocelli (simple light detectors), or from wind-sensitive head hairs excite premotor interneurons that are simultaneously rhythmically activated by the central circuits. Thus, the course deviation signal is gated through these premotor interneurons and transmitted to the motor neurons at the appropriate phase for an effective change in the motor pattern. Much of the asymmetry in the form of the wingbeat that generates steering torques occurs during the downstroke (the power stroke) of the wings, while the upstroke remains relatively symmetrical. The

DEAFFERENTED

INTACT

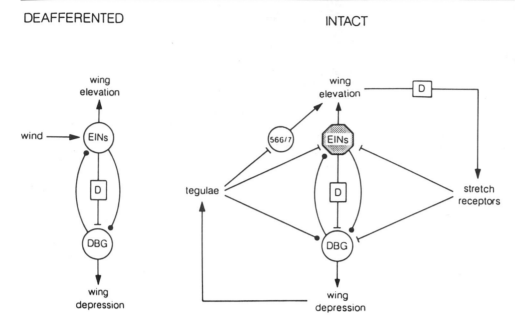

Figure 2. Models illustrating the role of proprioceptive feedback in generating the flight motor pattern. The deafferented system (left) consists of a depressor burst generator (DBG) with reciprocal inhibitory interactions with elevator interneurons (EINs). These interneurons also pass excitation from wind input to the DBG through a delay (D) pathway. In intact animals (right), feedback from proprioceptors can recruit interneurons (e.g., 566/7), as well as interact with the elements of the central rhythm generator. Stippling indicates that the activity pattern of the EINs is altered by the feedback, solid circles indicate inhibitory connections, and "T" bars represent excitatory connections. (From Pearson, K. G., and Ramirez, J.-M., 1992, Parallels with other invertebrate and vertebrate motor systems, in *Dynamic Biological Networks* (R. M. Harris-Warrick, E. Marder, A. I. Selverston, and M. Moulins, Eds.), Cambridge, MA: MIT Press, pp. 263–281; reprinted with permission.)

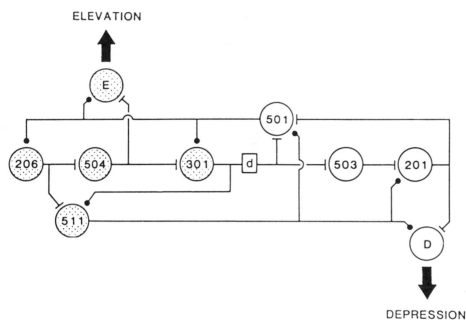

Figure 3. Model illustrating some of the connections between flight interneurons that may contribute to generating the central flight rhythm. Elevator interneurons are stippled. Interneuron 206 receives excitation from wind input. Note the similarity to the deafferented model in Figure 2. The heart of the circuit is delayed excitation (301 to 501) and feedback inhibition (501 to 301). Note also that the interneurons in this diagram are damaged by the transection and hemisection experiments described in the text. (From Robertson, R. M., 1986, Neuronal circuits controlling flight in the locust: Central generation of the rhythm, *Trends Neurosci.*, 9:278–280.)

deviation signal can be gated so that it affects only those motoneurons involved in controlling the form of the downstroke, and it need not interact directly with the central rhythm generator. Another reason for passing the signal through a bursting premotor interneuron may simply be to convert it to a rhythmical signal. This could be more effective in altering the phase of activity of the motoneuron.

Neuromodulation and Plasticity

The neural networks which control motor patterns are not static entities. Apart from the long-term plastic changes which are necessary to refine a coordinated motor performance, short-term dynamic alterations have been demonstrated which enable a network to modify its output under different condi-

tions (see NEUROMODULATION IN INVERTEBRATE NERVOUS SYSTEMS). The particular set of components and characteristics at any one time is controlled by the mix of circulating transmitters and neuromodulators. The neural networks in the locust flight system are similarly plastic.

Octopamine has a multifaceted role in the control and coordination of locust flight (Orchard, Ramirez, and Lange, 1993). From mobilization of energy resources to the modification of flight muscle properties, octopamine has influences throughout the locust, enabling it to fly efficiently and to respond to the metabolic demands of flight. Local injection of octopamine at specific sites in the thoracic ganglia is sufficient to release flight-like activity from the nervous system. Octopamine can induce intrinsic bursting properties (plateau potentials) in identified flight interneurons, and there are good reasons for supposing

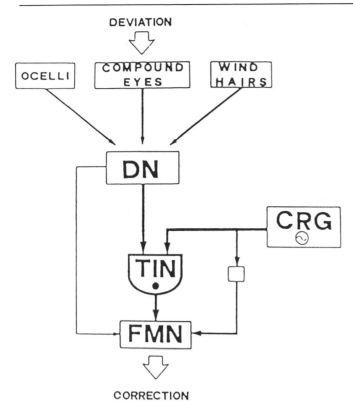

Figure 4. Model of the circuitry responsible for integrating exteroceptive information with the flight motor pattern. Sensory information traveling in descending interneurons (DN) passes through thoracic interneurons (TIN) to the flight motor neurons (FMN). The TINs are also driven rhythmically by the central rhythm generator (CRG) and thus can gate the sensory signal (the TIN is shown here as a logical AND-gate) to particular phases of the motor pattern. (From Rowell, H. F., and Reichert, H., 1985, Compensatory steering in locusts: The integration of non-phase locked input with a rhythmic motor output, in *Insect Locomotion* (M. Geweke and G. Wendler, Eds.), Berlin: Paul Parey Press, pp. 175 182; reprinted with permission.)

that this experimental manipulation reflects a physiological role for octopamine in the generation of normal flight motor patterns.

A short-term plasticity in the output of the flight system has recently been described and ascribed to associations between muscle-specific proprioceptive input and exteroceptive input signaling deviation from course (Möhl, 1993). The interesting proposal is that the central circuits provide a motor framework which is subsequently sculpted by the immediately preceding flight experience to provide the output that is most effective in controlling flight. Thus, the operating circuit can be tailored to the current condition of the animal and its flight system. It has been proposed that the temporal association of pattern-stabilizing proprioceptive feedback and an exteroceptive course deviation signal at synapses on a premotor neuron (like the thoracic interneurons illustrated in Figure 4) will result in a Hebb-like strengthening of the exteroceptive-to-premotor synapse. In this way, the afferent connections to those premotor interneurons with the greatest influence on the motor output can be strengthened. The specific synaptic mechanisms underlying this plasticity remain to be determined.

In contrast, there is some information on the synaptic mechanisms involved in the longer-term plasticity underlying functional recovery after a specific deafferentation (Büschges et al.,

1992). Ablation of the hindwing tegulae (see Figures 1 and 2) impairs the operation of the flight system, as reflected in a reduced wingbeat frequency. However, the system recovers by the forewing tegulae taking over the function of the ablated hindwing tegulae. New connections are made between the afferents and specific flight interneurons, accompanied by sprouting and growth of both the axonal branches of afferent fibers and the dendritic arbors of the interneurons.

Modeling the Locust Flight System

Initial attempts to model the locust flight system used electronic Lewis "neuromimes" which simulated the behavior of neural membranes and could be connected into networks (Wilson and Waldron, 1968). An arrangement of neuromimes into positively coupled subsets interconnnected with reciprocal inhibition successfully mimicked several features of the flight motor pattern. This bears little relation to current ideas of pattern generation in the flight system, but must be considered a success insofar as it could generate the pattern and did not conflict with the few experimental observations then available. Unfortunately, this early success has not been followed by successive families of detailed models. What mostly exists in the literature are conceptual models of the type illustrated in Figures 2 through 4. These are little more than summaries of experimental results, and have not been rigorously tested by attempts to implement them. Nevertheless, they do represent most of the current state of knowledge of the flight motor. One obvious omission is any indication of endogenous mechanisms of burst generation in Figure 3. However, although the existence of plateau properties has been established, their distribution among the interneuronal population has not been mapped. Recently, a computer simulation, using the program BioSim 3.0, has shown that the known properties and connections of flight neurons are sufficient to explain the oscillatory behavior of the flight circuitry (Grimm and Sauer, 1994).

Discussion

Locust flight motor patterns are generated by an interactive mixture of the intrinsic properties of flight neurons, the operation of complex circuits, and phase-specific proprioceptive input. These mechanisms are subject to the concentrations of circulating neuromodulators, and the patterns are also modulated according to the demands of a constantly changing sensory environment that requires adaptive behavior. The system is flexible and plastic in both the short and long term, able to operate despite severe ablations, and subsequently able to recover from these lesions.

Models always benefit from a more extensive database from which to work, and there is still much to be learned about how the flight motor operates. Nevertheless, knowledge of the flight system is currently at a stage at which detailed models would be beneficial. Would a change in the parameters of the model circuit (Grimm and Sauer, 1994), according to the known effects of temperature on conduction velocities and synaptic interactions, replicate the observed effects of temperature on the motor patterns? Can proprioceptive feedback be added to increase the frequency of the output? Is it possible to generate a model that mimics the coordination of the four wings using a single depressor burst generator located primarily in the metathoracic ganglion? Alternatively, can a set (or pair) of coupled oscillators be made to behave like the output from the thoracic ganglia under the onslaught of hemisections and transections? Can the effect of specific ablations and recoveries be accurately modeled? These are important, testable questions which have

560 Part III: Articles

arisen from the data generated by intracellular investigations in dissected organisms.

Without doubt, the neural processes involved in higher brain functions will not be first described in the locust. However, the basis for these higher functions is likely related both to the generation of patterns of electrical activity in time and space and to the modulation of these patterns by the extracellular environment, by the periphery, and by experience. The control of such spatiotemporal patterning can profitably be investigated in the locust flight system.

Acknowledgments. I thank the Natural Sciences and Research Council of Canada, and the Faculty of Graduate Studies and Research at Queen's University, for financially supporting my research.

Road Map: Motor Pattern Generators and Neuroethology
Background: I.1. Introducing the Neuron
Related Reading: Crustacean Stomatogastric System; Locomotion, Invertebrate; Spinal Cord of Lamprey: Generation of Locomotor Patterns

References

Büschges, A., Ramirez, J.-M., Driesang, R., and Pearson, K. G., 1992, Connections of the forewing tegulae in the locust flight system and their modification following partial deafferentation, *J. Neurobiol.*, 23:44–60.

Grimm, K., and Sauer, A. E., 1994, The high number of neurons contributes to the robustness of the locust flight-CPG against parameter variation, *Biol. Cybern.* (in press).

Kammer, A. E., 1985, Flying, in *Comprehensive Insect Physiology, Biochemistry and Pharmacology* (G. A. Kerkut and L. I. Gilbert, Eds.), New York: Pergamon, pp. 491–552. ◆

Möhl, B., 1993, The role of proprioception for motor learning in locust flight, *J. Comp. Physiol.*, 172:325–332.

Orchard, I., Ramirez, J.-M., and Lange, A. B., 1993, A multifunctional role for octopamine in locust flight, *Annu. Rev. Entomol.*, 38:227–249. ◆

Pearson, K. G., and Ramirez, J.-M., 1990, Influence of input from the forewing stretch receptors on motoneurones in flying locusts, *J. Exp. Biol.*, 151:317–340.

Reichert, H., Rowell, C. H. F., and Griss, C., 1985, Course correction circuitry translates feature detection into behavioural action in locusts, *Nature*, 315:142–144.

Robertson, R. M., 1989, Idiosyncratic computational units generating innate motor patterns: Neurones and circuits in the locust flight system, in *The Computing Neurone* (R. Durbin, R. C. Miall, and G. Mitchison, Eds.), London: Addison-Wesley, pp. 262–277. ◆

Robertson, R. M., and Pearson, K. G., 1985, Neural circuits in the flight system of the locust, *J. Neurophysiol.*, 53:110–128.

Robertson, R. M., Pearson, K. G., and Reichert, H., 1982, Flight interneurons in the locust and the origin of insect wings, *Science*, 217:177–179.

Ronacher, B., Wolf, H., and Reichert, H., 1988, Locust flight behavior after hemisection of individual thoracic ganglia: Evidence for hemiganglionic premotor centers, *J. Comp. Physiol.*, 163:749–759.

Rowell, C. H. F., 1988, Mechanisms of flight steering in locusts, *Experientia*, 44:389–395. ◆

Wilson, D. M., and Waldron, I., 1968, Models for the generation of the motor output pattern in flying locusts, *Proc. IEEE*, 56:1058–1064.

Long-Term Depression in the Cerebellum

F. Crepel, N. Hemart, D. Jaillard, and H. Daniel

Introduction

The participation of the cerebellum in motor learning (see CEREBELLUM AND MOTOR CONTROL) was postulated by Brindley as early as 1964 and was formalized by Marr (1969) and Albus (1971). According to the theory, the gain of excitatory synaptic transmission between parallel fibers (PFs) and Purkinje cells (PCs) (Eccles, Ito, and Szentagothai, 1967) is modified during motor learning by the other excitatory afferents to PCs—the climbing fibers (CFs)—to adjust the cerebellar output to the desired motor command. In this scheme, conjunctive activation of PCs by CFs and PFs leads to long-term depression (LTD) of synaptic transmission at PF–PC synapses (Figure 1).

From a neurochemical viewpoint, the cerebellum of mammals is also an interesting region of the brain to study synaptic plasticity. Indeed, like most excitatory afferents in the brain, PFs and CFs are likely to use glutamate as a neurotransmitter (reference cited in Crepel and Audinat, 1991). Now, in marked contrast to most other neuronal cell types, the postsynaptic ionotropic receptors of PCs to glutamate do not include *N*-methyl-D-aspartate (NMDA) receptors, but only α-amino-3 hydroxy-5-methyl-isoxazole-4-propionate (AMPA) receptors, these receptors being termed on a pharmacological basis by the names of their selective agonists NMDA and AMPA, respectively. (For these and other studies not cited explicitly, see Crepel and Audinat, 1991.) Because NMDA receptors normally play a crucial role in the induction of long-term changes in synaptic efficacy in the hippocampus as well as in neocortical areas (see NMDA RECEPTORS: SYNAPTIC, CELLULAR, AND NETWORK MODELS), the cerebellum appears particularly interesting for the study of the cellular mechanisms of synaptic plasticity operating in their absence. On the other hand, AMPA receptors have interesting desensitizing properties, i.e., their response to glutamate fades (desensitizes) within a few milliseconds in the presence of the agonist, which may thus contribute to the rapid decay of excitatory synaptic potentials mediated by these receptors. We will see later that LTD is likely to involve some sort of desensitization of the AMPA receptors located at PF–PC synapses.

This review deals with main experimental data, gathered over the last 10 years, on the participation of excitatory amino acid receptors of PCs in synaptic plasticity as a possible cellular basis of motor learning in the cerebellum.

Discovery of Long-Term Depression

Ito and co-workers were the first to demonstrate *in vivo* (Ito, Sakurai, and Tongroach, 1982) that, in rabbit cerebellum, conjunctive stimulation of PFs and CFs leads to LTD of synaptic transmission at PF–PC synapses. The fact that only those PF–PC synapses activated in conjunction with CFs are affected indicates that the changes in synaptic strength are restricted to the activated synapses.

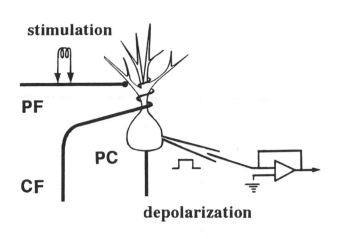

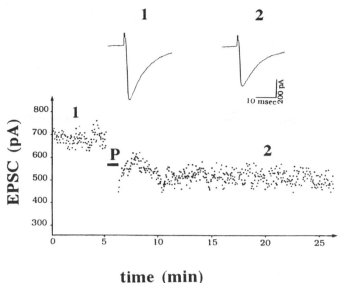

time (min)

Figure 1. Long-term depression (LTD) in cerebellar slices. An example of LTD induced by pairing (P) PF-mediated EPSCs with depolarization of PCs, giving rise to a calcium spike (duration of 1 minute). EPSC amplitudes are plotted against time before and after the pairing. The insets display control EPSC (1) and EPSC 15 minutes after the end of the pairing period (2). Diagrams of cerebellar circuits and experimental arrangements are shown on the left.

Role of Calcium in Long-Term Depression of Synaptic Transmission

Because CFs are known to strongly depolarize PCs, thus leading to calcium influx into these cells through the activation of their voltage-gated calcium channels, Ito and co-workers proposed that calcium plays a major role in LTD induction. Accordingly, LTD of PF-mediated EPSPs is consistently induced in slices when calcium spikes, induced in the postsynaptic cell by depolarizing current pulses, are substituted for CF activation during the pairing protocol with PF stimulation (see Figure 1). In contrast, when weaker depolarizing pulses are given during the pairing protocol, so that only sodium spikes are induced in PCs, LTD is no longer observed and is replaced by long-term potentiation (LTP) of PF-mediated EPSPs (see the section that follows). Similarly, in cultured PCs, Linden and Connor (1991) have shown that LTD of AMPA-mediated currents is induced by conjunctive ionophoretic glutamate pulses and PC depolarization sufficient to produce calcium entry through voltage-gated calcium channels. Finally, Konnerth and co-workers (1992) have shown that, in pairing experiments, a transient rise in internal calcium is sufficient to induce LTD.

Which Glutamate Receptors Participate in LTD Induction?

In slices, no LTD could be induced by a pairing protocol with calcium spikes in the presence of the AMPA receptor antagonist 6-cyano-7-nitroquinoxaline-2,3-dione (CNQX) (F. Crepel et al., unpublished data). Similarly, in cultured PCs, Linden and co-workers (1991) also showed that no LTD occurs unless AMPA receptors are activated during its induction phase. This is, of course, in keeping with the fact that LTD induction requires that PF–PC synapses are stimulated in conjunction with CF activation of PCs (see previous section). Now, it is also well established that, besides AMPA receptors mediating fast excitatory synaptic transmission at PF–PC synapses, the dendritic spines of PCs also bear the so-called metabotropic receptors to glutamate (Baude et al., 1993). These receptors are not coupled with ionic channels like AMPA receptors, but with enzymatic cascades leading to calcium release from internal stores and activation of protein kinase C (PKC), the role of which in LTD is emphasized in the following discussion. Therefore, this led several groups to test the possible role of metabotropic receptors in LTD. Ito and Karachot showed that coactivation of AMPA and metabotropic glutamate receptors of PCs by AMPA and by the selective metabotropic receptor agonist trans-1-amino-cyclopentyl-1,3-dicarboxylate (trans-ACPD), respectively, is sufficient to induce a long-lasting desensitization of AMPA receptors of these cells, suggesting that activation of metabotropic receptors of PCs also plays an important role in LTD. This view has been supported by experiments by Linden and co-workers (1991), who demonstrated that, in cultured PCs, induction of LTD by conjunctive depolarization and stimulation of glutamate receptors requires the activation of both their AMPA and metabotropic receptor subtypes. Finally, we have shown, in cerebellar slices, that pairing-induced LTD was strongly potentiated by activation of metabotropic receptors of PCs by trans-ACPD (Daniel et al., 1992). All of these experiments point out a role for both AMPA and metabotropic receptors in the induction of LTD (Figure 2). However, confirmation of the role of metabotropic receptors in physiologically induced LTD will only be achieved when selective antagonists of this class of receptors become available. Finally, numerous metabotropic receptors are also located close to CF–PC synapses (Baude et al., 1993), which raises the question of their possible contribution to LTD induction.

Second-Messenger Cascades Involved in LTD

As early as 1988, and as illustrated in Figure 2, we hypothesized that the cascade of events leading to LTD involves a coactivation of PKC in PCs by calcium entry through voltage-gated ionic channels and by diacylglycerol (DAG) produced by the activation of metabotropic receptors by glutamate released by PFs during the pairing protocol. It should be emphasized that this process, like the activation of AMPA receptors, is

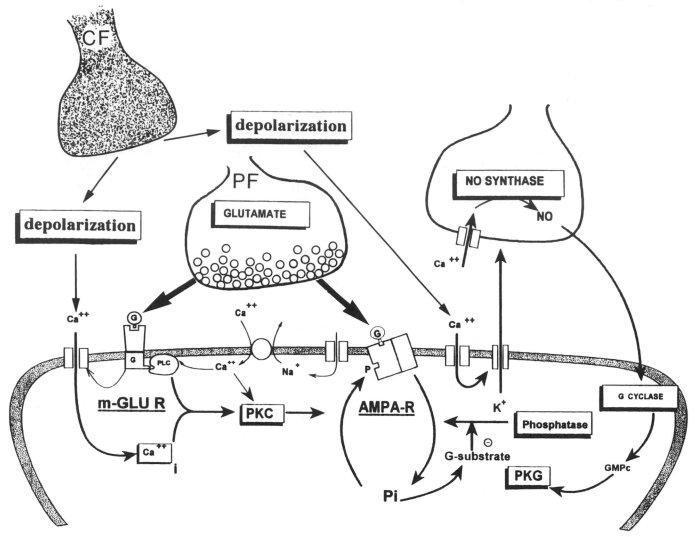

Figure 2. Putative cellular mechanisms involved in LTD. Schematic diagram of the signal transduction processes, involving the PKC and NO pathways, which are presumed to underlie LTD. In this scheme, we have illustrated the hypothesis according to which NO synthase involved in LTD is located outside PCs. Abbreviations: G, glutamate; m-GluR, metabotropic receptor; AMPA-R, AMPA receptor; G, protein; PLC, phospholipase C; cGMP, cyclic guanosine monophosphate; Pi, inorganic phosphate; P, phosphorylation.

synapse-specific, as required by theories of motor learning in the cerebellum. Indeed, a selective LTD of the responsiveness of PCs to glutamate was obtained by direct activation of PKC by phorbol esters, whereas LTD of PF-mediated EPSPs following their pairing with calcium spikes was nearly totally prevented by PKC inhibitors (reference cited in Crepel and Audinat, 1991). These results were recently confirmed and extended in cultured PCs (Linden and Connor, 1991).

There is now experimental evidence that LTD also depends on another cascade of events beyond calcium entry into PCs. Indeed, it is known that calcium can induce the formation of nitric oxide (NO) from arginine by activating a calmodulin-dependent NO synthase (reference cited in Garthwaite, Charles, and Chess-Williams, 1988). Since NO is a potent activator of soluble guanylate cyclase, this cascade of events leads to an increase in cGMP concentration and can thus activate a cGMP-dependent protein kinase—PKG—an enzyme particularly abundant in PCs (reference cited in Crepel and Audinat, 1991). Furthermore, NO is highly diffusible (although across a still largely unknown distance in nervous tissues; probably a

few tens of microns). It can, therefore, exert its effects not only in cells where it is produced, but also, probably, in neighboring cells. In principle at least, NO might thus create some cooperativity among neighboring synapses, or even among neighboring PCs, thus leading to the appearance of assemblies of neurons related to a given learning task. By contrast, synapse specificity would primarily be afforded by the PKC route described earlier.

Therefore, it is of prime importance that inhibitors of NO synthase partially prevent the induction of LTD, whereas NO donors and cGMP are able to reproduce an LTD-like phenomenon when they are directly injected into PCs (Crepel and Jaillard, 1990; Daniel et al., 1993). In 1990, Ito and Karachot, using the grease-gap method, also established that desensitization of AMPA receptors of PCs was likely to require the production of cGMP via the production of NO. Subsequently, Shibuki and Okada (1991) showed that protocols known to induce LTD did, indeed, lead to the production of NO in cerebellar slices. Therefore, these results show that NO also plays a role in LTD induction, and that its site of action is probably the soluble guanylate cyclase of PCs. According to Ito and Karachot (1992), cGMP, in turn, activates a PKG, thereby allowing phosphorylation of its specific substrate (G-substrate, a potent inhibitor of phosphatases). In this scheme (see Figure 2), LTD would involve both phosphorylation of AMPA receptors of PCs by PKC, and inhibition of their dephosphorylation by the NO route. As stated before, the PKC pathway would be responsible for synapse specificity of LTD, whereas the NO pathway might also be involved in some cooperation among neighboring synapses or even cells.

However, in the cerebellum, NO synthase has not been identified in PCs, but only in neighboring elements. Thus, one hypothesis is that NO might be produced by the NO synthase located in PFs and/or in basket cells when these neurons are activated by both PF stimulation (Eccles, Ito, and Szentágothai, 1967) and by the large efflux of potassium which follows the entry of calcium into PCs during pairing experiments. This possibility is illustrated in Figure 2. However, it is still possible that the NO synthase involved in the induction of LTD is located in the PCs themselves, but is different from the already-known forms of the enzyme, thus explaining why it has not been visualized so far.

Does LTD Involve Desensitization of AMPA Receptors of PCs?

The fact that coactivation of PCs by CFs and by direct application of glutamate in their dendritic fields leads to a persistent decrease in their responsiveness to this agonist (Ito, Sakurai, and Tongroach, 1982) led Ito and co-workers to propose that induction of LTD might ultimately lead to long-term desensitization of ionotropic glutamate receptors of PCs, as well as to the observed decrease in synaptic efficacy. Accordingly, in 1988, we showed that, in acute slices, pairing ionophoretic application of glutamate and calcium spike firing of PCs induced LTD of their responsiveness to this agonist. Similar results were obtained in dissociated cultures (Linden and Connor, 1991). However, the observed decrease in the efficacy of glutamate in activating PCs might be attributable to causes other than a true desensitization of glutamate receptors.

Recently, the nootropic compound aniracetam has been shown to markedly reduce desensitization of AMPA receptors (reference cited in Vyklicky, Patneau, and Mayer, 1991). It was, therefore, an important finding to show that, in whole cell clamped PCs in acute slices, aniracetam has a larger potentiating effect on PF-mediated excitatory postsynaptic currents (EPSCs) during expression of LTD than normally, and that this compound also significantly blocks the induction of LTD. Indeed, these data strongly support the view that this change in synaptic efficacy involves a genuine change in the kinetics or desensitization properties of these receptors, or both. Furthermore, and in keeping with the previous observations of Ito and Karachot (1992), these results also suggest that induction of LTD requires an initial agonist-dependent desensitization of AMPA receptors of PCs.

Discussion

In conclusion, LTD induction and expression can be tentatively explained in the following way (see Figure 2): when AMPA receptors of PCs are activated by glutamate released by PFs, their agonist-dependent desensitization leads to some conformational change which, for instance, might expose a putative phosphorylation site to PKC. As, during the pairing protocol, PKC is activated by calcium entry through voltage-gated

calcium channels and by DAG produced by activation of metabotropic receptors at PF–PC synapses, this kinase can link inorganic phosphorus to the desensitized AMPA receptors, thereby phosphorylating them. On the other hand, the NO route ultimately inhibits phosphatases, thus preventing removal of phosphorus from AMPA receptors, thereby allowing their phosphorylated state to be maintained. As a consequence of this phosphorylation, the kinetics involved in opening and closing AMPA-receptor coupled channels are likely to be affected, or a larger fraction of glutamate receptors than normal is stabilized in a desensitized state at rest, or both, thus explaining the maintenance of LTD. The presence of these biochemical cascades of events leading to LTD might well explain that the phenomenon develops rather slowly, within 10 to 15 minutes following the end of the pairing protocol.

Road Map: Mechanisms of Neural Plasticity
Background: Ion Channels: Keys to Neuronal Specialization
Related Reading: Cerebellum and Conditioning

References

Albus, J. S., 1971, A theory of cerebellar function, *Math. Biosci.*, 10: 25–61.
Baude, A., Nusser, Z., Roberts, J. D., Mulvihill, E., McIlhinney, R. A., and Somogyi, P., 1993, The metabotropic glutamate receptor (metabotropicR1a) is concentrated at perisynaptic membrane of neuronal subpopulations as detected by immunogold reaction, *Neuron*, 11:771–787.
Crepel, F., and Audinat, E., 1991, Excitatory amino acid receptors of cerebellar Purkinje cells: Development and plasticity, *Prog. Biophys. Mol. Biol.*, 55:31–46. ◆
Crepel, F., and Jaillard, D., 1990, Protein kinases, nitric oxide and long-term depression of synapses in the cerebellum, *NeuroReport*, 1:133–136.
Daniel, H., Hemart, N., Jaillard, D., and Crepel, F., 1992, Coactivation of metabotropic glutamate receptors and of voltage-gated calcium channels induces long-term depression in cerebellar Purkinje cells in vitro, *Exp. Brain Res.*, 90:327–331.
Daniel, H., Hemart, N., Jaillard, D., and Crepel, F., 1993, Long-term depression requires nitric oxide and guanosine 3′:5′ cyclic monophosphate production in rat cerebellar Purkinje cells, *Eur. J. Neurosci.*, 5:1079–1082.
Eccles, J. C., Ito, M., and Szentágothai, J., 1967, *The Cerebellum as a Neuronal Machine*, Berlin: Springer-Verlag.
Garthwaite, J., Charles, S. L., and Chess-Williams, R., 1988, Endothelium-derived relaxing factor release on activation of NMDA receptors suggests role as intercellular messenger in the brain, *Nature*, 336:385–388.
Ito, M., and Karachot, L., 1992, Protein kinases and phosphatase inhibitors mediating long-term desensitization of glutamate receptors in cerebellar Purkinje cells, *Neurosci. Res.*, 14:27–38.
Ito, M., Sakurai, M., and Tongroach, P., 1982, Climbing fibre induced depression of both mossy fibre responsiveness and glutamate sensitivity of cerebellar Purkinje cells, *J. Physiol. (Lond.)*, 324:113–134.
Konnerth, A., Dreesen, J., and Augustine, G. J., 1992, Brief dendritic calcium signals initiate long-lasting synaptic depression in cerebellar Purkinje cells, *Proc. Natl. Acad. Sci. USA*, 89:7051–7055.
Linden, D. J., and Connor, J. A., 1991, Participation of postsynaptic PKC in cerebellar long-term depression in culture, *Science*, 254: 1656–1659.
Linden, D. J., Dickinson, M. H., Smeyne, M., and Connor, J. A., 1991, A long term depression of AMPA currents in cultured cerebellar Purkinje neurons, *Neuron*, 7:81–89.
Marr, D., 1969, A theory of cerebellar cortex, *J. Physiol.*, 202:437–470.
Shibuki, K., and Okada, D., 1991, Endogenous nitric oxide release required for long-term synaptic depression in the cerebellum, *Nature*, 349:326–328.
Vyklicky, L., Patneau, D. K., and Mayer, M. L., 1991, Modulation of excitatory synaptic transmission by drugs that reduce desensitization at AMPA/kaïnate receptors, *Neuron*, 7:971–984.

Markov Random Field Models in Image Processing

Anand Rangarajan and Rama Chellappa

Introduction

Markov random field models have become useful in several areas of image processing. The success of Markov random fields (MRFs) can be attributed to the fact that they give rise to good, flexible, stochastic image models. The goal of image modeling is to find an adequate representation of the intensity distribution of a given image. What is adequate often depends on the task at hand, and MRF image models have been versatile enough to be applied in the areas of image and texture synthesis (Chellappa and Kashyap, 1985), image compression, restoration (Geman and Geman, 1984), tomographic reconstruction (Geman and Graffigne, 1987), image and texture segmentation (Rangarajan, Chellappa, and Manjunath, 1991), texture classification, (Derin and Elliott, 1987), and surface reconstruction (Geiger and Girosi, 1991). Our aim is to highlight the central ideas of this field using illustrative examples and to provide pointers to the many applications.

A guiding insight underlying most of the work on MRFs in image processing is that the information contained in the local, physical structure of images is sufficient to obtain a good, global image representation. This notion is captured by means of a local, *conditional* probability distribution. Here, the image intensity at a particular location depends only on a *neighborhood* of pixels. The conditional distribution is called an MRF. For example, a typical MRF model assumes that the image is locally smooth except for relatively few intensity gradient discontinuities corresponding to region boundaries or edges. The MRF image models are defined on the image intensities and on a further set of *hidden* attributes (edges, texture, and region labels). The observed quantities are usually noisy, blurred images, feature vectors, or projection data in the case of emission tomography. The intensity image underlying the observations is needed in applications like restoration and tomographic reconstruction, whereas region, boundary, and texture labels are sought in applications like texture segmentation.

Once the local, conditional probability distribution of the MRF is specified, there are five remaining steps involved. First, the joint distribution of the MRF is obtained. In this way, the image is represented in one global, joint probability distribution. Next, the process by which the observations are generated from the image is captured in a *degradation* probability distribution. In image restoration, for example, the degradation corresponds to a (typically uniform) blur. Then, Bayes's theorem is invoked to obtain the posterior probability distribution of the image given the observations. The posterior distribution gives us the probability that an image (with smooth regions and sharp region boundaries, for example) could have been degraded to obtain the particular observed noisy, blurred image. Once the posterior probability distribution is obtained, we can associate a cost with each configuration in the posterior. For example, if only the true underlying image will do, the cost penalizes all other images equally. The cost is formulated, keeping in mind the task at hand. A measure of the cost is minimized with respect to the image intensities (in image recovery tasks) or image attributes (in labeling tasks). Finally, since the MRFs are specified with model parameters, these are estimated from a training set (if one exists) or adaptively along with the cost minimization phase alluded to earlier. The overall MRF framework fits well within a Bayesian estimation/inference paradigm. In the next section, we step through all five phases of MRF modeling.

A Framework for Estimation and Inference

As mentioned in the Introduction, MRF image models represent knowledge in terms of "local" probability distributions. Specifically, the kinds of probability distributions generated by MRFs have a local neighborhood structure. Neighborhood systems commonly used by MRFs are depicted in Figure 1*A*.

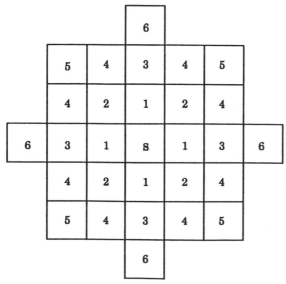

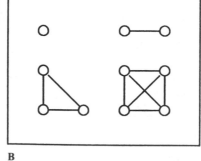

A   B

Figure 1. *A*, Neighborhood systems for MRFs. *B*, Cliques in MRFs.

In our exposition, we have adopted much of the notation found in Geman and Graffigne (1987).

Let us associate an image with a random process X whose typical element is X_s, where $s \in S$ refers to a site in the image. The local conditional distribution can be written as follows:

$$\Pr(X_s = x_s | X_t = x_t, t \neq s, t \in S) = \Pr(X_s = x_s | X_t = x_t, t \in G_s) \tag{1}$$

where X and x denote the random field and a particular realization, respectively, and G_s is the local neighborhood at site s. Note that in general G_s can be large or small, but it is usually a local neighborhood in keeping with the spirit of MRF modeling.

Let s be the site (i, j), and let the local neighborhood be a first-order neighborhood ($G(s)$ is the collection $(i, j+1)$, $(i, j-1)$, $(i+1, j)$, $(i-1, j)$). Then, let the conditional density take the form

$$p(X_s = x_s | X_t, t \in G_s) = \frac{1}{\sqrt{2\pi}} \exp\left[-\frac{1}{2}\left(x_{ij} - \frac{1}{4}[x_{i, j+1} + x_{i, j-1} + x_{i+1, j} + x_{i-1, j}] \right)^2 \right] \tag{2}$$

This is a very simple, special case of the first-order Gauss-Markov model (Woods, 1972; Besag, 1974). The Gauss-Markov model has been widely used in image processing tasks (Dubes and Jain, 1989).

The MRF model consists of a set of *cliques*. A clique is a collection of sites such that any two sites are neighbors. Different orders of cliques are shown in Figure 1B. The *order* of a clique refers to the number of distinct sites that appear multiplicatively. We now calculate the clique energies involving the site x_{ij} by expanding the conditional probability density and collecting the terms. There are cliques of order one and two. They are

$$\frac{x_{ij}^2}{2} \qquad -\frac{x_{ij} x_{i, j+1}}{4} \quad \text{and} \quad -\frac{x_{ij} x_{i+1, j}}{4} \tag{3}$$

The first term in Equation 3 is of order one, and the latter two terms are of order two.

MRF-Gibbs Equivalence

We now ask the following question: given the conditional probability structure $\Pr(X_s = x_s | X_t = x_t, t \in G_s)$, what is the joint probability distribution $\Pr(X = x)$? This is of utmost importance since it is the joint probability distribution and not the conditional distribution that contains the complete image representation.

Before relating the conditional and joint distributions, we introduce the concept of a Gibbs distribution, which will turn out to be crucial in specifying the relationship. A Gibbs distribution is specified by an *energy function* $E(x)$ and can be written as

$$\Pr(X = x) = \frac{1}{Z} \exp(-E(x)) \tag{4}$$

where the *partition function*

$$Z = \sum_x \exp(-E(x)) \tag{5}$$

is a normalizing constant and involves a summation over all possible configurations of X. Energy functions have been widely used in spin-glass models of statistical physics. The min-imum energy configuration corresponds to an ordered system of spins. $E(x)$ cannot take infinite values.

Our detour into Gibbs distributions is justified for the following reason. The Hammersley-Clifford theorem (Besag, 1974; Geman and Geman, 1984) states that any conditional distribution has a joint distribution which is Gibbs (Kinderman and Snell, 1980; Dubes and Jain, 1989) if the following conditions hold.

1. *Positivity*: $\Pr(X = x) > 0$.
2. *Locality*: $\Pr(X_s = x_s | X_t = x_t, t \neq s, t \in S) = \Pr(X_s = x_s | X_t = x_t, t \in G_s)$.
3. *Homogeneity*: $\Pr(X_s = x_s | X_t = x_t, t \in G_s)$ is the same for all sites s.

The locality condition is the same as the Markov property (Equation 1). The Hammersley-Clifford theorem allows us to shuttle between the conditional probability structure in Equation 1 and the joint probability in Equation 4.

The recipe for obtaining the joint density function is as follows: (1) assemble the different clique energies from the conditional probability, and (2) compute the energy function by adding up the clique energies.

We calculate the energy function for the simple first-order Gauss-Markov model:

$$E(x) = \frac{1}{2}\left(\sum_{ij} \left[x_{ij}^2 - \frac{x_{ij} x_{i, j+1}}{2} - \frac{x_{ij} x_{i+1, j}}{2} \right] \right)$$
$$= \frac{1}{8} \sum_{ij} \left[(x_{ij} - x_{i, j+1})^2 + (x_{ij} - x_{i+1, j})^2 \right] \tag{6}$$

It can be seen from the energy function $E(x)$ and the conditional density that the essence of the Hammersley-Clifford theorem lies in the clique energies. We examined the conditional density and teased apart the different orders of cliques (first and second order) and the associated clique energies. Then, all clique energies were summed (taking care to count each clique only once), yielding the energy function $E(x)$. Our presentation has been quite terse, and further details on cliques and the transition from the conditional to the joint probability distribution can be found in other studies (Besag, 1974; Geman and Geman, 1984; Kinderman and Snell, 1980).

The Prior and Degradation Models

Naturally, we are not content with merely obtaining MRF-Gibbs image models. These models can be used in a variety of image processing and pattern recognition tasks. As mentioned previously, MRF modeling fits perfectly into a Bayesian estimation/inference paradigm. A Bayesian setup consists of two ingredients—the prior and the degradation model. The prior model is defined on the set of image attributes X that are of interest. In edge-preserving image restoration for example, X includes the set of image intensities and a further set of binary-valued edge labels. In texture segmentation, X includes the image intensities and a set of texture labels at each location. The degradation model is a model of the physical process by which the observations are generated. Usually, we are faced with noisy and incomplete observations. Denote the set of observations by Y, and let the degradation model also be a Gibbs-Markov distribution:

$$\Pr(Y = y | X = x) = \frac{1}{Z_D(x)} \exp(-E_D(x, y)) \tag{7}$$

where

$$Z_D(x) = \sum_y \exp(-E_D(x, y)) \qquad (8)$$

In general, the partition function $Z_D(x)$ is a function of the image attributes x. $E_D(x, y)$ is the energy function corresponding to the degradation model. For example,

$$E_D(x, y) = \frac{1}{2} \sum_s \left(y_s - \sum_t h_{st} x_t \right)^2$$

yields a Gaussian degradation model wherein Y is obtained by blurring X with a *point spread function h* and adding additive Gaussian noise at each site s. This type of degradation model routinely occurs in image restoration and tomographic reconstruction.

A Bayesian Posterior Energy Function

Given the degradation and prior models, Bayesian estimation/ inference proceeds as follows. The posterior distribution $\Pr(X = x \mid Y = y)$ is obtained by using Bayes's theorem:

$$\Pr(X = x \mid Y = y) = \frac{\Pr(Y = y \mid X = x) \Pr(X = x)}{\Pr(Y = y)} \qquad (9)$$

Once the posterior distribution is obtained, an estimate ($\hat{X}$) of X is found by minimizing the expected cost, which is a measure of distance between the true and estimated values.

$$C = \sum_x C(x, x^*) \Pr(X = x \mid Y = y) \qquad (10)$$

where x^* is the true value. When the familiar squared-error cost is used, the *minimum mean square error* (MMSE) estimator turns out to be the conditional mean $\mathscr{E}(X \mid Y = y)$ ($\mathscr{E}$ denotes the expectation operator). If the cost penalizes all x different from x^* ($C(x, x^*) = \delta_{x, x^*}$), the *maximum a posteriori (MAP)* estimator results.

When the degradation and prior models are Gibbs, the posterior is Gibbs as well. To see this, assume a prior energy function $E_P(x)$ giving $\Pr(X = x) = (1/Z_P) \exp(-E_P(x))$. The posterior distribution (using Equation 9) is

$$\Pr(X = x \mid Y = y) = \frac{\exp(-E_D(x, y) - \log(Z_D(x)) - E_P(x))}{\sum_x \exp(-E_D(x, y) - \log(Z_D(x)) - E_P(x))} \qquad (11)$$

The posterior energy function $E(x) = E_D(x, y) + \log(Z_D(x)) + E_P(x)$. In the case of the MAP estimate, the entire Bayesian estimation engine reduces to minimizing just this posterior energy function $E(x)$, since the partition function of the posterior is independent of x. However, when the MMSE estimate is desired, the expected value of X in the posterior distribution needs to be computed. This computation is usually intractable, since it involves computing the partition function of the posterior distribution.

MAP Estimation

Restricting our focus to MAP estimation, we observe that MAP estimation reduces to minimizing the posterior energy function $E(x)$. This minimization involves the different kinds of processes which make up X. For example, in edge-preserving image restoration (Geman and Geman, 1984), the process X includes both continuous image intensities and binary-valued edge variables. Consequently, the minimization of the posterior objective function is a difficult problem due to the presence of nontrivial local minima. A general technique for finding

global minima is SIMULATED ANNEALING (q.v.) but it is usually computationally very intensive. Recently, a lot of effort has been expended in obtaining good suboptimal solutions to the MAP estimation problem (Geiger and Girosi, 1991; Lee et al., 1993). Deterministic annealing is a general method that has emerged recently. Deterministic annealing methods begin with a modified posterior:

$$\Pr(X = x \mid Y = y) = \frac{1}{Z(\beta)} \exp(-\beta E(x)) \qquad (12)$$

where $\beta > 0$ is the inverse temperature. Note that the partition function is now a function of the inverse temperature. The terminology is inherited from statistical mechanics (see STATISTICAL MECHANICS OF NEURAL NETWORKS). The idea of cooling a system slowly to reach a minimum energy configuration has a computational parallel in MRFs. The basic idea is to embed the posterior in a β exponentiated manner and to track the maximum of this posterior through gradual increase of β. In this manner, the posterior energy function is increasingly closely approximated by a sequence of smooth, continuous energy functions.

The main reason for doing this is based on the following statistical mechanics identity:

$$F(\beta) \stackrel{\text{def}}{=} -\frac{1}{\beta} \log Z(\beta) = \mathscr{E}(E(x)) - \frac{1}{\beta} S(\beta) \qquad (13)$$

where S is the entropy (defined as $-\Sigma_x \Pr(X = x \mid Y = y) \times \log(\Pr(X = x \mid Y = y))$). The entropy is proportional to the logarithm of the total number of configurations, and as the temperature is reduced (and fewer configurations become likely), it gradually goes to zero. Also, the expected value of the posterior energy goes to the minimum value of the energy. The key idea in deterministic annealing is to minimize the *free energy F* instead of $E(x)$ while reducing the temperature to zero. The free energy (at low β) is a smooth approximation to the original nonconvex energy function and approaches $E(x)$ as β tends to infinity. However, the free energy involves the logarithm of the partition function, which is intractable. An approximation to the free energy (usually called the naive mean field approximation) is minimized instead. While details are beyond the scope of this presentation (see Geiger and Girosi, 1991; Lee et al., 1993; and CONSTRAINED OPTIMIZATION AND THE ELASTIC NET), we present an example illustrating the method. Let the energy function contain only binary-valued variables and take the following form:

$$E(x) = \sum_{ij} T_{ij} x_i x_j + \sum_i h_i x_i \qquad x_i \in \{0, 1\} \qquad (14)$$

The free energy F is given by

$$F(v) = \sum_{ij} T_{ij} v_i v_j + \sum_i h_i v_i$$
$$+ \frac{1}{\beta} \sum_i [v_i \log(v_i) + (1 - v_i) \log(1 - v_i)] \qquad (15)$$

where $v_i \in [0, 1]$. The free energy consists of two terms. The first term can be seen as an approximation to the expected value of the energy once the identification $v_i \approx \mathscr{E}(x_i)$ is made. Now

$$\mathscr{E}(E(x)) = \sum_{ij} T_{ij} \mathscr{E}(x_i x_j) + \sum_i h_i \mathscr{E}(x_i) \qquad (16)$$

When the expected value of the product $x_i x_j$ is replaced by the product of the expected values ($v_i v_j$), the naive mean field approximation results. The third term in Equation 15 is an approximation to the entropy. At each setting of β, Equation 15 is minimized with respect to v, after which β is increased. In this manner, a deterministic relaxation network is obtained. There are questions regarding the choice of annealing schedules and the quality of the minima obtained, etc., and for the most part, except for very specific posterior energy functions, there is a dearth of analytical results in this area. However, the method is quite general and has been applied with varying degrees of success in a variety of image processing tasks like restoration, tomographic reconstruction, flow field segmentation, and surface reconstruction.

Parameter Estimation

So far we have concentrated on estimating X given the noisy observations Y. We have emphasized that Gibbs-Markov models are specified by local clique energies (from which the global distribution can be obtained). Consider a prior distribution

$$\Pr(X = x \mid \theta) = \frac{1}{Z(\theta)} \exp\left(-\frac{1}{2} \sum_k \sum_{\langle s,t \rangle_k} \theta_k (x_s - x_t)^2 \right) \quad (17)$$

where θ_k is a parameter associated with clique $\langle s, t \rangle$. Since pairwise interactions are used, a clique between pixels s and t is denoted by $\langle s, t \rangle$. This is the general form of the Gauss-Markov model. The model is a generalization of our earlier model (Equation 6) since it has the same clique form, albeit with a more general neighborhood structure. The partition function involves a sum over the configurations of X and is a function of θ. In addition to the estimation/inference problem, we are also saddled with the problem of parameter estimation.

The parameters can be estimated by maximizing the joint probability of X with respect to the unknown parameters. In most cases, this computation is intractable in its pure form and approximations have to be devised. An interesting alternative is to maximize the "pseudo-likelihood" with respect to the unknown parameters (Besag, 1977). The pseudo-likelihood takes advantage of the local conditional probability structure of MRFs. The parameters are now estimated by maximizing the product of the conditional distributions at each site s with respect to the parameters. The availability of a suitable training set is critical to both likelihood and pseudo-likelihood parameter estimation. When a training set is not available, parameter estimation and cost minimization proceed in lockstep. There are also issues of consistency and efficiency of these estimates, and more details can be found in Kashyap and Chellappa (1983).

Discussion

In sum, the MRF framework is well suited to a wide variety of image processing problems. Our exposition has been brief, and we have ignored important issues like validation, choice of the order of MRF models, and size of training sets. Validation, for example, takes us into the bias/variance dilemma (Geman, Bienenstock, and Doursat, 1992). MRF models, being parametric, introduce a certain kind of bias into the image representation. This seems to be the right kind of bias (in terms of reducing variance) for tasks like image restoration, tomographic reconstruction, and texture segmentation. However, if the order of the chosen model is incorrect, high bias could result. It is in bias/variance terms that MRF image models should be compared with "mechanical" (as opposed to probabilistic) models

like splines, generic representations like radial basis functions, and *tabula rasa*, feedforward neural networks. Also, there are interesting similarities between Gauss-Markov models and thin-plate splines (Wahba, 1990; see also GENERALIZATION AND REGULARIZATION IN NONLINEAR LEARNING SYSTEMS). For example, the simple case of the first-order Gauss-Markov model with the parameters $\theta_1 = \theta_2 = \frac{1}{4}$ is identical to the discrete membrane (first-order thin-plate spline in two dimensions). Correspondences of this sort should be expected since MRF models, splines, and RBFs impose local smoothness constraints, albeit in different ways. Finally, there are interesting relationships between MRFs and recurrent neural networks at both computational and algorithmic levels (Rangarajan et al., 1991).

Road Map: Cooperative Phenomena
Related Reading: Cellular Automata; Regularization Theory and Low-Level Vision; Textured Images: Modeling and Segmentation

References

Besag, J., 1974, Spatial interaction and the statistical analysis of lattice systems, *J. R. Statist. Soc. Ser. B*, 36:192–236. ◆

Besag, J., 1977, Efficiency of pseudo-likelihood estimation for simple Gaussian fields, *Biometrika*, 64:616–618.

Chellappa, R., and Kashyap, R. L., 1985, Texture synthesis using spatial interaction models, *IEEE Trans. Acoust. Speech Signal. Process*, 33:194–203.

Derin, H., and Elliott, H., 1987, Modeling and segmentation of noisy and textured images using Gibbs random fields, *IEEE Trans. Pattern Analysis Machine Intell.*, 9:39–55.

Dubes, R. C., and Jain, A. K., 1989, Random field models in image analysis, *J. Appl. Stat.*, 16:131–164. ◆

Geiger, D., and Girosi, F., 1991, Parallel and deterministic algorithms from MRFs: Surface reconstruction, *IEEE Trans. Pattern Analysis Machine Intell.*, 13:401–412.

Geman, S., Bienenstock, E., and Doursat, R., 1992, Neural networks and the bias/variance dilemma, *Neural Computat.*, 4:1–58.

Geman, S., and Geman, D., 1984, Stochastic relaxation, Gibbs distributions and the Bayesian restoration of images, *IEEE Trans. Pattern Analysis Machine Intell.*, 6:721–741.

Geman, S., and Graffigne, C., 1987, Markov random fields image models and their application to computer vision, in *Proceedings of the International Congress of Mathematicians, 1986* (A. M. Gleason, Ed.), Providence, RI: American Mathematical Society. ◆

Kashyap, R. L., and Chellappa, R., 1983, Estimation and choice of neighbors in spatial interaction models of images, *IEEE Trans. Inform. Theory*, 29:60–72.

Kinderman, R., and Snell, J. L., 1980, *Markov Random Fields and Their Applications*, Providence, RI: American Mathematical Society. ◆

Lee, M., Rangarajan, A., Zubal, I. G., and Gindi, G., 1993, A continuation method for emission tomography, *IEEE Trans. Nuclear Sci.*, 40:2049–2058.

Rangarajan, A., Chellappa, R., and Manjunath, B. S., 1991, Markov random fields and neural networks with applications to early vision, in *Artificial Neural Networks and Statistical Pattern Recognition: Old and New Connections* (I. K. Sethi and A. K. Jain, Eds.), New York: Elsevier Science. ◆

Wahba, G., 1990, *Spline Models for Observational Data*, CBMS-NSF Regional Conference Series in Applied Mathematics, vol. 59, Philadelphia, PA: SIAM.

Woods, J. W., 1972, Two-dimensional discrete Markovian fields, *IEEE Trans. Inform. Theory*, 18:101–109.

Memory-Based Reasoning

David L. Waltz

Introduction

Memory-Based Reasoning (MBR; see Waltz, 1990) refers to a family of nearest-neighbor-like methods (Dasarthy, 1991) for making decisions or classifications. Memory-based reasoning differs from other nearest-neighbor methods primarily in its metrics for computing the distance between examples, and in MBR's suitability for use with symbolic-valued features. Nearest-neighbor methods generally use a simple overlap distance metric, which defines distance as the number of mismatched features between two instances. MBR uses metrics related to the Value Distance Metric (VDM), introduced in Stanfill and Waltz (1986). Considerable effort on representing training cases is generally required in order to use neural nets on symbolic problems. MBR is specifically designed to handle such cases.

How Does MBR Work?

Every MBR system requires a similarity metric for judging the distance between an item to be classified and all items in the example database. If the MBR system uses k nearest neighbors, then a scheme for combining the information from k examples is also needed. We can illustrate the idea with a simple metric, the Modified Value Difference Metric (MVDM) (Cost and Salzberg, 1993), which is similar to the ones that have been most commonly used in MBR applications. Assume the database is relational with n-tuples consisting of $n-1$ predictor fields $(p_1, p_2, \ldots, p_{n-1})$ plus a goal field G. (This is not the fully general case; but it is the simplest and most common one, and serves well for purposes of illustration.) Each case in the database is a vector of predictor values $(a_1, a_2, \ldots, a_{n-1})$, plus a goal which can take on any of a finite set of values. The goal field G is an element of $\{g_1, g_2, \ldots, g_m\}$. Then the distance between a novel situation $B = (b_1, b_2, \ldots, b_{n-1})$ and a case A from the database, $A = (a_1, a_2, \ldots, a_{n-1})\}$, is

$$\sum_j \left[\sum_i |\text{Prob}(g_j|b_i) - \text{Prob}(g_j|a_i)| \right]$$

where j ranges over all possible values for the goal field, and i ranges over all predictor fields. This metric basically compares the distribution of cases for each pair of field values of A and B. The distance between A and B is small if the distributions of classifications associated with each of the field values are similar.

To keep computation tractable, the set of goal fields may be limited by, for example, only indexing over goals that correspond to database instances where at least one element of the predictor field exactly matches the corresponding element of the item to be classified. Such metrics are called *non-zero overlap metrics*.

The result of applying an MBR metric is a rank-ordered list of cases from the database, with distance values for each. The classification proposed for the novel situation is then the goal field value of the case with the minimum distance (single-neighbor case), or the goal field value whose weighted sum of distances over the k closest cases is the smallest (k-nearest-neighbor case).

Comparisons of MBR and Neural Nets

Several projects (described in more detail under following subheadings) have compared MBR with backpropagation neural nets. In terms of decision accuracy, MBR has often outperformed neural nets, and in some cases all the other learning methods with which it has been compared (Zhang, Mesirov, and Waltz, 1992; Cost and Salzberg, 1993; Rachlin et al., 1994); results for MBR are generally comparable with the best other learning methods. Memory-based reasoning has also been applied to cases which are beyond the representational reach of current neural net methods, such as classification of free-text examples of arbitrary lengths (Creecy et al., 1992; Linoff, Masand, and Waltz, 1992) or object recognition. Like neural nets, MBR systems can be easily and quickly built, with very little programming required. Unlike neural nets, no learning phase is necessary. Updating is therefore simple, requiring only additions, deletions, and modifications to MBR's database of examples, and decisions are always based on all known data, even the most recent. Also unlike neural nets, MBR does not generally require elaborate re-representation schemes. Memory-based reasoning provides "justifications" for decisions, namely the nearest example(s) from the database, whereas considerable analysis effort is generally required to understand why a neural net system behaves as it does. Thus debugging and tuning is far easier with MBR. And finally, MBR methods can provide confidence levels.

Memory-based reasoning does have a major disadvantage: while it does not require a training phase, MBR's decision phase is computationally expensive, and MBR systems have high memory requirements—the entire database of examples is kept—whereas with neural nets, the total storage requirements for a trained net are generally much smaller than the training set used to create it. As a result, MBR systems have typically been implemented on massively parallel computers or on special-purpose hardware (Atkeson, 1990). Moreover, neural nets can extract features to form compact descriptions of decision surfaces; MBR cannot.

NETtalk Task

The first MBR system, MBRtalk (Stanfill and Waltz, 1986), used as its main example the NETtalk database, and compared its performance with NETtalk (Sejnowski and Rosenberg, 1987). The NETtalk task is to produce pronunciations for all English words, based on a small (700-word) training set. For this task, NETtalk was trained with approximately 4000 seven-letter-wide examples, generated by moving the window across the 700-word training sample. The goal was to pronounce the central letter of the window. NETtalk, a backpropagation system (with some special output processing), achieved a 78% letter-by-letter generalization performance (i.e., 78% correct pronunciations of the central letter in a seven-letter-wide window from a novel word). Mooney et al. (1989) attempted to reproduce these results, but with a somewhat different training set (808 words) and a much larger test set; they achieved a correctness of 60.7%. MBRtalk produced a 78% letter-by-letter generalization performance on the original NETtalk database, the same as that reported by Sejnowski and Rosenberg (1987). MBRtalk was able to demonstrate 93% correct generalization after training with a 16,000-word corpus, a task that has not been attempted with a neural net system, and one that would probably require very long training times. Memory-based reasoning does not require training, though it is possible to do some precalculation in order to make the MBR decision phase run faster.

Protein Structure Prediction

For several years, the best protein secondary structure prediction systems were based on neural nets (Qian and Sejnowski, 1988). After considerable experimentation and tuning, Zhang et al. (1992) showed that MBR outperforms backpropagation on this task, albeit by a very small margin—64.5% correct for MBR versus 63.5% for a three-layer backpropagation system, and 64.0% for a cascaded backpropagation system (Qian and Sejnowski, 1988), all using eight-way cross-validation on the same 19,861-residue database. Interestingly, MBR and neural nets agreed with each other only about 80% of the time, and both methods agreed with a statistical system (with 63.5% performance) about as often. The outputs of the three methods were combined in a hybrid system using a backpropagation net, yielding a performance of 66.4%—a small improvement that is nonetheless better than any of the others alone with high statistical significance.

Tests on UCI Repository of Machine-Learning Databases

Several papers have reported results on data sets from the University of California Irvine (UCI) repository that allow us to compare MBR, specifically the PEBLS system (Cost and Salzberg, 1993), with backpropagation neural nets. (See Weiss and Kulikowski, 1991, for a summary of backpropagation performance on these data sets, along with comparison with learning methods.) The results reported below are from different papers that did not directly compare these two methods, so it is not absolutely certain that the results can be meaningfully compared.

The soybean disease database is a small one, with 50 features, 289 examples, and 17 categories. Backpropagation net performance was about 94.5% on this dataset (Mooney et al., 1989), and 93.2% for MBR (Rachlin et al., 1994).

The iris database (consisting of features and classifications of iris species) is even smaller, with 4 features, 10 values per feature, 3 classes, and 150 examples. Using the leave-one-out training method, the best backpropagation net reported in Weiss and Kapouleas (1989) achieved 96.7% correct, while PEBLS achieved 95.3%.

MBR as a Neural Model?

It has been suggested that the cerebellum stores many examples of motor movements that can then be interpolated to provide smooth motor movements (Albus, 1981; Atkeson, 1990). Memory-based reasoning, as a variant of case-based reasoning (CBR), has been proposed as an associative memory implementation. A number of researchers have argued for associative memory as a general model of the cortex. Memory-based reasoning—using either one cell per record with a winner-take-all network, or an n-nearest-neighbor memory plus a combining procedure for making final decisions—naturally implements associative memory.

Discussion

In general, MBR offers accuracies comparable to neural networks for classification tasks, at the price of greater computational and memory costs for each decision, but with no need for a separate learning or update phase. Memory-based reasoning can handle some problems that have not proved amenable to neural net solutions to date (see following paragraphs). On the other hand, MBR methods, unlike neural nets, do not extract features that can be used to produce compact descriptions of a task domain.

While their overall accuracies seem generally similar, MBR and neural nets form decision surfaces differently, and so will perform differently on particular examples. The decision surfaces of MBR for single-nearest-neighbor systems are jagged piecewise linear line segments that lie equidistantly between pairs of close but differently classified cases in the database. These surfaces become smoother for n-nearest-neighbor schemes. Memory-based reasoning can become arbitrarily accurate if large numbers of cases are available, and if these cases are well-behaved and properly categorized. Neural nets cannot respond well to isolated cases, but tend to be good at smooth extrapolation. To give an example, consider training a NETtalk system to pronounce the letter p and assume that all p's are either pronounced P (as in pig) or F (as in $photo$) except for one example where p is silent ($psychology$). Given that there are many P and F examples, these will statistically dominate the hidden units for a backpropagation net, and it is very unlikely that words beginning with ps will be correctly pronounced. But MBR is able to pronounce a ps word correctly, even with only a single near example.

In some cases MBR can be used for tasks outside the competence areas of neural nets using presently known training methods. In Linoff, Masand, and Waltz (1992), for example, a text-similarity metric ("vector similarity"—basically normalized weighted word overlap) allowed an MBR system to assign key words to news articles, based on a database of previously hand-labeled news stories. Each news story was simply a body of text with one or more human-generated key words attached. In effect, the input space was n-dimensional vector with binary inputs, where n is the number of distinct words or word pairs in the overall database (n was about 200,000 in this prototype system). The goal was a set of five to ten key words, selected from a total of about 400 possible key words. Such a task is beyond current neural net training methods, for which the encoding of free-text input presents a difficult challenge.

Road Map: Artificial Intelligence and Neural Networks

References

Albus, J., 1981, *Brain, Behavior, and Robotics*, New York: BYTE Books.

Atkeson, C., 1990, Using local models to control movement, in *Advances in Neural Information Processing Systems 2* (D. S. Touretzky, Ed.), San Mateo, CA: Morgan Kaufmann, pp. 316–323.

Cost, S., and Salzberg, S., 1993, A weighted nearest neighbor algorithm for learning with symbolic features, *Machine Learn.*, 10:57–78. ◆

Creecy, R., Masand, B., Smith, S., and Waltz, D., 1992, Trading MIPS and memory for knowledge engineering, *Commun. ACM*, 35:48–64.

Dasarthy, B., 1991, *Nearest Neighbor (NN) Norms*, Washington, DC: IEEE Computer Society Press.

Linoff, G., Masand, B., and Waltz, D., 1992, Classifying news stories using memory based reasoning, in *Proceedings of the SIGIR Conference*, New York: ACM, pp. 59–65.

Mooney, R., Shavlik, J., Towell, G., and Gove, A., 1989, An experimental comparison of symbolic and connectionist learning algorithms, in *Proceedings of the Eleventh International Joint Conference on Artificial Intelligence*, San Mateo, CA: Morgan Kaufmann, pp. 775–780.

Qian, N., and Sejnowski, T., 1988, Predicting the secondary structure of globular proteins using neural network models, *J. Mol. Biol.*, 202: 865–884.

Rachlin, J., Kasif, S., Salzberg, S., and Aha, D., 1994, Towards a better understanding of memory-based and Bayesian classifiers, in *Proceedings of the Eleventh International Conference on Machine Learning*, New Brunswick, NJ, pp. 242–250. ◆

Sejnowski, T., and Rosenberg, C., 1987, Parallel networks that learn to pronounce English text, *Complex Systems*, 1:145–168.

Stanfill, C., and Waltz, D., 1986, Toward memory-based reasoning, *Commun. ACM*, 29:1213–1228. ◆

Waltz, D. 1990, Memory-based reasoning, in *Natural and Artificial Parallel Computation* (M. Arbib and J. Robinson, Eds.), Cambridge, MA: MIT Press, pp. 251–276. ◆

Weiss, S. M., and Kapouleas, I., 1989, An empirical comparison of pattern recognition, neural nets, and machine learning classification methods, in *Proceedings of the Eleventh International Joint Conference on Artificial Intelligence*, San Mateo, CA: Morgan Kaufmann, pp. 781–787.

Weiss, S., and Kulikowski, C., 1991, *Computer Systems That Learn*, San Mateo, CA: Morgan Kaufmann. ◆

Zhang, X., Mesirov, J., and Waltz, D., 1992, A hybrid method for protein secondary structure prediction, *J. Mol. Biol.*, 225:1049–1063.

Mental Arithmetic Using Neural Networks

James A. Anderson

Introduction

Sometime during the first years of elementary school, students are supposed to learn a few hundred arithmetic facts along with a few simple algorithms. Many students find the facts and algorithms difficult to learn and easy to forget. Even adults under mild time pressure make over 7% errors on elementary arithmetic facts (Graham, 1987). The more complex algorithmic parts of arithmetic build on a foundation of learned information and elementary operations, but they also display extensions which bear directly on the most complex aspects of human cognition. Arithmetic forms a well-defined and well-circumscribed model of parts of higher-level cognition.

Neural Network Arithmetic Models

Neural networks often act as associators (see ASSOCIATIVE NETWORKS), coupling an input pattern with an output pattern. The major problem with association as a mechanism for arithmetic learning is that the structure of elementary arithmetic lends itself to severe *associative interference*. For example, a set of simple facts such as $6 \times 3 = 18$, $6 \times 4 = 24$, $6 \times 5 = 30$, and so on, has a multiplicand "6" associatively linked to several different products. Many errors in human performance can be shown to be of this type, and associative models do a good job at duplicating human error patterns (Campbell, 1987).

Traditional arithmetic algorithms are associative in that one step follows another. The earliest neural network models for arithmetic used this pattern. Rumelhart, Smolensky, McClelland, and Hinton (1986) used three-digit multiplication as an illustration of how to generate an iterative serial process using an associative neural network. The physical appearance of the problem in standard form on the paper or blackboard triggers the first partial multiply, which is then physically written down. This new pattern triggers the next step, until the problem is completed. This associative model for arithmetic is compatible with classic S → R (Stimulus → Response) association. An abstract algorithm is realized by a series of associatively linked steps with the neural networks used to provide a linkage mechanism. An external medium is used to write partial results.

Cottrell and T'sung (1991) developed a more detailed neural network of arithmetic simulation based on association. Their network simultaneously learned both the arithmetic facts themselves and the procedural operations on them. The output representation contained an *action* field which determined the next step: carry, write, shift to the next column, or finish. The arithmetic computation continued until the network signaled that it was finished, so indefinite length numbers could be added. The network used a version of recurrent backpropagation, a modification of backpropagation that incorporates previous states into the current input state (see RECURRENT NETWORKS: SUPERVISED LEARNING). The simulation was restricted to base 4 arithmetic. Even so, learning times were long, requiring thousands of presentations to achieve acceptable accuracy.

Number Representations

Proper data representation is usually the most critical part of neural network design. The three best developed connectionist models for arithmetic fact learning use similar representations for number. Entities such as number are psychologically much more complex than they first appear to be. Suppose a subject is asked whether 85 is bigger than 86, or whether 85 is bigger than 14. The response time is long when the numbers compared are close together and short when the numbers compared are far apart (Link, 1990). Another place where such response time patterns are seen is in comparisons of *sensory* magnitudes, that is, when a subject has to decide which of two weights is heavier, or which of two lights is brighter. Response times are longer if the two lights are of nearly equal intensity. This pattern suggests that the representation of number contains a part which acts like a weight or a light intensity. The conclusion is that humans use a rich coding of number which contains powerful sensory components, and there is a sensory magnitude attached to what might be considered an abstract quantity.

The importance of "sensory" components in mathematics in general comes as no surprise to those who have studied the thought processes of mathematicians (Davis and Anderson, 1979). In *The Psychology of Invention in the Mathematical Field* (1945), Jacques Hadamard, a famous mathematician himself, interviewed his peers about how they did mathematics. Most of those he talked to said they did not reason abstractly, but instead used visualization, or kinesthetic imagery with imagined muscle motions, or even auditory imagery. Language-based,

Bar Code Magnitude Representation

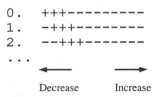

```
0.    +++---------
1.    -+++--------
2.    --+++-------
...
```

Decrease   Increase

Hybrid Representation

```
0.    Zero+++---------
1.    One -+++--------
2.    Two --+++-------
...
```

"symbolic"   "analog"

Figure 1. Representation for number used by Dallaway and by McCloskey and Lindemann (top) and by Anderson, Bennett, and Spoehr (bottom). The "spelled-out" letters in the hybrid representation indicate an essentially random vector, different for each number.

and formal, abstract reasoning were conspicuous by their rarity.

One way to incorporate a "sensory" part into the representation of number is to construct a topographic map of number magnitude, a technique inspired by the multitude of topographic maps found in cerebral cortex. A moving "bar" of activation on a topographic scale formed of individual elements would code the magnitude of the number, just as location of activity on the cortical surface roughly codes visual position or frequency of a sound (Figure 1). This representation seems to capture at least a few of the similarity properties detected in human experimental data and is also implicit in the number line model for the real numbers.

Arithmetic Fact Learning

The three most detailed neural network models for arithmetic fact learning use data representations containing a topographic component (Anderson, Bennett, and Spoehr, 1994; Dallaway, 1994; McCloskey and Lindemann, 1992).

MATHNET

McCloskey and Lindemann (1992) developed MATHNET, a network model for arithmetic fact learning. An earlier version of their basic network used a standard backpropagation learning rule. MATHNET in its final version used a network implementing simulated annealing. Number magnitudes were represented as bar codes. Two bar-coded numbers were used as input, and the output also contained two bar-coded numbers, one for the 10s place and one for the 1s place. There was one layer of hidden units.

Dallaway's Model

A connectionist model of arithmetic learning and performance was proposed by Dallaway (1994). One part of the model learned the arithmetic facts using a data representation much like MATHNET, with a topographic map of number magni-

tude, in this case involving an exponential decay of activation from the central location of the digit. Learning used backpropagation with the addition of an output activation equation to model response times.

A Hybrid Model

Anderson, Bennett, and Spoehr (1994) used a two-part number representation that had both an arbitrary, "symbolic" part, sometimes a random vector and sometimes a coding of the number name, and a bar code for magnitude. One suggested virtue of such a hybrid code is that it allows movement from an "analog" world to a "symbolic" world, since arbitrary patterns become associated with magnitudes. The network used in the Anderson, Bennett, and Spoehr model was the Brain State in a Box (BSB) model, an attractor network model where the dynamics of the network move the state vector to a stable low-energy attractor state (Anderson, 1995).

Results

The overall pattern of results of the simulations were consistent: First, there were often errors. A BSB simulation which has learned as well as it can reconstructs the correct answers about 70% of the time; that is, the network is a "C" arithmetic student. McCloskey and Lindemann (1992) reported similar error rates when MATHNET used a shortened annealing schedule, though it was possible to have error-free performance at other times. Second, errors were not random. Errors were "close" to the correct answer. It is clearly the analog component in the representation that causes this error pattern. Such an error pattern is a consistent feature of human performance for elementary multiplication (Graham, 1987), and is also seen in all of the models described in this section.

The human arithmetic algorithm seems to be something like, "The answer to a multiplication is (1) familiar, that is, a product for some elementary problem, and (2) about the "right size." Arithmetic fact learning and retrieval is a process that combines memory with estimation.

Each of these three models displayed reasonable fits for several of the well-established experimental findings for arithmetic. The learning rules used were different, yet many of the qualitative aspects of model behavior were similar. It seems that the data representation is responsible for most of the effects seen, and the exact learning algorithm used is of less importance.

Knowing More Than You Learned

Although arithmetic fact learning is of interest by itself, we also want to know more about other aspects of arithmetic and mathematics. Clearly, mathematics is a place where algorithms and symbol processing must play an important role. One of the most striking aspects of human cognition is its flexibility; that is, the same data can be used and reused in many different ways.

Consider the arithmetic relationships "bigger" and "smaller." There are approximately 100 single-digit comparisons, for example, "Is 7 bigger than 5?" But there are approximately 10,000 two-digit comparisons, 1,000,000 three-digit comparisons, and so on. It would not be possible for a neural net to see and learn all potential comparisons. Certainly, children do not learn this way.

Anderson, Bennett, and Spoehr (1994) suggested one way of doing number comparisons based on the topographic data representation for number. The analog part of the number representation gave information about size. An abstract problem could be spatialized by using the data representation, and a network solution could be obtained. The use of specialized units to learn algorithmic operations by Cottrell and T'sung (1991) and by Dallaway (1994) seemed less successful. A good "data representation" to use for the algorithmic aspects of arithmetic is unclear but is unlikely to be "grandmother cells" (see LOCALIZED VERSUS DISTRIBUTED REPRESENTATIONS for this concept).

Discussion

Both experimental data and the success of network models with using a "sensory" component suggest that flexible cognitive computation may be determined to a significant extent by physical brain architecture and the way it influences data representation. Mental computation need not be general to be useful. The great power of abstraction is its flexibility, universality, and power. However, a less universal data representation that is partially sensory is also powerful. The sensory part of a hybrid code can make contact with the large, highly developed sensory processing systems and can use many of its subroutines and its special-purpose hardware. For example, it can make use of the scaling, renormalizing, and other operations found in sensory systems. It is not much harder to compare 1 and 2 and 1379 and 2995, a scale change.

If arithmetic is assumed to be at all typical of complex high-level cognition, it seems to involve a fascinating hybrid computation: partially abstract and partially sensory based, the two parts enhancing and supporting each other. Perhaps true understanding of a complex system occurs when the abstract and the sensory-based components cooperate and can check each other's results.

Road Map: Connectionist Psychology
Related Reading: Connectionist and Symbolic Representations

References

Anderson, J. A., 1995, *An Introduction to Neural Networks*, Cambridge, MA: MIT Press. ◆

Anderson, J. A., Bennett, D. A., and Spoehr, K. T., 1994, A study in numerical perversity: Teaching arithmetic to a neural network, in *Neural Networks for Knowledge Representation and Inference* (D. S. Levine and M. Aparicio, Eds.), Hillsdale, NJ: Erlbaum, pp. 311–335.

Campbell, J. I. D., 1987, The roll of associative interference in learning and retrieving arithmetic facts, in *Cognitive Processes in Mathematics* (J. A. Sloboda and D. Rogers, Eds.), Oxford: Oxford University Press, pp. 107–122.

Cottrell, G. W., and T'sung, F. S., 1991, Learning simple arithmetic procedures, in *Advances in Connectionist and Neural Computation Theory*, vol. 1, *High Level Connectionist Models* (J. A. Barnden and J. B. Pollack, Eds.), Norwood, NJ: Ablex, pp. 305–321.

Dallaway, R., 1994, Dynamics of arithmetic: A connectionist view of arithmetic skills, Cognitive Science Research Papers 306, Brighton, UK: University of Sussex.

Davis, P. J., and Anderson, J. A., 1979, Nonanalytic aspects of mathematics and their implication for research and education, *SIAM Rev.*, 21:112–127. ◆

Graham, D. J., 1987, An associative retrieval model of arithmetic memory: How children learn to multiply, in *Cognitive Processes in Mathematics* (J. A. Sloboda and D. Rogers, Eds.), Oxford: Oxford University Press, pp. 123–141.

Hadamard, J., 1945, *The Psychology of Invention in the Mathematical Field*, Princeton, NJ: Princeton University Press. ◆

Link, S., 1990, Modeling imageless thought: The relative judgement theory of numerical comparisons, *J. Math. Psychol.*, 34:2–41.

McCloskey, M., and Lindemann, A. M., 1992, MATHNET: Preliminary results from a distributed model of arithmetic fact retrieval, in *The Nature and Origin of Mathematical Skills* (J. I. D. Campbell, Ed.), Amsterdam: North Holland, pp. 365–409.

Rumelhart, D. E., Smolensky, P., McClelland, J. L., and Hinton, G. E., 1986, Schemata and sequential thought processes in PDP models, in *Parallel Distributed Processing: Explorations in the Microstructure of Cognition* (D. E. Rumelhart, J. L. McClelland, and PDP Research Group, Eds.), vol. 2, *Psychological and Biological Models*, Cambridge, MA: MIT Press, pp. 7–57.

Minimum Description Length Analysis

Richard S. Zemel

Introduction

In this article, we review a variety of ways in which ideas relating to minimum description length (MDL) have been applied to neural networks. We begin with a brief introduction to the historical roots of MDL, and then describe the direct relationship between MDL and Bayesian model selection methods. We divide the applications of MDL to neural networks into two categories corresponding to the two main classes of learning in networks: supervised and unsupervised.

Historical Background

The underlying approach in MDL—applying coding theory to determine simplicity—grew out of work from the mid-1960s, when Kolmogorov, Solomonoff, and Chaitin introduced theories concerning the information content of an object. Instead of relating information to probabilities as Shannon had (e.g., Shannon, 1948), they adopted a computational approach. They defined the information in a binary string to be the length of the shortest program with which a general-purpose computer

can generate the string. The resulting *algorithmic theory of information* has had significant implications in many areas, but it has not had much impact in the practical construction of programs, because the proposed form of information is not computable in its pure form. However, a number of learning techniques (including MDL) have been derived by making approximations to this information measure. The interested reader should consult Li and Vitanyi (1993) for a detailed mathematical review of the history of MDL and its relationship to many other current inductive inferencing techniques.

Defining the Minimum Description Length Principle

Minimum description length can be seen as a principled version of Occam's razor, where the goal is to find the simplest accurate description of a set of data. An informal definition of the MDL principle (Rissanen, 1989) is as follows:

The best model to explain a set of data is the one which minimizes the summed length, in bits, of:

1. The description of the model, and
2. The description of the data, when encoded with respect to the model.

In algorithmic information theory, the model is a general computational model, i.e., a Turing machine. The MDL approach makes the problem tractable by considering particular classes of models. For example, if the goal is to infer decision trees, then the model is a decision tree, while if the goal is to learn the weights of a neural network, the model may be a network with a particular architecture. In this article, a model refers to various aspects of a network, such as its weights and activities.

It is useful to formulate MDL based on a communication protocol in which these terms are unified into a single encoded message that must be decoded to reproduce the data. The sender transmits a message, encoded in the description language $\mathscr{L}$, that conveys both the model M and the data D with respect to the model; this second term can be seen as the *residuals*, i.e., aspects of the data not predicted by the model. The standard goal of inferring an optimal M from the data is then equivalent to minimizing the length of this encoded message:

$$|\mathscr{L}(M, D)| = |\mathscr{L}(M)| + |\mathscr{L}(D \text{ using } M)| \qquad (1)$$

The notion of comparing models based on simplicity can equivalently be expressed from a Bayesian perspective. The goal is to infer a model M from a set of observations D. Bayes's theorem states that the posterior probability of a model is:

$$p(M \mid D) = \frac{p(M)p(D \mid M)}{p(D)} \qquad (2)$$

The most plausible model is then inferred by comparing these posterior probabilities:

$$\text{argmax}_M[p(M)p(D \mid M)]$$
$$= \text{argmax}_M[\log p(M) + \log p(D \mid M)] \qquad (3)$$

The tradeoff inherent in both of these approaches between simpler, more constrained networks and more complex, general networks echoes the *bias-variance dilemma* in statistics: introducing many parameters incurs high variance, while restricting the number of parameters incurs high bias in the set of possible solutions (Geman, Bienenstock, and Doursat, 1992). MDL and Bayesian analysis (see BAYESIAN METHODS FOR SUPERVISED NEURAL NETWORKS) offer an approach to this dilemma by formalizing the Occam's razor idea—a complex network is only preferred when its predictions are sufficiently more accurate—as an inference rule.

The link between the two objectives (Equations 1 and 3) is provided by the *optimal coding theorem* (Shannon, 1948), which states that x can be communicated at a cost that is bounded below by $-\log_2 p(x)$ bits. Applying this theorem produces the general MDL equation:

$$-\log p(M, D) = -\log p(M) - \log p(D \mid M) \qquad (4)$$

Shannon's theorem describes the optimal code if the true probability distribution for a set of discrete alternatives is known. In general, however, one does not know the true distribution, so because coding from a description language based on the wrong probability distribution will always take more bits on average, selecting an appropriate probability distribution for the codes is a key aspect of MDL applications. MDL provides a method of comparing these choices based on the resulting code lengths.

The distribution must be chosen to suit the nature of the task. For example, in a classification task, the data given the

model consist of a number of discrete alternatives, each with some probability of occurrence. Here we can save bits by simply communicating the fact that the model output is correct on the correctly classified examples. When the information to be encoded takes on real values, a continuous distribution is required. A coding distribution that is often (implicitly) selected is a Gaussian. In this case, if we assume that the residuals (the second term in Equation 1) are independent and have a zero mean, fixed-variance Gaussian distribution, then if the values are encoded to some fixed accuracy, the code length is the familiar summed-squared-error cost function.

An Example of Applying MDL to Neural Networks

One of the standard approaches to improving generalization in neural networks can be formulated as an MDL technique. This approach adds an extra term to the error function which penalizes the complexity of the network, so that the objective function used to train the network involves a tradeoff between the data misfit and the network complexity:

$$\text{Cost} = \alpha \cdot \text{Complexity} + \text{Error}$$

If we regard each possible weight vector of the network as a potential model, then this complexity term is simply the cost of specifying the model in the definition of MDL above.

Applying Shannon's theorem then equates the complexity of a network to the negative log probability of its weights. Thus, the critical question becomes the choice of encoding scheme, or prior distribution on the weights. A simple prior is a radially symmetric, mean-zero Gaussian. Setting the variance of this Gaussian to $1/\alpha$ yields an often-used complexity term—the sum of the squares of the weights, $\Sigma_j w_j^2$. Differentiating this measure produces simple *weight decay* in the learning rule, which forces weights with small gradients from the error term to decay away, leaving only the required weights for the task. This example points out the key role of probability distributions in MDL; now we consider other encoding distributions and see how this choice affects the models that MDL favors.

Applications of MDL

The MDL principle has been applied in a wide variety of areas over the last decade. For example, a range of recent applications was presented at the 1990 American Association for Artificial Intelligence symposium entitled "The Theory and Application of Minimal-Length Encoding."

With respect to neural networks, the duality between Bayesian analysis and MDL means that a range of applications of Bayesian techniques to neural networks also may be expressed in MDL terms. Few neural network methods have referred directly to minimal length encoding, but this relationship to Bayesian techniques makes many network methods relevant to this article.

We separate MDL applications into two classes, involving fundamentally different formulations in terms of the communication protocol. In supervised learning, each data item consists of an input-output pair. The sender and receiver both have access to the inputs. The task of the sender is to succinctly communicate the desired outputs for each input. The network is a *generative* model of the output given the input; this model includes the network architecture and weights, while the data are the residuals.

In unsupervised learning, the receiver does not have access to the input, and the sender must provide enough information to allow an accurate input reconstruction. For example, in a clustering task, the sender first communicates the cluster centers (i.e., the weights of a competitive learning network). Then he or she only needs to say which cluster each input belongs to, and the residual error, or the distance to the cluster center. Given this information, the receiver can recover the actual input. Thus, the network here is a generative model of the input itself.

Note that this communication protocol formulation is only a device to derive an MDL objective function which can be used to train a neural network. The algorithms described in later sections are not actually interested in sending the message, but rather in developing good models of a dataset.

Supervised Learning

In supervised learning, the primary application of MDL techniques has been to improve the generalization performance of networks. Good generalization requires that the amount of information required to specify the output vectors of the training cases must be considerably larger than the number of independent parameters of the network. When only a small amount of labeled training data is available, a large network thus will not readily produce a good solution.

A range of techniques have been proposed to address this problem. These include weight sharing, weight/unit pruning (see NEUROSMITHING: IMPROVING NEURAL NETWORK LEARNING), and cross-validation training. MDL techniques offer an alternative approach to this problem. In this section, we highlight two types of techniques. The first class assumes that the network architecture is given and uses MDL to limit the complexity of the network weights. The second class uses MDL to select between potential network architectures.

Using MDL to Determine Network Weights

The standard neural network training problem of finding the appropriate set of weights can be usefully formulated in MDL terms. We showed that a radially symmetric Gaussian prior on the weights produces a weight decay term in the learning rule. Many different types of priors have been discussed in the literature. MacKay (1992) compared several priors for a single learning problem. He shows that for this problem, using the simple weight-decay prior with a different α for separate weight classes—weights into hidden units, hidden unit biases, and weights into output units—achieves better generalization than using a single α for all the weights.

Buntine and Weigend (1991) proposed a more complicated prior on the weights. This prior can be viewed as a mixture of two distributions: a uniform distribution and a narrow zero-mean Gaussian. The Gaussian encourages small weight values to approach zero; the uniform distribution takes responsibility for larger weights and provides little pressure to change these values. The combined effect is to simplify the network by eliminating small weights.

Nowlan and Hinton (1992) proposed a more sophisticated form of weight decay term where the prior is adapted to the data. For many problems, such as translation-invariant recognition, improved network performance can be achieved by constraining particular subsets of the weights to share the same value. Nowlan and Hinton (1992) accomplished this by fitting a mixture of Gaussians to the weights, allowing the network to decide which weights should be tied together.

Finally, Hinton and van Camp (1993) extended this work to consider the general MDL objective where the cost function is the sum of two encoding costs: the weights of the network and the output error residuals. On a sample problem, coding the weights using an adaptive mixture-of-Gaussians prior allowed the network to find three sharp clusters for the weights. Dis-

covering this structure avoided overfitting the data, because the network was able to generalize even though the number of training cases was less than the dimensionality of the input vector.

Note that the MDL objectives just described involve several *hyperparameters*, such as the regularization constant α controlling the tradeoff between the error and complexity terms. Several methods have been proposed for determining hyperparameters (e.g., Buntine and Weigend, 1991; MacKay, 1992).

Using MDL to Determine a Network Architecture

Minimum description length methods have also been applied to evaluate network architectures (see BAYESIAN METHODS FOR SUPERVISED NEURAL NETWORKS). The weights for various network architectures are learned using an MDL objective, and then these networks are compared based on the same data-fit/complexity tradeoff. MacKay (1992) demonstrates how this MDL/Bayesian procedure accurately predicts the appropriate number of hidden units on a small interpolation problem, where the target is determined by examining how well the various architectures generalize to an unseen test set.

Kendall and Hall (1993) proposed an MDL approach to network construction where the model and the data misfit are encoded over a discrete space. They compute the code length of the network parameters from a histogram of the weight values; since the task is assumed to be classification, the data code length is simply the cost of specifying which training cases are incorrectly predicted by the model. The authors find that using a genetic optimization algorithm to minimize the total description length succeeds in finding optimal network architectures on some simple problems.

Minimum description length methods have also been applied to architecture selection in other areas relevant to neural networks. For example, Stolcke and Omohundro (1993) described an algorithm for inferring the structure of a Hidden Markov Model which involves a Bayesian/MDL approach. The algorithm begins with a model that directly encodes the training data, and then successively merges states based on a description length criterion which penalizes models according to the number of transitions and output values at each state.

Unsupervised Learning

For unsupervised learning networks, in which the goal of learning can be viewed as a form of probability density estimation, MDL techniques have been applied in several ways. The most popular approach is to use MDL to learn the number of distributions in the estimate, as well as the parameters of those distributions. A second approach involves applying MDL methods not only to the weights of the network, but also to the activities of the hidden units.

MDL and Finite Mixture Models

A standard statistical technique for unsupervised classification can be formulated in an MDL framework. In finite mixture models, each datum in a training set is assumed to be drawn from one of J different classes. While learning the class parameters, we can also determine the optimal J by maximizing the posterior distribution of this number given the data set. This search for J resembles the architecture selection approach just described.

AutoClass (Cheeseman et al., 1988) is an unsupervised classification system based on this approach. It determines J by starting with more classes than are believed to be present and iteratively eliminating them. A prior that makes classes accounting for little data improbable can be seen as a description length prior on the model complexity. Similar approaches have been used in clustering algorithms, where a complexity term penalizing complex clusterings (based on measures such as their summed entropy) is added to create an MDL-style objective trading off data fit for model simplicity (see DATA CLUSTERING AND LEARNING).

MDL and Autoencoders

In the unsupervised schemes just described, the model cost was some function of the number of underlying components in the clustering or mixture model algorithm, while the data cost was simply the summed-squared error. This represents one type of MDL objective function. If we adopt a more general viewpoint, we see that a wide range of other objective functions is possible.

In particular, we can view unsupervised learning in terms of an autoencoder (i.e., a network which attempts to reproduce its inputs on its outputs), where the MDL objective is to minimize the total cost of communicating the input vectors to a receiver. There are three terms in the description length:

- The *representation cost* is the number of bits required to communicate the representation (the hidden unit activities) that the algorithm assigns to each input vector.
- The *model cost* is the number of bits required to specify the hidden-to-output weights of the network, which provide an estimate of the input from its representation.
- The *reconstruction cost* is the number of bits required to fix up errors in the estimate of the input.

The sum of these three terms provides an objective function for training the autoencoder.

We can view many unsupervised algorithms in terms of this framework by understanding how they encode each of these three terms. For example, in competitive learning, the representation is the identity of the winning hidden unit, so the average representation cost is at most the log of the number of units, while the reconstruction cost is proportional to the squared difference between the winner's weight vector and the input. Standard competitive learning algorithms minimize this latter cost, while algorithms that attempt to limit the number of clusters can be seen as trading off representation cost for reconstruction cost. PRINCIPAL COMPONENT ANALYSIS (q.v.) can be viewed as a version of MDL in which we limit the representation cost by only having a few hidden units and we ignore the model cost and the accuracy with which the hidden activities must be coded.

Many new algorithms can be derived by adopting new assumptions about the structure of the data and using them to formulate different methods of encoding the three terms in this MDL cost function. Any method that communicates each hidden activity independently tends to lead to *factorial* representations because any mutual information between hidden units causes redundancy in the communicated message, so the pressure to keep the message short squeezes out the redundancy. Zemel (1993) and Hinton and Zemel (1994) describe algorithms derived from this MDL approach for learning factorial representations. Zemel and Hinton (1994) describe how this MDL approach can also be used to develop population codes in which the activities of hidden units are locally correlated so as to form a topographical map.

Discussion

Minimum description length methods have been applied in a variety of neural network training paradigms, for learning both the weights and the architecture. These methods involve selecting a class of models and an encoding scheme for the two terms in the description: the model and the data. Given these elements, an appropriate objective function can be constructed for either an unsupervised or supervised learning problem. Because of a common underlying framework, many Bayesian inferencing methods can also be viewed in MDL terms.

Many important issues remain to be explored in this area. One key issue concerns when it is better to use other methods of improving generalization, such as stopping training based on performance on a validation set, versus using an MDL-based regularization. In the unsupervised learning area, an open problem concerns formulating appropriate priors for learning hierarchical representations. Finally, an area that is ripe for MDL applications is temporal learning, in which MDL can be used to develop concise models that can accurately predict sequential events.

Acknowledgment.   I thank the ONR for support.

Road Map: Learning in Artificial Neural Networks, Statistical
Background: I.3. Dynamics and Adaptation in Neural Networks
Related Reading: Competitive Learning; Unsupervised Learning with Global Objective Functions

References

Buntine, W., and Weigend, A., 1991, Bayesian back-propagation, *Complex Systems*, 5:603–643.

Cheeseman, P., Kelly, J., Self, M., Stutz, J., Taylor, W., and Freeman, D., 1988, AutoClass: A Bayesian classification system, in *Proceedings of the Fifth International Conference on Machine Learning*, pp. 54–62.

Geman, S., Bienenstock, E., and Doursat, R., 1992, Neural networks and the bias/variance dilemma, *Neural Computat.*, 4:1–58. ◆

Hinton, G. E., and van Camp, D., 1993, Keeping neural networks simple by minimizing the description length of the weights, in *Sixth ACM Conference on Computational Learning Theory*, Santa Cruz, CA.

Hinton, G. E., and Zemel, R. S., 1994, Autoencoders, minimum description length, and Helmholtz free energy, in *Advances in Neural Information Processing Systems 6* (J. D. Cowan, G. Tesauro, and J. Alspector, Eds.), San Mateo, CA: Morgan Kaufmann, pp. 3–10.

Kendall, G., and Hall, T., 1993, Optimal network construction by minimum description length, *Neural Computat.*, 5:210–212.

Li, M., and Vitanyi, P. M. B., 1993, *An Introduction to Kolmogorov Complexity and Its Applications*, Reading, MA: Addison-Wesley. ◆

MacKay, D., 1992, A practical Bayesian framework for backpropagation networks, *Neural Computat.*, 4:448–472.

Nowlan, S. J., and Hinton, G. E., 1992, Simplifying neural networks by soft weight-sharing, *Neural Computat.*, 4:473–493.

Rissanen, J., 1989, *Stochastic Complexity in Statistical Inquiry*, Singapore: World Scientific. ◆

Shannon, C. E., 1948, A mathematical theory of communication, *Bell System Tech. J.*, 27:379–423, 623–656.

Stolcke, A., and Omohundro, S., 1993, Hidden Markov Model induction by Bayesian model merging, in *Advances in Neural Information Processing Systems 5* (S. J. Hanson, J. D. Cowan, and C. L. Giles, Eds.), San Mateo, CA: Morgan Kaufmann, pp. 11–18.

Zemel, R. S., 1993, A minimum description length framework for unsupervised learning, PhD thesis, University of Toronto. ◆

Zemel, R. S., and Hinton, G. E., 1994, Developing population codes by minimizing description length, in *Advances in Neural Information Processing Systems 6* (J. D. Cowan, G. Tesauro, and J. Alspector, Eds.), San Mateo, CA: Morgan Kaufmann, pp. 11–18.

Model-Reference Adaptive Control

Petros Ioannou

Introduction

The control of plants (i.e., physical systems) with unknown and continuously changing dynamics has been the main challenge of control theorists since the early designs of dynamical systems. Feedback and modern control techniques were developed to meet this challenge but only to a certain degree. It was soon realized that feedback controllers with fixed parameters or gains may not meet the performance requirements when the plant parameters change over time in a significant way. In the late 1950s this observation led to a strong interest in adaptive control techniques. With adaptivity the feedback controller parameters are adjusted based on the plant input-output information in an effort to compensate for the changes in the plant parameters and preserve performance. A typical example of an adaptive control system is a car driver. The visual field of the driver provides estimates about the relative distance and speed between his car and the vehicle ahead and about the position of his car in the lane. Based on these estimates or "measurements" the driver decides about the inputs to be given to the steering wheel and gas or brake pedals. The driver can perform the same function, i.e., safe driving for a family of cars of completely different dynamics, by modifying, i.e., adapting, his actions to accommodate the difference in the dynamics of the various vehicles.

One of the main approaches to adaptive control, introduced in the 1950s, is model-reference adaptive control (MRAC) (Landau, 1979). The basic structure of MRAC is shown in Figure 1. Given a plant with unknown or partially known parameters and the performance requirements it has to meet, a reference model that describes the desired behavior of the plant is first developed. The controller structure $C(\theta)$ is chosen to be the same or similar to the one that could be used if the plant parameters were fixed and known. Since the plant parameters denoted by θ_p are unknown, the controller parameters denoted by θ which are functions of θ_p cannot be calculated. The purpose of the adaptation mechanism is to generate estimates of θ by processing the plant input u and output y that carry information about θ_p. The objective of the controller and adaptation mechanism that form the MRAC scheme is to guarantee stability and convergence of the tracking error e (which is the measure of the deviation of the plant output y from the desired output given by y_m) to zero for any given reference input signal r. This objective is possible provided a controller $C(\theta)$ exists

that can force the input-output plant operator from r to y to be equal to that of the reference model from r to y_m.

In the case of the car driver the reference model may represent the desired behavior of ideal driving that the driver tries to imitate. The model is stored in his brain and is developed during driving lessons and from experience. The controller $C(\theta)$ and adaptation mechanism represent the actions of the human driver that generate the input u in an effort to force the vehicle to behave as the reference model by driving any mismatch, i.e., the error e, as close to zero as possible. The reference signal may represent inputs from the roadway such as lane geometry, speed of leading vehicle, and so on. In the context of neural networks, the reference model may represent the behavior of an excellent driver, and $C(\theta)$ a neural controller whose weights θ are adjusted by the adaptation mechanism. The adaptation mechanism could be a backpropagation algorithm or a similar algorithm.

The existence and stability of an MRAC depends very much on the form and properties of the plant to be controlled. When the plant is linear time invariant (LTI), i.e., when it is described by a linear differential equation whose parameters are constant, the design of MRAC to meet the objective of reference model following is possible provided the zeros of the plant or the inverse of the plant is stable. This restriction arises from the fact that the controller has to cancel the dynamics of the plant and replace them with those of the reference model. Cancellation of unstable plant dynamics may lead to unbounded signals. Therefore, in an effort to maintain well-behaved signals, the inverse of the plant is assumed to be stable (Narendra and Annaswamy, 1989; Astrom and Wittenmark, 1989; Sastry and Bodson, 1989; Ioannou and Sun, 1995).

MRAC may be divided into two major classes: direct MRAC and indirect MRAC. In direct MRAC, the adaptation mechanism is a parameter estimator (also called an adaptive law) that generates direct estimates of the controller parameters. In indirect MRAC, the adaptive mechanism consists of a parameter estimator for estimating the plant parameters and an algebraic equation for calculating the controller parameters. Both direct and indirect MRAC use the same assumptions about the plant and reference model and guarantee similar performance.

MRAC: Scalar Example

Consider the following first-order plant:

$$\dot{x} = ax + bu$$
$$y = x \tag{1}$$

where a and b are unknown parameters but the sign of b is known. The control objective is to choose an appropriate control law u such that all signals in the closed-loop plant are bounded and the plant state x tracks the state x_m of the reference model given by

$$\dot{x}_m = -a_m x_m + b_m r \tag{2}$$

for any bounded piecewise continuous reference signal $r(t)$ where $a_m > 0$, b_m are known, and $x_m(t)$ and $r(t)$ are measured at each time t. It is assumed that a_m, b_m, and r are chosen so that x_m represents the desired output response of the plant.

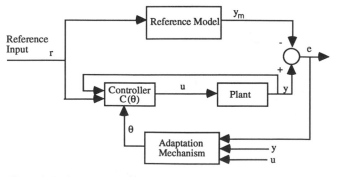

Figure 1. Basic structure of MRAC.

Control Law

In order for x to track x_m for any reference input signal $r(t)$, the control law should be chosen so that the closed-loop plant transfer function from the input r to output $y = x$ is equal to that of the reference model. We therefore propose

$$u = -k^*x + l^*r \tag{3}$$

where k^* and l^* are calculated so that $x(s)/r(s) = x_m(s)/r(s)$, which implies that

$$l^* = \frac{b_m}{b} \qquad k^* = \frac{a_m + a}{b} \tag{4}$$

provided of course that $b \neq 0$, i.e., that the plant is controllable. The control law (Equations 3 and 4) guarantees that the transfer function of the closed-loop plant, i.e., $x(s)/r(s)$, is equal to that of the reference model. Such a transfer function matching guarantees that $x(t) = x_m(t)$, for all $t \geq 0$ when $x(0) = x_m(0)$ or $|x(t) - x_m(t)| \to 0$ as $t \to \infty$ exponentially fast when $x(0) \neq x_m(0)$, for any reference signal $r(t)$ of the class considered.

When the plant parameters a and b are unknown, Equation 3 cannot be implemented. In this case we propose the control law

$$u = -k(t)x + l(t)r \tag{5}$$

where $k(t)$ and $l(t)$ are the estimates of k^* and l^*, respectively, at time t, and we search for a parameter estimator or adaptive law to generate $k(t)$ and $l(t)$ on-line.

Adaptive Law

We view the design of the adaptive law as an on-line parameter estimation problem for the unknown constants k^* and l^*. We start with the plant equation (Equation 1), which we express in terms of k^* and l^* by adding and subtracting the desired input term $-bk^*x + bl^*r$ to obtain

$$\dot{x} = -a_m x + b_m r + b(k^*x - l^*r + u) \tag{6}$$

that together with Equation 5 and the definition of the tracking error $e = x - x_m$ lead to

$$e = \frac{b}{s + a_m}(k^*x - l^*r + u) \tag{7}$$

Substituting $u = -k(t)x + l(t)r$ in Equation 7 and defining the parameter error $\tilde{k} = k - k^*$, $\tilde{l} = l - l^*$, we have

$$e = \frac{b}{s + a_m}(-\tilde{k}x + \tilde{l}r)$$

or

$$\dot{e} = -a_m e + b(-\tilde{k}x + \tilde{l}r) \tag{8}$$

The development of this differential equation relating the estimation or tracking error e with the parameter error is a significant step in deriving the adaptive laws for updating $k(t)$ and $l(t)$. We assume that the structure of the adaptive law is given by

$$\dot{k} = f_1(e, x, r, u)$$
$$\dot{l} = f_2(e, x, r, u) \tag{9}$$

where the functions f_1 and f_2 are to be designed, and we propose the function

$$V(e_1, \tilde{k}, \tilde{l}) = \frac{e^2}{2} + \frac{\tilde{k}^2}{2\gamma_1}|b| + \frac{\tilde{l}^2}{2\gamma_2}|b| \tag{10}$$

where γ_1 and $\gamma_2 > 0$, often referred to as a Lyapunov candidate for the system of Equations 8 and 9. Then the time derivative $\dot{V}$ along any trajectory of Equations 8 and 9 is given by

$$\dot{V} = -a_m e^2 - b\tilde{k}ex + b\tilde{l}er + \frac{|b|\tilde{k}}{\gamma_1}f_1 + \frac{|b|\tilde{l}}{\gamma_2}f_2 \tag{11}$$

Since $|b| = b \operatorname{sgn}(b)$, the indefinite terms in Equation 11 disappear by choosing $f_1 = \gamma_1 ex \operatorname{sgn}(b)$ and $f_2 = -\gamma_2 er \operatorname{sgn}(b)$; i.e., the adaptive laws (Equation 9) become

$$\dot{k} = \gamma_1 ex \operatorname{sgn}(b)$$
$$\dot{l} = -\gamma_2 er \operatorname{sgn}(b) \tag{12}$$

and

$$\dot{V} = -a_m e^2 \tag{13}$$

Analysis

Treating $x_m(t)$ and $r(t)$ in Equation 8 as bounded functions of time, it follows from Equations 10 and 13 that V is a Lyapunov function, and the equilibrium $e_e = 0$, $\tilde{k}_e = 0$, $\tilde{l}_e = 0$ is uniformly stable. Furthermore, e, $\tilde{k}$, and $\tilde{l}$ are bounded, i.e., they belong to L_∞, and e is square integratable, i.e., it belongs to L_2. Since $e = x - x_m$ and $x_m \in L_\infty$, we also have $x \in L_\infty$ and $u \in L_\infty$, and therefore all signals in the closed loop are bounded. Now from Equation 8 we have $\dot{e} \in L_\infty$, which together with $e \in L_2$ implies that $e(t) \to 0$ as $t \to \infty$.

We have established that the control law of Equation 5 together with the adaptive law of Equation 12 guarantees boundedness for all signals in the closed-loop plant and in addition that the plant state $x(t)$ tracks the state of the reference model x_m asymptotically with time for any reference input signal which is bounded and piecewise continuous. These results do not imply that $k(t) \to k^*$ and $l(t) \to l^*$ as $t \to \infty$, i.e., that in the limit as $t \to \infty$ the transfer function of the closed-loop plant approaches that of the reference model. In order to achieve such a result, the reference input r has to be sufficiently rich, i.e., to contain at least one non-zero frequency. For example, $r(t) = \sin \omega t$ for some $\omega \neq 0$ guarantees the exponential convergence of $x(t)$ to $x_m(t)$ and of $k(t)$ and $l(t)$ to k^* and l^*, respectively. In general, the design of $r(t)$ to be sufficiently rich may not be desirable, especially in cases where the control objective involves tracking of signals that are not rich in frequencies. The MRAC scheme given by Equations 5 and 12 is referred to as a direct scheme, since k and l are adjusted directly without estimating the plant parameters. (For further development of these ideas, see Ioannou and Datta, 1991; Ioannou and Sun, 1995.)

In the preceding examples we assumed that the plant parameters a and b are fixed constants. When a and b are changing with time, these MRAC schemes have to be modified to handle the variations as shown in Tsakalis and Ioannou (1993).

Nonlinear MRAC

The preceding examples illustrate the design procedure and analysis of simple MRAC applied to LTI plants, and they can be extended to a special class of nonlinear systems whose nonlinear elements are known (see Isidori, 1989; Kokotovic, 1991). In the case of complex systems that appear in such areas as engineering and biology, where the plant cannot be approximated by an LTI plant or its nonlinear elements are unknown, the concept of MRAC in conjunction with neural network techniques may be used. Figures 2 and 3 illustrate two different neural network (NN) MRAC schemes, a direct one in Figure 2

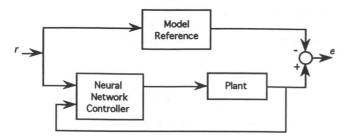

Figure 2. Direct NN-MRAC.

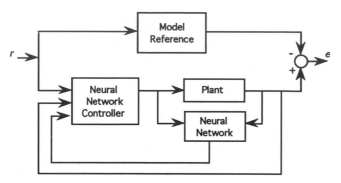

Figure 3. Indirect NN-MRAC.

and an indirect one in Figure 3. The reference model could be an LTI or a neural network that has already been trained to represent the desired behavior of the plant in response to some input commands or reference inputs r. As an example, consider the design of an autopilot for an aircraft. A very experienced and specially gifted pilot flies the aircraft by performing various normal as well as complicated maneuvers. The data collected are used to train a neural network. The trained neural network now models the behavior of this very special pilot and can be used as a reference model whose behavior is to be matched by the autopilot. In the direct NN-MRAC scheme (see Figure 2) the controller is another neural network whose weights are adjusted to minimize the error e between the plant and reference model output. In the indirect NN-MRAC scheme (see Figure 3) a neural network is used to learn the behavior of the nonlinear plant. The controller is designed based on the neural network that approximates the plant, or it could also be another neural network. The adjustment of the signals of the neural networks are based on the minimization of the error signal e. For more details about the various forms of NN-MRAC, see Narendra and Parthasarathy (1990).

As an example consider the nonlinear plant

$$\dot{x} = f(x) + g(x)u \tag{14}$$

where the nonlinear functions $f(x)$ and $g(x)$ are unknown. The input u is to be chosen so that x follows the state x_m of the LTI reference model

$$\dot{x} = -a_m x_m + b_m r$$

In this case the following control law may be attempted:

$$u = -\frac{1}{\hat{g}(x, \theta_y)} [\hat{f}(x, \theta_f) + a_m x - b_m r] \tag{15}$$

where $\hat{g}(x, \theta_y)$ and $\hat{f}(x, \theta_f)$ are the estimated nonlinearities generated by a neural network or other universal approximators, and θ_y and θ_f are adjustable parameters such as the weights of the neural network generated by an adaptive law. The details of the design of the adaptive laws and the stability of Equation 15 are given in Polycarpou and Ioannou (1991).

The NN-MRAC (Equation 15) is an indirect one, since it is based on the estimated or approximated plant model

$$\dot{x} = \hat{f}(x, \theta_f) + \hat{g}(x, \theta_y)u$$

that is generated by the neural network. The generalization of this approach to high-order nonlinear plants whose full state is not available for measurement is a topic for research.

Discussion

In general, the use of neural networks for MRAC for nonlinear plants poses several stability problems that have not been resolved yet. The difficulty is partly due to the lack of control laws and strong stability results even in the case of known nonlinearities. Another difficulty is the fact that the neural network that approximates the nonlinear plant may also be complicated to the point that the choice of the control law is not clear. Single-layer neural networks with a special class of nonlinearities may help alleviate this latter difficulty.

Despite the lack of strong stability results, researchers have been able to motivate the use of neural networks in many control applications by demonstrating their usefulness with computer simulations.

It is the opinion of this author that the lack of stability results will limit considerably the widespread application of neural networks for controlling dynamical systems.

Road Map: Control Theory and Robotics
Background: Motor Control, Biological and Theoretical
Related Reading: Adaptive Control: General Methodology; Adaptive Control: Neural Network Applications; Process Control

References

Astrom, K. J., and Wittenmark, B., 1989, *Adaptive Control*, New York: Addison-Wesley.
Ioannou, P. A., and Datta, A., 1991, Robust adaptive control: A unified approach, *Proc. IEEE*, 79:1736–1768. ◆
Ioannou, P. A., and Sun, J., 1995, *Robust Adaptive Control*, Englewood Cliffs, NJ: Prentice Hall. ◆
Isidori, A., 1989, *Nonlinear Control Systems*, 2nd ed., New York: Springer-Verlag.
Kokotovic, P. V., 1991, *Foundations of Adaptive Control*, New York: Springer-Verlag.
Landau, Y. D., 1979, *Adaptive Control: The Model Reference Approach*, New York: Marcel Dekker.
Narendra, K. S., and Annaswamy, A. M., 1989, *Stable Adaptive Systems*, Englewood Cliffs, NJ: Prentice Hall.
Narendra, K. S., and Parthasarathy, K., 1990, Identification and control of dynamical systems using neural networks, *IEEE Trans. Neural Netw.*, 1:4–28.
Polycarpou, M., and Ioannou, P. A., *Identification and Control of Nonlinear Systems Using Neural Network Models: Design and Stability Analysis*, University of Southern California, Electrical Engineering Systems, Report 91–09–01.
Sastry, S., and Bodson, M., 1989, *Adaptive Control: Stability, Convergence, and Robustness*, Englewood Cliffs, NJ: Prentice Hall.
Tsakalis, K. S., and Ioannou, P. A., 1993, *Linear Time-Varying Systems: Control and Adaptation*, Englewood Cliffs, NJ: Prentice Hall.

Modular and Hierarchical Learning Systems

Michael I. Jordan and Robert A. Jacobs

Introduction

Modular and hierarchical systems allow complex learning problems to be solved by dividing the problem into a set of subproblems, each of which may be simpler to solve than the original problem. Within the context of supervised learning—our focus in this article—modular architectures arise when we assume that the data can be well described by a collection of functions, each of which is defined over a relatively local region of the input space. A modular architecture can model such data by allocating different modules to different regions of the space. Hierarchical architectures arise when we assume that the data are well described by a multiresolution model—a model in which regions are divided recursively into subregions.

Modular and hierarchical systems present an interesting credit assignment problem: it is generally the case that the learner is not provided with prior knowledge of the partitioning of the input space (but cf. Hampshire and Waibel, 1989). Knowledge of the partition would correspond to being given *labels* specifying how to allocate modules to data points. The assumption we make is that such labels are absent. The situation is reminiscent of the unsupervised clustering problem in which a classification rule must be inferred from a data set in which the class labels are absent. Indeed, the connection to clustering has played an important role in the development of the supervised learning algorithms that we present here (Nowlan, 1990).

The learning algorithms that we describe solve the credit assignment problem by computing a set of values—*posterior probabilities*—that can be thought of as estimates of the missing "labels." These posterior probabilities are based on a probabilistic model of the input-output characteristics of each of the network modules. This approach to learning in modular systems was developed by Jacobs et al. (1991). Jordan and Jacobs (1994) extended the modular system to a hierarchical system, made links to the statistical literature on classification and regression trees (Breiman et al., 1984), and developed an Expectation Maximization (EM) algorithm for the architecture. We describe these developments in the remainder of the article, emphasizing the probabilistic framework.

The Mixture of Experts (ME) Architecture

The modular architecture that we consider is shown in Figure 1. The architecture is composed of N modules referred to as *expert networks*, each of which implements a parameterized function $\mu_i = f(\mathbf{x}, \boldsymbol{\theta}_i)$ from inputs $\mathbf{x}$ to outputs μ_i, where $\boldsymbol{\theta}_i$ is a parameter vector. We attach a probabilistic interpretation to each of the expert networks by assuming that the experts generate outputs $\mathbf{y}$ with probability $P(\mathbf{y}|\mathbf{x}, \boldsymbol{\theta}_i)$, where μ_i is the conditional mean associated with the density P.

Because we assume that different expert networks are appropriate in different regions of the input space, the architecture requires a mechanism that identifies, for any given input $\mathbf{x}$, that expert or blend of experts that is most likely to produce the correct output. This is accomplished via an auxiliary network, known as a *gating network*, that produces as output a set of scalar coefficients g_i that serve to weight the contributions of the various experts. These coefficients are not fixed constants but vary as a function of the input $\mathbf{x}$.

The probabilistic interpretation of the gating network is as a *classifier*, a system that maps an input $\mathbf{x}$ into the probabilities

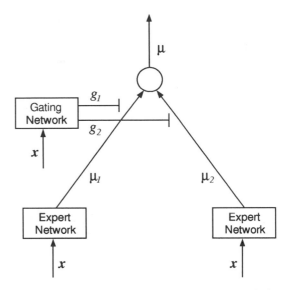

Figure 1. A mixture of experts architecture. The output μ is the conditional mean of $\mathbf{y}$ given $\mathbf{x}$ (see text).

that the various experts will be able to generate the desired output (based on knowledge of $\mathbf{x}$ alone). These probabilities (the g_i) are constrained to be non-negative and sum to one (for each $\mathbf{x}$).

There are many ways to enforce the probabilistic constraints on the g_i. One approach is to utilize the *softmax* function (Jacobs et al., 1991; see also WINNER-TAKE-ALL MECHANISMS). Define an intermediate set of variables ξ_i as parameterized functions of the input $\mathbf{x}$:

$$\xi_i = \xi_i(\mathbf{x}, \boldsymbol{\eta}) \tag{1}$$

where $\boldsymbol{\eta}$ is a parameter vector, and define the outputs g_i in terms of the ξ_i as follows:

$$g_i = \frac{e^{\xi_i}}{\sum_j e^{\xi_j}} \tag{2}$$

It is readily verified that the g_i are non-negative and sum to one for each $\mathbf{x}$. This approach has the virtue of having a simple probabilistic interpretation: the ξ_i are the discriminant surfaces for a classification problem involving the exponential family of probability distributions (Jordan and Jacobs, 1994).

The Mixture Model

Let us now specify the probabilistic model underlying the mixture of experts architecture more precisely. We assume that the training set $\mathcal{X} = \{\mathbf{x}^{(l)}, \mathbf{y}^{(l)}\}_{l=1}^{L}$ is generated in the following way. Given the choice of an input $\mathbf{x}$, a label i is chosen with probability $P(i|\mathbf{x}, \boldsymbol{\eta}^0)$ (where the superscript 0 distinguishes the parameter values in the model from the parameter estimates in the network). Given the choice of the label and given the input, the target output $\mathbf{y}$ is assumed to be generated with probability $P(\mathbf{y}|\mathbf{x}, \boldsymbol{\theta}_i^0)$. Each such data point is assumed to be generated independently in this manner.

Note that a given output $\mathbf{y}$ can be generated in N different ways, corresponding to the N different choices of the label i.

Thus the total probability of generating **y** from **x** is given by the sum over i:

$$P(\mathbf{y}|\mathbf{x}, \Theta^0) = \sum_i P(i|\mathbf{x}, \eta^0) P(\mathbf{y}|\mathbf{x}, \theta_i^0) \quad (3)$$

where Θ^0 denotes the vector of all of the parameters ($\Theta^0 = [\theta_1^0, \theta_2^0, \ldots, \theta_N^0, \eta^0]^T$). The density in Equation 3 is known as a *mixture density*. It is a mixture density in the output space, conditional on the choice of input.

It is the task of the gating network to model the probabilities $P(i|\mathbf{x}, \eta^0)$, which can be construed as class probabilities in a multiway classification of the input **x**. We parameterize these probabilities via Equations 1 and 2, identifying the gating network outputs g_i with $P(i|\mathbf{x}, \eta)$.

It is straightforward to compute moments of the mixture density. For example, the conditional mean $\mu = E(\mathbf{y}|\mathbf{x}, \Theta)$ is readily obtained by taking the expected value of Equation 3:

$$\mu = \sum_i g_i \mu_i$$

where μ_i is the conditional mean associated with the probability distribution $P(\mathbf{y}|\mathbf{x}, \theta_i^0)$. This convex combination of the outputs of the experts is a natural choice for the output of the modular architecture (cf. Figure 1), although it can also be advantageous to make fuller use of the entire mixture density.

A Gradient-Based Learning Algorithm

To develop an algorithm for estimating the parameters of a mixture of experts architecture, we make use of the maximum likelihood (ML) principle. That is, we choose parameters for which the probability of the training set given the parameters (a function known as the *likelihood*) is largest. As is commonly the case in statistics, it is actually more convenient to work with the logarithm of the likelihood than with the likelihood itself. Taking the logarithm of the product of N densities of the form of Equation 3 yields the following log likelihood:

$$l(\mathcal{X}, \Theta) = \sum_l \log \sum_i P(i|\mathbf{x}^{(l)}, \eta) P(\mathbf{y}^{(l)}|\mathbf{x}^{(l)}, \theta_i) \quad (4)$$

One approach to maximizing the log likelihood is to use gradient ascent (a better approach is to use the EM algorithm, as discussed in a later section). Computing the gradient of l with respect to μ_i and ξ_i yields

$$\frac{\partial l}{\partial \mu_i} = \sum_l h_i^{(l)} \frac{\partial}{\partial \mu_i} \log P(\mathbf{y}^{(l)}|\mathbf{x}^{(l)}, \theta_i) \quad (5)$$

and

$$\frac{\partial l}{\partial \xi_i} = \sum_l (h_i^{(l)} - g_i^{(l)}) \quad (6)$$

where $h_i^{(l)}$ is defined as $P(i|\mathbf{x}^{(l)}, \mathbf{y}^{(l)})$. In deriving this result, we have used Bayes's rule:

$$P(i|\mathbf{x}^{(l)}, \mathbf{y}^{(l)}) = \frac{P(i|\mathbf{x}^{(l)}) P(\mathbf{y}^{(l)}|\mathbf{x}^{(l)}, i)}{\sum_j P(j|\mathbf{x}^{(l)}) P(\mathbf{y}^{(l)}|\mathbf{x}^{(l)}, j)}$$

This suggests that we define $h_i^{(l)}$ as the *posterior probability* of the ith label, conditional on the input $\mathbf{x}^{(l)}$ and the output $\mathbf{y}^{(l)}$. Similarly, the probability $g_i^{(l)}$ can be interpreted as the *prior probability* $P(i|\mathbf{x}^{(l)})$, the probability of the ith label given only the input $\mathbf{x}^{(l)}$. Given these definitions, Equation 6 has the natural interpretation of moving the prior probabilities toward the posterior probabilities.

An interesting special case is an architecture in which the expert networks and the gating network are linear and the probability density associated with the experts is a Gaussian

with identity covariance matrix. In this case, Equations 5 and 6 yield the following on-line learning algorithm (*on-line* meaning that we have dropped the summation across l):

$$\Delta \theta_i = \rho h_i^{(l)} (\mathbf{y}^{(l)} - \mu_i^{(l)}) \mathbf{x}^{(l)T} \quad (7)$$

and

$$\Delta \eta_i = \rho (h_i^{(l)} - g_i^{(l)}) \mathbf{x}^{(l)T} \quad (8)$$

where ρ is a learning rate. Note that both of these equations have the form of the classical LMS rule (*delta algorithm*), with the updates for the experts in Equation 7 being modulated by their posterior probabilities.

It is also of interest to examine the expression for the posterior probability in the Gaussian case:

$$h_i^{(l)} = \frac{g_i^{(l)} e^{-1/2 (\mathbf{y}^{(l)} - \mu_i^{(l)})^T (\mathbf{y}^{(l)} - \mu_i^{(l)})}}{\sum_j g_j^{(l)} e^{-1/2 (\mathbf{y}^{(l)} - \mu_j^{(l)})^T (\mathbf{y}^{(l)} - \mu_j^{(l)})}} \quad (9)$$

This is a normalized distance measure that reflects the relative magnitudes of the residuals $\mathbf{y}^{(l)} - \mu_i^{(l)}$. If the residuals for expert i are small relative to those of the other experts, then $h_i^{(l)}$ is large; otherwise, $h_i^{(l)}$ is small. Note, moreover, that the $h_i^{(l)}$ are positive and sum to one for each $\mathbf{x}^{(l)}$; this implies that credit is distributed to the experts in a competitive manner.

It is straightforward to utilize other members of the exponential family of densities as component densities for the experts, to allow dispersion (e.g., covariance) parameters to be incorporated in the model, and to estimate the dispersion parameters via the learning algorithm (Jordan and Jacobs, 1994; Jordan and Xu, in press).

The Hierarchical Mixture of Experts (HME) Architecture

The ME architecture solves complex function approximation problems by allocating different modules to different regions of the input space. This approach can have advantages for problems in which the modules are simpler than the large network that would be required to solve the problem as a whole. If we now inquire about the internal structure of a module, however, we see that the same argument can be repeated. Perhaps it is better to split a module into simpler submodules rather than to use a single module to fit the data in a region. This suggests a thoroughgoing divide-and-conquer approach to supervised learning in which a tree-structured architecture is used to perform multiple, nested splits of the input space (Figure 2). The splitting process terminates in a set of expert networks at the leaves of the tree, which, because they are defined over relatively small regions of the input space, can fit simple (e.g., linear) functions to the data. This hierarchical architecture, suggested by Jordan and Jacobs (1994), has close ties to the classification and regression tree models in statistics and machine learning (e.g., Breiman et al., 1984). Indeed, the architecture can be viewed as a probabilistic variant of such models.

The mathematical framework underlying the HME architecture is essentially the same as that underlying the ME architecture. We simply extend the probability model to allow nested sequences of labels to be chosen, corresponding to the nested sequence of regions needed to specify a leaf of the tree. The probability model for a two-level tree is as follows:

$$P(\mathbf{y}|\mathbf{x}, \Theta) = \sum_i P(i|\mathbf{x}, \eta) \sum_j P(j|i, \mathbf{x}, v_i) P(\mathbf{y}|\mathbf{x}, \theta_{ij}) \quad (10)$$

which corresponds to a choice of label i with probability $P(i|\mathbf{x}, \eta)$, followed by a conditional choice of label j with probability $P(j|i, \mathbf{x}, v_i)$. This probability model yields the following log likelihood function:

Figure 2. A two-level binary hierarchical architecture. The top-level gating network produces coefficients g_i that effectively split the input space into regions, and the lower-level gating networks produce coefficients $g_{j|i}$ that effectively split these regions into subregions. The expert networks fit surfaces within these nested regions. Deeper trees are formed by expanding the expert networks recursively into additional gating networks and subexperts.

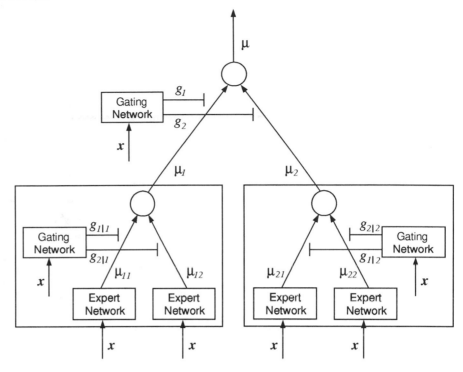

$$l(\mathscr{X}, \Theta) = \sum_l \log \sum_i P(i|\mathbf{x}^{(l)}, \boldsymbol{\eta}) \sum_j P(j|i, \mathbf{x}^{(l)}, \mathbf{v}_i) P(\mathbf{y}^{(l)}|\mathbf{x}^{(l)}, \boldsymbol{\theta}_{ij}) \tag{11}$$

where, as in the one-level case, the prior probabilities $g_i^{(l)} = P(i|\mathbf{x}^{(l)}, \boldsymbol{\eta})$ and $g_{j|i}^{(l)} = P(j|i, \mathbf{x}^{(l)}, \mathbf{v}_i)$ are defined in terms of underlying variables ξ_i and ξ_{ij} using the softmax function (cf. Equation 2). We also use Bayes's rule to define posterior probabilities in the obvious way:

$$h_i^{(l)} = \frac{g_i^{(l)} \sum_j g_{j|i}^{(l)} P(\mathbf{y}^{(l)}|\mathbf{x}^{(l)}, \boldsymbol{\theta}_{ij})}{\sum_j g_j^{(l)} \sum_k g_{k|j}^{(l)} P(\mathbf{y}^{(l)}|\mathbf{x}^{(l)}, \boldsymbol{\theta}_{jk})} \tag{12}$$

and

$$h_{j|i}^{(l)} = \frac{g_{j|i}^{(l)} P(\mathbf{y}^{(l)}|\mathbf{x}^{(l)}, \boldsymbol{\theta}_{ij})}{\sum_j g_{j|i}^{(l)} P(\mathbf{y}^{(l)}|\mathbf{x}^{(l)}, \boldsymbol{\theta}_{ij})} \tag{13}$$

The posterior probability $h_i^{(l)}$ can be viewed as the credit assigned to the ith nonterminal in the tree, and the posterior probability $h_{j|i}^{(l)}$ is the credit assigned to the branches below the nonterminals. The product $h_i^{(l)} h_{j|i}^{(l)}$ is therefore the credit assigned to expert (i, j).

A recursive relationship is available to compute the posterior probabilities efficiently in deep trees. The recursion proceeds upward in the tree, passing the denominator of the conditional posterior upward, multiplying by the priors, and normalizing (cf. the computation of h_i from $h_{j|i}$ in Equations 12 and 13).

We obtain a gradient ascent learning algorithm by computing the partial derivatives of l. If, as in the preceding section, we assume linear experts and a linear gating network, as well as Gaussian probabilities for the experts, we obtain the following LMS-like learning algorithm:

$$\Delta \boldsymbol{\theta}_{ij} = \rho h_i^{(l)} h_{j|i}^{(l)} (\mathbf{y}^{(l)} - \boldsymbol{\mu}_{ij}^{(l)}) \mathbf{x}^{(l)\mathsf{T}} \tag{14}$$

$$\Delta \boldsymbol{\eta}_i = \rho (h_i^{(l)} - g_i^{(l)}) \mathbf{x}^{(l)\mathsf{T}} \tag{15}$$

and

$$\Delta \mathbf{v}_{ij} = \rho h_i^{(l)} (h_{j|i}^{(l)} - g_{j|i}^{(l)}) \mathbf{x}^{(l)\mathsf{T}} \tag{16}$$

Each of these partial derivatives has a natural interpretation in terms of credit assignment. Credit is assigned to an expert by taking the product of the posterior probabilities along the path from the root of the tree to the expert (cf. Equation 14). The updates for the gating networks move the prior probabilities at a nonterminal toward the corresponding posterior probabilities, weighting these updates by the product of the posterior probabilities along the path from the root of the tree to the nonterminal in question (cf. Equation 16).

An EM Algorithm

Jordan and Jacobs (1994) have derived an Expectation Maximization (EM) algorithm for estimating the parameters of the ME and HME architectures. This algorithm, an alternative to gradient methods, is particularly useful for models in which the expert networks and gating networks have simple parametric forms. Each iteration of the algorithm consists of two phases: (1) a recursive propagation upward and downward in the tree to compute posterior probabilities (the "E step"), and (2) solution of a set of local weighted maximum likelihood problems at the nonterminals and terminals of the tree (the "M step"). Jordan and Jacobs (1994) tested this algorithm on a nonlinear system identification problem (the forward dynamics of a four-degree-of-freedom robot arm), and they report that it converges rapidly, converging nearly two orders of magnitude faster than backpropagation in a comparable multilayer perceptron network.

Discussion

Supervised learning can be viewed as the problem of approximating a nonlinear function from noisy samples of the function. Function approximation in neural networks has traditionally been based on superpositions of simple basis functions (typically either logistic functions or Gaussians). The work described in this article can be viewed as an alternative approach to function approximation that is based on the principle of divide-and-conquer rather than the principle of superposition. It is important to emphasize that neither divide-and-conquer nor superimposition is likely to work best in all cases; rather, the goal should be to characterize those classes of problems for which the different approaches are most appropriate. The use of probabilistic models can be helpful in this regard, clarifying the implicit assumptions made by the different forms of function approximator. Finally, in many problems the issue of learning speed is as important as the issue of approximation accuracy. In the area of learning speed, divide-and-conquer approaches appear to have a natural advantage.

We conclude with a brief list of suggestions for additional reading. The ME has been applied to control problems (Gomi and Kawato, 1992; Jacobs and Jordan, 1993), where it takes the form of an adaptive gain scheduling controller. In related work, a variant of the ME architecture has been applied by Cacciatore and Nowlan (1994) to the control of jump-linear systems. There have been applications of the HME architecture to speech coding (Waterhouse and Robinson, 1995; see also Hampshire and Waibel, 1989). A theoretical analysis of the ME and HME architectures has been provided by Jordan and Xu (in press); these authors present results on the convergence rates of the EM algorithm. Jordan (1994) discusses the model selection problem for HME architectures. Biological applications of the ME architecture have been presented by Jacobs and Kosslyn (in press) and Nowlan and Sejnowski (1993).

Acknowledgment.   Preparation of this paper was supported in part by a grant from ATR Human Information Processing Laboratories.

Road Maps: Dynamic Systems and Optimization; Learning in Artificial Neural Networks, Statistical
Background: I.3. Dynamics and Adaptation in Neural Networks
Related Reading: Averaging/Modular Techniques for Neural Networks; Bayesian Methods for Supervised Neural Networks; Modular Neural Net Systems, Training of

References

Breiman, L., Friedman, J. H., Olshen, R. A., and Stone, C. J., 1984, *Classification and Regression Trees*, Belmont, CA: Wadsworth International.

Cacciatore, T., and Nowlan, S. J., 1994, Mixtures of controllers for jump linear and nonlinear plants, in *Advances in Neural Information Processing Systems 6* (J. D. Cowan, G. Tesauro, and J. Alspector, Eds.), San Mateo, CA: Morgan Kaufmann.

Gomi, H., and Kawato, M., 1992, Recognition of manipulated objects by motor learning, in *Advances in Neural Information Processing Systems 4* (J. Moody, S. Hanson, and R. Lippmann, Eds.), San Mateo, CA: Morgan Kaufmann.

Hampshire, J., and Waibel, A., 1989, *The Meta-Pi network: Building distributed knowledge representations for robust pattern recognition*, Technical Report CMU–CS–89–166, Pittsburgh, PA: Carnegie Mellon University.

Jacobs, R. A., and Jordan, M. I., 1993, Learning piecewise control strategies in a modular neural network architecture, *IEEE Trans. Syst. Man Cybern.*, 23:337–345.

Jacobs, R. A., Jordan, M. I., Nowlan, S. J., and Hinton, G. E., 1991, Adaptive mixtures of local experts, *Neural Computat.*, 3:79–87. ◆

Jacobs, R. A., and Kosslyn, S. M., in press, Encoding shape and spatial relations: The role of receptive field size in coordinating complementary representations, *Cognit. Sci.*

Jordan, M. I., 1994, A statistical approach to decision tree modeling, in *Proceedings of the Seventh Annual ACM Conference on Computational Learning Theory* (M. Warmuth, Ed.), New York: ACM Press. ◆

Jordan, M. I., and Jacobs, R. A., 1994, Hierarchical mixtures of experts and the EM algorithm, *Neural Computat.*, 6:181–214. ◆

Jordan, M. I., and Xu, L., in press, Convergence properties of the EM approach to learning in mixture-of-experts architectures, *Neural Netw.*

Nowlan, S. J., 1990, Maximum likelihood competitive learning, in *Advances in Neural Information Processing Systems 2* (D. Touretzky, Ed.), San Mateo, CA: Morgan Kaufmann.

Nowlan, S. J., and Sejnowski, T., 1993, Filter selection model for generating visual motion signals, in *Advances in Neural Information Processing Systems 5* (S. J. Hanson, J. D. Cowan, and C. L. Giles, Eds.), San Mateo, CA: Morgan Kaufmann.

Waterhouse, S. R., and Robinson, A. J., 1995, Nonlinear prediction of acoustic vectors using hierarchical mixtures of experts, in *Advances in Neural Information Processing Systems 7* (G. Tesauro, D. Touretzky, and J. Alspector, Eds.), San Mateo, CA: Morgan Kaufmann, pp. 177–186.

Modular Neural Net Systems, Training of

P. Gallinari

Introduction

Since the mid-1980s, neural networks (NNs) have been used for many applications in different areas. After early success on relatively simple tasks, it appeared rapidly that using a single NN to build complex and high-level systems was a dead end in many cases. Although NNs are theoretically capable of learning complex functions, many problems usually forbid the use of a single large NN in practical applications or for very complex tasks, e.g., perception or distributed decision making.

Important achievements with NNs—for example, in speech or image processing—have been accomplished by designing systems in which several NN modules cooperate with each other or with other techniques to complete a global task. We will use the generic term *modular systems* for these different architectures, and the term *hybrids* will refer to heterogeneous systems.

Up to now, modular systems have mostly been developed as ad hoc solutions for a given task. There exist several basic theoretical and practical problems which remain to be solved. We will focus here on the development of modular systems for application purposes. In the next section, we discuss the need for such systems. Then we present a classification of these systems and give some examples. In the last section, we present some solutions for their design and training.

Why Modular Systems?

Different problems may lead to the need for modular systems; some of them can be described as follows:

- *Reducing model complexity.* The complexity of global NN systems usually drastically increases with the task size or difficulty. Modular systems allow us to keep this complexity proportional to the task. Sophisticated systems have been designed in several areas, ranging from speech applications (Bennani and Gallinari, 1992) to motor control (Gomi and Kawato, 1993).
- *Incorporating knowledge.* The system architecture may incorporate a priori knowledge when there exists an intuitive or a mathematical understanding of problem decomposition; for example, the current state of knowledge in image or visual processing has led to the design of hierarchical systems. In comparison to a global NN, the number of parameters is kept small and the search space and training time are reduced. Such knowledge acts as a constraint on the learning problem and allows for better generalization performances.
- *Data fusion and prediction averaging.* Modular systems allow us to take into account data from different sources and nature. Combining or averaging the output of several modules may lead to increased robustness and better generalization.
- *Hybrid systems.* Heterogeneous systems allow us to combine different techniques to perform successive tasks, ranging, e.g., from signal to symbolic processing.
- *Learning different tasks simultaneously.* Specialized modules may be shared among several systems trained to perform different tasks.
- *Robustness and incrementality.* Cooperative systems perform better and are more robust than single-stage systems (de Bollivier, Gallinari, and Thiria, 1991; Fogelman, Lamy, and Viennet, 1993). They are easily modified and extended.

In many other cases—e.g., for cognitive modeling, computational neuroscience, or hardware implementations—modularity is a basic requirement. Other articles in this *Handbook* cover recent developments in these areas.

Classes of Modular Systems

Although the motivations we have discussed may lead to a large variety of modular systems, it is possible to identify basic principles. We propose below a classification and give some examples. These have been chosen from the literature to illustrate different situations and are not intended to be exhaustive.

Partitioning of the input space.   Instead of training a large system on the whole input data space, it is often more efficient to partition the latter into several subspaces and to train a simple system to fit the local data onto each subspace. Partitioning may be performed once or recursively, as in trees (see MODULAR AND HIERARCHICAL LEARNING SYSTEMS). Partition-based modular systems, by far the most frequent in the NN literature, have been used for classification or regression. For example, Bennani and Gallinari (1992) describe a speaker identification system in which speakers are first classified into broad categories before their fine identification by specialized modules, the outputs of which are then combined (see SPEAKER IDENTIFICATION). Gomi and Kawato (1993) propose two cooperative architectures for the recognition and control of manipulated objects. Here, expert networks specialize either in portions of the input space or in problem-specific functions. For speech recognition, in order to model the nonstationarity of speech signals, multistate systems (Iso and Watanabe, 1991) are trained to locally model specific segments of speech. Several models of neural trees inspired by the statistical and artificial intelligence literature on classification trees have been proposed. Data

fusion also falls into this category; the partition of the input space corresponds in this case to the different sensors.

Successive processing.   Here, the global problem is decomposed into successive tasks, each of them being carried out by a specialized module. Thiria et al. (1992) describe a system for a large application in remote sensing. It is composed of three successive layers of NNs, each layer corresponding to a specific processing of the data and built from several NNs which operate in parallel. Combination of successive classifiers is discussed in de Bollivier et al. (1991) and Fogelman et al. (1993), where it is shown to increase the performances compared to individual classifiers. Similar ideas have been proposed for prediction in time-series applications (Ginzburg and Horn, 1994). Most hybrid systems also fall into this category. In the connectionist community, the most extensive and advanced use of hybrids has been for speech processing. Many systems have been proposed based on the cooperation of NN and Hidden Markov Models (HMM) or dynamic programming (see SPEECH RECOGNITION: PATTERN MATCHING). In these systems, NNs are used for speech segment modeling and local scoring, whereas segmentation is performed by dynamic programming techniques. Early examples may be found in Bourlard and Wellekens (1990), Iso and Watanabe (1991), Driancourt and Gallinari (1992), and Bengio et al. (1992).

Combining decisions.   One way to increase the performances of individual systems is to process the data by several independent modules and to combine their outputs. Such systems may cover a larger range of situations and have lower prediction errors than individual modules. Recently, combination of predictions has emerged as an alternative to model selection. This approach has been presented to the NN community by Battiti and Colla (1994) and Wolpert (1994). (See also AVERAGING-MODULAR TECHNIQUES FOR NEURAL NETWORKS).

This classification is not unique and is merely intended to illustrate the variety of modular systems. Many systems do combine these different aspects.

Designing and Training Modular Systems

The need for modular systems is not a specificity of NNs. The design and analysis of cooperative systems have been the subject of intensive research over the last 20 years. Cooperation has been formalized extensively by system theory in the 1970s and analyzed from different points of view, e.g., pattern recognition or control. We will focus here on NN modular systems. Attempts to formalize the design and training of these systems are recent, and it is still questionable whether a general theory of these systems would be of any help. In any case, it is useful to identify and formalize the main problems of cooperative systems and to present some techniques for deciding when to use and how to implement cooperation.

Designing a modular system amounts to performing the right *decomposition*—using the right technique at the right place and, when possible, estimating the parameters optimally according to a global goal. Modules will adjust their parameters according to local goals defined by external signals. Global training will set up these local goals so that the global computation is optimal in some sense.

Cooperatively training heterogeneous modules, each described with its own formalism (adaptive optimization, differential equations, heuristic rules, etc.), becomes very rapidly impossible. The only way to bypass this limitation is to define a unified formalism for the subsystems. The modules will thus share common basic functional principles which will allow co-operation. In order to design an ideal system, we then have to solve successively three basic problems which will be addressed in the following sections:

- Define a unified formalism for connectionist algorithms.
- Perform optimal training.
- Perform task decomposition.

Unified Formalism

A large number of connectionist algorithms are expressed as discrete adaptive learning rules of the form

$$w(n) = w(n - 1) - \varepsilon(n)F(x(n), w(n - 1)) \qquad (1)$$

where w are the system parameters (e.g., weights) and x is the concept to be learned. F may be the derivative of some cost function (e.g., multilayer perceptron, radial basis functions, adaline, perceptron). In many cases (e.g., learning vector quantization, topological maps, competitive learning), F corresponds to a heuristic weight updating rule. It appears that most of the heuristic rules proposed in the NN literature may also be shown to optimize some local analytical goal J (Bottou and Gallinari, 1992), so that Equation 1 may be replaced in almost any case by some adaptive gradient algorithm. For simplicity, we will use the simple form

$$w(n) = w(n - 1) - \gamma(n)\nabla_w J(x(n), w(n - 1)) \qquad (2)$$

where J is a distance or information measure defined on the system output for pattern x, and ∇_w is its gradient. This form is more amenable to formal derivations than Equation 1, but more sophisticated algorithms do exist.

Adaptive algorithms like Equation 2 have arisen in the field of stochastic approximations. In the context of learning systems, they have been extensively studied by Tsypkin (1973) as a means of optimizing a global cost function C (usually the expectation of the *local cost* function J over the space X of the concept to be learned; see Equation 3). When C is identifiable, it thus represents explicitly the *goal of learning*.

$$C(w) = \int_X J(x, w)p(x)\, dx = E_X\{J(x, w)\} \qquad (3)$$

Using the probabilistic formalism for learning proposed in Tsypkin (1973), this problem is addressed for NN algorithms in Bottou and Gallinari (1992), where an analytical goal is derived for many NN algorithms, including heuristic learning rules. Although inferring a local cost function J from updating rules has often been easy, it is not always so simple to infer the global cost C. In such cases, modifications of original algorithms have been proposed, allowing us to make the goal of learning explicit for these algorithms.

This formalism applies as well to many classical techniques in the field of speech and signal processing, prediction, approximation, and control. It unifies techniques from different domains and thus allows us to train heterogeneous modules simultaneously in hybrid systems.

Of course, adaptive or stochastic techniques are not the only viable formalism for connectionist algorithms. For many applications, the adaptive requirement which has been essential in the derivation of many original NN algorithms is no longer a necessity. In this case, classical optimization techniques may be used. However, the formalism presented here is more general; it can describe a larger number of NN algorithms than nonstochastic methods and has been found more efficient for large problems.

Training Algorithms

In most modular systems, both in the neural and nonneural literature, modules are trained independently or sequentially according to a local, user-defined goal (e.g., Bennani and Gallinari, 1992; Thiria et al., 1992). For these systems, neither the task decomposition nor the estimation of module parameters is optimal for the global computation. Sometimes it is possible to improve upon this situation, optimizing the system as a whole according to a global goal. This is possible only if the different modules have been expressed using a unified formalism such as that of the preceding section.

Bottou and Gallinari (1991) have proposed a general adaptive algorithm for global training. A cost $J(x, w)$ for the whole system may be written as the composition of several functions which will either measure the quality of the system through the local errors of modules or define the decomposition of the parametric relationship between the inputs and outputs of the different modules. Training the system amounts to optimizing J with respect to constraints which define respectively the *intra*- and *intermodule* architecture.

Global training of modular systems has led to improved performances for different applications in signal processing (de Bollivier et al., 1991) and image processing (Fogelman et al., 1993). For speech, Bengio et al. (1992) discuss extensively the problem of global training for MLP-HMM hybrids. Global tuning also allows use of smaller systems than does local training (Driancourt and Gallinari, 1992).

Task Decomposition

Task decomposition may be either performed explicitly or learned. It can apply to one level—e.g., when using competition among modules or for fusion applications. It can also be used for building hierarchies which handle successive parts of the whole processing. The techniques discussed in this article may be used at different stages of the design, but the global system structure is up to the user.

Explicit decomposition will be performed by a designer with a strong understanding of the problem. Several current large modular systems have been designed according to domain specialists (see, for example, Thiria et al., 1992). Attempts to integrate symbolic and numerical paradigms have also motivated the use of expert knowledge together with numerical optimization (see LEARNING BY SYMBOLIC AND NEURAL METHODS).

Automatic decomposition is most useful when, because of the nature or the complexity of the task, expert knowledge is missing. Jordan and co-workers (see MODULAR AND HIERARCHICAL LEARNING SYSTEMS) have proposed a series of algorithms for automatic task decomposition. In this model, expert networks learn to specialize onto subtasks, space regions, or function parts, and cooperate via a gating net. Neural trees provide another way to perform automatic decomposition by recursively partitioning the input space into successive regions. This decomposition usually aims at optimizing information criteria. These two approaches share many similarities, but operate at different levels of granularity.

Discussion

Several problems are inherent to the design and training of modular systems. Except for tree-like approaches, automatic design is still an open problem. Global training may be plagued with local minima which may be avoided by using careful initialization, constraints, or regularization. Different optimization techniques may be used if gradient descent is found inefficient. Many other problems remain to be resolved for the development and understanding of modular NN systems. These systems have, however, already been found to be of considerable utility in several domains.

Road Map: Learning in Artificial Neural Networks, Deterministic
Background: I.3. Dynamics and Adaptation in Neural Networks
Related Reading: Stochastic Approximation and Neural Network Learning

References

Battiti, R., and Colla, A. M., 1994, Democracy in neural nets: Voting schemes for classification, *Neural Netw.*, 7:691–707.

Bengio, Y., de Mori, R., Flammia, G., and Kompe, R., 1992, Global optimization of a neural network—Hidden Markov Model hybrid, *IEEE Trans. Neural Netw.*, 3:252–269.

Bennani, Y., and Gallinari, P., 1992, Task decomposition through a modular connectionist architecture: A Talker Identification System, in *Third International Conference on Artificial Neural Networks* (I. Aleksander and J. Taylor, Eds.), Amsterdam: North-Holland, vol. 1, pp. 783–786.

Bottou, L., and Gallinari, P., 1991, A framework for the cooperation of learning algorithms, in *Advances in Neural Information Processing Systems 3* (R. P. Lippmann, J. E. Moody, and D. S. Touretzky, Eds.), San Mateo, CA: Morgan Kaufmann, pp. 781–788.

Bottou, L., and Gallinari, P., 1992, A unified formalism for NN training algorithms, in *Proceedings of the International Joint Conference on Neural Networks*, Seattle: IEEE, vol. 4, pp. 7–12.

Bourlard, H., and Wellekens, C., 1990, Links between Markov models and multilayer perceptrons, *IEEE PAMI*, 12:1167–1178.

de Bollivier, M., Gallinari P., and Thiria, S., 1991, Neural nets and task decomposition, in *Proceedings of the International Joint Conference on Neural Networks*, Seattle: IEEE, vol. 2, 573–576.

Driancourt, X., and Gallinari, P., 1992, A speech recognizer optimally combining learning vector quantization, dynamic programming and multilayer perceptron, in *International Conference on Acoustic Speech and Signal Processing*, Toronto: IEEE, pp. 609–612.

Fogelman, F., Lamy, B., and Viennet, E., 1993, Multimodular neural network architectures for pattern recognition: Applications in optical character recognition and human faces recognition, *Int. J. Pattern Recognition Artif. Intell.*, 7(4).

Ginzburg, I., and Horn, D., 1994, Combining neural networks for time series analysis, in *Advances in Neural Information Processing Systems 6* (J. D. Cowan, G. Tesauro, and J. Alspector, Eds.), San Mateo, CA: Morgan Kaufmann, pp. 216–225.

Gomi, H., and Kawato, M., 1993, Recognition of manipulated objects by motor learning with modular architecture networks, *Neural Netw.*, 6:485–497.

Iso, K., and Watanabe, T., 1991, Large vocabulary speech recognition using neural prediction model, in *International Conference on Acoustic Speech and Signal Processing*, Toronto: IEEE, pp. 57–60.

Thiria, S., Mejia, C., Badran, F., and Crépon, M., 1992, Multimodular architecture for remote sensing operations, in *Advances in Neural Information Processing Systems 4* (J. E. Moody, S. J. Hanson, and R. P. Lippmann, Eds.), San Mateo, CA: Morgan Kaufmann, pp. 675–682.

Tsypkin, Y. A. Z., 1973, *Foundations of the Theory of Learning Systems*, New York: Academic Press. ◆

Wolpert, D. H., 1994, Stacked generalization, *Neural Netw.*, 5:241–260.

Motion Perception

Ellen C. Hildreth and Constance S. Royden

Introduction

When object surfaces move relative to an observer, a continually changing pattern of light intensity is projected onto his retina. Measurement of the movement of features in this changing image allows the observer to track objects, segment the scene into multiple surfaces, recover the three-dimensional (3D) shape and motion of objects, and recover the 3D movement of the observer through space.

Models proposed for motion analysis in biological vision systems typically divide motion processing into two stages. First, one must measure the direction and speed of movement of features in the two-dimensional (2D) image. Second, one must interpret these measurements in terms of the 3D structure of the scene and the motion of the observer or objects in the scene. Figure 1 illustrates these two stages. In Figure 1*A*, a stationary observer views a scene that contains a rotating cylinder and translating square in front of a plane that is moving toward the observer. Figure 1*B* shows the resulting *optic flow*, which is the pattern of image velocities that is projected onto the retina. The dashed lines indicate the borders of the cylinder and square, which must be inferred from this flow pattern. Along these borders, a sudden change in the direction and speed of motion signals their presence. Over the surface of the cylinder, the smooth variation in velocity conveys its 3D shape. On the background plane, features move radially outward from a central location called the *focus of expansion* that corresponds to the direction of motion of the plane relative to the observer. The two stages of motion measurement and interpretation aim at constructing representations of information such as that contained in Figures 1*B* and 1*A*. This article discusses models for performing these two stages in biological systems.

Models for 2D Motion Measurement

Temporal variations of image intensity provide the only information available for deriving the optic flow, or pattern of image velocities, shown in Figure 1*B*. For biological systems, the first stage in analyzing these temporal variations uses neural mechanisms with limited spatial and temporal receptive fields. Such mechanisms provide information limited by the *aperture problem* illustrated in Figure 2*A*. For the moving edge **E**, viewed within the aperture **A**, each of the velocities shown by the arrows yields the same temporal variation of image intensity. A motion detector with such a limited field of view can only measure the component of velocity in the direction perpendicular to the orientation of this edge. To compute the overall pattern of 2D image velocities, the outputs of multiple motion detectors within an extended image region must be integrated.

Models for the initial detection of image movement are presented in DIRECTIONAL SELECTIVITY IN THE CORTEX and DIRECTIONAL SELECTIVITY IN THE RETINA. Most motion detection models require that the image intensity function varies continuously with time. For these models, if the image sequence consists of discrete images, the spatial and temporal displacements

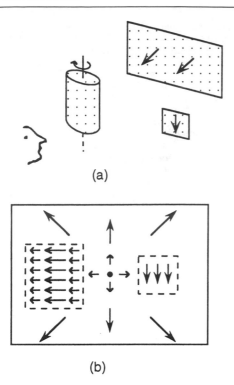

(a)

(b)

Figure 1. *A*, Observer viewing a moving 3D scene. *B*, The resulting pattern of image velocities.

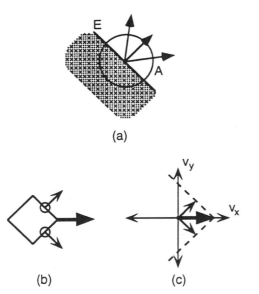

(a)

(b)   (c)

Figure 2. *A*, The aperture problem. Only the component of velocity in the direction perpendicular to the orientation of the edge **E** can be recovered from information viewed within the aperture **A**. *B*, The thin arrows represent the components of velocity in the direction perpendicular to the edges of a diamond translating to the right. *C*, The motion measurements from part *B* are replotted in velocity space, where the axes represent the horizontal and vertical components of velocity. Each measurement is consistent with a family of velocities that terminate along the corresponding dashed line. Only the velocity shown by the thick arrow is consistent with both measurements.

of features between two successive images must be small. This type of motion is often referred to as *short-range* motion. In addition to processing short-range motion, the human visual system has the ability to infer motion when the spatial and temporal displacements between discrete image frames are large. To analyze this *long-range* motion, a computational mechanism must establish a *correspondence* between features in successive images. That is, the corresponding projections of a single physical feature in two successive frames must be identified. Models of motion correspondence have been proposed to analyze short-range and long-range motion. Reviews of models of motion correspondence can be found elsewhere (Hildreth and Koch, 1987; Aggarwal and Nandhakumar, 1988; Martin and Aggarwal, 1988; Mitiche and Bouthemy, in press; see STEREO CORRESPONDENCE AND NEURAL NETWORKS for the analogous problem for stereo vision, and REGULARIZATION THEORY AND LOW-LEVEL VISION for a general theoretical framework). The remainder of this section addresses models that integrate short-range motion measurements to compute an image velocity field.

Motion Integration

Because of the aperture problem, the image $I(x, y, t)$ alone does not uniquely determine the pattern of 2D velocities. A single measurement of motion within a limited image region can be consistent with a family of 2D velocities, as suggested in Figure 2*A*. The solution to the aperture problem requires the integration of multiple motion measurements over an extended image region. To obtain a unique solution, this integration requires additional assumptions about the expected structure of the 2D velocity field.

Most models of motion integration assume that velocity is constant within limited image regions. One scheme uses the components of velocity derived from two features at different orientations in the image to define this 2D velocity uniquely. This is accomplished by computing the intersection of the constraints imposed by these two velocity components, as shown in Figures 2*B* and 2*C*. Movshon et al. (1985) proposed a mechanism based on this intersection of constraints. Movshon et al. also provide physiological evidence suggesting that, in the macaque monkey, motion-selective neurons in area V1 (primary visual cortex) and in the earlier stages of processing in the middle temporal area MT (layers 4 and 6) respond only to the component of velocity in the direction perpendicular to a moving, oriented grating pattern. In layers 2, 3, and 5 of area MT, a population of neurons was found to respond to the overall 2D direction of motion of a moving pattern containing two gratings at different orientations. Movshon et al. proposed that this latter population may serve to integrate measurements of the components of motion to solve the aperture problem. Sereno (1993) presents a neural network model that learns how to solve the aperture problem in the way that Movshon et al. suggest.

Another approach to the motion integration stage uses the fact that a single 2D velocity is represented by a plane in a 3D frequency space whose axes are the horizontal, vertical, and temporal frequencies. Thus, spatiotemporal energy mechanisms that are selective for a range of different spatial and temporal frequencies give maximal responses when their preferred frequencies are located along this plane. Models proposed by Heeger (1987) and Grzywacz and Yuille (1990) rely on the detection of this plane of maximal motion energy (see DIRECTIONAL SELECTIVITY IN THE CORTEX).

Finally, models have been proposed that use a more general assumption that the pattern of image velocities varies smoothly within limited image regions. These models compute a 2D velocity field that minimizes the variation in velocity over the regions (for review, see Hildreth and Koch, 1987). Wang et al. (1989) show how one such model, proposed by Horn and Schunck, can be implemented in a neural network. In this implementation, the first stage of processing, which could be performed in the retina, yields measurements of the spatial and temporal gradients of the changing visual image. These measurements are combined to yield orientation- and direction-selective measurements of local motion components. Finally, a class of neurons computes a 2D velocity for each image region that is consistent with the local direction-selective motion measurements. Interaction between neighboring velocity units yields an overall pattern of velocities with minimal variation over the image.

These models of motion integration generally do not perform well in regions where the velocity field is discontinuous, for example, in the vicinity of boundaries between object surfaces undergoing different motions. Several approaches to the detection of motion discontinuities have been considered (for review, see Hildreth and Koch, 1987; Martin and Aggarwal, 1988; and Mitiche and Bouthemy, in press). These approaches include: (1) the design of center-surround mechanisms to detect large spatial changes in the velocity field; (2) the detection of image regions in which the initial motion components are not consistent with a single 2D velocity; and (3) the detection of steep variations in the computed velocity field.

Models for 3D Structure and Motion Recovery

The variations in 2D velocity across the image can be used to recover the 3D structure of object surfaces and the 3D motion of the observer and objects in the environment. The problems of recovering 3D structure and 3D motion are closely related theoretically. The main problem faced by both is that the pattern of image velocities does not uniquely define the 3D motion and structure, so that a unique interpretation requires additional assumptions. Most models assume *rigidity*; that is, they attempt to interpret the moving image as the projection of object surfaces that move rigidly over time. Some models only recover the 3D shape of moving objects, which may be represented by the depth or surface orientation at discrete image locations. Other models also recover the parameters of motion of the observer or objects. These parameters typically include three parameters of translation and three parameters of rotation, allowing the representation of general 3D rigid motion between an observer and object surface.

Models for recovering 3D motion and structure fall into two main classes, depending on whether they rely on discrete or continuous image motion measurements. In the discrete approach, isolated image features are tracked over time. Their sequence of positions forms the input to a system of equations, the solution of which depends on the parameters of 3D structure, the motion of the observer, and the motions of object surfaces. In the continuous approach, 3D structure and motion parameters are derived from 2D velocities, sometimes together with their spatial and temporal derivatives, at one or more instants of time.

The discrete approach has yielded theoretical results concerning the minimal number of motion measurements required to compute 3D structure and motion uniquely (for review, see Aggarwal and Nandhakumar, 1988). For example, for orthographic projection, one can compute a unique 3D structure for as few as four points given their image positions at three different times. This requires that the points are not coplanar and move rigidly in space. Results of this sort indicate that, in prin-

ciple, a unique 3D structure and motion can be recovered by integrating 2D motion information over a limited spatial region and temporal window. Models based on the direct application of these theoretical results, however, do not perform well in the presence of error in the image motion measurements. This suggests that these models may not reflect the mechanisms underlying the human recovery of 3D structure from motion.

Some recent models compute 3D structure and motion by integrating 2D motion information over an extended image region or over extended time. These models perform better in the presence of motion measurement error than those that are limited in space and time. For example, Ullman (see Martin and Aggarwal, 1988) proposed a model for the human recovery of structure from motion that builds up 3D structure over time. The image is assumed to consist of a set of discrete points that are the projections of features moving in 3D space. The 3D structure of these points is represented by X, Y, Z coordinates (i.e., measured image coordinates and computed depths), which change over time. The model makes incremental improvements in the estimated structure as it considers each new view of a set of moving features. This model allows the computed 3D structure of the object to undergo a minimal change from one moment to the next, thus relaxing the rigidity assumption. The formulation of a biologically plausible implementation of Ullman's model has been addressed in Ando (1993).

Other discrete models integrate 2D motion information over an extended time, using a technique based on recursive estimation theory known as Kalman filtering (for review, see Heeger and Jepson, 1992; see also ADAPTIVE SIGNAL PROCESSING). This technique sequentially predicts and updates current estimates of 3D motion and structure as new information about the 2D motion of image features becomes available. The Kalman filter has been used to estimate the depths or relative depths of moving features and the six parameters describing the 3D translation and rotation of an observer relative to a stationary surface. The biological plausibility of this approach has not been addressed.

A variety of continuous models have been proposed for recovering 3D motion and structure (for references, see Aggarwal and Nandhakumar, 1988; Martin and Aggarwal, 1988; Heeger and Jepson, 1992; and Perrone, 1992). Koenderink and van Doorn suggested using local changes in image structure (e.g., divergence, curl and shear) to measure properties of motion and surface shape. Longuet-Higgins and Prazdny proposed an algorithm that uses measures of local velocity changes to derive the parameters of shape and 3D motion of planar surfaces.

Similar to the early discrete methods that integrate 2D motion over limited spatial and temporal windows, the model of Longuet-Higgins and Prazdny proved to be very sensitive to error in the image velocity measurements. Approaches that integrate 2D motion information over more extended spatial regions have yielded more reliable behavior (see Heeger and Jepson, 1992). One such model relies on the special case of a moving plane, which gives rise to an image velocity field that can be represented by second-order polynomials. Waxman and his colleagues proposed models for deriving the parameters of this second-order flow field. In another approach, Bruss and Horn developed a model that uses least-squared error minimization to compute the depth of the 3D surface and the parameters of the observer's translation and rotation that best account for the measured 2D image velocities. This model also integrates velocity information over extended image regions.

Several continuous models divide the motion recovery problem into separate computations of scene structure, observer translation, and observer rotation (for review, see Heeger and

Jepson, 1992; Perrone, 1992). In terms of Figure 1, a representation of scene structure includes the 3D shape of the cylinder, locations of the boundaries of the cylinder and square, and the depth of the cylinder and square relative to the background plane. Translation of the observer relative to the scene yields an expanding pattern of motion, as illustrated for the background plane in Figure 1B. Rotation of the observer yields a pattern of velocities whose directions and speeds are roughly constant over a large part of the visual field.

Perrone's model, for example, first recovers the parameters of the observer's rotation with filters that find the best common motion vector over the full 2D velocity field. It then uses these rotational parameters to construct a bank of 2D filters, each of which is optimally tuned to detect a possible focus of expansion at a particular location in the visual field. After the model recovers the observer's 3D translation, it uses this information to recover the depth, within a scale factor, to each surface location. Perrone's model also uses biologically motivated input motion measurements, such as the outputs of spatiotemporal motion energy filters. Hatsopoulos and Warren (1991) present a simple, linear neural network that can determine an observer's direction of motion from the optic flow field, assuming that the observer undergoes a pure translation relative to a rigid environment. The models of Perrone (1992) and Hatsopoulos and Warren (1991) are both intended to reflect known physiological behavior of neurons in areas MT and MST (middle superior temporal cortex) in primate visual cortex.

In a different continuous model, Heeger and Jepson (1992) derive a set of algebraic equations that relate the observer's direction of translation to the 2D image velocities, independent of the observer's rotation and scene structure. This analysis forms the basis for a model that first recovers the translational parameters and then independently recovers the rotational parameters and scene structure. Lappe and Rauschecker (1993) presented a neural network implementation of this model. One final approach uses the fact that image regions of significant depth variation cause sudden changes in 2D velocity over space that can be used directly to recover the location of the focus of expansion that corresponds to the observer's direction of translation. Models based on this observation have been proposed by Longuet-Higgins and Prazdny and by Rieger and Lawton (for review, see Hildreth, 1992).

Discussion

This article has focused on models for measuring 2D image motion and for recovering the 3D structure and motion of the observer and scene. Superficially, most of the models cited here exhibit general biological plausibility, in that they can be implemented by a network of simple, local processing mechanisms operating in parallel. Some models have been implemented in artificial neural networks (for example, Sereno, 1993; Hatsopoulos and Warren, 1991), but such networks often embody a restricted formulation of the motion analysis problem. As noted throughout this article, physiological observations provide some indication of where these computations take place and general properties of the representation of 2D and 3D motion information. It remains a challenge to incorporate all of the important aspects of motion analysis, including the measurement of short-range and long-range motion, detection of motion discontinuities, recovery of 3D structure and observer motion, into a neuronal model that exhibits a broad range of human behavior.

Acknowledgments.   Dr. Hildreth is supported by the National Science Foundation (IRI 9196196). Dr. Royden is supported by the Science Scholars Fellowship Program of the Bunting Institute of Radcliffe College.

Road Map: Vision
Related Reading: Active Vision; Motion Perception: Self-Organization; Perception of Three-Dimensional Structure

References

Aggarwal, J. K., and Nandhakumar, N., 1988, On the computation of motion from sequences of images: A review, *Proc. IEEE*, 76:917–935. ◆

Ando, H., 1993, Dynamic reconstruction and integration of 3D structure information, PhD Thesis, MIT, Department of Brain and Cognitive Sciences.

Grzywacz, N. M., and Yuille, A. L., 1990, A model for the estimate of local image velocity by cells in the visual cortex, *Proc. R. Soc. Lond. B Biol. Sci.*, 3:15–44.

Hatsopoulos, N. G., and Warren, W. H., 1991, Visual navigation with a neural network, *Neural Netw.*, 4:303–317.

Heeger, D. J., 1987, Model for the extraction of image flow, *J. Opt. Soc. Am. A*, 4:1455–1471.

Heeger, D. H., and Jepson, A. D., 1992, Subspace methods for recovering rigid motion, I: Algorithm and implementation, *Int. J. Comput. Vis.*, 7:95–117.

Hildreth, E. C., 1992, Recovering heading for visually-guided navigation, *Vis. Res.*, 32:1177–1192.

Hildreth, E. C., and Koch, C., 1987, The analysis of visual motion: From computational theory to neuronal mechanisms, *Annu. Rev. Neurosci.*, 10:477–533. ◆

Lappe, M., and Rauschecker, J. P., 1993, A neural network for the processing of optic flow from ego-motion in man and higher mammals, *Neural Comp.*, 5:374–391.

Martin, W. N., and Aggarwal, J. K., Eds., 1988, *Motion Understanding: Robot and Human Vision*, Boston: Kluwer.

Mitiche, A., and Bouthemy, P., in press, Computation and analysis of visual motion: A review, *Int. J. Comput. Vis.* ◆

Movshon, J. A., Adelson, E. H., Gizzi, M. S., and Newsome, W. T., 1985, The analysis of moving visual patterns, in *Pattern Recognition Mechanisms* (C. Chagas, R. Gattas, and C. G. Gross, Eds.), Rome: Vatican Press, pp. 117–151.

Perrone, J., 1992, Model for the computation of self-motion in biological systems, *J. Opt. Soc. Am. A*, 9:177–194.

Sereno, M, 1993, *Neural Computation of Pattern Motion*, Cambridge, MA: MIT Press.

Wang, H. T., Mathur, B., Hsu, A., and Koch, C., 1989, Computing optical flow in the primate visual system: Linking computational theory with perception and physiology, in *The Computing Neuron* (R. Durbin, C. Miall, and G. Mitchinson, Eds.), Reading, MA: Addison-Wesley.

Motion Perception: Self-Organization

Jonathan A. Marshall

Introduction: Motion Perception and Learning

Visual motion information has several obvious uses, including *segmenting* objects or image regions based on optic flow discontinuities; *navigating* and determining heading from wide-field optic flow; *tracking* moving objects; and computing *depth* from motion parallax. However, this article concerns a less obvious use: the special role of visual motion in helping visual systems *learn* about the structure of the visual world.

Motion information is particularly useful in helping visual systems learn because it discloses some important *invariances* in the visual world. As a visual object (e.g., a chair) moves, its image *appearance* (location, size, shape, color/highlights, etc.) changes, yet the object itself usually keeps its perceptual *identity* (it stays represented as the same chair). By observing moving objects and assuming that their perceptual identity is invariant, a visual system can learn about the *dimensions* along which a moving object's appearance may vary. The key research question within the scope of this article is: How can a visual system learn to detect and represent both the invariant visual identity and the variant appearance of objects, in a manner consistent with the structure of the visual world?

Motion detection can be broken conceptually into two stages: (1) extraction of basic local motion signals from image sequences, and (2) integration of multiple motion signals across the images to create globally consistent, coherent representations of object motion. This article briefly describes issues of self-organization in both stages.

Self-Organization of Basic Motion Detectors

Many neurons in visual cortex respond selectively to motion direction, speed, and location, in the changing light patterns projected on the retina (see MOTION PERCEPTION; DIRECTIONAL SELECTIVITY IN THE CORTEX; and VISUAL CORTEX CELL TYPES AND CONNECTIONS). How can a complete set of motion detector neurons, with a spectrum of preferred directions, speeds, and locations, self-organize? For simplicity, this discussion assumes the existence of a collection of feature detector neurons, with receptive fields distributed retinotopically around the image (although this assumption is not strictly necessary).

Light activates the feature detector neurons in sequence over time. When a visual feature sweeps by, the feature detector neurons that it activates excite many neurons in the next stage, which are initially nonspecific or uncommitted.

Local correlations across time exist between the activation of pairs of nearby feature detector neurons, even when many moving objects are present simultaneously. Standard Hebbian-type learning rules, in a competitive learning framework (see COMPETITIVE LEARNING), can use such correlations (plus a range of signal transmission time delays) to shape the receptive fields of the neurons in the higher stage (Fredericksen, 1993). Each active neuron in the next stage becomes more selective for one motion sequence and less selective for other motion sequences. Thus, these neurons tend to become local motion detectors, selective for a range of motion sequences.

Strong lateral inhibition gives winner-take-all behavior; it prevents more than one local motion detector neuron within a neighborhood from becoming simultaneously active. Thus, all the local motion detector neurons tend to learn to respond to different motion sequences. Since the correlations are strongest

at short spatial and temporal ranges, this process leads to the development of a short-range motion detection network. However, with strong lateral inhibition, the stage of local motion detectors can represent motion of only one visual feature (or just a few features, as long as they do not come too close together).

This model is thus inadequate. To produce usable representations of object motion, other motion processing stages must combine the outputs of these short-range motion-detector neurons. The local two-dimensional motion information gathered across the image must be combined to produce representations of three-dimensional object motion.

Learning About Motion Transparency and Element Integrity Constraints

In "motion transparency," dots moving in two directions (for example) within an image region are perceived as two overlaid moving sheets. This phenomenon suggests that the winner-take-all strategy is wrong as a model of learning in vision. How can a neural network simultaneously represent multiple transparently overlaid motions within an image region? Moreover, how can it learn to do so? Essentially the same issue arises near occlusion boundaries, where multiple motions may be present within a local neighborhood.

One self-organizing solution is to allow the lateral inhibitory connection weights, as well as the bottom-up excitatory weights, to vary according to a learning rule (Hubbard and Marshall, 1994; Marshall, 1990, 1991; Marshall, Alley, and Hubbard, 1994). An "anti-Hebbian" inhibitory learning rule (see POST-HEBBIAN LEARNING RULES) can cause the lateral inhibitory connections to become selective in a way that lets motion transparency be represented. Such a rule strengthens the inhibitory connections between local motion detector neurons that represent similar motions and weakens the inhibitory connections between local motion detector neurons that represent dissimilar motions. The selectively weakened inhibition lets certain combinations of neurons (those that represent different motions, even within the same image region) become simultaneously active in response to image sequences containing multiple motions. Thus, with the inhibitory learning rule, the network is no longer winner-take-all, and multiple motions, even motions that are transparently overlaid, can be simultaneously represented. The same inhibitory learning rule also lets a visual system learn a network structure (Hubbard and Marshall, 1994) that enforces two *element integrity* constraints for the internal consistency of the representations: "[1] The splitting of one element into parts during movement, or [2] the fusing together of different elements into one, should be penalized" (Dawson, 1991).

Another way to represent motion transparency is to have a system of multiple subnetworks, each tuned to a particular motion, plus a set of controller networks that decide which of the subnetworks best accounts for the visual input (Nowlan and Sejnowski, 1994). When multiple transparently overlaid motions are present, the controller networks can choose to activate multiple subnetworks that code the motions. Such a system can be trained to give the "correct" responses, using a supervised learning algorithm (Nowlan and Sejnowski, 1994). It may be possible to develop an unsupervised learning rule that would allow a similar structure to form via self-organization.

Learning About Motion Propagation

Psychophysical studies on motion perception suggest that human visual systems perform a *trajectory-specific* propagation of computed moving stimulus information to successive image locations where a stimulus is predicted to appear. For example, a visual stimulus appears to have *inertia* (Anstis and Ramachandran, 1987); its motion at one instant generates an expectation, or bias, or priming, for it to be represented as continuing to move along the same trajectory. Other phenomena, such as kinetic subjective contours (Kellman and Cohen, 1984), also suggest similar trajectory-specific interactions (Marshall, 1991). The propagation lets the visual system gather information about an object even while it moves from location to location, instead of having to recompute de novo all data about the object at each location.

These propagating trajectory-specific priming signals could be carried by long-range excitatory horizontal intrinsic connections (LEHICs) (Gabbott, Martin, and Whitteridge, 1987; Hubbard and Marshall, 1994; Marshall, 1991), discovered in visual cortex of several animal species. LEHICs often span great distances across the cortex, corresponding to long ranges in image space. They most strongly interconnect regions of *like specificity* with regard to certain receptive field attributes, e.g., stimulus orientation. Such connectivity patterns are consistent with the hypothesis that LEHICs support trajectory-specific propagation. However, further neurobiological data are needed to confirm the hypothesis, e.g., the precise timing and spatial relationships between the LEHIC-interconnected neurons. Trajectory-specific time-delayed LEHIC structures that carry the priming signals can be formed by simple Hebbian-type learning rules (Marshall, 1990, 1991; Hubbard and Marshall, 1994).

Learning About Occlusion Events

Another interesting question is how a neural network might self-organize to detect visual occlusion events, and thereby represent the depth relations between moving objects. Frost (1993) described neurons in pigeon optic tectum that respond selectively to visual occlusion or disocclusion events. Assad and Maunsell (1994) have identified neurons (in the posterior parietal cortex of macaque monkeys) that respond to the "inferred" motion of an invisible stimulus. Psychophysical evidence—e.g., the continuous perception of a unitary illusory figure in the kinetic subjective contours display (Kellman and Cohen, 1984) —suggests that human visual systems maintain representations of moving visual objects even while the objects are temporarily occluded and invisible (Marshall, 1991).

Can a neural network self-organize to detect and represent moving visual objects even while they are temporarily invisible? Marshall, Alley, and Hubbard (1994) have shown how a neural network, using a new *disinhibition* rule, can self-organize both an On channel, which represents objects while they are visible, and an opponent Off channel, which temporarily maintains and propagates a representation of objects after they have become invisible. The resulting network can predict the disappearance and *reappearance* of multiple moving objects as they engage in occlusion relationships with one another. A key benefit is that such a network can *learn* the relationship between depth and visibility during occlusion events. If human visual systems operate according to similar rules, then they can learn about depth relations from occlusion events.

Learning About Invariance Under Transformations

Földiák (1991) described a simple learning mechanism whereby visual neurons can become selective for an invariant property, such as orientation, merely by remaining active for an extended time during the motion of objects. The extended activation lets a neuron learn to respond to orientation alone, regardless of stimulus location. Földiák's method elicits the self-organization of neurons that would be part of a "What" subsystem, invariant with respect to where a given orientation is present. A complementary "Where" subsystem would respond to the *variant* portion of the moving oriented stimulus: its changing location (see VISUAL SCENE PERCEPTION: NEUROPHYSIOLOGY). The Where subsystem might be invariant with respect to features like the orientation of the object. The What and Where subsystems would need to interact so that the location (Where) and the feature (What) can be grouped or bound together, i.e., be represented as referring to the same visual object. An open research question is how a Where subsystem could also self-organize.

Learning About Motion Grouping and Binding

A major question in motion perception is how locally measured motion signals can *cohere* into global, consistent representations of object motion, in a way that can then be read out usably for behaviors, e.g., motor control, navigation, tracking, etc. Two tasks are involved: *deciding* which local features belong together as parts of an object and *representing* the parts of an object as coherent.

One basis on which visual systems might decide which features belong together in a coherent grouping is *common motion*, or common fate. Sereno (1989) and Zhang, Sereno, and Sereno (1993) have described how a simple Hebbian-type rule can let neurons learn to respond selectively to large-field combinations of local motions in the "aperture problem" and in rotations and dilations (see MOTION PERCEPTION; DIRECTIONAL SELECTIVITY IN THE CORTEX).

A visual system must be able to represent all possible groupings or bindings that link features into coherent objects. Yet it cannot do so by allocating one neuron for each possible binding, or else a combinatorially large number of neurons would be needed. Two alternative methods have been proposed recently. In the first method (see SYNCHRONIZATION OF NEURONAL RESPONSES AS A PUTATIVE BINDING MECHANISM), binding of features is represented by temporally synchronous oscillatory activations of the neurons that code the features. An open research question is how the capability to represent binding through synchrony might self-organize. In the second method (Nigrin, 1993), binding can be represented by self-organized axo-axonal gates that selectively enable connections between the representations of features and the representations of objects.

Discussion: Learning as Unification

The research just described has shown that visual systems can self-organize to represent moving images in terms of several invariance properties: local motion, transparency, element integrity, propagation, grouping, and occlusion. Self-organizing neural networks are therefore very helpful in our understanding of the underlying structure of neural circuits for vision. However, artificial vision systems for practical applications may benefit from this understanding even if they do not self-organize; the final learned circuitry of a self-organizing neural network can simply be hardwired into such artificial systems.

This article has made three main points. First, motion is particularly informative in showing both the variabilities and the invariances in visual objects. Second, simple unsupervised learning rules can cause a visual system receiving motion information to self-organize the capability to detect and represent visual data in terms of these variabilities and invariances.

Third, several challenging open research questions in motion perception can be addressed fruitfully through self-organization methods.

An additional point bears mention. The same learning rules that describe the formation of motion processing mechanisms may also be able to describe the formation of *many* diverse visual mechanisms. Such rules have been used to describe the development of neural mechanisms for stereopsis, length and orientation selectivity, contrast detection, and other capabilities. The advantage of describing many visual processing mechanisms as the products of a common set of learning rules or developmental principles is *unification* (Marshall, 1991; Marshall, Alley, and Hubbard, 1994): these diverse mechanisms then become part of a single theory of vision. Such a unified theory should be a key goal of research on self-organizing neural networks and visual perception.

Acknowledgments. This work was supported in part by the Office of Naval Research (N00014–93–1–0208) and the National Eye Institute (EY09669).

Road Map: Development and Regeneration of Neural Networks
Background: I.3. Dynamics and Adaptation in Neural Networks
Related Reading: Perception of Three-Dimensional Structure; Perceptual Grouping; Stereo Correspondence and Neural Networks

References

Anstis, S. M., and Ramachandran, V. S., 1987, Visual inertia in apparent motion, *Vis. Res.*, 27:755–764.

Assad, J. A., and Maunsell, J. H. R., 1994, Neuronal correlates of inferred motion in macaque posterior parietal cortex, *Invest. Ophthalmol. Vis. Sci.*, 35:1663.

Dawson, M. R. W., 1991, The how and why of what went where in apparent motion: Modeling solutions to the motion correspondence problem, *Psychol. Rev.*, 98:569–603. ◆

Földiák, P., 1991, Learning invariance from transformation sequences, *Neural Computat.*, 3:194–200.

Fredericksen, R. E., 1993, The biological computation of visual motion, PhD Dissertation, TR 93–036, University of North Carolina at Chapel Hill, Department of Computer Science.

Frost, B. J., 1993, Time to collision sensitive neurons in nucleus rotundus of pigeons, in *Visual Motion and Its Role in the Stabilization of Gaze* (F. A. Miles and J. Wallman, Eds.), Amsterdam: Elsevier Science.

Gabbott, P. L. A., Martin, K. A. C., and Whitteridge, D., 1987, Connections between pyramidal neurons in layer 5 of cat visual cortex (area 17), *J. Comp. Neurol.*, 259:364–381.

Hubbard, R. S., and Marshall, J. A., 1994, Self-organizing neural network model of the visual inertia phenomenon in motion perception, TR 94–001, University of North Carolina at Chapel Hill, Department of Computer Science.

Kellman, P. J., and Cohen, M. H., 1984, Kinetic subjective contours, *Percept. & Psychophys.*, 35:237–244.

Marshall, J. A., 1990, Self-organizing neural networks for perception of visual motion, *Neural Netw.*, 3:45–74. ◆

Marshall, J. A., 1991, Challenges of vision theory: Self-organization of neural mechanisms for stable steering of object-grouping data in visual motion perception, in *Stochastic and Neural Methods in Signal Processing, Image Processing, and Computer Vision* (S.-S. Chen, Ed.), Proceedings of the SPIE 1569, San Diego, pp. 200–215. ◆

Marshall, J. A., Alley, R. K., and Hubbard, R. S., 1994, Learning to represent visual depth from occlusion events (submitted for publication). ◆

Nigrin, A., 1993, *Neural Networks for Pattern Recognition*, Cambridge, MA: MIT Press. ◆

Nowlan, S. J., and Sejnowski, T. J., 1994, A selection model for motion processing in area MT of primates (submitted for publication).

Sereno, M. I., 1989, Learning the solution to the aperture problem for pattern motion with a Hebb rule, in *Advances in Neural Information Processing Systems 1* (D. Touretzky, Ed.), San Mateo, CA: Morgan Kaufmann, pp. 468–476.

Zhang, K., Sereno, M. I., and Sereno, M. E., 1993, Emergence of position-independent detectors of sense of rotation and dilation with Hebbian learning: An analysis, *Neural Computat.*, 5:597–612.

Motivation

Clark Dorman and Paolo Gaudiano

Introduction

The ability of humans and animals to survive in a constantly changing environment is a testament to the power of biological processes. At any moment in our lives, we are faced with many sensory stimuli, and we can typically generate a large number of behaviors. How do we learn to ignore irrelevant information and suppress inappropriate behavior so that we may function in a complex environment?

In this article we discuss *motivation*, the internal force that produces actions on the basis of the momentary balance between our needs and the demands of our environment. We first describe motivation and how it is studied, focusing on behavioral and physiological studies. We then discuss the role of motivation in behavioral theories and neural network modeling.

Although the word *motivation* is common in everyday language, it is not easy to define rigorously in a scientific context. The concept of motivation is related to, but distinct from, other concepts such as *instincts*, *drives*, and *reflexes*. Motivated behavior is usually goal oriented; the goal may be associated with a drive, such as hunger or thirst (called *primary motivation*). However, motivation is also closely tied to sensory stimuli: an animal will not usually exhibit eating behavior unless food is presented. Unlike instinctive behavior, motivation depends on *affect* (emotional state). Finally, motivation can be learned (in which case it is called *secondary motivation*) and typically elicits more complex behaviors than simple reflexes.

An animal is always performing some activity, even when that activity is sleep. At any given time, the environment offers the opportunity to carry out many different behaviors, such as exploratory or consummatory behaviors, but an animal typically performs a single voluntary activity at a time. The study of motivation is concerned with which activity the animal performs in a given environment and how the animal maintains a given activity or changes between different activities as a function of environmental events and internal needs.

Motivation is typically studied using two approaches: psychological studies manipulate environmental events and monitor the resulting patterns of motivated behavior; physiological studies attempt to clarify the neural or endocrine origin of motivation. For instance, psychological studies might examine how an animal is able to maintain a constant goal-oriented activity as the surrounding stimuli change, or how an animal is able to switch spontaneously between behaviors as its needs change. Physiological studies attempt to identify physiological variables and neural regions that are related to motivated behavior.

Psychological Studies of Motivation

Motivation figured prominently in the earliest studies of animal psychology around the end of the nineteenth century. The improvements in our knowledge of physiology fostered a significant increase in physiological and psychological studies of motivation around the middle of the twentieth century. A library search on *motivation* will uncover numerous writings published in the 1940s and 1950s. Motivation played a significant role in many theories of behavior, especially Hull's theory (described later in this article).

Behavioral studies of motivation frequently focus on basic functions related to survival, such as eating, drinking, and avoiding harmful stimuli. Other motivated behaviors that have been studied, such as sexual behavior or social interactions, do not seem as closely related to immediate survival.

Hunger has frequently been studied in psychological studies of motivation because the food intake of the animal can be easily controlled. The motivation to eat is not directly controlled by feelings of hunger; when presented with the opportunity to eat, animals eat in anticipation of hunger and continue to eat after satiation to maintain themselves until the next meal. Motivation is also influenced by the subjective value assigned to the rewards arising from motivated behavior; in turn, this subjective value can be influenced by learning. In an elegant experiment, Crespi (1942) demonstrated that rats' motivation to obtain food, measured as the speed with which the rats ran down an alley toward food, can be altered not only by changing the absolute magnitude of the reward (the amount of food), but also by changing the amount of reward relative to what the rat expects to find at the end of the alley. In Crespi's experiment, three groups of rats were trained to run down an alley to receive 1, 16, or 256 food pellets. Motivation was measured as the speed with which the rats approached the food. Initially, the running speed was proportional to the size of the reward, with the rats receiving 256 pellets showing the greatest speed. In the second part of the experiment, all three groups of rats were given 16 pellets at the end of the alley. The rats that were switched from 256 to 16 pellets exhibited less motivation (ran slower) than those that had always received 16 pellets, while the rats that were switched from 1 pellet to 16 pellets ran significantly faster. We can sympathize with the rat's behavior by imagining how differently we would react if our salary was cut from a high level to some lower level *x*, as opposed to its being raised to *x* from an initially lower level.

Physiological Studies of Motivation

Research on motivation has focused on the physiological basis for hunger, thirst, and other biological drives (see review by Grossman, 1988). Animals and humans possess complex mechanisms for *homeostasis*, that is, for maintaining an efficient balance between internal needs and environmental affordances to satisfy these needs. Taking for example the need for food, the mechanisms involved in maintaining blood glucose level encompass neural, endocrine, and other physical and chemical mechanisms whose purpose is to monitor continuously the internal need for energy and whose state affects motivated behavior aimed at finding and consuming food.

A significant amount of motivation-related neural circuitry appears to be located in the hypothalamus (see chap. 48 by Kupfermann in Kandel, Schwartz, and Jessell, 1991). In partic-

ular, there appear to be discrete hypothalamic areas that play significant roles in the control of homeostatic signals relating to feeding, drinking, and temperature regulation. Most of these areas are organized in *opponent pairs*, that is, areas having opposite effects on the function they regulate. For example, the control of body temperature is jointly regulated by the anterior hypothalamus, which is responsible for the generation of temperature-lowering behaviors such as dilation of skin blood vessels, and the posterior hypothalamus, which is responsible for the generation of temperature-increasing behaviors such as shivering. Electrical stimulation of these areas leads to an enhancement of the corresponding behavior, while lesion of each area leads to a suppression of the corresponding behavior. For example, electrical stimulation of the anterior hypothalamus produces panting, while lesions in the same area lead to chronic hyperthermia.

The control of homeostasis and motivated behavior is not relegated to hypothalamic areas. For one thing, many brain areas are involved with the control of motivated behavior, so that, for example, feeding behaviors may be disrupted by stimulation or lesion of areas outside of the hypothalamus. In a similar vein, animals that are subjected to hypothalamic lesions sometimes exhibit gradual but marked recovery of the functions that were disrupted by the lesions, suggesting the existence of other neural centers that are capable of performing regulatory tasks. These observations are not surprising given the complexity of a seemingly simple behavior such as feeding, which requires the ability to seek out, identify, and consume food, all tasks that involve the coordination of sensory, cognitive, and motor skills.

Hull's Behavioral Theory

Hull's theory provides a framework within which motivated behavior can be analyzed. Hull (1943) proposed that "the initiation of learned, or habitual, patterns of movement or behavior is called motivation." In addition, Hull proposed a distinction between *primary motivation*, the evocation of action in relation to primary needs, and *secondary motivation*, the evocation of action in relation to secondary reinforcing stimuli or incentives.

Primary motivation is the cornerstone of Hull's *drive reduction theory*. According to Hull, events that threaten survival give rise to internal drive states, and behaviors that reduce the drive are thus rewarding. For instance, lack of food causes an increase in the hunger drive, and the consumption of food is rewarding because it leads to a reduction in the hunger drive.

A stimulus that is repeatedly associated with the onset of a drive state can become an *acquired drive*. Once developed, an acquired drive can motivate behavior on subsequent occasions, even in the absence of cues that elicit the original drive state. Stimuli with this property become *incentives*, and their ability to evoke behaviors is known as *secondary*, or *incentive motivation*. For instance, throughout our lives, we learn to associate the sight of food with the impending act of consuming food, so we feel hungry when we see food.

Motivated behavior requires both drives and appropriate stimuli. Hull's theory captures this relationship by proposing that the *behavior potential* for a given action is the product of the drive strength and incentive level associated with that action:

$$_sE_R = D \times V \times K \times {_sH_R} \tag{1}$$

$_sE_R$ is called the *reaction potential*, the likelihood that a given behavior will be emitted. D represents the drive level, V is proportional to the stimulus intensity, K is the *incentive motivation* associated with stimuli present in the environment, and $_sH_R$ is the *habit strength* associated with the behavior. The multiplicative relationship between all of the variables suggests that all of these factors must work synergistically in order for a behavior to have a large reaction potential, that is, in order for the behavior to have a high probability of being emitted.

In our daily experiences, we are faced with a continuously fluctuating combination of drives and incentives. Somehow we must be able to select the behavior that is most appropriate in a given situation while suppressing other, less adequate behaviors. Thus, motivated behavior requires a form of *competition*. According to Hull, at any given time, the behavior with the greatest potential to reduce a given drive is released. If the drive persists, that behavior is inhibited, and the second strongest response in the *drive hierarchy* is released, and so on.

Hull's theory fell out of favor for a number of reasons (e.g., Klein, 1991). For one thing, his theory (Equation 1) predicts that behaviors should not be emitted in the absence of motivation, because in that case $K = 0$ and the reaction potential is likewise zero. However, a simple experiment by Sheffield and Roby (1950, see discussion in Barker, 1994) showed that rats could learn to perform a behavior to obtain saccharin-flavored water even though they were not hungry and even though saccharin has no nutritive value. A more significant challenge to Hull's theory came from experiments showing that in some instances drive *induction* can be motivating: In 1954, Olds and Milner (see Barker, 1994) discovered that electrical stimulation of a brain region called the *medial forebrain bundle* is rewarding for rats and that rats will learn to perform tasks that lead to electrical stimulation as a reward. Later studies have shown that electrical stimulation is not only rewarding but also is a direct source of motivation, that is, it can cause the release of behaviors in the absence of appropriate stimuli or homeostatic cues. These findings suggest that brain stimulation is motivating because it *induces* a drive. The concept of reward through drive induction was used by Mowrer (1960), and it plays a role in some of the neural network theories described below.

Another reason for the limited success of Hull's theory was that his mathematical approach was different from typical qualitative, descriptive learning theories. In that sense, it was ahead of its time. Current studies of computational neuroscience and neural networks are promoting an increased role for mathematical models in the study of brain function.

Neural Networks of Motivated Behavior

As reviewed in various articles in this *Handbook*, many studies of neural networks suppose that learning involves correlation between input and output, or requires the presence of an explicit error signal paired with each input. However, these networks learn without reference to the internal state of the network or the external state of the environment. In other words, neural network learning frequently lacks a parallel to the idea of motivation (but see REINFORECEMENT LEARNING). The idea of motivation has been used explicitly only by a handful of neural network researchers. The work of Grossberg and colleagues (see Grossberg, 1982, 1986, 1989), whose efforts to model animal and human behavior with dynamic neural networks span the past three decades, provides a computational neural framework within which it is possible to give a natural interpretation to the concept of motivation and to the role of drives and incentives in the generation of purposive behavior.

Grossberg (1971) proposed a neural model of *instrumental* and *classical* conditioning (see CONDITIONING; EMOTION AND COMPUTATIONAL NEUROSCIENCE) that embodies many of the concepts discussed in this article. Grossberg's model simulates neurons that represent sensory stimuli from the environment as well as neurons that represent internal drive signals. Reinforcement focuses attention on relevant environmental stimuli and allows the organism to learn what stimuli have value as reinforcers. In his later work, Grossberg expanded the notion of drive neurons to what he termed a *sensory-drive heterarchy*, in which both appetitive and aversive drives combine with sensory stimuli and compete to determine which behavior will be emitted in response to a given combination of internal needs and environmental stimuli.

The joint action of drives and reinforcers in Grossberg's network embodies Hull's intuition that drives and incentives combine in a multiplicative fashion (Equation 1). However, Grossberg's model extends Hull's ideas by including both drive induction and drive reduction and by describing dynamic aspects of behavior and learning, rather than static relationships. A detailed discussion of the relationship between Hull's drive reduction theory and Grossberg's neural theory of conditioning can be found elsewhere (see chap. 1 in Grossberg, 1986).

The idea of drive neurons that modulate learning is found in several other neural network models. The models of Klopf (chap. 7 in Byrne and Berry, 1989) and Sutton and Barto (1990) explicitly incorporate the idea of a drive neuron. Klopf's *drive-reinforcement theory* suggests that changes in drive level have reinforcing properties. In this case, however, Klopf suggests that organisms seek stimulation and that reward comes from *increases* in drive level, as suggested in drive induction theories. Stimuli that occur in contiguity with increases in drive are associated with the behaviors that caused the change in drive level. These stimuli can then energize the behavior of the animal. Sutton and Barto (1990) proposed the existence of an *eligibility trace* that determines when learning can occur. This level of control is important when a system must improve its performance on the basis of only general information about its success or failure, which occurs after a potentially long sequence of actions has been performed. The ability to assign credit or blame to elements of a network for events that took place in the past is known as the *temporal credit assignment problem*. In addition to making interesting predictions about conditioning phenomena and temporal learning, the work of Sutton, Barto, and colleagues has led to a number of useful applications in robotics and control.

Strong support for the existence of drive neurons has come from experimental and modeling work on both vertebrates and invertebrates. They require convergence of sensory and drive inputs to become active. In vertebrates, some cells in the hypothalamus respond only when they receive convergent input from internal drive signals and relevant external sensory stimuli (see chap. 48 in Kandel et al., 1991). Many invertebrates are also capable of sophisticated forms of learning and motivated behavior (Colgan, 1989). Alkon (chap. 1 in Byrne and Berry, 1989), Hawkins (chap. 5 in Byrne and Berry, 1989), and Byrne and colleagues (Buonomano, Baxter, and Byrne, 1990) have found evidence of *facilitator neurons* in the mollusks *Aplysia* and *Hermissenda*. While the location and specific action of facilitator neurons vary in different preparations, in all cases the facilitator neuron plays a role similar to that of drive neurons: it is closely linked to fundamental aspects of the animal's life, such as the onset of shock, and it modulates learning at associative synapses.

Discussion

We have described motivation as the internal force that energizes behaviors and determines which particular behavior will be emitted in response to internal needs and a given set of environmental stimuli. Motivation is a complex topic of research that has been studied from many different approaches. We have summarized some of the psychological and physiological experiments that probe the role of motivation in the behavior of humans and animals. We reviewed Hull's drive reduction theory, one of the most influential and rigorous behavioral theories from the field of psychology. We have also looked at neural network models that directly or indirectly use the concept of motivation or related concepts, such as drives and homeostasis.

Motivation is a concept that is difficult to describe quantitatively. Perhaps for this reason, it is largely unused by neural network modelers. However, we believe that the study of motivation can be useful for students of brain theory in two ways. First, several areas of brain research are closely related to motivation, for instance, the study of conditioning and reinforcement learning (see CEREBELLUM AND CONDITIONING; COGNITIVE MAPS; CONDITIONING; EMOTION AND COMPUTATIONAL NEUROSCIENCE; REINFORCEMENT LEARNING IN MOTOR CONTROL). Second, neural network models that consider complex goal-oriented behavior may find motivation essential. In particular, motivation (or the lack thereof) is perhaps the main difference between learning in typical neural networks and learning in humans and animals. Motivation allows humans and animals to take into account internal needs and external stimuli in deciding what should and should not be learned in a particular situation. The inclusion of concepts such as drives and motivation will be particularly important in neural network applications that involve interactions between simulated organisms and a realistic environment.

Acknowledgments. C. Dorman is supported by AFOSR: F49620–92–J–0334. P. Gaudiano is supported by a Sloan Fellowship (BR–3122) and AFOSR:F49620–92–J–0499.

Road Map: Connectionist Psychology

References

Barker, L. M., 1994, *Learning and Behavior: A Psychobiological Perspective*, New York: Macmillan. ◆
Buonomano, D. V., Baxter, D. A., and Byrne, J. H., 1990, Small networks of empirically derived adaptive elements simulate some higher-order features of classical conditioning, *Neural Netw.*, 3:507–523.
Byrne, J. H., and Berry, W. O., Eds., 1989, *Neural Models of Plasticity*, San Diego: Academic Press.
Colgan, P., 1989, *Animal Motivation*, New York: Chapman and Hall. ◆
Crespi, L. P., 1942, Quantitative variation of incentive and performance in the white rat, *Am. Psychol.*, 55:467–517.
Grossberg, S., 1971, On the dynamics of operant conditioning, *J. Theor. Biol.*, 33:225–255.
Grossberg, S., Ed., 1982, *Studies of Mind and Brain: Neural Principles of Learning, Perception, Development, Cognition and Motor Control*, Boston: Reidel. ◆
Grossberg, S., Ed., 1986, *The Adaptive Brain*, vol. 1, *Cognition, Learning, Reinforcement, and Rhythm*, Amsterdam: Elsevier/North-Holland.
Grossberg, S., Ed., 1989, *Neural Networks and Natural Intelligence*, Cambridge, MA: MIT Press.
Grossman, S. P., 1988, Motivation, in *Encyclopedia of Neuroscience* (G. Adelman, Ed.), Boston: Birkhäuser, pp. 60–65. ◆
Hull, C. L., 1943, *Principles of Behavior*, New York: Appleton-Century. ◆
Kandel, E. R., Schwartz, J. H., and Jessell, T. M., Eds., 1991, *Principles of Neural Science*, 3rd ed., New York: Elsevier. ◆
Klein, S. B., 1991, *Learning: Principles and Applications*, 2nd ed., New York: McGraw-Hill. ◆
Mowrer, O. H., 1960, *Learning Theory and Behavior*, New York: Wiley.
Sutton, R. S., and Barto, A. G., 1990, Time-derivative models of Pavlovian reinforcement, in *Learning and Computational Neuroscience* (J. W. Moore and M. Gabriel, Eds.), Cambridge, MA: MIT Press.

Motoneuron Recruitment

Daniel Bullock

Introduction

Motoneurons are neurons that directly innervate muscle fibers, and motoneuron discharges cause muscle fibers to contract. Such contractions generate the forces that produce active accelerations and decelerations of limb segments as well as the forces that oppose static loads. Moreover, co-contractions of opposing muscles allow us to stiffen joints and thereby maintain desired postures despite perturbations of unexpected magnitude and direction. Because of the direct anatomical link between motoneurons and contractile fibers, there is a close relationship between motoneuron activity and force production. A motoneuron and the contractile fibers that it innervates constitute a *motor unit*.

The range of forces producible by one motor unit is small. To make it possible to generate large forces, motor units must be combined into larger aggregates. The results of such aggregation are the muscles. Immediately associated with each muscle is a population, or pool, of motoneurons. Muscles are therefore composite structures whose force-generating components, the motor units, are typically heterogeneous and partly neuronal. When simultaneously excited, such aggregates can

produce much larger forces than a single motor unit. How are these heterogeneous aggregates of force-generating elements recruited in the service of reflexes, voluntary movement, and posture? Such task-dependent recruitment is achieved by a combination of specialized neurons and specialized neural networks.

Consider the simple question of control of force magnitude. If any excitatory input were sufficient to cause simultaneous excitation of all motor units, then the minimum force produced by the aggregate would be much too large for many purposes. To produce accurate movements, forces must be finely graded in response to the input to the motoneuron pool. The fine grading of forces required for accuracy favors a design that allows both partial activation of the motoneuron-fiber pool and finely graded changes, up or down, from preexisting states of activation.

Such force grading by a large cell-fiber aggregate is the context in which the *size principle* of motoneuron recruitment is usually discussed (Henneman and Mendell, 1981). The size principle encompasses a number of related aspects of the design of motoneuron pools and their embedding within the sensory-motor system. In this design, an excitatory input reaches all elements of the motoneuron pool at the same time and with equal magnitude. However, elements of the motoneuron pool differ in their activation thresholds. Because the distribution of threshold values varies from small to large, the larger the excitatory input to the pool, the more elements become active. This variability permits a continuously changing input signal to produce a graded force response from the muscle. The size principle states that as the excitatory input to the pool increases, motoneurons are recruited in order by size from smallest to largest because motoneurons with larger somatic volumes also have higher thresholds. Similarly, as excitatory input declines or as inhibitory input increases, motoneurons are de-recruited in order by size, in this case, from largest to smallest.

The grading of force by recruitment, which is necessarily quantal, is supplemented by finer grading through firing rate modulation of individual cells because each cell's firing rate is sensitive to input fluctuations in its suprathreshold range. This design permits finely graded increments and decrements in force over the entire range, from the very small force produced by a single motor unit to the large force produced by simultaneous maximal excitation of perhaps hundreds of motor units forming a muscle.

It might appear that the size principle makes each spinomuscular force generator a fixed-gain, near-linear, amplifier of excitatory inputs. However, several factors complicate the situation. First, the gain is not fixed because muscle force can become decoupled from motoneuron pool activation if a contraction-opposing load causes muscle yielding or if the muscle fatigues. Second, the amplification function is often faster than linear because motoneurons with larger cell bodies and thus higher recruitment thresholds typically project by larger, faster-conducting axons to more muscle fibers, each of which exhibits shorter twitch contraction times. Third, the description given above of a motor unit, although in agreement with common usage, is also arbitrary. Several other closely linked neural and sensory constituents appear in most muscle control systems of most vertebrates, and they appear to be part of the basic apparatus for force generation. For example, before exiting the spinal cord, the axons of most alpha-motoneurons emit collaterals that excite Renshaw cells, which in turn inhibit those same alpha-motoneurons. Fourth, the net torque developed at a joint depends on the balance of forces created by groups of muscles arranged into mutually antagonistic sets. To ensure that opponent muscles strike the right balance, motoneuron recruitment must be regulated by opponent neural interactions. Each of these considerations reveals a need for network control of recruitment.

Compensations for Fatigue and Yielding

Muscle fatigue and yielding make the functional relation between pool activation and force inherently variable, and several features of the biological network appear to provide compensations that reduce the variability in this linkage. Nichols and Houk (1973) argued that two feedbacks from muscle receptors to spinal motoneuron pools interact to reduce variability in *muscle stiffness*: the ratio of muscle force changes to muscle length changes. Muscle yielding, which reduces stiffness, also leads to two consequences for muscle receptors: increased activity by stretch-sensitive receptors, the spindles, and decreased activity by tension-sensitive receptors, the Golgi-tendon organs (GTOs). Because spindle feedback directly excites alpha-motoneurons through type Ia sensory fibers, whereas GTOs inhibit motoneurons through Ib interneurons, both feedbacks are compensatory. Kirsch and Rymer (1987) proposed that

GTO feedback also had appropriate characteristics to compensate for muscle fatigue. Bullock and Grossberg (1989) argued that the known covariation of motor unit size and contraction rate could also be seen as a contributor to yielding compensation.

Linearization or Equalization of Pool Responses

The covariation of recruitment threshold, number of fibers contacted, and fiber contraction rates with motoneuron size would produce a faster-than-linear relationship between excitatory input to the motoneuron pool and the force output of the muscle, at least when the system is not approaching saturation. Akazawa and Kato (1990) and Bullock and Grossberg (1989) independently proposed that Renshaw feedback could modify this transduction. The analysis by Akazawa and Kato treated a single motor unit pool and showed that inhibitory Renshaw feedback may be able to linearize the relationship between excitatory inputs and force outputs. The FLETE model (Figure 1) analyzed by Bullock and Grossberg encompassed a lumped pair of motor unit pools associated with biomechanically opposed muscles. These authors sought to explain how spinal circuitry enabled the higher brain to achieve independent control of joint angle and joint stiffness. By *F*actoring the *LE*ngth and *TE*nsion properties of muscle, the FLETE network allows a descending co-contraction command to stiffen and thereby stabilize the joint at any desired angle. Available data (Humphrey and Reed, 1983) indicate that voluntary stiffness adjustments are achieved by varying an excitatory signal sent to both opponent motor unit pools. Bullock and Grossberg showed that in the absence of Renshaw feedback, a descending co-contractive signal would generally be unequally amplified by recruitment events within opposing motor unit pools. Such unequal amplification would lead to an undesired joint rotation as well as to a change in joint stiffness. They then showed that Renshaw-mediated feedback could help to guarantee independent control of joint stiffness and joint angle by equalizing

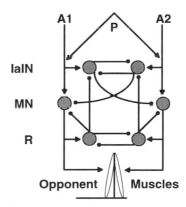

Figure 1. Partial connectivity of the FLETE model for independent control of joint angle and joint stiffness. To set the desired joint angle, the higher brain reciprocally adjusts descending signals A1 and A2 directed to two opposing alpha-motoneuron (MN) pools that project to opposing muscles. Descending signal P to both motoneuron pools adjusts joint stiffness without modifying joint angle if increments in P lead to equal increments in the force outputs of the two opposing muscles. Renshaw (R) cell feedbacks, among others, compensate for nonlinearities in the motoneuron response function and thereby help to assure equal force increments in the two muscles affected by P. Renshaw feedback disinhibits opponent MNs through the Ia interneurons (IaIN). In this figure and in Figure 2, arrow and dot line endings, respectively, indicate excitatory and inhibitory synapses.

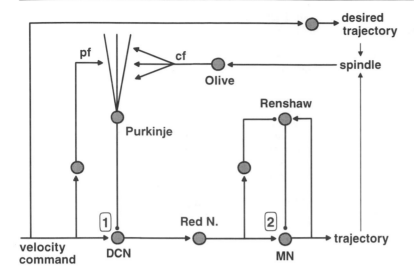

Figure 2. Network model incorporating two sites for controlling the gain of a velocity command by release from inhibition. Before learning, a velocity command directed toward a muscle through the deep cerebellar nuclear (DCN) pathway will have a negligible effect as a result of Purkinje cell inhibition of DCN sites and Renshaw cell inhibition of alpha-motoneurons (MN). However, trajectory errors detected by muscle spindles activate the inferior olive, whose climbing fibers (cf) reach the dendrites of Purkinje cells. Climbing fiber activity causes long-term depression of coactive parallel fiber (pf) synapses that excite Purkinje cells. Depression of Purkinje cell excitation causes disinhibition of DCN sites and opens the gate for passage of the velocity command to the red nucleus (Red N.). The red nucleus both excites alpha-motoneurons and inhibits Renshaw cells.

the two pools' amplifications of the co-contractive signal. This equalization, which need not involve global linearization of recruitment, is achieved by a local circuit that incorporates mutual inhibition between opponent Renshaw pools (Ryall, 1970) and between type Ia reciprocal inhibitory interneurons, which, like alpha-motoneurons, are inhibited by Renshaw cells (Hultborn, Jankowska, and Lindstrom, 1971). Subsequent extensions (e.g., Bullock and Grossberg, 1991; Bullock and Contreras-Vidal, 1993; Bullock, Contreras-Vidal, and Grossberg, 1993) showed that the FLETE model is applicable to a multijoint system with both mono- and bi-articular muscles and that the independent control property is enhanced by the incorporation of additional known pathways, such as the spindle (Ia) and GTO feedbacks mentioned above. Multiphasic motoneuron recruitment, as shown by the triphasic electromyogram bursts that are characteristic of rapid joint rotations, emerged within a FLETE network that includes velocity-sensitive muscle spindles.

Adaptive Control of Motoneuron Gain

Renshaw cells also mediate descending modulation of the motoneuron recruitment process. Hultborn, Lindstrom, and Wigstrom (1979) proposed that the gain of the monosynaptic stretch reflex could be controlled by descending inputs to Renshaw cells. Henatsch et al. (1986) later discovered that stimulation in the red nucleus (RN), which receives potent inputs from the nucleus interpositus of the cerebellum and which projects to spinal pools through the rubrospinal pathway, enhances the monosynaptic stretch reflex by inhibiting Renshaw cells, thereby releasing alpha-motoneurons from Renshaw inhibition. The same RN stimulation also excites motoneurons. Bullock and Grossberg (1989) proposed that the implied bivalent rubral projection to Renshaw cells and alpha-motoneurons afforded adaptive, or learning-based, control of gain of movement commands directed to motoneuron pools. Bullock, Contreras-Vidal, and Grossberg (1993) introduced a neural network comprising a central trajectory generator, the FLETE model, and a model cerebellar network capable of learning to modulate motoneuron recruitment through a bivalent output to Renshaw cells and alpha-motoneurons. Simulations of the circuit (Figure 2) showed that if the cerebellar circuit received both a desired velocity command and an error feedback routed from spindle to cerebellum through the inferior olive, then a learning-

adjusted cerebellar transduction substantially enhanced the dynamic tracking characteristics of the limb by transiently exciting and removing inhibition from the agonist motoneuron pool (see Figure 2). This theory is broadly consistent with other models of cerebellar learning (e.g., Albus, 1971; Grossberg and Kuperstein, 1986; Ito, 1984; Kawato, Furukawa, and Suzuki, 1987) and with observations of phasic RN and interpositus activity during learned movements (Martin and Ghez, 1991).

Discussion

Neural network analyses have begun to clarify how local spinal circuits cooperate with central adaptive circuits for task-dependent control of motoneuron recruitment, but many basic questions remain. Network models must be expanded to accommodate the connectivities that govern recruitment in different species, which differ dramatically in both biomechanics and behavioral specializations. An especially promising area for research is adaptively timed cerebellar modulation of motoneuron recruitment spanning many muscle synergists because such timed parallel modulation appears to be critical for the production of many types of coordinated movement.

Acknowledgment.   Preparation of this article was supported by the Office of Naval Research (ONR N00014–92–J–1309 and N00014–93–1–1364).

Road Map: Biological Motor Control
Related Reading: Cerebellum and Motor Control; Equilibrium Point
    Hypothesis; Muscle Models

References

Albus, J. S., 1971, A theory of cerebellar function, *Math. Biosci.*, 10: 25–61.
Akazawa, K., and Kato, K., 1990, Neural network for control of muscle force based on the size principle of motor unit, *Proc. IEEE*, 78: 1531–1535.
Bullock, D., and Contreras-Vidal, J. L., 1993, How spinal neural networks reduce discrepancies between motor intention and motor realization, in *Motor Control and Variability* (K. Newell and D. Corcos, Eds.), Champaign, IL: Human Kinetics Press, pp. 183–221. ◆
Bullock, D., Contreras-Vidal, J. L., and Grossberg, S., 1993, Speed scaling and adaptive cerebellar control of Renshaw cell and motoneuron gain, *Abstr. Soc. Neurosci.*, 19:1594.

Bullock, D., and Grossberg, S., 1989, VITE and FLETE: Neural modules for trajectory formation and postural control, in *Volitional Action* (W. A. Hershberger, Ed.), Amsterdam: Elsevier North-Holland, pp. 253–298.

Bullock, D., and Grossberg, S., 1991, Adaptive neural networks for control of movement trajectories invariant under speed and force rescaling, *Hum. Mov. Sci.*, 10:3–53. ◆

Grossberg, S., and Kuperstein, M., 1986, *Neural Dynamics of Adaptive Sensory-Motor Control: Ballistic Eye Movements*, Amsterdam: Elsevier.

Henatsch, H. D., Meyer-Lohmann, J., Windhorst, U., and Schmidt, J., 1986, Differential effects of stimulation of the cat's red nucleus on lumbar alpha motoneurons and their Renshaw cells, *Exp. Brain Res.*, 62:161–174.

Henneman, E., and Mendell, L. M., 1981, Functional organization of motoneuron pool and its inputs, in *Handbook of Physiology*, sect. 1, vol. II, *Motor Control* (V. Brooks, Ed.), Bethesda, MD: American Physiological Society, pp. 423–507. ◆

Hultborn, H. M., Jankowska, E., and Lindstrom, S., 1971, Recurrent inhibition of interneurons monosynaptically activated from group Ia afferents, *J. Physiol. (Lond.)*, 215:623–636.

Hultborn, H. M., Lindstrom, S., and Wigstrom, H. 1979, On the function of recurrent inhibition in the spinal cord, *Exp. Brain Res.*, 37:399–403.

Humphrey, D. R., and Reed, D. J., 1983, Separate cortical systems for control of joint movement and joint stiffness: Reciprocal activation and coactivation of antagonist muscles, in *Motor Control Mechanisms in Health and Disease* (J. Desmedt, Ed.), New York: Raven, pp. 347–372. ◆

Ito, M., 1984, *The Cerebellum and Neural Control*, New York: Raven.

Kawato, M., Furukawa, K., and Suzuki, R., 1987, A hierarchical neural-network model for control and learning of voluntary movement, *Biol. Cybern.*, 57:169–185.

Kirsch, R. F., and Rymer, W. Z., 1987, Neural compensation for muscular fatigue: Evidence for significant force regulation in man, *J. Neurophysiol.*, 57:1893–1910.

Martin, J. H., and Ghez, C., 1991, Task-related coding of stimulus and response in cat red nucleus, *Exp. Brain Res.*, 85:373–388.

Nichols, T. R., and Houk, J. C., 1973, Reflex compensation for variations in the mechanical properties of a muscle, *Science*, 181:182–184.

Ryall, R. W., 1970, Renshaw cell mediated inhibition of Renshaw cells: Patterns of excitation and inhibition from impulses in motor axon collaterals, *J. Neurophysiol.*, 33:257–270.

Motor Control, Biological and Theoretical

R. C. Miall

Introduction

Biological motor control can be characterized as a problem of controlling nonlinear systems whose states are monitored with slow, sometimes low-quality, sensors. In response to changing sensory inputs, internal goals, or motor errors, the motor system must solve several basic problems: selection of an appropriate action and transformation of control signals from sensory to motor coordinate frameworks; coordination of the selected movement with other ongoing behaviors and with postural reflexes; and monitoring the movement to ensure its accuracy. These stages may be interrelated, so that separation of any one particular problem into these individual stages may not be possible. This article describes some of the ways we think that the motor system solves these tasks, based on principles (and terminology) whose origins are in engineering.

A motor control system acts by sending motor commands to a controlled object, often referred to as "the plant," which in turn acts on the local environment (Figure 1). The plant or the environment has one or more variables which the controller attempts to regulate: either to maintain them at a steady level in the face of disturbances (when the controller is known as a regulator) or to follow some changing reference value (a servo controller). The controller may make use of sensory signals from the environment, from its reference inputs, and possibly also from the plant to determine what actions are required. The sensory inputs from the plant can provide information about the "state" of the controlled object, where state can be considered as all the relevant variables that adequately describe the controlled object. If the controller bases its actions on signals which are not affected by the plant output, it is said to be a feedforward controller: the control path is the thick line from left to right in Figure 1*A*, which does not require return signals. If the controller bases its actions on a comparison between the reference values and the controlled variables (via a comparator, Figure 1*B*), it is a feedback controller. One can add more complex control strategies to these simple systems (Figure 1*C*, 1*D*), which are described in more detail in the following sections.

Feedforward control schemes may be grouped as those based on direct control and those based on indirect control using internal models. Direct control is taken here to mean control without explicit knowledge of the behavior of the plant. "Endpoint" or equilibrium point control (Bizzi, Mussa-Ivaldi, and Giszter, 1991; see EQUILIBRIUM POINT HYPOTHESIS) makes use of the springlike properties of muscles. For any set of springs pulling across the joints of a limb, there are one or more stable positions into which the limb passively settles. So, if the central nervous system (CNS) can define appropriate endpoint muscle tensions, the limb will move into the desired position without the controller knowing either its starting position or its behavior during the movement. An alternative direct scheme is to generate the appropriate commands—muscle torques—but again without any explicit knowledge of the plant. In the limit, one could use a vast memorized lookup table to store appropriate motor commands for every possible input state (e.g., Atkeson and Reinkensmeyer, 1991). However, the numbers grow explosively: for the classical pole-balancing task, the state-space can be partitioned into just three cart positions, six pole angles, three cart velocities, and three pole angular velocities (Barto, Sutton, and Anderson, 1983), but this still gives 162 possible states. For a 6-degrees-of-freedom arm with 10% resolution of joint angles and velocities, the state-space totals 10^{12}.

The advantage of feedforward control is that it can, in the ideal case, give perfect performance with no error between the reference and the controlled variable. The main disadvantages in biological systems are the practical difficulties in developing an accurate controller and the lack of corrections for unexpected disturbances. If the controller is not accurate, output errors go unchecked. Because no biological system can be both perfectly accurate and also free from external disturbances, there is usually a need for error correction. Most biological

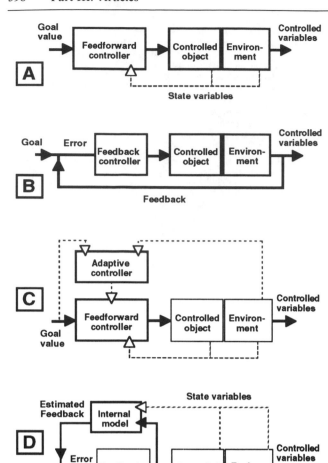

Figure 1. *A*, Feedforward control. *B*, Negative feedback control. *C*, Adaptive controllers using "off-line" feedback can adjust parameters of the feedforward controller. *D*, An internal model of the controlled object can replace the feedback loop with a rapid feedback estimate.

examples of feedforward control are therefore coupled with a feedback controller.

In contrast, the major advantage of negative feedback control is that it is a very simple, robust strategy. The controller drives the plant so as to cancel the feedback error signaled by the comparator. Because it constantly seeks to cancel the error, it operates well without exact knowledge of the controlled object, and despite internal or external disturbances.

The design criteria for negative feedback control are dominated by the "closed-loop gain." Gain is defined as the ratio of a system's output to its input; for a servo controller, the gain should be close to unity, so that a given input (the reference value) evokes an output of equal magnitude. In a feedback circuit (Figure 1*B*), one can define both open-loop and closed-loop gains. The closed-loop gain K_c is given by the ratio of response to reference amplitudes. The open-loop gain K_o is given by the ratio of the response to the error; it gives the response expected if the feedback path shown in Figure 1*B* is cut, thus "opening" the loop. The feedback gain is determined by the open-loop gain where $K_c = K_o/(1 + K_o)$. For ideal control, the closed-loop gain should be unity under all conditions; thus, the open-loop gain should be nearly infinite, ensuring that K_c approaches unity. In practice, the open-loop gain is usually

frequency dependent and can never reach infinity, and so the closed-loop gain is also frequency dependent and less than unity. Notice that the comparison of the reference value and the controlled variable, to calculate the error signal, is affected by the dynamics of the motor control system. If an action is issued by the controller, its effects are not immediately apparent to the comparator, but are delayed by the plant dynamics, by feedback dynamics, and by transport delays on both the forward and feedback paths. Hence, there is a frequency at which these delays combine to impart a 180° phase lag to the feedback signal; the open-loop gain at that frequency now only needs to be unity for the closed-loop gain to become infinite, forcing the system into instability. Any trivially small error or disturbance will be overcorrected and result in even bigger errors, leading to yet bigger corrections. In biological systems, where delays are inevitable, the comparison is always out of date. Thus, the principal disadvantages of feedback control are that the closed-loop gain falls below unity (which means that errors are never completely corrected) and that the speed of responses must be kept low with respect to the feedback loop delay to avoid instability.

Two more complex aspects of motor control indicated by Figure 1 are those of adaptive control and of internal models. Adaptive control relies on monitoring performance over a longer time scale than that used by negative-feedback control, so that one gains a measure of the average performance, rather than the moment-to-moment errors. The adaptive controller is then used to adjust the motor responses, either by modulating the feedforward controller as indicated in Figure 1*C*, or by modulating the open-loop gain of a feedback controller. The advantage of ADAPTIVE CONTROL (see both articles under that heading) is that it can compensate for gradual changes in the motor performance of the controlled object. More complex adaptive controllers can also be designed to predict changes in the reference value.

Internal Models

Two forms of internal model must be distinguished. An ideal feedforward controller ensures that the plant output (the controlled variable) is always identical to the goal or reference value. Thus, it inputs the reference value (and often also the state signals, Figure 1*A*), and it outputs a motor command; the motor command shifts the plant to a new state, which should equal the reference value. Hence, one can describe the ideal feedforward controller as an *inverse* of the plant: the plant translates commands into states; the inverse controller translates desired states into commands. If the transfer function of the plant is represented as P, its inverse is P^{-1}, and the transfer function of the complete system (from reference value to controlled variable) is $P \cdot P^{-1} = 1 \cdot 0$. Note again that this implies that the perfect system has a gain of unity. Inverse modeling is covered in more detail in Jordan (in press); see also SENSORIMOTOR LEARNING.

The alternative type of internal model is known as a direct or forward model of the plant (Figure 1*D*). It inputs a copy of the motor command being sent to the plant and the current plant state, and it outputs an estimate of the next state of the plant or of the controlled variables. This estimate is available to the feedback controller more rapidly than actual feedback. Thus, one can replace the external feedback loop with an internal loop, avoiding some or all of the delays just above. A negative-feedback loop with negligible delay and with a high open-loop gain rapidly and accurately drives its plant in the direction to minimize the comparator error. Thus, a fast internal loop through a forward model is functionally equivalent to an inverse dynamic model. The disadvantage is that, viewed from outside the loop, it functions as a feedforward controller: it disregards the actual feedback and therefore is no longer error correcting.

Physiological Control of Movement

Feedback Control

Feedback controllers are found throughout physiology; I mention only two examples from vertebrate motor systems. The major tension-producing fibers of the vertebrate muscle, known as extrafusal fibers, contract after excitation of the muscle by alpha motoneurons. However, the amount of tension produced by the muscle in response to a motor command varies with the length of the muscle, its speed of contraction, its fatigue, and so on. The muscles therefore are provided with numerous sensory structures known as muscle spindles and Golgi tendon organs (GTO) that signal back to the CNS the muscle length and muscle tension, respectively. The spindles are complex sensorimotor structures, combining contractile elements (intrafusal fibers, excited by specialized gamma motoneurons) with a central sensory region responding to stretch. Their afferent fibers monosynaptically excite alpha motoneurons serving the same muscle and synergistic muscles. This circuit (the stretch reflex) is a feedback controller for muscle length: if the muscle is passively stretched, the spindles respond, excite the alpha motoneurons, and the resulting reflex contraction of the extrafusal fibers restores the muscle to its original length, silencing the spindles again. Thus, the spindles signal a deviation from their regulated length, and the controller (the alpha motoneuron) acts to cancel the error. The GTOs also act in a feedback manner. They are attached to the tendons of muscle and respond to increased tension in the tendon. They excite interneurons, which inhibit motoneurons of that muscle and other muscles acting around the same joint. If the tension in the tendon increases, for example, because of an external load, the GTOs are activated, and through the inhibitory interneuron inhibit the motoneurons, causing the muscle to relax. This relaxation then reduces the tension, and thus the negative-feedback loop serves to maintain a controlled level of tension. This description of the spindle and GTO is a vast oversimplification, but emphasizes their basic control properties. Together they act to maintain a muscle in its current state: changes in length or in tension are automatically opposed.

Feedforward Control

Motor systems can make use of the spindles and GTOs to ensure that actions occur as planned. By sending motor commands both to the alpha and the gamma motoneurons, both the force-producing extrafusal fibers of the muscle and the much weaker intrafusal fibers of the spindle co-contract. If the joint fails to move fast enough, because of an unexpected load, then the spindle contracts within the main muscle, its sensory region is self-stimulated, and additional excitatory drive is reflexively added to the alpha motoneurons to overcome the load. This theory, while incomplete, is a very elegant example of servo-control. Thus, for the supraspinal centers driving the movement, the spinal circuits can be treated as a feedforward controller, autonomously regulating the muscles without the need for feedback to these higher centers.

Another example of a feedforward controller is found in the oculomotor system (see COLLICULAR VISUOMOTOR TRANSFORMATIONS FOR SACCADES). Although the human eye muscles have muscle spindles, they do not seem to have a functional stretch reflex; passive movements of the eyes are not reflexively ad-justed, and even seem to be ignored, so that if one pushes on the side of the eye, the resulting movement is reported by the visual system as movement of the external world. The reason the oculomotor system may be able to operate in feedforward mode is that the load (the spherical eyeball) is constant, unaffected by external weights or by gravity, and therefore more easily controlled than a multijointed limb. The feedforward controller may be an inverse model like that shown in Figure 1C (Krauzlis and Lisberger, 1989); an alternative proposal suggests an internal forward model like in Figure 1D (Robinson, 1975). Functionally, of course, there is powerful *visual* feedback: if the eyes drift, the error is reported as slip of the visual image over the retina.

Discussion

Internal Models

Although this section concentrates on problems of voluntary motor control, the arguments proposed here apply equally to other physiological systems. Visual guidance of the human arm is based on sensory information from the visual system with processing delays of up to 100 ms, and from proprioceptors with delays of perhaps 50–100 ms. Motor commands issued by the CNS may take 50–100 ms to start muscle contraction. So feedback signals from the environment lag significantly behind the issue of each motor command. Despite this, we control our limbs skillfully and accurately with movement durations of well under half a second. Thus, our motor control cannot be based entirely on feedback signals; we also employ feedforward control. However, the primate arm has variable nonlinear dynamics, changing with growth, age, fatigue, and orientation to the gravity field. These complexities mean that it is likely (although not yet certain) that control is based on an internal representation of the motor system—an internal model. This could be an inverse or a forward model of the limb. Both can transform a desired limb motion or position signal into the corresponding muscle torques.

How could we identify such an internal model? It should receive as inputs either the motor goal or an efferent copy of the motor command and also should receive proprioceptive information about the current state of the body. There must be a mechanism to allow the model to adapt to predict accurately the behavior of the limb, i.e., a neural learning mechanism. And the output of the model must form either the motor command or a sensory prediction of the action outcome. The cerebellum is a strong contender for these representations (Ito, 1984; Kawato and Gomi, 1992; Miall et al., 1993; CEREBELLUM AND MOTOR CONTROL). Other possible sites are the motor cortex (Kalaska et al., 1989), parietal cortex (Kalaska and Crammond, 1992), and the spinal cord (Bizzi et al., 1991), although a spinal representation would probably be more closely related to individual muscles.

There are strong connections from the posterior parietal cortex to the lateral hemispheres of the cerebellum, and from there to premotor and motor cortices. This cerebro-cerebellar pathway is a major route by which visual information reaches the cortical motor areas for the guidance of the limbs (Stein and Glickstein, 1992). Thus, for limb control, the input to the lateral cerebellum would be the current error in hand position or the desired movement of the hand. Spinocerebellar tracts provide a large array of proprioceptive signals, updating the cerebellum on the current state of the limb. As mentioned in the Introduction, visual and proprioceptive signals are in different coordinates, and need to be brought into a common format to control the arm; this may be another important task for the cerebellum. For adaptation, we know that coincident activity in climbing fiber and parallel fiber inputs to Purkinje cells results in a sustained change in the strength of the parallel fiber: Purkinje cell synapse (see LONG-TERM DEPRESSION IN THE CEREBELLUM). The cerebellum is therefore thought by many to act as an adaptive inverse model on the feedforward control pathway (see Figure 1C; Ito, 1984; Kawato and Gomi, 1992). The alternative model (Figure 1D) is equally valid, and my belief is that the cerebellum represents an adaptive forward model on a feedback pathway (Miall et al., 1993).

Problems with Theories of Internal Models

It is difficult to distinguish the two control strategies based on internal models unless one can get access to their internal structure (for example, blocking the internal feedback loop needed for a forward model). An accurate inverse model is preferable to an accurate forward model, because any physiological loop must have some delay, so that the fast internal feedback (see Figure 1D) cannot approach infinite open-loop gain. In their favor, forward models allow the brain to directly explore the consequences of proposed actions, testing them out on the model before executing them in the real world. However, while both forward and inverse models are well suited to feedforward control of physiological systems, they are still dependent on sensory feedback to ensure that their parameters remain accurate. Thus, they potentially suffer the difficulties of any feedback system with delayed sensory signals. An adaptation of a control scheme known as a "Smith Predictor" can resolve this problem by including an estimate of the feedback delays within the motor system (Miall et al., 1993).

Other difficulties with these models lie in strong assumptions about the relationship between engineering box diagrams like Figure 1 and biology. Thus, predictions based on these models may not be satisfied by brain lesions or by neurophysiological recordings. These difficulties may well be resolved by weakening our assumptions, acknowledging the differences between linear engineering models and the highly nonlinear realities, and treating inputs and outputs as parallel, distributed signals.

Acknowledgments.   I thank the Wellcome Trust and MRC & McDonnell-Pew Centres, Oxford, for support.

Road Maps: Control Theory and Robotics; Biological Motor Control

References

Atkeson, C. G., and Reinkensmeyer, D. J., 1991, Using associative content-addressable memories to control robots, in *Neural Networks for Control* (W. T. Miller III, R. S. Sutton, and P. J. Werbos, Eds.), Cambridge, MA: MIT Press, pp. 255–285.
Barto, A. G., Sutton, R. S., and Anderson, C. W., 1983, Neuronlike adaptive elements that can solve difficult learning control problems, *IEEE Trans. Syst. Man Cybern.*, 13:834–846.
Bizzi, E., Mussa-Ivaldi, F. A., and Giszter, S., 1991, Computations underlying the execution of movement: A biological perspective, *Science*, 253:287–291.
Ito, M., 1984, *The Cerebellum and Neural Control*, New York: Raven.
Jordan, M. I., in press, Computational aspects of motor control and motor learning, in *Handbook of Perception and Action: Motor Skills* (H. Heuer and S. Keele, Eds.), New York: Academic Press. ◆
Kalaska, J. F., Cohen, D. A. D., Hyde, M. L., and Prud'homme, M., 1989, A comparison of movement direction–related versus load direction–related activity in primate motor cortex, using a two-dimensional reaching task, *J. Neurosci.*, 9:2080–2102.
Kalaska, J. F., and Crammond, D. J., 1992, Cerebral cortical mechanisms of reaching movements, *Science*, 255:1517–1523.
Kawato, M., and Gomi, H., 1992, A computational role of four regions of the cerebellum based on feedback-error-learning, *Biol. Cybern.*, 68:95–103.
Krauzlis, R. J., and Lisberger, S., 1989, A control systems model of smooth pursuit eye movements with realistic emergent properties, *Neural Computat.*, 1:116–122.
Miall, R. C., Weir, D. J., Wolpert, D. M., and Stein, J. F., 1993, Is the cerebellum a Smith Predictor? *J. Motor Behav.*, 25:203–216.
Robinson, D. A., 1975, Oculomotor control signals, in *Basic Mechanisms of Ocular Motility and Their Clinical Implications* (G. Lennerstrand and P. Bach-y-Rita, Eds.), Oxford: Pergamon, pp. 337–374.
Stein, J. F., and Glickstein, M. G., 1992, Role of the cerebellum in visual guidance of movement, *Physiol. Rev.*, 72:967–1017. ◆

Motor Pattern Generation

Jeffrey Dean and Holk Cruse

Introduction

Whereas traditional ethology focuses on what David Marr called the computational theory of behavior (what an animal does and why), computational neuroscience—like neuroethology—is particularly interested in control algorithms and parameter representations (how a system does what it does) and implementations. Neuroethology (see NEUROETHOLOGY, COMPUTATIONAL) considers biological networks and proceeds inductively from behavior and physiology to general principles. Computational neuroscience seeks similar principles, testing in simulations hypotheses derived from various theoretical and experimental approaches.

A useful start is to consider what questions a theory of motor control should answer. First and foremost, how is a particular motor pattern generated? Which system properties are necessary? Which are sufficient? In neuroscience, behavior is often studied as patterns of neural activity, but the complex transformation in going from neural activity to muscle contraction and behavior cannot be ignored. A hierarchical description of motor function (e.g., Figure 1) depicts this transformation simply as effectors—usually muscles—transforming into movement a plan represented in neural activity, but this neglects two interactions. The physical plant can significantly influence movement, and its characteristics may already be incorporated into activity generated by higher centers.

Important questions at other levels are: What turns motor patterns on and off? What factors affect their form, amplitude, and speed? Again, physiological results show that the separation of function and the specific pattern of communication between levels implied in a hierarchical flow chart do not necessarily hold for the neural hardware. Individual neurons may contribute to, and communication links may cross, several levels (Figure 1).

Several questions address the system as a whole. Is the output stable with respect to external disturbances, and how can

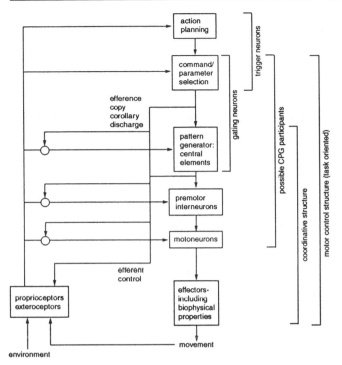

Figure 1. A hierarchical description of motor pattern generation, showing interactions between efferent, afferent, and central elements. Here, "motor control structure" designates the set of elements together with their states and interactions which produce patterned motor output appropriate to a particular task. Sensory information can modify activity at all levels, but may itself be modified by ongoing central activity (efference control, corollary discharge). The hierarchy of functions does not necessarily correspond to separate anatomical levels. Single neurons may participate in several functions, as indicated by the first three brackets at the right, and their interactions can spread over intervening levels.

speed, form, and amplitude be adapted to task demands? How are external and internal variables represented?

For practical reasons, experimental and theoretical work has focused on simple stereotyped movements, exemplified in escape reactions and target movements, and on rhythmic behaviors such as breathing and feeding, locomotion and communication, and patterns of activity and rest. Such basic processes are essential for survival. As a result, evolution has produced robust systems amenable to many experimental techniques. Rhythmic activity also provides an obvious marker of elements associated with the control system. Regular, repetitive behavior facilitates both statistical analysis to discriminate random variation from functional characteristics as well as quantitative stability analysis.

Motor systems also share phenomenological properties which invite modeling in various ways. Unfortunately, many models only demonstrate particular algorithms for how natural motor patterns might be generated, but few discriminate among alternatives. Taking physics as a paradigm, one might hope that accumulating comparative data on diverse animals would lead to a general, unified theory of biological motor control. While such a theory might be useful for engineers designing control systems, one must remember that the constraints of biological evolution may mean that each biological motor pattern generator (MPG) may only be one which was adequate and possible given the components in the animal's ancestors. Thus, performance and computations are often similar and

some algorithms may be shared, but the hardware is apt to be different for all but phylogenetically close relatives.

Metaphors and Tools for Studying Motor Behavior

Movement and its control interest scientists with quite different orientations. Biologically oriented researchers are primarily interested in how animals function. Technically oriented workers are primarily interested in building machines. Ideas have passed in both directions, but during most of the growth in neuroscience, technical sciences have provided models and tools for neuroscientists.

Since Descartes reduced animals to machines, biologists have used metaphors from contemporary technology to create theories of the brain and motor systems. As technology advanced, metaphors evolved from pneumatic actuators and card readers through magnetic tape to the computer-based metaphors of the present. For example, the orchestration of complex muscle activity was attributed to a central motor score, which could be stored and replayed like magnetic tape: Hoyle envisioned both a motor tape for motoneuron activation and a sensory tape for the sensory signals expected during undisturbed execution.

The current dominant metaphor is *motor program*, analogous to the set of instructions for a computer (Keele, Cohen, and Ivry, 1990). Varying usage led to alternative suggestions (*coordinated control program*, Arbib, 1981; *motor control structure*, Cruse et al., 1990). However, *motor program* remains entrenched in the literature. It is still apt if *program* is understood in a quite general sense as one that can generate complex output in the absence of input, vary its output depending on inputs, and modify itself. In other words, it represents a state established in the nervous system which produces coherent, task-related patterns of intrinsic activity and responses to external stimuli.

Both physiologists and psychologists adopted analytical tools developed in other fields. One important tool is cybernetics, the formal analysis of the role of feedback in a system's input-output characteristics, whereby systems can range from single neurons to whole animals (see MOTOR CONTROL, BIOLOGICAL AND THEORETICAL). The theory of dynamic systems is another conceptual tool suited to motor control and to rhythmic behavior in particular (see COOPERATIVE PHENOMENA). Motivated by the emergent properties of complex systems, it seeks collective variables summarizing important features of a system and control parameters which influence its behavior.

Now, as this volume illustrates, neuroscience provides important models for technical fields, as the latter develop systems inspired by the massively parallel, inherently redundant and self-organizing structure of nervous systems. However, the wish for specific solutions is confounded by the incredible diversity of biological adaptation.

Elements of Motor Behavior

The basic elements are the sensory systems which signal the state of the organism and its surroundings, the muscles which produce force, and the nervous system which links the two and contributes its own intrinsic activity. All possess some characteristics that make them unattractive to engineers. Muscles are relatively slow and the force they provide depends in a complex manner on muscle length, its rate of change, and the pattern of activation (see MUSCLE MODELS). Sense organs typically combine information about the value of a parameter and its rate of change in a complex way. Most neurons transmit information slowly, using a pulse code with limited bandwidth.

Together, transmission delays, component variability, and limited coding precision can cause feedback-induced oscillations. Variability in muscle responses requires feedback supervision, but inherent conduction delays can destabilize any loop with a high gain. (Conceivably, such oscillations might be useful; but as a rule, reflexes are configured to avoid instability and biological rhythms are actively generated.)

The contrast between the apparent inadequacies of the components and the exquisite motor performance of many animals leaves one feeling there must be a trick somewhere. Some limitations may actually simplify problems. For example, the spring-like properties of muscles could reduce movement planning to specifying endpoints (Bizzi et al., 1992). The highly parallel organization of biological control systems is a more obvious source of compensation.

A Brief History of Theories of Motor Pattern Generation

When physiological study of the nervous system began in earnest, the most accessible parts—the peripheral sensory and motor elements and their combination in elementary reflexes—were naturally investigated first. Because it is also natural to build theories on what is known, reflexes assumed a central role, making animals into reactive, reflex-driven machines. In biology, motor patterns were explained as chains of reflexes in which each movement creates appropriate exteroceptive and proprioceptive stimuli to elicit the next movement, leading to complex behavioral sequences or, if the chain is closed, rhythmic behaviors. For example, timing and continuation of leg movements in walking were attributed to reflexes based on changes in leg position, loading, and equilibrium signals during the step. However, such conceptual models were never simulated quantitatively. In psychology, the analogous theory was called *response chaining*. (Such reaction-based formulations were extended in ethology to interactions between individuals and in psychology to behaviorism.)

The rigidity of reflex or response chain theories caused their downfall (e.g., Keele et al., 1990; Lashley, 1951). Both had to postulate subtle differences in stimulus constellations to account for different behaviors under seemingly identical conditions and were unable to explain rule-based behavior in novel situations. Response chains based on peripheral, sensory loops cannot explain rapid sequences when event intervals are shorter than minimum response latencies. Chains based on central loops have difficulty explaining concurrency of preparatory movements and their dependence on antecedent *and* subsequent actions.

A new theory was needed, one incorporating generalized versions of behaviors—plans or schemas (Arbib, 1981; Lashley, 1951)—which are modifiable by external and internal factors. In psychology, neurological findings implicate particular structures in planning (see BASAL GANGLIA; CEREBELLUM AND MOTOR CONTROL), but details are unclear. In biology, motor programs took on concrete form, beginning with the realization that the central nervous system (CNS) alone can produce organized activity. T. Graham Brown demonstrated that cats deprived of sensory inputs could still produce coordinated stepping and envisioned a central pattern generator (CPG)—i.e., a set of elements in the CNS able, in the absence of patterned sensory inputs, to produce patterned motor output related to natural behaviors (see HALF-CENTER OSCILLATORS UNDERLYING RHYTHMIC MOVEMENTS). ("Neural oscillator" would be more precise for the many CPGs involved in rhythmic behaviors.) Both his experimental approach, deafferentation, and his conceptual model remain current. Subsequently, von Holst's (1969)

quantitative analyses of rhythmic behaviors were best explained by multiple rhythm generators interacting in different ways. With improved physiological techniques, motor patterns generated within the CNS could be traced sometimes to single neurons and sometimes to networks of neurons.

For a time, the role of CPGs was overemphasized at the expense of peripheral elements. Wilson's demonstration of a basic flight rhythm in deafferented locusts was particularly influential (see LOCUST FLIGHT: COMPONENTS AND MECHANISMS IN THE MOTOR). Sensory influences were relegated to providing tonic excitation rather than cycle-by-cycle modulation. Initial descriptions of neural oscillators reinforced this trend, especially when evidence was found for neural oscillators in virtually all rhythmic behaviors (Delcomyn, 1980). For nonrhythmic behaviors, an analogous self-sufficiency of central circuits is evident in the ability of the deafferented CNS to specify limb configurations (Bizzi et al., 1992). In general, however, the overemphasis on central mechanisms also reflected a willingness to overlook differences between intrinsically generated activity and natural motor patterns.

The necessary correction occurred when it was shown that sensory inputs could influence the frequency, timing, and form of rhythmic activity and fulfill criteria for inclusion in the pattern generator: rhythmic activity correlated with the behavior and the ability to shift or "reset" the phase of the rhythm when appropriately stimulated. In locusts, for example, actual wing movement and loading, visual stimuli, and changes in air-flow past the head modulate and stabilize wing movement on a cycle-by-cycle basis (Wendler, 1985; see LOCUST FLIGHT: COMPONENTS AND MECHANISMS IN THE MOTOR). Similar entrainment by sensory inputs occurs in other systems with well-developed neural oscillators. For nonrhythmic movements, sensory inputs may be essential for signaling when necessary preparations are completed (Pearson, 1981).

Hence, it appears that most MPGs incorporate both central and peripheral elements. Motor patterns represent an interaction between intrinsic CNS activity (CPGs) and peripheral influences reflecting biomechanical characteristics, reflexes, and general afferent activity (Pearson, 1981; Wendler, 1985). The relative importance of central and peripheral contributions is subject to evolutionary selection depending upon the predictability of both the external surroundings and the result of efferent commands. With respect to survival, these two elements are analogous to adaptation via genetic programs and learning or developmental plasticity, respectively.

Central Pattern Generators

Understanding of CPGs has increased most for neural oscillators involved in rhythmic behavior. Early proposals were conceptual models to explain simple two-phase rhythms—e.g., alternation between stance and swing. Half-center oscillator models like that of Brown (see Camhi, 1984) contain two functional units connected by reciprocal inhibition and subject to a common excitatory drive. (A functional unit can be either one neuron or groups of neurons, synchronized by reciprocal, chemical or electrical excitation.) Turning this bistable circuit into an oscillator requires the inhibition to decay with time or a functionally equivalent change to occur. (Several alternative two-phase oscillators are possible; see Camhi, 1984.) George Szekely demonstrated that more activity phases can be generated by neurons or groups forming a ring with recurrent inhibition (Camhi, 1984).

Early findings distinguished two kinds of neural oscillators: in pacemaker cells ("cellular oscillators," e.g., heart Purkinje cells or bursting cells of *Aplysia*), rhythms depend only on membrane properties and continue when isolated from the nervous system, whereas in "network oscillators," they depend upon the connectivity of cells with simpler properties. Network oscillators were found in several invertebrate and vertebrate preparations (e.g., Cohen, Grillner, and Rossignol, 1988; see also CRUSTACEAN STOMATOGASTRIC SYSTEM). Initial interest focused on connectivity because it was easier to characterize than cellular properties and because several networks contained reciprocal or recurrent inhibition, providing a satisfying agreement between experiment and theoretical simulations using simple model neurons.

Subsequent results showed this dichotomy to be oversimplified (Getting, 1989). First, the time delays in simple half-center models require nonlinear membrane properties. Second, actual connectivity is much more complex than simple recurrent inhibition. For example, the five-neuron, recurrent ring in the swim CPG of the leech also contains reciprocal inhibitory connections within and diagonally across the ring (see Camhi, 1984). Third, some neurons within putative network oscillators possess nonlinear membrane properties allowing them to oscillate or switch between inactivity and high, prolonged activity (see OSCILLATORY AND BURSTING PROPERTIES OF NEURONS). Such nonlinearities occur even at the level of motoneurons, the final common path to the muscles. Fourth, synaptic connections can be equally complex, changing in strength or even sign depending on time or on the relative potentials of sender and recipient.

These complications bedevil study of even "simple" invertebrate networks with a few identifiable neurons. As complexity increases, attributing function to particular elements becomes more difficult. Network simulations and physiological data show that systems can perform appropriately even when some neurons have nonfunctional or even dysfunctional connections (Robinson, 1992). Since evolutionary selection works most directly on behavior, eliminating all elements whose contribution opposes the overall performance may not be possible or necessary. Such rogues (Robinson, 1992) occur even in simple resistance reflexes.

For biologists studying a particular nervous system, simulations using realistic synaptic interactions and membrane characteristics are absolutely necessary (Getting and Dekin, 1985). The practical difficulty is the ever-increasing list of membrane and synaptic properties which must be measured in experiments and computed in simulations. Calculation of synaptic effects becomes still more formidable when the complex geometry of real neurons is considered (see DENDRITIC PROCESSING). This geometry influences responses according to both the timing and location of inputs. Thus, simulation of even a single real neuron, let alone a network, is a formidable problem (see SINGLE-CELL MODELS). In summary, the optimism following early successful simulations using simplistic neurons has dissipated, but newer models have become much more realistic.

Conventional artificial neural networks can generate temporally patterned output if they are recurrent or receive signals representing a kind of timing pulse. A simple step rhythm can be produced by a four-unit, recurrent network in which two units represent the states and two represent position criteria for changing states. The recurrent connections cause each movement to continue until its endpoint is reached. More complicated networks can generate movement trajectories (e.g., SENSORIMOTOR LEARNING). Such networks produce complex, context-dependent outputs, features attributed to schemas, and thus suggest directions for future biological research. However, they are still very simplistic compared to biological networks.

Connections typically are still homogeneous and unit properties are much simpler.

Switching Patterns On and Off

Functionally (Figure 1), pattern generators are controlled by a switch. In the half-center oscillator, this switch is the tonic excitation. The corresponding neural elements for both rhythmic and nonrhythmic behaviors were labeled *command neurons* (Kupfermann and Weiss, 1978). Activity in such a neuron should be both necessary and sufficient for producing the behavior. If action potentials are required, the neuron's threshold determines the behavioral threshold. In practice, very few neurons fulfill this strict criterion (Camhi, 1984). More often, control is distributed among many neurons, another example of redundancy in biological systems. As a result, the concept has been extended to "command systems" containing multiple "command elements" (see COMMAND NEURONS AND COMMAND SYSTEMS).

Two kinds of switches have been identified. Some behaviors continue only as long as an appropriate stimulus is present; others continue after it ends. A similar distinction applies to command neurons (Kupfermann and Weiss, 1978): activity in "trigger neurons" elicits a longer-lasting behavior, while that in "gating neurons" determines the duration of the behavior.

Command and pattern generation are not always clearly distinct in the neural hardware. Activity in some gating neurons is not constant, like a simple on-off command, but modulated in time with the output rhythm. Moreover, some can reset the rhythm and thus fulfill criteria for inclusion in the CPG. Integration of command and pattern generation occurs to different degrees. Some leech neurons are essentially outside the CPG, and the recurrent, excitatory paths modulating their activity simply help prolong this activity. (In contrast, trigger neurons often receive recurrent inhibition to prevent reactivation.) In other cases (e.g., the *Tritonia* escape system), the command signal depends on membrane properties of CPG neurons: depending on stimulus strength, the network produces either a single, longitudinal contraction or multicycle, dorsoventral flexions. Thus, a "polymorphic," anatomical network can produce different outputs (Getting and Dekin, 1985).

Nonlinear effects of neural activity (see ACTIVITY-DEPENDENT REGULATION OF NEURONAL CONDUCTANCES) and modulatory transmitters or hormones (see NEUROMODULATION IN INVERTEBRATE NERVOUS SYSTEMS) which modify membrane properties (Harris-Warrick and Marder, 1991) can produce qualitative changes in outputs or even functional reconfigurations of networks. Such plasticity makes hardware simulations that much more difficult. Thus, biological MPGs show a protean variability compared to artificial neural networks, where unit properties are usually static and output changes are controlled by inputs or by learned transitions.

Peripheral Factors Affecting Motor Patterns

Outputs of CPGs are usually subject to feedforward and feedback modification depending on sensory information about the surroundings and the performance. Formally, feedback depends on the consequences of motor activity and acts to reduce deviations ("errors") from a reference value, as in stretch reflexes or optomotor responses. Feedforward pathways are involved in selecting actions or setting parameters in advance—e.g., specifying the size of a saccade or an escape turn. However, in a natural, continuous stream of behavior, the distinction between error-correcting feedback and parameter-setting feedforward mechanisms is not always sharp (Cruse et al., 1990).

Relative contributions of central and peripheral components vary considerably. For example, central elements appear weaker in walking than in flying and swimming, where the substrate is more forgiving. In flying, variability in wing movements affecting lift and steering can be corrected over several subsequent cycles, whereas in walking, not finding a foothold or tripping over an obstruction must be corrected immediately. When stick insects and crayfish walk, the movement of the leg itself, as signaled by sensory inputs, actually determines the state of the pattern generator (Baessler, 1986a; Cruse, 1990). Intrinsic activity is slow, variable, and fragmentary.

Contributions of sensory inputs can be modified at various levels (Figure 1). Sensory activity can be modulated peripherally through efferent control or centrally at the output synapses of the primary afferents. Reflexes occurring in one situation can be reversed or replaced by wholly new responses. In invertebrates (Baessler, 1986b), negative feedback, resistance reflexes, which oppose leg displacement during posture, are replaced by assistance reflexes as stance begins to boost propulsion. At other times, they remain active, opposing deviations from the normal trajectory. Phase-dependent reflex modulation may be a natural consequence of thresholds in neural activation functions (see LOCUST FLIGHT: COMPONENTS AND MECHANISMS IN THE MOTOR).

Motor behaviors are often studied based on neural activity, but the transformation between this activity and movement is not trivial. Different neural rhythms in the crayfish CPG controlling chewing were characterized long before their functional significance was determined (see CRUSTACEAN STOMATO-GASTRIC SYSTEM). Even motoneuron activity is difficult to relate to movement, partly due to the sophistication with which animals use passive and elastic properties of their muscles and skeletons. For example, treating biped walkers simply as a system of passive, damped pendulums explains many characteristics, suggesting that muscle activity is only occasionally necessary to maintain the pendular motion or correct disturbances. Similarly, according to the endpoint control hypothesis (see EQUILIBRIUM POINT HYPOTHESIS), muscle activity is not related to movement as such, but to the target, the allowable deviation, and an estimate of possible errors. In summary, the control system need not drive movements in a rigid way (see also CERE-BELLUM AND MOTOR CONTROL). Moreover, approximate algorithms and models may suffice, especially when animals are small or compliances are high so impacts or stresses arising from inaccuracies can be absorbed without injury.

Coordinating Multiple MPGs

Many motor behaviors contain distinct subunits which may be active concurrently or sequentially. Movements of different body parts may need to be coordinated spatially and temporally (e.g., EYE-HAND COORDINATION IN REACHING MOVEMENTS; LOCOMOTION, INVERTEBRATE). Behavioral sequences may require that one action be completed before another begins. For this purpose, several MPGs can be arranged hierarchically or in parallel, depending upon task requirements. For speaking and typing, the types of errors and time delays, the concurrence of preparatory and execution phases, and the modulation of elements depending on preceding and following elements indicate a hierarchical arrangement, whereas relative coordination of concurrent rhythms (von Holst, 1969) indicates a parallel arrangement.

Both examples suggest that interactions among different MPGs can occur at many levels, ranging from overall planning

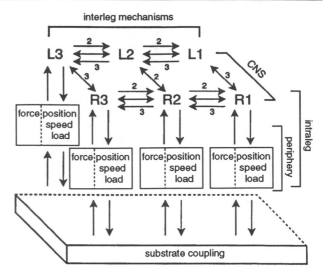

Figure 2. Redundancy in central and peripheral mechanisms of step coordination. In the stick insect, multiple, redundant mechanisms within the CNS and intraleg mechanisms affect leg coordination. Numbers beside an arrow indicate more than one mechanism. The periphery includes the skeletomuscular and sensory elements. (Boxes for left front and middle legs, L1 and L2, are omitted for clarity.)

through parameter setting and the timing and form of activity in different CPGs to peripheral loops involving mechanical coupling and reafference. Network models provide a natural way of simulating these interactions.

Coordination of rhythms has been studied extensively in several systems (see CRUSTACEAN STOMATOGASTRIC SYSTEM; CHAINS OF COUPLED OSCILLATORS). MPGs are often highly redundant and thus robust. In insect walking, multiple, centrally mediated coordination mechanisms are augmented by multiple locally mediated (intraleg) mechanisms sensitive to leg position, state, and load (Cruse, 1990; Figure 2).

Modeling the periphery is particularly important when it strongly affects movement. Algorithms for step coordination, demonstrated using a simplified representation of leg position, will be unable to control a machine with real legs if they try to place legs in unreachable positions. Behavioral data on limb interactions, the basis of many coordination models, actually represent the total effect of internal activity and any reflexes. Dynamic systems theory, using collective variables like phase, can sometimes provide a succinct encapsulation of this system behavior.

Discussion

For the near future, interactions between pure and applied researchers or between experimentalists and neural engineers may continue to be fruitful but, to some extent, unsatisfying. Neuroscience clearly has a place for computational neuroscientists using simulations of biological networks to represent acquired knowledge and test its completeness. Artificial neural networks provide new tools and, more important, reemphasize several significant concepts: (1) the common currency in neural networks is neuronal activity; (2) this activity need not be simply related to physical parameters; (3) processing is distributed and parallel; and (4) approximations may suffice. These conclusions have important implications for interpreting physiological experiments (Robinson, 1992).

Neural engineers naturally look to animals for inspiration in improving technical systems, hoping for a concise list of biological solutions for particular computational problems. Because animals are so diverse, neuroscientists can provide only rudimentary guidelines. Each species has its own answer to a very complex and specialized problem, an important part of which is making more of its kind. Animals are bound by their genes and developmental constraints to change in small steps that are themselves adaptive or at least not seriously detrimental to the individual's survival. In contrast, technological innovation can create machines or animats which break radically with the past. Prototypes hopelessly outperformed by existing technologies are not doomed to extinction. Nevertheless, the very diversity of biological solutions enhances what biology has to offer. Better understanding of biological computations should help select algorithms for technical applications, especially as artificial networks approach the complexity of biological networks, not merely in numbers of units or connectivity but also in unit properties and plasticity. Optimization techniques based on biological learning or evolution have already proved useful in adapting algorithms to new problems (see LOCOMOTION, INVERTEBRATE).

Acknowledgments. The authors were supported by the Deutsche Forschungsgemeinschaft (Cr 58/8) and the Bundesministerium fuer Forschung und Technik (01 IN 104B/1).

Road Map: Motor Pattern Generators and Neuroethology
Related Reading: Corollary Discharge in Visuomotor Coordination; Human Movement: A System-Level Approach

References

Arbib, M., 1981, Perceptual structures and distributed motor control, in *Handbook of Physiology—The Nervous System*, vol. 2, *Motor Control* (V. B. Brooks, Ed.), Bethesda, MD: American Physiological Society, pp. 1449–1480. ◆

Baessler, U., 1986a, On the definition of central pattern generator and its sensory control, *Biol. Cybern.*, 54:65–69. ◆

Baessler, U., 1986b, Afferent control of walking movements in the stick insect *Cuniculina impigra*, II: Reflex reversal and the release of the swing phase in the restrained foreleg, *J. Comp. Physiol. A*, 158:351–362.

Bizzi, E., Hogan, N., Mussa-Ivaldi, F. A., and Giszter, S., 1992, Does the nervous system use equilibrium-point control to guide single and multiple joint movements? *Behav. Brain Sci.*, 15:603–613. ◆

Camhi, J. M., 1984, *Neuroethology: Nerve Cells and the Natural Behavior of Animals*, Sunderland, MA: Sinauer. ◆

Cohen, A. H., Grillner, S., and Rossignol, S., 1988, *Neural Control of Rhythmic Movements in Vertebrates*, New York: Wiley.

Cruse, H., 1990, What mechanisms coordinate leg movement in walking arthropods? *Trends Neurosci.*, 13:15–21.

Cruse, H., Dean, J., Heuer, H., and Schmidt, R. A., 1990, Utilization of sensory information for motor control, in *Relationships Between Perception and Action: Current Approaches* (O. Neumann and W. Prinz, Eds.), Berlin: Springer-Verlag, pp. 43–79. ◆

Delcomyn, F., 1980, Neural basis of rhythmic behavior in animals, *Science*, 210:492–498.

Getting, P. A., 1989, Emerging principles governing the operation of neural networks, *Annu. Rev. Neurosci.*, 12:185–204. ◆

Getting, P. A., and Dekin, M. S., 1985, *Tritonia* swimming: A model system for integration within rhythmic motor systems, in *Model Neural Networks and Behavior* (A. E. Selverston, Ed.), New York: Plenum, pp. 3–20.

Harris-Warrick, R. M., and Marder, E., 1991, Modulation of neural networks for behavior, *Annu. Rev. Neurosci.*, 14:39–57. ◆

Keele, S. W., Cohen, A., and Ivry, R., 1990, Motor programs: Concepts and issues, in *Attention and Performance XIII: Motor Representation and Control* (M. Jeannerod, Ed.), Hillsdale, NJ: Erlbaum, pp. 77–110. ◆

Kupfermann, I., and Weiss, K. R., 1978, The command neuron concept, *Behav. Brain Sci.*, 1:3–39. ◆

Lashley, K. S., 1951, The problem of serial order in behavior, in *Cerebral Mechanisms in Behavior: The Hixon Symposium* (L. A. Jeffress, Ed.), New York: Hafner, pp. 112–136.

Pearson, K. G., 1981, Function of sensory input in insect motor systems, *Can. J. Physiol. Pharmacol.*, 59:660–666.

Robinson, D. A., 1992, Implications of neural networks for how we think about brain function, *Behav. Brain Sci.*, 15:644–655. ◆

von Holst, E., 1969, *Zur Verhaltensphysiologie bei Tieren und Menschen*, Munich: R. Piper & Co.

Wendler, G., 1985, Insect locomotory systems: Control by proprioceptive and exteroceptive inputs, in *Insect Locomotion* (M. Gewecke and G. Wendler, Eds.), Hamburg: Verlag Paul Parey, pp. 245–254.

Multiprocessor Simulation of Neural Networks

Hélène Paugam-Moisy

Introduction

Parallel Aspects of Cognitive and Neural Processing

Parallel aspects often emerge in cognitive and neural processing. The action of typing a letter, with the help of text processing software, may be addressed for illustrating the parallel use of multiple memory processing. Functional memory allows the fingers to move toward the right keys on the keyboard; short-term visual memory helps the typist to visualize the group of words or the sentence he is writing; long-term memory interposes in order to check the spelling of words; and anticipatory memory is working for the author of the text to forecast the next sentence from the content of later sentences, and to control the global significance of the whole text. All these forms of memory work concurrently, with constant interleaving, contributing to a unique result.

Several other parallel aspects have been recently pointed out by neurobiologists and psychologists (see Kosslyn and Koenig, 1992). One of them is the frequent association of several perception processes which enhance each other, like the conjunction and the reinforcement of visual and auditory acuities for improving pattern recognition. The brain's style of processing information appears to be a model of parallel processing. More precisely, the brain consists of multiple sets of various kinds of neurons, and it functions as a distributed system. This notion can be enlightening in making a distinction between distributed computation, in which a set of distinct types of computation are distributed across possibly different processors, and parallel computation, in which a given function is broken down into similar computations on a multiplicity of processors.

Basic Principles of Parallel Computation

Parallel computers have been conceived for reducing time for information processing. The first of them were vector processors, the parallelism of which was oriented toward accelerating calculation. More and more, different models arose, which can be classified into shared memory and distributed memory parallel computers, or into SIMD (Single Instruction flow, Multiple Data flow) and MIMD (Multiple Instruction flow, Multiple Data flow) machines (Hwang, 1993). These machines aim to be universal computers, but they are mainly used for improving the speed of massive numerical computation (Quinn, 1987).

Parallel programming is a rather new exercise which requires much more dexterity and abstraction than sequential programming. Parallel algorithms are often derived from sequential ones, according to several classical principles like divide and conquer, pointer jumping, partitioning and pipelining (Jàjà, 1992). Most of them deal with classical algorithms such as the sorting problem or linear algebra routines. Only in a few cases are they conceived directly, starting from the features of the problem which they have to solve (Carey, 1989).

Parallel Aspects of Artificial Neural Networks

Historically (Quinlan, 1991), the main interest of Warren McCulloch and Walter Pitts was to show how artificial neurons might compute primitive logical operations, and how complex logical processes might be carried out by networks of neurons. In the same way, some of the predicates studied by Marvin Minsky and Seymour Papert were fundamental to binary arithmetic. Minsky and Papert showed that the simple perceptron (no hidden layer or feedback) is not a universal computing machine. This first stage of neural network research was poorly related to parallel computation, which is quite natural in the computational context of the 1940s and 1960s, where calculators and serial Von Neumann computers were dominant (Arbib and Robinson, 1990).

Several features of parallel computation emerged in artificial neural network behavior with the new wave of connectionist research in the mid-1980s. The concept of neural networks was then revisited in a radically new context: parallel computers actually existed and parallel computation was a stand-alone domain of research and development. Furthermore, artificial intelligence and logicosymbolic methods showed their limitations. Artificial neural networks appeared as a new way of processing information by learning, without deterministic programming. The most original characteristics of connectionist models are their distributed way of storing information, their ability for fast retrieval, their strong self-organizing power, and their fault-tolerance properties. Since some of these features are similar to those of parallel and distributed computation, artificial neural networks and connectionist research are often named *parallel distributed processing*, which is the title of a best-seller in the domain (Rumelhart, McClelland, and PDP Research Group, 1986).

Attempts to Capture Neural Parallelism on Computers

Mapping Neural Networks on Parallel Computers

Most of the present connectionist models are derived from the formal neuron of McCulloch and Pitts, and from assembling such elementary units, with some variants like sigmoid transition functions or real valued states and weights. This point is an obvious parallel aspect of neural networks: mapping one neuron per processor seems to be a good idea. It seems to be realizable on SIMD multiprocessors, which currently possess several thousand processors. But the main drawback of this method comes from the prohibitive number of communications thus involved. Unless the network is sparsely connected,

the order of magnitude of the connection number is the square of the unit number. What is worse, the neural connections are neither local nor regular. Hence they involve long-distance communications. These characteristics do not fit a typical SIMD architecture very well. If one is considering such an implementation on an MIMD parallel computer, then the main limitation is that the number of available processors is usually much smaller than the number of artificial neurons in a connectionist network. Therefore, the granularity of the parallelization requires more subtle attention and appears to depend on the nature of the target parallel machine.

Some isolated works have taken this approach. Most of them concern only peculiar neural networks and precise parallel computers (optimizing communication load balancing on a hypercube of transputers, mapping a multilayer neural network with a constant number of neurons per layer, etc.). The drawback of this approach is that no general methods can emerge. A wider study has been carried out for the MIMD model of parallel computers, and for almost any model of neural networks. Ghosh and Hwang (1989) define the "connectivity" of a set of units as the ratio of the number of actual connections among these units to the number of connections in the fully connected case. They indicate how to arrange the units in remote regions, influence regions, and core regions. Hence, each processor can be associated to a core region, the core regions belonging to the same influence region being mapped on neighboring processors, in order to minimize the global communication cost. These remarks argue in favor of modular neural networks, as confirmed by a more recent analysis of implementation constraints of neural networks on transputer networks (Murre, 1993).

Nevertheless, the architecture of neural networks poses difficulties for capturing this form of parallelism because of the irregular connectivity of the current models. The concept of parallel computers and those of neural networks differ in their management of connection links (see also DIGITAL VLSI FOR NEURAL NETWORKS).

Parallelizing the Matrix Product Computations

Another parallel aspect comes from the way neural networks process. The chief calculations of every artificial neuron, which are repeated thousands of times, are matrix-vector products. They appear both in network processing, for the computation of the neural states from the weighted sums of inputs, and in many learning algorithms, for the computation of the weight updates. Hence, exploiting parallel linear algebra algorithms, which are well developed on current parallel computers, seems to be a good manner of improving neural computation.

However, except for Hopfield networks and a few other models, neural networks are not fully connected. For instance, in a feedforward multilayer network, without links skipping layers, the weight matrix is composed of successive blocks under the diagonal. In general cases, weight matrices are sparse, but they present regular architectures. Hence, neither general parallel algorithms for matrix-vector products nor special algorithms for randomly sparse matrices are suitable for the required computation. In some cases, good speed-up can be performed in cautiously applying adaptive mapping of non-zero coefficients on the processors, in order to balance the communications (Pétrowski and Paugam-Moisy, 1993). This method is more easily implemented on MIMD computers than on SIMD computers. Contrary to mapping the neurons on the processors, the most suitable neural architectures are fully connected networks (which provide dense matrices) or multilayer networks, the layers of which are numerous and have nearly the same number of units. Nevertheless, one drawback of this approach is similar to those of the previous method—the loss of generality.

For making use of classical techniques in parallel computation, this method of parallelizing weight update computations can be coupled with pipelining the examples to be learned by the neural network. If several examples are simultaneously (though not synchronously) in process, the succession of matrix-vector products can be seen as a matrix-matrix product. The new matrix is full because its components are the coordinates of a set of examples. However, this new point of view may imply some disturbances in the neural computation because a block of examples is now processed with the same weight matrix. This difficulty will be addressed in the next section.

In conclusion, since the connections between neural units are irregular and sparse, extracting parallel computation from neural computation is a complex problem. It is not straightforward to efficiently apply algebraic parallel algorithms, either for computing neural network states or for applying learning rules.

Parallelizing by Blocks of Examples

A better way to avoid communications consists in exploiting the large number of repetitive example presentations required by most of the learning rules. One solution consists in duplicating the neural network, one copy per processor. For instance, backpropagation learning can be computed on a two-dimensional torus of P processors, each implementing a multilayer network. Locally, every processor π accumulates the weight updates $\Delta W_\pi = \Sigma_{1 \leq i \leq K} \Delta W(x_i(\pi))$, which it computes for its packet of K examples. Then an all-to-all communication allows the P processors to exchange their local weight updates in order to actualize their weight matrix in adding the total weight update: $W_{\text{new}} = W_{\text{old}} + \Delta W$, where $\Delta W = \Sigma_{1 \leq \pi \leq P} \Delta W_\pi$. All the copies of the neural network are identical at every time. The weights are updated periodically and synchronously on every processor.

This method can only be implemented on distributed memory parallel computers, with sufficient local memory on each processor. Then the parallel aspect is found in the numerous patterns or examples that are presented to the neural network, and the amount of required memory is balanced by a gain of time. Usually, patterns are sequentially processed, but just as the brain has to process a lot of simultaneous stimuli, why not capture this opportunity for modeling by presenting several patterns in parallel to artificial neural network simulations?

When the network is fixed with invariable weights—for the generalization phase, for instance—this way of parallelizing neural processing is the optimal technique. The speed-up thus obtained is almost linear. It means that using P processors nearly divides the running time by a factor P (Paugam-Moisy, 1993). The problem becomes more complex for the learning phase, because this method implies processing the examples by blocks, which may involve disturbances to the convergence of learning algorithms, as mentioned above. This difficulty will be next addressed in cases of the most widely used connectionist models: backpropagation learning, topological feature maps, and an incremental neural network.

For multilayer neural networks, if K examples are processed by each of the P processors, a block of $B = KP$ examples is presented between two successive weight updating, as explained above. However, the size of the local data ΔW_π to be communicated is the same as the size of the weight matrix W, whatever the number K of processed examples. Consequently, the larger the block size B, the lower the number of data exchanges be-

tween the processors per epoch, and the smaller the communication time (an "epoch" designs one complete presentation of all the examples of the training set). However, presenting the examples by blocks alters the convergence of backpropagation. If the weight updating is deferred, the number of epochs required for reaching a given success in learning increases immoderately. Though these two effects cancel out, acceptable speed-up is still achievable in two ways:

1. Carefully adapting the learning rate, according to the block size, and optimizing both the block size and the number of processors (Paugam-Moisy, 1993).
2. Immediately applying local weight corrections, after each example, and only exchanging all the ΔW_π after a rather large block B. The various copies of the neural network are temporarily desynchronized (the weight matrices W_π differ when the K examples are processed, and become equal again just after the weight update exchange), but this strategy improves the speed of learning convergence (Girard and Paugam-Moisy, 1994).

In the case of topological feature maps, the difficulty is slightly different. The Kohonen learning rule implies only local weight updates, in a neighborhood of the unit which is closest to the presented example. Hence, the different units of the map have to be cleverly distributed on the processors in order to assure a good work balancing (see Pétrowski and Paugam-Moisy, 1993). Performances can be improved by processing examples by blocks. The optimal condition is that the distribution of examples in the training set is uniform with respect to the features to be discriminated, so that neighborhoods of examples in the same block do not overlap. In cases of overlapping, heuristics are proposed for managing the weight updating (Demian and Mignot, 1993).

The difficulty is the same as for multilayered networks: the larger the block size B, the smaller the communication time, but presenting examples by blocks alters the convergence of learning. Assuming that the distribution is uniform, an optimal block size can be statistically estimated (Demian and Mignot, 1993). But since topological feature maps are a nonsupervised connectionist model, the uniform distribution of examples is not straightforward. The user can adapt his strategy of presentation only if the features of the examples are known.

The problem is again different with incremental neural models. Incremental neural networks are characterized by an architecture which is dependent on the examples and adaptive through the learning phase. Take, for example, an incremental neural classifier which changes or creates its prototypes according to some learning rule. Local processing of example blocks again induces disturbances in the learning behavior. For instance, a new prototype could be created on a processor, though it already exists on another one. What is worse is that the exchange of modified or new prototypes involves an amount of data which increases with the block size. This drawback can be overcome in supervised learning phase if examples belonging to different classes are presented to different processors. Hence, the problem becomes nearly the same as for topological feature maps. For reaching good speed-up in parallelizing incremental neural networks by blocks of examples, it is important to finely tune the strategy of example presentation (Azcarraga, Paugam-Moisy, and Puzenat, 1994).

A conclusion would be that the coarse grain of MIMD computers better fits the parallel aspects of neural networks, especially in generalization phases, but is not yet well adapted to the learning algorithms of the present connectionist models.

Other Links Between Neural Networks and Parallel Computation

Parallel Optimization of Neural Networks

This topic addresses a meta-level of neural network parallelization. Instead of implementing one neural network on a processor network, a population of neural networks can be implemented. The purpose of such a device is to accelerate the tuning of neural architecture and parameters when applying a connectionist model to a given problem. As a matter of fact, the scarceness of constructive theoretical results for designing performing neural networks is an actual difficulty for the developer of applications. This gap is at least as important as the prohibitive running time of some learning algorithms on real-world size applications. Hence the contribution of parallel computation to this problem is worth considering.

A solution consists in developing an optimization algorithm on a supervisor process which governs a farm of different neural network processes. The function to be optimized can be the learning or the generalizing performance of the neural networks. The optimizer process can compute a hill-climbing-like algorithm or a genetic algorithm (Paugam-Moisy, 1993).

Optimization Problems Solved by Neural Networks

A radically different way of considering neural networks and computation consists in using neural networks as parallel devices for solving hard computational problems (see PROBLEM SOLVING, CONNECTIONIST, and NEURAL OPTIMIZATION).

Connectionist methods can then be applied to solve problems which are directly related to parallel computation and distributed systems, such as processor network designing, crossbar switch scheduling, channel routing, etc.

Discussion: Toward Another Form of Parallel Computation

Whereas the preceding section points out that neural networks and parallel computation can help each other in solving their respective problems, an earlier section emphasized the difficulties of capturing parallel aspects of artificial neural networks on current parallel computers. Distributing artificial neurons or connections on processors involves mapping difficulties and the risk of creating communication bottlenecks which drastically slow down the performances in running time. Neither the granularity of SIMD computers nor those of MIMD machines is well suited to the inherent parallelism of artificial neural networks. Distributing calculations of state or weight updating comes up against the difficulty of irregular connectivity and sparse matrices. The risk of generating bottlenecks arises again in applying this method. Distributing the examples by packets on several copies of the neural network seems to be the most promising way of parallelizing neural networks today. This method presents the advantage of being more general and could be successfully applied to any connectionist model. The most suitable computers are then distributed-memory MIMD machines.

Nevertheless, several parallel features are actually present in neural network architectures as well as in neural computation (learning and generalization phases). But these aspects are more complex and less regular than a plain network of identical elementary processors. As brain theory tells us, a biological neural network looks more like a multitude of intricate and distinctive subsystems, which are always working concurrently and which present a large variety of granularities in their way of processing information. Emulating the brain, neural parallel computing would only draw strength from concurrent diversity and modularity. Timorous evolutions arise in this direction, such as parallel computing on networks of dissimilar stand-alone computers, or modular neural network building with their own learning rules. The concept of modularity has to be promoted. Solutions have to be sought both in computing science and in cognitive science. In the coming years several difficulties could be resolved by large-scale cooperation of these two communities.

Present points of view on both parallel computation and neural networks are much too distinct. Moreover, recent results have shown that neural networks with real weights are more powerful than traditional models of computation (see PARALLEL COMPUTATIONAL MODELS). On the one hand, current use of neural networks today does not take advantage of all their intrinsic properties. On the other hand, the intrinsic properties of parallel computers, such as asynchronous behavior, are not yet fully understood and exploited. These problems give rise to a great challenge for future research, both in cognitive science and in computer science.

Road Map: Implementation of Neural Networks
Background: Backpropagation: Basics and New Developments; Self-Organizing Feature Maps: Kohonen Maps
Related Reading: Programmable Neurocomputing Systems

References

Arbib, M. A., and Robinson, J. A., 1990, *Natural and Artificial Parallel Computation*, Cambridge, MA: MIT Press. ◆

Azcarraga, A., Paugam-Moisy, H., and Puzenat, D., 1994, A parallel incremental neural classifier on a distributed-memory MIMD computer, in *Applications in Parallel and Distributed Computing*, Amsterdam: North-Holland, pp. 13–22.

Carey, G. F., 1989, *Parallel Supercomputing: Methods, Algorithms and Applications*, New York: Wiley.

Demian, V., and Mignot, J.-C., 1993, Optimization of the self-organizing feature map on parallel computers, in *Proceedings of the International Joint Conference on Neural Networks*, Piscataway, NJ: IEEE, pp. 483–486.

Ghosh, J., and Hwang, K., 1989, Mapping neural networks onto message-passing multicomputers, *J. Parallel Distrib. Comput.*, 6:291–330.

Girard, D., and Paugam-Moisy, H., 1994, Strategies of weight updating for parallel back-propagation, in *Applications in Parallel and Distributed Computing*, Amsterdam: North-Holland, pp. 335–336.

Hwang, K., 1993, *Advanced Computer Architecture*, New York: McGraw-Hill.

Jàjà, J., 1992, *An Introduction to Parallel Algorithms Reading*, MA: Addison-Wesley. ◆

Kosslyn, S., and Koenig, O., 1992, *Wet Mind: The New Cognitive Neuroscience*, New York: Free Press.

Murre, J. M., 1993, Transputers and neural networks: An analysis of implementation constraints and performance, *IEEE Trans. Neural Netw.*, 4:284–292.

Paugam-Moisy, H., 1993, Parallel neural computing based on network duplicating, in *Parallel Algorithms for Digital Image Processing, Computer Vision and Neural Networks* (I. Pitas, Ed.), New York: Wiley, pp. 305–340.

Pétrowski, A., and Paugam-Moisy, H., 1993, Parallel neural computation based on algebraic partitioning, in *Parallel Algorithms for Digital Image Processing, Computer Vision and Neural Networks* (I. Pitas, Ed.), New York: Wiley, pp. 259–304.

Quinlan, P. T., 1991, *Connectionism and Psychology*, Hemel Hempstead, Eng.: Harvester Wheatsheaf. ◆

Quinn, M. J., 1987, *Designing Efficient Algorithms for Parallel Computers*, New York: McGraw-Hill.

Rumelhart, D. E., McClelland, J. L., and PDP Research Group, Eds., 1986, *Parallel Distributed Processing: Explorations in the Microstructure of Cognition*, Cambridge, MA: MIT Press.

Muscle Models

Andrew M. Krylow, Thomas G. Sandercock, and W. Zev Rymer

Introduction

Muscle is a remarkable mechanical actuator which transduces chemical energy into force and motion, thereby providing power to move the skeleton. Because of the complexity of this transduction, and the intricacies of muscle microstructure and architecture, no comprehensive models are yet able to predict muscle performance completely. For this reason, muscle models are widely used to fulfill a variety of more narrowly defined objectives, ranging from attempts to promote understanding at the molecular level to more practical simulations of whole muscle behavior as part of a broader study of basic musculoskeletal biomechanics or issues of neural control.

Muscle models can be usefully classified in order of increasing complexity. The more elaborate models generally have a wider range of application and can give more accurate results. However, this increase in fidelity is achieved at the cost of an increase in the mathematical complexity, involving parameters that are often not known initially and are not readily measured. Aside from the mechanisms of force generation, many other processes such as activation, potentiation, and fatigue also play an important role. In the absence of a single model that captures all these features, models are usually simplified to include only those behaviors that are deemed of interest in a particular application. This article outlines three major model classes and provides guidelines for their application.

1. *Input-output models.* The simplest of models are "black box" input-output models that attempt to capture very specific behavior over a restricted range of operation. Such models commonly use linear transfer function descriptors to transform neural excitation into force.

2. *Lumped parameter mechanical models.* The next level in complexity is typified by lumped parameter mechanical models. These are often composed of combinations of linear mechanical elements such as springs and dashpots to create fairly simple viscoelastic analogs of muscle. Nonlinear relations representing hyperbolic force-velocity behavior and tendon properties can also be incorporated; such models are termed "Hill models." The parameters characterizing the elements of these models are usually directly measurable by experiments. Model inputs may be neural excitation or length and force perturbations, while outputs may include muscle force, stiffness, and the time course of muscle length changes.

3. *Cross-bridge models.* More sophisticated "cross-bridge" models attempt to reproduce the dynamics of molecular processes that are responsible for force generation in muscle. These models incorporate mathematical descriptions of the dynamics of cross-bridge populations, their driving chemical reactions, and the resulting mechanical consequences. Such models usually require knowledge of numerous parameters and rate functions for the underlying reactions. Most of these parameters are not directly measurable, making the fitting of the model to a specific muscle or experimental situation difficult. Again, inputs can consist of neural excitation pulses or mechanical perturbations, while various outputs can be obtained from such a model, ranging from mechanical variables to thermodynamic information.

We next summarize briefly the relevant basic physiology of muscle. See McMahon (1984) for more details.

Muscle Architecture

A muscle is composed of many long thin cells, or "fibers," arranged parallel to each other. Most fibers terminate in microtendons, which merge to form a common tendon that connects to the skeleton. Because of this parallel organization, the total force a muscle can produce is proportional to the summed cross-sectional area of all the fibers. The fibers are, in turn, composed of several thousand parallel *myofibrils*. Each myofibril is composed of repeating microscopic units (2–3 μm in length) called sarcomeres, which are the basic contractile units of muscle. Since sarcomeres within a fiber are linked in series and contract together, many key muscle properties, such as the maximum speed at which a muscle can shorten, are proportional to the length of the fiber. For this reason, muscle contractile properties are often normalized by both the muscle cross-sectional area and the fiber length.

Muscles come in an array of sizes and shapes, reflecting differences in fiber length, fiber number, and fiber orientation. There are also systematic differences in biochemistry and metabolic properties. See Alexander (1981) for a discussion of muscle and tendon architecture and Burke (1981) for a discussion of muscle fiber and motor unit specialization.

Input-Output (System) Models

At the simplest level, muscle may be treated as a linear system with a single input and output. In fact, since muscle has many simultaneous inputs—primarily neural activation, but also length, force, temperature, etc.—additional constraints on the system are usually required to enable linear systems approaches to be applied.

For example, Mannard and Stein (1973) modeled isometric cat soleus muscle as a linear system, with a neural pulse train input and a force output. Here, motor axons were driven by a random electrical pulse train, and the resulting force data were well fitted by a critically damped, linear, second-order system. Essentially, this muscle acted as a low pass filter, with a cutoff of 5 Hz. Difficulties with the linear systems approach became apparent when slight changes were made in the experiment. By changing the amplitude of the input (mean stimulus rate to the ventral roots), system gain changed by more than a factor of 4, and the cutoff frequency ranged from 8 to 2 Hz. Changing muscle length also changed the system parameters. Thus, the linear approximation assumed by this model holds true only for closely specified conditions.

There are several other nonlinearities in muscle that limit the usefulness of the linear systems approach. First, active muscle behaves quite differently when it is shortening than when it is lengthening—behavior inconsistent with linear system properties. If active muscle is stretched rapidly, force may drop precipitously after an initial region of high stiffness, giving rise to muscle "yield." Second, muscle force shows marked hysteresis when measured during increasing neural activation compared with decreasing activation.

Because a linear system model is at best an approximation, the model must be identified for the specific application for which it is used. Attempts to identify a more broadly applicable model by using nonlinear system techniques generally fail because muscle is quite nonstationary, and the system changes before it can be fully characterized. Nonetheless, the linear systems approach remains attractive because of its well-developed theoretical background, which allows relatively simple system identification techniques to be implemented.

Lumped Parameter (Hill) Models

The earliest experimentally based descriptions of muscle resulted in muscle models that were composed of viscoelastic elements. The most widely applicable of these models is that of A. V. Hill and can be described as follows: Muscle is composed operationally of three elements: (1) a contractile element (CE) that acts as an active force generator, (2) an elastic element (SE) that represents the combined stiffness of tendon and crossbridges in series with the force generator, and (3) a second elasticity in parallel with the previous two elements (PE) that represents the passive tissue contributions to muscle force (Figure 1A).

Hill (1938) characterized the CE by applying a series of constant force inputs (i.e., an "isotonic load") to active muscle. Muscle responds to such an input by shortening with an initially constant velocity. Larger loads result in smaller velocities. Plotting a number of such force-velocity pairs demonstrates this tradeoff (Figure 1B), which can be fitted well with a hyperbola of the form

$$V_{CE} = \frac{b(P_0 - F)}{F + a} \qquad F \leq P_0 \tag{1}$$

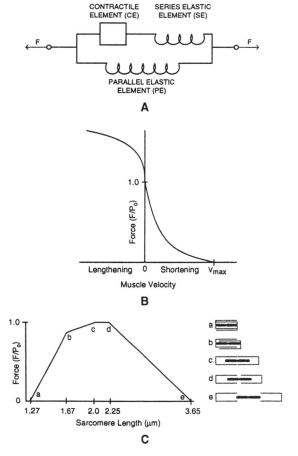

A

B

C

Figure 1. Schematic representations of (A) Hill model structure, (B) the force-velocity relation for both concentric and eccentric regions, and (C) isometric force-length relation, and the corresponding sarcomere geometry responsible for this effect.

where F is the applied isotonic force, V_{CE} is the resulting initial velocity of shortening, P_0 is the maximum isometric force, and a and b are empirical constants. Equation 1 has been found to describe the steady-state force-velocity behavior of a wide variety of skeletal muscles during shortening. The constitutive relation for the CE embodies this hyperbolic force-velocity tradeoff.

The series elasticity is usually a purely linear spring of stiffness k, and the passive elasticity often takes the form of an exponential function that increases with extension. The total muscle length is the sum of the lengths of the CE and SE. The contribution of the PE depends on the precise muscle geometry, but is often important only at long muscle lengths, and is thus frequently neglected in practice. When passive tension is neglected, a single, first-order, ordinary differential equation expresses the dynamics of this model with force, F, as an output:

$$\frac{dF}{dt} = k_{SE}(V_M - V_{CE}) \tag{2}$$

where V_M is the velocity of the end of the whole muscle, V_{CE} is the contractile element velocity from Equation 1, and k_{SE} is the series elastic stiffness. The CE property described by Equation 1 applies only to shortening muscle, and only when the muscle is operating in a length range at which maximal isometric force is produced. Modifications to this standard model extend its applicability to situations where large length excursions occur. For example, length dependence can be incorporated into the force-velocity relation (Equation 1) by changing P_0 to reflect the isometric force available at the current muscle length and scaling the parameters a and b by this factor as well. A length-dependent nonlinear SE stiffness can also be included.

In contrast, it is necessary to define an entirely new F-V relation to describe behavior during lengthening contraction, i.e., when the muscle is forced to lengthen by an external load that exceeds its active force-generating capacity. One such relation (Mashima et al., 1972) is given by

$$V_{CE} = \frac{b'(P_0 - F)}{2P_0 - F + a'} \qquad F > P_0 \tag{3}$$

where a' and b' are empirical constants. Such an extended model has been shown to perform well for complex motions involving both eccentric and concentric contractions (Krylow and Sandercock, in press).

Cross-Bridge Models

The main components of the sarcomere (Squire, 1981) are two sets of interdigitating protein filaments called thin filaments (made up partly of actin) and thick filaments (made up largely of myosin). Molecular projections (cross-bridges) on the thick filaments interact with receptor sites on the actin, when suitably activated by calcium ions and in the presence of adenosine triphosphate (ATP), to produce force and relative motion between the two sets of filaments. Each cross-bridge is believed to act independently, interacting cyclically with successive actin sites to produce a "ratchet-like" action. The forces thus produced between the two sets of filaments are in a direction to cause each sarcomere to shorten. The actin and myosin filaments are essentially inextensible, and sarcomere shortening occurs because of the relative sliding of the filaments past each other (sliding filament theory).

Muscle force exhibits a pronounced length dependence which can be explained by the sliding filament theory. As muscle length changes, the relative overlap of the actin and myosin filaments in each sarcomere changes because of telescoping of the sarcomere structure, and this overlap determines the maximum number of available cross-bridges at any given muscle length. Figure 1C shows the idealized length-tension curve measured during steady-state isometric contraction when the muscle is fully active.

In contrast to lumped parameter models, which try to reproduce macroscopic behavior with discrete mechanical elements, cross-bridge models strive to incorporate the known microstructure of the sarcomere together with metabolic kinetics of muscle to predict macroscopic variables such as whole muscle force, stiffness, shortening velocity, energy consumption, heat liberation, etc. The prototypical scheme for this type of model (Huxley, 1957) idealizes the interaction of actin and myosin as consisting of two possible cross-bridge states—either bound or unbound—and derives equations that describe the evolution of the distribution of bond lengths for the population of cross-bridges in the bound state. The cross-bridges are assumed to act independently as linear springs, producing force in proportion to their extension, when bound to actin. It is necessary to define simplified chemical reactions that describe the transitions between the states, and the form of the rate functions that determine the extent and directions of these reactions. Since the chosen rate functions depend explicitly on the cross-bridge length, the reactions are coupled directly to mechanical events of the cross-bridge cycle as well as to external perturbations imposed on the muscle by loading conditions. See McMahon (1984) and Zahalak (1981, 1992) for more details.

Tendon Properties

An idealized structure is often assumed for the tendon, in which muscle is connected to a linear series elastic element with very high stiffness. In fact, the tendon is far from being simply an inextensible link, and its mechanics modify muscle output significantly. For example, under some conditions the muscle fibers can shorten while the complete muscle-tendon structure lengthens. The tendon can be used to store energy, as was demonstrated in the Achilles tendon of the wallaby (Alexander, 1981), where it plays an integral role in efficient jumping locomotion. Although tendon is often simply modeled as an ideal spring, the mechanical properties of tendon are more complex.

Its force-length curve is often described as exponential, exponential-linear, or a quadratic function of length. When a tendon is stretched and released, the measured force shows a pronounced hysteresis which is a complex function of its history. Nonlinear tendons have been incorporated as the series elastic elements in Hill-type models, providing some improvement in model accuracy. However, these models have other inaccuracies which in most applications make the addition of nonlinear tendon properties unnecessary.

The cross-bridges that generate force in a muscle are themselves often modeled as ideal springs. They can be lumped together with tendon to define a global muscle stiffness. For example, in fully active cat soleus muscle, which is considered to have a short, stiff tendon, half of its compliance (the reciprocal of stiffness) is attributed to the tendon. In muscles with longer tendons, the compliance of the tendon predominates. At lesser activation levels, the cross-bridge compliance is the primary source of elasticity in the muscle.

Models of Activation

Muscle receives neural input in the form of discrete action potentials. Through a complex sequence of events, each action potential results in a release of calcium from stores in the sarcoplasmic reticulum. This calcium binds to regulatory proteins and allows cross-bridge cycling to proceed. The calcium is quickly sequestered, deactivating the actin-binding sites and allowing the muscle to relax. The frequency of action potential arrival at the muscle determines the degree of muscle activation. At low frequencies, muscle responds with discrete force transients (twitches). At high frequencies, isometric force fuses into a smooth contraction (tetanus) that rises to maximal levels (Figure 2).

The muscle models presented earlier need to be coupled with a model of activation to be fully comprehensive. These models fall into two basic categories: (1) models based on an estimated mean level of neural excitation to the muscle; and (2) models that translate a sequence of discrete action potentials into muscle activation. When simulating voluntary movement, models based on estimated mean excitation are preferable because the

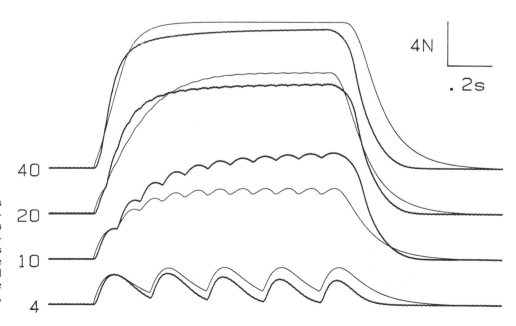

Figure 2. Cat soleus force responses (thick lines) for increasing rates of synchronous stimulation. At low stimulus rates, force pulses are responses to individual stimulus pulses. Force pulses fuse to produce smooth traces as the rate approaches 40 Hz. Simple Hill model predictions (thin lines), using the activation scheme given by Equation 4, show qualitatively similar behavior.

true excitation is never known, and little is accomplished by trying to estimate the action potential sequence to all motor units in the muscle. However, when muscle is stimulated by a known action potential train, modeling of the discrete action potentials is necessary to reproduce the ripple occurring in an unfused tetanus.

The most widely used method to estimate the neural drive to a muscle is the rectified and filtered electromyogram (REMG) (Deluca, 1979). The electromyogram signal is recorded using either surface or intramuscular electrodes and results from the complex summation of electric fields produced by each muscle fiber action potential. The REMG provides a measure of the total number of action potentials to the muscle. It often has a linear or exponential relation to isometric force and can be used as the input to a low pass filter model or as the input to a Hill-type model. Unfortunately, the relationship between the REMG and force varies in different muscles, for different recording electrodes, or even with the placement of the electrodes. Furthermore, because of its stochastic properties, an ensemble average of the REMG is needed for rapidly changing levels of excitation. At best, REMG is a crude approximation of neural drive.

The second approach to modeling activation is to approximate the physiological events after the arrival of an action potential at the muscle. Unfortunately, the complexity of the events precludes an accurate and simple model. In addition, activation is strongly influenced by muscle length. Well-documented activation-related phenomena can more than double or half the force measured from a muscle stimulated by identical pulse trains. See Burke (1981) for discussions of potentiation, doublets, sag, and fatigue.

A simple model to transform a time sequence of action potentials into activation for a Hill-type model is described by

$$x(t) = \sum_n \delta(t - t_n)$$

$$\dot{r}(t) = -C_1 r(t) + C_2 x(t)$$

$$\dot{y}(t) = \begin{cases} -C_3 y(t) + r(t) & y(t) \le 1 \\ -C_3 y(t) & y(t) > 1 \end{cases} \tag{4}$$

where $x(t)$ is the input, $\delta(t)$ are unit impulses representing action potentials at times t_n, and C_1, C_2, and C_3 are constants. Activation, $y(t)$, is used to scale the force-velocity relationship —P_0, a, b, a', and b' in Equations 1 and 3 are multiplied by $y(t)$. The results of the model applied to cat soleus (Figure 2) show fair agreement with the experimental data measured during isometric conditions. The most critical error in this model occurs if the muscle shortens or lengthens just before or during relaxation. Apparently the cross-bridges break more quickly during shortening and remain attached longer during lengthening. There is no mechanism to represent this in a Hill-type model, where activation dynamics and cross-bridge dynamics are confounded in the force-velocity curve. More sophisticated activation models coupled with a cross-bridge model (Zahalak, 1992) address this problem.

Discussion

As outlined in the Introduction, the choice of a particular type of model is determined by the intended use of the resulting information. Linear system–type muscle models can provide intuitive insights in the frequency domain that are not easily obtained from the other methods, but muscle's inherent nonlinearities make such models locally applicable at best. Cross-bridge–type models are essentially the only choice to study molecular mechanisms. To study whole muscle or multiple muscle systems, both cross-bridge– and Hill-type models offer possibilities, although Hill-type models are much more accessible. Cross-bridge models are capable of a wider range of behaviors, but they pay the price by needing more parameters. Because of the difficulty in estimating these parameters, cross-bridge models are rarely used to study control and multimuscle movement. Hill-type models are by far the most widely used.

A Hill model extended to include eccentric contraction and large muscle length changes does a good job predicting muscle behavior, and is the most practical solution for many requirements, but has at least two major weaknesses. First, the extensions to Hill-type models for lengthening contractions fit well only under limited conditions and cannot predict muscle "yield." Second, when activation is coupled with the Hill model, the model has no mechanism to handle varying cross-bridge persistence observed with different movement histories. Systematic methods to identify the parameters in a simplified cross-bridge–type model could make it the method of choice.

It is not known how accurate a muscle model must be to effectively study control of movement. Since muscle is a nonstationary system, with crucial time- and history-dependent properties, it is difficult, even under carefully controlled laboratory conditions, to get the identical response to the same input. Perhaps neural control systems make such differences unimportant. Conversely, Lehman (1990) has shown that predicted neural control signals are very sensitive to the muscle model structure. Here, activations necessary to reproduce experimental wrist motions were calculated using three different muscle models: a linear viscoelastic model, a Hill model with constant SE stiffness, and a Hill model with activation-dependent stiffness. The most complex muscle model predicted control signals that most closely resembled the actual EMG signals.

The nervous system embodies a great deal of processing capability to handle the difficult task of controlling multiarticular limb movements. A large part of this processing power is likely to be devoted to compensating for the variation of muscle properties.

Road Map: Biological Motor Control
Related Reading: Human Movement: A System-Level Approach; Motoneuron Recruitment

References

Alexander, R. M., 1981, Mechanics of skeleton and tendons, in *Handbook of Physiology, Section 1: The Nervous System*, vol. 2, *Motor Control*, Part 1 (V. B. Brooks, Ed.), Bethesda, MD: American Physiological Society, pp. 17–42.

Burke, R. E., 1981, Motor units: Anatomy, physiology, and functional organization, in *Handbook of Physiology, Section 1: The Nervous System*, vol. 2, *Motor Control*, Part 1 (V. B. Brooks, Ed.), Bethesda, MD: American Physiological Society, pp. 345–422.

Deluca, C. J., 1979, Physiology and mathematics of myoelectric signals, *IEEE Trans. Biomed. Eng.*, BME-26:313–326.

Hill, A. V., 1938, The heat of shortening and the dynamic constants of muscle, *Proc. R. Soc. B*, 126:136–195.

Huxley, A. F., 1957, Muscle structure and theories of contraction, *Prog. Biophys.*, 7:255–318.

Krylow, A. M., and Sandercock, T. G., in press, Test of a modified Hill model in reproducing force responses involving eccentric contraction, *J. Biomech.*

Lehman, S. L., 1990, Input identification depends on model complexity, in *Multiple Muscle Systems: Biomechanics and Movement Organization* (J. M. Winters and S. L.-Y. Woo, Eds.), New York: Springer-Verlag, pp. 94–100.

McMahon, T. A., 1984, *Muscles, Reflexes, and Locomotion*, Princeton, NJ: Princeton University Press. ◆

Mannard, A., and Stein, R. B., 1973, Determination of the frequency response of isometric soleus muscle in the cat using random nerve stimulation, *J. Physiol.*, 229:275–296.

Mashima, H., Akazawa, K., Kushima, H., and Fujii, K., 1972, The force-load-velocity relation and the viscous-like force in the frog skeletal muscle, *Jpn. J. Physiol.*, 22:103–120.

Squire, J., 1981, *The Structural Basis of Muscular Contraction*, New York: Plenum.

Zahalak, G. I., 1981, A distribution-moment approximation for kinetic theories of muscular contraction, *Math. Biosci.*, 55:89–114.

Zahalak, G. I., 1992, An overview of muscle modeling, in *Neural Prostheses: Replacing Motor Function After Disease or Disability* (R. B. Stein, P. Hunter Peckham, and D. B. Popovic, Eds.), New York: Oxford University Press, pp. 17–57. ◆

Neocognitron: A Model for Visual Pattern Recognition

Kunihiko Fukushima

Introduction

The *neocognitron* (Fukushima, 1980, 1988b, 1991) is a neural network model for deformation-resistant visual pattern recognition.

In primary visual cortex, neurons respond selectively to local features of a visual pattern, such as lines or edges in particular orientations. In the inferotemporal cortex, cells exist that respond selectively to certain figures such as circles, triangles, or squares, or even human faces. Thus, the visual system seems to have a hierarchical architecture, in which simple features are first extracted from a stimulus pattern, then integrated into more complicated ones. In this hierarchy, a cell in a higher stage generally has a larger receptive field and is more insensitive to the position of the stimulus. This kind of physiological evidence suggested the network architecture for the neocognitron.

The neocognitron is a hierarchical network consisting of many layers of neuron-like cells. There are forward connections between cells in adjoining layers. Some of these connections are variable and can be modified by learning. The neocognitron can acquire the ability to recognize patterns by learning. Since it has a large power of generalization, presentation of only a few typical examples of deformed patterns (or features) is enough for the learning process to be successful. It is not necessary to present all of the deformed versions of the patterns which might appear in the future. After learning, the neocognitron can recognize input patterns robustly, with little effect from deformation, changes in size, or shifts in position. It is even able to correctly recognize a pattern which has not been presented before, provided it resembles one of the training patterns.

The principle of the neocognitron can be used in various kinds of pattern recognition systems, such as systems recognizing handwritten characters (Fukushima, 1988b; Fukushima and Wake, 1991).

The Network Architecture

The neocognitron has a multilayered architecture, as shown in Figure 1, in which each rectangle represents a two-dimensional array of cells. Each cell receives its input connections from only a limited number of cells situated in a small area on the preceding layer. The density of cells in each layer is designed to decrease with the order of the stage.

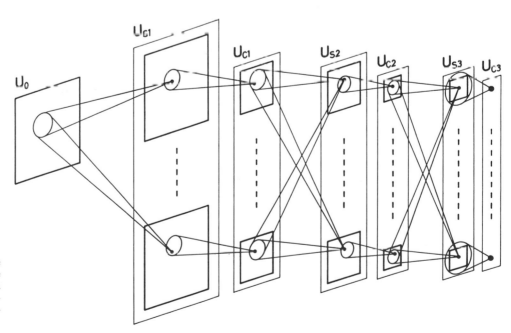

Figure 1. The network architecture of the neocognitron. Each rectangle drawn with heavy lines represents a "cell-plane." The cells in each cell-plane are arranged in a two-dimensional array.

The lowest stage of the hierarchical network is an input layer U_0, consisting of a two-dimensional array of receptor cells. Each succeeding stage has a layer U_S consisting of "S-cells," followed by another layer U_C consisting of "C-cells." Thus, in the whole network, layers of S-cells and C-cells are arranged alternately.

Each layer of S-cells or C-cells is divided into subgroups, called "cell-planes," according to the features to which they respond. The cells in each cell-plane are arranged in a two-dimensional array. Each rectangle drawn with heavy lines in Figure 1 represents a cell-plane. The connections converging to the cells in a cell-plane are homogeneous and topographically ordered. In other words, the connections have a translational symmetry such that each of the cells of a cell-plane shares the same set of input connections. This condition of translational symmetry holds for both fixed and variable connections. The modification of variable connections is always done under this condition.

S-cells are feature-extracting cells. They resemble simple cells in the visual cortex in their response. Connections converging to these cells may be modified by learning. After learning, S-cells are able to extract features from input patterns. In other words, an S-cell is activated only when a particular feature is presented in its receptive field. The features extracted by the S-cells are determined during the learning process. Generally speaking, local features, such as lines in particular orientations, are extracted in the lower stages. More "global" features, such as parts of a training pattern, are extracted in higher stages.

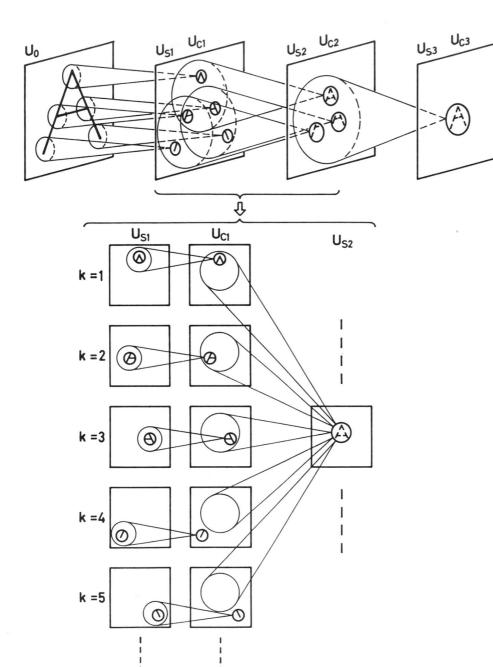

Figure 2. Illustration of the process of pattern recognition in the neocognitron (Fukushima, 1980). As shown in the upper half of the figure, local features extracted in lower stages are gradually integrated into more "global" features. The lower half of the figure is an enlarged illustration of a part of the network. The cell-plane with $k = 1$ in layer U_{S1} consists of S-cells which extract ∧-shaped features. Since the stimulus pattern "A" contains the ∧-shaped feature at the top, an S-cell near the top of this cell-plane is active. A C-cell in the succeeding cell-plane ($k = 1$) in U_{C1} has excitatory input connections from S-cells situated in the circle, and is activated if one of these S-cells is active. Only one cell-plane is shown in U_{S2} in this enlarged illustration. Each S-cell in this cell-plane detects the existence of features $k = 1, 2, 3$ in U_{C1}, and at the same time the absence of features $k = 4, 5$.

C-cells, which resembles complex cells in the visual cortex, are inserted in the network to allow for positional errors in the features of the stimulus. The connections from S-cells to C-cells are fixed and invariable. Each C-cell receives signals from a group of S-cells that extract the same feature, but from slightly different positions (Figure 2). The C-cell is activated if at least one of these S-cells is active. Even if the stimulus feature is shifted in position and another S-cell is activated instead of the first one, the same C-cell keeps responding. Hence, the C-cell's response is less sensitive to shifts in the position of the input pattern.

The layer of C-cells at the highest stage is the recognition layer: the response of the cells in this layer is the final result of pattern recognition by the neocognitron.

Principles of Deformation-Resistant Recognition

In the whole network, with its alternate layers of S-cells and C-cells, the process of feature extraction by the S-cells and toleration of positional shift by the C-cells is repeated. During this process, local features extracted in lower stages are gradually integrated into more global features. Finally, each C-cell of the recognition layer at the highest stage integrates all the information of the input pattern and responds only to one specific pattern. Figure 2 illustrates this situation schematically.

Tolerating positional error a little at a time at each stage, rather than all in one step, plays an important role in endowing the network with the ability to recognize even distorted patterns. Figure 3 illustrates this situation. Let an S-cell in an intermediate stage of the network have already been trained to extract a global feature consisting of three local features of a training pattern "A," as shown in Figure 3A. The cell tolerates a positional error of each local feature if the deviation falls within the dotted circle. Hence, the S-cell responds to any of the deformed patterns shown in Figure 3B. The toleration of positional errors should not be too large at this stage. If large errors are tolerated at any one step, the network may come to respond erroneously, such as by recognizing a stimulus like Figure 3C as an "A" pattern.

Since errors in the relative position of local features are thus tolerated in the process of extracting and integrating features, the same C-cell responds in the recognition layer at the highest stage, even if the input pattern is deformed, changed in size, or shifted in position.

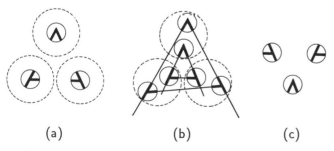

(a)   (b)   (c)

Figure 3. Illustration of the principle for recognizing deformed patterns (Fukushima, 1988a). An S-cell, which has already been trained to extract a global feature consisting of three local features as shown in part A, tolerates a positional error of each local feature if the deviation falls within the dotted circle. Hence, the S-cell responds to any of the deformed patterns shown in part B. The toleration of positional errors should not be too large at this stage. If large errors are tolerated at any one step, the network may come to respond erroneously, such as by recognizing a stimulus like the one in part C as an "A" pattern.

Self-Organization of the Neocognitron

The neocognitron can be trained to recognize patterns through either unsupervised or supervised learning. Various training methods have been proposed, and this section introduces two of them.

In the case of unsupervised learning, the self-organization of the network is performed using two principles. The first principle is a kind of "winner-take-all" rule (see WINNER-TAKE-ALL MECHANISMS): among the cells situated in a certain small area, only the one responding most strongly has its input connections reinforced. The change of each input connection to this maximum-output cell is proportional to the intensity of the response of the cell from which the relevant connection leads.

Figure 4 illustrates this process, showing only the connections converging to an S-cell. The S-cell receives variable excitatory connections from a group of C-cells of the preceding stage. The cell also receives a variable inhibitory connection from an inhibitory cell, called a V-cell. The V-cell receives fixed excitatory connections from the same group of C-cells as does the S-cell, and always responds with the average intensity of the output of the C-cells.

The initial strength of the variable connections is very weak and nearly zero (Figure 4A). Suppose the S-cell responds most strongly of the S-cells in its vicinity when a training stimulus is presented (Figure 4B). According to the winner-take-all rule just described, variable connections leading from activated C- and V-cells are reinforced, as shown in Figure 4C. The variable excitatory connections to the S-cell grow into a "template" that exactly matches the spatial distribution of the response of the cells in the preceding layer. The inhibitory variable connection from the V-cell is also increased at the same time, but not strongly, because the output of the V-cell is not as large.

After the learning, the S-cell acquires the ability to extract a feature of the stimulus presented during the learning period. Through the excitatory connections, the S-cell receives signals indicating the existence of the relevant feature to be extracted. If an irrelevant feature is presented, the inhibitory signal from the V-cell becomes stronger than the direct excitatory signals from the C-cells, and the response of the S-cell is suppressed (Fukushima, 1989).

Once an S-cell is thus selected and reinforced to respond to a feature, the cell usually loses its responsiveness to other features. When a different feature is presented, a different cell usually yields the maximum output and has its input connections reinforced. Thus, a "division of labor" among the cells occurs automatically.

The second principle for the learning is introduced in order that the connections being modified always preserve translational symmetry. The maximum-output cell not only grows by itself, but also controls the growth of neighboring cells, working, so to speak, like a seed in crystal growth. To be more specific, all of the other S-cells in the cell-plane, from which the "seed cell" is selected, follow the seed cell, and have their input connections reinforced by having the same spatial distribution as those of the seed cell.

Although the neocognitron can thus be trained by unsupervised learning, supervised learning is still useful when we want to train a system to recognize, for instance, handwritten characters, which should be classified not only on the basis of similarity in shape but also on the basis of certain conventions. In the case of supervised learning, the "teacher" presents training patterns to the network and points out the positions of the features which should be extracted. The cells whose receptive field centers coincide with the positions of the features take the

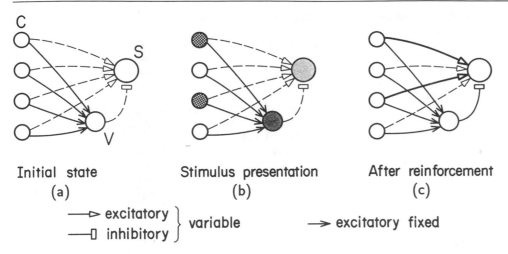

Figure 4. The process of reinforcement of the forward connections converging to a feature-extracting S-cell (Fukushima, 1988a). The density of the shadow in the circle represents the intensity of the response of the cell. *A,* The initial state before training. *B,* Stimulus presentation during the training. *C,* The connections after reinforcement.

place of the "maximum-output cells" and become seed cells. The other process of reinforcement is identical to that of the unsupervised learning and occurs automatically.

It is another advantage of the neocognitron that these learning methods, both supervised and unsupervised, require extremely short training times compared with other learning algorithms such as backpropagation. In an extreme case of unsupervised learning, for example, three presentations of a training set consisting of one training pattern from each category was sufficient to train the network to recognize 10 numeric characters robustly (Fukushima and Wake, 1992).

The optimal scale of the neocognitron changes depending on the set of patterns to be recognized. If the complexity of the patterns is high, the total number of stages in the hierarchical network needs to be large. Conversely, the necessary number of cell-planes in each stage of the network increases with the number of categories of patterns to be recognized. However, the increase in scale is not proportional. For example, if we compare a system recognizing 35 alphanumeric characters with a system recognizing 10 numerals, the number of characters to be recognized increases 3.5 times, but the number of cells increases only 1.9 times (Fukushima and Wake, 1991). This is a result of a fact that the local features extracted in the lower stages are common, and they usually are contained in many patterns of different categories. Although the number of cells is large in these systems, the number of parameters required to describe the network is quite small, because all of the cells in each cell-plane share the same set of input connections.

Selective Attention Model (SAM)

Although the neocognitron has considerable ability to recognize deformed patterns, it does not always recognize patterns correctly when two or more patterns are presented simultaneously. The *selective attention model* (SAM) has been proposed to eliminate these defects (Fukushima, 1986, 1987, 1988a). In the SAM, backward (i.e., top-down) connections were added to the conventional neocognitron-type network, which had only forward (i.e., bottom-up) connections.

When a composite stimulus, consisting of two patterns or more, is presented, the SAM focuses its attention selectively on one of the patterns, segments it from the rest, and recognizes it. After the identification of the first segment, the SAM switches its attention to recognize another pattern. The SAM also has the function of associative recall. Even if noise or defects affect the stimulus pattern, the SAM can recognize it and recall the

complete pattern from which the noise has been eliminated and defects corrected. These functions can be successfully performed even for deformed versions of training patterns, which have not been presented during learning.

The SAM has some similarity to the ADAPTIVE RESONANCE THEORY (ART) model (q.v.; see also Carpenter and Grossberg, 1987), but the most important difference between the two is the fact that the SAM has the ability to accept patterns deformed in shape and shifted in position, while ART does not have, in principle, such functions. With the SAM, not only the recognition of the patterns, but also the filling-in process for defective parts of imperfect input patterns works on the deformed and shifted patterns themselves. The SAM can repair the deformed pattern without changing the basic shape and location of the deformed input pattern. The deformed patterns themselves can be repaired at their original locations, thus preserving their deformation.

The principles of the SAM can be extended to be used for several applications: for example, the recognition and segmentation of connected characters in cursive handwriting of English words (Fukushima and Imagawa, 1993), and the recognition of Chinese characters (Fukushima, Imagawa, and Ashida, 1991).

Road Map: Learning in Artificial Neural Networks, Deterministic
Background: I.3. Dynamics and Adaptation in Neural Networks
Related Reading: Convolutional Networks for Images, Speech, and Time Series; Selective Visual Attention; Visual Cortex Cell Types and Connections

References

Carpenter, G. A., and Grossberg, S., 1987, ART 2: Self-organization of stable category recognition codes for analog input patterns, *Appl. Opt.*, 26:4919–4930.
Fukushima, K., 1980, Neocognitron: A self-organizing neural network model for a mechanism of pattern recognition unaffected by shift in position, *Biol. Cybern.*, 36:193–202.
Fukushima, K., 1986, A neural network model for selective attention in visual pattern recognition, *Biol. Cybern.*, 55:5–15.
Fukushima, K., 1987, A neural network model for selective attention in visual pattern recognition and associative recall, *Appl. Opt.*, 26:4985–4992.
Fukushima, K., 1988a, A neural network for visual pattern recognition, *IEEE Computer*, 21(3):65–75. ◆
Fukushima, K., 1988b, Neocognitron: A hierarchical neural network capable of visual pattern recognition, *Neural Netw.*, 1:119–130.

Fukushima, K., 1989, Analysis of the process of visual pattern recognition by the neocognitron, *Neural Netw.*, 2:413–420.

Fukushima, K., 1991, Neural networks for visual pattern recognition, *IEICE Trans.*, E74:179–190. ◆

Fukushima, K., and Imagawa, T., 1993, Recognition and segmentation of connected characters with selective attention, *Neural Netw.*, 6:33–41.

Fukushima, K., Imagawa, T., and Ashida, E., 1991, Character recognition with selective attention, in *Proceedings of the International Joint Conference on Neural Networks, 1991*, vol. 1, New York: IEEE, pp. 593–598.

Fukushima, K., and Wake, N., 1991, Handwritten alphanumeric character recognition by the neocognitron, *IEEE Trans. Neural Netw.*, 2: 355–365.

Fukushima, K., and Wake, N., 1992, Improved neocognitron with bend-detecting cells, in *Proceedings of the International Joint Conference on Neural Networks, 1992*, vol. 4, New York: IEEE, pp. 190–195.

Neural Optimization

Carsten Peterson and Bo Söderberg

Introduction

Many combinatorial optimization problems require exhaustive searches to achieve exact solutions, with the computational effort growing exponentially or worse with system size N. Different kinds of heuristic methods are therefore often used to find reasonably good solutions. The artificial neural network (ANN) approach falls within this category. The ANN method in the optimization domain really brings something new "to the table." In contrast to existing search and heuristics methods, the ANN approach does not fully or partly explore the different possible configurations. Rather it "feels" its way in a fuzzy manner toward good solutions. This procedure is done in a way that allows for a statistical interpretation of the results. The key element in this approach is the mean-field (MF) approximation (Hopfield and Tank, 1985; Peterson and Söderberg, 1989).

The ANN method gives high-quality solutions to a variety of problems. In addition, it has the appealing feature that the MF equations are isomorphic to equations which can be directly implemented in VLSI (see ANALOG VLSI FOR NEURAL NETWORKS), facilitating hardware implementations. Such tight bonds between algorithms and hardware are quite unique.

Recurrent Networks

Recurrent networks appear in the context of associative memories (Hopfield, 1982) and difficult optimization problems (Hopfield and Tank, 1985; Peterson and Söderberg, 1989). Simple models for magnetic systems ("spin glasses") have a lot in common with recurrent networks—with an atomic spin seen as analogous to the "firing" state of a neuron—and have therefore been the source of much inspiration for neural network studies.

The Hopfield model (Hopfield, 1982) is based on the energy function

$$E = -\frac{1}{2} \sum_{i \neq j} w_{ij} s_i s_j \tag{1}$$

with symmetric weights w_{ij} and binary variables (*Ising* neurons) $s_i = \pm 1$ (or 0, 1). By appropriate choice of w_{ij} corresponding to the stored patterns, the model serves as an associative memory, with an asynchronous dynamics that locally minimizes E (see COMPUTING WITH ATTRACTORS).

$$s_i(t+1) = \text{sgn}\left(\sum_{j \neq i} w_{ij} s_j(t) \right) \tag{2}$$

In ANN optimization solvers, one also uses Hopfield-type energy functions onto which the problems are mapped with a clever choice of w_{ij} representing the problem specification. The neurons s_i should then with a suitable method settle into a stable state, where the solution to the problem is given by the configuration $(s_1, s_2, \ldots)$. The key problem here is to reach the global minimum or at least a very low-lying local minimum.

There are two families of ANN algorithms for optimization problems:

- *The "pure" neural approach* based on either Ising (Hopfield and Tank, 1985) or K-state (Potts) (Peterson and Söderberg, 1989) neurons with MF equations for the dynamics.
- *Deformable templates* (Durbin and Willshaw, 1987; see CONSTRAINED OPTIMIZATION AND THE ELASTIC NET), where the neural degrees of freedom have been integrated out and one is left with coordinates for possible solutions.

The latter pathway is preferable for low-dimensional geometrical problems like the traveling-salesman problem (TSP), whereas the former is the only possibility for high-dimensional or nongeometric problems like scheduling. In this article we focus on the "pure" neural approach.

Optimization with Ising Neural Networks

We start with binary neuron systems, illustrating all steps in some detail with the graph bisection (GB) problem.

The Graph Bisection Problem

The neural approach is particularly transparent for graph bisection because of its binary nature. The problem is defined as follows. Consider a graph of N nodes to be partitioned into two halves such that the connectivity (cut size) between the two halves is minimized (Figure 1A). This problem is mapped onto the Hopfield energy function (cf. Equation 1) by assigning to each node i a binary neuron $s_i = \pm 1$, and to each pair of vertices i, j $(i \neq j)$ a value $w_{ij} = 1$ if they are connected, and $w_{ij} = 0$ if they are not. In terms of Figure 1A, $s_i = \pm 1$ represents whether node i is in the left or in the right partition. With this notation $w_{ij} s_i s_j \neq 0$ only for a connected pair i, j, being $+1$ if they are in the same partition and -1 if not. In addition, one needs to add a "constraint term" to the right-hand side of Equation 1 that penalizes situations where the nodes are not equally partitioned. Since $\Sigma s_i = 0$ only for a balanced partition, a term proportional to $(\Sigma s_i)^2$ will subsequently do the trick. Our energy function for graph bisection thus takes the form

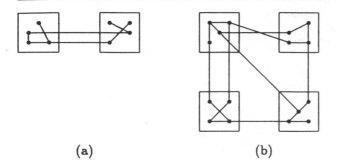

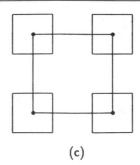

Figure 1. *A*, A graph bisection problem. *B*, A $K = 4$ graph partition problem. *C*, An $N = 4$ TSP problem. (Reprinted with permission from Peterson, C., and Söderberg, B., 1989, A new method for mapping optimization problems onto neural networks, *Int. J. Neural Syst.*, 1:3–22.)

(a)   (b)   (c)

$$E = -\frac{1}{2} \sum_{ij} w_{ij} s_i s_j + \frac{\alpha}{2} \left(\sum_i s_i \right)^2 \quad (3)$$

where the *constraint coefficient* (imbalance parameter) α sets the relative strength between the cut size and the balancing term. This balancing term represents a *global constraint*. The generic form of Equation 3 is

$$E = \text{Cost} + \text{Global constraint} \quad (4)$$

which is typical when casting combinatorial optimization problems onto neural networks. The origin of the difficulty inherent in this kind of problem is very transparent here; the problem is frustrated because of the competition between the two constraints (cost and global constraint). This often leads to the appearance of many local minima.

The next step is to find an efficient procedure for minimizing Equation 3, such that local minima are avoided as much as possible.

The Mean-Field Equations

If one minimizes Equation 3 according to a local optimization rule (cf. Equation 2), the system will very likely end up in the local minimum closest to the starting point, which is not desired. Thus, a stochastic algorithm is called for that allows for uphill moves. One such method is SIMULATED ANNEALING (SA) (Kirkpatrick, Gelatt, and Vecchi, 1983), in which configurations according to the Boltzmann distribution

$$P[s] = \frac{1}{Z} e^{-E[s]/T} \quad (5)$$

are generated with neighborhood search methods. In Equation 5, Z is the *partition function*

$$Z = \sum_{[s]} e^{-E[s]/T} \quad (6)$$

and T is the "temperature" representing the noise level of the system. For $T \to 0$ the Boltzmann distribution collapses into a delta function around the configuration minimizing E. If one generates configurations at successively lower T (annealing), these are less likely to get stuck in local minima than if $T = 0$ from the start. Needless to say, such a procedure can be very CPU consuming. The MF approach aims at approximating the stochastic SA method with a set of deterministic equations.

To this end, embed the spins s_i in a linear space (in this case $\mathbb{R}$), introduce a new set of variables v_i living in this space, and set them equal to the spins with a Dirac delta function. Then Z takes the form

$$Z = \sum_{[s]} \int d[v] e^{-E[v]/T} \prod_i \delta(s_i - v_i) \quad (7)$$

Rewriting the delta functions in terms of a new set of variables u_i gives

$$Z = \sum_{[s]} \int d[v] \int d[u] e^{-E[v]/T} \prod_i e^{u_i(s_i - v_i)} \quad (8)$$

Then carry out the original sum over $[s]$, and write the product as a sum in the exponent:

$$Z \propto \int d[v] \int d[u] e^{-E[v]/T - \sum_i u_i v_i + \sum_i \log \cosh u_i} \quad (9)$$

The original partition function is now rewritten entirely in terms of the new variables $[u, v]$, with an effective energy in the exponent. So far no approximation has been made. We next approximate Z in Equation 8 by the extremal value of the integrand, obtaining

$$v_i = \tanh \left(-\frac{\partial E[v]}{\partial v_i} \middle/ T \right) \quad (10)$$

The mean-field variables v_i can be seen as approximations to the thermal averages $\langle s_i \rangle_T$ of the original binary spins. What we have obtained is a set of deterministic equations emulating the stochastic behavior. High temperatures correspond to very smooth sigmoids, and the low-temperature limit is given by a step function of Equation 2. The MF equations are solved iteratively, either synchronously or asynchronously, under annealing in T.

Mean-Field Dynamics

The dynamics of recurrent ANNs typically exhibits a behavior with two phases (Figure 2): At large enough temperatures ($T \to \infty$), the system relaxes into the trivial fixed point $v_i^{(0)} = 0$. As the temperature is lowered, a phase transition is passed at $T = T_c$; and as $T \to 0$, fixed points $v_i^{(*)} = \pm 1$ emerge representing a specific decision made as to the solution to the problems in question (Figure 3). The position of T_c, which here depends

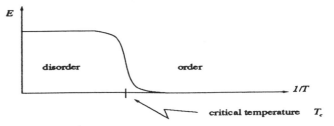

Figure 2. E as a function of T illustrating a phase transition. [From Peterson, C., and Söderberg, B., 1993, Artificial neural networks, in *Modern Heuristic Techniques for Combinatorial Problems* (C. Reeves, Ed.), Oxford: Blackwell, fig. 5.5; reprinted with permission from Alfred Waller Limited.]

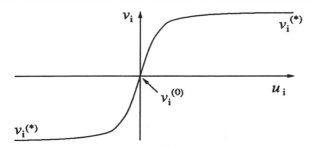

Figure 3. Fixed points in $v_i = \tanh(u_i)$. (Reprinted with permission from Peterson, C., and Söderberg, B., 1989, A new method for mapping optimization problems onto neural networks, *Int. J. Neural Syst.*, 1:3–22.)

on w_{ij} and α, can be estimated by linearizing the sigmoid function (tanh) around $v_i^{(0)} = 0$, i.e., linearizing Equation 10. For synchronous updating it is clear that if one of the eigenvalues of the matrix ω/T is > 1 in absolute value, the fixed point becomes unstable, and the solutions will wander away into the nonlinear region. In the case of serial updating, the philosophy is the same but the analysis slightly more complicated (Peterson and Söderberg, 1989). Prior estimation of the largest eigenvalues is simple to do. Also, this analysis is important for avoiding oscillatory behavior, which appears for eigenvalues < -1.

Results

Very good numerical results were obtained for the graph bisection problem (see Peterson and Söderberg, 1989, for references) for a wide range of problem sizes. The solutions are comparable in quality to those of the SA method. CPU time consumption is lower than for any other known method of comparable performance. The approach, of course, becomes even more competitive with respect to time consumption if the intrinsic parallelism is exploited on dedicated hardware.

Optimization with Potts Neural Networks

For GB and many other optimization problems, an encoding in terms of binary elementary variables is natural. However, there are many problems where the natural elementary decisions are of the type one-of-K with $K > 2$.

Early attempts to approach such problems by neural network methods used *neuron multiplexing* (Hopfield and Tank, 1985), where for each elementary K-fold decision, K binary 0/1-neurons were used, with the additional constraint that precisely one of them be on ($= 1$). These *syntax* constraints were implemented in a soft way as penalty terms. As it turned out in the original work on the traveling salesman problem (Hopfield and Tank, 1985), as well as in subsequent investigations for the graph partition problem (Peterson and Söderberg, 1989), this approach does not give rise to high-quality solutions in a parameter-robust way. An alternative encoding is to use *Potts neurons* with the syntax constraint built in. In this way the dynamics is confined to relevant parts of the solution space (Figure 4), leading to better performance.

Potts Spins

A K-state Potts spin is a variable that has K possible values (states). For our purposes, the best representation is in terms of a vector in the Euclidean space $\mathscr{E}_K$. Thus, denoting a spin vari-

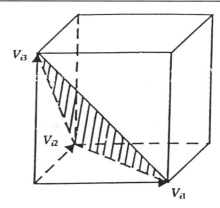

Figure 4. The volume of solutions corresponding to the neuron multiplexing encoding for $K = 3$. The shaded plane corresponds to the solution space of the corresponding Potts encoding. (Reprinted with permission from Peterson, C., and Söderberg, B., 1989, A new method for mapping optimization problems onto neural networks, *Int. J. Neural Syst.*, 1:3–22.)

able by $\mathbf{s} = (s_1, s_2, \ldots, s_K)$, the ath possible state is given by the ath principal unit vector, defined by $s_a = 1$, $s_b = 0$ for $b \neq a$. These vectors point to the corners of a regular K-simplex (see Figure 4 for the case of $K = 3$). They are all normalized and mutually orthogonal, and fulfill in addition the syntax constraint $\Sigma_a s_a = 1$.

The MF equations for a system of Potts spins s_i with an energy $E(s)$ are derived following the same path as in the Ising neuron case: Rewrite the partition function as an integral over $\mathbf{u}_i$ and $\mathbf{v}_i$, and approximate it with the maximum value of the integrand. One obtains

$$- u_{ia} = -\frac{\partial E(\mathbf{v})}{\partial v_{ia}}/T \tag{11}$$

$$v_{ia} = \frac{e^{u_{ia}}}{\sum_b e^{u_{ib}}} \tag{12}$$

from which it follows that the *Potts neurons* $\mathbf{v}_i$, which approximate the thermal average of $\mathbf{s}_i$, satisfy

$$v_{ia} > 0 \qquad \sum_a v_{ia} = 1 \tag{13}$$

One can thus think of the neuron component v_{ia} as the probability for the ith Potts spin to be in state a. For $K = 2$ one recovers the formalism of the Ising case in a slightly disguised form.

Again one can analyze the linearized dynamics, as in the earlier section on mean-field dynamics, in order to estimate the critical temperature T_c.

The Graph Partition Problem

A generalization of the graph bisection problem is *graph partitioning* (GP): An N-node graph, defined by a symmetric connection matrix $w_{ij} = 0, 1$, $i \neq j = 1, \ldots, N$, is to be partitioned into K subsets of N/K nodes each, while minimizing the cut size, i.e., the number of connected node pairs winding up in different subsets (see Figure 1B).

This problem is naturally encoded in terms of K-state Potts spins as follows: For each node $i = 1, \ldots, N$, a Potts spin variable $\mathbf{s}_i = (s_{i1}, \ldots, s_{iK})$ is assigned, such that the spin component s_{ia} takes the value 1 or 0 depending on whether node i belongs to subset a or not. A suitable energy function is then given by (cf. Equation 3)

$$E = -\frac{1}{2} \sum_{i,j=1}^{N} w_{ij} \mathbf{s}_i \cdot \mathbf{s}_j + \frac{\alpha}{2}\left(\left(\sum_{i=1}^{N} \mathbf{s}_i\right)^2 - \sum_{i=1}^{N} \mathbf{s}_i^2\right) \quad (14)$$

where the first term is a cost term (cut size), while the second is a penalty term with a minimum when the nodes are equally partitioned into the K subsets. Note that the diagonal contributions are subtracted in the second term. These are irrelevant for the encoding but affect the dynamics.

The Traveling Salesman Problem

In the traveling salesman problem (TSP) the coordinates $\mathbf{x}_i \in \mathbb{R}^2$ of a set of N cities are given. A closed tour of minimal total length is to be chosen such that each city is visited exactly once. This problem is somewhat reminiscent of (the trivial) $K = N$ graph partition (see Figure 1C). We define an N-state Potts neuron $\mathbf{s}_i$ for each city $i = 1, \ldots, N$, such that the component s_{ia} ($a = 1, \ldots, N$) is 1 if city i has the tour number a and 0 otherwise. Let d_{ij} be the distance between city i and j. Then a suitable energy function is given by

$$E = \sum_{i,j=1}^{N} d_{ij} \sum_{a=1}^{N} s_{ia} s_{j(a+1)} + \frac{\alpha}{2}\left(\left(\sum_{i=1}^{N} \mathbf{s}_i\right)^2 - \sum_{i=1}^{N} \mathbf{s}_i^2\right) \quad (15)$$

where the first term is a cost term, and the second a soft constraint term penalizing configurations where two cities are assigned the same tour number.

Scheduling Problems

Scheduling problems have a natural formulation in terms of Potts neurons. In its purest form, a scheduling problem consists entirely in fulfilling a set of basic constraints, each of which can be encoded as a *penalty term* that will vanish when the constraint is obeyed. In many applications, however, there exist additional preferences within the set of legal schedules, that lead to the appearence also of *cost terms*.

Consider a simplified scheduling problem, where N_p teachers lecture N_q classes in N_x classrooms at N_t time slots. For the solutions it is required that every teacher p give a lecture to each of the classes q, using the available rooms x and the available time-slots t, with no conflicts in space (classrooms) or time. This statement defines the *basic constraints* that have to be satisfied. In addition, various *preferences* regarding continuity in classrooms and so on might be present. The basic constraints are as follows:

1. An event (p, q) should occupy precisely one space-time slot (x, t).
2. Different events (p_1, q_1) and (p_2, q_2) must not occupy the same space-time slot (x, t).
3. A teacher p must not take part in more than one event at a time.
4. A class q must not take part in more than one event at a time.

A schedule fulfilling all the basic constraints is said to be *legal*.

The obvious encoding is in terms of Potts spins $\mathbf{s}_{pq}$; the component $s_{pq;xt}$ is defined to be 1 if the event (p, q) takes place in the space-time slot (x, t) and 0 otherwise. Thus we need $N_p N_q$ distinct K-state Potts spins, with $K = N_x N_t$. Then the first constraint is trivially satisfied through the usual Potts condition

$$\sum_{x,t} s_{pq;xt} = 1 \quad (16)$$

for each event (p, q). The other three constraints are implemented using energy penalty terms as follows:

$$E = \frac{1}{2}\sum_{x,t}\left(\sum_{p,q} s_{pq;xt}\right)^2 + \frac{1}{2}\sum_{p,t}\left(\sum_{q,x} s_{pq;xt}\right)^2 + \frac{1}{2}\sum_{q,t}\left(\sum_{p,x} s_{pq;xt}\right)^2 \quad (17)$$

Again, mean-field variables $v_{pq;xt} \sim \langle s_{pq;xt}\rangle_T$ are introduced, and solving the corresponding MF equations (cf. Equations 11 and 12) with annealing results in a very efficient algorithm.

An important simplification can be made, since it turns out that with the above encoding the MF equations give rise to two separate phase transitions: one in x and one in t. It is therefore economical to implement this factorization already at the encoding level by replacing each spin $\mathbf{s}_{pq}$ by the direct product of two spins: an N_x-state Potts spin $\mathbf{s}_{pq}^{(X)}$ for assigning classrooms, and an N_t-state Potts spin $\mathbf{s}_{pq}^{(T)}$ for assigning time slots, with separate Potts conditions replacing Equation 16. This method reduces the dimensionality from $N_p N_q N_x N_t$ to $N_p N_q (N_x + N_t)$, and as a consequence the sequential CPU execution time goes down; the solution quality is not affected.

The preceding synthetic scheduling problem contains several simplifications as compared to realistic problems. In Gislén, Söderberg, and Peterson (1992b) real-world problems were successfully dealt with; a somewhat extended formalism was required to handle all details.

Results

Extensive numerical tests on the Potts approach for the three different problem areas have been conducted. In all cases convergence is achieved after a very modest number (of the order of 50 to 100) of iterations. In fact, the number of iterations needed is found empirically to be independent of problem size (Peterson and Söderberg, 1989; Gislén, Söderberg, and Peterson, 1992b), partly because of the prior estimate of T_c. In most cases the MF algorithm performs better than other methods with respect to quality and speed.

For traveling-salesman problems the quality of the solutions is also very good (Peterson and Söderberg, 1989; Peterson, 1990). However, the "pure" Potts approach requires an excessive number of degrees of freedom, and it is therefore often advantageous to employ the deformable templates approach (see next section).

Deformable Templates

The preceding optimization problems were all treated as *pure assignment* problems—all variables are logical (neurons). In some areas it is advantageous to take a *parametric assignment* approach, which results in a hybrid picture where MF Potts decision variables are supplemented by geometric *template* variables. TSP (Durbin and Willshaw, 1987) and particle physics track finding (Ohlsson, Peterson, and Yuille, 1992; Gyulassy and Harlander, 1991) are examples where it pays off to use the deformable templates approach.

The Knapsack Problem

The graph partition problem is characterized by an *equality* constraint, which is implemented with a polynomial penalty term. However, in many optimization problems, in particular those of resource allocation type, one has to deal with *inequalities*. One such problem category is the knapsack problem, where one has a set of N items i with associated utilities c_i and loads a_{ki}. The goal is to fill a "knapsack" with a subset of the items such that their total utility,

$$U = \sum_{i=1}^{N} c_i s_i \quad (18)$$

is maximized, subject to a set of M load constraints,

$$\sum_{i=1}^{N} a_{ki}s_i \leq b_k \qquad k = 1, \ldots, M \qquad (19)$$

defined by load capacities $b_k > 0$. The encoding is in terms of binary decision variables (spins) $s_i \in \{1, 0\}$, representing whether or not item i goes into the knapsack.

We consider a class of problems, where a_{ki} and c_i are random numbers $\in [0, 1]$ and $b_k = b$ (fixed). The most difficult case $b = N/4$ is picked. The expected number of used items in an optimal solution is then about $N/2$, and an exact solution becomes inaccessible for large N. The *set covering* problem is a special case with random $a_{ki} \in \{0, 1\}$, and $b_k = 1$, which is a relatively simple class according to the preceding discussion.

In the optimal solution to such a problem, there will be a strong correlation between the value of c_i and the probability for s_i to be 1. With a simple heuristic based on this observation, one can often obtain near-optimal solutions very fast. We will therefore also consider a class of harder problems with more narrow c_i distributions—*homogeneous* problems.

A suitable neural energy function for the problem defined in Equation 18 is (Ohlsson, Peterson, and Söderberg, 1993)

$$E = -\sum_{i=1}^{N} c_i s_i + \alpha \sum_{k=1}^{M} \Phi\left(\sum_{i=1}^{N} a_{ki}s_i - b_k\right) \qquad (20)$$

where Φ is a penalty function ensuring that the constraint in Equation 19 is fulfilled. The coefficient α governs the relative strength between the utility and constraint terms. An appropriate choice of $\Phi(x)$ is $x\Theta(x)$. (Here, $\Theta(x) = 1$ if $x > 0$ and 0 otherwise.) Minimizing Equation 20 is done with the MF equations. Because of the nonpolynomial form of the constraint, the derivative $\partial E/\partial v_i$ in Equation 10 is replaced by a difference

$$\frac{\partial E}{\partial v_i} \rightarrow -c_i + \alpha \sum_{k=1}^{M} \Bigg[\Phi\left(\sum_{j=1}^{N} a_{kj}v_j - b_k\right)\Big|_{v_i=1}$$
$$- \Phi\left(\sum_{j=1}^{N} a_{kj}v_j - b_k\right)\Big|_{v_i=0} \Bigg] \qquad (21)$$

The MF equations are again solved iteratively by annealing in T.

In Ohlsson, Peterson, and Söderberg (1993) this ANN approach is compared with other approaches. For small problem sizes it is feasible to use an exact algorithm, branch and bound, for comparison. For larger problem sizes, one is confined to other approximate methods, such as simulated annealing, greedy heuristics, and simplex based on linear programming.

As expected, linear programming and in particular greedy heuristics benefit from nonhomogeneity in terms of both quality and CPU capacity, while for homogeneous problems the neural network algorithm is the winner. For larger problem sizes, it is not feasible to use the exact branch and bound algorithm. The best we can do is to compare the different approximate approaches, neural network, simulated annealing, and linear programming. The conclusions from problem sizes ranging from 50 to 500 are the same. The real strength in the neural network approach is best exploited for more homogeneous problems.

Discussion

There are two alternative paths within the ANN paradigm for finding good solutions to combinatorial optimization problems —the pure ANN and the deformable templates approaches. For generic combinatorial optimization problems, such as scheduling or assignment problems, one uses the pure neural approach in which the basic steps are as follows:

- Map the problem onto a neural network by a suitable encoding of the solution space and an appropriate choice of energy function. Where applicable, use Potts encoding rather than Ising neuron multiplexing.
- Utilize prior knowledge about phase transition properties from analyzing the linearized dynamics.
- While annealing, solve the corresponding mean-field equations iteratively.
- When the MF equations have converged, the solutions are checked with respect to "legality"—whether they satisfy the basic constraints. If not, one either performs a simple corrective postprocessing or re-anneals the system (possibly with modified constraint coefficients).

For low-dimensional geometrical problems like traveling salesman and track-finding problems, it is often advantageous to use a hybrid procedure, the deformable templates method (see CONSTRAINED OPTIMIZATION AND THE ELASTIC NET).

A binary (Ising) neuron can be considered as a vector living on a "sphere" in one dimension. The MF approach can be generalized to variables defined on spheres in higher dimensions. Such *rotor* neurons may be used in geometrical optimization problems with angular variables (Gislén, Söderberg, and Peterson, 1992a).

The MF approximation can be viewed as a variational approach—one approximates the true energy $E(s_i)$ with a trial one $E_0(s_i; u_i)$ and optimizes with respect to the variational parameters u_i. This procedure can be used in a wide range of situations not necessarily confined to discrete optimization problems (Jönsson, Peterson, and Söderberg, 1993).

Road Map: Dynamic Systems and Optimization
Background: I.3. Dynamics and Adaptation in Neural Networks
Related Reading: Fractal Strategies for Neural Network Scaling

References

Durbin, R., and Willshaw, D., 1987, An analog approach to the traveling salesman problem using an elastic net method, *Nature*, 326:689–691.

Gislén, L., Söderberg, B., and Peterson, C., 1992a, Rotor neurons: Basic formalism and dynamics, *Neural Computat.*, 4:737–745.

Gislén, L., Söderberg, B., and Peterson, C., 1992b, Complex scheduling with Potts neural networks, *Neural Computat.*, 4:805–831.

Gyulassy, M., and Harlander, H., 1991, Elastic tracking and neural network algorithms for complex pattern recognition, *Comput. Phys. Commun.*, 66:31–46.

Hopfield, J. J., 1982, Neural networks and physical systems with emergent collective computational abilities, *Proc. Natl. Acad. Sci. USA*, 79:2554–2558.

Hopfield, J. J., and Tank, D. W., 1985, Neural computation of decisions in optimization problems, *Biol. Cybern.*, 52:141–152.

Jönsson, B., Peterson, C., and Söderberg, B., 1993, A variational approach to correlations in polymers, *Phys. Rev. Lett.*, 71:376–379.

Kirkpatrick, S., Gelatt, C. D., and Vecchi, M. P., 1983, Optimization by simulated annealing, *Science*, 220:671–680.

Ohlsson, M., Peterson, C., and Söderberg, B., 1993, Neural networks for optimization problems with inequality constraints: The knapsack problem, *Neural Computat.*, 5:331–339.

Ohlsson, M., Peterson, C., and Yuille, A., 1992, Track finding with deformable templates: The elastic arms approach, *Comput. Phys. Commun.*, 71:77–98.

Peterson, C., 1990, Parallel distributed approaches to combinatorial optimization problems: Benchmark studies on TSP, *Neural Computat.*, 2:261–269.

Peterson, C., and Söderberg, B., 1989, A new method for mapping optimization problems onto neural networks, *Int. J. Neural Syst.*, 1:3–22.

Neuroanatomy in a Computational Perspective

Almut Schüz

Introduction

This article is intended to help the modeler get a feel for real brains. It provides an introduction to the basic principles of mammalian brain organization and offers an impression of the variability of brain structures and of the number of neuronal components of various parts of the brain. By focusing especially on the cerebral cortex, the largest part of the brain, the article shows how quantitative neuroanatomy contributes to brain theory.

The Role of Neuroanatomy in Brain Theory

In simple organisms in which the sensory organs and effectors (muscles, glands) are connected by direct pathways, tracing such pathways may provide a sufficient explanation of behavior. However, in the course of evolution, the number of neurons has increased considerably and the pathways between input and output have become less and less direct. Some figures suggest the orders of magnitude involved in seeking an explanation of higher brain functions.

The human brain contains between 7×10^{10} and 8×10^{10} neurons (Haug, 1986). Most of them belong to the cerebellar cortex (approximately 5×10^{10}) and the cerebral cortex (approximately 1.5×10^{10}). Many other parts of the human brain, such as the first relay station of the optic nerve in the thalamus, have approximately 10^6 neurons. The corpus callosum, the main fiber bundle which connects the two hemispheres, has approximately 10^8 fibers. For a comprehensive collection of figures, see Blinkov and Glezer (1968).

In small mammals, the numbers are still very large, although they are reduced by a factor of 1000: the cerebral cortex of the mouse, for example, contains 10^7 neurons, and the corpus callosum of the mouse contains 10^5 fibers.

However, the high number of neurons is not the main obstacle to understanding the neural mechanisms underlying higher brain functions. The situation is complicated by the fact that most connections in the brain are not one to one. Many neurons receive input from thousands of other neurons and distribute their output to as many. In such nerve networks, it is beyond practical means to trace the anatomical pathways in detail. Even if a direct anatomical connection between two neurons were demonstrated, whether the activity of neuron A leads to the activation of neuron B largely depends on what else is going on in the network (Aertsen and Preißl, 1991).

Thus, in complex nerve networks, the task of neuroanatomy is not so much to study all of the connections in detail, but rather to show the typical structural properties of a specific part of the brain. These properties may provide clues to the understanding of its specific function, as will be shown later for the cerebral cortex.

White and Gray Matter: Projection Versus Computation

One general principle of construction in the vertebrate brain is the separation of neuronal pieces that are responsible for the long-range connections projecting from one part of the brain to another from those performing the actual information processing. This distinction can even be seen macroscopically on sections through the brain which reveal regions of different shading, the white and gray matter. Apart from blood vessels and glia cells, the white matter is composed exclusively of axons, many of which are myelinated. On the other hand, the gray matter contains all components of neurons: cell bodies, dendrites, and axons, most of which are unmyelinated.

Because the gray matter also contains the synapses, it is the place where neurons interact. Within the gray matter, a neuron can reach other neurons within a radius of a few hundred microns, up to a few millimeters at most, while through the white matter it can reach cells located centimeters or even a meter away.

The largest mass of white matter in the human brain is in the cerebral cortex. The white matter of the hemispheres consists largely of fibers which connect different parts of the cortex to each other. The internal capsule, which contains fibers connecting the cortex with other parts of the brain, represents only a small fraction of this area. Therefore, one of the principles of cortical connectivity is to connect the cortex to itself. In comparison, the thin sheet of white matter accompanying the cerebellar cortex indicates that local computation plays a major role in this part of the brain. For this type of analysis of brain structures, see Braitenberg (1977).

White Matter, Brain Size, and Connectedness

Large brains have comparatively more cortical white matter than small brains. For example, 42% of the human neocortex (white and gray matter together) is white matter, while in the hedgehog it is only 13% (Frahm, Stephan, and Stephan, 1982). Clearly, if one wanted to connect a large cortex in a similar style as one with fewer neurons, the fiber mass would have to increase more than proportionately to the number of neurons. The question is whether the increase in fiber mass in larger brains suggests a higher, a lower, or an equal degree of interconnectedness. The answer is either lower or equal, depending on the definition of *interconnectedness*. If this term indicates the percentage of neurons of the cortex reached by an individual neuron, there is clearly a lower interconnectedness in larger cortices. However, if one defines comparable cortical compartments in different species, as indicated, for example, by the diameter of the largest dendritic trees, and if one postulates a complete set of connections between all of the compartments, the relative increase in white matter might be sufficient (Braitenberg, 1978).

Not only the number of connections but also the time delays involved play a role in the interconnectedness of various parts of a network. In a large brain, distant elements may not be able to collaborate efficiently because of the delays in the transmission from one point to the other. Since conduction velocity depends on the thickness of an axon, this problem could theoretically be solved by an appropriate increase in the diameter of the longer axons in larger brains. However, this remedy is self-defeating since thicker fibers would further increase brain size and therefore increase time delays, which would in turn require thicker fibers, and so on. Starting with a brain as large as the human one, such a series would converge at a volume which is approximately 50% larger if one wanted interhemispheric signals to travel twice as fast as they do (Ringo et al., 1994). The way out of this dilemma has been proposed to be the higher degree of functional specialization of cortical regions in larger brains (Ringo et al., 1994).

A similar argument has been put forward by Mitchison (1992). He showed that if the neurons in the cortex were not organized into spatially distinct areas connected through the white matter, but merged into one huge piece of gray matter, the volume of the human cortex would have to increase by a factor of 10 to maintain the same degree of connectedness.

Overall Connectivity of the Brain

There are no known isolated parts of the brain; that is, anatomical connections can be traced from any part of the brain to any other, although sometimes over many intermediate stations.

However, the brain as a whole is highly structured. Some parts interact directly with each other by way of reciprocal connections, such as the cerebral cortex and the thalamus. Other parts may be arranged in loops. Two examples are: (1) the loop cerebral cortex—basal ganglia—thalamus—cerebral cortex, and (2) the loop cerebral cortex—pontine nuclei—cerebellum—thalamus—cerebral cortex (see Còté and Crutcher, 1991).

Such loops may be neatly separated when they run through the same part of the brain, for example, the thalamus in the case of the two loops just mentioned (Asanuma, Thach, and Jones, 1983). At other places, cross-talk may be assumed. For example, the cortical regions involved in these two loops partly overlap. Thus, even if we know the main pathways between various parts of the brain, it may be difficult to predict the route an individual signal will take. In addition, shortcuts to the main stream may exist, part of a loop may have parallel lines over further relay stations, or subloops may be added.

Depending on the route a signal takes, it can be fed back either negatively or positively onto its parent structure. Positive feedback is sometimes transmitted by way of disinhibition, when two inhibitory stations are connected in a series. This type of positive feedback characterizes the corticocortical loop through the basal ganglia.

The projections between the various parts of the brain usually suggest parallel processing in that large regions of one part are connected to large regions of the other through thousands of fibers rather than through one or a few input and output lines.

The projections from one part of the brain to another often form special patterns which reflect various kinds of computation. For example, the projections may be organized such that neighborhood relationships are maintained (e.g., the retinotopic projections in the visual system). In other cases, complex, patchy divergent or convergent patterns may occur, suggesting either separation or combination of inputs from different sources (e.g., projections from the various regions of the cerebral cortex to the basal ganglia).

The projections can also differ in that they may be point to point (sometimes one point to several points) or that their terminal arbors may be smeared over large regions of the target structure. To a certain degree, this anatomical distinction corresponds to an important functional difference. Relatively restricted terminal arbors are typical for pathways which are involved in information processing, such as those between the specific thalamic nuclei and the cerebral cortex. In contrast, the neurons of some small nuclei in the brainstem (especially those of the locus coeruleus) can form huge terminal arbors which extend over large portions of their target structure. These nuclei seem to be involved in the global regulation of the level of activity, providing the background on which the information processing takes place.

Geometry of the Gray Matter

Some of the typical structural features of the various parts of the brain can easily be recognized at both the macroscopic and

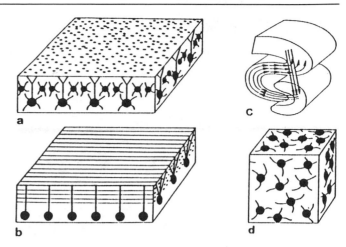

Figure 1. Different types of geometry in the gray matter. *A*, The structure in the horizontal plane is isotropic and different from that in the vertical plane (as in the cerebral cortex). *B*, All three planes of sectioning are different from each other (as in the cerebellar cortex). *C*, In the hippocampus, the horizontal (folded) plane is isotropic, but overlaid by a subpopulation of fibers which run in one direction only. These are, furthermore, arranged such that a cyclic operation is suggested. *D*, All three dimensions are equal. [Modified from Braitenberg, V., and Schüz, A., 1993, Allgemeine Neuroanatomie, in *Neuro- und Sinnesphysiologie* (R. F. Schmidt, Ed.), Berlin: Springer.]

the microscopic level. The gray matter can either form lumps (often called *nuclei*) or show a two-dimensional, layered arrangement as in the so-called cortices. With respect to function, the latter type suggests that the same kind of operation is performed over the whole surface. One would expect such an arrangement, for example, in the processing of two-dimensional pictures. Indeed, this type of arrangement is found in visual processing in all vertebrates as well as in diverse invertebrates.

Among the cortices, further distinctions can be made (Figure 1). In the cerebral cortex, the plane is isotropic in the sense that different directions cannot be distinguished in the histological picture. In contrast, in the cerebellar cortex, two perpendicular dimensions of the cortical plane are organized in completely different ways. Most of the axons run in a laterolateral direction, while most of the dendritic trees are flattened perpendicularly. This arrangement suggests that in the cerebellar cortex, a computation occurs along quasi-one-dimensional lines (Braitenberg, 1977).

A mixture of these two types also exists. In the hippocampus, the dendrites and some axonal systems are spread in all directions of the plane, while other axonal systems (mossy fibers, Schaffer collaterals) superimposed onto these run in one direction only. Furthermore, the arrangement of the latter is suggestive of cyclic operation (Figure 1*C*).

In contrast to cortices, in many nuclei geometry does not seem to play an important role. Their fine structure appears to be isotropic in all three dimensions (Figure 1*D*).

Histology and Connectivity

The various histological methods reveal further structural details. Different parts of the brain may differ with respect to density and size of neurons, shape of axonal and dendritic trees, or arrangement in layers. Layers can be distinct, as in the cerebellar cortex, or more vague, as in the cerebral cortex.

However, despite detailed knowledge of structure down to the electron microscopical level, it is often difficult to grasp the underlying principles of connectivity. What are the rules, for example, behind the felt of axons stained in Figure 2? To how

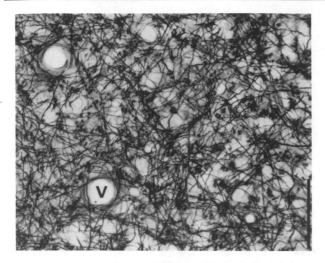

Figure 2. Axonal net. Light micrograph of the visual cortex of a monkey (area 17). Horizontal section is shown through layer IV. Only axons are stained. A blood vessel (V) is shown. The bar equals 50 μm.

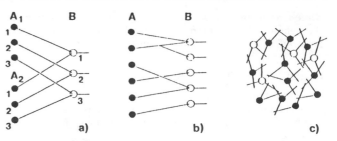

Figure 3. Three networks illustrating various degrees of specificity. *A*, All connections are determined. *B*, The connections are specified only with respect to type, not with respect to individual neurons. *C*, There is no specificity. The neurons connect to whatever they happen to meet. [From Schüz, A., 1992, Randomness and constraints in the cortical neuropil, in *Information Processing in the Cortex* (A. Aertsen, and V. Braitenberg, Eds.), Berlin: Springer, fig. 1, p. 4. Reprinted with permission.]

many neurons do they belong, and where do they come from? Is what we see in Figure 2 an intermingling of parallel circuits, or are they all part of the same network?

Specificity Versus Randomness

One of the crucial questions which springs to mind when studying illustrations such as Figure 2 is the degree to which the target of an individual fiber is defined. The question is relevant to the anatomist in determining at what level of detail to analyze the structure. The theoretician, too, is faced with the question of whether to base a model on strictly predetermined connectivity or on a random network. For a comprehensive discussion, see Szenthágothai and Arbib (1974).

In Figure 3, three possibilities are depicted. In Figure 3*A*, neurons of type A connect to neurons of type B in a strictly defined manner. There is specificity between types (A, B) and between individual neurons ($A_{1/1}$ and $A_{2/1}$ to B_1, etc.). A prerequisite for this kind of connectivity is that the neurons can be labeled individually, either by their geometric arrangement or by some other means, for example, by a chemical marker.

On the other hand, the specificity could be restricted to types of neurons (e.g., A connects to B only) without further specification, as shown in Figure 3*B*. In this case, it is no longer crucial whether a certain neuron A connects to a particular neuron B rather than to its neighbor.

In the third case (Figure 3*C*), there is no specificity whatsoever. The neurons of both types are intermingled and connect to everything they happen to meet.

In comparative anatomy, all three types can be found. An example of the first kind is the visual system of the fly. There, the photoreceptors are arranged in strict geometrical order and connect to the first optic ganglion in a completely determined manner (Braitenberg, 1977).

The second kind of network is similar to the situation in the cerebellar cortex. There, it is determined how the various types of neurons are connected. The granule cells, for example, connect to the other four cell types of the cerebellar cortex, but not to other granular cells; basket cells contact only Purkinje cells, etc. In view of the large number of granule cells and their extremely high packing density, it is hard to believe that it should be crucial for one of the target cells (e.g., a Purkinje cell) to be contacted by one particular granule cell rather than by a neighboring one.

The network in Figure 3*C* may be similar to the situation in the cerebral cortex. Although type-specialized connections as seen in Figure 3*B* exist in the cortex, they account for only a minority of the connections there. In that case, the depiction in Figure 3*C* seems to be more to the point than the other two kinds, as will be shown below.

Of course, even in the network shown in Figure 3*C*, constraints to complete randomness are imposed by rules on a more macroscopic level. For example, the overall connectivity depends on the density of neurons, the way they ramify, and the distribution of the various cell types, although only in a probabilistic manner. These constraints, visible with a light microscope, are in part genetically determined and in part refined by activity-dependent self-organization (see SELF-ORGANIZATION AND THE BRAIN).

Statistical Neuroanatomy

One approach to the question of connectivity in complex nerve networks is a quantitative assessment of the various components shown by the different histological methods (cell bodies, synapses, axons, etc.). How such data can be used to determine the typical structural properties of a network and to constrain neural models has been shown in detail for the cerebral cortex (see Braitenberg and Schüz, 1991, and references there). The main points are briefly summarized here.

Basic Structure of the Cerebral Cortex

Some of the quantities which have been measured are shown in Table 1 (a–m). They refer to the mouse cortex and are corrected for tissue shrinkage. The letters n–s show further quantities which can be derived from those. A number of these quantities require some explanation.

The *relative density of axons* (Table 1, q) is defined as follows. Imagine a piece of cortex punched out perpendicularly to the cortical surface and just large enough to contain the whole axonal tree of an individual neuron within the gray matter (excluding possible secondary axonal arbors made at distant places). If the total length of this individual axon is divided by the sum of the lengths of all axons within this volume, the relative axonal density is obtained. It quantifies the axonal contribution of an individual neuron to the volume within which it ramifies and is a measure for the intermingling of axons belonging to different neurons. The *relative density of dendrites* is

Table 1. Characteristics of the Mouse Cerebral Cortex

Measured quantities

a:	Volume (iso- and allocortex)	2×87 mm^3
b:	No. of sensory input fibers	$<10^6$
c:	Density of neurons	9.2×10^4/mm^3
d:	Percent pyramidal cells	85%
e:	Density of synapses	7.5×10^8
f:	Percent type I synapses	89%
g:	Percent synapses on spines	75%
h:	Density of axons	3.6 km/mm^3
i:	Density of dendrites	0.4 km/mm^3
j:	Length of axonal tree	10–40 mm
k:	Length of dendritic tree	4 mm
l:	Range of axonal tree (pyramidal cell)	1 mm
m:	Range of dendritic tree	0.2 mm

Deduced quantities

n:	(a, c) Total no. of neurons	2×10^7
o:	(c, e) Synapses/neuron	8000
p:	(e, h) Synapses/length of axon	200/mm
q:	(h, j, l) Relative density of axons (pyramidal cells)	10^{-5}
r:	(i, k, m) Relative density of dendrites (pyramidal cells)	10^{-3}
s:	(p, q, r) Probability of synapses between 2 pyramidal cells 0.1 mm apart	0 synapses, $p = 0.9$ 1 synapse, $p = 0.09$ 2 synapses, $p = 0.004$

Conclusions

t:	(b, n) No. of neurons $\gg$ no. of input fibers
u:	(d, f, g) Most connections between neurons of one kind
v:	(f) Most connections excitatory
w:	(o, s) Great divergence and convergence
x:	(s) Connections very weak
y:	(u–x) Mixing machine
z:	(t, u, g) Memory rather than computation
z':	(y, z) Associative memory with formation of cell assemblies

Modified from Braitenberg and Schüz (1991). The numbers represent measurements on light and electron micrographs in the cortex of the mouse (a–m), quantities which can be deduced from those measurements (n–s), and conclusions which can be drawn for the connectivity (t–x), as well as the functional interpretation (y–z'). The letters in parentheses indicate from which other quantities the corresponding quantity or conclusion can be derived.

derived similarly by dividing the dendritic length of an individual neuron by the total dendritic length in the volume within which it ramifies.

The *probability of synapses between two pyramidal cells* is determined as follows. A prerequisite for a synapse between two neurons is that the axon of one touches a dendrite of the other. The probability of this occurrence increases with greater overlap of the axonal tree of the one neuron with the dendritic tree of the other. The probability also increases with higher relative axonal and dendritic density of the two neurons.

From these data, some properties of cortical connectivity can be inferred (Table 1, t–x). A functional interpretation is also given in Table 1 (y–z').

In short, the network of the cortex consists of one main type of cell, the pyramidal cells (including the spiny stellate cells), which are connected by excitatory synapses. Most of these synapses are located on dendritic spines. Since synapses on spines are assumed to be modifiable in strength, one may conclude that one of the main tasks of this network is storage of information.

Nonpyramidal cells (about 15%) are interspersed among the pyramidal cells. They are inhibitory. Both kinds of neurons connect onto both types of cells. The inhibitory neurons, however, contribute only about 11% of the synapses to the cortex.

The number of synapses contributed by input fibers is small compared with the number of synapses which connect cortical neurons to each other. Most pyramidal cells have a local axonal tree which connects them to other neurons in their neighborhood as well as a far-reaching axonal tree which, in most cases, connects them to another region of the cortex through the white matter.

The low relative axonal and dendritic densities have interesting implications. In the mouse, the local axonal tree of an individual pyramidal cell is interwoven with the dendrites of 10^5 other neurons and competes with axons from approximately 5×10^5 other neurons ramifying in the same region. These various ramifications do not generally align, but freely permeate each other. It is therefore compelling to assume that the 8000 synapses of a pyramidal cell connect to almost as many different neighbors. This assumption suggests that theories based on a probabilistic connectivity, as shown in Figure 3C, more accurately represent the structure of the cortex than others which rely on precise circuits.

This description of cortical structure fits Hebb's theory of cell assemblies remarkably well. This theory postulates that meaningful events are represented by groups of neurons which are connected more strongly to each other than to other neurons (see HEBBIAN SYNAPTIC PLASTICITY: COMPARATIVE AND DEVELOPMENTAL ASPECTS). These groups are formed through a learning process. The learning is assumed to strengthen the connections between neurons which are often activated together, a process known as *associative storage*.

For the realization of such cell assemblies, a large number of similar components must be connected into a network. Precisely predetermined connections are not required since, as a result of learning, the patterns of interactions between neurons are different for each brain. What is crucial, however, is an initial connectivity that is sufficiently rich to allow as many constellations of neuronal activity as possible to be detected and learned in the connections. The fact that the individual pyramidal cell seems to strive toward a large number of different synaptic neighbors, together with the large mass of cortico-cortical fibers, indicates that the cortex is well suited for this task. The excitatory and modifiable (plastic) synapses which are implicit in Hebb's theory are also present in the cortex. The fact that individual neurons are weakly connected through one or only a few synapses implies that only correlated activity of many neurons can activate another neuron. This implication is also in agreement with Hebb's theory.

The linkage of the theory of cell assemblies with the structure of the cortex has led to more precise formulations of this concept. It permits estimates of the storage capacity of the cortex and the size and internal structure of cell assemblies (Palm, 1993) and allows concrete ideas to develop about the regulation of their dynamics through the hippocampus (Miller, 1991) or the striatum (Miller and Wickens, 1991).

Diffuse Connections and Patchy Projections

The order in the structure of the cortex seems to be largely restricted to a more global level (layers, patterns of ramification, organization of input, etc.). For more detailed information about the orderly aspects of cortical structure, see White

(1989). I want to elaborate briefly on one such point, the phenomenon of patchy projections, since it can be directly related to the disorder described above. In contrast to the individual neuron, which tends to mix its processes as much as possible with those of other neurons, groups of neurons have a tendency to stick together when they project from one place to another.

If a pyramidal cell connects to all of its postsynaptic neighbors by one or only a few synapses, one may wonder how excitation can spread at all. A solution to this problem was found by Abeles (1990) in his theory of synfire chains.

The anatomical answer to this question seems to be the phenomenon of patchy projections. Weak connections do not adversely affect the process as long as enough neurons overlapping each other are activated by the same event. Thus, incoming sensory fibers which densely ramify in the primary sensory areas of the cortex may cause activity there, especially if the incoming sensory fibers are topographically organized. In the visual system, for example, one can assume that in most cases several neighboring input fibers are activated by a sensory event.

However, this activity has little chance of propagating within the cortex if the neurons from one point project diffusely all over the cortex. On the other hand, if axons from neighboring neurons tend to join into bundles on their way to other regions, they have a better chance of activating a new set of neurons there. The combination of diffuseness of connections at a given point in the cortex and the patchiness of projections from such a point seems to be the anatomical solution to the dilemma of achieving a high degree of connectivity on the one hand and getting a message across on the other.

Other Principles of Connectivity

The connectivity of the cortex contrasts with that of other networks realized in the brain. The cerebellum has already been discussed. It differs from the cerebral cortex primarily because of its strict geometrical order and complete lack of positive feedback connections within the cerebellar cortex.

Another contrasting network is the striatum. Although the cortex and the striatum have a number of features in common, a fundamental difference is that the cortex operates primarily on the basis of mutual excitation of neurons, while mutual inhibition seems to play an essential role in the striatum. There, more than 90% of neurons are inhibitory. Thus, while in the cortex the dominating principle seems to be cooperation, the connectivity of the striatum suggests competition between neurons, a principle which is also referred to as *winner-takes-all* (Miller and Wickens, 1991; Wickens, 1993).

Road Map: Biological Networks
Related Reading: Cortical Columns, Modules, and Hebbian Cell Assemblies; Visual Cortex Cell Types and Connections

References

Abeles, M., 1990, *Corticonics*, Cambridge, Eng.: Cambridge University Press.
Aertsen, A., and Preißl, H., 1991, Dynamics of activity and connectivity in physiological neuronal networks, in *Non-linear Dynamics and Neuronal Networks* (H. Schuster, Ed.), Weinheim, Ger.: VCH, pp. 281–301.
Asanuma, C., Thach, W. T., and Jones, E. G., 1983, Distribution of cerebellar terminations and their relation to other afferent terminations in the ventral lateral thalamic region of the monkey, *Brain Res. Rev.*, 5:237–265.
Blinkov, S. M., and Glezer, I. I., 1968, *The Human Brain in Figures and Tables: A Quantitative Handbook*, New York: Plenum. ◆
Braïtenberg, V., 1977, *On the Texture of Brains*, Berlin: Springer-Verlag. ◆
Braitenberg, V., 1978, Cortical architectonics: General and areal, in *Architectonics of the Cerebral Cortex* (M. A. B. Brazier and H. Petsche, Eds.), New York: Raven, pp. 443–465.
Braitenberg, V., and Schüz, A., 1991, *Anatomy of the Cortex: Statistics and Geometry*, Berlin: Springer-Verlag. ◆
Côté, L., and Crutcher, M. D., 1991, The basal ganglia, in *Principles of Neural Science* (E. R. Kandel, J. H. Schwartz, and T. M. Jessel, Eds.), East Norwalk, CT: Appleton and Lange, pp. 647–659. ◆
Frahm, H. D., Stephan, H., and Stephan, M., 1982, Comparison of brain structure volumes in insectivora and primates, I: Neocortex, *J. Hirnforsch.*, 23:375–389.
Haug, H., 1986, History of neuromorphometry, *J. Neurosci. Methods*, 18:1–17.
Miller, R., 1991, *Cortico-Hippocampal Interplay and the Representation of Contexts in the Brain*, Berlin: Springer.
Miller, R., and Wickens, J. R., 1991, Corticostriatal cell assemblies in selective attention and in representation of predictable and controllable events: A general statement of corticostriatal interplay and the role of striatal dopamine, *CINS* (*Concepts Neurosci.*), 2:65–95.
Mitchison, G., 1992, Axonal trees and cortical architecture, *Trends Neurosci.*, 15:122–126.
Palm, G., 1993, On the internal structure of cell assemblies, in *Brain Theory: Spatio-Temporal Aspects of Brain Function* (A. Aertsen, Ed.), Amsterdam: Elsevier, pp. 261–270.
Ringo, J. L., Doty, R. W., Demeter, S., and Simard, P. Y., 1994, Time is of the essence: A conjecture that hemispheric specialization arises from interhemispheric conduction delay, *Cereb. Cortex*, 4:331–343.
Szenthágothai, J., and Arbib, M. A., 1974, Conceptual models of neural organization, *Neurosci. Progr. Bull.*, 12:3. ◆
White, E. L., 1989, *Cortical Circuits: Synaptic Organization of the Cerebral Cortex—Structure, Function and Theory*, Boston: Birkhäuser. ◆
Wickens, J., 1993, *A Theory of the Striatum*, London: Pergamon.

Neuroethology, Computational

Dave Cliff

Introduction

In recent years, a number of neural network researchers have used the term *computational neuroethology* (CNE) to describe a specific approach to *neuroethology*, the intersection of neuroscience (the study of nervous systems) and ethology (the study of animal behavior). The definition of computational neuroethology is similar, but is not quite as dependent on studying animals; animals just happen to be biological *autonomous agents*. Nonbiological autonomous agents include some robots and some simulated agents operating in virtual worlds. Here, autonomous agents are self-governing systems capable of operating (i.e., perceiving and acting) in environments which are complex, uncertain, and dynamic. For the sake of brevity in the rest of the text, autonomous agents will be referred to simply as *agents*.

CNE can be distinguished from classical computational neuroscience by its increased emphasis on studying the neural control of behavior within the context of neural systems which are both embodied and situated within an environment.

Put most simply, CNE involves the use of computational modeling in attempting to understand the neural mechanisms responsible for generating useful behaviors in an agent. The word *useful* is imprecise; it is more common to talk of *adaptive* behaviors. In the ethology literature, an adaptive behavior is usually defined as a behavior which increases the likelihood that an animal will survive long enough to produce viable offspring. Often implicit in this definition is the assumption that the animal's environment is sufficiently unforgiving (or hostile) that if the animal does nothing, it will die before it can reproduce. In studying artificial agents, the utility of the behavior is evaluated by different criteria, such as computational or economic efficiency.

Neural networks that generate adaptive behavior should not be confused with *adaptive* neural networks, where connection strengths may alter as a result of experience. Adaptation or plasticity may itself give rise to new or improved adaptive behaviors, but many adaptive behaviors are genetically determined (e.g., hardwired behaviors such as reflexes and instincts).

In the context of adaptive behavior research, it is clear that the neural system is one component in the *action-perception cycle*, where actions allow the agent to perceive information concerning its environment, which may lead to changes in the agent's internal state, which may in turn affect further actions, which affect what information can be perceived, and so on. This notion has long been stressed by Arbib (1972:16):

> In speaking of human perception, we often talk as if a purely passive process of classification were involved—of being able, when shown an object, to respond by naming it correctly. However, for most of the perception of most animals and much of human behavior, it is more appropriate to say that the animal perceives its environment to the extent that it is *prepared* to *interact* with that environment in some reasonably structured fashion.

As defined above, CNE may not seem to be particularly distinguishable from most work in neural network research. After all, many people in computational neural network research might argue that their work may, ultimately, lead to an understanding of the neural mechanisms underlying the generation of (some) adaptive behaviors. For example, face recognition is an adaptive behavior in humans and could probably be classed as an adaptive behavior in, for example, a security robot. So why is a backpropagation network that learns to distinguish between photographs of human faces (for example) not classed as work in CNE?

Motivations

This section gives a brief overview of the motivations for the CNE approach. Further discussion of methodological issues is provided elsewhere (Beer, 1990; Cliff, 1990).

Typically, connectionist models employ homogeneous groups of highly idealized and simplified neuron models (called *units*), connected in a regular fashion, which exhibit some form of learning or adaptation.

Many connectionist models can be described, in essence, as mapping or transforming between representations: input data are presented to the network in a particular format, and the network is judged successful when its outputs can be interpreted as a correct representation of the results of performing the desired transformation. In almost all cases, the input and output representation formats are prespecified by the experimenter, although this situation is not entirely true of unsupervised learning, and a number of connectionist models draw inspiration from biological data in their choice of input and output representations. If such networks are to be employed in artificial agents or are to be of use in understanding biological agents, then there is a necessary (often unspoken) assumption that, eventually, it will be possible to assemble a pipeline of such input-output transducer networks which links sensory inputs to motor outputs and produces adaptive behavior. The most significant issue here is that there is a dependence on intermediate representations, which may not be justifiable: neural sensorimotor pathways generating adaptive behaviors might not be neatly partitioned into representation-transforming modules. When we "open up the black box," we may not find any patterns of activity identifiable as a representation in the conventional sense, and even if we do, there is no guarantee that they will be in strong accordance with representations chosen a priori by connectionist modelers.

This statement should not be mistaken for an argument against representation, nor for a denial of the vital role played by internal states in the generation of adaptive behaviors. It is simply an awareness of the dangers of being misled by a priori notions of representation. One of the safest ways to avoid these dangers is to model, as far as possible, *entire* sensorimotor pathways (i.e., the complete sequence of neural processing, from sensory input to motor output) involved in the generation of adaptive behavior. This modeling requires that the agent be studied while *embedded in an environment*. Most sensorimotor processing for adaptive behavior involves dynamic interaction with the environment; an embedded agent is part of a closed-loop system in that certain actions can affect subsequent sensory inputs. The sensorimotor pathway should not be viewed as a pipeline transforming a given input representation to a desired output representation, but rather as a link in the action-perception cycle.

When such an approach is adopted, the true nature of the representations and processing necessary for the generation of relatively complex adaptive behaviors is more likely to be revealed, and the validity of any a priori assumptions is clarified.

Naturally, it is beyond the state of the art to attempt to model complete sensorimotor pathways in humans or other large mammals, but experimental work in the neuroethology literature provides a wealth of data from less intellectually able animals, such as arthropods (the animal class which includes insects, spiders, and crustacea) and amphibians. Such animals are used as the domains of study in most CNE research, as will be seen in the discussion of current CNE projects later in this article.

Before that, the above arguments are illustrated by reference to a series of thought experiments. In his book *Vehicles: Experiments in Synthetic Psychology*, Braitenberg (1984) described specifications for a series of simple mobile vehicles operating in a world with simplified kinematics. The series of vehicles starts with an elementary device which performs primitive heat-seeking behavior; it progresses through vehicles that exhibit positive or negative taxes (i.e., orientation toward or away from a directional stimulus) and primitive forms of learning, pattern detection, and movement detection; and it culminates in vehicles which exhibit chaotic dynamics and predictive behavior. The internal control mechanisms of all of the vehicles are rigorously minimal: the simpler vehicles contain nothing more than wires connecting sensors to actuators, while the more advanced vehicles employ nonlinear threshold devices with delays and pseudo-Hebbian adaptation.

The key point of these thought experiments is that Braitenberg uses the psychological language indicative of intentional mental states to describe the observed behavior of the vehicles. He ascribes *fear, aggression, love, values and taste, rules, trains of thought, free will, foresight, egotism,* and *optimism* to his vehicles, and he demonstrates that while such terms may be useful at the level of description of an external observer, the internal causal mechanisms could be surprisingly simple.

Braitenberg's vehicles are strongly reminiscent of the simple electromechanical creature *Machina speculatrix*, designed and built by Walter (1953): *M. speculatrix* was built from a photoelectric cell, a touch sensor, two electronic "neurons," and assorted gear trains from old clocks and gas meters. Nevertheless, Walter noted that the patterns of behavior it produced could be reasonably described as exhibiting *speculation, discernment,* and *self-recognition* (Walter, 1953:113–114, 244).

While Braitenberg's vehicles are nothing more than thought experiments, they provide insight to possible organizational principles in natural and artificial creatures, and they demonstrate the limits of applicability of intentional terminology. As such, they are of relevance to the philosophy of cognitive science. In a commentary on Arbib's work (discussed further below), Lloyd (in Arbib, 1987:442–443) noted that there is generally a tradeoff between accuracy and completeness in cognitive modeling and argued that completeness is more desirable than accuracy for two reasons highlighted by Arbib's work:

Focusing only on components can lead one to overlook emergent effects of cooperative computation; and working with complete models, even if wrong in detail, nonetheless provides us with analytical tools applicable to future data and future, more accurate models. . . . I think the reasons to push for completeness go beyond these, however, and reveal a further source of the value of Arbib's work for cognitive science. . . .

[In 1978, Dennett] proposed that one approach the complexity of humans by looking at simpler systems, first solving cognitive problems as they arise in these "simple minds" and then bootstrapping towards increasingly complex and human like cognizers. Two sorts of systems appealed to Dennett as fruitful stepping stones: living systems and artifactual systems born of engineering imagination, "Martian three-wheeled iguanas" and the like. (Lloyd, in Arbib, 1987:443)

Thus, it can be seen that the CNE approach has some parentage in the philosophy literature. Indeed, the focus in CNE is on understanding what Dennett (in Ewert, 1987:373) calls the *wise wiring* underlying the generation of adaptive behavior. The presence of such ideas in the philosophy literature is due, at least in part, to prior arguments (such as those found in Arbib, 1972). At present, the focus in most CNE research is on adaptive behaviors which serve the "four Fs": feeding, fleeing, fighting, and reproduction (this is an old joke, often attributed to Paul McLean). These behavioral modes can be argued as underlying much of the more complex adaptive behaviors witnessed in higher animals, including the intelligent activities that inspire philosophers to posit the existence of a "language of thought."

To summarize, research in CNE can be characterized as placing increased emphasis on modeling entire adaptive-behavior-generating sensorimotor pathways in agents embedded in environments which supply sensorimotor feedback. Such an approach lessens the chances of making untenable assumptions concerning issues of representation and processing. To study such pathways where there is reliable biological data, it is necessary to focus attention on relatively simple animals, such as arthropods or amphibians.

It is important to note that there is a tradition of related work in the artificial neural network literature. Research in reinforcement learning for control tasks is the most closely related (see REINFORCEMENT LEARNING IN MOTOR CONTROL).

Some Current Research Projects

This section describes some CNE projects of direct relevance to the arguments summarized above: Arbib's work on visuomotor activity in frogs and toads; Beer's work on locomotion, guidance, and behavioral choice in cockroaches; Cliff's work on visual tracking in hoverflies; and Franceschini's work on equipping autonomous robots with fly-like compound eyes. All of these are ongoing projects, so the descriptions serve as snapshots of their current status, rather than as final reviews of completed research programs. Following the descriptions of these projects, some related work is discussed.

The Computational Frog

Probably the most advanced project in CNE is the work of Arbib and his students on an evolving family of models of visually mediated behavior in frogs and toads (for a review of the project with peer commentary, see Arbib, 1987). Arbib named his simulation model *Rana computatrix*, the computational frog, after Walter's *M. speculatrix*.

Arbib's computational modeling is accurately based on data from biological experimental work performed by the neuroethologist Ewert (1987). The *R. computatrix* models are faithful to the known biology, and there is an interplay between the experimental and theoretical work: Arbib constructs "an evolving set of model families to mediate flexible cooperation between theory and experiment" (Arbib, 1987:407).

Briefly, the evolving nature of Arbib's work is due to his use of *incremental* modeling. His models explore a variety of different connectivities and parameter settings within the overall paradigm of visuomotor brain function in the frog and the toad. An initial first approximation model was extended and refined in a number of stages, leading to a family of models for *R. computatrix*.

Arbib's approach involves the definition of a number of functional *schemas*. Schemas can be modeled by interacting layers of neuronlike elements or by networks of intermediate-level units. The network models can be related to experimental data concerning neural circuitry, and the development process iterates (Arbib, 1987:411; see SCHEMA THEORY).

The primary focus in the *R. computatrix* models has been on how frogs and toads use vision to detect and catch prey in environments that include obstacles and barriers. Arbib has evolved a series of schema-based models which account for depth perception as an interaction between accommodation and binocular clues. At the lowest level, the schemas are plausibly based on known details of the relevant neurological data (for further details, see VISUOMOTOR COORDINATION IN FROGS AND TOADS).

One of the more striking results from this work, with reference to Marr's (1982) well-known theory of vision, is the indication that (in frogs and toads at least) there are different perceptual mechanisms for different visual stimuli (i.e., the depths to prey and to barriers are extracted from the optic array by different processing channels and are integrated in the sensorimotor pathways much later than Marr's theory might suggest).

The Computational Cockroach

Beer's (1990) book contains methodological arguments for CNE and details of experimental work on a computational cockroach, *Periplaneta computatrix*, which is a simulated hexapod agent embedded in an environment, inspired by neuroethological studies of the cockroach *Periplaneta americana*. The real cockroach uses chemotaxis as one of several strategies to locate food sources. If its path along an odor gradient is blocked by an obstacle, then it performs stereotyped edgefollowing behavior. The artificial cockroach is controlled by a heterogeneous neural network which was inspired by biological data and has been used to study issues in locomotion, guidance, and behavioral choice.

The primary external sensory input was simulated chemosensory information: patches of food in the environment gave off odor gradients detectable under an inverse-square law relating distance to odor intensity. The neural networks also received mechanosensory input from, for example, proprioceptors in the limbs and tactile sensors which signal the presence of food under the mouth. The simulation model included elementary kinematics: if the artificial cockroach failed to adopt a stable position for a sufficient length of time, it fell down.

Results from the simulation sessions demonstrated behavior in the model that was highly similar to behavior in the real animal, and Beer subsequently performed lesion experiments by selectively deleting connections or units from the *P. computatrix* control network. Again, the results from the artificial system were in agreement with the biological data.

P. computatrix was inspired by biological data, but it was not intended as a biological model. The various behaviors were generated by heterogeneous neural networks. The neuron model employed by Beer was more faithful to biology than many of the formal neurons used in connectionism: the units involved differential equations modeling membrane potentials, which gave his model neural assemblies a rich intrinsic dynamics (for further details, see LOCOMOTION, INVERTEBRATE).

The primary focus was on designing architectures for such units that could act as controllers for the various behaviors that *P. computatrix* should exhibit. Thus, there was no treatment of learning in the initial body of work on the cockroach. More recently, Beer reported on work which extended the original *P. computatrix* simulation model, testing it by allowing it to control walking in a real hexapod robot (Beer et al., 1992).

In the physical implementation, the control network was still simulated (i.e., the units in the neural network were not realized physically), but the sensorimotor connections to the artificial neural network were interfaced to physical sensors and actuators by means of analog-digital and digital-analog converters. Beer reported that in all cases, the response of the physical robot was similar to that observed in simulation. However, the implementation revealed one problem in the controller which had not been examined in the simulation. This problem (involving disturbances in the cross-body phasing of the legs) was easily rectified, but it showed that simulation models cannot be trusted as perfectly replicating any physical implementation.

The Computational Hoverfly

In studying issues in active vision gaze control with spatially variant foveal sampling, Cliff (1992) constructed a simulated embedded agent whose environment and optical system were inspired by studies of the hoverfly *Syritta pipiens*. The computational hoverfly (known as *Syritta computatrix*, or SyCo) was a simulated agent existing in its own virtual reality.

At the behavioral level, SyCo replicated the visually guided tracking movements made when male *S. pipiens* pursue conspecific flies in the hope of finding a mate. The network processing model was based on previously untested models proposed in the biology literature by Collett and Land (1975).

In the SyCo simulator, a dynamic three-dimensional world model of the relative positions and orientations of SyCo and a number of target flies were used to synthesize visual input with an accurate model of the optical anatomy of male *S. pipiens*. Within the simulator, the visual input was passed through parallel image-processing networks which effected crude targetidentification mechanisms (such as bug detectors). The results of this process were fed to tracking networks based on proposals by Collett and Land. The output of the tracking networks could alter the position and orientation of SyCo within its simulated world. The positions and orientations of the target flies could also vary dynamically, and the positions and orientations of the objects in the model were further varied by perturbations which model noise in effectors and crosswinds or turbulence in the air.

The simulation studies revealed opportunities for correction and extension of the prior models: simulating a proposed model enforces a degree of mechanistic rigor which is highly likely to expose any shortcomings or discrepancies in the model. For example, constructing the SyCo simulation required a more accurate characterization of the optical data and of the fly's interaction with its environment (Cliff, 1992) than was previously available in the literature. Furthermore, experience with simulating the original proposed model suggested alternative, more parsimonious models which could account for similar behavior.

Further experimentation with real animals would be required to establish which of the alternative models comes closer to the mechanisms actually operating in *S. pipiens*. Thus, while (as with Beer's work) SyCo was not intended as a biological model, the findings from SyCo may inform future studies of the real biological system. The need in CNE for highly coupled modeling, theoretical analysis, and biological experimentation is manifest.

A Robot with Compound Eyes

The work of Franceschini and colleagues (see, e.g., Franceschini, Pichon, and Blanes, 1992) can be considered applied CNE. Franceschini's research background is in the neurophysiology of vision in flies (for further details, see VISUOMOTOR COORDINATION IN FLIES). He and his students developed a visually guided autonomous mobile robot that heads for a goal location while simultaneously avoiding nearby obstacles, without the use of three-dimensional world models or explicit representations of the robot's surroundings. The visuomotor controller for the robot is based on a custom-built massively parallel analog asynchrous network designed according to principles elucidated in neurophysiological studies of fly vision. The robot uses visual motion information generated by this circuitry (which is essentially a one-dimensional, 360° horizontal compound eye) to slalom through a cluttered environment toward the goal at a speed of $50 \text{ cm} \cdot \text{s}^{-1}$.

However, the robot is not only an engineering endeavor; its development has helped to further the understanding of neural processing of visual information for the control of action in flies. In constructing the robotic system, Franceschini and colleagues were forced to address issues, such as the effects of using low-tolerance components, which led to further understanding of the details of the fly's visuomotor nervous system (Franceschini, personal communication, 1993).

Related Work

Other relevant work includes the work of Brooks's group at the Massachusetts Institute of Technology (MIT), who have constructed a number of robots which they refer to as *creatures*. Brooks has argued that the study of insect-level behaviors is more likely to reveal the fundamental mechanisms of cognition than is the study of human-level intelligent activities. There is insufficient space here to discuss this work in the detail it deserves (for further details and a review of how this work fits into the history of artificial intelligence and cognitive modeling, see Brooks, 1991). Typically, the creature-robots are autonomous agents which wander around office-style environments (namely, areas of the MIT Artificial Intelligence Laboratory). Brooks and his colleagues have demonstrated that relatively complex adaptive behaviors (such as autonomous navigation by map building) arise from agents whose control systems are organized as layers of *behavior-generating* modules (for discussion of a particular example, a hexapod walking robot, see Brooks, 1989). Typically, the control architectures for such agents are built from "combinatorial circuits plus a little timing circuitry" (Brooks, 1991). The use of combinatorial circuits does not preclude such work from being classed as CNE: Brooks (1991) uses the term to describe some of his own work (i.e., Brooks, 1989). Further details of such robotics research are given elsewhere in this volume (see REACTIVE ROBOTIC SYSTEMS).

Significantly, biologists Altman and Kien (1989) have identified strong similarities between Brooks's control-architecture principles and recent models of motor control proposed as underlying the generation of behaviors in a number of phylogenetically diverse animals. The similarities are in the rejection of traditional notions of linear hierarchical control of motor output, with the execution of behavioral outputs governed by a centralized command center (e.g., a command neuron). Instead, distributed heterarchical decentralized control systems with inputs and outputs at many levels have been proposed as better accounting for the interaction between sensory input, central pattern generation, and behavioral output, in locusts, cats, and frogs.

Discussion

CNE studies neural mechanisms which generate adaptive behaviors. Hence, it requires that agents are studied within the context of their environmental and behavioral niches.

From the above descriptions, some patterns emerge: all of the CNE projects mentioned are dependent on the availability of fairly detailed neuroethological data. Such data invariably come from invasive in vivo experimentation, and the neuroanatomy of lower animals, such as arthropods, is particularly amenable to such techniques. Certain neurons performing particular functions are readily locatable in different individual animals of the same species. There are many obstacles preventing the collection of such data from human subjects. Furthermore, by definition, any truly general principles underlying the neural generation of adaptive behaviors are common to a number of species, so only cross-species studies will help to identify general principles (Cliff, 1990:37).

Most of the CNE projects have largely eschewed the study of learning (plasticity), postponing this study until sufficient knowledge of the architecture of primary sensory-motor pathways is known to clearly understand how plasticity might increase the capacity for generation of adaptive behavior (see Beer, 1990:62); so far, the design approach has had much to offer. Nevertheless, it seems reasonable to expect a clean transi-

tion from the study of fixed network connectivities to variable connection strengths. Furthermore, the use of genetic algorithms can allow for the study of evolutionary learning in networks with nonplastic connectivities. Fixed-weight networks can be specified by genotypes; over a number of generations, the average behavior of a population of such networks may improve as a result of the effects of mutation and recombination in reproduction, if coupled with an appropriate selection pressure and fitness evaluation function (for further discussion of such issues, see "GENOTYPES" FOR NEURAL NETWORKS). Such techniques have been employed to develop useful CNE models (e.g., Stork, Jackson, and Walker, 1992).

Road Map: Motor Pattern Generators and Neuroethology
Related Reading: Command Neurons and Command Systems; Echolocation: Creating Computational Maps; Frog Wiping Reflexes; Locust Flight: Components and Mechanisms in the Motor; Motor Pattern Generation; Visuomotor Coordination in Salamanders

References

Altman, J. S., and Kien, J., 1989, New models for motor control, *Neural Computat.*, 1:173–183.

Arbib, M. A., 1972, *The Metaphorical Brain: An Introduction to Cybernetics as Artificial Intelligence and Brain Theory*, New York: Wiley-Interscience. ◆

Arbib, M. A., 1987, Levels of modelling of mechanisms of visually guided behavior, *Behav. Brain Sci.*, 10:407–465.

Beer, R. D., 1990, *Intelligence as Adaptive Behavior: An Experiment in Computational Neuroethology*, San Diego: Academic Press. ◆

Beer, R. D., Chiel, H. J., Quinn, R. D., Espenschied, K., and Larsson, P., 1992, A distributed neural network architecture for hexapod robot locomotion, *Neural Computat.*, 4:356–365.

Braitenberg, V., 1984, *Vehicles: Experiments in Synthetic Psychology*, Cambridge, MA: MIT Press/Bradford Books. ◆

Brooks, R. A., 1989, A robot that walks: Emergent behaviors from a carefully evolved network, *Neural Computat.*, 1:253–262.

Brooks, R. A., 1991, Intelligence without reason, in *Proceedings of the Twelfth International Joint Conference on Artificial Intelligence (IJCAI-91)*, San Mateo, CA: Morgan Kaufmann, pp. 139–159. ◆

Cliff, D., 1990, Computational neuroethology: A provisional manifesto, in *From Animals to Animats: Proceedings of the First International Conference on Simulation of Adaptive Behavior (SAB90)* (J.-A. Meyer and S. W. Wilson, Eds.), Cambridge, MA: MIT Press, pp. 29–39. ◆

Cliff, D., 1992, Neural networks for visual tracking in an artificial fly, in *Towards a Practice of Autonomous Systems: Proceedings of the First European Conference on Artificial Life (ECAL91)* (F. J. Varela and P. Bourgine, Eds.), Cambridge, MA: MIT Press/Bradford Books, pp. 78–87. ◆

Collett, T. S., and Land, M. F., 1975, Visual control of flight behavior in the hoverfly, *Syritta pipiens* L., *J. Comp. Physiol.*, 99:1–66.

Ewert, J.-P., 1987, Neuroethology of releasing mechanisms: Prey-catching in toads. *Behav. Brain Sci.*, 10:337–405.

Franceschini, N., Pichon, J.-M., and Blanes, C., 1992, From insect vision to robot vision, *Philos. Trans. R. Soc. Lond. Biol.*, 337(1281): 283–294.

Marr, D., 1982, *Vision: A Computational Investigation into the Human Representation and Processing of Information*, San Francisco: Freeman.

Stork, D., Jackson, B., and Walker, S., 1992, "Non-optimality" via pre-adaptation in simple neural systems, in *Artificial Life II* (C. Langton, C. Taylor, J. D. Farmer, and S. Rasmussen, Eds.), Redwood City, CA: Addison-Wesley, pp. 409–429.

Walter, W. G., 1953, *The Living Brain*, London: Duckworth; reprinted by Pelican/Penguin, 1961. ◆

Neuromodulation in Invertebrate Nervous Systems

Patsy S. Dickinson

Introduction

Because animals live in changing environments, behavior and hence the output of nervous systems must be flexible. This flexibility manifests itself in several ways that are important to modeling and experimental studies of neural networks. First, neurons are not all alike; they show a rich variety of conductances that endow them with different functional properties. Second, these properties and hence the collective activity of interacting groups of neurons are not fixed, but are subject to modulation that can change their characteristics and output. Modulation as a result of both locally released neuromodulators and more widely acting hormones differs qualitatively from the moment-to-moment integration of excitation and inhibition that a neuron receives. The effects of neuromodulators are long lasting, affecting neurons or circuits for seconds, minutes, or even longer.

Receptor sensitivity, neuronal membrane properties and synapses, neuromuscular junctions, and muscle properties are all subject to modulation. Together, these modulations allow the same groups of neurons to generate wide arrays of behaviors and to respond appropriately to a wide variety of sensory stimuli. Frequently, the same modulators act at multiple levels to influence or bias motor output. Because of their relative simplicity and accessibility, invertebrate nervous systems have provided the clearest examples of modulation and its importance to neuronal output.

Control of Modulator Release

One factor that plays an important role in determining the effects of modulators and therefore in determining the best ways to incorporate neuromodulation into network models is the nature and timing of release. Some neuromodulators are hormones; thus, they are spatially widespread and show relatively slow temporal changes. At the other end of the spectrum are modulators that are released locally at defined neuropilar locations and are rapidly broken down. Because of the diversity of release patterns, it may not be possible to use the same techniques in formulating models for all networks that are modulated. When modulation occurs on a time scale of minutes to hours, it may be possible to have two or more models for a given neuron, one corresponding to the control state and the other(s) corresponding to the modulated state(s). However, when modulation persists for only seconds, more dynamic changes to the equations may be necessary because the characteristics of the neurons or synapses change so rapidly.

Modulation of Sensory Systems

The sensory information that an animal needs depends on a number of factors, including its activity patterns and motivational state. Thus, the sensitivities of many sensory receptors can be modulated, as is seen for a stretch receptor, the oval organ, in crustaceans. This organ contains three sensory afferents and provides proprioceptive feedback to the gill ventilatory system. Proctolin increases the amplitude of the receptor potential and hence the number of action potentials produced in two of these afferents (Y, Z), whereas both octopamine and serotonin decrease these responses. Interestingly, the dendrites within the oval organ itself contain proctolin, suggesting that

receptor activity might modulate receptor sensitivity in that increased activity would induce greater proctolin release, thereby increasing receptor gain (Pasztor, 1989). The role of proctolin in modulating this sensory response is even more intriguing because proctolin is also found in motor terminals that innervate the muscles of the gill, where it enhances contractions of that muscle and because proctolin centrally modulates the ventilatory rhythm (Pasztor, 1989).

Modulation at Central Levels

Modulators can activate, terminate, or modify rhythmic pattern-generating networks. The detailed outputs of many rhythmic patterns are highly variable, with the frequency, phase relationships within the pattern, and number of neurons participating in the pattern subject to change. Additionally, many neurons or muscles participate in more than one behavior. Thus, when modeling networks, the mechanisms by which modulators sculpt specific patterns of activity from more generalized pattern generator substrates must be considered. Getting and Dekin (1985), who first described such networks as *polymorphic*, showed that the same network could be reconfigured to produce both escape swimming and reflexive withdrawal in the nudibranch mollusk *Tritonia*. The circuit that underlies these behaviors is shown in Figure 1*A*. Because the monosynaptic excitation between the three dorsal swim interneurons (DSI) is paralleled by inhibition mediated by the I (inhibitory) cell, one cannot predict the motor output from the circuit without knowing the relative strengths of synaptic connections, strengths that can be modulated. In the resting state, the inhibitory connection dominates, resulting in the functional circuit underlying reflexive withdrawal (Figure 1*B*). When the animal begins to swim, the C2 neuron is activated, thereby inhibiting the I cell (Figure 1*C*). The excitatory connections dominate, and the circuit drives the rhythmic motor output that underlies swimming (Getting and Dekin, 1985). Additionally, studies by Katz, Getting, and Frost (1994) showed that the DSI neurons themselves modulate the system, enhancing the synaptic strength of the synapses from C2 onto its followers in the network. Models of the *Tritonia* swim network do not yet include modulation.

The effects of a modulator on a specific neuronal network depend in part on the state of the system when the modulator is applied. Modulators themselves may affect this state, thereby altering the responses of the network to other inputs. In the lobster stomatogastric system, for example, the peptide proctolin by itself has no effect on the cardiac sac pattern. However, if proctolin is applied shortly after red pigment-concentrating hormone (RPCH) is applied, it elicits strong cardiac sac bursting (Dickinson, 1989). In formulating models of such systems, it should be remembered that the response of a system to one modulator may not always be identical, but instead may be a longer-term function of the modulatory history of the animal.

Alteration of Intrinsic Properties of Neurons

Many neurons have voltage-dependent conductances that allow them to generate rhythmic bursts of action potentials in the absence of synaptic input (see OSCILLATORY AND BURSTING PROPERTIES OF NEURONS for further details). However, many of these conductances are activated only in the presence of an ap-

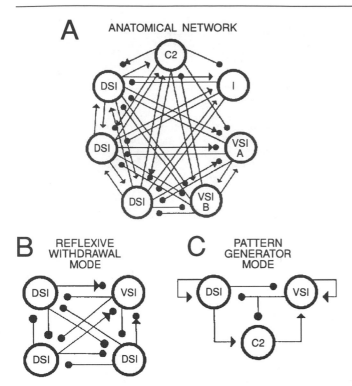

A ANATOMICAL NETWORK

B REFLEXIVE WITHDRAWAL MODE

C PATTERN GENERATOR MODE

Figure 1. The circuits underlying reflexive withdrawal and swimming in *Tritonia*. *A*, The basic circuit, showing the monosynaptic connections between the neurons of the network. Circles represent inhibitory synapses; triangles indicate excitatory synapses. When both symbols are present, the pathway is multicomponental. *B*, The functional circuit during reflexive withdrawal, when C2 is silent. *C*, The functional circuit during swimming, when C2 is active and dorsal swim interneurons (DSI) and ventricular swim interneurons (VSI) produce alternating bursts of impulses. (Reproduced, with permission, from Getting, P. A., 1989, Emerging principles governing the operation of neural networks, *Annual Review of Neuroscience*, Volume 12, p. 196. © 1989, by Annual Reviews Inc.)

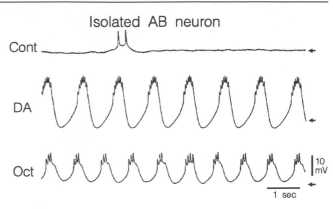

Isolated AB neuron

Cont

DA

Oct

Figure 2. Dopamine (DA) and octopamine (Oct) induce bursting in a synaptically isolated anterior burster (AB) neuron (of the lobster pyloric network). Each amine results in a different bursting pattern. Arrows indicate control membrane potential (-50 mV); "Cont" indicates the control. (Modified from Flamm, R. E., and Harris-Warrick, R. M., 1986, Aminergic modulation in the lobster stomatogastric ganglion, II: Target neurons of dopamine, octopamine and serotonin within the pyloric circuit, *J. Neurophysiol.*, 55:877. Reprinted with permission.)

propriate neuromodulator. For example, the anterior burster neuron (AB) in the lobster pyloric network does not oscillate when completely isolated from its network synaptic partners and modulatory inputs. However, the isolated neuron oscillates strongly when superfused with any of several neuromodulators (Figure 2; Flamm and Harris-Warrick, 1986). The characteristics of these oscillations (burst period, duration, amplitude) are different for each modulator. This diversity of responses results at least partly from the fact that bursting in this neuron can be driven by a number of different voltage-dependent currents. Each amine activates a different subset of these conductances, resulting in different bursting patterns (Harris-Warrick and Marder, 1991).

Modulators also alter the ability of neurons to generate plateau potentials. Neurons that can generate plateau potentials have two stable membrane potentials. At one, the cell is hyperpolarized and silent; at the other, it is depolarized and fires action potentials. Shifts between the two levels occur abruptly and regeneratively when the neuron's membrane potential crosses a threshold value in response to current injection or postsynaptic potentials. The abilities of neurons to generate plateau potentials can be enhanced or suppressed by modulatory inputs. For example, octopamine can induce plateau potentials as well as bursting in neurons that are involved in the

control of locust flight (Ramirez and Pearson, 1991). In the stomatogastric system, activity in the anterior pyloric modulator neuron enhances or suppresses the abilities of different neurons to generate plateau potentials. One effect of the changes in plateau capability is an altered sensitivity to synaptic input. When plateaus are generated, inputs strong enough to trigger the regenerative shift from one level to the other have greater effects than they would otherwise have, whereas those that are too weak to trigger the shift have little effect (Dickinson and Nagy, 1983). Once the threshold is reached, further increases in synaptic strength have little effect. In network computations, this characteristic effectively decreases the importance of synaptic strength (Marder, 1993). Consequently, the postsynaptic response becomes nonlinear. In addition, the effective duration of synaptic inputs is changed since, once shifted, the postsynaptic neuron remains at its new level; hence, the effect of inputs sufficient to induce a switch from one plateau level to another long outlasts the stimulus duration (Figure 3).

Changes in the abilities of neurons to generate plateau potentials can have far-reaching effects on network activity. When the plateau properties of a pyloric neuron, the ventricular dilator (VD), are suppressed, the VD no longer fires with the pyloric pattern. Instead, if the much slower cardiac sac pattern is active, the VD fires with this network (reviewed in Harris-Warrick and Marder, 1991).

The roles of different conductances in determining the firing patterns of nonoscillatory neurons have been examined in a model based on data from the lateral pyloric (LP) neuron of the stomatogastric system. The LP neuron is modulated by proctolin, and adding the proctolin current to the model alters both its activity and other membrane currents in ways that reflect the biological effects of proctolin. These models confirm that even small currents that produce minor changes in membrane potential may have profound effects (reviewed in Marder et al., 1993).

When modeling networks subject to neuromodulation, one must consider that, although a modulator may act on only a subset of neurons in the network, neurons that are not direct targets of the modulator can be affected and can affect the output of the system through their synapses within the network

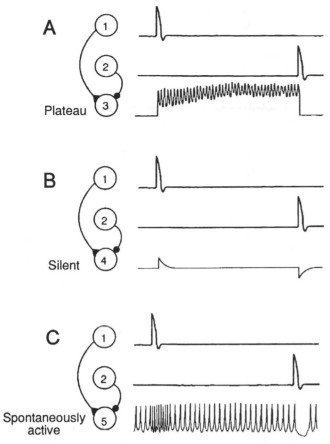

Figure 3. Plateau properties in a postsynaptic neuron increase the duration of its response to synaptic input. *A,* When the follower cell, 3, has plateau properties, an excitatory postsynaptic potential (from 1, triangle) triggers a shift to a depolarized plateau, and an inhibitory postsynaptic potential (from 2, circle) terminates the plateau. When the follower cell (4, 5) is silent (part *B*) or spontaneously active (part *C*), the effect of the same excitatory postsynaptic potential or inhibitory postsynaptic potential lasts for a short time after the input. [From Marder, E., 1993, Modulating membrane properties of neurons: Role in information processing, in *Exploring Brain Functions: Models in Neuroscience* (T. A. Poggio and D. A. Glaser, Eds.), Chichester, Eng.: Wiley, p. 30. Copyright 1993. Reprinted by permission of John Wiley & Sons, Ltd.]

(for further discussion, see CRUSTACEAN STOMATOGASTRIC SYSTEM).

Alteration of Synaptic Efficacy by Neuromodulators

The efficacy of both chemical and electrical synapses can be changed by the actions of neuromodulators. For chemical synapses, changes in the amount of transmitter released or in the responsiveness of postsynaptic neurons can contribute to modulation. In *Aplysia,* for example, changes in synaptic efficacy are largely responsible for the long-term changes that underlie learning (see INVERTEBRATE MODELS OF LEARNING: *APLYSIA* AND *HERMISSENDA*). In the stomatogastric system, for example, RPCH increases the efficacy of the synapses from a single presynaptic neuron onto its follower cells; this increase in synaptic efficacy is sufficient to cause a functional rewiring of the pattern generators and to provoke the generation of a novel rhythm (reviewed in Harris-Warrick and Marder, 1991).

Electrical synapses are likewise subject to modulation. Serotonin, octopamine, and dopamine alter electrical coupling in the lobster pyloric system, with coupling at some synapses increased while that at other synapses is decreased. A simple equivalent circuit model of these synapses enabled Johnson, Peck, and Harris-Warrick (1993) to test whether the experimentally recorded changes in input resistance are sufficient to account for the observed changes in electrical coupling. Their model suggests that, at some modulated synapses, the recorded changes in input resistance are insufficient, and a change in junctional resistance must also occur. Because the same modulators can change the efficacies of both chemical and electrical synapses between the same neurons, the effective sign of a synapse can be changed. Dopamine, for example, alters synaptic efficacy between several pairs of pyloric neurons that are connected by dual synapses. Under control conditions, the electrical component dominates, and the synapses are largely excitatory. In dopamine, however, the inhibitory chemical component dominates, and the synapses are largely inhibitory.

Modulation of Neuromuscular Junctions and Muscles

Neuromodulators in many systems exert effects on neuromuscular junctions or on muscles themselves. These effects are often consistent with the central or sensory effects of the same modulators. Modulators, which are released both from motor neurons and from exogenous sources, can change the amplitude, duration, or speed of muscle contraction. These effects result from changes at one or more of three levels: presynaptic effects resulting in altered transmitter release from motor terminals, electrical properties and excitability of the muscle fibers themselves, and excitation-contraction coupling (Calabrese, 1989).

Discussion

Modulation is prevalent in invertebrate as well as vertebrate nervous systems, and it occurs at all levels: sensory, central, and motor. It can affect the output of a system by altering the membrane properties of neurons, their synaptic interactions, and internal processes such as excitation-contraction coupling in muscles. The responses of a given system to a specific neuromodulator are not always the same and may depend on the state of the preparation when the modulator is applied or activated. Because many invertebrate systems are modulated by 10 or more substances, this variability has important consequences.

One implication of these studies for models of neuronal functioning is that neither the membrane properties nor the synaptic connectivity of neurons is static. Additionally, at least partly because of the dynamic nature and complexity of the systems, models may enable us to better understand both the functional roles and the mechanisms of modulation. Models can be used to test fundamental assumptions underlying neuronal and network function and modulation, such as whether specific changes in synaptic connectivity or membrane properties can account for the observed modulations. However, modulation is absent from almost all computer simulations, and thus far, techniques allowing modulation to be incorporated naturally into such simulations have not been developed. Given the importance of modulation in both invertebrate and vertebrate nervous systems, it is unlikely that models will completely fulfill their potential role as illuminators of nervous system function or as computing devices in their own right, until modulation is incorporated into them.

Acknowledgments. This work was supported in part by HFSP and NSF BNS–9310003.

Road Map: Biological Networks
Related Reading: Activity-Dependent Regulation of Neuronal Conductances; Ion Channels: Keys to Neuronal Specialization; Locust Flight: Components and Mechanisms in the Motor

References

Calabrese, R. L., 1989, Modulation of muscle and neuromuscular junctions in invertebrates, *Semin. Neurosci.*, 1:25–34. ◆

Dickinson, P. S., 1989, Modulation of simple motor patterns, *Semin. Neurosci.*, 1:15–24. ◆

Dickinson, P. S., and Nagy, F., 1983, Control of a central pattern generator by an identified modulatory interneurone in Crustacea, II: Induction and modification of plateau properties in pyloric neurones, *J. Exp. Biol.*, 105:59–82.

Flamm, R. E., and Harris-Warrick, R. M., 1986, Aminergic modulation in the lobster stomatogastric ganglion, II: Target neurons of dopamine, octopamine and serotonin within the pyloric circuit, *J. Neurophysiol.*, 55:866–881.

Getting, P. A., and Dekin, M. S., 1985, *Tritonia* swimming: A model

system for integration within rhythmic motor systems, in *Model Neural Networks and Behavior* (A. I. Selverston, Ed.), New York: Plenum, pp. 3–20. ◆

Harris-Warrick, R. M., and Marder, E., 1991, Modulation of neural networks for behavior, *Annu. Rev. Neurosci.*, 14:39–57. ◆

Johnson, B. R., Peck, J. H., and Harris-Warrick, R. M., 1993, Amine modulation of electrical coupling in the pyloric network of the lobster stomatogastric ganglion, *J. Comp. Physiol. [A]*, 172:715–732.

Katz, P. S., Getting, P. A., and Frost, W. N., 1994, Dynamic neuromodulation of synaptic strength intrinsic to a central pattern generator circuit, *Nature*, 367:729–731.

Marder, E., 1993, Modulating membrane properties of neurons: Role in information processing, in *Exploring Brain Functions: Models in Neuroscience* (T. A. Poggio and D. A. Glaser, Eds.), Chichester, Eng.: Wiley, pp. 27–42. ◆

Marder, E., Abbott, L. F., Bucholtz, F., Epstein, I. R., Golowasch, J., Hooper, S. L., and Kepler, T. B., 1993, Physiological insights from cellular and network models of the stomatogastric nervous system of lobsters and crabs, *Am. Zool.*, 33:29–39. ◆

Pasztor, V. M., 1989, Modulation of sensitivity in invertebrate sensory receptors, *Semin. Neurosci.*, 1:5–14. ◆

Ramirez, J. M., and Pearson, K. G., 1991, Octopamine induces bursting and plateau potentials in insect neurones, *Brain Res.*, 549:332–337.

Neurosimulators

Jacob M.J. Murre

Introduction

Since 1985, more than a hundred research groups have developed some form of simulator for neural networks. This article briefly reviews a selection of about 40 of these. A *neurosimulator* could be described as a software package created for the specific purpose of reducing the time and effort involved in solving a problem using neural networks. Apart from saving time, it can also increase the reliability of the simulations if these are based on standard neural network paradigms (e.g., variants of backpropagation). As we shall discuss below, when a completely new paradigm is developed, the advantages of using one of the existing neurosimulators are less clear.

A neural network simulation may be concerned with anything ranging from subcellular structures such as channels, synapses, etc., to extensive connectionist systems. Even though the range of problems is extremely diverse, most simulations share a common structure. The following sequence of subtasks, or phases, occurs in many simulation problems:

1. *Translation* of the problem into a neural network representation: This always involves deciding upon data representation, data presentation scheme, and model architecture (including network dynamics and parameter specification). When working with adaptive models, it may also include the selection of a set of training data that adequately represents the domain of the problem. If this set is too restricted, the learned behavior of the network will not generalize well.

2. *Testing* the validity of the decisions made in Phase 1: Are the data processed in the intended manner? Is the network architecture well constructed? Does the model behave roughly as expected?

3. The actual *running* of the simulations: This may also include lesioning of certain pathways, parameter tuning, and recording of various network outputs. Depending on the size of

the network, the complexity of the algorithms, and on the size of the training data (in the case of adaptive models), this may take up to several days or longer. Unsatisfactory results may take the user back to Phase 2.

4. *Analyzing* the results: How has the network achieved the results? How well does the system generalize? How closely does it approach the experimental data? How sensitive is the system to perturbations?

5. *Extending* the neural network paradigm used: This phase occurs frequently, especially if a completely new problem is tackled. Special-purpose extensions to the paradigm may be necessary, or a completely new paradigm may have to be developed. After Phase 5, the user must again return to Phase 2.

6. *Incorporating* the completed model in applications, either as a stand-alone program (e.g., for demonstrations) or as part of a larger system (e.g., a pattern recognizer as part of a manufacturing system): Even pure research cannot always escape this phase—for example, when one decides to interface a neural network with an expert system or with a genetic algorithm.

The field of neurosimulators is moving very fast. In the past couple of years, I have assembled references to about 100 neurosimulators. It is impossible to discuss a significant portion of these systems in any detail. Thus, the overview presented below is not intended to be a complete consumer's guide to current neurosimulators, but to discuss important issues in the choice and design of neurosimulators, and to give some hints about the choice of simulator. The paper is organized into four main sections: neurosimulators for novices; from academia; for single-neuron studies; and for business and industry. This organization represents the practical approach of this review: systems in the first two groups are cheap or free, whereas systems in the latter groups are more expensive. Single-neuron simulators have been treated separately, as most of them are special-purpose simulators. The review concludes with a discussion of future developments in neurosimulators.

In the lists and tables throughout this article, the following abbreviations are used to denote price categories:

n.c.   Free or nominal charge (+C means source code in C or C++ is included)
$ $20–200
$$ $200–2000
$$$ $2000–10,000

The lists and tables also generally include the name of the simulator; the developer or distributor; an e-mail address or phone/fax number; major hardware platforms supported; and an indication of the paradigms supported. For paradigms, *several* means that between two and seven paradigms are supported, and *many* means that more than seven are supported.

Neurosimulators for Novices

The best way to develop a feeling for neural networks is by running them on a computer and watching the parameters evolve on the screen. Neurosimulators are an excellent way to explore the many different paradigms available, such as backpropagation, Hopfield networks, or Kohonen networks (self-organizing maps). Most of the packages mentioned in the section on business and industry are well suited for novices. Some low-priced introductory packages have been developed to be used in conjunction with a textbook on neural networks. Two of the more interesting ones are:

• *Simulator CORTEX* is best used in conjunction with the book by I. Alexander and H. Morton (1990), *An Introduction to Neural Computing*, New York: Chapman and Hall. The simulator has to be bought separately and can also be used on its own; several paradigms; $. A new and more sophisticated version, *Cortex III*, is currently being developed by Michael Reiss of Unistat in the United Kingdom; tel.: +44 81-9641130; PC platform; many paradigms; $$$.

• The *PDP Simulator* is included in the book by J. L. McClelland and D. E. Rumelhart (1988), *Explorations in Parallel Distributed Processing*, Cambridge, MA: MIT Press. Mac, PC, and Unix platforms; several paradigms; $. A new, more general version of this system is being developed by McClelland's group at Carnegie Mellon University. The new *PDP Simulator* will be based on object-oriented design principles (C++).

Often low-priced packages are merely demonstration programs, but ones like the *PDP Simulator* are sufficiently powerful and easy to use to allow for more complicated simulations to be carried out. Users more experienced in computer programming and simulation might also try out some of the systems mentioned in the next section, especially those that have a well-developed graphical interface.

Neurosimulators Available from Academia

It seems that by now every university-based neural network group has developed its own neurosimulator. On the one hand, this is an advantageous situation since almost all of these systems are available for a nominal fee. On the other hand, it clearly illustrates that no particular simulator addresses the needs of significant sectors of the research community. It seems urgently necessary to somehow integrate the combined efforts of all these groups.

Possible drawbacks of using simulators from academia are that most of the neurosimulators developed there are suited for users who already have experience with simulation. Documentation of these systems may—but need not—be less extensive than of the more expensive commercial packages. The number of supported computer platforms is often limited (usually Unix) and user-interaction and user-support are usually much less developed than for commercial systems. On the positive side, they are likely to include the most recently developed methods in neural network research, source code is often available, and they are usually free.

Examples are (Table 1): *GRADSIM* (many backpropagation variants), *Mactivation* (excellent graphics), *MUME* (supports several computer platforms), *SNNS* (supports many neural

Table 1. Neurosimulators from Academia

Simulator	Developer or Distributor	E-mail Address or Fax	Hardware Platforms Supported	Paradigms Supported	Price
GRADSIM	University of Toronto	watrous@ai.toronto.edu	Sun, VAX, others	many	n.c., +C
Mactivation	University of Colorado at Boulder	mikek@boulder.colorado.edu	Mac	several	n.c.
MIRRORS/II	University of Maryland	mirrors@cs.umd.edu	Unix		n.c.
MUME 6.0	Sydney University, Australia	fax: +61 2/660-1228	Unix/Ultrix, PC, others	many	n.c., +C
NeuralShell	Ohio State University	sca@dopey.eng.ohio-state.edu	Unix	many	n.c., +C
NSL	Alfredo Weitzenfeld, University of Southern California	alfredo@usc.edu	Unix		n.c., +C
PlaNet	Yoshiro Miyata, University of Colorado at Boulder	miyata@boulder.colorado.edu	Unix/X-Windows/SUN Tools	many	n.c.
RCS (Rochester Connectionist Simulator)	University of Rochester	cs.rochester.edu	Unix, Mac		n.c.
SNNS (Stuttgart Neural Network Simulator)	University of Stuttgart, Germany	zell@informatik.uni-stuttgart.de	Unix/Ultrix/X-Windows	many	n.c., +C
XERION	Drew van Camp, University of Toronto	drew@cs.toronto.edu ftp: ftp.cs.toronto.edu (cd pub/xerion)	Unix/X-Windows	many	n.c., +C

See the article Introduction for an explanation of terms and abbreviations used for paradigms and price.

Table 2. Software Libraries and Code Generators from Academia

Library or Code Generator	Developer or Distributor	E-mail Address	Hardware Platforms Supported	Paradigms Supported	Price
Aspirin/ MIGRAINS	MITRE Corporation	connectionists-request@cs.cmu.edu ftp: pt.cs.cmu.edu (cd /afs/cs/project/ connect/code)	Unix/X-Windows; many others including Convex and Cray	many	n.c., +C
CALMLIB (Library for Categorizing And Learning Module)	Jacob Murre, Medical Research Council, Cambridge, UK	jaap.murre@mrc-apu.cam.ac.uk		CALM	n.c., +C
CasCorl (Cascade Correlation Simulator in Lisp and C) and *Quickprop*	Scott Fahlman	fahlman@cs.cmu.edu			n.c., +C/Lisp
SOM_PAK and *LVQ_PAK*	Teuvo Kohonen's group, Helsinki University of Technology	som@cochlea.hut.fi for SOM_PAK; lvq@cochlea.hut.fi for LVQ_PAK		self-organizing map, (SOM_PAK); learning vector quantization (LVQ_PAK)	n.c., +C
tlearn	University of California at San Diego	elman@ucsd.edu	Unix, PC, Mac	backpropagation and variants	n.c., +C

See the article Introduction for an explanation of terms and abbreviations used for paradigms and price.

network paradigms, including ART1 and ART2), *PlaNet*, *NeuralShell*, *RCS*, and *XERION*. Some groups base most user-interaction of their simulator on a neural network simulation language—for example, *MIRRORS/II* and *NSL*—although these systems also include graphical interaction. *NSL* is a system based on object-oriented programming techniques (see NSL: NEURAL SIMULATION LANGUAGE). The development environment includes tools for analyzing network behavior. It also allows importing of C-code to extend the simulator's functionality.

This is only a small selection of the available systems that seem to be most widely used at the moment. Several promising neurosimulators are currently under development (also see below), so that it is worthwhile to ask around whether any new systems are being made available.

For the experienced programmer, a viable option is to obtain one of the available neural network libraries (Table 2). The programs are usually written in portable C source code (see for example Blum, 1992, and Rao and Rao, 1993, for introductions to programming neural networks in C++). These libraries differ from the packages mentioned above in that the user is expected to develop most of the functionality of the simulation by writing additional source code. The advantage of this approach is that it is easy to alter and extend the algorithms. The disadvantage is usually that despite support of the libraries, programming can still be time-consuming. Several of the more extensive packages, however, can also be used by writing a program in some higher-level neural network language.

The MITRE Corporation, a nonprofit organization funded by the U.S. federal government, makes available a simulator called *Aspirin/MIGRAINES*. The main approach of their system is to read a network description file and to generate C code, which can then be compiled and run. Other systems provide network generators as well, but this system focuses exclusively on this method. The network specification language is called *Aspirin*. Another part of the system (called *MIGRAINES*) converts data into formats readable by existing visualization packages (e.g., Gnuplot). A drawback of using libraries or

code-generators is that the code has to be recompiled and linked every time a significant change in the network is made.

Single-Neuron Modeling

About ten neurosimulators have been developed for single-neuron modeling. These allow exploration of single neurons and of small networks. Real neurons are much more complex than the single-parameter nodes employed in most artificial neural networks (i.e., using a single activation value per node). In a real neuron, electrical activity spreads gradually through the dendritic branches. Its computational characteristics depend on the structure and electrochemical properties of these branches. To model this in biological simulators, a neuron is divided into a number of compartments. At all positions inside a single compartment, the activity is assumed to be equal. A complete neuron is built up out of many such isopotential compartments, like a stick-figure.

Table 3 lists seven single-neuron simulators, which are all based on such compartmental modeling. For a more thorough review of these simulators, the reader is referred to De Schutter (1992; also see Miller, 1990). In that review, the reader will find information about what aspects of the biology can be handled by each simulator (ion channels, synapse plasticity, etc.), as well as many other details. I shall here confine myself to mentioning a few general aspects.

A widely used neurosimulator is *NEURON*. This simulator seems to be especially favored by those who combine experimentation and modeling of biologically realistic networks of neurons. *SABER* and *SPICE* are commercial simulators that can also model other dynamical systems (*SPICE* is also available in a basic, noncommercial version). The user-interface of simulators *AXONTREE*, *GENESIS*, *NEMOSYS*, *NEURON*, and *NODUS* is supported by graphics or dialogs; the others rely on interaction through scripts or file input. *GENESIS* and *NEMOSYS* seem to offer the most flexible interaction in the testing phase of the simulation process. *GENESIS* has limited support for simulations with large neural networks as well. *NEURON* and *GENESIS* also support a parallel hardware

Table 3. Single-Neuron Simulators

Simulator	Developer or Distributor	E-mail Address or Phone	Hardware Platforms Supported	Price
AXONTREE	Y. Manor, Hebrew University, Jerusalem	idan@hujivms.bitnet	Unix/X-Windows	n.c., +C
GENESIS	Jim Bower, California Institute of Technology	genesis@caltech.bitnet	Unix/X-Windows	n.c., +C
NEMOSYS	University of California, Berkeley	eeckman@mozart.llnl.gov	Unix/X-Windows	n.c., +C
NEURON	Duke University Medical Center	ftp: neuron.neuro.duke.edu	Unix/X-Windows, PC (MS-Windows under development), Cray, others	n.c., +C
NODUS	Erik De Schutter, California Institute of Technology	erik@cns.caltech.edu	Mac	n.c.
SABER	Analogy Inc., USA	tel: +1 503/626-9700	Unix, VMS, PC, others	$$$
SPICE	University of California, Berkeley	tel: +1 510/643-6687	VMS, Unix, Mac, PC	n.c., +C

Table 4. Commercial Neurosimulators

Simulator	Developer or Distributor	E-mail Address or Phone/Fax	Hardware Platforms Supported	Paradigms Supported	Price
ANNE	Casey Bahr, George Mason University	king@gmuvax2.gmu.edu		Hypercube I and II, backpropagation, others; part of Hypercube's User's Library	
ANSim	SAIC, USA	tel: +1 619/546-6290	PC, others		$$
Brainmaker	California Scientific Software, USA	tel: +1 800/284-8112			$$
Cognitron	Cognitive Software, USA	tel: +1 317/577-4158	Mac	several	$$
DESIRE/NEUNET	Granino Korn, Industrial Consultants, USA	tel: +1 509/687-3390	PC, Sun		$$
MacBrain	Neurix, USA	tel: +1 617/577-1202	Mac	many	$$
Mimenice	Mimetics, France	tel: +33 1-40910990	Sun/X-Windows		$$$
Nestor NDS	Nestor, USA	tel: +1 401/331-9640	PC		$$
NEURAL DESK	Neural Computer Sciences, UK	tel: +44 703-667775	PC/Windows	several	$$
NeuralWorks Professional II/Plus	NeuralWare, USA	tel: +1 412/741-5959	PC, Mac, Sun, others	many algorithms	$$$
NeuroSoft	HNC, USA	tel: +1 619/546-8877	PC, SUN	many	$$
NeuroWindows	Ward Systems Group Inc., USA	fax: +1 301/662-5666	PC/Windows	several	$$
OWL	HyperLogic Corporation, USA	tel: +1 619/746-2765	[C library]	many	$$
Plexi	Lucid, USA	fax: +1 415/329-8480	Symbolics, SUN	several	$$
SN2.8	Neuristique, France	xd@bop.neuristique.fr	Unix	many	$$$

platform. Because running networks of many biological neurons takes a very large amount of computer time, parallel hardware is no luxury for this task. *SABER* and *SPICE* are also able to run networks of several neurons but lack parallel support. Important for possible extensions is whether source code is available, which is the case for *AXONTREE*, *GENESIS*, *NEMOSYS*, *NEURON*, and *SPICE*.

Business and Industry

A large proportion of the problems in business and industry fall in a limited number of categories, such as credit assessment, signal analysis, time-series prediction, pattern recognition, and process control. Because the solutions can often be based on standard neural network technology, neurosimulators may be particularly useful for these applications. An overview of simulators suitable for modeling investing and trading in the markets can be found in Caldwell (1993). Most commercial packages have excellent documentation. Special training courses are often available, and companies that market these systems can usually assist in every stage of the solution process. Not surprisingly, user-interaction of commercial systems tends to be better geared to the needs of someone working in business or industry. *NEURAL DESK*, for example, allows a neural network to be run from within a spreadsheet or other software package (e.g., Excel or Superbase), without having to leave the current program. Many commercial systems offer predefined example problems that can be amended to the problem at hand. Several of the better-known commercial neurosimulators are listed in Table 4.

NeuralWorks Professional II/Plus is a widely used system that offers a range of built-in network paradigms, display modes, and monitor options. It also includes possibilities for exchange with other popular programs (database and spreadsheets). New paradigms can be added using a special neural dynamics language or by dynamically linking user-supplied routines in C. With a separate utility, the *Designer Pack*, networks can be converted into C source code. Systems comparable to *NeuralWorks Professional II/Plus* are HNC *NeuroSoft* and SAIC *ANSim*. To extend the applicability of *NeuroSoft*, HNC sells a network description language called *AXON*. SAIC provides *ANSpec*, a network specification language based on concepts from concurrency. Though the latter utility allows smooth implementation on parallel machines (see discussion below), it may be difficult for someone unfamiliar with parallel processing.

Some simulators include a network specification language and a compiler. *NEUNET* is one example of this approach. It is derived from a package for dynamical systems called *DESIRE* (Direct Executing SImulation in REal time). A neural network simulation is programmed in a BASIC-like language with support for operations with arrays, matrices, and complex numbers. Like BASIC, the code is interpreted during the testing phase. For the running phase, a special run-time compiler is called to generate efficient machine code. The result of this is that execution can be fast, while extensibility is fully guaranteed. The disadvantage is that the user has to do extensive programming to make a simulation work. Built-in support for monitoring the network during a simulation is limited. Several other systems include higher network languages. Mimetics' *Mimenice* is a modular system based on a higher-level, C-like language called *G*. Mimetics also offers a utility to generate C-code. *Mimenice* focuses on just three paradigms: backpropagation, learning vector quantization, and topological maps. The philosophy behind this is that these networks cover a large range of useful applications, and users can always write their own learning schemes.

Three other companies share Mimetics' philosophy that only a few algorithms are useful for industrial applications. Nestor is one of the oldest neural network companies. It has been active in developing applications for industry since 1983. Their simulator is called the *Nestor NDS* (Nestor Development System). It allows the user to build networks using the learning algorithm developed by the founders of Nestor (this algorithm is protected by patents). Another single-paradigm simulator is *Brainmaker*, which limits the user to backpropagation. *NEURAL DESK* is a simulator that consists of several separate tools that allow easy interaction with existing standard software such as spreadsheets. The number of neural network paradigms is limited to four.

The interfaces of the above systems are mostly based on dialogs and edit-compile cycles. *MacBrain* and *Plexi* are two systems that offer excellent graphics-based direct manipulation of networks. The drawback of the extensive graphics is that these are implemented in such a way that the maximum network size is limited. The advantage is that direct-manipulation of network objects greatly improves user-interaction when specifying network architectures. *Plexi*'s range of paradigms can be extended by programming and recompiling (incrementally) using *Symbolic*'s object-oriented development environment. *MacBrain* can be extended a variety of ways, including importing external C routines. Neuristique's *SN2.8* also offers excellent graphical and statistical facilities. Its Lisp-based code can be extended or altered for high-level modeling. *Cognitron* is a system developed primarily for Macintosh computers, but with less emphasis on graphical interaction than *Plexi* or *MacBrain*. Like *SN2.8*, new activation and learning rules can be added by programming in Lisp. The system is also able to export Lisp code. *Cognitron* offers support for a variety of parallel machines, including transputers. In order to handle paradigms that require sequenced processing of layers like backpropagation, a special phasing option has been included that allows the user to specify the order in which layers must be processed.

ANNE is a simulator/library developed for Intel's Hypercube. It has now become part of the User Libraries. The simulator mainly focuses on backpropagation, but it can be extended to other paradigms. *OWL* is a good example of a commercially available library (for C), as is *NeuroWindows*. The latter can be linked dynamically under MS-Windows. It mainly provides a neural network engine (i.e., a set of calculation routines) for a limited number of paradigms.

Discussion

The tools available for neural network simulations still fall short of the standards achieved in more mature fields such as statistics or computer-aided design. The neurosimulators mentioned above are most useful for prototyping small to midsized standard applications. Especially, novice users and those interested in applying neural network technology to practical problems in business and industry can benefit significantly from using such a system. This was already concluded by Cohen (1989) and little has changed in the meantime. For the connectionist researcher, the tools discussed so far still fall short of what is required. There are about 300 researchers in European universities listed in the DEANNA database (a project of the European-Community-funded ESPRIT program). Only 34 researchers say they ever use a neurosimulator tool, and among these, many actually use a tool developed by their own group.

Why is it that more than 85% of these neural network researchers still write their own simulation software? The main problem is that designing a truly functional simulator is very hard. For research purposes, a neurosimulator requires at the very least 1. a highly developed interface; 2. a scalable design (i.e., through parallel hardware); and 3. extensibility with new neural network paradigms. An excellent interface is necessary to support simulation in the problem definition and testing phases. Scalability means that the system is able to handle even very large networks. The support of suitable parallel hardware will be necessary to achieve this. Extensibility implies that it is easy both to alter the paradigms provided with a system and to develop completely new paradigms. Unfortunately, the requirement of extensibility clashes with both the scalability requirement and with the interface requirement. It appears to be very difficult to design a system with a sophisticated interface in which new paradigms can be "implanted" without either major programming efforts (possibly involving recompiling the full source code of the package) or loss of full interface support (e.g., newly defined paradigms appear on the screen as white boxes and it is impossible to inspect what is going on inside them). Similarly, support of parallel hardware usually requires *explicit* programming of parallel routines for each neural network defined, which can result in parallel programs that are a nightmare to debug. It has so far proven to be impossible to carry out automatic parallelization of computer algorithms in general. The domain of neural networks, however, may be sufficiently constrained to make such an undertaking feasible. Moreover, neural networks may themselves be used as techniques to *optimize* the mapping. The resulting process would be a type of neural bootstrap. These new developments in neurosimulators will become available in the near future:

- *GALATEA*; Mimetics, Paris, France; tel: +33 1-40910990; Sun and dedicated hardware; many paradigms.
- *MetaNet*; Jaap Murre, Medical Research Council, Cambridge, UK; jaap.murre@mrc-apu.cam.ac.uk; PC platform; many paradigms.
- *SESAME*; Alexander Linden, GMD, Germany; sesame-request@gmd.de, ftp: ftp.gmd.de (cd gmd/as/sesame); Unix/ X-Windows platform; many paradigms. (An experimental version is already available.)

One of these systems, *GALATEA*, aims to solve the above three problems. Under development at Mimetics (also supported by ESPRIT), it combines a neural network environment with a higher-level language called *N* (see Marcadé, 1992; *N* is an extension of C++). With a "system architecture builder," the user can define new networks. Newly defined networks are represented in such a way that they can be mapped onto a variety of parallel hardware. The system will also include a silicon compiler for the generation of specifications for application-specific integrated circuits (ASICs). A similar approach to parallelism and extensibility is taken in the *MetaNet* system. *MetaNet* strongly emphasizes direct manipulation of neural network objects (Murre, 1992, appendix B5). Different users will have different needs for interaction with their networks. These specific needs are served by POINTs (problem-oriented interaction tools). Not only can a user define a new network algorithm, but also the way in which it appears and behaves on the screen. This allows each user to tailor the system to specific tasks (i.e., using different POINTs of view) and to manage the level of detail of the displays.

Both *GALATEA* and *MetaNet* are based on a building-block philosophy, also called *modularity*. This approach fits the design intuitions of most network developers. It also leads to well-organized systems that promote reuse of existing modules. Modularity at the interface level is usually mirrored by an object-oriented approach at the formal, internal level of a simulator. A system that, like *NSL*, takes a fully object-oriented approach to simulation is *SESAME*, which is under development at the German National Research Center for Computer Science. *SESAME*'s design is more akin to that of a general simulator. Building blocks can be defined in different classes: mathematical, utility (e.g., I/O functions), and graphical. An interesting feature of the system is that it allows autoconfiguration of parameters via a constraint satisfaction process (Linden et al., 1993).

An example of a general neurosimulator is *Simulink*, which is part of *MATLAB*:

- *Simulink/MATLAB/Neural Network Toolbox*; Math Works, USA; fax: +1 508/653-2997, info@mathworks.com; many hardware platforms, including Mac, PC, Unix, Convex, and Cray; $$.

A small *Neural Network Toolbox* is available for this system. In general simulators, a disadvantage is that the user may have to spend some efforts encoding the specific neural network algorithms used. But this drawback is usually outweighed by the excellent simulation support a general simulator may offer. Especially in cases where a neural network is tested on some simulated control tasks, such as a simulated robot arm (i.e., as opposed to a real arm), a general simulator may offer advantages over a specialized neural network simulator because the control dynamics of the arm can be programmed in the same system.

Conclusion

More efforts need to be undertaken to develop simulation systems that fulfill requirements of the research community, such as flexible user interaction, scalability, and extensibility. If a truly functional system were available at an affordable price, neural network research could be greatly enhanced worldwide. Not only could simulations be realized faster and in a more reliable manner, a great advantage would be that such a system would enforce a standard for neural network simulations. This in turn would enable a much more efficient spread of knowledge through exchange of models and simulations in a standard format. Computational modeling has become very prominent recently. Its continued success will depend on the development of the right tools to build and investigate models.

Acknowledgment. This work is supported by the Medical Research Council.

Road Map: Implementation of Neural Networks
Related Reading: Digital VLSI for Neural Networks; Multiprocessor Simulation of Neural Networks; Programmable Neurocomputing Systems

References

Blum, A., 1992, *Neural Networks in C++: An Object-Oriented Framework for Building Connectionist Systems*, New York: Wiley. ◆

Caldwell, R., 1993, Selecting the right neural network tool, *Neurove$t J.*, 1(2):21–24. (Address: P.O. Box 764, Haymarket, VA 22069, USA; e-mail: rbcaldwell@delphi.com.) ◆

Cohen, H., 1989, How useful are current neural network software tools? *Neural Netw. Rev.*, 3:102–113.

De Schutter, E., 1992, A consumer guide to neuronal modeling software, *Trends Neurosci.*, 15:462–464. ◆

Linden, A., Sudbrak, T., Tietz, C., and Weber, F., 1993, An object-oriented framework for the simulation of neural nets, in *Advances in Neural Information Processing Systems 5* (C. L. Giles, S. J. Hanson, and J. D. Cowan, Eds.), San Mateo, CA: Morgan Kaufmann.

Marcadé, E., 1992, *ESPRIT II Project 5293—GALATEA User Manual* (draft version), Paris: Mimetics (fax: +33 1-40919055).

Miller, J. P., 1990, Computer modelling at the single-neuron level, *Nature*, 347:783–784.

Murre, J. M. J., 1992, *Learning and Categorization in Modular Neural Networks*, Hemel Hempstead, Eng.: Harvester Wheatsheaf, and Hillsdale, NJ: Erlbaum.

Rao, V. B., and Rao, H. V., 1993, *C++, Neural Networks, and Fuzzy Logic*, New York: MIS Press. ◆

Neurosmithing: Improving Neural Network Learning

Russell Reed and Robert J. Marks II

Introduction

The goal of supervised training is to make the system output equal to a desired target function for any input. The standard approach is to (1) define an error function measuring the difference between the target and actual output functions, (2) determine how changes in parameters (network weights) affect the error, and (3) adjust parameters in a way that reduces the error. The backpropagation algorithm is the most commonly used technique for training multilayer perceptrons. Typically, the error function is the sum of squared differences between the desired targets $t(x_k)$ and the actual network outputs $y(x_k)$ summed over all training patterns k,

$$E(w) = \sum_k (t(x_k) - y(x_k))^2$$

Because the network output is a function of the weights, E is a function of w. If it could be plotted as a function of w, E might look like a rough landscape with hills and valleys, high where E is high and low where E is low. Backpropagation, as an approximation to gradient descent, could then be viewed as placing a marble at some random point on the landscape and letting it roll to the lowest point. The core of the algorithm is a repeated loop in which (1) the derivative chain rule is applied to determine how weight changes affect the error and (2) the weights are adjusted by small increments in the direction that reduces the error. In "batch" mode, every training pattern is considered before each weight change, and the algorithm approximates gradient descent when the step size is small enough. In "on-line" mode, a random subset of patterns (usually just one) are considered before each weight change. When the step size is small enough, this approximates stochastic gradient descent since the accumulated weight changes tend to average to the true (negative) gradient. Since weight updates are much more frequent, however, the error may decrease faster when the training data are highly redundant. An added benefit is that the randomness of the individual weight changes may help jostle the "marble" out of small "potholes" and thus help prevent convergence to shallow local minima.

In spite of its apparent simplicity, the algorithm has proven remarkably effective, and there are many examples of networks trained to implement relatively complex functions. This is not to say that difficulties never occur, however. Backpropagation training is often very time consuming, for example, and may converge to suboptimal solutions. In the following, some practical techniques are described that may be helpful to accelerate learning and avoid potential problems. Some of the remarks are very basic and may be viewed as a checklist of standard procedures. Others are more specific. Many of the remarks apply to any learning system, but unless otherwise stated, the focus is on supervised learning in feedforward networks such as multilayered sigmoid perceptrons.

Data Preparation

Neural networks are often trained from examples of a desired input-output relationship. Aside from possible constraints built into the architecture or training algorithm, the examples are the only information provided about the target function, so it is important that they adequately describe the function.

Distribution of the data. In general, larger data sets are desirable from the standpoint of statistical accuracy since sparse data may contain spurious correlations and miss significant features of the function. Since the data distribution provides information about the relative importance of different regions of the function, it should generally match the distribution of patterns that will occur in normal operation.

Redundant and irrelevant information. Conversely, since neural networks are often applied to tasks where little is known about the appropriate choice of variables and their relationship to the target function (indeed, other techniques might be used if more were known), there is a temptation to provide as much information as possible and let the network sort it out. This might be feasible when data are abundant and training times unimportant, but it may lead to poor generalization otherwise.

As a rule, any external knowledge about which variables are important and how they relate to the target function should be used to reduce the amount of irrelevant information presented to the network. Although (ideally) the system should learn to ignore redundant and irrelevant inputs, these make its task harder and, when training sets are small, there may not be enough information to demonstrate that extra inputs are actually irrelevant. If the input dimension exceeds the sample size, for example, the data can be fitted exactly by a linear equation which will probably generalize poorly.

Dimensionality-reducing preprocessing, such as principal components analysis, is often used to avoid this problem. An alternative is to place a bottleneck (narrow hidden layer) in the network structure, thereby forcing the system to eliminate redundancies. Since the representation formed at a bottleneck is related to the principal components, weight initialization from principal components information has been suggested (Georgiou and Koutsougeras, 1992).

A case in which redundant input variables may be desirable is when the data are noisy (but abundant), since they can be averaged to reduce the effective noise if the noises are independent.

Variable centering and normalization. Centering and normalization put variables with different ranges on an equal footing. Without normalization, a system modeling an electronic device with voltages from 0 to 10,000 V and currents from 0 to 0.01 A would probably need very small weights from the voltage inputs and large weights from the currents. The system is very poorly conditioned, and training times will probably be long. Since backpropagation weight changes are proportional to the signal magnitudes, a single learning rate would probably not work for both. If these were output targets (in a network with linear outputs), the network would almost surely ignore errors in the currents as long as voltage errors remain. A commonly used normalization is

$$X' = (X - \mu)/\sigma$$

where μ is the mean value of X, and σ is its standard deviation. Normalization based on minimum and maximum values is also common.

Known nonlinearities. In general, it helps to eliminate known nonlinearities. Conversion from Cartesian to cylindrical coordinates, for example, may simplify a problem. Functions involving products or ratios of positive inputs can be made linear by taking logarithms. Of course, these sorts of transformations are completely problem dependent.

Problem decomposition and modularization. Learning is almost always easier if a task can be broken into smaller noninteracting parts. Separate networks then can be trained independently for each subproblem and combined. The result is (1) shorter training times because each subnetwork is smaller and (2) better generalization because each subnetwork is better constrained by available examples. Assuming the subtasks are truly independent, a system which does both together cannot do better, and may do worse, since its task is more complex.

Realistic engineering applications almost always require some sort of high-level partitioning. Systems like a postal zip code reader, for example, usually have separate segmentation and recognition subsystems. The tasks can be partitioned because digit identity is basically independent of size, location, etc.

Problem decomposition is completely task dependent, of course. A problem may be broken down in many ways, and knowing how to partition a task is a large part of knowing how to solve it. When high-level human knowledge is unavailable but data are abundant, an alternative is to divide the input space into pieces, assigning relatively simple subnetworks to learn each piece. Clustering or vector quantization techniques (possibly implemented as neural networks) can form the necessary partition. An example of partitioning by self-organization is the "mixture-of-experts" model (Jacobs et al., 1991; see MODULAR AND HIERARCHICAL LEARNING SYSTEMS).

Architecture Selection

One of the central tasks in network design is selection of an architecture. The goal is to find a network powerful enough to solve the problem, yet simple enough to train easily and generalize well. An advantage of local representation systems such as radial basis functions, self-organizing maps, Adaptive Resonance Theory (ART), and others is that they usually train much faster than layered networks (trained by backpropagation). They tend to generalize less well from an equivalent amount of data, however, so they are best used when data are abundant.

Although much work has been done on selecting appropriate structures and sizes, it is still basically an art. An approach that often works well in practice is to guess an approximate initial size and then use node creation and pruning algorithms (see Topology-Modifying Neural Network Algorithms) to adjust the size during training, along with generalization-aiding techniques to suppress overfitting problems that may occur if the net is too large.

Pruning

Since the target function is unknown, it is often impossible to predict what size or configuration is appropriate. Although one can train a number of networks and choose the smallest/least complex one that learns the data, this can be inefficient if many networks have to be trained before an acceptable one is found. Even if the optimum size were known, the smallest adequate network might be difficult to train.

The pruning approach is to train a network that is larger than necessary and then remove unnecessary parts. The large initial size allows reasonably quick learning with less sensitivity to parameters, while the reduced complexity of the trimmed system favors improved generalization. In several studies, pruning techniques have produced small networks that generalize well where it was very difficult to obtain a solution by training the small network (obtained by pruning) from scratch with random weights (Sietsma and Dow, 1991).

Many pruning techniques have been suggested; a survey can be found in Reed (1993). Many of the algorithms fall into two broad groups. One group estimates the sensitivity of the error to removal of elements and removes those with the least effect. Another group adds terms to the error function that penalize unnecessarily complex solutions; many of these can be viewed as forms of regularization. In general, sensitivity methods modify a trained network; the network is trained, sensitivities are estimated, and then elements are removed. Penalty methods, however, modify the cost function so that optimization drives unnecessary weights to zero and, in effect, removes them during training. Even if the weights are not actually removed, the network acts like a smaller system. An advantage is that training and pruning are effectively done in parallel so the network can adapt to minimize errors introduced by pruning.

Although pruning and penalty term methods often may be faster than searching for and training a minimum-size network, they do not necessarily reduce training times; larger networks may take longer to train because of sheer size, and pruning takes some time itself. The goal, however, is improved generalization rather than faster training speed.

Constructive Methods

The opposite approach to pruning is to build the network incrementally by adding elements until a suitable configuration is found. The basic idea is to start with a small network, train until the error stops decreasing and then add a new node (or nodes) and resume training, repeating until an acceptable error is achieved. Algorithms differ in the network structures used, when new units are added, where they are placed, how they are initialized, etc.

In some cases, constructive methods can be faster than pruning methods since significant learning may occur while the network is still small. The approaches are not incompatible and are often used together. Since constructive methods, when used alone, sometimes create larger networks than necessary, a follow-up pruning phase can be useful to reduce the size.

It should be noted that pruning and constructive techniques are a means of adjusting network size rather than a way of deciding what size is appropriate. Other criteria are often useful to decide when to stop adding or removing elements.

Weight Initialization

The normal initialization procedure is to set weights to "small" random values. The randomness is intended to break symmetry, while "small" weights are chosen to avoid immediate saturation. Typically, weights are randomly selected from a range such as $(-A/\sqrt{N}, A/\sqrt{N})$, where N is the number of inputs to the node and A is between 2 and 3. More structured methods of initialization are discussed in Wessels and Barnard (1992) and Nguyen and Widrow (1990).

Initialization from a decision tree is considered in Sethi (1990). Since decision trees can be constructed very quickly, overall training time may be much shorter.

For problems in which the desired input-output relationship is well understood and expressible by a small set of rules, initialization based on a fuzzy logic implementation has been suggested.

Shortening Learning Times

Many heuristics have been developed in an attempt to shorten training times. The standard techniques of "on-line" training and momentum both tend to increase learning speed. "On-line" learning can be faster than batch learning since, with M training patterns, on-line learning will make M times as many weight updates in the same time. The effectiveness presumably implies redundancy in the data such that small samples give nearly as much information as the complete set. With *momentum*, a fraction of the previous weight change is added to the current weight change to give the system memory. This tends to stabilize the direction of movement by averaging opposing changes and often allows use of larger learning rates. In the analogy of the marble on the hilly surface, this gives the marble inertia, allowing it to coast over relatively flat areas and roll over small bumps.

Adaptive Learning Rates

The learning rate parameter has a direct effect on learning times. The "best" value, however, depends on the task to be solved and varies with local characteristics of the error surface, which change as the network learns. Different nodes in the net also may have different optimal rates, so there are no general rules for choosing a good fixed value a priori. With very small learning rates, learning is slower than necessary, and the system may settle in local minima, which it could easily escape otherwise. Very large rates, however, may send the system on wild jumps in essentially random directions or, in less extreme cases, cause it to oscillate around a solution instead of settling to a minimum.

An alternative to setting a fixed rate a priori is to change it dynamically. A typical approach is to start with a moderate value, reduce it when the error starts to oscillate, and increase it when the error is decreasing very slowly. A moderate initial rate allows the system to find a rough initial solution quickly; reduction to a small value then allows the system to settle to the minimum. One of the most cited references for automatic learning rate adjustment is Jacobs (1988).

More sophisticated optimization methods such as conjugate-gradient or quasi-Newton methods often converge much faster than simple gradient descent, but these generally assume the error surface is well approximated by a quadratic function and may not work well when the assumption is not valid. The approximation is usually valid near a minimum, though, and these techniques can speed up final convergence to the end-point after a rough solution is found by other methods.

Avoiding Paralysis

Paralysis occurs when nodes are driven into saturation. Since the tails of the sigmoid function are flat, the slope becomes very small when the node input is large. Consequently, weight updates are small, and learning is slow. One cause of saturation is large weights which amplify a normal activity pattern and create large signals that saturate nodes in following layers. Another cause is excessively high learning rates. The $E(w)$ graph often has a "stair-step" shape with large nearly flat regions separated by steep "cliffs" where E changes abruptly for small changes in w. (This is especially true for binary classification problems.) In using large learning rates to cross the plateaus quickly, there is a risk of taking a huge step in a wild direction on reaching a cliff. This may then create large weights and lead to saturation.

In simulations, code can be added to detect paralysis before it becomes serious and correct it by reducing the learning rate. Keeping a copy of the weights allows a step to be retracted and the step size reduced if saturation occurs suddenly.

Another guard against paralysis is the use of a nonsaturating node nonlinearity. Quickprop (Fahlman, 1988) uses a normal sigmoid nonlinearity but adds a small constant, e.g., 0.1, to the calculated derivative so it does not go to zero on saturation. This avoids paralysis but may make it more difficult for the network to settle to a solution.

Another technique is to reduce the sigmoid gain or, equivalently, to scale all node input weights by a factor less than 1 when saturation is detected in a node. This preserves the direction of the weight vector while reducing its magnitude. Weight decay tends to have a similar effect, since, when a node saturates, the decay term dominates other weight changes and reduces weights until the node comes out of saturation. Training with input noise sometimes has similar effects and also may help "jostle" the system out of saturation.

Hints

Another idea for accelerating learning and improving generalization is the use of "hint functions" (Suddarth, 1988; Yu and Simmons, 1990). Additional output nodes are appended to the network and trained to learn additional functions related to the function of interest but easier to express or learn. The hints may accelerate convergence by generating nonzero derivatives in regions where the original error function is flat and may aid generalization by providing additional constraints penalizing solutions which somehow match the original function on the

training samples but do not include intermediate concepts embedded in the hints. After training, the extra nodes may be removed. Yu and Simmons (1990) demonstrate accelerated learning and improved generalization for a five-bit parity function by the use of hint nodes that count the number of ON bits using a thermometer representation.

Improving Generalization

Although neural networks are trained to minimize errors on the training patterns, we usually want the system to generalize from the examples and learn the underlying function so it will do well on new examples from the same function. A rule of thumb for generalization is that small simple systems are preferable to large complex systems if they give equal performance. Poor generalization usually results when the response does extreme things (like oscillating wildly) just to fit the data points. Simple systems tend to generalize better because they have less power to do things that are not "supported by the data." That is, they have fewer degrees of freedom and are better constrained by the available data. Pruning is one of the major ways of reducing network size to favor generalization.

Size is not the only factor affecting generalization, however. A network which is too small will have insufficient power to fit the desired function and will perform poorly. Also, large networks can mimic smaller networks. In most cases, it is not obvious what size is sufficient, and there is the risk of choosing an overly complex system. The techniques summarized below are designed to prevent an overly powerful network from overfitting the data.

Early Stopping

A simple estimate of the true generalization performance can be obtained by measuring the error on a separate testing data set which the network does not see while training. Since this reduces the size of the training set and is subject to statistical errors, more sophisticated methods are sometimes used.

Typically the training and test set errors decrease together in early learning stages as the network learns major features of the target function. With an unnecessarily complex system, the test set error usually reaches a minimum at some point and then begins to increase as the network exploits idiosyncrasies of the training data. A simple way of avoiding overfitting is to monitor the test set error and stop when the minimum is detected. This technique can be used with most of the other generalization-aiding techniques.

Regularization

Most techniques for improving generalization work by imposing additional constraints on the solution. The idea behind regularization methods (see GENERALIZATION AND REGULARIZATION IN NONLINEAR LEARNING SYSTEMS) is that one of the least restrictive assumptions is that the target function is smooth, i.e., that small changes in the input do not cause large changes in the output. This bias is embedded in the learning algorithm by adding terms to the cost function that penalize nonsmooth solutions. A generic cost function is

$$E = \sum_k (t(x_k) - y(x_k))^2 + \lambda E(\text{complexity})$$

where $E(\text{complexity})$ measures the complexity of the solution and λ balances the tradeoff between smoothing and error reduction. The complexity term is often a differential operator measuring how much the output changes over the region of interest.

Although this provides a way of biasing the learning algorithm, success depends on selection of an appropriate value for λ to determine the strength of the bias. Most other generalization heuristics have a similar parameter balancing error reduction and other constraints. This is often chosen by cross-validation.

Constraints to discourage overfitting are usually most helpful in the final learning stages and may be harmful in early stages if they bias the solution too much before the network has "seen sufficient evidence." When initial weights are small, saturation and overfitting usually do not become problems until later. Thus, it is often useful to change λ dynamically, starting at 0 and increasing gradually once an acceptable error is achieved or when cross-validation indicates overfitting.

Weight Decay

Large weights tend to cause sharp transitions in the sigmoid functions, and thus large changes in the output result from small changes in the inputs. A simple way to obtain some of the benefits of pruning without complicating the learning algorithm much is to add a decay term like $-\beta w$ to the weight update rule. Nonessential weights then decay to zero and can be removed. Even if not removed, they have no effect on the output, so the network acts like a smaller system. This can be viewed as a form of regularization, since a $\beta\Sigma(w_i^2)$ regularizing term yields a $(-\beta w_i)$ decay term in the weight update rule. Several methods are compared in Hergert, Finnoff, and Zimmerman (1992).

A drawback of the $\Sigma(w_i^2)$ penalty term is that it favors weight vectors with many small components over ones with a few large components, even when this is an effective choice. An alternative is (Weigend, Rumelhart, and Huberman, 1991)

$$\lambda \sum_i w_i^2/(w_i^2 + w_o^2)$$

where w_o is a constant. For $|w_i| \ll w_o$, the cost of a weight is small but grows like w_i^2. For $|w_i| \gg w_o$, the cost saturates and approaches a constant λ, so the weight does not incur additional penalties.

Weight Sharing

"Soft weight sharing" (Nowlan and Hinton, 1992) is another method that allows large weights when they are effective by giving "preferred status" to several other weight values besides zero. This reduces system complexity by increasing correlation among weight values.

"Hard" weight sharing is commonly used in image processing networks where the same kernel is applied repeatedly at different positions in the image (see CONVOLUTIONAL NETWORKS FOR IMAGES, SPEECH, AND TIME SERIES). In a neural network, separate nodes could be used to apply the kernel at different locations, and the number of weights could be huge. Constraining nodes that compute the same kernel to have equal weights greatly reduces the number of independent parameters and makes an otherwise unmanageable problem tractable (LeCun et al., 1990).

Adding Noise to the Data

Many studies have noted that adding small amounts of input noise during training often helps generalization and fault tolerance. Since the network never sees the same pattern twice, it cannot simply memorize the training patterns. This is equivalent to imposing a smoothness assumption, since we are effectively telling the network that slightly different inputs give about the same output. The effective target function is obtained by convolving the noise density with the original function. This is a smoother version of the original and helps prevent overfitting because, although the original function may be known only at discrete sample points, the effective function is continuous over the entire input space. The network is forced to use excess degrees of freedom to approximate the smoothed function instead of forming an arbitrarily complex surface that may match the target only at the sample points. Even though the network may be large, it models a simpler system. A drawback of training with noise is that it can be very slow, and there is the question of how much noise to use.

Multiple Networks

Another idea for improving generalization is to combine the outputs of several systems that classify novel examples differently because of differences in architecture, randomness of initialization, variations in parameters, differences in training data, etc., or because completely different types of classifiers are used. With a mean-square-error function, the best generalization is expected when the system produces the expected value of all possible functions consistent with the examples, weighted by their probability of occurrence

$$f'(x) = \int f(x)p_f(f)\,df$$

Averaging outputs of different systems is a very simple approximation to this expected value and tends to damp out extreme behaviors not justified by the data. Combination of subsystems is also an issue in the "mixture-of-experts" model (see MODULAR AND HIERARCHICAL LEARNING SYSTEMS).

Discussion

A number of commonly used techniques for improving learning in multilayer perceptrons have been mentioned. Many are quite simple and easily implemented but can have a significant effect on the speed and probability of successful learning. This list is by no means complete, of course. Some of the most important factors affecting learning, e.g., representation of input-output variables, have not been considered.

Although neural networks are often said to learn solely from the examples, at the most basic level it is biases built into the network structure and training algorithms that determine what the network can learn from the data. The network designer implicitly manipulates biases at each stage of the process, from data preparation to final tuning of the training parameters. Many of the techniques mentioned here can be thought of as biases that guide the network by making some functions easier to learn while excluding others.

Road Map: Learning in Artificial Neural Networks, Deterministic
Background: Backpropagation: Basics and New Developments; Perceptrons, Adalines, and Backpropagation

References

Fahlman, S. E., 1988, Faster-learning variations of back-propagation: An empirical study, in *Proceedings of the 1988 Connectionist Models Summer School* (D. Touretzky, G. Hinton, and T. Sejnowski, Eds.), San Mateo, CA: Morgan Kaufmann, pp. 38–51.

Georgiou, G. M., and Koutsougeras, C., 1992, Embedding domain information in backpropagation, in *Proceedings of the SPIE Conference on Adaptive and Learning Systems*, Bellingham, WA: SPIE.

Hergert, F., Finnoff, W., and Zimmermann, H. G., 1992, A comparison of weight elimination methods for reducing complexity in neural networks, in *Proceedings of the International Joint Conference on Neural Networks*, vol. 3, Piscataway, NJ: IEEE, pp. 980–987.

Jacobs, R. A., 1988, Increased rates of convergence through learning rate adaptation, *Neural Netw.*, 1:295–307.

Jacobs, R. A., Jordan, M. I., Nowlan, S. J., and Hinton, G. E., 1991, Adaptive mixtures of local experts, *Neural Computat.*, 3:79–87.

LeCun, Y., et al., 1990, Handwritten digit recognition with a back-propagation network, in *Advances in Neural Information Processing 2* (D. S. Touretzky, Ed.), San Mateo, CA: Morgan Kaufmann, pp. 396–404.

Nguyen, D. H., and Widrow, B., 1990, Improving the learning speed of 2-layer neural networks by choosing initial values of the adaptive weights, in *Proceedings of the International Joint Conference on Neural Networks*, Piscataway, NJ: IEEE, pp. 211–226.

Nowlan, S. J., and Hinton, G. E., 1992, Simplifying neural networks by soft weight-sharing, *Neural Computat.*, 4:473–493.

Reed, R. D., 1993, Pruning algorithms: A survey, *IEEE Trans. Neural Netw.*, 4:740–744. ◆

Sethi, I. K., 1990, Entropy nets: From decision trees to neural networks, *Proc. IEEE*, 78:1605–1613.

Sietsma, J., and Dow, R. J. F., 1991, Creating artificial neural networks that generalize, *Neural Netw.*, 4:67–79.

Suddarth, S. C., 1988, The symbolic-neural method for creating models and control behaviors from examples, PhD thesis, University of Washington.

Weigend, A. S., Rumelhart, D. E., and Huberman, B. A., 1991, Generalization by weight-elimination applied to currency exchange rate prediction, in *Proceedings of the International Joint Conference on Neural Networks*, vol. 1, Piscataway, NJ: IEEE, pp. 837–841.

Wessels, L. F. A., and Barnard, E., 1992, Avoiding false local minima by proper initialization of connections, *IEEE Trans. Neural Netw.*, 3:899–905.

Yu, Y.-H., and Simmons, R. F., 1990, Extra output biased learning, in *Proceedings of the International Joint Conference on Neural Networks*, vol. 3, Piscataway, NJ: IEEE, pp. 161–166.

NMDA Receptors: Synaptic, Cellular, and Network Models

Jim-Shih Liaw, Theodore W. Berger, and Michel Baudry

Introduction

NMDA receptors are subtypes of receptors for the excitatory neurotransmitter glutamate. These receptors are selectively activated by the agonist *N*-methyl-D-aspartate (NMDA), a glutamate analog. They are involved in diverse physiological as well as pathological processes, such as visual perception, motor pattern generation, learning and memory, and epilepsy- or stroke-induced neuronal damage.

NMDA receptors have three unique properties that distinguish them from other ligand-gated channels (channels activated by molecules which bind to them). First, they mediate a relatively slow excitatory postsynaptic potential (EPSP). Second, their activation requires not only the binding of an agonist (a compound that mimics the effect of a neurotransmitter), but also the depolarization of the postsynaptic membrane to remove the voltage-dependent blockade of the channel by Mg^{2+} ions. As a consequence, NMDA receptors act as coincidence detectors of presynaptic and postsynaptic activity. Third, NMDA receptors are 10 times more permeable to Ca^{2+} ions than to Na^+ or K^+. They often are colocalized with other receptor channels that conduct fast EPSPs (e.g., the α-amino-3-hydroxy-5-methyl-4-isoxazole proprionic acid, or AMPA, receptor). The interaction between the slow NMDA-mediated and fast non-NMDA-mediated currents provides the basis for a range of interesting dynamic properties which contribute to a diversity of neuronal processes.

NMDA receptors have attracted a great deal of interest in neuroscience because of their role in learning and memory. Their properties of coincidence detectors make them an ideal molecular device for producing Hebbian synapses, i.e., synapses whose strength is modified depending on the correlation of presynaptic and postsynaptic activity (see HEBBIAN SYNAPTIC PLASTICITY). Furthermore, the influx of Ca^{2+} ions through NMDA receptor channels triggers a cascade of molecular processes that leads to various forms of synaptic plasticity, including short-term potentiation (STP), long-term potentiation (LTP), and long-term depression (LTD). As a result, they have been a subject of intense investigation by experimentalists interested in understanding the features of the protein which cause its functional properties and theorists and modelers attempting to incorporate NMDA receptors into models of synapses, neurons, or circuitries. We will briefly review data related to the biological characteristics of the receptor and models which have been used to describe its function in isolated membrane patches, in models of neurons, and in models of complex circuits.

Molecular Structure of NMDA Receptors

Cloning techniques have provided evidence for the existence of a multiplicity of NMDA receptor subunits. It is generally assumed that the NMDA receptors are multimeric entities (i.e., composed of several subunits), possibly pentameric proteins by analogy with the acetylcholine nicotinic receptor. Moreover, because different combinations of receptor subunits produce receptors with different physiological and pharmacological properties, it is likely that several functional classes of NMDA receptors exist in adult neurons. A major challenge will be to determine their cellular distribution and the mechanisms regulating subunit expression and receptor assembly and turnover.

NMDA receptor activation requires not only the presence of glutamate but also the presence of another amino acid, glycine, which thus has been called a co-agonist. Whereas it is generally agreed that there are at least two glutamate binding sites per molecule of receptor, the exact number of glycine binding sites remains a subject of controversy. Molecular biological techniques have further clarified the nature of the voltage-dependent magnesium blockade of the channel, the properties of the zinc binding site, and the domain of the proteins involved in the calcium permeability of the channel.

As with most ligand-gated ion channels, a major characteristic of the NMDA receptor is the existence of a rapid desensitization (a long period of inactivation) after agonist-induced activation. Although the precise mechanism underlying receptor desensitization is not well understood, most interpretations assume that receptors can exist in different states depending on the occupancy of the different sites and the open or closed state of the receptor-associated ion channel.

Kinetic Models of NMDA Receptors

The behavior of a receptor can be described by the kinetic scheme of its transition between various discrete states representing the occupancy of the different binding sites and the functional states of the channel. A simple example is $A + R \leftrightarrow AR \leftrightarrow AR^*$, where A, R, AR, and AR^* represent agonist, receptor, receptor-agonist complex in the closed state, and receptor-agonist complex in the open state, respectively. Several kinetic models of NMDA receptors have been developed to study various aspects of the nature of the receptor and the dynamics of its behavior.

Glycine Binding

Clements and Westbrook (1991) developed a kinetic model of the NMDA receptor to study the number of glycine binding sites and the effect of glycine binding on glutamate binding. They performed recordings in outside-out patches from cultured hippocampal neurons to measure the rate constants for agonist binding, open-close transitions, and desensitization of the NMDA receptor. First, two sets of models were constructed independently, one set for glutamate binding and the other for glycine binding. The measured rate constants were used in models assuming one, two, or three agonist binding sites. For both glutamate and glycine binding, a two-site model provided a superior fit for the time course of NMDA channel activation. The two-site models for the binding of glutamate and glycine were combined to evaluate their interaction during simultaneous application. Two hypotheses were tested: one assumed independent binding of glutamate and glycine, and the other assumed that both glycine sites must be occupied before glutamate can be bound. The independent binding model produced better fits than the sequential model under all test conditions. Finally, desensitization was interpreted as an interaction between the glutamate and the glycine site such that glycine binding produced a decreased affinity of glutamate for its binding site (Lester, Tong, and Jahr, 1993). These interpretations were challenged by Johnson and Ascher (1992), who best fitted their data with a single site for glycine and observed that the binding of glycine had no effect on glutamate binding. Binding experiments suggest a strong positive allosteric effect between glutamate and glycine binding (Marvizon and Baudry, 1993).

Number of States of the NMDA Receptor

Two models of NMDA receptor channel kinetics have been developed to address the issue of the number of states of the receptor (Jahr and Stevens, 1990; Chauvet and Berger, 1994). Jahr and Stevens developed a model which assumed that the NMDA receptor exists in three states, closed (C), open (O), and blocked (B). The kinetic behavior of the NMDA receptor channel was characterized by four experimentally measured quantities: open time (T_o), interruption time (T_i), number of interruptions (N), and burst length (T_b). The predictions made by the model did not match several key experimental findings, and these shortcomings suggested an extension to include a second blocked state in addition to the Mg^{2+} block. The four-state model described the NMDA receptor behavior under all conditions except the low-amplitude, second-exponential component in T_o, which occurred with low Mg^{2+} concentrations and positive voltage. A theory based on the four-state model postulates that the interruptions could be the result of a voltage-dependent conformational change which is facilitated by the binding of Mg ions to some sites on the NMDA receptor.

The Chauvet et al. model also assumes the existence of three channel states—closed (C), open (O), and desensitized (D)—

and binding sites for three ligands: glutamate, glycine, and Mg^{2+}. The model accounts for desensitization by allowing for multiple states to be entered after the conditions of ligand binding or unbinding have been met. NMDA receptor channel kinetics are determined by a total of 13 parameters, corresponding to six rate constants for the binding and unbinding of each of the three ligands, an additional two rate constants for the voltage-dependent component of Mg^{2+} binding, and five rate constants for the allowed transitions between channel states. More recently, the model has been expanded to allow rate constants for state transitions to vary as a function of the number of ligand molecules bound to the receptor (Chauvet and Berger, 1995).

Parameters of the model were optimized against recordings of NMDA receptor-mediated excitatory postsynaptic currents (EPSCs) obtained from patch-clamp experiments. The model successfully predicts the number of channels with non-zero conductances as a function of time for a variety of experimental conditions: (1) brief (5 ms) application of glutamate, during which desensitization would be minimal; (2) longer (750 ms) application of glutamate, during which desensitization would make a major contribution to EPSC decay; (3) paired pulse application of glutamate to estimate recovery from desensitization; and (4) application of glutamate in the presence of different concentrations of glycine to examine allosteric interactions between the two amino acids, among others. Additional studies in which the effects of AP5 (an antagonist for the NMDA receptor) and Mg^{2+} were compared confirmed that the long decay time of the EPSC response to glutamate is not caused by repeated binding of agonist to the receptor site, but is more likely the result of a series of channel openings which result from a single binding event. The model also found that including a second binding site for glutamate resulted in a significantly better fit to experimental data. Although a strong allosteric effect of glycine was observed on the rate of channel opening in response to glutamate, little difference was found between models that incorporated one or two binding sites for glycine.

NMDA Receptors in Models of Neurons

NMDA receptor models have been incorporated into several models of synapses and neurons that were developed to answer questions related to short-term as well as long-term synaptic plasticity and to mechanisms of synaptic integration. As mentioned in the introduction to this article, a major focus is on the role of the voltage-dependent Mg^{2+} blockade, which gives rise to the property of coincidence detection and Ca^{2+} influx through the NMDA receptor in initiating the molecular events leading to the changes in synaptic strength. Furthermore, the interplay between the slow NMDA-mediated and fast non-NMDA-mediated currents leads to subtle features in the timing of synaptic inputs in inducing LTP.

Short-Term Synaptic Plasticity

Short-term changes in the excitability of a hippocampal pyramidal neuron (in a region called CA1) was studied by Pongrácz et al. (1992) using a compartmental model. The pyramidal neuron was represented as a five-compartment model, including apical and basal dendrites, a compartment for the soma, and a layer representing extracellular K^+ concentration. The model included intrinsic membrane conductances and excitatory (NMDA and AMPA) and inhibitory ($GABA_A$ and $GABA_B$) synaptic conductances distributed in the compartments of the apical dendrite. The voltage dependency of the NMDA receptor (as a result of Mg^{2+} blockade) resulted in a weak conductance by single stimulation, but the conductance increased with increasing number and intensity of repeated stimulation. The model suggested that such cumulative activation of NMDA-mediated synaptic conductances contributed to the frequency-dependent EPSP potentiation, a form of short-term plasticity, of the hippocampal neurons.

Long-Term Synaptic Plasticity

Holmes and Levy (1990) developed a compartmental model of a granule cell from a hippocampal area (the dentate gyrus) to study the role of Ca^{2+} and the subsequent biochemical events involved in triggering LTP. Specifically, they were concerned with the mechanisms for amplifying the calcium signal and the relative timing of presynaptic and postsynaptic activation. An 11-compartment model was constructed to represent a spine and a small patch of the neighboring dendrite. Calcium dynamics, including Ca^{2+} influx, buffering, pumping, and diffusion were computed over this domain. One glutamate binding site and the voltage-dependent Mg^{2+} block were included in the NMDA receptor kinetics. The amplitude of peak intracellular free Ca^{2+} concentration was regarded as the indicator of the induction of LTP at a particular synapse. When few synapses were activated, Ca^{2+} influx was low, even with high input frequency. When many synapses were activated at the same time, a steep rise in Ca^{2+} influx was seen with increasing frequency because of the voltage dependency of the NMDA-mediated conductance. However, total Ca^{2+} influx never increased by more than fourfold, which is too small to account for the selective induction of LTP. The three- to four-fold increase could be amplified 20- to 30-fold by transient saturation of the fast Ca^{2+} buffering system. When a weak input was paired with a strong one, the largest increase in peak $[Ca^{2+}]_i$ was seen in cases in which the weak stimulation preceded the strong input by 1 to 8 ms, because of the slow rate constant of NMDA receptor kinetics. This model has been extended by De Schutter and Bower (1993) to evaluate the effect of Ca^{2+} permeability of the NMDA receptor channel. Maximum amplification of $[Ca^{2+}]_i$ was obtained at permeability close to the values reported in the literature. Amplification decreased significantly when permeability was reduced by more than 50%. Furthermore, simulations showed that $[Ca^{2+}]_i$ was as much as 80% higher at distal spines than at proximal ones.

Similar issues were studied by Brown et al. (1991), who constructed a 3000-compartment model of a hippocampal CA1 neuron based on anatomical data. The synapses were located on the dendritic spines represented by a biophysical model including NMDA receptor and AMPA receptor conductances. The model of an NMDA receptor channel assumed three-state channel kinetics (open, closed, and blocked). Calcium dynamics were computed, and the peak concentration of calcium-bound CamKII, $[CaM\text{-}Ca_4]$, was assumed to be the indicator of the magnitude of LTP. Their simulations concluded that: (1) changes in $[Ca^{2+}]$ and $[CaM\text{-}Ca_4]$ were restricted to spines which received synaptic stimulation, consistent with the experimental observation that LTP induction in CA1 is specific to synapses that satisfy the conjunction of pre- and postsynaptic activation; (2) as a result of the slow unbinding rate of the NMDA receptor, the requirement of synchronized pre- and postsynaptic activation could be relaxed as long as the latter occurs within 40 ms of the former; and (3) the spine created a microenvironment with a compartmentalized Ca^{2+} signal isolated from the dendrite, amplified second messenger signals, and steepened voltage-dependent processes underlying LTP induction.

Synaptic Integration

In addition to synaptic plasticity, NMDA receptors are also involved in the integration of spatiotemporal patterns of inputs by a neuron. Fox and Daw (1992) developed an electrical model of a neuron in area 17 of the visual cortex to study neuronal responses to visual stimuli. The model was composed of two compartments: a somatic component and a dendritic compartment with NMDA and non-NMDA receptors. Their model showed that instead of switching on only at a higher level of contrast, the NMDA receptor-mediated conductance contributed to the response to visual stimuli in a graded fashion, all the way from near-threshold to saturation. This property of the NMDA receptor could not be accounted for solely by its voltage dependency. In addition, the higher affinity (binding rate) of NMDA receptors for glutamate, compared with the non-NMDA receptor, was involved.

A compartmental model was developed to study the integrative behavior of a complex dendritic tree, with particular focus on the role of NMDA receptors in the generation of neuronal responses (Mel, 1993). An anatomically characterized cortical pyramidal cell was represented by a model consisting of 903 coupled electrical compartments. A glutamatergic synapse was placed on the distal end of a spine containing both AMPA and NMDA receptor channels. The major finding from the simulations of the model was that when the NMDA receptor channels constituted a large portion of the synapse, the neuron responded preferentially to spatially clustered, rather than random, distribution of a given number of activated synapses. This response was caused by the voltage dependency of the NMDA receptors such that they were more effective when activated in groups. As a result of activity-dependent synaptic modification, the synapses on a dendritic tree were organized so that stimuli which activated a similar set of synapses as those patterns presented in the learning period had a higher probability of eliciting a neuronal response. Therefore, the manipulation of the spatial ordering of afferent activation of a dendritic tree provides a biological strategy for storing and classifying patterned information.

NMDA Receptors in Models of Neuronal Circuits

The role of NMDA receptors in neuronal networks has been studied in various systems. In this section, we briefly review their role in generating oscillatory or epileptiform activity in neuronal networks. Here, the interaction between the excitatory and inhibitory synapses shapes the dynamics of the neural network.

Role of NMDA Receptors in Oscillators

A network of interneurons conforming to experimentally identified cell types was constructed to simulate the spinal locomotor pattern generation in lamprey (Trävén et al., 1993). Excitatory synapses displayed both NMDA and AMPA receptors, while the inhibitory synaptic transmission was glycinergic and was mediated by chloride. The NMDA receptor current was modeled as a product of channel conductance, the difference of the membrane potential and the equilibrium potential, and a state variable which accounted for the voltage-dependent Mg^{2+} block of the channel. Oscillatory bursts could be evoked in a postsynaptic cell driven by NMDA receptor-mediated synaptic current, but the presynaptic neuron had little effect on oscillation frequency. The presynaptic control of oscillation frequency increased when non-NMDA receptors were added. A continuous range of burst rate of the network could be produced by the NMDA and non-NMDA receptor–mediated conductances. The simulations suggest that the spinal locomotor network could be modulated by controlling the balance between NMDA and AMPA receptor–mediated synaptic input. The NMDA receptor–containing synapses primarily stabilized the rhythmic motor output, whereas the non-NMDA receptor–containing synapses provided direct phasic control of the burst pattern.

NMDA Receptors and Epileptiform Activity

Traub and colleagues (Traub, Miles, and Jefferys, 1993) developed a computer model of hippocampal region CA3 that consisted of pyramidal neurons and inhibitory interneurons. Each pyramidal neuron was composed of 19 compartments with six voltage-dependent ionic conductances. Each pyramidal neuron was randomly connected to 20 other pyramidal neurons through excitatory (NMDA and AMPA) receptors and to 20 interneurons through inhibitory (GABA$_A$ and GABA$_B$) receptors. The computation of NMDA receptor-mediated current involved a scaling factor, a synaptic conductance term with a slow decay time constant, and a term representing the voltage-dependent Mg^{2+} blockade. The simulation suggested the following conditions for the occurrence of population oscillations: (1) the strength of excitatory synapses falls within a limited range; (2) the after-hyperpolarization conductance is significantly reduced; (3) the inhibitory postsynaptic potentials (IPSPs) are blocked; and (4) the apical dendrites of the pyramidal neurons are depolarized. The NMDA receptor conductance was not necessary for the population oscillation. The model generated synchronized population bursts that resemble experimental data obtained from hippocampal slices perfused with a GABA$_A$ receptor blocker. The synchronized firing was blocked in the absence of AMPA receptor conductance, but persisted without NMDA receptor current. Introduction of GABA inhibition (most significantly GABA$_A$ IPSP) suppressed the synchronized bursts. The model predicted that dendritic calcium spikes occurred during each secondary burst which was generated by the AMPA receptor current. However, with sufficiently high NMDA receptor conductance, synchronized bursts could occur in the absence of AMPA receptor current.

Discussion

This review of the various models of NMDA receptors at the synaptic, cellular, and network levels illustrates that the three unique properties of these receptors (i.e., slow conductance, voltage and transmitter dependency, and calcium permeability) provide the basis for their involvement in STP, LTP, synaptic integration, motor pattern generation, and epileptiform activity.

It also shows the usefulness of this approach in investigating the contribution of different characteristics of the receptors at the functional level. Many issues remain unresolved, in part because of the limited knowledge concerning the exact number of binding sites for the various effectors of the receptors, the mechanisms underlying desensitization, and the anatomical localization of the receptors. Therefore, as understanding of the characteristics and properties of NMDA receptors improves, new models will be needed to capture these properties and evaluate their contributions to the computational properties of individual neurons as well as to complex circuitries.

Road Map: Mechanisms of Neural Plasticity
Background: Ion Channels: Keys to Neuronal Specialization
Related Reading: Hebbian Synaptic Plasticity: Comparative and Developmental Aspects; Scratch Reflex; Spinal Cord of Lamprey: Generation of Locomotor Patterns; Synaptic Currents, Neuromodulation, and Kinetic Models

References

Brown, T. H., Zador, A. M., Mainen, Z. F., and Claiborn, B. J., 1991, Hebbian modifications in hippocampal neurons, in *Long Term Potentiation: A Debate of Current Issues* (M. Baudry and J. Davis, Eds.), Cambridge, MA: MIT Press, pp. 357–389. ◆

Chauvet, G. A., and Berger, T. W., 1994, A hierarchical model derived from an n-level field theory to study LTP in the hippocampus, in *Long Term Potentiation* (J. Davis and M. Baudry, Eds.), Cambridge, MA: MIT Press, pp. 337–369.

Chauvet, G. A., and Berger, T. W., 1995, Kinetic model of the NMDA receptor-channel, submitted for publication.

Clements, J. D., and Westbrook, G. L., 1991, Activation kinetics reveal the number of glutamate and glycine binding sites on the N-methyl-D-aspartate receptor, *Neuron*, 7:605–613. ◆

De Schutter, E., and Bower, J. M., 1993, Sensitivity of synaptic plasticity to the Ca^{2+} permeability of NMDA channels: A model of long-term potentiation in hippocampal neurons, *Neural Computat.*, 5:681–694.

Fox, K., and Daw, N., 1992, A model for the action of NMDA conductances in the visual cortex, *Neural Computat.*, 4:59–83.

Holmes, W. R., and Levy, W., 1990, Insights into associative long-term potentiation from computational models of NMDA receptor-mediated calcium influx and intracellular calcium concentration changes, *J. Neurophysiol.*, 63:1148–1168.

Jahr, C. E., and Stevens, C. F., 1990, A quantitative description of NMDA receptor-channel kinetic behavior, *J. Neurosci.*, 10:1830–1837.

Johnson, J. W., and Ascher, P., 1992, Equilibrium and kinetic study of glycine action on the *N*-methyl-D-aspartate receptor in cultured mouse brain neurons, *J. Physiol. (Lond.)*, 455:339–365.

Lester, R. J. A., Tong, G., and Jahr, C. E., 1993, Interaction between the glycine and glutamate binding sites of the NMDA receptor, *J. Neurosci.*, 17:1088–1098.

Marvizon, J. C., and Baudry, M., 1993, Receptor activation by two agonists: Analysis by nonlinear regression and application to *N*-methyl-D-aspartate receptors, *Anal. Biochem.*, 213:3–11.

Mel, B. W., 1993, NMDA-based pattern discrimination in a modeled cortical neuron, *Neural Computat.*, 4:502–516.

Pongrácz, F., Poolos, N. P., Kocsis, J. D., and Shepherd, G. M., 1992, A model of NMDA receptor-mediated activity in dendrites of hippocampal CA1 pyramidal neurons, *J. Neurophysiol.*, 68:2248–2259.

Traub, R. D., Miles, R., and Jefferys, J. G. R., 1993, Synaptic and intrinsic conductances shape picrotoxin-induced synchronized afterdischarges in the guinea-pig hippocampal slice, *J. Physiol. (Lond.)*, 461:525–547. ◆

Träven, H., Brodin, L., Lansner, A., Ekeberg, Ö., Wallén, P., and Grillner, S., 1993, Computer simulations of NMDA and non-NMDA receptor-mediated synaptic drive: Sensory and supraspinal modulation of neurons and small network, *J. Neurophysiol.*, 70:695–709.

Noise Canceling and Channel Equalization

Bernard Widrow and Michael A. Lehr

Introduction

The fields of adaptive signal processing and adaptive neural networks have been developing independently but have the adaptive linear combiner (ALC) in common. With its inputs connected to a tapped delay line, the ALC becomes a key component of an adaptive filter. With its output connected to a quantizer, the ALC becomes an adaptive threshold element or adaptive neuron.

Adaptive filters have enjoyed great commercial success in the signal processing field. All high-speed modems now use adaptive equalization filters. Long-distance telephone and satellite communications links are being equipped with adaptive echo cancelers to filter out echo, allowing simultaneous two-way communications. Other applications include noise canceling and signal prediction.

Adaptive threshold elements, however, are the building blocks of neural networks. Today neural nets are the focus of widespread research interest. Although neural network systems have not yet had the commercial impact of adaptive filtering, they are already being used widely in industry, business, and science to solve problems in control, pattern recognition, prediction, and financial analysis.

The commonality of the ALC to adaptive signal processing and adaptive neural networks suggests that the two fields have much to share with each other. This article describes the manner in which the ALC can be used in practical adaptive noise canceling and channel equalization.

The Adaptive Linear Filter

The *adaptive linear combiner* is the basic building block for most adaptive systems. Its output is a linear combination of its inputs. At each sample time, this element receives an input signal vector or input pattern vector $\mathbf{X} = [x_0, x_1, x_2, \ldots, x_n]^T$, and a desired response d, a special input used to effect learning. The components of the input vector are weighted by a set of coefficients, the weight vector $\mathbf{W} = [w_0, w_1, w_2, \ldots, w_n]^T$. The sum of the weighted inputs is then computed, producing a linear output, the inner product $y = \mathbf{X}^T\mathbf{W}$. The output signal y is compared with desired response d, and the difference is the error signal, ε. To optimize performance, the ALC's weights are generally adjusted to minimize the mean square of the error signal. Of the many adaptive algorithms to adjust the weights automatically, the most popular is the Widrow-Hoff LMS (least mean square) algorithm devised in 1959 (Widrow and Hoff, 1960). For the weight update occurring at sample time k, this algorithm is given simply by

$$\mathbf{W}_{k+1} = \mathbf{W}_k + 2\mu\varepsilon_k\mathbf{X}_k, \tag{1}$$

where μ is a small constant which determines stability and learning speed. The LMS algorithm represents an efficient implementation of the method of gradient descent on the mean-square-error surface in weight space (Widrow and Stearns, 1985).

Digital signals used by *adaptive filters* generally originate from sampling continuous input signals by analog-to-digital conversion. Digital signals are often filtered by means of a tapped delay line or transversal filter, as shown in Figure 1. The sampled input signal is applied to a string of delay elements (denoted by z^{-1}), each delaying the signal by one sampling

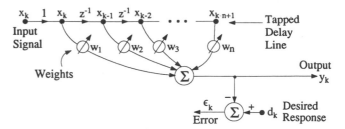

Figure 1. Linear adaptive transversal filter.

period. An ALC is seen connected to the taps between the delay elements. The filtered output is a linear combination of the current and past input signal samples. By varying the weights, the impulse response from input to output is directly controllable. Since the frequency response is the Fourier transform of the impulse response, controlling the impulse response controls the frequency response. The weights are usually adjusted so that the output signal provides the best least-squares match over time to the desired-response input signal.

The literature reports many other forms of adaptive filters (Widrow and Stearns, 1985; Haykin, 1991). Some filters contain a second tapped delay line which feeds the output of the filter back to the input through a second set of weights. This feedback results in a transfer function which contains both poles and zeros. Because it uses no signal feedback, the filter of Figure 1 realizes only zeros. Another variation is adaptive filters based on ladderlike architectures called lattice structures which achieve more rapid convergence under certain conditions. The simplest, most robust, and most widely used filter, however, is that of Figure 1, adapted by the LMS algorithm.

The adaptive filter of Figure 1 has an input signal and produces an output signal. The desired response is supplied during training. A question naturally arises: If the desired response were known and available, why would one need the adaptive filter? Put another way, how would one obtain the desired response in a practical application? There is no general answer to these questions, but studying successful examples provides some insight.

Noise Canceling

Separating a signal from additive noise is a common problem in signal processing. Figure 2*A* shows a classical approach to this problem using optimal Wiener or Kalman filtering. The purpose of the optimal filter is to pass the signal s without distortion while stopping the noise n_0. In general, this cannot be done perfectly. Even with the best filter, the signal is distorted, and some noise goes through to the output.

Figure 2*B* shows another approach to the problem using adaptive filtering. This approach is viable only when an additional "reference input" is available containing noise n_1, which is correlated with the original corrupting noise n_0. In Figure 2*B*, the adaptive filter receives the reference noise, filters it, and subtracts the result from the noisy "primary input," $s + n_0$. For this adaptive filter, the noisy input $s + n_0$ acts as the desired response. The "system output" acts as the error for the adaptive filter. Adaptive noise canceling generally performs

Figure 2. Separation of signal and noise. *A*, Classical approach. *B*, Adaptive noise-canceling approach.

(a)

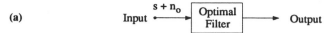

(b)

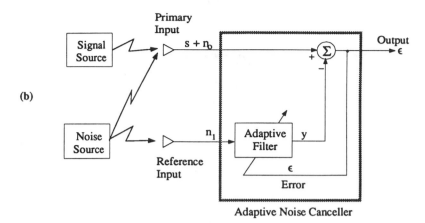

Adaptive Noise Canceller

much better than the classical approach, since the noise is subtracted out rather than filtered out.

One might think that some prior knowledge of the signal s or of the noises n_0 and n_1 would be necessary before the filter could adapt to produce the noise-canceling signal y. A simple argument shows, however, that little or no prior knowledge of s, n_0, n_1, or their interrelationships is required.

Assume that s, n_0, n_1, and y are statistically stationary and have zero means. Assume that s is uncorrelated with n_0 and n_1, and suppose that n_1 is correlated with n_0. The output is

$$\varepsilon = s + n_0 - y \tag{2}$$

Squaring, one obtains

$$\varepsilon^2 = s^2 + (n_0 - y)^2 + 2s(n_0 - y) \tag{3}$$

Taking expectations of both sides of Equation 3, and realizing that s is uncorrelated with n_0 and with y, yields

$$\mathrm{E}[\varepsilon^2] = \mathrm{E}[s^2] + \mathrm{E}[(n_0 - y)^2] + 2\mathrm{E}[s(n_0 - y)]$$
$$= \mathrm{E}[s^2] + \mathrm{E}[(n_0 - y)^2] \tag{4}$$

Adapting the filter to minimize $\mathrm{E}[\varepsilon^2]$ does not affect the signal power $\mathrm{E}[s^2]$. Accordingly, the minimum output power is

$$\mathrm{E}_{min}[\varepsilon^2] = \mathrm{E}[s^2] + \mathrm{E}_{min}[(n_0 - y)^2] \tag{5}$$

When the filter is adjusted so that $\mathrm{E}[\varepsilon^2]$ is minimized, $\mathrm{E}[(n_0 - y)^2]$ is therefore also minimized. The filter output y is then a best least-squares estimate of the primary noise n_0. Moreover, when $\mathrm{E}[(n_0 - y)^2]$ is minimized, $\mathrm{E}[(\varepsilon - s)^2]$ is also minimized, since, from Equation 2,

$$(\varepsilon - s) = (n_0 - y) \tag{6}$$

Adjusting or adapting the filter to minimize the total output power is tantamount to causing the output ε to be a best least-squares estimate of the signal s for the given structure and adjustability of the adaptive filter and for the given reference input.

There are many practical applications for adaptive noise canceling techniques. One involves canceling interference from the mother's heart when attempting to record clear fetal electrocardiograms (ECG). Figure 3 shows the location of the fetal and maternal hearts and the placement of the input leads. The abdominal leads provide the primary input (containing fetal ECG and interfering maternal ECG signals), and the chest leads provide the reference input (containing pure interference, the maternal ECG). Figure 4 shows the results. The maternal ECG from the chest leads was adaptively filtered and subtracted from the abdominal signal, leaving the fetal ECG.

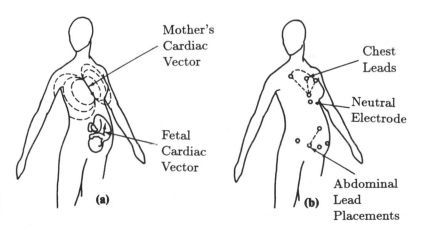

Figure 3. Canceling maternal heartbeat in fetal electrocardiography. *A*, Cardiac electric field vectors of mother and fetus. *B*, Placement of leads.

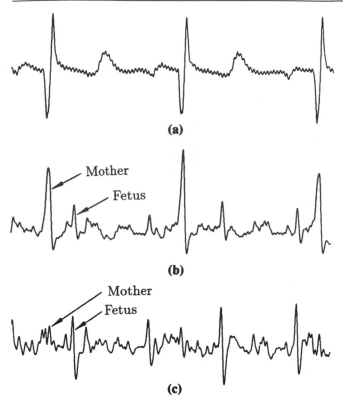

Figure 4. Result of fetal ECG experiment. *A*, Reference input (chest lead). *B*, Primary input (abdominal lead). *C*, Noise canceler output —the fetal ECG signal with maternal interference removed.

Channel Equalization

Telephone channels, radio channels, and even fiber-optic channels can have nonflat frequency responses and nonlinear phase responses in the signal passband. Sending digital data at high speed through these channels often results in a phenomenon called "intersymbol interference," caused by signal pulse smearing in the dispersive medium. Equalization in data modems combats this phenomenon by filtering incoming signals. A modem's adaptive filter, by adapting itself to become a channel inverse, can compensate for the irregularities in channel magnitude and phase response.

The *adaptive equalizer* in Figure 5 consists of a tapped delay line with an adaptive linear combiner connected to the taps. Deconvolved signal pulses appear at the weighted sum, which is quantized to provide a binary output corresponding to the original binary data transmitted through the channel. Any least-squares algorithm can adapt the weights, but the telecommunications industry uses the LMS algorithm almost exclusively.

In operation, the weight at a central tap is generally fixed at unit value. Initially, all other weights are set to zero so that the equalizer has a flat frequency response and a linear phase response. Without equalization, telephone channels can provide quantized binary outputs that reproduce the transmitted data stream with error rates of 10^{-1} or less. As such, the quantized binary output can be used as the desired response to train the neuron. It is a noisy desired response initially. Sporadic errors cause adaptation in the wrong direction, but on average, adaptation proceeds correctly. As the neuron learns, noise in the desired response diminishes. Once the adaptive equalizer converges, the error rate typically is 10^{-6} or less. The method,

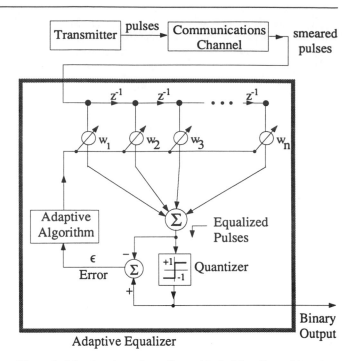

Figure 5. Adaptive channel equalizer with decision-directed learning.

called "decision-directed" learning, was invented by Robert W. Lucky of AT&T Bell Labs.

Using a modem with an adaptive equalizer enables transmitting approximately four times as much data through the same channel with the same reliability as without equalization.

Discussion

The simple concept of adapting the weights of a linear combiner to cause its output to approximate a desired response is the basis for the field of adaptive signal processing. In a large number of practical cases, it is possible to exploit this idea to solve difficult signal-processing problems with surprising accuracy, even when the statistics of the involved signals are unknown. Adaptive noise canceling and adaptive channel equalization are two examples which indicate the power of this approach. The burgeoning fields of neural networks, adaptive inverse control, and active noise control provide an indication of the generality and importance of methods based on this approach.

Acknowledgments. This work was sponsored by NSF under grant IRI 91–12531, by ONR under contract no N00014–92–J–1787, by EPRI under contract RP 8010–13, and by the U.S. Army under contract DAAK70–92–K–0003.

Road Map: Applications of Neural Networks
Related Reading: Adaptive Filtering; Adaptive Signal Processing; Perceptrons, Adalines, and Backpropagation

References

Haykin, S., 1991, *Adaptive Filter Theory*, 2nd ed., Englewood Cliffs, NJ: Prentice-Hall.
Widrow, B., and Hoff, M., 1960, Adaptive switching circuits, in *1960 IRE WESCON Convention Record*, Part 4, New York: IRE, pp. 96–104.
Widrow, B., and Stearns, S., 1985, *Adaptive Signal Processing*, Englewood Cliffs, NJ: Prentice-Hall. ◆

Nonmonotonic Neuron Associative Memory

Shuji Yoshizawa

Introduction

The history of the research on autocorrelation associative memory or content-addressable memories goes back to the early 1970s. K. Nakano, T. Kohonen, J. Anderson, and S. Amari independently proposed neural network models of associative memory in 1972. After almost a decade of silence, J. Hopfield (1982) introduced the concept of energy function into the dynamics of neural networks by analogy with spin glasses, and showed by computer simulation that the memory capacity of the associative memory is approximately $0.15n$, where n is the number of neurons. That is, the autocorrelation associative memory can recall up to $0.15n$ randomly generated patterns with a small percentage of errors. This is called the *relative capacity*.

It was proved that the *absolute capacity* is asymptotically $n/(2\log n)$ (Weisbuch, 1985; McEliece et al., 1987), where the absolute capacity is the maximum number of randomly generated patterns which are memorized as the equilibria of the network. On the other hand, Amit, Gutfreund, and Sompolinsky, (1985a, b) showed by using the well-known replica method of statistical mechanics that the relative capacity is about $0.138n$. Amari and Maginu (1988) studied the dynamical aspects of the recall process by using the statistical neurodynamical method, by which the relative capacity of $0.16n$ was obtained.

Through these studies it became clear that there are two flaws in the conventional associative memory model. The first one is that its absolute capacity of $n/(2\log n)$ is smaller than n in order, and its relative capacity of $0.15n$ is not large. The second one is the fact that (a) the network has many equilibria other than the memorized patterns, which are called spurious memories (Gardner, 1986), and (b) there is no way to decide whether a memorized pattern is recalled or not.

There are a number of modifications of the prototypes. They include, for example, the generalized-inverse memory matrix (Kohonen, Amari), the optimal capacity by a general memory matrix (Gardner), the sparsely encoded associative memory (Willshaw and associates, Palm, Amari, Meunier and associates), introduction of sparsity in connections (Yanai and associates), introduction of excitatory and inhibitory neurons (Shinomoto), and the chaos association memory (Aihara). In these modifications, elaborated structures of the memory or special patterns are used. See Yoshizawa, Morita, and Amari (1993) for specific references on these modifications.

On the other hand, M. Morita has shown that the performance of the conventional associative memory is improved remarkably by replacing the usual sigmoid or hard-limiter neuron with a nonmonotonic neuron (Morita, Yoshizawa, and Nakano, 1990; Morita, 1993). A nonmonotonic neuron is an analog device whose output is a nonmonotonic function of its internal potential which follows a first-order differential equation. One remarkable fact is that the absolute capacity is about $0.4n$, which is much greater than the absolute capacity $n/(2\log n)$ of the conventional model and is larger than even the relative capacity, $0.15n$. Moreover, the problem of spurious memories is also dissolved. When the network fails to recall a correct memorized pattern, the state shows a chaotic behavior instead of falling into a spurious memory.

In this article, we introduce some results of computer simulation of the nonmonotonic neuron associative memory and theoretical analysis of the simplified neuron model.

Autocorrelation Associative Memory

The neural network model of autocorrelation associative memory consists of n neurons. Memorized patterns are n-dimensional random vectors whose elements take the values ± 1 randomly with equal probability, and they are denoted by $\mathbf{s}^{(\mu)} = (s_1^{(\mu)} s_2^{(\mu)} \ldots s_n^{(\mu)})^{\mathrm{T}}$ $(\mu = 1, 2, \ldots, m)$, where T denotes the transpose operator.

The memory matrix W is constructed as follows:

$$w_{ij} = \frac{1}{n} \sum_{\mu=1}^{m} s_i^{(\mu)} s_j^{(\mu)} \qquad i \neq j, w_{ii} = 0$$

which is the result of Hebbian learning.

In matrix form, this relationship is expressed as follows:

$$W = \frac{1}{n} SS^{\mathrm{T}} - aE_n \qquad (1)$$

where

$$S = [\mathbf{s}^{(1)} \mathbf{s}^{(2)} \ldots \mathbf{s}^{(m)}] \qquad (2)$$

and a is the memory ratio defined by

$$a = \frac{m}{n} \qquad (3)$$

Without loss of generality, since we can "flip" $+1$ and -1 as needed, we can assume

$$\mathbf{s}^{(1)} = (1 \ldots 1)^{\mathrm{T}} \qquad (4)$$

and so it is enough to analyze the recalling dynamics of $\mathbf{s}^{(1)}$.

The recalling dynamics is given by

$$\frac{du_i}{dt} = -u_i + \sum_{j=1}^{n} w_{ij} f(u_j)$$

or in vector form

$$\frac{d\mathbf{u}}{dt} = -\mathbf{u} + Wf(\mathbf{u}) \qquad (5)$$

where $\mathbf{u} = (u_1 u_2 \ldots u_n)^{\mathrm{T}}$ is an n-dimensional vector which represents the internal potential of neurons, and f is the scalar output function of the neuron that operates on each element of the vector $\mathbf{u}$.

The recall process of the autocorrelative associative memory is as follows. For a given input pattern $\mathbf{u}_0$ we evaluate Equation 5 with initial condition $\mathbf{u}_0$, and if the solution converges to an equilibrium state $\mathbf{u}_e$, the recalled pattern is determined, not by $f(\mathbf{u}_e)$ but by the signum of $\mathbf{u}_e$: $\mathrm{sgn}(\mathbf{u}_e)$. If the solution does not converge, we cannot determine a recalled pattern, and we deem that the recall failed.

Comparison of Recall Processes of Conventional and Nonmonotonic Associative Memories

Conventional associative memories use a sigmoid output function. On the other hand, the Morita model uses a nonmonotonic function as shown in Figure 1 (Morita et al., 1990; Morita, 1993).

$$f(u) = \frac{1 - \exp(-cu)}{1 + \exp(-cu)} \times \frac{1 + \kappa \exp c'(|u| - h)}{1 + \exp c'(|u| - h)} \qquad (6)$$

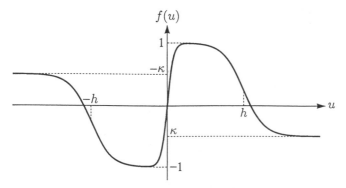

Figure 1. Nonmonotonic output function of the neuron model.

When the memory ratio a is small, the conventional and nonmonotonic neuron associative memories work correctly, and the memorized pattern $\mathbf{s}^{(1)}$ has its basin of attraction around it. The states of the associative memory converge to the memorized pattern $\mathbf{s}^{(1)}$ if the direction cosines $\sum_{i=1}^{n} x_i(0)s_i^{(1)}/n$ of the initial states $\mathbf{x}(0)$ are greater than a critical value, and they go away from $\mathbf{s}^{(1)}$ if the direction cosines are less than the critical value. We call this critical value the *critical direction cosine*.

When the memory ratio increases, the situation changes drastically. Figure 2 shows the time evolutions of the recall processes for a memory ratio of 0.32, where the abscissas and ordinates are the time and the direction cosine between the signum of $\mathbf{u}(t)$ and $\mathbf{s}^{(1)}$, respectively. The value 1.0 of the abscissa means that the state of the associative memory coincides

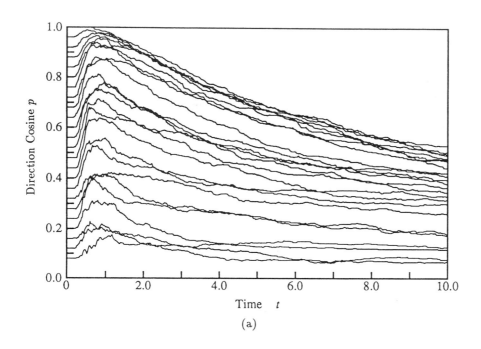

(a)

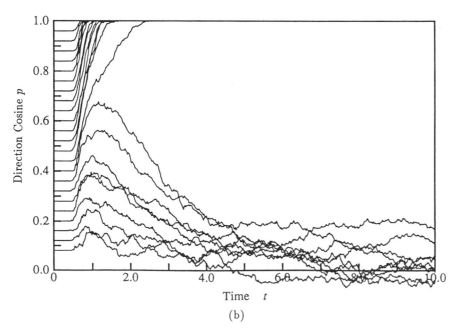

(b)

Figure 2. Time sequences of recall process ($n = 1000$; $a = 0.32$). *A*, In the case of a conventional neuron, the memorized pattern is unstable. *B*, In the case of a nonmonotonic neuron, a basin of attraction exists around the memorized pattern.

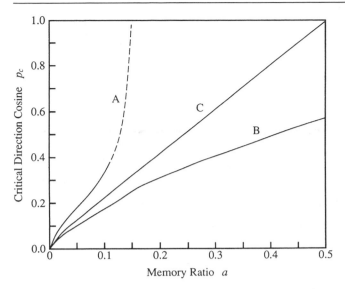

Figure 3. The relation between the memory ratio and the critical direction cosine. *A*, sigmoid neuron + autocorrelation matrix; *B*, sigmoid neuron + generalized inverse matrix; *C*, nonmonotonic neuron + autocorrelation matrix.

with the memorized pattern completely, and the value 0.0 implies that the state of the memory is orthogonal to it.

Figure 2*A* shows that the direction cosines for the conventional neuron converge to values far less than 1 even if their initial values are set close to 1. This means that the memorized pattern $s^{(1)}$ is unstable and cannot be recalled. For a nonmonotonic neuron (Figure 2*B*), on the other hand, it can be seen that there is a basin of attraction around $s^{(1)}$: If an initial pattern has direction cosine greater than the critical direction cosine (0.44 in this special example), the state of the network approaches to $s^{(1)}$, and if its direction cosine is less than the critical cosine, it goes away from $s^{(1)}$. The values of the parameters of the nonmonotonic function (see Figure 1) used in these simulations are $h = 0.5$, $\kappa = -1$, and the number of neurons is $n = 1000$. Figure 3 is a summary of the relation between the memory ratio and the critical direction cosine, from which it can be seen that the nonmonotonic model (curve *B*) is superior to the sigmoid model (curve *A*). For the benefit of comparison, a result for the sigmoid neuron with the generalized-inverse memory matrix is shown by curve *C*. The generalized inverse matrix for *S* is defined by

$$W = S(S^{\mathrm{T}}S)^{-1}S^{\mathrm{T}}$$

and for this memory matrix, every memorized pattern $s^{(\mu)}$ is an equilibrium state. It should be noted that recall is perfect in the cases of the nonmonotonic model (*B*) and the generalized-inverse memory matrix (*C*), while recalled patterns include a small percentage of errors in the conventional model (*A*).

A remarkable feature of Figure 2*A* is that the evolution of **u** does not converge to an equilibrium state but continues to oscillate (chaotically) when memorized pattern $s^{(1)}$ cannot be recalled. This shows that the failure of correct recall is detectable by the behavior of the recall process itself, and it also suggests that spurious memories are rare for the nonmonotonic neuron model.

The other remarkable feature of the nonmonotonic neuron associative memory is that its ability does not deteriorate even for correlated memorized patterns. Figure 4 shows the critical direction cosine for correlated patterns which are grouped into *k* clusters within which patterns $s^{(\mu\kappa)}$ have correlation *c* with

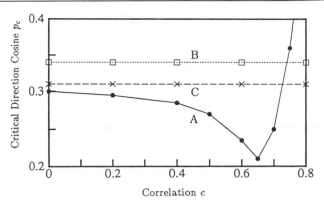

Figure 4. The critical direction cosine for the clustered patterns. *A*, nonmonotonic neuron + autocorrelation matrix; *B*, sigmoid neuron + generalized inverse matrix; *C*, nonmonotonic neuron + generalized inverse matrix.

each other:

$$c = \frac{1}{n}\sum_{i=1}^{n} s_i^{(\mu\kappa)} s_i^{(\nu\kappa)}$$

In the case of conventional associative memory, the correlation works as a noise which contaminates the correct recall pattern. Contrary to this, a correlation less than 0.7 helps the recall process of the nonmonotonic neuron memory and gives a bigger basin of attraction than the generalized inverse memory.

Theoretical Estimate of the Capacity of the Nonmonotonic Neuron Associative Memory

By approximating the nonmonotonic function shown in Figure 1 by the piecewise linear function $x(u)$ defined by the second line of Equation 7, we can obtain a theoretical estimate of the absolute capacity or the maximum memory ratio within which there exists a stable equilibrium of Equation 5 in the same quadrant as the memorized pattern (Yoshizawa et al., 1993).

The recall dynamics for the associative memory using the piecewise linear neuron is described by

$$\begin{cases} \dfrac{d\mathbf{u}}{dt} = -\mathbf{u} + W\mathbf{x}(\mathbf{u}) \\[2mm] \mathbf{x}(\mathbf{u}) = \mathrm{sgn}(\mathbf{u}) - k\mathbf{u} \qquad k = \dfrac{1}{a} \end{cases} \tag{7}$$

where sgn is the signum function. The condition that the probability of the existence of the equilibrium of Equation 7 is equal to 1 is given by

$$\frac{\pi}{2} - \sin^{-1}\exp\left\{-\left[\frac{v_0^2}{2\sigma^2} - \log\{1 - \Phi(v_0)\}\right]\right\} < \tan^{-1}\left\{\sqrt{\frac{1-a}{a}}\right\}$$

where v_0 is the value of v at which the function

$$\Psi(v,\sigma) = \frac{v^2}{2\sigma^2} - \log\{1 - \Phi(v)\}$$

takes its minimum, $\Phi(v)$ and σ being as follows:

$$\Phi(v) = \int_{-\infty}^{v} \frac{1}{\sqrt{2\pi}}\exp\left\{-\frac{u^2}{2}\right\}du$$

$$\sigma = \frac{\sqrt{1-2a}}{a}$$

Solving this inequality relation numerically, we have the following result.

Main result.   In the case of $k = a^{-1}$, the absolute capacity of the nonmonotonic neuron associative memory is estimated at $0.398n$. Thus, the total information $I = (\text{capacity}) \times (\text{information/pattern})$ storable in the memory is $0.398n^2$.

In numerical experiments with Equation 7 for $n = 1000$, the upper limit of a for correct recall was between 0.4 and 0.41.

Remark.   If the simple autocorrelation matrix is used, the best capacity and total information attainable by the sigmoid neuron associative memory are of the order $(n/\log n)^2$ and n^2, respectively. These values are obtained when random sparsely encoded patterns are used (Amari, 1989). In this sense the nonmonotonic neuron associative memory has the best ability.

Concluding Remarks

In the piecewise linear model, the stability of equilibrium state $\mathbf{x}_0$ is easily determined if its location is known:

1. Equilibrium state $\mathbf{x}_0$ is stable when $k < 1/a$.
2. Equilibrium state $\mathbf{x}_0$ is neutrally stable when $k = 1/a$.
3. Equilibrium state $\mathbf{x}_0$ is unstable when $k > 1/a$. (This implies that directions of nonmemorized vectors are unstable.)

Our consideration has been restricted to the existence of equilibrium solutions and local stability. For a complete understanding of the capability of the nonmonotonic neuron associative memory, we have to investigate further such problems as the size of the basin of attraction, the full sketch of spurious memories, and the behavior for clustered memorized patterns.

Road Map: Learning in Artificial Neural Networks, Deterministic
Background: I.3. Dynamics and Adaptation in Neural Networks
Related Reading: Statistical Mechanics of Learning

References

Amari, S., 1989, Characteristics of sparsely encoded associative memory. *Neural Netw.*, 2:451–457.
Amari, S., and Maginu, K., 1988, Statistical neurodynamics of associative memory, *Neural Netw.*, 1:63–73. ◆
Amit, D. J., Gutfreund, H., and Sompolinsky, H., 1985a, Spin-glass models of neural networks, *Phys. Rev. A*, 32:1007–1018.
Amit, D. J., Gutfreund, H., and Sompolinsky, H., 1985b, Storing infinite numbers of patterns in a spin-glass model of neural networks, *Phys. Rev. Lett.*, 55:1530–1533.
Gardner, E., 1986, Structure of metastable states in the Hopfield model, *J. Phys. A*, 19:1047–1052.
Hopfield, J. J., 1982, Neural network and physical systems with emergent collective computational abilities, *Proc. Natl. Acad. Sci. USA*, 79:2554–2558.
McEliece, R. J., Posner, E. C., Rodemich, E. R., and Venkatesh, S. S., 1987, The capacity of the Hopfield associative memory, *IEEE Trans. Inform. Theory*, IT-33:461–482. ◆
Morita, M., 1993, Associative memory with nonmonotone dynamics, *Neural Netw.*, 6:115–126.
Morita, M., Yoshizawa, S., and Nakano, K., 1990, Analysis and improvement of the dynamics of autocorrelation associative memory, *Trans. Inst. Electron., Inform. Commun. Engrs.*, J73-D-III:232–242.
Weisbuch, G., 1985, Scaling laws for the attractors of Hopfield networks, *J. Phys. Lett.*, 46:623–630.
Yoshizawa, S., Morita, M., and Amari, S. 1993, Capacity of associative memory using a nonmonotonic neuron model, *Neural Netw.*, 6:167–176. ◆

NSL: Neural Simulation Language

Alfredo Weitzenfeld

Introduction

In the area of neural network simulation many tools have been built to assist the scientist in the task of modeling and simulating neurons at different levels of detail. The more detailed neuronal models, such as the *Hodgkin-Huxley* model (Hodgkin and Huxley, 1952) and the *compartmental* model (Rall, 1959), permit only the modeling of a few neurons at a time, and are supported by simulation systems such as GENESIS or NEURON (De Schutter, 1992). The coarser neural models, such as the *leaky integrator* model (Arbib, 1989), permit the modeling of thousands of neurons, and are supported by simulation systems such as NSL (Weitzenfeld, 1991; Weitzenfeld and Arbib, in press). In this article we will focus on the simulation of large neural networks with NSL.

The NSL System

NSL, Neural Simulation Language, is a general-purpose simulation system providing a high-level language with many constructs and libraries developed to ease the specification of large neural networks. This system integrates *object-oriented* programming methodologies (Wegner, 1990) in its design and implementation, providing a simulation environment for users with little programming background, as well as those with more extensive programming expertise who can use C++ (Stroustrup, 1991) as an extension of NSL's modeling language. NSL is offered as public domain software (anonymous file transfer protocol, FTP, from usc.edu).

The system, whose architecture is shown in Figure 1, includes a command interpreter for interactive and batch processing, an X-Windows graphical interface, and temporal and spatial displays, including 2D and 3D graphics.

NSL Language

In order to model neural networks with NSL (NSL language compiler shown in Figure 1) it is necessary to describe (1) the neurons making up the network, (2) the neurons' interconnections, and (3) the dynamics of the neurons and their interconnections.

Neurons

The basic neural model in NSL is the single-compartment neuron, having one output and many inputs, as shown in Figure 2. The internal state of the neuron is described by a single scalar quantity, its membrane potential m, which depends on the neuron's inputs and its past history. The output is described by another single scalar quantity, its firing rate M, and may serve as input to many other neurons, including itself. As the input to a neuron varies, the membrane potential and firing rate also vary.

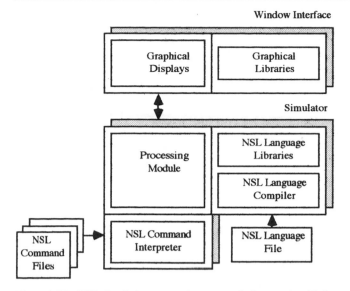

Figure 1. The NSL simulation system is composed of two units: (1) the simulator, containing the processing module, NSL language compiler, NSL language libraries, and NSL command interpreter; and (2) the Window interface, containing the graphical displays and the graphics libraries.

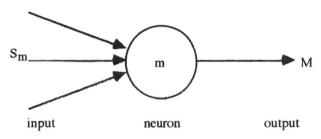

Figure 2. The single-compartment neuron model is represented by one value m corresponding to its membrane potential, and one value M corresponding to its firing rate. The symbol S_m represents the set of inputs to the neuron. There is a single output.

The *membrane potential* for m is described by the differential equation

$$\tau_m \frac{dm(t)}{dt} = f(S_m, m, t)$$

which depends on the neuron's input S_m, previous values of m, and time parameter t. The symbol τ_m is the time constant. The choice of f defines the particular neural model utilized. In particular, the *leaky integrator* model is described by $f(S_m, m, t) = -m(t) + S_m(t)$, or

$$\tau_m \frac{dm(t)}{dt} = -m(t) + S_m(t)$$

The *firing rate* M, the output of the neuron, is obtained by applying a *threshold function* to the neuron's membrane potential,

$$M(t) = \sigma(m(t))$$

where σ is usually a nonlinear function. Some of the most common threshold functions—*ramp*, *step*, *saturation*, and *sigmoidal*—are shown in Figure 3.

In NSL two **DATA** structures are required to represent a single neuron; one structure corresponds to the membrane potential and the other one to the firing rate. The notation is as follows (with a semicolon at the end of each statement):

$$\text{DATA m;}$$
$$\text{DATA M;}$$

The membrane potential m is represented by a differential equation

$$\text{DIFF}(m, \tau_m) = f(S_m, m);$$

where DIFF defines a first-order differential equation for m with time decay τ_m (time parameter t is implicit in the equation). The leaky integrator model corresponds to

$$\text{DIFF}(m, \tau_m) = -m + S_m;$$

The firing rate M is represented simply by

$$M = \sigma(m);$$

where σ represents the choice of threshold function.

Interconnections

When building neural networks, the output of a neuron serves as input to other neurons. Links among neurons carry a connection weight which describes how neurons affect each other. Links are excitatory or inhibitory depending on whether the weight is positive or negative. The most common formula for the input to a neuron v is

$$S_v = \sum_{i=1}^{n} w_i M_i(t)$$

where $M_i(t)$ is the firing rate of neuron m whose output is connected to the ith input of neuron v, and w_i is the weight on that link.

For example, a neural network architecture corresponding to the *Maximum Selector* model (see Arbib, 1989) is shown in Figure 4. The variables u and v represent membrane potential (analogous to m), and U and V represent firing rate (analogous to M).

The input to neuron v is given by

$$S_v = w_1 U_1 + w_2 U_2 + w_3 U_3 + \ldots + w_n U_n$$

while the input to the u_i neuron (there are n such equations) is

$$S_{ui} = w_m V + w_{ui} U_i + S_i$$

In NSL, these expressions describing interconnections among neurons in the neural network are described in a similar way. For example, the input to neuron v, represented by S_v, would be the summation of the outputs of all the neurons u multiplied by the corresponding connection weights w:

$$S_v = w_1 * U_1 + w_2 * U_2 + w_3 * U_3 + \ldots + w_n * U_n$$

The input to neuron u_i is represented by S_{ui} (there are n such equations)

$$S_{ui} = w_m * V + w_{ui} * U_i + S_i$$

Layers and Masks

When modeling thousands of neurons and their interconnections, it becomes extremely difficult to name every single one of them. Since in the brain we often find neural networks structured into two-dimensional homogeneous neural layers, with regular connection patterns between various layers, we extend the basic neuron abstraction into neural layers and *connection masks*.

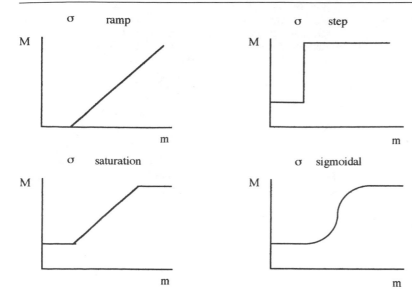

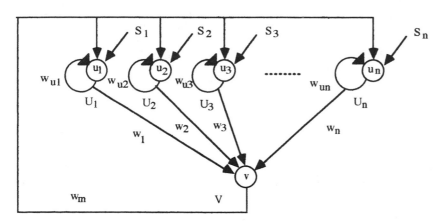

Figure 3. Some typical threshold functions.

Figure 4. The neural network shown corresponds to the architecture of the *Maximum Selector* model (see Arbib, 1989), where u_i and v represent membrane potentials, U_i and V represent firing rates, S_i represent inputs to the network, and w_i represent connection weights.

The computational advantage of introducing such concepts when describing a neural network is that neural layers and connection masks can then be concisely described as higher-level data structures. Instead of describing neurons on a one-by-one basis, a layer can be described as an array, and, similarly, the connections between layers can be described by a mask storing synaptic weights. An interconnection among neurons would then be processed by computing a spatial convolution of a mask and a layer. For example, as shown in Figure 5, if A represents an array of outputs from one layer of neurons, and B represents the array of inputs to another layer, and if the mask $W(k, l)$ (for $-d \le k, l \le d$) represents the synaptic weight from the $A(i + k, j + l)$ (for $-d \le k, l \le d$) elements to the $B(i, j)$ element for each i and j, we then have

$$B = \sum_{k=-d}^{d} \sum_{l=-d}^{d} W(k, l) A(i + k, j + l)$$

which can be described by a simple expression

$$B = W * A$$

In order to support layers and masks, the basic **DATA** structure in NSL is extended with two layer types, **VECTOR** and **MATRIX**, differing according to the number of dimensions they have. To simplify matters, masks, which may have any

rectangular shape, are also defined as layers whose values are interpreted in a different way.

For example, the layers of neurons shown in Figure 4 would be described by

```
VECTOR(S,n);
VECTOR(u,n);
VECTOR(U,n);
DATA(v);
DATA(V);
```

Sample Model

The complete set of equations describing the Maximum Selector model, shown in Figure 4, is

$$\tau_u \frac{du_i(t)}{dt} = -u_i + w_u f(u_i) - w_m g(v) - h_1 + s_i \qquad 1 \le i \le n$$

$$\tau_v \frac{dv}{dt} = -v + w_n \sum_{i=1}^{n} f(u_i) - h_2$$

where w_{ui} is the connection weight for the self-connection of

Figure 5. The symbol W represents the connection or convolution mask between layers A and B corresponding to the equation $B = W*A$. In this example W is a 3×3 mask which is overlapped over a window of A to obtain a single value in B.

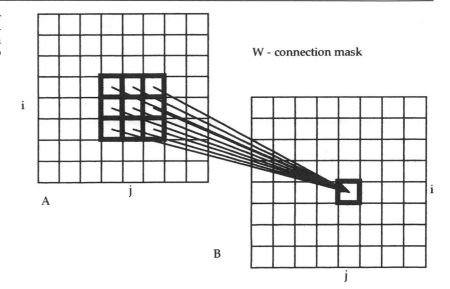

W - connection mask

u_i, $w_{u1} = w_{u2} = \ldots = w_{un} = w_u$ and $w_1 = w_2 = \ldots = w_n$ (in this particular model these connection weights are the same), h_1 and h_2 are constants, and the threshold functions are

$$f(u_i) = \begin{cases} 1 & u_i > 0 \\ 0 & u_i \leq 0 \end{cases} \qquad g(s_i) = \begin{cases} s_i & s_i > 0 \\ 0 & s_i \leq 0 \end{cases}$$

In NSL the description of the complete model is arranged into modules, the INIT_MODULE containing reinitialization statements, and the RUN_MODULE containing statements which are continuously executed as part of the simulation.

The above equations correspond to the following code arranged in two RUN_MODULEs:

```
RUN_MODULE(U)
{
    DIFF(u,τ_u) = −u + w_u*U − w_m*V − h_1 + S;
    U = step(u);
}

RUN_MODULE(V)
{
    DIFF(v,τ_v) = −v + SUM(w_n*U) − h_2;
    V = ramp(v);
}
```

Note that S, u, and U are vector layers and all operations are applied to the layer as a whole. SUM($w_n * U$) first multiplies the connection weight w_n by the firing rate U, and then returns a single value corresponding to the vector summation of the expression. This procedure is necessary since v is a single-element layer.

Discussion

NSL has been successfully utilized as a simulation tool for both biological and artificial neural networks, where various types of applications have been developed, such as the *visuomotor coordination* model (Arbib and Lee, 1993) and the *generation of saccades* model (Dominey and Arbib, 1992). The main challenge in the development of NSL, as well as with other simulation tools, is on one hand to provide a general-purpose user-friendly simulation environment, while on the other hand to be as efficient as possible in the time-consuming process of neural network simulation.

As NSL keeps on evolving, it will offer a distributed and parallel framework for the simulation of neural networks (Weitzenfeld and Arbib, 1991) integrating with *schema* models, as described in ASL, Abstract Schema Language (Weitzenfeld, 1993), to enable the development of hierarchical and distributed neural networks, such as are needed in robotics applications (Fagg et al., 1992).

Road Map: Implementation of Neural Networks
Background: I.1. Introducing the Neuron
Related Reading: Convolutional Networks for Images, Speech, and Time Series; Neurosimulators

References

Arbib, M. A., 1989, *The Metaphorical Brain 2: Neural Networks and Beyond*, New York: Wiley-Interscience. ◆

Arbib, M. A., and Lee, H. B., 1993, Anuran visuomotor coordination for detour behavior: From retina to motor schemas, in *From Animals to Animats 2: Proceedings of the Second International Conference on Simulation of Adaptive Behavior* (J.-A. Meyer, H. L. Roitblat, and S. Wilson, Eds.), Cambridge, MA: MIT Press, pp. 42–51.

De Schutter, E., 1992, A consumer guide to neuronal modeling software, *Trends Neurosci.*, 15:462–464. ◆

Dominey, P. F., and Arbib, M. A., 1992, A cortico-subcortical model for generation of spatially accurate sequential saccades, *Cereb. Cortex*, 2:153–175.

Fagg, A. H., King, I. K., Lewis, M. A., Liaw, J. S., and Weitzenfeld, A., 1992, A neural network based testbed for modeling sensorimotor integration in robotics applications, in *Proceedings of the International Joint Conference on Neural Networks*, New York: IEEE.

Hodgkin, A. L., and Huxley, A. F., 1952, A quantitative description of membrane current and its application to conduction and excitation in nerve, *J. Physiol. (Lond.)*, 117:500–544.

Rall, W., 1959, Branching dendritic trees and motoneuron membrane resistivity, *Exp. Neurol.*, 2:503–532.

Stroustrup, B., 1991, *The C++ Programming Language*, 2nd ed., Reading, MA: Addison-Wesley. ◆

Wegner, P., 1990, Concepts and paradigms of object-oriented programming, *SIGPLAN, OOPS Messenger*, 1:7–87. ◆

Weitzenfeld, A., 1991, *NSL: Neural Simulation Language*, Version 2.1, CNE-TR 91–05, Los Angeles, CA: Center for Neural Engineering, University of Southern California. ◆

Weitzenfeld, A., 1993, *A Hierarchical Computational Model for Distributed Heterogeneous Systems*, TR 93–02, Los Angeles, CA: Center for Neural Engineering, University of Southern California.

Weitzenfeld, A., and Arbib, M., 1991, A concurrent object-oriented framework for the simulation of neural networks, Proceedings of ECOOP/OOPSLA '90 Workshop on Object-Based Concurrent Programming, *SIGPLAN*, *OOPS Messenger*, 2:120–124.

Weitzenfeld, A., and Arbib, M. A., in press, NSL Neural Simulation Language, in *Neural Network Simulation Environments* (J. Skrzypek, Ed.), Norwell, MA: Kluwer Academic. ◆

Object Recognition

John E. Hummel

Introduction

A fundamental problem for theories of human object recognition is to explain how we recognize objects regardless of viewpoint (the position and size of the image on the retina, and the orientation in depth from which the object is viewed). Many neural network (NN) models of object recognition have been proposed, and in virtually all cases, a model's approach to the problem of *viewpoint invariance* (VI) plays the dominant role in determining its strengths and weaknesses as an account of human shape recognition. This article discusses some strategies NN researchers have employed in the attempt to model VI object recognition. There is a vast literature on pattern recognition in NNs, but this review is limited to models of VI object recognition as accounts of human object recognition.

Evaluating Models of Object Recognition

It is important to consider a few criteria for evaluating models of VI.

1. *Scaling.* Many modelers simplify the VI problem by considering only 2D objects (typically letters or words), and most models are tested with only a small number of objects (typically much fewer than 100). It is important to consider how the size and performance of a model changes with the number of objects it must recognize. In the case of 2D models, it is also critical to consider whether and how the model could be generalized to recognize 3D objects at different orientations in depth.

2. *Dealing with unfamiliarity.* People can easily appreciate the equivalence of two views of an unfamiliar object (Biederman and Gerhardstein, 1994). To account for this capacity, an NN must be able to activate a VI representation of an object's shape independent of any representation of the object's identity.

3. *Time course of recognition.* Behavioral data (e.g., Intraub, 1981) suggest that no stage of object recognition may require more than 100 ms. This figure is particularly important for models employing time-consuming procedures, such as transformations or iterative constraint satisfaction.

4. *Equivalence classes and errors.* A model of human object recognition must partition objects into natural equivalence classes. For example, people naturally classify both Camrys and Celicas as "cars." A model should show a similar capacity to generalize. Furthermore, a model should not only succeed where human recognition succeeds, but also fail where the human fails. For example, human recognition is largely unaffected by orientation in depth, but is adversely affected by rotations in the picture plane (e.g., when an object is upside down; cf. Jolicoeur, 1990).

Models of Object Recognition

Most NN models of VI can be characterized in terms of four approaches. These are: Direct Template Matching, Hierarchical Template Matching, Transform and Match, and Structural Description. These approaches are neither exhaustive nor mutually exclusive, but they provide a useful framework for understanding NN models of VI.

Direct Template Matching

Direct Template Matching (DTM) decomposes recognition into two stages. In the first, the to-be-recognized image is matched to specific views of each known object. In the second stage, VI is achieved by connecting different views of the same object to a single representation. A model by Poggio and Edelman (1990) provides an example. This model contains three layers of units. Units in the input layer respond to features at specific locations in the image. Units in the second layer respond to specific conjunctions of input units, each conjunction corresponding to one view of one object. In the second stage, VI is achieved by mapping the view units (Layer 2) to a single representation of the object in Layer 3.

The primary strength of DTM is that it is rapid, and likely possible within the 100 ms required for recognition by humans. Additionally, some DTM models simulate some aspects of human performance. However, DTM does not fare well on the other criteria. First, it does not scale well, storing separately all views of all known objects. More important, DTM can only match images to known views, so it can only treat two views of an object as equivalent if it has preexisting templates for each. Finally, DTM represents each view of an object in terms of the literal positions of the features available in that view, so it is probably too rigid to account for human capacity for generalization. For example, people have no difficulty recognizing a new model of car as a car, even if it is not similar to any known car in terms of the literal positions of its features.

Hierarchical Template Matching

Hierarchical Template Matching (HTM) achieves both pattern classification and VI gradually. An image is mapped through layers of units representing progressively more complex features at progressively less specific locations in the image. HTM is hierarchical in the sense that features in early layers are components of features in later layers; it is template-like in that the features detected in each layer are based on the literal positions of features in previous layers. Fukushima and Miyake's (1982) neocognitron (a model of character recognition rather than VI object recognition) is an example of HTM.

The NEOCOGNITRON (q.v.) starts with a pixel-based image of a numeral and achieves position-independent recognition of the numeral. It consists of several processing layers, each divided into *simple cells* (S-cells) and *complex cells* (C-cells). S-cells in a given layer detect features by responding to conjunctions of C-cells in the previous layer (in the first layer, S-cells respond to conjunctions of pixels). In the earliest layers, S-cells are highly location selective. For example, an S-cell in Layer 1 might respond to five pixels forming a particular horizontal line. Other S-cells in that layer would respond to horizontal lines at different locations in the image, and still others to lines at different orientations. C-cells receive excitation from small neighborhoods of S-cells representing a particular feature. For example, a given C-cell in Layer 1 might receive excitation from all S-cells in Layer 1 representing horizontal lines in a small region of the image. C-cells thus respond to the same features as S-cells, but with less location-specificity. By mapping an image through successive layers of S- and C-cells, the neocognitron gradually detects more complex features in less specific locations, eventually discarding location altogether. In the last layer, each C-cell responds to a complete numeral regardless of its position in the image.

HTM is attractive in its apparent neural plausibility. Like neurons in mammalian visual systems, units in HTM models have larger receptive fields in later layers. Bottom-up HTM is also consistent with the temporal constraint in that it performs no iterative procedures or transformations. Some HTM models also simulate some human perceptual phenomena, including capacity limitations, interference, and facilitation (cf. Mozer, 1991). Scaling is a concern with HTM. The neocognitron is quite large and recognizes only numerals. It is unclear how gracefully HTM will scale to recognize objects at different orientations in depth. A more serious weakness is that HTM cannot achieve VI for unfamiliar patterns because the representation of a pattern is only VI at the highest level of classification. For example, an unfamiliar pattern such as "$" might activate units in the neocognitron's intermediate layers, but it would not activate units in the highest (numeral) layer, and only the latter are completely indifferent to location.

Transform and Match

Transform and Match (TAM) models place images and stored representations into a common reference frame (e.g., via translation, scaling, and rotation) so that they may be directly compared (e.g., via DTM). TAM was first proposed by Pitts and McCulloch (1947) and has more recently been developed by Hinton (1981), Olshausen, Van Essen, and Anderson (1993; see ROUTING NETWORKS IN VISUAL CORTEX), and Seibert and Waxman (1992; see VISUAL PROCESSING OF OBJECT FORM AND ENVIRONMENT LAYOUT), among others. Many neurally plausible mechanisms exist for performing transformations. Hinton's model uses parallel constraint satisfaction to simultaneously settle on a transformation and an object interpretation; Seibert and Waxman's model employs log polar transformations to achieve invariance with translation, scale, and rotation in the picture plane, and links stored views together to recognize objects at different orientations in depth; and Olshausen et al.'s model uses multilayered feedforward networks to place images into correspondence with stored representations.

An advantage of TAM is that it achieves VI with fewer units than either DTM or HTM. Rather than encoding all views of an object (like DTM) or gradually achieving VI by means of a potentially large and complex feature hierarchy (like HTM), TAM discounts viewpoint by means of transformations. A disadvantage is that transformations take time. Time costs are well established for recognizing objects misoriented in the picture plane (cf. Jolicoeur, 1990), but there are no such costs for translation, size, or left-right reflection (cf. Cooper, Biederman, and Hummel, 1992). Another disadvantage of many TAM models is that they are adept at dealing with rotations in the picture plane but not rotations in depth, precisely the opposite of the pattern observed in humans. Finally, TAM models typically perform a type of template match on the transformed image, making them subject to the rigidity and intolerance-of-novelty limitations of DTM.

Structural Description

Structural Description (SD) models discard the literal positions of an object's features and instead code their interrelations. Hummel and Biederman's (1992) model is an example of this approach. This model consists of seven layers of units. Local interactions among units in the first two layers group image features into volumetric parts by synchronizing oscillations in their outputs. (See DYNAMIC LINK ARCHITECTURE for more about synchrony in NNs.) Layers 3 through 5 use the synchronized outputs of Layer 2 to derive a structural description of the object in terms of its parts and their interrelations. This description activates units in layers 6 and 7 representing the object category. Like human performance, the model's performance is invariant with the position and size of the image, largely invariant with orientation in depth and left-right reflection, and sensitive to rotations in the picture plane.

The critical difference between SD and the other approaches (DTM, HTM, and TAM) is that in SD, spatial relations are coded explicitly. For example, the Hummel and Biederman (H&B) model would represent that a lamp's shade is above its base by activating "above" (Layer 5) in response to the shade, and "below" in response to the base. Representing relations as units in the network introduces the need to bind relations to parts (e.g., to represent that "above" refers to the shade rather than the base). The H&B model uses the synchrony established in layers 1 and 2 to bind relations to parts. For example, units representing the shade would fire in synchrony with units representing "above" while units for the base fire in synchrony with "below."

SD models are attractive in their scaling properties and their capacity to account for findings in human object recognition. They are efficient because independently representing parts and relations obviates the need to represent all possible parts in all possible positions, sizes, and orientations. SD is also consistent with recent data showing that parts play an important role in object recognition and that transformations are often not required for object recognition. Finally, SD does not suffer the rigidity and inability to deal with novelty that characterize the template-based approaches (e.g., H&B's model can generate a VI description of an object regardless of whether the object is familiar). A disadvantage of SD is that explicitly binding parts to relations takes time and is subject to capacity limitations (cf. Hummel and Biederman, 1992). One weakness of H&B's model (but not of SD in general) is that, although it specifies which relations are bound to each part (e.g., the lamp shade is above), it does not specify the other parts to which those relations refer (e.g., *what* the shade is above). SD is also sensitive to how an image is segmented into parts; the same image, decomposed in different ways, can result in very different descriptions.

Discussion

Each approach to object recognition discussed here has strengths and weaknesses, both as a solution to the problem of

VI and as way to model human object recognition. Rather than a set of discrete categories into which NN models neatly fit, these approaches provide a way to anticipate a model's properties. Most models are hybrids, adopting techniques and properties from two or more approaches. For example, Seibert and Waxman's (1992) model is a hybrid of TAM and HTM.

It remains an open question which approaches will provide the best account of which aspects of human performance, but some speculation is possible. DTM has been widely applied to problems requiring sensitivity to tiny differences between objects (e.g., face recognition). HTM is suggestive in its similarity to the known organization of mammalian visual systems. TAM reduces the representational demands of DTM and HTM, and may constitute a link between recognition and visual tasks such as imagery. SD is promising in its capacity to explicitly code relations.

Road Maps: Artificial Intelligence and Neural Networks; Vision
Related Reading: Binding in the Visual System; Structured Connectionist Models

References

Biederman, I., and Gerhardstein, P. C., 1994, Recognizing depth-rotated objects: Evidence and conditions for 3-dimensional viewpoint invariance, *J. Exp. Psychol. Hum. Percept. Perform.*, 19:1162–1182.

Cooper, E. E., Biederman, I., and Hummel, J. E., 1992, Metric invariance in object recognition: A review and further evidence, *Can. J. Psychol.*, 46:1–24. ◆

Fukushima, K., and Miyake, S., 1982, Neocognitron: A new algorithm for pattern recognition tolerant of deformations and shifts in position, *Pattern Recognition*, 15:455–469.

Hinton, G. E., 1981, A parallel computation that assigns canonical object-based frames of reference, in *Proceedings of the Seventh International Joint Conference on Artificial Intelligence*, Vancouver, Canada, pp. 683–685.

Hummel, J. E., and Biederman, I., 1992, Dynamic binding in a neural network for shape recognition, *Psychol. Rev.*, 99:480–517.

Intraub, H., 1981, Identification and processing of briefly glimpsed visual scenes, in *Eye Movements: Cognition and Visual Perception* (D. Fisher, R. A. Monty, and J. W. Sender, Eds.), Hillsdale, NJ: Erlbaum, pp. 181–190.

Jolicoeur, P., 1990, Identification of disoriented objects: A dual systems theory, *Mind Lang.*, 5:387–410.

Mozer, M. C., 1991, *The Perception of Multiple Objects: A Connectionist Approach*, Cambridge, MA: MIT Press.

Olshausen, B., Van Essen, D., and Anderson, C., 1993, A neurobiological model of visual attention and invariant pattern recognition based on dynamic routing of information, *J. Neurosci.*, 13:4700–4719.

Pitts, W., and McCulloch, W. W., 1947, How we know universals, *Bull. Math. Biophys.*, 9:127–147.

Poggio, T., and Edelman, S., 1990, A neural network that learns to recognize three-dimensional objects, *Nature*, 317:314–319.

Seibert, M., and Waxman, A. M., 1992, Learning and recognizing 3D objects from multiple views in a neural system, in *Neural Networks for Perception*, vol. 1, *Human and Machine Perception* (H. Wechsler, Ed.), San Diego, CA: Academic Press, pp. 427–444.

Ocular Dominance and Orientation Columns

Kenneth D. Miller

Introduction

The classic example of activity-dependent neural development is the formation of ocular dominance columns in the primary visual cortex of the cat or monkey (reviewed in Miller and Stryker, 1990). The primary visual cortex (V1) receives signals from the lateral geniculate nucleus of the thalamus (LGN), which in turn receives input from the retinas of the two eyes (Figure 1).

To describe ocular dominance columns, several terms must be defined. First, the *receptive field* of a cortical cell refers to the area on the retinas in which appropriate light stimulation evokes a response in the cell, and also to the pattern of light stimulation that evokes such a response. Second, a *column* is defined as follows: V1 extends many millimeters in each of two "horizontal" dimensions. Receptive field positions vary continuously along these dimensions, forming a *retinotopic* map, a continuous map of the visual world. In the third, "vertical" dimension, the cortex is about 2 mm in depth, and consists of six layers. Receptive field positions do not significantly vary through this depth. Such organization, in which cortical properties are invariant through the vertical depth of cortex but vary horizontally, is called *columnar* organization and is a basic feature of cerebral cortex. Finally, *ocular dominance*, or eye preference, describes the degree to which a cortical cell's responses are better driven by stimulation of one eye or the other. Like retinotopy, ocular dominance has a columnar organiza-

tion: alternating stripes or patches of cortex are dominated throughout the cortical depth by a single eye, and are known as *ocular dominance columns*.

The anatomical basis for ocular dominance columns is the segregated pattern of termination of the LGN inputs to V1 (Figures 1 and 2A). Inputs serving a single eye terminate in alternating stripes or patches of cortex. This segregation arises early in development. Initially, LGN inputs project in an overlapping manner, without apparent distinction by eye represented. The terminal arbors of individual LGN inputs extend horizontally in V1 for distances as large as 2 mm (for comparison, a typical spacing between cortical cells is perhaps 20 μm). Subsequently, beginning either prenatally or shortly after birth depending on the species, the inputs representing each eye become confined to the alternating, approximately 0.5-mm-wide, ocular dominance patches.

This segregation results from an activity-dependent competition between the geniculate terminals serving the two eyes. The signal indicating that different terminals represent the same eye appears to be the correlations in their neural activities. These correlations exist because of both spontaneous activity, which is locally correlated within each retina, and visually induced activity, which correlates the activities of retinotopically nearby neurons within each eye and, to a lesser extent, between the eyes. The segregation process is competitive. If one eye is caused to have less activity than the other during a critical period in which the columns are forming, the more active eye

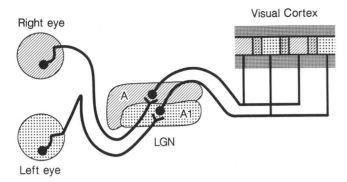

Figure 1. Schematic of the mature visual system. Retinal ganglion cells from the two eyes project to separate layers of the lateral geniculate nucleus (LGN). Neurons from these two layers project to separate patches or stripes within visual cortex (V1). Binocular regions (receiving input from both eyes) are depicted at the borders between the eye-specific patches. (Reprinted by permission from Miller, K. D., Keller, J. B., and Stryker, M. P., 1989, Ocular dominance column development: Analysis and simulation, *Science*, 245:605–615. © 1989 by the AAAS.)

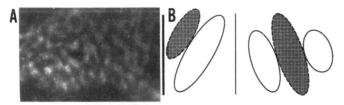

Figure 2. *A*, Ocular dominance columns from cat V1. A horizontal cut through the LGN recipient layer of V1 is shown. Terminals serving a single eye are labeled white. Dark regions at edges are out of plane containing LGN terminals. Region shown is 5.3 × 7.9 mm. Photograph generously supplied by Dr. Y. Hata. *B*, Two examples of simple cell receptive fields (RFs). Regions of the visual field from which a simple cell receives ON-center (white) or OFF-center (dark) input are shown. Note. Ocular dominance columns (*A*) represent an alternation, across cortex, in the type of input (left or right eye) received by different cortical cells, while a simple cell RF (*B*) represents an alternation across visual space in the type of input (ON or OFF center) received by a *single* cortical cell.

takes over most of the cortical territory; but the eye with reduced activity suffers no loss of projection strength in retinotopic regions in which it lacks competition from the other eye.

Orientation columns are another striking feature of visual cortical organization. Most V1 cells are orientation selective, responding selectively to light/dark edges over a narrow range of orientations. The preferred orientation of cortical cells varies regularly and periodically across the horizontal dimension of cortex, and is invariant in the vertical dimension. It has not yet been possible to study whether the initial development of orientation selectivity is activity dependent, although its maturation clearly is (see discussion in Miller, 1994). However, it has long been a popular notion that orientation selectivity, like ocular dominance, may develop by a process of activity-dependent synaptic competition.

The inputs from LGN to V1 serving each eye are of two types: ON-center and OFF-center. Both kinds of cells have circularly symmetric, orientation-insensitive receptive fields, and both respond to contrast rather than uniform luminance. ON-center cells respond to light against a dark background, or to light onset; OFF-center cells respond to dark against a light background, or to light offset. In the cat, the orientation-selective V1 cells that receive the bulk of LGN input are *simple cells*: cells with receptive fields consisting of alternating oriented subregions that receive exclusively ON-center or exclusively OFF-center input (Figure 2*B*). One theory for the development of orientation selectivity is that, like ocular dominance, it develops through a competition between two input populations: in this case, a competition between the ON-center and the OFF-center inputs (Miller, 1994).

Correlation-Based Models

To understand ocular dominance and orientation column formation, two processes must be understood: (1) The development of *receptive field structure*: Under what conditions do receptive fields become monocular (drivable only by a single eye) or orientation selective? (2) The development of *periodic cortical maps* of receptive field properties: What leads ocular dominance or preferred orientation to vary periodically across the horizontal dimensions of cortex, and what determines the periodic length scales of these maps? Typically, the problem is simplified by consideration of a two-dimensional model cortex, ignoring the third dimension in which properties such as ocular dominance and orientation are invariant.

One approach to addressing these problems is to begin with a hypothesized mechanism of synaptic plasticity and to study the outcome of cortical development under such a mechanism. Most commonly, theorists have considered a "Hebbian synapse" (see HEBBIAN SYNAPTIC PLASTICITY): a synapse whose strength is increased when pre- and postsynaptic firing are correlated, and possibly decreased when they are anticorrelated. Other mechanisms can lead to similar dynamics, in which synaptic plasticity depends on the correlations among the activities of competing inputs. Models based on such mechanisms are referred to as correlation-based models (Miller, 1990).

The Von Der Malsburg Model of V1 Development

Von der Malsburg (1973; von der Malsburg and Willshaw, 1976) first formulated a correlation-based model for the development of visual cortical receptive fields and maps. His model had two basic elements. First, synapses of LGN inputs onto cortical neurons were modified by a Hebbian rule that is *competitive*, so that some synapses were strengthened only at the expense of others. He enforced the competition by holding constant the total strength of synapses converging on each cortical cell (conservation rule). Second, cortical cells tended to be activated in *clusters*, due to intrinsic cortical connectivity—for example, short-range horizontal excitatory connections and longer-range horizontal inhibitory connections.

The conservation rule leads to competition among the inputs to a single target cell. Inputs that tend to be coactivated—that is, that have correlated activities—are mutually reinforcing, working together to activate the postsynaptic cells and thus to strengthen their own synapses. Different patterns that are mutually un- or anti-correlated compete, since strengthening of some synapses means weakening of others. Cortical cells eventually develop receptive fields responsive to a correlated pattern of inputs.

The clustered cortical activity patterns lead to competition between different groups of cortical cells. Each input pattern comes to be associated with a cortical cluster of activity. Overlapping cortical clusters contain many coactivated cortical cells, and thus become responsive to overlapping, correlated input patterns. Adjacent, nonoverlapping clusters contain many anti-correlated cortical cells, and thus become responsive to un- or anti-correlated input patterns. Thus, over distances on the scale of an activity cluster, cortical cells will have similar response properties; while, on the scale of the distance between nonoverlapping clusters, cortical cells will prefer un- or anti-correlated input patterns. This combination of local continuity and larger-scale heterogeneity leads to continuous, periodic cortical maps of receptive field properties.

In computer simulations, this model was applied to the development of orientation columns (von der Malsburg, 1973) and ocular dominance columns (von der Malsburg and Willshaw, 1976). For orientation columns, inputs were activated in oriented patterns of all possible orientations. Individual cortical cells then developed selective responses, preferring one such oriented pattern, with nearby cortical cells preferring nearby orientations. For ocular dominance columns, inputs were activated in monocular patterns consisting of a localized set of inputs from a single eye. Individual cortical cells came to be driven exclusively by a single eye, and clusters of cortical cells came to be driven by the same eye. The final cortical pattern consisted of alternating stripes of cortical cells preferring a single eye, with the width of a stripe approximately set by the diameter of an intrinsic cluster of cortical activity.

In summary, a competitive Hebbian rule leads individual receptive fields to become selective for a correlated pattern of inputs. Combined with the idea that the cortex is activated in intrinsic clusters, this process suggests an origin for cortical maps: coactivated cells in a cortical cluster tend to become selective for similar, coactivated patterns of inputs. These basic ideas are used in most subsequent models.

Mathematical Formulation

A typical correlation-based model is mathematically formulated as follows (von der Malsburg, 1973; Linsker, 1986; Miller, Keller, and Stryker, 1989). Let $x, y, \ldots$ represent retinotopic positions in V1, and let $\alpha, \beta, \ldots$ represent retinotopic positions in the LGN. Let $S^\mu(x, \alpha)$ be the synaptic strength of the connection from α to x of the LGN projection of type μ, where μ may signify left-eye, right-eye, ON-center, OFF-center, and so on. Let $B(x, y)$ represent the synaptic strength and sign of connection from the cortical cell at y to that at x. For simplicity, $B(x, y)$ is assumed to take different signs for a fixed y as x varies, but alternatively, separate excitatory-projecting and inhibitory-projecting cortical neurons may be used. Let $a(x)$ and $a^\mu(\alpha)$ represent the activity of a cortical or LGN cell, respectively.

The activity $a(x)$ of a cortical neuron is assumed to depend on a linear combination of its inputs:

$$a(x) = f_1\left(\sum_{\mu, \alpha} S^\mu(x, \alpha)a^\mu(\alpha) + \sum_y B(x, y)a(y)\right) \quad (1)$$

Here, f_1 is some monotonic function such as a sigmoid or linear threshold.

A Hebbian rule for the change in feedforward synapses can be expressed

$$\Delta S^\mu(x, \alpha) = A^\mu(x, \alpha)f_2[a(x)]f_3[a^\mu(\alpha)] \quad (2)$$

Here, $A(x, \alpha)$ is an "arbor function," expressing the number of synapses of each type from α to x; a minimal form is $A(x, \alpha) = 1$ if there is a connection from α to x, $A(x, \alpha) = 0$ otherwise. A typical form for the functions f_2 and f_3 is $f(a) = (a - \langle a \rangle)$, where $\langle a \rangle$ indicates an average of a over input patterns. This

function yields a *covariance rule*: synaptic change depends on the covariance of postsynaptic and presynaptic activity.

Next, the Hebbian rule must be made *competitive*. This purpose can be accomplished by conserving total synaptic strength over the postsynaptic cell (von der Malsburg, 1973), which in turn may be done either subtractively or multiplicatively (Miller and MacKay, 1994). The corresponding equations are

$$\frac{d}{dt}S^\mu(x, \alpha) = \Delta S^\mu(x, \alpha) - \varepsilon(x)A(x, \alpha) \quad \text{(Subtractive)} \quad (3)$$

$$\frac{d}{dt}S^\mu(x, \alpha) = \Delta S^\mu(x, \alpha) - \gamma(x)S^\mu(x, \alpha) \quad \text{(Multiplicative)}$$

$$(4)$$

where $\varepsilon(x) = [\Sigma_{\kappa, \alpha}\Delta S^\kappa(x, \alpha)]/[\Sigma_{\kappa, \alpha}A(x, \alpha)]$, and $\gamma(x) = [\Sigma_{\kappa, \alpha}\Delta S^\kappa(x, \alpha)]/[\Sigma_{\kappa, \alpha}S^\kappa(x, \alpha)]$. Either form of constraint ensures that $\Sigma_{\mu, \alpha}dS^\mu(x, \alpha)/dt = 0$. Alternative methods have been developed to force Hebbian rules to be competitive (see BCM THEORY OF VISUAL CORTICAL PLASTICITY; Miller and MacKay, 1994).

Finally, synaptic weights may be limited to a finite range, $s_{\min}A(x, \alpha) \leq S^\mu(x, \alpha) \leq s_{\max}A(x, \alpha)$. Typically, $s_{\min} = 0$ and $s_{\max}$ is some positive constant.

Semilinear Models

In semilinear models, the f's in Equations 1 and 2 are chosen to be linear. Then, after substituting for $a(x)$ from Equation 1 and averaging over input patterns (assuming that all inputs have identical mean activity, and that changes in synaptic weights are negligibly small over the averaging time), Equation 2 becomes

$$\Delta S^\mu(x, \alpha) = \lambda A(x, \alpha)\sum_{y, \beta, \kappa} I(x - y)[C^{\mu\kappa}(\alpha - \beta) - k_2]S^\kappa(y, \beta)$$
$$+ k_1 A(x, \alpha) \quad (5)$$

Here, $I(x - y)$ is an element of the intracortical interaction matrix $\mathbf{I} \equiv (1 - \mathbf{B})^{-1} = 1 + \mathbf{B} + \mathbf{B}^2 + \ldots$, where the matrix $\mathbf{B}$ is defined in Equation 1. This relationship summarizes intracortical synaptic influences including contributions by $0, 1, 2, \ldots$ synapses. The covariance matrix $C^{\mu\kappa}(\alpha - \beta) = \langle(a^\mu(\alpha) - \bar{a})(a^\kappa(\beta) - \bar{a})\rangle$ expresses the covariation of input activities. The factors λ, k_1, and k_2 are constants. Translation invariance has been assumed in both cortex and LGN.

When there are two competing input populations, Equation 5 can be further simplified by transforming to sum and difference variables: $S^S \equiv S^1 + S^2$, $S^D \equiv S^1 - S^2$. Assuming equivalence of the two populations (so that $C^{11} = C^{22}, C^{12} = C^{21}$), Equation 5 becomes

$$\Delta S^S(x, \alpha) = \lambda A(x, \alpha)\sum_{y, \beta} I(x - y)[C^S(\alpha - \beta) - 2k_2]S^S(y, \beta)$$
$$+ 2k_1 A(x, \alpha) \quad (6)$$

$$\Delta S^D(x, \alpha) = \lambda A(x, \alpha)\sum_{y, \beta} I(x - y)C^D(\alpha - \beta)S^D(y, \beta) \quad (7)$$

Here, $C^S \equiv C^{11} + C^{12}$, $C^D \equiv C^{11} - C^{12}$.

How Semilinear Models Behave

Linear equations like Equations 6 and 7 can be understood by finding the eigenvectors or "modes" of the operators on the right side of the equation (see PATTERN FORMATION, BIOLOGICAL). The eigenvectors are the synaptic weight patterns that grow independently and exponentially, each at its own rate. The fastest-growing eigenvectors typically dominate development and determine basic features of the final pattern, although the final pattern ultimately is stabilized by nonlinearities such as the limits on the range of synaptic weights or the nonlinearity involved in multiplicative renormalization (Equation 4).

I will focus on the behavior of Equation 7 for S^D. The variable S^D describes the difference in the strength of two competing input populations. Thus, it is the key variable describing the development of ocular dominance segregation, or development under an ON-center/OFF-center competition. For analysis of Equation 6, see MacKay and Miller (1990).

Equation 7 can be simply solved in the case of full connectivity from the LGN to the cortex, when $A(x, \alpha) \equiv 1$ for all x and α. Then modes of $S^D(x, \alpha)$ of the form $e^{ikx}e^{il\alpha}$ grow exponentially and independently, with rate proportional to $\tilde{I}(k)\tilde{C}^D(l)$, where $\tilde{I}$ and $\tilde{C}^D$ denote the Fourier transforms of I and C^D, respectively. The wave number k determines the wavelength $2\pi/|k|$ of an oscillation of S^D across cortical cells, while the wave number l determines the wavelength $2\pi/|l|$ of an oscillation of S^D across geniculate cells. The fastest-growing modes, which will dominate early development, are determined by the k and l that maximize $\tilde{I}(k)$ and $\tilde{C}^D(l)$, respectively. The peak of a function's Fourier transform corresponds to the cosine wave that best matches the function, and thus represents the "principal oscillation" in the function.

To understand these modes (Figure 3), consider first the set of inputs received by a single cortical cell, that is, the shape of the mode for a fixed cortical position x. This can be regarded as the "receptive field" of the cortical cell. Each receptive field oscillates with wave number l. This oscillation, of $S^D \equiv S^1 - S^2$, is an oscillation between receptive field subregions dominated by S^1 inputs and subregions dominated by S^2 inputs. Thus, in ocular dominance competition, monocular cells (cells whose entire receptive fields are dominated by a single eye) are formed only by modes with $l = 0$ (no oscillation). Monocular cells thus dominate development if the peak of the Fourier transform of the C^D governing left/right competition is at $l = 0$. Now instead consider an ON/OFF competition: S^1 and S^2 represent ON- and OFF-center inputs from a single eye. Then the receptive fields of modes with nonzero l resemble simple cells: they receive predominantly ON-center and predominantly OFF-center inputs from successive, alternating subregions of the visual world. Thus, simple cells can form if the C^D governing ON/OFF competition has its peak at a nonzero l.

Now consider the arborizations or "projective fields" projecting from a single geniculate point, that is, the shape of the mode for a fixed geniculate position α. These oscillate with wave number k. In ocular dominance competition, this means that left- and right-eye cells from α project to alternating patches of cortex. When monocular cells form ($l = 0$), these alternating patches of cortex are the ocular dominance columns: alternating patches of cortex receiving exclusively left-eye or exclusively right-eye input. Thus, the width of ocular dominance columns—the wavelength of alternation between right-eye- and left-eye-dominated cortical cells—is determined by the peak of the Fourier transform of the intracortical interaction function I. In ON/OFF competition, with $l \neq 0$, the identity of the cortical cells receiving the ON-center or OFF-center part of the projection varies as α varies, so individual cortical cells receive both ON- and OFF-center input, but from distinct subregions of the receptive field.

In summary, there is an oscillation across receptive fields, with wave number l determined by the peak of $\tilde{C}^D$, and an oscillation across arbors, with wave number k determined by the peak of $\tilde{I}$ (see Figure 3). These two oscillations are "knit together" to determine the overall pattern of synaptic connectivity. The receptive field oscillation, which matches the receptive field to the correlations, quantitatively describes von der Malsburg's finding that individual receptive fields become selective for a correlated pattern of inputs. Similarly, the arbor oscillation matches projective fields to the intracortical interactions, and thus to the patterns of cortical activity clusters. This pattern of oscillations quantitatively describes the relationship between activity clusters and maps. Note that the factor e^{ikx} can be regarded as inducing a phase shift, for varying x, in the structure of receptive fields. Thus, cortical cells that are nearby on the scale of the arbor oscillation have similar receptive fields, while cells $\frac{1}{2}$ wavelength apart have opposite receptive fields.

The competitive, renormalizing terms (Equations 3 and 4) do not substantially alter this picture, except that multiplicative renormalization can suppress ocular dominance development in some circumstances (Miller and MacKay, 1994). These results hold also for localized connectivity (finite arbors), and thus generally characterize the behavior of semilinear models (Miller, 1990; Miller and Stryker, 1990). The major difference in the case of localized connectivity is that, if k or l corresponds to a wavelength larger than the diameter of connectivity from or to a single cell, then it is equivalent to $k = 0$ or $l = 0$, respectively.

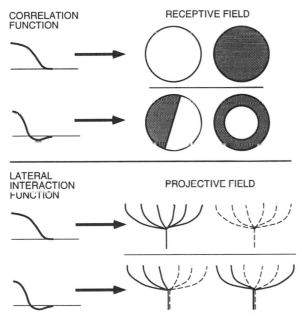

CORRELATION FUNCTION

RECEPTIVE FIELD

LATERAL INTERACTION FUNCTION

PROJECTIVE FIELD

Figure 3. Schematic of the outcome of semilinear correlation-based development. *Top:* The correlation function (C^D) determines the structure of receptive fields (RFs). White RF subregions indicate positive values of S^D; dark subregions, negative values. If C^D oscillates, there is a corresponding oscillation in the type of input received by individual cortical cells, as in simple cell RFs. Alternative RF structures could form, as shown; but oriented simple-cell-like outcomes predominate for reasonable parameters (Miller, 1994). When C^D does not oscillate, individual cortical cells receive only a single type of input, as in ocular dominance segregation. *Bottom:* The intracortical interactions (I) similarly determine the structure of projective fields. Here, solid lines indicate positive values of S^D; dotted lines indicate negative values.

Understanding Ocular Dominance and Orientation Columns with Semilinear Models

This understanding of semilinear models leads to simple models for the development of both ocular dominance columns

(Miller et al., 1989) and orientation columns (Miller, 1994), as follows.

Monocular cells develop through a competition of left- and right-eye inputs in a regime in which $\tilde{C}^D(l)$ is peaked at $l = 0$. The wavelength of ocular dominance column alternation is then determined by the peak of $\tilde{I}(k)$.

Orientation-selective simple cells develop through a competition of ON-center and OFF-center inputs in a regime in which $\tilde{C}^D(l)$ is peaked at $l \neq 0$. The mean wavelength of alternation of ON-center and OFF-center subregions in the simple cells' receptive fields is determined by the peak of $\tilde{C}^D(l)$. This wavelength corresponds to a cell's preferred spatial frequency under stimulation by sinusoidal luminance gratings. In individual modes, all cortical cells have the same preferred orientation, but their spatial phase varies periodically with cortical position. The mixing of such modes of all orientations leads to a periodic variation of preferred orientation across cortex.

This model of ocular dominance column formation is similar to that of von der Malsburg (von der Malsburg and Willshaw, 1976). The latter model assumed anti-correlation between the two eyes; this assumption was required because of the use of multiplicative renormalization (Equation 4). With subtractive renormalization (Equation 3), ocular dominance column formation can occur even with partial correlation of the two eyes (Miller and MacKay, 1994). The model can be compared to experiment, particularly through the prediction of the relation between intracortical connectivity and ocular dominance column width.

The model of orientation-selective cell development is quite different from that of von der Malsburg (1973). Von der Malsburg postulated that oriented input patterns lead to the development of orientation-selective cells. The ON/OFF model instead postulates that ON/OFF competition results in oriented receptive fields, in the absence of oriented input patterns; the circular symmetry of the input patterns is spontaneously broken. This symmetry-breaking potential of Hebbian development was first discovered by Linsker (1986). In all of these models, the continuity and periodic alternation of preferred orientation is due to the intracortical connectivity. The ON/OFF model can be compared to experiment most simply by the measurement of C^D, to determine whether it has the predicted oscillation.

Related Semilinear Models

Linsker (1986) proposed a model that was highly influential in two respects. First, he pointed out the potential of Hebbian rules to break symmetry spontaneously, yielding orientation-selective cells given approximately circularly symmetric input patterns. Second, he demonstrated that Hebbian rules could lead to segregation *within* receptive fields, so that a cell came to receive purely excitatory or purely inhibitory input in alternating subregions of the receptive field. This model has been thoroughly analyzed in MacKay and Miller (1990). Two factors underlay the results. One factor was that oscillations in a correlation function can induce oscillations in a receptive field, as described earlier. The other factor was a constraint in the model fixing the percentage of positive or negative synapses received by a cell; this forced an alternation of positive and negative subregions even when the correlation function did not oscillate.

Tanaka (1991; Miyashita and Tanaka, 1992) has independently formulated models of ocular dominance and orientation columns that are similar to those we have described. The major difference is that he works in a regime in which each cortical cell comes to receive only a single LGN input. Tanaka defines cortical receptive fields as the convolution of the input arrangement with the intracortical interaction function. This definition means that a cortical cell's receptive field is due to its single input from the LGN, plus its input from all other cortical cells within reach of the intracortical interaction function. Thus, orientation selectivity in this model arises from the breaking of circular symmetry in the pattern of inputs to different cortical cells, rather than to individual cortical cells.

The Problem of Map Structure

The preceding models account well for basic features of primary visual cortex. However, many details of real orientation maps are not replicated by these models (Erwin, Obermayer, and Schulten, in press; Wolf et al., 1994). One reason may be the simplicity of the model of cortex: the real cortex is three-dimensional, rather than two; has cell-specific connectivity, rather than connectivity that depends only on distance; and has plastic rather than fixed intracortical connections. Another reason is that the details of map structure inherently involve non-linearities, by which the fastest-growing modes interact and compete; whereas the semilinear framework only focuses on early pattern formation, in which the fastest-growing modes emerge and mix randomly without interacting.

Some simple models that focus on map development rather than receptive field development strikingly match the map structures observed in monkeys (Erwin et al., in press). One such model (Obermayer, Blasdel, and Schulten, 1992) uses the self-organizing map (SOM) of Kohonen (see SELF-ORGANIZING FEATURE MAPS: KOHONEN MAPS). In the SOM, only a single cluster of cortical cells is activated in response to a given input pattern. This is an abstraction of the idea that the cortex responds in localized activity clusters. Otherwise, the SOM is much like the correlation-based models. However, an abstract representation of the input is used. In correlation-based models, the input space may have thousands of dimensions, one for each input cell. In the SOM model of visual cortex, the input space instead has five dimensions: two represent retinotopic position, and one represents each of ocular dominance, orientation selectivity, and preferred orientation. Each cortical cell receives five "synapses," corresponding to these five "inputs." Assumptions are made as to the relative "size" of, or variance of the input ensemble along, each dimension. There is no obvious biological interpretation for this comparison between dimensions. Under certain such assumptions, Hebbian learning leads to maps of orientation and ocular dominance that are, in detail, remarkably like those seen in macaque monkeys (Erwin et al., in press; Obermayer et al., 1992).

The SOM and other models based on the "elastic net" algorithm (Durbin and Mitchison, 1990; see also CONSTRAINED OPTIMIZATION AND THE ELASTIC NET) lead to locally continuous mappings in which a constant distance across the cortex corresponds to a roughly constant distance in the reduced "input space." As a result, when one input feature is changing rapidly across cortex, the others are changing slowly. Thus, the models predict that orientation changes rapidly where ocular dominance changes slowly, and vice versa. It may be this feature that is key to replicating the details of macaque maps. A model that forces such a relationship to develop between ocular dominance and orientation, while assuring periodic representation of each, also gives a good match to primate visual maps (Swindale, 1992).

Cat orientation maps also have significant structure that could not arise from simple linear considerations (Wolf et al.,

1994). Analysis suggests that this map could result from a local "diffusion" of preferred orientations; it will be interesting to develop a biologically interpretable model of such a process.

Open Questions

Among the many open questions in the field are these: How can biologically interpretable models replicate the details of cortical maps? Might orientation selectivity arise from early oriented wave patterns of retinal activity or other mechanisms, rather than through ON/OFF competition? How might intracortical plasticity affect receptive field and map development (Sirosh and Mikkulainen, in press)? How might input correlations affect column size (Goodhill, 1993)? How will development be altered by incorporation of more realistic cortical connectivity rules and more realistic nonlinear learning rules? For example, might input correlations help determine the self-organization of plastic intracortical connections or the size of nonlinearly determined cortical activity clusters, each of which in turn would shape the pattern of input synapses? These and other questions may be answered soon.

Acknowledgments. K. D. Miller is supported by grants from the Lucille P. Markey Charitable Trust and the Searle Scholars' Program.

Road Map: Development and Regeneration of Neural Networks
Background: I.3. Dynamics and Adaptation in Neural Networks
Related Reading: Development and Regeneration of Eye-Brain Maps; Hebbian Synaptic Plasticity: Comparative and Developmental Aspects; Visual Cortex Cell Types and Connections

References

Durbin, R., and Mitchison, G., 1990, A dimension reduction framework for understanding cortical maps, *Nature*, 343:644–647.

Erwin, E., Obermayer, K., and Schulten, K., in press, Models of orientation and ocular dominance columns in the visual cortex: A critical comparison, *Neural Computat.*

Goodhill, G. J., 1993, Topography and ocular dominance: A model exploring positive correlations, *Biol. Cybern.*, 69:109–118.

Linsker, R., 1986, From basic network principles to neural architecture (series), *Proc. Natl. Acad. Sci. USA*, 83:7508–7512, 8390–8394, 8779–8783.

MacKay, D. J. C., and Miller, K. D., 1990, Analysis of Linsker's applications of Hebbian rules to linear networks, *Network*, 1:257–298.

Miller, K. D., 1990, Correlation-based models of neural development, in *Neuroscience and Connectionist Theory* (M. A. Gluck and D. E. Rumelhart, Eds.), Hillsdale, NJ: Erlbaum, pp. 267–353. ◆

Miller, K. D., 1994, A model for the development of simple cell receptive fields and the ordered arrangement of orientation columns through activity-dependent competition between ON- and OFF-center inputs, *J. Neurosci.*, 14:409–441.

Miller, K. D., Keller, J. B., and Stryker, M. P., 1989, Ocular dominance column development: Analysis and simulation, *Science*, 245:605–615.

Miller, K. D., and MacKay, D. J. C., 1994, The role of constraints in Hebbian learning, *Neural Computat.*, 6:98–124.

Miller, K. D., and Stryker, M. P., 1990, The development of ocular dominance columns: Mechanisms and models, in *Connectionist Modeling and Brain Function: The Developing Interface* (S. J. Hanson and C. R. Olson, Eds.), Cambridge, MA: MIT Press, pp. 255–350. ◆

Miyashita, M., and Tanaka, S., 1992, A mathematical model for the self-organization of orientation columns in visual cortex, *NeuroReport*, 3:69–72.

Obermayer, K., Blasdel, G. G., and Schulten, K., 1992, A statistical mechanical analysis of self-organization and pattern formation during the development of visual maps, *Phys. Rev. A*, 45:7568–7589.

Sirosh, J., and Mikkulainen, R., in press, A unified neural network model for the self-organization of topographic receptive fields and lateral interactions, *Neural Computat.*

Swindale, N. V., 1992, A model for the coordinated development of columnar systems in primate striate cortex, *Biol. Cybern.*, 66:217–230.

Tanaka, S., 1991, Theory of ocular dominance column formation: Mathematical basis and computer simulation, *Biol. Cybern.*, 64:263–272.

von der Malsburg, C., 1973, Self-organization of orientation selective cells in the striate cortex, *Kybernetik*, 14:85–100.

von der Malsburg, C., and Willshaw, D. J., 1976, A mechanism for producing continuous neural mappings: Ocularity dominance stripes and ordered retino-tectal projections, *Exp. Brain Res.*, supp. 1:463–469.

Wolf, F., Pawelzik, K., Geisel, T., Kim, D. S., and Bonhoeffer, T., 1994, Optimal smoothness of orientation preference maps, in *Computation in Neurons and Neural Systems*, Boston: Kluwer, pp. 97–102.

Olfactory Bulb

John S. Kauer, Joel White, and Gordon M. Shepherd

Introduction

The olfactory bulb is of special interest to neural modelers. It was one of the first regions of the brain for which compartmental models of neurons were constructed, and this work led to some of the first computer models of functional microcircuits. The aim of this article is to provide a brief orientation to the modeling studies that have focused on three main levels of organization: integrative processing within individual neurons, functional operations within microcircuits, and sensory processing in bulbar networks. Together with the article on the OLFACTORY CORTEX, this article provides an introduction to the nature of information processing in this system.

The Concept of a Basic Circuit

In a simply organized region such as the olfactory bulb, it has been possible to combine the results of anatomical, physiological, neurochemical, and computational studies into a consensual basic circuit. The basic circuit is a key concept for computational neuroscience in that it defines the irreducible minimum of neural components necessary to incorporate into neuronal and network models to capture the neural basis of the functions carried out by that region (see Shepherd, 1990).

The basic circuit for the olfactory bulb (Figure 1) consists of straight pathways that pass from olfactory nerves through mitral or tufted cells to their output axons, and two levels for

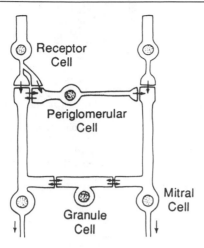

Figure 1. Basic circuit for the olfactory bulb. (From Shepherd, 1990.)

lateral interactions. The first level is the olfactory glomeruli, which are modular units containing the synaptic connections of olfactory axon terminals onto mitral, tufted, and periglomerular (short-axon cell) dendrites; intraglomerular dendrodendritic interactions between these neurons; and interglomerular interactions by means of periglomerular cell axons. This modular organization is reminiscent of modular columns and barrels in the cortex and patches in the striatum. This first level of synaptic interconnections involves processing of the afferent input, and it may be referred to as the level of *input processing*.

The second level consists of integrative mechanisms whose main function is to control the output from the olfactory bulb to the olfactory cortex; this level may be referred to as the level of *output control*. These mechanisms involve dendrodendritic synapses through anaxonal (granule) cells. The basic circuit allows one to identify these levels of organization within a given region and to compare them with other regions. For example, the basic circuit for the retina has similar levels for input processing and output control.

Compartmental Models of Olfactory Bulb Neurons

As detailed elsewhere in this book (see PERSPECTIVE ON NEURON MODEL COMPLEXITY), Rall introduced the compartmental method for computational modeling of neurons in 1964. Motoneurons in the spinal cord and mitral and granule cells in the olfactory bulb (Rall et al., 1966; Rall and Shepherd, 1968) were the first neurons to be modeled using this approach.

Mitral or tufted cells can be activated synaptically by the olfactory nerves and antidromically from the mitral cell axons (in the lateral olfactory tract). Activation is followed by long-lasting synaptic inhibition, both recurrent inhibition onto the same cell and lateral inhibition onto neighboring cells. Early, it was postulated that the granule cells might function as inhibitory interneurons mediating these effects, but the nature of this function was uncertain because, unlike Renshaw cells in the classical pathway for recurrent inhibition, the granule cells lack axons. The problem was attacked by constructing computational models of the mitral and granule cell populations and reconstructing their sequential activation. These models included a simplified Hodgkin-Huxley-type action potential model; passive or active dendritic membrane properties; synaptic excitation and inhibition; and background synaptic modulation. These properties were distributed in compartments scaled for the axon, soma, and dendrites of the mitral cells and the

dendrites and soma of the granule cells (Rall and Shepherd, 1968).

These studies have been extended by incorporating different active membrane conductances into the compartmental models and assessing their spatial distribution in the dendritic trees by systematic parameter searches (Bhalla and Bower, 1993). The mitral cell model predicted that Ca and slow K channels are concentrated in the glomerular dendritic tuft; sodium, fast K, K_A, and K_{Ca} channels in the soma; and lower concentrations of these channels in the secondary dendrites. The results support the concept of conditional signal propagation through the mitral cell dendritic tree. The granule cell model predicted active properties in both the soma and the dendrites. The authors emphasize that further experimental data are needed to constrain the models.

Microcircuits for Self-Inhibition and Lateral Inhibition

The original theoretical analysis suggested that the mitral cell dendrites would be the most likely source of the synaptic input to the granule cell spines and that the spines in turn would provide the synaptic input for inhibition of the mitral cell dendrites. Electron microscopical studies confirmed the presence of reciprocally oriented dendrodendritic synapses between mitral dendrites and granule spines that could mediate these interactions. It was an unusual instance in which theoretical models led to previously unknown neuronal structures, connections, and interactions.

The circuit model for mitral cell activation followed by self-inhibition and recurrent inhibition through granule cells is shown in Figure 2. The salient features for neural modelers

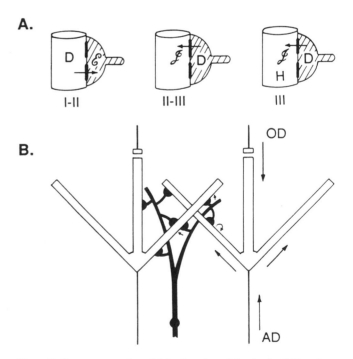

Figure 2. Compartmental model for function of dendrodendritic microcircuit. *A*, Impulse in the mitral cell soma spreads into the dendrites to activate excitatory output synapse onto the granule cell spine. *B*, The granule cell spine generates a depolarizing excitatory postsynaptic potential (EPSP), which then activates the reciprocal inhibitory synapse back onto the mitral cell dendrite. Abbreviations are as follows: D, depolarization; H, hyperpolarization; E, excitation; J, inhibition; OD, orthodromic; AD, antidromic. (From Rall and Shepherd, 1968.)

are as follows. First, the circuit includes only parts of the dendritic trees of the mitral and granule cells. This expresses the principle that neural circuits do not necessarily engage entire neurons, but may pass through only local parts of neurons. Second, impulse generation is not a necessary property for synaptic interactions to take place; graded changes in membrane potential and passive spread through dendritic trees may be sufficient. Third, the synaptic potentials in the granule spines (see DENDRITIC SPINES) are very large (because of the high input resistance and very fast because of the low capacitance and high conductance load). Their spread is controlled by the resistance of the spine stem and is relatively effective over the short distances within the granule dendritic tree. Fourth, the reciprocally interconnected dendrites and their synaptic interactions constitute a functional unit for specific operations: self-inhibition and lateral inhibition. From a computational perspective, one may refer to these as *microcircuits* for performing information-processing functions (see Shepherd, 1978).

These types of microcircuits are not unique to the olfactory bulb. Synaptic interconnections between dendrites have been found in a number of regions in the nervous system. They vary in pattern and extent, suggesting that they form microcircuits of different degrees of complexity for processing information at different levels of complexity (reviewed in Shepherd, 1990).

Discrete Component Network Models

We first describe network models constructed on the basis of simplified compartmental models of the neuronal types described above. A key principal of all olfactory bulb models is parallel processing of inputs through the olfactory glomeruli and bulbar elements (Kauer, 1991).

White et al. (1992) studied how membrane potential responses in different cell types may be shaped by microcircuit interactions in the salamander olfactory bulb. Their model simulates the membrane potentials of individual receptor, mitral, granule, and periglomerular cells after electrical or odor stimulation. Fewer cells are found in the model than in the real system (7,224 vs. several million), but the ratios between the populations are maintained. Mitral cells are modeled with three compartments (soma and two dendritic tufts) using standard cable equations. The other cell types are represented by single compartments, with connections between cells calculated as synaptic conductance changes over time.

This model has been tested with a variety of simulations that allow direct comparisons between the model's behavior and physiological recordings. To model responses to odor input, a spatially distributed pattern of activity was applied to the receptor cell population. Low-intensity stimulation of a small number of receptor cells causes only a few mitral cells to respond with a few impulses during the duration of the stimulus. With higher-intensity odor stimulation (Figure 3), the temporal properties of the responding mitral cells are more complex. The excitatory-inhibitory sequence is preceded in some mitral cells by a brief hyperpolarization similar to that seen in recordings from mitral or tufted cells in the salamander as well as from relay neurons in the insect antennal lobe. Long-lasting depolarizations are seen in granule cells and periglomerular cells, as in experimental recordings. Thus, each of the major types of response seen in salamander cells is reproduced by varying the spatial pattern of activity applied to the receptor cells.

Meredith (1992) has developed a model consisting of a 30 × 30 matrix of functional units representing glomeruli and their associated bulbar neurons. The degree to which individual cells interact is determined by the amount of dendritic overlap. This

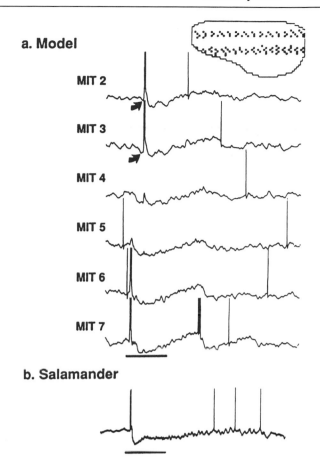

Figure 3. Odor stimulation of modeled mitral (MIT) cell responses (MIT 2–7) (*A*) compared with experimental recordings (*B*). Note the pattern of excited receptor cells in the receptor sheet in the inset above. Arrows in part *A* indicate a brief initial hyperpolarization which is also seen in experimental recordings. (From White et al., 1992.)

model has been used to investigate the possible role of lateral inhibition in producing the nonmonotonic stimulus intensity-response relationships observed experimentally in mitral or tufted cells.

The role of glomeruli as functional units in the coding of odor molecules has been the subject of several studies. The idea that odor recognition involves spatial pattern discrimination is supported by both experiment and theory (Kauer, 1991). Sejnowski, Kienker, and Shepherd (1985) showed that even crude simulations of the patterns of glomerular activation revealed by 2-deoxyglucose studies were sufficient to provide for discrimination of different odors by a neural network model.

Schild and Riedel (1990) examined the sequence of processing from receptor cells to mitral or tufted cells through a 10 × 10 glomerular array. Similar odor stimuli were mapped to similar glomerular activity patterns:

> Regions of the receptor cell class activity space with high information content are spread over relatively large areas on the glomerular layer while other regions with little information are only scarcely projected to the glomerul(i).

Each ensemble of activated glomeruli stimulates maximally its set of attached mitral or tufted cells; this set suppresses neighboring less active mitral or tufted cells through lateral inhibitory interactions to enhance discrimination. The authors note that if each glomerulus encodes only a single odor, the model could store only 100 odors, whereas if glomerular activities overlap, the number of possible glomerular patterns (e.g., stored odors) is increased "by many orders of magnitude." The model also takes account of receptor cell replacement and learning.

Encoding of odor concentration has been examined by Anton, Lynch, and Granger (1991) in a model representing 625 olfactory receptor cells providing input through a single glomerulus to 23 mitral or tufted cells, 90 inhibitory granule cells, and 15 periglomerular cells. The frequency of impulse firing in the mitral or tufted cells was relatively independent of the frequency of impulse firing in the receptor cells, but the number of activated cells was approximately linearly related. This finding supports the hypothesis that the glomerulus can function as a frequency-to-spatial encoder of odor concentration (i.e., concentration is encoded by the number of activated glomeruli and connected mitral or tufted cells).

Berkowicz, Mori, and Shepherd (1992) tested the widely held hypothesis that a given glomerulus receives predominant input from similarly tuned receptor cells and supplementary inputs from other receptor cell types. In their model, three sets of receptor cells send primary and secondary projections to three target glomeruli. The activity of mitral cells connected to each glomerular type is modeled using the GENESIS compartmental simulator (see NEUROSIMULATORS). The results show that the patterns of mitral cell activity most nearly resemble the experimental recordings of responses to closely related types of odor molecules when a glomerulus receives both primary and secondary receptor cell projections.

The universality of glomeruli in olfactory pathways has stimulated the investigation of invertebrate antennal lobes, where the populations of receptor cells, glomeruli, principal (mitral or tufted-like), and local interneurons can be counted precisely. In many insects, the male has specialist odor receptor cells that are narrowly tuned to female pheromones and project to a specialized macroglomerulus that functions as a labeled line for transmitting this species-specific information to higher brain centers (the mushroom bodies). By contrast, as in vertebrates, generalist receptor cells respond to the broad ranges of odor molecules involved in behaviors such as feeding. In contrast to the specialist system, it is believed that the generalist system requires learning for odor recognition.

Rospars and Fort (1994) have developed a model for generalist odor processing in which odor quality is encoded by sets of molecular properties acting independently on odor receptors contained in receptor cells. The receptor cells map odor quality by sending differential projections to spatially distributed sets of glomeruli. Each glomerulus in turn has its output through a single principal neuron which precisely reflects the activity of the glomerulus as a functional unit. The function of lateral inhibition is to enhance the differences in activity between different principal neurons. This model resembles the model of Schild and Riedel (1990; see above).

Parallel studies have been carried out on the olfactory pathway of the bee. In the model of Linster and Masson (1994), odor stimulation differentially activates populations of glomeruli; antagonistic interactions between glomeruli are mediated by local interneurons. These interactions generate resting oscillatory activity. Odor stimulation moves the oscillations to a new level. Simulation of central inputs from the brain makes the oscillatory patterns less noisy, suggesting that this activity may represent attentional mechanisms. The authors hypothesize that central input enhances fine-tuning underlying feature extraction at the glomerular level.

Generalized Population Models

At a more abstract level are network models that simulate the behavior of cell populations as manifested in electroencephalographic (EEG) recordings. Individual cells or modules are not modeled; *activity* is therefore an abstract property representing aggregate changes in membrane potential and impulse activity in cell populations.

Freeman (1987) and Yao and Freeman (1990) developed a model to investigate the oscillatory properties of olfactory circuitry and to study how patterned input may lead to patterned output. The basic configuration consists of coupled oscillators representing populations of neurons in the olfactory bulb, anterior olfactory nucleus (AON), and prepyriform (olfactory) cortex (PPC). In this model, the mitral cell population receives excitatory input from the olfactory receptor cell population and also locally from the periglomerular cell population. The mitral cell output excites AON and PPC oscillators, which in turn provide excitatory inputs back to the granule and periglomerular cell populations. This model appears to reproduce several properties of EEG recordings in the olfactory bulb, AON, and PPC. Once initiated with a brief receptor cell population input, the model exhibits continuous oscillations, similar to resting EEG recordings. Subsequent receptor cell input elicits bursts of oscillations that are also similar to EEG recordings after odor stimulation. Large oscillations similar to those seen during seizures could be elicited by manipulating the strength of connections between the model populations. Yao and Freeman (1990) extended this model by dividing the lumped olfactory bulb into as many as 32 parallel interconnected channels, each consisting of periglomerular, mitral, and granule cell populations, as before. The model includes modifiable synapses to explore the ability to reproduce output patterns from partial input patterns, as in associative memory. Associative memory mechanisms based on studies in the snail *Limax* have also been simulated by Gelperin, Tank, and Tesauro (1989).

Li and Hopfield (1989) investigated how input patterns may be processed by populations of olfactory bulb cells and how these processes may be influenced by centrifugal inputs. The model consisted of 10 mitral cell nodal elements and 10 granule cell nodal elements. Each input element presented an envelope of activity patterned after the spike firing of olfactory receptor cells in response to odor pulse stimulation. Some patterns of receptor cell activity generated oscillatory output of the mitral cell elements, reproducing some of the EEG patterns seen in the vertebrate olfactory bulb, while other patterns did not. In the case of mitral cell element oscillations, the patterns in the 10 mitral cell elements were different for different input patterns. Li continued this work, investigating the mechanisms by which central inputs, acting on granule cells, may modulate patterns of mitral cell element oscillation. Specific central inputs were found that could enhance or suppress the mitral cell element response pattern to a specific odor input. The influences of these central inputs on self- and cross-adaptation to different input patterns were also investigated.

Summary

The olfactory bulb has a simple structure that has made it attractive for analysis of microcircuit organization and models of neural circuits. For brain theorists, its organization offers examples of information processing without impulses and output functions of dendrites that have forced new concepts of the neuron as a complex computational unit. It provides many opportunities for correlating membrane and cellular properties with network functions, thus pointing the way toward a deeper understanding of the neural basis of network functions.

Acknowledgments. JSK is supported by grants from NIDCD and ORN. GMS is supported by grants from NIDCD, NIMH, and NASA (Human Brain Project).

Road Maps: Mammalian Brain Regions; Other Sensory Systems
Background: I.1. Introducing the Neuron

References

Anton, P. S., Lynch, G., and Granger, R., 1991, Computation of frequency-to-spatial transform by olfactory bulb glomeruli, *Biol. Cybern.*, 65:407–414.

Berkowicz, D. A., Mori, K., and Shepherd, G. M., 1992, A model to explore the relationship between olfactory receptor cell specificity and mitral cell response, *Assoc. Chemoreception Sci. Abstr.*, 14:157.

Bhalla, U. S., and Bower, J. M., 1993, Exploring parameter space in detailed single neuron models: Simulations of the mitral and granule cells of the olfactory bulb, *J. Neurophysiol.*, 69:1948–1965.

Freeman, W. J., 1987, How brains make chaos in order to make sense of the world, *Behav. Brain Sci.*, 10:161–195. ◆

Gelperin, A., Tank, D. W., and Tesauro, G., 1989, Olfactory processing and associative memory: Cellular and modeling studies, in *Neural Models of Plasticity: Experimental and Theoretical Approaches* (J. H. Byrne and W. O. Berry, Eds.), New York: Academic Press, pp. 133–149.

Kauer, J. S., 1991, Contributions of topography and parallel processing to odor coding in the vertebrate olfactory pathway, *Trends Neurosci.*, 14:79–85. ◆

Li, Z., and Hopfield, J. J., 1989, Modeling the olfactory bulb and its neural oscillatory processings, *Biol. Cybern.*, 61:379–392.

Linster, C., and Masson, C., 1994, Modeling odor processing in the bee antennal lobe: Modulation of glomerular dynamics by central feedback, *Eur. Chemosci. Res. Org. XI Abstr.*, 49.

Meredith, M., 1992, Neural circuit computation: Complex patterns in the olfactory bulb, *Brain Res. Bull.*, 29:111–117.

Rall, W., and Shepherd, G. M., 1968, Theoretical reconstruction of field potentials and dendro-dendritic synaptic interactions in olfactory bulb, *J. Neurophysiol.*, 31:884–915.

Rall, W., Shepherd, G. M., Reese, T. S., and Brightman, M. W., 1966, Dendro-dendritic synaptic pathway for inhibition in the olfactory bulb, *Exp. Neurol.*, 14:44–56.

Rospars, J.-P., and Fort, J.-C., 1994, Coding of odour quality: Roles of convergence and inhibition, *Network*, 5:121–145.

Schild, D., and Riedel, H., 1990, Monte Carlo generation of chemosensory maps in the olfactory bulb: Glomerular activation patterns, in *Chemosensory Information Processing* (D. Schild, Ed.), Berlin: Springer, pp. 359–369.

Sejnowski, T. J., Kienker, P. K., and Shepherd, G. M., 1985, Simple pattern recognition models of olfactory discrimination, *Soc. Neurosci. Abst.*, 11:970.

Shepherd, G. M., 1978, Microcircuits in the nervous system, *Sci. Am.*, 238(2):92–103. ◆

Shepherd, G. M., Ed., 1990, *The Synaptic Organization of the Brain*, New York: Oxford University Press. ◆

White, J., Hamilton, K. A., Neff, S. R., and Kauer, J. S., 1992, Emergent properties of odor information coding in a representational model of the salamander olfactory bulb, *J. Neurosci.*, 12:1772–1780.

Yao, Y., and Freeman, W. J., 1990, Model of biological pattern recognition with spatially chaotic dynamics, *Neural Netw.*, 3:153–170.

Olfactory Cortex

Matt Wilson and Gordon M. Shepherd

Introduction

The olfactory cortex has traditionally played an important role in theoretical studies of cortical function. It is the earliest cortical region to differentiate in the evolution of the vertebrate forebrain. It is the only region within the forebrain to receive direct sensory input. The olfactory input processed by the cortex dominates the behavior of most vertebrate species. Thus, the role of the olfactory cortex is critical for the evolution of much of vertebrate behavior. Finally, the olfactory cortex has the simplest organization among the main types of cerebral cortex. These features suggest that the olfactory cortex may serve as a model for understanding basic principles underlying cortical organization.

The olfactory pathway begins with the olfactory receptor neurons in the nose, which project their axons to the olfactory bulb. The function of the OLFACTORY BULB (q.v.) is to perform the initial stages of sensory processing of the olfactory signals before sending this information to the olfactory cortex. The olfactory cortex is defined as the region of the cerebral cortex that receives direct connections from the olfactory bulb (Figure 1). It is subdivided into several areas that share a basic organization but are distinct in terms of details of cell type, lamination, and site of output to the rest of the brain. The main area involved in olfactory perception is the piriform (also called prepyriform) cortex (see Figure 1), which projects to the mediodorsal thalamus, which in turn projects to the frontal neo-

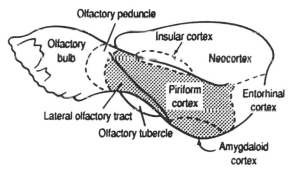

Figure 1. The relation of the olfactory cortex to the main components of the olfactory pathway. [From Haberly, L. B., 1990, Olfactory cortex, in *The Synaptic Organization of the Brain* (G. M. Shepherd, Ed.), New York: Oxford University Press.]

cortex. Prepyriform cortex, often regarded as the main olfactory cortex, will be the subject of this article.

Evolutionary Significance of the Olfactory Cortex

For brain theorists interested in the principles of cortical organization, the early appearance of the olfactory cortex in phylogeny deserves attention. The cerebral cortex first appears in vertebrate evolution in the fish as a simple structure composed of three layers: a superficial layer containing incoming nerve fibers, dendrites of intrinsic and output neurons, and scattered cell bodies of interneurons; a layer of the large cell bodies of output neurons; and a deep layer of interspersed input and output fibers and scattered cell bodies of interneurons. This arrangement is the classical three-layered cortex. The cortex on the ventrolateral surface that receives direct olfactory input from the olfactory bulb is termed the *palaeocortex*, which is the olfactory cortex as described above. On the medial surface is another part related to the septum, termed the *archicortex*; this part is the *anlage* of the hippocampus in higher vertebrates. On the dorsal surface is the dorsal cortex, generally believed to be the anlage of the neocortex.

During phylogeny, the palaeocortex and archicortex develop in extent and complexity, but retain their three-layered character (see also EVOLUTION OF THE ANCESTRAL VERTEBRATE BRAIN). The neocortex, however, emerges in mammals as a five- to six-layered structure. It is controversial among evolutionary neurobiologists whether the dorsal cortex in fact can be considered an early representation of the neocortex or whether it is more properly considered an *anlage* (i.e., a predecessor of the true neocortex). In reptiles, such as the turtle, this dorsal cortex has become sufficiently differentiated to serve as the visual cortex for visual input relayed from the thalamus. Whether this area can be regarded as a true primary visual cortex, homologous to the primary visual cortex in the mammal, or whether it is only a primitive anlage of the primary visual cortex is a matter for debate.

With the rise of modern studies of synaptic organization, it was hypothesized that comparisons between brain cortical regions in phylogeny should focus less on the numbers of layers and more on the particular types of circuits that are present and the functions that they mediate (Shepherd, 1974). We will

therefore identify the main types of circuits that are present in the olfactory cortex. We will then describe compartmental and network models and discuss the insights gained from these models into olfactory processing and their relevance for understanding the general properties of cortical networks.

Basic Circuit for Olfactory Cortex

As discussed elsewhere (see OLFACTORY BULB), the concept of a basic, or *canonical*, circuit is of critical importance for computational neuroscience and brain theory. The basic circuit combines the results of anatomical, physiological, neurochemical, and computational studies into a consensual representation of the main circuits in a particular region (Shepherd, 1974). This objective is facilitated by the extent to which the region in question has distinct layers, clearly differentiated cell types, and readily characterizable inputs. Of all cortical regions, the olfactory cortex best satisfies these criteria.

Our current understanding of the basic circuit for the olfactory cortex has arisen from a series of anatomical, physiological, and pharmacological studies (summarized in Haberly, 1990). The current model originated in studies by Haberly and Shepherd (1973) and Shepherd (1974) and is summarized in Figure 2. The main features of the basic circuit include the following:

The primary sensory input (through the lateral olfactory tract from the olfactory bulb) makes its synapses on the most distal parts of the apical dendrites of the pyramidal neurons. This process continues the pattern present in the earlier stages of the olfactory pathway, in which primary sensory input is delivered to the most distal parts of the dendrites of the sensory neurons and their axons in turn make synapses on the most distal dendrites of their targets, the mitral or tufted cells of the

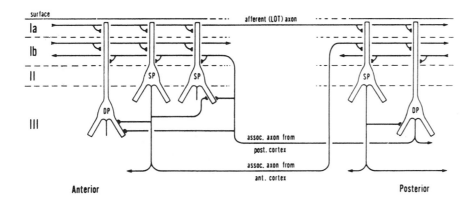

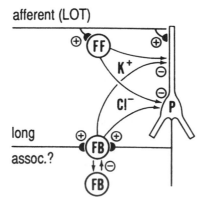

Figure 2. The basic circuit of the olfactory cortex. P indicates pyramidal neurons, which are further distinguished as superficial (SP) and deep (DP). LOT is the lateral olfactory tract. FF indicates feedforward; FB, feedback. [From Haberly, L. B., 1990, Olfactory cortex, in *The Synaptic Organization of the Brain* (G. M. Shepherd, Ed.), New York: Oxford University Press.]

olfactory bulb. Thus, distal dendrites, rather than being sites for weak background modulation of neuronal activity, are the preferred sites for rapid transmission of specific sensory information from neuron to neuron in this pathway. This fact is counterintuitive and against the lore in the field, but it is critically important for brain theorists and neural modelers because it means that the specific properties of distal dendrites must be included in network models to represent the mechanisms of processing in this pathway. The properties of dendrites are discussed elsewhere in this book (see DENDRITIC PROCESSING).

The distal inputs in the olfactory cortex are made exclusively onto dendritic spines of the apical dendrites. These are small branches, only a few microns in length and 0.1–0.2 μm in diameter, whose head (1–2 μm across) receives the excitatory input synapses. Spines are of considerable current interest as sites for activity-dependent mechanisms, such as long-term potentiation (LTP), that may underlie learning (see HEBBIAN SYNAPTIC PLASTICITY). Both the afferent excitatory inputs and the recurrent excitatory inputs are made onto spines, and both show properties of LTP.

The intrinsic cortical circuits for processing information consist of inhibitory and excitatory local circuits. The inhibitory circuits are of two types: those for feedforward inhibition and those for feedback (lateral) inhibition (see Figure 2). A given interneuron may be involved exclusively in one of these types, or it may be a node for convergence and integration of both types. The excitatory circuits provide not only for the excitation of the inhibitory interneurons in the feedback (lateral) pathway, but also for direct recurrent excitation of other pyramidal neurons. These intrinsic excitatory and inhibitory inputs are made to different regions and levels of the apical and basal dendritic trees of the pyramidal neurons. Thus, in the case of apical dendrites, these inputs can gate the transfer of the specific sensory responses in the distal dendrites to the soma.

Thus, the essential elements of the olfactory cortex basic circuit can be summarized as follows: (1) pyramidal output neurons, with apical and basal dendritic fields; (2) differentiation of pyramidal neurons into subtypes in sublayers; (3) reception of excitatory inputs by dendritic spines; (4) different modes of input driving (direct excitation; feedforward inhibition); (5) intrinsic recurrent axon collaterals for feedback and lateral inhibition; (6) intrinsic recurrent axon collaterals for feedback and lateral excitation; and (7) lamination of inputs to the dendritic trees of pyramidal neurons. Taken together, these constitute a unique set of circuit elements characteristic of the olfactory cortex and also shared with the other type of three-layered cortex, the dentate-hippocampal complex. Furthermore, these elements are embedded in most regions of the neocortex, where they are further elaborated into additional layers, additional subtypes of neurons and internal circuits, and additional types of inputs and outputs (see Shepherd and Koch, 1990).

Network Models of the Olfactory Cortex

The first suggestion that the olfactory cortex could serve as a simple model for learning and memory in cortical networks was made by Haberly (1985). In his landmark report, he described the features of the olfactory cortex outlined above and pointed out that this organization distributed in a broad sheet would subserve the functions of a cortex with content-addressable memory. The critical features were the widespread distribution of inputs by the input fibers and the presence of recurrent excitation providing for a wealth of combinatorial possibilities for activation and reactivation of the cortical circuits. He further pointed out the possible similarities between

processing of the olfactory input by the olfactory cortex and processing of complex visual stimuli in the face area of the neocortex.

These suggestions stimulated studies by several laboratories which have established the olfactory cortex as an attractive subject for network models of cortical functions. We will summarize briefly some of the main studies to date.

Lynch (1986) drew on the concept of the basic olfactory circuit to discuss the principles of a model of piriform (olfactory) cortex which would function as an associative memory network having the ability to identify conjunctions of odor components that make up complex odors. The role of the piriform cortex in olfactory memory was contrasted with the role of the hippocampus in maintaining or enabling the establishment of long-term olfactory associations. A reduced model of the basic olfactory circuit was described which incorporated properties of LTP and implemented a "combinatorial memory system." In this model, novel combinations of stimulus features during learning result in unique representations. Complex stimuli composed of previously experienced odor components produce a response that is biased toward the existing representations within the cortex. It was proposed that the piriform cortex could be regarded as a model for a general cortical memory representational system of this type.

In pursuing this model, Ambros-Ingerson, Granger, and Lynch (1990) obtained results that led to a proposal that interactions between the olfactory bulb and olfactory cortex, with synaptic modification in the input pathway, could result in a form of hierarchical clustering which could construct perceptual hierarchies used for storage and recognition of complex olfactory stimuli. This proposal was a departure from earlier models which had explored the role of intrinsic excitatory connections on associative memory function. In this model, the olfactory cortex selectively inhibits previously active olfactory bulbar neurons. The response to subsequent odor presentations leads to responses which reflect the differences between stimuli. While experimental work (Granger et al., 1991) reported a tendency for cortical response generalization after the presentation of a number of similar stimuli, supporting the type of clustering predicted by this model, this intriguing hypothesis awaits further experimental testing.

Building on the work of Haberly and collaborators, Wilson and Bower (1989) used the concept of the piriform cortex to approach the investigation of cortical function on two fronts. First, compartmental models of pyramidal neurons based on anatomical and physiological data, as reviewed above, were used to simulate intracellular potentials, extracellular field potentials, and ensemble impulse activity as recorded experimentally in response to orthodromic volleys in the lateral olfactory tract. Second, this network model, constrained by physiological response properties, was used for simulations which attempted to demonstrate computational properties which would underlie its functions as an associative memory. The results of this study showed the ability to store and retrieve patterned impulse information in a network which displayed the physiological responses and temporal dynamics of the real cortex using only a local Hebbian-type rule for modification of intrinsic excitatory synaptic interactions (Figure 3).

An interesting aspect of these simulations was the suggestion of a role for oscillatory phenomena, such as the prominent gamma-range (30–100 Hz) extracellular oscillations. It was proposed that these oscillations coordinate computational processes which directly underlie the associative memory function of the piriform cortex (Wilson and Bower, 1992).

Based on this physiologically based model, Liljenstrom (1991) drew on the work of Hopfield to develop further a

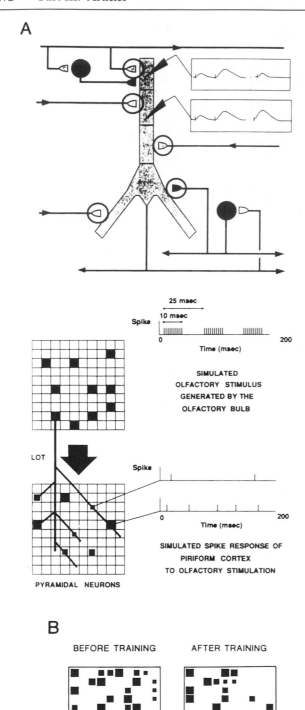

Figure 3. Autoassociation memory functions of a network model based on the basic circuit of Figure 2. *A* (top) Pyramidal neuron model with excitatory (open) and inhibitory (filled) synaptic inputs. Circles indicate synapses where Hebbian modification rules applied during learning. *A* (middle), Bulbar input patterns to the cortex (spike activity indicated by size of square); LOT is the lateral olfactory tract. *B*, Responses of pyramidal neurons before and after learning produced a stable activity pattern. [Modified from Hasselmo, M. E., Wilson, M. A., Anderson, B. P., and Bower, J. M., 1990, Associative memory function in piriform (olfactory) cortex: Computational modeling and neuropharmocology, *Cold Spring Harbor Symp. Quant. Biol.*, 55:599–610.]

model that used simplified sigmoidal output units. The local circuits for feedforward and feedback inhibition, together with the excitatory interactions between pyramidal neurons, were critically important for the input-output dynamics. The model demonstrated simultaneous slow theta and rapid gamma oscillations that are characteristic of the olfactory cortex. It reproduced the experimental effects of acetylcholine both on the modulation of these oscillations, by selectively increasing excitability and suppressing intrinsic synaptic transmission, and on associative memory functions.

Hasselmo developed a similar network model based on a simplified version of the basic circuit to investigate learning mechanisms (see Hasselmo, Anderson, and Bower, 1992). This model explored the effects of selective modification of both input and intrinsic excitatory connections. Associative memory performance was enhanced by the combination of suppression of intrinsic fiber transmission and increased excitability during learning, as seen with cholinergic modulation. The model could also exhibit self-organizing properties under these conditions. This work represents a synthesis of abstraction and physiological detail in modeling cortical function. It was proposed that this model could represent a basic unit for learning and memory in cortical circuits.

Nonlinear and chaotic properties of the olfactory cortex were studied by Freeman (1987), who showed that a simple cortical model could display the chaotic properties of electroencephalogram rhythms; he proposed that a feedback gain parameter Q sets the level of arousal of the cortex. It is this factor that appears to be sensitive to the modulatory actions of acetylcholine, as described above. Simulated actions of acetylcholine on this parameter in the simplified olfactory cortical model produce point attractor, limit cycle attractor, and strange chaotic or nonchaotic attractor behavior (Wu and Liljenstrom, 1994).

Thus far, the models of olfactory cortical function discussed here have been at the neuronal or circuit level. The behaviors at these levels depend in turn on properties at the membrane and synaptic level and on the differential properties of cell bodies and dendrites. An initial step toward assessing these properties has been made by constructing compartmental models of apical dendrites and their spines and then analyzing their responses to excitatory and inhibitory inputs. Basic logic operations of AND, OR, and NOT-AND were found to arise from spine interactions when active membrane properties were placed in the spines (Shepherd and Brayton, 1987) or the dendritic branch (Shepherd, Woolf, and Carnevale, 1989). These studies support the idea discussed earlier that distal dendrites of pyramidal neurons receive and process rapid, precise input information, rather than mediating only slow and weak background modulation. They emphasise the importance of including apical and dendritic properties in network models to represent the full complexity of cortical circuits and cortical functions.

Summary

The organization of the olfactory cortex can be summarized by a basic circuit composed of a unique set of elements, which may be embedded and elaborated in more complicated cortical regions. Models of the olfactory cortex emphasize the importance of cortical dynamics, including the interactions of intrinsic excitatory and inhibitory circuits and the role of oscillatory potentials in generating the computations performed by the cortex, replacing earlier interpretations of simple modulation or synchronization of activity. In this way, the olfactory cortex and neural networks based on it may be useful in the study of computations defined by cortical dynamics.

Acknowledgments. MW has been supported by grants from NSF and ONR. GMS has been supported by grants from NIDCD, NIMH, and NASA (Human Brain Project).

Road Maps: Mammalian Brain Regions; Other Sensory Systems
Background: I.1. Introducing the Neuron
Related Reading: Oscillatory Associative Memories

References

Ambros-Ingerson, J., Granger, R., and Lynch, G., 1990, Simulation of paleocortex performs hierarchical clustering, *Science*, 247:1344–1348.

Freeman, W. J., 1987, Simulation of chaotic EEG patterns with a dynamic model of the olfactory system, *Biol. Cybern.*, 56:139–150.

Granger, R., Staubli, U., Powers, H., Otto, T., Ambros-Ingerson, J., and Lynch, G., 1991, Behavioral tests of a prediction from a cortical network simulation, *Psychol. Sci.*, 2:116–118.

Haberly, L. B., 1985, Neuronal circuitry in olfactory cortex: Anatomy and functional implications, *Chem. Senses*, 10:219–238.

Haberly, L. B., 1990, Olfactory cortex, in *The Synaptic Organization of the Brain* (G. M. Shepherd, Ed.), New York: Oxford University Press. ◆

Haberly, L. B., and Shepherd, G. M., 1973, Current density analysis of summed evoked potentials in opossum prepyriform cortex, *J. Neurophysiol.*, 36:389–802.

Hasselmo, M. E., Anderson, B. P., and Bower, J. M., 1992, Cholinergic modulation of cortical associative memory function, *J. Neurophysiol.*, 67:1231–1246.

Hasselmo, M. E., Wilson, M. A., Anderson, B. P., and Bower, J. M., 1990, Associative memory function in piriform (olfactory) cortex: Computational modeling and neuropharmocology, *Cold Spring Harbor Symp. Quant. Biol.*, 55:599–610.

Liljenstrom, H., 1991, Modeling the dynamics of olfactory cortex using simplified network units and realistic architecture, *Int. J. Neural Syst.*, 2:1–15.

Lynch, G., 1986, *Synapses, Circuits and the Beginnings of Memory*, Cambridge, MA: MIT Press. ◆

Shepherd, G. M., 1974, *The Synaptic Organization of the Brain*, New York: Oxford University Press. ◆

Shepherd, G. M., and Brayton, R. K., 1987, Logic operations are properties of computer-simulated interactions between excitable dendritic spines, *Neuroscience*, 21:151–165.

Shepherd, G. M., and Koch, C., 1990, Introduction to synaptic circuits, in *The Synaptic Organization of the Brain*, 3rd ed. (G. M. Shepherd, Ed.), New York: Oxford University Press, pp. 3–31. ◆

Shepherd, G. M., Woolf, T. B., and Carnevale, N. T., 1989, Comparisons between active properties of distal dendritic branches and spines: Implications for neuronal computations, *J. Cognit. Neurosci.*, 1:273–286.

Wilson, M., and Bower, J. M., 1989, The simulation of large-scale neural networks, in *Methods of Neuronal Modeling: From Synapses to Networks* (C. Koch and I. Segev, Eds.), Cambridge, MA: MIT Press, pp. 291–334.

Wilson, M., and Bower, J. M., 1992, Cortical oscillations and temporal interactions in a computer simulation of piriform cortex, *J. Neurophysiol*, 67:981–995.

Wu, X., and Liljenstrom, H., 1994, Regulating the nonlinear dynamics of olfactory cortex, *Network*, 5:47–60.

Optical Architectures for Neural Network Implementations

B. Keith Jenkins and Armand R. Tanguay, Jr.

Introduction

A neural computing paradigm that differs markedly from the conventional digital computing paradigm is emerging from the collection of artificial neural network models. This neural computing paradigm promises to enable the design and manufacture of highly parallel, analog based computers for applications in sensor signal processing and fusion, pattern recognition, associative memory, and robotic control. An additional application may be the simulation of a number of aspects of *biological* neural networks, at significantly larger scales and lower simulation times than would be possible using conventional digital computers.

The basic requisite functions for a neural network technology base appear to be neuron unit response, weighted interconnections (fixed and variable), input/output, learning computation and weight update, and duplication capability (i.e., the capability for making a copy of a pretrained neural network). The hardware systems considered in this article make the following assumptions for artificial neural network functionality: First, the interconnections and synapses are functionally (and physically) distinct from the neuron units. Second, each synaptic weight is multiplicative and is stored as part of the interconnection. This article focuses on systems that use analog weight values and have the capability for their modification. Third, sets of postsynaptic signals are summed at the neuron unit input, and only a small number of such sets are allowed at each neuron unit input because of physical hardware limitations. (One or two sets are sufficient for most artificial neural network models.) For example, a large number of postsynaptic signals may be summed in two sets at each neuron

unit input: one set as positive inputs and the other set as negative inputs. Finally, in some systems the neuron unit functionality is quite general and can be designed according to need; in other systems it constitutes a fixed nonlinear response function.

While semiconductor-based very large scale integration (VLSI) technology is appropriate for immediate implementation of these functions in a relatively small-scale neural network (but see DIGITAL VLSI FOR NEURAL NETWORKS for a multiplexing technique), the incorporation of photonic technology will enable the future implementation of much larger-scale neural networks. For example, direct implementation of a fully connected network of N neuron units requires area $O(N^2)$ in VLSI, limiting such networks to less than approximately 1000 neuron units and 10^6 interconnections on a complementary metal oxide semiconductor (CMOS) chip (assuming 0.5 μm design rules), even without learning capability. By using optics to implement the interconnections in a third dimension, the electronic chip area required becomes just $O(N)$, and the resulting hardware module can potentially implement on the order of 10^5 neuron units and 10^{10} weighted interconnections (Jenkins and Tanguay, 1992). Because implementations of such *large-scale* neural networks stand to have the greatest impact in many application areas, this article considers implementation techniques that use both electronics and optics, together, to achieve this scalability. To efficiently combine optical and electronic hardware into the same system, *photonic* components, which integrate electronic and optical functionality on the same device, are used in conjunction with purely optical and in some cases purely electronic systems, to yield a photonic computing module for large-scale neural network implementation.

This article first defines the class of neural network models that will be considered, describes a common architectural framework that encompasses most optical implementations of large-scale neural network computers, and introduces a requisite set of photonic components. A companion article describes the operational principles of these photonic components (see OPTICAL COMPONENTS FOR NEURAL NETWORK IMPLEMENTATIONS). Next, a more specific example of a photonic neural network computing module that is based on a particular optical implementation technique is described herein. The article concludes with a discussion of past, current, and future research directions in photonic neural network computing and their implications in research on neural network models and applications.

Architectural Characteristics of Optical Implementations

Although a wide variety of optical and photonic components are used in optical neural network architectures, most proposed and demonstrated optical implementations have many characteristics in common. This section focuses on a general architectural framework that emphasizes these common characteristics and, in so doing, also illuminates some of the key fundamental differences.

Architecture Fundamentals

Optical and photonic implementations of neural networks typically use either 1D arrays of neuron units interconnected by 2D elements, or 2D arrays of neuron units interconnected by 2D or 3D elements. This article considers only the latter scenario and some of its variants, primarily because of their preferable scaling properties. Depending on the architectural layout of the system and the interconnection elements it uses, the system can be truly adaptive, can have weights that are programmable but are not modified in accordance with a learning algorithm, or can have weights that are precomputed but fixed in hardware. While this article concentrates on truly adaptive systems, the other types of systems use many of the same architectural techniques but have simpler hardware requirements. Finally, this article discusses the implementation of a single-layer neural network. If the capability for feedback is incorporated, such a single physical layer can be directly generalized to implement any number of neural network layers (Jenkins and Tanguay, 1992; Farhat, 1987).

We assume that the computation process of any one layer of an artificial neural network can be represented by

$$y_i = f\left[\sum_j w_{ij} x_j\right] \tag{1}$$

in which neuron unit j is situated at the input to the layer of interconnections, neuron unit i is situated at the output of the layer of interconnections, y_i is the output of neuron unit i, x_j is the output of neuron unit j, w_{ij} is the weight associated with the interconnection between them, and the function f represents the activation function of each neuron unit. The term inside the brackets, the activation potential, will be denoted by ρ_i. Note that this activation potential is a matrix-vector product between an interconnection weight matrix and an input vector. The function f then operates independently on each element of the resulting vector; this type of nonlinear function is called a *point nonlinearity*, and it lends itself to implementation with a spatial light modulator.

Most learning algorithms for artificial neural networks fall into one of a small number of classes. One such class is characterized by outer products between the input vector, x, and the training vector, δ, and can be specified by:

$$\Delta w_{ij} = \alpha \delta_i x_j - \beta w_{ij} \tag{2}$$

in which $\Delta w_{ij} = w_{ij}(k+1) - w_{ij}(k)$ is the weight update, k is the iteration index, α is a (positive) learning gain constant, and β is a (nonnegative) decay constant that is included primarily for hardware convenience. Suitable choices of δ give different learning algorithms, such as Hebbian, Widrow-Hoff, and single and multilayer least mean squares (LMS) (Jenkins and Tanguay, 1992; Kosko, 1992; see PERCEPTRONS, ADALINES, AND BACKPROPAGATION). As two different examples of the latter, optical architectures for implementation of error backpropagation have been described by Wagner and Psaltis (1987) and by Owechko (1993). For illustrative purposes, this article considers only this outer-product class of algorithms. Although other classes of learning algorithms can also be implemented using optical hardware (e.g., Wagner and Psaltis, 1993; Carpenter and Grossberg, 1987), research has focused primarily on the outer-product algorithms.

Photonic Components

Before discussing optical neural network systems, a brief introduction to their key optical and photonic components is in order. These components include detector arrays, modulator arrays, laser diode arrays, smart-pixel spatial light modulators, and volume holograms. (For more detail than is given below, see OPTICAL COMPONENTS FOR NEURAL NETWORK IMPLEMENTATIONS).

A *detector array* is used to convert a set of optical signals in parallel to a set of electronic signals. A key feature that is required for its use in high-performance photonic neural network systems is fully parallel output of the electrical signals over the 2D array. An example follows.

Each modulator in a *modulator array* is used to convert an electrical signal into an optical signal. A separate optical beam is provided by an external source (e.g., laser or LED), and upon passing through the modulator, has its intensity or phase modified in time in accordance with the electrical driving signal. A full 2D modulator array can then modulate a spatially expanded input beam independently over its 2D cross section. Similarly, a *laser diode array* can *generate* a set of beams in response to electrical driving signals.

Finally, to allow an information-carrying input beam to modulate (or generate) a separate readout beam, a detector array can be mated directly to a modulator (or laser diode) array. If the electrical connections between the two arrays are fully parallel, the resulting device can provide a large number of high-bandwidth channels. Such a device is spatially segmented into an array of pixels, with each pixel comprising one or more detectors, one or more modulators (or laser diodes), and a set of signal processing electronics. Such *smart-pixel* (so called because of the inclusion of signal processing electronics) *spatial light modulators* (so called because of an optical input effecting an optical output within each pixel) and *smart-pixel source arrays* (in the case of laser diodes) are the most promising sets of devices for implementation of 2D arrays of neuron units with optical input and optical output.

Optical Architectures

A conceptual diagram of an optical system for implementation of an artificial neural network is given in Figure 1. Two smart-pixel spatial light modulators are used, one each to implement a 2D array of training term generators (SLM_1), and a 2D array of neuron units (SLM_2). The interconnection output plane

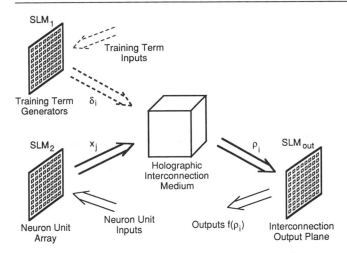

Figure 1. Conceptual diagram of an optical neural network system architecture, showing two layers of neuron units (SLM$_2$ and SLM$_{out}$) and one layer of interconnections (optical subsystem incorporating the holographic interconnection medium). Many variants of this geometry with similar properties are possible.

may consist of a detector array that couples into an electronic system, or the input detector plane of a spatial light modulator (SLM$_{out}$), as shown. In the latter case, the SLM may serve as input to subsequent signal processing stages, or the existing training-term or input neuron-unit-array SLM in the case of a system with feedback.

The key interconnection element includes a medium that stores volume holograms. If the holograms are properly recorded, then when only the x_j terms illuminate the holographic medium, the output beam from the medium gives the set of products between each x_j and the corresponding stored w_{ij}'s, in parallel, as needed during the computing phase of the neural network. In the learning phase, if the optics is arranged appropriately, the medium can record a set of weight updates in accordance with the outer product rule, when illuminated by both a set of x_j terms and a set of δ_i terms. A full 3D holographic medium is preferred for this function because of its high weight-storage capacity.

During hologram readout, by arranging the lenses and optical sources (not shown) appropriately, the beam from each input pixel x_j can illuminate the holographic medium at a unique position, angle, or wavelength. The selective nature of volume holograms then allows only the appropriate interconnection weights w_{ij} to be multiplied by each input beam, yielding the set of products $w_{ij}x_j$. The output beam optics is then set up so that the sum over j of these terms appears at the location of each output term ρ_i in the interconnection output plane.

During hologram recording, the training plane generates the δ_i terms needed for the learning phase. An exposure is made of the interference pattern between beams emanating from the two left-hand planes in Figure 1. With appropriate choices of parameters, this exposure can increment the value of each interconnection weight by an amount that is approximately proportional to $\delta_i x_j$, as given in Equation 2. It should be noted that, when designing an architecture that will generate and record these weight updates, care must be taken to insure that the appropriate interference terms are recorded and that not too much cross-talk is inadvertently created. In fact, one of the key differences in the various approaches to full-scale optical neural network architectures is the technique used to avoid such cross-talk (Owechko, 1993; Asthana et al., 1993, and references

therein). In addition, the actual weight update that physically occurs in holographic systems can be somewhat different from the idealized expression of Equation 2, and this depends on the particular architecture employed (Wagner and Psaltis, 1993).

In the computing phase of a neural network, the holographic interconnection time is limited only by the travel time of the light from input modulator to output detector, so the overall computing time is typically determined by other factors such as the SLM switching time. During learning, the holographic material sensitivity or response time typically limits the weight update rate, and for a given update rate puts an upper bound on the achievable learning gain constant; such holographic material response times depend on the material and vary over many orders of magnitude from one material to another. The decay rate or retention time of holograms in read/write holographic materials puts a lower bound on the achievable decay constant and is similarly material dependent.

Large-scale optical *nonadaptive* networks are important as well. In this case, the interconnection hologram does not have to be recorded in accordance with a specific learning algorithm. If the weights are known a priori, then any applicable recording technique can be used to prerecord the hologram. In many cases, however, the weights may not be known. A common scenario may involve the training of a "master" network; once it has been trained, multiple copies of the network would be produced, for example, in a production environment. If the network is large, and particularly if it uses volume holographic interconnections, then making direct copies of the volume hologram may be more practical than probing the values of all of the weights and then loading those weight values into a recording system. Thus, the capability of rapidly copying a multiplexed volume interconnection hologram may be important (see OPTICAL COMPONENTS FOR NEURAL NETWORK IMPLEMENTATIONS).

To summarize and extend the concepts described in this section, most optical and photonic neural network architectures of this class can be generalized in the following ways: First, they are modular, so that multiple "modules" can be cascaded while permitting parallel communication between modules. Second, they have the capability for implementing multiple layers, by adding feedback capability to the single layer described, or by cascading multiple modules. The former technique yields a simpler hardware implementation when lateral, feedforward, and feedback connections spanning multiple layers are required. Additionally, optical and photonic neural network architectures of the type just described have the following features: analog, weighted connections with analog neuron unit activation; simultaneous, parallel updates of all interconnection weights at each learning iteration; and fully parallel multiplication by all weights during each computing iteration. Other desirable features that are incorporated into these optical and photonic neural network architectures to varying degrees include capability for bipolar signals and weights; scalability to very large numbers of neuron units with high connectivity; generality, so that different neuron models, network models, and learning algorithms can be implemented within the same basic technology; compatibility of different components within a given architecture; overall feasibility of the proposed combination of algorithm, architecture, devices, and materials; high optical throughput; and low interconnection cross-talk.

A Sample Implementation

This section describes a specific photonic implementation technique that potentially satisfies all of these desirable features. This implementation technique is a particular manifestation of the general concept of Figure 1.

Physically, this implementation technique uses smart-pixel SLMs for the 2D neuron unit and training term planes. The structure of these SLMs depends substantially on the physical quantity used to represent the input and output signals in the neuron unit and training planes. Optical intensity is used in this implementation. Unlike all-electronic implementations, which typically represent signals by voltage or current, intensity is inherently a real, *nonnegative* value. As a result, to allow for the representation of *bipolar* signals, each neuron unit (and, typically, each training term generator) incorporates dual-channel encoding. One example of dual-channel encoding represents each bipolar signal by one positive channel and one negative channel. With dual-channel encoding, each SLM pixel comprises two integrated detectors, two modulators, and integrated analog signal processing electronics.

The interconnections within this implementation are based on a technique for multiplexed volume holography that uses double angular multiplexing and incoherent/coherent recording and readout (Jenkins and Tanguay, 1992; Asthana et al., 1993). This interconnection technique exhibits an advantageous combination of total number of channels, interchannel crosstalk, and total optical throughput efficiency. The remainder of this section describes this technique in more detail, along with other aspects of the photonic neural network implementation.

A key feature of this implementation technique is the use of a source array that consists of an array of lasers, each of which generates light independently of the others. Because of this independence, a pair of beams from *different* lasers is not capable of interfering and writing a hologram, whereas a pair of beams generated from the *same* laser can interfere and thereby write a hologram into a holographic medium. From this source array, a set of coherent beam pairs is formed, one pair from each laser, so that they can write holograms pairwise in a controlled fashion.

First we describe the recording and readout of a single hologram with this system, using only a single laser source; then we extend the description to the multiple-hologram, multiple-source case. In Figure 2, the second laser of the source array generates a beam (indicated by solid lines) that is subsequently split into an upper object beam path and a lower reference beam path. The reference beam passes through a single pixel of the neuron unit array (SLM_2), and the object beam path passes through the entire set of pixels in the training term plane (SLM_1). Because these two beams originated from the same laser, they interfere inside the holographic medium, and a hologram is written (or a previously written hologram is updated)

in the medium. This hologram can be recalled at any time thereafter by closing the shutter (to prevent further recording) and turning on the same source and reference beam SLM pixel, at which time the recorded data array $\{\delta_j\}$ (originally from SLM_1) emerges from the hologram and is imaged onto the detector array.

The concept of single-hologram recording and readout is extended in two ways to implement a complete neural network interconnection: First, a series of exposures can be made sequentially in time, in accordance with a learning algorithm. Each such exposure updates the set of weights that are stored in the holographic medium by $x_j\delta_i$ for a specific j and for all i. Second, all lasers S_j of the source array are turned on simultaneously to perform these operations over all input neuron units j. In this way, all weights (over all i and j) can be updated simultaneously during each neural learning step.

For the computing phase of a neural network, the shutter is closed, blocking the upper beam path. With all lasers of the $N \times N$ source array again simultaneously on, the $N \times N$ set of neuron unit outputs x_j are encoded onto the set of beams at SLM_2, and each beam (x_j) reads out its corresponding hologram (or $M \times M$ array of output data), scaled by x_j, yielding the array $w_{ij}x_j$ for all i at the detector array. Each of these holograms is physically distinct, although their positions may partially or fully overlap in the holographic medium. Because all N^2 sources are on, all such arrays (one from each source, or one for each value of j) are incident on the detector array, and because of the independence of the laser sources, the detected value is the array of sums

$$\rho_i = \sum_{j=1}^{N^2} w_{ij}x_j \tag{3}$$

in which i represents the array index and takes on M^2 values. As described in reference to Figure 1, this $M \times M$ array of data can then be used for output, can be the input of another SLM as the input to another physical layer, or can be fed back to either of the existing SLMs of Figure 2, depending on the artificial neural network model being implemented.

This specific angular and spatial multiplexing arrangement at the holographic medium provides a high degree of independence among the different holograms during readout. This in turn permits all of the weight multiplications in a large-scale system to occur simultaneously without compromising the detected postsynaptic signals given by Equation 3. In general, the holographic medium stores N^2 holograms, each consisting of M^2 weighted interconnection channels, thereby providing a total of N^2M^2 interconnection weights.

This volume holographic interconnection technique can now be incorporated into a full photonic neural network computing module. By adding a few features to the overall architecture of Figure 1, we can implement this module for the broad class of adaptive neural networks based on outer product learning. First, the array of training term generators may consist of a single smart-pixel SLM (as shown) or multiple SLMs in cascade, depending on the complexity of the training term computation, number of inputs, number of pixels, and available chip area. Second, the system output values $f(\rho_i)$, as well as their target values t_i if needed, can be directed to the inputs of the training term generators. The resultant computing module implements one layer of a feedforward neural network based on outer product learning.

As a third feature, the interconnection output potentials ρ_i can be fed back to a portion of the neuron unit array (SLM_2) inputs, to implement a multilayer network, lateral connections, and feedback connections. These output potentials can also be sent to the training term generator to implement potential-

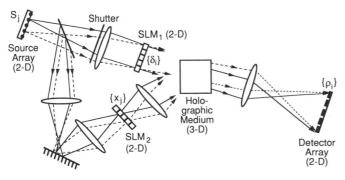

Figure 2. Example of a volume holographic interconnection architecture. The desired interconnection outputs (or training terms) are generated on SLM_1, and the interconnection input nodes (or neuron units) are implemented by SLM_2. Training term inputs to SLM_1 and neuron unit inputs to SLM_2 are not shown.

based learning algorithms; likewise, they can be sent in parallel to the neuron unit array input of another photonic computing module, providing complete cascadability.

Discussion

Research in optical and photonic neural network implementations has included the development and analysis of new architectures, analysis of limits in the scalability to large numbers, development of the technology base for near-optimal individual components, and experimental demonstration of neural networks for the cases of small numbers of neuron units or small numbers of independent stored holograms using currently available but nonoptimal components. This continuing research bodes well for the future implementation of full-scale photonic neural networks with high-bandwidth and high-capacity components.

Additional research and development directions that are key for the eventual realization of physically small, high-performance, reasonable-cost photonic neural networks include increased focus on the manufacturability of the photonic and optical components, development of packaging and miniaturization techniques for photonic computing systems, and increased design automation and flexibility. Work along these lines has begun in a number of groups; for example, substantial progress has already been made in the areas of photonic digital interconnection systems and components for early vision. Most of this work can be applied directly to photonic neural network computing modules.

Finally, the increasing likelihood of having a technology base to implement artificial neural networks with very large numbers of neuron units (e.g., 10^5 per module) and weighted interconnections (e.g., 10^{10} per module) puts increasing importance on the development of artificial neural network models specifically for operation at such large scales. It also opens up the potential for, and necessity of, the development of application areas for very large-scale artificial neural networks.

Acknowledgments. This work was supported in part by the Advanced Research Projects Agency and the Air Force Office of Scientific Research under grant AFOSR–89–0466, grant F49620–92–J–0472, and the University Research Initiative program.

Road Map: Implementation of Neural Networks
Related Reading: Programmable Neurocomputing Systems

References

Asthana, P., Nordin, G. P., Tanguay, A. R., Jr., and Jenkins, B. K., 1993, Analysis of weighted fan-out/fan-in volume holographic optical interconnections, *Appl. Opt.* 32:1441–1469.
Carpenter, G. A., and Grossberg, S., Eds., 1987, Feature issue on neural networks, *Appl. Opt.*, 26:4909–5111. ◆
Farhat, N. H., 1987, Optoelectronic analogs of self-programming neural nets: Architecture and methodologies for implementing fast stochastic learning by simulated annealing, *Appl. Opt.*, 26:5093–5103.
Jenkins, B. K., and Tanguay, A. R., Jr., 1992, Photonic implementations of neural networks, in *Neural Networks for Signal Processing* (B. Kosko, Ed.), Englewood Cliffs, NJ: Prentice Hall, pp. 287–382. ◆
Johnson, R. V., and Tanguay, A. R., Jr., 1989, Fundamental physical limitations of the photorefractive grating recording sensitivity, in *Optical Processing and Computing* (H. Arsenault, T. Szoplik, and B. Macukow, Eds.), New York: Academic Press, pp. 59–102.
Kosko, B., 1992, *Neural Networks and Fuzzy Systems: A Dynamical Systems Approach to Machine Intelligence*, Englewood Cliffs, NJ: Prentice Hall, chaps. 4 and 5, pp. 111–220. ◆
Mao, C. C., and Johnson, K. M., 1993, Optoelectronic array that computes error and weight modification for a bipolar optical neural network, *Appl. Opt.*, 32:1290–1296.
Owechko, Y., 1993, Cascaded-grating holography for artificial neural networks, *Appl. Opt.*, 32:1380–1398.
Wagner, K., and Psaltis, D., 1987, Multilayer optical learning networks, *Appl. Opt.*, 26:5061–5076.
Wagner, K., and Psaltis, D., Eds., 1993, Feature issue on optical implementations of neural networks, *Appl. Opt.*, 32:1249–1476. ◆

Optical Components for Neural Network Implementations

Armand R. Tanguay, Jr., and B. Keith Jenkins

Introduction

Technological implementations of neural networks in both software and hardware forms have been motivated to a greater or lesser extent by biological neural networks, but also include network topologies, synaptic interconnection rules, and neuron unit functionalities that combine to yield novel system-level properties. Software implementations of such networks are highly flexible in design and reconfiguration (see NEUROSIMULATORS), but are time-, power-, and computational-resource consumptive for even modest numbers of neuron units and interconnections. Both electronic and photonic neural network implementations in hardware are designed to circumvent these limitations in applications that require compact systems and high computational throughputs.

In fully electronic (primarily very large scale integration [VLSI]-based) neural networks, the neuron units and the weighted (synaptic) interconnection matrices are incorporated on a planar microelectronic chip (see DIGITAL VLSI FOR NEURAL NETWORKS, ANALOG VLSI FOR NEURAL NETWORKS, and SILICON NEURONS). An important advantage of the integrated circuit approach to neural network implementation is the capability for near-term technology insertion, with leverage provided by a well-established technology base. An equally important limitation, however, is the difficulty of scaling up neural chips to incorporate large numbers of neuron units in fully (or near fully) interconnected architectures. This limitation derives from the limited pinout, off-chip communication bandwidth, and on-chip interconnection density available in both current generation and projected chip designs.

In this article, we consider the photonic implementation of neural networks, in which optical (free-space) interconnection techniques in conjunction with photonic switching and modulating devices are used to expand the number of neuron units and the complexity of the interconnections by using the off-chip (third) dimension for synaptic communication (Jenkins and Tanguay, 1992; Wagner and Psaltis, 1993). This merging of optical and photonic devices with appropriately matched

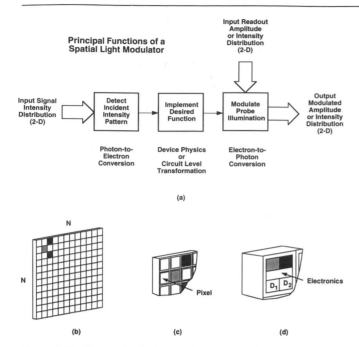

Principal Functions of a Spatial Light Modulator

(a)

(b)   **(c)**   **(d)**

Figure 1. Fundamental principles of spatial light modulator function. *A*, Block diagram of the principal functions of an optically addressed spatial light modulator, including detection, functional implementation, and modulation. *B*, Schematic diagram of an $N \times N$ array of spatial light modulator pixels, in which three pixels are shown in different transmission states. *C*, Expanded view of the pixel array. *D*, Expanded view of a single pixel within the array, illustrating one possible "smart" pixel configuration that incorporates two detector elements D_1 and D_2, control electronics for functional implementation, and two modulator elements. (After Jenkins and Tanguay, 1992.)

electronic circuitry provides novel features such as fully parallel weight updates and modular scalability, as well as both short- and long-term synaptic plasticity.

We focus herein on the operational principles of key photonic devices that constitute photonic neural networks; architectural and systems considerations are provided in a companion article (see OPTICAL ARCHITECTURES FOR NEURAL NETWORK IMPLEMENTATIONS). Such devices include surface-emitting semiconductor diode laser arrays that provide optical power for the network; "smart-pixel" spatial light modulators (Figure 1) that make up arrays of neuron units with optical inputs and outputs; and volume holographic optical elements that implement dense, weighted, high fan-in, high fan-out, and reconfigurable interconnections among neuron units (Jenkins and Tanguay, 1992; Jahns and Lee, 1994).

Sources and Source Arrays

Electrical power is relatively inexpensive, widely available, well characterized, and reasonably abundant. The situation in photonic technology, however, is quite different. Although coherent, monochromatic optical power can be provided by many different types of laser sources, including argon-ion, neodymium-YAG, helium-neon, helium-cadmium, dye, excimer, carbon-dioxide, and semiconductor diode lasers, perhaps only the last is compact, power efficient, and inexpensive enough for incorporation in commercially viable photonic neural networks. Even so, the range of output powers available from these sources (1–100 mW for single-element diode lasers) is rather

limited. Taking an upper bound (with continued development) of approximately a watt gives us a realistic estimate of the average coherent source power available for at least circuit-level implementations of photonic neural networks.

Semiconductor diode lasers (Bhattacharya, 1994) have a broad range of commercial applications, including compact-disc player recording and readout, fiber optical communications systems, merchandise optical scanners, and laser printers. The physical size of these lasers is small enough (approximately $0.3 \times 1 \times 5$ mm) to fit inside a standard transistor (or integrated circuit [IC]) package. Lasers with output powers of 1–10 mW are relatively inexpensive, costing a few tens of dollars in quantity on the average.

Most optoelectronic implementations of neuron unit arrays have photodetectors or modulation windows that are smaller in size than each individual pixel, as shown schematically in Figure 1. Hence, if a single high-intensity optical power source is to efficiently illuminate such neuron unit arrays, it must be combined with appropriate array generation optics. The optical source array generation problem has received considerable attention because of significant interest in optical interconnection systems. In one approach, computer-generated binary phase holograms (Jahns and Lee, 1994) are configured to form large grid patterns of regularly spaced illuminated spots. Using this technique, $N \times N$ arrays of spots can be generated that exhibit both high optical throughput efficiencies and low scattered light. The resulting light beamlets are *mutually coherent*, as they derive from the same laser source. This mutual coherence is important for the generation of independent holographic interconnection networks (Asthana et al., 1993; Jenkins and Tanguay, 1992).

An alternative to the generation of multiple optical sources from a *single* laser source is that of direct fabrication of *multiple-source arrays*. One striking example is the successful fabrication of more than 10^6 independent surface-emitting semiconductor diode lasers on a single gallium arsenide chip (Jewell et al., 1990), as shown in Figure 2. These lasers are not designed to be mutually coherent (phase locked). In fact, over time constants (milliseconds) typical of holographic recording in real-time materials such as photorefractive crystals, such arrays are effectively *mutually incoherent*. Arrays of individually coherent but mutually incoherent optical sources allow implementation of dense neural network interconnection

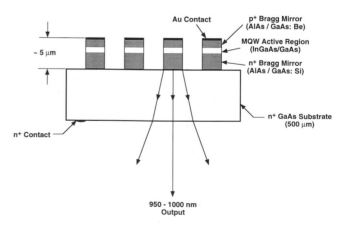

Figure 2. Cross-sectional view of a surface-emitting laser diode source array (after Jewell et al., 1990). In this example, the individual semiconductor diode lasers emit *through* the GaAs substrate.

systems with very high efficiency and very low interchannel crosstalk (Jenkins and Tanguay, 1992).

Another solid-state device that is capable of both single-source and source-array fabrication is the light-emitting diode (LED). Closely related to the semiconductor laser diode, the LED can be fabricated with considerably less processing complexity by elimination of the high-reflectivity mirrors that form the semiconductor laser cavity. Light-emitting diodes are relatively broadband (incoherent) sources, and as such are not usable as sources either for holographic recording applications or for readout of high-storage-capacity holograms. In addition, LEDs are relatively inefficient emitters, and hence are rather power consumptive.

Implementation of Neuron Unit Arrays

The switching function of providing an output that is nonlinearly dependent on one or more inputs is a principal distinguishing characteristic of neuron units. (We use the phrase *neuron units* here and throughout to differentiate from biological neurons.) Electronic circuits are well suited to the switching task as long as the number of inputs (representing the fan-in) and the number of outputs (representing the fan-out) are both kept relatively small (less than a few hundred for analog fan-in and fan-out). However, for purely electronic neural-network implementations that combine high connectivity with a concomitantly large number of neuron units, the gate count and area required for interconnection routing rapidly get out of hand.

Photonic implementations of neural networks take advantage of simple beam-combining mechanisms to multiplex inputs and outputs, and as such exhibit a much higher capacity for fan-in and fan-out than do typical electronic implementations. The use of optical rather than electronic interconnections for the fan-in and fan-out functions thus allows for a larger number of neuron units (10^5–10^6) and total number of interconnections (10^{10}–10^{12}) within a given module.

In photonic implementations, the neuron units are typically hybrid structures in which certain photonic devices (*detectors*) are used to provide for the conversion of optical inputs into one or more electrical input signals, electronic circuits are used to implement the nonlinear transfer function (*functional transformation*) between input and output, and other photonic devices (*emitters* or *modulators*) are used to convert the electrical output signals to optical outputs. Neuron unit arrays that incorporate modulators (*spatial light modulators*) and emitters (*active emitter arrays*), are discussed separately in the following sections.

Spatial Light Modulators

Optically addressed spatial light modulators alter the distribution of either the amplitude or phase of a two-dimensional readout beam in response to the information (local intensity or exposure) encoded on an input (writing or recording) beam. The simplest example of a spatial light modulator, albeit one that cannot operate at real-time frame rates, is photographic film.

In addition to the basic functions of optical detection and functional transformation, the third key function performed by a spatial light modulator is that of optical modulation, as shown schematically in Figure 1*A*. Optoelectronic spatial light modulators that allow for independent control of these three functions within each picture element ("pixel") are commonly

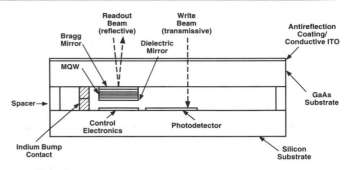

Figure 3. Cross-sectional view of a hybrid spatial light modulator, in which the photodetectors and control electronics are fabricated on a silicon substrate, and the MQW modulator elements are fabricated on a GaAs substrate. The two sets of devices (only one pixel within the array is shown) are bump contacted on a pixel-by-pixel basis to provide parallel electrical connections. (After Jenkins and Tanguay, 1992.)

referred to as "smart pixel" devices, in which each smart pixel incorporates sophisticated electronic circuitry to generate programmable nonlinear transfer functions (see Figure 1*B–D*).

A wide range of physical modulation mechanisms have been investigated for use in various types of spatial light modulators, including electro-optic, electroabsorptive, electrorefractive, magneto-optic, acousto-optic, surface deformation, photochromic, and photorefractive effects. The use of these physical modulation mechanisms has been addressed in a number of review articles (Tanguay, 1985; Warde and Fisher, 1987) and journal special issues (e.g., Lentine et al., 1994).

Both monolithic and hybrid approaches to the development of optoelectronic spatial light modulators have been employed. In the *monolithic* approach, the detectors, control circuitry, and modulation elements within each pixel are integrated within a single class of materials on a supporting substrate. An example of such an approach is the integration of phototransistors with metal-semiconductor field-effect-transistors (MESFETs) to drive multiple-quantum-well (MQW) optical modulators (Jahns and Lee, 1994), all fabricated on a common gallium arsenide (GaAs) substrate.

In the *hybrid* approach, however, certain of the device functions may be integrated on one substrate to optimize either their performance characteristics or their manufacturability, whereas others are integrated on a separate substrate. The two substrates are then interconnected such that mating pixels are in pairwise electrical contact, as shown in Figure 3 (Jenkins and Tanguay, 1992).

As an example of such a spatial light modulator that implements a two-dimensional neuron unit array, a silicon-based complementary metal oxide semiconductor logic (CMOS) chip has been designed and fabricated (Jenkins and Tanguay, 1992) that incorporates two input detectors, control circuitry, and two (optical modulator) output drivers within each 100×100-μm pixel, as shown in Figure 1*D*. The pixel layout allows for two 30×50-μm detectors, followed by a 15-transistor dual-input, dual-output differential amplifier that implements a sigmoidlike transfer function. (Other functions, such as time integration, time differentiation, logarithmic amplification, and sigmoidal derivatives, can be implemented just as easily.) Output pads are provided for hybrid bonding to an InGaAs/AlGaAs MQW modulator structure fabricated on a GaAs substrate, as shown in Figure 3. Using 2-μm CMOS design rules, the design allows for the integration of 10^4 pixels/cm^2, or 6×10^4 pixels/in.^{2} at a bandwidth in excess of 16 MHz/pixel.

Active Emitter Arrays

Neuron unit arrays with optical inputs and outputs can also be configured in such a manner that the outputs comprise individual sources rather than modulation elements, by monolithic or hybrid integration of detectors and control electronics with either light-emitting diodes or surface-emitting semiconductor diode lasers (Jahns and Lee, 1994). These approaches do not require a probe beam for readout; at the same time, they do not allow for mutual coherence of the light read out from the entire set of pixels within the array. As a result of the optical inefficiency of LEDs and of the threshold operating characteristics of surface-emitting semiconductor diode lasers, such devices tend to be far more power consumptive than modulator arrays at a given bandwidth and optical output power.

Photonic Interconnections

To implement dense, weighted optical interconnections among large two-dimensional arrays of neuron units, three types of optical components can be used, including planar holographic optical elements, volume holographic optical elements, and micro-optical elements such as gradient index (GRIN) lens arrays.

Holographic Optical Elements

The basic principle of using holographic optical elements for implementing interconnections among neuron units is shown in Figure 4. A given neuron unit at a location p_1 in the input plane can be connected with a second neuron unit in the output plane at a location p_2 by using a holographic optical element as an angle-to-angle encoder. Light from the input neuron unit is collimated by a lens at an angle ϑ_1 determined by the location p_1. As such, the lens acts as a position-to-angle encoder. Light incident on the holographic optical element at angle ϑ_1 diffracts from a stored diffraction grating at a new angle ϑ_2 determined by the orientation and spacing of the grating. The second lens performs an angle-to-position encoding function, focusing the light on the desired point p_2 in the output plane. The interconnection weight is related to the modulation depth (strength) of the recorded diffraction grating. Multiple points in the input plane can be interconnected to multiple points in the output plane with essentially arbitrary weights and arbitrary fan-out and fan-in by superimposing multiple holographic gratings to form a *multiplexed hologram*.

Planar holographic optical elements can be fabricated by optical recording in thin layers of photosensitive materials such as dichromated gelatin, or by the creation of multiple phase levels

in a substrate by using semiconductor processing (chemical or ion beam etching) techniques to generate a surface relief pattern capable of selective optical diffraction (Veldkamp, 1993; Nishihara and Marchand, 1990; Jahns and Lee, 1994). In the latter case, design of the holographic optical element is typically performed by numerical iteration, and the resulting element is referred to as a *computer-generated hologram*. Two-phase-level holograms are also called *binary phase gratings*. Computer-generated holograms have proven to be very useful for array generation functions (Jahns and Lee, 1994), and weighted fan-out of single sources in multiple plane architectures (Veldkamp, 1993).

The use of planar holographic optical elements for weighted interconnections is limited to relatively sparse interconnections among small sets of neuron units, since any recorded hologram that is designed to connect a single input point to one or more output points will in fact also connect *every* other input point to corresponding sets of output points. Planar holograms also exhibit diffraction artifacts (multiple orders), low diffraction efficiencies, and limited multiplexing capacity. However, multiple layers of planar holograms can be combined to form Stratified Volume Holographic Optical Elements (SVHOEs), which can emulate the interconnection properties of the *volume* holograms described in the next section (Nordin, Johnson, and Tanguay, 1992).

Volume Holographic Optical Elements

The solution to the weighted interconnection problem just outlined is to extend the holographic medium into the third dimension (the direction of light propagation), creating a volume holographic optical element (VHOE) to replace the thin planar holographic optical element. Two essential properties bear directly on the use of such elements in photonic interconnections. The first is that diffraction is limited to the first order (with no diffraction artifacts) if the holographic medium is made thick enough. The second is that of *angular selectivity*. Specifically, the range of input angles that can diffract from a given grating decreases as the thickness of the grating is increased. This property therefore eliminates the inadvertent connection of all input points pairwise to a matching set of output points and allows for the generation of the *independent, weighted* interconnections that are desired for neural network applications.

The extremely narrow angular alignment characteristics of volume diffraction gratings allow the simultaneous multiplexing of large numbers of such independent, weighted interconnections to be recorded (see Figure 4). In addition, the use of angular multiplexing allows for both fan-out from a given input point to a number of output points and fan-in from a number of input points to a single output point (Asthana et al., 1993).

Key characteristics of such volume holographic optical interconnections for neural network applications include the capacity for both short-term and long-term memory, real-time recording capability that can be used to implement learning (synaptic plasticity), and the capacity for single-step copying of the entire interconnection network (Piazzolla, Jenkins, and Tanguay, 1992).

The development of a viable photonic interconnection technology is based on the availability of appropriate photosensitive recording materials. The class of photosensitive recording materials that has been most extensively investigated for photonic interconnection applications has the capacity for sensitive holographic recording, is available in "thick" samples that allow for the formation of volume holographic gratings, and exhibits a high multiplexing capacity. These so-called photore-

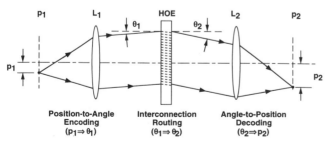

Figure 4. A point-to-point interconnection system, using a holographic optical element (HOE) for interconnection routing, and lenses as position-to-angle and angle-to-position encoders. The holographic optical element effectively performs an input-angle-to-output-angle transformation. (After Jenkins and Tanguay, 1992.)

fractive materials (Tanguay, 1985; Gunter and Huignard, 1988) include single crystals of semi-insulating optical materials such as bismuth silicon oxide and lithium niobate, as well as semi-insulating semiconductors such as GaAs and cadmium telluride.

Micro-optical Elements

Two-dimensional arrays of microlenses can be used to implement several neural network functions such as beam array generation and localized optical interconnections. Such lenses can be fabricated either by GRIN or by computer-generated holographic (CGH) techniques (Nishihara and Marchand, 1990; Jahns and Lee, 1994). Array sizes up to 256×256 are commercially available, with lens diameters in the range 100–250 µm.

Detectors and Detector Arrays

Detectors act as photon-to-electron converters by transforming incident optical intensity into electronic form, usually a voltage or a current. As such, they are important components for the photonic implementation of neural networks in at least two functional areas: (a) as input transducers for the optical detection function of optically addressed spatial light modulators, and (b) as output transducers for the translation of optically generated intermediate and final results to an appropriate electronic format for further processing or display.

Key *single-pixel* detectors (which may be integrated into independent arrays) include photoconductors, semiconductor *p-n* junction photodiodes, avalanche photodiodes, bipolar junction phototransistors, and metal-oxide-semiconductor field effect phototransistors (Sze, 1981). An example of an *interconnected detector array* (which typically includes some form of parallel-to-serial conversion) is the light-sensitive element in the CCD (charge-coupled-device) camera (Janesick, 1987).

Integration and Packaging

The integration of multiple planes of detectors with local, optical interconnections between adjacent planes potentially provides a photonic analog to the mammalian multilayer retina (Veldkamp, 1993). Several alternative approaches are possible, including the use of either modulator or emitter arrays for the local interconnection sources; binary phase holograms, computer-generated holograms, or microlens arrays for implementing the weighted local interconnection function; and any of the photodetection elements described in the previous section. Areal densities in excess of 10^5 pixels/cm^2 can be envisioned using such techniques. Applications of this technology span smart cameras, robotic vision systems, smart sensors, and visual representation generators for advanced electronic and photonic neural networks.

The integration and packaging of full-scale photonic neural networks is a technology in its infancy, as most photonic neural networks currently exist only in the breadboard stage of development. Packaging issues of concern include automated alignment of critical optical components; computer-aided design of compact optical systems that incorporate refractive, diffractive, emissive, and absorptive optical elements; stability under g-loading; and heat removal from active, power-dissipating components.

The Future of Neuro-optical Computation

The usefulness of computational systems is often determined by the maximum computational throughput capacity that can be squeezed into the smallest system volume within a tolerable energy-dissipation constraint or power budget. In the case of

photonic neural networks, the most important factors that influence the computational throughput capacity are the storage capacity of the VHOE and the operational bandwidth of the neuron units (nonlinear spatial light modulators).

The operational bandwidth of spatial light modulators that can be used as neuron unit arrays will prove to be orders of magnitude larger than the bandwidth characteristic of biological systems. Feasibility analyses and device characterization studies indicate no fundamental or technological barriers to operation of individual neuron units at bandwidths exceeding 100 MHz (Jenkins and Tanguay, 1992). Full duty-cycle power-dissipation effects may eventually force a modest lowering of this bandwidth. Nevertheless, the interesting question is whether photonic neural networks in comparison with biological neural networks can, at least for certain applications in which the system response time is at a premium, make up for a mismatch in total neuron count and interconnection complexity with excess bandwidth.

A fair question to ask is whether the physical volume of a neuro-optical computer module would be better off densely packed with silicon chips that emulate the same functionality. The preliminary evidence suggests that the answer shifts rather dramatically from an emphatic *yes* in the limit of small numbers of neurons and required interconnections, to a more tenuous *no* as the number of neurons and density of required interconnections increases. Perhaps the most interesting question for the future of neuro-optical computation is the clear identification of this performance boundary.

Acknowledgments. This work was supported in part by the Advanced Research Projects Agency and the Air Force Office of Scientific Research under grant AFOSR-89-0466, grant F49620-92-J-0472, and the University Research Initiative program.

Road Map: Implementation of Neural Networks
Related Reading: Biomaterials for Intelligent Systems

References

Asthana, P., Nordin, G. P., Tanguay, A. R., Jr., and Jenkins, B. K., 1993, Analysis of weighted fan-out/fan-in volume holographic optical interconnections, *Appl. Opt.*, 32:1441–1469.
Bhattacharya, P., 1994, *Semiconductor Optoelectronic Devices*, Englewood Cliffs, NJ: Prentice Hall. ◆
Gunter, P., and Huignard, J.-P., Eds., 1988, *Photorefractive Materials and Their Applications, I: Fundamental Phenomena*, New York: Springer-Verlag. ◆
Jahns, J., and Lee, S. H. (Eds.), 1994, *Optical Computing Hardware*, San Diego, CA: Academic Press. ◆
Janesick, J. R., Ed., 1987, Special issue on charge-coupled-device manufacture and application, *Opt. Engrg.*, 26:827–943. ◆
Jenkins, B. K., and Tanguay, A. R., Jr., 1992, Photonic implementations of neural networks, in *Neural Networks for Signal Processing* (B. Kosko, Ed.), Englewood Cliffs, NJ: Prentice Hall, chap. 9, pp. 287–382. ◆
Jewell, J. L., Lee, Y. H., Scherer, A., McCall, S. L., Olsson, N. A., Harbison, J. P., and Florez, L. T., 1990, Surface-emitting microlasers for photonic switching and interchip connections, *Opt. Engrg.*, 29:210–214.
Lentine, A. L., Lee, J. N., Lee, S. H., and Efron, U., Eds., 1994, Feature issue on spatial light modulators and their applications, *Appl. Opt.*, 33:2767–2860. ◆
Nishihara, H., and Marchand, E. W., Eds., 1990, Feature issue on microoptics, *Appl. Opt.*, 29:5049–5135. ◆
Nordin, G. P., Johnson, R. V., and Tanguay, A. R., Jr., 1992, Diffraction properties of stratified volume holographic optical elements, *J. Opt. Soc. Am. [A]*, 9:2206–2217.
Piazzolla, S., Jenkins, B. K., and Tanguay, A. R., Jr., 1992, Single-step

copying process for multiplexed volume holograms, *Opt. Lett.*, 17:676–678.

Sze, S. M., 1981, *Physics of Semiconductor Devices*, 2nd ed., New York: Wiley, pp. 743–789. ◆

Tanguay, A. R., Jr., 1985, Materials requirements for optical processing and computing devices, *Opt. Engrg.*, 24:002–018. ◆

Tanguay, A. R., Jr., 1988, Physical and technological limitations of optical information processing and computing, *Mater. Res. Bull.*, 13:36–40. ◆

Veldkamp, W. B., 1993, Wireless focal planes "on the road to amacronic sensors," *IEEE J. Quantum Electronics*, 29:801–813. ◆

Wagner, K., and Psaltis, D., 1993, Eds., Feature issue on optical implementations of neural networks, *Appl. Opt.*, 32:1249–1476. ◆

Warde, C., and Fisher, A. D., 1987, Spatial light modulators: Applications and functional capabilities, in *Optical Signal Processing* (J. L. Horner, Ed.), San Diego, CA: Academic Press, chap. 7.2, pp. 477–523. ◆

Optimization Principles in Motor Control

Tamar Flash and Neville Hogan

Introduction

Optimization theory has become an important research tool in our attempts to discover organizing principles that guide the generation of goal-directed motor behavior. It provides a convenient way to formulate a coarse-grained model of the underlying neural computation without requiring specific details of the way those computations are carried out. Generally speaking, this application of optimization theory consists of defining an objective function that quantifies what is to be regarded as optimum (i.e., best) performance and then applying the tools of variational calculus to identify the specific behavior that achieves that optimum. This forces us to make explicit, quantitative hypotheses about the goals of motor actions and allows us to articulate how those goals relate to observable behavior. Not all motor behaviors are necessarily optimal, but attempts to identify optimization principles can be useful for developing a taxonomy of motor behavior and gaining insight into the neural processes that produce motor behavior.

Ill-Posed Problems in Motor Behavior

Many optimization-based models in the literature have been developed to address the "excess degrees-of-freedom" problem (Bernstein, 1967). How does the motor system select the behavior it uses from the infinite number of possibilities open to it? In mathematical parlance, this is an "ill-posed" problem in the sense that many solutions are possible. For example, most limb segments are moved by a larger number of muscles than appear to be necessary. To reach for a cup of coffee the hand may move along many different paths. The same figural form (e.g., the letter *Z* or an ellipse) may be drawn using a wide variety of time profiles for the pen's position. The central question is how the nervous system chooses values for the large number of parameters that can be controlled. One appealing possibility is that the nervous system has evolved to select solutions that maximize the organism's fitness, i.e., that are optimal in some sense. More specifically, the hypothesis is that, in performing a motor task, the brain produces coordinated actions that minimize some measure of performance (such as effort, smoothness, etc.). In this article we review several studies in which the validity of such ideas was examined in the context of upper limb movements. Similar ideas have been explored in the context of other effector systems and motor actions such as whole body posture, gait, and various sporting activities, but they will not be considered here. The interested reader is referred to Winters and Woo (1990) for further information.

Arm Trajectory Formation

Our first topic is the kinematic aspects of movement. *Kinematics* refers to the time course of limb position, velocity, etc., while *dynamics* refers to variables such as forces and torques. In principle, even a single degree of freedom movement (e.g., elbow rotation) can be performed in many different ways. Thus, while the hand path is constrained to follow a circular arc, its speed along the path may follow many different time profiles. One way to gain insight into the processes responsible for the selection of specific limb trajectories is to experimentally observe human movements. Patterns or invariances in the observed behavior suggest hypotheses about the way these movements are organized. Optimization theory provides a mathematical tool for concisely formulating and testing these hypotheses. The key step is the identification of an *objective function* which defines a measure of performance by assigning a number to each member of the class of possible behaviors under study (e.g., arm trajectories). One member of this class (e.g., one trajectory) will then be selected to maximize or minimize that function. How the objective function is defined determines what aspects of the motor behavior are considered important.

Kinematic Versus Dynamic Objective Functions

In this article we will consider two different types of objective functions that have been proposed (out of the multitude of possibilities), as they reflect two major competing theories of how motor computations are organized. The first type of objective function is based solely on kinematic variables (e.g., limb position and its time derivatives). If a kinematic objective function can be found that leads to optimal trajectories that accurately reproduce the patterns of observed behavior, it implies that the brain ignores nonkinematic factors in selecting and producing that behavior. This would be consistent with a theory that neural computations to produce movement are organized hierarchically and executed by proceeding from the abstract (i.e., move to that light over there) to the particular (i.e., activate that set of motoneurons in this manner). The most compelling evidence supporting this idea is that similar kinematic patterns are observed even when widely different musculoskeletal systems are involved in producing motor behavior. One's signature on paper is equally as recognizable and distinctive as one's signature on a blackboard, despite the enormous differences in the mechanics and physiology of the body parts used to produce it. Nevertheless, a troubling aspect of this theory is that it seems to imply that, at least at the higher levels of the postulated hierarchy, the brain does not take into account *any* dynamic considerations such as the energy required, the loads on the limb segments, or the force and fatigue limitations of its peripheral neuromuscular system.

To circumvent this problem within the framework of optimization theory, a second type of objective function may be formulated based on dynamic variables (e.g., joint torques, muscle forces, etc., and their time derivatives). If a dynamic objective function can be found that leads to optimal trajectories that accurately reproduce the patterns of observed behavior, it implies that the brain considers dynamic factors in selecting and producing that behavior. It is also consistent with a theory that neural computations to produce movement are executed in parallel, taking all relevant factors (e.g., dynamics as well as kinematics) into account simultaneously.

Single-Joint Movements

As has been frequently observed, single-joint movements are characterized by single-peaked, bell-shaped speed profiles. This finding and the tendency of natural movements to be characteristically smooth and graceful led Hogan (1984) to suggest that motor coordination can be mathematically modeled by postulating that voluntary movements are made, at least in the absence of any other overriding concerns, to be as smooth as possible. For mathematical convenience (there are many other plausible measures of smoothness) maximizing smoothness was expressed as minimizing mean-squared average *jerk*, the third time derivative of position. In the single-joint case

$$C = \int_{t_0}^{t_f} \left(\frac{d^3\theta}{dt^3}\right)^2 dt$$

where $\theta(t)$ is the joint angle, and t_0 and t_f are the initial and final movement times, respectively. Using variational calculus, the unique time history of joint positions that minimizes this performance measure may be derived analytically. It is described by the following quintic polynomial in time:

$$\theta(t) = c_0 + c_1 t + c_2 t^2 + c_3 t^3 + c_4 t^4 + c_5 t^5$$

where c_i, $i = 0, \ldots, 5$, are unspecified coefficients whose values are determined by the conditions at the beginning and end of the movement (boundary conditions). Originally, Hogan (1984) analyzed movements that start and end at rest and therefore assumed zero initial and final velocities and accelerations. Consequently, the predicted trajectories were characterized by symmetrical bell-shaped speed profiles. For movements of different amplitudes and durations, the ratio of peak speed to average speed was invariant at 1.88. For a repetitive sequence of movements, speed profiles were again symmetrical, and this ratio was again invariant, but with a value of 1.57. These predictions appear to be in good agreement with observation. A constant ratio of peak speed to average speed has been reported by several researchers, with values between 1.60 and 1.90, depending on the conditions of measurement. However, a distinctive feature of these minimum-jerk movements is their symmetrical speed profile, and that is not always observed experimentally. For example, when enhanced accuracy of target acquisition is demanded, an asymmetrical speed profile is typically observed, with the peak speed occurring earlier in the movement. This indicates that the simple minimum-jerk theory may need to be modified. Hogan and Flash (1987) proposed to account for this asymmetry by adding to the objective function an additional term to minimize hand-to-target error integrated across the movement. An alternative is to modify the boundary condi-

tions. Wiegner and Wierzbicka (1992) reported that during single-joint elbow rotations the hand does not reach zero acceleration at the time that the velocity first becomes zero and that the movement continues in a more or less underdamped manner. To model movements generated over a large range of movement speeds, they used the minimum-jerk model with modified boundary conditions to account for the asymmetry observed for very slow or very rapid movements.

Another alternative is to use a dynamic objective function. This requires formulation of a model of neuromuscular and skeletal mechanics to relate dynamic variables (e.g., forces) to kinematics. Hasan (1986) proposed a minimum-effort theory of single-joint movement generation based on a model that described neuromuscular behavior as equivalent to a "spring-like" element driving the limb toward a neurally defined "equilibrium position," determined by simultaneous activation of agonist and antagonist muscle groups. Minimization of effort was expressed as follows:

$$C = \int_{t_0}^{t_f} \left(\sigma(t) \left(\frac{d\beta}{dt}\right)^2 \right) dt$$

where σ is the joint *stiffness* (describing the rate of change of the restoring force generated by the "spring-like" element with its displacement from equilibrium) and $d\beta/dt$ is the time derivative of the equilibrium position. Using this minimum-effort criterion, which penalizes unnecessarily high stiffnesses and limits excessive antagonist muscle co-contraction, Lan and Crago (1994) recently demonstrated that this model is capable of accounting not only for the asymmetry of the speed profiles but also for the characteristic three-burst pattern of muscle activity commonly observed in relatively fast single-joint movements. Thus, for single-joint movements, optimization theories using both kinematic and dynamic objective functions have been applied with success. A more telling test of these theories is found in multijoint movements.

Multijoint Movements and the Question of Coordinates

The kinematics of multijoint arm movements may be represented in a number of different ways, e.g., as a series of hand positions, joint angles, or muscle lengths. Each of these may be considered as alternative "coordinate frames" for describing the movement. The neural computations underlying multijoint arm movements may make use of any one (or even several) of these representations. Experimental observations of unconstrained human reaching movements are characterized by approximately straight hand paths and symmetrical bell-shaped speed profiles, which remain nearly invariant despite changes in movement direction, speed, and starting position (Morasso, 1981). Because these features are evident only in the motions of the hand and not in the movements of individual limb segments, it was proposed that the neural computations underlying movement production take place in terms of hand motion through extracorporeal space and not in terms of joint rotations.

Flash and Hogan (1985) showed that the maximum-smoothness theory reproduced all of these features, provided the objective function was expressed in terms of the Cartesian coordinates of the hand as follows:

$$C = \int_{t_0}^{t_f} \left(\left(\frac{d^3x}{dt^3}\right)^2 + \left(\frac{d^3y}{dt^3}\right)^2 \right) dt$$

where $x(t)$, $y(t)$ describe the hand position coordinates and t_f is the movement duration.

Minimizing this objective function yielded analytic expressions for the hand trajectories. For unrestrained point-to-point movements starting and ending at rest, the model predictions agreed closely with experimental data and successfully accounted for the invariance of hand trajectories under translation, rotation, amplitude, and speed scaling.

In more complex curved movements, patterns were again evident in hand kinematics but not in joint kinematics. When subjects were instructed to generate curved or obstacle-avoidance movements, although the hand paths appeared smooth, movement curvature was not uniform; the trajectories displayed two or more curvature maxima. The hand speed profiles also had two or more maxima, and the minima between adjacent peaks temporally corresponded to the peaks in curvature.

To describe curved and obstacle-avoidance movements, the maximum-smoothness model was extended by assuming that a small number of points along the path through which the hand should pass are specified (Flash and Hogan, 1985). The time of passage through those "via" points and the hand velocity at that time were not specified a priori but were predicted by the model. For the simplest case of one via point between the initial and final positions, the theory yielded explicit mathematical expressions for the hand motion (Flash and Hogan, 1985) that reproduced all the features of the experimental observations: distinct maxima in the speed profile with a minimum between them which coincided temporally with a curvature maximum; trajectory shape invariant under translation, rotation, amplitude, and time scaling; and nearly equal durations of movement from the initial position to the via point and from the via point to the final position. The latter observation was referred to as the *isochrony principle* (Viviani and Terzuolo, 1982)—the phenomenon that movement durations of large and small segments of a trajectory are roughly equal.

Minimum Torque Change Model

In contrast to the maximum-smoothness model, Uno, Kawato, and Suzuki (1989) postulated that movement selection optimizes the rate of change of actuator efforts, e.g., joint torques. Although minimizing jerk and minimizing the rate of change of joint torques appear conceptually similar (in a single-joint system with predominantly inertial dynamics they are proportional to one another, the scale factor being limb inertia), there are important differences. First, the objective function was based on dynamic variables (the time rate of change of torque); therefore the predicted motion depends sensitively on the (assumed) dynamic behavior of the musculoskeletal system. Second, the objective function was formulated in terms of joint torques (instead of other representations of actuator effort, e.g., hand force). This implies that motor computations are based on a joint-space representation of behavior. Although (as outlined above) kinematic patterns are most evident in hand motions in extracorporeal space, approaches based on either joint or muscle spaces (see Kawato, 1990) have the advantage that they can generate solutions to important aspects of the ill-posed motor-control problems, such as kinematic redundancy (the apparent excess degrees of freedom) or actuator redundancy (the apparent excess of muscles). The maximum-smoothness model expressed in hand coordinates does not address these issues.

Uno et al. (1989) reported that the performance of the minimum torque change model surpassed that of the maximum-smoothness model in that it better accounted for several characteristics of human arm trajectories. It predicted that, for movements performed in the horizontal plane, point-to-point hand paths are roughly straight, except for movements where the starting point of the arm is at the side of the body and the endpoint is in front of the body. Moreover, the hand paths and speed profiles of curved movements depend on whether they curve away from or toward the body. The authors reported experiments confirming all of these predictions, apparently disproving the maximum-smoothness theory. However, an independent study (Flash, 1990) showed that the trajectories predicted by the minimum torque change theory are sensitive to the model of upper limb dynamics that is needed in order to solve the optimization problem. For example, the straightness of the horizontal planar point-to-point movements predicted by Uno et al. (1989) are a consequence of unrealistic inertial parameters used to model the human arm. Using more realistic inertial parameters, the trajectories predicted by the minimum torque change model were found to be unrealistically curved, with double-peaked rather than single-peaked speed profiles.

Motor Adaptation Studies

The most critical comparison of these two models arises from their fundamental differences. According to kinematically based optimization models, neural computations specify intended motions independently of movement dynamics or external load conditions. In contrast, dynamically based optimization models imply that external loads profoundly influence intended motions. For example, according to the minimum torque change model, movements in the presence of elastic loads are more curved than unloaded movements, whereas the maximum-smoothness model predicts no effect.

Investigating motor adaptation to elastic loads, Uno et al. (1989) concluded that the behavior in the presence of the load was different from the unloaded case. Completely different results, however, were obtained by Gurevitch (1993) when static elastic loads were unexpectedly introduced during human reaching movements toward visual targets. The first few trials following load application movements were found to be misdirected and to miss the target, but following a small number of practice trials (5–7), the loaded movements tended to follow straight hand paths with symmetrical velocity profiles (Gurevich, 1993). In another recent study (Shadmehr and Mussa-Ivaldi, 1994), velocity-dependent force fields were used to perturb the motion, and the perturbed trajectories performed in the presence of the new force fields were again found to converge toward the ones seen in the unloaded case. In a third related study, Wolpert, Ghahramani, and Jordan (in press) used altered visual feedback conditions that caused an increase in the perceived curvature of aiming movements. This led to significant corrective adaptation of the movements actually produced; the hand movements became curved, thereby reducing the visually perceived curvature. These results support the notion that arm trajectories follow a kinematic plan formulated in extrinsic visual space independent of movement dynamics or external force conditions. They are incompatible with the assumptions of dynamically based optimization models formulated in terms of intrinsic coordinates.

Why Smoothness?

The kinematic and dynamic objective functions discussed above are based on different measures of smoothness in different coordinate frames. One obvious question is: why smoothness? Maximizing smoothness might be consistent with minimizing "wear and tear" on the system, but the theories are successful even for slow movements under light loads. Maximizing smoothness is equally compatible with increasing the predictability of the trajectory or reducing the amount of information needed to internally represent motion plans (Hogan, 1984; Flash and Hogan, 1985). Smoothness maximization and the superposition of elemental movements to generate more complicated arm trajectories are also closely related to regularization-based approaches to learning from examples. Those approaches view learning as equivalent to identifying a function from sparse and noisy data. The tradeoff between accurate data reproduction and "well-behavedness" of the mapping is achieved by maximizing the smoothness of the function. Poggio (1990) pointed out the relevance of regularization theory (see REGULARIZATION THEORY AND LOW-LEVEL VISION) to the arm trajectory planning problem and suggested that RADIAL BASIS FUNCTION NETWORKS (q.v.) might be a plausible implementation for the minimum-jerk model by composing the entire movement from elemental motion primitives. The idea that a minimum-jerk movement planner can be directly implemented by a radial basis function network is attractive from the biological perspective. Other neural implementation schemes for solving dynamics-based optimization problems for arm trajectory generation have been suggested by Kawato (1992).

Discussion

One of the exciting challenges of brain theory is the need to deal with reality at the level of whole, functioning systems. Traditionally, scientific endeavor has advanced our state of knowledge by delving into finer and finer details of isolated pieces of reality—the essence of the reductionist approach. But reductionism, despite its evident success, is based on a fallacy: the wishful belief that, once the fine details are known, understanding the behavior of objects "in the large" will follow. But it has become increasingly evident that this "bottom-up" approach is severely limited in its ability to describe whole systems. Large-scale, complex systems exhibit behavior that emerges primarily from interactions among their parts. To understand them, a "top-down" approach is far more effective, beginning at a coarse-grained macroscopic level, proceeding to finer levels of detail as their structure is discerned. Optimization theory provides a powerful set of mathematical tools that lend themselves well to a top-down approach to studying the brain. As we have reviewed above, optimization theory facilitates a rigorous approach, based on macroscopic observations of psychophysical behavior, to some fundamental and far-reaching questions about the structure of neural computations.

Acknowledgments.   Tamar Flash is supported in part by grant 3164-3-93 from the Israeli ministry of Science and by the Robert and Giampiero Alchadeff Award. Neville Hogan is supported in part by NIH grants AR 40029 and NS 09343.

Road Map: Biological Motor Control
Related Reading: Equilibrium Point Hypothesis; Eye-Hand Coordination in Reaching Movements; Geometrical Principles in Motor Control; Human Movement: A System-Level Approach; Limb Geometry: Neural Control

References

Bernstein, N. A., 1967, *The Coordination and Regulation of Movement*, Oxford: Pergamon. ◆

Flash, T., 1990, The organization of human arm trajectory control, in *Multiple Muscle Systems: Biomechanics and Movement Organization* (J. M. Winters and S. L. Y. Woo, Eds.), New York: Springer-Verlag, pp. 283–301. ◆

Flash, T., and Hogan, N., 1985, The coordination of the arm movements: An experimentally confirmed mathematical model, *J. Neurosci.*, 7:1688–1703.

Gurevich, I., 1993, Strategies of motor adaptation to external loads during planar two-joint arm movements, PhD thesis, Department of Applied Mathematics and Computer Science, Weizmann Institute of Science, Rehovot, Israel.

Hasan, Z., 1986, Optimized movement trajectories and joint stiffness in unperturbed inertial loaded movements, *Biol. Cybern.*, 53:373–382.

Hogan, N., 1984, An organizing principle for a class of voluntary movements, *J. Neurosci.*, 4:2745–2754.

Hogan, N., and Flash, T., 1987, Moving gracefully: Quantitative theories of motor coordination, *Trends Neurosci.*, 10:170–174. ◆

Kawato, M., 1992, Optimization and learning in neural network for formation and control of coordinated movement, in *Attention and Performance XIV: Synergies in Experimental Psychology, Artificial Intelligence, and Cognitive Science—A Silver Jubilee* (D. Mayer and D. S. Kornblum, Eds.), pp. 283–301. Cambridge, MA: MIT Press. ◆

Lan, N., and Crago, P. E., 1994, Optimal control of antagonist muscle stiffnesses during voluntary movements, *Biol. Cybern.*, pp. 397–405.

Morasso, P., 1981, Spatial control of arm movements, *Exp. Brain Res.*, 42:223–227.

Poggio, T., 1990, A theory of how the brain might work, in *Cold Spring Harbor Symp. Quant. Biol.*, 55:899–910. ◆

Shadmehr, R., and Mussa-Ivaldi, F. A., 1994, Adaptive representation of dynamics during learning of a motor task, *J. Neurosci.*, 5:3208–3224.

Uno, Y., Kawato, M., and Suzuki, R., 1989, Formation and control of optimal trajectory in human multijoint arm movement: Minimum torque-change model, *Biol. Cybern.*, 61:89–101.

Viviani, P., and Terzuolo, C., 1982, Trajectory determines movement dynamics, *J. Neurosci.*, 7:431–437.

Wiegner, A. W., and Wierzbicka, M., 1992, Kinematic models and human elbow flexion movements: Quantitative analysis, *Exp. Brain Res.*, 88:665–673.

Winters, J. M., and Woo, S. L. Y., 1990, *Multiple Muscle Systems: Biomechanics and Movement Organization*, New York: Springer-Verlag. ◆

Wolpert, D. M., Ghahramani, Z., and Jordan, M. I., in press, Are arm trajectories planned in kinematic or dynamic coordinates? An adaptation study, *Exp. Brain Res.*

Oscillatory and Bursting Properties of Neurons

Xiao-Jing Wang and John Rinzel

Introduction

Rhythmicity is a common feature of temporal organization in neuronal firing patterns. Historically, when recordings from *isolated* nerves became possible in the 1930s, systematic study of repetitive firing behaviors ensued. Arvanitaki (1939) and Hodgkin (1948) identified three categories of crustacean axons by their rhythmic discharge patterns: those that fire repetitively over a wide (I) or narrow (II) range of frequencies and those whose firing hardly repeats (III). Later, Arvanitaki also pioneered the *Aplysia* preparation and discovered *bursting* oscillations where impulse clusters occur periodically, separated by phases of quiescence.

Since then, many other stereotypical single-neuron patterns, including a fascinating variety of endogeneous oscillations, have been identified (Llinás, 1988; Connors and Gutnick, 1990). One wonders anew about categorizing neuronal firing modes and the criteria on which to base such a classification. Hodgkin and Huxley (1952) showed that many spiking properties can be explained in terms of various active ionic currents across the cell membrane. Today, many types of ion channels are known (see ION CHANNELS: KEYS TO NEURONAL SPECIALIZATION), and some specific neuronal rhythms have been linked to selected subsets of channels. However, membrane potential oscillations with apparently similar characteristics can be generated by dif-ferent ionic mechanisms and by other biophysical factors, such as cable properties. In addition, a given cell type may display several different firing patterns under different neuromodula-tory conditions. For these reasons, the visual appearance of particular voltage time courses and the presence of certain ion-ic mechanisms are insufficient bases for classification. A ratio-nal scheme should consider a cell's complete *repertoire* of dy-namical modes and the nature of transitions between modes.

Here we apply the mathematics of dynamical systems to de-scribe precisely the dynamical modes of neuronal firing and the transformations between them. The approach was pioneered by FitzHugh with his phase space analysis of nerve membrane excitability (FitzHugh, 1961). In this theoretical framework, membrane dynamics is described by coupled *differential equa-tions*, e.g., à la Hodgkin and Huxley (cf. Rinzel and Ermentrout, 1989), the behavior modes by *attractors*, and the transitions between modes by *bifurcations*. The rest state is represented by a time-independent *steady state* and repetitive firing by a *limit cycle*. The transition from resting to oscillating typically occurs either through a *Hopf bifurcation* or a *homoclinic bifurcation* (Figure 1); (e.g., see Rinzel and Ermentrout, 1989; and DY-NAMICS AND BIFURCATION OF NEURAL NETWORKS). The firing frequency versus applied current curves are qualitatively differ-ent in the two cases (minimum frequency being non-zero or zero, respectively), and they might subserve an abstract basis

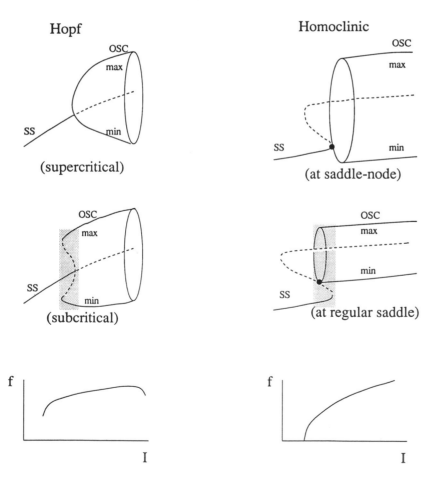

Figure 1. Schematic bifurcation diagrams from a steady state (SS) to an oscillatory firing state (OSC). The abscissa is a control parameter such as the applied current intensity. The ordinate corresponds to the membrane potential, the repetitive firing state being indicated by the maximal (max) and minimal (min) amplitudes of the oscillatory membrane potential. The solid curve indicates stable and the dashed curve un-stable. In the lowermost panels, the ordinate (f) is the frequency of repetitive firing and I is applied current. The left panels show Hopf bifurcation. At the onset of oscillation, the rhythmic amplitude is small and the frequency is finite. The bifurcation may be supercriti-cal, where the new oscillatory branch is stable; or sub-critical, where the new oscillatory branch is unstable and becomes stable at a turning point. The right panels show homoclinic bifurcation. It corresponds to the co-alescence of an oscillatory state with an unstable steady state. This steady state can be either of saddle-node or saddle type. As this bifurcation point is approached, the amplitude of oscillation remains finite, while the rhythmic frequency tends to zero (the period diverging to infinity). In the case of a subcritical Hopf bifurca-tion or a normal homoclinic bifurcation, there is a range of parameter values where a steady-state attrac-tor and an oscillatory attractor coexist (bistability, shaded region).

for the distinction between the Arvanitaki-Hodgkin type II and type I axons. Our review generalizes this theoretical methodology to characterize various *bursting* oscillations in single neurons, elaborating on a qualitative classification scheme for bursting mechanisms proposed by Rinzel (1987).

Neuronal Bursting: Examples

We summarize some qualitative features of observed bursting patterns and then relate these to our classification scheme. We briefly mention conductance mechanisms that are *sufficient* to produce some of these bursting oscillations. While network synaptic interactions and dendritic cable properties influence bursting behavior, for the most part, our discussion concerns an isolated, isopotential neuron. The main biophysical idea is that rhythmicity is generated by a depolarization process which is autocatalytic (positive feedback), followed by a *slower* repolarization process (negative feedback). These opposing processes may involve activation and slow inactivation of an inward ionic current or a fast inward current and a slower outward current. Such features underlie action potential generation, and for bursting, there is at least another *slower* negative feedback process.

The burst pattern shown in Figure 2A has a *square-wave* form, with abrupt periodic switching between rest (silent phase) and depolarized repetitive firing (active phase). Spiking here is primarily caused by a *high-threshold* fast calcium current and a Hodgkin-Huxley-like potassium current. A minimal biophysical mechanism for square-wave bursting involves a calcium-activated potassium current (see Rinzel, 1985, and citation there for Chay and Keizer, 1983). During the active phase, each calcium spike increases slightly $[Ca^{2+}]_i$, slowly turning on this current and eventually repolarizing the membrane to terminate the active phase. During the silent phase, the Ca^{2+} channels are closed, $[Ca^{2+}]_i$ decreases, and as the potassium conductance deactivates, the cell slowly depolarizes until the threshold for the next active phase is reached. Suggested alternative mechanisms for this type of bursting include slow inactivation by Ca^{2+}, and/or by voltage of the Ca^{2+} current (Cook, 1991). Here, if spikes are abolished by pharmacologically blocking the calcium current, bursting is lost.

While the dopamine-secreting neuron (Figure 2B) superficially appears to be a square-wave burster, we would not classify it as such. Its underlying slow wave persists even when action potentials are blocked. It appears to be of dendritic origin, and it drives somatic spiking through electrotonic interaction.

The bursting patterns of Figure 2C and 2D exhibit brief spike bursts riding on a slow *triangular wave*. Thalamocortical relay cells (see Figure 2C) burst at the delta-wave frequency (3 Hz) of quiet sleep (Steriade, McCormick, and Sejnowski, 1993), while the 5-Hz oscillation in inferior olivary cells (see Figure 2D) is probably involved with movement tremor (see Llinás and Yarom, 1986). Remarkably, in both cases, rhythmic bursting occurs for maintained hyperpolarizing rather than depolarizing stimuli. The underlying slow wave (as a result of a low-threshold calcium current) is unmasked when the fast action potentials are blocked and is sometimes seen for modest hyperpolarizing inputs, even without blocking spikes. The Ca^{2+} current activates rapidly below the voltage threshold for action potentials. Its inactivation by voltage, with a time scale like that of the triangular wave's depolarization, provides the slow negative feedback.

The *Aplysia* R15 neuron is the quintessential experimental model of an endogenous burster (Figure 2E) (Adams and Benson, 1985). The sodium spike rate during a burst first in-

creases and then decreases; hence, the term *parabolic* bursting. Blocking these spikes reveals an underlying quasi-sinusoidal slow wave that is generated primarily by a Ca^{2+} current. This current activates more slowly and at lower depolarizations than that associated with the square-wave bursting of Figure 2A. Its slow activation and the slower $[Ca^{2+}]_i$ that inactivates it provide the two variables for a minimal model of a parabolic burster's underlying slow oscillator (see the next section).

Parabolic burst-like features are seen in the 10-Hz oscillations of mammalian thalamic reticular neurons (Figure 2F) during the spindle waves of quiet sleep (Steriade et al., 1993). The oscillation depends on a low-threshold calcium current, such as that of triangular bursting. In addition to this current's slow inactivation, there is likely a second slow variable to support the parabolic pattern, e.g., $[Ca^{2+}]_i$ for activating a calcium-dependent potassium current in these cells.

A different kind of burst pattern consists of spike clusters interspersed with epochs of small-amplitude subthreshold oscillations (Figure 2G, 2H). The envelope of fast events slowly waxes and wanes, forming an approximate spindle or ellipse; hence, the term *elliptic bursting*. Here, the inactive phase is not totally silent, but often shows small oscillations. The frequency of intraburst spiking is comparable to that of the interburst subthreshold oscillations. Only recently has this bursting pattern been reported for mammalian neurons and associated with important functional roles, such as the limbic system's theta rhythm (not shown) and the gamma fast oscillations (approximately 40 Hz) that occur intermittently with increased alertness and focused attention (see Figure 2H). Experimental (Llinás, Grace, and Yarom, 1991) and computational (Wang, 1993) studies indicate that the 40-Hz elliptic bursts involve a persistent Na^+ conductance and a specific voltage-dependent transient K^+ conductance.

Some oscillations (Figure 2I, 2J) depend on the electrical cable properties of neuronal dendrites and intracellular sources of regenerative ion fluxes. The bursting behavior of some pyramidal neurons (Figure 2I) in the neocortex (Connors and Gutnick, 1990) and in the hippocampus depends on high-threshold calcium channels located on the distal dendrites, while the faster sodium spikes are generated primarily in the perisomatic region. Computer simulations suggest that a one-compartment description is inadequate and that electrotonically distinct compartments must be explicitly modeled and analyzed (e.g., Traub et al., 1991). Figure 2J displays the bursting pattern of a pituitary gonadotropin-releasing cell. While it resembles the square-wave form of Figure 2A, here the underlying slow rhythm is generated by a cytoplasmatic second messenger system that leads to nonlinear, time-dependent, calcium fluxes across the endoplasmic reticulum membrane (Stojilkovic and Catt, 1992) and to oscillations in $[Ca^{2+}]_i$.

Bursting Systems Analysis: Fast- and Slow-Phase Space Dynamics

Since different bursters may have qualitatively similar patterns, a qualitative classification should not depend on quantitative properties such as the rhythm's period or its precise biophysical bases. Our general framework involves a *geometrical* analysis of the bursting dynamics for a model's differential equations (Rinzel, 1985, 1987). The model for an isopotential neuron may be written as:

$$\frac{dX}{dt} = F(X, Y) \qquad (1)$$

$$\frac{dY}{dt} = G(X, Y) \qquad (2)$$

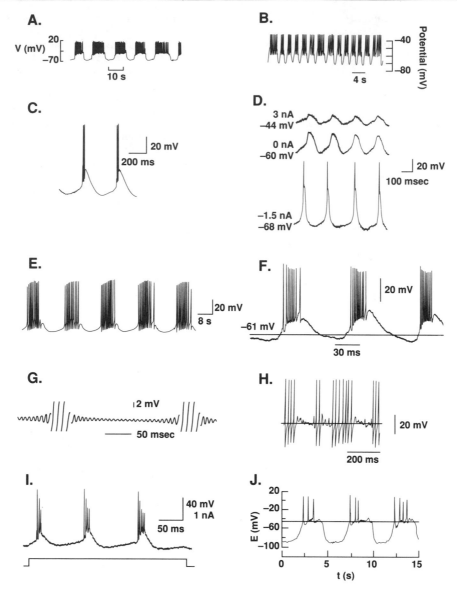

Figure 2. Examples of rhythmic bursting, showing the time courses of membrane potential, with the exception of *G*, which is extracellular voltage. See text for explanations. *A*, Pancreatic β-cell [From Sherman, A., Carroll, P., Santos, R. M., and Atwater, I., 1990, Glucose dose response of pancreatic beta-cells: Experimental and theoretical results, in *Transduction in Biological Systems* (C. Hidalgo et al. Eds.), New York: Plenum, p. 123; reprinted with permission.] *B*, Dopamine-containing neurons in the rat midbrain. (From Johnson, S. W., Seutin, V., and North, R. A., 1992, Burst firing in dopamine neurons induced by *N*-methyl-D-aspartate: Role of electrogenic sodium pump, *Science*, 258:665–667; reprinted with permission. Copyright 1992 by the AAAS.) *C*, Cat thalamocortical relay neuron. (From McCormick, D. A., and Pape, C.-H., 1991, Properties of a hyperpolarization-activated cation current and its role in rhythmic oscillation in thalamic relay neurons, *J. Physiol. Camb.*, 431:291–318; reprinted with permission.) *D*, Guinea pig inferior olivary neuron. (From Benardo, L., and Foster, R. E., 1986, Oscillatory behaviors in inferior olive neurons: Mechanism, modulation, cell aggregates, *Brain Res. Bull.*, 17:773–784; copyright 1986; reprinted with permission from Elsevier Science Ltd.) *E*, *Aplysia* R15 neuron. (From Lotshaw, D. P., Levitan, E. S., and Levitan, I. B., 1986, Fine tuning of neuronal electrical activity: Modulation of several ion channels by intracellular messengers in a single identified nerve cell,

J. Exp. Biol., 124:302–322; reprinted with permission of Company of Biologists Ltd.) *F*, Cat thalamic reticular neuron. (From Mulle, C., Madariaga, A., and Deschênes, M., 1986, Morphology and electro-physiological properties of reticularis thalami neurons in cat: In vivo study of a thalamic pacemaker, *J. Neurosci.*, 6:2134–2145; reprinted with permission of the Society for Neuroscience.) *G*, *Sepia* giant axon. (From Arvanitaki, A., 1939, Recherche sur la réponse oscillatoire locale de l'axone géant isolé de *Sepia*, *Arch. Int. Physiol.*, 49:209–256; reprinted with permission.) *H*, Rat thalamic reticular neuron. (From Pinault, D., and Deschênes, M., 1992, Voltage-dependent 40 Hz oscillations in rat reticular thalamic neurons in vivo, *Neuroscience*, 51:245–258; copyright 1992; reprinted with permission from Elsevier Science Ltd.) *I*, Mouse neocortical pyramidal neuron. (From Agmon, A., and Connors, B. W., 1989, Repetitive burst-firing neurons in the deep layers of mouse somatosensory cortex, *Neurosci. Lett.*, 99:137–141; reprinted with permission.) *J*, Rat pituitary gonadotropin-releasing cell. (From Tse, A., and Hille, B., 1993, Role of voltage-gated Na$^+$ and Ca^{2+} channels in gonadotropin-releasing hormone-induced membrane potential changes in identified rat gonadotropes, *Endocrinology*, 132(4):1475–1481; reprinted with permission. © The Endocrine Society.)

where the vectors X and Y represent the variables with fast and slow time scales, respectively. Typically, the membrane potential is a fast variable, so Equation 1 might be the membrane's current balance equation:

$$C_m \frac{dV}{dt} = -\sum_i I_i + I_{app}$$

The other dynamic variables include the gating variables for specific ionic channels plus relevant second-messenger variables and ionic concentrations. Here, we consider only one or two slow variables Y_k, which might be a slow voltage-dependent gating variable or $[Ca^{2+}]_i$, or both.

The fast- and slow-phase space dissection method (Rinzel, 1985, 1987) exploits the presence of two disparate time scales. For simplicity, suppose there is only one slow variable, Y. One first treats Y as a *control parameter* and considers the dynamics of Equation 1 as a function of Y. The fast subsystem's various behavioral states are then summarized in a bifurcation diagram, plotting response amplitude, for example V, versus Y, as in Figure 1, but where Y (instead of I) is the parameter. When the full system is considered, Y evolves slowly in time according to Equation 2, slowly sweeping through a range of values while the fast subsystem slowly tracks its stable states (*attractors*). For example, an oscillatory state of the fast subsystem corresponds to the repetitive firing of a burst's active phase. During a silent phase, the fast subsystem would be following a pseudo-steady state of hyperpolarized V. To complete the description, one must understand the slow dynamics from Equation 2 to know where on the fast subsystem's bifurcation diagram Y will be increasing or decreasing. When the full system, Equations 1 and 2, is integrated and the resulting burst trajectory is projected onto the (V, Y) plane, it coincides with portions of the bifurcation diagram. Through visualization of this geometrical representation, one can make predictions about the qualitative behavior of bursting and the effects of various parameter changes.

Square-wave bursting. The prototypical fast- and slow-phase plane (Figure 3A) was originally developed for the Chay-Keizer model of β-cell bursting (Rinzel, 1985), where $[Ca^{2+}]_i$ was the slow, negative-feedback variable (see the earlier section on neuronal bursting). For the fast-slow dissection, one first constructs the fast subsystem's bifurcation diagram by treating Y as a parameter. This construction yields the Z-shaped curve of steady states. The oscillatory state surrounding the upper branch corresponds to repetitive spiking of an active phase. It terminates by contacting the unstable middle steady-state branch at a homoclinic bifurcation. The Z-curve's lower branch represents a stable steady state of hyperpolarization as tracked during a burst's silent phase. In an intermediate range of Y values, there is bistability of the depolarized oscillation and the hyperpolarized steady state.

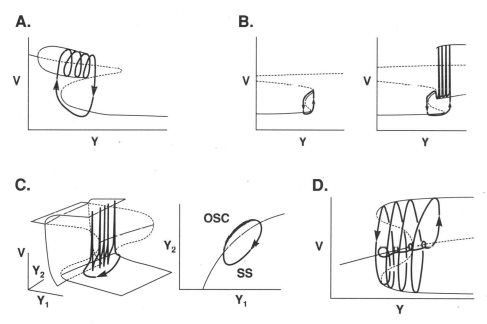

Figure 3. Fast- and slow-phase plot of bursting dynamics. The variable Y is a slow variable (there are two slow variables Y_1 and Y_2 in part C). In each case, the bifurcation diagram is computed for the fast subsystem, with Y treated as a parameter, and plotted in terms of membrane potential (V) behavior as a function of Y. The solid curve shows stable and the dashed curve unstable branches. The oscillatory state of repetitive firing is represented by its maximum and minimum of V (cf. Figure 1). The heavy curves with arrows are bursting trajectories of the full system plotted on the (V, Y) plane or the (V, Y_1, Y_2) space. *A,* Square-wave bursting is based on a bistability of a steady state and a repetitive firing state in the fast subsystem and periodic switching between the two, induced by the slow-variable dynamics. *B,* Triangular bursting has a similar phase plot as in part *A,* but the fast subsystem's steady-state curve is quintic rather than cubic, with two branches of stable steady states. Depending on whether the stable re-petitive firing state overlaps with the lowermost steady-state branch, oscillations of the full system may be either purely subthreshold (left panel) or bursting (right panel). For simplicity, the repetitive firing state is shown only on the right panel, not on the left panel. *C,* Parabolic bursting is generated by an oscillation in a two-variable (Y_1 and Y_2) slow subsystem (right panel) that induces smooth periodic switching between a steady state (SS) and a repetitive spiking state (OSC) (which do not overlap) of the fast subsystem. *D,* Elliptic bursting involves a subcritical Hopf bifurcation in the fast subsystem. Bursting involves slow switching between a steady state and a repetitive firing state that are bistable in the fast subsystem. The silent phase exhibits damped or growing small oscillations as its trajectory passes through the Hopf bifurcation point. (Parts *A* and *C–D* are adapted from Rinzel, 1987; Part *B* from Rush, M., and Rinzel J., 1994, *Biol. Cybern.,* 71:281–291.)

Next, Y is allowed to vary according to its kinetics. Bursting occurs if the slow kinetics dictate that Y increases (decreases) when the fast spike-generating subsystem is in its upper (lower) state, where the voltage-dependent channels are (are not) activated. The slow Y modulation induces abrupt switching between the two coexisting states, and thus temporal alternation occurs between a train of spikes and a resting phase, as seen in Figure 2A.

Triangular bursting. Figure 3B shows fast- and slow-phase planes associated with triangular bursting. A minimal model has one slow variable, and its fast subsystem has regimes of bistability, as with square waves. Here, however, the steady-state curve has five branches composed of *two* S-shaped portions in different V ranges. These S-curves correspond to the two sets of regenerative currents active in the subthreshold voltage ranges, such as in thalamic relay or inferior olivary cells. The depolarized oscillatory state (repetitive spiking) joins the middle steady-state branch at its right knee (a saddle-node homoclinic bifurcation). Different oscillation patterns occur depending on whether the right knee of the lower S extends rightward beyond that of the right knee of the upper S. Otherwise (see Figure 3B, left), a slow subthreshold oscillation without fast spikes may occur. The alternative case (see Figure 3B, right) corresponds to more intense hyperpolarizing input, when triangular bursting arises (see Figure 2D). The term *triangular* refers to the gradually falling V time course of the active phase, related to the middle branch's steep slope (see Figure 3B).

Parabolic bursting. This bursting type has a smooth, underlying slow-subthreshold wave. Its generation requires at least two slow variables, one for positive feedback and the other for negative feedback. The minimal fast- and slow-phase plot has three dimensions: V and the two slow variables (Figure 3C, originally constructed for a model of the *Aplysia* R15 neuron; see Rinzel, 1987). Steady states of the fast subsystem are now represented by a *Z-surface*. Similarly, a surface describes the fast oscillatory (repetitive spiking) attractors. These periodic solutions disappear through homoclinic bifurcation as they contact the Z-surface precisely at its lower knee, forming a saddle-node coalescence (see Figure 1). Here, the fast subsystem is monostable. The slow-variable-phase plane is divided into two nonoverlapping regions: one for the resting steady state and the other for the repetitive spiking regime of the fast subsystem.

When the slow variables are allowed to vary, an oscillation may occur in this two-variable slow system (see Figure 3C, right). If the slow oscillatory trajectory visits both of the fast subsystem's regimes, bursting occurs, with repetitive smooth switching between the resting and spiking states. As a burst begins and ends, its trajectory crosses a homoclinic bifurcation of the fast subsystem and spike frequency drops dramatically; hence, the parabolic nature.

Elliptic bursting. A minimal model has only one slow variable (Figure 3D, originally constructed for a modified FitzHugh-type model; see Rinzel, 1987). The fast subsystem has bistability because of a *subcritical* Hopf bifurcation (see Figure 1) of periodic solutions from a monotonic steady-state curve. As with square waves, during bursting, the full system operates in the (V, Y) regime of bistability, repetitively switching between the steady state and the spiking state. A distinguishing feature, however, is that the silent phase, when the fast subsystem operates near its steady state, is no longer truly silent: it can display small oscillations which damp and then grow as the trajectory slowly passes through the Hopf bifurcation point, where the steady state is a spiral-type fixed point.

Complex bursting. The theoretical study of certain bursting types (see Figure 2I, 2J) is relatively recent, and mathematical understanding of their mechanisms is just emerging. For analyzing the case of Figure 2I, one requires a minimal model of at least two electrotonically separated compartments. Electrical coupling might introduce another, possibly intermediate, time scale. As for Figure 2J, one must take into account the interaction between second-messenger-mediated calcium fluxes from intracellular pools and voltage-dependent plasma membrane calcium currents.

The classification discussed here is based on various fast- and slow-phase plots. Although consistent with some of the waveform phenomenology, the two may sometimes disagree. For instance, a system with the fast- and slow-phase plot of Figure 3C may burst with a slow wave that is less sinusoidal and more rectangular if one slow variable is much slower than the other. However, in contrast to a square-wave burster (see Figures 2A and 3A), its slow wave may persist, even with the fast action potentials blocked.

Discussion

We have reviewed various neuronal bursting oscillations and, by using notions and analytic tools from the mathematics of dynamical systems, we discussed how these bursting patterns might be theoretically described and classified. Our examples are minimal for these categories. Indeed, one can imagine subcategories based on differences in the fast subsystem's bifurcation diagram. In summary, bursting in a single-compartment model typically involves some slow processes which induce repetitive switching between a relatively quiescent state and an active state of repetitive spiking of a faster system. In the cases of square-wave, triangular, and elliptic bursting, one slow variable is sufficient, and the fast subsystem must be bistable. In the case of parabolic bursting, bistability in the fast subsystem is not necessary, and two slow variables are required.

The geometrical analysis by fast and slow dissection illustrates how novel and powerful theoretical approaches can emerge from fruitful interactions between neurobiology and the science of dynamical systems. Possible extensions might consider cable-like distributed systems with local burst-generating dynamics, systems with many slow variables, or systems with complicated bifurcation diagrams, perhaps involving chaotic attractors. One can expect that dynamical systems methods, including fast and slow dissection, may also play a role in our understanding of neural networks with many synaptically coupled neurons, as long as there are disparate time scales in the system.

Road Map: Biological Neurons
Background: Dynamics and Bifurcation of Neural Networks; Ion Channels: Keys to Neuronal Specialization
Related Reading: Half-Center Oscillators Underlying Rhythmic Movements; Thalamocortical Oscillations in Sleep and Wakefulness

References

Adams, W. B., and Benson, J. A., 1985, The generation and modulation of endogenous rhythmicity in the *Aplysia* bursting pacemaker neurone R15, *Prog. Biophys. Mol. Biol.*, 46:1–49. ◆
Arvanitaki, A., 1939, *Les Variations graduées de la polarization des systèmes excitables*, Paris: Hermann. ◆

Connors, B. W., and Gutnick, M. J., 1990, Intrinsic firing patterns of diverse neocortical neurons, *Trends Neurosci.*, 13:99–104. ◆

Cook, L. D., Satin, L. S., and Hopkins, W. F., 1991, Pancreatic B cells are bursting, but how? *Trends Neurosci.*, 14:411–414. ◆

FitzHugh, R., 1961, Impulses and physiological states in models of nerve membrane, *Biophys. J.*, 1:445–466.

Hodgkin, A. L., 1948, The local electric changes associated with repetitive action in a non-medullated axon, *J. Physiol. (Lond.)*, 107:165–181.

Hodgkin, A. L., and Huxley, A. F., 1952, A quantitative description of membrane current and its application to conduction and excitation in nerve, *J. Physiol. (Lond.)*, 117:500–544.

Llinás, R., 1988, The intrinsic electrophysiological properties of mammalian neurons: Insights into central nervous system function, *Science*, 242:1654–1664. ◆

Llinás, R. R., Grace, T., and Yarom, Y., 1991, *In vitro* neurons in mammalian cortical layer 4 exhibit intrinsic oscillatory activity in the 10- to 50 Hz frequency range, *Proc. Natl. Acad. Sci. USA*, 88:897–901.

Llinás, R. R., and Yarom, Y., 1986, Oscillatory properties of guinea-pig inferior olivary neurones and their pharmacological modulation: An *in vitro* study, *J. Physiol. (Lond.)*, 376:163–182.

Rinzel, J., 1985, Bursting oscillations in an excitable membrane model, in *Ordinary and Partial Differential Equations: Proceedings of the 8th Dundee Conference* (B. D. Sleeman and R. J. Jarvis, Eds.), Lecture Notes in Mathematics, 1151, New York: Springer, 304–316.

Rinzel, J., 1987, A formal classification of bursting mechanisms in excitable systems, in *Proceedings of the International Congress of Mathematicians* (A. M. Gleason, Ed.), Providence, RI: American Mathematical Society, pp. 1578–1594.

Rinzel, J., and Ermentrout, G. B., 1989, Analysis of neural excitability and oscillations, in *Methods in Neuronal Modeling: From Synapses to Networks* (C. Koch and I. Segev, Eds.), Cambridge, MA: MIT Press, 135–169. ◆

Steriade, M., McCormick, D. A., and Sejnowski, T. J., 1993, Thalamo-cortical oscillations in the sleep and aroused brain, *Science*, 262:679–685. ◆

Stojilkovic, S. S., and Catt, K. J., 1992: Calcium oscillations in anterior pituitary cells, *Endocr. Rev.*, 13:256–280. ◆

Traub, R., Wong, R., Miles, R., and Michelson, H., 1991, A model of a CA3 hippocampal pyramidal neuron incorporating voltage-clamp data on intrinsic conductances, *J. Neurophysiol.*, 66:635–649.

Wang, X.-J., 1993, Ionic basis for intrinsic 40 Hz neuronal oscillations, *Neuroreport*, 5:221–224.

Oscillatory Associative Memories

Joachim M. Buhmann

Introduction

Assemblies of cooperating binary units have been postulated as the basic building blocks of associative memories. Artificial neural networks with a simple dynamics evolve from initial states to predefined fixed points, the memory traces. Storage and retrieval in these neural networks with fixed-point attractors are limited to one pattern at a time. Biological and functional considerations, especially in olfactory cortex, motivate oscillatory neural network dynamics which model experimentally observed neural activity patterns in associative recall tasks. Binding features to stored patterns when several memory traces are coactivated requires a network dynamics with oscillatory or even chaotic activity patterns.

Associative Storage in Dynamical Systems

Associative recall and completion of information is one of the astonishing abilities of intelligent living beings. The search for mechanisms which produce this ability of associative memory has yielded a class of computational systems composed of many neuron-like, nonlinear units. The basic principle of associative information recall is a network dynamic which maps initial network states to a subset of final states. According to dynamical systems theory, the asymptotic dynamics of such an assembly of neurons can be a fixed point, a limit cycle, or a chaotic attractor. These attractors of the neural dynamics are identified with memorized items, the memory traces (see COMPUTING WITH ATTRACTORS).

The best understood artificial neural networks with associative abilities are networks with fixed-point dynamics (Amari, 1972). The prototypical model of this class, suggested by Hopfield (1982), comprises a network of n binary units which are connected in a symmetric fashion and typically show a distributed activity pattern. The neural dynamics with synchronous or asynchronous (stochastic) update of neuron states drives the network into a stationary state, a fixed point of the neural dynamics. These fixed points are supposed to be identical

or at least very similar to a set of p random patterns $\{\xi^v \mid v = 1, \ldots, p\}$, $\xi_i^v \in \{1, 0\}$, $i = 1, \ldots, n$, which represent the information content of the associative memory. A connectivity pattern which guarantees the closeness of pattern states and fixed points for a moderate number of patterns ($p/n < 0.14$) is defined by a version of Hebb's rule

$$W_{ik} = \frac{1}{N} \sum_{v=1}^{p} (\xi_i^v - \tfrac{1}{2})(\xi_k^v - \tfrac{1}{2}) \tag{1}$$

i.e., correlated activity of presynaptic and postsynaptic units ($\xi_i^v = \xi_k^v$) increases synaptic strength, and uncorrelated activity ($\xi_i^v \neq \xi_k^v$) decreases it. The standard Hopfield model and variants of it were analyzed by statistical physicists in great detail (Amit, 1989; Hertz, Krogh, and Palmer, 1991; see STATISTICAL MECHANICS OF LEARNING).

What are the computational limits of fixed-point associative memories? Any stationary activity pattern of the network is interpreted as a stored memory trace. Suppose now that a fixed-point network is initialized in a state which corresponds to a superposition of several stored patterns, none of them being dominant. A classification of the input as one of the components is inappropriate, since all the information of the admixed and equally important patterns has to be suppressed and, therefore, is lost for further processing. Most likely, the associative memory relaxes into a fixed-point state which corresponds to a superposition of patterns. The mixture is treated as a new, composite pattern. Unfortunately, the information is lost as to which part of the network response belongs to which component pattern of the mixture. This information loss has been called the superposition catastrophe (von der Malsburg, 1981); it limits associative memories with fixed-point dynamics to sequential associative recall of one pattern at a time.

The observation of nonstationary neural activity in visual cortex (Gray et al., 1989) and olfactory cortex (Freeman, 1991) has stimulated interest in associative memory models with an oscillatory or chaotic dynamics. Neural modeling of associative memory has either aimed at rendering biological behavior as faithfully as possible or at exploring new associative recall principles with segmentation capability which promise to make technical storage devices more versatile. We will discuss three different models of oscillatory associative memory which represent both research philosophies: (1) an artificial neural network with adaptive threshold control, (2) a minimal model of olfactory cortex, and (3) a model for pattern segmentation in the time domain.

A Neural Assembly with Dynamic Thresholds

Starting with the standard Hopfield model of a conventional associative memory, Horn and Usher (1991) introduced a dynamic threshold to model fatigue effects in neuronal response. The neuron states are described by the binary variables $S_i \in \{-1, 1\}$. Neurons with a high activity level raise their threshold and thereby lower their sensitivity. The neurons are sequentially updated in random order. A probabilistic update rule, parameterized by a computational temperature T, mimics the stochastic influences which are abundant in biological systems (see SIMULATED ANNEALING). The equation of motion of the neuron i is given by

$$S_i(t + 1) = F_T\left(\sum_{k \neq i} W_{ik}S_k(t) - \theta_i(t)\right) \tag{2}$$

where $F_T(x) = \pm 1$ with probability $(1 + \exp(\mp 2x/T))^{-1}$. The condition $T = 0$ reduces Equation 2 to a deterministic threshold rule. The qualitative behavior of the network at finite T can also be described by n deterministic equations for the expectation value of the stochastic neuronal states S_i (Horn and Usher, 1991). The dynamic threshold $\theta_i(t)$ integrates the past activity of neuron i with an exponential decay, i.e.,

$$\theta_i(t + 1) = \frac{\theta_i(t)}{c} + bS_i(t + 1) \tag{3}$$

Permanently active or inactive neurons $[S_i(t) = S_i(t - 1) = \ldots = \pm 1]$ saturate their threshold at the values $\theta_i(t) = \pm bc/(c - 1)$, respectively. Obviously, appropriate values of b and c destabilize a stationary pattern state and force the network out of the present metastable state into another one. A Hopfield model with adaptive threshold control exhibits oscillatory pattern activity for a small number of patterns stored. Information recall takes place during intermittent periods of minimal fluctuations in network activity. A readout network which is sensitive to changes in neuron states $[S_i(t + 1) \neq S_i(t)]$ integrates the neural activities of Horn and Usher's oscillatory associative memory and retrieves a stored pattern during periods of quasi-stationarity. The network has settled in a quasi-stationary state when only few neurons change their neural state per time interval. The length of these metastable recall periods is controlled by the parameter c, with an increase in c prolonging the recall periods.

Associative memories with dynamic thresholds emulate fatigue effects in neural networks which prevent the network dynamics from being trapped for an indefinite time in a stable attractor state, regardless of the input pattern that is presented. An internal renewal mechanism of this sort is necessary for associative memories which are expected to correct changing input patterns. The dynamically adjusted thresholds increase the network's sensitivity to new stimuli, thereby bringing associative memory models closer to biological reality.

A Model for Olfactory Cortex

A brain region which apparently has anatomical structure and dynamics similar to associative memory models is the *olfactory cortex*. Preliminary evidence for this hypothesis comes from the sparse but distributed activity responses of olfactory neurons to similar odors (Haberly and Bower, 1989; Freeman, 1987). The olfactory cortex, whose dynamics is strongly dominated by an inhibitory feedback loop between pyramidal cells and inhibitory interneurons, is a brain region with pronounced activity oscillations. How might oscillatory activity enhance the associative abilities of the network compared to associative memories with fixed-point attractors? Oscillations in associative recall (according to Freeman's view) provide an unbiased access to the memory traces and facilitate transitions from one pattern to others. Freeman's very detailed olfactory model (Freeman, 1987) even shows deterministic chaotic behavior, stimulating speculation on the possible role of chaos in the brain (Freeman, 1991).

Baird has proposed an oscillatory associative memory which models the architecture of olfactory cortex in a simplified fashion (Baird, 1990). The model comprises excitatory neurons with long-range connections and local inhibitory neurons which are only connected to nearby excitatory cells. The excitatory layer of neurons implements the associative recall dynamics with its mutual cooperation among cells belonging to the same assembly and its competition between cells which represent different patterns. Inhibitory cells are connected to their excitatory counterparts and serve as the inhibitory feedback element to stabilize a limit cycle in the neural activity. A schematic drawing of the model is shown in Figure 1. The specific neural dynamics suggested by Baird is defined by n pairs of leaky integrator equations for the membrane potentials $x_i(t)$, $y_i(t)$, $i = 1, \ldots, n$, of excitatory, inhibitory cells, respectively, i.e.,

$$\frac{dx_i}{dt} = \tau x_i - hy_i + \sum_{j=1}^{p} W_{ij}x_j - \sum_{j,k,l=1}^{p} W_{ijkl}x_jx_kx_l + I_i(t) \tag{4}$$

$$\frac{dy_i}{dt} = -\tau y_i + gx_i \tag{5}$$

The rate constant τ permits gauging the simulation time scale to the physiological one. The term $I_i(t)$ abbreviates a time-dependent input to unit i. The pairwise connections W_{ij} between the excitatory units are designed to establish a set of desired patterns χ^v, $v = 1, \ldots, p$ as limit cycles of the dynamics. A stable oscillatory dynamics is guaranteed for $p \leq (n/2)$. The fourth-order synapses W_{ijkl}, which are omitted in Figure 1 for reasons of clarity, stabilize these limit cycles and suppress spurious attractor states by competition. The competition mechanism resembles winner-take-all networks in its structure and functional role. The connections are chosen according to the Hebb rule

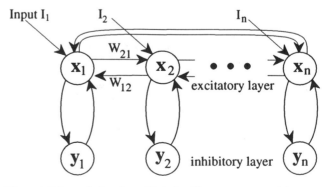

Figure 1. Schematic drawing of Baird's olfactory cortex model.

$$W_{ij} = \sum_{v=1}^{p} \alpha^v \chi_i^v \chi_j^v \qquad (6)$$

and the multiple outer product rule for the fourth-order weights

$$W_{ijkl} = c\delta_{ij}\delta_{kl} - d \sum_{v=1}^{p} \chi_i^v \chi_j^v \chi_k^v \chi_l^v \qquad (7)$$

The requirement $c > d$ has to be satisfied for stable limit cycles (Baird, 1990). With such a choice of weights, the network exhibits for asymptotically long times ($t \to \infty$) oscillatory activity of the form

$$\begin{pmatrix} x_k^v(t) \\ y_k^v(t) \end{pmatrix} = \begin{pmatrix} |\chi^v| \exp(i\theta_x^v + i\omega^v t) \\ \sqrt{g/h}|\chi^v| \exp(i\theta_y^v + i\omega^v t) \end{pmatrix} \qquad (8)$$

Equation 8 describes the recall of a single activity pattern χ^v. Baird's network is not designed to recall multiple patterns as in Horn and Usher's associative memory, but to provide a mathematically tractable sketch of olfactory cortex. The phases θ_x, θ_y are identical for all excitatory and inhibitory neurons, respectively, a consequence of the simplicity of the model. The phase lag $\theta_x - \theta_y$ between the two populations is nearly 90° for a broad parameter range, which fits nicely the experimentally measured dynamics of olfactory activity.

Baird has shown that the network (Equations 4 and 5) is capable of storing Fourier components and suggested that complex motor patterns might be memorized in neural structures with oscillatory response. This hypothesis still has to be substantiated by experimental results from motor cortex. Recently an oscillatory network has successfully been used to recognize handwritten characters by the temporal signature of pen movement (Baird et al., 1991).

Pattern Segmentation in Oscillatory Associative Memory

Obviously, the time domain endows an associative network with an additional degree of freedom which is not available to fixed-point attractor networks. Time can be used to label and segment patterns in neural networks which would avoid the superposition catastrophe of fixed-point memories. Binding and segmentation of pieces of information is not only a problem in associative recall in olfactory cortex, but the question of feature binding appears also in early vision where no associative recall takes place. Neurophysiological findings of oscillatory activity in primary visual cortex (Gray et al., 1989) support speculations that temporal correlations form the basic mechanism for information binding in cortex (von der Malsburg, 1981; see SYNCHRONIZATION OF NEURONAL RESPONSES AS A PUTATIVE BINDING MECHANISM).

Wang, Buhmann, and von der Malsburg (1990) have demonstrated that such an oscillatory neural network is able to separate several simultaneously presented patterns on a microscopic time scale of milliseconds but to preserve the information about the patterns on a psychological time scale of a fraction of a second. The network relaxes to a limit cycle representing one pattern and switches to another limit cycle after a brief period of time. Such a switching mechanism based on oscillatory activity might explain the ability of rats to discriminate different odors in a mixture without suppressing all except one.

Wang et al. used population equations of the leaky integrator type for pools of excitatory and inhibitory neurons:

$$\tau_x \frac{dx_i}{dt} = -x_i + G_x \left(T_{xx}x_i - T_{xy}F(y_i) + \sum_{k \neq i} W_{ik}x_k(t) + I_i - H_i \right) \qquad (9)$$

$$\tau_y \frac{dy_i}{dt} = -y_i + G_y(-T_{yy}y_i + T_{yx}x_i) \qquad (10)$$

$$H_i = \alpha \int_0^t x_i(\tau) \exp(-\beta(t - \tau)) \, d\tau \qquad (11)$$

According to these equations, the variables x_i and y_i are restricted to the interval [0, 1]. They describe the average firing rate of excitatory or inhibitory neuron pools, respectively. The parameters τ_x and τ_y are the time constants of the excitatory and inhibitory activity of the oscillators. The architecture of the network corresponds again to Figure 1 and is similar to Baird's model. The symbols G_x and G_y represent sigmoid gain functions of the form

$$G_r(v) = [1 + \exp(-(v - \theta_r)/\lambda_r)]^{-1} \qquad r \in \{x, y\} \qquad (12)$$

with thresholds θ_x or θ_y and gain parameters $1/\lambda_x$ or $1/\lambda_y$. The inhibitory neuron pools influence the excitatory neuron pools in a nonlinear way described by the quadratic function $F(y_i) = (1 - \eta)(y_i/\bar{y}) + \eta(y_i/\bar{y})^2$, ($0 \leq \eta \leq 1$). The tuning parameter $\bar{y}$ is used to adjust the average inhibitory activity. The delayed self-inhibition term H_i is analogous to a dynamic threshold. In addition to the interaction between excitatory units x_i and inhibitory units y_i, parameterized by the weights T_{rs}, $r, s \in \{x, y\}$, an excitatory unit x_i receives time-dependent external input $I_i(t)$ from a sensory area or from other networks, and internal input $\sum_{k \neq i} W_{ik}x_k(t)$ from other oscillators.

Let us now study how the oscillatory network performs the segmentation task: p sparsely coded, random N-bit words $\{\xi^v | v = 1, \ldots, p\}$, $\xi_i^v \in \{0, 1\}$, $i = 1, \ldots, n$ are stored in the network. The probability that a bit equals 1 is a, typically $a < 0.2$. As in Horn and Usher's and Baird's models, the synapses are chosen according to a Hebbian correlation rule

$$W_{ik} = \frac{1}{aN} \sum_v (\xi_i^v - a)(\xi_k^v - a) \qquad (13)$$

In a simulation study, 50 oscillators were used and 8 patterns (each with 8 active units) were stored in the memory. A superposition of the first three patterns is offered to the network for associative restoration and segmentation. In Figure 2, the activity of five representative oscillators is monitored, and three sample time slots are marked in which the component patterns appear in isolation. All three components are recognized as stored patterns. The exact number of patterns that can be represented simultaneously depends on details of implementation, but a reasonable estimate seems to be the 7 ± 2 which is often cited as the storage capacity of short-term memory.

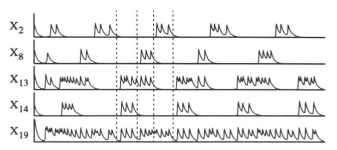

Figure 2. Activity of an oscillatory associative memory for pattern segmentation. The activity of five representative oscillators is monitored during pattern recall. The oscillators x_2, x_8, and x_{14} are exclusively active in patterns 1, 2, and 3, respectively. Oscillator x_{13} is active in patterns 2 and 3; oscillator x_{19} is part of all three patterns.

An alternative, but quite novel, proposal for olfactory pattern segmentation has been suggested by Hopfield (1991). The spatial location of odors and the individual odor qualities are identified and segmented on the basis of temporal fluctuations. Coactivation of receptors is exploited to strengthen or weaken synapses on a rapid time scale, enabling the network to disambiguate mixture components. Information about odor mixtures and their spatial distribution is encoded in the weights of rapidly changing synapses (von der Malsburg, 1981), thereby forming neural assemblies to represent odor sources for a short time period. The architecture of the model is adapted from a caricature of the olfactory bulb (Li and Hopfield, 1989), an oscillatory neural filter with sensitivity control for odor enhancement or suppression.

Discussion

Oscillatory associative memories are motivated by biological findings in olfactory and visual cortex. The merit of these networks as parallel computing devices is determined by their ability to exploit the time-varying neural activity for information processing beyond associative pattern recall and pattern completion. The capability of oscillatory associative memories to segment superimposed patterns in time clearly enhances the functionality of conventional associative memories. It also demonstrates a plausible mechanism by which neural assemblies might dynamically bind pieces of information together by activity correlations which possibly might trigger rapid synaptic changes. It remains to be seen if chaotic network dynamics provides additional network capabilities beyond the computational power of oscillatory associative memories.

Road Map: Self-Organization in Neural Networks
Background: I.3. Dynamics and Adaptation in Neural Networks
Related Reading: Spatiotemporal Association in Neural Networks

References

Amari, Shun-Ichi, 1972, Learning patterns and pattern sequences by self-organizing nets of threshold elements, *IEEE Trans. Comput.*, C-21:1197–1206.
Amit, D., 1989, *Modelling Brain Function*, Cambridge, Eng.: Cambridge University Press. ◆
Baird, B., 1990, Bifurcation and learning in models of oscillating cortex, *Physica D*, 42:365–384.
Baird, B., Freeman, W., Eeckman, F., and Yao, Y., 1991, Applications of chaotic neurodynamics in pattern recognition, *Proc. Soc. Photo-Opt. Instrum. Eng.*, 1469:12–23.
Freeman, W. J., 1987, Simulation of chaotic EEG patterns with a dynamic model of the olfactory system, *Biol. Cybern.*, 56:139–150.
Freeman, W. J., 1991, The physiology of perception. *Sci. Am.*, 264:78–87. ◆
Gray, C. M., König, P., Engel, A. K., and Singer, W., 1989, Synchronization which reflects global stimulus properties, *Nature*, 338:334–337.
Haberly, L. B., and Bower, J. M., 1989, Olfactory cortex: Model circuit for study of associative memory? *Trends Neurosci.*, 12:258–264.
Hertz, J., Krogh, A., and Palmer, R. G., 1991, *Introduction to the Theory of Neural Computation*, New York: Addison-Wesley. ◆
Hopfield, J. J., 1982, Neural networks and physical systems with emergent collective computational abilities, *Proc. Natl. Acad. Sci. USA*, 79:2554.
Hopfield, J. J., 1991, Olfactory computation and object perception, *Proc. Natl. Acad. Sci. USA*, 88:6462–6466.
Horn, D., and Usher, M., 1991, Parallel activation of memories in an oscillatory neural network, *Neural Computat.*, 3:31–43.
Li, Z., and Hopfield, J. J., 1989, Modeling the olfactory bulb and its neural, oscillatory processings, *Biol., Cybern.*, 61:379–392.
von der Malsburg, C., 1981, *The Correlation Theory of Brain Function*, Internal Report, Max-Planck-Institut für Biophysikalische Chemie, reprinted in *Models of Neural Networks, 2* (K. Schulten and H.-J. van Hemmen, Eds.), 1994, New York: Springer-Verlag.
Wang, D. L., Buhmann, J. M., and von der Malsburg, C., 1990, Pattern segmentation in associative memory, *Neural Computat.*, 2:94–106. ◆

PAC Learning and Neural Networks

Martin Anthony and Norman Biggs

Introduction

In this article, we discuss the *probably approximately correct* (PAC) learning paradigm as it applies to artificial neural networks. The PAC learning model is a probabilistic framework for the study of learning and generalization. It is useful not only for neural classification problems, but also for learning problems more often associated with mainstream artificial intelligence, such as the inference of Boolean functions. In PAC theory, the notion of succesful learning is formally defined using probability theory. Very roughly speaking, if a large enough sample of randomly drawn training examples is presented, then it should be likely that, after learning, the neural network will classify most other randomly drawn examples correctly. The PAC model formalizes the terms *likely* and *most*. Furthermore, the learning algorithm must be expected to act quickly, since otherwise it may be of little use in practice.

There are thus two main emphases in PAC learning theory. First, there is the issue of how many training examples should be presented. Second, there is the question of whether learning can be achieved using a fast algorithm. These are known, respectively, as the *sample complexity* and *computational com-*plexity problems. This article provides a brief introduction to these. We highlight the importance of the Vapnik-Chervonenkis dimension, a combinatorial parameter which measures the *expressive power* of a neural network, and describe how this parameter quantifies fairly precisely the sample complexity of PAC learning. In discussing the computational complexity of PAC learning, we shall present a result which illustrates that in some cases the problem of PAC learning is inherently intractable.

PAC Learning

Basic Definitions

In this section, we describe the basic *probably approximately correct* (PAC) model of learning introduced by Valiant (1984). This model is applicable to neural networks with one output unit which outputs either the value 0 or 1; thus, it applies to *classification* problems. In the PAC model, it is assumed that the neural network receives a sequence of *examples x*, each labeled with the value $t(x)$ of the particular *target function* which is being *learned*. A fundamental assumption of this model is that these examples are presented independently and at random according to some fixed (but unknown) probability distribution on the set of all examples.

We first explain how to formalize the notion of generalization. Suppose that the set of all possible examples is $X = \mathbb{R}^n$ or $X = \{0, 1\}^n$, where n is the number of inputs to the network, and that the target function t can be computed by the neural network in some state. A *training sample* for t of length m is an element $\mathbf{s}$ of $(X \times \{0, 1\})^m$, of the form

$$\mathbf{s} = ((x_1, t(x_1)), (x_2, t(x_2)), \ldots, (x_m, t(x_m)))$$

We shall denote by $S(m, t)$ the set of all training samples of length m for t. The learning algorithm accepts the training sample $\mathbf{s}$ and alters the state of the network in some way in response to the information provided by the sample. It is desired that the resulting state is an approximation to the target function.

Probability and Approximation

If $L(\mathbf{s})$ is the function computed by the network after training sample $\mathbf{s} \in S(m, t)$ has been presented and learning algorithm L has been applied, one way in which to assess the success of the learning process is to measure how close $L(\mathbf{s})$ is to t. Since there is assumed to be some probability distribution P on the set of all examples, and since t takes only the values 0 or 1, we may define the *error* of a function h (with respect to t), $\mathrm{er}_P(h, t)$, to be the P-probability that a randomly chosen example is classified incorrectly by h. In other words,

$$\mathrm{er}_P(h, t) = P(\{x \in X : h(x) \neq t(x)\})$$

The aim is to ensure that the error of $L(\mathbf{s})$ is *usually small*. Since each of the m examples in the training sample is drawn randomly and independently according to P, the sample vector $\mathbf{x}$ is drawn randomly from X^m according to the product probability distribution P^m. Thus, more formally, we want it to be true that with high P^m-probability the sample $\mathbf{s}$ arising from $\mathbf{x}$ is such that the function $L(\mathbf{s})$ computed after training has small error with respect to t. This leads us to the following formal definition of PAC learning.

The learning algorithm L is a *PAC-learning algorithm* for the network if *for any* given $\delta, \varepsilon > 0$, there is a sample length $m_0(\delta, \varepsilon)$ such that *for all* target functions t computable by the network and *for all* probability distributions P on the set of examples, we have

$$m \geq m_0(\delta, \varepsilon) \Rightarrow P^m(\{\mathbf{s} \in S(m, t) : \mathrm{er}_P(L(\mathbf{s}), t) > \varepsilon\}) < \delta$$

In other words, provided the sample has length at least $m_0(\delta, \varepsilon)$, then it is "probably" the case that after training on that sample, the function computed by the network is "approximately" correct. (We should note that the product probability distribution P^m is really defined not on subsets of $S(m, t)$ but on sets of vectors $\mathbf{x} \in X^m$. However, this abuse of notation is convenient and is unambiguous: for a fixed t, there is a clear one-to-one correspondence between vectors $\mathbf{x} \in X^m$ and training samples $\mathbf{s} \in S(m, t)$.) Note that the probability distribution P occurs twice in the definition: first in the requirement that the P^m-probability of a sample be small and second through the fact that the error of $L(\mathbf{s})$ is measured with reference to P. The crucial feature of the definition is that we require that the sample length $m_0(\delta, \varepsilon)$ be independent of P and of t. It is not immediately clear that this is possible, but the following informal arguments explain why it can be done. If a particular example has not been seen in a large sample $\mathbf{s}$, the chances are that this example has low probability (with respect to P) and therefore misclassification of that example contributes little to the error

of the function $L(\mathbf{s})$. In other words, the penalty paid for misclassification of a particular example is its probability, and, very loosely speaking, the two occurrences of the probability distribution in the definition can therefore balance or cancel each other.

The Finite Case

We shall show that if the network computes only a finite number of functions (for example, when the weights of a neural network are restricted to a finite set of allowed values), then there is a PAC learning algorithm for the network.

We say that the learning algorithm L is *consistent* if, given any training sample $\mathbf{s} = ((x_1, t(x_1)), (x_2, t(x_2)), \ldots, (x_m, t(x_m)))$, the functions $L(\mathbf{s})$ and t agree on x_i, for each i between 1 and m. Such a condition seems quite natural. We should note, however, that neither the standard on-line perceptron learning algorithm nor the on-line backpropagation algorithm is, in general, consistent. But the batch versions of these algorithms, in which one repeatedly cycles through the training sample until no further changes are required, *are* consistent algorithms.

Suppose that the network is capable of computing a total of M different functions and let t be any one of these. If h is computable by the network and has error $\varepsilon_h \geq \varepsilon$ with respect to t and P, then the probability (with respect to the product distribution P^m) that h agrees with t on a random sample is clearly at most $(1 - \varepsilon_h)^m$. This is at most $\exp(-\varepsilon_h m)$, using a standard approximation. Thus, since there are certainly at most M such functions h, the probability that *some* function computable by the network has error at least ε *and* is consistent with a randomly chosen sample $\mathbf{s}$ is at most $M \exp(-\varepsilon m)$. For any fixed positive δ, this probability is less than δ provided

$$m \geq m_0(\delta, \varepsilon) = \frac{1}{\varepsilon} \log\left(\frac{M}{\delta}\right)$$

This bound is independent of both the distribution and the target function.

This analysis shows that if a network only computes a finite number of functions, then there is a PAC learning algorithm for the network and, moreover, *any* consistent learning algorithm for the network is a PAC learning algorithm. The argument fails if the network in question computes infinitely many functions, and it is not immediately clear that PAC learning is possible in such circumstances. In the next section, we present a theory which shows that, in many such cases, it is possible.

PAC Learning and the Vapnik-Chervonenkis Dimension

The Vapnik-Chervonenkis Dimension

In this section, we show how the problem of PAC learning can be addressed by means of a combinatorial parameter known as the Vapnik-Chervonenkis dimension (henceforth called the VC-dimension). Suppose $\mathcal{N}$ is a neural network which outputs 0 or 1 and suppose $\mathcal{N}$ accepts examples from a set X (for example, $X = \mathbb{R}^n$, where n is the number of inputs). We say that a set T of examples is *shattered* by $\mathcal{N}$ if for each of the $2^{|T|}$ possible ways of dividing T into two disjoint sets T_1 and T_0, there is *some* function f computable by $\mathcal{N}$ such that $f(x) = 1$ if $x \in T_1$ and $f(x) = 0$ if $x \in T_0$. In what follows, it is sometimes convenient to say that x is a positive (respectively, negative) example of f if $f(x) = 1$ (respectively, $f(x) = 0$). The *VC-dimension* of $\mathcal{N}$, denoted VC-dimension$(\mathcal{N})$, is defined to be the largest size of a set of examples shattered by $\mathcal{N}$. The VC-dimension may be thought of as a measure of the *expressive power* of the network, although Vapnik and Chervonenkis (1971) defined this parameter in a more general context and not specifically in the context of neural networks (see VAPNIK-CHERVONENKIS DIMENSION OF NEURAL NETWORKS). It should be noted that the notion of Vapnik-Chervonenkis dimension is, in a sense, an extension to that of linear (or vector-space) dimension. Dudley (1978) proved that if $\mathcal{F}$ is a vector space of real functions defined on a set X and if, for $f \in \mathcal{F}$, we define $f_+ : X \to \{0, 1\}$ by $f_+(x) = 1 \Leftrightarrow f(x) > 0$, then the VC-dimension of $\{f_+ : f \in \mathcal{F}\}$ is the linear dimension of $\mathcal{F}$.

It is instructive at this stage to determine the VC-dimension of the simplest neural network, the *simple real perceptron* $\mathcal{P}_n$ on n inputs. This network consists of n real-valued inputs, each of which is connected by a weighted connection to the single, linear threshold, output unit. (The weights can be any real numbers.) It is clear that, for functions computable by $\mathcal{P}_n$, the sets of positive examples and negative examples are separated by a hyperplane.

Theorem 1. For any positive integer n, let $\mathcal{P}_n$ be the simple real perceptron with n inputs. Then

$$\text{VC-dimension}(\mathcal{P}_n) = n + 1$$

Proof. Let T be *any* set of $n + 2$ examples. It can be shown that there is a nonempty subset T_1 of T such that, if $T_0 = T \setminus T_1$, then $\text{conv}(T_1) \cap \text{conv}(T_0) \neq \varnothing$, where $\text{conv}(A)$ denotes the convex hull of A. (This follows from Radon's theorem, which may be found in Grunbaum, 1967, for instance.) It follows immediately that the sets T_1 and T_0 cannot be separated by a hyperplane; in other words, there can be no function f computable by $\mathcal{P}_n$ such that $f(x) = 1$ if $x \in T_1$ and $f(x) = 0$ if $x \in T_0$. Therefore T is not shattered and the VC-dimension of $\mathcal{P}_n$ must be at most $n + 1$. It remains to prove the reverse inequality. Let o denote the origin of $\mathbb{R}^n$ and, for $1 \leq i \leq n$, let e_i be the point with a 1 in the ith coordinate and all other coordinates 0. Then $\mathcal{P}_n$ shatters the set $T = \{0, e_1, e_2, \ldots, e_n\}$ of $n + 1$ examples. To see this, suppose that $T_1 \subseteq T$. For $i = 1, 2, \ldots, n$, let α_i be 1 if $e_i \in T_1$ and -1 otherwise, and let θ be $-\frac{1}{2}$ if $0 \in T_1$, $\frac{1}{2}$ otherwise. Then it is straightforward to verify that if h is the function computed by the perceptron when the threshold is θ and the weights are $\alpha_1, \alpha_2, \ldots, \alpha_n$, then $h(x) = 1$ if $x \in T_1$ and $h(x) = 0$ if $x \in T_0$. Therefore, T is shattered and, consequently, VC-dimension$(\mathcal{P}_n) \geq n + 1$. $\square$

Finite VC-Dimension Characterizes PAC Learning

We have observed that if $\mathcal{N}$ computes only a finite number of functions, then any consistent learning algorithm is a PAC algorithm, and a value of $m_0(\delta, \varepsilon)$ involving the number of computable functions can be determined. It turns out that, as far as PAC learning is concerned, it is not the size of the set of computable functions which is crucial, but the *VC-dimension* of the network. More precisely, we have the following key result, due to Blumer et al. (1989) and Ehrenfeucht et al. (1989).

Theorem 2. If a neural network $\mathcal{N}$ has finite VC-dimension $d \geq 1$, then any consistent learning algorithm L for $\mathcal{N}$ is a PAC learning algorithm. Moreover, there is a constant K such that a sufficient sample length $m_0(\delta, \varepsilon)$ for any such algorithm is

$$K\varepsilon^{-1}(d \ln(\varepsilon^{-1}) + \ln(\delta^{-1}))$$

On the other hand, there is a constant c such that for any PAC learning algorithm for $\mathcal{N}$, the sufficient sample length $m_0(\delta, \varepsilon)$ must be at least $c\varepsilon^{-1}(d + \ln(\delta^{-1}))$ for all $\varepsilon \leq \frac{1}{8}$ and $\delta \leq \frac{1}{100}$.

In fact, an analog of Theorem 2 holds for general classes of $\{0, 1\}$-valued functions, and not simply those computable by neural networks.

VC-Dimension of Neural Networks

We now discuss some results on the VC-dimensions of certain types of networks. A more detailed treatment of this topic may be found in the article VAPNIK-CHERVONENKIS DIMENSION OF NEURAL NETWORKS. First, we start with the feedforward linear threshold network. The first part of the following result is due to Baum and Haussler (1989) and the second part is due to Maass (1993a).

Theorem 3. There is $K > 0$ such that, if $\mathcal{N}$ is any feedforward linear threshold network having W variable weights and thresholds and N threshold units, then VC-dimension$(\mathcal{N}) \leq KW \log N$. Furthermore, there is $c > 0$ such that some feedforward linear threshold networks having W weights and N threshold units have VC-dimension at least $cW \log N$; in other words, the upper bound is tight to within a constant.

Anthony and Holden (1993) have investigated the VC-dimension of a general type of network, first introduced in the 1960s, which includes simple RADIAL BASIS FUNCTION NETWORKS and "polynomial discriminators." A *linearly weighted neural network*, with n real inputs and a single Boolean output, is defined by a fixed set $\phi_1, \phi_2, \ldots, \phi_k$ of *basis* functions, each of which maps $\mathbb{R}^n$ to $\mathbb{R}$. The state of the network is determined by a variable weight vector $\mathbf{w} = (w_1, w_2, \ldots, w_k)$. The output corresponding to example $x \in \mathbb{R}^n$ and state $\mathbf{w}$ is 1 or 0 according to whether the weighted sum $\Sigma_{i=1}^{k} w_i \phi_i(x)$ is positive or not. For example, a *polynomial discriminator* is obtained when all the functions ϕ_i are monomials, that is, products of the components of x, such as $x_1^2 x_3 x_n^3$. The *degree* d of a polynomial discriminator is the maximum total degree of any ϕ_i, in the usual sense. Anthony and Holden (1993) proved that the VC-dimension of a polynomial discriminator of degree d is at most $\binom{n+d}{d}$, and that if one only permits $\{0, 1\}$-valued inputs, the VC-dimension is at most $\Sigma_{i=1}^{d} \binom{n}{i}$. The *radial basis function networks* are another important class of linearly weighted networks. Here each ϕ_i is defined in terms of the distance of the example from a fixed "center" y_i: $\phi_i(x) = \phi(\|x - y_i\|)$, where ϕ is a fixed function and $\|z\|$ is the usual Euclidean norm of z. Anthony and Holden showed that the VC-dimension of a radial basis function network of this form is equal to k, the number of basis functions, if $\phi(r)$ takes any one of the forms r, $\exp(-cr^2)$, $(r^2 + c^2)^\alpha$, $(r^2 + c^2)^{-\beta}$, in which α and β are positive constants with $\beta < 1$.

Other work on the VC-dimension of neural networks includes that of MacIntyre and Sontag (1993) and Goldberg and Jerrum (1993). In both of these papers, techniques from logic are used to study the VC-dimension of neural networks of certain types. The first paper proves, among other things, the finiteness of the VC-dimension of feedforward networks in which the output unit is a linear threshold and all other computational units have the standard sigmoid activation function $f(x) = 1/(1 + e^{-x})$. However, explicit bounds on the VC-dimension are not given. In the second paper, upper bounds are obtained for the VC-dimension of networks similar to that just described, but in which the activation functions are not the standard sigmoid ones.

The Computational Complexity of PAC Learning

Efficiency with Respect to Accuracy, Example Size, and Sample Length

Thus far, a learning algorithm has been defined as a function which maps training samples into hypotheses. We shall now be more specific about the computational effectiveness of this function. If the process of PAC learning by an algorithm L is to be of practical value, it should be possible to implement the algorithm "quickly." We wish to quantify the behavior of a learning algorithm for a particular neural network architecture with respect to the size of the network. In particular, we wish to consider how the running time of the algorithm varies with the number n of inputs to the network: for a learning algorithm to be efficient, this running time should increase polynomially with n. However, there is another important consideration in any discussion of efficiency. Until now, we have regarded the accuracy parameter ε as fixed but arbitrary. It is clear that decreasing this parameter makes the learning task more difficult, and therefore the time taken to produce a probably approximately correct output should be constrained in some appropriate way as ε decreases; the appropriate condition is that the running time must be polynomial in $1/\varepsilon$. Formally, we say that a learning algorithm L is *efficient with respect to accuracy ε, example size n, and sample length m* if its running time is polynomial in the length m of the training sample and if there is a value of $m_0(\delta, \varepsilon)$ sufficient for PAC learning which is polynomial in n and ε^{-1}.

Hardness Results

In complexity theory, two important classes of problems, RP and NP, are defined. The class RP is the class of all problems which can be solved by "randomized" algorithms in polynomial time, while NP is the class of problems which can be solved by nondeterministic Turing machines in polynomial time. (We refer the reader to the book by Cormen, Leiserson, and Rivest, 1990.) It is conjectured, and widely believed, that these classes are not the same; more precisely, it is believed that RP is a strict subset of NP. This is known as the "RP $\neq$ NP" conjecture. For fixed k, for each n, let $\mathscr{P}_n^k$ be the neural network which consists of k linear threshold units, each connected to all of n inputs, the outputs of these threshold networks then being combined together by a hardwired AND gate. Thus, the network outputs 1 if and only if all k threshold units output 1. Blum and Rivest (1988) proved (essentially) the following result. (See also Anthony and Biggs, 1992.)

Theorem 4. Let $\mathscr{P}_n^k$ be as described, where $k \geq 2$. If there is a PAC learning algorithm for $\mathscr{P}_n^k$ which is efficient with respect to accuracy, example size, and number of inputs, then the "RP $\neq$ NP" conjecture is false.

Thus it is extremely unlikely that there is an efficient PAC learning algorithm for this surprisingly simple class of neural networks.

Discussion

We have considered basic PAC learning as it applies to learning in artificial neural networks. There are two distinct aspects: the length of training sample to be used and the efficiency of learning. In other words, we have the *sample complexity* problem and the *computational complexity* problem. The Vapnik-Chervonenkis dimension of a neural network determines in a fairly precise way the length of sample sufficient for PAC learning. This dimension can, in many cases, be related to the structure of the network, as in the examples presented here. Techniques from computational complexity theory can be applied to show that in a number of cases, *efficient* algorithmic PAC learning is impossible unless the NP $\neq$ RP conjecture is false.

There are several recent important extensions and generalizations of the PAC model which can be applied to artificial neural networks. We refer the reader to the survey of Anthony and Biggs (1993) and to the references cited there. One very important recent paper is that of Haussler (1992). This paper studies ways in which the PAC model may be extended in order to discuss the learning of functions whose values are other than simply 0 or 1. In particular, Haussler bounds the length of training sample which should be used for valid learning in neural networks with real-valued outputs. There have also been new approaches to and new results on the computational complexity problem; see Maass (1993b), for example.

Road Map: Computability and Complexity
Background: I.3. Dynamics and Adaptation in Neural Networks
Related Reading: Learning and Generalization: Theoretical Bounds; Parallel Computational Models; Time Complexity of Learning

References

Anthony, M., and Biggs, N., 1992, *Computational Learning Theory: An Introduction*, Cambridge, UK: Cambridge University Press. ◆

Anthony, M., and Biggs, N., 1993, Computational learning theory for artificial neural networks, in *Mathematical Approaches to Neural Networks* (J. G. Taylor, Ed.), North-Holland Mathematical Library, Amsterdam: North-Holland, pp. 25–63. ◆

Anthony, M., and Holden, S. B., 1993, On the power of polynomial discriminators and radial basis function networks, in *Proceedings of the Sixth Annual ACM Conference on Computational Learning Theory*, New York: ACM Press, pp. 158–164.

Baum, E. B., and Haussler, D., 1989, What size net gives valid generalization? *Neural Computat.*, 1:151–160.

Blum, A., and Rivest, R. L., 1988, Training a 3-node neural network is NP-complete, in *Proceedings of the 1988 Workshop on Computational Learning Theory*, San Mateo, CA: Morgan Kaufmann, pp. 9–18. (See also *Neural Netw.*, 1992, 5:117–127.)

Blumer, A., Ehrenfeucht, A., Haussler, D., and Warmuth, M. K., 1989, Learnability and the Vapnik-Chervonenkis dimension, *J. ACM*, 36:929–965.

Cormen, T. H., Leiserson, C. E., and Rivest, R. L., 1990, *Introduction to Algorithms*, Cambridge, MA: MIT Press. ◆

Dudley, R., 1978, Central limit theorems for empirical measures, *Ann. Probab.*, 6:899–929.

Ehrenfeucht, A., Haussler, D., Kearns, M., and Valiant, L., 1989, A general lower bound on the number of examples needed for learning, *Inform. and Computat.*, 82:247–261.

Goldberg, P., and Jerrum, M., 1993, Bounding the Vapnik-Chervonenkis dimension of concept classes parameterized by real numbers, in *Proceedings of the Sixth Annual ACM Conference on Computational Learning Theory*, New York: ACM Press, pp. 361–369.

Grunbaum, B., 1967, *Convex Polytopes*, London: Wiley.

Haussler, D., 1992, Decision theoretic generalizations of the PAC model for neural net and other learning applications, *Inform. and Computat.*, 100:78–150.

Maass, W., 1993a, Bounds on the computational power and learning complexity of analog neural nets, in *Proceedings of the Twenty-fifth Annual ACM Symposium on the Theory of Computing*, New York: ACM Press, pp. 335–344. (See also Maass, W., 1994, Neural nets with superlinear VC-dimension, *Neural Computat.*, 6:877–884.)

Maass, W., 1993b, Agnostic PAC-learning of functions on analog neural nets, to appear in *Neural Computat*. [Extended abstract appears in *Advances in Neural Information Processing Systems 6* (J. D. Cowan, G. Tesauro, and J. Alspector, Eds.), San Mateo, CA: Morgan Kaufmann, 1994.]

MacIntyre, A., and Sontag, E. D., 1993, Finiteness results for sigmoidal "neural" networks, in *Proceedings of the Twenty-fifth Annual ACM Symposium on the Theory of Computing*, New York: ACM Press, pp. 325–334.

Valiant, L. G., 1984, A theory of the learnable, *Commun. ACM*, 27:1134–1142.

Vapnik, V. N., and Chervonenkis, A. Y., 1971, On the uniform convergence of relative frequencies of events to their probabilities, *Theory Probab. Appl.*, 16:264–280.

Pain Networks

Marshall Devor

Introduction

The pain system encodes information on the intensity, location, and dynamics of strong, tissue-threatening stimuli. This sensory-discriminative function is common to all sensory systems. Pain differs from the others in the ability of various extrinsic, or stimulus-generated, and intrinsic, or central nervous system (CNS)–generated, state variables to modulate the relation between stimulus and felt response. At one extreme is *allodynia*, a state in which the slightest touch is agonizing. At the other is the apparent analgesia of the maimed soldier pulling his buddy out of the line of fire. By all reports, it is not that the soldier overcomes his pain by dint of heroism, but that, temporarily, the pain is simply not felt. Perceptual modulation makes good evolutionary sense. In times of danger, it is presumably adaptive to turn off pain, even in the presence of injury, so as not to impede fight or flight. At other times, it is adaptive to amplify pain to reduce the use of an injured limb and facilitate healing. Even under everyday conditions, normal, rational people display wide person-to-person and trial-to-trial variability in the amount of pain reported after the administration of calibrated noxious stimuli. For this reason, pain professionals usually avoid speaking of *pain stimuli* (or *pain receptors*). The preferred term is *noxious stimuli* (or *nociceptors*) because these stimuli may or may not evoke pain, depending on the current emotional-motivational and cognitive-evaluative context (Wall and Jones, 1991). In this article, I will review the fundamentals of pain signal processing and modulation with an eye toward problem areas that might be advanced with computational methods.

The Pain System

Receptor Encoding

Primary somatosensory neurons (*primary afferents*) are located in the paraspinal dorsal root ganglia (DRGs) (see Figure 1*B*). Each has a peripheral axon that travels in a nerve and terminates in a sensory transducer ending in skin, muscle, viscera, etc., and a central axon that travels in a dorsal root and ends synaptically in the spinal cord and/or the brainstem. Nerve impulses travel from the receptor ending, past the DRG, and on into the CNS without pause. As in other sensory systems, somatosensory primary afferents show modality specificity. Some, low-threshold afferents, respond to specific weak stimuli such as touch and vibration. Others, nociceptors, respond only to strong stimuli and either display submodality specificity (e.g., pinch, hot, cold) or are polymodal. Low-threshold afferents mostly have heavily myelinated, fast-conducting Aβ axons. All nociceptors have slowly conducting axons with thin or no myelin (Aδ- and C-axons, respectively).

The firing frequency of each primary afferent type encodes only a narrow range of stimulus intensities. As stimulus intensity increases from touch to pinch, there is a progressive handoff of the encoding function from low-threshold afferents to nociceptors. This process is the somatosensory system's solution to the problem of coding more than six decades of discriminable stimulus strengths using neurons that can vary their firing rate over no more than approximately two decades. Normally, stimulation of low-threshold afferents, even at maximal frequencies, evokes a sensation of pressure or rapid vibration, but not pain. I stress the word *normally* because there are system states in which stimulation of low-threshold afferents does evoke pain (see below).

Central Convergence

The specificity of primary afferents formed the basis for the classical belief that each somatosensory modality, including pain, remains separate and specific all the way to consciousness. Acceptance of this *Specificity Theory of Pain* was only partly shaken by the discovery in the 1960s that most neurons in the spinal cord and brainstem that receive synaptic input from nociceptive afferents also receive low-threshold input. That is, they are modality-convergent wide-dynamic-range (WDR) neurons. The identification of a small proportion of spinal neurons that are normally nociceptive selective left room for the belief that these were the exclusive painsignaling neurons. However, it has been established that most of these nociceptive-selective neurons send their ascending axons through a restricted part of the dorsolateral spinal white matter and that cutting this tract does not reduce pain sensation. On the other hand, pain is eliminated, at least for a time, by cutting the ascending pathways that carry the axons of WDR neurons. For this and other reasons, the existence of a pain-specific pathway is unlikely.

But how can WDR neurons transmit both touch and pain information? A number of schemes have been advanced to answer this question. For example, the Comparator Model holds that activity in WDR neurons is interpreted as pain only if nociceptive-selective neurons are also active, while the Spike Pattern Model posits that pain is encoded in a particular pattern of WDR discharge. I favor the Population Vector Model, in which touch is felt when a few WDR neurons fire slowly and pain is felt when many fire rapidly. Precision coding over many decades of stimulus discriminability is achieved through the progressive recruitment of WDR neurons that have sequentially overlapping encoding functions. Thus, any given stimulus, weak or strong, produces a unique aggregate of impulse activity viewed across the entire population of involved neurons.

Modulation and Gate Control

Modulation in the periphery. Sensory modulation is obvious from everyday experience. For example, in sunburned skin, gentle touch evokes pain. This phenomenon was explained early on by *peripheral sensitization*, the fact that tissue inflammation can increase the gain of nociceptive endings. Many of the details of the resulting primary hyperalgesia, including the identity of chemical mediators involved, are now known (Levine and Taiwo, 1994).

Modulation at the first central synapse. The idea of central modulation was brought to center stage with Melzack and Wall's (1965) Gate Control Theory of Pain. This theory was based on the observation that afferent activity in low-threshold Aβ afferents inhibits the response of postsynaptic WDR neurons to nociceptive input; hence, the idea of closing a gate on pain. An everyday example is the relief obtained from gently rubbing bruised skin. Melzack and Wall attributed this effect primarily to presynaptic inhibition of nociceptive primary afferent terminals by spinal interneurons (see SG in Figure 1*A*)

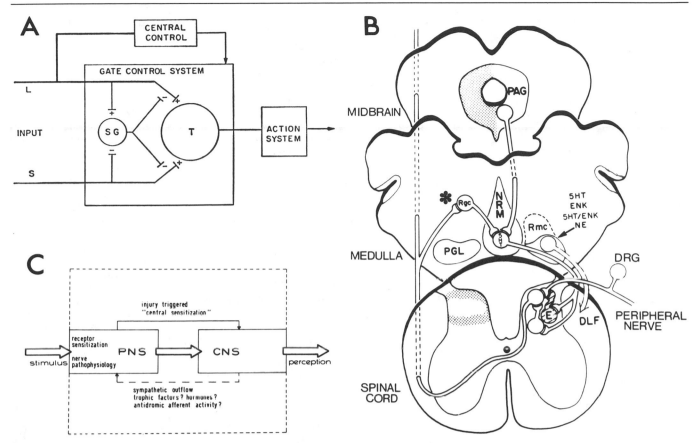

Figure 1. Three circuits for the modulation of pain signals. *A*, The original Gate Control system of Melzack and Wall (1965). Input from low-threshold (L) afferents and nociceptors (S) activates wide-dynamic-range transmission (T) neurons in the dorsal horn of the spinal cord. The former, but not the latter, activate substantia gelatinosa (SG) interneurons which presynaptically inhibit the nociceptive input. Central control is also noted. *B*, Subsequently discovered details of the central control system (from Basbaum and Fields, 1984). Activity in the midbrain periaqueductal gray (PAG), relayed through specific medullary nuclei, including the nucleus raphe magnus (NRM) and the reticular magnocellular nucleus (Rmc), evokes synaptic inhibition on ascending pain-signaling neurons in the spinal cord via enkephalinergic

(E) spinal interneurons. This pain-control pathway is composed of fine unmyelinated axons which descend in the dorsolateral funiculus (DLF) of the spinal white matter and use as neurotransmitters 5-hydroxytryptamine (5HT), enkephalin (ENK), and/or norepinephrine (NE). A collateral branch of the ascending pain-signaling neuron contributes to the descending inhibition via the reticular gigantocellular nucleus (Rgc, asterisk), completing a negative feedback loop. *C*, Pain amplification mechanisms associated with tissue and nerve damage. The pain signal is modulated in the peripheral nervous system (PNS) and central nervous system (CNS) by local processes (e.g., ectopic firing and peripheral sensitization) and by a combination of feedback and feedforward processes. (From Devor et al., 1991.)

activated by low-threshold Aβ afferents. The main evidence was that Aβ input, in contrast to Aδ or C input, generates subthreshold depolarization of the central terminals of neighboring afferents. This primary afferent depolarization (PAD) reduces the amount of neurotransmitter released from the neighbors and thus inhibits their ability to drive postsynaptic neurons.

Over the years, the gate control theory has undergone revision. For example, it is now known that low-threshold input generates postsynaptic inhibition in addition to presynaptic inhibition. Likewise, there remains controversy concerning the theory's proposal that nociceptor input selectively disinhibits PAD, thereby generating primary afferent hyperpolarization and *opening* spinal gates. Such details aside, the fundamental concept of signal modulation introduced by Melzack and Wall has been richly vindicated.

Modulation by descending inhibition. Central control was noted in the gate control model (see Figure 1*A*), but the details emerged later (Basbaum and Fields, 1984). The midbrain peri-

aqueductal gray matter (PAG) (see Figure 1*B*) is a nodal point for a descending inhibitory control circuit through which the brain gates ascending nociceptive information. Electrical stimulation of the PAG, or microinjection there of opiates, activates hindbrain nuclei with cells containing the neurotransmitters serotonin (e.g., see NRM in Figure 1*B*) and norepinephrine. These cell groups send impulses along a descending spinal tract (see DLF in Figure 1*B*), the effect of which is to inhibit the response of dorsal horn WDR neurons to noxious inputs. Responses to innocuous inputs are largely unaffected. This midbrain-medullo-spinal inhibitory circuit appears to be largely responsible for the analgesia obtained from systemic injection of morphine and related narcotics. Morphine excites PAG neurons, initiating the descending inhibition of noxious inflow from the spinal cord.

There is evidence that at least some types of psychogenic analgesia may be mediated by emotional or cognitive (limbic or cortical) drive of PAG neurons. For example, in experimental animals, intermittent unavoidable electrical foot shocks and defeat by a competitor evoke *stress-induced analgesia*. This analgesia may be blocked by opiate receptor antagonists, such as naloxone, implicating the involvement of one of the endogenous opiate-like neurotransmitters (enkephalin, endorphin, or dynorphin; Tricklebank and Curzon, 1984). A similar process could explain the absence of pain in the soldier mentioned in the example above. It has even been proposed that the placebo effect is mediated by this circuit. Placebos are drugs or manipulations that have no direct pharmacological effect on pain, but provide analgesia when the patient is led to *believe* that they will. The cerebral network that subserves belief may trigger enkephalin release in the PAG.

Pain inhibits pain: Ascending feedback on the descending inhibitory circuit, diffuse noxious inhibitory control, and propriospinal inhibition.   The reticular core of the hindbrain, and indeed of the entire neuraxis, is dominated by large neurons with widely spread dendrites. Neurons in this *isodendritic core of the brainstem* (Ramon-Moliner and Nauta, 1966) receive diffuse inputs from many sources and modalities, including high-threshold (nociceptive) drive. The brainstem neurons whose descending axons inhibit pain (e.g., see NRM in Figure 1*B*) are a part of this isodendritic core. Therefore, descending inhibition can be turned on by ascending noxious drive (see asterisk in Figure 1*B*). LeBars et al. (1984) noted a related pain-inhibits-pain negative feedback loop, nicknamed DNIC (diffuse noxious inhibitory control), whereby noxious stimuli applied over widespread parts of the body yield diffuse noxious inhibition. Finally, polysynaptic inhibitory networks associated with the isodendritic core of the spinal cord itself yield regional pinch-evoked inhibition of WDR neurons independent of the brain. This process is propriospinal inhibition.

Central sensitization.   Pain modulation involves excitation, not just inhibition. I noted above that inflammation sensitizes nociceptive endings locally at sites of trauma, yielding primary hyperalgesia. However, tenderness also spreads to surrounding skin where there is no inflammation. It has long been suspected that this secondary hyperalgesia is caused by impulses entering the CNS along Aβ touch afferents. Solid evidence of this effect has been obtained (Meyer, Campbell, and Raja, 1994). *Central sensitization*, the term adopted for this state of Aβ afferent-mediated pain, can be triggered by a momentary noxious input. Once triggered, it normally persists for tens of minutes before fading. Continuous noxious input and perhaps neural injury can apparently maintain it indefinitely (Gracely, Lynch, and Bennett, 1992). The existence of Aβ pain is a dramatic violation of the classical dogma that Aβ input exclusively mediates touch and vibration sense, while Aδ or C input is necessary for pain and temperature sense. A working model of central sensitization, based on neuropeptide enablement of glutamate receptors of the *N*-methyl-D-aspartate (NMDA) type, has been proposed by Woolf (1991).

Neuropathic Pain

Particularly bizarre and intractable pain states are associated with damage to peripheral nerves, spinal roots, or the CNS. Such neuropathic pains are doubly paradoxical. Not only is pain felt with no noxious stimulus present, but the conducting channel itself is compromised. Sensation in this situation should be reduced, not augmented! Certainly, the most dramatic example of neuropathic pain is phantom limb pain in amputees, but this category also includes various common conditions such as diabetic neuropathy, much low-back pain (e.g., radiculopathy with sciatica), and many instances of cancer

pain. Factors contributing to neuropathic pain are summarized in Figure 1*C*.

Nerve Pathophysiology and Ectopia

A key to understanding neuropathic pain is the observation that some afferent neurons become hyperexcitable when their axon has been injured (Devor, 1994). The result is ongoing and stimulus-evoked firing that originates at abnormal (ectopic) sites, usually the region of injury or the sensory cell body in the DRG. The most likely cause of ectopic hyperexcitability is remodeling of the axon membrane's normal electrical properties (Figure 2*A*) after injury. Specifically, there is evidence of the accumulation of excess voltage-sensitive Na$^+$ channels in the terminal swellings (end-bulbs) and sprouts that develop in the region of injury (Figure 2*B*). Na$^+$ channel accumulation lowers the local repetitive firing threshold (Figure 2*C*), thus priming the axon to fire in response to previously ineffective generator depolarizations (stimuli), and even to fire spontaneously.

Ectopic firing (ectopia) in injured afferents contributes to neuropathic pain in two ways. First, it injects an abnormal afferent impulse barrage into the CNS. Second, it may initiate and maintain a state of central sensitization in which weak cutaneous stimulation, and ectopia in Aβ afferents, evoke pain (Gracely et al., 1992; Devor, 1994).

Neuron-to-Neuron Cross-Excitation in the Nerves, Skin, and DRG

An additional contribution to pathophysiological discharge after nerve injury derives from a breakdown in the normal compartmentalization of neighboring afferent neurons in the periphery. Activity in one group of neurons is then able to excite others, amplifying and distorting natural and ectopic signals. Several forms of neuropathic cross-excitation have been identified (Devor, 1994). These include ephaptic (electrical) crosstalk; crossed afterdischarge, a novel form of nonsynaptic cross-excitation among afferents, apparently mediated by K$^+$ ions or neurotransmitters; and nonsynaptic coupling between sympathetic efferent neurons and afferents, mediated by norepinephrine release. Sympathetic-sensory coupling appears to be responsible for sympathetic-related pain states, such as causalgia and reflex sympathetic dystrophy.

Long-Term Spatial Reorganization in the CNS

Injury to peripheral nerves may also trigger long-term changes in CNS somatosensory processing that are presumably independent of nociceptor-driven central sensitization. In a typical experiment, a microelectrode is inserted into the dorsomedial part of the lower lumbar spinal cord. This is the location of cells that respond to stimuli applied to the toes and foot. If the nerves of the foot are cut acutely, these cells cease to respond. However, if the animal is investigated a few days (for rats) or weeks (for cats) later, the cells are found to have acquired a new receptive field on the nearest adjacent innervated skin (i.e., thigh or lower back). This shift reflects spreading of the thigh map into the zone formerly occupied by the foot (Devor and Wall, 1981).

As expected, spatial remapping in the spinal cord and brainstem is reflected in corresponding remapping in the cerebral cortex. (see SOMATOTOPY: PLASTICITY OF SENSORY MAPS). Indeed, the predicted perceptual correlates are well known. For example, upper-limb amputees often feel stimuli to the shoulder or cheek as originating in the phantom hand. If such func-

Figure 2. Nerve pathophysiology as a result of membrane channel remodeling. *A,* Stimulus transduction and encoding at normal sensory endings depends on the precise regulation of membrane electrical properties in the sensory ending. *B,* The membrane channels (-ch), receptors (-r), and other proteins responsible for electrogenesis are synthesized in the cell body and transported down the axon. These include K^+, Na^+, and Ca^{++} ion channels, mechanosensitive stretch-activated (SA) channels, and alpha-adrenoreceptors (α-r). In the presence of nerve injury, these channels and receptors dam up in the axon membrane at the cut nerve end (right), rendering it hyperexcitable and a source of ectopic impulse generation and neuropathic pain. *C,* Numerical simulation demonstrating that the accumulation of voltage-sensitive Na^+ channels (gNa^+_{max}) sharply reduces the threshold for rhythmic firing, but has less effect on the threshold for evoking individual nerve impulses. (From Matzner and Devor, 1992.)

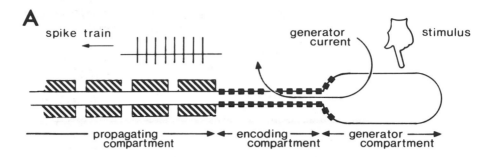

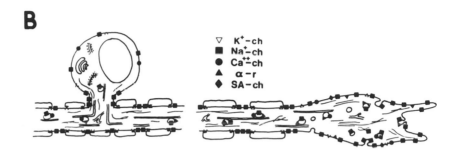

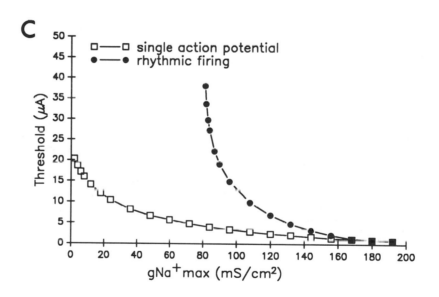

tional remapping occurs in the modality domain as it does in the spatial domain, one of the expected consequences would be pain in response to light touch.

Although the detailed mechanism of somatotopic remapping has not yet been established, it almost certainly involves modifications of synaptic efficacy in the somatosensory network. One particularly attractive possibility is that the trigger involves sustained alterations in sensory input after the injury as a result of either ectopic firing or modified patterns of limb use. Thus, for example, in the absence of an arm, the shoulder may be stimulated abnormally, inducing its expansion into the now silenced arm map. Evidence often cited in support of this process is the reported sensation in phantom limbs of pre-amputation sores, bunions, and other long-standing pre-amputation pains (Katz and Melzack, 1990). Strong and persistent stimuli of this sort are said to be *burned into* the CNS. Unfortunately, it is difficult to know whether these anecdotes reflect true centralization of preexisting sensory patterns or new sensations associated with ectopic activity in the injured peripheral nerve.

Where Is Pain?

I have referred to specific neural circuits and processes which modulate the conscious experience of pain. But in what circuit does pain sensation actually occur? The answer is not known, of course, but there are hints. First, it is safe to exclude the peripheral nervous system and spinal cord on the grounds that quadriplegics can experience pain, even pain referred to anesthetic parts of the body (phantom body pain). Interestingly, the cerebral cortex also appears to play only a supporting role, even though some regions respond to noxious stimuli (e.g., postcentral gyrus, cingulate gyrus). Extensive cortical lesions do not preclude pain sensation, even in parts of the body whose cortical representation has been destroyed. Furthermore, electrical stimulation of the cortex only rarely evokes pain. The most likely seat of pain sensation is in the brainstem and subcortical forebrain regions that are selectively activated by noxious stimuli. Some of these belong to the nonspecific (open) systems, such as the reticular formation and its midbrain and thalamic extensions (Ramon-Moliner and Nauta, 1966). Others belong to the classical specific pain pathways or to the limbic system (Chen, 1993).

Perspective

Pain, particularly persistent pain, remains a medical health problem of the first order. Witness the prominence of alternative approaches built largely on the therapeutic failures of conventional medicine. It is also a remarkable basic science challenge. Situated at the interface of body and mind, in the pain system, only a few synapses intervene between the biophysics of stimulus transduction and the mysteries of perception, affect, and belief. At each level of analysis are problems that could be fruitfully approached using computational methods. For example, in the periphery, the fine diameter of C-fiber endings precludes direct electrophysiological measurement. Testable hypotheses concerning the ionic mechanisms of transduction, encoding, and sensitization (e.g., during inflammation) could be provided by theoretical analysis. At the level of spinal processing are issues such as the modes of convergence that go into synthesizing natural receptive fields, the problem of how WDR neurons encode specific sensations, and mechanisms of functional plasticity (e.g., central sensitization and somatotopic remapping). Finally, theoretical analysis might provide insights into the higher-level functions that control descending pain inhibition.

Acknowledgments. The author's work is supported primarily by the US-Israel Binational Science Foundation (BSF), the German-Israel Foundation for Research and Development (GIF), and the Israel Science Foundation.

Road Map: Other Sensory Systems
Related Reading: Electrolocation; Somatosensory System

References

Basbaum, A., and Fields, H. L., 1984, Endogenous pain control systems: Brain-stem spinal pathways and endorphin circuitry, *Annu. Rev. Neurosci.*, 7:309–338.
Chen, A. C. N., 1993, Human brain measures of clinical pain: A review, II: Tomographic imagings, *Pain*, 54:133–144.
Devor, M., 1994, The pathophysiology of damaged peripheral nerve, in *Textbook of Pain*, 3rd ed. (P. D. Wall and R. Melzack, Eds.), London: Churchill-Livingstone, pp. 79–100. ◆
Devor, M., Basbaum, A. I., Bennett, G. J., Blumberg, H., Campbell, J. N., Dembowsky, K. P., Guilbaud, G., Janig, W., Koltzenberg, M., Levine, J. D., Otten, U. H., and Portenoy, R. K., 1991, Mechanisms of neuropathic pain following peripheral injury, in *Towards a New Pharmacology of Pain* (A. I. Basbaum and J.-M. Besson, Eds.), Dahlem Konferenzen, Chichester, Eng.: Wiley, pp. 417–440. ◆
Devor, M., and Wall, P. D., 1981, The effect of peripheral nerve injury on receptive fields of cells in the cat spinal cord, *J. Comp. Neurol.*, 199:277–291.
Gracely, R. H., Lynch, S. A., and Bennett, G. J., 1992, Painful neuropathy: Altered central processing, maintained dynamically by peripheral input, *Pain*, 51:175–194.
Katz, J., and Melzack, R., 1990, Pain "memories" in phantom limbs: Review and clinical observations, *Pain*, 43:319–336. ◆
LeBars, D., Calvino, B., Villanueva, L., and Cadden, S., 1984, Physiological approaches to counter-irritation phenomena, in *Stress Induced Analgesia* (M. D. Tricklebank and G. Curzon, Eds.), Chichester, Eng.: Wiley, pp. 67–101.
Levine, J., and Taiwo, Y., 1994, Inflammatory pain, in *Textbook of Pain*, 3rd ed. (P. D. Wall and R. Melzack, Eds.), London: Churchill-Livingstone, pp. 45–56. ◆
Matzner, O., and Devor, M., 1992, Na$^+$ conductance and the threshold for repetitive neuronal firing, *Brain Res.*, 597:92–98.
Melzack, R., and Wall, P. D., 1965, Pain mechanisms: A new theory, *Science*, 150:971–979.
Meyer, R. A., Campbell, J. N., and Raja, S. N., 1994, Peripheral neural mechanisms of nociception, in *Textbook of Pain*, 3rd ed. (P. D. Wall and R. Melzack, Eds.), London: Churchill-Livingstone, pp. 13–44. ◆
Ramon-Moliner, E., and Nauta, W. J. H., 1966, The isodendritic core of the brainstem, *J. Comp. Neurol.*, 126:311–335.
Tricklebank, M. D., and Curzon, G., Eds., 1984, *Stress Induced Analgesia*, Chichester, Eng.: Wiley, pp. 1–194.
Wall, P. D., and Jones, M., 1991, *Defeating Pain*, New York: Plenum, pp. 1–285. ◆
Woolf, C. J., 1991, Excitability changes in central neurons following peripheral damage, in *Hyperalgesia and Allodynia* (W. D. Willis Jr., Ed.), New York: Raven, pp. 221–243.

Parallel Computational Models

Michel Cosnard

Introduction

Neural networks are studied mainly for their learning capabilities and their abilities to process information in parallel. This article deals mainly with the latter. In the 1930s, the theory of effective computation was developed, showing that Turing machines could apparently compute any effectively computable function. In 1943, McCulloch and Pitts showed that a network of neurons could provide the "control box" for any Turing machine. It was not until the 1960s that attention shifted from the question, "Do two classes of machines have equivalent computing power?" to questions of complexity, which asked how that computing power varied as a function of some parameter, such as the length of tape used by a Turing machine and its computation, or the number of steps required. Similarly, for neural nets, one can look at complexity as a function of the number of neurons or the depth of the circuit required to solve a class of problems. The aim of the current article, then, is to place considerations of neural networks within this general framework of computational complexity. From the computing viewpoint, several parallel machine models have been introduced, and their relations in term of complexity have been analyzed (for reviews, see van Emde Boas, 1990; and Balcazar, Diaz, and Gabarro, 1988). However, the analysis of the relations between standard models of parallel computation and neural networks is still in its infancy.

Reasonable sequential machine models include the Turing machine and the random access machine. The *Turing machine* is the reference model for defining computability. Its main ingredients are a tape, i.e., a one-dimensional sequence of boxes containing symbols which can be read or written, and a cell computing a transition function (finite automaton; see AUTOMATA AND NEURAL NETWORKS). The input of the machine is written on the tape, the number of symbols of this input defining the size of the problem. The cell points to the current box, reads its content, changes its state, writes the result of the transition into the current box, and points to the box one position to the left or to the right. The answer can be found on the tape when the machine stops. The number of steps is called the computation time. If it is a polynomial (exponential) with respect to the size of the input, the problem is said to be polynomial (respectively, exponential). Such a model is difficult to use for constructing algorithms for high-level problems, particularly the constraint of accessing the data one bit after the other on a tape. The *Random Access Machine* (RAM) has been introduced to overcome these drawbacks. The RAM is composed of a computing cell and a set of registers capable of storing integers. In one unit of time the cell reads the integer stored in one register (at any position, thus the name of the machine), computes a function, and stores the result in a register. The two models are strongly related in the sense that any problem that can be solved on a Turing machine in polynomial time can also be solved on a RAM in polynomial time. The same is true for polynomial space (number of registers).

Parallel models of computation are defined in relation to these reasonable sequential models. A productive idea is to define models in which any problem that can be solved in polynomial space on a reasonable sequential machine model can be solved in polynomial time, and vice versa. Various models, including vector machines, tree machines, parallel RAMs (known as PRAMs; see the later section "Parallel Random Access Machines"), uniform circuits, and systolic arrays, satisfy this last requirement and hence can reasonably be called parallel. One can ask whether there are models of parallel computations that do not belong to this class and particularly what the status of neural networks might be.

Neural networks with real weights are often used even for computing Boolean functions. These networks have been proposed as models for parallel analog machines. The use of real numbers in computations can be discussed. However, one can argue that there exist physical or biological systems in which real quantities which may not be measurable can have a nonnegligible effect on the behavior of the system, as for example chaotic systems (see CHAOS IN NEURAL SYSTEMS). A tentative model of such systems would consist of abstract machines using real numbers for internal computations but using discrete, although unbounded, inputs and outputs. Neural networks appear to be ideal candidates.

Finite Parallel Abstract Machines

Boolean Circuits

A Boolean circuit is like a neural network, but without loops, and with the "neuron" at each node taking the form of a Boolean gate (i.e., all values are binary, 0 or 1) labeled as an input, output, constant, AND, OR, or NOT node. Each node has fan-in and fan-out bounded by a constant. A Boolean network with n input nodes and one output node computes a Boolean function. The value of an input is the value of the corresponding variable. The output value of a gate is obtained by applying the function labeling the gate to its input values. The output of a circuit is the value of the output gate. Allowing a circuit to have several output nodes makes it possible to have circuits computing elaborate functions (addition, multiplication, sorting, etc.). The size of the network is the number of gates; the depth is the length of the longest path in the network.

Boolean circuits are parallel by definition, since if we assign each gate to a different processor, we get a data-flow parallel abstract machine. In this model, a node computes its output values as soon as the input values are ready.

We are often interested in defining circuits for generic functions depending on a given parameter (the size of the input, for instance). An n-input, one-output circuit can only compute a Boolean function restricted to input length n. Hence, a parameterized function can be computed by a sequence of circuits (C_n). If there is a sequence (C_n) computing f, and a polynomial p, such that the size of C_n is bounded by $p(n)$, then we say that f is computed by polynomial-size circuits. We are interested in families that can be described easily. A family is said to be uniform if the nth circuit can be constructed using a Turing machine working in logarithmic space.

Boolean circuits also define a class of problems that can be solved efficiently in parallel. Define NC^k as the set of Boolean functions computed by uniform families of polynomial-size circuits such that the depth of C_n is polylogarithmic, that is, proportional to $\log^k n$ (recall that $\log^k n = \log n \cdot \log^{k-1} n$ and so $\log^k n > \log^{k-1} n$). The set NC is the union of all sets NC^k. Call P the set of problems that can be solved in polynomial time on a Turing machine. A challenging question in parallel complexity theory is to decide whether $NC = P$ (the strict inclusion is likely to occur). Moreover, the sets NC^k form a hierarchical sequence by inclusion (NC^k is clearly included in NC^{k+1}). However, it is still unknown if this hierarchy is strict (is NC^k strictly included in NC^{k+1}?). Many problems have been shown to be in NC. Moreover, several problems in P have been isolated as possible candidates for not being in NC. These problems can be shown to be equivalent in the sense that if one of them belongs to NC, then $NC = P$. They form a set called P-complete. More than 100 problems have been proved to be P-complete.

Threshold Circuits

By *threshold circuits* we mean discrete neural networks using a loop-free interconnection network. We reserve the name *neural nets* for recurrent and continuous networks in general. A Boolean threshold gate g of fan-in n is associated with some weights $w = w_1, \ldots, w_n$ and a threshold θ. For any input $x = x_1, \ldots, x_n$ the dot product $w \cdot x$ is computed. g is equal to 1 if $w \cdot x \geq \theta$ and equal to 0 otherwise (i.e., it is just like a McCulloch-Pitts neuron). While we might allow the weights to take arbitrary real values, this is unnecessary. Since the set of inputs is finite, we can assume (for the purpose of mathematical analysis as distinct from electronic implementation) that the weights and thresholds are integers.

A threshold circuit is a circuit in which all the gates are threshold gates. The weight of the circuit is the maximum weight of any gate. Some problems can be solved by using threshold circuits with large weights (exponential in the size of the problem); others, by using circuits with small weights (logarithmic in the size of the problem). For general surveys of the relations between complexity theory and threshold circuits, see Orponen (1992) and Parberry (1990); see also STRUCTURAL COMPLEXITY AND DISCRETE NEURAL NETWORKS.

A threshold gate can be computed by an NC Boolean circuit. AND and OR gates are special cases of threshold gates. Thus, the general parallel complexity classes induced by Boolean circuits are not changed. More precisely, in close analogy to the definition of NC^k, let us call TC^k the set of functions computable by threshold ciruits of polynomial size and depth $O(\log^k n)$. The following hierarchy relates TC^k and NC^k:

$$TC^k \subseteq NC^{k+1}$$

None of these inclusions are known to be strict.

Parallel Random Access Machines

We shall now consider parallel models of computation that correspond to the classical von Neumann model. In this model a program is a sequence of instructions to be applied to a set of data, both stored in a memory. A control unit takes the instructions one after the other from the memory and executes them on the data, producing results that are also stored in the memory. This model is sequential since the instructions are ordered. A rather natural extension is to perform the same operation on several pieces of data. A Parallel Random Access Machine (PRAM; Karp and Ramachandran, 1990) consists of several independent sequential processors, each with its own private memory, communicating with one another through a global memory. In one cycle time, each processor reads one global or local memory location, executes a single RAM operation, and writes into one global or local memory location.

The power of the PRAM is usually defined by means of uniform families of Boolean circuits (see Karp and Ramachandran, 1990, for a detailed presentation). Concerning the polynomial problem class, the two models are equivalent.

Infinite Parallel Abstract Machines

We shall now introduce more elaborate models of parallel computation. These models capture the complexity of infinite dimensional computations in various senses (either in the space of representations or in the space of computations).

Cellular Automata

We briefly define CELLULAR AUTOMATA (q.v.) for the one-dimensional case. A one-dimensional cellular space is a collection of cells associated to every integer point of $\mathbb{Z}$. In a *cellular automaton*, each cell is a finite automaton with a finite set of states Q and a local transition function δ, depending on the cell itself and some neighboring cells. The transition function and the neighborhoods are the same for all the cells (uniform neighborhoods). In one unit of time, each cell computes its new state using the transition function. Cellular automata have an infinite memory with an infinite number of identical finite control units working in parallel and synchronously. Von Neumann introduced cellular automata (see SELF-REPRODUCING AUTOMATA), and was thus the founder of the study of non–von-Neumann computers!

The cellular automata constitute an intrinsically uniform and parallel model of computation. If we are interested in finite computations, we shall only consider cellular automata with bounded support. Martin (1993) has proved that bounded-support cellular automata can efficiently simulate PRAMs.

Neural Networks

As an immediate generalization of cellular automata, consider that the underlying net is no longer homogeneous but an arbitrary digraph and that the local transition functions can vary from cell to cell. Discrete neural nets provide one class of examples. However, mainly for biological analogy reasons, in the classical discrete neural nets the local transition functions first compute weighted sums of inputs from neighboring cells, so that the cells are unable to detect where specific inputs come from.

Following Siegelmann and Sontag (1992), a *neural net* is a quadruple $(Q, X, U, \{f_i\})$ consisting of an *activation set* $Q \subseteq \mathbb{R}$, a *weighted digraph* X (i.e., a set of vertices connected by directed edges that are labeled with numbers), a set U of *input nodes*, and a family of *activation functions* $\{f_i\}$ from $\mathbb{R}$ into Q, one for each vertex i in X. The local dynamics is defined by

$$x_i(t + 1) = f_i(\textstyle\sum_j w_{ij} x_j(t) + \sum_j b_{ij} u_j(t) + \theta_i)$$

where the first summation is taken over all neighbor cells j of cell i, and the second is taken over all inputs. The activation function is often taken as a sigmoid (in some cases we shall consider the saturated-linear function). In the remaining discussion we take $Q = [0, 1]$ and fix a subset O of X to consist of the output cells which communicate the outputs of the network to the environment. These networks are often called *recurrent first-order neural nets*. They are called discrete if the inputs and outputs are Boolean values, and *real* if the inputs and outputs are real.

There exists an important difference between cellular automata and neural nets: a neural net has a finite number of nodes and computes with real numbers; a cellular automaton has an infinite support, and each cell is a finite-state automaton.

Computability Properties of Neural Networks

Blum, Shub, and Smale (1989) have introduced a powerful model of computation over the real numbers corresponding to the Turing machine. A RealRAM is simply a RAM in which we allow computations on real numbers. It was used by Cucker (1992) and Cosnard and Koiran (1992) to study the complexity of some problems in the BSS (Blum, Shub, and Smale) models. To get a model of parallel machines which operate on real numbers, it is sufficient to generalize the RealRAM to the RealPRAM. For this, take the earlier definition of the PRAM and replace the phrase "executes a single RAM operation" with "executes a single RealRAM operation."

Siegelmann and Sontag (1992) have shown that if one restricts consideration to nets whose weights are rational numbers, then one obtains a model that can simulate a multitape Turing machine in linear time (see also AUTOMATA AND NEURAL NETWORKS). It was already proved by von Neumann that cellular automata could simulate Turing machines. There are no direct simulation results between neural networks and cellular automata. However, it is reasonable to conjecture that rational (and also real) neural nets can be simulated by cellular automata, but that the reciprocal is false for cellular automata with infinite support.

In Siegelmann and Sontag (1992), the computational properties of real neural networks (RNNs) are analyzed. They prove that RNNs can solve in polynomial time discrete problems that cannot be solved in polynomial time by Turing machines without some extra information. Moreover, RNNs can solve all discrete problems, including noncomputable problems, in exponential time. The proofs of these results are based on the fact that, in polynomial time, real neural networks and non-uniform Boolean circuits compute the same class of functions, $P/poly$.

Is the RealRAM more powerful than an RNN for discrete problems? The problem is still open, but Koiran (1993) shows that any function in P/poly can be computed in polynomial time using a RealRAM. Clearly the NC class of problems that can be solved on a PRAM in polylogarithmic time using a polynomial number of processors (NC_D) can be extended to RealPRAM. Any discrete problem in NC_D belongs to NC_R. However, contrary to the discrete case, Cucker (1992) has shown that

$$NC_R \neq P_R$$

The detailed study of the relations between real neural networks and realPRAM is still a largely open problem.

Discussion

Recent results have shown that neural networks with real weights are more powerful than traditional models of computation. This property, combined with their intrinsic parallelism, make these networks very appealing. However, it is still a technological open problem to prove that full use can be made of these properties (see MULTIPROCESSOR SIMULATION OF NEURAL NETWORKS). The comeback of real numbers in complexity theory has produced an interesting renewal of this theory and has introduced many open questions. The combination of recent results with computational learning theory will certainly strengthen the foundations of neural network theory.

Acknowledgments. This work was supported by the Programmes de Recherche C3 and Cognisciences.

Road Map: Computability and Complexity
Background: Automata and Neural Networks
Related Reading: Structural Complexity and Discrete Neural Networks

References

Balcazar, J. L., Diaz, J., and Gabarro, J., 1988, *Structural Complexity I*, EATCS Monographs on Theoretical Computer Science, New York: Springer-Verlag.

Blum, L., Shub, M., and Smale, S., 1989, On a theory of computation and complexity over the real numbers: NP-completeness, recursive functions and universal machines, *Bull. Am. Math. Soc.*, 21:1–46.

Cosnard, M., Garzon, M., and Koiran, P., 1993, Computability properties of low-dimensional systems, in *STACS'93*, Lecture Notes in Computer Science, New York: Springer-Verlag.

Cosnard, M., and Koiran, P., 1992, Relations between models of parallel abstract machines, in *Heinz Nixdorf Seminar*, Lecture Notes in Computer Science, New York: Springer-Verlag.

Cucker, F., 1992, $NC_R \neq P_R$, *J. Complexity*, 8:230–238.

Karp, R. M., and Ramachandran, V., 1990, Parallel algorithms for shared-memory machines, in *Handbook of Theoretical Computer Science*, vol. A, *Algorithms and Complexity*, Amsterdam: North-Holland, pp. 870–941. ◆

Koiran, P., 1993, Puissance de calcul des réseaux de neurones artificiels, Doctoral Thesis, École Normale Supérieure de Lyon, France.

Martin, B., 1993, A uniform universal CREW PRAM, in *MFCS 93*, Lecture Notes in Computer Science, New York: Springer-Verlag (to appear).

Orponen, P., 1992, Neural networks and complexity theory, in *MFCS 92*, Lecture Notes in Computer Science, New York: Springer-Verlag, pp. 50–61.

Parberry, I., 1990, A primer on the complexity theory of neural networks, in *Formal Techniques in Artificial Intelligence: A Sourcebook*, New York: Elsevier. ◆

Siegelmann, H. T., and Sontag, E. D., 1992, Neural networks with real weights: Analog computational complexity, in COLT92.

van Emde Boas, P., 1990, Machine models and simulations, in *Handbook of Theoretical Computer Science*, vol. A, *Algorithms and Complexity*, Amsterdam: North-Holland, pp. 3–66.

Pattern Formation, Biological

J. Cook and J. D. Murray

Introduction

Spatial pattern formation is of interest in many disciplines, including developmental biology, physiology, neurobiology, epidemiology and ecology. In population biology, patchiness in population densities is the norm rather than the exception. In developmental biology, groups of previously identical cells follow different developmental pathways, depending on their position. The rich spectrum of mammalian coat patterns and the patterns found on fish, reptiles, molluscs, and butterflies reflect developmental processes that are still incompletely understood. Stationary patterns, as well as a wide variety of waves, have been observed in chemical reactions. Ocular dominance stripes reflect patterns in the connectivity of the visual cortex, while hallucination patterns can be partially explained as activity patterns in the visual cortex.

Although these patterns occur on a wide range of spatial scales, spanning the molecular, cellular, individual, and population levels, a common feature is that macroscopic patterns result from microscopic interactions. Mathematical models have been proposed for the mechanisms of biological pattern formation based on this principle (Figure 1). We shall describe two of the main classes, reaction-diffusion and neural field, and we shall mention others. Each model can exhibit spontaneous pattern formation, that is, patterns develop in homogenous

environments without particular initial conditions, boundary conditions, or other external forces to drive them. The patterns are therefore self-organizing and symmetry breaking.

Our goal is to take a mechanism and to determine (1) the range of parameters in which pattern formation is expected, (2) the nature of the pattern (steady, oscillating, or moving through space), and (3) the scale of the pattern. Our strategy is as follows. First, a suggested mechanism is translated into a set of mathematical equations (the model). We have found that an appreciation of the pattern formation potential of existing models is invaluable here, and we hope that this survey will be useful in that regard. Once the model has been specified, we determine a homogenous steady state and use linear stability analysis to determine whether perturbations to such an unpatterned state will grow or decay. For parameters supporting pattern formation, we isolate unstable modes and use the dominant mode to predict the scale of the pattern.

Reaction-Diffusion Models

Reaction-diffusion models are the most widely studied of the models we shall discuss. They have been applied in developmental biology, population biology, epidemiology, neurophysiology, chemistry, and physics (see Meinhardt, 1982; Murray,

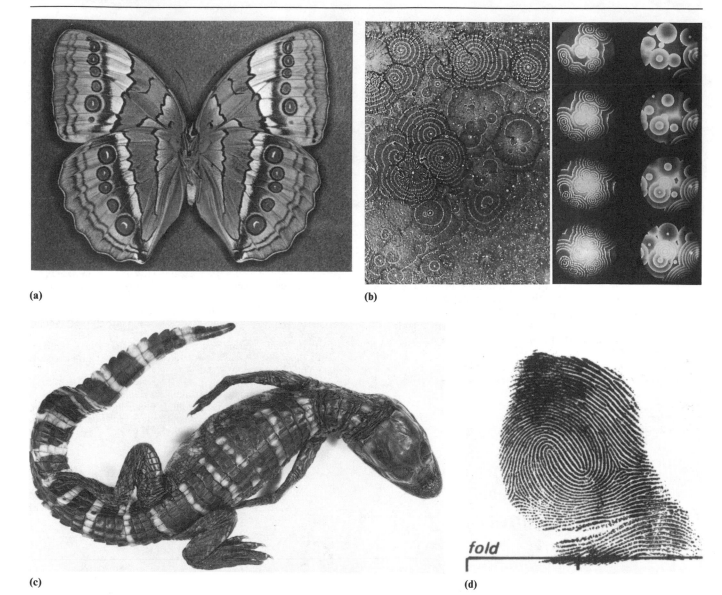

(a)

(b)

(c)

(d)

Figure 1. A small selection of the diverse spatial patterns for which model mechanisms have been proposed. (a) The butterfly (*Stichophthalama camadeva*) shown exhibits most of the basic pattern elements observed in butterfly wings (photograph courtesy of H. F. Nijhout). (b) Example of moving and stationary bands of amoebae of the slime mold *Dictyostelium discoideum* (photograph courtesy of P. C. Newell). (c) Stripes on an alligator (*Alligator mississipiensis*) (photograph courtesy of M. W. J. Ferguson). (d) Typical human fingerprint.

1989). We refer to the variables, which depend on time and space, as *species*, recognizing that a species may be a type of molecule or cell. Species disperse and react, and these two processes are independent. Developing expressions for local interactions between species and for flux, and invoking conservation laws, we obtain the general form

$$\frac{\partial n}{\partial t} = f(n) + D\nabla^2 n \qquad (1)$$

where $n(x, t)$ is the vector of species densities, f is the vector of reaction terms, and D is the diffusion coefficient. Initial conditions and boundary conditions must also be specified.

The patterns in which we are most interested are stable, stationary ($\partial n/\partial t = 0$), inhomogenous solutions to Equation 1. For a single species in a single spatial dimension, it can be shown that a homogenous steady state cannot be destabilized by diffusion. Interestingly, this result is not true for all two-dimensional domains, for example, a peanut-shaped domain (Murray, 1989, chap. 14). Two-species models are more interesting still.

Two-Species Models

In 1952 Alan Turing (of Turing machine fame) suggested that differential diffusion of two interacting species could act to destabilize a steady state. Since diffusion is usually thought of as a stabilizing (or *smoothing*) force, this was a startlingly original idea. It has since been supported both mathematically (Murray, 1989, chap. 14), and experimentally (Ouyang and Swinney, 1991). Murray (1989, chap. 15) and Meinhardt (1982) describe many applications.

Consider the two-species system given by

$$\frac{\partial A}{\partial t} = f(A, B) + d_A \left(\frac{\partial^2 A}{\partial x^2}\right) \qquad \frac{\partial B}{\partial t} = g(A, B) + d_B \left(\frac{\partial^2 B}{\partial x^2}\right) \qquad (2)$$

in which a steady state exists at $(A_0, B_0)(f(A_0, B_0) = g(A_0, B_0) = 0)$. Linearizing around (A_0, B_0), one obtains

$$\frac{\partial A}{\partial t} = f_A A + f_B B + d \left(\frac{\partial^2 A}{\partial x^2}\right) \qquad \frac{\partial B}{\partial t} = g_A A + g_B B + d_B \left(\frac{\partial^2 B}{\partial x^2}\right)$$
$$(3)$$

where $f_A = \partial f / \partial A$ is evaluated at (A_0, B_0), etc. We assume that the steady state is stable in the absence of spatial interaction (diffusion) and therefore that $g_B < 0$, without loss of generality.

We look for solutions of the form $e^{\lambda t + ikx}$ and generate a *dispersion relation* relating eigenvalues, λ, to modes, k (the growth rate of mode k is $Re(\lambda(k))$). The dispersion relation provides conditions for unstable modes to exist (see Murray, 1989, chap. 14):

$$f_A + g_B < 0 \qquad \Delta = f_A g_B - f_B g_A > 0$$
$$(g_B < 0 \text{ without loss of generality})$$
$$d_A > 0 \qquad \delta f_A + g_B > 0 \qquad (\delta f_A + g_B)^2 > 4\delta\Delta$$
$$\text{where } \delta = d_B / d_A$$
$$(4)$$

Some necessary conditions for Turing instability are (1) the self-inhibiting species, B ($g_B < 0$), must diffuse at the higher rate ($\delta > 1$), and (2) A must be self-activating ($f_A > 0$). Also, f_B and g_A must have opposite signs. It is usual to call the species which promotes growth of the other the *activator* and the other species the *inhibitor*. The two possible cases are illustrated schematically in Figure 2. In case 1, A is the activator, which is also self-activating, while the inhibitor, B, diffuses at a higher rate

and inhibits not only A but also itself. In case 2, B is the activator, is again self-inhibiting, and again diffuses at a higher rate. It can be shown that in case 1 the two species occur at high or low density together (Figure 2c), whereas in case 2, A is at high density where B is low, and vice versa (Figure 2d).

We now give two analogies for the mechanisms underlying Turing instabilities. Consider case 1, and refer to Figure 3a. Let A be prey to a predator, B. How can patterns arise as in Figure 2c when predators disperse more rapidly than their prey? Suppose there were an area of increased prey density. In the absence of diffusion this would be damped out after a temporary increase in both populations. However, with high predator dispersal it is possible that the local increase in predators partially disperses and hence is not strong enough to push the prey population back toward equilibrium. Furthermore, when predators disperse they lower the prey density in neighboring regions and cause the opposite effect. It is thus possible to have alternating clumps of high and low population density of both species.

Consider the second type of dynamics (Figure 2b, 2d; Figure 3b). Suppose now that A is a slowly dispersing, *autocatalytic* ($f_A > 0$) predator, and B is its prey. In an area of high prey density, without diffusion predator numbers would increase at the expense of the prey, and eventually both populations would return to the steady state. However, there is a transient increase in the predator population and a reduction in the prey population to below its steady-state value. The resulting net influx of rapidly dispersing prey from neighboring regions would cause the predator population to drop in those regions while prey flourish. A pattern can become established in which areas of few predators and many prey supply with prey to areas which contain few prey and large numbers of predators.

The dispersion relation also indicates the scale on which a pattern occurs, through the wavelength of the fastest growing

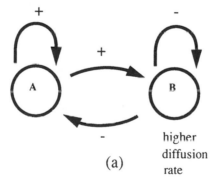

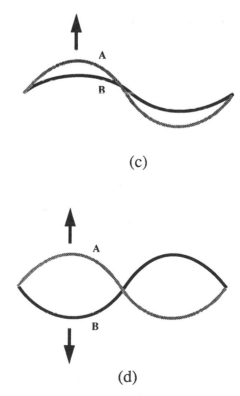

(a)   higher diffusion rate

(c)

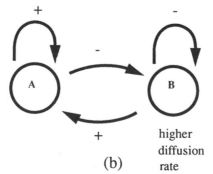

(b)   higher diffusion rate

(d)

Figure 2. Turing instabilities. Some interactions support diffusion-driven (Turing) instabilities. In part (a), self-activating A also activates B, which inhibits both species. The resulting spatial pattern is shown schematically in part (c). In part (b), self-activating A now inhibits B but is itself activated by B. The resulting pattern is shown in part (d). Corresponding reaction phase planes are shown in Figure 3.

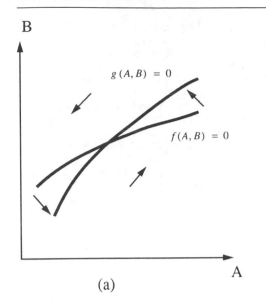

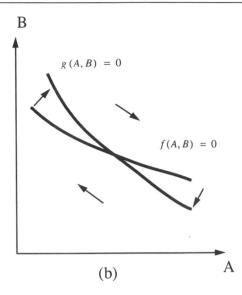

(a)   (b)

Figure 3. Reaction phase planes for Turing instabilities. Part (a) shows the phase plane corresponding to (a) and (c) of Figure 2. Part (b) shows the phase plane corresponding to (b) and (d) of Figure 2. The steady state is at the intersection of the two null-clines. Arrows represent the direction of change due to local species interaction.

mode:

$$l_c = \frac{2\pi(d_B - d_A)^{1/2}}{[(\delta + 1)((-f_B g_A)/\delta)^{1/2} - f_A + g_B]^{1/2}} \quad \text{where } \delta = \frac{d_B}{d_A} \quad (5)$$

(see Murray, 1989, chap. 14). Equation 6 indicates that Turing instabilities can occur on a broad range of spatial scales. For large δ, $l_c \approx (2\pi)\sqrt[4]{(d_A d_B)/(-f_B g_A)}$.

Mathematical analysis can also provide insight into the effect of boundary size and shape on pattern formation. In models for animal coat patterns (Murray, 1989, chap. 15) one finds that only crosswise stripes can occur in long narrow domains, while spots can occur on wider domains. This is a possible explanation for why animals that have spots over most of their bodies (e.g., leopards) tend to have hooped patterns, or no pattern at all, on their tails. The qualitative form of the pattern is governed by the size and shape of the animal at the time at which pattern is determined.

Neural Field Models

In reaction-diffusion models, all of the action is local. However, in some applications, direct instantaneous interactions occur between distant points. For example, neurons activate (excite) or inhibit distant neurons and, in population biology, individuals interact via common global resources. Neural field models incorporate long-range interaction explicitly (Murray, 1989, chap. 16; see also LAYERED COMPUTATION IN NEURAL NETWORKS).

Murray (1989, chap. 16) describes three applications of neural field models. In a model presented by Swindale (1980) for the formation of ocular dominance stripes, the equations describe the evolution of the densities of synapses associated with each eye (see also OCULAR DOMINANCE AND ORIENTATION COLUMNS). A model for hallucination patterns presented by Ermentrout and Cowan (1979) describes the patterns of activity in a network of excitatory and inhibitory neurons. Finally, there is a model for pattern formation on the shells of molluscs (Ermentrout, Campbell, and Oster, 1986) in which neural activity influences the sequential laying down of pigmentation. The earliest model of this type was proposed by Wilson and Cowan (1973). Our exposition is based closely on the paper by Ermentrout and Cowan (1980) and that by Ermentrout in the volume edited by Amari and Arbib (1980).

Single Cell Type

We first develop a model based on a single cell type or layer which will later be extended. Let V_k be the membrane potential associated with the kth neuron and let $I_j = S(V_j)$ be the output from another neuron, j. The postsynaptic potential, ϕ_{jk}, depends on the input history via a function $h(t)$:

$$\phi_{jk} = \int_{-\infty}^{t} h(t - \tau)\alpha_{jk} I_j(\tau)\, d\tau = \alpha_{jk}(h * I_j) \quad (6)$$

The asterisk refers to temporal convolution, and α_{jk} are constants related to the network connectivity. For simplicity, we shall assume that $h(\tau) = \exp(-\tau)$. The membrane potential is the sum of the postsynaptic potentials:

$$V_k = \sum_j \phi_{jk} = \sum_j \alpha_{jk}(h * S(V_j)) \quad (7)$$

Next we assume that neurons are identical and have identical connectivity (relative to their location). This allows us to pass to a continuum limit:

$$V(x) = \int_{-\infty}^{\infty} w(x - y)(h * S(V(y)))\, dy = w \otimes h * S(V) \quad (8)$$

where $w \otimes S$ signifies the spatial convolution, with $w(y)$ reflecting how activation and inhibition depend on the distance between neurons. By our choice of h, we can differentiate and remove the temporal convolution:

$$\frac{\partial V}{\partial t} = -V + w \otimes S(V) \quad (9)$$

We follow Ermentrout and Cowan (1980) and define a new variable, Y, which is a *coarse-grained* (time-averaged) measure of activity, $Y = h * S(V)$, in contrast to the instantaneous firing rate, I. Equation 9 becomes

$$\frac{\partial Y}{\partial t} = -Y + S(w \otimes Y) \quad (10)$$

Consider an example in which both excitatory and inhibitory interactions occur. The combined excitation/inhibition kernel, w, can be constructed as a difference between two Gaussians (Figure 4a). The example shown exhibits short-range activation and long-range inhibition.

Figure 4. Spatial interaction kernel. (a) A kernel $w(s)$ exhibiting long-range inhibition and short-range activation can be constructed from a sum of Gaussian distributions. (b) Fourier transform, $W(k)$, of the kernel in part (a).

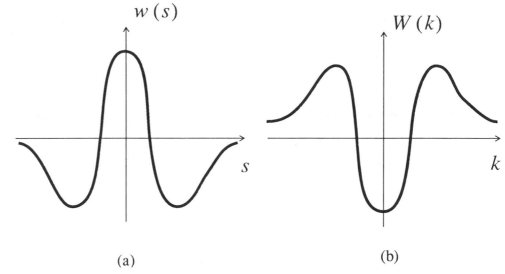

(a) (b)

We linearize about the homogenous steady state, $Y \equiv 0$, and look for solutions of the form $Y \propto e^{\lambda t + ikx}$. This leads to

$$\lambda = -1 + S'(0) W(k) \qquad (11)$$

where $W(k)$ is the Fourier transform of w shown in Figure 4b, and we assume $S'(0) > 0$. A mode of wave number k will grow if $\lambda(k) > 0$, and the potential for pattern and its scale are predicted by the value and the location of the maximum of $W(k)$, respectively.

Two Cell Types

A model with two cell types—excitatory and inhibitory—was used by Ermentrout and Cowan (1979) to explain hallucinogenic patterns in the visual cortex. Let coarse-grained activity in an excitatory layer be represented by E, and activity in an inhibitory layer by I. The extension of the previous example is

$$\frac{\partial E}{\partial t} = -E + S_E[(E \otimes w_{EE}) - (I \otimes w_{IE})]$$

$$\frac{\partial I}{\partial t} = -I + S_I[(E \otimes w_{EI}) - (I \otimes w_{II})] \qquad (12)$$

Linearizing around the steady state $E \equiv I \equiv 0$, and looking for solutions proportional to $\exp(\lambda t + ikx)$, we find

$$\lambda^2 + R_1(k)\lambda + R_2(k) = 0 \qquad (13a)$$

where $R_1(k) = 2 - d_E W_{EE} + d_I W_{II}$, and

$$R_2(k) = 1 - d_E W_{EE} + d_I W_{II} + d_I d_E (W_{IE} W_{EI} - W_{EE} W_{II}) \qquad (13b)$$

where the d_j's are positive constants ($d_j = S_j'(0)$) and $W_{ij}(k)$ is the Fourier transform of $w_{ij}(x)$.

A mode is unstable if $R_1(k) < 0$ or $R_2(k) < 0$. We insist that the system is stable to homogenous perturbations, i.e., that $R_1(0) > 0$ and $R_2(0) > 0$. Small wave number perturbations cannot grow since $W_{ij}(k) \to 0$ as $k \to \infty$. However, either R_1 or R_2 (or both) can be negative for some intermediate values of k. There are thus two possible types of bifurcation (see DYNAMICS AND BIFURCATION OF NEURAL NETWORKS): a Hopf-type bifurcation, resulting in an oscillating pattern (when $R_1 < 0$ for some k and $R_2 > 0$ for all k), and a bifurcation to steady stationary patterns ($R_2 < 0$ for some k and $R_1 > 0$ of all k).

Ermentrout and Cowan (1979) showed how repeating stripes and lattices can appear as solutions in two dimensions. They suggested that hallucinogenic drugs could destabilize an otherwise stable homogenous visual cortex. Under the experimentally supported logarithmic mapping from the retina to the visual cortex such patterns would be perceived as tunnels, spirals, etc., as have been reported in the literature (see Ermentrout and Cowan, 1979, or Murray, 1989, chap. 16).

Discussion

We have seen how two fundamentally different types of model can give rise to patterns. We now briefly mention other mechanisms that have been studied.

Mechanical Models

It is known that cells of the mesenchyme, and others, can exert significant *traction* on the extracellular matrix. Furthermore, these cells exhibit *haptotaxis*, whereby they move up gradients of adhesion. Mechanical models for pattern formation were first introduced by Oster, Murray, and co-workers (Murray, 1989, chap. 17). Harris, Stopak, and Warner (1984) presented graphic experimental confirmation of the predictions. Example equations, corresponding to conservation of cells (n), extracellular matrix (ρ), and force balance, are

$$\frac{\partial n}{\partial t} + \frac{\partial}{\partial x}(nv) = rn(1 - n) - \alpha \frac{\partial}{\partial x}\left(n \frac{\partial \rho}{\partial x}\right) + D \frac{\partial^2 n}{\partial x^2}$$

$$\frac{\partial \rho}{\partial t} + \frac{\partial}{\partial x}(\rho v) = 0 \qquad (14)$$

$$\frac{\partial}{\partial x}\left[E \frac{\partial u}{\partial x} + \tau n\left(\rho + \gamma \frac{\partial^2 \rho}{\partial x^2}\right)\right] = 0$$

where u is the displacement and α is the haptotactic constant. Cells and the matrix convect as a single phase with material velocity $v = Du/Dt$. E is the modulus of elasticity of the matrix, τ measures the strength of cell traction, and $\sqrt{\gamma}$ is a measure of the scale of direct cell-matrix interactions.

Two fundamentally different types of instability can occur. Firstly, traction can be large enough to overcome matrix resistance: a *mechanical instability*. Second, even a stiff matrix subject to traction from cells which are strongly haptotactic can support cell clumping: a *haptotaxis-induced instability*.

Negative and Long-Range Diffusion

If, in a reaction-diffusion equation, the diffusion coefficient were negative, this would intuitively cause clumping (imagine viewing a movie of diffusion in reverse). Though such a problem is ill-posed mathematically, this can be rescued by adding a biharmonic term, as in the Cahn-Hilliard equation (Cahn, 1968). Murray (1989, chap. 9) shows that a biharmonic term is a natural modification when fluxes have a *long-range* component (they depend on densities in a *neighborhood* of the reference point).

Chemotaxis

Chemotaxis is the name given to the process whereby cells move up or down a chemical (chemoattractant or chemorepellent) gradient. The original formulation of a mathematical model for chemotaxis is due to Keller and Segel (1970). Chemoattractant (c) is secreted and degraded by cells (n), and cells respond to gradients in with a convection speed of $\chi(\partial c/\partial x)$:

$$\frac{\partial n}{\partial t} = D_n \frac{\partial^2 n}{\partial x^2} - \chi \frac{\partial}{\partial x}\left[n \frac{\partial c}{\partial x} \right] \qquad \frac{\partial c}{\partial t} = f(n, c) + D_c \frac{\partial^2 c}{\partial x^2} \quad (15)$$

For large clumps to form, cells must diffuse at the lower rate and the chemoattractive (destabilizing) force must be only slightly stronger than the diffusive (stabilizing) force.

Cross-Taxis

Two species can exhibit taxis with respect to one other:

$$\frac{\partial A}{\partial t} = D_1 \frac{\partial^2 A}{\partial x^2} - \chi_1 \frac{\partial}{\partial x}\left[A \frac{\partial B}{\partial x} \right] \qquad \frac{\partial B}{\partial t} = D_2 \frac{\partial^2 B}{\partial x^2} - \chi_2 \frac{\partial}{\partial x}\left[B \frac{\partial A}{\partial x} \right]$$
$$(16)$$

(if $\chi_1 > 0$ and $\chi_2 > 0$, for example, the two species move up each other gradients). Though not greatly studied, such models are known to be susceptible to blow up. See Murray (1989, chap. 9) for references.

Convection-Driven Instabilities

Differential convection can cause instabilities in parameter ranges where Turing instabilities do not occur. Equation 2 becomes

$$\frac{\partial A}{\partial t} = f(A, B) + c_A \frac{\partial A}{\partial x} + d_A \frac{\partial^2 A}{\partial x^2}$$
$$\frac{\partial B}{\partial t} = g(A, B) + c_B \frac{\partial B}{\partial x} + d_B \frac{\partial^2 B}{\partial x^2} \qquad (17)$$

where c_A and c_B are the convection speeds. The predicted pattern is a series of peaks and valleys that move through space. Rovinsky and Menzinger (1993) recently confirmed these predictions experimentally.

Discrete-Time and Delay Models

All models discussed so far have involved rates of change which depend only on current values. Yet time lags can be natural elements of many models, and they can significantly affect pattern formation. For example, a single species neural field model with time lags can support oscillating patterns under conditions of local inhibition and long-range excitation.

Mathematical Techniques

We have barely scratched the surface of the mathematical analysis of pattern formation. Close to bifurcation (loss of stability), it is possible to analyze the behavior of small amplitude patterns. For example, one can determine whether the two basic patterns, hexagonal and striped, are stable with respect to each other (Murray, 1989, chap. 15). We have only briefly confronted the fact that pattern formation generally takes place on a finite domain, perhaps with a particular geometry. At the very least, this reduces the number of modes which can occur. We refer the reader to Murray's (1989, chap. 15) modeling of mammalian coat pattern formation.

Acknowledgments. J. Cook acknowledges the support of the Engineering and Physical Science Research Council of Great Britain. This work was in part supported by grant DMS–9106848 from the National Science Foundation and PHS/NIH 2P41 RR01243–12 from the National Institutes of Health.

Road Maps: Dynamic Systems and Optimization; Self-Organization in Neural Networks
Related Reading: Cellular Automata; Cooperative Phenomena; Development and Regeneration of Eye-Brain Maps; Phase-Plane Analysis of Neural Activity

References

Amari, S., and Arbib, M. A., Eds., 1980, *Competition and Cooperation in Neural Nets*, New York: Springer-Verlag.

Cahn, J. W., 1968, Spinodal decomposition: The 1967 Institute of Metals Lecture, *Trans. Metall. Soc. AIME*, 242:167–180.

Ermentrout, G. B., Campbell, J., and Oster, G., 1986, A model for shell patterns based on neural activity, *Veliger*, 28:369–388.

Ermentrout, G. B., and Cowan, J., 1979, A mathematical theory of visual hallucination patterns, *Biol. Cybern.*, 34:137–150.

Ermentrout, G. B., and Cowan, J., 1980, Large scale spatially organizing activity in neural nets, *SIAM J. Appl. Math.*, 38:1–21.

Harris, A. K., Stopak, D., and Warner, P., 1984, Generation of spatially periodic patterns by a mechanical instability: A mechanical alternative to the Turing model, *J. Embryol. Exp. Morphol.*, 80:1–20.

Keller, E. F., and Segel, L. A., 1971, Travelling bands of chemotactic bacteria: A theoretical analysis, *J. Theor. Biol.*, 30:235–248.

Meinhardt, H., 1982, *Models of Biological Pattern Formation*, London: Academic Press. ◆

Murray, J. D., 1989, *Mathematical Biology*, New York: Springer-Verlag. ◆

Ouyang, Q., and Swinney, H. L., 1991, Transition from a uniform state to hexagonal and striped Turing patterns, *Nature*, 352:610–612.

Rovinsky, A. B., and Menzinger, M., 1993, Self-organization induced by the differential flow of activator and inhibitor, *Phys. Rev. Lett.*, 70:778–781.

Swindale, N. V., 1980, A model for the formation of ocular dominance stripes, *Proc. R. Soc. Lond. B Biol. Sci.*, 208:243–264.

Turing, A. M., 1952, The chemical basis of morphogenesis, *Philos. Trans. R. Soc. Lond. B Biol. Sci.*, 237:37–72.

Wilson, H. R., and Cowan, J. D., 1973, A mathematical theory of the functional dynamics of cortical and thalamic nervous tissue, *Kybernetik*, 13:55–80.

Pattern Recognition

Yann LeCun and Yoshua Bengio

Introduction

Pattern Recognition (PR) addresses the problem of classifying objects, often represented as vectors or as strings of symbols, into categories. The difficulty is to synthesize, and then to efficiently compute, the *classification function* that maps objects to categories, given that objects in a category can have widely varying input representations. In most instances, the task is known to the designer through a set of example patterns whose categories are known, and through general, a priori knowledge about the task, such as "the category of an object is not changed when the object is slightly translated or rotated in space."

Historically, the field of PR started with the early efforts in neural networks (perceptrons, adalines, etc.: see PERCEPTRONS, ADALINES, AND BACKPROPAGATION). While neural networks (NNs) have sometimes played the role of an outsider in PR, the recent progress in learning algorithms and the availability of powerful hardware have made them the method of choice for many PR applications.

Because most PR problems are too complex to be solved entirely by handcrafted algorithms, machine learning has always played a central role in PR. Learning automatically synthesizes a classification function from a set of labeled examples. Unfortunately, no learning algorithm can be expected to succeed unless it is guided by prior knowledge. The traditional way of incorporating knowledge about the task is to divide the recognizer into a feature extractor and a classifier. Since most learning algorithms work better in low-dimensional spaces with easily separable patterns, the role of the feature extractor is to transform the input patterns so that they can be represented by low-dimensional vectors, or short strings of symbols, that (a) can be easily compared or matched, and (b) are relatively invariant to transformations that do not change the nature of the input objects. The feature extractor contains most of the prior knowledge and is rather specific to the task. It also requires most of the design effort because it is often handcrafted, although unsupervised learning methods such as PRINCIPAL COMPONENT ANALYSIS (q.v.) can sometimes be used. The classifier, on the other hand, is often general purpose and trainable. One of the main problems with this approach is that the recognition accuracy is largely determined by the ability of the designer to come up with an appropriate set of features. This turns out to be a daunting task which, unfortunately, must be redone for each new problem.

One of the main contributions of neural networks to PR has been to provide an alternative to this design: properly designed multilayer networks can learn complex mappings in high-dimensional spaces without requiring complicated handcrafted feature extractors. Networks containing hundreds of inputs and tens of thousands of parameters can be trained on databases containing several 100,000 examples. This allows designers to rely more on learning and less on detailed engineering of feature extractors. Crucial to success is the ability to tailor the network architecture to the task, which allows incorporating prior knowledge and, therefore, learning complex tasks without requiring excessively large networks and training sets.

The success of multilayer networks relies on one surprising fact: gradient-based minimization techniques can be used to learn very complex nonlinear mappings. Generalizations of the concept of gradient-based learning have allowed one to view many PR techniques, neural and nonneural, in a unified way, including not only traditional multilayer feedforward nets with sigmoid units and dot products, but also many other structures such as Radial Basis Functions (RBFs), Hidden Markov Models (HMMs), vector quantizers, etc. (see RADIAL BASIS FUNCTION NETWORKS; SPEECH RECOGNITION: PATTERN MATCHING; LEARNING VECTOR QUANTIZATION). Many recent efforts have been directed at combining adaptive modules of different types into a single system and training them cooperatively by propagating gradients through them, particularly for recognizing composite objects such as handwritten or spoken words.

Learning and Generalization

Due to the presence of noise, the high dimension of the input, and the complexity of the mapping to be learned, PR applications create some of the most challenging problems in machine learning. Most learning methods are trained by minimizing a *cost function* computed over a set of training examples. The cost function is generally of the form

$$C(W) = \sum_X Q(X, F(X, W)) + H(W) \qquad (1)$$

where X is a training example, $F(X, W)$ is the recognizer output for pattern X and "parameters" W, $Q(X, F(X, W))$ is a single-pattern cost function (the training error), and $H(W)$ is a measure of "capacity" of the recognizer (the *regularizer*; see GENERALIZATION AND REGULARIZATION IN NONLINEAR LEARNING SYSTEMS). Such cost functions attempt to model the real measure of performance, i.e., the *testing error* (error rate on a test set disjoint from the training set). (See also LEARNING AND GENERALIZATION: THEORETICAL BOUNDS).

System designers have to strike the right balance between learning the training set (by using powerful learning architectures) and minimizing the difference between the training error and the test error (by limiting the capacity of the machine). Large machines can learn the training set but may perform poorly if the training set is not large enough, a problem known as overparameterization, or overfitting. On the other hand, too little capacity yields underfitting—i.e., large error on both training and test sets.

Most adaptive recognizers stand between two extremes of a continous spectrum. At one end, parameter-based methods, in which a set of learned parameters determines the input-output relation, put the emphasis on minimizing the first term in Equation 1 with a fixed H (e.g., multilayer neural networks). At the other end, memory-based methods, which rely on matching, or comparing, the incoming pattern with a set of learned or stored prototypes, keep the first term close to zero and attempt to minimize the regularizer (e.g., nearest-neighbor algorithms).

Although, in principle, any appropriate functional form for F, Q, and H can be used, the choice is largely determined by (a) the belief that it is well suited to the task, and (b) the efficiency of the available minimization algorithms. There is a strong incentive to choose smooth and well-behaved functions whose gradient can be computed easily, so that gradient-based minimization algorithms can be used, as opposed to inefficient combinatorial search methods. Preferably, F will be a smooth real-valued function (e.g., layers of sigmoid units), rather than a discrete function (e.g., layers of threshold units); Q is often chosen to be the mean square error between the actual output and a target, rather than the number of misclassified patterns, which would be more relevant, but which is practically impossible to minimize.

A Few Basic Classification Methods

Linear and Polynomial Classifiers

A linear classifier is essentially a single neuron. An elementary two-class discrimination is performed by comparing the output to a threshold (multiple classes use multiple neurons). Training algorithms for linear classifiers are well studied (see PERCEPTRONS, ADALINES, AND BACKPROPAGATION). Their limitations are well known: the likelihood that a partition of P vectors of dimension N be computable by a linear classifier decreases very quickly as P increases beyond N (Duda and Hart, 1973). One method to ensure separability is to represent the patterns by high-dimensional vectors (large N). If necessary, the dimension of original input vectors can be enlarged using a set of basis functions ϕ_i:

$$F(X, W) = \sum_i w_i \phi_i(X) \qquad (2)$$

A simple example is when the basis functions are cross products of K or fewer coordinates of the input vector X (F is a polynomial of degree K). Such polynomial classifiers have been studied since the early 1960s, and have been "renamed" in the context of NNs as sigma-pi units or high-order nets. Unfortunately polynomial classifiers are often impractical because the number of features scales like N^K. Nevertheless, feature selection methods can be used to reduce the number of product terms, or to reduce the number of original input variables.

Local Basis Functions

Another popular kind of space expansion (Equation 2) uses *local* basis functions, that are activated within a small area of the input space. A popular family are the Radial Basis Functions (RBFs): $\phi_i(X) = e^{-(X-P_i)^2}$, where the P_i are a set of appropriately chosen "prototypes." Methods based on such expansions can cover the full spectrum between parameter-based and purely memory-based methods by varying the number of prototypes, the way they are computed, and the classifier that follows the expansion (which can be more complex than a simple weighted sum). At one extreme, each training sample is used as a prototype, to which the sample's label is attached. In the K-nearest neighbors algorithms, the K nearest prototypes to an unknown pattern vote for its label. In the Parzen windows method, the normalized sum of all the $\phi_i(X)$ associated with a particular class is interpreted as the conditional probability that X belongs to that class (Duda and Hart, 1973). In the RBF method, the output is a (learned) linear combination of the outputs of the basis functions. Associating a prototype with each training sample can be very inefficient, and increases the "complexity" term. Therefore, several methods have been proposed to *learn* the prototypes. One way is to use unsupervised clustering techniques such as K-means to put prototypes in regions of high sample density, but supervised methods can also be used (see RADIAL BASIS FUNCTION NETWORKS). An important one is LVQ2, in which prototypes that are near a training sample are moved away from it if its assigned class differs from the sample's, and moved toward it if its class is equal to the sample's (see LEARNING VECTOR QUANTIZATION). Another important supervised method for RBF networks is simply gradient descent: the partial derivatives of the cost function with respect to the parameters of the basis functions (the prototype vectors) can be computed using a form of backpropagation: the same way gradients can be backpropagated through sigmoids and dot products, they can be backpropagated through exponentials and Euclidean distances. The parameters can then be adjusted using the gradient. It has been argued that the local property leads to faster learning than standard multilayer nets, and good rejection properties (Lee, 1991). Several authors enhance the power of prototype-based systems by using distance measures that are more complex than just Euclidean distance between the prototypes and the input patterns (such as general bilinear forms with learned coefficients). Methods that add prototypes as needed have also been proposed, notably the RCE algorithm (see COULOMB POTENTIAL LEARNING).

Maximum Margin Classifiers

A recently proposed elegant way of avoiding the curse of dimensionality in polynomial and local classifiers rests on the fact that if the w_i in Equation 2 are computed to maximize the *margin* (the minimum distance between training points and the classification surface), the W obtained after training can be written as a linear combination of a small subset of the expanded training examples (Boser, Guyon, and Vapnik, 1992). Points in this subset are called *support points*. This leads to a surprisingly simple way of evaluating high-degree polynomials in high-dimensional spaces without having to explicitly compute all the terms of the polynomial. For example, maximum-margin polynomials of degree K can be computed using

$$F(X) = \sum_{j \in S} \alpha_j (X \cdot P_j + 1)^K \qquad (3)$$

where the P_j are the support points (subset of the training set), and the α_j are coefficients that uniquely determine the weights W. Learning the α_j amounts to solving a quadratic programming problem with linear inequality constraints. Excellent results on handwritten digit images have been obtained with a fourth-degree polynomial computed with this method (Bottou et al., 1994). The number of multiply-adds per recognition was a few 100,000s, much less than the $O(400^4)$ multiply-adds required to directly evaluate the polynomial.

Complex Distance Measures

Although many memory-based methods use simple distance measures (Euclidean distance) and large collections of prototypes, some applications can take advantage of more complex, problem-dependent, distance measures and use fewer prototypes. Ideal distance measures should be invariant with respect to transformations of the patterns that do not change their nature (e.g., translations and distortions for characters, time or pitch distortion for speech). With invariant distances, a single prototype can potentially represent many possible instances of a category, reducing the number of necessary prototypes. An important family of invariant distance measures is *elastic matching*. Elastic matching comes down to finding the point closest to the input pattern on the surface of all possible deformations of the prototype. Naturally, the exhaustive search approach is prohibitively expensive in general. However, if the surface is smooth, better search techniques can be used: gradient descent (Burr, 1983), or conjugate gradient (Hinton, Williams, and Revow, 1992). If the deformations are along one dimension (as in speech), dynamic programming can find the best solution efficiently. In an interesting technique, recently proposed in Simard, LeCun, and Denker (1993), the surface of a deformed prototype is approximated by its tangent plane at the prototype. The matching problem reduces to finding the minimum distance between a point and a plane, which can be done efficiently. This has been applied to handwritten character recognition with great success.

Multilayer Networks, Gradient-Based Learning

The vast majority of applications of NNs to PR are based on multilayer feedforward networks trained with backpropagation. At first, it seems almost magical that an algorithm as simple as gradient descent works at all to learn complex nonlinear mappings (nonconvex, ill-conditioned error surfaces). Minsky and Selfridge's warning about the limitations of "hill-climbing" methods for machine learning in 1961 (see LEARNING AS HILL-CLIMBING IN WEIGHT SPACE) is an indication of the general belief that it could not work. Surprisingly, experiments show that local minima are rarely a problem with large networks. As evidence of the success of backpropagation, all but two of the entries in the last NIST character recognition competition used some form of backpropagation network.

PR problems are often characterized by large and redundant training sets with high-dimensional inputs, which translates into large networks, and long learning times. Much effort has been devoted to speeding up training using refined nonlinear optimization methods (conjugate gradient, quasi-Newton methods, etc.). These are essentially batch methods (the weights are updated after a complete pass through the training set), which can rarely compete with "carefully tuned" stochastic (on-line) gradient descent (where the weights are updated after each pattern presentation). This is due to the presence of redundancy in large, natural training sets. On typical large-scale image or speech recognition tasks, stochastic gradient descent converges in one to a few dozen epochs. To avoid overlearning, a validation set should be set aside, and training should be stopped when the error rate on the validation set stops decreasing. An important limitation to the popularity of NN techniques for PR is that certain simple tricks must be used and many common pitfalls must be avoided that are part of the "oral culture" rather than scientific facts (LeCun, 1989).

Once backpropagation with feedforward networks of sigmoid units and dot products established the value of gradient-based learning, it seemed natural to extend the idea to other structures. Minimizing a cost function through gradient-based learning can be seen as the unifying principle behind many methods: radial basis functions or mixtures of Gaussians, learning vector quantization, HMMs, and many prototype-based methods using various distance measures. Experiments have shown the advantage of using different types of modules in different parts of a learning system. In particular, sigmoids and dot products seem better for processing large amounts of high-dimensional and low-level information (early feature extraction), while RBF or other more local modules seem better suited for final classification, a more memory-intensive task. With the gradient-based learning framework, modules of different types can be connected in any configuration, and trained cooperatively by backpropagating gradients through them. To achieve this, one only needs to be able to compute the partial derivatives of each output of a module with respect to each input and each parameter of the module (see MODULAR NEURAL NET SYSTEMS, TRAINING OF). In addition, many cost functions can be considered as just another module (with a scalar output) through which gradients can be backpropagated. Examples include the mean square error, modified LVQ cost functions, maximum likelihood, maximum mutual information, cross entropy, classification figure of merit, and several types of statistical post-processors.

Local/Global and Modular Methods

It has recently been suggested that good PR systems should behave differently in different parts of the input space. For example, parts of the input space may be very sparsely populated, requiring a low-capacity learner, while denser areas may require a more complex one. A simple idea is to use a collection of modules, each of which is activated when the input lies in a particular region. A separate module, called a *gater*, decides which module should be activated. When the gater is differentiable, the whole system (modules plus gater) can be trained cooperatively (see MODULAR AND HIERARCHICAL LEARNING SYSTEMS). In such multimodular systems, parameters are relatively decoupled across modules, which is believed to allow for faster training (or better scaling of training time). In another interesting "semilocal" method, a simple network (e.g., single layer) is trained each time a new test pattern is presented, using training patterns in the neighborhood of this test pattern; training is done "on demand" during recognition (Bottou and Vapnik, 1992).

In general, local methods learn fast, but they are expensive at run-time in terms of memory and, often, of computation. In addition, they may not be appropriate for problems with high-dimensional inputs. Global methods, such as multilayer networks, take longer to train, but they are quite compact, and they execute quickly. They can handle high-dimensional inputs, particularly when specialized architectures are used.

Specialized Architectures, Convolutional Networks

The great hope that multilayer networks brought with them was the possibility of eliminating the need for a separate handcrafted feature extractor, relying on the first layers to automatically learn the right set of features. Although fully connected networks fed with "raw" character images (or speech spectra) have very large numbers of free parameters, they have been applied with some success (Martin and Pittman, 1991). This can be explained as follows. With small initial weights, a multilayer network is almost equivalent to a single-layer network (each layer is quasi-linear). As incremental learning proceeds, the weights gradually increase, thereby progressively increasing the effective capacity of the system (to the authors' knowledge, this explanation was first suggested by Léon Bottou in 1988).

Nevertheless, using a specialized network architecture, instead of a fully connected net, can reduce the number of free parameters and facilitate the learning of invariances. In certain applications, the need for a separate handcrafted feature extractor can be eliminated by wiring the first few layers of the network in a way that forces it to learn relevant features and eliminate irrelevant variability. CONVOLUTIONAL NETWORKS (q.v.), including Time-Delay Neural Networks (TDNNs), are an important class of specialized architectures, well suited for dealing with 1D or 2D signals such as time series, images, or speech. Convolutional networks use the techniques of local receptive fields, shared weights, and subsampling (loosely based on the architecture of the visual cortex) to ensure that the first few layers extract and combine local features in a distortion-invariant way. Although the wiring of the convolutional layers is designed by hand, the values of all the coefficients are learned with a variant of the backpropagation algorithm. The main advantage of this approach is that the feature extractor is totally integrated into the classifier, and is produced by the learning process rather than by the hand of the designer (LeCun et al., 1990). Due to the weight-sharing technique, the number of free parameters in a convolutional network is much less than in a fully connected network of comparable power, which has the effect of reducing the complexity term in Equation 1 and improving the generalization. The success of convolutional nets of various types has had a major impact on several application domains: speech recognition, character recognition, object spotting. On handwriting recognition tasks, they compare favorably with other techniques (Bottou et al., 1994) in terms of accuracy, speed, and memory requirements. Character recognizers using convolutional nets have been deployed in commercial applications. A very promising feature of convolutional nets is that they can be efficiently replicated, or scanned, over large input fields, resulting in the so-called Space Displacement Neural Net (SDNN) architecture (see below, and CONVOLUTIONAL NETWORKS).

Networks with recurrent connections can be used to map input sequences to output sequences, while taking long-term context into account. The main advantage of recurrent networks over TDNNs for analyzing sequences is that the span of the temporal context that the network can take into account is not hard-wired within a fixed temporal window by the architectural choices, but can be learned by the network. However, theoretical and practical hurdles (Bengio et al., 1994) limit the span of long-term dependencies that can be learned efficiently.

Recognition of Composite Objects

In many real applications, the difficulty is not only to recognize individual objects but also to separate them from context or background. For example, one approach to handwritten word recognition is to segment the characters out of their surrounding and recognize them in isolation. A typical handwritten word recognizer uses heuristics to form multiple, possibly overlapping, character candidates by cutting the word or by joining nearby strokes. Then the recognizer must either classify each candidate as a character or reject it as a noncharacter. In many applications, such as cursive handwriting or continuous speech, it is difficult, or even impossible, to devise robust segmentation heuristics. One approach to avoid explicit segmentation is to simply scan the recognizer over all possible locations on the input (character string or spoken sentence) and collect the sequence of corresponding recognizer outputs. Although this is computationally very expensive in general, replicated convolutional networks (SDNN or TDNN) can be used to do that very efficiently. In the case of handwriting recognition, an SDNN output will contain a well-identified label when centered on a character. Between characters, the output should indicate a reject. However, combinations of off-center characters may cause ambiguous outputs (e.g., "cl" labeled as "d"). Since both methods (explicit segmentation and scanning) generate many extraneous candidates, a post-processor is required to resolve ambiguities and pull out the most consistent interpretation, retaining genuine characters and rejecting erroneous stroke combinations, possibly taking linguistic constraints into account (a lexicon or grammar). For this to succeed, the recognizer must be trained not only to classify characters, but also to reject noncharacters. The search for the best interpretation is easily done within the framework of hidden Markov models. A graph is built in which each path corresponds to a possible interpretation of the input and in which each node is given probabilities of matching recognizer outputs. Dynamic programming can be used to find the path of highest probability, which yields the most likely interpretation.

Multimodule Architectures and Cooperative Training

Such combinations (Figure 1) of neural networks and HMMs (or other graph-based post-processors) have been proposed by several authors, mostly for speech recognition (see SPEECH RECOGNITION: PATTERN MATCHING), but also for handwriting recognition (see HAND WRITTEN DIGIT STRING RECOGNITION and CONVOLUTIONAL NETWORKS).

The main technical difficulty is in training such hybrid systems. Training the recognizer exclusively on presegmented characters is neither sufficient nor always possible, since (a) the recognizer must be trained to reject noncharacters, and (b) in many cases, such as cursive handwriting, segmented characters are not available, only whole words are. The solution is to train the recognizer and the post-processor simultaneously to minimize an error measure at the word level. This means being able to backpropagate gradients through the HMM down to the recognizer, or to generate desired outputs for the recognizer using the best path in the graph (see SPEECH RECOGNITION: PATTERN MATCHING; and Franzini, Lee, and Waibel, 1990). Simultaneous training of such hybrids has been reported to yield large reductions in error rates over independent training of the modules in speech recognition (for TDNN/dynamic time warping, see Driancourt, Bottou, and Gallinari, 1991, and Haffner, Franzini, and Waibel, 1991; for TDNN/HMM, see Bengio et al., 1992). Similar reductions of error rates have been reported for on-line handwriting recognition (see Bengio, LeCun, and Henderson, 1994, for an SDNN/HMM approach).

Discussion

Neural networks, particularly multilayer backpropagation NNs, provide simple yet powerful and general methods for synthesizing classifiers with minimal effort. However, most practical systems combine NNs with other techniques for pre- and post-processing. On isolated character recognition tasks, multilayer nets trained with variants of backpropagation have approached human accuracy, at speeds of about 1000 characters per second using NN hardware. NNs have allowed workers to minimize the role of detailed engineering and maximize the role of learning. Despite the recent advances in multimodule architectures and gradient-based learning, several key questions are still unanswered, and many problems are still out of reach. How much has to be built into the system, and how much can be learned? How to achieve true transformation-invariant perception with NNs? Convolutional nets are a step in the right direction, but new concepts will be required for a complete solution (see DYNAMIC LINK ARCHITECTURE). How to recognize compound objects in their context? The accuracy of the best NN/HMM hybrids for written or spoken sentences

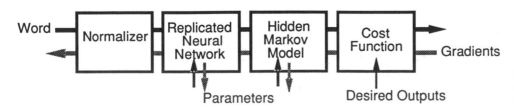

Figure 1. A multimodule architecture combining a convolutional NN with an HMM post-processor.

cannot even be compared with human performance. Topics such as the recognition of 3D objects in complex scenes are totally out of reach. Human-like accuracy on complex PR tasks such as handwriting and speech recognition may not be achieved without a drastic increase in the available computing power. Several important questions may simply resolve themselves with the availability of more powerful hardware, allowing the use of brute-force methods and very large networks.

Road Map: Learning in Artificial Neural Networks, Deterministic
Background: I.3. Dynamics and Adaptation in Neural Networks
Related Reading: Concept Learning; Neurosmithing: Improving Neural Network Learning; Object Recognition; Stochastic Approximation and Neural Network Learning

References

Bengio, Y., LeCun, Y., and Henderson, D., 1994, Globally trained handwritten word recognizer using spatial representation, space displacement neural networks and hidden Markov models, in *Advances in Neural Information Processing Systems 6* (J. Cowan, G. Tesauro, and J. Alspector, Eds.), San Mateo, CA: Morgan Kaufmann, pp. 937–944. ◆

Bengio, Y., Mori, R. D., Flammia, G., and Kompe, R., 1992, Global optimization of a neural network-hidden Markov model hybrid, *IEEE Trans. Neural Netw.*, 3:252–259. ◆

Bengio, Y., Simard, P., and Frasconi, P., 1994, Learning long-term dependencies with gradient descent is difficult, *IEEE Trans. Neural Netw.*, Special Issue on Recurrent Neural Networks, 5:157–166.

Boser, B., Guyon, I., and Vapnik, V., 1992, An algorithm for optimal margin classifiers, in *Fifth Annual Workshop on Computational Learning Theory*, Pittsburgh, pp. 144–152.

Bottou, L., Cortes, C., Denker, J., Drucker, H., Guyon, I., Jackel, L., LeCun, Y., Muller, U., Sackinger, E., Simard, P., and Vapnik, V., 1994, Comparison of classifier methods: A case study in handwritten digit recognition, in *International Conference on Pattern Recognition*, Jerusalem, Israel.

Bottou, L., and Vapnik, V., 1992, Local learning algorithms, *Neural Computat.*, 4:888–900.

Burr, D., 1983, Designing a handwriting reader, *IEEE Trans. Pattern Analysis Machine Intell.*, 5:554–559.

Denker, J., and LeCun, Y., 1991, Transforming neural-net output levels to probability distributions, in *Advances in Neural Information Processing Systems 3* (R. P. Lippman, R. Moody, and D. S. Touretzky, Eds.), San Mateo, CA: Morgan Kaufmann, pp. 853–859.

Driancourt, X., Bottou, L., and Gallinari, P., 1991, Learning vector quantization, multi layer perceptron and dynamic programming: Comparison and cooperation, in *Proceedings of the International Joint Conference on Neural Networks, 1991*, vol. 2, New York: IEEE, pp. 815–819.

Duda, R., and Hart, P., 1973, *Pattern Classification and Scene Analysis*, New York: Wiley. ◆

Franzini, M., Lee, K., and Waibel, A., 1990, Connectionist Viterbi training: A new hybrid method for continuous speech recognition, in *International Conference on Acoustics, Speech and Signal Processing*, Albuquerque, NM, pp. 425–428.

Haffner, P., Franzini, M., and Waibel, A., 1991, Integrating time alignment and neural networks for high performance continuous speech recognition, in *International Conference on Acoustics, Speech and Signal Processing*, Toronto, pp. 105–108.

Hinton, G., Williams, C., and Revow, M., 1992, Adaptive elastic models for hand-printed character recognition, in *Advances in Neural Information Processing Systems 4* (J. Moody, S. Hanson, and R. Lipmann, Eds.), San Mateo, CA: Morgan Kaufmann, pp. 512–519. ◆

LeCun, Y., 1989, *Generalization and Network Design Strategies*, Technical Report CRG-TR-89-4, University of Toronto, Department of Computer Science; short version in *Connectionism in Perspective* (R. Pfeifer, Z. Schreter, F. Fogelman, and L. Steels, Eds.), New York: Elsevier, 1989. ◆

LeCun, Y., Boser, B., Denker, J., Henderson, D., Howard, R., Hubbard, W., and Jackel, L., 1990, Handwritten digit recognition with a back-propagation network, in *Advances in Neural Information Processing Systems 2* (D. Touretzky, Ed.), San Mateo, CA: Morgan Kaufmann, pp. 396–404. ◆

Lee, Y., 1991, Handwritten digit recognition using K nearest neighbor, radial-basis function, and backpropagation neural network, *Neural Computat.*, 3:441–449.

Martin, G., and Pittman, J., 1991, Recognizing hand-printed letters and digits using back-propagation learning, *Neural Computat.*, 3(2): 258–267.

Simard, P., LeCun, Y., and Denker, J., 1993, Efficient pattern recognition using a new transformation distance, in *Advances in Neural Information Processing Systems 5* (S. J. Hanson, J. D. Cowan, and C. L. Giles, Eds.), San Mateo, CA: Morgan Kaufmann, pp. 50–58. ◆

Perception of Three-Dimensional Structure

James S. Tittle and James T. Todd

Introduction

One of the most perplexing phenomena in the study of human perception is the ability of observers to determine the layout and three-dimensional (3D) structure of objects in the environment from the two-dimensional (2D) patterns of light that project onto the retina. Over the past 25 years, there have been numerous neural-based theories of how the biological systems perform low-level visual processes such as edge detection, the computation of 2D motion, or determining which points correspond to one another in stereoscopically presented pairs of images. However, relatively few models address the issue of how these 2D image features are used to compute an appropriate neural representation of 3D structure (notable exceptions include Grossberg and Mingolla, 1987; and Pentland, 1989).

There are two important issues that need to be considered in evaluating any computational model of 3D form perception.

The first involves how 3D structure is perceptually represented. After all, to compute an object's shape from visual information, one must first define precisely what *shape* is. There are numerous attributes of 3D structure that potentially could be represented by the visual system (e.g., curvature, relative depth, or local orientation), and the relative computational difficulty of analyzing these different attributes can vary dramatically. It is much more difficult, for example, to determine the precise Euclidean distance between a pair of visible points than merely to assess which point is closer in depth (e.g., see Todd and Reichel, 1989).

A second related issue to consider in evaluating different computational models is the plausibility of their underlying assumptions. Because there is a one-to-many mapping between the structure of light at a point of observation and the structure of the environment, all computational analyses of 3D form perception must restrict the set of possible interpretations by assuming the existence of environmental constraints. Unfortunately, however, many of the constraints that have been employed for this purpose seem to have been adopted more for their mathematical convenience than for their ecological validity. The problem with the adoption of such constraints is that the resulting analyses of 3D form may function effectively only within narrowly defined contexts, which have a small probability of occurrence in the natural environments of real biological organisms.

In the remainder of this article, we review various computational models that have been proposed for analyzing an object's 3D structure from different types of optical information, such as shading, texture, motion, and stereo. We also examine how the performance of these models compares with the capabilities and limitations of actual human observers in judging different aspects of 3D structure under varying viewing conditions. Our goal is to identify the specific representations and computational mechanisms by which 3D form is perceptually analyzed within the human visual system.

Structure from Shading

The most basic type of information available to any visual system is the amount of light that reflects onto different regions of the retina from illuminated surfaces in the environment. Smooth gradations in surface luminance are usually referred to as *shading*. Although there exist many examples from both painting and photography that image shading can be a perceptually salient source of information about 3D structure, it is not at all obvious how the human visual system makes use of this information, or even what attributes of an object's structure are perceptually specified. The analysis of image shading is difficult because the luminance of any visible surface region can depend on the positions and spectral composition of its sources of illumination, the local reflectance and orientation of the surface, and the position of the observer. To compute shape from shading, it is necessary to somehow decompose these different factors.

The first computational analyses of image shading were developed by Horn and his co-workers (Horn and Brooks, 1989) to determine the local orientation of a visible surface region from the intensity of its reflected light. To constrain the solution, these models assume (1) that the surface is smooth; (2) that it has a known uniform reflectance function, usually Lambertian (i.e., a surface with reflectance that depends only on the angle of illumination), with no specular components; and (3) that it has a uniform pattern of illumination with a known direction and spectral composition. Subsequent analyses have also been developed which do not require prior knowledge about the direction of illumination. These models use the gradient of image intensity to compute surface orientation, such that the magnitude of the gradient is used to estimate slant, and the direction of the gradient is used to estimate tilt. This method assumes that the observed surface region is locally spherical.

Most models of shape from shading are designed to represent shape using an orientation map, but there are a number of notable exceptions to this rule. For example, Pentland (1989) has proposed a biologically plausible neural mechanism for computing relative surface depths up to a indeterminate scale factor for Lambertian surfaces whose directions of illumination are greater than 30° from the observer's line of sight. Other analyses have been proposed (e.g., Koenderink and van Doorn, 1980) for determining the sign of Gaussian curvature from singular points within the overall pattern of image isointensity contours. A possible neural mechanism for obtaining the field of isointensity contours has been developed by Grossberg and Mingolla (1987). These authors have argued that a surface's 3D form can be neurally represented by the global pattern of activity generated by these contours at different spatial scales. It is important to keep in mind, however, that the pattern of image isointensity contours does not remain invariant over changes in viewing position or the direction of illumination, and some higher-order mechanism therefore would be required with this type of representation to achieve shape constancy.

A fundamental characteristic of almost all of the models described is that they are designed to compute some particular property of local surface structure (e.g., orientation, curvature, or depth), based on some degree of prior knowledge about the surface reflectance function and/or the pattern of illumination. There is a growing amount of evidence to suggest, however, that the perception of shape from shading by actual human observers does not share this characteristic. For example, one important limitation of all current computational models is that they cannot cope with the occurrence of indirect illumination, shadows, transparencies, or specular highlights, yet human observers seem to have little difficulty in dealing with these phenomena (e.g., see Mingolla and Todd, 1986). Other research suggests, moreover, that human observers are surprisingly insensitive to the properties of relative local depth or orientation, which most of these models are designed to compute. On the basis of this evidence, Todd and Reichel (1989) have argued that perception of shape from shading may largely consist of sensitivity to ordinal depth relations among neighboring surface regions.

Structure from Surface Texture and Contour

The next level of optical structure used for perceiving 3D shape results from the discontinuities in the image intensity field that are seen as texture elements and contours on a surface. Optical texture provides information about 3D structure because (1) the projected size and distance between elements decreases with increasing distance from the observer, and (2) the projected shape of texture elements is increasingly compressed (i.e., foreshortened) as the angle between surface orientation and viewing direction increases. Because of this relationship between texture size, density, and compression and the 3D layout, a visual system could potentially recover information about surface shape from the systematic variations in the projection of texture elements.

Gibson (1950) was the first to identify a quantitative relationship between the optical gradients of both texture size and compression and the slant of a planar surface. Extensions of Gibson's analysis have been recently developed to recover the 3D structure of curved surfaces. For example, Witkin (1981) showed that it is possible to recover the orientation of curved surfaces from local measurements of texture element compression observed under parallel projection. Similarly, it is also possible to recover changes in surface depth under polar projection from the relative sizes or lengths of individual texture elements (Todd and Akerstrom, 1987). One drawback of these approaches is that they assume all texture elements are isotropic (no orientation bias) or have a circular symmetry. In addition, they also require that texture elements are homogeneously distributed across a surface (i.e., the texture all across the surface has comparable distributions of size, shape, and density), but recent psychophysical results (Todd and Akerstrom, 1987) indicate that human observers can perceive shape from texture with nonhomogeneously distributed elements.

According to these theories, human observers should be able to perceive the 3D structure of surfaces from two distinct sources of texture information: changes in texture (1) size or length and (2) compression. However, Todd and Akerstrom (1987) found that observers could perceive curved surfaces viewed with either polar or parallel projection, and when projected texture elements had constant area, and compression. Grossberg and Mingolla (1987) have partially implemented an analysis of this result, based on the projected widths of texture elements, using their neural net model of early visual processing.

Another source of 3D shape information comes from the occluding contour (i.e., the locus of points on a surface for which the surface normal is perpendicular to the viewing direction) of objects and surfaces. The occluding contour is similar to the silhouette of an object, but it is more general because there can be internal contours caused by object self-occlusion. Koenderink and van Doorn (1982) analyzed occluding contours and showed that they provide qualitative information (sign of Gaussian curvature) about the 3D structure of surfaces. Specifically, their analysis indicates that convex occluding contours result from elliptic (positive Gaussian curvature), surface regions and concave occluding contours result from hyperbolic (negative Gaussian curvature) surface regions. Koenderink and van Doorn have further shown that, as we move along a surface away from an occlusion point, the ordinal depth must decrease monotonically until a local depth minimum is reached. Thus, important information about the ordinal structure of surfaces is contained in the region near an occluding contour, and Todd and Reichel (1989) have presented psychophysical evidence that human observers make use of this information to perceive the 3D structure of surfaces.

Structure from Motion

All of the previously described optical information is contained in a single static image, but when either the observer or an object in the environment moves, the pattern of relative motion in the image provides additional information about 3D structure. Recent theoretical analyses of structure-from-motion (SFM) tend to be based either on discrete points and views or the spatial derivatives of optic flow fields. In both cases, the trend has been to provide a precise description of the minimum information (points and views or spatial derivatives of flow fields) needed to uniquely determine an object's structure and motion. However, recent psychophysical evidence (Todd and Bressan, 1991) indicates that human observers may not recover the unique solution provided by these analyses.

The earliest of these minimum point and view analyses was presented by Ullman (1979), who showed that three views of four non-coplanar points was the minimum amount of information needed to uniquely (up to reflection about the image plane) specify 3D structure from motion. This analysis assumes orthographic projection and is based on the assumption that the points are moving as a rigid configuration. Numerous other analyses using a rigidity constraint were subsequently presented based on both optic flow and discrete points and views approaches.

Although the SFM theories described vary in their input representations (discrete frames or velocity/acceleration fields) and their assumptions or constraints (e.g., rigidity or fixed axis rotation), they all demonstrate that it is mathematically possible to recover the correct Euclidean metric structure of a configuration of moving points. However, recent empirical evidence suggests that humans cannot make use of this informa-

tion and instead recover the affine 3D structure available from only two views of a motion sequence (Todd and Bressan, 1990). These results are consistent with recent theoretical analyses describing how two views allow for the recovery of 3D structure up to a one-parameter family defined by an affine stretching transformation along the line of sight (e.g., Todd and Bressan, 1990).

Structure from Binocular Disparity

Although both Euclid and Leonardo da Vinci observed that the right and left eyes project different images of the same 3D scene, it was not until the nineteenth century that Wheatstone provided the first experimental evidence that depth is perceived when disparate images are presented separately to each eye. Since that time, binocular disparity (the angular difference between the images of corresponding features in the right and left eyes) has been considered one of the most powerful sources of 3D shape information. The goal of this visual process has usually been described as the recovery of 3D metric surface properties (i.e., depth, slant, or curvature). However, some researchers have argued that stereopsis may only reliably provide ordinal or even topological information about 3D structure. The primary difficulty in perceiving 3D shape from binocular disparity arises because the relationship between depth and horizontal binocular disparity is not one-to-one: the binocular disparity for a fixed depth interval varies inversely with the squared distance to fixation. Thus, to achieve veridical depth perception, horizontal binocular disparities must be scaled either by the squared viewing distance or vertical disparity, and many studies have been conducted to determine if the visual system does in fact do such a scaling. However, the results from this line of research have not been consistent.

Some early research indicated that, at close viewing distances (less than 2 m), disparities are correctly scaled by viewing distance. However, one problem with almost all of this research is that it was done with simple stereograms consisting of just a few isolated elements rather than a continuous disparity field such as would result from the binocular viewing of any smooth surface. Recently, several researchers have examined the influence of viewing distance on the perception of shape for smooth surfaces and found that perceived depth from binocular disparities was not scaled correctly (Johnston, 1991).

One reason observers may have had difficulty obtaining veridical depth perception in the studies mentioned is because of the absence of vertical disparities. When binocularly viewing a 3D scene, there are both horizontal disparities (caused by the horizontal separation of the two eyes) and vertical disparities (a perspective effect that occurs when part of the scene is closer to one eye than to the other). Although the vertical disparities are quite small relative to the horizontal disparities, it has been shown mathematically that horizontal and vertical disparities taken together uniquely determine the depth of points in the environment (Mayhew and Longuet-Higgins, 1982). Consequently, there has been much recent interest in determining whether human observers make use of vertical disparities. The initial evidence indicates that observers are sensitive to vertical disparities, but only for large visual angle (approximately 60°) displays.

Structure from Multiple Sources

In a natural viewing situation, the available sources of optical information for 3D structure provide related but not necessarily redundant descriptions of the environment. (e.g., shading may specify surface orientation while binocular disparity specifies relative depth.) Furthermore, the relevance or salience of these different aspects of 3D structure may vary depending on the tasks being performed by the observer. Thus, there are many ways in which the visual system could combine these related but nonredundant structural descriptions.

The most recent theoretical approaches for integrating multiple sources of information about 3D structure are the Bayesian models (e.g., Bülthoff and Yuille, 1991) and the weighted linear combination model of Maloney and Landy (1989). Although both approaches could adequately explain results indicating a linear combination of multiple sources, only the Bayesian theory can account for those studies that have found evidence for a nonlinear (i.e., facilitatory) interaction (e.g., Tittle and Braunstein, 1993). The key difference between these theories can be characterized by a distinction between weak and strong fusion models. Specifically, a *weak* fusion model is one in which representations of 3D structure are computed independently from each source and then combined, and a *strong* fusion model is one in which nonlinear interactions occur during the computation of 3D structure from each source.

The theoretical and psychophysical study of how the visual system integrates 3D structure from multiple sources is a relatively recent development, and it is not yet clear whether the strong or weak fusion models best characterize our perceptions. In fact, current psychophysical results strongly suggest that there is no one combination rule that describes how multiple sources interact in all situations. Instead, it appears that the combination of multiple sources of 3D structure is an adaptive process that depends on the particular stimulus configuration under observation. Given this state of affairs, it seems likely that the strong fusion approach, characterized by Bayesian models of integration, will be more successful in explaining how the perception of 3D structure from multiple sources varies depending on such factors as the specific task demands confronting an observer, or the relative amount of noise a stimulus configuration contains for each source of information.

Discussion

It is interesting to note when reviewing the literature on 3D form perception that the capabilities and limitations of existing computational models often have surprisingly little overlap with the empirical data obtained from actual human observers. Many of these models are designed to generate precise metrical representations of 3D structure, but they are only able to achieve this precision by making dubious assumptions about environmental constraints that seem to be adopted more for their mathematical convenience than for their ecological validity. As a result, these models can only function effectively within narrowly constrained contexts that would seldom be encountered in a natural environment. There is a growing amount of evidence to suggest, however, that the human visual system does not share these characteristics. Observers' perceptions of Euclidean metric structure are surprisingly inaccurate and imprecise, yet their abilities to successfully recognize and interact with objects does not seem to depend on any specific set of viewing conditions. To the extent that there is a tradeoff between metrical precision and constancy in visual form perception, the evidence suggests that the forces of evolution must favor constancy.

There is one important caveat that should be noted with respect to these conclusions. In most psychophysical experiments on visual form perception, observers are asked to make judgments based on the phenomenal appearance of an object's structure, but there is some evidence to suggest that there may be aspects of perceptual knowledge involved with motor interactions (e.g., grasping an object) that are not directly accessible to our conscious awareness (see DISSOCIATIONS BETWEEN VISUAL PROCESSING MODES). Consider, for example, the common tasks of avoiding obstacles during high-speed locomotion, braking a moving vehicle with the correct force to prevent an impending collision, or running to intercept a moving projectile. How could observers perform such activities without having accurate knowledge at some level about the Euclidean metric structure of the environment? It is interesting to note in this regard that all of these tasks have been theoretically analyzed to show how they could be performed by monitoring simple aspects of optical stimulation with no knowledge whatsoever of Euclidean distance relations in 3D space (e.g., see Lee, 1976). Because similar behaviors are performed by animals such as flies and frogs with minimal neural capacity, it seems reasonable to conclude that they are accomplished with relatively simple mechanisms. What exactly must be computed about an object's shape to successfully grasp it is a fundamental question that has yet to be answered.

Road Map: Vision
Related Reading: Motion Perception; Stereo Correspondence and Neural Networks; Visual Processing of Object Form and Environment Layout; Visuomotor Coordination in Flies

References

Bülthoff, H., and Yuille, A. L., 1991, Bayesian models for seeing shapes and depth, *Comments Theor. Biol.*, 2:283–314.

Gibson, J. J., 1950, *The Perception of the Visual World*, Boston: Houghton Mifflin. ◆

Grossberg, S., and Mingolla, E., 1987, Neural dynamics of surface perception: Boundary-webs, illuminants, and shape from shading, *Commun. Vis. Graph. Image Proc.*, 37:116–165.

Horn, B. K. P., and Brooks, M. J., 1989, *Shape from Shading*, Cambridge, MA: MIT Press. ◆

Johnston, E. B., 1991, Systematic distortions of shape from stereopsis, *Vis. Res.* 31:1351–1360.

Koenderink, J. J., and van Doorn, A. J., 1980, Photometric invariants related to solid shape, *Opt. Acta*, 27:981–996.

Koenderink, J. J., and van Doorn, A. J., 1982, The shape of smooth objects and the way contours end, *Perception*, 11:129–137.

Lee, D. N., 1976, A theory of visual control of braking based on time-to-collision, *Perception*, 5:437–459.

Maloney, L., and Landy, M., 1989, A statistical framework for robust fusion of depth information, in *Visual Communication and Image Processing IV* (W. A. Perlman, Ed.), *Proc. Soc. Photo-Opt. Instrum. Eng.*, 1199:1154–1163.

Mayhew, J. E., and Longuet-Higgins, H. C., 1982, A computational model of binocular depth perception, *Nature*, 297:376–378.

Mingolla, E., and Todd, J. T., 1986, Perception of solid shape from shading, *Biol. Cybern.*, 53:137–151.

Pentland, A. P., 1989, A possible neural mechanism for computing shape from shading, *Neural Computat.*, 1:208–217.

Tittle, J. S., and Braunstein, M. L., 1993, Recovery of 3-D shape from binocular disparity and structure from motion, *Percept. & Psychophys.*, 54:157–169.

Todd, J. T., and Akerstrom, R. A., 1987, Perception of three-dimensional form from patterns of optical texture, *Percept. & Psychophys.*, 13:242–255.

Todd, J., and Bressan, P., 1990, The perception of 3-dimensional affine structure from minimal apparent motion sequences, *Percept. & Psychophys.*, 48:419–430.

Todd, J. T., and Reichel, F. D., 1989, Ordinal structure in the visual perception and cognition of smoothly curved surfaces, *Psychol. Rev.*, 96:643–657.

Ullman, S., 1979, *The Interpretation of Visual Motion*, Cambridge, MA: MIT Press. ◆

Witkin, A. P., 1981, Recovering surface shape and orientation from texture, *Artif. Intell.*, 17:17–45.

Perceptrons, Adalines, and Backpropagation

Bernard Widrow and Michael A. Lehr

Introduction

The field of neural networks has enjoyed major advances since 1960, a year which saw the introduction of two of the earliest feedforward neural network algorithms: the perceptron rule (Rosenblatt, 1962) and the LMS algorithm (Widrow and Hoff, 1960). Around 1961, Widrow and his students devised Madaline Rule I (MRI), the earliest learning rule for feedforward networks with multiple adaptive elements. The major extension of the feedforward neural network beyond Madaline I took place in 1971 when Paul Werbos developed a backpropagation algorithm for training multilayer neural networks. He first published his findings in 1974 in his doctoral dissertation (see BACKPROPAGATION: BASICS AND NEW DEVELOPMENTS). Werbos's work remained almost unknown in the scientific community until 1986, when Rumelhart, Hinton, and Williams (1986) rediscovered the technique and, within a clear framework, succeeded in making the method widely known.

The development of backpropagation has made it possible to attack problems requiring neural networks with high degrees of nonlinearity and precision (Widrow and Lehr, 1990; Widrow, Rumelhart, and Lehr, 1994). Backpropagation networks with fewer than 150 neural elements have been successfully applied to vehicular control simulations, speech generation, and undersea mine detection. Small networks have also been used successfully in airport explosive detection, expert systems, and scores of other applications. Furthermore, efforts to develop parallel neural network hardware are advancing rapidly, and these systems are now becoming available for attacking more difficult problems such as continuous speech recognition.

The networks used to solve the above applications varied widely in size and topology. A basic component of nearly all neural networks, however, is the adaptive linear combiner.

The Adaptive Linear Combiner

The adaptive linear combiner has as output a linear combination of its inputs. In a digital implementation, this element receives at time k an input signal vector or input pattern vector $\mathbf{X}_k = [x_0, x_{1_k}, x_{2_k}, \ldots, x_{n_k}]^T$, and a desired response d_k, a special input used to effect learning. The components of the input vector are weighted by a set of coefficients, the weight vector $\mathbf{W}_k = [w_{0_k}, w_{1_k}, w_{2_k}, \ldots, w_{n_k}]^T$. The sum of the weighted inputs is then computed, producing a linear output, the inner product $s_k = \mathbf{X}_k^T \mathbf{W}_k$. The components of $\mathbf{X}_k$ may be either continuous analog values or binary values. The weights are essentially continuously variable and can take on negative as well as positive values.

During the training process, input patterns and corresponding desired responses are presented to the linear combiner. An adaptation algorithm automatically adjusts the weights so the output responses to the input patterns will be as close as possible to their respective desired responses. In signal processing applications, the most popular method for adapting the weights is the simple LMS (least mean square) algorithm (Widrow and Hoff, 1960), often called the Widrow-Hoff Delta Rule (Rumelhart, Hinton, and Williams, 1986). This algorithm minimizes the sum of squares of the linear errors over the training set. The linear error ε_k is defined to be the difference between the desired response d_k and the linear output s_k during presentation k. Having this error signal is necessary for adapting the weights. Both the LMS rule and Rosenblatt's perceptron rule will be detailed in later sections.

An important element used in many neural networks is the "ADAptive LInear NEuron," or *adaline* (Widrow and Hoff, 1960). In the neural network literature, such elements are often referred to as *adaptive neurons*. The adaline is an adaptive threshold logic element. It consists of an adaptive linear combiner cascaded with a hard-limiting quantizer which is used to produce a binary ± 1 output, $y_k = \mathrm{sgn}(s_k)$. A bias weight, or *threshold*, w_{0_k}, which is connected to a constant input, $x_0 = +1$, effectively controls the threshold level of the quantizer. Such an element may be seen as a McCulloch-Pitts neuron augmented with a learning rule for adjusting its weights.

In single-element neural networks, the weights are often trained to classify binary patterns using binary desired responses. Once training is complete, the responses of the trained element can be tested by applying various input patterns. If the adaline responds correctly with high probability to input patterns that were not included in the training set, it is said that generalization has taken place. Learning and generalization are among the most useful attributes of adalines and neural networks.

With n binary inputs and one binary output, a single adaline is capable of implementing certain logic functions. There are 2^n possible input patterns. A general logic implementation would be capable of classifying each pattern as either $+1$ or -1, in accordance with the desired response. Thus, there are 2^{2^n} possible logic functions connecting n inputs to a single binary output. A single adaline is capable of realizing only a small subset of these functions, known as the linearly separable logic functions or threshold logic functions. These are the set of logic functions that can be obtained with all possible weight variations. With two inputs, a single adaline can realize 14 of the 16 possible binary logic functions. The two it cannot learn are exclusive OR and exclusive NOR functions. With many inputs, however, only a small fraction of all possible logic functions are realizable, i.e., linearly separable. Combinations of elements or networks of elements can be used to realize functions which are not linearly separable.

Nonlinear Neural Networks

One of the earliest trainable layered neural networks with multiple adaptive elements was the *Madaline I* structure of Widrow and Hoff. In the early 1960s, a 1000-weight Madaline I was built out of hardware and used in pattern recognition research (Widrow and Lehr, 1990). The weights in this machine were memistors, electrically variable resistors developed by Widrow and Hoff which are adjusted by electroplating a resistive link in a sealed cell containing copper sulfate and sulfuric acid.

Madaline I was configured in the following way. Retinal inputs were connected to a layer of adaptive adaline elements, the outputs of which were connected to a fixed logic device that generated the system output. Methods for adapting such systems were developed at that time. An example of this kind of network is shown in Figure 1. Two adalines are connected to an AND logic device to provide an output. With weights suitably chosen, the separating boundary in pattern space for the system can implement any of the 16 two-input binary logic functions, including the exclusive OR and exclusive NOR functions.

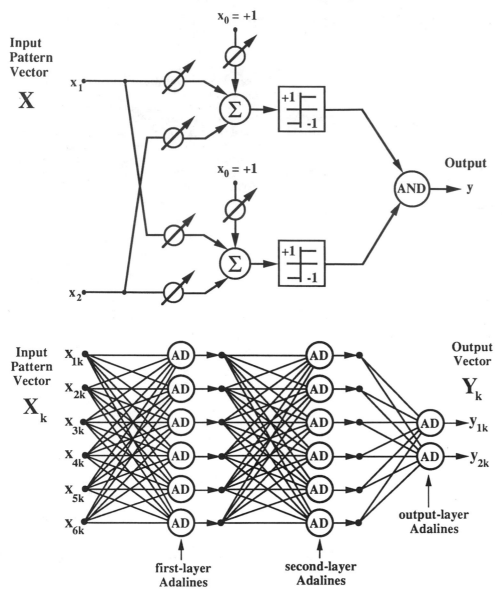

Figure 1. A two-adaline form of madaline.

Figure 2. A three-layer adaptive neural network.

Madalines were constructed with many more inputs, with many more adaline elements in the first layer, and with various fixed logic devices such as AND, OR, and majority vote-taker elements in the second layer. Those three functions are all threshold logic functions.

Multilayer Networks

The madaline networks of the 1960s had an adaptive first layer and a fixed threshold function in the second (output) layer (Widrow and Lehr, 1990). The feedforward neural networks of today often have many layers, all of which are usually adaptive. The backpropagation networks of Rumelhart et al. (1986) are perhaps the best-known examples of multilayer networks. A three-layer feedforward adaptive network is illustrated in Figure 2. It is "fully connected" in the sense that each adaline receives inputs from every output in the preceding layer.

During training, the responses of the output elements in the network are compared with a corresponding set of desired responses. Error signals associated with the elements of the output layer are thus readily computed, so adaptation of the output layer is straightforward. The fundamental difficulty associated with adapting a layered network lies in obtaining *error signals* for hidden layer adalines, that is, for adalines in layers other than the output layer. The backpropagation algorithm provides a method for establishing these error signals.

Learning Algorithms

The iterative algorithms described here are all designed in accord with the *Principle of Minimal Disturbance: Adapt to reduce the output error for the current training pattern, with minimal disturbance to responses already learned*. Unless this principle is practiced, it is difficult to simultaneously store the required pattern responses. The minimal disturbance principle is intuitive. It was the motivating idea that led to the discovery of the LMS algorithm and the madaline rules. In fact, the LMS algorithm had existed for several months as an error reduction rule before it was discovered that the algorithm uses an instantaneous gradient to follow the path of steepest descent and minimizes the mean square error of the training set. It was then given the name LMS (least mean square) algorithm.

The LMS Algorithm

The objective of adaptation for a feedforward neural network is usually to reduce the error between the desired response and the network's actual response. The most common error function is the mean square error (MSE), averaged over the training set. The most popular approaches to mean-square-error reduction in both single-element and multielement networks are based on the method of gradient descent.

Adaptation of a network by gradient descent starts with an arbitrary initial value $\mathbf{W}_0$ for the system's weight vector. The gradient of the mean-square-error function is measured and the weight vector is altered in the direction opposite to the measured gradient. This procedure is repeated, causing the MSE to be successively reduced on average and causing the weight vector to approach a locally optimal value.

The method of gradient descent can be described by the relation

$$\mathbf{W}_{k+1} = \mathbf{W}_k + \mu(-\nabla_k) \tag{1}$$

where μ is a parameter that controls stability and rate of convergence and ∇_k is the value of the gradient at a point on the MSE surface corresponding to $\mathbf{W} = \mathbf{W}_k$.

The LMS algorithm works by performing approximate steepest descent on the mean-square-error surface in weight space. This surface is a quadratic function of the weights and is therefore convex and has a unique (global) minimum. An instantaneous gradient based on the square of the instantaneous error is

$$\hat{\nabla}_k = \frac{\partial \varepsilon_k^2}{\partial \mathbf{W}_k} = \left\{ \begin{array}{c} \dfrac{\partial \varepsilon_k^2}{\partial w_{0k}} \\ \vdots \\ \dfrac{\partial \varepsilon_k^2}{\partial w_{nk}} \end{array} \right\} \tag{2}$$

LMS works by using this crude gradient estimate in place of the true gradient ∇_k. Making this replacement into Equation 1 yields

$$\mathbf{W}_{k+1} = \mathbf{W}_k = \mu(-\hat{\nabla}_k) = \mathbf{W}_k - \mu \frac{\partial \varepsilon_k^2}{\partial \mathbf{W}_k} \tag{3}$$

The instantaneous gradient is used because (a) it is an unbiased estimate of the true gradient (Widrow and Stearns, 1985), and (b) it is easily computed from single data samples. The true gradient is generally difficult to obtain. Computing it would involve averaging the instantaneous gradients associated with all patterns in the training set. This is usually impractical and almost always inefficient.

The present error or *linear* error ε_k is defined to be the difference between the desired response d_k and the linear output $s_k = \mathbf{W}_k^T \mathbf{X}_k$ before adaptation:

$$\varepsilon_k \triangleq d_k - \mathbf{W}_k^T \mathbf{X}_k \tag{4}$$

Performing the differentiation in Equation 3 and replacing the linear error by the definition in Equation 4 gives

$$\mathbf{W}_{k+1} = \mathbf{W}_k - 2\mu\varepsilon_k \frac{\partial(d_k - \mathbf{W}_k^T \mathbf{X}_k)}{\partial \mathbf{W}_k} \tag{5}$$

Noting that d_k and $\mathbf{X}_k$ are independent of $\mathbf{W}_k$ yields

$$\mathbf{W}_{k+1} = \mathbf{W}_k + 2\mu\varepsilon_k \mathbf{X}_k \tag{6}$$

This is the LMS algorithm. The learning constant μ determines stability and convergence rate (Widrow and Stearns, 1985).

The Perceptron Learning Rule

The Rosenblatt α-perceptron (Rosenblatt, 1962), diagrammed in Figure 3, processed input patterns with a first layer of sparse, randomly connected, fixed-logic devices. The outputs of the fixed first layer fed a second layer which consisted of a single adaptive linear threshold element. Other than the convention that its input signals and its output signal were $\{1, 0\}$ binary, and that no bias weight was included, this element was equivalent to the adaline element. The learning rule for the α-perceptron was very similar to LMS, but its behavior was in fact quite different.

Adapting with the perceptron rule makes use of the *quantizer error* $\tilde{\varepsilon}_k$, defined to be the difference between the desired re-

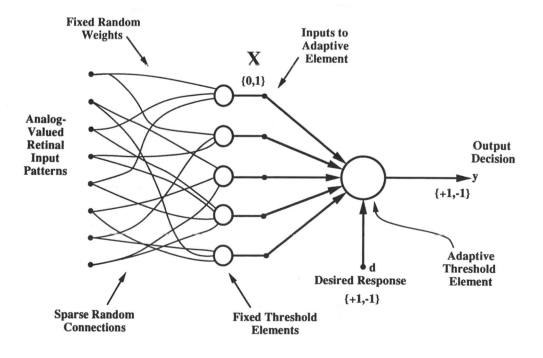

Figure 3. Rosenblatt's α-perceptron.

sponse and the output of the quantizer

$$\tilde{\tilde{\varepsilon}}_k \triangleq d_k - y_k \tag{7}$$

The perceptron rule, sometimes called the *perceptron convergence procedure*, does not adapt the weights if the output decision y_k is correct, i.e., if $\tilde{\tilde{\varepsilon}}_k = 0$. If the output decision disagrees with the binary desired response d_k, however, adaptation is effected by adding the input vector to the weight vector when the error $\tilde{\tilde{\varepsilon}}_k$ is positive, or subtracting the input vector from the weight vector when the error $\tilde{\tilde{\varepsilon}}_k$ is negative. Note that the quantizer error $\tilde{\tilde{\varepsilon}}_k$ is always equal to either $+1$, -1, or 0. Thus, the product of the input vector and the quantizer error $\tilde{\tilde{\varepsilon}}_k$ is added to the weight vector. The perceptron rule is identical to the LMS algorithm, except that with the perceptron rule, one-half of the quantizer error, $\tilde{\tilde{\varepsilon}}_k/4$, is used in place of the linear error ε_k of the LMS rule. The perceptron rule is nonlinear, in contrast to the LMS rule, which is linear. Nonetheless, it can be written in a form which is very similar to the LMS rule of Equation 6:

$$\mathbf{W}_{k+1} = \mathbf{W}_k + 2\mu \frac{\tilde{\tilde{\varepsilon}}_k}{2} \mathbf{X}_k \tag{8}$$

Rosenblatt normally set μ to one. In contrast to LMS, the choice of μ does not affect the stability of the perceptron algorithm, and it affects convergence time only if the initial weight vector is non-zero. Also, while LMS can be used with either analog or binary desired responses, Rosenblatt's rule can be used only with binary desired responses.

The perceptron rule stops adapting when the training patterns are correctly separated. There is no restraining force controlling the magnitude of the weights, however. The direction of the weight vector, not its magnitude, determines the decision function. The perceptron rule has been proven capable of separating any linearly separable set of training patterns (Rosenblatt, 1962; Nilsson, 1965). If the training patterns are not linearly separable, the perceptron algorithm goes on forever, and in most cases the weight vector gravitates toward zero. As a result, on problems which are not linearly separable, the perceptron often does not yield a low-error solution, even if one exists.

This behavior is very different from that of the LMS algorithm. Continued use of LMS does not lead to an unreasonable weight solution if the pattern set is not linearly separable. Nor, however, is this algorithm guaranteed to separate any linearly separable pattern set. LMS typically comes close to achieving such separation, but its objective is different, i.e., error reduction at the linear output of the adaptive element.

"Backpropagation" for the Sigmoid Adaline

A *sigmoid adaline* element incorporates a sigmoidal nonlinearity. The input-output relation of the sigmoid can be denoted by $y_k = \text{sgm}(s_k)$. A typical sigmoid function is the hyperbolic tangent

$$y_k = \tanh(s_k) = \left(\frac{1 - e^{-2s_k}}{1 + e^{-2s_k}} \right). \tag{9}$$

We shall adapt this adaline with the objective of minimizing the mean square of the *sigmoid error* $\tilde{\varepsilon}_k$, defined as

$$\tilde{\varepsilon}_k \triangleq d_k - y_k = d_k - \text{sgm}(s_k) \tag{10}$$

The method of gradient descent is used to adapt the weight vector. By following the same line of reasoning used to develop LMS, the instantaneous gradient estimate obtained during presentation of the kth input vector $\mathbf{X}_k$ can be found to be

$$\hat{\nabla}_k = \frac{\partial(\tilde{\varepsilon}_k)^2}{\partial \mathbf{W}_k} = 2\tilde{\varepsilon}_k \frac{\partial \tilde{\varepsilon}_k}{\partial \mathbf{W}_k} = -2\tilde{\varepsilon}_k \text{sgm}'(s_k)\mathbf{X}_k \tag{11}$$

Using this gradient estimate with the method of gradient descent provides a means for minimizing the mean square error even after the summed signal s_k goes through the nonlinear sigmoid. The algorithm is

$$\mathbf{W}_{k+1} = \mathbf{W}_k + \mu(-\hat{\nabla}_k) = \mathbf{W}_k + 2\mu\delta_k\mathbf{X}_k \tag{12}$$

where δ_k denotes $\tilde{\varepsilon}_k \text{sgm}'(s_k)$. The algorithm of Equation 12 is the *backpropagation* algorithm for the single adaline element, though the backpropagation name only makes sense when the algorithm is utilized in a layered network, which will be studied later.

If the sigmoid is chosen to be the hyperbolic tangent function (Equation 9), then the derivative $\text{sgm}'(s_k)$ is given by

$$\text{sgm}'(s_k) = \frac{\partial(\tanh(s_k))}{\partial s_k}$$
$$= 1 - (\tanh(s_k))^2 = 1 - y_k^2 \tag{13}$$

Accordingly, Equation 12 becomes

$$\mathbf{W}_{k+1} = \mathbf{W}_k + 2\mu\tilde{\varepsilon}_k(1 - y_k^2)\mathbf{X}_k \tag{14}$$

The single sigmoid adaline trained by backpropagation shares some advantages with both the adaline trained by LMS and the perceptron trained by Rosenblatt's perceptron rule. If a pattern set is linearly separable, the objective function of the sigmoid adaline, the mean square error, is minimized only when the pattern set is separated. This is because, as the weights of the sigmoid adaline grow large, its response approximates that of a perceptron with weights in the same direction. The sigmoid adaline trained by backpropagation however, also shares the advantage of the adaline trained by LMS: it tends to give reasonable results even if the training set is not separable.

Backpropagation training of the sigmoid adaline does have one drawback, however. Unlike the linear error of the adaline, the output error of the sigmoid adaline is a nonlinear function of the weights. As a result, its mean square error surface is not quadratic, and may have local minima in addition to the optimal solution. Thus, unlike the perceptron rule, it cannot be guaranteed that backpropagation training of the sigmoid adaline will successfully separate a linearly separable training set. Nonetheless, the single sigmoid adaline performs admirably in many filtering and pattern classification applications. Its most important role, however, occurs in multilayer networks, to which we now turn.

Backpropagation for Networks

The backpropagation technique is a substantial generalization of the single sigmoid adaline case discussed in the previous section. When applied to multilayer feedforward networks, the backpropagation technique adjusts the weights in the direction opposite to the instantaneous gradient of the sum square error in weight space. Derivations of the algorithm are widely available in the literature (Rumelhart, Hinton, and Williams, 1986; Widrow and Lehr, 1990). Here we provide only a brief summary of the result.

The instantaneous sum square error ε_k^2 is the sum of the squares of the errors at each of the N_y outputs of the network. Thus

$$\varepsilon_k^2 = \sum_{i=1}^{N_y} \varepsilon_{ik}^2 \tag{15}$$

In its simplest form, backpropagation training begins by presenting an input pattern vector $\mathbf{X}$ to the network, sweeping forward through the system to generate an output response vector **Y**, and computing the errors at each output. We continue by sweeping the effects of the errors backward through the network to associate a *square error derivative* δ with each adaline, computing a gradient from each δ, and finally updating the weights of each adaline based on the corresponding gradient. A new pattern is then presented and the process is repeated. The initial weight values are normally set to small random values. The algorithm will not work properly with multilayer networks if the initial weights are either zero or poorly chosen non-zero values.

The δ's in the output layer are computed just as they are for the sigmoid adaline element. For a given output adaline,

$$\delta = \tilde{\varepsilon}\, \text{sgm}'(s) \qquad (16)$$

where $\tilde{\varepsilon}$ is the error at the output of the adaline and s is the summing junction output of the same unit.

Hidden layer calculations, however, are more complicated. The procedure for finding the value of $\delta^{(l)}$, the value of δ associated with a given adaline in hidden layer l, involves respectively multiplying each derivative $\delta^{(l+1)}$ associated with each element in the layer immediately downstream from the given adaline by the weight connecting it to the given adaline. These weighted square error derivatives are then added together, producing an error term $\varepsilon^{(l)}$, which in turn is multiplied by $\text{sgm}'(s^{(l)})$, the derivative of the given adaline's sigmoid function at its current operating point. Thus, the δ corresponding to adaline j in hidden layer l is given by

$$\delta_j^{(l)} = \text{sgm}'(s_j^{(l)}) \sum_{i \in N^{(l+1)}} \delta_i^{(l+1)} w_{ij}^{(l+1)} \qquad (17)$$

where $N^{(l+1)}$ is a set containing the indices of all adalines in layer $l+1$ and $w_{ij}^{(l+1)}$ is the weight connecting adaline i in layer $l+1$ to the output of adaline j in layer l.

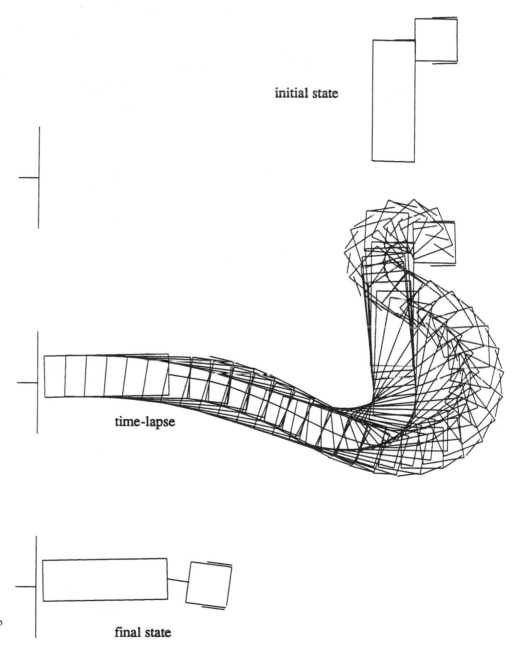

initial state

time-lapse

final state

Figure 4. Example of a truck backup sequence.

Updating the weights of the adaline element using the method of gradient descent with the instantaneous gradient is a process represented by

$$\mathbf{W}_{k+1} = \mathbf{W}_k + \mu(-\hat{\nabla}_k) = \mathbf{W}_k + 2\mu\delta_k\mathbf{X}_k \qquad (18)$$

where $\mathbf{W}$ is the adaline's weight vector and $\mathbf{X}$ is the vector of inputs to the adaline. Thus, after backpropagating all square error derivatives, we complete a backpropagation iteration by adding to each weight vector the corresponding input vector scaled by the associated square error derivative. Equations 16, 17, and 18 comprise the general weight update rule of the backpropagation algorithm for layered neural networks.

Many useful techniques based on the backpropagation algorithm have been developed. One popular method, called *backpropagation through time*, allows dynamical recurrent networks to be trained. Essentially, this is accomplished by running the recurrent neural network for several time steps and then "unrolling" the network in time. This results in a virtual network with a number of layers equal to the product of the original number of layers and the number of time steps. The ordinary backpropagation algorithm is then applied to this virtual network and the result is used to update the weights of the original network. This approach was used by Nguyen and Widrow (1989) to enable a neural network to learn without a teacher how to back up a computer-simulated trailer truck to a loading dock (Figure 4). This is a complicated and highly nonlinear steering task. Nevertheless, with just six inputs providing information about the current position of the truck, a two-layer neural network with only 26 sigmoid adalines was able to learn of its own accord to solve this problem. Once trained, the network could successfully back up the truck from any initial position and orientation in front of the loading dock.

Discussion

Although this article has focused on pattern classification issues, nonlinear neural networks are equally useful for such tasks as interpolation, system modeling, state estimation, adaptive filtering, and nonlinear control. Unlike their linear counterparts which have a long track record of success, nonlinear neural networks have only recently begun proving themselves in commercial applications. The capabilities of multielement neural networks have improved markedly since the introduction of Madaline Rule I. This has resulted largely from development of the backpropagation algorithm, easily the most useful and popular neural network training algorithm currently available. As we have seen, backpropagation is a generalization of LMS which allows complex networks of sigmoid adalines to be efficiently adapted. Backpropagation and related algorithms are in a large part responsible for the dramatic growth the field of neural networks is currently experiencing.

The timing of the current boom in the field of neural networks is also due to the rapid advance of computer and microprocessor performance which continues to improve the feasibility and cost-effectiveness of computationally expensive techniques in relation to classical approaches of engineering and statistics. Although single-element linear adaptive filters are still used more extensively than nonlinear multielement neural networks, the latter are potentially applicable to a much wider range of problems. Furthermore, the applications for which multielement neural networks are best suited often involve complicated nonlinear relationships for which classical solutions are either ineffective or unavailable. The continued advancement of neural network algorithms and techniques, in conjunction with improvements in the special and general purpose computer hardware used to implement them, sets the stage for a future in which neural networks will play an increasing role in commercial and industrial applications.

Acknowledgments.   This work was sponsored by NSF under grant IRI 91–12531, by ONR under contract N00014–92–J–1787, by EPRI under contract RP 8010–13, and by the U.S. Army under contract DAAK70–92–K–0003.

Road Map: Learning in Artificial Neural Networks, Deterministic
Background: I.3. Dynamics and Adaptation in Neural Networks
Related Reading: Adaptive Control: Neural Network Applications; Adaptive Filtering; Learning as Hill-Climbing in Weight Space

References

Nilsson, N., 1965, *Learning Machines*, New York: McGraw-Hill. ◆

Nguyen, D., and Widrow, B., 1989, The truck backer-upper: An example of self-learning in neural networks, in *Proceedings of the International Joint Conference on Neural Networks*, vol. 2, New York: IEEE, pp. 357–363.

Rumelhart, D. E., Hinton, G. E., and Williams, R. J., 1986, Learning internal representations by error propagation, in *Parallel Distributed Processing: Explorations in the Microstructure of Cognition* (D. E. Rumelhart, J. L. McClelland, and PDP Research Group, Eds.), vol. 1, *Foundations*, Cambridge, MA: MIT Press, chap. 8. ◆

Rosenblatt, F., 1962, *Principles of Neurodynamics: Perceptrons and the Theory of Brain Mechanisms*, Washington, DC: Spartan.

Widrow, B., and Hoff, M. E., Jr., 1960, Adaptive switching circuits, in *1960 IRE WESCON Convention Record*, Part 4, New York: IRE, pp. 96–104.

Widrow, B., and Lehr, M. A., 1990, 30 years of adaptive neural networks: Perceptron, madaline, and backpropagation, *Proc. IEEE*, 78: 1415–1442. ◆

Widrow, B., Rumelhart, D., and Lehr, M. A., 1994, Neural networks: Applications in industry, business, and science, *Commun. ACM*, 37(3):93–105.

Widrow, B., and Stearns, S. D., 1985, *Adaptive Signal Processing*, Englewood Cliffs, NJ: Prentice-Hall. ◆

Perceptual Grouping

Steven W. Zucker

Introduction

While gazing at the skies, the ancients saw more than just stars; they grouped the stars into clusters and identified the clusters with objects both real and mythological. This illustrates that vision is a process by which descriptions of the world are inferred from images and, as such, indicates that such inferences must encompass two distinct functions: (i) the agglomeration of "local" structures into more "global" ones; and (ii) the abstraction of these agglomerated structures into a single, encompassing one.

While star gazing illustrates both of these functions, more concrete examples abound. An example of the first type is the process of curve inference, in which a collection of (local) "edge elements" are inferred from an image and then grouped together to form a (global) contour. This type of grouping is illustrated by a process of "walking" along the edge elements, and turning only when it is necessary. Thus, edge element agglomeration involves a form of curve continuation.

Another example of grouping agglomerates more complex objects—e.g., two curves instead of two local edge elements—into a single, more abstract one. This occurs when curves are interpreted as "bounding contours" of an object, and provides the basis for separating a "figure" from "background."

These examples indicate the role that perceptual grouping plays in vision, a role that is bounded from below by edge detection and from above by object recognition. The bounds further evoke the different tensions shaping grouping processes. From below is the tension introduced by computational efficiency: since the number of elements in an image is so enormous, vision systems, both biological and artificial, strive toward parallel processing by decomposing complex global computations into networks of local, parallel ones. Grouping coordinates these local processes and puts the local results together in a bottom-up, or data-directed, fashion, as in curve detection. The tension from above derives from object recognition. Because objects in the world are often obscured by projection into images, identification may depend on the proper interpretation of a few fragments. Grouping links these fragments together, but in a manner that now is model driven (or top down). Thus, grouping is intimately related to the segmentation problem, and its role is to bind elements in a manner such that, as the Gestalt psychologists put it: *The whole is greater than the sum of its parts* (Koffka, 1935).

Perceptual grouping thus fills the gulf between early vision (edge detection) and high-level vision (object recognition). This gulf is clearly enormous, which suggests that it is not filled by a single process. In this article, we survey the range of processes that have been implicated in perceptual grouping, as well as the different functional roles attributed to them. In the end, we show that perceptual grouping involves inferences over different types of objects, ranging from point sets indicating (projected) locations of interest on solid objects through curvelike objects to surface coverings.

Gestalt Principles of Perceptual Grouping

Stars in the sky are one example of the perception of dot patterns on a homogeneous ground; a figure is readily seen, and not just a sum of dots. This elementary but insightful observation was made by Wertheimer in 1923. It shows immediately that there must be certain principles of organization; just consider all of the mathematical possibilities for forming clusters of size $2, \ldots, (n-1)$, within a field of n dots. All other aspects of the stimulus being equal, Wertheimer claimed that those elements of a pattern are grouped together that are most proximal spatially; most proximal temporally; most similar geometrically; part of the most continuous pattern; part of the most closed pattern; arranged in uniform density; evolving with common fate; most symmetric; exhibiting a common orientation; or optimizing an intuitive measure of "figural goodness" or *pragnanz*.

The difficulty in reading this sentence illustrates the difficulty in operationalizing the Gestalt principles. How should the "elements" be determined, and what criterion provides the ordering behind "most" (see also Kanisza, 1979; Kubovy and Pomerantz, 1981)? Since groups can be formed of groups, it is usually assumed that grouping is a distinct binding process that is recursively applied to more abstract entities. Marr (1982), e.g., postulated grouping to be those processes that write from the "primal sketch" (an explicit representation of the "important information about the two-dimensional image, primarily the intensity changes and their geometrical organization") back into it. In physiology, there is a view that grouping is accomplished by the synchronous firing of neurons (Singer, 1990; see SYNCHRONIZATION OF NEURONAL RESPONSES AS A PUTATIVE BINDING MECHANISM). While this recursive view of grouping is attractive in the simplicity of its abstraction and the uniformity of its implementation, unfortunately, it does not elaborate the Gestalt principles. It is from the details of the different grouping processes that functionality emerges.

Operationalizing the Gestalt Principles

We illustrate the subtlety in building a computational theory of the Gestalt principles with the task of inferring a curve from an image. This shows how spatial proximity, common fate, and common orientation can all interact in a form of orientation good continuation.

Imagine an arrangement of dots in the plane. Grouping the dots into a curve involves determining which ones should be linked together. The children's game of "connect the dots" comes with numbers indicating the order; perceptual grouping does not. Clearly the task would be simplified if pairings were given, e.g., as a vector that points from one dot to its neighbor. Such a vector defines an "orientation" for each dot, which can be viewed as the tangent to the curve that passes through the dots. Inferring such tangents, and then finding an integral curve that passes through them, is precisely how the problem is approached computationally (Figure 1). Two distinct stages arise, in which local grouping problems are distinct from global ones. In particular, estimates of orientation formed purely locally are noisy and incomplete (Figure 1*B*) and can be refined by networks that require nearby tangents to be consistent according to curvature (Figure 1*C*). Curvature arises naturally in this case, just as steering corrections are necessary when driving along a road. Neurobiologically, tangents can be represented by orientation-selective cells, and curvature by the end-stopped property (Dobbins, Zucker, and Cynader, 1987). Consistency in responses can coordinate activities between cells in adjacent hypercolumns, leading to a form of synchronicity in firing. The global requirements are different, however, because spatial proximity may not be available to determine topological con-

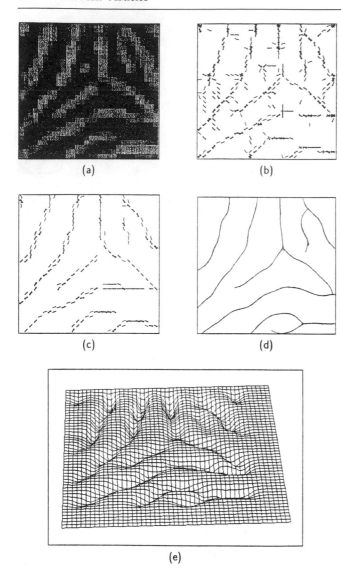

(a)

(b)

(c)

(d)

(e)

Figure 1. Illustration of an algorithm for curve detection. Part *A* shows an image of a portion of a fingerprint, rich in curves. The algorithm consists of two stages. Local representation: initial estimates of orientation (part *B*), followed by refined estimates based on a form of orientation good continuation (part *C*). Global representation: a potential distribution of possible curve positions (part *E*), the most likely of which are shown in part *D*; see Zucker et al. (1989).

nectivity; for discussion, see Zucker, Dobbins, and Iverson (1989). Such schemes may not work for sparse patterns; see the following sections.

The Functional Roles of Perceptual Grouping

The process of curve inferencing illustrates the abstraction of increasingly more global elements (local orientations, curvatures, then global curves) from an image. Differential geometry (Koenderink, 1990) provided the model. Other types of elements can also emerge, leading to a notion of *figure*. The mathematical notion of dimensionality classifies the types:

Zero-dimensional configurations are point clusters that indicate either (i) sparse background noise, e.g., the uniform distributions of stars in the sky, or rocks in a field; or (ii) key locations on an object, e.g., the stars indicating Orion

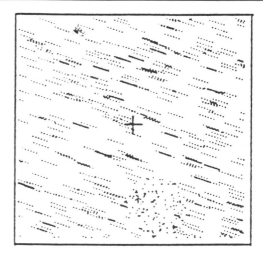

Figure 2. Examples of two-dimensional groupings: an oriented texture flow pattern with a patch of unoriented texture field inserted.

or the extremal points of high-curvature and corners in a line drawing.

One-dimensional configurations are clusters that organize into curves, e.g., the grouping of points into a line (Orion's belt) or the grouping of edge elements into the bounding contour of an object (see Figure 1).

Two-dimensional configurations are clusters that organize into surface coverings (Figure 2), e.g., hair, fur, and parallel pinstripes on a shirt; regular patterns of dots or other elements, as in a honeycomb or wallpaper. Such surface coverings project into regions in images, and differential geometry again plays a role, both with the curvilinear organization along flow patterns (functionally exhibiting growth) and with gradients (functionally exhibiting depth relationships; see Gibson, 1950).

Edge fragments group into long contours, and surface markings group into textures. These represent surfaces and their boundaries in many cases. Building on this, once representations of contours and textures become available, their emergent properties can be used to support further groupings with still more abstract functional interpretations. For example, *endpoints* are an important emergent property from curves, because they can indicate an occlusion; this is especially compelling when endpoints further align along another curve (the occluder's boundary) (Kanisza, 1979). Such emergent organizations provide a segmentation of the image that is related to objects in the scene; geometrically they can be interpreted as intersurface relationships.

Contours and textures are interrelated. Texture discontinuities arise when a surface covered in one texture occludes another, and this discontinuity contour constrains the possible surface shape (Koenderink, 1990). In general, contours define intersurface relationships, while textures define intrasurface properties. It should therefore be possible to combine different cues to make a richer whole, in the same way that different glances must be combined to form a full percept of a large object; Witkin and Tenenbaum (1983) review early computer vision algorithms.

As contour fragments align into closed boundaries around regions of texture, the notion of figure as being separate from ground emerges (Elder and Zucker, 1993; see FIGURE-GROUND SEPARATION). The Gestalt psychologists again identified intuitive principles, such as figures forming from smaller (enclosed) rather than larger (enclosing) portions of the scene; and again they are difficult to operationalize. One useful concept is non-accidentalness (Witkin and Tenenbaum, 1983): e.g., random points thrown onto a plane are unlikely to lie along a straight line; two contours are unlikely to be symmetric but probabilistically independent (Lowe, 1985); and 50 moving dots are unlikely to follow paths consistent with their lying on a rigid object. At a more structural level, consider the bounding contour of a dog behind a tree; "matched-Ts," or aligned, T-shaped contour fragments, indicate occlusion boundaries of figures, an observation that has been used in computer vision programs (Binford, 1981).

The final class of patterns is perhaps most typical of those indicating constellations from the Introduction. This class consists in those arrangements of elements whose positions are determined by locational features on the object rather than by specific geometric constraints. An example might be as simple as three points indicating the vertices of a triangle, or as complex as one of the stellar constellations.

The triangle example indicates the importance of boundary discontinuities as key points to trigger an object percept; Attneave (1954) stressed the importance of points of high curvature, and Koenderink (1990) reviews other geometric points along contours and on surfaces. Johansson (1975) constructed "moving light displays" by attaching point sources to joints on people, and found that, although the points looked random when the subject was stationary, their coherent movement was rapidly recognizable; even sex could be determined. Another example are the trigger features for the Kanizsa subjective triangle (Figure 3; see ILLUSORY CONTOUR FORMATION).

We refer to such patterns as zero-dimensional; there is little structure to capture them from within the image alone. Rather,

they are relatively isolated patterns of points that take on their structure from semantics. These abilities may well be exercising the tremendous top-down influences in vision. Perhaps this ability derives from the need to recognize an object that is highly obscured; e.g., a predator hiding behind a tree.

Discussion

The diversity of perceptual grouping was illustrated with a sequence of examples, ranging from the earliest forms (grouping edge elements into curves) through interactions between curves and textures to emergent features. The result was an intuitive Gestalt concept of figures separating from background and ultimately supporting object recognition. Mathematical dimension classified the elements, and different mechanisms are needed for different tasks. The physiology of early grouping derives from an embedding of differential-geometric constructs into the visual cortex, but higher levels of grouping are elusive.

Acknowledgments. This work was supported by NSERC, AFOSR, the Newton Institute for Mathematical Sciences, University of Cambridge, and the Canadian Institute for Advanced Research.

Road Map: Vision
Related Reading: Perception of Three-Dimensional Structure; Visual Cortex Cell Types and Connections

References

Attneave, F., 1954, Informational aspects of visual perception, *Psych. Rev.*, 61:183–193.
Binford, T., 1981, Inferring surfaces from images, *Artif. Intell.*, 17:205–244.
Dobbins, A., Zucker, S. W., and Cynader, M., 1987, Endstopped neurons in the visual cortex as a substrate for calculating curvature, *Nature*, 329:438–441.
Elder, J., and Zucker, S. W., 1993, Contour closure and the perception of shape, *Vis. Res.*, 33:981–991.
Gibson, J., 1950, *The Perception of the Physical World*, Boston: Houghton Mifflin. ◆
Johansson, G., 1975, Visual motion perception, *Sci. Am.*, 232:76–88. ◆
Kanisza, G., 1979, *Organization in Vision*, New York: Praeger.
Koenderink, J., 1990, *Solid Shape*, Cambridge, MA: MIT Press. ◆
Koffka, K., 1935, *Principles of Gestalt Psychology*, New York: Harcourt Brace Jovanovich. ◆
Kubovy, M., and Pomerantz, J. R., Eds., 1981, *Perceptual Organization*, Hillsdale, NJ: Erlbaum.
Lowe, D., 1985, *Perceptual Organization and Visual Recognition*, Norwell, MA: Kluwer.
Marr, D., 1982, *Vision*, San Francisco: Freeman.
Singer, W., 1990, The formation of cooperative cell assemblies in the visual cortex, *J. Exp. Biol.*, 153:177–197.
Wertheimer, M., 1923, Laws of organization in perceptual forms, *Psych. Forsch.*, 4:301–350; trans. in Ellis, W., 1938, *A Source Book of Gestalt Psychology*, London: Routledge and Kegan Paul, pp. 71–88.
Witkin, A., and Tenenbaum, J. M., 1983, On the role of structure in vision, in *Human and Machine Vision* (J. Beck, B. Hope, and A. Rosenfeld, Eds.), New York: Academic Press, pp. 481–544.
Zucker, S. W., Dobbins, A., and Iverson, L., 1989, Two stages of curve detection suggest two styles of visual computation, *Neural Computat.*, 1:68–81.

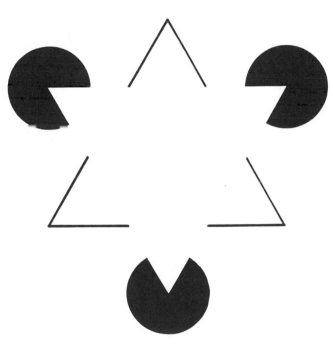

Figure 3. Illustration of the emergence of figure from background. A Kanizsa (1975) subjective triangle illustrates the completion of a figure from separate, localized trigger elements.

Perspective on Neuron Model Complexity

Wilfrid Rall

Introduction

There is a wide range of choice in model complexity, from very simple to rather complex neuron models. Which model to choose depends, in each case, on the context. How much information do we already have about the neurons under consideration? What questions do we wish to explore?

Sometimes we wish to model a particular biological neuron whose anatomy and physiology are known in considerable experimental detail. In such cases, we may choose to specify a model that includes at least some of the dendritic branching of the neuron, because synapses from one source may be distributed preferentially to either a distal or a proximal dendritic location, while synapses from another source may end mainly at the soma, or on a different dendritic tree of the same neuron. Also, there may be a functionally significant nonuniformity in the distribution of channel densities of several ion channel types over the surface of the soma and dendrites. How much detail is needed depends on the biological experiments to be simulated and the questions asked.

Conversely, many network modelers are not constrained by anatomical or physiological data. For some network modeling, this is partly justified by a paucity of available data. However, more often, network modelers are constrained by their mathematical methods, which lead them to focus on abstract networks composed of extremely simple units. The simplest units are two-state, binary units, analagous to atomic spin, previously studied for condensed matter physics (see, e.g., NEURAL OPTIMIZATION). Such binary units do not resemble neurons, but they do have a strong appeal for nerve-net modelers, who have generated an extensive literature. That literature lies outside the scope of the current article.

When simple binary units are compared with a dendritic neuron model (especially with nonuniform distributions of synapses and ion channels), it becomes apparent that one dendritic model neuron can perform tasks that would require a network of many simple units to duplicate. For the purpose of machine design, it may seem quite appropriate to consider the tradeoffs in cost and flexibility (between the one realistic model and the many binary units), but for functional insights and understanding of biological nervous systems, I freely state my bias for the more realistic neuron models. I do not choose the most complex, in the sense of including all known anatomical and physiological details; I favor an intermediate level of complexity, which preserves the most significant distinctions between regions (soma, proximal dendritic, distal dendritic, different trees) especially when further justified by nonuniform distributions of synapses and ion channels; see also Segev (1992).

The claim is sometimes made that network properties depend primarily on the connectivity between the units, and not on the properties of the units. While this may be true for some gross network properties, I do not believe this to be true for many of the actual biological networks which perform important, complicated tasks. I regard it as a worthwhile challenge for like-minded neural modelers to provide interesting demonstrations in support of this belief. The challenge is to demonstrate a useful computation or discrimination that can be accomplished with a dendritic neuron model, or a network composed of such models, and then show that this useful capacity is lost when all of the dendritic membrane is lumped with the soma, and all of the inputs to each neuron are now delivered to that lumped membrane. There are valuable examples that already meet this challenge; several of these are presented in three later sections of this article; other examples can be found in a recent review by Borst and Egelhaaf (1994; see also VISUOMOTOR COORDINATION IN FLIES).

Brief Historical Notes

Neurons are biological cells, and their electrical properties depend on ions and the cell membrane, in a manner brilliantly elucidated by Hodgkin, Huxley, and Katz during the period 1948–1952. It is a fascinating historical coincidence that two seeds of their important insights can be found in a single 1902 volume of *Pfluegers Archiv*, in pioneering articles by Bernstein and by Overton. Following the earlier theoretical insights of Nernst and Planck, Bernstein recognized the importance of the potassium ion concentration difference across the membrane in determining a non-zero resting potential; he regarded excitation as a brief breakdown of the membrane, a concept that prevailed until 1948, when Hodgkin and Katz showed that the key is a sudden overwhelming increase in membrane permeability to sodium ions. Overton's 1902 paper had correctly emphasized the importance of the external sodium ion concentration to the excitability properties of nerve, but no one put these ideas together in 1902. Between 1900 and 1914, several investigators, including Hermann, Lucas, and Lapique, recognized the importance of membrane capacitance; the concept of nerve membrane as a leaky integrator, with a threshold for an action potential, was used to understand the strength-duration curve for a threshold stimulus. During the 1930s, several investigators, including Rashevsky, Hill, and Monnier, developed mathematical models of excitation and inhibition; Rashevsky's textbook *Mathematical Biophysics* (1948) includes many examples of network modeling by himself, by Householder, Landahl, and others, and by McCulloch and Pitts, whose famous 1943 paper arose in the context of Rashevsky's research seminars at the University of Chicago; see also historical notes in Schwartz (1990). Ever since that time, many neuron modelers have been content with the leaky integrator neuron model which reduces a neuron to a single node which integrates synaptic excitation (+) and synaptic inhibition (−) delivered to it by other neurons. Several errors caused by these oversimple assumptions were demonstrated by compartmental computations in 1962; see Rall's chapter in Reiss (1964) or in Segev, Rinzel, and Shepherd (1995). Other chapters in Reiss (1964) also provide several interesting early perspectives on neural modeling. The mathematical modeling of nonlinear membrane properties has been presented and discussed in an outstanding early review by FitzHugh (1969), and in a chapter by Rinzel and Ermentrout that appears in Koch and Segev (1989).

The concept of a nerve axon as an extended core conductor (i.e., membrane cylinder with ionic conducting media inside and outside) goes back to the 1870s, when it was treated mathematically by Hermann and Weber; both the concept of passive electrotonus in membrane cylinders and the mathematics (of passive cable theory) were explored over the years, culminating in classic papers by Hodgkin and Rushton, and by Davis and Lorente de Nó, both around 1946–1947; see references in Rall (1977). Before 1900, neuroanatomical studies by Ramón y Cajal demonstrated the extensiveness of dendritic branching for most neuron types; this was confirmed by many anatomists, and later (in the 1950s), use of the electron microscope made it possible to verify the existence of very many synapses on the dendritic branches and on the dendritic spines of neurons. These anatomical facts, together with the introduction of intracellulalar microelectrode recording from single dendritic neurons (in the 1950s), made it urgent to extend cable theory to the dendrites of individual neurons. This was begun in the late 1950s and carried forward into the 1960s and 1970s; for review, see Jack, Noble, and Tsien (1975) or Rall (1977); see also Koch and Segev (1989), McKenna, Davis, and Zornetzer (1992), Rall et al. (1992), Segev et al. (1995), and DENDRITIC PROCESSING.

Dendritic Neuron Model Complexity: Geometric Versus Membrane Complexity

The concept of complexity in dendritic neuron models can be explored quite efficiently by making a two-dimensional chart. One dimension would be membrane complexity, ranging from the simple case of a passive linear membrane to that of postsynaptic membrane models with time-varying ion permeability (or conductance), and then to excitable membrane models with voltage-dependent ion conductances as described by Hodgkin and Huxley (see AXONAL MODELING), or as now described with increasing detail in terms of many different species of ion channels whose voltage and time dependence are currently being characterized (see ION CHANNELS: KEYS TO NEURONAL SPECIALIZATION). The other dimension would be geometric complexity, ranging from the simple case of an isopotential region of membrane (a soma, or a space-clamped section of a cylinder) to that of a uniform membrane cylinder with two sealed ends (or with one end voltage clamped, or current clamped), and then to several dendritic trees attached to a soma (with or without an axon), where the soma may be shunted and the branching of the trees may be specified to varying degrees of arbitrariness. The most complicated geometric case, with arbitrary branching and shunted soma, has recently been solved analytically (for transients, assuming passive membrane) in a mathematical tour de force by Major, Evans, and Jack (1993); see also Holmes, Segev, and Rall (1992). The less complicated, but illuminating, case of idealized branching with a point soma was solved earlier by Rall and Rinzel; see the 1973 and 1974 papers reprinted in Segev et al. (1995). However, these analytical methods do depend on the assumption of linear membrane properties. When nonlinear membrane complexity is combined with geometric complexity, the transient solutions can be obtained computationally by using compartmental models; see 1964 and 1968 papers reprinted in Segev et al. (1995); see also DENDRITIC PROCESSING and several chapters in Koch and Segev (1989) and in McKenna et al. (1992).

Dendritic Model Can Provide Spatiotemporal Discrimination

Figure 1 summarizes a demonstration of how a dendritic neuron model could perform a discrimination between two contrasting spatiotemporal patterns of synaptic input (i.e., possible movement detection); this discrimination is lost if the compartments and inputs are lumped together. A neuron is represented by a chain of 10 compartments; compartment 1 represents the soma, while compartments 2 to 10 represent dendritic membrane of the same neuron, with increasing cable distance from the soma. One spatiotemporal input sequence, A-B-C-D, has the proximal dendritic input first, followed in time by progressively more distal dendritic input locations. The other input pattern, namely, D-C-B-A, is opposite in having the most distal input first, followed in time by progressively more proximal input locations. Comparison of the resulting computed voltage transients (EPSP at the soma), shown in Figure 1, reveals that

input sequence D-C-B-A yields a significantly larger voltage amplitude than does input sequence A-B-C-D. Intuitive understanding of this computed result is obtained by noting that the delayed proximal input builds on membrane depolarization which has spread to the soma (with delay) from the distal dendrites (that were activated earlier). If the voltage threshold for spiking at the soma were tuned between these two peak amplitudes, a spike would be produced by sequence D-C-B-A, but a spike would not be produced by sequence A-B-C-D; this would constitute a discrimination between these two sequences. The dashed curve in the figure shows the computed result when the compartments are lumped together; either sequence of input synapse activation then produces the same intermediate result, and no discrimination would be possible.

Models for Mitral and Granule Cell Populations in Olfactory Bulb

A rather different example is provided by the neuron models used for the mitral cell and granule cell populations in simulating experiments on the OLFACTORY BULB (q.v.) of rabbit; see the 1968 paper of Rall and Shepherd, in Segev et al. (1995); or see figures 2.11 and 2.12 in Koch and Segev (1989). Here, the task was to model and compute extracellular field potentials that matched those observed experimentally in olfactory bulb when the mitral cell population was activated in near synchrony by means of an antidromic volley. Compartmental models were used; a nine-compartment model (three axonal, one somatic, and five dendritic) was used to simulate antidromic activation of a mitral cell, while a 10-compartment model was used to simulate nonspiking activity in the dendrites of an axonless granule cell. The dendritic compartments were absolutely essential for the computation of electric current flow between different dendritic regions of each granule cell and between the dendrites and soma of each mitral cell; without these currents, it would have been impossible to compute the field potentials generated by the synchronously activated neuron populations. Also, this modeling led to a critically important distinction in the depth distribution of the two fields: the larger, longer-lasting field potentials generated by the very large population of granule cells extended to significantly greater depth in the olfactory bulb than did the earlier, smaller, briefer field potentials generated by the mitral cell population. The difference between these two fields was such that neither population could have generated the other field. This provided the key to our prediction of (and the functional interpretation of subsequent electron microscopic evidence for) dendrodendritic synaptic interactions between the mitral secondary dendrites and the distal dendrites of the granule cells, which are intermingled in the external plexiform layer of the bulb. If these cells had been modeled as lumped somas, without dendrites, neither the successful simulation of the experimental field potentials nor the exciting new insights about a dendrodendritic pathway for recurrent inhibition would have been possible.

Similarly, for the earlier simulations and insights obtained for motoneurons of cat spinal cord, we found that observations made at the soma seemed paradoxical until they were understood in terms of synaptic events that occur in distal dendrites; see the 1967 paper in Segev et al. (1995); these results and insights would not have been possible without dendritic compartments in the neuron model.

Comment on Functional Aspect of Dendrodendritic Interactions

To highlight an important functional difference, note first that motoneurons do exhibit the classical functional polarity envis-

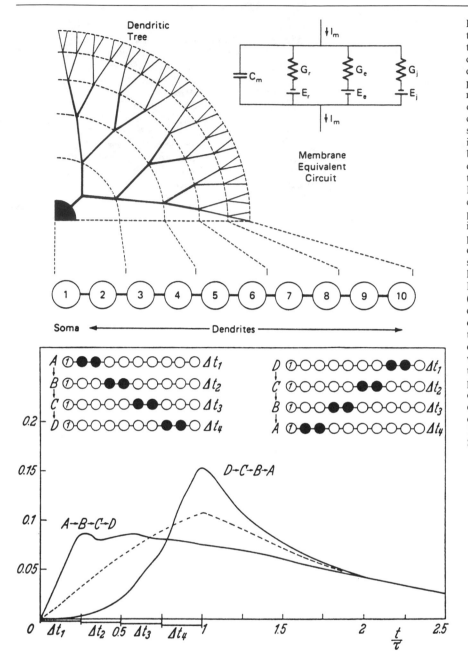

Figure 1. Effect of spatiotemporal dendritic pattern of synaptic input on the computed EPSP at the soma, for a 10-compartment model. Upper diagram indicates the mapping of a soma and dendritic tree into a chain of 10 equal compartments. Compartment 1 represents the soma membrane, while compartments 2 to 10 represent dendritic membrane, from proximal to distal locations. The middle diagram (at left) shows the synaptic input sequence A-B-C-D, meaning proximal dendritic input location active first, followed by successive activation at increasingly more distal input locations; this input pattern produced the soma voltage transient (computed composite EPSP) labeled A-B-C-D at lower left. The middle diagram (at right) shows the opposite synaptic input sequence, D-C-B-A, meaning distal dendritic input location first, followed by successively more proximal input locations; this input pattern produced a significantly different soma voltage transient (computed composite EPSP), having a delayed rise to a larger peak amplitude, labeled D-C-B-A. In both cases, each input compartment (shown in black) received a synaptic excitatory conductance pulse ($G_\varepsilon = G_r$, for a duration, 0.25τ) during one of the four labeled periods. The same total amount of synaptic input produced the dashed curve when the spatiotemporal pattern was eliminated by smearing the synaptic conductance in space and time ($G_\varepsilon = 0.25G_r$, in eight compartments (compartments 2 to 9) for the full time duration from $t = 0$ to $t = \tau$. The membrane equivalent circuit (upper right) holds for each compartment. Further details can be found in the 1964 chapter by Rall in Reiss (1964), reprinted in Segev et al. (1995).

aged by Ramón y Cajal and Sherrington (as well as most modelers). The dendrites receive inputs from many sources (their effects converge on the soma); the output (when spike threshold is exceeded) is an all-or-nothing action potential propagated by the axon to muscle units which may be quite distant; i.e., classically, input is received by the dendrites, and the output is delivered by the axon. In contrast, the dendrites of both the mitral cells and the granule cells are functionally different, because they both send as well as receive synaptic information, locally. The mitral secondary dendrites, which are smooth and spineless, send nonspiking synaptic excitatory output which is received as input by the spines (see DENDRITIC SPINES) of the adjacent granule cells. The granule cells have no axons and perhaps no action potentials; their spines receive graded synaptic excitatory input and then send graded synaptic output which is inhibitory to the adjacent mitral cell dendrites.

It is important to emphasize that this is not a rare anomaly found only in the olfactory bulb; evidence for dendrodendritic synapses and for graded local synaptic interactions is now found in many parts of the brain (e.g., retina and inferior olive). In 1965, when we (Rall, Shepherd, Reese, and Brightman; see 1966 and 1968 papers reprinted in Segev et al., 1995) first presented our interpretations of dendrites that send as well as receive, some critics resisted this concept as heretical; however, our functional interpretation of these dendrodendritic synapses is now widely accepted by physiologists and anatomists. This kind of graded two-way synaptic interaction is very different from the classical functional polarity just described for motoneurons; it provides graded functional coupling between neurons (without axonal impulses). The implications have so far hardly been explored in theoretical networks. Such exploration will require explicit modeling of dendritic compartments; a point neuron model would be useless for this. Note also that computational exploration of localized plastic changes at synapses and at dendritic spines depends on neuron models that include dendritic compartments.

Network Rhythmogenesis Using the Traub Model and a Reduced Model

A 19-compartment cable model for the pyramidal cells of the CA3 region of guinea pig hippocampus was developed by Traub et al. (1991); see also the chapter by Traub and Miles in McKenna et al. (1992). Based on experimental measurements, parameters were chosen for each compartment, using up to six active ionic conductances, and controlled by 10-channel gating variables. They succeeded in finding a set of physiologically reasonable parameters for which the network of model neurons could simulate several important aspects of the experimental repertoire of the slightly disinhibited hippocampal slice preparation. Although Traub et al. recognized that their successful simulations of network behavior depended on specifying significantly different ion channel densities for the soma and for the dendrites, the critical importance of this difference was made starkly clear by the modeling of Pinsky and Rinzel (1994); they obtained essentially the same behavioral repertoire by using a network composed of a severely reduced neuron model consisting of only two compartments per pyramidal cell. One compartment represented the soma and proximal dendrites, while the other compartment represented the distal dendrites. To be more specific, the ion channels for fast-spiking currents (inward sodium, and delayed rectifier) were restricted to the soma-like compartment, and the ionic channels for the slower calcium currents (calcium-inward and calcium-modulated currents) were restricted to the dendrite-like compartment. I hasten to add that these results also show that at least two compartments are needed for simulations of this behavior; a single lumped compartment, with all of the ion channels in parallel, could not produce the same behavior, especially the rhythm which basically involves an alternating flow of current between the two coupled compartments. A special advantage of the reduced neuron model is that much simpler computations can explore how much the interesting behavior depends on the values of key parameters, especially the parameter which defines the tightness of coupling between the two compartments. Also, the behavior of very large networks can be explored more efficiently using such a reduced neuron model. Further study may show that the two-compartment model cannot match the fuller model in certain important tests, but, in any case, these findings so far represent a very satisfying example that illustrates the thesis of this article.

Discussion

In an earlier essay offering perspective on neural modeling (a chapter in Binder and Mendell, 1990), I provided a completely different set of examples. One of these provided a detailed consideration of the number of degrees of freedom to be found in a neuron model composed of a thousand compartments. Such models exist today because of tremendous improvements in anatomical methods and in computation facilities now available to experimental investigators. Because they have the morphological data and a computer, why not put everything into the model? The answer is that you can if you wish to, but you should be aware of the huge number of degrees of freedom implied by the large number of parameters that must be specified; as someone once pointed out: given enough free parameters, he could fit an elephant. Is the membrane uniform, or do we know the density of every channel species in every membrane compartment? How are the inputs distributed to the many compartments? Today, the data needed for such detailed specifications are largely missing; however, such data are beginning to become at least partly available for some neurons. Where the data are not available, the modeler must make reasonable guesses. If it seems reasonable to assign the same parameter values to many neighboring compartments, one should consider lumping those compartments together to produce a simpler model with fewer compartments. Nevertheless, one important merit of the larger model is that it can be used to test whether it can perform some interesting task that cannot be performed by the reduced model.

As stated earlier, my preference is for intermediate levels of complexity; I vote for the smallest number of compartments that can preserve what one judges to be the functionally important differences between dendritic regions with regard to ion channel densities and to distributions of synapses from different sources. If a five-compartment model can provide a good approximation of the interesting properties of a 1000-compartment model, I would prefer the smaller model for two important reasons: (i) it helps sharpen our intuitive understanding about what is essential to obtaining the behavior of interest, and (ii) it can greatly facilitate computations with networks composed of such neuron models. I expect modeling of this kind will continue to be particularly fruitful in the near future; see also discussion by Segev (1992).

Concluding Comment

As when drawing, painting, sculpting, or composing music, so too, when deeply engaged in neural modeling, I believe that much of the fun and satisfaction comes from interactions between my conscious mind and my subconscious sources of creativity. It seems that preliminary sketching serves to plant seeds in the subconscious, where they can grow, if nurtured. Conscious pursuit of the problem can then stimulate differentiation and development in the subconscious and may produce fruits that can reach conscious awareness (popping up like mushrooms produced by an underground mycelium). Such fruits may provide exciting new insights for the conscious mind. Indeed, the pleasure of such creative discovery can become almost addictive for those fortunate enough to have both the interest and the opportunity for creative activity. I hasten to add that a lot of hard work is usually required to test and polish before one can produce a finished product. Pioneering in dendritic neuron modeling provided me with such an opportunity; now, with retirement upon me, I hope to persist by sculpting, painting, and by designing a house for a natural, mountain setting.

Road Map: Biological Neurons
Background: I.1. Introducing the Neuron

References

Binder, M. D., and Mendell, L. M., 1990, *The Segmental Motor System*, Oxford: Oxford University Press.
Borst, A., and Egelhaaf, M., 1994, Dendritic processing of synaptic information by sensory interneurons, *Trends Neurosci.*, 17:257–263.
FitzHugh, R., 1969, Mathematical models of excitation and propagation in nerve, in *Biological Engineering* (H. P. Schwann, Ed.), New York: McGraw-Hill. ◆
Holmes, W. R., Segev, I., and Rall, W., 1992, Interpretation of time constant and electrotonic length estimates in multi-cylinder or branched neuronal structures, *J. Neurophysiol.*, 68:1401–1420.

Jack, J. J. B., Noble, D., and Tsien, R. W., 1975, *Electric Current Flow in Excitable Cells*, Oxford: Oxford University Press. ◆

Koch, C., and Segev, I., 1989, *Methods in Neuronal Modeling: From Synapses to Networks*, Cambridge, MA: MIT Press. ◆

Major, G., Evans, J. D., and Jack, J. J. B., 1993, Solutions for transients in arbitrarily branching cables: I, Voltage recording with a somatic shunt, *Biophys. J.*, 65:423–449.

McKenna, T., Davis, J., and Zornetzer, S. F., 1992, *Single Neuron Computation*, San Diego, CA: Academic Press. ◆

Pinsky, P. F., and Rinzel, J., 1994, Intrinsic and network rhythmogenesis in a reduced Traub model for CA3 neurons, *J. Computat. Neurosci.*, 1:39–60.

Rall, W., 1977, Core conductor theory and cable properties of neurons, in *Handbook of Physiology: The Nervous System: Cellular Biology of Neurons*, sect. 1, vol. I, part 1, chap. 3, Bethesda, MD: American Physiological Society, pp. 39–97. ◆

Rall, W., Burke, R. E., Holmes, W. R., Jack, J. J. B., Redman, S. J.,

and Segev, I., 1992, Matching dendritic neuron models to experimental data, *Physiol. Rev.*, 72:S159–S186. ◆

Rashevsky, N., 1948, *Mathematical Biophysics*, Chicago: University of Chicago Press; reissued, 1960, New York: Dover.

Reiss, R., Ed., 1964, *Neural Theory and Modeling*, Stanford, CA: Stanford University Press.

Schwartz, E. L., 1990, *Computational Neuroscience*, Cambridge, MA: MIT Press. ◆

Segev, I., 1992, Single neurone models: Oversimple, complex and reduced, *Trends Neurosci.*, 15:414–421. ◆

Segev, I., Rinzel, J., and Shepherd, G. M., Eds., 1995, *The Theoretical Foundation of Dendritic Function: Selected Papers of Wilfrid Rall with Commentaries*, Cambridge, MA: MIT Press. ◆

Traub, R., Wong, R., Miles, R., and Michelson, H., 1991, A model of a CA3 hippocampal pyramidal neuron incorporating voltage-clamp data on intrinsic conductances. *J. Neurophysiol.*, 66:635–649.

Phase-Plane Analysis of Neural Activity

Bard Ermentrout

Introduction

Models of neural networks often involve the solutions to differential equations which govern the time evolution of these complex systems. The dynamical behavior of these networks ranges from the convergence to an equilibrium (generally desired in connectionist applications) to oscillatory behavior (in models of central pattern generators and bursting) through possibly chaotic behavior (see COMPUTING WITH ATTRACTORS). There are many ways to analyze these models; the most used techniques employ simulation. In this article I will give an overview of an alternative technique for studying the *qualitative* behavior of small systems of interacting neural networks. One form that the models take is

$$\tau_i \frac{dx_i}{dt} = -x_i + f_i\left(\sum_{j=1}^{n} w_{ij}x_j + s_i\right) \qquad i = 1, \ldots, n \qquad (1)$$

where x_i represents the activity or firing rate of the ith neuron, τ_i is the time constant, w_{ij} are the connection weights, s_i are inputs, and f_i are typically saturating nonlinear functions that have the form shown in Figure 1. That is, the nonlinear functions are increasing and bounded. Some typical examples are

$$f(x) = \tanh(x) \qquad (2)$$

$$f(x) = \tan^{-1}(x) \qquad (3)$$

$$f(x) = \frac{1}{1 + \exp(-x)} \qquad (4)$$

Often, a slightly different form of Equation 1 is chosen where the nonlinearities are inside the sums. The tranformation from one to the other is elementary, and all of the following holds for either type of model.

A complete analysis of networks of the form given in Equation 1 is obviously impossible. However, if $n = 2$, then a fairly complete description of Equation 1 can be given. Thus, the goal of this article is to introduce the reader to the qualitative theory of differential equations in the plane. In particular, I will analyze two neuron networks that consist of (1) two excitatory cells, (2) two inhibitory cells, and (3) an excitatory and inhibitory cell. The advantages of restricting the analysis to these

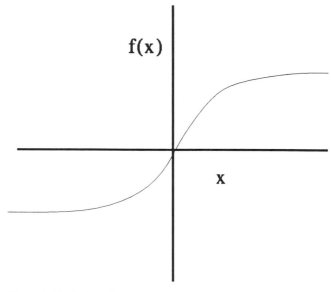

Figure 1. Typical nonlinear input-output function of a single model neuron.

small networks are the special topology of the plane, the completeness of the analysis possible, and finally the ease of exposition. Indeed, an overview of nonlinear dynamics can be obtained through these simple examples.

The approach of this article is not restricted to neural networks and can be applied to a variety of other systems such as positive-feedback biochemical models (Segel, 1984), activator-inhibitor systems (Edelstein-Keshet, 1988), hormonal models (Sherman and Rinzel, 1989), population and disease models (Murray, 1989), and membrane models of the action potential (Rinzel and Ermentrout, 1989; Morris and Lecar, 1981). The techniques are powerful and provide insights into the behavior of these systems that would otherwise only be accessible through simulation. Computational methods are a very powerful adjunct to this type of analysis, and together with the qualitative analysis of this article, they enable the researcher to understand his or her system.

In the following section, I will describe a pair of neurons coupled with mutual inhibition and mutual excitation. The penultimate section is devoted to a summary of the rich behavior of an excitatory-inhibitory pair. Finally, some comments on numerical methods and software close the article.

Two Coupled Cells of the Same Type

In this section, we analyze the behavior of two cells that act by means of mutual inhibition or mutual excitation. I will use phase-plane analysis to draw a complete picture of the phase space.

General Considerations

Before analyzing the two-component neural network, I will first give a brief description of phase-plane techniques in general. Consider a planar differential equation:

$$x' = f(x, y) \qquad (5)$$

$$y' = g(x, y) \qquad (6)$$

At each point (x_0, y_0) there is a solution $(x(t), y(t))$ such that $(x(0), y(0)) = (x_0, y_0)$ and such that the tangent to the trajectory is $(f(x_0, y_0), g(x_0, y_0))$. Thus, at any point in the plane, one can easily determine the direction of the trajectory by simply evaluating f and g at that point. These directions enable one to paint a qualitative picture of the dynamics of Equations 5 and 6; that is, we can determine where x and y are increasing and decreasing with time. The most crucial points are those values of x and y at which the direction of the trajectory changes. Thus, setting $f(x, y) = 0$ defines a curve in the plane where x does not change and breaks the plane into regions where x is either increasing or decreasing. This curve is called the x-nullcline. The curve $g(x, y) = 0$ defines the y-nullcline. The two curves together break the plane into regions of four distinct types: (1) x and y are increasing, (2) x and y are decreasing, (3) x increases and y decreases, and (4) x decreases and y increases. The intersection of the two nullclines occurs at points where both x and y are not changing—that is, equilibria or rest states of Equations 5 and 6.

The behavior of trajectories away from equilibria is straightforward and is found by simply looking at the signs of f and g. Near the equilibria, one can look at the linearization of (f, g) about the equilibrium. This search results in a two-by-two matrix called the Jacobian:

$$A = \begin{pmatrix} a & b \\ c & d \end{pmatrix} \qquad (7)$$

The eigenvalues of A determine the behavior of the equilibria. If they both have negative real parts, the equilibrium is stable, and if any have positive real parts, the equilibrium is unstable. If both are real and of the same sign, the point is called a node. Nodes consist of infinitely many trajectories emanating from (unstable) or entering (stable) the equilibrium. If both eigenvalues are complex, the rest state is a vortex, and trajectories spiral into (stable) or out of (unstable) the rest point. If the eigenvalues have opposite signs, the rest state is a saddle point. Then a single pair of trajectories, stable manifold or set, enter the rest point, and a pair of trajectories, the unstable manifold, leave the equilibrium. When the determinant of A is negative, the rest point is a saddle; if it is positive and the trace is nonzero, the equilibrium is a node or vortex. Cases for which the

real part is zero do not persist for small changes in the parameters and often indicate the appearance of new equilibria or periodic solutions. A simple necessary and sufficient criterion for linear stability is that the trace $a + d$ be negative and the determinant of A, $ad - bc$, be positive. A complete description of phase-plane methods can be found in Edelstein-Keshet (1988), as well as most texts on ordinary differential equations.

Crossed Excitatory and Inhibitory Networks

The first result I want to establish in systems that have mutual coupling of the same sign is that periodic solutions are impossible. Once this is established, then a complete characterization can be made by simply studying the intersections of the nullclines.

Theorem 1. Suppose that $w_{21} w_{12} \geq 0$. Then there are no periodic solutions to

$$\tau_1 x_1' = -x_1 + f(w_{11} x_1 + w_{12} x_2 + s_1) \qquad (8)$$

$$\tau_2 x_2' = -x_2 + f(w_{21} x_1 + w_{22} x_2 + s_2) \qquad (9)$$

Since this theorem does not appear in the literature, I will prove it.

Proof. First, note that if either $w_{12} = 0$ or $w_{21} = 0$, then the network is decoupled and so no periodic solutions exist. Thus, I will assume that both are non-zero. If a periodic solution exists, then there must be a time t_0 such that $x_1'(t_0) = 0$ and $x_2'(t_0) \neq 0$. Suppose that $x_2'(t_0) > 0$. The opposite case follows in the same manner. Next, suppose that w_{12} and w_{21} are positive. Differentiating Equation 8 with respect to t and evaluating at t_0, one obtains

$$\tau_1 x_1'' = f'(w_{11} x_1 + w_{12} x_2 + s_1) w_{12} x_2' > 0$$

so that x_1' will be positive for t slightly larger than t_0. Now both x_1' and x_2' are positive. If at some later time one of these derivatives vanishes, then the preceding argument shows that at slightly later times, the derivatives are again positive. Thus, neither derivative can ever change sign, and so there are no periodic solutions. If w_{21} and w_{12} are negative, then a similar argument shows that x_1' and x_2' have opposite signs and can never switch signs for all time. Thus, no periodic solutions exist. □

All solutions to Equations 8 and 9 are bounded, and Theorem 1 implies that trajectories are monotone, and therefore all solutions tend to equilibria. This fact in turn means that the time constants can be set to 1 without loss of generality as the dynamics is completely trivial. The intersections of the nullclines and some observations on the signs of the coefficients in the linearized matrix based on the nullclines allow one to completely determine the number and stability type of the equilibria.

Recall that f is increasing and bounded. Without loss of generality, one can assume the minimum of f is 0 and the maximum is 1. The function f is invertible, and the inverse is also monotone with asymptotes at 0 and 1. The formula for the x_1-nullcline is

$$x_2 = (-w_{11} x_1 - s_1 + f^{-1}(x_1))/w_{12} \qquad (10)$$

The x_2-nullcline satisfies

$$x_1 = (-w_{22} x_2 - s_2 + f^{-1}(x_2))/w_{21} \qquad (11)$$

If $h(x) = (-w_s x - s + f^{-1}(x))/w_c$, then h is monotone if w_s is either positive or small and negative. However, if w_s is large

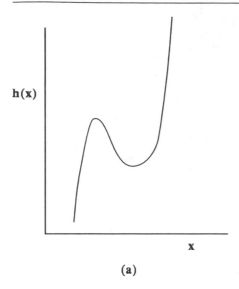

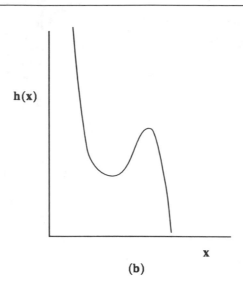

Figure 2. Nullcline shape for (A) mutual excitation and (B) mutual inhibition.

enough, h develops a kink and is "cubic" shaped (Figure 2). If w_c is positive (mutual excitation), then $h \to -\infty$ as $x \to 0$ and $h \to \infty$ as $x \to 1$ (Figure 2A.) When w_c is negative (mutual inhibition), the asymptotes are switched (Figure 2B.) Finally, the stimulus parameter s merely translates the nullclines up and down in the case of the x_1-nullcline and left-right for the x_2-nullcline. The phase-plane is easy to construct with these observations.

In both cases, there can be up to nine different equilibria, and there is always at least one. Figure 3 shows some typical configurations for mutually inhibitory interactions. To assess the stability of the equilibria, one need only look at the positions of the nullclines at the equilibria. Referring to Equation 7, I will use the nullclines to determine the signs and relative magnitudes of the entries in this matrix. For mutually inhibiting cells, the following are necessary and sufficient for stability:

1. The slope of both nullclines is negative through an equilibrium.
2. The slope of the x_1-nullcline is steeper than the x_2-nullcline through the equilibrium.

If either of these is violated, the equilibrium is unstable.

For mutually excitatory cells, the conditions for stability are these:

1. The slope of both nullclines is positive through an equilibrium.
2. The slope of the x_1-nullcline is steeper than the x_2-nullcline through the equilibrium.

Tangential intersections are saddle nodes, and as some parameter varies these intersections will lead to either two new equilibria or the disappearance of the pair. The matrix A has a zero eigenvalue when there are tangential crossings, for then the slopes of the nullclines are the same. That is, $-a/b = -c/d$, so $ad - bc = 0$. A bit of counting shows that when there are nine equilibria, four are stable nodes, four saddles, and one unstable node. As the parameters vary, a pair of equilibria is lost, including a saddle point and either a stable node or the unstable node, leaving seven equilibria. Further losses of equilibria (or gains, up to a maximum of nine) are obtained as the

parameters vary, ending in the minimum of a single globally stable equilibrium.

When there are several stable equilibria, it is important to determine what initial conditions lead to which of the equilibria. The set of all initial data that tend to a particular equilibrium point is called the *basin of attraction* of the equilibrium point. For the present networks, this is very easy to determine geometrically. Figure 3 depicts a network of mutually inhibitory cells that has five equilibria labeled a through e. The preceding discussion allows one to conclude that a, c, and e are stable nodes, and b and d are saddle points. Each saddle point has associated with it one positive eigenvalue and one negative eigenvalue. Corresponding to this negative eigenvalue is the very stable manifold for the saddle point, and it consists of the set of all initial conditions that tend to the equilibrium point as $t \to \infty$. For two-dimensional neural nets, this is a one-dimensional set. I have drawn it for each of the two saddle points in Figure 3 as the dashed lines pointing into b and d. These curves divide the two-dimensional plane into three regions, labeled A, C, and E. All initial data in A tend to equilbrium point a and so on. Thus, although the saddle points are unstable, their stable manifolds provide the boundaries which determine the final states of the network given the initial state. From this description the reader should be able to construct complete qualitative pictures for other configurations of the nullclines for mutually excitatory or inhibitory nets.

Summarizing, I have used phase-plane analysis to show that for a pair of coupled neurons with mutual excitation or inhibition, the only stable solutions are equilibria. The stable manifolds of the saddle-points divide the plane into domains of attraction for each of the stable equilibria. All equilibria are approached monotonically, and there can be up to four stable steady states.

A Pair of Excitatory and Inhibitory Cells

In many regions of cortex and in fact throughout the central nervous system, many of the coupled excitatory and inhibitory cells comprise a local neural network. These networks have been the subject of numerous mathematical and computational investigations (Ermentrout and Cowan, 1979; Kyriazi and Simons, 1993; Ellias and Grossberg, 1975; Wilson and Cowan, 1972). One can view such systems as either two neurons acting

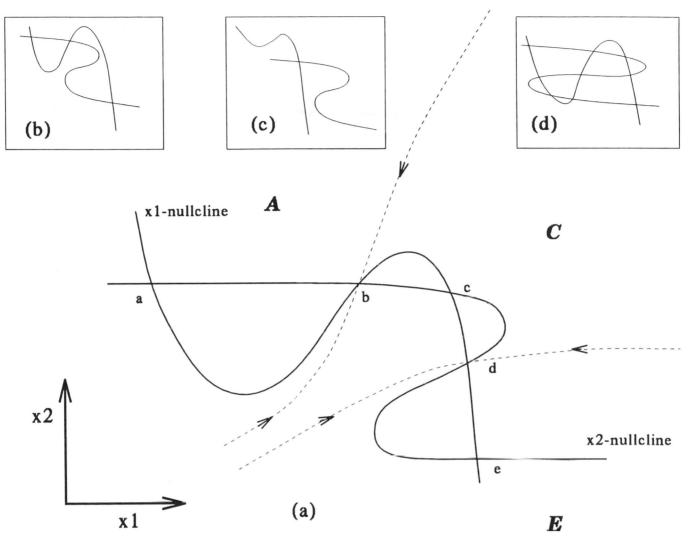

Figure 3. Phase-plane for two mutually inhibitory neurons. Nullclines are solid lines, and the stable manifolds of the saddle points b and d are shown dashed. The points a, c, and e are stable nodes with domains of attraction A, C, and E, respectively. Insets show some other possible nullcline configurations.

in isolation (a difficult experiment to imagine) or more reasonably as a two-layer network with spatially homogeneous activity. Then each component is the activity of a *pool* of cells rather than the activity of a single cell.

I will consider a network of the form

$$x_1' = -x_1 + f(w_{11}x_1 - w_{12}x_2 + s_1) \tag{12}$$

$$x_2' = (-x_2 + f(w_{21}x_1 - w_{22}x_2 + s_2))/\tau \tag{13}$$

where all of the weights are nonnegative. I have introduced a time constant for the inhibitory neurons because one cannot expect them to have the same temporal behavior as the excitatory cells. The Jacobian matrix $A = [\partial x_i'/\partial x_j]$ at an equilibrium point $(\bar{x}_1, \bar{x}_2)$ has coefficients

$$a = -1 + w_{11}f'(w_{11}\bar{x}_1 - w_{12}\bar{x}_2 + s_1) \tag{14}$$

$$b = -w_{12}f'(w_{11}\bar{x}_1 - w_{12}\bar{x}_2 + s_1) < 0 \tag{15}$$

$$c = (w_{21}f'(w_{21}\bar{x}_1 - w_{22}\bar{x}_2 + s_2))/\tau > 0 \tag{16}$$

$$d = (-1 - w_{22}f'(w_{21}\bar{x}_1 - w_{22}\bar{x}_2 + s_2))/\tau < 0 \tag{17}$$

It is clear that all of the coefficients except for a have a fixed sign independent of the parameters. If $w_{11}f' > 1$, then $a > 0$ and the system is called an *activator-inhibitor* system, since x_1 activates both itself and x_2 while x_2 inhibits everthing to which it connects. Activator-inhibitor models occur ubiquitously in biology, and their dynamics is rich and varied. Oscillations, excitability, and multiple steady states are among the possible behaviors of these networks. Since a very complete analysis of these systems as applied to neural excitation is given in Rinzel and Ermentrout (1989), I only sketch some of the dynamic behavior possible for this network.

The qualitative behavior of any planar model can be understood by combining nullcline analysis with local stability analysis of the equilibria which depends on the coefficients of the Jacobian A. The neural model studied in Rinzel and Ermentrout (1989) has exactly the same nullcline structure and has a Jacobian matrix with the same structure as the neural net model.

It is instructive to first consider the effects of parameters on the shapes of the nullclines. A typical nullcline configuration is

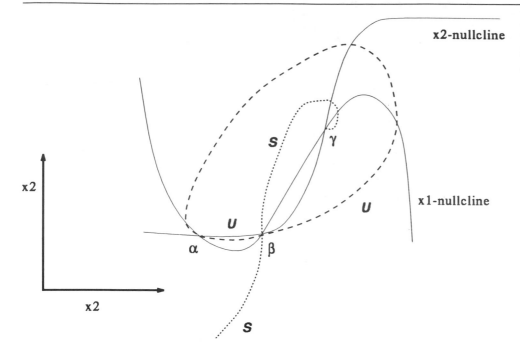

x2-nullcline

x1-nullcline

x2

x2

Figure 4. Typical phase-plane for an excitatory-inhibitory pair. Nullclines and a typical trajectory are shown. There is a unique globally attracting equilibrium point.

shown in Figure 4 for Equations 12 and 13. The x_2-nullcline is always monotonically increasing; w_{21} sharpens it while w_{22} makes it shallower and s_2 shifts it left and right. As described earlier, the effect of w_{11} is to kink the x_1-nullcline while w_{12} makes it less kinked. The variable s_1 shifts it up and down. Finally, the parameter τ has no effect on the nullclines but dramatically alters the dynamics and stability of the equilibria. Changing τ has no effects on the determinant of A (so a saddle point cannot become a node) but can switch the sign of the trace of A and so change a point from a stable node to an unstable node.

The positions of the nullclines make it clear that there can be up to five equilibria and at least one equilibrium point. Furthermore, any equilibria that occur on the "unkinked" part of the x_1-nullcline are necessarily asymptotically stable, since then $a < 0$ in Equation 14. Thus, the trace $a + d < 0$, and the determinant $ad - bc > 0$. If w_{11} is sufficiently small so that the x_1-nullcline is monotone, then there is only one equilibrium point, and it is globally stable. This statement follows from the fact that all solutions are bounded and from an application of Bendixson's negative criterion (Edelstein-Keshet, 1988) which eliminates periodic orbits when $a + d < 0$. Any time the inhibitory nullcline has a lesser slope than the excitatory nullcline, the equilibrium is a saddle point. These considerations along with the preceding discussion show how the parameters affect the local existence and stability of various rest states. The global dynamics is much more complicated, since one cannot eliminate the possibility of limit cycle solutions.

Excitability

One important difference between networks consisting of one excitatory and one inhibitory layer and the networks described under "Two Coupled Cells of the Same Type" is the possibility of excitable dynamics. As was shown, trajectories of the activity of cells are necessarily monotone. Thus if, say, x_1 is increasing, then it can never decrease again. However, in mixed networks, no such restriction occurs, and it is possible for x_1 to initially increase before decreasing again. In particular, a net-

work is said to be *excitable* if there is a *unique globally stable* rest state with the following property. Small perturbations from rest decay monotoncally back to rest, but perturbations larger than some *threshold* continue to grow before decaying back to the stable rest state (Figure 5). There are at least two qualitatively different types of excitability for networks with the present structure. In type I excitability, there are three equilibria, while in type II, there is one. These two cases are described in Rinzel and Ermentrout (1989). In the context of neural networks, this type of behavior has been called an *active transient*. It can be viewed as a transient excitatory activity due to a stimulus that is eventually quelled by the inhibitory interneuronal feedback.

Periodic Solutions

Periodic solutions occur generally (although not strictly) when there is a single rest state on the middle branch and it is unstable. This point must necessarily be a node, and the boundedness of solutions thus implies that a limit cycle exists. If some parameter (say τ) is varied in such a way as to make the unique equilibrium go from a stable point to an unstable point (without introducing any new rest states), then a *Hopf bifurcation* generically occurs, and its existence implies that a periodic solution exists near the rest state. For planar systems, easily checked necessary conditions for a Hopf bifurcation are that the determinant of A remain positive and the trace change from negative to positive as the parameter is varied. If this new limit cycle is unstable, then there can be regimes in parameter space where there are two stable behaviors: (1) a stable rest state and (2) a stable *large-amplitude* periodic solution (Figure 6). This is known as *bistability*.

Other Behavior

In addition to excitability, multiple equilibria, oscillations, and bistability, other types of dynamic behavior can be found in these simple models. Infinite period oscillations and homoclinic trajectories can be obtained in some parameter regimes. (A

Figure 5. Excitable dynamics. The value $\bar{x}_1$ is the globally stable rest state, and x_1^* is the threshold. Trajectory a is subthreshold, and b is superthreshold.

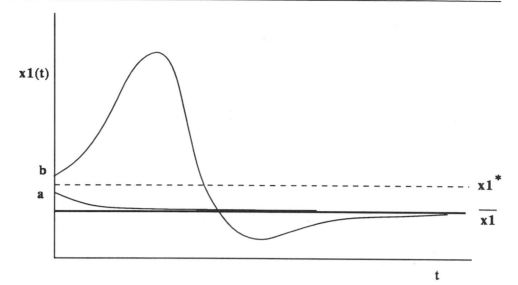

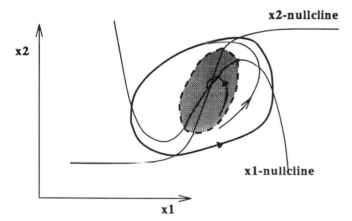

Figure 6. Phase-plane for bistable regime of parameters. Nullclines are shown as well as the stable periodic orbit (dark line), the unstable periodic orbit (dashed line), and representative trajectories (thin lines). The gray area denotes the domain of attraction for the fixed point. The rest of the plane is attracted to the stable periodic orbit.

homoclinic trajectory is one that leaves a saddle point from one side and enters it from another.) Homoclinics are important, since they separate qualitatively different types of behavior. Furthermore, when one periodically stimulates a system with homoclinics, it is possible to obtain complex irregular behavior called *chaos* which is impossible in a simple planar system (see Guckenheimer and Holmes, 1983).

There are many other pictures possible with this simple model, and I urge the reader to explore the phase-plane dynamics of this excitatory inhibitory net. Phase-plane methods provide a powerful analytic and qualitative tool for studying small neural networks. When combined with sophisticated numerical tools, these methods make possible a complete understanding of the global dynamics.

In systems with more than two components, it is difficult to make any general comments on behavior. For symmetrically coupled networks with no self-connections, a complete analysis can be given (Hopfield, 1984; see COMPUTING WITH ATTRACTORS). Weakly coupled systems of intrinsically oscillatory networks can be analyzed with the techniques described in the article CHAINS OF COUPLED OSCILLATORS.

Numerical Methods

Computers are a valuable adjunct in the exploration of systems of differential equations. For this article, I have used a program called PHASEPLANE that is available for both MS-DOS computers and Unix workstations. The latter software, called xpp, is available by anonymous file transfer protocol (FTP) to mthsn4.math.pitt.edu in the /pub subdirectory. PHASEPLANE allows the user to plot trajectories, determine stability, and draw nullclines for systems of differential equations. There are several other packages that do similar tasks, notably PHASER (MS-DOS only, Kocak, 1989), DSTOOL (available from Cornell for Unix), and GRIND (available on many platforms from R. DeBoer at Los Alamos). To get *global* pictures of the dynamics as one or two parameters are varied, a powerful numerical package written by Doedel (1981) called AUTO can be used.

The main method used for integrating the neural net equations such as those in this article is the fourth-order Runge-Kutta algorithm. Standard Newton algorithms are used for computing equilibria. Stability is determined by numerically computing the Jacobian and then using the QR algorithm to compute eigenvalues.

Road Map: Dynamic Systems and Optimization
Related Reading: Pattern Formation, Biological

References

Doedel, E. J., 1981, AUTO: A program for the automatic bifurcation and analysis of autonomous systems, *Cong. Num.*, 30:265–285.

Edelstein-Keshet, L., 1988, *Mathematical Models in Biology*, New York: Random House. ◆

Ellias, S. A., and Grossberg, S., 1975, Pattern formation, contrast control, and oscillations in the short-term memory of shunting on-center off-center surround networks, *Biol. Cybern.*, 20:69–98.

Ermentrout, G. B., and Cowan, J. D., 1979, Temporal oscillations in neuronal nets, *J. Math. Biol.*, 7:265–280.

Guckenheimer, J., and Holmes, P. J., 1983, *Nonlinear Oscillations, Dynamical Systems, and Bifurcations of Vector Fields*, Heidelberg: Springer-Verlag. ◆

Hopfield, J. J., 1984, Neurons with graded responses have collective computational properties like those of two-state neurons, *Proc. Natl. Acad. Sci. USA*, 81:3088–3092.

Kocak, H., 1989, *Differential and Difference Equations Through Computer Experiments*, Heidelberg: Springer-Verlag. ◆

Kyriazi, H., and Simons, D., 1993, Thalamocortical response transformations in simulated whisker barrels, *J. Neurosci.*, 13:1601–1615.

Morris, C., and Lecar, H., 1981, Voltage oscillations in the barnacle giant muscle fiber, *Biophys. J.*, 35:193–213.

Murray, J. D., 1989, *Mathematical Biology*, Heidelberg: Springer-Verlag. ◆

Rinzel, J., and Ermentrout, G. B., 1989, Analysis of neural excitability and oscillations, in *Methods of Neuronal Modeling: From Synapses to Networks* (C. Koch and I. Segev, Eds.), Cambridge, MA: MIT Press. ◆

Segel, L. A., 1984, *Modeling Dynamic Phenomena in Molecular and Cellular Biology*, New York: Cambridge University Press. ◆

Shepherd, G. M., 1990, *The Synaptic Organization of the Brain*, Oxford, Eng.: Oxford University Press. ◆

Sherman, A., and Rinzel, J. M., 1989, Collective properties of insulin-secreting cells, in *Cell to Cell Signaling* (A. Goldbeter, Ed.), New York: Academic Press.

Wilson, H. R., and Cowan, J. D., 1972, Excitatory and inhibitory interactions in localized populations of model neurons, *Biophys. J.*, 12:1–24.

Philosophical Issues in Brain Theory and Connectionism

Andy Clark

Introduction

"What are the philosophical issues raised by such and such a scientific development?" The question is difficult because it is itself a philosophical one. How we conceive the spread of philosophical issues arising out of brain theory and connectionism depends, to a large extent, on how we view the task of philosophy. In this article, I sketch the central issues as they appear to someone committed to viewing the philosopher as an active participant in an interdisciplinary venture. As such, I shall not restrict the philosophical issues to the a priori investigation of what must be true about minds (if anything), nor to the role of connectionism and brain science in illuminating old philosophical questions, nor to the conceptual analysis of new terms and locutions. Instead, I view a major task of the philosopher as trying to achieve a slightly wider view of the shape and significance of developments in a multidisciplinary endeavor and of contributing to that endeavor by seeking to understand how the pieces of the new jigsaw puzzle are supposed to fit together to yield a consistent and unified view of mind and its place in nature. In this spirit, I highlight three questions: (1) Does thought need syntax? (2) What constrains a good model? (3) What is missing so far?

The first question concerns the nature of the connectionist vision of mind: in what way, if any, does such a vision constitute a genuine alternative to the *rule and symbol* vision associated with classical artificial intelligence? The second question concerns the crucial methodological issue of how results emerging from the various brain sciences can help to constrain connectionist models to promote biological relevance. Finally, the third question focuses attention on three major problems which must be solved before the union of connectionism and brain science can really bear fruit. The problems center on the role of innate knowledge, the multiple usability of information, and the difficulties of generalizing solutions which work for small artificial networks to the larger arena of biologically realistic neural resources.

Does Thought Need Syntax?

"Does thought need syntax" is the biggest, most difficult, and least well defined of our three issues. But it gets to the heart of what appears to be the single most philosophically interesting feature of many, but not all, connectionist models. The intuition is that connectionism opens up the possibility of understanding the mind in terms which are divorced from what may be termed the *syntactic image*. Many writers have noted some such possibility (see Churchland, P. M., 1989; Churchland, P. S., and Sejnowski, 1990; Clark, 1989). The difficulty lies in stating exactly what the alleged difference amounts to.

As a first pass, all of the above authors agree that the syntactic image (Clark, 1993) involves depicting all or most of human cognition as involving something like logical operations applied to something like linguistic (sentential) structures. The prime philosophical exponent of the identification of genuine cognitive processes with such operations on quasi-sentential entities is Fodor (see Fodor, 1987; Fodor and Pylyshyn, 1988). Fodor argues in favor of an innate symbolic code (the *Language of Thought*) and of mental processes as involving operations defined over the syntactically structured strings of that code. The underlying image is of an inner economy in which symbol strings are operated on by procedures sensitive to the structure of the string.

On the face of it, many connectionist models are profoundly different from the syntactic-image approach. In a typical trained-up network, we do not find grammatical strings nor, a fortiori, processing operations sensitive to the structure of such strings. Instead, we find knowledge organized around prototypical complexes of properties represented in a high-dimensional space (see Churchland, 1989). The space is highly organized in the sense that data items which need to be treated in closely related ways become encoded in neighboring regions of the space. It is this *semantic metric* (see Clark, 1993) which allows the network to generalize and to respond well under conditions of noise, etc. But this systematic organization of the encoded knowledge does not, on the face of it, amount to the provision of a genuine syntax. One way to see this difference is to ask what rules of combination of represented elements apply, and in what systematic ways we can operate on complexes of represented items. As Fodor and Pylyshyn (1988) point out, we cannot, for instance, simply take the data points in such an encoding and reorganize them into new complex representations at will. In that sense, there is no analog to the logical operations of detaching an element from one string (complex representation) and adding it to another. Instead, the dynamics of connectionist processing look different (see COMPOSITIONALITY IN NEURAL SYSTEMS).

Caution is needed, however, because the ideas of a quasi-linguistic encoding and a quasi-logical operation are inherently vague. It may well be that the systematic organization of information in networks should count as providing some genuine, although unfamiliar, type of syntax. The completely general question "Is there a syntax there or not?" is probably of little ultimate value. We will do better to concentrate on whatever concrete differences we can fix.

To illustrate, one such difference may lie not in whether connectionist systems can support some degree of compositional structuring of representations, but in what kind of structure they support. A variety of connectionist techniques have been developed to allow for structure-sensitive processing (processing that is sensitive to the individual components of a complex representation), but such techniques have been nicely described (Van Gelder, 1990) as providing *functional*, as opposed to *concatenative*, compositional structure. A complex representation has concatenative structure if it embeds the individual constitutive elements unaltered within it. It has functional compositional structure if such components are usable or retrievable, but the complex expression does not itself embed unaltered tokens of these parts. Many connectionist schemes for dealing with compositional structure are only functionally compositional (e.g., RAAM architectures, tensor product encodings; for a review, see CONNECTIONIST AND SYMBOLIC REPRESENTATIONS). Here, then, is one example of a well-defined difference relative to which connectionist encodings may be genuinely less "text-like" than those exploited by classical syntactic engines.

A major benefit of exploring the space of connectionist cognitive models is thus that it may help us expand our sense of the possible nature of internal representation and hence better understand what is essential to notions such as *structure*, *syntax*, and *complex representation*. Doing so, we may discover which aspects of our models are simply artifacts of our familiarity with one representational format, viz., the format of atomic elements and grammar common to language and logic.

What Constrains a Good Model?

If connectionism is to be taken seriously as a new, nonsentential paradigm for understanding cognition, important methodological issues must be addressed. Connectionist techniques of representation and processing are powerful, and they can be used to approximate nearly any function or performance profile the theorist desires. The mere fact that some input-output pattern P is found in human cognition and can be mimicked using some connectionist model is, in itself, of only marginal psychological interest. Hence, the question, what additional factors constrain a good model?

Such additional constraints come largely from two directions: *outward*, from work in disciplines such as psychology and psycholinguistics, and *upward*, from work in neuroscience and brain theory. From psychology and psycholinguistics we can extract vast bodies of constraining data which go way beyond the mere specification of a task-specific input-output mapping. Such data can concern, for example, the relative difficulty of parsing certain sentences or solving certain problems, the time course of problem solving, the developmental profile of skill acquisition, and the way in which new and old knowledge interacts in the context of new learning (for detailed examples, see Karmiloff-Smith, 1992; Harris and Coltheart, 1986; DEVELOPMENTAL DISORDERS).

For current purposes, however, it is the upward constraints which I seek to highlight. The question here concerns the proper relation between connectionist computational modeling and the detailed constraints emerging from the various brain sciences. Such sciences include neuroanatomy, neurochemistry,

lesion studies, and research on the single cell, circuit, and systems level. It seems clear that any acceptable model of human information processing must respect the results of such studies. To do so, some intelligible relation must exist between the theories put forward by, for example, connectionist computational modeling and the entities and lawful interactions studied by the brain sciences. It is a duty sadly neglected by both classical artificial intelligence and a great deal of connectionist work to make some effort to display the precise nature of such relations.

Such a task is complicated by the variety of levels of interest which may characterize the brain sciences. These include the levels of biochemical specification: single cells, circuits, subsystems, and networks of subsystems. One model of the explanatory strategy of artificial intelligence which fails to take account of such diversity is Marr's (1982) image of the three levels at which a machine carrying out an information processing task needs to be understood. According to Marr, there is a top level (level 1, computational theory) at which we merely specify a function to be computed. Below that (level 2, representation and algorithm), we describe a set of computational steps designed to apply to a particular representation of the inputs and outputs to compute that function. Below that, and independent of it, another level (level 3, implementation) merely concerns how the algorithm and representations can be instantiated in some real-world device, such as the brain or a digital computer.

The idea of a single level of implementation detail is clearly an oversimplification. Likewise is the idea of a single level of algorithmic specification. Each of the distinct neuroscientific interests mentioned above (single cell, circuit, system, etc.) requires understanding in terms of its own computational profile and implementation. At the very least, Marr's image needs to be multiplied. In fact, however, the situation is worse, since the underlying idea that studies at each level can be independently pursued is itself highly dubious. For example, our top-level decomposition of a task into subtasks apt for computational modeling may be challenged once we become familiar with the distribution of information processing resources in the brain. What we originally thought of as two distinct functions may actually share circuitry in the brain (see Arbib, 1993:278). Such a result will not be devoid of psychological significance, since it will figure in an explanation of the breakdown profile as revealed by, for example, lesion studies of the system. What is mere implementation detail relative to one set of explanatory interests may be crucial algorithmic detail relative to another.

For this reason, I concur with both Arbib (1993) and Churchland and Sejnowski (1990) that Marr's three-level analysis, especially its methodological moral concerning the independence of the levels, does not afford a useful conceptualization of the relations between studies of the brain and computational models.

How, then, should we conceive the bridge between idealized artificial intelligence models and brain theory? Arbib (1993) pursues the quest for a single bridging level, which he calls SCHEMA THEORY (q.v.). Schemas are defined by a store of knowledge and an application procedure (the two being conceived of as intimately combined). Such functionally defined schemas are not required to map onto brain regions in a one-to-one manner. However, the schema framework aims to do justice to the distributed, cooperative style of computation in the brain by insisting that schemas interact without executive control and that each schema contains only a partial representation of the world. The interaction of the totality of such schemas constitutes the distributed analog to the central logicist representations posited by classical artificial intelligence. In addition, the theorist's allocation of functions to schemas and schemas to circuitry may be subsequently revised to better fit with neurological data (e.g., lesion studies).

The general methodological style of the schema theory approach is clearly an advance over less reflective attempts to use connectionist modeling. In explicitly addressing the question of the relation between its computational posits and real neurological structures, schema theory fulfills a major requirement of good cognitive modeling. It is too soon, however, to decide whether the specific high-level ontology of schemas and partial representations is adequate for all cases. It may well be that schema theory embodies just one bridging notion among many that we will need to explain the wide variety of cognitive operations and the multiplicity of levels of neuroscientific interest. What is clearly correct in Arbib's approach is the insistence that effective bridging requires real neuroscientific knowledge on the part of the computational theorist and that one role of such knowledge is to help to suggest abstract constructs which (like those of schema theory) have a foot in both camps. Conversely, as neuroscientists gain greater expertise in computational modeling, the constructs of the neurosciences may themselves become more closely adapted to the explanatory environment of artificial intelligence (see Churchland and Sejnowski, 1992).

In short, the key to a fruitful and mutually constraining union between brain theory and, for example, connectionist modeling is mutual knowledge and the recognition of the multiple types of explanatory interest which characterize both domains. No model can be expected to do justice to all aspects of its target. What we can and should expect is a clear statement of what aspects of the target phenomenon are supposed to be explained, and (if it is a computational model) at what level, if any, the computational story is intended to capture real neurophysiological facts.

What Is Missing So Far?

This question is the riskiest issue to address simply because what is missing today may not be missing tomorrow. However, I see three major stumbling blocks for the unfolding scientific understanding of the mind.

First, we need to pay much more attention to the issue of innate structure and knowledge. A key lesson which emerges from attempts to use neural networks to solve complex problems is that the setting of initial parameters makes all the difference. Such parameters include gross structural items (e.g., groupings of densely connected units with sparser links to other such groupings); computational resources (e.g., the nature and variety of learning rules, procedures for adding and deleting units and connections); and actual weight settings. Since connectionist learning is highly sensitive to the order in which problems are solved (see Elman, 1991), the nature of the products of early learning will also heavily influence later processing. For a variety of reasons, it is therefore imperative that connectionists forge stronger links with developmental psychology (see Karmiloff-Smith, 1992; Clark, 1993; COGNITIVE DEVELOPMENT; LANGUAGE ACQUISITION) and developmental neurobiology.

Second, a problem concerning the flexible use of achieved knowledge must be addressed. Current connectionist systems too often wed their achieved knowledge to a single task. However, one of the keys to advanced cognition apparently lies in our ability to use the knowledge achieved in solving one problem to reduce the effort required to solve some other, but related, problem. This ability enables us to use the representational fruits of one problem-solving episode to reduce future computational labors. Connectionism should likewise profit from such economies. Increased representational conservatism must therefore be placed high on the future agenda (for a discussion of the problem of the flexible use of achieved knowledge, see Clark, 1993).

Part of the solution to this problem of understanding flexible information use will probably involve greater attention to the distribution of the elements of a problem solution among various interacting resources in the brain. That is, one source of such flexibility may involve the brain's exploitation of various kinds of modular decomposition (see Jacobs, Jordan, and Barto, 1991). To the extent that this process is in effect, we will need to focus increased attention on the properties of what are in effect networks of networks, i.e., we will need to develop tools for understanding the interaction of multiple distributed sources of knowledge and skills (see DISTRIBUTED ARTIFICIAL INTELLIGENCE).

Finally and most seriously, there is a problem of scaling up. Most existing networks are tiny compared with the networks found in the brain. A single cortical column may have 200,000 neurons. Most of the artificial networks studied today comprise fewer than 5000 units. It is not clear that the principles of training and processing which work for these small networks will continue to work as we approach the big league of neural similitude. In particular, existing techniques may not yield results in real time when applied to such astronomically large systems. The real data will be messier and much more varied than in the past; inputs will not come ready sorted into training regimes for task X, training regimes for task Y, and so on. In addition, there is the important challenge of dealing with the temporal dimension. The target patterns in real input data are extended over time, and the target outputs are themselves importantly temporally sequenced. Scaling up thus involves learning how to deal with multiple, mixed, noisy, temporally patterned inputs using the resources of large, but highly structured, neural organizations. The gap between this goal and our current computational understanding is great, but the potential rewards are greater still.

Discussion

We have seen both the promise and the problems associated with the use of connectionist models for understanding the nature of the mind. The promise is nothing less than a new vision of mental processes, a vision which replaces logical operations defined over quasi-linguistic strings with alternative types of structure-sensitive processing. We also saw how such models can and should be constrained by results emerging from the various brain sciences to provide a realistic account of biological cognition. We also noted some serious stumbling blocks which remain, including the need to accommodate innate knowledge and structure and to address the *scaling problem*.

Thus, the philosophical task is to monitor and help investigate the relations which bind the three moving parts of a tripartite puzzle: *Mind/Model/Brain*. What is most exciting about the connectionist unpacking of the middle term is that it does not model the brain too closely on an unreflective, intuitive picture of the mind. Thus, we often depict our own minds as storehouses of propositional knowledge (that the sky is blue, the earth is round, etc.). Given that characterization, it is all too easy to imagine that what nature put inside our heads amounts to a device for storing and processing just such sentence-like strings of symbols. However, this step is not compulsory. Knowledge may be (correctly) propositionally specified and yet may not be stored in any quasi-sentential fashion. Connectionism encourages us to explore at least one nonsentential model in concrete detail. However, connectionist models ultimately have to answer to the third part of the puzzle, the brain. If such models are to function as scientifically valuable hypotheses concerning the type of processing involved in biological cognition, they must be carefully specified and evaluated. The specification involves identifying what level of neural detail is supposed to be replicated. The evaluation hinges on biological and psychological data appropriate to that level of detail. These tasks require the knowledge both of brain theory and of artificial neural networks that this *Handbook* provides. It is at this crucial intersection that much exciting work remains to be done.

Road Map: Connectionist Psychology
Related Reading: Artificial Intelligence and Neural Networks; Consciousness, Theories of; Perspective on Neuron Model Complexity; Structured Connectionist Models

References

Arbib, M., 1993, Review of A. Newell, *Unified Theories of Cognition*, *Artif. Intell.*, 58:265–283.
Churchland, P. M., 1989, *A Neurocomputational Perspective*, Cambridge, MA: MIT Press. ◆
Churchland, P. S., and Sejnowski, T., 1990, Neural representation and neural computation, in *Mind and Cognition: A Reader* (W. Lycan, Ed.), Oxford: Blackwell, pp. 224–251.
Churchland, P. S., and Sejnowski, T., 1992, *The Computational Brain*, Cambridge, MA: MIT Press. ◆
Clark, A., 1989, *Microcognition: Philosophy, Cognitive Science and Parallel Distributed Processing*, Cambridge, MA: MIT Press. ◆
Clark, A., 1993, *Associative Engines: Connectionism, Concepts and Representational Change*, Cambridge, MA: MIT Press. ◆
Elman, J., 1991, *Incremental Learning or the Importance of Starting Small*, Technical Report 9101, San Diego, CA: University of California, Center for Research in Language.
Fodor, J., 1987, *Psychosemantics: The Problem of Meaning in the Philosophy of Mind*, Cambridge, MA: MIT Press. ◆
Fodor, J., and Pylyshyn, Z., 1988, Connectionism and cognitive architecture: A critical analysis, *Cognition*, 28:3–71.
Jacobs, R., Jordan, M., and Barto, A., 1991, Task decomposition through competition in a modular connectionist architecture: The what and where visual tasks, *Cognit. Sci.*, 15:219–250.
Harris, M., and Coltheart, M., 1986, *Language Processing in Children and Adults*, London: Routledge and Kegan Paul. ◆
Karmiloff-Smith, A., 1992, *Beyond Modularity: A Developmental Perspective on Cognitive Science*, Cambridge, MA: MIT Press. ◆
Marr, D., 1982, *Vision*, New York: Freeman.
Van Gelder, T., 1990, Compositionality: A connectionist variation on a classical theme, *Cognit. Sci.*, 14:355–384.

Planning, Connectionist

Bartlett W. Mel

Introduction

Planning has played a central role in many accounts of intelligent behavior. In simple terms, planning may be viewed as "thinking to organize action." In the words of Holland et al. (1983):

> [We] believe that cognitive systems construct models of the problem space that are then mentally "run," or manipulated to produce expectations about the environment. . . . The cognitive system attempts to plan a sequence of actions that will transform the initial problematic state into a goal-satisfying state. An adequate mental model can accomplish this task by mimicking the environment up to an acceptable level of approximation.

We adopt this perspective so that, for the current discussion, planning involves (1) a *goal*, (2) an *internal model* of the agent or external world, and (3) a *search mechanism* that exercises the mental model to generate and evaluate hypothetical action sequences for purposes of *goal reduction* (i.e., reaching the goal, or reducing the search by reaching a subgoal).

Most work on planning has been carried out within the fields of robotics and traditional artificial intelligence (AI). Robot motion planning has most often involved explicit geometric techniques for representing, and then efficiently searching, the multidimensional configuration space of a robot to find collision-free paths to a goal (Latombe, 1991). In some cases, this search may be carried out using gradient descent on a potential field, specially constructed with a global minimum at the goal configuration and with local maxima placed at illegal or undesirable robot configurations (see POTENTIAL FIELDS AND NEURAL NETWORKS).

In the field of AI, planners have often focused on high-level problem-solving tasks, such as how to get to Aunt Nellie's house from somewhere far away. Such systems have generally assumed a preexisting model of the world that specifies which actions have what consequences and used highly abstracted representations of actions and world states. For example, Newell and Simon's General Problem Solver (GPS) (see Winston, 1984) might plan the trip to Aunt Nellie's as follows: need to take airplane to get there, but need to be at airport, but need to drive car to get there, but need to be at car to drive it, so walk to car, drive to airport, get on plane, but now need to get from airport to Aunt Nellie's house, but need car, so rent car, drive car to Aunt Nellie's, register success.

The benefits of planning were summarized in *The Handbook of Artificial Intelligence* (Cohen and Feigenbaum, 1982, vol. 3, chap. XV) as "reducing search, resolving goal conflicts, and providing a basis for error recovery." Some have argued that, despite these benefits, the primary function of the brain is not to think and plan, but to act and react (Albus, 1979; see also Brooks, 1991); Agre and Horswill (1993) point out that even humans frequently avoid the need to plan through the use of cultural systems and artifacts. But planning cannot *always* be avoided; several examples are considered in the following sections in which planning or proto-planning functionality has been modeled using connectionist approaches (see also REACTIVE ROBOTIC SYSTEMS).

Connectionist Planning: Five Examples

In keeping with the current working definition of planning, it is often possible to distinguish two architectural components of a planner. These components are (1) the *forward model*, i.e., that which allows the planner to predict the consequences of its own actions, and (2) that which exercises the forward model during search. One or both of these components can be implemented within a connectionist framework, as described in the following sections.

Simulating Tic-Tac-Toe

Rumelhart et al. (1986) were among the first connectionists to underscore the potential benefits of simulation within a mental model, suggesting that internal simulation is the basis for sequential thought processes in general, including thinking, language, and problem solving. To make their ideas concrete, they handwired a simple network capable of generating good moves in the game of tic-tac-toe. They then connected two such networks together, inputs-to-outputs, provided an initial blank board configuration, and then allowed the doublet network to produce a sequence of moves and countermoves as each network responded to the other's output. The authors argued that both networks could be contained within the "mind" of a single player—with one of the two networks acting as a mental model of the player's opponent. The sequence of moves so generated could then be viewed as an internal simulation of tic-tac-toe play. While this study was among the earliest demonstrations that a connectionist model could simulate the future in a sequential decision process, it did not address the more general modes of sequential processing needed for planning per se, such as backtracking when undesirable outcomes

are anticipated, or subgoal management, nor did it incorporate learning.

Mental Rotation as Visual Planning

Mel (1986) presented a connectionist model to demonstrate that a retinotopic map of the visual field could be used as the basic data structure underlying mental rotation and other smooth transformations of three-dimensional (3D) object images. The system, called VIPS, was organized as a retinotopically mapped array of binocular visual units, and a population of motor units that specified VIPS' state of motion (e.g., forward, backward, sideways, clockwise-circling, etc.). VIPS learned by moving through its simple simulated world while fixating a variety of 3D wireframe objects. For each type of motion, a distinct set of excitatory lateral interconnections was formed among the visual-field units according to a Hebbian learning rule. The learned pattern of lateral connections encoded the systematic way in which neural activity "flowed" around the retinotopic map as 3D objects were transformed in front of the eyes. This system of lateral connections was thus a crude mental model that allowed VIPS to simulate visual transformations of 3D objects, as controlled by its internally represented motor state. The mental model was, in and of itself, capable of sequential behavior via the lateral feedback pathway internal to the visual representation. However, no connectionist mechanism was proposed in this work that could exer-

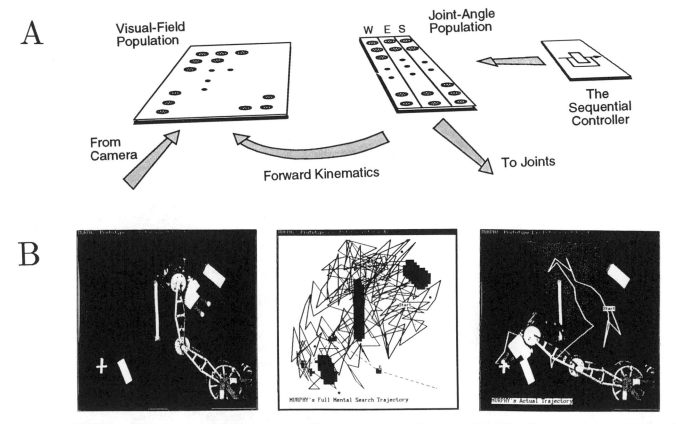

Figure 1. *A*, MURPHY's architecture, consisting of a camerotopically mapped array of visual units, a three-part map of coarsely tuned joint angle units for the wrist, elbow, and shoulder, and a nonconnectionist sequential controller mechanism. *B*, An example reaching problem. Target is white cross; other white objects are obstacles. Left panel shows initial configuration. Center panel shows trace of hand position during internal search. Right panel shows physically executed trajectory en route to target. (From Mel, 1990.)

cise the mental model in more sophisticated ways, such as for viewpoint invariant object recognition.

Planning Trajectories for a Multi-Link Arm

In a related visual-motor domain, a connectionist system called MURPHY planned collision-free trajectories for a multi-link arm (Mel, 1990). The physical setup consisted of a video camera and a robot arm with three joints that moved in the image plane of a video camera. MURPHY first learned the kinematic relationship between joint angle commands and the resulting pose of the arm in the visual field. This was done by coarsely sampling the 3D joint space during an initial "flailing" period. Each arm configuration generated one training exemplar, used to train the association between joint angle units, and visual units organized in a 2D camerotopic array (Figure 1A). This forward kinematic map was trained using a single-layer network of radial basis functions (see RADIAL BASIS FUNCTION NETWORKS), implemented as sigma-pi units. (A sigma-pi unit computes a sum of product terms over its input lines.) After training, MURPHY could generate "mental" images of his arm in an arbitrary pose by disabling the connections to the robot motors, setting up a hypothetical joint command, and allowing the learned kinematic map to generate an internal image of the arm. The connectionist portion of MURPHY's architecture thus provided the mental model that was used to search heuristically for collision-free paths in the presence of obstacles. As in Mel (1986), the search process itself was driven by an architecturally indeterminate outside module that used MURPHY's connectionist representations to guide the search process. Examples of internal search and physically executed reaching trajectories are shown in Figure 1B.

Improving Plans Using Gradient Descent

Thrun, Moller, and Linden (1991) presented a connectionist architecture that planned using a supervised error-correction procedure to iteratively optimize a sequence of actions aimed at a specific goal. Their approach used a backpropagation learning scheme that attempted to minimize predicted *future* error, in addition to current error, where the future error signals were provided by looking ahead in the internal model of the world. The authors demonstrated their approach for the problem of reaching for a rolling ball. They pointed out that when the current distance between hand and ball is minimized at every step, which can be achieved by backpropagating the ball-hand mismatch through the robot kinematics to adjust the joint commands, the resulting reach trajectory is an inefficient arc that promptly falls behind the rolling ball (Figure 2A). Thus, rather than adjust the joint commands based on current arm and ball locations only, forward models are used to advance both the ball and the hand into the future by one or more steps, so that future ball-hand errors could contribute to the current hand adjustment. When adjustments are made in this way, hand trajectories begin to reflect where the ball *will* be rather than where it is (Figure 2B).

Combining Planning with Reinforcement Learning

REINFORCEMENT LEARNING (q.v.) is often discussed in the context of connectionist computation. This is true in part because the state transition graphs underlying reinforcement learning lend themselves to relatively direct connectionist implementation. Reinforcement learning usually involves acquisition of an optimal control "policy" for a sequential decision task, where

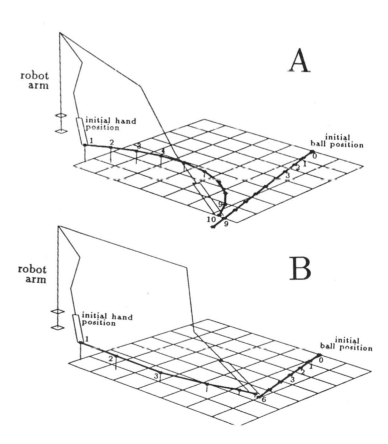

Figure 2. Improving plans by minimizing future error. A connectionist architecture (not shown) first learns to predict ball and hand trajectories. *A,* Reaching trajectory when current ball position is used to target hand movement, i.e., without planning ahead. *B,* Reaching trajectory when models of ball position and hand position are iterated into the future, so that future position errors can contribute to current targeting of the hand. (From Thrun, Moller, and Linden, 1991.)

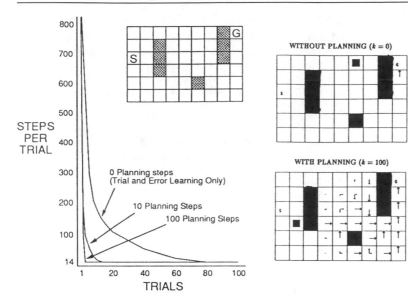

Figure 3. Dyna architecture that includes planning. At left, results for learning with and without planning, i.e., with 0, 10, and 100 lookahead runs per trial in a simple maze problem. At right, policies found by planning and nonplanning versions of Dyna system, by middle of second trial. The black square is the current location of the system. The arrows indicate action probabilities (excess over smallest) for each direction of movement. (From Sutton, 1991.)

an optimal policy is essentially a table specifying which action to perform in every situation to maximize expected rewards. Most current reinforcement learning methods use dynamic programming in some form to iteratively improve the stored policy. In such cases, it is usually assumed that the learning system has a complete (though perhaps probabilistic) model of the world in advance. The distinction between canonical planning and canonical reinforcement learning may be reduced to a difference in emphasis. The emphasis in the case of canonical planning is to search with a mental model for a good single path to a relatively ephemeral goal. One emphasis in reinforcement learning is to build an adaptive critic which provides "signposts" throughout the state space for the optimal path from each state to a relatively persistent goal.

Sutton (1991) has pointed out that the dynamic programming step of a reinforcement learning algorithm is equivalent to many one-step search steps that update the "values" of each state based on the values of the states that can be reached by applying a single action. Sutton makes the further point that in cases in which a model of the world is not available in advance to such a system, but is incrementally acquired through physical practice in the environment, the internal search steps associated with the dynamic programming algorithm are equivalent to planning. The use of planning in this form can greatly amplify the value of physical experience in a reinforcement learning context, as shown in Figure 3.

Other Work

Neurobiological Models of Planning

A handful of modeling efforts have been aimed at explaining the neurobiological basis for motion planning. Albus (1979) proposed an elaborate framework for understanding hierarchical task decomposition for planning, based in part on his CMAC model for the cerebellum (see CEREBELLUM AND MOTOR CONTROL). Arbib and House (1987) discuss the proto-planning activities of the frog during prey-catching in the presence of barriers, and present a preliminary model based on potential fields that could help to explain the frog's ability to follow a curved, barrier-avoiding path to its prey target. Mel (1991) used the connectionist components in MURPHY to provide a

framework for interpreting the complex neurobiological database relevant to the neural substrates for the planning of reaching movements. Droulez and Berthoz (1991) presented a model in which an efference copy signal in the eye movement system continuously updated a retinotopic memory representation of the visual field to compensate for retinal movement during saccades and other eye movements (see DYNAMIC REMAPPING). The shifting of neural activity within their "dynamic memory" structure is in essence an internal simulation and is similar in spirit to the motor-induced internal shifts of neural activity previously discussed in the context of mental rotation (Mel, 1986).

Discussion

The basic structure of planning as a computation has been well understood for decades; success or failure in particular cases is most likely determined by the appropriateness of internal representations chosen for the agent's sensations and actions, for the structure of the world, and for the plans themselves. The fact that a planner's internal model of the world or its sequential control components can in principle be represented as connectionist networks is unsurprising. While the standard advantages of neural network computation may motivate a connectionist implementation in one case or another, it remains an important open question whether connectionist computation has anything fundamental to add to the understanding of planning, beyond straightforward representational considerations.

Perhaps the most intellectually fertile areas for future work that involves "connectionist planning" lie in the field of computational neuroscience. A few questions of great interest concern how real, living neural tissue carries off the mental modeling and internal search procedures involved in planning, including the ability to represent multiple competing behavioral options, where such "planning" systems are located in the brain, and what experimental programs might be pursued to gather neurobiological data relevant to these questions.

Acknowledgments.   The author is supported by a McDonnell-Pew Fellowship and by the Office of Naval Research.

Road Map: Artificial Intelligence and Neural Networks
Related Reading: Cerebellum and Motor Control; Distributed Artificial Intelligence; Expert Systems and Decision Systems Using Neural Networks; Hippocampus: Spatial Models; Problem Solving, Connectionist; Reinforcement Learning in Motor Control; Sensorimotor Learning

References

Agre, P., and Horswill, I., 1993, Cultural support for improvisation, in *Proceedings of the AAAI*, Menlo Park, CA: AAAI Press, pp. 363–368.

Albus, J., 1979, Mechanisms of planning and problem solving in the brain, *Math. Biosci.*, 45:247–293.

Arbib, M. and House, D., 1987, Depth and detours: An essay on visually guided behavior, in *Vision, Brain, and Cooperative Computation* (M. Arbib and A. Hanson, Eds.), Cambridge, MA: MIT Press, pp. 129–163.

Brooks, R., 1991, New approaches to robotics, *Science*, 253:1227–1232.

Cohen, P. R., and Feigenbaum, E. A., 1982, *The Handbook of Artificial Intelligence*, vol. 3, Los Altos, CA: William Kaufmann.

Droulez, J., and Berthoz, A., 1991, The concept of dynamic memory in sensorimotor control, in *Motor Control: Concepts and Issues* (D. Humphrey and H.-J. Freund, Eds.), New York: Wiley, pp. 137–161.

Holland, J., Holyoak, K., Nisbett, K., and Thagard, P., 1983, *Induction: Processes of Inference, Learning, and Discovery*, Cambridge, MA: MIT Press.

Latombe, J.-C., 1991, *Robot Motion Planning*, Boston, MA: Kluwer. ◆

Mel, B., 1986, A connectionist learning model for 3-d mental rotation, zoom, and pan, in *Proceedings of the 8th Annual Conference of the Cognitive Science Society*, pp. 562–571.

Mel, B., 1990, *Connectionist Robot Motion Planning*, Cambridge, MA: Academic Press. ◆

Mel, B., 1991, A connectionist model may shed light on neural mechanisms for visually-guided reaching, *J. Cognit. Neurosci.*, 3:273–292.

Rumelhart, D., Smolensky, P., McClelland, J., and Hinton, G., 1986, Schemata and sequential thought processes in PDP models, in *Parallel Distributed Processing: Explorations in the Microstructure of Cognition* (D. E. Rumelhart, J. L. McClelland, and PDP Research Group, Eds.), vol. 2, *Psychological and Biological Models*, Cambridge, MA: MIT Press, pp. 7–57.

Sutton, R., 1991, Integrated modeling and control based on reinforcement learning and dynamic programming, in *Advances in Neural Information Processing Systems 3* (R. Lippman, J. Moody, and D. Touretzky, Eds.), San Mateo, CA: Morgan Kaufmann, pp. 471–478.

Thrun, S., Moller, K., and Linden, A., 1991, Planning with an adaptive world model, in *Advances in Neural Information Processing Systems 3* (R. Lippman, J. Moody, and D. Touretzky, Eds.), San Mateo, CA: Morgan Kaufmann, pp. 450–456.

Winston, P., 1984, *Artificial Intelligence*, Reading, MA: Addison-Wesley. ◆

Post-Hebbian Learning Rules

Harel Z. Shouval and Michael P. Perrone

Introduction

The cornerstone of all unsupervised learning rules is the Hebb rule (Hebb, 1949) (see, e.g., HEBBIAN SYNAPTIC PLASTICITY). However, the Hebb rule in its original form is unstable, i.e., the synaptic weights will all be driven to their maximal value.

Numerous learning rules have been proposed which overcome this difficulty (von der Malsburg, 1973; Oja, 1982; Sanger, 1989). The outcome of learning in all of these models depends only on first- and second-order statistics of their inputs. Therefore, in order to understand these algorithms, despite their differences, it is first necessary to know what the principal components of the inputs are. When this is understood it becomes easy to account for the differences as well.

Some of these models have been used to explain properties of learning in the visual system. In order to do this, additional assumptions about the relevant visual environment and the architecture of the network have to be made. We will survey these different assumptions and try to understand the sources of the differing results: Do they arise from the different assumptions about the environment, the architecture, or the learning rule?

Hebb's Rule and Its Instability

The original Hebb rule states how synapse efficacies are strengthened (Hebb, 1949):

> When an axon in cell A is near enough to excite cell B and repeatedly and persistently takes part in firing it, some growth process or metabolic change takes place in one or both cells such that A's efficiency in firing B, is increased.

Later, it was suggested that synapse strengths may decrease when the presynaptic cell fails to fire the postsynaptic cell. If this is taken in a slightly more general context it implies that synapse strengths change as a function of the correlations between the pre- and postsynaptic neurons, i.e., they increase when the activities of the neurons are correlated and decrease when they are anticorrelated. This generalized Hebbian rule is the cornerstone of many unsupervised learning schemes (von der Malsburg, 1973; Rochester et al., 1956; Linsker, 1986; Oja, 1982).

A mathematically simple form, which will nevertheless be general enough for understanding the basic properties of such learning rules, is the linear rule (Linsker, 1986)

$$\Delta w_i = -\eta(x_i - x_0)(y - y_0) \tag{1}$$

where x_i are the activities of presynaptic neurons, y is the activity of the postsynaptic neuron, w_i is the value of the synaptic efficacy between presynaptic neuron i and the postsynaptic neuron, and η, x_0, and y_0 are constants.

This concept is easiest to analyze in terms of single postsynaptic linear neurons, i.e., when the neuron's activity is given by

$$y = \sum_i w_i x_i \tag{2}$$

For this case, we can calculate the average of the synaptic efficacies over the input probability distribution. We denote this average by $E[\cdot]$. This average weight change is of interest to us since we define the fixed points as those values of w_i for which the average weight change is zero. When y is replaced by x_i, we write w_{ij} for the strength of its synaptic connection from x_j.

Inserting Equation 2 into Equation 1, averaging the result, and assuming the same probability distribution for each input, i.e., $E[x_j] = E[x_i] = \mu$, we obtain, for the vector $\mathbf{W}$ of synaptic weights w_{ij},

$$E[\Delta \mathbf{W}] = -\eta(Q - k_2 J)\mathbf{W} + \eta k_1 \hat{e} \quad (3)$$

in which $Q_{ij} = E[(x_i - \mu)(x_j - \mu)]$ is the correlation function of the input patterns, J is the matrix $J_{ij} = 1$, $k_1 = y_0(x_0 - \mu)$, $k_2 = \mu(x_0 - \mu)$, and $\hat{e}$ is the vector $e_i = 1$. It is important to notice that k_1 and k_2 are not free parameters which can be set arbitrarily. For instance, if $x_0 = \mu$ then $k_1 = k_2 = 0$.

A fundamental problem with this learning rule is that it has no stable fixed points (MacKay and Miller, 1990). The vector $\mathbf{W}$ will asymptotically become parallel to the eigenvector with the largest eigenvalue of the matrix $Q - k_2 J$, but its magnitude will diverge to infinity. For example, in the case where $k_2 = 0$, the weight vector $\mathbf{W}$ would asymptotically become parallel to the principal component of the input (which is the largest eigenvector of the correlation matrix in the case), but its magnitude would approach infinity. This instability is not biologically plausible since it implies that the synaptic strengths would grow arbitrarily large.

In order to avoid this unbiological behavior, one must go beyond the basic Hebb rule by imposing additional constraints. One approach is to assume that there exists a biologically imposed saturation limit which the synapses cannot exceed. This approach, although conceptually simple and easy to implement, does not preserve one of the basic properties of the Hebb rule: the ability to identify the principal component of the data.

Before discussing how this problem can be corrected without appealing to saturation limits, we note that some researchers have avoided this problem by assuming that the neuron learns slowly enough for the receptive field to become approximately parallel to the principal component and that learning stops quickly enough for the synapses to avoid saturation (Linsker, 1986; MacKay and Miller, 1990; Miller, Keller, and Striker, 1989).

Stabilized Hebb Rule

The absence of a bounded fixed point for the Hebb rule without saturation was noted by Rochester (Rochester et al., 1956): "If no additional rule were made the Hebb rule would cause synapse values to rise without bound." Normalization procedures to bound synapse strengths by supplementing the Hebb rule with additional constraints have been proposed (von der Malsburg, 1973; Kohonen, 1982). The method chosen by von der Malsburg was to renormalize after each Hebbian update by dividing the strength of each synapse by the sum of the synapse strengths of the neuron. Kohonen chose to divide by the sum of the squares of synapse strengths. Dividing by the sum of squares always keeps the weights bounded, whereas dividing by the sum of the weights can fail if the principal component has a sum of weights equal to zero.

These stabilization schemes seem biologically implausible since they are nonlocal. Each synapse needs to "know" the weights of all the other synapses terminating on the same neuron. A mechanism for propagating this nonlocal information is not known. However, this type of stabilization can be implemented in another way (Oja, 1982). Instead of dividing after each Hebbian update, an extra term can be added to the update rule. Setting for simplicity, $\mu = x_0$ which implies that, $k_1 = k_2 = 0$, this rule takes the form

$$\Delta w_i = x_i y - \beta(\mathbf{W}) w_i \quad (4)$$

where $\beta(\mathbf{W})$ is a scalar function of the vector $\mathbf{W}$. To see what is

then the fixed point of the learning rule, it is convenient to go back to the correlational formulation of the learning rule. In this formulation the fixed points of Equation 3 (i.e., the points where we have $E[\Delta \mathbf{W}] = 0$) are given by

$$Q\mathbf{W} = \lambda \mathbf{W} \quad (5)$$

where $\lambda = \beta(\mathbf{W})$. Thus the fixed points are the eigenvectors of the correlation matrix Q, and the function $\beta(\mathbf{W})$ sets the normalization of these vectors. Oja chooses $\beta(\mathbf{W}) = y^2$, implying that $E[\beta(\mathbf{W})] = \mathbf{W}^T Q \mathbf{W}$. Therefore the weights will be normalized such that $\mathbf{W}^2 = 1$. In fact, the only stable fixed point is the principal component (see PRINCIPAL COMPONENT ANALYSIS). Computationally, the importance of the principal component is that it is the projection which maximizes the variance and in a Gaussian channel carries more information than any other projection (see INFORMATION THEORY AND VISUAL PLASTICITY).

When posed in this way, the learning rule seems more biologically plausible, since the stabilization requires that each synapse have information of its own efficacy and of the square of the postsynaptic potential.

If instead of using Oja's choice of β we choose $\beta = \hat{e} \cdot xy$, this would, when averaged, result in $\hat{e}Q\mathbf{W}$. Thus an eigenvector with eigenvalue λ is normalized such that $\hat{e}\mathbf{W} = 1$, which is equivalent to the normalization proposed by von der Malsburg (1973). This type of normalization function β is problematic in the cases where $\hat{e}\mathbf{W} = 0$ and normalization is not possible.

Finding Multiple Principal Components

Learning rules which extract only the largest principal component generally capture only a fraction of the information in high-dimensional data. In order to reduce the dimensionality of the data yet have a relatively small reconstruction error, it is useful to extract several principal components. This can be accomplished using a network with lateral inhibition. Different forms of lateral inhibition will find different features. If we assume that there are P output neurons, where P is smaller than the input dimensionality, then the resulting receptive fields will span the same subspace as the first P principal components but in general will be different from the principal components themselves. However, by adding additional constraints, one can force the features to extract the principal components (Oja, 1992; Sanger, 1989). A heuristic way of understanding these methods is that the lateral inhibition between the first and second neurons acts effectively to subtract from the input to the second neuron those parts of the input which result from the first neuron's receptive field, and so on for successive neurons.

Predicting the Receptive Fields of Hebbian Neurons

Once this stable family of learning rules has been proposed, the question which naturally arises is: Does the neural machinery in our brain implement or approximate any of these learning rules? There are several ways to approach such a question, the most obvious one being the direct experimental approach, which means examining what the learning rule is in real live tissue using electrophysiology. Although this method is ultimately the best, it is currently unable to give a conclusive answer to this question. In this section, we will discuss a different approach which tries to predict what the receptive fields of neurons using this family of learning rules should be. These receptive fields can then be compared to the real biological receptive fields. If they are similar, then the learning rule used in the brain has more chance of being similar as well; if they are not, then we probably have to examine other types of learning rules. The results of these predictions depend on assumptions about the nature of the environment to which these neurons are exposed. We will survey several different results which vary in both the assumptions about the environment and in the methods with which they were obtained.

The first investigations into these questions were usually performed with simplified model environments (von der Malsburg, 1973; Kohonen, 1982). An environment composed of bars was used by von der Malsburg (1973) to train a network of neurons. The network he used had a lateral Mexican hat inhibition between the output neurons, and the sum of the weights was kept a constant. The neuron's receptive field in these simulations became parallel to one of the training patterns, and therefore resembles at least qualitatively the receptive fields of cortical neurons, which indeed are highly selective to bars of certain orientations. What these simulations showed is that this kind of learning rule can produce neurons which are selective to one of their training patterns. However, it is not clear that a discrete set of bars is a good model for a realistic environment, although one can argue heuristically that when we view the environment through a sufficiently small window, the boundaries between objects resemble straight edges.

Another approach is to assume a correlation structure of the incoming inputs to the visual cortical cells (Linsker, 1986; Miller et al., 1989). This approach is aimed at modeling prenatal development and assumes that, due to the structure of the input channels (i.e., the retina and the LGN), there exist correlations between neighboring inputs to the cortical neurons. Since the outcome of these types of learning rules depends on the second-order correlations in the inputs, the resulting receptive field structures depend critically on the assumed correlations in the input and on the details of the neurocircuitry in the input channels. Another interesting aspect of these models is that the correlation functions assumed are typically radially symmetrical, but this symmetry has to be broken in order for the receptive fields to be orientation selective.

In our context it is interesting to examine what is the origin of those correlation functions which produce receptive fields similar to those obtained in the brain. In the model produced by Linsker (1986), this is the correlation function between neurons in the first layer, not neurons in the input layer. Due to the learning rule used by Linsker, the relevant correlation function is $Q - k_2 J$. Furthermore, the correlation function in this layer depends also on the arbor functions between two consecutive layers, which implies a different hard cutoff for different synapses. This hard cutoff is necessary since the rule used is not stabilized. Since this arbor function has the form $A(\mathbf{r}) = \exp[-(r^2/\sigma^2)]$, it turns out that, after learning, Q has the form $Q(\mathbf{r} - \mathbf{r}') = \exp[-(r - r')^2/\sigma^2]$. Given this correlation function, the structure of the receptive fields which evolve can be examined analytically (MacKay and Miller, 1990). It has therefore been shown that the principal component of this correlation function is orientation selective only when $k_2 < -3$. The principal component of Q is radially symmetrical and thus not orientation selective. However, when the component $k_2 J$ (MacKay and Miller, 1990) is added, it affects only the radially symmetrical solutions, reducing their eigenvalues and thus making the orientation selective solution the highest eigenvalue solution. In order to examine if this is a plausible explanation, we need to see whether such negative values of k_2 are plausible. Since $k_2 = \mu(x_0 - \mu)$, these values could be achieved only if spontaneous activity levels $\approx \mu$ typically induce considerable depotentiation. If, on the other hand, they induce no change $(x_0 = \mu)$, then $k_2 = 0$, which is clearly outside the orientation selective domain.

In Miller's models (Miller et al., 1989; Miller, 1992), the relevant correlations exist in the input layer and result from the correlations present in the dark activity in the retina. Since there is no conclusive evidence as to what these correlations indeed are, the resulting receptive fields depend crucially on the assumptions made about these correlations. In Miller's first work on the subject, which addressed mostly the questions of binocularity, the assumed correlation decreased with length similar to the correlation functions which evolved in Linsker's first layer, and since in his model $k_2 = 0$, no orientation selective cells were produced. In order to achieve orientation selective cells, another type of correlation function must be chosen. Miller (1992) chose a Mexican hat correlation function and

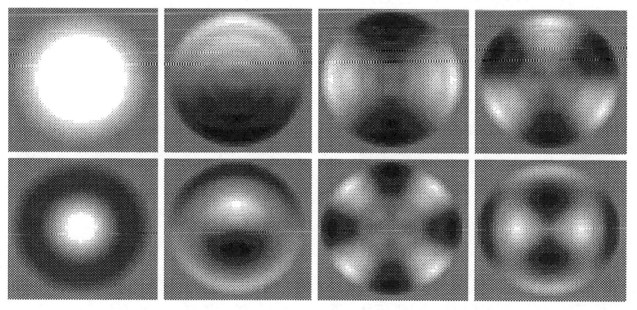

Figure 1. The shapes of the first eight principal components ($b = 0$, $a = 1$, and $\phi_{mi} = 0$), ordered from left to right, top to bottom.

indeed obtained orientation selective receptive fields. The assumption that correlations exist in the dark activity in the retina comes from experiments performed on dark activity in adult retinas (Mastronarde, 1989). If selectivity indeed evolves because of these correlations, it stands to reason that selectivity would evolve in dark reared animals as well, while it is well known that in cats selectivity of animals reared in the dark eventually drops off (Imbert and Buisseret, 1975).

Another type of approach is to create a model environment which is as realistic as possible for an animal with open eyes. This type of approach was pursued by Hancock, Buddeley, and Smith (1992), who trained a network of the type proposed by Sanger (1989) to extract principal components from a set of natural scenes. These scenes were viewed through a small Gaussian window whose centers were chosen at random. The results they obtained were a set of well-defined receptive fields. It seems from these results that here too a nearly radially symmetrical environment is used, since the different receptive fields seem to be composed of a radial part multiplied by an angular part, in the form $\psi(\mathbf{r}) \propto f(r) \cos(m\theta_\phi)r$, which is what would be predicted for a radially symmetrical environment. It is also interesting to notice that the first principal component predicted in this experiment is a radially symmetrical one, which has no resemblance to receptive fields of neurons in the visual cortex. These types of simulations can be performed with a hardbound window, rather than a soft Gaussian window. The results obtained can be explained analytically (Liu and Shouval, 1994) if we use Field's result (Field, 1987) that the power spectrum of natural scenes is assumed to be scale invariant, i.e., $\tilde{C}(k) = 1/K^2$, and radially symmetrical. This spectrum corresponds to $C(\mathbf{r}) = c(\log(r/a) + b)$, in which the constant b is undetermined and needs to be measured directly.

In Figure 1 we display the first eight localized principal components of natural scenes as predicted by this analysis, assuming $b = 0$. The radial symmetry assumption implies that the angular part of the receptive fields has the form $\cos(l\theta + \phi)$ or $\sin(l\theta + \phi)$. Both of these will exist for $l > 0$, but they have the same eigenvalue.

These results are in good agreement with simulation results using natural scenes as training patterns. In the simulations, the phase of the angular part of the solution, i.e., ϕ in $\cos(l\theta + \phi)$, is always the same, and the sine and cosine solutions have consistently different eigenvalues. This cannot be explained by a completely radially symmetrical correlation function, but is a result of a small nonradially symmetrical component of the correlation function (Liu and Shouval, 1994).

Discussion

Many of the proposed learning rules, which stem from the Hebbian rule, can be shown to belong to the same class, i.e., rules which are sensitive only to the first and second moments of their inputs. There are two major differences between the different learning rules: the use of small additions to the correlation function (Linsker, 1986) and the method with which they stabilize the weights and prevent them from approaching infinity. What is common to all of these methods is that the final state of the network is in general determined by the principal components of the correlation or altered correlation function.

Different assumptions have been made about the environment. It seems that for many reasonable choices of radially symmetrical correlation functions, the major principal component is radially symmetrical as well, which stands in contrast to what is found in the visual cortex. It seems that whenever the prominent receptive field is orientation selective, it arises from an effective correlation function which has a prominent anti-correlation region (Linsker, 1986; Miller, 1992). In order to achieve such correlation functions, strong assumptions about the learning rule (Linsker, 1986) or about the correlation functions in the dark retina (Miller, 1992) must be made. These assumptions should be tested experimentally.

Another possible difference between these learning rules and biological neurons is that, from a computational point of view, only two receptive fields of each angular type are needed, and these should differ by a phase of $\pi/2$ in terms of which any orientation may be expressed as a linear combination. This mathematical simplification is clearly in contrast to the continuum of orientations found in the cortex. If we do not assume the special types of correlation functions which elicit orientation selective cells, the above models typically predict a prominent radially symmetrical receptive field, which is also nonbiological. Thus we can conjecture that the difference between the receptive field structures in the cortex and that found in the models could stem either from the wrong learning rule being employed in the models or from the real inputs to the visual cortex, which arrive there after being preprocessed in the retina and LGN, not being radially symmetrical.

Road Map: Mechanisms of Neural Plasticity
Background: I.3. Dynamics and Adaptation in Neural Networks
Related Reading: BCM Theory of Visual Cortical Plasticity; Hebbian Synaptic Plasticity: Comparative and Developmental Aspects; Ocular Dominance and Orientation Columns; Visual Coding, Redundancy, and "Feature Detection"

References

Field, D. J., 1987, Relations between the statistics of natural images and the response properties of cortical cells, *J. Opt. Soc. Am. A*, 4:2379–2394. ◆
Hancock, P. J., Baddeley, R. J., and Smith, L. S., 1992, The principal components of natural images, *Network*, 3:61–70.
Hebb, D. O., 1949, *The Organization of Behavior*, New York: Wiley. ◆
Imbert, M., and Buisseret, P., 1975, Receptive field characteristics and plastic properties of visual cortical cells in kittens reared with or without visual experience, *Exp. Brain Res.*, 22:25–36.
Kohonen, T., 1982, Self-organization of topologically correct feature maps, *Biol. Cybern.*, 43:59–69.
Linsker, R., 1986, From basic network principles to neural architecture, *Proc. Natl. Acad. Sci. USA*, 83:7508–7512, 8390–8394, 8779–8783.
Liu, Y., and Shouval, H., 1994, Localized principal components of natural images: An analytic solution, *Network*, 5:317–324.
MacKay, D. J., and Miller, K. D., 1990, Analysis of Linsker's simulations of Hebbian rules to linear networks, *Network*, 1:257–297.
Mastronarde, D. N., 1989, Correlated firing in cat retinal ganglion cells, *Trends Neurosci.*, 12:75–80.
Miller, K. D., 1992, Development of orientation columns via competition between on- and off-center inputs, *NeuroReport*, 3:73–76.
Miller, K. D., Keller, J. B., and Striker, M. P., 1989, Ocular dominance column development: Analysis and simulation, *Science*, 245:605–615.
Oja, E., 1982, A simplified neuron model as a principle component analyzer, *J. Math. Biol.*, 15:267–273.
Oja, E., 1992, Principal components, minor components, and linear neural networks, *Neural Netw.*, 5:927–935.
Rochester, N., Holland, J., Haibt, L., and Duda, W., 1956, Tests on a cell assembly theory of the action of the brain, using a large scale digital computer, *IRE Trans. Inform. Theory*, IT-2:80–93.
Sanger, T. D., 1989, Optimal unsupervised learning in a single-layer linear feedforward neural network, *Neural Netw.*, 2:459–473.
von der Malsburg, C., 1973, Self-organization of orientation sensitive cells in striate cortex, *Kybernetik*, 14:85–100.

Potential Fields and Neural Networks

Ashraf A. Kassim and B. V. K. Vijaya Kumar

Introduction

The ability to automatically generate paths is crucial for robots. A survey of such motion-planning methodologies is found in Latombe (1991). Stated in simple terms, the objective of path planning is to find a path for the robot from an origin or starting point to a destination or goal without colliding with any other obstacles present in its physical workspace environment. Robots come in a variety of shapes and sizes, thereby complicating the development of generalized solutions to the path-planning problem. Path-planning methodologies are applied either in the robot's *coordinate* (world) *space* or *configuration space* (Cspace) (Lozano-Perez, 1983). The Cspace is the set of all possible configurations of the robot. A robot's configuration is a set of independent parameters which completely specifies the position of every point on the robot. The robot is thus represented as a point in the Cspace, while the obstacles and physical workspace boundaries are transformed into forbidden regions called *Cspace obstacles*.

In this article, we examine the *potential field approach* to path planning (Khatib, 1986) and how *artificial potential fields* (APFs) useful for path planning can be developed by artificial neural networks.

Potential Field Approach to Path Planning

In the potential field approach to path planning, a mathematically postulated potential energy is evaluated over the robot's Cspace. The resulting energy topology or APF forms a distributed representation of the robot's environment and can be used to plan paths for the robot by evaluating the associated gradient or "force." In the potential field approach to path planning, the robot is represented as a point moving under the influence of an *artificial potential field* (APF) from the robot's *initial configuration* (i.e., initial positions and orientations) to the desired *goal configuration*. The APF is produced as some combination of an *attractive potential* which attracts the robot toward the goal configuration and *repulsive potentials* which push it away from the Cspace obstacles. Therefore, the APF forms a distributed representation of the topological structure of the robot's environment. There are a variety of APF models for path planning, and most are described by analytical functions which are usually difficult to synthesize over arbitrary configuration spaces (Latombe, 1991). Also, many of these fields *approximate* the shapes of the Cspace obstacles to reduce the computational complexity.

A simple and often used attractive potential is the *parabolic well* (Khatib, 1986), U_{attr}, which is given by:

$$U_{\text{attr}}(C) = \tfrac{1}{2}\xi\|C - C_{\text{goal}}\|^2 \tag{1}$$

where ξ is a positive scaling factor and $\|C - C_{\text{goal}}\|$ is the Euclidean distance from the configuration C to the goal configuration, C_{goal}. An artificial attractive force is obtained by evaluating the derivative of U_{attr}.

Repulsive potentials provide a potential barrier around Cspace obstacles with some specific region of influence such that there is little or no influence beyond this region. A possible repulsive potential function is as follows:

$$U_{\text{rep}}(C) = \begin{cases} \dfrac{1}{2}\eta\left(\dfrac{1}{\varphi(C)} - \dfrac{1}{\varphi(0)}\right)^2 & \text{if } \varphi(C) \le C_0 \\[2ex] 0 & \text{if } \varphi(C) > C_0 \end{cases} \tag{2}$$

where η is a positive scaling factor, $\varphi(C)$ is the distance from the Cspace obstacle, and C_0 is called the distance of influence of the Cspace obstacle. The APF is formed as some combination of attractive and repulsive potentials. Thus, it is possible for the APFs to contain *saddle points* or *local minima* which can potentially trap the gradient-based path-finding processes. However, it is possible to construct APFs which have few or no local minima (Latombe, 1991).

Unlike the APFs, which are described by analytical functions, the construction of a numerical potential field over a discretized representation of the Cspace is much easier. Such *grid potentials* can be easily implemented and operate directly on discretized environments. They can also be made to be free of local minima. The distributed nature of grid potential computations enable them to be performed massively in parallel (Barraquand and Latombe, 1991; Lemmon, 1991a; Prassler and Milios, 1991), and the process is reminiscent of *wavefront expansion operations* (Latombe, 1991). The realization that APFs for path planning can be implemented by parallel and distributed processing systems has created interest in the application of artificial neural networks to the problem of generating these APFs.

The main advantage of the potential field approach to path planning is the ease of constructing motion planners based on the approach, which are quite efficient, reasonably reliable, and particularly fast.

A Neural Network Which Develops APFs for Path Planning

The *Wave Expansion Neural Network*, or WENN (Kassim and Vijaya Kumar, 1992, 1994), is capable of implementing grid potentials. The WENN is a single layer of locally interconnected neurons organized in the form of overlapping neighborhoods denoted by v. Each WENN neuron has a unique label i and is characterized by its activation x_i and connection weights w_{ji} associated with each v-neighbor neuron j (collectively represented by the vector W_{v_i}). Figure 1 shows a WENN neuron.

The evolution of the WENN neuron activation and connection weights are governed by the WENN *activation* and *learning equations*, respectively. The general form of the WENN activation is given by:

$$x_i^+ = x_i^- + F(I(x_i^-, \alpha_{v_i}) + E(\beta_{v_i}) + P(X_{v_i}, W_{v_i})) \tag{3}$$

while that of the learning equations for each of the n connection weights associated with each neighbor neuron $j \in v_i$ are given by:

$$w_{ji}^+ = w_{ji}^- + G_k(I(x_i, \alpha_{v_i}) + V_k(X_{v_i})) \tag{4}$$

where $j = i_k$ and $k \in [1, n]$. The various functions of the activation and learning equations are based on the particular APF that is desired. The inhibitory function $I(\cdot)$ provides the inhibition which prevents the neuron from activating. The *investigatory functions* $V_k(\cdot)$ in the learning equations enable the adaption of the connection weights in such a way that only *one* is activated while the rest are inhibited from activation. Activity

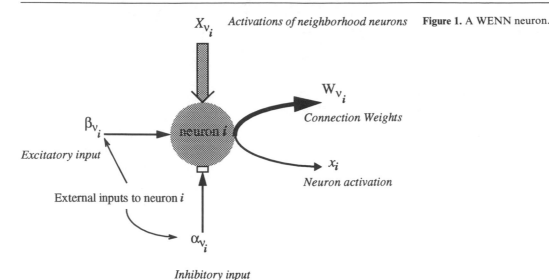

X_{v_i} *Activations of neighborhood neurons*   **Figure 1.** A WENN neuron.

W_{v_i}

Connection Weights

β_{v_i}

Excitatory input

x_i

Neuron activation

External inputs to neuron i

α_{v_i}

Inhibitory input

in the WENN neural field initiates at uninhibited neurons which receive external excitation through the excitatory inputs in the excitatory function $E(\cdot)$. Other uninhibited neurons activate in response to the activations of their v neighbors through the propagatory function $P(\cdot)$. Thus, the WENN neural activities propagate in *waves* originating from externally excited neurons. Specific choices of the investigatory functions determine the pattern of propagation of WENN neural activity. The activity is allowed to propagate until there are no further neural evolutions in the WENN neural field, and the resulting neural activity distribution is the desired APF.

Attractive potentials have been developed by the WENN. Such a WENN corresponds in size and structure to that of the discretized Cspace over which an attractive potential needs to be developed, and its v neighborhoods consist of the nearest neighbor neurons. The Cspace is mapped through the WENN's *inhibitory inputs*, α_i, such that only the Cspace obstacle neurons are inhibited. Only the C_{goal} neuron receives external excitation through its *excitatory input* β_i, while other neurons receive no excitation. Inhibition in the activation and learning equations occurs when the neuron activation or inhibitory input is active (i.e., non-zero), which inhibits further change of the neuron activation and the connection weights.

Initially, the WENN neural field is at rest with all neuron states identically zero. The C_{goal} neuron is the only one in the neural field to be agitated through its excitatory input. Neural activity therefore initiates at the C_{goal} neuron, which becomes active with a unit-value activation. This activity propagates to the uninhibited v-neighbors of the C_{goal} neuron. The activation equation increments the incoming activity at each of C_{goal}'s uninhibited v-neighbors so that they subsequently become active with an activation value of 2. The Cspace obstacle neurons are inhibited, and neural activity cannot propagate through them. The process continues as activity propagates in the form of *wavefronts* away from C_{goal} and around Cspace obstacles. An *activity wavefront* is the set of all neurons which activate simultaneously (i.e., have the same activation values) and are thus analogous to equipotential surfaces. The neural activity propagates throughout the WENN neural field until all C_{free} neurons are activated, i.e., *equilibrium state* is achieved. The WENN neuron activity distribution then represents the attractive potential.

An interesting property of the attractive potential developed by the WENN is the *absence of local minima*. This is because as activity is propagated from one neuron to the next, the activation of the subsequent neuron is *always* greater than that of the previous neuron in the sequence. A path to C_{goal} is computed by following the maximum negative gradient in the WENN neuron activity distribution from any point, C_{init}, in the free Cspace. The next neuron in the path is the one with the smallest *non-zero* activation value which forms the steepest gradient with respect to C_{init} among all of C_{init} neuron's active neighbors. The process continues until C_{goal} is reached. The path is optimal in the L_1 metric and is a set of adjacent C_{free} points associated with a set of WENN neurons beginning with C_{init}. All Cspace points corresponding to the neurons identified by the path-planning procedure form the physical path for a point object (i.e., Cspace is the Cartesian space). For larger objects, the Cspace points correspond to a set of configurations that the object would need to translate through from C_{init} to reach C_{goal}.

Figure 2 shows a path for a point robot in a two-dimensional (2D) physical workspace which was determined using the WENN neural activity distribution. The discretized 2D workspace of a point robot shown in Figure 2 was mapped onto the WENN, and the neuron corresponding to C_{goal} was activated. Figure 3 shows the resulting WENN neural activity distribution after the network had reached the equilibrium state. The activity distribution which represents the attractive potential is used to evaluate the path from C_{init} to reach C_{goal} as described earlier. The gradient descent procedure produces a piecewise linear path which can be smoothened before being used by the point robot.

The WENN has also been used to develop other APFs which are useful for path planning (Kassim and Vijaya Kumar, 1994). Repulsive potentials, for example, are developed by the WENN by spreading activation from the surfaces of the C-obstacles. The WENN activity wavefronts originate at the C-obstacle surfaces and spread until they meet at points which are equidistant between C-obstacles. Figure 4 shows the WENN neural activity distribution corresponding to the repulsive potential developed over the workspace of Figure 2. The maxima of the resulting neuron activity distribution (i.e., repulsive potential) are a form of *Voronoi diagram* (Canny and Donald, 1988), which is also a useful tool in path planning.

Figure 2. A randomly generated 2D workspace with the resulting path found using the WENN neuron activity distribution developed over the workspace.

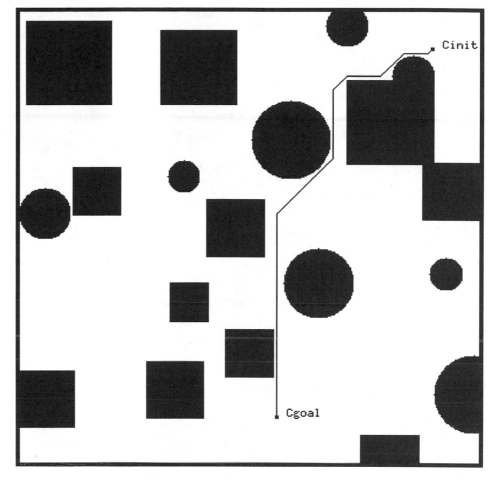

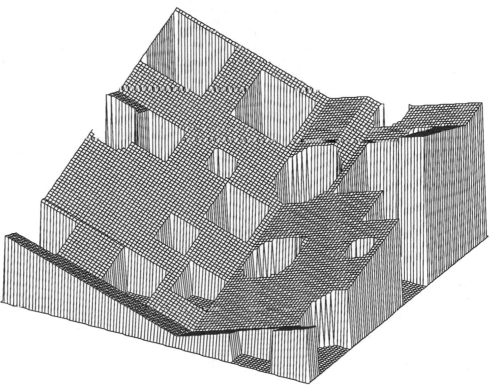

Figure 3. A 3D view of the WENN neuron activity distribution (i.e., attractive potential) developed over the workspace of Figure 2.

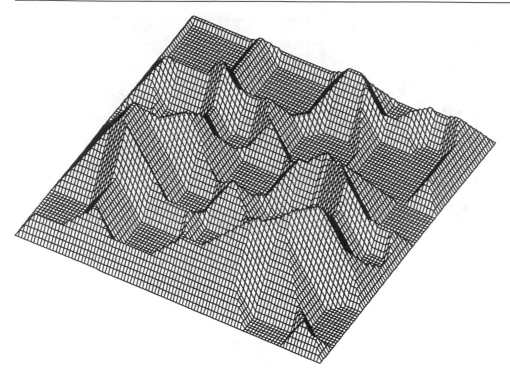

Figure 4. A 3D view of the WENN neural activity distribution which represents the distance from the obstacle surface (i.e., repulsive potential) developed over the workspace of Figure 2.

Other Neural Networks Which Are Capable of Developing APFs Useful for Path Planning

There are several other neural networks which are also capable of developing APFs useful for path planning. Park and Lee (1991) used feedforward neural networks to develop *collision penalty functions* over workspaces of robots with polyhedral obstacles. The collision penalty function is developed by using a three-layer feedforward neural network for each obstacle in the robot's workspace environment. The number of neurons in the input layer corresponds to the dimensionality of the environment. The connection weights and thresholds of the neurons in each neural network correspond to the respective set of inequality constraints which describe the particular polyhedral object. The neurons in the input layer represent the coordinates of a point. The collision penalty function is a function of the distance from the obstacles such that it is maximum at the obstacles and diminishes with increasing distance, becoming zero at a short distance from the obstacle. The collision penalty function is used by the path-planning algorithm as a repulsive potential which repels the path away from obstacles. A collision-free path is evaluated in the course of minimization of an energy function which is related to the length of the path and the collision penalties at points along the path due to the obstacles. The minimization proceeds from an initial arbitrary choice of a path between the initial and desired goal positions. The procedure can be extended to larger polyhedral objects by evaluating the collision penalties at points within the object.

Lemmon (1991a) developed a cooperative neural field with locally interconnected fixed weight connections based on the *Competitively Inhibited Neural Network* (CINN) (Lemmon, 1991b). The neural network's resulting oscillatory behavior was used to develop attractive potentials over two-dimensional workspaces of point robots. The neural network is called the *Oscillatory Neural Field* (ONF), and it is a two-dimensional field of locally interconnected neurons whose size corresponds with that of the discretized workspace over which APFs are to be developed. The ONF neuron is basically characterized by its *short-term activity* (STA) and *long-term activity* (LTA). The point robot's discretized workspace is mapped to the ONF, and points corresponding to the obstacles are inhibited from activation. Only the neuron corresponding to C_{goal} receives external excitation. In the initial state of the ONF, all STA and LTA states are zero. The first neuron to become active is the C_{goal} neuron. This activity then spreads to its *uninhibited* neighbors, at which point the C_{goal} neuron deactivates or *switches off* because of negative feedback. Subsequently, the uninhibited neighbors of the C_{goal} neuron's neighbors and the C_{goal} neuron become active while the C_{goal} neuron's neighbors switch off. Thus, the uninhibited neurons oscillate, and the LTA state counts the number of times the output changes. The activity is allowed to propagate throughout the ONF around the obstacle neurons until the neuron corresponding to the robot's initial position is activated. A path from the robot's initial position to its desired goal position is found by moving in the direction of increasing LTA state. The LTA state distribution is thus a form of an attractive potential field with the attractor being located at the goal position. Paths produced using the ONF are similar to those produced using the WENN attractive potentials.

Discussion

The neural networks discussed in this article are different in many respects, but they all serve as platforms for developing APFs useful for planning paths. While the WENN is capable of developing a variety of APFs useful for path planning, the other neural networks are capable of developing one type of APF or another. Park and Lee's feedforward neural networks operate on algebraic descriptions of the workspace, while the other neural networks develop APFs directly over discretized representations of the robot's workspace or Cspace. Thus, the feedforward neural networks very much depend on the particular workspace or Cspace, unlike the ONF and WENN.

Although the ONF and WENN are similar in some respects, there are significant differences between them. These differences enable the WENN to develop a variety of APFs, including the attractive potential field. The ability of the ONF and WENN to work with the discretized image of a robot's workspace makes them suitable for direct use with the robot's visual information-gathering system. The process of developing potentials by spreading activation by the ONF and WENN neural networks are somewhat similar to biological detour behavior models (Arbib and House, 1987). These models use spread functions to develop depth fields which provide information about both the distance and direction of objects in the field of view of animals.

Neural networks are highly parallel structures and therefore can be simulated on parallel computers. The WENN can be possibly implemented on digital hardware (Kassim and Vijaya Kumar, 1994), making it a highly attractive solution to the real-time path-planning problem. However, the process of developing the Cspace, especially that of the higher degree of freedom robots, is computationally intensive. Thus far, APFs developed by neural networks have been restricted to the physical workspaces of robots or the Cspaces of low degree of freedom robots.

Acknowledgments. The authors thank the Air Force Office of Scientific Research (grant no. 89–0551) and the National University of Singapore for support of this research.

Road Map: Control Theory and Robotics
Related Reading: Exploration in Active Learning; Planning, Connectionist; Reactive Robotic Systems

References

Arbib, M. A., and House, D. H., 1987, Depth and detours: An essay on visually guided behavior, in *Vision, Brain and Cooperative Competition* (M. A. Arbib and A. R. Hanson, Eds.), Cambridge, MA: MIT Press, pp. 129–163. ◆

Barraquand, J., and Latombe, J.-C., 1991, Robot motion planning: A distributed representation approach, *Int. J. Robotics Res.*, 10:628–649. ◆

Canny, J. F., and Donald, B. R., 1988, Simplified Voronoi diagrams, *Discrete Comput. Geom.*, 3:219–236.

Kassim, A. A., and Vijaya Kumar, B. V. K., 1992, A neural network architecture for generating potential fields for path planning, *Proc. Soc. Photo-Opt. Instrum. Eng.*, 1766:94–105. ◆

Kassim, A. A., and Vijaya Kumar, B. V. K., 1994, Using the wave expansion neural network for path generation, *Proc. Soc. Photo-Opt. Instrum. Eng.*, 2243:406–419.

Khatib, O., 1986, Real-time obstacle avoidance for manipulators and mobile robots, *Int. J. Robotics Res.*, 5:90–98. ◆

Latombe, J.-C., 1991, *Robot Motion Planning*, Norwell, MA: Kluwer.

Lemmon, M. D., 1991a, 2-Degree-of-freedom robot path planning using Cooperative Neural Fields, *Neural Computat.*, 3:350–362.

Lemmon, M. D., 1991b, *Competitively Inhibited Neural Networks for Adaptive Parameter Estimation*, Norwell, MA: Kluwer.

Lozano-Perez, T., 1983, Spatial planning: A configuration approach, *IEEE Trans. Comput.*, C-32:108–120.

Park, J., and Lee, S., 1991, Neural computation for collision free path planning, *J. Intelligent Manufacturing*, 2:315–326.

Prassler, E., and Milios, E., 1991, Parallel distributed robot navigation in the presence of obstacles, in *Proceedings of the IEEE International Conference on Robotics and Automation*, Sacramento, CA, pp. 475–478.

Principal Component Analysis

Erkki Oja

Introduction

Principal Component Analysis (PCA) and the closely related Karhunen-Loève transform, or the Hotelling transform, are standard techniques in feature extraction and data compression (see, e.g., Devijver and Kittler, 1982; Oja, 1983). As an example, take a sequence of 8 × 8 pixel windows from a digital image. They are first scanned into vectors x whose elements are the gray levels of the 64 pixels in the window. In real-time digital video transmission, it is essential to reduce these data as much as possible without losing too much of the visual quality, because the total amount of data is very large. Using the PCA, a compressed representation y is obtained from x, which can be stored or transmitted. Typically, y can have as few as 10 elements, and a good replica of the original 8 × 8 image window can still be reconstructed from it.

In general terms, the input vectors are random vectors x with K elements. No assumptions about the probability density of the vectors are needed, as long as their first- and second-order statistics are known or can be estimated from a sample. Typically the elements of x are measurements like pixel gray levels or values of a signal at different time instants. They are mutually correlated.

In the PCA transform, vector x is linearly transformed to another vector y with N elements, $N < K$, so that the redundancy induced by the correlations is removed. This transformation is achieved by finding a rotated coordinate system such that the elements of x in the new coordinates become uncorrelated. For instance, if x has a Gaussian density that is constant over ellipsoidal surfaces, then the rotated coordinate system coincides with the principal axes of the ellipsoid. In addition to achieving uncorrelated components, the variances of the elements of y will be strongly decreasing in most applications, with a considerable number of the elements so small that they can be discarded altogether. The elements that remain constitute vector y.

In mathematical terms, consider a linear combination

$$y_1 = \sum_{k=1}^{K} w_{k1} x_k = w_1^T x$$

of the elements $x_1, \ldots, x_K$ of vector x, where $w_{11}, \ldots, w_{K1}$ are scalar coefficients or weights, elements of a K-dimensional vector w_1, and w_1^T denotes the transpose of w_1. Usually it is assumed that x has zero mean; if not, then the mean vector is estimated separately and subtracted from x to obtain a zero mean vector.

The factor y_1 is the *first principal component* of x if the variance of y_1 is maximally large under the constraint that the norm of w_1 is constant (see, e.g., Devijver and Kittler, 1982). Then the weight vector w_1 maximizes the PCA criterion

$$J_1^{PCA}(w_1) = E\{y_1^2\} = E\{(w_1^T x)^2\} = w_1^T C w_1 \qquad \|w_1\| = 1 \qquad (1)$$

where $E\{\cdot\}$ is the expectation over the density of input vector

x, and the norm of w_1 is defined as

$$\|w_1\| = (w_1^T w_1)^{1/2} = \left[\sum_{k=1}^{K} w_{k1}^2 \right]^{1/2}$$

The matrix C in Equation 1 is the $K \times K$ covariance matrix defined by

$$C = E\{xx^T\} \qquad (2)$$

The solution is given in terms of the unit-length eigenvectors $c_1, \ldots, c_K$ of the matrix C. With $\lambda_1, \ldots, \lambda_K$ the corresponding eigenvalues in decreasing order (or nonincreasing in case of multiple eigenvalues), the solution is given by

$$w_1 = c_1$$

The criterion J_1^{PCA} in Equation 1 can be generalized to N principal components, with N any number between 1 and K. Denoting the nth ($1 \le n \le N$) principal component by $y_n = w_n^T x$, with w_n the corresponding weight vector, the variance of y_n is maximized under the constraints

$$\|w_n\| = 1 \qquad w_n^T w_m = 0 \qquad m < n \qquad (3)$$

Note that, compared to the first principal component, there is now the extra constraint that the weight vector w_n must be orthonormal with all the previous weight vectors. The solution is

$$w_n = c_n$$

and it follows that

$$E\{y_n^2\} = c_n^T C c_n = \lambda_n$$

This equation can often be used in advance to determine N, if the eigenvalues are known. The eigenvalue sequence $\lambda_1, \lambda_2, \ldots$ is usually sharply decreasing, and it is possible to set a limit below which the eigenvalues, hence principal components, are insignificantly small. This limit determines how many principal components are used.

Another possible extension of Equation 1 is

$$J_N^{PCA}(w_1, \ldots, w_N) = E\left\{ \sum_{n=1}^{N} y_n^2 \right\} = E\left\{ \sum_{n=1}^{N} (w_n^T x)^2 \right\}$$

$$= \sum_{n=1}^{N} w_n^T C w_n = \max \qquad (4)$$

$$w_m^T w_n = \delta_{mn} \qquad (5)$$

This criterion determines the subspace spanned by vectors $w_1, \ldots, w_N$ in a unique way as the subspace spanned by the N first eigenvectors $c_1, \ldots, c_N$, but does not specify the basis of this subspace at all.

If the constraint in Equation 5 is changed to

$$w_m^T w_n = \omega_n \delta_{mn} \qquad (6)$$

where all the numbers ω_n are positive and different, then this problem will have a unique solution given by scaled eigenvectors (Oja, Ogawa, and Wangviwattana, 1992).

To use the closed-form solutions given here, the eigenvectors of the covariance matrix C must be known. These are rarely known in practice. In an on-line data-compression application like image or speech coding, it is usually not possible to solve the eigenvector-eigenvalue problem for computational reasons. The PCA solution is then replaced by suboptimal standard transformations. Another alternative is to derive gradient ascent algorithms for the maximization problems that we have discussed. The algorithms will then converge to the solutions of the problems, i.e., to the eigenvectors. This approach is the basis of the neural network learning rules.

PCA Learning Algorithms and Neural Networks

Neural networks provide a novel way for parallel on-line computation of the PCA expansion. The Principal Component Analysis (PCA) network (Oja, 1992) is a layer of parallel linear artificial neurons shown in Figure 1. The output of the nth unit ($n = 1, \ldots, N$) is $y_n = w_n^T x$, with x denoting the K-dimensional input vector of the network and w_n denoting the weight vector of the nth unit. The number of units N will determine how many principal components the network will compute. Sometimes this number can be determined in advance for typical inputs, or N can be equal to K if all principal components are required.

The PCA network learns the principal components by unsupervised learning rules, by which the weight vectors are gradually updated until they become orthonormal and tend to the theoretically correct eigenvectors. The network also has the ability to track slowly varying statistics in the input data, maintaining its optimality when the statistical properties of the inputs do not stay constant. Because of their parallelism and adaptivity to input data, such learning algorithms and their implementations in neural networks are potentially useful in feature-detection and data-compression tasks.

Several basic learning algorithms are listed here. In the following, k denotes discrete time, thus $x(k)$ is a stream of input data vectors (e.g., image windows or segments of a time-varying signal) entering the learning neural network. The learning weights are $w_j(k), j = 1, \ldots, N$.

The Stochastic Gradient Ascent (SGA) Algorithm

The Stochastic Gradient Ascent (SGA) algorithm (Oja, 1983) is obtained from Equation 1 by taking the gradients with respect to weight vector w_j and using the normalization constraints. Denoting

$$\Delta w_j(k - 1) = w_j(k) - w_j(k - 1) \qquad (7)$$

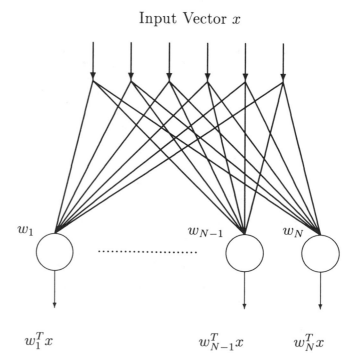

Input Vector x

w_1 w_{N-1} w_N

$w_1^T x$ $w_{N-1}^T x$ $w_N^T x$

Figure 1. The basic linear PCA layer.

the learning rule is

$$\Delta w_j(k-1)$$

$$= \gamma(k)y_j(k)\left[x(k) - y_j(k)w_j(k-1) - 2\sum_{i<j}y_i(k)w_i(k-1)\right] \tag{8}$$

where $\gamma(k)$ are the step sizes in the gradient ascent, typically a sequence of small numbers tending slowly to zero.

The first term on the right contains the product $y_j(k)x(k)$, which is a Hebbian term, and the other terms are implicit orthonormality constraints. The case $j = 1$ gives the constrained Hebbian learning rule of the basic PCA neuron introduced by Oja (1982). The convergence of the vectors $w_1(k), \ldots, w_N(k)$ to the eigenvectors $c_1, \ldots, c_N$ was established by Oja (1983). A modification called the Generalized Hebbian Algorithm (GHA) was presented by Sanger (1989), who also applied it to image coding, texture segmentation, and the development of receptive fields.

The algorithm may have significance in hierarchical clustering of learned cues in the cerebral cortex. Ambros-Ingerson, Granger, and Lynch (1990) performed simulations on the olfactory paleocortex, receiving inputs from the olfactory bulb. They used a network model resembling the SGA algorithm in which the first neuron (in their case, a competitive subnet) learns the input as such, and consequent neurons (subnets) learn progressively masked versions of the input. Masking corresponds to subtracting from the input the previous weight vectors as in Equation 8. The simulation revealed how perceptual hierarchies may arise for recognizing environmental cues.

The Subspace Network Learning Algorithm

The Subspace Network learning algorithm is as follows (Oja, 1983; Williams, 1985):

$$\Delta w_j(k-1) = \gamma(k)y_j(k)\left[x(k) - \sum_{i=1}^{N}y_i(k)w_i(k-1)\right] \tag{9}$$

It is obtained as a gradient ascent maximization of Equation 4. The network implementation is analogous to the SGA algorithm but still simpler because the feedback term, depending on the other weight vectors, is the same for all neuron units. Thus learning at an individual connection weight w_{ji} is local as it only depends on y_i, x_j, and the feedback term, all of which are easily accessible at that position in a hardware network. The convergence was studied by Williams (1985), who showed that the weight vectors $w_1(k), \ldots, w_N(k)$ will not tend to the eigenvectors $c_1, \ldots, c_N$ but only to a rotated basis in the subspace spanned by them, by analogy with the subspace criterion presented in the introductory section.

The Weighted Subspace Algorithm

The Weighted Subspace algorithm, proposed by Oja et al. (1992), is derived from the optimization criterion (Equation 4) with the constraints of Equation 6:

$$\Delta w_j(k-1) = \gamma(k)y_j(k)\left[x(k) - \theta_j\sum_{i=1}^{N}y_i(k)w_i(k-1)\right] \tag{10}$$

This algorithm is similar to the Subspace Network algorithm except for the scalar parameters $\theta_1, \ldots, \theta_N$, which are inverses of the parameters $\omega_1, \ldots, \omega_N$ in Equation 6. If all of them are chosen different and positive, then it has been shown by Oja et al. (1992) that the vectors $w_1(k), \ldots, w_N(k)$ will tend to the true PCA eigenvectors $c_1, \ldots, c_N$ multiplied by scalars. The algo-

rithm is appealing because it produces the true eigenvectors but can be computed in a fully parallel way in a homogeneous network.

Also, minor components defined by the eigenvectors corresponding to the smallest eigenvalues can be computed by similar algorithms (see Oja, 1992). A recent overview of these and related neural network realizations of signal-processing algorithms is given by Cichocki and Unbehauen (1993).

Learning PCA by Backpropagation Learning

Another possibility for PCA computation in neural networks is the multilayer perceptron network, which learns by the backpropagation algorithm in unsupervised autoassociative mode. This network was suggested for data compression by Cottrell, Munro, and Zipser (1987), and it was shown to be closely connected to the theoretical PCA by Bourlard and Kamp (1988) and Baldi and Hornik (1989).

The three-layer perceptron net is shown in Figure 2, in which the input and output layer have K units and the one hidden layer has $N < K$ units. The outputs of the hidden layer are given by

$$h = S(W_1 x + w_1) \tag{11}$$

where W_1 is the input-to-hidden-layer weight matrix, w_1 is the corresponding bias vector, and S is the usually nonlinear activation function, to be applied element-wise. The output y is an affine linear function of hidden layer outputs:

$$y = W_2 h + w_2 \tag{12}$$

with obvious notation. In autoassociative mode, the same vectors x are used both as inputs and as desired outputs, and thus y should be as close to x as possible in the least-mean-square sense.

For the linear net, backpropagation learning is especially feasible because, as was shown by Baldi and Hornik (1989), the "energy" function has no local minima. The three-layer net has been used for image compression, e.g., by Cottrell et al. (1987).

A network with an extra hidden layer after the input layer and another before the output layer, i.e., a five-layer perceptron net, can also be used for data compression, although due to the nonlinearities, it will not compute the PCA. With some general activation functions such a net is theoretically able to perform better than a three-layer net or a PCA net. A five-layer fully nonlinear net may be problematic to train by backpropagation, especially as the second and fourth layers may have to

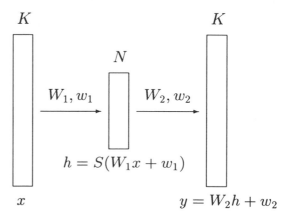

Figure 2. The three-layer multilayer perceptron network in autoassociative mode.

be large. Also the generalization ability may not be as good as that obtained by the linear PCA (Wang and Oja, 1993).

Discussion

The algorithms reviewed in this article are typical learning rules for the adaptive PCA extraction problem, and they are especially suitable for neural network implementations. In numerical analysis and signal processing, many other algorithms have been reported for different computing hardware. A review is Comon and Golub (1990). Experimental results on the PCA algorithms both for finding the eigenvectors of stationary training sets and for tracking the slowly changing eigenvectors of nonstationary input data streams have been reported by Oja (1983) and Sanger (1989). An obvious extension of the PCA neural networks would be to use nonlinear units, e.g., perceptrons, instead of the linear units. This method allows nontrivial cascading of neural layers to more complicated networks. They will then optimize some other criteria, related but not equivalent to the PCA; as an example, see Xu (1991).

Road Map: Dynamic Systems and Optimization
Background: I.3. Dynamics and Adaptation in Neural Networks; Motor Control, Biological and Theoretical
Related Reading: Associative Networks; Face Recognition; Post-Hebbian Learning Rules

References

Ambros-Ingerson, J., Granger, R., and Lynch, G., 1990, Simulation of paleocortex performs hierarchical clustering, *Science*, 247:1344–1348.
Baldi, P., and Hornik, K., 1989, Neural networks and principal components analysis: Learning from examples without local minima, *Neural Netw.*, 2:52–58.
Bourlard, H., and Kamp, Y., 1988, Auto-association by multilayer perceptrons and singular value decomposition, *Biol. Cybern.*, 59:291–294.
Cichocki, A., and Unbehauen, R., 1993, *Neural Networks for Optimization and Signal Processing*, New York: Wiley. ◆
Comon, P., and Golub, G., 1990, Tracking a few extreme singular values and vectors in signal processing, *Proc. IEEE*, 78:1327–1343. ◆
Cottrell, G. W., Munro, P. W., and Zipser, D., 1987, Learning internal representations from gray-scale images: An example of extensional programming, in *Proceedings of the Ninth Annual Conference of the Cognitive Science Society*, Hillsdale, NJ: Erlbaum, pp. 462–473.
Devijver, P. A., and Kittler, J., 1982, *Pattern Recognition: A Statistical Approach*, London: Prentice-Hall. ◆
Oja, E., 1982, A simplified neuron model as a principal components analyzer, *J. Math. Biol.*, 15:267–273.
Oja, E., 1983, *Subspace Methods of Pattern Recognition*, Letchworth, Eng.: Research Studies Press and J. Wiley. ◆
Oja, E., Ogawa, H., and Wangviwattana, J., 1992, Principal Component Analysis by homogeneous neural networks, Part I: The Weighted Subspace criterion, *IEICE Trans. Inform. Syst.*, E75-D:366–375.
Oja, E., 1992, Principal components, minor components, and linear neural networks, *Neural Netw.*, 5:927–935. ◆
Sanger, T. D., 1989, Optimal unsupervised learning in a single-layer linear feedforward network, *Neural Netw.*, 2:459–473.
Wang, L., and Oja, E., 1993, Image compression by MLP and PCA neural networks, in *Proceedings of the Eighth Scandinavian Conference on Image Analysis*, Tromso, Norway, pp. 1317–1324.
Williams, R., 1985, *Feature Discovery Through Error-Correcting Learning*, Technical Report 8501, University of California at San Diego, Institute of Cognitive Science.
Xu, L., 1991, Least Mean Square error reconstruction principle for self-organizing neural nets, *Neural Netw.*, 6:627–648.

Problem Solving, Connectionist

Steven Hampson

Introduction

The task addressed here is the construction of connectionist models that learn to produce sequences of actions (operator applications) that can move from any given current state to a state that satisfies a positive primary goal, while simultaneously avoiding negative goal states. No built-in knowledge other than an ability to recognize states that satisfy the primary goals is provided. Three behavior models, derived from animal learning theory, are considered.

The models are primarily developed in a Boolean (binary) state-space, although continuous inputs are also possible. Specifically, the world is modeled as a series of discrete "states" which can be described with a collection of binary (True/False, 0/1) features. Likewise the model outputs are a collection of binary operators which are capable of changing the world by changing the values of the features. Maze running is used as a concrete example of adaptive problem solving. A particular location in the maze is chosen as the goal state, and various other locations are identified as bad states. The general approach is applicable to problem solving in any state-space, but maze examples provide easily generated test cases that can exercise many important issues of problem solving.

An important issue in problem solving involves the tradeoffs between "procedural" and "declarative" knowledge, the basic issue being what is learned and stored for direct application versus what is inferred or computed at the time of behavior. Two extreme approaches to problem solving are considered which highlight this difference—one based on a Stimulus-Response + Stimulus-Evaluation (S-R/S-E) model of behavior in which state evaluations and appropriate responses are learned, and the other on a Stimulus-Stimulus (S-S) model, which simply records the observed relations between input stimuli and, when behavior is required, computes an appropriate response based on these accumulated "facts." In animal learning theory, the S-R/S-E model is viewed as procedural, while the S-S model is considered declarative. Both are "complete" in the sense that they can assemble arbitrarily long sequences of arbitrary actions in order to achieve a final goal, but they vary considerably in their time/space characteristics. In particular, the S-R/S-E model generally is slow to learn, but space efficient (in the number of nodes needed), while the S-S model has the opposite characteristics. A third (Stimulus, Response)-Stimulus (S, R)-S model which learns the effects of operators is also developed. It provides a look-ahead capability that is generally intermediate in behavioral characteristics between the S-R/S-E and S-S models. The general characteristics of these three models are considered with particular attention to the issue of multiple goals of variable strength.

Connectionist implementation of these models offers some advantages over more "symbolic" approaches, since the necessary continuously adjustable learning rates and continuous input and output functions are naturally provided by connectionist representations and learning schemes.

S-R/S-E Problem Solving

This model of CONDITIONING (q.v.) has a long history in animal learning theory (Bower and Hilgard, 1981, chap. 2) and is the simplest model of instrumental conditioning. It involves the learning of two functions: stimulus-response and stimulus-evaluation mappings.

Training an S-R Network Using Evaluation

Assuming that only one operator can be applied at a time, and that accurate evaluation is always available (for the contrary case, see REINFORCEMENT LEARNING), it is easy to assign credit or blame to individual operators, and modify behavior accordingly.

1. If an operator fired and things got worse, then it was wrong and should be off.
2. If an operator fired and things got better, then it was right and should be on and (perhaps) all other operators should be off.
3. If no operator fired and things got worse, then an operator should have fired so all increase their output.
4. If no operator fired and things got better, then none should have fired, and none did.

This trial-and-error strategy will cycle through the operators, repeating some, until an acceptable response is made. The system can then stabilize on correct output for that input pattern. With an appropriately chosen learning rate (i.e., not too fast) the system will eventually stabilize on correct output for all inputs to which it has been exposed. A standard two-level connectionist network can be used to learn and represent the S-R mapping.

The basic strategy is to produce behavior variation by some mechanism, and to evaluate the results of the executed action. The evaluation is used to increase the frequency of "good" responses and decrease the frequency of "bad" responses.

Learned Evaluation

Given an S-R network that can be trained with a good/bad evaluation signal, the next problem is to produce such a signal. The specific problem is how sequences of actions can be learned if only the final goal can be identified and evaluated. This problem of temporal credit assignment can be addressed with the introduction of learned, secondary evaluation. Primary (innate) evaluation identifies a particular goal state (with fixed evaluation = 1.0), and secondary (learned) evaluation is a heuristic measure of how desirable a state is with respect to achieving the primary goal. What is generally desired is an evaluation function that monotonically increases, or monotonically decreases with a state's problem-solving "distance" from the primary goal.

In an unknown environment, the Stimulus-Evaluation (S-E) function must be learned. A learning rule that can achieve this is:

$$\text{Eval} := \text{Eval} + (C * \text{Next\_Eval} - \text{Eval}) * rt$$

where C and rt are constants between 0 and 1, Eval is the evaluation of the input state at time t, and Next_Eval is the evaluation of the next input state at time $t + 1$ (see REINFORCEMENT LEARNING for a somewhat different formulation). The

evaluation function is adjusted on each state transition. With this learning rule, the Eval of a state approaches $C * \text{Next\_Eval}$, the rate of change depending on rt. If $rt = 1$, Eval is immediately adjusted to match $C * \text{Next\_Eval}$. An rt value of 0.1 is generally used. A standard two-level network can be used to learn the desired evaluation function.

In effect, this adjustment rule says that the learned evaluation of a state should predict the evaluations of the states that generally succeed it. This learning strategy was used in Samuel's (1963) classic checkers-playing program, and it is very similar to Rescorla's (1977) model of higher-order (multistep) conditioning in animals. Sutton (1988) and Werbos (1990) provide further discussion of this general learning strategy. With this learning rule, a state's learned evaluation indicates whether the goal can be reached, and, if $C < 1.0$, how far the goal is. Specifically, with a primary evaluation of 1.0 for the goal state, and a constant C for all transitions, a state's learned evaluation will approach and stay in the neighborhood of C^n, where n is the average number of state transitions to the primary goal. If the number of steps to the goal from a particular state is always exactly n, the state's evaluation is stable at exactly C^n. A C value of 0.9 is generally used. More generally, C could vary to reflect the "cost" of each transition. In this model, a state's evaluation decreases as its problem-solving distance to the goal increases. Other monotonic functions are possible, but an optimum evaluation of C^n is used in the problem-solving models considered here.

Assuming that the primary goal state can be identified and evaluated, this S-E learning rule, plus the previous S-R learning rule, permits goal-seeking operator sequences to be assembled. For example, in an m-state sequence, the transition to the final goal state, S_m (primary evaluation = 1.0), is immediately rewarding, so an operator can be learned that achieves it. State S_{m-1} is (almost) as good as the goal state once the correct operator for the $S_{m-1} \Rightarrow S_m$ transition is learned, and its learned evaluation approaches $C * 1.0 = C$. The learned evaluation of S_{m-1} then makes the transition from S_{m-2} to S_{m-1} rewarding, so an appropriate operator for that transition can be learned, and so on, in the manner of dynamic programming. Action sequences are incrementally assembled backwards from the goal, producing a gradient of evaluation leading to the final goal state. Similar S-R/S-E models have been developed by several researchers (Barto, Sutton, and Anderson, 1983; Jordan and Jacobs, 1990; Klopf, Morgan, and Weaver, 1993).

Only a single positive goal has been considered so far. However, it is a simple matter to extend the model to include innately bad states (negative primary evaluation), and negative secondary evaluation of places that generally lead to primary bad states. For example, in maze running, dead-end passages can be identified as innately bad, even when forward progress cannot yet be measured. A simple method of combining separate positive and negative evaluations is to add them. This is generally adequate for simple mazelike examples, but the problem of goal interaction is an important and difficult issue that needs to be addressed in any general model of behavior. This issue will be considered in somewhat greater detail later in this article.

Lookahead/S-E Problem Solving

The S-R/S-E model is purely reactive in the sense that a response is directly triggered by the stimulus and that no planning, or sequential consideration of alternative lines of actions, is required or possible. In this basic model, evaluation is used only to train the operators; it does not in itself directly influence the choice of operators. If, however, some form of internal simulated search were possible, evaluation alone could be used to choose the best next state, and the appropriate course of action, without the need for S-R heuristics. This section considers (S, R)-S operator models as one possible approach to simulated search. A certain amount of simulated search before action is used in essentially all artificial intelligence (AI) game-playing programs. (See any introductory AI text for a discussion of search techniques.) One-step lookahead is of interest as a limited form of search which is found in many models of instrumental conditioning.

Two methods of internal search will be considered: operator models and stimulus-stimulus associations. The first approach is considered in this section. The second approach will be summarized in the next section. The two approaches can be characterized as (S, R)-S, and S-S associations, respectively. Both can be viewed as providing simple "world models," which organisms can use to generate expectations of possible future states. In the first case, given a stimulus and a possible response (operator), the state resulting from the application of the operator can be predicted. In the second case, the possible neighboring states can be directly predicted without reference to intervening operators; a separate system is used to choose an appropriate operator once the desired next state has been chosen.

In artificial domains (e.g., checkers), operator models are generally given, but in real-world conditions, a connectionist model or biological organism might require specialized networks to learn and represent the expected transitions or effects of its operators (Colwill and Rescorla, 1986). Perhaps the most obvious (S, R)-S function to learn is: $(S_1$ and $R_x) \Rightarrow S_2$, which can be read as "if you are in state S_1 and perform response R_x, then expect to go to state S_2." This function can be learned by a standard two-level network, taking a current state description and a possible response as input, and producing an expected state description as output.

Unlike learning S-R operator heuristics, operator models can be trained with an explicit, perfect teacher on every overt transition. Specifically, every transition provides a (S, R)-S training example. If the predicted (S, R)-S transition is different from the observed transition, it can be immediately corrected. Learning the expected effects of an operator contrasts with training S-R heuristics in which the problem is to decide when the operators should be applied. Evaluation learning is the same as in the S-R/S-E model.

The most time-expensive step of this approach is the sequential consideration of all possible transitions from the current state. In maze problems, it may be possible to consider each alternative; but in more general problem-solving behavior, the number of neighboring states may be too large to search exhaustively. However, considering a few transitions is presumably better than considering none at all, and the benefits can be maximized by searching the right few. If learned S-R heuristics were used to suggest an ordering of candidate transitions, the result would be quite similar to Mowrer's (1960) model of instrumental conditioning.

Goal Representation and Utilization

The current-state "stimulus" is from the external environment, but internally generated goal features can also be included, which significantly extends the behavioral capabilities of the models. In this section, the models are extended to include explicit goal representation and utilization. The S-S model is also briefly described. Two main topics are considered: whether goal magnitude is used as a *desired value* for a feature or as a *measure of importance* of achieving the goal. Goal features (e.g., "alleviate thirst," or "maze goal at location (x, y)") can be either binary [0, 1] or continuous (0, 1).

Goals as Desired Values

With the inclusion of goal features as part of the input stimulus, an operator can implement a servomechanism. A great deal of intelligent behavior and its underlying neural machinery can be analyzed in terms of hierarchies of continuously valued servomechanisms (e.g., Gallistel, 1980, chap. 6; Albus, 1981, chap. 5). Minimally, a servomechanism takes two inputs, the current value of some feature, and the desired value for that feature. The output of the unit can affect the value of the feature, and is used to adjust the actual value toward the goal value. A servomechanism can thus be viewed as a (Stimulus + Goal) ⇒ Response, or (S, G)-R, operator. This mapping can be learned by a standard connectionist network, so learning (S, G)-R associations requires no modification of the previously discussed S-R learning mechanism. Goal features can simply be added as undistinguished features to the S-R network's inputs.

Goals of Variable Importance

Another important issue in problem solving is the possibility of multiple primary goals with independently variable importance or strength. It is useful in this context to consider a Stimulus-Stimulus (S-S) model of problem solving (Deutsch, 1960). Its strengths and weaknesses provide a point of comparison, and it suggests a simple approach to handling multiple goals of variable strength. It can be summarized as follows: In the S-S model, each observed state of the world is represented by a unique node, and all neighboring states are connected with a link strength of C. These next-state links can be learned during interaction with the environment by observing which transitions are possible or impossible. Each node computes its output as the maximum of its inputs. With this simple "world model," any state can be activated as a primary goal (fixed output $= G$), and by spreading activation, all other states are activated with a subgoal strength of $C^n * G$, where n is their minimum distance to the goal state. Consequently, for a primary goal strength of $G = 1.0$, each state's computed subgoal strength is equal to its optimum learned evaluation in the S-E model. Any number of primary goals can be activated with any goal strength. Since each node outputs the maximum of its inputs, strong goals simply overwhelm weak goals. The output of bad states is held at zero so that the goal gradient is forced to spread around them. Behavior is generated by moving to the neighboring state with the greatest goal strength in the internal model (see Cognitive Maps).

An advantage of the S-S model is that it is guaranteed to take the shortest path from any current state to any single goal, and that it easily handles multiple goals of variable strength. A disadvantage is that, unlike the other two models, it requires a complete decoding of the input space (one node per state). In this section, the S-R/S-E and lookahead/S-E models are extended so that they can also deal with multiple goals of variable strength in a manner that parallels the S-S model.

For simulation purposes, each primary goal is represented by a unique goal feature that indicates when the goal is active and how important it is. Its "goal strength" (G) is a continuous value between 0 and 1. Since the evaluation (E) of an input state with respect to a particular primary goal is inversely proportional to the estimated distance or cost to achieve the goal from that state, the product EG seems a reasonable measure of a state's "total evaluation" that takes into consideration both the distance and current desirability of the goal. As previously observed, for a perfect evaluation function, this is equal to the subgoal strength of the corresponding state node in the S-S model.

Learning evaluation functions for multiple primary goals is straightforward if one-step lookahead is possible. All states that are one step away from the current state are considered and evaluated, and each primary goal is adjusted toward the best of its next-state evaluations. "Bad" next states are simply ignored. This is a straightforward generalization of the single-goal case.

With lookahead, choosing an appropriate next state to actually move to is equally easy. Simply move to the neighboring state that has the single largest $E \cdot G$ value over all primary goals. Given optimum evaluation functions, this mimics the behavior of the S-S model. However, because a state that satisfies a number of weaker goals might be preferable to a state that satisfies a single strong goal, an alternative strategy of going to the state with the largest sum of $E \cdot G$ rather than the single largest $E \cdot G$ has some appeal. Unfortunately, this strategy runs the risk of being trapped at $E \cdot G$ maxima that do not satisfy any goals. Like a frog snapping at the "average" position of several flies, the model can get stuck if it tries to work on more than one goal at a time (see VISUOMOTOR COORDINATION IN FROGS AND TOADS).

A similar problem of goal interaction can occur if multiple desired-value goals are simultaneously active in the input to the (S, G)-R operator network. A simple solution to this is to limit the active goal input to the (S, G)-R system to a single goal at a time. With lookahead, the best next-state $E \cdot G$ can be used to choose the best single goal. Without lookahead, this next-state information is not available, but the $E \cdot G$ values of the current state provide an estimate of the best next-state $E \cdot G$ for each goal. Accordingly, the primary goal with the largest $E \cdot G$ for the current state can be chosen as the single goal to be worked on, and it alone is input to the (S, G)-R system with a strength of 1.0.

Training the (S, G)-R/S-E model for multiple goals is only slightly more complex than the original single-goal case. For the chosen goal, operator and evaluation training can be treated exactly as in the single goal case. That is, for the chosen goal, Eval is adjusted toward $C * \text{Next\_Eval}$. Also as before, if Next_Eval > Eval, the total $(S + G)$ input pattern is a positive training example for the applied operator, and if Next_Eval < Eval, it is a negative example.

The (S, G)-R/S-E system can also learn about nonchosen goals. In particular, if Next_Eval $* C >$ Eval for *any* goal, the achieved transition is desirable with respect to that goal, even if that goal wasn't the one chosen to work on. Consequently, if Next_Eval $* C >$ Eval for any primary goal, that goal's evaluation can also be adjusted. With this training rule, a goal's evaluation function is more like the *optimum* cost to the goal rather than the *expected* cost. With one-step lookahead, the two are the same, since optimum evaluation immediately leads to optimum behavior; but with S-R heuristics, the operators must separately learn the optimum transitions that correspond to the optimum evaluation.

Simulation Results

The three models were tested on 10×10 mazes. Specifically, the world is defined as a 10×10 grid with a prespecialized feature for each of the 100 input patterns. Four operators are provided to move left, right, up, and down. The input space is continually cycled through, randomizing the order after each cycle, until evaluation and/or behavior is stable. See Hampson (1990, 1994) for simulation details; only the most general results are summarized here.

The behavior of the S-S model is easy to predict. Link learning is fast since a transition, or lack of one, has to be observed only once to be incorporated into the net. Behavior is always perfect in the sense of taking the shortest route to the best goal.

With one-step lookahead, learning is slower since learned evaluation must incrementally back up one step at a time. Consequently, learning time increases with the longest optimum path in the maze. However, once evaluation is learned, behavior is still optimum. Operator selection time increases with the number of possible transitions from the current state.

Learning in the S-R model is slower still, since the S-R system can try only one transition per pattern presentation, rather than considering all neighboring states by internal simulation. More specifically, learning time increases with both the length of the longest optimum path and the number of operators that can be applied per state. In addition, the system can stabilize on suboptimum evaluation and behavior. On the other hand, operator selection time is essentially independent of the number of possible transitions.

In order to investigate the effects of multiple goals, an additional set of 100 internal goal features, one per possible state, was included in the system.

Evaluation learning time for the lookahead model is basically unaffected by the addition of multiple goals, and as before, once correct evaluation is learned, behavior is identical to the S-S model.

As before, evaluation learning time for the (S, G)-R/S-E model is slower than the lookahead/S-E model. In addition, because training on one (S, G)-R function partially disrupts (S, G)-R associations for other goals, (S, G)-R learning time increases dramatically with the number of goals. Appropriate (S, G)-R associations can eventually be learned, but learning was so slow that it was impractical to train the (S, G)-R/S-E network over all possible $(S + G)$ inputs.

Finally, the presence of obstacles (bad states) poses no further problems for the S-S or lookahead/S-E models, but further complicates the learning task for the (S, G)-R/S-E model. Specifically, with a single goal, the (S, G)-R/S-E system could easily deal with obstacles, but with multiple goals and obstacles, the necessary (S, G)-R operator functions are more complex and the network becomes impractically expensive in the number of nodes needed. In order to produce reasonable convergence characteristics, something approaching a complete decoding would be required having a unique node for all possible (S, G) pairs.

Discussion

The three models considered here have very different time/space requirements, consequently their relative merits are highly domain dependent. However, the excessive training time and complexity of the (S, G)-R/S-E system resulting from multiple goals and obstacles highlights an issue that is directly addressed in the S-S model. The complete S-S model (Deutsch, 1960) has two separate components: the previously described goal-setting S-S network, and a goal-achieving (S, G)-R network which takes the output of the S-S net as its G input. Neither component is trained by reinforcement learning; they simply observe and memorize the possible one-step transitions in the world, and the one-step effects of the individual operators, respectively. Only immediately adjacent (S, G) pairs are learned by the (S, G)-R operators, and at most $S * O$ nodes are needed, where O is the number of operators. Using S nodes and at most S network updates, the S-S net computes the hard part of the overall (S, G)-R mapping by setting appropriate one-step subgoals, which are then given to the one-step operators.

This explicit division of labor seems a reasonable strategy; while any problem-solving task can always, in principle, be solved with a sufficiently large set of (S, G)-R heuristics, it is also possible to learn the effects of the operators and the layout of the world as independent processes, which are only combined at the time of actual problem solving. By treating them as separate processes, more specialized, high-level techniques can be utilized for each task. In general, the ability to plan ahead and establish appropriate subgoals which are then fed to a simple "motor" system seems almost unavoidable. At some point, higher-level planning systems are necessary to augment the basic (S, G)-R/S-E system. The ability of the S-S system to set appropriate subgoals is one possible approach to that problem.

Thus, while a single, general mechanism can, in principle, learn correct behavior, the overall problem-solving task can also be approached as a number of more specialized tasks, each of which permits more specialized learning and representation mechanisms. By identifying specific aspects of the overall task and addressing them with specialized systems, performance can often be greatly improved. The complexity of the nervous system clearly reflects the variety of storage and processing tasks that can be productively addressed with specialized structures.

Road Map: Artificial Intelligence and Neural Networks
Related Reading: Planning, Connectionist; Reinforcement Learning in Motor Control

References

Albus, J. S., 1981, *Brains, Behavior and Robotics*, Peterborough, NH: BYTE/McGraw-Hill.

Barto, A. G., Sutton, R. S., and Anderson, C. W., 1983, Neuron-like adaptive elements that can solve difficult learning control problems, *IEEE Trans. Syst. Man Cybern.*, SMC-13:834–846.

Bower, G. H., and Hilgard, E. R., 1981, *Theories of Learning*, Englewood Cliffs, NJ: Prentice Hall.

Colwill, R. M., and Rescorla, R. A., 1986, Associative structures in instrumental learning, in *The Psychology of Learning and Motivation*, vol. 20 (G. H. Bower, Ed.), New York: Academic Press.

Deutsch, J. A., 1960, *The Structural Basis of Behavior*, Chicago: University of Chicago Press.

Gallistel, C. R., 1980, *The Organization of Action*, Hillsdale, NJ: Erlbaum.

Hampson, S. E., 1990, *Connectionistic Problem Solving: Computational Aspects of Biological Learning*, Boston: Birkhäuser.

Hampson, S. E., 1994, Problem-solving in artificial neural networks, *Progr. Neurobiol.*, 42:229–281.

Jordan, M. I., and Jacobs, R. A., 1990, Learning to control an unstable system with forward modeling, in *Advances in Neural Information Processing Systems 2* (D. S. Touretzky, Ed.), San Mateo, CA: Morgan Kaufmann.

Klopf, A. H., Morgan, J. S., and Weaver, S. E., 1993, A hierarchical network of control systems that learn: Modeling nervous system functions during classical and instrumental conditioning, *Adapt. Behav.*, 1:263–319.

Mowrer, O. H., 1960, *Learning Theory and Behavior*, New York: Wiley.

Rescorla, R. A., 1977, Pavlovian second-order conditioning: Some implications for instrumental behavior, in *Operant Pavlovian Interactions* (H. Davis and H. M. B. Hurwitz, Eds.), Hillsdale, NJ: Erlbaum.

Samuel, A. L., 1963, Some studies in machine learning using the game of checkers, in *Computers and Thought* (E.A. Feigenbaum and J. Feldman, Eds.), New York: McGraw-Hill.

Sutton, R. S., 1988, Learning to predict by the methods of temporal differences, *Machine Learn.*, 3:9–44.

Werbos, P. J., 1990, Consistency of HDP applied to a simple reinforcement learning problem, *Neural Netw.*, 3:179–189.

Process Control

Lyle H. Ungar

Introduction

Neural networks are being used throughout the chemical-processing and related industries such as production of polymers, steel, and composites. Similar techniques are used in production of aircraft and many other industries. This article covers the use of neural networks for process control: not just control as it is typically defined (see ADAPTIVE CONTROL: NEURAL NETWORK APPLICATIONS and ADAPTIVE CONTROL: GENERAL METHODOLOGY) but a host of related applications such as virtual sensors and fault detection and diagnosis. It concentrates on examples in the control of chemical processes, but it should be of interest to people in a wide variety of industries with similar problems.

Neural networks have been used in a vast variety of different control structures and applications, serving as controllers and as process models or parts of process models (e.g., as virtual sensors). They have been used to recognize and forecast disturbances, to detect and diagnose faults, to combine data from partially redundant sensors, to perform statistical quality control, and to adaptively tune conventional controllers such as PIDs (proportional, integral, and derivative controllers). Many different structures of neural networks have been used, with feedforward networks, radial basis functions, and recurrent networks being the most popular. Most of the examples cited here use standard feedforward networks; RADIAL BASIS FUNCTION NETWORKS (q.v.) are more popular in adaptive control, since they are linear in the coefficients of the basis functions (Sanner and Slotine, 1992).

The different uses of neural networks covered in this article can be characterized by a set of mappings as shown in Table 1. Neural networks may be used in direct control, where they provide a mapping from the current state (or, more commonly, current and recent sensor readings as in an ARMA model—see FORECASTING for definitions and details) and the desired next state of the plant to the control action to be taken. In indirect

Table 1. Input-Output Mappings for Different Uses of Neural Networks

Use	Input-Output Mapping
Direct control	Current state + Desired next state → Control action
Indirect control	Current state + Control action → Next state
Virtual sensors	Easy-to-measure properties → Hard-to-measure properties
Fault diagnosis	Sensor readings → Faults

control, the networks are used as a nonlinear model of the plant: given the current state of the plant and a proposed control action, they predict the next state of the plant. This model can then be used in a variety of controllers, such as model predictive control (MPC).

Neural networks can also be used as a piece of a larger model of the plant. Often one wishes to control variables such as concentration or viscosity, which are relatively hard to measure accurately. Other variables may be much easier to measure and be related to the variables of interest in a deterministic but highly nonlinear way. For example, a variety of spectral techniques such as infrared (IR) measurements can be used to predict concentrations of chemical species and hence quality properties such as viscosities. Neural networks can be trained on the basis of data collected in the laboratory to learn the mapping between the easier-to-measure and the harder-to-measure variables and then used as "virtual sensors." For example, it is often inconvenient and expensive to measure the exact chemical composition of the gases going up a stack, yet it is important to be able to monitor and control them to meet environmental regulations. Cheaper and more reliable sensors can be used to measure related variables, and then neural networks used to estimate the concentrations.

Finally, neural networks are being used to detect and diagnose faults and to compensate for sensor failures or drifts. These applications are closely related to control, serving either as preprocessors (filters) for controller inputs or as components in supervisory control schemes running above conventional process controllers.

This article covers three of the most important uses of neural networks in chemical process control: model predictive control, virtual sensors, and fault detection and diagnosis.

Model Predictive Control with Neural Networks

As mentioned previously, there are a number of ways that neural networks can be used in process controllers. In almost all cases, the strengths and weaknesses of the control scheme remain unchanged when a neural network is used in place of ARMA or mechanistic models. Many different neural network architectures can be used in neurocontrollers. The most popular are backpropagation networks and radial basis functions. In both of these cases, an ARMA style model is typically used in which the inputs to the neural network include lagged values of the relevant measured variables. Alternatively, when the process has variable time delays, it may be more advantageous to use internally RECURRENT NETWORKS (q.v.) and input just the current measured variable.

The *Internal Model Control* (IMC) framework (Garcia and Morari, 1982; Nahas, Henson, and Seborg, 1992) provides a good example of how neural networks can be incorporated into controllers (Figure 1). In conventional IMC, a model of the plant, typically linear, is partially inverted to determine a control action. Neural networks can be used as the controller (for direct control), as the plant model (to be "inverted" for indirect control), or as both (Psichogios and Ungar, 1991).

For indirect control the model of the plant is learned by training a neural network so that its output $\hat{y}_t$ approximates the function $y_{t+1} = f(y_t, u_t)$, where y_t is a set of plant measurements at time t, and u_t contains control actions. Lagged values of y and u may also be given as inputs to the network. Finding a control action requires partially inverting this model to find the control action u_t to minimize the difference $\| \hat{y}_{t+1} - y_{t+1}^{\mathrm{sp}} \|$ between the plant state predicted at time $t + 1$, $\hat{y}_{t+1}$, and that specified by the set point y_{t+1}^{sp}. If the function f were linear, the minimum would be easy to find analytically. However, neural networks are nonlinear, and therefore Newton's method or similar nonlinear equation-solving methods must be used to find the optimal control action. Fortunately, the Jacobian (the derivative of f with respect to each of the inputs or, equivalently, the linearization of f) required for such methods is generally available, since it is required when using the backpropagation or conjugate gradient algorithms to determine the network weights.

As an alternative to the computationally burdensome partial inversion of the plant model, a neural network can be trained to directly approximate the desired control function:

$$u_t = g(y_t, y_{t+1}^{\mathrm{sp}}) \tag{1}$$

This relationship provides a control law for a direct feedforward controller. In general, the model (or inverse model) learned will be imperfect, and feedback must be used to stabilize the system. IMC provides one framework for doing so, as shown in Figure 1. Consider the case where the model gives an estimate $\hat{y}_{t+1}$ of the plant output which is higher than the actual output y_{t+1}. The difference $y_{t+1} - \hat{y}_{t+1}$, which is the estimation error, is subtracted from the set point y_{t+1}^{sp} so that the controller tries to achieve a lower target y. If the error in the model is roughly constant, this method will cancel the error due to model inaccuracy.

Under certain conditions, one can train the controller network directly from past plant data, but if the controller is used in an IMC framework, one must be sure it is an accurate inverse of the plant model, since although IMC compensates for approximate plant models, it does not compensate for inaccuracies in inverting the model. (If the plant model is not invertible, this approach will, of course, be hard to use.) One technique often used is to propagate errors back through the plant model to the controller (Jordan and Rumelhart, 1992; see SENSORIMOTOR LEARNING).

Although it is relatively easy to implement and widely used, IMC has several disadvantages. It is typically used as a one-

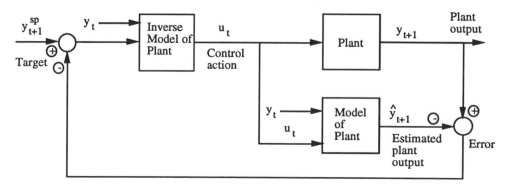

Figure 1. IMC control structure.

step-ahead predictor, and as such will not work on unstable plants or when there is a non-minimum phase response which requires looking more than one time step into the future to determine the correct answer. (For example, increasing the concentration of a reactant will often initially decrease the temperature and then later increase it; one-step-ahead controllers fare badly on such problems.) Also, IMC lends itself poorly to including constraints on the actuators (e.g., valves cannot open beyond some maximum) or on the controlled variables (e.g., temperature should not exceed some value). Thus it is more common to use neural networks in model predictive controllers.

The most widely used method of incorporating neural networks into controllers is within the framework of model predictive control (MPC), where the neural network is used as a process model (Psichogios and Ungar, 1991; Hunt et al., 1992). In this case, the method of MPC and all of its advantages and disadvantages remain unchanged; it simply becomes easier to obtain the model.

In a typical MPC architecture, one uses an optimizer to pick a sequence of control actions u to minimize the difference between the target y^{sp} and the actual value $\hat{y}$ over the next N time steps. More formally, one wishes to pick u to find the minimum:

$$\min \sum_{i=1}^{N} a_i[y_i^{sp} - (\hat{y}_i - d)]^2 \qquad (2)$$

where d is an estimate of persistent disturbances, used to provide a form of feedback to account for modeling errors, measurement errors, and process disturbances, and a_i is a weighting, typically giving more weight to errors in the near future.

Usually, one sets the control action constant over the last portion of the time horizon ($j = M$ to N).

$$u_j = u_{j+1} \qquad j = M, \ldots, N-1 \text{ for } N > M \qquad (3)$$

As mentioned earlier, MPC allows one to set bounds on the output y, e.g., to keep a temperature in some range

$$y_{min} \leq y_i \leq y_{max} \qquad i = 1, 2, \ldots, N \qquad (4)$$

MPC also allows one to set bounds on the control action u_i,

$$u_{min} \leq u_i \leq u_{max} \qquad i = 1, 2, \ldots, N \qquad (5)$$

and on the rate of change of the control action,

$$|u_i - u_{i+1}| \leq \Delta u_{max} \qquad i = 1, 2, \ldots, M-1 \qquad (6)$$

One then uses an optimization technique such as Sequential Quadratic Programming (SQP) to minimize Equation 2 subject to the constraints of Equations 3–6 and the neural network model of the plant,

$$\hat{y}_{t+1} = f(y_t, y_{t-1}, \ldots, u_t) \qquad (7)$$

Such optimization methods typically use gradient descent using the derivatives of the objective function (Equation 2) with respect to the control variables u_i, with Lagrange multipliers added to handle constraints. Note that if one is going to optimize over many steps, it is important to train the neural network to minimize the multistep prediction error rather than the one-step-ahead prediction error (see FORECASTING).

Neural networks are often offered as a panacea for the nonlinear models needed for MPC. After all, if one can take data from previous plant operation, fit a model, and hand it over to an optimizer, one should get a "perfect" controller. Unfortunately, this assumption is not true. Plant data are often inadequate to generate good models. In particular, data taken during closed-loop operation tend to provide a model of the controller rather than the plant. For example, if a plant has a

temperature controller on it, then when the plant temperature is above the set point, the controller will increase the flow of cooling water. Plant data will thus show that high cooling-water flow rates are correlated with high temperatures. A controller built using such a model will, of course, give disastrous performances. Similarly, but less spectacularly, if a given input has remained constant in the past, a neural network model (like any other regression-style model) will "show" that the input has no effect on the output. This result, too, will not provide a good basis for control.

Another problem with neural network models is that they are not necessarily accurate; insufficient data can produce models which are aphysical, even when an adequate network structure is built which includes the necessary delays of inputs. For example, they may not satisfy conservation of mass or energy, particularly when they are extrapolating or interpolating in regions where few data were available for training. Networks may give accurate predictions one time step in the future, but the predictions many time steps in the future used by the model predictive controller may still be inaccurate (see FORECASTING). More subtly, the models may be reasonably accurate in making predictions in the time domain but may give poor control because they are inaccurate in the frequency domain. Accurate models in the time domain (where neural networks are almost exclusively used) do not guarantee accurate models in the frequency domain, and frequency domain accuracy can be far more important for accurate control of dynamic systems. (Stability of controllers on a plant is determined by their dynamic characteristics in key frequency ranges, which may not be obvious when looking at standard plots of plant state over time.)

It bears repeating that although neural networks ease the task of building accurate multivariable nonlinear models, they do not eliminate the problems and difficulties involved in developing accurate and robust controllers. Although neural networks give good nonlinear models (and are much harder to analyze than linear models), their use is subject to all the usual limitations and dangers of conventional models. One must still determine appropriate control structures based on the plant and disturbance structure, including time delays and frequency response characteristics. IMC structures are not appropriate for controlling unstable systems, and accurate plant models are of no use if the disturbances dominate process behavior or if the plant is inherently hard to control because of poor design.

Hybrid Networks, Virtual Sensors, and Fault Diagnosis

Neural networks can also be used in conjunction with first principles models (Psichogios and Ungar, 1992) or as virtual sensors (Piovoso and Owens, 1991). Often one has a reasonable partial theoretical model of a plant but lacks certain parts of the model such as the dependence of viscosity or reaction kinetics on temperature and concentration. Neural networks can be used to learn that portion of the model. The overall hybrid model contains both the theoretical equations such as mass and energy balances and the neural networks, and can be used in an MPC scheme as described in the preceding section.

More formally, plants can often be described in the form

$$dx/dt = f(x, u, p) \qquad (8)$$

$$p = g(x, u) \qquad (9)$$

where the function f is known (e.g., kinematics or mass and energy conservation) but contains "parameters" p which depend on the plant state x and possibly the control action u in unknown ways. Observations of x and u can then be used to train a neural network to approximate the unknown function

Figure 2. Plant and controller using estimator as virtual sensor for z.

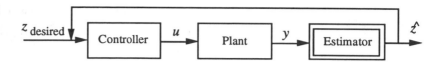

z desired → Controller → u → Plant → y → Estimator → $\hat{z}$

g using an extension of the backpropagation algorithm even though the parameters p cannot be measured. The resulting plant model consisting of Equations 8 and 9 can be trained more accurately, using less data, and extrapolates better than pure neural network plant models which do not incorporate the known form of f (Psichogios and Ungar, 1992).

A similar situation arises when one cannot easily measure some quality variables (e.g., viscosity, photodegradation, or even concentrations of certain chemical species). For example, consider the control of the viscosity of a polymer product, where viscosity cannot be measured on-line. Temperatures y in the reactor can easily be measured, and the measurements can perhaps be supplemented with measurement of near-IR absorption. On-line measurement of viscosity is difficult. Often companies take samples, which are then sent to a laboratory for analysis. It can take from a half hour up to several hours to get an analysis from the laboratory. A neural network trained with product samples whose viscosity z has been measured in the lab can be trained to predict z as a function of variables y such as temperature and pressure which are measured on line:

$$z = f(y) \qquad (10)$$

Such neural networks can be used as *virtual sensors* to give on-line estimations of viscosity for a controller.

Willis et al. (1992) give a nice example of the use of virtual sensors for inferential control of penicillin production. To optimally control penicillin production, one would like to have accurate on-line estimates of biomass concentration during fermentation. Such measurements are unfortunately not available. Willis et al. use a neural network to estimate biomass from carbon dioxide evolution rate, batch age, and the rates that the two primary substrates are fed to the reactor. These plant measurements (y in Figure 2) are then used to estimate the biomass z, which is the controlled variable.

Neural networks have also been used to minimize the effect of sensor noise and drift (data reconciliation), as elements of nonlinear statistical process controllers (SPC), and for fault diagnosis, sometimes as a prelude to reconfiguring control loops to cope with faults such as sensor or actuator failure. Neural networks can be trained which take a set of sensor readings as inputs and predict the same sensor readings as outputs (Kramer, 1992). When fewer hidden nodes are used than inputs, this system has the effect of nonlinearly projecting the sensor readings onto a smaller dimensional space and then reconstructing the sensor readings, thus reducing the noise in the signal. Such rectified sensor readings take advantage of the correlation structure of the sensor readings and, like other forms of filtered signals, can reduce the effect of sensor noise on control.

Neural networks have also been used extensively for fault diagnosis. In theory, it is relatively simple to train a network on a set of data where the inputs to the network are sensor readings just after a fault and the outputs of the network indicate what the fault was (Venkatasubramanian and Chan, 1989). A typical output-encoding scheme is to have one output for each fault, and have 1 represent the presence of a fault and 0 represent the absence of a fault. This design works well on simulated faults, but it can be more problematic in the real world. Apart from the usual questions of how much past (lagged) data to

include, in many plants there is not a sufficient quantity of data recorded from real faults, so one is forced to rely on less realistic data from simulators. Faults are often of different magnitudes (e.g., a leak may be of different sizes) and may occur when the plant is in different states (e.g., running at capacity or during start-up), and thus extensive fault data are required. Most data sets are for the case of single faults. It is only sometimes the case that a network trained on individual faults will be able to diagnosis combinations of simultaneous faults.

Discussion

The field of neural networks for process control is much too extensive to be fully summarized here: hundreds of papers are published each year in the area, and several commercial vendors offer control systems using neural networks. It is clear that neural networks are attractive models to use when processes are nonlinear and good first principles (mechanistic) models are not available, either for entire processes or for parts of the process. Most control schemes using neural networks are identical to those that do not use neural networks; the "neurocontrollers" just use neural networks as process models (e.g., in MPC), as virtual sensors, or as controllers. Neural networks can also be used to forecast disturbances (see FORECASTING) as part of a feedforward control system. Often, neural controllers are used as a feedforward component of a more complex control architecture in which a standard feedback controller rejects disturbances and gives adequate control while the neural network is being trained (see, e.g., Lee and Park, 1992, and citations there to Kawato and others). Many other uses of neural networks are possible, including their use as data preprocessors, as fault detectors (possibly coupled to a system which reconfigures the control loops when faults are discovered), and as automatic tuners for conventional controllers. Looking farther into the future, neural networks will be used to provide continuous optimization of plants, perhaps using REINFORCEMENT LEARNING (q.v.; see also Barto, 1990).

Road Map: Applications of Neural Networks
Background: I.3. Dynamics and Adaptation in Neural Networks; Motor Control, Biological and Theoretical
Related Reading: Applications of Neural Networks; Model-Reference Adaptive Control; Reinforcement Learning in Motor Control

References

Barto, A. G., 1990, Connectionist learning for control, in *Neural Networks for Control* (W. T. Miller, R. S. Sutton, and P. J. Werbos, Eds.), Cambridge, MA: MIT Press, pp. 5–58. ◆

Bhat, N., and McAvoy, T. J., 1990, Use of neural nets for dynamic modeling and control of chemical processes, *Comput. Chem. Engrg.*, 14:573–583. ◆

Garcia, C. E., and Morari, M., 1982, Internal model control, 1: A unifying review and some new results, *Ind. Eng. Chem. Process. Des. Dev.*, 21:308–323.

Hunt, K. J., Sbarbaro, D., Zbikowski, R., and Gawthrop, P. J., 1992, Neural networks for control systems—A survey, *Automatica*, 28: 1083–1112. ◆

Jordan, M. I., and Rumelhart, D. E., 1992, Forward models: Supervised learning with a distal teacher, *Cognit. Sci.*, 16:307–354.

Kramer, M. A., 1992, Autoassociative neural networks, *Comput. Chem. Engrg.*, 16:313–328.

Lee, M., and Park, S., 1992, A new scheme combining neural feedforward control with model-predictive control, *Am. Inst. Chem. Eng. J.*, 38:193–200.

Nahas, E. P., Henson, M. A., and Seborg, D. E., 1992, Nonlinear internal model control strategy for neural network models, *Comput. Chem. Engrg.*, 16:1039–1057.

Narendra, K. S., and Parthasarathy, K., 1990, Identification and control of dynamical systems using neural networks, *IEEE Trans. Neural Netw.*, 1:4–27.

Piovoso, M. J., and Owens, A. J., 1991, Sensor data analysis using neural networks, in *Chemical Process Control: CPC IV*, Proceedings of the Fourth International Conference on Chemical Process Control, New York: AICHE, pp. 101–118.

Psichogios, D. C., and Ungar, L. H., 1991, Direct and indirect model based control using artificial neural networks, *Ind. Engrg. Chem. Res.*, 30:2564–2573.

Psichogios, D. C., and Ungar, L. H., 1992, A hybrid neural network: First principles approach to process modeling, *Am. Inst. Chem. Eng. J.*, 38:1499–1511.

Sanner, R. M., and Slotine, J.-J. E., 1992, Gaussian networks for direct adaptive control, *IEEE Trans. Neural Netw.*, 3:837–863.

Venkatasubramanian, V., and Chan, K., 1989, A neural network methodology for process fault diagnosis, *Am. Inst. Chem. Eng. J.*, 35:1993.

Willis, M. J., Montague, G. A., Di Massimo, C., Tham, M. T., and Morris, A. J., 1992, Artificial neural networks in process estimation and control, *Automatica*, 28:1181–1187.

Programmable Neurocomputing Systems

Nelson Morgan

Introduction

Neurocomputing is processing in which information is represented in the pattern of activity and connections between elements that perform simple operations on their inputs (e.g., a saturating nonlinearity of a weighted sum of the inputs). One of the attractions of neurocomputing is that it is often quite naturally parallelized. For this reason, much recent research in neurocomputing hardware has focused on efficient implementation of the most common neural algorithms, such as Hopfield networks, self-organizing feature maps, and backpropagation learning for multilayer perceptrons. Typically, matrix operations are at the core of these algorithms, and neural hardware is often customized for their efficient implementation. This can result in chips that are cost-effective solutions to specific applications, assuming that the design is also optimized for memory and I/O requirements.

Other research in neurocomputing is focused on developing new algorithms. For most of this work, specialized neurocomputing chips are not sufficient, and programmable computers are required. Workstations provide the necessary programming flexibility, but they are much slower than the best specialized chips. It is likely that specialization will always provide a speed advantage over the fully general approach.

Fortunately, computers can be more general than a single-function engine yet more specialized than a general-purpose computer. Architectures can be designed to be optimal for a class of applications and still have greater flexibility than a fixed-function implementation. Some common characteristics of these computers are:

- Single-task machine (no multi-user operating system)
- Physical memory only (no virtual memory)
- Arithmetic elements, interconnection, and memory paths designed to be efficient for a given class of algorithms
- Relatively short wordlengths (often fixed-point or integer) that are sufficient for the target tasks
- Some programmability

Neurocomputing hardware will nearly always include a fast dot-product engine that will be optimized for fixed-point or single precision floating-point operations. This is in contrast to both conventional supercomputers requiring double-precision floating-point, and experimental (e.g., analog) neural chips using low-precision computational mechanisms. Otherwise, neurocomputers vary widely in the extent of specialization for neural computation. All can be programmed, but some with great difficulty, since the focus is on accelerating a limited form of computation. For instance, the matrix/vector operations that characterize much neural computation can often only be modified by writing microcode or nonstandard assembly language routines. In other systems, higher-level languages and commercial assemblers can be used for most modifications. The latter case needs additional resources for the control required for generality. However, a simple Reduced Instruction Set Computing (RISC) CPU can be fabricated on a compact piece of silicon, and can provide general-purpose programming capabilities. In some way, neurocomputers must permit users to program non-neural pieces of a full task.

This article provides a 1994 snapshot of programmable neurocomputers. Factors affecting utility and performance of these machines are discussed, with a look toward future machines. In fact, neurocomputer design is not fundamentally different from mainstream computer design; the ideas presented here should look familiar to the reader versed in conventional computer architecture (see Hennessy and Patterson, 1990, for a good text on this topic).

The focus in this article is on computing systems used for research and development of neural algorithms, as opposed to systems for embedded applications of production-ready networks, or research into network mappings onto the physics of electronics or optics. In this context, the word *programmable* refers to the ability of the neurocomputer to run programs, as opposed to the mutability of neural parameters such as weights. Finally, the applications focus is on network training for cases in which workstation technology is insufficient, for instance, with large amounts of training data and a large number of trained parameters.

System Requirements

Neurocomputation

While other operations are also necessary, fast computation of matrix products is the primary requirement for neurocomputing. Fundamental equations for search, learning, and recall primarily consist of matrix/matrix multiplications (including matrix/vector or vector/matrix multiplication), as described in Ramacher and Ruckert (1991).

In neurocomputing systems for algorithmic research, data streams are most commonly digital, and the functionality must be programmable. Therefore, digital arithmetic elements are currently used. In some cases, these elements implement floating-point arithmetic. This is flexible, but for efficiency many neurocomputing machines use fixed-point arithmetic. Several studies have reported that 16 bits of resolution for the weights and 8 bits for activations are sufficient, as long as longer words are used for intermediate calculations. These studies were typically done for backpropagation, and other kinds of neurocomputation may require different precision. For algorithmic research, hardware should permit multiple precision for further flexibility.

Memory bandwidth is important for neurocomputation, as for many other domains. In some sense, memory bandwidth dominates the design problem as very large scale integration (VLSI) has progressed. Roughly speaking, chip computation is proportional to die area, while memory bandwidth is proportional to the chip perimeter. Although neural algorithms have been developed that reduce memory bandwidth, they often slow convergence, yielding no net savings.

Caching schemes are more useful for instructions than for weights in common learning paradigms, since weights are typically not reused in per-pattern update schemes. Instruction caching can be very important; without a cache, off-chip memory bandwidth can be strongly impacted by the need to read in instructions from external memory.

Other Computational Requirements

Requirements for full neurocomputing applications can be complex. Users express desired operations in a standard programming language, and execution should be efficient on the hardware despite a mix of non-neural operations. Even for neurocomputation, matrix sparsity, conditional execution, and weight sharing can complicate the data flow considerably. Neurocomputer design is simplified by implicit assumptions about these factors.

A fast computational chip is not enough to make a neurocomputer useful. System design includes consideration of testing and debugging, memory bandwidth and latency, communication between processors, I/O, and software. Even when programmability is not a design goal, there must at least be an interface to a programmable host's software. Without these considerations, the hardware may either lose performance or never be fully debugged.

Traditional computer science continues to be important for neurocomputer design; for instance, the rule of thumb called Amdahl's Law (named after Gene Amdahl, who published a related statement in 1967; see Hennessy and Patterson, 1990, for extended discussions on the topic and its history). The gist of this law is that the improvement of overall system performance attributable to the speeding up of one part of the system (typically through parallelism) is limited by the fraction of the job that is not speeded up (e.g., serial code). For instance, if 90% of a task is sped up infinitely, the overall task is done only 10 times as quickly.

Amdahl's law holds for neurocomputers as well. Stated in the context of this paradigm, Amdahl's law (for neurocomputers) would be: *A neurocomputing accelerator can at best speed up an application by a factor of 1/(fraction of non-neural computation).*

Thus, a neurocomputing capability of a billion connections per second coupled with a conventional capability of 10 million arithmetic or logical operations per second suggests a ratio of 100 to 1 between neural and non-neural operations (to avoid the non-neural operations dominating).

Of course, a good neurocomputer should provide hardware to assist in the most common non-neural operations, such as relevant address calculations. However, a broad interpretation of Amdahl's law should hold for any real system. In such a system, balance is required not only between classes of instructions, but also between computation and I/O, interprocessor communication costs, and, perhaps most importantly, ease of use for problems of interest. The statistic most interesting to the neurocomputer user is problems solved per week, rather than connections per second. If the software is insufficient to the task, then the desired algorithms will never be translated satisfactorily to the machine, and the hardware will be useless. This is not idle speculation; in fact, the majority of all parallel research machines have never been used outside of machine validation and toy applications.

The cost of non-multiply-accumulate operations (e.g., disk I/O, interprocessor communication, and logical data operations) may not be negligible. For $O(N^2)$ connections, $O(N)$ inputs and outputs, and P processors, most of the run time is attributable to the non-neural component if it takes t_{nc} steps per input or output unit, with

$$t_{nc} > N/P$$

For example, if $N = 100$ for a net with one hidden layer and uniform layer size, there are 20,000 connections and 200 inputs and outputs. Assume 100 processors. If each multiply-accumulate takes a single cycle, a forward pass will take 200 cycles. If the per-unit non-parallelized non-dot-product overhead is greater than 1 cycle, it will take most of the execution time.

Some major causes for this overhead are:

- *Instruction balance*: Preprocessing of inputs and postprocessing of outputs of networks increase execution time for a real task. Possible solutions include: using more general chips; adding some general capability on an otherwise specialized chip; modifying the algorithm to be more neural; and building specialized hardware to speed up the non-neural operations.

- *Interprocessor communications*: Since interchip communication is typically slower than on-chip communication, data movement between chips can impact performance. Systolic or assembly-line data movement patterns can reduce this effect but are not always appropriate for full tasks. Generally, off-chip communication is made as fast as possible, while the problem is mapped to maximize local (on-chip) data movement.

- *I/O*: Interaction with the host computer or disk is often orders of magnitude slower than arithmetic. While neural algorithms are notorious for their poor cache behavior, disk access in large sequential blocks is still a reasonable policy for most algorithms. Once the blocks are in memory, random access may be used without the large penalty paid for random disk access. For some designs and tasks, however, the system can be I/O bound.

Software

If a neurocomputer is not flexible, the theoretical speed is useless for algorithm development. Unfortunately, flexibility and speed are often contradictory requirements. The user needs a general-purpose programming environment. Often one must port existing serial code. In general, compilers will not take advantage of the special capabilities of the neurocomputer.

A common compromise is to use a general programming language (e.g., C) augmented by hand-coded library routines. These routines generally perform such common operations as matrix/vector multiplies. When all code (including the non-library sections) is implemented on the neurocomputer, there is no overhead for remote procedure calls between host and neurocomputing engine. When part of the code is implemented on the host machine, the overhead for such calls and returns degrades performance.

The most common library routines are designed to be very efficient. For this purpose, routines are written in microcode or assembly code. The hardware may require programming with specialized coprocessor (e.g., vector) instructions. Communications primitives are sometimes provided (e.g., for efficient broadcast capabilities).

In addition to compilers, debuggers and other programming tools can be important in practice. System-level libraries, e.g., for file I/O, are also necessary for complete programs.

A neural network simulator can improve system ease of use. It is difficult, however, to permit broad flexibility of expression while preserving the high performance that is obtained by using the efficient library routines. In practice, this almost always results in some compromises; many changes to the algorithm will not easily map to existing libraries.

Software development must include routines for testing and debugging. Some debugging can be done using serial versions of the basic routines on a workstation. Subtle problems in library code are more difficult to detect, suggesting that greater effort must be made to check these routines at design time.

We briefly note the requirement for system debugging hardware and software. For instance, the data network can jam so that the system cannot return to a neutral state. Diagnostic hardware and software permit resetting and examination of the system state.

A Brief Survey of Implementations

Hardware accelerators for neurocomputing have been built for decades. Fully programmable neurocomputers are more recent, profiting from progress in VLSI and computer system design.

Parallel computers are commonly Single Instruction stream, Multiple Data stream (SIMD) machines. SIMD computers execute each instruction on multiple processing elements. Alternatively, computers can be Multiple Instruction stream, Multiple Data stream (MIMD). For an intermediate computational model called Single Program Multiple Data stream (SPMD), the same program is run on all processing nodes, but with no constraint to execute each instruction simultaneously. The SPMD model has much of the simplicity of SIMD without requiring tight ("lock-step") instruction-level node synchronization.

In this section, some major neurocomputer types are described, along with some examples. While far from a complete list of neurocomputing systems, these are chosen to illustrate each design class.

Designs Based on Commercial DSP Chips

Many neurocomputers use Digital Signal Processing (DSP) chips from Texas Instruments, Motorola, AT&T, and others. These chips provide fast dot-product arithmetic, general-purpose programming support, convenient addressing modes, and

loop support. Other neurocomputers have used general-purpose microprocessors. To some extent, the distinction between general CPUs and DSP chips is disappearing, since the former are becoming much better at signal processing, and the latter are adopting industry-wide standards such as the IEEE 754 floating-point format.

Machines based on DSP chips are generally flexible, since general-purpose programming languages can be used for most routines, and many operations can be sped up through parallelism. Conversely, raw peak performance is often much less than what is available from a more specialized system, since much of the resources (e.g., silicon area, output pins) is devoted to providing general programmability. For instance, the Texas Instruments TMS320C30 has a single multiplier-accumulate data path and uses 32-bit floating-point for all operations. A similar-sized chip developed by LSI Logic for an image-processing application some years ago performed 64 8×8 multiplies per cycle; however, it was not programmable.

DSP-based neurocomputers usually have between 4 and 64 DSPs, each with local memory, and a ring-style interconnection between processors. For the size and regularity of common neurocomputing tasks, this simple approach is often sufficient.

Two examples are:

- *RAP*: This system (Ring Array Processor) was developed in Berkeley (Morgan et al., 1992). Implementations used between 4 and 40 TMS320C30 DSP chips. The largest RAP has a peak capability of 1.28 Gflops. The maximum network recall speed for this machine is 640 million connections per second (MCPS), assuming two floating-point operations/connection. One test actually measured 570 MCPS. A global ring runs at the DSP clock speed of 16 MHz. SPMD programming is used, so that nodes need not be in lock-step synchrony, but only a single program is run on all nodes. A matrix and vector library was written in TI assembler. More general processing is done in C and C++, and object-oriented environments also have been written. Large multilayer perceptrons have been trained for speech recognition, including one with 342 inputs, 4000 hidden units, and 61 phonetic outputs (1.6 million free parameters). This was trained on 7 million acoustic vectors to do speaker-independent continuous speech recognition for a 5000-word vocabulary. RAP machines have also been used for numerical integration of coupled differential equations for modeling neural masses and also for recurrent networks. (See DIGITAL VLSI FOR NEURAL NETWORKS.)
- *Music*: This system (MUltiprocessor System with Intelligent Communication) uses up to 63 Motorola 96002 DSP chips, each implementing IEEE single-precision arithmetic with a peak performance of 60 Mflops (Muller et al., 1992). A global 5 MHz ring connects nodes, and global communication can be overlapped with computation. As with RAP, SPMD is used, and memory is distributed. Most programming is done in C, but time-critical sections are written in assembler. Peak performance for the largest system is 3.8 Gflops, or 1900 MCPS. A number of algorithmic studies have been performed making use of the computational capabilities of the machine.

Designs Based on Systolic MAC Chips

Much of the silicon resources in DSP-based neurocomputers are devoted to general-purpose capabilities, since each node can run complete programs. If nearly all computation is assumed to be matrix oriented, the computation can be accelerated by a chip that is focused almost exclusively on the Multiply-Accumulate (MAC) operation, and programmability can be relegated to the board or host level. A common class of systems that rely on this assumption are systolic systems, so called because the data are rhythmically computed by a set of pipelined arithmetic elements. These systems commonly rely on SIMD operation of many nodes in a ring or mesh topology.

Two examples are:

- *SYNAPSE* (SYnthesis of Neural Algorithms on a Parallel Systolic Engine): This SIMD machine (Ramacher and Ruckert, 1991) was designed to accelerate several matrix equations that apply to most neural models. A Motorola MC68040 is used for non–compute-intensive operations, and a systolic array of neural signal processors is used for computationally intensive operations. The signal processor, called an MA16, has 16 fixed-point multipliers per chip, with peak performance of 800 MCPS.
- *SNAP*: This SIMD machine uses a chip with four processing elements implementing IEEE 32-bit floating-point arithmetic (Means and Lisenbee, 1991). A linear (ring) systolic array is used, and an eight-chip system has a peak performance of 640 MCPS. Each processing element includes a floating-point multiplier, a floating-point adder, and an integer arithmetic logic unit (ALU). A microinstruction can implement any logical or arithmetic operation, though the architecture is optimized for multiply-accumulate–based operations. C programs can call efficient matrix/vector library routines, or run on an Intel i860 host. A proprietary microassembly language can also be used.

Designs Based on Nonsystolic MAC Chips

A bus is a reasonable structure for the broadcast operations that typify some neural computation, but has electrical and communications limitations for large numbers of processors. There is only one major bus-based neurocomputing engine in this category that we know of (see DIGITAL VLSI FOR NEURAL NETWORKS).

- *CNAPS* (Connected Node Architecture for Parallel Systems): This system is based on a chip with more than 11 million transistors (Hammerstrom et al., 1993, in Przytula and Prasanna, 1993). For such a large chip, it is unlikely that all arithmetic units will operate properly, so redundancy techniques are used to guarantee 64 functional processors on each chip. Each processor performs fixed-point arithmetic and uses a common bus. A typical system uses four computational chips, with a peak computational throughput of 6.4 billion CPS. The programming model is SIMD, although conditional execution based on a flag value is provided for each processing element. A specialized control processor can be programmed in a subset of ANSI C, using library functions where possible.

Designs Based an RISC + Coprocessor Model

Ideally, the programming interface to a neurocomputer would appear to be the same as for a workstation but would produce efficient code for neurocomputation. We do not know how to do this. However, calling efficient library functions from C has been a reasonable approach. It has often been difficult to extend libraries because of the specializations in the assembler or microcode.

For greater ease of programming, designers can use a standard RISC instruction set that is augmented with vector instructions. We have been developing a fixed-point vector MIMD machine in Berkeley that follows this approach, called the Connectionist Network Simulator (CNS-1) (Asanović et al., 1994). This machine consists of 4 to 512 nodes connected in a mesh that is end-connected on one dimension (forming a cylinder). Some characteristics of the node design are:

- MIPS-II standard instruction set architecture, including coprocessor interface.
- Specialized coprocessor implementing fixed-point vector instructions.
- Multiple (8) arithmetic units working simultaneously on a vector instruction.
- Multiple (4) registers for each vector element to provide temporal parallelism (so that a single vector instruction can operate on multiple elements).
- Multiple (3) pipelines for each data element so that multiplication, addition, and memory access can all be done simultaneously.
- A fast network interface.

The standard instruction set permits the use of commercial software tools, augmented to accommodate the vector instructions. Users program in assembly language for the computationally intensive pieces (the library routines), but mostly program in a high-level language, maintaining high efficiency for target applications. The gap between speed on scalar code (on a node) and vector code is only an order of magnitude, which is small for a machine of this power. Similarly, the gap between the bandwidth for local memory references (within node) and off-node references is roughly one order of magnitude. Finally, I/O capabilities for input from disks or other off-machine sources (such as video cameras) are being designed to run at network speeds.

CNS-1 is currently being designed, and projected peak performance of a 128-node system is 50–100 billion CPS. Smaller systems will also be built with 4 to 16 CPUs, and a paper design will be done for a larger system with 512 nodes.

Discussion

As with other computers, good system design is important for neurocomputers. In particular, attention must be paid to scalar and other non-neural operations, communication costs, I/O between the host/disk subsystem(s), and the neurocomputer, software, and diagnostic capabilities. Careful attention to these factors can result in a system that provides good sustained performance for many neurocomputing problems and flexible programming capabilities for users.

In this article, I have focused on operations other than dot-product computation. This was deliberate, since matrix arithmetic is the most obvious system design requirement and is dealt with adequately in many papers (for instance in S. Y. Kung's chapter in Przytula and Prasanna, 1993). However, even this operation is not handled trivially. For instance, memory access to large memories must keep up with the computational engine. Often a design will assume re-use to take advantage of larger, slower memories (for instance, evaluating network outputs for several input patterns at a time), but more generally, memory bandwidth should match computational throughput. Memory speeds (particularly for DRAM) have not kept pace with CPU clock rates. However, new developments of high-speed DRAM technology should help. These include synchronous DRAM, high-speed serial DRAM interface (Ramlink), and packet-based DRAM interface (Rambus).

Finally, most neurocomputer designs implicitly assume densely connected networks. As networks grow larger (as they always do once users have the capability), they tend to be more sparsely connected. Networks with a million neurons are unlikely to be fully connected, even if the memory for such a behemoth is affordable, as the data required for training so many weights will not be available. Therefore, a major concern for future neurocomputers must be efficient implementation of sparsely interconnected networks.

Road Map: Implementation of Neural Networks
Related Reading: Multiprocessor Simulation of Neural Networks; Neurosimulators; Silicon Neurons

References

Asanović, K., Beck, J., Feldman, J., Morgan, N., and Wawrzynek, J., 1994, A supercomputer for neural computation, in *Proceedings of the IEEE International Conference on Neural Networks*, Piscataway, NJ: IEEE Press, vol. 1, pp. 5–9.

Graf, H., Janow, R., Henderson, D., and Lee, R., 1991, Reconfigurable neural net chip with 32K connections, in *Advances in Neural Information Processing Systems 3* (R. P. Lippman,, R. Moody, and D. S. Touretzky, Eds.), San Mateo, CA: Morgan Kaufmann, pp. 1032–1038.

Hennessy, J., and Patterson, D., 1990, *Computer Architecture: A Quantitative Approach*, San Mateo, CA: Morgan Kaufmann. ◆

Kato, H., 1990, A parallel neurocomputer architecture toward billion connection updates per second, in *Proceedings of the International Joint Conference on Neural Networks*, vol. 2, New York: IEEE, pp. 47–50.

Means, R., and Lisenbee, L., 1991, Extensible linear floating point SIMD neurocomputer array processor, in *Proceedings of the International Joint Conference on Neural Networks*, vol. 1, New York: IEEE, pp. 587–592.

Morgan, N., Beck, J., Kohn, P., Bilmes, J., Allman, E., and Beer, J., 1992, The Ring Array Processor (RAP): A multiprocessing peripheral for connectionist applications, *J. Parallel Distrib. Comput.*, 14:248–259.

Muller, U., Baumle, B., Scott, W., Kohler, P., Gunzinger, A., and Guggenbuhl, W., 1992, Achieving supercomputer performance for neural net simulation with an array of digital signal processors, *IEEE Micro Mag.*, 12(5):55–65.

Przytula, K. W., and Prasanna, V. K., 1993, *Parallel Digital Implementations of Neural Networks*, Englewood Cliffs, NJ: Prentice Hall. ◆

Ramacher, U., and Ruckert, U., 1991, *VLSI Design of Neural Networks*, Dordrecht: Kluwer. ◆

Wawrzynek, J., Asanovic, K., and Morgan, N., 1993, The design of a neuro-microprocessor, *IEEE Trans. Neural Netw.*, 4:394–399.

Prosthetics, Neural

Gerald E. Loeb

Introduction

Two types of physical systems are known to be capable of real-time information processing: (1) electronic circuits, in which information is carried by electrons in metal conductors, and (2) neural circuits, in which information is carried by ions in water. Much current research is concerned with discovering or exploiting common principles of information processing in these two systems. Thus, it is natural that real-time interfaces between these systems have been developed so that each can be used to study the other.

Neural prosthetics are clinical applications of neural control interfaces whereby information may be exchanged between neural and electronic circuits. The electronic technology to date has been derived from cardiac pacemakers, which themselves have evolved from the fixed-rate, single-channel stimulators of the 1950s to programmable and adaptive systems equipped with sensors and sophisticated data processing capabilities. Prostheses that interact directly with neurons are now used routinely to treat sensorineural deafness (cochlear prosthesis) and some types of respiratory insufficiency (phrenic nerve stimulator). Feasibility studies are in progress regarding the treatment of incontinence and impotence and the restoration of useful vision. There is active clinical research on the reanimation of paralyzed limbs with functional neuromuscular stimulation (FNS) in patients who have had a stroke or spinal cord injury and on the treatment of various dystonias, such as spasticity and tremor.

This article is not concerned with the current design and performance of specific devices; these devices have been described elsewhere (Loeb, 1989; Agnew and McCreery, 1990). Rather, it considers these applications from the perspective of the following synergistic relationships between basic and applied research in neural control:

1. The clinical and commercial value of neural prostheses justifies the development of technology that is also useful in basic research.
2. The implantation of sophisticated neural control interfaces in sentient observers creates unique opportunities for a new class of psychophysical research into naturally occurring neural circuitry.
3. The development of functional replacement parts for the nervous system forces researchers to examine and test theories of neural computing more rigorously than they might do otherwise.
4. The development of neural prosthetic controllers that can deal successfully with the exigencies of daily life will almost certainly require principles and methods of neural networks and other forms of adaptive control.

Hardware Interfaces

In principle, information could be transferred into and out of the nervous system by any of several means, including chemical, magnetic, optical, and ultrasonic. In practice, neural prostheses require temporospatial resolution and physical portability that has been achieved only with the types of electrical signals that are familiar to most neurophysiologists. Thus, the future of neural prosthetics depends on the well-developed biophysical principles of excitable membranes and on the development of technology that can approach physical limits that are readily predictable from those principles.

Stimulation

The enormous information-carrying capacity of the nervous system is achieved by having great numbers of parallel channels, each of which has a low bandwidth by modern electronic standards. In most parts of the nervous system, these channels are packed closely together, often not in any particular geometrical relationship. Electrical currents induced in the extracellular fluids surrounding these neural channels tend to activate them according to the strength of the voltage gradient in time and space and the geometric size and orientation of the neural processes (Ranck, 1975). Because the tissues of the body are moderately good volume conductors, electrical current tends to spread radially from a point source, and the voltage gradient falls as the square of the distance from the source. In general, the amount of information that can be transferred into the nervous system by a neural control interface depends on finding a location in which the target neurons are spread as widely as possible in some geometrically organized way and in which relatively large numbers of closely spaced electrodes can be positioned close to those neurons without compromising their viability.

Recording

The close packing of parallel channels in a volume-conductive medium makes it even more difficult to record signals from the nervous system than to inject signals into it. The volume conductivity surrounding the neurons quickly dissipates their action potentials so that it is difficult to detect an extracellular action potential with an amplitude of greater than 0.1 mV, even when the recording electrode is within a few microns of the neuron (Rall, 1962). Such an electrode must, of course, have a tiny exposed contact area. At low voltages, the metal-electrolyte junction resembles a small capacitor in series with the access resistance of the immediately adjacent fluid, presenting a relatively high impedance (on the order of 1 mΩ at 1 kHz) and its associated thermal noise (on the order of 0.01 mV). It is likely that such an electrode will record action potentials from several of the closely packed neurons, differing only slightly in amplitude and shape and subject to large changes with the tiniest shift of position of the electrode.

Given the relatively low signal-to-noise level and mechanical lability of microelectrode signals, it is not surprising that their use is confined to basic research. There is, however, considerable interest in recording such signals, particularly in patients with spinal cord injury, where it would be most useful to record sensory information from intact somatosensory afferents in the paralyzed limb and command information from the disconnected, but functional, motor cortex.

Cochlear Prostheses

Current Technology

Cochlear prostheses use direct electrical stimulation of auditory nerve cells to bypass absent or defective hair cells that normally transduce acoustic vibrations into neural activity. They are the most sophisticated and successful neural prostheses to date, and they are still evolving rapidly. The currently available devices all use multicontact electrodes inserted into the scala tympani of the cochlea so that they can differentially activate auditory neurons that normally encode different pitches of sound (Loeb, 1990). An external, wearable control unit determines a pattern of multichannel electrical stimulation according to the spectral content in the signal from a microphone and a previously stored map of auditory sensations that can be elicited by electrical stimulation of each contact.

There are several competing algorithms for converting the acoustical information in speech sounds into electrical stimuli;

they are sometimes used in combinations, making it difficult to summarize the alternatives. One pair of options is to distribute the electrical stimulation among the parallel channels according to the raw spectral energy distribution itself or to extract the various frequencies associated with the fundamental vibration of the vocal cords and with the clusters of harmonic frequencies (called *formants*) produced by the vocal tract. Another set of options involves the electrical waveforms applied to the electrodes, which may be band-filtered versions of the continuous analog signal detected by the microphone or brief, biphasic waveforms applied sequentially to the various electrodes at repetition frequencies that may encode low-frequency information (*rate pitch*) or may be at arbitrarily fast rates (Wilson et al., 1991).

Research Questions

Is there an optimal strategy for presenting information to all postlinguistically deafened patients?   Patient variability is perhaps the major obstacle to the widespread use of cochlear prostheses in the many patients who cannot effectively use an acoustic hearing aid. Profoundly deaf patients, with essentially no detectable hearing, benefit from even poor results with a cochlear prosthesis (e.g., by gaining an awareness of environmental sounds and improving their lip-reading skills). Lesser, but more common, impairments are defined in terms of unaided sound thresholds that can be amplified acoustically for detectability. However, there may not be sufficient resolution for comprehension of speech. Patients with cochlear implants obtain results that range from virtually complete rehabilitation (able to understand unstructured conversation over the telephone) to only general awareness of environmental sounds. In some cases, high levels of performance occur almost immediately after the initial adjustment of the speech processor, while other patients show gradual improvement over a period of years. It remains unclear how much of this variability stems from the condition of the remaining auditory nerve, the adequacy of tailoring of the speech processor strategy to the patient, or the inherent psychoacoustic capabilities or plasticity of the higher central nervous system centers, which are known to vary widely, even among normal hearing subjects.

The next phase of research on cochlear prostheses will focus more on predicting or resolving poor performance than on achieving occasional triumphs. This focus will require a concerted, multidisciplinary approach to understanding the details of the electrical fields produced by various electrode geometries, their interactions with the unusual biophysical properties of spiral ganglion cells, and the neural circuits responsible for extracting spectral information from neural activity according to spatial distribution, phase-locked repetition rates, and interneuronal phasing. Modeling at the level of single neurons and feature-detecting networks will contribute to this effort, but it is complicated by the relative importance of the precortical relays of the auditory brainstem. These relays are a maze of highly specialized neurons and connection patterns associated with the many aspects of acoustical signal processing; their rules for forming and modulating connections are completely unknown.

How does neural plasticity interact with cochlear prosthetic strategies in prelinguistically deafened children?   In the visual system, repair of congenital problems such as corneal opacities and strabismus must be undertaken in the first few years of life to be effective because the developing brain has a limited period during which it learns to process sensory information. Attempts to apply cochlear prostheses to adults who lost hearing before acquiring speech have been unsuccessful, presumably because they have missed a similar critical period in the development of the auditory nervous system. Conversely, cochlear prostheses applied to deaf children before or during the presumed critical period for the development of hearing may be even more successful than those applied to postlinguistically deafened adults. The temporospatial patterns of neural activity produced by even the best cochlear prosthesis are likely to represent a great distortion from those that are remembered by the adult patient, whereas the naive nervous system of a child may learn to cope with the noise and may even develop neural algorithms to extract information that are unlike those seen in the adult. Animal studies have shown that chronic electrical stimulation of the deafened immature nervous system tends to reduce the atrophic changes that deafness usually produces in various levels of the auditory brainstem (Leake et al., 1991; Moore and Kowalchuck, 1988). Clinical studies in children have demonstrated surprisingly good results, even with relatively crude prostheses (Miyamoto et al., 1994), but this research is hampered by its necessarily long-term and poorly controllable nature. Within the next few years, however, there will be substantial numbers of mature patients whose auditory nervous systems have developed under the exclusive influence of a wide range of prostheses based on different speech processing algorithms. Some are likely to be given differently designed cochlear prostheses in the contralateral ear after reaching maturity. Detailed psychoacoustic testing of their various capabilities and limitations should provide important insights about the rules for information processing in auditory neural networks.

Visual Prostheses

Current Technology

Attempts to provide useful visual sensations in the blind by direct electrical stimulation of the visual cerebral cortex go back more than 25 years (Brindley and Lewin, 1968), but there is no active clinical research on a functional visual prosthesis. The initial devices used arrays of small electrodes (approximately 1 mm in diameter) on the pial surface. Relatively high stimulus currents (approximately 1 mA for a 200 µs pulse) were required to activate neurons whose cell bodies tend to lie about 2 mm deep in the cortex. Because of current spread by volume conduction, such stimulation presented to a single electrode presumably recruits neurons scattered over many adjacent cortical columns, but the surround inhibitory mechanisms actually result in surprisingly well-formed sensations of a single dot of light called a *phosphene*. This finding seems to suggest that a complete, if coarse-grained, picture could be built up from a sufficient number of such phosphenes. The problem is that the processes responsible for focusing operate slowly, so stimulus trains presented concurrently but interleaved between even two such sites produce unpredictable, nonlinear interactions (Girvin, 1988).

More recently, intracortical microelectrodes have been employed successfully to create similar phosphenes with stimulus currents (10–20 µA) that would tend to recruit only a few neurons within the immediate vicinity of the electrode tip (Bak et al., 1990). When two sites spaced less than 1 mm apart are stimulated concurrently, their phosphenes seem to combine and fuse in a predictable and desirable manner. Recent advances in silicon fabrication (Wise and Najafi, 1991) actually make it feasible to build dense arrays with hundreds of contacts and associated electronic circuitry that are probably safe to implant and operate continuously for long periods.

Researchers are starting to examine the details of the sensations that can be evoked by intracortical microstimulation to determine whether they provide a promising basis set for a functional visual prosthesis (Cha, Horch, and Normann, 1992). Such a prosthesis would be most useful as a mobility aid, an open-set pattern recognition problem that shows little sign of yielding to artificial intelligence approaches such as optical character recognition, which now offers a satisfactory reading tool for the blind. Mobility aids based on sensory substitution (e.g., sonar ranging devices that provide auditory or tactile feedback) have the disadvantage that they interfere with the normal function of the substituted modality, on which blind subjects are particularly dependent.

Research Questions

How does temporal patterning of neural activity affect perception in the primary visual cortex? Many current theories of visual perception involve temporal modulation and synchronization of neural activity (e.g., Engel et al., 1992). While distinctive patterns of neural activity have been recorded at various levels of the nervous system, it is not known whether these patterns represent critical components of the perceptual process of "feature binding" or are merely epiphenomena in highly cross-coupled networks of elements with broadly similar preferred firing rates (see SYNCHRONIZATION OF NEURONAL RESPONSES AS A PUTATIVE BINDING MECHANISM). One direct way to answer this question would be to ask a knowledgeable observer to describe the visual sensations evoked by different temporospatial patterns of activity induced directly in the visual cortex. The results of such an experiment are likely to have substantial implications for the design of a functional visual prosthesis, perhaps extending greatly the range of percepts that can be evoked by a limited set of electrodes or perhaps complicating greatly the preprocessing of the visual image into electrical stimuli that will produce the desired percepts regardless of context.

What is the role of active gaze control in the perception and interpretation of complex visual scenes? Most neurophysiological experiments in the visual cortex are performed on paralyzed and partially anesthetized animals; most theories of perception are based on the passive analysis of a visual scene presented to a retina fixed in space. However, it has long been known that a subject confronted by even a relatively simple visual object immediately engages in a complex series of highly targeted saccadic gaze shifts. Presumably, this process helps to compensate for the limited field of high-resolution foveal vision, but it also raises the question of how the myriad snapshots of the real world are assembled into a coherent mind's-eye montage. It should be instructive to study the eye movements made by subjects looking at virtual images created from electrically induced cortical phosphenes and at the eye movements that they make in response to such images in the absence of retinotectal input.

Does highly patterned, chronic activation of cortical subpopulations induce plastic changes in the responses of these and adjacent, chronically deprived subpopulations? There have been dramatic demonstrations of remapping of the somatosensory maps and motor representations in the primary cortex in response to various surgical, electrical, and behavioral modifications of cortical input (Merzenich and Grajski, 1990; see SOMATOTOPY: PLASTICITY OF SENSORY MAPS). Presumably, the subjects who are implanted with intracortical stimulation electrodes have been blind for some time and may have had to cope with progressively deteriorating vision for a greater or lesser period. It will be important to monitor perceptual changes over time as a function of the history of stimulation of various sites.

Sensorimotor Control

Current Technology

Most of the work to date on sensorimotor control has been preoccupied with the efferent interface, the stimulation part of FNS, whereby many muscles can be independently recruited to produce smoothly graded, predictable levels of force output. This problem is made particularly difficult by the fact that even a simple behavioral task usually requires the coordinated recruitment of dozens of muscles distributed over one or more entire limbs. Ultraminiature, wireless technology seems likely to overcome many of these problems by permitting a single, amplitude-modulated signal transmitted from an external antenna to control precisely a virtually unlimited number of separate stimulation sites (Loeb et al., 1991). This capability will permit attention to be directed to the more serious problems of sensorimotor control. In the intact musculoskeletal system, the amount of signal traffic related to proprioceptive feedback is at least an order of magnitude greater than that related to efferent control of motor units. Developing artificial sensors or means for recording afferent signals from the natural sensors (which are usually still functional) is another technological challenge, but one that can probably be met with existing technology. Which sensors are needed, how much resolution they need, and how they will be used in control systems are largely unresolved questions.

Research Questions

Control theory for dynamic mechanical systems is still crude, as can be seen from the performance of robots in coping with complex external loads and unexpected perturbations. For the purely sensory prostheses for vision and hearing described above, the problem is basically to provide raw information in as complete and natural a form as possible to the nervous system, which then applies cognitive processes to extract the information that is currently meaningful to the observer. In the case of FNS, the prosthesis itself must process relatively simple command signals into a complex output that depends on the current posture and mechanical constraints that obtain in the patient. For naturally controlled movements, this complex output includes not only the temporal pattern of muscle recruitment but also a matrix of gain values between each sensor signal and each muscle (Loeb, Levine, and He, 1990). It remains unclear whether FNS control systems can or should emulate such distributed control strategies or whether they should rely on more discrete, state-dependent switching among simple open-loop or single-input servo-controlled outputs for each of the various muscles required to produce the desired trajectory of the limb (Prochazka, 1993; Chizeck, 1992; Popovic, 1993).

At first glance, adaptive neural networks might seem to be unpromising candidates for the control of FNS prostheses. After all, the vastly more sophisticated set of neural networks in the infant's brain requires a couple of years of flailing about before even the simplest motor tasks can be performed at all successfully. The properly designed prosthetic system, however, can use the collective expertise of the mature patient and therapists, much as an adaptive computer software package and its users simultaneously modify their behaviors to converge rapidly on efficient and effective results. Even relatively simple adaptive logic networks have been used successfully to recognize a variety of patient-specific cues in the correct anticipation of the next action required in sequences of sitting, standing, walking, and stair climbing. The input signals come from simple sensors, such as foot-pressure switches, and the outputs trigger sequences of muscle activation that are appropriate for the next phase of the activity that the patient apparently intends. Adaptive neural networks have also been applied to the related problems of myoelectrically controlled, powered orthoses for amputees, in which the voluntarily controlled electromyogram signals from a few remaining muscles in the stump of a limb are used to infer the mechanical actions required from a motorized artificial limb.

The FNS system must also contend with principles of mechanics that are independent of the patient and that greatly limit the set of motor programs that are musculoskeletally achievable, neuromuscularly stable, and biomechanically safe. Currently, much of this information is provided subjectively by expert clinicians and engineers and is incorporated during the lengthy, trial-and-error adjustment of various motor programs that are stored in the patient's wearable control unit. As the motor behaviors become more complex and as FNS systems emerge from the research and development laboratory into the general clinical environment, it will be essential to develop more formal models of muscloskeletal mechanics and more quantitative kinetic analyses of observed kinematic patterns.

Finally, if it becomes possible to chronically record motor cortical activity as command signals for FNS prostheses, we will face serious limitations in our understanding of the encoding of these command signals, their dependence on sensory feedback, and their interaction with extrapyramidal systems, such as the cerebellum and basal ganglia. It will be particularly instructive to study the ability of subjects to learn to produce and modify patterns of neural activity in regions of the cortex that have been chronically deprived of their usual somatosensory input and output convergence with other motor circuits.

Conclusions

The growing clinical application of neural prosthetics should provide a major catalyst for the expansion of basic knowledge about the nervous system. The devices themselves provide unique opportunities for psychophysical testing of current theories of neural computing and immediate incentives for improving those theories when their limitations are revealed. The technology that is being developed to build these prostheses has considerable spin-off potential as neurophysiological research tools. Conversely, the nervous system embodies proven solutions to computational problems that have resisted the conventional algorithmic approaches of robotics and artificial intelligence. It is difficult to imagine a more appropriate application of electronic neural networks than in the repair of the biological systems that have inspired them.

Road Map: Applications of Neural Networks
Related Reading: Auditory Periphery and Cochlear Nucleus; BCM Theory of Visual Cortical Plasticity; Human Movement: A System-Level Approach; Silicon Neurons; Walking

References

Agnew, W. F., and McCreery, D. B., 1990, *Neural Prostheses: Fundamental Studies*, Englewood Cliffs, NJ: Prentice Hall. ◆
Bak, M., Girvin, J. P., Hambrecht, F. T., Kufta, C. V., Loeb, G. E., and Schmidt, E. M., 1990, Visual sensations produced by intracortical microstimulation of the human occipital cortex, *Med. Biol. Eng. Comput.*, 28:257–259.

Brindley, G. S., and Lewin, W. S., 1968, The sensations produced by electrical stimulation of the visual cortex, *J. Physiol. (Lond.)*, 196:479–493.

Cha, K., Horch, K., and Normann, R. A., 1992, Mobility of performance with a pixelized vision system, *Vis. Res.*, 32:1367–1372.

Chizeck, H. J., 1992, Adaptive and nonlinear control methods for neuroprostheses, in *Neural Prostheses: Replacing Motor Function After Disease or Disability* (R. B. Stein, H. P. Peckham, and D. Popovic, Eds.), New York: Oxford University Press, pp. 298–328.

Engel, A., Konig, P., Kreiter, A. K., Schillen, T. B., and Singer, W., 1992, Temporal coding in the visual cortex: New vistas on integration in the nervous system, *Trends Neurosci.*, 15:218–226.

Girvin, J. P., 1988, Current status of artificial vision by electrocortical stimulation, *Neuroscience*, 15:58–62. ◆

Leake, P. A., Hrdaek, G. T., Rebscher, S. J., and Snyder, R. L., 1991, Chronic intracochlear electrical stimulation induces selective survival of spiral ganglion neurons in neonatally deafened cats, *Hear. Res.*, 54:251–271.

Loeb, G. E., 1989, Neural prosthetic interfaces with the nervous system, *Trends Neurosci.*, 12:195–201. ◆

Loeb, G. E., 1990, Cochlear prosthetics, *Annu. Rev. Neurosci.*, 13:357–371.

Loeb, G. E., Levine, W. S., and He, J., 1990, Understanding sensorimotor feedback through optimal control, *Cold Spring Harbor Symp. Quant. Biol.*, 55:791–803.

Loeb, G. E., Zamin, C. J., Schulman, J. H., and Troyk, P. R., 1991,

Injectable microstimulator for functional electrical stimulation, *Med. Biol. Eng. Comput.*, 29:NS13–NS19.

Merzenich, M. M., and Grajski, K., 1990, Cortical network changes underlying representational plasticity, *Cold Spring Harbor Symp. Quant. Biol.*, 55:873–887.

Miyamoto, R. T., Osberger, J. M., Robbins, A. M., Myres, M. Y., and Kessler, K., 1994, Prelingually deafened children's performance with a nucleus multichannel cochlear implant, *Am. J. Otol.*, 14:437–445.

Moore, D. R., and Kowalchuck, N. E., 1988, Auditory brainstem of the ferret: Effects of unilateral cochlear lesions on cochlear nucleus volume and projections to the inferior colliculus, *J. Comp Neurol.*, 272:503–515.

Popovic, D., 1993, Finite state model of locomotion for functional electrical stimulation systems, *Prog. Brain Res.*, 97:397–407.

Prochazka, A., 1993, Comparison of natural and artificial control of movement, *IEEE Trans. Biomed. Eng.*, 1:7–17. ◆

Rall, W., 1962, Electrophysiology of a dendritic neuron model, *Biophys. J.*, 2:145–167.

Ranck, J. B., Jr., 1975, Which elements are excited in electrical stimulation of mammalian central nervous system: A review, *Brain Res.*, 98:417–440. ◆

Wilson, B., Finley, C., Lawson, D., Wolford, R., Eddington, D., and Rabinowitz, W., 1991, Better speech recognition with cochlear implants, *Nature*, 352:236–238.

Wise, K. D., and Najafi, K., 1991, Microfabrication techniques for integrated sensors and microsystems, *Science*, 254:1335–1342.

Protein Structure Prediction

Burkhard Rost and Chris Sander

Introduction

What is a protein?   The information for life is stored by a four-letter alphabet in the genes. Proteins perform most important tasks in organisms, such as catalysis of biochemical reactions, transport of nutrients, recognition and transmission of signals. Proteins are formed by joining amino acids into a long, stretched chain, the protein sequence. Proteins differ in the number (from 30 to 30,000) and in the arrangement of the amino acids (called *residues*, when joined in proteins). In water, the chain folds up to a unique three-dimensional (3D) structure. The main driving force is the need to pack residues for which a contact with water is energetically unfavorable into the interior of the molecule. This is only possible if the protein forms regular patterns of a macroscopic substructure called *secondary structure* (Figure 1; see Bränden and Tooze, 1991).

What determines protein function and structure?   The 3D structure of a protein determines its function. The 3D structure is uniquely determined by the sequence. Can the code be deciphered—i.e., can 3D structure be predicted from sequence? In principle, yes; but the computer time required to predict 3D structure from first principles is many orders of magnitude beyond today's possibilities. However, one reason to want to know the structure is rational drug design.

Why not simply look by microscope at the 3D structure?   The techniques to experimentally determine 3D structure of a protein are rather complicated. Today, the sequence is known for some 36,000 proteins, but only for 2000 has the 3D structure been determined by experiment. Large gene sequencing projects increase the sequence-structure gap further. The most accurate way to predict 3D structure from sequence is by

homology modeling—i.e., search for a protein with similar sequence that has a known 3D structure and then model the 3D structure of the unknown protein in analogy to the known one. Such techniques lead to a reduction of the sequence-structure gap by some 9000 proteins.

Why can homology modeling be successful?   The exchange of a few residues can already destabilize a protein. This implies that the majority of the 20^N possible sequences of length N form different structures. But has evolution created such an immense variety? The evolutionary pressure to conserve function and the discontinuity of the universe of structures have the result that structure is more conserved than sequence. Evolution has produced pairs of proteins which have the same 3D structure with only 25% identical residues. For such pairs, 3D structure can be predicted rather accurately by homology.

Can the egg be unboiled?   When an egg is boiled, the proteins it contains unfold. Can this procedure be reversed in theory? Or, can the encrypted code of protein folding be deciphered from sequence? Current tools to predict 3D structure from sequence are rather limited (Rost and Sander, 1994b). The problem has to be simplified. One extreme simplification is to predict one-dimensional (1D) strings of secondary structure assignment (Figure 1).

How can neural networks predict protein structure?   In practice, the most successful predictions are based on an analysis of common features in the data bank of known 3D structures. Artificial neural networks are well suited for pattern classification. Here, we shall attempt to show how neural networks can be used to predict protein structure. First, we give examples of

1D sequence
secondary structure

MQIFVKTLTGKTITLEVEPSDTIENVKAKIQDKEGIPPDQQRLIFAGKQLEDGRTLSDYNIQKESTLHLVLRLRGG
 EEEEEE EEEEE   HHHHHHHHHHHH HHHEEEEE EE   HHH   EEEEEE

2D contact map   **3D** schematic trace

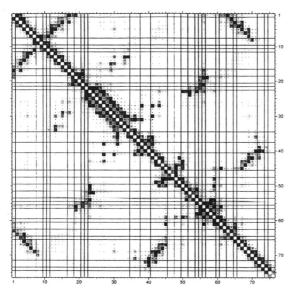

Figure 1. Hierarchy of protein structure. *1D*, The amino acid sequence determines the formation of 3D structure. Here, the chain of ubiquitin (1ubq) is shown. The 3D structure can be projected onto a 1D string of repetitive patterns: the secondary structure (*H*, α-helix; *E* [extended], β-strand; and *blank*, loop). *2D*, The 3D structure can be projected onto a 2D matrix: the entry at position *ij* of the matrix gives the contact between residue *i* and residue *j* (plot by *Conan*: M. Scharf, 1989,

Analyse von Paarwechselwirkungen in Proteinen, University of Heidelberg, Department of Physics). *3D*, The trace of the protein chain in 3D is plotted schematically. *C* gives the end and *N* the beginning of the protein. The longer helix is on the right-hand side, while three of the strands are indicated by arrows (plot by *Molscript*: P. Kraulis, 1991, MOLSCRIPT: A program to produce both detailed and schematic plots of protein structures, *J. Appl. Crystallogr.*, 24:946–950).

how the data bank of known 3D structures can be used to predict secondary structure and, following that, other structural features. Finally, we briefly review attempts to predict entire 3D structures.

Prediction of Secondary Structure

Presenting the Protein to the Network

The usual goal of secondary structure prediction methods is to classify a pattern of adjacent residues as either H (α-helix), E (for extended β-strand), or L (for loop = all others). Sequences are translated into patterns by shifting a window of adjacent residues through the protein and looking up the secondary structure for the central residue (Figure 2).

Networks used for secondary structure prediction are multi-layer feedforward networks (Figure 2). The network error is given by the difference between actual network output (uniquely determined by the choice of connections) and desired output (looked up from data bank). *Training* or *learning* means changing the connections such that the error decreases for the given examples (gradient descent; see, e.g., BACKPROPAGATION: BASICS AND NEW DEVELOPMENTS and LEARNING AS HILL-CLIMBING IN WEIGHT SPACE). If training is successful, the patterns are correctly classified. But how can new patterns be classified correctly? The hope is that the network extracts general rules by the classification of the training patterns. The generalization ability is checked by another set of test samples for

which the mapping of sequence window to secondary structure is known as well. Sufficient testing is crucial (Rost and Sander, 1994a).

Prediction Performance of Simple Neural Networks

Networks of the type described reach values for three-state overall prediction accuracy of around 60%. This is comparable to the performance of non-network methods. In the five years following the first application of neural networks to the prediction of secondary structure (Qian and Sejnowski, 1988), more than 20 groups have followed (Hirst and Sternberg, 1992; Rost and Sander, 1994a). Prediction accuracy was not improved significantly without using biological expertise, as we shall see in the next section.

Using Evolutionary Information of Multiple Sequences

Some residues can be replaced by others without changing the structure. But not every amino acid can be replaced by any other. On the contrary, the residue substitution patterns are very specific for a certain 3D structure. Can this information be used to improve the prediction accuracy for neural networks? Indeed, using evolutionary information as derived from a database of proteins with homologous 3D structure improves the performance accuracy by about 10 percentage points to >72% (Rost and Sander, 1994a). The basic procedure is as follows: First, sequences of proteins which are similar enough in se-

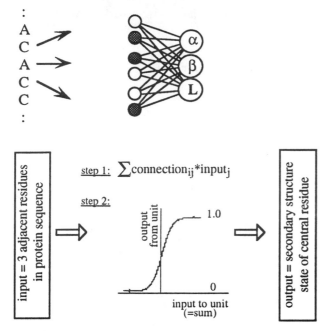

Figure 2. Neural network for secondary structure prediction. Here, for simplicity, a window of three residues is used as input (for actual applications, 10 to 30 residues are used). For each position in the window, two input units (for simplicity, two instead of 20 amino acids) are used: the first is set to one (shaded dark) if the amino acid is an A, the second if it is a C. The signal is transmitted via the connections between input and output layer. At each output unit, two computational steps are performed. First, the input signal is multiplied with the connection, and this product is summed over all input units. Second, from this sum, the output of each output unit is compiled according to the trigger function shown.

quence to know that they are similar as well in 3D structure are aligned (optimal fit). Second, for each residue position, it is counted how often any of the 20 amino acids occur in the alignment at that position. Third, the counts are used as input to a network.

Prediction of Structural or Functional Protein Class

According to the relative content in secondary structure, proteins can be classified into *structural classes*. From the predicted secondary structure, the relative content of secondary structure of a protein can be calculated to predict structural class (Rost and Sander, 1994a). An alternative approach is to directly predict the secondary structure content of a protein by a neural network that uses as input a vector of 20 components giving the frequency with which each amino acid occurs in a particular protein (Muskal and Kim, 1992).

A different task is to predict that two proteins are similar in terms of function or 3D structure. Methods have been based (1) on multiple feedforward networks (Frishman and Argos, 1992), using proteins of similar sequences as input; (2) on simple feedforward networks, using different amino acid features as input (Wu et al., 1992); and (3) on Kohonen maps (see SELF-ORGANIZING FEATURE MAPS: KOHONEN MAPS), using residue pair frequencies as input (Ferrán and Ferrara, 1992). Whereas feedforward networks are useful to learn a classification into known features (secondary structure, structural class), the Kohonen maps have been applied to render a general classification scheme (e.g., *A* and *B* are similar, and *A* is more

similar to *C* than *B*). Such a classification is in general a priori not evident (and in itself provides a controversial research area, attempting to answer questions like "Are we more similar to an orangutan than to a pig?").

Prediction of Other Structural or Functional Features

Most applications of neural networks use a similar sliding window input as described above. Approaches address the predictions of surface exposure, disulfide bonds, and function-specific sequence motifs.

Surface exposure. A simple feature of 3D structure that is also of interest for molecular biology is the extent to which a residue is exposed to the solvent. Holbrook, Muskal, and Kim (1990) used a network to classify amino acid residues as either buried or exposed. The result was evaluated on too small a data set, yielding some 70% accuracy in two states (buried/exposed).

Disulfide bonds. Disulfide bonds between cysteine residues (one of the 20 amino acids is cysteine) are often of functional and structural importance. Muskal, Holbrook, and Kim (1990) used a single-layer feedforward network to predict the existence or absence of disulfide bonds.

Function-specific sequence motifs. Often, function depends on a rather short (5 to 10 residues) sequence motif (unique pattern of adjacent amino acids). Residues that are associated with particular functions were also subject to neural network predictions. Examples are: (1) sequence motifs that reveal binding of energy storage molecules (Hirst and Sternberg, 1992); (2) sequence motifs specific for particular proteins—e.g., the immunoglobulins (Bengio and Pouliot, in Hirst and Sternberg, 1992); and (3) signal peptide motifs in sequences (Ladunga et al., 1991).

Aiming at Prediction in 3D

Distance Constraints

The projection of the 3D structure onto a two-dimensional distance matrix (Figure 1) could be an important step on the way to predicting 3D structure. This enterprise was undertaken by Bohr et al. (1990), who used a neural network to predict residues which are closer than 8 angstroms ($=8 \cdot 10^{-10}$ m) to any of the 30 residues adjacent in sequence. The predicted fragments of the distance matrix were used for a simple steepest descent energy minimization procedure. The training set comprised 13 proteins. The method was tested on only one protein that has sufficient sequence identity to proteins used for training. (The prediction was worse than the one that could have been obtained by homology modeling.)

Spin-Glass Models for Proteins

The putative analogies of the energy landscapes of spin glasses and proteins led to a multitude of models attempting to describe protein folding with the formalism known from spin-glass theory (Elber, 1993). Such models have been used for attempts to predict 3D structure (Goldstein, Luthey-Schulten, and Wolynes, 1992). The principal idea is to define an effective energy function from a database of known 3D structures that is flexible enough to enable description of a large class of structures and simple enough to surpass the multiple minima problem of conventional energy minimization calculations for proteins. The analogy to spin-glass theory consists in constructing an energy function based on pairwise interactions (between residues). For some cases, such methods are comparable to non-network methods which use statistically derived energy functions (Rost and Sander, 1994b).

Discussion

Neural networks can be used for predicting structural features of proteins. There were at least 50 articles on the application of neural networks for protein structure prediction until 1993. One message of the literature is convincing: neural networks can be used to predict secondary structure, structural class, family relations, surface exposure, functional motifs, distance matrices, and even the 3D structure of proteins.

Neural network methods are seldom superior to non-network approaches. The second message of the literature is that networks are superior to alternative techniques, but this answer is not convincing! The general problem is a lack of rigor in evaluating results. A common example is the allowance of significant sequence identity between test and training set. Any evaluation that allows for sequence identity has to be compared to homology modeling. And in this comparison, all prediction methods are clearly inferior. The conclusion is that neural network applications have almost never yielded significant improvements over current techniques (Hirst and Sternberg, 1992). An exception is a network that uses evolutionary information to predict secondary structure (Rost and Sander, 1994a). So far, this is the only example for a neural network prediction of protein structure being clearly superior to alternative techniques.

Neural network predictions have not been made sufficiently available to biochemists. Unfortunately, the tendency to overestimate the performance accuracy of network prediction has not contributed much to their acceptance by biochemists. Another problem is that almost none of the network methods is publicly available to those researchers who need predictions.

Neural network techniques will continue to be useful for the prediction of protein structure. First, the problem of predicting protein structure is far from solved. For a sequence of unknown 3D structure for which no homology to a known fold can be detected, the best one can achieve today is a more or less reliable prediction of secondary structure, surface exposure, or functional class. Second, the constantly growing data banks provide an increasing body of information about protein structure. Chances are that methods based on data bank analysis will be the first to practically solve the prediction of protein 3D structure. Third, neural networks might be well suited for appropriately incorporating the increased information. Using

evolutionary information will be one way to improve predictions by networks. Neural network applications can become increasingly important for the research of tomorrow's molecular biology, provided that testing is done with care and that methods become available to potential users.

Road Map: Applications of Neural Networks
Background: I.3. Dynamics and Adaptation in Neural Networks

References

Bohr, H., Bohr, J., Brunak, S., Fredholm, H., Lautrup, B., and Petersen, S. B., 1990, A novel approach to prediction of the 3-dimensional structures of protein backbones by neural networks, *FEBS Lett.*, 261:43–46.

Bränden, C., and Tooze, J., 1991, *Introduction to Protein Structure*, New York, London: Garland. ◆

Elber, R., 1993, New simulation methods for proteins and DNA, *Curr. Opin. Struct. Biol.*, 3:260–264.

Ferrán, E., and Ferrara, P., 1992, Clustering proteins into families using artificial neural networks, *Comput. Appl. Biosci.*, 8:39–44.

Frishman, D., and Argos, P., 1992, Recognition of distantly related protein sequences using conserved motifs and neural networks, *J. Mol. Biol.* 228:951–962.

Goldstein, R. A., Luthey-Schulten, Z. A., and Wolynes, P. G., 1992, Protein tertiary structure recognition using optimized Hamiltonians with local interactions, *Proc. Natl. Acad. Sci. USA*, 89:9029–9033.

Hirst, J. D., and Sternberg, M. J. E., 1992, Prediction of structural and functional features of protein and nucleic acid sequences by artificial neural networks, *Biochemistry*, 31:615–623.

Holbrook, S. R., Muskal, S. M., and Kim, S.-H., 1990, Predicting surface exposure of amino acids from protein sequence, *Protein Eng.*, 3:659–665.

Ladunga, I., Czakó, F., Csabai, I., and Geszti, T., 1991, Improving signal peptide prediction accuracy by simulated neural network, *Comput. Appl. Biosci.*, 7:485–487.

Muskal, S. M., Holbrook, S. R., and Kim, S.-H., 1990, Prediction of the disulfide-bonding state of cysteine in proteins, *Protein Eng.*, 3:667–672.

Muskal, S. M., and Kim, S.-H., 1992, Predicting protein secondary structure content: A tandem neural network approach, *J. Mol. Biol.*, 225:713–727.

Qian, N., and Sejnowski, T. J., 1988, Predicting the secondary structure of globular proteins using neural network models, *J. Mol. Biol.*, 202:865–884.

Rost, B., and Sander, C., 1994a, Combining evolutionary information and neural networks to predict protein secondary structure, *Proteins*, 19:55–72.

Rost, B., and Sander, C., 1994b, Structure prediction of proteins—Where are we now? *Curr. Opin. Biotechnol.*, 5:372–380. ◆

Wu, C., Whitson, G., McLarty, J., Ermongkonchai, A., and Chang, T.-C., 1992, Protein classification artificial neural system, *Protein Sci.*, 1:667–677.

Pursuit Eye Movements

Richard J. Krauzlis

Introduction

When viewing objects, monkeys and humans use a combination of saccadic and pursuit eye movements to keep the retinal image of the object of regard within the high-acuity region near the fovea. While these movements mix seamlessly in normal behavior, their properties and origins are quite distinct. Saccades are ballistic movements that quickly direct the eyes to-

ward a visual target, thereby translating the image of the target from an eccentric retinal location to the fovea. In contrast, pursuit is a continuous movement that slowly rotates the eyes to compensate for any motion of the visual target, minimizing the drift of the target's image across the retina that might otherwise compromise visual acuity. While other mammalian species can generate smooth optokinetic eye movements—which track the motion of the entire visual surround—only primates can smoothly pursue a single element of a complex visual scene, regardless of the motion this causes elsewhere on the retina. Pursuit eye movements therefore represent a specialization of the primate central nervous system.

Basic Features of Pursuit Behavior

The basic features of pursuit can be illustrated by considering the *ramp paradigm* (Figure 1), in which a target initially at rest moves at a constant speed. The onset of target motion is often accompanied by a step, eliminating the need for a catch-up saccade (Rashbass, 1961). The eye velocity records obtained with this paradigm can be divided into distinct phases. During the latent phase (1), the target is moving, but the eyes have not yet begun to move. During the initiation of pursuit (2), the eye accelerates at a nearly constant rate related to the image speed experienced during the latent phase. This is followed by a transition phase (3), as eye velocity continues to increase and often overshoots target velocity slightly. During sustained pursuit (4), eye velocity either settles to a steady-state value or oscillates around a value near target velocity.

The ramp paradigm also indicates several of the constraints associated with the pursuit system. First, because the retina is part of the eye, there is a reciprocal relationship between the motion of the target's retinal image and the motion of the eyes. During the latent phase, *image velocity* (the difference between target and eye velocities) is equal to target velocity. After the latent phase, image velocity decreases and then remains near zero during sustained pursuit. Pursuit is therefore organized as

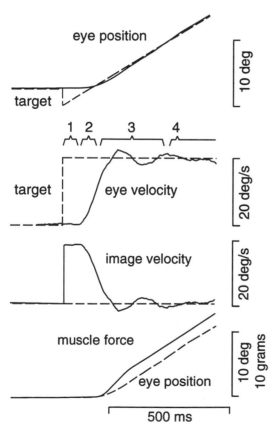

Figure 1. Basic features of pursuit can be seen with the ramp paradigm. The target jumps to a slightly eccentric position and moves at a constant speed of 20 degrees per second that is matched by the subject's eye movement after a few hundred milliseconds.

a negative-feedback system: the eye movement output of the system acts to reduce the visual motion input to the system (Robinson, 1981).

Another constraint is imposed by the delay (approximately 100 ms) associated with sensory and motor processing. Combined with negative feedback, this delay could make the system unstable; in fact, under certain conditions, pursuit does exhibit large amplitude oscillations. To compensate for this constraint, the pursuit system uses a combination of short- and long-term predictive mechanisms. For example, it is believed that the motor pathways for pursuit include a circuit that retains the current value of pursuit eye speed. This *velocity integrator* represents a form of short-term prediction and can maintain pursuit eye speed in the absence of vision. Visual inputs (like image velocity in Figure 1) indicate how eye speed should change, and are best related to eye acceleration during pursuit.

Finally, the pursuit system must provide a steadily increasing *muscle force* (Figure 1) to produce a constant-speed eye movement. The required muscle force increases in parallel with eye position as a function of time and can be approximated by taking the mathematical integral of desired eye speed. This process of integration is believed to be common to all eye movements and is accomplished by a *position integrator* contained within the brainstem (Cannon and Robinson, 1987). In addition, the pursuit system compensates for the mechanical aspects of the oculomotor *plant*—the inertial mass of the eye and the visco-elastic properties of the eye muscles. As indicated by the offset between muscle force and eye position, the applied force begins with an additional increment to overcome the sluggish dynamics of the eye (Robinson, 1965). Without this pre-emphasis, it would take three to four times as long for eye speed to match target speed. The neurons innervating the eye muscles during pursuit therefore provide an "inverse dynamics" version of the eye movement command—a signal which, after it is transformed by the eye plant, produces the desired eye motion.

Models of Pursuit

Current models of pursuit vary in their organization and in the features of pursuit that they are designed to reproduce. However, they are similar in that they are concerned with describing pursuit behavior without explicit reference to the neural structures that might be responsible.

Image motion models (Figure 2A) focus on replicating in detail the initial increase in eye velocity at the initiation of pursuit, the overshoot in eye velocity during the transition to sustained pursuit, and the oscillations observed during sustained pursuit (Krauzlis and Lisberger, 1989). The key feature of these models is the presence of multiple visual inputs, which are designed to reflect the complexity of the visual signals used to drive pursuit. In these models, the temporal features of pursuit eye movements (see Figure 1) are matched primarily by adjusting the dynamics of the visual inputs, while the premotor processing is simplified.

Target velocity models (Figure 2B) also can replicate the profile of eye velocity as a subject initiates and maintains pursuit, but they use a different structure (Robinson, Gordon, and Gordon, 1986). The key feature of these models is the construction of an estimate of target motion by adding a copy of the eye velocity output to the visual motion input. In these models, the processing of visual inputs is greatly simplified, and the characteristic features of pursuit are accomplished by the properties of the premotor processing.

Despite the differences between these two classes of models, they accomplish similar transformations of the input signal

Figure 2. Three classes of pursuit models. The input to each model is target speed; the output is eye speed. The dashed lines indicate the physical coupling of the eye and retina. *A, Image motion* models include complex processing of visual motion inputs for pursuit. *B,* In *target velocity* models, a positive-feedback loop is used to construct an internal estimate of target velocity. *C,* In *predictive* models, the output from a long-range predictive mechanism provides an additional input for pursuit.

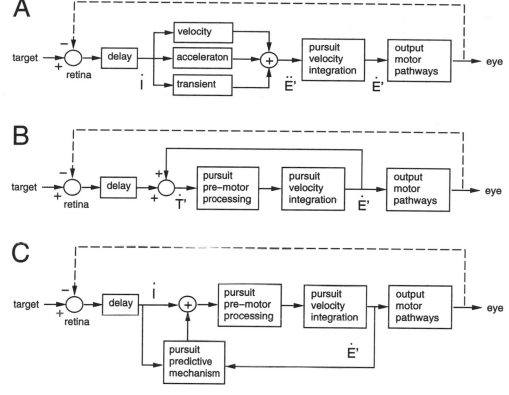

Table 1. Summary of Physiological Studies of Pursuit

Structure	A. Lesions	B. Electrical Stimulation	C. Single-Unit Recording
1. V1	Deficits in saccades and pursuit		
2. Extrafoveal MT	Transient retinotopic deficits in the initiation of pursuit		Visual responses tuned for direction/ speed of small stimuli
3. Foveal MT	Deficits in initiating pursuit; deficits for ipsilateral sustained pursuit	Ipsilateral eye acceleration if applied during sustained pursuit	Visual responses to motion of small stimuli
4. MST	Deficits in initiating pursuit; deficits for ipsilateral sustained pursuit	Ipsilateral eye acceleration if applied during sustained pursuit	Visual responses to motion of small or large stimuli; nonvisual responses
5. 7a, VIP			Visual responses to stimulus motion; nonvisual responses
6. FEF	Deficits in sustained and predictive or anticipatory pursuit	Eye acceleration, often ipsilateral, if applied during fixation or pursuit	Visual responses to stimulus motion responses during tracking
7. DLPN	Deficits in initiating pursuit; deficits for ipsilateral sustained pursuit	Ipsilateral eye acceleration if applied during sustained pursuit	Visual responses best for moving large stimuli; nonvisual responses
8. DMPN, NRTP	Deficits in pursuit		Visual responses to large stimuli
9. NOT	Deficits in ipsilateral tracking	Ipsilateral eye acceleration	Visual responses to large stimuli
10. LTN			Visual responses to large stimuli
11. Ventral paraflocculus	Long-lasting deficit in pursuit	Ipsilateral eye acceleration if applied during fixation or pursuit	Responses to eye and head velocity; visual responses during pursuit
12. Oculomotor vermis	Long-lasting deficit in pursuit	Evokes saccadic eye movements	Responses to eye and head velocity, passive visual responses
13. VN, FN, NPH	Sustained deficits in pursuit and saccadic eye movements		

Details of experimental findings can be found in several longer reviews (Eckmiller, 1987; Keller and Heinen, 1991; Krauzlis, 1994; Lisberger, Morris, and Tychsen, 1987).

and, with certain simplifying assumptions, can be shown to be formally equivalent (Deno, Keller, and Crandall, 1989). However, structural differences between the models do have implications. For example, it has been observed that image motion models can account for the altered pursuit found when the delay in the visual feedback is changed, while target velocity models cannot (Goldreich, Krauzlis, and Lisberger, 1992).

Predictive models (Figure 2C) address the role of prediction in pursuit. These models cover a wide range of approaches, but they share the feature of asserting that mechanisms other than immediate processing of visual inputs are required to replicate all features of pursuit behavior (Barnes, 1993). These mechanisms often involve extracting and subsequently recognizing patterns of target motion. This additional information can supersede the effects of visual feedback and produce movements that are not simply visual reflexes.

The Neural Pathways for Pursuit

The importance of both visual areas of the cerebral cortex and oculomotor regions of the cerebellum have been clearly demonstrated by experimental lesions (Table 1). These results suggest that the cerebral cortex provides sensory inputs that the cerebellum and premotor nuclei, in turn, convert into commands for pursuit. However, several observations argue that the conveyance of activity along the cortico-ponto-cerebellar pathways does not constitute a straightforward progression of visual signals toward commands for pursuit. For example, some of the signals provided by the cerebral cortex are not purely visual. Nonvisual signals have been demonstrated by recording the activity of isolated units as a monkey continuously tracks a target moving in the units' preferred direction. When the target is briefly turned off, the response of neurons in several regions continues unabated. Conversely, visual signals have been recorded at the level of the cerebellum, in addition to "motor" signals related to eye movements. Furthermore, electrical stimulation of some sites produces pursuitlike movements only if the subject is already engaged in pursuit, while stimulation of other sites produces smooth movements whether the subject is pursuing or fixating. These effects suggest that the pursuit pathways contain a "switch" that governs the transmission of sensory and motor signals, an idea consistent with the behavioral distinctions that can be drawn between pursuit and other eye movements (Luebke and Robinson, 1988).

Perspectives on the Neural Pathways for Pursuit

To illustrate the difficulties in drawing conclusions about the overall organization of the neural pathways for pursuit, we next consider how three particular structures—the floccular region, the oculomotor vermis, and extrastriate cortex—may contribute to pursuit.

During sustained pursuit, floccular Purkinje cells (the cerebellar output neurons) show a continuous increase or decrease in firing rate. This sustained activity is suggested to result from the reciprocal connnections between the floccular region and the premotor nuclei (Figure 3A). The loop formed by these connections could provide the neural substrate for the pursuit velocity integrator included in the pursuit models (see Figure 2). Furthermore, these Purkinje cells display transient responses to moving stimuli that are the object of a pursuit eye movement, but only modest responses to stimuli presented during fixation. These transient responses at the initiation of pursuit may reflect visual inputs to the floccular region that drive the initial acceleration of the eye ($\ddot{E}$) and are then incorporated through feedback into the sustained eye velocity command ($\dot{E}$), analogous

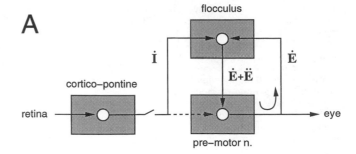

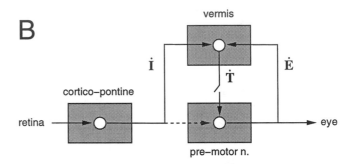

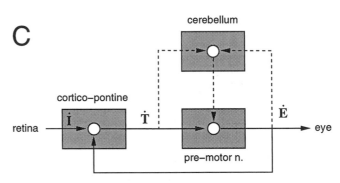

Figure 3. Three perspectives on the neural substrates for pursuit. *A,* The output of the floccular region may represent a command for pursuit eye velocity ($\dot{E}$) and eye acceleration ($\ddot{E}$). The broken line at the output of the corticopontine pathways indicates that the pursuit "switch" appears to lie upstream. *B,* The output of the vermis may represent a neural reconstruction of target velocity. The pursuit "switch" appears to lie downstream of the vermis. *C,* The output of the corticopontine pathways may provide a signal that already encodes target velocity and can be used directly by the premotor nuclei for pursuit.

to the flow of signals in the image motion models (see Figure 2A). Another similarity is that this scheme does not include an internal estimate of target motion.

Like the floccular region, the oculomotor vermis receives a combination of visual and eye motion inputs (Figure 3B). However, vermal Purkinje cells respond to moving stimuli presented during fixation or pursuit. This conditional linkage between activity in the vermis and pursuit eye movements, indicated by the "switch" in Figure 3B, suggests that the vermis may be part of a premotor circuit that provides a set of candidate signals for pursuit. As in the target velocity models (see Figure 2B), the combination of visual and eye motion signals may be used to construct an internal estimate of target motion.

In extrastriate cortex, there are several sites where eye movement signals have been found in conjunction with visual motion signals, such as the medial superior temporal area (MST), the frontal eye fields (FEF), and the posterior parietal cortex. Ablation of these areas leads to deficits in both initiating and sustaining pursuit. These results suggest that cortical inputs may be sufficient to drive pursuit and that the critical neural pathway may be a direct link between the cerebral cortex and the brainstem (Figure 3C). If the critical computations for pursuit were accomplished in the cerebral cortex, the pursuit deficits resulting from cerebellar lesions might be viewed as one aspect of a general deficit in oculomotor control. For example, the cerebellum might be important for compensating for the mechanical properties of the eye plant, a function associated with the output motor pathways in the pursuit models (see Figure 2), while the cerebral cortex might compute the internal estimates of target motion that are used to drive pursuit, a function associated with the predictive or premotor elements in the pursuit models.

Discussion

The aim of this article has been to draw parallels between behavioral, modeling, and physiological approaches to the study of pursuit. The current challenge is to bridge the gap between mathematical models and physiological data. Toward this end, models of pursuit need to become more biomorphic; their organization needs to conform more closely to known anatomy, and their components need to resemble more nearly actual neurons. Conversely, physiological studies of pursuit should explicitly recognize the conceptual models underlying their design and produce quantitative tests of those models.

Road Map: Primate Motor Control
Background: Motor Control, Biological and Theoretical
Related Reading: Cerebellum and Motor Control; Collicular Visuomotor Transformations for Saccades

References

Barnes, G. R., 1993, Visual-vestibular interaction in the control of head and eye movement: The role of visual feedback and predictive mechanisms, *Prog. Neurobiol.*, 41:435–472. ◆
Cannon, S. C., and Robinson, D. A., 1987, Loss of the neural integrator of the oculomotor system from brain stem lesions in monkey, *J. Neurophysiol.*, 57:1383–1409.
Deno, D. C., Keller, E. L., and Crandall, W. F., 1989, Dynamic neural network organization of the visual pursuit system, *IEEE Trans. Biomed. Eng.*, BME36:85–92.
Eckmiller, R., 1987, Neural control of pursuit eye movements, *Physiol. Rev.*, 67:797–857. ◆
Goldreich, D., Krauzlis, R. J., and Lisberger, S. G., 1992, Effect of changing feedback delay on spontaneous oscillations in smooth pursuit eye movements of monkeys, *J. Neurophysiol.*, 67:625–638.
Keller, E. L., and Heinen, S. J., 1991, Generation of smooth-pursuit eye movements: Neuronal mechanisms and pathways, *Neurosci. Res.*, 11:79–107. ◆
Krauzlis, R. J., 1994, The visual drive for smooth eye movements, in *Visual Detection of Motion* (A. T. Smith and R. J. Snowden, Eds.), London: Academic Press, pp. 437–473. ◆
Krauzlis, R. J., and Lisberger, S. G., 1989, A control systems model of smooth pursuit eye movements with realistic emergent properties, *Neural Comp.*, 1:116–122.
Lisberger, S. G., Morris, E. J., and Tychsen, L., 1987, Visual motion processing and sensory-motor integration for smooth pursuit eye movements, *Annu. Rev. Neurosci.*, 10:97–129. ◆
Luebke, A. E., and Robinson, D. A., 1988, Transition dynamics between pursuit and fixation suggest different systems, *Vis. Res.*, 28:941–946.
Rashbass, C., 1961, The relationship between saccadic and smooth tracking eye movements, *J. Physiol. (Lond.)*, 159:326–338.
Robinson, D. A., 1965, The mechanics of human smooth pursuit eye movement, *J. Physiol. (Lond.)*, 180:569–591.
Robinson, D. A., 1981, The use of control systems analysis in the neurophysiology of eye movements, *Annu. Rev. Neurosci.*, 4:463–503. ◆
Robinson, D. A., Gordon, J. L., and Gordon, S. E., 1986, A model of the smooth pursuit eye movement system, *Biol. Cybern.*, 55:43–57.

Radial Basis Function Networks

David Lowe

Introduction

The Radial Basis Function Network (RBFN) is conceptually a very simple and yet intrinsically powerful network structure. The radial basis function network constructs global approximations to functions using combinations of "basis" functions "centered" around weight vectors (Figure 1), whereas a multilayer perceptron constructs an architecture out of separating hyperplanes. An extra distinction is that the radial basis function employs a distance function to convert the vector input pattern into a scalar at the hidden layer, as opposed to a vector dot product. The network's strength derives from a rich interpretational foundation, since it lies at the confluence of a variety of "established" scientific disciplines. Thus, although the original motivation of this particular network structure was in terms of functional approximation techniques, the network may be "derived" on the basis of statistical pattern processing theory, regression and regularization, biological pattern formation, mapping in the presence of noisy data, and so on. However, in addition to exhibiting a range of useful theoretical properties, it is above all a practically useful construct as it

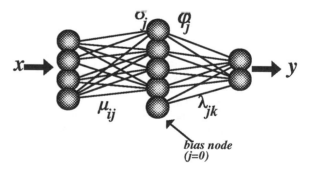

Figure 1. The basic radial basis function structure. There is a nonlinear basis function $\phi_j(\dots)$ centered around each hidden node weight vector μ which also has a (possibly) adaptable "range of influence" σ_j. The output of the hidden node j, h_j is given as a *radial* function of the distance between each pattern vector and each hidden node weight vector, $h_j = \phi_j(\|x - \mu_j\|/\sigma_j)$. This is the main difference from a multilayer perceptron. The network outputs are evaluated by a traditional scalar product between the vector of hidden node outputs and the weight vector attached to output node k, as $o_k = h \cdot \lambda_k$.

may be applied to problem domains in discrimination (see, e.g., Niranjan and Fallside, 1990, for a speech classification example), time series prediction (see articles in Rao Vemuri and Rogers, 1994, for financial and other examples) and other mapping problems, and feature extraction/topographic mapping problem domains (e.g., see Lowe, 1993, for an example of chemical odor concentration coding).

The Basic RBFN Structure

The radial basis function network is a single-hidden-layer feedforward network with linear output transfer functions and nonlinear transfer functions $\phi_j(\ldots)$ on the hidden layer nodes. Many types of nonlinearities may be used. There is also typically a bias on each output node. The primary adjustable parameters (see Figure 1) are the final layer weights $\{\lambda_{jk}\}$ connecting the jth hidden node to the kth output node. There are also weights $\{\mu_{ij}\}$ connecting the ith input node with the jth hidden node and occasionally a "smoothing" factor matrix $\{\Sigma_j\}$.

The mathematical form of the RBFN is as follows. The kth component of the output vector y_p corresponding to the pth input pattern x_p is expressed as

$$[y(x_p)]_k = \sum_{j=0}^{h} \lambda_{jk} \phi_j(\|x_p - \mu_j\|; \Sigma_j) \qquad (1)$$

where $\phi_j(\ldots)$ denotes the nonlinear transfer function of hidden node j [$\phi_0(\ldots) \equiv 1$ is the bias node], and the possible dependence upon a "smoothing" matrix is left explicit. The most common example of the smoothing factor is in the use of a general Gaussian transfer function, i.e., $\phi(z) \approx \exp - [z^T \Sigma^{-1} z]$. Since the general expression is an analytic function of the variables corresponding to the basis function positions and smoothing factors, it is possible to adapt them by a full nonlinear least squares process if required (Lowe, 1989; Moody and Darken, 1989). This is usually not necessary. As can be seen from Equation 1, the main difference from a multilayer perceptron is that the output of the hidden node j, h_j is given as a *radial* function of the distance between each pattern vector and each hidden node weight vector, $h_j = \phi_j(\|x - \mu_j\|)$, rather than a scalar product, $h_j^{\mathrm{MLP}} = \phi_j(x \cdot \mu_j)$.

One of the advantages of the RBFN is that the first layer weights $\{\mu_j, \Sigma_j; j = 1, \ldots, h\}$ may often be determined or specified by a judicious use of prior knowledge, or adapted by simple techniques. Early work (Broomhead and Lowe, 1988) found it sufficient to position the basis functions at data points sampled randomly according to the distribution of the data. This method ensured that network resources were concentrated in regions of higher data density. Another early technique (Moody and Darken, 1989) was to position the centers of the basis functions according to a K-means clustering process on the data points and then set the smoothing parameters of the assumed Gaussian basis functions to be the average distance between cluster centers. Therefore, once the weights associated with the first layer have been specified, the major problem in "training" an RBFN is focused upon the determination of the final layer weights. Since the RBFN is typically employed to perform a *supervised* discrimination or prediction task such as time series forecasting, this training usually takes the form of the optimization of a cost function requiring the outputs of the network to somehow closely approximate a set of known target values. It is common to attempt to minimize a standard residual sum-of-squares cost function, though other cost functions may be employed. Since this is a linear optimization process (the parameters $\{\lambda_{jk}\}$ occur linearly when minimizing the residual sum squared error measure), the radial basis function network is computationally more attractive in applications than

a multilayer perceptron even though they are both computationally universal architectures (Park and Sandberg, 1991).

RBFNs for Classification

We begin by discussing the use of RBFNs for discrimination and classification tasks, such as in static speech recognition experiments (e.g., Niranjan and Fallside, 1990).

Figure 2 illustrates a simple classification example where the distribution of data points exhibits a simple clustering. There are primarily two ways to separate these clusters. One is by a segregation of the space into polygonal cells. The straight lines in the figure illustrate this decomposition of the pattern space into regions as would be obtained by a simple multilayer perceptron where the lines represent the class boundaries. An alternative is to describe the clusters of data themselves as if they were generated according to an underlying probability density function, modeled here in the figure by elliptical distributions. Thus one method concentrates upon class boundaries, and the other focuses upon regions where the data density is highest. These are complementary approaches with their respective disadvantages and advantages. The latter alternative is the radial basis function approach, and it may be motivated from the perspective of kernel-based density estimation (Lowe, 1991; Tråvén, 1991).

In classification we are primarily interested in the posterior, $p(c|x)$, the probability that class c is present given the observation x. However, it is easier to model other related aspects of the data such as the unconditional distribution of the data $p(x)$ and the likelihood of the data, $p(x|c)$, which is the probability that the data were generated given that they came from a specific class c. We can then recreate the posterior from these quantities according to Bayes's theorem, $p(c_i|x) = p(c_i)p(x|c_i)/p(x)$ (see BAYESIAN METHODS FOR SUPERVISED NEURAL NETWORKS). The distribution of the data is modeled as if it were generated by a mixture distribution, i.e., a linear combination of parameterized state, or basis functions such as Gaussians. Since individual data clusters for each class are not likely to be approximated by a single Gaussian distribution, we need several basis functions per cluster. We assume that the likelihood and the

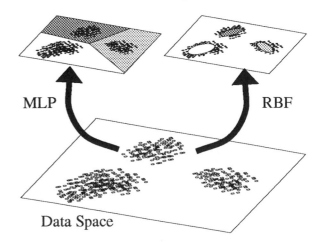

Figure 2. Dissection of pattern space by clusters and hyperplanes. The multilayer perceptron exploits the logistic nonlinearity to create combinations of hyperplanes to dissect pattern space into separable regions. Subsequent layers combine these regions to allow the formation of nonconvex class boundaries. The radial basis function dissects pattern space by modeling *clusters* of data directly and so is more concerned with data distributions.

unconditional distribution can both be modeled by the same set of distributions, $q(x|s)$ but with different mixing coefficients, i.e., $p(x) = \Sigma_s \hat{p}(s)q(x|s)$ and $p(x|c_i) = \Sigma_s p(s;i)q(x|s)$. Then the quantity we are interested in, $p(c_i|x) = p(c_i)p(x|c_i)/p(x)$, is given by

$$p(c_i|x) = \sum_s \frac{p(c_i)p(s;i)}{\hat{p}(s)} \cdot \frac{\hat{p}(s)q(x|s)}{\sum_{s'} \hat{p}(s')q(x|s')} \equiv \sum_j \lambda_{ij}\phi(x|j)$$

where $\lambda_{ij} = p(c_i)p(j;i)/\hat{p}(j)$ relates the overall significance of state j to class i, and $\phi(x|j)$ is a normalized basis function, $\hat{p}(j)q(x|j)/\Sigma_j \hat{p}(j)q(x|j)$.

This formula gives an RBF architecture. For a total of h functions used to approximate the likelihood and the unconditional density, there are h hidden nodes corresponding to the normalized basis functions, and the final layer weights relate the significance of the hidden nodes to the c output class nodes, providing the class conditional information. Of course, the positions and possibly also the ranges of influence of each of these basis functions need to be specified/adapted to allow an adequate model of each data cluster. This can be achieved by unsupervised clustering techniques.

In this manner the RBFN is an ideal network to be applied to classification problems. Note that the architecture of RBFNs for density estimation is more general than the preceding motivational discussion. In particular, note that it is not essential that each basis function itself should be a probability density function.

RBFNs for Prediction

The previous section motivated the radial basis function network by a statistical interpretation of data distributions. In that case the underlying generator of the data (the probability density function) was sampled stochastically. However, the original formulation of the RBFN (Broomhead and Lowe, 1988) was developed in order to produce a deterministic mapping of data by exploiting links with traditional function approximation. This approach attempted to introduce the notion that the training of neural networks could be described as curve fitting. Consequently, "generalization" has a natural interpretation as interpolating along this fitting surface.

The basic idea was as follows. Assume that we have a set of input/output pairs of input/target patterns representing data from an unknown but smooth surface in $\mathbb{R}^n \otimes \mathbb{R}^c$. As a simple example, consider a set of (x, y) pairs generated according to $y = x^2$. In this approach, the problem is to choose a function $y: \mathbb{R}^n \to \mathbb{R}^c$ which satisfies the interpolation conditions $y(x_p) = t_p, p = 1, 2, \ldots, P$. This is strict interpolation in which the function is constrained to pass through all the known data points. The strategy in interpolation theory was to construct a linear function space spanned by a set of nonorthogonal basis functions which depended on the positions of the known data points. The radial basis function expansion mapping to one dimension was originally expressed as $y(x) = \Sigma_{j=1}^P \lambda_j \phi(\|x - x_j\|)$. By using the interpolation conditions, the fitting parameters λ may be determined by matrix inverse methods. The approach was generalized to higher dimensional mappings, to incorporate bias terms, and to account for the fact that strict interpolation is not a good strategy for real-world noisy data, leading to the feedforward neural network topology discussed already.

In the case of the simple $y = x^2$ example mentioned earlier, the inputs are x values, the targets are specific y values, and the RBFN is constructed so as to produce a fitting surface to the parabola $y = x^2$ which is a surface in $\mathbb{R}^1 \otimes \mathbb{R}^1 = \mathbb{R}^2$. Note that this is curve fitting to the parabola in the product space,

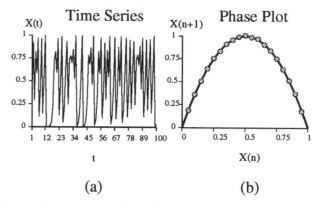

Figure 3. Time series prediction of the quadratic map. Part A shows the original time series, and part B shows the predicted and actual manifolds which generated the data. The predicted values are indicated by the circles, and the actual map (given by $X(n + 1) = 4X(n)[1 - X(n)]$) by the solid line. Part B indicates that the generating function of the time series is being approximated by the network.

not the data samples themselves. This parabola may be interpreted as the generator of the observed data as locations on the parabola "produce" the (x, y) input/output pairs.

Applied in the context of deterministic time series prediction (see, e.g., articles in Rao Vemuri and Rogers, 1994) we assume that there exists an underlying generating function producing the observed time series samples. However, the unknown "surface" producing the data is not likely to be a simple parabola, but is a stationary dynamical system characterized by an attracting manifold. This manifold is embedded as a surface in the high-dimensional product space, just as the simple parabola was embedded in $\mathbb{R}^2$. It is this underlying manifold which is the primary quantity of interest that the RBF network attempts to approximate. Hence it is also assumed that in addition to stationarity, the manifold has some degree of smoothness, just as it was necesssary that the parabola was "smooth" to allow a fitting surface to be produced. Note that it is not important that the raw time series itself should be smooth. This observation explains why networks can be successful in predicting deterministically chaotic time series, since although the time series themselves may exhibit apparent randomness, the underlying map which produces the samples is itself usually very smooth. This point is illustrated in Figure 3 by using an RBFN to predict the one-step-ahead values of the time series generated by the quadratic map. Although the example is artificial, the basic principle and underlying assumptions carry over to the prediction of real-world time series.

Miscellaneous Topics

There are many topics related to radial basis function networks that we have been unable to discuss in this brief introductory article. Among them are the issues of how many centers to use, what types of nonlinearities may be employed, how to choose the smoothing parameters, how to optimize the various weights, and how to assist generalization through regularization. Some of these topics are briefly mentioned in the following paragraphs.

Choice of Kernel Function

The theory of statistical density estimation produces many recommended bounded kernels which may or may not be density functions themselves. Examples of $\phi(z)$ are the Epanechnikov ($\frac{3}{4}[1 - z^2/5]/\sqrt{5}$ for $|z| < \sqrt{5}$, and 0 otherwise), the triangular ($1 - |z|$ for $|z| < 1$ and 0 otherwise), and of course the Gaussian. From interpolation theory the following choices are common: cubic (z^3), thin plate spline ($z^2 \log z$), inverse multi-quadric ($[z^2 + c^2]^{-1/2}$), multiquadric ($[z^2 + c^2]^{1/2}$), and again the Gaussian. Note that these functions do not have finite support, and indeed some of the choices are *unbounded* functions, contrary to intuition and common folklore that the network basis functions are localized. However, it is correct that the parameters of the network may be chosen such that $y(x) \to 0$ as $x \to \infty$ so that the network as a whole achieves a localized response.

Preventing Overfitting: Optimization Under Constraints, Regularization

Various schemes have emerged to discourage overfitting of the training data points. Such options include (1) choosing a small set of initial centers and adapting the positions and spreads to best describe the data in some sense (Moody and Darken, 1989, Lowe, 1989); (2) regularization—having a center located at each data point but adding a smoothing term which is an effective constraint on the possible weight values, the magnitude of this extra term governing the amount of smoothing applied to the fitting surface (Bishop, 1991; Poggio and Girosi, 1990); and (3) selecting centers over a subset of the data points in the training set incrementally to maximize the descriptive power of the data variance obtained by adding each new basis function (Chen, Cowan, and Grant, 1991).

We choose finally to discuss regularization. Extensive and useful reviews of this approach along with further references can be found in Haykin (1994), Girosi, Jones, and Poggio (in press), or GENERALIZATION AND REGULARIZATION IN NONLINEAR LEARNING SYSTEMS. Basically we wish to interpolate a finite data set using an approximation $y(x)$. Of all possible approximators, we wish to choose the one that minimizes the augmented functional

$$H[y] = \sum_{p=1}^{P} (t_p - y(x_p))^2 + \eta \| \hat{O}y \|^2$$

where $\hat{O}$ is an operator such as $\partial/\partial x$ which embodies the constraints of our prior knowledge on desired "smoothness" constraints. The variable η is the regularization parameter and usually embodies the degree to which the constraint should dominate the data. Interestingly, the functional which formally minimizes $H[y]$ takes the form of a radial basis function network; i.e., $y(x) = \Sigma_{p=1}^{P} \lambda_p G^{\dagger}(x, x_p)$. Here, $G^{\dagger}(x, x_p)$ denotes a Green's function which is a solution of an equation determined by the regularization operator. If the operator $\hat{O}^{\dagger}\hat{O}$ is rotationally and translationally invariant, then the Green's function is only a function of the radial differences of its arguments, i.e., $G(\|x - x_p\|)$. Thus once again the form of the radial basis function may be derived, but this time from the perspective of preventing overfitting by regularization. As in the previous interpolation case, the weighting coefficients λ_p may be determined by the solution of a linear equation. Again, strictly speaking, this approach requires a center at each data point, and overfitting is avoided by imposing the smoothing constraint. However, in practice the number of centers may also be chosen to vary.

Discussion

This brief introduction has discussed the motivation and application of the radial basis function network from a variety of perspectives. We have chosen to concentrate upon contrasting a statistical pattern-processing perspective with a deterministic dynamical systems viewpoint. However, both perspectives had the common philosophical basis that the aim of the network is to approximate the underlying structure which generated the observed data, rather than the data itself. This is actually how a multilayer perceptron operates also; however, the RBFN was introduced to make this link with curve fitting and interpolation explicit.

We have tried to illustrate that the RBFN may be employed in classification tasks and time series prediction and other mapping tasks. Combined with its computational tractability the RBFN is of great use in practical applications. But above all, its strength and utility derives from its simplicity and a close relationship with other areas of signal and pattern processing, as well as other neural network architectures.

Road Map: Learning in Artificial Neural Networks, Statistical
Background: I.3. Dynamics and Adaptation in Neural Networks
Related Reading: Coulomb Potential Learning; Data Clustering and Learning; Modular Neural Net Systems, Training of; Pattern Recognition; Process Control

References

Bishop, C., 1991, Improving the generalization properties of radial basis function neural networks, *Neural Computat.*, 3:579–588.

Broomhead, D. S., and Lowe, D., 1988, Multivariable functional interpolation and adaptive networks, *Complex Systems*, 2:321–355.

Chen, S., Cowan, C. F. N., and Grant, P. M., 1991, Orthogonal least squares learning algorithm for radial basis function networks, *IEEE Trans. Neural Netw.*, 2:302–309.

Girosi, F., Jones, M., and Poggio, T., in press, Regularization theory and neural network architectures, *Neural Computat.*

Haykin, S., 1994, Radial basis function networks, in *Neural Networks: A Comprehensive Foundation*, New York: Macmillan, chap. 7. ◆

Lowe, D., 1989, Adaptive radial basis function nonlinearities and the problem of generalisation, in *First IEE International Conference on Artificial Neural Networks*, Conference Publication 313, London: Institute of Electrical Engineers, pp. 171–175.

Lowe, D., 1991, On the iterative inversion of RBF networks: A statistical interpretation, in *Second IEE International Conference on Artificial Neural Networks*, Conference Publication 349, London: Institute of Electrical Engineers, pp. 29–33.

Lowe, D., 1993, Novel "topographic" nonlinear feature extraction using radial basis functions for concentration coding in the "artificial nose," in *Third IEE International Conference on Artificial Neural Networks*, Conference Publication 372, London: Institute of Electrical Engineers, pp. 95–99.

Moody, J., and Darken, C., 1989, Fast learning in networks of locally tuned processing units, *Neural Computat.*, 1:281–294.

Niranjan, M., and Fallside, F., 1990, Neural networks and radial basis functions in classifying static speech patterns, *Comput. Speech Lang.*, 4:275–289.

Park, J., and Sandberg, I. W., 1991, Universal approximation using Radial Basis Function networks, *Neural Computat.*, 3:246–257.

Poggio, T., and Girosi, F., 1990, Networks for approximation and learning, *Proc. IEEE*, 78:1481–1497.

Rao Vemuri, V., and Rogers, R. D., 1994, *Artificial Neural Networks: Forecasting Time Series*, Los Alamitos, CA: IEEE Computer Society Press.

Tråvén, H. G. C., 1991, A neural network approach to statistical pattern classification by "semiparametric" estimation of probability density functions, *IEEE Trans. Neural Netw.*, 2:366–377.

Reaching: Coding in Motor Cortex

Apostolos P. Georgopoulos

Introduction

A common and behaviorally meaningful movement is reaching to targets in space. Reaching involves well-coordinated motion about the shoulder and elbow joints for transporting the hand in space and bringing it to a desired location. A reaching movement can be regarded as a vector, from its origin to its target, with direction and amplitude. The results of several studies support the view that these two parameters reflect separate processing constraints (Georgopoulos, 1991). First, accuracy of pointing is much better for direction than for amplitude; second, when subjects are forced to make a motor response at a time shorter than the usual reaction time, the direction and amplitude of the motor trajectory are affected differently; and third, peripheral sensory neuropathy affects differentially the direction and amplitude of the movement (Ghez et al., 1990). Moreover, the generation of arm movements is not a stereotypic process but seems to involve processing that is subject to interference by distracting sensory and cognitive loads (Frens and Erkelens, 1991); this is in contrast to visually evoked saccades which are unaffected under such conditions. This susceptibility of the arm movement–generating process to distracting (or competing) processes underscores the complexity of the central nervous processes that are involved in this function.

Neural Coding of Reaching Movements

Coding by Single Cells

A relation between the direction of reaching and changes in neuronal activity has been established for several brain areas, including the motor cortex, the premotor cortex, area 5, the cerebellar cortex, and the deep cerebellar nuclei (Georgopoulos, 1991). This relation is a broad, tuning function, the peak of which denotes the "preferred" direction of the cell, that is, the direction of movement for which the cell's activity would be highest. Typically, cell activity varies as a linear function of the cosine of the angle formed between the preferred direction of the cell and the direction of reaching (Figure 1):

$$d_i(M) = b_i + k_i \cos \Theta_{C_i M} \qquad (1)$$

where d_i is the discharge rate of the ith cell with movement in direction M, b_i and k_i are regression coefficients, and $\Theta_{C_i M}$ is the angle formed between the direction of movement M and the cell's preferred direction C_i. (Although other functions could fit the data, the cosine function is a simple one that explains a good percentage of variation in cell activity.) Equation 1 holds both for two-dimensional (2D) reaching movements performed on a plane and for free three-dimensional (3D) reaching movements (Georgopoulos, 1990). Preferred directions of single cells range throughout the directional continuum and are multiply represented in the motor cortex (Georgopoulos, Taira, and Lukashin, 1993). Finally, preferred directions tend to shift in the horizontal plane with different starting points of the movement (Caminiti and Johnson, 1992).

In contrast to the clear relations obtained between single cell activity and direction of movement, the relations between neuronal activity at the single cell level and movement amplitude have been elusive (Georgopoulos, 1990). These relations have been recently elucidated (Fu, Suarez, and Ebner, 1993). In these studies, monkeys moved a handle over a planar working surface in eight directions (0–360° in 45° intervals) and six am-plitudes (1.4–5.4 cm in 0.8-cm increments) in a pseudorandom order. The activity of cells in the motor and premotor cortex was directionally tuned in a cosine fashion, as described previously (Georgopoulos, 1990); the preferred direction was very similar for movements of different amplitudes. Cell activity increased with movement amplitude. Two aspects of this latter finding are noteworthy: first, the highest increase of neuronal activity with movement amplitude was not always along the cell's preferred direction; and second, the best relations with movement amplitude were observed for cell activity during but not before the movement; in the latter case, the direction of movement is the most important factor. These findings indicate that the motor cortex is involved primarily in the *specification* of movement direction, because the movement is planned during the reaction time, and in *monitoring* movement amplitude, because the movement evolves during the movement time. Since the rate of change of velocity and of force scale with amplitude from the beginning of the onset of the motor response, these findings raise interesting questions concerning the potential involvement of other structures in setting the scaled initial motor parameters.

The findings of the studies just reviewed regard discrete reaching movements. Schwartz (1993) studied the neural mechanisms of continuous, drawing movements by recording the activity of single cells in the motor cortex while the monkey traced on a touch screen sinusoids of various amplitudes and spatial frequencies; under these conditions, the direction and speed of movement changed continuously in time. In another task, monkeys made equal-amplitude movements from a central point to peripheral targets (center → out task). The following were found: (i) In the center → out task, cell activity varied in a cosine fashion with the direction of the movement, as found previously (Georgopoulos et al., 1993); (ii) in the tracing task, the ongoing direction of movement explained most of the variance in ongoing cell activity; and (iii) a good proportion of the remaining, nondirectional variance could be accounted for by the ongoing speed of the movement. This relation to speed was best observed for movements near the cell's preferred direction.

Coding by Neuronal Populations

The broad directional tuning of single cell activity indicates that a given cell participates in movements of various directions; from this result, and from the fact that preferred directions range widely, it follows that a movement in a particular direction will engage a whole population of cells. A unique code for the direction of movement (Georgopoulos, Schwartz, and Kettner, 1986) regarded this population as an ensemble of vectors. Each vector represents the contribution of a directionally tuned cell: it points in the cell's preferred direction and is weighted (i.e., has length) according to the change in cell activity associated with a particular movement direction. The weighted vector sum of these neuronal contributions is the "population vector":

$$P(M) = \sum_i^N w_i(M) C_i \qquad (2)$$

The population vector points in the direction of the movement for discrete movements in 2D and 3D space (Figure 2; see also Georgopoulos et al., 1993).

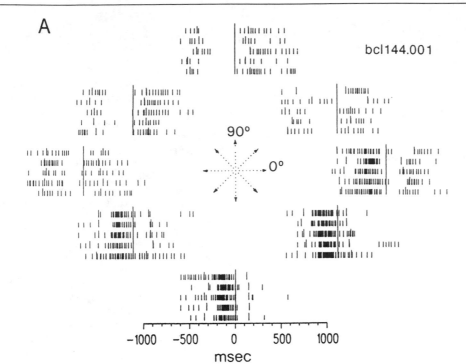

A

bcl144.001

90°

0°

−1000   −500   0   500   1000

msec

Figure 1. Discharge patterns during the center → out task. *A*, Rasters are arranged schematically at each target location around the center start position. Each raster is aligned to the exit from the center start position ($T = 0$). The first long tick mark of each trial is the target onset time, the second is the time of movement onset, and the third is the time of target acquisition. *B*, The cosine tuning function was derived from the average rate of discharge between target onset and target acquisition for each movement (circles). The vertical line through each circle is the standard deviation of that average rate. A cosine tuning function was fitted to these data: $D = b_0 + k\cos(\Theta - \Theta_0)$, $b_0 = 37.9$, $k = 36.1$ (the units are spikes/s), $r^2 = 0.95$. (From Schwartz, A. B., 1992, Motor cortical activity during drawing movements: Single-unit activity during sinusoid tracing, *J. Neurophysiol.*, 68:528–541, fig. 4; reproduced with permission.)

B

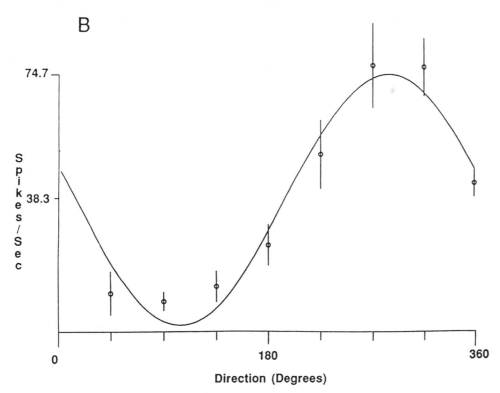

74.7

S
p
i
k
e
s
/
S
e
c

38.3

0   180   360

Direction (Degrees)

The length of the population vector is proportional to the instantaneous speed of the movement. This was indicated initially for 3D reaching movements (Georgopoulos, Kettner, and Schwartz, 1988) but was shown decisively for continuous, tracing movements (Schwartz, 1993). In the latter study, population vectors calculated during the trajectory were added successively tip-to-tail, resulting in a "neural" trajectory that predicted well the ensuing trajectory of the actual movement by an average time lead of approximately 120 ms (Schwartz, 1993). Therefore, the population vector carries information concerning the unfolding movement trajectory. Finally, the population vector is an unbiased predictor of the direction of movement even when the movement begins from different points in space (Georgopoulos, 1990; Caminiti and Johnson, 1992).

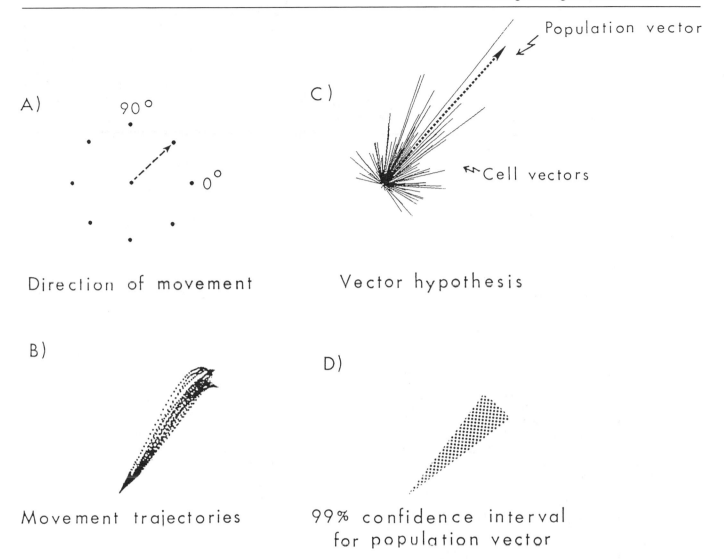

A) 90° 0°

Direction of movement

C) Population vector / Cell vectors

Vector hypothesis

B)

Movement trajectories

D)

99% confidence interval
for population vector

Figure 2. Neuronal population coding of movement direction illustrated for a motor cortical population ($N = 241$ cells) and movement direction toward 12 o'clock in the 2D working surface. *A*, Movement direction. *B*, Family of trajectories made by a well-trained monkey. *C*, Vectorial contributions of single cells (continuous lines) add to yield population vector (broken line), which is in the direction of the move-ment. *D*, Ninety-nine percent confidence interval for the population vector. [Modified from Georgopoulos, A. P., et al., 1984, The representation of movement direction in the motor cortex: Single cell and population studies, in *Dynamic Aspects of Neocortical Function* (G. M. Edelman, W. M. Cowan, and W. E. Gall, Eds.), New York: Wiley, pp. 501–524; reproduced with permission.]

Neural Events Signaling Motor Intention

When a movement is produced as soon as a stimulus appears, some time intervenes between the occurrence of the stimulus and the beginning of the movement, which is the traditional reaction time. This time varies depending on the sensory modality of the stimulus and any imposed constraints on the movement, but it usually takes 200–300 ms. In other cases, a delay can be imposed so that the movement will be initiated after a period of waiting, while the stimulus is still present. These *instructed delay paradigms* probe a step further the representation of intended movements, in the sense that there is not an immediate motor output while the representation is being kept active. A specific case of delayed tasks involves movements that have to be produced on the basis of information kept in *memory*. The difference from the instructed delay task is that now the stimulus defining the direction of the movement is turned off after a short period of presentation, and the movement is triggered after a delay by a separate "go" signal. Thus, information concerning the intended movement has to be retained during the memorized delay.

In all three cases, the representation of information about the intended movement can be studied under different conditions which impose different constraints on the system. It would be interesting to know whether this representation could be identified and visualized during the reaction time, the instructed delay, and the memorized delay periods. Since the information assumed to be represented is about direction, the neuronal population vector could be a useful tool by which to identify this representation. Indeed, the population vector computed at short intervals (e.g., every 20 ms) pointed in the direction of the intended movement during the reaction time, an instructed delay period, or during a memorized delay period (Smyrnis et al., 1992; Figure 3). These findings (i) underscore the usefulness of this analysis as a tool for visualizing representations of the intended movement, and (ii) show that in the

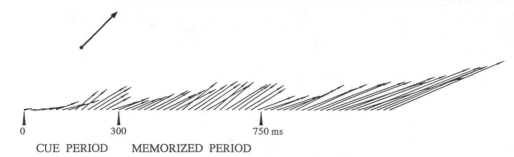

Figure 3. Population vectors in the memorized delay task for the direction indicated are plotted every 20 ms. The arrow on top indicates the direction of the *cue* signal present during the first 300 ms of the delay period. (From Smyrnis, N., et al., 1992, Motor cortical activity in a memorized delay task, *Exp. Brain Res.*, 92:139–151; reproduced with permission.)

presence or absence of an immediate motor output, as well as when the directional information has to be kept in memory, the direction of the intended movement is represented in a dynamic form at the ensemble level. These results also document the involvement of the motor cortex in the representation of intended movements under various behavioral conditions.

Neural Correlates of Motor Cognitive Transformations

In the delayed movement tasks just described, the movement to be made is unequivocally defined in the sense that its direction is determined by the location of a stimulus relative to the starting point of the movement. In that situation, the visual information concerning direction is used to generate the appropriate motor command to implement a movement in that direction. This direction has to be generated and kept available during the delay period, but it is defined from the beginning; therefore, the direction of the intended movement is the same throughout the various times considered. In a different experiment (Georgopoulos et al., 1993), the direction of the movement had to be determined freshly at every trial according to a certain rule, namely, that the movement direction be at an angle (counterclockwise, CCW, or clockwise, CW) from the stimulus direction. In this experiment, the motor intention is not fixed but has to be derived as the solution to the problem. One possible strategy to solve this problem would be to mentally rotate the stimulus direction in the instructed departure (CCW or CW) by an amount equal to the required angular shift. This hypothesis predicts an increase of the reaction time with the angle; indeed, the results of the experiments in human subjects (Georgopoulos and Massey, 1987) showed an increase of the reaction time with the angle and therefore supported the mental rotation hypothesis. The neural mechanisms underlying the process of mental rotation in the movement domain were investigated by training monkeys to perform a task in which they made a movement in a direction 90° CCW from a stimulus direction: if a mental rotation of an imagined vector was taking place, the neuronal population vector would reveal it. Indeed, the population vector rotated during the reaction time from the stimulus to the movement direction through the 90° CCW angle (see Figure 4; Georgopoulos et al., 1993).

Discussion

The results of these experiments indicate that the motor cortical representation of movement is not obligatorily connected with the production of movement; that is, its presence does not necessarily lead to motor output. Concerning downstream points where motor cortical activity could be gated, segmental and propriospinal (Lundberg, 1979) levels in the spinal cord are good candidates, given the extensive convergence of several supraspinal inputs on these interneuronal systems. Therefore, engagement of the motor cortex is not a sufficient condition for

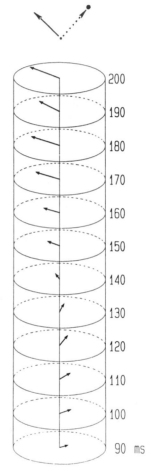

Figure 4. Rotation of the population vector for a different set of rotation trials. The stimulus and movement directions are indicated by the interrupted and continuous lines at the top. The population vector in the 2D space is shown for successive time frames beginning 90 ms after stimulus onset. Notice its rotation counterclockwise from the direction of the stimulus to the direction of the movement. (From Georgopoulos, A. P., et al., 1989, Mental rotation of the neuronal population vector, *Science*, 243:234–235, ©AAAS. Reproduced with permission.)

triggering the movement. Conversely, motor cortical activation seems to be necessary for appropriate planning of the movement, as suggested by the disturbed reaching movements produced by reversible inactivation of the motor cortex (Martin and Ghez, 1993). Another point concerns the nature of the information represented. It may not be appropriate to assign all of this information to the upcoming movement, for it may reflect processes subserving the translation of visual or memorized information to motor output. The complexity of potential explanatory factors (e.g., muscular activity, position of the hand, direction of movement, etc.) for motor cortical activity in behavioral tasks is suggested by the results of studies where such factors were dissociated (Georgopoulos, 1991), but also by the results of the directional transformation study (Georgopoulos et al., 1993), which showed that motor cortical activity does reflect a process involved in mental rotation. It is important to realize that, unlike primary sensory cortices, motor cortex is the site of convergence from a large number of other areas, both cortical and subcortical. Therefore, the discharge patterns of motor cortical cells are generated through this convergence rather than being the outcome of a faithful transmission through sensory lines. Moreover, the motor cortex is not the "final" motor path from the cerebral cortex, since several premotor areas possess direct and dense projections to the spinal cord. It is possible that the motor and premotor areas might be concerned with different but overlapping aspects of motor control and that a particular movement might be the result of this parallel processing. These findings have an important implication, among others, and that is that the spinal motor mechanisms involved in the production of voluntary movement can be properly understood only if one takes into account (a) the convergent pattern of influences from the motor and premotor cortical areas, (b) influences from subcortical structures such as the red nucleus and the reticular formation, and (c) the organization and dynamic interplay of spinal interneuronal circuits involved in the transmission of central commands, the generation of stereotypic motor patterns, and the control of afferent input from the moving limb. Conversely, the patterns of activity of precentral corticospinal neurons will have to be understood in the light of their influences on the spinal mechanisms. In that respect, most of the attention in behaving primates has been focused on those motor cortical neurons that are presumably monosynaptically projecting onto motoneurons, and practically no attention has been paid to the cortical influences on spinal interneuronal mechanisms, in spite of the fact that it is through the latter that a major, and indeed exclusive in some species (e.g., the cat), cortical effect is exerted. There is little doubt that understanding the interactions among the various motor areas, and in particular those between the motor cortex and the spinal cord, is now the biggest challenge in deciphering the "natural intelligence" of the motor system. In that respect, the results of recent studies in the spinal cord of the frog (Bizzi, Mussa-Ivaldi, and Giszter, 1991; see FROG WIPING REFLEXES) provide an important advance in understanding the motor spinal mechanisms at the ensemble level and form a base for understanding the interactions between central motor areas and spinal interneuronal systems intercalated in the translation of central commands to muscle output.

Acknowledgments.   This work was supported by United States Public Health Service grants NS17413 and PSMH48185, and by an Office of Naval Research contract.

Road Map: Primate Motor Control
Related Reading: Eye-Hand Coordination in Reaching Movements; Gaze Coding in the Posterior Parietal Cortex; Grasping Movements: Visuomotor Transformations; Reaching Movements: Implications of Connectionist Models; Short-Term Memory

References

Bizzi, E., Mussa-Ivaldi, F. A., and Giszter, S., 1991, Computations underlying the execution of movement: A biological perspective, *Science*, 253:287–291.

Caminiti, R., and Johnson, P. B., 1992, Internal representation of movement in the cerebral cortex as revealed by the analysis of reaching, *Cereb. Cortex*, 2:269–276.

Frens, M. A., and Erkelens, C. J., 1991, Coordination of hand movements and saccades: Evidence for a common and a separate pathway, *Exp. Brain Res.*, 85:682–690.

Fu, Q. G., Suarez, J. I., and Ebner, T. J., 1993, Neuronal specification of direction and distance during reaching movements in the superior precentral premotor area and primary motor cortex of monkeys, *J. Neurophysiol.*, 70:2097–2116.

Georgopoulos, A. P., 1990, Neurophysiology of reaching, in *Attention and Performance XIII* (M. Jeannerod, Ed.), Hillsdale, NJ: Erlbaum, pp. 227–263.

Georgopoulos, A. P., 1991, Higher order motor control, *Annu. Rev. Neurosci.*, 14:361–377.

Georgopoulos, A. P., Kalaska, J. F., Crutcher, M. D., Caminiti, R., and Massey, J. T., 1984, The representation of movement direction in the motor cortex: Single cell and population studies, in *Dynamic Aspects of Neocortical Function* (G. M. Edelman, W. M. Cowan, and W. E. Gall, Eds.), New York: Wiley, pp. 501–524.

Georgopoulos, A. P., Kettner, R. E., and Schwartz, A. B., 1988, Primate motor cortex and free arm movements to visual targets in three-dimensional space, II: Coding of the direction of movement by a neuronal population, *J. Neurosci.*, 8:2928–2937.

Georgopoulos, A. P., Lurito, J. T., Petrides, M., Schwartz, A. B., and Massey, J. T., 1989, Mental rotation of the neuronal population vector, *Science*, 243:234 235.

Georgopoulos, A. P., and Massey, J. T., 1987, Cognitive spatial-motor processes, 1: The making of movements at various angles from a stimulus direction, *Exp. Brain Res.*, 65:361–370.

Georgopoulos, A. P., Schwartz, A. B., and Kettner, R. E., 1986, Neuronal population coding of movement direction, *Science*, 233:1416–1419.

Georgopoulos, A. P., Taira, M., and Lukashin, A., 1993, Cognitive neurophysiology of the motor cortex, *Science*, 260:47–52.

Ghez, C., Gordon, J., Ghiraldi, M. F., Christakos, C. N., and Cooper, C. N., 1990, Roles of proprioceptive input in the programming of arm trajectories, *Cold Spring Harbor Symp. Quant. Biol.*, 55:837–847.

Lundberg, A., 1979, Integration in a propriospinal motor centre controlling the forelimb in the cat, in *Integration in the Nervous System* (H. Asanuma and V. J. Wilson, Eds.), Tokyo: Igaku-Shoin, pp. 47–64.

Martin, J. H., and Ghez, C., 1993, Differential impairments in reaching and grasping produced by local inactivation within the forelimb representation of the motor cortex of the cat, *Exp. Brain Res.*, 94:429–443.

Schwartz, A. B., 1992, Motor cortical activity during drawing movements: Single-unit activity during sinusoid tracing, *J. Neurophysiol.*, 68:528–541.

Schwartz, A. B., 1993, Motor cortical activity during drawing movements: Population representation during sinusoid tracing, *J. Neurophysiol.*, 70:28–36.

Smyrnis, N., Taira, M., Ashe, J., and Georgopoulos, A. P., 1992, Motor cortical activity in a memorized delay task, *Exp. Brain Res.*, 92:139–151.

Reaching Movements: Implications of Connectionist Models

John F. Kalaska

Introduction

Visually guided reaching movements are deceptively simple: move the hand to the spatial location of a target. Yet how the central nervous system (CNS) accomplishes this task is a deep problem that addresses many motor control issues, including the cognitive events underlying response selection, the serial ordering of motor acts, multiarticular coordination, sensorimotor transformations, and the excess-degrees-of-freedom and inverse-transform problems. This article considers the implications of connectionist solutions to many of these problems for the overall functional organization of the cerebral cortical components of the motor system.

Reaching to a Target Appears to Involve a Sequence of Information-Processing Events

One can arbitrarily divide motor control into two stages (Figure 1): (1) *response selection*: specification of a motor plan in task-space parameters; and (2) *response implementation*: transformation of the motor plan into actuator-space parameters. Separation of planning from execution is advantageous, because the form of the motor act can then be defined in terms that are independent of the details of its implementation, but the neural reality of this separation is debatable.

Response selection and implementation have been further divided into a hierarchical sequence of processing stages. For reaching movements, this is often described as a sequence of transformations between representations of the movement in different reference frames in task and actuator space (Georgopoulos, 1991; Kalaska, 1991; Kalaska and Crammond, 1992; Soechting and Flanders, 1992). Details of the models differ. However, most recognize three major levels of representation in one form or another (Figure 1): (1) extrinsic kinematics, such as motion of the hand through space; (2) intrinsic kinematics, such as joint motions; and (3) dynamics, such as the causal forces that produce the movement. Terms such as kinematics and dynamics are being used here to distinguish neuronal representations of the spatiotemporal form of movement from those of the causal means of its implementation.

These models predict that the CNS will contain several serial representations of reaching movements in different reference frames, but there is still no consensus on their nature. Moreover, not all of the steps in Figure 1 need be planned and imposed by supraspinal mechanisms (Georgopoulos, 1991; Kalaska and Crammond, 1992). Instead, some may result from

the inherent pattern-generating abilities of spinal cord circuits and the mechanical properties of the peripheral musculoskeletal system. This is the fundamental assumption of equilibrium point models (see EQUILIBRIUM POINT HYPOTHESIS).

Physiological Studies Do Not Support a Simple Serial Control Process

Many lines of evidence have implicated a number of separate areas of the cerebral cortex in the control of arm movements (Figure 2; Kalaska, 1991; Humphrey and Tanji, 1991; Kalaska and Crammond, 1992). However, anatomical data show that information flow among these multiple arm movement representations is not unidirectional and therefore not strictly hierarchical. Instead, they are interconnected by a complex web of serial, parallel, convergent, divergent, feedforward, and feedback projections.

Neurophysiological studies have described the movement-related patterns of cell activity in different cortical areas but have not resolved what are the *parameters* or *coordinate frames* for reaching in each area (Georgopoulos, 1991; Humphrey and Tanji, 1991; Kalaska, 1991; Kalaska and Crammond, 1992). Single-cell discharge covaries with many potential parameters at different putative stages in the control hierarchy in different areas. The correlation with a given movement parameter typically accounts for some of the variance of a cell's discharge, but rarely for all or none, as if single cells do not uniquely "encode" a specific movement parameter. Instead, they behave as if they are transmitting partial information about several different parameters of movement at different representational levels in the putative control hierarchy (Fetz, 1992; Kalaska, 1991; Robinson, 1992). Furthermore, the cell population within a cortical area, rather than being homogeneous, shows a wide range of partial correlations to different movement parameters, as if processing different combinations of movement-related information. Although the populations in different areas are distinguished by different combinations of response properties, there is also an overlap in the range of responses across areas. There is also extensive temporal overlap in the activation of cells across areas before movement.

In summary, these studies do not support a strict hierarchy of cortical motor areas. The gradual changes of cell responses in different areas and their sequential recruitment imply some degree of serial organization. In contrast, the distributed representation of different parameters across many cortical areas,

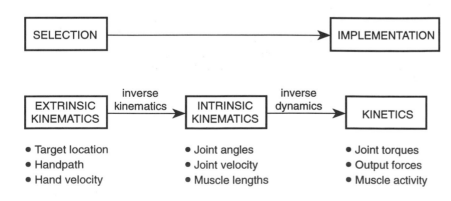

Figure 1. A putative three-step hierarchy for the control of reaching movements, based on the formalisms of mechanics. Arrows symbolize coordinate transformations. Terms below brackets indicate potential movement parameters in the corresponding coordinate frameworks.

Figure 2. Distribution of cerebral cortical neuronal populations activated during reaching movements. MI: primary motor cortex; SMA: supplementary motor area; PMd and PMa: dorsal and arcuate premotor cortex; PF and PFd: prefrontal and dorsal prefrontal cortex; PA5 and PA7b: parietal cortex areas 5 and 7b; CMAd, CMAr, and CMAv: dorsal, rostral, and ventral cingulate motor areas. (Adapted from Kalaska and Crammond, 1992.)

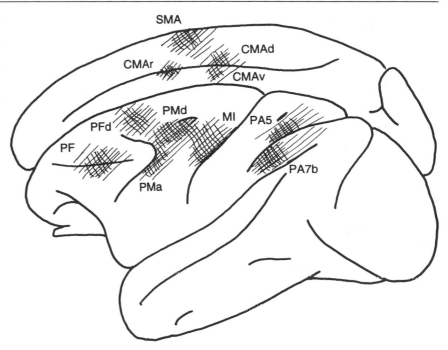

the coexistence of several levels of representation in any one area, and their extensive temporal coactivation during movement favor a more parallel organization.

Connectionist Models of Reaching Movements

A number of connectionist models have been developed for reaching movements. Although most were not intended to be "physiological," they illustrate the range of solutions that are competent to solve the computational problems underlying reaching movements.

For instance, the robotic controller INFANT (Figure 3A) learns the inverse kinematics transform between target spatial location and the arm posture required to place the hand at the target (Kuperstein, 1988). Arrays of elements encode the visual field input from a stereo camera, the angle of gaze of the cameras, and "muscle activity" (the last array actually codes joint angles as muscle lengths). The gaze and visual-field maps are linked to the muscle map by arrays of adjustable weights. During an initial learning period, an external signal generates random arm-muscle map activation patterns that displace the robot arm to different postures (*motor babbling*). The stereo camera foveates the hand after each babble. The resultant visual-field and gaze-angle inputs pass through the weight maps to generate a predicted arm-muscle activation vector. The difference between the predicted and externally generated activation vectors is used to modify the weight maps. After many random babbles, the external and predicted vectors converge, at which point INFANT can reach accurately to foveated objects in its workspace. The adjustable-weight maps represent in distributed manner the transformation between all combinations of gaze angle and visual field input and the requisite final muscle lengths. INFANT is essentially a static feedforward *equilibrium point* model that learns only the mapping between target location and final arm posture; no trajectory or time-varying signals are planned.

In contrast, the robotic controller MURPHY (Figure 3B; Mel, 1991; see PLANNING, CONNECTIONIST) can plan a trajectory. It contains two parallel subsystems. A visual-field array

encodes the location of objects, including MURPHY's arm. A visual hand-velocity array signals the direction of hand motion. Joint-angle and joint-velocity maps signal the angle and rate of change of angle of the three joints in the robot arm. During an initial "learning by doing" period, a forward kinematics transform from joint angles to arm posture is learned. Random external signals onto the joint-angle map displace the arm to different postures, viewed by a camera. This activates the visual-field map, which learns the visual-field activation (i.e., arm spatial location) corresponding to each joint-angle map activation pattern. An inverse differential kinematics transform from hand velocity to joint velocity is learned at the same time, by using the current joint angles of the arm and the desired direction of motion of the hand toward the target to calculate the required joint-velocity map signals. Once trained, MURPHY approaches a target by an incremental trajectory of small steps. Each step is selected by passing many joint-angle activation vectors through the forward kinematics network to generate "mental images" of the position of the arm in space on the visual-field array. Processing modules outside of the network select the "best" of the tested next incremental postures that moves the arm toward the target while avoiding obstacles.

The DIRECT model (Figure 4A; Bullock, Grossberg, and Guenther, 1993) also learns the inverse differential kinematics transform from hand motions and joint motions by motor babbling, but its implementation is quite different. A set of arrays signals target location in body-centered space. Another array uses visual and joint angle inputs to encode the spatial position of the robot arm. The difference between these two spatial signals specifies the desired direction of hand movement toward the target. A subsequent array determines the requisite joint rotations, given the hand movement direction and current arm posture (inverse differential kinematics). Integration converts the joint rotation signals into joint angle signals that are sent to the actuators to move the limb.

Jordan (1990; see also SENSORIMOTOR LEARNING) devised a multilayer network (Figure 4B) with two input vectors, plan and state. The plan vector defines the desired motor act in task

A "INFANT"

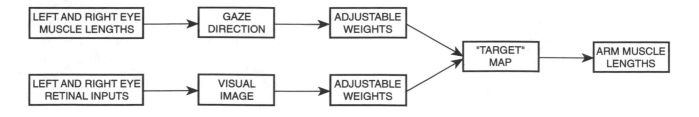

B "MURPHY"

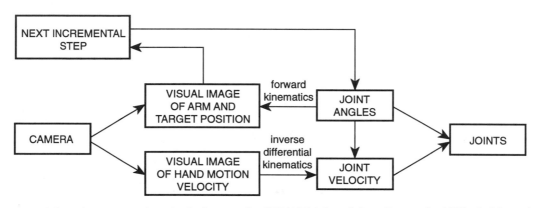

Figure 3. *A*, Schematic representation of robotic controller INFANT (adapted from Kuperstein, 1988). *B*, Schematic representation of robotic controller MURPHY (adapted from Mel, 1991).

space. The state vector reflects the current status of the actuator system. Both vectors project onto a layer of hidden units. The model first learns the forward kinematics transform from actuator unit activity to task space motions. The forward network is then coupled to the inverse kinematics network. Because the combination of forward and inverse models is the identity transform, a plan vector input should generate an identical output vector. Differences between the two are used to train the inverse kinematics network.

These models learn the transformation from extrinsic to intrinsic kinematics. Biological systems also must deal with the inverse dynamics transformation from desired motions to causal forces (although equilibrium-point models argue that this problem is finessed, not solved). A hierarchical network by Kawato et al. (1990) can compute joint torques from the movement trajectory by solving inverse kinematics and inverse dynamics simultaneously, rather than sequentially. This model is a serial cascade of identical four-layer modules that each compute the change in trajectory and torque during a specific finite time interval along the trajectory, determined by its position in the cascade. The output of each module is the input to the next module in the cascade. It is trained by putting joint torque signals into both the control network and the actuators and using the difference between the actual trajectory and the trajectory predicted by the network to adjust network connection weights.

Most of these models were based on control theory formalisms. A few network models have been largely inspired by

physiological data. For instance, Burnod et al. (1992) examined how the CNS might learn to reach to any target position from any initial arm posture. Their model has three arrays of modules called matching, synergy, and motor units, the latter representing spinal motor pools. Each matching and synergy unit is a three-layer network. Synergy unit output activates a weighted set of motor units, displacing the hand in space. The network must learn how the direction of motion caused by activation of each synergy unit changes with starting arm posture. Each synergy unit receives lateral inputs from other synergy units and a feedforward projection from matching units. Matching units learn the inverse kinematics transform from hand space to actuator space. They receive a visual input signaling the direction of movement of the hand in space and an efferent copy of synergy unit output. Both matching and synergy units also receive a signal about arm position. During a motor-babbling period, random patterns of synergy unit output generate arm movements and activate matching units. Visual input from the resulting hand motion converges on the matching units, which learn which synergy units are activated at a particular initial arm posture to produce a particular hand motion. Conversely, synergy units learn which matching units are activated by the motion generated by its output in different initial arm positions, that is, how the motion of the hand in space caused by its output changes with arm initial posture. Once trained, a signal about intended hand movement direction projected onto the matching units activates the appropriate synergy units as a function of starting arm posture. Al-

A   "DIRECT"

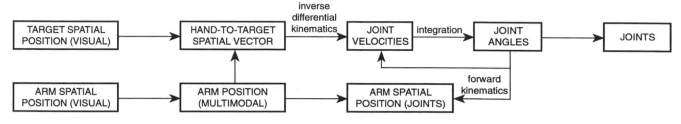

B   JORDAN

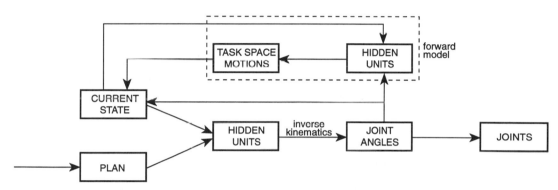

Figure 4. *A*, Schematic representation of robotic controller DIRECT (adapted from Bullock et al., 1993). *B*, Schematic representation of robotic controller proposed by Jordan (adapted from Jordan, 1990).

though synergy output is described in terms of muscle activity, the model only learns the inverse kinematics transform.

Networks and Neurophysiology

The network models just described, rather than revealing how the CNS controls reaching movements, demonstrate that many schemes are capable of doing so. Nevertheless, they provide some insights on a number of biological motor control issues.

Hierarchy, Heterarchy, or Anarchy?

Most of the networks reviewed perform a sequence of coordinate transformations. Moreover, their behavioral capacity is usually limited. They make reflexlike reaching responses to targets, but they cannot formulate more general or flexible plans. This presumably requires still further hierarchically higher planning levels. For instance, Jordan's model can implement a movement sequence, but it cannot determine the sequence itself. That information is provided by the plan vector input, which is in essence the output from putative higher-level response-selection processes. Similarly, MURPHY organizes reaching trajectories by virtue of modules superimposed on the network that select the next incremental step to implement.

Some networks are strict feedforward serial hierarchies (see Figure 3*A*), but others possess more complex architectures with multiple connections among layers that begin to resemble the complexity of biological motor systems. The multiple pathways allow for variable patterns of information flow among the components of the network—a heterarchy, not a hierarchy.

One important potential benefit of this architecture is flexible context-dependent control. Movements are made in many behavioral contexts with different control constraints. For instance, a critical control factor in tool use is the output forces at the hand, and the biomechanical form of its implementation is often unimportant. In contrast, a control constraint in dance is the graceful coordination of the sequence and form of limb movements, whereas during walking on a tightrope, maintenance of balance is paramount, and all else, including gracefulness, is secondary. The motor system must be able to adapt to these different contexts. The control constraints and priorities of a simple hierarchy are inflexible because information flow is unidirectional, and the processing in each serial stage is completely subservient to the preceding stages.

One should not confuse a heterarchy with anarchy. In many contexts, a heterarchy could even behave serially. For instance, the abrupt appearance of a target initiates a flow of activity throughout the network, successively recruiting different neuronal populations. This produces exactly the pattern of staggered recruitment but extensive temporal co-activation of different cortical areas that has been seen repeatedly in neurophysiological studies.

An important question is whether each transformation or movement representation within the heterarchy corresponds to a distinct anatomical structure, such as a cortical area, or takes some other form. In most network models, each layer is a discrete functional unit with unique properties and a specific role. Some models claim that certain network components represent specific CNS structures (Burnod et al., 1992; Kawato et al., 1990; Mel, 1991). Alternatively, others suggest that the analog of each network layer is a set of interconnected modules distributed across several anatomical structures (Alexander, DeLong, and Crutcher, 1992; see also BASAL GANGLIA). The motor system would possess many such distributed networks in parallel, and each cortical area would contain components of several distributed representations.

Movement Is a Real-Time Process Requiring Time-Varying Signals

Many networks are static—a constant input signal generates a constant output. However, movements unfold in time and require the generation of time-varying outputs. Several models suggest that the capacity to generate time-varying signals emerges from the inherent dynamical properties of networks with recurrent connections between layers. Multiple recurrent connections are a prominent characteristic of the motor system.

For instance, the Jordan network has a recurrent loop from its actuator output back onto the input state units. This loop is not a servocontrol feedback error signal; its role is to alter the input signal as a function of the current output signal. As a result, the network output varies incrementally with each iteration of the loop, even though the plan input vector remains constant. The DIRECT model also contains several feedback loops. For instance, a recurrent copy of the output signal about current joint angles projects to the arrays doing the inverse differential transform to ensure that the joint rotations account for current limb geometry during movement, and it also updates the signal specifying the location of the hand in space, which in turn gradually drives the spatial direction vector to zero as the hand approaches the target. Even Kawato's cascade network is equivalent to a recurrent loop, with each iterative cycle of the loop "unrolled" as a separate cascade module. Admittedly, the cascade is an engineering convenience and not a realistic model of any part of the motor system.

Sensorimotor Transformations

Network models show that sensorimotor transformations can be made implicitly by the convergence of input signals in different reference frames onto a population of neurons. The neurons learn the association between the input signals that produces the desired output in a new reference frame. In both MURPHY and DIRECT models, for instance, many array elements respond selectively to combinations of joint and visual inputs to affect sensorimotor transformations. The CNS may likewise perform sensorimotor transformations implicitly by bringing together input signals in appropriate reference frames onto neurons through the extensive interconnections between different components of the motor system.

Movement and the Single Neuron

This mechanism for transformations suggests that single cells do not signal a single movement parameter in a specific reference frame. Instead, they learn the cause-and-effect associations between input and output signals during repeated movements. This association is expressed by discharge that covaries with more than one movement parameter and need not conform to any given formalism. Nevertheless, the musculoskeletal system must obey the universal laws of motion, and so the internal mechanisms of the motor system must reflect those laws in one way or another. As a result, the discharge of a motor cortex neuron may show a good correlation to output

forces or torques, but that does not mean that the cell is explicitly coding either parameter. Indeed, it does not know a newton-meter from a Fig Newton.

This may have computational advantages. One can think of sensorimotor transformations as complex surfaces in a multidimensional space of interdependent movement parameters. The discharge of a single cell is a vector in that parameter space, with different weightings for different parameters, determined by convergent inputs from different parts of the parameter space. Each cell can only make a partial contribution to the definition of the transformation surface. Its complete representation is distributed across many cells, so that global movement parameters in a particular reference frame can only be seen in the discharge of cell populations. Distributed coding of movement parameters has been observed in several cortical areas (Georgopoulos, 1991; Kalaska and Crammond, 1992; see also REACHING: CODING IN MOTOR CORTEX). Population coding may overcome the computational limitations imposed by the fact that single cells are stochastic processors with limited channel capacity (see SENSORY CODING AND INFORMATION THEORY). Furthermore, if cell activity reflects the covariation of different movement parameters, that is, the local shape of the transformation surface, it implicitly accounts for the complex interactions between movement parameters defined by the laws of motion. Therefore, the multivariate partial correlations shown by cells may be not just an artifact of task design, biomechanics, or the laws of motion, but rather a result of the computational mechanisms used by the system.

Neural Networks: Nattering Nabobs of Negativism?

This discussion might lead one to despair of our ability ever to understand the neuronal mechanisms of motor control. At the extreme, it suggests that the brain is a multilayer network of hidden units on which fragments of information are scattered in a chaotic and undecipherable manner. As a result, attempting to make sense of single-cell activity recorded in behaving animals is a fool's game (Fetz, 1992; Robinson, 1992; for a more positive view by Fetz, see DYNAMIC MODELS OF NEUROPHYSIOLOGICAL SYSTEMS).

Such pessimism is most certainly unwarranted. It has become fashionable, for instance, to stress the widespread distribution of similar types of cell responses in different cortical areas. However, although cell populations in different areas are not totally different, they are nevertheless not the same. The arm representation in each area has its own unique character, with different combinations of response properties that can be seen repeatedly across individuals. This consistency cannot be happenstance. Just as the general architecture of any network model has to be carefully crafted because the combination of inputs converging on each layer has a specific purpose, the design of the limb control system also is purposeful.

Much Ado About Networks

Networks are just the latest in a long line of brain models. Their shortcomings as physiological models are well known. It must also be recognized that network models have been applied successfully to motor control in part because of their power to solve exactly the types of nonlinear transformations and ill-posed problems that predominate the field. In that sense, they are competent because they are the right tool applied to the appropriate problem.

One should not take any current network models as literal models of the motor system. Instead, they should be viewed only as metaphors of how the motor system processes information and as tools to generate and test hypotheses about the nature of information processing within the motor system. These models can be used profitably as an adjunct to neurophysiological studies. There are a vast range of network architectures and learning rules that are competent to solve motor-control problems, and no amount of simulation will prove any given network as a good physiological model of brain function. Any motor control model with pretences of biological validity will only be validated by neurophysiological studies.

Road Map: Primate Motor Control
Background: I.3. Dynamics and Adaptation in Neural Networks
Related Reading: Cerebellum and Motor Control; Eye-Hand Coordination in Reaching Movements; Motor Control, Biological and Theoretical; Recurrent Networks: Supervised Learning

References

Alexander, G. E., DeLong, M. R., and Crutcher, M. D., 1992, Do cortical and basal ganglionic motor areas use "motor programs" to control movement? *Behav. Brain Sci.*, 15:656–665. ◆

Bullock, D., Grossberg, S., and Guenther, F. H., 1993, A self-organizing neural model of motor equivalent reaching and tool use by a multijoint arm, *J. Cognit. Neurosci.*, 5:408–435.

Burnod, Y., Grandguillaume, P., Otto, I., Ferraina, S., Johnson, P. B., and Caminiti, R., 1992, Visuomotor transformations underlying arm movements toward visual targets: A neural network model of cerebral cortical operations, *J. Neurosci.*, 12:1435–1453.

Fetz, E. E., 1992, Are movement parameters recognizably coded in the activity of single neurons? *Behav. Brain Sci.*, 15:679–690. ◆

Georgopoulos, A. P., 1991, Higher order motor control, *Annu. Rev. Neurosci.*, 14:361–377. ◆

Humphrey, D. R., and Tanji, J., 1991, What features of voluntary motor control are encoded in the neuronal discharge of different cortical areas?, in *Motor Control: Concepts and Issues* (D. R. Humphrey and H. J. Freund, Eds.), Chichester, Eng.: Wiley, pp. 413–443. ◆

Jordan, M. I., 1990, Motor learning and the degrees of freedom problem, in *Attention and Performance XIII: Motor Representation and Control* (M. Jeannerod, Ed.), Hillsdale, NJ: Erlbaum, pp. 796–836. ◆

Kalaska, J. F., 1991, Reaching movements to visual targets: Neuronal representations of sensorimotor transformations, *Semin. Neurosci.*, 3:67–80. ◆

Kalaska J. F., and Crammond D. J., 1992, Cerebral cortical mechanism of reaching movements, *Science*, 255:1517–1523. ◆

Kawato, M., Maeda, Y., Uno, Y., and Suzuki, R., 1990, Trajectory formation of arm movement by cascade neural network model based on minimum torque-change criterion, *Biol. Cybern.*, 62:275–288.

Kuperstein, M., 1988, Neural model of adaptive hand-eye coordination for single postures, *Science*, 239:1308–1311.

Mel, B. W., 1991, A connectionist model may shed light on neural mechanisms for visually guided reaching, *J. Cognit. Neurosci.*, 3: 273–292.

Robinson, D. A., 1992, Implications of neural networks for how we think about brain function, *Behav. Brain Sci.*, 15:644–655. ◆

Soechting, J. F., and Flanders, M., 1992, Moving in three-dimensional space: Frames of reference, vectors, and coordinate systems, *Annu. Rev. Neurosci.*, 15:167–191. ◆

Reactive Robotic Systems

Ronald C. Arkin

Introduction

Reactive systems are a relatively recent development in robotics that has redirected artificial intelligence research. This new approach grew out of a dissatisfaction with existing methods for producing intelligent robotic response and a growing awareness of the importance of looking at biological systems as a basis for constructing intelligent behavior. Reactive robots are also referred to as behavior-based robots—they are instructed to perform through the activation of a collection of low-level primitive behaviors. Complex physical behavior emerges through the interaction of the behavioral set and the complexities of the environment in which the robot finds itself. This methodology provides more rapid and flexible response than is attainable through traditional methods of robotic control.

Some of the hallmark characteristics of purely reactive robotic systems include:

1. *Behaviors are basic building blocks.* A behavior in these systems is usually a simple sensorimotor pair, where sensory activity consists of providing necessary information to support low-level reactive motor response, such as avoiding obstacles, escaping from predators, being attracted to goals, etc.

2. *Abstract representational knowledge is avoided.* Creating and maintaining accurate representations of the world is a time-consuming, error-prone process. Purely reactive systems do not maintain world models, instead reacting directly to the stimuli the world presents. This is particularly useful in highly dynamic and hazardous worlds, where the environment is unpredictable and potentially hostile.

3. *Animal models of behavior are often used as a basis for these systems.* Models from neuroscience, cognitive psychology, and ethology are used to capture the nature of the behaviors that are necessary for a robot's safe interaction with a hostile world.

4. *Demonstrable robotic results have been achieved.* These techniques have been applied to a wide range of robots, including six-legged walking robots, pipe-crawling robots, robots for indoor/outdoor activities, mobile manipulators, dextrous hands, and entire herds of mobile robots. Because these systems are highly modular, they can be constructed incrementally from the bottom up by adding new behaviors to an existing repertoire. From an engineering perspective, this is quite desirable because it facilitates the growth and application of existing software and hardware systems to new domains.

Even more recently, hybrid reactive/deliberative robotic architectures have emerged which combine aspects of more traditional artificial intelligence (AI) symbolic methods and use of abstract representational knowledge with the responsiveness, robustness, and flexibility of purely reactive systems. Both purely reactive and hybrid architectures are discussed within this article.

Biological Basis for Reactive Robotic Systems

Many of the designers of reactive systems look to biology as a source of models for use in robots. Although the diversity of these efforts is significant, ranging from traditionally engineered systems to those that dedicate themselves to faithfully replicating biological behavior, this article reports on a few exemplars that have affected reactive and hybrid system design.

Action-oriented perception.   Neuroscientists and psychologists (especially the cognitive and ecological communities) have provided models for the relationships between perceptual activities and behaviors required for a particular task. One excellent example is presented in Arbib (1972). His model of action-oriented perception shows that what an agent needs to perceive is based on its needs to act. This is a primary guiding principle in the design of reactive robots. The traditional computer vision community often views perception as a disembodied perceiver that interprets images without consideration of what the knowing agent needs to do. In contrast, the strong coupling between action and perception is one of the hallmarks of purely reactive robotic systems (see also ACTIVE VISION). Neisser has further developed these ideas in the context of cognitive psychology (see Arkin, 1990a, for a review of those aspects relevant to robotic systems).

Ethological studies.   A pressing question for reactive robotic system designers is just what behaviors are necessary or sufficient for a particular task and environment. Many of these researchers have turned to ethological studies as a source for behaviors that are relevant in certain circumstances. Specific models used in reactive robotic systems have been quite varied, including bird flocking, ant foraging, fish schooling, and cockroach escape, among others. One example involving toad detour behavior (Arbib and House, 1987) provided motivation and justification for the use of vector fields in reactive schema-based robot navigation (Arkin, 1990a; see also POTENTIAL FIELDS AND NEURAL NETWORKS).

Co-existence of parallel planning and execution systems (hybrid systems).   Norman and Shallice (1986) have modeled the co-existence of two distinct systems concerned with controlling human behavior. One system models "automatic" behavior and is closely aligned with reactive systems. This system handles automatic action execution without awareness, starts without attention, and consists of independent parallel activity threads (schemas). The second system controls "willed" behavior and expresses an interface between deliberate conscious control and the automatic system.

While purely reactive robotic systems are compatible with the modeled automatic system (e.g., Brooks, 1986), most hybrid robotic systems (e.g., Arkin, 1990b; Gat, 1992) incorporate both willed (deliberative) and automatic (reactive) components in a manner somewhat consistent with this model.

One problem confronting the reactive robotic systems designer is that much of the data reported by biological scientists is often presented statistically. While this may be useful within the context of their home disciplines, it is important for process models to be constructed whenever possible to facilitate the adoption of this work into intelligent robotic systems (see NEUROETHOLOGY, COMPUTATIONAL).

Purely Reactive Robotic Systems

Reactive robotic systems originate in the cybernetic movement of the 1940s. Grey Walter (1953) developed an electromechanical "tortoise" capable of moving about the world, avoiding perceived threats, and being attracted to certain goals. Of special interest was the inclusion of changing goals regarding the robot's recharging station. When power was low, the tortoise was attracted to and docked with the recharger. When sufficient energy was acquired, it lost its "appetite" (charger attraction) and was repelled by it. There was no use of abstract representational constructs as found in traditional AI; perception directly controlled motor action. Simple behaviors were created: head toward weak light, back away from strong light, and turn-and-push to avoid obstacles.

Braitenberg (1984) revived interest in this class of creatures. He demonstrated, using simple analog circuitry, that "creatures" could be built that manifested behaviors comparable to those found in animals, e.g., cowardice, aggression, love, exploration, and logic. These thought experiments in "synthetic psychology" showed that seemingly complex behavior could result from a collection of simple sensorimotor transformations.

Brooks (1986) was an early leader of the purely reactive robotic paradigm. His group pushed this approach with the development of the subsumption architecture. He articulated the departure from classical AI and broke away from the *sense-plan-act* paradigm that dominated AI in the 1970s and 1980s, as typified by robots like Shakey that used resolution theorem proving as their primary reasoning mechanism. This new position brought into question the role of representational knowledge in AI altogether. The subsumption architecture was biologically motivated only in the behaviorist sense, since it produced overt results that resembled certain insect systems but was unconcerned about the underlying biological mechanisms that produced them.

At about the same time the subsumption architecture appeared, other researchers were interested in pursuing parallels in biological and mechanical systems. A sort of cybernetics revival occurred. Studies produced by ethologists, neuroscientists, and others provided models that were used within reactive robotic systems. These researchers' goals varied. For example, Arkin (1990a) exploited these models with the purpose of constructing intelligent robotic systems, using interacting schemas as a basis for reactive robotic control systems design (see SCHEMA THEORY). Beer, alternatively, used robotic systems to demonstrate the fidelity of neuroscientific models (see LOCOMOTION, INVERTEBRATE). Significant conferences are now dedicated to animal and computational systems relationships (see, for example, Cliff et al., 1994).

Figure 1 presents a simple reactive control system example. A robot controlled by this system wanders around avoiding collisions until it finds a path, which it then follows until it locates its goal. This consists of four behaviors: *avoid-obstacle* prevents the robot from colliding with anything; *wander* ensures movement in the absence of goal or path attraction; *stay-on-path* guides the robot down a hall or road to find the goal near the path's end; *move-to-goal* attracts the robot to the final goal. The perceptual strategies for each behavior are also depicted. The behavior coordination mechanism can be of several forms. Arbitration or action-selection mechanisms are typically found in subsumption-style architectures where only one behavior is active at any given time. This action-selection mechanism can be complex, involving extensive connections between behaviors for inhibition/suppression. The schematic representation of this mechanism is greatly simplified in this figure.

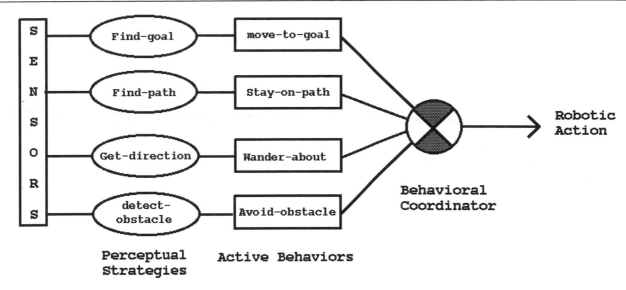

Figure 1. Example of a reactive control system.

Other coordinators may involve blending, as in schema-based reactive control systems, where all active behaviors contribute somewhat to the overall coordinated motion.

Hybrid Reactive/Deliberative Robotic Systems

Hybrid architectures permit reconfiguration of reactive control systems based on available world knowledge, adding considerable flexibility over purely reactive systems. Dynamically reconfiguring the control system based on deliberation (reasoning over world models) is an important addition to the overall competence of general-purpose robots.

It should be recognized that purely reactive robotic systems are not appropriate for every robotic application. In situations where the world can be accurately modeled, where there is restricted uncertainty, and there exists some guarantee of virtually no change in the world during execution (such as an engineered assembly workcell), deliberative methods are often preferred, since a plan can, most likely, be effectively carried out. In the real world, in which biological agents function, these prerequisites for purely deliberative planners do not exist. If roboticists hope to have their machines functioning in the same environments that we do, methods like reactive control are required. Many believe that hybrid systems capable of incorporating both deliberative reasoning and reactive execution are needed to deliver the full potential of robotic systems.

Arkin was among the first to advocate the use of both deliberative (hierarchical) and reactive (schema-based) control systems within the autonomous robot architecture. Incorporating a traditional planner that could reason over a flexible and modular reactive control system, specific robotic configurations could be constructed that integrated behavioral, perceptual, and a priori environmental knowledge (Arkin, 1990b). This system was tested on a wide range of applications, both inside and outdoors.

Gat (1992) proposed a three-level hybrid system (Atlantis) incorporating a Lisp-based deliberator, a sequencer that handled failures of the reactive system, and a reactive controller. This system was fielded and tested successfully on Mars rover prototypes.

Perception and Reactivity

A fundamental guiding principle for purely reactive systems is that perceptual activities should always be viewed on the basis of motor needs (i.e., *a need-to-know basis*). A large body of mainstream computer vision research is concerned with the abstract task of image understanding which typically is independent of a particular agent's needs. Proponents of purely reactive control advocate that perception serves motor action, and image interpretation algorithms must take this into account. Sensing strategies should be constructed taking advantage of the knowledge of underlying behavioral requirements. This eliminates the need to construct global representations of the world, an activity avoided in purely reactive robotic systems. By creating perceptual algorithms that extract only relevant information and that exploit expectations of what is necessary and sufficient to be perceived, efficient sensor processing is a natural consequence.

Hybrid approaches, nonetheless, are more consistent with the views of neuroscientists (e.g., Mishkin, Ungerleider, and Macko, 1983) on "what" and "where" visual systems which account for the maintenance of spatial relationships in a more than purely reactive manner (see VISUAL SCENE PERCEPTION: NEUROPHYSIOLOGY).

There are three ways in which reactive systems can use perceptual information: perceptual channeling (sensor fission), action-oriented sensor fusion, and perceptual sequencing. Perceptual channeling is straightforward: a motor behavior requires a particular stimulus for it to be invoked, so a single sensor system is created. A simple sensorimotor circuit results. There are numerous examples (e.g., Brooks, 1991; Maes, 1990).

Action-oriented sensor fusion (Arkin, 1993) permits the construction of representations (percepts) which are local to individual behaviors. Restricting the representation to the requirements of a particular behavior allows the benefits of reactive control to remain while permitting more than one sensor to provide input, resulting in increased robustness.

Sometimes fixed action patterns require varying stimuli to support them over time and space. As a behavioral response unfolds, it may be modulated by different sensors or different views of the world. Perceptual sequencing provides for the coordination of multiple perceptual algorithms over time in support of a single behavioral activity. Perceptual algorithms are phased in and out based on the needs of the agent and the environmental context in which it is situated.

Discussion

Space prevents an extensive survey of the wide range of reactive robotic systems; the reader is referred to Maes (1990), Efken and Shaw (1993), Brooks (1991), and Lyons and Hendriks (1992) for additional information. These methods have gained dramatically in popularity and utility since the mid-1980s and are being applied to robotic systems throughout the world.

Hybrid reactive/deliberative architectures have been created to address several of the potential shortcomings of purely reactive systems. They permit the incorporation of world knowledge and the construction of global representations, yet preserve the strength of reactive execution and responsiveness to environmental change.

Road Map: Control Theory and Robotics
Background: I.2. Levels and Styles of Analysis
Related Reading: Planning, Connectionist

References

Arbib, M. A., 1972, *The Metaphorical Brain: An Introduction to Cybernetics as Artificial Intelligence and Brain Theory*, New York: Wiley.

Arbib, M. A., and House, D., 1987, Depth and detours: An essay on visually guided behavior, in *Vision, Brain, and Cooperative Computation* (M. Arbib and A. Hanson, Eds.), Cambridge, MA: MIT Press, pp. 139–163.

Arkin, R. C., 1990a, The impact of cybernetics on the design of a mobile robot system: A case study, *IEEE Trans. Syst. Man Cybern.*, 20:1245–1257.

Arkin, R. C., 1990b, Integrating behavioral, perceptual, and world knowledge in reactive navigation, *Robot. Auton. Syst.*, 6:105–122.

Arkin, R. C., 1993, Modeling neural function at the schema level: Implications and results for robotic control, in *Biological Neural Networks in Invertebrate Neuroethology and Robotics* (R. Beer, R. Ritzmann, and T. McKenna, Eds.), San Diego: Academic Press, pp. 383–410.

Braitenberg, V., 1984, *Vehicles: Experiments in Synthetic Psychology*, Cambridge, MA: MIT Press.

Brooks, R, 1986, A robust layered control system for a mobile robot, *IEEE J. Robot. Automat.*, 2:14–23.

Brooks, R., 1991, New approaches to robotics, *Science*, 13 Sept., pp. 1227–1232. ◆

Cliff, D., Husbands, P., Meyer, J.-A., and Wilson, S. W., Eds., 1994, *From Animals to Animats 3: Proceedings of the Third International Conference on Simulation of Adaptive Behavior*, Cambridge, MA: MIT Press.

Efken, J., and Shaw, R., 1993, Ecological perspectives on the new artificial intelligence, *Ecol. Psychol.*, 4:247–270. ◆

Gat, E., 1992, Integrating planning and reacting in a heterogeneous asynchronous architecture for controlling real-world mobile robots, in *Proceedings of the Tenth National Conference on Artificial Intelligence*, San Jose, CA, pp. 809–815.

Lyons, D., and Hendriks, A., 1992, Reactive planning, in *Encyclopedia of Artificial Intelligence*, 2nd ed., (S. Shapiro, Ed.), New York: Wiley. ◆

Maes, P., Ed., 1990, *Designing Autonomous Agents*, Cambridge, MA: MIT Press/Elsevier, 1990. ◆

Mishkin, M., Ungerleider, L. G., and Macko, K. A., 1983, Object vision and spatial vision: Two cortical pathways, *Trends Neurosci.*, 6:414–417.

Norman, D., and Shallice, T., 1986, Attention to action: Willed and automatic control of behavior, in *Consciousness and Self-Regulation: Advances in Research and Theory*, vol. 4 (R. Davidson, G. Schwartz, and D. Shapiro, Eds.), New York: Plenum, pp. 1–17.

Walter, W. G., 1953, *The Living Brain*, New York: Norton.

Recurrent Networks: Supervised Learning

Kenji Doya

Introduction

The backpropagation algorithm for feedforward networks (Rumelhart, Hinton, and Williams, 1986) has been successfully applied to a wide range of problems from neuroscience to consumer electronics (see BACKPROPAGATION and APPLICATIONS OF NEURAL NETWORKS). However, what can be implemented by a feedforward network is just a static mapping of the input vectors. Needless to say, our brain is not a stateless input-output system but a high-dimensional nonlinear dynamical system. In order to model dynamical functions of the brain or to design a machine that performs as well as a brain does, it is essential to utilize a system that is capable of storing internal states and implementing complex dynamics.

This is why learning algorithms for *recurrent neural networks*, which have feedback connections and time delays, have been widely studied. In a recurrent network, the state of the system can be encoded in the activity pattern of the units, and a wide variety of dynamical behaviors can be programmed by the connection weights.

A popular subclass of recurrent networks comprises those with symmetric connection weights. In this case, the network dynamics is guaranteed to converge to a minimum of the "energy" function (see ENERGY FUNCTIONS FOR NEURAL NETWORKS). Typical examples are associative memory networks (see ASSOCIATIVE NETWORKS), optimization networks (see CONSTRAINED OPTIMIZATION AND THE ELASTIC NET), and winner-take-all networks (see WINNER-TAKE-ALL MECHANISMS).

However, steady-state solutions are only a limited portion of the capabilities of recurrent networks. A recurrent network can serve as a sequence recognition system (see LANGUAGE PROCESSING) or as a sequential pattern generator (see DYNAMICS AND BIFURCATION OF NEURAL NETWORKS). More generally, it is capable of transforming an input sequence into some other output sequence. It can be used as as a nonlinear filter (see ADAPTIVE SIGNAL PROCESSING), a nonlinear controller (see ADAPTIVE CONTROL: NEURAL NETWORK APPLICATIONS), or a finite state machine (Giles et al., 1992).

The main subject of this article is the supervised learning algorithms for training recurrent networks to perform temporal tasks. The problem setup is similar to the case of feedforward networks; a network is given a desired output for an input. An error function is defined, and its gradient with respect to the weights is derived. However, the major difference is that the input and output are not static vectors but *time sequences*.

In early studies, the standard backpropagation algorithm was adopted for training recurrent networks by regarding them as feedforward networks. Popular cases are Jordan's sequence-generating network (Jordan, 1986) and Elman's sequence prediction network (Elman, 1990). However, these approximate learning schemes take into account the network dynamics for only one step back into time.

In a recurrent network, a change in a weight can affect the future behavior of the entire network. Learning algorithms that take into account this recurrent effect have been obtained for both discrete-time models (Rumelhart et al., 1986; Williams and Zipser, 1989) and continuous-time models (Pearlmutter, 1989; Doya and Yoshizawa, 1989; Rowat and Selverston, 1991). The basic principle is to run a linearized version of the network dynamics and estimate the effect of a small change in a weight onto the error function.

There are two ways for doing this sensitivity analysis; one is to run the linearized system *forward in time*, and the other is to run its "adjoint" system *backward in time*. Each has its own strengths and weaknesses. In the following sections, we will formulate these algorithms for both discrete-time and continuous-time models and then discuss technical problems in using them.

Discrete-Time Models

First, we start with a discrete-time model. We denote the state of the ith unit by y_i and the connection weight from the jth to the ith unit by w_{ij}. Both external inputs u_j and recurrent inputs y_j are represented as z_j for convenience.

$$y_i(t+1) = f\left(\sum_{j=1}^{n+m} w_{ij}z_j(t)\right) \qquad i = 1, \ldots, n$$

$$z_j(t) = \begin{cases} y_j(t) & j \leq n \\ u_{j-n} & j > n \end{cases} \qquad (1)$$

The output nonlinearity $f(\cdot)$ is usually a squashing function such as $f(x) = 1/(1 + e^{-x})$ or $f(x) = \tanh x$. Note that their derivatives are conveniently given by $f'(x) = f(x)(1 - f(x))$ and $f'(x) = 1 - f(x)^2$, respectively. We can introduce a bias parameter by assuming that one of the inputs u_j is constant.

The goal of learning is to set the parameters w_{ij} so that the output trajectory $(y_1(t), \ldots, y_n(t))$ follows a desired trajectory $(d_1(t), \ldots, d_n(t))$ $(t = 1, \ldots, T)$ with a given initial state $(y_1(0), \ldots, y_n(0))$ and an input sequence $(u_1(t), \ldots, u_m(t))$ $(t = 0, \ldots, T-1)$. We define the error function

$$E = \sum_{t=1}^{T} \sum_{i=1}^{n} \mu_i(t)\tfrac{1}{2}(y_i(t) - d_i(t))^2 \qquad (2)$$

and perform gradient descent on E with respect to the weights w_{ij}. The masking function $\mu_i(t)$ specifies which components of the trajectory are to be supervised at what time. In a typical case, $\mu_i(t) \equiv 1$ for output units, and $\mu_i(t) \equiv 0$ for hidden units. When only the end point of the trajectory is specified, $\mu_i(T) = 1$ and $\mu_i(t) = 0$ for $t < T$.

Real-Time Recurrent Learning

The effect of weight change on the network dynamics can be seen by simply differentiating the network dynamics equation (Equation 1) by a weight w_{kl} (Williams and Zipser, 1989).

$$\frac{\partial y_i(t+1)}{\partial w_{kl}} = f'(x_i(t))\left[\sum_{j=1}^{n} w_{ij}\frac{\partial y_j(t)}{\partial w_{kl}} + \delta_{ik}z_l(t)\right] \qquad i = 1, \ldots, n \qquad (3)$$

where $x_i(t) = \sum_{j=1}^{n+m} w_{ij}z_j(t)$ is the net input to the unit, and δ_{ik} is Kronecker's delta ($\delta_{ik} = 1$ if $i = k$ and 0 otherwise). The term $\delta_{ik}z_l(t)$ represents an *explicit* effect of the weight w_{kl} onto the unit k, and the term $\sum_{j=1}^{n} w_{ij}(\partial y_j(t)/\partial w_{kl})$ represents an *implicit* effect onto all the units due to network dynamics.

Equation 3 for each unit $i = 1, \ldots, n$ constitutes an n-dimensional linear dynamical system (with time-varying coefficients), where $(\partial y_1/\partial w_{kl}, \ldots, \partial y_n/\partial w_{kl})$ is taken as a dynamical variable. Since the initial state $y_i(0)$ of the network is independent of the connection weights, the appropriate initial condition for Equation 3 is

$$\frac{\partial y_i(0)}{\partial w_{kl}} = 0 \qquad i = 1, \ldots, n$$

Thus we can compute $\partial y_i(t)/\partial w_{kl}$ *forward in time* by iterating Equation 3 simultaneously with the network dynamics (Equation 1). From this solution, we can calculate the error gradient as follows:

$$\frac{\partial E}{\partial w_{kl}} = \sum_{t=1}^{T} \sum_{i=1}^{n} \mu_i(t)(y_i(t) - d_i(t))\frac{\partial y_i(t)}{\partial w_{kl}} \qquad (4)$$

A standard *batch* gradient descent algorithm is to accumulate the error gradient by Equation 4 and update each weight w_{kl} by

$$w_{kl} := w_{kl} - \varepsilon\frac{\partial E}{\partial w_{kl}} \qquad (5)$$

where $\varepsilon > 0$ is a learning rate parameter.

An alternative update scheme is the gradient descent of *current* output error $\sum_{i=1}^{n}\tfrac{1}{2}\mu_i(t)(y_i(t) - d_i(t))$ at each time step, namely,

$$w_{kl}(t+1) = w_{kl}(t) - \varepsilon\sum_{i=1}^{n} \mu_i(t)(y_i(t) - d_i(t))\frac{\partial y_i(t)}{\partial w_{kl}} \qquad (6)$$

Note that we assumed that w_{kl} is a constant, not a dynamical variable, in deriving Equation 3, so we have to keep the learning rate ε small enough. However, this *on-line* update scheme was shown to be effective in a number of temporal learning tasks (Williams and Zipser, 1989), and it is often called *real-time recurrent learning*.

A drawback of this error gradient calculation *forward in time* is that we have to solve an n-dimensional system (Equation 3) for *each* of the weights w_{kl} ($i = 1, \ldots, n$; $t = 1, \ldots, T$). It requires $O(n^3)$ memories and $O(n^4)$ computations.

Backpropagation Through Time

Another learning algorithm for a discrete-time model can be derived by "unfolding" a recurrent network into a multilayer network (Rumelhart et al., 1986). In this scheme, T-step iteration of a recurrent network is regarded as one sweep of operation in a T-layered feedforward network with identical connection weights w_{ij} between successive layers. The error gradient can be derived in a same way as in the standard backpropagation, except that the output errors are not only given in the last layer but also added in each layer.

$$\frac{\partial E}{\partial y_i(t)} = \sum_{j=1}^{n} \frac{\partial E}{\partial y_j(t+1)}f'(x_j(t))w_{ji} + \mu_i(t)(y_i(t) - d_i(t))$$

$$i = 1, \ldots, n \qquad (7)$$

Since the error E is independent of the state at $t > T$, the boundary condition for Equation 7 is given at the final time step as

$$\frac{\partial E}{\partial y_i(T+1)} = 0 \qquad i = 1, \ldots, n$$

Thus, the learning equation (Equation 7) can be iterated *backward in time* from $t = T$ to 1.

From the solution $\partial E/\partial y_i$, the error gradients are given by

$$\frac{\partial E}{\partial w_{ij}} = \sum_{t=1}^{T} \frac{\partial E}{\partial y_i(t)} f'(x_i(t-1)) z_j(t-1) \tag{8}$$

and the weights are updated in a *batch* using Equation 5.

The advantage of this algorithm is that we have to solve only *one* n-dimensional system (Equation 7) for adjusting all the weights. Therefore only $O(n^2)$ computations are required. However, since the learning equation (Equation 7) has to be solved *backward* in time, we cannot update the weights *on-line* and have to store the history of the network state $y_i(t)$ ($i = 1, \ldots, n; t = 1, \ldots, T$), which requires $O(nT)$ memories.

Continuous-Time Models

A continuous-time model is a natural choice for modeling systems that are governed by differential equations. Time constants of continuous-time models are convenient parameters for setting local memory spans for individual units. They can also be adjusted by learning, as we will discuss.

Slightly different versions of continuous-time models have been studied. Here, we focus on the following model (Pineda, 1988; Pearlmutter, 1989):

$$\tau_i \dot{y}_i(t) = -y_i(t) + f\left(\sum_{j=1}^{n+m} w_{ij} z_j(t) \right) \qquad i = 1, \ldots, n$$

$$z_j(t) = \begin{cases} y_j(t) & j \le n \\ u_{j-n} & j > n \end{cases} \tag{9}$$

However, similar derivations apply to other models as well (Doya and Yoshizawa, 1989; Rowat and Selverston, 1991).

We define an error integral

$$E = \int_0^T \sum_{i=1}^{n} \mu_i(t) \tfrac{1}{2}(y_i(t) - d_i(t))^2 \, dt \tag{10}$$

and derive a gradient descent algorithm for minimizing E for a desired trajectory $(d_1(t), \ldots, d_n(t))$ $(0 \le t \le T)$ with a given initial state $(y_1(0), \ldots, y_n(0))$ and an input sequence $(u_1(t), \ldots, u_m(t))$.

Variation Method

The effect of a change in a weight w_{kl} on the state $y_i(t)$ can be estimated by differentiating the network dynamics equation (Equation 9) as follows (Rowat and Selverston, 1991):

$$\tau_i \frac{d}{dt}\left(\frac{\partial y_i}{\partial w_{kl}} \right) = -\frac{\partial y_i}{\partial w_{kl}} + f'(x_i(t)) \left[\sum_{j=1}^{n} w_{ij} \frac{\partial y_j}{\partial w_{kl}} + \delta_{ik} z_l(t) \right]$$

$$i = 1, \ldots, n \tag{11}$$

This equation forms an n-dimensional linear differential equation system with the state variable $(\partial y_1/\partial w_{kl}, \ldots, \partial y_n/\partial w_{kl})$ and is called a *variation system* of the network dynamics (Equation 9). The initial condition for this system is given by

$$\frac{\partial y_i(0)}{\partial w_{kl}} = 0 \qquad i = 1, \ldots, n$$

because the initial state of the network is independent of the weights. We can numerically integrate Equation 11 *forward in time* concurrently with the network dynamics (Equation 9).

From the solution $\partial y_i(t)/\partial w_{kl}$ $(0 \le t \le T)$, the error gradient is given by

$$\frac{\partial E}{\partial w_{ij}} = \int_0^T \sum_{i=1}^{n} \mu_i(t)(y_i(t) - d_i(t)) \frac{\partial y_i(t)}{\partial w_{kl}} \, dt \tag{12}$$

We can use either the *batch* update scheme (Equation 5) at the end of a sequence, or the *on-line* update scheme

$$\dot{w}_{kl} = -\varepsilon \sum_{i=1}^{n} \mu_i(t)(y_i(t) - d_i(t)) \frac{\partial y_i(t)}{\partial w_{kl}} \tag{13}$$

with sufficiently small learning rate $\varepsilon > 0$.

The error gradient for a time constant τ_k is given by the following variation equation:

$$\tau_i \frac{d}{dt}\left(\frac{\partial y_i}{\partial \tau_k} \right) = -\frac{\partial y_i}{\partial \tau_k} + f'(x_i(t)) \left[\sum_{j=1}^{n} w_{ij} \frac{\partial y_j}{\partial \tau_k} - \delta_{ik} \dot{y}_k(t) \right]$$

$$i = 1, \ldots, n \tag{14}$$

Adjoint Method

The *backward* algorithm for a continuous-time model can be derived in several ways, for example, by finite difference approximation (Pearlmutter, 1989). Here we derive the algorithm as an "adjoint" system of the *forward* learning equation (Equation 11).

A pair of n-dimensional linear systems

$$\dot{p} = A(t)p + b(t)$$

$$\dot{q} = -A^*(t)q - c(t)$$

is called *adjoint* to each other when A^* is the transpose of matrix A. A useful property of adjoint systems is that their solutions satisfy the following *Green's equality* (Hartman, 1982):

$$\int_0^T q(t) \cdot b(t) \, dt - \int_0^T c(t) \cdot p(t) \, dt = q(T) \cdot p(T) - q(0) \cdot p(0)$$

We can actually compose an adjoint system of the variation equation (Equation 11):

$$\dot{q}_i = \frac{q_i(t)}{\tau_i} - \sum_{j=1}^{n} \frac{f'(x_j(t))}{\tau_j} w_{ji} q_j(t) - \mu_i(t)(y_i(t) - d_i(t)) \tag{15}$$

where we put $p_i = \partial y_i/\partial w_{kl}$, $A_{ij}(t) = [f'(x_i(t))/\tau_i] w_{ij} - \delta_{ij}/\tau_i$, $b_i(t) = [f'(x_i(t))/\tau_i] \delta_{ik} y_l(t)$, and $c_i(t) = \mu_i(t)(y_i(t) - d_i(t))$. With the boundary conditions $p_i(0) = \partial y_i(0)/\partial w_{kl} = 0$ and $q_i(T) = 0$, Green's equality becomes

$$\int_0^T \sum_{i=1}^{n} q_i(t) \frac{f'(x_i(t))}{\tau_i} \delta_{ik} y_l(t) \, dt$$

$$= \int_0^T \sum_{i=1}^{n} \mu_i(t)(y_i(t) - d_i(t)) \frac{\partial y_i}{\partial w_{kl}} \, dt \tag{16}$$

Note that the right-hand side is identical to the error gradient (Equation 12). Thus, we have an alternative form of the error gradient

$$\frac{\partial E}{\partial w_{kl}} = \int_0^T q_k(t) \frac{f'(x_k(t))}{\tau_k} z_l(t) \, dt \tag{17}$$

Similarly, the error gradient for a time constant is given by

$$\frac{\partial E}{\partial \tau_k} = \int_0^T q_k(t) \frac{f'(x_k(t))}{\tau_k} (-\dot{y}_k(t)) \, dt \tag{18}$$

As in the discrete-time case, we first run the network dynamics (Equation 9) *forward in time* and then run the adjoint system (Equation 15) *backward in time* with the terminal condition $q_i(T) = 0$. The weights are updated in batch by (Equation 5).

Technical Remarks

Forward or Backward

A forward algorithm requires $O(n^4)$ computations. Therefore, it is not suitable for a fully connected network with tens or hundreds of units. However, for a small-sized network or a network with only local connections, on-line weight update can be an advantage.

In order to allow on-line weight update with the efficiency of the backward algorithm, a truncated version of the backpropagation-through-time algorithm has been proposed (Williams and Zipser, 1990).

Teacher Forcing

The so-called *teacher forcing* technique has been shown to be helpful, especially in training a network into an autonomous dynamical system (Pineda, 1988; Doya and Yoshizawa, 1989; Williams and Zipser, 1989). In this scheme, the desired output $d_i(t)$ is used to drive the network dynamics in place of the feedback of its actual output $y_i(t)$.

The reasons for the need of teacher forcing are as follows:

- The state of the network is assigned to the desired one of many attractor domains (Pineda, 1988).
- In learning oscillatory patterns, unless the phase of the network output is synchronized to the teacher signal, there will be an apparently large error (Doya and Yoshizawa, 1989).
- It will avoid a local minimum solution of static output at the mean value of the dynamic teacher signal (Williams and Zipser, 1989).
- The linearized equation for a limit cycle trajectory is not asymptotically stable if the system is running autonomously (Doya, 1992).

One problem with this technique is that the trajectory learned with teacher forcing may not be stable when the network is run autonomously after learning. Several heuristics have been proposed for enhancing the stability of the non-forced trajectory:

Noisy forcing: Add some noise to the forcing input.
Partial forcing: Use a mixed input $z_i(t) = y_i(t) + \alpha(d_i(t) - y_i(t))$ with $0 < \alpha < 1$ (Williams and Zipser, 1990), and decrease the forcing rate α with the progress of learning.
Part-time forcing: Turn on forcing to synchronize the network to the teacher, and then turn off forcing to train the autonomous trajectory.

Bifurcation Boundaries

In many learning tasks, the goal is not only to replicate particular sample trajectories but to reconstruct some "attractors" in the state space, such as fixed points, limit cycles, and chaotic attractors.

For example, when a network is trained as a finite state machine, it must have distinct attractors in order to represent discrete states. For another example, when a network is trained as a periodic oscillator, it must have a limit cycle attractor. When we gradually change network parameters, we expect the shape and location of attractors to change continuously. However, that expectation is not always fulfilled. At some points in the parameter space, attractors can emerge, disappear, or change their stability. Such a phenomenon is known as *bifurcation* in nonlinear systems theory (Wiggins, 1990; see DYNAMICS AND BIFURCATION OF NEURAL NETWORKS and CHAOS IN NEURAL SYSTEMS).

At a bifurcation point, the linearized equations that are used for gradient computation can lose asymptotic stability. Accordingly, when the network goes through a bifurcation point, the solution of the learning equation can grow rapidly, and the gradient descent algorithm can become unstable (Doya, 1992).

Although this might sound like a rare, pathetic situation, bifurcation is actually an inevitable step in many learning tasks (Pineda, 1988; Doya, 1992). If the connection weights w_{ij} are initialized with small random values, the network dynamics has a single global attractor point. In order to have multiple attractor domains or a limit cycle, the network must go through some bifurcation boundary. Conversely, until the network goes through an appropriate bifurcation, even a simple memory task can be very difficult because of exponential decay of the error gradient (Bengio, Simard, and Frasconi, 1994).

Incremental Training

It has been reported that gradual increase of the complexity of training examples is critical for successfully training a network as a finite state machine (Giles et al., 1992). A possible reason is that a network can acquire memory mechanisms only gradually, by going through bifurcation boundaries. If we impose examples that require many internal states with long time delay from the beginning, we might have little chance of correctly training the network. This problem of "developmental" capability of recurrent networks needs further examination.

Discussion

A fully connected recurrent neural network can potentially be a very powerful system for temporal information processing. It can be shown, as a corollary to the universal approximation theorem for three-layered networks (Hornik, 1991; see KOLMOGOROV'S THEOREM), that a recurrent network can, with enough units, approximate any vector field or map. However, it does not mean that such an approximation can be readily achieved by error gradient descent learning.

As mentioned earlier, the error gradient can decay or expand exponentially in time, thus making gradient descent more difficult than in the case of feedforward networks. Convergence of learning depends critically on the choice of network topology, initial weights, and the choice of training samples. These are some of the reasons why networks with specialized architectures have been crafted for specific problems, for example, networks with tapped delay lines or local recurrent loops.

One direction of future study is to find more efficient algorithms and heuristics for reliable learning. An alternative direction is to find out a new principle for exploiting the emergent properties of nonlinear dynamical systems (see DYNAMICS AND BIFURCATION OF NEURAL NETWORKS; CHAOS IN NEURAL SYSTEMS; and COLLECTIVE BEHAVIOR OF COUPLED OSCILLATORS).

Road Map: Learning in Artificial Neural Networks, Deterministic
Background: I.3. Dynamics and Adaptation in Neural Networks
Related Reading: Dynamic Models of Neurophysiological Systems; Self-Organization in the Time Domain; Spatiotemporal Association in Neural Networks

References

Bengio, Y., Simard, P., and Frasconi, P., 1994, Learning long-term dependencies with gradient descent is difficult, *IEEE Trans. Neural Netw.*, 5:157–166.

Doya, K., 1992, Bifurcations in the learning of recurrent neural networks, in *Proceedings of 1992 IEEE International Symposium on Circuits and Systems*, New York: IEEE, vol. 6, pp. 2777–2780.

Doya, K., and Yoshizawa, S., 1989, Adaptive neural oscillator using continuous-time back-propagation learning, *Neural Netw.*, 2:375–386.

Elman, J. L., 1990, Finding structure in time, *Cognit. Sci.*, 14:179–211.

Giles, C. L., Miller, C. B., Cheng, D., Chen, H. H., Sun, G. Z., and Lee, Y. C., 1992, Learning and extracting finite state automata with second-order recurrent neural networks, *Neural Computat.* 4:393–405.

Hartman, P., 1982, *Ordinary Differential Equations*, Boston: Birkhäuser.

Hornik, K., 1991, Approximation capabilities of multilayer feedforward networks, *Neural Netw.*, 4:251–257.

Jordan, M. I., 1986, Attractor dynamics and parallelism in a connectionist sequential machine, in *Proceedings of the Eighth Annual Conference of the Cognitive Science Society*, Hillsdale, NJ: Erlbaum, pp. 531–546; reprinted, IEEE Tutorials Series, New York: IEEE, 1990.

Pearlmutter, B. A., 1989, Learning state space trajectories in recurrent neural networks, *Neural Comput.*, 1:263–269.

Pineda, F. J., 1988, Dynamics and architecture for neural computation, *J. Complexity*, 4:216–245.

Rowat, P. F., and Selverston, A. I., 1991, Learning algorithms for oscillatory networks with gap junctions and membrane currents, *Network*, 2:17–41.

Rumelhart, D. E., Hinton, G. E., and Williams, R. J., 1986, Learning representations by back-propagating errors, *Nature*, 323:533–536.

Wiggins, S., 1990, *Introduction to Applied Nonlinear Dynamical Systems and Chaos*, New York: Springer-Verlag.

Williams, R. J., and Zipser, D., 1989, A learning algorithm for continually running fully recurrent neural networks, *Neural Comput.*, 1: 270–280.

Williams, R. J., and Zipser, D., 1990, *Gradient Based Learning Algorithms for Recurrent Connectionist Networks*, Technical Report NU-CCS-90-9, Boston: Northeastern University, College of Computer Science.

Regularization Theory and Low-Level Vision

Jose L Marroquin

Introduction

Current research in computational vision follows two main paradigms: in the first one, it is considered that the first task that a visual system has to solve consists in reconstructing, from the set of images that constitute the sensory input, a set of fields that represent, on one hand, the physical properties of the three-dimensional surfaces around the viewer, and on the other, the boundaries between patches that "belong together" in some sense, and thus, that may correspond to the outlines of plausible physical objects in the scene. This process, which is usually called early or low-level vision, is supposed to be performed in natural systems by a set of loosely coupled neural networks (computational modules), each one of which specializes in the reconstruction of a particular field. Thus, specific modules have been proposed for the computation of brightness edges; depth from stereo and shading; color, lightness and albedo; velocity and optical flow; spatial and spatiotemporal interpolation and approximation; and so on.

In the second view, it is noted that many of the problems that have to be solved using vision do not need a complete reconstruction of the three-dimensional world; for a given task, it may be possible to feed the raw sensory data to a network (such as a multilayer perceptron) which directly generates the desired control commands. The plausibility of this approach is illustrated, for example, in Pomerleau 1991; see VISION FOR ROBOT DRIVING), where such a network is used for an autonomous navigation task. In this case, however, it is also necessary to determine a set of fields defined on the same lattice as the observations: these fields represent the weights that indicate the relative importance of each pixel value for the subnetwork of the corresponding hidden unit.

In both cases, the determination of the corresponding fields exhibits an important common characteristic: due to the loss of information inherent to the imaging and sensory transduction processes and, in the second case, to the fact that one usually has a limited number of available "examples" to train the network, the values of the fields are constrained by the data, but not determined in a unique and stable way (i.e., the reconstruction problems are mathematically ill-posed). As a result, the networks that implement the solutions must incorporate in their structure prior knowledge about the reconstructed fields.

For the sake of clarity, this article focuses on the reconstruction (multimodule) paradigm (although most of the results may be extended to the action-oriented case as well). The general problem that we consider is then the following:

Suppose that we are given sensory measurements in the form of a set of observed fields g at the nodes of a regular lattice L (usually a square lattice is assumed, although other arrangements are possible). From these measurements, one wishes to reconstruct a field $f = \{ f_i, i \in L \}$, given the "direct" equations that model g in terms of f and some noise process n:

$$\phi(g, f, n) = 0$$

The simplest instance of this problem is image filtering: here, g consists of a single field (the noisy observed image); f is the desired reconstructed image; and the observation model is

$$g - f - n = 0 \tag{1}$$

Another example is the recovery of depth from stereoscopic pairs of images (Grimson, 1982; see STEREO CORRESPONDENCE AND NEURAL NETWORKS). Here, the observations $g = (g_L, g_R)$ are the gray levels measured in the left and right retinas, respectively, and f is the associated disparity between pairs of corresponding points (if this "correspondence problem" is solved, and if the geometry of the sensors is known, the actual recovery of depth is a matter of simple geometric computations). If the sites of the lattice are identified by a two-dimensional index $i = (i_x, i_y)$, and assuming horizontal epipolar lines, a simplified direct equation is

$$g_L(i_x, i_y) - g_R(i_x + f_i, i_y) - n_i = 0$$

for each $i \in L$.

In the first example, the field f is underconstrained, because the noise field is not known. In the second one, even in the absence of noise, the field f is not uniquely determined, because there may be many points in the right image with the same gray level as a given point in the left one. Similar ambiguous situations arise in other early vision problems for different reasons, and in all these cases it is necessary to introduce additional prior constraints.

In this article we present systematic ways for introducing these constraints, and for embedding the solution algorithms in suitable networks.

Standard Regularization

The classical way of finding solutions to ill-posed problems is found in the so-called regularization methods (Tikhonov and Arsenin, 1977; Bertero, Poggio, and Torre, 1988). In the standard case, the observation model ϕ is assumed to be of the form

$$g - Af = 0$$

where A is a (noninvertible) linear operator. The regularized solution is obtained by incorporating a prior constraint on the smoothness of f; this solution is defined as the minimizer (over f) of the functional

$$H(f) = \|Af - g\|^2 + \lambda \|Pf\|^2 \qquad (2)$$

where $\|\cdot\|$ is the L_2 norm; P is a linear operator chosen in such a way that the stabilizing functional $\|Pf\|^2$ is a measure of the overall smoothness of f (standard choices for P are, for example, the magnitude of the gradient or the two-dimensional Laplacian); and λ is a parameter that controls the tradeoff between smoothness of the solution and fidelity to the data.

If, as in our case, f and g are defined only at the sites of a lattice, this functional can be written in terms of the variables $\{f_i, g_i, i \in L\}$: each term is just the sum, over all sites of L, of the squares of the finite difference approximations to the corresponding differential operators.

Standard regularization methods have been applied to solve a variety of early-vision problems (see, for example Marroquin, Mitter, and Poggio, 1987, and references contained therein, and also Poggio and Girosi, 1990).

A fundamental limitation of this approach is that it can only implement global smoothness constraints on the solution; to handle the most interesting case of piecewise smoothness, a more general approach is needed. This is discussed in the next section.

Probabilistic Regularization

In the probabilistic regularization approach, the value of the field and the observation at each location are regarded as random variables (i.e., f and g are considered realizations of the random fields F and G), so that the reconstruction of f is understood as an estimation problem. The prior knowledge about the solution is expressed in the form of a joint probability distribution for the field F that specifies the desired dependencies between values at neighboring sites. In this way, one may specify not only global smoothness constraints (so that standard regularization becomes a particular case), but also piecewise smoothness, as well as constraints on the shape of the discontinuities.

The basic tool in this approach is Bayes's rule, which specifies the way in which prior information (i.e., the prior distribution P_F) is to be combined with the constraints generated by the observations (i.e., the conditional distribution $P_{G|F}$) to generate the posterior distribution $P_{F|G}$:

$$P_{F|G}(f; g) = \frac{P_F(f) P_{G|F}(f; g)}{P_G(g)}$$

Note that since the observations g are given, $P_G(g)$ is a constant (f denotes the collection of variables that are to be inferred). The optimal estimator $\hat{f}^*$ is then obtained as the minimizer of the expected value (taken with respect to the posterior distribution) of an appropriate cost function $C(f, \hat{f})$.

This approach, then, requires the specification of three basic components (besides the cost function): the observation model $P_{G|F}$, the prior distribution P_F, and the network that will effect the reconstruction. We will now analyze them in detail.

The Observation Model

The form of the constraints that sensor measurements impose on the reconstructed field depends upon the particular assumptions that are made about the image formation process. In the simplest case, these constraints take the form

$$g = Af + n$$

where A is a noninvertible operator, and n is a field of independent random variables, each with probability distribution P_N. In this case, the conditional distribution is simply

$$P_{G|F}(f; g) \prod_{i \in L} P_N(g_i - (Af)_i) \qquad (3)$$

which can also be written in the general form

$$P_{G|F}(f; g) = \exp\left[\sum_{i \in L} - \Phi_i(f, g)\right] \qquad (4)$$

Note that, for a particular problem, the definition of the conditional distribution is not unique, and different definitions may lead to reconstructing networks with very different performances, regardless of the prior term (see, for example, Poggio, Yang, and Torre, 1989, for an analysis of the different possible definitions of optical flow).

Prior Distribution

The success of the Bayesian approach depends on the specification of a probability distribution $P_F(f)$ that models the desired behavior of the solution. In particular, one would like to be able to specify a distribution in which fields where neighboring sites exhibit the appropriate dependencies are more probable than those in which these local constraints are violated. A general way of constructing such distributions is by defining an "energy" function $U(f)$, which is formed by a sum of terms that measure the violation of the local constraints. The probability distribution of the field is then given by the Gibbs measure:

$$P_F(f) = \frac{1}{Z}\exp[-U(f)] \qquad (5)$$

where Z is a normalizing constant.

More precisely, if we define a neighborhood system $\{N_i, i \in L\}$, that is, a collection of subsets of sites indexed by the sites of L: $\{N_i \subset L, i \in L\}$ with the properties

$$i \notin N_i$$

$$i \in N_j \Leftrightarrow j \in N_i$$

its *cliques* consist of either single sites or subsets of sites such that any two of them belonging to the same clique are neighbors of each other. With this definition, the internal energy may be written as

$$U(f) = \sum_C V_c(f) \qquad (6)$$

where C ranges over all the cliques of the neighborhood system, and each "potential function" V_c depends only on $\{f_i, i \in C\}$.

A random field F whose probability distribution is given by Equations 5 and 6 is called a *Markov random field* on L (see MARKOV RANDOM FIELD MODELS IN IMAGE PROCESSING).

The potential functions represent the *user interface* of the model, since through them one may specify the desired characteristics of the sample fields. Although they may be arbitrarily specified, there are three basic types that are generally used, depending on the characteristics of the desired reconstruction:

1. *Piecewise constant fields.* Here, each f_i can only take a finite (usually small) number of values. These fields are mostly used in classification problems (e.g., texture segmentation). The most widely used potential is the generalized Ising potential for cliques of size 2:

$$V_C(f_i, f_j) = -\beta \quad \text{if } f_i = f_j$$
$$= \beta \quad \text{otherwise}$$

2. *Globally smooth fields.* This case is analogous to the one discussed in standard regularization; the prior energy U corresponds to the stabilizing functional, so that the potentials may be obtained by discretizing and squaring the corresponding differential operators; for example, if the operator P is the gradient magnitude, the potentials take the form of a first-order quadratic model, which in physical terms corresponds to the deformation energy of a membrane:

$$V_C(f_i, f_j) = (f_i - f_j)^2 \quad (7)$$

where i and j denote a pair of nearest neighbor sites in the lattice. The second-order model corresponds to the bending energy of a thin plate, and the neighborhood system has cliques that consist of sets of three neighboring sites i, j, k lying on a horizontal or vertical straight line, and of sets of four sites p, q, r, s lying at the corners of a square whose side equals the lattice spacing. The corresponding potentials are

$$V_{C_3}(f) = (-f_i + 2f_j - f_k)^2 \quad (8)$$

and

$$V_{C_4}(f) = \tfrac{1}{4}(-f_r + f_s + f_p - f_q)^2 \quad (9)$$

where (r, q) and (s, p) lie at opposite corners of the square.

If one adopts the observation model (Equation 3), and assumes that P_N is a zero-mean Gaussian distribution, the posterior energy becomes equivalent to the discretized functional (Equation 2) of standard regularization, and its (unique) maximizer corresponds to the MAP estimator [see item 3(b)]. The

networks corresponding to the first- and second-order reconstruction models are shown in Figure 1.

3. *Piecewise smooth fields.* This is the most important and general case. There are two basic approaches for the construction of the potentials:

(a) The discontinuities of the field are explicitly modeled by means of an auxiliary "line field" s (originally introduced by Geman and Geman, 1984), which is defined on a "dual" lattice whose sites are between each pair of (horizontal or vertical) neighboring sites of L (s is thus indexed by a pair of indices corresponding to sites of L). Each line element s_{ij} may take values on the set $\{0, 1\}$, indicating the presence or absence of a line (discontinuity), respectively (in some models, s is allowed to take other integer values to encode the orientation of the line as well).

The prior energy takes the form

$$U(f, s) = \sum_C V_C(f) B_C(s) + \sum_D W_D(s) \quad (10)$$

where the potentials V_C are convex (e.g., quadratic) functions similar to those of the globally smooth case. The coupling potential $B_C(s)$ equals 0 if for at least one pair of neighboring sites $i, j \in C$ the corresponding line element s_{ij} is different from zero; otherwise, $B_C(s) = 1$.

The line potentials $W_D(s)$ assign penalties to different local line configurations. They are summed over the cliques D of a neighborhood system defined on the dual lattice, and they are used to favor, for example, piecewise smooth lines, and to prevent the formation of smooth patches that are too thin or too small.

(b) The discontinuities are implicitly modeled by nonconvex potentials: if the cliques D of the dual neighborhood system consist only of single line sites (i.e., if the line field is noninteracting), the term $\Sigma_D W_D(s)$ in Equation 10 will be proportional to the total number of active line elements. In this case, it is possible to show that an equivalent energy function may be obtained as a function of the f field only, if the convex potentials V_C are replaced by appropriate nonconvex ones. Since the resulting energy depends only on the real-valued variables $\{f_i\}$, it can be minimized, for example, by gradient descent methods.

The simplest nonconvex potentials are based on truncated quadratic functions (Blake and Zisserman, 1987); others commonly used are of the form

$$\frac{1}{1 + kx^2} \quad \text{or} \quad \frac{1}{1 + k|x|}$$

where x stands for a linear combination of the values of the field at the sites of the clique.

The best model to use depends on the particular application; as a simple example, consider the problem of filtering noise-corrupted images (i.e., an observation model of the form of Equation 1), without blurring their main edges (Figure 2). This may be accomplished using a prior with truncated quadratic first- or second-order potentials (Equation 7 or 8 and 9). The resulting optimal (MAP) estimators are shown in parts C and D of Figure 2. Note that the first-order model (part C) tends to produce piecewise constant images, so that the second-order one may be better for reconstructing smooth gradients (part D), at the expense of increased computational complexity.

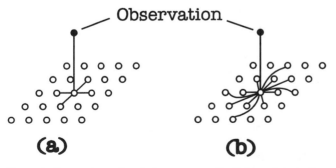

Figure 1. Connections of typical processors of cellular automata that implement the piecewise smooth reconstruction models discussed in the text. (The filled circles represent the observed image values at the corresponding pixel.) *A*, First-order model. *B*, Second-order model.

Other examples of the application of this approach to a variety of problems, as well as extensions and theoretical results, may be found in Chellapa and Jain (1993).

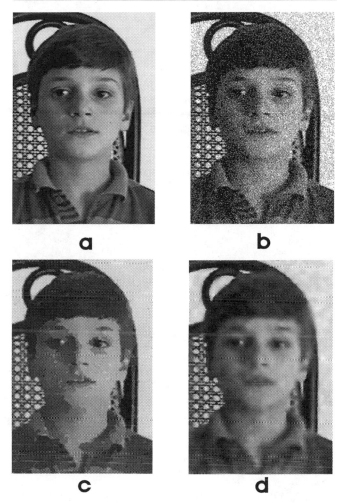

Figure 2. Piecewise smooth reconstruction of the image (*A*) from the observations (*B*) corrupted with zero-mean, additive white Gaussian noise ($\sigma = 40$). Parts *C* and *D* show the reconstructions obtained with the first- and second-order prior models, respectively, discussed in the text.

Networks

Since the reconstruction is needed at the sites of the pixel lattice L, it is very natural to model the reconstructing network as a cellular automaton that consists of an array of processors or cells located also at the sites of L. The state of these processors at a given time t is denoted by $\xi^{(t)} = \{\xi_i^{(t)}, i \in L\}$. The interconnection pattern between processors is specified by the defined neighborhood system. The state of each processor changes from time to time with a rule that depends on its own state and that of its neighbors:

$$\xi_i^{(t+1)} = R(\xi_j^{(t)}, j \in N_i \cup \{i\})$$

Cellular automata (CA) may be deterministic (DCA) or stochastic (SCA), depending on the nature of the rule R.

Given this model for the architecture of a computational module, the important question is how to specify R, so that in the deterministic case, the DCA has a fixed point and the reconstructed field f is obtained from it, and in the stochastic case, the automaton is regular and f is obtained from time averages of functions of its state.

In the case of globally smooth reconstructions, the energy function is usually convex, and the best estimator is obtained

by minimizing this energy. The reconstructing networks are in this case equivalent to iterative methods for matrix inversion (Ortega and Rheinholdt, 1970). They may also be implemented analogically with pure resistor networks (see Marroquin, Mitter, and Poggio, 1987).

In the case of piecewise smooth or piecewise constant fields, however, the energy becomes highly nonconvex, and more powerful methods are required. The most general are based on SCA: a stationary SCA is mathematically a Markov chain with transition probabilities specified by R. If this chain is regular (see Kemeny and Snell, 1960), its states will have a unique invariant distribution π. In this case, the law of large numbers for regular chains establishes that the average of any function of the state $Y(\xi)$ taken with respect to π may be approximated arbitrarily well by the time average of $Y(\xi^{(t)})$ obtained by observing the evolution of the automaton. If this invariant measure coincides with $P_{F|G}$, one may use this property for estimating, for example, the posterior mean (by taking the time average of the state at each site) or the posterior marginals (by counting the number of times a given cell is in each state), from which the optimal estimators may be obtained.

The most commonly used SCAs are the Metropolis and the Gibbs sampler algorithms (Geman and Geman, 1984); the general procedures for constructing an SCA with a given invariant Gibbsian measure and different convergence properties are given in Marroquin and Ramirez (1991). Also see Besag (1974) for methods for constructing related deterministic networks.

Discussion

The techniques we have presented allow one to design networks that perform the reconstruction tasks associated with each early-vision module in an independent way. It is an accepted fact, however, that both the accuracy and the computational efficiency would improve if these modules were allowed to interact during the reconstruction process. This interaction may be implemented directly within the probabilistic regularization framework by defining a combined prior energy for several modules, which includes terms that constrain the location of the discontinuities of the associated fields so that they coincide in most places; in this direct approach, however, the computational complexity of the corresponding algorithms soon becomes exceedingly high.

Alternatively, a model has been proposed (Marroquin, 1992) in which each module first constructs an intermediate multilayered representation where a relative probability is assigned to every valid value of the field at each site (that is, each module computes a *measure field*). Then, in a second stage, these measure fields interact with each other to produce the final reconstruction. This approach allows for rich interactions with reasonable computational loads, and it appears to work well in simple synthetic examples. The problem of modeling module interactions in more realistic situations, however, still remains very much open.

Road Map: Vision
Related Reading: Active Vision; Generalization and Regularization in Nonlinear Learning Systems; Motion Perception; Perception of Three-Dimensional Structure

References

Bertero, M., Poggio, T., and Torre, V., 1988, Ill-posed problems in early vision, *Proc. IEEE*, 76:869–889. ◆
Besag, J., 1974, Spatial interaction and the statistical analysis of lattice systems, *J. R. Statist. Soc. Ser. B.*, 36:192–326.

Blake, A., and Zisserman, A., 1987, *Visual Reconstruction*, Cambridge, MA: MIT Press.

Chellapa, R., and Jain, A., Eds., 1993, *Markov Random Fields: Theory and Practice*, Boston: Academic Press. ◆

Geman, S., and Geman, D., 1984, Stochastic relaxation, Gibbs distributions and the Bayesian restoration of images, *IEEE Trans. Pattern Analysis Machine Intell.*, 6:721–741.

Grimson, W. E. L., 1982, A computer implementation of a theory of stereo vision, *Philos. Trans. R. Soc. Lond. B Biol. Sci.*, 298.

Kemeny, J. G., and Snell, J. L., 1960, *Finite Markov Chains and Their Applications*, New York: Van Nostrand.

Marroquin, J. L., 1992, Random measure fields and the integration of visual information, *IEEE Trans. Syst. Man Cybern.*, 22:705–716.

Marroquin, J. L., Mitter, S., and Poggio, T., 1987, Probabilistic solution of ill-posed problems in computational vision, *J. Am. Statist. Assoc.*, 82:76–89.

Marroquin, J. L., and Ramirez, A., 1991, Stochastic cellular automata with invariant Gibbsian measures, *IEEE Trans. Inform. Theory*, 37: 541–551.

Ortega, J. M., and Rheinholdt, W. C., 1970, *Iterative Solution of Nonlinear Equations in Several Variables*, New York: Academic Press.

Poggio, T., and Girosi, F., 1990, Regularization algorithms for learning that are equivalent to multilayer networks, *Science*, 247:978–992.

Poggio, T., Yang, W., and Torre, V., 1989, Optical flow: Computational properties and networks, biological and analog, in *The Computing Neuron* (R. Durbin, C. Miall, and G. Mitchison, Eds.), Reading, MA: Addison-Wesley.

Pomerleau, D. A., 1991, Efficient training of artificial neural networks for autonomous navigation, *Neural Computat.*, 3:88–97.

Tikhonov, A. N., and Arsenin, V. Y., 1977, *Solutions of Ill-Posed Problems*, Washington, DC: Winston and Sons.

Reinforcement Learning

Andrew G. Barto

Introduction

The term *reinforcement* comes from studies of animal learning in experimental psychology, where it refers to the occurrence of an event, in the proper relation to a response, that tends to increase the probability that the response will occur again in the same situation. Although not used by psychologists, the term *reinforcement learning* has been widely adopted by theorists in engineering and artificial intelligence to refer to a class of learning tasks and algorithms based on this principle of reinforcement. The simplest reinforcement learning methods are based on the common-sense idea that if an action is followed by a satisfactory state of affairs or an improvement in the state of affairs, then the tendency to produce that action is strengthened, i.e., reinforced.

Reinforcement learning is usually formulated as an *optimization problem* with the objective of finding an action or a strategy for producing actions that is optimal, or best, in some well-defined way. Although in practice it is more important that a reinforcement learning system continue to improve than it is for it to actually achieve optimal behavior, optimality objectives provide a useful categorization of reinforcement learning into three basic types, in order of increasing complexity: *nonassociative*, *associative*, and *sequential*. All these types of reinforcement learning differ from the more commonly studied paradigm of supervised learning, or "learning with a teacher," in significant ways that we discuss in the course of this article. The article REINFORCEMENT LEARNING IN MOTOR CONTROL contains additional information. For more detailed treatments, the reader should consult Barto (1992) and Sutton (1992).

Nonassociative and Associative Reinforcement Learning

Figure 1 shows the basic components of both nonassociative and associative reinforcement learning. The learning system's actions influence the behavior of some process. A *critic* sends the learning system a *reinforcement signal* whose value at any time is a measure of the "goodness" of the current process behavior. Using this information, the learning system updates its action-generation rule, generates another action, and the process repeats. In nonassociative reinforcement learning, the only input to the learning system is the reinforcement signal, whereas in the associative case, the learning system also re-

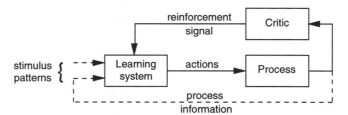

Figure 1. Nonassociative and associative reinforcement learning. A critic evaluates the actions' immediate consequences on the process and sends the learning system a reinforcement signal. In the associative case, in contrast to the nonassociative case, stimulus patterns (dashed lines) are available to the learning system in addition to the reinforcement signal.

ceives stimulus patterns that provide information about the process and possibly other information as well (the dashed lines in Figure 1). Thus, whereas the objective of nonassociative reinforcement learning is to find the optimal action, the objective in the associative case is to learn an associative mapping that produces the optimal action on any trial as a function of the stimulus pattern present on that trial.

An example of a nonassociative reinforcement learning problem has been extensively studied by learning automata theorists (Narendra and Thathachar, 1989). Suppose the learning system has m actions $a_1, a_2, \ldots, a_m$, and that the reinforcement signal simply indicates "success" or "failure." Further, assume that the influence of the learning system's actions on the reinforcement signal can be modeled as a collection of success probabilities $d_1, d_2, \ldots, d_m$, where d_i is the probability of success given that the learning system has generated a_i. The d_i's do not have to sum to one. With no initial knowledge of these values, the learning system's objective is to eventually maximize the probability of receiving "success," which is accomplished when it always performs the action a_j such that $d_j = \max\{d_i | i = 1, \ldots, m\}$.

One class of learning systems for this problem consists of *stochastic learning automata* (Narendra and Thathachar, 1989). Suppose that on each trial, or time step, t, the learning system selects an action $a(t)$ from its set of m actions according to a probability vector $(p_1(t), \ldots, p_m(t))$, where $p_i(t) = Pr\{a(t) = a_i\}$.

A stochastic learning automaton implements a common-sense notion of reinforcement learning: if action a_i is chosen on trial t and the critic's feedback is "success," then $p_i(t)$ is increased and the probabilities of the other actions are decreased; whereas if the critic indicates "failure," then $p_i(t)$ is decreased and the probabilities of the other actions are appropriately adjusted. Many methods for adjusting action probabilities have been studied, and numerous theorems have been proven about how they perform.

One can generalize this nonassociative problem to illustrate an associative reinforcement learning problem. Suppose that on trial t the learning system senses stimulus pattern $x(t)$ and selects an action $a(t) = a_i$ through a process that can depend on $x(t)$. After this action is executed, the critic signals success with probability $d_i(x(t))$ and failure with probability $1 - d_i(x(t))$. The objective of learning is to maximize success probability, achieved when on each trial t the learning system executes the action $a(t) = a_j$, where a_j is the action such that $d_j(x(t)) = \max\{d_i(x(t))|i = 1, \ldots, m\}$. Unlike supervised learning, examples of optimal actions are not provided during training; they have to be *discovered* through exploration by the learning sytem. Learning tasks like this are related to instrumental, or cued operant, tasks used by animal learning theorists, and the stimulus patterns correspond to discriminative stimuli.

Following are key observations about both nonassociative and associative reinforcement learning:

1. *Uncertainty* plays a key role in reinforcement learning. For example, if the critic in the preceding example evaluated actions deterministically (i.e., $d_i = 1$ or 0 for each i), then the problem would be a much simpler optimization problem.

2. The critic is an abstract model of any process that evaluates the learning system's actions. It need not have direct access to the actions or have any knowledge about the interior workings of the process influenced by those actions.

3. The reinforcement signal can be any signal evaluating the learning system's actions, and not just the success/failure signal described earlier. Often it takes on real values, and the objective of learning is to maximize its expected value. Moreover, the critic can use a variety of criteria in evaluating actions, which it can combine in various ways to form the reinforcement signal.

4. The critic's signal does not directly tell the learning system what action is best; it only evaluates the action taken. The critic also does not directly tell the learning system how to change its actions. These are key features distinguishing reinforcement learning from supervised learning, and we will discuss them further.

5. Reinforcement learning algorithms are *selectional* processes. There must be *variety* in the action-generation process so that the consequences of alternative actions can be compared to select the best. Behavioral variety is called *exploration*; it is often generated through randomness (as in stochastic learning automata), but it need not be.

6. Reinforcement learning involves a conflict between *exploitation* and *exploration*. In deciding which action to take, the learning system has to balance two conflicting objectives: it has to exploit what it has already learned to obtain high evaluations, and it has to behave in new ways—explore—to learn more. Because these needs ordinarily conflict, reinforcement learning systems have to somehow balance them. In control engineering, this is known as the conflict between control and identification (see IDENTIFICATION AND CONTROL). It is absent from supervised and unsupervised learning unless the learning system is also engaged in influencing

which training examples it sees (see EXPLORATION IN ACTIVE LEARNING).

Associative Reinforcement Learning Rules

Several associative reinforcement learning rules for neuron-like units have been studied. Consider a neuron-like unit receiving a stimulus pattern as input in addition to the critic's reinforcement signal. Let $x(t)$, $w(t)$, $a(t)$, and $r(t)$, respectively, denote the stimulus vector, weight vector, action, and value of the reinforcement signal at time t. Let $s(t)$ denote the weighted sum of the stimulus components at time t:

$$s(t) = \sum_{i=1}^{n} w_i(t)x_i(t)$$

where $w_i(t)$ and $x_i(t)$, respectively, are the ith components of the weight and stimulus vectors.

Associative Search Unit

One simple associative reinforcement learning rule is an extension of the Hebbian correlation learning rule. This rule, called the *associative search rule*, was motivated by Klopf's (1982) theory of the self-interested neuron. To exhibit variety in its behavior, the unit's output is a random variable depending on the activation level:

$$a(t) = \begin{cases} 1 & \text{with probability } p(t) \\ 0 & \text{with probability } 1 - p(t) \end{cases} \qquad (1)$$

where $p(t)$, which must be between 0 and 1, is an increasing function (such as the logistic function) of $s(t)$. Thus, as the weighted sum increases (decreases), the unit becomes more (less) likely to fire (i.e., to produce an output of 1). If the critic takes time t to evaluate an action, the weights are updated according to the following rule:

$$\Delta w(t) = \eta r(t)a(t - \tau)x(t - \tau) \qquad (2)$$

where $r(t)$ is $+1$ (success) or -1 (failure), and $\eta > 0$ is the learning rate parameter.

This is basically the Hebbian correlation rule with the reinforcement signal acting as an additional modulatory factor. Thus, if the unit fires in the presence of an input x, possibly just by chance, and this action is followed by "success," the weights change so that the unit will be more likely to fire in the presence of x and inputs similar to x in the future. A failure signal makes it less likely to fire under these conditions. This rule makes clear the three factors minimally required for associative reinforcement learning: a stimulus signal, x; the action produced in its presence, a; and the consequent evaluation, r.

Selective Bootstrap and Associative Reward-Penalty Units

Widrow, Gupta, and Maitra (1973) extended the Widrow-Hoff, or LMS (least-mean-square), learning rule so that it could be used in associative reinforcement learning problems. They called their extension of LMS the *selective bootstrap* rule. A selective bootstrap unit's output, $a(t)$, is either 0 or 1, computed as the deterministic threshold of the weighted sum, $s(t)$. In supervised learning, an LMS unit receives a training signal, $z(t)$, that directly specifies the desired action at trial t and updates its weights as follows:

$$\Delta w(t) = \eta[z(t) - s(t)]x(t) \qquad (3)$$

In contrast, a selective bootstrap unit receives a reinforcement signal, $r(t)$, and updates its weights according to this rule:

$$\Delta w(t) = \begin{cases} \eta[a(t) - s(t)]x(t) & \text{if } r(t) = \text{success} \\ \eta[1 - a(t) - s(t)]x(t) & \text{if } r(t) = \text{failure} \end{cases}$$

where it is understood that $r(t)$ evaluates $a(t)$. Thus, if $a(t)$ produces "success," the LMS rule is applied with $a(t)$ playing the role of the desired action. Widrow et al. (1973) called this *positive bootstrap adaptation*: weights are updated as if the output actually produced was in fact the desired action. On the other hand, if $a(t)$ leads to "failure," the desired action is $1 - a(t)$, i.e., the action that was *not* produced. This is *negative bootstrap adaptation*. The reinforcement signal switches the unit between positive and negative bootstrap adaptation, motivating the term *selective bootstrap adaptation*.

A closely related unit is the *associative reward-penalty* (A_{R-P}) unit of Barto and Anandan (1985). It differs from the selective bootstrap algorithm in two ways. First, the unit's output is a random variable like that of the associative search unit (Equation 1). Second, its weight-update rule is an *asymmetric* version of the selective bootstrap rule:

$$\Delta w(t) = \begin{cases} \eta[a(t) - s(t)]x(t) & \text{if } r(t) = \text{success} \\ \lambda\eta[1 - a(t) - s(t)]x(t) & \text{if } r(t) = \text{failure} \end{cases}$$

where $0 \le \lambda \le 1$ and $\eta > 0$. This rule's asymmetry is important because its asymptotic performance improves as λ approaches zero, but $\lambda = 0$ is, in general, *not* optimal.

We can see from the selective bootstrap and A_{R-P} units that a reinforcement signal is less informative than a signal specifying a desired action. It is also less informative than the error $z(t) - a(t)$ used by the LMS rule. Because this error is a signed quantity, it tells the unit *how*, i.e., in what direction, it should change its action. A reinforcement signal—by itself—does not convey this information. If the learner has only two actions, it is easy to deduce, or estimate, the desired action from the reinforcement signal and the actual action. However, if there are more than two actions, the situation is more difficult because the reinforcement signal does not provide information about actions that were not taken. One way a neuron-like unit with more than two actions can perform associative reinforcement learning is illustrated by the *Stochastic Real-Valued* (SRV) unit of Gullapalli described in REINFORCEMENT LEARNING IN MOTOR CONTROL.

Weight Perturbation

For the units described in the preceding section (except the selective bootstrap unit), behavioral variability is achieved by including random variation in the unit's output. Another approach is to randomly vary the weights. Following Alspector et al. (1993), let δw be a vector of small perturbations, one for each weight, which are independently selected from some probability distribution. Letting $\mathscr{E}$ denote the function evaluating the system's behavior, the weights are updated as follows:

$$\Delta w = -\eta\left[\frac{\mathscr{E}(w + \delta w) - \mathscr{E}(w)}{\delta w}\right] \tag{4}$$

where $\eta > 0$ is a learning rate parameter. This is a gradient descent learning rule that changes weights according to an estimate of the gradient of $\mathscr{E}$ with respect to the weights. Alspector et al. (1993) say that the method *measures* the gradient instead of *calculating* it as the LMS and error backpropagation algorithms do. This approach has been proposed by several researchers for updating the weights of a unit, or of a network, during supervised learning, where $\mathscr{E}$ gives the error over the training examples. However, $\mathscr{E}$ can be any function evaluating

the unit's behavior, including a reinforcement function (in which case, the sign of the learning rule would be changed to make it a gradient *ascent* rule).

Reinforcement Learning Networks

Networks of A_{R-P} units have been used successfully in both supervised and associative reinforcement learning tasks (Barto, 1985; Barto and Jordan, 1987), although only with feedforward connection patterns. For supervised learning, the output units learn just as they do in error backpropagtion, but the hidden units learn according to the A_{R-P} rule. The reinforcement signal, which is defined to increase as the output error decreases, is simply *broadcast* to all the hidden units, which learn simultaneously. If the network as a whole faces an associative reinforcement learning task, all the units are A_{R-P} units, to which the reinforcement signal is uniformly broadcast (Barto, 1985). Another way to use reinforcement learning units in networks is to use them only as output units, with hidden units being trained via backpropagation. Weight changes of the output units determine the quantities that are backpropagated. An example is provided by a network for robot peg-in-hole insertion described in REINFORCEMENT LEARNING IN MOTOR CONTROL.

The error backpropagation algorithm can be used in another way in associative reinforcement learning problems. It is possible to train a multilayer network to form a model of the process by which the critic evaluates actions. After this model is trained sufficiently, it is possible to estimate the gradient of the reinforcement signal with respect to each component of the action vector by analytically differentiating the model's output with respect to its action inputs (which can be done efficiently by backpropagation). This gradient estimate is then used to update the parameters of an action-generation component. Jordan and Jacobs (1990) illustrate this approach.

The weight perturbation approach carries over directly to networks by simply letting w in Equation 4 be the vector consisting all the network's weights. A number of researchers have achieved success using this approach in supervised learning problems. In these cases, one can think of each weight as facing a reinforcement learning task (which is in fact nonassociative), even though the network as a whole faces a supervised learning task. An advantage of this approach is that it applies to networks with arbitrary connection patterns, not just to feedforward networks.

It should be clear from this discussion of reinforcement learning networks that there are many different approaches to solving reinforcement learning problems. Furthermore, although reinforcement learning *tasks* can be clearly distinguished from supervised and unsupervised learning tasks, it is more difficult to precisely define a class of reinforcement learning *algorithms*.

Sequential Reinforcement Learning

Sequential reinforcement requires improving the long-term consequences of an action, or of a strategy for performing actions, in addition to short-term consequences. In these problems, it can make sense to forgo short-term performance in order to achieve better performance over the long term. Tasks having these properties are examples of *optimal control problems*, sometimes called *sequential decision problems* when formulated in discrete time (see ADAPTIVE CONTROL: NEURAL NETWORK APPLICATIONS).

Figure 1, which shows the components of an associative reinforcement learning system, also applies to sequential reinforcement learning, where the box labeled "Process" is a system being controlled. A sequential reinforcement learning system tries to influence the behavior of the process in order to maximize a measure of the total amount of reinforcement that will be received over time. In the simplest case, this measure is the sum of the future reinforcement values, and the objective is to learn an associative mapping that at each time step t selects, as a function of the stimulus pattern $x(t)$, an action $a(t)$ that maximizes

$$\sum_{k=0}^{\infty} r(t + k)$$

where $r(t + k)$ is the reinforcement signal at step $t + k$. Such an associative mapping is called a *policy*.

Because this sum might be infinite in some problems, and because the learning system usually has control only over its expected value, researchers often consider the following *expected discounted sum* instead:

$$E\{r(t) + \gamma r(t + 1) + \gamma^2 r(t + 2) + \cdots\} = E\left\{\sum_{k=0}^{\infty} \gamma^k r(t + k)\right\}$$

(5)

where E is the expectation over all possible future behavior patterns of the process. The discount factor γ determines the present value of future reinforcement: a reinforcement value received k time steps in the future is worth γ^k times what it would be worth if it were received now. If $0 \leq \gamma < 1$, this infinite discounted sum is finite as long as the reinforcement values are bounded. If $\gamma = 0$, the robot is "myopic" in being only concerned with maximizing immediate reinforcement; this is the associative reinforcement learning problem discussed earlier. As γ approaches 1, the objective explicitly takes future reinforcement into account: the robot becomes more farsighted.

An important special case of this problem occurs when there is no immediate reinforcement until a goal state is reached. This is a *delayed reward* problem in which the learning system has to learn how to make the process enter a goal state. Sometimes the objective is to make it enter a goal state as quickly as possible. A key difficulty in these problems has been called the *temporal credit-assignment problem*: When a goal state is finally reached, which of the decisions made earlier deserve credit for the resulting reinforcement? (See also REINFORCEMENT LEARNING IN MOTOR CONTROL.) A widely studied approach to this problem is to learn an *internal evaluation function* that is more informative than the evaluation function implemented by the external critic. An *adaptive critic* is a system that learns such an internal evaluation function.

Samuel's Checkers Player

Samuel's (1959) checkers-playing program has been a major influence on adaptive critic methods. The checkers player uses an evaluation function to assign a score to each board configuration; and the system makes the move expected to lead to the configuration with the highest score. Samuel used a method to improve the evaluation function through a process that compared the score of the current board position with the score of a board position likely to arise later in the game. As a result of this process of "backing up" board evaluations, the evaluation function improved in its ability to evaluate the long-term consequences of moves. If the evaluation function can be made to score each board configuration according to its true promise of eventually leading to a win, then the best strategy for playing is to myopically select each move so that the next board configuration is the most highly scored. If the evaluation function is optimal in this sense, then it already takes into account all the possible future courses of play. Methods such as Samuel's that attempt to adjust the evaluation function toward this ideal optimal evaluation function are of great utility.

Adaptive Critic Unit and Temporal Difference Methods

An adaptive critic unit is a neuron-like unit that implements a method similar to Samuel's. The unit is a neuron-like unit whose output at time step t is $P(t) = \Sigma_{i=1}^n w_i(t)x_i(t)$, so denoted because it is a *prediction* of the discounted sum of future reinforcement defined by Equation 5. The adaptive critic learning rule rests on noting that correct predictions must satisfy a consistency condition relating predictions at adjacent time steps. Suppose that the predictions at any two successive time steps, say steps t and $t + 1$, are correct. This assumption means that

$$P(t) = E\{r(t) + \gamma r(t + 1) + \gamma^2 r(t + 2) + \cdots\}$$
$$P(t + 1) = E\{r(t + 1) + \gamma r(t + 2) + \gamma^2 r(t + 3) + \cdots\}$$

Now notice that we can rewrite $P(t)$ as follows:

$$P(t) = E\{r(t) + \gamma[r(t + 1) + \gamma r(t + 2) + \cdots]\}$$

But this is exactly the same as

$$P(t) = E\{r(t)\} + \gamma P(t + 1)$$

An estimate of the error by which any two adjacent predictions fail to satisfy this consistency condition is called the *temporal difference (TD) error* (Sutton, 1988):

$$r(t) + \gamma P(t + 1) - P(t)$$

(6)

where $r(t)$ is used as an unbiased estimate of $E\{r(t)\}$. The term *temporal difference* comes from the fact that this error essentially depends on the difference between the critic's predictions at successive time steps.

The adaptive critic unit adjusts its weights according to the following learning rule:

$$\Delta w(t) = \eta[r(t) + \gamma P(t + 1) - P(t)]x(t)$$

(7)

A subtlety here is that $P(t + 1)$ should be computed using the weight vector $w(t)$, not $w(t + 1)$. This rule changes the weights to decrease the magnitude of the TD error. Note that if $\gamma = 0$, it is equal to the LMS learning rule (Equation 3). By analogy with the LMS rule, we can think of $r(t) + \gamma P(t + 1)$ as the prediction target: it is the quantity that each $P(t)$ should match. The adaptive critic is therefore trying to predict the next reinforcement, $r(t)$, *plus its own next (discounted) prediction*, $\gamma P(t + 1)$. It is similar to Samuel's learning method in adjusting weights to make current predictions closer to later predictions.

Although this method is very simple computationally, it actually converges to the correct predictions of the expected discounted sum of future reinforcement if these correct predictions can be computed by a linear unit. This finding is shown by Sutton (1988), who discusses a more general class of methods, called *TD methods*, that include Equation 7 as a special case. It is also possible to learn nonlinear predictions using, for example, multilayer networks trained by back propagating the TD error. Using this approach, Tesauro (1992) produced a system that learned how to play expert-level backgammon.

Actor-Critic Architectures

In an actor-critic architecture, the predictions formed by an adaptive critic act as reinforcement for an associative reinforcement learning component, called the *actor* (Figure 2). To distinguish the adaptive critic's signal from the reinforcement signal supplied by the original, nonadaptive critic, we

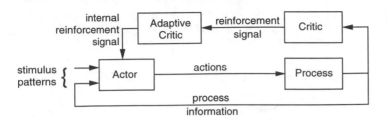

Figure 2. Actor-critic architecture. An adaptive critic provides an internal reinforcement signal to an *actor* which learns a policy for controlling the process.

call it the *internal reinforcement signal*. The actor tries to maximize the *immediate* internal reinforcement signal, while the adaptive critic tries to predict total future reinforcement. To the extent that the adaptive critic's predictions of total future reinforcement are correct given the actor's current policy, the actor actually learns to increase the total amount of future reinforcement.

Barto, Sutton, and Anderson (1983) used this architecture for learning to balance a simulated pole mounted on a cart. The actor had two actions: application of a force of a fixed magnitude to the cart in the plus or minus directions. The nonadaptive critic only provided a signal of failure when the pole fell past a certain angle or the cart hit the end of the track. The stimulus patterns were vectors representing the state of the cart-pole system. The actor was essentially an associative search unit as described above whose weights were modulated by the internal reinforcement signal.

Q-Learning

Another approach to sequential reinforcement learning combines the actor and adaptive critic into a single component that learns separate predictions for each action. At each time step the action with the largest prediction is selected, except for an exploration factor that causes other actions to be selected occasionally. An algorithm for learning predictions of future reinforcement for each action, called the *Q-learning* algorithm, was proposed by Watkins (1989), who proved that it converges to the correct predictions under certain conditions. Although the Q-learning convergence theorem requires lookup-table storage (and therefore finite state and action sets), many researchers have heuristically adapted Q-learning to more general forms of storage, including multilayer neural networks trained by backpropagating the Q-learning error.

Dynamic Programming

Sequential reinforcement learning problems (in fact, all reinforcement learning problems) are examples of stochastic optimal control problems. Among the traditional methods for solving these problems are dynamic programming (DP) algorithms. As applied to optimal control, DP consists of methods for successively approximating optimal evaluation functions and optimal policies. Bertsekas (1987) provides a good treatment of these methods. A basic operation in all DP algorithms is "backing up" evaluations in a manner similar to the operation used in Samuel's method and in the adaptive critic.

Recent reinforcement learning theory exploits connections with DP algorithms while emphasizing important differences. Following is a summary of key observations:

1. Because conventional DP algorithms require multiple exhaustive "sweeps" of the process state set (or a discretized approximation of it), they are not practical for problems with very large finite state sets or high-dimensional continuous state spaces. Sequential reinforcement learning algorithms *approximate* DP algorithms in ways designed to reduce this computational complexity.
2. Instead of requiring exhaustive sweeps, sequential reinforcement learning algorithms operate on states as they occur in actual or simulated experiences in controlling the process. It is appropriate to view them as *Monte Carlo* DP algorithms.
3. Whereas conventional DP algorithms require a complete and accurate model of the process to be controlled, sequential reinforcement learning algorithms do not require such a model. Instead of computing the required quantities (such as state evaluations) from a model, they estimate these quantities from experience. However, reinforcement learning methods can also take advantage of models to improve their efficiency.

It is therefore accurate to view sequential reinforcement learning as a collection of heuristic methods providing computationally feasible approximations of DP solutions to stochastic optimal control problems.

Discussion

The increasing interest in reinforcement learning is due to its applicability to learning by autonomous robotic agents. Although both supervised and unsupervised learning can play essential roles in reinforcement learning systems, these paradigms by themselves are not general enough for learning while acting in a dynamic and uncertain environment. Among the topics being addressed by current reinforcement learning research are these: extending the theory of sequential reinforcement learning to include generalizing function approximation methods; understanding how exploratory behavior is best introduced and controlled; sequential reinforcement learning when the process state cannot be observed; how problem-specific knowledge can be effectively incorporated into reinforcement learning systems; the design of modular and hierarchical architectures; and the relationship to brain reward mechanisms.

Road Map: Learning in Artificial Neural Networks, Deterministic
Background: I.3. Dynamics and Adaptation in Neural Networks
Related Reading: Planning, Connectionist; Problem Solving, Connectionist

References

Alspector, J., Meir, R., Yuhas, B., Jayakumar, A., and Lippe, D., 1993, A parallel gradient descent method for learning in analog VLSI neural networks, in *Advances in Neural Information Processing Systems 5* (S. J. Hanson, J. D. Cowan, and C. L. Giles, Eds.), San Mateo, CA: Morgan Kaufmann, pp. 836–844.
Barto, A. G., 1985, Learning by statistical cooperation of self-interested neuron-like computing elements, *Hum. Neurobiol.*, 4:229–256.
Barto, A. G., 1992, Reinforcement learning and adaptive critic methods, in *Handbook of Intelligent Control: Neural, Fuzzy, and Adaptive*

Approaches (D. A. White and D. A. Sofge, Eds.), New York: Van Nostrand Reinhold, pp. 469–491. ◆

Barto, A. G., and Anandan, P., 1985, Pattern recognizing stochastic learning automata, *IEEE Trans. Syst. Man Cybern.*, 15:360–375.

Barto, A. G., and Jordan, M. I., 1987, Gradient following without back-propagation in layered networks, in *Proceedings of the IEEE First Annual Conference on Neural Networks* (M. Caudill and C. Butler, Eds.), San Diego: IEEE, pp. 11629–11636.

Barto, A. G., Sutton, R. S., and Anderson, C. W., 1983, Neuronlike elements that can solve difficult learning control problems, *IEEE Trans. Syst. Man Cybern.*, 13:835–846. Reprinted in *Neurocomputing: Foundations of Research* (J. A. Anderson and E. Rosenfeld, Eds.), Cambridge, MA: MIT Press, 1988.

Bertsekas, D. P., 1987, *Dynamic Programming: Deterministic and Stochastic Models*, Englewood Cliffs, NJ: Prentice Hall. ◆

Jordan, M. I., and Jacobs, R. A., 1990, Learning to control an unstable system with forward modeling, in *Advances in Neural Information Processing Systems 2* (D. S. Touretzky, Ed.), San Mateo, CA: Morgan Kaufmann, pp. 324–331.

Jordan, M. I., and Rumelhart, D. E., 1992, Supervised learning with a distal teacher, *Cognit. Sci.*, 16:307–354.

Klopf, A. H., 1982, *The Hedonistic Neuron: A Theory of Memory, Learning, and Intelligence*, Washington, DC: Hemisphere.

Narendra, K., and Thathachar, M. A. L., 1989, *Learning Automata: An Introduction*, Englewood Cliffs, NJ: Prentice Hall. ◆

Samuel, A. L., 1959, Some studies in machine learning using the game of checkers, *IBM J. Res. Develop.*, 3:210–229. Reprinted in *Computers and Thought* (E. A. Feigenbaum and J. Feldman, Eds.), New York: McGraw-Hill, 1963, pp. 71–105.

Sutton, R. S., 1988, Learning to predict by the method of temporal differences, *Machine Learn.*, 3:9–44.

Sutton, R. S., Ed., 1992, *A Special Issue of Machine Learning on Reinforcement Learning, Machine Learn.*, 8. Also published as *Reinforcement Learning*, Boston: Kluwer Academic, 1992. ◆

Tesauro, G. J., 1992, Practical issues in temporal difference learning, *Machine Learn.*, 8:257–277.

Watkins, C. J. C. H., 1989, Learning from delayed rewards, PhD Thesis, Cambridge University, Cambridge, UK.

Widrow, B., Gupta, N. K., and Maitra, S., 1973, Punish/reward: Learning with a critic in adaptive threshold systems, *IEEE Trans. Syst. Man Cybern.*, 5:455–465.

Reinforcement Learning in Motor Control

Andrew G. Barto

Introduction

How do we learn motor skills such as reaching, walking, swimming, or riding a bicycle? Although there is a large literature on motor skill acquisition which is full of controversies (for a recent introduction to human motor control, see Rosenbaum, 1991), there is general agreement that motor learning requires the learner, human or not, to receive response-produced feedback through various senses providing information about performance. Careful consideration of the nature of the feedback used in learning is important for understanding the role of reinforcement learning in motor control (see REINFORCEMENT LEARNING). One function of feedback is to guide the performance of movements. This is the kind of feedback with which we are familiar from control theory, where it is the basis of servomechanisms, although its role in guiding animal movement is more complex. Another function of feedback is to provide information useful for improving *subsequent* movement. Feedback having this function has been called *learning feedback*. Note that this functional distinction between feedback for control and for learning does not mean that the signals or channels serving these functions need to be different.

Learning Feedback

When motor skills are acquired without the help of an explicit teacher or trainer, learning feedback must consist of information automatically generated by the movement and its consequences on the environment. This has been called *intrinsic feedback* (Schmidt, 1982). The "feel" of a successfully completed movement and the sight of a basketball going through the hoop are examples of intrinsic learning feedback. A teacher or trainer can augment intrinsic feedback by providing *extrinsic feedback* (Schmidt, 1982) consisting of extra information added for training purposes, such as a buzzer indicating that a movement was on target, a word of praise or encouragement, or an indication that a certain kind of error was made.

Most research in the field of artificial neural networks has focused on the learning paradigm called *supervised learning*, which emphasizes the role of training information in the form of desired, or *target*, network responses for a set of training inputs (see PERCEPTRONS, ADALINES, AND BACKPROPAGATION). The aspect of real training that corresponds most closely to the supervised learning paradigm is the trainer's role in telling or showing the learner what to do, or explicitly guiding his or her movements. These activities provide standards of correctness that the learner can try to match as closely as possible by reducing the error between its behavior and the standard. Supervised learning can also be relevant to motor learning when there is no trainer because it can use intrinsic feedback to construct various kinds of *models* that are useful for learning. Barto (1990) and Jordan and Rumelhart (1992) discuss some of the uses of models in learning control.

In contrast to supervised learning, reinforcement learning emphasizes learning feedback that *evaluates* the learner's performance without providing standards of correctness in the form of behavioral targets. Evaluative feedback tells the learner whether, and possibly by how much, its behavior has improved; or it provides a measure of the "goodness" of the behavior; or it just provides an indication of success or failure. Evaluative feedback does not directly tell the learner what it *should* have done, and although it sometimes provides the *magnitude* of an error, it does not include *directional* information telling the learner how to change its behavior, as does the error feedback of supervised learning. Although the most obvious evaluative feedback is extrinsic feedback provided by a trainer, most evaluative feedback is probably intrinsic, being derived by the learner from sensations generated by a movement and its consequences on the environment: the kinesthetic and tactile feel of a successful grasp or the swish of a basketball through the hoop. Instead of trying to match a standard of correctness, a reinforcement learning system tries to maximize the goodness of behavior as indicated by evaluative feedback. To do this, it has to actively try alternatives, compare the resulting evaluations, and use some kind of selection mechanism to guide behavior toward the better alternatives. Although evaluative feedback is often called *reinforcement* feedback, it need not involve pleasure or pain.

Motor learning involves feedback carrying many different kinds of information. Consequently, it is incorrect to view motor learning strictly in terms of either supervised, reinforcement, or any other learning paradigms that have been formulated for theoretical study. Aspects of all of these paradigms play interlocking roles. However, reinforcement learning may be an essential component of motor learning.

Learning from Consequences

The simplest reinforcement learning algorithms are based on the commonsense idea that if an action is followed by a satisfactory state of affairs, or an improvement in the state of affairs, then the tendency to produce that action is strengthened, i.e., reinforced. This basic idea follows Thorndike's classical *Law of Effect* (Thorndike, 1911). Although this principle has generated considerable controversy over the years, it remains influential because its general idea is supported by many experiments and it makes such good intuitive sense (e.g., Glazer, 1971).

To illustrate how this principle applies to motor learning, we first discuss it within the general context of control. Then we describe several special cases related to motor control. Figure 1*A*, is a variation of the classical control system diagram. A controller provides control signals to a controlled system. The behavior of the controlled system is influenced by disturbances, and feedback from the controlled system to the controller provides information on which the control signals can depend. Commands to the controller specify aspects of the control task's objective.

In Figure 1*B*, the control loop is augmented with another feedback loop that provides learning feedback to the controller. In accordance with common practice in reinforcement learning, a *critic* is included that generates evaluative learning feedback on the basis of observing the control signals and their consequences on the behavior of the controlled system. The critic also needs to know the command to the controller be-

cause its evaluations must be different depending on what the controller should be trying to do. The critic is an abstraction of whatever process supplies evaluative learning feedback, both intrinsic and extrinsic, to the learning system. It is often said that the critic provides a *reinforcement signal* to the learning system. In most artificial reinforcement learning systems, the critic's output at any time is a number that scores the controller's behavior: the higher the number, the better the behavior. Assume for the moment that the behavior being scored is the immediately preceding behavior. We discuss more complex temporal relationships in later sections. For this process to work, there must be some *variability* in the controller's behavior so that the critic can evaluate many alternatives. A learning mechanism can then adjust the controller's behavior so that it tends toward behavior that is favored by the critic.

A learning rule particularly suited to reinforcement learning control systems implemented as artificial neural networks was developed by Gullapalli (1990) in the form of what he called a Stochastic Real-Valued (SRV) unit. An SRV unit's output is produced by adding a random number to the weighted sum of the components of its input pattern. The random number is drawn from a zero-mean Gaussian distribution. This random component provides the unit with the variability necessary for it to "explore" its activity space. When the reinforcement signal indicates that something good happened just after the unit emitted a particular output value in the presence of some input pattern, the unit's weights are adjusted to move the activation in the direction in which it was perturbed by the random number. This has the effect of increasing the probability that future outputs generated for that input pattern (and similar input patterns) will be closer to the output value just emitted. If the reinforcement signal indicates that something bad happened, the weights are adjusted to move future output values away from the value just emitted. Another part of the SRV learning rule decreases the variance of the Gaussian distribution as learning proceeds. This decreases the variability of the unit's behavior, with the goal of making it eventually stick (i.e., become deterministic) at the best output value for each input pattern. Using this learning rule, an SRV unit learns to produce the best output in response to each input pattern (given appropriate assumptions). Unlike more familiar supervised

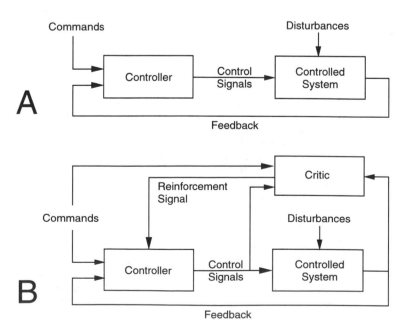

Figure 1. *A*, A basic control loop. A controller provides control signals to a controlled system, whose behavior is influenced by disturbances. Feedback from the controlled system to the controller provides information on which the control signals can depend. Commands to the controller specify aspects of the control task's objective. *B*, A control system with learning feedback. A *critic* provides the controller with a reinforcement signal evaluating its success in achieving the control objectives.

learning units, it is never given target outputs; it has to discover what outputs are best through an active exploration process.

Overcoming the Distal Error Problem

To see how reinforcement learning might work in motor learning, consider the problem of learning the control signals required to move the end of an arm from its initial position to a desired final position. This problem combines aspects of the inverse kinematics and inverse dynamics problems in robotics, where the end of the arm is called the end-effector. Many researchers have used artificial neural networks for these problems, but relatively few have employed reinforcement learning. Lipitkas et al. (1993) provide a simple example of a reinforcement learning approach. To avoid the complexity of transforming each desired arm trajectory into corresponding joint trajectories, and then computing the required time functions of forces, they propose storing prototypical force time functions in memory. The prototypes are modified on playback according to the demands of each movement. Their controller is a network receiving inputs coding the starting location of the end-effector as well as command inputs giving the Cartesian coordinates of the target end-effector position (Figure 2). The six outputs of the network provide parameters to a torque generator which generates time-varying signals for driving the joint actuators of a dynamic arm model. The time-varying signals are parameterized by six numbers determining characteristics of their wavelike shapes (e.g., giving the magnitudes and relative timing of the half-waves). During each movement, the controller operates in open-loop mode, generating the torque time functions without the aid of sensory feedback. The learning problem for the network, then, is to learn a function associating each pair of end-effector starting and target positions to the values of the six parameters that will accomplish the movement.

A straightforward application of supervised learning is not possible here because the required training examples are not available: It is not known what parameters will work for any pair of starting and target positions (except possibly the trivial cases in which the starting position is already the target position, but these are not useful as training examples). This is an instance of what has been called the *distal error problem* (Jordan and Rumelhart, 1992) for supervised learning. This problem is present whenever the standard of correctness required for supervised learning is available in a coordinate system that is different from the one in which the learning system's activity must be specified for learning. In the case of learning how to move an arm from a starting position to a target position, the standard of correctness is the target position, but what must be learned are the control signals to the joint actuators. The resulting end-effector position error is distal to the output of the

controller that is to be learned. Although a nonzero distal error vector indicates that the controller made an error, it does not tell the controller how it should change its behavior to reduce the error.

The distal error problem can be solved by using a model of the controller's influence on the arm's movement (possibly learned through supervised learning) to translate distal error vectors into error vectors required for supervised learning (e.g., Jordan and Rumelhart, 1992). Another approach is to learn an inverse model of the controller's influence on the arm's movement (also discussed by Jordan and Rumelhart, 1992). Reinforcement learning offers yet another way to overcome the distal error problem because learning feedback in the form of error vectors is not required. Continuing with the arm movement example, Lipitkas et al. (1993) defined a reinforcement signal that attains a maximum value of 1 if the arm reaches the desired position and stops there. The signal decreases depending on the distance between the end-effector's final position and the target position and on its tangential velocity as it passes the target position. The reinforcement signal could include other criteria of successful movements as well. With inputs coding starting and target end-effector positions, the network employs SRV units to generate six parameter values using its current weights. The torque generator generates a movement using these parameter values. When the movement is completed, it is scored by the reinforcement signal, and the network's weights are changed according to Gullapalli's SRV learning rule. After a few thousand movements with different starting and target end-effector positions, the system could move with reasonable accuracy for new pairs of starting and target positions as well as for the pairs on which it was trained. This amount of practice is required because the system effectively has to search the six-dimensional parameter space for each starting and target position.

Gaining an understanding of the relative advantages and disadvantages of reinforcement learning and model-based approaches to the distal error problem is a topic of current research. It is clear that reinforcement learning approaches are much simpler, but reinforcement learning can be slower in terms of the amount of experience required for learning. This is true because reinforcement learning methods tend to extract less information from each experience than do the model-based approaches. However, in some problems, reinforcement learning can significantly outperform model-based approaches (e.g., Gullapalli, 1991; Markey and Mozer, 1992). This occurs when it is easier to learn the right actions than it is to model their effects on a complicated process. Hybrid learning architectures using both approaches are promising alternatives.

Other examples have been developed in which reinforcement learning is used to address the distal error problem. Gullapalli, Barto, and Grupen (1994) devised a reinforcement learning

Figure 2. Block diagram of a reinforcement learning controller of an arm (after Figure 1 of Lipitkas et al., 1993). Given inputs coding the starting and target positions of the end effector, the network controller learns to provide correct parameters to a torque generator which generates, in open-loop mode, time-varying torque signals to the arm. The reinforcement signal evaluates the success of each movement after its completion.

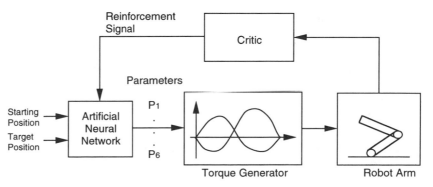
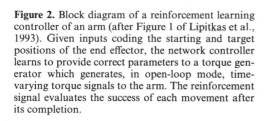

network by which a real robot arm learns how to perform a peg-in-hole insertion task. Although this task is important in industrial robotics, it is also suggestive about animal skill acquisition. Unlike the open-loop example of Lipitkas et al. (1993) just described, the peg-in-hole network learns a closed-loop control rule that guides the robot arm using sensory feedback during peg insertion. Learning occurs throughout each peg-insertion attempt, with the reinforcement signal derived from assessments of the progress of the peg toward the desired inserted position and of the forces generated at the robot's wrist. The robot learns to perform insertion tasks even when the clearance between the hole and the peg is many times less than the amount of uncertainty in the robot's sensory feedback.

Another reinforcement learning system was developed by Fagg and Arbib (1992) which models the role of the primate premotor cortex in triggering movements on the basis of visual stimuli. In contrast to the systems just mentioned, this model learns to *select* from previously learned motor behaviors on the basis of sensory cues. The reinforcement signal is $+1$ when the model selects the correct behavior for a given sensory cue, and -1 otherwise. Although the reinforcement value -1 indicates that an error was made, it does not tell the model which behavior would have been correct or give a clue as to how the model should change its behavior. (The minus sign is not really giving directional information.) This model produces performance curves that are qualitatively similar to those observed in animal experiments.

Credit Assignment

Although reinforcement learning systems do not suffer from the distal error problem, they do suffer from the related *credit assignment problem*. A scalar evaluation of a complex mechanism's behavior does not indicate which of its many action components, both internal and external, were responsible for the evaluation. Thus, it is difficult to determine which of these components deserve the credit (or the blame) for the evaluation. One approach is to assign credit equally to *all* the components so that, through a process of averaging over many variations of the behavior, the components that are key in producing laudable behavior are most strongly reinforced, while inappropriate components are weakened. This approach is commonly taken in reinforcement learning systems, but learning by this process can be very slow in complex systems, such as those involved in motor control. This problem is sometimes referred to as the *structural* credit assignment problem: How is credit assigned to the internal workings of a complex structure? The backpropagation algorithm (e.g., Rumelhart, Hinton, and Williams, 1986) addresses structural credit assignment for artificial neural networks by means of its backpropagation process, and this technique can also be used by reinforcement learning systems (e.g., Gullapalli et al., 1994). However, understanding structural credit assignment mechanisms that are more plausible for biological systems is a frontier of current research.

Reinforcement learning principles lead to a number of alternatives to the backpropagation method for structural assigning credit in complex neural networks. In these methods (see also REINFORCEMENT LEARNING), a single reinforcement signal is uniformly *broadcast* to all the sites of learning, either neurons or individual synapses. Computational studies provide ample evidence that any task that can be learned by error backpropagation can also be learned using this approach, although possibly more slowly. Moreover, these network learning methods are consistent with anatomical and physiological evidence

showing the existence of diffusely projecting neural pathways by which neuromodulators (see NEUROMODULATION IN INVERTEBRATE NERVOUS SYSTEMS) can be widely and nonspecifically distributed. It has been suggested that some of these pathways may play a role in reward-mediated learning. A specific hypothesis is that dopamine mediates synaptic enhancement in the corticostriatal pathway in the manner of a broadcast reinforcement signal (Wickens, 1990; see BASAL GANGLIA). Although hypotheses like this are far from being proven, they are much more appealing to many neuroscientists than hypotheses about how error backpropagation might be implemented in the nervous system.

Another aspect of the credit assignment problem occurs when the temporal relationship between a system's behavior and evaluations of that behavior is not as simple as assumed in the previous discussion. How can reinforcement learning work when the learner's behavior is temporally extended and evaluations occur at varying and unpredictable times? Under these more realistic conditions, it is not always clear what events are being evaluated. This has been called the *temporal* credit assignment problem. It is especially relevant in motor control because movements extend over time and evaluative feedback may become available, for example, only after the end of a movement. An approach to this problem that is receiving considerable attention is the use of methods by which the critic itself can learn to provide useful evaluative feedback immediately after the evaluated event. According to this approach, reinforcement learning is not only the process of improving behavior according to given evaluative feedback; it also includes learning how to improve the evaluative feedback itself. These methods have been called *adaptive critic methods* (see REINFORCEMENT LEARNING and Barto, 1992).

Discussion

Motor learning is too complex to view strictly in terms of either supervised learning or reinforcement learning. Feedback used in motor learning ranges from specific standards of correctness to nonspecific evaluative information, and many learning mechanisms with differing characteristics probably interact to produce the motor learning capabilities of animals. However, reinforcement learning principles may be indispensable for motor learning because they seem necessary for improving motor performance beyond the standards of correctness required by supervised learning.

Road Maps: Control Theory and Robotics; Learning in Biological Systems; Biological Motor Control
Background: I.3. Dynamics and Adaptation in Neural Networks
Related Reading: Problem Solving, Connectionist; Sensorimotor Learning

References

Barto, A. G., 1990, Connectionist learning for control: An overview, in *Neural Networks for Control* (T. Miller, R. S. Sutton, and P. J. Werbos, Eds.), Cambridge, MA: MIT Press, pp. 5–58. ◆
Barto, A. G., 1992, Reinforcement learning and adaptive critic methods, in *Handbook of Intelligent Control: Neural, Fuzzy, and Adaptive Approaches* (D. A. White and D. A. Sofge, Eds.), New York: Van Nostrand Reinhold, pp. 469–491. ◆
Fagg, A. H., and Arbib, M. A., 1992, A model of primate visual-motor conditional learning, *Adapt. Behav.*, 1:3–37.
Glazer, R., 1971, *The Nature of Reinforcement*, New York: Academic Press.
Gullapalli, V., 1990, A stochastic reinforcement algorithm for learning real-valued functions, *Neural Netw.*, 3:671–692.

Gullapalli, V., 1991, A comparison of supervised and reinforcement learning methods on a reinforcement learning task, in *Proceedings of the 1991 IEEE International Symposium on Intelligent Control*, Los Alamitos, CA: IEEE Computer Society Press, pp. 394–399.

Gullapalli, V., Barto, A. G., and Grupen, R. A., 1994, Learning admittance mappings for force-guided assembly, in *Proceedings of the 1994 International Conference on Robotics and Automation*, Los Alamitos, CA: IEEE Computer Society Press, pp. 2633–2638.

Jordan, M. I., and Rumelhart, D. E., 1992, Supervised learning with a distal teacher, *Cognit. Sci.*, 16:307–354.

Lipitkas, J., D'Eleuterio, G. M. T., Bock, O., and Grodski, J. J., 1993, Reinforcement learning and the parametric motor control hypothesis applied to robotic arm movements, in *Proceedings of the Knowledge-Based Systems and Robotics Workshop*, Gloucester, Ont.: Business Intelligence Systems, pp. 101–106.

Markey, K. L., and Mozer, M. C., 1992, Performance comparison of reinforcement learning algorithms on discrete functions, *Proceedings of the International Joint Conference on Neural Networks*, vol. I, San Diego, CA: IEEE Publishing Services, pp. 853–859.

Rosenbaum, D. A., 1991, *Human Motor Control*, San Diego: Academic Press. ◆

Rumelhart, D. E., Hinton, G. E., and Williams, R. J., 1986, Learning internal representations by error propagation, in *Parallel Distributed Processing: Explorations in the Microstructure of Cognition*, vol. 1, *Foundations*, (D. E. Rumelhart, J. L. McClelland, and PDP Research Group, Eds.), Cambridge, MA: MIT Press.

Schmidt, R. A., 1982, *Motor Control and Learning*, Champaign, IL: Human Kinetics. ◆

Thorndike, E. L., 1911, *Animal Intelligence*, Darien, CT: Hafner.

Wickens, J., 1990, Striatal dopamine in motor activation and reward-mediated learning: Steps towards a unifying model, *J. Neural Transm.*, 80:9–31.

Respiratory Rhythm Generation

John E. Lewis

Introduction

Breathing in mammals is a complex neural and muscular process, consisting of two alternating phases, inspiration and expiration. It allows the exchange of oxygen and carbon dioxide between internal and external environments, and hence the rate and depth of breathing regulate blood levels of these important gases. Because of the many feedback control mechanisms (e.g., those mediated by chemoreceptors and lung stretch receptors), the respiratory rhythm is robust, flexible, and generally stable to perturbations. Although some of the control mechanisms have been described in detail, the neural mechanisms involved in generating the rhythm—the respiratory rhythm generator (RRG)—are not well understood (Cohen, 1981; Euler, 1986). This article briefly describes two theories of the RRG and then discusses the role of network models in the investigation of the RRG.

The neuronal networks constituting the RRG are located in the brainstem region of the medulla. Within the medulla, distinct populations of respiratory-related neurons have been classified according to their spiking patterns and anatomical location (Euler, 1986; Ezure, 1990). Some respiratory-related neurons fire specifically during inspiration (*I*-neurons) or expiration (*E* neurons), while others fire during the transitions between these two phases (phase-spanning neurons). Classifying neurons in this way has led to the description of many additional populations and subpopulations. These different neural populations give rise to the respiratory rhythm, and they control respiration by exciting different motor neuron pools innervating the respiratory musculature (Ezure, 1990). The motor neurons themselves are not thought to participate in rhythm generation. The RRG's output is usually monitored from the activity of motor neurons contained in the phrenic nerve, which innervates the diaphragm (the principle inspiratory-related muscle).

Theories of Respiratory Rhythm Generation

The RRG is generally thought to consist of a *central oscillator*, responsible for the timing of the rhythm, and a *pattern-forming network* that shapes the neural output to the respiratory muscles. Two theories for the central oscillator of the RRG have been proposed: a *network theory* requiring stereotyped synaptic connections between distinct populations of respiratory neurons (Richter, Ballanyi, and Schwarzacher, 1992), and a *pacemaker theory* involving endogenously active pacemaker neurons (Feldman, Smith, and Ellenberger, 1990). Both theories consider that the central oscillator requires a form of tonic excitatory drive (e.g., chemoreceptor activation). In many systems, network interactions and pacemaker neurons probably coexist, with the relative importance of each varying.

A Network Theory

Richter and co-workers (1992) proposed a three-phase theory of the RRG that is based on network interactions between several neural populations. This theory separates the respiratory rhythm into three distinct neural phases: inspiratory (*I*), postinspiratory (*pI*; or stage 1 expiratory), and expiratory (*E*; or stage 2 expiratory). These phases can be associated with the patterned output of phrenic nerve activity.

The proposed network consists principally of five populations of neurons within the three-phase framework: early I (*eI*), ramp I (*I_R*), and late I (*LI*) groups; a *pI* group; and an *E* group (Figure 1). The respiratory cycle is outlined as follows: The rapid onset of *eI* activity marks the start of inspiration, with the I_R neurons increasing in firing frequency because of regenerative self-excitation. This activity excites *LI* neurons whose firing begins during the latter part of inspiration and peaks near the *I-E* transition. *LI* neurons are thought to be responsible for the reversible *I* off-switch, where *I* can be transiently inhibited. Decreasing *eI* activity releases the *pI* neurons from inhibition, which then exhibit a rapid onset of activity and terminate inspiration (irreversible *I* off-switch); the *pI* firing frequency decreases during the remainder of the *pI* phase. The decrease in *pI* activity allows *E* neurons to progressively increase their firing frequency. When expiration ends, *E* neurons are inhibited, and the next cycle begins. The reciprocal inhibition between the *eI* and *pI* populations is considered the primary source of the rhythm.

A Pacemaker Theory

Feldman and co-workers (1990) have outlined a theory in which the central oscillator in the RRG consists of pacemaker

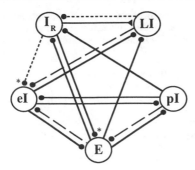

Figure 1. Schematic representation of the connectivity of the Botros-Bruce and the Ogilvie-Gottschalk network models of the RRG. The dotted line denotes connections found only in the Botros-Bruce model, the long-dashed line denotes those found only in the Ogilvie-Gottschalk model, and the solid line denotes connections found in both models. For comparison, the Richter model incorporates all connections shown except for the two marked by an asterisk (*). Inhibitory and excitatory connections are represented by circles and triangles, respectively.

neurons. In preparations of isolated neonatal brainstem and spinal cord (Feldman, Smith, and Liu, 1991), large reductions in inhibitory synaptic transmission did not affect respiratory pattern timing. Because the authors considered inhibition necessary for network-driven oscillations, they argued that this observation indicated the existence of pacemaker neurons. In adult animals, however, blocking inhibition abolishes the respiratory rhythm. This contradiction has led to the suggestion that the importance of pacemaker properties decreases during development, and synaptic interactions dominate in the adult (Richter et al., 1992).

Network Models of the RRG

Several quantitative models of the RRG have been proposed. These models are almost exclusively based on network interactions, predominantly inhibitory, between distinct neural populations. In addition, each population usually receives constant excitatory drive. Generally, the activity of each population, usually representing some average over the population, is described by an equation of the form:

$$\tau_i \frac{dx_i}{dt} + x_i = S_i + \sum_{j=1}^{n} w_{ij} G(x_j) \qquad (1)$$

where x_i is the activity of the ith population of n total populations, τ_i is a time constant, S_i is the tonic drive, w_{ij} are the connection strengths from the jth population to the ith population, and G is an activation function, typically sigmoidal. In some cases, the neural populations have been given the property of frequency adaptation. This type of adaptation (different from learning in adaptive neural networks) can be viewed as a negative feedback within the population and results in an intrinsic decay of activity. An additional equation describes this property so that each population requires two equations:

$$\tau_i \frac{dx_i}{dt} + x_i = S_i + \sum_{j=1}^{n} w_{ij} G(x_j) - A_i y_i$$
$$\alpha_i \frac{dy_i}{dt} + y_i = F(x_i) \qquad (2)$$

where y_i represents adaptation of the ith population, A_i and α_i are time constants of adaptation, and F is similar to G.

Two recently proposed models of the RRG are based on the Richter three-phase theory and are formulated with five primary neural populations: eI, I_R, LI, pI, and E. The Ogilvie-Gottschalk model (Ogilvie et al., 1992; Gottschalk et al., 1994) includes adaptation (Equation 2), whereas the Botros-Bruce model (Botros and Bruce, 1990) does not (Equation 1). The connectivity differs slightly in the respective networks (Figure 1). In addition, a dynamic variable describing tonic drive is explicitly included in the Ogilvie-Gottschalk model.

Model Evaluation

Different criteria are used to evaluate current network models of the RRG. The models usually produce an oscillatory output similar to that observed experimentally because of the liberty one has in choosing parameter values. Because many different networks may produce similar outputs, it is important to understand the general network properties required for a given output. Presently, this level of understanding has only been achieved in simple two-element networks, e.g., reciprocal inhibition. Thus, the periodic trajectory in phase space (the limit cycle) can be similar between model and experimental systems. However, the nature of the phase spaces in each system must be compared in other ways.

One approach is to determine how changing system parameters influences network output. For example, Botros and Bruce (1990) investigated the effects of increasing tonic input to the neural populations in their model. They found that as long as the LI population received only small increases relative to the others, the duration of the inspiratory phase changed very little. In contrast, the peak activity level of each population increased with increases in tonic input. These observations are similar to those observed experimentally when CO_2 levels are increased. Thus, the model predicts that the tonic input attributable to CO_2 might influence the LI population less than other populations. Similarly, Ogilvie et al. (1992) showed that certain changes in the tonic input to pI and E populations in their model could result in multiple bursts of pI activity for single I_R bursts. Again such patterns have been observed experimentally in some conditions. Another possibility is to describe the different activity patterns as the level of synaptic inhibition is globally lowered, thus mimicking the Feldman experiments. Because a quantitative comparison of the inhibition level between model and experiment is not yet possible, looking carefully at the qualitative transitions in pattern may disclose similarities in the two. Analyses of models in these different ways can suggest what parameters may be important for rhythm generation in the experimental system, as well as develop a context for previously unexplained and future experimental observations. In addition, the mechanisms involved in the abnormal rhythms observed in such investigations may help interpret different respiratory pathologies (Cherniack and Longobardo, 1986) like Cheyne-Stokes breathing (where the rhythm waxes and wanes) and Biot's breathing (where long periods of apnea separate similar breaths).

Another approach is to investigate the response of the model and experimental system to transient perturbations. One commonly used protocol for delivering perturbations is *phase resetting*. Phase resetting experiments consist of perturbing an intrinsically oscillating system at different times during its cycle, with the hope of better understanding the system phase space and the mechanisms producing the oscillation (Winfree, 1980; Glass and Mackey, 1988). In further discussion, the cophase θ_i refers to the effect of a given perturbation on the oscillation, and is defined as the latency from the end of perturbation to the onset of the ith following cycle ($i = 1, 2, \ldots$) normalized to the control cycle duration. Cophase plots are constructed by plotting θ_i versus the phase ϕ of perturbation.

Winfree (1980) made specific predictions concerning phase resetting of nonlinear oscillators. Given that two topologically distinct types of resetting occur, namely type 1 or weak (cophase plot with average slope of −1) and type 0 or strong (cophase plot with average slope of 0), Winfree predicted that a *singularity* exists in the response to perturbation. Theoretically, a singularity means that a critical stimulus (of intermediate strength at a specific phase) can stop the oscillation. Experimentally, resetting with the critical stimulus would result in a random latency in the return of the oscillation, i.e., random cophase. This has important implications for biological oscillations. For example, the possibility that the respiratory rhythm could be stopped with a single perturbation represents a potentially life-threatening vulnerability.

Classification of phase resetting in this way provides a context for comparing models with experiments. The respiratory rhythm has been perturbed experimentally in a variety of ways, including lung inflation, brainstem stimulation, and stimulation of the vagus, carotid sinus, and superior laryngeal nerves. Although the general effects of such perturbations have been documented, only a few studies have systematically investigated phase resetting of the respiratory rhythm as outlined by Winfree (1980). One group has found evidence of a singularity in the RRG with superior laryngeal nerve (SLN) stimulation (Paydarfar, Eldridge, and Kiley, 1986; for an alternative interpretation see Lewis et al., 1990).

We have investigated the phase resetting of a simplified three-phase network model configured to mimic the activities of *I*, *pI*, and *E* populations (Lewis et al., 1992). Cophase plots for the model showed both type 1 and type 0 resetting and compared favorably with those obtained in previous experiments involving SLN stimulation of the respiratory rhythm (Lewis et al., 1990). Gottschalk et al. (1994) showed cophase plots for their model that were comparable to the experimental data of Paydarfar. Models like these, showing limit cycle oscillations, necessarily have a singularity and thus lend credibility to Paydarfar's prediction. However, because many different models can show similar cophase plots (Figure 2), phase resetting cannot be considered a sensitive method of evaluating RRG models. An additional consideration is the nature of the singularity. Gottschalk et al. showed that the singularity in their model was an unstable point, and thus functionally irrelevant because it would be impossible to stop the oscillation with a single stimulus. However, a perturbation close to an unstable steady state can result in a long recovery time to the control oscillation even though the oscillation is not stopped. Experimentally, this may be considered functionally similar to a stable nonoscillatory state.

In addition to phase resetting, other stimulation protocols can be used to investigate oscillating systems and evaluate different models. Periodic stimulation, in which stimuli are delivered at a fixed rate, can be very useful in this context (Glass and Mackey, 1988). We have investigated another protocol called *fixed-delay stimulation* (Lewis et al., 1992). This protocol consists of delivering stimuli repeatedly at a constant delay from the onset of an oscillator's cycle. The response to fixed-delay stimulation can be complex in cases where a single perturbation produces effects on the oscillator that last longer than the cycle in which it is given. In the case of the respiratory rhythm, SLN stimulation is such a perturbation. We previously used fixed-delay stimulation to study the RRG, with SLN input, and then compared the results with the response of the simple three-phase model mentioned earlier (Lewis et al., 1992). This comparison showed that the aftereffects of stimulation in the model were shorter in duration than in the experimental system, a difference not seen with phase resetting. Periodic stimulation is expected to show similar differences. Because the Botros-Bruce and Ogilvie-Gottschalk models are formulated in a more realistic context, an investigation of these models using such protocols may suggest a locus for the long-duration aftereffects seen in the experiments.

Discussion

The previous sections have outlined selected topics in the study of respiratory rhythm generation. The fundamental question remains, given a large collection of brainstem neurons with different firing characteristics, how is the rhythm produced and subsequently controlled by the various feedback mechanisms? This question is also relevant to many neural pattern generators (see LOCOMOTION, INVERTEBRATE; CRUSTACEAN STOMATOGASTRIC SYSTEM; LOCUST FLIGHT; SPINAL CORD OF LAMPREY).

The mechanisms of respiratory rhythm generation have been investigated through anatomical and phase resetting studies. Mathematical models have played a limited role. The complex nature of the respiratory system demands multiple approaches of study, including different levels of modeling. The network models discussed in this article are based on interactions between neurons having similar connectivity and firing patterns within a population. Because phase resetting may not be able to distinguish different models in this class, other evaluation techniques are required. Nonetheless, to the extent that these models explain experimental results, we can consider the possibility that the overall behavior does not rely on the details of individual neurons, but on the activities of distinct neural populations. Other levels of modeling also may be instructive.

Figure 2. A comparison of cophase plots for a stimulus of intermediate magnitude. The *i*th cophase θ_i is plotted twice versus the phase of stimulation ϕ (Winfree, 1980). Experimental data are from Lewis et al. (1990). The simple three-phase model is a cyclic inhibitory network of three elements: *I*, *pI*, *E* (Lewis et al., 1992). Parameter values for the Botros-Bruce model are as in table 1 from Botros and Bruce (1990). SLN nerve stimulation was mimicked by a transient increase in *pI* activity for the three-phase model (point stimulus) and in the tonic drive to the *LI* population for the Botros-Bruce model (duration was 2% of the control cycle length to approximate the experimental stimulus). Stimulus magnitudes were 0.06 and 14.0 for each model, respectively (note that these are arbitrary units corresponding to the original models).

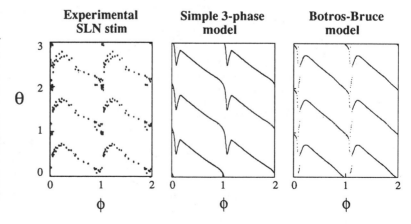

Experiments are showing the characteristics of individual respiratory neurons, providing the possibility that networks of realistic model neurons may be constructed. Feldman and co-workers (1990) have argued that such modeling is required to understand the mechanisms of rhythm generation. Conversely, Euler (1986) has argued that to fully understand respiratory rhythm generation, a detailed description of the various feedback control mechanisms must be obtained. This may be true of many neural pattern generators. The respiratory system provides a useful model for investigating this issue because pathologies such as Cheyne-Stokes and Biot's breathing are naturally occurring experiments involving respiratory control.

Until we have a sufficiently detailed experimental description of the RRG and its control mechanisms, networks consisting of many realistic model neurons may be underconstrained and difficult to evaluate. As illustrated in Figure 2, even simple models can replicate some experimental observations (see also Glass and Mackey, 1988). Thus, one approach is to study the simplest model that is consistent with a particular observation, in the hope of distilling the essential features for a good model. Building on such simple models will ensure a directed search for more accurate and realistic models.

Acknowledgment.   The author thanks the National Institute of Mental Health for a predoctoral fellowship (MH–10677).

Road Map: Motor Pattern Generators and Neuroethology
Related Reading: Chaos in Neural Systems; Half-Center Oscillators Underlying Rhythmic Movements; Motor Pattern Generation

References

Botros, S. M., and Bruce, E. N., 1990, Neural network implementation of a three-phase model of respiratory rhythm generation. *Biol. Cybern.*, 63:143–153.

Cherniack, N. S., and Longobardo, G. S., 1986, Abnormalities in respiratory rhythm, in *Handbook of Physiology: The Respiratory System II*, Bethesda, MD: American Physiological Society, pp. 729–749. ◆

Cohen, M. I., 1981, Central determinants of respiratory rhythm, *Annu. Rev. Physiol.*, 43:91–104. ◆

Euler, C. von, 1986, Brain stem mechanisms for generation and control of breathing pattern, in *Handbook of Physiology: The Respiratory System II*, Bethesda, MD: American Physiological Society, pp. 1–67. ◆

Ezure, K., 1990, Synaptic connections between medullary respiratory neurons and considerations on the genesis of respiratory rhythm, *Prog. Neurobiol.*, 35:429–450. ◆

Feldman, J. L., Smith, J. C., Ellenberger, H. H., Connelly, C. A., Liu, G., Greer, J. J., Lindsay, A. D., and Otto, M. R., 1990, Neurogenesis of respiratory rhythm and pattern: Emerging concepts, *Am. J. Physiol.*, 28:R879–R886. ◆

Feldman, J. L., Smith, J. C., and Liu, G., 1991, Respiratory pattern generation in mammals: *In vitro* en bloc analyses, *Curr. Opin. Neurobiol.*, 1:590–594. ◆

Glass, L., and Mackey, M. C., 1988, *From Clocks to Chaos: The Rhythms of Life*, Princeton, NJ: Princeton University Press. ◆

Gottschalk, A., Ogilvie, M. D., Richter, D. W., and Pack, A. I., 1994, Computational aspects of the respiratory pattern generator. *Neural Comp.*, 6:56–68. ◆

Lewis, J. E., Bachoo, M., Polosa, C., and Glass, L., 1990, The effects of superior laryngeal nerve stimulation on the respiratory rhythm: Phase resetting and aftereffects, *Brain Res.*, 517:44–50.

Lewis, J. E., Glass, L., Bachoo, M., and Polosa, C., 1992, Phase resetting and fixed-delay stimulation of a simple model of respiratory rhythm generation, *J. Theor. Biol.*, 159:491–506.

Ogilvie, M. D., Gottschalk, A., Anders, K., Richter, D. W., and Pack, A. I., 1992, A network model of respiratory rhythmogenesis, *Am. J. Physiol.*, 263:R962–R975.

Paydarfar, D., Eldridge, F. L., and Kiley, J. P., 1986, Resetting of mammalian respiratory rhythm: Existence of a phase-singularity, *Am. J. Physiol.*, 250:R721–R727.

Richter, D. W., Ballanyi, K., and Schwarzacher, S., 1992, Mechanisms of respiratory rhythm generation, *Curr. Opin. Neurobiol.*, 2:788–793. ◆

Winfree, A. T., 1980, *The Geometry of Biological Time*, New York: Springer-Verlag. ◆

Retina

Robert G. Smith

Introduction

At the most basic level, the retina transduces light intensity signals over space and time and transmits them to the brain. However, the signal transmitted by the retina does not code intensity directly. Instead, a variety of neuron types transform visual signals in a multitude of ways to code properties of the visual world, such as contrast and motion. This article describes a conceptual theory, based on the actual problems faced by neurons, to explain why the retina codes visual signals and how the structure of the retina is related to its coding function.

The vertebrate retina reliably responds to light contrast as low as 1% (Shapley and Enroth-Cugell, 1984). However, as the delicate visual signal is amplified as it passes through the retina from photoreceptors to the ganglion cell, the biological limitations of neural processing add distortion and noise with every neuron. The ease with which we see fine gradations in the presence of such biological limitations suggests the hypothesis that a major function of intricate retinal circuits is to maintain the signal's quality in spite of the limitations. This hypothesis would predict that much of the retina's signal coding and structural detail is derived from the need (intrinsic to the retina) to optimally amplify the signal and eliminate noise.

Structure

Layers and Cell Classes

The retina is a thin (100–200 μm) tissue at the rear surface of the eye consisting of three layers of neurons and glial cells (Figure 1) (see Dowling, 1987; Rodieck, 1988; Sterling, 1990). Neurons in the *outer nuclear layer* (ONL) are exclusively photoreceptors. The *inner nuclear layer* (INL) (the middle layer) contains the cell bodies of *horizontal cells*, *bipolar cells*, and *amacrine cells*. Between these two layers lies the *outer plexiform layer* (OPL), in which bipolar and horizontal cells extend dendritic processes laterally to receive synaptic contacts from photoreceptors. The innermost cell layer, the *ganglion cell layer* (GCL), contains cell bodies of ganglion cells and amacrine cells. Between the INL and GCL lies the *inner plexiform layer*

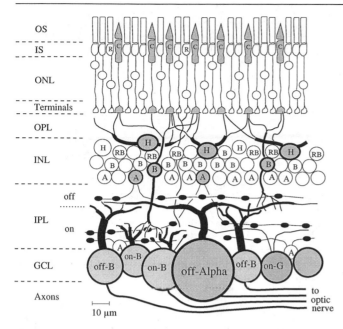

OS
IS
ONL
Terminals
OPL
INL
off
IPL
on
GCL
Axons

10 μm

to optic nerve

Figure 1. Structure of the retina, showing the outer segment (OS), inner segment (IS), outer nuclear layer (ONL), outer plexiform layer (OPL), inner nuclear layer (INL), inner plexiform layer (IPL), ganglion cell layer (GCL), horizontal cells (H), bipolar cells (B), amacrine (A), and rod bipolar (RB) cells.

(IPL), where bipolar, amacrine, and ganglion cells are synaptically connected. Ganglion cells send their output to the brain through axons that lie on the inner surface of the retina.

Cell Types

Each class of neuron described above comprises several cell types, and overall, the retina comprises several dozen types (Sterling, 1990). A cell type is defined by a distinctive morphology, distribution, and synaptic connection pattern (Rodieck, 1988) or a distinctive physiology or immunocytochemical staining pattern. Although the retina of one species may contain cell types that are not present in another, the same five retinal cell classes exist in all vertebrate species (Kuffler, Nicholls, and Martin, 1984; Dowling, 1987; Sterling, 1990). Therefore, all vertebrates likely share similar neural circuit organization.

Receptive Fields and Connectivity

To understand the physiological function of a retinal neuron, investigators often measure its *receptive field* (the region in space and time over which it responds to light with a change in voltage). The receptive fields of retinal neurons consist of a circular region in visual space to which the neuron is most sensitive, called the *center*, and a larger, weaker antagonistic region concentric with the center, called the *surround* (Kuffler et al., 1984). The receptive field of a neuron is determined by the receptive fields of neurons presynaptic to it and the signal-filtering properties of its input synapses as well as the signal processing performed inside the neuron. For example, a ganglion cell's receptive field must reflect not only its own morphology and membrane channels but also the receptive field properties of the bipolar cells that contact it. These, in turn, must originate to some extent in the receptive field properties

of all of the photoreceptors, horizontal cells, and amacrine cells presynaptic to the bipolar cells.

While receptive field analysis is a powerful method for studying the function of a neural circuit (see Kuffler et al., 1984), the origin of a receptive field's components is difficult to grasp in a circuit that includes several layers of neurons, even when studied by modern engineering analysis methods (Shapley and Enroth-Cugell, 1984). The problem seems to be the difficulty of relating components of a physiologically based model to the details of cell morphology, synaptic connectivity, and membrane channels (e.g., see DIRECTIONAL SELECTIVITY IN THE RETINA). However, by computationally simulating these biophysical details, it is possible to test specific hypotheses about neural circuit connectivity with only partial knowledge of the circuit's detail (Teeters and Arbib, 1991; Freed, Smith, and Sterling, 1992). The analysis of retinal circuitry has also been extended to an information-theoretic approach (Atick and Redlich, 1992; Laughlin, Howard, and Blakeslee, 1987).

Functional Modules

Since neurons of each cell type are spaced in a regular array across the retina (see Figure 1), one might hope to decompose the arrays of different types into a repeating structure that would correspond to a *functional module* (i.e., a small neural circuit duplicated over the retina that contains several neuron types and performs a specific signal-processing function, such as generating a ganglion cell's receptive field). The problem is that each cell type is distributed with a different density, so individual neurons in one array cannot easily be grouped with those of other arrays on the basis of their proximity (Rodieck, 1988; Sterling, 1990). The key to identifying functional modules is, of course, their synaptic connections. By tracing each neuron's presynaptic circuit in a series of electron micrographs, an essentially complete circuit of the cat beta ganglion cell was reconstructed (Sterling, 1990).

Functional Circuits

Photoreceptors and Adaptation

The outer segment of a vertebrate photoreceptor transduces light through a multistep biochemical cascade (Liebman, Parker, and Dratz, 1987) into an electrical signal that is conducted through the photoreceptor's axon to its terminal in the OPL. In response to a flash of light, the photoreceptor closes channels in its outer segment (i.e., it hyperpolarizes). Over a limited range of stimulus intensity, the change in the light-modulated signal is proportional to the change in the light stimulus. The advantage of this coding function is that a photoreceptor responds well to small differences in intensity (i.e., low-contrast signals) that are common in the visual world. The disadvantage is that at intensities outside this limited range, the photoreceptor responds poorly. At low intensities, the photoreceptor's transduction gain (i.e., proportion of change in its output signal to a change in input) is insufficient, and at high intensities, the photoreceptor's response saturates. To solve such saturation problems, a photoreceptor continuously adjusts the intensity range to which it responds. This process is called *adaptation*. In some species, adaptation in a photoreceptor can modulate transduction gain by up to four log units.

There are two classes of photoreceptors, rods and cones (see Rodieck, 1988). Rods are sensitive to single photons and are bleached by daylight (Rodieck, 1988). Cones are approximately two log units less sensitive and can regenerate their pigment in daylight (photopic intensity range). At twilight (the mesopic intensity range), cones do not respond well, so rods are coupled through gap junctions to their neighboring cones, causing the rod signal in twilight to pass directly into the cones, where it is carried by the lower-gain cone pathway (Daw, Jensen, and Brunken, 1990). For the low-intensity range encountered at night (scotopic intensity range), a special *rod bipolar* pathway (see Figure 1) carries quantal single-photon signals, removes dark noise, and adapts over an extra three log units of intensity (Sterling, 1990).

Outer Plexiform Layer

The axon terminal of a cone transmits its signal to bipolar cells with a chemical synapse which increases signal gain at the cost of adding extra noise and reducing the intensity range over which the bipolar cell can respond (Laughlin et al., 1987). Here we consider how the function for the OPL circuit might be related to the synapse's limitations.

To reduce the tendency of the cone signal to saturate its synapse, the OPL filters the signal both spatially and temporally (Laughlin et al., 1987). The filter consists of two components: a spatial low-pass filter constructed from lateral electrical connections and a spatiotemporal high-pass filter constructed from horizonal cells. The low-pass filter is constructed by electrically coupling a cone directly to its neighbors with gap junctions. This coupling tends to remove uncorrelated noise from the cone's response. The coupling also causes *neural blur*, which is useful for providing an anti-aliasing filter for the next stage of processing in the IPL.

The high-pass filter is constructed by subtracting a local average from the cone by negative feedback. Horizontal cells sum the inputs from many cone terminals and provide negative feedback to each through a GABAergic synapse (Sterling, 1990). The synaptic structure that performs this function, called a *triad*, is complex because it is both feedforward and feedback and thus is termed a *reciprocal* feedback connection (Dowling, 1987). This type of coding has been termed *predictive* (Srinivasan, Laughlin, and Dubs, 1982) because the ideal signal to subtract from each cone would be a local mean signal produced by averaging signals from the immediately surrounding cones over a short time interval.

Synaptic Function and Noise

The glutamatergic synapse that transmits a cone's signal to bipolar and horizontal cells adds noise that originates in the random fluctuation of the synaptic vesicle release rate (Kuffler et al., 1984; Sterling, 1990). To reduce the amount of noise relative to the signal, the vesicle release rate must be high. Therefore, the chemical synapse of the photoreceptor contains a special structure, called a *ribbon*, that functions as a docking site and reservoir for vesicles (Kuffler et al., 1984; Rodieck, 1988; Sterling, 1990), allowing a high vesicle release rate.

IPL: Bipolar and Amacrine Circuits

The mammalian retina contains approximately 10 types of bipolar and more than 20 types of amacrine cells (Sterling, 1990; Kolb, Nelson, and Mariani, 1981). The dendritic trees of bipolar cells arborize in the OPL to receive contacts from multiple photoreceptors, and their axons terminate in the IPL. In contrast, amacrine cells make no connections in the OPL and extend their dendrites in the IPL to contact bipolar cells, other amacrine cells, and ganglion cells.

Bipolar cells respond to light as photoreceptors do, with a voltage proportional to the intensity change, but their response range is narrower and they adapt over a wider range of stimuli. Bipolar cells contact ganglion cells with glutamatergic synapses, so they employ the same synaptic ribbon structure found in the photoreceptor to allow high release rates and reduce noise. A bipolar cell may contact several ganglion cell types, each with different numbers of synapses (from two to several dozen), suggesting a specific coding of the bipolar signal (Sterling, 1990; Teeters and Arbib, 1991).

The need for several bipolar cell types may be related to a need for different coding requirements by different types of ganglion cells. For example, some bipolar cells have a transient response of 20–50 ms at the beginning of a light flash, while others respond more slowly. The circuitry unique to a bipolar cell type therefore might define a special coding property, such as chromatic, velocity, or direction sensitivity. This specialization increases the signal-to-noise ratio and reduces distortion, which is an advantage for a visual signal that is destined to pass through another synapse.

Function of Amacrine Cells

Amacrine cells are diverse both morphologically (Kolb et al., 1981) and neurochemically (Masland, 1988; Rodieck, 1988). Many amacrine cells have a large (1–2 mm), but sparse dendritic field with fine dendritic processes (0.2 μm diameter) that resemble axons (Kolb et al., 1981; Dowling, 1987). In contrast to the generally passive membrane properties of bipolar cells, these amacrine cells fire action potentials. This firing allows them to transmit signals laterally over the extent of their large dendritic fields (Masland, 1988).

Amacrine cells are generally either GABAergic (Kuffler et al., 1984; Rodieck, 1988) or glycinergic, which implies that they perform subtractive or shunting control functions. Some (e.g., the cholinergic starburst amacrine) are involved in temporal processing and respond transiently to light (Masland, 1988). Amacrine circuitry is thought to be responsible for directional selectivity in ganglion cells (see DIRECTIONAL SELECTIVITY IN THE RETINA), excitatory transient and peripheral effects, and several types of gain control (Shapley and Enroth-Cugell, 1984; Dowling, 1987).

Amacrine cells receive synaptic contacts from bipolar cells at a structure called a *dyad*, where a bipolar ribbon synapse contacts two postsynaptic cells, either a ganglion cell and an amacrine cell, or two amacrine or ganglion cells (Rodieck, 1988). The similarity between the synaptic dyad in the IPL and the triad in the OPL is striking. Both contain synaptic ribbons, and both include reciprocal feedback from a lateral neuron. The reason may be the problem of noise. The reciprocal feedback from an amacrine varicosity to the bipolar cell that feeds it can spatiotemporally filter the bipolar signal, reducing the signal's range before transmission to ganglion cells occurs (Masland, 1988; Dowling, 1988).

Ganglion Cell Coding

The salient feature of a ganglion cell's response to light is its exquisite sensitivity to low-contrast stimuli over a wide range of light intensity (Kuffler et al., 1984). Ganglion cells are specialized into a diverse set of cell types that code different properties of the visual world (see DIRECTIONAL SELECTIVITY IN THE RETINA; Kolb et al., 1981; Maturana, Lettvin, and McCulloch, 1960; Rodieck, 1988). Some give a tonic response to stationary stimuli (e.g., the X cell of cat retina), and others give a more phasic response to signal the presence of flashing or moving stimuli (e.g., the Y cell of cat retina). Many species (lower vertebrates, but also mammals) possess ganglion cells with more complex receptive fields, for example, they respond only to small moving objects that could be insects (Maturana et al., 1960; Teeters and Arbib, 1991). In some species (e.g., primates, turtles), color-opponent ganglion cells provide excellent color coding (Rodieck, 1988; Dowling, 1987). In most cases, summation of inputs by a ganglion cell is linear, so a nonlinear ganglion cell is dependent on its presynaptic amacrine and bipolar circuitry to produce complex properties (Freed et al., 1992).

To transmit a signal to the brain, the ganglion cell codes its intracellular voltage as a firing rate of action potentials along its axon (Kuffler et al., 1984). This process is limited by noise and dynamic range in a manner similar to synaptic coding. Noise in the spike generator of ganglion cells causes the action potential frequency to vary, which can obscure the ganglion cell's signal at low firing rates. To cope with the problem of noise, the retina contains two subclasses of ganglion cells, called *on* and *off*. These cells respond with opposite polarity to a light stimulus, the on cell increasing its firing rate and the off cell reducing its rate. In many species, the responses of on- and off-ganglion cells are symmetric, which allows the retina to code bright and dark objects without much distortion.

IPL: On- and Off-Sublaminae

To supply on- and off-ganglion cells with appropriate signals, the IPL is organized into on- and off-sublaminae. To split the light signal into two symmetrically opposite responses, two bipolar cell subclasses respond oppositely to the neurotransmitter (glutamate) released by the cones. The bipolar and amacrine cell types are divided roughly equally between on- and off-sublaminae, although some arborize (e.g., an on-off amacrine) in both.

An on-bipolar cell contains in its dendritic membrane metabotropic receptors. When bound by glutamate released by a photoreceptor, these receptors signal a cytoplasmic second-messenger system after a short delay to turn off the synapse's ionic channels (Sterling, 1990; Dowling, 1987). Thus, the on-bipolar depolarizes when the photoreceptor decreases its glutamate release (i.e., in response to light). An off-bipolar contains ionic channels that open when glutamate binds to the channel's receptor site. The on- and off-bipolar cells therefore have the proper responses to directly excite their respective ganglion cells (i.e., the on-bipolar depolarizes to a bright spot, and the off-bipolar depolarizes to a dark spot).

Discussion

Diversity of Coding

There are several reasons for the diversity of retinal circuitry. Although all ganglion cell circuits receive the same information from the OPL, by discarding part of the information, a ganglion cell can specialize in coding one or more properties of the signal (i.e., contrast, motion, bright, dark, or colored light flashes). The exact details of coding are probably related to the ecological niche occupied by the organism. Because of their quantal nature, rod signals are qualitatively different from cone signals over a range of more than three log units, so there is an advantage in having a separate rod pathway. Such specialization in coding increases the signal-to-noise ratio and makes better use of the limited dynamic range of neurons, synapses, and the spike train in the ganglion cell axon (Srinivasan et al., 1982). Specialization in coding also simplifies the task of brain circuitry in visual segmentation (Atick and Redlich, 1992;

Kuffler et al., 1984), so it appears that the brain's need for special coding contributes to the retina's circuit structure.

Local Processing in Retinal Circuits

The receptive fields of many retinal neurons, ganglion cells in particular, share important properties: their center-surround organization, high contrast sensitivity, and wide-ranging adaptation. The high sensitivity of the retina is achieved at the cost of complexity: to the extent that each retinal circuit amplifies the signal, it must adapt to reduce the signal's dynamic range, and this adaptation implies circuit complexity. For example, the net effect of the OPL circuit is to create for the photoreceptor a receptive field with a broad center region and a wide antagonistic surround (Sterling, 1990) that adapts temporally and spatially. By removing information about absolute light intensity, the OPL circuit transmits what is left (i.e., information about contrast). In turn, the IPL circuit removes more information about absolute light intensity, accentuating the center-surround receptive field in bipolar and amacrine cells. In the process, it regulates its gain to prevent saturation at high contrast (Shapley and Enroth-Cugell, 1984). Thus, it appears that circuits along the retinal pathway all contribute to the ganglion cell's receptive field properties for a similar reason: to prevent noise or saturation from degrading the signal.

Function of the Ganglion Cell Receptive Field

The well-known antagonistic center-surround and adaptation properties of the ganglion cell receptive field seem less designed for the brain's ulterior purposes than to preserve the quality of retinal signals. The circuitry of both OPL and IPL increases the lateral extent of center and surround, but the need for high visual acuity mandates that OPL and IPL circuits not extend too far laterally. Thus, the retina is shaped by a compromise between spatial acuity and accuracy of coding in the need to compensate for its limitations.

Testing the Theory

Although knowledge of the biophysical components of retinal circuitry and its receptive fields is progressing rapidly, such knowledge does not guarantee a useful theory. For example, the biophysical properties and visual responses of bipolar and amacrine cells that are presynaptic to a ganglion cell are currently being measured with whole-cell patch recordings. These presynaptic responses presumably correspond to components of the ganglion cell's receptive field. However, such knowledge alone cannot answer the question of function in design (i.e., why the individual components exist). The answer can be derived only from synthetic models that integrate details of the retina's neural circuitry with the noise and dynamic range limitations inherent to neural biology.

Computational modeling promises to help find the answers (Teeters and Arbib, 1991; Freed et al., 1992). Once the basic signal flow and function in a retinal circuit have been established, a series of simulations can determine what biological limitations are most significant for the circuit under different conditions. The effect of synaptic noise on the retina's performance can be tested by simulating noise of various types to the signal pathway and comparing the resulting signal-to-noise ratios as a measure of signal quality.

Acknowledgment. This work was supported by MH48168.

Road Maps: Mammalian Brain Regions; Vision

References

Atick, J. J., and Redlich, A. N., 1992, What does the retina know about natural scenes? *Neural Comp.*, 4:196–210.

Daw, N., Jensen, R. J., and Brunken, W. J., 1990, Rod pathways in mammalian retinae, *Trends Neurosci.*, 13:110–115. ◆

Dowling, J. E., 1987, *The Retina: An Approachable Part of the Brain*, Cambridge, MA: Harvard University Press. ◆

Freed, M. A., Smith, R. G., and Sterling, P., 1992, Computational model of the on-alpha ganglion cell receptive field based on bipolar cell circuitry, *Proc. Natl. Acad. Sci. USA*, 89:236–240.

Kolb, H., Nelson, R., and Mariani, A., 1981, Amacrine cells, bipolar cells, and ganglion cells of the cat retina: A Golgi study, *Vis. Res.*, 21:1081–1114.

Kuffler, S. W., Nicholls, J. G., and Martin, R. A., 1984, *From Neuron to Brain: A Cellular Approach to the Function of the Nervous System*, Sunderland, MA: Sinauer Associates. ◆

Laughlin, S. B., Howard, J., and Blakeslee, B., 1987, Synaptic limitations to contrast coding in the retina of the blowfly *Calliphora*, *Proc. R. Soc. Lond. B Biol. Sci.*, 231:437–467.

Liebman, P. A., Parker, K. R., and Dratz, E. A., 1987, The molecular mechanism of visual excitation and its relation to the structure and composition of the rod outer segment, *Annu. Rev. Physiol.*, 49:765–791. ◆

Masland, R. H., 1988, Amacrine cells, *Trends Neurosci.*, 11:405–410. ◆

Maturana, H. R., Lettvin, J. Y., and McCulloch, W. S., 1960, Anatomy and physiology of vision in the frog (*Rana pipiens*), *J. Gen. Physiol.*, 43:129–175.

Rodieck, R. W., 1988, The primate retina, *Comp. Primate Biol.*, 4:203–278. ◆

Shapley, R. M., and Enroth-Cugell, C., 1984, Visual adaptation and retinal gain controls, *Prog. Retinal Res.*, 3:263–346. ◆

Srinivasan, M. V., Laughlin, S. B., and Dubs, A., 1982, Predictive coding: A fresh view of inhibition in the retina, *Proc. R. Soc. Lond. B Biol. Sci.*, 216:427–459.

Sterling, P., 1990, Retina, in *The Synaptic Organization of the Brain*, 3rd ed. (Gordon M. Shepherd, Ed.), New York: Oxford University Press. ◆

Teeters, J. L., and Arbib, M. A., 1991, A model of anuran retina relating interneurons to ganglion cell responses, *Biol. Cybern.*, 64:197–207.

Robot Control

Carme Torras

Introduction

A robot is a multifunctional and reprogrammable mechanism able to move in a given environment. Three broad classes of robots can be distinguished on the basis of their mobility: *Robot manipulators* have a fixed base, and their mobility comes from their articulated structure, thus operating on a bounded three-dimensional (3D) workspace. *Robot vehicles* move on two-dimensional (2D) surfaces by using wheels or other similar continuous traction elements. *Walking robots* are designed to move through rough terrains by using articulated legs (see WALKING). Of course, mixed possibilities also exist, such as robot manipulators mounted on a wheeled vehicle.

For robots to be of practical use, their mechanical structure has to be augmented with a controller that permits commanding them to perform the desired tasks. Controllers are usually hierarchically structured from the lowest level of servomotors to the highest levels of trajectory generation and task supervision. The activity taking place at all these levels is conceptually the same: an actual motion (of a single joint, the end-effector, or the entire robot) is made to adjust as closely as possible to a commanded motion through the use of sensory feedback.

Robot tasks are usually specified in world coordinates (or, alternatively, in terms of sensor readings), while robot moves are governed by their actuator's variables. Therefore, the availability of precise mappings from physical space or sensor space to joint space or motor space is a crucial issue in robot control. The problem is that these mappings are often highly nonlinear and it is difficult (when not impossible) to derive them analytically. Furthermore, because of environmental changes or robot wear-and-tear, the mappings may vary in time; in that case, one would like the control structure to adapt to these variations. It goes without saying that only a few of these mappings can be characterized uniquely in terms of inputs and outputs, most of them being instead dependent on state variables (or the short-term history of inputs).

Classical control methods usually rely on a reference model, whose discrepancy with the real system may lead to large errors. Moreover, model estimation techniques assume that the system structure is known, and only the parameters need adjustment. Adaptive control methods, i.e. those not based on a model, have until recently achieved a considerable development only in the case of linear systems. However, since neural networks are essentially general procedures for the learning of nonlinear mappings, they constitute a promising approach to adaptive nonlinear control (Miller, Sutton, and Werbos, 1990b; see ADAPTIVE CONTROL: NEURAL NETWORK APPLICATIONS). Moreover, the representation of history-dependent behavior can be accomplished very naturally by using either recurrent networks or feedforward networks with tapped delays in the inputs.

Neural Network Approaches to Control

The output of a system is a function of both its current state and its input. Thus, a controller can be thought of as an inverse model of a system in that, given a desired output and the current state, the controller has to generate the input that will produce that output.

The most straightforward neural control approach, *direct inverse modeling*, uses the system itself to generate input-output pairs and trains the inverse model directly by reversing the roles of inputs and outputs. The applicability of this approach is restricted to systems characterized by one-to-one mappings (otherwise, the inverse is a one-to-many mapping) and its success depends on the quality of the sampling (the inputs have to be selected so that the induced outputs cover adequately the output space).

The *forward modeling* approach proceeds in two stages. In the first stage, a forward model of the system is learned from input-output pairs. The second stage consists of composing the obtained forward model with another network and training the composition of the two to approximate the identity mapping. The weights of the forward model are held fixed in this second stage, while the weights of the controller network undergo adaptation. In the case that the forward mapping is many-to-one, this approach can be biased to find a particular inverse with certain desired properties.

The first stage can be obviated if the Jacobian matrix of the system is known. This is the matrix of partial derivatives of outputs with respect to inputs. In the case of a robot manipulator, the Jacobian permits deriving the linear and angular velocity of the end-effector from the joint velocities (Paul, 1981). By a straightforward application of the chain rule, the Jacobian can be used to derive the input errors as a function of the output errors (Kröse and van der Smagt, 1993), which is precisely the purpose of the forward model in the forward modeling approach.

Finally, the *feedback error learning* approach requires having a conventional feedback controller linked to the system, and the role of the neural controller is to make the feedback error signal tend to zero. One interesting characteristic of this approach is that there is no need of a separate training phase, but instead the system is trained during operation.

Jordan (1993) provides a detailed treatment of the concepts presented in this section. In Chapter 9 of Miller et al. (1990b), Kawato compares the three approaches.

Neural Learning Procedures Used in Robot Control

The above control approaches can be applied under both supervised and unsupervised (or self-supervised) training modes and through the use of correlational, reinforcement, or error-minimization learning procedures (Torras, 1989). This distinction among learning procedures is made on the basis of the type of problem information they use.

Correlational procedures use no problem information; their goal is to carry out feature discovery or clustering. In a robot control setting, these procedures are often used to represent a given state space in a compact and topology-preserving manner (see the later subsections "Self-Organizing Topologic Maps" and "CMAC Network"). The correlational procedures most widely used for robot control are SELF-ORGANIZING FEATURE MAPS and ADAPTIVE RESONANCE THEORY (q.v.).

Error-minimization procedures require complete target information—in the form of input/output pairs—and their goal is to build a mapping from inputs to outputs that generalizes adequately. Two such procedures, namely the LMS rule and backpropagation, have been applied to robot control as described below.

Reinforcement-based procedures lie between both extremes. They make use only of a reward/penalty signal, whose goal is to build a mapping that maximizes reward (see REINFORCEMENT LEARNING). These procedures have been applied to sensorimotor integration as described later in this article.

Mappings Underlying Robot Control

As explained in the Introduction, robot control critically depends on the availability of accurate mappings from physical space or sensor space to joint space or motor space.

One such mapping relates the world coordinates of a workspace to the joint coordinates of a robot arm. This is called *inverse kinematics*, because the natural (direct) map relates the values of the joint coordinates defining an arm configuration to the position and orientation of its end-effector (hand, gripper,...) in the workspace.

Another mapping relates the desired trajectory of the end-effector to the forces and torques that need to be exerted at the different joints to realize such a trajectory. This is called *inverse*

dynamics, again due to the fact that the natural (direct) map is going the other way around.

A generic mapping encompassing many particular instances is the so-called *sensorimotor mapping*, which relates sensory patterns to appropriate motor commands.

In the following sections, several neural models proposed to carry out these different mappings are reviewed.

Inverse Kinematics

Using neural networks to learn the inverse kinematics of robot arms is of particular interest when a precise model of some joints is lacking or when, due to the operation conditions of the robot (in space, underwater, etc.), it is hardly possible to recalibrate it.

Feedforward Networks

The simplest way to tackle the learning of inverse kinematics is to apply the *direct inverse modeling approach* to a feedforward network using backpropagation. The one-to-many problem mentioned previously is obviated by designing the training set in a way that the resulting function is one-to-one (by using robot configurations with the links always in the same half-spaces).

Jordan and Rumelhart (1992) have applied the *forward modeling approach* to learning the inverse kinematics of a three-link robot in the plane. Although in this case the inverse mapping is two-to-three, the controller network managed to converge toward a particular inverse kinematic function. The authors point out that a minimum-norm constraint or temporal smoothness constraints can be easily incorporated into the learning procedure to bias the choice of the particular inverse function obtained.

The conclusion reached after extensive experimentation with feedforward networks under both approaches is that a coarse mapping can be obtained quickly, but an accurate representation of the true mapping is often not feasible or is extremely difficult. The reason for this seems to be the global effect that every connection weight has obtained on the final approximation (Kröse and van der Smagt, 1993).

Self-Organizing Topological Maps

An obvious way to avoid the aforementioned global effect is using local representations, so that every part of the network is responsible for a small subspace of the total input space.

Ritter, Martinetz, and Schulten (1992) have used a 3D self-organizing feature map together with the LMS rule to learn the inverse kinematics of a robot arm with three degrees of freedom in 3D space. A direct inverse modeling approach under a completely unsupervised training mode is used. The inputs to each neuron are the retinal coordinates of a point in the workspace, and the outputs (after correct learning) are the joint angles and the Jacobian corresponding to that point.

Extensive experimentation has shown that the network self-organizes into a reasonable representation of the workspace in about 30,000 learning cycles. This should be taken as an experimental demonstration of the powerful learning capabilities of this approach, because the conditions in which it is made to operate are the worst possible ones: no a priori knowledge of the robot model, random weight initialization, and random sampling of the workspace during training.

In a practical setting, this approach could be used to adapt a nominal inverse kinematics to the actual working conditions of a robot after some wear-and-tear has occurred.

Inverse Dynamics

The dynamics case differs from the kinematics in that some sort of fixed robot controller is needed to generate the training data.

CMAC Network

Miller et al. (1990a) have combined the table look-up facilities provided by the Cerebellar Model Articulation Controller (CMAC) developed by Albus (see CEREBELLUM AND MOTOR CONTROL) with an error-correction scheme similar to the LMS rule (see PERCEPTRONS, ADALINES, AND BACKPROPAGATION) to accomplish the dynamic control of a robot manipulator with five degrees of freedom.

The idea underlying this combination is similar to that of enlarging self-organizing maps with the LMS rule, as described earlier. Here, CMAC is used to represent the state space in a compact and localized manner, as self-organizing maps were used there to cover the robot workspace.

The task consists of teaching a robot to follow a given trajectory. This is done by supplying successive points along the trajectory to both the neural network and a fixed-gain controller and then adding up their responses to command the robot. After each cycle, the actual command given to the robot, together with its current state, are used as an input-output pair to train the neural network following a direct inverse modeling approach.

As learning progresses, the CMAC network approximates the inverse dynamics mapping, so that the difference between the current and desired states tends to zero, and consequently the neural network takes over control from the fixed-gain controller. The network converges to a low error (between one and two position encoder units) within 10 trials, provided enough weight vectors are used.

Feedback Error Learning

The same trajectory learning task above has been tackled by Kawato et al. (1987) by using a feedback error learning approach. The robot they use has three degrees of freedom and the neural network consists of only three neurons (one per joint) whose inputs are 13 nonlinear functions of the joint velocities and accelerations. Therefore, 39 weights undergo adaptation using an error-correction rule similar to the LMS one.

The teaching scheme is the same described in the preceding subsection, the two approaches differing in the error signal used to modify the weights. Kawato et al. do not generate input-output pairs, but use the output of the feedback controller directly as the error signal. This error measure is less accurate than that used by Miller et al., but has the advantage of being directly available in the control loop.

The authors report that, after training the robot to follow a trajectory lasting six seconds for 300 trials, the average feedback torque decreased from a few hundreds to just a few units, demonstrating that the neural network had taken over control from the fixed-gain controller. Moreover, the mean square error in the joint angles decreased steadily 1.5 orders of magnitude.

Sensorimotor Integration

Visual Servoing

The basic step to perform vision-based robot control is that of moving a camera so that the image captured matches a given reference image. The target is thus no longer a position in space but a desired sensory pattern. Envisaged applications include visual inspection and grasping of parts that cannot be precisely placed (think of underwater or space settings).

The classical way of tackling this task consists of defining a set of image features and then deriving an interaction matrix relating 2D shifts of these features in the image to 3D movements of the camera (Samson, LeBorgne, and Espiau, 1990). Hashimoto et al. (1992) have used backpropagation to learn the interaction matrix. The training procedure consists of moving the camera from the reference position to random positions and then using the displacement in image features together with the motion performed as input-output pairs. The system thus follows a direct inverse modeling approach. In operation, the robot is commanded to execute the inverse of the motion that the network has associated to the given input.

The results obtained through simulation show that, after 30,000 backpropagation iterations, pixel errors range from −1 to 1. These results seem reasonable for this kind of task. However, more research is needed to include a system like this in a control loop.

Fine Manipulation

Insertion of components with small clearance is an example of a task for which it is extremely difficult to devise a detailed force control strategy that performs correctly in all possible situations while being subjected to real-world conditions of uncertainty and noise. Therefore, the possibility of using neural networks to learn the action to apply in response to each force pattern (i.e., the appropriate sensorimotor mapping) looks very attractive.

Gullapalli, Grupen, and Barto (1992) have used an *associative reinforcement learning system* to learn active compliant control for 2D peg-in-hole insertion (see REINFORCEMENT LEARNING IN MOTOR CONTROL). Their system takes the position of the peg as well as the force and moment sensations as inputs, and produces a velocity command as output. The reinforcement signal depends on the discrepancy between the sensed and the desired position of the peg, with a penalty term being activated whenever the sensed forces on the peg exceed a preset maximum.

The training runs start with the peg at a random position and orientation with respect to the hole, and end when either the peg is successfully inserted or 100 time steps have elapsed. Experimentation carried out on a real robot show that, after 150 trials, the robot is always able to complete the insertion. Moreover, the time to insertion decreases continuously from 100 to 20 time steps over 500 training runs.

Mobile Robot Navigation

Along the same line described in the preceding subsection, Millán and Torras (1992) have developed a *reinforcement connectionist learning system* able to find safe paths for a mobile robot in a 2D environment. The input is an attraction force exerted by the goal and repulsion forces exerted by the obstacles. The output represents a robot action—i.e., a step length and orientation. A reinforcement signal assesses how good the robot move is in response to the given sensorial situation, according to the optimization goal pursued, which is a compromise between minimizing path length and maximizing clearance. All signals are coded as real numbers.

After 75,000 learning steps, the system has learned a reasonable sensorimotor map, leading to the generation of safe paths from almost all initial placements in the workspace. The sensorimotor map exhibits a remarkable noise tolerance (it is almost insensitive to 20% white noise), good generalization abilities in front of goal changes and new obstacles, and appropriate handling of dynamic goals and obstacles.

Recently, the navigation system has been considerably enhanced and installed in a wheeled cylindrical platform of the NOMAD 200 family (Millán and Torras, 1994). The codification of inputs has been adapted to the two rings of infrared and ultrasound sensors mounted on the robot, and the learning time has been considerably reduced by using a modular network and initializing it with built-in reflexes.

Discussion

Neural robot controllers are now at the prototyping stage, having demonstrated their ability to elegantly cope with a variety of simplified control tasks. Common simplifications include working in simulation, considering a reduced number of degrees of freedom, assuming perfect measurement of robot position within its workspace, dealing with an idealized robot shape, and lowering precision requirements.

The next stage of development will not only have to avoid such simplifications, but also address issues such as robustness in front of environmental disturbances, learning speed, and the development of a rationale for parameter assignment. To attain the last stage of being widely used in industry and services, neural controllers must also be made easy to integrate with geometric and symbolic systems, which can exploit the a priori knowledge to gain efficiency (see REACTIVE ROBOTIC SYSTEMS; Torras, 1993).

Acknowledgments.   The author acknowledges support from the ESPRIT III Program of the European Community under contracts no. 6715 (project "CONNY: Robot Control Based on Neural Network Systems") and no. 7274 (project "B-LEARN II: Behavioural Learning: Combining Sensing and Action").

Road Map: Control Theory and Robotics
Background: I.3. Dynamics and Adaptation in Neural Networks
Related Reading: Sensorimotor Learning

References

Gullapalli, V., Grupen, R., and Barto, A., 1992, Learning reactive admittance control, in *Proceedings of the IEEE International Conference on Robotics and Automation*, Los Alamitos, CA: IEEE Computer Society Press, vol. 2, pp. 1475–1480.

Hashimoto, H., Takashi, K., Kudou, M., and Harashima, F., 1992, Self-organizing visual servo system based on neural networks, *IEEE Control Sys.*, April:31–36.

Jordan, M. I., 1993, Computational aspects of motor control and motor learning, in *Handbook of Perception and Action: Motor Skills* (H. Heuer and S. Keele, Eds.), New York: Academic Press. ◆

Jordan, M. I., and Rumelhart, D. E., 1992, Forward models: Supervised learning with a distal teacher, *Cognit. Sci.*, 16:307–354.

Kawato, M., Uno, Y., Isobe, M., and Suzuki, R., 1987, A hierarchical model of voluntary movement and its application to robotics, in *Proceedings of the IEEE First International Conference on Neural Networks*, Piscataway, NJ: IEEE Press, vol. 4, pp. 573–582.

Kröse, B. J. A., and van der Smagt, P. P., 1993, *An Introduction to Neural Networks*, 5th ed., Amsterdam: University of Amsterdam, chap. 7. ◆

Millán, J. del R., and Torras, C., 1992, A reinforcement connectionist approach to robot path finding in non-maze-like environments, *Machine Learn.*, 8:363–395.

Millán, J. del R., and Torras, C., 1994, Efficient reinforcement learning of navigation strategies in an autonomous robot, in *Proceedings of the IEEE/RSJ/GI International Conference on Intelligent Robots and Systems (IROS'94)*, Piscataway, NJ: IEEE Press, vol. 1, pp. 15–22.

Miller, W. T., Hewes, R. P., Glanz, F. H., and Kraft, L. G., 1990a, Real-time dynamic control of an industrial manipulator using a neural-network-based learning controller, *IEEE Trans. Robot. Automat.*, 6:1–9.

Miller, W. T., Sutton, R. S., and Werbos, P. J., Eds., 1990b, *Neural Networks for Control*, Cambridge, MA: MIT Press.

Paul, R. P., 1981, *Robot Manipulators: Mathematics, Programming and Control*, Cambridge, MA: MIT Press. ◆

Ritter, H., Martinetz, T., and Schulten, K., 1992, *Neural Computation and Self-Organizing Maps*, New York: Addison-Wesley. ◆

Samson, C., LeBorgne, M., and Espiau, B., 1990, *Robot Control: The Task Function Approach*, Oxford Engineering Science Series 22, Oxford: Oxford Science Publications. ◆

Torras, C., 1989, Relaxation and neural learning: Points of convergence and divergence, *J. Parallel Distrib. Comput.*, 6:217–244.

Torras, C., 1993, From geometric motion planning to neural motor control in robotics, *AI Commun.*, 6:3–17. ◆

Routing Networks in Visual Cortex

Charles H. Anderson, Bruno A. Olshausen, and David Van Essen

Introduction

What are the mechanisms that allow us to attend to a selected region of visual space? How can we develop a motor plan for limb movement at a cognitive level and then selectively direct that plan to a specific limb? These questions deal with the basic issue of controlling the flow of information from one location to another, which is of central importance to the design and programming of all electronic computers. How this is accomplished in neurobiological systems is largely unresolved (Van Essen, Anderson, and Olshausen, 1994).

The visual system of primates, with its highly specialized ability to recognize objects over an extreme range of scales, provides a prime example of why the control of information flow is important in biological systems. Our perception of the overall shape of a hand is similar whether viewed from a distance of 20 cm, where it subtends most of the visual field, or from a distance of 20 meters, where it subtends only a fraction of the fovea. How does the recognition system get access to the visual information from the central part of the fovea at one instant in time and then get access to most of the visual field at another? Lashley (1942) regarded this issue of perceptual invariance to be the most elementary problem of cerebral function and doubted that any progress could be made in understanding the brain at a neuronal level until this problem was solved.

Historically, Pitts and McCulloch (1947) were the first to propose a neurobiological mechanism that addressed the issue of shift and scale invariance. Although their entire proposal cannot be reconciled with our current understanding of the

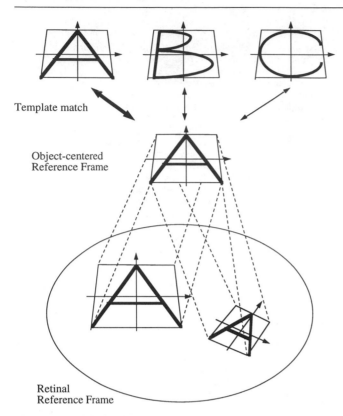

Template match

Object-centered
Reference Frame

Retinal
Reference Frame

Figure 1. Model of reference frame transformation. The process of attending to an object places it into a canonical, or object-based, reference frame. This representation is then suitable for matching to a set of previously stored objects.

brain, it is noteworthy that their model incorporated a switching mechanism for shifting and rescaling sensory input data. Switches and the means to control them are the central elements required of any solution to the invariance question, but surprisingly, they have been largely neglected by neurobiologists who model the brain.

Palmer (1983) suggested that *visual attention* plays a key role in forming invariant object representations. In his view, the process of attending to an object places it into a canonical, or object-based, reference frame, as illustrated in Figure 1. The recognition system in this approach then would consist of two major components: A low-level circuit rescales, repositions, and otherwise reformats the visual information into a standard format, while a high-level circuit actually carries out the labeling or recognition process on the information presented in the standard format. This division of task responsibility has many computational advantages, as demonstrated by machine vision systems that use it (Burt, 1988). There is also support for the idea in the psychology and neurophysiology literature (Olshausen, Anderson, and Van Essen, 1993).

Routing Circuit Model

The goal of our model is to provide a neurobiologically plausible mechanism for shifting and rescaling the representation of an object from its retinal frame of reference into an object-centered one while maintaining local spatial relationships. This requires a mechanism for dynamically accessing cortical areas, which cannot be achieved by modifying the physical arrangement of axons, dendrites, and their synaptic connections. In

our routing circuit, the switching is proposed to take place using multiplicative interactions on the dendrites (Koch and Poggio, 1992), using a mechanism similar to the coincidence detectors observed in the auditory spatial localization systems of the barn owl (Konishi, 1991; see SOUND LOCALIZATION AND BINAURAL PROCESSING) and bat (Suga, 1990; see ECHOLOCATION: CREATING COMPUTATIONAL MAPS). The gating could be carried out on neuronal bodies, but hundreds of times more neurons would be required to construct an equivalent circuit. Our model addresses many of the same issues raised by von der Malsburg and Bienenstock (1986), differing primarily from their pairwise synchronous coupling by the introduction of a third set of neurons to dynamically control the structure of the network. This introduction of a third set of neurons separates the processing of the parameters of scale and position from the statistically independent ones dealing with what an object is.

The basic elements of a routing circuit are illustrated in Figure 2 (Olshausen et al., 1993). This consists of an input layer of 33 nodes, an output layer of five nodes, and two layers in between. Additionally, a set of control units make multiplicative contacts onto the feedforward pathways to change connection strengths. This network has been constructed so that (a) the fan-in (number of inputs) on any node is the same—in this case, five; (b) the spacing between inputs doubles at each successive stage; and (c) the number of nodes within a layer is such that the spread of its total input field just covers the layer below. This connection scheme has the attractive property of keeping the fan-in on any node fixed to a relatively low number while allowing the nodes in the output layer access to any part of the input layer. This property is important in scaling up the model, since real neurons are limited in the number of synapses and hence in the degree of convergence and divergence at each stage.

The biological two-dimensional equivalent to this circuit in the macaque monkey visual system would have an input width of 600 nodes and an output width of 30 nodes, with each node fanning out to connect to 1000 nodes in the next higher level. This would require five levels whose biological substrates for the recognition pathway might correspond to visual areas V1a, V1b, V2, V4, and IT (V1 is subdivided into two levels because of its greater thickness). Receptive field sizes and the extent of neuronal connections through these cortical areas generally follow a pattern of doubling in size at each stage, as suggested by the model. Control may be subserved by neurons in the pulvinar, which make extensive connections with all areas in the visual cortex, and possibly by neurons in layer 6 of the cortex as well (Van Essen and Anderson, 1990).

Each layer of the hierarchy illustrated in Figure 2 is computed as a weighted sum of small patches:

$$I^1[m, n] = \sum_k c_k^l I_k^l[m, n] \tag{1}$$

where the nodes are labeled by the indices m, n, which specify their location within a layer. The weighting factor, c_k^l, is determined by the firing rate of the kth control neuron of level l. Each patch, $I_k^l[m, n]$, is derived from the previous layer by performing a small, fixed shift and rescaling operation, which can be approximated by a linear affine transformation:

$$I^k[m, n] = \sum_{i, j} \Gamma_{ijk}^{l-1} I^{l-1}[m - i, n - j] \tag{2}$$

The weights Γ_{ijk}^{l-1} (see Figure 3) describe the fixed shift and rescale operations on the kth patch. The use of small patches reduces the number of control neurons and imposes the spatial relationship constraint on a local scale. The consistency of the transformation on a global scale, namely that adjacent areas of the map need to shift and rescale roughly in the same manner

Figure 2. A simple, one-dimensional dynamic routing circuit. Connections are shown for the leftmost node in each layer. The connections for the other nodes are the same, but merely shifted. N denotes the number of nodes within each layer, and l denotes the layer number. A set of control units (not explicitly shown) provide the necessary signals for modulating connection strengths so that the image within the window of attention in the input is mapped onto the output nodes.

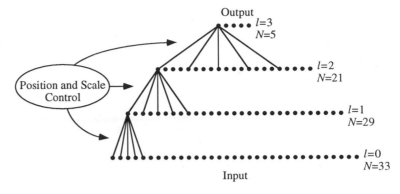

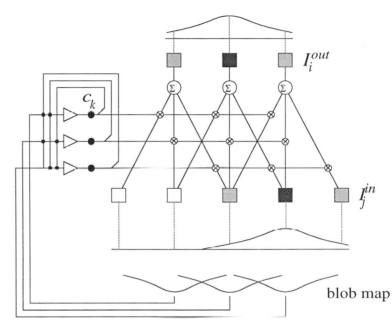

Figure 3. Autonomous control circuitry for tracking a blob in the input. Each control unit corresponds to a different position of the window of attention: left (c_0), center (c_1), or right (c_2). For example, to accomplish the remapping shown, the values on the control units should be $c_2 = 1$ and $c_0 = c_1 = 0$. Each control unit has a Gaussian receptive field in the input layer. The control units compete among each other, via negatively weighted interconnections, so that only the control unit corresponding to the strongest blob in the input prevails.

at any given time, is imposed by mutual excitation and inhibition among the control neurons. The state of the control neurons then provides explicit information about the scale and position of the attentional window. This circuit can be scaled up to provide a model of preattentive vision, which in general has as its input a saliency map, such as gradients in color or texture (Olshausen et al., 1993).

Discussion

The most important neurophysiological prediction of the routing circuit model is that the receptive fields of cortical neurons should be dynamic, shifting and rescaling with attention. Partial evidence for such dynamic effects has been observed in areas V4 and IT (Moran and Desimone, 1985; Connor, Gallant, and Van Essen, 1993). One would also expect to find neurons in the pulvinar and cortex whose activity is related to the control of visual attention.

The type of switching circuits discussed in this section are inherently more complex than performing coordinate transformations on the vector location of a single point, as addressed by Andersen et al. (1993). Groh and Sparks (1992) provide a more detailed discussion of the differences between these two types of models and offer possible alternatives. They also emphasize, as did von der Malsburg and Bienenstock (1986), that it is quite reasonable to suppose neurons can manipulate hun-

dreds of dynamic variables using distributed transistorlike behavior within their complex dendritic trees, which are hundreds of times more powerful than equivalent circuits that allow modulation only at the level of single neurons (see PERSPECTIVE ON NEURON MODEL COMPLEXITY).

More generally, the introduction of mechanisms to store and retrieve the state of the control neurons suggests a neural implementation for pointers, which suggests how symbolic processing might be implemented in the brain (see STRUCTURED CONNECTIONIST MODELS). Finally, these circuits provide insight into possible mechanisms for controlling the flow of many forms of information in the brain, as needed to enable general attentional access to various sensory and cognitive resources, as well as enabling selective execution of motor outputs that are planned at a high level.

Road Map: Vision
Related Reading: Collicular Visuomotor Transformations for Saccades; Dynamic Link Architecture; Gaze Coding in the Posterior Parietal Cortex; Selective Visual Attention

References

Andersen, R. A., Snyder, L. H., Chiang-Shan, L., and Stricanne, B., 1993, Coordinate transformations in the representation of spatial information, *Curr. Opin. Neurobiol.*, 3:171–176. ◆

Burt, P. J., 1988, Attention mechanisms for vision in a dynamic world, *IEEE Trans. Pattern Analysis Machine Intell.*, 11:1217–1222.

Connor, C. E., Gallant, J. L., and Van Essen, D. C., 1993, Effects of focal attention on receptive field profiles in area V4, *Soc. Neurosci. Abst.*, 19:974.

Groh, J. M., and Sparks, D. L., 1992, Two models for transforming auditory signals from head-centered to eye-centered coordinates, *Biol. Cybern.*, 67:291–302.

Koch, C., and Poggio, T., 1992, *Multiplying with Synapses and Neurons: Single Neuron Computation, Neural Nets: Foundations to Applications*, San Diego: Academic Press. ◆

Konishi, M., 1991, Deciphering the brain's codes, *Neural Computat.*, 3:1–18.

Lashley, K., 1942, The problem of cerebral organization in vision, *Biol. Symp.*, 7:301–322.

Moran, J., and Desimone, R., 1985, Selective attention gates visual processing in the extrastiate cortex, *Science*, 229:782–784.

Olshausen, B., Anderson, C., and Van Essen, D., 1993, A neurobiological model of visual attention and invariant pattern recognition based on dynamic routing of information, *J. Neurosci.*, 13:4700–4719.

Palmer, S. E., 1983, The psychology of perceptual organization: A transformational approach, in *Human and Machine Vision* (J. Beck, B. Hope, and A. Rosenfeld, Eds.), New York: Academic Press. ◆

Pitts, W., and McCulloch, W. S., 1947, How we know universals: The perception of auditory and visual forms, *Bull. Math. Biophys.*, 9:127–147.

Suga, N., 1990, Cortical computational maps for auditory imaging, *Neural Netw.*, 3:3–21, and *Cold Spring Harbor Symp.*, 55:585–597.

Van Essen, D. C., and Anderson, C. H., 1990, Information processing strategies and pathways in the primate retina and visual cortex, in *An Introduction to Neural and Electronic Networks* (S. F. Zornetzer, D. L. Davis, and C. Lau, Eds.), New York: Academic Press, pp. 43–72. ◆

Van Essen, D. C., Anderson, C. H., and Olshausen, B., 1994, Dynamic routing strategies in sensory, motor and cognitive processing, in *Large Scale Neuronal Theories of the Brain* (C. Koch and J. Davis, Eds.), Cambridge, MA: MIT Press, pp. 271–299. ◆

von der Malsburg, C., and Bienenstock, E., 1986, Statistical coding and short-term synaptic plasticity: A scheme for knowledge representation in the brain, in *Disordered Systems and Biological Organization* (E. Bienenstock et al., Eds.), NATO ASI Series, vol. F20, Berlin: Springer-Verlag.

Saccades and Listing's Law

Klaus Hepp

Introduction

In humans and monkeys, the saccadic system is the best studied and most fully understood voluntary motor system. In its most constrained action, with the head upright and fixed and the eyes looking at distant targets, its function is to direct the fovea of the *cyclopean eye* to the most interesting part of the visual field. For this purpose, it is tightly coupled to the visual system to replace by successive *foveations* the high demands on neural circuitry, which a high-acuity panoramic vision would otherwise require. In the rhesus monkey, the area devoted to para-foveal vision (1° radius) in the primary visual cortex is approximately 5 cm². If nature had attributed the same space to the full visual field of 90° radius, then the entire cortex of the monkey would already be used for the low-order visual processing of area V1. A spatiotemporal tradeoff using saccades is dictated by hardware requirements as well as by the necessity of reducing the inflow of information for visual attention (see SELECTIVE VISUAL ATTENTION).

In this article we characterize the saccadic system on the functional and kinematic level, describe the neurophysiology of the rapid eye movement (REM) generator in the reticular formation, and discuss the contribution of the superior colliculus (SC) to REM in three dimensions. In the Discussion section, we speculate about the neural implementation of Listing's law. Anatomy (Büttner-Ennever, 1989) and the saccadic system (Wurtz and Goldberg, 1989; Sparks and Mays, 1990; Fukushima, Kaneko, and Fuchs, 1992; Moshovakis and Highstein, 1994; Delgado-Garcia, Godeaux, and Vidal, 1994) have been recently reviewed.

Kinematics of Saccades and Listing's Law

Saccades (French: brusque, irregular movements) of the eye are a class of rapid voluntary movements which we execute during alertness one to three times per second. They are distinguished by a high initial acceleration and deceleration (in the monkey, of as great as 50,000 deg/s²), by a peak velocity increasing with amplitude up to approximately 1000 deg/s, and by a duration which increases linearly with amplitude.

The kinematics of eye movements are non-Euclidean. The eye is approximately a center-fixed sphere with three degrees of freedom of rotation. Every eye position within the oculomotor range is uniquely characterized by the rotation R between an arbitrarily chosen head- and eye-fixed coordinate system. Rotations are intrinsically specified by their axis e (unit vector) and nonnegative angle $\rho \geq 0$ (using the right-hand rule). Unless otherwise stated, we shall discuss only *conjugate* eye movements, where both eyes are rotated together. *Listing's law* states that all eye positions during fixation with the head upright and stationary have their rotation axes in a plane called *Listing's plane*. It is convenient to combine e and ρ into *rotation vectors* $r = \tan(\rho/2)e$. Then the rotation vector $r_1 * r_2$ for the (noncommutative) product $R_1 * R_2$ of two rotations (first R_2, then R_1) is $(r_1 + r_2 + r_1 \times r_2)/(1 - r_1 \cdot r_2)$, and the inverse R^{-1} corresponds to $-r$. It is always possible to choose a head-fixed coordinate system in which the *primary direction* is the x-axis, orthogonal to Listing's plane and along the direction of sight, when the eye is in primary position $\rho = 0$. In monkeys and humans, the primary direction lies close to the midplane. Then the y-axis for vertical eye rotations can be taken along the interocular line, and r_x, r_y, and r_z are the *torsional*, *vertical*, and *horizontal* components of eye position, respectively.

To a good approximation, saccades from r_1 to r_2 are fixed-axis rotations or, equivalently, their trajectories are straight lines between r_1 and r_2. The *difference vector* $d = r_2 - r_1$, the three-dimensional (3D) generalization of the classical saccade vector, lies in Listing's plane, but not the *quotient vector* $q = r_2 * (-r_1)$, unless r_1 and r_2 are collinear. The term q characterizes the rotation Q of the eye from R_1 to R_2, $Q * R_1 = R_2$. The direction of q is the rotation axis, and its length is approximately half of the rotation angle of the eye during the saccade. When the head is upright and stationary, the entire saccade trajectory lies in Listing's plane. The same is true for smooth-pursuit eye movements. In the monkey, Listing's plane has a standard deviation of only 0.5° relative to a torsional oculomotor range of more than $\pm 10°$. Listing's plane in the head is stable over many months, and it is maintained by neural control, since it thickens to approximately 3° standard deviation during light sleep.

There are different theories about the functional importance of Listing's law. They all start from Donders' law, which states that the torsion of the eye is uniquely determined by the direction of sight or that the motor system of the eye, a redundant manipulator without obstacles and external forces, reduces the dimensionality of its control space from three to two, whenever possible.

A simple motor proof of Listing's law uses the fact that fixed-axis rotations are geodesics on the 3D rotation group to show that an oculomotor space constrained by Donders' law and spanned by saccades is a plane. More profound is Helmholtz's *visual* proof from the principle of *easiest orientation*, which is motivated by the desire for seeing a stationary figure at rest, when it is moved around by saccades. For Listing's law, the average torsional error is the absolute minimum if the oculomotor range is nearly circular. Recently, attempts have been made to generalize Listing's law to convergent eye movements, for instance, in terms of a *cyclopean* and a *vergence* (Listing's) plane $\{c_x = v_y = 0\}$ for the rotation vectors of the right and left eye, $r = v * c$, $l = (-v) * c$ (Minken, Gielen, and van Gisbergen, 1994). With $d = r*(-l) = \tan(\alpha/2)n$, $n = d/|d|$, and $v = \tan(\alpha/4)n$, one can relate this parameterization to Helmholtz's principle (Hepp, 1995).

Saccades in Listing's plane are only a small subset of all conjugate REMs. Fast phases of nystagmus evoked by passive head rotation are convenient for exploring the REM generator in three dimensions as well as the stability of Listing's plane. In the alert monkey, fast nystagmus phases in the light are kinematically similar to saccades, except that torsional REM components are small.

The Saccade Generator in the Brainstem

The static force of a muscle is a function of its length and innervation. The final common pathway for all control signals, by which the brain can move the eyes, is through the innervation of the extraocular muscles. Hence, the firing patterns of the extraocular motoneurons (MNs) during fixation provide a view of the statics of the oculomotor *plant* (eye, muscles, and orbital tissue) from within the brain. The on-directions of the horizontal recti (LR, MR) have essentially no vertical components and an angle of approximately 15° relative to the horizontal axis in the ipsitorsional direction (pairing MR with the contralateral LR). The vertical recti (SR, IR) and obliques (SO, IO) act almost diagonally in the vertical-torsional plane (Suzuki et al., 1994). Figure 1 shows the burst-tonic innervation pattern of a right abducens (LR) MN for two saccades with rightward components. Typically, for fixed-axis rotations of the eye, the velocities of the horizontal and vertical components of the saccade trajectory are coupled. For leftward saccades, this MN would be inhibited, and the firing rate after the saccade would be lower.

Robinson found for the firing rate $f(t)$ of an MN in Listing's plane a linear relation $f(t - \tau) = f_0 + ke \cdot r(t) + le \cdot r'(t) + me \cdot r''(t)$, if $f > 0$, where $'$ denotes d/dt, $\tau \approx 5$ ms is a time delay, e is the static on-direction, m is significant only at saccadic velocities, and l/k is of the order of the time constant (approximately 100 ms) of the plant. This relation is a refinement of Henneman's law of MOTONEURON RECRUITMENT (q.v.), and it motivates the *velocity command* hypothesis: all oculomotor

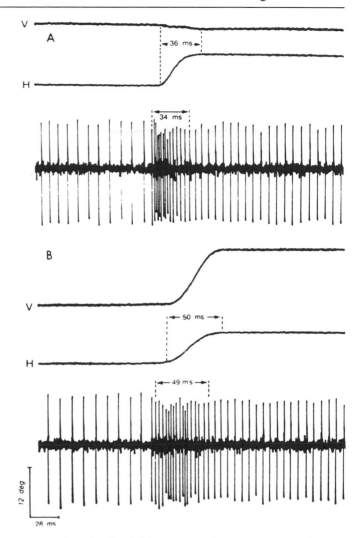

Figure 1. Saccade-related firing pattern of a motoneuron (from King, Lisberger, and Fuchs, 1986).

programs act at the level of the MNs by a direct velocity input, which also changes the activity in common integrators for holding the eyes in all tasks. From the MN on-directions, one expects to find a horizontal and a vertical-torsional integrator. Both should be mirror-symmetric relative to the midplane, should operate in a push-pull manner, and should be orthogonal in Listing's plane.

The horizontal integrator depends on a bilateral network in the left and right nucleus prepositus hypoglossi (NPH), where many neurons have firing patterns similar to those of horizontal MNs. The vertical-torsional integrator depends on the INC on both sides of the midplane. Here, the neurons have ipsitorsional on-direction (as the MN on the same side, which all generate positive torsion on the right side of the brainstem). After inactivation of an integrator, saccades can still be made, but the eyes return to a horizontal or vertical-torsional equilibrium position with approximately the plant time constant. For the saccadic system, neural integrators are necessary to keep the eyes on target as well as for feedback control during the saccade.

There are populations of excitatory and inhibitory short-lead burst neurons (SBNs) in four disjoint areas, which carry the velocity command for all REMs in a half-space. Horizontal

REMs in the ipsilateral direction are implemented in the right and left caudal parapontine reticular formation (cPPRF), and upward and downward REMs are implemented with ipsitorsional components (clockwise on the right side) in the right and left rostral medial longitudinal fasciculus (riMLF). Lesions of one of these structures permanently abolish all REMs into the horizontal or torsional half-space, while leaving the other components intact. Microstimulation moves the eye into the corresponding direction, where it is held by the integrator. The firing patterns of the SBNs in these areas also have a clear geometrical interpretation: the average firing rate in the velocity-related burst, which precedes the firing pattern in the MN by approximately 4 ms (see Figure 1), is approximately cosine tuned around on-directions, which are predominantly ipsilateral in the PPRF and vertical-ipsitorsional in the riMLF.

In a midline strip between the right and left PPRF lie the omnipause neurons (OPNs), which fire regularly with 150–200 Hz, except for a pause during all REMs. These OPNs inhibit the SBNs and are responsible for the high velocity and acceleration of saccades. At the sleep-wake phase transition between alertness and light sleep, the OPNs are silenced in parallel with changes in the electroencephalogram reading, and then there is no longer a clear distinction between rapid and slow eye movements. In neural terms, a saccade is characterized by a pause of the OPNs. These neurons gate all preparatory activities in higher saccade-related areas. Microstimulation of OPNs during a targeting saccade stops the eye in midflight. Afterward, even in total darkness, REM brings the eye to the remembered target, accompanied by the appropriate activity patterns of the SBN.

This work and other experiments on perturbations of saccades show that saccades are generated by feedback circuits. In the Robinson *position feedback* model, saccades are driven by SBNs, which are excited by a desired eye position command and inhibited by internal eye position feedback from the integrator. In the reticular formation, a neural representation of the desired eye position in the head (at a time preceeding the onset of the saccadic burst in the SBN) has never been found. In fact, there are two main streams with velocity input to the REM generator: one from the cortical eye fields (the frontal eye fields, FEF; and the lateral intraparietal area, LIP) and the motor layers of the superior colliculus (SC) for visually and memory-guided saccades; and the other from the vestibular nuclei for fast phases of nystagmus.

These commands polysynaptically activate the horizontal SBNs in the cPPRF. An intermediate area with input from the SC and FEF and direct output to the cPPRF combines the rostral PPRF (rPPRF) and the caudal nucleus reticularis tegmenti pontis (cNRTP). Here one finds predominantly horizontal long-lead burst neurons (LBN). Microstimulation evokes (with latencies between those in the motor SC and those in the cPPRF) horizontal saccades, and microlesions produce ipsilateral hypometric saccades. Another excitatory input population to the horizontal excitatory burst neurons (EBNs) are the burster-driving neurons around the NPH. These LBNs are generators of the fast phases of nystagmus, but they are also active during visually evoked saccades.

Many LBNs in the brainstem carry a vestibular or visual signal, even if no eye movements are made, so they are neither strictly *sensory* nor *motor*. LBNs could be involved in *velocity feedback* circuits of the saccadic system by comparing eye displacement commands with eye velocity generated by the SBN.

Role of the SC for Saccades in Listing's Plane

On each side of the midline, the *upper* layers of the SC have a representation of the contralateral visual field in retinocentric coordinates. Corresponding to this visual map, the *deeper* layers contain a motor map of saccade-related burst neurons, which can bring the fovea to a target on the visual map. The motor SC lies within the feedback loop of the saccade generator. When one interrupts a visually evoked saccade in midflight by stimulation of the OPN area, presaccadic burst neurons in the motor map in the SC (in register with the visual target and projecting into the PPRF) interrupt their firing and resume it during the continuation of the saccade (Keller and Edelman, 1993).

There is a logarithmic transformation between retinal polar coordinates (R, ϕ) and SC surface coordinates (u, v) such that a selected visual target at (R, ϕ) generates in the contralateral motor SC a "hill" of activity around (u, v) that is approximately circular and translation invariant. The question arises how these coordinates are related to the 3D description of REM in terms of rotation vectors.

When the eye is in r_1 and a target relative to the fovea is seen in the direction s (unit vector), which could be foveated with the eye in r_2, then $d = r_2 - r_1 \approx (0, -s_z/2, s_y/2)$, if r_1 and r_2 are in Listing's plane. The *vector* model postulates that the motor map in the SC is two dimensional and that a hill of activity around $d \approx (u, v)$ or a microstimulation at d, when the eye is in r_1, would generate a saccade which *translates* the eye to $r_2 = r_1 + d$. Then r_2 would be in Listing's plane if r_1 were. The *quotient* model assumes that a point on the SC motor map corresponds to a quotient vector q such that microstimulation at q with the eye at r_1 *rotates* the eye to a position $r_2 = q * r_1$. Here the motor map of the SC is the 3D set of all $q = r_2 * (-r_1)$ with r_1 and r_2 in Listing's plane. The experimental data are much better explained by the vector model than by the quotient model (Hepp et al., 1993). Hence, the output neurons of the motor SC do not "know" the rotation axis of the impending saccade, and structures downstream of or parallel to the SC are needed for generating saccades in Listing's plane. This observation confirms the conclusion from adaptation experiments that neurons in the motor SC have activity which reflects target location in retinocentric coordinates, and not the metric of the actual saccade (Goldberg et al., 1993).

It has been shown that the motor map of the SC is at least four dimensional (4D). Many neurons have eye-position-dependent planar gain fields in Listing's plane (van Opstal, in Delgado-Garcia et al., 1994). These gain fields have slopes in all directions. Therefore, the motor SC also encodes, as its input from area LIP (Andersen et al., 1990), a distributed representation of desired target position in head coordinates that are active at saccadic onset and could control the reticular saccade generator by *dual feedback*. Desired eye displacement and end position can be read out simultaneously from this 4D map by summing the activity of all neurons representing either different d vectors or different initial eye positions and generating in parallel a velocity (default mode) and an end-position (backup mode) feedback.

Discussion

Three mechanisms are needed for the implementation of Listing's law: (1) a neural specification of Listing's plane as a set of equilibrium positions of the oculomotor system; (2) controls that ensure that the velocity commands from the premotor circuits maintain the eyes in Listing's plane if possible; and (3) a stabilization mechanism which returns the eye to Listing's plane after it has been displaced, for example, by a head movement.

The static equilibrium positions of both eyes depend on the direction of sight of the cyclopean eye, the convergence angle, and the head position relative to gravity. The constraints of

Listing's law are lost at the transition from alertness to sleep, when the eyes start to drift. The backward transition from sleep to alertness occurs often during a saccade, and then the firing rate of every MN reverts to its rate-position hypersurface. Similarly, if the eye is driven into the torsional direction by a head rotation, then it remains there until it is reset to a new position in Listing's plane by the next REM (Crawford and Vilis, 1991). It is not known how and where the equilibrium innervations of the two eyes are defined and how significant deviations from equilibrium can trigger corrective saccades or fast phases of nystagmus when the head is moving. For the orientation of Listing's plane relative to gravity, the cerebellum seems to play an important role since lesions of the nodulouvular lobe lead to oscillations of Listing's plane in the torsional direction (Angelaki and Hess in Delgado-Garcia et al., 1994).

One understands slightly better how the REM generator transforms a 2D motor command d from the SC, when the eye is initially in the Listing position r_1, into a fixed-axis rotation to $r_2 = r_1 + d$. Mathematically, this transformation is trivial. The SC activity around d has to generate a rotation vector velocity signal $\{r'(t), t_1 \le t \le t_2\}$, which is linearly integrated. There are many SBNs in the cPPRF and riMLF which code rotation vector velocity and acceleration as a population. They often show an excellent linear relation between the number of spikes N in the burst and the component $d \cdot e$ of d in their on-direction e and between the average firing rate and e (see Figure 2 and Hepp et al., in Delgado-Garcia et al., 1994). Since N is related to the firing rate $f(t)$ by $N = \int f(t)\,dt$ and since $r_2 - r_1 = \int r'(t)\,dt$, these SBNs carry a velocity command. If one assumes that their signal is linearly integrated by the NPH and INC to change the firing rate in the MN from r_1 to r_2, then this process could implement a control to move the eye from one Listing position to another.

However, these *neural population vectors* are only internal representations of saccades, and to move the eye, they must generate *torques $m(t)$* by the innervation of the eye muscles. If one assumes that the plant dynamics satisfies the second-order equation $Ir'' + Br' + Kr = m$, with m linearly related to the neuronal trajectory, then the eye trajectory would remain in Listing's plane. However, for a rigid body, r' and r'' should be replaced by eye *angular velocity* $\omega = 2(r' + r \times r')/(1 + r \cdot r)$ and its derivative ω'. Such a model predicts for realistic choices of I, B, K (compatible with the Robinson equation) torsional "blips" of more than $5°$ for large $r_1 \times r_2$ and a return to List-

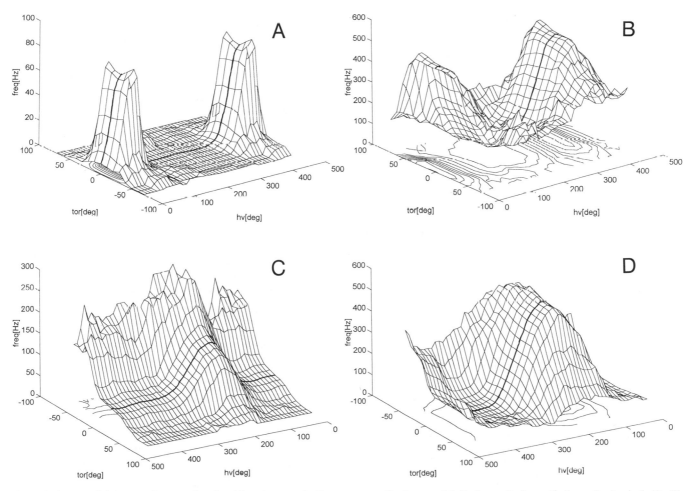

Figure 2. Average firing rate of a saccade-related burst neuron in the right SC (*A*); of an almost purely downward SBN (*B*) and a purely torsional SBN (*C*) from the right and left riMLF; and of a right SR MN (*D*) as a function of saccade direction e during REM in all directions. Here $e = d/|d|$, $d = r_2 - r_1$, and the torsional (tor) and horizontal-vertical (hv) coordinates between $0°$ and $450°$ ($0°$ being down, $90°$ left, and so on) are the azimuth and elevation of e in polar coordinates $e = (\sin(\text{tor}), \cos(\text{tor}) * \cos(\text{hv}), \cos(\text{tor}) * \sin(\text{hv}))$. The thickest line marks the tuning curves in Listing's plane. Note the narrow tuning of the SC neuron, which only fires for d vectors around Listing's plane, and the ipsitorsional on-direction of the SR MN (with soma on the left side) intermediate between those of the two riMLF SBNs.

ing's plane with the time constant of the plant (Schnabolk and Raphan, 1994). Such blips have not been found in monkeys and humans (Tweed, Misslisch, and Fetter, 1994). Hence, either important neural control signals are missing, the plant model is unrealistic, or both.

The recent experience with neural network models shows the power of highly simplified neurons with adaptive couplings and also their counterintuitive *firing patterns*. It is hopeless to trace the complex dynamics of the plant and of the sensory transducers inside mammalian sensorimotor circuits. However, the example of the SC shows that in the highly optimized interaction between vision and saccades, the direct pathways respect the geometry of the outside physical world and allow nontrivial predictive models.

Acknowledgment.   This work was supported by Swiss National Science Foundation grant 31.28008.89.

Road Map: Primate Motor Control
Related Reading: Collicular Visuomotor Transformations for Saccades; Head Movements: Multidimensional Modeling

References

Andersen, R. A., Bracewell, R. M., Barash, S., Gnadt, J. W., and Fogassi, L., 1990, Eye position effects on visual, memory, and saccade-related activity in areas LIP and 7a of macaque, *J. Neurosci.*, 10:1176–1196.

Büttner-Ennever, J. A., 1989, *Neuroanatomy of the Oculomotor System*, Amsterdam: Elsevier.

Crawford, J. D., and Vilis, T., 1991, Axes of eye rotation and Listing's law during rotations of the head, *J. Neurophysiol.*, 65:407–423.

Delgado-Garcia, J. M., Godeaux, E., and Vidal, P. P., 1994, *Information Processing Underlying Gaze Control*, London: Pergamon.

Fukushima, K., Kaneko, C. R. S., and Fuchs, A. F., 1992, The neural substrate of integration in the oculomotor system, *Prog. Neurobiol.*, 39:606–639.

Goldberg, M. E., Musil, S. Y., Fitzgibbon, E. J., Smith, M., and Olson, C. R., 1993, The role of the cerebellum in the control of saccadic eye movements, in *Role of the Cerebellum and Basal Ganglia in Voluntary Movement* (N. Mano, I. Hamada, and M. R. DeLong, Eds.), Amsterdam: Elsevier, pp. 203–211.

Hepp, K., 1995, Mathematical proofs of Listing's law and their implication for binocular vision, *Vision Res.*, in press.

Hepp, K., van Opstal, A. J., Straumann, D., Hess, B. J. M., and Henn, V., 1993, Monkey superior colliculus represents rapid eye movements in a two-dimensional motor map, *J. Neurophysiol.*, 69:965–979.

Keller, E. L., and Edelman, J. A., 1993, Effects of electrical stimulation in the omnipause region on cells in the superior colliculus of the monkey, *Soc. Neurosci. Abst.*, 19:354.3.

King, W. M., Lisberger, S. G., and Fuchs, A. F., 1986, Oblique saccadic eye movements of primates, *J. Neurophysiol.*, 56:769–784.

Minken, A. H. W., Gielen, C. C. A. M., and van Gisbergen, J. A. M., 1994, An alternative 3D interpretation of Hering's equal-innervation law for vergence and version eye movements, *Vision Res.*, 34, in press.

Moshovakis, A. K., and Highstein, S. M., 1994, The anatomy and physiology of primate neurons that control rapid eye movements, *Annu. Rev. Neurosci.*, 17:465–488. ◆

Schnabolk, C., and Raphan, T., 1994, Modeling three-dimensional velocity-to-position transformation in oculomotor control, *J. Neurophysiol.*, 71:623–638.

Sparks, D. L., and Mays, L. E., 1990, Signal transformations required for the generation of saccadic eye movements, *Annu. Rev. Neurosci.*, 13:309–336. ◆

Suzuki, Y., Straumann, D., Hepp, K., and Henn, V., 1994, Three-dimensional representation of position coding in extra-ocular motoneurons in the rhesus monkey, in *Contemporary Ocular Motor and Vestibular Research: A Tribute to David A. Robinson* (A. F. Fuchs, T. Brandt, U. Büttner, and D. S. Zee, Eds.), Stuttgart: Thieme Verlag, in press. ◆

Tweed, D., Misslisch, H., and Fetter, M., 1994, Testing models of the oculomotor velocity to position transformation, *J. Neurophysiol.*, in press.

Wurtz, R. H., and Goldberg, M. E., 1989, *The Neurobiology of Saccadic Eye Movements*, Amsterdam: Elsevier.

Schema Theory

Michael A. Arbib

Introduction

Schema theory provides a language, at a relatively high level of abstraction, for brain theory, cognitive psychology, and distributed artificial intelligence (including robotics). A *schema* is an active entity such as that involved in driving a car or recognizing a tree. In many cases it can be studied apart from its implementation in terms of more detailed processes, such as a neural network. Moreover, the notion of schema as defined here is *recursive*: a schema defined functionally may be refined in terms of "smaller" schemas, and so on until such time as a secure foundation of neural localization or technological implementation is attained. Conversely, when we learn a new task, such as driving a car, we may quickly approximate the skill by marshaling a stock of existing schemas and then tune the resultant assemblage through experience to emerge with a new schema for skilled performance of the task. Note that, when the terminology of schemas is used in a psychological description of human behavior, there is no a priori claim as to whether or not the use of that schema in directing behavior requires conscious knowledge of the schema.

A schema is both a store of knowledge and the description of a process for applying that knowledge. A schema may be instantiated to form multiple *schema instances* as active copies of the process to apply that knowledge. For instance, given a schema for perceiving a chair, we may need several active instances of the schema to subserve our perception of a room containing several chairs. The network of interacting schemas provides processes for going from a particular situation and a particular structure of goals and tasks to a suitable course of action. This action may involve passing of messages, changes of state, instantiation to add new schema instances to the network, or deinstantiation to remove instances; it may also involve self-modification and self-organization.

Schema theory has been widely used for informal modeling of phenomena in both cognitive psychology (e.g., Mandler, 1985) and neuropsychology (e.g., Shallice, 1988), while Arbib and Hesse (1986) have developed epistemological analyses linking the schemas of the individual to the social construction of reality (including an account of consciousness and free will). The present article is consistent with these approaches, but goes further to suggest how schema theory may be viewed as a distributed model of computation compatible with, but not restricted to, neural computation. In this light, schema theory not only provides a rigorous analysis of behavior which requires no prior commitment to hypotheses on the localization of each unit of the analysis, but can also provide a functional analysis which is linkable to a structural analysis as this becomes appropriate. Although, on occasion, a schema may be identified with the function of a single module, the two notions must be kept distinct. *Modules* are *structural* entities, whereas schemas are *functional* entities. In general, a single schema may require several modules (which might, for example, be neural networks) for its implementation, whereas a single module in a brain or distributed computing system may be involved in the implementation of a variety of different schemas.

A schema network need not have a top-level executive since schema instances can combine their effects by distributed processes of competition and cooperation (i.e., interactions which, respectively, decrease and increase the activity levels of these instances), rather than by the operation of an inference engine on a passive store of knowledge. This distributed processing without executive control may lead to apparently emergent behavior. A corollary to this view is that knowledge receives a distributed representation in the brain. A multiplicity of different representations must be linked into an integrated whole, but such linkage may be mediated by distributed interactions.

Schemas in Neuroscience, Cognitive Psychology, and Artificial Intelligence

Schema theory posits an active and selective process of cognition which in some sense constructs reality as much as it embodies it. Head and Holmes (1911) spoke of a "body schema" in the parietal lobe to explain why a person with damage to that brain lobe on one side might lose awareness of the body on the opposite side as being her own. Bartlett (1932) carried the idea of the schema into psychology: when people try to recall a story they have heard, they reconstitute the story in relation to a set of familiar schemas, rather than by rote memorization of arbitrary details. Craik (1943) furthered these ideas with his observation that the brain creates a model of the world, forming expectations on which actions can be based adaptively.

Quite separately, Piaget (e.g., 1971) used schemas in his study of COGNITIVE DEVELOPMENT (q.v.). *Assimilation* is the process of making sense of the situation in terms of the available stock of schemas, and *accommodation* is the development of new schemas to the extent that mismatches arise. This formulation provides a point of contact between schema theory and the use of learning rules to allow a neural network to adapt itself to some specified input-output behavior.

Schema theory provides a knowledge representation protocol which is related to the frames (Minsky, 1975) and scripts of artificial intelligence (AI); a schema is generally smaller than a frame but larger than a neuron. However, a schema is more like a molecule than an atom in that schemas may well be linked to others to provide yet more comprehensive schemas. As the name "frame" suggests, frames tend to "build in" from the overall framework, while schema theory is more generative. Rather than represent a birthday party by a single frame with specific slots to be filled in, schema theory would form such a representation as an assemblage of schemas for salient objects and actions.

In the HEARSAY system for speech understanding (see DISTRIBUTED ARTIFICIAL INTELLIGENCE), various hypotheses as to phonemes, words, and phrases occurring in the speech stream are generated as the interpretation of the input proceeds and are placed on a multilevel data structure called the *blackboard*.

Computing agents called knowledge sources (e.g., the lexicon, processes embodying grammatical knowledge, etc.) act upon hypotheses at one level to try to come up with a hypothesis at another level. Knowledge sources are brought in one at a time by a serial scheduler in HEARSAY. By contrast, schemas describe units in a network of active processes communicating with each other.

The strand of schema theory most related to brain theory and robotics owes much to the work of Warren McCulloch. McCulloch and Pitts showed that any finite automaton could be implemented by a formal neural network (see AUTOMATA AND NEURAL NETWORKS). Pitts and McCulloch (1947) studied neural networks for pattern recognition to show how visual input could control motor output via the distributed activity of a layered neural network without the intervention of executive control. Kilmer, McCulloch, and Blum (1969) showed how activity in a brain region could set the organism's overall mode of behavior through competition and cooperation (again, no executive control) of modules which aggregated the activity of many neurons.

Much work in brain theory and AI contributes to schema theory, even though the scientists involved do not use this term. The subsumption architecture of Brooks (1986) controls robots with layers made up of asynchronous modules akin to schemas. This work shares with schema theory the point that no single, central, logical representation of the world need link perception and action. It also shares with the work of Grey Walter and Valentino Braitenberg (see REACTIVE ROBOTIC SYSTEMS) the study of the "evolution" of simple "creatures" with increasingly sophisticated sensorimotor capacities.

Rana computatrix, a family of models of VISUOMOTOR COORDINATION IN FROGS AND TOADS (q.v.), provides studies which integrate action and perception in distributed systems. This, plus the work of Nikolai Bernstein on "synergies" as units of motor control, led to the analysis of visual perception and motor control in terms of "perceptual schemas" and "motor schemas" (Arbib, 1981), useful both in models of brain function (see COMMAND NEURONS AND COMMAND SYSTEMS; GRASPING MOVEMENTS: VISUOMOTOR TRANSFORMATIONS; and Jeannerod et al., in press) and in robotics (see REACTIVE ROBOTIC SYSTEMS).

There is a strong resemblance between schemas and "agents" with the ability to communicate with other agents, and whose functionality is specified by some behavior (see DISTRIBUTED ARTIFICIAL INTELLIGENCE). Whereas much of schema theory was motivated by analysis of individual cognition or by analogies with the functions of interacting brain regions, much work in distributed artificial intelligence (DAI) was motivated by a social analogy. For example, Minsky (1985) viewed the mind as a "society" whose "members," agents, are analogous to schemas. In future, we will not only need to understand the integration of schemas and/or neural networks within the processors of the distributed network that constitutes a single brain or machine; we will also have to understand how vast problems (such as ecological control) can only be handled by human-machine networks where agents may be schemas, machines, people, or groups of these.

Schemas Characterized

Schema theory provides, *inter alia*, a language for the study of action-oriented perception (see REACTIVE ROBOTIC SYSTEMS). A *perceptual schema* embodies processes for recognizing a given domain of interaction, with various parameters representing properties such as size, location, and motion. An *assemblage* of perceptual schemas provides an estimate of environmental state with a representation of goals and needs. New sensory input as well as internal processes update the schema assemblage. The internal state is also updated by knowledge of the state of execution of current plans made up of *motor schemas* akin to control systems. Motor schemas can be combined with perceptual schemas to form *coordinated control programs* which control the phasing in and out of patterns of co-activation, with mechanisms for the passing of control parameters from perceptual to motor schemas.

The *activity level* of an instance of a perceptual schema represents a "confidence level" that the object represented by the schema is indeed present; that of a motor schema may signal its "degree of readiness" to control some course of action. Schema instances can become *activated* in response to certain patterns of input from sensory stimuli (in which case we say they are data driven) or in response to other schema instances that are already active (hypothesis driven). The activity level of a schema may be but one of many parameters that characterize it (a schema for "ball" might include parameters for its size, color, and velocity). It is thus important to distinguish "activity level" as a particular parameter of a schema from the "neural activity" which will vary with different neural implementations of the schema.

Schema theory is a learning theory, too. In a general setting, there is no fixed repertoire of basic schemas. Rather, new schemas may be formed as assemblages of old schemas; but, once formed, a schema may be tuned by some adaptive mechanism. This tunability of schema assemblages allows them to start as composite but emerge as primitive, much as a skill is honed into a unified whole from constituent pieces. When used in conjunction with neural networks, schema theory offers a means of providing a functional/structural decomposition, and it is to be contrasted with models which employ a learning rule to train an otherwise undifferentiated network to respond as specified by a training set. An influential study of schemas in a connectionist framework was provided by Rumelhart et al. (1986), who showed that many properties of schemas used in the cognitive psychology literature may be seen as emergent properties of adaptive, connectionist networks. However, their approach does not address the issue of how new schemas may be assembled from old, and the synthesis of the two viewpoints remains an open challenge (for some related issues see, e.g., COMPOSITIONALITY IN NEURAL SYSTEMS).

Cooperative computation, a shorthand for "computation based on the competition and cooperation of concurrently active agents," is the style of schema-based computation. Cooperation yields a pattern of "strengthened alliances" between mutually consistent schema instances that allows them to achieve high activity levels to constitute the overall solution of a problem (as perceptual schemas become part of the current short-term model of the environment, or motor schemas contribute to the current course of action). As a result of competition, instances which do not meet the evolving (data-guided) consensus lose activity, and thus are not part of this solution (though their continuing subthreshold activity may well affect later behavior). Draper et al. (1989) demonstrate this style of schema interaction in a machine vision system; see also VISUAL SCHEMAS IN OBJECT RECOGNITION AND SCENE ANALYSIS.

Just as thermodynamics is a free-standing branch of physics but statistical mechanics relates it to the more microscopic dynamics of atoms and molecules, so is schema theory a free-standing branch of AI or cognitive psychology but brain theory relates it to the more microscopic dynamics of neurons and neural networks (see SELF-ORGANIZATION AND THE BRAIN). A given schema may have many different implementations, either biologically or technologically. In brain theory, a given schema, defined functionally, may be distributed across more than one brain region; conversely, a given brain region may be involved in many schemas. Hypotheses about the localization of schemas in the brain may be tested by lesion experiments or functional imaging, with possible modification of the model (e.g., replacing one schema by several interacting schemas with different localizations) and further testing. Given hypotheses about the neural localization of schemas, we may then model a brain region by seeing if its known neural circuitry can indeed be shown to implement the posited schema. When the model involves properties of the circuitry that have not yet been tested, it lays the ground for new experiments. In DAI, individual schemas may be implemented by artificial neural networks, or in some programming language on a "standard" (possibly distributed) computer. In providing an account of the development (or evolution) of schemas, we find that new schemas often arise as "modulators" of existing schemas, rather than as new systems with independent functional roles. Thus, for example, schemas for control of dextrous hand movements serve to modulate less specific schemas for reaching with an undifferentiated grasp unadapted to the shape or planned use of an object.

Schemas as a Programming Methodology for Neural Computation

It has become a truism in the computer industry that software is at least as important as hardware in the advance of computer technology. But some have argued that, with the use of the learning and self-organization principles of artificial neural networks, programming will no longer be necessary. However, if we consider that the human brain is a giga-gigaflop machine (10^{15} synapses running at a millisecond "clock rate") and that it takes 25 years for such a system to learn the skills for which we grant a Ph.D., it can be seen that more efficient techniques will be required for even the fastest and most adaptive of neural computers. In short, we will still require sophisticated programming to reap the full benefits of neural computing. This requires a methodology which allows an overall task to be decomposed into subtasks, with further stepwise refinement (and various processes of iteration up and down the levels of complexity) being required before subtasks are derived for which implementation is a straightforward process. Schema theory provides the beginnings of such a methodology.

The schema-based style of cooperative computation is far removed from serial computation and the symbol-based ideas that have dominated conventional AI. The RS (Robot Schema) language (Lyons and Arbib, 1989) formalizes each schema instance as a *port automaton*, i.e., an automaton with a set of input and output ports through which it can communicate with other instances. As action and perception progress, certain schema instances need no longer be active (they are *deinstantiated*), while new ones are added (*instantiated*) as new objects are perceived and new plans of action are elaborated. A basic schema definition includes a $(C++)$-like behavior specification making explicit how an instance of the schema behaves, but this basic behavior could also be implemented in terms of a neural network (see NSL: NEURAL SIMULATION LANGUAGE).

Schemas may be formed hierarchically as assemblages, which also have ports but have their behavior defined through the interactions of instances. An assemblage may itself be considered a schema for further processes of assemblage formation; and the network itself will be dynamic, growing and shrinking as various instantiations and deinstantiations occur. The RS syntax for an assemblage tells us how to put schemas together in a way which does not depend on how the behavior specification is given. It is thus simple to extend RS to make it possible, in the basic schemas, to define the behavior directly in terms of a neural network as well as by a C-like program.

This new style of computation will include programming in silicon; programming in network protocols; the use of mechatronics for integrated design of robotic subsystems; and principles of *adaptive programming* that incorporate lessons from brain theory and connectionism. Such adaptation will include development of user models that allow the computer to adapt itself to each individual user.

Classically, something is computable if it can be mapped into a recursive function $f: \mathbb{N} \to \mathbb{N}$. Such processes can, more generally, be captured by a finite (but expandable) graph of finite connectivity with a finite set of neighborhood functions. This includes the "two von Neumanns" of stored programs and CELLULAR AUTOMATA (q.v.). We get a starker generalization if we embrace the "two Turings" of the classic Turing machine of serial computation and the "stripe machine" which explains pattern formation through processes of reaction and diffusion of morphogens in a discrete set of cells (see PATTERN FORMATION, BIOLOGICAL). The suggestion, then, is that neural computation will still need a programming language for the formation of discrete assemblages (schema theory) but that we will now allow building blocks to be continuous, generalizing a discrete finite basis to a continuous basis from which overall programs may be built. Ingredients for such building blocks include leaky integrator neurons (see Section I.1 of Part I), dynamic systems (see COMPUTING WITH ATTRACTORS), and optimization (see NEURAL OPTIMIZATION), with modulation between the subsystems in the style of competition and cooperation.

The bit no longer has to be the building block. It wasn't anyway; AI uses symbols, not bits, as the grounding level. But as we turn to both analog VLSI inspired by the design of neural networks (see SILICON NEURONS) and optical computing (see OPTICAL ARCHITECTURES FOR NEURAL NETWORK IMPLEMENTATIONS) as implementation technologies, analog computations become available as practical building blocks. In the process, the line between computation and control disappears. Biological control theory (see MOTOR CONTROL, BIOLOGICAL AND THEORETICAL) usually studies neural circuitry specialized for the control of a specific function, be it the stretch reflex or the vestibulo-ocular reflex. Yet most behavior involves complex sequences of coordinated activity of a number of control systems. The notion of a coordinated control program, introduced earlier, combines control theory and the computer scientist's notion of a program. These coordinated control programs can control the time-varying interaction of a number of control systems.

However, just as we emphasized that neural computing demands a programming methodology, so is it clear that symbolic coding remains a key issue. We will need to develop a whole new theory of computation to understand, for example, when analog "searches" (e.g., SIMULATED ANNEALING) are more efficient than symbolic searches. The proper treatment of decision making under uncertainty will demand judicious blends of continuous processes and discrete graph structures, as we are beginning to see in the attempts to build large but tractable expert systems. The DYNAMIC LINK ARCHITECTURE (q.v.) shows that *fast* synaptic changes may offer a new approach to the problem of graph matching, providing a way to solve the binding problem of linking items in a neurally based architecture without using the explicit addressing of conventional computers to build pointers as in symbolic AI (see STRUCTURED CONNECTIONIST MODELS).

A classic form of program specification uses the notation $\{P\}S\{Q\}$ to specify that S be implemented by a function $f: X \to Y$ such that if $P(x)$ is true and $f(x)$ is defined, then $Q(f(x))$ is also true. To extend this to schemas, we have to relate the predicate logic style of program specification to the learning approach to neural net specification (which is in turn closely related to system identification: see IDENTIFICATION AND CONTROL). In training a neural network with a finite training set, we may write $\{x\}N\{y\}$ to require that the function f of the net N satisfy $f(x_i) \approx y_i$ for each i, in some suitable sense of approximation. Here we have a finite set of requirements, rather than the global specification provided by the predicates P and Q of $\{P\}S\{Q\}$.

Where computability theory has until now stressed the Turing machine halting problem, this new methodology stresses the convergence problem for optimization techniques subject to constraints appropriate to the current task (see CONSTRAINED OPTIMIZATION AND THE ELASTIC NET). We need to understand how symbolic structures may be seen as a limiting case of continuous activity so that recursive definitions of program correctness may be related to recursive proofs of near optimality. In this regard, it may be encouraging that Manes and Arbib (1986) developed a general canonical fixpoint theory that applies to the recursive definition of functions in programming languages when taken in a partially additive category, yet also applies to more analytic problems when applied in a category of metric spaces.

Many studies of adaptive neural networks have been in the context of learning strategies for pattern recognition and/or motor control. In such cases (see, e.g., REINFORCEMENT LEARNING IN MOTOR CONTROL), the specification of performance need be neither in terms of symbolic predicates nor in terms of numerical optimization; rather, the performance specification rests on relating schemas to "world semantics," with success based on sensory and motor criteria for interaction with the rest of the world.

Road Map: Connectionist Psychology
Related Reading: I.2. Levels and Styles of Analysis; Developmental Disorders; Motor Pattern Generation; Neuroethology, Computational; Philosophical Issues in Brain Theory and Connectionism; Sensor Fusion; Sensorimotor Learning; Time Perception: Problems of Representation and Processing

References

Arbib, M. A., 1981, Perceptual structures and distributed motor control, in *Handbook of Physiology—The Nervous System II. Motor Control* (V. B. Brooks, Ed.), Bethesda, MD: American Physiological Society, pp. 1449–1480. ◆
Arbib, M. A., and Hesse, M. B., 1986, *The Construction of Reality*, Cambridge, Eng.: Cambridge University Press.
Bartlett, F. C., 1932, *Remembering*, Cambridge, Eng.: Cambridge University Press.
Brooks, R. A., 1986, A robust layered control system for a mobile robot, *IEEE J. Robot. Automat.*, RA-2:14–23.
Craik, K. J. W., 1943, The Nature of Explanation, Cambridge, Eng.: Cambridge University Press.
Draper, B. A., Collins, R. T., Brolio, J., Hanson, A. R., and Riseman, E. M., 1989, The schema system, *Int. J. Comput. Vis.*, 2:209–250.
Head, H., and Holmes, G., 1911, Sensory disturbances from cerebral lesions, *Brain*, 34:102–254.
Jeannerod, M., Arbib, M. A., Rizzolatti, G., and Sakata, H., in press, The neural mechanisms of grasping, *Trends Neurosci.*
Kilmer, W. L., McCulloch, W. S., and Blum, J., 1969, A model of the

vertebrate central command system, *Int. J. Man-Mach. Stud.*, 1:279–309.

Lyons, D. M., and Arbib, M. A., 1989, A formal model of computation for sensory-based robotics, *IEEE Trans. Robot. Automat.*, 5:280–293.

Mandler, G., 1985, *Cognitive Psychology: An Essay in Cognitive Science*, Hillsdale, NJ: Erlbaum.

Manes, E. G., and Arbib, M. A., 1986, *Algebraic Approaches to Program Semantics*, New York: Springer-Verlag.

Minsky, M. L., 1975, A framework for representing knowledge, in *The Psychology of Computer Vision* (P. H. Winston, Ed.), New York: McGraw-Hill, pp. 211–277.

Minsky, M. L., 1985, *The Society of Mind*, New York: Simon and Schuster.

Piaget, J., 1971, *Biology and Knowledge*, Edinburgh: Edinburgh University Press.

Pitts, W. H., and McCulloch, W. S., 1947, How we know universals: The perception of auditory and visual forms, *Bull. Math. Biophys.*, 9:127–147.

Rumelhart, D. E., Smolensky, P., McClelland, J. L., and Hinton, G. E., 1986, Schemata and sequential thought processes in PDP models, in *Parallel Distributed Processing: Explorations in the Microstructure of Cognition*, vol. 2, *Psychological and Biological Models* (J. L. McClelland, D. E. Rumelhart, and PDP Research Group, Eds.), Cambridge, MA: MIT Press, chap. 14. ◆

Shallice, T., 1988, *From Neuropsychology to Mental Structure*, Cambridge, Eng.: Cambridge University Press. ◆

Scratch Reflex

Paul S.G. Stein

Introduction

An organism can scratch itself in response to a mechanical stimulus that generates a force at a site on the body surface (Stein, 1983). During successful scratching, a nearby limb moves toward and rubs against the site. If a different site is stimulated, different limb movements are required for a successful scratch.

Some organisms do not require the entire central nervous system to produce a successful scratch reflex. Some vertebrates with a complete spinal cord transection at the level of the neck or upper back, termed spinal vertebrates, can perform a successful hindlimb scratch in response to a mechanical stimulus delivered to a site on the body surface posterior to the complete transection. This stimulus excites neural networks in the spinal cord posterior to the complete transection (Stein, 1983). Scratching has been demonstrated in the following spinal vertebrates: dog (Sherrington, 1906), cat (Arshavsky, Gelfand, and Orlovsky, 1986), turtle (Stein, Mortin, and Robertson, 1986; Stein, 1989), and frog. In the frog, scratch reflex is termed *wiping reflex* (Berkinblit, Feldman, and Fukson, 1986, 1989; see FROG WIPING REFLEXES).

Strategies of Scratching: The Forms of a Scratch

The set of all successful scratches is constrained by the physical construction of the organism's limbs and body, i.e., its biomechanics. For some organisms, there may be a set of sites on the body surface, e.g., sites on the middle of the back of a turtle or a human, that cannot be rubbed directly by a limb of that organism. There may be some sites that can be scratched using only one strategy of movement, e.g., a human can scratch some sites on the upper back using only a strategy in which the elbow is placed over the shoulder. These sites belong to a set termed a *pure-form domain*. While the same motor strategy is used to rub against each site within a pure-form domain, parametric adjustment of limb movement is required to reach each specific site. There may be other sites that can be scratched using either of several strategies, e.g., a human can scratch a site on the side of the thorax using either the hand or the elbow. These sites belong to a set termed a *transition zone*. Each scratch movement strategy is termed a *form* of the scratch (Stein, Mortin, and Robertson, 1986). The concept of movement form can be applied to other motor acts. There are several forms of locomotion in a horse, e.g., walk, trot, and gallop (see GAIT TRANSI-

TIONS). Forward stepping and backward stepping are among the forms of stepping produced by humans.

Several scratch strategies are produced by the spinal turtle. In each strategy, a distinct portion of the hindlimb is used to exert force against the stimulated site (Stein, Mortin, and Robertson, 1986). The turtle uses the dorsum of the foot for *rostral scratching*; it uses the side of the knee for *pocket scratching*; it uses the side of the foot or heel for *caudal scratching*. Biomechanical constraints play a key role for each of these movement strategies. The rostral strategy is the only strategy that can be used to place a portion of the hindlimb against a site on the region that connects the upper shell and the lower shell in the middle of the body. The foot cannot reach sites in the pocket region just anterior to the turtle's hip; only the side of the knee can be used to generate force against a site in the pocket region. Thus, the spinal turtle can select the biomechanically appropriate form that produces successful scratches; selection of the proper scratch strategy does not require supraspinal structures, i.e., the brainstem and the brain. Experiments described in a later section establish that selection is an intrinsic property of spinal cord neural networks.

Coordinate System Transformations in the Scratch Reflex

Several transformations occur during scratching. First, a mechanical-to-sensory transformation takes place on the body surface when the tactile stimulus activates cutaneous primary afferent neurons. Second, a sensory-to-motor transformation takes place within the central nervous system (CNS). Third, a motor-to-mechanical transformation takes place in the limb. It is possible to examine all three transformations at the same time during actual scratching. For neural network studies, it is useful to examine the first and second transformations in the absence of the third transformation, i.e., in the absence of actual movements. Experiments that examine the response to a tactile stimulus that elicits scratch motor output while neuromuscular synapses are blocked are described in later sections.

Several coordinate systems help describe the different transformations. First, the rectilinear orthogonal Cartesian coordinate system is useful for describing the site on the body surface that receives the sensory stimulus as well as for describing the position in space of the portion of the limb that rubs against the site. Second, the muscle/motor-pool coordinate system is useful for describing the output of the CNS. Third, the body degree-of-freedom, also termed joint-angle, coordinate system is useful for describing the movements of the multijointed limb. Each of these coordinate systems is useful for describing a particular moment; thus, the additional dimension of time must be added to each of these coordinate systems to describe the temporal history of an episode of scratching.

Cartesian Coordinates

Cartesian coordinates are useful for the description of an organism's personal space at the body surface and its extrapersonal space in the immediate vicinity of the body surface. When a limb rubs against a site on the body surface, both the site and the portion of the limb contacting the site are at the same coordinate location in a Cartesian space.

Muscle/Motor-Pool Coordinates

The set of motor neurons that synaptically activate a given muscle is termed a motor pool. Each muscle of the body or its motor pool can be viewed as a dimension of a coordinate system. The amplitude of each dimension is determined by the intensity of activation of each muscle/motor-pool (see MOTONEURON RECRUITMENT). Note that this coordinate system is different from the other coordinate systems based on traditional geometries. During each movement, there is a distinct "motor pattern" of muscle/motor-pool activation that occupies a region of an abstract space whose dimensions are muscles/motor-pools and time.

Body Degree-of-Freedom Coordinates

Each degree of freedom of movement in a body is a dimension of a coordinate system. Some body joints, e.g., knee, have a single degree of freedom and constitute only a single dimension. Other body joints, e.g., hip, have several degrees of freedom and therefore require several dimensions of this coordinate system. The description of the transformation from muscle/motor-pool space to body degree-of-freedom space requires considerable mathematics and extensive information about the geometry and physics of the musculoskeletal system.

Coordinate System Analyses of the Turtle Scratch Reflex

Data obtained from studies of the scratch reflex in the spinal turtle (Stein, 1989) allow application of the concepts outlined in the previous section. These data are described in this section. Other data obtained from frog (Berkinblit, Feldman, and Fukson, 1986, 1989; see FROG WIPING REFLEXES) and from cat (Arshavsky, Gelfand, and Orlovsky, 1986) are also consistent with these concepts.

Cartesian Coordinate Description of Receptive Fields

If a stimulus applied to a site on the body surface elicits a specific form of scratch reflex in which a nearby limb reaches toward and rubs against the stimulated site, then that site is a member of the receptive field for that form of scratch. In the spinal turtle, there is a receptive field for rostral scratch, a receptive field for pocket scratch, and a receptive field for caudal scratch.

Stimulation of most sites in one form's receptive field elicits only scratches of that form; these sites constitute the pure-form domain of that form's receptive field. The *pure-form domain* for each scratch form is the set of sites in which only one scratch form is biomechanically possible. There is also a set of transition-zone sites located between the pure-form domain for one form and the pure-form domain for another form. The *transition zone* is the set of sites in which more than one scratch form is biomechanically possible. Stimulation of a site in this transition zone can elicit either one scratch form, or the other scratch form, or a blend response of both scratch forms. There are two types of blends, the switch response and the hybrid response. In a *switch response*, several cycles of one form are followed smoothly by several cycles of the other form. In a *hybrid response*, each of several successive cycles has two rubs per cycle; one rub uses one scratch form, and the other rub uses the other scratch form.

The occurrence of blends supports the notion that there is shared neural circuitry between the neural network that generates one form of scratch and the neural network that generates another form of scratch. In particular, detailed analyses of blends support the concept that interneurons controlling the rhythm of hip movements are shared among the networks responsible for each of several forms of scratching.

The receptive field for turtle hindlimb scratch is a continuous surface on the body that contains three pure-form domains and two transition zones. Each transition zone is a space that separates two pure-form domains; this fact has important implications for the understanding of the biological bases of motor strategy selection (Stein, Mortin, and Robertson, 1986).

Muscle/Motor-Pool Coordinate Description of Motor Patterns

The electromyographic (EMG) activity of individual muscles that play critical roles for each scratch form may be recorded during scratching in the spinal turtle. The monoarticular knee extensor muscle is active during the rub against the stimulated site for all three forms of scratch. Scratching in the turtle is rhythmic; all three forms of scratching display rhythmic alternation between hip flexor muscle activity and hip extensor muscle activity. Timing of the monoarticular knee extensor muscle is distinct for each scratch form. The monoarticular knee extensor muscle is active (1) during the latter part of hip flexor muscle activity in a rostral scratch, (2) during hip extensor muscle activity in a pocket scratch, and (3) after the burst of hip extensor muscle activity in a caudal scratch. Thus, all three muscles are active in each of the three forms of the scratch. The motor pattern of muscle/motor-pool activation is distinct for each scratch form.

Body Degree-of-Freedom Coordination Description of Movement

The time course of hip angle (angle of hip flexion/extension) and knee angle (angle of knee flexion/extension) has been studied during each of the three forms of turtle scratch reflex. Rhythmic alternation between hip flexion and hip extension occurs for all three scratch forms. Timing of knee extension in the cycle of hip flexion and extension is distinct for each scratch form. The knee extends during the latter part of hip flexion in a rostral scratch; the knee extends during hip extension in a pocket scratch; the knee extends after hip extension is completed in a caudal scratch. A specific "movement pattern" in the joint-angle coordinate system is distinct for each form, i.e., there is regulated timing of knee extension in the cycle of hip movement.

This movement pattern for each form in joint-angle space is similar to the motor pattern for each form measured in muscle/motor-pool space. In both spaces for each form of the scratch, there is a regulated timing of the knee with respect to the cycle of the hip. Similar changes of timing of a distal joint in the cycle of a proximal joint have been observed for other behaviors, e.g., forward versus backward stepping in humans.

Spinal Cord Networks for Scratch Reflex

The neuronal networks responsible for producing the scratch reflex in the cat (Arshavsky, Gelfand, and Orlovsky, 1986) and in the turtle (Stein, 1989) are only partially understood. Additional experimental work is necessary for a more complete understanding of these networks. The neuronal networks for chewing in lobster are examples of networks that are understood in greater detail (see CRUSTACEAN STOMATOGASTRIC SYSTEM). Our current understanding of scratching neuronal networks relies on experiments that demonstrate the ability of the spinal cord to produce a motor pattern in the absence of actual movements. Movements are prevented by blockade of muscle acetylcholine receptors with a specific antagonist, e.g., curare. The motor pattern is measured as the electroneurographic (ENG) activities of specific motor pools in response to a stimulation of a site in a scratch receptive field. The ENG motor patterns recorded in the absence of "real" movements are termed *fictive* motor patterns.

The ENG motor pattern for each scratch form in the spinal immobilized turtle is generated in response to stimulation of a site in the receptive field for that scratch form; the ENG motor pattern is an excellent replica of the EMG motor pattern recorded during actual movements. Each motor pool monitored in the immobilized turtle using ENG recording techniques innervates a muscle that was previously monitored using EMG recording techniques during actual movements. These results establish that spinal cord neuronal networks can select the appropriate scratch motor pattern in response to stimulation of a specific site on the body surface; thus, motor strategy selection is a property of spinal cord neuronal networks. These rhythmic scratch motor patterns are produced in an "open-loop" condition without benefit of timing cues from movement-related sensory feedback; thus, rhythmic motor pattern generation is an intrinsic property of spinal cord neuronal networks.

Motor patterns produced in the absence of movement-related feedback are termed *central motor patterns*. The neuronal network responsible for generating a central motor pattern for a behavior is termed a *Central Pattern Generator (CPG)* for that behavior. A goal of current research is to disclose the properties of the CPG for each scratch form. It is possible that the CPG for one scratch form shares no neural circuitry with the CPG for another scratch form; such a lack of overlap in circuitry is not likely to occur, however. Recent single-neuron recordings in the turtle support the hypothesis that the CPG for the rostral scratch may share many neural elements with the CPG for the pocket scratch (Berkowitz and Stein, 1994a, 1994b).

The scratch motor pattern is not independent of movement-related sensory input, however. Motor patterns are subject to important modulations caused by sensory input (see MOTOR PATTERN GENERATION). For example, EMG recordings during actual scratching in the cat demonstrate amplitude and phase modulations of the motor pattern caused by sensory feedback during paw contact with the stimulated site (Kuhta and Smith, 1990); similar modulations of the EMG motor pattern are also seen in the spinal turtle when the foot catches against the fire-polished glass rod used to deliver the mechanical stimulus (Stein, 1983).

Localization and Distribution of Spinal Cord Neuronal Networks

The spinal cord is a segmental structure. Each segment receives sensory input from a specific region of the body surface, termed the *dermatome* of that spinal segment. Each segment contains the cell bodies of motor neurons that innervate a specific set of muscles. The hindlimb enlargement is the set of spinal segments that contain the cell bodies of motor neurons that innervate hindlimb muscles. In both the turtle and the cat, there are five spinal segments that constitute the hindlimb enlargement.

The anterior segments of the hindlimb enlargement play an important role in scratch rhythm generation in the cat (Arshavsky, Gelfand, and Orlovsky, 1986) and in the turtle (Mortin and Stein, 1989). In all limbed vertebrates, the anterior portion of the hindlimb enlargement contains hip flexor motor neurons and knee extensor motor neurons. A scratch motor rhythm is produced by the most anterior segment of the turtle hindlimb enlargement in response to stimulation of a site in that segment's dermatome. A rhythmic pocket scratch motor pattern is produced by the three most anterior segments of the turtle hindlimb enlargement in response to stimulation of a site in the dermatome of the most anterior segment. The spinal segment just anterior to the hindlimb enlargement also contributes to rhythmogenesis. Thus, neuronal networks for scratching are contained in and distributed among a set of spinal segments.

Multisecond Excitability Changes in Scratch Neuronal Networks

The turtle scratch motor response can continue for several seconds after the cessation of sensory stimulation (Currie and Stein, 1988). For an additional several seconds after the cessation of motor neuron activity, there is an increased excitability of spinal cord neuronal networks. This afterexcitability is form specific and is a physiological measure of spinal cord selection processes. NMDA RECEPTORS (q.v.) contribute to this afterexcitability (Currie and Stein, 1992). The long time constant of NMDA receptor activation is well suited for multisecond excitability changes (Daw, Stein, and Fox, 1993).

Spinal cord neurons, termed long-afterdischarge interneurons, are activated by stimulation in a region of a scratch receptive field and are active for many seconds after the cessation of stimulation (Currie and Stein, 1990). Long-afterdischarge interneurons may play a role in motor pattern selection in scratch neuronal networks. NMDA receptors contribute to the excitability of long-afterdischarge interneurons (Currie and Stein, 1992).

Broad Tuning of Interneurons in Neuronal Networks for Scratching

There is broad tuning in the responses of individual turtle interneurons activated by stimulation of sites in the receptive fields for the rostral scratch and for the pocket scratch (Berkowitz and Stein, 1994a, 1994b). Some of these interneurons are activated throughout the entire region of the scratch receptive fields for each of several forms. Many of these interneurons may be members of both the rostral scratch CPG and the pocket scratch CPG. For each interneuron, there is usually a site whose stimulation results in the highest frequency of action potentials; stimulation of other sites usually results in firing frequencies of the interneuron that decrease as the distance from the site that evokes the largest response increases. These data are consistent with the hypothesis that motor pattern selection results from the summed activities of a population of broadly tuned interneurons that are shared by several CPGs.

Conclusions

Scratching can be used to uncover important characteristics of neuronal networks that perform sensory-to-motor transformations. Future experiments are required for a more complete understanding of the properties of these neuronal networks.

Acknowledgment. The author's research is supported by NIH grant NS30786.

Road Map: Motor Pattern Generators and Neuroethology
Related Reading: Geometrical Principles in Motor Control; Limb Geometry: Neural Control

References

Arshavsky, Y. I., Gelfand, I. M., and Orlovsky, G. N., 1986, *Cerebellum and Rhythmical Movements*, Berlin: Springer-Verlag. ◆

Berkinblit, M. B., Feldman, A. G., and Fukson, O. I., 1986, Adaptability of innate motor patterns and motor control mechanisms, *Behav. Brain Sci.*, 9:585–599.

Berkinblit, M. B., Feldman, A. G., and Fukson, O. I., 1989, Wiping reflex in the frog: Movement patterns, receptive fields, and blends, in *Visuomotor Coordination* (J.-P. Ewert and M. A. Arbib, Eds.), New York: Plenum, pp. 615–629. ◆

Berkowitz, A., and Stein, P. S. G., 1994a, Activity of descending propriospinal axons in the turtle hindlimb enlargement during two forms of fictive scratching: Broad tuning to regions of the body surface, *J. Neurosci.*, 14:5089–5104.

Berkowitz, A., and Stein, P. S. G., 1994b, Activity of descending propriospinal axons in the turtle hindlimb enlargement during two forms of fictive scratching: Phase analyses, *J. Neurosci.*, 14:5105–5119.

Currie, S. N., and Stein, P. S. G., 1988, Electrical activation of the pocket scratch central pattern generator in the turtle, *J. Neurophysiol.*, 60:2122–2137.

Currie, S. N., and Stein, P. S. G., 1990, Cutaneous stimulation evokes long-lasting excitation of spinal interneurons in the turtle, *J. Neurophysiol.*, 64:1134–1148.

Currie, S. N., and Stein, P. S. G., 1992, Glutamate antagonists applied to midbody spinal cord segments reduce the excitability of the fictive rostral scratch reflex in the turtle, *Brain Res.*, 581:91–100.

Daw, N., Stein, P. S. G., and Fox, K., 1993, The role of NMDA receptors in information processing, *Annu. Rev. Neurosci.*, 16:207–222. ◆

Kuhta, P. C., and Smith, J. L., 1990, Scratch responses in normal cats: Hindlimb kinetics and muscle synergies, *J. Neurophysiol.*, 64:1653–1667.

Mortin, L. I., and Stein, P. S. G., 1989, Spinal cord segments containing key elements of the central pattern generators for three forms of scratch reflex in the turtle, *J. Neurosci.*, 9:2285–2296.

Sherrington, C. S., 1906, *The Integrative Action of the Nervous System*, New Haven: Yale University Press.

Stein, P. S. G., 1983, The vertebrate scratch reflex, *Symp. Soc. Exp. Biol.*, 37:383–403. ◆

Stein, P. S. G., 1989, Spinal cord circuits for motor pattern selection in the turtle, *Ann. NY Acad. Sci.*, 563:1–10.

Stein, P. S. G., Mortin, L. I., and Robertson, G. A., 1986, The forms of a task and their blends, in *Neurobiology of Vertebrate Locomotion* (S. Grillner, P. S. G. Stein, D. G. Stuart, H. Forssberg, and R. M. Herman, Eds.), London: Macmillan, pp. 201–216. ◆

Selective Visual Attention

Bruno A. Olshausen and Christof Koch

Introduction

It may seem ironic that biological vision systems would employ a serial attentional strategy, since one usually thinks of such systems as natural examples of massively parallel computation. However, in any physical computational system, processing resources are limited, which leads to bottlenecks similar to those faced by the von Neumann architecture. Nowhere is this more evident than in the human visual system, where the amount of information provided by the optic nerve—estimated to be in the range of 10^8–10^9 bits per second—far exceeds what the brain is capable of fully processing and assimilating into conscious experience. The strategy that nature has devised for dealing with this bottleneck is to select certain portions of the input to be processed preferentially, shifting the processing focus from one location to another in a serial fashion. This strategy, commonly referred to as *selective visual attention*, is employed by a wide variety of biological vision systems, from jumping spiders to humans. Here, we shall review what has been learned of visual attention from studies in psychophysics, neurophysiology, and computational modeling.

Psychophysics

Spotlight Metaphor

A metaphor commonly employed in the psychophysics community is that visual attention acts as a *spotlight*, enhancing information within a selected region of the image (or alternatively, filtering out information outside of the spotlight). Importantly, this spotlight can move about the scene independent of eye movements. An athlete, for example, may attend to other players on his team without moving the eyes so as not to reveal to his opponents where he is actually "looking." This form of attention is usually referred to as *covert attention* to distinguish it from the overt attentional shifts that take place with eye movements (see COLLICULAR VISUOMOTOR TRANSFORMATIONS FOR SACCADES). This article focuses primarily on covert attention.

An early study of covert attentional shifts (Posner, 1980) showed that the efficiency with which an observer can detect signal events occurring at various locations in the visual field—while keeping the position of the eyes fixed—changes as a function of where the observer is instructed to attend. Generally, those events occurring within the focus of attention yield faster or more accurate detection than those events occurring outside the focus. Thus, information within the spotlight of attention is somehow processed preferentially. Another interesting finding of Posner's study was that detection efficiency also increased at those locations in the visual field that were to be the target of an eye movement, which would seem to indicate that eye movements are preceded by movements of covert attention.

Feature Integration

Treisman and Gelade (1980), on the basis of visual search studies, have proposed that attention serves to glue together the various features of an object, such as color, form, and motion. Their *conjunctive search* experiments show that when one is required to find a target item that differs from distracting items only by a conjunction of features—for example, a green

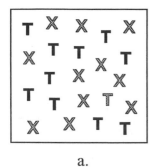
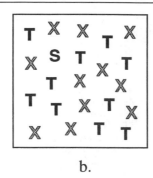

a.   b.

▓▓ green
▬ red

Figure 1. Demonstration of visual search. *A*, The green *T* differs from the distracting items (red *T*s and green *X*s) only by a conjunction of features (i.e., green and *T*). Finding it among the distractors requires progressively longer search times as the number of distractors is increased. *B*, The red *S* differs from the distractors by a single feature (its shape), thus resulting in short search times that are independent of the number of distractors.

T among green *X*s and red *T*s (Figure 1*A*)—the search time increases linearly with the number of distractors. Conversely, searching for an item that can be distinguished by a single feature—for example, an *S* among green *X*s and red *T*s (Figure 1*B*)—can be accomplished in an amount of time independent of the number of distractors. It appears from these results that the conjoining of features requires a serial process that analyzes each object one at a time, whereas differences in a single feature "pop out" immediately. Treisman's original theory thus proposed that the various attributes of an object are encoded in separate representations in the brain which are computed in parallel, and that attention is required to dynamically bind (or integrate) together all those attributes belonging to the same object. Dynamic binding is accomplished by selectively gating each of the separate feature maps so that only those features lying within the attentional spotlight are passed on to higher areas for analysis and recognition. (For an alternative view, see SYNCHRONIZATION OF NEURONAL RESPONSES AS A PUTATIVE BINDING MECHANISM.)

More recently, Treisman's feature integration theory has been modified, since it has been shown that many kinds of conjunctions (e.g., motion and stereoscopic disparity) can be detected in an amount of time that is independent of the number of distractors. Furthermore, conjunction searches can yield reaction times that range from roughly 0 msec per item to 20 or more milliseconds per item, confounding the simple distinction between either "parallel" or "serial" search (see Treisman, 1988, for a review). The modified theory proposes that the attentional spotlight need not necessarily be constrained to the two spatial dimensions, but that it can act along feature dimensions as well. For example, if one were searching for a red object, then all red objects in the scene may be enhanced (or the non-red objects filtered out) so that a conjunction search would be restricted just to red objects. The most recent evidence indicates that attention may be allocated on the basis of fairly complex attributes as well, such as three-dimensional surfaces (He and Nakayama, 1993).

Dynamics

There is evidence for at least two different dynamical forms of attention—one that is fast and transient and another that is slower and sustained (Nakayama and Mackeben, 1989). The transient component is driven involuntarily (for example, by a moving object, or a flickering light), while the sustained component is under voluntary control (such as the way an athlete directs covert attention to track other players). More recently it has been discovered that shifts in the transient form of attention can be unusually rapid when a temporal gap is placed between the disappearance of the fixation mark (i.e., the currently attended location at the center of the screen) and the appearance of the target to be attended (Mackeben and Nakayama,

1993). It is hypothesized that the removal of the fixation mark allows attention to disengage from the fovea and thus be deployed more rapidly to the peripheral target. Interestingly, this speeded form of covert attention appears to underlie an analogous form of very fast eye movements, termed *express saccades*.

How the attentional spotlight moves from one location to another within a scene is an unresolved issue. Some studies support the notion that attention moves continuously between locations (i.e., it traverses all intervening locations), whereas other studies are consistent with discrete, saltatory movements. The speed with which the attentional spotlight can move from one location to the next appears to be on the order of 30–50 msec (Saarinen and Julesz, 1991), making attentional shifts four to six times faster than eye movements. A recent study by Duncan, Ward, and Shapiro (1994), using stimuli displayed at randomly spaced intervals, indicates that attentional shifts may be considerably slower; however, the relatively long masking times used in this experiment may have impeded the disengagement of attention (cf. Mackeben and Nakayama, 1993), leading to slower shifts.

Size

In addition to changing position within the visual field, the attentional spotlight can also change size. A number of studies suggest that the bandwidth of the attentional window is limited, so that increasing the size of the window does not necessarily allow more information to be processed, but merely spreads the same amount of processing resources over a larger area. This effect has yet to be demonstrated conclusively or characterized quantitatively, however.

Neurophysiology

Brain Areas

Neurophysiological studies have disclosed a wide variety of brain areas involved in visual attention. Cortex, superior colliculus, and pulvinar all seem to be involved in various aspects of visual attention, as discerned by single-cell, lesion, and imaging studies. Figure 2 illustrates the neuroanatomical relationships among these areas. Roughly, visual cortex can be subdivided into two processing streams: an occipitotemporal stream that is mainly concerned with object recognition, and an occipitoparietal stream that is mainly concerned with spatial relationships among objects and how one acts on them through reaching and grasping (see VISUAL SCENE PERCEPTION: NEUROPHYSIOLOGY). The superior colliculus is a structure of the midbrain that is involved in controlling eye movements. It receives its input from the retina and cortex and projects to the pulvinar (but has no direct projection to the cortex). The pulvinar is a large subcortical structure that is part of the thalamus and is itself

Figure 2. Major brain areas involved in selective visual attention. To avoid clutter, some connection pathways (e.g., pulvinar-MT) are not explicitly shown.

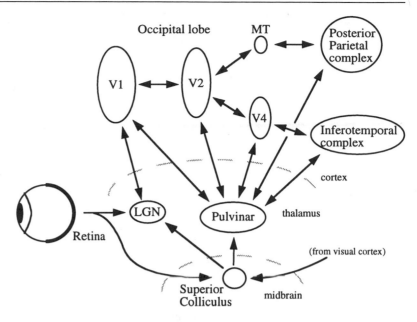

divided into at least four nuclei. It is heavily interconnected with visual cortical areas.

Lesion Studies

Studies of humans with damage to specific brain areas have shown a number of areas to be involved in the control of attention. Posner and Petersen (1990) have proposed on the basis of these studies that the parietal cortex is responsible for disengaging attention; the superior colliculus, for moving attention; and the pulvinar, for engaging attention. The latter conclusion is consistent with lesion studies in primates, which show that deactivation of the pulvinar decreases an animal's ability to filter out irrelevant stimuli from a scene.

Imaging Studies

Imaging studies using positron emission tomography (PET) have disclosed a number of brain areas that are differentially activated as a function of one's attentional state. Most notably, Corbetta et al. (1990) have shown that attending to different visual attributes (e.g., color, form, or motion) yields differential amounts of activity in the brain areas specialized for processing each of these attributes. Similar experiments showed that the pulvinar is differentially activated during a task requiring the filtering out of irrelevant stimuli.

Single-Cell Studies

Single-cell studies in awake, behaving monkeys have been able to show more precisely the effect of visual attention on neural processing. For example, the receptive fields of neurons in V4 and IT have been shown to shift toward attended stimuli (Wise and Desimone, 1988; Connor, Gallant, and Van Essen, 1993; Desimone and Duncan, 1995), providing evidence for the selective filtering of form information through the occipitotemporal pathway. Neurons in the posterior parietal complex and in the dorsomedial portion of the pulvinar show an enhancement or suppression in firing rate for attended stimuli within their receptive fields, supporting the conjecture that these areas are involved in encoding the saliency of visual objects or controlling information flow (Robinson and Petersen, 1992).

Computational Models

The Computational Perspective

Visual attention can be viewed as a means for reducing the amount of incoming visual information to a manageable size so that it can be dealt with by the limited computational resources of the brain. A number of models have been proposed that attempt to explain psychophysical and neurophysiological findings with this perspective in mind, and to generate predictions for future experiments. Two common features of these models are a "saliency map," for registering potentially interesting areas of the input, and a mechanism for gating, or dynamically routing, information flow.

Saliency Map

Given that the purpose of visual attention is to focus computational resources on a specific, "conspicuous" or "salient" region within a scene, it has been proposed that the control structure underlying visual attention needs to represent such locations within a topographic *saliency map* (Koch and Ullman, 1985). The control of attention is thought to be carried out through one or more of such maps that do not code for particular features (e.g., the color or velocity of a stimulus), but rather for how different or how salient a particular stimulus is relative to its neighborhood (i.e., a red object among many green ones). A winner-take-all mechanism then selects the currently most salient feature in the map and directs attention to its location through a gating mechanism (Figure 3). An alternative to this scheme, proposed by Desimone and Duncan (1995), is that the control of attention is highly distributed throughout the cortex, without any explicit saliency map.

Dynamic Routing

A neurobiological model for the dynamic routing of visual information has been proposed by Olshausen, Van Essen, and Anderson (1993). This model was motivated by the need to form position- and size-invariant representations of objects for recognition (see ROUTING NETWORKS IN VISUAL CORTEX). A set of *control neurons* dynamically modify the synaptic strengths of intracortical connections so that information from an attended

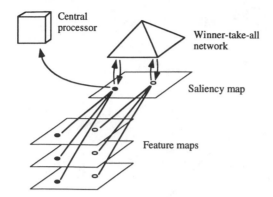

Figure 3. According to the theory of Koch and Ullman (1985), conspicuous regions within a scene are represented within a topographic *saliency map*. A winner-take-all mechanism selects the currently most salient location in the map (denoted by the black node) and directs attention to its location. (Adapted from Koch and Ullman, 1985.)

region of primary visual cortex (V1) is selectively routed to higher cortical areas. The control neurons are hypothesized to reside in the pulvinar and deep layers of cortex and modulate information flow through multiplicative synaptic gating mechanisms (e.g., through the NMDA-receptor channel). The control neurons in turn are driven by a saliency map that may be represented in either the superior colliculus or parietal cortex. A specific prediction of this model is that receptive fields of cortical neurons should shift with attention, which is in accordance with the observed dynamic changes in the receptive fields of V4 and IT neurons.

Discussion

From the combination of psychophysical, neurobiological, and computational studies reviewed here, we can conclude that visual attention acts as a short-term binding mechanism for bringing together a select portion of visual information for higher-level analyses. Importantly, this dynamic aspect of attention relieves the brain of having to create a plethora of hardwired neural representations for each of the feature combinations that occur in the visual world at every possible position and size, since complex spatial relationships among features need only be encoded within the window of attention. While

this strategy has the possible disadvantage of requiring a serial analysis of a scene, it results in a tremendous savings in computational resources.

Road Map: Vision
Related Reading: Dissociations Between Visual Processing Modes; Figure-Ground Separation; Thalamus

References

Connor, C. E., Gallant, J. L., and Van Essen, D. C., 1993, Effects of focal attention on receptive field profiles in area V4, *Soc. Neurosci. Abst.*, 19:974.
Corbetta, M., Miezin, F. M., Dobmeyer, S., Shulman, G. L., and Petersen, S. E., 1990, Selective and divided attention during visual discrimination of shape, color, and speed: Functional anatomy by positron emission tomography, *J. Neurosci.*, 11:2383–2402.
Desimone, R., and Duncan, J., 1995, Neural mechanisms of selective visual attention, *Annu. Rev. Neurosci.*, 18:193–222. ◆
Duncan, J., Ward, R., and Shapiro, K., 1994, Direct measurement of attentional dwell time in human vision, *Nature*, 369:313–315.
He, Z. J., and Nakayama, K., 1993, Common surface rather than common depth determines attention in 3-D search task, *Soc. Neurosci. Abst.*, 19:773.
Koch, C., and Ullman, S., 1985, Shifts in selective visual attention: Towards the underlying neural circuitry, *Hum. Neurobiol.*, 4:219–227.
Mackeben, M., and Nakayama, K., 1993, Express attentional shifts, *Vis. Res.*, 33:85–90.
Nakayama, K., and Mackeben, M., 1989, Sustained and transient components of focal visual attention, *Vis. Res.*, 29:1631–1647.
Olshausen, B. A., Van Essen, D. C., and Anderson, C. H., 1993, A neurobiological model of visual attention and invariant pattern recognition based on dynamic routing of information, *J. Neurosci.*, 13:4700–4719.
Posner, M. I., 1980, Orientation of attention, *Q. J. Exp. Psychol.*, 32:3–25.
Posner, M. I., and Petersen, S. E., 1990, The attention system of the human brain, *Annu. Rev. Neurosci.*, 13:25–42. ◆
Robinson, D. L., and Petersen, S. E., 1992, The pulvinar and visual salience, *Trends Neurosci.*, 15:127–132. ◆
Saarinen, J., and Julesz, B., 1991, The speed of attentional shifts in the visual field, *Proc. Natl. Acad. Sci. USA*, 88:1812–1814.
Treisman, A., and Gelade, G., 1980, A feature integration theory of attention, *Cognit. Psychol.*, 12:97–136.
Treisman, A., 1988, Features and objects: The fourteenth Bartlett Memorial Lecture, *Q. J. Exp. Psychol. [A]*, 40:201–237. ◆
Wise, S. P., and Desimone, R., 1988, Behavioral neurophysiology: Insights into seeing and grasping, *Science*, 242:736–741. ◆

Self-Organization and the Brain

Christoph von der Malsburg

Introduction

The process of network self-organization is fundamental to the organization of the brain. It takes place on several temporal scales: the ontogenetic/learning time scale of hours, days, and years; and probably also the functional time scale of fractions of a second to minutes (see DYNAMIC LINK ARCHITECTURE). The basic concepts of network self-organization are discussed here. More extended introductions to the topic of self-organization are found in Prigogine and Stengers (1984) and Murray (1993), for instance.

One often speaks of some structural trait of an organism as being "genetically determined." This seems to imply that the genes contain a blueprint describing the organism in full detail. However, all the stages of brain organization (not just evolution) more or less strongly involve an element of self-organization and creativity. It has often been emphasized that the genes cannot, in any naive sense, contain the full information necessary to describe the brain. The cerebral cortex alone contains at least on the order of 10^{14} synapses. Forgetting considerations of genome size, one can hardly imagine how ontogeny could select the correct wiring diagram out of all of the alternatives if all were equally likely. Besides, judging from the variability of the vertebrate brain structure, the precision of the ontogenetic process is not sufficient to specify individual connections.

The conclusion one must draw is that ontogeny makes use of self-organization, that is, of general rules to generate neural structure and of principles of error correction. Above all, ontogenesis can only produce structures with a high degree of regularity—for example, homogeneity, repetitivity, or continuity. Knowing the mechanism of ontogeny is of extreme importance: one cannot understand the function of the brain without knowing its structure, and one cannot know the structure of the brain without knowing the principles of its ontogenesis.

Abstract Scheme of Organization

There are well-studied paradigms of pattern formation, especially in physics, physical chemistry, and astronomy: convection, crystallization (or more generally, phase transitions), reaction-diffusion systems (the emergence of spatial and temporal chemical patterns, e.g., in the Zhabotinski-Belusov reaction), and star and galaxy formation. I will attempt to give here a general description of the basic mechanisms of organization by using the important example of convective pattern formation, the so-called Bénard problem (see Prigogine and Stengers, 1984).

Organization takes place in systems consisting of a large number of interacting elements. These could be atoms in a liquid or crystal, or small subvolumes of liquid in convection currents, in a reaction-diffusion system or in an evolving star system—or, in the application that is of interest here, these would be synapses in nerve networks. Initially, self-organizing systems are in a relatively undifferentiated state: atoms move randomly and all subvolumes of the liquid are in the same state of motion or have the same chemical composition. Then, some small, typically random deviations from that state arise; for example, some convective fluid motion sets in. To stress the random nature of typical small deviations, they are called *fluctuations*.

In the prime example, the Bénard phenomenon, a flat vessel is filled with liquid and its bottom is homogeneously heated. As long as the temperature gradient is below a certain threshold, heat is conducted from the lower to the upper surface without bulk movement of the liquid. However, above that threshold, the warmer, lighter liquid near the bottom rises and cooler liquid from the top flows down. Under homogeneous conditions, this flow pattern is very regular and has the form of hexagons or rolls.

From this and many other organizing systems, the following three principles may be abstracted:

1. *Fluctuations self-amplify.* This self-amplification is analogous to (asexual) reproduction in Darwinian evolution. In the Bénard system, fluctuations are created by thermal motion. If a small column of liquid moves upward, more warm liquid is drawn in from the bottom, the column becomes less dense, and its upward movement is accelerated. Downward movement accelerates analogously.
2. *Limitation of resources leads to competition among fluctuations and to the selection of the most vigorously growing (the "fittest") at the expense of the others.* In the Bénard system, upward movement in one place requires downward movement in other places. The columns with the least density will win and rise.
3. *Fluctuations cooperate.* The presence of a fluctuation can enhance the fitness of some of the others, in spite of the overall competition in the field. (In many systems, the

"fitness" of a fluctuation is identical with the degree of cooperation with other fluctuations.) The liquid near a column of rising liquid is dragged up by viscosity.

The identification of these three principles with features of a concrete system is sometimes ambiguous. In the Bénard system, competition in terms of upward movement might also be seen as cooperation between upward movement occurring in one place and downward movement occurring in another place. Whole coherent patterns of movement, again, compete as long as there is local contradiction between them: liquid cannot move up and down at the same place.

A fundamental and very important observation about organizing systems is the fact that global order can arise from local interactions. Many originally random local fluctuations can coalesce into a globally ordered pattern of deviations from the original state. The intermolecular forces acting within a volume of liquid are of extremely short range, yet the patterns of convective movement they give rise to may be coherent and ordered on a large scale. This fact will be one of extreme importance to the brain, in which local interactions between neighboring cellular elements create states of global order, ultimately leading to coherent behavior.

The stage for the organization of a pattern is set by the forces between elements and by initial and boundary conditions. In the Bénard system, these forces are the hydrodynamic interactions, gravity, thermal conduction, and expansion. Boundary conditions are set by temperatures at the upper and lower boundary and by the form of the vessel. In the nervous system, the stage for the generation of connection patterns is ultimately set by prespecified rules for the interaction of cellular processes and signals, and by the environment. Because nerve cells are connected by long axons, there is an important and exciting difference between the nervous system and most other examples studied so far. Neural interactions are not necessarily topologically arranged; connected cells are "neighbors" although they may be located at different ends of the brain. This gives rise to genuinely new phenomena. Some of the ordered structures within the nervous system may not "look" ordered to our eye, which relies essentially on spatial continuity. However, in the concrete cases considered here, ordinary space will still play a dominant role.

An organizing system may contain a symmetry such that there are several equivalent organized patterns. These compete with each other during organization. In the Bénard system, if set up in a circular pan, any organized pattern could be rotated around the center of the pan by an arbitrary angle to obtain another valid pattern. One of these has to be spontaneously selected during pattern formation, a process that is called *spontaneous symmetry breaking.* When the boundary or initial conditions are slightly deformed, so that the original symmetry is destroyed, one organized pattern is favored. In general, self-organizing systems react very sensitively to symmetry-breaking influences.

Neural Network Organization

Two types of variables are relevant to network organization: signals and interconnections. Signals are the action potentials that are propagated down the axonal trees of neurons. Connections control neural interactions and are characterized by weight variables. These measure the size of the effect exerted on the postsynaptic membrane by arriving nervous impulses. Correspondingly, organization takes place on two levels: activity and connectivity.

On the ontogenetic time scale, one is interested mainly in network self-organization, which has the following general form. Assume that previous processes have already set up a primitive network. This network, together with input signals, creates activity patterns, and these activity patterns in turn modify connections by synaptic plasticity. The feedback loop between changes in synaptic strengths and changes in activity patterns must be positive, so that coherent deviations from the undifferentiated state self-amplify, conforming to the first of the principles previously formulated. The process is constrained by the requirement that modifications in a synaptic connection have to be based on locally available signals. These are the presynaptic signals, the postsynaptic signal, and possibly modulatory signals that are broadcast by central structures. The postsynaptic signal could be a local dendritic signal or the outgoing axonal signal.

The requirements of self-reinforcement and locality suffice to specify the mechanism of synaptic plasticity in excitatory synapses: A strong synapse leads to coincidences of pre- and postsynaptic signals which, in turn, increase the strength of the synapse. Hebb (1949) gave this formulation:

> When an axon of cell A is near enough to excite cell B and repeatedly or persistently takes part in firing it, some growth process or metabolic change takes place in one or both cells such that A's efficiency, as one of the cells firing B, is increased.

This rule is referred to as "Hebbian plasticity" (see HEBBIAN SYNAPTIC PLASTICITY). The corresponding rule for inhibitory synapses would have a synapse strengthened if it was successful in inhibiting the postsynaptic element. At present, however, most authors consider inhibition as a rigid service system that does not take part in network self-organization.

Hebb's rule corresponds to the "self-reproduction" of the general scheme of organization. To stabilize the system, some competition for limited "resources" has to be introduced. Most likely, there is a mechanism of isostasy by which each cell keeps the temporal average of its activity (taken over the span of some hours) constant. As a consequence, the increase in strength in some synapses must be compensated for by a decrease in others. Only the more successful synapses can grow; the less successful ones weaken and eventually disappear. For technical reasons, some models discuss a simpler competition rule for synapses, in which the sum of the synaptic weights of all synapses converging on a cell is kept constant. This rule leads to certain functional deficits and is probably not realistic. Synaptic plasticity, constrained by competition, implements organizing principles 1 and 2.

One synapse on its own cannot efficiently produce favorable events. For that it needs the cooperation of other synapses that converge onto the same postsynaptic neuron and that carry coincident signals. This implements the third organizing principle. In order for such coincidences to occur consistently, there must be a causal connection between presynaptic cells. Synaptic plasticity is the means by which the nervous system detects such causal connections. Coincidences may result from excitatory links between presynaptic neurons. They may, however, also be caused by simultaneous stimulation of sensory cells, in which case they point to the existence of causal connections in the external world.

The rules of cooperation and competition act on a local scale. The phenomenon of self-organization is the emergence of globally ordered states, as discussed in context with the emergence of global convection patterns in the Bénard phenomenon. The term *global order* is used for configurations that bring the local rules into a state of optimal mutual consistency with each other. The fact that the external world takes part in the game leads to the adaptation of the nervous system to it.

The rules for the adjustment of synaptic weights that have been introduced are able to produce ordered connection patterns. However, they do not necessarily organize the nervous system for optimal biological utility. For this, two types of controls are necessary: (1) genetic control of boundary conditions and interaction rules to favor certain useful connection patterns; and (2) control by central structures that are able to evaluate the degree of biological desirability of activity states. If a state proves to be useful, a gating signal is sent to all of the brain, or to an appropriate part of it, to authorize synaptic plasticity. That state is thereby stabilized, and the likelihood for its future appearance is increased.

Central control as the *only* criterion for growth or decay of synapses is not sufficient. Assume our nervous system evaluates the usefulness of its state once per second. It then could create less than 3×10^9 bits of information in our lifetime, for that is about the maximum number of seconds given to us. This certainly is not sufficient to regulate the strengths of all of the 10^{14} synapses of our cerebral cortex. On the other hand, this amount of information may be sufficient to select from among the relatively small universe of ordered connectivity patterns that can be created by rules of local cooperation and competition under predetermined constraints.

Conclusion

The last two or three decades have seen a revolution in thinking about organization and the origin of structures. This revolution has swept across all fields of human thinking and is also deeply affecting our view of brain and mind. In the prerevolutionary view, organization is the result of a preexisting plan, which has arisen outside the field of study in some separate agent. This plan is realized with the help of some explicit mechanism, and the study of this mechanism is the study of the process of organization. You may refer to this scheme by the word *hetero-organization*. In the post-revolutionary view, there is no preexisting plan. But there is a dynamical system of interacting elements which spontaneously fall into globally ordered patterns. Thus, the "plan" can only be read off the final product of the process of organization and is in general not accessible ahead of time.

There is a number of mathematical tools to study self-organization in a systematic way. Among them are systems of nonlinear differential equations and their analysis with the help of analytical methods, stability analysis and bifurcation theory, and numerical simulation. Some of these methods are introduced in Murray (1993; see also PATTERN FORMATION, BIOLOGICAL). An altogether different approach is represented by the methods of statistical mechanics and phase transition physics; for a general introduction, see Prigogine and Stengers (1984).

The general principles of self-organization in neural networks and other systems presented here are applied and illustrated in various articles of this *Handbook*. Prime examples are the formation of cortical domains (stripes or blobs or barrels) and the establishment of retinotopic mappings or of orderly arranged orientation sensitivity in the visual cortex (see DEVELOPMENT AND REGENERATION OF EYE-BRAIN MAPS AND OCULAR DOMINANCE AND ORIENTATION COLUMNS). Great challenges still lie ahead, especially a more detailed understanding of the ontogenesis of the brain and nervous system, the ontogenetic construction of areas and their connective architecture, the construction of the functioning brain by growth and learning, and the rapid construction of the states of mind with which we represent and deal with our environment.

Road Map: Self-Organization in Neural Networks
Related Reading: Cooperative Phenomena; "Genotypes" for Neural Networks; Hebbian Synaptic Plasticity: Comparative and Developmental Aspects; Statistical Mechanics of Learning

References

Hebb, D. O., 1949, *The Organization of Behavior*, New York: Wiley. ◆

Murray, J., 1993, *Mathematical Biology*, 2nd ed., New York: Springer-Verlag. ◆

Prigogine, I., and Stengers, I., 1984, *Order out of Chaos: Man's New Dialogue with Nature*, Toronto and New York: Bantam. ◆

Self-Organization in the Time Domain

John G. Taylor

Introduction

Besides space, time is a crucial aspect of incoming information. There are many situations in the animal world where the ability to process, store, recognize, or recall temporal sequences of patterns has great survival value to the animal. At the human level, language has allowed for the efficient handling of high-level concepts. In this article we will not consider hardwired delay-line systems, but restrict ourselves to adaptive neural networks able to learn the temporal features on inputs.

In order to achieve learning of such temporal structures, there may be an ability, in either the net itself or in an auxiliary structure, for the temporal character of the input to be buffered temporarily. This buffering might be achieved directly by reasonably long time constant neurons on a "history" net, as in the net L of Figure 1.

Leaky integrator neurons (LINs) have activity on their surface at a given time which dies away with a certain time constant. This decay of past activity corresponds to the effect of new activity being reduced by a factor (less than one) at each subsequent time step. The decay time constant is related to this reduction factor in a simple manner. Experimental data on cortical neurons show that biologically realistic values of this

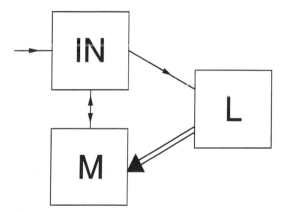

Figure 1. The architecture of the learning system for predictive self-learning. The first box, at the top left-hand side, accepts the inputs *IN*. These are then relayed in a 1 : 1 manner to the buffer net *L*. This is composed of a set of LINs, with different rates of leakage, so as to store the pattern history in differing degrees on the various neurons. The "crumbling history" held on the buffer net *L* is then transmitted to the neurons in the net *M*. These have adaptive weights on their inputs from *L*, which are modified so as to produce the next input of the sequence in *IN*. Retrieval of the sequence is achieved by connecting the output of the net *M* to the input net and letting the successive patterns of the sequence generate themselves from a suitable initial seed.

decay time for a simple neuron will have a maximum of about 50 milliseconds. However, the presence of integrating circuits with a time constant of up to 20 seconds for the vestibulo-ocular reflex indicate that model neurons with a similarly large time constant may be taken as a first approximation to more complex underlying microcircuitry.

It is important that these model LINs have a range of time constants, so that a history of the past inputs is almost in a one-to-one correspondence with the activity stored in the buffering net *L* of Figure 1. This activity is then stored and used to guide output of the further network *M* of Figure 1 to produce the *next* input (Reiss and Taylor, 1991), leading to the method of predictive self-learning, which will be discussed in more detail after the more complete description of temporal neurons in the next section. Alternatively, the changing nature of the patterns can be learned directly by comparison of new with ongoing net activity. The adaptation can be done using Hebbian learning in the case, say, of a recurrent Hopfield net (Amit, 1989) to learn the transition between sequences of patterns. This might be called the temporal Hopfield map. One may also use a Mexican-hat style of lateral inhibition to produce a topographic map, but with temporal neurons holding a history of past activity. This method is appropriately termed the temporal Kohonen map. This approach can also be improved by storage of histories of inputs on separate synapses leading to traces and improved sequence storage (Critchley, 1994). The method and these extra features will be discussed in the section "Temporal Topographic Maps."

There have been numerous other approaches to temporal sequence storage and generation using temporal neurons, for example, the outstar avalanche approach of Grossberg. This does not use a range of time constants for the neurons, so may have problems with the disambiguation of sequences which have repetitions in them (such as the sequence ABCABD). There is also the fact that loss of the feedback structure of hippocampus causes amnesia for episodic memory (see SHORT-TERM MEMORY); neurobiological realism therefore supports the use of feedback architectures for some aspects of temporal sequence learning, rather than simple feedforward systems.

Temporal Neurons

Leaky integrator neurons act by storing activity coming onto their surface, this activity $A(t)$ at time t dying away with certain time constant. Thus in discrete time

$$A(t + 1) = (1 - d)A(t) + I(t) \qquad (1)$$

where $(1 - d)$ is the reduction factor experienced at each time step, and $I(t)$ is the input at that time. Further temporality could be obtained by including nontrivial geometry for the neurons, but we will not pursue that topic here. A simple way of looking at what the LIN type of node can achieve (Bressloff and Taylor, 1992) is to consider the task of learning a given binary classification of a set of sequences $S^m(r)$ (r denoting time, m the pattern label). For LINs on each of the input lines of a single-layer perceptron, their activity at time t, on iteration of the right-hand side of Equation 1 successively, with all the LINs having the same time constant, will be

$$\sum_{r=0}^{t} \tilde{S}^m(r)(1 - d)^{t-r} \qquad (2)$$

which will be denoted as $\tilde{S}^m(t)$.

Thus if $S^m(r)$ for $0 < r < t$ is applied to the input LINs, then the output at time t is the value $\tilde{S}^m(t)$. The classification problem, for which the outputs of some of the input patterns are to be 1 and for the remaining patterns 0, can be solved by a single-layer perceptron provided the set of patterns $\tilde{S}^m(r)$ is linearly separable. This is a different problem from that of the linear separability of the set of patterns $S^m(r)$. Consider, for example, the two-dimensional vectors

$$A = (1, 0)^T \qquad B = (0, 1)^T$$

and consider the task of learning the mappings $AB \to 1, 0$; $BA \to 0, 0$. This is an ordering problem, since the pattern A produces output 1 or 0 according to whether it is before or after the pattern B; it could not be solved by a standard perceptron. But it can be learned for the pattern $\tilde{S}$ formed by Equation 2, since (with $k = 1 - d$)

$$\tilde{A}(1) = (1, 0)^T \qquad \tilde{B}(1) = (0, 1)^T$$
$$\tilde{A}(2) = (k, 1)^T \qquad \tilde{B}(2) = (1, k)$$

and the sets $\tilde{A}(1)$ and $\tilde{A}(2)$, $\tilde{B}(1)$, $\tilde{B}(2)$ are linearly separable in $\mathbb{R}^2$. We should add that choice of different values for d on the different input LINs adds greater flexibility. For example, the mapping $AAA \to 101$, on two input LINs with $A = (1, 1)^T$, is not linearly separable if $d_1 = d_2$ (where d_1 and d_2 are the decay constants of the two LINs) but is so if $d_1 \neq d_2$. Learning the correct set of d's to make a task linearly separable is an interesting problem.

Finally, input channel dynamics can be included by replacing $I(t)$ by channel variable $C_2(t)$, which is coupled to another channel variable C_1 by an equation like Equation 1, while C_1 instead of A is driven by $I(t)$ by Equation 1. The result is equivalent to convoluting the input with the *alpha function* $\alpha^2 t e^{-\alpha t}$ before adding it on the right of Equation 1. This procedure gives an effect of delaying the maximum of the input till a time that is a fraction $1/\alpha$ of its original arrival at the synapse. It is relevant to note that some papers on biological memory (e.g., Bliss et al., in Baudry and Davies, 1991) consider long-term potentiation (LTP; see HEBBIAN SYNAPTIC PLASTICITY) as a reduction in the latency $1/\alpha$, so that it should be added to the biologically realistic adaptive parameters of the net (unlike the time constant).

Predictive Self-Learning

From temporal sequence classification let us turn to temporal sequence storage (TSS), using what is termed *predictive self-learning*. The basic idea behind this approach (Reiss and Taylor, 1991) is to use a single-layer perceptron net (in general with sigmoidal output) to produce the next input $P(t + 1)$ of a pattern sequence, the input being a stored history of the earlier patterns by means of a net of LINs. This may be achieved by the architecture of Figure 1, where the input IN is fed in a 1 : 1 fashion into the net L of LINs with a range of time constant

and channel variables. The output of these LINs, suitably thresholded, is fed onto the SLP net M. The weights of these latter inputs are modified in the teaching phase, so that the output of M is identical to the input on IN at one later time step (by the delta learning rule; see PERCEPTRONS, ADALINES, AND BACKPROPAGATION). On feeding the output of M back into IN then, the stored sequence (provided it has been correctly learned) will be generated from the net M from that time.

A great deal of simulation of the predictive self-learning approach (Reiss and Taylor, 1991) led to a general understanding of the parameters needed to achieve effective response. The range of time constants must cover the length of the sequences, in steps able to separate their main patterns. Such a criterion is to be expected from the need to be able to store the patterns without loss of information. The number of output neurons must be equal to that of the inputs (which is fixed by the feedback architecture). Finally the number of adaptive weights on each of the output neurons must be at least equal to the number of distinct patterns to be stored in the sequence (a feature arising from the capacity of the system, as will be discussed shortly).

Improvement of pattern storage has been achieved (Reiss and Taylor, 1991) by adding to the architecture of Figure 1 lateral connections to the net M so as to increase the size of basins of attraction for each of the patterns. The method of TSS seems very effective and was explored by Reiss and Taylor (1991) for its dependence on the various channel and neuron parameters. In particular the storage capacity was shown by simulation to be of $O(N)$, where N is the number of neurons in L. This result was also shown more recently by replica symmetry using statistical mechanical techniques.

The tools of statistical mechanics allow averaged properties of single-layer neural networks, with binary decision nodes, to be analyzed very efficiently (see STATISTICAL MECHANICS OF LEARNING). The crucial quantity for such nets is their partition function, from which the volume of the weight space of nets able to solve a given input-output task may be determined. The vanishing of this volume corresponds to the net reaching maximum capacity. Calculation of the partition function for a single-layered perceptron, using replica symmetry (in which many copies are used, with a suitable symmetry between them) is well known (Amit, 1989). These methods were extended to the case of LINs in the output (Bressloff and Taylor, 1992), leading to agreement with the simulation results noted in the preceding paragraph.

A slightly different version of TSS (Bressloff and Taylor, 1992), still using the architecture of Figure 1, guarantees convergence of the learning algorithm for suitable temporal sequences. This approach uses linear outputs of the LIN net L, and the nodes of M are thresholded, which becomes a threshold SLP. In terms of the transformed pattern space $\tilde{S}^m(t)$ one can use the perceptron learning algorithm for the change of the weight from the ith LIN to the jth neuron in M, with guarantee of convergence provided the set of patterns $\tilde{S}^m(t)$ is linearly separable. Moreover, an extension can be given to the optimal learning algorithm of Gardner (1988) with a nonzero stability parameter K (which guarantees recall of degraded patterns with a noise level determined by K). This model has not yet been explored by simulation. The learning of a function of time can also be obtained by these approaches.

The Temporal Hopfield Net

The Hopfield net (Hopfield, 1982; see COMPUTING WITH ATTRACTORS) used a set of neurons laterally coupled through weights, $w_{ij} = w_{ji}$ (where, i, j label the neurons with no self-coupling, so $w_{ii} = 0$). Storage of a set of P patterns $\{u_i^\mu\}$, $\mu = 1, \ldots, P$ was found to be effective by means of the pseudo-Hebbian correlation rule

$$w_{ij}^{(1)} = \frac{1}{N} \sum_{\mu=1}^{P} u_i^\mu u_j^\mu \qquad (3)$$

This rule works well for recall of noisy patterns provided $P < 0.14N$; more patterns can be stored by using, for example, the inverse of the pattern correlation matrix for correlated and biased patterns (Coombes and Taylor, 1994). To store temporal sequences it was proposed (Sompolinsky and Kanter, 1986; Kleinfeld, 1986) that the transition overlap between patterns, if ordered temporally by index μ with exponentially smoothed incoming activity on each neuron, be effective

$$w_{ij}^{(2)} = \frac{\lambda}{N} \sum u_i^{(\mu+1)} u_j^{(\mu)} \qquad (4)$$

$$h_i^{(2)}(t) = \int_{-\infty}^{t} dt^1 w(t - t^1) \sum w_{ij}^{(2)} u_j(t^1) \qquad (5)$$

Here $h_i^{(2)}$ is the activity at time t, to which the activity coming from the effect of $w_{ij}^{(1)}$ of Equation 3 in the connection matrix must be added, in order to determine the update of activity on the ith neuron. The resulting network proves effective in storage of sequences, up to the capacity limit of $0.14N$, as for the static net. It has been extended in various ways, to be more robust and faithful in recall to the initial pattern temporal structure. However, there are problems of the creation of spurious states and of a low maximal capacity.

Temporal Topographic Maps

We now turn to an alternate approach to the storage of temporal sequences which is based on some biological realism for its foundation, the Kohonen topographic map (see SELF-ORGANIZING FEATURE MAPS: KOHONEN MAPS). It is proposed to extend this to take complete account of temporal features of neurons by replacing the nontemporal neurons in Kohonen maps by the LINs discussed in the previous section. In this way it may be possible to include contextual content in a biologically realistic manner, in comparison to the 15,000–20,000 expert system rules apparently needed for the phonetic typewriter (see SPEECH RECOGNITION: A HYBRID APPROACH). Simulations of the temporal topographic map approach (Chappell and Taylor, 1993) have already shown the validity of the idea; our purpose here is to describe the system and why it is expected to work in a qualitative manner.

Some constraint must be put on weight vectors in order for competitive learning to produce a topographic map. Alternatively, there must be a modification of the input activity as a linearly weighted sum, by including quadratic terms in the weights and inputs. Taking the latter approach, we replace the update rule (Equation 1) for a given neuron, with synaptic weight vector w, by

$$A(t + 1) = (1 - d)V(t) - (1/2)\|I(t) - w\|^2 \qquad (6)$$

where I and w are vectors of the same size. We will limit our discussion to binary inputs only, corresponding to biologically realistic spike train communication. We also restrict to one dimension for the moment. Consider only trying to store sequences of length k, although sequences of arbitrary length will be considered in the training. Binary sequences of length k, may be enumerated, say, by the integer n. Then, maximizing the activity at any one time t in Equation 6 for the sequence n leads to a value for the winning weight. This may be shown,

from Equation 6, to have the value

$$w(n) = S_k(n) + \text{Noise} \qquad (7)$$

The first term in Equation 7 is that arising from summing the binary input sequences of length k by an LIN, as in the section "Temporal Neurons." This term S_k is therefore the sum of powers of d for each of the spikes in the input sequence, while the noise, for sequences of arbitrary length (and assuming the training has been done), has the value of the remaining terms in such sums for a LIN, and so has the form

$$\text{Noise} = (1 - d)^k d \cdot z$$

where z is a random variable uniformly distributed over $(0, 1)$ [as may be shown by using the expansion of z to the base $(1 - d)^{-1}$]. For $d > 1/2$ the values $S_k(n)$ are a set of 2^k numbers in $(0, 1)$ ordered by n. With the same condition on d, the noise terms do not cause the values of $w(n)$ to lose the ordering given by n. Thus applying the usual training rule of rotating the weight of the winning node of each input, and those of its neighbors, to be closer to the input at that time, there will result a linearly ordered topographic map along the line of the 2^k input sequences of length k. The neighborhood function must be chosen to have a width ultimately smaller than the noise term in Equation 7, and so be bounded by $(1 - d)^k d$. To obtain the initial convergence from a random initial set of weight values is not necessarily simple, however, and careful choice is needed of the neighborhood function. Simulations, in both one and two dimensions, bear out this analysis. Moreover, they indicate a more powerful method of inclusion of context by means of following the whole winning trajectory during the input period, instead of solely using the final winner. Care must be taken to avoid ambiguity if two sequences have many common elements, but any system (without some attempt at chunking) will find such a task difficult. Reference should also be made to Wang and Arbib (1990), who use LINs to create sequence-detecting neurons, although without any spatial topography in the resulting net (see TEMPORAL PATTERN PROCESSING).

A more effective way to store sequences of arbitrary length is to use traces on input lines (Critchley, 1994). Thus a trace x_j on the jth input line is defined by the update rule

$$x_j(t) = (1 - \delta)x_j(t - 1) + I_j(t) \qquad (8)$$

where $I_j(t)$ is the binary input at time t on the jth input line. Thus a history of the inputs on a particular axon is stored at the synapse and does not intermingle with the net membrane potential. The traces are used in place of the actual inputs in the neural dynamics. If the separate histories of inputs are held on appropriate dendritic spines (which may be valid for distal dendrites at great electrotonic length from the soma) then more precise processing appears possible. Thus binary input sequences of arbitrary length are mapped uniquely by Equation 8 into points in the interval $[0, \delta]$. Thus the standard Kohonen learning rule applied to the resulting traces $\{\bar{x}_j\}$ leads to a topographic representation of the input sequences. This approach is currently being analyzed for its applications to speech and other temporal sequence problems.

Discussion

All the methods presented in this article have some non-zero degree of biological reality. The looped structure of the hippocampus and the loop of the thalamus, frontal lobe, and basal ganglia may be sites of implementation of predictive self-learning. There may also be a more local form of this learning (Reiss and Taylor, 1992). The hippocampus, especially the field CA3, with its extensive axon collaterals, is appropriate to consider for the temporal Hopfield map. It has been suggested that similar processing is occurring in temporal lobe pattern storage (Amit, 1993). This memory system is not expected to be topographic in structure, especially since there is now accepted to be a complete lack of topography in the ordering in CA3 of cells responsive to particular external cues (the so-called place cells; see O'Keefe and Nadel, 1975; see also HIPPOCAMPUS: SPATIAL MODELS, where the hippocampus is analyzed in spatial, rather than temporal, terms). However, all of these maps are undoubtedly self-organizing, with no external teacher. Finally, the temporal Kohonen map may be regarded as a refinement of the Kohonen map, allowing for some sort of topographic representation, say, of spoken or written words. However, it appears necessary to include the relational structure between the words to give any sense of meaning to them, say, by the ideas associated with action-based scripts and schemes.

Road Map: Self-Organization in Neural Networks
Background: I.3. Dynamics and Adaptation in Neural Networks
Related Reading: Spatiotemporal Association in Neural Networks

References

Amit, D., 1989, *Modelling of Brain Function*, Cambridge, Eng.: Cambridge University Press.
Amit, D., 1993, Cognitive neuro-psychology: An empirical basis for neural modelling and cognitive psychology, in *Neural Computing: Research and Applications* (G. Orchard, Ed.), Bristol, Eng.: Adam Hilger, pp. 21–28.
Baudry, M., and Davies, J. L., Eds., 1991, *Long-Term Potentiation*, Cambridge, MA: MIT Press.
Bressloff, P. C., and Taylor, J. G., 1992, Temporal Sequence Storage capacity of time-summating neural networks, *J. Phys. A*, 25:833.
Chappell, G. J., and Taylor, J. G., 1993, The temporal Kohonen map, *Neural Netw.*, 6:441–445.
Coombes, S., and Taylor, J. G., 1994, Using generalized principal component analysis to achieve associative memory in a Hopfield net, *Network*, 5:75–88.
Critchley, D., 1994, The self-organizing map, PhD Thesis, University of London.
Gardner, E., 1988, The space of interactions in neural models, *J. Phys. A*, 21:257–270.
Hopfield, J., 1982, Neural networks and physical systems with emergent collective computational abilities, *Proc. Natl. Acad. Sci. USA*, 79:2554–2558.
Kleinfeld, D., 1986, Sequential state generation by model neural networks, *Proc. Natl. Acad. Sci. USA*, 83:9469–9473.
O'Keefe, J., and Nadel, L., 1975, *The Hippocampus as a Cognitive Map*, Oxford: Oxford University Press.
Reiss, M., and Taylor, J. G., 1991, Storing temporal sequences, *Neural Netw.*, 4:773–787.
Reiss, M., and Taylor, J. G., 1992, Does the hippocampus store temporal sequences? *Neural Netw. World*, 3:365–384.
Sompolinsky, H., and Kanter, I., 1986, Temporal association in asymmetric neural networks, *Phys. Rev. Lett.*, 57:2861–2864.
Wang, D., and Arbib, M., 1990, Complex temporal sequence learning based on short-term memory, *Proc. IEEE*, 78:1536–1543.

Self-Organizing Feature Maps: Kohonen Maps

Helge Ritter

Introduction

A first and very important step in many pattern recognition and information-processing tasks is the identification or construction of a reasonably small set of important features in which the essential information for the task is concentrated.

The *self-organizing feature map* is an approach by which such features can be obtained by means of an unsupervised learning process. The feature map is a nonlinear method that is based on a layer of adaptive units which gradually develop into an array of feature detectors that is spatially organized in such a way that the location of the excited units becomes indicative of statistically important features of the input signals. In this process, the importance of features is derived from the statistical distribution of the input signals (*stimulus density*). In particular, clusters of frequently occurring input stimuli will become represented by a larger area in the map than clusters of more seldom occurring stimuli. The resulting correspondence between signal features and response locations in the layer has many properties of a nonlinear, compressed image or "map" of the original signal space and resembles very much the *topographic feature maps* found in many brain areas. As a neural model, the feature map provides a bridge between microscopic adaptation rules postulated at the single-neuron or synapse level, and the formation of experimentally better, accessible, macroscopic patterns of feature selectivity in neural layers. From the viewpoint of applied neural computation, the feature map provides a nonlinear generalization of *principal component analysis* and has proven valuable in many different application contexts, ranging from pattern recognition and optimization to robotics (Kohonen, 1984; Ritter, Martinetz, and Schulten, 1992).

The Basic Feature Map Algorithm

The neurons of a feature map form a two-dimensional sheet and are connected to a common bundle of input fibers, schematically depicted in Figure 1. Any input pattern is encoded as an activity pattern on the input fibers and gives rise to excitation of some local group of neurons. For self-organization of a

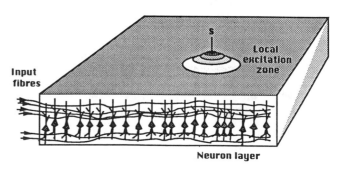

Figure 1. Schematic representation of a feature map. Nerve fibers providing the input signal excite the neurons by way of synaptic connections. Lateral interactions restrict the neural responses and the synaptic adaptation to a local excitation zone. Each possible position *s* of the excitation zone can be viewed as a compressed image in a two-dimensional space of the original stimulus features.

feature map, this excitation must cause suitable local changes in the response properties of these neurons. These changes must be such that after a learning phase the spatial positions of the excited groups specify a mapping of the input patterns onto the two-dimensional sheet with the property of a topographic map; i.e., the distance relations in the high-dimensional space of the input signals must be approximately represented as distance relationships on the two-dimensional neural sheet. The main requirements for such self-organization are that (1) the neurons are exposed to a sufficient number of different inputs, (2) for each input, only the synaptic input connections to the excited group are affected, (3) similar updating is imposed on many adjacent neurons, and (4) the resulting adjustment is such that it enhances the same responses to a subsequent, sufficiently similar input (Kohonen, 1982).

Mathematically, the neural sheet is represented in a discretized form by a (usually) two-dimensional lattice A of formal neurons. The input pattern is described by an input vector $\mathbf{x}$ from some pattern space V. The responsiveness of a neuron at a site $r \in A$ is measured by the dot product $\mathbf{x} \cdot \mathbf{w}_r$, where $\mathbf{w}_r$ is the vector of the neuron's synaptic efficacies (the dimensionality of $\mathbf{x}$ and of the $\mathbf{w}_r$ is equal to the number n of input lines). The input vectors are normalized to unit length (the weights $\mathbf{w}_r$ need not be normalized explicitly because sooner or later the process will normalize them automatically). The connections to neurons will be modified if they are close to the site s for which $\mathbf{x} \cdot \mathbf{w}_s$ is maximal. A function h_{rs} that takes larger values for sites r close to the chosen s is used to describe this distance-dependent modulation of the adaptive changes. A rather realistic modeling choice for h_{rs} is a Gaussian

$$h_{rs} = \exp\left(-\frac{\|r - s\|^2}{\sigma^2}\right) \tag{1}$$

whose variance $\sigma^2/2$ will control the radius of the group. The adjustments corresponding to the input $\mathbf{x}$ are then given by

$$\mathbf{w}_r^{(\text{new})} = (1 - \varepsilon \cdot h_{rs})\mathbf{w}_r^{(\text{old})} + \varepsilon \cdot h_{rs} \cdot \mathbf{x} \tag{2}$$

Equation 2 can be justified by assuming the traditional Hebbian law for synaptic modification together with an additional nonlinear, "active" forgetting process for the synaptic strengths (Kohonen, 1984).

An important role in this process is played by the *neighborhood kernel* h_{rs}. Its effect is to focus adaptation to an entire *neighborhood zone* of the layer A, but to a varying degree, decreasing the amount of adaptation with distance from the respective adaptation center s chosen at each step (*neighborhood cooperation*; it is essential that the distance dependence of h_{rs} be based on *distances in the space of the lattice A*, not on distances in the original signal space V). Consequently, neurons that are close neighbors in A will tend to specialize on similar patterns. After learning, this specialization is used to define the mapping from the space V of patterns onto the (discretized) space A: each pattern vector $\mathbf{x} \in V$ is mapped to one of the discretized locations of A that are represented by the formal neurons. The image of $\mathbf{x}$ under this mapping is defined to be the location $s = s(\mathbf{x})$ associated with the neuron for which $\mathbf{x} \cdot \mathbf{w}_s$ is largest (the *winner neuron*).

A frequently useful modification of this basic algorithm is to determine the site $s(\mathbf{x})$ by *minimizing the Euclidean difference* $\|\mathbf{x} - \mathbf{w}_s\|$. This approach is equivalent to the previous method of maximizing the scalar product $\mathbf{w}_s \cdot \mathbf{x}$ when the vectors are normalized. However, in most cases this modification also works very well when normalization is omitted. It is often somewhat simpler to use, and most theoretical aspects of the method are easier to consider in this setting.

Visualization of Feature Maps

There are two main ways to visualize a feature map. In the first approach, one labels each neuron by the test pattern that excites this neuron maximally (*best stimulus*). This procedure resembles the experimental procedure by which sites in a brain area are labeled by those stimulus features that are most effective in exciting neurons at this site. The labeling produces a partitioning of the lattice A into a number of coherent regions (for a well-ordered map), each of which contains neurons that are specialized for the same pattern. In the example of Figure 2, each training and test pattern was a coarse description of one of 16 animals (using a data vector of 13 simple binary-valued features; Ritter and Kohonen, 1989). Evidently, in this case a topographic map that exhibits the similarity relationships among the 16 animals has been formed.

In the second approach, the feature map is visualized as a *virtual net* in the original pattern space V. The virtual net is the set of weight vectors $\mathbf{w}_r$ displayed as points in the pattern space V, together with lines that connect those pairs $(\mathbf{w}_r, \mathbf{w}_s)$, for which the associated neuron sites (r, s) are nearest neighbors in the lattice A. The virtual net is very well suited to display the topological ordering of the map. Unfortunately, its use is limited to continuous and at most three-dimensional spaces. Figures 3A and 3B show the development of the virtual net of a two-dimensional 20×20-lattice A from a disordered initial state (Figure 3A) into an ordered final state (Figure 3B) when the stimulus density is concentrated in the vicinity of the surface $z = x \cdot y$ in the cube V given by $-1 < x, y, z < 1$.

Discussion

The main characteristics of the feature map algorithm. Geometrically, the adaptive process (Equation 1) can be viewed as a sequence of *local deformations* of the virtual net in the space of input patterns. As a result, the virtual net gradually becomes deformed in such a way that it *approximates the shape of the stimulus density* $P(\mathbf{x})$ in the space V. If the topology of the virtual net (which is the same as the topology of the lattice A) and the topology of the stimulus density are the same (as, e.g., in Figures 3A, 3B), there will be a smooth deformation of the

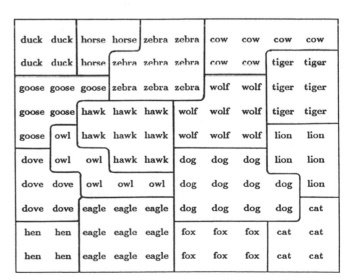

Figure 2. Visualization of a 10×10 feature map for a set of pattern vectors describing binary features of 16 animal species. The spatial arrangement of the labeled map regions reflects the similarity relationships between the animals.

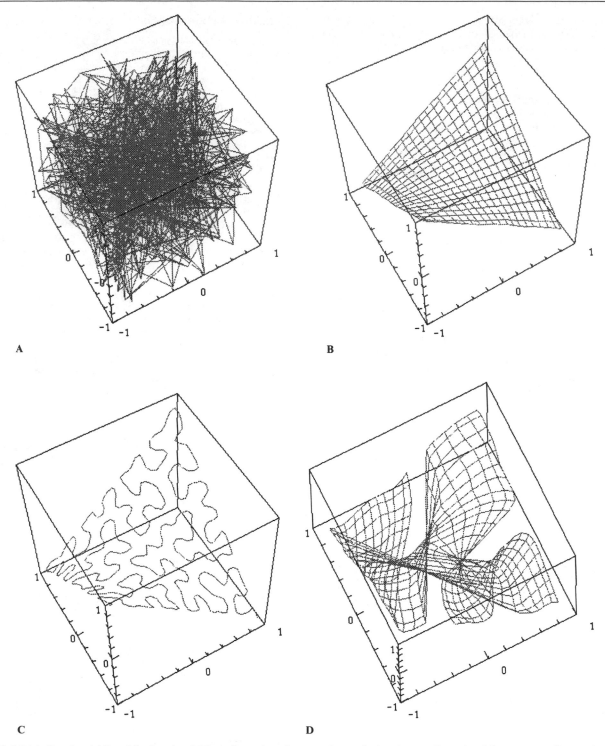

A

B

C

D

Figure 3. Initial, disordered (*A*) and final, ordered (*B*) configuration of the *virtual net* for a two-dimensional feature map, developing under the influence of a two-dimensional stimulus density that is embedded in a three-dimensional signal space. *Dimension conflict* is shown in part *C*:

when replacing the two-dimensional feature map of parts *A* and *B* by a one-dimensional chain, a space-filling fractal curve results. A *topological defect* (*D*) is characterized by several patches of globally conflicting local orderings.

virtual net so that it matches the stimulus density precisely, and the feature map algorithm will find this deformation in the case of successful convergence. If both topologies differ, e.g., if the dimensionalities of the stimulus manifold and the virtual net are different, no such deformation will exist, and the resulting approximation will be a compromise between the then conflicting goals of matching spatially close points of the stimulus manifold to points that are neighbors in the virtual net. An example is depicted in Figure 3C, where the map manifold is one-dimensional (a chain of 400 units), the stimulus manifold is two-dimensional (the surface $z = x \cdot y$), and the embedding space V is three-dimensional (the cube $-1 < x, y, z < 1$). In this case, the dimensionality of the virtual net is lower than the dimensionality of the stimulus manifold (this is the typical situation), and the resulting approximation resembles a "space-filling" fractal curve.

Properties of the features in a feature map.   The geometric interpretation of the previous section suggests that a good approximation of the stimulus density by the virtual net requires that the virtual net is oriented tangentially at each point of the stimulus manifold. Therefore, a d-dimensional feature map will select a (possibly locally varying) subset of d independent features that capture as much of the variation of the stimulus distribution as possible. This is an important property that is also shared by the method of PRINCIPAL COMPONENT ANALYSIS (q.v.). An important difference, however, is that the selection of the "best" features can vary smoothly across the feature map and can be optimized locally. Therefore, the feature map can be viewed as a nonlinear extension of principal component analysis.

Relations of the feature map with more traditional approaches.   The feature map is also related to the method of *vector quantization* (see LEARNING VECTOR QUANTIZATION). The aim of vector quantization is to achieve data compression by mapping the members of a large (possibly continuous) set of different input signals to a much smaller, discrete set (the *codebook*) of code labels.

This codebook can be identified with the set of weight vectors $\mathbf{w}_r$ of a feature map. Then the determination of the site $s(\mathbf{x})$ of the winner neuron can be considered as assignment of a code s to a data vector $\mathbf{x}$. From this code, $\mathbf{x}$ can be reconstructed approximately by taking $\mathbf{w}_{s(\mathbf{x})}$ as its reconstruction. This procedure results in an average (mean square) reconstruction error

$$E = \int P(\mathbf{x})(\mathbf{x} - \mathbf{w}_{s(\mathbf{x})})^2 d^d\mathbf{x} \qquad (3)$$

and it can be seen that in the special case $h_{rs} = \delta_{rs}$ (absence of neighborhood cooperation) the adaptation rule (Equation 2) is a stochastic minimization procedure for E, and becomes equivalent to a standard algorithm for vector quantization. However, a further source of reconstruction errors may be the confusion of two codes s', s'' with some probability $p(s', s'')$. This situation requires the minimization of a more general error measure, now given by

$$E = \sum_{s', s''} p(s', s'') \int_{s(\mathbf{x}) = s'} P(\mathbf{x})(\mathbf{x} - \mathbf{w}_{s'})^2 d^d\mathbf{x} \qquad (4)$$

It can be shown that the minimization of this error measure, and, therefore, the construction of an optimal code for the new situation, is closely related to the feature map formation, provided one takes $h_{ss'} = p(s, s')$, i.e., the neighborhood cooperation in the map is derived from the "confusion matrix" $p(s, s')$ (Luttrell, 1990; Kohonen, 1991).

Characterizing the ordering achieved with a feature map.   In the general case, a straightforward quantitative characterization of the nature of the *ordering* produced by the feature map turns out to be difficult, except for the case when both V and A are one-dimensional. For higher-dimensional cases, no comparably clear answer is known. In many practical applications, one may not even have a good guess for the "intrinsic" dimensionality of the signal manifold. Using a two-dimensional feature map then must be considered as an attempt to find out how well a two-dimensional map can approximate the unknown data topology, or one uses approaches where the lattice A can dynamically adjust its topology to the data (Martinetz and Schulten, 1994). A useful measure to compare the "faithfulness" of feature maps of different dimensionalities is the *wavering product* (Bauer and Pawelzik, 1992). Another possibility is to seek for a formulation of the feature map algorithm in terms of an optimization problem and take the cost function of this problem, e.g., the function E in Equation 4, as a measure of ordering (Kohonen, 1991). It has been shown, however, that there is no "energy function" (i.e., the adaptation rule of Equation 2 cannot be derived from the gradient of some "order measure") for the feature map algorithm in general (Erwin, Obermayer, and Schulten, 1992).

Stationary states of the feature map algorithm.   The issue of ordering is closely related to the issue of classifying the possible *stationary states* of the feature map algorithm. A stationary state is a configuration for which the average change of weights per adaptation step vanishes. A stationary state is stable if all sufficiently small perturbations from it decay on the average. However, for a given stable stationary state this definition still leaves the possibility that there may exist "very unlikely" sequences of adaptation steps which lead to a different stationary state. If such sequences exist, a stable stationary state is called *metastable*; otherwise it is called *absorbing*. Only the absorbing states are the "true" final states of the time evolution of a feature map. If the system, however, is driven into a metastable state, it may become "trapped" for such a long time that the state cannot be distinguished from a true absorbing state within realistic observation times. Usually, the metastable states are those that occur as only partially ordered, while those states that intuitively appear as fully ordered seem to have the property of being absorbing (Erwin et al., 1992).

Speed and reliability of the ordering process in relation to the various parameters of the model.   The convergence process of a feature map can be roughly subdivided into a first ordering phase, in which the correct topological order is produced, and a subsequent fine-tuning phase. So far, rigorous convergence proofs for the full process have only been obtained for the one-dimensional case and for the two-dimensional case when the correct topological order is specified along the border of A (Cottrell and Fort, 1986; Erwin et al., 1992).

A very important role for the first phase is played by the neighborhood kernel h_{rs}. Usually, h_{rs} is chosen as a function of the distance $\|r - s\|$; i.e., h_{rs} is translation invariant. The algorithm will work for a wide range of different choices (locally constant, Gaussian, exponential decay) for h_{rs}, but for fast ordering the function h_{rs} should be convex over most of its support. Otherwise, the system may get trapped in partially ordered, metastable states, and the resulting map will exhibit "topological defects," i.e., conflicts between several locally ordered patches (Figure 3D shows a typical example) (Erwin et al., 1992).

Another important parameter is the distance up to which h_{rs} is significantly different from zero. This range sets the radius of the adaptation zone and the length scale over which the response properties of the neurons are kept correlated. The smaller this range, the larger the effective number of degrees of freedom of the network and, correspondingly, the harder the ordering task from a completely disordered state. Conversely, if this range is large, ordering is easy, but finer details are averaged out in the map. In addition, for Gaussian neighborhood kernels, a sufficiently large range also makes the function h_{rs} convex over the entire network. Therefore, formation of an ordered map from a very disordered initial state is favored by a large initial range (a sizable fraction of the linear dimensions of the map) of h_{rs}, which then should decay slowly to a small final value (on the order of a single lattice spacing or less). Although statistical considerations seem to dictate a $1/t$ decay law (Heskes and Kappen, 1993), the faster exponential decay is suitable in many cases.

The relationship between stimulus density and weight vector density. During the second, fine-tuning phase of the map formation process, the density of the weight vectors becomes matched to the signal distribution. Regions with high stimulus density in V lead to the specialization of more neurons than regions with lower stimulus density. As a consequence, such regions appear magnified on the map; i.e., the map exhibits a locally varying *magnification factor*. In the limit of no neighborhood, the asymptotic density of the weight vectors is proportional to a power $P(\mathbf{x})^\alpha$ of the signal density $P(\mathbf{x})$ with exponent $\alpha = d/(d + 2)$. For a nonvanishing neighborhood, the power law remains valid in the one-dimensional case, but with a different exponent α that now depends on the neighborhood function (Ritter, 1991). For higher dimensions, the relation between signal and weight vector distribution is more complicated, but the monotonic relationship between local magnification factor and stimulus density seems to hold in all cases investigated so far.

The effect of a "dimension conflict." In many cases of interest, the stimulus manifold in the space V is of higher dimensionality than the map manifold A, and, as a consequence, the feature map will display the features that have the largest variance in the stimulus manifold. These features have also been termed *primary*. However, under suitable conditions further, *secondary*, features may become expressed in the map. The representation of these features is in the form of repeated patches, each representing a full "miniature map" of the secondary feature set. The spatial organization of these patches is correlated with the variation of the primary features over the map: the gradient of the primary features has larger values at the boundaries separating adjacent patches, while it tends to be small within each patch. The conditions for the occurrence of secondary features have been analyzed for simplified situations (Ritter et al., 1992; Obermayer, Blasdel, and Schulten, 1992). These results show that the stability of a two-dimensional map with only two primary features requires that the ratio of the variance of the primary features across the range of the neighborhood function to the variance of the signal manifold perpendicular to the directions of the primary features be below a certain threshold. If the signal distribution is such that no such map configuration exists, additional features will become expressed in the map, with the role of the secondary features assigned to the features with the lower variance.

Modeling the properties of observed brain maps. Many regions in the brain are known to be topographic representations of sensory surfaces. It has been shown that the qualitative structure, including, e.g., the spatially varying magnification factor and certain experimentally induced reorganization phenomena, can be reproduced with the Kohonen feature map algorithm (Ritter et al., 1992). A more stringent test is provided by the more complex maps that are found in the visual cortex (see OCULAR DOMINANCE AND ORIENTATION COLUMNS). In V1, there is a hierarchical representation of the features retinal position, orientation, and ocular dominance, with retinal position acting as the primary feature (pair). The secondary features orientation and ocular dominance form two correlated spatial structures: (1) a system of alternating bands of binocular preference and (2) a system of regions of orientation-selective neurons arranged in parallel iso-orientation stripes such that orientation angle changes monotonically in the perpendicular direction. The iso-orientation stripes are correlated with the binocular bands such that both band systems tend to intersect perpendicularly. In addition, the orientation map exhibits several types of singularities that tend to cluster along the monocular regions between adjacent stripes of opposite binocularity. It turns out that all these features, including even many quantitative aspects, can be remarkably well reproduced by the Kohonen feature map algorithm (Obermayer et al., 1992; Blasdel and Obermayer, 1994). One may conclude that, despite its computational simplicity, the Kohonen feature map algorithm can successfully model a striking range of features of observed brain maps.

Applications of the feature map algorithm and local linear maps. Artificial feature maps have proven useful in many pattern recognition applications. The "classic example" is the *Neural Typewriter* of Kohonen, where a feature map is used to create a map of phoneme space for subsequent speech recognition (see SPEECH RECOGNITION: A HYBRID APPROACH). Subsequent work has demonstrated the possibility of creating feature maps of language data that are ordered according to higher-level, semantic categories (Ritter and Kohonen, 1989). Other applications of the feature map include process control, image compression, time series prediction, optimization, generation of noise-resistant codes, synthesis of digital systems, and robot learning (Kohonen, 1990).

For some of these applications, an extension of the basic feature map to produce a vectorial output value has been proven very useful. To this end, each formal neuron is considered as computing a *locally valid linear map* (Ritter et al., 1992):

$$\mathbf{y}_r(\mathbf{x}) = \mathbf{w}_r^{(out)} + \mathbf{A}_r(\mathbf{x} - \mathbf{w}_r^{(in)}) \tag{5}$$

of the input pattern $\mathbf{x}$. Selection of a winner neuron s and adaptation of the synaptic weights (now denoted by $\mathbf{w}_r^{(in)}$) proceeds as in Equations 1 and 2. The output of the network can be taken as $\mathbf{y}^{(net)}(\mathbf{x}) = \mathbf{y}_{s(\mathbf{x})}(\mathbf{x})$, i.e., as the output of the local map of the winner neuron $s(\mathbf{x})$. A better choice that avoids discontinuities when the winner changes is to use the map for a smooth "gating" of the contributions of all neurons such that the winner neuron s makes the strongest contribution and the other neurons r contribute more weakly with increasing distance between r and s.

The additional quantities (output weights $\mathbf{w}_r^{(out)}$ and matrices $\mathbf{A}_r$) can be adjusted by a learning rule of the perceptron type:

$$\Delta\mathbf{w}_r^{(out)} = \varepsilon^{(out)} h_{rs}(\mathbf{y}^{(target)} - \mathbf{y}^{(net)}) + \mathbf{A}_r\Delta\mathbf{w}_r^{(in)} \tag{6}$$

$$\Delta\mathbf{A}_r = \varepsilon^A h_{rs}(\mathbf{y}^{(target)} - \mathbf{y}^{(net)})(\mathbf{x} - \mathbf{w}_r^{(in)})^T/\|\mathbf{x} - \mathbf{w}_r^{(in)}\|^2 \tag{7}$$

This approach assumes that for each input $\mathbf{x}$ a desired target value $\mathbf{y}^{(target)}$ is known; i.e., training of the local linear maps is supervised ($\varepsilon^{(out)}$ and ε^A are two new learning rates).

Road Map: Self-Organization in Neural Networks
Background: I.3. Dynamics and Adaptation in Neural Networks
Related Reading: Data Clustering and Learning; Development and Regeneration of Eye-Brain Maps; Stochastic Approximation and Neural Network Learning

References

Bauer, H. U., and Pawelzik, K., 1992, Quantifying the neighborhood preservation of self-organizing feature maps, *IEEE Trans. Neural Netw.*, 3:570–579.

Blasdel, G. G., and Obermayer, K., 1994, Putative strategies of scene segmentation in monkey visual cortex, *Neural Netw.*, 7:865–881.

Cottrell, M., and Fort, J. C., 1986, A stochastic model of retinotopy: A self-organizing process, *Biol. Cybern.*, 53:405–411.

Erwin, E., Obermayer, K., and Schulten, K., 1992, I: Self-organizing maps: Stationary states, metastability and convergence rate; II: Self-organizing maps: Ordering, convergence properties and energy functions, *Biol. Cybern.*, 67:35–45, 47–55.

Heskes, T., and Kappen, B., 1993, On-line learning process in artificial neural networks, in *Mathematical Approaches to Neural Networks* (J. G. Taylor, Ed.), Amsterdam: Elsevier Science, pp. 199–233.

Kohonen, T., 1982, Self-organized formation of topologically correct feature maps, *Biol. Cybern.*, 43:59–69.

Kohonen, T., 1984, *Self-Organization and Associative Memory*, Series in Information Sciences 8, Heidelberg: Springer-Verlag. ◆

Kohonen, T., 1990, The self-organizing map, *Proc. IEEE*, 78:1464–1480. ◆

Kohonen, T., 1991, Self-organizing maps: Optimization approaches, in *Artificial Neural Networks* (T. Kohonen, K. Mäkisara, O. Simula, and J. Kangas, Eds.), Amsterdam: Elsevier Science, vol. 2, pp. 981–990.

Luttrell, S. P., 1990, Derivation of a class of training algorithms, *IEEE Trans. Neural Netw.*, 1:229–232.

Martinetz, T., and Schulten, K., 1994, Topology representing networks, *Neural Netw.*, 7:507–522.

Obermayer, K., Blasdel, G., and Schulten, K., 1992, A statistical mechanical analysis of self-organization and pattern formation during the development of visual maps, *Phys. Rev. A*, 45:7568–7589.

Ritter, H., 1991, Asymptotic level density for a class of vector quantization processes, *IEEE Trans. Neural Netw.*, 2:173–175.

Ritter, H., and Kohonen, T., 1989, Self-organizing semantic maps, *Biol. Cybern.*, 61:241–254.

Ritter, H., Martinetz, T., and Schulten, K., 1992, *Neural Computation and Self-Organizing Maps*, Reading, MA: Addison-Wesley. ◆

Self-Reproducing Automata

Arthur W. Burks

Origin of the Subject

Self-reproducing automata are computer-logical models of self-reproduction introduced by John von Neumann. These models were part of the theory of automata he started in 1944, to assist in the design of more powerful electronic computers (Burks, 1970, 1987; von Neumann, 1966).

From 1936 to 1944, work by John V. Atanasoff, J. Presper Eckert, John W. Mauchly, and others (including the present writer) had led to successful high-speed electronic computer circuits, and also to a serial electronic memory capable of storing 1000 words. Von Neumann then adapted the logical neuron elements of McCulloch and Pitts (see SINGLE-CELL MODELS) to represent the known basic electronic switching and storage circuits. Each logical neuron element consisted of a simple threshold switch (possibly with inhibition) followed by a one-pulse-time delay line.

Using this logical net symbolism, von Neumann worked out the logical design of the Electronic Discrete Variable Automatic Computer (EDVAC), the first electronic computer that could store its program in replaceable rather than read-only memory. He also designed the programming language for this machine—the first programming language capable of modifying itself, and thus the first modern programming language. (Earlier computer programs were all stored in read-only memory and so could not change themselves within the course of a problem run.) Von Neumann next conceived of the random-access memory. He then worked out the logical design and programming system for the first computer with such a memory, assisted by Herman Goldstine and the present writer. The organization of that computer is now called the *von Neumann architecture*.

Thus, von Neumann had invented a method for representing the switching and memory structure of a modern electronic computer in abstraction from the very much more complicated details of electrical engineering and physics needed to achieve high-speed reliable digital computing. His next step was to extend this method of logical abstraction to the computing aspects of living systems, including evolution. He believed that a theory of automata of this kind would be useful in designing more powerful computers.

He first conceived a robot model of self-reproduction and then a cellular automaton model. He planned to move onto automaton models of evolution, but died prematurely in 1957.

Robot Model of Self-Reproduction

We explain his robot model first. Imagine a class of automata made of robot parts and operating in an environment of such parts. There are computer parts (switches, memory elements, wires), input-output parts (sensing elements, display elements), action parts (grasping and moving elements, joining and cutting elements), and straight bars (to maintain structure and for constructing a storage tape). There are also energy sources that enable the robots to operate and move around. These five categories of parts are sufficient for the construction of robots that can make complexes of various kinds, including other robots.

These parts suffice for making a robot version of any *finite automaton*—every electronic computer is a finite automaton. Sensing and acting parts can be added to this robot so that it can make an indefinitely expandable *storage tape* from straight bars. A "blank tape" consists of bars joined in sequence, and the robot stores information on this tape by attaching a bar or not at each junction. Thus, a robot could be made with the theoretical computing capacity of any electronic computer.

At this point, von Neumann made use of the theoretical machine concept of Alan Turing (Davis, 1965). Turing defined a computing machine as a central device interacting with an indefinitely extendable storage tape. He gave no indication of how to build the central device, but specified it by a *state transition table*. Such a table gives, for each state of the central device and the symbol on the tape square under scan, the next state of the central device, what is to be written on the tape, and how the tape is to be moved. Turing showed that there exists a universal computing machine that could carry out the computation of any Turing machine whose state table is written on the tape of the universal machine.

Von Neumann used this theorem to show that there is a universal computing robot. A universal computing robot can be augmented to form a universal constructing robot—a robot that can construct any robot when given its description. A *self-reproducing robot* then results from applying the universal constructing robot to itself. It is not easy to depict this because robot automata are three-dimensional. Instead, we refer the reader to the descriptions of universal construction and self-reproduction given below for the (two-dimensional) cellular automaton case (Figures 2, 3, and 4).

Cellular Automaton Model of Self-Reproduction

Cellular Automata

When von Neumann conceived his robot model of self-reproduction, it was not feasible to work out the details. At Stanislaw Ulam's suggestion, however, he began to develop a cellular automaton model of self-reproduction (see CELLULAR AUTOMATA). He first thought of working in a three-dimensional cellular space built up of cubes, and then decided to simplify the problem by making the construction in two-dimensional space, an infinite checkerboard or "space" of cells, each cell containing the same 29-state finite automaton, each automaton being connected directly to its four immediate neighbors. (John Conway's well-known game of LIFE uses a nine-cell immediate neighborhood but only two states per cell.)

We can think of the finite automaton in each cell as a simple structure of switches and unit delay elements. The whole system is clocked, operating over successive times $t = 0, 1, 2, 3, \ldots$. At each moment t, a delay element either emits a pulse (representing "1") or does not (representing "0"). Thus, the state of the whole cellular automaton at time t will be followed by another state (usually a different state) at time $t + 1$.

Von Neumann designed the automaton that was in each cell so that many different kinds of automata could be constructed or embedded in the system as contiguous complexes of cells, including: the equivalent of any finite automaton (and hence any electronic computer), a universal Turing machine, a universal constructing automaton, and a self-reproducing automaton.

To explain how these automata worked, we will partition the set of 29 states of which each cell is capable into five subsets:

1. The *blank state U* represents empty space. Initially, and hence at any subsequent time step, only a finite number of cells can be in states other than U, the rest of the infinite space being empty. Since a universal computing machine must have an indefinitely long tape, it is natural to use a cell in state U as a blank square of a storage tape.

2. An *ordinary transmission element*, written as a single arrow, represents a disjunctive switch followed by a unit delay. Its output goes to the cell to which it points, and its inputs are the other three sides of its own cell.

3. The *confluent element C* represents a conjunction followed by two units of delay. Its inputs are the arrows pointed toward it; the output of a C goes to all the arrow inputs that are contiguous to it, so the confluent element is also a branching element.

Figure 1 shows a simple finite automaton embedded in 24 cells. Suppose that both input i_1 and input i_2 are stimulated by pulses at time t. These pulses impinge on the conjunctive state C at the next moment of time, and this element sends a pulse into the transmission element to its right at time $t + 3$. This

Figure 1. A simple finite cellular automaton. After inputs i_1 and i_2 are stimulated simultaneously, a pulse repeatedly cycles in the interior and produces a periodic output at o_1.

pulse then cycles around the rectangular path repeatedly, and on each cycle also goes to the output o_1 of the figure. Thus, pulses are emitted from the output of the automaton at times $t + 8, t + 17, t + 26, \ldots$.

At any time t, the output of a delay element will be either a pulse or the absence of a pulse. Hence, each arrow (ordinary transmission element) represents two states, and each confluent element C represents four states; thus, the cells of Figure 1 illustrate 13 different states, including the blank state U.

4. Von Neumann's cellular automaton system also has four *special transmission elements* (eight states), which are used for destruction (reducing a cell to U). Negation is accomplished by destruction. Special transmission elements are represented as double arrows (see Figure 2). The two kinds of transmission elements play dual roles in destruction: a pulse from an ordinary transmission element into a cell with a special transmission element will change that cell back to U; a pulse from a special transmission element into a cell with an ordinary transmission element or a confluent element will change that cell back to U.

5. Finally, there are eight *transient states*, used for construction. Construction is accomplished when a transmission element sends a timed sequence of pulses and absences thereof into a cell that is initially in state U. Each such pulse sequence transforms the cell from state U through a sequence of transient states into an inactive transmission or confluent element.

Construction and destruction operations are used for building new automata in blank areas of cells, and also for erasing information from devices. For example, any sequence that is cycling around in Figure 1 can be erased by sending the appropriate sequence of destruction and construction signals into one of the ordinary transmission states along the bottom of the automaton. Sequences of destructive and constructive signals play essential roles in the operation of the "constructing arm," a device that von Neumann used in two ways. He used a one-dimensional form of it for accessing an indefinitely long storage tape (Figures 2, 3, and 4) and a two-dimensional form of it for constructing automata in blank regions of the cellular space (Figures 3 and 4).

Universal Computer and Universal Constructor

We next show how computers, constructing systems, and self-reproducing automata can be designed in this cellular automaton system.

A *universal computer* (universal Turing machine) is an idealized object consisting of a finite automaton M_T (named after Alan Turing) and a potentially infinite *tape unit*, illustrated in Figure 2. Each cell x_n of the infinite row of storage cells is either in state U (for "0") or has an ordinary transmission state pointing down (for "1"). The automaton M_T reads the cell x_n under scan by sending a specific pulse, no-pulse sequence into i_3. There are two possible outcomes: (1) If cell x_n is in the state

Figure 2. Universal computer.

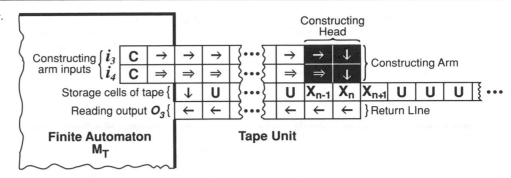

single-arrow-pointed-down, the reading sequence returns to the reading output o_3 unchanged; but (2) if x_n is in state U, the reading sequence will change the cell to the single-arrow-pointed-down state, and a single pulse will appear at the reading output o_3. (Note that this read-out process is destructive—cell x_n is left in the state that represents "1" in either case.)

The internal program of cellular automaton M_T will specify the next tape operation to be carried out: whether the constructing arm and return line are to be extended, contracted, or left where they are; and whether cell x_n is to be left in state 0 or in state 1. M_T achieves these changes by sending appropriate sequences of temporally spaced pulses into the inputs i_3 and i_4 of the *constructing arm*. These pulse, no-pulse sequences travel to the constructing head and cause it to perform various destructive and constructive operations on neighboring cells, thereby extending or contracting the construction arm and the return line. For example, the first step in this process is for M_T to send a sequence of pulses to the special transmission element of the constructing head that will kill the ordinary transmission element located on its right and then build a different element in its place.

The constructing arm of a universal computer (Figure 2) moves only forward and backward. But von Neumann designed the constructing head so that a constructing arm can operate in two dimensions. When the arm is fed the proper sequences of signals on its inputs, it will extend itself so that its head is in a blank area of the cellular space, lay down a cellular automaton in that area, and then pull back to the finite automaton that is supplying its input signals. A two-dimensional constructing arm is used in the *universal constructor*, shown in Figure 3.

The finite automaton M_{vN} (named after von Neumann) can operate both a tape unit and a constructing arm. Suppose a suitably coded description $D\{m\}$ of some finite automaton is placed on the tape of M_{vN} along with the tape contents $T\{m\}$ that are to be stored on the tape of m. When M_{vN} is started, it reads and interprets $D\{m\}$ and sends the appropriate signals into its construction arm so that the arm moves to the designated area, constructs automaton m and its tape, and copies $T\{m\}$ onto this tape. Then M_{vN} withdraws its constructing arm.

Thus, the universal constructor can construct *any* automaton m with any initial tape contents $T\{m\}$. Note that a universal constructing machine is a two-dimensional generalization of a universal computing machine.

As in the robot model, cellular automaton self-reproduction results from applying the universal constructor to itself; see Figure 4. M_{vN} is designed so that when it finds a period after using $D\{m\}$ to construct the machine m, it goes back and copies $D\{m\}$ onto the tape that it has attached to the new machine m. Consequently, automaton M_{vN} with the description "$D\{M_{vN}\}$." on its tape uses that description twice: to construct a new automaton M_{vN} and to copy the description $D\{M_{vN}\}$ onto the

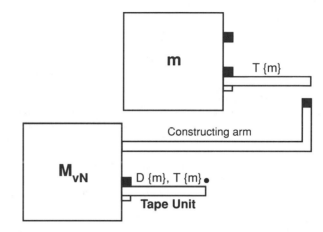

Figure 3. Universal constructor.

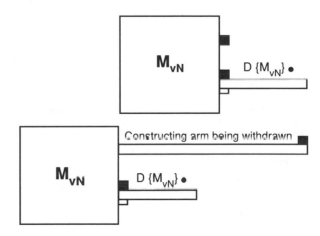

Figure 4. Self-reproduction.

tape of the new automaton. In this case, the constructed automaton is the same as the constructing automaton. This is *automaton self-reproduction*.

Conclusion

Von Neumann's robot and cellular automaton models of self-reproduction give a formal or logical representation of the activities of computation, communication, construction, and destruction, as well as of self-reproduction. The two models constitute an in-principle proof that self-reproductive intelligent robots could be constructed. A self-reproducing automaton describes itself and so belongs to the general class of grammatical self-referential statements, which includes Kurt Gödel's famous undecidable formula of formal arithmetic (Davis, 1965), Turing's universal computer simulating itself, and von Neumann's computer programs that modify themselves. In contrast, statements that refer to their own truth-status may lead to contradiction (e.g., "This sentence is false" is both true and false) and are not logically grammatical.

The subject of cellular automata has developed into a theory (pursued mathematically) and a practice (pursued by simulation). A cellular automaton is a kind of model universe, with a space of any dimensionality divided into cells. This universe has a local "law of nature," which may be deterministic or probabilistic. Each local law of a cellular automaton universe yields a global "law of nature." In the deterministic case, a starting state yields a unique history of nature, and in the probabilistic case, a starting state yields a statistical distribution of histories.

Cellular automata are used to study models of physical processes (e.g., gases and liquids, chaotic processes) and models of living processes, including evolution and the emergence of various phenomena. Investigators also explore artificial systems that are suggested by real systems. These artificial physical and living systems are not designed as models, but are pursued for their own interest. Powerful computers are used to carry out such studies interactively.

Von Neumann did not actually complete his cellular automaton model but left manuscripts from which the present writer completed it (von Neumann, 1966). My introductions to the parts of Aspray and Burks (1987) give an account of von Neumann's contributions to computer design and summarize his theory of automata. The book also contains a list of publications that build on von Neumann's work. Von Neumann had planned a third and a fourth model of self-reproduction (counting his sketch of a robot model as the first).

As we saw earlier, he had used highly simplified neuron-like elements for his logical design of the first modern electronic computer (EDVAC), each such neuron-like element consisting of a simple threshold switch followed by a one-microsecond delay line. The 29-state automaton in each cell of his cellular automaton would be a simple network of such elements. For his third model, von Neumann would have replaced the simplified neurons of this design by neurons with realistic characteristics of excitation, threshold response, and fatigue.

His fourth model of self-reproduction was to be a continuous model. He planned to base it on a system of nonlinear partial differential equations of the diffusion type (see PATTERN FORMATION, BIOLOGICAL). Presumably these would be the equations describing the functioning of a realistic network of neurons. Von Neumann also suggested an automaton model of evolution that would incorporate mutation and natural selection, and that would explain how complex organic systems evolved from physico-chemical systems.

All these studies and questions were part of his plan to develop a theory of automata, both artificial and natural, that would study principles of organization, control, information, and language and be applicable to the design of computers. (Today some parallel computers embody a cellular automaton structure.) Automata theory would be similar to Norbert Wiener's cybernetics, but would emphasize digital computers over analog computers, and logic and programming over continuous mathematics.

Von Neumann's use of idealized neuron-like elements in automata theory led him to write *The Computer and the Brain* (Yale University Press, 1958). The following, expressed in terms of today's software, illustrates the kind of analogy he was interested in. A person may communicate with a computer in a high-level visual language, but the hardware executes the commands in a low-level machine language. Similarly, a mathematician reasons in a natural language with symbols and images, but the central nervous system "executes" these thoughts in a language of chemical and electrical signals.

Acknowledgment.   This article is an extension of my definition of *self-reproducing automata* in *The Cambridge Dictionary of Philosophy* (Robert Audi, Ed.), Cambridge, Eng.: Cambridge University Press, 1995. The material is adapted with permission.

Road Map: Dynamic Systems and Optimization
Background: I.1. Introducing the Neuron
Related Reading: I.2. Levels and Styles of Analysis; Automata and Neural Networks; "Genotypes" for Neural Networks; Parallel Computational Models

References

Aspray, W., and Burks, A. W., Eds., 1987, *Papers of John von Neumann on Computers and Computer Theory*, Cambridge, MA: MIT Press.

Burks, A. W., Ed., 1970, *Essays on Cellular Automata*, Urbana, IL: University of Illinois Press.

Burks, A. W., 1980, From the ENIAC to the stored program computer: Two revolutions in computers, in *A History of Computing in the Twentieth Century* (N. Metropolis, J. Howlett, and G.-C. Rota, Eds.), New York: Academic Press, pp. 311–344. ◆

Burks, A. W., 1987, Von Neumann's self-reproducing automata, in *Papers of John von Neumann on Computers and Computer Theory* (W. Aspray and A. W. Burks, Eds.), Cambridge, MA: MIT Press, pp. 491–552. ◆

Davis, M., Ed., 1965, *The Undecidable*, Hewlett, NY: Raven.

von Neumann, J., 1966, *Theory of Self-Reproducing Automata*, edited and completed by A. W. Burks, Urbana: University of Illinois Press.

Semantic Networks

John A. Barnden

Introduction

Semantic networks (SNs) are a way of representing information abstractly. They are commonly used in symbolic cognitive science. There are interesting similarities and differences between SNs and neural networks (NNs). Some attempts have been made to implement or emulate SNs in NNs and to form hybrid SN-NN systems.

SNs were originally developed for couching "semantic" information, either in the psychologist's sense of static information about concepts or in the semanticist's sense of the meanings of natural language sentences. However, they are also used as a general knowledge representation tool. The more elaborate types of SN are similar in their representational abilities to sophisticated forms of symbolic logic. (For more information on SNs, see Findler, 1979; Barr and Feigenbaum, 1981, section

III.3C; Sowa, 1984; Brachman and Levesque, 1985, including the well-known critique by Woods; Rich and Knight, 1991, chaps. 4, 9–11; Lehmann, 1992).

The Nature of Semantic Networks

As with NNs, there are many different styles of SN. However, all SNs share two general features:

- The nodes are to be interpreted (by us) as representing physical or nonphysical entities in the world, classes or kinds of entity, relationships, or concepts.
- The links, which are almost always directed, encode specific relationships between entities, concepts, etc., where the type of relationship is specified by a symbolic label on the link.

We roughly characterize SNs as either *restricted* or *general*, although there are cases in between. The restricted class is typified by the small fragment shown in Figure 1. Nodes represent kinds, activities, or individuals (all shown in the diagram by lower-case labels). IS-A is the subkind relationship, whereas INST is the instance (i.e., individual-to-kind) relationship. IS-A links are often labeled as A-KIND-OF or AKO links. The individuals in the fragment are Canny and Osten.

The central purpose of a restricted network is to organize concepts, kinds, or classes into a taxonomy, which may or may not be strictly hierarchical, and to state properties of the included entities. A general principle is that

- Information general to a class, kind, or concept should be held at the node representing it, and not repeated at nodes for subclasses, subkinds, or subconcepts (or particular instances of them); instead, these subentities implicitly *inherit* the information at the former node.

An allied fundamental principle is that

- Information attached to a node can contradict information attached to an ancestor node; in that case, the former information overrides the latter.

Here, an ancestor node of a node N is a node that can be reached from N by a succession of IS-A links, or by an INST link and then possibly some IS-A links. Thus, the fragment in Figure 1 says that birds (in general) have flying as their main method of locomotion, but that for ostriches (in general), the main method of locomotion is running. On the other hand, canaries in general, and Canny in particular, implicitly inherit the flying because no node between those nodes and the bird node has a MAIN-LOCOMOTION link contradicting the flying. Osten inherits the running of ostriches in general as well as their love of sand.

The general type of SN is typified by the fragment shown in Figure 2. The q node represents the proposition that every pig loves rain. Alternatively, it represents the situation of every pig loving rain. The fragment shown is a direct analog of the logic formula

$$(\forall x)\, \text{is-pig}(x) \Rightarrow \text{loves}(x, \text{rain})$$

Although general SNs often contain taxonomic information, this information is less important to the overall purpose of the network. Rather, the focus is on representing a much wider variety of types of information than is practical to represent in restricted SNs. In contrast to restricted networks, general networks usually have a small selection of link labels, typified by those in Figure 2. Often, labels have a close connection to deep case relationships in natural language semantic theories (see the AGENT and OBJECT links in Figure 2). Relationships such as loving are now represented by nodes (see the *loves* node in Figure 2) rather than by link labels (see the LOVE link in Figure 1). Also, the more elaborate general SNs have facilities analogous to the connectives and quantifiers of formal logic (see the *implication* and *forall* nodes in Figure 2. The QUANT link to the *forall* node says that the q node represents a universally quantified proposition. The VAR link points to the x node, playing the role of the variable x in the formula above. The BODY link points to the body of the proposition, which has the form of an implication proposition (i node). ANTE stands for antecedent, and CONSE represents consequent. PRED links point from nodes representing simple propositions to the nodes for the predicates in the propositions. ARG represents argument.

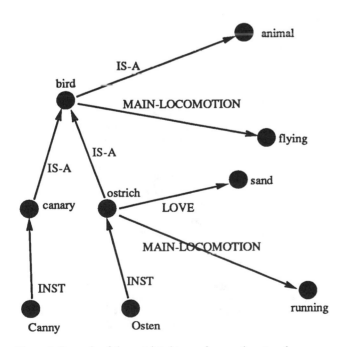

Figure 1. Example of the restricted type of semantic network.

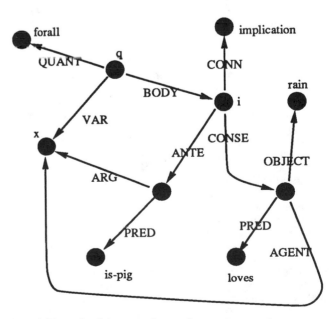

Figure 2. Example of the general type of semantic network.

Thus, formal logic expressions can be recast as SNs. However, the two forms of representation are not equivalent, partly because SNs implicitly involve implementational assumptions that facilitate some particularly important operations (see below), whereas formal logic is devoid of implementational assumptions. Also, to parallel the inheritance-blocking features of SNs in logic, one has to turn to special, advanced forms of logic (for an introduction, see Rich and Knight, 1991, chap. 7).

Processing in Semantic Networks

One common and basic type of processing in SNs is *spreading activation*. During the course of spreading activation, if one node is active at some moment, then the nodes directly connected to it can become activated. A given episode of spreading activation might only pass activation across links of specific types. Activation here is often a yes-no matter as opposed to the graded activation typically used in NNs. However, graded activation can be used in SNs.

A common variant of spreading activation is *marker passing*. Markers are symbolic objects, usually of several different types. The marker(s) that a node sends out depends in some possibly complex way on markers that it receives from other nodes. Markers are often simple symbolic objects that bear no information except their type. However, more complex markers are used in some SNs; for instance, a marker can contain information on its node of origin or the nodes or links it has passed through. Complex markers are used in the system of Charniak (1986), for example. A marker can have a graded energy level that affects its ability to cause a receiving node to produce a marker in response (see the system of Hendler in Barnden and Pollack, 1991).

The main motivation for activation spread or marker passing is to allow *intersection search*. For instance, consider an SN system used to understand natural language text. Marker passing might start with nodes representing the word senses of various words in the text. The collision of markers at nodes means that paths have been found between word senses. This information can be used to help to disambiguate the words (see Yu and Simmons, 1990). Marker passing can also be used to manage simple inheritance reasoning. For instance, in Figure 1, Osten's main method of locomotion could be found by sending out a marker from the Osten node and constraining it to travel only along a chain of IS-A links, possibly preceded by a single INST link, and followed by one MAIN-LOCOMOTION link. If the system prefers answers found first to answers found later, the *running* answer is preferred over the *flying* answer, thereby effecting the desired handling of exceptions (blocking of inheritance).

Marker passing is important, but is generally only used for restricted purposes. In general networks that handle more elaborate reasoning, the brunt of the inferencing is typically done by subnetwork matching and construction processes (see Barr and Feigenbaum, 1981, section III.C3, for a brief account; see Cravo and Martins, 1993, for an advanced implemented system).

In any SN with INST and IS-A links, inheritance reasoning is simpler and quicker than it would otherwise be. In SNs without the special links, or in ordinary logical frameworks, traversal of taxonomic relationships is a more elaborate process. These observations rely on certain strong implementational assumptions that are generally made about SNs, although usually only tacitly. The assumptions apply to computer implementations of SNs as well as to hypothesized realizations of SNs in the brain. One important assumption is that, if a processing mechanism is attending to a given node N, then it can efficiently transfer attention to nodes connected by single links to N.

Finally, SNs can have *attached procedures or rules*. These algorithms, expressed in some symbolic format, are attached to nodes or links, and they typically have localized effects supporting particular inference functions (for a brief introduction to this topic, see Rich and Knight, 1991, chap. 4 and section 10.3.2).

Contrasts to Neural Networks

In NNs, in contrast to general SNs, activation spread is almost always the only mechanism for short-term computation (as opposed to long-term computation, such as slow adaptation). Moreover, the types of activation spread and marker passing in SNs are typically more complex than activation spread in NNs, especially when complex markers are used. In strong contrast to almost all NNs, the topology of an SN, especially a general SN, can be changed by processing mechanisms in arbitrarily extensive and complex ways in the short term.

NN links do not have type labels. As a result, NNs generally do not treat different input links differently. Instead, a weighted sum of the input values is formed, submerging the identities of different links. The most salient exception to this rule is the differentiation of input links into excitatory and inhibitory links, but this distinction is crude compared with the link differentiation by labels in SNs. Nevertheless, NN links that impinge on other links and modulate them (see NEUROMODULATION IN INVERTEBRATE NERVOUS SYSTEMS) can be used to obtain many of the effects of link typing in SNs.

When an NN link is directed, activation can spread in only one direction across the link. It is less common to impose this constraint in SNs. It is therefore easier in SNs to express asymmetric relationships between nodes without consequently constraining the accessibility of the nodes from each other. As an example of the problem raised for NNs, consider the use of symmetric links in some artificial NNs, notably Hopfield nets. These allow activation spread in both directions, but do not directly support conceptually asymmetric relationships. More indirect methods are needed to handle such relationships (see Barnden and Srinivas, 1991).

As a result of these differences, SNs cope much more readily with many of the representational and processing needs that arise in high-level cognitive activities, such as common-sense reasoning and natural language understanding. The needs in question are listed in ARTIFICIAL INTELLIGENCE AND NEURAL NETWORKS (q.v.). NNs have difficulty in rapidly creating new, complex bodies of information and in structurally matching complex, and possibly highly temporary, bodies of information. Certainly, any direct parallel of subnetwork matching and subnetwork creation in SNs is difficult to achieve in typical NNs (but see DYNAMIC LINK ARCHITECTURE). However, some NN systems have been developed that can directly implement SNs (see the next section). They depart from mainstream NNs in that they have more elaborate and specialized structure.

On the other hand, NNs appear to have advantages with respect to other needs posed by high-level cognition (see ARTIFICIAL INTELLIGENCE AND NEURAL NETWORKS). These other needs include context sensitivity, graceful degradation, and automatic adaptation. A contrastive factor here is that link weights are fundamental in NNs, but uncommon in SNs. This difference confers a graded associativity quality on NNs that has no direct parallel in most SNs, although the length of a path between two SN nodes has often been assumed to encode the strength of association between them: the shorter the path, the stronger the association.

Bringing Semantic and Neural Networks Together

Several researchers have overtly used NNs to implement or approximately emulate SNs. Other work has aimed at mixed systems that combine NNs with SNs or are compromises between the two. Major examples of the implementation-emulation type of work can be found in Hinton (1989), Sumida (1991), and the chapters by Barnden, Bookman and Alterman, J Diederich, Dyer, and Shastri in Barnden and Pollack (1991). See Touretzky (1990) for a framework that is not aimed specifically at SNs but that could readily implement them. Because of the complexity of the systems that implement or emulate SNs, we omit description of them here.

For examples of combined SN-NN systems, see the chapters by Hendler and Lehnert in Barnden and Pollack (1991). In the Hendler case, nodes in the SN can also serve as input-output nodes in an NN. Markers passed to such SN nodes cause them to inject neural activation into the NN, and vice versa. The symbolic markers have a numeric strength, and this strength affects the level of neural activation instigated by a marker, and vice versa. The NN acts as an intermediary between SN nodes, allowing an SN node representing one concept to stimulate an SN node encoding a similar concept. The NN is trained by backpropagation to associate concepts with sets of low-level features. Concepts are thereby viewed as similar according to the extent to which they share features. Hendler's system makes symbolic reasoning less rigid by enriching it with similarity-based reasoning.

For compromises between SNs and NNs, see the chapters by Eskridge and Kokinov in Holyoak and Barnden (1994). Unlike Hendler's system, these systems are not divided into separate neural and semantic parts. In Eskridge's system, nodes communicate by means of numeric activation signals, much as in NNs. However, symbolic markers travel along with the numeric activation, the links have labels, activation can be constrained to spread only over links with specified labels, links can be dynamically created, and specialized SN processing rules can be invoked. The system performs complex analogical processing, with the symbolic aspects coping with complex structure manipulation and the neural aspects helping with retrieval of source analogs and weighing of alternatives.

Discussion

NNs and SNs have much in common and should be regarded as two points in a rich, quasi-continuous space of computational architectures rather than as radically different types of network. There are important differences in the nature and usage of links and in the degree to which computation can be thought of as local to individual nodes (although in restricted SNs, the computation can be as local as it is in NNs). There are various ways of implementing or emulating SNs in NNs and of forming hybrid SN-NN systems.

Road Map: Artificial Intelligence and Neural Networks
Related Reading: Bayesian Networks; Compositionality in Neural Systems; Structured Connectionist Models

References

Barnden, J. A., and Pollack, J. B., Eds., 1991, *Advances in Connectionist and Neural Computation Theory*, vol. 1, *High Level Connectionist Models*, Norwood, NJ: Ablex.

Barnden, J. A., and Srinivas, K., 1991, Encoding techniques for complex information structures in connectionist systems, *Connection Sci.*, 3:263–309.

Barr, A., and Feigenbaum, E. A., Eds., 1981, *The Handbook of Artificial Intelligence*, vol. I, Los Altos, CA: Morgan Kaufmann. ◆

Brachman, R. J., and Levesque, H. J., 1985, *Readings in Knowledge Representation*, Los Altos, CA: Morgan Kaufmann.

Charniak, E., 1986, A neat theory of marker passing, in *Proceedings of the Fifth National Conference on Artificial Intelligence (AAAI-86)*, Los Altos, CA: Morgan Kaufmann, pp. 584–588.

Cravo, M. R., and Martins, J. P., 1993, SNePSwD: A newcomer to the SNePS family, *J. Exp. Theor. Artif. Intell.*, 5:135–148.

Findler, N. V., Ed., 1979, *Associative Networks: Representation and Use of Knowledge by Computers*, New York: Academic Press.

Hinton, G. E., 1989, Implementing semantic networks in parallel hardware, in *Parallel Models of Associative Memory*, updated ed. (G. E. Hinton and J. A. Anderson, Eds.), Hillsdale, NJ: Erlbaum, pp. 191–221.

Holyoak, K. J., and Barnden, J. A., Eds., 1994, *Advances in Connectionist and Neural Computation Theory*, vol. 2, *Analogical Connections*, Norwood, NJ: Ablex.

Lehmann, F. W., 1992, *Semantic Networks in Artificial Intelligence*, New York: Pergamon.

Rich, E., and Knight, K., 1991, *Artificial Intelligence*, 2nd ed., New York: McGraw-Hill. ◆

Sowa, J. F., 1984, *Conceptual Structures: Information Processing in Mind and Machine*, Reading, MA: Addison-Wesley.

Sumida, R. A., 1991, Dynamic inferencing in parallel distributed semantic networks, in *Proceedings of the Thirteenth Annual Conference of the Cognitive Science Society*, Hillsdale, NJ: Erlbaum, pp. 913–917.

Touretzky, D. S., 1990, BoltzCONS: Dynamic symbol structures in a connectionist network, *Artif. Intell.*, 46:5–46.

Yu, Y.-H., and Simmons, R. F., 1990, Truly parallel understanding of text, in *Proceedings of the Eighth National Conference on Artificial Intelligence (AAAI-90)*, Menlo Park, CA: AAAI Press, pp. 996–1001.

Sensor Fusion

Robin R. Murphy

Introduction

An intelligent agent uses sensor fusion to overcome the limitations of individual senses for a particular task. When one sense cannot provide all of the necessary information, *complementary* observations may be provided by another sense. For example, haptic sensing complements vision in placing a peg in a hole when the effector occludes the agent's view. Also, senses may offer *competing* observations, such as the competition between vision and the vestibular system in maintaining balance and its occasional side effect of seasickness. Another type of interplay between senses is the use of information extracted by one sense to focus the attention of another sense, *coordinating* the two. Audition cueing vision is a common instance of temporal cooperation between senses.

A model of sensor fusion must account for a diverse set of phenomena. At a minimum, it must detail how observations which are asynchronous or noisy are combined into a coherent percept. These observations may be both competing and complementary; for example, one sense may provide the focus of attention for another sensor while providing additional complementary information. The information may be quantitative (the agent is 10 feet from object A) or qualitative (the agent is to the right of object A). Furthermore, the perceptual information must be represented in a way that permits it to be shared by multiple tasks. For example, vision may be simultaneously used for navigation (heading for the front door) and recognition (where are the car keys?). Such a representation must also support the integration of perception with motor control and other aspects of the entire perceptual process, including assimilation into a single long-term-memory world model.

Unfortunately, sensor fusion is not well understood by either psychologists or computer scientists. Much of the published work in the psychology of perception has been concerned with defining the *characteristics* of sensor fusion through experiments rather than proposing *models* to explain these characteristics. This article reviews the cognitive and behavioral literature deriving two insights into sensor fusion. First, the process of sensor fusion can be represented as a perceptual schema. This representation gives researchers a mechanism for investigating the interplay between the senses as well as defining the role of sensor fusion in terms of other processes. Second, sensor fusion involves both combining measurements and reasoning about the evidential contributions of those measurements, bridging the neurophysiological and cognitive science viewpoints. Companion research activity in sensor fusion from an artificial intelligence perspective is also presented. Recent efforts in sensor fusion for autonomous mobile robots reinforce the utility of these insights.

Sensor Fusion as a Perceptual Schema

SCHEMA THEORY (q.v.) is likely to be a powerful tool in exploring sensor fusion, as seen by three different sources: the work of psychophysicist Marks (1978) on the unity of the senses; neurophysiological studies on the merging of senses in cats by Stein and Meredith (1993); and studies on perceptual modes by the cognitive psychologists Pick and Saltzman (1978). These characterizations of sensor fusion are consistent with its encapsulation as a *perceptual schema*.

To review, a *perceptual schema* is a knowledge representation expressing the generic knowledge about a percept (e.g., a model) and the basic pattern of activity for accomplishing that perception (e.g., how to combine observations from different senses). The basic pattern of activity is reactive; it is adaptable (within limits) to changes in the perceptual context. Perception in schema theory is action oriented. For example, a perceptual schema may be instantiated by a motor schema to supply perception for the motor schema's task. Therefore, if there is a motor schema to manipulate an object (e.g., a coffee cup), the supporting perceptual schema may be based on a three-dimensional description of the object. If another motor schema is to merely locate the object, a separate perceptual schema based on a reduced representation may be used instead. Since, as in the above example, multiple schemas may use the same input source, the features relevant to a task serve as submodalities. Also, a perceptual schema may trigger a motor schema, such as when a loud noise coupled with a sudden movement leads to an orientation or investigatory behavior. Perceptual schemas are modular with respect to information content and are also distributed, allowing a perceptual schema to be constructed of subschemas tailored to each sense. These subschemas may compete and cooperate with each other in their attempt to accomplish a perceptual task. This situation implies a common format for communication between the senses.

Neural Models

Marks (1978) offers a theory of sensory correspondence based on psychophysical studies. These studies confirm the existence of a set of attributes common to all senses. Consider that sensory stimuli can be measured in terms of intensity of input and duration. Likewise, attributes such as number, location, and size of objects are independent of modality. Given this commonality, Marks theorizes that all sensory input is at some point in the neural processing converted to a shared representation and fusion performed by a small set of combination mechanisms, the choice of which is determined by context.

The neurological model developed by Stein and Meredith (1993) agrees with the conclusions made by Marks and answers many questions about how sensor fusion is accomplished. Their work demonstrates that stimulus to different senses in cats is initially segregated; each sense has dedicated neurons and a unique layout of its receptive field. However, these sensory signals all converge on the superior colliculus. Neurons in this structure can be clearly defined as *multisensory*; they respond to stimuli from more than one sensor. The response of these multisensor neurons depends on the characteristics of the receptive fields of the individual senses, the overlap in time between the activity patterns, and the strength of the stimuli. Multisensory integration appears to be an accrual process: a stronger response is evoked from weak stimuli to multiple sources than from a strong stimulus from an individual sense.

Stein and Meredith estimate that nearly 75% of the descending efferent projections from the superior colliculus (which control behavior) are multisensory. Furthermore, signals from the cerebral cortex descend as well to the superior colliculus, and studies suggest that they influence behavior through these multisensor neurons based on context and experience.

Perceptual Modes

In one of the few overviews of sensory integration from a cognitive vantage, Pick and Saltzman (1978) segregate intersensory integration into schema-like *perceptual modes*. These modes embody different outcomes of processing, with those outcomes determined by the classes of tasks to be performed by the agent. As each perceptual mode is expected to be implemented by a separate neural mechanism, modes appear to involve recruitment of a neuron into different circuits.

Modes explain why different percepts can result from the same stimulus. The visual stimulus of a room is the same, but what is seen when passing through the room versus searching the room for missing car keys is different. The functional difference between percepts is the task to be supported. Vision is used for locomotion in the first instance and for focused search in the second. The presence of modes may indicate that *intrasensory* integration of submodalities of a sensor may use a subset of the mechanisms for intersensory fusion.

Evidence in Sensor Fusion

The second insight from the literature is that sensor fusion is both a measurement-driven process and an evidential process; that is, the senses provide evidence as well as measurements for fusion. Studies with sensory integration have been somewhat contradictory.

On one hand, some neuroethological studies indicate that sensor fusion is a measurement-driven process, where the measurements supplied by each sensor are combined mathematically according to a formula specific to each percept. The combination takes place without regard to any indications that the measurements may be incorrect. The classic example is the vertical orientation of angelfish. Studies by von Holst, reported in Schöne (1984), indicate that angelfish maintain their vertical orientation relative to two different sources of perceived information, gravity and the horizon line (water surface). Under normal circumstances, angelfish swim upright, both perpendicular to the horizon line and parallel to gravity. If the horizon line is tilted, the fish will swim sideways, compromising midway between the stimulus from the horizon line and gravity. Clearly, this level of sensory integration blindly integrates the contributions from each source without any mechanism for detecting discordances between the modalities.

However, there are many instances in which sensor fusion appears to be evidence driven. For example, in the Roelof illusion, where the subject sees an object under a prism which distorts location, the visual evidence of where the hand is contradicts the proprioceptive evidence of where the hand is believed to be. Rather than staying in the fixed relationship specified by a measurement-driven interpretation of sensor fusion, the senses of adult subjects are able to recalibrate and eventually grasp the object as it is moved to new locations under the prism. When the prism is removed, the subjects have difficulty reaching the object until they adjust once more. Rieser et al. (in press) performed a variation on this illusion in which, based on visual cues, subjects adjust their internal sense of how much distance is covered while walking. Also, studies with dominant modalities (Lee, 1978; Mack, 1978; Pick and Saltzman, 1978) show that the evidence from one sense may be weighted more than evidence from others, if available. The handling of evidence may be unconscious, as in the variations on the Roelof illusion, or it may be deliberate. An example of conscious reinterpretation of the senses is the response of subjects to illusions in which the floor appears flat but is actually tilted. The subjects resolve these contradictions by suppressing vision and relying on the vestibular modality (Johansson, 1974).

Bower (1974) addresses the dispute between measurement and evidence by proposing a four-level framework of sensory integration which encompasses both. Each perceptual process (or schema) falls into one of four levels of sensor fusion. The levels are based on the amount of information processing required in integration. The taxonomy represents implied stages in perceptual evolution, from simple systems that integrate tightly coupled features (i.e., measurement driven) to complex systems which adjust sensing according to the evidence afforded by the situation.

These levels are not hierarchical except in the sense that the higher the level, the more information processing is required. Freedman and Rekosh (1968) also support the basic idea that compensation mechanisms are based on the detection and resolution of incongruities among different features rather than on an unmediated response from the sensorimotor system. They maintain that the sensorimotor processes are rearranged in response to discordances. Bower's taxonomy provides an interesting decomposition which is compatible with the modularity of schema theory. The decomposition accounts for blind integration in level I as well as the use of dominant sense modalities in levels II and III. Unfortunately, no cognitive studies directly evaluating Bower's taxonomy have been conducted. However, the taxonomy offers further insight into the possible interac-

tions between the cerebral cortex and the superior colliculus in Stein and Meredith's model.

Companion Efforts in Artificial Intelligence and Artificial Neural Networks

Sensor fusion is also of interest to the artificial intelligence (AI) and artificial neural networks (ANN) communities, especially to the members who work with autonomous mobile robots. Issues of interest are how to use the information from one sensor to focus the attention of another (cross-modality matching) and how to combine the information from multiple sensors to improve measurement accuracy or confidence in recognition. The demands of the open world, with its unpredictable devastating failures, such as dirt being kicked up onto a planetary rover's camera lens, also necessitate the use of multiple sensors to provide robustness.

Artificial neural network efforts have concentrated on correlating sensors for object recognition, especially target acquisition. The result has produced new techniques for training a neural network for a specific object, but no new insights into sensor fusion. The close coupling of perception and action, along with the role of cognitive influences on perception, is largely ignored in these systems, although some work has been done in ANNs for fault tolerance.

Researchers working with autonomous mobile robots have been more concerned with developing a general model of sensor fusion which captures the broad nature of biological sensory integration. Murphy (1992) surveyed architectures developed at Carnegie Mellon, the University of Pennsylvania, Georgia Tech, and in Europe which present systems for robotic sensor fusion. These architectures almost exclusively focus on how sensors influence and dominate each other in tasks involving object recognition. The majority treat sensor observations as a probability density function capturing the confidence in the object classification. The probabilities are typically combined with a variant of Bayes's rule. Many other researchers have also produced notable results in aspects of AI sensor fusion, particularly adaptations of logic and Bayesian theory to evidential reasoning. Overall, these approaches have one or more of the following shortcomings: the flavor or complexity of sensing interactions is not adequately captured; the architectures are restricted to a few sensing modalities and cannot be readily extended; and the sensor fusion process has been treated separately from the context of the needs of the motor control scheme as well as other perceptual processes, such as active perception.

Forays into sensor fusion by the AI community have largely ignored cognitive and behavioral psychology. Of the references cited, only the work of Luo and Kay (1989) and Murphy (1992) is explicitly motivated by cognitive models of sensor fusion. In particular, Murphy's Sensor Fusion Effects (SFX) architecture is strongly cognitively based. It uses computational processes called *perceptual schemas*, which fuse sensor observations according to *fusion state*. These fusion states correspond to the first three levels of Bower's hypothesized taxonomy. $State_1$: *complete sensor fusion* combines evidence from each sensor according to a percept-specific weighting function. The weighting function expresses any sensor dominance effects. $State_2$: *fusion with the possibility of recalibration* provides feedback to a suspect sensing modality, bringing it closer to the consensus of the majority of the sensors. $State_3$: *fusion with the possibility of suppression* filters out observations from a suspect sensor that are inconsistent with a highly certain consensus from other sensors. SFX used Dempster-Shafer theory to combine and propagate evidence; unlike Bayesian schemes, evidence in Dempster-Shafer theory accrues, in keeping with the biological neural network model of sensor fusion advocated by Stein and Meredith (1993).

Experiments with SFX implemented on a mobile security robot operating in a cluttered environment demonstrate the utility of these states. The use of a State$_2$ configuration allowed the thermal camera to be periodically recalibrated as its internal temperature reference drifted, and a State$_3$ configuration permitted erroneous readings from ultrasonic sensors to be ignored. The percept model used for recognition embeds spatial and temporal constraints between sensors. The results favor a model of sensor fusion that is driven by both measurement and evidence and in which the balance between the two depends on the task and the particular sensing configuration.

Discussion

No complete theory of sensor fusion has been presented in the cognitive and biological literature which explains how sensors influence and dominate each other while producing more accurate or confident perception. Sensor fusion admittedly covers many functions, so one conclusion is that such a theory either does not exist or may be too abstract to be useful. However, the work of Stein and Meredith (1993) and Marks (1978) suggests that deriving such a theory is possible and will rely on a small set of organizing principles which will direct companion efforts in AI and ANNs.

The current literature can be interpreted as offering two insights into the organization of sensor fusion. First, sensor fusion can be expressed within a perceptual schema format, especially in the coupling of action to perception (e.g., multisensor neurons as the major type of efferent neuron). The ramifications indicate that despite the large number of ways in which senses can interact, the mechanisms for expressing those relations appear to be small in number and general in purpose. It also implies that there is a representation common to all senses which is used to share and coordinate multiple senses (e.g., multisensor neurons in the superior colliculus). Second, sensor fusion may require differing levels of evidential reasoning, depending on the perceptual requirements of the task, the characteristics of the senses being used, and the impact of the environment on the reliability of those senses (e.g., levels of sensory integration).

Road Map: Other Sensory Systems
Related Reading: Collicular Visuomotor Transformations for Saccades; Dynamic Remapping; Reactive Robotic Systems; Sound Localization and Binaural Processing

References

Bower, T. G. R., 1974, The evolution of sensory systems, in *Perception: Essays in Honor of James J. Gibson* (R. B. MacLeod and H. L. Pick, Jr., Eds.), Ithaca, NY: Cornell University Press, pp. 141–153. ◆

Freedman, S. J., and Rekosh, J. H., 1968, The functional integrity of spatial behavior, in *The Neuropsychology of Spatially Oriented Behavior* (S. J. Freedman, Ed.), Homewood, IL: Dorsey, pp. 153–162.

Johansson, G., 1974, Projective transformations as determining visual space perception, in *Perception: Essays in Honor of James J. Gibson* (R. B. MacLeod and H. L. Pick, Jr., Eds.), Ithaca, NY: Cornell University Press, pp. 117–138.

Lee, D., 1978, The functions of vision, in *Modes of Perceiving and Processing Information* (H. L. Pick, Jr., and E. Saltzman, Eds.), New York: Wiley, pp. 159–170. ◆

Luo, R. C., and Kay, M. G., 1989, Multisensor integration and fusion in intelligent systems, *IEEE Trans. Syst. Man Cybern.*, vol. SMC-19:901–931.

Mack, A., 1978, Modes of perceiving and processing information, in *Modes of Perceiving and Processing Information* (H. L. Pick, Jr., and E. Saltzman, Eds.), New York: Wiley, pp. 171–186.

Marks, L. E., 1978, *The Unity of the Senses: Interrelations Among the Modalities*, New York: Academic Press. ◆

Murphy, R. R., 1992, *An Architecture for Intelligent Robotic Sensor Fusion*, Technical Report GIT-ICS-92/42, Atlanta: Georgia Institute of Technology, College of Computing.

Pick, H. L., and Saltzman, E., 1978, Modes of perceiving and processing information, in *Modes of Perceiving and Processing Information* (H. L. Pick, Jr., and E. Saltzman, Eds.), New York: Wiley, pp. 1–20.

Rieser, J. J., Pick, H. L., Ashmead, D. H., and Garing, A. E., in press, The calibration of human locomotion and models of perceptual-motor organization, *J. Exp. Psychol. Hum. Learn. Mem.*

Schöne, H., 1984, *Spatial Orientation: The Spatial Control of Behavior in Animals and Man* (C. Strausfeld, Trans.), Princeton, NJ: Princeton University Press. ◆

Stein, B., and Meredith, M. A., 1993, *The Merging of the Senses*, Cambridge, MA: MIT Press. ◆

Sensorimotor Learning

Lina L.E. Massone

Introduction

A sensorimotor transformation maps signals from various sensory modalities into an appropriate set of efferent motor commands to skeletal muscles or to robotic actuators. Sensorimotor learning refers to the process of tuning the internal parameters of the various structures of the central nervous system (CNS) or of some processing architecture in such a way that a satisfactory motor performance will result.

Sensorimotor transformations and sensorimotor learning can be viewed as complex parallel distributed information-processing tasks. A number of different components contribute to the tasks' complexity. On the sensory side, signals from various sources (vision, hearing, touch, pain, proprioception) need to be integrated and interpreted (Stein and Meredith, 1993; see SENSOR FUSION). In particular, they need to be translated into a form that can be used for motor purposes because the coordinates of afferent sensory signals are different from the coordinates of the movements they are guiding. This convergent process, from many parallel sensory signals to an intermediate internal representation, is referred to as the early stages in a sensorimotor transformation. A widely accepted concept that describes the computed internal representation is the so-called motor program: a specification of the important parameters of the movement to be executed (Keele and Summers, 1976). Bernstein used the term *motor program* to denote a prototype of a planned movement described with an abstract central language that encodes specific features of the movement itself.

This idea was then elaborated by Schmidt (1988) and modified by Arbib (1990), who introduced the concept of *coordinated control program*: a combination of perceptual and motor functional units called schemas, whose dynamic interaction causes the emergence of behavior. A different approach was proposed by Schoener and Kelso (1988), who redefined motor programs in dynamic terms as transitions between the equilibrium states of a dynamical system. Interestingly, there is some experimental evidence that some of the parameters of the motor programs could be actuator-independent, i.e., unrelated to the particular plant that will carry out the movements: Bernstein (1967), for example, observed in his classic experiments the presence of similar patterns in cursive writing executed with different joints or with different parts of the body. Motor programs, however, are likely to be task-dependent entities and to exhibit different structures and different coordinate frames for different classes of movements, like, for example, reaching, grasping, and manipulating (see GRASPING MOVEMENTS: VISUO-MOTOR TRANSFORMATIONS).

A quite different process is needed in order to translate the computed internal representations into a set of commands to the actuators (muscles/motors). This process is referred to as the late stages in a sensorimotor transformation. An important feature of the late stages is their intrinsic divergence: it is a mapping from a lower to a higher number of degrees of freedom. There are in fact many more muscles in, say, an arm than there are independent variables in a representation of movement expressed, say, in a body-centered coordinate frame. The redundancy of the motor system is an important and desirable property: it allows the same motion to be carried out with different muscle combinations under different task-related and environmental circumstances. Hence it provides the basis for the organism to be able to cope, for example, with fatigue and fast reaction. But, most importantly, redundancy underlies the smoothness of body movements: without a redundant system we would be moving like puppets. It is important to understand that redundancy is exploited by the CNS in order to coordinate body movements efficiently. Consequently, any model of sensorimotor tranformations and sensorimotor learning should be based on the same constructive approach—exploit redundancy in order to enrich the motor repertoire—rather than try to eliminate redundancy—a destructive approach.

One additional degree of complexity in the transformation from sensory to motor is the fact that such transformations occur continuously over time but with basic events that occur at various different time scales and various different spatial resolutions, and hence require different grains of parallelism. Moreover, not only does the information need to be transformed between different coordinate frames, but also a translation is often required from signals that are spatially coded to signals that are temporally coded and vice versa—the so-called spatiotemporal and temporospatial transformations. Interesting examples of such transformations can be found in the saccadic/gaze system (Massone, 1994; Massone and Khoshaba, 1994; see COLLICULAR VISUOMOTOR TRANSFORMATIONS FOR SACCADES).

How is this tremendous complexity handled by the CNS, and how does the CNS learn what the correct architectural and processing choices are for the large ensemble of neurons involved? What are the important aspects that need to be taken into account when trying to model such a complex process?

Computational Approaches

From a computational perspective, it is important to point out that many aspects of sensorimotor transformations and learn-

ing can be captured by the network-like style of computation of connectionist models. While the correspondence between such a computational style and brain mechanisms remains an open and controversial issue, it is true that some of the basic properties of connectionist models match very well the problems that one must face when dealing with sensorimotor transformations and learning.

Coordinate Transformation

The transformation f that occurs between an n-dimensional input layer and an m-dimensional output layer of a network

$$f: \mathbb{R}^n \to \mathbb{R}^m, \quad y = f(x) \qquad (1)$$

can be viewed as a very general, nonlinear coordinate transformation process. This view becomes particularly important when one uses it in association with error correction mechanisms and learning algorithms. The ability of connectionist learning algorithms to solve the credit assignment problem constitutes an efficient mechanism for translating errors expressed in one particular coordinate frame ($\mathbb{R}^m$) into errors expressed in a different coordinate frame ($\mathbb{R}^n$). The relevance of this approach for motor control purposes lies in the fact that often errors are available, through some measuring system, only in task or task-related coordinates and not in the coordinate frame of the controller. One example related to arm movements is that of a hand-coordinate error measured by the visual system that needs to be translated into an error in muscle-activation space—error that is not directly available. Another example involving a different system, the vestibulo-ocular reflex, is the translation of the retinal slip signal measured by the accessory optic system into gain and phase errors. These two examples, although quite different in nature, share a common general structure, based on two primary components: (1) a controlled object that translates a motor command into an actual movement, i.e., into task coordinates, and (2) a controller that generates the motor commands for the controlled object. The computation of the motor commands and their translation into movements (task coordinates) is referred to as the forward path. The opposite computation, from task coordinates to motor commands, is referred to as the inverse path. In particular, the inverse path is what is needed in order to learn an appropriate controller for a given controlled object. The next subsection will introduce three different approaches that have been proposed in the literature for this particular learning task. A more detailed account of this topic can be found, for example, in Jordan and Rumelhart (1992) and Kawato (1990). See also MOTOR CONTROL, BIOLOGICAL AND THEORETICAL.

Forward and Inverse Models

A forward model is a representation of the transformation from motor commands to movements, in other words, a model of the controlled object. An inverse model is a representation of the transformation from desired movements to motor commands. Hence an inverse model can be used as the controller for the controlled object: the cascade of the two things implements an identity mapping as shown in Figure 1. How can the inverse model of a controlled object be learned?

A first method is the so-called direct inverse modeling. The approach (Figure 2) consists of activating the controlled object (or a forward model of it) with some motor commands generated, for example, in a random fashion. The output produced by the controlled object is used as the input to the inverse model, which in turn computes an estimated motor command.

Figure 1. An identity mapping realized as the cascade of an inverse model and the corresponding controlled object.

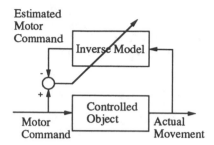

Figure 2. Direct inverse modeling.

The estimated motor command is then compared to the actual motor command, and an error is computed. The error can be used to modify the inverse model. If the inverse model is, for example, a multilayer network with randomly initialized weights, the error can be used to update the network's weights with supervised learning techniques. If this process (generation of random motor command, error computation, weights update) is repeated long enough, the network will learn, over time, to behave like the inverse of the controlled object. The trained network can now be used as a controller as shown in Figure 1. This learning scheme, although conceptually simple, has, in practice, a number of drawbacks as pointed out by Jordan (1990), most notably the fact that it cannot properly control nonlinear redundant objects (i.e., nonlinear objects with a lower number of degrees of freedom in task space than in motor command space) because it cannot solve for the inherent ambiguity of the inverse transformation. It is worth pointing out that the inability of handling redundancy is a very serious drawback, because redundancy is, as emphasized in the introductory section, an intrinsic, extensive, and important property of the musculoskeletal system.

A different and more effective approach is the distal supervised learning approach of Jordan and Rumelhart (1992) that consists of two phases depicted in Figure 3. During the first phase (see Figure 3A) the forward model is learned by activating the controlled object and the forward model with the same motor command, again randomly generated. The error between the output of the forward model and the output of the controlled object is then computed and used to update the weights of the network that represents the forward model. Over a number of learning trials the forward model learns to emulate the controlled object. The second phase consists of training the controller. This is achieved with the scheme shown in Figure 3B: A desired trajectory is inputted to the controller, the actual movement is computed at the output of the controlled object, and an error between the two is computed. The error is then propagated backward through the forward model and transformed into a motor command error, which is in turn used to update the weights of the controller. As learning progresses, the controller becomes an inverse model of the controlled object. The two phases can also be merged, so that the forward model and the controller can be learned in parallel. The forward model approach presents significant advantages over direct inverse modeling. First of all, because of its goal-oriented

character, it can properly cope with nonlinear redundant objects. In addition, it can be used simultaneously for learning and for control, while direct inverse modeling cannot. Another interesting feature of the forward model approach is that the same scheme can be used to learn an input-output transformation that is not necessarily an identity mapping. This ability would be most useful, for example, in the case of a controller that transforms a sensory stimulus or an internal representation of it into motor commands, as in the late stages of a sensorimotor transformation.

A third approach, called feedback-error learning and proposed by Kawato (1990), is depicted in Figure 4. In this scheme an error is computed between a desired trajectory and the actual trajectory executed by the plant. A feedback controller, which is a linear approximation of the inverse model, transforms the trajectory error into a motor-command error, which is used to train the inverse model and to compute the actual motor command that activates the plant. The latter is computed as the sum of the motor-command error plus the motor command computed by the inverse model. As learning progresses and the error decreases, the feedback controller play a lesser and lesser role in the computation until the plant is completely controlled by its inverse model. Like the forward model approach, feedback-error learning is able to control successfully nonlinear redundant plants. For a biological interpretation of feedback-error learning see CEREBELLUM AND MOTOR CONTROL.

Optimization and Learning

A computational approach to error transformation based on distributed learning schemes must be able to deal with the redundancy present at the various stages of sensorimotor transformations. Because the pathway for the transformation of a motor program expressed in task coordinates into patterns of muscle activation is a divergent one, there is redundancy in the inverse tranformation from an error expressed in task coordinates to an error expressed in muscle coordinates: there are infinitely many possible combinations of muscle errors that would lead to the same task error. In order to solve for the redundancy problem in a way that is meaningful from the standpoint of the tasks being carried out, one must make use of optimization techniques so that the error transformation process will make specific choices for the solutions, rather than arbitrary ones. One way of tying optimization to learning is to incorporate optimization principles in the learning algorithms themselves, i.e., to constrain the learning procedure to follow specific paths. Most connectionist learning algorithms are formulated as the minimization of a cost function. Consequently, one possible approach to learning with extra degrees of freedom is to include in the cost function additional terms (besides the usual quadratic error terms) that represent optimization criteria in some analytical form.

Consider, for example, a model of the late stages of the sensorimotor transformation of a planar reaching movement executed by a 3-joint, 8-muscle arm, and assume (Figure 5) that in such a model the controller maps an internal representation of the movement target into a time-varying pattern of activity of the eight muscles of the arm. The controlled object is the arm itself or a model of its biomechanical properties; it transforms the activation of the muscles into joint angles (forward dynamics) and then the joint angles into Cartesian coordinates of the hand (forward kinematics). In this model the number of degrees of freedom in muscle space is higher than the number of degrees of freedom in joint space, which is in turn higher than the number of degrees of freedom in hand space. Hence

Figure 3. *A*, Learning the forward model. *B*, Learning the controller.

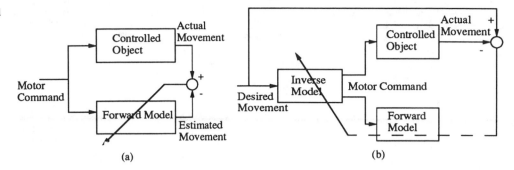

(a)   (b)

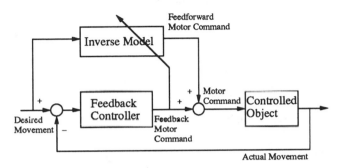

Figure 4. Feedback-error learning.

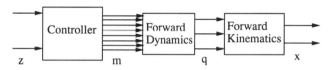

Figure 5. A sensorimotor transformation from an internal representation of a sensory stimulus (z) to hand coordinates (x); m: muscle activations; q: joint angles.

the controlled object presents both a kinematic and a dynamic redundancy: there is an infinite number of ways in which an error in hand coordinates maps into an error in joint coordinates, and an infinite number of ways in which an error in joint coordinates maps into an error in muscle coordinates. Let us now adopt the forward model approach to learning the controller, which requires us to propagate the hand error backward in order to compute an error in muscle space—a highly ill-posed process in our example. If we choose the cost function J that the learning algorithm minimizes as a simple quadratic form of the hand error

$$J = \frac{1}{2}(x^* - x)^{\mathrm{T}}(x^* - x) \tag{2}$$

with corresponding learning rule for the network weights w

$$\Delta w = -\alpha \nabla_w J \tag{3}$$

$$\nabla_w J = -(x^* - x)^{\mathrm{T}} \frac{\partial x}{\partial w} \tag{4}$$

then the resulting error in muscle space will be picked at random among the infinite possible configurations—a solution that is unlikely to bear any physiological significance. The cost function, however, can be enriched to include terms that force the learning process to compute the error in muscle space according to a specific criterion. The following cost function, for example, realizes an optimization criterion of minimum change

of the muscles activation (see also Kawato, 1992). It is the sum of the quadratic cost function (Equation 2) plus a term that penalizes the change over time of the vector m.

$$J = \frac{1}{2}(x^* - x)^{\mathrm{T}}(x^* - x) + \frac{1}{2} \sum_{k=1}^{N} (m_k - m_{k-1})^{\mathrm{T}}(m_k - m_{k-1})$$

$$\tag{5}$$

In the second part of the equation, k represents the time index and N the number of time steps in the reaching movement. The corresponding learning rule is given by

$$\Delta w = -\alpha \nabla_w J \tag{6}$$

$$\nabla_w J = -(x^* - x)^{\mathrm{T}} \frac{\partial x}{\partial w} + \sum_{k=1}^{N} (m_k - m_{k-1})^{\mathrm{T}} \frac{\partial m_k}{\partial w}$$

$$- (m_k - m_{k-1})^{\mathrm{T}} \frac{\partial m_{k-1}}{\partial w} \tag{7}$$

By substituting q for m in Equations 5 to 7, one can obtain a learning rule that solves for the redundancy of the arm by minimizing the change over time of the joint angles, i.e., the angular velocity.

The use of modified cost functions to learn with extra degrees of freedom was first introduced by Jordan (1990) in the context of arm movements and speech: he introduced smoothness constraints (of which Equation 5 is an example) as well as their opposite, distinctiveness constraints, and rest-configuration constraints. Alternative and successful ways of optimizing the learning process have been proposed by Kawato (1992). Massone and Bizzi (1989) utilized a minimum-potential-energy criterion to construct the training off line. See also OPTIMIZATION PRINCIPLES IN MOTOR CONTROL.

Representing Information

Another aspect that plays a major role in the computation underlying sensorimotor learning is how the information is represented. A design principle widely used by the CNS is that of coarse coding (Hinton, McClelland, and Rumelhart, 1986), which uses large and partially overlapped receptive fields. Through this principle the CNS achieves greater sensitivity and resolution, higher signal-to-noise ratio, wider dynamic range, better resistance to damage due to loss or independent drift of single neurons and for reduced demand for precision in ontogenesis and for stability of performance. Coarse coding turns out to be also crucial in the design of computational systems capable of learning sensorimotor mappings. Massone and Bizzi (1990) provided a comparison between the performances of a learning algorithm for an arm controller when different types of representation were used for the sensory information at the network input layer. The comparison showed that a coarse-coding scheme is superior, in terms of both number of epochs of the learning algorithm and the ability of the controller to generalize, to other more local schemes. The reason for the latter effect is that a coarse-coding scheme requires a lower number of units to represent a certain amount of information than a local scheme does, and hence the corresponding network exhibits a lower number of computational degrees of freedom. The reason why coarse coding is also likely to require a lower number of epochs to learn a certain set of patterns is that the receptive fields of the input units overlap. This overlapping causes an input unit to become active more often than in a local scheme. Consequently, the weights that originate from each input unit are trained, in the case of coarse coding, more often than in the case of local coding.

Acknowledgment.   This work was supported by National Science Foundation grant BCS–9113455 to Lina L.E. Massone.

Road Maps: Biological Motor Control; Control Theory and Robotics; Learning in Biological Systems
Background: I.3. Dynamics and Adaptation in Neural Networks
Related Reading: Eye-Hand Coordination in Reaching Movements; Identification and Control; Reinforcement Learning in Motor Control; Robot Control; Schema Theory

References

Arbib, M. A., 1990, Programs, schemas and neural networks for control of hand movements: Beyond the RS framework, in *Attention and Performance XIII* (M. Jeannerod, Ed.), Hillsdale, NJ: Erlbaum, pp. 111–138.
Bernstein, N. A., 1967, *The Coordination and Regulation of Movements*, Oxford: Pergamon. ◆
Hinton, G. E., McClelland, J. L., and Rumelhart, D. E., 1986, Distributed representations, in *Parallel Distributed Processing: Explorations*
in the Microstructure of Cognition (D. E. Rumelhart, J. L. McClelland, and PDP Research Group, Eds.), vol. 1, *Foundations*, Cambridge, MA: MIT Press, pp. 77–109.
Jordan, M. I., 1990, Motor learning and the degrees of freedom problem, in *Attention and Performance XIII* (M. Jeannerod, Ed.), Hillsdale, NJ: Erlbaum, pp. 796–836. ◆
Jordan, M. I., and Rumelhart, D. E., 1992, Forward models: Supervised learning with a distal teacher, *Cognit. Sci.*, 16:307–354.
Kawato, M., 1990, Computational schemes and neural network models for formation and control of multijoint arm trajectory, in *Neural Networks for Control* (W. T. Miller, R. S. Sutton, and P. J. Werbos, Eds.), Cambridge, MA: MIT Press, pp. 197–228.
Kawato, M., 1992, Optimization and learning in neural networks for formation and control of coordinated movement, in *Attention and Performance XIV* (D. Meyer, Ed.), Hillsdale, NJ: Erlbaum, pp. 821–849. ◆
Keele, S. W., and Summers, J. J., 1976, The structure of motor programs, in *Motor Control—Issues and Trends* (G. E. Stelmach, Ed.), San Diego: Academic Press, pp. 109–142. ◆
Massone, L., 1994, A neural network system for control of eye movements: Basic mechanisms, *Biol. Cybern.*, 71:293–305.
Massone, L., and Bizzi, E., 1989, A neural network model for limb trajectory formation, *Biol. Cybern.*, 61:417–425.
Massone, L., and Bizzi, E., 1990, On the role of input representations in sensorimotor mapping, in *Proceedings of the International Joint Conference on Neural Networks*, vol. 1, New York: IEEE, pp. 173–176.
Massone, L., and Khoshaba, T., 1994, Local dynamic interactions in the collicular motor map: A neural network model (submitted for publication).
Schmidt, R. A., 1988, *Motor Control and Learning: A Behavioral Emphasis*, Champaign, IL: Human Kinetics. ◆
Schoener, G., and Kelso, J. A. S., 1988, Dynamic pattern generation in behavioral and neural systems, *Science*, 239:1513–1520.
Stein, B. E., and Meredith, M. A., 1993, *The Merging of the Senses*, Cambridge, MA: MIT Press. ◆

Sensory Coding and Information Theory

John Hertz

Introduction

The brain is an information-processing machine. Although this statement is commonplace, most neurobiologists only use the word *information* in the informal, qualitative sense. However, there has been a school of thought, dating from the 1950s, that has taken the statement seriously in the formal sense of Shannon information theory (Shannon, 1948). In recent years, quantitative studies of neuronal information transmission have yielded unexpected results which compel us to reexamine conventional views of how the brain works and how it can be modeled theoretically.

Only a few years after Shannon's invention of information theory, MacKay and McCulloch (1952) attempted to estimate the information transmission capacity of single spiking neurons. Assuming that it is limited only by the refractory time (1 ms) and the discriminability of successive spikes, one easily obtains an upper bound on the transmission rate of the order of 1000 bits/s. A more restrictive bound is obtained by computing the actual entropy of spike trains. This is less than the above limit, because neurons spike less than half of the time and different 1-ms intervals are not independent. Still, numbers on the order of 500 bits/s are found.

Now, if neurons are intrinsically very noisy devices, reliable transmission will require a high degree of redundancy, and the actual rate at which they convey information will be correspondingly lower. It has been commonplace to suppose that this is the case, but neural firing may appear noisy to us only because we do not understand it. To achieve even the beginning of an understanding of how the brain works, it is necessary to measure the rate at which neurons actually carry information.

In the 1970s, Eckhorn and Pöpel (1974, 1975) laid the foundation for much recent work by measuring the rate at which neurons in the cat lateral geniculate nucleus transmitted information about a random train of visual flashes, independent of a priori assumptions about how it was coded. They found rates from about 10 to as high as 60 bits/s, depending on how fast the stimulus was flashed.

Here, we review two more recent sets of investigations. The first, by Bialek and his collaborators, deals with vision in flies, hearing in frogs, and mechanoreception in crickets. The second, by Optican and Richmond and their co-workers, treats spatial pattern vision in monkeys.

Information-Theoretic Background

Whatever system we are interested in, the formal characterization of the problem is the same. The animal is presented a stimulus s from some set S, and the response of a neuron (or several neurons) is measured. For spiking neurons, the response can be represented generally in the following way. Time is divided into intervals (typically 1 ms) small enough that there is never more than one spike per interval. The response can then be described by a vector **r** with one component per interval, equal to a 1 or 0 according to whether or not the neuron fired a spike in that interval. (For present purposes, we will assume that submillisecond timing of neuronal spikes does not carry information.)

The average information gained from the observation of a response **r** about which stimulus s (out of a set S) was presented is

$$I(S; R) = \left\langle \sum_s P(s|\mathbf{r}) \log_2 \left(\frac{P(s|\mathbf{r})}{P(s)} \right) \right\rangle_\mathbf{r} \qquad (1)$$

where $P(s|\mathbf{r})$ is the conditional probability that the stimulus was s, given that the response was **r**; $P(s)$ is the probability of stimulus s without any knowledge about the response; and the average is over the response distribution $P(\mathbf{r})$. Since $P(s) = \int P(s|\mathbf{r})P(\mathbf{r}) d\mathbf{r} = \langle P(s|\mathbf{r}) \rangle_\mathbf{r}$, we can always find $P(s)$ if we know $P(s|\mathbf{r})$ and $P(\mathbf{r})$. Therefore, the calculation of the transmitted information requires simply the accurate estimation of the conditional probabilities $P(s|\mathbf{r})$. However, performing the calculation is generally a difficult task because of the high dimensionality (typically at least several hundred) of **r**. The two groups of investigators use different methods to deal with this problem.

Insects and Amphibians

In the method adopted by Bialek and his collaborators (Bialek et al., 1991; Rieke, Warland, and Bialek, 1993), the estimate of the stimulus $s(t)$ is expanded in a Taylor series in **r**:

$$s_{\text{est}}(t) = \int_{-\infty}^{t} dt' F_1(t - t') r_{t'}$$
$$+ \int_{-\infty}^{t} dt' \, dt'' F_2(t - t', t - t'') r_{t'} r_{t''} + \cdots \qquad (2)$$

where $r_{t'}$ is the spike number recorded at time t'. In their fly experiments, the cell recorded from (called H1) is one which is sensitive to the overall horizontal movement of the visual field. The stimulus is the angular velocity of a randomly moving rigid random pattern. For the cricket it is a white noise audio signal, while for the bullfrog it is either that or a noisy simulated frog call. In all cases, the stimulus distribution is Gaussian. In these animals, the cells recorded from respond to these signals almost linearly; that is, the quadratic term in the fit in Equation 2 is relatively small. In this discussion, we will ignore it.

Estimating the decoding kernel F_1 is a straightforward task. In neural network terms, it amounts to training a linear unit with inputs equal to the immediate past response values r_{t-1}, $r_{t-2}, \ldots, r_{t-T}$ (for a suitable window size T) and target equal to the present (known) stimulus $s(t)$. Once this is done, one can compare the estimated stimulus with the true one and determine the distribution of noise in the system. Since we are working with a linear, time-invariant filter, it is easiest to do this after Fourier transforming both the true stimulus $s(t)$ and the estimated one $s_{\text{est}}(t)$. The time interval over which the Fourier transform is computed must be long enough to contain any significant temporal structure in the stimulus or the response. In these calculations, 1-second segments are used. Thus, for each such segment, one has Fourier components tilde $\tilde{s}(\omega)$ of the actual stimulus and tilde $\tilde{s}_{\text{est}}(\omega)$ of the estimated one for many different frequencies. These are plotted against each other on a scatter plot. The deviations from a straight line fitted to these data represent a noise $\tilde{\eta}(\omega)$. Plotting the distribution of

$\tilde{\eta}(\omega)$ by sampling over thousands of 1-second intervals, one finds that it is Gaussian to a high degree of accuracy. Since by design of the experiment the a priori stimulus distribution was also Gaussian, all the probability densities in Equation 1 are Gaussians, and one can perform the integrals analytically. The result is

$$I(S; R) = \frac{1}{2} \int_{-\infty}^{\infty} \frac{d\omega}{2\pi} \log \left[1 + \frac{\langle |\tilde{s}(\omega)|^2 \rangle}{\langle |\tilde{\eta}(\omega)|^2 \rangle} \right] \qquad (3)$$

Since the signal and noise powers $\langle |\tilde{s}(\omega)|^2 \rangle$ and $\langle |\tilde{\eta}(\omega)|^2 \rangle$ have been measured at each frequency in the process described above for separating signal and noise, it is simple to calculate the information transmission rate. For the H1 neuron in the blowfly, the result rate is 64 ± 1 bits/s. For the bullfrog vibratory receptor, it is 155 ± 3 bits/s, and for the cricket filiform hair, it is 294.6 ± 6 bits/s. (These numbers include small corrections—a few percent—which come from including the quadratic term in the expansion Equation 2 in the calculation.) These rates are about half the absolute upper bounds set by the spike train entropies. Thus, these neurons are not nearly as noisy as we might have believed; a large fraction of the variability in their firing patterns is actually used to convey information.

How do these neurons code the information they transmit? For the blowfly, de Ruyter van Steveninck and Bialek (1988) found, in a detailed analysis of short segments (up to three spikes) of the responses that the timing of the spikes was significant: the information carried by a pair of spikes and the interval between them was several times that carried by a single spike. On the other hand, *exact* knowledge of the spike times was not necessary to make an optimal decoding of the response. The code proved robust against errors in spike times of the order of several milliseconds.

Primate Pattern Vision

The second set of measurements of information transmission that we discuss here have been performed by Richmond and Optican and their co-workers, who have made electrophysiological recordings in the visual system of rhesus monkeys (Optican and Richmond, 1987; Richmond and Optican, 1990). The purpose of the experiments was to investigate quantitatively how these animals, whose visual systems are very similar to those of humans, process information about spatial structure in patterns that they look at.

The stimulus patterns used in these experiments were based on Walsh functions (Figure 1). These patterns contain many local spatial features (edges, bars, corners, etc.) a functioning visual system would have to analyze, and their spectra are spread over many spatial frequencies and orientations. The stimuli were presented to the animal for several hundred milliseconds. The spike trains emitted by neurons during or immediately after this period, measured with 1-ms resolution, were then analyzed for information conveyed about the stimuli.

Rather than following the indirect route taken by Bialek and his collaborators, Richmond and Optican chose to estimate the conditional probabilities in Equation 1 directly. The high dimensionality of the response vector **r** necessitated some preprocessing of the data. They therefore performed a two-step compression of their reponses before constructing their models for $P(s|\mathbf{r})$. In the first step, the spikes were smeared into a continuous signal by convolving them with a Gaussian kernel. (Typically, the half-width of this kernel was 3 to 5 ms.) The resulting function was then sampled at 4- or 5-ms intervals to give a lower-dimensional representation of **r**. A principal com-

Figure 1. Walsh patterns used as stimuli in the experiments of Richmond and Optican.

ponent analysis was then performed on these data, rotating to a new coordinate frame in which the largest variance is in direction number 1, the next largest is in direction number 2, and so on. The rotated response vectors were then truncated, retaining only the first N principal components. Crudely speaking, this preprocessing amounted to a weakly time-dependent low-pass filtering of the response. Other response representations, such as spike times or interspike intervals, can be employed in place of the principal components, but they have not been found to contain any information beyond that carried in the first five principal components.

Once the responses are represented this way, the main task of fitting the conditional probabilities $P(s|\mathbf{r})$ can be tackled. The most effective way found so far to do this is with a feedforward neural network (Hertz et al., 1992; Kjær, Hertz, and Richmond, 1994). That is, $P(s|\mathbf{r})$ is fit by the output of such a network, parametrized by its weights and biases. Different models, using different numbers of inputs (i.e., principal components) and hidden units, are compared on the basis of test set error, averaged over different splits of the data into training and test sets. The result of this model comparison is that three to five principal components and around six hidden units are almost always nearly optimal, although one can exceed these numbers by factors of two without significantly worsening the fit quality.

Once the fit is made, the network outputs $P(s|\mathbf{r})$ can be used to estimate the transmitted information Equation 1 by sampling over the data set. Typically, the results are between $\frac{1}{2}$ bit and 1 bit. The responses analyzed in these experiments were 320 ms long, so the apparent average transmission rate was about 2 bits/s, dramatically lower than that found by Bialek and his co-workers in much more primitive nervous systems. However, performing the above kind of analysis on different portions of the response period reveals that most of the information is actually transmitted in the first 30 ms (Tovee et al., 1993). The peak transmission rate, achieved in this initial burst, is about 30 bits/s. It is plausible that these neurons maintain this rate under normal operating conditions with dynamic stimuli. In fact, the transmission rates which Eckhorn and Pöpel (1975) found for lateral geniculate neurons in the cat are of this magnitude.

So far, for both sets of experiments, we have focused solely on rates of information transmission, not on the content of the messages transmitted. It is possible to analyze this aspect of the code by performing the calculations described above for subsets of the full stimulus set, selected for particular values of average spatial frequency and of the spread of spatial frequencies within them. The dependence of the transmitted information on these variables tells us both about what spatial frequencies a neuron transmits information most effectively and about its sensitivity to differences in spatial frequency. Such an analysis (Kjær et al., 1994) reveals a surprising richness and complexity: these neurons (complex cells; see VISUAL CORTEX CELL TYPES AND CONNECTIONS) do not act as simple detectors of oriented bars and edges. It appears that to the extent that they detect such features, they tend to do so in a way that is insensitive to (i.e., generalizes across) their internal texture.

The present analysis has been restricted to information about static spatial structure. Recent work by McClurkin, Zarbock, and Optican (1994), using the same Walsh stimuli but also varying them in color and luminance, finds that the information carried about these attributes is typically several times that carried about spatial pattern. Thus, a total peak rate comparable to the 64 bits/s reported by Bialek et al. (1991) for the H1 neuron in the blowfly is likely.

Discussion

These examples demonstrate that quantitative measurements of neuronal information transmission can be carried out. Already, the studies of insect and frog sensory neurons have yielded the important result that information about the stimulus is carried continuously at a reasonable fraction of the maximum possible rate consistent with the overall spike statistics. The results for primate visual cortical neurons suggest that comparable rates are also achieved there, though only in an initial firing burst of around 30 ms. What these neurons are doing with their bandwidth the rest of the time remains unknown. One should remember, however, that cortical neurons are not simply transducers of external stimuli. Though they do perform some kind of transduction, they do so in the context of inputs from other cells in V1 and in other areas of the brain. Indeed, there are generally many more afferents to a given cortical neuron from other cortical neurons than from sensory input pathways, so it is natural to expect that a large fraction of the neuronal activity is devoted to such internal cortical communication. It might be possible to analyze this kind of information transmission by studying systematically the mutual information between signals obtained from different neurons in multiunit recordings.

As we have seen, systematic study of the content of neuronal messages is also possible, but such investigations are only beginning to be carried out. They will be important in connection with an interesting line of theoretical work seeking to understand cortical organization and neural processing in terms of efficient coding (see VISUAL CODING, REDUNDANCY, AND "FEATURE DETECTION").

Road Map: Biological Neurons
Related Reading: Information Theory and Visual Plasticity; Synaptic Coding of Spike Trains

References

Bialek, W., Rieke, F., de Ruyter van Steveninck, R., and Warland, D., 1991, Reading a neural code, *Science*, 252:1854–1857. ◆

Eckhorn, R., and Pöpel, B., 1974, Rigorous and extended application of information theory to the afferent visual system of the cat, I: Basic concepts, *Kybernetik*, 16:191–200.

Eckhorn, R., and Pöpel, B., 1975, Rigorous and extended application of information theory to the afferent visual system of the cat, II: Experimental results, *Biol. Cybern.*, 17:7–17.

Hertz, J. A., Kjær, T. W., Eskandar, E. N., and Richmond, B. J., 1992, Measuring natural neural processing with artificial neural networks, *Int. J. Neural Syst.*, 3(supp):91–103. ◆

Kjær, T. W., Hertz, J. A., and Richmond, B. J., 1994, Decoding cortical neuronal signals: Network models, information estimation, and spatial tuning, *J. Computat. Neurosci.*, 1:109–139.

McClurkin, J. W., Zarbock, J. A., and Optican, L. M., 1994, Temporal codes for colors, patterns and memories, in *Visual Cortex of Primates* (A. Peters and K. S. Rockland, Eds.), New York: Plenum, pp. 443–467. ◆

MacKay, D. M., and McCulloch, W. S., 1952, The limiting information capacity of a neuronal link, *Bull. Math. Biophys.*, 14:127–135.

Optican, L. M., and Richmond, B. J., 1987, Temporal encoding of two-dimensional patterns by single units in primate inferior temporal cortex, III: Information theoretic analysis, *J. Neurophysiol.*, 57:162–178.

Richmond, B. J., and Optican, L. M., 1990, Temporal encoding of two-dimensional patterns by single units in primate primary visual cortex, II: Information transmission, *J. Neurophysiol.*, 64:370–380.

Rieke, F., Warland, D., and Bialek, W., 1993, Coding efficiency and information rates in sensory neurons, *Europhys. Lett.*, 22:151–156.

de Ruyter van Steveninck, R., and Bialek, W., 1988, Real-time performance of a movement-sensitive neuron in the blowfly visual system: Coding and information transfer in short spike sequences, *Proc. R. Soc. Lond. B Biol. Sci.*, 234:379–414.

Shannon, C. E., 1948, A mathematical theory of communication, *Bell Syst. Tech. J.*, 27:379–423, 623–653. ◆

Tovee, M. J., Rolls, E. T., Treves, A., and Bellis, R. P., 1993, Information encoding and the responses of single neurons in the primate temporal visual cortex, *J. Neurophysiol.*, 70:650–654.

Short-Term Memory

Emmanuel Guigon and Yves Burnod

Introduction

It is now generally agreed that temporally distinct neural processes contribute to the acquisition and expression of brain functions. Transient variations of membrane potential (neuronal activity), with a time scale of milliseconds, reflect the flow of information from neuron to neuron and define the function of neuronal networks. These variations can result in long-lasting (and maybe permanent) alterations in neuronal operations—for instance, through activity-dependent changes in synaptic transmission.

There is now strong evidence for a complementary process, acting over an intermediate time scale (short-term memory, STM). This process is involved in performing tasks requiring temporary storage and manipulation of information to guide appropriate actions (Goldman-Rakic, 1987; Baddeley, 1992). Two main issues should be addressed when studying STM: (1) How is neural information selected for storage and temporarily stored in STM for future use in a temporal sequence of sensorimotor events, and how is a large amount of information buffered when its future use is not known? (2) How can a long-term memory (LTM) representation of temporal sequences of events be constructed, and how can information selected by STM process be transferred to LTM?

Figure 1 outlines a general scheme which allows models including short-term and long-term memory to be compared. A network of units processes spatial and temporal information, represented by their short-term (milliseconds to a few seconds) activities (x_i, y_j). The units store information by changing their synaptic weights (W), with long-lasting effects (days, years). Information can be stored over intermediate time scales of seconds, minutes, or hours by short-term memory elements, represented by units (z_k). At the neural level (Figure 1A), STM appears to be an intermediate step between neuron activity and LTM within single neurons or simple circuits. At the system level (Figure 1B), two brain structures have been extensively studied for their role in STM processes, the hippocampus (Squire, Knowlton, and Musen, 1993) and the prefrontal cortex (Fuster, 1989).

This article discusses the relationships between neuronal activity, STM, and LTM at the neural and system levels, and then presents a neural network model, which illustrates the properties of short-term memory in the prefrontal cortex.

Short-Term Memory at the Neural Level

This section focuses on the temporal patterns generated by single neurons, simple circuits or networks, which may be responsible for short-term memory.

Biological Neurons and Simple Circuits

From activity to STM. A wide variety of temporal patterns of activity are actively generated by neurons and local circuits of neurons, such as transforming transient inputs into long-

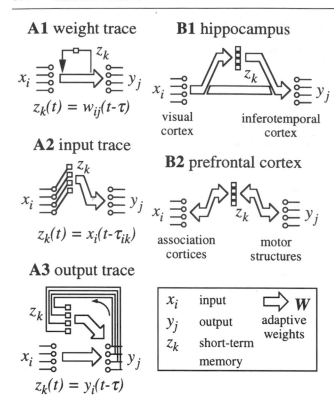

A1 weight trace

$$z_k(t) = w_{ij}(t-\tau)$$

A2 input trace

$$z_k(t) = x_i(t-\tau_{ik})$$

A3 output trace

$$z_k(t) = y_i(t-\tau)$$

B1 hippocampus

visual
cortex   inferotemporal
cortex

B2 prefrontal cortex

association
cortices   motor
structures

x_i	input	
y_j	output	W adaptive weights
z_k	short-term	
	memory	

Figure 1. Architecture for short-term and long-term memory at two levels of brain organization. Each network has an input pathway (x_i), an output pathway (y_j), a short-term memory pathway (z_k), and a set of adaptive weights (W). A, The neural level: A1, STM units store transient variations in synaptic weights; A2, STM units store synaptic inputs in delay lines; A3, STM units store recent history of the activity in the network provided by recurrent connections. B, The system level: B1, STM units in the hippocampus contribute to the transfer of information from STM to LTM; B2, STM units in the prefrontal cortex are the temporal link between sensory and motor events.

lasting sustained or oscillatory activity. Experimental studies in invertebrates (Harris-Warrick and Marder, 1991) have demonstrated that such temporal patterns produce motor programs and are generated both by the molecular properties of each neuron, and the connectivity of the local network.

It is now well established that, in vertebrates, long-lasting activities are neural correlates of transient memory processes (mainly in the frontal lobe of the cerebral cortex). This pattern of activity allows past events to be represented and behavioral reaction to future, predictable events to be prepared in what is called *working memory* or *active memory* (Goldman-Rakic, 1987; Fuster, 1989).

Several models have addressed the question of how to maintain such a sustained activity. One popular model is a network with reverberating excitatory feedback loops forming a *cell assembly* (Hebb, 1949). A model of reciprocally inhibitory neurons, described by a system of nonlinear differential equations, can generate bistable activities (Kirillov, Myre, and Woodward, 1993). Such a model describes the conditions of stability of the two states, the role of noise, and the input commands for transitions between states. Zipser et al. (1993) provided direct evidence for bistability of cortical neurons in a recurrent neural network trained to mimic the input-output characteristics of an active memory module.

The intrinsic properties of a single neuron can also be responsible for generating bistable activities, via a set of ionic channels (for example, persistent sodium currents in the spinal cord): one state is silent, and the other is continuous activity that can be triggered by a transient synaptic input (see NEUROMODULATION IN INVERTEBRATE NERVOUS SYSTEMS).

From STM to LTM.   A cascade of molecular events occurs in neurons after synaptic activation; these include the activation/inactivation of the various types of sodium and potassium channels with different time constants, a calcium influx and second messenger cascades, short-term changes in the probability of transmitter release, and short-term potentiation (STP). These events define memory traces which outlast the duration of synaptic events. They also constitute initial steps for the formation of longer memory traces, such as long-term potentiation (LTP) which can last for hours (see HEBBIAN SYNAPTIC PLASTICITY and NMDA RECEPTORS: SYNAPTIC, CELLULAR, AND NETWORK MODELS).

STM in Neural Network Models

From a computational point of view, a simple way to implement short-term memory is to consider that neuronal variables have an effect which outlasts their duration. This effect can concern synaptic weights, synaptic inputs, or membrane potential as illustrated in Figure 1, A1–A3.

Short-term memory can appear to be an intermediate step in the learning process at the level of each synapse, in the same way as STP and LTP (Figure 1, A1). The strength of the synapse is transiently modulated by the successive events in a sequence. Since associative learning rules, such as the wide classes of Hebbian rules, are based on the temporal coincidence of two events, transient synaptic modifications allow the time overlap to be increased, and thus association between temporally separate events to be learned. Sutton and Barto (1981), proposed a model in which the time scale of neuronal activations (milliseconds to hundreds of milliseconds) is extended to the time scale of temporal correlations between successive sensory and motor events during classical conditioning (seconds to minutes).

Temporal sequences of events can be turned into spatial patterns (Figure 1, A2). In time-delay neural networks (Waibel, 1989), a sequence is represented by a vector in such a way that the position in the vector corresponds to the temporal order of events. Thus, the events occurring in a preselected time-window can be learned as a single spatial pattern.

Time can also be processed when units are linked by recurrent connections (Figure 1, A3). In recurrent networks, a subset of input neurons represents the trace of activity of output neurons that is the result of the computation performed by the network in the previous time step (see LANGUAGE PROCESSING). A new input is thus interpreted in a specific context, which can be learned from the previous computations performed by the network. In this way, the neuronal operations at a given time are modulated by the recent history of the network. In these models, memory results from sequential transitions between states rather than from the buffering of specific items.

Short-Term Memory at the System Level

Short-term memory may also be a property of large-scale neural architecture involving cortical and subcortical regions of the brain (Figure 1, B1 and B2).

Hippocampus

The hippocampus is important for STM to LTM transfer, but it does not appear to be the substrate of LTM (Squire et al.,

1993). Monkeys with hippocampal lesions are severely impaired at remembering recently learned objects, but perform correctly with objects learned long ago. The hippocampus is needed for some time after learning. Permanent memory develops in adjacent cortical areas of the temporal lobe, to which the hippocampus is connected in parallel (see Figure 1, *B1*). The neural correlates of the long-term memory storage of new patterns in these temporal regions are now well documented: for example, in recognition of visual patterns, neural activities are selective for "prototypes," invariant in size and orientation, and sustained activities represent the temporal links between these prototypes.

Memory processes in the hippocampus appear to be based on three different forms of plasticity within a serially organized anatomical circuit that comprises the corticohippocampal pathways (entorhinal cortex → dentate → CA3 → CA1). Experimental data indicate that synaptic potentiation can persist for hours in mossy fibers (between dentate and CA3), for several days in cortical projections to the dentate gyrus, for several weeks in the CA1 (see HEBBIAN SYNAPTIC PLASTICITY).

The hippocampus can be viewed as a control system between STM and LTM in temporal cortical areas (see ADAPTIVE RESONANCE THEORY). This "search and orienting" system is able to determine whether an input is a new example of a previously stored prototype or a new prototype. A "resonant state" appears when low-level inputs and high-level expectancies are matched. During this state, the input example can be stored. When there is a mismatch, the hippocampal control system triggers a memory search for a better category by activating a new high-level expectancy (a new "hypothesis"). If the input is too different from any previously learned prototype, the hippocampal control system selects an uncommitted population of high-level neurons to store a new category. Once a memory is formed, the hippocampus is not needed for retention or retrieval: familiar events have a direct access to their recognition code.

Prefrontal Cortex

The prefrontal cortex is involved in integration of temporally separate events into purposive behavioral structures (Goldman-Rakic, 1987; Fuster, 1989). This function is reflected in the performance of tasks using temporal delays for structuring behavioral reactions to environmental stimuli. The paradigmatic test is the delayed response task, which requires a subject to memorize an instruction stimulus and to wait for a go signal before responding to it. This task is typically impaired after lesions of the prefrontal cortex (Fuster, 1989).

Prefrontal neurons recorded during delayed response tasks in monkeys display patterns of sustained activity which reflects the short-term mnemonic aspects related to instruction cues, the expectation of forthcoming signals, and the preparation of the behavioral reaction. Sustained activities have three important characteristics, which define their cardinal role in the learning and execution of behavioral tasks. First, whatever the modalities used (visual or auditory cues, arm or eye movement responses), they occur during the delay between an instruction cue and the final permission to use the information contained therein to produce a response. Second, the duration of the activity is linked to the duration of the delay. Increasing the delay's length leads to a prolonging of the activity. Third, these activities are a product of learning and there appears to be a relationship between the amount of delay activation and the level of performance (Fuster, 1989).

Goldman-Rakic (1987) has proposed that the prefrontal cortex is necessary for expression of behaviors guided by representation or internalized models and not when the behavior is guided by external stimuli. The mechanism of the prefrontal cortex is related to a distributed system of interconnected neural networks. Specific functions would thus come from dynamics of the system and interactions between independent networks, rather than from a strictly hierarchical processing based on the convergence through association regions (Figure 1, *B2*). In a such a system, the "working memory," defined as the formation of selective memory traces of relevant events, appears as a relevant concept which characterizes the specificity of prefrontal functions (Goldman-Rakic, 1987).

Short-term memory in the prefrontal cortex appears to be subserved by sustained activities. Furthermore, these activities may also be involved in the formation of permanent memory (Fuster, 1989). Thus, the same mechanism is likely to participate in formation and retention of memories. This property is in contrast with the hippocampus, in which different mechanisms contribute to the formation and the retention of memories.

A Model of Short-Term Working Memory

We have built a computational model of prefrontal circuits to illustrate a strategy for implementing short-term memory in a neural network (Guigon et al., in press). Based on the principles of organization and operations in the prefrontal cortex, this model shows that short-term memory in a neural network can be obtained by processing units which switch between two stable states of activity (bistable behavior) in response to synaptic inputs. The sustained activity of a given neuron represents a temporal link between two sensory or motor events. It also shows that long-term representation of tasks requiring short-term memory can result from activity-dependent changes in the synaptic transmission controlling the bistable behavior. After learning, the sustained activity of a given neuron represents both the selective memorization of a past event and the selective anticipation of a future event.

Description of the Model

The neural network model, designed according to the principles of organization of prefrontal connections, was trained to execute a delayed response task (Figure 2). The architecture of the network is shown in Figure 2A. Each sensory event is coded by the all-or-none activation of a specific unit in the sensory layer, and movements toward the levers are coded in the motor layer. Matching units model neurons in the associative sensory and motor areas connected to the prefrontal cortex. These units implement sensorimotor relations, such as a direct relation between the position of the lever and movement toward the lever. Bistable units model prefrontal neurons. Each bistable unit is reciprocally connected to one matching unit and receives nonreciprocal projections from some other matching units. This connectivity defines multiple interactions between matching and bistable units, but does not correspond to an a priori representation of particular functions. Bistable units are connected to a drive pathway *d* (thirst), which is made active at the beginning of each behavior of the network, and to a reinforcement pathway *r* (receipt of liquid), which is activated when a correct behavior is produced by the network.

The function of the network is defined by the dynamics of processing units and by the adjustable connection coefficients between processing units. Neural processing function of matching units is modeled by a nonlinear interaction between inputs, which reflects the modulation of sensory inputs and motor outputs by memorized conditions. Matching units respond to the coactivation of sensory and bistable inputs, but

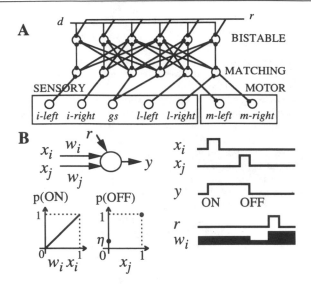

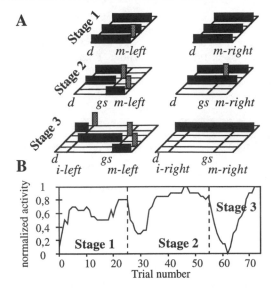

Figure 2. *A*, Architecture of the network for learning a delayed-response task. The task involves two lights mounted above two horizontally arranged levers and a trigger light. At each trial, one light (instruction stimulus) comes on for a short period; a few seconds later, the trigger light (go signal) comes on and the animal touches the lever indicated by the instruction: it receives a reward. Notations: *l-left* and *l-right*, positions of left and right levers; *m-left* and *m-right*, movements toward the levers; *gs*, go signal; *i-left* and *i-right*, instruction stimuli; *d* and *r*, drive and reinforcement. Black dots indicate synapses. *B*, Dynamics of bistable units. The unit has two weighted input pathways x_i (w_i) and x_j (w_j), a reinforcement pathway r, and an output pathway y. Variables are binary. Weights vary in [0, 1]. Qualitative variations in the activity y and the synaptic weight w_i when input and reinforcement pathways are activated as shown in the tracings. Unlike classical neural automata, which display transient responses to transient inputs, the present neuron remains activated (state ON) for some time after the input i. The neuron then returns to rest after the second input j (state OFF). Transition to the ON state follows a classical law used to model the stochastic behavior of neurons (graph p(ON)): the probability of transition is proportional to the summed inputs. Transition to the OFF state has two components (graph p(OFF)): a spontaneous transition with a fixed probability η (effect of noise) and an unconditional transition following subsequent inputs. Only the transition to the ON state is controlled by a synaptic weight.

Figure 3. Computer simulations. *A*, Activities in three bistable units (black) and three matching units (shaded) are qualitatively displayed for each training stage and for left and right trials. The task events are those described in Figure 2. Note the gradual changes in the relationships between neuronal activity and task events and the differentiation for left versus right trials. *B*, Variations in the level of activity of a bistable unit during the training period. The graph is constructed from the activity during reinforced left trials. Each horizontal division corresponds to a trial. Vertical dashed lines indicate the transitions between training stages. Note the combination of increasing and decreasing activity: activity decreases at the transition between two stages and increases after the transition.

not to activation of either input alone. We postulate that prefrontal neurons have two stable states of activity (bistable), and that transitions between these states are elicited by synaptic inputs (Figure 2*B*; Guigon et al., in press). We also postulate that this bistable behavior is controlled by learning and allows sensorimotor sequences to be built up under the control of a reinforcement signal.

Bistable Units Implement Short-Term Memory

Computer simulations of the neural network in Figure 2 were used to train it to execute a delayed response task in three successive stages: (1) movement, reward; (2) go signal, movement, reward; (3) instruction stimulus, go signal, movement, reward. The rationale for this protocol is that the training protocols used with animals are progressive, stage-by-stage procedures. The contribution of bistable units to the execution of the delayed response task is illustrated in Figure 3*A*. Each graph qualitatively displays the activity of three bistable units at a given training stage. During execution of the task (stage 3), bistable units display different patterns of activity defined by the temporal relationship between task events and peaks of

activity. Each unit is active between two successive task events. The most interesting pattern is the differential delay activity. This is a sustained activity between the onset of the instruction stimulus and the onset of the go signal specific for right versus left trials. All these patterns have been described in the prefrontal cortex during the delayed response task (Fuster, 1989).

At each training stage, bistable units play a complementary role in encoding the temporal structure of the task. Individual units are selective for a specific sequence of events, but the set of units is able to bridge all the gaps between the events of the current task. Matching units displayed transient activity that was time-locked to sensory or motor events and that was correlated with the end of activity in bistable units. They signal the occurrence of a specific sensory or motor event in the context of a specific behavior.

Long-Term Changes in Bistable Units

Variations in the activity of bistable units are correlated with the changes in reinforcement contingency, depending on variations in the reinforcement rate (Figure 3*B*). Two behaviors are alternatively performed by the network when changing from stage 1 to stage 2: one is the previously correct behavior (self-initiated movements) and the other is the new correct behavior (stimulus-triggered movements). The mean activity during reinforced trials increases for leftward self-initiated movements during the first stage. During the transition from stage 1 to stage 2, activity first decreases and then increases with the increase in the performance rate. The same phenomenon occurs between stage 2 and stage 3 (Figure 3*B*). The experience gained at each trial in the learning period is thus transferred to a long-term representation of the task.

Discussion

In most neural networks, information stored into long-term memory reflects correlation between transient neuronal activities. This form of storage is efficient for encoding long-term memory of objects, but less efficient for encoding temporal sequences of events (Sutton and Barto, 1981). We have described different strategies, which use short-term memory mechanisms to link temporally separate events. In some cases, memory is an implicit consequence of neural network architecture and neuronal dynamics. In others, an explicit neural correlate of short-term memory is observed (theta rhythm, sustained activity). We have presented a simple model in order to illustrate a possible mechanism of short-term working memory in the prefrontal cortex. The model has shown the dual role of sustained activity in the short-term retention of relevant cues and in the formation of long-term memory of simple sequential behavior. Further study of these strategies should provide initial cues for the understanding of complex behaviors involving planning, reasoning, and language.

Road Map: Learning in Biological Systems
Background: I.3. Dynamics and Adaptation in Neural Networks
Related Reading: Disease: Neural Network Models; Dynamic Models of Neurophysiological Systems; Hippocampus: Spatial Models; Recurrent Networks: Supervised Learning; Temporal Pattern Processing

References

Baddeley, A. D., 1992, Working memory, *Science*, 255:556–559. ◆

Fuster, J. M., 1989, *The Prefrontal Cortex*, 2nd ed., New York: Raven. ◆

Goldman-Rakic, P. S., 1987, Circuitry of primate prefrontal cortex and regulation of behavior by representational memory, in *Handbook of Physiology: The Nervous System, Higher Functions of the Brain*, vol. 5 (F. Plum, Ed.), Bethesda, MD: American Physiological Society, pp. 373–417. ◆

Guigon, E., Dorizzi, B., Burnod, Y., and Schultz, W., in press, Neural correlates of learning in the prefrontal cortex of the monkey: A predictive model, *Cereb. Cortex*.

Harris-Warrick, R. M., and Marder, E., 1991, Modulation of neural networks for behavior, *Annu. Rev. Neurosci.*, 14:39–57. ◆

Hebb, D. O., 1949, *The Organization of Behaviour*, New York: Wiley. ◆

Kirillov, A. B., Myre, C. D., and Woodward, D. J., 1993, Bistability, switches and working memory in a two-neuron inhibitory-feedback model, *Biol. Cybern.*, 68:441–449.

Squire, L. R., Knowlton, B., and Musen, G., 1993, The structure and organization of memory, *Annu. Rev. Psychol.*, 44:453–495. ◆

Sutton, R. S., and Barto, A. G., 1981, Toward a modern theory of adaptive networks: Expectation and prediction, *Psychol. Rev.*, 88: 135–170.

Waibel, A., 1989, Modular construction of time-delay neural networks for speech recognition, *Neural Comp.*, 1:39–46.

Zipser, D., Kehoe, B., Littlewort, G., and Fuster, J., 1993, A spiking network model of short-term active memory, *J. Neurosci.*, 13:3406–3420.

Silicon Neurons

Rodney Douglas and Misha Mahowald

Introduction

Silicon neurons are analog electronic circuits fabricated in Complementary Metal Oxide Semiconductor (CMOS) medium using Very Large Scale Integration (VLSI) methods. CMOS is a medium for manufacturing transistors whose conductivity can be altered by an applied electric field. This technology is commonly used to construct the digital circuits found in general-purpose computers, but the silicon neurons discussed in this article are not a kind of digital computer. Instead, the same CMOS VLSI technology is used to construct analog circuits whose physics is analogous to the physics of membrane conductivity. This analogy permits the circuits to emulate the electrophysiological behavior of biological neurons in real time, while the high component density offered by VLSI technology provides a means of fabricating large networks of silicon neurons.

Neuronal systems are difficult to model because they are composed of large numbers of nonlinear elements and have a wide range of time constants. Consequently, their mathematical behavior can rarely be solved analytically. The usual approach is to simulate these problems on a general-purpose digital computer (Koch and Segev, 1989). But for any given computer, the speed of these simulations is limited by the shortest time constant in the problem. Furthermore, the simulation time slows dramatically as the number and coupling of elements increase. By contrast, silicon neurons operate in real time, and the speed of the network is independent of the number of neurons or their coupling. Thus, networks of silicon neurons are especially suited to the investigation of questions that arise from real-time interaction of the system with its environment. Nevertheless, the design of special-purpose hardware is a significant investment, particularly if it is analog hardware, since analog VLSI (aVLSI) design is still very much an art form.

Analog VLSI has a controversial role in the study of neural computation. This controversy arises out of a debate over the role of precision in computation. Digital computation is guaranteed precise to the number of bits used in the computation. However, the most compact analog circuits have low precision caused by uncertain calibration between transistors. Some proponents of neuromorphic aVLSI design claim that these circuits provide a natural route for exploring the principles of biological computing, which must also make do with low precision (Mead, 1989; Mead, 1990). Unlike the ideal components used in conventional computers, real neurons are not homogenous. Even within a morphological class such as the pyramidal cells of the cerebral cortex, they show a wide range of behavior. They are poorly insulated conductors; they have a variety of nonlinear conductance elements, and large amounts of stray capacitance; they are sensitive to environmental changes; and a significant fraction malfunction or stop during the operational life of the system. Nevertheless, the smallest vertebrate brain is vastly more competent at interaction with the real world than our most elaborate supercomputers. From the level of conductances, through synapses to neurons and networks, the nervous system elements obtain precision, speed, and computational power using imperfect elements. Understanding the architectures and adaptive processes that allow the nervous system to extract precise information from a noisy and ambiguous environment with uncalibrated components is a central problem of computational neuroscience (see FAULT TOLERANCE for a "digital" perspective). Analog VLSI circuits have similar intrinsic variability, and so synthesis of silicon neurons is a method of exploring the principles of computations that must use unreliable components. The philosophy of neuromorphic engineering is that the medium of computation is an intrinsic part of the computation itself.

Mapping Neurons into CMOS

The strategy for mapping neurons into CMOS varies between research groups, and this article does not review the full range of options that are being explored. Instead, we focus on a few examples which are representative of the various types. The feature that distinguishes a silicon neuron from, say, a discrete-time or leaky integrator neuron is that it includes a large number of time constants. The silicon neuron's dynamical complexity implies that the input–output relationship of silicon neurons cannot be encapsulated by a single sigmoidal function; instead, a given synaptic input has a different effect on the output, depending on where and when it is applied.

Conceptually, the neuron is often divided into three parts: the dendrite that receives inputs, the soma that translates the

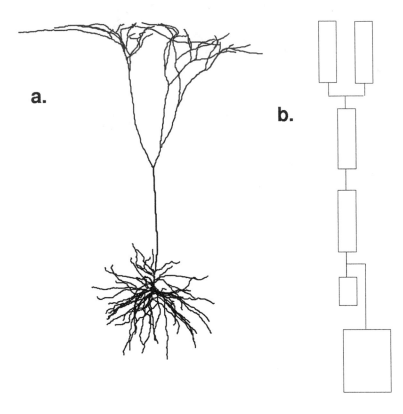

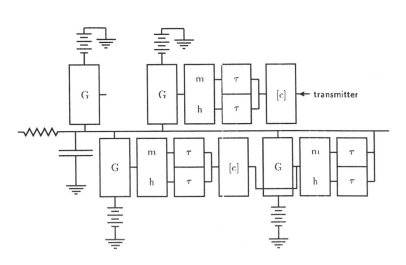

Figure 1. Mapping neurons in CMOS aVLSI circuits. *A*, Biophysical characteristics and morphology of neurons are obtained from intracellular recordings and single cell labeling, followed by three-dimensional reconstruction. This example shows a layer 5 pyramidal neuron in cat visual cortex reconstructed in three dimensions. Scale bar, 250 μm. *B*, To construct a silicon neuron, the detailed pyramidal data must be simplified into a compartmental model. The degree of simplification is a compromise that depends on the problem being addressed. In this prototype, the pyramidal cell has been reduced to six compartments (rectangles) that represent (bottom to top) the basal dendrites, soma, trunk of the apical dendrite (two compartments), and the branched apical tuft (two compartments). *C*, Each compartment comprises an axial conductance, which connects adjoining compartments, a capacitor and a leak conductance, which represent the passive cell membrane, and circuit modules that emulate various active ionic conductances. The modules are variations on the three basic types shown here: voltage-sensitive, ligand- (transmitter-) sensitive, and ion-concentration sensitive (usually calcium). The modules have activation, *m*, and an inactivation, *h*, subcircuits, each of which are governed by time constants, τ. The lower concentration, [c], element is shown connected to a voltage-sensitive conductance, because the concentration, *c*, of the ion is affected by the voltage-regulated flow of the ion into the cell. Once a repertoire of ionic conductance modules has been designed, any desired profile of currents can be inserted into the compartments of the silicon neuron.

inputs to an output, and the axon that distributes the outputs. At a more physical level, traditional modelers of detailed activity of biological neurons have divided the continuous neuronal membrane of the dendrites and soma into a series of compartments to facilitate numerical computation (Koch and Segev, 1989; Traub and Miles, 1991; see DENDRITIC PROCESSING). Each compartment is considered to be isopotential and spatially uniform in its properties. The connectivity of the compartments mirrors the spatial morphology of the modeled cell (Figure 1). Elias (1993) has constructed neuromorphic VLSI neurons with 112 passive compartments that model the leakiness of the cellular membrane and the axial resistance of the intracellular medium, using space-efficient switched capacitors to implement resistances. Each compartment provides temporal filtering of the inputs. The resolution of the segmentation is a compromise between the questions that must be addressed by the model, the resources required by each compartment, and error tolerance. For example, neurons with between 5 and 30 compartments are a common compromise for digital simulations of cortical and hippocampal circuits (Douglas and Martin, 1993; Traub and Miles, 1991).

In addition to the passive properties of the lipid membrane, the biological neuronal membrane contains active ionic channels (see ION CHANNELS: KEYS TO NEURONAL SPECIALIZATION). In the silicon neurons of Douglas and Mahowald (1994), the compartments are populated by modular subcircuits, each of which emulates the physics of a particular ionic conductance (Figure 1C). They have designed and tested a number of circuit modules. These modules emulate the sodium and potassium spike currents, persistent sodium current, various calcium currents, calcium-dependent potassium current, potassium A-current, nonspecific leak current, exogenous (electrode) current source, excitatory synapse, potassium-mediated inhibitory synapse, and chloride-mediated (shunting) inhibitory synapse. The prototypical circuits are modified in various ways to emulate the particular properties of a desired ion conductance. For example, some conductances are sensitive to calcium concentration rather than membrane voltage and require a separate voltage variable representing free calcium concentration. Synaptic conductances are sensitive to ligand concentrations, and these circuits require a voltage variable representing neurotransmitter concentration. The dynamics of the neurotransmitter concentration in the cleft is governed by additional time constant circuits. This array of ionic conductances gives rise to state-dependent dynamics within the compartments.

The dynamics of the silicon neuron is determined by the compartment model that is implemented. Different implementations emulate more or less faithfully the dynamics of real neurons (Etienne-Cummings et al., 1994; Elias, 1993). The dynamics of the somatic compartment of the Douglas and Mahowald neuron is illustrated in Figure 2. This silicon neuron mimics the electrophysiology and temporal pattern of action-potential discharge of real neurons. Action potentials are the output signals that are transmitted through the axon to this neuron's postsynaptic targets.

Biological neurons communicate with one another through axons that ramify widely to make connections with many target neurons. It is impractical to hardwire every network configuration that one might wish to emulate. Instead, the connectivity of the emulator should be reconfigurable. One approach is to provide direct connections between neurons that can be configured by sets of switches that are under digital control (Van der Spiegel et al., 1994). This approach requires a large silicon area for the switches, but it has the advantage of continuous communication between neurons. It is appropriate when the neuron output is encoded as a continuous analog

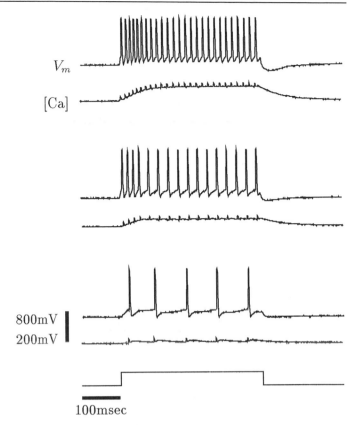

V_m

$[Ca]$

800mV

200mV

100msec

Figure 2. Real-time response of a silicon neuron emulating the response of a cortical pyramidal neuron to three different levels of intrasomatic current injection. Time course of the rectangular current injections is shown below. Each pair of response traces consists of the somatic membrane potential, V_m (800-mV scale bar), and a voltage that is proportional to the intrasomatic free calcium concentration, $[Ca]$ (200-mV scale bar). Notice the adaptation of discharge and after-train hyperpolarization. (Quantization noise on traces was caused by the recording device.) The striking aspect of this silicon neuron emulation is the degree to which qualitatively realistic neuronal behavior arises out of the interaction of circuits that are analogs of the physics of neurons, rather than explicit and precise mathematical models. This is not only true for particular parameter settings. In general, adjustments of the parameters cause the behavior of the neuron to change smoothly, in ways that are consistent with our understanding of neuronal biophysics.

variable. An alternative approach is to use the high speed of CMOS wires to *multiplex* slow neuronal signals.

Action-potential representations of neuronal output are compatible with multiplexing because the output of the neuron is active only during the action potential. Furthermore, the action potential is a digital amplitude signal that can be robustly transmitted between chips. Digital amplitude signals are robust against noise and interchip variability and have been used to advantage in VLSI neural networks (Murray and Tarassenko, 1994). Event-based digital data-encoding methods, such as the Address-Event Representation (AER) (Mahowald, 1994) or Virtual Wires (Elias, 1993), broadcast action potential events occurring in neurons, one at a time, onto a common data bus (Figure 3). Many silicon neurons can share the same bus because switching times in CMOS are much faster than the switching times of neurons. Events generated by silicon neurons can be broadcast and removed from the data bus at frequencies of more than a megahertz. Therefore, more than 1000 address-events could be transmitted in the time it takes one

a.

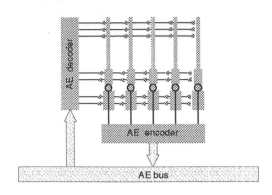

b.

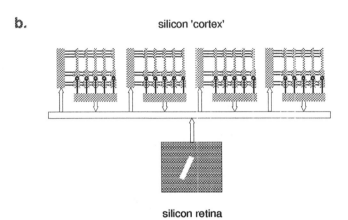

Figure 3. Proposed connections between silicon neurons based on the address-event representation. This figure shows a system that is under development. Although various aspects of the AER have been tested successfully (Mahowald, 1994), full networks of silicon neurons have not yet been fabricated. *A*, Multineuron chip attached to an address-event bus. Action potentials generated by neurons (gray broomsticks) are detected by an on-chip AE encoder that broadcasts the binary addresses of the source neurons on the AE bus. AE decoders activate the destination synapses "connected" to the source neurons. *B*, A silicon retina (Mead, 1989) and a number of multineuron chips communicate through a common AE bus.

neuron to complete a single action potential. Multiplexing strategies are most effective if, like their biological counterparts, only a small fraction of the silicon neurons embedded in a network are active at any time.

Event-based digital encoding methods facilitate network reconfigurability. These digital multiplexing schemes work by placing on the common communications bus the identity (a digital address) of the neuron generating an action potential. In one type of implementation, the bus broadcasts this address to all synapses, which decode the addresses. In this way, those synapses that are "connected" to the source neuron detect that it has generated an action potential, and they initiate a synaptic input on the dendrite to which they are attached. In a more elaborate implementation, the event-address can be translated from a postsynaptic bus to a presynaptic bus through a programmable lookup table that maps the addresses of source neurons to their destination synapses. The topology of the network is defined by the mapping of source neurons to recipient synapses.

Working with Analog Silicon

Custom CMOS circuits are created by an iterative process of design, fabrication, and experiment. In the design phase, the correspondence between elements of the analog circuit and those of the neural system are established, and the variable parameters are identified. Computer simulations and mathematical analyses of the proposed circuit subunits are useful at this stage. The electronic circuit design then is transposed into a layout design that expresses the circuit as a sandwich of layers in the silicon chip. The layout is drawn with specialized computer-aided design (CAD) software on a workstation or personal computer. The final layout instructions are used by the silicon foundry to fabricates the chip. The MOSIS service at the University of Southern California's Information Sciences Institute, which is available to academic researchers in the United States, accepts layout by electronic mail and returns a fabricated chip in approximately 10 weeks. Through the MOSIS service, fabrication costs range from approximately $500 for four pieces of a small 2.2 × 2.2 mm chip suitable for prototyping a few neurons, to $12,000 for 32 pieces of a large 7.9 × 9.2 mm chip suitable for fabricating a retina or a network of silicon neurons.

It is important to consider the range of desired behaviors when designing the circuits. Within certain limits, the dynamics of the compartmental models can be varied parametrically when the neuron is in use. Often, the behavior of analog circuits can be controlled by the voltages applied to the gates of their various transistors. In the Douglas and Mahowald neurons, these parameters determine, for example, the temporal dynamics of activation and inactivation, the voltages at which they occur, and the maximum conductance that can be obtained when fully activated. The effect of changing these parameters is immediate. Thus, the electrophysiological "personality" of the silicon neuron can be switched rapidly, for example, from a regular adapting to a bursting pyramidal cell. Of course, only the parameters that were incorporated into the design at the time of fabrication are available for reconfiguring the performance of the neuron. If additional properties are required, another silicon neuron with different morphology, or different types of channels, can be fabricated using variations of the basic circuit modules.

To produce compact circuits, it may be necessary to make approximations in the design of the circuit modules. For example, the analog conductances may saturate (Douglas and Mahowald, 1994), and hence these neurons would not perform like real neurons if they were clamped at voltages very far away from the resting potential. These errors are insignificant when the cell is operating in the physiological range because the linear regions of the circuits are arranged in such a way that the deviation from true neuronal behavior is minimized when the cell is operating there (see Figure 2).

The circuits constituting the neuron must be arranged spatially on the surface of silicon chips. One possibility is to distribute the dendritic compartments of individual neurons across multiple chips (Van der Spiegel et al., 1994). This approach is useful because the number of compartments, and hence the number of synaptic inputs to a particular neuron, is not limited by the chip boundaries. However, this division requires a method of transmitting accurately between multiple chips the analog voltages and currents at the compartment boundaries. The alternative approach is to fabricate entire silicon neurons (Elias, 1993) and networks of neurons on the same chip. The single-chip solution is appropriate for networks of simple neurons that have only local connectivity mediated by graded synapses, such as the outer layers of the retina

(Mahowald, 1994). It is also appropriate for networks of spiking neurons distributed across multiple chips, where only robust action-potential events need be transmitted between chips. However, the number of neurons that can be fabricated on a single chip is limited, and depends on the complexity of the neuron. A large development chip has an area of roughly 100 mm^2. Using available fabrication technology, a retinal chip can accommodate a 64×64 array of simplified photoreceptors, horizontal and bipolar cells, and retinal ganglion cells. For more comprehensive neurons and more general connectivity, a reasonable target would be a linear array of approximately 100 neurons, each having approximately 10 dendritic compartments. The number of synapses (a few 10s to 100) depends on the degree of biological realism incorporated in the synapse. There are ways to improve the number of synaptic inputs. One possibility is to map multiple presynaptic inputs into single postsynaptic circuits. This strategy raises the effective number of inputs by an order of magnitude. A further consideration is that evolving aVLSI technology is expected to bring a 10-fold improvement of scale during the next decade. Furthermore, a wafer-scale technology may be developed if fault-tolerant design strategies can be devised. Nevertheless, an important focus of future work must be the development of more compact synapses and also of learning synapses.

Once the chip has been fabricated, its performance is explored using similar experimental methods to those used in a real neurophysiological preparation, except that many more variables can be observed. For example, the response of the analog chip to stimulation is measured in real time, with an oscilloscope. Except for the variable parameters included in the design, the circuits cannot be altered after fabrication, and so errors in the specification of the neuron cannot be corrected as easily as in software simulations. Also, care must be taken to plan experiments before the chip is fabricated so that instrumentation circuitry can be included to observe the state of interesting analog variables. The designs of circuits evolve with understanding gained by experiment.

Discussion

A number of groups are currently investigating the properties of analog silicon neurons in networks. At the time of this writing, the scale of these networks ranges from tens of neurons (Elias, 1993) to 1000 neurons (Van der Spiegel et al., 1994). The next five years should see an increase in the number of neurons by a factor of 10, attributable both to an improvement in the basic fabrication technology and to improved implementations. The optimal degree of biological realism is an open question and is likely to be task dependent. Furthermore, the cost of a computational element depends ultimately on the device physics of the computational primitives, so that more effective methods for performing a computation may become available as time goes on. The most promising path for the development of these networks is interfacing them to sensors and effectors that can interact dynamically with the real world. (See Douglas, Mahowald, and Mead, 1995, for a review.) For example,

Etienne-Cummings et al. (1994) have used a silicon retina to provide real-time edge-enhanced input to a neural emulator to do visual motion estimation. Analog VLSI is not the only way to build such systems, but it does have some striking advantages. Analog emulation is inherently parallel, the circuits are extremely compact by comparison with a digital circuit performing an equivalent computation, and the power consumption is often a few orders of magnitude less than their digital equivalents. These properties lend themselves to the construction of small, autonomous neuromorphic systems that could interact directly with the world and so provide a platform for studying animal behaviors by emulation.

Acknowledgments. We acknowledge the support of the Office of Naval Research and the fabrication facilities provided by MOSIS.

Road Maps: Biological Neurons; Implementation of Neural Networks
Background: Perspective on Neuron Model Complexity
Related Reading: Analog VLSI for Neural Networks; Digital VLSI for Neural Networks

References

Douglas, R., and Mahowald, M., 1994, A constructor set for silicon neurons, in *An Introduction to Neural and Electronic Networks*, 2nd ed. (S. F. Zornetzer, J. L. Davis, C. Lau, and T. McKenna, Eds.), San Diego: Academic Press, pp. 227–296.

Douglas, R., Mahowald, M., and Mead, C., 1995, Neuromorphic analogue VLSI, *Annu. Rev. Neurosci.*, 18:255–281. ◆

Douglas, R. J., and Martin, K. A., 1993, Exploring cortical microcircuits: A combined anatomical, physiological, and computational approach, in *Single Neuron Computation* (J. D. T. McKenna and S. Zornetzer, Eds.), Orlando, FL: Academic Press, pp. 381–412.

Elias, J. G., 1993, Artificial dendritic trees, *Neural Computat.*, 5:648–664.

Etienne-Cummings, R., Donham, C., Van der Spiegel, J., and Mueller, P., 1994, Spatiotemporal computation with a general purpose analog neural computer: Real-time visual motion estimation, in *Proceedings of the International Conference on Neural Networks*, vol. III, Orlando, FL: IEEE, pp. 1836–1841.

Koch, C., and Segev, I., 1989, *Methods in Neuronal Modelling: From Synapses to Networks*, Cambridge, MA: MIT Press.

Mahowald, M., 1994, *An Analog VLSI System for Stereoscopic Vision*, Boston: Kluwer.

Mead, C., 1989, *Analog VLSI and Neural Systems*, Reading, MA: Addison-Wesley. ◆

Mead, C., 1990, Neuromorphic electronic systems, *Proc. IEEE*, 78: 1629–1636.

Murray, A., and Tarassenko, L., 1994, *Analogue Neural VLSI*, London: Chapman and Hall.

Traub, R. D., and Miles, R., 1991, *Neuronal Networks of the Hippocampus*, Cambridge, UK: Cambridge University Press.

Van der Spiegel, J., Donham, C., Etienne-Cummings, R., Fernando, S., Mueller, P., and Blackman, D., 1994, Large scale analog neural computer with programmable architecture and programmable time constants for temporal pattern analysis, in *Proceedings of the International Conference on Neural Networks*, vol. III, Orlando, FL: IEEE, pp. 1830–1835.

Simulated Annealing

Scott Kirkpatrick and Gregory B. Sorkin

Introduction

Simulated annealing is a general framework for optimizing the behavior of complex systems of many parameters, as well as a new way of thinking about such systems. It was introduced in the early 1980s in response to the need for computer algorithms to automatically design very large scale integration (VLSI) electronic chips and computing hardware built out of such chips. Even after the circuitry for such hardware has been designed, the physical layout of the thousands to millions of logical components and their interconnecting wires is beyond human abilities. Physical design involves optimizing some combination of power consumption, performance, manufacturing yield, and cost, or optimizing some of these subject to constraints on others.

The first step in automating this process is to introduce an explicit cost function, C, which quantifies the different characteristics to be optimized and scales them so that they can be combined into a single number. C is extremely problem specific; for example, different cost functions are required for layouts in different VLSI circuit technologies with different speeds, maximum wire densities, and so forth. Given a description of a "problem instance," e.g., a particular logic design to be realized in a given VLSI technology, C computes the cost of a *configuration* (or *solution*, or *state*), e.g., the locations of all the components and the routes of the wires.

The task is difficult for two reasons: first, because the number of parameters describing a configuration is so large, and second (and more fundamentally), because the objectives, e.g., high performance and low cost, are in conflict. It is generally found that solutions offering good tradeoffs, while more numerous than solutions optimizing a single objective, are not described by simple rules or symmetries. They must be found by more expensive, explicit search through configurations.

This is generally done with a *local search* strategy, in which a small portion of a proposed design is modified to create a new one (Papadimitriou and Steiglitz, 1982). In the simplest strategy, *iterative improvement*, the new configuration is adopted as the current working state if its cost is lower than that of the old one. For example, one could select two circuits of the same size on a VLSI chip, interchange them and re-route the wires connected to each, and then see if the new chip runs faster and still meets technology constraints. This sort of search terminates when no further improvement is possible (a *local minimum* has been reached), or improvements have become so rare and so small that it is not worth continuing.

The *moves*, i.e., the rules for a single modification, are crucial. A single move should allow quick reevaluation of the cost (perhaps by changing only a few parameters), and a few moves should suffice to randomize the configuration. (For more on the second, *rapid mixing*, property, see Sinclair, 1992.) It is easy to quantify both of these requirements and usually easy to satisfy them. A finer art is the choice of a move which perturbs the cost only slightly: the locality of the move must have some correlation with the cost function so that the small modifications will gradually home in on a good solution.

As the number of variables, or *problem size*, grows, problems of scale rapidly become overwhelming. Techniques such as conjugate gradient minimization, which prescribe effective ways of tracing what can be tortuous paths to a local mini-

mum, can address some of the difficulties. But the essential difficulty is that "frustrated" problems, with competing objectives, have many local minima with high barriers between them.

Typically, a local minimum can be found "in polynomial time," i.e., with a computational effort that scales as some small power of the number of variables, N. The catch is that for many classic problems in combinatorial optimization, notably the *NP-complete* problems, the effort required to find the global optimum is believed to scale superpolynomially, e.g., as $\exp(N)$ (Papadimitriou and Steiglitz, 1982). For some such problems there are *approximation algorithms* which give nearly optimal results in polynomial time, but others are provably hard to approximate. Even when a heuristic solves a classical problem well, it may not be applicable to a more constrained, less elegant, variant of the same problem encountered in real engineering practice.

The strength of simple iterative improvement is its adaptability to these real-world problems; its Achilles heel is that it tends to get stuck at a local minimum, never reaching the global minimum. An obvious tactic to improve this or other heuristics is to start from each of many randomly chosen initial states and keep the best solution obtained. (Or, if the heuristic is randomized, it can be re-run with different "coin-tosses" from the same initial state.) While this can help, the number of trials is of necessity a tiny fraction of the size of the state space (typically exponential in N), and the improvement is rarely enough to make a bad heuristic into a good one. Another tactic applicable to local search is to give the system some sort of a "kick" uphill when it reaches a local minimum, or even to add small amounts of noise at each step in the search. This can help by permitting more extended search in the neighborhoods of local optima, but details of its use are very problem specific.

Simulated Annealing

Simulated annealing prescribes a controlled way to introduce noise for more robust iterative search. The control parameter is called *temperature*, by analogy with physical systems such as alloys, where the best crystals are grown by annealing out their defects: first heating or even melting the material, then cooling slowly to allow the system to find its state of lowest energy, or optimum arrangement. Having the right *annealing schedule*, the sequence of temperatures and length of time spent at each, is vital in crystal growing. Cool too rapidly, and a glass or polycrystalline composite forms instead of the perfect single crystal desired. If we make an equivalence between the cost in an optimization problem and the energy in a physical system, computational techniques used to simulate the annealing of a physical system can be applied to search for an optimization problem's lowest-cost solution, or *ground state* of lowest energy. The earliest journal articles describing simulated annealing for this purpose are by Kirkpatrick, Gelatt, and Vecchi (1983), and Černy (1985).

The Metropolis algorithm (Metropolis et al., 1953) is the simplest of many techniques used to generate a population of configurations representative of a physical system with energy given by C in equilibrium at a temperature T. The basic idea of simulated annealing is to generate a series of configurations which at first span a wide range of energies but eventually come to include only the lowest values of C. Simulated annealing is described by the following pseudo-code:

```
OUTER LOOP: choose a temperature, T;
INNER LOOP (Metropolis algorithm):
   until (system reaches equilibrium) do
      from state X, propose a local rearrangement X';
      evaluate ΔC = C(X) − C(X');
      if ΔC ≤ 0 or exp(−ΔC/T) > random(0, 1) then
         accept;
      if accept then update X to X'
   end INNER LOOP;
   if (system is "frozen") then leave
end OUTER LOOP
```

For each temperature, this process generates a series of accepted configurations that converges to a distribution in which the probability of a state X is proportional to $\exp(-C(X)/T)$, known in statistical mechanics as the Boltzmann factor. The test for equilibrium is usually performed by watching for a running average of C to stabilize, although poorly understood theoretical (and practical) issues lurk under this rug. When the control parameter T (which must have the same units as C) is large, most proposed rearrangements are accepted, and the system is said to be *melted*. At intermediate values of T, the rearrangements explored organize the system on larger scales. When T is small, only rearrangements that lead to improvements are accepted; these quickly become few, and the system is said to be *frozen*. While it is possible to keep track of the best solution found at any point during the search, the final configuration is almost always equally good and is more amenable to theoretical analysis.

To take a concrete example, let C be the cost function for a circuit placement problem, incorporating contributions from the estimated length of wire required to connect the circuits, density penalties for overcrowding the spaces available for wires or circuits, and so forth. *Infeasible* states, such as those where circuits overlap or wires are overcrowded, are allowed but are penalized in the cost function. This is not only easier than constructing a move set that never violates any constraints, but it may also facilitate the search by allowing passage between good configurations through infeasible ones. As the temperature is reduced, the cost penalties become more significant, and the constraint violations are gradually removed.

Figure 1 shows the different stages of annealing in a real application. The logic from six chips in an older semiconductor technology was combined, and using simulated annealing, placed on a single chip in a newer, denser technology. The same objectives that originally led certain circuits to be grouped on the same chips should now lead them to be grouped in the same region of the combined chip. With the circuits color-coded (gray-scaled, in this reproduction) to reflect the chips they came from, they are seen to be randomly mixed when the design is melted (A). As cooling starts (B), circuits with related functions coalesce into loose blobs or clusters. At lower temperatures, in stage (C), the clusters are taking on their final shapes. (The diagonal boundaries which now appear are favored over vertical or horizontal boundaries because they expose both vertical and horizontal channels of wiring, thereby increasing communication between clusters.) Finally, at the lowest temperature (D), the last few mistakes have been found and eliminated: six distinct regions, corresponding to the original six chips, are visible. The solution is a very good local optimum.

Developing an annealing schedule that will work robustly for a whole class of problems of similar type is still an open problem. The procedure described by Kirkpatrick, Gelatt, and Vecchi (1983), increasing the temperature in steps with a constant ratio until the system melts, then decreasing the temperature in constant ratio steps until it freezes, is robust but expensive. An improvement attributable to White (1984), estimating the melting temperature as the standard deviation in C found by preliminary random rearrangements, is often used to start annealing.

The Metropolis algorithm is only one variant of computing with random numbers, or *Monte Carlo methods*, employed in statistical physics. For a thorough review, see Binder (1978). For algorithms which explore the Boltzmann distribution in systems with continuous-state spaces, see the more general discussion in Whitman and Kalos (1982).

Simulated annealing poses serious challenges to the theorist to understand and improve the technique; to determine when convergence can be assured; to find annealing schedules and move sets that are robust and efficient; and to characterize problems for which the method works well. It is easy to prove that the distribution of values of C obtained by running the Metropolis algorithm at a single temperature converges to the Boltzmann distribution. Convergence to the global minimum also occurs if the temperature as a function of time step, n, is decreased no faster than in proportion to $C_{max}/\ln(1+n)$, where C_{max} is the height of the lowest "mountain pass" separating two distant local minima (Hajek, 1988). Of course, this infinite-time schedule can only be approximated in reality, and, empirically, truncated logarithmic schedules are slower than the geometric schedules effective in practice, for the same expected quality of result. More detail on these topics can be found in any good textbook on annealing, such as Laarhoven and Aarts (1988).

Stronger results are possible for restricted problem classes. For example, Jerrum and Sorkin (1993) prove that "annealing" at a single appropriate temperature efficiently finds the optimum bisection for a class of random graphs. In the interesting phase of the algorithm's execution, a bisection only slightly better than average is improved to a nearly optimal one in just linear time; the proof relies on the randomness inherent in annealing to "smooth out" the irregularities of the chosen random graph.

Discussion

Only qualitative understanding has been achieved on the question of when simulated annealing is better than downhill search by iterative improvement. It seems reasonable to assume that roughness of the configuration space makes downhill search difficult and annealing appropriate; this might occur if the moves are sparse and paths between good states must climb over energy mountains. At the opposite extreme, a system in which any configuration can be transformed into any other has no energy barriers; but in this case, no information about a configuration is implied by the energies of its neighbors, so local search is ineffective. Finally, while the constant temperature (INNER LOOP) algorithm has been proven superior to simple descent on specific problems by several authors, the superiority of the annealing (OUTER LOOP) portion of the approach with varying temperatures still rests on artificial mathematical models, physical intuition, appealing pictures, and other anecdotal evidence. However, numerical work (e.g., Strenski and Kirkpatrick, 1991) has shown that, when a finite computing resource is available, varying the temperature produces the best final energies on average.

Applications of simulated annealing continue to accumulate. Descendants of the original circuit placement packages are still in regular use in many semiconductor companies. And an-

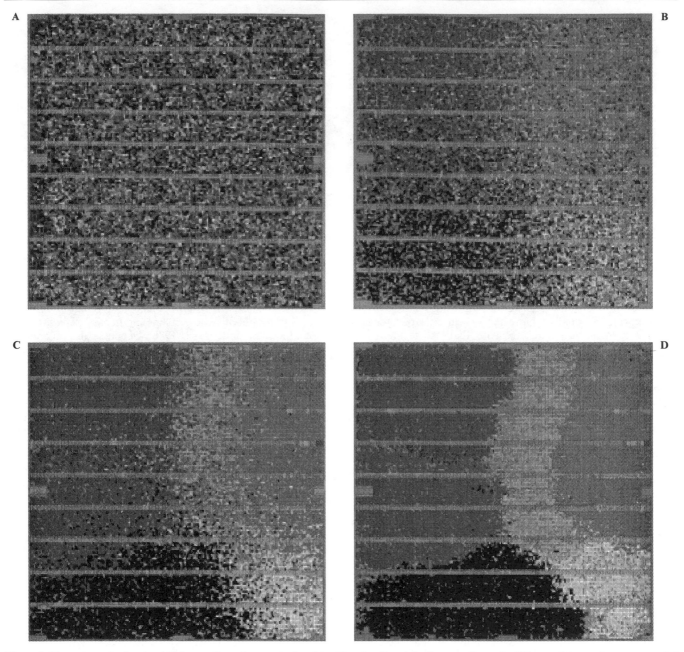

Figure 1. Four stages in the simulated annealing placement of a chip. The circuits originally came from six different chips, and are gray-scaled accordingly. They are randomly jumbled at high temperature (*A*); but at lower temperatures (*B* and *C*), circuits originally from the same chip begin to coalesce. At the lowest temperature (*D*), six distinct regions are evident.

nealing's applications are wide ranging: for example, large inverse problems in seismic and tomographic imaging, molecular structure determination, complex layouts of factory machinery, and digital and analog circuit design.

The relevance of annealing to explicitly neural computation was first identified by Geman and Geman (1984) in their model of stochastic optimization as a basic characteristic of parallel low-level visual perception. Network computational structures with and without any explicit use of temperature both exist: *Boltzmann machines* and standard *backpropagation*, respectively (see BOLTZMANN MACHINES and BACKPROPAGATION: BASICS AND NEW DEVELOPMENTS). It is generally believed that weight update as performed in backpropagation provides a highly connected, smooth solution space, and thus is amenable to simple gradient descent optimization. However, it is not uncommon for workers using complex multilayered networks to report that convergence of their algorithms was improved by adding a little *thermal* noise to the weight increments while training.

Does the brain use randomized computational strategies, or even "anneal" to extract a memory from an associative store? Given the large amounts of noise accompanying neural signals, these are attractive speculations, but they do not seem to be testable at our current level of knowledge.

Road Map: Cooperative Phenomena
Related Reading: Constrained Optimization and the Elastic Net; Fractal Strategies for Neural Network Scaling; Markov Random Field Models in Image Processing

References

Binder, K., 1978, *Monte Carlo Methods in Statistical Physics*, New York: Springer. ◆
Černy, V., 1985, Thermodynamic approach to the Travelling Salesman Problem, *J. Optim. Theory Appl.*, 45:41–51.
Geman, S., and Geman, D., 1984, Stochastic relaxation, Gibbs distributions, and the Bayesian restoration of images, *IEEE Trans. Pattern Analysis Machine Intell.*, 6:721–741. ◆
Hajek, B., 1988, Cooling schedules for optimal annealing, *Math. Oper. Res.*, 13:311–329.
Jerrum, M., and Sorkin, G. B., 1993, Simulated annealing for graph bisection, in *Proceedings of the 34th Annual IEEE Symposium on Foundations of Computer Science*, pp. 94–103.
Kirkpatrick, S., Gelatt, C. D., Jr., and Vecchi, M. P., 1983, Optimization by simulated annealing, *Science*, 220:671–680. ◆
Laarhoven, P. J. M. van, and Aarts, E. H. L., 1988, *Simulated Annealing: Theory and Applications*, Boston: Kluwer. ◆
Metropolis, N., Rosenbluth, A. W., Rosenbluth, M. N., Teller, A. H., and Teller, E., 1953, Equation of state calculations by fast computing machines, *J. Chem. Phys.*, 21:1087–1092.
Papadimitriou, C. H., and Steiglitz, K., 1982, *Combinatorial Optimization, Algorithms and Complexity*, New York: Prentice-Hall.
Sinclair, A., 1992, Improved bounds for mixing rates of Markov chains and multicommodity flow, *Combin. Probab. Comput.*, 1:351–370.
Strenski, P., and Kirkpatrick, S., 1991, Analysis of finite-length annealing schedules, *Algorithmica*, 6:346–366.
White, S. R., 1984, Concepts of scale in simulated annealing, in *Proceedings of the IEEE International Conference on Circuit Design*, Silver Spring, MD: IEEE Computer Society Press, pp. 646–651.
Whitman, P., and Kalos, M., 1982, *Monte Carlo Methods*, New York: Springer.

Single-Cell Models

William Softky and Christof Koch

Introduction

Most of the roughly 10^{10} neurons in the human cerebral cortex are tiny, membrane-bound bags of salt water, shaped like trees (including roots). Each is surrounded by more salt water and by other cells, many of which are neurons to which it is connected. Most of the neurons communicate by means of brief, all-or-none pulses (called spikes, or action potentials), each lasting about a millisecond. But we still do not understand many aspects of those pulses.

Researchers often want to distill the complex shape and behavior of a real neuron into a simpler model, either to guide a neurobiological experiment or to construct a functional network. But choosing which neuronal properties to keep and which to ignore is heavily influenced by how one interprets the pulses.

Most theories assume that information is carried in the average rate of pulses over a time much longer than a typical pulse-width, so that the occurrence times of particular pulses simply represent jitter in an averaged analog signal. A neural model in such a theory might be a mathematical function which produces a real-valued output from its many real-valued inputs; that function could be linear or nonlinear, static or adaptive, and might be instantiated in analog silicon circuits or in digital software. Alternatively, the neural model might perform its analog computation by producing an output pulse rate which depends monotonically on the average of its input pulse rates.

But a few theories assume that each single neural pulse carries reliable, precisely timed information. A neural model in such a theory must fire only on the exact coincidence of several input pulses, and must quickly "forget" when it last fired, so that it is always ready to fire on another coincidence. Whether real neurons operate in this regime or in the slower average-rate regime awaits further neurobiological experimentation. Both types of codes and single-neuron models have special features and advantages; understanding the models touches issues of bandwidth, nonlinearity, and the fundamental precision and function of single nerve cells.

Formal Models

It is easiest to understand and analyze the models which are the least like real neurons. Virtually all such models share two features in common. Firstly, each model neuron combines many inputs, both *excitatory* and *inhibitory*, into a single output. And each neuron has at least one internal state variable (conceptually corresponding to the cell's membrane potential), which increases monotonically with the total amount of excitatory input and decreases with inhibitory input. So the neuron is constrained to "adding up" (in a rough sense) its positive and negative inputs, and *cannot* independently assign an arbitrary value to each of its myriad possible input combinations.

McCulloch-Pitts Model

The McCulloch-Pitts model neuron celebrated its fiftieth anniversary in 1993, and myriads of its progeny are present in digital circuits in the form of logic gates. The explicit assumptions of this model are that each binary "pulse" represents a logical statement (i.e., *true* or *false*), and that each neuron performs an exact, noise-free, synchronous computation on its input pulses.

If any one of the model neuron's inhibitory inputs is active, the output is shut off or inactive. Otherwise, all the active excitatory inputs x_i are multiplied by their *synaptic weights* w_i and then added. However, only if this activity level exceeds a preset "threshold" θ is the output active:

$$Y = \begin{cases} 1 & \text{if } \sum_i w_i x_i > \theta \text{ and no inhibition} \\ 0 & \text{otherwise} \end{cases} \tag{1}$$

McCulloch and Pitts showed that enough of such units, with weights and connections set properly and operating synchronously, could in principle perform any possible computation (Arbib, 1987).

An even simpler model is the *linear* neuron, which just adds up its inputs and delivers the sum as output, with no thresholds or other nonlinearities. The neuron's real-valued output is the sum of its inputs x_i, weighted by real-valued coefficients w_i:

$$Y = \sum_i w_i x_i \qquad (2)$$

Networks of linear neurons can be treated analytically, using well-established matrix methods. But unlike the spikes and rates from real neurons, outputs from the linear model can become negative or arbitrarily large.

Perceptron Model

Rosenblatt's perceptron is formally similar to the McCulloch-Pitts model, having synchronous inputs and producing outputs between 0 and 1. But the perceptron creates a real-valued (not binary) output, representing the *average firing rate* of the cell. As with the linear model, the internal variable V of a perceptron is the weighted sum of its inputs:

$$V = \sum_i w_i x_i \qquad (3)$$

A *threshold* or *bias* θ is subtracted from V and is then passed through a continuous and monotonically increasing function g:

$$Y = g(V - \theta) \qquad (4)$$

The nonlinear function g is *sigmoidal*: it asymptotes 0 as $V \ll \theta$ and saturates at 1 for $V \gg \theta$. This function mimics the biological relationship between the cell's input current and its firing rate in several ways: the output is non-negative, it is very small below θ, it monotonically increases with input, and it has an upper bound.

Hopfield Neurons

In Hopfield's binary model, the output of neuron i is the step function of V_i and the threshold θ,

$$Y_i = \begin{cases} 0 & \text{if } V_i < \theta \\ 1 & \text{if } V_i \geq \theta \end{cases} \qquad (5)$$

Unlike the McCulloch-Pitts or the perceptron model, each Hopfield neuron updates its state at a *random* time, independently of any other neurons.

Both Hopfield's binary and continuous-valued models (Figure 1) are similar to perceptrons in isolation but can act as associative memories in highly interconnected networks.

Polynomial Neurons

The appeal of models like those just described is that only a very simple function of the inputs—a weighted sum—is neces-

sary for them to work. But such a sum cannot tell apart the individual contributions to it. For the neuron to respond strongly to correlations among particular input pairs or groups, one must include multiplicative terms and then sum over the products. Such a *sigma-pi* (Σ = sum, Π = product) neuron computes its internal state as the sum of contributions from a set of monomials

$$V = a_1 + b_1 x_1 + b_2 x_2 + c_1 x_1^2 + c_2 x_1 x_2 + \ldots \qquad (6)$$

This state variable then can be passed through the usual nonlinear function g. It is clear that such "neurons" are computationally richer than linear or threshold units—just one can implement parity, exclusive-or, or *lookup table* functions. Furthermore, such models also better represent the operations of real neurons containing highly branched dendrites with voltage-dependent membrane conductances (Mel, 1994).

Biophysical Models

While many crucial properties of real neurons remain unknown, biophysical neural models at least attempt to incorporate some known properties of neural tissue. Like real neurons, these models produce spikes rather than continuous-valued outputs.

The Hodgkin-Huxley Model of Squid Axon

In its quiescent state, the inside of a typical neuron has a negative voltage or *resting potential* (relative to the extracellular fluid). The cell membrane—a lipid bilayer approximately 50 Å thick—acts like a capacitor. The electrical charge carriers (various species of ions such as Na^+, K^+, Cl^-, and Ca^{2+}) pass through special pores or *ionic channels* embedded in the membrane (Hille, 1992). Although each individual channel is either open or closed (in a partially stochastic manner), the current through many channels in parallel is well approximated by continuous, deterministic equations, much as the laws of electrical current describe averages over many electrons.

In the simplest case, the channels' collective behavior is like an ohmic (or *passive*) resistor across the membrane. The combination of the resistance and the capacitance creates a membrane time-constant τ, characterizing the $1/e$ falloff time of a small, brief voltage (usually τ is between 5 and 50 ms).

Other ion channels are nonlinear: their conductance *depends on* voltage. For instance, an action potential occurs when the membrane potential becomes high enough that voltage-controlled sodium channels open, initiating the fast positive-feedback event of a spike (Figure 2). One spike lasts between $\frac{1}{2}$ and 1 millisecond, and is followed by a few milliseconds of *refractory period* during which it is difficult or impossible to fire another spike.

This process was described by Hodgkin and Huxley in 1952 in one of the most successful of all models in neurobiology: a four-dimensional set of coupled, nonlinear, partial differential equations. Those equations describe the initiation and propagation of action potentials in axons well enough that they are often treated as "gospel truth," although they are technically imperfect phenomenological fits rather than expressions derived from first principles.

Simplified versions—the so-called FitzHugh-Nagumo and van der Pol oscillator equations—yield qualitatively the same kind of subthreshold behavior and limit-cycle oscillations as the original Hodgkin-Huxley equations, but their reduced parameters cannot be interpreted biophysically.

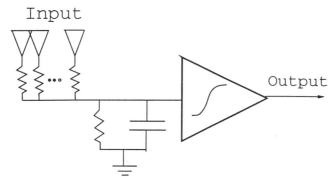

Figure 1. The most popular neural model takes real-valued inputs (analogous to slowly varying electrical currents), adds them up linearly, and produces as output a sigmoidal function of that sum.

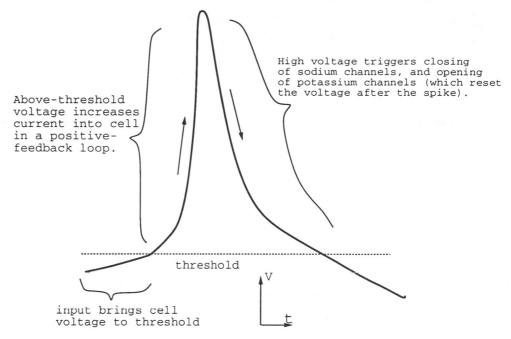

Figure 2. A "spike"—the fundamental form of neural output—occurs when input current brings the cell voltage above a certain threshold. The voltage allows positively charged sodium ions to enter the cell, further increasing the voltage and sodium current. This process halts within a millisecond, and is reversed when the elevated voltages allow potassium ions to flow out of the cell, thereby bringing the cell voltage back below threshold.

High voltage triggers closing of sodium channels, and opening of potassium channels (which reset the voltage after the spike).

Above-threshold voltage increases current into cell in a positive-feedback loop.

threshold

input brings cell voltage to threshold

V

t

Integrate-and-Fire Models

A very different approach is to divide the membrane behavior conceptually into two distinct and discontinuous regimes: a prolonged period of linear *integration* (adding up of inputs), and a sudden *firing*. This integrate-and-fire model relaxes the requirement that a single set of continuous differential equations describe the cell's two very different regimens. The cell voltage starts from zero, increasing or decreasing according to the synaptic input. When the voltage reaches a certain threshold V_{thr}, the cell instantly fires an output pulse and resets the voltage. After a refractory period—a brief "dead time" during which the cell cannot fire at all—the unit is ready to fire again.

The simplest type of integrator model is a leak-free capacitance. With steady DC input current, it acts like a relaxation oscillator or a current-to-frequency converter, producing regular output pulses at a rate depending on the input current.

If the input instead arrives in brief excitatory pulses (e.g., because of spikes from other cells), so that N pulses are necessary to reach threshold, then this model acts like a divide-by-N counter, firing on every Nth input pulse. Its output rate depends on the *average* of the overall rate of the inputs. For small, random input pulses, the output firing is fairly regular as the input randomness is averaged out. The fact that such a neuron can smooth out input noise is one of its great advantages.

The addition of a leak resistance in parallel to the capacitance makes a *leaky integrate-and-fire neuron*, which only fires if the excitatory input is strong enough to overcome the leak. The time-constant $\tau = RC$ divides the model's operation into two qualitatively distinct regimes: temporal integration and fluctuation-detection. When τ is larger than the mean time between output spikes, the leak is insignificant, and the model temporally integrates its input (Figure 3). When τ is much smaller than the average output interval, then production of a spike depends not on the *average* input but on input *fluctuations*: only a rare fluctuation brings the voltage above threshold. Here, the output represents a precisely timed threshold-crossing computation with a binary output—in this regime the neuron is neither linear nor analog.

Integrate-and-fire models can approximately account for the average firing rates of real neurons, but those models have serious problems reproducing the observed *variability* of pulse output (Softky and Koch, 1993).

Modified Single-Point Models

More realistic neural models must account for nature's rich array of nonlinear currents and additional internal variables, only some of which are understood.

Internal variables. Most of the simple models just outlined have only a single internal variable: the membrane potential. An additional variable, such as the concentration of free, intracellular calcium, can give a wider array of functions. Calcium concentration roughly represents a running average of the cell's recent activity. This temporal averaging—along with the ability of calcium ions to remain trapped near synapses—makes it a candidate for modulating synaptic strength. Calcium also participates in *adaptation*, a hysteresis effect in which the calcium accumulated from past spikes makes it more difficult to fire new ones.

Additional ionic currents. Most neurons typically contain a dozen or more *nonlinear* ion channels, whose conductance depends on the cell voltage (see ION CHANNELS: KEYS TO NEURONAL SPECIALIZATION). There are slow positive-feedback currents, such as calcium and persistent sodium currents, which tend to amplify large voltage excursions. There are also negative-feedback currents like those found for potassium, which tend to hyperpolarize the cell, acting like a kind of active inhibition or adaptation. These "active" currents can strongly influence a cell's response to input, but their strengths in real cells are often unknown.

Cable and Compartmental Models

The *single-point* models just described assume that neurons do not have any significant spatial extent. But most real neurons

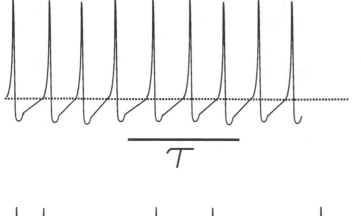

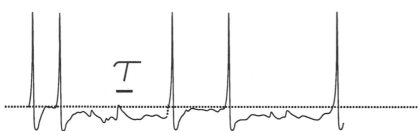

Figure 3. A schematic of the two distinct operating regimes of an integrate-and-fire model. The top trace shows how current input ramps the neuron's internal "voltage" up to produce regular spikes. The input current determines the *average slope* of the voltage between spikes. This can only occur if the interval between spikes is less than the neuron's time constant, so that the spiking frequency reliably reflects the average input current. A strikingly different situation is shown in the lower trace, where a relatively shorter time constant causes the neuron to *forget* when its last spike occurred. Now the input current determines the *average voltage* between spikes, which is nearly constant and below spiking threshold. Here a spike is only generated by a brief fluctuation in input current rather than a slower average over it, so that each spike can be interpreted as the distinct binary output of a fast, multiplicative computation, reporting the precise coincidence of several contributing inputs.

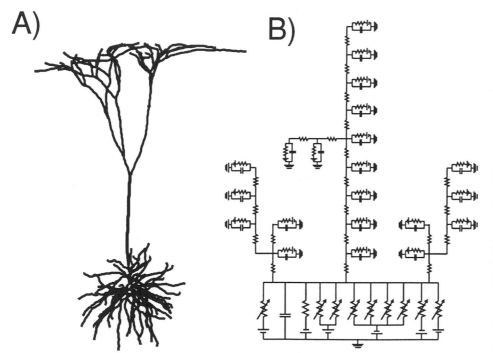

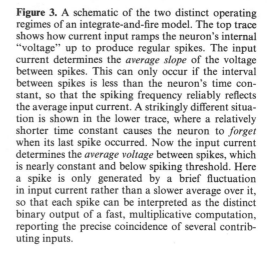

Figure 4. The most realistic form of single-neuron model numerically simulates the electrical properties of a branched cell membrane. First, the observed shape of the cell (*A*) is approximated by a collection of connected cylinders of the appropriate length and diameter. Then, each cylinder is simulated as a single electrical unit composed of an axial resistance and membrane capacitances and resistances (*B*). While computationally intensive, this numerical approach can treat the whole variety of cell shapes and nonlinear electrical properties which are ignored by the traditional single-compartment models.

have intricate dendritic trees (where the synaptic input impinges), as well as an axon and its branches (where the output spike is carried away). The unique shapes of those dendrites and axons can distinguish between various cell classes.

Axons are usually thought of as delay-lines without any significant information-processing ability (see AXONAL MODELING). But dendrites display a much richer repertoire of information-processing operations. The simplest "passive" dendrite model, pioneered by Rall, is a single capacitive, resistive cable.

Its voltage is characterized by the *cable equation*, which has two main parameters: τ is the membrane time constant, and λ, the *electrotonic space-constant*, is a characteristic distance over which a steady-state voltage attenuates. Signals always attenuate and temporally smooth as they spread from their sources in such a dendritic cable (Rall, 1989; Jack et al., 1983).

Because dendrites with branches are not as easily analyzed, modelers resort to discretizing the cable equation (like decomposing the dendritic tree into hundreds of simple electrical compartments connected by ohmic resistors, as in Figure 4), and then solving the compartments' equations numerically (Segev, 1992; see DENDRITIC PROCESSING).

Synaptic Conductances

Real synapses—whether slow or fast, inhibitory or excitatory, passive or voltage-gated—are best modeled as *conductances* in the cell membrane in series with driving potentials. They have many computational possibilities, especially in dendrites.

For example, if some of the synapses are inhibitory (with the driving potential close to the resting potential), then the inhibition can veto excitation *only* if the inhibition lies on the path between the location of the excitatory synapse and the cell body (Koch, Poggio, and Torre, 1982).

Also, when many synapses are simultaneously active, their increased conductance further attenuates distant input (by reducing λ) and makes the cell sensitive to fluctuations at a faster timescale (by reducing τ).

Experiments have shown that dendrites contain not only passive membranes but nonlinear, *active* ones. Active membranes on dendrites and their even tinier DENDRITIC SPINES (q.v.) can amplify synaptic input (so that synapses near and far from the cell body are equally effective) and can support nonlinear sigma-pi computation (Mel, 1994). They also can make the cell very sensitive to events at fast time-scales (Softky, 1994).

Silicon Neurons

Very Large Scale Integrated circuits (VLSI) of analog transistors represent a natural way to emulate neuronal functions (Mead, 1989). The silicon medium has many features in common with nervous tissue: it must use processing strategies that are fast and reliable, robust against noise and component variability, and can be easily interconnected. This approach works best when the circuits' structure reflects the computations they carry out, as occurs in the *silicon retina*.

Such transistor "neurons" typically operate much faster than their biological counterparts, but still orders of magnitude slower than transistors in digital computers. And wiring large numbers of them to each other is difficult in the layered, two-dimensional integrated circuits currently used.

More recently, Mahowald and Douglas (see Mahowald et al., 1992) fabricated SILICON NEURONS (q.v.) in analog VLSI. By exploiting transistor current-voltage properties similar to those of voltage-dependent ionic channels (Hille, 1992), they built subcircuits corresponding to the channels. A "neuron" containing these properties shows very similar behavior to a cortical pyramidal cell, including identical discharge curves and adaptation in the firing frequency in response to constant input; the scheme might be extended to model complex dendritic structures.

Discussion

After several decades of research, we still do not understand some of the most fundamental functions of nerve cells.

Learning

The most remarkable feature of the nervous system is its ability to change its internal structure in response to previous input—that is, to learn. Theoretical aspects of learning are well covered elsewhere (see the three road maps on learning in Part II of the *Handbook*). But at the single-cell level, there are many issues left to resolve.

Strengthening or weakening the weight of a single synapse w_{ij} seems to involve complex events, including calcium accumulation on both sides of the connection. But there are several different types of increase and decrease, occurring under different circumstances and lasting from seconds to days. These mechanisms are not yet well understood, especially in cerebral cortex.

A more potent (but even more poorly understood) type of learning could occur as axonal branches and connections "die off," while other branches form elsewhere and connect to other cells. If such structural plasticity participates in learning, then at least as much information could be stored in the existence/nonexistence of the synaptic connections as in their current-pulse amplitudes or "strengths."

The Dendrites

While most input to cortical cells arrives on intricately branched dendritic trees, we do not understand how a tree processes the input. Whether those tiny dendrites smooth out brief, localized input fluctuations—or instead amplify them—depends on the type of nonlinear membrane properties to be found there. Nonlinear dendrites can make a single "neuron" function as a large collection of distinct, multiplicative subunits.

Slow Analog Versus Fast Digital

There are two distinct, self-consistent interpretations of single-neuron function which are diametrically opposed, but there is evidence for both.

In the most popular interpretation, the fundamental computation is like that of a perceptron or a Hopfield neuron: a slow real-valued output resulting from slowly varying real-valued inputs, in which the timing of single spikes is inconsequential. (In fact, decades of experiments on most parts of the brain have found that only average rates—and not individual spike times—correlate with simple stimuli.)

This corresponds to an integrate-and-fire neuron gathering many small inputs, smoothing out their irregularities in the dendritic tree, summing the results in the cell body, and producing a firing rate as output. Here the best internal variable is the *average current* into the cell body, and the output firing rate can be interpreted as current into the *next* cell (the cell voltage is just a repeating ramp whose phase has no significance). This form of computation is most effective when the fluctuations in current are small, so that the cell fires in response to the mean current.

An alternative is McCulloch and Pitts' original interpretation of their binary neuron: the *active* state corresponds to a *single spike* rather than to a prolonged firing rate, so that every spike carries some kind of independent message. Although such a model can in principle transmit information much faster than an analog neuron (Stein, 1967), there are four fundamental criticisms of it: Are binary computations more appropriate than analog ones? Is there any need for such temporal precision and the broader bandwidth it confers? Can that temporal precision be implemented in a realistic cell? And how sensitive is such a cell to the noise which exists in cortex?

Other neural systems—such as the fly's visual system—can indeed carry significant information by single spikes (Bialek, 1992; see SENSORY CODING AND INFORMATION THEORY). And there is a need for *some* improved neural bandwidth to solve complex perceptual problems, like segmentation and binding. But there is so far no solid evidence that most cortical areas need such fast temporal resolution.

This single-spike regime corresponds to an integrate-and-fire type cell which fires according to its instantaneous (rather than average) voltage, so it is sensitive to input fluctuations. But for the cell to respond reliably to fluctuations, the average current must not dominate them—otherwise the mean current will ramp up the voltage and fire the cell, fluctuations or not.

Noise

A major attraction of analog neural models is their ability to withstand noise, and even to make use of it. This robust situation stands in marked contrast to that of digital computers, which are vulnerable to even single-bit errors.

Many neural algorithms actually work better with noise added, because noise tends to smooth out nonlinearities in transfer functions. For example, one can linearize the highly peaked and cusped frequency-response curve of a leaky-integrator neuron by convolving the input with noise, so that the transfer function samples a wider frequency range of input and hence produces a less narrowly peaked output (Knight, 1972).

While such noise can be useful in theoretical models, we do not yet understand its origin in the cortex. We do know that many sensory and motor neurons usually fire quite regularly, i.e., with low timing noise in their spike trains. Conversely, cortical cells—which most neural networks purport to model —seem to fire as irregularly as possible (Softky and Koch, 1993). But even in vitro cortical cells can fire very regulary when injected with DC current, suggesting that their irregularity comes from synaptic sources interacting with deterministic membrane properties, rather than from any intrinsically "stochastic" spiking mechanism.

If single neurons are deterministic, then the possibility remains that the irregular spike trains contain densely packed information, such as precise temporal patterns or hidden synchrony (Abeles, 1990). The mean firing rate is certainly crucial in many nervous systems and in many parts of the brain. But if in some areas, under some circumstances, the exact timing of spikes also matters, then the fast, binary aspects of the original McCulloch-Pitts model neuron may be in for a revival.

Road Map: Biological Neurons
Related Reading: Analog VLSI for Neural Networks; Perspective on Neuron Model Complexity; Synaptic Coding of Spike Trains

References

Abeles, M., 1990, *Corticonics*, Cambridge, Eng.: Cambridge University Press. ◆
Arbib, M., 1987, *Brains, Machines, and Mathematics*, 2nd ed., New York: Springer-Verlag. ◆
Bialek, W., 1992, Optimal signal processing in the nervous system, in *Princeton Lectures on Biophysics*, New York: World Scientific, pp. 321–401. ◆
Hille, B., 1992, *Ionic Channels of Excitable Membranes*, 2nd ed., New York: Sinauer. ◆
Jack, J., Noble, D., and Tsien, R., 1983, *Electric Current Flow in Excitable Cells*, Oxford: Oxford University Press. ◆
Knight, B., 1972, Dynamics of encoding in a population of neurons, *J. Gen. Phys.*, 59:734–766.
Koch, C., Poggio, T., and Torre, V., 1982, Retinal ganglion cells: A functional interpretation of dendritic morphology, *Philos. Trans. R. Soc. Lond. B Biol. Sci.*, 298:227–263.
Mahowald, M., Douglas, R., LeMoncheck, J., and Mead, C., 1992, An introduction to silicon neural analogs, *Semin. Neurosci.*, 4:83–92. ◆
Mead, C., 1989, *Analog VLSI and Neural Systems*, Menlo Park, CA: Addison-Wesley. ◆
Mel, B., 1994, Information processing in dendritic trees, *Neural Computat.*, 6:1031–1085. ◆
Rall, W., 1989, Cable theory for dendritic neurons, in *Methods in Neuronal Modelling* (C. Koch and I. Segev, Eds.), Cambridge, MA: MIT Press, pp. 9–62.
Segev, I., 1992, Single neurone models: Oversimple, complex, and reduced, *Trends Neurosci.*, 15:414–421. ◆
Softky, W., 1994, Sub-millisecond coincidence detection in active dendritic trees, *Neuroscience*, 58:15–41.
Softky, W., and Koch, C., 1993, The highly irregular firing of cortical cells is inconsistent with temporal integration of random epsp's, *J. Neurosci.*, 13:334–350. ◆
Stein, R., 1967, Some models of neuronal variability, *Biophys. J.*, 7:37–68.

Somatosensory System

Oleg V. Favorov and Douglas G. Kelly

Introduction

Diverse theoretical and modeling studies of the somatosensory system, primarily at the level of the cerebral cortex, have contributed to a progressively comprehensive understanding of somatosensory information processing. The prevailing view of this processing is that tactile stimulus representation in the nervous system is changed from an original form (more or less isomorphic to the stimulus itself) to a completely distributed form (underlying perception) in a series of partial transformations in successive subcortical and cortical networks. As demonstrated, for example, by Johnson et al. (1991), at the level of peripheral afferents, the representation of a scanned tactile form, such as a letter of the alphabet embossed on a rotating drum, is an isomorphic neural image of that form because of the topography of the receptor sheet and the small receptive fields (RFs) of those afferents. In contrast, at the level of the primary somatosensory cortex (SI), consisting of cytoarchitectural areas 3a, 3b, 1, and 2, the neural image of the stimulus already has become much more complex, as evidenced by the complexity and heterogeneity of the responses of cortical neurons, which are already sensitive to the shape and temporal features of peripheral stimuli, rather than simply reflecting the overall intensity of stimulation of their RFs.

Transformation from Periphery to SI

The modeling study of Bankman, Hsiao, and Johnson (1990) sought to determine how much of this transformation from the periphery to the SI can be attributed to the most general known principles of organization of somatosensory pathways: (1) ascending connections are locally divergent so that any given site in one network receives a convergent input from somatotopically adjacent regions of the preceding network; (2) each network consists of repeating circuits; (3) lateral inhibitory connections within each network are responsible for the emergence of complexly organized RFs; and (4) there is little, if any, loss of spatial information from one network to the next. These principles were implemented in a multilayer, feedforward network in which each unit in a given layer computed either the sum or the difference of the activity levels of homologous units of adjacent modules in the preceding layer (effecting a two-dimensional Walsh-Hadamard transform). The first layer of the network simulated the spatiotemporal patterns of activation expected to take place in the receptor sheet in response to stimuli in the form of letters swept along the skin (letters were chosen as an example of small, but spatially complex stimuli). Actual spike trains recorded from monkey cutaneous mechanoreceptors in response to the same letters were used to generate these patterns. The activity patterns evoked in some of the units in different layers of the model were similar to the activity patterns, evoked by the same stimuli, that have been observed in a substantial number (37%) of SI neurons. The units whose responses matched those of the cortical neurons closely had RFs with elongated excitatory and inhibitory subregions, similar to the RFs of simple cells of the visual cortex. Whether such RFs are present in the SI is not well documented, principally because the three-dimensional shape of the skin surface presents serious technical difficulties in the control and delivery of elongated stimuli. However, the behavior of some of the units with more complex RFs was not observed in neurophysiological studies, and the responses of some SI neurons could not be matched by the network.

The partial success of the network that embodies the organizational principles stated above suggests that those principles are not fully constraining and points to the need for greater attention to biological details, as in the model of Kyriazi and Simons (1993). That model is similar to the model of Bankman et al. (1990) in that (1) it also addresses the nature of neural image transformation (in this case, transformation from the thalamus to cortical layer 4) and (2) it uses spike trains recorded from real thalamic neurons (in response to three stimulation paradigms) for input and compares the responses of the model neurons with those recorded from neurons in cortical layer 4. The network was designed to model the organization of a cortical *barrel*, a discrete neuronal aggregate in layer 4 of the rodent SI that receives its principal input from one of the facial whiskers. The network consists of a layer of 100 thalamic cells feeding to a layer of 70 excitatory and 30 inhibitory cortical cells. The structural and functional details of the network were chosen to reflect accurately those of real barrels. The crucial features of the cortical layer include the following: (1) cells have nonlinear activation properties; (2) inhibitory cells are more responsive than excitatory cells; (3) each excitatory and inhibitory cell receives connections from multiple thalamic cells; and (4) all cells in the cortical layer are strongly interconnected. Under these conditions, the model accurately simulated the cortical response to all three types of whisker stimulations studied (brief or prolonged deflection of a single whisker or sequential deflection of two whiskers in a condition-test paradigm), and it reproduced the known differences between the response properties of thalamic and cortical neurons. Specifically, excitatory cells in the cortical layer of the model shared a greater signal-to-noise ratio than cells in the thalamic layer, and they had RFs with more tightly focused excitatory centers and stronger inhibitory surrounds than did thalamic cells. Traditionally, focusing of RF properties has been assumed to be achieved by lateral inhibitory interactions among neighboring cortical columns. One of the major lessons of the modeling work of Kyriazi and Simons (1993) is the demonstration of a different mechanism of RF focusing, one used by the rodent SI and based on the greater responsiveness of inhibitory neurons in a cortical column to weak stimuli applied at the RF margins and the resulting suppression of the responses of excitatory neurons within the *same* column.

Development of SI Topography

The topographical organization of thalamocortical connections is one of the principal determinants of how representations of peripheral stimuli are transformed in the SI. How the topography is developed and maintained in the SI has been the subject of a number of theoretical and modeling studies (see SOMATOTOPY: PLASTICITY OF SENSORY MAPS). For example, Senft and Woolsey (1991) addressed the question of how the barrels of rodent SI are formed. They found that the barrels in a cortical field closely approximate Dirichlet domains (which subdivide a field into convex polygons, each having a center point and consisting of all points that are closer to this center point than to those of other polygons). Senft and Woolsey propose that thalamocortical fibers that carry information from any particular facial whisker initially (around the day of birth) terminate in the SI in an approximately Gaussian distribution, partially overlapping the projections from neighboring whiskers. Each such whisker-based group of thalamocortical fibers thus provides the predominant afferent input to some region of the cortex. These regions (dominated by a single whisker) form a Dirichlet-like pattern in the SI, with the center point of each Dirichlet domain coinciding with the peak of the Gaussian distribution of a particular whisker. Within each Dirichlet domain, the coherently acting thalamocortical fibers of the dominant whisker-based group cooperate with each other and compete with fibers from other whisker-based groups. As a result, the fibers of the dominant group arborize extensively within their domain, while those from other groups withdraw from it, thus producing the typical discrete clustering of afferents characteristic of the barrel cortex.

How the adult barrel cortex adapts to changes in sensory inputs was modeled by Benuskova, Diamond, and Ebner, (1994). To account for the results of their physiological study of the adult rat barrel cortex after trimming of all but two whiskers, they built a model of a single representative neuron from a barrel whose principal whisker was left intact. This neuron received two afferent inputs: (1) short-latency thalamocortical input from the principal whisker and much weaker input from the neighboring whiskers, and (2) long-latency corticocortical input from other cells within the same barrel and from the adjacent barrels. Experimentally, after the whiskers were trimmed, the short-latency input from the thalamus gradually increased, the long-latency corticocortical input from the trimmed whiskers gradually decreased, and the long-latency inputs from the uncut whiskers initially increased, but gradually returned to their starting levels. Benuskova et al. (1994) demonstrated with their model that this behavior is suggestive of synaptic plasticity as proposed by Bienenstock, Cooper, and Munro (1982; see BCM THEORY OF VISUAL CORTICAL PLASTICITY): active synapses strengthen whenever the postsynaptic activity is greater than a certain synaptic modification threshold and weaken when the postsynaptic activity is less than that threshold, and the threshold itself varies as a function of the postsynaptic activity, averaged over the recent past. The transient increase in the strength of long-latency corticocortical input from the nontrimmed whiskers is explained as the result of a transient decrease in average postsynaptic activity after whisker cutting, resulting in a transient lowering of the synaptic modification threshold.

The barrel cortex is a specialized somatosensory network in that it receives input from discrete sensory structures. Pearson, Finkel, and Edelman (1987) developed a model of a more prototypical SI network that receives input from the skin of the hand. Their aim was to account for the topographical reorganization of the adult primate SI in response to experimentally induced manipulations of peripheral input. The model is a two-dimensional network of locally interconnected excitatory and inhibitory cells that receive topographical projections from two receptor sheets, corresponding to the glabrous and dorsal surfaces of the hand. The intrinsic connections are arranged to produce a Mexican-hat distribution of shorter-range excitatory and longer-range inhibitory interactions within the network. Both intrinsic and extrinsic excitatory connections undergo activity-dependent modifications according to a Hebbian-like synaptic rule. Although this network is originally given somatotopically distributed afferent connections and homogeneously organized intrinsic connections, under repeated sensory stimulation, it spontaneously resolves into a mosaic of small, discrete regions. The excitatory intrinsic connections among neurons within each such region strengthen greatly, whereas connections among different regions weaken profoundly. Afferent connections also reorganize, and neurons within each region acquire identical or similar focal RFs, whereas neurons in adjacent regions acquire nonoverlapping RFs. As a result, across the network, RFs do not shift continuously, but in abrupt steps at the borders between these regions. After repetitive preferential stimulation of a particular local skin region or after local transection of afferent connections, there are en masse shifts in the RFs, leading to dramatic changes in the topographical map of the network. These changes correspond closely to those seen in the monkey cortex after similar perturbations.

Grajski and Merzenich (1990) expanded on the work of Pearson et al. by showing that Hebbian plasticity in combination with lateral inhibition is sufficient to account for the inverse relationship that exists in the cortex between RF size and cortical magnification (i.e., the size of the cortical area representing a unit area of skin). Using a three-layer network (skin, subcortical, and cortical layers) with a Mexican-hat pattern of intrinsic connections and Hebbian plasticity of afferent and intrinsic connectivity, they successfully simulated two biological studies. In one, preferential stimulation of a particular restricted skin region resulted in an increase of its cortical representation coupled with a decrease in RF size. In the second study, cortical magnification was reduced by lesioning a restricted cortical region, which led to an increase in RF size in cortical territories around the lesion.

Segregates, Minicolumns, and Macrocolumns

Although it was not known at the time, Pearson et al.'s model predicts accurately that the forelimb region of the SI is organized as a mosaic of discrete columns, similar to barrels in the whisker region of the rodent SI. Favorov and colleagues (e.g., Favorov and Diamond, 1990) have shown that the forelimb region of the SI in cats and monkeys is made up of columns, 0.3–0.6 mm in diameter, separated from each other by sharp boundaries that compartmentalize the SI into a honeycomb-like mosaic. The defining characteristic of these columns (they were named *segregates*) is that, as in the Pearson model, the *strongest* afferent input to local clusters of neurons *throughout* a given column comes from *the same* focal skin site (the *segregate RF center*). At the borders between segregates, the site of the strongest input shifts abruptly to a new, prominently displaced skin locus.

The similarity between the discrete regions described by Pearson and colleagues and SI segregates would be complete except for one notable difference: whereas the regions in the Pearson model have a uniform RF composition, an SI segregate is made up of neurons with a large variety of RF sizes, shapes, and positions on the skin. In fact, segregates are composed of smaller, 0.05-mm-diameter columns called *minicolumns*; neurons located in the same minicolumn have similar RFs, but neurons located in different, even adjacent, minicolumns typically have RFs that differ significantly in size and shape and frequently overlap only minimally on the skin, having in common only the segregate RF center (e.g., Favorov and Diamond, 1990; Tommerdahl et al., 1993).

Favorov and Kelly (1994a, 1994b) developed a model of a typical SI segregate to evaluate a hypothesis that the diversity in the RFs among minicolumns in a segregate is an outcome of cortical network self-organization during perinatal development. In general, local RF variability is an ignored feature of cortical topographical organization. Instead, it is typical (and convenient) to view cortical topographical organization as locally uniform (i.e., to view neighboring neurons as having very similar RFs and to view RFs as becoming progressively more different as the cortical distance separating two neurons increases). This view has strongly influenced the design of existing models of cortical organization. It led to the use of a Mexican-hat pattern of lateral connections in network models because this arrangement readily generates (through Hebbian self-organization) locally uniform topographical maps in neural networks. However, there is ample experimental evidence that neurons near each other frequently have RFs that vary extensively in size, shape, and position (cited in Favorov and Kelly, 1994b). However, the Mexican-hat pattern of lateral connections cannot generate neural networks exhibiting such local RF variability. What is needed to achieve the local RF variations characteristic of living sensory networks and to preserve topographical orderliness on a more global scale is a connection pattern in which near neighbors have inhibitory lateral interactions and more distant neighbors have excitatory interactions. The living cortical network possesses such a pattern of lateral connectivity; it is provided by the axons of inhibitory *double-bouquet* cells and excitatory *spiny stellate* cells (reviewed in Favorov and Kelly, 1994a). The model of SI segregate as a group of 61 minicolumns developed by Favorov and Kelly was built around such a connectional pattern. Each minicolumn in it consists of three cells representative of different cell types in the cortex: an input spiny stellate cell (which receives thalamic input and distributes it to all other cells of the same minicolumn and, to a lesser degree, to other nearby minicolumns), an intrinsic double-bouquet cell (which inhibits neighboring minicolumns), and an output *pyramidal* cell. Connections from the thalamus to minicolumns were plastic; they were allowed to self-organize in accordance with a Hebbian rule during a developmental period in which the network was driven by punctate skin stimuli.

Control by the segregate's minicolumns of their own thalamic connections is the central property of this model. That is, the thalamic connections to minicolumns are not prescribed explicitly according to some preconceived idea about the mapping relationship between skin and cortex, but are selected by the model itself, driven by the history of skin stimulation. As a result, through self-organization, a segregate acquires a complex, richly detailed pattern of thalamic connections in which neighboring minicolumns choose connections from partially shifted groups of thalamic neurons, reproducing experimentally observed features of the termination patterns of thalamocortical axons. The model also produces local RF variability by reshuffling the positions of RFs among neighboring minicolumns within the modeled segregate, thus producing complex patterns of RFs across the segregate that are similar to those observed experimentally in microelectrode penetrations through SI segregates (see Favorov and Diamond, 1990). Furthermore, this self-organized network has complex, nonlinear functional properties which caused its individual neurons to be sensitive to the shape and temporal features of peripheral stimuli, rather than simply reflecting in their responses the overall intensity of stimulation of their RFs. In particular, even though the network was exposed only to stationary point stimuli during self-organization, its neurons acquired the ability to discriminate the direction of moving stimuli as well as the orientation of bar stimuli. Different stimulus directions and orientations were represented by different neurons in the network, and the maps of neurons with these preferences had many features in common with maps in the visual cortex.

Discussion

What picture of sensory information processing in the SI emerges out of these studies? Most fundamentally, one sees that the processing is modular on at least two different scales, macrocolumnar and minicolumnar. The studies of Pearson et al. (1987) and Senft and Woolsey (1991) suggest that discrete macrocolumns (i.e., barrels or segregates) are small, sharply delineated cortical areas that, during perinatal development, are innervated by a selected group of thalamic neurons, sharing similar RFs, whose axons all terminate extensively *throughout* the territory of that macrocolumn. According to Favorov and Kelly (1994a), within a segregate, thalamocortical axons connect to the minicolumns selectively, not uniformly, so each minicolumn receives afferent connections from a unique subset of the thalamic neurons that project to that segregate. The differences in afferent inputs to neighboring minicolumns in a segregate are further amplified by lateral inhibitory interactions among adjacent minicolumns. As a result, in response to a tactile stimulus, even a simple punctate stimulus, an activated segregate generates a complex, spatially heterogeneous pattern of activity that consists of active minicolumns interdigitated throughout the segregate with other, much less active minicolumns. Furthermore, tactile stimuli, even the most restricted ones, usually activate not a single segregate, but a local group of segregates (Tommerdahl et al., 1993). Each segregate in such an active group generates its own pattern of minicolumnar activation. Thus, the SI response to a peripheral stimulus takes the form of a patchwork of active minicolumns that extends across a number of segregates. The richly detailed minicolumnar patterns generated by the model are remarkably stimulus specific, being sensitive to stimulus location, shape, and temporal characteristics (such as direction of motion for moving stimuli). Information about stimulus properties, such as location, motion, and geometric detail, is encoded by the distribution of activities among minicolumns in the activated cortical field in a way that allows easy extraction of information about various stimulus features independent of each other (e.g., spatial patterns can be recognized independently of their positions on the skin).

To complete the picture of stimulus representation in the SI, according to Whitsel et al. (1991), an initial cortical response to a stimulus application undergoes substantial modifications with a continuous or repetitive exposure to that stimulus. Those authors proposed a conceptual model of cortical dynamics that showed how such modifications could result from activity-dependent increases in extracellular concentrations of potassium and their modulating effects on the state of *N*-methyl-D-aspartate (NMDA) receptors and neuromodulators, such as acetylcholine, and on the development of long-lasting afterhyperpolarization in active neurons. These changes in the network state translate into changes in pericolumnar inhibition, leading to progressive changes in the spatial pattern of cortical response. These changes make this pattern progressively more stimulus specific.

The theoretical and modeling work on the somatosensory system has focused on stimulus representation at the level of the input (middle) layers of the primary SI (i.e., issues of topographical organization of the SI). However, the delivery of peripheral information to the cortical network is only one of the basic questions regarding cortical information processing. In particular, it remains to be explored what novel functional processes operate outside the input cortical layers (i.e., in the upper and deep layers). In addition, the functional properties of networks comprising more than one sensory cortical area have not been modeled. There is also a need to extend investigations to include more realistic kinds of tactile sensory experiences.

Acknowledgments. This work was supported in part by NIMH R01 grant MH48654 and NINDS R01 grant NS30686.

Road Maps: Mammalian Brain Regions; Other Sensory Systems
Background: I.3. Dynamics and Adaptation in Neural Networks
Related Reading: Cortical Columns, Modules, and Hebbian Cell Assemblies; Ocular Dominance and Orientation Columns

References

Bankman, I. N., Hsiao, S. S., and Johnson, K. O., 1990, Neural image transformation in the somatosensory system of the monkey: Comparison of neurophysiological observations with responses in a neural network model, *Cold Spring Harbor Symp. Quant. Biol.*, 55:611–620.

Benuskova, L., Diamond, M. E., and Ebner, F. F., 1994, Dynamic synaptic modification threshold: Computational model of experience-dependent plasticity in adult rat barrel cortex, *Proc. Natl. Acad. Sci. USA*, 91:4791–4795.

Bienenstock, E. L., Cooper, L. N., and Munro, P. W., 1982, Theory for the development of neuron selectivity: Orientation specificity and binocular interaction in visual cortex, *Proc. Natl. Acad. Sci. USA*, 79:2082–2086.

Favorov, O. V., and Diamond, M. E., 1990, Demonstration of discrete place-defined columns—segregates—in the cat SI, *J. Comp. Neurol.*, 298:97–112.

Favorov, O. V., and Kelly, D. G., 1994a, Minicolumnar organization within somatosensory cortical segregates, I: Development of afferent connections, *Cereb. Cortex*, 4:408–427.

Favorov, O. V., and Kelly, D. G., 1994b, Minicolumnar organization within somatosensory cortical segregates, II: Emergent functional properties, *Cereb. Cortex*, 4:428–442.

Grajski, K. A., and Merzenich, M. M., 1990, Hebb-type dynamics is sufficient to account for the inverse magnification rule in cortical somatotopy, *Neural Computat.*, 2:77–84.

Johnson, K. O., Phillips, J. R., Hsiao, S. S., and Bankman, I. N., 1991, Tactile pattern recognition, in *Information Processing in the Somatosensory System* (O. Franzen and J. Westman, Eds.), New York: Stockton, pp. 305–318.

Kyriazi, H. T., and Simons, D. J., 1993, Thalamocortical response transformation in simulated whisker barrels, *J. Neurosci.*, 13:1601–1615.

Pearson, J. C., Finkel, L. M., and Edelman, G. M., 1987, Plasticity in the organization of adult cerebral cortical maps: A computer simulation based on neuronal group selection, *J. Neurosci.*, 7:4209–4223.

Senft, S. L., and Woolsey, T. A., 1991, Mouse barrel cortex viewed as Dirichlet domains, *Cereb. Cortex*, 1:348–363.

Tommerdahl, M., Favorov, O. V., Whitsel, B. L., Nakhle, B., and Gonchar, Y. A., 1993, Minicolumnar activation patterns in cat and monkey SI cortex, *Cereb. Cortex*, 3:399–411.

Whitsel, B. L., Favorov, O. V., Kelly, D. G., and Tommerdahl, M., 1991, Mechanisms of dynamic peri- and intracolumnar interactions in somatosensory cortex: Stimulus-specific contrast enhancement by NMDA receptor activation, in *Information Processing in the Somatosensory System* (O. Franzen and J. Westman, Eds.), New York: Stockton, pp. 353–369.

Somatotopy: Plasticity of Sensory Maps

Sherre L. Florence and Jon H. Kaas

Introduction

Somatotopy, a dominant feature of subdivisions of the somatosensory system, is defined by a topographic representation, or map, in the brain of sensory receptors on the body surface (see SOMATOSENSORY SYSTEM for a review of cortical somatosensory receptive fields and their development). In mammals, there are orderly representations of cutaneous receptors at various levels: in the spinal cord, lower brainstem, thalamus, and neocortex. In the past, these sensory maps were considered relatively permanent, point-to-point representations, with little or no changes in internal organization once the mature patterns had developed. During the last decade, however, evidence for plastic changes at all levels of the adult somatosensory system has accumulated in a wide range of mammalian species (reviewed by Kaas, 1991), and it has become clear that the somatotopic organization of sensory maps represents both the peripheral distribution of receptors and dynamic aspects of brain function. Changes in the relative levels of sensory stimulation as a result of experience or injury produce modifications in sensory maps. The current review deals predominantly with the changes that occur in cortical maps after peripheral sensory denervations. The three following issues are discussed: (1) What features of somatotopic maps change, and under what conditions are the changes produced? (2) What mechanisms account for these changes? (3) What are the functional consequences of sensory map changes? This review focuses on studies of plasticity of the primary somatosensory cortex in primates, but plastic changes also occur at subcortical levels, in other systems, and in other species (see Kaas, 1991; Merzenich and Jenkins, 1993).

Features of Somatotopic Plasticity

The preponderance of evidence for plastic changes in the sensory systems of adult mammals has come from studies of the organization of the primary somatosensory cortex in which the somatotopic organization of the body representation has been well defined (e.g., Kaas, 1991). In primates, the large representation of the hand in the somatosensory cortex is particularly suitable for these types of investigation since some of the plastic changes involve small topographic shifts that would be harder to detect in more compact representations. The approach typically is to crush or cut a sensory nerve to eliminate the relay of sensory information from a certain portion of the skin surface to the brain. The effects of the deprivation on the somatosensory pathway can be studied minutes, weeks, or even years later, using microelectrode mapping techniques which allow large regions of the nervous system to be sampled in each animal in a relatively limited span of time. In one of the first clear demonstrations of adult plasticity, Merzenich and colleagues (for review, see Kaas, 1991; Merzenich and Jenkins, 1993) cut the median nerve that innervates the glabrous (palmar) surface of the thumb, index, and long fingers, and adjacent palmar pads in monkeys. Immediately after the nerve was cut, the majority of neurons in the large cortical zone where median nerve skin is typically represented were unresponsive. However, at a small number of recording sites, neurons had acquired new receptive fields which were on the dorsal rather than the glabrous surfaces of the digits. Because of the rapidity of these changes, the mechanism that underlies this

reorganization must be an unmasking of previously impotent connections (see below for more discussion).

Given a period of weeks to months after the peripheral denervation, other types of reorganization are found in the primary somatosensory cortex in addition to the synaptic unmasking that occurs immediately after the manipulation. One commonly observed change in map organization that results from peripheral denervation is expansion of the representations of nearby innervated skin of the hand, such as the preserved digits and palmar pads, into cortical regions where the receptive fields of the neurons, before the denervation, were on the median-nerve-innervated skin. This kind of somatotopic

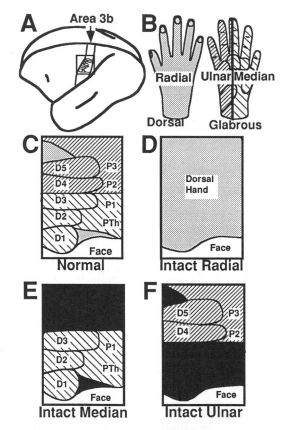

Figure 1. The reorganization of the primary somatosensory cortex area 3b in adult owl monkeys after cutting and ligation of combinations of nerves that innervate the skin of the hand. Part *A* shows the location of the hand representation in area 3b on a dorsolateral view of an owl monkey brain. Part *B* shows a schematic representation of the nerve territories of the skin of the hand. Part *C* shows the cortical organization of the hand representation in normal animals, including digits (D) and palmar pads (P). Medial is to the top and rostral is to the left. Part *D* shows the representation after cutting and ligation of the median and ulnar nerves to denervate the entire palmar surface of the hand. The large representation of the denervated hand becomes completely reactivated by the representation of the dorsal surface of the hand. In contrast, when the radial nerve is cut in combination with either the ulnar (part *E*) or median (part *F*) nerve, the deprived zone in area 3b is only minimally reactivated (regions that could not be activated by cutaneous stimulation are shown in black).

expansion involves only a portion of the overall map changes. The most extensive reorganizations demonstrated to date result from a takeover of the deactivated cortex such that neurons in the deprived zone become responsive to stimulation at novel skin locations. For example, after denervation of part or all of the glabrous surface of the hand, the deprived cortex is activated by inputs from the hairy surface of the hand (Figure 1) (see also Garraghty and Kaas, 1991). Even more dramatic changes in the topographic organization of area 3b were reported in monkeys after the forearm had been denervated by lesioning the cervical dorsal root ganglia (called *dorsal root rhizotomy*) to remove all sensory inputs. The entire block of cortex in which the hand typically is represented, an area of 70 mm² or more, was taken over by the representation of the face (Pons et al., 1991).

In the types of sensory denervations mentioned above, the region of deprived cortex ultimately is completely reactivated; however, a persisting or permanent deactivation of at least a portion of the deprived cortex follows other types of injuries. For example, Merzenich and colleagues (reviewed in Merzenich and Jenkins, 1993) found that amputation of a single digit resulted in takeover of the small zone of deprived cortex by the representations of the adjacent digits and pads. However, after amputation of two or more adjacent digits, the representation of the intact glabrous hand did not completely take over the deprived region, leaving a zone of deactivated cortex. In another experiment, Garraghty et al. (1994) found that much of the deprived cortex remains unresponsive to tactile stimulation after combined transection of the radial nerve with either the median or ulnar nerves. This manipulation mimics amputation since entire digits are denervated (see Figure 1).

Clues from these exceptions may be particularly useful for unraveling the mechanisms that contribute to the sensory map changes. For example, permanent cortical deactivations typically follow complete denervation of both the glabrous and hairy surfaces of the hand. However, the large cortical representations of the glabrous surface of the hand can be completely reactivated if the inputs from the hairy surface remain intact (Figure 1; see also Garraghty and Kaas, 1991). We propose that this pattern of reactivation is a reflection of small changes in the primary sensory inputs to the first-order somatosensory relay (the cuneate nucleus of the brainstem), where inputs from the dorsal surface of the hand are directly adjacent to or partially overlapping inputs from the glabrous surface of the hand. After denervation of the glabrous surface, presumably, the inputs from the hairy skin come to activate the neurons in the cuneate nucleus that once responded preferentially to glabrous inputs. This change occurs through a change in the distribution or effectiveness of the projections from the hairy skin.

Further evidence of the impact of subcortical connections on cortical maps comes from our study of the effects of long-standing hand amputation in monkeys. As shown in Figure 2, the region where the hand normally is represented in area 3b becomes reactivated to a large extent by the skin of the forearm adjacent to the amputation. In the same animal, the distribution of the primary sensory afferents from the skin of the forearm adjacent to the amputation extends well beyond its normal limits, into the region previously occupied by inputs from the hand. Presumably, this expanded projection is the source of reactivation of the neurons in the cortex.

The cortical changes described above all result from a loss of sensory-driven activity, but similar changes have been found after an increase in activity or a change in activity patterns. These changes have been demonstrated with a diversity of imaginative surgical and behavioral paradigms in monkeys (re-

Area 3b

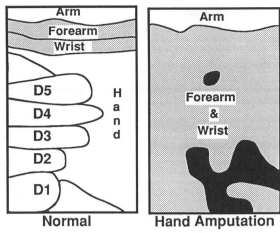

Cuneate Nucleus

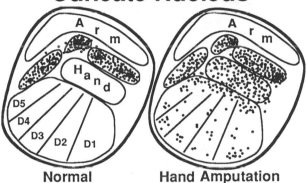

Figure 2. Schematic depiction of the reorganization of the primary somatosensory cortex area 3b and the expansion of sensory afferents in the cuneate nucleus of the brainstem in an adult owl monkey after amputation of the hand. The locations of the wrist and forearm representations (shading) in area 3b of normal monkeys are shown at the top left. After amputation of the hand, the representations of the wrist and forearm (shading) expand and take over most of the hand representation (top right). At a few recording sites, neurons did not respond to light, tactile stimuli (shown in black). Medial is to the top in area 3b, and rostral is to the left. In schematics of the cuneate nucleus (bottom), the normal distribution of sensory afferent terminations from the wrist and forearm (dots) is shown on the left, and the expansion of the projection from the wrist and forearm into the region that was denervated after amputation of the hand is shown on the right. Medial is to the left in the cuneate nucleus, and dorsal is to the top.

viewed by Merzenich and Jenkins, 1993) and also have been reported in humans. For example, in blind subjects who read using the Braille method, recordings of somatosensory evoked potentials demonstrated that the area of activation in the somatosensory cortex for the reading (right index) finger is larger than that for other fingers (Pascual-Leone and Torres, 1993). Additionally, in patients with syndactyly (webbed fingers), the topographic organization of the digit representations changed within 1 week of surgical separation of the syndactyl digits (Mogliner et al., 1993). Before the surgical correction, the pattern of cortical activation that resulted from tactile stimulation of the hand was nontopographic; in contrast, after surgery, representations of the individual digits in the somatosensory cortex became distinct and assumed a somatotopic progression.

Mechanisms Underlying Map Plasticity

Since sensory deprivations produce both immediate, small-scale map changes and large-scale adjustments that evolve over a period of weeks or months, there must be multiple mechanisms underlying the recovery. Immediate changes have been attributed to an unmasking of existing connections through a reduction of activity-driven activation of inhibitory neurons and to a disruption of the normal balance between inhibitory and excitatory influences on neurons in the sensory map. The potential for immediate changes after sensory deprivation ultimately is limited by the amount of overlap of the central projections from different skin regions; however, small changes at each level of the primary somatosensory cortex could amount to significant immediate cortical reorganization. Changes that take place over a longer time course are likely to result from a combination of events, with synaptic unmasking accounting for only a small proportion of the reorganization.

Some of the long-term changes are presumed to involve Hebbian-type mechanisms. A model of cortical plasticity based on Hebb-like synaptic modifications can replicate most of the more limited somatotopic map changes, including changes in cortical organization after peripheral nerve injury and regeneration, the expansion of representations after the removal of competing inputs (such as after digit amputation), and expansions after overstimulation of the inputs to a digit representation (Sklar, 1991). This model is based on the theory that a neuron possesses a synaptic modification threshold that dictates whether the efficacy of synapses impinging on the neuron is strengthened or weakened. The threshold changes in relation to the neuron's recent average postsynaptic activity so that synapses potentiate when postsynaptic activity is greater than the synaptic modification threshold and weaken when activity is less than the threshold. A similar model can explain visually driven changes in the developing visual cortex (Bienenstock, Cooper, and Munro, 1982; see BCM THEORY OF VISUAL CORTICAL PLASTICITY).

Currently, the strongest candidate for the neural mechanism that underlies the Hebbian-type modification is the N-methyl-D-aspartate (NMDA) receptor mechanism (reviewed by Bear, Cooper, and Ebner, 1987; see NMDA RECEPTORS: SYNAPTIC, CELLULAR, AND NETWORK MODELS). The NMDA receptor is thought to coexist with other receptors for excitatory amino acids and to mediate synaptic transmission jointly. In the developing visual system, lasting changes in synaptic effectiveness that occur with experience are absent when the NMDA receptors are selectively blocked (reviewed by Rauschecker, 1991). Similarly, in the mature somatosensory system, the expansion of intact representations into denervated zones that is typically found after dorsal root rhizotomies in the cat somatosensory cortex was disrupted when NMDA receptors were chronically blocked by the infusion of NMDA antagonists (Kano, Lino, and Kano, 1991).

Additional mechanisms for long-term map reorganization may involve major changes in gene expression. For example, injured neurons in the rat spinal cord show increased production of the growth associated protein GAP-43 (for review, see Mendell and Lewin, 1992); the increased levels of GAP-43 could result in a host of subcellular changes, including neuronal growth. Under most circumstances, in the adult nervous system, new growth is likely to be limited to the formation of new synapses or the local extension of dendrites and axon arbors. However, the large-scale changes in map organization may be mediated by greater amounts of neural growth.

Long-distance growth of axons has been demonstrated in adult rats and monkeys. In the rat spinal cord, McMahon and Kett-White (reviewed in Mendell and Lewin, 1992) demonstrated that the central axon processes of the sciatic nerve could be induced to grow into the spinal cord territory of other sensory afferents after a manipulation including lesioning the dorsal roots adjacent to the sciatic innervation territory to denervate the spinal cord and subsequently crushing the sciatic nerve to induce a central as well as a peripheral growth state in the regenerating neurons. When both of these criteria were met, the sciatic nerve afferents sprouted well beyond the normal sciatic nerve territory, into spinal cord zones denervated by the dorsal root lesions. Less extensive sprouting of injured peripheral nerve afferents was demonstrated in the adult monkey spinal cord after median nerve crush (Florence et al., 1993). The central projections from individual median-nerve-innervated digits had extended into the territory of adjacent digits and pads that were also innervated by the median nerve. Presumably, the central arbors of injured sensory nerves expanded within the region deprived of normal activation patterns. Finally, in monkeys with long-standing hand amputation, the expansion of the peripheral nerve projections from the skin of the forearm proximal to the amputation into the denervated portions of the spinal cord and cuneate nucleus in the brainstem (see Figure 2) may reflect new growth of the central arbors of sensory nerves that were injured by the amputation or may reflect sprouting of the intact, uninjured projections.

Functional Consequences of Sensory Map Changes

It is commonly assumed that large sensory representations relate to superior sensory abilities since skin zones with good tactile acuity, such as the hands and lips, have large representations in the somatosensory pathway (magnification factor). To some extent, these large representations reflect a high density of peripheral receptors; however, it appears that adding neurons to a processing circuit may enhance the performance of that circuit. For example, after monkeys were trained in a tactile discrimination task using a specific finger, the representation of that finger in the cortex was expanded, and performance improved (reviewed in Merzenich and Jenkins, 1993). Also, human amputees often demonstrate increased tactile acuity for skin zones adjacent to the site of amputation, presumably the same skin whose cortical representation is expanded. In contrast, no improvement in the sensory performance of the reading finger was detected in blind subjects who read using the Braille method (Pascual-Leone and Torres, 1993). Thus, the consequence of the expansion of the area of activation in the somatosensory cortex for the reading finger is unclear.

Cortical expansions may also lead to sensory mislocalizations. Ramachandran, Stewart, and Rogers-Ramachandran (1992) have shown that amputation can lead to perceived errors in stimulus location. Some patients with forelimb amputations reported tactile stimuli on the face as being both on the face and on the missing forelimb. These mislocalizations could be a result of takeover of the denervated forelimb representation by inputs from the face, much like that reported by Pons et al. (1991) after forelimb deafferentation in monkeys, and they indicate that, at least in this case, expanded representations do not lead to better performance. Thus, there must be limits to the extent to which processing circuits can benefit from increasing neuron number. Beyond that limit, additional neurons appear to be a detriment to sensory abilities.

For such profound changes in the topographic organization of the somatosensory cortex to affect the behavior of the individual, the use-driven alterations almost certainly must be relayed through higher-order sensory cortical areas into the motor pathway. Indeed, studies of the motor cortex in which electrical stimulation was used to evoke muscular contractions indicate that motor maps can be significantly altered with repeated stimulation of a cortical zone using electrical currents at or below those necessary to initiate a muscular contraction. The movement that originally is evoked at the stimulation site has an expanded representation after repeated stimulation (reviewed by Merzenich and Jenkins, 1993). This type of adaptive process leads to the aquisition of new skills and recovery from brain injury.

Discussion

The reviewed findings show that both altered sensory experience and peripheral injury result in modifications of the body representation in the somatosensory cortex of primates. Current models of cortical function can account for many of the observed small-scale changes in somatotopic organization, but no models exist to account for the more large-scale reorganizations that occur after long-term sensory deprivation. To successfully explain the types of topographic changes that have been found with large-scale reorganizations, models may need to include somatotopic reorganization at subcortical levels of the somatosensory pathway based predominantly on synaptic weight changes, but also perhaps involving limited new growth of axonal connections.

Acknowledgments. This work was supported by NIH grant NS16446 (JHK) and by the John F. Kennedy Center for Research on Education and Human Development (SLF).

Road Map: Development and Regeneration of Neural Networks
Related Reading: Development and Regeneration of Eye-Brain Maps; Pain Networks

References

Bear, M. F., Cooper, L. N., and Ebner, F. F., 1987, A physiological basis for a theory of synapse modification, *Science*, 237:42–48. ◆

Bienenstock, E. L., Cooper, L. N., and Munro, P. W., 1982, Theory for the development of neuron selectivity: Orientation specificity and binocular interaction in visual cortex, *J. Neurosci.*, 2:32.

Florence, S. L., Garraghty, P. E., Carlson, M., and Kaas, J. H., 1993, Sprouting of peripheral nerve axons in the spinal cord of monkeys, *Brain Res.*, 601:343–348.

Garraghty, P. E., and Kaas, J. H., 1991, Large-scale functional reorganization in adult monkey cortex after peripheral nerve injury, *Proc. Natl. Acad. Sci. USA*, 88:6976–6980.

Garraghty, P. E., Hanes, D. P., Florence, S. L., and Kaas, J. H., 1994, Pattern of peripheral deafferentation predicts reorganizational limits in adult primate somatosensory cortex, *Somatosens. Mot. Res.*, 11: 109–117.

Kaas, J. H., 1991, Plasticity of sensory and motor maps in adult mammals, *Annu. Rev. Neurosci.*, 14:137–167. ◆

Kano, M., Lino, K., and Kano, M., 1991, Functional reorganization of adult cat somatosensory cortex is dependent on NMDA receptors, *NeuroReport*, 2:77–80.

Mendell, L. M., and Lewin, G. R., 1992, Removing constraints on neural sprouting, *Curr. Biol.*, 2:259–261. ◆

Merzenich, M. M., and Jenkins, W. M., 1993, Reorganization of cortical representations of the hand following alterations of skin inputs induced by nerve injury, skin island transfers, and experience, *J. Hand Ther.*, 6:89–104. ◆

Mogliner, A., Grossman, J. A. I., Ribary, R., Joliot, M., Volkmann, J., Rapaport, D., Beasley, R. W., and Llinás, R. R., 1993, Somatosensory cortical plasticity in adult humans revealed by magnetoencephalography, *Proc. Natl. Acad. Sci. USA*, 90:3593–3597.

Pascual-Leone, A., and Torres, F., 1993, Plasticity of sensorimotor cortex representation of the reading finger in Braille readers, *Brain*, 116:39–52.

Pons, T. P., Garraghty, P. E., Ommaya, A. K., Kaas, J. H., Taub, E., and Mishkin, M., 1991, Massive cortical reorganization after sensory deafferentation in adult macaques, *Science*, 252:1857–1860.

Ramachandran, V. S., Stewart, M., and Rogers-Ramachandran, D. C., 1992, Perceptual correlates of massive cortical reorganization, *NeuroReport*, 3:583–587.

Rauschecker, J. P., 1991, Mechanisms of visual plasticity: Hebb synapses, NMDA receptors and beyond, *Physiol. Rev.*, 71:587–615. ◆

Sklar, E., 1991, A simulation of somatosensory cortical map plasticity, *Proceedings of the International Joint Conference on Neural Networks, 1990*, vol. 3, New York: IEEE, pp. 727–732.

Sound Localization and Binaural Processing

Yehuda Albeck

Introduction

In the most general sense, the term *sound localization* means the vivid perception of the location of an object that emits sound. In the more restricted context of psychophysics and neurophysiology, the term denotes the precision with which a subject can point at a sound source and the computational algorithm that the brain uses to complete such tasks. Psychologists can infer the characteristics of the human sound localization system by tuning their models to describe the performance of human subjects. These models emphasize performance: how well people localize sounds and what parameters affect this ability.

Neurobiologists study the underlying neural processing, focusing on neural characteristics and patterns of connectivity such as tuning curves, axonal conduction latencies, and inhibitory or excitatory projections. They use behavioral studies in normal and lesioned animals to study the role and significance of different groups of neurons for sound localization. The two disciplines produce similar, but distinct models.

In this article, I concentrate on one aspect of sound localization: the use of the interaural time difference (ITD) to estimate the azimuthal angle of a sound source. I describe one biological model—the ITD detection in the barn owl's brainstem—and two psychological models—the cross-correlation (CRC) model extended by inhibition and monaural processors and the straightness model. These models have much in common. The biological model is well understood in terms of information processing. The psychological models are defined in realistic biological terms, and both contain elements that are present in the biological system. The binaural time pathway in the cat is also well understood (for reviews and references to more biological models, see Yin and Chan, 1988; Casseday and Covey, 1987).

More specifically, these models have a common theoretical foundation, the CRC model. This model suggests that the brain attempts to match the sounds in the two ears by shifting one sound relative to the other. To estimate the match, the brain averages the multiplication of the instantaneous amplitudes. This operation is similar to computing the CRC function of the sounds in the two ears. The shift that produces the best match is assumed to be the one that just balances the *real ITD*. However, the models differ in the detailed composition of the binaural cross-correlator and the way it is interpreted by higher systems. By comparing the models, I intend to highlight the merits of each approach and advocate the need for a unified theory that can satisfy the requirements of both fields.

Models of binaural interaction are founded on a simple logic. The mathematical description, however, is tedious. Thus, the description is only a qualitative introduction. Detailed mathematical formulation, simulation procedures, and experimental paradigms can be found in the original articles (for review, see Yin and Chan, 1988; Yost and Hafter, 1987; and other chapters in these books).

The Cross-Correlation Model

Figure 1 shows a possible implementation of the CRC model (Jeffress, 1948). The ears decompose the sounds to their spectral components using a bank of bandpass filters. Each output line conveys the waveform in a specific spectral channel. Each spectral channel interacts with the corresponding channel from the other ear. The representation of information in distinct spectral channels is called *tonotopic organization*. Each binaural neuron multiplies the signals from both sides and averages the signal over the recent past. The inputs from the ears to each binaural neuron are identical; only the relative time shift is different. The signal from the left side reaches the leftmost unit before the signal from the right side does. The opposite is true for the neuron in the right border of the processor. Between these two extremes, the signals are shifted relative to each other by many intermediate values.

The model uses three important features. First, the input signal must represent the temporal structure of the sound. This representation is achieved by phase locking the spikes in the auditory nerve to the signal. In other words, the instantaneous likelihood of observing a spike in the auditory nerve is approximately proportional to the amplitude of the half-wave rectified sound at that time. The barn owl's auditory nerve phase locks to frequencies as high as 9 kHz. Second, the signal has to be accurately delayed to each binaural neuron by selecting the length of the axons from each side to provide the exact amount of conduction delay. The relevant range of ITDs for barn owls is ± 170 μs (the size of the head divided by the speed of sound). Finally, neurons in the binaural processor must effectively multiply the signals. The neural system employs coincidence detectors to mimic the multiplication. These neurons respond only

when spikes from the two sides arrive simultaneously, or at least within a fraction of a millisecond. Mathematically, this operation is equivalent to multiplication because the likelihood of a coincidence is the product of the likelihoods of the single independent events (see review in Yin and Chan, 1988). The biophysics of the coincidence detection are not clear (see discussion in Yin and Chan, 1993).

This binaural processor contains a map that *place codes* the azimuth of the target. The neuron for which the axonal interaural delay balances the acoustic interaural delay becomes most active. Its place along the array codes the real ITD. The ITD response of a single cell is usually wider than the behavioral acuity; thus, the brain must use global coding to read the map. The nature of this mechanism is not known.

The Time Pathway in Barn Owls

Barn owls use hearing to locate their prey. When an owl hears a sound, it quickly turns its head to gaze directly at the sound source. The head-turning behavior involves auditory, visual, and motor systems. The auditory system's role is to provide a two-dimensional map of the space. In this map, the location of the target is coded by the firing of a group of neurons. The neurons in the map are organized in a two-dimensional array that resides in the external nucleus of the inferior colliculus. The location of the activity along one axis codes the azimuthal angle of the target. Along the perpendicular axis, the *interaural intensity difference* is represented, which codes the elevation of the target. This map projects to a motor map that controls the owl's neck muscles.

The creation of the ITD map occurs in four computational steps. The brain divides this task among at least eight structures on each side of the brain. A block diagram of the network is shown in Figure 2. The sound is bandpass filtered and transduced to a series of phase-locked spikes in the ears. These series are relayed through the auditory nerve to the nucleus magnocellularis, one of the cochlear nuclei. Each nucleus magnocellularis projects to a nucleus laminaris on both sides. Each nucleus laminaris, in turn, projects to the core of the contralateral inferior colliculus. All of these nuclei are tonotopically organized. The contralateral inferior colliculus projects to the lateral shell of the inferior colliculus on the contralateral side, and the lateral shell projects to the homolateral inferior colliculus. Note that the pathway crosses the midline back and forth three times, but binaural interactions take place only in the nucleus laminaris (see review in Konishi et al., 1988).

Typical ITD response curves of neurons from the nucleus laminaris, lateral shell, and inferior colliculus are plotted in

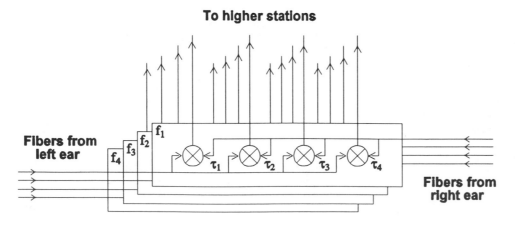

To higher stations

Fibers from left ear

f_1 f_2 f_3 f_4

τ_1 τ_2 τ_3 τ_4

Fibers from right ear

Figure 1. Schematic diagram of the cross-correlation model. The binaural processor has two dimensions, frequency and interaural time difference. In this example, each of the four spectral channels, f_1–f_4, has four coincidence detectors. The signals from each ear pass along delay lines on the way along the nucleus toward the contralateral side. Each detector provides an estimate of the cross-correlation of the signals at a specific band f_i for a specific time shift τ_i.

Figure 2. Block diagram of the barn owl's binaural time pathway. The nucleus magnocellularis (NM) relays the auditory nerve (AN) spikes to the nucleus laminaris (NL) on both sides. The NL projects to the core of the inferior colliculus (IC) on the contralateral side. The IC innervates the lateral shell (LS) of the IC, and the LS in turn projects to the inferior colliculus (ICx). Other nuclei, such as the superior olive (SO) and ventral lateral anterior lemniscus (VLVa), also participate, but are not discussed here. Tonotopic organized nuclei are shown in a cascade.

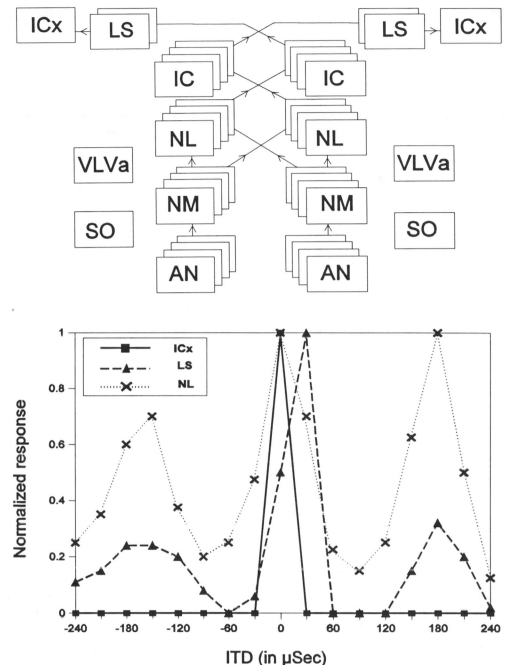

Figure 3. Normalized interaural time difference (ITD) response curves from the barn owl's brainstem. The nucleus laminaris (NL) response is phase ambiguous and has a high baseline. In the lateral shell (LS), the secondary peaks are suppressed. In the external nucleus of the inferior colliculus (ICx), only a narrow range of ITD evokes some response.

Figure 3. Recordings in the nucleus laminaris confirm the predictions of the CRC model. The nucleus contains a systematic map of frequencies and ITDs (Carr and Konishi, 1990). In the nucleus laminaris and the contralateral inferior colliculus, the curve is periodic; this phenomenon is called *phase ambiguity*. A typical ITD response in the lateral shell shows a central dominant peak and smaller secondary peaks. In many inferior colliculus neurons, the secondary peaks disappear. This type of neuron is called *space specific*. Note also that the width of the ITD response decreases from approximately 70 μs in the nucleus laminaris to less than 30 μs in the inferior colliculus.

How does the phase-ambiguous curve transform into a space-specific response? It takes a combination of two processes, which are illustrated in Figure 4. The tonotopically or-

ganized nucleus laminaris is depicted as a matrix of binaural neurons. Each neuron has its own characteristic frequency and ITD. All neurons that have the same ITD, from all frequency channels, converge in the lateral shell. The convergence creates a composed ITD response curve, with partial suppression of secondary peaks. The real ITD appears in the same location across all of the spectral channels. The secondary peaks are separated from the primary peak by one period of the characteristic frequency; therefore, they have different locations for different frequencies. Thus, the contribution of the secondary peaks is spread out and cannot create a sharp, distinct peak in the composed ITD curve. Consequently, space-specific neurons cannot localize pure-tone stimuli (Takahashi and Konishi, 1986).

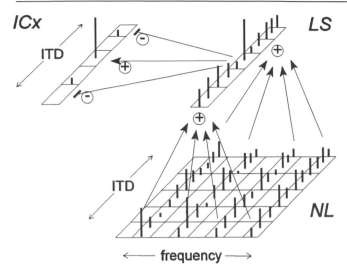

Figure 4. The nucleus laminaris (NL) can be viewed as a matrix of neurons. Each row in this matrix represents a group of neurons with the same characteristic frequency and different ITDs. Columns represent neurons with the same ITD and different characteristic frequency. All the neurons in a specific row excite a single neuron in the lateral shell (LS). The excitation is denoted by an arrow and a plus sign. This projection transforms the NL matrix into a one-dimensional array of ITD-sensitive neurons and destroys the frequency tuning. LS neurons supply excitation to inferior colliculus (ICx) neurons with the same ITD and inhibition to other ICx neurons with ITDs larger or smaller than their own. The inhibition is denoted by an arrow with a minus sign. These ICx neurons show a sharp tuning to the sound location and a broad tuning to frequency.

The second step involves inhibition (Fujita and Konishi, 1991). Neurons in the lateral shell project to inferior colliculus neurons with the same ITD. Simultaneously, they inhibit other inferior colliculus neurons with different ITDs. The inhibition, along with a high threshold, suppresses the activity in all neurons except those that receive a projection from the central ITD peak.

Psychophysical Models

The CRC model provides a basic mechanism that explains how a neural network can extract the ITD of a binaural signal. Early theoretical approaches used this mechanism to calculate the theoretical limits on the network's performance derived from the amount of noise in the system (see review in Colburn and Durlach, 1978).

More recent studies focused on how the binaural processor is used to estimate sound locations and enhance detection in a noisy environment (see review in Stern and Trahiotis, in press). The basic CRC mechanism explains the lateralization experiments. *Lateralization* refers to the perception of an image inside the head which occurs in earphone experiments. *Localization* refers to external perception. The model does not explain how and why the interaural intensity difference can affect lateralization (see, e.g., Yost and Hafter, 1987). This effect is called *time-intensity trading*. The interaural intensity difference cannot change the CRC predictions because the signals from both sides are simply multiplied. Lindemann (1986) suggested that the signal traveling along the binaural processor has two components. The sound-evoked signal attenuates as it travels along the delay line, and it is replaced by a second independent signal. If the sound in the right ear is faint, the binaural units on the right of the processor will multiply it by the left signal

and produce a small response. On the other hand, the units on the left multiply the strong signal from the left ear by the independent signal and produce a higher response. The peak of the response is then shifted leftward, and a time-intensity trade results. In addition, Lindemann suggested that each unit in the binaural processor inhibits the input to the other units (for more details, including the distinction between static and dynamic inhibition, see Lindemann, 1986). This inhibition sharpens the ITD peaks and enhances the stronger peak when another peak appears.

These extensions to the CRC model elegantly explain the time-intensity interaction. The model does not explain the interaction between different spectral channels. A pure 500-Hz tone with an ITD of −1.5 ms sounds as if it has an ITD of +0.5 ms. Both solutions are identical for a pure CRC model, but human subjects always perceive the ITD which is smaller in absolute value. As the spectrum of the signal broadens, the perception crosses the midline and moves toward the location that corresponds to an ITD of −1.5 ms. To explain this finding and similar experiments, Stern, Zeiberg, and Trahiotis (1988) stated the *straightness principle*. They suggested that an intermediate mechanism assigns greater weight to units that are active in coincidence with units of different spectral channels, but of the same ITD. The weighted image is further processed to lateralize the sound. Neither model provides a perfect fit for every psychophysical experiment (see the review on lateralization in Yost and Hafter, 1987). Each model captures some aspects and probably reflects some characteristics of the biological system.

Discussion

It is instructive to compare the biological and psychophysical approaches. The most striking characteristic of the biological system is its modular structure. Each nucleus performs a limited task. For example, the nucleus laminaris seems to pay little attention to the interaural intensity difference. This feature is taken care of by another nucleus. Also, there is no evidence of monaural processing in the nucleus laminaris. A special pathway is devoted to the preservation of monaural characteristics. The separate pathways converge only in the inferior colliculus. Similar mechanisms exist in other animals (Casseday and Covey, 1987).

The models differ in a more subtle way in across-frequency interaction. The owl employs a two-steps process, involving excitation and inhibition over the coincidence mechanism suggested in the straightness model. From a computational point of view, the owl's process is a realization of the straightness principle. Lateral inhibition in the inferior colliculus is probable, but in a different context than that suggested by Lindemann.

An important distinction between the models is the purpose of the network. The owl's network is responsible for head turning toward a sound source. Psychophysical models deal with perception, which may involve different brain structures (see DISSOCIATIONS BETWEEN VISUAL PROCESSING MODES for a distinction between *what* and *where* systems in vision). I suggest that models of perception should integrate several modalities and abandon the effort to construct a single, super-binaural processor that accounts for all of the psychophysical data in a single step. A more realistic model should have two separate modules for the processing of ITD and interaural intensity difference as well as a third decision mechanism that weighs both. The decision mechanism should be able to change its strategy based on monaural information, such as spectral composition and envelope structure.

The CRC model is sufficiently detailed to allow a successful silicon implementation of sound localization (Lazzaro and Mead, 1989) that captures the essential features of the biological system. On the other hand, the details of the ITD detection in the cellular level are still not known. Intracellular recording in the nucleus laminaris and in analog mammalian structures is difficult; therefore, little is known about its biophysics and pharmacology.

Other biological systems possess some similarity to the owl's time pathway. The jamming-avoiding behavior of the weak electric fish and the echolocation of the bat are well documented (see review in Carr, 1993). However, these models provide little information about the internal representation of the external environment in the brain. Usually, more information is extracted by studying neural pathways that lead to a motor response, such as head turning in owls or frequency adjustments in electric fish and bats. A more holistic theory of sound localization should unify the biological and psychological models. Such a theory may explain psychophysical phenomena in terms of realistic biological structures and provide testable predictions for experimentalists in both fields.

Acknowledgment. This research was supported by a fellowship from the International Human Frontier Science Program Organization.

Road Map: Other Sensory Systems
Related Reading: Collicular Visuomotor Transformations for Saccades; Echolocation: Creating Computational Maps; Electrolocation

References

Carr, C. E., 1993, Processing of temporal information in the brain, *Annu. Rev. Neurosci.*, 16:223–243. ◆
Carr, C. E., and Konishi, M., 1990, A circuit for detection of interaural time differences in the brain stem of the barn owl, *J. Neurosci.*, 10:3227–3246.
Casseday, J. H., and Covey, E., 1987, Central auditory pathways in directional hearing, in *Directional Hearing* (W. A. Yost and G. Gourevitch, Eds.), New York: Springer-Verlag, pp. 109–145. ◆
Colburn, H. S., and Durlach, N. I., 1978, Models of binaural interaction, in *Handbook of Perception*, vol. 4 (E. C. Carterette and M. P. Friedman, Eds.), New York: Academic Press, pp. 467–518. ◆
Fujita, I., and Konishi, M., 1991, The role of GABAergic inhibition in processing of interaural time difference in the owl's auditory system, *J. Neurosci.*, 11:722–739.
Jeffress, L. A., 1948, A place theory of sound localization, *J. Comp. Physiol. Psychol.*, 41:35–39.
Konishi, M., Takahashi, T. T., Wagner, H., and Carr, C. E., 1988, Neurophysiological and anatomical substrates of sound localization in the owl, in *Auditory Function* (G. M. Edelman, W. E. Gall, and W. M. Cowan, Eds.), New York: Wiley, pp. 721–745. ◆
Lazzaro, J., and Mead, C., 1989, A silicon model of auditory localization, *Neural Computat.*, 1:47–57.
Lindemann, W., 1986, Extension of a binaural cross-correlation model by contralateral inhibition, I: Simulation of lateralization for stationary signals, *J. Acoust. Soc. Am.*, 80:1608–1622.
Stern, R. M., Zeiberg, A. S., and Trahiotis, C., 1988, Lateralization of complex binaural stimuli: A weighted-image model, *J. Acoust. Soc. Am.*, 84:156–165.
Stern, R. M., and Trahiotis, C., in press, Models of binaural interaction, in *Handbook of Perception and Cognition*, vol. 6 (B. C. J. Moore, Ed.), San Diego: Academic Press.
Takahashi, T. T., and Konishi, M., 1986, Selectivity for interaural time difference in the owl's midbrain, *J. Neurosci.*, 6:3413–3422.
Yin, T. C. T., and Chan, J. C. K., 1988, Neural mechanism underlying interaural time sensitivity to tones and noise, in *Auditory Function* (G. M. Edelman, W. E. Gall, and W. M. Cowan, Eds.), New York: Wiley, pp. 385–430. ◆
Yin, T. C. T., and Chan, J. C. K., 1993, Interaural time sensitivity in medial superior olive of cat, *J. Neurophysiol.*, 64:465–488.
Yost, W. E., and Hafter, E. R., 1987, Lateralization, in *Directional Hearing* (W. A. Yost and G. Gourevitch, Eds.), New York: Springer-Verlag, pp. 49–84. ◆

Sparse Coding in the Primate Cortex

Peter Földiák and Malcolm P. Young

Introduction

A central goal in the study of cortical function is to understand how states of the environment are represented by firing patterns of cortical neurons. Electrophysiological recordings from single cells have revealed a remarkably close relationship among stimuli, neural activity, and perceptual states. The nature of this relationship and interpretations of experimental results are fiercely debated. Is sensory information represented by the activity of single, individually meaningful cells, or is it only the global activity pattern across a whole cell population that corresponds to interpretable states? There are now strong theoretical reasons and experimental evidence suggesting that the brain adopts a compromise between these extremes which is often referred to as *sparse coding*.

Sparse Coding

The brain must encode the state of the environment by the firing pattern of a large but fixed set of neurons. Consider coding by units that are either "active" or "passive" where the code assigns states to subsets of active units. An important characteristic of such a code is the *activity ratio*, the fraction of active neurons at any one time. At its lowest value is *local representation*, where each state is represented by a single active unit from a pool in which all other units are silent, e.g., letters on a typewriter keyboard are locally encoded. In *dense distributed representations*, each state is represented on average by about half of the units being active. Examples of this are the binary (ASCII) encoding of characters used in computers or the coding of visual images by the retinal photoreceptor array. Codes with low activity ratios are called *sparse codes*.

The activity ratio affects several aspects of information processing such as the architecture and robustness of networks, the number of distinct states that can be represented and stored, generalization properties, and the speed and rules of learning (Table 1).

The representational capacity of local codes is small: they can represent only as many states as the number of units in the pool, which is insufficient for any but the most trivial tasks. Even when the number of units is as high as that in the primate cortex, the number of discriminable states well exceeds this number. Making associations between a locally encoded item and an output, however, is easy and fast. Single-layer networks

Table 1. Properties of Coding Schemes

	Representational Capacity	Memory Capacity	Speed of Learning	Generalization	Interference	Fault Tolerance	Simultaneous Items
Local	Very low	Limited	Very fast	None	None	None	Unlimited
Sparse	High	High	Fast	Good	Controlled	High	Several
Dense	Very high	Low	Slow	Good	Strong	High	One

can learn any output association in a single trial by local, Hebbian strengthening of connections between active representation and output units, and the linear separability problem does not arise. In such a *lookup table*, there is no interference between associations to other discriminable states, and learning information about new states does not interfere with old associations. This, however, also means that there will be no generalization to other discriminable states—which is a fundamental flaw, as we can expect a system never to experience precisely the same pattern of stimulation twice.

Dense distributed, or "holographic," codes, on the other hand, can represent a very high number ($\sim 2^N$) of different states by combinatorial use of units. In fact, this power is largely superfluous, as the number of patterns ever experienced by the system will never approach this capacity, and therefore dense codes usually have high statistical redundancy. The price to pay for the potential (but unused) high information content of each pattern is that the number of such patterns that an associative memory can store is unnecessarily low. The mapping between a dense representation and an output can be complex (a linearly nonseparable function), therefore requiring multilayer networks and learning algorithms that are hard to implement biologically. Even efficient supervised algorithms are prohibitively slow, requiring many training trials and large amounts of the kind of training data that is labeled with either an appropriate output or reinforcement. Such data is often too risky, time consuming, or expensive to obtain. Distributed representations in intermediate layers of such networks ensure a kind of automatic generalization (Hinton, McClelland, and Rumelhart, 1986), however, this often manifests itself as unwanted interference between patterns. A further serious problem is that no new associations can be added without retraining the network with the complete training set.

Sparse codes combine advantages of local and dense codes while avoiding most of their drawbacks. Codes with small activity ratios can still have sufficiently high representational capacity, while the number of input-output pairs that can be stored in an associative memory is far greater for sparse than for dense patterns (see SPARSELY CODED NEURAL NETWORKS). This is achieved by decreasing the amount of information in the representation of any individual stored pattern. As a much larger fraction of all input-output functions are linearly separable using sparse coding, a single supervised layer with simple learning rules, following perhaps several unsupervised layers, is more likely to be sufficient for learning target outputs, avoiding problems associated with supervised training in multilayer networks. As generalization takes place only between overlapping patterns, new associations will not interfere with previous associations to nonoverlapping patterns.

Distributed representations are tolerant to damage. However, redundancy far smaller than that in dense codes is sufficient to produce robust behavior. By simply duplicating units with 99% reliability (assuming independent failures), reliability increases to 99.99%. Sparse representations can be even more tolerant to damage than dense ones if high accuracy is required or if the units are highly unreliable. Arguments supporting dense coding may also be challenged by recent studies on

Alzheimer's disease (Hodges, Salmon, and Butters, 1992) suggesting that patients may irreversibly lose specificity, and even whole concepts, independently for individual objects.

Sparseness can also be defined with respect to components. A scene may be encoded in a distributed representation while, at the same time, object features may be represented locally. The number of simultaneously presented items decreases as activity ratio increases because the addition of active units eventually results in activation of "ghost" subsets, corresponding to items that were not intended to be activated.

To utilize the favorable properties of sparse representations, densely coded inputs must be transformed into sparse form. As the representational capacity of sparse codes is smaller, this cannot be achieved perfectly for all possible patterns on the same number of units. Information loss can be minimized by increasing the number of representation units or by losing resolution—but only in parts of pattern space that are usually not used. Both measures seem to be taken in the cortex. First, the number of neurons in the primary visual cortex is about two orders of magnitude higher than the number of optic nerve fibers that indirectly provides its input. Second, the natural sensory environment consists of patterns that occupy only a small fraction of pattern space; that is, it has large statistical redundancy. Barlow (1972) suggested that it is the nonaccidental conjunctions, "suspicious coincidences" of features, or "sensory clichés" that must be extracted that give good discrimination in populated regions of pattern space. By making explicit representations for commonly occurring features of the natural environment, such as facial features, our visual system is much better at discriminating natural images than, for instance, random dot patterns. As events are linked to the causes of sensory stimuli in the environment, such as objects, rather than arbitrary combinations of receptor signals, associations can be made more efficiently, based on such direct representations (Barlow, 1991).

An unsupervised algorithm for learning such representations in a nonlinear network using local learning rules has been proposed (Földiák, 1990) which uses Hebbian forward connections to detect nonaccidental features, an adaptive threshold to keep the activity ratio low, and anti-Hebbian, decorrelating lateral connections to keep redundancy low. Simulations suggest that these three constraints force the network to implement a sparse code with only little information loss. Another interesting effect can be observed: high probability ("known") patterns are represented on fewer units while new or low probability patterns get encoded by combinations of larger numbers of features. This algorithm is not limited to detecting only second-order correlations, so it seems suitable for multilayer applications. Its theoretical treatment, however, is difficult as it does not explicitly minimize well-defined objective functions.

Sparse Coding in the Cortex

It is easy to measure sparseness in network models, where the responses of all units can be observed. An idealized "wavelet" filter model of simple cell responses in primary visual cortex has shown that wavelet coefficients of natural images show high kurtosis; that is, for natural images, most wavelet units have outputs near zero and only a small subset of units gives large responses, there being a different subset for each image (Field, 1994; see VISUAL CODING, REDUNDANCY, AND "FEATURE DETECTION"). Evaluating the sparseness of coding in brains, however, is difficult: it is hard to record a set of neurons simultaneously across which sparseness could be measured. New techniques, such as optical recording and multiple electrode recording, may eventually yield data on the density of coding, but there are presently formidable technical difficulties to overcome. We have more information about neurons' breadth of tuning across various stimulus sets than about sparseness per se. Coding across stimuli and across cells are, however, closely related. What evidence is there for sparse coding from single unit recordings in sensory cortex?

The most immediate observation during physiological experiments is the difficulty of finding effective stimuli for neurons in most cortical areas. Each neuron appears to have specific response properties, typically being tuned to several stimulus parameters. In primary visual cortex, many neurons only respond strongly when an elongated stimulus, such as a line, edge, or grating, is presented within a small part of the visual field, and then only if other parameters, including orientation, spatial frequency (width), stereoscopic disparity, and perhaps color or length fall within a fairly narrow range. This suggests that at any moment during the animal's life, only a small fraction of these neurons will be strongly activated by natural stimuli. The problem of finding the preferences of cells becomes severe in higher visual areas, such as area V4, and especially in inferotemporal cortex (IT). Cells' preferences in IT are often difficult to account for by reference to simple stimulus features, such as orientation, motion, position, or color, and they appear to lie in the domain of shape (Gross, Rocha-Miranda, and Bender, 1972; Tanaka et al., 1991). Cells here show selectivity for complex visual patterns and objects, such as faces, hands, complex geometrical shapes, and fractal patterns, and the responses are usually not predictable from responses to simple stimuli. Cells responding to faces but not to a large collection of control stimuli could be considered, on the one hand, to be very tightly tuned cells in the space of all possible stimuli. On the other hand, they may have quite broad tuning and show graded responses to stimuli within the specific categories for which they show selectivity. To estimate, therefore, how often these cells are activated in behaving animals would require much more accurate knowledge of the animals' visual environment and their behavior, or access to the cell's response during natural behavior.

Cells with apparent selectivity for faces might be selective for the full configural and textural information present in a preferred face stimulus, or be triggered simply by the presence of two roughly collinear bars (most faces have eyebrows), or a colored ovoid. Two approaches have been taken to explore IT cells' preferences. One, which has been widely employed (Gross et al., 1972) but which has recently been applied as systematically as possible by Tanaka and his colleagues (Fujita et al., 1992; Tanaka et al., 1991), has been to try to determine preferred features of cells by simplifying the stimuli that excite them. This method begins by presenting many objects to the monkey while recording from a neuron to find objects that excite it. Component features of effective stimuli, as judged by the experimenters, are then presented singly or in combination. By assessing the cell's firing rate during presentation of each simplified stimulus, the protocol attempts to find the simplest feature combination that maximally excites the cell. This approach suffers from the problem that even simple objects contain a rich combination of color, orientation, depth, curvature, texture, specular reflections, shading, shape, and other features that may not be obvious to the experimenters. As any feature combination may be close enough to the preferences of a cell for it to become excited, the simplified stimuli that are actually presented are only a small subset of all possible combinations, selected according to the experimenter's intuitions. Hence, it is not possible to conclude that the best simplified stimulus found using this method is optimal for the cell, only that it was the best of those presented. It cannot even be assumed that cells code only one optimal set of features, since it is possible that they could exhibit two or more maxima, corresponding to quite different feature combinations (Young, 1993).

According to the results of this method, IT cells show preferences for patterns that are simpler than real visual objects. One interpretation of these results is that IT might consist of a large number of detectors of pattern "partials," which together might constitute an "alphabet" (Stryker, 1992). The detection of such partials would seem to suggest that these cells will have broader tuning than cells with selectivity for the full configuration. The idea that an IT cell reliably signals the presence of the particular pattern "partial" seems not to be supported by results of Tanaka et al. (1991), who showed that the presence of other visual features can disrupt the cell's response to its "partial," a result which is inconsistent with the visual alphabet concept. Hence, the simplification approach captures neither necessary nor sufficient descriptions of the behavior of IT cells, and does not yet present a clear message on the sparseness of representation.

A second systematic approach to the issue of how IT cells participate in recognition has been to quantify cells' responses to a stimulus set by making a numerical model. This model can then be compared by regressionlike methods to other numerical models that capture various ways in which the stimuli differ, to find which stimulus dimensions are good predictors of physiological responses. This statistical modeling approach has been applied to both single cell responses (Yamane, Kaji, and Kawano, 1988), and responses of populations of cells (Young and Yamane, 1992, 1993) in IT cortex. In essence, this approach involves examining internal relations between a set of stimuli and the responses to them, to determine whether the cells' activities are sufficient to carry information about the stimuli.

In an analysis of population processing in IT along these lines (Young and Yamane, 1992), responses to 27 faces were examined. A quantitative numerical model of population responses to the faces was made by applying multidimensional scaling (MDS). This analysis produced a two-dimensional configuration that accounted for 70% of the variance in the data, suggesting that coding was redundant, at least for this set of 27 stimuli. The stimuli in this study had been extensively quantified previously (Yamane et al., 1988). This was exploited to find whether identifiable information was carried at the level of population responses, as would be expected if either sparse or dense population coding underlay the responses. These cells' responses were more similar the more similar the faces were, where similarity was computed based on the physical face measurements. In addition, measurement variables encoding relations between the eyes and the hairline were significantly related to the population model. No other models were significantly related. Hence, these analyses suggest that the IT population may have been coding general physical properties of the faces, with a particular emphasis on the face's upper part (Young and Yamane, 1992, 1993). These results suggest that neurons responsive to faces in IT share dimensions of specificity and that these shared dimensions correspond to physical properties of faces.

It was also possible to show that these cells formed an interpretable population code that could sometimes identify individual faces using the population vector technique (Georgopoulos et al., 1982; see REACHING: CODING IN MOTOR CORTEX). The direction of population vectors in this space derived by MDS was, in general, close to the stimulus vectors' direction, despite the fact that responses from only about 40 cells entered the analysis. This was interpreted as evidence for a sparse population code (Young and Yamane, 1992, 1993). These results suggest that cells' activity in IT is best predicted by quite complex combinations of facial cues rather than by simple stimulus features, and that neurons participate in recognition by signaling in the form of a sparse code. This approach, however, itself suffers from limitations. It may only be applied to stimuli, like faces, for which there is good information on physical and psychological differences, and where the physical similarity space can be estimated. Information about dimensions in which most objects that primates can recognize differ is lacking, and so this approach cannot at present bear on the issue of the representation's sparseness in the general case.

An alternative interpretation of responses of arbitrary neurons may, however, be applicable in the more general case (Földiák, 1993). By recording a neuron's responses to a set of stimuli, we are sampling the conditional probability distribution of responses given the particular stimulus being presented, $P(\text{response}|\text{stimulus})$. By repeated presentation, this conditional probability distribution can be estimated, and Bayes's rule can be applied to calculate the posterior distribution:

$$P(\text{stimulus}|\text{response})$$

$$= P(\text{response}|\text{stimulus})P(\text{stimulus})/P(\text{response})$$

This probability distribution of the stimuli given a particular response constitutes an interpretation of the neural response, a "reading of the neural code," allowing an optimal guess of the stimulus by an ideal statistical observer of the neurons (the "ideal homunculus"). Such a hypothetical observer does not tell us how the response is actually used in the brain (neither do other approaches), but it does provide a theoretical upper bound on performance and a baseline for the comparison of any neural mechanism's efficiency. This approach can also be used to analyze responses of a set of cells. The Bayesian combination of cell responses extracts more information than the population vector approach, as has been demonstrated in an analysis of responses of neurons in cat primary visual cortex (Földiák, 1993). It has been shown that the identity of the stimulus from a fixed set can be reliably estimated based on responses of only a small number of cells (approximately eight), a number far smaller than that required by the population vector method applied to the same data. Furthermore, the Bayesian approach is also applicable in situations where there is no obvious space in which to define population vectors.

Finally, we note a difficulty for all attempts to measure sparseness in the cortex. In the extreme case, a cell with tuning so precise that it responds only to a single object will sustain its firing near its background rate when shown anything else. Researchers have only limited time and stimuli available to explore the cell's preferences during an experiment, and invariably go on to the next unit if they cannot determine what it is that the cell prefers, which strongly biases estimates of the specificity distribution. So whether there are any cells with extremely high specificity (approaching the specificity of "grandmother cells"), we cannot expect to find them experimentally using current methods. On the other hand, a cell that appears to respond only to a very limited number of a set of stimuli, as for example some human medial temporal lobe cells shown in

Heit, Smith, and Halgren (1988) and the very tightly tuned cell from monkey temporal cortex shown in Young and Yamane (1993), cannot be interpreted as conclusive evidence for extremely narrow tuning because of uncertainty about their responses to untested stimuli.

Discussion

The theoretical reasons and experimental evidence discussed here support the hypothesis that sparse coding is used in cortical computations, while the degree of sparseness is still a subject for future research. The full description of high-level cells will require far more detailed knowledge of their anatomical connectivity and better understanding of the lower-level sensory neurons out of which their responses are constructed.

Road Map: Vision
Background: I.3. Dynamics and Adaptation in Neural Networks
Related Reading: Bayesian Methods for Supervised Neural Networks; Binding in the Visual System; Connectionist and Symbolic Representations; Localized Versus Distributed Representations

References

Barlow, H. B., 1972, Single units and sensation, *Perception*, 1:371–394. ◆

Barlow, H. B., 1991, Vision tells you more than "what is where," in *Representations of Vision* (A. Gorea, Ed.), Cambridge, Eng.: Cambridge University Press, pp. 319–329.

Field, D., 1994, What is the goal of sensory coding? *Neural Computat.*, 5:559–601.

Földiák, P., 1990, Forming sparse representations by local anti-Hebbian learning, *Biol. Cybern.*, 64:165–170.

Földiák, P., 1993, The "Ideal Homunculus": Statistical inference from neural population responses, in *Computation and Neural Systems* (F. Eeckman and J. Bower, Eds.), Norwell, MA: Kluwer, pp. 55–60.

Fujita, I., Tanaka, K., Ito, M., and Cheng, K., 1992, Columns for visual features of objects in monkey inferotemporal cortex, *Nature*, 360:343–346.

Georgopoulos, A., Kalaska, J., Caminiti, R., and Massey, J., 1982, On the relations between the direction of two-dimensional arm movements and cell discharge in primate motor cortex, *J. Neurosci.*, 2:1527–1537.

Gross, C. G., Rocha-Miranda, C., and Bender, D., 1972, Visual properties of neurons in the inferotemporal cortex of the macaque, *J. Neurophysiol.*, 35:96–111.

Heit, G., Smith, M., and Halgren E., 1988, Neural encoding of individual words and faces by the human hippocampus and amygdala, *Nature*, 333:773–775.

Hinton, G., McClelland, J., and Rumelhart, D., 1986, Distributed representations, in *Parallel Distributed Processing: Explorations in the Microstructure of Cognition* (D. Rumelhart, J. McClelland, and PDP Research Group, Eds.), vol. 1, *Foundations*, Cambridge, MA: MIT Press, pp. 77–109. ◆

Hodges, J., Salmon, D., and Butters, N., 1992, Semantic memory impairments in Alzheimer's disease, *Neuropsychologia*, 30:301–314.

Stryker, M., 1992, Elements of visual perception, *Nature*, 360:301–302.

Tanaka, K., Saito, H., Fukada, Y., and Moriya, M., 1991, Coding visual images of objects in the inferotemporal cortex of the macaque monkey, *J. Neurosci.*, 6:134–144. ◆

Yamane, S., Kaji, S., and Kawano, K., 1988, What facial features activate face neurons in the inferotemporal cortex of the monkey? *Exp. Brain Res.*, 73:209–214.

Young, M. P., 1993, Visual cortex: Modules for pattern recognition, *Curr. Biol.*, 3:44–46.

Young, M. P., and Yamane, S., 1992, Sparse population coding of faces in the inferotemporal cortex, *Science*, 256:1327–1331.

Young, M. P., and Yamane, S., 1993, An analysis at the population level of the processing of faces in the inferotemporal cortex, in *Brain Mechanisms of Perception and Memory* (T. Ono, L. Squire, D. Perrett, and M. Fukuda, Eds.), New York: Oxford University Press, pp. 47–70.

Sparsely Coded Neural Networks

C. Meunier and J.-P. Nadal

What Is Sparse Coding?

Hopfield's seminal paper on associative memories (Hopfield, 1982) aroused a wave of interest in networks of formal neurons in the community of physicists. Hopfield considered a fully connected network of binary units with real-valued connection weights that become active when the input from the rest of the network exceeds a given threshold. Though retaining only the most basic aspects of biological neurons, this network revived a paradigm of distributed memory with good robustness properties, making it amenable to exact analytical studies (Amit, Gutfreund, and Sompolinsky, 1987a; see also STATISTICAL MECHANICS OF NEURAL NETWORKS), thanks to a fruitful analogy with disordered magnetic systems (spin glasses). The storage of specific configurations, called patterns, relied on a Hebbian learning rule where the weight of the connection between any two neurons is increased or decreased depending on whether these neurons are simultaneously active or not (see HEBBIAN SYNAPTIC PLASTICITY: COMPARATIVE AND DEVELOPMENTAL ASPECTS).

In the Hopfield model, the patterns were chosen randomly with half of the neurons active in the average. Such a situation is referred to as *dense coding*. However, it soon proved interesting to investigate the opposite situation of *sparse coding*, where only a small fraction of neurons are active in each pattern, for the following reasons:

- The maximum number of patterns that can be stored in a network as fixed points of its dynamics increases with the degree of sparseness. Moreover, the optimal results (estimated by E. Gardner for an arbitrary network) are almost obtained for the so-called Willshaw model (Willshaw, Buneman, and Longuet-Higgins, 1969) where connections are binary, a rather striking fact that needed to be understood.
- Sparse coding is more realistic from a biological viewpoint, as only a small fraction of neurons are active at a given time in the central nervous system (CNS). It has been advocated, for instance, that the hippocampus might behave in some respects as a sparsely encoded associative memory (Rolls, 1989). However, in simple sparsely coded networks, the active neurons fire at maximum frequency, in contradiction with the biological situation where firing rates are generally low. This problem of low firing rates has recently induced a new line of work unrelated to sparse coding.

Two types of architecture have been investigated for sparsely coded networks:

- Autoassociative memories where the neurons are densely interconnected (diluted situations have also been studied); the dynamics then converges toward attractor states.
- Heteroassociative memories where the network consists of several layers linked by feedforward connections; the network then implements an input-output relationship without any real dynamics. For such layered networks, sparse coding corresponds to the situation where both input and ouput patterns are sparse.

All studies, to our knowledge, rely on simple two-state neurons. The learning rules are generally Hebbian, with modifications introduced to take into account the correlations between patterns and to constrain the dynamics of the network to states of low activity. Versions where connections have discrete or bounded efficacies have also been considered.

We should also make precise the very definition of sparse coding. As a rule, situations where the level of activity f (i.e., the fraction of active neurons) is low are generally referred to as *biased networks*. The term *sparse coding* is preferred in the limit of vanishing activity, when the activity f goes to 0 at least as $\ln(N)/N$ when the size N of the network goes to infinity. Most studies have been devoted to the case of a $\ln(N)/N$ scaling. However, more general situations where the activity decays as a power law have also been considered (Meunier, Yanai, and Amari, 1991).

Does Sparse Coding Really Enhance the Storage Capacity?

The Pattern Capacity

One of the limitations of the original Hopfield network is the relatively low number of patterns that can be stored: the ratio α of the number p of patterns to the number N of neurons of the network cannot exceed the critical value $\alpha_c \simeq 0.14$ (Amit et al., 1987a; Hopfield, 1982). This estimate holds for fully random unbiased patterns. Note that it is conventional to use binary ($S_i = 0, 1$) or spinlike ($S_i = \pm 1$) variables to describe the state of activity of neurons. However, other representations of neurons have also been used: shifting from one representation to the other is equivalent to redefining the coupling constants and threshold, and the learning rule accordingly.

When generalizing the original Hopfield model to biased patterns, one finds that the critical capacity strongly increases with the bias (Amit et al., 1987b). Therefore, one expects a strong enhancement of the storage capacity for sparsely encoded associative memories (that is, f going to zero). E. Gardner derived optimal capacities for both dense and sparse codings (Gardner, 1988). Her viewpoint relied on a statistical physics approach in the space of coupling constants: the optimal capacity corresponds to the vanishing of the volumic fraction of couplings that stabilize a given set of patterns. For dense coding ($f = \frac{1}{2}$), she found that $\alpha_c = 2$, to be compared to the value $\alpha_c = 0.14$ obtained for the regular Hopfield model. This brings to light the poor performance obtained with Hebbian learning for dense coding. For sparse coding, her calculation yields

$$\alpha_c = -\frac{1}{2f \ln(f)} \tag{1}$$

where $f \sim \ln(N)/N$ is the mean activity of patterns. Remarkably, as opposed to dense coding, this optimal capacity can be reached or approached with Hebbian learning rules (Dayan and Willshaw, 1991; Horner, 1989; Nadal and Toulouse, 1990; Palm and Sommer, 1992; Perez-Vicente and Amit, 1989; Tsodyks and Feigelman, 1988), in sharp contrast with the case of dense coding (see above). Particularly striking is the fact that the optimal capacity is almost reached [$\alpha_c = -\ln(2)^2/(f \ln(f))$] by the original Willshaw model (Willshaw et al., 1969; see also below). The Willshaw model uses a clipped Hebb rule where connections are 0 or 1. A connection is set to 1 whenever the two neurons it links are simultaneously active in at least one pattern. All other models consider autoassociative memories with an unclipped learning rule of the form

$$J_{i,j} = \sum_{\mu} (\xi_i^{\mu} - a)(\xi_j^{\mu} - a) - b \qquad (2)$$

where $J_{i,j}$ is the strength of the connection between neurons i and j, ξ_i^{μ} denotes the state 0 or 1 of neuron number i in the pattern ξ^{μ}, and the summation is over the whole set of patterns. This generalizes the Hopfield model for which $f = a = \frac{1}{2}$, and $b = 0$. Many authors have calculated a maximum capacity by optimizing over the parameters a, b and the activation threshold of neurons. Some authors have adopted a distinct but equivalent viewpoint where optimization is performed over the representation of neurons (Perez-Vicente and Amit, 1989; Horner, 1989; Horner et al., 1989; Dayan and Willshaw, 1991). All these models yield a capacity equal to the optimal capacity $\alpha_c = -1/(2f\ln(f))$ up to logarithmic corrections. The speed of convergence to the limit when f goes to 0 has also been investigated by H. Horner et al., who found a very slow convergence rate to the asymptotics of α_c, given by

$$-\frac{1}{2f\ln(f)}\left\{1 + \sqrt{\frac{\ln(\ln(|f|))}{\ln(|f|)}}\right\} \qquad (3)$$

The Information Capacity

The above estimates show that the critical capacity diverges as f goes to 0. However, this result should be qualified by the fact that the amount of information contained in any pattern goes to 0 in this limit. Therefore, a better criterion of the performance of the network is the maximum amount of information that can be stored. This leads to the notion of *information capacity* (per connection) (Palm and Sommer, 1992; Willshaw et al., 1969; Gardner, 1988; Nadal and Toulouse, 1990). For perfect storage it reads

$$I_c = \alpha_c S(f) \qquad (4)$$

where $S(f)$ denotes the information content of a pattern with activity f expressed in bits per connection:

$$S(f) = \frac{f\ln(f) + (1-f)\ln(1-f)}{\ln 2} \qquad (5)$$

One can understand the scaling of α_c with N or f by writing that the information capacity is a finite quantity: α_c is then of order $1/S(f)$. Note also that for $f = \frac{1}{2}$, the information capacity I_c and the pattern capacity α_c coincide. It was shown by Gardner (1988) that the optimal information capacity decreases to a non-zero value (equal to $1/(2\ln 2) \approx 0.72$) when f goes to zero. This value is about one third of the information capacity obtained with the dense coding scheme and is reached for simple Hebb-like learning rules. One should also note that in practical cases—in particular when dealing with computer science applications of neural nets—most people use the pattern capacity instead of the information capacity. This is justified by the difficulty of estimating the information content of the data in practice.

Refinements

Several remarks will clarify the above picture and reveal some subtleties:

- The capacity of an associative memory requires the prior definition of a quality criterion. One generally asks that patterns be exactly stored or at least that the level of errors be vanishingly low. Capacities can also be defined for a given error rate. These different definitions may lead to quite different estimates of the capacity, as shown by Nadal (1991) for the Willshaw model.

- In addition, two different notions of activity level f are commonly used: the exact fraction of active neurons in every pattern, or the average number of active neurons. Nadal (1991) analyzed the implications of the definition of f in the framework of the Willshaw model. For a fixed number of active neurons, the information capacity, as computed by Willshaw et al., is equal to $\ln 2 \simeq 0.693$, whereas for a fluctuating number of active neurons (equal to f in average) it decreases to 0.236. One should note that the optimal result of Gardner was obtained for an average f and should be compared to the latter of the estimates above.

- In view of the above, the capacity of the Willshaw model should be compared to the optimal information capacity for models with binary connections $(0, 1)$ and an adjustable threshold. Gutfreund and Stein (1990) showed that this optimal value was close to 0.29.

- The information capacity can still be defined in the overloading regime—that is, when α is greater than α_c—so that the patterns are stored with errors. It was shown (Nadal and Toulouse, 1990) for the covariant rule that in the limit where the number of patterns grows to infinity (the error rate then goes to $\frac{1}{2}$), the information capacity is finite and equal to $1/(\pi \ln 2) \approx 0.46$ (bits/connection).

- For autoassociative memories, Palm considers an alternative choice of the quality criterion, which requires not only the perfect storage of patterns but also the absence of spurious fixed points with the same mean activity f. This leads to reduction of the pattern capacity (and information capacity) by a factor of two (Palm and Sommer, 1992).

- The above results have been generalized to sparser scalings such as $f \sim N^{-s}$ with $0 < s < 1$ (Amari, 1989; Meunier et al., 1991).

Methods

Different analytical approaches were used to derive these results. All of them rely on taking the large N limit. This limit is not only relevant in view of the large number of neurons that may be involved, in the cortex, in a memory task, but is also of special interest from the theoretical point of view. Indeed, sharp transitions between different behaviors (analogous to phase transitions in statistical mechanics) occur only in this limit.

- The most commonly used method is signal-to-noise analysis (valid for both auto- and heteroassociative memories). This method consists in writing under which conditions a pattern is stable. It is equivalent to a mean field analysis where one checks the stability of any pattern with respect to the random fluctuations induced by the storage of the other patterns.

- For autoassociative memories and symmetric couplings, one can define an energy function (see ENERGY FUNCTIONS FOR NEURAL NETWORKS) and use statistical mechanics to compute the equilibrium properties of the network (Amit et al., 1987a).

- Numerical simulations on biased networks have been performed by most authors in order to check theoretical predictions. Large-scale simulations are necessary in the regime of interest, especially for autoassociative networks, since sparse coding requires $1 \ll \ln N \ll N$ (N should be at least equal to 10^4).

The Dynamics of Sparsely Coded Networks

For an associative memory, the storage capacity is only half of the story: one is also interested in the retrieval dynamics, in particular in the nature of attractors (spurious states versus prototype attractors) and in the size and shape of basins of attraction.

The dynamical properties of a recurrent network are not readily amenable to theoretical study, except for diluted networks where a large fraction of connections have been randomly cut. For a connectivity C (i.e., a mean number of connections per neuron) of the order of $\ln N$, the dynamics can be solved exactly as the temporal correlations between neurons are destroyed by the spatial dilution. This method has been used for the study of sparse coding in Tsodyks and Feigelman (1988), yielding the same scaling behavior for the capacity as in Equation 2 with here $\alpha = p/C$. The dynamic mean field theory is a more general approach not relying on dilution. It is generally used to derive the stationary states of neural networks. Horner et al. (1989) have studied the time dependent mean field theory in the case of sparse coding. As the mean field equations are difficult to solve, Horner et al. relied on an approximation scheme that interpolates between the short time dynamics and the large time limit (that yields the stationary states). Some aspects of the dynamics can also be analyzed when computing the equilibrium properties (Amit et al., 1987b). The main results concern the existence and stability of spurious states.

The main outcome of all these studies is that one must constrain the dynamics of the network to maintain a constant mean activity equal to the bias of the stored patterns. This is necessary to avoid both the decay of the dynamics to a fully quiescent state, and in the overloading regime, an ever-increasing activity leading to a fully active state (Meunier et al., 1991). Moreover, if no constraint is imposed, many spurious states appear. In particular, the symmetric mixtures of an even number of prototypes become stable for sparsely encoded networks, whereas only odd mixtures are stable for the Hopfield network (Amit et al., 1987b). Still worse, these spurious states are global minima of the energy. A constrained dynamics suppresses the spurious states. It yields a correct retrieval for states with activity comparable to the activity of the stored patterns (Amit et al., 1987b; Horner et al., 1989). One should note that for the Hopfield model, some noise may be introduced in the dynamics by adopting a stochastic update rule for the neurons. Mixture states then disappear when increasing the noise level while the desired patterns are not affected. This useful role of the noise is not encountered in sparsely encoded networks. As a rule the (rigid) constraint is imposed via a global inhibitory term, proportional to the activity of the network. This is equivalent to choosing properly the constant threshold of each neuron, so that the constraint is automatically taken into account in the calculation of the optimal capacity. An alternative (soft) constraint has been considered in Amit et al. (1987b), and consists in adding a quadratic cost term in the energy which favors a given mean activity. When the relative weight of this additional term goes to infinity, one recovers the previous case of rigid constraint.

The size of the basins of attraction has been estimated (Meunier et al., 1991; Horner et al., 1989), as well as the maximal amount of information that can be retrieved (Horner et al., 1989; Amit et al., 1987b). It will not come as a surprise that the basins of attraction are made of configurations that strongly overlap one of the prototypes and, hence, of similar mean activity.

The Lessons of Sparse Coding

How rewarding was the study of sparse coding? First, we gained some insight into the notion of capacity, with a strong emphasis on information quantities. This was possible because exact results could be derived in the sparse coding limit without relying on the involved statistical mechanics methods used for the Hopfield model (Amit et al., 1987a, 1987b). Next, the notion of information capacity made it natural to consider the overloading regime. In particular, one can show that for the Willshaw model the maximal storage of information can be achieved in the overloading regime (Nadal and Toulouse, 1990). This, again, stresses the remarkable properties of the model introduced by Willshaw, Buneman, and Longuet-Higgins in the 1960s. It is still far from clear why this simple model achieves such a high capacity as compared to the optimal capacity estimated for binary connections. Other open problems and perspectives should be mentioned. Although the storage properties have been analyzed in detail, the specificity of sparse coding in the ability of the network to generalize has not been considered. This is probably because more attention has been devoted to associative networks. Finally, as low activity in biological networks exhibits the two aspects of sparse coding and low firing rate, it would be very interesting to study networks that combine these two aspects.

Road Map: Cooperative Phenomena
Background: I.3. Dynamics and Adaptation in Neural Networks
Related Reading: Neural Optimization; Nonmonotonic Neuron Associative Memory; Sparse Coding in the Primate Cortex; Statistical Mechanics of Learning

References

Amari, S., 1989, Characteristics of sparsely encoded associative memory, *Neural Netw.*, 2:451–457.
Amit, D. J., Gutfreund, H., and Sompolinsky, H., 1987a, Statistical mechanics of neural networks near saturation, *Ann. Physics*, 173:30–67.
Amit, D. J., Gutfreund, H., and Sompolinsky, H., 1987b, Information storage in neural networks with low levels of activity, *Phys. Rev. A*, 35:2293–2303.
Dayan, P., and Willshaw, D., 1991, Optimizing synaptic learning in linear associative memories, *Biol. Cybern.*, 65:253–265.
Gardner, E., 1988, The space of interactions in neural network models, *J. Phys. A*, 21:257–270.
Gutfreund, H., and Stein, Y., 1990, Capacity of neural networks with discrete synaptic couplings, *J. Phys. A*, 23:2613.
Hopfield, J. J., 1982, Neural networks and physical systems with emergent collective computational abilities, *Proc. Natl. Acad. Sci. USA*, 79:2554–2558.
Horner, H., 1989, Neural networks with low levels of activity: Ising vs. McCulloch-Pitts neurons, *Z. Phys. B*, 75:133–136.
Horner, H., Bormann, D., Frick, M., Kinzelbach, H., and Schmidt, A., 1989, Transients and basins of attraction in neural networks models, *Z. Phys. B*, 76:381–398.
Meunier, C., Yanai, H., and Amari, S., 1991, Sparsely coded associative memories: Capacity and dynamical properties, *Network*, 2:469–487.
Nadal, J.-P., 1991, Associative memory: On the (puzzling) sparse coding limit, *J. Phys. A*, 24:1093–1101.
Nadal, J.-P., and Toulouse, G., 1990, Information storage in sparsely coded memory nets, *Network*, 1:61–74.
Palm, G., and Sommer, F. T., 1992, Information capacity in recurrent McCulloch-Pitts networks with sparsely coded memory states, *Network*, 3:177–186.
Perez-Vicente, C. J., and Amit, D. J., 1989, Optimized network for sparsely coded patterns, *J. Phys. A*, 22:559.
Rolls, E. T., 1989, The representation and storage of information in neuronal networks in the primate cerebral cortex and hippocampus, in *The Computing Neuron* (R. Durbin, C. Miall, and G. Mitchison, Eds.), Wokingham: Addison-Wesley, pp. 125–159.
Tsodyks, M. V., and Feigelman, M. V., 1988, The enhanced storage capacity in neural networks with low activity level, *Europhys. Lett.*, 6:101–105.
Willshaw, D. J., Buneman, O. P., and Longuet-Higgins, H. C., 1969, Non-holographic associative memory, *Nature*, 222:960–962.

Spatiotemporal Association in Neural Networks

Andreas V.M. Herz

Introduction

The capability to process spatiotemporal information is a pre-requisite for any action in or reaction to a natural, that is, time-varying, environment. This simple fact explains why biological organisms have developed highly sophisticated mechanisms to recognize, generate, and learn pattern sequences (see TEMPORAL PATTERN PROCESSING and MOTOR PATTERN GENERATION).

A closed theory of the neural processes underlying spatio-temporal associations is far beyond present knowledge. But even complicated pattern sequences consist in general of simpler spatiotemporal building blocks with a duration of up to a few hundred milliseconds. Hardwired or learned in a reliable manner, those building blocks are stored in dedicated brain regions and facilitate a faithful replay or recognition of the entire spatiotemporal object.

How could elementary pattern sequences be represented in neural structures at a low architectural and computational cost? What are possible mechanisms to memorize spatiotemporal associations in a robust fashion within model neural networks? Is it possible to understand the global computation on a qualitative and quantitative level? What is the relevance of these models for biological systems?

The present article tries to answer these questions. It focuses on formal neural networks whose dynamics can be analyzed using methods from nonlinear dynamics and statistical mechanics. Such networks are necessarily caricatures of biological structures. Still, they may capture aspects that are important for more elaborate approaches and real neural systems.

Signal Delays

Signal delays are omnipresent in the brain. Time lags of a few milliseconds are characteristic for axonal propagation of action potentials, synaptic transmission, and dendritic transport processes. The delay times are of the same order of magnitude or only slightly smaller than the typical time scale of various important neurobiological phenomena—for example, mean interspike intervals or periods of neural oscillations. The incorporation of signal delays into theoretical models of neural dynamics is thus mandatory, especially if one allows for a distribution of time lags.

Kleinfeld (1986) and Sompolinsky and Kanter (1986) proposed models for temporal associations, using a single delay line between each pair of neurons. Tank and Hopfield (1987) presented a feedforward architecture for sequence recognition based on multiple delays, encoding information relative to the very end of a given sequence. Elements of both approaches have been combined to construct a class of feedback networks with a *broad* distribution of transmission lines (Coolen and Gielen, 1988; Herz et al., 1989; Kerszberg and Zippelius, 1990). This article focuses on a special class of delay networks with discrete-time dynamics and synchronous updating. Systems with (random) sequential dynamics are discussed in Amit (1989) and Kühn and van Hemmen (1991). Work related to the present topic has also been reported by Bauer and Krey (1991), who extended Gardner's analysis (see STATISTICAL MECHANICS OF LEARNING; SPARSELY CODED NEURAL NETWORKS) to spatiotemporal associations, and Bressloff (1991), who studied the dynamics of "time-summating" networks with stochastic updating. Further references can be found in Kühn and van Hemmen (1991).

Neural Dynamics

Throughout what follows, a neural network will be described as a collection of N two-state neurons with activities $S_i = 1$ for a firing cell and $S_i = -1$ for a quiescent one. The neurons are connected by synapses with modifiable efficacies $J_{ij}(\tau)$, where τ denotes the delay for the information transport from j to i. For simplicity, a model will be considered where each pair of neurons is linked by *several* delay lines with time lags $0 \leq \tau \leq \tau_{max}$, which are integer multiples of a small unit time step $\Delta t = 1$. External stimuli are fed into the system through receptor neurons $\sigma_i = \pm 1$ with normalized input sensitivity γ, $0 \leq \gamma \leq 1$. The postsynaptic potentials h_i are then given by

$$h_i(t) = (1 - \gamma) \sum_{j=1}^{N} \sum_{\tau=0}^{\tau_{max}} J_{ij}(\tau) S_j(t - \tau) + \gamma \sigma_i(t) \qquad (1)$$

The network dynamics are assumed to be synchronous; that is, all neurons are updated in parallel. A spike is generated if the postsynaptic potential exceeds the firing threshold. In what follows, this threshold is set to zero for simplicity so that

$$S_i(t + 1) = \text{sign}[h_i(t)] \qquad (2)$$

If one takes synaptic noise into account, the postsynaptic potential becomes a fluctuating quantity $h_i + v_i$, where v_i denotes stochastic contributions due to the probabilistic nature of neurotransmitter release. A careful analysis of synaptic transmission reveals that under the assumption of linear dendritic processing, the variable v_i is distributed according to a Gaussian probability distribution (Amit, 1989). The probability for spike generation may then be approximated by the stochastic dynamics

$$\text{Prob}[S_i(t + 1) = \pm 1] = \tfrac{1}{2}\{1 \pm \tanh[T^{-1} h_i(t)]\} \qquad (3)$$

where Prob denotes probability and T represents the noise level. Within the physics literature, the update rule (Equation 3) is known as Glauber dynamics. In the limit $T \to 0$, one recovers the deterministic description (Equation 2).

Hebbian Learning

According to Hebb's neurophysiological postulate for learning (Hebb, 1949), information presented to a neural network is physically embedded through an alteration of the network structure: "When an axon of cell A is near enough to excite cell B and *repeatedly* or *persistently* takes part in firing it, some growth process or metabolic change takes place in one or both cells such that A's efficiency, as one of the cells firing B is increased" (see HEBBIAN SYNAPTIC PLASTICITY).

How should this postulate be implemented in a formal neural network with transmission delays?

Let us focus on a connection with delay τ between neurons j and i. According to Hebb's postulate, the corresponding synaptic strength $J_{ij}(\tau)$ will be increased if cell j takes part in *firing* cell i. (In its physiological context, the postulate was formulated for excitatory synapses only, but for simplicity, it will be applied to all synapses of the model network.) Due to the delay τ and the parallel dynamics, it takes $\tau + 1$ time steps until neuron j actually influences the *state* S_i of neuron i: τ time steps for the signal propagation (Equation 1) and one further time step to determine the cell's new firing state given the postsynaptic potential (Equation 2). Following Hebb's rule, $J_{ij}(\tau)$ should therefore be altered at time $t + 1$ by some function of

$S_j(t - \tau)$ and $S_i(t + 1)$—most simply by their product. Starting with a *tabula rasa*, $J_{ij}(\tau) = 0$, one thus obtains after P learning sessions, labeled by μ and each of duration D_μ,

$$J_{ij}(\tau) = \varepsilon(\tau)N^{-1} \sum_{\mu=1}^{P} \sum_{t_\mu=1}^{D_\mu} S_i(t_\mu + 1)S_j(t_\mu - \tau) \equiv \varepsilon(\tau)\tilde{J}_{ij}(\tau) \quad (4)$$

The parameters $\varepsilon(\tau)$ model morphological characteristics of the delay lines; N^{-1} is a scaling factor useful for the theoretical analysis. By Equation 4, synapses act as microscopic feature detectors during the learning sessions: they measure and store correlations of the taught sequences in both space (i,j) and time (τ). This process leads to a resonance phenomenon: connections with delays that approximately match the time course of the external input receive maximum strength. Note that these connections are also the ones that would support a stable sequence of the same duration. Thus, due to a subtle interplay between external stimulus and internal architecture (distribution of τ's), the Hebb rule (Equation 4), which *prima facie* appears to be instructive in character, exhibits in fact also pronounced selective aspects.

Furthermore, an external stimulus encoded in a network with a *broad* distribution of transmission delays enjoys a rather multifaceted representation. According to Equation 4, synaptic couplings with delays that are short compared to the typical time scale of single patterns within the taught sequence are almost symmetric in the sense that $J_{ij}(\tau) \approx J_{ji}(\tau)$. They encode the individual patterns of the sequence as *unrelated static objects*. On the other hand, synapses with transmission delays that approximately match the duration of single patterns of the sequence are able to detect the transitions between patterns. The corresponding synaptic efficacies are asymmetric and establish various temporal relations between the patterns, thereby representing the complete sequence as *one dynamic object*.

Once the learning sessions are over, the $J_{ij}(\tau)$ are kept fixed. The retrieval process (Equations 1–3), operating with the very same delays as the synaptic dynamics (Equation 4), is then able to extract the spatiotemporal information encoded in the $J_{ij}(\tau)$. Retrieval is therefore extremely robust, as shown in numerous simulations (Herz et al., 1989).

It should be noted that this mechanism requires delay times that are of the same order as the duration of *single* quasi-static elements of a temporal sequence. The length of the *entire* sequence, however, remains unconstrained. Let me also remark that a number of authors have discussed the interplay between neural and synaptic dynamics and, in particular, the role of transmission delays (e.g., Hebb, 1949; Caianiello, 1961). However, the full consequences, discussed in this article, have been explored only recently.

Global Analysis

Equations 1–4 describe a "double dynamics" where both neurons and synapses change in time. In general, such a scenario cannot be analyzed mathematically. The situation becomes simplified to some extent if teaching is performed in a "clamped" fashion where the system evolves strictly according to the external stimuli, $S_i(t_\mu) = \sigma_i(t_\mu - 1)$, so that the synapses measure the true correlations of the external stimuli without any interference by the internal dynamics. Clamped learning is achieved by setting $T = 0$ in Equation 3 and $\gamma = 1$ in Equation 1 during learning sessions. Numerical experiments demonstrate that this condition can be relaxed to a certain degree without any significant changes in the emergent network behavior.

One may derive approximate equations of motion for macroscopic network quantities that characterize how closely the retrieval sequences resemble the taught sequences (for a detailed discussion, see Kühn and van Hemmen, 1991). These equations are nonlinear because of the nonlinear signal processing of single neurons (Equations 2 and 3) and contain previous activities because of the memory effects modeled by Equation 1. The solutions agree well with numerical simulations, but an exact analysis is in general not possible.

The mathematics becomes tractable in the special case where all input sequences $\sigma_i(t_\mu)$ are cyclic with equal periods $D_\mu = D$. If one defines patterns $\xi_{ia}^{\mu 0}$ by $\xi_{ia}^{\mu 0} \equiv \sigma_i(t_\mu = a)$ for $0 \le a < D$, one obtains

$$\tilde{J}_{ij}(\tau) = N^{-1} \sum_{\mu=1}^{P} \sum_{a=0}^{D-1} \xi_{i,a+1}^{\mu 0} \xi_{j,a-\tau}^{\mu 0} \quad (5)$$

Note that the synaptic strengths are in general still asymmetric in the sense that $\tilde{J}_{ij}(\tau) \neq \tilde{J}_{ji}(\tau)$. They do, however, obey the symmetry $\tilde{J}_{ij}(\tau) = \tilde{J}_{ji}(D - (2 + \tau))$ for $\tau < D - 1$ and $\tilde{J}_{ij}(D - 1) = \tilde{J}_{ji}(D - 1)$. For all networks whose a priori weights $\varepsilon(\tau)$ satisfy $\varepsilon(\tau) = \varepsilon(D - (2 + \tau))$ for $\tau < D - 1$, one has thus found an "extended synaptic symmetry,"

$$J_{ij}(\tau) = J_{ji}((D - (2 + \tau)) \, (\text{modulo } D)) \quad (6)$$

generalizing Hopfield's symmetry assumption $J_{ij} = J_{ji}$ (see COMPUTING WITH ATTRACTORS) in a natural way to the temporal domain. The symmetry (Equation 6) suggests that one may be able to find a Lyapunov (or "energy") function (see ENERGY FUNCTIONS FOR NEURAL NETWORKS) for the noiseless retrieval dynamics (Equation 2). This is indeed the case if $\varepsilon(D - 1) = 0$ (Herz, Li, and van Hemmen, 1991). One takes $\gamma = 0$ in Equation 1 and defines

$$H(t) \equiv -\frac{1}{2} \sum_{i,j=1}^{N} \sum_{a,\tau=0}^{D-1} J_{ij}(\tau)S_i(t - a)S_j(t - (a + \tau + 1))$$

$$(\text{modulo } D)) \quad (7)$$

By Equations 1, 2, and 6, the difference $\Delta H(t) \equiv H(t) - H(t - 1)$ is

$$\Delta H(t) = -\sum_{i=1}^{N} [S_i(t) - S_i(t - D)]h_i(t - 1) \quad (8)$$

As a finite sum of finite terms, H is bounded. The function $H(t)$ is nonincreasing because the right-hand side of Equation 8 is nonpositive: the term $S_i(t)h_i(t - 1)$ equals $|h_i(t - 1)|$ because of the dynamics (Equation 2) and is therefore larger than or at least equal to the product $S_i(t - D)h_i(t - 1)$, which is $+h_i(t - 1)$. Consequently, $\Delta H(t)$ has to vanish as $t \to \infty$. This result is possible only if the system settles into a state with $S_i(t) = S_i(t - D)$ for all i.

The analysis has exposed two important facts: (a) the retrieval process of certain delay networks is governed by a Lyapunov function. The time evolution during retrieval sessions can thus be understood as a downhill march in an abstract spatiotemporal energy landscape. (b) The networks relax to a static state or a limit cycle with $S_i(t) = S_i(t - D)$—oscillatory solutions with the same period as that of the taught cycles or a period which is equal to an integer fraction of D.

Stepping back for an overview, one notices that H is a Lyapunov function for all networks which exhibit an "extended synaptic symmetry" (Equation 6) and for which the matrix $\mathbf{J}(D - 1)$ vanishes. The Hebbian synapses (Equation 4) are one important special case and will be the main subject of the further discussion.

Statistical Mechanics

This section will show that certain delay networks relax to fixed points or oscillatory solutions with well-determined periods.

Are there limit cycles that closely resemble the taught sequences? How many sequences can be stored in a given network?

The answers are obtained in a two-step process. First, it is demonstrated that networks with cyclic temporal associations and deterministic dynamics can be mapped onto symmetric Hopfield-like systems without delays. In a second step, one shows that this correspondence holds for the stochastic Glauber dynamics (Equation 3) as well. One may then apply equilibrium statistical mechanics to derive quantitative results. A detailed discussion of the technical issues involved can be found in Herz et al. (1991). Let us briefly sketch the main ideas.

D-periodic oscillatory solutions of the retrieval dynamics can be interpreted as static states in a fictitious "D-plicated" system with N rows and D columns of cells with activities S_{ia} where $1 \leq i \leq N$ and $0 \leq a < D$. The parallel dynamics of the original system are reproduced by the update rule

$$S_{ia}(t+1) = \begin{cases} \text{sign}[\sum_{j=1}^{N} \sum_{b=0}^{D-1} J_{ij}^{ab} S_{jb}(t)] & \text{if } a = t \text{ (modulo D)} \\ S_{ia}(t) & \text{otherwise} \end{cases}$$

(9)

In terms of the original synaptic efficacies $J_{ij}(\tau)$, the couplings J_{ij}^{ab} are given by $J_{ij}^{ab} = J_{ij}((b-a-1)(\text{modulo } D))$. Because of Equation 6, they are symmetric, that is, $J_{ij}^{ab} = J_{ji}^{ba}$.

The time evolution of the new network has a pseudosequential characteristic: synchronous within single columns and sequentially ordered with respect to these columns. The interpretation of the retrieval process has changed significantly: a *limit cycle* of period D in the original network corresponds to a *fixed point* of the new system of size ND. Storing one cycle $\sigma_i(t_\mu) = \xi_{ia}^{\mu 0}$ in the delay network thus corresponds to memorizing D shifted duplicates $\xi_{ia}^{\mu\nu}$, $0 \leq \nu \leq D$, in the equivalent system. This correspondence reflects the fact that a cycle with period D can be retrieved in D different time-shifted versions in the original network.

Let us now turn to quantitative results. They were obtained within a replica-symmetric theory (see STATISTICAL MECHANICS OF GENERALIZATION) for the case where each of the P learning sessions corresponds to teaching a (different) cycle of D unbiased random patterns $\xi_{ia}^{\mu 0}$, each lasting for one time step.

The retrieval quality for a given cycle μ is best described in terms of the generalized "overlap" $m^\mu = \max_\nu \{N^{-1}D^{-1}\Sigma_{i,a}\xi_{ia}^{\mu\nu} S_{ia}\}$: this order parameter equals one for perfect recall and zero if the network state is uncorrelated with the μth cycle. Retrieval is possible as long as the network dynamics admits solutions with one large m^μ. Owing to the interference between different memories, such solutions are possible only if the number P of stored sequences remains below a critical number P_c. The analysis shows that P_c scales linearly with N for large N, $P_c = \alpha_c N$. The factor α_c is called the storage capacity of the network. As in the Hopfield model, the retrieval quality m^μ drops suddenly as P is increased beyond $\alpha_c N$—a first-order phase transition (see COOPERATIVE PHENOMENA).

It should be noted that each cycle consists of D patterns so that the storage capacity for *single* patterns is bar $\bar{\alpha}_c = D\alpha_c$. During the recognition process, however, each of them will trigger the cycle it belongs to and cannot be retrieved as a static pattern. For systems with a delay distribution that vanishes for $\tau = D - 1$ and is uniform for $\tau < D - 1$, that is, $\varepsilon(\tau) = (D-1)^{-1}(1 - \delta_{\tau,D-1})$, one obtains

$$\begin{array}{cccccc} D & 2 & 3 & 4 & 5 & \infty \\ \alpha_c & 0.100 & 0.110 & 0.116 & 0.120 & 0.138 \end{array}$$

(10)

The findings agree well with estimates from a finite-size analysis of data from numerical simulations. The results demonstrate that the storage capacity for temporal associations is comparable to that for static memories. As an example, take $D = 2$. In the limit of large N, $0.100 \cdot N$ two-cycles of the form $\xi_{i0}^{\mu 0} \rightleftharpoons \xi_{i1}^{\mu 0}$ may be recalled as compared to $0.138 \cdot N$ static patterns (Amit, 1989); since $2 \times 1.00/1.38 \approx 1.45$, this result amounts to a 1.45-fold increase of the information content per synapse.

The influence of the weight distribution on the network behavior is demonstrated by some choices of $\varepsilon(\tau)$ for $D = 4$:

$$\begin{array}{ccccccc} \tau = & 0 & 1 & 2 & 3 & \alpha_c & m_c \\ \varepsilon(\tau) = & \frac{1}{3} & \frac{1}{3} & \frac{1}{3} & 0 & 0.116 & 0.96 \\ \varepsilon(\tau) = & \frac{1}{2} & 0 & \frac{1}{2} & 0 & 0.100 & 0.93 \\ \varepsilon(\tau) = & 0 & 1 & 0 & 0 & 0.050 & 0.93 \end{array}$$

(11)

The storage capacity decreases with a decreasing number of delay lines, but measured *per synapse*, it increases. However, networks with only a few delays are less fault-tolerant, as is well known from numerical simulations.

Discussion

Learning schemes can only be successful if the structure of the learning task is compatible with both the network architecture and the learning algorithm. In the present context, the task is to store simple temporal associations. It can be accomplished in neural networks with a broad distribution of signal delays and Hebbian synapses which, during learning periods, operate as microscopic feature detectors for spatiotemporal correlations within the external stimuli. The retrieval process utilizes the very same delays and synapses, and is therefore rather robust.

Let me emphasize again that the Hebbian mechanism discussed here does *not* limit the overall length of a pattern sequence stored in a neural network with transmission delays. It is the duration of a single pattern within a temporal association that is constrained and cannot exceed the time lags provided by transmission delays.

Quantitative results for the storage capacity and retrieval quality can be obtained for a certain class of delay networks using a global Lyapunov function and techniques from statistical mechanics. Unlike the general case, here the total length of a pattern sequence has to approximately match the maximum time lag to allow for the extended synaptic symmetry (Equation 6). The analytical findings prove that an extensive number of temporal associations can be stored as spatiotemporal attractors for the retrieval dynamics. Numerical simulations with pattern sequences that are long with respect to the available signal delays show a qualitatively similar picture. Both results indicate that dynamical systems with delayed interactions can be programmed in an efficient manner to perform associative computations in the space-time domain.

The approach sketched in this article has also been extended to neural networks with continuous units that store cycles of correlated real-valued pattern sequences (Herz, 1991). Numerical studies have been performed for low-dimensional trajectories (small N) with high numbers of data points (large D). For many examples, good retrieval has been obtained without any need for highly time-consuming supervised learning schemes. However, algorithms of the latter kind may eventually be necessary to solve more sophisticated real-world associations. Here, once again, the existence of Lyapunov functions is of great help, since it allows for the application of powerful techniques from statistical mechanics to a wide class of supervised learning strategies such as spatiotemporal extensions of the concept of BOLTZMANN MACHINES (q.v.) or more-general contrastive-learning schemes (Baldi and Pineda, 1991).

Delays add a new dimension—time—to *any* learning mechanism based on correlation measurements of pre- and post-synaptic events. This extra dimension naturally extends the structure of objects representable in the synaptic code generated by such a learning mechanism, thereby considerably increasing its potential power. Transmission delays, therefore, do not induce a loss of the associative capabilities of neural networks as one might have feared. On the contrary, if properly included in the learning process, they provide a physical structure to perform spatiotemporal computations at low architectural cost. Nature may have opted to make constructive use of them, as is also indicated by the success of more elaborate models with delayed interactions.

Acknowledgments. This work has been supported by the Deutsche Forschungsgemeinschaft and the Beckman Institute.

Road Map: Self-Organization in Neural Networks
Background: I.3. Dynamics and Adaptation in Neural Networks
Related Reading: Oscillatory Associative Memories; Self-Organization in the Time Domain

References

Amit, D. J., 1989, *Modeling Brain Function: The World of Attractor Neural Networks*, Cambridge, Eng.: Cambridge University Press. ◆

Baldi, P., and Pineda, F., 1991, Contrastive learning and neural oscillations, *Neural Computat.*, 3:526–545.

Bauer, K., and Krey, U., 1991, On the storage capacity for temporal pattern sequences in networks with delays, *Z. Phys. B*, 84:131–141.

Bressloff, P. C., 1991, Stochastic dynamics of time-summating binary neural networks, *Phys. Rev. A*, 44:4005–4016.

Caianiello, E., 1961, Outline of a theory of thought processes and thinking machines, *J. Theor. Biol.*, 1:204–235.

Coolen, A. C. C., and Gielen, C. C. A. M., 1988, Delays in neural networks, *Europhys. Lett.*, 7:281–285.

Hebb, D. O., 1949, *The Organization of Behavior*, New York: Wiley.

Herz, A. V. M., 1991, Global analysis of parallel analog networks with retarded feedback, *Phys. Rev. E*, 44:1415–1418.

Herz, A. V. M., Li, Z., and van Hemmen, J. L., 1991, Statistical mechanics of temporal association in neural networks with transmission delays, *Phys. Rev. Lett.*, 66:1370–1373.

Herz, A. V. M., Sulzer, B., Kühn, R., and van Hemmen, J. L., 1989, Hebbian learning reconsidered: Representation of static and dynamic objects in associative neural nets, *Biol. Cybern.*, 60:457–467.

Kerszberg., M., and Zippelius, A., 1990, Synchronization in neural assemblies, *Phys. Scr.*, 33:54–64.

Kleinfeld, D., 1986, Sequential state generation by model neural networks, *Proc. Natl. Acad. Sci. USA*, 83:9469–9473.

Kühn, R., and van Hemmen, J. L., 1991, Temporal association, in *Models of Neural Networks* (E. Domany, J. L. van Hemmen, and K. Schulten, Eds.), Berlin: Springer-Verlag. ◆

Sompolinsky, H., and Kanter, I., 1986, Temporal association in asymmetric neural networks, *Phys. Rev. Lett.*, 57:2861–2864.

Tank, D., and Hopfield, J. J., 1987, Neural computation by concentrating information in time, *Proc. Natl. Acad. Sci. USA*, 84:1896–1900.

Speaker Identification

Younès Bennani

Introduction

Connectionist systems have gained widespread acceptance for tackling problems where the relationship between input and desired output is highly complex and nonlinear. Unfortunately, the required connectionist system often has a large number of parameters, and the amount of training data is frequently very limited. This places a serious constraint on the ability of the connectionist system to correctly generalize. Two ways to handle this problem are: attempt to reduce the system's complexity, and incorporate a priori knowledge into its architecture. Since a complex problem can often be decomposed into a series of much simpler subproblems, decomposing the single connectionist system into a set of modules which tackle each of these subproblems, while cooperating together to solve the global problem, is a powerful method to both reduce complexity and incorporate a priori knowledge about the problem. In this article, a modular connectionist system is developed for the problem of text independent speaker identification (TISI).

Naturally, the time needed to train such a system grows with the number of speakers to identify. An additional problem is that, for real speaker identification systems, the amount of training data is often limited to a few utterances, while the range of possible test utterances for a text independent system is essentially unlimited.

Thus, TISI suffers from the twin problems of limited training data and more-than-linear growth of training time with the number of speakers. A review on connectionist approaches for speaker recognition can be found in Bennani and Gallinari (1994).

In our work with TISI, we noticed that certain vocal characteristics allow for division of the population of speakers into homogenous and separable classes (Bennani, 1992). We can thus design a modular multiexpert system which has some analogy with Jacobs et al. (1991; see MODULAR AND HIERARCHICAL LEARNING SYSTEMS); however, our architecture and gating network are very different from those introduced there. This article develops a modular TISI system, based on Time-Delay Neural Networks (TDNNs), and its performance is evaluated against a system based on Multivariate Auto-Regressive Models (MARMs) (Grenier, 1980).

A Multiexpert Connectionist Architecture for TISI

We have used a significant part of the TIMIT database (Fisher et al., 1987) containing 33 females and 69 males. We have chosen, for each speaker, five sentences for training and five for testing.

Within the classes of females and males, it may be possible to distinguish many subclasses, which group together speakers with similar vocal characteristics. We have found that precisely such a subdivision is possible by using a k-means clustering technique labeled by a majority vote on the set of speech vectors formed by the training data. In essence, this subdivision of the population of speakers reflects an underlying structure in the problem, which is a form of a priori knowledge. We will refer to each of these subgroups as a typology, and our proposed system is based on using a separate connectionist module for each typology.

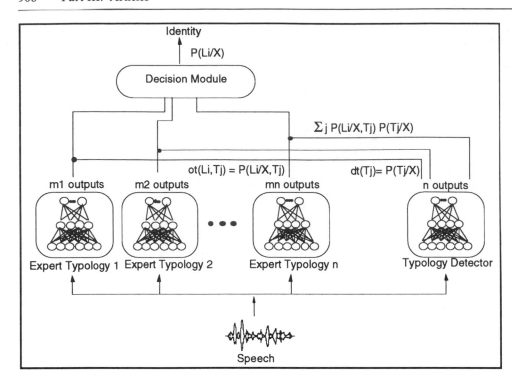

Figure 1. Architecture of the system. The speech coefficients X enter the system and are fed to each typology expert module and to the typology detector module. Each expert module outputs a probability per speaker, $P(L_i|X, T_j)$, given that speaker i belongs to that typology j. The typology detector module outputs a probability, $P(T_j|X)$, that the speech X belongs to a particular typology j. The decision module combines these probabilities to produce a final probability per speaker, $P(L_i|X)$, given by $\Sigma_{j=1\ldots n}P(L_i|X, T_j)P(T_j|X)$.

The system illustrated in Figure 1 consists of two types of networks: a typology detector, and expert modules. The speech data for all speakers are clustered by a k-means technique, and these clusters are then used to form the speaker typologies. For the population used in our evaluation, we have found that 16 typologies produced a balanced set of clusters. The typology detection module is trained to classify the input speech according to the typology label found by the k-means technique. Each expert module of the system is dedicated to the discrimination between speakers within the same typology.

In the identification phase, all frames of a sentence are presented to the system, as *win* successive windows of 25 acoustic vectors. At time t when presented with a window W_t, the system produces m activations $a_t(L_i)$, $i = 1 \ldots m$, where m is the total number of speakers.

To compute a_t, the system proceeds in the following fashion: at time t, output d_t of the typology detector is used as a weighting factor for the outputs o_t of the expert modules. We assume that the final output of the system is a combination of the outputs of the expert modules, with the typology detector network determining the proportion of each expert module's output in the combination. The successive activations of the system are accumulated over the duration of the sentence to give the final activation for each speaker.

We have noticed in our experiments that, after training, a set of three successive windows was sufficient for a perfect identification (*win* = 3). If this is compared with existing systems, these three windows correspond to an extremely short utterance (less than a second) for a text-independent identification. The majority of systems use utterances of the order of several seconds.

Results and Comparisons with a Conventional System

Multivariate Auto-Regressive Models (MARMs) are a well-known technique for speech processing. MARMs have been used for speaker identification by Grenier (1980) and more recently by Bimbot et al. (1992) and Montacie and Le Floch (1992). In order to analyze and compare our connectionist ap-

Table 1. Comparison of the Two Approaches: Connectionist and MARM

System	Identification Time	Test Duration	95% Confidence Interval
Typology detector	< 1 s	≈ 0.5 s	[99.9%, 100%]
Global connectionist system	< 1 s	≈ 0.75 s	[99.9%, 100%]
MARM system	minutes	≈ 2.5 s	[91.4%, 95.6%]

proach, we have built a MARM-based system and performed tests with the two systems on the same database. In our evaluation, identification performances are given as a 95% confidence interval.

The results in Table 1 show the superiority of the connectionist multimodule approach in comparison to the MARM technique. Although the MARM performances are somewhat lower, they are much easier to train. The connectionist method is discriminant and very fast during the identification phase: less than one second is effectively real time. Moreover, only a short duration of the speech signal is required.

The MARM method has the advantage that the characteristics are computed in one step from the whole training set. However, they are not discriminant enough, and this leads to confusion when the number of speakers increases.

Discussion

Modularity is an indispensable tool in the design and analysis of complex systems, and the notion of modularity has been found to be of considerable utility in the connectionist domain. The essential idea behind modularity is that if a task can be decomposed into subtasks, each of which has its own idiosyncratic properties, then a system designed to solve it should itself be decomposable into distinct expert modules, each allocated to a distinct subtask. In addition, a priori domain knowledge can be used in suggesting an appropriate decomposition of the global task. We have performed a set of experiments which have demonstrated the validity of the modular connectionist approach for text independent speaker recognition. Finally, although the main emphasis of this work has been on speaker identification, the ideas presented are also directly applicable to speaker verification and many other speech recognition problems.

Road Map: Applications of Neural Networks
Related Reading: Modular Neural Net Systems, Training of; Speech Recognition: A Hybrid Approach

References

Bennani, Y., 1992, Speaker identification through a modular connectionist architecture: Evaluation on the TIMIT database, in *Proceedings of the International Conference on Spoken Language Processing*, S4.4, Banff, Canada, pp. 607–610.

Bennani, Y., and Gallinari, P., 1994, Connectionist approaches for automatic speaker recognition, in *Proceedings of the European Speech Communication Association Workshop*, Martigny, Switzerland.

Bimbot, F., Mathan, L., De Lima, A., and Chollet, G., 1992, Standard and target driven AR-vector models for speech analysis and speaker recognition, in *Proceedings of the International Conference on Signal and Speech Processing*, San Francisco, vol. 2, pp. 5–8.

Fisher, W., Zue, V., Bernstein, J., and Pallett, D., 1987, An acoustic-phonetic database, *J. Acoust. Soc. Amer.*, suppl. A, 81:S92. ◆

Grenier, Y., 1980, Utilisation de la prediction lineaire en reconnaissance et adaptation au locuteur, in *Proceedings of XI Journées d'Etudes sur la Parole*, Strasbourg, France, pp. 163–171.

Jacobs, R. A., Jordan, M. I., Nowlan, S. J., and Hinton, G. E., 1991, Adaptive mixtures of local experts, *Neural Computat.*, 3:79–87. ◆

Montacie, C., and Le Floch, J. L., 1992, AR-vector models for free-text speaker recognition, in *Proceedings of the International Conference on Spoken Language Processing*, Banff, Canada, pp. 475–478.

Speech Recognition: A Hybrid Approach

Kari Torkkola and Teuvo Kohonen

Introduction

The difficulty of speech recognition in general stems from the high degree of variability of the acoustic speech signals that should be mapped to the same linguistic units, for instance, the phonemes as discussed here. Being linguistic abstractions, phonemes do not directly correspond to clearly distinct acoustic segments. They have varying lengths, they may overlap partly, and their acoustic appearances vary according to their context, as well as across speakers. Even the same speaker is not able to produce exactly the same acoustic signal for the same utterance.

We describe an approach to automatic speech recognition, the aim of which is to transcribe carefully spoken utterances into phoneme sequences and eventually to transform them into orthographically correct written text. In this article we present our modular hybrid approach, concentrating on those aspects that are most "neural," namely, phonemic classification using two methods that are instances of artificial neural networks (ANNs): the Self-Organizing Map (SOM) and Learning Vector Quantization (LVQ) (see SELF-ORGANIZING FEATURE MAPS: KOHONEN MAPS; and LEARNING VECTOR QUANTIZATION). Either of these methods is used to classify speech segments into phoneme classes time-synchronously. The results are decoded into phoneme strings using Hidden Markov Models (HMMs), a statistical methodology for representing and recognizing stochastic time series (see SPEECH RECOGNITION: PATTERN MATCHING). This is currently the dominant approach in automatic speech recognition (Rabiner, 1989). Our approach differs from conventional HMM methodology in the following way: by using ANNs as a component, we emphasize *discrimination* between phonemes, whereas the aim of HMM is *representation*.

We have chosen the phonemes for the basic recognition units in our system because of the following characteristics of the language we are working with, namely Finnish: (1) it contains only 21 phonemes (except for loan words); (2) the written text symbols (graphemes) and the phonemes correspond to each other almost uniquely; and (3) the words are highly inflected—for example, a single verb can have over 1000 different combinations of inflections, in which the base form of the word can also change. Our main objective has been to write out text from *arbitrary but carefully articulated dictation*, whereby the

prosodic features and coarticulation effects can be kept to a minimum.

It must be emphasized that the division of the problem into subtasks may be very language dependent. For instance, for English, some dictation machines are currently commercially available, and several are at the experimental stage (Baker, 1989; Averbuch et al., 1987). Because of the relatively simple morphology of English, each word form can be a separate entry in the lexicon. This kind of a clearly defined "word level" in the recognition process would be much more complicated for Finnish.

This presentation is structured as follows: The next section gives some details of the SOM and the LVQ algorithms. The section after that describes the general structure of our "phonetic typewriter," concentrating on the application of the SOM and the LVQ to classification of speech feature vectors. The final section is devoted to discussion.

The SOM and LVQ Methods

We have used *three* slightly different adaptive vector quantization methods for the classification of speech feature vectors, all of them belonging to the category of *supervised learning* methods. In the context of classification, supervised learning means that since the correct class of each training sample is known, the classifier is trained to produce this class label as its output. In unsupervised learning, not even the number of classes may be known; the underlying structure of the data directs the training.

The oldest of these methods was the *learning subspace method* (Kohonen et al., 1979). The newest of them is called *learning vector quantization* (Kohonen, 1990a). The third of these methods is a modification of the *self-organizing map* (Kohonen, 1990b), which in its original form is an *unsupervised* learning algorithm, but it can be made to improve its classification accuracy in a supervised way, as described later in this article.

The LVQ approach is tuned toward *pattern classification* tasks, whereas SOM is more appropriate for constructing new *representations* of the input data. In line with the vector quantization literature, the set of weight vectors can be called a *code-*

book. We will first recapitulate the main features of the training algorithms of both the basic unsupervised SOM and the LVQ.

The Self-Organizing Map

The SOM belongs to the category of *competitive-learning* algorithms. In its simplest form the corresponding "neural network" consists of a planar layer of neural cells, each one having an adaptive weight vector $m_i \in \mathbb{R}^n$. A common input vector $x \in \mathbb{R}^n$ is connected in parallel to all the cells. An essential feature in the SOM is the *winner-take-all* (*WTA*) function, by which *that cell whose m_i matches best with x in some metric and also its topological neighbors in the network are activated for learning.* It seems that this function is implemented in biological networks by lateral connectivity, and several theoretical modeling approaches to it exist (for a review, cf., e.g. Kohonen, 1993; and WINNER-TAKE-ALL MECHANISMS). The "winner" m_c should satisfy

$$d(x, m_c) = \min_i \{d(x, m_i)\} \qquad (1)$$

where $d(x, m_i)$ is some distance function applied to x and m_i. Thereafter, the degree of matching of the m_i with x is *increased* adaptively (in discrete time steps t) according to

$$m_i(t + 1) = m_i(t) - \lambda_{ci}(t) \cdot \nabla_{m_i(t)} d[x(t), m_i(t)] \qquad (2)$$

Here $\lambda_{ci}(t)$ is the so-called *neighborhood function*, which defines how strongly cell c controls the learning rate of cell i; normally this rate decreases with spatial distance of cells c and i in the neural network, as well as with time.

One of the simplest distance measures is Euclidean, for which Equation 2 reads (with d the square of the vectorial distance)

$$m_i(t + 1) = m_i(t) + h_{ci}(t)[x(t) - m_i(t)] \qquad (3)$$

where $h_{ci}(t)$ is the same as $\lambda_{ci}(t)$, up to a multiplicative scalar constant.

Learning Vector Quantization

In the basic LVQ, $h_{ci}(t) = \alpha(t)\delta_{ci}$, where δ_{ci} is the Kronecker delta, and $\alpha(t)$ is a scalar learning rate. Supervised learning in LVQ then means that if x is first drawn from a training set of samples for which the classification is known, the correction must be taken with a plus sign as in Equation 3 if the classes of x and m_c agree, but with a minus sign if the classes disagree. This way the decision borders become better defined, and the average rate of misclassification errors is minimized (i.e., discrimination is improved). Other versions of the LVQ algorithms also exist (Kohonen, 1990a).

The Neural Phonetic Typewriter

History

Speech recognition research at Helsinki University of Technology underwent six phases between 1975 and 1993. The equipment completed in 1985 was built around a coprocessor board of our own construction, which had enough capacity to perform the classification of short-time speech spectra by the SOM, as well as to apply symbolic postprocessing to correct for coarticulation errors. It was able to transcribe *unlimited* Finnish and Japanese dictation into text at a *letter accuracy* exceeding 92%. This system became known as the *Neural Phonetic Typewriter* (Kohonen, 1988).

"Neural" Phonemic Classification

The "neural" principle applied in this design up to 1986 was a special *supervised* SOM (Kohonen, Mäkisara, and Saramäki, 1984). To this end the input vectors were formed of two parts x_s and x_u, where x_s was a 15-component short-time spectrum vector, and x_u was supposed to correspond to a unit vector with its components assigned to different phonemic classes. In reality, during training, a value of 0.1 was used for each "1" position of the "unit" vector. During recognition, x_u was not considered.

The original Finnish language has only 21 phonemes, of which the three unvoiced plosives were grouped into one class, resulting in a 19-component x_u. The concatenated 34-dimensional vectors $x = [x_s^T, x_u^T]^T$ were then used as inputs to the SOM. Notice that since x_u is the same for vectors of the same class but different for different classes, the clustering of the vectors x along with the classes is enhanced, leading to improved class separation. The weight vectors $m_i \in \mathbb{R}^{34}$ then also tend to approximate to the density of the concatenated x, not of the signals x_s.

Supervised learning here means that since the classification of each x_s in the training set is known, the corresponding x_u value must be used during training. During recognition of an unknown x, only its x_s part is compared with the corresponding part of the weight vectors.

This scheme was later replaced by the LVQ algorithms, although the supervised SOM yielded recognition scores almost as good. It may be useful to summarize here the differences between these methods. The unsupervised SOM constructs a topology-preserving representation of the statistical distribution of all of the input data. The supervised SOM tunes this representation more toward discrimination of pattern classes, whereas the LVQ is completely inclined toward this task. In the LVQ, the topological order of the weight vectors is no longer considered, and one can therefore not refer to any "map" when using LVQ. The weight vectors of the cells directly define decision borders between classes.

The Architecture of the System

We will now briefly describe the architecture of the system (Torkkola et al., 1991), which is delineated in Figure 1.

As mentioned, our principal task is classification of short-time feature vectors into phonemic classes. As feature vectors we use groups of (say, three) adjacent 20-component mel-scale cepstra. To allocate more resources to cases that are hard to separate (like unvoiced plosives), we are actually using two different LVQ classifiers (or codebooks): one trained for all phonemes, whereby the closure parts of /k/, /p/, and /t/ are regarded as one class /#/, and another trained only by examples of the glottal stop and the bursts of /k/, /p/, and /t/. These codebooks now produce two parallel streams of classification results. These streams consist of phoneme labels every 10 milliseconds, and they are called quasi-phoneme sequences.

The quasi-phoneme sequences are decoded into phonemic transcriptions using fairly standard multiple-codebook HMM techniques (Rabiner, 1989). The main difference from normal discrete HMMs is that we are not using codebooks that aim to *represent* the underlying signal faithfully, but LVQ codebooks that are tuned to *discriminate* between phonemes.

The phonemic transcriptions obtained as a result of decoding by HMMs may contain coarticulation effects and other systematic errors. These can be corrected partly by the *Dynamically Expanding Context* (DEC) algorithm (Kohonen, 1986). The same stage also transforms phoneme sequences into text. The

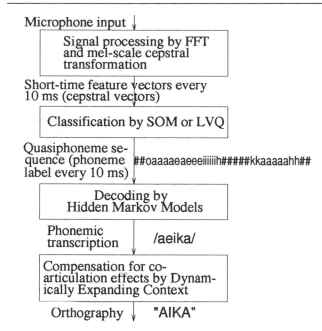

Microphone input

Signal processing by FFT and mel-scale cepstral transformation

Short-time feature vectors every 10 ms (cepstral vectors)

Classification by SOM or LVQ

Quasiphoneme sequence (phoneme label every 10 ms)   ##oaaaaeaeeeiiiiiih#####kkaaaaahh##

Decoding by Hidden Markov Models

Phonemic transcription   /aeika/

Compensation for co-articulation effects by Dynamically Expanding Context

Orthography   "AIKA"

Figure 1. Block diagram of the Phonetic Typewriter.

central idea underlying the DEC is to derive unique symbol-to-symbol(s) mapping rules from two exemplary streams of symbols: the phonemic transcription produced by the LVQ/HMM system (the source) and the orthographic form of the utterance (the target). The DEC constructs rules that are most general in the sense of an explicitly defined specificity hierarchy.

The algorithm starts at the most general level, by creating initial context-free transformation rules, which in this case are phoneme-to-letter rewrite rules. Whenever two rules with the same condition part have different productions, the conflict is resolved by taking more phoneme symbols around the initial condition part to make the rules unique. In this way, the generality of the rules will be maximal, while all the conflicts within the training cases are resolved. The resulting rule set can be interpreted as a set of decision trees, where the decisions are made according to the context of the phoneme in question.

The rule set can then be used to correct errors similar to those appearing in the training examples. Each symbol in the transcription to be corrected is searched for in the respective rule tree according to its context. The corrected result will be the concatenation of the found productions. In speech recognition experiments the DEC has been able to correct 50–75% of the errors remaining in phonemic transcriptions. Using the described configuration, letter accuracies around 94–95% can be obtained.

Discussion

We have described a hybrid approach to develop an automatic dictation machine, the Phonetic Typewriter, partly based on ANNs. Our approach consists of three main parts: (1) classification of speech feature vectors into phoneme classes using SOM or LVQ networks, (2) phoneme sequence modeling and recognition using HMMs, and (3) mapping from erroneous phonemic transcriptions to orthography using a learning grammar.

One advantage of neural networks for speech recognition is their discriminative ability. However, the temporal nature of the speech signal poses difficulties to current ANN architec-

tures, whereas HMMs are well known for their ability to model temporal structures. Thus, a hybrid of these two methodologies seems to be desirable. For examples of other kinds of ANN/HMM hybrids, we could mention Bourlard, Morgan, and Renals (1992) and Le Cerf, Ma, and Van Compernolle (1994).

Our approach appears to work well with languages whose orthography and phonemic transcriptions have a simple correspondence. This finding has been verified by experiments with Finnish and romanized Japanese. Other languages similar in this sense are Italian and Spanish. However, we do not claim that our approach would be usable as such with languages very different in this aspect, such as English, French, or Chinese. For these languages, an acceptable solution to an automatic dictation machine might look totally different. The system ought to have knowledge about the dictionary, since the mapping from phonemic transcriptions to orthography is one-to-many.

Since a corresponding system would not be useful as such for many major languages of the world (like English), it is hard to compare this architecture to others. However, it is possible to construct a similar system using HMM techniques only. Compared to such a system, ours has the advantage of improved discrimination between phonemes by the virtue of LVQ. This is exactly what we need, because in Finnish recognizing phonemes is already almost recognizing letters.

The Phonetic Typewriter implements a pure bottom-up (signal-processing) approach, which does not apply much knowledge about the language. The two partial tasks where some linguistic knowledge is involved are decoding by the HMM, where conditional phoneme sequence probabilities are estimated from text, and postprocessing by the DEC, where the morphology of the training vocabulary is implicitly involved. Were it possible to integrate more human linguistic knowledge, improved performance might be the result. However, this possibility is still an open problem facing all speech recognition research.

Road Map: Applications of Neural Networks
Background: I.3. Dynamics and Adaptation in Neural Networks
Related Reading: Speaker Identification

References

Averbuch, A., et al., 1987, Experiments with the TANGORA 20000 word speech recognizer, in *Proceedings of the International Conference on Acoustics, Speech, and Signal Processing*, Piscataway, NJ: IEEE, pp. 701–704.

Baker, J. M., 1989, DragonDictate™–30K. Natural language speech recognition with 30,000 words, in *Proceedings of the European Conference on Speech Communication and Technology*, Edinburgh: CEP Consultants, pp. 161–163.

Bourlard, H., Morgan, N., and Renals, S., 1992, Neural nets and hidden Markov models: Review and generalizations, *Speech Commun.*, 11:237–246.

Kohonen, T., 1986, Dynamically Expanding Context, with application to the correction of symbol strings in the recognition of continuous speech, in *Proceedings of the 8th International Conference on Pattern Recognition*, Washington, DC: IEEE Computer Society Press, pp. 1148–1151.

Kohonen, T., 1988, The "Neural" Phonetic Typewriter, *IEEE Computer*, 21:11–22.

Kohonen, T., 1990a, Improved versions of learning vector quantization, in *Proceedings of the International Joint Conference on Neural Networks*, Ann Arbor, MI: IEEE Neural Networks Council, pp. 545–550.

Kohonen, T., 1990b, The self-organizing map, *Proc. IEEE*, 78:1464–1480. ◆

Kohonen, T., 1993, Physiological interpretation of the self-organizing map algorithm, *Neural Netw.*, 6:895–905.

Kohonen, T., Mäkisara, K., and Saramäki, T., 1984, Phonotopic maps —Insightful representation of phonological features for speech recognition, in *Proceedings of the 7th International Conference on Pattern Recognition*, Silver Spring, MD: IEEE Computer Society Press, pp. 182–185.

Kohonen, T., Németh, G., Bry, K.-J., Jalanko, M., and Riittinen, H., 1979, Spectral classification of phonemes by learning subspaces, in *Proceedings of the International Conference on Acoustics, Speech, and Signal Processing*, Piscataway, NJ: IEEE, pp. 97–100.

Le Cerf, P., Ma, W., and Van Compernolle, D., 1994, Multilayer perceptrons as labelers for hidden Markov models, *IEEE Trans. Speech Audio Process.*, 2:185–193.

Rabiner, L. R., 1989, A tutorial on Hidden Markov Models and selected applications in speech recognition, *Proc. IEEE*, 77:257–286. ◆

Torkkola, K., Kangas, J., Utela, P., Kaski, S., Kokkonen, M., Kurimo, M., and Kohonen, T., 1991, Status report of the Finnish phonetic typewriter project, in *Proceedings of the International Conference on Artificial Neural Networks*, Amsterdam: Elsevier, pp. 771–776.

Speech Recognition: Feature Extraction

Nelson Morgan and Hervé Bourlard

Introduction

Given the complexity of the problem, the *Automatic Speech Recognition* (ASR) formalism is generally split into four major components:

1. A signal processing and feature extraction module that transforms the speech signal into a sequence of acoustic feature vectors. This is the issue that is addressed in this paper.
2. A local hypothesis generator that may either produce a label or some phonetic hypotheses about a speech segment (associated with one or more acoustic vectors). This is usually based on speech unit models (typically word or phoneme models) that are trained on a large amount of speech data containing many occurrences of the speech units in many different contexts.
3. A time alignment and pattern-matching module that transforms the local hypothesis into a global decision for word or sentence recognition.
4. A language module that should interact with the pattern-matching module to help the recognizer incorporate syntactic, semantic, and pragmatic constraints. For more information on language modeling by neural networks, see, e.g., Morgan and Scofield (1991).

Much of the work in ASR focuses on implementing steps 2 and 3 with statistical speech modeling, using either Hidden Markov Models (HMMs), Multilayer Perceptrons (MLPs), or a combination of both, to enhance recognition performance. (These topics are discussed in the companion article, (SPEECH RECOGNITION: PATTERN MATCHING.) However, neither of these approaches can entirely compensate for the effects of choosing a poor representation (i.e., feature extraction) in the first place.

Feature Extraction

Feature extraction, in its general formulation, consists of transforming the speech waveform into a sequence of acoustic feature vectors that is robust to acoustic variations, but is representative of the lexical content, i.e., the words and phonemes that are pronounced.

This front-end analysis is generally based on some form of spectral analysis performed at regular intervals—for instance, every 10 ms. For each such analysis, a fixed-length (20 to 30 ms) window or "frame" over which the signal is assumed to be stationary is used for the spectral estimate. Many variants of spectral analysis have been used, including Linear Predictive Coding (LPC), Perceptual Linear Prediction (PLP) (Hermansky, 1990), and log power spectral or cepstral coefficients (which are the Fourier transform of the log spectrum) computed from a spectrum with "mel scale" spacing, which roughly corresponds to auditory "critical bands" (Davis and Mermelstein, 1980). A critical band is the range of frequencies passed by the equivalent auditory filter used to model masking phenomena (i.e., the covering of the perception of a tone by a noise with similar frequency content). For a good overview of the basic speech analysis techniques that underlie these approaches, we refer the reader to Schaefer and Rabiner (1975). Most of these techniques are motivated by Fourier analysis and suffer from several potential weaknesses when used for speech recognition, including:

- Because the same type of feature extraction is used for the entire recognition process, the analysis generates features that are not optimally adapted to the inherent variability among speech sounds.
- By performing frame-by-frame analysis (at a constant frame rate), the time domain information and the dynamics of the signal are poorly modeled. To alleviate this problem, static feature vectors are usually augmented by their first (and sometimes second) time derivatives.

Partly as a consequence of these limitations of the front-end analysis, the performance and the robustness of state-of-the-art speech recognizers is still deficient. Even with an infinite amount of training data and unlimited CPU power, it is unclear whether these recognizers would achieve human performance.

Another major problem of existing speech analysis techniques is their lack of robustness to adverse conditions, including: (1) stationary and nonstationary errors such as additive noise; (2) distorted acoustics and speech correlated noise, including nonlinear distortions; and (3) channel (microphone and filter characteristics) variations, including stationary or slowly varying convolutional error. Robust feature extraction should be capable of handling these problems.

Also, typically, most feature extraction methods apply the same transformation regardless of the quality or content of the speech at each time; i.e., training or adaptation is not used. This frequently results in significant degradations.

In this article, we briefly discuss some of the approaches that have been used to alleviate these limitations, as well as to get more discriminant (and class-dependent) features. In keeping with the focus of this *Handbook*, we will restrict ourselves to those approaches that have been interpreted (however loosely) as "neural," e.g., due to the modeling of some function of large numbers of neurons or to the use of popular approaches based on gross models of neuron behavior.

Auditory Models

The mammalian auditory nerve appears to retain much time domain information, although there is little doubt that some sort of spectral analysis occurs (Greenberg, 1988). This suggests that a conventional power spectrum estimate may not be the appropriate representation for speech (or other auditory signals), particularly in the presence of noise. A number of researchers have proposed models to better simulate the auditory nerve response characteristics. In some cases, these have been shown to yield improved speech recognition in noisy environments. For a good review of these physiologically motivated preprocessors, see Greenberg, (1988).

Physiologically based front ends are computationally intensive. While general purpose computing capabilities are improving at an exponential rate, some researchers have worked to develop efficient VLSI realizations of auditory models. The current implementations have primarily been continuous time analog CMOS designs, although some recent work has been done in digital interfaces so that the analog models can be used as part of a conventional computing system. One such design merges a cochlear filter bank, simplified inner hair cell models, and an arbitration tree to communicate the time and location of firing events (filter and hair cell) off the chip (Lazzaro et al., 1993). This chip, like most of the others in this class, are designed to consume very little power, and thus may be suitable for portable applications.

All of these approaches are based on physiological models. Other, more engineering-oriented representations are currently in use or under investigation, such as mel-scaled features (Davis and Mermelstein, 1980), the auditory spectrum of PLP (Hermansky, 1990), and the newer RelAtive SpecTral Analysis (RASTA) approaches to overcoming additive and convolutional noise (Hermansky, Morgan, and Hirsch, 1993). These engineering-oriented approaches are based on simple properties of psychological or psychoacoustic models of the auditory system. For instance, PLP was designed to be more robust to interspeaker acoustic variabilities, in particular between speakers with very different vocal tracts (e.g., between adults and children). RASTA approaches provide some robustness to spectral changes that are much slower (or faster) than phonemic rates, performing a bandpass filtering on spectral trajectories reminiscent of forward temporal masking. Most of these psychological models tend to be less demanding computationally than the physiologically based models.

Some auditory representations, such as the correlogram of Lyon and Slaney (Lyon, 1991) fall somewhere in between these two categories, being partially based on physiology and partly on observed behavior in auditory experiments.

A Noise-Reduction Network

In Tamura and Waibel (1988), an MLP was trained, using the backpropagation (BP) algorithm, to extract clean speech from a noisy input signal (for both stationary and nonstationary noise). In this case, 60 sample points of noisy speech were presented at the input of the network and 60 points of the original clean signal were used as the desired output. The training was performed on a phonetically balanced set of utterances to permit the network to discriminate noise from actual speech. Good performance has been reported on this task, particularly

in terms of listener preference, in comparison with spectral subtraction. In a related approach, several researchers have experimented with such mappings for feature vectors rather than the speech waveform. In one such experiment (Trompf, 1992), linear predictive cepstra (LPC) were mapped from an additive-noise condition to a clean condition. In this case, an isolated word task was strongly improved by such processing, and the use of a moderate amount of additive noise in the training provided a fair amount of robustness to a wide range of noise amplitudes. Preliminary cross-noise tests (between computer-room and printer-room noise) showed some small degree of degradation, except for the case of extremely high noise levels (in which case the degradation was much larger). It should be noted that many other kinds of noise might be far more dissimilar and may not be handled well by this method. Another interesting note is that a purely linear network mapping did moderately well at the mapping, although performance was significantly better when additional nonlinear elements were employed. None of these systems are adaptive (after the initial training), and thus are unable to adjust to slowly varying noise conditions.

Self-Organized Feature Maps

Many researchers have experimented with variants on the ideas proposed by T. Kohonen for the unsupervised learning of spatially related features (Kohonen, 1988). In this approach, reference feature vectors are used to partition a two-dimensional space (referred to as a *self-organized*, or *topological*, *map*) that the *n*-dimensional feature vectors are mapped to while preserving the neighborhood relations of the initial space. In their application of this approach to speech recognition, Kohonen and his colleagues constructed what they called a Neural Phonetic Typewriter (see SPEECH RECOGNITION: A HYBRID APPROACH). A mapping was found between the clusters discovered by the self-organized map and the phonemes of Finnish (and also of Japanese in a second implementation). Because both languages have an unambiguous mapping between the orthographic representation (letters) and the phonemic transcription, recognition at the word level was then possible. However, discrimination between similar phonemes was evidently not reliable from these mappings alone, requiring Kohonen to develop another map to analyze other information such as transient spectra.

A number of researchers have also experimented with a related method that directly incorporates supervisory information in the formation of the reference vectors. These approaches are generally called Learning Vector Quantization (LVQ), as they basically consist of training a topological map using supervisory class membership information to adjust the map to discriminate between classes. There are a number of variants, most notably LVQ2 (McDermott and Katagiri, 1989). At least for small speech classification tasks such as phoneme recognition, these approaches appear to be competitive with backpropagation.

Autoassociative MLPs

Feature extraction can be implemented in several different ways. In the case of class-dependent feature extraction, some form of supervision is usually required. For instance, an MLP can be trained to discriminate between classes. This is a form of nonlinear discriminant analysis, where a linear discriminant computes a weighted sum of terms which are nonlinear functions of the MLP inputs and compares this sum to a threshold, where the weights and threshold are optimized to discriminate between classes. In both cases, the weighted sum can be interpreted in terms of feature extraction or feature transformation.

However, it is sometimes desirable to do unsupervised nonlinear feature extraction, in particular to reduce dimensionality while preserving or enhancing salient features in the data. MLPs have been applied to this task. Because MLPs usually require supervision for training, the most efficient way to do unsupervised feature extraction with MLPs is to use teaching signals that are identical to the input, as this avoids explicit segmentation and labeling of the signal. In this case, the MLP is trained to give an output pattern that is the same as the pattern presented at the input via one or several layers of hidden units where the middle hidden layer would have a dimension lower than the input dimension and would encode the feature vector extracted from the input. This use of an MLP as a trainable nonlinear feature extractor was systematically investigated in Elman and Zipser (1988) for speech coding and in Cottrell, Munro, and Zipser (1988) for image compression. For this particular mode of operation, known as autoassociation or identity mapping, the output layer does not generally contain any nonlinear function (at least for real valued inputs) since the desired output pattern is identical to the input pattern.

However, it has been shown in Bourlard and Kamp (1988) that this kind of autoassociative MLP with linear output units (and a single hidden layer containing $p < n$ hidden units, where n is the number of input units) is, at best, an indirect way of performing data compression via a Karhunen-Loève transform. More precisely, it was shown that the optimal weight values can be efficiently evaluated by the standard linear algebra commonly used for Singular Value Decomposition (SVD), also known as PRINCIPAL COMPONENT ANALYSIS (q.v.), and that, in this case, the nonlinearity in the hidden units is theoretically of no help. The optimal solution can thus be obtained directly from linear algebra (no local minima, no need of gradient procedure). Additionally, the computation for the direct SVD method is considerably more efficient.

Though this conclusion may sound pessimistic, there are still some potential benefits of the MLP-based approach. It provides an efficient parallel implementation of the SVD algorithm that can be easily integrated in a general neural network framework. Also, the MLP solution typically requires only moderately precise arithmetic for weights and activations (as opposed to the high precision that is required for the direct SVD), and so may be more suitable for hardware neural network implementation.

We should note here that, in theory, multiple hidden layers with more hidden units than input and output units ($p > n$) could provide a nonlinear expansion prior to the SVD-equivalent mappings. In this case, the overall transformation could potentially be improved over SVD.

Ignorance-Based Feature Extraction

A number of researchers have experimented with the automatic learning of features relevant to particular classification tasks; work of this nature goes back at least as far as Rosenblatt and his students at Cornell in the late 1950s. For speech classification problems, a layered network approach was used in 1981 to select relevant features for voiced-unvoiced classification (Gevins and Morgan, 1984). In such schemes, domain-specific knowledge was used to guide the choice of candidate feature sets, and the candidates were evaluated on the basis of classification performance. Many other researchers have used separation criteria such as interclass variance versus intraclass variance to develop features using Linear Discriminant Analysis

(LDA). Similarly, the hidden layer of a trained MLP classifier can be viewed as a feature extractor in which the signal transformation has been optimized by a discriminant criterion.

All of these techniques have in common the idea of a mixture between knowledge-based and ignorance-based methods. In the former class of approaches, domain-specific knowledge must be used to reduce the search space of possible features, as well as to reduce the noise or irrelevant contributions from features that are unlikely to represent discriminant properties of the population. In contrast, the latter family of approaches is automatic and data-driven, and is characterized by searches for combinations and weightings of system parameters to optimize some simple criteria. Such approaches make sense for searches in data spaces that are insufficiently determined by human knowledge. However, ignorance-based approaches alone often perform poorly. For instance, a number of researchers have experimented with using raw speech waveforms to train an MLP to learn speech categories. However, in practice, the speech waveform is a notoriously ambiguous predictor of the speech class, and the imperfect training of the MLP leads to much poorer speech classification than can be achieved with even the simplest power spectral measure. This simple piece of domain-specific knowledge can have an enormous effect on speech recognition performance.

Therefore, domain knowledge (which, as we have noted above, can include representations of physiological or psychological models of hearing) can be used to limit the search for relevant features. Using ignorance-based approaches such as MLP estimators is necessary, however, to map plausible inputs to the most discriminant representations, i.e., to representations that are most useful for discriminating between pattern classes. For classification or recognition tasks, there are major advantages to using mappings chosen to discriminate between the potential classes. For this reason, dimensionality reduction techniques such as autoassociation or SVD typically are not sufficient to provide a good speech representation; components accounting for small amounts of the signal variance can often be very important for discrimination. Thus, for classification or recognition purposes, linear or nonlinear discriminant approaches are probably to be preferred for the ignorance-based aspect of feature selection and extraction.

Discussion

As with many other subject areas, the descriptor *neural* can mean a variety of different things for speech recognition feature extraction. Models of auditory function on a number of levels have been investigated, sometimes based on detailed neurophysiological data and sometimes based on behavioral (psychoacoustic) models. Purely mathematical approaches, sometimes implemented with sigmoidal dot-product units, as commonly used now in MLPs, have also been used to reduce dimensionality from a raw short-term spectral representation. Finally, both variance-based and discriminant approaches to dimensionality reduction can be evaluated directly in terms of recognition performance (an ignorance-based approach) by constraining the feature choices using domain-specific knowledge (a knowledge-based approach). In such a fashion, features that have been explicitly designed for robust representation of speech in the presence of stationary convolutional or additive error (Hermansky et al., 1993) (thus using domain specific knowledge, i.e., that additive and convolutional error is common for the speech application) can be combined to optimize phonemic classification by an MLP trained with an ignorance-based procedure such as BP.

Road Map: Applications of Neural Networks
Background: I.3. Dynamics and Adaptation in Neural Networks
Related Reading: Auditory Periphery and Cochlear Nucleus; Prosthetics, Neural; Speaker Identification

References

Bourlard, H., and Kamp, Y., 1988, Auto-association by multilayer perceptrons and singular value decomposition, *Biol. Cybern.*, 59: 291–294.

Cottrell, G., Munro, P., and Zipser, D., 1988, Image compression by backpropagation: A demonstration of extensional programming, in *Models of Cognition: A Review of Cognitive Science*, vol. 1 (N. E. Sharkey, Ed.), Norwood, NJ: Ablex, pp. 208–240.

Davis, S., and Mermelstein, P., 1980, Comparison of parametric representations of monosyllabic word recognition in continuously spoken sentences, *IEEE Trans. Acoust. Speech Signal Process.*, 28:357–366.

Elman, J., and Zipser, D., 1988, Learning the hidden structure of speech, *J. Acoust. Soc. Am.*, 83:615–626.

Gevins, A., and Morgan, N., 1984, Ignorance-based systems, in *Proceedings of the IEEE International Conference on Acoustics, Speech, and Signal Processing*, New York: IEEE, pp. 39A.5.1–39A.5.4.

Greenberg, S., 1988, The ear as a speech analyzer, *J. Phonetics*, 16:139–149. ◆

Hermansky, H., 1990, Perceptual linear predictive (PLP) analysis of speech, *J. Acoust. Soc. Am.*, 87:1738–1752.

Hermansky, H., Morgan, N., and Hirsch, H., 1993, Recognition of speech in additive and convolutional noise based on RASTA spectral processing, in *Proceedings of the IEEE International Conference on Acoustics, Speech, and Signal Processing*, New York: IEEE, vol. 2, pp. 83–86.

Kohonen, T., 1988, The "Neural" Phonetic Typewriter, *IEEE Computer*, 21:11–22.

Lazzaro, J., Wawrzynek, J., Mahowald, M., Sivilotti, M., and Gillespie, D., 1993, Silicon auditory processor as computer peripherals, *IEEE Trans. Neural Netw.*, 4:523–528.

Lyon, R., 1991, CCD correlators for auditory models, in *Proceedings of the IEEE Asilomar Conference on Signals, Systems, and Computers*, Los Alamitos, CA: IEEE Computer Society Press, pp. 785–788.

McDermott, E., and Katagiri, S., 1989, Shift-invariant, multi-category phoneme recognition using Kohonen's LVQ2, in *Proceedings of the IEEE International Conference on Acoustics, Speech, and Signal Processing*, New York: IEEE, pp. 81–84.

Morgan, D. P., and Scofield, C. L., 1991, *Neural Networks and Speech Processing*, Norwell, MA: Kluwer Academic.

Schaefer, R., and Rabiner, L., 1975, Digital representations of speech signals, *Proc. IEEE*, 63:662–667. ◆

Tamura, S., and Waibel, A., 1988, Noise reduction using connectionist models, in *Proceedings of the IEEE International Conference on Acoustics, Speech, and Signal Processing*, New York: IEEE, pp. 553–556.

Trompf, M., 1992, Neural network development for noise reduction in robust speech recognition, in *Proceedings of the International Joint Conference on Neural Networks*, New York: IEEE, vol. 4, pp. 722–727.

Speech Recognition: Pattern Matching

Hervé Bourlard and Nelson Morgan

Introduction

In the companion article (see SPEECH RECOGNITION: FEATURE EXTRACTION), the overall formalism generally used for Automatic Speech Recognition (ASR) was presented, and the first step of this process, referred to as *feature extraction*, was discussed. It was shown that the general goal of feature extraction is to transform the sampled and digitized speech signal into a sequence of acoustic vectors $X = \{x_1, \ldots, x_n, \ldots, x_N\}$, where x_n is a d-dimensional vector typically representing the nth 10-ms speech frame.

In this paper, we will mainly focus on the steps following feature extraction, i.e., *local hypothesis generation* and *pattern matching*. In the sections that follow, we will describe

- The basic principle of Hidden Markov Models (HMMs), the speech recognition formalism most commonly used today (Rabiner, 1989; Lee, 1989).
- Artificial Neural Network (ANN) approaches that have been investigated to classify temporal patterns like speech. However, we will see that we do not currently know how to solve the whole problem of ASR solely with ANNs.
- HMM-motivated neural networks, i.e., networks that implement basic HMM algorithms.
- Hybrid HMM/ANN approaches, which, we believe, have the potential to improve over state-of-the-art, HMM-based ASR systems.

For brevity's sake, this article will not address the problem of language modeling. However, this is also an important topic of research for both standard and ANN approaches. For more information on language modeling and neural networks, see, e.g., Morgan and Scofield (1991). For a good book on speech processing in general and speech recognition in particular, see Deller, Proakis, and Hansen (1993).

Hidden Markov Models (HMMs)

General Description

One of the greatest difficulties in speech recognition is to model the inherent statistical variations in speaking rate and pronunciation. The most efficient approach developed for this problem consists in modeling each speech unit (e.g., words, phonemes, triphones, or syllables) by an HMM. Recently, a number of researchers demonstrated (for some limited tasks) several accurate, large-vocabulary, speaker-independent, continuous speech recognition systems based on this approach. For good overviews of the fundamentals of HMMs, see Rabiner, (1989) and Lee (1989).

Although speech is a nonstationary process, HMMs assume that the sequence of feature vectors is a piecewise stationary process. That is, an utterance $X = \{x_1, \ldots, x_n, \ldots, x_N\}$ is modeled as a succession of discrete stationary states $Q = \{q_1, \ldots, q_k, \ldots, q_K\}$, $K < N$, with instantaneous transitions between these states. In this case, an HMM is defined as a stochastic finite state automaton with a particular (generally strictly left-to-right, given the sequential nature of speech) topology. An example of a simple HMM is given in Figure 1.

This could be the model of a word or phoneme which is assumed to be composed of two stationary parts (respectively associated with the beginning and the end of the word or the

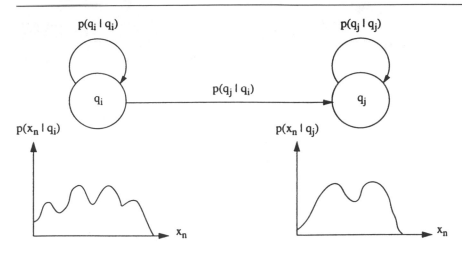

Figure 1. Schematic of a two-state, left-to-right hidden Markov model (HMM).

phoneme). Thus, there are two concurrent stochastic processes: the sequence of HMM states modeling the temporal structure of speech, and a set of state output processes modeling the (locally) stationary character of the speech signal. The HMM is called *hidden* because there is an underlying stochastic process (i.e., the sequence of states) that is not observable, but affects the observed sequence of events. Ideally, there should be one HMM for every word or sentence allowed in the recognition task. Since this is generally not feasible for large lexicon and/or continuous speech, a hierarchical scheme is usually adopted where sentences are modeled as sequences of words and where words are modeled as sequences of subword units (usually phonemes). In this case, each subword unit is represented by its own HMM.

Once the topology of the HMMs has been defined (usually arbitrarily!), the main criterion used for training and decoding is based on the "likelihood" $P(X|M)$, i.e., the probability of an acoustic vector sequence X given a Markov model M. It can be shown that, provided several assumptions are met (Bourlard and Morgan, 1993), the (global) probability $P(X|M)$ can be expressed and computed in terms of (local) probabilities $p(x_n|q_k)$ and $p(q_l|q_k)$, respectively referred to as emission and transition probabilities. Usually, emission probabilities $p(x_n|q_k)$ are estimated by assuming that they are described as (mixtures of) multivariate Gaussian densities.

Given this formalism, there are then two solutions to the "estimation problem," i.e., the calculation of $P(X|M)$. The first one, usually referred to as the *forward-backward algorithm*, computes the actual likelihood $P(X|M)$ by summing the probabilities of all possible paths through the HMM that can generate the acoustic vector sequence X. In other words, if $S = \{s_1, \ldots, s_n, \ldots, s_N\}$ is a path of length N (N = number of acoustic vectors in X), in which $s_n \in Q$ is the state visited at time n, the forward-backward algorithm computes, in a very efficient way, the global likelihood

$$P(X|M) = \sum_{\{S\}} P(X|S) \qquad (1)$$

where $P(X|S) = \prod_{n=1}^{N} p(x_n|s_n)p(s_n|s_{n-1})$, and where the sum extends over all possible paths S of length N in M.

Another criterion, usually referred to as the *Viterbi criterion*, approximates the actual likelihood as the likelihood of only the best path through the HMM, i.e.,

$$\overline{P}(X|M) = \max_{\{S\}} P(X|S) \qquad (2)$$

In this case, the solution to this problem can be obtained by

using a dynamic programming algorithm (also referred to as Dynamic Time Warping, or DTW; see Ney, 1984) that finds the shortest path through a graph (the HMM), i.e., the best segmentation of X in terms of the statistical densities associated with each HMM state. The dynamic programming recurrence then has the following form:

$$P(X_1^{n+1}|S_1^{n+1}) = p(x_{n+1}|s_{n+1}) \max_{\{s_n\}} \{P(X_1^n|S_1^n)p(s_{n+1}|s_n)\} \qquad (3)$$

where X_1^n is the partial acoustic vector sequence $\{x_1, \ldots, x_n\}$, S_1^n is the optimal state sequence associated with X_1^n, and $\{s_n\}$ is the set of possible predecessor states of s_{n+1} (given by the topology of the HMM). In this case, $\overline{P}(X|M)$ is obtained at the end of the sequence and is equal to $\max_{\{s_N\}} P(X_1^N|S_1^N)$, where $\{s_N\}$ represents the set of possible final states of M.

During training, the HMMs' parameters have to be calculated to maximize the likelihood of each training utterance given the associated (and known during training) Markov models; this is usually referred to as *maximum likelihood* training. Powerful iterative training procedures exist for both criteria and have been proved to converge to a local minimum. Also, efficient decoding procedures (based on the estimation algorithms) that find the (sequence of) HMM(s) maximizing the full likelihood or the Viterbi criterion associated with an unknown utterance X are known. Segmentation of the utterances in terms of the speech units results from the Viterbi training and decoding.

For good reviews of these training and decoding algorithms, see Rabiner (1989); Bourlard and Morgan (1993); and Deller, Proakis, and Hansen (1993).

HMM Advantages and Drawbacks

HMMs have been very useful for speech recognition. Over the last few years, a number of laboratories have demonstrated large-vocabulary (at least 1000 words), speaker-independent, continuous speech recognition systems based on HMMs. HMMs are models that can deal efficiently with the temporal aspect of speech (time warping) as well as with frequency distortion. They also benefit from powerful and efficient training and decoding algorithms. For training, only the transcription in terms of the speech units which are trained is necessary and no explicit segmentation of the training material is required. Also, HMMs can easily be extended to include phonological and syntactical rules (at least when they use the same statistical formalism).

These assumptions permit efficient optimization of statistical models. However, they can also limit generality. As a consequence, they also suffer from several drawbacks, including:

- Poor discrimination due to the training algorithm, which maximizes likelihoods instead of a posteriori probabilities $P(M|X)$ (i.e., the HMM associated with each speech unit is trained independently of the other models).
- A priori choice of model topology and statistical distributions, e.g., assuming that the probability density functions associated with the HMM state can be described as (mixtures of) multivariate Gaussian densities, each with a diagonal-only covariance matrix (i.e., correlations between components of the acoustic vectors are disregarded).
- Assumption that the state sequences are first-order Markov chains. ·
- Typically, no acoustical context is used, so that possible correlations between successive acoustic vectors is overlooked.

Most of the research done so far on ANNs for ASR has been aimed at avoiding some or all of these drawbacks, or at defining acoustic features that are less sensitive to noise or speaker dependencies.

ANNs and Time Sequence Matching

Both HMMs and ANNs can learn. Since they have been the most extensively studied in the framework of speech recognition, approaches using Multilayer Perceptrons (MLPs) (see PERCEPTRONS, ADALINES, AND BACKPROPAGATION), a particular form of ANNs, will be the primary consideration here. MLPs have several advantages that make them particularly attractive for ASR, e.g.:

- They provide discriminant-based learning; that is, when trained for classification—e.g., using a Least Mean Square (LMS) criterion—the parameters of the MLP output classes (which could be associated with speech units or, as shown later on, with HMM states) are trained to minimize the error rate while maximizing the discrimination between the correct output class and the rival ones. In other words, MLPs not only train and optimize the parameters of each class on the data belonging to that class, but also have the goal of rejecting as much as possible the data belonging to other (rival) classes.
- Because ANNs are capable of incorporating multiple constraints and finding optimal combinations of constraints for classification, features do not need to be treated as independent. More generally, there is no need for strong assumptions about the statistical distributions of the input features (as is usually required in standard HMMs).
- They have a very flexible architecture which easily accommodates contextual inputs and feedback.
- ANNs are typically highly parallel and regular structures, which makes them especially amenable to high-performance architectures and hardware implementations.

The pattern matching problem in continuous speech recognition can be formulated in this simple manner: How can an input sequence (e.g., X) be properly explained in terms of an output sequence (e.g., a sequence of phonemes or words) when the two sequences are not synchronous (since there are multiple acoustic vectors associated with each pronounced word or phoneme)? However, a connectionist formalism is not very well suited to solve such a problem; most previous applications

of ANNs to speech recognition have depended on severe simplifying assumptions (e.g., small vocabulary, isolated words, known word, or phoneme boundaries). In this section, several ANNs that have been used to solve subtasks of the ASR problem will be reviewed. However, these approaches do not address the problem of time alignment (and segmentation, when required), which is efficiently solved by dynamic programming in HMMs. Neural networks that have been designed to implement some aspect of HMM functionality will be discussed in the next section.

For a good earlier review of the different approaches using neural networks for speech recognition, see Lippmann (1989). For a good overview of ANNs for speech processing in general, see Morgan and Scofield (1991).

Static Networks

The simplest way to perform sequence recognition is to turn the temporal pattern into a spatial pattern at the input layer of an MLP for which the standard backpropagation (BP) algorithm (see BACKPROPAGATION: BASICS AND NEW DEVELOPMENTS) can be used for training. In this case, the entire input sequence to be trained or recognized is stored in a buffer at the input of the MLP (Lippmann, 1989) and the possible speech units of the lexicon (e.g., isolated words or presegmented phonemes) are associated with the output units of the MLP. In some cases, such tapped delay MLPs are used in conjunction with conventional time alignment techniques.

Although this general approach to sequence recognition has been shown to lead to similar or slightly better performance than standard HMMs, it has several drawbacks:

- The input buffer must be large enough to accommodate the longest possible input sequence, which increases the number of parameters and, consequently, the number of required training examples.
- The network is not automatically shift and distortion invariant; to have such a property, it is often necessary to train it on a large number of utterances for each output class and to shift them everywhere through the input layer.
- This approach does not seem to be appropriate for continuous speech recognition since such a network is used with "static" input patterns associated with only one output class.

Recurrent Networks

Ideally, ANNs used for speech recognition should accept input vectors sequentially, which requires some kind of recurrent internal state that would be a function of the current input and the previous internal state (Kuhn, Watrous, and Ladendorf, 1990; Bridle, 1990; Robinson, 1994). Various recurrent networks (see RECURRENT NETWORKS: SUPERVISED LEARNING) using time-step delayed recurrent loops on the hidden and/or output units of a feedforward network have been proposed and tried. For sequences with a small maximum length T, we can turn these recurrent networks into equivalent feedforward networks (by "unfolding" them over the time period T) that can be trained by a slightly modified form of BP, referred to as BP through time, in which

- All copies of the "unfolded" weights remain identical during training, and
- The desired outputs are functions of time, and errors are computed (and backpropagated) for every copy of the output layer. Another solution is to define the target function

(and to backpropagate the error) only at certain times (i.e., on certain copies of the output units) corresponding, for instance, to the end of the words or phonemes.

An implementation of BP through time for speech recognition was proposed in Watrous and Shastri (1987), where sequential processing was performed with the "temporal flow model." In this case, delayed self-loops are added to each hidden unit (for a single hidden layer) and to each output unit of an MLP.

Partially recurrent networks primarily use feedforward connections except for a restricted subset of feedback connections (Jordan, 1989). These networks are usually implemented by extending the input field with additional "feedback units" containing the hidden or output unit values generated by the preceding input. The feedback units remember some aspects of the past, and so the state of the whole network at any time depends on the aggregate of previous states as well as on the current input.

Approximating Recurrent Networks by Feedforward MLPs

An intermediate solution between the spatial input model and the recurrent model is to approximate the recurrent network over a finite time period by a feedforward network in which the loops are replaced by the explicit use of several preceding activation values. This kind of network, usually referred to as a Time-Delay Neural Network (TDNN) (Lang, Waibel, and Hinton, 1990), can be trained to recognize short time sequences (e.g., acoustic vector sequences associated with phonemes). In this case, the activations in each layer are computed from the current and multiple delayed values of the preceding layer. If the delayed activation values are not constrained to have the same set of weights, this kind of network is in one sense more general than a standard recurrent network since it has more parameters. However, it is also more restrictive since it is constrained to look over a finite time period.

Discussion

All of these models have been shown to yield good performance (sometimes better than HMMs) on short isolated speech units. By their recurrent aspect and their implicit or explicit temporal memory, they can perform some kind of integration over time. However, neural networks by themselves have not been shown to be effective for large-scale recognition of continuous speech. There is at least one fundamental difficulty with supervised training of a connectionist network for continuous speech recognition: a target function must be defined, even though the training is done for connected speech units where the segmentation is generally unknown. With recurrent neural networks, even for training isolated units, there is no principled method for selecting this target function. This is not a problem for HMM-based training, which only requires the sequence of speech units and not their temporal segmentations. For recognition, HMMs not only tackle the variability of speech pronunciation, but are also efficient tools for connected speech recognition and segmentation. This property seems to be difficult to achieve using a connectionist architecture by itself.

ANN Models of HMMs

Because of the success of HMM algorithms for speech recognition problems, several neural network implementations of HMMs have been studied. Although these are generally merely different implementations of the same formalism, they can help in understanding the relationships and limitations of each system.

The Viterbi Network

The Viterbi network (Lippmann, 1989) emulates the function of the Viterbi algorithm and, more specifically, of the dynamic programming recursion (Equation 3). In this case, each HMM is associated with a neural network in which each output unit corresponds to an HMM state. In the case of Gaussian HMMs, the first hidden layer computes the set of Gaussian outputs for the input vectors that are presented sequentially to the network. In fact, it can be shown that the logarithm of these Gaussians can be implemented using the perceptron structure. Each output node is also complemented by time-delayed connections between the different output units to represent the topology of the underlying HMM followed by a comparator subnetwork to compute the minimum of the activation values of output nodes at the previous time step. For this network, the outputs are roughly equivalent to the negative logarithm of output probabilities for the original HMM.

However, while interesting, this ANN implementation of Equation 3 does not change the basic data movement requirements of a practical implementation of the Viterbi algorithm which, in the case of continuous speech recognition, can be dominated by pointer bookkeeping (i.e., the information that has to be stored during dynamic programming to keep track of the best path and, consequently, the word or phoneme sequence associated with the best scoring path obtained at the end of the utterance). Also, it does not overcome the limitations of standard HMMs.

The Alpha-Net

In Bridle (1990), a recurrent neural network was introduced that emulates the formulation of HMMs using the full likelihood criterion. In this case, the units in the recurrent loop are linear and the acoustic vectors enter the loop via a multiplication to simulate the operation of a standard HMM node. The training of this network is done by a backpropagation through time that takes exactly the same form as the "forward-backward" training algorithm (Rabiner, 1989).

Given this equivalence between HMMs and the Alpha-Nets, some of the HMM constraints can be relaxed. For example, the constraint that probabilities sum to one and are positive could be dropped in the neural network implementation. Also, this approach can easily be generalized to other, more interesting (i.e., discriminant) training criteria like maximum mutual information.

Combining HMMs and ANNs

The idea of combining HMMs and ANNs was motivated by the observation that HMMs and ANNs had complementary properties: (1) HMMs are clearly dynamic and very well suited to temporal data, but several assumptions limit their generality; (2) ANNs can approximate any kind of nonlinear discriminant functions, are very flexible, and do not need strong assumptions about the distribution of the input data, but they cannot properly handle time sequences. However, HMMs are based on strict formalisms, making them difficult to interface with other modules in a heterogeneous system. Still, two important links between HMMs and MLPs have been studied recently and used successfully in ASR systems.

Hybrid HMM/MLP Approaches

In these approaches, MLPs are used to compute the emission probabilities required in HMM systems (Bourlard and Morgan, 1993). It has indeed been shown that if each output unit of an MLP is associated with a state q_k of the set of states $Q = \{q_1, q_2, \ldots, q_K\}$ on which the HMMs are defined, it is possible to train the MLP to generate a posteriori probabilities of the output classes conditioned on the input.

Assume a set of acoustic vectors $X = \{x_1, \ldots, x_n, \ldots, x_N\}$, where each of these vectors is labeled in terms of the MLP output classes (i.e., the q_k's corresponding to phonemes or HMM-states). This labeling could be obtained from a presegmentation or the segmentation resulting from a standard Viterbi training; it will be shown below that this segmentation can also be obtained iteratively from the hybrid HMM/MLP approach itself. Assume further that the MLP is trained with one of several common cost functions (e.g., least mean square, entropy, or relative entropy). Finally, assume that the network is trained for classification, i.e., that only one of the output units is "on" during training. In this case, it can be proved that, if the MLP contains enough parameters and if the training does not get stuck at a local minimum, the optimal output values of the MLP are estimates of posterior probabilities $p(q_k|x_n)$, when x_n is presented at the input of the MLP. The true values of these probabilities are known to lead to the optimal classification; they are discriminant by nature and minimize the classification error rate. It has been confirmed experimentally that MLP outputs do approximate posterior probabilities (Bourlard and Morgan, 1993). In a hybrid system, then, the MLP can be used to compute the emission probabilities required by the HMMs, while the HMMs take care of time warping properties. Also, the iterative Viterbi training used in standard HMMs remains valid for the MLP estimation case. In this case, the segmentation provides targets to train the MLP. Once the MLP is trained, its output values can be used as probabilities for dynamic programming to find a better segmentation, which provides a new segmentation and new targets to train the MLP further. Again, convergence of this iterative process (still to a local optimum) can be proved. See Bourlard and Morgan (1993) for a full description of this, together with the (numerous) modifications required to turn this basic scheme into a state-of-the-art system.

This hybrid HMM/MLP approach may provide more discriminant estimates of the emission probabilities needed for HMMs, without requiring strong hypotheses about the statistical distribution of the data. Since this result still holds with modified MLP architectures, the approach has been extended in a number of ways, including:

- Extending the input field to accommodate not only the current input vector but also its right and left contexts, leading to HMM systems that take into account the correlation between acoustic vectors (Bourlard and Morgan, 1993).
- Partially recurrent MLPs (Robinson, 1994) feeding back previous activation vectors on the hidden or output units, leading to a higher-order HMM.

In recent years, the performance of these hybrid approaches to continuous speech recognition has often been comparable to some of the best HMM-based systems. It also appears that in fair comparisons (i.e., using the same feature set, underlying HMM, and language model), this approach consistently shows significant improvements over systems using conventional "maximum likelihood" approaches (Bourlard and Morgan, 1993). A partially recurrent approach to this same hybrid system has also done very well on the same large vocabulary tasks (Robinson, 1994).

Initial work in using global optimization methods for continuous speech recognition by hybrid HMM/MLP approaches has also been done. For example, in Bengio et al. (1992), a similar approach is used to optimize the input parameters via either a linear or nonlinear transformation, training the parameters of the HMM according to a maximum likelihood criterion.

Predictive Neural Networks

For signal processing and speech recognition, MLPs have also been used for regression and prediction. In this case, given the previous p samples or frames of speech presented at the input of an MLP, we may train this MLP to predict the next sample or frame (by minimizing the squared error between the present sample or frame and its predicted value). This results in a kind of nonlinear autoregressive (AR) model, which is usually referred to as a predictive MLP (Levin, 1993). Since linear AR models (Rabiner, 1989) can be used to estimate emission probabilities in a particular form of HMM (usually called autoregressive HMM), predictive MLPs can also be embedded in a Markov process (Levin, 1993) to give a piecewise stationary model of speech dynamics.

If each HMM state is associated with its own predictive MLP, it can be shown that the prediction errors of both linear and nonlinear (MLP based) AR processes can be interpreted as logarithms of emission probabilities of the form $p(x_n|q_k, \{x_{n-1}, \ldots, x_{n-p}\})$ which can be used in dynamic programming of HMMs. An advantage of using this type of model is that, rather than estimating posterior likelihoods $p(x_n|q_k)$ as used in standard HMMs, it explicitly addresses observation independence by modeling the observations as an AR process. However, predictive MLPs do not have the discriminative character of direct MLP probability estimators.

Discussion

ANN systems have been developed to approach the pattern matching and search aspects of speech recognition from a number of perspectives. Research is currently dominated by recurrent and feedforward networks, each trained with some form of backpropagation. For continuous speech recognition, functional ANN systems require integration into a larger recognizer to provide the global model for matching a complete utterance to a sequence of words. This structure is currently most often provided by HMMs. Ultimately, it is possible that statistical models, possibly based on neural networks, could replace HMMs with a structure that is not prone to as many restrictive assumptions; approaches based on models of perception or articulation are plausible alternatives. However, as of this writing, it is certainly too early to proclaim the ultimate success of neural network approaches. Nonetheless, speech researchers are already absorbing ANN approaches into their "tool kits" as potential solutions to difficult subproblems in speech recognition.

Road Map: Applications of Neural Networks
Background: I.3. Dynamics and Adaptation in Neural Networks

References

Bengio, Y., de Mori, R., Flammia, G., and Kompe, R., 1992, Global optimization of a neural network: Hidden Markov model hybrid, *IEEE Trans. Neural Netw.*, 3:252–259.

Bourlard, H., and Morgan, N., 1993, *Connectionist Speech Recognition: A Hybrid Approach*, Norwell, MA: Kluwer. ◆

Bridle, J., 1990, Alpha-Nets: A recurrent neural network architecture with a hidden Markov model interpretation, *Speech Commun.*, 9:83–92.

Deller, J. R., Proakis, J. G., and Hansen, J. H., 1993, *Discrete-Time Processing of Speech Signals*, New York: Macmillan. ◆

Jordan, M., 1989, Serial order: A parallel distributed processing approach, in *Advances in Connectionist Theory: Speech* (J. L. Elman and D. E. Rumelhart, Eds.), Hillsdale, NJ: Erlbaum.

Kuhn, G., Watrous, R. L., and Ladendorf, D., 1990, Connected recognition with a recurrent network, *Speech Commun.*, 9:41–48.

Lang, K. J., Waibel, A. H., and Hinton, G. E., 1990, A time-delay neural network architecture for isolated word recognition, *Neural Netw.*, 3:23–43.

Lee, K. F., 1989, *Automatic Speech Recognition: The Development of the Sphinx System*, Norwell, MA: Kluwer. ◆

Levin, E., 1993, Hidden control neural architecture modeling of non-linear time varying systems and its applications, *IEEE Trans. Neural Netw.*, 4:109–116.

Lippmann, R. P., 1989, Review of neural networks for speech recognition, *Neural Computat.*, 1:1–38. ◆

Morgan, D. P., and Scofield, C. L., 1991, *Neural Networks and Speech Processing*, Norwell, MA: Kluwer. ◆

Ney, H., 1984, The use of a one-stage dynamic programming algorithm for connected word recognition, *IEEE Trans. Acoust. Speech Signal Process.*, 32:263–271. ◆

Rabiner, L. R., 1989, A tutorial on hidden Markov models and selected applications in speech recognition, *Proc. IEEE*, 77:257–286. ◆

Robinson, A. J., 1994, An application of recurrent nets to phone probability estimation, *IEEE Trans. Neural Netw.*, 5:298–305.

Watrous, R. L., and Shastri, L., 1987, Learning phonetic features using connectionist networks: An experiment in speech recognition, in *Proceedings of the First IEEE International Conference on Neural Networks*, San Diego, vol. 2, pp. 619–627.

Spinal Cord of Lamprey: Generation of Locomotor Patterns

Thelma L. Williams and Karen A. Sigvardt

Introduction

The patterns of muscle activation giving rise to locomotion in the lamprey can be generated by the isolated spinal cord. Progress in the understanding of the underlying mechanisms has been made by modeling at several different levels, from the detailed biophysical modeling of small collections of cells to the application of the mathematics of coupled oscillators. In this article, we highlight some of the questions addressed by the modeling, and how the answers have provided insights for biological function.

Locomotion in the lamprey is relatively simple compared with limbed vertebrates. Swimming is produced by a wave of curvature passing down the body that pushes the animal forward. This wave is produced by alternating activation of the muscles on the left and right sides of the body with a head-to-tail delay. In faster swimming, the cycle period decreases, with a proportionate decrease in the head-to-tail delay, such that the intersegmental phase lag (time delay per segment divided by cycle period) remains constant. This phase lag shows no systematic variation along the length of the body, being approximately equal to 1% per segment. Since the lamprey has approximately 100 body segments, this ensures a full wavelength on the body at all swimming speeds. The task, then, is to understand how the spinal cord circuits produce both the rhythmic alternation and the intersegmental phase delay independent of position along the body and independent of frequency.

The lamprey locomotor network can be studied in a piece of spinal cord in vitro. When treated with drugs that mimic the action of excitatory amino acid transmitters, such a preparation produces *fictive locomotion*, the same pattern of ventral root activity as during swimming in the intact lamprey (Figure 1*A*). This occurs without the descending activity from the brain that normally initiates and modulates locomotion, and without the input from peripheral receptors that provides movement-related feedback (see MOTOR PATTERN GENERATION). The basic rhythm is thus produced by rhythm-generating circuitry within the spinal cord. In this article, we discuss the modeling of this neuronal network.

In early studies it was shown that rhythmic activity alternating on the left and right sides could be produced by as few as one-and-one-half segments taken from anywhere along the length of the spinal cord. Furthermore, a piece containing many segments produces coordinated motor activity, with the appropriate intersegmental phase delay. For these reasons, the lamprey spinal cord has been viewed as a chain of oscillators coupled together in some way that ensures the appropriate phase delays. It is not yet clear whether the thousands of cells in the spinal cord are functionally grouped into unit oscillators or whether the system more nearly approximates a continuum. All the models to be considered here begin with the assumption of discrete unit oscillators. The consequences of a more continuous distribution of rhythm-generating capacity are yet to be explored.

Modeling the Segmental Oscillator

A model of the segmental neuronal network for basic rhythm generation (Figure 1*B*) was proposed by Buchanan and Grillner in 1987 and shown by computer simulation to generate rhythmic left-right alternating activity (Grillner et al., 1991). This model is based on intracellular recordings of the activity patterns and synaptic connections between spinal neurons. The three "neurons" in each half of the network represent three heterogeneous neuron classes that are *broadly defined* (rather than *identified* as is possible in many invertebrate networks). Excitatory interneurons (EINs) have ipsilateral axons that excite their synaptic targets, crossed caudal interneurons (CCs) have crossed axons that inhibit their targets, and lateral interneurons (LINs) are large neurons with ipsilateral axons that inhibit their targets. Neurons in all these classes have axons extending over more than one segment, the EINs over a few, the CCs caudally up to 20 segments, and the LINs caudally up to 50 segments; the distribution of axonal length among the neurons of each class has not been investigated. During fictive locomotion, the activity of neurons in these classes is approximately in phase with the ipsilateral motoneurons, which receive

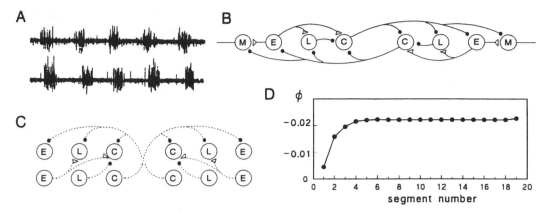

Figure 1. *A*, Rhythmic activity recorded in vitro (fictive locomotion) from two different ventral roots on the same side, separated by 15 segments. Here rostral-caudal delay is approximately 18% of the cycle duration. *B*, Network model (Buchanan and Grillner, 1987) for the unit oscillator of the lamprey central pattern generator, consisting of identified cell classes: crossed caudal interneurons (C), excitatory interneurons (E), lateral interneurons (L). Motoneurons (M) are output elements only, not participating in rhythm generation. Filled circles: inhibitory synapses; open triangles: excitatory synapses. *C*, Model of coupling between adjacent segments consisting of a repeat of the synaptic connections within a unit oscillator but with decreased synaptic strength. Only ascending coupling shown (Williams, 1992b). *D*, Intersegmental phase lags for simulated chain of 20 identical oscillators, each with the structure shown in part *B* and both ascending and descending coupling as shown in *C*. Synaptic strength in ascending coupling was 0.05 times the strength within a segment; for descending coupling, 0.02. Measurements on a pair of coupled oscillators gave $r = 0.4$, $\phi_A = -0.023$, $\phi_D = 0.023$. Data points: measured intersegmental phase lags of the simulated chain; solid line: Equation 2 (Williams, 1992a). Negative phase lag represents rostral-to-caudal delay, as seen in part *A*.

excitatory synaptic inputs during the active portion of the cycle (when the muscles on the same side contract) and inhibitory inputs during the interburst interval. All but one of the synaptic connections in the model have been confirmed by paired cell recordings. However, multiple postsynaptic targets of an individual neuron are rarely identified (the experiments are very difficult) but instead are inferred for a class of neurons on the basis of recordings between pairs of individual neurons. For example, it is inferred from three kinds of paired recordings that EIN of the model forms excitatory connections with ipsilateral LIN, CC, and MN; whether an individual EIN makes connections with all three has not been tested. Connections from a CC to three of its four targets, contralateral LIN, MN, and CC, have been demonstrated by Buchanan in two cases, although the target neurons have not been in the same segment. Another difference between the model and spinal cord biology is the absence of a class of small inhibitory interneurons which are active in phase with the ventral root bursts and have been shown to have a powerful effect on locomotor patterns (Buchanan and Grillner, 1988).

The function of the simple network of Figure 1*B* can be deduced intuitively from the proposed connections: all neurons receive tonic excitation (not shown) that simulates the effect of tonic activity in descending brainstem systems in the intact animal or bath application of excitatory amino acids in the in vitro preparation. If the neurons on one side are active, those on the other side are inhibited by CC. To get alternation, there must be a mechanism to terminate the activity of CC on the active side, thus releasing the contralateral neurons from inhibition. Several mechanisms for burst termination have been suggested. CC activity could be terminated by inhibition from ipsilateral LIN. This requires that LIN becomes active later in the excitatory phase than CC, as has been demonstrated in the lamprey. This mechanism has been shown to be sufficient, by connectionist simulations (Buchanan, 1991). Other possible mechanisms, which depend on particular cellular properties, include (1) spike frequency adaptation attributable to summation of the after-hyperpolarization in CC and (2) repolarization by outward currents activated by the influx of calcium ions through the NMDA-activated channel. Similar mechanisms have been found to contribute significantly to the generation of rhythmicity in simple invertebrate networks (see HALF-CENTER OSCILLATORS UNDERLYING RHYTHMIC MOVEMENTS).

To investigate the contributions of cellular properties, Grillner and his colleagues have developed a biophysical simulation of this network model, containing Hodgkin-Huxley kinetics and dendritic compartments (see AXONAL MODELING and DENDRITIC PROCESSING). The values of most of the parameters used in the simulations have not been measured in lamprey neurons; many of the parameters were, however, adjusted to make the membrane potential trajectories of each of the model neurons resemble those recorded from lamprey neurons of that class (Grillner et al., 1991).

Because of the large parameter space, it is difficult to deduce intuitively what the consequence of changes in various network parameters might be; insights are gained by experimenting with the model. For example, Brodin et al. (1991) have shown that incorporating currents through simulated NMDA-receptor–activated channels can extend the range of frequencies produced, corresponding to results in the in vitro preparation (see NMDA RECEPTORS: SYNAPTIC, CELLULAR, AND NETWORK MODELS). Insight into burst termination mechanisms is provided by the finding that simulations without the LIN neurons can still produce rhythmic output, although the rhythm is slower and more irregular (Hellgren, Grillner, and Lansner, 1992).

Thus, the biophysical simulations allow investigation of the relative contributions of various cellular properties to the function of the model, while the connectionist simulations, with orders of magnitude fewer parameters, allow investigation of the role of cellular connectivity. Whether this network is in fact responsible for rhythm generation in the lamprey is still not established, because of the lack of detailed knowledge of the spinal cord circuitry and the biophysical properties of different cell types in the lamprey. What is needed for further progress is the design of critical experiments to test predictions made by studies of the model.

Population Models

Vertebrate nervous systems consist not of individually identifiable neurons, but of populations of neuron types. Each cell in the circuit of Figure 1*B* can be taken to represent a population of cells with similar but not identical properties. Consequences for network stability of having multiple neurons of each class have been investigated by Ferrar, Williams, and Bowtell (1993). These authors incorporated random fluctuations in the parameters of the network, using connectionist simulations. The number of cells representing each cell type was then increased, with a corresponding decrease in synaptic strengths. They found that doubling the number of neurons reduced the standard deviation of the cycle duration by a factor of approximately 4. In addition, the doubled network was more robust, sustaining rhythmic activity in the face of much higher levels of random fluctuation in simulated membrane potentials.

In an independent study, Hellgren, Grillner, and Lansner (1992) increased the number of cells in biophysical simulations and randomly assigned slightly different parameters to each cell within a class. The network then produced a larger range of burst frequencies in response to changes in tonic excitation.

Modeling Intersegmental Coordination

The mechanisms whereby the spinal cord produces a constant intersegmental phase lag of approximately 1% are not known, nor is the identity of the cells involved. Intracellular recordings and neuroanatomical studies have shown that many neurons, mostly unidentified, project to other segments in both ascending and descending directions. Progress has been made by viewing the system as a chain of coupled oscillators. The mathematical analysis developed by Kopell and Ermentrout (see CHAINS OF COUPLED OSCILLATORS) allows predictions for the behavior of such a chain in response to experimental manipulation. In the next section we will review application of the theory to experiments on the lamprey spinal cord in vitro. In this section, we hope to provide a bridge between mathematics and biology by describing application of the mathematical analysis to simulated chains of neural oscillators with the structure shown in Figure 1*B*.

Each oscillator in the chain is assumed to have its own intrinsic frequency, which it would exhibit if uncoupled from its neighbors. Through coupling, oscillators can speed each other up or slow each other down, depending on the phase lag between them. When the oscillators in a chain have entrained each other, their common frequency is given by the sum of an oscillator's intrinsic frequency and the change in frequency caused by the coupling from its rostral and from its caudal neighbor. These are the basic assumptions of the Kopell-Ermentrout analysis, which are incorporated in the following set of equations. For the kth oscillator in a chain of n coupled oscillators all cycling at the same frequency Ω (and thus phase-locked),

$$
\begin{aligned}
\Omega &= \omega_1 + C_A(\phi_1) & k = 1 \\
\Omega &= \omega_k + C_A(\phi_k) + C_D(-\phi_{k-1}) & 1 < k < n \quad (1) \\
\Omega &= \omega_n + C_D(-\phi_{n-1}) & k = n
\end{aligned}
$$

where ω_k is the intrinsic frequency of oscillator k, ϕ_k is the difference in phase between oscillator $k + 1$ and oscillator k,

$C_A(\phi_k)$ represents the frequency change produced in oscillator k by the coupling from oscillator $k + 1$, and $C_D(-\phi_{k-1})$ the change produced through descending coupling from oscillator $k - 1$. (See CHAINS OF COUPLED OSCILLATORS for a comparison of these empirically derived C-functions with the mathematically defined H-functions of that chapter.)

If each C-function in Equation 1 is replaced by a linear approximation in the region near its intersection with the ϕ-axis, these equations can be solved explicitly to give the predicted phase lag of each segment in the chain as:

$$
\phi_k = \phi_A \frac{(1 - r^k)}{(1 - r^n)} + \phi_D \frac{(r^k - r^n)}{(1 - r^n)} \quad (2)
$$

where k is the segment number, n is the number of oscillators in the chain, r is the ratio of the slopes of the (linearized) C-functions, and ϕ_A and ϕ_D are the ϕ-axis intercepts of the ascending and descending coupling functions, respectively (Williams, 1992a).

From this equation, it can be shown that to have uniform phase lag over most of the chain, ascending and descending coupling functions must not have equal slopes; the coupling with the steeper slope is "dominant" and determines the phase lag over most of the chain. If $r < 1$, ascending coupling is dominant; if $r > 1$, descending coupling is dominant. The value of ϕ for all k lies between ϕ_D and ϕ_A, being closest to ϕ_D at the head end (small k) and closest to ϕ_A at the tail end (large k). Unless r is very nearly equal to 1 (symmetric coupling), the transition from ϕ_D to ϕ_A is restricted to a boundary region at one end (Figure 1*D*), the head end if ascending coupling is dominant ($r < 1$), and the tail end if descending coupling is dominant ($r > 1$). This is also the behavior predicted by the full nonlinear equations (see CHAINS OF COUPLED OSCILLATORS).

For computer-simulated oscillators, the C-function of a particular form of coupling can be determined by coupling a pair of oscillators together in one direction only. Finding the phase lag which develops in the steady-state when the pair have a given difference in intrinsic frequency is equivalent to measuring $\phi(C)$. By making measurements at a range of frequency differences, $C(\phi)$ can be found. From the intercepts and slopes of the C-functions obtained for a particular choice of ascending and descending coupling, the parameters of Equation 2 can be determined, and the behavior of a chain of oscillators with coupling in both directions then can be predicted (Williams, 1992a).

Using the connectionist simulation developed by Buchanan (1991), C-functions were derived for oscillators with the structure shown in Figure 1*B*. Coupling between segments consisted of the same synaptic connections occurring within the unit oscillator, extended to neighboring oscillators with reduced strength (Figure 1*C*). The solid line in Figure 1*D* is drawn from Equation 2 with parameters derived from such C-functions, with ascending coupling dominant. The data points in Figure 1*D* show the results of a simulation of a chain of oscillators with such coupling. It can be seen that the phase lags along the chain are nearly equal to the values predicted by Equation 2. These simulated oscillators thus behave as predicted by the mathematical analysis, providing some assurance that the abstract analysis is applicable to neuronal circuits.

Since the intersegmental phase lags (except for those in a small boundary region) are equal to the ϕ-axis intercept of the dominant coupling (ϕ_A or ϕ_D), phase lag is independent of frequency if this ϕ-axis intercept is independent of frequency. Within a region of the parameter space of the simulated oscillators, such frequency-independence was seen (Williams, 1992b). In this way, the occurrence of a constant wavelength of curvature on the lamprey body at all swimming speeds can be achieved.

Predictions and Experiments

Three types of biological experiments have been performed to test predictions of the mathematical analysis and to gain insights into the intersegmental coordinating system (for review, see Sigvardt, 1993). All three have indicated that in the lamprey, ascending coupling is dominant. First, the existence of a rostral boundary region such as occurs in Figure 1D has been demonstrated (Williams and Sigvardt, 1994). Second, it is possible to entrain the spinal cord rhythm mechanically over a larger range of frequencies from the caudal end than from the rostral end (Williams et al., 1990). The third type of experiment concerns the effects on intersegmental phase lags of variations in the intrinsic frequencies of the oscillators along the length of the chain. According to theory (see CHAINS OF COUPLED OSCILLATORS), for the chain to produce coordinated activity in the face of such differences, the phase lags must change in such a way that the coupling functions compensate for the differences in intrinsic frequency. Mathematical analysis and simulations predict that, for a moderate-amplitude step change in intrinsic frequency midway down the chain, the phase lags should change on one side of the step but not the other, depending on whether ascending or descending coupling is dominant. In experiments in which all segments of a spinal cord preparation are bathed in the same concentration of amino acid, the phase lags are not significantly different in the rostral and caudal halves. If, however, the two halves are perfused with different concentrations, it can be assumed that the oscillators in the two halves have different intrinsic frequencies, which should result in phase differences. The results (Sigvardt, 1993) were as predicted for ascending coupling dominant: statistical analysis of the data indicated that the phase lags in the caudal compartment were not changed, whereas those in the rostral compartment increased when the concentration was higher there and decreased when it was lower. These three sets of experiments indicate that the mathematical analysis of the lamprey spinal cord as a chain of coupled oscillators is valid, and that ascending coupling is dominant in the lamprey and thus sets the intersegmental phase lag.

One of the problems encountered in the theoretical interpretation of the results of experiments has been the intrinsic variability in biological data, both within a given experiment and between different experiments. Measurements spanning 20 segments or more in the swimming lamprey yield intersegmental delays with coefficients of variation (standard deviation divided by mean) of typically less than 0.20. In the in vitro preparation, however, which is artificially activated and devoid of the stabilizing action of sensory feedback, the standard deviations tend to be larger. With recordings from electrodes separated by many segments, and careful selection of sections of record containing stable activity, the standard deviations are of similar magnitude to the recordings from intact animals (Wallén and Williams, 1984). For unedited data they may be an order of magnitude larger and can have quite different mean values (Matsushima and Grillner, 1992). Thus it is important to design and analyze experiments carefully and to test hypotheses statistically.

Conclusions

The lamprey central pattern generator (CPG) for locomotion provides a clear illustration of the value of different levels of modeling. For questions about the importance of particular cellular properties to function, such as the mechanism by which locomotor bursts may be terminated, biophysical modeling is essential. For discovering the rules under which coupled oscillators operate, mathematical theory can be powerful. And for investigating whether the mathematical analysis can be successfully applied to particular neuronal circuits, connectionist modeling, with its more modest computational demands and smaller parameter space, has proved valuable. At all levels, the most powerful approach is a tight interaction between the models and the biology: biological data are used to build a model; investigation of the model then leads to predictions for biological experiments. The ideal goal is to design critical experiments which can provide clear-cut answers to simple questions.

Road Map: Motor Pattern Generators and Neuroethology
Background: I.1. Introducing the Neuron
Related Reading: Oscillatory and Bursting Properties of Neurons

References

Brodin, L., Tråvén, H. G. C., Lansner, A., Wallén, P., Ekeberg, Ö., and Grillner, S., 1991, Computer simulations of N-methyl-D-aspartate receptor-induced membrane properties in a neuron model, *J. Neurophysiol.*, 66:473–484.

Buchanan, J. T., 1991, Neural network simulations of coupled locomotor oscillators in the lamprey spinal cord, *Biol. Cybern.*, 66:367–374.

Buchanan, J. T., and Grillner, S., 1987, Newly identified "glutamate interneurones" and their role in locomotion in the lamprey spinal cord, *Science*, 4799:312–314.

Buchanan, J. T., and Grillner, S., 1988, A new class of small inhibitory interneurons in the lamprey spinal cord, *Brain Res.*, 438:404–407.

Cohen, A. H., Ermentrout, B. E., Kiemel, T., Kopell, N., Mellen, N., Sigvardt, K. A., and Williams, T. L., 1992, Modelling of intersegmental coordination in the lamprey central pattern generator for locomotion, *Trends Neurosci.*, 15:434–438. ◆

Ferrar, C. H., Williams, T. L., and Bowtell, G., 1993, Effects on variability of duplicating the cells in a pattern generating network, *Neural Computat.*, 5:587–596.

Grillner, S., Wallén, P., Brodin, L., and Lansner, A., 1991, Neuronal network generating locomotor behavior in lamprey: Circuitry, transmitters, membrane properties, and simulation, *Annu. Rev. Neurosci.*, 14:169–199. ◆

Hellgren, J., Grillner, S., and Lansner, A., 1992, Computer simulation of the segmental neural network generating locomotion in lamprey by using populations of network interneurons, *Biol. Cybern.*, 68:1–13.

Matsushima, T., and Grillner, S., 1992, Neural mechanisms of intersegmental coordination in lamprey: Local excitability changes modify the phase coupling along the spinal cord, *J. Neurophysiol.*, 67:373–388.

Sigvardt, K. A., 1993, Intersegmental coordination in the lamprey central pattern generator for locomotion, *Semin. Neurosci.*, 5:3–15. ◆

Sigvardt, K. A., and Williams, T. L., 1992, Models of central pattern generators as oscillators: Mathematical analysis and simulations of the lamprey locomotor CPG, *Semin. Neurosci.*, 4:37–46. ◆

Wallén, P., and Williams, T. L., 1984, Fictive locomotion in the lamprey spinal cord in vitro compared with swimming in the intact and spinal animal, *J. Physiol.*, 347:225–239.

Williams, T. L., 1992a, Phase coupling in simulated chains of coupled oscillators representing the lamprey spinal cord, *Neural Computat.*, 4:546–558.

Williams, T. L., 1992b, Phase coupling by synaptic spread in systems of coupled neuronal oscillators, *Science*, 258:662–665.

Williams, T. L., and Sigvardt, K. A., 1994, Intersegmental phase lags in the lamprey spinal cord: Experimental confirmation of the existence of a boundary region, *J. Computat. Neurosci.*, 1:61–67.

Williams, T. L., Sigvardt, K. A., Kopell, N., Ermentrout, G. B., and Remler, M. P., 1990, Forcing of coupled nonlinear oscillators: Studies of intersegmental coordination in the lamprey locomotor central pattern generator, *J. Neurophysiol.*, 64:862–871.

Statistical Mechanics of Generalization

Manfred Opper

Introduction

The theory of learning in artificial neural networks has benefited from various fields of research. Among these, statistical mechanics has become an important tool for understanding a neural network's ability to generalize from examples. It is the aim of this article to explain some of the basic principles and ideas of this approach.

In the following, we assume a feedforward network of N input nodes, receiving real-valued inputs, summarized by the vector $\mathbf{x} = (x(1), \ldots, x(N))$. The configuration of the network is described by its weights and will be abbreviated by a vector of parameters $\mathbf{w}$. Using $\mathbf{w}$, the network computes a function $F_{\mathbf{w}}$ of the inputs $\mathbf{x}$ and returns $\sigma = F_{\mathbf{w}}(\mathbf{x})$ as its output.

In the simplest case, a neural network should learn a binary classification task. That is, it should decide that a given input $\mathbf{x}$ belongs to a certain class of objects and respond with the output $F_{\mathbf{w}}(\mathbf{x}) = +1$, or, if the input does not belong, it should answer with $\sigma = -1$. To learn the underlying classification rule, the network is trained on a set of m inputs $\mathbf{x}^m = \{\mathbf{x}_1, \ldots, \mathbf{x}_m\}$ together with the classification labels $\sigma^m = \{\sigma_1, \ldots, \sigma_m\}$, which are provided by a trainer or *teacher*. Using a *learning algorithm*, the network is adapted to this *training set* $(\sigma^m, \mathbf{x}^m)$ by adjusting its parameters $\mathbf{w}$ such that it responds correctly on the m examples.

How well will the trained network be able to classify an input that it has not seen before? In order to give a quantitative answer to this question, a common model assumes that all inputs, those from the training set and the new one, are produced independently at *random* with the same probability density from the network's environment. Fixing the training set for a moment, the *probability* that the network will make a *mistake* on the new input defines the generalization error $\varepsilon(\sigma^m, \mathbf{x}^m)$. Its *average*, ε, over many realizations of the training set, as a function of the number of examples, gives the so-called *learning curve*. This will be our main interest in the following.

Clearly, ε also depends on the specific algorithm that was used during the training. Thus, the calculation of ε requires the knowledge of the network weights generated by the learning process. In general, these weights will be complicated functions of the examples, and an explicit form will not be available in most cases.

The methods of statistical mechanics provide an approach to this problem, which often enables an *exact* calculation of learning curves in the limit of a very large network, i.e., for $N \to \infty$. At first glance it may seem surprising that a problem will simplify when the number of its parameters is increased. However, this phenomenon is well known for physical systems like gases or liquids which consists of a huge number of molecules. Clearly, there is no chance of estimating the complete *microscopic* state of the system, which is described by the rapidly fluctuating positions and velocities of all particles. On the other hand, the description of the *macroscopic* state of a gas requires only a few parameters like density, temperature, and pressure. It was one of the major achievements of statistical mechanics to show that such quantities can be calculated by suitably *averaging* over a whole ensemble of microscopic states that are compatible with macroscopic constraints.

Applying similar ideas to neural network learning, the problems which arise from specifying the details of a concrete learning algorithm can be avoided. In the statistical mechanics approach, one studies the ensemble of *all* networks which implement the same set of input/output examples to a given accuracy. It is believed that in this way the typical generalization behavior of a neural network (in contrast to the worst or optimal behavior) is described. From a less formal viewpoint, we may consider that an ensemble is realized by a stochastic training algorithm.

The Perceptron

In this section I will explain this approach for one of the simplest types of networks, the *single-layer perceptron* (see PERCEPTRONS, ADALINES, AND BACKPROPAGATION). This machine, for which a great variety of results have been obtained, is far from being a toy model. Since the single-layer architecture is a substructure of multilayer networks, many of the steps in the subsequent calculations also appear in the analysis of more complex networks.

The adjustable parameters of the perceptron are the N weights $\mathbf{w} = (w(1), \ldots, w(N))$. The output is a weighted sum

$$\sigma = F_{\mathbf{w}}(\mathbf{x}) = \text{sign}\left(\sum_{i=1}^{N} w(i)x(i)\right) = \text{sign}(\mathbf{w} \cdot \mathbf{x}) \qquad (1)$$

of the input values, where $\text{sign}(y) = +1$ if $y > 0$, and is otherwise -1. Since the length of $\mathbf{w}$ can be normalized without changing the performance, we choose $\|\mathbf{w}\|^2 = N$.

The input/output relation (Equation 1) has a simple geometric interpretation: Consider the *hyperplane* $\mathbf{w} \cdot \mathbf{x} = 0$ in the N-dimensional space of inputs. All inputs that are on the same side as $\mathbf{w}$ are mapped onto $+1$, those on the other side onto -1. Perceptrons realize *linearly separable* classification problems. In the following, we assume that the rule to be learned belongs to this class so that it can be exactly implemented by a perceptron. We may think that the classification labels σ_k, which come from an ideal expert, are being generated by some other perceptron with weights $\mathbf{w}_t$, the "teacher" perceptron.

The geometric picture immediately gives us an expression for the generalization error. A misclassification of a new input $\mathbf{x}$ by a "student" perceptron $\mathbf{w}_s$ occurs only if $\mathbf{x}$ is between the separating planes defined by $\mathbf{w}_s$ and $\mathbf{w}_t$. If the inputs are drawn randomly from a spherical distribution, the generalization error is proportional to the angle between $\mathbf{w}_s$ and $\mathbf{w}_t$. We obtain

$$\varepsilon(\sigma^m, \mathbf{x}^m) = \frac{1}{\pi} \arccos(N^{-1}\mathbf{w}_s \cdot \mathbf{w}_t) \equiv \frac{1}{\pi} \arccos(R) \qquad (2)$$

The *overlap* $R = N^{-1}\mathbf{w}_s \cdot \mathbf{w}_t$ measures the similarity between student and teacher.

Following the pioneering work of Elizabeth Gardner (1988), we will not concentrate on a specific $\mathbf{w}_s$, which is the result of a concrete learning algorithm. We will rather assume that $\mathbf{w}_s$ was chosen *at random* from the ensemble of all student perceptrons that are consistent with the training set. This space of consistent vectors is usually termed the *version space*. It turns out, that the generalization error for such a *typical* student can be calculated from the volume V of the version space which is defined by

$$V(\sigma^m, \mathbf{x}^m) = \int d\mathbf{w} \prod_{k=1}^{m} \Theta(\sigma_k \mathbf{w} \cdot \mathbf{x}_k) \qquad (3)$$

Here, the Heaviside step function $\Theta(x)$ equals 1 if x is positive, and zero otherwise. Thus, only coupling vectors for which the outputs σ_k are correct, i.e., $\sigma_k \mathbf{w}_s \cdot \mathbf{x}_k > 0$, contribute.

The volume V is a measure for our uncertainty on the weights of the unknown teacher. In the language of statistical mechanics, such a degree of uncertainty of the state of a physical system defines the *entropy* $\mathscr{S} = \ln V(\sigma^m, \mathbf{x}^m)$. The learning of an increasing number of labeled examples reduces the set of consistent vectors $\mathbf{w}_s$ and leads to a decrease of the volume (Equation 3). The decrease $\Delta\mathscr{S}$ of the entropy equals the *amount of information* that is gained on the unknown teacher $\mathbf{w}_t$.

As we will see in the following, by calculating the entropy one will get the generalization error ε for free.

Entropy and the Replica Method

A priori, it is not clear how to get a useful estimate of the entropy $\mathscr{S}$: the expression $V(\sigma^m, \mathbf{x}^m)$ is a *random variable* and fluctuates with the random training set, so an averaging process is necessary. As is typical for a volume, V scales like $\simeq L^N$, where N is the dimension of the version space and L its typical diameter. Thus, for large N, its fluctuations will be over many orders of magnitude, and simply averaging V will put much weight on untypical events.

On the other hand, $\mathscr{S} = \ln(V) \simeq N \ln L$ grows linearly with N, which makes it plausible that the fluctuations of $N^{-1}\mathscr{S}$ will be averaged out by the additive, random contributions of very many degrees of freedom. The same argument applies to the overlap $R = N^{-1}\mathbf{w}_s \cdot \mathbf{w}_t$. We conclude that $N^{-1}\mathscr{S}$ and R are *self-averaging* quantities which can be safely replaced by their average values.

The calculation of the averaged entropy of the examples is a nontrivial problem, which can be solved by a tool of statistical physics, the *replica method*. This is based on the identity

$$\mathscr{S}_{av} = \langle\langle \ln V \rangle\rangle = \lim_{n \to 0} n^{-1} \ln(\langle\langle V^n \rangle\rangle - 1) \qquad (4)$$

Here, the brackets denote the average over the examples. Often, the average of V^n, which is the phase-space volume of the *n-fold* replicated system, can be calculated for *all integers n*. At the end of the calculation, an appropriate analytical continuation to real n is necessary. For integer n, we have

$$\langle\langle V^n \rangle\rangle = \int \prod_{a=1}^{n} d\mathbf{w}_a \exp\left[\alpha N \ln \left\langle\left\langle \prod_{a=1}^{n} \Theta(\sigma \mathbf{w}_a \cdot \mathbf{x}) \right\rangle\right\rangle\right] \qquad (5)$$

Here, we have scaled the number of inputs like $m = \alpha N$, keeping α fixed to obtain a nontrivial limit for $N \to \infty$, and used the fact that all examples are *statistically independent* and identically distributed.

The calculation of the high-dimensional integrals in Equation 5 is made possible by two ideas: The first utilizes the symmetry of the *spherical distribution*. Since the labels σ were produced by the teacher perceptron $\mathbf{w}_t$, the inner average in Equation 5 will depend *only* on the $n(n + 1)/2$ relative angles between the vectors $\mathbf{w}_a, a = 1, \ldots, n$, and $\mathbf{w}_t$. Thus, $\ln\langle\langle \ldots \rangle\rangle = \mathscr{G}_1(n, \{q_{ab}, R_a\})$ is a function of the overlaps

$$
\begin{aligned}
q_{ab} &= N^{-1}\mathbf{w}_a \cdot \mathbf{w}_b \qquad a < b \\
R_a &= N^{-1}\mathbf{w}_a \cdot \mathbf{w}_t
\end{aligned} \qquad (6)
$$

By introducing the measure $e^{N\mathscr{G}_2(n, \{q_{ab}, R_a\})}$ of all vectors $\mathbf{w}_a$ that are constrained to fixed overlaps $\{q_{ab}, R_a\}$, Equation 5 can be converted into an expression that contains only integrations over R_a and q_{ab}

$$\langle\langle V^n \rangle\rangle = \int \prod_a dR_a \prod_{a<b} dq_{ab} \exp[N\mathscr{G}(n, \{q_{ab}, R_a\})] \qquad (7)$$

with $\mathscr{G} = \alpha\mathscr{G}_1 + \mathscr{G}_2$. The explicit form of $\mathscr{G}$ has been given in Györgyi and Tishby, (1990).

The limit $N \to \infty$ provides a second simplification: The integrals in Equation 7 are dominated by values $R_a(n)$ and $q_{ab}(n)$, for which the exponent $\mathscr{G}(n, \{q_{ab}, R_a\})$ is maximal. Other values have an exponentially (in N) smaller weight.

It can be shown that these most probable values can be continued to noninteger n and have limits $R = \lim_{n\to 0} R_a(n)$ and $q = \lim_{n\to 0} q_{ab}(n)$, where R coincides with the average teacher-student overlap needed for the generalization error (Equation 2). The variable q gives the average overlap between two random student vectors in the version space. In statistical mechanics q and R are called *order parameters* (see COOPERATIVE PHENOMENA). This name is justified when we look at the ordering of the student vectors, when more and more examples are learned. When q and R are small, the student vectors typically point in arbitrary directions, showing few similarities with each other and with the teacher. On the other hand, for a small version space, i.e., when many examples have been presented, all students closely resemble the teacher, and q and R approach their maximal value 1.

The formal continuation of Equation 7 to noninteger $n \simeq 0$ is far from trivial: The symmetry of $\mathscr{G}(n, \{q_{ab}, R_a\})$ under permutation of replica indices a, b, suggests the *replica symmetric assumption* $q_{ab}(n) = q(n)$ and $R_a(n) = R(n)$, which is correct for the present perceptron problem. However, a more complicated scheme for continuing the matrices q_{ab} to noninteger dimensions, which allows for a *replica symmetry breaking* (Mézard, Parisi, and Virasoro, 1987), must be applied if the version space of the learning problem is sufficiently complex.

Within replica symmetry, R and q are obtained as solutions of the optimization problem

$$N^{-1}\mathscr{S}_{av} = \text{extr}_{\{q, R\}} \lim_{n \to 0} n^{-1}(\mathscr{G}(n, q, R) - 1) \qquad (8)$$

This expression yields, from Equation 2, the desired learning curve ε as a function of the relative number of examples α, which is shown in Figure 1 (solid line). For a small size of the training set ($\alpha \to 0$), q and R are close to zero, and the generalization error $\varepsilon \approx \frac{1}{2}$, which is not better than a random guessing of the output. To ensure good generalization, m, the size of the training set, must significantly exceed N, the number of couplings. Finally, when the ratio $\alpha = m/N$ grows large, q and R approach 1, and the error decreases slowly to 0 as $\varepsilon \simeq 0.62\alpha^{-1}$.

The shrinking of the space of network couplings resembles a similar result obtained for the learning in attractor neural networks as presented in the article STATISTICAL MECHANICS OF LEARNING. For the latter case, however, the output bits of the corresponding perceptron are completely random (given by the random patterns to be stored), instead of being defined by a teacher network. As the number of patterns grows, the volume of couplings already decreases to zero at a finite critical capacity $\alpha = 2$.

So far, we have discussed the *typical* generalization ability of a perceptron learning a linear separable rule. Is it possible to generalize faster, by using more sophisticated learning strategies? The answer is, Not much, if we are restricted to random input examples. Using the replica approach, the learning curve of an optimal Bayes classifier (Opper and Kinzel, in press; Watkin, Rau, and Biehl, 1993) for linear separable rules has been calculated, yielding $\varepsilon \simeq 0.44\alpha^{-1}$. Studies of multilayer networks indicate that the asymptotic α^{-1} decay is generic for networks with continuous weights and learnable problems.

The situation changes if the learner is free to ask the teacher questions (Watkin et al., 1993), i.e., if she can choose highly

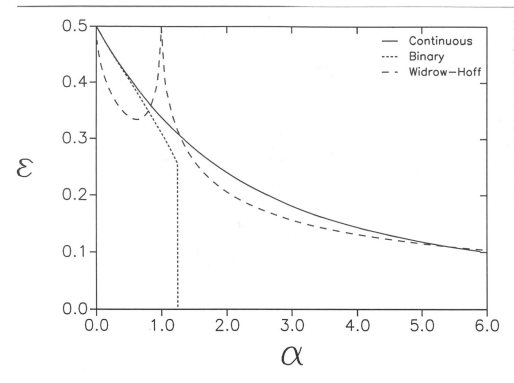

Figure 1. Generalization errors ε for a typical continuous perceptron (solid curve) and a typical binary perceptron (dotted curve) as a function of the relative size $\alpha = m/N$ of the training set. For $\alpha = 1.24$, the version space of the binary perceptron consists of the teacher only, and the generalization error drops discontinuously to zero. The dashed curve refers to a perceptron trained with the Widrow-Hoff algorithm. For $\alpha \approx 1$, the mismatch between the nonlinear teacher and the linear student becomes apparent: Although all examples are perfectly learned, generalization becomes impossible ($\varepsilon \approx \frac{1}{2}$). This overfitting phenomenon diappears for $\alpha > 1$, when the Widrow algorithm learns with training errors.

informative inputs. Then the decrease of the generalization error ε can be exponentially fast in α.

Discontinuous Learning

The method of the preceding section will even provide information about generalization abilities, when (at present) no efficient learning algorithm is known. This is the case when the network weights are constrained to binary values $w(j) \in \{+1, -1\}$. Such a choice may give a crude model for the effects of a finite weight precision in digital network implementations.

For a binary perceptron, perfect learning is equivalent to a hard combinatorial optimization problem (integer linear programming), which in the worst case is believed to require a learning time that grows exponentially with N. Using the replica method, the dotted learning curve in Figure 1 is obtained. For sufficiently small α, the discreteness of the version space has only minor effects. However, since there is a minimal volume of the version space when only one coupling vector is left, the generalization error ε drops to zero at a finite value $\alpha_c = 1.24$. Remarkably, this transition is discontinuous. Therefore, for α slightly below α_c, the few coupling vectors $\mathbf{w}$ which are consistent with all examples typically differ in a finite fraction of bits.

The discreteness of couplings is not the only source of phase transitions in neural networks. Nonsmooth learning curves will occur for continuous weights in multilayer nets if the architecture allows for symmetries that can be spontaneously broken. These include the permutation symmetry between different hidden units in the *committee machine* and the inversion symmetry for a *parity machine*.

Algorithms and Overfitting

The statistical ensemble approach can in many cases also be applied to the performance of concrete algorithms, if the task of the algorithm is to minimize *training energy*. A famous

example for such a learning strategy is *backpropagation* (see BACKPROPAGATION: BASICS AND NEW DEVELOPMENTS), where the quadratic deviation

$$E(\mathbf{w}_s|\sigma^m, \mathbf{x}^m) = \sum_{k=1}^{m} (\sigma_k - F_{\mathbf{w}_s}(\mathbf{x}_k))^2 \qquad (9)$$

between the network's and the teacher's outputs is minimized by a gradient descent of E.

Generalizing the method of the previous sections, we will abandon the assumption that the student network is able to implement the learning task perfectly. The ensemble of students is now defined by all vectors $\mathbf{w}_s$ which achieve a certain accuracy in learning, i.e., which have a fixed training energy. For actual calculations, it is simpler to fix the *average* energy, allowing for small fluctuations which can be neglected for a large network. Statistical mechanics provides a solution to this problem by a probability density p in the space of networks which weights each student according to

$$p(w_s|\sigma^m, \mathbf{x}^m) \propto e^{-\beta E(\mathbf{w}_s|\sigma^m, \mathbf{x}^m)} \qquad (10)$$

The parameter β has to be adjusted such that the average energy achieves the desired value. This so-called *Gibbs distribution* has its origin in the the theory of physical systems which are in thermal equilibrium with their environment. There, $T = 1/\beta$ plays the role of the temperature.

Of special interest is the value $\beta = \infty$, i.e., zero temperature. Here the probability distribution p is entirely concentrated at the vector $\mathbf{w}_s$ for which E is minimal. This distribution is produced by a learning algorithm that finds the total minimum of the training energy. Using the replica method in a suitable way, the generalization error can be calculated.

Let us briefly illustrate the results of this method for a single-layer perceptron, where during the training phase, the student is replaced by a simple *linear* function

$$F_{\mathbf{w}_s}(\mathbf{x}) = \mathbf{w}_s \cdot \mathbf{x}$$

in Equation 9. The backpropagation algorithm, then, reduces

to the so-called *Widrow-Hoff* rule (see PERCEPTRONS, ADALINES, AND BACKPROPAGATION). For a teacher of the same linear type, the classification rule is learned completely with $m = N$ examples. A rather different behavior occurs if the teacher is the *nonlinear* rule (Equation 1), and for generalization also the student's output is given by Equation 1. Although all examples are still perfectly learned up to $\alpha = m/N = 1$, the generalization error increases to the random guessing value $\varepsilon = \frac{1}{2}$ (see Figure 1, dashed line), a phenomenon termed *overfitting*.

If $m > N$, the minimal training error E is *greater than zero*. Nevertheless, ε decreases again and approaches 0 asymptotically for $\alpha \to \infty$. This result shows that one can achieve good generalization with algorithms that allow for learning errors.

The introduction of a temperature into learning theory is not merely a formal trick. Stochastic learning with a nonzero temperature may be useful to escape from local minima of the training energy, enabling better learning of the training set. Surprisingly, it can lead to *better generalization abilities* if the classification rule is not completely learnable by the net. In the simplest case, that outcome occurs when the rule contains a degree of randomness or noise (Györgyi and Tishby, 1990; Opper and Kinzel, in press).

Discussion

The statistical mechanics approach to learning allows us to understand the typical generalization behavior of large neural networks. Its major tool, the replica method, has been illustrated for the single-layer perceptron. Even for this simple network, interesting phenomena like discontinuous learning and overfitting can be observed. Since the field is rapidly developing, these topics are only a small selection of learning problems that have recently been attacked by statistical mechanics. The more complex problem of learning in multilayer networks is of current interest. Besides the replica method, approximations like the annealed theory and the high-temperature limit enable a calculation of the richly structured learning curves. For perceptron learning, new problems which aim to bring the theory closer to reality have been investigated recently. These include time-dependent rules, the influence of input distributions, intelligent pruning of network weights, and unsupervised learning. Finally, a unification of the statistical mechanics methods with ideas coming from other fields of research like mathematical statistics is a challenging problem (see LEARNING AND GENERALIZATION: THEORETICAL BOUNDS).

A more detailed review of the techniques, models, and further references can be found in the longer review articles and books listed in the references.

Road Maps: Cooperative Phenomena; Learning in Artificial Neural Networks, Statistical
Background: I.3. Dynamics and Adaptation in Neural Networks
Related Reading: Sparsely Coded Neural Networks; Spatiotemporal Association in Neural Networks; Statistical Mechanics of Neural Networks

References

Gardner, E., 1988, The space of interactions in neural network models, *J. Phys. A*, 21:257–270.
Györgyi, G., and Tishby, N., 1990, Statistical theory of learning a rule, in *Neural Networks and Spin Glasses* (W. K. Theumann and R. Koeberle, Eds.), Singapore: World Scientific, pp. 3–36.
Hertz, J. A., Krogh, A., and Palmer, R. G., 1991, *Introduction to the Theory of Neural Computation*, Redwood City, CA: Addison-Wesley. ◆
Mézard, M., Parisi, G., and Virasoro, M. A., 1987, *Spin Glass Theory and Beyond*, Singapore: World Scientific.
Opper, M., and Kinzel, W., in press, Statistical mechanics of generalization, in *Physics of Neural Networks II* (J. L. van Hemmen, E. Domany, and K. Schulten, Eds.), New York: Springer-Verlag.
Seung, H. S., Sompolinsky, H., and Tishby, N., 1992, Statistical mechanics of learning from examples, *Phys. Rev. A*, 45:6056–6091.
Watkin, T. L. H., Rau, A., and Biehl, M., 1993, The statistical mechanics of learning a rule, *Rev. Modern Phys.*, 65:499–556.

Statistical Mechanics of Learning

Andreas Engel and Annette Zippelius

Introduction

Methods from statistical mechanics can be used to analyze collective properties for information processing emerging in large systems of interacting neurons. In such model networks, both the activity of the neurons (denoted by S_i, $i = 1, \ldots, N$) and the couplings between the neurons (denoted by J_{ij}, $i, j = 1, \ldots, N$) evolve in time. This coupled dynamics can be rather complex, and for a first understanding it is useful to separate the two kinds of dynamic variables. One is then led to consider either the time evolution of the neuronal activity $\{S_i(t)\}$ for fixed values of the couplings or the gradual adjustment of the interaction strengths $J_{ij}(t)$ to a fixed pattern set of neural activity configurations. The former case, usually referred to as retrieval dynamics, is described in detail in the article STATISTICAL MECHANICS OF NEURAL NETWORKS, together with an introductory discussion of the relevant concepts of statistical mechanics. Here we analyze the second case of *learning* dynamics.

The central aim of both biological and technical neural networks is to perform a specific task of information processing by their retrieval dynamics. The success of the neuron dynamics in accomplishing this task depends crucially on the appropriate choice of the couplings J_{ij}. In technical applications and for simple problems it may be possible to properly design the J_{ij} a priori. More interesting and more relevant for complex problems and biological situations, however, is the idea that given the conditions and constraints of the problem, the system adapts itself to the task. Interesting questions concerning this type of learning dynamics are these:

- Do fixed points of the coupling dynamics exist? That is, are there networks able to perform the desired task? If there are several solutions, how are they related to each other?
- Are these fixed points attractive? That is, does the learning process converge?
- How long does it take the system to reach the solution?

Statistical mechanics can provide answers to these questions in the case of large networks with simple architecture processing

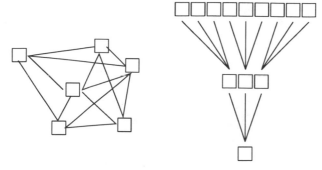

Figure 1. Schematic structure of an attractor neural network (left) and a feedforward neural network.

random patterns. For simplicity we will consider in this article only networks made of McCulloch-Pitts neurons, although the generalization to somewhat more realistic model neurons is possible. We hence deal with a network of N binary neurons with activity $S_i = \pm 1$ connected by real-valued couplings J_{ij} and obeying the dynamics

$$S_i(t + 1) = \text{sign}\left(\sum_{j=1}^{N} J_{ij} S_j(t)\right) \tag{1}$$

where $\text{sign}(x) = +1$ for $x > 0$, and is otherwise -1. We can interpret this equation in two related but slightly different ways (Figure 1):

1. On the one hand, Equation 1 defines for $i, j = 1, \ldots, N$ the retrieval dynamics of a densely connected attractor neural network. Since we are interested here in the learning dynamics only, all we need from this dynamical rule for the neurons is the condition for a fixed point, i.e., for an activity configuration $\xi^\mu = \{\xi_i^\mu\}$ invariant under Equation 1. Obviously we must have

$$\xi_i^\mu = \text{sign}\left(\sum_{j=1}^{N} J_{ij} \xi_j^\mu\right) \tag{2}$$

or equivalently

$$\xi_i^\mu \sum_{j=1}^{N} J_{ij} \xi_j^\mu \geq 0 \tag{3}$$

for all i. Attractor neural networks can function as associative memories if they are able to stabilize a whole set of patterns $\{\xi^\mu\}$, $\mu = 1, \ldots, p$ as fixed points of their dynamics (Hopfield, 1982; see COMPUTING WITH ATTRACTORS). We are interested in characterizing the general properties of coupling matrices J_{ij} satisfying Equation 3 for all $\mu = 1, \ldots, p$. Note that the equations for different values of i are independent of each other if we do not impose additional requirements on the couplings J_{ij} (e.g., the symmetry requirement $J_{ij} = J_{ji}$).

2. On the other hand, Equation 1 describes for $i = 0$, $j = 1, \ldots, N$ a simple feedforward network with N input bits and one output bit. Such an architecture is called a perceptron, and every choice of the coupling vector $\mathbf{J}$ prescribes a binary classification ($\xi_0^\mu = \pm 1$) of the patterns ξ^μ. Given p input-output pairs $\{\xi^\mu, \xi_0^\mu\}$, the relevant question is to find a coupling vector $\mathbf{J} = \{J_{01}, \ldots, J_{0N}\}$ implementing all these mappings, i.e., again to find solutions of Equation 3. Hence the storage problem for an attractor neural network is mathematically equivalent to the classification problem for a perceptron.

Clearly, with J_{ij}, all λJ_{ij} with $\lambda > 0$ are also solutions of Equation 3. In order to suppress this trivial multiplicity, one imposes the normalization condition

$$\sum_{j=1}^{N} J_{ij}^2 = N \tag{4}$$

for all i. Moreover it is useful to require instead of Equation 3 the somewhat stronger condition

$$\frac{\xi_i^\mu}{\sqrt{N}} \sum_{j=1}^{N} J_{ij} \xi_j^\mu \geq \kappa > 0 \tag{5}$$

with the stability parameter κ. The idea behind this is the following. Consider again an attractor neural network. A pattern ξ^μ is a stable fixed point of the neuron dynamics if Equation 3 holds. This property might, however, not be very robust. If only a tiny fraction of neurons deviate from their prescribed values ξ_i^μ, they may in turn destabilize others, and the neuron dynamics will move away from ξ^μ. If, on the other hand, the stronger condition (Equation 5) is fulfilled, a small fraction of misaligned neurons, though still reducing the postsynaptic potential $\sum_j J_{ij} \xi_j^\mu$, will not immediately change its sign. Hence the fixed point will be accompanied by some non-zero basin of attraction, a very desirable property for an associative memory. Roughly, κ gives the size of this basin of attraction, and the prefactor in Equation 5 ensures that this relation is independent of N if N is large. For the perceptron problem, a non-zero κ implies similarly that slightly distorted patterns are also classified with the original.

In the following we will discuss how one can answer the three questions we have posed, both for the perceptron and the attractor neural network, i.e., for the solutions of Equation 5 if the patterns are generated at random. Studying random patterns is the best compromise so far between nontrivial pattern structure and mathematical feasibility of the problem. Moreover, as is characteristic for statistical mechanics, we will consider large networks by taking advantage of the thermodynamic limit $N \to \infty$, relevant to many technical and biological neural networks. For definiteness, the discussion will always be in terms of the perceptron problem.

Statics of Perceptron Learning

Let us first investigate whether solutions of Equation 5 exist at all. Qualitatively, the situation is rather clear (Figure 2). For $p = 0$, there is no condition at all, and the whole surface of the N-dimensional sphere defined by Equation 4 is at our disposal for choosing $\mathbf{J}$. Prescribing just a single input-output mapping $\{\xi^1, \xi_0^1\}$ introduces a hyperplane (with normal ξ^1) into the picture, and only $\mathbf{J}$ vectors on the correct side (given by ξ_0^1) of this hyperplane remain solutions. Every new pattern introduces a new hyperplane, and the volume containing the remaining admissible solutions shrinks with increasing pattern set size p. It is then conceivable that for some number p_c of patterns this volume is reduced to a single point representing the unique perceptron still able to classify all patterns correctly and that no solution will remain if even more patterns are added. This threshold p_c is called the *critical storage capacity*.

In order to analyze this scenario quantitatively we introduce the function

$$\chi(\mathbf{J}; \xi^\mu, \xi_0^\mu) = \prod_{\mu=1}^{p} \theta\left(\frac{\xi_0^\mu}{\sqrt{N}} \sum_{j=1}^{N} J_j \xi_j^\mu - \kappa\right) \tag{6}$$

where $\theta(x)$ denotes the Heaviside function: $\theta(x) = 1$ if $x \geq 0$, and $\theta(x) = 0$ if $x < 0$. Obviously $\chi(\mathbf{J}; \xi^\mu, \xi_0^\mu)$ is 1 for perceptrons $\mathbf{J}$ which implement all the desired input-output mappings correctly, and zero otherwise. Hence

$$V(\kappa; \xi^\mu, \xi_0^\mu) = \frac{\int \prod_j dJ_j \delta(\sum_{j=1}^{N} J_j^2 - N) \chi(\mathbf{J}, \xi^\mu, \xi_0^\mu)}{\int \prod_j dJ_j \delta(\sum_{j=1}^{N} J_j^2 - N)} \tag{7}$$

Figure 2. With increasing number p of patterns to be stored, the volume of admissible coupling vectors (shaded) shrinks (see text).

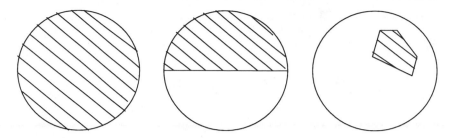

is the fractional volume of perceptrons able to realize the prescribed classification. Here the δ functions enforce the normalization (Equation 4). The fractional volume is 1 for $p = 0$ and will decrease with increasing p and κ. If it is zero, no perceptron that can perform the required task exists at all.

It is impossible to directly calculate $V(\kappa; \xi^\mu, \xi_0^\mu)$, since it still depends on the details of the pattern set $\{\xi^\mu, \xi_0^\mu\}$. In the limit $N \to \infty$, $p \to \infty$, $\alpha = p/N = \text{const}$, however, one can determine its typical value $V_{\text{typ}}(\alpha, \kappa)$ by an appropriate average over the patterns. As in related physical problems analyzed using statistical mechanics, the fluctuations are assumed to tend to zero in this limit, and $V(\kappa; \xi^\mu, \xi_0^\mu) = V_{\text{typ}}(\alpha, \kappa)$ for almost all random pattern sets of size $p = \alpha N$.

The calculation of $V_{\text{typ}}(\alpha, \kappa)$ is an elegant application of the methods of statistical mechanics to this problem. Details of this calculation can be found in Gardner (1988). For random patterns with independent bits $\xi_i^\mu = \pm 1$ with equal probability, the final result is

$$V_{\text{typ}}(\alpha, \kappa) \sim \exp\left[N \min_q \left[\frac{1}{2}\log(1-q) + \frac{q}{2(1-q)} + \alpha \int \frac{dt}{\sqrt{2\pi}} e^{-t^2/2} \log \int_{(\kappa - \sqrt{q}t)/\sqrt{1-q}}^{\infty} \frac{dz}{\sqrt{2\pi}} e^{-z^2/2}\right]\right] \quad (8)$$

The value q_{min} of the auxiliary parameter q minimizing the expression in the exponent is to be found numerically. It turns out that q_{min} gives the cosine of the typical angle between two *different* solutions $\mathbf{J}^{(1)}$ and $\mathbf{J}^{(2)}$ and hence characterizes the size of the solution space. There is a threshold value $\alpha_c(\kappa)$ of the storage ratio α for which $q_{\text{min}} = 1$, i.e., for which the typical angle between two solutions becomes zero. This value indicates that the solution space has shrunk to a point and that no solution will survive if α is further increased. Taking the limit $q_{\text{min}} \to 1$ one finds from Equation 8 the simple result

$$\alpha_c = \left[\int_{-\kappa}^{\infty} \frac{dt}{\sqrt{2\pi}} e^{-t^2/2}(t + \kappa)^2\right]^{-1} \quad (9)$$

This dependence of α_c on κ is shown in Figure 3. As expected, α_c decreases as the stability parameter decreases, since there is a tradeoff between storage capacity and basin of attraction. The maximal possible storage capacity is hence given by $\alpha_c(\kappa = 0) = 2$. This result can also be obtained using different techniques (Cover, 1965), which are, however, not easily generalized to $\kappa > 0$.

Note that this calculation does not provide us with the explicit coupling vector $\mathbf{J}$ realizing the task for a definite pattern set but merely proves (or disproves) the *existence* of a solution. We come back to this point in the next section. Explicit learning rules, e.g., the Hebb rule and the pseudo-inverse rule, give an explicit prescription on how to determine the coupling vector $\mathbf{J}$ for a particular pattern set $\{\xi^\mu, \xi_0^\mu\}$. However, their storage capacity (0.14 and 1.0, respectively) is significantly lower than the optimal one. In fact, the strength of the previously described determination of α_c is that it gives the maximal possible value for *any* learning algorithm. This value depends only on the general setup of the problem (architecture of the network, required stability κ) and the probability distribution according to which the pattern set was generated.

The formalism sketched here allows several interesting generalizations. We mention a few in the following paragraphs.

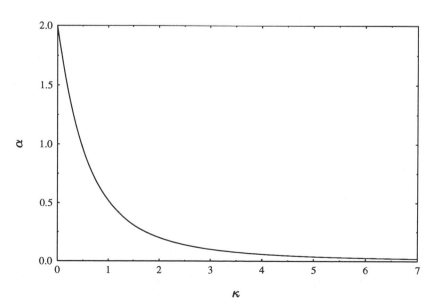

Figure 3. Dependence of the storage capacity α_c on the stability parameter κ as given by Equation 8.

Correlated patterns. Independent random patterns composed in turn of independent random bits are a very crude approximation to real-life situations of pattern processing. They can be refined by allowing for correlations inside the pattern set. Two different types of correlations can occur: those between the same bits in different patterns $\langle\langle\xi_i^\mu\xi_j^\nu\rangle\rangle = \delta_{ij}C^{\mu\nu}$ and those between different bits of one pattern $\langle\langle\xi_i^\mu\xi_j^\nu\rangle\rangle = \delta^{\mu\nu}C_{ij}$. Both types can increase the storage capacity α_c. On the other hand, both types reduce the information content *per pattern*. One can show that the tradeoff between these two tendencies always decreases the total amount of information stored in the network. An important example of the first type of correlations is given by patterns with low level of activity a (i.e., a low proportion of active neurons; see SPARSELY CODED NEURAL NETWORKS). The maximal possible storage capacity diverges if $a \to \pm 1$. Also, more structured data—for example, pattern hierarchies where the patterns are organized in a tree of correlated classes—can be dealt with.

Other constraints on the couplings. Using continuous variables for the couplings J_j requires considerable precision in a practical setup of the network. This is disadvantageous in technical systems and implausible in biological ones. It is hence of interest to study networks with discrete valued couplings. In the extreme case of binary couplings $J_j = \pm 1$, one finds $\alpha_c \cong 0.83$. In view of the severe restriction imposed on the set of allowed coupling vectors, this is a surprisingly modest reduction of α_c. Note also that α_c must be smaller than 1 in this case, since the system cannot store more information than is contained in its own structure.

Error tolerance. For $\alpha > \alpha_c$ no $\mathbf{J}$ vector can be found classifying all the patterns correctly. It is then of interest to determine the minimal possible fraction of misclassified patterns. To this end, one has to generalize the volume $V(\kappa; \xi^\mu, \xi_0^\mu)$ defined in Equation 7 to a quantity that takes into account also $\mathbf{J}$ vectors with a small fraction of errors. The error rate of a coupling vector $\mathbf{J}$ is given by

$$E(\alpha, \mathbf{J}) = \frac{1}{\alpha N}\sum_{\mu=1}^{\alpha N}\theta\left(\kappa - \frac{\xi_0^\mu}{\sqrt{N}}\sum_{j=1}^{N}J_j\xi_j^\mu\right) \qquad (10)$$

and the suitable generalization of $V_{\text{typ}}(\alpha, \kappa)$ (Equation 8) is to calculate the averaged free energy density

$$f(\alpha, T) = -\lim_{N\to\infty}\frac{T}{N}\left\langle\left\langle\log\int\prod dJ_i\delta\left(\sum_{j=1}^{N}J_j^2 - N\right)\right.\right.$$
$$\left.\left.\times\exp\left(-\frac{E(\alpha, \mathbf{J})}{T}\right)\right\rangle\right\rangle_{\xi^\mu} \qquad (11)$$

where $\langle\langle\ldots\rangle\rangle_{\xi^\mu}$ denotes the average over the distribution of the patterns. The parameter T controls the range of error rates substantially contributing to f. By varying it one can "zoom in" on the region of error rates one is interested in (see the equivalent procedure in optimization theory as described in SIMULATED ANNEALING). In particular for $T \to 0$, only $\mathbf{J}$ vectors with very small error rates play a role in Equation 11, and accordingly $f_{\min}(\alpha) = \lim_{T\to 0}f(\alpha, T)$ gives the minimal fraction of misclassified patterns at storage level α. Of course, one finds $f_{\min}(\alpha) = 0$ if $\alpha \le \alpha_c$. For $\alpha \ge \alpha_c$, $f_{\min}(\alpha)$ monotonically increases with α.

Learning from examples: generalization. A very interesting situation is given by a perceptron that stores input-output mappings that are not random but generated by a special rule. In the simplest case, that rule could be provided by another,

unspecified perceptron $\mathbf{T}$. The storage capacity is now trivially infinite, since the choice $\mathbf{J} = \mathbf{T}$ classifies all patterns correctly, by definition. However, a new interesting question arises, namely, given that the perceptron $\mathbf{J}$ has learned to classify p randomly chosen patterns in accordance with the rule $\mathbf{T}$, what is the probability that it will also classify a new, so far unseen, pattern in the same way as the reference perceptron $\mathbf{T}$? It turns out that for large N, $p = \alpha N$ training patterns are sufficient to make this probability nearly 1; i.e., the perceptron $\mathbf{J}$ has learned the underlying rule $\mathbf{T}$ on the basis of a rather small subset of all the 2^N possible input-output mappings (see STATISTICAL MECHANICS OF GENERALIZATION).

Dynamics of Perceptron Learning

We now turn to the question of how to determine the couplings $\mathbf{J}$ that realize the optimal networks considered previously. No explicit learning rules $\mathbf{J} = \mathbf{J}(\xi^\mu)$ are known that yield the maximal storage capacity α_c, and hence one has to consider iterative schemes that successively modify the couplings until a solution is reached. This procedure is conveniently described as an optimization process. One first defines a cost function $E(\mathbf{J})$ that is minimal for the desired coupling vector $\mathbf{J}$. A simple choice is just the fraction of errors as given by Equation 10. Because it is a stepwise constant function, however, it is not well suited for minimization. One therefore prefers the generalized form

$$E(\mathbf{J}) = \sum_\mu\left(\kappa - \frac{\xi_0^\mu}{\sqrt{N}}\sum_{j=1}^{N}J_j\xi_j^\mu\right)^a\theta\left(\kappa - \frac{\xi_0^\mu}{\sqrt{N}}\sum_{j=1}^{N}J_j\xi_j^\mu\right) \qquad a > 0$$
$$(12)$$

which additionally takes into account how severely the storage conditions are violated for every misclassified pattern. Learning now consists in the minimization of this cost function. A suitable method employs updates of the form

$$\delta J_k^{(\mu)} \sim \left(\kappa - \frac{\xi_0^\mu}{\sqrt{N}}\sum_{j=1}^{N}J_j\xi_j^\mu\right)^{a-1}\xi_0^\mu\xi_k^\mu\theta\left(\kappa - \frac{\xi_0^\mu}{\sqrt{N}}\sum_{j=1}^{N}J_j\xi_j^\mu\right)$$
$$(13)$$

done sequentially with respect to the patterns which successively bias the couplings by increments proportional to $\xi_0^\mu\xi_k^\mu$ toward those patterns μ which are not yet stored with the desired stability. With $a = 1$, Equation 13 is a variant of the famous perceptron learning rule (Rosenblatt, 1962); $a = 2$ corresponds to the so-called adatron algorithm, a rather fast method to generate the perceptron with optimal stability (Kinzel and Opper, 1991).

The crucial question concerning such iterative prescriptions is whether they converge, i.e., whether they indeed eventually find the solution of the problem. For the learning rules (Equation 13) with $a = 1, 2$ this convergence can be proven provided a solution *exists at all*. This fact underlines again the importance of the results described in the preceding section on the existence of solutions. Given the parameters of the problem (here α and κ) one can first decide whether the problem is solvable by a perceptron at all. If so, the convergence proof ensures that the learning algorithm will indeed find a solution.

Finally we mention that in the comparatively simple case of the perceptron, the third question concerning the time necessary for the learning process to converge can also be determined using methods from statistical mechanics. Typically one finds that the number t of necessary updates (Equation 13) of the couplings increases with the storage ratio α and diverges as an inverse power $t \sim (\alpha_c - \alpha)^{-b}$ when α approaches α_c. This topic is discussed in detail in Kinzel and Opper (1991).

Multilayer Networks

The perceptron is the simplest feedforward network having just an input layer and an output layer. Accordingly, the class of Boolean functions between input and output one can realize with a perceptron is rather limited (it is the class of linearly separable functions). Simple but very useful Boolean functions turn out not to be in this class, the most prominent example being the *exclusive or* (XOR). Hence the possible technical applications of the perceptron are rather limited.

It turns out that by adding to the network a third layer between the input and output layers containing sufficiently many neurons (Figure 4), one can realize *any* Boolean function between input and output. It is hence very interesting and important for technical applications to analyze the properties of such more powerful networks.

However, few of the results obtained for the perceptron carry over to this more complicated case of multilayer feedforward nets. So far, neither general results on the storage capacity nor learning algorithms with convergence proofs are available. Some progress has been made for nets with one hidden layer, far fewer elements in this layer than in the input layer, and a fixed Boolean function from the hidden layer to the output. Most prominent among those are the *parity machine*, giving the product of the values of the hidden units as output, and the *committee machine*, giving the majority vote of the hidden units as output. Their complete input-output behavior is hence given by

$$\xi_0^\mu = \prod_{k=1}^{K} \text{sign}\left(\sum_{j=1}^{N} J_{kj}\xi_j^\mu\right) \tag{14}$$

and

$$\xi_0^\mu = \text{sign}\left(\sum_{k=1}^{K} \text{sign}\left(\sum_{j=1}^{N} J_{kj}\xi_j^\mu\right)\right) \tag{15}$$

respectively. In both cases the storage capacity is likely to diverge with the number K of hidden units as $\alpha_c \sim K \log K$ for $K \to \infty$ (but always $K \ll N$). The main stumbling block for analytical progress is that, owing to the possibility of different internal representations of the patterns by the hidden neurons, the solution space is no longer connected, giving rise to difficult mathematical problems.

Many special learning algorithms have been devised for multilayer networks, but their efficiency seems to vary widely with the problem at hand, and their performance can only be studied numerically. An interesting new aspect of learning in multilayer networks is given by the possibility of modifying the architecture. A simple procedure of this kind is to successively add hidden units until the desired task can be realized (see TOPOLOGY-MODIFYING NEURAL NETWORK ALGORITHMS).

Discussion

Aspects of the learning process in neural network models can be effectively studied with methods from statistical mechanics. Detailed results have been obtained for the simplest feed forward network, the perceptron, when it has very many input bits and is processing random patterns. This problem is mathematically equivalent to the design of an attractor neural network functioning as associative memory.

We have given a short account of some of the basic ideas in this field. A more detailed introduction can be found in the textbooks by Hertz, Krogh, and Palmer (1991) and by Müller and Reinhard (1991). The book edited by Domany, van Hemmen, and Schulten (1991), and the Elizabeth Gardner memorial issue of the *Journal of Physics A: Mathematical and General*, vol. 22, no. 12 (1989), provide an interesting collection of research papers on the subject.

Current research topics include generalizations to multilayer networks, learning static, noisy and time-varying rules from examples, and the relation of the statistical mechanics results to those obtained on the same problems in mathematical statistics and computer science.

Road Maps: Cooperative Phenomena; Learning in Artificial Neural Networks, Statistical
Background: Computing with Attractors
Related Reading: Fault Tolerance; Spatiotemporal Association in Neural Networks

References

Cover, T. M., 1965, Geometrical and statistical properties of systems of linear inequalities with applications in pattern recognition, *IEEE Trans. Electron. Comput.*, EC-14:326–334.

Domany, E., van Hemmen, J. L., and Schulten, K., 1991, *Physics of Neural Networks*, Berlin: Springer-Verlag.

Gardner, E., 1988, The space of interactions in neural network models, *J. Phys. A*, 21:257–270.

Hertz, J. A., Krogh, A., and Palmer, R. G., 1991, *Introduction to the Theory of Neural Computation*, Redwood City, CA: Addison-Wesley. ◆

Hopfield, J. J., 1982, Neural networks and physical systems with emergent collective computational abilities, *Proc. Natl. Acad. Sci. USA*, 79:2554–2558.

Kinzel, W., and Opper, M., 1991, Dynamics of learning, in *Physics of Neural Networks* (E. Domany, J. L. van Hemmen, and K. Schulten, Eds.), Berlin: Springer-Verlag.

Müller, B., and Reinhard, J., 1991, *Neural Networks: An Introduction*, Berlin: Springer-Verlag. ◆

Rosenblatt, F., 1962, *Principles of Neurodynamics*, New York: Spartan.

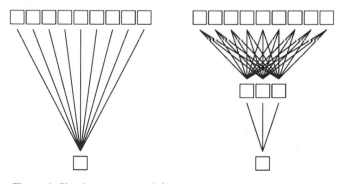

Figure 4. Simple perceptron (left) and multilayer network with one hidden layer.

Statistical Mechanics of Neural Networks

Annette Zippelius and Andreas Engel

Introduction

Statistical mechanics describes the collective properties of systems composed of very many interacting elements on the basis of the individual behavior and mutual interaction of these elements. The term *collective* refers here to properties which result substantially from the interaction of the constituents and which show up only if the whole system contains sufficiently many elements. In physics, statistical mechanics in this way bridges the gap between microscopic structure and macroscopic properties of matter. A basic example is the calculation of the equation of state for a gas—i.e., the law according to which pressure changes with temperature—from the equation of motion governing the dynamics of the molecules. Numerous detailed investigations of physical systems have revealed two remarkable features of this approach:

1. The collective properties of a system can be more complex than the simplicity of its constituents seems to allow. This feature is a consequence of the phenomenon of spontaneous *symmetry breaking*, which means that a macroscopic system can well be in a state of lesser symmetry (i.e., higher complexity) than the underlying microscopic dynamics. For example, the statistical distribution of atomic *spins* in an unmagnetized bar of iron is symmetrical, but once the iron is magnetized the mean direction of the spins is constrained to lie in a specific direction—the symmetry has been broken. Such symmetry breaking is often the sign of a *phase transition* when matter changes from one state of organization to another.

2. Very few of the details characterizing the elements of a macroscopic system are essential for the qualitative collective properties. In the example of the gas, e.g., an attractive interaction between the molecules results in a condensation point, such that for temperatures below this point there is a phase transition from gas to liquid. Practically any law of mutual attraction would give qualitatively the same transition, just changing the specific value for the transition temperature. The same is true for the characterization of spins in models of magnetization. Consequently, extremely simplified models for the microscopic elements can be used. To find out which microscopic properties are *relevant* is, of course, the first step of the program, usually a nontrivial and decisive one.

Neural networks, biological as well as technical ones, are (in many, but not all, cases) composed of a large number of elements (neurons or electronic circuits) strongly interacting through connections (synapses or wires). They can perform information processing of amazing complexity. Although the elements are themselves already rather complicated systems which, at least in the case of biological neurons, are still far from being understood completely, it is generally believed that these abilities for information processing are far beyond the possibilities of a single or a few units and are therefore in some sense collective properties. It is then tempting to assume that, similarly to complex physical systems, these collective properties are again fairly independent of many of the microscopic details. The aim of a statistical mechanics of neural network models is hence to investigate to what extent the methods of physics can be applied to show how, in a large system of simple interacting units reminiscent of biological neurons, there can

emerge collective properties for complex information processing comparable to those found in real neural networks.

Within the approach of statistical mechanics, we define a neural network as an ensemble of N neurons of activity S_i, $i = 1, \ldots, N$, which interact through synaptic couplings J_{ij}, $i, j = 1, \ldots, N$. Both the activity and the couplings can be real-valued or restricted to a discrete set and in general evolve dynamically. However, most studies assume a separation of time scales, such that either

- the neurons evolve dynamically, $S_i = S_i(t)$, for a fixed set of couplings J_{ij}, or
- the couplings evolve dynamically, $J_{ij} = J_{ij}(t)$, for a given activity state $\{S_i\}$ or a set of p such states $\{\xi_i^\nu\}$, numbered by $\nu = 1, \ldots, p$.

In the first case, the focus is on processing of information. Problems of this kind will be discussed in this article. In the second case, the important question is how to adapt the couplings to a particular processing task. Questions concerning such types of learning dynamics are the subject of the companion article STATISTICAL MECHANICS OF LEARNING.

Statistical mechanics provides a quantitative analysis of the properties of a very large network (mathematically in the limit $N \to \infty$). A large class of models can be solved exactly. In this article we treat the Hopfield model and diluted networks, while the single-layer perceptron is treated in STATISTICAL MECHANICS OF LEARNING.

Many of the information-processing tasks a neural network can perform involve a set of special configurations $\{\xi_i^\nu\}$ of neuron activity, called patterns. Generally the performance depends on the detailed realization of this pattern set. The focus of statistical mechanics is then on the *typical* behavior of the network; i.e., one wants to know how it performs typically for a given problem rather than to pinpoint the properties specific to one particular pattern set $\{\xi_i^\nu\}$. An effective way to describe such a typical performance mathematically is to generate the pattern set at random according to a given probability distribution (e.g., $\xi_i^\mu = \pm 1$ with equal probability) and to determine averages of the quantities of interest. As it turns out, many interesting quantities are self-averaging, meaning that for $N \to \infty$ the performance on a special random pattern set is identical to the average performance. By calculating the average of the relevant quantities, one can hence make reliable predictions for the results of a real or numerical experiment involving just one realization of the pattern set. This approach has been very successful in the theory of disordered physical systems, and many of the mathematical techniques developed there have been proven to be adequate for the statistical mechanics of neural networks.

Neuron Dynamics

An extremely simplified model of the individual neurons suitable for a statistical mechanics analysis was introduced in 1943 by Warren McCulloch and Walter Pitts. They devised a neuron as a bistable threshold unit which is either active (firing) or passive (quiescent). The state of neuron i can therefore be described by a binary variable $S_i = 0, 1$, or equivalently, $S_i = \pm 1$. The former choice was used by McCulloch and Pitts, whereas the latter became standard in statistical mechanics because of its formal similarity to magnetic ("spin") systems, many of which have been studied by physicists under the general heading of *spin glasses*. The interaction between N such neurons takes place through real-valued couplings J_{ij}. More precisely, a neuron S_j contributes to the (postsynaptic) potential of neuron S_i, with the value $J_{ij}S_j$ depending on its own activity state. Neuron S_i in turn sums up all these contributions and compares the sum with a neuron-specific threshold θ_i. If the accumulated potential $\Sigma_j J_{ij}S_j$ exceeds this threshold, neuron S_i becomes (or remains) active; otherwise, it remains (or becomes) passive. The discrete time dynamics of the whole network is therefore given by

$$S_i(t + 1) = \text{sign}(h_i(t)) = \text{sign}\left(\sum_j J_{ij}S_j(t) - \theta_i\right) \quad (1)$$

where $\text{sign}(x) = +1$ for $x > 0$, and $\text{sign}(x) = -1$ otherwise, and where $t = 0, 1, 2, 3, \dots$ denotes the discrete time steps.

The collective behavior of an ensemble of such neurons is often understood in the context of associative memory, i.e., the recall of information on the basis of partial knowledge of its content. The use of neural networks as associative memories is built on three main principles (see COMPUTING WITH ATTRACTORS):

- The information which is to be stored and retrieved in the system is represented as one or many configurations of neuronal activity, called patterns $\{\xi_i^\nu\}$ $(i = 1, \dots, N; \nu = 1, \dots, p)$.
- The couplings are designed to make the patterns attractors of the neuron dynamics.
- Association is the dynamic evolution of an arbitrary initial state into one of these attractors.

In this section we discuss the performance of the network for a given set of couplings which we assume to be adapted to a set of patterns $\{\xi_i^\nu\}$. Some of the relevant questions are these (of which we address the first two in some detail):

- What are the attractors of the neuron dynamics, and how well are they correlated with the patterns?
- What is the capacity? That is, what is the maximal number of patterns which can be stored and retrieved?
- What are the basins of attraction? That is, what is the maximal amount of noise permissible in the initial stimulus?
- How long does it take the network to reach an attractor?

Of particular interest, therefore, are the stationary states of the dynamics, for which $S_i h_i > 0$ for all $i = 1, \dots, N$ (see Equation 1). If the coupling constants are symmetric, i.e., $J_{ij} = J_{ji}$, and if there are no self-interactions, i.e., $J_{ii} = 0$, then an *energy function*

$$E(\{S_i\}) = -\frac{1}{2}\sum_{i \neq j} J_{ij}S_iS_j + \sum_i \theta_iS_i \quad (2)$$

exists and is monotonically decreasing for the *asynchronous* dynamics of Equation 1, i.e., when a single neuron is chosen randomly at each time t for updating. The stationary states of the dynamics must then coincide with the minima of this energy, which in turn can be analyzed with the methods of equilibrium statistical mechanics.

Before embarking on that analysis, however, it is very useful to slightly generalize this approach. By describing the neurons in terms of binary activity variables we ignored almost all microscopic details of the network. The influence of the neglected microscopic degrees of freedom on the dynamics of the activity states can be modeled by a random-noise term, changing the deterministic updating process (Equation 1) into a probabilistic one:

$$S_i(t + 1) = \text{sign}(h_i(t) + \varphi_i(t)) \quad (3)$$

The probability distribution $P(\varphi)$ of the noise is not very critical; a simple choice would be a Gaussian $P(\varphi) = 1/\sqrt{\pi T} \times \exp(-\varphi^2/T)$ which would imply that neuron i will be firing (quiescent) at time $(t + 1)$ with probability

$$p_\pm = \frac{1}{2}\left(1 \pm \frac{2}{\sqrt{\pi}}\int_0^{h_i(t)/T} dx\, e^{-x^2}\right) \quad (4)$$

The strength of the noise is denoted by T, such that for $T = 0$ we recover the deterministic dynamics of Equation 1. By analogy with physical systems, we will often refer to T as the temperature, keeping in mind, however, that it is usally different from the (thermodynamic) temperature of the neural network under consideration. The stochastic process (Equation 3) is also the basic dynamics of Monte Carlo simulations. In that context a different choice

$$P(\psi) = \frac{1}{2T}\left[\cosh\left(\frac{\varphi}{T}\right)\right]^{-2} \quad (5)$$

is quite common. It gives rise to a firing probability $p_+ = \frac{1}{2}[1 \pm \tanh(h_i(t)/T)]$, which differs only slightly (at most 1%) from Equation 4. Both firing probabilities are in accordance with the important physical principle of detailed balance, which ensures that the system will ultimately converge to a steady-state distribution (see, e.g., Reif, 1982, for a textbook account of relevant results from "ordinary" statistical mechanics).

In deterministic neural networks (see COMPUTING WITH ATTRACTORS), the state of the neural network will tend toward a single attractor. Once noise is included, no such determinism is possible. Instead, statistical mechanics provides tools to find an *equilibrium distribution* in which each state has a probability of occurrence for which the probability of leaving a state is exactly balanced by the probability of entering it. If we choose the noise statistics of Equation 5, we end up in the equilibrium distribution in which states are distributed according to the *Boltzmann distribution* in which the probability of state $\{S_i\}$ is given by

$$P(\{S_i\}) = \exp(-\beta E(\{S_i\}))/Z \quad (6)$$

where $\beta = 1/T$ and Z is a normalization factor, $Z = \Sigma_{\{S_i\}}\exp(-\beta E(\{S_i\}))$. All equilibrium expectation values can then be calculated by summing over all configurations, weighted with their Boltzmann factor. For example, the mean activity of neuron k at equilibrium is given by

$$\langle S_k \rangle_{th} = \frac{\sum_{\{S_i\}} S_k \exp(-\beta E(\{S_i\}))}{\sum_{\{S_i\}} \exp(-\beta E(\{S_i\}))} \quad (7)$$

where the notation $\langle \cdot \rangle_{th}$ means "average over all states at thermal equilibrium." This depends on $\beta = 1/T$. Note that equivalently we could write

$$\langle S_k \rangle_{th} = \frac{1}{\beta}\frac{\partial F}{\partial \theta_k} \quad (8)$$

with the free energy $F(\beta, \theta_i)$ defined by

$$F(\beta, \theta_i) = -\frac{1}{\beta}\log\sum_{\{S_i\}}\exp(-\beta E(\{S_i\})) \quad (9)$$

It is a standard result of statistical mechanics that all averages with the Boltzmann distribution, i.e., all macroscopic quantities, can be calculated similarly by appropriate derivatives of the free energy. The main task within a statistical mechanics analysis, therefore, is the calculation of the free energy.

The inclusion of noise has the effect that many configurations—and not just the minima of E—contribute to the expectation values. In other words, a much larger part of configuration space can be explored. A similar approach has been put forward in optimization theory to include nearly optimal solutions (see SIMULATED ANNEALING and NEURAL OPTIMIZATION).

In statistical mechanics powerful methods have been developed to calculate the free energy (Equation 9). Most of them, however, only apply in the thermodynamic limit $N \to \infty$. In physics we have abundant experimental evidence that this limit exists. Quantities like the energy (Equation 2) and the free energy (Equation 9) are expected to be *extensive* variables, increasing linearly with N. Hence the corresponding intensive quantities, e.g., the free energy per neuron $f = F/N$, have a well-defined limit as $N \to \infty$. Furthermore, fluctuations of extensive quantities are generally negligible in the thermodynamic limit. Typically the variance relative to the mean decreases like $N^{-1/2}$. Therefore, the results obtained theoretically employing the limit $N \to \infty$ very accurately describe real systems with finite but large N. This is a characteristic feature of statistical distributions involving very many degrees of freedom.

Statics of the Hopfield Model

The following choice of couplings was used in early studies of ASSOCIATIVE NETWORKS (q.v.), by Hopfield (1982), and extensively thereafter (see Amit, Gutfreund, and Sompolinsky, 1985):

$$J_{ij} = \frac{1}{N} \sum_{v=1}^{p} \xi_i^v \xi_j^v \qquad (10)$$

The patterns are random variables, and the synaptic couplings of Equation 10 correspond to one realization of the random process. In the simplest case for statistical analysis, all ξ_i^v are statistically independent, identically distributed, and equally likely to be ± 1. To further simplify the analysis we set $\theta_i = 0$.

There is no limit on the range of interaction for the coupling matrix of Equation 10. Instead, each neuron interacts with every other neuron. Hence the *local field* (i.e., the "force" exerted on a single neuron by the activity of all other neurons)

$$h_i = \sum_{j=1}^{N} J_{ij} S_j = \frac{1}{N} \sum_{j=1}^{N} \sum_{v=1}^{p} \xi_i^v \xi_j^v S_j \qquad (11)$$

is a large sum of many random numbers, which are only weakly correlated. As long as the number p of patterns remains finite as $N \to \infty$, we can neglect these correlations and replace h_i by its thermal average:

$$h_i \cong \sum_{v=1}^{p} \xi_i^v \frac{1}{N} \sum_{j=1}^{N} \xi_j^v \langle S_j \rangle_{th} \qquad (12)$$

The most important characteristic of an associative memory is the overlap

$$m^v = \frac{1}{N} \sum_{j=1}^{N} \xi_j^v \langle S_j \rangle_{th} \qquad (13)$$

It describes the average alignment of the stationary states with the patterns. The overlap (as well as the free energy) is self-averaging; i.e., its fluctuations from pattern set to pattern set go to zero in the thermodynamic limit. It is hence equal to its average over the pattern statistics with probability 1: $m^v = \langle\langle \xi_i^v \langle S_j \rangle_{th} \rangle\rangle_\xi$. In terms of the local field h_i (Equation 12), the activity is given by $\langle S_i \rangle_{th} = \tanh(\beta h_i)$ so that the overlap vector $\mathbf{m} = (m^1, m^2, \ldots, m^p)$ is the solution of

$$\mathbf{m} = \langle\langle \xi \tanh \beta \mathbf{m} \cdot \xi \rangle\rangle_\xi \qquad (14)$$

In the context of associative memory we are interested in states $\{S_i\}$ which are correlated with one pattern only, i.e., $m^v = m \delta_{vv_0}$. Such states are called *Mattis* or *retrieval states* and are indeed solutions of Equation 14 below a critical noise level $T_c = 1$. These retrieval states are the global minima of E for $T \simeq T_c$, for $T \to 0$, and presumably for all T in between. The failure of this property for $T > T_c$ is an example of a *phase transition* in neural networks (compare the dramatic change in the properties of water at the critical temperature 100°C). Equation 14 also has solutions with several non-zero components of $\mathbf{m}$. They correspond to network configurations which resemble mixtures of stored patterns. Such states clearly disturb the retrieval process. One of the surprising results of the statistical analysis is the fact that noise actually improves the efficiency of the network as an associative memory by eliminating these *spurious states* for $T \geq 0.5$.

It is also possible to store an extensive number of patterns, i.e., $p \to \infty$ with $N \to \infty$, such that $\alpha = p/N$ finite (Amit, Gutfreund, and Sompolinsky, 1987). This approach gives rise to a *static* random contribution z in the local field h

$$\langle S_i \rangle_{th} = \tanh \beta h(z) \qquad (15)$$

$$h(z) = \mathbf{m} \cdot \xi + \sqrt{\alpha} r z$$

The overlap $\mathbf{m}$ and the variance r of the new noise term have to be calculated from the coupled equations

$$\mathbf{m} = \int \frac{dz}{\sqrt{2\pi}} e^{-z^2/2} \langle\langle \xi_i \langle S_i \rangle_{th} \rangle\rangle_\xi$$

$$q = \int \frac{dz}{\sqrt{2\pi}} e^{-z^2/2} \langle\langle\langle S_i \rangle_{th}^2 \rangle\rangle_\xi \qquad (16)$$

$$r = \frac{q}{(1 - \beta(1 - q))^2}$$

Thermal noise (T) and the static noise (cross-talk, z) resulting from many other stored patterns destroy the retrieval states above a critical line in the (T, α) plane. In the absence of thermal noise ($T = 0$), retrieval states can only exist up to $\alpha_c \simeq 0.14$. This result is in very good agreement with extensive computer simulations of the model. The transition is discontinuous with a sudden jump of the overlap from $m(\alpha \to \alpha_c) \simeq 0.97$ to $m = 0$ for $\alpha > \alpha_c$. Hence, when the network is overloading, the retrieval quality abruptly changes.

For $\alpha < \alpha_c$, retrieval states coexist with so-called spin-glass states. In the latter the neurons build up static random configurations almost uncorrelated with the stored patterns. They can be understood as descendants of the mixture states encountered for finite p now mixing very many patterns. The retrieval states are global minima of the energy E for $\alpha \leq 0.05$ only.

Numerous extensions and generalizations of the Hopfield model have been discussed. We mention just a few of them briefly.

Robustness. It has been shown that the Hopfield model is surprisingly robust with respect to various perturbations, like synaptic noise or synaptic dilution (defined prior to Equation 21 in the next section). Even synaptic clipping, i.e., retaining only the sign of J_{ij}, just reduces the capacity from $\alpha_c \sim 0.14$ to $\alpha_c \sim 0.1$.

Forgetting and unlearning. In the Hopfield model all memories are destroyed completely upon over-loading the network. This destruction can be avoided by using special learning rules that weigh patterns differently, thereby forgetting older items in favor of newer ones. It has also been suggested that "unlearning"—a procedure to weaken bonds in a specific way—can improve the performance of the network.

Other pattern statistics.   The assumption of uncorrelated patterns in the Hopfield model does not apply to many situations of practical interest. An explicit construction of the coupling matrix for patterns with arbitrary correlations is provided by the pseudo-inverse rule. If the patterns are linearly independent, it is given by

$$J_{ij} = \frac{1}{N} \sum_{\nu,\mu}^{p} \xi_i^{\nu} (C^{-1})_{\nu\mu} \xi_j^{\mu} \quad \text{with} \quad C_{\nu\mu} = \frac{1}{N} \sum_{i=1}^{N} \xi_i^{\nu} \xi_i^{\mu} \quad (17)$$

This coupling matrix is nothing but a projector $\Sigma_j J_{ij} \xi_j^{\nu} = \xi_i^{\nu}$, which implies that the patterns are stationary states because

$$h_i \xi_i^{\nu} = \xi_i^{\nu} \sum_j J_{ij} \xi_j^{\nu} = 1 \quad (18)$$

This matrix has been widely used in the context of self-organizing maps (Kohonen, 1984) as well as attractor neural networks. The storage capacity is $\alpha_c = 1$, and a local learning rule was given, to construct the couplings in an iterative scheme. Special cases of correlated patterns include hierarchical correlations, which are important for the organization of human memory, and biased patterns, which have non-zero average mean activity (see SPARSELY CODED NEURAL NETWORKS).

Dynamical Properties

Several interesting questions concern the dynamic behavior of neural networks during retrieval, for example:

- How long does it take the system to retrieve a pattern?
- Which initial conditions are iterated into a particular attractor?

If the couplings are asymmetric $J_{ij} \neq J_{ji}$, as in the majority of biological systems, then a neural network is only defined as a dynamic process (Equation 1). Such models have in general no energy function, and complex, time-dependent attractors may occur. Even if one is only interested in stationary states or distributions, these can in general only be obtained as the long time limit of the dynamics. Since the methods of nonequilibrium dynamics are much less developed than those of equilibrium statistical mechanics, it is no surprise that genuine dynamic problems, like the basins of attraction, are much less understood than the stationary properties of neural networks. Nevertheless, nontrivial dynamic behavior is of great importance in biological as well as technical neural networks.

One motivation to study cyclic attractors, for example, is the abundant appearance of temporal sequences in cognition and of rhythmic motion in motor control tasks. Hopfield suggested a network which was designed to store and retrieve a sequence $\xi^1, \xi^2, \ldots, \xi^q$, such that the system, if initialized in pattern ξ^1, makes a collective transition to ξ^2, then from ξ^2 to ξ^3, and so on.

The following set of couplings

$$J_{ij} = \frac{1}{N} \sum_{\nu=1}^{q} \xi_i^{\nu} \xi_j^{\nu} + \frac{\lambda}{N} \sum_{\nu=1}^{q-1} \xi_i^{\nu+1} \xi_j^{\nu} \quad \lambda > 1 \quad (19)$$

generate such a temporal sequence of states, provided the neurons are updated synchronously and without any noise, either in the dynamics or in the initial state. Noise tolerance or recall of sequences with asynchronous dynamics can be achieved with delayed synapses, which induce transitions only after the system has been stabilized for a time τ in one of the patterns. The synaptic delay does not have to match the period of the sequence exactly. In fact, a broad distribution of synaptic delays is well suited for learning of sequences whose relevant time scale is represented in the spectrum (see SPATIOTEMPORAL ASSOCIATION IN NEURAL NETWORKS). Connections which approximately match the time scale of the sequence will be enforced, a process which can be seen as learning by selection.

Complex dynamical behavior naturally occurs in asymmetrically *diluted networks*, i.e., networks in which only a small fraction of possible synaptic couplings are non-zero. The assumption of complete connectivity in the Hopfield model is implausible for biological systems and unacceptable for technical implementations. Hence we are led to consider diluted networks, e.g., a diluted Hopfield model

$$J_{ij} = \frac{c_{ij}}{K} \sum_{\nu=1}^{p} \xi_i^{\nu} \xi_j^{\nu} \quad (20)$$

The $c_{ij} \in \{0, 1\}$ are independent random variables, and $c_{ij} = 1$ with probability K/N. In the limit of strong dilution $\log K \ll \log N$, the dynamic model defined by Equations 1 and 20 is exactly solvable (Derrida, Gardner, and Zippelius, 1987). Specializing to retrieval states with overlap $m(t)$ with, e.g., pattern ξ^1, one finds for synchronous updating

$$m(t+1) = f(m(t)) = \int \frac{dy}{\sqrt{2\pi}} e^{-y^2/2} \tanh \beta(m(t) - \sqrt{\alpha}y) \quad (21)$$

Here we have taken the limit $K \to \infty$ (after $N \to \infty$) and anticipated a finite capacity $\alpha = P/K$ per synapse. Retrieval states correspond to stable fixed points $m_s^* = f(m_s^*) \neq 0$. They exist up to $\alpha \leq \alpha_c = 2/\pi$. As $\alpha \to \alpha_c$, m_s^* decreases continuously to zero, in marked contrast to the discontinuous behavior of m at α_c found in the preceding section for the fully connected Hopfield model. Note also that the retrieval states $m_s^* \neq 0$ do not correspond to fixed points in phase space. Although the overlap m is constant in time, there is substantial dynamic activity on the microscopic level. This can be seen from the fact that two states with the same overlap have in general a Hamming distance different from zero, and the time-delayed auto-correlation function exhibits dynamic noise even at zero temperature.

The size of the basins of attraction is determined by an unstable fixed point m_u^* of Equation 21 with $0 \leq m_u^* < m_s^*$. If the system is started in a configuration with initial overlap $m(t = 0)$ larger than m_u^*, it will be iterated into m_s^*. If $m(t = 0) < m_u^*$, the dynamics will evolve the state toward $m^* = 0$. The basins of attraction have been calculated in this way for a variety of learning rules in the limit of strong dilution.

Discussion

In this article we have tried to elucidate how the concepts and methods of statistical mechanics can be successfully applied to understand certain aspects of large neural networks. We have emphasized the goals of statistical mechanics more than the abundance of results. A specific example, namely, the Hopfield model for associative memory, has been discussed in detail to show the kind of quantitative result that can be expected from a statistical mechanical analysis. To put this model in a more general context we have indicated how it can be extended to more complex tasks. We have focused here on the dynamics of neurons in the presence of fixed synaptic couplings. The inverse problem—the adaptation of the couplings to given neuronal activity states—has also been analyzed very sucessfully with the methods of statistical mechanics (see STATISTICAL MECHANICS OF LEARNING). Statistical mechanics of neural networks is a rapidly developing field with widespread research activity. For more information, the interested reader is referred to textbooks (Amit, 1989; Domany, van Hemmen, and Schulten, 1991; Hertz, Krogh, and Palmer, 1991) and the articles in this *Handbook*.

Road Map: Cooperative Phenomena
Background: Computing with Attractors
Related Reading: Constrained Optimization and the Elastic Net; Cooperative Phenomena; Dynamics and Bifurcation of Neural Networks; Statistical Mechanics of Generalization

References

Amit, D. J., 1989, *Modeling Brain Function: The World of Attractor Neural Networks*, New York: Cambridge University Press. ◆
Amit, D. J., Gutfreund, H., and Sompolinsky, H., 1985, Spin-glass models of neural networks, *Phys. Rev. A*, 32:1007–1018.
Amit, D. J., Gutfreund, H., and Sompolinsky, H., 1987, Statistical mechanics of neural networks near saturation, *Ann. Physics*, 173:30–67.
Derrida, B., Gardner, E., and Zippelius, A., 1987, An exactly solvable asymmetric neural network model, *Europhys. Lett.*, 4:167–173.
Domany, E., van Hemmen, J. L., and Schulten, K., 1991, *Physics of Neural Networks*, Berlin: Springer-Verlag. ◆
Hertz, J. A., Krogh, A., and Palmer, R. G., 1991, *Introduction to the Theory of Neural Computation*, Redwood City, CA: Addison-Wesley. ◆
Hopfield, J. J., 1982, Neural networks and physical systems with emergent collective computational abilities, *Proc. Natl. Acad. Sci. USA*, 79:2554–2558.
Kohonen, T., 1984, *Self-Organization and Associative Memory*, Berlin: Springer-Verlag. ◆
Reif, F., 1982, *Fundamentals of Statistical and Thermal Physics*, New York: McGraw-Hill. ◆

Steelmaking

William E. Staib and James N. McNames

Introduction

The steel industry contains an abundance of neural network applications. The absence of accurate analysis tools and the noise inherent to steelmaking processes make traditional (non-neural) systems difficult to design and suboptimal, thereby requiring lengthy development times. Conversely, neural network systems are able to generalize from noisy data and require less design, which leads to faster and more accurate solutions (van der Walt, van Deventer, and Barnard, 1991; Cilliers, 1991; Thibault, Flament, and Hodouin, 1991).

Neural networks can also automate tasks where automation was not previously possible. For example, the ability of neural networks to identify and classify complicated patterns makes them capable of applications such as early warning and detection, quality inspection, and defect classification, tasks previously performed exclusively by human operators.

Figure 1 shows some examples of how neural networks are presently applied throughout the steelmaking process: initially a neural controlled furnace melts scrap metal; molten steel is then shaped using a continuous casting process controlled by a neural system; a neural optical character classification system is used to ensure readability of stenciled identification codes; finally, neural networks detect and classify defects in the produced steel. Detailed descriptions of these processes are given below.

Intelligent Arc Furnace Controller

Electric arc furnaces are used by steel plants throughout the world to melt scrap metal. Each furnace contains three large suspended high-voltage electrodes that can be lowered or raised to control the amount of electrical current transferred to the scrap located in the bottom of the furnace.

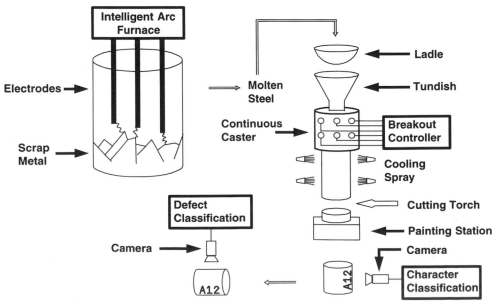

Figure 1. Examples of neural networks applied to steelmaking processes.

Figure 2. The Intelligent Arc Furnace™ neural network structure.

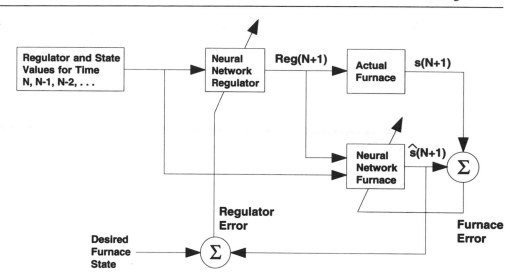

Since the scrap's electrical characteristics are constantly changing as it melts, steady, continuous currents are very difficult to obtain. The Intelligent Arc Furnace Controller (IAF), created by Neural Applications Corporation, uses neural networks to efficiently control electrode position to maintain amp setpoints (Staib and Staib, 1992).

The IAF uses two cascaded neural networks: one neural network, the regulator, is trained to control the positions of all three electrodes; and a second neural network, the model, is trained to predict the next state of the furnace given the regulator's output (Figure 2). Both neural networks are trained continuously on-line. The IAF also uses two adaptive expert systems: the first is used to detect abnormal operating conditions; the second is used to determine the best current setpoints for the regulator.

Since 1991, the IAF has been installed at over 13 different locations including Nucor-Yamato Steel and Birmingham Steel Corporation. In some cases, the IAF has led to an estimated savings of more than $1 million per year.

Breakout Prediction

Once the molten steel is available from the arc furnace (or blast furnace), it is often molded by a continuous casting process. This process begins with pouring molten steel into a large funnel, called a *tundish*, which feeds one end of the mold (see Figure 1). The steel closest to the mold walls cools (and solidifies) most quickly and forms a solid shell that surrounds the still-molten core. The shell usually forms near the entrance of the mold, but the core does not solidify until after the steel has exited the mold.

Occasionally this solid steel shell sticks to the mold wall near the entrance. This can cause the shell to break and create a hole of molten steel next to the mold wall (Figure 3). The break may travel the length of the mold and spill molten steel as it exits the mold. This is called a *breakout* and can result in expensive down time.

A break can be detected by observing subtle temperature changes along the outside wall of the mold. Once a break is

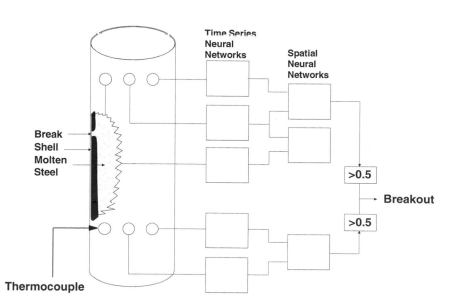

Figure 3. Continuous caster breakout prediction.

detected, a breakout can be prevented by slowing the rate at which steel is drawn through the mold. However, similar temperature variations occur during normal operation that do not result in breakouts, which makes detection of breaks difficult. To solve this problem, Fujitsu Limited teamed up with Nippon Steel Corporation to develop a neural network break-detection system for the prevention of breakouts (Tanaka et al., 1991; Hammerstrom, 1993).

To measure the temperatures along the mold, two sets of thermocouples were placed on the mold's outside wall: one set was placed around the perimeter of the mold near the entrance; a second set was placed around the perimeter closer to the exit of the mold (Figure 3). This two-set structure allows the progression of a break through the mold to be detectable.

The temperatures given by the thermocouples are the inputs to multiple neural networks organized into two layers. The first layer of neural networks, called *time-series neural networks*, were trained to recognize breaks. Each thermocouple is connected to a separate time-series network which receives 10 time-sampled temperature changes. A second layer of neural networks, called *spatial networks*, are connected to processed outputs of adjacent time-series neural networks. These spatial networks are used to detect breaks that may pass in between thermocouples and might only be partially detected by each time-series neural network.

Both time-series and spatial neural networks were fully connected and trained using backpropagation. The system was tested at the plant, where it predicted all breakouts and had one-eighth of the false predictions of the conventional system during its first year of operation. The system is still in use and has been operating since 1990.

Optical Character Classification

After steel exits the mold, it is water-cooled until the core has solidified, and then it is cut into slabs. Since identification of these slabs is crucial for accurate and prompt delivery, Kawasaki Steel Corporation paints each slab with an alphanumeric code (Asano et al., 1991). Occasionally, the paint becomes blurred or marred, making the code unreadable, or the characters are overpainted or not painted at all, making identification of the slab impossible.

To detect this problem, Kawasaki Steel Corporation has developed a neural-network-based system to ensure the quality of the identification codes (Asano et al., 1991). In most optical character recognition (OCR) tasks, the quality of the optical image is high and the goal is recognition. In this case, the quality of the image may be highly degraded and the goal is to determine the readability of the character.

Processed images from a line scan camera were used as inputs to the neural classification system. If a character was judged to be unreadable, then an alarm was sounded and an operator made the final decision on whether the code was readable.

The system has been applied and tested on more than 30,000 slabs since August 1989. The correct classification rate is greater than 99%. All unreadable characters have been detected. Readable characters create false alarms 3% of the time, and thus the amount of time in which operators are required for the process has been drastically reduced.

Rolling Mills

After the continuous caster (see Figure 1), steel is often rolled into sheets instead of going through a painting station. The average sheet thickness is controlled by adjusting the pressure that the rolls exert on each sheet. Traditional linear controllers are often slow to react and cannot always incorporate all of the process information available or achieve the desired tolerances. Preliminary studies indicate that neural control systems can eliminate these difficulties by incorporating nonlinear predictive neural models. In process simulations, neural control systems consistently out-perform traditional linear controllers (Sbarbaro-Hofer, Neumerkel, and Hunt, 1993). Prototype systems are currently under development.

Rolling mills are also plagued by uneven roll wear, which can create sheets with nonuniform thickness and can cause ripples. Usually, skilled operators manually compensate for these problems. Prototype neural network systems have been developed that completely automate this process and improve performance (Hattori, Nakajima, and Morooka, 1993).

Defect Recognition and Classification

Surface inspection of strip and sheet metal products is a crucial step at many stages of the manufacturing process. Traditionally, human inspectors have been used due to the lack of automated alternatives. There are many problems associated with using human inspectors, including: fluctuation over time, varied defect criteria among inspectors, difficulty in obtaining people to do boring work, and limited inspection rates. Neural network inspection systems could alleviate these problems.

Although neural network solutions are not yet sufficiently robust for industrial implementations, preliminary studies demonstrate that neural systems make automation possible with improved performance (Neubauer, 1991; Haataja et al., 1991).

Discussion

Although there are some exceptions, standard feedforward neural networks are dominant in steelmaking applications. A variety of training algorithms may be employed for this class of neural networks, but for most applications that do not utilize on-line training, the simple backpropagation algorithm is adequate and most widely used.

The studies and applications discussed have demonstrated the advantages of neural networks in the steel industry. Noisy data, insufficient knowledge, and inaccurate mathematical models enhance the advantages of neural networks. As these advantages become publicly known and understood, neural applications will become increasingly preferred and widespread in steelmaking.

Road Map: Applications of Neural Networks
Background: I.3. Dynamics and Adaptation in Neural Networks
Related Reading: Adaptive Control: Neural Network Applications; Process Control

References

Asano, K., Tateno, J., Maruyama, S., Arai, K., Ibaragi, M., and Shibata, M., 1991, Neural network model for recognition of characters stenciled on slabs, in *Expert Systems in Mineral and Metal Processing: Proceedings of the IFAC Workshop*, Oxford: Pergamon, pp. 147–153.

Cilliers, J. J., 1991, Neural networks for steady-state process modeling and fault diagnosis, in *Expert Systems in Mineral and Metal Processing: Proceedings of the IFAC Workshop*, Oxford: Pergamon, pp. 161–165.

Haataja, R., Kerttula, M., Piironen, T., and Laitinen, T., 1991, Expert systems for the automatic surface inspection of steel strip, in *Expert Systems in Mineral and Metal Processing: Proceedings of the IFAC Workshop*, Oxford: Pergamon, pp. 70–78.

Hammerstrom, D., 1993, Neural networks at work, *IEEE Spectrum*, June, pp. 26–32.

Hattori, S., Nakajima, M., and Morooka, Y., 1993, Application of pattern recognition and control techniques to shape control of rolling mills, *Hitachi Rev.*, 42:165–170.

Neubauer, C., 1991, Fast detection and classification of defects on treated metal surfaces using back propagation neural network, in *Proceedings of the 1991 IEEE International Joint Conference on Neural Networks*, New York: IEEE, vol. 2, pp. 1148–1153.

Sbarbaro-Hofer, D., Neumerkel, D., and Hunt, K., 1993, Neural control of a steel rolling mill, *IEEE Control Syst. Mag.*, 13(3):69–75.

Staib, W. E., and Staib, R. B., 1992, The intelligent arc furnace controller: A neural network electrode position optimization system for the electric arc furnace, in *Proceedings of the International Joint Conference on Neural Networks*, New York: IEEE, vol. 3, pp. 1–9.

Tanaka, T., Endo, H., Kamada, N., Naito, S., and Kominami, H., 1991, Trouble forecasting system by multi-neural network continuous casting process of steel production, in *Artificial Neural Networks: Proceedings of the 1991 International Conference, ICANN-91*, Amsterdam: North-Holland, vol. 1, pp. 835–840.

Thibault, J., Flament, F., and Hodouin, D., 1991, Modeling and control of mineral processing plants using neural networks, in *Expert Systems in Mineral and Metal Processing: Proceedings of the IFAC Workshop*, Oxford: Pergamon, pp. 25–30.

van der Walt, T. J., van Deventer, and Barnard, E., 1991, The simulation of ill-defined metallurgical processes using a neural net training program based on conjugate-gradient optimization, in *Expert Systems in Mineral and Metal Processing: Proceedings of the IFAC Workshop*, Oxford: Pergamon, pp. 179–184.

Stereo Correspondence and Neural Networks

John P. Frisby

Introduction

Stereoscopic vision exploits the fact that points in a three-dimensional (3D) scene in general project to different positions in the images formed in the left and right eyes. These differences are termed *disparities*. The stereo correspondence problem is: how can a pair of stereo images be mapped into a single representation, called a disparity map, that makes explicit the disparities of all points common to both images? A great deal of computer vision research has addressed this problem because a disparity map is a useful first step toward building a representation of 3D scene structure.

The stereo correspondence problem is illustrated in Figure 1. The image locations of both correct and false matches fall on geometrical constructs called epipolar lines. The importance of this geometrical constraint is that it makes the stereo-matching problem a one-dimensional search. Without it, the stereo problem becomes formally identical to the two-dimensional motion correspondence problem.

Any claims made in this article which do not receive explicit citations are supported by references in Frisby and Pollard (1991), whose review of the computational literature forms the basis for this article; Howard and Rogers (1995); and Regan (1991), which provides a wide introduction to psychophysical and neurophysiological work on binocular vision.

Matching Primitives

The first issue to be resolved in designing any stereo algorithm is, what image features should serve as the points to be matched on the epipolar lines of Figure 1? Edge points are the most common choice because they usually stand in reasonably close correspondence to scene entities, and hence they seem a sensible choice because the ultimate goal is to deliver 3D information about scene structures (Marr and Poggio, 1976). That property is not possessed by image point intensities because a given scene entity often produces different pixel/receptor values in the left and right images. This fact is usually expressed by saying that stereo projections do not preserve photometric invariance. Other possible matching primitives that have been tried include image regions, for example, in stereo algorithms relying on area cross-correlation, and "corner points," usually defined as image features possessing a sharply peaked auto-correlation function.

Edge and corner points produce sparse disparity maps. It is possible to interpolate a dense disparity surface between the data points in such maps, using processes that readily lend themselves to implementations in neural nets (Blake and Zisserman, 1987). Texture boundary cues help shape such interpolation processes in human vision.

Human stereo vision uses a diverse set of matching primitives, including dots, edges, regions, lines, and texture boundaries. It is clear that stereo matching can be achieved without recourse to matching primitives at the level of "objects" (chairs, faces, etc.) because human stereo vision can find correctly matching points even when no high-level objects are discernible in each stereo half-image. This is demonstrated by random dot stereograms, made by taking two copies of a noise field, one for each stereo half-image, and then shifting chosen regions to create disparate zones, but hiding these shifts monocularly by filling in the gaps so generated with new random noise texture (Julesz, 1971).

Exploiting the Epipolar Constraint

To take advantage of the one-dimensional search characteristic of the stereo correspondence problem requires knowing the location of the epipolar lines for a given primitive. This in turn requires a solution to the nontrivial problem of knowing the spatial relationships between the two cameras/eyes, as well as facts about focal length, etc. This is the business of camera calibration, as it is termed in the computational literature. For a stereo camera rig, it typically involves solving at least 12 free parameters. The usual approach for a fixed geometry rig is to infer most of them at startup time from stereo images of an accurately measured test object. The need for a calibrated test object can be avoided by using "corner" features from natural images as the inputs for an optimal least-squares fit of the various camera parameters to the matched corner data. Dispensing with a test object is obviously desirable in modeling biological systems.

Retinal disparities have both horizontal and vertical components (epipolar lines are tilted unless the fixation point is very distant). Horizontal disparities contain information about 3D scene structure, but the size of retinal horizontal disparity arising from any given scene interval is also a function of the positions of the eyes and the viewing distance. Hence horizontal

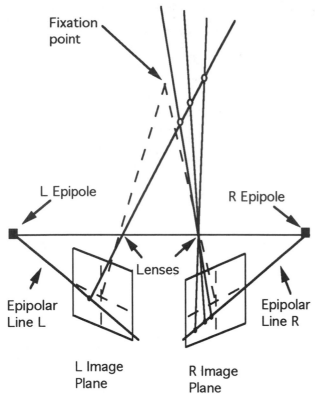

Figure 1. Epipolar geometry and the stereo correspondence problem. L, left; R, right. A line of sight is shown extending from an image point (solid circle) in the L image through the optic center of the lens out into the world. A sample of the possible locations in the world that could have given rise to that point are shown as unfilled circles. These are imaged in the R image on a geometrical construct known as an epipolar line, thereby reducing to a single dimension the stereo correspondence problem, which is the task of deciding which points in the L and R images are to be matched. Each epipolar line intersects a line passing through the optical centers of the lenses at a point called the epipole.

disparities need to be *scaled* or *calibrated* if they are to be used to recover metric measures of scene structure, a process sometimes referred to as *stereoscopic depth constancy*. The information required for calibration can be recovered from vertical disparities which to first order are determined only by the prevailing gaze and elevation angles of the eyes and by the distance to the fixation point (Mayhew and Longuet-Higgins, 1982). There is clear psychophysical evidence that human vision uses vertical disparities for scaling horizontal disparities during its recovery of metric 3D scene structure if the field of view is sufficiently large (Howard and Rogers, 1995). Information on eye positions from oculomotor mechanisms is also used for scaling horizontal disparities. The recovery of qualitative 3D surface descriptions (such as planarity, collinearity, etc.) from horizontal disparities imposes less stringent requirements on the information to be obtained from vertical disparities and can be achieved with smaller fields of view (Gårding et al., in press).

The fact that we can adjust, given time, to perturbations imposed by various distorting lenses shows that the *eye/head calibration* mechanisms in human vision are adaptive in character, suggesting that processes providing continual recalibration are at work. Such processes are of obvious benefit during maturation in coping with the growth of the head.

Constraints for Resolving the Matching Problem

The epipolar constraint is not sufficient for solving the stereo correspondence problem because several matching primitives typically fall on any given epipolar line. Hence other constraints are required, derived from aspects of the viewed world and its projection into stereo images. From these constraints can be inferred binocular matching rules capable of eliminating false matches while preserving correct ones. The main constraints that have been used in neural networks models for stereo correspondence are:

The Constraint of Compatibility of Matching Primitives

It has often been suggested that the size of the ambiguity problem can be substantially reduced, if not altogether eliminated, by matching features of similar shape and contrast. This strategy has already been referred to in connection with choosing matching primitives such as "corner" points, but even using those distinctive primitives is not always enough for achieving disambiguation, because similar "corner" points often exist in natural images. Marr's (1982) way of expressing this constraint, which he dubbed the *compatibility constraint*, was: match only left and right features which could have arisen from the same scene entity. To implement this constraint requires careful attention to how scene features appear in stereo projections. This is necessary to determine bounds on the ranges of, for example, size (spatial frequency) and orientation differences allowed between left and right image features if they are to form potential matches.

The Cohesivity Constraint

Marr and Poggio (1976) argued that it is reasonable to assume the visual world is made up of matter separated into objects whose surfaces are generally smooth compared with their overall distances from the viewer. In other words, the visual world is not usually made up only of clouds of dust particles or snowflakes. They termed this the cohesivity constraint, and they used it to underpin their continuity binocular matching rule: prefer possible matches that could have arisen from smooth surfaces. Poggio, Torre, and Koch (1985) show how the use of a smoothness constraint in stereo can be seen as just one instance of its more general use as a regularizer for solving ill-posed vision problems (see REGULARIZATION THEORY AND LOW-LEVEL VISION), so termed because they are either underdetermined or overdetermined.

The Uniqueness Constraint

Marr and Poggio (1976) also noted that a scene entity cannot be in two places at the same time, from which they derived the binocular matching rule: each matching primitive from each image may be assigned, at most, one disparity value.

The Figural Continuity Constraint

Mayhew and Frisby (1981) used the cohesivity constraint to justify a different matching rule: cohesive objects generate surface edges and surface markings that are spatially continuous, therefore prefer matches which preserve figural continuity. That is, give preference to point-for-point matches which are part of an edge whose other component points also have matches, such that the whole forms a figurally continuous structure of matched points. Mayhew and Frisby found psychophysical evidence suggesting that human stereovision exploits this constraint and developed a correspondence algorithm, STEREOEDGE, which gave preference to matches that could be linked in continuous long binocular strings.

The Ordering Constraint

If opaque surfaces can be assumed, then it is possible to exploit the fact that the order of primitives along an epipolar line in one image is preserved in the other image. It seems that human stereo vision never breaks this constraint.

Neural Network Stereo Correspondence Algorithms

Much research has attempted to solve the stereo correspondence problem by implementing one or more of these constraints with processes of excitation and inhibition in networks of elements that represent competing (usually edge-based) matches. Few have taken seriously the problem of camera calibration, often seeking to solve the (often trivially easy) problems posed by random dot stereograms for which rasters are assumed to be the appropriate epipolar lines. The algorithms differ both in the constraints they try to implement and in the

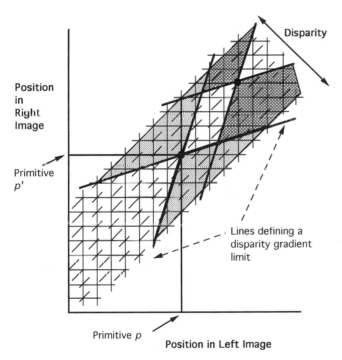

Figure 2. The axes represent the locations of matching primitives on a corresponding pair of left and right epipolar lines. The lines perpendicular to each axis represent lines of sight extending from each primitive, and their intersections represent potential matches. Primitives with zero disparity fall on the diagonal passing through the origin. Other disparity planes lie parallel to that diagonal. The potential match pp' formed from the primitives p and p' and lying in a nonzero disparity plane is shown at the intersection of two bold lines whose slope reflects a particular disparity gradient. The shaded areas on one side of these lines depict regions in which neighbors of pp' would have too steep a disparity gradient with pp' to be allowed to support that match in the PMF stereo algorithm. Gradient limit lines extending from a second match are also shown to emphasize that all matches have their own region of support in disparity space.

details of how they do so. They are described with reference to the network shown in Figure 2.

Dev's Stereo Neural Network

Dev was the first to describe (in 1975) a stereo algorithm using a neural network in which excitation was exchanged (giving mutual support) between elements representing potential matches with the same disparity (hence lying on the diagonals of Figure 2). The support window was in fact two-dimensional (not shown in Figure 2), because excitation was exchanged between potential matches of the same disparity lying on different epipolar lines within a local neighborhood. This use of excitation implements the cohesivity constraint because it favors matches with the same depth. Inhibition was passed between elements lying on lines orthogonal to the diagonals (lines not shown in Figure 2). That way of using inhibition was severely criticized by Marr (1982), who argued that it amounts to a constraint of forbidding double matches along radial lines out from the viewer, for which there is no justification.

Marr and Poggio's Stereo Neural Network

Marr and Poggio (1976) described a similar algorithm to that of Dev. Again, the cohesivity constraint was implemented by excitation exchanged between elements lying on the diagonals of Figure 2. The difference lay in the inhibitory connections. These were between elements lying on the same lines of sight, shown as the vertical and horizontal lines passing through each potential match in Figure 2. This scheme implements the uniqueness constraint properly. Mathematical analyses of the algorithm showed that states satisfying the cohesivity and uniqueness constraints were stable states of the network, and that the network converges for a wide range of parameter values (Marr, 1982).

PMF and the Disparity Gradient Limit Constraint

A limitation of the Dev and Marr/Poggio algorithms is that they impose a very restricted version of the cohesivity constraint, essentially one that favors frontoparallel surfaces. Yet it is evident from inspection of table tops, floors, etc., that human stereo vision copes readily with scenes that are full of a wide variety of slants. Even locally frontoparallel surface patches are far from characteristic of our visual world. The key idea in Pollard, Mayhew, and Frisby's (1985; see Frisby and Pollard, 1991) stereo algorithm, called PMF, is that neighboring potential matches exchange support (mutual excitation) if their relative disparities are similar, not just if they are the same. Similarity is defined in terms of the disparity between potential matches not exceeding a disparity gradient (DG) limit. The butterfly shapes in Figure 2 illustrate the bounds on neighboring matches falling within and outside this limit. DG is defined in PMF as the difference in disparities divided by feature separation.

The use of a disparity gradient limit in PMF was prompted by human psychophysical experiments by Burt and Julesz which showed that if two disparate dots are brought closer and closer together while their disparity is kept unaltered, there comes a point (DG about 1.0) at which binocular fusion breaks down into diplopia.

The Marr/Poggio algorithm essentially imposed a DG limit of zero, by allowing only matches with the same disparity to exchange support. Using a DG limit greater than zero is much less restrictive: DGs of 1.0 arise from a planar surface with a slant of 84° for a viewing distance of 65 cm and interocular separation of 6.5 cm. Hence, one way to view the use of a DG limit is as a means of "parameterizing" the binocular matching rule of seeking matches which preserve surface smoothness, or perhaps it is better to say surface jaggedness. This is because potential matches falling within the DG limit do not need to arise from "smooth" surfaces, if by that is meant adopting a selection procedure equivalent to fitting locally planar or curved surfaces. If the DG limit is set close to zero, then the disambiguating power is great but the range of surfaces that can be dealt with is correspondingly small; *mutatis mutandis* if the DG limit is increased toward the theoretical limit of 2.0, which is the value obtained when the viewing direction from one eye to an opaque surface lies in that surface.

The matching strength of each potential match in PMF is computed as the sum of the strengths of all potential matches in a local image neighborhood that satisfy a moderate DG limit with respect to it (the limit used varies as the algorithm proceeds, from initially prudent, 0.5, to finally generous, 1.5). Because the probability of a neighboring match falling within the DG limit by chance increases (almost linearly) with its distance away from the match under consideration, the contribution of each match in the neighborhood is weighted inversely by its distance away. Prazdny, in a similar algorithm, has used a Gaussian weighting function. The uniqueness constraint is exploited in PMF by selecting the best-supported matches using a form of winner-take-all discrete relaxation. Also, to the basic PMF algorithm have now been added other procedures implementing the figural continuity constraint, the use on startup of a set of strong seed point matches to speed up selection of subsequent matches, and an explicit use of the ordering constraint at a late stage to resolve remaining ambiguities (Frisby and Pollard, 1991).

Surface Discontinuities and Uncorrelated Points

Even if the cohesivity constraint is used in a form that treats the world as locally "not very jagged," it is nevertheless true that many common scenes contain surface discontinuities that are very jagged indeed. For example, tufts of grass or hair are full of discontinuities in depth that exceed a DG limit of 1.0 in almost all directions, and yet they give excellent stereo percepts. Any stereo algorithm that purports to model human stereo competence must be able to cope with such objects.

Within the framework of the neural net models being reviewed here, one means for doing this is to compute only the quantity of *support* that exists for a particular match, and not to exact a penalty for any neighboring matches that do not satisfy the particular definition of "smooth" surface being implemented. For example, the extent to which the DG limit is offended by other potential matches in the neighborhood of a candidate match does not directly affect the selection procedure of PMF. The advantage of this design feature can be appreciated by considering a large step between two planar surfaces. This creates steep disparity gradients across it but not necessarily along it. Within-DG-limit support exchanged between neighboring matches picks out matches lying on one or other side of the step. This helps their correct selection without introducing inappropriate evidence from the other surface, as would happen if penalties were imposed from neighbors lying outside the DG limit.

Discontinuities from opaque surfaces usually create features in one image that do not appear in the other image because they are hidden from view. Human vision is remarkably fast at detecting uncorrelated image regions, which suggests that they

are used in some way, not just treated as noise left over after stereo matching has been completed.

The Problem of Transparency

Human stereo vision can deal successfully with points arising from overlapping transparent surfaces. The difficulty they pose is avoiding computations associated with points from one surface becoming entangled with those from other surfaces. Neural network algorithms can solve this problem in the way just described for dealing with discontinuities, namely by computing the support for any given potential match without imposing a penalty for neighbors which do not satisfy the support rule. In this way, points arising from each transparent surface can "discover" their mutually supporting neighbors and achieve selection. Pollard and Frisby (1990) demonstrate the efficacy of this procedure and point out that the ability of human vision to perceive matches lying in multiple transparent depth planes need not be interpreted as a violation of the uniqueness constraint.

Neurophysiological Studies of Disparity Mechanisms

Disparity-tuned neurons have been studied in a variety of species. Their properties vary considerably, but it is customary to classify them into three broad classes defined in terms of whether they are tuned to disparities close to zero, to "far," or to "near" disparities with respect to the fixation point (Poggio, 1991). The possible roles of these cells in solving the stereo correspondence problem is unclear, although some are sensitive to targets in random-dot stereograms and hence must be at a representational level at or after the stage at which this problem is solved. Lehky and Sejnowski (1990) have used these neurophysiological data to constrain the design of a disparity channels model capable of generating the psychophysical disparity discrimination threshold function which shows greatest sensitivity around zero disparity. Agreement between the model and the data was found to be possible only when 20–200 channels were used for each patch of the visual field, with the breadth of the disparity tuning curve for each channel increasing with peak sensitivity (broadest channels furthest from zero disparity) and all channels having their steepest change in sensitivity crossing zero disparity. They extended their model to incorporate short- and long-range inhibitory interactions between cells at nearby spatial locations, with the result that it could model psychophysical data for detection of discontinuities and transparency.

Discussion

This article has reviewed approaches to solving the stereo correspondence problem. The main point to be emphasized is that neural network approaches to this problem (as indeed any others) need to be firmly grounded in an analysis of the constraints to be implemented in the network. It is also worth emphasizing that solving the stereo correspondence problem as it has been posed here produces a disparity map that is simply another iconic (imagelike) representation which needs itself to be interpreted if the information it contains is to be used, for example, for controlling visually guided behavior. Thus, there is much more to stereo vision than solving the stereo correspondence problem. A further limitation is that the neural networks described here have been handcrafted by their originators. A topic of considerable interest is how to create networks able to learn their own constraints. O'Toole (1989) described a stereo network model that developed connection weights implementing the smoothness constraint after associative learning of a large number of example mappings from disparity data to surface depth data. Becker and Hinton (1992) described a network that replaces an external teacher with internally derived teaching signals generated by the assumption that neighboring parts of the input have common causes in the external world. Small modules look at separate but related parts of the input and discover these common causes by striving to produce outputs that agree with each other, thereby learning to exploit the smoothness constraint. A further limitation of the models described here is that they do not meet the challenges posed by eye movements. A current research issue is how to keep track of epipolar geometry as the eyes move (Mayhew, Zheng, and Cornell, 1993).

Road Map: Vision
Related Reading: Motion Perception; Perception of Three-Dimensional Structure; Thalamus

References

Becker, S., and Hinton, G. E., 1992, Self-organising neural network that discovers surfaces in random-dot stereograms, *Nature*, 355:161–163.

Blake, A., and Zisserman, A., 1987, *Visual Reconstruction*, Cambridge, MA: MIT Press. ◆

Dev, P., 1975, Perception of depth surfaces in random dot stereograms: A neural model, *Int. J. Man-Mach. Stud.*, 7:511–528.

Frisby, J. P., and Pollard, S. B., 1991, Computational issues in solving the stereo correspondence problem, in *Computational Models of Visual Processing* (M. Landy and J. A. Movshon, Eds.), Cambridge, MA: MIT Press, pp. 331–358.

Gårding, J., Porrill, J., Mayhew, J. E. W., and Frisby, J. P., in press, Stereopsis, vertical disparity and relief transformations, *Vis. Res.*

Howard, I. P., and Rogers, B. J., 1995, *Binocular Vision and Stereopsis*, Oxford: Oxford University Press. ◆

Julesz, B. J., 1971, *Foundations of Cyclopean Perception*, Chicago: University of Chicago Press. ◆

Lehky, S. R., and Sejnowski, T. J., 1990, Neural model of stereoacuity and depth interpolation based on a distributed representation of stereo disparity, *J. Neurosci.*, 10:2281–2299. ◆

Marr, D., 1982, *Vision*, San Francisco: Freeman. ◆

Marr, D., and Poggio, T., 1976, Cooperative computation of stereo disparity, *Science*, 194:283–287.

Mayhew, J. E. W., and Frisby, J. P., 1981, Psychophysical and computational studies towards a theory of human stereopsis, *Artif. Intell.*, 17:349–387. ◆

Mayhew, J. E. W., and Longuet-Higgins, H. C., 1982, A computational model of binocular depth perception, *Nature*, 297:376–379.

Mayhew, J. E. W., Zheng, Y., and Cornell, S., 1993, The adaptive control of a four-degrees-of-freedom stereo camera head, *Philos. Trans. R. Soc. Lond. B Biol. Sci.*, 337:315–326.

O'Toole, A. J., 1989, Structure from stereo by associative learning of the constraints, *Perception*, 6:767–782.

Poggio, G. F., 1991, Physiological basis of stereoscopic vision, in *Binocular Vision* (D. Regan, Ed.), vol. 9 of *Vision and Visual Dysfunction* (J. Cronly-Dillon, General Ed.), London: Macmillan, pp. 224–238. ◆

Poggio, T., Torre, V., and Koch, C., 1985, Computational vision and regularization theory, *Nature*, 317:314–319.

Pollard, S. B., and Frisby, J. P., 1990, Transparency and the uniqueness constraint in human and computer stereo vision, *Nature*, 347:553–556.

Regan, D., Ed., 1991, *Binocular Vision*, vol. 9 of *Vision and Visual Dysfunction* (J. Cronly-Dillon, General Ed.), London: Macmillan.

Stochastic Approximation and Neural Network Learning

Christian J. Darken

Introduction

Stochastic approximation (SA) is the study of a particular class of stochastic processes reminiscent of gradual learning. These processes can be interpreted as sequentially finding a minimum of a cost function, most often by performing a noisy version of gradient descent. They have been used as models of biological processes (Hebbian and reinforcement learning) as well as for the study and design of neural networks and other systems for signal processing and control applications. LMS (least-mean-square) adaptive filters (see ADAPTIVE FILTERING), on-line backpropagation for neural networks (see BACKPROPAGATION: BASICS AND NEW DEVELOPMENTS), and *k*-means clustering are all examples of SA algorithms in active use. On-line principal-component and reinforcement-learning algorithms are also SA or closely related.

To the engineering community, acquaintance with SA is a valuable tool for the design of efficient algorithms. Engineering algorithms which are SA can be greatly enhanced in some settings by choosing learning rate schedules based on SA theory. Furthermore, it often pays to design new algorithms to be SA. In addition to good convergence behavior, SA algorithms typically have the following advantages:

- Low computational complexity for each update (as compared to approaches calculating the true gradient by averaging over large quantities of data).
- Robustness (they can perform even when higher derivatives of the cost function, required for other approaches, fail to exist).
- Simplicity of implementation in software and hardware.

Finally, new algorithms developed in ignorance of SA may nonetheless be SA or analyzable by tools developed for SA.

We will define the class of SA processes and then describe what is known about various subclasses. We emphasize guarantees of convergence, convergence speed, and the computational complexity of implementing the various processes, with occasional comments on the quality (depth) of the minimum typically found (there are often several minima).

Background

Stochastic approximation had its origin in a 1951 paper by Herbert Robbins and Sutton Monro concerning a recursive method for finding the zero of an unknown monotonic function when evaluations of the function are contaminated by noise. Since then, the scope of the subject has expanded enormously. There is no standard definition of stochastic approximation, but we will hazard a rough one in order to convey an idea of the subject under discussion. Any stochastic process which can be interpreted as minimizing a cost function based on noisy gradient measurements in a sequential, recursive manner can reasonably lay claim to being stochastic approximation. *Sequential* means that each estimate of the location of a minimum is used to make a new observation (another gradient measurement) which in turn immediately leads to a new estimate. *Recursive* means that the estimates are not allowed to depend arbitrarily on all past gradient measurements (which would then have to be stored if the process were implemented as an algorithm). Instead, the estimates are allowed to depend on previous measurements only through a fixed number of scalar statistics, here labeled ϕ_1 through ϕ_m, which are themselves recursive. Let $X_n \in \mathbb{R}^d$ be the nth estimate of the location of a minimum of the cost function h. The value $G_n(X_n)$ is a noisy but unbiased estimate of the true gradient $g(X_n) = \nabla h(X_n)$, i.e.,

$$E[G_n(X_n)|X_1, \ldots, X_n, G_1(X_1), \ldots, G_{n-1}(X_{n-1})] = g(X_n) \quad (1)$$

The ϕ_i are updated as

$$\phi_{i,n} = \Lambda_i(n, X_n, G_n(X_n), \phi_{1,n-1}, \ldots, \phi_{m,n-1}) \quad (2)$$

Then we can write a fairly general SA process as

$$X_{n+1} = X_n - a_n(X_n, G_n(X_n), \phi_{1,n}, \ldots, \phi_{m,n})G_n(X_n) \quad (3)$$

where a_n is a scalar sometimes called the *learning rate schedule* because it affects how much will be "learned" from the nth gradient measurement. The proper choice of learning rate schedule is important for getting good performance from an SA algorithm. Since each step is in the direction of the noisy gradient measurement, this process may be called stochastic gradient descent, and it may be considered to be deterministic gradient descent in the cost function with the addition of some zero-mean noise to the updates. The noise threatens to keep stochastic gradient descent from converging to a minimum, but often also has the beneficial effect of dislodging X_n from shallow minima (cf. Simulated Annealing).

The process in Equation 3 is general enough to contain backpropagation, LMS, and k-means clustering, for example, but requires generalizing to apply to more complex "higher-order" algorithms, such as stochastic variants of Newton's method.

Even this simplified equation is a nonlinear, time-varying, stochastic difference equation. No explicit solutions to such equations can be expected. Instead, one studies various properties of the equation. In particular, the convergence rate of the *misadjustment error*, defined as $E\|X_n - X^*\|^2$ where X^* is the location of a minimum, is of particular interest.

Backpropagation as Stochastic Approximation

As a specific example of an SA process, we study the gradient descent associated with on-line backpropagation as typically used. The algorithm is

$$X_{n+1} = X_n - \eta[f(X_n, V_n) - Z_n]\nabla_X f(X_n, V_n) + \alpha(X_n - X_{n-1}) \quad (4)$$

The variable X_n is the vector of *weights* of the network. The (V_n, Z_n) are the *exemplars* (V_n is an *input*, Z_n is the *desired output*), which are randomly and independently drawn from a database $\mathscr{E}$ with replacement. The gradient is taken with respect to the first argument of f. The momentum term, $\alpha(X_n - X_{n-1})$, is briefly discussed in the subsection "Other First-Order Meth-

ods." For now we choose the value zero for the momentum parameter α. Take $a_n = \eta$ and the cost function to be

$$h(X) = \frac{1}{|\mathscr{E}|} \sum_{(v,z) \in \mathscr{E}} \frac{1}{2}[f(X, v) - z]^2 = E_{V,z}\frac{1}{2}[f(X, V) - Z]^2 \quad (5)$$

Then

$$g(X) \equiv \nabla_X h(X) = E_{V,z}\{[f(X, V) - Z]\nabla_X f(X, V)\} \quad (6)$$

Taking $G_n(X_n) \equiv [f(X_n, V_n) - Z_n]\nabla_X f(X_n, V_n)$ and $a_n \equiv \eta$, note that Equation 4 has the form of Equation 3. Since the X_i are independent,

$$E[G_n(X_n)|X_1, \ldots, X_n, G_1(X_1), \ldots, G_{n-1}(X_{n-1})] \quad (7)$$

$$= E[G_n(X_n)|X_n] \quad (8)$$

$$= g(X_n) \quad (9)$$

which meets the condition of Equation 1. Thus backpropagation can be interpreted as an SA algorithm. Convergence theory specific to backpropagation on neural networks is given in White (1989).

Similar arguments can be made for LMS and certain clustering (competitive learning) algorithms. In fact, LMS can be considered a simple form of backpropagation where the function f is linear in both X_n and V_n. Clustering algorithms are quite different, however. Instead of approximating an input-output mapping, clustering algorithms attempt to represent the distribution of the inputs, usually by adjusting the location of a small number of representative points (the *cluster centers*).

Regularity Conditions

Even the simpler SA process defined in Equation 3 is too general for rigorous study. In order to draw conclusions which are strong enough to merit interest, two kinds of additional conditions are needed: conditions on the shape of h, the function we are optimizing, and conditions on the stochastic behavior of the noise. Many different sets of conditions have been studied theoretically. It is an exaggeration, but only slightly, to say that there is one set of conditions for every paper on the subject, so we must leave it to the reader to see which theory applies in his or her case of interest. Nevertheless, the SA literature is large and considers many of the issues pertinent to neural networks. In what follows, though we refer to "the minimum," much of the theory extends to the multiple minimum case, in which case one can substitute "that minimum to which we are converging." For this treatment, we ignore the case where minima are hyperplanes or other manifolds rather than isolated points. The theory also extends to various kinds of dependent noises.

Power Law Schedules

In this section we discuss learning rate schedules of the form $a_n = cn^{-p}$, where $\frac{1}{2} < p \leq 1$ and c is a positive scalar. A special case is the harmonic schedule, c/n. Stochastic approximation with power law schedules is the most thoroughly studied class of SA processes. These schedules are both useful in themselves and important reference points for later work. References to these results may be found in the general works and survey articles listed at the end of this article.

Stability

Even when the learning rate schedule decreases to zero, the possibility that the variance of the parameter vector will increase to infinity still exists, because the parameter vector can be pushed diminishingly small distances (by the noise) which nonetheless sum to infinity. Also, the parameter vector may fail to converge to a minimum with probability greater than zero. However, there are numerous sets of sufficient conditions for convergence to a minimum with probability one or in mean square. Stochastic approximation can converge to the *global* minimum of the cost function (but only in probability) if a_n is asymptotic to $1/\log n$ and Gaussian noise is added to the gradient (in addition to the noise which is on the gradient originally) (Kushner, 1987). As in SIMULATED ANNEALING (q.v.), the parameter vector must have a significant probability of taking even sharply uphill steps in order to escape from arbitrarily deep and wide local minima.

Asymptotic Distribution

For some choices of learning rate schedule, a_n, we know that a normalized version of the signed error converges weakly to a random variable, i.e., for some $p > 0$, $n^p(X_n - X^*) \rightsquigarrow Z$. The $\rightsquigarrow$ notation stands for convergence in distribution, which is also known as *weak convergence*. One intuition as to the meaning of this result is as follows. If we imagine many simultaneous, but independent, runs of the process, an envelope containing some fixed fraction of the runs at each n would go to zero like n^{-p}. Any of the processes could enter and leave this envelope an infinite number of times. The classic result for c/n learning rates is that if $c > c^*$, where c^* is some task-dependent number, then $\sqrt{n}(X_n - X^*) \rightsquigarrow Z$, where Z is normal with mean zero and variance $c^2\sigma^2/(c/c^* - 1)$, and σ^2 is the variance of the gradient noise at the minimum. The critical value is given by $c^* \equiv 1/2\lambda$, where λ is the smallest eigenvalue of the Hessian of the cost function h at the minimum. In geometrical terms, λ is the curvature of the objective function in the direction of least curvature at the minimum. The *Hessian* of a function $f(w_1, \ldots, w_m)$ is the matrix with elements $H_{ij} = \partial^2 f/\partial w_i \partial w_j$. This matrix is positive definite at any minimum of f, and may therefore be diagonalized by a rotation of the coordinate system. The eigenvalue λ is thus the smallest diagonal element of the matrix in the rotated coordinates, and thus may be written $\partial^2 h/\partial^2 w'$, where w' is one of the new coordinates. If the learning rate schedule is ct^{-p} with $\frac{1}{2} < p < 1$, however, then $n^{p/2}(X_n - X^*) \rightsquigarrow Z$, where Z is normal. It was later shown that if $c = c^*$, $(n/\log n)^{1/2}(X_n - X^*) \rightsquigarrow Z$, where Z is normal. Furthermore, if $c < c^*$, $n^{c/2c^*}(X_n - X^*) \rightsquigarrow Z$, for some Z not generally normal. See the general references for pointers to specific papers.

Asymptotic Trajectories

What do these results tell us about what we will see X_n doing on any particular run? Some possibilities are ruled out by the asymptotic distribution results, but we still do not know the shape of the sample paths. We need stronger results from the theories of weak convergence and strong approximation to get this information. See the general references for descriptions of these subjects.

Table 1. Asymptotic Behavior of the Misadjustment, $M = E\|X - X^*\|^2$, as a Function of p and c

$p < 1$	any c	$M \propto n^{-p}$
$p = 1$	$c < c^*$	$M \propto n^{-c/c^*}$
$p = 1$	$c > c^*$	$M \propto n^{-1}$

Note: $a_n = c/n^p$, and c^* is the inverse of twice the smallest eigenvalue of the Hessian of the cost function at the minimum.

Transient Properties

Even in the fortunate case where the noise is small compared to the size of the gradient, power law learning rate schedules do not achieve exponential convergence. For instance, if $a_n \propto n^{-1}$ for a problem where the noise has been completely removed, the misadjustment error will still only reduce as n^{-1}.

Asymptotically Optimal Power Law Schedule

Which power law schedule is the best to use from asymptotic performance considerations? We summarize the asymptotic behavior of the misadjustment in Table 1.

It is apparent that it is best to take a_n to go as c/n asymptotically, where $c > c^*$. If the misadjustment for such a schedule is $M_*(n)$ and that for a schedule with $c < c^*$ or $p < 1$ is $M(n)$, then by Table 1, it must be the case that

$$\lim_{n \to \infty} \frac{M(n)}{M_*(n)} = \infty \tag{10}$$

The penalty for using a suboptimal schedule (one with $c < c^*$ or $p < 1$) becomes increasingly severe for large times. However, if M_1 is the misadjustment for the schedule $c_1 \neq n$ and M_2 that for c_2/n, and if c_1 and c_2 are both greater than c^*, then

$$\lim_{n \to \infty} \frac{M_1(n)}{M_2(n)} < \infty \tag{11}$$

i.e., the misadjustments stay some fixed multiple apart.

Moreover, using the asymptotic distribution result given above, we can also find a best value for c. The expected misadjustment is asymptotically proportional to $c^2/n(c/c^* - 1)$. For any n, the minimum of this expression is achieved by $c = 2c^*$.

Constant Schedules

Taking $a_n = \eta$ to be constant is the usual choice made for LMS and backpropagation, and is sometimes used for clustering (competitive learning) as well. In the following subsections we explore the consequences of the constant schedule for the asymptotic behavior of the misadjustment error.

Stability

The parameter vector will not generally converge when a constant schedule is used. In fact, it is possible for the variance of the parameter vector to blow up. However, we do have some asymptotic knowledge about the process in the limit of small η which we present in the next subsection.

Asymptotic Distribution

Assuming that the second-order statistics of the parameter vector X are converging, we will give a form for $E(X - X^*) \times (X - X^*)^{\mathrm{T}}$ valid for all SA processes with constant learning rate in the limit of small learning rates and large times. Let X^* be the location of the minimum of the cost function h, to which we are converging. Let Σ be the covariance matrix of the gradient noise at X^*. Likewise, define G as $E\nabla h(X^*)[\nabla h(X^*)]^{\mathrm{T}}$ and H as the Hessian matrix of h at X^*. Then, for small η and large n,

$$E(X_n - X^*)(X_n - X^*)^{\mathrm{T}} \approx \frac{\eta}{2} H^{-1}(G + \Sigma) \tag{12}$$

(Kushner and Huang, 1981).

Two conclusions to draw are that the size of the residual misadjustment ($\mathrm{Tr}\, E(X_n - X^*)(X_n - X^*)^{\mathrm{T}}$) is proportional to η, and also to a term which depends upon details of the noise and curvature of h. (These are usually unknown in applications.)

Transient Properties

In a region in which the noise is of small magnitude and the cost function is quadratic, convergence can be exponential. That is, the behavior is just as in deterministic gradient descent. The usual illustration is to take the loss function to be exactly quadratic and the magnitude of the noise to be zero. The corresponding recursion is $X_{n+1} = X_n - \eta X_n = (1 - \eta)X_n$ which has solution $X_n = (1 - \eta)^n X_0$.

The constant learning rate schedule is often observed to escape from a bad (i.e., high-lying) local minimum and transit to a better one. The chances of such an occurrence are enhanced with a constant schedule as opposed to a schedule which decreases to zero, because the parameter vector travels further with each step, enhancing the likelihood of escaping from the current attractor because of a lucky combination of gradient noises. Of course, transitions from a good minimum to a bad one are also possible, but these are much less commonly observed in practice. Note that the only known guaranteed way of escaping *all* local minima requires the addition of more noise to the gradient, as discussed previously.

Other Varieties of Stochastic Approximation

In this section we discuss a few learning rate schedules and other first-order and higher-order (in the complexity of updates) recursive optimization techniques. A larger, but still far from comprehensive, bibliography of methods is given in Darken (1993).

Nonclassical Learning Rate Schedules

There are many learning rate schedules in the statistics and engineering literature other than the so-called classical constant and power law schedules. Kesten's method adjusts the learning rate on the basis of the inner product $G_n(X_n) \cdot G_{n+1}(X_{n+1})$. When this quantity is positive, indicating two successive steps in roughly the same direction, the learning rate is kept constant. Otherwise it is lowered. This technique has been proven (Zhulenev and Medovyi, 1978) to speed learning for some tasks, but it slows them for others. A related group of more complex techniques (for example, Sutton, 1992) allow the learning rate to be raised as well as lowered. These approaches are often justified as attempts to find a learning rate which is optimal for some local region of the parameter space. The class of tasks for which there exist locally optimal learning rates is unknown, however. This class of schedules typically leads to nonconvergent behavior.

Search-then-converge schedules attempt to obtain both the good transient properties of constant schedules and the good asymptotic properties of the optimal power law schedule (Darken and Moody, 1991). An example of such a schedule is

$$a_n = \eta \frac{1 + \dfrac{c}{\eta}\dfrac{n}{\tau^2}}{1 + \dfrac{c}{\eta}\dfrac{n}{\tau^2} + \tau^2 \dfrac{n^2}{\tau^4}} \tag{13}$$

This function is approximately constant with value η at times small compared to τ (the *search phase*). At times large compared with τ (the *converge phase*), the function decreases as c/n.

Furthermore, given that one is using a c/n schedule, it is possible to determine on-line whether $c > c^*$ or not by using appropriate statistics. An adaptive search-then-converge schedule which increases c on-line when necessary has also been studied (Darken, 1993).

Other First-Order Methods

Other recursive optimization techniques are first order [i.e., require only $O(d)$ computations per update where d is the dimension of X_n] but cannot be put in the form of Equation 3. The simplest of these techniques is the addition of a momentum term $\alpha(X_n - X_{n-1})$ to the update equation (Equation 3). Momentum can speed convergence, but if the size of the momentum parameter is too large, the system may converge more slowly, or even become unstable (Tugay and Tanik, 1989). The appropriate size for the momentum parameter depends upon the details of the task.

Another family of techniques is that of stochastic analogs of conjugate gradient optimization. Conjugate gradient is a deterministic optimization algorithm which converges quickly and has an implementation requiring only $O(d)$ computations per parameter vector update. See Kushner and Gavin (1973), Møller (1993), and references therein.

The Polyak-Ruppert technique is a fascinatingly simple method for achieving $1/n$ convergence of the misadjustment (Polyak and Juditsky, 1991). This method uses a power law learning rate schedule cn^{-p}, $p < 1$, to produce a slowly convergent process which is then averaged. The misadjustment of the average converges at the $1/n$ rate. This technique should work well for many signal-processing tasks, but it could be problematic for control applications where penalties apply for evaluating the gradient (i.e., operating the system) far from the minimum. Additionally, this estimator may suffer from sluggish transient performance (because of the averaging).

Second-Order Methods

Higher-order recursive stochastic optimization techniques include signal orthogonalizers and several Newton-like techniques. One very thoroughly studied technique is that of Ljung and Söderström (1983). The techniques achieve an $O(1/n)$ convergence rate with a constant typically better than that achieved by first-order techniques. However, the update complexity of these methods is quadratic or worse in the number of system parameters (the dimensionality of X_n).

Stopping Rules

Finally, we note that obtaining the fastest asymptotic convergence rate is not the only question of interest. It is also possible to imagine that each recursive step of our learning algorithm is costly, and we would like to turn off our learning machine as soon as it has found an answer that is "good enough," i.e., an estimate of the minimum which is close enough to the true location of a minimum. This is the issue of finding an efficient *stopping rule* for stochastic approximations (Ljung, Pflug, and Walk, 1992:II§9).

Discussion

Applying stochastic approximation theory is useful both to obtain guarantees of finding a minimum and to facilitate converging quickly to a minimum once one has been found. However, the difficult part of some neural network optimization problems may be finding the location of a minimum in the first place. For such tasks, choosing the best optimization technique to use remains something of an art.

For further reading, Benveniste, Métivier, and Priouret (1990) provides a general reference and text which targets the engineering community and includes numerous applications as well as theory. Ruppert (1991) is a thorough survey article, and Ljung, Pflug, and Walk (1992) is a useful monograph.

Road Map: Learning in Artificial Neural Networks, Statistical
Background: I.3. Dynamics and Adaptation in Neural Networks
Related Reading: Boltzmann Machines; Data Clustering and Learning; Unsupervised Learning with Global Objective Functions

References

Benveniste, A., Métivier, M., and Priouret, P., 1990, *Adaptive Algorithms and Stochastic Approximations*, Berlin: Springer-Verlag. ◆

Darken, C., 1993, Learning rate schedules for stochastic gradient algorithms, PhD Dissertation, Yale University.

Darken, C., and Moody, J., 1991, Note on learning rate schedules for stochastic optimization, in *Advances in Neural Information Processing Systems 3* (R. P. Lippmann, J. E. Moody, and D. S. Touretzky, Eds.), San Mateo, CA: Morgan Kaufmann, pp. 832–838.

Kushner, H., 1987, Asymptotic global behavior for stochastic approximation and diffusions with slowly decreasing noise effects: Global minimization via Monte Carlo, *SIAM J. Appl. Math.*, 47:169.

Kushner, H., and Gavin, T., 1973, Extensions of Kesten's adaptive stochastic approximation method, *Ann. Statist.* 1:851.

Kushner, H., and Huang, H., 1981, Asymptotic properties of stochastic approximations with constant coefficients, *SIAM J. Control Optim.*, 19:87.

Ljung, L., Pflug, G., and Walk, H., 1992, *Stochastic Approximation and Optimization of Random Systems*, Basel: Birkhäuser. ◆

Ljung, L., and Söderström, T., 1983, *Theory and Practice of Recursive System Identification*, Cambridge, MA: MIT Press.

Møller, M., 1993, Supervised learning on large redundant training sets, *Int. J. Neural Syst.*, 4:15–25.

Polyak, B., and Juditsky, A., 1991, Acceleration of stochastic approximation by averaging, *SIAM J. Control Optim.*, 29:378–389.

Robbins, H., and Monro, S., 1951, A stochastic approximation method, *Ann. Math. Statist.*, 22:400–407.

Ruppert, D., 1991, Stochastic approximation, in *Handbook of Sequential Analysis* (B. Ghosh and P. Sen, Eds.), New York: Marcel Dekker. ◆

Sutton, R., 1992, Adapting bias by gradient descent: An incremental version of Delta-Bar-Delta, in *Proceedings of the Tenth National Conference on Artificial Intelligence (AAAI-92)*, Menlo Park, CA: AAAI Press, pp. 171–176.

Tugay, M., and Tanik, Y., 1989, Properties of the momentum LMS algorithm, *Signal Processing*, 18:117–127.

White, H., 1989, Some asymptotic results for learning in single hidden layer feedforward networks, *J. Am. Statist. Assoc.*, 84:1008–1013.

Zhulenev, S., and Medovyi, V., 1978, The strong law of large numbers and normality of Kesten's procedure, *Theory Probab. Appl.*, 23:615–621.

Structural Complexity and Discrete Neural Networks

Ian Parberry

Introduction

It has been conjectured that neural networks can perform tasks that are extremely difficult for conventional computers. One thing that characterizes these tasks is the large amount of input data. The typical neural network research project tackles a scaled-down version using only a small amount of input, while the more challenging problem of scaling up the solution to real-world input sizes is optimistically left as a subject for future research (and funding).

A connectionist neural network model is usually a finite circuit using simple computational elements that interact to perform a computation. The normal approach to circuit design is to start with a circuit that solves a smaller version of the problem. Scaling up the circuit will likely involve adding more inputs; hence more neurons, and possibly more layers, will be needed to integrate the additional data into the computation. It is reasonable to conjecture that future research breakthroughs and advances in hardware technology will allow us to build circuits with successively more neurons and layers. The usefulness of a neural network design will depend on how the numbers of neurons and layers scale as the number of inputs increases. If its resource requirements grow too quickly, then it is unlikely that technology will be able to keep pace with demands.

Current technology allows us to build a finite circuit with the following components:

- 100 inputs
- 10^6 processing elements (called *gates* or *processors*)
- 2 inputs to each processor (we will use the terminology *fan-in* 2)
- Processors that compute Boolean conjunction, disjunction, and complement

It is likely that a series of future technological jumps will allow us to increase some or all of these values. That is, we will produce a family of circuits scaling from current technological limits up to brain-scale circuits, which appear to have

- 10^7 inputs
- 10^{10} processors (neurons)
- Fan-in 10^3
- Processors that compute complicated functions

A useful abstraction of the scaling process is to imagine an infinite series of finite circuits, one for each possible input size (Figure 1).

Since our aim is to investigate how the resource usage of various circuit designs will scale up on very simple problems, we consider here only problems that are less grandiose than the type of computation that is typically exhibited as something that brains do well and computers do poorly. Two important circuit resources are the *size*, defined to be the number of gates used, and the *depth*, defined to be the number of gates on the longest path from an input to an output.

The main body of this article is divided into two sections. The first section describes conventional structural complexity theory, and the second compares and contrasts this with a neu-

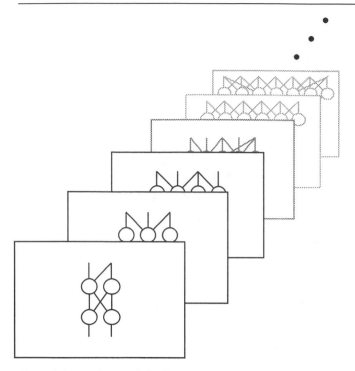

Figure 1. A neural network family.

ral network complexity theory. In the interest of brevity, proofs have been omitted, and only a few representative references are given. For more details the reader may consult Parberry (1990, 1994, in press) and reference lists therein.

The symbol $\mathbb{B}$ denotes the Boolean set $\{0, 1\}$, $\mathbb{N}$ denotes the natural numbers, and $\mathbb{R}$ denotes the real numbers. If $x \in \mathbb{R}$ is positive, $\lfloor x \rfloor$ denotes the largest natural number not exceeding x. All logarithms are to base two. The expression $\log^k n$ is used as shorthand for $(\log n)^k$. *Polylog* means bounded above by $\log^k n$ for some $k \in \mathbb{N}$.

If $f, g: \mathbb{N} \rightarrow \mathbb{N}$, $f(n)$ is said to be $O(g(n))$ if there exists c, $n_0 \in \mathbb{N}$ such that for all $n \geq n_0$, $f(n) \leq c \cdot g(n)$. If S is a set, S^n denotes the n-fold Cartesian product of S,

$$\underbrace{S \times S \times \cdots \times S}_{n \text{ times}} = \{(s_1, \ldots, s_n) | s_i \in S \text{ for } 1 \leq i \leq n\}$$

Structural Complexity Theory

Computational complexity theory is the study of resource-bounded computation. *Computation* can be defined using many different machine models. Conventional computation can be studied using the *Turing machine*, or the *random-access machine* (usually abbreviated to RAM) which models the conventional von Neumann–style computer, or the circuit family. Parallel computers (see PARALLEL COMPUTATIONAL MODELS) can be modeled by the PRAM, which is a collection of RAMs which communicate through a shared memory, or the circuit family (under different resource constraints than for conventional computers). Shared memory access for a PRAM can be either CREW (concurrent read, exclusive write), in which simultaneous reads from single memory cells by many processors is allowed but simultaneous writes are not, or CRCW (concurrent read, concurrent write), in which both simultaneous reads and writes are permitted.

The term *resource* means any finite commodity that is needed for a computation (such as time, memory, and hardware). Resources are measured as a function of input size. A prime tenet of computational complexity theory is that resource usage that grows more than polynomially with input size is bad. A *complexity* class is defined to be the set of problems that can be solved within a particular resource bound on a particular model of computation. *Structural complexity theory* is the study of relationships between complexity classes. Of most interest are the *robust* complexity classes: those that can be defined in a natural and elegant way using several different machine models and resources.

Figure 2 shows some important complexity classes, described briefly in the following list. For more information and references, see Garey and Johnson (1979), Johnson (1990), Parberry (1994, in press), and van Emde Boas (1990). All circuit families mentioned here use gates that compute Boolean conjunction, disjunction, and complement.

1. $\mathcal{AC}^0$ is the set of decision problems that can be solved in constant depth on a circuit family using gates of arbitrary fan-in. An alternative formulation is constant time on a CRCW PRAM with polynomially many processors.
2. $\mathcal{NC}^1$ is the set of decision problems that can be solved in logarithmic depth on a circuit family using gates of fan-in at most two. An alternative formulation is logarithmic time on a CREW PRAM with polynomially many processors. Note that $\mathcal{AC}^0 \subset \mathcal{NC}^1$, since the parity problem is in $\mathcal{NC}^1$ but not in $\mathcal{AC}^0$.
3. $\mathcal{NC}$ is the set of decision problems that can be solved by a small, fast parallel computer. Alternative characterizations include these:

   - Polylog time on a PRAM with polynomially many processors
   - Polylog depth on a polynomial size circuit family
   - Polylog reversals and polynomial time on a deterministic Turing machine

   Clearly, $\mathcal{NC}^1 \subseteq \mathcal{NC}$, but it is not known whether the containment is proper.
4. $\mathcal{P}$ is the set of decision problems that can be solved in polynomial time on any of a wide range of machine models, including the RAM and the deterministic Turing machine. It can also be characterized as the set of problems that can be solved by a polynomial-size family of circuits. Although $\mathcal{NC} \subseteq \mathcal{P}$, it is not known whether $\mathcal{P} = \mathcal{NC}$ (it is popularly conjectured otherwise; see item 5).
5. $\mathcal{P}$-complete problems are the hardest problems in $\mathcal{P}$ in the sense that if one of them is in $\mathcal{NC}$, then $\mathcal{P} = \mathcal{NC}$. $\mathcal{P}$-completeness is an analog for parallel computation of the more well-known $\mathcal{NP}$-completeness (see item 7). The standard interpretation of the $\mathcal{P}$-complete problems is that they are *inherently sequential*; that is, they can be solved fast on a sequential computer but probably not much faster on a parallel computer.
6. $\mathcal{NP}$ is the class of problems that can be solved in polynomial time by a nondeterministic Turing machine. It can also be characterized as the set of existential decision problems whose witnesses can be verified in polynomial time. More precisely, problems in $\mathcal{NP}$ have the following form: On input x, is

$$\exists_p y P(x, y)$$

true, where P is a predicate in $\mathcal{P}$? The subscript p on any quantifier indicates that it is a *polynomially bounded quan-*

Figure 2. Some of the important conventional complexity classes and their expected relationships.

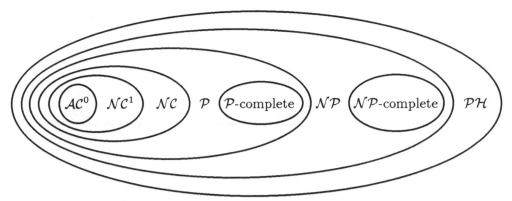

tifier; that is, it ranges over the set of objects that are no larger than a polynomial in the size of x. So, "$\exists_p y$" is shorthand for "There exists y with size no more than polynomially larger than the size of x." Although $\mathcal{P} \subseteq \mathcal{NP}$, it is not known whether $\mathcal{P} = \mathcal{NP}$ (it is popularly conjectured otherwise; see item 7).

7. $\mathcal{NP}$-complete problems are the hardest problems in $\mathcal{NP}$ in the sense that if one of them is in $\mathcal{P}$, then $\mathcal{NP} = \mathcal{P}$.
8. The *polynomial-time hierarchy* is defined as follows. Let $\Sigma_p^0 = \Pi_p^0 = \mathcal{P}$. Then, for all $i \geq 1$, Σ_p^i consists of the problems of the following form: On input x, is

$$\exists_p y\, P(x, y)$$

true, where P is a predicate in Π_p^{i-1}? Similarly, Π_p^i consists of the problems of the following form: On input x, is

$$\forall_p y\, P(x, y)$$

true, where P is a predicate in Σ_p^{i-1}? Hence, for example, $\Sigma_p^1 = \mathcal{NP}$, and Π_p^1 is the set of complements of problems in $\mathcal{NP}$, commonly known as co-$\mathcal{NP}$. The *polynomial hierarchy* is then defined to be

$$\mathcal{PH} = \bigcup_{i \geq 1} \left(\Sigma_p^{i-1} \cup \Pi_p^{i-1} \right)$$

Threshold Circuits

Many neural network models differ from the circuit models in the preceding section in that the gates compute linear threshold functions. We also allow the values 0 and 1 to be used as constants in any place in the circuit. This technologically plausible assumption is often used for convenience in conventional circuit complexity classes. This section is divided into two subsections. The first studies linear threshold functions, and the second threshold complexity classes.

Threshold Logic

A *linear threshold function* $f: \mathbb{B}^n \to \mathbb{B}$ is defined as follows:

$$f(x_1, \ldots, x_n) = \begin{cases} 1 & \text{if } \sum_{i=1}^n w_i x_i \geq h \\ 0 & \text{otherwise} \end{cases}$$

for some $w_1, \ldots, w_n, h \in \mathbb{R}$. The values $w_1, \ldots, w_n$ are called *weights*, and the value h is called the *threshold*. There are many different normal forms for linear threshold functions. One may assume without loss of generality that any one of the following holds:

1. The threshold $h = 0$.
2. The weights are integers.

3. The thresholding inequality is strict, that is, f is defined by

$$f(x_1, \ldots, x_n) = \begin{cases} 1 & \text{if } \sum_{i=1}^n w_i x_i > h \\ 0 & \text{if } \sum_{i=1}^n w_i x_i < h \end{cases}$$

(and the third case of $\sum_{i=1}^n w_i x_i = h$ never occurs).
4. The weights are positive integers, assuming that hardware for performing Boolean complementation is available.

Muroga, Toda, and Takasu showed in 1961 that the weights of any linear threshold function can be made positive integers bounded above by $n^n/2^n$. Johan Håstad has recently found a linear threshold function that requires integer weights at least $n^{n/2-O(n)}$ (for a proof see Parberry, 1994).

Threshold Complexity Classes

The standard complexity classes described in the section "Structural Complexity Theory" are based on the Boolean operations conjunction, disjunction, and complement. What happens if we replace these operations by linear threshold functions? Since conjunction, disjunction, and complement are linear threshold functions, we obtain complexity classes that include the originals. The study of threshold complexity classes and their relationships to conventional complexity classes was initiated by Parberry and Schnitger, and independently by Hajnal et al. The journal versions of these papers appear in Hajnal et al. (1993) and Parberry and Schnitger (1988, 1989). Many of the results described here are from these articles.

1. Existential and universal quantifiers are analogs of the Boolean functions OR and AND respectively. It is possible to define threshold quantifiers which are similarly analogous to linear threshold functions (the definitions get quite technical). Replacing the quantifiers in the definition of the polynomial-time hierarchy by these new threshold quantifiers gives a larger complexity class called the *threshold hierarchy*, which is identical to the *counting hierarchy* of Jacobo Torán and Klaus Wagner.
2. Replacing the existential quantifier in the definition of $\mathcal{NP}$ by a threshold quantifier gives a new complexity class at the first level of the threshold hierarchy. This class is known as $\mathcal{PP}$, for *probabilistic polynomial time*.
3. Since linear threshold functions can be computed in polynomial time, $\mathcal{P}$ remains invariant if the gates used in the circuit characterization are replaced by linear threshold gates.
4. Since linear threshold functions are in $\mathcal{NC}$, $\mathcal{NC}$ remains invariant if the gates used in the circuit characterization are replaced by linear threshold gates.

5. Since linear threshold functions are in $\mathcal{NC}^1$, $\mathcal{NC}^1$ remains invariant if the gates used in the circuit characterization are replaced by linear threshold gates.
6. Let $\mathcal{TC}^0$ be the analog of $\mathcal{AC}^0$ using linear threshold gates. $\mathcal{TC}^0$ is by far the most interesting threshold complexity class. The facts that we have observed so far imply that $\mathcal{AC}^0 \subseteq \mathcal{TC}^0 \subseteq \mathcal{NC}^1$. All symmetric functions (those that remain invariant under any permutation of their inputs) are in $\mathcal{TC}^0$. Hence, parity is in $\mathcal{TC}^0$. Since it is well-known that parity is not in $\mathcal{AC}^0$ (Furst, Saxe, and Sipser, 1984), this result implies that $\mathcal{AC}^0 \neq \mathcal{TC}^0$.

The sequence of sets

$$\mathcal{TC}_1^0 \subseteq \mathcal{TC}_2^0 \subseteq \mathcal{TC}_3^0 \subseteq \cdots$$

is called the $\mathcal{TC}^0$ hierarchy. The $\mathcal{TC}^0$ hierarchy is less robust than $\mathcal{TC}^0$. It is an open question whether $\mathcal{TC}_d^0 \neq \mathcal{TC}_{d+1}^0$ for all $d \geq 1$. This would be called *separating the $\mathcal{TC}^0$ hierarchy*. The parity function separates $\mathcal{TC}_1^0$ from $\mathcal{TC}_2^0$, and Hajnal et al. (1993) give a function that separates $\mathcal{TC}_2^0$ from $\mathcal{TC}_3^0$. The $\mathcal{TC}^0$ hierarchy has not, to date, been separated above depth 3. The corresponding hierarchy for *monotone $\mathcal{TC}^0$* ($\mathcal{TC}^0$ without negative weights or Boolean negations) has been separated by Andrew Yao. A conjecture slightly weaker than the collapse of the $\mathcal{TC}^0$ hierarchy is that $\mathcal{AC}^0$ is contained in $\mathcal{TC}_3^0$; however, all that is known is that every function in $\mathcal{AC}^0$ can be computed by threshold circuits of depth 3 and size $n^{\text{polylog}(n)}$, as shown by Eric Allender.

The class $\mathcal{TC}^0$ is fairly robust, for example:

- $\mathcal{TC}^0$ remains invariant when all weights are restricted to ± 1. Mikael Goldmann has shown that the size increases polynomially, and the depth increases by 1.
- $\mathcal{TC}^0$ is robust to restrictions on fan-in. That is, fan-in can be reduced by a polynomial amount with only a polynomial increase in size and a constant-multiple in depth.
- $\mathcal{TC}^0$ is robust to randomness. A probabilistic variant of threshold circuits can be defined using *random gates*, each of which outputs a Boolean value with some fixed probability. If the error in the circuit is bounded away from 0.5, then the randomness can be removed using a sampling technique developed by Len Adleman.
- $\mathcal{TC}^0$ is robust to cycles in the network. That is, polynomial-size synchronous cyclic threshold circuits running in constant time compute exactly the functions in $\mathcal{TC}^0$. The proof of this claim uses an "unrolling" technique that first appeared in Savage (1972).

Discussion

The *family of circuits* is a plausible abstract model of scalable neural networks. If a neural network is to be considered feasibly scalable, the depth should remain constant, and the size must grow polynomially. One of the differences between conventional computation and neural network computation is that the former is based on conjunction and disjunction, while the latter is based on linear threshold functions. Generalizing standard complexity classes to include computation based on linear threshold functions gives new complexity classes at high levels (above $\mathcal{P}$) and at low levels (below $\mathcal{NC}^1$). The low-level complexity classes are the most interesting because they model the observed ability of neural systems to compute with only a few layers of neurons, and they are a source of rich and mathematically deep open problems. Many complexity classes (neural network or otherwise) are robust; that is, many details of the neural network model are relatively unimportant in the analysis of scalability. More research is needed to determine which features of both natural and artificial neural networks are important and which are not. Exponential lower bounds for even the simplest neural network model, the threshold circuit, are very difficult to obtain.

Acknowledgments. The research described in this paper was supported by the National Science Foundation under grant number CCR–9302917 and by the Air Force Office of Scientific Research, Air Force Systems Command, USAF, under grant number F49620–93–1–0100.

Road Map: Computability and Complexity
Related Reading: Automata and Neural Networks; Time Complexity of Learning

References

Furst, M., Saxe, J. B., and Sipser, M., 1984, Parity, circuits and the polynomial time hierarchy, *Math. Syst. Theory*, 17:13–27.
Garey, M. R., and Johnson, D. S., 1979, *Computers and Intractability: A Guide to the Theory of NP-Completeness*, San Francisco: W. H. Freeman.
Hajnal, A., Maass, W., Pudlák, P., Szegedy, M., and Turán, G., 1993, Threshold circuits of bounded depth, *J. Comput. Syst. Sci.*, 46:129–154.
Johnson, D. S., 1990, A catalog of complexity classes, in *Algorithms and Complexity* (J. van Leeuwen, Ed.), vol. A of *Handbook of Theoretical Computer Science*, Amsterdam and Cambridge, MA: Elsevier and MIT Press, pp. 67–161. ◆
Muroga, S., Toda, I., and Takasu, S., 1961, Theory of majority decision elements, *J. Franklin Inst.*, 271:376–418.
Parberry, I., 1990, A primer on the complexity theory of neural networks, in *Formal Techniques in Artificial Intelligence: A Sourcebook* (R. Banerji, Ed.), vol. 6 of *Studies in Computer Science and Artificial Intelligence*, Amsterdam: North-Holland, pp. 217–268.
Parberry, I., 1994, *Circuit Complexity and Neural Networks*, Cambridge, MA: MIT Press.
Parberry, I., in press, Circuit complexity and feedforward neural networks, in *Mathematical Perspectives on Neural Networks* (P. Smolensky, M. Mozer, and D. Rumelhart, Eds.), Developments in Connectionist Theory, Hillsdale, NJ: Erlbaum.
Parberry, I., and Schnitger, G., 1988, Parallel computation with threshold functions, *J. Comput. Syst. Sci.*, 36:278–302.
Parberry, I., and Schnitger, G., 1989, Relating Boltzmann machines to conventional models of computation, *Neural Netw.*, 2:59–67.
Rumelhart, D. E., Hinton, G. E., and McClelland, J. L., 1986, A general framework for parallel distributed processing, in *Parallel Distributed Processing: Explorations in the Microstructure of Cognition* (D. E. Rumelhart, J. L. McClelland, and PDP Research Group, Eds.), vol. 1, *Foundations*, Cambridge, MA: MIT Press, pp. 282–317.
Savage, J. E., 1972, Computational work and time on finite machines, *J. ACM*, 19:660–674.
van Emde Boas, P., 1990, Machine models and simulations, in *Algorithms and Complexity* (J. van Leeuwen, Ed.), vol. A of *Handbook of Theoretical Computer Science*, Amsterdam and Cambridge, MA: Elsevier and MIT Press, pp. 1–66. ◆

Structured Connectionist Models

Lokendra Shastri

Introduction

Artificial intelligence (AI) and cognitive science have made considerable advances over the past four decades, but it is widely believed that solutions resulting from the classical approach to these disciplines lack scalability, gradedness, robustness, and flexibility. Consider scalability. Although existing AI systems may perform credibly within restricted domains, they do not scale up; as the domain grows larger, a system's performance degrades drastically, and it can no longer solve interesting problems in acceptable time scales. Consider gradedness. Research in AI and cognitive science has made it apparent that the solution of a cognitive task emerges as a result of rich context-sensitive interactions among a large number of graded factors. Classical models, rooted in the von Neumann architecture for serial computation, are not suited for articulating this view of computation (for a discussion of robustness and flexibility, see CONNECTIONIST AND SYMBOLIC REPRESENTATIONS).

Research in connectionism is motivated by the belief that to address the limitations mentioned above, one must pay attention to the computational characteristics of the brain. After all, the brain is the only physical system that exhibits the requisite attributes, and it seems reasonable to expect that identifying neurally motivated constraints, even at an abstract computational level, and incorporating them into our models would lead to novel and critical insights.

In addition to recognizing the importance of neurally motivated constraints, the *structured connectionist* approach (Feldman, Fanty, and Goddard, 1988) also recognizes that a number of insights acquired by disciplines such as computer science, AI, psychology, linguistics, and learning theory will have to be leveraged in developing solutions to difficult problems in AI and cognitive science. These insights pertain to recognizing the power of problem decomposition, hierarchical processing, and structured representations; the need for representational and inferential adequacy; and the role of complexity analysis.

A key difference between the structured connectionist approaches and the *distributed* approach is as follows: The fully distributed approach assumes that each *item* (concept or mental object) is represented as a pattern of activity distributed over a *common* pool of nodes (van Gelder, 1992). This notion of representation has several fundamental limitations. Consider the representation of *John and Mary*. If *John* and *Mary* are represented as patterns of activity over the entire network such that each node in the network has a specific value in the patterns for *John* and *Mary*, respectively, then how can the network represent *John* and *Mary* at the same time? The situation becomes more complex if the system must represent relations such as *John loves Mary*, or *John loves Mary but Tom loves Susan*. In contrast to the distributed approach, the structured approach holds that small clusters of nodes can have distinct representational status. For simplicity, structured connectionist models often equate a small cluster of nodes with a single *idealized* node. In particular, there are small clusters of nodes that act as *focal* nodes or *handles* of *learned* concepts and provide access to more elaborate node structures which make up the detailed encoding of concepts. Such a detailed encoding might include various features of the concept as well as its relationship to other concepts (see Feldman, 1989, and the article by Shastri in Barnden and Pollack, 1991). The fully distributed view is also inconsistent with the continually emerging data about the localization of function in the brain. However, many studies take advantage of distributed coding without requiring a global distribution of activity for each encoding, so that the *fully* distributed view is something of a straw man.

The structured approach is often incorrectly equated with the *grandmother cell* approach, which assumes that each concept is represented by a distinct node. This misunderstanding stems from an incorrect interpretation of the representational role of focal nodes.

A second difference between the structured and *fully* distributed approaches concerns learning. The latter approach underplays the importance of structure and assumes that essentially all of the required structure emerges as a result of general purpose learning processes operating on relatively unstructured hidden layers. The structured approach holds that such a tabula rasa view is untenable on the grounds of computational complexity; training unstructured networks with general purpose learning techniques is not a feasible way to obtain scalable solutions to complex problems. The structured approach (and many approaches using locally distributed coding in modular neural networks) emphasizes the importance of prior structure for effective learning and requires the initial design of network models (e.g., the broad representational significance of nodes, the number of representational levels in the network, and the network interconnection pattern) to reflect the structure of the problem.

Some Neural Constraints on Cognitive Models

Representational Constraints

With more than 10^{11} computing elements and 10^{15} interconnections, the human brain's capacity for encoding, communicating, and processing information seems awesome. However, if the brain is powerful, it is also limited. First, neurons are slow computing devices. Second, although the spatiotemporal integration of inputs performed by neurons is complex, it is relatively undifferentiated in terms of the needs of symbolic computation. Third, neurons communicate through relatively simple messages that can encode only a few bits of information. Hence, a neuron's output cannot be expected to encode names, pointers, or complex structures.

A specific limitation of neurally plausible systems is that they have difficulty representing composite structures in a dynamic fashion (see COMPOSITIONALITY IN NEURAL SYSTEMS). Consider the representation of the fact *give(John, Mary, Book1)*. This fact cannot be represented dynamically by simply activating the roles *giver*, *recipient*, and *give-object* and the constituents *John*, *Mary*, and *Book1*. Such a representation would be indistinguishable from the representation of *give(Mary, John, Book1)*. The problem is that representing a fact requires representing the appropriate *bindings* between roles and their fillers. It is easy to represent static (long-term) bindings with dedicated nodes and links. For example, one could posit a separate binder node for each role-filler pair to represent role-filler bindings. Such a scheme is adequate for representing long-term knowledge because the required binder nodes may be recruited over time. However, this scheme is implausible for representing dynamic bindings that arise during language understanding and visual processing since it is unlikely that mechanisms exist for establishing new links within such time scales. The alternative that interconnections between *all possible* pairs of roles and fillers already exist and the appropriate ones become active temporarily to represent dynamic bindings is also excluded because of the prohibitively large number of such role-filler bindings.

Scalability in Time

We can visually recognize items from a potential pool of 100,000 common items in approximately 100 ms and can understand language at the rate of several words per second, even though doing so involves perceptual processing, lexical access, parsing, and reasoning. Therefore, we can perform a wide range of visual, linguistic, and inferential tasks within a few hundred milliseconds. This observation provides a powerful constraint that can inform our search for cognitive models (Feldman and Ballard, 1982).

Scalability in Space

Although the number of neurons in the brain is large, it is not too large compared with the magnitude of the problems the brain must solve. Consider vision and reasoning. The retinal output consists of 1 million signals, and our common-sense knowledge base may contain more than 100,000 items. This observation suggests that any model of vision or reasoning whose node requirement grows quadratically or higher with respect to the size of the problem may not be neurally plausible.

From Constraints to Predictions

While cognitive agents solve a wide range of tasks efficiently, their cognitive ability is limited in a number of ways. Examples abound in vision, language, and short-term memory. It is believed that structured connectionist models that incorporate the representational and scalability constraints discussed above would help in understanding and explaining the strengths and limitations of human cognition.

Some Structured Connectionist Models

Early Work

One of the earliest examples of a structured connectionist model is the interactive activation model for letter perception proposed by McClelland and Rumelhart (1981). The model consisted of three layers of nodes corresponding to visual letter features, letters, and words. Nodes representing mutually exclusive hypotheses within the letter and word layers inhibited each other. For example, since only one letter may exist in a given letter position, all nodes representing letters in the same position inhibited each other. A node in the feature layer was connected through excitatory connections to nodes in the letter layer representing letters that contained that feature. Similarly, a node in the letter layer was connected through excitatory connections to nodes in the word layer representing words that contained that letter in the appropriate position. Additionally, there were reciprocal connections from the word layer to the letter layer. The interconnection pattern allowed bottom-up perceptual processing to be guided by top-down expectations. The model could explain a number of psychological findings about the preference of words and pronounceable nonwords over other nonwords and isolated letters.

Other examples of early structured connectionist models were word sense disambiguation models developed by Cottrell and Small (1983) and Waltz and Pollack (1985). Most words have multiple senses, but we use contextual and syntactic information to rapidly disambiguate the meanings of words. These models demonstrated how such disambiguation might occur. Cottrell and Small's model consisted of a three-level network that included the lexical (word) level, the word sense level, and the case level. There were inhibitory links between different noun senses of the same word and between different predicate senses of the same word. A node at the lexical level was connected to all of its senses at the word sense level. Connections between the word sense level and the case level expressed all feasible bindings between predicates and objects. As a sentence was inputted by activating the appropriate lexical items in a sequence, activation flowed through the network and the combination of lexical items, word senses, and case assignment that best fit the input formed a stable coalition of active nodes.

Another example of a structured connectionist model is the connectionist semantic network model (CSN) (Shastri, 1988). CSN viewed memory as a collection of concepts organized in an IS-A hierarchy (e.g., "bird IS-A animal") and allowed the attachment of property values to concepts. Unlike traditional Semantic Networks (q.v.), the property-value attachment in CSN consisted of distributional information indicating how members of a concept were distributed with respect to the different values of the property. CSN could answer inheritance queries, i.e., infer the *most likely* value of a specified property for a given concept and also recognition queries, i.e., given a description consisting of property-value pairs, find the concept that *best* matched the given description. CSN found answers to queries by combining information encoded in the network in accordance with an evidential formalization based on the principle of maximum entropy. In particular, CSN could use distributional information to deal with exceptional and conflicting information in a principled manner and disambiguate *multiple inheritance* situations that could not be dealt with by extant formulations of inheritance in AI.

CSN encoded concepts, properties, and values with focal nodes. The IS-A relations were encoded as links, and property values were attached to concepts by connecting the appropriate property, value, and concept nodes through binder nodes. The weights on links between concept, value, and binder nodes captured the distributional information associated with a property-value attachment. A query was posed by activating the appropriate nodes. Thereafter, CSN performed the required inferences automatically by propagating graded activations and combining activations with appropriate activation combination rules.

Recent Models of Memory and Reasoning

The models discussed above made significant contributions, but were limited in their expressive power and inferential ability. One of their key limitations was that they did not address the dynamic binding problem. For example, the McClelland and Rumelhart model required *n*-fold repetition of letter and feature layers to deal with words of length *n*; it could not dynamically bind a letter to a position in a word. The Cottrell and Small and Waltz and Pollack systems prewired all possible bindings using dedicated nodes and links. Recently, there has been significant progress in solving this problem. New models include the CONPOSIT system (see article by Barnden and Srinivas in Barnden and Pollack, 1991) the ROBIN system (Lange and Dyer, 1989), the SHRUTI system (Shastri and Ajjanagadde, 1993), and the CONSYDERR system (Sun, 1992). We give a brief overview of SHRUTI, which shares a number of representational and functional features with ROB-

Figure 1. Encoding of predicates, concepts, and the rules $\forall x, y, z$ [$give(x, y, z)$ $\Rightarrow$ $own(y, z)$], $\forall x, y$ [$buy(x, y)$ $\Rightarrow$ $own(x, y)$], and $\forall x, y$[$own(x, y)$ $\Rightarrow$ $can\text{-}sell(x, y)$]. Abbreviations are as follows: recip: recipient; g-obj: give-object; b-obj; buy-object; o-obj: own-object; p-seller: possible-seller; cs-obj: can-sell-object.

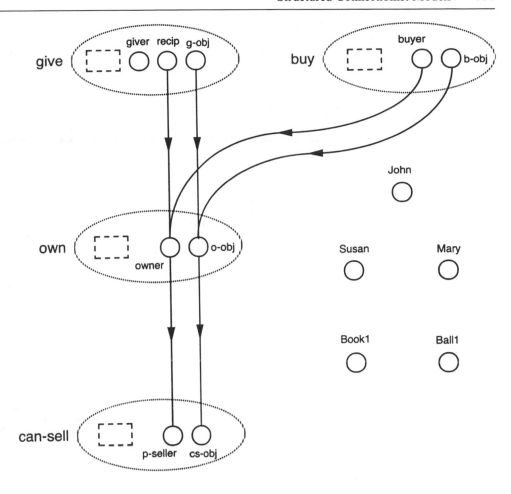

IN, but uses a different mechanism for representing dynamic bindings.

SHRUTI can encode a large number of specific facts, general rules, and IS-A relations between concepts and can perform a broad class of reasoning with extreme efficiency. SHRUTI encodes an n-ary predicate as a cluster of nodes which includes n role nodes (Figure 1). Nodes such as *John* and *Mary* correspond to focal nodes of the complete representations of the individuals John and Mary. A rule is encoded by linking the roles of the antecedent and consequent predicates in accordance with the correspondence between roles specified in the rule. SHRUTI represents dynamic bindings using *synchronous* firing of the appropriate argument and concept nodes. For example, the dynamic fact *give(John, Mary, Book1)* is represented by a rhythmic pattern of activity wherein the focal nodes *John*, *Mary*, and *Book1* are firing in synchrony with the role nodes *giver*, *recipient*, and *give-object*, respectively. By virtue of the interconnections between role nodes of the predicates *give*, *own*, and *can-sell*, this state of activation evolves so that (i) *owner* and, in turn, *p-seller* start firing in synchrony with *recipient* and, hence, *Mary*; and (ii) *own-object* and, in turn, *can-sell-object* start firing in synchrony with *give-object* and, hence, *Book1*. The resulting firing pattern corresponds to the dynamic facts *give(John, Mary, Book1)*, *own(Mary, Book1)*, and *can-sell(Mary, Book1)*. The key assumption here is that if nodes A and B are linked, the firing of A leads to a synchronous firing of B. For more on the role of synchrony and the dynamic binding problem, see COMPOSITIONALITY IN NEURAL SYSTEMS.

SHRUTI can encode "long-term" facts and a bounded number of instantiations of each predicate and concept. The latter

feature allows it to deal with reasoning involving *bounded recursion*. SHRUTI can also represent a type (IS-A) hierarchy, and it allows categories as well as instances in rules, facts, and queries. By using appropriate weights on links, the system can also encode soft, or evidential, rules. The time SHRUTI takes to generate a chain of inference is independent of the total number of rules and facts and is equal to $l * \alpha$, where l is the number of steps in the chain of inference and α is the time required for connected nodes to synchronize. If we assume α to be approximately 100 ms, SHRUTI demonstrates that a system of simple computing elements can encode millions of items and draw interesting inferences in a few hundred milliseconds. An implementation of the system on a CM-5 encodes more than 300,000 items and responds to queries with derivation lengths of up to 8 in less than a second.

Instead of using synchronous firing of nodes to represent and propagate bindings, ROBIN and CONSYDERR assign a distinct *signature* to each concept and propagate these codes to establish bindings. A signature may take the form of a unique activation value or a pattern of activity. CONPOSIT creates bindings by virtue of the relative position of active nodes and the similarity of patterns. The use of temporal synchrony in SHRUTI leads to a number of predictions about the capacity of the *working memory* underlying rapid reasoning. Thus, SHRUTI predicts that a large number of facts may be coactive in working memory and a large number of rules may fire simultaneously as long as the maximum number of distinct entities that can occur as role fillers in the dynamic facts is small (at most 10) and only a small number of instances of each predicate (approximately 3) may be coactive at the same time. The temporal approach also predicts that the depth to which an agent may reason rapidly, but systematically, is bounded. All of these constraints are motivated by biological considerations. For example, each entity participating in dynamic bindings occupies a distinct phase. Hence, the number of distinct entities that can occur as role fillers in dynamic facts cannot exceed $\lfloor \pi_{max}/\omega \rfloor$, where π_{max} is the maximum delay between consecutive firings of synchronous cell clusters and ω equals the allowable jitter in the firing times of synchronous cell clusters.

Henderson (1994) developed an on-line parser for English using a SHRUTI-like architecture. The parser's speed is independent of the size of the grammar, and it can recover the structure of arbitrarily long sentences as long as the dynamic state required to parse the sentence does not exceed the capacity of the parser's working memory. The parser shows that the constraints on the working memory help to explain several properties of human parsing involving long-distance dependencies, garden path effects, and our limited ability to deal with center embedding.

Significance of Structure

The representational and inferential power of structured connectionist systems such as SHRUTI and their ability to draw inferences in parallel is directly attributable to their use of structured representations. Any system that uses *fully* distributed representations will be incapable of representing multiple dynamic facts and applying multiple rules simultaneously. Attempts to develop distributed systems to handle relations invariably end up positing several distinct banks (one for each role), thereby stepping away from a fully distributed mode, or revert to seriality. It is not surprising that distributed systems such as DCPS (Touretzky and Hinton, 1988) have a limited capacity for encoding dynamic structures and are serial at the level of rule application.

Learning in Structured Networks

The models discussed thus far did not address the issue of learning in detail, although an outline of how rule learning might occur in a SHRUTI-like system appears in Shastri and Ajjanagadde (1993). Regier's (1992) model for learning the lexical semantics of natural language spatial terms provides a concrete example of learning within the structured connectionist paradigm. The model observes movies of simple two-dimensional objects moving relative to one another, where each movie is labeled as an example of some spatial term from a natural language, and learns the association between the label (word) and the event or relation it describes. The model successfully learned several spatial terms for diverse natural languages. The model includes structured network components that reflect prior constraints about the task as well as the usual *hidden layers*, and it demonstrates how structured connectionist networks can incorporate flexible learning ability and leverage prior structure to achieve tractability.

In addition to incremental learning driven by repeated exposure to a large body of training data, structured models have also made use of one-shot learning using *recruitment* learning schemes (e.g., see Shastri, 1988, and the article by Diederich in Barnden and Pollack, 1991).

Discussion

Structured connectionism offers a rich framework for developing models of cognition that are guided by biological, behavioral, and computational constraints. The approach has been productive and has resulted in a number of models that are informed by insights from diverse disciplines, such as computer science, AI, psychology, linguistics, and neuroscience. Now that some difficult representational problems have been resolved, the focus of work is shifting toward the study of structured adaptive networks that are grounded in perception and action.

Acknowledgment. The preparation of this paper was funded by ONR grant N00014–93–1–1149.

Road Map: Artificial Intelligence and Neural Networks
Background: I.2. Levels and Styles of Analysis
Related Reading: Connectionist and Symbolic Representations; Language Processing; Localized Versus Distributed Representations; Oscillatory Associative Memories; Sparse Coding in the Primate Cortex

References

Barnden, J., and Pollack, L., 1991, Eds., *Advances in Connectionist and Neural Computation Theory*, vol. 1, Norwood, NJ: Ablex. ◆

Cottrell, G. W., and Small, S. L., 1983, A connectionist scheme for modeling word sense disambiguation, *Cognition and Brain Theory*, 6:89–120.

Feldman, J. A., 1989, Neural representation of conceptual knowledge, in *Neural Connections, Mental Computation* (L. Nadel, L. A. Cooper, P. Culicover, and R. M. Harnish, Eds.), Cambridge, MA: MIT Press. ◆

Feldman, J. A., and Ballard, D. H., 1982, Connectionist models and their properties, *Cognit. Sci.*, 6:205–254. ◆

Feldman, J. A., Fanty, M. A., and Goddard, N. H., 1988, Computing with structured neural networks, *IEEE Computer*, March, pp. 91–103. ◆

Henderson, J., 1994, Connectionist syntactic parsing using temporal variable binding, *J. Psycholinguist. Res.*, 23:353–379.

Lange, T. E., and Dyer, M. G., 1989, High-level inferencing in a connectionist network, *Connection Sci.*, 1:181–217.

McClelland, J. L., and Rumelhart, D. E., 1981, An interactive activation model of context effects in letter perception, Part 1: An account of basic findings, *Psychol. Rev.*, 88:375–407.

Regier, T., 1992, The acquisition of lexical semantics for spatial terms: A connectionist model of perceptual categorization, PhD Thesis, University of California, Berkeley, available as TR-92-062, International Computer Science Institute.

Shastri, L., 1988, *Semantic Networks: An Evidential Formulation and Its Connectionist Realization*, London: Pitman. ◆

Shastri, L., and Ajjanagadde, V., 1993, From simple associations to systematic reasoning: A connectionist encoding of rules, variables and dynamic bindings using temporal synchrony, *Behav. Brain Sci.*, 16:417–494. ◆

Sun, R., 1992, On variable binding in connectionist networks, *Connection Sci.*, 4:93–124.

Touretzky, D. S., and Hinton, G. E., 1988, A distributed connectionist production system, *Cognit. Sci.*, 12:423–466.

van Gelder, T., 1992, Defining "distributed representation," *Connection Sci.*, 4:175–191.

Waltz, D. L., and Pollack, J. B., 1985, Massively parallel parsing: A strongly interactive model of natural language interpretation, *Cognit. Sci.*, 9:51–74.

Synaptic Coding of Spike Trains

J. P. Segundo, M. Stiber, and J.-F. Vibert

Introduction

At every synapse, a relation, or *coding*, between corresponding spike trains is established. Physiologists recognized early the importance of whether neuronal spikes were close or far apart, arose regularly or irregularly, etc. This work implied studying the instants t_i ($i = \ldots, -1, 0, 1, 2, \ldots$) when spikes occurred; i.e., *assimilating discharges to point processes* composed of points along a continuum. Hagiwara, Katsuki, Gerstein, and their collaborators were pioneers (Segundo, Stiber, and Vibert, 1993). The train's *timing* is equivalently the t_i and the sets of intervals are $T_i = t_i - t_{i-1}$ or the instantaneous rates $1/T_i$, whose statistics are related straightforwardly. Trains have interval and instantaneous rate *averages* and *patterns*. Patterns reflect their *pacemaker*, *bursty*, etc., character, implying dispersions, correlations, and so forth. Simultaneous trains imply cross-intervals and their statistics.

Experimental trains are bounded. With samples involving n spikes, the averages of interval $(T_2 + \ldots + T_n)/(n - 1) = S/(n - 1)$ and rate $(n - 1)/S$ are reciprocals. With samples involving a span S, average rates are n/S. Average intervals cannot be estimated under identical conditions (e.g., $S/(n - 1)$ ignores the ends, and $S/(n + 1)$ exceeds S). These average rates differ from average instantaneous rates.

Descriptions in this article, illustrated by electrical records and interspike interval displays, will be qualitative. Rigorous analyses demand time and frequency domain statistics derived from point process and nonlinear mathematics (e.g., Rosenberg et al., 1989; Segundo, Stiber, and Vibert, 1993).

In oscillator studies, *entrainment* exists when an oscillator is driven at frequencies similar to its natural one (without extraneous influences) and exhibits only that frequency or divisors (Torra i Genís, 1985). Here, entrainment means simply the carrying along and how corresponding modulations relate.

The rationale for assimilating series of events to point processes is that their timing says something about the underlying dynamics. The rates versus intervals opposition is false, their theoretical equivalence implicit in assimilations to point processes. Practically, however, there may be reasons for preferring one.

Experimental Data

Presynaptic discharges are invariant or fluctuate, becoming faster or slower and occurring gradually or quickly, once or repeatedly, or irregularly or regularly. *Faster* (accelerating) and *slower* mean, respectively, increasing and decreasing rates. Sporadic, steady-state, modulated, and transient trains are described; they imply different correspondences and entrainments (Perkel and Bullock, 1965).

All discharge parameters are influential. Generally, driven postsynaptic discharges are faster or slower than natural ones if excitatory postsynaptic potentials (EPSPs) or inhibitory postsynaptic potentials (IPSPs) are involved, respectively. Significant postsynaptic changes require minimal presynaptic changes.

Data involve moderately powerful IPSPs, intermediate postsynaptic rates, and stationary epochs collected in a prototype from crayfish stretch receptor organs involving GABAergic IPSPs, A-like receptors, Cl^-, and a pacemaker postsynaptic neuron. A pacemaker has practically invariant interspike intervals. Conclusions are applicable to other situations, including different presynaptic spikes, central synapses, etc.

Sporadic Discharges

Presynaptic intervals are long and irregular. Each arrival is unaffected by earlier ones, and each has invariant consequences. Most postsynaptic intervals do not contain IPSPs and are natural; a few are lengthened irregularly.

The sequence *primary effects, ringing, undisturbed epoch* follows each arrival. Right after the inhibitory spike, the postsynaptic discharge shows a pause and a burst. They compose the *primary synaptic effects*, slowing and acceleration with IPSPs and acceleration and slowing with EPSPs. This observation reflects refractoriness and the opposing consequences at the ON and OFF of stimulating currents. *Ringing* follows, involving equally spaced pauses and bursts that attenuate, reflecting triggered intrinsic postsynaptic periodicities. An *undisturbed epoch* of natural intervals follows.

Parts *A*, *B*, and *C* of Figure 1 show discharges at an inhibitory synapse. Approximately one dozen sweeps are superimposed from the presynaptic train (1) and from the concomitant postsynaptic train (2). A presynaptic spike is present in all sweeps at the vertical arrow; the positions of additional spikes vary from one sweep to another, depending on the presynaptic timing and entrainment. With sporadic discharges (*A*), the additional presynaptic spikes are irregular and sparse. Preceding the arrivals at the arrow (to its left), postsynaptic timings (2) vary unsystematically between sweeps: superimposed, they generate a uniform distribution. In contrast, following the arrow, all postsynaptic discharges show the same slowings (gaps) and accelerations (clusters); superposition accentuates them, revealing primary effects and ringings.

Steady-State Presynaptic Discharge Patterns

Nonsporadic trains with closely packed spikes are common. Arrivals are conditioned by facilitation, refractoriness, and so

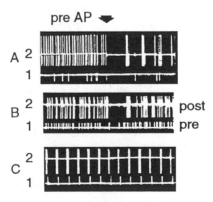

Figure 1. Electrical records. The presynaptic discharges are sporadic (*A*), Poisson (*B*), or pacemaker (*C*). In each part, row 1 shows superimposed sweeps of the presynaptic discharges; all sweeps have a presynaptic spike at the vertical arrow. Row 2 shows concomitant postsynaptic discharges. Compare the distributions of spikes to the left and the right of the arrow (before and after the reference presynaptic spike). Data are from the inhibitory synapse in crayfish stretch receptor organs.

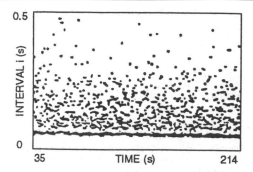

Figure 2. Postsynaptic interspike intervals across time for a Poisson presynaptic pattern. The few points seen in a sparse horizontal cluster represent natural intervals without inhibition. Most intervals, however, are lengthened by inhibition, as shown by the irregularly dispersed points. Presynaptic intervals (not shown) were dispersed irregularly.

forth caused by earlier arrivals. Therefore, postsynaptic discharges are not simple superpositions of the consequences of sporadic arrivals. Nonsporadic trains are steady state, modulated, or transient. Steady-state presynaptic discharges are stationary over few and over many spikes.

Poisson discharges approximate Poisson processes; patterns are highly irregular [see Figure 1*B*(1)], with intervals of many durations appearing randomly.

Poisson patterns elicit irregular postsynaptic discharges. Only a few postsynaptic intervals are natural; they are represented by a sparse, horizontal cluster in Figure 2. Most intervals contain arrivals and are lengthened, as shown by the irregularly dispersed points in the figure. Higher presynaptic rates are associated with fewer natural and more lengthened intervals; corresponding discharges become similar. The primary effects remain relatively unchanged [see Figure 1*B*(2), right of arrow]. Ringings and undisturbed epochs are recognizable only within long interarrival intervals and are unclear otherwise.

Pacemaker patterns have invariant intervals [see Figure 1*C*(1)]. Entrainment depends on the natural postsynaptic pattern; if Poisson, it resembles those with irregular arrivals. If it is a pacemaker, it resembles those of nonlinear oscillators, and different discharges occur (Segundo, Stiber, and Vibert, 1993). The *p : q* lockings are *p : q* entrainments, and *p* presynaptic and *q* postsynaptic intervals repeat in the same order. The most common is 1 : 1 [see Figure 1*C*(2) and Figure 3*A*]. *Intermittent* discharges are locked most of the time, but destabilize briefly and irregularly. *Messy* forms, which are complicated and unpredictable, are *erratic*, with low presynaptic rates and dispersed intervals (Figure 3*B*), or *stammerings*, with high presynaptic rates and intervals that are multiples of the presynaptic interval. Chaotic issues are demonstrable for erratic forms in living preparations and for all messy forms in simulations. Forms are staggered characteristically along the presynaptic rate scale. The presynaptic periodicity of pacemaker arrivals always imposes its mark, but maximally with lockings.

Presynaptic discharges with *intermediate patterns* between Poisson and pacemaker elicit intermediate consequences. A resemblance to pacemaker driving persists, even with substantial irregularity.

Different overall average rates or intervals are associated with dissimilar postsynaptic trains. Plots of presynaptic versus postsynaptic averages always increase overall with EPSPs or decrease with IPSPs. Locally, they are pattern dependent. They are monotonic Poisson driven, but when pacemaker driven, they alternately increase and decrease, with each rate producing a specific form (i.e., locked, intermittent, or messy) located

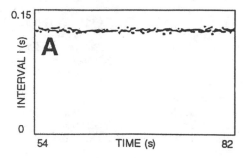

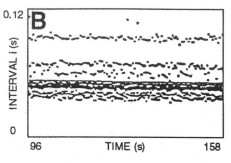

Figure 3. Postsynaptic interspike intervals across time for presynaptic pacemaker patterns with different average rates and different postsynaptic forms. *A*, Locked (1 : 1) form: all postsynaptic intervals are almost equal; the pacemaker discharges at the same rate. *B*, Messy, erratic form, with dispersed interval categories.

characteristically on the presynaptic rate scale. Through intermediate patterns, postsynaptic versus presynaptic plots change from monotonic to zigzag as discharges regularize.

Negligible consequences of varying presynaptic averages around extremes (i.e., saturations) are common and pattern dependent. High IPSP rates stop postsynaptic firing, and pacemaker discharges are more effective.

Periodically Modulated Presynaptic Discharges

Discharges of neurons that participate in periodic functions are periodic and influence other neurons (Segundo, Stiber, and Vibert, 1993). Numerous physiologically relevant modulations are transferred. Overall, presynaptic and postsynaptic discharges accelerate and slow together if EPSPs are involved or change in opposition if IPSPs are involved. Local distortions (e.g., joint accelerations with IPSPs, heterogeneous timings) are significant, however.

Timings, although with trends, resemble a succession of pacemaker-driven forms, and they are related similarly to driver rates. Locked, intermittent, and messy forms may therefore be building blocks for synaptic transfer. Similarity is clear with low modulation frequencies, but it decreases as frequencies increase because of hysteresis. Hysteresis reflects the postsynaptic cell's sensitivity to how fast the presynaptic discharge changes. It is greater to accelerations than to slowings. Coding depends on every modulation parameter.

Distortions are reduced when slow presynaptic modulations are perturbed by fast variabilities or noise. This fidelity increase is counterintuitive.

Presynaptic Transients

Everyday life shows pervasive transients. Presynaptic transients determine postsynaptic discharges that differ from them, and they depend on all of the transient's parameters (Segundo, Stiber, and Vibert, 1993). Transients, if steplike, cause marked shifts, with slow adaptations to stationary states; half-modulated cycles code as periodically modulated. Consequences extend beyond the transient's termination (e.g., OFF rebounds).

The following situation does not involve more than one synapse, but because it is critical for brain operation, it must be included. Sets of individually weak terminals converging on a single neuron are ubiquitous (e.g., motor and Purkinje neurons), although they are neglected because of technical difficulties (Segundo, 1970; Segundo, Stiber, and Vibert, 1993). The central limit theorem may apply, but the required similarities and independence often are absent. Correlations between discharges are critical for how converging terminals drive the shared postsynaptic neuron. The tendency to fire simultaneously or in a particular order favors timings with higher spike-eliciting or spike-preventing probabilities.

The supporting argument, which is partly conjectural, follows. When terminals are uncorrelated, their joint influence is noiselike, depending on synaptic strengths and rates, but not on patterns (e.g., pacemaker). As terminals become weaker and more numerous, the noise average changes little, but its dispersion decreases and approaches a DC bias. Correspondingly, postsynaptic discharges tend to follow a pacemaker pattern if EPSPs are involved, or slow without changing pattern with IPSPs. When terminals tend to fire synchronously, the joint influence approaches that of a single stronger terminal, and the distortions of stronger terminal coding become clearer as synchrony gets tighter.

Distortions of low-frequency signals imposed by a set of weak correlated terminals or by a single strong one are attenuated by fast noises imposed by a separate set of independent weak terminals. The same weak terminals may provide both signal and noise if their discharges are coherent at low frequencies, but independent at higher ones. Discharge correlation is a meaningful parameter that is indispensable for a full understanding of synaptic coding. Demonstrations that correlations exist and vary under physiological circumstances abound (Villa, 1992).

Formal Issues

Work by Bialek and Rieke (1992), Bullock (1961, 1970), and Perkel and Bullock (1965) is especially relevant here.

The Nerve Cell as an Analyzer of Arriving Discharges

Presynaptic changes associated with postsynaptic changes are said to be *read* by the postsynaptic cell; otherwise, they are *not read*. For example, IPSP accelerations that are large enough result in slowings, and are read. This observation implies the role of the nerve cell as *analyzer of arriving discharges*, critical in networks in which changes that are not read become operational dead ends. *Reliability of neurons* means that their behavior is the same when conditions are the same. Both fundamental notions were introduced by Bullock (1970).

Investigators read discharges a posteriori, identifying prolonged or multiple stationary epochs and characterizing them by overall statistics. Using conventional methods, this practice has provided useful insights, and is legitimate.

The organism, on the other hand, must use readouts over shorter spans with fewer spikes. A neuron reads arriving trains as a decision maker that unceasingly resolves the *fire/not fire* alternative. Its decision emerges from matching on the run its own excitability with all active drives. Matching is within a window, the *integration period*, which covers a bounded recent epoch and slides unceasingly along time. It reflects jointly the time constants of PSP summations, membranes, encoder excitability, and so forth. Presynaptic spikes and postsynaptic spikes within integration periods, called *influential*, favor or hinder spiking at the period's endpoint. Their averages and patterns are critical.

Synaptic Coding, Probabilities, Uncertainty, and Information

Synaptic coding is the relation between corresponding pre- and postsynaptic activities. The rules that summarize it constitute the *synaptic code*. Each synapse has its code; separate synapses usually have different codes, even when incoming or outgoing on the same cell. Here we discuss spike trains, but any activity can participate in a code.

The main conclusion from the experimental data is that synaptic coding involves several situations, each with varied rules. For example, one situation involves moderately powerful IPSPs, pacemakers, steady-state arrivals, and rules which allude to patterns and rates. The full code must cover exhaustively synapse and neuron types, discharge features, and so on (e.g., EPSPs, silent cells, transients, and spectra). Systematizations of the synaptic code, even at elementary levels, must be long and complicated if they are to be faithful.

Each neuron X or Y participates in complementary coding schemes. A topographic relation matches each to the set of cells, S_X or S_Y, it acts on; this is *spatial coding*. Next, at each synapse, a functional relation matches corresponding spike trains; this is *temporal coding*. A conceptual relation R matches all presynaptic trains M with all postsynaptic trains N. Subsets of that relation are r_L arising during the individual's natural life and r_E observed experimentally; both overlap partially. R, r_L, and r_E can be approached prospectively, taking a particular presynaptic train m^* and identifying its postsynaptic associates, or retrospectively, taking a postsynaptic train n^{**} and identifying its presynaptic associates.

Relations have probabilistic aspects implying uncertainties. Possible inputs are the timings of influential presynaptic spikes, possible outputs, and spike/no spike. Each train has an a priori probability, called *generating* if it is presynaptic. One observer predicts prospectively the forthcoming output, and another infers retrospectively a recent input. A priori uncertainties are decreased for the prospective observer by knowledge of the recent input and for the retrospective observer by knowledge of the resulting output. Input and output provide information about each other, and this information is quantifiable. Physiological implications are especially clear retrospectively for sensory afferents or prospectively for motoneurons, but are harder to discern elsewhere.

More uncertainty is removed (i.e., more information is provided) by averages and patterns jointly than by either alone. To make reliable statements, and because of the situation's formal character, observers need a minimum of a priori knowledge, particularly about the presynaptic generating probabilities. This approach allows quantification of, for example, the influences of presynaptic spike summation rules and of refractoriness.

Discussion

The key points can be listed as follows:

1. Synaptic coding is varied and complicated, involving numerous unexpected distortions plus some predictable features (e.g., overall accelerations by EPSPs).

2. Coding depends on numerous variables (e.g., synaptic type, discharges involved). Discharge variabilities have simplifying influences.
3. In addition to discharge assimilation to point processes, formal issues with physiological implications include, the neuron's analyzer role, the definition of coding, and probabilistic and informational aspects.

Linking this article to brain theory are questions regarding generalizations to other discharges and synapses, long-term changes (neuromodulation, plasticity), participating membrane and molecular mechanisms, and the roles of deterministic issues and noise. Coding fidelity is useful biologically in some sensory or motor situations (although not always); its compensation poses an interesting problem.

Behaviors are reproduced satisfactorily by models, mainly formal and incorporating variables recognized in living preparations that refer to presynaptic spikes, thresholds, currents, conductances, and so forth. Models ignore issues such as the neuron's regional inhomogeneity and the individual synapse's complexity.

Circuit-oriented concepts are indispensable for understanding nervous systems. Synapses are operational units therein, but claiming their hegemony is similar to claiming that the atom holds the key that unravels the universe. However, the relative significance of the whole should not be exaggerated. Synaptic coding and neuronal entrainments, although not sufficient, are indispensable for comprehending neural networks and the nervous systems that they are intended to mimic. Networks with simple units reveal surprisingly complex behaviors; those with more realistic units show great promise and will add important dimensions to the body of knowledge (Aihara, Takabe, and Toyoda, 1993).

Road Map: Biological Neurons
Background: I.1. Introducing the Neuron
Related Reading: Diffusion Models of Neuron Activity; Sensory Coding and Information Theory; Spinal Cord of Lamprey: Generation of Locomotor Patterns

References

Aihara, K., Takabe, T., and Toyoda, M., 1990, Chaotic neural networks, *Phys. Lett. A*, 144:333–340.
Bialek, W., and Rieke, F., 1992, Reliability and information transmission in spiking neurons, *Trends Neurosci.*, 15:428–434.
Bullock, T. H., 1961, The problem of recognition in an analyzer made of neurons, in *Sensory Communication* (W. A. Rosemblith, Ed.), New York: Wiley, pp. 717–724.
Bullock, T. H., 1970, The reliability of neurons, *J. Gen. Physiol.*, 355: 556–684.
Perkel, D. H., and Bullock, T. H., 1965, Neural coding, *Neurosci. Res. Program Bull.*, 6:221–348.
Rosenberg, J., Amjad, A. M., Breeze, P., Brillinger, D. R., and Halliday, D. M., 1989, The Fourier approach to the identification of functional coupling between neuronal discharges, *Prog. Biophys. Mol. Biol.*, 53:1–31.
Segundo, J. P., 1970, Communication and coding by nerve cells, in *The Neurosciences: Second Study Program* (G. C. Quarton, T. Melnechuk, and F. O. Schmitt, Eds.), New York: Rockefeller University Press, pp. 569–586.
Segundo, J. P., Stiber, M., and Vibert, J.-F., 1993, Synaptic coding by discharges, in *Tutorial Texts: International Joint Conference on Neural Networks*, October 25–29, Nagoya, Japan, pp. 7–21.
Torra i Genís, C., 1985, Temporal-pattern learning in neural models, in *Lecture Notes in Biomathematics* (S. Levin, Ed.), New York: Springer-Verlag.
Villa, A. E. P., 1992, Les catastrophes cachées du cerveau, *Le Nouveau Golem*, 1:33–64.

Synaptic Currents, Neuromodulation, and Kinetic Models

Alain Destexhe, Zachary F. Mainen, and Terrence J. Sejnowski

Introduction

Synaptic interactions are essential to neural network models of all levels of complexity. Synaptic interactions in "realistic" network models pose a particular challenge, since the aim is not only to capture the essence of synaptic mechanisms, but also to do so in a computationally efficient manner to facilitate simulations of large networks. In this article, we review several types of models which address these goals.

Synaptic currents are mediated by ion channels activated by neurotransmitter released from presynaptic terminals. *Kinetic models* are a powerful formalism for the description of channel behavior, and are therefore well suited to the description of synaptic interactions, both traditional and neuromodulatory. Although full representation of the molecular details of the synapse generally requires highly complex kinetic models, we focus on simpler kinetic models which are very efficient to compute. We show how these models capture the time courses of several types of synaptic responses as well as the important phenomena of summation, saturation, and desensitization.

Models of Synaptic Currents

For neural models that do not include action potentials, synaptic currents are typically modeled as a direct function of some presynaptic activity measure. In the simplest case, synaptic interactions are described by a sigmoid function, and presynaptic activity is interpreted as the average firing rate of the afferent neuron. Alternatively, the postsynaptic currents can be described by a first-order differential equation in which one term depends on the presynaptic membrane potential through a sigmoid function (Wang and Rinzel, 1992). Another possibility is to intepret the activity level as the fraction of neurons active per unit of time, thus representing the interaction between neural populations rather than single neurons (Wilson and Cowan, 1973).

For spiking neurons, a popular model of postsynaptic currents (PSCs) is the alpha function

$$r(t - t_0) = \frac{(t - t_0)}{\tau_1} \exp[-(t - t_0)/\tau_1] \qquad (1)$$

(Rall, 1967), where $r(t)$ resembles the time course of experimentally recorded postsynaptic potentials (PSPs) with a time constant τ_1. The alpha function and its double-exponential generalization can be used to approximate most synaptic currents with a small number of parameters and, if implemented properly, at low computation and storage requirements (Srinivasan and Chiel, 1993). Other types of template functions were also proposed for spiking neurons (Tsodyks, Mitkov, and Sompolinsky, 1990; Traub and Miles, 1991). The disadvantages of the alpha-function, or related approaches, include the lack of correspondence to a plausible biophysical mechanism and the absence of a natural method for handling the summation of successive PSCs from a train of presynaptic impulses.

The most fundamental way to model synaptic currents is based on the kinetic properties of the underlying synaptic ion channels. The kinetic approach is closely related to the well-known model of Hodgkin and Huxley (1952) for voltage-dependent ion channels (reviewed in Armstrong, 1992; see AXONAL MODELING). Kinetic models are powerful enough to describe in great detail the properties of synaptic ion channels and can be integrated coherently with chemical kinetic models for enzymatic cascades underlying signal transduction and neuromodulation. The drawback of kinetic models is that they are often complex, with several coupled differential equations, making them too costly to be used in simulations involving large populations of neurons. We show how these limitations can be ameliorated.

The Kinetic Description

Ion channels are proteins that have distinct conformational states, some of which are *open* and conduct ionic current and some of which are *closed*, *inactivated*, or *desensitized* and do not conduct. Single-channel recording techniques have demonstrated that the transitions between conformational states occurs both rapidly and randomly or stochastically (reviewed in Sakmann and Neher, 1983). It has furthermore been shown that the behavior of single-ion channels is well described by *Markov models*, a class of stochastic model in which transitions between states occurs with a time-independent probability.

It is straightforward to move from a microscopic description of single-channel behavior to a macroscopic description of a population of similar channels. In the limit of large numbers, the stochastic behavior of individual channels can be described by a set of continuous differential equations analogous to ordinary chemical reaction kinetics. The kinetic analog of Markov models posits the existence of a group of conformational states $S_1 \dots S_n$ linked by a set of transitions

$$\begin{array}{ccc} S_1 \rightleftarrows S_2 \rightleftarrows S_4 \rightleftarrows \dots \\ \downarrow\uparrow \qquad \downarrow\uparrow \\ S_3 \rightleftarrows \dots \end{array} \qquad (2)$$

Define s_i as the fraction of channels in state S_i and r_{ij} as the rate constant of the transition

$$S_i \underset{r_{ji}}{\overset{r_{ij}}{\rightleftarrows}} S_j \qquad (3)$$

which obeys the kinetic equation

$$\frac{ds_i}{dt} = \sum_{j=1}^{n} s_j r_{ji} - \sum_{j=1}^{n} s_i r_{ij} \qquad (4)$$

The wide range of interesting behavior exhibited by channels arises from the dependence of certain transitions on factors extrinsic to the channel, primarily either the binding of another molecule to the protein or the electric field across the cell membrane. These influences are referred to as *ligand-gating* and *voltage-gating*, respectively. Ligand-gating is typified by synaptic receptors, which are ion channels that are gated by neurotransmitter molecules. Other channels are gated by molecules inside the cell, most prominently the so-called second-messengers such as calcium ions or G-proteins.

In the case of voltage-dependent ion channels, the transition between two states S_i and S_j occurs with rate constants that are dependent on voltage, such as

$$S_i \underset{r_{ji}(V)}{\overset{r_{ij}(V)}{\rightleftarrows}} S_j \qquad (5)$$

The functional form of the voltage dependence can be obtained from single-channel recordings (see Sakmann and Neher, 1983). The kinetics-based description of the voltage dependence of channels is quite general. In particular, the well-known model of Hodgkin and Huxley (1952) for the fast sodium channel and the delayed-rectifier potassium channel can be written in a kinetic form which is equivalent to the original Hodgkin-Huxley equations.

In the case of ligand-gated ion channels, the transition between two states S_i and S_j can depend on the binding of a ligand L:

$$L + S_i \underset{r_{ji}}{\overset{r_{ij}}{\rightleftarrows}} S_j \qquad (6)$$

which can be rewritten as

$$S_i \underset{r_{ji}}{\overset{r_{ij}([L])}{\rightleftarrows}} S_j \qquad (7)$$

where $r_{ij}([L]) = [L]r_{ij}$ and $[L]$ is the concentration of ligand. The functional dependence of the rate constants is linear in the ligand concentration, and in some cases, can also depend on the voltage.

Ligand-Gated Channels: AMPA, NMDA, and GABA-A Receptors

The most common types of ligand-gated channels are the excitatory AMPA and NMDA types of glutamate receptor and the inhibitory GABA-A receptor. Many kinetic models have been constructed. For example, an accurate model of the AMPA receptor is

$$D_2 \underset{r_9([L])}{\overset{r_{10}}{\rightleftarrows}} D_1 \underset{r_7([L])}{\overset{r_8}{\rightleftarrows}} C \underset{r_2}{\overset{r_1([L])}{\rightleftarrows}} C_1 \underset{r_4}{\overset{r_3([L])}{\rightleftarrows}} C_2 \underset{r_6}{\overset{r_5}{\rightleftarrows}} O \quad (8)$$

(Standley, Ramsey, and Usherwood, 1993), where C is the unbound closed state, C_1 and C_2 are, respectively, the singly and doubly bound closed states, O is the open state, and D_1 and D_2 are the desensitized singly and doubly bound states, respectively. r_1 through r_{10} are the associated rate constants, and $[L]$ is the concentration of neurotransmitter in the synaptic cleft.

The six states of this AMPA model are required to account for some of the subtle dynamical properties of these receptors, yet simplified schemes with far fewer states and transitions are often very good approximations for the time course and the dynamic behavior of most synaptic currents (see Destexhe et al., 1994c). In particular, consider the simplest kinetic schemes involving two states

$$C \underset{r_2}{\overset{r_1([L])}{\rightleftarrows}} O \qquad (9)$$

or three states

$$C \underset{r_2}{\overset{r_1([L])}{\rightleftarrows}} O \atop r_5 \underset{r_6([L])}{\searrow} \nearrow r_4 \; D \; \searrow r_3 \qquad (10)$$

In these two schemes, C and O represent the closed and open states of the channel, D represents the desensitized state, and $r_1 \dots r_6$ are the associated rate constants. Not only are these simple schemes easier to compute than more complex schemes, but the time course of the current can be obtained analytically (Destexhe et al., 1994b, 1994c).

Another useful means to simplify the model is suggested by experiments using artificial application of neurotransmitter, where it has been seen that synaptic currents with a time course very similar to intact synapses can be produced using very brief pulses of agonist (Colquhoun, Jonas, and Sakmann, 1992).

These data suggest that a model for the AMPA synapse does not require a detailed kinetic model for transmitter release, as the response time course is dominated by the postsynaptic kinetics rather than the time course of the neurotransmitter concentration. Hence, we can assume that the neurotransmitter is delivered as a brief (≈ 1 ms) pulse triggered at the time of each presynaptic spike.

Simplified kinetic schemes for the AMPA response can be compared with detailed kinetic models to judge the quality of the approximation (Figure 1A–D). Both simple and detailed synaptic responses first require a trigger event, corresponding

to the release of neurotransmitter in the synaptic cleft. In simulations of the detailed kinetics, the time course of neurotransmitter was derived using a model which included presynaptic action potentials, calcium-dependent fusion of presynaptic vesicles, and clearance of neurotransmitter. Figure 1A–C shows the AMPA response resulting from a high-frequency train of presynaptic spikes. The amplitude of successive PSCs decreased progressively because of an increasing fraction of receptors in desensitized states. One of the simplified schemes gave a good fit both to the time course of the AMPA current and to the response desensitization that occurs during multiple successive events (Figure 1D). Alpha functions, in contrast, did not match the summation behavior of the synaptic current (Figure 1E).

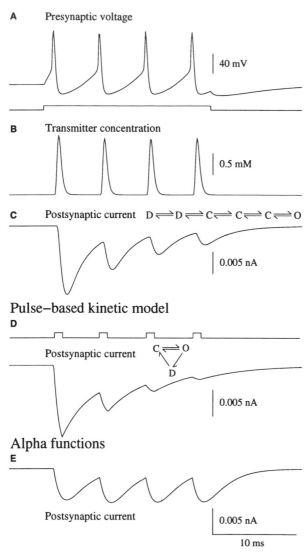

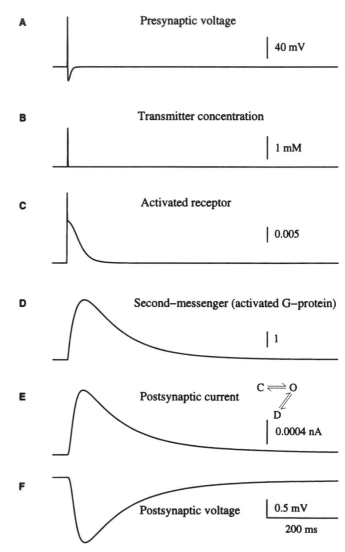

Figure 1. Comparison of three models for AMPA receptors. *A*, Presynaptic train of action potentials elicited by current injection. *B*, Corresponding glutamate release in the synaptic cleft obtained using a kinetic model for transmitter release. *C*, Postsynaptic current from AMPA receptors modeled by a six-state Markov model (Standley et al., 1993). *D*, Same simulation with AMPA receptors modeled by a simpler three-state kinetic scheme and transmitter time course approximated by spike-triggered pulse (above current trace). *E*, Postsynaptic current modeled by summated alpha functions. (Modified from Destexhe et al., 1994c.)

Figure 2. Kinetic model of synaptic currents acting through second messengers. *A*, Presynaptic action potential elicited by current injection. *B*, Time course of transmitter in the synaptic cleft obtained by a kinetic model for transmitter release. *C*, Activated GABA-B receptors after binding with transmitter. The activated receptor catalyzes the formation of a second messenger, a G-protein subunit. *D*, Time course of activated G-protein. *E*, Postsynaptic current produced by the gating of K$^+$ channels by G-protein. *F*, Inhibitory postsynaptic potential. (Modified from Destexhe et al., 1994c.)

Procedures similar to those applied to the AMPA response can be used to obtain simple kinetic models for other types of ligand-gated synaptic channels, including the NMDA and GABA-A receptors. Two-state and three-state models provide good fits of averaged whole-cell recordings of the corresponding PSCs (see Destexhe et al., 1994c, for more details).

Second Messenger-Gated Channels: GABA-B and Neuromodulation

Some neurotransmitters do not bind directly to the ion channel but act through an intracellular second messenger, which links the activated receptor to the opening or closing of an ion channel. This type of synaptic interaction occurs at a slower time scale as ligand-gated channels and is therefore distinguished as *neuromodulation*. Examples of neuromodulators such as GABA (GABA-B), acetylcholine (M2), noradrenaline (alpha-2), serotonin (5HT-1), dopamine (D2), and others gate a K^+ channel through the direct action of a G-protein subunit, G_α (Brown, 1990; Brown and Birnbaumer, 1990). We have developed a kinetic model of the G-protein–mediated slow intracellular response (Destexhe et al., 1994c) that can be applied to any of these transmitters.

A detailed kinetic model of the GABA-B response (Figure 2) was compared with a two-state model, where it was assumed that the time course of the activated G-protein occurs as a pulse of 80–100 ms duration. As in the case of ligand-gated channels, elementary kinetic schemes capture the essential dynamics of more detailed models; the equations are the same as ligand-gated channels, but with [L] representing the second messenger. The slow time course of the neuromodulators is reflected in the low values of the rate constants in Equations 9 and 10. These elementary schemes provide excellent fits to whole-cell recordings of GABA-B PSCs, and are also fast to compute.

The neuromodulators just listed share a very similar G-protein–mediated intracellular response. The basic method applied to the GABA-B response can thus be used to model these currents, with rate constants adjusted to fit the time courses reported for the particular responses. Details on the rate constants obtained from fitting different kinetic schemes to GABA-B PSC and other neuromodulators are given in Destexhe et al. (1994c).

Discussion

Although it has been possible to develop remarkably detailed models of the synapse (Bartol et al., 1991), substantial simplification is necessary for large-scale network simulations involving thousands of synapses. A variety of abstract representations of the synapse are available. We advocate a class of model based directly on kinetics of the ion channel molecules mediating synaptic responses. Simplified kinetic models can be implemented with minimal computational expense, while still capturing both the time course of individual synaptic and neuromodulatory events and also the interactions between successive events (summation, saturation, desensitization), which may be critical when neurons interact through bursts of action potentials (Destexhe et al., 1994a).

Programs that simulate the models of this article using the NEURON simulator are available by anonymous ftp to salk.edu in/pub/alain.

Acknowledgments. This research was supported by the Howard Hughes Medical Institute, the Office for Naval Research, and the National Institutes of Health.

Road Map: Biological Neurons
Background: Ion Channels: Keys to Neuronal Specialization
Related Reading: Dynamic Clamp: Computer-Neural Hybrids; NMDA Receptors: Synaptic, Cellular, and Network Models; Oscillatory and Bursting Properties of Neurons

References

Armstrong, C. M., 1992, Voltage-dependent ion channels and their gating, *Physiol. Rev.*, 72:S5–S13.

Bartol, T. M., Jr., Land, B. R., Salpeter, E. E., and Salpeter, M. M., 1991, Monte Carlo simulation of miniature endplate current generation in the vertebrate neuromuscular junction, *Biophys. J.*, 59:1290–1307.

Brown, A. M., and Birnbaumer, L., 1990, Ionic channels and their regulation by G-protein subunits, *Annu. Rev. Physiol.*, 52:197–213.

Brown, D. A., 1990, G-proteins and potassium currents in neurons, *Annu. Rev. Physiol.*, 52:215–242.

Colquhoun, D., Jonas, P., and Sakmann, B., 1992, Action of brief pulses of glutamate on AMPA receptors in patches from different neurons of rat hippocampal slices, *J. Physiol. (Lond.)*, 458:261–287.

Destexhe, A., Contreras, C., Sejnowski, T. J., and Steriade, M., 1994a, A model of spindle rhythmicity in the isolated thalamic reticular nucleus, *J. Neurophysiol.*, 72:803–818.

Destexhe, A., Mainen, Z., and Sejnowski, T. J., 1994b, An efficient method for computing synaptic conductances based on a kinetic model of receptor binding, *Neural Computat.*, 6:14–18.

Destexhe, A., Mainen, Z., and Sejnowski, T. J., 1994c, Synthesis of models for excitable membranes, synaptic transmission and neuromodulation using a common kinetic formalism, *J. Computat. Neurosci.*, 1:195–230.

Hodgkin, A. L., and Huxley, A. F., 1952, A quantitative description of membrane current and its application to conduction and excitation in nerve, *J. Physiol. (Lond.)*, 117:500–544.

Rall, W., 1967, Distinguishing theoretical synaptic potentials computed for different soma-dendritic distributions of synaptic inputs, *J. Neurophysiol.*, 30:1138–1168.

Sakmann, B., and Neher, E., Eds., 1983, *Single-Channel Recording*, New York: Plenum.

Srinivasan, R., and Chiel, H. J., 1993, Fast calculation of synaptic conductances, *Neural Computat.*, 5:200–204.

Standley, C., Ramsey, R. L., and Usherwood, P. N. R., 1993, Gating kinetics of the quisqualate-sensitive glutamate receptor of locust muscle studied using agonist concentration jumps and computer simulations, *Biophys. J.*, 65:1379–1386.

Traub, R. D., and Miles, R., 1991, *Neuronal Networks of the Hippocampus*, Cambridge, Eng.: Cambridge University Press.

Tsodyks, M., Mitkov, I., and Sompolinsky, H., 1993, Pattern of synchrony in inhomogeneous networks of oscillators with pulse interactions, *Phys. Rev. Lett.*, 71:1280–1283.

Wang, X. J., and Rinzel, J., 1992, Alternating and synchronous rhythms in reciprocally inhibitory model neurons, *Neural Computat.*, 4:84–97.

Wilson, H. R., and Cowan, J. D., 1973, A mathematical theory of the functional dynamics of nervous tissue, *Kybernetik*, 13:55–80.

Synchronization of Neuronal Responses as a Putative Binding Mechanism

Wolf Singer

Introduction

Numerous sensory and motor functions of the nervous system require that subsets of distributed neuronal responses are selected and bound together for further joint processing.

Such *binding* operations are indispensable when specific sensory or motor patterns are represented by the graded activity of populations of neurons rather than by the highly specialized responses of individual cells. The main feature of population, or assembly, coding is that it uses the information contained in the spatiotemporal constellation of distributed responses. Thus, a given cell can participate at different times in the representation of different patterns. This capability reduces substantially the number of cells required for the representation of different patterns and allows for considerably more flexibility in the generation of new representations. However, whenever such population codes are used, problems arise that are commonly addressed as *binding problems*. For the evaluation of population codes, it is necessary to identify in an unambiguous way the responses which participate in the representation of a particular content. In the case of coarse coding of elementary stimulus features, responses representing the same feature must be associated selectively and therefore must be distinguished from responses to other features. Similar distinctions are required for PERCEPTUAL GROUPING (q.v.). Responses of cells participating in the representation of the same object must be identified and segregated from responses to other objects.

If only a single feature is present, binding problems do not arise because all of the responses can be associated with each other indiscriminately. However, the environment of the organism is usually crowded with different perceptual objects and features. Hence, a large number of neurons will be active simultaneously, and because of broad tuning, many of them will be coactivated by different features that may even belong to different perceptual objects. For a successful association of responses coding for the same feature, for the selective association of features belonging to the same object, and for the segregation of different objects from one another and from the background, it is crucial to have a mechanism which selects from the many simultaneous responses those which can be related to one another in a meaningful way to avoid false conjunctions. Similar problems arise in motor coordination. Groups of cells controlling particular muscles must become associated with one another in a highly flexible but unambiguous way to generate a specific motor pattern. What is required then is an unambiguous way to identify neuronal responses that must be bound together.

An effective way to select among neuronal responses is to enhance their saliency, for example, by increasing their rate. This change leads to more effective temporal summation in the respective target cells. However, selecting neurons solely on the basis of enhanced discharge rates has two disadvantages. First, it precludes the option of encoding information about features or constellations of features in the *graded* responses of distributed populations of neurons. Second, it limits the number of populations that can be enhanced simultaneously without becoming confounded. The only populations that would remain segregatable are those that are clearly defined by a place code.

Therefore, it has been proposed that the synchronization of responses on a time scale of milliseconds is a more efficient mechanism for response selection and binding of population responses (von der Malsburg, 1985; DYNAMIC LINK ARCHITECTURE; see also Milner, 1974, for a related proposal). Synchronization also increases the saliency of responses because it allows for effective spatial summation in the population of neurons receiving convergent input from synchronized input cells. In addition, synchronization expresses unambiguous relations among input neurons because it enhances selectively and with high temporal precision only the saliency of responses that are synchronous. Thus, response synchronization is an effective means for selecting response constellations for further joint processing. Moreover, it has the required flexibility since different assemblies can be organized in rapid temporal succession with a multiplexing strategy.

The assumption is that the discharges of neurons undergo a specific temporal patterning so that cells participating in the encoding of related contents eventually come to discharge in synchrony. This patterning is thought to be based on a self-organizing process that is mediated by a highly selective network of reentrant connections. Thus, neurons having joined into an assembly coding for the same feature or, at higher levels, for the same perceptual object or for a particular movement trajectory, would be identifiable as members of the assembly because their responses would contain episodes during which their discharges are synchronous.

Predictions

If an assembly of cells coding for a common feature, a common perceptual object, or a particular motor act is distinguished by the temporal coherence of the responses of the constituting neurons, predictions can be derived which are accessible to experimental testing.

1. Spatially segregated neurons should exhibit synchronized response episodes if they are activated by a single stimulus or by stimuli that can be grouped into a single perceptual object.
2. Synchronization should be frequent among neurons within a particular cortical area, but it should also occur between cells distributed across different cortical areas if these cells respond to a common feature or perceptual object.
3. The probability that neurons synchronize their responses both within a particular area and across areas should reflect some of the Gestalt criteria used for perceptual grouping.
4. Individual cells must be able to rapidly change the partners with which they synchronize their responses if stimulus configurations change and require new associations.
5. If more than one object is present in a scene, several distinct assemblies should form. Cells belonging to the same assembly should exhibit synchronous response episodes, while no consistent temporal relations should exist between the discharges of neurons belonging to different assemblies.

6. Synchronization should occur as the result of a self-organizing process that is based on mutual and parallel interactions between distributed cortical cells.

7. The connections that determine synchronization probabilities should be highly specific because the criteria according to which distributed responses are bound together reside in the functional architecture of these connections.

8. The synchronizing connections should allow for interactions at levels of processing where responses of neurons already express some feature selectivity to permit feature-specific associations. Therefore, corticocortical connections should contribute to synchronization.

9. The synchronizing connections should have adaptive synapses that allow for use-dependent long-term modifications of synaptic gain to permit the acquisition of new grouping criteria when new object representations are to be installed during perceptual learning.

10. These use-dependent synaptic modifications should follow a correlation rule whereby synaptic connections should strengthen if pre- and postsynaptic activity is often correlated, and they should weaken if there is no correlation. This activity is required to enhance the grouping of cells which code for features that often occur in consistent relations, as for features constituting a particular object.

11. These grouping operations should occur over multiple processing stages because the search for meaningful groupings must be performed at different spatial scales and according to different feature domains. This task could be achieved by distributing the grouping operations over different cortical areas since neighborhood relations differ in different areas because of remapping of input connections.

Experimental Testing of Predictions

All predictions have received experimental support, with most of the data obtained in the mammalian visual cortex. Neurons recorded simultaneously with a single electrode transiently engaged in synchronous discharges when activated with a single stimulus (Gray and Singer, 1987). In multiunit recordings, these locally synchronous discharges often appear as clusters of spikes that follow one another at rather regular intervals of 15–30 ms. Accordingly, autocorrelograms computed from such response epochs often exhibit periodic modulation (Gray and Singer, 1987; Eckhorn et al., 1988).

This phenomenon of local response synchronization has been observed with multiunit and field potential recordings in several independent studies in different areas of the visual cortex of anesthetized cats (areas 17, 18, 19, and PMLS), in area 17 of awake cats, in the optic tectum of awake pigeons, and in various areas of the visual cortex of anesthetized and awake-behaving monkeys. Similar synchronization phenomena have also been observed in the somatosensory and motor cortex of awake monkeys (for review, see Singer, 1993; Singer and Gray, 1995).

Multielectrode recordings showed that similar response synchronization can occur between spatially segregated cell groups within the same visual area (Gray et al., 1989, 1992; Engel et al., 1990), between different cortical areas (Eckhorn et al., 1988; Engel et al., 1991a; Murthy and Fetz, 1992; Nelson et al., 1992), and even across hemispheres (Engel et al., 1991b). Feature-dependent response synchronization has been observed in the visual thalamus, and it could be demonstrated that it is induced by feedback connections from the visual cortex (Sillito et al., 1994). Thus, synchronization seems to be used for response selection at processing levels as peripheral as the thalamus. These single-unit data are complemented by a large body of evidence derived from field potential and electroencephalogram recordings, which all indicate that distributed groups of neurons can engage in synchronous activity (for a review of the extensive literature, see Singer, 1993; Singer and Gray, 1995).

The Dependence of Response Synchronization on Stimulus Configuration

As outlined above, the hypothesis of temporally coded assemblies requires that the probabilities with which distributed cells synchronize their responses should reflect some of the Gestalt criteria applied in perceptual grouping. A clear dependence of synchronization probability could be established with respect to the criteria of vicinity, continuity, and common fate (Gray et al., 1989; Engel et al., 1990).

In related experiments, Engel et al. (1991a, 1991b) demonstrated in the cat that the synchronization of activity between cells in area 17 and a motion-sensitive area in the suprasylvian sulcus (PMLS) and between area 17 in the two hemispheres exhibits a similar dependence on the properties of the visual stimulus (Figure 1).

Experiments have also been performed to test the prediction that simultaneously presented but different contours should lead to the organization of two independently synchronized assemblies of cells (Engel et al., 1991c). If groups of cells with overlapping receptive fields but different orientation preferences are activated with a single moving light bar, they synchronize their responses (Engel et al., 1990, 1991c). This finding agrees with the postulate derived from the hypothesis of coarse coding that all responses of cells participating in the representation of a stimulus should be bound together. However, if such a set of groups is stimulated with two independent, spatially overlapping stimuli which move in different directions, the activated cells split into two independently synchronized assemblies. Cells whose feature preferences match better with stimulus 1 form one synchronously active assembly, and those matching better with stimulus 2 form the other (Figure 2). Thus, although the two stimuli evoke graded responses in all of the recorded groups, cells representing the same stimulus remain distinguishable because their responses exhibit synchronized response epochs while showing no consistent correlations with the responses of cells activated by different stimuli. To extract this information, a read-out mechanism is required which is capable of evaluating coincident firing at a millisecond time scale. An analysis of the integrative properties of cortical pyramidal cells suggests that in these cells, the window for effective temporal summation may be as short as a few milliseconds (Softky and Koch, 1993).

Another important feature of the experiments with changing stimulus configurations is the demonstration that individual cells can actually change the partners with which they synchronize. Cell groups that engaged in synchronous response episodes when activated with a single stimulus no longer did so when activated with two stimuli, but then synchronized with other groups.

Experience-Dependent Development of Synchronizing Connections

The theory of assembly coding implies that the criteria according to which particular features, rather than others, are grouped reside in the functional architecture of the assembly

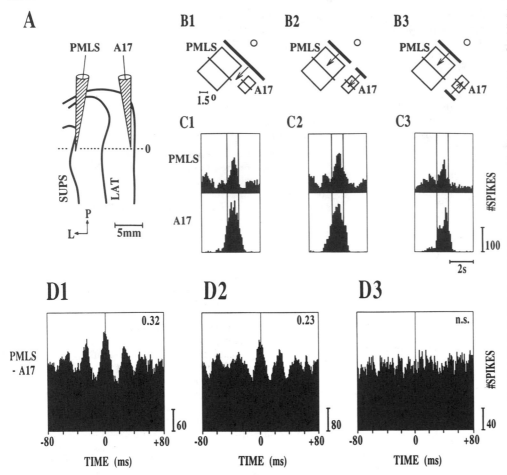

Figure 1. Interareal synchronization is sensitive to global stimulus features. Part *A* shows the position of the recording electrodes with respect to area 17 (A17), the posteriomedial lateral suprasylvian sulcus (PMLA), the lateral sulcus (LAT), and the suprasylvian sulcus (SUPS); posterior (P) and lateral (L) are indicated. Parts *B1*, *B2*, and *B3* show plots of the receptive fields of the PMLS and A17 recording. The diagrams show the three stimulus conditions tested. The circle indicates the visual field center. Parts *C1*, *C2*, and *C3* show peristimulus time histograms for the three stimulus conditions. The vertical lines indicate 1-s windows for which autocorrelograms and cross-correlograms were computed. Parts *D1*, *D2*, and *D3* show cross-correlograms computed for the three stimulus conditions. The number in the upper right corner is the relative modulation amplitude of each correlogram. The strongest correlogram modulation is obtained with the continuous stimulus. The cross-correlogram is less regular and has a lower modulation amplitude when two comoving light bars are used as stimuli, and there is no significant modulation (n.s.) with two light bars moving in opposite directions. (From Engel et al., 1991a.)

forming coupling connections. Therefore, it is of particular interest to identify the connections responsible for synchronization, to study their development, to identify the rules according to which they are selected, to establish correlations between their architecture and synchronization probabilities, and if possible, to relate these neuronal properties to perceptual functions.

Examination of interhemispheric response synchronization revealed that responses of neurons in area 17 of the two hemispheres synchronize in much the same way as responses within the same hemisphere if evoked by coherently moving stimuli (Engel et al., 1991b). This observation agrees with the postulate that contours extending across the midline of the visual field should be bound by the same mechanism as contours located within the same hemifield. Sectioning the corpus callosum abolished response synchronization (Engel et al., 1991b). This finding identifies corticocortical connections as the substrate for synchronization and proves that zero-phase synchronization can be the result of reciprocal coupling connections. In mammals, corticocortical connections develop mainly postnatally, and they attain their final specificity through an activity-dependent selection process. When strabismus is induced in three-week-old kittens, it leads to a profound rearrangement of corticocortical connections. Normally, these connections link cortical territories irrespective of whether they are dominated by the same or by different eyes. In the strabismics kittens, by contrast, the tangential intracortical connections come to link with high selectivity only territories served by the same eye (Löwel and Singer, 1992). The functional correlate of these

changes in the architecture of corticocortical connections is a modification of synchronization probabilities. In the strabismic animals, response synchronization no longer occurs between cell groups connected to different eyes, while it is normal between cell groups connected to the same eye.

The results from strabismic animals have several implications. First, they further support the notion that tangential intracortical connections contribute to response synchronization. Second, these results confirm the prediction that the assembly forming connections should be susceptible to use-dependent modifications and should be selected according to a correlation rule. Third, the reduced synchrony among cells driven by different eyes supports the hypothesis that synchronization acts as a binding mechanism. Strabismic subjects become unable to fuse signals conveyed by different eyes into coherent percepts, even if these signals are made retinotopically contiguous by optical compensation of the squint angle. Thus, in strabismic subjects, binding mechanisms appear to be abnormal or missing between cells driven by different eyes. The lack of corticocortical connections and response synchronization could contribute to this deficit in addition to the loss of binocular neurons.

These correlations are, at the least, compatible with the view that the architecture of corticocortical connections, by determining the probability of response synchronization, could set the criteria for perceptual grouping. Since this architecture is shaped by experience, it is possible that some of the binding and segmentation criteria are acquired or modified by experience.

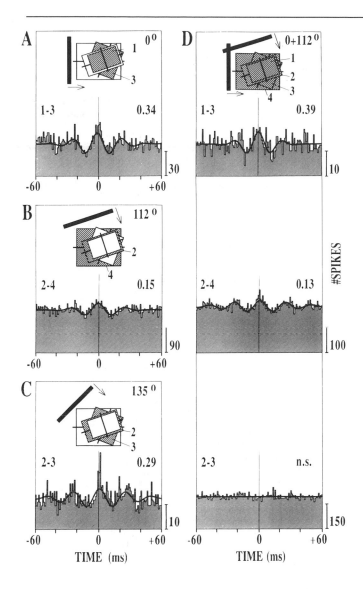

Figure 2. Stimulus dependence of short-range interactions. Multiunit activity was recorded from four different orientation columns of area 17 of the cat visual cortex separated by 0.4 mm. The four cell groups had overlapping receptive fields and orientation preferences of 22° (group 1), 112° (group 2), 157° (group 3), and 90° (group 4), as indicated by the thick line drawn across each receptive field in parts *A*, *B*, *C*, and *D*. Responses to stimulation with single moving light bars of varying orientation (left) are compared with responses to the combined presentation of two superimposed light bars (right). For each stimulus condition, the shading of the receptive fields indicates the responding cell groups. Stimulation with a single light bar yielded a synchronization between all cells activated by the respective orientation. Thus, groups 1 and 3 responded synchronously to a vertically orientated (O°) light bar (*A*), groups 2 and 4 responded synchronously to a light bar at an orientation of 112° (*B*), and groups 2 and 3 responded synchronously to a light bar of intermediate orientation (*C*). Simultaneous presentation of two stimuli with orientations of O° and 112°, respectively, activated all four groups (*D*). However, in this case, the groups segregated into two distinct assemblies, depending on which stimulus was closer to the preferred orientation of each group. Thus, responses were synchronized between groups 1 and 3, which preferred the vertical stimulus, and between groups 2 and 4, which preferred the stimulus oriented at 112°. The two assemblies were desynchronized with respect to each other, so there was no significant synchronization between groups 2 and 3. The cross-correlograms between groups 1 and 2, 1 and 4, and 3 and 4 were also flat (not shown). The correlograms are shown superimposed with their fitted Gabor function. The number at the upper right of each correlogram indicates the correlation strength, which is expressed as the relative amplitude of the center peak. Scale bars indicate the number of spikes. N.S. stands for "not significant." (From Engel et al., 1991c.)

ified amblyopia revealed highly significant differences in the synchronization behavior of cells driven by the normal and the amblyopic eye, respectively. The responses to single moving bars that were recorded simultaneously from spatially segregated neurons connected to the amblyopic eye were much less well synchronized with one another than the responses recorded from neuron pairs driven by the normal eye. This difference was even more pronounced for responses elicited by gratings of different spatial frequency. For responses of cell pairs activated through the normal eye, the strength of synchronization tended to increase with increasing spatial frequency and tended to decrease further for cell pairs activated through the amblyopic eye.

Apart from these highly significant differences between the synchronization behavior of cells driven by the normal and the amblyopic eye, no other differences were found in the commonly determined response properties of these cells. Thus, cells connected to the amblyopic eye continued to respond vigorously to gratings whose spatial frequency was too high to be discriminated with the amblyopic eye in the preceding behavioral tests. These results suggest that disturbed temporal coordination of responses, such as reduced synchrony, may be one of the neuronal correlates of the amblyopic deficit. Indeed, if synchronization of responses at a millisecond time scale is used by the system to tag and identify the responses of cells which code for the same feature or contour, disturbance of this temporal patterning could be the cause of the crowding phenomenon. If responses evoked by nearby contours can no longer be associated unambiguously, but become confounded, perceptual deficits are expected that closely resemble the crowding phenomenon. As a possible reason for the reduced synchronization among cells driven by the amblyopic eye, one might consider abnormalities in the network of corticocortical connections linking cell groups dominated by this eye. It is conceivable that the continuous suppression of the signals provided from

Impaired Response Synchronization Correlates with Perceptual Disturbances

Further indications of a relation between experience-dependent modifications of synchronization probabilities and functional deficits come from a recent study of strabismic cats who had amblyopia. Strabismus, when induced early in life, not only abolishes binocular fusion and stereopsis, but also may lead to amblyopia of one eye. This condition develops when the subjects solve the problem of double vision, not by alternating the use of the two eyes, but by constantly suppressing the signals from the deviated eye. The amblyopic deficit usually consists of reduced spatial resolution and distorted and blurred perception of patterns. A characteristic phenomenon in amblyopia is crowding, the drastic impairment of the ability to discriminate and recognize figures if they are surrounded with other contours. The identification of neuronal correlates of these deficits in animal models of amblyopia has remained inconclusive because the contrast sensitivity and spatial resolution capacity of neurons in the retina and lateral geniculate nucleus were normal. In the visual cortex, identification of neurons with reduced spatial resolution or otherwise abnormal receptive field properties remained controversial. However, multielectrode recordings from the striate cortex of cats exhibiting behaviorally ver-

this eye impedes the experience-dependent specification of the respective intracortical synchronizing connections. Because reduced synchrony also impairs the transmission of responses, and hence their saliency, the current results can further account for the fact that amblyopic patients have difficulty attending to the signals conveyed by the amblyopic eye when both eyes are open. In that case, signals from the amblyopic eye are usually eliminated from further processing and are not perceived.

Discussion

The experimental results reviewed in this article are compatible with predictions derived from the hypothesis that synchronization of neuronal responses at a time scale of milliseconds may be used in cortical processing. In a distributed network such as the neocortex, where any given cell contacts any other cell with only a few synapses, but individual cells receive converging input from many thousands of cells, synchronization of discharges is a particularly efficient mechanism to increase the saliency of responses. In addition, synchronization establishes unambiguous relations between responses because it enhances with great selectivity the saliency of only those response episodes that contain coincident discharges. Hence, in principle, synchronization can be used to select with high spatial and temporal precision those constellations of responses that should be considered for further processing. The proposal is that this selection is achieved in a distributed and highly parallel operation by the system of corticocortical association fibers. Their function then would consist essentially of adjusting the *timing* of discharges rather than modulating discharge rates. Eventually, however, synchronization will also influence firing rates because synchronous inputs are more effective in driving cells than asynchronous inputs. Hence, rate codes and synchronization codes can coexist in the same network. The prediction is that rate coding should prevail at levels of processing that are close to sensory and motor output, while the intermediate computations should be based predominantly on the shifting of temporal relations. It will be interesting to determine whether the performance of neuronal network models can be improved by combining rate and synchronization codes for the selection and grouping of responses and by using response synchronization to define assemblies.

Road Map: Vision
Related Reading: Binding in the Visual System; Cortical Columns, Modules, and Hebbian Cell Assemblies; Hebbian Synaptic Plasticity: Comparative and Developmental Aspects; Oscillatory Associative Memories; Structured Connectionist Models

References

Eckhorn, R., Bauer, R., Jordan, W., Brosch, M., Kruse, W., Munk, M., and Reitboeck, H. J., 1988, Coherent oscillations: A mechanism for feature linking in the visual cortex? *Biol. Cybern.*, 60:121–130.

Engel, A. K., König, P., Gray, C. M., and Singer, W., 1990, Stimulus-dependent neuronal oscillations in cat visual cortex: Inter-columnar interaction as determined by cross-correlation analysis, *Eur. J. Neurosci.*, 2:588–606.

Engel, A. K., Kreiter, A. K., König, P., and Singer, W., 1991a, Synchronization of oscillatory neuronal responses between striate and extrastriate visual cortical areas of the cat, *Proc. Natl. Acad. Sci. USA*, 88:6048–6052.

Engel, A. K., König, P., Kreiter, A. K., and Singer, W., 1991b, Interhemispheric synchronization of oscillatory neuronal responses in cat visual cortex, *Science*, 252:1177–1179.

Engel, A. K., König, P., and Singer, W., 1991c, Direct physiological evidence for scene segmentation by temporal coding, *Proc. Natl. Acad. Sci. USA*, 88:9136–9140.

Gray, C. M., Engel, A. K., König, P., and Singer, W., 1992, Synchronization of oscillatory neuronal responses in cat striate cortex: Temporal properties, *Vis. Neurosci.*, 8:337–347.

Gray, C. M., König, P., Engel, A. K., and Singer, W., 1989, Oscillatory responses in cat visual cortex exhibit inter-columnar synchronization which reflects global stimulus properties, *Nature*, 338:334–337.

Gray, C. M., and Singer, W., 1987, Stimulus-specific neuronal oscillations in the cat visual cortex: A cortical functional unit, *Soc. Neurosci. Abst.*, 13:404.3.

Löwel, S., and Singer, W., 1992, Selection of intrinsic horizontal connections in the visual cortex by correlated neuronal activity, *Science*, 255:209–212.

Milner, P. M., 1974, A model for visual shape recognition, *Psychol. Review*, 81:521–535.

Murthy, V. N., and Fetz, E. E., 1992, Coherent 25- to 35-Hz oscillations in the sensorimotor cortex of awake behaving moneys, *Proc. Natl. Acad. Sci. USA*, 89:5670–5674.

Nelson, J. I., Salin, P. A., Munk, M. H. J., Arzi, M., and Bullier, J., 1992, Spatial and temporal coherence in cortico-cortical connections: A cross-correlation study in areas 17 and 18 in the cat, *Vis. Neurosci.*, 9:21–38.

Sillito, A. M., Jones, H. E., Gerstein, G. L., and West, D. C., 1994, Feature-linked synchronization of thalamic relay cell firing induced by feedback from the visual cortex, *Nature*, 369:479–482.

Singer, W., 1993, Synchronization of cortical activity and its putative role in information processing and learning, *Annu. Rev. Physiol.*, 55:349–374. ◆

Singer, W., and Gray, C. M., 1995, Visual feature integration and the temporal correlation hypothesis, *Annu. Rev. Neurosci.*, 18:555–586. ◆

Softky, W. R., and Koch, C., 1993, The highly irregular firing of cortical cells is inconsistent with temporal integration of random EPSPs, *J. Neurosci.*, 13:334–350.

von der Malsburg, C., 1985, Nervous structures with dynamical links, *Ber. Bunsenges. Phys. Chem.*, 89:703–710.

Telecommunications

Anandkumar and Harold Szu

Introduction

The rapidly changing world of telecommunications requires new adaptive techniques to enhance its speed and usefulness. Neural networks, as a result of their inherent adaptive/learning qualities, can contribute greatly to this explosive area of technology. Just as higher levels of the nervous system are built on primitive subsystems, today's information networks are built on existing cable systems and T1 lines (T1 is a long-haul cable system which allows for the connection of remote local area networks with typical bandwidths of 56 kilobits per second), using well-defined protocols like ISDN (Integrated Services Digital Networks) and TCP/IP (Transfer Control Protocol/Internet Protocol, a protocol stack which allows for end-to-end transport communication over an Internet including error checking, flow control, and sequencing processes). Communication systems are organizations with their own needs for survivability and are subject to economic competition. Telecommunication networks of the future, on the other hand, will need both adaptivity and "intelligence" for many reasons—reliability, information filtering, speech recognition, fault tolerance, congestion control, network management, resource allocation and market prediction. For reasons of adaptivity, fast response, parallelism, and intelligence, neural networks could significantly contribute to the dynamic field of telecommunications (Alspector, Goodman, and Timothy, 1993).

Some recent developments and findings are summarized in the following sections, particularly in the areas of maximum-likelihood sequence estimation, packet routing, code division multiple access (CDMA), asynchronous transfer mode (ATM), and predictive coding.

Sequence Estimation

It is well known that channel impairments introduced during signal propagation gravely affect the performance of a digital communication system over band-limited multipath fading channels (Proakis, 1993). Maximum Likelihood Sequence Estimation (MLSE) is an optimum method of detecting digital data over time-varying and time-dispersive channels in the presence of additive white Gaussian noise. Although it has error rate performance superior to that of its linear counterparts, it is extremely impractical owing to its complexity and computational intensity. The performance achieved by MLSE is possible only if channel characteristics are precisely known and if the sampling rates are higher than symbol rates. Furthermore, the noise statistics for the receiver needs to be white Gaussian.

The cost function to be minimized/optimized in the MLSE has the same form as the Lyapunov energy function associated with the Hopfield network. The cost function is the selection of a sequence as the best estimate of the transmitted sequence, such that the selection maximizes the conditional a posteriori probabilities. Recent work on a four-layer perceptron has shown, via simulation and experimentation, that the performance achieved by the artificial neural network is superior to those of MLSE since neither estimation of the channel characteristics nor a sampling rate higher than the symbol rate is required (Bang, Sheu, and Choi, 1993). Input to the artificial neural network is the transmitted signal added to white Gaussian noise. The output is a binary value. As training symbols increase, the decision boundary closely approximates that of the optimum receiver. Training of the network was performed by the Kalman filtering algorithm. The complexity of the Hopfield-based MLSE receiver is proportional to $n * l$, where n is the number of symbols in a given sequence and l is the channel memory. This is far superior to the exponential complexity exhibited by commonly used MLSE algorithms like the Viterbi algorithm (Bang et al., 1993).

Packet Routing

The routing of packets in a network with irregular topology and unpredictable usage patterns calls for an algorithm which monitors its own performance and adapts to unforeseen situations. The algorithm has to learn to minimize the number of "hops" a packet will take, especially in congested routes (Rudin, 1976). A variant of the REINFORCEMENT LEARNING (q.v.) algorithm, called Q-learning (Watkins, 1989), has been successfully used on a distributive network which relies on lo-

cal information (conforming to ANN principles) at each node to keep accurate statistics on which routing policies lead to minimal routing times (Littman and Boyan, 1993). This study was conducted on an irregularly connected network, and proved that the self-adjusting algorithm could discover a more efficient routing policy than that provided for by the precomputed shortest path approach. The learning algorithm did not know the network topology or traffic patterns a priori and did not need any centralized routing control system.

Code Division Multiple Access (CDMA)

The Code Division Multiple Access (CDMA) technique is in vogue due to cellular communications. It is being used for multiple point-to-point digital communication networks. In this technique, each transmitter sends messages by modulating its own characteristic "signature" waveform which is known a priori to the receiver. The receiver (a conventional detector) passes the received signal through a bank of matched filters and retrieves information based on the sign of the output. A major limitation of this conventional detector has been its performance degradation when the powers of the transmitting users are dissimilar (near-far resistant).

Optimum near-far resistant multiuser-demodulation is an NP-complete problem and is equivalent to the maximization of a quadratic function. In an Optimal Multiuser Detector/ Demodulator (OMD), the objective function is the suppression of additive white Gaussian noise and multiple access interference. There is a lot of ongoing research toward developing suboptimal receivers that are near-far resistant, computationally practical, and have a performance that is comparable to that of an optimal receiver. The traditional suboptimal receivers have a major drawback in that they have hardware complexities that are exponential in the number of users (Kechriotis and Manolakos, 1993).

A novel implementation of a CDMA optimal multiuser detector using a Hopfield network has shown results that outperform conventional detectors at much lower computational rates. The equation that represents the OMD objective function is quite like the Hopfield energy function (with a few transformations). The input to the Hopfield network is a transmitted sequence corrupted with noise. The output is a voltage that represents the detected bits. The proposed Hopfield network has a linear hardware complexity and computational complexity that is constant irrespective of the number of users. Based on factors like the signal-to-noise ratio and bit error figures, the Hopfield network outperforms a conventional detector and compares to the OMD at a much lower computational cost (Kechriotis and Manolakos, 1993). The Hopfield network's fast convergence time makes it particularly attractive for use in high-speed data communications.

Asynchronous Transfer Mode (ATM)

Virtually all communication services in the future will be supported by ATM, which is a packet and connection-oriented switching and multiplexing technique. The ATM switch is a statistical multiplexer that uses backbones to transfer voice, data, and images on the same physical connection by utilizing many virtual connections within the same physical connection. It allows customers on different virtual circuits to negotiate a quality-of-service (QOS) contract based upon their specific traffic types and needs. ATM allows the network to statistically multiplex user data in order to maximize bandwidth efficiency.

A fundamental principle of broadband services, in which ATM will be used, is to prevent congestion rather than to react to it. Therefore, connection admission control (CAC) traffic enforcement and traffic shaping will be basic in ATM traffic control. A recent study on the performance evaluation of a single ATM multiplexer compared traditional techniques to hybrid ones that incorporated a multilayer perceptron, and proved via numerical results that the neural network approach improved bandwidth efficiency. The traditional approach is based on analytical approximations that could lead to an over-controlled network. The hybrid method is composed of a supervised Multilayer Perceptron (MLP) in combination with a variant of the fluid flow performance formula. (The total traffic stream to an ATM multiplexer is usually analyzed in two time scales, namely, cell scale and burst scale. Cell scale analyzes queue buildups caused by simultaneous cell arrivals. Burst scale considers queue buildups that are due to fluctuations in the total arrival rate that may exceed the link capacity at times. The fluid flow model is a very accurate tool that is used to evaluate burst scale performance.) The MLP was trained using knowledge derived from different fluid flow performance methods (Kvols, 1992; Nordstrom, 1993). In another study, neural networks were successfully used to characterize and predict the statistical variations of the packet arrival process and the complex nonrenewal process (Nordstrom, 1993).

Usage Parameter Control (UPC) is the structure that defines the virtual connection and maps it into a QOS contract. One study concluded that neural networks have great potential as traffic descriptors for the UPC algorithm which forms the basis of ATM traffic control. Related work in this field of ATM has shown that a window-based scheduling algorithm (for input queuing), implemented using continuous Hopfield-type neural networks, increases the maximum throughput with lower cell loss probabilities and buffer sizes than with traditional approaches (Park et al., 1993).

Predictive Coding

Predictive coding is an established technique that is used to represent speech signals using lower rates than the standard Pulse Code Modulation (PCM) techniques. To accomplish a reduction in the bit rate, a combination of adaptive predictive encoding and adaptive quantization are used. The resulting technique is commonly referred to as Adaptive Differential Pulse Code Modulation (ADPCM). Conventional ADPCM systems use linear predictors, although it is well known that the biological mechanism responsible for speech generation is non-linear. The conventional ADPCM systems use the Least Mean Square (LMS) algorithm for its adaptation. Lack of stability and mistracking are known problems with a linear predictor. This shortcoming was addressed in a recent study, wherein a predictor was constructed using a nonlinear subsection that consisted of real-time recurrent neural networks and a linear subsection whose weight update algorithm was influenced by the neural network. The neural network had to perform its learning "on-line" as speech signals were continuously being processed. Based on experimental research, the study concluded that the overall performance of 16 Kb/s nonlinear ADPCM system was better than that of a 32 Kb/s linear ADPCM, despite the fact that the linear system had the advantage of operating at twice the bit rate (Haykin et al., 1993).

Discussion

Based on experimental results, there is now ample evidence that artificial neural networks (ANNs) will be used in several realms of telecommunications to solve previously intractable issues. Most current research and development in the use of ANNs in telecommunications has focused on problems associated with ATM and cellular communications. Simplicity of architecture, training/learning procedures, relative ease of hardware implementation (both analog and digital VLSI), innate nonlinearity and fault tolerance make ANNs an integral part of future telecommunication systems.

Road Map: Applications of Neural Networks
Background: I.3. Dynamics and Adaptation in Neural Networks
Related Reading: Adaptive Signal Processing; Noise Canceling and Channel Equalization

References

Alspector, J., Goodman, R., and Timothy, B. X., 1993, *Proceedings of the International Workshop on Applications of Neural Networks to Telecommunications*, IEEE Communications Society, LEA Publishers.

Bang, S. H., Sheu, B. J., and Choi, J., 1993, Programmable VLSI neural network processors for equalization of digital communication channels, in *Proceedings of the International Workshop on Applications of Neural Networks to Telecommunications*, IEEE Communications Society, LEA Publishers.

Haykin, S., et al., 1993, 16 kbps adaptive differential pulse code modulation of speech, in *Proceedings of the International Workshop on Applications of Neural Networks to Telecommunications*, IEEE Communications Society, LEA Publishers.

Kechriotis, G. I., and Manolakos, E. S., 1993, Implementing optimal CDMA multiuser detector with Hopfield neural networks, in *Proceedings of the International Workshop on Applications of Neural Networks to Telecommunications*, IEEE Communications Society, LEA Publishers.

Kvols, K., 1992, Bounds and approximations for the periodic on/off queue with applications to ATM traffic control, in *INFOCOM '92*, May, Italy.

Littman, M., and Boyan, J., 1993, A distributed reinforcement learning scheme for network routing, in *Proceedings of the International Workshop on Applications of Neural Networks to Telecommunications*, IEEE Communications Society, LEA Publishers.

Nordstrom, E., 1993, A hybrid admission control scheme for broadband ATM traffic, in *Proceedings of the International Workshop on Applications of Neural Networks to Telecommunications*, IEEE Communications Society, LEA Publishers.

Park, Y. K., et al., 1993, ATM cell scheduling for broadband switching by neural network, in *Proceedings of the International Workshop on Applications of Neural Networks to Telecommunications*, IEEE Communications Society, LEA Publishers.

Proakis, J. G., 1993, *Digital Communications*, New York: McGraw-Hill.

Rudin, H., 1976, On routing and delta routing: A taxonomy and performance comparison of techniques for packet-switched networks, *IEEE Transcripts on Communications*, COM-24(1):43–59.

Taraff, A. A., et al., 1993, Neural networks for ATM multimedia traffic prediction, in *Proceedings of the International Workshop on Applications of Neural Networks to Telecommunications*, IEEE Communications Society, LEA Publishers.

Verdu, S., 1989, Computational complexity of Optimum Multiuser Detection, *Algorithmica*, 4:303–312.

Watkins, C., 1989, Learning from delayed rewards, PhD Thesis, Kings College, University of Cambridge.

Temporal Pattern Processing

DeLiang Wang

Introduction

Temporal pattern processing underlies various intelligent behaviors, including hearing, speech, and vision. Because we live in an ever-changing environment, an intelligent system, whether it be a human being or a robot, must be able to encode patterns over time and to recognize and produce the temporal patterns. Time can be embodied in a temporal pattern in two different ways:

- *Temporal order* refers to the ordering among the components of a sequence. For example, the sequence *A-B-C* is different from *C-B-A*. Temporal order may also refer to a syntactic structure, such as subject-verb-object, where each component may be any of a number of possible symbols.
- *Time duration* can play a critical role for temporal processing. In speech recognition, for example, we want rate invariance while distinguishing relative durations of the vowel /i:/ (as in b*ee*t) and /i/ (as in b*i*t).

A temporal pattern may be continuous. In this case, it is usually sampled into a discrete pattern before processing. Thus, in the following discussion, we assume discrete patterns. Following Wang and Arbib (1990), a sequence is defined as *complex* if it contains repetitions of the same subsequence like *A-B-C-D-B-C-E*; otherwise, as *simple*. For production of complex sequences, the correct successor can be determined only by knowing symbols prior to the current one. We refer to the prior subsequence required to determine the current component as the *context* of the symbol, and the length of this context as the *degree* of the component. The *degree of a sequence* is defined as the maximum degree of its components. Thus, a simple sequence is a degree 1 sequence.

There is no doubt that temporal pattern processing should be of central importance for neural network research. However, only a relatively small body of literature deals with this topic, partly because the two most influential neural architectures, associative memories and multilayer perceptrons with backpropagation (see ASSOCIATIVE NETWORKS and BACKPROPAGATION: BASICS AND NEW DEVELOPMENTS), deal only with static patterns. Another reason, perhaps, is that temporal pattern processing is particularly challenging because the information needed is embedded in time (thus is inherently dynamic), not simultaneously available. Nonetheless, this topic has been studied by a number of investigators. Most of the models proposed so far try to extend the two classic architectures one way or another.

Fundamentally different from static pattern processing, temporal processing requires that a neural network have a capacity of short-term memory (STM) in order to maintain a component for some time. This capacity is necessary because a temporal pattern stretches over a certain time period. Thus, how to encode STM becomes one of the criteria for classifying neural networks for temporal processing (see also Mozer, 1993). This article will provide an outline of temporal pattern processing and, in the end, point out several outstanding problems yet to be addressed.

STM Models

Delay Lines

This simplest form of STM uses a fixed-length buffer to maintain the $N + 1$ most recent input items. Figure 1 shows two possible implementations by either a shift register or an array of different delay lines. The delay-line STM transforms a temporal pattern into a spatial one where time forms another dimension. The idea forms the basis of many neural network models for recognition (e.g., Waibel et al., 1989).

Decay Traces

An item in STM decays in time, corresponding to the decay theory of forgetting in STM. The decay usually takes on the exponential form. Theoretically, time information can be precisely recovered from the current value of $x_i(t)$. But because of rapid decay and noise, only a limited number of the most recent items can be reliably discerned from STM. This form has been used by Jordan (1986), Mozer (1989, see Mozer, 1993), and Wang and Arbib (1990), among others. Figure 2*A* shows a typical decay trace.

This model uses only one unit to represent a symbol, rather than $N + 1$ units as in delay-line STM. However, in its simplest form, this model is not adequate to represent complex sequences, since it cannot tell whether the symbol occurs more than once. Wang and Arbib (1990) proposed the idea of a pushdown buffer to keep multiple occurrences of a symbol. Since the pushdown buffer is used only for multiple occurrences of the same symbol, it requires a buffer whose size is the same as the degree of the sequence, as opposed to the buffer size as large as the sequence length required by delay-line STM.

Exponential Kernels

Tank and Hopfield (1987) proposed a set of normalized exponential kernels to sample the history, described as

$$f_k(t) = \left(\frac{t}{k}\right)^{\alpha} e^{\alpha(1-t/k)} \qquad \text{for } k = 1, \ldots, K \qquad (1)$$

where α regulates the width of each kernel. Notice that $f_k(t) = 1$ if $t = k$. Figure 2*B* shows a set of four kernels. There are K

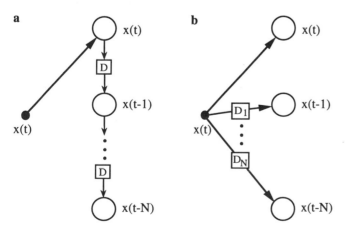

Figure 1. Two delay-line STM diagrams. The letter *D* in a box represents a delay interval. *A*, A shift register. *B*, An array of different delays, where $D_1 < D_2 < \ldots < D_N$.

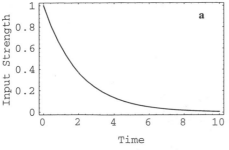

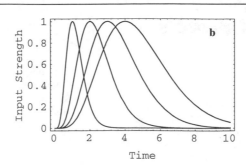

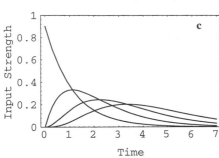

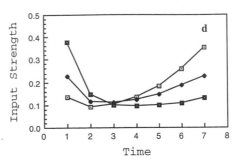

Figure 2. STM traces. *A*, Exponential decay. *B*, Normalized exponential kernels; $\alpha = 5.0$, and $k = 1, \ldots, 4$ for the curves from left to right, respectively. *C*, Gamma kernels; $\mu = 0.9$, and $k = 1, \ldots, 4$ for the curves from left to right, respectively. *D*, The STORE model; $\beta = 0.5$ for the empty square bow, 0.3 for the diamond bow, and 0.15 for the filled square bow. Seven items are kept in STM.

units to represent each symbol. Unlike delay-line STM, where each unit samples a symbol at a specific time step, each unit in this model samples a symbol within a certain period peaked at a specific time step ($t = k$).

Along similar lines, de Vries and Principe (1992) proposed the so-called *gamma* model, which uses a set of gamma kernels (integrands of Γ-functions, hence the name)

$$g_k(t) = \frac{\mu^k}{(k-1)!} t^{k-1} e^{-\mu t} \qquad \text{for } k = 1, \ldots, K \qquad (2)$$

where μ is a parameter between 0 and 1. The value K is called the order of the memory, and there are K units for storing S in STM. Figure 2C shows a set of four kernels. Since g_k has a maximum value at $t = (k - 1)/\mu$, μ determines the depth of the peak of each kernel in STM. Thus, unlike normalized exponential kernels, an N-step history may be sampled by less than N gamma kernels. Another advantage is that the kernel functions can be computed recursively, whereas in normalized exponential kernels the convolution must be computed between the kernel functions and the activity history of S.

Interactive Models

The STM models examined so far are autonomous, for the trace of each symbol is fully independent of other input symbols in STM. A basic property of human STM is that it has limited capacity (7 ± 2), so that whether and how long an item is held in STM critically depends on other inputs entering STM. The following two models address interactions between the items in STM.

Based on the interference theory of forgetting, Wang and Arbib (1993) proposed an STM model in which an input item stays in STM as long as the number of later items does not exceed the value T (the capacity). More specifically,

$$x_i(t) = \begin{cases} T & \text{if } I_i(t) = 1 \\ x_i(t-1) - 1 & \text{if } x_i(t-1) > 0, y(t) = 1 \\ x_i(t-1) & \text{otherwise} \end{cases} \qquad (3)$$

where $y(t)$ detects whether there is a new input entering STM. Once $x_i(t)$ receives an external input I_i, its activation value is

brought to T. This value decrements when later inputs come in. This function was implemented as a mutually inhibitory network. This *interference-based* STM model has flexible time traces for storing input symbols, depending on how frequently later inputs enter STM. According to Equation 3, more recent items have greater activations, thus showing a recency factor.

The study of human retention of sequences shows that in addition to recency, there is also a primacy factor whereby the beginning items of a sequence are less prone to forgetting, thus exhibiting the bowing effect. With an interactive STM model, called STORE, Bradski, Carpenter, and Grossberg (1992) showed that both recency and primacy can be exhibited by the following model using a pair of units for storing S:

$$x_i(t + 1) = x_i(t) + [\beta I_i(t) + y_i(t) - x_i(t)x(t)]I(t) \qquad (4a)$$

$$y_i(t + 1) = y_i(t) + [x_i(t) - y_i(t)][1 - I(t)] \qquad (4b)$$

where $x(t) = \Sigma_j x_j(t)$ and $I(t) = \Sigma_j I_j(t)$. The symbol β is the only parameter in the model. The behavior of STORE can be interpreted by two elements. The first is the global inhibition term in Equation 4a, which reduces the value of x_i in favor of new items. The second element is the excitatory loop between x_i and y_i, which clearly favors old items in STM. Combined together, they are able to produce the bowing shape for a sequence of items. Figure 2D shows three different bows generated with different values of β. The STORE model currently cannot handle complex sequences.

Temporal Pattern Recognition

The shared goal of all STM models is to make input history somehow available simultaneously when recognition takes place. With an STM model in place, recognition is no different from the recognition of static patterns.

Template Matching Using Hebbian Learning

The architecture for this type of recognition is simply a two-layer network: the input layer which incorporates STM, and the sequence-recognition layer where each unit encodes each individual sequence. Figure 3 shows the diagram of this archi-

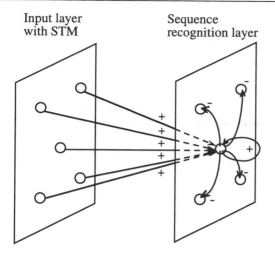

Figure 3. Architecture for temporal pattern recognition based on template matching. The plus sign indicates a positive connection, and the minus sign a negative one.

tecture. The recognition scheme is essentially temporal matching, where templates are formed through Hebbian learning

$$W_{0i}(t) = W_{0i}(t-1) + Cs_0(t)(x_i(t) - W_{0i}(t-1)) \quad (5)$$

where W_{0i} is the connection weight between unit x_i in the input layer and the sequence recognizer s_0 of the recognition layer. The sequence recognizer s_0 is selected either by self-organization, e.g., winner-take-all, or the system. The parameter C regulates learning rate. Hebbian learning is applied after the presentation of the entire sequence has been completed. The templates thus formed can be used to recognize specific input sequences. The recurrent connections within the second layer are used for selecting a winner during training or recognition.

Kohonen (1990) demonstrated an architecture called the *Phonetic Typewriter* for phoneme recognition (see SPEECH RECOGNITION: A HYBRID APPROACH). The Phonetic Typewriter extracts a vector of frequency components using the Fast Fourier transform. After this step, his algorithm of feature mapping is applied for recognition, where winner-take-all is applied to both training (selecting s_0 of Equation 5) and recognition. The Phonetic Typewriter has been applied to recognize Finnish and Japanese phonemes (Kohonen, 1990). Wang and Arbib (1990, 1993) adopted a learning method very similar to Equation 5, and showed that the recognition algorithm plus either decay trace or interference-based STM can recognize any complex sequence. Furthermore, the interference-based STM model was extended to multiple layers so that the algorithm can recognize hierarchical sequences with much greater lengths, reminiscent of information chunking.

The Associative Memory Approach

The dynamics of Hopfield's associative memory model (see COMPUTING WITH ATTRACTORS) can be characterized as evolving toward the memory state most similar to the current input pattern. If one views each memory state as a category, the Hopfield net performs pattern recognition: the recalled category is the recognized pattern. This process of dynamic evolution can also be viewed as an optimization process which minimizes a cost function until equilibrium is reached.

With the normalized exponential kernel STM, Tank and Hopfield (1987) described a recognition network based on associative memory dynamics. Like the architecture of Figure 3,

a layer of sequence recognizers receives inputs from the STM model. Each recognizer encodes a different template sequence by its unique weight vector acting upon the inputs in STM. In addition, recognizers inhibit each other, thus forming a competitive network. The recognition process uses the current input sequence (evidence) to bias a minimization process so that the most similar template wins the competition, thus activating its corresponding recognizer. They demonstrated that, because of the exponential kernels, recognition is fairly robust with respect to *time warping*, distortions in durations. A similar architecture is applied to speaker-independent spoken digit recognition (Unnikrishnan, Hopfield, and Tank, 1992).

Multilayer Perceptrons

Perhaps the most popular approach to temporal pattern learning is multilayer perceptrons (MLP). These have been demonstrated to be very effective for static pattern recognition. It is natural to combine MLPs with an STM model to do temporal pattern recognition. Using delay-line STM as the input layer, Waibel et al. (1989) reported an architecture called Time-Delay Neural Networks (TDNN) for spoken phoneme recognition. Besides the input layer, TDNN uses two hidden layers and an output layer where each unit encodes one phoneme. The feedforward connections converge from the input layer to each successive layer so that each unit in a specific layer receives inputs within a limited time window from the previous layer. After very slow training, they demonstrated impressive recognition performance. For the three stop consonants /b/, /d/, and /g/, the accuracy of speaker-dependent recognition reached 98.5%. According to their comparative experiments, this performance was favorably comparable with hidden Markov models, one of the most successful engineering approaches to speech recognition.

Mozer in 1989 proposed the *focused backpropagation algorithm* for sequence recognition (see Mozer 1993). A self-connected context layer, which realizes decay STM, is added to a typical three-layer MLP. A sequence is viewed as a set of fixed-length context-component pairs, called *Wickelphones*. Because each Wickelphone is supposed to be unique, the temporal sequence to be recognized is transformed into a simple sequence without recurring Wickelphones. Thus, the time dimension in TDNN can be eliminated, resulting in a much simpler architecture to train. Mozer argued that this architecture should scale better with respect to sequence length. It appears that determining how to decide the length of a Wickelphone remains an issue. A Wickelphone that is too small may lead to an ambiguous representation, and one that is too large significantly increases the size of the network.

Temporal Pattern Production

One of the early models of sequence production is the *outstar avalanche* of Grossberg (1969), which is composed of n sequential outstars. Each outstar $\mathbf{M}_i$ stores a static pattern and is activated by a signal in the vertex v_i. These vertices are connected as $v_1 \rightarrow v_2 \rightarrow \ldots \rightarrow v_n$, and a signal from v_i arrives with some delay at v_{i+1}. So an initial signal at v_1 can produce sequentially the spatial patterns stored in $\mathbf{M}_1, \mathbf{M}_2, \ldots, \mathbf{M}_n$, respectively. In the last several years, synchronized with the resurgence of neural networks, a number of solutions have been proposed for temporal pattern generation.

The Associative Memory Approach

Since associative memory studies how to associate one pattern with another, its mechanism can be readily extended to storing and producing a sequence. A sequence is treated as a set of pairs between consecutive components, and these pairs are stored into an associative memory. Hence, after the first component of the sequence is presented, the next component will be activated from the memory shortly, which further activates the third one, etc. This approach has been investigated by Sompolinsky and Kanter (1986) and Buhmann and Schulten (1987), among others. Unlike other approaches to sequence production, it seems to permit direct biological interpretation.

This scheme, however, leads to ambiguity when producing a complex sequence, where one symbol may be followed by different symbols. Several investigators have proposed using high-order networks to fix the problem. In a kth-order network, the input to each unit is the weighted sum of k-tuples, instead of individual units, and each k-tuple is a product of k units. In such a network, one component in a sequence is associated by a prior subsequence of length k. This idea is essentially the same as using Wickelphones for sequence recognition. Thus, a sequence of degree k can be produced without ambiguity by a kth-order associative memory. Guyon et al. (1988) described general high-order networks for producing complex sequences. The major problem with high-order networks is the immense number of connections, which grows exponentially with the order of the network.

The Multilayer Perceptron Approach

Jordan (1986) described the first MLP architecture with recurrent connections for sequence production, shown in Figure 4. The input layer has two parts. The first part, composed of plan units, receives external input, representing the identity of the sequence. The second part consists of state units which receive one-to-one projections from the output layer, forming decay trace STM. After a sequence is stored into the network by backpropagation training, it can be reproduced by an external input representing the name of the sequence. This input activates the first component of the sequence in the output layer. This component will flow back to the input layer and, together with the external input, activate the second component, and so on. A particular component of a sequence is produced by the part of the sequence prior to the component, earlier components having smaller roles due to exponential decay. Elman (1990) later modified Jordan's architecture by having the hid-

den layer connect to a part of the input layer, called the context layer. The context layer simply duplicates the activation values of the hidden layer at the previous time step. Elman used this architecture for learning a set of individual sequences satisfying a syntactic description, and found that the network exhibits a kind of recognition of the syntax itself (see LANGUAGE PROCESSING). This result suggests a promising way of learning high-level structures, crucial for natural language parsing. In both models, since the history is somehow encoded, certain complex sequences can be produced. On the other hand, because the entire history is blended into a single state, different subsequences cannot be uniquely represented.

Temporal sequence prediction is a task where the neural network is required to predict a sequence based on the past history in a specific domain, e.g., the financial markets. Since the network has no way knowing what is going to happen, it has to acquire the regularity hidden in the history and use it to generate the future behavior. This task has been studied using the MLP approach, and several encouraging results have been reported. The interested reader is referred to *Predicting the Future and Understanding the Past*, edited by A. Weigend and N. Gershenfeld (Redwood City, CA: Addison-Wesley, 1993).

Self-Organization of Context Learning

By now it should be clear that the production of complex sequences is a major issue. The approaches described so far rely on either fixed-degree contexts which are not very practical because of implementation overhead, or a composite vector recording the traces of the history which seems prone to ambiguity. In order to ensure disambiguation, the idea of encoding fixed-degree contexts must use a degree not smaller than the degree of the sequence, which is usually greater than the degrees of most components. This analysis calls for a mechanism of self-organization, where each component in a sequence can learn the degree of its own context, thus avoiding using a preset degree.

Recently, Wang and Arbib (1993) proposed such a mechanism. The model was based on the idea of using anticipation for disambiguation. Figure 5 shows the architecture of the model, where the input layer is an interference-based STM model and a unit in the detector layer encodes the context for each component in a sequence. The feedback connections from the detector layer to the input layer associate context detectors with their corresponding components. Each context detector learns to recognize a specific subsequence in the same way as for sequence recognition, and it has a parameter denoting the

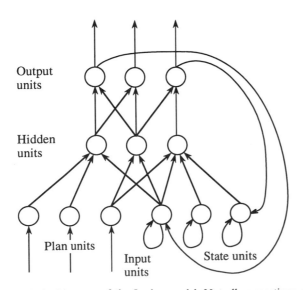

Figure 4. Architecture of the Jordan model. Not all connections are shown.

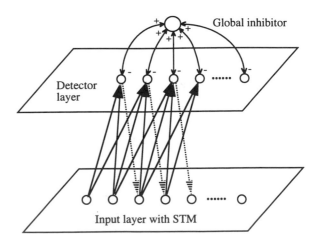

Figure 5. Architecture for self-organization of context learning.

degree of the context to be encoded, which is initially set to 1. If the sequence is a simple one, it can be easily learned. Otherwise, the following situation is bound to occur: A subsequence in STM will anticipate more than one following component, i.e., activate more than one unit in the second layer (potential ambiguity). The global inhibitor (Figure 5) will detect this situation, and its firing will then increment the degree parameters of the units currently active in the detector layer so that these units will detect longer contexts next time. This process continues during training until no ambiguity exists. Wang and Arbib (1993) showed that this algorithm of self-organization optimally identifies the context of each component in an arbitrary sequence. For instance, the algorithm found the degrees $\{1, 2, 3, 1, 1, 2, 3, 4, 1, 1, 2, 3, 4, 1, 2, 2, 3, 4, 2\}$ for all but the first component of the sequence *J-B-A-C-D-A-B-A-E-F-A-B-A-G-H-A-B-A-H-I*, respectively.

Discussion

Although the topic of neural network processing of temporal patterns has been attended to just over the last several years, much progress has been made both in recognition and in production. There are, however, many questions yet to be answered despite the progress. The following are, I think, three major aspects:

1. *Rate invariance and time warping.* Humans show rate invariance to a certain extent in recognizing a temporal pattern. *Rate invariance* is different from what I call *interval invariance*; the former is invariance to only global scaling in the durations of components, and the latter is invariance to all changes in the durations. Interval invariance is exhibited in several network models, but not rate invariance. One must be careful about time-warp invariance. We would like to have invariance over limited warping, but dramatic change in relative duration must be recognized differently (see the introductory section).
2. *Temporal pattern segmentation.* Temporal pattern recognition, including speech recognition, could perform well after presegmentation has been done. Segmentation of simultaneous temporal signals is a tremendous challenge, which has hardly been addressed at all.
3. *Top-down coding.* So far, all models deal with only input-oriented, bottom-up processing. However, long-term memory of the past sequences can dramatically influence the outcome of temporal processing. Top-down coding must be incorporated into future neural network models.

The neural network approach is not handicapped in any way toward solving these problems. On the contrary, these challenges, I believe, pave the way to an exciting future.

Acknowledgments. The preparation of this paper was supported in part by NSF grant IRI–9211419 and ONR grant N00014–93–1–0335.

Road Map: Self-Organization in Neural Networks
Background: I.3. Dynamics and Adaptation in Neural Networks
Related Reading: Recurrent Networks: Supervised Learning; Self-Organization in the Time Domain; Short-Term Memory; Spatiotemporal Association in Neural Networks; Speech Recognition: Pattern Matching

References

Bradski, G., Carpenter, G. A., and Grossberg, S., 1992, Working memory networks for learning temporal order with application to three-dimensional visual object recognition, *Neural Computat.*, 4:270–286.
Buhmann, J., and Schulten, K., 1987, Noise-driven temporal association in neural networks, *Europhys. Lett.*, 4:1205–1209.
de Vries, B., and Principe, J. C., 1992, The gamma model—A new neural model for temporal processing, *Neural Netw.*, 5:565–576.
Elman, J. L., 1990, Finding structure in time, *Cognit. Sci.*, 14:179–211.
Grossberg, S., 1969, Some networks that can learn, remember, and reproduce any number of complicated space-time patterns, I, *J. Math. Mech.*, 19:53–91.
Guyon, I., Personnaz, L., Nadal, J. P., and Dreyfus, G., 1988, Storage and retrieval of complex sequences in neural networks, *Phys. Rev. A*, 38:6365–6372.
Jordan, M. I., 1986, Attractor dynamics and parallelism in a connectionist sequential machine, in *Proceedings of the Eighth Annual Conference of the Cognitive Science Society*, Hillsdale, NJ: Erlbaum, pp. 531–546.
Kohonen, T., 1990, The self-organizing map, *Proc. IEEE*, 78:1464–1480. ◆
Mozer, M. C., 1993, Neural net architectures for temporal sequence processing, in *Predicting the Future and Understanding the Past* (A. Weigend and N. Gershenfeld, Eds.), Redwood City, CA: Addison-Wesley, pp. 243–264. ◆
Sompolinsky, H., and Kanter, I., 1986, Temporal association in asymmetric neural networks, *Phys. Rev. Lett.*, 57:2861–2864.
Tank, D. W., and Hopfield, J. J., 1987, Neural computation by concentrating information in time, *Proc. Natl. Acad. Sci. USA*, 84:1896–1900.
Unnikrishnan, K. P., Hopfield, J. J., and Tank, D. W., 1992, Speaker-independent digit recognition using a neural network with time-delayed connections, *Neural Computat.*, 4:108–119.
Waibel, A., Hanazawa, T., Hinton, G. E., Shikano, K., and Lang, K. J., 1989, Phoneme recognition using time-delay neural networks, *IEEE Trans. Acoust. Speech Signal Process.*, 37:328–339.
Wang, D. L., and Arbib, M. A., 1990, Complex temporal sequence learning based on short-term memory, *Proc. IEEE*, 78:1536–1543.
Wang, D. L., and Arbib, M. A., 1993, Timing and chunking in processing temporal order, *IEEE Trans. Syst. Man Cybern.*, 23:993–1009.

Textured Images: Modeling and Segmentation

Jenq-Neng Hwang

Markov Random Field Modeling

Textures have been considered as stochastic two-dimensional image fields. Given a sample input texture, the goal of texture modeling is to analyze the parameters associated with the stochastic description of the texture and then generate the corresponding textured image that both resembles the sample input texture visually and matches it closely from a statistical point of view (Cross and Jain, 1983). Once stochastic texture models are established, segmentation of several textures mixed in an image can also be carried out in a stochastic manner (Derin and Elliott, 1987).

A human looking at two different types of four-gray-level artificial textures (Figure 1) can tell them apart by observing that the first texture has balanced small black/white stripes of all orientations (horizontal, vertical, and both diagonals), while

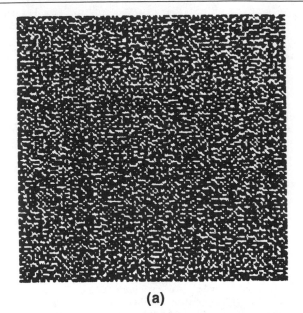

(a)

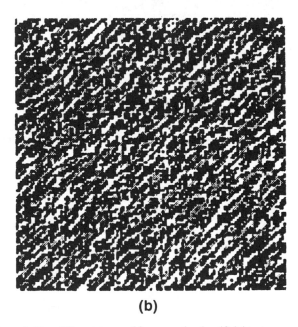

(b)

Figure 1. Two different types of four-gray-level artificial textures.

the second texture possesses larger black/white stripes along all directions except one diagonal direction (i.e., upper-left corner to lower-right corner). This observation can be interpreted in another manner if we assume the gray level of a pixel in the textured image is highly dependent on those of neighboring pixels. More specifically, the pixel gray levels of the first texture have comparable correlation with those of all neighboring pixels, while the pixel gray levels of the second texture have comparable correlation with those of most neighboring pixels except with the upper-left and lower-right pixels. This notion of near-neighbor dependence motivates the research of Markov random field modeling of textures (Besag, 1974; Cross and Jain, 1983; Geman and Geman, 1984; Chellappa and Jain, 1993). We review this approach briefly (see also MARKOV RANDOM FIELD MODELS IN IMAGE PROCESSING), but the thrust of

this article will be to advocate an alternative approach based on neural networks.

Markov random field (MRF) image modeling parametrically and stochastically represents the gray levels of images as random variables. More specifically, an MRF model builds the joint distribution $P(\mathbf{X})$ of a textured image (random field) $\mathbf{X} = \{x_{ij}\}$ of size $N \times N$ based on a local conditional distribution formulation, $\{P(x_{ij}|\eta_{ij})\}$, of all pixels x_{ij}, where η_{ij} is the predefined set of neighboring pixels surrounding x_{ij}. For example, the first-order neighborhood system η_{ij}^1 contains the nearest four pixels surrounding x_{ij}, and the second-order neighborhood system η_{ij}^2 contains the nearest eight pixels surrounding x_{ij}, and so on. Even though the joint distribution $P(\mathbf{X})$ is not readily available given all local conditional distributions, $\{P(x_{ij}|\eta_{ij})\}$, the synthesis (realization) of the texture can be directly carried out using solely the local conditional distributions, $\{P(x_{ij}|\eta_{ij})\}$, and the so-called *Gibbs sampler* algorithm (Geman and Geman, 1984; Gelfand et al., 1990).

The Gibbs sampler updating scheme proceeds as follows. Given an arbitrary starting set of values $\mathbf{X}^{(0)} = (x_1^{(0)}, \ldots, x_{N^2}^{(0)})$, we update a single component $x_1^{(0)}$ to $x_1^{(1)}$ according to $P(x_1|x_1^{(0)}, \ldots, x_{N^2}^{(0)})$ in a stochastic manner (i.e., based on the resulting conditional distribution value and a uniform distributed random number). The Gibbs sampler procedure then updates $x_2^{(0)}$ to $x_2^{(1)}$ according to $P(x_2|x_1^{(1)}, x_3^{(0)}, \ldots, x_{N^2}^{(0)})$ and continuous up to the updating of $x_{N^2}^{(0)}$ according to $P(x_{N^2}|x_1^{(1)}, x_3^{(1)}, \ldots, x_{N^2-1}^{(1)})$ to complete one iteration. Several iterations are required for the updating to converge (Geman and Geman, 1984), and the resulting $\mathbf{X}^{(t)}$ can be regarded as a simulated observation from (near) maximum(s) of joint distribution $P(\mathbf{X})$ as $t \to \infty$. In a two-dimensional MRF, the conditional distributions, $\{P(x_{ij}|x_{mn}), (m, n) \neq (i, j)\}$, are further replaced by the local conditional distributions, $\{P(x_{ij}|\eta_{ij})\}$. Moreover, the pixel visiting mechanism need not be sequentially ordered; it can be random or deterministic (e.g., raster scan).

As observed previously, the visual perception of a texture is usually characterized by the correlation (interaction) of the pixels with their neighboring pixels. To allow a computationally tractable formulation of the local conditional distributions, $\{P(x_{ij}|\eta_{ij})\}$, in an MRF setting, most researchers adopt the *Gibbs distribution* (GD; see MARKOV RANDOM FIELD MODELS IN IMAGE PROCESSING). More specifically, we first form several subsets of pixels (*clique primitives*) inside the neighborhood whose mutual gray-level interaction has contributed to the overall characteristics of textures. For example, a regular and periodic pair of horizontally adjacent pixels with comparably large gray levels may contribute to long horizontal stripe of textures. We then have to quantitatively formulate the interaction of the pixels in each primitive based on numerical functions (*clique functions*). Finally, for each defined clique function, a weighting coefficient (*clique parameter*) is assigned to indicate its relative importance. The GD formulation imposes a very restricted choice of the functional form (e.g., the exponential with linearly weighted exponents) of the clique functions defined over a set of clique primitives associated with a specific pixel x_{ij}. Each clique primitive, which more or less characterizes the black/white stripes seen in Figure 1, contains the specific pixel x_{ij} and a subset of the neighboring pixels η_{ij} (Cross and Jain, 1983).

In addition, the clique parameters, which specify the degree of contribution of the clique functions to the linearly weighted sums of exponents that generate the local conditional probabilities, require tedious and laborious estimation procedures. Various estimation techniques have been proposed for clique parameter estimation: e.g., maximum likelihood methods (Besag, 1974; Kashyap and Chellappa, 1983), maximum pseudo-likelihood (Besag, 1986), linear least-squares with histogramming (Derin and Elliott, 1987), and multistart maximum likelihood (Won and Derin, 1988). In spite of the efforts in creating a computationally tractable functional form in a GD formulation, these parameter estimation methods sometimes generate unreliable or inconsistent solutions because of the requirements of either the formulation of a set of nonlinear equations that are difficult and cumbersome to solve (Besag, 1974; Besag, 1986; Kashyap and Chellappa, 1983), the least square formulation to a big set of linear equations based on very ad hoc clique functions (Derin and Elliott, 1987), or a nonlinear optimization problem subject to hundreds of thousands of inequality constraints (Won and Derin, 1988). Moreover, most real-world textured images may possess approximately local neighboring pixel interactions but not necessarily follow the assumptions of a GD, where the global distribution of the image is constituted from the local distribution in an exponential form based on the linearly weighted sums of subjectively defined clique functions.

Neural Network Modeling

In a classification application, it is normally assumed that the input vector, $\mathbf{z} \in \mathbb{R}^n$, belongs to one of M classes, $\mathscr{g}_m$, $1 \le m \le M$. The main objective of a classification task is to decide to which of the M classes the vector $\mathbf{z}$ belongs. The decision can be made based on some form of deterministic discriminant function, e.g., the Euclidean distance measure. A more general decision rule is based on the probabilistic decision, such as the *maximum a posteriori* (MAP) approach which guarantees the minimum classification error (Duda and Hart, 1973). In a MAP approach, for each of the classes one is required to estimate the posterior probability, $P(\mathscr{g}_m|\mathbf{z}) \propto P(\mathbf{z}|\mathscr{g}_m)P(\mathscr{g}_m)$, which is usually computed via Bayes's rule. Since the a priori probability, $P(\mathscr{g}_m)$, is relatively easier to compute, most conventional pattern recognition approaches focus on estimating the likelihood $P(\mathbf{z}|\mathscr{g}_m)$.

Neural Network as a MAP Classifier

Conversely, when a backpropagation neural network (BPNN), i.e., a multilayer feedforward network trained by backpropagation, is used for this classification task, there is usually an input layer of n neurons corresponding to the n-dimensional input vector $\mathbf{z}$, one or two layers of "appropriately chosen" hidden neurons, and one output layer of M neurons with each one representative of one of the M different classes (e.g., the desired binary output vector for class one is $\mathbf{t} = [1, 0, 0, \ldots, 0]$). It has been shown that the continuous-valued output activations, $\mathbf{y} = (y_1, y_2, \ldots, y_M)$, of a BPNN trained by the standard backpropagation learning, which minimizes the mean squared error (MSE) between the actual outputs $\mathbf{y}$ and the desired binary targets $\mathbf{t}$, can be directly interpreted as a least squares estimate of the posterior probabilities $\{P(\mathscr{g}_m|\mathbf{z}); m = 1, \ldots, M\}$ (Richard and Lippmann, 1991). Therefore, neural network classifiers bypass the stage of estimating the likelihood and directly estimate the posterior probability in a classification task.

To ensure the BPNN's outputs to be valid probabilities (i.e., their values should be nonnegative and sum to 1), additional care is required. For example, the positivity can be guaranteed by the use of exponential or sigmoid activation functions in the output layer. However, to make network outputs sum to 1 without further normalization, an additional (fixed weights) layer of neurons can be added on top of the output layer to impose this constraint. Instead of using the minimum MSE

criterion, the Kullback-Leibler (KL) criterion (see LEARNING AND STATISTICAL INFERENCE), which was derived from information theory, may be preferred.

The MSE criterion tends to give more accurate estimates at large outputs than at small outputs (Makhoul, 1991). In classification applications, where a winner is chosen based on the largest output, this MSE criterion may not cause any adverse effect. On the other hand, accurate approximation of the local conditional distributions $\{P(x_{ij}|\eta_{ij})\}$, regardless of their magnitudes, is critical to the success of texture modeling. The performance of texture modeling can be severely degraded by the deficiency created by the MSE criterion. The KL criterion, which overcomes this deficiency and provides a better probability estimate, can thus be adopted. The KL criterion uses the relative entropy E_{KL} as the cost function to be minimized and serves as a better measure of the difference between the desired probabilities $\{t_m = P(\mathscr{g}_m|\mathbf{z})\}$ and the actual estimated probabilities $\{y_m = \hat{P}(\mathscr{g}_m|\mathbf{z})\}$ determined by the activation value of output neurons (Wittner and Denker, 1988). More specifically, when the desired outputs are binary valued and given N training vectors $\{\mathbf{z}^{(n)}, n = 1, \ldots, N\}$, E_{KL} can be approximated by:

$$E_{KL} = -\sum_{m=1}^{M}\sum_{n=1}^{N}[t_m^{(n)}\log y_m^{(n)} + (1 - t_m^{(n)})\log(1 - y_m^{(n)})]$$

where $y_m^{(n)} = y_m^{(n)}(\mathbf{z}^{(n)}, \mathbf{W})$ is a function of inputs $\mathbf{z}^{(n)}$ and all the BPNN interconnection weights $\mathbf{W}$. The relative entropy cost function differs from MSE in that it weights errors more heavily and in a more balanced manner when actual outputs are near zero and one. Although the introduction of the additional layer and the new KL training criterion complicates the learning, the chain rule involved in backpropagation learning can still be worked out through the fixed-weight layer to get all the gradients for updating the weights. For the remainder of this article, all the BPNNs for texture modeling are trained by the KL criterion.

The BPNN Probabilistic Modeling

In an MRF setting, each clique type associated with the assumed neighborhood system should be well defined. Since only the local conditional distributions are required in a GD formulation of an MRF, we are interested in the distribution of a center pixel x_{ij} conditioned on its neighbors η_{ij}, i.e., $P(x_{ij}|\eta_{ij})$. Instead of using the restricted parametric formulation defined in a GD for representing $P(x_{ij}|\eta_{ij})$, we can establish this local conditional probability easily in a BPNN setting for classification problems (Hwang and Chen, 1993).

In BPNN probabilistic modeling, training is constructed by simply feeding the BPNN with neighborhood configuration η_{ij} of each pixel, x_{ij}, and assigning the gray level of x_{ij} to be the target $\mathbf{t} = \mathscr{g}_m$ in a classification formulation (i.e., each gray level of x_{ij} is assigned as one class). More specifically, consider for example the second-order neighborhood system shown in Figure 2. The BPNN contains eight inputs to represent eight neighboring pixels, η_{ij}, an appropriately chosen number of hidden neurons, and M output neurons to represent M gray levels. The desired outputs are set so that only the mth output neuron is 1 and the rest are 0 if the gray level of x_{ij} is equal to m. After training using some pairs of training data obtained from training textured images, the BPNN responds to a new neighborhood configuration of inputs by providing the continuous-valued local conditional distribution (or the classifier's posterior probability), $\{P(x_{ij} = m|\eta_{ij}), m = 1, \ldots, M\}$, as outputs for the center site being assigned to all gray levels; i.e., the BPNN represents a model of the underlying field which is conventionally modeled by GD in a parametric form for $P(x_{ij}|\eta_{ij})$.

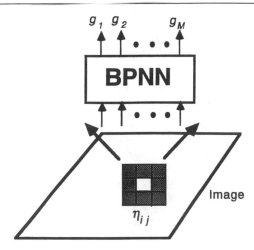

Figure 2. A BPNN is trained to provide continuous-valued local conditional distributions of an image of M gray levels under a second-order neighborhood system.

Comparative Simulation Studies

To support the claims of unreliable and inconsistent estimation of clique parameters based on the existing MRF approaches and to demonstrate the usefulness of the proposed BPNN probabilistic modeling, application of the existing MRF approaches and the BPNN method to model artificial and real-world textures is presented. It is our belief that the textured image modeling is a much more difficult task than the textured image segmentation, which is presented in the later section.

We studied two different types of four-gray-level artificial textures used in Derin and Elliott (1987). These two artificial textures are created from two second-order MRFs with the clique parameters being $[1, 1, -1, 1, 0, 0, 0, 0, 0]$ for type 1 and $[2, 2, -1, -1, 0, 0, 0, 0, 0]$ for type 2. Note that there should be nine contributing clique parameters in total for a second-order MRF, and only four of them are set to be non-zero.

Based on the original artificial textured images of sizes 128×128, the clique parameters were estimated using all 128×128 pixels based on the popular linear least-squares MRF method following the same clique function definitions as in Derin and Elliott (1987). Our results showed that the estimation results based on existing MRF approaches for artificial textures are quite unreliable and very sensitive to the number of parameters to be estimated and the selection of alternative clique functions, e.g., autologistic, autobinomial, autopoisson, etc. (Besag, 1974).

We then applied the BPNN probabilistic modeling of these two artificial textures. The BPNN contains eight inputs to represent eight neighboring pixels, 10 hidden neurons, and four output neurons to represent four gray levels. All the 128×128 pixels were again used as training data, and 1000 training iterations were used. After training, the BPNN responds to the neighborhood configuration inputs by providing the local conditional distributions, $\{P(x_{ij} = m|\eta_{ij})\}$, as outputs for x_{ij} being assigned to any of the four gray levels ($m = 1, 2, 3, 4$).

Compared with those synthesized images using linear least-squares MRF approaches, we found the BPNN modeling gave desired textures more consistently. Moreover, we also noticed that the Gibbs sampling used in iteratively updating and synthesizing the textured images required a much smaller number of iterations (approximately 10) when using the local conditional distribution generated by the BPNN model than when using those generated by the MRF model (approximately 100

iterations). This is another indication of a more effective approximation of the local conditional distribution in representing the random field. Note that the interconnection weights of the trained BPNNs do not correspond directly with the desired clique parameters of artificial textures. This fact does not affect the applicability of BPNN texture modeling since the real-world textures to be modeled do not follow the GD formulation. Therefore, the clique parameters are only useful for intuitive interpretation of the texture pixel interactions in the local regions (if they can be reliably estimated).

This BPNN modeling also can be used to discover textures that it has not been trained on. More specifically, by sending one-by-one the neighboring pixel blocks, $\{\eta_{ij}\}$, of an unknown texture patch to a trained BPNN texture model, if the accumulated posterior probabilities given by the winning output neuron is large enough, then a match of texture can be declared; otherwise, a new texture is discovered.

Application to Texture Segmentation

The proposed BPNN probabilistic modeling, when incorporated with the Gibbs sampling procedure, can be employed in a hierarchical manner for Bayesian textured image segmentation (Derin and Elliott, 1987). In this application, the Gibbs sampler focuses on estimation of the global MAP configuration through a sequential and iterative updating of the labels of all the pixels (Besag, 1986).

A mixed-textured image (Figure 3A) that consists of the two textures shown in Figure 1 was to be segmented. These two textures were first modeled by two BPNNs as discussed earlier. The segmentation mask is also modeled by a second-order Gibbs distribution as shown in Figure 3B. The segmentation result after 150 iterations of the hierarchical Gibbs sampling is shown in Figure 3C. The segmentation result can be further smoothed and improved by using a simple 3×3 median filter over the segmented label image (Figure 3D).

MRF texture modeling parametrically and probabilistically builds the joint distribution $P(\mathbf{X})$ of a textured image based on a local conditional distribution formulation $\{P(x_{ij}|\eta_{ij})\}$. This modeling enables a clean mathematical description of the random field, and thus makes feasible various applications involving textures, e.g., texture coding, texture synthesis, classification, segmentation, etc. Unfortunately, to maintain such clean mathematical description while calling for a computationally tractable functional form (i.e., a GD formulation), most parameter estimation methods cannot generate reliable or consistent solutions (Besag, 1974; Besag, 1986; Kashyap and Chellappa, 1983; Derin and Elliott, 1987; Won and Derin, 1988). Moreover, most real-world textured images may involve more complicated neighboring-pixel interactions which do not necessarily follow the clean formulation of a GD, where the global distribution of the image is constituted from the local distribution in an exponential form based on the linearly weighted sums of subjectively defined clique functions.

Neural networks, which are powerful and nonparametric approaches without very restricted assumptions of functional form, have thus been applied to texture classification and segmentation tasks (Visa, 1990; Greenspan, Goodman, and Chellappa, 1991; Kollias and Sukissian, 1992; Muhamad and Deravi, 1993; Schumacher and Zhang, 1994). Unlike our BPNN approach to probabilistic texture modeling, which attempts to approximate the unified local conditional distribution formulation $\{P(x_{ij}|\eta_{ij})\}$ without subjectively defined clique functions and the subsequent laborious clique parameter estimations, all these methods aim at using neural networks to sort out the unknown and complicated neighboring-pixel interaction, so

Figure 3. *A*, A mixed textured image of two four-gray-level artificial textures. *B*, The segmentation mask is modeled by a second-order Gibbs distribution. *C*, The segmentation result after 150 iterations of the hierarchical Gibbs sampling. *D*, The smoothed segmentation result using a simple 3 × 3 median filter.

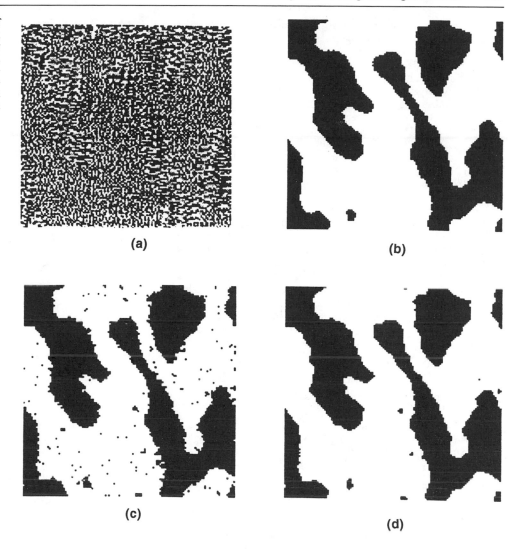

(a)

(b)

(c)

(d)

that the texture classification and segmentation tasks can avoid the tedious parameter estimations. All these methods, with each output neuron representing each texture class, focus on using neural networks to delineate the classification boundaries between several texture classes. Therefore, they can be used only for texture classification and segmentation. On the other hand, the proposed BPNN texture modeling, with each output neuron representing each gray level of one specific texture class, focuses on using neural networks to regress the unknown nonlinear local conditional distributions.

Visa (1990) classified an unknown texture of size 512 × 512 by first encoding the neighboring-pixel interaction based on 14 features (such as energy, entropy, correlation, homogeneity, and inertia) derived from the gray tone co-occurrence matrix (GTCM) representation. These encoded features were then classified by a self-organizing neural network trained with learning vector quantization (LVQ). Because of the way features are derived, this approach can be used only for texture classification, not segmentation.

Greenspan et al. (1991) classified a windowed block of unknown texture by first preprocessing this block using a set of Garbor filters. These filters encoded the neighboring-pixel interaction by decomposing them into multiple spatial frequency and orientation channels. Instead of directly sending these encoded continuous-valued features into a first-order Bayesian classifier, a Kohonen feature map (see SELF-ORGANIZING FEATURE MAPS: KOHONEN MAPS) first reduced the 256 gray levels to 6 gray levels for computing the smaller-size GTCM which could be directly used as input features for texture classification by a standard feedforward multilayer neural network.

With growing use of wavelets for efficient image representation (see WAVELET DYNAMICS), Schumacher and Zhang (1994) described the neighboring-pixel interaction based on the discrete wavelet transform (DWT), which could be a good multiresolution representation of textures commonly illustrated by quasi-periodic patterns of edges and contours. To efficiently train a feedforward multilayer neural network classifier based on the DWT representation, four subnetworks were trained separately with DWT coefficients of different resolution, and a voting scheme was adopted to generate the texture classification result.

Discussion

A BPNN trained with the KL criterion can be a powerful mechanism for probabilistic modeling of the local interactions in a random field. The comparative simulation results show that a BPNN can provide a fairly good estimate of Bayesian a posteriori probabilities required in texture modeling. While conventional MRF modeling relies heavily on restricted parametric representation and tedious parameter estimation, the BPNN probabilistic modeling gives a more relaxed and more reliable solution, as evidenced by successful application to textured image modeling and segmentation of both artificial and real-world textures.

It can be argued that the proposed BPNN modeling approach requires more parameters (interconnection weights of the BPNN) than the number of clique parameters used in a conventional MRF modeling. Even though this challenge is valid, there is no way to increase the number of clique parameters for a better performance without increasing the order of the neighborhood system under the very restricted MRF formulation. More importantly, when the order of the neighborhood system increases in an MRF model, the number of clique parameters increases exponentially, and the parameter estimation performance degrades rapidly. Conversely, the BPNN probabilistic modeling increases its number of interconnection weights linearly (i.e., the input dimension increases), with slight performance degradation observed from our experience.

In spite of its seemingly superior performance based on a limited set of simulations, the proposed BPNN texture modeling can possibly suffer performance degradation in the presence of texture rotation, texture scaling, and also larger number of gray levels. All these difficulties are yet to be overcome before the BPNN texture modeling can be of practical use.

Acknowledgments. This research was partially supported by the National Science Foundation under Grant No. ECS-9014243, and by NASA under Contract No. NAGW-1702.

Road Map: Vision
Background: Markov Random Field Models in Image Processing
Related Reading: Figure-Ground Separation; Regularization Theory and Low-Level Vision

References

Besag, J., 1974, Spatial interaction and the statistical analysis of lattice systems (with discussion), *J. R. Statist. Soc. Ser. B*, 36:192–326.

Besag, J., 1986, On the statistical analysis of dirty pictures, *J. R. Statist. Soc. Ser. B*, 48:259–302.

Chellappa, R., and Jain, A. K., 1993, *Markov Random Fields: Theory and Application*, Orlando, FL: Academic Press.

Cross, G. R., and Jain, A. K., 1983, Markov random field texture modeling, *IEEE Trans. PAMI*, 5:149–163.

Derin, H., and Elliott, H., 1987, Modeling and segmentation of noisy and textured images using Gibbs random fields, *IEEE Trans. Pattern Analysis Machine Intell.*, 9:39–55.

Duda, R. O., and Hart, P. E., 1973, *Pattern Classification and Scene Analysis*, New York: Wiley.

Gelfand, A., Hills, S. E., Racine-Poon, A., and Smith, A. F. M., 1990, Illustration of Bayesian inference in normal data models using Gibbs Sampling, *J. Am. Statist. Assoc.*, 85:972–985.

Geman, S., and Geman, D., 1984, Stochastic relaxation, Gibbs distribution, and the Bayesian restoration of images, *IEEE Trans. Pattern Analysis Machine Intell.*, 6:721–741.

Greenspan, H., Goodman, R., and Chellappa, R., 1991, Texture analysis via unsupervised and supervised learning, in *Proceedings of the International Joint Conference on Neural Networks*, Piscataway, NJ: IEEE, vol. 1, pp. 639–644.

Hwang, J. N., and Chen, T. Y., 1993, Textured image segmentation via neural network probabilistic modeling, in *Proceedings of the International Conference on Neural Networks*, Piscataway, NJ: IEEE, vol. 3, pp. 1702–1707.

Kashyap, R., and Chellappa, R., 1983, Estimation and choice of neighbors in spatial interaction model of images, *IEEE Trans. Inform. Theory*, 29:60–72.

Kollias, S., and Sukissian, L., 1992, Adaptive segmentation of textured images using linear prediction and neural networks, in *Proceedings, Neural Networks for Signal Processing*, Piscataway, NJ: IEEE, pp. 401–410.

Makhoul, J., 1991, Pattern recognition properties of neural networks, in *Proceedings of the IEEE Workshop on Neural Networks for Signal Processing*, Piscataway, NJ: IEEE, pp. 173–186.

Muhamad, A. K., and Deravi, F., 1993, Neural network texture classifiers using direct input coocurrence matrices, in *Proceedings of the International Conference on Acoustics, Speech and Signal Processing*, Piscataway, NJ: IEEE, vol. 5, pp. 117–120.

Richard, M. D., and Lippmann, R. P., 1991, Neural network classifiers estimate Bayesian a posteriori probabilities, *Neural Computation*, 3:461–483.

Schumacher, P., and Zhang, J., 1994, Texture classification using neural networks and discrete wavelet transform, in *Proceedings of the International Conference on Image Processing*, Piscataway, NJ: IEEE, vol. 3, pp. 903–907.

Visa, A., 1990, A texture classifier based on neural network principles, in *Proceedings of the International Joint Conference on Neural Networks*, Piscataway, NJ: IEEE, vol. 1, pp. 491–496.

Won, C. S., and Derin, H., 1988, Maximum likelihood estimation of gaussian Markov random field parameters, in *Proceedings of the International Conference on Acoustics, Speech and Signal Processing*, Piscataway, NJ: IEEE, pp. 1040–1043.

Thalamocortical Oscillations in Sleep and Wakefulness

Terrence J. Sejnowski, David A. McCormick, and Mircea Steriade

Introduction

The brain spontaneously generates complex patterns of neural activity. As the brain enters slow-wave (quiescent) sleep, the rapid patterns characteristic of the aroused state are replaced by low-frequency, synchronized rhythms of neuronal activity. At the same time, electroencephalographic (EEG) recordings shift from low-amplitude, high-frequency rhythms to large-amplitude, slow oscillations. In what follows, we concentrate primarily on this slow-wave sleep, rather than rapid eye movement (REM) sleep, whose oscillatory properties resemble those of wakefulness. Thus, "sleep" without further qualification will mean "quiescent sleep."

The dramatic reduction in forebrain responsiveness during sleep, the pervasiveness of these changes, and the discovery of the underlying specific cellular mechanisms suggest that sleep oscillations are highly orchestrated and highly regulated. Experimental and modeling studies have shown how sleep rhythms emerge from an interaction between the intrinsic properties of these neurons and the networks through which they interact.

The thalamus and cerebral cortex are intimately linked by means of reciprocal projections. The thalamus is the major gateway for the flow of information toward the cerebral cortex and is the first station at which incoming signals can be blocked by synaptic inhibition during sleep. This mechanism contributes to the shift that the brain undergoes as it changes from an aroused state, open to signals from the outside world, to the closed state of sleep. The early stage of quiescent sleep is associated with EEG spindle waves, which occur at a frequency of 7–14 Hz; as sleep deepens, waves with slower frequencies (0.1–4 Hz) appear on the EEG recording. This review of these thalamocortical oscillations is adapted from Steriade, McCormick, and Sejnowski (1993).

Delta Oscillations

Delta waves (1–4 Hz) were initially shown to arise between cortical layers 2–3 and 5 (Steriade, Jones, and Llinás, 1990). Intracellular recordings in vivo and in vitro indicate that the thalamus is also involved in the generation of this rhythm (Figure 1A, 1B). A delta-frequency rhythm can be generated in single cells by the interplay of two intrinsic currents of thalamocortical neurons: the hyperpolarization-activated cation current (I_h) and the transient low-threshold Ca^{2+} current (I_t). A wide variety of other ionic currents (see ION CHANNELS: KEYS TO NEURONAL SPECIALIZATION for a general view of such channels), with different voltage dependencies and kinetics of activation and inactivation, contribute to the shaping of the amplitude and time course of each burst of action potentials, as revealed through biological experiments and computational modeling (Figure 1) (McCormick and Huguenard, 1992; Lytton and Sejnowski, 1992).

The hyperpolarization of thalamocortical cells is a critical factor for the interplay between I_h and I_t that generates delta oscillation. At the normal resting level in vivo, I_t is inactivated, but a hyperpolarization of 10 mV can lead to spontaneous, self-sustained delta oscillation (Figure 1A). The dependence of

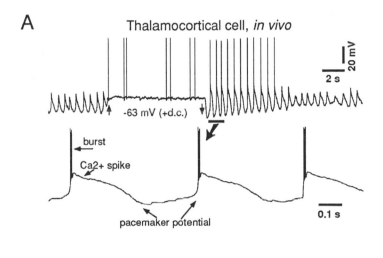

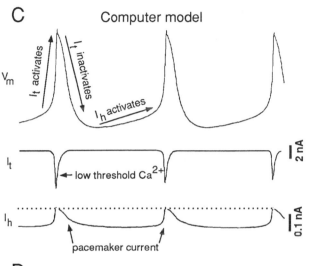

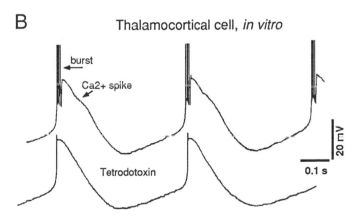

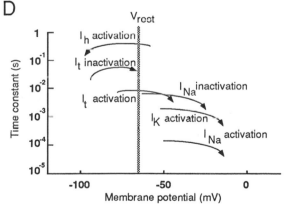

Figure 1. Intrinsic cellular mechanisms of thalamic delta oscillation. A, Voltage-dependency of delta oscillation. Intracellular recording in vivo of the lateroposterior thalamocortical neuron after decortication of areas projecting to that nucleus in an anesthetized cat is shown. The cell oscillated spontaneously at 1.7 Hz. A 0.5-nA depolarizing current pulse (between arrows), bringing the membrane potential to −63 mV, prevented the oscillation, and its removal set the cell back into the oscillatory mode. Three cycles marked by the horizontal bar in the upper trace are expanded below. B, Spontaneous rhythmic burst firing in a cat lateral geniculate relay cell recorded in vitro before and after block of voltage-dependent Na^+ conductances with application of the Na^+ channel blocker tetrodotoxin. C, Computational model of rhythmic generation of I_t as a consequence of interplay between I_t and the pacemaker current I_h. As the membrane becomes depolarized by I_h

from hyperpolarized levels, the threshold for I_t is reached, leading to a Ca^{2+} spike. D, Diagram of activation and inactivation for the primary ionic currents in thalamocortical cells. Each arc represents the time constant for activation or inactivation of a voltage-dependent current. Most currents begin to activate (or inactivate) on the left side of the arc and are fully activated (or inactivated) on the right side. One exception is the cation current I_h, which activates with hyperpolarization and does not inactivate. Different combinations of currents are active at different membrane potentials. The voltage-dependent Na^+ current, I_{Na}, and the delayed rectifier K^+ current, I_K, are responsible for the fast action potentials; V_{rest} is the resting potential. (Reprinted with permission from Steriade, M., McCormick, D. A., and Sejnowski, T. J., 1993, Thalamocortical oscillations in the sleeping and aroused brain, *Science*, 262:679–685; © AAAS.)

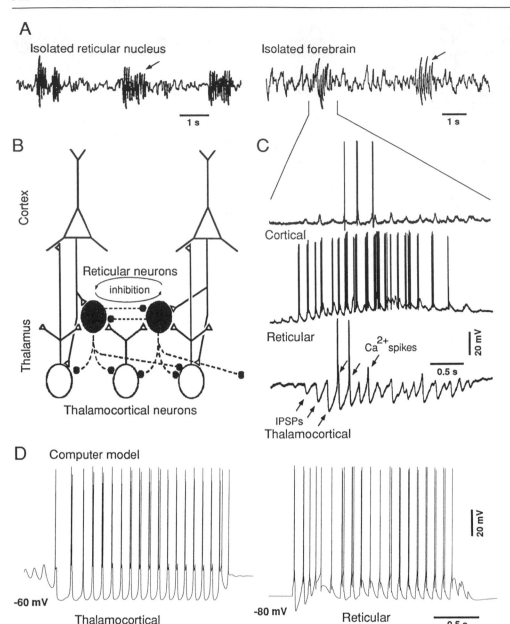

Figure 2. Sleep spindles. *A*, left: Field potentials recorded in vivo through a microelectrode inserted in the deafferented reticular thalamic nucleus of a cat. The arrow indicates one spindle sequence. *A*, right: Spindles recorded in vivo in the intralaminar centrolateral thalamic nucleus of a cat in an isolated forebrain preparation. Two spindle sequences are shown (the second marked by an arrow), and between them are lower-frequency (delta) waves. *B*, Schematic diagram of neuronal connections involved in spindling. *C*, Intracellular recordings of one spindle sequence in three neuronal types (cortical, reticular thalamic, and thalamocortical) of cats in vivo. *D*, Computer model of 8–10-Hz spindling in a pair of interconnected thalamocortical and reticular neurons. A burst of spikes in the thalamocortical cell excites the reticular thalamic cell, which in turn hyperpolarizes and produces a rebound burst in the thalamocortical neuron (as in vivo; compare with part *C*). (Reprinted with permission from Steriade, M., McCormick, D. A., and Sejnowski, T. J., 1993, Thalamocortical oscillations in the sleeping and aroused brain, *Science*, 262:679–685; © AAAS.)

delta oscillation on membrane hyperpolarization can also be demonstrated in simulations of thalamic neurons based on Hodgkin-Huxley-like kinetic models (see SYNAPTIC CURRENTS, NEUROMODULATION, AND KINETIC MODELS) of the ionic currents (Figure 1*C*).

Corticothalamic volleys potentiate and synchronize the delta oscillations of simultaneously recorded thalamic cells. In simulations of thalamocortical cells oscillating in the bursting mode at delta frequency, depolarizing cortical inputs are easily able to reset the cell to a new phase of its rhythm. Thalamic synchronization can also be induced by stimulating cortical foci that are not directly connected to the thalamic nuclei where the recordings are performed; this recruitment of thalamic cells may be achieved through the reticular thalamic nucleus, which receives collaterals of layer 6 corticothalamic cells and thalamic neurons that project to the cortex. The reticular cells are exclusively inhibitory and project back to the thalamus, but not to the cerebral cortex, and also innervate other cells of the reticu-

lar thalamic nucleus (Figure 2*B*). The reticular nucleus is uniquely positioned to influence the flow of information between the thalamus and the cerebral cortex (see THALAMUS).

Spindle Waves

Spindle oscillations consist of 7–14-Hz waxing and waning field potentials, grouped in sequences that last for 1–3 s and recur once every 3–10 s (see Figure 2*A*). The EEG spindles are the epitome of brain electrical synchronization at sleep onset, an electrographic landmark for the transition from waking to sleep that is associated with loss of perceptual awareness. These oscillations are generated in the thalamus as the result of synaptic interactions and intrinsic membrane properties of inhibitory neurons of the reticular thalamic nucleus and excitatory thalamocortical cells, and their interaction with cortical pyramidal neurons (see Figure 2*B*).

In intracellular recordings of reticular and thalamocortical cells as well as from computational modeling, these two neuronal classes behave inversely during spindles (Figure 2C). In reticular cells, rhythmic (7–14 Hz) bursts are generated by low-threshold Ca^{2+} spikes and are superimposed on a slowly rising and decaying depolarizing envelope. The bursts of reticular cells inhibit large numbers of thalamocortical cells through their divergent GABAergic axons, leading to the appearance of rhythmic (7–14 Hz) inhibitory postsynaptic potentials (IPSPs) in thalamocortical neurons (Figure 2C). Some of these IPSPs result in enough removal of inactivation of the low-threshold Ca^{2+} current to be followed by a rebound Ca^{2+} spike and an associated burst of action potentials (Figure 2C). These periodic bursts in thalamocortical cells converge onto reticular neurons and facilitate their rhythmic oscillation.

A simple model consisting of a thalamocortical cell reciprocally interacting with a reticular cell already demonstrates the essential features of spindling (Destexhe, McCormick, and Sejnowski, 1993). The waxing and waning of the spindling in this two-neuron model is controlled by the intracellular calcium level in the thalamocortical neuron, which increases with each Ca^{2+} spike; calcium binding to the I_h channels changes their voltage dependence and eventually terminates the spindle, as shown in Figure 2D (Destexhe, Babloyantz, and Sejnowski, 1993).

Isolation of the reticular nucleus from the rest of the thalamus and cerebral cortex abolishes spindle oscillations in thalamocortical systems, but the deafferented reticular thalamic nucleus can generate oscillations at spindle frequencies (Steriade et al., 1987). Axonal and, in some species, dendrodendritic interconnections between reticular cells may allow the coupling and interaction of these endogenous oscillators, thereby generating oscillations in an isolated nucleus. Models of simplified reticular thalamic neurons with full connectivity and slow mutual inhibition exhibit synchronous oscillatory activity, but the frequency is below the range of the spindling rhythm (Wang and Rinzel, 1993; Destexhe et al., 1994a). An array of model reticular neurons with fast inhibition between locally connected neurons exhibits 8–10-Hz oscillations in the local field potential in the model (based on the average membrane potential for a cluster of nearby neurons) that wax and wane in a fashion similar to what has been observed in vivo (Destexhe et al., 1994a).

Spindling has been observed in thalamic slice preparations (von Krosigk, Bal, and McCormick, 1993). However, when the reticular cells were isolated from the thalamocortical cells, spindling was abolished. The modeling suggests that that may occur because a larger and more intact collection of reticular thalamic cells is needed to generate spindle waves autonomously. Another possible reason is that the presence of neuromodulators in vivo keeps the resting levels of reticular cells more depolarized than in vitro; in the model, the oscillations in the reticular network are abolished at resting levels that are too hyperpolarized (Destexhe et al., 1994b).

Traveling spindle waves have been observed in vitro (McCormick, unpublished data) and in thalamic models based on sheets of interacting thalamocortical and reticular neurons (Destexhe and Sejnowski, unpublished modeling).

Absence Seizures

The spindles of natural sleep are related to the development of a peculiar pattern of oscillatory activity, the spike-and-wave EEG complexes, which are associated with absence (petit mal) epileptic seizures. Because the reticular thalamic nucleus is central to the genesis of spindles, decreasing or abolishing the in-

hibitory efficacy of reticular neurons on thalamocortical cells would also decrease the incidence of epileptic spike-and-wave discharges. This hypothesis is supported by recent experiments showing that, in animals with genetic absence epilepsy, thalamic injections of a selective agonist of $GABA_B$ receptors increase the incidence of spike-and-wave discharges, whereas injections of a $GABA_B$ antagonist decrease these seizures in a dose-dependent manner.

The activation of $GABA_B$ receptors in thalamocortical neurons produces a slow increase in K^+ conductance and a deep hyperpolarization and also enhances the removal of inactivation of the low-threshold Ca^{2+} spike. As a consequence, there is a larger than usual rebound burst discharge in a greater than usual proportion of thalamocortical cells. These facilitated rebound bursts further excite reticular cells, quickly resulting in the generalization of paroxysmal activity. Further support for the $GABA_B$ hypothesis derives from a model of spindling in which the frequency of spindling could be shifted from 8–10 Hz to 2–4 Hz by slowing the kinetics of the inhibitory synaptic potentials from that of $GABA_A$ (5–25 ms) to that of $GABA_B$ (100–250 ms) (von Krosigk, Bal, and McCormick, 1993; Destexhe, McCormick, and Sejnowski, 1993).

Arousal

Electrical activation of certain brainstem and hypothalamic regions, including the reticular activating system, causes a variety of neurotransmitters, including acetylcholine (ACh), norepinephrine (NE), serotonin (5-HT), histamine (HA), and glutamate to be released though diffuse ascending axonal arborizations. These neuromodulators mimic arousal by suppressing sleep spindles, delta waves, and slow cellular rhythms and by replacing these low-frequency oscillations with activity similar to that of the awake, attentive animal. In cortical pyramidal neurons, ACh, NE, 5-HT, HA, and glutamate can reduce three distinct K^+ currents, thereby resulting in a significantly enhanced responsiveness to depolarizing inputs and changes in the neuronal firing mode (McCormick, 1992). Adenosine and GABA can reduce excitability by increasing membrane K^+ conductance.

These neurotransmitter systems abolish the low-frequency rhythms in thalamocortical systems during waking and rapid eye movement (REM) sleep and also promote more tonic activity or the appearance of high-frequency oscillation. The changes in firing between sleep and arousal in thalamic neurons are accomplished by depolarization of the membrane potential by 5–20 mV, which inactivates the low-threshold Ca^{2+} current and therefore inhibits burst firing. These results have been simulated in models of thalamocortical and reticular neurons.

High-Frequency Oscillations

Changes in the activity pattern generated by cortical neurons and circuits are less stereotyped than those of thalamic cells and circuits, although some common features exist. The low-frequency oscillations of the cortical EEG disappear on arousal and are replaced by higher-frequency (20–80 Hz, mainly around 40 Hz) rhythms. As in the thalamus, these alterations in cortical activity occur, at least in part, through the depolarization of pyramidal cells, presumably through the reduction of specialized K^+ conductances by ACh, NE, and other neuromodulators.

The high-frequency (20–80 Hz) oscillations in the EEG occur during some behaviors, such as immobility during hunting and focused attention to stimuli during complex sensory or motor tasks. Neurons throughout the nervous system (e.g., the retina, lateral geniculate nucleus, and cortex) have the ability to generate repetitive trains of action potentials in the frequency range of 20–80 Hz, although the synchronization of this activity into behaviorally relevant subgroups of widely spaced neurons has only been demonstrated in the cerebral cortex (Gray, 1994).

The diversity of cortical cells and their complex interactions make it difficult to model cortical networks with the same confidence with which thalamic networks have been modeled. However, it is not difficult to generate oscillatory activity in the 20–80-Hz range with networks of simplified neurons (Koch, 1993). These models reveal the need to regulate the tendency of recurrent networks to oscillate. The excitability of neurons can be controlled by inhibition. However, inhibition is also an efficient mechanism for synchronizing large populations of pyramidal neurons because of voltage-dependent mechanisms in their somas and the strategic location of inhibitory boutons on the somas and the initial segments of axons, where action potentials are initiated (Lytton and Sejnowski, 1991). Realistic simulations of cortical neurons show that sparse excitatory connectivity between distant populations of neurons can produce synchronization within one or two cycles, but only if the long-range connections are made on inhibitory as well as excitatory neurons (Bush and Sejnowski, in press).

Discussion

This article has focused on the events that occur during the transition from wakefulness to sleep and on the rhythms of deep, slow-wave sleep. Dreams occur during another sleep state, REM sleep. This sleep state is characterized by an abolition of low-frequency oscillations and an increase in cellular excitability, much like wakefulness, although motor output is significantly inhibited. Despite great interest, there is no generally accepted function for dreams or, for that matter, for the sleep state itself.

During spindling and slow-wave sleep, the thalamus excites the cortex with patterns of activity that are more spatially and temporally coherent than normally would be encountered in the awake state. Depolarizing pulses of Ca^{2+} that enter the thalamic and cortical neurons may influence enzyme cascades and regulate gene expression, homeostatically adjusting the balance of ionic currents and regulatory mechanisms. This widespread activity could be used to reorganize cortical networks after learning occurs during the awake state (Wilson and McNaughton, 1994).

Inhibitory neurons in the thalamus and cortex are of particular importance in producing the synchrony and controlling the spatial extent of the coherent populations. Synchrony and other network properties could be used to control the flow of information between brain areas and to decide where to store important information. Synchronization enhances the strength of signals, but also reduces the amount of information that can be encoded.

The ascending neuromodulatory transmitter systems delicately tune the state and excitability of the different parts of the nervous system so that it is appropriate for the analysis of sensory information, the cognitive processing and storage of this information, and the subsequent performance of the appropriate neuronal and behavioral responses. Uncovering and modeling the cellular mechanisms of these dynamic changes may provide important clues to long-standing questions ranging from the functional role of sleep to the nature of cognitive representations.

Road Map: Biological Networks
Background: Ion Channels: Keys to Neuronal Specialization
Related Reading: Neuromodulation in Invertebrate Nervous Systems; Oscillatory and Bursting Properties of Neurons; Synchronization of Neuronal Responses as a Putative Binding Mechanism

References

Bush, P., and Sejnowski, T. J., in press, Inhibition synchronizes sparsely connected cortical neurons within and between columns of realistic network models, *J. Computat. Neurosci.*

Destexhe, A., Babloyantz, A., and Sejnowski, T. J., 1993, Ionic mechanisms for intrinsic slow oscillations in thalamic relay neurons, *Biophys. J.*, 65:1538–1552.

Destexhe, A., McCormick, D. A., and Sejnowski, T. J., 1993, A model for 8–10 Hz spindling in interconnected thalamic relay and reticularis neurons, *Biophys. J.*, 65:2473–2477.

Destexhe, A., Contreras, D., Sejnowski, T. J., and Steriade, M., 1994a, A model of spindle rhythmicity in the isolated thalamic reticular nucleus, *J. Neurophysiol.*, 83:803–818.

Destexhe, A., Contreras, D., Sejnowski, T. J., and Steriade, M., 1994b, Modeling the control of reticular thalamic oscillations by neuromodulators, *NeuroReport*, 5:2217–2220.

Gray, C., 1994, Synchronous oscillations in neuoronal systems: Mechanisms and functions, *J. Computat. Neurosci.*, 1:11–38.

Koch, C., 1993, Computational approaches to cognition: The bottom-up view, *Curr. Opin. Neurobiol.*, 3:203–208.

Lytton, W. W., and Sejnowski, T. J., 1991, Simulations of cortical pyramidal neurons synchronized by inhibitory interneurons, *J. Neurophysiol.*, 66:1059–1079.

Lytton, W. W., and Sejnowski, T. J., 1992, Computer model of ethosuximide's effect on a thalamic neuron, *Ann. Neurol.*, 32:131–139.

McCormick, D. A., 1992, Neurotransmitter actions in the thalamus and cerebral cortex and their role in neuromodulation of thalamocortical activity, *Prog. Neurobiol.*, 39:337–388.

McCormick, D. A., and Huguenard, J. R., 1992, A model of the electrophysiological properties of thalamocortical relay neurons, *J. Neurophysiol.*, 68:1384–1400.

Steriade, M., Domich, L., Oakson, G., and Deschênes, M., 1987, The deafferented reticular thalamic nucleus generates spindle rhythmicity, *J. Neurophysiol.*, 57:260–273.

Steriade, M., Jones, E. G., and Llinás, R. R., 1990, *Thalamic Oscillations and Signaling*, New York: Wiley-Interscience.

Steriade, M., McCormick, D. A., and Sejnowski, T. J., 1993, Thalamocortical oscillations in the sleeping and aroused brain, *Science*, 262:679–685.

von Krosigk, M., Bal, T., and McCormick, D. A., 1993, Cellular mechanisms of a synchronized oscillation in the thalamus, *Science*, 261:361–364.

Wang, X.-J., and Rinzel, J., 1993, Spindle rhythmicity in the reticularis thalami nucleus: Synchronization among inhibitory neurons, *Neuroscience*, 53:899–904.

Wilson, M., and McNaughton, B., 1994, Reactivation of hippocampal ensemble memories during sleep, *Science*, 265:676–679.

Thalamus

David Mumford

Introduction

The thalamus is a subdivision of the brain of all mammals. It is situated at the top of the brainstem, in the middle of the inverted bowl formed by the cerebral hemispheres. It is shaped roughly like a pair of small eggs, oriented on the posterior-anterior axis and side by side, one in each hemisphere. Its cell count is approximately 2%–4% of the cortex.

The thalamus has a striking position in the flowchart of data in the brain: *essentially all input to the cortex is relayed through the thalamus.* The main exceptions are the diffuse projections of several brainstem nuclei carrying neuromodulators such as acetylcholine, but presumably not carrying detailed information-bearing signals such as those encoding sensory or motor data; the lateral olfactory tract that conveys the sense of smell to the cortex; and a connection from the amygdala to the prefrontal cortex, which duplicates a thalamic connection. The majority of the input to the cortex—visual, auditory, and somatosensory information; planning-tuning-motor output of the basal ganglia and cerebellum; and emotional-motivational output of the mammillary body—reaches the cortex exclusively through the thalamus. The thalamus is thus the principal gateway to the cortex, the cortex's window on the world at every level. What is equally striking is that the thalamocortical pathways are reciprocated by feedback pathways from the cortex back to the thalamus, forming a massive system of local loops between the thalamus and the entire cortex. For instance, Sherman and Koch (1986) estimate that in the cat, there are approximately 10^6 fibers from the *lateral geniculate nucleus* (LGN), the visual area of the thalamus, to the visual cortex but 10^7 fibers in the reverse direction. In other words, there are approximately 10 times more feedback than feedforward paths.

As knowledge of the anatomy and connections of the thalamus developed, the initial belief was that the principal function of the thalamus was simply to relay information from subcortical structures to the cortex. This view was reinforced by the simplicity of the internal circuitry of the thalamus and the apparent faithfulness of transmission shown by neurophysiological studies. Most of the cells in the thalamus are excited by subcortical input and send their output directly to the cortex, with no collaterals to other thalamic cells (see below for a more detailed description). This finding suggests that the thalamus is one of the simplest structures in the brain, and one that hardly requires modeling.

The central question concerning the thalamus is, however, what is the use of the massive feedback pathways from every area of the brain to its thalamic input nucleus? As Jones (1985:819) puts it: "Can it be that such a highly organized and numerically dense projection has virtually no functional significance? One doubts it. The very anatomical precision speaks against it. Every dorsal thalamic nucleus receives fibers back from all cortical areas to which it projects." Jones noted the central puzzle of the thalamus from a modeling perspective. The existence of these feedback pathways makes it evident that the thalamus plays an essential cognitive role, engaging in a dialogue with the cortex in which some information is being computed or combined, but what information?

Anatomy of the Thalamus

The thalamus is not a homogeneous mass of neurons, but a collection of smaller nuclei. In humans, there are approximately 50 of these nuclei in each hemisphere, with some subdivisions much clearer than others. An exhaustive survey of our knowledge of the nuclei and their connections (through 1984) is provided by Jones (1985).

Most of the nuclei are called *specific nuclei.* These nuclei connect in an ordered topological pattern, one nucleus to one area of the cortex. The topography of the projection from these nuclei to their cortical target area varies somewhat, but tends to follow what Jones calls a *rod-to-column* pattern of projection. In other words, a thalamic nucleus is divided into a family of disjoint *rods,* each of which projects to a column of cortical tissue slicing perpendicularly through the cortical plate from layer 1 to layer 6 (cf. Jones, 1985: figure 3.20, p. 126; figure 3.22, p. 129; and p. 811). This tendency has many variations, but specific thalamic relay cells seem to be constrained to synapse in specific cortical columns to preserve the relationship of their data with parallel data streams. The best known example, on which the largest amount of research has been done, is the projection of the LGN in the thalamus to cortical area V1, the primary visual area. This projection preserves the two-dimensional retinal layout of the visual image, through the LGN and onto the cortical surface, preserving in addition the separation of the signals from the two eyes onto distinct *ocular dominance columns* in area V1.

In addition to the specific nuclei, there are also *nonspecific* nuclei which project diffusely, often to the entire cortex. They play various kinds of regulatory roles and will not be discussed here. These specific and nonspecific nuclei make up the dorsal thalamus. In addition, there are several structures called the *ventral thalamus,* including the reticular thalamic nucleus (RE) and perigeniculate nucleus. These structures form a thin layer of cells covering the anterior, dorsal, and lateral surfaces of the thalamus (the perigeniculate over the LGN) through which all thalamocortical and corticothalamic fibers must pass and which sends inhibitory projections back to the thalamus. From an anatomical point of view, Crick (1984) states that, "If the thalamus is the gateway to the cortex, the RE might be described as the guardian of the gateway."

The principal cell type in the specific nuclei of the thalamus is the medium to large excitatory cells known as relay cells. They make up approximately 65%–80% of all cells. Their axons go directly to the cortex, giving off no local collaterals, except on cells in the reticular nucleus as they pass through this structure. These axons synapse principally in the cortex in layer 4 or deep layer 3, the standard input layers of the cortex. Some reports show a small group of small excitatory cells which project more diffusely, possibly to several areas of the cortex, synapsing principally in layer 1 as well as in layer 6 (Jones, 1985: 97, 158). The remaining cells are inhibitory GABAergic interneurons which provide the only intrathalamic circuitry (for cell counts, see Jones, 1985:166–167). They synapse on the relay cells and on each other. Figure 1 shows a synopsis of these circuits.

The thalamic relay cells do not *always* relay information faithfully as it comes in. In drowsiness, in non–rapid eye movement (REM) sleep, or after the administration of various laboratory preparations, these cells go into an oscillatory mode in which they alternate between short, high-frequency bursts and extended periods of hyperpolarization, repeating at a frequency of 7–14 Hz. This oscillatory mode is a key property of thalamic relay cells, but it is not clear whether it has any cogni-

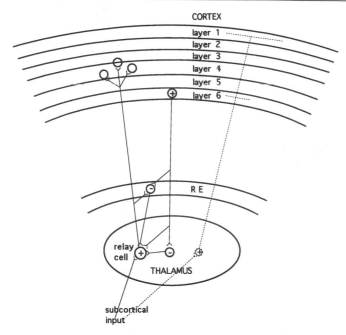

Figure 1. Simplified diagram of the neurons of the thalamus and their principal connections. The signs indicate which neurons are excitatory and which are inhibitory.

tive significance (see THALAMOCORTICAL OSCILLATIONS IN SLEEP AND WAKEFULNESS).

Gating and Selective Attention Through Feedback

Perhaps the most widespread belief about the role of feedback is that cortical feedback gates thalamic transmission of subcortical data; hence, it allows the cortex to attend to part of these data selectively. Such ideas were suggested by Singer (1977), Crick (1984), Sherman and Koch (1986), Koch (1987), and Desimone et al. (1990), among others.

To distinguish this model from others, it is useful to describe it mathematically. Consider the visual pathway from the LGN to area V1, and let $I(x, y, t)$ represent the visual signal incident on the retina as a function of coordinates on the retina and time. The ganglion cell output of the retina has been modeled as a filtered and rectified function of I: $J_\alpha = (I * F_\alpha)_+$, for a set of filters F_α, one for each class of ganglion cells. If the LGN was a faithful relay station, J_α would also model the activity K_α of the LGN relay cells. The gating hypothesis is, roughly, that instead, the relay cell activity is modeled by $K_\alpha(x, y, t) = w_\alpha(x, y, t) \cdot J_\alpha(x, y, t)$ where w_α represents weights that selectively enhance or suppress parts of the signal. The idea is that weights w_α which gate the strength of the signal as it passes through the relay cells can be set up either (1) by excitation of the LGN relay cells on the distal parts of their dendrites by direct corticothalamic feedback, possibly using N-methyl-D-aspartate (NMDA) channel mechanisms (Koch, 1987), or (2) by inhibition through an intermediate inhibitory cell in RE or LGN. These mechanisms should be contrasted with possible, much more complex, transformations of the signal $J_\alpha \to K_\alpha$ that the LGN might perform as a result of cortical feedback (possibly after several cycles of sending signals to the cortex, then back, then to the cortex again, etc.).

Evidence for such gating was discovered by Singer and Schmielau (1976) in their study of LGN relay cell responses to binocular stimuli. The LGN relay cells are, to first approxima-

tion, monocular cells. Different layers of the LGN respond to different eyes. In general, there is an interaction between LGN relay cells whose receptive fields overlap to inhibit each other, e.g., center-surround cells with adjacent receptive fields, *on* and *off* cells with the same receptive fields, and cells in the sustained versus the transient pathway inhibit each other. This observation is interpreted by Singer (1977) as a way of increasing the signal-to-noise ratio in the relaying process. When signals from opposite eyes are compared, the signals will differ by a left-right *disparity shift* whose size depends on the distance between the plane of fixation and the visible surface in that direction. What Singer and Schmielau discovered was that relay cells in laminae A and A1 of the cat LGN, responding to ispilateral and contralateral retinas, inhibit each other as noted above unless the visible surface lies in the fixation plane. In that case, their responses are enhanced. Moreover, whereas this inhibition between left and right responsive relay cells is mediated by LGN inhibitory cells, the suppression of the inhibition and its replacement by an enhancement when the visible surface is part of the fixation plane is caused by *corticofugal* signals. The effect is to highlight objects on the fixation plane and suppress nearer and farther objects whose binocular signals are out of registration. Their article contains a proposal for circuitry underlying this effect.

An influential model incorporating this idea of gating was proposed by Crick (1984). To explain the latency of low-level visual responses, Anne Treisman and others proposed that some tasks can be performed in parallel on the entire visual field, while others can be performed only in one small *window of attention* at a time. In Crick's model, these windows are created by the reticular nucleus RE suppressing the relay cells, except for those within the window of attention. He proposes that temporary cell assemblies are then created, like a buffer for the subimage "in the *searchlight*," within which further processing will occur. The mechanisms for such gating of visual signals are discussed in Sherman and Koch (1986).

Active Roles for the Thalamic Buffer

Two objections to the hypothesis described in the last section are that (a) simply suppressing or enhancing different parts of the subcortical signal would not seem to require such a massive feedback pathway, and (b) some thalamic nuclei do not receive major subcortical input, yet they still are reciprocally connected to the cortex with massive pathways.

To my knowledge, the first person to propose a more complex role for the LGN was Harth. Beginning in 1974, he developed his ALOPEX neural net theory for visual processing and especially for corticothalamic feedback. As described in Harth, Unnikrishnan, and Pandya (1987): "A model is proposed in which the feedback pathways serve to modify afferent sensory stimuli in ways that enhance and complete sensory input patterns, suppress irrelevant features, and generate quasi-sensory patterns when afferent stimulation is weak or absent." Although formulated as a time-varying gate, as in the previous section, this theory is an iterative one in which many signals traverse the thalamocortical loop, optimizing by SIMULATED ANNEALING (q.v.) an objective function which seeks to enhance remembered patterns partially or noisily present in the input. Mathematically, the key point is that if we break up the time interval within each fixation into a sequence of times $\{t_n\}$, then

$$K_\alpha(x, y, t_n) = F(J_\alpha(x, y, t_n), \dots, K_\beta(u, v, t_{n-1}), \dots)$$

where F represents the iteration of some algorithm combining retinal input with feedback. The idea that an iterative algorithm is carried out in the thalamocortical loop has received interesting experimental confirmation in the oscillations observed in Ribary et al. (1991).

Experimental evidence that the LGN is doing much more than gating comes from Murphy and Sillito (1987). They report that most relay cells in the cat LGN are *end-stopped*, i.e., they respond to moving bars of a certain length, but their response drops off markedly when the length of the bar is increased. However, if cortical feedback is removed by destruction of those visual areas projecting to the LGN, the end stopping ceases. This observation is puzzling because the cells in the cortex which project to the LGN do not show end stopping, but the result is that the feedback has strong, complex inhibitory effects. One possible interpretation is that, under cortical feedback, the LGN cells become more narrowly tuned to specific types of local features, e.g., bars of a particular length or curved bars.

To clarify the various models for active roles of the thalamus, I would like to distinguish two distinct ideas in Harth's model. One is the concept of generating completed sensory patterns from memory when the actual stimulus is noisy and incomplete. The roots of this idea go back at least as far as MacKay (1956), who proposed that feedback in general may represent a process of actively creating from memory synthetic patterns which try to match as closely as possible the current stimulus. According to this theory, the feedback signal was the pattern synthesized from memory, and the feedforward signal contained features of the stimulus or of the difference between the stimulus and the feedback (the *residual*). MacKay's ideas were developed from a Bayesian statistical perspective in the *Pattern Theory* of Grenander (1976–81, 1994). An influential neural net model of this type is the ADAPTIVE RESONANCE THEORY (ART) (q.v.) of Carpenter and Grossberg (1987). Pece (1992) developed a related theory in which thalamocortical feedback implements MacKay's ideas. He proposes that the area V1 to LGN pathway carries *negative* feedback so that after iteration, area V1 converges to a pattern of activity whose feedback closely matches the most salient visual patterns in the stimulus and then cancels the retinal traces in the LGN. We call this type of algorithm *feedback-pattern-synthesis*.

A second computational idea in Harth's theory is that the LGN is an internal sketchpad, or an *active blackboard*, on which various patterns can be written, misleading patterns can be suppressed, and a best reconstruction can be generated. Evidence for this hypothesis comes from Sillito et al. (1994). They find that cortical feedback synchronizes the firing of specific sets of LGN relay cells, namely those responding to a common feature, like parts of the same edge. This type of function is a kind of enhanced image processing, which can be considered independently of the idea of feedback-pattern-synthesis. The concept of a blackboard was first introduced in computer science in the HEARSAY speech project at Carnegie Mellon University, in the 1970s; it refers to a small shared common memory on which multiple experts can read and write, possibly combining their results by the convergence of weak pieces of evidence or one expert vetoing another if their conclusions conflict (see DISTRIBUTED ARTIFICIAL INTELLIGENCE).

In Mumford (1991, 1992, 1994), an integrated theory for the corticothalamic and corticocortical feedback loops was proposed in which the active blackboard image processing role is assigned to the thalamocortical feedback loop, while feedback-pattern-synthesis is the role assigned to the corticocortical feedback loops. The activity of each specific nucleus in the thalamus is assumed to act as a blackboard in representing the current view of the world for those areas of the cortex to which it is connected. The thalamic buffers connected to primary and secondary sensory areas and to associational and multimodal ar-

eas will contain progressively more abstract representations of some aspect of the world. Each view is based on data coming from connections to the external world through subcortical pathways and on data computed in the cortex and written on the thalamus by cortical feedback. These top-down data may be used to enhance the bottom-up signals, to reconstruct missing data, or to externalize for further processing views of the world created purely by mental imagery. For instance, the actual retinal signals are both noisy and complex, with multiple physical effects creating a highly coded, but incomplete view of three-dimensional objects and their illumination. The cortex must disentangle the effects of lighting, texture, shape, and depth. The hypothesis is that, instead of copying the image into successive buffers in a feedforward architecture as various remembered patterns are identified and used to construct the world scene behind the viewed image, small numbers of buffers in the thalamic nuclei are used to combine the reconstructions made by various cortical experts. Many cortical experts can search independently for a large variety of patterns in the image, sending them all to the thalamus, where a kind of voting takes place by summation in the dendritic arbors of the relay cells and by inhibition through the interneurons. Thus, the active thalamic blackboard can be used to decide which representation is the most successful, rejecting the weaker matches. The resulting pattern of activity is then sent back to the cortex as an enhanced view of the world.

The function of corticothalamic feedback remains a matter of speculation. It is hoped, however, that these speculations will stimulate another generation of more sophisticated experiments, with more complex stimuli, that will enable us to see further.

Road Maps: Mammalian Brain Regions; Vision
Related Reading: Electrolocation; Selective Visual Attention; Stereo Correspondence and Neural Networks; Visual Schemas in Object Recognition and Scene Analysis

References

Carpenter, G., and Grossberg, S., 1987, A massively parallel architecture for a self-organizing neural pattern recognition machine, *Comput. Vis. Graph. Image Proc.*, 37:54–115.

Crick, F., 1984, Function of the thalamic reticular complex: The searchlight hypothesis, *Proc. Natl. Acad. Sci. USA*, 81:4586–4590.

Desimone, R., Wessinger, M., Thomas, L., and Schneider, W., 1990, Attentional control of visual perception: Cortical and subcortical mechanisms, *Cold Spring Harbor Symp. Quant. Biol.*, 55:963–971.

Grenander, U., 1976–81, *Lectures in Pattern Theory I–III*, New York: Springer-Verlag.

Grenander, U., 1994, *Pattern Theory*, New York: Oxford University Press.

Harth, E., Unnikrishnan, K. P., and Pandya, A. S., 1987, The inversion of sensory processing by feedback pathways: A model of visual cognitive functions, *Science*, 1987:184–187.

Jones, E. G., 1985, *The Thalamus*, New York: Plenum. ◆

Koch, C., 1987, The action of the corticofugal pathway on sensory thalamic nuclei: A hypothesis, *Brain Res.*, 23:399–406.

MacKay, D., 1956, The epistemological problem for automata, in *Automata Studies* (C. E. Shannon and J. McCarthy, Eds.), Princeton, NJ: Princeton University Press, pp. 235–251.

Mumford, D., 1991, On the computational architecture of the neocortex, pt. I, The role of the thalamo-cortical loop, *Biol. Cybern.*, 65:135–145.

Mumford, D., 1992, On the computational architecture of the neocortex, pt. II, The role of the cortico-cortical loop, *Biol. Cybern.*, 66:241–251.

Mumford, D., 1994, Neuronal architectures for pattern-theoretic problems, in *Large Scale Neuronal Models of the Brain* (C. Koch, Ed.), Cambridge, MA: MIT Press, pp. 125–152.

Murphy, P., and Sillito, A., 1987, Corticofugal feedback influences the generation of length tuning in the visual pathway, *Nature*, 329:727–729.

Pece, A. E. C., 1992, Redundancy reduction of a Gabor representation: A possible computational role for feedback from primary visual cortex to lateral geniculate nucleus, in *Artificial Neural Nets* (I. Aleksander and J. Taylor, Eds.), Amsterdam: Elsevier Science.

Ribary, U., Ioannides, A., Singh, K., Hasson, R., Bolton, J., Lado, F., Mogilner, A., and Llinás, R., 1991, Magnetic field tomography of coherent thalamocortical 40-Hz oscillations in humans, *Proc. Natl. Acad. Sci. USA*, 88:11037–11041.

Sherman, M., and Koch, C., 1986, The control of retinogeniculate transmission in the mammalian LGN, *Exp. Brain Res.*, 63:1–20.

Sillito, A., Jones, H., Gerstein, G., and West, D., 1994, Feature-linked synchronization of thalamic relay cell firing induced by feedback from the visual cortex, *Nature*, 369:479–482.

Singer, W., 1977, Control of thalamic transmission by corticofugal and ascending reticular pathways in the visual system, *Physiol. Rev.*, 57: 386–419.

Singer, W., and Schmielau, F., 1976, The effect of reticular stimulation on binocular inhibition in the cat LGN, *Exp. Brain Res.*, 14:210–226.

Time Complexity of Learning

J. Stephen Judd

Introduction

This article delves into questions about learning, specifically the number of computing steps required to support simple associative memory in neural networks, and includes comments on mistake bounds.

Unfortunately, many of the learning algorithms reported in the literature are often very slow even for the small network sizes (tens or hundreds of nodes) used in experiments; they have all been unacceptably slow in large networks. It is clear that we need to be able to scale up our applications to much bigger networks, and so we need to understand the cause of these learning difficulties and how to deal with large sizes.

The type of learning investigated here is known as supervised learning. In this paradigm, input patterns (called *stimuli*) are presented to a machine paired with their desired output patterns (called *responses*). The object of the learning machine is to remember the associations presented during a training phase so that in a future retrieval phase the machine will be able to emit the associated response for a given stimulus.

It is assumed that the networks can change their behavior (by changing their weights); in the work reported here, this change does not involve altering the connectivity structure.

A set of (stimulus, response) *pairs*, or SRpairs, will herein be called a *task*. When a small task is drawn randomly from a large set of possible pairs, the literature usually calls it a *sample*.

An *architecture* specifies the input lines, the connectivity from each node to others, and which nodes will be network outputs. It includes all data about a circuit except what functions the nodes perform. In most of this article, we consider only feedforward networks for which a stimulus fields a unique response.

Each node in a network is designed to compute one of a certain family of *node functions*. Typical examples pass the weighted sum of inputs through a step function or sigmoid function.

A *configuration*, $F = \{f_1, f_2, \ldots, f_n\}$, of a network is an assignment of one function from the node function set to each node in the architecture, to specify what that node computes. An architecture, A, and a configuration, F, together define a mapping from the space of stimuli to the space of responses. This mapping describes how the network will behave during retrieval.

A goal of neural network learning research has been to find a "learning rule" that each network node can follow to adjust its weights, i.e., to find a configuration such that the retrieval behavior of the whole network eventually implements some desired mapping from stimuli to responses. It was hoped that a learning rule, and especially some biologically plausible learning rule, would work for any network design. Many researchers have developed candidates for such a learning algorithm, as this book attests, but much of this article reports studies where the biological nature is sacrificed, in recognition of the fact that the form of the learned representations may be as important as how they are obtained.

There are several measures of the difficulty of learning: one is the amount of time it takes to learn, another is the amount of data it takes to learn, and another is the number of mistakes that will be made during a learning process. For each of these questions, there are other issues that have to be specified: how "neural" the algorithms are, how exactly correct they need to be, how dependable they need to be, how they get their data, what they are allowed to manipulate, and how helpful the teacher is. This article deals with some questions regarding time and mistakes; see VAPNIK-CHERVONENKIS DIMENSION OF NEURAL NETWORKS for some data complexity issues.

Neural Algorithms

The original perceptron had the impact it did because its learning rule was deemed to be "biologically plausible." This attribute is still a hallmark of many learning schemes. Typically, a sample of data is collected and then repeatedly presented to the machine while it incrementally alters its hypothesis toward the correct one.

Rosenblatt (1961) and others proved a theorem stating that the various perceptron learning rules eventually converge to correct weights if such weights do exist (i.e., if the task is linearly separable). This development demonstrated that the perceptron would learn in finite time, but it gave no scaling information. The scaling issues are with respect to s, the number of input lines in the stimulus vector.

Muroga (1965) showed that there are linearly separable functions whose weights are approximately as large as 2^s. Thus, even when the function is performable, it will take the various perceptron learning rules $\Omega(2^s)$ adjustments before getting acceptable weights. Hampson and Volper (1986) extended the argument to the average case (as opposed to the worst case) and derived a bound of $\Omega(1.4^s)$. (The Ω notation is like the O notation in that it makes no claim about the constant multiplier, but whereas $O(f(n))$ says that the scaling is no worse than $f(n)$, $\Omega(f(n))$ claims that the scaling is at least as bad as $f(n)$.)

Tesauro (1987) studied time-scaling issues in some simple families of multilayered networks and measured learning time as a function of the size of the task. He used three networks of a particular style, one particular algorithm (backpropagation), and one particular function from which he draws t random SRpairs to make up a task. He then plotted learning time as a function of t and claimed it to be the sum of a polynomial and an exponential. The polynomial dominated in the low ranges but, after a certain point, the exponential dominated.

Hampson and Volper (1986) explored several algorithms and learning situations for the single perceptron to see how it behaves as the number of input lines, s, is scaled up. They report exponential times for all but a few simple cases. When the additional stimulus bits are irrelevant or redundant, or when the task being learned is an OR or AND, then algorithms exist with running times that are low polynomials in s.

Some attempts have been made to analyze the behavior of these learning algorithms in the context of composite networks. David Rumelhart, Geoffrey Hinton, and Ronald Williams showed that when the generalized delta rule is used in an arbitrary feedforward network for making weight updates, the net has a gradient-descent behavior (see PERCEPTRONS, ADALINES, AND BACKPROPAGATION). This is typical of "neural" algorithms, but because the error surface in weight space is in general multimodal, the algorithm may descend into a local minimum and thereby never discover fully correct responses.

Tesauro and Janssens (1988) report empirical results studying the learning time in a series of (network, task) pairs parameterized by a single quantity q. The net has q inputs, $2q$ nodes in the first layer (fully connected to each input), and a single output node (fully connected to each node in the first layer). The task is a complete listing of the $t = 2^q$ SRpairs for the parity function on q bits. When trained using backpropagation, they observe learning times of approximately 4^q. Since the task has size 2^q, this means the training time is $4^q/2^q = 2^q$ times the amount of data to be learned. This result might also be reinterpreted as evidence that the learning time scaled exponentially in the size of the network for *this* problem.

When we relax the requirement for the networks to be trained by neural algorithms and instead look at the problem of using *any* type of computer and *any* type of algorithm available, many other types of analysis are available; the rest of this article surveys results that are oblivious to neural desiderata.

Mistake Bounds

These results all have to do with time costs. This short section introduces the situation where mistakes are the real issue and time is not a primary consideration. (Sometimes they amount to the same thing, if, for instance, the input space is sampled randomly.) This mistake measure is quite relevant when considering the plight of animals behaving (and learning) in their environment. As long as they make appropriate actions for their situations, the animals are successful; only when they do something wrong is there any penalty for their incomplete understanding of the world.

On each of many cycles, a machine is shown only the stimulus, and it must make a prediction as to what the response is. Only after it makes its prediction does it get access to the real answer. What is the fewest number of mistakes that an algorithm could make in learning a given class of functions?

The classic perceptron training algorithm has a mistake bound inversely proportional to a polynomial in the separation between the classes. (The *separation* is the maximum distance between two parallel hyperplanes which separate the classes and have no points between them.) Unfortunately, the bound is exponential in s.

Littlestone (1987) found polynomial on-line mistake bounds for a perceptron learning a variety of classes of functions. For the case where the target function is a simple disjunction of some subset of the input bits, he gave an algorithm that makes $O(k \log s)$ mistakes, k being the size of the relevant subset. When learning k-DNF expressions (i.e., disjunctive normal forms which are the disjunction of many terms, each of which is the conjunction of no more than k Boolean literals), his algorithm has an upper bound of $O(kl \log s)$ mistakes. (l is the length of the expression learned, and s essentially measures the number of irrelevant input bits.) This is remarkable both for being linear in k and for being logarithmic in s.

Maass and Turan (1994) show that for binary input vectors of length s, there is an algorithm to learn linear threshold functions using only $O(s^2 \log s)$ mistakes. Furthermore, it runs in time polynomial in s and n, where n is the number of bits in the required weights. They also show (1993) that to learn the intersection of two half-spaces on $\{0, 1, \ldots, 2^n - 1\}^2$ (i.e., the AND of two perceptrons, each of which has just two inputs), the mistake bound is $\Omega(2^n)$.

Loading

Judd contributed a formalization of the neural network learning problem suitable for applying classical time-complexity tools of analysis. He allows an architecture and a task to be selected arbitrarily (even by an adversary), but then the machine is allowed to see all of it at once and is given a time budget scaled by the total size of those things. In complexity-theoretic form, the *loading problem* is phrased as follows:

Input: An architecture A and a task T.
Output: A configuration F for A such that every stimulus is mapped to its desired response, or a message that there is no such configuration.

This formalization also quashes the concern for biological plausibility but gets at the fundamental computational issues arising from the desire to capture the training data in the form of a neural circuit.

For the special case of architectures that are simple linear threshold devices, the loading problem can be solved in polynomial time by linear programming. (Each weight is treated as a variable, and each item in the task is translated into a linear constraint on the vector of weights. The algorithms are polynomial in the number of variables, the number of constraints, and the number of bits of precision in the constraint equations.)

For the general case, however, Judd (1990) provided a theoretical result that fits well with the empirical observations of exponentiality: Loading is *NP*-complete. This implies that no algorithm for use in arbitrary architectures can guarantee to load any given performable task in polynomial time. (Caveats, terminology, and the related complexity-theoretical concepts of *NP*-completeness are explained thoroughly in Garey and Johnson, 1979; for the basic definition, see STRUCTURAL COMPLEXITY AND DISCRETE NEURAL NETWORKS.)

Blum and Rivest (1989) found a different proof of the *NP*-completeness that differs from Judd's in that different parameters are scaled up. Judd scales up the number of nodes and the number of outputs while keeping fixed the number of SRpairs (at 3), the fan-in to individual nodes (at 3), and the number of inputs (at 2). Blum and Rivest's proof keeps the number of nodes fixed (at 3) and the number of outputs constant (at 1) while scaling up the number of SRpairs and the number of inputs.

These results provide a variety of pointers as to where the intrinsic constraints lie; the game now is to try to thread our way through these landmines and find specialized systems that will work in spite of the limits on the general case. Essentially, the goal that has been formulated is to find an algorithm that is *guaranteed* to load *any* (performable) task in *any* conceivable architecture. The *NP*-completeness theorems imply this is not always feasible. But there are several ways to constrain the problem in such a way that some special regularity in it might facilitate its solution. Such constraints would involve restrictions on architectural design, restrictions on tasks restrictions, or different criteria of success.

Shallow neural networks were defined in Judd (1990); the definition effectively limits the depth of networks while allowing the width to grow arbitrarily, and it is used as a model of neurological tissue like cortex where neurons are arranged in arrays tens of millions of neurons wide but only tens of neurons deep. One justification for the definition is that certain parts of the brain (e.g., visual cortex) are quite shallow compared with their great width, and the direction of information flow is predominantly local and unidirectional along the shallow axis, though thalamocortical and corticocortical loops (see THALAMUS) take us beyond this into the class of recurrent networks.

Most of the shallow families studied have *NP*-complete loading problems, with the exception of a family of networks which are not only shallow but "one-dimensional" as well—these networks are allowed to extend arbitrarily in width. In these families, the complexity of loading is linear in the size of the network.

All of the results just reviewed are for feedforward networks only. Often, networks with cycles in them are defined to issue their responses when the net activity stabilizes. The loading of such networks has not had as much attention, but presumably part of what a loading algorithm would need to do would be to decide if the configuration would ever stabilize. There are plenty of indications that intractability lurks in these waters too; Porat (1987) proves that, in such a system, the problem of deciding if a configured network stabilizes or cycles is *NP*-hard. Wiklicky shows that when the weights are unbounded (i.e., of infinite precision, something not physically possible), the loading problem in recurrent networks is not even decidable (see AUTOMATA AND NEURAL NETWORKS).

Average Cases and Approximations

NP-completeness is a worst-case analysis and can sometimes be too pessimistic. Average case analyses can be much more relevant (and optimistic), provided the definition of *average case* is reasonable. As mentioned earlier, Tesauro (1987) and Hampson and Volper (1986) both examine average cases for simple machines, while, as discussed in the preceding section, the worst-case complexity of loading shallow one-dimensional neural networks is linear in the size of the network. However, when the network has a huge number of units (as cortex has), even linear time might be unacceptable. (Furthermore, the algorithm that was given to achieve this time was based on a single serial processor and was biologically implausible.) Judd (1991) considered the more biological *parallel* model of processing and demonstrated an expected-time complexity that is *constant* (i.e., independent of the width of the network). This holds even when internode communication channels are short and local, thus adhering to further neurological and VLSI constraints. The expectation was over tasks.

The loading problem has a very exacting criterion of success in training: either the machine performs perfectly or it does not. If the criterion were more lenient, the problem might be much easier. Some probabilistic or approximate criterion of learning might be more appropriate.

Amaldi and Kann (1993) study the problem of doing the best possible approximation of loading (i.e., getting an optimal

number of data points correctly classified) in a simple perceptron. They show that a polynomial-time algorithm can easily get half of the optimal number correct, but no polynomial time algorithm can always find the weights that get 100% of the optimal number correct.

Loading arbitrary networks is *NP*-complete even when only 67% of the responses are required to be retrieved correctly (Judd, 1990).

PAC Learning

The loading problem requires a network to remember any particular task, even if it has arisen in some peculiar (and malicious) way. Suppose instead that the machine is to learn some function, *f*, over its *whole* stimulus space, but that some distribution, *D*, over the stimulus space is given, and that the machine is allowed to sample the value of the function at *random* points in the space. This model of data selection is relieved of the vagaries inherent in preselecting a task.

Suppose further that we weaken the fidelity required of the learner; instead of demanding that all SRpairs are recalled exactly, we ask merely that most of the answers be correct. An acceptable configuration is one that will know the right answer for a fraction $1 - \varepsilon$ of the stimulus space (weighted by *D*).

Suppose further that we weaken the dependability of the learner; instead of requiring that an acceptable configuration be found *every* time the algorithm is run, we ask merely that one be found with probability $1 - \delta$.

The criterion called "PAC-learnability" (see PAC LEARNING AND NEURAL NETWORKS) holds when there is an algorithm that runs in time polynomial in $1/\delta$ and $1/\varepsilon$ (and some size parameters) that meets these fidelity and dependability requirements.

Golea, Hancock, and Marchand (1992) show how a particular type of architecture (ones where all nodes have fan-out of 1, a definition equivalent to a system of non-overlapping perceptrons) can PAC-learn if the SRpairs are chosen from a *uniform* distribution.

Maass (1993:theorem 4.3) shows that PAC-learnability holds for the same class of architectures of fan-out 1 when outfitted with any node functions that are just piecewise linear (which includes threshold gates). It is *precision* that is the size parameter being scaled up here. The run time increases polynomially with the number of bits in the input variables and the number of bits in the weights; the number of nodes and input lines are fixed.

Maass extends this result to *any* architecture by enlarging the learning machine. For any network $\mathcal{N}$, he constructs another network $\hat{\mathcal{N}}$ that is an expansion of that network and which is of the fan-out 1 type used in the previous description, but only a polynomial expansion of it. Then he proves that any function that can be performed by $\mathcal{N}$ can be PAC-learned by $\hat{\mathcal{N}}$. The node functions employed are also the more general class of piecewise polynomial functions.

Discussion

The simple problem of associating a set of stimuli with their responses is computationally trivial if a table lookup data structure can be selected for the job. Essentially, the intractability results just discussed show that for neural networks the problem is computationally quite different: when the data structure is required to be some fixed and arbitrary architecture, this simple associative problem is intractable. To avoid this, researchers have been scurrying in many directions: narrowing the classes of functions to be learned, relaxing the criteria for success in learning, improving the quality of teaching, searching for average cases or probable performance, tinkering with new algorithms, making more creative definitions of neural network classes, and constantly delineating what can be achieved under these new conditions. (One activity that was not covered in this article is the attempt to develop algorithms that will construct the architecture as well as specify weights for it. Such studies will eventually connect the much older field of circuit design to the neural network community.)

The scale-up problem will not be solved without a deeper understanding of the issues and tradeoffs involved in learning. As time passes, definitions will get sharper and more focused on exactly those classes of functions that are learnable with reasonable neural networks, and it will be clearer what our engineering ambitions should rightly be.

Road Map: Computability and Complexity
Background: I.3. Dynamics and Adaptation in Neural Networks
Related Reading: Learning and Generalization: Theoretical Bounds; Parallel Computational Models

References

Amaldi, E., and Kann, V., 1993, *The Complexity and Approximability of Finding Maximal Feasible Subsystems of Linear Relations*, Technical Report ORWP 93/11, Lausanne: Department of Mathematics, EPFL.

Blum, A., and Rivest, R. L., 1989, Training a 3-node neural net is *NP*-complete, in *Advances in Neural Information Processing Systems 1* (D. S. Touretzky, Ed.), San Mateo, CA: Morgan Kaufmann, pp. 494–501.

Garey, M. R., and Johnson, D. S., 1979, *Computers and Intractability: A Guide to the Theory of NP-Completeness*, San Francisco: Freeman.

Golea, M., Hancock, T., and Marchand, M., 1992, On learning mu-perceptron networks with binary weights, in *Neural Information Processing Systems 6*.

Hampson, S. E., and Volper, D. J., 1986, Linear function neurons: Structure and training, *Biol. Cybern.*, 53:203–217.

Judd, J. S., 1990, *Neural Network Design and the Complexity of Learning*, Cambridge, MA: MIT Press. ◆

Judd, J. S., 1991, Constant time loading of shallow 1-dimensional networks, in *Neural Information Processing Systems 4*, pp. 863–870.

Littlestone, N., 1987, Learning quickly when irrelevant attributes abound: A new linear-threshold algorithm, in *28th Symposium on Foundations of Computer Science*, IEEE, pp. 68–77.

Maass, W., 1993, Bounds for the computational power and learning complexity of analog neural nets, in *Proceedings of the 25th ACM Symposium on the Theory of Computing*.

Maass, W., and Turan, G., 1993, Algorithms and lower bounds for on-line learning of geometrical concepts, *Machine Learn.* (in press).

Maass, W., and Turan, G., 1994, How fast can a threshold gate learn?, in *Computational Learning Theory and Natural Learning Systems: Constraints and Prospects* (D. Hanson and R. Rivest, Eds.), Cambridge, MA: MIT Press (in press).

Muroga, S., 1965, Lower bounds of the number of threshold functions and a maximum weight, *Trans. Electron. Comput.*, 14:136–148.

Porat, S., 1987, *Stability and Looping in Connectionist Models with Asymmetric Weights*, Technical Report TR 210, Rochester, NY: Computer Science Deptartment, University of Rochester.

Rosenblatt, F., 1961, *Principles of Neurodynamics: Perceptrons and the Theory of Brain Mechanisms*, Washington, DC: Spartan.

Tesauro, G., 1987, Scaling relationships in back-propagation learning: Dependence on training set size, *Complex Systems*, 1:367–372.

Tesauro, G., and Janssens, R., 1988, Scaling relationships in back-propagation learning: Dependence on predicate order, *Complex Systems*, 2:39–44.

Time Perception: Problems of Representation and Processing

Ernst Pöppel and Kerstin Schill

Introduction

"Time perception" is unique. Time must be conceived of as being reconstructed on the basis of events perceived. Thus the question is: How are events made available by brain processes? To approach this question, we believe that it is necessary (a) to look at the spatial mode of functional representation, and (b) to provide a catalog of elementary temporal experiences that have to be explained under the topic *time perception*.

Anatomical studies indicate the segregation of elementary functions, i.e., their modular representation in the brain (e.g., Leise, 1990). However, theoretical considerations suggest that each mental act is characterized by simultaneous neuronal activity in different brain areas. Take, for instance, spontaneous speech: To speak a sentence requires that lexical, syntactical, semantic, phonetic, and prosodic competences are all operative (Pöppel, 1988, 1994). Because of the suspected distributed mode of representation of mental acts, the brain has to deal with the following logistical problem: How are distributed activities related to each other to represent integrated processes? (Compare the problem of BINDING IN THE VISUAL SYSTEM.)

With respect to the catalog comprising human time perception, the following elementary temporal experiences can be distinguished. The most basic phenomenon is the experience of *simultaneity* versus *nonsimultaneity* of stimuli. Nonsimultaneity is a necessary but not sufficient condition for the experience of temporal order or *successiveness* of stimuli. If different stimuli can be related to each other with respect to the before-after relationship, they are defined here as representing discrete primordial *events*. Successive events are integrated up to approximately two to three seconds to set up an operational *temporal window*. This temporal window can also be referred to as the *subjective present*. The elementary temporal experience of *duration* is based on the amount of information processed or events perceived within basic integration units of a few seconds. Subjective time within this hierarchical taxonomy of elementary temporal experiences (Pöppel, 1988) is analyzed by looking at and looking for neuronal mechanisms of information processing.

Prerequisites for the Representation of Temporal Structure

What are necessary prerequisites for the neural representation of the temporal structure of the environment? This is an essential question for understanding MOTION PERCEPTION (q.v.) or in general the perception of dynamic vision. Current approaches for the modeling of dynamic vision are dominated by the view that sensory data are processed in a continuous mode and that the internal representation is nothing but a processed version of the spatiotemporal input modified according to a "continuously" operating algorithm. We will call this the *collinear* concept of the relation between the external spatiotemporal structure and its internal representation. This collinear concept is a property of all "filter" schemas of temporal processing (for review, see Watson, 1986), in which the internal stream of information is seen as a modified (filtered) version of the external stream of data.

The transient/sustained dichotomy, for example, suggests a processing of spatial information by the temporal low-pass filtering sustained system and a detection of temporal changes by the high-pass filtering transient system (Watson, 1986). A more refined concept, the "optic flow" paradigm (see MOTION PERCEPTION), assumes the internal representation of a vector field which assigns, at a given instant of time, a velocity vector to each spatial point of an image. The computation of this vector can be based on a comparison of the input from two points in time and space (Reichardt, 1957) or on higher-order approximations (e.g., Arbib, 1989, Section 7.2; Zetzsche and Barth, 1991). Common to these approaches is that the internal signals change in accordance with the variation of external signals, and that at any chosen instant of external time, a collinearly related instant of time of the input sequence is available internally.

In contrast to these *collinear* concepts, Schill and Zetzsche (in press) have suggested a spatiotemporal memory model based on what they call an *orthogonal* representation of temporal information. Their concept assumes that at any given instant of external time, features belonging to a past sequence are internally available—i.e., there exists an internal representation of time which is orthogonal to the physical time. The orthogonal concept is not merely a metaphor of representation; it allows experimental predictions like those on "iconic memory," and it resolves inconsistencies arising within the collinear view.

The orthogonal representation can be achieved by the mapping of time into simultaneously available, spatially distributed properties (such as the instantaneous spike rates of an array of neurons). A consequence of the mapping of time into space is the resulting discreteness of the internal *representation* of time —i.e., internal time is spatially "quantized."

There are two basic approaches to implementing the spatiotemporal model in neural or technical hardware: as a shift register or as an ordered or cascaded delay-line structure. Both approaches lead to a similar type of representation of spatiotemporal information, but they have to be differentiated with respect to their modes of processing. The shift-register solution can be thought of as a stack of frames, each frame representing the spatial input pattern at a certain moment in time. In this case, a kind of clock has to be assumed which initiates the periodic shifts, resulting in a discrete mode of processing. In contrast to this, the delay-line solutions imply a continuous mode of information processing (cf. Uttley, 1954).

Simultaneity, Nonsimultaneity, and Succession

The previous section suggests that the processing of sensory data is of a discrete nature. This hypothesis is supported by many experimental observations of hierarchically related elementary temporal experiences. To be perceived as nonsimultaneous, two discrete physical stimuli must be separated by a minimal temporal interval. This interval (the *fusion threshold*) is different for the different sensory systems. For example, two auditory stimuli have to be separated by 1 to 2 ms to be heard as nonsimultaneous, but visual or tactile stimuli separated by

such intervals are still perceived as simultaneous. The auditory system has the best, the visual system the worst temporal acuity.

Temporal resolving power must be broken down into at least two measures. One is the interval that separates two stimuli so that they are perceived as nonsimultaneous rather than simultaneous. The second measure refers to the phenomenon of successiveness—i.e., whether the temporal order of stimuli has to be indicated.

In such a situation, these stimuli must be distinct so that they can be temporally labeled differently. Stimuli must be separated in the 40-Hz domain when their order has to be indicated. This order threshold value seems to be the same for different sensory modalities (vision, hearing, touch). Order thresholds reflect a time-organizing system that is independent of peripheral sensory mechanisms. If two stimuli can be put into a temporal order, it is necessary that they be defined as independent entities. We call these basic entities *primordial events*. They provide the raw material for further cognitive processing.

Oscillations as an Organizational Principle

A system which operates in a discrete mode for both representation and processing of time requires a specialized organizing program. Time-organizing programs have been suggested under various headings, such as biological clocks, subjective time quanta, processing units, perceptual moments, central oscillations, excitability cycles, and system states. The existence of such system states should be reflected in approximately the same numerical value for their duration if different experimental paradigms are used. This is in fact the case. Through many studies, values close to 30 ms have been suggested (Pöppel, 1978, 1994). For instance, histograms of simple or choice reaction time gathered under strictly stationary conditions often are multimodal, with a 30–40 ms temporal interval between adjacent modes. Similar observations have been made for oculomotor behavior; histograms for the latencies of pursuit eye movements show intermodal distances of 30–40 ms. Thus, the initiation of a movement with respect to a stimulus is of a noncontinuous nature.

These phenomena (order thresholds, multimodal response histograms for reaction times or latencies of eye movements, etc.) can be explained on the basis of excitability cycles. Suprathreshold stimuli instantaneously entrain or resynchronize a neuronal relaxation oscillation. Each period of this oscillation hypothetically sets up the temporal frame for a singular system state and, thus, a primordial event as defined here. Studies that report oscillations in single-cell activity (e.g., Gray et al., 1989) are not related to issues discussed here; in those studies (see SYNCHRONIZATION OF NEURONAL RESPONSES AS A PUTATIVE BINDING MECHANISM), synchronized activities of oscillatory responses of spatially distributed cells are interpreted as coding binding (i.e., time is used only as a representational medium).

Sequencing of Events

The discreteness of events is a necessary but not sufficient condition for time tags being related to such events. An additional neuronal mechanism has to be assumed which delivers *time tags*. On the basis of neuropsychological observations, special mechanisms of the left prefrontal cortex may be related to time tagging and, thus, sequencing. In general, patients with left hemispheric lesions seem to do worse than controls in reproducing sequences of stimuli. The dissociation of memory for events and the times at which the event happened is typical for patients suffering Korsakoff psychosis. These and other observations suggest that the time tags defining the position of events in a sequence can be selectively destroyed, pointing to a separate neuronal mechanism.

The experiments on order threshold only allow us to infer a mechanism that codes the before-after relationship of *two* events. Such binary order information which stresses the discreteness of information processing may, however, be used as input by a further mechanism to set up longer sequences. Modeling the storage and recall of sequential information with neural network approaches also seems to require input sequences with a discrete nature in the time domain (Bradski, Carpenter, and Gross, 1992).

Automatic Temporal Binding: The Subjective Present

Given a sequence of events, one has to ask how a phenomenal continuity on the basis of separate events is possible. Is such temporal linking context-driven, i.e., dependent on what is processed? Is it determined by the sequence of time tags, or is temporal linking presemantic? A large set of experimental observations indicates that the linking mechanism of primordial events is presemantic and automatic. A temporal binding process, which is independent of what is processed, links successive information up to approximately 3 seconds (Pöppel, 1994). The elementary temporal experience that is mediated by this temporal integration can be referred to as the *subjective present*.

What is the experimental basis for assuming a temporally limited integration? Evidence comes from many different areas. If, for instance, the duration of temporal intervals has to be reproduced, subjects estimate these intervals veridically up to approximately 2 to 3 s and underestimate longer intervals considerably. These and other results imply that information can be stored without loss concerning its intensity for intervals up to 2 to 3 s.

Further support for the hypothesis of an integration with a limit of 2 to 3 s comes from psycholinguistic and psychomotor research. Spontaneous speech in different languages is apparently organized in such a way that verbal utterances usually last 2 to 3 s. And human behavior appears to be structured in such a way that short movement episodes (like intentional hand or arm movements) are embedded within 3-second windows. This observation is related to the temporal limits of anticipation: Regularly appearing stimuli can be anticipated by controlled movements if the interstimulus interval is not longer than approximately 3 s; if this interval gets longer, one is no longer able to program a temporally precise movement.

The 3-second integration provides temporal windows for conscious activity (Pöppel, 1988). How is the succession of these discrete temporal windows linked together on a hierarchically higher level? How does the brain produce subjective continuity, the flow of time, if representation is discontinuous? It is suggested that continuity is made possible by semantic binding of what is represented in successive 3-second temporal windows (i.e., the contents of what is represented are linked). The observation that continuity can break down, as in the case of some schizophrenic states or under intoxication, implies that under normal circumstances a specific neuronal program is responsible for the effectiveness of semantic binding.

Duration Estimation

The subjective flow of time is sometimes slowed down, sometimes accelerated. *Duration estimation* refers to the ability to indicate how much time has elapsed, usually by referring to conventional time units (like seconds or minutes). The subjective flow of time is influenced by the number of events represented and processed. For the estimation of time on the basis of events, special mechanisms of integration or summation have to be assumed. This summation process appears to be the basis for duration estimation. It is suggested that duration estimation is based on system states with a period of approximately 30 ms and presemantic temporal integration processes of approximately 3 s duration. Not every system state may deliver an event; thus, a situation devoid of interesting information results in a small number of events per integration period. A summation process, being dependent on a variable number of events and a rather fixed integration period, allows duration estimation without an external clock.

This selective summation process that underlies duration estimation has to be distinguished from a mechanism that averages temporal information. For this mechanism, which seems to be operative for longer time intervals (many seconds to minutes), it is assumed that processed information generates a personal reference system of subjective time. To set up such a reference system by learning, clock time may be useful—i.e., one can refer to intervals with conventional time units. Duration estimates in this case are made by comparing the storage size produced by the events within a particular interval with a subjective frame of reference gradually developed by averaging the durations of different time intervals experienced in the past. The input to this temporal reference system can be conceived of being provided by the temporal summation process referred to earlier. Duration estimates of longer intervals appear, thus, to be dependent on two independent mechanisms which are hierarchically related: a selective summation process and a temporal reference system individually set up by averaging temporal information.

Discussion

The prerequisites for the representation of temporal structure suggest an orthogonal mapping of time into space, which implies a spatially "discrete" representation of time. Furthermore, the discussion of elementary temporal experiences shows that temporally discrete system states are observed at different levels of information processing.

The ecological usefulness of temporally discrete processing of information by using successive system states becomes plausible if one considers the problem of intermodal integration. Because of different transduction times of the sensory systems—the visual system being the slowest—the central availability of stimuli characterizing one object or one event through different modalities will practically never be simultaneous. In addition, physical distance of an object is a critical variable for the integration of acoustic and optic data because of the enormous difference between sound and light velocity; identifying an object with varying distances both visually and aurally thus requires temporal tolerance zones.

Potential "temporal diplopia" is even true for a single channel: the central availability of different regions of an object characterized by different brightness can differ by tens of ms, because transduction time is flux-dependent. An *atemporal* system state within which all neuronal information is treated as *cotemporal* can theoretically overcome the temporal unpredictability of sensory information, allowing binding operations by special neuronal programs throughout distributed brain areas (Pöppel, 1994). The problem mentioned here refers also to a question of guidance control in autonomous robots, because unequivocal trajectories have to be programmed on the basis of sensory input which is unpredictable in time. It is suggested that the atemporal system states that define temporal tolerance zones are also the kernels of primordial events.

The alternative mode to discrete temporal processing, obviously, is continuous information processing. One possible way to solve the problem of different central availability of stimuli in a continuous mode of processing is to assume that the generation of discrete categories occurs on a "higher" level, thus allowing continuous processing at a lower level. However, the generation of discrete categories as higher-level system states would necessarily introduce a semantic domain; centrally available schemas would be checked in a top-down manner against afferent information, and objects would be perceived if the neuronal information supported the schema.

The assumption of a top-down control is also made by Arbib (1989). He suggests a strategy for an economical description of events over time characterized by a slide-box metaphor. Slides correspond to the different entities perceived within the current scene. Depending on how long entities are relevant to ongoing behavior, they may be maintained for any period of time; but during this period, various parameters of different slides may be adjusted to relate them to the changing situation of the observer in the environment. Such a top-down control is complemented by the data-driven addition and deletion of slides to meet the "real time" requirements of temporal information processing (cf. the notion of *schema assemblage* discussed in Schema Theory).

We want to stress that it is essential, in modeling time perception, to distinguish between the formal operational modes (the "syntax" of temporal processing) which are reflected in the different elementary experiences and the information processed within these operational modes (the "semantics" implemented at the different hierarchically related levels of processing). The syntax is reflected in different neuronal processes that give rise to the catalog of elementary temporal experiences, and this syntax has to be thought of as being presemantic. It provides a hierarchical network that allows a bottom-up processing of sensory information in a predetermined temporal structure.

Road Map: Other Sensory Systems
Related Reading: Spatiotemporal Association in Neural Networks; Temporal Pattern Processing

References

Arbib, M. A., 1989, *The Metaphorical Brain 2: Neural Networks and Beyond*, New York: Wiley.
Bradski, G., Carpenter, G. A., and Grossberg, S., 1992, Working memory networks for learning temporal order with application to 3-D visual object recognition, *Neural Comp.*, 4:270–286.
Gray, C. M., König, P., Engel, A. K., and Singer, W., 1989, Oscillatory responses in cat visual cortex exhibit inter-columnar synchronization which reflects global stimulus properties, *Nature*, 338:334–337.
Leise, E. M., 1990, Modular constructions of nervous systems: A basic principle of design for invertebrates and vertebrates, *Brain. Res. Rev.*, 15:1–23.
Pöppel, E., 1978, Time perception, in *Handbook of Sensory Physiology* (R. Held, H. Leibowitz, and H.-L. Teuber, Eds.), Berlin: Springer-Verlag, pp. 713–729.
Pöppel, E., 1988, *Mindworks: Time and Conscious Experience*, Boston: Harcourt Brace Jovanovich. ◆
Pöppel, E., 1994, Temporal mechanisms in perception, *Int. Rev. Neurobiol.*, 37:185–202. ◆
Reichardt, W., 1957, Autokorrelations-Auswertung als Funktionsprinzip des Zentralnervensystems, *Z. Naturforsch.*, 12b:448–457.
Schill, K., and Zetzsche, C., in press, A model of visual spatiotemporal memory: The icon revisited, *Psychol. Res.* ◆
Uttley, A. M., 1954, The classification of signals in the nervous system, *Electroencephalogr. Clin. Neurophysiol.*, 6:479–494.
Watson, A., 1986, Temporal sensitivity, in *Handbook of Perception and Human Performance* (K. R. Boff, L. Kaufman, and J. P. Thomas, Eds.),vol. 1, *Sensory Processes and Perception*, New York: Wiley, chap. 6.
Zetzsche, C., and Barth, E., 1991, Direct detection of flow discontinuities by 3D curvature operators, *Pattern Recognition Lett.*, 12:771–779.

Topology-Modifying Neural Network Algorithms

Timur Ash and Garrison Cottrell

Introduction

Neural network performance is fundamentally tied to the topology of the network. The capacity and accuracy of a network mapping is determined by the number of free parameters (typically weights) in the network. Networks that are too small cannot accurately approximate the desired input-to-output mapping. Networks that are too large do not generalize well and require longer training times. It is therefore desirable for best performance to find the smallest network that will "fit" the training data.

This article reviews algorithms which adjust the network's topology, as well as other parameters, to arrive at a compact network representation. We emphasize those algorithms which have been most often applied or which are representative of a broader class of specific methods. The intent is to convey their general flavor and unique mechanisms. A more complete survey with extensive references is available in Ash and Cottrell (1994).

Taxonomy

Algorithms that modify network topologies can be categorized into two broad approaches: growing (*constructive*) and pruning (*destructive*). Growing approaches start with networks that are too small to solve a problem and add elements during the training process. Growing methods offer a natural starting point (the smallest initial network possible) in the search for an architecture. Growing methods also benefit from initially working with smaller networks, thus minimizing training expense on serial hardware. However, the rate of new element growth needs to be controlled properly, or the network may grow without bound. Growing methods can also result in a wide variance of final network sizes, based on the control parameters.

Pruning methods start with a large network initially and trim it until the error becomes unacceptable. These methods have the advantage that the proper mapping can easily be learned by the network (i.e., there is little chance that the network is too simple to model the mapping). Pruning approaches also offer a natural stopping criterion. If the further-pruned network cannot be retrained to the required error tolerances, the last valid architecture is kept. There are some shortcomings to pruning approaches. Since the majority of the training time is spent with a network that is larger than necessary, these methods are computationally wasteful on serial machines. Since the pruning approach starts with a large network, it may get stuck in one of the intermediate-sized solutions, because of the shape of the error surface, and never find the smallest.

Growing Algorithms

Growing algorithms start with a small network and add weights or nodes to their architectures until the problem has been solved. The following subsections will examine three key growing algorithms in detail.

Dynamic Node Creation (DNC)

There are two basic ways to construct feedforward networks that grow additional nodes or weights dynamically. Either one can allow retraining of existing network weights when new elements are added, or one can freeze their values. The Dynamic Node Creation (DNC) method (Ash, 1989) adds fully connected nodes to the hidden layer of a three-layer backpropagation (BP) network during training and allows all weights to be retrained. No direct weights between the input and output layers are used.

Although DNC does not freeze nodes after adding them to the network, the already-trained weights will tend to remain near their original values, so a newly added node will generally have to find some smaller component of the error that is not yet covered. A similar process can be seen in projection pursuit regression (Hwang et al., 1992; Zhao and Atkeson, 1992) where the largest remaining component of the error is modeled.

The network starts with a single hidden node. Regular BP training takes place until the desired mapping is learned or until another hidden node needs to be added to the network. The procedure is repeated until user-specified stopping criteria (based on acceptable-average and worst-case error cutoffs) are met.

A sliding time history window of width w is kept for the network's average output node error (E). The drop in error since the creation of the last hidden node is calculated. If the error has not decreased quickly enough over the width of the history window (based on a user defined *trigger slope* parameter Δ_T), a new node is added to the hidden layer. That is, a new node can be added after w epochs if

$$\frac{E_t - E_{t-w}}{E_{t_*}} < \Delta_T$$

where t represents the current training epoch, and t_* is the epoch after which the last hidden node was added.

This method has been tested on the encoder problem, symmetry, parity, and binary addition with carry. In *every* run, a solution to the problem at hand was discovered. Although not every run resulted in a minimal topology, most resulted in networks that were within two nodes of the known best solutions. Computational expense for this method is competitive (in terms of total arithmetic operations required) with learning using BP networks that have the "correct" number of hidden nodes (DNC was 44% more efficient on the problems tested). Since expense grows geometrically with increasing network size, DNC benefits by initially working with smaller networks.

It is important to understand the effect of training the smaller networks on the way to finding the larger solution. This can be evaluated by considering how many epochs (complete presentations of the data set) the DNC network spends with its largest architecture before the mapping is learned. In every case tested, DNC networks spent less time training at the maximum size than their BP counterparts. The initial training in lower-dimensional space helps DNC find a solution.

Cascade Correlation

The Cascade Correlation (CC) learning architecture (Fahlman and Lebiere, 1990) has proven to be one of the more robust and useful methods for growing network architectures. The system begins with direct connections from inputs to outputs. New hidden nodes are added one at a time. Each new node's inputs are connected to all of the original inputs, as well as all previously created intermediate nodes.

Learning in the network proceeds in two phases. In the first, the existing network is trained using *quickprop* (a fast variation on BP learning). When the error has not been reduced significantly after a number of training cycles (controlled by a user-set "patience" parameter), a new node is added to the network. If the current error is low enough, the algorithm terminates.

The new node comes from a group of candidate nodes that are trained to maximize their output's correlation with the residual errors on the network outputs. The one that maximizes this correlation is chosen and inserted into the network. At this stage, the node has learned a feature that correlates highly with the residual error, and its input weights are frozen. Only network weights that connect to output nodes are then retrained (again using quickprop) in order to incorporate the contribution of the new intermediate node.

The goal of the candidate node training is to maximize S, the magnitude of the correlation between the candidate node's value and the residual output error observed at node o, summed over all output nodes. Thus, S is defined as

$$S = \sum_o \sum_p (V_p - \overline{V})(E_{p,o} - \overline{E}_o)$$

where o and p index the output nodes and patterns, respectively, V is the output of the candidate node, and E is the error.

In order to maximize S, $\partial S/\partial w_i$ is computed (the partial derivative of S with respect to each of the candidate node's incoming weights w_i). The expanded and differentiated formula for S is

$$\partial S/\partial w_i = \sum_{p,o} \sigma_o (E_{p,o} - \overline{E}_o) f'_p I_{i,p}$$

where σ_o is the sign of the correlation between the candidate's value and the output o, f'_p is the derivative for pattern p of the candidate node's activation function with respect to its inputs, and $I_{i,p}$ is the input the candidate node receives from node i for pattern p.

The CC technique was tried on the " two-spirals" problem and the n-input parity problem. The two-spirals problem involves the classification of a point in a plane as belonging to one of two interlocking concentric spirals. This problem is known to be difficult for BP to solve. In both cases, CC consistently builds networks that were able to solve the problems. However, the performance of CC on the two-spirals problem is not as good as that of the projection pursuit network described in Hwang et al. (1993; this reference also discusses limitations of CC for applications where smooth function interpolation is desired).

Meiosis (A Local Growing Method)

Hanson's Meiosis networks (Hanson, 1990) use local rules for splitting nodes when a mapping cannot be achieved with the current network. This purpose is achieved by letting the network error drive the splitting process. Nodes that receive more error split into two nodes. The splitting is done by using stochastic weights. The weights are drawn from a normal distribution on every feedforward pass through the network:

$$P(w_{ij} = w_{ij}^*) = N(\mu_{w_{ij}}, \sigma_{w_{ij}})$$

Network learning is accomplished by modifying the mean μ_{w_i} and standard deviation σ_{w_i} for each weight distribution based on the error gradient using the sampled w^*. The mean is modified using

$$\mu_{w_{ij}}(n+1) = \alpha\left(-\frac{\partial E}{\partial w_{ij}^*}\right) + \mu_{w_{ij}}(n)$$

The variances are updated similarly. However, instead of using the signed gradient, its absolute value is used. This has the effect of increasing variance whenever there is *any* error:

$$\sigma_{w_{ij}}(n+1) = \varsigma\left(\beta\left|-\frac{\partial E}{\partial w_{ij}^*}\right| + \sigma_{w_{ij}}(n)\right)$$

The decay parameter (ς) is strictly less than 1, and it guarantees that as the network settles on a solution, its behavior will become deterministic. For higher damping rates (smaller values of ς), the system becomes deterministic quickly. Lower damping rates allow the system to jump around the solution space more before settling. Thus, this *stochastic delta rule* implements a local, adaptive simulated annealing.

A Meiosis network starts with random initial mean and variance values. Initially, the network only has one hidden node. Meiosis training then proceeds as follows:

1. A forward stochastic pass is made to produce an output.
2. The output is compared to the target, and the errors are used to update the weight distribution means and variances.
3. The variance-to-mean ratios are computed for each hidden node's input and output weights. If these ratios are greater than a threshold, the node will be split.
4. Those that are candidates for Meiosis are split. For each new node, weights are assigned half the variance of the old nodes and a new mean consisting of a "jittered" value centered at the old mean.

New node creation stops based on the network's error and stage of the annealing process.

Meiosis seems to be an effective method for the problems tried. It yields reasonable results and is not computationally expensive. Since node splitting is a local phenomenon, it can be implemented on parallel hardware. The work could easily be extended to more hidden layers.

Although the computation actually took a little less time than standard BP on the small parity problems examined, simulated annealing is typically slower on larger problems. Also, the author notes a sensitivity to the value of the decay parameter selected. Low values do not allow any node splitting, while high ones result in continuous splitting without regard to the problem being modeled. The value used for the example problems was handpicked.

Pruning Algorithms

Pruning methods fall into two general classes. One group of methods continuously seeks to bias the learning of the network during training in order to minimize its size. These *on-line* methods usually add other criteria to the goal of minimizing the error, such as reducing weight magnitudes or minimizing the variance of their distributions. Examples of on-line methods are the various *weight decay* techniques that add a term to the error function being minimized that penalizes weights in certain magnitude ranges. The objective (cost) function being minimized in such approaches is composed of two terms: an error term and a bias term that is used to keep weights small:

$$C = E + B$$

In essence, weight decay methods seek to drive weight values to insignificance (near zero) for later removal. Since the error and magnitude changes are computed locally for each weight, these methods can be implemented on parallel hardware. One of the earliest versions of such methods was proposed by Rumelhart (1987) and subsequently examined in Weigend, Rumelhart, and Huberman (1991) (see next subsection). Hanson and Pratt (1989) examined similar approaches.

Other pruning approaches are *posttraining* methods that measure a weight's importance and remove the least important weights. Many methods can be used to determine the importance of the particular weights in the network. Typically the resulting networks need to be further trained to fine-tune their performance after weight removal.

Rumelhart's Weight Decay (An On-Line Pruning Method)

David Rumelhart, in a research talk at the University of California, San Diego, in 1987, proposed a method for decaying network weights by adding a weight size penalty term to the error term being minimized and terms to keep weight distributions small by preferring incoming weights to be the same and outgoing weights to be the same. A simplification of this approach has subsequently been applied by Weigend et al. (1991). The function being minimized is composed of two main terms:

$$C = \sum_{i,p}(t_{ip} - o_{ip})^2 + \lambda\sum_{i,j}^{w}\frac{w_{ij}^2/w_0^2}{1 + w_{ij}^2/w_0^2}$$

The first term is the standard sum of squared errors over the set of training examples. The second term is the penalty term for each weight's magnitude. The variable w_0 is a scaling term (typically set to 1 for the standard zero-to-one sigmoid activation range). The variable λ is a weighting term describing the relative importance of the error-versus-weight sizes. In practice, λ is set to zero until the error is minimized and then slowly increased until performance declines.

The method was tried on a time-series prediction task. The weight elimination method reduced the network from the eight initial hidden nodes to three, with performance comparable to the best eight-hidden-node network found without weight decay. However, the network required four times as much training time as the regular BP network. No comparison was made to a BP network trained from scratch with three hidden nodes. The weight decay network performance was significantly better than the threshold auto-regressive model to which it was compared. The weight decay factor (λ) described in the method essentially had to be hand-tuned for best performance. A fixed training time-out was the only stopping criterion used. Whether implementation of this method can be automated is still an open question.

Optimal Brain Surgeon

LeCun, Denker, and Solla (1990) propose a method called Optimal Brain Damage (OBD). This is a posttraining method that removes selected weights in order, based on a notion of *saliency*. The OBD method calculates an approximation to the actual effect on error of deleting each weight from the network, to derive a measure of saliency. This is to be contrasted with Rumelhart's approach which simply penalizes small weights (a heuristic assumption). The effect on the global error of removing a weight from a network can be approximated by a Taylor series. OBD makes a series of simplifying assumptions (including the calculation of only diagonal terms in the Hessian matrix of error with respect to the weights) in order to rank-order all weights by saliency.

The Optimal Brain Surgeon (OBS) method (Hassibi and Stork, 1993) improves on OBD by calculating the off-diagonal terms of the Hessian to determine saliency more accurately. These terms turn out to be significant in all test problems, resulting in different weights being removed. In addition, OBD only deletes weights, while OBS also adjusts for a weight's deletion by properly adjusting the values of all remaining weights.

The procedure works as follows:

1. Train a "reasonably large" network to minimum error.
2. Compute $\mathbf{H}^{-1}$ (the inverse Hessian for the weights).
3. Find the weight (indexed by q) that gives the smallest saliency (rise in network error after removing the weight):

$$L_q = \frac{w_q^2}{2[\mathbf{H}^{-1}]_{qq}}$$

If this candidate error increase is smaller than a user-chosen cutoff, then delete the candidate weight and proceed to step 4; otherwise proceed to step 5.
4. Update all remaining weights in the network according to

$$\delta w = -\frac{w_q}{[\mathbf{H}^{-1}]_{qq}} \mathbf{H}^{-1} \cdot \mathbf{e}_q$$

where $\mathbf{e}_q$ is the unit vector (in weight space which corresponds to the scalar weight w_q).
5. Continue until no more weights can be deleted without a large increase in network error. (At this point it may be desirable to retrain the network and repeat the procedure.)

OBS yields excellent results on the test problems examined, often resulting in better performance than existing solutions with significantly fewer final weights. For example, an OBS-pruned network for the NETtalk problem resulted in better generalization with 1560 weights (compared to 18,000 weights in the original network).

Summary

This article provides a taxonomy of neural network methods that modify network topologies. One clear result of this survey is the need for a set of benchmark problems and consistent reporting procedures. Without actually implementing these algorithms and running them on the same test suites, quantitative comparisons are difficult. However, it is still possible to draw some conclusions about the effectiveness of the differing approaches.

In general, the most effective growing methods use a greedy (global) approach to minimizing error (i.e., the largest sources of error are minimized first). Methods like DNC (Ash, 1989) have shown improved training times and consistently small network sizes with good convergence. Cascade Correlation (Fahlman and Lebiere, 1990) and later work on projection pursuit learning networks (Hwang et al., 1993) are robust methods capable of modeling complex mappings consistently. Local growing methods such as Meiosis (Hanson, 1990) also work well, but should be tested on larger problems.

Among the pruning methods, on-line approaches such as weight decay do not provide consistent results. The addition of a bias term to the error function being minimized seems to slow down training or prevent convergence of the network (Hanson and Pratt, 1989). In addition, the relative weighting of the bias versus the error-minimization term, as well as selection of a proper weight decay constant, are problematic, often requiring manual tuning (Weigend et al., 1991). Off-line pruning methods such as Optimal Brain Damage (LeCun et al., 1990) show promise. Optimal Brain Surgeon (Hassibi and Stork, 1993), enhanced by the use of more complete second-derivative information, appears to be the best pruning method currently.

Road Map: Learning in Artificial Neural Networks, Deterministic
Background: I.3. Dynamics and Adaptation in Neural Networks
Related Reading: Neurosmithing: Improving Neural Network Learning

References

Ash, T., 1989, Dynamic node creation in backpropagation networks, *Connection Sci.*, 1:365–375.

Ash, T., and Cottrell, G., 1994, *A Review of Learning Algorithms That Modify Network Topologies*, Technical Report CS94-348, San Diego: Department of Computer Science and Engineering, University of California at San Diego, March. ◆

Fahlman, S. E., and Lebiere, C., 1990, The Cascade-Correlation learning architecture, in *Advances in Neural Information Processing Systems 2* (D. S. Touretzky, Ed.), San Mateo, CA: Morgan Kaufmann, pp. 524–532.

Hanson, S. J., 1990, Meiosis networks, in *Advances in Neural Information Processing Systems 2* (D. S. Touretzky, Ed.), San Mateo, CA: Morgan Kaufmann, pp. 533–541.

Hanson, S. J., and Pratt, L. Y., 1989, Comparing biases for minimal network construction with back-propagation, in *Advances in Neural Information Processing Systems 1* (D. S. Touretzky, Ed.), Palo Alto, CA: Morgan Kaufmann.

Hassibi, B., and Stork, D. G., 1993, Second order derivatives for network pruning: Optimal Brain Surgeon, in *Advances in Neural Information Processing Systems 5* (S. J. Hanson, J. D. Cowan, and C. L. Giles, Eds.), San Mateo, CA: Morgan Kaufmann. ◆

Hwang, J.-N., Li, H., Maechler, M., Martin, R. D., and Schimert, J., 1992, A comparison of projection pursuit and neural network regression modeling, in *Advances in Neural Information Processing Systems 4* (J. E. Moody, S. J. Hanson, and R. P. Lippmann, Eds.), San Mateo, CA: Morgan Kaufmann, pp. 1159–1166.

Hwang, J.-N., You, S.-S., Lay, S.-R., and Jou, I-C., 1993, What's wrong with a Cascaded Correlation learning network: A projection pursuit learning perspective, in *Neuroprose* computer archive (ftp archive.cis.ohio-state.edu, cd pub/neuroprose/hwang.cclppl.ps.Z), Ohio State University, 1993. ◆

LeCun, Y., Denker, J. S., and Solla, S. A., 1990, Optimal Brain Damage, in *Advances in Neural Information Processing Systems 2* (D. S. Touretzky, Ed.), San Mateo, CA: Morgan Kaufmann, pp. 598–605.

Weigend, A. S., Rumelhart, D. E., and Huberman, B. A., 1991, Back-propagation, weight-elimination and time series prediction, in *Connectionist Models: Proceedings of the 1990 Summer School* (D. S. Touretzky, Ed.), San Mateo, CA: Morgan Kaufmann, pp. 105–116.

Zhao, Y., and Atkeson, C. G., 1992, Some approximation properties of projection pursuit learning networks, in *Advances in Neural Information Processing Systems 4* (J. E. Moody, S. H. Hanson, and R. P. Lippmann, Eds.), San Mateo, CA: Morgan Kaufmann, pp. 936–943.

Traveling Activity Waves

John Milton, Trevor Mundel, Uwe an der Heiden, Jean-Paul Spire, and Jack Cowan

Introduction

Waves of neural activity traveling with velocities of 10–90 cm/sec have been recorded with microelectrodes and electrode grids from hippocampal slices (Miles, Traub, and Wong, 1988) and with velocities of 20–30 cm/sec from the cortical surface (Rosenblueth and Cannon, 1942). Rotating waves have also been observed (Shevelev et al., 1992). The organization of large masses of neurons into synchronized waves of activity lies at the basis of phenomena such as the electroencephalogram (EEG) and evoked potentials.

One mechanism for the spread of neural activity involves the diffusion of K^+ ions. This produces waves of spreading depression that travel at much lower velocities than those cited earlier, i.e., at approximately 1–3 mm/min (see WAVE PROPAGATION IN CARDIAC MUSCLE AND IN NERVE NETWORKS). Here we focus on activity waves which spread by nerve conduction. First we define the possible states of neurons in a network. Next, we briefly review wave phenomena in neural networks. In particular, we draw attention to the important role played by the relative refractory state of a neuron in shaping the spatiotemporal dynamics. Following that, we demonstrate how a relative refractory state can be incorporated into a continuum model for a neural network, and we extend this formulation to a network model. One example of an interesting wave phenomenon which can arise in such networks—a migrating "traveling" wave pattern—is also mentioned.

The Relative Refractory State

Consider a slab or slice of tissue with neural packing density $\rho(\underline{x})$, where ρ is the number of neurons per unit volume of tissue. At any time t, neurons are in one of three states: quiescent (sensitive), activated, or relative refractory. The duration Δ_a of the active state is commonly referred to as the "absolute refractory period" and lasts about 1–3 ms. The absolute refractory period is followed by a period of relative refractoriness of approximately 5–200 ms, denoted by Δ_r, during which the neuron can be made to fire; however, a higher input is required than when the neuron is at rest. This relative refractory period arises because of the interplay of two processes: (1) the rapid decay of the threshold to its resting value (complete within 3–5 ms); and (2) the slower decay of the membrane hyperpolarization to its resting potential (complete within 60–200 ms).

All neurons in the central nervous system (CNS) exhibit a relative refractory state. It can be anticipated that this will have a nontrivial influence on network dynamics since the threshold depends on the time elapsed since the neuron last fired, and hence on the recent history of the neuron.

Wave Phenomena

Early computational studies of excitable networks drew attention to the role of two factors for producing traveling waves of excitation: (1) the interconnectivity of network elements, and (2) the recovery dynamics of the elements after excitation (Beurle, 1956, 1962; Farley, 1965). In neural tissue, it is most natural to focus on the interconnectivity of elements with fixed and uniform relative refractory properties. Here we briefly discuss the wave phenomena which arise in neural networks in which the probability of connectivity, $P(r)$, is an exponentially decreasing function of interneuronal distance, r, i.e.,

$$P(r) \approx \beta \exp\left[-\frac{|r|}{\sigma} \right] \tag{1}$$

where β and σ are positive constants (Sholl, 1956). The relationship between neural connectivity and propagation velocity has recently become a topic of active interest (Chu, Milton, and Cowan, 1994; Ermentrout and McLeod, 1993; Idiart and Abbott, 1993).

One-Dimensional Lines

Most models of wave phenomena in neural networks neglect the relative refractory state. An early example is the Wilson-Cowan model describing wave propagation in a one-dimensional (1D) line composed of excitatory and inhibitory neurons (Amari, 1977; Wilson and Cowan, 1973). When the inhibition is weak, a propagating wave can be initiated by a brief narrow pulse that gives rise to a pair of waves traveling in opposite directions from the region of stimulation. Figure 1 shows such an effect. Wave pairs are generated once per cycle as long as the stimulus pulse $P(x, t)$ persists, and each wave travels away from the locus of stimulation without attenuation. It follows that a very brief stimulus generates a single wave pair, whereas a stimulus of longer duration generates a succession of such pairs. The propagation velocity is a function of the connectivity parameters β and σ, whereas the wavelength depends on P. For $\sigma = 50\ \mu m$, the propagation velocity is 4 cm/s.

Two-Dimensional Excitatory Slabs

The first authors to draw attention to the importance of a relative refractory effect in shaping the spatiotemporal dynamics of

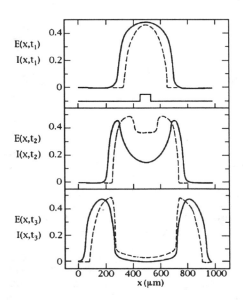

Figure 1. Generation of traveling wave pairs. $E(x, t)$ (solid lines), $I(x, t)$ (dashed lines). The region initially stimulated is indicated in the top graph. $t_1 = 20$ ms, $t_2 = 30$ ms, $t_3 = 30$ ms. (Redrawn from Wilson and Cowan, 1973.)

neural networks were Beurle and Farley and Clark (for reviews, see Beurle, 1962; Farley, 1965; Chu et al., 1994; Milton, Chu, and Cowan, 1993). The first analytical formulation was given by Wilson and Cowan (1972). Beurle and Farley and Clark's early simulations composed of excitatory integrate-and-fire neurons (see SINGLE-CELL MODELS) subject to point stimulation demonstrated the close relationship between the compactness of interneuronal connectivity and the time course of the relative refractory state in shaping wave properties.

When the network is more "loosely connected" and the recovery dynamics are adjusted so that refiring in the refractory trough becomes possible, then it is possible for the whole network to undergo synchronous bulk oscillations (Farley, 1965). In some cases, these bulk oscillations can continue; in others, they cease spontaneously. In addition to these large-amplitude bulk oscillations, migrating localized oscillatory bursts of activity can occur. As the network becomes tighter and the relative refractory state more prolonged, the network produces both traveling circular and spiral waves (Beurle, 1962; Chu et al., 1994; Farley, 1965; Milton et al., 1993). Figure 2 shows, for example, a traveling spiral wave generated in response to prolonged point stimulation. The importance of the relative refractory period is emphasized by the fact that spiral waves do not arise in such models when this period is neglected.

For physiologically plausible choices of the parameters (i.e., a σ of approximately 250 μm), these traveling waves have velocities of 20–25 cm/s with wavelengths of 1–4 mm. The probability that a spiral occurs in such networks increases with the tightness of the connectivity (Chu et al., 1994; Farley, 1965). Such networks have been shown to possess three stable states (Milton et al., 1993): a state in which all neurons are at rest, and two self-maintaining states—one associated with spirals and another with disorganized spatial patterns. It has been suggested that the spiral waves arise through an exchange of stability in a manner similar to a subcritical Hopf bifurcation (for more on bifurcations, see DYNAMICS AND BIFURCATION OF NEURAL NETWORKS). Moreover, in such networks, spirals may arise from certain initial conditions and not others. Finally, spirals do not appear to arise from central point stimulation if $P(r)$ is not a monotone decreasing function (Chu et al., 1994).

In 1D and 2D networks composed of both excitatory and inhibitory neurons, the wave properties depend on the relative spatial extent of the two neural populations (see for example, an der Heiden, 1979; Wilson and Cowan, 1973). In general, standing wave patterns predominate when the spatial range of

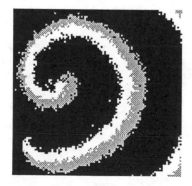

Figure 2. Generation of a traveling spiral in response to prolonged point stimulation. Gray depicts quiescent; black, relative refractory; and white, active.

inhibition exceeds that of excitation, whereas traveling wave phenomena arise in the converse case.

Neural Continuum Models

Our ultimate goal is to derive a network equation that includes a relative refractory state as a first step toward understanding the dynamical behavior shown in Figure 2. We can derive such a model by making discrete the appropriate continuum model. In this section, we describe such a continuum model (more details can be found in Wilson and Cowan, 1972).

Consider a slab or slice of tissue with neural packing density $\rho(\mathbf{x})$, where ρ is the number of neurons per unit volume of tissue, and $\mathbf{x}$ is the position coordinate. At any time t, neurons are in one of three states: quiescent (sensitive), activated, or relative refractory. Let the relative proportions of neurons in active, quiescent, and relative refractory states in such a sheet or slab be denoted, respectively, by $q(\mathbf{x}, t)$, $a(\mathbf{x}, t)$ and $r(\mathbf{x}, t)$, so that $q + a + r = 1$. Let $e(\mathbf{x}, t)\, dt$ be the proportion of neurons activated at the point $\mathbf{x}$, in the interval dt, and $r(\mathbf{x}, t, t')\, dt'$ be the proportion of neurons which were activated during the interval $(t', t' + dt')$ which are still relative refractory. Then

$$q = 1 - \int_{t-\Delta_a}^{t} e(\mathbf{x}, t')\, dt' - \int_{t-\Delta_r}^{t} r(\mathbf{x}, t, t')\, dt'$$

It can be shown that the proportion of neurons in such a sheet or slab which are activated at the point $\mathbf{x}$, in the interval dt, is given by

$$e(\mathbf{x}, t + \Delta)\, dt = \left[1 - \int_{t-\Delta_a}^{t} e(\mathbf{x}, t')\, dt' \right.$$
$$\left. - \int_{t-\Delta_r}^{t} r(\mathbf{x}, t, t')\, dt' \right] \varphi[v - \theta_q]\, dt$$
$$+ \int_{t-\Delta_r}^{t} \varphi[v - \theta(t - t')] r(\mathbf{x}, t, t')\, dt'\, dt \quad (2)$$

where Δ is the synaptic delay, θ_q is the resting threshold, $\theta(t - t')$ is the relative refractory threshold at the time $t - t'$ after the last neural firing at the point $\mathbf{x}$, and $\varphi[v]$ is the proportion of neurons receiving at least threshold excitation per unit time, expressed as a function of the mean voltage built up at the point $\mathbf{x}$ relative to the threshold there. This voltage can itself be written in terms of the activity of neurons coupled to those at the point $\mathbf{x}$ via the weighting function $w(\mathbf{x} - \mathbf{x}')$, the strength or weight of connections at the point $\mathbf{x}$ from all neurons a distance $|\mathbf{x} - \mathbf{x}'|$ away. Finally, the proportion $r(\mathbf{x}, t, t')\, dt'$ can itself be expressed in terms of a differential equation (Wilson and Cowan, 1972)

$$\frac{dr(\mathbf{x}, t, t')}{dt} = -\varphi[v - \theta(t - t')] r(\mathbf{x}, t, t') \quad (3)$$

subject to the initial condition $r(\mathbf{x}, t, t) = e(\mathbf{x}, t)$, the solution of which is:

$$r(\mathbf{x}, t, t') = e(\mathbf{x}, t') \exp\left[-\int_{t'}^{t} \varphi[v(s) - \theta(s - t')]\, ds \right]$$

Given that the membrane time constant $\tau \approx 3$–5 ms, it is clear that any fluctuations faster than approximately 100 Hz will be filtered out by neural membrane impedances. Thus it is appropriate to define new variables, $\hat{y}$, as

$$\hat{y}(t) = \frac{1}{\tau} \int_{-\infty}^{t} \exp\left[-\frac{(t-s)}{\tau} \right] y(\mathbf{x}, s)\, ds \approx \frac{1}{\tau} \int_{t-\tau}^{t} y(\mathbf{x}, s)\, ds$$

Thus the "time coarse-grained" evolution equation is:

$$\tau \frac{dE(\mathbf{x}, t)}{dt} = -E(\mathbf{x}, t) + \left[1 - \Delta_a - \int_{t-\Delta_r}^{t} r(\mathbf{x}, t, t') \, dt' \right]$$
$$+ \int_{t-\Delta_r}^{t} \varphi[v - \theta(t - t')] r(\mathbf{x}, t, t') \, dt' \quad (4)$$

where $E(\mathbf{x}, t) = (1/\tau) \int_{-\infty}^{t} \exp[-(t - t')/t] e(\mathbf{x}, t') \, dt'$ is a time coarse-grained version of $e(\mathbf{x}, t)$. Equations 3 and 4, together with the voltage equation, are the fundamental ones which govern the origin and propagation of fronts and waves in a slab of tissue.

Network Equations

The network formulation can be obtained from the continuum model described in the preceding section by using the relation

$$\rho(\mathbf{x}) = \sum_{j=1}^{N} \delta(\mathbf{x} - \mathbf{x}_j) \quad (5)$$

where $\delta(\mathbf{x})$ is the multidimensional Dirac delta function. If we rewrite Equation 4 in integral form, neglect the relative refractory effect, and make use of Equation 5 we obtain the equation

$$\tau \frac{dE_i(t)}{dt} = -E_i(t) + [1 - \Delta_a E_i(t)] \varphi[v_i - \theta_q] \quad (6)$$

where $E_i(t)$ is the fraction of time in a long time interval, during which the ith neuron is active, i.e., the mean *firing rate*, and where $v_i(t)$, its membrane potential, is given as

$$v_i(t) = \sum_{j=1}^{N} w_{ij} E_j \left(t - \frac{|i - j|}{v} \right) \quad (7)$$

plus any external stimulus that may exist.

In going from Equation 4 to a network equation incorporating relative refractoriness, it is not reasonable to time coarse-grain over the relative refractory period, since many of the important dynamical effects occur on comparable time scales (i.e., most EEG rhythms of clinical relevance are in the frequency range of 1–40 Hz). Thus, in contrast to the absolute refractory period, a relative refractory period has a major influence on network dynamics. However, reducing Equation 4 directly to a discrete network form is not straightforward. The difficulty arises because of the form of $\theta(t - t')$, which depends on the time elapsed since the neuron last fired, and hence on the history of the neuron.

A mathematically simpler way to do this is to assume that the neural membrane potential depends on one (or more) recovery or state variables, w^k, $k = 1, \ldots, n$. We can then incorporate a relative refractory state into the dynamics of a neuron by modifying Equation 7 (Stein et al., 1974), e.g.:

$$v_i(t) = \sum_{j=1}^{N} w_{ij} E_j \left(t - \frac{|i - j|}{v} \right) - w_i(t)$$
$$\dot{w}_i(t) = -\lambda w_i(t) + \mu v_i(t) \quad (8)$$

where λ and μ are constants.

Neural Activity Waves

Numerical simulations of networks described by Equation 8 show that it is sufficient to capture the dynamics of the integrate-and-fire models described earlier when $k = 1$. More complex dynamics, such as bursting patterns, become possible when $k \geq 2$.

As is clear from inspection of Equation 8, networks comprising neurons with relative refractory as well as absolute refractory states necessarily evolve dynamically in a higher-dimensional space. A single first-order ODE (ordinary differential equation) has a 0-dimensional invariant set, a point, corresponding to a stationary state, and a system of two first-order ODEs can have, in addition, a one-dimensional invariant set, a closed contour in its 2D phase space, corresponding to a cycle. Similarly, a system of three first-order ODEs can have a two-dimensional invariant set, a surface, corresponding to almost periodic motion or chaos, and so on. It follows from this that a network of excitatory cells without refractoriness can only switch on or off locally; thus, only propagating fronts are to be expected in such a network. However, a network of excitatory cells with relative refractoriness can cycle locally, so propagating waves and spirals can be supported. Similar phenomena are to be expected in networks comprising both excitatory and inhibitory cells without relative refractoriness, since inhibition can itself be thought of as a kind of nonlocal refractoriness. In more complex networks comprising both excitatory and inhibitory cells with relative refractoriness, the local dynamics are four-dimensional. Thus, both oscillatory and chaotic behavior are to be expected, either stationary or traveling, depending on the relative spatial ranges of excitation and inhibition. One example which we have observed is a migrating standing wave pattern that resembles a moving checkerboard (Mundel, Cowan, and Milton, in preparation).

Discussion

Since neurons are excitable cells, it can be anticipated that some of the wave phenomena described for excitable media with reaction-diffusion–type kinetics (see, for example, WAVE PROPAGATION IN CARDIAC MUSCLE AND IN NERVE NETWORKS) will have their counterparts in the dynamics of neural networks. However, neural networks possess a number of features which distinguish them from the simpler excitable media of chemical and cardiac systems. Most importantly, (a) neural networks contain both inhibitory and excitatory connections; (b) interneuronal connectivity extends beyond nearest neighbors; (c) there can be significant delays in the conduction of activity between two neurons; and (d) neurons have both absolute and relative refractory periods. However, a major limiting factor to progress in the study of neural waves is the scarcity of experimental methods with sufficient spatiotemporal resolution. From elementary considerations, the minimum spatial and temporal resolution would be, respectively, approximately 0.5–1 mm and approximately 0.08–2.5 ms. Continued study of the properties of these networks as well as improved monitoring techniques will be required before the functional significance, if any, of standing, traveling, and rotating neural activity waves becomes resolved.

Acknowledgments.   This work is supported in part by NATO, the National Institutes of Mental Health, and the Brain Research Foundation of the University of Chicago.

Road Map: Biological Networks
Background: I.1. Introducing the Neuron
Related Reading: EEG Analysis; Epilepsy: Network Models of Generation; Layered Computation in Neural Networks

References

Amari, S.-I., 1977, Dynamics of pattern formation in lateral-inhibition type neural fields, *Biol. Cybern.*, 27:77–87.
Beurle, R. L., 1956, Properties of a mass of cells capable of regenerating pulses, *Philos. Trans. R. Soc. Lond. B Biol. Sci.*, 669:55–94.

Beurle, R. L., 1962, Functional organization in random networks, in *Principles of Self-Organization* (H. v. Foerster and G. W. Zopf, Eds.), New York: Pergamon, pp. 291–314.

Chu, P. H., Milton, J. G., and Cowan, J. D., 1994, Connectivity and the dynamics of integrate-and-fire networks, *Int. J. Bifur. Chaos*, 4:237–243.

Ermentrout, G. B., and McLeod, J. B., 1993, Existence and uniqueness of travelling waves for a neural network, *Proc. R. Soc. Edinb. Sect. A (Math. Phys. Sci.)*, 123:461–478. ◆

Farley, B. G., 1965, A neuronal network model and the "slow potentials" of electro-physiology, *Comput. Biomed. Res.*, 2:265–294.

an der Heiden, U., 1979, *Analysis of Neural Networks*, New York: Springer-Verlag.

Idiart, M. A. P., and Abbott, L. F., 1993, Propagation of excitation in neural network models, *Network*, 4:285–294.

Miles, R., Traub, R. D., and Wong, R. K. S., 1988, Spread of synchronous firing in longitudinal slices from the CA3 region of the hippocampus, *J. Neurophysiol.*, 60:1481–1496.

Milton, J. G., Chu, P. H., and Cowan, J. D., 1993, Spiral waves in integrate-and-fire neural networks, in *Advances in Neural Information Processing Systems 5* (S. J. Hanson, J. D. Cowan, and C. L. Giles, Eds.), San Mateo, CA: Morgan Kaufmann, pp. 1001–1007.

Rosenblueth, A., and Cannon, W. B., 1942, Cortical responses to electrical stimulation, *Am. J. Physiol.*, 135:690–741.

Shevelev, I. A., Tsicalov, E. N., Gorbach, A. M., Budko, K. P., and Sharaev, G. A., 1992, Temperature tomography of the brain cortex: Thermoencephalography, *J. Neurosci. Methods*, 46:49–57.

Sholl, D. A., 1956, *The Organization of the Cerebral Cortex*, London: Methuen.

Stein, R. B., Leung, K. V., Mangeron, D., and Oguztoreli, M. N., 1974, Improved neuronal models for studying neural networks, *Kybernetik*, 15:1–9.

Wilson, H. R., and Cowan, J. D., 1972, Excitatory and inhibitory interactions in localized populations of model neurons, *Biophys. J.*, 12:1–24.

Wilson, H. R., and Cowan, J. D., 1973, A theory of the functional dynamics of cortical and thalamic nervous tissue, *Kybernetik*, 13:55–79.

Unsupervised Learning with Global Objective Functions

Suzanna Becker

Introduction

In this article, we review three types of neural network learning procedures which can be considered *unsupervised*: information-preserving algorithms, density estimation techniques, and invariance-based learning procedures. This decomposition does not imply three strictly nonoverlapping classes, but rather is meant to emphasize the different underlying principles that motivated each algorithm's development. We will use the term *unsupervised* to refer to those algorithms for which there is no externally derived teaching signal informing the network as to whether or not it has produced the correct response for each input pattern. Invariably, though, for each unsupervised learning procedure there is an implicit *internally derived* training signal; this training signal may be based on the network's ability to predict its own input, or on some more general measure of the quality of its internal representation.

Global Objective Functions or Synaptic Learning Rules?

Since our concern is with unsupervised learning in *networks* and their global behavior, we will focus on algorithms based upon globally defined objective functions, rather than synaptic learning rules. By performing gradient descent in a global objective function we can reduce a global algorithm into synaptic-level steps (weight changes), but the converse is not necessarily true; i.e., a given synaptic learning rule may not correspond to the derivative of any global objective function. Many advantages are afforded by the "global approach." It allows us to understand the operation of the network in an information-processing sense, i.e., in terms of what sort of transformation the network applies to the input; such an understanding can be elusive if we begin with a synaptic learning rule and then try to predict its global behavior. The global approach also adheres to the principles of good algorithm design well-known to the computer scientist: we start with a conceptual specification of what the learning is meant to accomplish; this is translated into a computational specification—the objective function, which is then refined into detailed computational steps—the synaptic learning rules. This top-down approach allows us to explore different implementations of the same learning algorithm, such as batch versus on-line versions. Finally, the global objective function provides a quantitative measure of the success of the learning procedure, and we can (usually) detect its convergence.

In contrast to this top-down approach, the earliest computational models of learning were based on Hebb's synaptic learning principle; Hebb postulated that synaptic efficacy should increase whenever two pre- and postsynaptic neurons are coactive. Many computational models have built upon this principle (see POST-HEBBIAN LEARNING RULES). It has also gained popularity among neurobiologists as a plausible candidate for a cortical synaptic learning mechanism. It is therefore of interest to computational modelers to try to translate their global learning procedures into local, biologically plausible learning rules such as Hebbian learning.

Self-Organization in Perceptual Systems

One of the major motivations for studying unsupervised learning is to discover the general computational principles underlying brain self-organization. Evidence of experience-dependent plasticity has been reported in a wide variety of brain areas. Perhaps the most startling evidence comes from a series of studies by Sur and colleagues (reviewed in Sur, 1989), who found that when primary visual cortical input pathways are artificially rerouted to the auditory cortex in ferrets, the "auditory" cortical cells develop responses to visual stimuli and exhibit typical visual cortical receptive fields. According to Asanuma (1991:217), "The long-held belief that the cortical representation of the sensory periphery is hard wired in adults has become less and less tenable." It seems that the brain has a dynamic restructuring capacity which is not restricted to primary sensory areas (see, e.g., SOMATOTOPY: PLASTICITY OF SENSORY MAPS), but may be a ubiquitous property of the adult neocortex (Asanuma, 1991). This possibility raises a number of questions: Are there any general, unsupervised organizing principles which predict cortical reorganization, and can they be expressed computationally, as global objective functions for learning? Is more than one such principle required? What architectural constraints are necessary for successful learning, and how do they interact with the choice of objective function? It is these sorts of questions that unsupervised learning research is concerned with.

Information-Preserving Algorithms

Since there is no external teaching signal for unsupervised learning, the goal of the learning must be stated solely in terms of some transformation on the input which will preserve the interesting structure. The first task then is to define what constitutes interesting structure. The most general possible goal is to try to preserve *all* of the information by simply memorizing the input patterns. Pattern-associators (see ASSOCIATIVE NETWORKS) can be used as such by operating in autoassociative mode, i.e., by storing each input pattern associated with itself. All of these models suffer capacity limitations: only a limited number of patterns can be stored and perfectly recalled by a network of fixed size.

Minimizing Reconstruction Error

Given the limited ability of networks to store a set of patterns exactly, a better strategy might be to try to find a *compressed* representation of the patterns. This may be helpful for preprocessing noisy data and for modeling early stages of perceptual processing. A standard data compression technique is principal component analysis (PCA) (see PRINCIPAL COMPONENT ANALYSIS). Several learning procedures (reviewed in Becker and Plumbley, in press) have been developed which converge to the first N principal directions of the input distribution. These methods are optimal with respect to minimizing the mean squared reconstruction error for linear networks. However, there is no guarantee that a linear method like PCA will capture the interesting structure in arbitrary input distributions.

A more general method for finding a compressed representation that minimizes reconstruction error is to use a nonlinear backpropagation network as an auto-encoder (Hinton, 1989) by making the desired states of the N output units identical to the states of the N input units on each case. Data compression can be achieved by making the number of hidden units $M < N$. Further, the features discovered by the hidden units may be useful for subsequent stages of processing such as classification. However, with complicated input patterns containing multiple features, it may not be possible to relate the activities of individual hidden units to specific features. One way to constrain the hidden-unit representation is to add extra penalty terms to the objective function. For example, Saund (1989) added a constraint that caused hidden units to represent high-dimensional data as single points on a lower-dimensional constraint surface, by penalizing activation patterns that deviated from unimodal distributions. This technique encourages units to represent a single scalar dimension that best characterizes the input. Zemel and Hinton (see MINIMUM DESCRIPTION LENGTH ANALYSIS) generalized this idea by imposing an MDL-based penalty term on hidden-unit activities.

Direct Minimization of Information Loss

Another approach to ensuring that the important information in the input is preserved in the output is to use concepts from information theory. Many learning procedures have been proposed which minimize the information loss in a network, subject to processing constraints (reviewed in Becker and Plumbley, in press). The common feature of these methods is the preservation of mutual information (Shannon, 1948)

between the input vector **x** and output vector **y**:

$$I_{x;y} = H(\mathbf{x}) - H(\mathbf{x}|\mathbf{y}) \tag{1}$$

where $H(\mathbf{x}) = -\int_x p(\mathbf{x}) \log p(\mathbf{x}) \, dx$ is the entropy of random variable x with probability distribution $p(x)$, and $H(\mathbf{x}|\mathbf{y}) = -\int_{x,y} p(\mathbf{x},\mathbf{y}) \log p(\mathbf{x}|\mathbf{y}) \, dx \, dy$ is the entropy of the conditional distribution of **x** given **y**. This measure tells us the amount of information (uncertainty) in **x** less the uncertainty remaining in **x** when **y** is known. Thus, $I_{x;y}$ is high when **x** is difficult to predict a priori but becomes much easier to predict after being told **y**.

If the network is free of processing noise and has enough units, its output layer can convey all the information contained in the input simply by copying the input. Linsker (1988) proposed applying the *Infomax principle* in the presence of Gaussian processing noise at the output layer for linear networks; when the input distribution is Gaussian, the information is

$$I = 0.5 \log\left(\frac{|\mathbf{Q^y}|}{V(n)}\right) \tag{2}$$

where $|\mathbf{Q^y}|$ is the determinant of the covariance matrix of the output vector **y** (the signal plus noise), and $V(n)$ is the noise variance. Maximizing this quantity results in a tradeoff between maximizing the variances of the outputs, and decorrelating them, depending on the noise level. For a single output unit, this leads to a simple Hebb-like learning rule.

An alternative optimality criterion proposed by Barlow (1989) is to find a minimally *redundant* encoding of the sensory input vector into an n-element feature vector, which should facilitate subsequent learning. If the n features are statistically independent, then the formation of new associations with some event V (assuming the features are also approximately independent conditioned on V) only requires knowledge of the conditional probabilities of V given each feature y_i, rather than complete knowledge of the probabilities of events given each of the 2^m possible sensory inputs. Barlow proposes that one could achieve featural independence by finding a *minimum entropy encoding*: an invertible code which minimizes the sum of the feature entropies (see VISUAL CODING, REDUNDANCY, AND "FEATURE DETECTION").

Several approximate solutions to Barlow's model in the linear case are reviewed by Becker (1991). The nonlinear case is of course much more difficult to learn, requiring a much stronger result of statistically independent, rather than just decorrelated, outputs. In general this is an intractable problem; that is, to verify the statistical independence of n items requires the enumeration of on the order of n^n statistics. Thus, tractable approximations to this objective function are needed.

Density Estimation Techniques

Rather than trying to retain all the information in the input, we could try to characterize its underlying probability distribution by developing a more abstract representation. Many standard statistical methods fall under the category of density estimation techniques (for a good introduction, see Silverman, 1986), and several unsupervised learning procedures can be viewed in this way. The general approach is to assume a priori a class of models which constrains the general form of the probability density function, then search for the particular model parameters defining the density function most likely to have generated the observed data. This can be cast as an unsupervised learning problem by treating the network weights as the model parameters, and the overall function can be computed by the network as being directly related to the density function.

Mixture Models and Competitive Learning

One possible choice of prior model is a mixture of Gaussians. The prior assumption in this case is that each data point was actually generated by one of n Gaussians having different means μ_i, variances σ_i^2, and prior probabilities π_i. Fixing the model parameters μ_i, σ_i, and π_i, we can compute the probability of a given data point $\mathbf{x}$ under a mixture-of-Gaussians model as follows:

$$p(\mathbf{x}|\{\mu_i\},\{\sigma_i\},\{\pi_i\}) = \sum_{i=1}^{n} \pi_i P_i(\mathbf{x}, \mu_i, \sigma_i) \qquad (3)$$

where $P_i(\mathbf{x}, \mu_i, \sigma_i)$ is the probability of $\mathbf{x}$ under the ith Gaussian. Applying Bayes's rule, we can also compute the probability that any one of the Gaussians generated the data point $\mathbf{x}$:

$$p(i|\mathbf{x},\{\mu_j\},\{\sigma_j\},\{\pi_j\}) = \frac{\pi_i P_i(\mathbf{x}, \mu_i, \sigma_i)}{\sum_{j=1}^{n} \pi_j P_j(\mathbf{x}, \mu_j, \sigma_j)} \qquad (4)$$

Given these probabilities, we can now use as a cost function the log likelihood of the data given the model:

$$\log(L) = \sum_x \log(p(\mathbf{x}|\{\mu_i\},\{\sigma_i\},\{\pi_i\})) \qquad (5)$$

By maximizing this function, we can approximate the true probability distribution of the data, given our prior model assumptions. Note that by taking the log of L, we obtain a cost function which is a sum (rather than a product) of probabilities for each input pattern. The model parameters can then be adapted by performing gradient ascent in $\log(L)$. The Expectation Maximization (EM) algorithm (Dempster, Laird, and Rubin, 1977) alternately applies Equation 3 (the expectation step) and adapts the model parameters (the maximization step) to converge on the maximum likelihood mixture model of the data.

COMPETITIVE LEARNING (q.v.) procedures perform a discrete approximation to density estimation. The general idea is that units compete to respond (e.g., by a winner-take-all activation function or lateral inhibition), so that only the winning unit in each competitive cluster is active. Only this unit learns on each case, by moving its weight vector closer to the current input pattern. Hence, each unit minimizes the squared distance between its weight vector and the patterns nearest to it, as in standard k-means clustering. This version of competitive learning is closely related to fitting a mixture-of-Gaussians model with equal priors π_i and equal fixed variances σ_i^2. Using the EM algorithm, every unit (not just the winner) moves its mean closer to the current input vector, in proportion to the probability that its Gaussian model accounts for the current input (Equation 4). Competitive learning approximates this step by making a binary decision as to which unit accounts for the input. Thus, the same learning rule applies, except that the proportional weighting is replaced by an all-or-none decision.

Nowlan (1991) proposed a "soft competitive learning" model for neural networks. Rather than only allowing the winner to adapt, each unit adapts its weights for every input case, in proportion to how strongly it responds on a given case. This is an on-line version of the EM algorithm for Gaussian densities with equal priors, and adaptive means and variances. Nowlan found this method to be superior to the traditional "hard competitive learning models" on several classification tasks.

Combinatorial Representations

A major limitation of mixture models and competitive learning is that they employ a 1-of-n encoding, in which a single unit or model is assumed to have generated the data. A *multiple-causes* model is more appropriate when the most compact data description consists of several independent parameters (e.g., color, shape, size). Several examples of this approach are reviewed in Becker and Plumbley (in press). For example, Neal's (1992) multilayer "connectionist belief networks" resemble stochastic BOLTZMANN MACHINES (q.v.), but they are strictly feedforward. Output states are clamped to patterns selected from the environment, while the hidden-unit state space is randomly explored. The weights are adjusted so as to increase the probability of the hidden units generating the clamped output patterns. The network thereby learns to represent features in the hidden layer which explain correlations in the pattern set.

Invariance-Based Learning

The methods discussed so far try to extract useful structure from raw data, assuming minimal prior knowledge. How can unsupervised learning be applied beyond these preprocessing stages, to extract higher-order features and build more abstract representations? One approach is to restrict our search to particular kinds of structure. We can make constraining assumptions about the structure we are looking for and build these constraints into the network's architecture or objective function to develop more efficient, specialized learning procedures.

Spatially and Temporally Coherent Features

Becker and Hinton's (1992) Imax learning procedure discovers properties of the input that are coherent across space and time, by maximizing the mutual information between the *outputs*, y_a and y_b, of network modules that receive input from different parts of the sensory input (e.g., different modalities, or different spatial or temporal samples). Note how this objective function differs from the Infomax principle; the latter tries to retain *all* of the information in the input by maximizing the mutual information between inputs and outputs, whereas Imax tries to extract only those features common to two or more distinct parts of the input.

Under Gaussian assumptions about the signal and noise, Becker and Hinton derived the following objective function for the learning:

$$I = 0.5 \log \frac{V(v_a + v_b)}{V(y_a - y_b)} \qquad (6)$$

This measure tells how much information the average of y_a and y_b conveys about the common underlying signal, i.e., the feature which is coherent across the two input samples. When applied to networks composed of multilayer modules that receive input from adjacent, nonoverlapping regions of the input, Imax discovered higher-order image features (i.e., features not learnable by single-layer or linear networks) such as stereo disparity in random dot stereograms. One way to apply Imax to more than two modules is to have each module make a prediction about a linear combination of several neighboring modules' outputs. Becker and Hinton showed that a layer of linear units can thereby interpolate surface depth by learning to optimally combine local depth measurements. Note that Imax requires backpropagation of derivatives to train the weights to the hidden units, and the storage of several statistics on each link to compute the mutual information derivatives. Thus, a more biologically plausible approximation is needed.

Discussion

We have argued in favor of the global objective function approach to modeling unsupervised learning processes, and explored several powerful learning procedures based on this approach. These methods have had success in modeling early perceptual processing. With the incorporation of highly constraining prior models, unsupervised learning procedures can form even more abstract representations of data and extract higher-order features. A major direction for future research is to find tractable instantiations of these learning procedures and to apply them in multiple learning stages to form a diversity of representational levels. Additionally, in order to remain within the realm of biological plausibility, many of these learning models must be extended to yield simple, local synaptic learning rules.

Acknowledgments. The author acknowledges support from the McDonnell-Pew Program in Cognitive Neuroscience, research grant 92–40, and the Natural Sciences and Engineering Research Council of Canada.

Road Map: Learning in Artificial Neural Networks, Statistical
Background: I.3. Dynamics and Adaptation in Neural Networks
Related Reading: Data Clustering and Learning; Hebbian Synaptic Plasticity: Comparative and Developmental Aspects; Information Theory and Visual Plasticity

References

Asanuma, C., 1991, Mapping movements within a moving motor map, *Trends Neurosci.*, 14:217–218.
Barlow, H. B., 1989, Unsupervised learning, *Neural Computat.*, 1:295–311.
Becker, S., 1991, Unsupervised learning procedures for neural networks, *Int. J. Neural Syst.*, 2:17–33. ◆
Becker, S., and Hinton, G. E., 1992, A self-organizing neural network that discovers surfaces in random-dot stereograms, *Nature*, 355:161–163.
Becker, S., and Plumbley, M., in press, Unsupervised neural network learning procedures for feature extraction and classification, *Int. J. Appl. Intell.*, special issue on Applications of Neural Networks (F. Pineda, Ed.). ◆
Dempster, A. P., Laird, N. M., and Rubin, D. B., 1977, Maximum likelihood from incomplete data via the EM algorithm, *Proc. R. Statist. Soc.*, B-39:1–38.
Hinton, G. E., 1989, Connectionist learning procedures, *Artif. Intell.*, 40:185–234. ◆
Linsker, R., 1988, Self-organization in a perceptual network, *IEEE Computer*, 21:105–117.
Neal, R. M., 1992, Connectionist learning of belief networks, *Artif. Intell.*, 56:71–113.
Nowlan, S. J., 1991, Maximum likelihood competitive learning, in *Advances in Neural Information Processing Systems 2* (D. S. Touretzky, Ed.), San Mateo, CA: Morgan Kaufmann, pp. 574–582.
Saund, E., 1989, Dimensionality-reduction using connectionist networks, *IEEE Trans. Pattern Analysis Machine Intell.*, 11:304–314.
Shannon, C. E., 1948, A mathematical theory of communication, *Bell System Tech. J.*, 27:379–423, 623–656.
Silverman, B., 1986, *Density Estimation for Statistics and Data Analysis*, London: Chapman and Hall. ◆
Sur, M., 1989, Visual plasticity in the auditory pathway: Visual inputs induced into auditory thalamus and cortex illustrate principles of adaptive organization in sensory systems, in *Dynamic Interactions in Neural Networks: Models and Data* (M. A. Arbib and S.-I. Amari, Eds.), New York: Springer-Verlag, pp. 35–51.

Vapnik-Chervonenkis Dimension of Neural Networks

Wolfgang Maass

Introduction

Let $\mathcal{N}$ be some arbitrary feedforward neural net with w weights from some weight space W (e.g., $W = \mathbb{N}, \mathbb{Q},$ or $\mathbb{R}$). If $\mathcal{N}$ has n input-nodes, and if the output gate has range $\{0, 1\}$, then $\mathcal{N}$ computes for any weight assignment $\alpha \in W^w$ a function $\mathcal{N}^\alpha$ from some n-dimensional domain X (e.g., $X = \mathbb{N}^n, \mathbb{Q}^n, \mathbb{R}^n$) into $\{0, 1\}$.

One says that a subset S of the domain X is *shattered* by $\mathcal{N}$ if every function $g: S \to \{0, 1\}$ can be computed on $\mathcal{N}$, i.e., for all $g: S \to \{0, 1\}$ there is an $\alpha \in W^w$ such that $g(x) = \mathcal{N}^\alpha(x)$ for all $x \in S$.

The *Vapnik-Chervonenkis dimension* of $\mathcal{N}$ (abbreviated VC-dimension$(\mathcal{N})$; see Cover, 1968, for an equivalent definition) is defined as the maximal size of a set $S \subseteq X$ that is shattered by $\mathcal{N}$, i.e.,

VC-dimension$(\mathcal{N}) := \max\{|S|: S \subseteq X \text{ is shattered by } \mathcal{N}\}$

Intuitively one may view the VC-dimension of a neural net $\mathcal{N}$ as the number of *degrees of freedom* that one has in specifying the input/output behavior of $\mathcal{N}$. Of course one can define the dimension more generally without reference to neural nets. For *any* class $\mathcal{F}$ of functions $f: X \to \{0, 1\}$ (see Vapnik and Chervonenkis, 1971), the VC-dimension of $\mathcal{F}$ is defined by

VC-dimension$(\mathcal{F}) := \max\{|S|: S \subseteq X \text{ and for all } g: S \to \{0, 1\}$

there exists $f \in \mathcal{F}$ such that $g(x) = f(x)$ for all $x \in S\}$

Thus our preceding definition of the VC-dimension of a neural net $\mathcal{N}$ is just a special case for the function class $\mathcal{F} := \{f: X \to \{0, 1\}:$ there is an $\alpha \in W^w$ such that $f(x) = \mathcal{N}^\alpha(x)$ for all $x \in X\}$. The relevance of the VC-dimension for the training of a neural net can be traced back to the following rather simple mathematical result:

Theorem 1. ("Sauer's lemma"; see appendix A2 of Blumer et al., 1989.) Let $\mathcal{F}$ be any class of functions from some finite set X into $\{0, 1\}$, and set $d := \text{VC-dimension}(\mathcal{F})$. Then $\mathcal{F}$ contains at most $\Sigma_{i=0}^d (|X|/i) \leq |X|^d + 1$ different functions.

With the help of this result one can establish theoretical results about the *generalization abilities* of a neural net. More precisely one can prove relationships between the *apparent error* of a neural net $\mathcal{N}^\alpha$ on a randomly drawn training set T, and the *true error* of $\mathcal{N}^\alpha$ for new examples drawn from the same distribution. This relationship is discussed in PAC LEARNING AND NEURAL NETWORKS (q.v.) for the idealized setting of the classical PAC-learning model, where one assumes that there exists some assignment α^* to the weights of $\mathcal{N}$ such that $\mathcal{N}^{\alpha^*}$ has true error 0. We will discuss in the fourth section of this article the corresponding results for the more realistic setting of *agnostic PAC learning*, where no unrealistic a priori assumption is required.

For either version of the PAC model, one can roughly say that the expected deviation of the *true error* of a trained neural net $\mathcal{N}^\alpha$ from the *apparent error* of $\mathcal{N}^\alpha$ on the training set T depends on the size of T relative to the VC-dimension of $\mathcal{N}$.

Hence it has become of considerable interest to derive estimates for the VC-dimension of various types of neural nets.

We will survey in this article the most important known bounds for the VC-dimension of neural nets that consist of linear threshold gates (next section) and for the case of neural nets with real-valued activation functions (third section). In the fourth section we discuss a generalization of the VC-dimension for neural nets with non-Boolean network output. With regard to a discussion of the VC-dimension of models for networks of *spiking neurons*, we refer to Maass (in press).

For comparing the asymptotic behavior of two functions $f, g: \mathbb{N} \to \mathbb{R}^+$ we use the customary notation $f(n) = O(g(n))$ $[f(n) = \Omega(g(n))]$ if there exists some constant $c > 0$ such that $f(n) \leq c \cdot g(n)$ [resp. $f(n) \geq c \cdot g(n)$] for all sufficiently large n, and $f(n) = \Theta(g(n))$ if both $f(n) = O(g(n))$ and $f(n) = \Omega(g(n))$.

VC-Dimension of Neural Nets with Linear Threshold Gates

A *linear threshold gate* with n inputs computes for given weights $\alpha = \langle \alpha_0, \alpha_1, \ldots, \alpha_n \rangle \in W^{n+1}$ the function

$$T^\alpha(x_1, \ldots, x_n) = \begin{cases} 1 & \text{if } \sum_{i=1}^n \alpha_i x_i + \alpha_0 \geq 0 \\ 0 & \text{otherwise} \end{cases}$$

from $\mathbb{R}^n$ into $\{0, 1\}$.

Theorem 2. (Wenocur and Dudley.) VC-dimension($\mathcal{N}$) = $n + 1$ if $\mathcal{N}$ consists of a single linear threshold gate with n inputs.

Sketch of the proof. One can easily verify that the set $S := \{\mathbf{0}\} \cup \{\mathbf{e}_i : i \in \{1, \ldots, n\}\}$ is shattered by $\mathcal{N}$ (where $\mathbf{e}_i \in \{0, 1\}^n$ denotes the ith unit vector). Hence VC-dimension($\mathcal{N}$) $\geq n + 1$.

The lower bound follows from Radon's theorem, which states that any set S of $\geq n + 2$ points in $\mathbb{R}^n$ can be partitioned into sets S_0 and S_1 such that the convex hulls of S_0 and S_1 intersect. Obviously such sets S_0 and S_1 cannot be separated by any hyperplane, hence not by any threshold gate. □

With the help of Theorem 1, one can derive from Theorem 2:

Corollary. A linear threshold gate with n inputs can compute at most $|X|^{n+1} + 1$ different functions from any set $X \subseteq \mathbb{R}^n$ into $\{0, 1\}$.

Theorem 3. (Cover, 1968; see also Baum and Haussler, 1989.) Let $\mathcal{N}$ be an arbitrary feedforward neural net with w weights that consists of linear threshold gates. Then VC-dimension($\mathcal{N}$) $= O(w \cdot \log w)$.

Sketch of the proof. Let S be some arbitrary set of m input-vectors for $\mathcal{N}$. By the Corollary to Theorem 2 a gate g in $\mathcal{N}$ can compute at most $|X|^{\text{fan-in}(g)+1} + 1$ different functions from any finite set $X \subseteq \mathbb{R}^{\text{fan-in}(g)}$ into $\{0, 1\}$ (fan-in(g) denotes the number of inputs of gate g). Hence $\mathcal{N}$ can compute at most $\Pi_{g \text{ gate in } \mathcal{N}} (m^{\text{fan-in}(g)+1} + 1) \leq m^{2w}$ different functions from S into $\{0, 1\}$. If S is shattered by $\mathcal{N}$, then $\mathcal{N}$ can compute all 2^m functions from S into $\{0, 1\}$. In this case the preceding implies that $2^m \leq m^{2w}$; hence $m \leq 2w \cdot \log m$. It follows that $\log m = O(\log w)$; thus $m = O(w \cdot \log w)$. □

It is tempting to conjecture that the VC-dimension of a neural net $\mathcal{N}$ cannot be larger than the total number of weights of all gates in $\mathcal{N}$, which is equal to the sum of the VC-dimensions of the individual gates in $\mathcal{N}$. In view of Theorem 2 this conjecture would imply an upper bound $O(w)$ for VC-dimension($\mathcal{N}$) in Theorem 3. However, the following result (whose proof re-

quires rather complex techniques from circuit theory) shows that the superlinear upper bound of Theorem 3 is in fact asymptotically optimal. Hence with regard to the VC-dimension it is fair to say that a neural net can be "more than the sum of its parts."

Theorem 4. (Maass, 1993, 1994b.) Assume that $(\mathcal{N}_n)_{n \in \mathbb{N}}$ is any sequence of neural nets with at least 2 hidden layers, where $\mathcal{N}_n$ has n Boolean input nodes and $O(n)$ gates. Furthermore assume that $\mathcal{N}_n$ has $\Omega(n)$ gates on the first hidden layer, and at least $4 \log n$ gates on the second hidden layer. We also assume that $\mathcal{N}_n$ is fully connected between any two successive layers (hence $\mathcal{N}_n$ has $\Theta(n^2)$ weights), and that the gates of $\mathcal{N}_n$ are linear threshold gates (or gates with the sigmoid activation function $\sigma(y) = 1/(1 + e^{-y})$, with round-off at the network output). Then

$$\text{VC-dimension}(\mathcal{N}_n) = \Theta(n^2 \cdot \log n)$$

Subsequently, Sakurai (1993) has shown that if one allows *real-valued* network inputs, then the lower bound of Theorem 4 also holds for certain neural nets with *one* hidden layer. In addition he has shown that for the case of real-valued inputs one can determine exactly the constant factor in these bounds.

VC-Dimension of Analog Neural Nets

We consider in this section and the next the case where some gates in $\mathcal{N}$ employ activation functions f with non-Boolean output, such as $\sigma(y) = 1/(1 + e^{-y})$ (a gate of fan-in m with activation function f and weights $\alpha_0, \ldots, \alpha_m$ computes the function $\langle y_1, \ldots, y_m \rangle \mapsto f(\Sigma_{i=1}^m \alpha_i y_i + \alpha_0)$). We first consider the case where the network output of $\mathcal{N}$ is nevertheless Boolean valued (e.g., because the output gate of $\mathcal{N}$ is a linear threshold gate).

It turns out that, in order to get upper bounds for the VC-dimension of such neural nets, it does not suffice to assume that the analog activation functions in $\mathcal{N}$ are *very smooth squashing functions*. Sontag (1992) has shown that for the real-analytic function $\Psi(y) := (1/\pi)\arctan(y) + (\cos y)/7(1 + y^2) + \frac{1}{2}$ a neural net with 2 real valued inputs, 2 hidden units with activation function Ψ, and a linear threshold gate as output gate has *infinite* VC-dimension (compare results in AUTOMATA AND NEURAL NETWORKS). Note that this function Ψ is strictly increasing and has limits $1, 0$ at $\pm \infty$ (hence it is a squashing function). For the case of neural nets with n Boolean inputs, Sontag constructed activation functions with the same analytic properties as the function Ψ, such that the neural net with the same architecture as above has the maximal possible VC-dimension 2^n.

Thus one cannot hope to prove significant upper bounds for the VC-dimension of an analog neural net if one only knows that its non-Boolean activation functions are very smooth, strictly increasing squashing functions. More subtle mathematical properties of the activation functions turn out to be crucial, such as the maximal possible number of zeros of any function that is definable from these activation functions with a specific set of operations.

The first upper bound for the VC-dimension of a neural net whose gates employ the activation function $\sigma(y) = 1(1 + e^{-y})$ is due to Macintyre and Sontag (1993). By using a sophisticated result from mathematical logic (order minimality of the elementary theory L of real numbers with the basic algebraic operations and exponentiation), they have shown that the VC-dimension of any finite feedforward neural net with this activation function is finite. Very recently, Karpinski and Macintyre have applied very complicated techniques from differential topology in order to achieve the following upper bound, which is *polynomial* in the number w of weights:

Theorem 5. (Karpinski and Macintyre, 1994.) The VC-dimension of any feedforward neural net with the sigmoid activation function σ is bounded by $O(w^4)$, where w is the total number of weights in the neural net. The same upper bound holds for a large class of activation functions that satisfy a certain Pfaffian differential equation.

For the case of neural nets of arbitrary constant depth with n Boolean inputs and polynomially in n many gates with piecewise polynomial activation functions and arbitrary real weights, it was shown in Maass (1993) that such circuits can be simulated by polynomial size neural nets that consist entirely of linear threshold gates. Hence a polynomial upper bound for the VC-dimension of such neural nets follows immediately from Theorem 3. Subsequently Goldberg and Jerrum have shown that with the help of *Milnor's theorem* from algebraic geometry one can prove directly a polynomial upper bound for arbitrary polynomial-size neural nets with piecewise polynomial activation functions. (In fact, their argument also applies to the case of piecewise rational activation functions.)

Theorem 6. (Goldberg and Jerrum, 1993.) Let $\mathcal{N}$ be any neural net with piecewise polynomial activation functions (with $O(1)$ pieces each), arbitrary real inputs and weights, and Boolean output. Then the VC-dimension of $\mathcal{N}$ is at most $O(w^2)$, where w is the total number of weights in $\mathcal{N}$.

It is an open problem whether this upper bound can be improved to $O(w \log w)$. The best known *lower bound* for the VC-dimension of an analog neural net with piecewise polynomial activation functions (or the activation function σ) is the same bound $\Omega(w \cdot \log w)$ as for the case of neural nets with linear threshold gates (see Theorem 4).

Generalization of the VC-Dimension for Neural Nets with Real-Valued Output

We consider here the case of an analog neural net $\mathcal{N}$ where the range of the activation function of the output gate is *not* Boolean-valued. If for example $\mathcal{N}$ has an output gate whose activation function is $\sigma(y) = 1/(1 + e^{-y})$, then it cannot compute *any* function with range $\{0, 1\}$. Thus its VC-dimension (as defined above) would be 0. Consequently, one has to consider for such neural nets $\mathcal{N}$ a more general notion of a "dimension" in order to give an upper bound for the number of training examples that are needed to train $\mathcal{N}$. A suitable generalization is provided by the notion of a *pseudodimension*.

In order to define the pseudodimension of a neural net $\mathcal{N}$ one has to specify a loss function l that is used to measure for any example $\langle x, y \rangle \in X \times Y$ the deviation $l(\mathcal{N}^{\alpha}(x), y)$ of the prediction $\mathcal{N}^{\alpha}(x)$ of the neural net from the target value y. Popular choices for l are $l(z, y) = |z - y|$, or $l(z, y) = (z - y)^2$. One then considers the class $\mathcal{F}_{\mathcal{N}, l}$ of all functions from $X \times Y$ into $\mathbb{R}$ of the form $\langle x, y \rangle \mapsto l(\mathcal{N}^{\alpha}(x), y)$ for some weight assignment $\alpha \in W^w$.

One would like to be able to say that $\mathcal{F}_{\mathcal{N}, l}$ "shatters" a certain subset S of its domain $X \times Y$. However, to do so one has to generalize the corresponding definition in the introductory section, since the functions in $\mathcal{F}_{\mathcal{N}, l}$ may assume other real values besides 0 or 1. This problem is solved by allowing an arbitrary "threshold" $t(\langle x, y \rangle)$ for each element $\langle x, y \rangle$ of the shattered set $S \subseteq X \times Y$ so that for any $f \in \mathcal{F}_{\mathcal{N}, l}$ one "rounds off" $f(\langle x, y \rangle)$ to 1 if $f(\langle x, y \rangle) \geq t(\langle x, y \rangle)$, and to 0 otherwise.

Definition. The pseudodimension $\dim_P^l(\mathcal{N})$ of $\mathcal{N}$ with respect to the loss function l is defined as the maximal size of a set

$S \subseteq X \times Y$ which is shattered by $\mathcal{N}$ in the sense that there exists some $t: S \to \mathbb{R}$ such that

> For all $g: S \to \{0, 1\}$ there exists $f \in \mathcal{F}_{\mathcal{N}, l}$ such that
>
> $g(\langle x, y \rangle) = 1 \Leftrightarrow f(\langle x, y \rangle) \geq t(\langle x, y \rangle)$ for all $\langle x, y \rangle \in S$

Remark. For the special case of the binary range $Y = \{0, 1\}$ and the discrete loss function l_D (where $l_D(z, y) = 0$ if $z = y$, and $l_D(z, y) = 1$ if $z \neq y$), the pseudodimension of a neural net $\mathcal{N}$ coincides with its VC-dimension.

If the size m of a training set $T = (\langle x_i, y_i \rangle_{i \leq m})$ (which is randomly drawn according to some arbitrary distribution D over $X \times Y$) is relatively large in comparison with the pseudodimension of $\mathcal{N}$, then the "apparent error" $(1/m) \sum_{i=1}^m l(\mathcal{N}^{\alpha}(x_i), y_i)$ of $\mathcal{N}^{\alpha}$ is (with high probability) close to the "true error" $E_{\langle x, y \rangle \in D}[l(\mathcal{N}^{\alpha}(x), y)]$ of $\mathcal{N}^{\alpha}$, provided that the range of the values $l(\mathcal{N}^{\alpha}(x), y)$ is bounded. This relationship is made more precise by the following result:

Theorem 7. (Haussler, 1992.) Assume that $\mathcal{F}_{\mathcal{N}, l}$ is a permissible class of functions from $X \times Y$ into some arbitrary bounded interval $[0, B]$ (the "permissibility" of $\mathcal{F}_{\mathcal{N}, l}$ is a somewhat technical measurability assumption, which is always satisfied if the weight space W is countable, e.g., for $W \subseteq \mathbb{Q}$). Then for any distribution D over $X \times Y$ and any sample $T = (\langle x_i, y_i \rangle)_{i \leq m}$ of m randomly drawn "training examples" (which are drawn independently according to distribution D), one has for any given $\varepsilon, \delta > 0$ and any sample size $m \geq (64B^2/\varepsilon^2)(2 \cdot \dim_P^l(\mathcal{N}) \ln(16eB/\varepsilon) + \ln(8/\delta))$ that with probability $\geq 1 - \delta$:

$$\text{For all } \alpha \in W^w \left(\left| \frac{1}{m} \sum_{i=1}^m l(\mathcal{N}^{\alpha}(x_i), y_i) \right. \right.$$

$$\left. \left. - E_{\langle x, y \rangle \in D}[l(\mathcal{N}^{\alpha}(x), y)] \right| \leq \varepsilon \right)$$

For neural nets $\mathcal{N}$ with the sigmoid activation function $\sigma(y) = 1/(1 + e^{-y})$ the only known upper bounds for the pseudodimension are given by a corresponding generalization of Theorem 5 (Karpinski and Macintyre, 1994) and by the following result:

Theorem 8. (Bartlett and Williamson, 1993.) Let $\mathcal{N}$ be a neural net with one hidden layer. Assume that the gates of the hidden layer use the activation function σ (alternatively these gates may compute a radial basis function $\mathbf{y} = \langle y_1, \ldots, y_m \rangle \mapsto e^{-\|\mathbf{y} - \mathbf{c}\|}$ with "weights" $\mathbf{c} \in \mathbb{R}^m$), and that the output gate outputs a weighted sum of its inputs. Then for discrete inputs from $\{-K, \ldots, K\}^n$ the pseudodimension of $\mathcal{N}$ is at most $8w \log_2(11 \cdot wK)$, where w denotes the total number of weights in $\mathcal{N}$.

The proof uses an exponential parameter transformation in order to transform the function that is computed by $\mathcal{N}$ into one that is polynomial in its parameters. One can then apply Milnor's theorem in a similar fashion to that in Theorem 6. □

For neural nets $\mathcal{N}$ with piecewise polynomial activation functions one can give the following upper bound (Maass, 1994a), which is shown with the help of Milnor's theorem in the same way as Theorem 6.

Theorem 9. $\dim_P^l(\mathcal{N}) = O(w^2)$ for arbitrary neural nets $\mathcal{N}$ with w real-valued weights and arbitrary piecewise polynomial activation functions that consist of $O(1)$ pieces of degree $O(1)$.

It is obvious from Theorem 4 that the pseudodimension of a neural net $\mathcal{N}$ with w weights can be as large as $\Omega(w \cdot \log w)$. It is an open problem whether the pseudodimension of a neural net with piecewise polynomial activation functions that consist of $O(1)$ pieces each (or with any other common activation function such as $\sigma(y) = 1/(1 + e^{-y})$) can be any larger.

Discussion

The VC-dimension of a neural net with Boolean output measures the "expressiveness" of such a neural net. The related notion of a pseudodimension provides a similar tool for the analysis of neural nets with real-valued output. The derivation of bounds for the VC-dimension and the pseudodimension of neural nets has turned out to be a rather challenging but quite interesting chapter in the mathematical investigation of neural nets. This work has brought a number of sophisticated mathematical tools into this research area, which have subsequently turned out to be also useful for the solution of a variety of other problems regarding the complexity of computing and learning on neural nets (see Roychowdhury, Siu, and Orlitsky, 1994, for an overview of the current state of affairs).

Bounds for the VC-dimension or pseudodimension of a neural net $\mathcal{N}$ provide estimates for the number of random examples that are needed to train $\mathcal{N}$ so that it has good generalization properties (i.e., so that the error of $\mathcal{N}$ on new examples from the same distribution is at most ε, with probability $\geq 1 - \delta$). From the point of view of a single application problem, these bounds tend to be too large, since they provide such generalization guarantees simultaneously for *any* probability distribution on the examples and for *any* training algorithm that minimizes disagreement on the training examples. For some special distributions and specific training algorithms, one has achieved tighter bounds with the help of heuristic arguments (replica techniques) from statistical mechanics.

Road Map: Computability and Complexity
Related Reading: Learning and Generalization: Theoretical Bounds; Statistical Mechanics of Generalization

References

Bartlett, P. L., and Williamson, R. C., 1993, The VC-dimension and pseudodimension of two-layer neural networks with discrete inputs, Technical Report, Department of Systems Engineering, Australian National University.

Baum, E. B., and Haussler, D., 1989, What size net gives valid generalization? *Neural Computat.*, 1:151–160.

Blumer, A., Ehrenfeucht, A., Haussler, D., and Warmuth, M. K., 1989, Learnability and the Vapnik-Chervonenkis dimension, *J. ACM*, 36:929–965.

Cover, T. M., 1968, Capacity problems for linear machines, in *Pattern Recognition* (L. Kanal, Ed.), Thompson, pp. 283–289.

Goldberg, P., and Jerrum, M., 1993, Bounding the Vapnik-Chervonenkis dimension of concept classes parameterized by real numbers, in *Proceedings of the 6th Annual ACM Conference on Computational Learning Theory*, New York: ACM Press, pp. 361–369.

Haussler, D., 1992, Decision theoretic generalizations of the PAC model for neural nets and other learning applications, *Inform. and Computat.*, 100:78–150.

Karpinski, M., and Macintyre, A., 1994, Polynomial bounds for VC-dimension of sigmoidal neural networks, Research Report 85116-CS, University of Bonn; extended abstract to appear in *Proceedings of EuroCOLT '95*, Lecture Notes in Computer Science, Berlin: Springer.

Maass, W., 1993, Bounds for the computational power and learning complexity of analog neural nets; extended abstract in *Proceedings of the 25th Annual ACM Symposium on the Theory of Computing*, New York: ACM Press, pp. 335–344.

Maass, W., 1994a, Agnostic PAC-learning of functions on analog neural nets, in *Advances in Neural Information Processing Systems* (J. Cowan, G. Tesauro, and J. Alspector, Eds.), San Mateo, CA: Morgan Kaufmann, pp. 311–318. Journal version appears in *Neural Computat.*

Maass, W., 1994b, Neural nets with superlinear VC-dimension, *Neural Computat.*, 6:875–882.

Maass, W., in press, On the computational complexity of networks of spiking neurons, in *Advances in Neural Information Processing Systems 7* (G. Tesauro, D. Touretzky, and J. Alspector, Eds.), San Mateo, CA: Morgan Kaufmann.

Macintyre, M., and Sontag, E. D., 1993, Finiteness results for sigmoidal "neural" networks, in *Proceedings of the 25th Annual ACM Symposium on the Theory of Computing*, New York: ACM Press, pp. 325–334.

Roychowdhury, V. P., Siu, K. Y., and Orlitsky, A., 1994, *Theoretical Advances in Neural Computation and Learning*, Boston: Kluwer Academic. ◆

Sakurai, A., 1993, Tighter bounds of the VC-dimension of three-layer networks, in *Proceedings of the 1993 World Congress on Neural Networks*, vol. 3, Hillsdale, NJ: Erlbaum, pp. 540–543.

Sontag, E. D., 1992, Feedforward nets for interpolation and classification, *J. Comput. Syst. Sci.*, 45:20–48.

Vapnik, V. N., and Chervonenkis, A. Y., 1971, On the uniform convergence of relative frequencies of events to their probabilities, *Theory Probab. Appl.*, 16:264–280.

Vestibulo-Ocular Reflex: Performance and Plasticity

Thomas J. Anastasio

Introduction

The vestibulo-ocular reflex (VOR) is one of the most frequently modeled systems in the nervous system. The function of the VOR is simply to stabilize the retinal image by producing eye rotations that counterbalance head rotations (Wilson and Melvill Jones, 1979). The VOR receives information about head rotation from the vestibular, semicircular canal receptors. It is mediated by interneurons in the vestibular nuclei (VN) of the brainstem, that relay head rotation signals from canal sensory afferent neurons to the motoneurons of the extraocular muscles. The interesting aspect of VOR neurophysiology is that vestibular nuclei neurons (VNNs) are much more than a simple relay. Among the functions performed by VNNs include multimodality integration, temporal signal processing, and adaptive plasticity. Theory and modeling have been employed in the study of all of these functions.

VOR Neurophysiology

The horizontal VOR is schematized in Figure 1. The VOR is bilaterally symmetric. Head rotation is sensed by excitation of the canal receptor on one side of the head, and inhibition of the

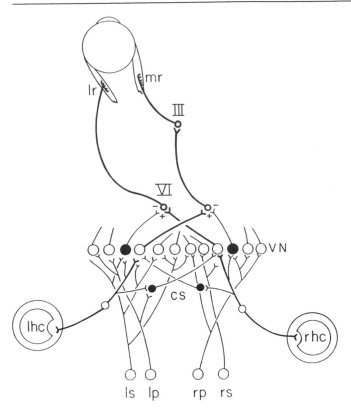

Figure 1. Schematic diagram of the bilateral, horizontal vestibulo-ocular reflex: *lhc, rhc*, left and right horizontal canal receptors and afferents; *lp, rp*, left and right pursuit inputs; *ls, rs*, left and right saccadic inputs; *cs*, commissures; *VN*, vestibular nuclei and neurons; *lr, mr*, lateral and medial rectus muscles of the left eye and their motoneurons; *VI*, abducens nuclei; *III*, oculomotor nuclei; *open circles*, excitatory neurons; *filled circles*, inhibitory neurons. (Redrawn from Anastasio, T. J., and Robinson, D. A., 1989, The distributed representation of vestibulo-oculomotor signals by brain-stem neurons, *Biological Cybernetics*, 61:79–88; used with permission.)

canal on the other side (*lhc* and *rhc* indicate the left and right horizontal canals). Similarly, eye rotations are produced by coordinated contraction and relaxation of the muscles on either side of the eye (*lr* and *mr*, the lateral and medial rectus muscles). In addition to vestibular signals, the VN utilizes information from other oculomotor sources to rotate the eyes. These include signals from the pursuit (*lp* and *rp*) and saccadic (*ls* and *rs*) systems. All of these signals converge and intermix on VNNs (Scudder and Fuchs, 1992), which represent them in a complex, seemingly random way.

Semicircular Canals and Canal Afferents

The semicircular canals are fluid-filled tori, carved out of the bone of the skull, with a membrane spanning the lumen at one end. Rotation of the head causes the fluid to flow, and this produces deflection of the membrane. Classical analysis of canal dynamics involved a consideration of the balance of forces associated with head rotational acceleration, fluid inertia and viscosity, and membrane elasticity (Wilson and Melvill Jones, 1979). This analysis showed that the semicircular canals are leaky integrating accelerometers, and that membrane deflection is mostly proportional to head rotational velocity. The response of the canal to an impulse head rotational acceleration would be an instantaneous deflection of the membrane

followed by an exponential decay back to resting position, with a time constant on the order of 5 seconds.

Hair cell receptors that are sensitive to canal membrane deflection make synaptic contact with canal afferents, which in turn transmit head rotation information to the VN. Canal afferents have a spontaneous discharge rate that is modulated up and down for on- and off-direction head rotations, respectively. As expected from the dynamic analysis of the canal, the response of the canal afferents to an impulse head rotational acceleration is primarily an exponential decay with a time constant of about 5 seconds (Wilson and Melvill Jones, 1979).

The Overall Vestibulo-Ocular Reflex

The VOR is driven by the canal afferent head rotational velocity signal, which is essentially inverted to produce an equal and opposite eye rotational velocity command. The afferent signal is mathematically integrated by central neural circuitry, in a lossy or leaky way, in order to make the VOR eye rotation command more effective. One type of leaky integration, velocity storage (Raphan, Matsuo, and Cohen, 1979), involves a lengthening of the time constant of the afferent head velocity signal from 5 to 20 seconds (approximately). This prolongs the period over which the VOR can produce counterbalancing eye rotations. Another type of leaky integration, velocity-position (Robinson, 1981), converts the VOR eye velocity command into an eye position command. Both velocity and position commands are needed to control eye rotation.

Vestibular Nuclei Neurons

Both forms of leaky integration, velocity storage and velocity-position, occur at the level of the VN (Robinson, 1981). The time constant of the VNN eye velocity command is as long as that of the VOR (about 20 seconds). Many VNNs also carry an eye position signal (Baker, Evinger, and McCrea, 1981). Both types of leaky integration may be mediated by an inhibitory commissural system (*cs* in Figure 1) that interconnects the VNNs bilaterally. Probably the most salient features of VNNs are that, in comparison with canal afferents, they have lower spontaneous rates and higher sensitivities (Baker et al., 1981). These factors combine to give VNNs a strong tendency to cut off (rectify) in response to head rotations in their off-directions.

Plasticity in the Vestibulo-Ocular Reflex

The VOR is well known for its ability to be adaptively modified. The process of vestibular compensation involves the partial restoration of VOR function following removal of the vestibular receptors from one side of the head (Smith and Curthoys, 1989). The VOR can also be modified by altering its visual input in such a way that the normal relationship between head rotation and apparent visual surround rotation is disrupted (Wilson and Melvill Jones, 1979). Models have shown how adaptive modification of VOR could be brought about by VNN synaptic plasticity. Models have also shed light on other aspects of VOR neurophysiology, like the seemingly random way in which VNNs represent vestibulo-oculomotor signals, the mechanisms by which VNNs produce velocity storage and velocity-position leaky integration, and VNN rectification.

VOR Models

Unilateral VOR Models

The first VOR models sought to explain velocity storage, and treated the VOR as if it were unilateral. The velocity storage

Figure 2. Head rotational acceleration impulse responses of some elements in the feedback control theory model of the VOR. *A: solid line*, head rotational velocity; *dashed curve*, canal afferent response; *dotted curve*, motoneuron eye velocity command. *B: solid curve*, leaky integrated canal signal; *dashed curve*, direct canal signal; *dotted curve*, eye velocity command. (Redrawn from Anastasio, T. J., 1993, Modeling vestibulo-ocular reflex dynamics: From classical analysis to neural networks, in *Neural Systems: Analysis and Modeling*, ed. F. Eeckman; used by permission of Kluwer Academic Publishers.)

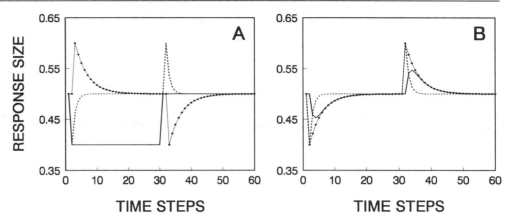

transformation in this one-sided VOR is illustrated in Figure 2*A*. The dashed curve shows the canal response to an impulse head rotational acceleration to one side, followed by another to the opposite side. These accelerations would produce a constant head rotational velocity in the direction of the first acceleration impulse, which would be terminated by the second, oppositely directed impulse (solid line). The dotted curve shows the corresponding motoneuron eye velocity command, which is equal but opposite in amplitude to the canal afferent signal. Also, the time constant of the motoneuron command is four times longer than that of the canal response, reflecting velocity storage. This prolongs the period over which the VOR can continue to produce an eye rotational velocity that is opposite to the imposed head rotational velocity.

The models that were proposed to account for velocity storage included a leaky integrator that was inserted into the VOR pathway. It represented a second leaky integration of head rotational acceleration, after the first by the canals themselves. In the model illustrated in Figure 3, the second leaky integrator (LI2) was configured to feedback onto VNNs (Robinson, 1981). The signal coming from LI2 is shown in Figure 2*B* (solid curve), and the canal signal is reproduced there for reference (dashed curve). It is easily seen that the canal and LI2 signals add up to give an exponential decay with the long VOR time constant. Since this summation occurs at the VN in the feedback model (Figure 3), the VNN eye velocity command would also have the longer VOR time constant (Figure 2*B*, dotted curve), as observed experimentally.

Bilateral VOR Models

Subsequent VOR models were bilateral. They showed how leaky integration, like velocity storage or velocity-position (LI2 and LI3 in Figure 3, respectively), could be produced (Robinson, 1989) or enhanced (Galiana, 1985) by bilateral interactions between VNNs. The bilateral models made use of the fact that VNNs are connected over the midline by inhibitory commissural fibers (Wilson and Melvill Jones, 1979). The feature of mutual inhibition of bilaterally organized VNNs was essential for VOR neural integration, since it allowed VNNs to integrate the push-pull modulations of the canal afferent discharge, but not their spontaneous discharge. Thus, when the bilateral canal inputs to VN are firing at a spontaneous rate that is approximately equal on both sides, the bilateral VNNs inhibit each other equally, and they also maintain a constant spontaneous discharge rate. But during head rotation, when the canal inputs are modulated in push-pull, the VNNs are activated differentially and exert net positive feedback on themselves, thereby producing leaky integration of their inputs.

Adaptive VOR Models

While most VOR models have fixed parameters, some have focused on plastic phenomena. In these adaptive models, the plastic elements correspond to synaptic weights between neural elements. In a bilateral VOR model, it was demonstrated that compensation for unilateral loss of the vestibular receptors

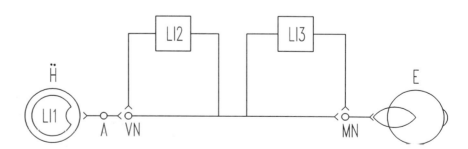

Figure 3. Schematic diagram of the feedback control theory model of the VOR. *Ḧ*, head rotational acceleration; *E*, eye rotational position; *A*, canal afferents; *VN*, vestibular nuclei; *MN*, extraocular motoneurons. (Adapted from Robinson, 1981.)

LI1 – canal leaky integrating accelerometer
LI2 – velocity storage leaky integrator
LI3 – velocity-position leaky integrator

could be accomplished at the level of the VNNs by changes in commissural weights alone (Galiana, 1985). A linear adaptive filter model was used to simulate the plastic changes that occur in the VOR under conditions where optical manipulation is used to alter the normal relationship between head rotation and apparent rotation of the visual surround. This model produced a close fit to the data, and demonstrated that plastic changes in VOR follow the gradient of error (Fujita, 1982). This lends support to the use of error-based learning algorithms to program neural network models of VOR.

Static Neural Network Models of VOR

Some of the most recent models of VOR have taken the form of adaptive neural networks. A generic architecture for VOR neural network models is shown in Figure 4. The networks have three layers, in which the input, hidden, and output units represent canal afferents, VNNs, and motoneurons, respective-

ly. Inputs can also include other modalities such as pursuit or bias. The number of model VNNs could vary from 2 to 40. The recurrent connections shown between VNNs are used only for dynamic neural networks (see below).

The static neural networks were used to study how vestibulo-oculomotor signals could be represented by relatively large numbers of VNNs. These networks contained 40 model VNNs, and the bilateral inputs represented contributions from both the vestibular and pursuit systems (Anastasio and Robinson, 1989). The networks were trained using backpropagation (Rumelhart, Hinton, and Williams, 1986; see also BACKPROPAGATION: BASICS AND NEW DEVELOPMENTS) to generate motoneuron activation patterns that would be appropriate for the patterns received as inputs. After training, the activation patterns of the hidden units (i.e., model VNNs) were tested, and the results for a typical run are shown in Figure 5. The motoneurons (plus signs) have vestibular and pursuit sensitivities that are equal and opposite, as demanded by training. But those of the VNNs vary over a broad range, and this matches the randomness in vestibulo-oculomotor patterns actually observed for VNNs.

Dynamic Neural Network Models of VOR

The dynamic neural network models of VOR, i.e., those with recurrent connections (Figure 4), resemble the linear, bilateral models of VOR in their connectivity. However, the dynamic networks were adaptive, and were trained using the recurrent backpropagation algorithm (Williams and Zipser, 1989; see RECURRENT NETWORKS: SUPERVISED LEARNING). Also, as in the static networks, units in the dynamic networks were nonlinear. They computed the weighted sum of their inputs, and bounded the result sigmoidally between zero and one. The dynamic neural networks were trained to reproduce the VOR with velocity storage (Anastasio, 1991, 1993), and the time-varying inputs and outputs were bilateral, push-pull versions of the unilateral

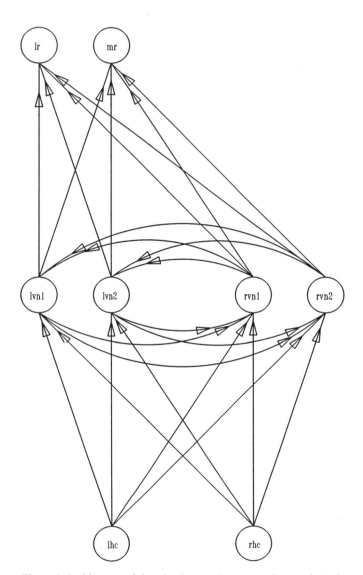

Figure 4. Architecture of the adaptive, nonlinear neural network models of the VOR: *lvn, rvn,* left and right vestibular nuclei neurons; other abbreviations as in Figure 1. (Redrawn from Anastasio, T. J., 1991, Neural network models of velocity storage in the horizontal vestibulo-ocular reflex, *Biological Cybernetics,* 64:187–196; used with permission.)

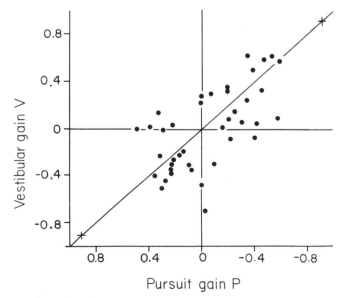

Figure 5. Variability in vestibular and pursuit related activity of 40 model vestibular nuclei neurons from an adaptive, static neural network model of the VOR. (Redrawn from Anastasio, T. J., and Robinson, D. A., 1989, The distributed representation of vestibulo-oculomotor signals by brain-stem neurons, *Biological Cybernetics,* 61: 79–88; used with permission.)

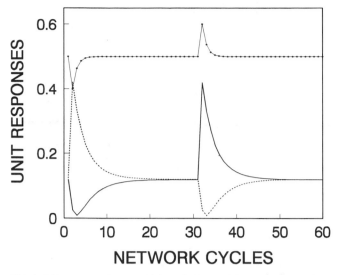

Figure 6. Responses of two model vestibular nuclei neurons in an adaptive, dynamic neural network model of velocity storage in the VOR: *solid and dashed curves,* vestibular nuclei neuron responses; *dotted curve,* canal afferent response. (Redrawn from Anastasio, T. J., 1993, Modeling vestibulo-ocular reflex dynamics: From classical analysis to neural networks, in *Neural Systems: Analysis and Modeling,* ed. F. Eeckman; used by permission of Kluwer Academic Publishers.)

signals shown in Figure 2*A*. The canal afferent signal from one side is reproduced in Figure 6 (dotted curve), in which are also shown the VNN responses from a network having only two VNNs, one on each side (solid and dashed curves). It is apparent from the figure that the model VNNs have lower spontaneous rates and higher sensitivities than the canal afferents, and this causes them to rectify, just as it does for real VNNs.

Why the learning algorithm produces rectifying VNN responses can be appreciated by considering the nature of the velocity storage transformation, which consists in adding the direct, sharply peaked canal afferent response to a leaky integrated version of it (see Figure 2*B*). The VNNs learn to mutually inhibit each other. This allows them to produce a leaky integration of the canal afferent signal, just as in the linear, bilateral VOR models discussed above. But since the two VNNs in the recurrent neural network are nonlinear, the same VNNs can also pass the sharp peak of the canal afferent signal. Thus, at peak, when the off-direction VNN rectifies, the feedback loop temporarily opens, allowing the direct canal afferent signal to pass unintegrated. These results suggest that VNNs rectify because this nonlinearity allows them to selectively pass the sharp peak of the canal afferent response and also integrate its rapidly decaying tail.

Dynamic neural network models of vestibular compensation can reproduce the time course of this adaptive process (Anastasio, 1992). In these models, fully trained recurrent networks are retrained following removal of the canal input from one side. They demonstrate that stages of compensation bringing more error reduction are reached sooner, suggesting that the compensatory process also follows the error gradient. The network models reproduce the behavior patterns of VNNs following compensation, but only if synapses at the motoneurons, as well as at the VNNs, are plastic. The prediction that synapses onto motoneurons also change in strength during compensation provides a challenge for experimental verification.

Road Map: Primate Motor Control
Background: I.3. Dynamics and Adaptation in Neural Networks
Related Reading: Dynamic Models of Neurophysiological Systems; Gaze Coding in the Posterior Parietal Cortex; Head Movements: Multidimensional Modeling; Pursuit Eye Movements; Saccades and Listing's Law

References

Anastasio, T. J., 1991, Neural network models of velocity storage in the horizontal vestibulo-ocular reflex, *Biol. Cybern.,* 64:187–196.
Anastasio, T. J., 1992, Simulating vestibular compensation using recurrent back-propagation, *Biol. Cybern.,* 66:389–397.
Anastasio, T. J., 1993, Modeling vestibulo-ocular reflex dynamics: From classical analysis to neural networks, in *Neural Systems: Analysis and Modeling* (F. Eeckman, Ed.), Norwell, MA: Kluwer, pp. 407–430. ◆
Anastasio, T. J., and Robinson, D. A., 1989, The distributed representation of vestibulo-oculomotor signals by brain-stem neurons, *Biol. Cybern.,* 61:79–88.
Baker, R., Evinger, C., and McCrea, R. A., 1981, Some thoughts about the three neurons in the vestibular ocular reflex, in *Vestibular and Oculomotor Physiology* (B. Cohen, Ed.), New York: New York Academy of Sciences, part IV, pp. 171–188. ◆
Fujita, M., 1982, Simulation of adaptive modification of the vestibulo-ocular reflex with an adaptive filter model of the cerebellum, *Biol. Cybern.,* 45:207–214.
Galiana, H. L., 1985, Commissural vestibular nuclear coupling: A powerful putative site for producing adaptive change, in *Adaptive Mechanisms in Gaze Control: Facts and Theories* (A. Berthoz and G. Melvill Jones, Eds.), Amsterdam: Elsevier Science, pp. 327–339.
Raphan, T., Matsuo, V., and Cohen, B., 1979, Velocity storage in the vestibulo-ocular arc (VOR), *Exp. Brain Res.,* 35:229–248.
Robinson, D. A., 1981, The use of control systems analysis in the neurophysiology of eye movements, *Annu. Rev. Neurosci.,* 4:463–503. ◆
Robinson, D. A., 1989, Integrating with neurons, *Annu. Rev. Neurosci.,* 12:33–45. ◆
Rumelhart, D. E., Hinton, G. E., and Williams, R. J., 1986, Learning internal representations by error propagation, in *Parallel Distributed Processing: Explorations in the Microstructure of Cognition,* vol. 1, *Foundations* (D. E. Rumelhart, J. L. McClelland, and PDP Research Group, Eds.), Cambridge, MA: MIT Press, pp. 318–362.
Scudder, C. A., and Fuchs, A. F., 1992, Physiology and behavioral identification of vestibular nucleus neurons mediating the horizontal vestibuloocular reflex in trained rhesus monkeys, *J. Neurophysiol.,* 68:244–264.
Smith, P. F., and Curthoys, I. S., 1989, Mechanisms of recovery following unilateral labyrinthectomy: A review, *Brain Res. Rev.,* 14:155–180. ◆
Williams, R. J., and Zipser, D., 1989, A learning algorithm for continually running fully recurrent networks, *Neural Computat.,* 1:270–280.
Wilson, V. J., and Melvill Jones, G., 1979, *Mammalian Vestibular Physiology,* New York: Plenum. ◆

Vision for Robot Driving

Dean A. Pomerleau

Task Description

An application domain which has proven quite amenable to artificial neural networks is vision-based autonomous driving. In this domain, the objective is to steer a robot vehicle, like the Navlab shown in Figure 1, based on input from an onboard video camera. The vehicle is equipped with motors on the steering wheel, brake, and accelerator pedal, enabling computer control of the vehicle's trajectory. A typical autonomous driving task which has been addressed using artificial neural networks is road following. For this task, the input consists of images from the video camera, and the output is a steering command which will keep the vehicle on the road.

Neural Network Model

The connectionist model for autonomous road following used in the ALVINN system (Pomerleau, 1991, 1993) is the feedforward multilayer perceptron shown in Figure 2. The input

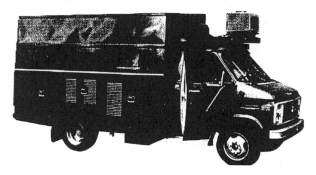

Figure 1. The CMU Navlab autonomous navigation testbed vehicle.

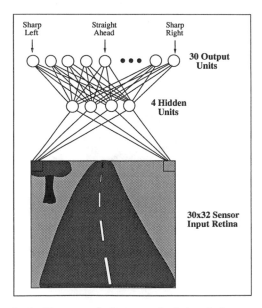

Figure 2. Architecture of the network designed for autonomous driving.

layer consists of a single 30×32 unit "retina" onto which a video image is projected. Each of the 960 input units is fully connected to the four-unit hidden layer, which is in turn fully connected to the output layer. Each of the 30 output units represents a different possible steering direction. The center-most output unit represents the "travel straight ahead" condition, while units to the left and right of center represent successively sharper left and right turns.

To drive the Navlab, an image from the video camera is reduced to 30×32 pixels and projected onto the input layer. After propagating activation through the network, the output layer's activation profile is translated into a vehicle steering command. Instead of simply selecting the output with the highest activation level, as is often done in classification tasks like character recognition and speech recognition (see WINNER-TAKE-ALL MECHANISMS), the steering direction dictated by the network is taken to be the center of mass of the "hill" of activation surrounding the output unit with the highest activation level (an example of *coarse coding*; see LOCALIZED VERSUS DISTRIBUTED REPRESENTATIONS). Using the center of mass of activation, instead of the most active output unit, to determine the direction to steer permits finer steering corrections, thus improving ALVINN's driving accuracy. This distributed output representation also makes generalization to similar situations easier for the network, since slight shifts in the road position in the input image lead to only slight changes in any single output unit's target activation.

Training "On-the-Fly"

The most interesting and novel aspect of the ALVINN system is the method used to train it. In this technique, called training "on-the-fly," the network is taught to imitate the driving reactions of a person. As a person drives, the network is trained with backpropagation (see BACKPROPAGATION: BASICS AND NEW DEVELOPMENTS) using the latest video image as input, and the person's steering direction as the desired output.

To facilitate generalization to new situations, variety is added to the training set by shifting and rotating the original camera image in software to make it appear that the vehicle is situated differently relative to the road ahead. The correct steering direction for each of these transformed images is created by altering the person's steering direction for the original image to account for the altered vehicle placement. So, for instance, if the person were steering straight ahead, and the image were transformed to make it appear the vehicle is off to the right side of the road, the correct steering direction for this new image would be to steer toward the left in order to bring the vehicle back to the road center. Adding these transformed patterns to the training set teaches the network to recover from driving mistakes, without requiring the human trainer to explicitly stray from the road center and then return.

ALVINN Driving Performance

Running on two Sun Sparcstations onboard the Navlab, training on-the-fly requires about two minutes during which a person drives over about a 1/4- to 1/2-mile stretch of training road. During this training phase, the network typically is presented with approximately 50 real images, each of which is transformed 15 times to create a training set of 750 images.

Figure 3. Video images taken on three of the roads ALVINN has been trained to handle.

Once it has learned, the network can accurately traverse the length of the road used for training, and also generalize to drive along parts of the road not encountered during training under a variety of weather and lighting conditions. In addition, since determining the steering direction from the input image merely involves a forward sweep through the network, the system is able to process 10 images per second and drive at up to 55 mph. This is over five times as fast as any nonconnectionist system has driven using comparable hardware (Crisman and Thorpe, 1990; Kluge and Thorpe, 1990).

The flexibility provided by the neural network has allowed ALVINN to learn to drive in a wide variety of situations. Individual networks have been trained to drive on single-lane dirt and paved roads, two-lane suburban and city streets, and multilane divided highways. Images taken from three of these domains are shown in Figure 3. On the highway, ALVINN has driven for up to 90 miles without human intervention.

For each road type, the feature detectors the network develops are slightly different. By examining the hidden unit representation, we find that for dirt roads the network develops "rut" detectors. For unlined paved roads, the network develops edge detectors and detectors for trapezoidal-shaped road regions. For lined highways, the network develops detectors for the lane markings. This ability to adapt its processing to fit the situation makes ALVINN more flexible than other autonomous driving systems (Crisman and Thorpe, 1990; Kluge and Thorpe, 1990).

Analysis

There are a number of factors that contribute to the success of neural networks in autonomous driving. The first is the fact that the training set is augmented with additional patterns by shifting and rotating each live video image. If the network is trained exclusively on live camera images, when driving autonomously it soon makes a steering mistake from which it has not learned to recover, and as a result drives off the road.

Perhaps a more fundamental characteristic of the task that makes autonomous driving easier than tasks such as speech and character recognition is the fact that spatial invariance is not required in this domain. In fact, it is the positions of features like lane markings in the input image which determines the correct steering response. Since spatial invariance is difficult to achieve using artificial neural networks, tasks which do not require this property are easier to learn.

A third factor contributing to ALVINN's success has been the care taken in task decomposition. Instead of training a single network to drive in all situations, separate networks are trained for different road types. For example, different networks are trained to drive on city streets and on divided highways. Currently, the choice of which network to use for a particular situation, and when a new network needs to be trained, is made manually. Open questions in the area of neural-network-based autonomous driving are (1) how to automatically decompose the driving task into manageable subtasks, each of which can be handled by a single network; and (2) how to automatically select the appropriate pretrained network for the current situation (see MODULAR AND HIERARCHICAL LEARNING SYSTEMS).

Acknowledgments. The principal support for the Navlab has come from DARPA, under contracts DACA76–85–C–0019, DACA76–85–C–0003, and DACA76–85–C–0002. This research was also funded in part by a grant from Fujitsu Corporation.

Road Map: Control Theory and Robotics
Background: I.3. Dynamics and Adaptation in Neural Networks
Related Reading: Active Vision; Reactive Robotic Systems

References

Crisman, J. D., and Thorpe, C. E., 1990, Color vision for road following, in *Vision and Navigation: The CMU Navlab* (C. E. Thorpe, Ed.), Boston: Kluwer, pp. 9–24.
Kluge, K., and Thorpe, C. E., 1990, Explicit models for robot road following, in *Vision and Navigation: The CMU Navlab* (C. E. Thorpe, Ed.), Boston: Kluwer, pp. 25–38.
Pomerleau, D. A, 1991, Efficient training of artificial neural networks for autonomous navigation, *Neural Computat.*, 3:88–97.
Pomerleau, D. A., 1993, *Neural Network Perception for Mobile Robot Guidance*, Boston: Kluwer.

Vision: Hyperacuity

Shimon Edelman and Yair Weiss

Introduction

What Is Hyperacuity?

The term *visual acuity* refers to the ability of an observer to resolve fine spatial details in a scene. The observer's acuity can be estimated using the familiar eyechart in an optometrist's office. Alternatively, it can be measured by determining the observer's discrimination threshold in a simple visual task, similar to those shown in Figure 1. In the latter case, the acuity estimate is found to be highly dependent on the task itself.

While in some tasks (e.g., in telling apart two nearby dots) thresholds are in the range of 30–60 arcsec, in other tasks such as the vernier, the threshold may be as low as 5 arcsec. A threshold of 5 arcsec means that the observer reliably resolves features that are less than 0.02 mm at a 1-m distance, or the size of a quarter viewed at 1 km!

One can better appreciate the astonishing precision of this performance by considering the optical properties of the eye. In the spatially most sensitive region of the retina, the fovea, the diameter of the photoreceptors is in the range of 30–60 arcsec,

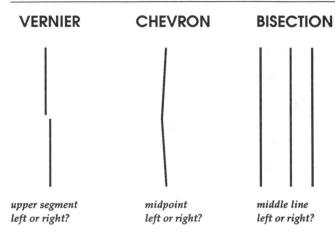

VERNIER **CHEVRON** **BISECTION**

upper segment *midpoint* *middle line*
left or right? *left or right?* *left or right?*

Figure 1. Three examples of stimuli (vernier, chevron, and bisection) used for testing spatial acuity of the visual system. The task is to tell the sense of the offset of some parts of the stimulus with respect to the others.

and the sizes of the receptive fields of the retinal ganglion cells may be even larger. Thus, humans can resolve detail with an accuracy of better than one-fifth the size of the most sensitive photoreceptor. G. Westheimer coined the term *hyperacuity* to describe such performance (Westheimer, 1981).

Why Is Hyperacuity Interesting?

Visual hyperacuity has been studied by experimentalists for more than 100 years, the first report of vernier hyperacuity dating back to 1892. By now, probably the most interesting point about hyperacuity is that it should not come as a surprise, if considered in the context of the computational neuroscience of the visual system. First, because of the point spread function of the eye's optics, a very small dot of light projected onto the retina may activate as many as 40 different photoreceptors. Thus, there is no a priori reason to expect a simple relation between the acuity exhibited by the system and the spacing of adjacent photoreceptors. Second, there is no reason to expect identical thresholds for different tasks, unless it is assumed that the first stage of vision, common to all subsequent processing, amounts to an internal reconstruction of the outside world in some unique and veridical way (for an alternative view, see ACTIVE VISION).

Computational Characterization of Hyperacuity

Models of visual information processing can be thought of as lying between two extremes. There is the notion that the goal of the system is to reconstruct the visual scene and display the result for the benefit of a decision module (it is difficult to resist calling this module anything but a homunculus). Conversely, there is the approach, related to the notion of direct perception, according to which vision is a process of transduction into an abstract feature space followed by classification in this space. Among the models of hyperacuity, one can find instances both of the reconstructionist and of the feature-space approaches.

Reconstructionist Models

In a classic paper entitled "Reconstructing the Visual Image in Space and Time," H. Barlow pointed out that the phenomenon of visual hyperacuity is analogous to our ability to perceive films at the cinema (Barlow, 1979). In a film, shown as a series of discrete frames, we perceive a continuous motion, presumably through a process of temporal interpolation. Likewise, in a hyperacuity task, although the photoreceptors provide a discrete sampling of the retinal image, we perceive detail smaller than the sampling rate, a process of spatial interpolation. Barlow suggested that the first stages of visual processing may be concerned with such interpolation (see also Crick, Marr, and Poggio, 1981). The output of the spatial interpolation process, he wrote, would be a reconstruction of the retinal image on a finer grid, which may correspond to the granular cells of the cortical layer $4c\beta$ found in the macaque's primary visual area. Such fine-grid reconstruction of the stimulus would then support hyperacuity-level performance by allowing a subsequent stage to base its decision on the activity of individual units, an impossible feat as far as the raw retinal image is concerned.

The Shannon-Nyquist sampling theorem states that if the signal is bandlimited, a sampling rate higher than twice the maximal frequency would enable perfect interpolation. Because of the optical point spread function, the retinal image is indeed bandlimited to spatial frequencies less than approximately 60 cycles/degree; hence, a sampling rate of 30 arcsec should suffice to reconstruct the retinal image. However, in many experiments, the sampling rate does not satisfy the requirements of the sampling theorem. For example, the subjects of Fahle and Poggio (1981) were shown discrete frames of the motion of a vernier target drifting at a constant velocity. Resolving the misalignment of the vernier thus entailed both spatial and temporal interpolation, with the sampling rates Δx, Δt imposed on the system. Fahle and Poggio found that humans can still solve the vernier task, at thresholds significantly lower than photoreceptor spacing, for a wide range of sampling rates. They concluded that an additional computational constraint is used by the visual system for spatiotemporal interpolation, namely, the assumption of constant velocity, and that this assumption may explain why the system opts for interpolation in the first place. Extending Barlow's suggestion, they proposed that the goal of interpolation may be broader than simple subreceptor precision, and that it involves some correction in the visual representation by enforcing *continuity* of the visual field in space and time (see REGULARIZATION THEORY AND LOW-LEVEL VISION).

Feature-Space Models

In 1986, H. Wilson showed that the responses of filters, similar in properties to the psychophysically defined spatial frequency channels and to the oriented receptive fields of cells in the primary visual cortex, can serve as the feature space for solving hyperacuity tasks. In Wilson's model, stimuli are represented as points in a multidimensional filter response space. The basic prediction of the model is that the threshold for discriminating two patterns is monotonically related to the Euclidean distance between the local representations of the two patterns, $\|\Delta R\|$. The term *local* here refers to the assumption that not all filter responses contribute to the calculation of ΔR: for each value of orientation and size, only filters overlapping the stimulus, as well as their spatial nearest neighbors, are used. Thus, a stimulus is coded by the response of a population of cells. See LOCALIZED VERSUS DISTRIBUTED REPRESENTATIONS for more on such *coarse coding*.

To test this prediction, Wilson computed the threshold of the model for a wide range of hyperacuity tasks and obtained a quantitative fit with psychophysical data, with the parameters of the filters determined independently by previous experiments. Wilson's model also provided an intuitive explanation of how widely tuned, discretely spaced filters can support the resolution of small features. The most informative filter, Wilson observed, is not necessarily the one which is most active. For example, in the vernier task, the most informative filters are the two oriented filters flanking the stimulus, which are significantly less active than the filter oriented at the same orientation as the stimulus.

A possible architecture for classification in the filter space was suggested by Poggio, Fahle, and Edelman (1992). In their model, the stimulus is transduced into $\mathbb{R}^n$ by a convolution with a bank of widely tuned, circularly symmetric Gaussian filters. Classification, or judging the offset of the vernier, is then carried out using radial basis function approximation (Poggio and Girosi, 1990; see RADIAL BASIS FUNCTION NETWORKS) in $\mathbb{R}^n$. Poggio et al. found that this model can resolve verniers with offsets significantly smaller than the spacing between the Gaussian filters, again demonstrating the importance of pooling of information between units. Subsequently, it was shown that if Wilson's filters are used as transducers into $\mathbb{R}^n$, then the classification can be performed using a simple perceptron (Weiss, Edelman, and Fahle, 1993).

Comparison Between Models

It seems that the reconstructionist models of hyperacuity were motivated equally by the desire to explain the performance of the visual system in acuity tasks, and by the desire to find a computational rationale for the existence of the dense sheet of cells in layer $4C\beta$ of the macaque primary visual cortex. The appeal of such models was somewhat diminished by the absence of a comparable dense representation area in cats, given the cat's ability to solve the vernier task at a hyperacuity level (Swindale and Cynader, 1986). Specifically, at least in this case, Barlow's fine-grid reconstruction scheme just outlined has no fine grid to rely on, and an alternative explanation seems to be required. Still, many of the insights on spatiotemporal interpolation remain valid (e.g., the utility of interpolation in avoiding motion smear and in filling in gaps in the retinal input caused by local occlusion), even when the interpolation is assumed somehow to be mixed with the decision procedure, rather than merely leading to an explicit reconstruction of the input on a finer grid.

The Neural Basis of Hyperacuity

What neurobiological mechanisms in the brain support hyperacuity-level performance? To obtain a comprehensive answer to this question, one must relate data from anatomy and physiology (e.g., dendritic tree shapes and functional tuning curves of neurons) to behavioral findings (e.g., thresholds) measured by psychophysicists.

Parker and Hawken (1985) suggested that this relation can be elucidated by measuring the psychometric function of a single unit, called the *neurometric function*. Rather than averaging the firing rate of a neuron over many presentations, this function captures the ability of a neuron to respond correctly to a single stimulus presentation. A response is defined to be correct if it agrees with the unit's behavior in the long run. For example, a neuron tuned to vertical bars is said to respond correctly in an orientation discrimination trial if its firing rate exceeded a certain value, given that the stimulus in that trial was vertical.

Once the neurometric curve (a plot of the percentage of correct responses versus the difference between the two patterns) is plotted, the behavioral threshold of the neuron can be determined by standard regression methods such as those used in psychophysics.

Applying this technique to neurons in the visual cortex of an anesthetized monkey, Parker and Hawken found that some neurons had thresholds significantly below photoreceptor spacing. Motivated by Barlow's suggestion that hyperacuity is linked to nonoriented neurons in cortical layer $4C\beta$, they compared the performance of these neurons with that of orientationally selective cells. They found that the nonoriented neurons performed worse in localization tasks. Using a similar technique, Swindale and Cynader (1986) measured the vernier threshold of cells with oriented receptive fields in cortical area 17 of the cat. They found cells with vernier thresholds as low as 6 arcmin.

The significance of these experimental results depends on the assumption that the threshold of a single cell limits the behavioral threshold of the organism. If an organism indeed can only perform as well as its best-tuned single neurons, then the finding of neurons with exceptionally low thresholds would constitute a step toward unraveling the neural basis of hyperacuity. If, however, information sufficient for solving the task at a hyperacuity level is available in the coarse-coded population response of only marginally selective units, then the response of each such unit does not tell us much about the emergent capability of the entire system.

A computational study by Snippe and Koenderink (1992) suggests that the population coding scenario may be more relevant to the understanding of hyperacuity than the single-unit scenario. Snippe and Koenderink determined the discrimination threshold of an ideal observer given the activities in an array of noise-perturbed receptors, which were assumed to have a Gaussian response profile, with an unknown width. These authors derived a formula for the threshold as a function of the width of the Gaussian, the distance between two neighboring receptors, and the functional dependence between the noise and signal in each receptor. They found that the threshold as compared with the tuning width can be made arbitrarily small if the individual receptors show high fidelity, or if there is considerable overlap between the receptors. Thus, an ensemble of cells with large *single-cell* thresholds may in fact support hyperacuity-level *behavioral* thresholds.

Perceptual Learning in Hyperacuity

Human performance in a vernier task improves with practice (that is, the threshold decreases), by as much as 40% over 2000 trials (McKee and Westheimer, 1978). As in many other cases of perceptual learning, improvement is specific for stimulus parameters such as orientation and the eye to which the stimulus is presented. Interestingly, learning does not require that the subjects be informed as to the correctness of their responses.

Poggio et al. (1992) proposed that improvement in performance with practice may result from synthesizing a perceptual module for the task by acquiring examples of the input-output relationship and by subsequent optimization of the use of the stored examples with practice. They demonstrated that a model based on this principle can learn to solve hyperacuity tasks starting with a *tabula rasa* state and continuously improving its performance. The basic prediction of their model is that generalization (supported by conservative interpolation among the stored examples) should be limited to stimuli that are similar to the previously seen ones, as indeed is observed in psychophysical experiments.

The model suggested by Poggio et al. requires feedback for parameter update. Weiss et al. (1993) developed a model that operates without feedback, in which learning is achieved not by synthesizing a new perceptual module, but rather by improving the performance of an existing one, based on a stimulus-driven, feedback-independent amplification of unit responses. They proposed a simple feedback-independent rule for synaptic modification and showed that this rule can improve the signal-to-noise ratio in an existing network module, as a result of repeated exposure to the stimulus. The psychophysical data on the time course of learning indicates that it can be divided into an initial fast and a subsequent slow and prolonged components. This finding suggests that at least two different learning strategies may coexist in the visual system, perhaps corresponding to the two rules just outlined.

Discussion

To conclude, we point out that the two most intriguing aspects of hyperacuity (the surprisingly high resolution exhibited by a "sloppily built" system, and the dependence of the resolution on the exact nature of the task) may be found in many biological systems, both in vision and in other natural computation domains, ranging from bat echolocation to jamming avoidance response of electric fish (Altes, 1988). The study of this phenomenon indicates that issues vitally important for the understanding of how the brain works—representation, processing, and learning—are frequently best addressed with the simplest of stimuli.

Road Map: Vision
Related Reading: Echolocation: Creating Computational Maps; Electrolocation; Visual Coding, Redundancy, and "Feature Detection"; Visual Cortex Cell Types and Connections

References

Altes, R. A., 1988, Ubiquity of hyperacuity, *J. Acoust. Soc. Am.*, 85: 943–952.

Barlow, H. B., 1979, Reconstructing the visual image in space and time, *Nature*, 279:189–190.

Crick, F. H. C., Marr, D. C., and Poggio, T., 1981, An information-processing approach to understanding the visual cortex, in *The Organization of the Cerebral Cortex* (F. Schmitt, Ed.), Cambridge, MA: MIT Press.

Fahle, M. W., and Poggio, T., 1981, Visual hyperacuity: Spatiotemporal interpolation in human vision, *Proc. R. Soc. Lond. B Biol. Sci.*, 213:451–477.

McKee, S. P., and Westheimer, G., 1978, Improvement in vernier acuity with practice, *Percept. & Psychophys.*, 24:258–262.

Parker, A. J., and Hawken, M. J., 1985, Capabilities of monkey cortical cells in spatial resolution tasks, *J. Opt. Soc. Am.*, 2:1101–1114.

Poggio, T., Fahle, M., and Edelman, S., 1992, Fast perceptual learning in visual hyperacuity, *Science*, 256:1018–1021.

Poggio, T., and Girosi, F., 1990, Regularization algorithms for learning that are equivalent to multilayer networks, *Science*, 247:978–982.

Snippe, H. P., and Koenderink, J. J., 1992, Discrimination thresholds for channel-coded systems, *Biol. Cybern.*, 66:543–551.

Swindale, N. V., and Cynader, M. S., 1986, Vernier acuity of neurones in cat visual cortex, *Nature*, 319:591–593.

Weiss, Y., Edelman, S., and Fahle, M., 1993, Models of perceptual learning in vernier hyperacuity, *Neural Computat.*, 5:695–718.

Westheimer, G., 1981, Visual hyperacuity, *Prog. Sens. Physiol.*, 1:1–37.

Visual Coding, Redundancy, and "Feature Detection"

David J. Field

Introduction

This article looks into the question of why cells in the visual pathway have the response properties that they do. The emphasis of this article will be on information processing approaches that consider the redundancy of the signal and the transformation of the signal as it passes along the visual pathway. We will consider three different information processing strategies. It will be proposed that each strategy is employed by the visual system to handle particular types of redundancy. However, before we consider these strategies we need to understand the notion of a "feature detector." It is a concept that is widespread in the literature and is often put forward as an opposing approach to an information processing strategy.

In every animal with a visual system, one can find cells that are selective to particular properties of the animal's visual environment. When early single unit recordings found cells responding to what seemed to be a meaningful stimulus (e.g., a neuron in the retina of the frog responding to the presence of a moving spot), many researchers were quick to attribute function (e.g., they were fly detectors). The notion that single cells signal the presence of particular features is widespread. Cells in the mammalian visual cortex are selective to a limited band of spatial frequencies. This led numerous researchers to suggest that these cells coded the presence of particular Fourier coefficients. These same cells are selective to the orientation and location of edges, and this led other researchers to suggest that these cells were "edge detectors."

The danger of describing a particular cell as an "X" detector is that it implies (1) that the cell responds only when that particular feature is present (e.g., an edge detector responds only in the presence of an edge) and (2) that it signals only the presence of that feature (e.g., signals the presence of an edge when the cell responds). It will be argued in this article that this kind of feature-specific coding is not typical of any of the cells in the early visual system [i.e., retina, lateral geniculate nucleus (LGN), or visual cortex] and therefore describing them as "detectors" of any type of feature is both misleading and inaccurate.

Simple cells in primary visual cortex are not edge detectors, any more than long wavelength selective cones can be described as "red spot detectors." These cells respond to and are certainly involved in the representation of stimuli other than edges and red spots. The debate regarding whether a cortical cell is a grating detector, an edge detector, or even a "Gabor function detector" is misplaced. For a cell to act as a feature detector, the cell must show a high degree of spatial nonlinearity that restricts its response to a very specific stimulus. As will be argued, this may be true of cells high in the visual pathway (e.g., inferotemporal cortex). However, for cells early in the visual pathway, it is argued that one must understand how the signal as a whole is transformed. We will return to the discussion of feature detection in the final section, where we will attempt to bridge the gap between information processing and cells selective to the particular features in an image.

Information Processing

Statistically speaking, the visual world is a very special place. Because of the physics of how objects and surfaces reflect light, the images that are projected onto our retinas are highly constrained and redundant. Following from the classic work of Shannon, *information processing* concentrates on the redundancy in the signal and the transmission of the signal through a *channel* that has limited bandwidth or is subject to noise. It is important to recognize that this approach does not concentrate on the "meaning" or interpretation of the signal. Rather, the emphasis is on the transform that allows for an accurate transmission of the signal given the limitations of the system.

This article is not intended as an introduction to the basics of information theory. This can be found in numerous textbooks and some papers (e.g., see Cover and Thomas, 1991; Atick, 1992; and INFORMATION THEORY AND VISUAL PLASTICITY). However, some general terms need to be noted. *Entropy*, for example, refers to the range of possible states that a signal is likely to have: With an image as the signal, we can consider the range of possible images that a code is likely to encounter. If the image class is large (e.g., natural scenes), determining the probability of each image is impossible. Rather, one determines the size of the population from the conditional relations among the input vectors (e.g., the correlations between pixels). The *redundancy* of the signal refers to these conditional relations. The higher the redundancy, the lower the range of possible signals, and the lower the entropy.

It is common to consider redundancy in terms of some *nth*-order conditional probability among the set of symbols used to represent the input. *First-order* statistics typically refer to the probabilities that individual symbols are used. *Second-order* refers to the pairwise conditional probabilities among the symbols (e.g., correlations). *Third-order* refers to conditional probabilities defined among triplets of symbols, etc. Unfortunately, most introductions to information theory use letter frequencies in language as examples of these different forms of redundancy. Letters are binary (present or not present), so the letter frequency is a complete description of the first-order statistics. However, when the symbols (e.g., the cells) have a continuous response distribution, there are two components of the first-order statistics: the variance of the response distribution and the shape of the response distribution. Given a fixed variance, a normal distribution in responses has the highest entropy and lowest redundancy (i.e., the normal distribution is most random). The shape of the response distribution will become quite important in the discussions that follow.

Redundancy is often discussed in terms of bit rates. However, this can become extremely complex when discussing continuous distributions. To understand how codes transform continuous distributions, it can be more useful to consider redundancy in terms of the state-space of possible inputs (e.g., Field, 1994). The state-space represents the space of all possible inputs and captures all of these high-order conditional probabilities in terms of the geometry of the space. Consider the case of an image consisting of three pixels. All possible images can be represented by a three-dimensional space where the coordinate axes represent the amplitudes of the three pixels. Random data (e.g., white noise) have maximum entropy and will fill this space uniformly. If there is any redundancy in the data, then the distribution of inputs must "clump" in some way. Although high-dimensional data require a high-dimensional state-space, as we will see, a number of general principles can be applied.

The other advantage of using the state-space is that many transforms are described by simple manipulations of the coordinate system. For example, if we use the pixel values as the coordinates of the initial state-space (defined as the *basis vectors*), then an orthonormal transform (e.g., a Fourier transform) is represented by a rotation of the coordinate axes. Each of the new basis vectors is represented by a linear sum of the old basis vectors. They remain orthogonal and of "normal" length. From this point of view, information processing strategies are interpreted in terms of the way they rotate, transform, or otherwise distort the state-space of probable inputs. Indeed, it will be argued in the following sections that a "good" information processing strategy transforms higher-order redundancy (conditional relations between vectors) into first-order redundancy (i.e., into changes in the variance or shape of the response distributions).

The information processing approach requires that one understand the redundancy inherent in the signal. To apply this approach to the visual system, one must understand the statistics of the visual environment. Until recently, few studies have tried to measure statistical redundancy in natural environments. There appears to be an implicit belief that the natural environment is quite random, but that is a misconception. Images with random independent pixel values have no redundancy. However, images of the natural environment (natural scenes) are mathematically quite unique. The following three subsections discuss three techniques for dealing with this highly redundant signal.

Redundancy Reduction: Compact Codes

When information processing strategies are applied to sensory systems, the most common approach is to consider how the sensory code removes redundancy. This was Attneave's (1954) and Barlow's (1961) basic proposal, and it has served as the basis of numerous image-processing strategies and neural networks (e.g., Atick, 1992; van Hateren, 1992). The general intent of this approach is to reduce the set of symbols (i.e., basis vectors) coding the signal, without losing information about the signal. For example, in a three-dimensional state-space, if all the data fall in plane, then it will be possible to apply a transform such that all the information in the signal is represented with only two vectors. The description of the stimulus with only these two vectors has less redundancy (i.e., the data in the two-dimensional subspace are more uniformly distributed than the data in the three-dimensional space).

If there exists a subspace that contains most of the data, then one can find that subspace using a technique called PRINCIPAL COMPONENT ANALYSIS (q.v.). The analysis depends on only the pairwise correlations in the input. If the data are highly correlated, then there will exist a subspace that will describe most of the variance in the signal. The code that represents this subspace has been called a *compact code* (Field, 1994).

Trichromacy is one example of a compact code. It has been noted that the chromatic spectra of most naturally occurring surfaces are relatively smooth (e.g., Maloney, 1986). That is, the spectra do not typically contain spikes. The strong correlations in spectra imply that there exists a subspace that can account for most of the variation in different spectra. Indeed, principal components analysis of natural spectra shows that most of the variance is described by the first three principal components. However, it is important to recognize that the principal components are not equivalent to the response profiles of the three cones. In general, the principal components are useful for identifying any subspace that is capable of describing the data. However, there can be many ways to describe that particular subspace and the principal components may not (and probably will not) provide the optimal vectors to describe it.

In applying redundancy reduction to the spatial domain, one must first understand the concept of stationarity. *Stationarity* means that across the populations of inputs, the statistics at one location are no different from any other (e.g., the statistics do not change as one moves around the image). When the statistics are stationary, then the principal components are described by the amplitude spectra of the data (i.e., the amplitudes of the Fourier coefficients will be uncorrelated). Natural scenes can largely be described as having stationary statistics. Therefore, the average amplitude spectrum will describe the principal components.

The scale invariance of natural scenes produces Fourier spectra with amplitudes that fall with spatial frequency (f) as approximately $1/f$ (Field, 1987). This means that to capture most of the variance with the fewest number of vectors, one should choose the low spatial frequencies. In terms of the spatial properties of natural scenes, a redundancy reduction strategy will simply remove the high spatial frequencies. We see evidence of this strategy in the visual system in the transformation from the retina to the optic nerve. In the human, there are over 100 million photoreceptors and only 1 million optic fibers. The compression is primarily achieved by throwing out high-frequency information in the periphery. This captures most of the variance in the image with the smallest number of cells.

A variety of neural networks implicitly or explicitly perform compact coding. If a network forces data through a bottleneck (i.e., a reduced dimensionality) and attempts to solve a problem that requires a large proportion of the variance in the signal, then the network will implicitly be performing compact coding. Under the right constraints, Hebbian learning has been shown to be capable of producing the principal components (see PRINCIPAL COMPONENT ANALYSIS). However, it must be reemphasized that the principal components are not likely to be particularly useful other than in identifying the subspace that can account for most of the variation in the stimulus (i.e., the entropy). Once one has selected the subspace (e.g., thrown out the high frequencies beyond the acuity limit), the directions of the principal components are probably not particularly important. To decide how to efficiently code that subspace, other factors need to be considered.

Redundancy Transformation: Sparse Codes

In the compact codes described above, the goal is to find a subset of vectors which account for most of the information across the population of inputs. In this section, we want to consider codes which distribute the information across all the vectors, but which use each vector relatively rarely. These codes are described as *sparse codes* because for any given input, most of the units are not responding. Sparse codes are possible only when the redundancy of the data has the correct form— i.e., when each input can be described by a small number of basis vectors but where a larger number of vectors are required to describe all the inputs. To give a simple example, if all the inputs in a three-dimensional set fell in three orthogonal planes, then any given input could be described with only two vectors while three would be required for the entire population.

Sparse codes have response histograms (i.e., first-order statistics) with high redundancy. As was noted, a Gaussian histogram has the highest entropy (lowest redundancy) of any distribution given a fixed variance (as opposed to a flat histogram which has the highest entropy given a fixed range). For most data sets (e.g., natural scenes) if one chooses a random vector (i.e., creates a random receptive field), the response histogram over the population of inputs is likely to be Gaussian. Relative to a Gaussian, sparse codes have histograms that show a high probability of no response and an increased probability of a large response. This change can be described in terms of a statistic called *kurtosis*, which represents the fourth moment of the distribution (Field, 1994). Because the response histograms deviate from Gaussian, they have less entropy (more redundancy). Sparse codes increase this first-order redundancy (the response histograms) by decreasing the higher-order redundancy (e.g., the relations among pixels). This transformation of redundancy has been proposed to account for the principal spatial properties of cells in the mammalian visual system and to explain why the wavelet transform (see WAVELET DYNAMICS) has proven so popular in applied mathematics (Field, 1993).

The *wavelet transform* consists of basis vectors that are all scaled versions of each other (i.e., only differ by translations, dilations, or rotations). The transform has been applied to a variety of naturally occurring data sets from turbulence to earthquakes (e.g., Farge, Hunt, and Vassilicos, 1993). It also captures the principal spatial response properties of cells in the visual cortex (oriented, localized self-similar receptive fields). It has been proposed that natural scenes have the type of redundancy that allows a wavelet transform to produce sparse response histograms (Field, 1987). In a given natural scene, edges occur at only a small number of locations and scales, but across all natural scenes they are likely to occur at all possible locations and scales. Because of this structure, only a subset of the wavelet basis vectors are required to code a given scene, but the full set is required to code all natural scenes. It has been demonstrated that when wavelet transforms are applied to digitized natural scenes, transforms with the properties of the cells in primary visual cortex are near to optimal for converting high-order to first-order redundancy (Field, 1987, 1993, 1994). That is, when coding natural scenes, artificial image transforms produce the most sparse responses when the bandwidths and spatial distributions of receptive fields are like that found in the visual cortex.

A number of studies have suggested that sparse codes could be useful to an organism (see SPARSE CODING IN THE PRIMATE CORTEX, and Field, 1994, for a review). However, it must be emphasized one cannot choose to perform a sparse code simply because one considers it to be useful. As with compact codes, the data must have the appropriate form of redundancy. Unlike compact codes, sparse coding does not depend on the correlations in the data and does not depend on the principal components. For natural scenes, or any signal with stationary statistics, the information that allows sparse coding is found using other statistics, which in Fourier terms is described by the phase spectrum.

Sparse coding represents one method of minimizing the relations between the basis vectors by converting high-order redundancy to first-order redundancy. Images or objects within images are represented by a relatively small number of active cells where the activity of a cell provides a high degree of information about the local structure. We will come back to this point. The final information processing strategy considers codes that make use of highly nonlinear vectors and conforms more closely to the notion of a feature detector.

Redundancy Specialization: Combinatorial Codes

A code is described as a *factorial code* (e.g., see Schmidhuber, 1992) when the response probabilities of all the vectors are independent of one another, given a particular set of inputs (i.e., the response of each vector provides no information about the responses of the other vectors). Both redundancy reduction and redundancy transformation increase the independence of the vectors of the code. Indeed, the cells in primary visual cortex may be as independent as possible given the statistics of the environment and the number of cells available. However, this cortical representation is by no means a factorial code. Because of the many forms of structure in the natural environment (e.g., continuity of borders, predictable relations of features within objects), the responses of these cells will not be independent.

It may be possible to capture the lack of independence using a specific type of nonlinearity. These codes will be described as *combinatorial codes* in line with the combination coding described by Tanaka et al. (1991). These codes show similarities to the sparse codes discussed above, but as described below, require a specific type of nonlinearity and are realistic only when they are used for a portion of the total entropy.

For these codes, each cell responds only when all of a constituent set of input cells respond. That is, the cell responds only in the presence of a particular combination of features and does not respond when that combination is not completely present. This nonlinear summation is described as an AND operation: it only responds if unit A responds *and* unit B responds (e.g., Zetzsche and Barth, 1990). The difficulty with this approach is that it requires an exponentially large number of cells to describe all the possible combinations. For a code with m inputs, the total number of pairwise combinations is $(m^2 - m)/2 \cong m^2$. In general, the total number of nth-order combinations is approximately m^n. If we consider the roughly 1 million optic fibers as the number of coordinates in V1, and we want to consider all possible second-order relations, we are talking about 10^{12} pairwise combinations. This is clearly an unreasonable strategy for coding complex scenes, and with more complex combinations (i.e., fifth order), complete codes become impossible, given the number of cells in the brain.

There are two solutions to this exponential explosion. First, one can limit this type of coding to a very specific subset of combinations that occur with relatively high probability or to those that are particularly meaningful. Faces, for example, represent a combination of features that occur with much higher probability than predicted from the probabilities of each of the individual features. Faces are also meaningful. So by either criterion, faces represent one possible direction for a combinatorial code. And there is considerable evidence for such detectors in the inferotemporal of primates (see SPARSE CODING IN THE PRIMATE CORTEX). There is also evidence of a variety of other types of combinatorial units (Tanaka et al., 1991).

However, even with the restriction of this coding process to relatively probable or meaningful combinations, the exponential explosion may still be too great. A face detector at every place in the visual field at all the scales that the face might occur requires as many detectors as found in the optic nerve. Anderson, Olshausen, and Van Essen (see ROUTING NETWORKS IN VISUAL CORTEX) have recently suggested an alternative strategy. They propose a biologically plausible "shifter circuit," which allows the image to be scaled and shifted to a normalized location. This process would vastly reduce the number of cells

required to code an object, but it does require that the system perform sufficient processing to identify the size and location of the object so that the shifter circuit can determine how to normalize the image. Biederman (1987; see OBJECT RECOGNITION) has suggested that human observers have rather specific spatial representations of objects that may number as high as 30,000. Clearly, building detectors at every possible scale and position is not realistic. However, if each object can be scaled and shifted to a normalized location and a characteristic view, 30,000 object detectors may not be unrealistic.

Combinatorial codes would be useful when there remain cases of high redundancy after the optimal sparse code has been determined, i.e., when the probability of a particular combination of vectors is greater than that predicted from the individual probabilities, as when $P(V_1 \text{ and } V_2 \text{ and } V_3) \gg P(V_1) \cdot P(V_2) \cdot P(V_3)$.

Combinatorial codes are necessarily nonlinear, requiring all of the constituent units to respond for the combined unit to respond. Such units should be largely silent but give a strong response when the appropriate combination of features is present. Like sparse codes, the responses should be rare. Indeed, such object-specific detectors have been described as sparse codes (see SPARSE CODING IN THE PRIMATE CORTEX). However, it must be emphasized that unlike the sparse coding described above, combinatorial codes must be nonlinear.

Combinatorial codes could theoretically preserve information, but this is unlikely considering the number of units required. Realistically, these codes are likely to represent only a small fraction of the total entropy of the input and be applied only when the probability of a combination is considerably higher than the probability predicted by independence. If this is the case, then these codes will be "blind" to low probability events. To allow the perception of novel combinations, later stages of the system would require access to the cells in the visual cortex and be pliable enough to build new configurations. By this scenario, the system as a whole will have access to both the cells in the primary visual cortex as well as later stages involved in combinatorial coding. Combinatorial codes, therefore, provide a means of coding specific objects and configurations that are common in the environment. However, one should not assume that all information in our visual environment is processed using these "feature detectors."

Discussion

In one sense, each of the strategies discussed above represents a method of "betting" as to what sort of redundancy is likely to occur in the typical environment. Each strategy provides a means for reducing the complex relationships that are typically found between cells. After these codes are applied, the particular relationships between cells will describe what is unique to that input rather than what is common across all inputs.

These information processing strategies concentrate on the statistics of the environment; for the most part, they do not focus on how important particular information might be to an animal. Across mammalian species, the visual coding of spatial information follows very similar lines. This may be mostly due to strong similarities in spatial statistics across different environments. However, it is unlikely that all differences in the visual systems of different species can be accounted for in terms of the statistics of their environment (e.g., the hawk's environment is not likely to allow higher acuity than the environment of a dog). A complete account will certainly require some consideration of species-specific codes based on the specific tasks that the animal must face.

Finally, a brief comment should be made regarding neural networks and learning rules. As was noted, there are various techniques for finding compact codes. There also have been some recent suggestions on how networks might develop sparse representations (Földiák, 1990; see SPARSE CODING IN THE PRIMATE CORTEX). However, it may be difficult to find a single learning rule that can achieve all three information processing tasks. One of the main points of this article is that although each information processing strategy increases the independence of the units, they depend on different forms of redundancy. We may discover that the three information processing strategies require a combination of learning rules rather than a single rule.

Acknowledgment.   This work was supported by NIH grant MH50588.

Road Map: Vision
Related Reading: Gabor Wavelets for Statistical Pattern Recognition; Localized Versus Distributed Representations; Unsupervised Learning with Global Objective Functions; Vision: Hyperacuity; Visual Cortex Cell Types and Connections

References

Atick, J. J., 1992 Could information theory provide an ecological theory of sensory processing? *Network*, 3:213–251. ◆

Attneave, F., 1954, Some informational aspects of visual perception, *Psychol. Rev.*, 61:183–193.

Barlow, H. B., 1961, The coding of sensory messages, in *Current Problems in Animal Behavior* (W. H. Thorpe and O. L. Zangwill, Eds.), Cambridge, Eng.: Cambridge University Press, pp. 330–360.

Biederman, I., 1987, Recognition by components: A theory of human image understanding, *Psychol. Rev.*, 94:115–147. ◆

Cover, T. M., and Thomas, J. A., 1991, *Elements of Information Theory*, New York: Wiley.

Farge, M., Hunt, J., and Vassilicos, J. C., Eds., 1993, *Wavelets, Fractals and Fourier Transforms: New Developments and New Applications*, Oxford: Clarendon.

Field, D. J., 1987, Relations between the statistics of natural images and the response properties of cortical cells, *J. Opt. Soc. Am.*, 4:2379–2394.

Field, D. J., 1993, Scale-invariance and self-similar 'wavelet' transforms: An analysis of natural scenes and mammalian visual systems, in *Wavelets, Fractals and Fourier Transforms: New Developments and New Applications* (M. Farge, J. Hunt, and J. C. Vassilicos, Eds.), Oxford: Clarendon, pp. 151–193.

Field, D. J., 1994, What is the goal of sensory coding? *Neural Computat.* 6:559–601. ◆

Földiák, P., 1990, Forming sparse representations by local anti-Hebbian learning, *Biol. Cybern.*, 64:165–170.

Maloney, L. T., 1986, Evaluation of linear models of surface spectral reflectance with small numbers of parameters, *J. Opt. Soc. Am. A*, 3:1673–1683.

Schmidhuber, J., 1992, Learning factorial codes by predictability minimization, *Neural Computat.*, 4:863–879.

Tanaka, K., Saito, H., Fukada, Y., and Moriya, M., 1991, Coding visual images of objects in the inferotemporal cortex of the macaque monkey, *J. Neurophysiol.*, 66:170–189.

van Hateren, J. H., 1992, A theory of maximizing sensory information, *Biol. Cybern.*, 68:23–29.

Zetzsche, C., and Barth, E., 1990, Fundamental limits of linear filters in the visual processing of two-dimensional signals, *Vis. Res.*, 30:1111–1117.

Visual Cortex Cell Types and Connections

J. S. Lund, Q. Wu, and J. B. Levitt

Introduction

Cortical visual function is one of the most exciting areas of computational neuroscience, where models can be constructed based on neuroanatomical and neurophysiological findings and can be used to simulate cortical functional properties (for a review, see chap. 4 in Churchland and Sejnowski, 1992). However, it has been disappointing that the models rarely predict the existence of anatomical features that could be looked for in the real cortex. It would be invaluable for the anatomists to have some indicators as to what architectural features among so many possibilities could be crucial determinants of visual cortical function.

We provide a brief review of some features of the microcircuitry of the primary visual cortex, area V1, and the physiological properties of cells in its different laminae. We then outline several hypotheses as to how the anatomical structure and connections might serve the functional organization of the region. We outline these hypotheses as simple models to illustrate how theorists might help experimentalists to resolve important questions and formulate new experimental approaches.

Overview of the Anatomy of Area V1

The anatomy of the visual system has been studied extensively in the macaque monkey (for an introduction, see Hubel, 1988), whose visual system is believed to be very similar to that of humans. Two main groups of neurons are present within the cortex; the most numerous group (approximately 80%) are excitatory in their influence on other cells and are characterized morphologically by having small spines on their dendritic surfaces that act as specialized sites for synaptic contact. These spiny neurons either have dendritic processes of equal length—and are called *stellate neurons*—or have one dendrite on the pial aspect of the cell that is greatly extended compared to the others—and are called *pyramidal neurons*. The less numerous (20%) group of cortical neurons is composed of cells which are generally stellate in morphology, but largely lack dendritic spines. These cells usually contain γ-aminobutyric acid (GABA) as their synaptic transmitter and are viewed as generally inhibitory in their output to other cells. These cells are called *local circuit neurons*, or *interneurons*.

There are two major processing streams in the primate visual system, usually referred to as P and M pathways, with different functional properties, originating from the retina of the eye and relaying through the parvocellular (P) and magnocellular (M) layers of the lateral geniculate nucleus (LGN) of the thalamus to area V1 (Figure 1). Approximately 80% of retinal ganglion cells are β cells which project to the P layers of the LGN, whereas 10% are α cells which project to the M layers of the LGN. In area V1, thalamic axons from the P and M divisions of the LGN terminate mainly in the lower half (β division) and upper half (α division), respectively, of layer 4C in middle depth of the V1 cortex. The P pathway has an additional zone

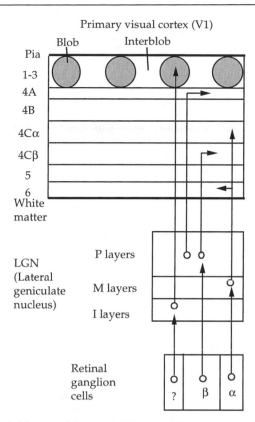

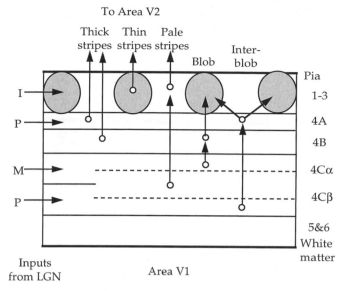

Figure 2. Diagram showing the intrinsic relays of excitatory spiny stellate neurons of layer 4C of area V1 to the superficial layers as well as the origins of four distinct pyramidal neuron relays to three interleaved stripe-like territories in visual association area V2.

Figure 1. Diagram of the visual pathways of the macaque monkey from the retina to V1. The route from α ganglion cells through the magnocellular layers (M layers) of the lateral geniculate nucleus (LGN) to layers 4Cα and 6 is known as the M pathway. The route from β ganglion cells through the parvocellular LGN layers (P layers) to layers 4Cβ and 4A is called the P pathway. The nature of the ganglion cells serving the interlaminar regions (I layers) of the LGN is uncertain. These interlaminar regions relay to punctate zones in the superficial layers that are called blobs.

of termination in a more superficial stratum called layer 4A, while the M pathway also contributes some terminals to layer 6, where pyramidal cells relaying back to the LGN are located. A third channel of input, whose origin in the retina is not yet known, relays through narrow, cell-sparse laminae (see I layers in Figure 1) inserted between the main M and P layers of the LGN to regularly arranged patches of neurons (known as blobs) in layers 2–3 of area V1. Physiological recordings indicate that cells in the P and M pathways differ in both chromatic and achromatic sensitivity as well as in spatial and temporal frequency tuning. These differences are described more fully in DeYoe and Van Essen (1988).

The primary thalamic recipient neurons (spiny stellate in morphology) in layer 4C of area V1 have staggered dendritic overlap through the depth of the layer. They send short, rising intrinsic axon fibers to three different target regions of the more superficial cortex, depending on their position in layer 4C (Figure 2). Accompanying the spiny stellate neurons in layer 4C are several morphological varieties of GABAergic stellate neurons, but little is known of the specific synaptic relationships or sources of input of most of these local circuit neurons (see Mize, Marc, and Sillito, 1992).

In the superficial layers 2–3 of area V1, pyramidal cells rather than spiny stellate cells are the principal excitatory spine-bearing neurons. The pyramidal neurons have intrinsic

axon projections, making long, horizontal, patchy, connections within the superficial layers 1–3 (Lund, Yoshioka, and Levitt, 1993). The pyramidal neuron axons also make efferent projections out of area V1 to other visual cortical areas, giving off horizontal collaterals to layer 5 before exiting area V1. It has been shown that pyramidal neurons in the blobs and in the interblob territories of layers 2–3 and pyramidal neurons in layers 4A and 4B project to three interleaved territorial segments of the adjacent visual association cortex, area V2 (for reviews, see Lund, 1988, 1990).

The organization of local circuit interneurons (largely GABAergic and inhibitory) in the primary visual cortex of the macaque monkey was summarized by Lund, Yoshioka, and Levitt (1994). The features used to describe these neurons are their laminar location, their general morphological characteristics, and especially the intra- and interlaminar patterns of projection of their axon arbors. Many of the patterns of their axon projections seem to relate to the projection patterns of the excitatory spine-bearing neurons in the same layers. Given that the population of GABAergic cells in all layers of the cortex is approximately 20% of the total cell population, any one variety of such cells is likely to have a fairly economical distribution; this sparse distribution is interesting in that the mosaic of inhibitory neuron dendritic surface and axon distribution may impose particular constraints on the way in which these neurons sample excitatory inputs and impinge on their postsynaptic targets. This geometry may determine specific response characteristics in the larger population of excitatory spiny neurons with heavily overlapped dendrites on which the interneurons synapse. The different patterns of axon projections within and between layers by each of the interneuron classes suggest that a different function is fulfilled by each class.

Anatomical and Functional Divisions in Layer 4C

As mentioned above, the thalamic input from M and P pathways ends principally in layers 4Cα and 4Cβ, respectively, of area V1. However, in terms of the intrinsic circuitry within area V1, layer 4C can also be considered as tripartite in that it has three outflows (see Figure 2; see also Lund, 1990; Lund, Yoshioka, and Levitt, 1994). Neurons in the middle of layer 4C provide a distinct source of relays out of the layer as well as projections to different destinations leaving from layers 4Cα and 4Cβ. This puzzling split into three channels instead of the incoming two M and P channels raises the question of how such a parcellation is achieved and what function the three outputs subserve.

Physiological measurements of field size and contrast sensitivity, which are distinctly different in the M and P inputs from the LGN, vary continuously as cells are recorded sequentially down through the depth of layer 4C (Blasdel and Fitzpatrick, 1984). The answer to how this functional gradient is achieved may lie in the way the spiny stellate neurons sample the incoming M and P inputs. As the position of spiny stellate neurons in layer 4C shifts from the top of the layer to its base, the neurons change the degree of their dendritic overlap into the α and β terminal zones of M and P LGN axons, since cells near the middle of layer 4C have dendrites that freely cross the junction between these two divisions. Thus, their relative sampling of M and P inputs may change as their dendrites extend to a greater or lesser degree into the M and P thalamic axon territories. However, the mature synaptic loading is the same for spiny stellate neurons at any depth in layer 4C. This finding suggests that the absolute number of contacts each cell will carry at maturity is fixed, but that the source of the contacts may vary. The cells seem to compute a response characteristic on the basis of the simple weight of M versus P inputs, and their position in depth may be the sole determinant of the ratio of synapses they eventually receive from each channel.

Based on this feature of dendritic overlap between the α and β compartments, the division of layer 4C into three partially overlapped zones with different targets of outputs appears to reflect neurons with the following characteristics. (1) They have access to M inputs and lie in the α and upper β layers; these neurons project to layer 4B. (2) They have access to P inputs and lie in the β and lower α layers; these neurons project to layer 4A. (3) They have shared M and P inputs and lie in the upper β and lower α layers; these neurons project to layer 3B interblob territories (see Figure 2). We are uncertain whether single neurons contribute both to the last option as well as to either the first or second option.

The neurons of the three territories targeted by the projections of different divisions of layer 4C (see Figure 2) share response characteristics and also show receptive field properties unique to their zone. Neurons in the blobs are monocularly driven and show weak orientation specificity, whereas neurons in the interblob regions show strong orientation specificity and are binocular. Blob cells may show particularly strong specificity to color, although interblob cells also can be color specific. Neurons of layer 4B and the layer 2–3 interblob territories have orientation specificity and binocularity in common, but the neurons of layer 4B are also directionally selective for moving stimuli. The neurons of layer 4A lack orientation specificity, are monocular, and have color specificity (Blasdel and Fitzpatrick, 1984). Neurons of both layers 4B and 4A send efferent projections to the extrastriate cortex and also send intrinsic relays to the blob regions in layers 2–3.

Origins of Orientation Specificities

Orientation preference first appears in the visual pathway of the macaque monkey in monocular simple cells located in the middle to upper region of layer 4C, but it is not shown by all of the neurons in this region (Blasdel and Fitzpatrick, 1984). (A simple cell fires only when a line or an edge of preferred orientation falls within a particular location of the cell's receptive field, while a complex cell fires wherever such a stimulus falls into its receptive field; see Hubel, 1988. Complex cells only begin to emerge beyond layer 4C in area V1.) The origin of orientation specificity is one of the most debated questions in visual neuroscience, and a variety of models have been offered to explain its generation, as reviewed by Ferster and Koch (1987; see also OCULAR DOMINANCE AND ORIENTATION COLUMNS). These models rely most heavily on physiological data, and they are generally structured around anatomical data from the cat visual cortex, but the macaque monkey primary visual cortex differs anatomically from area V1 of the cat. The precise retinotopic mapping found in layer 4C of the monkey appears to argue against Hubel and Wiesel's (1977) model of direct convergence of thalamic axons arising from rows of spatially offset geniculate cells which would provide an elongated combined receptive field. Models requiring cross-orientation inhibition would appear to require orientations to have been generated already, whereas most neurons in monkey layer 4C are nonoriented.

Models requiring offset inhibition would appear to require inhibitory axons to run laterally for some distance, which is not a feature of the main body of layer 4C, where orientation first appears. We have, however, noted the presence of short lateral projections (approximately 300–500 μm long) made by the axons of spiny stellate neurons of layer 4C, both within mid-layer 4C and off of their rising trunks in the interblob territories of layer 3B. These projections, arising from single points of layer 4C, appear to be limited to one or two axes across the retinotopic map of layers 4C and 3B. Figure 3 suggests a model with a pattern of connectivity that could generate two preferred orthogonal orientations, e.g., the horizontal and vertical orientations, in mid-layer 4C neurons. Since there is psychophysical as well as physiological evidence that vertical and horizontal orientations are dominant in at least the foveal region, this proposition is not unlikely. The identification of the targets of these layer 4C lateral projections (currently unknown) is crucial since either direct excitation of offset populations of spiny stellate neurons (as shown in Figure 3) or inhibition through contacts onto local circuit interneurons could be used to generate orientation specificity within layer 4C. A model for the circuitry that could generate orientation specificity in layer 4C, rather than in the superficial layers, is of particular importance since orientation specificity begins to appear first in this layer. In the next section, we will argue that it could be the projections of layer 4C neurons already showing orthogonal orientation specificities which may form the foundation for the long sequences of gradually changing orientation preference seen in the superficial layers.

Patterns of Lateral Connections in Superficial Layers

As mentioned earlier, the axons of pyramidal neurons in the superficial layers of the cortex make long-range intra-areal projections with regularly spaced clusters of terminals along their length. These regularly spaced terminal fields are a striking general feature of intrinsic cortical organization (Lund, Yoshioka, and Levitt, 1993), and they appear to arise from every point across the cortex rather than being a single, fixed-place lattice with locally connected cells in its lacunae. This lattice-like connectivity is observed in the primary visual cortex of the macaque monkey, and the function of the lattices of connections is an intriguing issue for the neurobiologist. These horizontal connections link cells with similar properties in the same cortical area (for a review, see Gilbert, 1992), and it is noticeable that functions of the same kind repeat across the cortex at a similar center-to-center distance as that between adjacent ter-

Figure 3. *A*, Diagram of the observed projection patterns of axon collaterals arising from spiny stellate neurons in mid-layer 4C of area V1. Left: The main axon trunk passes vertically to terminate in layer 3 over the cell of origin; axon collaterals also pass laterally to either side of the cell along a common axis, terminating in two spatially offset zones in layer 4; in addition, collaterals emerge from the rising axon trunk to terminate in layer 3 over the two offset terminal zones in layer 4. Center: The main axon trunk rises to layer 3 without emitting any laterally spreading collaterals. Right: The rising axon gives off lateral collaterals terminating just to one side of the parent cell in layer 4 and in layer 3. *B*, Diagram of a tangential view of layer 4C with hypothetical patterns of axon collateral projections for four spiny stellate neurons. *C*, If a horizontal line stimulus crosses the receptive fields of neurons *a* and *b*, both will respond, but the response of neuron *b* will be enhanced by input from concurrently active neuron *a*. If a vertical line stimulus crosses the fields of neurons *a* and *c*, the response of neuron *a* will be enhanced by concurrent input from neuron *c*. The model requires that input between spiny stellate neurons would enhance the activity of their neighbors to thalamic input, but it would not be sufficient to drive their neighbors' activity.

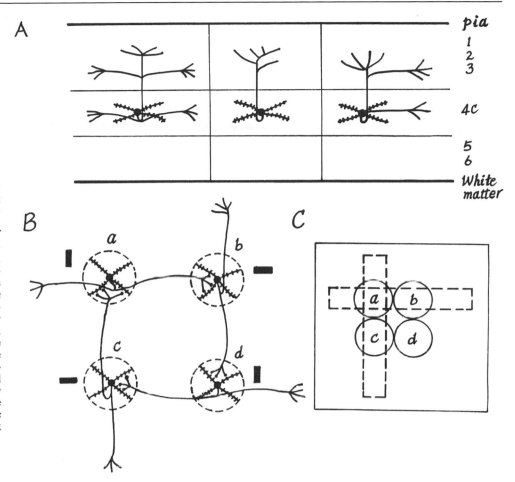

minal patches in the connectional lattice. It is interesting in this regard to note that neural network models in which excitatory connections link similar feature detectors have been proposed, and such connections were suggested to be useful for visual scene segmentation and depth perception (see Dev, 1975, and STEREO CORRESPONDENCE AND NEURAL NETWORKS). Our experiments with a combination of functional imaging and anatomical tracing, however, indicate that a proportion (one-third on average) of the links of single points also fall into regions of unlike functions. The implications of having both like and unlike connections is of interest in that it is consistent with the observation that cortical neurons may be influenced subliminally by, and perhaps even capable of responding to, stimulus features that do not normally appear to be a primary drive.

There is evidence that during the early postnatal period, there is no regular lattice connectivity in the visual cortex, but laterally spreading pyramidal neuron axon trunks extend for long distances across the superficial layers of the cortex (for a review, see Katz and Callaway, 1992). As postnatal visual stimulation begins to drive cortical activity, terminal collaterals develop from the long trunks, and the patchy pattern of connectivity begins to form. These connections are evidently constrained by the patterns of activity in the neurons since the connections can be driven into monocular domains if the animal is reared with alternating monocular vision or with strabismus, but the repetitive geometry of the connections remains unaltered.

In our anatomical investigations of these lattice connections across a number of cortical areas in the primate and the visual cortex of other species, we noted that the size of the clusters of terminals and the intervening uninnervated gaps between them are of equal width. Moreover, this dimension is closely matched to the width of the dendritic tree of single pyramidal neurons making the lattice connections. On the basis of these observations, we have proposed a model to explain the geometry of functional repeat and patterned connections. The model is based on the hypothesis that a local field of inhibition surrounds each active point across the cortex. As a basis for generating this inhibitory field, we propose that afferents to the superficial layers drive both pyramidal and local circuit basket neurons at any point. Since the axons of basket neurons in the superficial layers of the visual cortex spread three times the diameter of the local pyramidal neuron dendritic fields (Lund, Yoshioka, and Levitt, 1993), their terminals, which are known to target pyramidal neurons, will inhibit the activity of pyramids within reach of their axon. This inhibition will prevent the axons of colocalized, simultaneously activated pyramidal neurons from establishing terminals on other pyramidal neurons within this zone, since, using a Hebbian rule to govern the establishment of connections, they will not encounter simultaneously active cells within the basket neuron's inhibitory field. Figure 4 shows how such inhibition could lead to a more or less hexagonal pattern of patchy connectivity, much as is seen in the superficial cortical layers.

The constraints on connectivity that we hypothesized above would force afferent connections of different physiological properties entering the superficial layers to distribute their terminals in a discontinuous, nonoverlapping fashion, but afferents of similar property would terminate in patches at the same scale as the continuously overlapped intrinsic lattice connec-

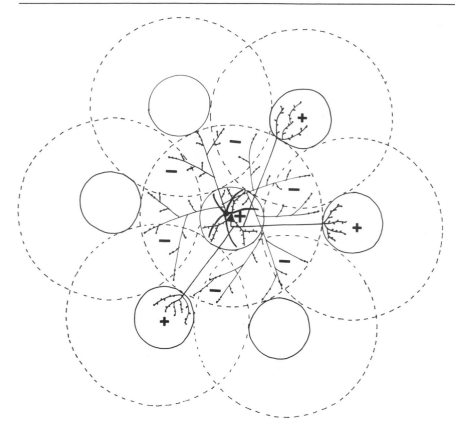

Figure 4. Diagram of intra-areal cortical connectivity suggested to explain the offset patchy distribution of pyramidal neuron axon terminals. The cortex is viewed from the surface, and one pyramidal neuron is indicated spatially colocalized with an inhibitory basket neuron. The basket neuron axon spreads over a region limited by the innermost hatched circle (indicated by minus sign). Coactivation of both pyramid and basket neurons would drive inhibition within the inner hatched circle, thereby making it less likely that the pyramidal neuron would find any other simultaneously active pyramidal neurons with that region and thus retarding the establishment of synaptic connectivity between different pyramids in this zone during development. The pyramidal neuron makes connections to zones (small circles marked by plus signs) outside the range of the coactive inhibitory axon field. However, every other pyramidal neuron it contacts will have a similar inhibitory surround from basket neurons colocalized with them (outer dashed circles), and the excitatory connectivity is restricted to a series of six points in hexagonal form across the cortex. These constraints are the same for any point, so the same hexagonal connectional matrix would be found around any single pyramidal cell, thus forming a continuum across the cortex. (From Lund et al., 1993.)

tions. Indeed, thalamic afferents to the blobs appear to show this distribution. Pyramidal neurons provide a continuum of dendritic surface across the parcellated zones of afferent input. Therefore, we propose that each neuron builds its response properties from sampling from the parcellated afferents to the degree that its dendrites enter into the afferent terminal territories. It is noticeable that in recording in tangential penetrations through the superficial cortical layers, functions usually change gradually (e.g., neurons gradually lose orientation specificity as the recording electrode enters a blob zone). This observation is what might be expected from our dendritic sampling model.

With regard to the generation of gradients of smooth change in orientation specificity, we suggested that the mid-layer 4C neurons that project to the interblob territory might generate two orthogonal orientation specificities (see the preceding section). Since these orientation specificities are two opposite functional extremes, we propose that they project to nonoverlapping territories in the interblob territories in the superficial cortex. Following our model, dendritic overlap could then generate a gradient of pyramidal neuron orientation specificities between these segregated extremes. Since several functions may distribute their gradients across the cortex, the many attributes of any one stimulus would be expected to alert only a specific set of lattice connections where the combined gradients for each attribute coincide at optimal levels for the stimulus attributes.

Discussion

In recent years, there has been a surge of interest in neural network simulations of visual cortex function. However, we have found that models based on the real anatomy of the cor-

tex are rare. One interesting example of a theoretical model, based on real anatomical and physiological observations and allowing testable predictions for the organization of laterally spreading intra-areal connections in the tree shrew visual cortex, is that of Mitchison and Crick (1982). In this article, we have given a brief overview of features of the anatomy of the primary visual cortex which we find interesting, and we have outlined several issues together with our own rudimentary models for how information processing may be carried out by this area. We hope that this discussion will stimulate theorists to take a step further along the line of biologically realistic modeling of visual cortex information processing. In particular, we hope that theorists will construct models that predict features of anatomy and physiology, helping to steer the neurobiologist toward important questions in data acquisition.

Acknowledgments. The research described here was supported by MRC grant G9203679N and NIH-NEI grant EY10021 awarded to J.S.L.

Road Maps: Mammalian Brain Regions; Vision
Related Reading: Gabor Wavelets for Statistical Pattern Recognition; Vision: Hyperacuity; Visual Scene Perception: Neurophysiology

References

Blasdel, G. G., and Fitzpatrick, D., 1984, Physiological organization of layer 4 in macaque striate cortex, *J. Neurosci.*, 12:3141–3163.
Churchland, P. S., and Sejnowski, T. J., 1992, *The Computational Brain*, Cambridge, MA: MIT Press. ◆
Dev, P., 1975, Perception of depth surfaces in random-dot stereograms: A neural model, *Int. J. Man-Mach. Stud.*, 7:511–528.
DeYoe, E. A., and Van Essen, D. C., 1988, Concurrent processing streams in monkey visual cortex, *Trends Neurosci.*, 11:219–226.

Ferster, D., and Koch, C., 1987, Neuronal connections underlying orientation selectivity in cat visual cortex, *Trends Neurosci.*, 10:487–497. ◆

Gilbert, C. D., 1992, Horizontal integration and cortical dynamics, *Neuron*, 9:1–13. ◆

Hubel, D. H., 1988, *Eye, Brain and Vision*, New York: Freeman. ◆

Hubel, D. H., and Wiesel, T. N., 1977, Receptive fields and functional architecture of monkey striate cortex, *J. Physiol. (Lond.)*, 195:215–245.

Katz, L. C., and Callaway, E. M., 1992, Development of local circuits in mammalian visual cortex, *Annu. Rev. Neurosci.*, 15:31–56.

Lund, J. S., 1988, Anatomical organization of macaque monkey striate visual cortex, *Annu. Rev. Neurosci.*, 11:253–288.

Lund, J. S., 1990, Excitatory and inhibitory circuitry and laminar mapping strategies in primary visual cortex of the monkey, in *Signal and Sense: Local and Global Order in Perceptual Maps* (G. M. Edelman, W. E. Gall, and W. M. Cowan, Eds.), New York: Wiley, pp. 51–66.

Lund, J. S., Yoshioka, T., and Levitt, J. B., 1993, Comparison of intrinsic connectivity in different areas of macaque monkey cerebral cortex, *Cereb. Cortex*, 3:148–162.

Lund, J. S., Yoshioka, T., and Levitt, J. B., 1994, Substrates for interlaminar connections in area V1 of macaque monkey cerebral cortex, in *Cerebral Cortex 10: Primary Visual Cortex in Primates* (A. A. Peters and K. S. Rockland, Eds.), New York: Plenum, pp. 37–60. ◆

Mitchison, G., and Crick, F., 1982, Long axons within striate cortex: Their distribution, orientation, and patterns of connection, *Proc. Natl. Acad. Sci. USA*, 79:3661–3665.

Mize, R. R., Marc, R. E., and Sillito, A. M., Eds., 1992, *GABA in the Retina and Central Visual System*, Amsterdam: Elsevier. ◆

Visual Processing of Object Form and Environment Layout

Allen M. Waxman, Michael Seibert, and Ivan A. Bachelder

Introduction

Visual sensing provides a rich source of information that is useful for a wide range of applications in both static and dynamic environments. In this article, we focus on two related types of visual information processing: (1) learning and recognition of three-dimensional (3D) objects and (2) qualitative mapping of two-dimensional (2D) environments defined by distributions of 3D visual landmarks. We show that mapping follows as a direct generalization of our object learning paradigm. Although our focus is the neural processing of visual imagery, we have applied similar concepts and methods to other sensing modalities, including multispectral infrared imagery and synthetic aperture radar imagery (Waxman et al., 1993).

The processing of object form relates to the *parvocellular* pathway in the visual cortex (see VISUAL CORTEX CELL TYPES AND CONNECTIONS). The learning and recognition of *objects* and their *spatial layout* relates to the *what* and *where* processing streams which project from the primary visual cortex to the temporal and parietal cortices, respectively (see VISUAL SCENE PERCEPTION: NEUROPHYSIOLOGY), and manifest themselves as qualitative maps of an environment expressed in the hippocampus (see HIPPOCAMPUS: SPATIAL MODELS). Thus, the pathway from the visual cortex to the temporal lobe to the hippocampus may reflect an evolution of reference frames, from retinotopic to egocentric to allocentric.

The advantage of a neurocomputational approach to these visual processing applications is twofold. From a modeling standpoint, we can probe the brains of animals to suggest computational network architectures. From a technology standpoint, we can implement these neural networks and systems in structured analog and digital VLSI (Very Large Scale Integration) circuits in conjunction with the sensors themselves, with the potential of achieving lightweight, low-power, smart sensors.

Learning and Recognition of Three-Dimensional Objects

An important alternative to the object-centered, model-based approach to object recognition is the view-based approach to learning and recognition of 3D objects. Motivation stems from physiological studies of shape processing in the macaque monkey temporal lobe and the superior temporal sulcus, an area in which single cells have been found that are highly tuned to particular views of faces and, in some cases, to the identity of particular individuals (Perrett et al., 1989). These cells display significant invariance properties, responding to particular face views invariant to lighting conditions and face position, size, orientation around the optical axis, and small deformations away from the preferred view. In addition, cells have been found that respond to rotations (i.e., transitions) between face views, such as those observed during natural inspection of an object. By pooling these viewer-centered cells in hierarchy, one obtains a single cell that may represent the object as a whole, independent of view. However, whether these cells are genuine *grandmother cells* coding for a unique face, or are rather part of a sparse coding system, remains open to debate (see LOCALIZED VERSUS DISTRIBUTED REPRESENTATIONS and SPARSE CODING IN THE PRIMATE CORTEX). In any case, the invariant view-based approach separates the problem of object recognition from that of determining precisely the object's pose, which is a major source of complexity in the model-based approach. Moreover, the need for a model of each object of interest is rendered moot by the ability to learn the required representations directly from the sensor data. Next, we summarize the neural system approach of Seibert and Waxman (1989, 1992; Waxman et al., 1993) and review other related studies.

In general, a viewer explores the appearance of a 3D object by manipulating it, moving around it, or tracking it as the object moves past the viewer. In each case, the viewer experiences a sequence of views, many of which are similar, but many of which are different. That is, the viewer experiences a set of characteristic views and view transitions, a natural unfolding of visual events over time. In our approach, each image is processed by multiple neural nets; the system architecture is shown in Figure 1.

Networks first perform feature extraction on segmented imagery (edge enhancement followed by high-curvature point detection) and then transform this illumination-invariant feature pattern into a new pattern which is invariant to object position (through tracking), scale and orientation (through log-polar mapping and centroiding), and small deformation (through overlapping Gaussian receptors) on the visual field.

Next, an *ART network* [see ADAPTIVE RESONANCE THEORY (ART)] takes as input these invariant patterns and clusters them into view categories, or *aspects*, in an unsupervised manner. Along with each category, a template is established or modified for the prototype feature code corresponding to that aspect. Figure 2 shows the result of ART-2 categorization, at a vigilance threshold of 0.93, for three model aircraft (F18, F16, HK-1). In each case, more than 400 views have been compressed into fewer than 30 categories, and we observe the emergence of aspects in the form of *category islands* on the viewing sphere around each object (i.e., extended viewing regions to which the invariant feature pattern is sufficiently similar that it is classified as the same category). These representations can be

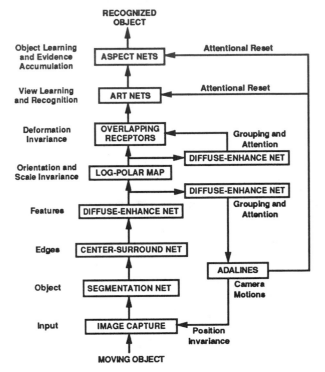

Figure 1. Modular neural system for view-based learning and recognition of 3D objects. Each module is a type of neural network whose function is also indicated in the diagram.

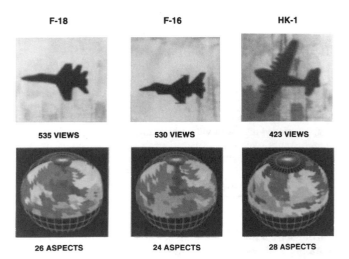

Figure 2. Learned aspect category spheres for three aircraft. More than 400 views in each case are quantized into approximately 25 categories (shown as shaded viewing islands) whose learned templates represent the object.

further consolidated without bias to ordering of the original data by using the learned template patterns as input to a new ART module at a lower vigilance setting.

Finally, the categories emerging from each ART network associated with each 3D object are connected to an *aspect network* which uses modified Hebbian learning to develop associations between categories triggered in sequence (i.e., it learns *aspect transitions*). In addition, each aspect network accumulates activity from the ART category nodes over time, combining object evidence across an image sequence and competing with one another to determine the instantaneous *winner*.

For each object, the learned aspect categories can be thought of as the nodes of a graph. The learned category transitions or neighbor relations on the category sphere then correspond to the arcs of the graph. This graph structure, which emerges within the aspect network in our system, is analogous to the *aspect graph* concept of Koenderink and van Doorn (1979).

An alternative approach to view-based learning was developed by Edelman and Weinshall (1991). It treats the mapping from feature patterns to objects as a problem in continuous function approximation, using *radial basis functions* (see RADIAL BASIS FUNCTION NETWORKS) to map previously labeled vertices of an observed wire-frame object into a winning category node. An alternative to our aspect network for exploiting category sequences was suggested by Bradski, Carpenter, and Grossberg (1992), whereby a fading temporal working memory stores category activations and then inputs such sequences into another ART network to learn and recognize aspect sequences of variable length. Fielding and colleagues at the Air Force Institute of Technology have adopted our conceptual approach, replacing our ART network for view categorization with an LBG vector quantizer and our aspect network for evidence accumulation with a hidden Markov model formalism (Fielding et al., in press). It is being applied to both visible and infrared imagery of military vehicles. A survey of neural network technology as applied to AUTOMATIC TARGET RECOGNITION (q.v.) has been presented by Roth (1990).

Learning Qualitative Maps of Environments

Another important application of visual information processing concerns mapping and navigating environments defined by spatially distributed landmarks. This area has received much attention in the artificial intelligence community (Kuipers and Levitt, 1988), and it has recently gained attention in the neurocomputing community (Bachelder and Waxman, 1993; Schmajuk and Blair, 1993; Wu and Penna, 1993; see COGNITIVE MAPS). In many respects, learning qualitative maps of a visual environment is analogous to learning representations of 3D objects; whereas one explores an object from vantage points *around* the object and quantizes these many views into a finite set of aspect categories, one explores an environment from vantage points *within* the environment and can quantize the spatial patterns of visual landmarks into a finite set of place categories. Thus, the combination of learning and recognition of landmarks and patterns of landmarks enables a view-based approach to represent both objects and environments. The analogy can be taken further by considering the learning of place category transitions in environments, yielding a graphlike map, as in the world graph of Arbib and Lieblich (1977).

In terms of the known neural pathways in the brain, recognition of visual landmarks is realized in the temporal lobe, whereas recognition of spatial relations is realized in the parietal lobe. These lobes interact through direct axonal projections as well as indirectly through the hippocampus (see VISUAL SCENE PERCEPTION: NEUROPHYSIOLOGY). The hippocampus is

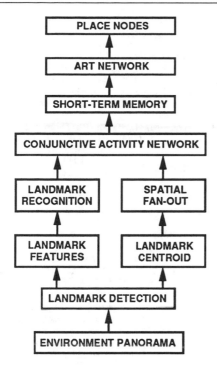

Figure 3. System architecture for learning environmental maps in the form of place fields. The system has been implemented in real time on the mobile robot MAVIN.

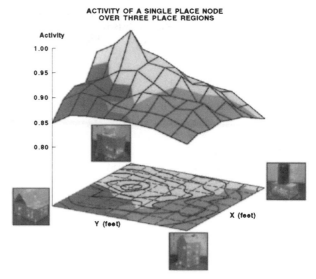

ACTIVITY OF A SINGLE PLACE NODE
OVER THREE PLACE REGIONS

Figure 4. Qualitative map of learned place fields and the activity pattern of one learned place node. A 9- × 8-foot area, with four landmark objects at its corners (insets), is explored by MAVIN. The area is automatically parcelled into three place fields (shaded areas on the lower surface). The activities of the place nodes (as shown on the upper surface) vary with position, decaying rapidly away from their maxima at place centers. The equi-activity contours of the place nodes strongly resemble measurements made of place cell activities in the rat hippocampus.

known to support the representation of environments in terms of *place cells* which form a *cognitive map* (O'Keefe and Nadel, 1978). These hippocampal place cells, found in the rat brain, are maximally active whenever the rat is located in (or about to enter) a small area (*place field*) of a known environment, such as a maze. In some experiments, these place cells are *heading sensitive* and are believed to be coding a *scene view*. (Heading-sensitive and heading-invariant cells may reflect the existence of hierarchy in the construction of place cells.) The rat can qualitatively map the entire maze in terms of a small collection of place cells, and somewhat more precise locations can be defined in terms of the relative activities of neighboring place cells. In fact, these place field maps are analogous to the aspect category spheres we introduced for 3D object learning.

Bachelder and Waxman (1993, 1994) extended the object learning approach of Seibert and Waxman (1992) and developed a neurocomputational system to learn place field maps (summarized in Figure 3). This system was implemented in real time on the mobile robot MAVIN (Mobile Adaptive VIsual Navigator) to map an area with four visual landmarks at its corners. The robot's field of view searches the environmental panorama, finding each already learned landmark in succession. The extracted features are passed along a *what* stream to the vision system described above for landmark recognition, and the feature centroid is passed along a *where* stream through a Gaussian spatial fan-out to a *conjunctive activity network*. This network brings together coarse location information and identity information preserved across all landmarks and then codes these conjunctions in *token nodes*. These tokens, in turn, activate a short-term memory storage, forming a pattern which represents the likelihood of *what landmark is approximately where*. This pattern serves as input to another ART network, whose learned category nodes can be interpreted as *place cells* in the system. If we restrict the robot's field of view so that only a subset of the landmarks is visible, the learned category nodes

would be better interpreted as *view cells*. Then, through repeated explorations of the same area from different headings, view cells would become hierarchically linked to a place cell through associative learning.

A sample environmental map composed of learned place fields, constructed by MAVIN, is shown on the lower part of Figure 4. The four landmark objects shown were located in the four corners of the laboratory area, and the robot collected panoramic views as it explored the area in 1-foot increments. The 9- × 8-foot area was learned as only three categories, shown as separately shaded place fields on this qualitative map. Increasing the vigilance level of the ART network shown in Figure 4 increases the number of emergent place fields. The three place node activities vary across each place field and overlap one another, as illustrated for one node on the upper part of Figure 4, and can be used to localize the robot within place fields (e.g., to localize a goal). This system is being extended to incorporate the learning of heading-sensitive place nodes and action-induced transitions between place nodes, which support a variety of route-planning tasks.

Discussion

Neural network approaches to visual information processing are rapidly progressing as a result of the significant guidance provided by animal models and psychophysical experiments. In this article, we described two modular neural systems constructed from a variety of neural networks, specifically, systems which learn objects and environments. These new techniques are now finding their way into a variety of applications, and their circuit-like architectures make them amenable to compact, low-power implementation in VLSI. We expect the combination of analog VLSI and structured digital implementations to make smart visual sensors a reality by the turn of the century.

Acknowledgments. We wish to acknowledge support for our research from the Advanced Research Projects Agency, the Air Force Office of Scientific Research, and the Office of Naval Research.

Road Map: Applications of Neural Networks
Related Reading: Analog VLSI for Neural Networks; Digital VLSI for Neural Networks; Exploration in Active Learning; Perception of Three-Dimensional Structure

References

Arbib, M. A., and Lieblich, I., 1977, Motivational learning of spatial behavior, in *Systems Neuroscience* (J. Metzler, Ed.), New York: Academic Press, pp. 221–239.

Bachelder, I. A., and Waxman, A. M., 1993, A neural system for mobile robot visual place learning and recognition, in *Proceedings of the 1993 World Congress on Neural Networks*, vol. 1, Hillsdale, NJ: Erlbaum, pp. 512–517.

Bachelder, I. A., and Waxman, A. M., 1994, Mobile robot visual mapping and localization: A view-based neurocomputational architecture that emulates hippocampal place learning, *Neural Netw.*, 7:1083–1099. ◆

Bradski, G., Carpenter, G. A., and Grossberg, S., 1992, Working memory networks for learning temporal order with application to 3-D visual object recognition, *Neural Computat.*, 4:270–286.

Edelman, S., and Weinshall, D., 1991, A self-organizing multi-view representation of 3D objects, *Biol. Cybern.*, 64:209–219.

Fielding, K. H., Ruck, D. W., Rogers, S. K., Welsh, B. M., and Oxley, M. E., in press, Spatio-temporal pattern recognition using hidden Markov models, *IEEE Trans. Aerospace Electronic Systems*.

Koenderink, J. J., and van Doorn, A. J., 1979, The internal representation of solid shape with respect to vision, *Biol. Cybern.*, 32:211–216.

Kuipers, B. J., and Levitt, T. S., 1988, Navigation and mapping in large-scale space, *AI Magazine*, 9:25–43. ◆

O'Keefe, J., and Nadel, L., 1978, *The Hippocampus as a Cognitive Map*, Oxford: Clarendon.

Perrett, D. I., Harries, M. H., Bevan, R., Thomas, S., Benson, P. J., Mistlin, A. J., Chitty, A. J., Hietanen, J. K., and Ortega, J. E., 1989, Frameworks of analysis for the neural representation of animate objects and actions, *J. Exp. Biol.*, 146:87–113.

Roth, M. W., 1990, Survey of neural network technology for automatic target recognition, *IEEE Trans. Neural Netw.*, 1:28–43. ◆

Schmajuk, N. A., and Blair, H. T., 1993, Place learning and the dynamics of spatial navigation: A neural network approach, *Adapt. Behav.*, 1:353–385.

Seibert, M., and Waxman, A. M., 1989, Spreading activation layers, visual saccades, and invariant representations for neural pattern recognition systems, *Neural Netw.*, 2:9–27.

Seibert, M., and Waxman, A. M., 1992, Adaptive 3D object recognition from multiple views, *IEEE Trans. Pattern Analysis Machine Intell.*, 14:107–124.

Waxman, A. M., Seibert, M., Bernardon, A. M., and Fay, D. A., 1993, Neural systems for automatic target learning and recognition, *Lincoln Lab. J.*, 6:77–116.

Wu, J., and Penna, M. A., 1993, Models for map building and navigation, *IEEE Trans. Syst. Man Cybern.*, 23:1276–1301. ◆

Visual Scene Perception: Neurophysiology

J. I. Nelson

Introduction

The cortex divides complex scene perception into two tasks executed by two streams of many cortical areas each. The *object vision* stream segments the attended, fixated object from its background and processes it with great precision, but lacks positional and contextual information that is available in the *place vision* stream. These differences in information processing require information exchange between the streams. We have the anatomical substrate for this exchange, but not the mechanisms. The futility of visual scene analysis when one stream loses access to the other is illustrated with clinical examples from strabismus and visual agnosia. The attributes of a cortical circuit which could provide massively parallel distributed processing among large clusters of areas are also described.

Object and Place Vision Streams

Anatomy

Lesion studies in the monkey have established that the parietal areas are concerned with perceiving and learning the spatial arrangement of objects, while object recognition depends heavily on temporal areas (Ungerleider and Mishkin, 1982). The cortical anatomy supports the separation of *where versus what* perception and memory into two streams of cortical areas for *place versus object vision*. The streams are shown in Figure 1. Area V4 (the fourth visual area, important for color perception) is an important gateway from the early visual areas to the temporal (object) stream, while area MT (the middle temporal area, or area V5, important for motion) is a gateway to the parietal (place) stream. In Figure 1, dotted lines in the parietal stream indicate that cells encoding the peripheral parts of the visual field are relatively more numerous and enjoy relatively more interarea information exchange compared with temporal stream areas, which exchange more information about whatever has attracted the observer's fixation. Everything is *not* connected to everything else, the connections are not all the same, and the labels have a logic and a history (e.g., TE is the Eth area in the temporal lobe). When the work of the temporal areas is done, the most significant task remaining is memory formation (Mishkin, 1993). The work of the parietal areas leads to motor action (below), so the two streams may be regarded equally as *WHAT-it-is* and *HOW-to-use-it* streams (Milner and Goodale, 1993).

Object Versus Place Binding

Activity in many areas must be able to modulate the response and functional coupling of a single neuron. Such massively parallel information exchange is needed to bring an object and its attributes together (*moving, spotted dog*), thus achieving binding (see BINDING IN THE VISUAL SYSTEM). Position information, which viewpoint-invariant models of object recognition discard (see OBJECT RECOGNITION), is also essential for perceiving the complex visual world and our place in it. It must also be bound. I will term the binding of visual attributes to an object *object binding* and the assembly of large objects or complex scenes in their proper topography using positional information *place binding*. I regard object binding as a task per-

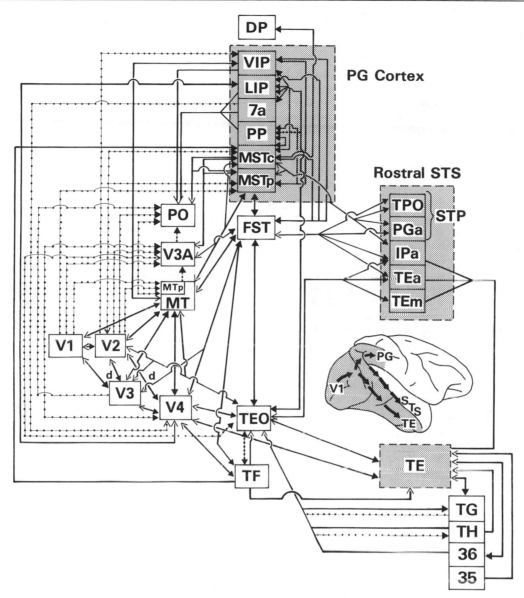

Figure 1. Wiring diagram for the primate visual system. The visual world is represented many times in approximately 27 different cortical areas (boxes). Object vision areas lie in the temporal lobe and have names beginning mostly with T. Spatial vision areas are parietal and often have P in their names. Connections between areas are usually reciprocal (double arrow heads). Lamination patterns specify most connections as upward or downward (solid and open arrowheads, respectively), so that a hierarchy may be defined for the visual system as a whole. Areas located the same number of steps above the first area (V1) in this hierarchy may also be interconnected (double solid arrowheads). In some cases, a pathway is demonstrated, but lamination data are lacking (no arrowheads). The terminal (highest) areas in each of the three streams are shaded; shading is repeated in the inset sketch of the brain. Segregation in the connectivity pattern supports the two-stream concept. V4 is a gateway to the temporal stream, and MT is a gateway to the parietal stream. The rostral (forward) side of the superior temporal sulcus (STS) (see inset) may be viewed as the culmination of a third motion stream. Connections between peripheral visual field representations (dotted lines) are more prominent in the parietal stream. A bridge for information exchange between the temporal and parietal streams is provided by connections between several higher areas, particularly temporal area TEO and parietal area FST, as well as by areas in the rostral STS. Today's terminology is mixed. The numerical terminology of Brodmann from the beginning of this century survives in the earliest areas whose specialized lamination makes them distinguishable under the microscope. Arbitrary Brodmann numbers (17, 18, 19) were renamed visual areas 1, 2, 3, and so on (V1, V2, V3, etc.) by Woolsey in the 1960s. Von Economo and later von Bonin and Bailey lettered lobes alphabetically (temporal areas E, F, G; i.e., TE, TF, TG). More areas made more letters; TEO is split off the temporal area E on the occipital side. Positional naming became common with the work of Allman and Kaas in the 1970s; e.g., in the intraparietal sulcus, LIP lies in the lateral wall. More functional knowledge has produced more hybrid functional and positional names; e.g., STP is a polysensory area in the superior temporal sulcus. (From Distler, C., Boussaoud, D., Desimone, R., and Ungerleider, L. G., 1993, Cortical connections of inferior temporal area TEO in macaque monkeys, *Journal of Comparative Neurology*, 334:125–150. Copyright 1993, John Wiley & Sons, Inc. Reprinted by permission of John Wiley & Sons, Inc.)

formed chiefly by temporal areas, and place binding as a task of the parietal areas.

Clinical Examples

Distortions in an Early Map

Patients with strabismus (i.e., crossed eyes) show crowding effects during monocular acuity testing (multiple letters become jumbled; isolated ones are fine) or more general distortions (striped patterns appear wavy or jagged). Sometimes it is difficult for the subject to point accurately to the location of a target or straight ahead. A physiological conjecture with considerable empirical support is that the representation of visual space in early cortical areas becomes disturbed when the eyes are misaligned and their signals are out of register. In the cortex, the two retinal images are in part combined by binocular neurons with out-of-register receptive fields matched to the ocular misregistration. Cells which possess these abnormal disparity tunings provide a substrate for anomalous retinal correspondence (Nelson, 1988). However, there is no synthesis for other, monocular cells, and so three maps arise in one cortical area: a map evoked by input from the left eye, a right eye map in register with the first, and a map with anomalous spatial values which is binocular, but also partially accessible by one eye alone. Not surprisingly, the topographical representation of visual space is discontinuous, and regular striped patterns appear jagged.

Place binding also depends on maps which are more abstract than this retinotopic one. Maps can be egocentric, and thus compensate for eye and head rotations, or allocentric. Allocentric maps are entities separate from our body (e.g., our mental map of how to drive across the town where Grandma lives). A disturbed oculomotor system defines strabismus and deprives egocentric map making of needed eye position information (see GAZE CODING IN THE POSTERIOR PARIETAL CORTEX). The strabismic patient who cannot point straight ahead has lost place binding at the second map level.

Visual Agnosia

Visual agnosia (reviewed by Farah, 1990) is seeing without understanding. The visual periphery is intact, yet the patient cannot interpret what is seen. A drawing can be copied (often in a slavish, line-by-line way), but the patient may be unable to name or describe the use of what he has drawn. The smallest obstacle, such as a printing error or the partial occlusion of a picture, derails perception. Patients seem unable to process part-whole relationships to form a larger perceptual construct.

Animal neurophysiology and human clinical observation can be described with consistent terminology if we order observations along two dimensions: the stream and the hierarchical level. The distinction between dorsal and ventral streams has gained currency in the clinical literature. It is expressed in the terms *dorsal* and *ventral simultanagnosia* (the subject cannot see two objects at once). I will generalize these terms slightly to dorsal and ventral agnosia. The term *dorsal agnosia* is meant to embrace dorsal simultanagnosia and also, for example, the more abstract inability to draw ground plans or follow a mapped route. In humans, the term *ventral simultanagnosia* is often applied to left hemisphere damage accompanying language as well as visual deficits. I use the term *ventral agnosia* to describe temporal stream deficits which are linked neither to hemispheric lateralization nor to hierarchical level. Examples in different streams at different hierarchical levels are shown in Table 1. Next I discuss the different functions of the two streams, beginning with object binding in the ventral stream.

Table 1. Binding Hierarchy

Early Visual System	Parietal Stream	Temporal Stream
Disturbed Topography in Strabismus	Disturbed Binding of Object and Topographical Information	Disturbed Binding of Object and Features
Lowest Level		
Stripes of grating appear wavy; real-life scenes look better	Cannot copy a line drawing; gross distortion of form	Poor contour synthesis; cannot perceive figure presented as dotted line
1 letter OK; multiple letters jumbled	Cannot count dots; cannot segregate ones done already	Printing defects erroneously bound to figure
Suppression of (parts of) one eye's input	Simultanagnosia; jumbling or competitive extinction of different objects which should be in different places	Parts not bound to whole; cowcatcher on locomotive seen as second car
Subject cannot point to straight ahead	Isolated objects OK, but pictures do not tell a story	Cannot recognize or name objects (associative agnosia/anomia); cannot imagine what they are used for (ideational apraxia)
	Trouble reaching for an object; gets lost on ward and in hometown	Cannot recognize own face, own car
Highest Level		

Failure of Object Binding

We synthesize scattered responses to an object's attributes so that the parts add up and a unified object representation arises. When this binding is impaired at a relatively low level, we have *apperceptive* visual agnosia. The patient cannot synthesize a line from dots, has more difficulty seeing curvy-contour objects than objects delineated by straight lines, and may be unable to distinguish a key ring from a coin. In the last example, it is as if, when the edge discontinuity lacks figurality, cortical processing to determine whether the space enclosed by the edge is empty or solidly filled is not invoked. The subject with *associative* agnosia sees, but cannot recognize. For example, a drawing of an automobile can be copied accurately, but the patient cannot convey by any means how it is used. He may use the name *woman*, but not recognize his wife.

The subject with associative visual agnosia cannot bind parts to the whole (Rubens and Bensen, 1971). Normally, a figure will be put together from its parts by an active process requiring computation and information exchange among several cortical areas. When this process is not possible, object recognition is not possible. Unable to recognize his own face in a mirror, a prosopagnosic patient complains, "I can see the eyes, nose, and mouth quite clearly, but they just don't add up. They all seem chalked in, like on a blackboard" (Pallis, 1955).

Central Versus Peripheral Visual Field Differences

The great differences in what object and spatial vision streams accomplish requires close cooperation between them. As it processes objects fixation by fixation, the only thing the ventral stream can achieve on its own is visual chaos. This observation is shown in the clinical examples below, but it is clear from the neurophysiology alone where the salient issues are central-peripheral visual field weighting and access to the motor system.

The Ungerleider group has consistently documented differences in central versus peripheral visual field weighting in the pathways of ventral and dorsal streams (Distler et al., 1993; see Figure 1). The central visual field is well represented in the ventral stream areas and in the connections between them. This system is concerned with identifying an object that is fixated in central vision and segmented from its background, which itself is largely discarded from processing. There is little regard for the object's relation to other objects in the visual field or to the observer (see FIGURE-GROUND SEPARATION).

The peripheral visual field is heavily represented in the parietal stream. Parietal areas with motion sensitivity are concerned with attracting our attention to an object that is not yet well identified, not yet even fixated. These areas monitor the context well beyond the point of regard. By virtue of their concern with motion perception, the areas in this stream must be concerned with visual processing which is extended in time and across space. Furthermore, access to the motor system enables the parietal areas to construct and maintain egocentric and allocentric maps (see GAZE CODING IN THE POSTERIOR PARIETAL CORTEX). This system takes into account eye and head position information to update maps in real time. The maps in turn are available to direct a saccadic eye movement or a stroll to the corner store. These are also the maps onto which the object output from successive fixations can be plotted. What happens when we lose these maps?

Failure of Place Binding

Dorsal simultanagnosia may involve difficulty in maintaining or shifting attention between objects, but the core deficit is visuospatial. Subjects cannot bind an object to a place. They cannot point to the location of a seen object with their eyes or hands unless it makes a noise (optic ataxia), and they cannot describe the relative position of two objects (visual disorientation). Multiple elements are jumbled, and complex scenes cannot be comprehended. With no landmarks, these patients stumble across previously inspected components again and again; counting is impossible. Without landmarks, placing the parts of a scene into their correct arrangement is a trying experience when copying a drawing. "As soon as I lift my pencil from the paper, I can't see what I have drawn and have to guess where to put the next detail" (Godwin-Austen, 1965. 455). When they are not bound to separate places, two different objects cannot be seen clearly at the same time. There is competitive extinction (Williams, 1970:61–62).

A sixty-eight-year-old patient studied by the author had difficulty finding his way around because "he couldn't see properly." It was found that if two objects (e.g., pencils) were held in front of him at the same time, he could see only one of them, whether they were held side by side, one above the other, or one behind the other. Further testing showed that single stimuli representing objects or faces could be identified correctly and even recognized when shown again, whether simple or complex. . . . If stimuli included more than one object, one only would be identified at one time, though the other would sometimes "come into focus" as the first went out.

In normal observers, unequal stimuli which lie on top of each other cause rivalry, masking, or suppression. The brain's response to a mismatch is to turn off one visual signal using competitive, WINNER-TAKE-ALL MECHANISMS (q.v.). Suppression of diplopia in strabismus seems relatively simple: the signal can be identified by the eye of origin. The brain's dilemma in choosing what to suppress in simultanagnosia is more fascinating, suggesting that these two forms of suppression occur at different levels. Meaningful scenes, presumably processed with higher areas, *reduce* distortions in the case of impaired vision with misaligned eyes (strabismic amblyopia; see Sireteanu, Lagreze, and Constantinescu, 1993), while such cognitive complexity only makes the situation worse for the agnosic patient.

In summary, suppression occurs in strabismus when objects which are mapped to the same tissue patch cannot be combined. In simultanagnosia, objects which should have been mapped to different tissue patches are combined instead.

Interaction Between Streams: Synthesis of Large Scenes

Our eyes dart here and there, and the retinal image changes. The stable, unified perceptual world we inhabit would be unthinkable without intense, two-way interactions between object and spatial vision subsystems. Several higher visual areas in both streams are reciprocally connected, and there are additional bridges through areas in the anterior bank of the superior temporal sulcus (see Figure 1). I will presume that an intense stream-to-stream exchange of information occurs in perception; for visuospatial *memory* formation, circuits involving temporal lobe limbic structures are also important. For scene perception, according to my proposal, the ventral *what* stream performs figure synthesis and object recognition on one fixated object after another and feeds the output of this processing to the dorsal stream, where objects are assembled onto a stable array. Otherwise, we have simultanagnosia. With most of its tissue concentrated on the central visual field, this process is not something that ventral stream areas are well prepared to do. These areas also lack the parietal stream's ready access to motor area information required for map stabilization.

We bind features to one object, bind a group of objects to a spatial arrangement, and perceive an overall scene (see VISUAL SCHEMAS IN OBJECT RECOGNITION AND SCENE ANALYSIS). It must be possible to dynamically shift the level of binding within dorsal and ventral hierarchies as recognition proceeds. There are no known biological mechanisms for accomplishing this shift. An exception comes from the seminal discovery that, within the larger receptive fields typical of neurons in higher cortical areas, the response to one of two or more small stimuli can be enhanced by directing a monkey's attention to it (reviewed in Desimone and Duncan, 1995). This information gives us at least one mechanism for changing the granularity of the analysis (for related modeling, see SELECTIVE VISUAL ATTENTION and ROUTING NETWORKS IN VISUAL CORTEX). What happens to patients with dorsal agnosia when *they* shift the granularity of their analysis?

If the patient is presented with a pattern of six dots arranged to form a rectangle, he can easily perceive and name it. But if the patient be then instructed to *count* the component dots, he experiences very considerable difficulty. The new task destroys immediate awareness of the configuration; it is now the separate elements which become the object of analysis. Under these conditions, the patient becomes unable to see clearly any dot other than that which he is immediately fixating. (Luria, 1959:446)

The injury of the patient described by Luria was a gunshot wound in the left occipitoparietal lobe, complicated by surgical removal of the bullet from the symmetrical position in the right hemisphere.

On its own, without access to an intact parietal stream, the ventral stream cannot construct percepts of an extended object or a complete visual environment. When patients with dorsal agnosia shift the granularity of their analysis, they have no place to store the previous snapshot along with its positional information. The closer these patients look, the less they see.

Assembly-Based Transmission

The degree of interaction needed between object and spatial vision streams can be accomplished only with mechanisms for massively parallel, distributed information exchange among multiple cortical areas. These mechanisms do not yet exist. New principles of spike transmission will have to emerge from cortical neuroscience. The form which the needed circuits must take is described below.

The dependence of object and spatial vision subsystems on each other requires an active neuron in one cortical area to share information with, and to be modulated by, the activity of neurons in many other areas. Object binding must also tap and combine attributes from *memory* (see Damasio's convergence zones, 1990) if an object is to be detected, perceived, understood, named, and reacted to (e.g., grasped and used in a task). The terms for failure of these five achievements are blindness, apperceptive visual agnosia, associative visual agnosia, anomia, and apraxia, respectively. Plans and purposes modulate object percepts, presumably through the rich connections which exist between the frontal cortex and the temporal lobe. For example, a person who is looking through a doorway at a chair on a table may want to move the chair to sit on it, may want to preserve the total configuration to change a lightbulb, or simply may want to avoid walking into obstacles, with no concern for details. We do not wander aimlessly through our maps. We get our bearings and set off because we need something and pursue a plan to obtain it. The motor system stabilizes egocentric and allocentric maps, and we use the maps (perhaps through the strong projection from the parietal to the premotor cortex) to drive the motor system to enable us to walk toward a cup of coffee and pick it up without spilling it. Using recalled sensory experiences, we construct plans in the frontal and prefrontal cortex. Planning means laying out a map in time, not just space. The dorsal stream is open ended because spatial maps lead to maps of imagined places and planned actions, and there can be any number of these. For a mammal with a cortex, day-to-day life requires exchanging activity among many maps.

What question should be addressed to provide a better basis for spike transmission and multiarea information exchange? I point to massive divergence in the pyramidal cell network of the cortex. One pyramidal cell synapses with 500–2000 other pyramidal target cells, leaving few synapses per target cell. With the mechanisms we know of for spike transmission, it is not possible to understand how one pyramidal cell, on its own, can fire any one of its target cells. How can we deal with this sparse connectivity problem? Parameters which would fire one target, e.g., unusually large excitatory postsynaptic potentials (EPSPs) or very long membrane time constants, will fire them all. If all targets fire, we have reflexive pathways or, at worst, epilepsy. Our task instead is to understand how the cortex can drive one target cell at one moment and another later. The key to cognition lies in this freedom of target choice. To hold the key, we must invoke an assembly of cells to participate in transmission events at a particular cell's target. The additional EPSPs from the additional participants will depolarize the target sufficiently to fire it, provided there is synchrony (see SYNCHRONIZATION OF NEURONAL RESPONSES AS A PUTATIVE BINDING MECHANISM). Different assemblies enable transmission to different targets so

that we can shift functional links dynamically within the envelope of all possibilities given by the anatomy, as called for in the article DYNAMIC LINK ARCHITECTURE. As I have argued here, each assembly must embrace multiple areas. Information which these areas find consistent will receive the boost needed for transmission. We must find circuits, synchronize them, and embed them somehow in the reciprocal projection systems between one area and multiple others.

Discussion

Shall we go down only this stream or that one? How long before we come out the other end? Is there time for top-down modulation or only for bottom-up processing? With single-cell transmission principles, multiple areas are a liability, not an asset, and next year there will be more of them. Until we have a marriage of assembly-based transmission principles and classic principles of cell-to-cell transmission, neuroscience cannot provide a basis for massively parallel, distributed information exchange. Without assembly-based transmission principles in the underlying neuroscience, cognitive neuroscience remains a castle in the air.

Road Maps: Mammalian Brain Regions; Vision
Related Reading: Dissociations Between Visual Processing Modes; Grasping Movements: Visuomotor Transformations; Planning, Connectionist; Visual Processing of Object Form and Environment Layout

References

Damasio, A. R., 1990, Synchronous activation in multiple cortical regions: A mechanism for recall, *Semin. Neurosci.*, 2:287–296. ◆
Desimone, R., and Duncan, J., 1995, Neural mechanisms of selective visual attention, *Annu. Rev. Neurosci.*, 18:193–222. ◆
Distler, C., Boussaoud, D., Desimone, R., and Ungerleider, L. G., 1993, Cortical connections of inferior temporal area TEO in macaque monkeys, *J. Comp. Neurol.*, 334:125–150.
Farah, M. J., 1990, *Visual Agnosia: Disorders of Object Recognition and What They Tell Us About Normal Vision*, Cambridge, MA: MIT Press. ◆
Godwin-Austen, R. B., 1965, A case of visual disorientation, *J. Neurol. Neurosurg. Psychiatry*, 28:453–458.
Luria, A. R., 1959, Disorders of "simultaneous perception" in a case of bilateral occipitoparietal brain injury, *Brain*, 82:437–449.
Milner, A. D., and Goodale, M. A., 1993, Visual pathways to perception and action, *Prog. Brain Res.*, 95:317–337. ◆
Mishkin, M., 1993, Cerebral memory circuits, in *Exploring Brain Functions: Models in Neuroscience* (T. A. Poggio and D. A. Glaser, Eds.), London: Wiley, pp. 113–125. ◆
Nelson, J. I., 1988, Binocular vision: Disparity detection and anomalous correspondence, in *Textbook of Optometry* (K. Edwards and R. Llewellyn, Eds.), London: Butterworths, pp. 217–237. ◆
Pallis, C. A., 1955, Impaired identification of faces and places with agnosia for colors, *J. Neurol. Neurosurg. Psychiatry*, 18:218–224.
Rubens, A. B., and Benson, D. F., 1971, Associative visual agnosia, *Arch. Neurol.*, 24:305–316.
Sireteanu, R., Lagreze, W. D., and Constantinescu, D. H., 1993, Distortions in 2-dimensional visual space perception in strabismic observers, *Vision Res.*, 33:677–690.
Ungerleider, L. G., and Mishkin, M., 1982, Two cortical visual systems, in *Analysis of Visual Behaviour* (D. G. Ingle, M. A. Goodale, and R. J. W. Mansfield, Eds.), Cambridge, MA: MIT Press, pp. 459–486. ◆
Williams, M., 1970, *Brain Damage and the Mind*, Baltimore: Penguin, pp. 61–62. ◆

Visual Schemas in Object Recognition and Scene Analysis

Risto Miikkulainen and Wee Kheng Leow

Introduction

Humans have the ability to rapidly and accurately recognize objects in a scene. We perform this task by matching visual inputs with object representations in our memory. These representations are not simply raw images in terms of light and dark pixels, but describe the spatial structure of objects. In many computational models, such representations are implemented as *visual schemas*, which are active functional units that cooperate and compete to determine which representation best matches the object. This article focuses on how visual schemas can be implemented in neural networks and how they can be used to model human object recognition and scene analysis.

Basic Concepts

Visual schemas describe objects in terms of the physical properties and the spatial arrangements of their components. Consider the schema for a deer, for example. The real-world deer has a variety of parts, such as a head, a neck, a body, and four legs. Each part has a characteristic shape and size: the head is a small triangular block, the neck a long cylinder, the body a large rectangular block, and the legs are long and slim cylinders. The parts are arranged in more or less specific locations. One end of the neck is attached to the head, and its other end is attached to the body. The four legs are attached to the bottom of the body to support it. All such information must be represented in the schema for a deer.

Recent psychological theories of human object recognition (Biederman, 1987; see OBJECT RECOGNITION) suggest that the shapes of the object components can be classified into about 36 different categories, called *geons*. Geons are similar to the generalized cylinders used in machine vision for describing three-dimensional (3D) shapes (Marr, 1982). Whereas a generalized cylinder is modeled mathematically by sweeping a two-dimensional (2D) cross section along a curvilinear axis, a geon is identified by a set of characteristics such as the shape of the cross section, whether the size of the cross section is uniform or expanding along the axis, whether the axis of symmetry is straight or curved, and so on. The sizes need not be very accurately measured, and they can be quantized into discrete intervals. Although the geon theory does not address certain aspects of human visual perception such as the recognition of faces and natural scenes, it captures many aspects of human object recognition and serves as psychological motivation for modeling object recognition in terms of schema systems.

Much of SCHEMA THEORY (q.v.) has been developed in the symbolic framework (Arbib, 1989; Rumelhart, 1980). Based on the experience with symbolic schema systems such as VISIONS (Draper et al., 1989; Hanson and Riseman, 1978), it is possible to summarize the three main functional characteristics that visual schema implementations should have:

1. Visual schemas are organized into a *schema hierarchy*, with scene schemas at the topmost level, object schemas at the next level, and so on.
2. Schema instances representing the components of an object *cooperate* to support the instantiation of the object schema.
3. Schema instances representing different objects (or scenes) *compete* to determine which one best matches the inputs.

It turns out that these characteristics are very naturally implemented in neural networks. Moreover, the standard learning mechanisms of neural networks make it possible to build systems that learn schemas from examples, which is difficult to do in the symbolic framework.

Visual Schemas in Neural Networks

Schemas are inherently structured representations, and representing structure in general is difficult with neural networks (see STRUCTURED CONNECTIONIST MODELS). There are three key ideas that allow us to approach the problem of representing visual schemas. First, part-whole relationships between an object and its parts can be represented by connections between localist units (Hinton, 1988). When a unit representing a part of an object is activated, its activity propagates to the unit representing the whole object. This process corresponds to bottom-up input. Conversely, through feedback connections from the object unit to the part units, the object can activate its parts, corresponding to top-down expectation.

Second, cooperative and competitive relationships among the schemas can be represented by connections among the part and schema units. This idea is adopted in the distributed schema system of Rumelhart et al. (1986). There are parts that can be found in several high-level schemas, while others belong only to a single schema. For example, both dining room and kitchen may contain a table, but a bed typically exists only in a bedroom. Part units that belong to the same schema are connected with positive weights and cooperate to support the schema. Part units that never exist in the same schema are connected with negative weights and compete, trying to activate different schemas. Given that certain components, say a table and a bed, are activated by the input, the activity propagates through the connections and eventually stabilizes. The resulting activity over the network represents the activation of the best matching schema, such as the bedroom schema. Since the network contains only part units and has no distinct units for the different schemas, only one schema can be distinctly activated at any one time in this model.

The third idea concerns the so-called *binding problem* which occurs when a part unit is connected to several schema units that contain the same component. Several distinct objects in the scene may contain different instances of the same part. When the part unit is activated, the system has to determine to which object the part instance belongs. The model described in the next section addresses the binding problem by focusing attention (see SELECTIVE VISUAL ATTENTION) on one component of an object at a time (Didday and Arbib, 1975) and cumulating the recognition results in the object schema hierarchy; for other issues, see BINDING IN THE VISUAL SYSTEM.

Implementation in VISOR

VISOR (Leow, 1994; Leow and Miikkulainen, 1994), a concrete implementation of these three ideas in the domain of object recognition and scene analysis, is one of the few explicitly schema-based neural network vision systems built to date (see also Arbib, 1989; Feldman, 1985; Hummel and Biederman, 1992). It consists of three main modules: the Low-Level Visual Module, the Schema Module, and the Response Module. The

Low-Level Visual Module (LLVM, currently simulated procedurally) focuses attention on one component of an object (such as the triangular roof of the arch) at a time, extracts its shape (what) and relative position (where), and sends this information to the Schema Module. The schema that best matches the input suggests shifting attention to a location where another component (such as the left pillar) is expected. The LLVM then shifts attention to the new position, and the process repeats until VISOR has looked at all the components in the scene.

In the Schema Module, schema representations are organized in two levels (Figure 1). The topmost level consists of scene schemas that receive inputs from lower-level object schemas. Object schemas in turn receive inputs from shape units that represent rough categories of shape and size modeled after Biederman's theory. The spatial structure of an object, such as an arch (Figure 1B), is represented in a 2D array of units called the Subschema Activity Map (SAM). Each component of the object is represented by a SAM unit at the corresponding position. The connections between the shape units and the SAM units encode what VISOR expects to find at each SAM position. For example, the arch, which consists of a triangle on top of two rectangles, is encoded in a 3 × 3 SAM: the top-center unit is strongly connected to the triangle-sensitive shape unit, and the two units on either side are most strongly activated by the same rectangle-sensitive shape unit.

At the scene level, spatial structure is often less rigid. For example, in a park scene, there may be several instances of the tree, and they may appear anywhere in the scene. The schema for such a scene consists of a column of SAM units without spatial relationships (Figure 1). The connection from an object

schema to a SAM unit indicates that several instances of the object may appear anywhere in the scene. A rough spatial structure, such as objects appearing at the top, center, or bottom sections of the scene, can be represented by multiple spatially organized columns.

In analyzing a park scene that contains, e.g., an arch surrounded by two trees, VISOR may begin by focusing at the center of the image—that is, at the triangular roof of the arch. The LLVM finds this component to be located at the top-center position of the object and enables the object schemas' top-center SAM units. It also activates the triangle-sensitive shape unit (Figure 1A). Because the arch schema expects a triangle at that relative position, its top-center SAM unit becomes highly activated. As VISOR looks at other components in the scene, the corresponding SAM units' activities are updated, and indicate how likely the other components of the schema are to be present in the scene. The schema's output unit sums up the component activities and indicates how well the entire schema matches the input. In other words, the components cooperate in supporting the schema activation. The output unit then sends activation to the SAM units of higher-level schemas, indicating, for example, that finding an arch in the scene suggests that the entire scene might depict a park.

Different schemas may share identical or similar parts. For instance, the roof of an arch may look like that of a house. In this case, the triangle-sensitive shape unit has a strong connection to SAM units in both the arch and the house schemas (Figure 1). If the triangles appear in the same relative position, as is the case with the arch and house, then the activation of the triangle-sensitive unit propagates to both arch and house SAMs. This way, whenever VISOR focuses on a new location, all schemas that match the input at that location are simultaneously activated. VISOR keeps shifting attention to other positions and accumulating activation in its schema hierarchy until it has seen all of the important inputs in the scene. In the arch example, the arch schema eventually develops a larger output activity because it matches the input object better than the house schema. It also inhibits the house schema through inhibitory connections between their output units, enhancing the difference. Thus, the schemas compete to determine which one best matches the focused object.

VISOR can also learn to encode schemas from examples. Initially, all SAM weights have small positive random values, indicating no specific spatial structure. Given an input object such as a house, the schemas become randomly activated in the scene analysis process. The schema that happens to match the input best modifies its weights through variations of Hebbian rules, and gradually learns to encode the spatial structure of the house. At the same time, the Response Module learns to associate the current schema hierarchy activation with the house label provided by the environment. In case that best-matching schema happens to already encode a different object, such as the arch, the Response Module will produce an incorrect label. The environment delivers a punishment signal which suppresses the activation of the arch schema, and another schema will become most active. If it is a new schema, no incorrect label will be produced, and learning proceeds as above. This punishment signal is analogous to the mismatch-reset signal in the ART network (Carpenter and Grossberg, 1987; see ADAPTIVE RESONANCE THEORY). It tells VISOR to find a different schema to encode the house without specifying which one.

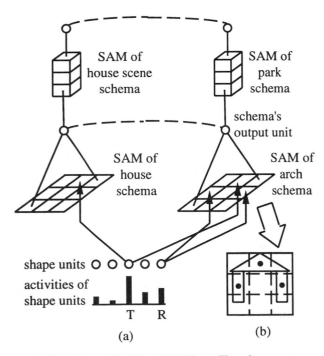

Figure 1. The Schema Module of VISOR. *A*, The schemas are organized into two levels: objects and scenes. Arrows represent one-way connections from low-level inputs; the solid lines represent both the bottom-up and top-down connections (which are different) in the schema hierarchy; and dashed lines indicate inhibition. The shape unit marked *T* is sensitive to flat triangles; the one marked *R*, to vertical rectangles. *B*, The arch image encoded by the arch schema. The grid represents the Subschema Activity Map (SAM). The black dots denote those SAM units that represent the arch components.

Discussion

The VISOR implementation of visual schema representation, application, and learning can give a computational account to several phenomena in human object recognition and scene analysis (Leow and Miikkulainen, 1994), but it is still a long way from capturing the full variety and complexity of real-world object recognition and scene analysis. Some objects have flexible or movable components that can appear in different spatial relations with each other, such as the limbs and body of a human reaching up or picking up something from the ground. For such objects, topological relationships such as "connected-to" would be more appropriate than rigid spatial relationships (Biederman, 1987). Many objects would need to be represented in 3D rather than as 2D projections, and it should be possible to recognize them from different viewpoints and also in different scales and orientations (Hummel and Biederman, 1992; Leow, 1994; Olshausen, Anderson, and Van Essen, 1993). Also, segmentation of scenes to their components and separation from the background is perhaps not possible strictly bottom-up as VISOR currently assumes, especially when the objects can be occluded. So far, it has been possible to give only partial answers to some of these questions, and others remain wide open.

Road Map: Vision
Related Reading: Routing Networks in Visual Cortex; Thalamus; Visual Scene Perception: Neurophysiology

References

Arbib, M. A., 1989, *The Metaphorical Brain 2: Neural Networks and Beyond*, New York: Wiley. ◆

Biederman, I., 1987, Recognition-by-components: A theory of human image understanding, *Psychol. Rev.*, 94:115–147.

Carpenter, G. A., and Grossberg, S., 1987, A massively parallel architecture for a self-organizing neural pattern recognition machine, *Comput. Vis. Graph. Image Proc.*, 37:54–115.

Didday, R. L., and Arbib, M. A., 1975, Eye movements and visual perception: A "two visual system model," *Int. J. Man-Mach. Stud.*, 7:547–569.

Draper, B. A., Collins, R. T., Brolio, J., Hanson, A. R., and Riseman, E. M., 1989, The schema system, *Int. J. Comput. Vis.*, 2:209–250. ◆

Feldman, J. A., 1985, Four frames suffice: A provisional model of vision and space, *Behav. Brain Sci.*, 8:265–313.

Hanson, A. R., and Riseman, E. M., 1978, VISIONS: A computer system for interpreting scenes, in *Computer Vision Systems* (A. R. Hanson and E. M. Riseman, Eds.), New York: Academic Press.

Hinton, G. E., 1988, Representing part-whole hierarchies in connectionist networks, in *Proceedings of the 10th Annual Conference of the Cognitive Science Society*, Hillsdale, NJ: Erlbaum, pp. 48–54.

Hummel, J. E., and Biederman, I., 1992, Dynamic bindings in a neural network for shape recognition, *Psychol. Rev.*, 99:480–517. ◆

Leow, W. K., 1994, *VISOR: A Neural Network System That Learns to Represent Schemas for Object Recognition and Scene Analysis*, Technical Report AI94-219, University of Texas at Austin, Deptartment of Computer Sciences.

Leow, W. K., and Miikkulainen, R., 1994, Priming, perceptual reversal, and circular reaction in a neural network model of schema-based vision, in *Proceedings of the 16th Annual Conference of the Cognitive Science Society*, Hillsdale, NJ: Erlbaum, pp. 560–565.

Marr, D., 1982, *Vision*, San Francisco: Freeman. ◆

Olshausen, B. A., Anderson, C. H., and Van Essen, D. C., 1993, A neurobiological model of visual attention and invariant pattern recognition based on dynamic routing of information, *Neuroscience*, 13:4700–4719.

Rumelhart, D. E., 1980, Schemata: The building blocks of cognition, in *Theoretical Issues in Reading Comprehension* (R. J. Spiro, B. C. Bruce, and W. F. Brewer, Eds.), New York: Wiley.

Rumelhart, D. E., Smolensky, P., McClelland, J. L., and Hinton, G. E., 1986, Schemata and sequential thought processing in PDP models, in *Parallel Distributed Processing: Explorations in the Microstructure of Cognition* (D. E. Rumelhart, J. L. McClelland, and PDP Research Group, Eds.), vol. 2, *Psychological and Biological Models*, Cambridge, MA: MIT Press, pp. 7–57. ◆

Visuomotor Coordination in Flies

Alexander Borst and Martin Egelhaaf

Introduction

Motion information plays a prominent role in the visual orientation of many animal species because the retinal images are continually displaced during self-motion. The resulting retinal motion patterns depend characteristically on the trajectory described by the animal as well as on the particular three-dimensional structure of the visual environment.

Consider, for instance, three common situations. (1) When an animal deviates from its course, the retinal image of the entire visual environment is coherently displaced in the opposite direction. (2) In contrast, the approach toward an obstacle leads to an expansion of the retinal image. In other words, all elements in the image move centrifugally away from the point that the animal is approaching. (3) When the animal passes a nearby object in front of a more distant background, the retinal images of the object and the background move at different velocities. This difference leads to discontinuities in the motion pattern. All of these retinal motion patterns induced by self-motion are particularly pronounced in fast-moving animals and especially in flying animals.

If evaluated appropriately, these different types of motion patterns can be used to guide visual orientation. (1) Rotatory large-field motion may signal to the animal unintended deviations from its course; thus, a system extracting this type of motion pattern could be an integral part of an autopilot which compensates for these deviations by corrective steering maneuvers (*optomotor response*). (2) Image expansion signals that the animal approaches an obstacle; a system evaluating this type of motion pattern may initiate deceleration and control extension of the legs to avoid crash landing (*landing response*). (3) Finally, discontinuities in the retinal motion field and small-field motion indicate nearby stationary or moving objects; therefore, mechanisms sensitive to this type of motion pattern may be part of a fixation system which induces turning reactions toward objects (*object response*).

The mechanisms underlying the extraction of these retinal motion patterns and their transformation into the appropriate motor activity have been analyzed extensively in the fly. Since the fly's orientation behavior relies heavily on motion information, its visual system was found to be highly specialized with respect to motion vision. Moreover, the fly's nervous system is amenable to an analysis on the basis of nerve cells which can be identified individually in each animal. This feature is a great advantage for an analysis of biological information processing in terms of neuronal circuits.

The Visuomotor System of the Fly

The transformation of the retinal images into appropriate motor activities is the consequence of specific biophysical properties of nerve cells and their connection patterns. Figure 1 summarizes those features of the fly's nervous system which are most relevant in the current context. The retinal images are initially transformed by a sequence of successive retinotopically organized layers of columnar nerve cells. At the level of the lobula plate, large-field elements are found which integrate the output signals of columnar neurons. These large-field cells connect to other brain areas or, through descending neurons, to the motor control centers. All of these large-field cells respond selectively to motion in a particular direction and play an important role in extracting the different types of retinal motion patterns. Because of their extraordinary structural constancy and highly invariant physiological properties, they can be identified individually in each animal (Strausfeld, 1989; Hausen and Egelhaaf, 1989). The motor output consists of indirect power muscles, which keep the thorax and the wings oscillating during flight, and a system of small steering muscles, which insert on various sclerites at the wing base and influence the wing posture during up- and downstrokes. While visual afferences to the power muscles mainly affect lift and thrust reactions of the fly, visual input to the steering muscles is important for the control of different kinds of flight maneuvers, such as torque (Heide, 1983).

Studies on Visual Orientation of the Fly

Although fly visual orientation has been studied also under free-flight conditions (Wagner, 1985), free-flight behavior is often too complex for a systematic analysis. In a more reductionist approach, therefore, most behavioral experiments were done on tethered flying animals under well-defined and sufficiently simple visual stimulus conditions which made it possible to establish stimulus-response relationships quantitatively. Visually induced turning responses were monitored by mounting the tethered flying fly to a torque meter and stimulating it with patterns of different size which could be moved with various velocities into different directions (Buchner, 1984; Heisenberg and Wolf, 1984). Essentially the same stimulus conditions were employed to investigate the visual release mechanisms of landing. In these experiments, the extension of the fly's forelegs was monitored by means of a light barrier. The leg extension is a fixed-action pattern which characterizes the initial phase of the landing response and is accompanied by deceleration of the animal (Borst, 1990).

From this sort of black box analysis, models of the underlying mechanisms could be derived. These models allowed us to design the appropriate visual stimuli for the identification of those neurons which may correspond to the different model elements (Hausen and Egelhaaf, 1989; Borst, 1990; Egelhaaf and Borst, 1993b). The outcome of this analysis is summarized below. However, it should be kept in mind that the analysis builds on knowledge of only a few dozen neurons (mainly the large ones) among the approximately 300,000 nerve cells of a fly brain. It also considers mainly those behaviors which can be observed in the laboratory during tethered flight. Thus, the scope can only be to explain the control of some behavioral components through well-defined retinal motion patterns, and not to understand the subject of fly vision and control of aerobatic flight maneuvers in its entirety.

Stages of Motion Computation

The only information available to the visual system is given by the time-dependent brightness values of the retinal image as sensed by the photoreceptors. The retinal input is transformed into the corresponding behavioral output in three principal steps: (1) Motion in the different parts of the retinal image

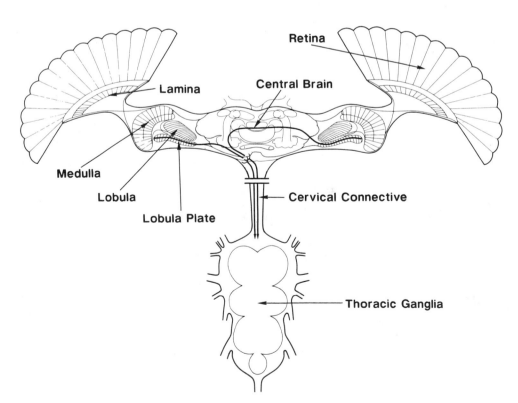

Figure 1. Schematic horizontal cross-section through the nervous system of the fly, with the retina; the three visual ganglia (lamina, medulla, and lobula complex, which is subdivided into the anterior lobula and the posterior lobula plate); the central brain; and the thoracic ganglia, with the motor control centers. (Modified from Hausen, 1984.)

is computed in parallel by two-dimensional retinotopic arrays of local movement detectors. (2) From their signals, various retinal motion patterns are extracted by spatial integration over arrays of appropriately directed local movement detectors. (3) The dynamical properties of these representations of retinal motion are tuned by temporal filtering to the needs of the fly in free flight.

Local Movement Detection

Motion in different parts of the visual field usually does not have the same direction and velocity (Figure 2, top). As a first step of motion analysis, therefore, a local representation of the different motion vectors must be computed. This process is done in parallel by local movement detectors. These are organized in two-dimensional retinotopic arrays which cover the entire visual field (Figure 2, center). On the basis of a behavioral analysis, many years ago, W. Reichardt and B. Hassen-

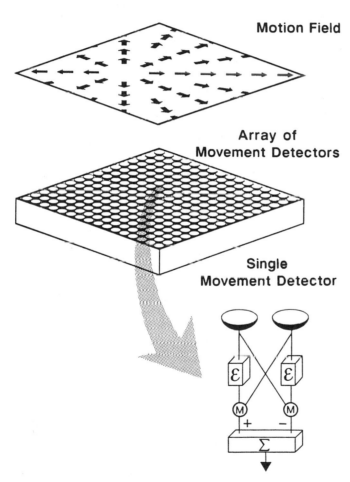

Motion Field

Array of Movement Detectors

Single Movement Detector

Figure 2. In the first step of motion analysis, motion in different parts of the visual field is computed in parallel by two-dimensional arrays of local movement detectors. The upper diagram shows an example of a retinal motion pattern. The middle diagram shows a two-dimensional array of movement detectors. The bottom diagram shows a correlation-type movement detector as a representative of a local motion detection mechanism. In its simplest form, its input is given by the light intensities as measured at two points in space. In each subunit, the detector input signals are multiplied (*M*) with each other after one of them has been delayed by some sort of temporal filter (*ε*). The outputs of both mirror-symmetrical subunits are then subtracted to give the final output signal of the detector.

stein proposed a formal model of such a local movement detector, the *correlation-type movement detector* (Figure 2, bottom) (Reichardt, 1961, 1987; Buchner, 1984; Borst and Egelhaaf, 1993). It consists of two mirror-symmetrical subunits. Their output signals are subtracted from each other. Each subunit has two input channels which become multiplied after one of the signals has been delayed by some sort of temporal filtering. Interestingly, this type of motion detector does not provide an exact measurement of the local pattern velocity. Its response also depends characteristically on the structure of the stimulus pattern. As a result, the instantaneous activity profile of a two-dimensional array of correlation-type motion detectors does not encode faithfully the geometrically calculated retinal motion pattern. Thus, additional processing steps are required to gain meaningful motion information from the output of local movement detectors and, in particular, useful representations of the different retinal motion patterns.

Extraction of Different Retinal Motion Patterns

The different types of motion patterns are characterized by different directions and velocities in the different parts of the visual field. Hence, specific information about the different retinal motion patterns can be extracted by intra- and interocular spatial integration over appropriately oriented local movement detectors. In the fly, this process occurs mainly in the lobula plate (see Figure 1), where motion-sensitive neurons with large dendrites are found. All of these neurons are activated by motion in a particular direction and are inhibited by motion in the respective opposite direction (Hausen, 1984; Hausen and Egelhaaf, 1989). Two functional classes of output elements of the lobula plate are particularly important in the current context, the *horizontal cells* (HS cells) and *figure detection cells* (FD cells) (Hausen and Egelhaaf, 1989; Egelhaaf and Borst, 1993b).

There are three HS cells which cover the dorsal, medial, and ventral parts of the lobula plate (Figure 3*A*). They are excited by motion from the front to the back in the dorsal, medial, and ventral parts of the ipsilateral visual field, respectively. Their responses increase, although not linearly, with increasing size of the stimulus pattern (Figure 3*B*). As a result of synaptic input from another identified large-field element of the contralateral lobula plate, part of the HS cells also respond to motion from the back to the front in the contralateral visual field. This input organization makes the HS cells particularly sensitive to coherent rotatory large-field motion around the animal's vertical axis. In studies where the HS cells were ablated from the circuit by microsurgical, laser, or genetic techniques, the large-field optomotor response was severely impaired (Heisenberg and Wolf, 1984; Hausen and Egelhaaf, 1989). Thus, the HS cells are likely to be part of the fly's autopilot, which compensates for unintended turns of the fly from its course.

The FD cells are a group of at least four output elements of the lobula plate which are all selectively tuned to small-field motion. For example, the FD1 cell is shown in Figure 3*A*. Its response is greatest during stimulation with a small pattern and declines when the stimulus pattern becomes larger (see Figure 3*C*). In contrast to many small-field cells found in other animals which have spatially separated excitatory and inhibitory subregions of their receptive field, the FD cells are excited by a small moving object anywhere within their receptive field of the ipsilateral eye. Ablation experiments demonstrated that the small-field tuning of the FD1 cell is mediated through inhibition by a single identified neuron, the ventral centrifugal horizontal (VCH) cell: After photoinactivation of the VCH cell, the small-field tuning of the FD1 cell was selectively lost (Egelhaaf

a

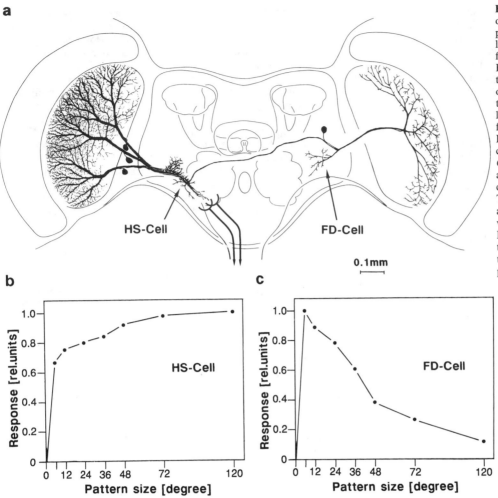

HS-Cell   FD-Cell

0.1mm

b

HS-Cell

Response [rel.units] — vertical axis (0, 0.2, 0.4, 0.6, 0.8, 1.0)
Pattern size [degree] — horizontal axis (0, 12, 24, 36, 48, 72, 120)

c

FD-Cell

Response [rel.units] — vertical axis (0, 0.2, 0.4, 0.6, 0.8, 1.0)
Pattern size [degree] — horizontal axis (0, 12, 24, 36, 48, 72, 120)

Figure 3. Spatial integration properties of neural elements in the fly's lobula plate extracting coherent rotatory large-field motion (HS cells) and small-field motion (FD cells), respectively. *A*, Frontal projection of arborizations of the three HS cells (shown in the left optic lobe) and one of the FD cells, the FD1 cell (shown in the right optic lobe). The HS cells were reconstructed from cobalt stainings (courtesy Klaus Hausen) and the FD1 cell from intracellular Lucifer yellow staining. *B* and *C*, Dependence of the mean response amplitude of an HS cell and an FD1 cell on the size of the stimulus pattern. The pattern was a random texture, the angular horizontal extent of which was varied. Whereas the response of the HS cell reaches its maximum for motion of large patterns, the response of the FD cell is strongest when a small pattern is moved in its receptive field.

and Borst, 1993b). The similarity between the functional properties of the FD cells and the dependence of visually induced turning responses on the different stimulus parameters suggests that the FD cells may control turns of the animal toward objects.

Visual interneurons extracting some representation of image expansion from the activity profiles of the retinotopic array of movement detectors have only been found in the cervical connective (see Figure 1). These cells' responses are strongest when the animal approaches an obstacle or a potential landing site. Their responses to different motion stimuli correlate well with changes in landing responses seen in behavioral experiments, suggesting that they are part of the neuronal circuit initiating landing behavior (Borst, 1990).

Temporal Tuning

The different types of motion patterns are characterized by their specific geometrical properties and also have specific dynamic features, as reflected in the temporal tuning of the respective control systems.

At the level of the lobula plate, the different motion-sensitive neurons still have similar dynamical properties (Egelhaaf and Borst, 1993a, 1993b). In the pathway tuned to coherent rotatory large-field motion, these signals become temporally low-pass filtered somewhere between the lobula plate and the steer-ing muscles which mediate the compensatory turning responses (Egelhaaf, 1991). Since active turns of the fly in free flight are brief and rapid (Wagner, 1985), the resulting retinal large-field motion consequently is also characterized by fast changes in its direction. Because of the dynamical tuning of the fly's autopilot, these deviations from course are not well compensated for by corrective steering maneuvers. Hence, the temporal tuning might be a simple computational means to prevent the visual consequences of active turns from being compensated for by the autopilot.

Some sort of low-pass filtering is also taking place in the fly's landing system, as suggested by experiments where the latency of the leg extension which accompanies the initiation of landing was found to vary with the stimulus strength in a graded way (Borst, 1990). The low-pass filter, in combination with a threshold device, converts the stimulus strength into a wide range of response latencies. A weak stimulus, such as a slowly expanding pattern, leads only to a slow increase in the temporally integrated signal and thus to a large response latency, whereas a strong stimulus, such as a pattern expanding with a higher velocity, leads to a rapid increase in the temporally integrated signal and thus needs less time to reach threshold. This capability ensures that the fly initiates landing earlier when it approaches a potential landing site with a higher velocity than when it reaches a site with a lower velocity and, thus, leads to a safe landing.

Local Motion Detection

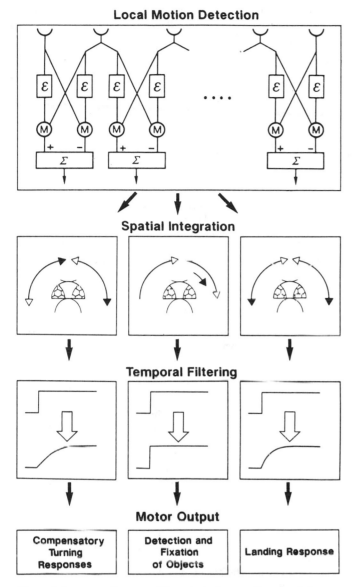

Spatial Integration

Temporal Filtering

Motor Output

| Compensatory Turning Responses | Detection and Fixation of Objects | Landing Response |

Figure 4. Summary diagram of the principal steps of motion information processing in the fly visual system. The top part illustrates detection of local motion by large retinotopic arrays of movement detectors. Only three detectors are shown (for abbreviations, see the legend of Figure 2). Next, the initial representation of local motion segregates into three pathways which extract different retinal motion patterns: rotatory large-field motion (left), relative motion of an object and its background (middle), and pattern expansion (right). The arrows indicate the direction of pattern motion; filled and open arrowheads indicate excitation and inhibition, and long and small arrows indicate large-field and small-field motion, respectively. The third part of the diagram shows temporal tuning of the representations of rotatory large-field motion and pattern expansion by some sort of low-pass filters with different time constants. The signals before and after the filters are shown in response to an onset of motion. The different pathways are involved in mediating different types of orientation behavior, as illustrated by the final part of the diagram.

Discussion

The features of the different retinal motion patterns which are important for flight control are extracted in a series of processing steps (Figure 4). (1) The initial explicit representation of motion is computed in parallel by retinotopic arrays of local motion detectors. (2) This representation is then segregated into different pathways which are selectively tuned to the different retinal motion patterns and feed different control systems of visual orientation behavior. This spatial tuning is achieved by appropriate inter- and intraocular spatial integration. (3) Before exerting their influence on the motor control centers, the signals are tuned by appropriate temporal filtering to the characteristic dynamical properties of the retinal motion fields as induced during different flight maneuvers.

Despite the relative simplicity of the computations underlying the different control systems, the problems they solve are of widespread relevance. There is good evidence from electrophysiological, behavioral, and psychophysical studies of similar mechanisms in such phylogenetically distant animals as insects and mammals, including humans (Egelhaaf and Borst, 1993a). These similarities hint at a convergent evolution and thus to some degree of optimal adaptive value of the mechanisms of motion processing exploited by biological systems. In contrast to technical systems, these mechanisms are not exact in a mathematical sense, but are fast, reliable, and robust against noisy signals. These features could make them attractive models for low-level processors in artificial seeing systems for robots (e.g., Franceschini, Pichon, and Blanes, 1992).

Road Map: Motor Pattern Generators and Neuroethology
Related Reading: Active Vision; Directional Selectivity in the Retina; Locust Flight: Components and Mechanisms in the Motor; Motion Perception; Reactive Robotic Systems

References

Borst, A., 1990, How do flies land? From behavior to neuronal circuits, *Bioscience.*, 40:292–299.
Borst, A., and Egelhaaf, M., 1993, Detecting visual motion. Theory and models, in *Visual Motion and Its Role in the Stabilization of Gaze* (F. A. Miles and J. Wallman, Eds), Amsterdam: Elsevier, pp. 3–27. ◆
Buchner, E., 1984, Behavioural analysis of spatial vision in insects, in *Photoreception and Vision in Invertebrates* (M. A. Ali, Ed.), New York: Plenum, pp. 561–621.
Egelhaaf, M., 1991, How do flies use visual motion information to control their course? *Zool. Jahrb. Physiol.*, 95:287–296.
Egelhaaf, M., and Borst, A., 1993a, Detecting visual motion: Invertebrate neurophysiology, in *Visual Motion and Its Role in the Stabilization of Gaze* (F. A. Miles and J. Wallman, Eds.), Amsterdam: Elsevier, pp. 53–77.
Egelhaaf, M., and Borst, A., 1993b, A look into the cockpit of the fly: Visual orientation, algorithms and identified neurons, *J. Neurosci.*, 13:4563–4574. ◆
Franceschini, N., Pichon, J. M., and Blanes, C., 1992, From insect vision to robot vision, *Philos. Trans. R. Soc. Lond. B Biol. Sci.*, 337:283–294.
Hausen, K., 1984, The lobula complex of the fly: Structure, function and significance in visual behavior, in *Photoreception and Vision in Invertebrates* (M. A. Ali, Ed.), New York: Plenum, pp. 523–559.
Hausen, K., and Egelhaaf, M., 1989, Neural mechanisms of visual course control in insects, in *Facets of Vision* (D. Stavenga and R Hardie, Eds.), Berlin: Springer-Verlag, pp. 391–424.
Heide, G., 1983, Neural mechanism of flight control in Diptera, in *Biona Report 2* (W. Nachtigall, Ed.), Stuttgart: G. Fischer-Verlag, pp. 35–52.
Heisenberg, M., and Wolf, R., 1984, *Vision in Drosophila*, Berlin: Springer-Verlag.
Reichardt, W., 1961, Autocorrelation: A principle for the evaluation of

sensory information by the central nervous system, in *Sensory Communication* (W. A. Rosenblith, Ed.), New York: MIT Press/Wiley, pp. 303–317.

Reichardt, W., 1987, Evaluation of optical motion information by movement detectors, *J. Comp. Physiol. [A]*, 161:533–547.

Strausfeld, N. J., 1989, Beneath the compound eye: Neuroanatomical analysis and physiological correlates in the study of insect vision, in *Facets of Vision* (D. Stavenga and R. Hardie, Eds.), Berlin: Springer-Verlag, pp. 317–359.

Wagner, H., 1985, Aspects of the free flight behavior of houseflies (*Musca domestica*), in *Insect Locomotion* (M. Gewecke and G. Wendlcr, Eds.), Berlin: Paul Parey Verlag, pp. 223–232.

Visuomotor Coordination in Frogs and Toads

Francisco Cervantes-Pérez

Introduction

The study of visually guided behaviors in amphibians, as in many other animals, has become significant for scientists working in a variety of fields. In neurobiology (see Llinás and Precht, 1976; Fite, 1976; Vanegas, 1984) and computational neuroscience (see articles by Arbib, Betts, Cervantes-Pérez, an der Heiden, and Lara in Ewert and Arbib, 1989), scientists are pursuing the quest for the understanding of how the animal's central nervous system (CNS) integrates the processing of sensory information to control motor behavior. Scientists working in artificial intelligence and robotics use functional principles generated in the study of living animals (e.g., amphibians) as models to build computer systems to control automata that display complex sensorimotor behaviors (see article by Arkin in Ewert and Arbib, 1989; NEUROETHOLOGY, COMPUTATIONAL; and REACTIVE ROBOTIC SYSTEMS).

Although anuran amphibians (frogs and toads) live in a three-dimensional world rich in multiple modes of sensory signals (e.g., visual, tactile, and olfactory), their behavior is guided primarily by visual information. Therefore, they have been studied mostly under visuomotor coordination paradigms. In these animals, visuomotor integration implies a complex transformation of sensory data, since the same locus of retinal activation might release different types of behavior, some directed toward the stimulus (e.g., prey catching) and others directed to an opposite part of the visual field (e.g., predator avoidance). Furthermore, the efficacy of visual stimuli to release a response (i.e., type of behavior, intensity, and frequency) is determined by many factors: (1) the stimulus situation (e.g., form, size, velocity, geometric configuration with respect to the direction of motion, spatiotemporal relationship with the animal); (2) the current state of internal variables of the organism, especially those related to motivational changes (e.g., season of the year, food deprivation); (3) previous experience with the stimulus (e.g., learning, conditioning, habituation); and (4) the physical condition of the animal's CNS (e.g., brain lesions). Thus, to define the next behavior required to interact with a specific external situation, other sensory signals (e.g., kinesthetic information about the muscles and joints intervening in the execution of the behavior), as well as signals indicating the state of motivational variables and of learning processes, must be integrated with those signals encoded in the locus and level of retinal activation.

In searching for the neural mechanisms responsible for connecting the presentation of a visual stimulus with the elicitation of a specific behavior, experimental and theoretical studies have generated and analyzed a series of hypotheses pointing at the interactions among neural elements of the retina, the optic tectum, and the thalamic-pretectal region, or *pretectum*, as the underlying mechanisms. For example, it has been proposed that the locus of activity within the optic tectum and pretectum defines the spatial location of visual stimuli and the direction of prey-catching and predator-avoidance behaviors (see Lara, Cromarty, and Arbib, 1982; Cervantes-Pérez, Lara, and Arbib, 1985; Liaw and Arbib, 1993; articles by Ewert and Grobstein in Ewert and Arbib, 1989; article by Cobas and Arbib in Arbib and Ewert, 1991).

Visually Guided Behaviors in Amphibians

Ethological results show that amphibians are capable of interacting with a wide variety of moving and nonmoving visual stimuli by presenting motor actions that can be classified into different behavioral patterns.

Responses to nonmoving stimuli. Diverse stationary objects may influence the animal's next action. In general, frogs move toward zones in the visual field where blue is preponderant, probably because this situation might be associated with the location of a pond (a proper habitat) and other stimuli whose presence is important for the frog's survival (e.g., prey to eat, water to maintain its body humidity) (see article by Grüsser and Grüsser-Cornhels in Llinás and Precht, 1976). When an obstacle, such as a barrier or chasm, is placed between the amphibian and potential prey, toads display *detour behavior*, following predictable routes to avoid the obstacle and catch the prey (see article by Arbib in Ewert and Arbib, 1989).

Mating. During the spring mating season, the presence of a female in a male's visual field may elicit: (1) an *orienting* response toward the female if she appears in the peripheral part of the visual field; (2) an *approaching* action to reduce the distance to the female when she is far afield in the binocular field; and (3) a *clasping* behavior if the female is within reaching distance in the frontal part of the visual field (Kondrashev, 1987).

Predator avoidance. Large stimuli at close distances may yield one of several avoidance behaviors in the frog and toad, depending on its parametric composition. A flying stimulus close to a frog releases a *ducking* behavior, but when it is far afield, the frog *orients* toward the opposite direction and *jumps* or *runs away*. In the presence of a ground predator, toads display a stiff-legged posture, puffing up and *orienting* toward the predator, and *tilting* the body. Then, they *sidestep* or *jump away* from the predator's location (see Liaw and Arbib, 1993; articles by Ingle and Ewert in Fite, 1976; Ewert in Vanegas, 1984; Cobas and Arbib in Arbib and Ewert, 1991).

Prey acquisition. The presence of potential prey may elicit an action from the following repertoire: (1) amphibians *orient* the head and body toward prey appearing in the peripheral part of the visual field, bringing the stimulus into the binocular zone; (2) they *stalk* (toad) or *jump toward* (frog) potential prey located far afield in the frontal part of the visual field, *approaching* the prey to bring it closer; (3) then *binocular fixation* occurs to properly locate the prey; (4) if the prey is within the snapping zone, amphibians *flick the tongue* to catch it and bring it into the mouth; (5) the mechanical stimulation produced by the prey inside the mouth triggers a *gulping reflex*; and (6) amphibians *wipe the snout* with their forelimbs (for a review, see Ewert in Vanegas, 1984).

Neural Basis for Visuomotor Integration

Structural Organization

Neuroanatomical studies have been used to delineate the structural organization of the visuomotor pathway in amphibians, which involves both serial and highly parallel interconnections. Visual information enters the system through the eye, activating the receptors to produce electrical changes that modify the level of activity displayed by the neural elements of the retina. The efferent fibers of retinal ganglion cells *distribute* neural signals, encoding the presence of moving and nonmoving stimuli to different structures in the animal's CNS. These structures include the optic tectum, the pretectum, the anterior thalamus, and the basal optic nucleus. A direct visuomotor pathway has been defined as follows. The optic tectum and pretectum receive signals from various types of retinal ganglion cells (R2, R3, and R4 project to the tectum, and R3 and R4 project to the pretectum), send a topographic projection to each other, and send efferent fibers that *converge* in the mesencephalic tegmentum and the reticular formation. These latter structures are reciprocally connected and have been postulated to be involved in generating motor patterns that are sent in turn to motoneuron nuclei in the medulla oblongata and spinal cord. The elements of these motoneuron nuclei exert spatiotemporal control on the activation of specific sets of muscles during the elicitation of a motor action (for review, see articles by Ewert, Grüsser-Cornehls, and Lázár in Vanegas, 1984; articles by Ewert, Lázár, and Weerasuriya in Ewert and Arbib, 1989).

Functional Complexity

Neurophysiological studies reinforce the anatomical hypothesis that visuomotor integration is subserved by a parallel distributed neural information processing system. The retina distributes visual information to different structures and has a great deal of parallelism within itself. Retinal ganglion cells are organized into different classes according to the size of their receptive fields, all of which extract information from the complete visual field. Class R2 cells have an excitatory receptive field (ERF) of 4°, whereas class R3 cells have an ERF of 8°, and class R4 cells have an ERF of 12–16° (see Ewert in Ewert and Arbib, 1989). In the same fashion, the tectum and pretectum have different types of neurons. Applying various physiological criteria, Ewert and co-workers (for review, see Ewert in Ewert and Arbib, 1989) offered a classification of ten cell types in the pretectum (named TH1–TH10) and nine classes in the tectum (named T1–T9). Taking into account their different sensitivities to changes of dynamic configural stimulus features, Ewert and colleagues identified several subclasses in some of the tectal cells. For example, T5 cells were divided into four subclasses (T5.1–T5.4) depending on the way they are activated by wormlike objects (i.e., a black rectangle moving against a white background, with its longest axis parallel to the direction of motion), anti-wormlike objects (same object, but with the longest axis moving perpendicular to the direction of

movement), and square objects. It has been postulated that these different types of tectal and pretectal neurons form a *command-releasing system* (see COMMAND NEURONS AND COMMAND SYSTEMS); that is, the activation of a motor pattern generator requires the simultaneous activation of appropriate combinations of specialized neurons. For example, the activation of T5.2 (prey-feature selecting unit) and T4 (arousal unit) may yield a prey-catching orienting response.

Visuomotor Coordination and the Theory-Experiment Cycle

In addition, neuroethological data on fixed action patterns and the analysis of schema-theoretical and neural network models point to the tectal-pretectal interconnections as the neural basis for the amphibian's ability to interact with a variety of moving visual stimuli. First, it was suggested that, in frogs, retinal ganglion cells with small receptive fields (class R2) were tuned as *bug detectors* (Lettvin et al., 1959); then, in a more distributed view, the *integrative properties* found in the optic tectum were emphasized as the basis for detecting prey (Grüsser and Grüsser-Cornehls in Llinás and Precht, 1976). Later, prey-predator discrimination was described as the outcome of an *interactive process* among the projection of ganglion cells of various classes to the tectum (classes R2, R3, and R4) and to the pretectum (classes R3 and R4), in combination with an inhibitory pretectal effect on the tectum (see Cervantes-Pérez et al., 1985; Ewert in Fite, 1976; Ewert in Vanegas, 1984; Arbib, Ewert, Betts, and Cervantes-Pérez in Ewert and Arbib, 1989; Cobas and Arbib in Arbib and Ewert, 1991).

Some hypotheses raised by experimental data address questions as to what information may be stored in the retino-tectal-pretectal interactions, how it is retrieved, and what dynamic processes are activated by the presence of specific stimuli. These questions have been analyzed with different kinds of theoretical models.

Ewert and von Seelen (1974) analyzed prey-predator discrimination phenomena by considering the optic tectum and the pretectum as *filters* that could be activated in parallel by signals coming from the retina, which also acts as a filter (Figure 1). The tectum has a first filter that is sensitive to wormlike objects, whereas the pretectum is modeled as a filter that is sensitive to anti-wormlike objects. Finally, a second filter in the tectum receives excitatory input from the first tectal filter and inhibition from the pretectal filter. The outcome of the processing in the filter-type model reproduced Ewert's physiological data. The level of tectal response has been correlated with the efficacy of visual stimuli to produce prey-catching behavior in which a wormlike object produces a strong tectal response, an anti-worm object elicits little or no response, and a square yields an intermediate response. However, the model is lumped in both space and time, and it describes neither the locus at which the toads directs its response nor the time at which the response is activated.

An der Heiden and Roth (in Ewert and Arbib, 1989) studied the same phenomenon by proposing an alternative model in which the activation dynamics of tectal cells (i.e., T5.1 and T5.2) are explained as the result of the spatiotemporal summation of retinal signals (i.e., ganglion cells R2 and R3) integrated with recurrent lateral inhibition among tectal neurons, that is, without pretectal influence. This model allowed the authors to reproduce important properties of the worm-anti-worm discrimination, but it did not address the issues of tectal disinhibition after pretectal lesion or intrinsic tectal geometry.

Our group has developed a family of models (*Rana computatrix*) to analyze issues of the possible role played by retino-

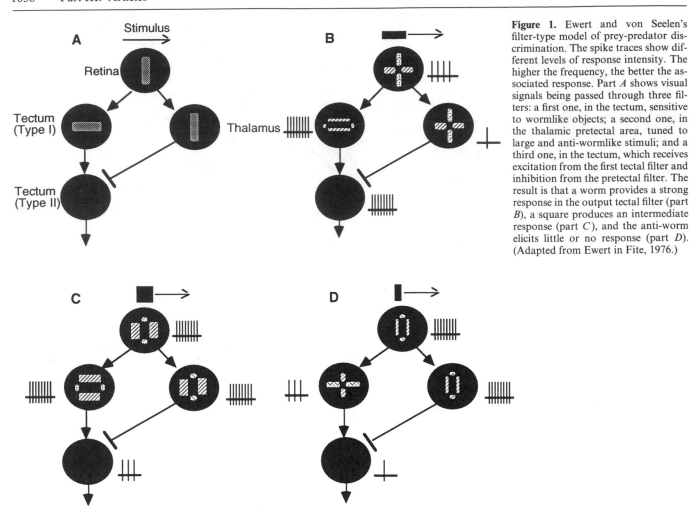

Figure 1. Ewert and von Seelen's filter-type model of prey-predator discrimination. The spike traces show different levels of response intensity. The higher the frequency, the better the associated response. Part *A* shows visual signals being passed through three filters: a first one, in the tectum, sensitive to wormlike objects; a second one, in the thalamic pretectal area, tuned to large and anti-wormlike stimuli; and a third one, in the tectum, which receives excitation from the first tectal filter and inhibition from the pretectal filter. The result is that a worm provides a strong response in the output tectal filter (part *B*), a square produces an intermediate response (part *C*), and the anti-worm elicits little or no response (part *D*). (Adapted from Ewert in Fite, 1976.)

tectal-pretectal interactions on the processing of sensory signals and their translation into commands for the control of prey-catching and predator-avoidance behaviors. Lara and co-workers (1982), assuming that retino-tectal-pretectal neural circuits are formed by arrays of functional units, proposed a layered neural network model of a single unit called a *Facilitation Tectal Column* (FTC). They suggested that short-term memory processes subserving prey-catching facilitation are encoded in the reverberatory activity of tectal self-exciting neural loops rather than on synaptic strengthening (see also Cervantes-Pérez and Arbib, 1990). Building on these modeling efforts, Cervantes-Pérez et al. (1985) presented an extended model consisting of an array of 8 × 8 FTC units including the effects of pretectal cell TH3 and ganglion cells R3 and R4 (Figure 2). They used this model to study whether hypotheses generated from anatomical data (i.e., retino-tectal-pretectal interactions, the presence of positive and negative feedback loops within the optic tectum) might subserve hypotheses about the functionality found in neuroethological studies (i.e., pretectal inhibition of the tectum underlying prey-predator discrimination, correlation of the physiological activity of tectal cell T5.2 (PY in Figure 2) with the elicitation of a prey-orienting response). These authors offered a neural network model that describes the spatiotemporal distribution of the mechanisms responsible for the activation of the animal's behavior. They extended the analysis to suggest that: (1) the movement direction invariance of the efficacy of a visual stimulus to elicit prey-

catching behavior is a consequence of the tectal architecture, and (2) if pretectal inhibition could be modulated by changes in motivational factors, then the optic tectum may act as a *dynamic filter*, showing that the same stimulus may yield different levels of tectal activation, depending on the amount of pretectal inhibition.

Although the FTC model was abstracted somewhat crudely from anatomical data (Székely and Lázár in Llinás and Precht, 1976), it has been useful to analyze how tectal circuitry involving *cooperation-competition* among positive and negative feedback loops works as a dynamic filter. The reverberatory circuits serve as mechanisms for short-term memory processes (prey-catching facilitation) and also as a *content-addressable memory* that stores information related to the probability of a stimulus fitting the prey category. This information is *retrieved* by the presence of a stimulus in the visual field (Cervantes-Pérez et al., 1985; Cervantes-Pérez and Arbib, 1990).

In light of new, ethological, physiological, and anatomical data (see Ewert, Grobstein, Matsumoto, and Lázár in Ewert and Arbib, 1989), the FTC model has been updated, especially to address the following questions. How is the level of tectal-pretectal activation used by motor pattern generators to produce the proper behavior? How can these levels of activation be modified by motivation or by learning processes? Both the tectum and the pretectum send efferent fibers to motor-pattern-generating networks (i.e., reticular formation, mesencephalic tegmentum). They contain neurons whose level of activation

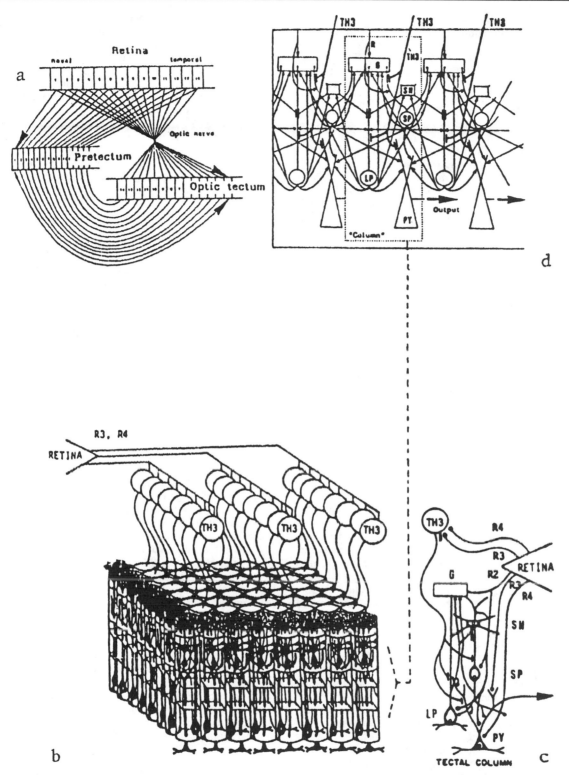

Figure 2. Neural network interactions among retina, optic tectum, and pretectum. *A*, The retina sends fibers retinotopically to both tectum (ganglion cells R2, R3, and R4) and pretectum (classes R3 and R4). *B*, There is a topographic projection from pretectal TH3 neurons to the optic tectum. *C*, Structural organization of the Facilitation Tectal Col- umn model: TH3 cell inhibits intrinsic (LP and SP) as well as the efferent tectal neurons corresponding to its topographic projection. *D*, Interactions among elements of neighboring tectal columns; arrows indicate excitatory influences and lines with bars inhibitory ones. (From Cervantes-Pérez in Ewert and Arbib, 1989.)

form a *command-releasing system*. That is, the simultaneous activation of appropiate combinations of tectal and pretectal cell populations may be closely correlated to the efficacy of a visual target to produce a motor response toward *prey* (e.g., tectal cells T1.3, T3, T4, and T5.2), *predator* (e.g., tectal cell T5.1, pretectal cells TH3 and TH4), or *mate* (e.g., tectal cells T4 and T5.1, during the spring) (Ewert in Ewert and Arbib, 1989). Furthermore, it has been proposed that the locus of activity within the optic tectum and pretectum defines the spatial location of visual stimuli and the direction of prey-catching and predator-avoidance behaviors. Cervantes-Pérez et al. (1985) proposed that the activation of tectal pyramidal neurons (i.e., Ewert's T5.2 cells) defines the stimulus spatial location as well as the direction of the corresponding *prey-catching orienting behavior* when the stimulus appears in the lateral visual field. These cells integrate excitatory signals from retinal ganglion cells R2, R3, and R4 as well as from intrinsic tectal neurons (i.e., large and small pear cells), with an inhibitory effect from TH3 neurons. When several pyramidal cells are active, the target location of a response is determined by the average of their target loci as weighted by their level of activation. These authors analyzed how pretectal inhibition on the tectum might account for prey-predator discrimination, assuming a hierarchical relationship. High pretectal activity, closely correlated with the appearance of possible predators in the visual field, should shut off the prey-catching system.

Analyzing experimental data on *direction-selective avoidance behavior*, Liaw and Arbib (1993) offered a neural network model of the retino-tectal-pretectal interactions to account for the anuran's ability to avoid looming objects (i.e., potential predators). A visual stimulus activates ganglion cells R3 and R4, which project to tectal cells T3 and T6. The former is activated by moving looming objects, and the latter by the presence of a stimulus in the upper part of the visual field. These authors suggested that a spatial map of the stimulus location is given by the population of T3 neurons, where: (1) the stimulus position is determined by localizing its center, encoded in the peak of activation produced by the stimulus on the T3 neuron population, and (2) the looming stimulus direction of motion is determined by monitoring the shift of the peak of neuronal activity within this population. In addition, the signals generated by these cells are integrated, along with depth information, by pretectal neuron TH6 to indicate whether the stimulus is a looming threat. Their model includes other tectal (T2, postulated to respond to stimuli with temporo-nasal movement) and pretectal (TH3, sensitive to larger stimuli) neurons in such a way that (1) the spatial signals conveyed by T2 and T3 neurons are integrated and transformed in the tegmentum to generate a motor heading map that specifies the escape direction, and (2) the appropriate motor response selection (e.g., jumping away, ducking) depends on the stimulus size (including TH6 and TH3 signals) and elevation (T6).

Other hypotheses explored under the theory-experiment cycle paradigm refer to the possibility of having the same tectal circuitry subserving different visually guided behaviors. Betts (in Ewert and Arbib, 1989) analyzed a dynamic neural network model of tectal-pretectal interactions, postulating that T5 neurons form a unique population with the same basic properties and that pretectal inhibition from TH3 neurons modulates the functioning of those properties to achieve specific characteristics (subclass T5.1, T5.2, or T5.3) for an individual T5 neuron at a given time. Betts suggested that, under these conditions, T5 cells have the potential to participate as feature analyzers during the execution of different motor behaviors. For example, in the absence of pretectal modulation, T5 neurons are strongly activated by the least prey-like objects. Thus, he postulated

that, in males, during the spring, T5 neurons exchange their role in prey recognition to participate in mate recognition. He incorporates the notion that pretectal inhibition might be lowered by hormonal influences during the mating season to allow this change of role. If the hypotheses embedded in Bett's model are correct, then the same population of tectal neurons (T5) must drive the activation of different motor pattern generators (e.g., prey catching and mate clasping). One possible explanation would be that the T5 population sends efferent fibers to both motor pattern generators, which requires the existence of proper mechanisms to disable the motor pattern generator for prey catching and simultaneously to enable the motor pattern generator for mating. No experimental data are available to posit hypotheses about what these mechanisms might be. However, Arbib (in Arbib and Ewert, 1991) discussed the possible role of hormonal effects (having high androgen levels to change the effector part of the system). Another plausible hypothesis could be that sensory information is encoded in the spatiotemporal response of the T5 neuron population (i.e., how many neurons are activated and in what temporal sequence) as well as in the time course of individual T5 neuron responses.

Neurobiological Operational Principles

Some neurobiological operational principles have been identified from the study of retino-tectal-pretectal interactions as the neural basis for visuomotor coordination. These principles may be of interest for scientists concerned with analyzing the structural and functional complexity of the amphibian's CNS and also for those interested in the synthesis of complex automata to carry out difficult tasks involving sensorimotor integration. Two examples are discussed in the following subsections.

Cooperation and Competition in Neural Networks

Ingle (1968) studied a *prey-selection phenomenon*: the snapping behavior of frogs confronted with two fly-like stimuli, each with enough prey-like characteristics to elicit a snapping response (i.e., size, form, type of motion, and distance at which the fly is located in the frontal visual field, or *snapping zone*). Frogs exhibited one of three reactions: (1) they snapped at one of the flies, (2) they did not snap at all, or (3) they snapped at the *average* fly. Didday (1976) developed a distributed network model of the optic tectum that explains this choice behavior in terms of a cooperation-competition process between two input signals. This model permits the network to act as a dynamic filter whose outcome is to transmit only the information encoded in the strongest signal. This process is one of the earliest examples of the *winner-take-all* algorithm (see WINNER-TAKE-ALL MECHANISMS). Didday's model offered an architecture whereby different layers in the optic tectum, organized in retinotopic correspondence, compete in such a way that the most active region in the *foodness* and *newness* cell layers, which receive direct input from the retina, eventually suppresses all other inputs and yields an above-threshold response in the *relative foodness* layer. This process is accomplished through the integration of foodness and newness cell layers activity with a lateral inhibition effect provided by the *sameness* cell layer. It was hypothesized that the level of activation in the foodness layer represents the input to the motor circuitry and encodes information to elicit a snapping response at the corresponding point in space. That is, tectal circuitry has the structural and functional properties to process information about multiple prey-like objects and to direct the animal's action toward one of them. This model was placed in a general theoretical perspective by Amari and Arbib (1977).

Time-Varying Scale Modulation

The anuran's prey-catching response can be influenced by changes in motivation (Cervantes-Pérez et al. and Merkel-Harff and Ewert in Arbib and Ewert, 1991; Cervantes-Pérez et al. in Rudomín et al., 1993) or by conditioning or learning processes (see Finkestädt in Ewert and Arbib, 1989). Neuro-anatomical studies have shown that the tectum and pretectum form closed loops with structures in the telencephalon and diencephalon. They have been analyzed through electrophysiological recordings and simulations of theoretical models as the underlying mechanisms of these modulatory processes. For example, the optic tectum projects to the anterior thalamus, which also receives input from the retina and connects to the medial pallium which, in turn, sends efferent fibers to the dorsal hypothalamus, a structure which directly connects to the optic tectum, closing the loop that has been postulated to subserve *stimulus-specific habituation* (Cervantes-Pérez et al. and Wang et al. in Arbib and Ewert, 1991; Lara in Ewert and Arbib, 1989). Cervantes-Pérez and collegues (in Arbib and Ewert, 1991) studied how the intensity and duration of the toad's prey-catching response is modulated when it is stimulated repetitively with a visual wormlike dummy. They observed that the toad's response gradually decreases as the experiment progresses, until complete inhibition is achieved, and that it is a stimulus-specific process in that there is almost no analogical generalization. These authors also studied how the dynamics displayed by a simplified model of the FTC, whose architecture involves cooperation-competition processes between positive and negative feedback loops, determine computational properties that may subserve the elicitation of prey-catching behaviors recorded during those stimulus-specific habituation experiments. They suggested that when the stimulus remains in the visual field, despite the animal's efforts to catch it, the processing of information within the modulatory path is carried out on a time scale slower than that of the activation dynamics. The result is increased inhibition in the tectum. A decrease in tectal activity reduces the probability of the current stimulus fitting the prey category. That is, Cervantes-Pérez and co-workers postulated that the activity of the *modulatory path* causes a change in the *activation dynamics* produced by the visual stimulus on the neural network (i.e., tectal-pretectal interactions) subserving visual stimuli classification, rather than changes in the *weight dynamics* related to its structural organization.

These functional principles might become a useful tool in neural computing, especially in cases in which: (1) various stimuli are presented at a time, and one must be selected for interaction; (2) the synchronization of events (e.g., stimuli presentation) separated in time is required to accomplish a given task; or (3) in a given neural network, trained with many cases to perform a complex task, the activation dynamics produced by few specific inputs must be modified, preferably without having to train the network from scratch.

Discussion

Experimental and theoretical studies have been used to establish how the underlying neural mechanisms of visuomotor integration in amphibians have developed into a *parallel distributed neural processing system* in which neural networks conformed by tectal-pretectal interactions represent more than a visual (sensory) map. Rather, they are the *site of integration* of signals coming from the retina and from brain regions involved in the processing of information related to those factors, identified by ethological studies, that have been postulated to modulate animal behavior. Therefore, the amphibian's response toward, or away from, visual stimuli could be explained in terms of the *integration* of neural signals generated by dynamic systems working at different time scales: (1) In *activation dynamics*, signals are activated in the CNS by the presence of a visual stimulus in the animal's receptive field (e.g., a prey-catching response may take hundreds to thousands of milliseconds). (2) In *motivational dynamics*, some changes affecting the animal's motivation occur at longer intervals (e.g., the daytime effect on the efficacy of the visual stimulus to yield the execution of motor programs might take place over a period of several hours) (see Cervantes-Pérez et al. in Rudomín et al., 1993). (3) In *time-varying scale dynamics*, learning processes require at least the occurrence of one activation dynamic process to modify future interactions with the same stimulus (e.g., a bee sting on a toad's tongue is enough to prevent the toad from catching bees in the future) (see Lara in Ewert and Arbib, 1989), whereas some learning paradigms require longer training (e.g., stimulus-specific habituation is accomplished in a period ranging from a few minutes to an hour and a half) (see articles by Ewert and Finkestädt in Ewert and Arbib, 1989; article by Cervantes-Pérez et al. in Arbib and Ewert, 1991).

Finally, it must be emphasized that the study of retino-tectal-pretectal interactions as the neural basis for amphibian visuomotor integration has become a good example of how brain theory connects the aims of neuroscience and neural engineering. It contributes to neuroscience by analyzing formal models to test hypotheses posed by experimental studies and to generate new hypotheses that stimulate new experimental and theoretical work. These modeling studies contribute, in turn, to neural engineering by identifying functional principles (e.g., the cooperation-competition in neural networks subserving the winner-take-all algorithm for object selection) as suitable models to build machines capable of carrying out difficult tasks when interacting with a complex environment.

Road Map: Motor Pattern Generators and Neuroethology
Related Reading: Active Vision; Frog Wiping Reflexes; Habituation; Visual Coding, Redundancy, and "Feature Detection"; Visuomotor Coordination in Salamanders

References

Amari, S. I., and Arbib, M. A., 1977, Competition and cooperation in neural nets, in *Systems Neuroscience* (J. Metzler, Ed.), New York: Academic Press, pp. 119–165.

Arbib, M. A., and Ewert, J. P., 1991, *Visual Structures and Integrated Functions*, Research Notes in Neural Computing, vol. 3, New York: Springer-Verlag. ◆

Cervantes-Pérez, F., Lara, R., and Arbib, M. A., 1985, A neural model of interactions subserving prey-predator discrimination and size preference in anuran amphibia, *J. Theor. Biol.*, 113:117–152.

Cervantes-Pérez, F., and Arbib, M. A., 1990, Stability and parameter dependency analyses of a Facilitation Tectal Column (FTC) model, *J. Math. Biol.*, 29:1–32.

Didday, R. L., 1976, A model of visuomotor mechanisms in the frog optic tectum, *Math. Biosci.*, 30:169–180.

Ewert, J. P., and Arbib, M. A., 1989, *Visuomotor Coordination: Amphibians, Comparisons, Models, and Robots*, New York: Plenum. ◆

Ewert, J. P., and von Seelen, W., 1974, Neurobiologie und system: Theorie eines visuellen Muster-Erkennungsmechanismus bei Kroten, *Kybernetik*, 14:105–114.

Fite, K. V., 1976, *The Amphibian Visual System: A Multidisciplinary Approach*, New York: Academic Press. ◆

Ingle, D., 1968, Visual releasers of prey catching behavior in frogs and toads, *Brain Behav. Evol.*, 1:500–518.

Kondrashev, S. L., 1987, Neuroethology and color vision in amphibians, *Behav. Brain Sci.*, 10:385.

Lettvin, J. Y., Maturana, H., McCulloch, W. S., and Pitts, W. H., 1959, What the frog's eye tells the frog brain, *Proc. IRE*, 47:1940–1951.

Lara, R., Cromarty, A., and Arbib, M. A., 1982, The role of the tectal column in facilitation of amphibian prey-catching behavior: A neural model, *J. Neurosci.*, 2:521–530.

Liaw, J. S., and Arbib, M. A., 1993, Neural mechanisms underlying direction-sensitive avoidance behavior, *Adapt. Behav.*, 1:227–261.

Llinás, R., and Precht, W., 1976, *Frog Neurobiology*, Berlin: Springer-Verlag.

Rudomín, P., Arbib, M. A., Cervantes-Pérez, F., and Romo, R., 1993, *Neuroscience: From Neural Networks to Artificial Intelligence*, Research Notes in Neural Computing, vol. 4, New York: Springer-Verlag.

Vancgas, II., 1984, *Comparative Neurology of the Optic Tectum*, New York: Plenum.

Visuomotor Coordination in Salamanders

Gerhard Manteuffel

Introduction

Visually guided behavior is the observable result of visuomotor coordination in the central nervous system. Although in an experimental situation, various behaviors can be produced and analyzed separately, in a natural environment, this separation cannot always be performed easily.

If an amphibian, for example, a salamander, is hunting a fly sitting on a leaf of grass, various sensory and motor programs are activated simultaneously. For instance, the movement of the image of the prey on the retinas may be caused by movement of the fly, movement of the leaf, or by self-motion of the approaching salamander. Thus, to be successful, our hunter must stabilize its gaze with respect to the prey to avoid a blurred image, then approach and determine whether the prey is close enough to capture.

Amphibians are supposed to be relatively simple vertebrates, and their behaviors have been studied for many decades. Early studies concentrated on the visuomotor aspects of prey capture and gaze stabilization. Electrophysiological experiments have been performed in the central nervous system at the levels of the retina and the optic tectum. In addition, the anatomy of the retinotectal system has been investigated with light and electron microscopy.

As shown in Figure 1, the centers involved in visuomotor coordination are frequently mutually interconnected. Thus, virtually all of these centers are involved in the control of all visual behaviors, which leads to an apparent lack of clear-cut functional "boxes," even in the brains of amphibians. Nevertheless, there are some centers which seem to be *mainly* responsible for the proper performance of particular behaviors, as discussed below.

Brain Areas

Retina

Efferent retinal projections to the midbrain and the mesodiencephalic junction terminate in the optic tectum, the pretectum, and the accessory optic system. The signals transferred to these targets differ considerably, however. The tectum receives largely crossed input from retinal ganglion cells with small excitatory receptive fields (ERFs) in the range of 3°–20° (Roth, 1987). In contrast, the accessory optic system and the pretectum receive input from large-field ganglion cells (ERF of 30–90°). The retinal input to the accessory optic system is completely crossed, while a well-developed ipsilateral retinal projection to the pretectum exists in addition (Manteuffel, 1989).

Thalamus

The thalamus of the diencephalon contains two visual areas (Wicht and Himstedt, 1988). The anterior thalamus is supplied by direct retinal afferents and projects bilaterally to the pallium

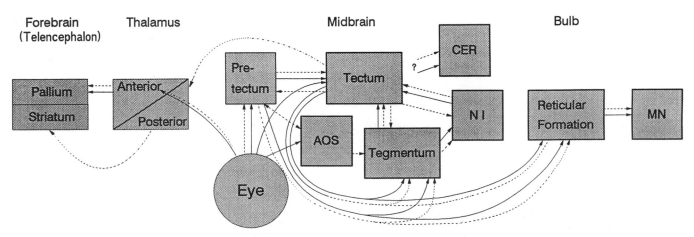

Figure 1. Major connections in the salamander brain. Broken lines indicate uncrossed and solid lines, crossing connections. The accessory optic system (AOS), cerebellum (CER), motoneurons (MN), and nucleus isthmi (NI) are shown. The midbrain centers, including the pretectum and the nucleus isthmi, represent the main sensorimotor interface, which is modulated by various descending signals from the forebrain (pathways not shown). The nucleus isthmi is a relay for the tectotectal transfer of visual signals. In the reticular formation, the descending signals are further processed to obtain the final format needed by the motoneurons.

of the telencephalon. The posterior thalamus is indirectly supplied with visual signals through the tectum.

Telencephalon

The telencephalon receives afferents to its pallial areas from the rostral thalamus and to the striatum from the posterior dorsal thalamus (Wicht and Himstedt, 1988). Both of these pathways may carry the visual signals which can be recorded from the dorsolateral pallium (Hechler and Himstedt, 1990). According to Finkenstädt (1981), lesions in the pallial area lead to uncontrolled prey-catching behavior and significant loss of habituation to prey stimuli. Retrograde visual amnesia has also been noted.

Pretectum

In the pretectum, most cells are binocular and are sensitive to optokinetic stimuli moving slowly (optimum range 1–10 deg/s) in a temporonasal direction with respect to the contralateral eye and in a nasotemporal direction with respect to the ipsilateral eye. Thus, they are most sensitive to stimuli turning around a vertical axis (Manteuffel, 1989). Pretectal output occurs with bilaterally descending fibers terminating in various tegmental and bulbar regions (Naujoks-Manteuffel and Manteuffel, 1988).

Accessory Optic System

The accessory optic system is exclusively supplied by contralateral retinal afferents. The cells project to the pretectal nucleus, where they match the retinal fibers from the contralateral eye, and to the nucleus of the medial longitudinal fasciculus. The units are sensitive to large-field movements with vertical components (Manteuffel, 1989).

Mesencephalic Tectum

The mesencephalic (optic) tectum receives various nonvisual afferents in addition to its optic input. Somatosensory afferents from the spinal levels are mostly relayed in the bulbar reticular formation. Vestibular and acoustic signals are relayed in the dorsal tegmentum (Manteuffel and Naujoks-Manteuffel, 1990) that projects to the tectum bilaterally. The vestibular input is thought to mediate a self-motion signal to the tectum which is synergistic to visual self-motion signals from the pretectum.

The retinal inputs from the contralateral eye establish nearly a Cartesian map of the surrounding visual space on the tectum. The nasal visual field is represented rostrally and the temporal field caudally; the superior visual field is represented close to the dorsal midline, whereas the ventrolateral margin of the tectum is stimulated by objects in the lower part of the visual field (Roth, 1987).

Each tectal hemisphere also receives an input from the ipsilateral eye, but with a reversed retinotopy that is point symmetrical to the contralateral one. This ipsilateral projection is achieved, at least in part, by an intertectal transfer (Manteuffel, Fox, and Roth, 1989).

The anatomy of the tectum in salamanders has been described in detail by Roth, Naujoks-Manteuffel, and Grunwald (1990). There are three main classes of neurons: pear-shaped, pyramidal, and ganglionic cells. Most output cells belong to the last two classes. Tectal cells show different responsiveness to various stimulus shapes. As summarized by Roth (1987), they respond in a much more complex fashion to visual stimuli than do retinal ganglion cells.

The output of the tectum has been analyzed by neuroanatomical tracing (Naujoks-Manteuffel and Manteuffel, 1988). Output cells are generally more numerous on the ipsilateral than on the contralateral side, but on both sides, cell density increases from the rostral pole of the tectum up to an isthmic level, and then decreases rapidly. The distribution is also nonhomogeneous in a mediolateral direction.

Behavior

Prey Catching

Prey catching in salamanders has been studied intensively by Himstedt and colleagues and Roth and associates. An overview is given in Roth (1987). To summarize the results of various authors, salamanders show a slight increase in the frequency of prey-directed saccades (i.e., head turning toward a prey dummy; saccadic eye movements seem to be absent in the context of prey catching) if the object becomes longer in the direction of movement. Thus, the shape of an object has some influence on the probability with which a saccade may occur, but it was shown that additional features, such as the type of movement or early experience, may be more significant.

Salamanders are able to judge the distance of prey. To that end, they apparently use monocular and binocular mechanisms such as accommodation (Roth, 1987), motion parallax, and disparity (Manteuffel et al., 1989).

Few data are available on the exact metrics of prey-directed saccades. They are largely ballistic movements, and they do not differ significantly under open- or closed-loop conditions. At larger angles ($> 30°$), the saccades tend to fall short. Therefore, two or more saccades are often needed to bring a target into the center of the visual field. A saccade is typically terminated within two seconds, independent of its amplitude. Thus, the velocity of the head movement is linearly related to the turning angle. Maximum values are less than $90°/s$ (Werner and Himstedt, 1985).

In monocular salamanders, saccades are directed away from distant prey. On approach, the saccades then reverse their direction and correct the gaze error. This process results in a strange curved path of approach in these animals (Roth, 1987).

Modeling of Prey-Directed Saccades

The metric properties of salamander saccades are largely a result of the particular visual and motor maps of the mesencephalic tectum. This observation was shown in a computational model designed on the basis of the anatomical and electrophysiological data from the tectum and the brainstem (Manteuffel and Roth, 1993). In that model (Figure 2), a continuous mapping from the retinas of both eyes to the motor layers of both tectal hemispheres was applied. The specific mapping was derived from empirical results; in addition, it was assumed that the visual layer of the tectum recruits the premotor neurons at the location of the excitation on the visual map. Inspired by the bimodal distribution of the efferent tectal neurons (Naujoks-Manteuffel and Manteuffel, 1988), it was further assumed that there are distinct populations of tectal neurons that are premotoric to the epaxial and hypaxial muscles.

The transfer from the visual map to the premotor map is controlled by a threshold which can be adjusted by motivational afferents, probably coming from forebrain sources. Therefore, any visual object may or may not elicit a saccade, depending on motivational or learned configurational properties.

The simulation of the model was performed in a three-dimensional space. The program consisted of eight basic steps.

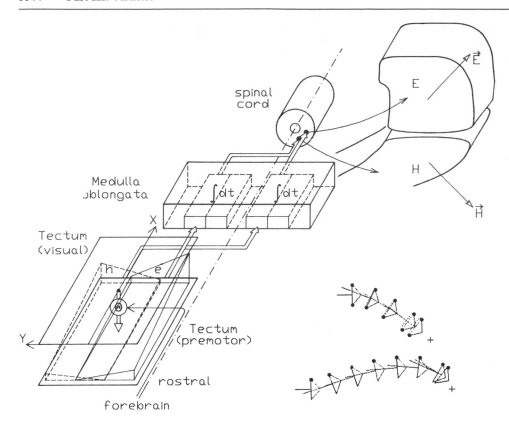

Figure 2. Schematic drawing of the visuomotor system involved in the saccade generation of salamanders and taken as a basis for the model of Manteuffel and Roth (1993). The visual tectal map is an orthogonal projection of the retina. At a given location, the visual excitation triggers the respective amount of premotor neurons through a threshold adjusted by forebrain inputs. The local numbers of premotor neurons that activate the epaxial and hypaxial systems (e, h) represent the motor map. The outputs reach the medial portion of the reticular nucleus in the medulla oblongata (bulb) through a crossed pathway and the lateral portion through an uncrossed pathway. There, the signals are integrated to drive the motoneurons of the epaxial and hypaxial muscles E and H, which move the head obliquely up and down. Simulation results under binocular (above) and monocular conditions are shown at the lower right. Crosses represent the targets. Monocular stimulation leads to a devious approach, as is found in the behavior of real animals.

1. Calculate the object position within the coordinate systems of the eyes (object on retinal maps). Stop the program if the object is not represented in either of the two maps.
2. Transfer to the tectal coordinates (bilateral tectal maps).
3. Calculate the tectal coordinate on the other hemisphere if the stimulus is monocular (tectotectal transfer).
4. Calculate the recruitment strength according to the empirically grounded functions on the motor map at the location beneath the visual map coordinate.
5. Find the winner of the antagonistic pairs of muscle recruitments.
6. Calculate the compound vector of head movement.
7. Create new graphics.
8. Stop the program if the object is reached, or calculate the next forward step of the model salamander in the direction of the previous head angle relative to body axis, and continue.

The simulated saccadic behavior (followed by approach) is similar to that observed in salamanders under binocular or monocular conditions. Saccades are correctly produced in all directions and fall short at larger angles (as happens in the animal) such that two or three saccades are necessary to center a target in the lateral visual field. Monocular stimulation yields the same curved approach as was found in real behavior. This effect is greatest in the simulations with large eye convergences and wide binocular visual fields. The same result is found in real salamanders.

The model is robust with respect to eye position. Without changing any other parameter of the model, the simulated salamander reaches (residual error <5°) its target at frontal eye convergences of 10°–70° and dorsal convergences of 0°–40°.

The model is also relatively robust with respect to the distribution of premotor cells (i.e., the recruitment parameters). The model's ability to hit a target is hardly affected by a change in the distribution of premotor neurons by some 20% with respect to the empirically found value. The rostrocaudal recruitment function can likewise be changed without dramatic effects as long as some increase in recruitment strength from rostral to caudal remains. However, the empirically found distribution seems to be near optimum.

The robust properties of the system have three primary sources. First, the system contains a feedback loop that is closed by way of the environment. Second, the intertectal transfer enables the system to keep functioning, even if it is occasionally monocular. Finally, the efferent recruitment system does not critically depend on the exact location of excitation on the sensory map since the recruitment parameters do not change dramatically with variation in excitation site. Thus, fairly correct saccades are produced under a large variety of input vectors which may be corrected by an additional saccade, if necessary.

Interestingly, an abolishment of the intertectal transfer (which has not yet been performed in real salamanders) does not change the ability of the binocular model to reach a target. However, dramatic effects occur under monocular conditions. Whereas with an intact intertectal transfer the target is eventually reached (although with a curved approach) (see Figure 2), interruption of the transfer prevents the monocular model salamander from reaching the target. Instead, it turns completely away from the target toward the side of the seeing eye until the object is out of sight. This outcome results from merely unilateral tectal stimulation, which leads to an exclusively unilateral motor reaction to the side of the seeing eye and away from the target.

The good performance of the model, which is tolerant of a change in many features, such as eye position or head size, is largely an effect of the kind of mapping and the continuous feedback, and it depends relatively little on the specific parameter values chosen. An important side effect of this mechanism may be that it allows evolutionary changes of the periphery without major change of the central structures.

In a similar approach, but using a three-layered neuronal network with 300 discrete neurons per layer instead of continuous mapping, Eurich et al. (1993) showed that few neurons, as typically found in salamander tecta, are able to produce the behavioral accuracy of prey localization.

Empirical Data and Modeling of Gaze Stabilization

The amphibian nystagmus of the head is similar in all species investigated to date. It does not reach a gain of 1, but maximally reaches a gain of 0.8 during optokinetic or vestibular stimulation. It is nonlinear since the gain during sinusoidal stimulation depends on stimulus amplitude, and the shape of the response is distorted with respect to an input sine wave at higher frequencies. Optokinetic gain reaches its maximum value at stimulus frequencies of approximately 0.03 Hz and then decreases for higher frequencies, whereas the gain of the vestibulocollic reflex increases for higher frequencies, reaching its maximum at approximately 0.5 Hz. Both add linearly and result in a flat gain curve when combined (Manteuffel, 1989).

The features of the optokinetic reflex in salamanders can be reproduced in a computer model under both constant velocity and sinusoidal input conditions (Manteuffel, 1989), taking into consideration the bell-shaped velocity tuning curves of pretectal neurons and assuming that these signals are fed to the ipsilateral neck motoneurons through a salamander-specific velocity storage system and a linear velocity-to-position integrator. This model was able to demonstrate that peculiar features of the optokinetic reflex of salamanders, such as oscillations in the low-stimulus velocity range and specific nonlinearities, are the results of a system that is simpler than that in higher vertebrates. It especially lacks an effective velocity storage element that serves a more continuous output. This finding was corroborated by behavioral studies that showed only a weak optokinetic afternystagmus in these species.

Discussion

The models discussed above show how low-level mechanisms add up to produce sometimes complicated behaviors, such as the devious approach of monocular salamanders. As one would expect from physiologically plausible models, they offer testable predictions of the behavioral effects of central manipulations.

Thus, the models enable us to make the important proposition that given a set of experimental data and assuming some precisely formulated additional conditions, a behavior can result that is indistinguishable from or similar to the behavior which is found in the animal. Hence, only modeling can tell us precisely what emerges from the empirical results gained at the level of single neurons and connections.

Road Map: Motor Pattern Generators and Neuroethology
Related Reading: Head Movements: Multidimensional Modeling; Neuroethology, Computational; Vestibulo-Ocular Reflex: Performance and Plasticity; Visuomotor Coordination in Frogs and Toads

References

Eurich, C., Roth, G., Schwegler, H., and Wiggers, W., 1993, Simulander: A neural network model for the orientation movement of salamanders, in *Genes, Brain, Behavior* (N. Elsner and M. Heisenberg, Eds.), Stuttgart: Thieme Verlag, p. 91.
Finkenstädt, T., 1981, Effects of forebrain lesions on visual discrimination in *Salamandra salamandra*, *Naturwissenschaften*, 68:268.
Hechler, N., and Himstedt, W., 1990, Visually evoked potentials in the telencephalon of the newt *Triturus alpestris*, in *Brain, Perception, Cognition* (N. Elsner and G. Roth, Eds.), Stuttgart: Thieme Verlag, p. 227.
Manteuffel, G., 1989, Compensation of visual background motion in salamanders, in *Visuomotor Coordination: Amphibians, Comparisons, Models, and Robots* (J. P. Ewert and M. A. Arbib, Eds.), New York: Plenum, pp. 311–340. ◆
Manteuffel, G., Fox, B., and Roth, G., 1989, Topographic relationships of ipsi- and contralateral visual inputs to the rostral tectum opticum in the salamander *Plethodon jordani* indicate the presence of a horopter, *Neurosci. Lett.*, 107:105–109.
Manteuffel, G., and Naujoks-Manteuffel, C., 1990, Anatomical connections and electrophysiological properties of toral and dorsal tegmental neurons in the terrestrial urodele *Salamandra salamandra*, *J. Hirnforsch.*, 31:65–76.
Manteuffel, G., and Roth, G., 1993, A model of the saccadic sensorimotor system of salamanders, *Biol. Cybern.*, 68:431–440. ◆
Naujoks-Manteuffel, C., and Manteuffel, G., 1988, The origins of descending projections to the medulla oblongata and rostral medulla spinalis in the urodele *Salmandra salamandra* (Amphibia), *J. Comp. Neurol.*, 273:187–206.
Roth, G., 1987, *Visual Behavior in Salamanders*, Studies of Brain Function, vol. 14, Berlin: Springer-Verlag. ◆
Roth, G., Naujoks-Manteuffel, C., and Grunwald, W., 1990, Cytoarchitecture of the tectum mesencephali in salamanders: A Golgi and HRP study, *J. Comp. Neurol.*, 291:27–42.
Werner, C., and Himstedt, W., 1985, Mechanism of head orientation during prey capture in salamander (*Salamandra salamandra L.*), *Zool. Jahrb. Physiol.*, 89:359–368.
Wicht, H., and Himstedt, W., 1988, Topologic and connectional analysis of the dorsal thalamus of *Triturus alpestris* (Amphibia, Urodela, Salamandridae), *J. Comp. Neurol.*, 267:545–561.

Walking

George A. Bekey

Introduction

Walking is a method of locomotion employed by legged land animals and humans, as well as legged robots. Neural mechanisms provide the coordination of numerous muscles to provide forward progression while maintaining a stable posture. This remarkable series of coordinated actions requires the integration of central control with spinal feedback and large numbers of local control and feedback systems. Locomotion at constant speed requires neural networks which stimulate systematic periodic sequences of leg movements. Neural mechanisms also provide for the changes of gait pattern exhibited by both humans and animals.

At slow velocities, stable walking is characterized by *static stability*, where the center of mass of the body remains within the polygon of support formed by the legs in contact with the ground. During motion, the support forces of the legs, momentum, and inertial forces are summed to produce *dynamic stability*. The requirement for dynamic stability also arises in fast gaits of quadrupeds. This situation is more complex in humans, since their bodies sway and they shift weight even during standing. Walking in humans is sometimes described as a pattern of controlled falling.

In both humans and animals, each leg alternates between *stance*, the time its end point (foot, paw, or hoof) is in contact with the ground, and *swing*, the time the limb spends in forward motion. During stance, muscular action propels the body forward. As the velocity of locomotion increases, the duration of the stance phase is reduced while that of the swing phase remains nearly fixed.

Human Gait Biomechanics

Stable human locomotion requires a periodic motion of the legs, resulting in velocities from about 40 m/min (a very slow walk) to about 82 m/min (normal level walking) and 120 m/min (fast walking). At speeds higher than about 130 m/min, humans switch to running, a different gait pattern (Perry, 1992). Stance begins when the heel touches the ground and ends when the toes leave the ground (Figure 1). Normal walking involves coordinated motion at the hip, knee, and ankle, none of which are simple pin or hinge joints. Muscles attach at various points along the rigid skeletal supporting structure, and often perform more than one function—e.g., a muscle may be responsible for deflection of the toes as well as ankle rotation. This complex architecture makes mathematical modeling and analysis extremely difficult, but also provides the redundancy which makes walking possible (though perhaps difficult) following disease or injury.

During normal human gait, the time spent in stance is approximately 60% of the gait cycle. The 40/60 ratio of swing to stance is nearly independent of the specific limb dimensions of the walkers.

Energy consumption in normal walking increases linearly with speed. Moreover, the major portion of the energy expended during walking is required for the vertical motion of the body center of gravity (CG), with the lesser portion being used for forward propulsion at normal walking speeds (McMahon, 1984). The ability of humans to walk long distances is related to their remarkable ability to coordinate ankle,

knee, and hip motion in order to minimize the vertical displacement of the CG. If no compensatory mechanisms were present, the change in body height between the points of single and double limb support (see Figure 1) would be about 9.5 cm. A mixture of adjustments known as the *determinants of gait* reduce this displacement by about 50%, to just 4.6 cm (Perry, 1992; McMahon, 1984), thus greatly reducing energy expenditure.

Quadruped Gait Biomechanics

The locomotion of quadrupeds differs from that of humans in a number of significant ways. Static stability is enhanced by the increased number of support points and by the horizontal posture of the spine, which decreases the potential moment of the upper portion of the body about its CG. In addition, quadruped locomotion is characterized by a number of different gaits —such as crawl, walk, trot, and gallop—which differ in the sequence in which the legs contact the ground (see GAIT TRANSITIONS). The transition from one gait pattern to another is related to speed and efficiency (energy consumption per unit distance traveled). Also, the ground contact area of the paws or hooves of many quadrupeds is small in relation to their body weight, when compared with human feet. The net effect of these properties is that quadruped locomotion is highly variable, and highly dependent on the species in question. Energy consumption is nearly linear with velocity, in both quadrupeds and humans (McMahon, 1984).

The study of gait patterns began in the late nineteenth century, with a remarkable series of stop-motion photographs obtained by Eadweard Muybridge, showing that during trot all four of a horse's legs could be off the ground at the same time (reported by Raibert, 1986). Muybridge then photographed walking and running of many other mammals, including humans, elephants, and cats.

Neural Control of Locomotion

The nervous system locations and the mechanisms of control of the rhythmic limb movements arising during locomotion have been the subject of intensive study. Major attention has been devoted to the study of leg coordination during walking of the stick insect (e.g., Bässler, 1983), the swimming movements of the lamprey (primarily by Grillner and his group, e.g., Grillner, et al., 1988) and movement in other animals (see Grillner, 1981,

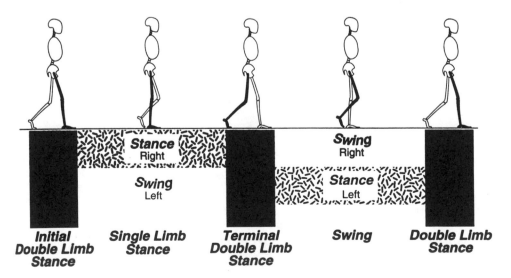

Figure 1. Stance and swing phases in normal human gait. Vertical dark bars are the periods of double-limb stance (right and left feet). Horizontal shaded bar is single-limb support (single stance). Total stance includes three intervals: the initial double stance, single-limb support, and the next (terminal) double stance. Swing is the clear bar that follows terminal double stance. Note that right single-limb support has the same time interval as left swing. During right swing there is left single-limb support. The third vertical bar (double stance) begins the next gait cycle. (From Perry, J., 1992, *Gait Analysis: Normal and Pathological Function*, Thorofare, NJ: Slack, p. 5; reprinted with permission of the publisher.)

and Cohen, 1992, for a review of the literature). The associated mechanisms and neural connections are very complex and widely distributed in the CNS. There are networks of neurons within the nervous system capable of producing the periodic discharges associated with walking or various running gaits; they are usually referred to as *Central Pattern Generators* (CPGs). A number of interneurons believed to be components of the CPG have been identified in several insects, but the situation in higher vertebrates is less clear. Hence, vertebrate CPGs remain primarily conceptual at the time of this writing. Nevertheless, the concept is so widely accepted that it is used extensively in simulation models of animal locomotion. The level of activity of the CPGs is controlled by higher centers and influenced by sensory feedback from peripheral receptors in the limbs, but considerable controversy surrounds the relative roles of peripheral and central mechanisms in walking. The cerebellum is responsible for fine control of locomotion; the loss of cerebellar function does not prevent locomotion, but the movements become less coordinated. The decision to initiate locomotion is made somewhere else in the brain, probably in one or more brainstem locations. However, decerebrate cats placed on a treadmill are capable of normal-appearing walking. Basically, the function of the CPGs is to activate alpha-motoneurons in reciprocal patterns, producing alternating sequences of flexion and extension in the various limb muscles.

Coordination between limbs probably involves some form of reciprocal inhibition between pairs of CPGs, phase delays in coupling between generators, and a variety of connections between generators. These mechanisms have been modeled in invertebrates (see LOCOMOTION, INVERTEBRATE). The control of locomotion is strongly influenced by feedback from peripheral receptors. Such factors as speed, load, and limb position provide afferent inputs which affect the CPGs. The interaction of these input and feedback signals in the control of locomotion is illustrated in Figure 2.

The above paragraphs have emphasized the neural control of locomotion in steady state, i.e., at a constant speed, while moving in a straight line and on level terrain. Changing terrain or a desired change in speed require major adaptations in the patterns of walking or running. Moving in a curved path requires the leg or legs on the outside of the curve to turn during swing and to traverse longer distances than the inside leg(s), so that the stride length of the two legs will no longer be equal. Walking uphill or downhill requires a change in body posture to ensure static stability. For quadrupeds, it also requires the rear legs to extend further back than during level walking. Walking along a constant elevation contour on the side of a hill requires dramatic adjustments of leg and shoulder or hip position to maintain stability.

Changes in speed also require complex neural mechanisms. Bipeds have only three modes of locomotion: walking, running, and jumping. Quadrupeds can exhibit a variety of gaits. It now appears that the switch from one gait pattern to another occurs, at least in part, by mechanisms which attempt to keep

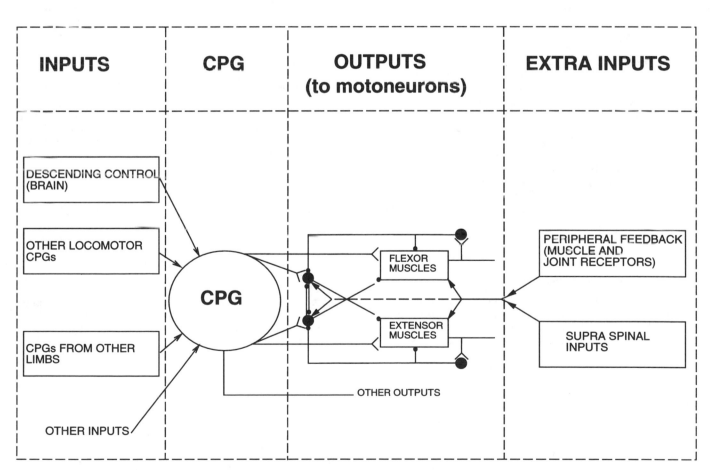

Figure 2. Organization of locomotor control (adapted from Grillner, 1981). Note the Central Pattern Generator (CPG) with inputs from higher centers, peripheral feedback, and outputs to motoneurons of flexors and extensors. The symbol ● indicates inhibitory synapses, ⟨ indicates excitatory synapses.

the muscular forces and joint loads in the legs within particular limits.

A further function of neural control is to allow locomotion in the presence of obstacles. Both animals and humans use vision to step around obstacles. However, for quadrupeds this applies only to their front legs, while an exquisite form of coordination between front and back legs ensures that the latter also avoid the obstacles. Clearly, a very complex problem in inverse kinematics is being solved by the nervous system in this process.

Pathological gait in humans following disease or injury is the subject of much contemporary research (e.g., Perry, 1992). The variety of potential disturbances to the neural control of locomotion is enormous, ranging from damage to motor neurons controlling a single muscle to the global effects of cerebral palsy. Pathology gives rise to remarkable compensatory mechanisms to allow for locomotion, often with unusual body postures or the use of muscles for unaccustomed functions.

Walking Robots

The construction of robot walking machines is an attempt to isolate a set of locomotion properties sufficient to produce stable forward motion with a minimum of hardware and software. Such an abstraction of properties from those developed by evolution is not an easy process, and many walking machines have not been highly successful.

Walking machines are basically of two types: They either concentrate on some aspect of kinematics and dynamics, including the design and fabrication of mechanical substitutes for skeletal, joint, and muscle function (e.g., Waldron et al., 1984) or they devote major attention to the control of the leg movement sequence, attempting to imitate animal gait patterns.

The first autonomous quadruped robot in the U.S. was constructed in the 1960s by Frank and McGhee (McGhee, 1966). This robot was controlled by a finite state machine using sensory feedback on the state of its joints, without any internal model of its kinematics or dynamics. The machine was capable of emulating a number of quadruped gait patterns, including crawl, walk, and trot. Other remarkable walking machines include the hexapods constructed by Waldron et al. (1984) at Ohio State University, the quadrupeds of Hirose and his collaborators in Japan (Hirose, 1984), the one-, two-, and four-legged machines fabricated by Raibert (Raibert, 1986), the menagerie of artificial insects of Brooks and his students (Brooks, 1992), and the insects simulated (and later fabricated) by Beer (1990) (see LOCOMOTION, INVERTEBRATE).

Control of Robot Gait Patterns

Much of the successful research in the field has concentrated on simplified controllers, based on models of information processing which only approximate certain aspects of the nervous system. Beginning with the work of Tomovic and McGhee (1966), the complexity of control has been reduced by representing the legs as *finite state machines* (FSMs). Locomotion then becomes a sequence of events rather than a continuous dynamical process, a gait pattern can be described as a sequence of states, and the control requires only a finite algorithm. At the simplest level, the stance and swing portions of a leg can be regarded as two states and used to describe the timing of various gait patterns. However, a two-state description is insufficient for a control algorithm which enables representation of hip, knee, and other joint movements. Tomovic and McGhee (1966) proposed the representation of joint motions by *cybernetic actuators*, which transform discrete inputs into continuous outputs. The

important contribution of these early investigations lies in the simplification which they produce, by transforming a problem of nonlinear dynamics into one of sequential decisions, triggered by sensory events. For the Frank quadruped, these events were hip and knee angle limits, which triggered the transitions to new states. The finite state approach has also been applied to the study of human walking and the development of rehabilitation devices (Bekey and Tomovic, 1986).

The finite state approach to control of walking machines can be viewed as an idealized neural network, which translates sensory inputs to motor outputs. Brooks (1992) added more complex input and output circuitry to the FSM, producing augmented FSMs, which are interconnected to provide for a variety of behaviors. This multilayer structure is known as the *subsumption architecture*. Raibert's quadrupeds are also controlled by FSMs (Raibert, 1986).

Learning to Walk

The gait patterns of walking robots are obtained by preprogrammed coordination between legs and the selection of appropriate stance and swing times (e.g., Beer, 1990; McGhee, 1967), requiring only an appropriate selection signal to switch between gaits. In living animals, the neural generators of particular patterns may also be "prewired." In an attempt to study how such "wiring" might arise, simulated evolution of gait patterns has been studied using genetic algorithms (GAs). GAs are basically optimization algorithms which use a process of variation and selection to search a parameter space while optimizing an appropriate "fitness function." They have been applied to locomotion in a hexapod robot by Lewis, Fagg, and Bekey (1994). The fitness function was a measure of stable and sustained periodic leg movements and forward progression along the robot's body axis. A robot was constructed, using simple leaky integrator model neurons for control of each leg, and the network gains were obtained from a genetic algorithm. The initial gains were random. Following a learning period consisting of evolution through some 50 generations, the robot developed a stable tripod gait, characteristic of insect locomotion (see LOCOMOTION, INVERTEBRATE).

Discussion

The neural mechanisms responsible for walking in humans and animals are complex and not completely understood. They are responsible for both forward progression and postural stability, energy minimization and gait control. The construction of walking machines requires models of periodic leg movement, gait control, and stability. The study of walking in animals and robots may lead to fruitful insights in both directions. As we learn more about the neural control of locomotion, we will be able to build better walking machines. Conversely, the construction and study of walking machines and their control may provide new hypotheses for the study of neural control mechanisms.

Road Map: Motor Pattern Generators and Neuroethology
Related Reading: Human Movement: A System-Level Approach; Motor Pattern Generation

References

Bässler, U., 1983, *Neural Basis of Elementary Behavior in Stick Insects*, New York: Springer-Verlag.
Beer, R. D., 1990, *Intelligence as Adaptive Behavior: An Experiment in Computational Neuroethology*, San Diego: Academic Press.

Bekey, G. A., and Tomovic, R., 1986, Robot control by reflex actions, in *Proceedings of the IEEE International Conference on Robotics and Automation*, Los Alamitos, CA: IEEE Computer Society Press, pp. 240–247.

Brooks, R. A., 1992, A robot that walks, in *Biological Neural Networks in Invertebrate Neuroethology and Robotics* (R. D. Beer, R. E. Ritzmann, and T. McKenna, Eds.), San Diego: Academic Press, pp. 319–354.

Cohen, A. V., 1992, The role of heterarchical control in the evolution of Central Pattern Generators, *Brain Behav. Evol.*, 40:112–124.

Grillner, S., 1981, Control of locomotion in bipeds, tetrapods and fish, in *Handbook of Physiology*, sec. 1, vol. 2, pt. 2, pp. 1179–1236.

Grillner, S., Buchanan, J. T., Wallen, P., and Brodin, L., 1988, Neural control of locomotion in lower vertebrates, in *Neural Control of Rhythmic Movements in Vertebrates* (A. H. Cohen, S. Rossignol, and S. Grillner, Eds.), New York: Wiley, pp. 1–40.

Hirose, S., 1984, A study of design and control of a quadruped walking vehicle, *Int. J. Robotics Res.*, 3:113–133.

Lewis, M. A., Fagg, A. H., and Bekey, G. A., 1994, Genetic algorithms for gait synthesis in a hexapod robot, in *Advanced Mobile Robots* (Y. F. Zheng, Ed.), New York: World, pp. 317–331.

McGhee, R. B., 1967, Some finite state aspects of legged locomotion, *Math. Biosci.*, 2:67–84.

McMahon, T. A., 1984, *Muscles, Reflexes and Locomotion*, Princeton, NJ: Princeton University Press. ◆

Perry, J., 1992, *Gait Analysis: Normal and Pathological Function*, Thorofare, NJ: Slack.

Raibert, R. A., 1986, *Legged Robots That Balance*, Cambridge, MA: MIT Press.

Tomovic, R., and McGhee, R. B., 1966, A finite state approach to the synthesis of bioengineering control systems, *IEEE Trans. Hum. Factors Electron.*, HFE-7:65–69.

Waldron, K. J., Vohnout, V. J., Pery, A., and McGhee, R. B., 1984, Configuration design of the adaptive suspension vehicle, *Int. J. Robotics Res.*, 3:37–48.

Wavelet Dynamics

Harold Szu and Brian Telfer

Introduction: Continuous WT on a Church Organ

Classical acoustic engineers have known for a long time how to design church organs that please human ears. Only recently has the underlying mathematics become formalized as the wavelet transform (WT). The organ has a very pleasant vibrato because of a fixed percentage of frequency spread Q. Each sound is not a pure tone, but rather a special group wave of constant Q which is called in short a *wavelet*. For example, the 100 Hz wind pipe generates from 99 Hz to 101 Hz having 2 Hz bandwidth, which gives the fidelity $Q \equiv \Delta f/f = (101 - 99)/100 = 2\%$. Then, an octave-higher sound with the same $Q = 2\Delta f/2f = 4/200 = 2\%$ must have a factor 2 bigger bandwidth $2\Delta f = 4$ Hz having inversely a half pipe length. This size or window reduction at high frequency is related to the wavelet interference. The larger the bandwidth, the easier the out-of-sync destructive interference. Thus, the larger the bandwidth Δf, the smaller the support Δt. This is the wavelet uncertainty principle—$\Delta t \Delta f \approx$ constant—of wave mechanics. In fact, a pure tone having zero spread $\Delta f = 0$ implies an infinite support $\Delta t \approx \text{constant}/\Delta f = \infty$, because the sinusoidal wave is periodically forever, while an infinite spread $\Delta f \approx \text{constant}/\Delta t = \infty$ represents a Dirac delta function of zero support:

$$\delta(t) = \int_{-\infty}^{\infty} df \exp(-2\pi i f t). \tag{1}$$

The acoustic designer has long known these two principles: the octave-higher node, which must have $2\Delta f$ according to the constant fidelity principle $Q = 2f/2\Delta f = f/\Delta f$, must have a half pipe length: $\Delta t/2 \approx \text{constant}/2\Delta f$, according to the wavelet uncertainty principle. Indeed, in church, the higher the pitch, the shorter the pipe.

To ensure an identical number of wavelet oscillations, if the 100 ± 1 wavelet $\psi(t)$ is recorded on music tape and played back faster in time by a factor of 2, it will produce the 200 ± 2 wavelet $\psi(2t)$ precisely. Since each wavelet, being a physical sound, must have a finite energy, we can use a $\sqrt{a}$ scale factor to normalize each *daughter wavelet* ψ_{ab} (i.e., each wavelet "descended" from the original wavelet ψ) by

$$\psi_{ab}(t) \equiv \psi((t-b)/a)/\sqrt{a} \equiv \psi(t')/\sqrt{a} \tag{2}$$

so that scale $a = \frac{1}{2}$ is the compressed octave version and their energies are identical:

$$\int_{-\infty}^{\infty} dt \, |\psi_{ab}(t)|^2 = \int_{-\infty}^{\infty} dt' \, |\psi(t')|^2 = \text{constant} \tag{3}$$

This gives the square-integrable Hilbert space to generalize the complete set from the affine time-scale transform: $t \to t' = (t-b)/a$, where a is the inverse of frequency f (namely wavelength) and b is the shift necessary to compensate the locality limitation. We can introduce mathematically the continuous wavelet transform (WT):

$$S(a,b) \equiv \int_{-\infty}^{\infty} dt' \, \psi_{ab}(t')^* s(t') \tag{4}$$

The original signal $s(t)$ is recovered by integrating the coefficient basis, $S(a,b)\psi_{ab}(t)$, over the dimensionless mode number: $dt \, df \approx -db \, d(1/a) = db \, da/a^2$,

$$s(t) = (1/c) \int_{-\infty}^{\infty} da/a^2 \int_{-\infty}^{\infty} db \left(\int_{-\infty}^{\infty} dt' \, \psi_{ab}(t')^* s(t') \right) \psi_{ab}(t) \tag{5}$$

For this, the following completeness condition must be observed:

$$\delta(t'-t) = (1/c) \int_{-\infty}^{\infty} da/a2 \int_{-\infty}^{\infty} db \, \psi_{ab}(t')^* \psi_{ab}(t) \tag{6}$$

which, when rewritten in the Fourier domain using $\psi_{ab}(t) = \int_{-\infty}^{\infty} df \exp(2\pi i f t)\Psi_{ab}(f)$ and $\Psi_{ab}(f) = \exp(i2\pi f b)\Psi(af)$, yields the condition for an admissible mother wavelet; i.e., the condition that the daughter wavelets (Equation 2) can give rise to an invertible transform via Equations 4 and 5.

$$\int_{-\infty}^{\infty} df |\Psi(f)|^2/|f| = \text{constant} \tag{7}$$

Note that the division of zero frequency implies for the purpose of convergence

$$\Psi(0) = \int_{-\infty}^{\infty} dt \exp(-i2\pi f t)|_{f=0}\psi(t) = 0 \tag{8}$$

Thus an admissible mother wavelet has no zero frequency (d.c.)

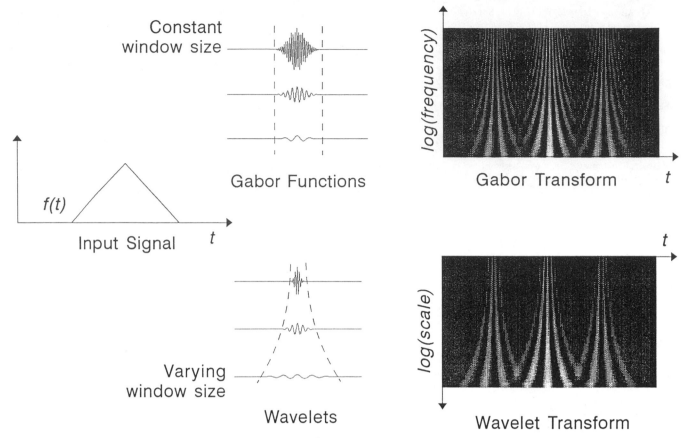

Constant
window size

Gabor Functions

Gabor Transform   t

f(t)

Input Signal   t

Varying
window size

Wavelets

log(frequency)

log(scale)

Wavelet Transform

Figure 1. Continuous WT: A short-time Fourier transform that has a fixed Gaussian window is called the Gabor transform. Morlet wavelet transform is better in locating a rooftop seismic image intensity discontinuity.

component and so must have a zero integrated area; namely, a small wave has a finite extent and the positive amplitude is always canceled by the negative amplitude. Since without the d.c. the WT nullifies any constant signal, it is a multiple-scale change detection similar to that performed in the visual system. Current interest in WT applications has been stimulated by the seismic imaging used in French oil exploration. In order to localize a sharp discontinuity better, the effect of fixed window FT versus Morlet WT is shown in Figure 1.

An adaptive WT, given in the section "Adaptive Wavelet Transforms," can mimic the human auditory system for phoneme analysis, a few features being used to classify the signal adequately. Moreover, each term suffers a less global noise contamination due to its local compact support. Such a multiple resolution WT (Mallat, 1989) yields parallel inputs for brainstyle computing for pattern recognition (Pati and Krishnaprasad, 1993, Sheng et al., 1993).

Discrete WT: The Haar Transform

In 1910, Alfred Haar suggested an orthogonal coordinate system satisfying a scaling relationship, known nowadays as the discrete WT. He followed the orthogonal Cartesian coordinate viewpoint, where in 4D: $x \equiv (1, 0, 0, 0)$, $y \equiv (0, 1, 0, 0)$, $z \equiv (0, 0, 1, 0)$, and $z' \equiv (0, 0, 0, 1)$ satisfies the property that the inner product of any pair will be zero, implying orthogonality. However, Haar chose even-odd orthogonality, namely, an even function $\phi(t) = (+, +, +, +)$ pointing in the first orthant of four positive axes; an odd function $y(t) = (+, +, -1, -1)$

pointing in the lower orthant of positive x, y; and negative z and z'. Their inner product likewise gives orthogonality in 4D. Furthermore, this scaling property repeats in a half interval, if $\psi(2t) = (+, -, 0, 0)$ is chosen pointing in the plane bounded by $x \geq 0$ and $y \leq 0$, and $\psi(2t - 1) = (0, 0, +, -)$ pointing in the plane bounded by $z \geq 0$ and $z' \leq 0$. For example, any 4D vector, e.g. a ramp $\mathbf{g}$, can be expanded in Haar coordinates

$$\mathbf{g} \equiv \begin{bmatrix} 2 \\ 4 \\ 6 \\ 8 \end{bmatrix} = \begin{bmatrix} + & + & + & 0 \\ + & + & - & 0 \\ + & - & 0 & + \\ + & - & 0 & - \end{bmatrix} \begin{bmatrix} 5 \\ -2 \\ -1 \\ -1 \end{bmatrix} \equiv [H4]\mathbf{G} \quad (9)$$

where the expansion coefficient $\mathbf{G}$ is easily computed because of the orthogonal property; i.e., the inverse matrix is the transpose matrix (superscript T indicates the matrix transpose operation if use is made of the basis normalization $(4, 4, 2, 2)$).

$$\mathbf{G} \equiv \{5, -2, -1, -1\}^T = [H4]^{-1}\mathbf{g} = [H4]^T\mathbf{g} \quad (10)$$

corresponding to $G(a, b) = \langle h_{ab}(t), g(t) \rangle$ from Equation 7. The Haar transform has $O(N)$ computational complexity due to the matrix reduction (Strang, 1994):

$$[H4] = \begin{bmatrix} [H2] & & 0 & 0 \\ & & 0 & 0 \\ 0 & 0 & [H2] & \\ 0 & 0 & & \end{bmatrix} \begin{bmatrix} 1 & 0 & 0 & 0 \\ 0 & 0 & 1 & 0 \\ 0 & 1 & 0 & 0 \\ 0 & 0 & 0 & 1 \end{bmatrix} \begin{bmatrix} [H2] & & 0 & 0 \\ & & 0 & 0 \\ 0 & 0 & 1 & 0 \\ 0 & 0 & 0 & 1 \end{bmatrix} \quad (11)$$

in terms of three uses of the reduced matrix:

$$[H2] = \begin{bmatrix} + & + \\ + & - \end{bmatrix} \qquad (12)$$

Similarly, $[H8]$ is reduced to $[H4]$. Zero entries reduce the dense matrix of $O(N^2)$ to $O(N)$ for large N data.

Subband Coding from a Digital Filter Bank Viewpoint

Equation 12 reveals also a base-2 subband nature as follows. The low-pass L and high-pass H filters define the scaling function ϕ and wavelet ψ:

$$L\{c_k\} = \{+, +\} \leftrightarrow \phi(t) = \phi(2t) + \phi(2t - 1) = \sum_k c_k \phi(2t - k) \qquad (13)$$

$$H\{d_k\} = \{+, -\} \leftrightarrow \psi(t) = \phi(2t) - \phi(2t - 1) = \sum_k d_k \phi(2t - k) \qquad (14)$$

The scaling function (Equation 13) of Haar is satisfied by $\text{rect}(t) = \text{rect}(2t) + \text{rect}(2t - 1)$, which has $\{1, 1\}$ coefficients that are equivalent to averaging two neighborhood pixels via the low-pass filter L in Equation 13. Likewise, a Haar wavelet $H = \{+, -\}$ (Equation 14) is a high-pass filter in differencing two neighborhood pixels. Given an arbitrary input signal, say $\{1, 1\}$ which is fed in parallel to low-pass $L = \{+, +\}$, and to high-pass $H = \{+, -\}$ (cf. Figure 2 analysis), the convolution product gives $\{1, 1\} \times \{+, +\} = \{1, 2, 1\}$ and $\{1, 1\} \times \{+, -\} = \{1, 0, -1\}$. Because of the smoothing, a smaller number of pixels is needed, and a down-sampling removes every other pixel in both channels giving $\{1, 2, 1\} \rightarrow \{1, 1\}$; $\{1, 0, -1\} \rightarrow \{1, -1\}$. The receiver for reconstruction takes the up-sampling (defined by interpolation fill-in zeros) giving $\{1, 1\} \rightarrow \{1, 0, 1\}$; $\{1, -1\} \rightarrow \{1, 0, -1\}$. Since Haar synthesis filters $L^+ = \{-, -\}$ and $H^+ = \{+, -\}$ are identical to those of analysis, the reconstruction gives $\{1, 0, 1\} \times \{-, -\} = \{-1, -1, -1, -1\}$ and $\{1, 0, -1\} \times \{+, -\} = \{1, -1, -1, 1\}$, which are added: $\{-1, -1, -1, -1\} + \{1, -1, -1, 1\} = \{-2, -2\} = \{1, 1\} \times (-2)$ to reproduce the original $\{1, 1\}$ within a factor. Without the down- and up-sampling, the system remains lossless, as easily verified with the above example. Such a discrete WT via subband coding was known in digital filter theory as lossless quadrature mirror filters. Only the high-pass outputs called detail signals are sent through the channel, and the low-pass outputs called reference signals are usually fed back to the input to produce the next level high-pass detail signal to be sent through recursively, as shown in Figure 2. Since the down-sampling is by 2 at every level of recursion, the net number of pixels remains the same $N + N/2 + N/4 + \ldots = 2N$. Otherwise, the amount of data to be stored in the original (Laplacian pyramid) algorithm would prohibitively increase. Such a down-sampling, however, sensitizes the quantization or alignment errors. Daubechies (1992) was the first to go beyond Haar's system, by assuming analysis filters L and H equal to synthesis

filters L^+ and H^+, and moreover made the simplifying assumption:

$$L = (c_0, c_1, c_2, c_3); \qquad H = (c_3, -c_2, c_1, -c_0) \qquad (15)$$

which reduced eight unknowns c's and d's to four unknowns c's, so that the self-adjoint system forms a unitary matrix (called paraunitary by Vetterli & Herley, 1992):

$$\begin{bmatrix} c_0 & c_1 & c_2 & c_3 \\ c_3 & -c_2 & c_1 & -c_0 \end{bmatrix} \begin{bmatrix} c_0 & c_3 \\ c_1 & -c_2 \\ c_2 & c_1 \\ c_3 & -c_0 \end{bmatrix} = \begin{bmatrix} \sum_i c_i^2 & 0 \\ 0 & \sum_i c_i^2 \end{bmatrix} \qquad (16)$$

Only four equations are needed to determine four c's: (1) Low-pass scaling (Equation 13) is normalized by $\int dt\, \phi(t) = 1$, which yields $c_1 + c_2 + c_3 + c_4 = 2$. (2) To extract the change, the high-pass wavelet filter H, Equation 14 should nullify a constant $(1, 1, 1, 1)$: $c_3 - c_2 + c_1 - c_0 = 0$, and also (3) nullify a linear ramp $(1, 2, 3, 4)$: $c_3 - 2c_2 + 3c_1 - 4c_0 = 0$. (4) Two daughter wavelets should be orthogonal in a shift by 2: $(c_3, -c_2, c_1, -c_0, 0, 0)(0, 0, c_3, -c_2, c_1, -c_0)^t = c_1 c_3 + c_0 c_2 = 0$, giving four filter coefficients in Table 1.

A contribution of WT to the filter bank theory was the discovery of the importance of the regularity condition which means that solving the scaling function $\phi(t)$ (Equation 13) by recursion should converge stably, and then a shifted version $\phi(2t - k)$ gives the wavelet (Equation 14). Then, a continuous WT: $\langle g(t), \psi(t) \rangle$ can be approximated by a discrete WT: filter bank convolution product: $g * H$. Note that a regularity condition is necessary but not sufficient, for it can still yield an unsymmetric scaling function that is fractal-like, which is less suited for image compressions, because the image value at a boundary is usually extrapolated by assuming a mirror-symmetric value. A symmetric scaling function is still unitary in a biorthogonal system $L \neq L^+$; $H \neq H^+$ (see Table 1):

$$\begin{bmatrix} 1, & \alpha, & \alpha, & 1 \\ 1, & \alpha, & -\alpha, & -1 \end{bmatrix} \begin{bmatrix} -1 & 1 \\ \alpha & -\alpha \\ \alpha & \alpha \\ -1 & -1 \end{bmatrix} = \begin{bmatrix} 2\alpha^2 - 2 & 0 \\ 0 & -2\alpha^2 + 2 \end{bmatrix}$$

Image Compression. Since the discrete WT is linear, it merely rearranges the information content. Data compression must result from dynamic range compression by exploiting the fact that a filtered image which has less detail requires fewer bits per pixel. In so doing, the convergence of recursive filtering depends crucially on the stability of the filter under iteration, the so-called *regularity property*. Since, for a biorthogonal system analysis, $L \neq$ synthesis L^+, both must be regular. Unfortunately, the simple four-tap L^+ is not regular. In Table 1, only the Spline $S_{9,7}$ system is regular: its L of nine coefficients and adjoint L^+ of seven coefficients satisfy the regularity conditions. Therefore, it is suited for image compression, as demonstrated by Antonini et al. (1992). In fact, using it, the FBI

Table 1. Analysis/Decomposition and Synthesis/Reconstruction Filters

System	L	H	L^+	H^+
Haar H_2	1 1	1 -1	-1 -1	1 -1
Hat H_3	1 2 1	1 2 -6 2 1	-1 2 6 2 -1	1 -2 1
Spline S_4	1 α α 1	1 α $-\alpha$ -1	-1 α α -1	1 $-\alpha$ α -1
Spline $S_{9,7}$	$.04, -.02, -.1, .4, \underline{.9}, .4, -.1, -.02, .04$	$-1 \times$ even L	$-.06, -.04, .4, \underline{.8}, .4, -.04, -.06$	$-1 \times$ odd H
Daubechies	$1 + \sqrt{3}, 3 + \sqrt{3}, 3 - \sqrt{3}, 1 - \sqrt{3}$	$1 - \sqrt{3}, -3 + \sqrt{3}, 3 + \sqrt{3}, -1 - \sqrt{3}$	Identical	Identical
Pyramid	$1 + 2a, 1, 4a, 1, 1 - 2a$	Difference image	Identical	Identical

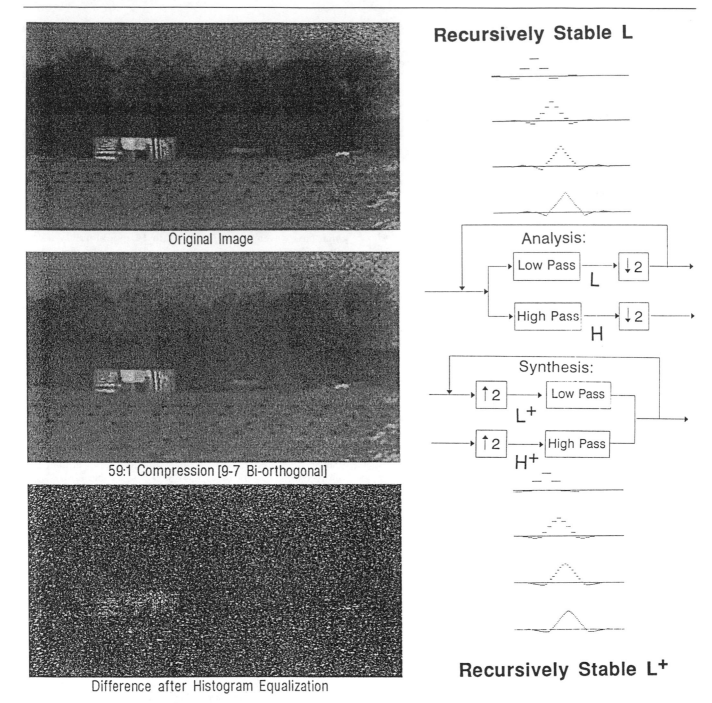

Recursively Stable L

Original Image

59:1 Compression [9-7 Bi-orthogonal]

Difference after Histogram Equalization

Recursively Stable L⁺

Figure 2. Discrete WT: Applications of iterative biorthogonal subband coding adopted by the FBI (shown in the right column) with recursively stable (low-pass filter) L with nine coefficients in scaling Equation 13 for the analysis stage; and the adjoint system L^+ with seven coefficients for the synthesis stage (see $S_{9,7}$ of Table 1) to multispectral infrared imagery to achieve simultaneously the sensor denoise and data compression ratio of 59 : 1.

(Hopper, 1994) has demonstrated a "lossless" 20 : 1 fingerprint compression which is believed to be able to yield improved identification of criminals. As a by-product, one can achieve noise suppression or clutter rejection by means of selective reconstruction. In Figure 2 we have applied $S_{9,7}$ recursively 4 levels to a multiple spectral band infrared imagery and shown that when the detail output at the first level was eliminated, the nonuniform infrared sensor noise was eliminated while preserving the target features at a 59 : 1 compression ratio.

Adaptive Wavelet Transforms

It is not so important for recognition that the wavelet features be orthogonal, as it is that they separate the classes of training data. Since classifiers normally require relatively lengthy off-line training, adapting wavelet features together with the classifier during training is an attractive approach to minimize the misclassification rate. A wedge-shaped filter in the a, b domain was used as the neural network preprocessing by Szu et al.

(1993) for scale invariant classification. Moreover, we wished to match the signal class better by adaptively constructing a superposition of admissible mothers called the *super-mother wavelet*. In so doing, the number of terms needed to represent the signals is greatly reduced and each becomes more robust in the sense of redundancy and stability under perturbation. The adaptive WT theorem of Szu and Telfer (1994) states that *the linear superposition of admissible mothers is also admissible.* This theorem allows us to use both noisy exemplars and a top-down performance measure similar to a supervised training strategy used in artificial neural networks (ANN), to construct by backpropagation the super-mother for each phoneme (Szu, Telfer, and Kadambe, 1992).

Another interesting example was given by Zhang and Benveniste (1992). The *cocktail party effect* is that one can hear one's own name no matter at what pitch it is whispered during a noisy drinking party. In that case, one's name is a wideband transient mother wavelet $h(t)$. The architecture is similar to that of the standard radial basis ANN (see RADIAL BASIS FUNCTION NETWORKS), except its radial basis functions are daughter wavelets called the "wave-nons" that are related to one another by the affine scale transform. The wave-nons can match in parallel the input sound, the name plus noise, with specific daughter neurons, and when it happens the mechanism of a winner-take-all (see WINNER-TAKE-ALL MECHANISMS) sets in, in order to suppress all the other weak SNR contributions from other daughter neurons. This eliminates other less-matched daughter neurons having more noise contribution than signal, and thus the wave net enhances the relative SNR, explaining the cocktail party effect (Szu, Sheng, and Chen, 1992). Recently, Mallat and Zhang (1993) have considered a fast algorithm of $O(N \log N)$ to construct a set of redundant waveforms to match a signal class using matching pursuit.

Discussion

Adaptive WTs that mimic aspects of the human auditory and visual systems can overcome the mismatch between instrumental data outputs and the multiple-resolution input (pyramids, preprocessing cones) of brain-style computing and reduce the amount of data using wavelet features. There are several technological implementation issues:

1. We need a fast discrete WT with an order (N) algorithm that is less sensitive to digitization error or pixel registration error. This requires a wavelet sampling theorem (Szu and Telfer, 1994).
2. Two-dimensional wavelet chips (cf. 1D digital chips: Szu et al., 1994) are needed that are programmable, modular, and plug-in.
3. We need more ways to base neural networks on user-friendly adaptive WT algorithms.

A real-time compression and decompression capability could have powerful applications in the commercial area of picturephones and in global markets for high-definition television, fingerprints, medical image transmission, and so on. A fast and discrete WT in software-hardware will become important to surveillance and communication applications. The synergism of wavelet chips and neurochips can together push the frontier of automation technology. In short, real-world information processing may be characterized as a *wideband transient* (which has the same acronym WT as its efficient tool, the wavelet transform). The freedom to adapt wavelet kernels to better mimic human sensory processing might open up new research and development in multiresolution neurodynamics, chaos, and fuzzy logic.

Acknowledgment. Support from the Independent Research Fund of the Naval Surface Warfare Center, Dahlgren Division, is appreciated.

Road Map: Vision
Related Reading: Gabor Wavelets for Statistical Pattern Recognition; Visual Coding, Redundancy, and "Feature Detection"

References

Antonini, M., Barlaud, M., Mathieu, P., and Daubechies, I., 1992, Image coding using wavelet transform, *IEEE Trans. Image Proc.* 1(2):205–220.

Daubechies, I., 1992, *Ten Lectures on Wavelets*, Philadelphia: SIAM.

Haar, A., 1910, Zur Theorie der orthogonalen Funktionen-systeme, *Math. Annal.* 69:331–371.

Hopper, T., 1994, Compression of grey-scale fingerprint images, In *Wavelet Applications* (ed. H. Szu), *Proc. SPIE*, 2242:180–187, April 5–8, Orlando (SPIE, WA). ◆

Mallat, S., 1989, A theory of multiresolution signal decomposition: The wavelet representation, *IEEE Trans. Pattern Anal. Machine Intell.*, PAMI-11:674–693. ◆

Mallat, S., and Zhang, Z., 1993, Matching pursuit with time-frequency dictionaries, *IEEE Trans. Signal Proc.*, 41(12):3397–3415.

Pati, Y. C., and Krishnaprasad, P. S., 1993, Analysis and synthesis of feedforward neural networks using discrete affine wavelet transforms, *IEEE Trans. Neural Netw.*, 4:73–85.

Sheng, Y., Roberge, D., Szu, H., and Lu, T., 1993, Optical wavelet matched filters for shift-invariant pattern recognition, *Opt. Lett.*, 18(2):209–301.

Strang, G., 1994, Wavelets, *Am. Sci.*, 82:150–255. ◆

Szu, H., Hsu, C., Thaker, P., and Zaghloul, M., 1994, Image wavelet transforms implemented by discrete wavelet chips, *Opt. Engrg.*, 33(7):2310–2325.

Szu, H., Sheng, Y., and Chen, J., 1992, Wavelet transform as a bank of matched filters, *Appl. Opt.*, 31(6):3267–3277. ◆

Szu, H., and Telfer, B., 1994, Mathematics of adaptive wavelet transforms: Relating continuous with discrete transforms, *Opt. Engrg.*, 33(7):2111–2124.

Szu, H., Telfer, B., and Kadambe, S., 1992, Neural network adaptive wavelets for signal representation and classification, *Opt. Engrg.*, 31(9):1907–1916. ◆

Szu, H., Yang, X.-Y., Telfer, B., and Sheng, Y., 1993, Neural network and wavelet transform for scale-invariant data classification, *Phys. Rev. E*, 48(2):1497–1501.

Vetterli, M., and Herley, C., 1992, Wavelets and filter bank: Theory and design, *IEEE Trans. Signal Proc.*, 40(9):2207–2232.

Zhang, Q., and Benveniste, A., 1992, Wavelet networks, *IEEE Trans. Neural Netw.*, 3:889–898.

Wave Propagation in Cardiac Muscle and in Nerve Networks

A. T. Winfree

Introduction

Like neuroglia, excitable cells in heart muscle connect to their immediate neighbors through electrically conductive gap junctions; there is nothing like arborization to contact remote neighbors and nothing like chemical synapses, synaptic delay, or choice of excitation or inhibition. Adjacent cells are typically quite similar in structure and function. Excitation conducts through a volume of such connected cells much as it does across a uniform sheet of excitable membrane. This mechanism is described by the "cable equation" augmented with detailed kinetics of the ionic channels involved—for example, in the Hodgkin-Huxley equation (see AXONAL MODELING) or its descendants adapted to the peculiarities on heart cell membranes. Such models are *reaction-diffusion* partial differential equations. The "reaction" part describes local ionic channel responses to changing membrane potential; the "diffusion" part is the cable equation, which can be viewed as diffusion of electric potential, scaled by a coefficient compounded of passive capacitance and resistivity terms.

This is all quite different from neuronal activity in the central nervous system, with the possible exception of relatively slow signaling mediated by chemical diffusion as in spreading depression, calcium waves, and nitric oxide oscillations. Although these various propagating excitations have completely different ionic mechanisms, they also share a few common principles. All involve local membrane events that propagate by diffusion: of electric potential in myocardial activation, of ions and neurotransmitters in the other cases. The local membrane depolarization constitutes excitability, whether of postsynaptic or nonsynaptic membrane, and the recovery process may or may not involve active transport. The pertinent equations are similar in form, so analogies are expected between these several phenomena. This article focuses on the best understood case, the cardiac activation wave. The sections that follow present the essential form of the nonlinear equation summarizing the cellular mechanisms of action potential propagation in myocardium; illustrate vortex solutions, the "rotors" in myocardium; present fibrillation in myocardium as a rotor-catalyzed turbulent mode of propagation; and sketch a scaling equivalence among rotors in physically diverse media. Finally, the Discussion section speculates on (limited) analogies in the central nervous system.

The Continuum Approximation

On scales exceeding 1 mm (its passive electrical space constant), healthy myocardium resembles continuous anisotropic media. Its mechanical motions are a delayed response to electrophysiological activation of cell membranes to admit calcium ions. To understand this controlling process, it suffices to understand propagated depolarization and local repolarization in a motionless medium. The basic electrophysiological mechanism in an electrically continuous medium (no synapses) has the form

$$\partial V_m(x, y, z, t)/\partial t = -J_m/C_m + D_x\partial^2 V_m/\partial x^2$$
$$+ D_y\partial^2 V_m/\partial y^2 + D_z\partial^2 V_m/\partial z$$

where t is time, V_m is local membrane potential, and J_m is a sum of (voltage- and time-dependent) local ion-channel current densities of the form

$$\sum(V_m - V_i)g_i$$

The sum is taken over the several kinds of ionic channel embedded in the cell membrane. V_i is the constant and uniform *Nernst potential* for each ionic species, proportional to the logarithm of its inside/outside concentration ratio; g_i is the membrane's conductance toward that ion, usually curve-fitted to indicate time and voltage dependencies of many components of that conductivity. These intrinsic properties g_i vary only slightly from cell to cell or regionally in heterogeneous tissues. C_m is a constant membrane-specific capacitance, and D_i are constant diffusion coefficients in this reaction-diffusion equation. The reciprocal of each is a local average of the product C_m times the cell surface/volume ratio times a directionally anisotropic resistivity that includes cytoplasm and a dozen gap junctions per cell, fudged with some tortuosity. In the fast direction, it is normally about 1 cm^2/s along fibers (and less, transversely or in ischemic tissue).

In a fibrous medium like myocardium, if we take xy as the epicardial plane, including x as the long axis of local fiber, $D_x > D_y = D_z$, resistivity is an order of magnitude smaller, and wave speed consequently several-fold greater, along axis x. We substitute rescaled distances $y' = y\sqrt{(D_x/D_y)}$ and $z' = z\sqrt{(D_x/D_z)}$ to recover the isotropic case with coefficients $D_{y'} = D_{z'} = D_x$. We call it D henceforth:

$$\partial V_m/\partial t = -J_m/C_m + D\nabla^2 V_m \tag{1}$$

In other words, solutions for the uniformly anisotropic continuum are exactly those of the isotropic case, appropriately expanded (including boundary conditions) in both directions transverse to the fast axis. Thus, no temporal aspect of the solution is affected at all by uniform anisotropy. This theorem is strictly true only of continua; in otherwise uniform media which are discretely cellular—i.e., grainy on a scale exceeding wavefront thickness—it is an approximation whose domain of adequacy needs to be determined experimentally in each case (Keener in Jalife, 1990; Glass, Hunter, and McCulloch, 1991). In another respect also, model Equation 1 presents less than the full physics of current flow in myocardium. A more correct (bidomain) model would attribute to each point in the three-dimensional continuum *two* potentials, their difference being V_m: an intracellular potential V_i, directing current flow through longitudinal and transverse cellular resistivities, and an interstitial potential V_e, directing current flow through longitudinal and transverse extracellular resistivities (Plonsey, 1989). Nonetheless, these and other corrections to the simple ideas underlying Equation 1 seem but little to alter the basic qualitative conclusions it supports regarding waves of bulk depolarization spanning many millimeters.

Systems like Equation 1 support flat wave propagation at a speed of about $\sqrt{(D/T)}$, where T is the nominal rise time of space-clamped excitation. Thus, in myocardium with D about 1 cm^2/s in the fiber long direction and T about 1 ms, action potential speed is about 30 cm/s. The corresponding thickness of the excitation front is speed times rise time, thus $\sqrt{(DT)}$: about 0.03 cm, which is a few cell lengths, so this tissue is not too grainy for the continuum approximation. It is useful to note that the product of activation front thickness by speed estimates D, and their ratio estimates T.

Propagation speed in systems like Equation 1 diminishes if the wavefront is curved; for small curvature, the decrement of speed is $-D/$(radius of curvature). Carelessly extrapolating to large curvature, speed falls to 0 at or before a radius of about $\sqrt{(DT)}$, comparable to front thickness. Thus, a smaller ball of excited cells constitutes less than a critical nucleus for spread of activation. After rescaling to isotropy as above, radius $\sqrt{(DT)} = 0.03$ cm in myocardium, so a minimum of 100 to 200 cells is estimated. This is a strict minimum: in less excitable media, propagation fails at larger radii, while speed is still positive.

Propagation speed in systems like Equation 1 also falls off as the interval between excitations shortens so that the medium has less fully recovered when excited again. This speed effect is not dramatic in myocardium but there is a minimum interval near 70 ms below which propagation is completely unstable. Between about 70 ms and about 125 ms, various modes of beat-skipping and alternation occur; at longer intervals, 1 : 1 propagation is stable. In myocardium, the concept of *refractory period* proves less useful than hoped, as its value depends on the interval since prior repolarization, and various instabilities dominate tissue response.

Vortex-like Propagation

Implicit in Equation 1 is an additional rotating vortex-like action potential called a *rotor*, which drifts freely through the medium like a tornado. This *re-entrant* propagation plays an essential role in cardiac arrhythmias leading to fibrillation and sudden cardiac death. Sudden cardiac death in human beings is usually the result of ventricular tachycardia, i.e., continuous beating of the ventricular muscle at intervals of less than a nominal 240 ms. *Fibrillation*, the fastest tachycardia, is incompatible with coherent muscular contraction and pumping of blood. The spatial pattern of electrical activation during fibrillation is complex, changes rapidly, and has never been well described, but it commonly originates from *re-entrant* action potentials resembling a vortex (in two dimensions) or a vortex line (in three dimensions) (Winfree, 1987; Jalife, 1990). The effective radius, r, of the rotor for purposes of substantial interaction with boundaries or other rotors in a typical excitable medium (excepting special cases such as marginally excitable media) is roughly the diffusion distance of the propagator species (membrane potential, in this case) during one rotation period, $\tau_0: r \sim \sqrt{(2D\tau_0)}$. Excitation propagates $2\pi r$ around the rotor in time τ_0 at speed roughly $\sqrt{(D/T)}$ so $\tau_0 \sim 8\pi^2 * T \sim \frac{1}{10}$ s, and $r \sim \sqrt{(2 * \frac{1}{10}\text{ s} * 1\text{ cm}^2/\text{s})} \sim \frac{1}{2}$ cm.

To test whether real myocardium does support such rotors, electrical stimuli were contrived that should evoke rotors of opposite-handedness some centimeters apart. According to theory, a rotor will arise in a two-dimensional (or three-dimensional) continuum along the intersection of two critical contours (Winfree, 1987; Winfree in Glass et al., 1991). One is an iso-stimulus contour, a locus of uniform current density $S = S^*$, believed to exceed several-fold the 4 mA/cm^2 required for pacing; this surrounds the electrode tip like a sphere. The other contour is an iso-phase contour along which, at the moment the stimulus is applied, cells are all at a certain phase $T = T^*$ near the transition from excitation to recovery. Such a contour line (or plane) thus rides several centimeters behind the activation front. Planes and spheres typically intersect in rings, or, if we restrict attention to the epicardial surface, lines and circles intersect in pairs of points, so rotors should appear on the surface in mirror-image counter-rotating pairs. Unless they are at least two rotor radii apart ($2r \sim 1$ cm), they overlap substantially, combine, and vanish. Marginal instigation of rotors (and of waxing/waning tachycardia at the rotor's period) must then involve intersections just that far apart. Critical values T^* and S^* were determined experimentally: $T^* = 170$ ms or 5 cm after local activation, and S^* is 5 V/cm or 20 mA/cm^2 (Frazier et al., 1989). Rotors in fact arose only where these contours crossed, and they turned at roughly the expected period in the expected directions. They can be located anywhere, depending only on the stimulus; they commonly drift through the medium at $<10\%$ of wave speed.

In normal hearts, this tachycardia promptly degenerates to fibrillation, so the minimum total current, $I = I^*$, required to instigate rotors also estimates the electrical threshold for ventricular fibrillation. Instigation requires the ring $S = S^*$ to be 1 cm in diameter so as to intersect the instantaneous T^* contour at points 1 cm apart. Around a unipolar electrode resting on the exposed surface of an isotropic medium, a hemisphere of radius $\frac{1}{2}$ cm would be maintained at current density $S^* = 20$ mA/cm^2 by a current $I = 2\pi(\frac{1}{2})^2 20$ mA/cm$^2 = 31$ mA, which in fact falls among scattered measurements of the least single DC impulse needed to instigate fibrillation.

Turbulent Propagation

The concurrence of the ventricular fibrillation threshold with conditions required for creating rotors by a single electrical pulse suggests that rotors (and other comparably short-period sources) catalyze the transition from orderly tachycardia. But no quantitative understanding of fibrillation, nor even a quantitative theory of its instigation, has yet commanded widespread assent. The classical cellular automaton "wandering wavelets" model (Moe, Rheinboldt, and Abildskov, 1964) supposes that atrial (not ventricular) fibrillation comes from irregularities of the substrate. In this model, any cell fires soon after recovering from prior excitation if an immediate neighbor then fires; then, if and only if some cells intrinsically recover more quickly than others, and given sufficient area or volume, appropriate stimuli can evoke persistent turbulent activity. Fibrillation has long been recognized as a spatially heterogeneous process, so it has been easy to suppose that its indispensable cause may be some kind of permanent physiological heterogeneity intrinsic to a grainy tissue. So it comes as a surprise to discover that in the presence of such heterogeneity as is normally supposed to be responsible for transition from tachycardia to fibrillation, no transition occurs in ventricular tissue unless it is functionally three-dimensional, i.e., thicker than the vortex core's diameter assayed above (Winfree, 1994). The significance of this thickness in other excitable media is that it allows a vortex line to wander freely within the bulk medium, rather than consisting only of stationary short segments perpendicular to the surfaces. It is also surprising that in detailed ionic models of ventricular membrane in the format of Equation 1, something like fibrillation develops from rotors in two-dimensional continua *devoid* of any heterogeneity. This has recently turned out to be true even of seemingly trivial modifications of the atrial model in which substantial heterogeneity was taken to be a necessary condition for turbulence. Thus, it seems necessary now to ask whether the originally conceived heterogeneity per se is the cause, or possibly only an incidental catalyst and modifier, of the transition to fibrillation in normal ventricular myocardium and possibly in diverse other excitable media.

Vortex Size and Period Versus Membrane Parameters

In the limit of small rise time/refractory period, in reaction-diffusion mechanisms like Equation 1, the rotor's size, period, and wave speed scale according to simple power laws of the diffusion coefficient, the rise time of excitation of the space-clamped system, and its refractory period. Comparing two similar media, let *rise* be the ratio of their rise times; *refr* the ratio of their refractory periods (both times being evaluated by an arbitrary criterion, but identically in both systems); and *diff* the ratio of their diffusion coefficients. Then the corresponding vortex core radii and (multiplying by 2π) wavelengths are expected to differ by the factor $(rise^1 * refr^2 * diff^3)^{1/6}$ while rotation periods differ by the factor $(rise^2 * refr^4)^{1/6}$, and wave speeds correspondingly differ by the factor $(rise^{-1} * refr^{-2} * diff^3)^{1/6}$. This supposes the two are similar FitzHugh-Nagumo-like systems, differing only in diffusion coefficient and rate multipliers of activation and of recovery kinetics, activation being very much faster than recovery. The modeling is simple-minded, but it provides a starting place for estimation of rotor periods and wavelengths in systems where they might be expected by analogy. For example, we can try using myocardial rotors to "foresee" those in the Belousov-Zhabotinsky (BZ) chemical reaction: rise is about $(1 \text{ s})/(1 \text{ ms}) = 1000$, refr is about $(20 \text{ s})/(250 \text{ ms}) = 80$, and diff is about $(2 \times 10^{-5} \text{ cm}^2/\text{s})/(1 \text{ cm}^2/\text{s}) = 2 \times 10^{-5}$; so spiral wavelength (or rotor radius) in BZ solutions might be 3 cm (or $\frac{1}{2}$ cm) $\times (1000^1 \times 80^2 \times 0.00002^3)^{1/6} = 1.8$ mm (or 0.6 mm), while rotation period becomes $0.1\text{s} \times (1000^2 \times 80^4)^{1/6} = 19\text{s}$, so propagation speed becomes 0.1 mm/s = 6 mm/min—all close to familiar values. Or for rotors in spreading depression: rise is about $(3 \text{ s})/(1 \text{ ms}) = 3000$, refr is about $(5 \text{ min})/(250 \text{ ms}) = 1200$, and diff is again 2×10^{-5}, thus roundly 5 mm wavelength (1 mm radius), 3 min rotations, and 2 mm/min propagation, all near observed values (Gorelova and Bures, 1983).

Discussion

Many excitable media are susceptible to turbulent propagation induced by rotors or other comparably short-period sources. Rotors and related propagation in myocardium and in chemically excitable media are quantitatively understandable in terms of the basic principles embodied in Equation 1. Similar principles seem to underlie (only) diffusion-based signaling in and between cortical cells—for example, spreading depression (Tuckwell, 1981). Calcium waves propagate intracellularly with diffusion coefficient somewhat less than $10^{-6} \text{ cm}^2/\text{s}$ through gap junctions among astrocytes cultured from the hippocampus and from the suprachiasmatic nucleus (Cooper, 1995). Their rotors have been observed in the *Xenopus* egg (Atri et al., 1993) and (with Ca rise time an order of magnitude shorter) in single cardiac myocytes (Lipp and Niggli, 1993). Velocities, periods, and wavelengths are quantitatively compatible with Equation 1 and its scaling laws (only intracellularly: front thickness is only

tens of microns). Waves of nitric oxide release in *Limax* olfactory bulb suggest a locally oscillatory excitable process coupled across space by $10^{-5} \text{ cm}^2/\text{s}$ extracellular molecular diffusion with activation front spanning many neurons (Delaney et al., 1994).

For the present, any analogy from cardiac waves to bulk propagation in the central nervous system stops about here, i.e., short of involvement in seizure propagation or normal high-speed neural processing.

Acknowledgment.   My work has been supported by the U.S. National Science Foundation since 1967.

Road Map: Biological Networks
Background: Axonal Modeling
Related Reading: Traveling Activity Waves

References

Atri, A., Amundson, J., Clapham, D., and Sneyd, J., 1993, A single-pool model for intracellular calcium oscillations and waves in the *Xenopus laevis* oocyte, *Biophys. J.*, 65:1727–1739.

Cooper, M. S., 1995, Intracellular signaling in neuronal-glial networks, *Biosystems*, 34:65–85. ◆

Delaney, K. R., Gelperin, A., Fee, M. S., Flores, J. A., Gervais, R., Tank, D. W., and Kleinfeld, D., 1994, Waves and stimulus-modulated dynamics in an oscillating olfactory network, *Proc. Natl. Acad. Sci. USA*, 91:669–673.

Frazier, D. W., Wolf, P. D., Wharton, J. M., Tang, A. S. L., Smith, W. M., and Ideker, R. E., 1989, Stimulus-induced critical point: Mechanism of electrical induction of reentry in normal canine myocardium, *J. Clin. Invest.*, 83:1039–1052.

Glass, L., Hunter, P., and McCulloch, A., Eds., 1991, *Theory of Heart*, New York: Springer-Verlag. ◆

Gorelova, N. A., and Bures, J., 1983, Spiral waves of spreading depression in the isolated chicken retina, *J. Neurobiol.*, 14:353–363.

Jalife, J., Ed., 1990, *Mathematical Approaches to Cardiac Arrhythmias*, Ann. NY Acad. Sci., 591. ◆

Lipp, P., and Niggli, E., 1993, Microscopic spiral waves reveal positive feedback in subcellular calcium signalling, *Biophys. J.*, 65:2272–2276.

Moe, G., Rheinboldt, W. C., and Abildskov, J. A., 1964, A computer model of atrial fibrillation, *Am. Heart J.*, 67:200–220.

Plonsey, R., 1989, The use of a bidomain model for the study of excitable media, *Lect. Math. Life Sci.*, 21:123–149. ◆

Tuckwell, H., 1981, Simplified reaction-diffusion equations for potassium and calcium ion concentrations during spreading cortical depression, *Int. J. Neurosci.*, 12:85–135.

Winfree, A. T., 1987, *When Time Breaks Down: The Three-Dimensional Dynamics of Electrochemical Waves and Cardiac Arrhythmias*, Princeton, NJ: Princeton University Press. ◆

Winfree, A. T., 1994, Electrical turbulence in three-dimensional heart muscle, *Science*, 266:1003–1006. ◆

Winner-Take-All Mechanisms

Alan L. Yuille and Davi Geiger

Introduction

A winner-take-all mechanism is a device that determines the identity, and sometimes the amplitude, of its largest input. Such mechanisms are necessary in network models for enforcing competition between different possible outputs of the network (Amari and Arbib, 1977). Classic examples include (i) associative memory models, (Hertz, Krogh, and Palmer, 1991),

where different stored memories compete to explain the input data, (ii) cooperative models of binocular stereo where competition is required to ensure that each feature has a unique match (Dev, 1975; Marr and Poggio, 1977; see STEREO CORRESPONDENCE AND NEURAL NETWORKS), and (iii) Fukushima's neocognitron (see NEOCOGNITRON: A MODEL FOR VISUAL PATTERN RECOGNITION) for feature extraction. In addition, such mechanisms are often required for combinatorial optimization problems (see CONSTRAINED OPTIMIZATION AND THE ELASTIC NET).

More recently, a softened version of winner-take-all has been proposed. This variant, known as *softmax*, consists of assigning each input a weight so that all weights sum to 1 and the largest input receives the biggest weight. Winner-take-all can be considered a limiting case of softmax when the biggest weight is set equal to 1. Softmax has been applied to problems in speech recognition (Bridle, 1989) and learning by mixtures of experts (Jordan and Jacobs, 1992). Softmax also appears from a statistical mechanics analysis of the winner-take-all problem.

This article first describes softmax and shows how winner-take-all can be derived as a limiting case. We then describe how they can both be derived from probabilistic, or energy function, formulations. This leads naturally to several algorithms for calculating them and their generalizations to more complicated systems. We then discuss very large scale integration (VLSI) and biological mechanisms for implementing these algorithms.

Softmax

The basic idea of softmax is very simple. Suppose we have inputs $I = \{I_a: a = 1, \ldots, N\}$ and let β be a positive parameter. We define a weight $w_a(I; \beta)$ for each input I_a by

$$w_a(I; \beta) = \frac{e^{\beta I_a}}{\sum_b e^{\beta I_b}} \qquad (1)$$

Clearly, because the exponential function is monotically increasing, the largest weight corresponds to the largest input. Moreover, the size ordering of the weights corresponds to the size ordering of the inputs. The weights sum up to 1, $\sum_a w_a(I; \beta) = 1$. In the limit as $\beta \to \infty$, the weight of the largest input tends to 1, all other weights tending to zero, and we obtain the classic winner-take-all. At the other extreme, as $\beta \to 0$, the weights all tend to $1/N$. Thus, the parameter β controls the sharpness of the softmax.

The variables $w_a(I; \beta)$ in Equation 1 preferentially weight the largest inputs. They can also be used to calculate a weighted average of the inputs to yield

$$O(I; \beta) = \sum_a w_a(I; \beta) I_a \qquad (2)$$

As $\beta \to \infty$, the output $O(I; \beta)$ selects the largest input value.

We cannot make β arbitrarily large in Equation 1. So how large must it be to ensure that we are close to the true winner-take-all solution? If we require that the largest weight differs from 1 by at most ε, then it can be shown that we need to have $\beta \geq (1/|\Delta I|) \log\{(N-1)(1-\varepsilon)/\varepsilon\}$, where $|\Delta I|$ is the difference between the largest and second-largest inputs.

Statistical Derivation of Softmax

The winner-take-all problem can be posed as follows: given a set of N inputs $\{I_a: a = 1, \ldots, N\}$, how can we index each input I_a with a binary variable V_a, such that $V_{a*} = 1$ for the largest input I_{a*}, and is zero otherwise? We can transform the winner-take-all problem into an optimization problem by introducing the energy function

$$E[V] = -\sum_a V_a I_a \qquad (3)$$

Minimizing Equation 3 with the constraint that $\sum_a V_a = 1$ will select the largest input.

The associated Gibbs distribution (for justification of this distribution, see CONSTRAINED OPTIMIZATION AND THE ELASTIC NET) is

$$P[V; \beta] = \frac{e^{-\beta E[V]}}{Z} \qquad (4)$$

where the constant $\beta = 1/T$. T is called the *temperature* of the system. Z is a normalization factor also known as the *partition function*. Hence, $Z = \sum_V e^{-\beta E[V]}$ where we sum over all allowable configurations of V. Because of the global constraint, these configurations are of the form $V_k = 1$, $V_j = 0$, for all $j \neq k$. Hence we find that

$$Z = \sum_j e^{\beta I_j} \qquad (5)$$

We can now directly evaluate the probabilities of the allowable configurations. These are

$$P[V_k = 1; V_j = 0, j \neq k] = \frac{e^{\beta I_k}}{\sum_j e^{\beta I_j}} \quad \text{for all } k \qquad (6)$$

This is exactly the softmax formula. So we can interpret the weights given by softmax as the probabilities given by the Gibbs distribution. It is straightforward to design a physical system whose states are specified by the Gibbs distribution, i.e., so that the probability of each unit firing is given by Equation 6.

Alternatively, we can reinterpret the softmax weights as the mean values of the output units (for a binary variable, the means are the same as the probabilities of firing). The means can be calculated (see CONSTRAINED OPTIMIZATION AND THE ELASTIC NET) by minimizing an effective energy:

$$E_{\text{eff}}[w, P; \beta] = -\sum_a w_a I_a + (1/\beta) \sum_a w_a \log w_a + P\left(\sum_a w_a - 1\right) \qquad (7)$$

where P is a Lagrange multiplier to impose the constraint $\sum_a w_a = 1$. The second term can be interpreted (see CONSTRAINED OPTIMIZATION AND THE ELASTIC NET) as an entropy term. It is straightforward to verify that E_{eff} is a convex energy function which is bounded below and hence has a unique minimum. This minimum can be obtained by extremizing E_{eff} with respect to w and P, yielding:

$$\frac{\partial E_{\text{eff}}}{\partial w_a} - I_a + (1/\beta)\{\log w_a + 1\} + P = 0$$
$$\frac{\partial E_{\text{eff}}}{\partial P_a} = \sum_a w_a = 1 \qquad (8)$$

We can solve these equations, eliminating P, to obtain the softmax formula (Equation 1). Moreover, the effective energy can be used (see the next section) to design algorithms for calculating softmax.

Dynamical Systems for Computing Softmax

There are many dynamical systems that converge to the softmax solution. One possibility is to do a variant of gradient descent which is modified to ensure that the constraint $\sum_a w_a = 1$ is satisfied. This corresponds to removing the component of the gradient in the direction which violates the constraints. This yields:

$$\frac{dw_a}{dt} = -\frac{\partial E_{\text{eff}}}{\partial w_a} + \frac{1}{N}\sum_b \frac{\partial E_{\text{eff}}}{\partial w_b} \qquad (9)$$

Substituting from Equation 7 gives:

$$\frac{dw_a}{dt} = I_a - (1/\beta)\log w_a - (1/N)\sum_b I_b + (1/N\beta)\sum_b \log w_b \qquad (10)$$

This update equation ensures that $d/dt(\sum_a w_a) = 0$, so the constraints are always satisfied (provided the initial conditions are chosen to satisfy them). It can be shown that this equation causes the energy to monotonically increase to the minimum of the effective energy and hence yield softmax as a solution.

Two other dynamical systems can be adapted from Waugh and Westervelt (1993). Their formulation applies to more general networks such as systems of competitive memories (see next section). Both systems converge to the softmax solution. Winner-take-all can be achieved by making β sufficiently large or by thresholding the softmax solution.

The first is a discrete-time dynamical system:

$$w_i(t + 1) = \exp\{\beta I_i + P(t)\} \tag{11}$$

where $P(t)$ is chosen to ensure that at any time $\Sigma_i w_i(t + 1) = 1$. This requires that $P(t) = -\log\{\Sigma_i e^{\beta I_i}\}$. Thus, the algorithm converges to the softmax solution $w_i(1) = e^{\beta I_i}/\{\Sigma_j e^{\beta I_j}\}$ in a single iteration. The simplicity of this algorithm, and its rapid convergence, reflects the simplicity of the winner-take-all/ softmax problem. The algorithm is usually applied (see the next section) to more complex problems.

The second is a continuous-time dynamical system:

$$\frac{dw_i(t)}{dt} = -w_i(t) + \exp\{\beta I_i + P(t)\} \tag{12}$$

where, as before, $P(t)$ is chosen to ensure that $\Sigma_i w_i = 1$. It is straightforward to verify that $\lim_{t \to \infty} w_i(t) = e^{\beta I_i}/\Sigma_k e^{\beta I_k}$, recalling that E_{eff} has a unique minimum and observing that it acts as a *Lyapunov* function for this system (Waugh and Westervelt, 1993). Note that this is only a special case of Waugh and Westervelt's formalism. In the general case, the exponential functions in Equations 11 and 12 could be replaced by other monotonic differentiable functions.

For additional algorithms for this problem, and detailed descriptions of the relationship of winner-take-all to effective energies, we highly recommend Elfadel (1993).

Generalizations

Winner-take-all mechanisms are usually part of more complicated systems. Here we show how they can be generalized to systems of competitive memories. The system converges to the "best" memory rather than to the "best" input.

Following Waugh and Westervelt (1993, see also Elfadel, 1993), we now generalize the previous system to more complicated models which consist of neurons grouped into clusters.

The variables are $\{w_{ia}\}$, where the index i labels the cluster and a the element in the cluster. We impose the constraint $\Sigma_a w_{ia} = R_i$, for all i to ensure that in the ith cluster only R_i neurons are firing. The input to the iath neuron is $h_{ia} = \Sigma_{jb} T_{ijab} w_{jb} + I_{ia}$, where I_{ia} are the external inputs and the T_{ijab} are connection strengths. Then Equations 11 and 12 are generalized to

$$w_{ia}(t + 1) = \exp\{\beta(h_{ia}(t) + B_i(t))\} \quad \text{for all } a, i \tag{13}$$

and

$$\frac{dw_{ia}(t)}{dt} = -w_{ia}(t) + \exp\{\beta(h_{ia}(t) + B_i(t))\} \quad \text{for all } a, i \tag{14}$$

where the $B_i(t)$ are chosen to enforce the constraints $\Sigma_a w_{ia} = R_i$, for all i.

These dynamical systems are guaranteed to converge (Waugh and Westervelt, 1993). Indeed, they will converge to a local minimum of an effective energy (Elfadel, 1993), generalized from Equation 7, given by

$$E_{\text{eff}}[w, P; \beta] = -\frac{1}{2} \sum_{ijab} T_{ijab} w_{ia} w_{jb} - \sum_{ia} w_{ia} I_{ia}$$

$$+ (1/\beta) \sum_{ia} w_{ia} \log w_{ia} + \sum_i P_i \left(\sum_a w_{ia} - R_i \right) \tag{15}$$

where the $\{P_i\}$ are Lagrange multipliers.

Waugh and Westervelt's formulation is more general than this. It can be obtained by replacing the exponentials in Equations 13 and 14 by monotonic differential functions F_{ia} and replacing the term $(1/\beta)\Sigma_{ia} w_{ia} \log w_{ia}$ in Equation 15 by $(1/\beta)\Sigma_{ia} \int_{w_0}^{w_{ia}} F_{ia}^{-1}(w) \, dw$.

Networks similar to Equations 13 and 14, and effective energies similar to Equation 15, can also be used to solve optimization problems (see CONSTRAINED OPTIMIZATION AND THE ELASTIC NET). They also can be related to the Potts models in statistical physics (see NEURAL OPTIMIZATION and Elfadel, 1993).

VLSI Analog Implementations

There have been several attempts to design analog VLSI circuits to compute winner-take-all.

Lazzaro et al. (1989) developed a winner-take-all circuit that was used for applications to auditory localization and visual stereopsis. This network was demonstrated to work effectively, and perturbation analysis was performed to demonstrate local

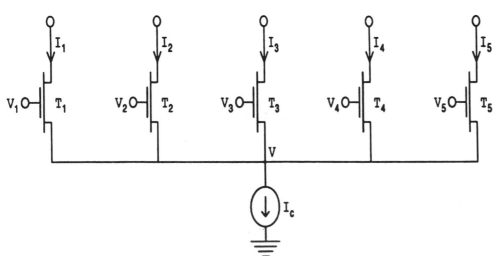

Figure 1. A VLSI circuit for winner-take-all proposed by Elfadel (1993). The circuit takes voltage as input and outputs current (see text).

stability. Yet, to our knowledge, no global analysis of the network has been performed, and convergence to the optimal solution has not been proven. The circuit takes current as input and outputs voltage.

A second circuit, from Waugh and Westervelt (1993), implements Equation 12. It is guaranteed to converge to the correct solution, provided the components of the circuit correctly implement the equations. In contrast to Lazzaro et al. (1989), the input is gate voltage, and the output is current.

A final circuit, shown in Figure 1, implements Elfadel's winner-take-all by a circuit containing transistors. For every one of the transistors, we can write the input currents as

$$I_m = I_0 e^{(\kappa V_m - V)/V_0} \qquad 1 \le m \le 5 \qquad (16)$$

where I_0 is a parameter to be determined, and κ is a parameter which plays the role of β. Applying Kirchoff's laws gives $\Sigma_{m=1}^5 I_m = I_c$, where I_c is the control current source. Substituting from Equation 16 allows us to solve for I_0. Substituting back into Equation 16 gives:

$$I_m = I_c \frac{e^{\kappa V_m/V_0}}{\sum_{p=1}^5 e^{\kappa V_p/V_0}} \qquad (17)$$

Thus, the input site with largest voltage V_M will have the largest current. Provided κ/V_0 is large enough, this current will tend to the control current source I_c, and the other currents will tend to zero.

Care must be taken when drawing conclusions from the circuit equations describing these networks. Such equations are idealizations and assume that the networks are built of perfect components. This is unrealistic, and so the design of the network must be robust to imperfections in the components. Analog networks of this type are somewhat fragile and not as robust as simple discrete methods for winner-take-all using analog comparators and digital latches. It seems, however,

that latch circuits are only capable of implementing pure winner-take-all and, unlike the analog circuits, cannot perform softmax.

Biological Models

Biological considerations have motivated some winner-take-all models (Elias and Grossberg, 1975). Here we discuss a model presented by Yuille and Grzywacz (1989). The network is updated by the following equation:

$$\tau \frac{dx_i}{dt} = -x_i + I_i K(x_1, x_2, \ldots, x_{i-1}, x_{i+1}, \ldots, x_N) \qquad (18)$$

where $x_i(t)$ is the state of the ith network element, τ is constant, and the function K is symmetric in all its variables, decreases (or remains constant) as they increase, and tends to zero when any of them goes to infinity. The initial values of the x_i are set to the inputs I_i, which are constrained to be positive. The first term on the right-hand side of Equation 18 corresponds to a time decay, and the second contains inhibition between the network elements (Figure 2).

Lateral inhibition is a biologically plausible mechanism which is often used in winner-take-all networks to ensure that only one neuron is firing (Amari and Arbib, 1977). An unusual aspect of this model (but see Didday, 1976) is that the output of each element feeds back to inhibit the inputs to other elements (see Figure 2). We consider a special choice of K, which is motivated by a biological mechanism called *shunting inhibition* (Yuille and Grzywacz, 1989). Let

$$K(x_1, x_2, \ldots, x_{i-1}, x_{i+1}, \ldots, x_N) = e^{-\lambda \sum_{j \ne i} x_j} \qquad (19)$$

where λ is a constant which plays a similar role to the parameter β in softmax.

By using the coordinate transformation $z_i = e^{-\lambda x_i}$, Equation 18 can be transformed into

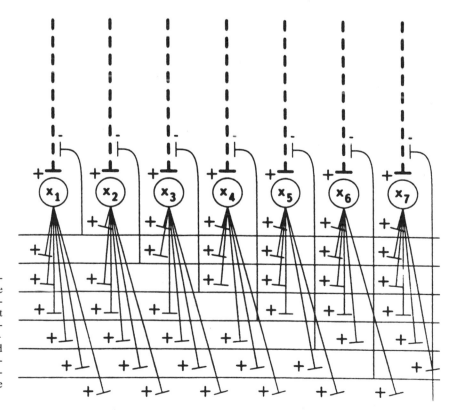

Figure 2. A biologically plausible circuit for winner-take-all. The dashed lines represent the inputs to the individual network elements. Each element excites inhibitory elements (similar to interneurons), which act on the presynaptic inputs of the other elements. Excitatory and inhibitory synapses are labeled with + and − signs, respectively. (From Yuille, A. L., and Grzywacz, N. M., 1989, A winner-take-all mechanism based on presynaptic inhibition, *Neural Computation*, 1:334–347; reprinted with permission of The MIT Press.)

$$\frac{dz_i}{dt} = -\frac{\lambda I_i z_i}{\tau} \frac{\partial E[z]}{\partial z_i} \qquad (20)$$

where

$$E[z] = \frac{1}{\lambda} \sum_j \left(\frac{z_j \log z_j - z_j}{I_j} \right) + \prod_j z_j \qquad (21)$$

satisfies the properties of a Lyapunov function; it is bounded from below and always decreases with time when using the dynamics specified by Equation 20. This can be seen by observing that $dE/dt = -\Sigma_i (\lambda I_i z_i / \tau)(\partial E/\partial z_i)^2$, and by recalling that the z's are constrained to be positive by Equation 18. Hence the system converges to a minimum of E.

To understand the global convergence of the system, we examine the Hessian of E, the matrix with components $H_{ij} = \partial^2 E/(\partial z_i \partial z_j)$. We see that on the diagonal we have $H_{ii} = 1/(\lambda I_i z_i)$ and off the diagonal $H_{jk} = \prod_{i \neq j,k} z_i$. From Equation 18 we see that the x_i are always positive and so the z_i lie in the range $[0, 1]$. Thus the diagonal elements are all greater than $1/(\lambda I_{max})$ and the off-diagonal elements are all less than 1. By making $\lambda \geq (N - 1)/I_{max}$ we can ensure that the Hessian is positive definite; hence, E is convex, and so there is a single solution to which the system converges. It was shown, in the large λ limit, that this corresponds to the winner-take-all solution (Yuille and Grzywacz, 1989).

Discussion

Winner-take-all is a special case of softmax. Both problems can be formulated in terms of energy minimization, and both can be solved by a number of continuous-time and discrete-time dynamical systems. Some of these systems can be implemented by VLSI circuits or by biologically plausible mechanisms.

These systems can be generalized in a straightforward way to systems of competitive memories or optimization problems. In these cases, only convergence to locally optimal solutions is guaranteed.

Acknowledgments. We would like to thank ARPA and the Air Force for support under contract F49620–92–J–0466.

Road Map: Dynamic Systems and Optimization
Background: I.3. Dynamics and Adaptation in Neural Networks
Related Reading: Learning Vector Quantization; Modular and Hierarchical Learning Systems; Visuomotor Coordination in Frogs and Toads

References

Amari, S., and Arbib, M., 1977, Competition and cooperation in neural nets, in *Systems Neuroscience* (J. Metzler, Ed.), San Diego: Academic Press, pp. 119–165. ◆

Bridle, J., 1989, Probabilistic interpretation of feedforward classification network outputs, with relationships to statistical pattern recognition, in *Neuro-computing: Algorithms, Architectures* (F. Fogelman-Soulié and J. Hérault, Eds.), New York: Springer-Verlag.

Dev, P., 1975, Perception of depth surfaces in random-dot stereograms, *Int. J. Man-Mach. Stud.*, 7:511–528.

Didday, R. L., 1976, A model of visuomotor mechanisms in the frog optic tectum, *Math. Biosci.*, 30:169–180.

Elias, S. A., and Grossberg, S., 1975, Pattern formation, contrast control, and oscillations in the short term memory of shunting on-center off-surround networks, *Biol. Cybern.*, 20:69–98.

Elfadel, I. M., 1993, *From Random Fields to Networks*, Research Laboratory of Electronics Technical Report No. 579, Cambridge: Massachusetts Institute of Technology. ◆

Hertz, J., Krogh, A., and Palmer, R. G., 1991, *Introduction to the Theory of Neural Computation*, Redwood City, CA: Addison-Wesley. ◆

Jordan, M. I., and Jacobs, R. A., 1992, Hierarchies of adaptive experts, in *Advances in Neural Information Processing Systems 4* (J. Moody, S. Hanson, and R. Lippmann, Eds.), San Mateo, CA: Morgan Kaufmann, pp. 985–993.

Lazzaro, J., Ryckebusch, S., Mahowald, M. A., and Mead, C. A., 1989, Winner-take-all networks of O(N) complexity, in *Advances in Neural Information Processing Systems* (D. S. Touretzky, Ed.), San Mateo, CA: Morgan Kaufmann, pp. 703–711.

Marr, D., and Poggio, T., 1977, Cooperative computation of stereo disparity, *Science*, 195:283–328.

Waugh, F., and Westervelt, R., 1993, Analog neural networks with local competition, I: Dynamics and stability, *Phys. Rev. E*, 47:4524–4536.

Yuille, A. L., and Grzywacz, N. M., 1989, A winner-take-all mechanism based on presynaptic inhibition, *Neural Computat.*, 1:334–347.

Editorial Advisory Board

Contributors

Emile H. L. Aarts   Phillips Research Laboratories, Eindhoven, The Netherlands
E-mail: aarts@prl.philips.nl
Boltzmann Machines

Larry F. Abbott   Center for Complex Systems, Brandeis University, Waltham, Massachusetts
E-mail: abbott@binah.cc.brandeis.edu
Activity-Dependent Regulation of Neuronal Conductances
Dynamic Clamp: Computer-Neural Hybrids

Hervé Abdi   School of Human Development, The University of Texas at Dallas, Richardson, Texas
Face Recognition

Kazuyuki Aihara   Department of Mathematical Engineering and Information Physics, Faculty of Engineering, The University of Tokyo, Tokyo, Japan
E-mail: aihara@sat.t.u-tokyo.ac.jp
Chaos in Axons

Yehuda Albeck   Division of Biology, California Institute of Technology, Pasadena, California
E-mail: albeck@asterix.bbb.caltech.edu
Sound Localization and Binaural Processing

Garrett E. Alexander   Department of Neurology, Emory University School of Medicine, Atlanta, Georgia
Basal Ganglia

Shun-ichi Amari   Faculty of Engineering, University of Tokyo, Tokyo, Japan
E-mail: amari@sat.t.u-tokyo.ac.jp
Learning and Statistical Inference

Franklin R. Amthor   Department of Psychology and Neurobiology Research Center, University of Alabama at Birmingham, Birmingham, Alabama
Directional Selectivity in the Retina

Uwe an der Heiden   University of Witten at Herdecke, Herdecke, Germany
Traveling Activity Waves

Anandkumar   Hughes Network Systems, Germantown, Maryland
E-mail: akumar@hns.com
Telecommunications

Thomas J. Anastasio   University of Illinois, Beckman Institute, Urbana, Illinois
E-mail: tstasio@uiuc.edu
Vestibulo-Ocular Reflex: Performance and Plasticity

Richard A. Andersen   Division of Biology, California Institute of Technology, Pasadena, California
E-mail: andersen@vis.caltech.edu
Gaze Coding in the Posterior Parietal Cortex

Charles H. Anderson   Department of Anatomy and Neurobiology, Washington University School of Medicine, St. Louis, Missouri
E-mail: cha@shifter.wustl.edu
Routing Networks in Visual Cortex

James A. Anderson   Department of Cognitive and Linguistic Sciences, Brown University, Providence, Rhode Island
E-mail: james_anderson@brown.edu
Associative Networks
Mental Arithmetic Using Neural Networks

Martin Anthony   Department of Mathematics, London School of Economics and Political Science, London, United Kingdom
E-mail: anthony@vax.lse.ac.uk
PAC Learning and Neural Networks

Michael A. Arbib   Center for Neural Engineering, University of Southern California, Los Angeles, California
E-mail: arbib@pollux.usc.edu
Part I; Part II; Schema Theory

Ronald C. Arkin   College of Computing, Georgia Institute of Technology, Atlanta, Georgia
E-mail: arkin@cc.gatech.edu
Reactive Robotic Systems

Tim Ash   University of California at San Diego, San Diego, California
E-mail: ash@cs.ucsd.edu
Topology-Modifying Neural Network Algorithms

Karl J. Åström   Department of Automatic Control, Lund Institute of Technology, Lund, Sweden
E-mail: kja@control.lth.se
Adaptive Control: General Methodology

Jean-Pierre Aubin   Centre de Recherche Mathematiques de la Decision; Université de Paris—Dauphine, Paris, France
E-mail: aubin@frulm63.bitnet
Learning as Adaptive Control of Synaptic Matrices

Ivan A. Bachelder   Machine Intelligence Technology Group, MIT Lincoln Laboratory, Lexington, Massachusetts
Visual Processing of Object Form and Environment Layout

José Bargas   Departamento de Neurociencias, Instituto de Fisiologia Celular, UNAM, Mexico City D. F., Mexico
E-mail: jbargas@ifcsun1.ifisiol.unam.mx
Ion Channels: Keys to Neuronal Specialization

John A. Barnden   Computing Research Laboratory, New Mexico State University, Las Cruces, New Mexico
E-mail: jbarnden@nmsu.edu
Artificial Intelligence and Neural Networks
Semantic Networks

Gabor T. Bartha   Wright Laboratory, Wright-Patterson Air Force Base, Dayton, Ohio
E-mail: bartha@corsair.aa.wpafb.af.mil
Cerebellum and Conditioning

Andrew G. Barto   Computer Science Department, University of Massachusetts, Amherst, Massachusetts
E-mail: barto@cs.umass.edu
Learning as Hill-Climbing in Weight Space
Reinforcement Learning
Reinforcement Learning in Motor Control

Joseph Bastian   Department of Zoology, University of Oklahoma, Norman, Oklahoma
E-mail: joe@ell.bio.uoknor.edu
Electrolocation

Michel Baudry   Department of Biological Sciences, University of Southern California, Los Angeles, California
E-mail: baudry@neuro.usc.edu
NMDA Receptors: Synaptic, Cellular, and Network Models

Suzanna Becker   Department of Psychology, McMaster University, Hamilton, Ontario, Canada
E-mail: becker@mcmaster.ca
Unsupervised Learning with Global Objective Functions

Mark Beeman Rush Neuroscience Institute, Chicago, Illinois
E-mail: mbeeman@rpslmc.edu
Emotion-Cognition Interactions

Randall D. Beer Department of Computer Engineering and Science, Case Western Reserve University, Cleveland, Ohio
E-mail: beer@alpha.ces.cwru.edu
Locomotion, Invertebrate

George A. Bekey Department of Computer Science, University of Southern California, Los Angeles, California
E-mail: bekey@robotics.usc.edu
Walking

Yoshua Bengio Dept. Informatique et Recherche Operationnelle, Université de Montréal, Montréal, Québec, Canada
E-mail: bengioy@iro.umontreal.ca
Convolutional Networks for Images, Speech, and Time Series Pattern Recognition

Younès Bennani CNRS, LIPN, URA-1507, University of Paris-Nord, Villetaneuse, France
E-mail: younes@lipn.univ-paris13.fr
Speaker Identification

Theodore W. Berger Department of Biomedical Engineering and Program in Neuroscience, University of Southern California, Los Angeles, California
E-mail: berger@bmsrs.usc.edu
NMDA Receptors: Synaptic, Cellular, and Network Models

Öjvind Bernander Computation and Neural Systems Program, California Institute of Technology, Pasadena, California
E-mail: ojvind@klab.caltech.edu
Axonal Modeling

Elie Bienenstock Division of Applied Mathematics, Brown University, Providence, Rhode Island; CNRS, Paris, France
E-mail: elie@dam.brown.edu
Compositionality in Neural Systems

Norman Biggs Department of Mathematics, London School of Economics and Political Science, London, United Kingdom
PAC Learning and Neural Networks

Andrew Blake Department of Engineering Science, University of Oxford, Oxford, United Kingdom
E-mail: ab@robots.oxford.ac.uk
Active Vision

Edward K. Blum Department of Mathematics, University of Southern California, Los Angeles, California
E-mail: blum@pollux.usc.edu
Dynamics and Bifurcation of Neural Networks

Alexander Borst Friedrich-Miescher-Laboratory of the Max-Planck-Society, Tübingen, Germany
E-mail: borst@sunwan.mpik-tueb.mpg.de
Visuomotor Coordination in Flies

Hervé Bourlard Faculté Polytechnique de Mons, TCTS, Mons, Belgium
E-mail: bourlard@fpms.fpms.ac.be
Speech Recognition: Feature Extraction
Speech Recognition: Pattern Matching

Bruce Bridgeman Departments of Psychology and Psychobiology, University of California, Santa Cruz, Santa Cruz, California
E-mail: bruceb@cats.ucsc.edu
Dissociations Between Visual Processing Modes

Thomas H. Brown Department of Psychology, Yale University, New Haven, Connecticut
E-mail: brown@james.psych.yale.edu
Hebbian Synaptic Plasticity

Joachim M. Buhmann Universität Bonn, Institut für Informatik III, Bonn, Germany
E-mail: jb@cs.bonn.edu
     jb@informatik.uni-bonn.de
Data Clustering and Learning
Oscillatory Associative Memories

Daniel Bullock Cognitive and Neural Systems Department, Boston University, Boston, Massachusetts
E-mail: danb@cms.bu.edu
Motoneuron Recruitment

C. J. C. Burges AT&T Bell Laboratories, Holmdel, New Jersey
E-mail: cjcb@big.att.com.
Handwritten Digit String Recognition

Neil Burgess Department of Anatomy, University College London, London, United Kingdom
E-mail: n.burgess@ucl.ac.uk
Hippocampus: Spatial Models

Arthur W. Burks University of Michigan, Ann Arbor, Michigan
E-mail: arthur_burks@um.cc.umich.edu
Self-Reproducing Automata

Yves Burnod Université Pierre et Marie Curie, Paris, France
Short-Term Memory

John H. Byrne Department of Neurobiology and Anatomy, University of Texas Medical School at Houston, Houston, Texas
E-mail: jbyrne@nba19.med.uth.tmc.edu
Invertebrate Models of Learning: Aplysia and Hermissenda

Ronald L. Calabrese Department of Biology, Emory University, Atlanta, Georgia
E-mail: rcalabre@biology.emory.edu
Half-Center Oscillators Underlying Rhythmic Movements

William H. Calvin University of Washington, Seattle, Washington
E-mail: wcalvin@u.washington.edu
Cortical Columns, Modules, and Hebbian Cell Assemblies

Nicholas T. Carnevale Center for Theoretical and Applied Neuroscience, Yale University, New Haven, Connecticut
E-mail: hines-michael@yale.edu
Computer Modeling Methods for Neurons

Gail A. Carpenter Department of Cognitive and Neural Systems, Boston University, Boston, Massachusetts
E-mail: gail@cns.bu.edu
Adaptive Resonance Theory (ART)

Francisco Cervantes-Pérez Departamento Académico de Computación, Instituto Tecnológico Autónomo de México, México D. F., Mexico
E-mail: cervante@lamport.rhon.itam.mx
Visuomotor Coordination in Frogs and Toads

Sumantra Chattarji Department of Psychology, Yale University, New Haven, Connecticut
E-mail: shona@james.psych.yale.edu
Hebbian Synaptic Plasticity

Rama Chellappa Department of Electrical Engineering and Center for Automation Research, University of Maryland, College Park, Maryland
E-mail: chella@eng.umd.edu
Markov Random Field Models in Image Processing

Hillel J. Chiel Departments of Biology and Neuroscience, Case Western Reserve University, Cleveland, Ohio
Locomotion, Invertebrate

Andy Clark Department of Philosophy, Washington University, St. Louis, Missouri
E-mail: andy@twinearth.wustl.edu
Philosophical Issues in Brain Theory and Connectionism

David Cliff University of Sussex, School of Cognitive and Computing Sciences, Brighton, England, United Kingdom
E-mail: davec@cogs.susx.ac.uk
Neuroethology, Computational

J. J. Collins Boston University, Neuromuscular Research Center, Boston, Massachusetts
E-mail: collins@buenga.bu.edu
Gait Transitions

Michael Conrad Department of Computer Science, Wayne State University, Detroit, Michigan
E-mail: conrad@cs.wayne.edu
Biomaterials for Intelligent Systems

Julian Cook Biomathematics Department, UCLA School of Medicine, Los Angeles, California
Pattern Formation, Biological

Leon N. Cooper Physics Department and Institute for Brain and Neural Systems, Brown University, Providence, Rhode Island
BCM Theory of Visual Cortical Plasticity
Coulomb Potential Learning

Michel Cosnard Laboratoire de l'Informatique du Parallélisme—CNRS, Ecole Normale Supérieure de Lyon, Lyon, France
E-mail: cosnard@lip.ens-lyon.fr
Parallel Computational Models

Gary Cottrell Departments of Computer Science and Engineering, University of California at San Diego, La Jolla, California
E-mail: gary@cs.ucsd.edu
Topology-Modifying Neural Network Algorithms

Jack D. Cowan Department of Mathematics, University of Chicago, Chicago, Illinois
E-mail: cowan@neuro.uchicago.edu
Development and Regeneration of Eye-Brain Maps
Fault Tolerance
Traveling Activity Waves

Francis Crepel Laboratoire de Neurobiologie et Neuropharmacologie du Développement, Université Paris-Sud, Orsay, France
Long-Term Depression in the Cerebellum

Terry Crow Department of Neurobiology and Anatomy, University of Texas Medical School at Houston, Houston, Texas
Invertebrate Models of Learning: Aplysia and Hermissenda

Holk Cruse Abteilung für Biokybernetik und Theoretische Biologie, Universität Bielefeld, Bielefeld, Germany
Motor Pattern Generation

H. Daniel Laboratoire de Neurobiologie et Neuropharmacologie du Développement, Université Paris-Sud, Orsay, France
Long-Term Depression in the Cerebellum

Christian J. Darken Siemens Corporate Research, Princeton, New Jersey
E-mail: darken@learning.siemens.com
Stochastic Approximation and Neural Network Learning

John Daugman The Computer Laboratory, University of Cambridge, Cambridge, England
E-mail: john.daugman@cl.cam.ac.uk
Gabor Wavelets for Statistical Pattern Recognition

Jules Davidoff Department of Psychology, University of Essex, Colchester, Essex, United Kingdom
E-mail: jdavid@uk.ac.sx
Color Perception

Jeffrey Dean Abteilung für Biokybernetik und Theoretische Biologie, Universität Bielefeld, Bielefeld, Germany
E-mail: jeff@bio128.uni-bielefeld.de
Motor Pattern Generation

Bruce Denby INFN Sezione di Pisa, Italy
E-mail: denby@pisa.infn.it
High-Energy Physics

Alain Destexhe The Salk Institute, Computational Neurobiology Lab, La Jolla, California
E-mail: alain@salk.edu
Synaptic Currents, Neuromodulation, and Kinetic Models

Marshall Devor Department of Cell and Animal Biology, Life Sciences Institute, Hebrew University, Jerusalem, Israel
E-mail: marshlu@vms.huji.ac.il
Pain Networks

Patsy S. Dickinson Department of Biology, Bowdoin College, Brunswick, Maine
E-mail: pdickins@polar.bowdoin.edu
Neuromodulation in Invertebrate Nervous Systems

Peter F. Dominey Vision et Motricité, INSERM Unité 94, Bron, France
E-mail: dominey@frmop11.bitnet
Eye-Hand Coordination in Reaching Movements

Clark Dorman Boston University, Department of Cognitive and Neural Systems, Boston, Massachusetts
Motivation

Rodney Douglas MRC Anatomical Neuropharmacology Unit, Oxford, England
E-mail: rjd@vax.ox.ac.uk
Silicon Neurons

Cathryn Downing The Computer Laboratory, Cambridge University, Cambridge, United Kingdom
Gabor Wavelets for Statistical Pattern Recognition

Kenji Doya ATR Human Information Processing Research Laboratories, Kyoto, Japan
E-mail: doya@hip.atr.co.jp
Recurrent Networks; Supervised Learning

Edmund H. Durfee Department of Electrical Engineering and Computer Science, University of Michigan, Ann Arbor, Michigan
E-mail: durfee@engin.umich.edu
Distributed Artificial Intelligence

Shimon Edelman Department of Applied Mathematics and Computer Sciences, The Weizmann Institute of Science, Rehovot, Israel
E-mail: edelman@wisdom.weizmann.ac.il
Vision: Hyperacuity

Martin Egelhaaf Centre for Visual Sciences, Research School of Biological Sciences, Australian National University, Canberra, Australia
Visuomotor Coordination in Flies

Jeffrey L. Elman Department of Cognitive Science, University of California at San Diego, La Jolla, California
E-mail: elman@crl.ucsd.edu
Language Processing

Andreas Engel Institut für Theoretische Physik, Otto-von-Guericke-Universität Magdeburg, Magdeburg, Germany
E-mail: engel@chaos.nat.uni-magdeburg.de
Statistical Mechanics of Learning
Statistical Mechanics of Neural Networks

G. Bard Ermentrout   Department of Mathematics, University of Pittsburgh, Pittsburgh, Pennsylvania
E-mail: bard@mthbard.math.pitt.edu
Phase-Plane Analysis of Neural Activity

Jorg-Peter Ewert   University of Kassel, Kassel, Germany
Command Neurons and Command Systems

Oleg V. Favorov   Department of Physiology, University of North Carolina at Chapel Hill, Chapel Hill, North Carolina
E-mail: kelly@stat.unc.edu
Somatosensory System

Jean-Marc Fellous   Center for Neural Engineering, University of Southern California, Los Angeles, California
E-mail: fellous@pollux.usc.edu
Emotion and Computational Neuroscience

Eberhard E. Fetz   Department of Physiology and Biophysics and Regional Primate Research Center, University of Washington, Seattle, Washington
E-mail: fetz@u.washington.edu
Dynamic Models of Neurophysiological Systems

David J. Field   Department of Psychology, Cornell University, Ithaca, New York
E-mail: djf3@cornell.edu
Visual Coding, Redundancy, and "Feature Detection"

Tamar Flash   Department of Applied Mathematics and Computer Science, The Weizmann Institute of Science, Rehovot, Israel
E-mail: tamar@wisdom.weizmann.ac.il
Optimization Principles in Motor Control

Sherre L. Florence   Department of Psychology, Vanderbilt University, Nashville, Tennessee
E-mail: florensl@ctrvax.vanderbilt.edu
Somatotopy: Plasticity of Sensory Maps

Françoise Fogelman-Soulié   SLIGOS/TAIC/RDF, Clamart, France
E-mail: fogelman@laforia.ibp.fr
Applications of Neural Networks

Peter Földiák   Psychological Laboratory, University of St. Andrews, St. Andrews, England
E-mail: peter.foldiak@st-and.ac.uk
Sparse Coding in the Primate Cortex

Yves Frégnac   Institut Alfred Fessard, CNRS, Gif sur Yvette, France
E-mail: fregnac@bobby.iaf.cnrs-gif.fr
Hebbian Synaptic Plasticity: Comparative and Developmental Aspects

A. Edward Friedman   Department of Neurology, University of Chicago, Chicago, Illinois
Development and Regeneration of Eye-Brain Maps

John P. Frisby   AI Vision Research Unit, University of Sheffield, Sheffield, United Kingdom
E-mail: j.p.frisby@sheffield.ac.uk
Stereo Correspondence and Neural Networks

Bernd Fritzsch   Department of Biomedical Sciences, Creighton University, Omaha, Nebraska
E-mail: berndfri@bif.creighton.edu
Evolution of the Ancestral Vertebrate Brain

Kunihiko Fukushima   Department of Biophysical Engineering, Osaka University, Osaka, Japan
E-mail: fukusima@bpe.es.osaka-u.ac.jp
Neocognitron: A Model for Visual Pattern Recognition

Elvira Galarraga   Departamento de Neurociencias, Instituto de Fisiologia Celular, UNAM, Mexico City D. F., Mexico
Ion Channels: Keys to Neuronal Specialization

Stephen I. Gallant   Belmont Research, Inc., Cambridge, Massachusetts
E-mail: sg@belmont.com
Expert Systems and Decision Systems Using Neural Networks

Patrick Gallinari   Institut Blaise Pascal, Université de Paris, Paris, France
E-mail: gallinari@laforia.ibp.fr
Modular Neural Net Systems, Training of

Paolo Gaudiano   Boston University, Department of Cognitive and Neural Systems, Boston, Massachusetts
E-mail: gaudiano@cns.bu.edu
Motivation

Davi Geiger   Courant Institute, New York University, New York, New York
E-mail: geiger@fred.cs.nyu.edu
Winner-Take-All Mechanisms

Stuart Geman   Division of Applied Mathematics, Brown University, Providence, Rhode Island
Compositionality in Neural Systems

Dedre Gentner   Department of Psychology, Northwestern University, Evanston, Illinois
E-mail: gentner@ils.nwu.edu
Analogy-Based Reasoning

Apostolos P. Georgopoulos   Brain Sciences Center, Veterans Affairs Medical Center, Minneapolis, Minnesota
E-mail: omega@maroon.tc.umn.edu
Reaching: Coding in Motor Cortex

Simon F. Giszter   Department of Anatomy and Neurobiology, Medical College of Pennsylvania and Hahnemann University, Philadelphia, Pennsylvania
E-mail: giszter@ai.mit.edu
Frog Wiping Reflexes

Leon Glass   Department of Physiology, McGill University, Montreal, Quebec, Canada
E-mail: glass@cnd.mcgill.ca
Chaos in Neural Systems

Eric Goles   Departamento de Ingeniería Matemática, Universidad de Chile, Santiago de Chile, Chile
E-mail: egoles@dim.uchile.cl
Energy Functions for Neural Networks

Stephen Grossberg   Center for Adaptive Systems and Department of Cognitive and Neural Systems, Boston University, Boston, Massachusetts
E-mail: diana@cns.bu.edu
Adaptive Resonance Theory (ART)
Figure-Ground Separation

Norberto M. Grzywacz   Smith-Kettlewell Institute, San Francisco, California
Email: nmg@skivs.ski.org
Directional Selectivity in the Cortex
Directional Selectivity in the Retina

Emmanuel Guigon   Université Pierre et Marie Curie, Paris, France
E-mail: guigon@ccr.jussieu.fr
Short-Term Memory

Hermann Haken   Institute for Theoretical Physics and Synergetics, Universitat Stuttgart, Stuttgart, Germany
E-mail: haken@theo1.physik.uni-stuttgart.de
Cooperative Phenomena

Dan Hammerstrom   Adaptive Solutions, Inc., Beaverton, Oregon
E-mail: strom@asi.com
Digital VLSI for Neural Networks

Steven Hampson   ICS Department, University of California at Irvine, Irvine, California
E-mail: hampson@ics.uci.edu
Problem Solving, Connectionist

Mary Hare   Center for Research in Language, University of California at San Diego, La Jolla, California
E-mail: hare@crl.ucsd.edu
Language Change

Simon Haykin   McMaster University, Communication Research Laboratory, Hamilton, Ontario, Canada
E-mail: haykin@mcmaster.ca
Adaptive Signal Processing

N. Hemart   Laboratoire de Neurobiologie et Neuropharmacologie du Développement, Université Paris-Sud, Orsay, France
Long-Term Depression in the Cerebellum

Klaus Hepp   Institut für Theoretische Physik, ETH Zurich, ETH-Honggerberg, Zurich, Switzerland
Saccades and Listing's Law

John Hertz   Nordita, Copenhagen, Denmark
E-mail: hertz@nordita.dk
Computing with Attractors
Sensory Coding and Information Theory

Andreas V. M. Herz   Department of Zoology, University of Oxford, Oxford, England
E-mail: andreas.herz@zoology.oxford.ac.uk
Spatiotemporal Association in Neural Networks

Ellen C. Hildreth   Department of Computer Science, Wellesley College, Wellesley, Massachusetts
E-mail: ehildreth@lucy.wellesley.edu
Motion Perception

Michael Hines   Center for Theoretical and Applied Neuroscience, Yale University, New Haven, Connecticut
E-mail: hines-michael@yale.edu
Computer Modeling Methods for Neurons

Neville Hogan   Departments of Mechanical Engineering and Brain and Cognitive Sciences, Massachusetts Institute of Technology, Cambridge, Massachusetts
Optimization Principles in Motor Control

William R. Holmes   Department of Biological Sciences, Ohio University, Athens, Ohio
E-mail: holmes@cneuro.zool.ohiou.edu
Dendritic Spines

Scott L. Hooper   Department of Biological Sciences, Ohio University, Athens, Ohio
E-mail: hooper@ouvaxa.cats.ohiou.edu
Crustacean Stomatogastric System

John E. Hummel   Department of Psychology, University of California at Los Angeles, Los Angeles, California
E-mail: jhummel@psych.ucla.edu
Object Recognition

Jenq-Neng Hwang   Information Processing Laboratory, Department of Electrical Engineering, University of Washington, Seattle, Washington
E-mail: hwang@ee.washington
Textured Images: Modeling and Segmentation

Nathan Intrator   Computer Science Department, Tel-Aviv University, Ramat-Aviv, Israel
E-mail: nin@math.tau.ac.il
BCM Theory of Visual Cortical Plasticity
Competitive Learning
Information Theory and Visual Plasticity

Petros Ioannou   Department of Electrical Engineering, University of Southern California, Los Angeles, California
E-mail: ioannou@bode.usc.edu
Model-Reference Adaptive Control

Robert A. Jacobs   Department of Psychology, University of Rochester, Rochester, New York
Modular and Hierarchical Learning Systems

D. Jaillard   Laboratoire de Neurobiologie et Neuropharmacologie du Développement, Université Paris-Sud, Orsay, France
Long-Term Depression in the Cerebellum

Marc Jeannerod   Vision et Motricité, INSERM Unité 94, Bron, France
E-mail: jeannerod@frmop11.bitnet
Corollary Discharge in Visuomotor Coordination

B. Keith Jenkins   Signal and Image Processing Institute, University of Southern California, Los Angeles, California
E-mail: jenkins@sipi.usc.edu
Optical Architectures for Neural Network Implementations
Optical Components for Neural Network Implementations

Michael I. Jordan   Department of Brain and Cognitive Sciences, Massachusetts Institute of Technology, Cambridge, Massachusetts
E-mail: jordan@psyche.mit.edu
Modular and Hierarchical Learning Systems

J. Stephen Judd   Siemens Corporate Research, Princeton, New Jersey
E-mail: judd@scr.siemens.com
Time Complexity of Learning

Jon H. Kaas   Department of Psychology, Vanderbilt University, Nashville, Tennessee
Somatotopy: Plasticity of Sensory Maps

John F. Kalaska   Département de Physiologie, Université de Montréal, Montreal, Quebec, Canada
E-mail: kalaskaj@ere.umontreal.ca
Reaching Movements: Implications of Connectionist Models

Kunihiko Kaneko   Department of Pure and Applied Sciences, University of Tokyo, Tokyo, Japan
E-mail: kaneko@cyber.c.u-tokyo.ac.jp
Cooperative Behavior in Networks of Chaotic Elements

Jagmeet S. Kanwal   Department of Biology, Washington University, St. Louis, Missouri
E-mail: kanwal@wustlb.wustl.edu
Echolocation: Creating Computational Maps

Annette Karmiloff-Smith   MRC Child Development Unit, London, England
E-mail: annette@cdu.ucl.ac.uk
Developmental Disorders

Ashraf A. Kassim   National University of Singapore, Singapore
E-mail: eleashra@leonis.nus.sg
Potential Fields and Neural Networks

John S. Kauer   Department of Neurosurgery, Tuft's University Medical School, New England Medical Center, Boston, Massachusetts
E-mail: jkauer@pearl.tufts.edu
Olfactory Bulb

Mitsuo Kawato   ATR Human Information Processing Research Laboratories, Kyoto, Japan
E-mail: kawato@hip.atr.co.jp
Cerebellum and Motor Control

Douglas G. Kelly   Department of Mathematics and Statistics, University of North Carolina, Chapel Hill, North Carolina
E-mail: kelly@stat.unc.edu
Somatosensory System

Scott Kirkpatrick   Thomas J. Watson Research Center, Yorktown Heights, New York
E-mail: kirk@watson.ibm.com
Simulated Annealing

Christof Koch   Computation and Neural Systems Program, California Institute of Technology, Pasadena, California
E-mail: koch@cns.caltech.edu
Axonal Modeling
Selective Visual Attention
Single-Cell Models

Teuvo Kohonen   Helsinki University of Technology, Neural Networks Research Centre, Espoo, Finland
E-mail: teuvo@hutmc.hut.fi
Learning Vector Quantization
Speech Recognition: A Hybrid Approach

Nancy Kopell   Department of Mathematics, Boston University, Boston, Massachusetts
E-mail: nk@math.bu.edu
Chains of Coupled Oscillators

Jan H. M. Korst   Philips Research Laboratories, Eindhoven, The Netherlands
Boltzmann Machines

Richard J. Krauzlis   Laboratory of Sensorimotor Research, National Institutes of Health, Bethesda, Maryland
E-mail: rjk@lsr.nei.nih.gov
Pursuit Eye Movements

Andrew M. Krylow   Department of Biomedical Engineering, Northwestern University, Evanston, Illinois
Muscle Models

B. V. K. Vijaya Kumar   Department of Electrical and Computer Engineering, Carnegie Mellon University, Pittsburgh, Pennsylvania
E-mail: kumar@ece.cmu.edu
Potential Fields and Neural Networks

Yoshiki Kuramoto   Department of Physics, Kyoto University, Kyoto, Japan
E-mail: kuramoto@ton.scphys.kyoto-u.ac.jp
Collective Behavior of Coupled Oscillators

Věra Kůrková   Institute of Computer Science, Czech Academy of Sciences, Prague, Czechia
E-mail: vera@uivt.cas.cz
Kolmogorov's Theorem

Francesco Lacquaniti   Istituto di Neuroscienze e Bioimmagini, Consiglio Nazionale delle Ricerche, Milan, Italy
Limb Geometry: Neural Control

Ofer Lahav   Institute of Astronomy, Cambridge, England, United Kingdom
E-mail: lahav@mail.ast.cam.ac.uk
        hlahav@wicc.weizmann.ac.il
Astronomy

Yann LeCun   AT&T Bell Laboratories, Holmdel, New Jersey
E-mail: yann@research.att.com
Convolutional Networks for Images, Speech, and Time Series
Pattern Recognition

Joseph E. LeDoux   Center for Neural Science, New York University, New York, New York
E-mail: ledoux@cns.nyu.edu
Emotion and Computational Neuroscience

Michael A. Lehr   Stanford University, Department of Electrical Engineering, Stanford, California
E-mail: lehr@isl.stanford.edu
Noise Canceling and Channel Equalization
Perceptrons, Adalines, and Backpropagation

Wee Kheng Leow   Department of Computer Sciences, The University of Texas at Austin, Austin, Texas
E-mail: leow@cs.utexas.edu
Visual Schemas in Object Recognition and Scene Analysis

Gregory W. Lesher   Enkidu Research, Ithaca, New York
Illusory Contour Formation

Daniel S. Levine   Department of Mathematics, University of Texas at Arlington, Arlington, Texas
E-mail: b344dsl@utarlg.uta.edu
Disease: Neural Network Models

J. B. Levitt   Department of Visual Science, Institute of Ophthalmology, London, England, United Kingdom
Visual Cortex Cell Types and Connections

John E. Lewis   Department of Biology, University of California at San Diego, La Jolla, California
E-mail: jlewis@ucsd.edu
Respiratory Rhythm Generation

Jim-Shih Liaw   Program in Neuroscience, University of Southern California, Los Angeles, California
E-mail: liaw@bmsrs.usc.edu
NMDA Receptors: Synaptic, Cellular, and Network Models

Raymond Lister   School of Computing Science, Queensland University of Technology, Brisbane, Australia
E-mail: raymond@fitmail.fit.qut.edu.au
Fractal Strategies for Neural Network Scaling

Gerald E. Loeb   Bio-Medical Engineering Unit, Queen's University, Kingston, Ontario, Canada
E-mail: loeb@biomed.queensu.ca
Prosthetics, Neural

Jason D. Lohn   Electrical Engineering Department, University of Maryland, College Park, Maryland
E-mail: jlohn@eng.umd.edu
Analog VLSI for Neural Networks

Fernando H. Lopes da Silva   Graduate School for the Neurosciences, University of Amsterdam, Amsterdam, The Netherlands
E-mail: silva@bio.uva.nl
EEG Analysis
Epilepsy: Network Models of Generation

David Lowe   Department of Computer Science and Applied Mathematics, Aston University, Birmingham, United Kingdom
E-mail: d.lowe@aston.ac.uk
Radial Basis Function Networks

Jennifer S. Lund   Department of Visual Science, Institute of Ophthalmology, London, United Kingdom
E-mail: smgxjsl@ucl.ac.uk
Visual Cortex Cell Types and Connections

Giuseppe Luppino   Istituto di Fisiologia Umana, Universitá di Parma, Parma, Italy
Grasping Movements: Visuomotor Transformations

Wolfgang Maass   Institute for Theoretical Computer Science, Technische Universität Graz, Graz, Austria
E-mail: maass@igi.tu-graz.ac.at
Vapnik-Chervonenkis Dimension of Neural Networks

David J. C. MacKay   Cavendish Laboratory, Cambridge, United Kingdom
E-mail: mackay@mrao.cam.ac.uk
Bayesian Methods for Supervised Neural Networks

Misha Mahowald MRC Anatomical Neuropharmacology Unit, Oxford University, Oxford, United Kingdom
E-mail: misha@zen.pharm.ox.ac.uk
Silicon Neurons

Zachary F. Mainen The Salk Institute, Computational Neurobiology Lab, La Jolla, California
Synaptic Currents, Neuromodulation, and Kinetic Models

Claudio Maioli Biomedical Sciences Department, Brescia University, Brescia, Italy
Limb Geometry: Neural Control

Hanspeter A. Mallot Max-Planck-Institut für biologische Kybernetik, Tübingen, Germany
E-mail: ham@.mpik-tueb.mpg.de
Layered Computation in Neural Networks

Gerhard Manteuffel FBN, Dummerstorf, Germany
E-mail: a12i@alf.zfn.uni-bremen.de
Visuomotor Coordination in Salamanders

Eve Marder Biology Department, Brandeis University, Waltham, Massachusetts
E-mail: marder@binah.cc.brandeis.edu
Activity-Dependent Regulation of Neuronal Conductances
Dynamic Clamp: Computer-Neural Hybrids

Arthur B. Markman Department of Psychology, Columbia University, New York, New York
E-mail: markman@psych.columbia.edu
Analogy-Based Reasoning

Robert J. Marks II Department of Electrical Engineering, University of Washington, Seattle, Washington
E-mail: marks@u.washington.edu
Neurosmithing: Improving Neural Network Learning

Jose L. Marroquin Centro de Investigacion en Matematicas, Guanajato, Mexico
E-mail: cimat@unamvml.bitnet
Regularization Theory and Low-Level Vision

Jonathan A. Marshall Department of Computer Science, University of North Carolina, Chapel Hill, North Carolina
E-mail: marshall@cs.unc.edu
Motion Perception: Self-Organization

Lina L. E. Massone Department of Electrical Engineering and Computer Science, Northwestern University, Evanston, Illinois
E-mail: massone@eecs.nwu.edu
Sensorimotor Learning

Massimo Matelli Istituto di Fisiologia Umana, Universitá di Parma, Parma, Italy
Grasping Movements: Visuomotor Transformations

Pietro Mazzoni Cambridge, Massachusetts
E-mail: pmazzoni@ai.mit.edu
Gaze Coding in the Posterior Parietal Cortex

James L. McClelland Department of Psychology, Carnegie Mellon University, Pittsburgh, Pennsylvania
E-mail: mcclelland+@cmu.edu
Cognitive Development

David A. McCormick Section of Neurobiology, Yale University School of Medicine, New Haven, Connecticut
Thalamocortical Oscillations in Sleep and Wakefulness

James N. McNames Neural Applications Corporation, Coralville, Iowa
Steelmaking

Bartlett W. Mel Department of Biomedical Engineering, University of Southern California, Los Angeles, California
E-mail: mel@quake.usc.edu
Planning, Connectionist

Jerry M. Mendel Signal and Image Processing Institute, Department of Electrical Engineering—Systems, University of Southern California, Los Angeles, California
E-mail: mendel@sipi.usc.edu
Fuzzy Logic Systems and Qualitative Knowledge

Janet Metcalfe Department of Psychology, Dartmouth College, Hanover, New Hampshire
E-mail: metcalfe@dartmouth.edu
Distortions in Human Memory

Claude Meunier Centre de Physique Theoretique, Ecole Polytechnique, Palaiseau, France
E-mail: meunier@orphee.polytechnique.fr
Sparsely Coded Neural Networks

R. Chris Miall Laboratory of Physiology, Oxford University, Oxford, United Kingdom
E-mail: rcm@physiol.ox.ac.uk
Motor Control, Biological and Theoretical

Risto Miikkulainen Department of Computer Sciences, The University of Texas at Austin, Austin, Texas
E-mail: risto@cs.utexas.edu
Visual Schemas in Object Recognition and Scene Analysis

Kenneth D. Miller Department of Physiology, University of California at San Francisco, San Francisco, California
E-mail: ken@phy.ucsf.edu
Ocular Dominance and Orientation Columns

John Milton Department of Neurology, University of Chicago Hospitals, Chicago, Illinois
E-mail: sp1ace@ace.bsd.uchicago.edu
Traveling Activity Waves

Ennio Mingolla Boston University, Center for Adaptive Systems and Department of Cognitive and Neural Systems, Boston, Massachusetts
E-mail: ennio@cns.bu.edu
Illusory Contour Formation

Laura A. Monti Rush Neuroscience Institute, Chicago, Illinois
Emotion-Cognition Interactions

Nelson Morgan International Computer Science Institute, Berkeley, California
E-mail: morgan@icsl.berkeley.edu
Programmable Neurocomputing Systems
Speech Recognition: Feature Extraction
Speech Recognition: Pattern Matching

David C. Mountain Department of Biomedical Engineering, Boston University, Boston, Massachusetts
E-mail: dcm@bu.edu
Auditory Periphery and Cochlear Nucleus

David Mumford Department of Mathematics, Harvard University, Cambridge, Massachusetts
E-mail:mumford@math.harvard.edu
Thalamus

Trevor Mundel Department of Mathematics, University of Chicago, Chicago, Illinois
Traveling Activity Waves

Bennet B. Murdock Department of Psychology, University of Toronto, Toronto, Ontario, Canada
E-mail: murdock@psych.toronto.edu
Classical Learning Theory and Neural Networks

Robin R. Murphy Department of Mathematical and Computer Sciences, Colorado School of Mines, Golden, Colorado
E-mail: rmurphy@mines.colorado.edu
Sensor Fusion

J. D. Murray Department of Applied Mathematics, University of Washington, Seattle, Washington
E-mail: murrayjd@amath.washington.edu
Pattern Formation, Biological

Jacob M. J. Murre MRC Applied Psychology Unit, Cambridge, England
E-mail: jaap.murre@mrc-apu.cam.ac.uk
Neurosimulators

Ferdinando A. Mussa-Ivaldi Department of Physiology, Northwestern University Medical School, Chicago, Illinois
E-mail: sandro@nwu.edu
Geometrical Principles in Motor Control

J.-P. Nadal Laboratoire de Physique Statistique, Ecole Normale Supérieure, Paris, France
E-mail: nadal@physique.ens.fr
Sparsely Coded Neural Networks

Kumpati S. Narendra Center for Systems Science, Department of Electrical Engineering, Yale University, New Haven, Connecticut
E-mail: narendra@koshy.eng.yale.edu
Adaptive Control: Neural Network Applications
Identification and Control

J. I. Nelson Laboratory of Neuropsychology, National Institute of Mental Health, Bethesda, Maryland
E-mail: jnelson@ln.nimh.nih.gov
Binding in the Visual System
Visual Scene Perception: Neurophysiology

Robert W. Newcomb Microsystems Laboratory, Electrical Engineering Department, University of Maryland, College Park, Maryland
E-mail: newcomb@eng.umd.edu
Analog VLSI for Neural Networks

Stefano Nolfi Institute of Psychology, National Research Council, Rome, Italy
E-mail: stefano@kant.irmkant.rm.cnr.it
"Genotypes" for Neural Networks

Anthony M. Norcia The Smith-Kettlewell Eye Research Institute, San Francisco, California
Directional Selectivity in the Cortex

Erkki Oja Laboratory of Computer and Information Science, Helsinki University of Technology, Espoo, Finland
E-mail: erkki.oja@hut.fi
Principal Component Analysis

John O'Keefe Department of Anatomy, University College London, London, England
E-mail: john@anat.ucl.ac.uk
Hippocampus: Spatial Models

Bruno A. Olshausen Department of Psychology, Cornell University, Ithaca, New York
E-mail: bao1@cornell.edu
Routing Networks in Visual Cortex
Selective Visual Attention

M. B. O'Neil Dyna-Quest Technologies, Inc., Sudbury, Massachusetts
Dynamic Clamp: Computer-Neural Hybrids

Manfred Opper Physikalisches Institut, Julius Maximilians Universität, Würzburg, Germany
E-mail: opper@physik.uni-wuerzburg.de
Statistical Mechanics of Generalization

Andrew Ortony Department of Psychology, Northwestern University, Evanston, Illinois
E-mail: ortony@aristotle.ils.nwu.edu
Emotion-Cognition Interactions

Alice J. O'Toole School of Human Development, The University of Texas at Dallas, Richardson, Texas
E-mail: otoole@utdallas.edu
Face Recognition

Ian Parberry Department of Computer Sciences, University of North Texas, Denton, Texas
E-mail: ian@ponder.csci.unt.edu
Structural Complexity and Discrete Neural Networks

Domenico Parisi Institute of Psychology, National Research Council, Rome, Italy
"Genotypes" for Neural Networks

Hélène Paugam-Moisy Ecole Normale Supérieure de Lyon, Lyon, France
E-mail: hpaugam@lip.ens-lyon.fr
Multiprocessor Simulation of Neural Networks

Judea Pearl Computer Science, University of California at Los Angeles, Los Angeles, California
E-mail: judea@cs.ucla.edu
Bayesian Networks

Denis Pélisson Vision et Motricité, INSERM Unité 94, Bron, France
Eye-Hand Coordination in Reaching Movements

Michael P. Perrone Thomas J. Watson Research Center, Yorktown Heights, New York
E-mail: mpp@watson.ibm.com
Averaging/Modular Techniques for Neural Networks
Coulomb Potential Learning
Post-Hebbian Learning Rules

Barry W. Peterson Northwestern University Medical School, Chicago, Illinois
E-mail: b-peterson2@nwu.edu
Head Movements: Multidimensional Modeling

Carsten Peterson Department of Theoretical Physics, University of Lund, Lund, Sweden
E-mail: carsten@thep.lu.se
Neural Optimization

Jan Pieter Pijn Instituut voor Epilepsiebestrijding Meer en Bosch, Heemstede, The Netherlands
EEG Analysis
Epilepsy: Network Models of Generation

David C. Plaut Department of Psychology, Carnegie Mellon University, Pittsburgh, Pennsylvania
E-mail: plaut@cmu.edu
Lesioned Attractor Networks as Models of Neuropsychological Deficits

Kim Plunkett Department of Psychology, University of Oxford, Oxford, England
E-mail: plunkett@dragon.psych.oxford.ac.uk
Cognitive Development
Language Acquisition

Dean A. Pomerleau School of Computer Science, Carnegie Mellon University, Pittsburgh, Pennsylvania
E-mail: dean.pomerleau@ius4.ius.cs.cmu.edu
Vision for Robot Driving

Ernst Pöppel Forschungszentrum Jülich, KFA, Jülich, Germany
Time Perception: Problems of Representation and Processing

Alexandre Pouget The Salk Institute, La Jolla, California
E-mail: alex@salk.edu
Dynamic Remapping

Claude Prablanc Vision et Motricité, INSERM Unité 94, Bron, France
E-mail: prablanc@frmop11.bitnet
Eye-Hand Coordination in Reaching Movements

Wilfrid Rall   Math. Research Br, NIDDK, Bethesda, Maryland
E-mail: wilrall@helix.nih.gov
Dendritic Spines
Perspective on Neuron Model Complexity

Anand Rangarajan   Department of Computer Science, Yale University, New Haven, Connecticut
E-mail: rangarajan-anand@cs.yale.edu
Markov Random Field Models in Image Processing

Michael Recce   Department of Anatomy, University College London, London, England
Hippocampus: Spatial Models

Russell Reed   Department of Electrical Engineering, University of Washington, Seattle, Washington
Neurosmithing: Improving Neural Network Learning

A. N. Refenes   Department of Decision Science, London Business School, London, England
E-mail: p.refenes@lbs.lon.ac.uk
Investment Management: Tactical Asset Allocation

Luigi M. Ricciardi   Dipartimento di Matematica e Applicazioni, Università degli Studi di Napoli 'Federico II,' Napoli, Italy
E-mail: ricciardi@matna1.dma.unina.it
Diffusion Models of Neuron Activity

John Rinzel   Mathematical Research Branch, National Institutes of Health, Bethesda, Maryland
Oscillatory and Bursting Properties of Neurons

Helge Ritter   Department of Information Science, Bielefeld University, Bielefeld, Germany
E-mail: helge@techfak.uni-bielefeld.de
Self-Organizing Feature Maps: Kohonen Maps

Giacomo Rizzolatti   Instituto di Fisiologia Umana, Universitá di Parma, Parma, Italy
Grasping Movements: Visuomotor Transformations

R. Meldrum Robertson   Department of Biology, Queen's University, Kingston, Ontario, Canada
E-mail: robertrm@biology.queensu.ca
Locust Flight: Components and Mechanisms in the Motor

Luc Rodet   Department of Psychology, University of Grenoble, Grenoble, France
Concept Learning

Burkhard Rost   EMBL Heidelberg, Protein Design Group, Heidelberg, Germany
E-mail: rost@embl-heidelberg.de
Protein Structure Prediction

Constance S. Royden   Department of Computer Science, Wellesley College, Wellesley, Massachusetts
Motion Perception

W. Zev Rymer   Sensory Motor Performance Program, Rehabilitation Institute of Chicago, Chicago, Illinois
E-mail: zevric@casbah.acns.nwu.edu
Muscle Models

Chris Sander   European Molecular Biology Organization, Protein Design Group, Heidelberg, Germany
Protein Structure Prediction

Thomas G. Sandercock   Department of Physiology, Northwestern University Medical School, Chicago, Illinois
Muscle Models

Kerstin Schill   Institut für Medizinische Psychologie, München, München, Germany
E-mail: kerstin@groucho.imp.med.uni-muenchen.de
Time Perception: Problems of Representation and Processing

Nestor A. Schmajuk   Department of Psychology: Experimental, Duke University, Durham, North Carolina
E-mail: nestor@acpub.duke.edu
Cognitive Maps
Conditioning

Almut Schüz   Max-Planck-Institut für Biologische Kybernetik, Tübingen, Germany
E-mail: schuez@sun04.mpik-tueb.mpg.de
Neuroanatomy in a Computational Perspective

Philippe G. Schyns   Department of Psychology, University of Montreal, Montreal, Quebec, Canada
E-mail: schyns@ai.mit.edu
Concept Learning

Idan Segev   Department of Neurobiology, Institute of Life Sciences and Center for Neural Computation, Hebrew University, Jerusalem, Israel
E-mail: idan@hujivms.huji.ac.il
Dendritic Processing

Jose P. Segundo   Department of Anatomy and Cell Biology, and Brain Research Institute, University of California, Los Angeles, California
E-mail: iaqfjps@mvs.oac.ucla.edu
Synaptic Coding of Spike Trains

Michael Seibert   Machine Intelligence Technology Group, MIT Lincoln Laboratory, Lexington, Massachusetts
Visual Processing of Object Form and Environment Layout

Mark S. Seidenberg   Neuroscience Program, University of Southern California, Los Angeles, California
E-mail: marks@neuro.usc.edu
Linguistic Morphology

Terrence J. Sejnowski   Salk Institute, San Diego, California
E-mail: terry@sdbio2.ucsd.edu
Dynamic Remapping
Synaptic Currents, Neuromodulation, and Kinetics Models
Thalamocortical Oscillations in Sleep and Wakefulness

Evelyne Sernagor   Smith-Kettlewell Institute, San Francisco, California
Directional Selectivity in the Retina

Reza Shadmehr   Department of Biomedical Engineering, Johns Hopkins University, Baltimore, Maryland
E-mail: reza@spindle.bme.jhu.edu
Equilibrium Point Hypothesis

Shihab A. Shamma   Department of Electrical Engineering and Institute for Systems Research, University of Maryland, College Park, Maryland
E-mail: sas@src.umd.edu
Auditory Cortex

Amanda J. C. Sharkey   Department of Computer Science, University of Sheffield, Sheffield, England
E-mail: a.sharkey@dcs.shef.ac.uk
Cognitive Modeling: Psychology and Connectionism

Noel E. Sharkey   Department of Computer Science, University of Sheffield, Sheffield, England
E-mail: n.sharkey@dcs.shef.ac.uk
Cognitive Modeling: Psychology and Connectionism

A. A. Sharp   Biology Department and Center for Complex Systems, Brandeis University, Waltham, Massachusetts
Dynamic Clamp: Computer-Neural Hybrids

Lokendra Shastri   International Computer Science Institute, Berkeley, California
E-mail: shastri@icsi.berkeley.edu
Structured Connectionist Models

Jude W. Shavlik   Computer Sciences Department, University of Wisconsin, Madison, Wisconsin
E-mail: shavlik@cs.wisc.edu
Learning by Symbolic and Neural Methods

Gordon M. Shepherd   Department of Neuroanatomy, Yale University, School of Medicine, New Haven, Connecticut
E-mail: gordon_shepherd@quickmail.cis.yale.edu
Olfactory Bulb
Olfactory Cortex

Harel Z. Shouval   Brown University, Providence, Rhode Island
E-mail: hzs@cns.brown.edu
Post-Hebbian Learning Rules

Larry E. Shupe   Department of Physiology and Biophysics and Regional Primate Research Center, University of Washington, Seattle, Washington
Dynamic Models of Neurophysiological Systems

John J. Shynk   University of California at Santa Barbara, Department of Electrical and Computer Engineering, Santa Barbara, California
E-mail: shynk@ece.ucsb.edu
Adaptive Filtering

Karen A. Sigvardt   Department of Neurology, University of California at Davis, Davis, California
Spinal Cord of Lamprey: Generation of Locomotor Patterns

Wolf Singer   Max-Planck-Institut für Hirnforschung, Frankfurt am Main, Germany
E-mail: singer@mpih-frankfurt.mpg.dbp.de
Synchronization of Neuronal Responses as a Putative Binding Mechanism

Robert G. Smith   Department of Neuroscience, University of Pennsylvania, Philadelphia, Pennsylvania
E-mail: rob@retina.anatomy.upenn.edu
Retina

Bo Söderberg   Department of Theoretical Physics, University of Lund, Lund, Sweden
Neural Optimization

William Softky   Mathematical Research Branch, NIDDK/NIH, Bethesda, Maryland
E-mail: bill@homer.niddk.nih.gov
Single-Cell Models

Eduardo D. Sontag   Rutgers Center for Systems and Control, Department of Mathematics, Rutgers University, New Brunswick, New Jersey
E-mail: sontag@hilbert.rutgers.edu
Automata and Neural Networks

Gregory B. Sorkin   Thomas J. Watson Research Center, Yorktown Heights, New York
Simulated Annealing

Jean-Paul Spire   Department of Neurology, University of Chicago, Chicago, Illinois
Traveling Activity Waves

William E. Staib   Neural Applications Corporation, Coralville, Iowa
E-mail: wstaib@neural.com
Steelmaking

Paul S. G. Stein   Department of Biology, Washington University, St. Louis, Missouri
E-mail: stein@wustlb.wustl.edu
Scratch Reflex

Mircea Steriade   Laboratoire de Neurophysiologie, Département de Physiologie, Faculté de Médecine Université Laval, Quebec, Ontario, Canada
Thalamocortical Oscillations in Sleep and Wakefulness

M. Stiber   Department of Computer Science, Hong Kong University of Science and Technology, Kowloon, Hong Kong
E-mail: stiber@cs.ust.hk
Synaptic Coding of Spike Trains

Michael C. Storrie-Lombardi   Institute of Astronomy, Cambridge, England
E-mail: mcsl@mail.ast.cam.ac.uk
Astronomy

Nobuo Suga   Department of Biology, Washington University, St. Louis, Missouri
E-mail: suga@batlab.wustl.edu
Echolocation: Creating Computational Maps

Harold Szu   Naval Surface Warfare Center, Silver Spring, Maryland
E-mail: hszu@ulysses.nswc.navy.mil
Automatic Target Recognition
Telecommunications
Wavelet Dynamics

Armand R. Tanguay, Jr.   Optical Materials and Devices Laboratory, University of Southern California, Los Angeles, California
E-mail: atanguay@mizar.usc.edu
Optical Architectures for Neural Network Implementations
Optical Components for Neural Network Implementations

John G. Taylor   Centre for Neural Networks, King's College London, London, England
E-mail: udah057@bay.cc.kcl.ac.uk
Self-Organization in the Time Domain

Brian Telfer   Naval Surface Warfare Center, Silver Spring, Maryland
Automatic Target Recognition
Wavelet Dynamics

Richard F. Thompson   University of Southern California, Los Angeles, California
E-mail: thompson@neuro.usc.edu
Cerebellum and Conditioning

Simon J. Thorpe   Centre de Recherche Cerveau & Cognition, Faculté de Medecine de Rangueil, Université Paul Sabatier, Toulouse, France
E-mail: thorpe@cix.cict.fr
Localized Versus Distributed Representations

Sebastian Thrun   Universität Bonn, Institut für Informatik III, Bonn, Germany
E-mail: thrun@carbon.cs.bonn.edu
Exploration in Active Learning

James S. Tittle   Department of Psychology, Ohio State University, Columbus, Ohio
E-mail: jtittle@magnus.acs.ohio-state.edu
Perception of Three-Dimensional Structure

James T. Todd   Department of Psychology, Ohio State University, Columbus, Ohio
E-mail: jtodd@magnus.acs.ohio-state.edu
Perception of Three-Dimensional Structure

Tommaso Toffoli   MIT Lab of Computer Science, Cambridge, Massachusetts
E-mail: tt@im.lcs.mit.edu
Cellular Automata

Kari Torkkola   Motorola, Inc., Tempe, Arizona
E-mail: torkkk@pcrl.sps.mot.com
Speech Recognition: A Hybrid Approach

Carme Torras   Institut de Cibernètica, Barcelona, Spain
E-mail: torras@ic.upc.es
Robot Control

David S. Touretzky   School of Computer Science, Carnegie Mellon University, Pittsburgh, Pennsylvania
E-mail: dst@cs.cmu.edu
Connectionist and Symbolic Representations

Lyle H. Ungar   Department of Chemical Engineering, University of Pennsylvania, Philadelphia, Pennsylvania
E-mail: ungar@central.cis.upenn.edu
Forecasting
Process Control

Dominique Valentin   School of Human Development, The University of Texas at Dallas, Richardson, Texas
Face Recognition

David Van Essen   Department of Anatomy and Neurobiology, Washington University School of Medicine, St. Louis, Missouri
Routing Networks in Visual Cortex

J. A. M. Van Gisbergen   Department of Medical Physics and Biophysics, University of Nijmegen, Nijmegen, The Netherlands
E-mail: admin@mbfys.kun.nl
Collicular Visuomotor Transformations for Saccades

A. J. Van Opstal   Department of Medical Physics and Biophysics, University of Nijmegen, Nijmegen, The Netherlands
Collicular Visuomotor Transformations for Saccades

Vladimir Vapnik   AT&T Bell Laboratories, Holmdel, New Jersey
E-mail: vlad@research.att.com
Learning and Generalization: Theoretical Bounds

Max Velmans   Department of Psychology, Goldsmiths College, University of London, London, England
E-mail: mlv@gold.ac.uk
Consciousness, Theories of

J.-F. Vibert   Biomathématiques, Biostatistique, Bioinformatique et Epidémiologie, INSERM U263, Faculté de Médecine Saint-Antoine, Paris, France
E-mail: vibert@b3c.jussieu.fr
Synaptic Coding of Spike Trains

Philippe Vindras   Faculte de Psychologie et de Sciences de l'Education, Carouge, Switzerland
Eye Hand Coordination in Reaching Movements

Christoph von der Malsburg   Institut für Neuroinformatik, Ruhr-Universität Bochum, Bochum, Germany; Departments of Computer Science and Neurobiology, University of Southern California, Los Angeles, California
E-mail: malsburg@usc.edu
Dynamic Link Architecture
Self-Organization and the Brain

Grace Wahba   Department of Statistics, University of Wisconsin at Madison, Madison, Wisconsin
E-mail: wahba@stat.wisc.edu
Generalization and Regularization in Nonlinear Learning Systems

David L. Waltz   NEC Research Institute, Princeton, New Jersey
E-mail: waltz@research.nj.nec.com
Memory-Based Reasoning

DeLiang Wang   Department of Computer and Information Science, Ohio State University, Columbus, Ohio
E-mail: dwang@cis.ohio-state.edu
Habituation
Temporal Pattern Processing

Xiao-Jing Wang   Department of Mathematics, University of Pittsburgh, Pittsburgh, Pennsylvania
E-mail: xjwang@math.pitt.edu
Oscillatory and Bursting Properties of Neurons

Xin Wang   Computer Science Department, University of California at Los Angeles, Los Angeles, California
E-mail: xwang@cs.ucla.edu
Dynamics and Bifurcation of Neural Networks

Allen M. Waxman   Machine Intelligence Technology Group, MIT Lincoln Laboratory, Lexington, Massachusetts
E-mail: waxman@ll.mit.edu
Visual Processing of Object Form and Environment Layout

Yair Weiss   Department of Brain and Cognitive Sciences, Massachusetts Institute of Technology, Cambridge, Massachusetts
E-mail: yweiss@media.mit.edu
Vision: Hyperacuity

Alfredo Weitzenfeld   Departamento Académico de Computación, Instituto Tecnológico Autónomo de México, México D.F., México
E-mail: alfredo@lamport.rhon.itam.mx
        alfredo@rana.usc.edu
NSL: Neural Simulation Language

Paul J. Werbos   National Science Foundation, Arlington, Virginia
E-mail: pwerbos@nsf.gov
Backpropagation: Basics and New Developments

Joel White   Program in Neuroscience, Tufts University School of Medicine, Boston, Massachusetts
E-mail: jwhite@opal.tufts.edu
Olfactory Bulb

Bernard Widrow   Department of Electrical Engineering, Stanford University, Stanford, California
E-mail: widrow@isl.stanford.edu
Noise Canceling and Channel Equalization
Perceptrons, Adalines, and Backpropagation

Thelma L. Williams   Physiology Department, St. George's Hospital Medical School, London, England
E-mail: twilliams@sghms.ac.uk
Spinal Cord of Lamprey: Generation of Locomotor Patterns

Matthew A. Wilson   Massachusetts Institute of Technology, Department of Brain and Cognitive Sciences, Cambridge, Massachusetts
E-mail: wilson@ai.mit.edu
Olfactory Cortex

A. T. Winfree   Department of Ecology and Evolutionary Biology, University of Arizona, Tucson, Arizona
E-mail: art@cochise.biosci.arizona.edu
Wave Propagation in Cardiac Muscle and in Nerve Networks

David A. Winter   Department of Kinesiology, University of Waterloo, Waterloo, Ontario, Canada
Human Movement: A System-Level Approach

Q. Wu   Department of Visual Science, Institute of Ophthalmology, London, England
E-mail: smgxqfw@ucl.ac.uk
Visual Cortex Cell Types and Connections

Shuji Yoshizawa   Department of Mechano-Informatics, University of Tokyo, Tokyo, Japan
E-mail: yoshi@bios.t.u-tokyo.ac.jp
Nonmonotonic Neuron Associative Memory

Malcolm P. Young   Department of Physiology, University of Oxford, Oxford, England
E-mail: mpy@physiol.ox.ac.uk
Sparse Coding in the Primate Cortex

Alan L. Yuille   Division of Applied Sciences, Harvard University, Cambridge, Massachusetts
E-mail: yuille@hrl.harvard.edu
Constrained Optimization and the Elastic Net
Winner-Take-All Mechanisms

A. D. Zapranis   London Business School, Department of Decision Science, London, England
E-mail: azaprani@neptune.lbs.lon.ac.uk
Investment Management: Tactical Asset Allocation

Klaus-Peter Zauner   Department of Computer Science, Wayne State University, Detroit, Michigan
Biomaterials for Intelligent Systems

Richard S. Zemel   Department of Psychology, Carnegie Mellon University, Pittsburgh, Pennsylvania
E-mail: zemel@cmu.edu
Minimum Description Length Analysis

Annette Zippelius   Institut für Theoretische Physik, Georg-August Universität Göttingen, Göttingen, Germany
E-mail: annette@theo-phys.gwdg.de
Statistical Mechanics of Learning
Statistical Mechanics of Neural Networks

Steven W. Zucker   Center for Intelligent Machines, McGill University, Montréal, Québec, Canada
E-mail: zucker@cim.mcgill.edu
Perceptual Grouping

Subject Index